DICTIONARY OF SCIENTIFIC PRINCIPLES

Stephen Marvin
West Chester University

WILEY

JOHN WILEY & SONS, INC., PUBLICATION

Library of Congress Cataloging-in-Publication Data:

Dictionary of scientific principles / by Stephen Marvin.
    p. cm.
  Includes bibliographical references.
  ISBN 978-0-470-14680-4 (cloth)
  1. Science–Dictionaries. I. Marvin, Stephen.
  Q123.D537 2010
  503–dc22
                            2010003426

  Printed in Singapore

10 9 8 7 6 5 4 3 2 1

To E. E. Barnes,
with long overdue thanks
for
sharing ideas, concepts, and challenges
in the way we seek and organize information

CONTENTS

PREFACE

The *Dictionary of Scientific Principles* is an attempt to compile the language of art used for various known rules or laws applied to a broad category of topics, including mathematics, medicine, sciences, psychology, management, and even philosophy and art. This project has taken over 6 years to develop to this point. I have consulted with scientists and colleagues on the development of this dictionary and had some help in organizing the files from an MS Excel spreadsheet. There are approximately 2000+ principles that form the language of art. Some are rewording of the same principle; For instance, the *principle of maximum entropy* is also listed as the *maximum entropy principle*. I exerted a great deal of effort to have this work prepared in time for the new millennium and to call it *Millennial Principles.* However, the myriad new discoveries in scientific and other disciplines necessitated the continual addition of new entries and cross-references to similar-context or related entries already listed in this volume. In creating this dictionary, I consulted many encyclopedias, dictionaries, books, indexes, and journal articles. There is no single source containing the breadth of coverage of all principles listed in this work. The references listed in footnotes are some of the many resources that I consulted. I hope that this will be an ongoing project, in order that new principles may be added in future editions or enhancements can be made in the applications listed. Many of the entries in this dictionary are excerpts from journal articles, summaries from other literature sources, or information obtained from unique Internet sources or available content definitions from patent files.

The *Dictionary of Scientific Principles* was prepared to provide information about basic fundamental properties, systems, activities, or phenomena that have become terms in common use, including eponyms, among various fields of study. It provides a brief description of the individual principle, a variety of definitions applied to the principle, and alternate names used to describe the principle in "see also" attachments to the name, with definitions of over 2000 terms, both current and historical. About 85% of these terms cannot be found in any other source such as dictionaries, encyclopedias, or other collected printed (hardcopy) or electronic works. The footnoted references are included to help the reader find further in-depth information as needed. The *Dictionary of Scientific Principles* neither attempts nor intends to exhaust the entire spectrum of meaning and potential intention with historical connections for each principle.

The principles included may be factual, historical, fictitious, or comical. Abbreviations are included [e.g., TNSTAAFL principle]. Some surname-based eponyms containing the term *law*, (e.g., *Newton's law*), are also described as *Newton's principle* and thus are included.

Principles have been included regardless of their frequency of use or the manner in which they were created. The polyuronid principle, for example, was found in only one single reference. Occasionally, names are in formative or transitional stages of development, which legitimately justifies the compiler's reasons for assigning different names to the same or very similar principles. The inclusion of a name as part of a term in no way depends on how well the person is or was known at the time, nor does it mean that this person will become well known in the future because of the principle with which she or he may be affiliated or associated. Many of the principles include names of famous persons, while a very large number include the names of people who were modest practitioners of their trades and who lived and died in anonymity. Such people could not be included in professional and membership directories, biographical listings, or national newspaper obituaries. Biographical information, as explained earlier, for many of the principles, is incomplete. Selection was made to include and focus on the principle, not the individual for whom it was named. Literary, historical, and mythological names are included. Many of the biographical resources on these names can be found in commonly available biographical sources.

A surname-based eponym contains both a proper noun (the name of the person after whom something is named) and a generic term. The eponym need not contain the person's real name (e.g., the Dilbert principle); a pseudonym can become an eponym, such as the *Tinkerbell principle*. Names may appear in multiple forms and they are included with cross-reference's to alternate forms including spelling variations. Associating names with specific individuals is often difficult since the names are coined not by the persons who first described the concept but by someone else, often many years later.

The entries in this *Dictionary of Scientific Principles* are arranged in alphabetic order with cross-references to alternate terms applied. The listing depends on the manner in which the principle was described. For a hypothetical example, the term *principle of XYZ* and its variation, *XYZ principle*, are both listed. Only usage dictates whether the name includes a possessive "s" (e.g., Einstein's theory of relativity). Principles containing more than one personal (e.g., *Borwein–Price principle*) name are followed by brief biographic notes regarding the people in the order to in which their names appear in the term.

ACKNOWLEDGMENTS

I want to thank my wife for being so supportive in encouraging me to continue. I want to thank my children for their interest and of apparent understanding my need to sacrifice some of our time together while this work was being completed. I wish to also thank Samantha Richardson for many patient days of editing and correcting spelling errors. Tracie Meloy helped with with organizing many text records into a single standard format. I must thank many members of the Philadelphia Chapter of the Special Library Association for their encouragement. I have to thank Barbara and Bruce, who thought I was crazy but persistent. I must thank the various companies who provided additional support, including the Dialog Corporation, NewsNet, MNIS and Telebase. I would like to express special appreciation for support and encouragement from friends at Penn State University Great Valley campus and West Chester University.

A very special note of thanks to those who contributed their subject expertise as collaborators to this work. With deep thanks for her dedication and effort despite her terminal illness, Jennifer Papin-Ramcharan, Librarian III, Engineering & Physical Sciences Division, The University of the West Indies, St. Augustine Campus, St. Augustine. Trinidad and Tobago, West Indies provided a very comprehensive review of mathematics and developed mathematical formulas to be included. She passed away September 9, 2009 and was a delightful tenacious supporter. She leaves to mourn, apart from her library family, her husband Oliver and four children, her mother and two sisters. A qualified engineer, University of Hong Kong and holder of B.Sc Math/Physics from the University of the West Indies, and Fulbright—LASPAU scholar, she served as the subject specialist for the Engineering and Physical Sciences Division. She received her M.L.S. from the University at Buffalo, State University of New York and continued to serve the UWI and the Library with distinction. Her memorial service was held at the St. Stephen's Anglican Church, High Street Princes with burial at the St. Nicholas Churchyard Cemetery.

I envy Gregory D. Mahlon, Science and Technology librarian, Penn State Mont Alto, Mont Alto, PA 17237-9799 and his steady, consistent, and well organized deliberations, comments, and humor regarding this project. There were many others who sent additions to be included and provided editing or content advise. Finally, I would like to personally thank the contributions from Eleanor Brown, Ph.D., Clinical Psychology, Assistant Professor, West Chester University of Pennsylvania and her student research assistant, Andrea Knorr. Ellie collaborated with colleagues and contributed several new entries from the field of psychology and medical related practice.

Very special thanks to E. E. Barnes for offering definitions and suggestions to the list.

NOTES TO THE READER

The Dictionary of Scientific Principles is an exercise in acquiring all known rules or laws commonly called *principles* and describing the language of art corresponding to usage. These principles cover all subjects ranging from science, to business, literature, philosophy, medicine, and society. Cross-references to other principles are listed with the definition. In addition to principle definitions, [denoted (D)], you will find applications [denotd (A)], which cover an equally broad field of multiple subject disciplines aiding in a search for principles as they relate to a certain subject.

PRINCIPLES–DEFINITIONS

A

AARON ANTONOVSKY'S COMFORT-THROUGH-DISCOMFORT PRINCIPLE [psychology] (AAron Antonovsky, 1923–1994, Israeli American Sociologist) Comfort, or well-being, arises through a process of making meaning out of discomfort, or distress, thereby arriving at a sense of coherence. When one's sense of coherence is strong, the stimuli that impinge on one are perceived as comprehensible, as being manageable, and as being meaningful, or challenges worth engaging in.* See also PRINCIPLE OF SOMATOMENTAL BALANCE; PRINCIPLE OF THE SALUTOGENETIC TRIAD; SALUTOGENESIS PRINCIPLE.

ABEL'S PRINCIPLE [mathematics] (Neils Henrik Abel, 1802–1829, Norwegian mathematician) (1) Also known as *Abel's theorems*, stating: (1) if $\sum_{n=0}^{\infty} a_n x^n$ converges for $|x| \prec R$ and for $x = R$, then the series converges uniformly on $0 \leq x \leq R$. (2) Abel's theorem of algebraic equations: For $n \geq 5$, the general equation of nth order cannot be solved by radicals.[†] (3) Abel's theorem for power series: If a power series in z converges for $z = a$, it converges absolutely for $|z| < |a|$. If the power series, $S(z) = \sum_{k=0}^{\infty} a_k (z - b)^k$, where a_k, b, z are complex numbers, converges at $z = z_0$, then it converges absolutely and uniformly within any disk $|z - b| \leq \rho$ of radius $\rho \prec |z_0 - b|$ and with center at b. It follows from the theorem that there exists a number $R \in [0, \infty]$ such that if $|z - b| \prec R$, the series is convergent, while if $|z - b| \succ R$, the series is divergent. The number R is called the *radius of convergence* of the series

$S(z)$, while the disk $|z - b| \prec R$ is known as the *disk of convergence* of the series. (4) If the three series with nth term a_n, b_n, and $c_n = a_0 b_n + a_1 b_{n-1} + \cdots + a_n b_0$, respectively, converge, then the third series equals the product of the first two series. (5) If a power series in z converges to $f(z)$ for $|z| \prec 1$ and to a for $z = 1$, then the limit of $f(z)$ as z approaches 1 equals a. Abel's continuity theorem: If the power series converges at point z_0 on the boundary of the disk of convergence, then it is a continuous function in any closed triangle T with vertices z_0, z_1, z_2, where z_1 and z_2 are located inside the disk of convergence. In particular, $\lim_{z \to z_0} S(z) = S(z_0), z \in T$. This limit always exists along the radius: The series converges uniformly along any radius of the disk of convergence joining points b and z_0. (6) Abel's theorem on Dirichlet series: If the Dirichlet series $\phi(s) = \sum_{n=1}^{\infty} a_n e^{-\lambda_n s}, s = \sigma + it, \lambda_n \succ 0$ converges at point $s_0 = \sigma_0 + it_0$, then it converges in the half-plane $\sigma \succ \sigma_0$ and converges uniformly inside any angle $|\arg(s - s_0)| \leq \theta \prec (\pi/2)$. It is a generalization of Abel's theorem on power series (take $\lambda_n = n$ and put $e^{-s} = z$).[‡]

ABSORPTION PRINCIPLE [physics, energy] (1) Light decreases exponentially with distance; fractional loss is the same for equal distances of penetration. Energy loss from the light appears as energy added to the medium.[§] (2) Penetration of a substance into the body of another.[¶] (3) Eigenfunction expansions for the self-adjoin operator governing the propagation of elastic waves in an unperturbed

*Suedfeld, P. (2005), Invulnerability, coping, salutogenesis, integration: Four phases of space psychology, *Av. Space Environ. Med.* **76**(6 suppl.), B61–B66.

[†]Bhatnagar, G. (1995), *Inverse Relations, Generalized Bibasic Series, and Their U(n) Extensions*, PhD thesis, Ohio State Univ., p. 105.

[‡]Chu, W. C. and Hsu, L. C. (1989), Some new applications of Gould-Hsu inversions, *J. Combin. Inform. Syst. Sci.* **14**, 1–4.

[§]Tsypkin, Ya. Z. (1997), Mowing approximation and absorption principle, *Dokl. Akad. Nauk* **357**(6), 750–751.

[¶]Schmitt, K. (1980), Gas heat pumps using the absorption principle, *VDI-Berichte* **353**, 79–86.

stratified media radiation.* See also LIMITING ABSORPTION PRINCIPLE; LIMITING AMPLITUDE PRINCIPLE.

ACCOUNTING PRINCIPLE [mathematics]   A collection of rules and procedures and conventions that define accepted accounting practice; includes broad guidelines as well as detailed procedures. Governs current accounting practice and that is used as a reference to determine the appropriate treatment of complex transactions.[†] See also GENERALLY ACCEPTED ACCOUNTING PRINCIPLE (GAAP).

ACETYLENE CYCLOADDITION AROMATIC HETEROCYCLE INERTIA PRINCIPLE [chemistry] See INERTIA PRINCIPLE.

ACKERMAN PRINCIPLE [engineering] (Rudolp Ackermann, 1764–1834, anglo-german inventor)   For any given corner, the outside wheel should have less turn angle than the inside one, because it is following a larger radius than the inside wheel. In order to minimize lateral skid while turning, the extensions of the center lines of the wheel axles must intersect at the center of the arc on which the vehicle turns.[‡] (2) When a vehicle is steered, it follows a path that is part of the circumference of its turning circle, which will have a centerpoint somewhere along a line extending from the axis of the fixed axle. The steered wheels must be angled so that they are both at a 90° angle to a line drawn from the circle center through the center of the wheel. Since the wheel on the outside of the turn will trace a larger circle than will the wheel on the inside, the wheels need to be set at different angles.[§]

*Zhu, L., Dong, X., Zhang, J., and Zhu, C. (2004), The principle of gas absorption and slag removal about liquid metal having through sprue and its application, *Zhuzao*, **53**(11), 901–904.
[†]Anon. (2000), *American Heritage Dictionary of the English Language*, 4th ed., Houghton Mifflin, New York.
[‡]Anon. (n.d.), *Total Vehicle Alignment*, section 6, p. 47; see http://www.goodyear.com/truck/pdf/radialretserv/Retread_S6_V.pdf.
[§]Anon. (2009), *Knowledge-Ackermann's Principle*, Northern Districts Rodders Club, Inc. (http://ndrc.org.au), vol. 22, p. 42.

ACKNOWLEDGMENT CHAINING PRINCIPLE [engineering]   The principle of acknowledgment chaining works by processes sending messages to the group of processes. Allows each message to be directly acknowledged only a few times, and through chains of acknowledgements, to be indirectly acknowledged by other processes. This leads to an efficient utilization of resources.[¶]

ACROPHONIC PRINCIPLE [linguistics]   (from Greek $\alpha K\rho o$-"tip" $+\phi\omega\nu\acute{\iota}a$, "voice," "the initial sound"; Dr. Richard Venezky, linguistics, USA)   Illuminating the nature of English writing by relating current spellings to the sounds, morphemic structure, and history of our language. As in telephone directory spelling, "A for apple," "B for . . . ," and so on.[∥]

ACTION PRINCIPLE [mathematics, physics] (1) An action principle is a method for reformulating differential equations of motion for a physical system as an equivalent integral equation. Although several variants have been defined, the most commonly used action principle is Hamilton's principle. An earlier, less informative action principle is Maupertuis' principle, which is sometimes called by its (less correct) historical name, the *principle of least action*. (Newton's second law is sometimes called an *action principle*.) Any force $\vec{F}$ acting on a body of mass m induces an acceleration $\vec{a}$ of that body, which is proportional to the force and in the same direction $\vec{F} = m\vec{a}$.[**] (2) Originally used to derive the equation of motion for a particle in classical mechanics, the *action* was defined by Hamilton to be the difference $L = T - U$, where T is the kinetic energy and U the potential energy of a mechanical system. Hamilton's principle (or the action principle) states that the

[¶]Malki, D. and Reiter, M. (1996), A high-throughput secure reliable multicast protocol, *Proc., 9th IEEE Computer Security Foundations Workshop*, AT&T Bell Labs., Murray Hill, NJ, June 10–12, 1996, pp. 13,14.
[∥]Venezky, R. L. (1975) *The American Way of Spelling: The Structure and Origins of American English Orthography*, Guilford Press, p. 1.
[**]Palacios, A. F. (2006), *The Hamilton-Type Principle in Fluid Dynamics: Fundamentals and Applications in Magnetohydrodynamics, Thermodynamics, and Astrophysics*, Springer, Vienna, p. 47.

motion of a mechanical system is such that the *action* integral $S[X(t)] = \int_{t=t_1}^{t=t_2} L(X,\dot{X})dt$ is *stationary* with respect to variations in path $X(t)$. The Euler–Lagrange equations for this system are known as Lagrange's equations $(d/dt)(\partial L/\partial \dot{X}) = (\partial L/\partial X)$, which are equivalent to Newton's equations of motion. In other words, a classical dynamical system evolves from t_1 to t_2 along a path $X(t)$ for which its action $S(X(t))$ is an extremum. The extremum is often a minimum. In such a case, nature is efficient; it spends the least amount of action. In these cases we speak of "the principle of least action." Most commonly, the term is used for a functional S that takes a function of time and (for fields) space, as input, and returns a scalar. Specifically, in classical mechanics, the input function is the evolution $q(t)$ of the system between two timepoints t_1 and t_2, where q represents the generalized coordinates. The *action* $S[q(t)]$ is defined as the integral of the Lagrangian L (in classical mechanics, the Lagrangian is defined as the kinetic energy T of the system minus its potential energy V) for an input evolution between the two timepoints $S[q(t)] = \int_{t_1}^{t_2} L[q(t), \dot{q}(t), t]dt$, where the endpoints of the evolution are fixed and defined as $q_1 = q(t_1)$ and $q_2 = q(t_2)$. According to Hamilton's principle, the true evolution $q_{\text{true}}(t)$ is an evolution for which the action $S[q(t)]$ is stationary (a minimum, a maximum, or a saddle point). This principle results in the equations of motion in what is known as Lagrangian mechanics.[*] See also HAMILTON PRINCIPLE; LEAST-ACTION PRINCIPLE; MAUPERTUIS' PRINCIPLE; PRINCIPLE OF LEAST ACTION.

ACTIVE   PRINCIPLE [pharmaceuticals]   (1) An ingredient giving a complex drug its chief therapeutic value. The portion of a pharmaceutical preparation producing the therapeutic action. By analogy, any substance providing the more significant value and thus producing the primary action. The substance in a preparation exerting an effect; as distinct from the substances, is also included.[†] (2) A constituent of a drug,

usually an alkaloid or glycoside, on which the characteristic therapeutic action of the substance largely depends.[‡]

ACYCLIC   DEPENDENCES   PRINCIPLE   (ADP) [computers] (Robert Cecil Martin) A package is a binary deliverable like a .jar file, or a dll as opposed to a namespace like a java package or a C++ namespace. Package metrics evaluate the structure of a system. The dependence graph of packages must have no cycles.[§]

ADAPTION-LEVEL PRINCIPLE [social psychology]   Our expectations of success and failure, satisfaction and dissatisfaction, even justice and injustice, are relative to our prior experience and to what we observe people like ourselves receiving. If our achievements rise above those expectations, we experience success and satisfaction; if they fall below, we feel dissatisfied and frustrated.[¶]

ADAPTIVE DIFFERENTIAL PULSE-CODE MODULATION (ADPCM) PRINCIPLE [computer science] (Developed by Alec H. Reeves, 1926) See PRINCIPLE OF PULSE-CODE MODULATION.

ADAPTIVE   PRINCIPLE [mathematics]   (1) Means for overall design of a precision measurement system by the correlation of the disturbance effect.[‖] (2) Analyzing a system and determining the compensation factors as

[*]Chen, Q., Zhang, J., and Zhu, L. (2005), Principles and performances of new-type air-floatation equipment, *Kuangye Gongcheng* **25**(1), 20–22.
[†]Anon. (2008), *Active Principle. A Dictionary of Nursing*, Oxford Univ. Press; see

Encyclopedia.com: (http://www.encyclopedia.com/doc/1O62-activeprinciple.html).
[‡]Anon. (2004), *Active Principle. The American Heritage Medical Dictionary*, Houghton Mifflin.
[§]Martin, R. C. (n.d.), *Dependency Inversion Group. Ten Commandments of OO Proramming* (http://www.butunclebob.com/ArticleS.UncleBob.PrinciplesOfOod).
[¶]Myers, D. and Ludwig, T. (1978), Let's cut the poortalk: Adaption level principle, *Saturday Rev.* **5**, 24.
[‖]Yina, C. and Guob, J. (2005), Novel comprehension of "common path" principle, *Optics Lasers Eng.* **43**(10), 1081–1095.

well as giving the possibility of correction.[*] Also referred to as *common-path principle*.

ADDITION PRINCIPLE [mathematics] (1) If two actions are mutually exclusive, and the first can be done in N_1 ways and the second in N_2 ways, then one action or the other can be done in $N_1 + N_2$ ways.[†] (2) If set A is a union of two mutually exclusive sets B and C, then $n(A) = n(B) + n(C)$, where $n(A)$ is the number of elements (objects) in set A, B and C are said to be mutually exclusive if B and C have no elements in common, i.e., $B \cap C = \emptyset$. In other words, if a set can be partitioned into disjoint subsets, then the number of objects in the set = the sum of the number of objects in each of its parts. The addition principle can be generalized to counting a set that is a union of several mutually exclusive sets.[‡] (3) Adding the same number to both sides of an equation does not change its solution set. (4) In probability, if $E_1, E_2, \ldots, E_s$ are all mutually exclusive events, then the probability some E_i occurs is the sum of the probabilities the individual E_i occur, i.e., $P(E_1 \cup E_2 \cup \cdots \cup E_s) = P(E_1) + P(E_2) + \cdots + P(E_s)$.[§] See also MULTIPLICATION PRINCIPLE; FUNDAMENTAL PRINCIPLE OF COUNTING.

ADDITIVITY PRINCIPLE [mathematics] (1) The value of a magnitude or property corresponding to a whole is equal to the sum of the values of the magnitudes or properties corresponding to its parts for any division of the whole into its parts.[¶] (2) The judged probabilities for complementary events should sum to unity.[‖] See also ADDITION PRINCIPLE.

ADIABATIC PRINCIPLE [physics, computer science] Correlates the capacity of a system to interact with the frequency of attempted interaction by another.[**] Also referred to as *ehrenfest adiabatic principle*.

ADJACENCY PRINCIPLE [geometry] Size cues between adjacent objects is more effective than the size cue between displaced objects in determining the perceived relative depth portion of objects.[††]

ADLER PINCIPLE [psychology] (Alfred Adler, 1870–1937, Austrian physicist) Theory placing emphasis on the individual's need to belong and to contribute, based on social equality and mutual respect. These sociopsychological concepts are integrated in Adlerian writings with concepts that bear on an individual's dynamics, such as the individual's goals and private logic.[‡‡]

ADLERIAN PRINCIPLES [psychology] (Alfred Adler, 1870–1937, Austrian physician) Emphasis on making the children aware of their goals in speech deficiency and misbehavior; leaving to them the decision to improve; encouraging their learning through mutual help within the group; and providing logical consequences for their behavior.[§§] (2) Applies four principles of conflict resolution: practicing mutual respect, pinpointing the real issue, changing the conflict agreement, and involving all concerned in decisionmaking.[¶¶]

[*]Zhang, Z. and Deshu, C. (1991), An adaptive approach in digital distance protection, *IEEE Trans. Power Deliv.* **6**(1), 135–142.

[†]Schay, G. (2007), *Introduction to Probability with Statistical Applications*, Springer, p. 16.

[‡]Breakbill, L. R. (2005), *Addition and Multiplication Principles*, Guided Lecture Notes (https://www.math.gatech.edu/academic/courses/core/math1711/html/1711-3.3.html).

[§]Brennan, J. W. (n.d.), *Understanding Algebra* (http://www.jamesbrennan.org).

[¶]Persoons, T., Hoefnagels, A., and Van den Bulck, E. (2006), Experimental validation of the addition principle for pulsating flow in close-coupled catalyst manifolds, *J. Fluids Eng.* **128**(4), 656–670.

[‖]Borho, B. W. (1982), On the Joseph-Small additivity principle for Goldie Ranks, *Compos. Math.* **47**(1), 3–29.

[**]Born, M. and Fock, V. (1928), Proof of the adiabatic principle, *Z. Phys.* **51**, 165–180.

[††]Gogel, W. C. (1978), The adjacency principle in visual perception, *Sci. Am.* **238**, 126.

[‡‡]Ferguson, E. D. (1996), Adlerian principles and methods apply to workplace problems, *Indiv. Psychol. J. Adler. Theory Res. Pract.* **52**(3), 270–287.

[§§]Vandette, J. (1964), Application of Adlerian principles to speech therapy, *J. Indiv. Psychol.* **20**(2), 213–218.

[¶¶]Dewey, E. A. (1985), Adlerian principles in conflict resolution, *Indiv. Psychol. J. Adler. Theory Res. Pract.* **4**(2), 237–242.

ADVANTAGE OF RARITY PRINCIPLE See CHAOTIC PRINCIPLE.

AFFINITY PRINCIPLE [chemistry]   (Julius Thomson and Marcellin Berthelot) Every simple or complex action of a purely chemical nature is accompanied by an evolution of heat. Energy difference between ionization of elements, combination of power, attraction, or magnetisim impacts the lowest state of the corresponding reaction.* See also THOMSON–BERHELOT PRINCIPLE.

AGGREGATION   PRINCIPLE [biology] A grouping or clustering of individuals within a population.† (2) The sum of a set of multiple measurements is a more stable and representative estimator than any single measurement.‡

AIR DISTRIBUTION PRINCIPLE [engineering] Arrangement by which gases may be patterned, ordered, confined, or dispersed in order to control the allotment, mixture, partitioning, circulation, or saturation for possible reaction.§

AIR FLOAT PRINCIPLE [engineering]   Drying by means of directing streams of a drying gas while at the same time floating or flowing air and carrying the surface according to the Coandă effect.¶

AIRFLOW PRINCIPLE [engineering] Measurement of the resistance to airflow through a

plug of fibers. This resistance can be directly related to fiber fineness.‖

ALARA PRINCIPLE [mathematics] Also referred to as *ALARP* or *as low as reasonably achievable*.

ALARP PRINCIPLE [mathematics] See AS LOW AS REASONABLY PRACTICAL PRINCIPLE. Also referred to as *ALARA* (as low as reasonably achievable).

ALEXANDER PRINCIPLE [psychology] (Frederick Matthias Alexander, 1869–1955, Australian Actor) Biofeedback and relaxation with methods employed in sport, physical education, and acting. The work of the athlete and the performances of the actor and the dancer are magnifications of skills demonstrated with all people at some level. Although the next-door neighbor may not dance, the upstairs resident may not act, the person across the road may never get out of the armchair, it is suggested that all ordinary people may benefit from the biofeedback principle and the Alexander principle as part of an education toward their full consummation of being.**

ALLEE'S PRINCIPLE [ecology] (Warder Clyde Allee, 1885–1955, American Zoologist) The tendency of individuals in some populations to flourish best at intermediate optimal population density.††

ALPHA PRINCIPLE [management]   (James Morgan, Allied Materials CEO) Once a company gains critical mass, it can accelerate like an alpha particle emitted from a radioactive atom and the company can experience exponential growth.‡‡

*Cropper, W. H. (2004), *Great Physicists: The Life and Times of Leading Physicists from Galileo to Hawking*, Oxford Univ. Press, p.128.
†DiMaggio, P. (1987), Classification in art, *Am. Sociol. Rev.* **52**(4), 440–455.
‡Rushton, J. P., Brainerd, C. J., and Pressley, M. (1983), Behavioral development and construct validity: The principle of aggregation, *Psychol. Bull.* **94**(1), 18–38.
§Ramakrishnan, V. and Sai, P. S. T. (1999), Optimizatoin of air distribution in direct reductin process in rotary kilns, *Can. J. Chem. Eng.* **77**(4), 775–783.
¶Chen, Q., Zhang, J., and Zhu, L. (2005), Principles and performances of new type air flotation equipment, *Kungye Gongcheng* **25**(1), 20–22.

‖Schorn, P., Voigt, H., and Noe, W. (2000), Sampling system for use in the analysis of biological processes. p.6. US Patent 6,085,602.
Hazelton, J. (1985), The purpose and relevance of nonclinical biofeedback relaxation and the Alexander principle, *Clin. Biofeed. Health Int. J.* **8(1), 52–67.
††Kennett, D. J. and Winterhalder, B. (2006), *Behavioral Ecology and the Transition to Agriculture*, Univ. California Press, p. 206.
‡‡Morris, K. (1995), The alpha principle: No guts, no glory. That's how Allied Materials plays the game. Take notes, *Finan. World.* **164**(24), 30.

ALPHABETIC PRINCIPLE [language] Two components of phonemic awareness involves recognition of phoneme identity across words and recognition of phonemic segmentation within words. Word identity can be equally easily taught using word-initial and word-final phonemes.*

ALTMAN PRINCIPLE [biology] (Sidney Altman, 1939-, Yale Univ., Nobel prize, Canadian molecular biologist) (1) Molecular biology studies made in the area of RNA processing. Ribonucleoprotein is a key enzyme in the biosynthesis of tRNA. RNase precursor is involved in processing all species of tRNA and is present in all cells and organelles that carry out tRNA synthesis.† (2) Used to prove a vector-valued equilibrium variant.‡

AMBARTSUMYAN'S INVARIANCE PRINCIPLE [physics, mathematics, chemistry] (Viktor Hambardzumyan, 1908–1996, [Viktor Ambartsumian-russified] Armenian physicist) (1) Density limit theorems for delays from limit theorems for queue lengths when studying queue-dependent arrival and or molecular orbits service completion rates.§ (2) Applied in demonstrating the existence of adaptive stabilizers and servomechanisms for a variety of nonlinear system classes.¶ (3) Prevention of nonlinear suppression of

signal in receiving and amplifying circuits.‖ See INVARIANCE PRINCIPLE.

AMINE CAPTURE PRINCIPLE [chemistry] Amino acid esters react with 4-methoxy-3-acyloxy-2-hydroxybenzaldehydes to form imines, which on reduction undergo intramolecular acyl transfer to form N-4-methoxy-2,3-dihydroxybenzyalmides, useful in peptide synthesis. The feasibility of peptide bond formation through a new principle of intramolecular acylation which is preceded by amine capture. Imine formation from salicylaldehydes occurs with unusually large rate and equilibrium constants.**

ANALYTICAL PRINCIPLE [chemistry] Elements used to make a determination toward a result such as the use of calorimetric analyzer limiting interference from chloride, temperature, or hydrogen ions.††

ANATOMOCLINICAL PRINCIPLES [medicine] Correlated with lesions of specific anatomic foci. Based on identified relations between structural brain lesions and behavioral disturbances.‡‡

ANCILLARITY PRINCIPLE [mathematics] A statistical experiment or a model M is defined as a triplet (χ, Ω, P), where $\chi = \{x\}$ is an abstract sample space, $\Omega = \{\theta\}$ is an abstract parameter space, and $P = \{P_\theta : \theta \in \Omega\}$ is a class of distributions on χ indexed by the parameter θ. It is assumed that χ and Ω are finite. The inference one can make on the basis of an observation x (in χ) given the experiment M can be denoted by $\inf F(\cdot | x, M)$.

*Byrne, B. and Fielding-Barnsley, R. (1990) Acquiring the alphabetic principle: A case for teaching recognition of phoneme identity. Journal of Educational Psychology. **82**(4), 805–812.

†Pedersona, T. and Politza, J. C. (2000), The nucleolus and the four ribonucleoproteins of translation, *J. Cell Biol.* **148**(6), 1091–1096.

‡Altmana, E., Boulognea, T., El-Azouzia, R., Jiménezb, T., and Wynterc, L. (2006), A survey on networking games in telecommunications, *Comput. Oper. Res.* **33**(2), 286–311.

§Remizovich, V. S. and Tishin, I. V. (1998), Use of the Ambartsumyan invariance principle in the problem of reflection of light ions, *Izvest. Akad. Nauk Ser. Fiz.* **62**(4), 770–777.

¶Ryanc, E. P. (1998), An integral invariance principle for differential inclusions with applications in adaptive control, *SIAM J. Control Optim.* (ftp://ftp.maths.bath.ac.uk/pub/preprints/maths9806.ps.Z).

‖Zakhar-Itkin, M. Kh. (1982), Application of the Ambartsumyan invariance principle to study of lengthy light guides with random inhomogeneities, *Radiophys. Quantum Electron.* **25**(11).

Kemp, D. S., Grattan, J. A., and Reczek, J. (1975), Letter: Peptide bond formation by the amine capture principle, *J. Org. Chem.* **40(23), 3465–3466.

††Ohls, K. (1980), ICP emission spectrometry—a complete analytical principle. 13th Spektrometertagung, 239–256.

‡‡Finitzo, T., Pool, K. D., and Chapman, S. B. (1991), Quantitative electroencephalography and anatomoclinical principles of aphasia. A validation study, *Ann. NY Acad. Sci.* **620**, 57–72.

The ancillarity principle states, that if P_θ is the same for all $\theta \in \Omega$, then $\inf F(\cdot|x, M)$ is the same for all x in χ. In other words, no inference about θ is possible on the basis of an observation x, under the experiment M.[*]

ANEMOMETER PRINCIPLE [engineering] Measure of low speeds in studies of air circulation. Rate of cooling being determined by the speed of the airflow.[†]

ANISOTROPIC PRINCIPLE [physics] Showing different properties as to velocity light transmission, conductivity of heat or elasticity, compressibility, and so on in different directions.[‡]

ANNA KARENINA PRINCIPLE [psychology] (Anna Karenina, Fictitious character, Leo Tolstoy) Derived from Leo Tolstoy's book *Anna Karenina*, which begins: "Happy families are all alike; every unhappy family is unhappy in its own way." An endeavor in which a deficiency in any one of a number of factors dooms it to failure. Consequently, a successful endeavor is one in which every last one of the possible deficiencies has been avoided.[§]

ANOKHIN–PARETO INTERACTION PRINCIPLE [mathematics] (Pyotr Kuzmich Anokhin, 1898–1974, Russian biologist) (Vilfredo Pareto, 1848–1923, Italian Sociologist) The control energy that is lost because the control actions.[¶]

[*]Chapman, S. B., Pool, K. D., Finitizo, T. and Hong, C.-T. (1989), Comparison of language profiles and electrocortical dysfunction in aphasia, *Proc. 18th Annual Conf. Clinical Aphasiology*, College Hill Press, p. 42.
[†]Laughlin, D. E. and Mahoney, R. P. (1972), New phonocardiographic transducers utilizing the hot-wire anemometer principle, *Med. Biol. Eng.* **10**(1), 43–55.
[‡]Fock, K. (2001), Static and dynamic modeling of magnetoelastic dynamometers, *Measurement* **30**(1), 75–84.
[§]Schneiderman, L., J., Gilmer, T., Teetzel, H. D., Dugan, D. O., Goodman-Crews, P., and Cohn, F. (2006), Dissatisfaction with the ethics consultations: The Anna Karenina principle, *Cambridge Quart. Healthcare Ethics CQ Int. J. Healthcare Ethics Committees* **15**(1), 101–106.
[¶]Kiforencko, B. N. and Kiforencko, S. I. (1993), The Anokhin-Pareto interaction principle in contol system theory, *J. Comput. Syst. Sci.* **31**(5), 165–170.

ANTENNA PRINCIPLE [physics] Focused reflected electromagnetic energy used to identify size, speed, and location of an object.[‖]

ANTHROPIC COSMOLOGICAL PRINCIPLE [physics] (John Barrow and Frank Tipler) We ought not be surprised at measuring a universe so finely tuned for life, for if it were different, we would not observe it.[**] See COSMOLOGICAL PRINCIPLE.

ANTHROPIC PRINCIPLE [astronomy, genetics] (Brandon Carter, b. 1942; Australian theoretical physicist, British mathematician) (1) The nature of the universe is constrained because of our presence as observers.[††] (2) Life, even if abundant on many worlds, is only an infinitesimal portion of the cosmos. The presence of intelligent life on Earth places limits on the many ways the universe could have developed and could have caused the prevailing conditions.[‡‡] See also BLACK HOLE PRINCIPLE; COPERNICAN PRINCIPLE; PRINCIPLE OF BLACK HOLE COMPLEMENTARITY; STELLAR PRINCIPLE; ANTHROPIC COSMOLOGICAL PRINCIPLE; COSMOLOGICAL PRINCIPLE; FINAL ANTHROPIC PRINCIPLE; STRONG ANTHROPIC PRINCIPLE; WEAK ANTHROPIC PRINCIPLE.

ANTHROPOMURPHIC PRINCIPLE [mathematics] Any universe built along conventional lines that contains intelligent polymorphs will conform to Murphy's law.[§§]

[‖]Zuber, H. (1987), Structural principles of the antenna system of photosynthetic organisms, *Prog. Photosynth. Res., Proc. 7th Int. Congr. Photosynthesis*, 1986, pp. 1–8.
[**]Williams, P. S. (2006), *The Big Bad Wolf, Theism and the Foundations of Intelligent Design. A Review of Richard Dawkins, the God Delusion*, Bantam.
[††]Walker, P. M. B. (1999), *Chambers Dictionary of Science and Technology*, Chambers, New York, p. 50.
[‡‡]Guillen, M. M. (1984), The center of attention (anthropic principle, scientific explanation based on human existence), *Psychol. Today* **1**, 74.
[§§]Stewart, L. (1995), The anthropomurphic principle (mathematical explanations for the dynamics of falling), *Sci. Am.* **273**, 106.

ANTIAGING PRINCIPLE [nutrition]   A 60% reduction in the typical normal daily caloric allotment will have an antiaging effect.[*]

ANTIARRHYTHMIC   PRINCIPLE [chemistry] (1) Channel modification is the response to the interaction with organic or inorganic molecules and causes repetitive activity by removal of inactivation.[†] (2) Sodium channel blocks controlling cardiac arrhythmias by the selective or isolated prolongation of repolarisation.[‡] (3) Pharmacologically induced removal of inactivator is kinetically indistinguishable from spontaneous failure on inactivation.[§]

ANTIBIOTIC PRINCIPLE [medicine]   (1) Characteristics show traits of heat resistance, time resistance, and a relationship between the pH of the medium generally with greater production around 7–8 days on media containing starch or carbohydrates.[¶] (2) Function, mode, and effect on the physiological ecology of animal habitats can be investigated to determine natural occurrences of antibiotic means to adjust to certain diet or plant growth interactions.[||]

ANTICOERCION   PRINCIPLE [psychology] (Aristotle) Everything (within the sphere of social conduct) forced is unjust (e.g., aggression is defined as the initiation of physical force, the threat of such, or fraud committed on persons or their property). Aristotle subscribed to two principles relating justice and nature: a positive principle linking the just and the natural.[**] Also termed *zero-aggression principle*; see NONAGGRESSION PRINCIPLE.

ANTIDEPRESSANT   PRINCIPLE [medicine] Increasing control over depression sympathetic activity by increasing use of exercise as an effective antidepressant or decreasing stress relieving the susceptibility toward depression. Exercise and stress have many opposing effects in the brain and are consistent with this hypothesis.[††]

ANTIDIURETIC   PRINCIPLE [psychology]   A peptide hormone limiting the amount of water excreted by the kidneys. Deficiencies of this hormone result in central diabetes insipidus. Excesses cause water retention and hyponatremia.[‡‡]

ANTIHISTAMINIC   PRINCIPLE [pharmacy] Tending to neutralize or antagonize the action of histamine or inhibit its production in the body.[§§]

ANTIINFLAMMATION   PRINCIPLE [medicine] Medical procedure or pharmaceuticals designed to prevent or inhibit coagulation, activation, thrombosis, tissue factor complex, heart disease, or biosynthesis.[¶¶] See also ANTIPHLOGISTIC PRINCIPLE; ANTITHROMBOTIC PRINCIPLE.

[*]Cousens, G. (2003), *Rainbow Green Live-Food Cuisine Tree of Life Café*, North Atlantic Books, p. 107.
[†]Kohlhardt, M. (1989), Elementary properties and the interaction of single myocardial Na+ channels with antiarrhythmic drugs, *Arzneimittelforschung* 39(1A), 126–129.
[‡]Advani, S. V. and Singh, B. N. (1995), Pharmacodynamic, pharmacokinetic and antiarrhythmic properties of d-sotalol, the dextro-isomer of sotalol, *Drugs* 49(5), 664–679.
[§]Pu, H.-L., Huang, X., Zhao, J.-H., and Hong, A. (2002), Bergenin is the antiarrhythmic principle of Fluggea virosa, *Planta Med.* 68(4), 372–374.
[¶]Brotzu, G. (1948), *Research on New Antibiotic*, Cagliari Institute of Hygiene (http://pacs.unica.it/brotzu/brotzuen.pdf).
[||]Sieburth, J. M. (1960), Acrylic acid, an antibiotic principle in phaeocystis blooms in antarctic waters, *Science* 132, 676.

[**]Kraut, R. (2005), *Aristotle's Politics: Critical Essays*, Rowman & Littlefield Publishers.
[††]Subarnas, A., Oshima, Y., and Ohizumi, Y. (1992), An antidepressant principle of Lobelia inflata L. (Campanulaceae), *J. Pharm. Sci.* 81(7), 620–621.
[‡‡]Bansi, H. W. and Olsen, J. M. (1959), Antidiuretic principle and osmotic conditions in the serum in obesity, *Acta Endocrinol.* 31, 426–432.
[§§]Antihistaminic (2007), *American Heritage Medical Dictionary*, Houghton Mifflin, Boston, (http://www.credoreference.com/entry/hmmedicaldict/antihistaminic).
[¶¶]Valente, A. (1952) Modern concepts on the antiinflammation effect of salicylates in ophthalmology, *Rev. Méd. Aéronaut.* 4(1), 7–22.

ANTIMAXIMUM PRINCIPLE [mathematics] A fundamental result in the theory of partial differential equations is the following maximum principle. According to it a function u on an interval $[a,b]$ that satisfies $-u'' \geqslant 0$ on $[a,b]$ achieves its *maximum* at a or b. If, additionally, $u(a) = u(b) = 0$, then $u \geqslant 0$ on $[a,b]$, and either $u \equiv 0$ on $[a,b]$ or $u \succ 0$ on $[a,b]$ with $u'(a) \succ 0$ and $u'(b) \prec 0$. In 1979 P. Clement and L. Peletier studied the classical Dirichlet problem $-\Delta u = \lambda u + f(x)$ in Ω, $u = 0$ on $\partial\Omega$, when the real parameter λ satisfies $\lambda \succ \lambda_1$ with suitable $\lambda_1 \succ 0$. They derived, for a certain value of λ, a conclusion that is opposite to the preceding one; namely, if $f \geqslant 0$ but not identically zero, then it implies $u \prec 0$.[*]

ANTIPERNICIOUS ANEMIA PRINCIPLE [medicine] (1) Theapeutic effect of liver in controlling the disease of pernicious anemia.[†] (2) Regeneration of red cells by the liver requiring the combination of an intrinsic factor, present in normal human gastric juice, with an extrinsic factor of vitamin B_{12}. Intrinsic factor is a specific B_{12} binding protein secreted by the stomach to enhance absorption of the vitamin.[‡]

ANTIPHLOGISTIC PRINCIPLE [pharmaceuticals] Medical procedure or pharmaceuticals designed to prevent or inhibit coagulation, activation, thrombosis, tissue factor complex, heart disease, or biosynthesis.[§] See also ANTIINFLAMMATION PRINCIPLE; ANTITHROMBOTIC PRINCIPLE.

[*]Godoy, T., Gossez, J. P., and Paczka, S. (2002), On the antimaximum principle for the p-Laplacian with indefinite weight, *Nonlinear Anal.* **51**(3), 449–467.
[†]Jacobs, H. R. (1937), The nature of the antipernicious anemia principle. II. Identification of the 5,6-quinone of dihydroindole-2-carboxylic acid in liver extract, *J. Lab. Clin. Med.* **22**, 890–892.
[‡]Schlesinger, A. (1933), Demonstration of the antipernicious-anemia principle in the gastric juice of a patient with the blood picture of pernicious anemia and stenosis of the small intestines, *Klin. Wochen.* **12**, 298–300.
[§]Wagner, H. and Flachsbarth, H. (1981), A new antiphlogistic principle from Sabal serrulata, I, *Planta Med.* **41**(3), 244–251.

ANTISYMMETRY PRINCIPLE [electronics, anatomy] In regard to electrode placement there is a fundamental neurological difference between antisymmetric and symmetric excitation, in which the skin polarization is respectively antisymmetric and symmetric with respect to the sagittal plane. In antisymmetric excitation, the weak frequency-modulated signals from the modulated afferents act antisymmetrically on the brain.[¶] See INVARIANCE PRINCIPLE; LINDELÖF PRINCIPLE; SCHWARZ REFLECTION PRINCIPLE; SYMMETRY PRINCIPLE.

ANTITHROMBOTIC PRINCIPLE [pharmacology] Medical procedure or pharmaceuticals designed to prevent or inhibit coagulation, activation, thrombosis, tissue factor complex, heart disease, or biosynthesis.[‖] See also ANTIINFLAMMATION PRINCIPLE; ANTIPHLOGISTIC PRINCIPLE.

ANTIVIRAL ACTIVE PRINCIPLE [pharmaceuticals] Method for quantization of the antiviral effect comprising the following steps: (1) transducing cells with a viral vector, containing all the genetic data required for infecting a cell with a type of target virus; (2) introducing a prospective antiviral into the cells; and (3) quantitatively analyzing the activity.[**] See also ACTIVE PRINCIPLE.

APPROXIMATION INDUCTION PRINCIPLE [mathematics, physics] See INDUCTION PRINCIPLE.

ARC-EXTINGUISHING PRINCIPLE [engineering] The high-temperature fluid near the arc striking part attributed to the thermal energy emission of the arc is caused by flowback into the gas storage chamber

[¶]Palting, P. (1995), On weak and strong conjugacy in the antisymmetry principle, *Int. J. Quantum Chem.* **54**(1), 19–26.
[‖]Lindahl, A. K. (1997), Tissue factor pathway inhibitor: From unknown coagulation inhibitor to major antithrombotic principle, *Cardiovasc. Res.* **33**(2), 286–291.
[**]Heinkelein, M., Jarmy, G., Jassoy, C., Rethwilm, A., and Weissbrich, B. (2001), *Method for Quantization of the Antiviral Effect of Antiviral Active Principles*, Patent WO/2001/007646 (http://www.wipo.int/pctdb/en/wo.jsp?wo=2001007646).

and mixed with the low-temperature fluid contained therein. The fluid pressure in the gas storage chamber is raised, whereby the fluid, at a temperature low enough to extinguish the arc, is blown against the arc to effect self-extinction in the process in which current decreases toward its zero point, while the low-temperature arc extinguishing fluid from the puffer chamber is forcibly blown against the arc to effect forcible extinction in the process in which the current decreases toward its zero point.[*] See also EXTINCTION PRINCIPLE; EXTINGUISHING PRINCIPLE.

ARCHIMEDEAN   PRINCIPLE [mathematics, physics] (Archimedes, c. 287–212 BC, Greek mathematician) (1) Also known as *Archimedean axiom*, originally formulated for segments, it states that if the smaller one of two given segments is marked off a sufficient number of times, it will always produce a segment larger than the larger one of the original two segments. This axiom can be applied in a similar manner for surfaces, volumes, positive numbers, etc. In general, the Archimedean axiom applies to a given quantity if for any two values C and D of this quantity such that $C \prec D$, it is always possible to find an integer n such that $Cn \succ D$. The axiom forms the basis of the process of successive division in arithmetic and in geometry.[†] (2) If a body is wholly or partially submerged in a fluid (liquid or gas), it experiences an upward force (upthrust) equal to the weight of fluid that it displaces. A body floating in a fluid displaces a weight of fluid equal to its own weight. Also known as the *principle of buoyancy*. A body wholly or partly immersed in a fluid will experience an upward thrust (upthrust) equal to the weight of fluid it displaces; the upthrust acts

[*]Ueyama, T., Ohnawa, T., Yamazaki, K., Tanaka, M., Ushio, M., and Nakata, K. (2005), High-speed welding of steel sheets by the tandem pulsed gas metal arc welding system, *Trans. JWRI* **34**(1), 11–18.
[†]Walker, P. M. B. (1999), *Chambers Dictionary of Science and Technology*, Chambers, New York, p. 60.

vertically through the center of gravity of the displaced fluid.[‡]

ARCHIMEDES'   PRINCIPLE [mathematics, physics] (Archimedes, c. 287–212 BC, Greek mathematician) (1) Predicting a ship's buoyancy by the distribution of weight for balance of heel and trim.[§] (2) Correlation between surface and volume of a sphere and its circumscribing cylinder.[¶] (3) A body immersed wholly or partially in a fluid is buoyed up by a force equal in magnitude to the weight of the volume of fluid it displaces.[‖] (4) Object immersed in a fluid has an upward force equal to the weight of the fluid displaced by the object. The Archimedes thrust $S = rgV$ ($S =$ force, $r =$ density, $g =$ weight, $V =$ volume) is generated by the resultant of all the forces that the fluid produces on the surface of the body by means of the hydrostatic or aerostatic pressure.[**] (5) The hydrostatic pressure a liquid produces because of the gravity force, depends on both the density and the height of the liquid inside a container.[††] (6) A hot-air balloon is subjected to an ascensional force Fa, which is given by the difference between the aerostatic thrust S and the weight P: $Fa = S - P = r_c Vg - r_h Vg$, where r_c and r_h are, respectively, the density of the external cool air and the one of the hot air inside the

[‡]Drenteln, N. S. (1904), Determination of the density of carbon dioxide according to the Archimedean principle, *Z. Phys. Chem. Unterr* **17**, 350–351.
[§]Yamamoto, S. (2002), Do not forget Archimedes' principle, *Kagaku to Kyoiku* **50**(10), 720–721.
[¶]Dettwiler, W., Ribordy, M., Donath, A., and Scherrer, J. R. (1978), Measurement of human body fat by means of gravimetry. Application of Archimedes' principle, *Schweizerische Med. Wochen.* **108**(48), 1914–1916.
[‖]Schultz, R. C., Dolezal, R. F., and Nolan, J. (1986), Further applications of Archimedes' principle in the correction of asymmetrical breasts, *Ann. Plast. Surg.* **16**(2), 98–101.
[**]Huerta, D. A., Sosa, V., Vargas, M. C., and Ruiz-Suarez, J. C. (2005), Archimedes' principle in fluidized granular systems, *Phys. Rev. E Stat. Nonlin. Soft Mat. Phys.* **72**(3, Pt. 1), 031307.
[††]Derjaguin, B. V. (1993), Amendment of Archimedes' principle, *Colloids Surf. A Physicochem. Eng. Aspects* **81**(1–3), 289–290.

balloon.* See also HYDROSTATIC PRINCIPLE, PRINCIPLE OF BUOYANCY.

AREA PRINCIPLE [mathematics] (1) (T. H. Gronwall, 1914) (theory of univalent functions) The area of the complement to the image of a domain under a mapping by a function regular in it is nonnegative. Let B be a domain of finite connectivity containing point ∞, and let $F(z) + b_0 + b_1/z, \dots$ be univalent in B. If B_F is the complementary set of $F(B)$ and $Q(w)$ is regular in an open set containing B_F, then the area principle simply states that the area of the set $Q(B_F)$ is nonnegative. If we take $B = \{z : |z| > 1\}$ and $Q(w) = w$, we obtain the classical area theorem: $\Sigma_{n=1}^{\infty} n|b_n|^2 \leq 1$.[†] (2) The sides of a triangle are equal to the ratio of the lengths depending on whether these segments have the same or opposite directions.[‡] (3) Exchange takes place across surfaces, and an increase of the ratio of surface area to volume leads to an increase in efficiency.[§] Also termed *area ratio principle*.

ARIOKHIN–PARETO INTERACTION PRINCIPLE [mathematics] Control energy lost due to mismatch in control actions with the system.[¶] Also known as PLUS/MINUS INTERACTION PRINCIPLE; INTERACTION EQUIVALENCE PRINCIPLE; GROUND SUPPORT INTERACTION PRINCIPLE; PRINCIPLE OF KUAN HSI; INTERACTION PRINCIPLE.

ARISTOTELIAN FIRST PRINCIPLES [psychology] (Aristotle, 384 BC–322 BC, Greek

philosopher) Earliest text on logic formulating the historical distinction of being called the first principle from the *Meta ta physica*, 1005b. "For the same (characteristic) simultaneously to belong and not belong to the same (object) in the same (way) is impossible." First expression of consistency in Western thought. Any defining and reasoning in any language on any topic assumes it a priori. It cannot be doubted, as all doubting is based on inconsistency, which assumes consistency a priori.[‖]

ARISTOTLE'S THEORY OF PRINCIPLES [psychology] (Aristotle, 384 BC–322 BC, Greek philosopher) Philosophy-based concepts on science related to induction/deduction, contradictions, or other sets of fundamental principles allowing for the development of deduction toward a coherent theory of knowledge.[**]

ARTIFICIAL IMMUNE PRINCIPLE [computer science] Simulated immune learning algorithm is used for determining the number and location of hidden layers by regarding the input data of network as antigens, and the centers of the hidden layer as antibodies.[††]

ARYABHATA'S RELATIVITY PRINCIPLE [astronomy] (Aryabhata, c. 476–550, Indian mathematician and astronomer) Ascribes the motion of the moon to Earth's rotation and developed an elliptical model of the heliocentric planetary system. Follows Gallilean relativity supporting Earth rotation, with possible underlying theory in which Earth (and the other solar system planets) orbits the sun, rather than the sun orbiting Earth.[‡‡]

*Beeget, A. (1909), *The Conquest of Aeronautis Aviation History*, Heinemann, London/Putnam, New York (http://www.archive.org/stream/conquestairaero00berggoog/conquestairaero00berggoog_djvu.txt).

[†]Gronwall, T. H. (1914), Some remarks on conformal representation, *Ann. Math. Ser. 2* **16**, 72–76.

[‡]Grünbaum, B. and Shepard, G. C. (1995), Ceva, Menelaus, and the area principle, *Math. Mag.* **68**, 254–268.

[§]Crawford, J. D., Terry, M. E., and Rourke, G. M. (1950), Simplification of drug dosage calculation by application of the surface area principle, *Pediatrics*, **5**(5), 783–790.

[¶]Kiforencko, B. N. and Kiforencko, S. I. (1993), Anokhin-Pareto interaction principle in control system theory, *J. Comput. Syst. Sci.* **31**(5), 165–170.

[‖]Cruz-Coke, R. (1994), Ethical principles in human scientific research, *Rev. Med. Chile* **122**(7), 819–824.

[**]van Deventer, Ch. M. (1927), The trace of an ancient theory in a modern principle, *Z. Phys. Chem.* **130**, 33–38.

[††]Bobo, Y., Shouwen, F., and Mingquan, S. (2007), Research on fault-tolerant controller for mobile robot based on artificial immune principle, *Proc. 3rd Int. Conf. Natural Computation*, ICNC, vol. 3, pp. 682–687.

[‡‡]Thurston, H. (1994), *Early Astronomy*, Springer-Verlag, New York.

AS LOW AS REASONABLY PRACTICAL (ALARP) PRINCIPLE [psychology, mathematics] (1) Applied in many areas to regulate the tolerable level of risk. Usually the principle is operationalized by assigning a value per fatality. A cost/benefit analysis is used to trade the expected value of lives saved with the costs of technical measures required to reduce risks. In sectors in which risks have been reduced over a period of years, it is difficult to pinpoint those areas in which further risk reduction might be sought. In this article we show that many different risk reduction mechanisms can be considered simultaneously in a decision analysis framework. Using influence diagrams it is straightforward to build mini–decision analysis models in which competing alternatives addressing the same risk can be compared. The minimodel decision alternatives are assembled into decision strategies representing the best possible combination of alternatives at different cost/benefit ratios. Disynergies between the different alternatives are highlighted through the model. The overall aim is to build a high-level model to explore the sensitivity of risk reduction measures to the value per fatality parameter. This enables decisionmakers to gain a better understanding of the cost of measures required to obtain a global reduction in risk.[*] (2) Specifies the boundary of tolerable risk at a level as low as reasonably practical, defined as falling between two limits. The top limit defines where operations should be for hidden and the bottom limit, the level below which risk is insignificant. If the process has the potential for inflicting an unacceptable level of risk, then steps must be taken to bring the risk level into the ALARP region.[†] Also known as *ALARA*; *ALARP*; *as low as reasonably achievable*.

ASSEMBLY SEQUENCE PRINCIPLE [engineering] Construction sequences should be planned in such a way as to allow the systems to work freely and have access to the site.[‡] See also STRONG-AXIS PRINCIPLE; SEVENTH JOINT PRINCIPLE; INTERFACE PRINCIPLE; STACKABILITY PRINCIPLE; PATH PRINCIPLE; DRILLED-CELL PRINCIPLE.

ASYMMETRIC INDUCTION PRINCIPLE [measurement, physics] Provides a simple and economical method of obtaining enantiomerically enriched products. Involves the reaction of organocopper reagents with enantiomerically enriched unsaturated esters. These esters, derived from scalemic alcohols and unsaturated carboxylic acids, react to give diastereomeric products that on hydrolysis yield enantiomers.[§] See INDUCTION PRINCIPLE.

ASYMPTOTIC MATCHING PRINCIPLE [engineering] (M. Van Dyke) The usual asymptotic expansion is called the *inner expansion*. The asymptotic expansion within certain powers is called the *outer expansion*. To obtain the inner expansion, a stretching transformation is introduced. The inner and outer expansions have a common region of validity and one can express the inner expansion of the outer expansion and the outer expansion of the inner expansion. The m-term inner expansion (of the n-term outer expansion) = the n-term outer expansion (of the m-term inner expansion, where m and n are any two integers.[¶] See also MATCHING PRINCIPLE.

AUFBAU PRINCIPLE [chemistry] (1) Governs the order in which the atomic orbitals are filled in elements of successive proton number.[‖] (2) A description of the buildup of elements in which the structure of each in sequence is obtained by simultaneously adding one positive charge (proton) to the nucleus of the atom and one negative charge (electron)

[*]Bedford, T. and Quigley, J. (2004), Risk reduction prioritization using decision analysis, *Risk Decision Policy* **9**(3), 223–236.

[†]Jackson, D. and Ramsey, M. (2003), SRP scientific meeting: ALARP: Principles and practices, *J. Radiol. Protect.* **23**(2), 235–238.

[‡]Nigg, E. A. (1992), Assembly and cell cycle dynamics of the nuclear lamina, *Semin. Cell Biol.* **3**(4), 245–253.

[§]Rossiter, B. E. and Swingle, N. M. (1992), Asymmetric conjugate addition, *Chem. Rev.* **92**(5), 771–806.

[¶]Irvine, T. F. and Hartnett, J. P. (1989), *Advances in Heat Transfer*, Academic Press, p. 7.

[‖]Daintith, J. (1999), *The Facts on File Dictionary of Chemistry*, 3rd ed., Facts on File, New York, p. 23.

to an atomic orbital.[*] (3) To achieve a multi-electron configuration, the required number of electrons must be added to the orbital one at a time, filling the most stable orbital first.[†] (4) Filling of successive electron shells around an atom, electron by electron. Each electron occupies the lowest energy level or atomic orbital available, so creating an electron structure for all the elements. Explains the structure of the periodic table.[‡] See also HUND'S PRINCIPLE OF MAXIMUM MULTICIPLICY; KITAIGORODSKII'S AUFBAU PRINCIPLE.

AUGUST KROGH PRINCIPLE [biology] (Schack August Steenberg Krogh, 1874–1949, Danish zoophysiologist) (Fians Krebs, in honor of August Krogh) (1) For many problems there is an animal on which it can be most conveniently studied.[§] (2) For a large number of problems there will be some animal of choice, on which a problem can be most conveniently studied.[¶]

AUSTERITY   PRINCIPLE [mathematics] Applied to holomorphic mapping of analytical quantities and applies to definite complex spaces as well as to immersions of meromorphic functions.[‖] See BOUNDARY PRINCIPLE.

AUTARKNESS   PRINCIPLE [psychology] See GENERATIVE PRINCIPLE OF WRITING.

AUTHORITY-LEVEL   PRINCIPLE [psychology] Maintenance of authority delegation requires that decisions within the authority level of an individual manager be made and not be referred upward in the organization structure.[**]

AUTOMATIC CONTROL PRINCIPLES [psychology, computers] (1) Instinctive attempt to avoid pain, discomfort, or unpleasant situations; desire to obtain maximum gratification with minimum effort.[††] (2) Applied to achieve the target steady operation state in which the control adjustments are determined according to the state parameters of the exchanger system.[‡‡] (3) Activities must be grouped to facilitate the accomplishment of goals, and the manager of each subdivision must have authority to coordinate its activities with the organization as a whole. The more clearly a position or a department defines the results expected, activities to be undertaken, organization authority delegated, and authority and informational relationships with other positions, the more adequately individuals responsible can contribute toward accomplishing enterprise objectives. If an activity is designed as a check on the activities of another department, the individual charged with such activity cannot adequately discharge her/his responsibility if s/he reports to the department whose activity s/he is expected to evaluate.[§§] See also PRINCIPLE OF DOMINANT SUBSYSTEMS; PRINCIPLE OF FUNDAMENTAL CHARACTERISTICS; PRINCIPLE OF OPERATION; EQUIPARTITION PRINCIPLE; PRINCIPLE OF CONSTRUCTION AND OPERATION; PRINCIPLE OF SMOOTH FIT.

AUTONOMOUS COOPERATION DISTRIBUTED MAXIMUM PRINCIPLE [psychology] Methodology for optimizing distributed systems by their autonomous activities by means

---

[*]Aufbau principle (2003), in *McGraw-Hill Dictionary of Scientific and Technical Terms*, McGraw-Hill, New York.

[†]Walker, P. M. B. (1999), *Chambers Dictionary of Science and Technology*, Chambers, New York, p. 76.

[‡]Scerri, E. R. (2003), Lowdin's remarks on the Aufbau principle and a philosopher's view of ab initio quantum chemistry, *Fund. World Quantum Chem.* **2**, 675–694.

[§]Wayne, R. and Staves, M. P. (1996), The August Krogh principle applies to plants, *Bioscience* **46**(5), 365–369.

[¶]Krebs, H. A. (1975), The August Krogh principle, *J. Exp. Zool.* **194**(1), 221–226.

[‖]Kozack, R. and Levin, F. S. (1986), Few-channel models of nuclear reactions: three-body model for deuteron elastic scattering and breakup, *Phys. Rev. C Nuclear Phys.* **34**(5), 1511–1519.

[**]Hancock, H. and Campbell, S. (1987), Impact of the leading an empowered organisation programme, *Nurs. Stand.* (Royal College of Nursing, UK) **20**(19), 41–48.

[††]Holzbock, W. G. (1958), *Automatic Control: Principles and Practice*. Reinhold, NY.

[‡‡]Li, K., Luo, X., Niemeyer B., and Li, M. (2004), The automatic optimal control process for the operation changeover of heat exchangers, *Proc. 8th Int. Conf. Advanced Computational Methods in Heat Transfer*, pp. 203–213.

[§§]Gulick, L. and Urwick, L. (1937), *Papers on the Science of Administration*, Inst. Public Administration, New York.

of the cooperative work among them.* See also MAXIMUM PRINCIPLE, MAXIMUM MODULUS PRINCIPLE, DISTRIBUTED MAXIMUM PRINCIPLE.

AUTOTELIC   PRINCIPLE [psychology] The balancing of skill and challenge should be the fundamental motivational driving force of the agent. This implies that (1) each component must be parameterized so that challenge levels can be self-adjusted by self-monitoring of performance, (2) each component must have the ability to increase skill to cope with new challenge, and (3) a complex agent could self-regulate buildup of skills and knowledge without the need for intervention of a designer to scaffold the environment, stage the reward functions, or bring resources progressively online in a maturational schedule.[†]

AUXILIARY   PRINCIPLE [biology]   (Willi Hennig, German dipterist, 1913–1976) Development of a coherent theory of the investigation and presentation of the relations that exist among species. See HENNIG–FARRIS AUXILIARY PRINCIPLE; HENNIG'S AUXILIARY PRINCIPLE; PRINCIPLE OF PARSIMONY; HENNIG'S AUXILIARY METHODOLOGICAL PRINCIPLE.

AVERAGING PRINCIPLE [mathematics] (Stefan Banach, 1892–1945) (1) The distribution of sample means possess less variability as shown by a smaller standard deviation of the sample means as $\sigma_x/n^{1/2}$, where σ_x is standard deviation and n is the size of the sample.[‡] (2) For measures corresponding to marginal conditional distributions of diffusion processes.[§] (2) Large deviations from

the law of large numbers for a sequence of independent and identically distributed random variables.[¶] (3) Reduction of noise received by a robot sensor by screening it over a period of time.[‖] (4) The averaging method is a classical tool for analyzing dynamical systems with fast angular variables; the idea is to average over the angles, to obtain an approximate evolution law for the slow variables.[**] See also GIBBS VARIATIONAL PRINCIPLE; CONTRACTION PRINCIPLE; LARGE-DEVIATION PRINCIPLE; CONTRACTION MAPPING PRINCIPLE.

AVOGADRO'S PRINCIPLE [chemistry] (Amedeo Avogadro) (1) Equal volumes of gases, at the same temperature and pressure, contain the same number of particles, or molecules.[††] (2) The number of molecules in a specific volume of gas is independent of the size or mass of the gas molecules.[‡‡] (3) The ideal-gas constant has the same value for all gases. One moles of an ideal gas occupies 22.4 liters (L) (dm^3) at standard temperature and pressure (STP). This volume is often referred to as the *molar volume of an ideal gas*. Real gases may deviate from this value.[§§] (4) The number of molecules in one mole is approximately 6.022×10^{23} particles per mole.[¶¶]

AVOIDANCE   PRINCIPLE [mathematics] Regards the antipolynomiality of the Hilbert–Samuel multiplicity of the graded components of the local cohomology modules of a finitely generated module over

*Ayoub, N., Seki, H., and Naka, Y. (1994), *Optimization of the Distributed Systems by Autonomous Cooperation-Distributed Maximum Principle*, DARS'94, Wakao, Japan., pp. 29–40.
[†]Steels, L. and Wellens, P. (2007), Scaffolding language emergence using the autotelic principle, *Proc. 2007 IEEE Symp. Artificial Life*, CI-ALife, pp. 325–332.
[‡]Burkey, M. L. (2006), *Basic Business Statistics*, 8th ed. (http://www.ncat.edu/~burkeym/Data/chap07%5B1%5D.ppt).
[§]Kir'yanov, V. A. and Chernenko, A. A. (1977), Principle of averaging in quantum mechanical problems of kinetics. General positions of the principle of averaging, *Elektrokhimiya* **13**n(2), 229–235.

[¶]Op. cit. Kir'yanov and Chernenko.
[‖]Op. cit. Kir'yanov and Chernenko.
[**]Muhler, R. and von Specht, H. (1999), Sorted averaging-principle and application to auditory brainstem responses, *Scand. Audiol.* **28**(3), 145–149.
[††]Walker, P. M. B. (1999), *Chambers Dictionary of Science and Technology* Chambers, New York, p. 83.
[‡‡]Gorin, G. (1997), Avogadro's idea: Is it a hypothesis, theory, law, principle, or what-have-you? *Book of Abstracts*, 213th ACS National Meeting, San Francisco, April 13–17, 1997.
[§§]Leduc, A. (1897), On the principle of Avogadro-Ampere considered as a limiting law, *C. R. Hebd. Seances Acad. Sci.* **124**, 285.
[¶¶]Kapustinskii, A. F. (1957), Avogadro's law and some of its recent developments, *Izvest. Akad. Nauk SSSR Ser. Khim.* p. 657–663.

a Noetherian homogeneous ring with two-dimensional local base ring.[*] (2) To teach a child to avoid a certain type of situation, simultaneously present to the child the situation to be avoided (or some representation of it) and some aversive conditon (or its representation).[†]

AXIOM OF ARCHIMEDES PRINCIPLE [mathematics, physics] (Archimedes, c. 287–212 BC, Greek mathematician)

(1) Archimedes' axiom, also known as the *continuity axiom* or *Archimedes' lemma*, survives in the writings of Eudoxus. The term was first coined by the Austrian mathematician Otto Stolz (1883). (1) Given two magnitudes having a ratio, one can find a multiple of either that will exceed the other. This principle was the basis for the method of exhaustion, which Archimedes invented to solve problems of area and volume.[‡] (2) In the introduction to his *Quadrature of the Parabola*, Archimedes formulated the following principle, since known as the *axiom of Archimedes*: "The more by which the greater of two unequal areas exceeds the less can, by being added to itself, be made to exceed any given area." This is equivalent to saying that, given any positive real numbers a and b such that $a \prec b$, there exists an integer n such that $na \succ b$. For two magnitudes a and d there exists an n such that $nd \geqslant a$. A system of magnitudes satisfying this is called and *Archimedean system*. Let x be any real number. Then there exists a natural number n such that $n > x$. Formally, the Archimedean property can be stated as follows: Let c, e be in R, the real numbers. Then the following two properties hold: (1) for any positive c, there exists a natural number n such that $n > c$; (2) for any positive e, there exists a natural number n, such that $1/n < e$. In simple terms, the Archimedean property can be regarded as either of the following two statements: (1) given any number, you can always pick another number that is larger than the original number; or (2) given any positive number, you can always pick another positive number that is less than the original number.[§]

AXIOM OF CHOICE PRINCIPLE [mathematics] (Ernst Friedrich Ferdinand Zermelo, 1871–1953, German mathematician)

One of the axioms in set theory. It is sometimes called *Zermelo's axiom of choice*. It makes it possible to form sets by choosing an element simultaneously from each member of an infinite collection of sets even when no algorithm exists for the selection. It states that for any family F of nonempty sets there exists a function f such that, for any set S from F, one has $f(S) \in S$ (f is called a *choice* or *selection function on* F). The set of values of f is called the *choice set*. A choice function for S may be regarded as selecting a member from each nonempty subset of S. For example, if $S = \{3, 4\}$, then nonempty subsets of S are $S_1 = \{3\}, S_2 = \{4\}$, and $S_3 = \{3, 4\}$. Two choice functions for S may then be defined: $f_1(S_1) = 3$, $f_1(S_2) = 4$, $f_1(X_3) = 3$, and $f_2(S_1) = 3, f_2(S_2) = 4, f_2(S_3) = 4$. Many postulates equivalent to the axiom of choice were subsequently discovered, including (1) *the well-ordering theorem*—on any set X there exists a total order $X \subseteq X \times X$ such that any nonempty set $U \subset X$ contains a least element in the sense of the relation R; (2) *the maximality principle (Zorn's lemma)*—if any totally ordered subset U of a partially ordered set X is bounded from above, X contains a maximal element; (3) any nontrivial lattice with a unit element has a maximal ideal; (4) the product of compact topological spaces is compact; and (5) any set X has the same cardinality as $X \times X$.[¶] See also WELL-ORDERING PRINCIPLE; ZORN'S MAXIMALITY PRINCIPLE; ZORN'S MAXIMUM PRINCIPLE.

[*]Brodmann, M., Kurmann, S., and Rohrer, F. (2007), An avoidance principle with an application to the asymptotic behaviour of graded local cohomology, *J. Pure Appl. Algebra* **210**(3), p. 639–643.

[†]Huitt, W. (1994), Principles for using behavior modification, in *Educational Psychology Interactive*, Valdosta State Univ., Valdosta, GA (http://chiron.valdosta.edu/whuitt/col/behsys/behmod.html).

[‡]Yonezawa, T. (2003), Experiment of measurement of Avogadro number using Archimedes' principle, *Kagaku to Kyoiku* **51**(7), 443–444.

[§]Burn, B. (2005), The vice: Some historically inspired and proof-generated steps to limits of sequences, *Educ. Stud. Math.* (Springer, Netherlands) **60**(3), 269–295.

[¶]Harper, J. M. and Rubin, J. E. (1976), Variations of Zorn's lemma, principles of cofinality, and Hausdorff's maximal principle, *Notre Dame J. Formal Logic*, **17**(4), 565–588.

B

BABINET'S PRINCIPLE [physics] (Jacques Babinet, 1794–1872, French physicist) (1) Used to solve diffraction problems in optics and electrodynamics. Sum of the waves diffracted by a certain boundary and by the complement of this boundary is equal to the undistorted wave. Given a diffracting hole of arbitrary shape, the conjugate to the hole produces an identical diffraction pattern.[*] (2) Diffraction patterns produced by complementary screens are identical; two screens are said to be complementary when the opaque parts of one correspond to transparent part of other.[†] (3) Radiation field beyond a screen that has apertures, added to that produced by a complementary screen, is identical to the field that would be produced by the unobstructed beam of radiation; thus the two diffraction patterns will also be complementary.[‡] See also COMPLEMENTARITY PRINCIPLE.

BABO'S PRINCIPLE [chemistry] (Lambert Heinrich von Babo, 1818–1899, German chemist) If a substance is dissolved in a liquid, the vapor pressure of the liquid is reduced; the amount of lowering is proportional to the amount of solute dissolved.[§] See also RAOULT'S PRINCIPLE; HENRY'S PRINCIPLE.

BACKWARD CHAINING PRINCIPLE [psychology, computers] An inference method whereby a system starts with a defined goal or an outcome to prove and tries to establish the facts to do so.[¶] See CHAINING PRINCIPLE; FORWARD CHAINING PRINCIPLE.

BALIAN–LOW UNCERTAINTY PRINCIPLE [mathematics] (1) If the Gabor family $G = \{e^{2\pi imt}g(t-n)\}_{m,n\in\mathbb{Z}}$ forms an orthonormal basis for $L^2(\square)$ then the variances $\int |t|^2|g(t)|^2\,dt$ and $\int |\xi|^2|\hat{g}(\xi)|^2\,d\xi$ cannot both be finite. A more general but less precise version states that if $\varepsilon \succ 0$ then $\int |t|^{p+\varepsilon}|g(t)|^2\,dt$ and $\int |\xi|^{q+\varepsilon}|\hat{g}(\xi)|^2\,d\xi$ cannot both converge when $1/p + 1/q = 1$. Here $\hat{g}$ is the Fourier transform of $g\epsilon L^2(\square)$.[‖] (2) $\mathcal{G}(f,1,1) = \{f_{m,n}\}_{m,n\in\mathbb{Z}} = \{e^{2\pi imt}f(t-n)\}_{m,n\in\mathbb{Z}}$, is an orthonormal basis for $L^2(\mathbb{R})$ then the strong uncertainty constraint $\Delta(f)\Delta(\hat{f}) = \infty$ must hold. The Balian-Low theorem depends crucially on the rigid structure of Gabor systems.[**]

BANACH CONTRACTION PRINCIPLE [mathematics] (Stefan Banach, 1892–1945, Polish mathematician) Every contraction in a complete metric space has a unique fixed point. Let (X,d) be a complete metric space; $f : X \to X$, a contraction. Then f has a unique fixed point in X.[††]

BANACH CONTRACTION PRINCIPLE IN A METRIC SPACE [mathematics] (Stefan Banach, 1892–1945, Polish mathematician)

[*]Walker, P. M. B. (1999), *Chambers Dictionary of Science and Technology*, Chambers, New York, p. 86.
[†]Konitz H. (1972), Babinet's principle in electron optics, *Microsc. Acta* **73**(1), 25–28.
[‡]Porai-Koshits, E. A. and Filipovich, V. N. (1955), Babinet's principle applied for the low-angle diffraction of x-rays by porous glasses, *Izvest. Akad. Nauk SSSR Ser. Khim.* pp. 21–30.
[§]Daintith, J. (1999), *The Facts on File Dictionary of Chemistry*, 3rd ed., Facts On File, New York, p. 25.

[¶]Walker, P. M. B. (1999), *Chambers Dictionary of Science and Technology*, Chambers, New York, p. 86.
[‖]Benedetto, J. J., Heil, C. and Walnut, D. F., (1995) Differentiation and the Balian-Low Theorem. J. Four. Anal. Appl. **1**, 355–402.
[**]Balian, R. (1981) Un princple d'incertitude fort en théerie du signal ou en mécanique quantique, Comptes-Rendus de l'Académie des Sciences. **292, 1357–1362**
[††]Palais, R. S. (2007), A simple proof of the banach contraction principle, *J. Fixed Point Theory Appl.* **2**(2), 221–223.

Underlies the Picard iteration method of solving differential equations numerically. A mapping $f : X \to X$ where X is a metric space, is a contraction if it decreases distances in the sense that there is a positive constant $\alpha \prec 1$ such that $d(f(x), f(y)) \leqslant \alpha . d(x, y)$ for all $x, y \in X$. If X is complete, then every contraction mapping has a unique fixed point: a point $a \in X$ such that $f(a) = a$.[*] (2) A metric space X will be said to be η-chainable if for every $a, b \in X$. there exists an n chain, that is, a finite set of points $a = x_0, x_i, \ldots, x\eta = b$ (η may depend on both a and b) such that $d(x_{i-1}, x_i) < \eta$ $(i = 1, 2, \ldots, \eta)$.[†]

BANACH FIXED-POINT PRINCIPLE [mathematics] (Stefan Banach, 1892–1945, Polish mathematician). (1) Let (X, d) be a complete metric space and $T : X \to X$ a contraction map. Then T has a unique fixed point $x_0 \in X$, i.e., $T(x_0) = x_0$. *Uniform spaces*: Every monotone function on a complete lattice has a least fixpoint (fixed point).[‡] (2) A fundamental tool to investigate fixpoint calculus.[§] See also KNASTER–TARSKI PRINCIPLE. Also known as *Caccioppoli–Banach principle* and *Banach fixed-point theorem*.

BANACH PRINCIPLE [mathematics] (Stefan Banach, 1892–1945, Polish mathematician) (1) Every nonexpansive mapping of a closed and bounded convex set, in a uniformly convex Banach space, into itself has a fixed point.,[¶] (2) *Uniform spaces:* Every monotone function

on a complete lattice has a least fixpoint.[‖] (3) Convergence of a sequence of operators to the finiteness of a maximal function.[**] Also known as *CACCIOPPOLI–BANACH PRINCIPLE*. See also KNASTER–TARSKI PRINCIPLE.

BANACH–STEINHAUS UNIFORM BOUNDEDNESS PRINCIPLE. See BOUNDEDNESS PRINCIPLE.

BANG-BANG PRINCIPLE [mathematics] (1) States constrained parabolic systems within boundary controls. Let $t \succ 0$ and suppose $x^0 \in C(t)$, for the system $\dot{x}(t) = Mx(t) + N\alpha(t)$. Then there exists a bang-bang control $\alpha(\cdot)$ that steers x^0 to 0 at time t, where a control $\alpha(\cdot) \in A$ is called *bang-bang* if for each time $t \geqslant 0$ and each index $i = 1, \ldots\ldots\ldots, m$, we have $|\alpha^i(t)| = 1$, where

$$\alpha(t) = \begin{bmatrix} \alpha^1(t) \\ \vdots \\ \alpha^m(t) \end{bmatrix}$$

and A is the cube $[-1, 1]^m$ in $\Re^m$. (2) Linear control systems in finite-dimensional space stated as if the system can be steered from a point to another point in a given time by a control and then transferred on by another control.[††] (3) A control system is being operated from a limited source of power; then, in order to move the system from one state to another in the shortest time, it is necessary to utilize the maximum power available.[‡‡]

BASE INVARIANCE PRINCIPLE [mathematics] See SCALE INVARIANCE PRINCIPLE.

[*]Anastassiou, G. A. (1995), Central limit theorem, weak law of large numbers for martingales in Banach spaces, and weak invariance principle. A quantitative study, *J. Multivar. Anal.* **52**, 158.

[†]Namsrai, K. (1991), Stochastic and quantum space-time metrics and the weak-field limit, *Int. J. Theor. Phys.* **30**(5), 587–710.

[‡]Karimov, A. K. and Mukhamedov, F. M. (2003), Banach principle in Jordan algebras and its application, *Dopovidi Natsion. Akad. Nauk Ukraini* (1), 22–24.

[§]Bellow, A. and Jones, R. L. (1996), A Banach principle for L^∞, *Adv. Math.* **120**(1), 155–172.

[¶]Edelstein, M. (1974), Fixed Point Theorems in Uniformly Convex Banach Spaces. *Proc. Am. Math. Soc.* **44**(2), 369–374.

[‖]Karimov, A. K. and Mukhamedov, F. M. (2003), Banach principle in Jordan algebras and its application, *Dopovidi Natsion. Akad. Nauk Ukraini* (1), 22–24.

[**]Palais, R. S. (2007), A simple proof of the Banach contraction principle, *J. Fixed Point Theory Appl.* 2(2), 221–223.

[††]LaSalle, J. P. (1960), The "bang-bang" principle. *Automatic and remote control, Proc. 1st Int. Congr. Int. Federation Automat. Control*, Moscow, vol. 1, pp. 493–497.

[‡‡]Wang, G. and Wang, L. (2007), The bang-bang principle of time optimal controls for the heat equation with internal controls, *Syst. Control Lett.* **56**(11–12), 709–713.

BASIC PRINCIPLE [law]   (1) Considered to be key, novel, essential, or rudimentary to the concept, property, or operation or a new or emerging technology.[*] (2) The basis from which other truths can be derived; "first you must learn the fundamentals"; "let's get down to basics."[†]

BATEMAN'S PRINCIPLE [biology] (Angus John Bateman, 1919–1996, British geneticist) (A. J. Bateman, geneticist) (1) Bateman's three principles is the key to understanding the operation of sexual selection in animals. "It can now be seen that the sex difference in variance of fertility, which is a sign of intr-masculine selection, is due to the effect of number of mates per fly on fertility" First principle is males show greater variance in number of offspring than do females. His second principle states: the higher variance, in males, of the number of mates per fly, a sign of intra-masculine selection. The third principle by Bateman states, "the stronger correlation in males, between number of mates and fertility.[‡] (2) The number of matings achieved is usually limited by male fitness, while the resources available for reproduction usually limit female fitness.[§]

BAYES PRINCIPLE [mathematics]   (Thomas Bayes, 1702–1761, British clergyman and mathematician)   (1) A method of calculating the probability of the outcome of mutually exclusive events on the basis of a calculation that considers the probability of each of the separate events. States exactly how the probability of a certain "cause" changes as different events actually occur.[¶] (2) If

$B_1, B_2, \ldots, B_n$ are a mutually exclusive and exhaustive set of events [i.e., a set of nonoverlapping events covering the set S of all possible outcomes of an experiment (i.e., the whole sample space)] and an event A is observed. The probability that the event B_j is the causal event giving rise to A, that is, the probability of B_j conditional on A, is given by Bayes' theorem:[‖]

$$\Pr(B_j|A) = \frac{\Pr(B_j)\Pr(A|B_j)}{\sum_i \Pr(B_i)\Pr(A|B_i)}$$

Also referred to as *Bayes theorem*; *law of inverse probability*.

BAYESIAN CLASSIFICATION PRINCIPLE [mathematics] (Thomas Bayes, 1702–1761, English mathematician)   A Bayesian classifier provides an algorithm in which a sample x is assigned to the class with a highest probability in accordance with the Bayesian decision rule. Bayesian decision processes are very theoretical or oriented toward business or sociological, rather than engineering, problems. Bayesian classifier requires an extensive database to determine the probability density functions for the particular classification application.[**]

BAYONET ATTACHMENT PRINCIPLE [engineering]   Where a part of a stator having lip ears or of a stator collar having lip ears is inserted into the stator connector and, with a twist of the inserted part, is attached to the stator connector. Where a suitable stator is inserted into the connector and, with a twist of the inserted stator, is attached to the connector. The stator can include a hollow tube housing having an open top and a closed bottom, with a wide lip extending outwardly and generally normal to the tube housing at the top

[*]Nagaya, T. (1996), Constraints programmed macro model of information selection. Limits of information processing and the basic principle of information behavior, in *Electronics and Communications in Japan*, Part III, *Fundamental Electronic Science*, vol. 79, p. 1.

[†]American Heritage (2002), BASIC. *American Heritage Dictionary of the English Language*, 4th ed., Houghton Mifflin, New York.

[‡]Arnold, S. J. (1994), Bateman's principles and the measurement of sexual selection in plants and animals, *Am. Natur.* **144**, S126–S149.

[§]Bateman, A. J., (1948), Intra-sexual selection in Drosophila, *Heredity*. **2**, 349–368.

[¶]Weise, L. (1970), Estimation of physical parameters with Bayes principle. Examples from the nuclear radiation measurement technique, *Atomkernenergie* **16**(3), 233–236.

[‖]Thompson, J. R. (1998), *Probability and Inference in What Is Probability* (http://www.stat.rice.edu/stat/FACULTY/courses/stat431/Bayes.pdf).

[**]Hu, J., Si, J., Olson, B. P., and He, J. (2004), Principle component feature detector for motor cortical control, *Proc.* Annual Int. Conf. IEEE Engineering in Medicine and Biology Society, vol. 6, pp. 4021–4024.

of the tube housing, and may have a stator retaining hole in the lip.*

BEAMSPLITTER PRINCIPLE [electronics] (1) Optical device for dividing a light beam into two or more paths.[†] (2) Optical wavelength division multiplex transmission system edge interference filter for separation or combination.[‡]

BEER–LAMBERT PRINCIPLE [physics] (August Beer, 1825–1863, German physicist, mathematician) (Johann Heinrich Lambert 1728–1777, Swiss mathematician physicist and astromoner) The fraction of the light absorbed by each layer of solution is the same. Describes the relationship between the proportions of light penetration.[§]

BEKESY TRAVELING WAVE PRINCIPLE [physics]   See TRAVELING WAVE PRINCIPLE.

BELL–EVANS–POLANYI PRINCIPLE [chemistry] (M. G. Evans) (Ronnie Bell, 1907–1966, British Chemist) (Polanyi Perby Mihály, 1891–1976, Hungarian British polymath) (Michael Polanyi, b. 3/12/1891, Northampton, England), Hungarian, British polymath) Linear relation between energy activation and enthalpy of reaction observed within a series of closely related reactions.[¶]

BELLAGIO PRINCIPLES [ecology]   Development of new ways to measure and assess progress toward sustainable development. In

response, significant efforts to assess performance have been made by corporations, nongovernment organizations, academics, communities, nations, and international organizations.[‖] See also HANNOVER PRINCIPLES; FOREST STEWARDSHIP COUNCIL (FSC) PRINCIPLES; GLOBAL SULLIVAN PRINCIPLES OF SOCIAL RESPONSIBILITY; MARINE STEWARDSHIP COUNCIL (MSC) PRINCIPLES; PERMACULTURE PRINCIPLES; EQUATOR PRINCIPLES; MELBOURNE PRINCIPLES; PRECAUTIONARY PRINCIPLE; SANBORN PRINCIPLES; TODDS' PRINCIPLES OF ECOLOGICAL DESIGN.

BELLMAN'S PRINCIPLE [mathematics] (Richard Ernest Bellman, 1920–1984, American mathematician) (1) Concerns the field of dynamic programming, which is a mathematical technique concerned with the optimization of multistage decision processes. In this technique, the problem is divided into small subproblems (stages), which are then solved successively, thus forming a sequence of decisions that leads to an optimal solution of the problem.[**] (2) Once all the functions involving a single variable have been combined, the size of the resulting interim function can be reduced by performing the global operation on the interim function. Only the values of the variable being removed that produce the best results relative to the global operation can then be carried forward for each combination of the remaining variables in the function.[††] (3) An optimal sequence of decisions in a multistage decision process problem has the property that whatever the initial state and decisions are, the remaining decisions must constitute an optimal policy with regard to the state resulting from the first decisions.[‡‡]

*Selby, T. W. (1996), *Stator Connector*, US Patent 5,681,985 (8/27/1996).

[†]Walker, P. M. B. (1999), *Chambers Dictionary of Science and Technology*, Chambers, New York, p. 102.

[‡]Zhang, W. and Sanders, B. C. (1994), Atomic beam splitter: Reflection and transmission by a laser beam, *J. Phys. B Atom. Mol. Opt. Phys.* **27**(4), 795–808.

[§]Rakhmankulova, M., Stavrou, S., W. Yuen, A. P., Raymond Zhou, R., Kessler, P., and Pevsner, P. H. (2008) Micropipette tips—the unsung heroes of mass spectrometry, *Rapid Commun. Mass Spectrom.* **22**(15), 2349–2354.

[¶]Anglada, J. M., Besalu, E., Bofill, J. M., and Crehuet, R. (1999), Prediction of approximate transition states by Bell-Evans-Polanyi principle: I, *J. Comput. Chem.* **20**(11), 1112–1129.

[‖]Hardi, P. and Zdan, T. (1997) Principles in Practice. International Institute for Sustainable Development. http://www.nssd.net/pdf/bellagio.pdf

[**]Garashchenko, F. G. and Pichkur, V. V. (1999), On a generalization of Bellman's principle and its application, *Dopovidi Natsion. Akad. Nauk Ukraini* **6**, 105–107.

[††]n.A (2003) Bellman's Principle. *McGraw-Hill Dictionary of Scientific and Technical Terms.* 10th edition.

[‡‡]Kwong, C. P. (1987), Development of the Schroedinger equation and Klein-Gordon equation via Bellman's principle of optimality, *Phys. Lett. A* **124**(4–5), 220–222.

BELLMAN'S PRINCIPLE OF OPTIMALITY [mathematics] (Richard Ernest Bellman, 1920–1984, American mathematician)   An optimal sequence of decisions in a multistage decision process problem has the property that whatever the initial state and decisions are, the remaining decisions must constitute an optimal policy with regard to the state resulting from the first decisions.* See also ORGANIZING PRINCIPLE.

BELMONT PRINCIPLE [psychology] (US Dept. Health, Education, and Welfare, 4/18/79, Belmont Conference Center)   Race, class, and ethics in research: Belmont principles to functional relevance. Explains three fundamental ethical principles must be followed for all scientific research involving human subjects: (1) respect for persons; (2) beneficence; and (3) justice.[†]

BELT FILTER PRESS PRINCIPLE [engineering] See FILTER PRESS PRINCIPLE.

BERGERON–FINDELSEN   PRINCIPLE [meteorology] (David J. Bergeron ???) (Wladyslaw Findelsen, 1926–Polish professor)   Initiation of precipitation in a cloud consisting mainly of supercooled water droplets is due to the presence of ice crystals that grow at the expense of the droplets because the vapor pressure with respect to ice is lower than that with respect to liquid water at the same temperature.[‡]

BERGMANN'S PRINCIPLE [evolution]   Carl Bergmann, 1821–1876, German Biologist An increase in the geographic latitude and depth of habitat (correlating mainly with lower temperatures) leads to an increased cell size,

lifespan of the animal, and, as a result, an increase in body size.[§]

BERGOFSKY PRINCIPLE [computer science] (Dan Brown, Digital Fortress) "If a computer tried enough keys, it was mathematically guaranteed to find the right one."[¶]

BERKELEIAN PRINCIPLES [philosophy] (George Berkeley, 1685–1753, Anglo-Irish philosopher)   The whole of the double-sided structure has its basis in something that is not merely of a physical order, and that the only context into which both can be fitted is that of a notion of the absolute where the two aspects can exist without contradiction.[‖]

BERMUDA PRINCIPLES [biology] (Bermuda, 1996).   Provides a basis for a free sharing of prepublished data on gene sequences among scientists. The three principles originally were automatic release of sequence assemblies larger than 1 kb (kilobase) (preferably within 24 h, immediate publication of finished annotated sequences, and an attempt to make the entire sequence freely available in the public domain for both research and development in order to maximize benefits to society.[**]

BERNOULLI'S   PRINCIPLE [mathematics] (Daniel Bernoulli, 1700–1782, Swiss mathematician)   (1) For a nonviscous incompressible fluid in steady flow, the sum of the pressure, potential and kinetic energies per unit volume is constant at any point.[††] (2) Law stating that the pressure of a fluid

[§]Timofeev, S. F. (2001), Bergmann's principle and deep-water gigantism in marine crustaceans, *Izvest. Akad. Nauk Ser. Biol. / Rossiisk. Akad. Nauk* **6**, 764–768.
[¶]Brown, D. (2003), *Digital Fortress*. St. Martin's Press.
[‖]Guru, N. (2008), *The Search for a Norm in Western Thought*, p. 29 (http://www.advaitavedanta.co.uk/content/10-the-search-for-a-norm-in-western-thought).
[**]Aled, E. (2008), Bermuda principles meet structural biology, *Nat. Struct. Mol. Biolo.* **15**(2), 116.
[††]Walker, P. M. B. (1999), *Chambers Dictionary of Science and Technology*, Chambers, New York, p. 107.

*Srivastava, U. K., Shenoy, G. V., and Sharma, S. C. (1991), *Quantitative Techniques for Managerial Decisions: Concepts, Illustrations and Problem*, New Age International, p. 732.
[†]Cassell, E. J. (2000), The principles of the Belmont report revisited. How have respect for persons, beneficence, and justice been applied to clinical medicine? *Hastings Center Report* **30**(4), 12–21.
[‡]Walker, P. M. B. (1999), *Chambers Dictionary of Science and Technology*, Chambers, New York, p. 107.

varies inversely with its velocity. An increase in fluid flow velocity produces a decrease in pressure and vice versa. The principle is expressed as $K = p + \frac{1}{2}\rho v^2 + h\rho g$, where K is a constant, p is the pressure of the fluid, ρ is the fluid density, v is the velocity of the fluid, h is the difference in elevation, and g is the acceleration due to gravity.[*] Also known as the *Bernoulli law; Bernoulli law of hydrodynamic pressure.*

BERNOULLI'S WEAK PRINCIPLE OF LARGE NUMBERS [mathematics] (Jacob Bernoulli, 1654–1705, Swiss mathematician) Commonly referred to as *Bernoulli's theorem.* The idea behind the law of large numbers is that if the size of a sample of statistically independent variables is increased indefinitely, good sample estimates of population parameters will tend to concentrate more and more closely about the true value; i.e., approximately in the long run, frequency settles down to probability. Bernoulli's theorem states that given a sequence of independent trials, in each one of which the probability of occurrence of a certain event A ("success") has the same value $p, 0 \prec p \prec 1$, then, if x_n is the random variable equal to the number of successful events in the first n trials and x_n/n is the frequency of occurrence, then

$$\lim_{n \to \infty} P\left(\left|\frac{x_n}{n} - p\right| \succ \varepsilon\right) = 0$$

for any epsilon,. $\varepsilon \succ 0$.[†]

BERTHELOT-THOMSEN PRINCIPLE [chemistry] (AKA Thomsen-Berthelot Principle, French Chemist) (Pierre Eugène Marcellin, Berthelot, 1827–1907) Of all chemical reactions possible, the one developing the greatest amount of heat will take place, with certain obvious exceptions such as changes of state.[‡] (Hons Peter Jorgen Julius Thomsen, 1826–1909, Danish chemist)

BETTI RECIPROCITY PRINCIPLE [mechanics] Defined by the appliction of forces $f_1(P)$ at P and $f_2(Q)$ at Q. The reciprocity theorem in this case is expressed by $f_1(P) - u_2(P) = f_2(Q) - u_1(Q)$, where $u_1(Q)$ is the displacement in state 1 measured at Q and $u_2(P)$ is the displacement in state 2 measured at P.[§] Also known as the Maxwell Betti Reciprocity Principle.

BEYOND THE PLEASURE PRINCIPLE [psychology] In the psychoanalytical theory of the mind we take it for granted that the course of mental processes is automatically regulated by the "pleasure principle"; that is to say, we believe that any given process originates in an unpleasant state of tension and thereupon determines for itself such a path that its ultimate issue coincides with a relaxation of this tension, i.e., with avoidance of "pain" or with production of pleasure. We know that the pleasure principle is adjusted to a primary mode of operation on the part of the psychic apparatus, and that for preservation of the organism amid the difficulties of the external world it is ab initio useless and indeed extremely dangerous. Under the influence of the instinct of the ego for self-preservation, it is replaced by the "reality principle," which, without giving up the intention of ultimately attaining pleasure, yet demands and enforces the postponement of satisfaction, the renunciation of manifold possibilities thereof, and the temporary endurance of pain on the long and circuitous road to pleasure. Replacement of the pleasure principle with the reality principle can account for only a small part, and not the most intense, of painful experiences. Another and no less regular source of pain proceeds from the conflicts and dissociations in the psychic apparatus during development of the ego toward a more highly

[*]Walker, J. (1988), Does convection or the Bernoulli principle make the shower curtain flutter inward? *Sci. Am.* **258**, 1.

[†]Föllmer, H. (1977), The Bernoulli principle and the Dirichlet problem. Mathematical economics and game theory, *Lecture Notes Econ. Math. Syst.* (Springer, Berlin) **141**, 208–216.

[‡]Briner, E. (1948), Chemical reactions for which the Berthelot-Thomson principle of maximum work is valid, *C. R. Acad. Sci.* **227**, 661–663.

[§]Constantinescu, A. and Korsunsky, A. (2007), *Elasticity with Mathematica: An Introduction to Continuum Mechanics and Linear Elasticity*, Cambridge Univ. Press, p. 100.

coordinated organization. The two sources of pain indicated here still do not nearly cover the majority of our painful experiences, but as to the rest one may say with a fair show of reason that their presence does not impugn the supremacy of the pleasure principle. Most of the pain we experience is of a perceptual order, perception of either the urge of unsatisfied instincts or something in the external world that may be painful in itself or may arouse painful anticipations in the psychic apparatus and is recognized by it as "danger." The reaction to these claims of impulse and these threats of danger, a reaction in which the real activity of the psychic apparatus is manifested, may be guided correctly by the pleasure principle or the reality principle that modifies this. It seems thus unnecessary to recognize a still more far-reaching limitation of the pleasure principle, and nevertheless it is precisely the investigation of the psychic reaction to external danger that may supply new material and new questions in regard to the problem treated here.* See also PLEASURE PRINCIPLE.

BIJECTION PRINCIPLE [mathematics]   Bijective proofs prove that two sets have the same number of elements by finding a bijective function (one-to-one correspondence) from one set to the other. Technique for finding a bijective function $f : A > B$ between two sets A and B and thus proves that both sets have the same number of elements: $|A| = |B|$.[†] See also COMBINATORIAL PRINCIPLE.

BIN LADEN PRINCIPLE [psychology] (Osama bin Mohammed bin Awad Awad bin Laden, 1957–al Qaeda leader) It is forbidden to co-operate or form any alliance with components because the lives have been contaminated

with corruption and heresy.[‡] Also referred to as *principle of Takfir*.

BIOMIMICRY PRINCIPLES. [ecology]   Animals based on the quadruped design fill a wide range of ecological niches.[§] See also HANNOVER PRINCIPLES; DEEP ECOLOGY'S BASIC PRINCIPLES; FOREST STEWARDSHIP COUNCIL (FSC) PRINCIPLES; GLOBAL SULLIVAN PRINCIPLES OF SOCIAL RESPONSIBILITY; MARINE STEWARDSHIP COUNCIL (MSC) PRINCIPLES; PERMACULTURE PRINCIPLES; THE BELLAGIO PRINCIPLES FOR ASSESSMENT; EQUATOR PRINCIPLES; MELBOURNE PRINCIPLES; PRECAUTIONARY PRINCIPLE; SANBORN PRINCIPLES; TODDS' PRINCIPLES OF ECOLOGICAL DESIGN.

BIOT PRINCIPLE [physics] (Jean Baptiste Biot, 1774–1862, French physicist) The rotation produced by optically active media is proportional to the length of the path, to the concentration, and to the inverse que of the wavelength of the light.[¶]

BIOTIC PRINCIPLE [biology]   The major biological principles considered are those indicated by the terms taxonomy, biogeography, ecology, morphology, physiology, reproduction, development, genetics, behavior, and evolution.[‖] See ANTHROPIC PRINCIPLE.

BIOT–SAVART PRINCIPLE [engineering] (Jean Baptiste Biot, 1774–1862, French physicist) (Félix Savart, 1791–1841, French physicist) (Jean-Baptiste Biot, 1774–1862 and Félix Savart, 1791–1841) (1) Expression for the intensity of magnetic flux density produced at a point a distance from a current-carrying

*Freud, S., Hubback, C. J. M., and Jones, E. (1922), *Beyond the Pleasure Principle*, The International Psycho-Analytical Press, London.

[†]Chen, C.-C. and Koh, K.-M. (1992), *Principles and Techniques in Combinatorics*, World Scientific, p. 27.

[‡]Schweitzer, Y. and Shay, S. (2008), *The Globalization of Terror: The Challenge of Al-Qaida and the Response of the International Community*, Transaction Publishers, New Brunswick, p. 14.

[§]Ng, S.K., Carter, S.J.B., and Bullen, F. (2006) A Biomimicry approach to automating visual road surveys, 22nd ARRB Conference—Research into Practice, Canberra Australia, http://rakan.jkr.gov.my/cawangan/Cjalan/documentation/seminar CourseNotes/arb/Papers/Ng.pdf.

[¶]Walker, P. M. B. (1999), *Chambers Dictionary of Science and Technology*, Chambers, New York, p. 115.

[‖]Stahnke, H. L. (1961), *Biotic Principle*. Ulus.. Charles E. Merrill Books.

conductor.* (2) A constant electric current carried by a wire gives the intensity for a magnetic field.[†]

BIOT'S VARIATIONAL PRINCIPLE (BVP) [engineering] (Jean-Baptiste Biot, 1774—1862, French physicist)   Conservation of energy for thermal conduction systems confined to special and dependent variables.[‡]

BIREFRINGE   BIREFRINGENT   PRINCIPLE [physics, engineering]   (1) Splitting of incident light into two rays vibrating at right angles to each other and causing two images to appear.[§] (2) Frequency doubling electrooptic modulation method.[¶]

BIRMAN–SCHWINGER   PRINCIPLE [computer science] (M. Sh. Birman, Russian Mathematician)   (Julian Seymour Schwinger, 1918–1994,   American   physicist)   See ORTHOGONAL PROJECTION PRINCIPLE.

BISMUT MAXIMUM PRINCIPLE. See   BISMUT PRINCIPLE.

BISMUT PRINCIPLE [mechanics] (Jean Michel Bismut, 1948-, French mathematician)   is a professor at Université Paris-Sud, France) (1) Fundamental importance to the semiconductor industry, which supplies solid-state components, such as integrated circuits, for the manufacture of electronic equipment.[∥] (2) Intense surface heat transfer associated with radiation, finite-element solutions display anomalous behaviors.[**] (3) Continuous differentiability of the functions of the equality constraints and a subdifferential regularity of the function of the inequality constraints.[††] (4) Modification of triangle-based adaptive stencils for the solution of scalar hyperbolic conservation laws.[‡‡] Also known as *maximum principle; boundary-point principle*; *bismut maximum principle*.

BISTABILITY PRINCIPLE [optics]   Two stable states of transmission generally obtained by controlling transmission by means of positive feedback from the optical output. Optical bistable devices are of two types: (1 intrinsic or all-optical type and (2) hybrid type.[§§]

BITTER PRINCIPLE [pharmacology]   Compound blocks the action of γ-aminobutyric acid (GABA), a possible presynaptic inhibitory transmitter.[¶¶]

BIVALENCE PRINCIPLE See PRINCIPLE OF BIVALENCE.

BLACK BOX PRINCIPLE [psychology]   An experimental process with measurable inputs and outputs can be used to obtain information about unknown instruments and

*Walker, P. M. B., (1999), *Chambers Dictionary of Science and Technology*, Chambers, New York, p.116
[†]van den Broek, S. P., Zhou, H., and Peters, M. J. (1996), Computation of neuromagnetic fields using finite-element method and Biot-Savart principle, *Med. Biol. Eng. Comput.* **34**(1), 21–26.
[‡]Vyrodov, I. P., Ermakova, N. G., and Matyugina, L. K. (1973), Use of Biot's principle in solving problems with moving boundaries along which the first boundary condition is an arbitrary time function, *Trudy Krasnod. Politekh. Inst.* **51**, 61–81.
[§]Walker, P. M. B. (1999), *Chambers Dictionary of Science and Technology*, Chambers, New York, p. 118.
[¶]Bartels, P. H. (1970), Principles of polarized light, In *Introd. Quant. Cytochem.*, International Tutorial on Quantitative Cytochemistry by Gunter F.Bahr, Editors George L. Wied, Guntur F. Bahrp Academic Press, NY pp. 519–538.

[∥]NASA (1994), *Bismuth-Tin Crystal Growth Monitored Using MEPHISTO Furnace*, Research and Technology, NASA Lewis Research Center, NASA TM-106764, p. 132.
[**]Delmarskii, Yu. K., Omel'chuk, A. A., and Zarubitskii, O. G. (1976), Principles of bismut anodic refining in a melt of potassium and zinc chlorides, *Ukrainskii Khim. Zh.* **42**(11), 1202–1204.
[††]Bismut, J. M. (1973), Conjugate convex functions in optimal stochastic control, *J. Math. Anal. Appl.* **44**, 384–404.
[‡‡]Serezhkin, V. N., Pushkin, D. V., and Serezhkina, L. B. (2006), Maximum filling principle and sublattice characteristics for the atoms of period 6 elements, *Russ. J. Coord. Chem.* **32**(11), 801–810.
[§§]Suhara, T. and Haruna, M. (1989), *Optical Integrated Circuits*, McGraw Hill, p. 331.
[¶¶]Curtis, D. B. and Watkins, J. C. (1965), The pharmacology of amino acids related to gamma-aminobutyric acid, *Pharmacol Rev.* **17**, 347–391.

mechanisms.* See also FEEDBACK PRINCI-
PLE; ISHIKAWA'S PRINCIPLE; INFORMATION-AND-
ENERGY PRINCIPLE; INTENTION-TO-TREAT PRIN-
CIPLE; OPEN–CLOSED SYSTEM PRINCIPLE.

BLACK HOLE PRINCIPLE [physics] (1) A
gravity so strong that no light can escape
and time itself is stretched.[†] (2) (Stephen
Hawking, British physicist, Cambridge
Univ.) A particle sucked into a black hole,
as a result of entropy, would eventually
have to exit the hole somewhere, sometime.[‡]
(3) The infinity state exists at the center of
black holes. This is known as the singularity,
an *unknowable* state. Like infinity, black
holes inherently cannot be perceived because
they are outside the spacetime continuum.
For reasons unknown, the undifferentiated
state begins to "slow down" (decelerate).
As it slows down, it spirals. This spiral is
in a configuration known as the *Fibonacci
series*, which is the hallmark of natural
patterns. As light spirals and slows down,
it enters the vibration of both negative
and positive spacetime regions. In this
way antimatter and matter are created.
The event horizon and inward collectively
represents the area of the black hole that
can no longer be perceived because it is
moving too rapidly. The antimatter region
(the dark matter found around black holes)
it is also responsible for the magnetic fields
present around a black hole. It is too fast
to see because it is in the realm of dark
matter. It is a creative entity that shapes the
matter region through gravitational force.
Arising from the region of black holes are
very rapid emissions of electrons that are
traveling at 95% of the speed of light. They
are traveling at this speed because they
are at the boundary between what can and
what cannot be perceived. These electrons
still must slow down to the speeds normally

encountered. The jets of matter that are
created continue to form galaxies, which are
shaped by the antimatter regions.[§] (4) The
universe is made up of a spinning fractal
pattern of light spinning from the light of
infinity into all the forces of the universe,
matter, and antimatter. The forces created
are reflections of the photon–photon electron
positron. Involves slowing of light from
infinity at the center of black holes, and at
event horizon, the movement from infinite,
undifferentiated light from the singularity
inside a black hole. This region exists outside
spacetime and slows as it moves in a spiral
motion to the zone within spacetime. The
movement creates the gravitational forces.[¶]
(5) [mathematics] Claims the worst-case
instances of an algorithmic problem lie in
a "black hole," which is a negligible set (in
rigorous mathematical sense).[‖] (6) [Liter-
ature] The absence of narrative attracts
a whole kaleidoscope of alien narrative
elements. In this case a scene, lacking a
plot, on a snippet of film, of little value by
itself creates such a strong magnetic field
that the plot winds around it like cotton
candy. However, a black hole is capable
of emanating intrigue only when someone
creates a situation necessary for it.[**] See
also STELLAR PRINCIPLE; PRINCIPLE OF BLACK
HOLE COMPLEMENTARITY; ANTHROPIC PRINCI-
PLE; REFLECTIONLESS APERTURE PRINCIPLE;
OPTICAL BLACK HOLE PRINCIPLE; HYDROGENIC
PRINCIPLE.

BLICHFELDT'S PRINCIPLE [mathematics]
(Hans Frederik Blichfeldt, 1873–1945),

*Meyer, R. D., Tamarapalli, J. R., and Lemons, J.
E. (1993), Arthroscopy training using a "black box"
technique, *Arthroscopy* **9**(3), 338–340.
[†]Barrow, J. D., and Tipler, F. J. (1986),
The Anthropic Cosmological Principle, Oxford
Univ. Press.
[‡]Hawking, S. W. (1976), Breakdown of predictabil-
ity in gravitational collapse, *Phys. Rev. D* **14**,
2460–2473.

[§]Bardeen, J. M., Carter, B., and Hawking, S. W.
(1973), The four laws of black hole mechanics, *J.
Commun. Math. Phys.* **31**(2), 161–170.
[¶]Argyrisa, J., Ciubotariu, C., and Matuttisa, H. G.
(2001), Fractal space, cosmic strings and spontaneous symmetry breaking, *Chaos Solitons Fractals*
12(1), 1–48.
[‖]Myasnikov, Alexei (2004) A new approach to
algorithmic problems in groups: stratification,
randomization, and black holes. Temple University
Mathematics Colloquium. http://math.temple.edu/
nqutierre/colloquium/myasnikovabtract.html.
[**]Adler, R. J., Chen, P., and Santiago, D. I. (2001),
The generalized uncertainty principle and black
hole remnants, *J. Gen. Relativ. Grav.* **33**(12),
2101–2108.

American mathematician) Any bounded planar region with positive area placed in any position of the unit square lattice can be translated so that the number of lattice points inside the region will be at least $A + 1$.[*] (2) Letting G be a group of linear transformations in n variables of finite order g, we find that if g is divisible by a prime p that is larger than $(n-1)(2n+1)$, then G contains a normal abelian p subgroup.[†] (3) Let the n space defined by rectangular coordinates $x_1, x_2, \ldots, x_n$ be divided into equal rectangular spaces by n systems of planes $x_1 = a_1 + b_1 t, \ldots, x_n = a_n + b_n t$, where $(t = 0, \pm 1, \pm 2, \ldots)$ and where $a_1, a_2, \ldots, a_n$ and $b_1, b_2, \ldots, b_n$ are given real numbers. These spaces are called *fundamental parallelepipeds*. In each of them let there be located, in an arbitrary manner, a given number of points, say, k, with none of them, however, lying on the boundaries of the parallelepipeds. These points shall be called *lattice points*. One lattice point $(k = 1)$ is at the center of each resulting parallelepiped. Assume that S represents any limited open n-dimensional continuum in the n space $x_1, x_2, \ldots, x_n$ having the (outer) volume V. By a suitable translation $x_i' = x_i + \delta_i$.$(i = 1, 2, \ldots, n)$, this continuum can be placed in such a position with reference to the fundamental parallelepipeds that the number of lattice points L contained in the continuum or lying as near as we please to its boundary is greater than Vk/W, where W represents the volume, and k is the number of lattice points of a fundamental parallelepiped.[‡]

BLOSSOMING PRINCIPLE [mathematics]   (L. Ramshaw)   (1) Provides a very simple interpretation of all the classical notions about polynomial functions. (2) Letting $F(u) =$

$f(\underbrace{u, \ldots, u}_{n})$, we can give its pth derivative by

$$\frac{d^p}{du^p} F(u) = \frac{n!}{(n-p)!} \underbrace{\Delta(1, \ldots, 1)}_{p} f(\underbrace{u, \ldots, u}_{n-p})$$

(3) Every polynomial has a unique polar form. Thus, given any polynomial $F : R \to R^d$ of degree n, there exists a uniquely defined symmetric n-affine mapping $f : R^n \to R$ with $f(t, \ldots, t) = F(t)$. The function f is called the "blossom" or *polar* form of F.[§]

BODE'S PRINCIPLE [astronomy] aka Titius Bode Principle (Johann Elert Bode, 1747–1826, German astronomer) (Johann Daniel Titius, 1729–1796, German astronomer)   (J. Titius, 1766; J. Bode, 1772) A numerical relationship linking the distances of planets from the sun.[¶]

BODY   SIZE   PRINCIPLE   (BSP) [biology] Predicts smaller animals, because of their greater surface–mass ratio, should engage in more frequent tick removal grooming than larger animals in order to compensate for higher costs of tick infestation.[||] See also OPTIMALITY PRINCIPLE; SELF-ORGANIZING PRINCIPLE; PRINCIPAL-COMPONENT ANALYSIS; ORGANIZING PRINCIPLE; SIZE PRINCIPLE.

BOGOLUTOV PRINCIPLE [measurement] [Also transliterated Bogolyubov] (Nikolay Nikolaevich Bogolyubov, 1909–1992, Russian mathematician) (Soviet mathematical physicist Nikolai Nikolaevich Bogolutov, 1909–, recognized for his many fundamental contributions in physics and mathematics) (1) A systematic formulation of the renormalization program for perturbative computations

[*]Grotstein, J. S. (1990), Nothingness, meaninglessness, chaos, and the "black hole" I—the importance of nothingness, meaninglessness, and chaos in psychoanalysis, *Contemp. Psychoanal.* **26**, 257–290.
[†]de Buda, R. (1989), Some optimal codes have structure, *IEEE J. Select. Areas Commun.* **7**(6), 893–899.
[‡]Op. cit. de Buda.

[§]Alt, L. (1993), Rational linear reparametrization of NURBs and the blossoming principle, *Comput. Aid. Geom. Design* **10**(5), 465–467.
[¶]Walker, P. M. B. (1999), *Chambers Dictionary of Science and Technology*, Chambers, New York, p.128.
[||]Olubayo, R. O., Jono, J., Orinda, G., Groothenhuis, J. G., and Hart, B. L. (2000), Comparative differences in densities of adult ticks as a function of body size on some East African antelopes, *Afr. J. Ecol.* **31**(1), 26–34.

of the S matrix. In mathematics, among his many important contributions we cite his work on nonlinear mechanics and the general theory of dynamical systems.[*] (2) A form of the fundamental variational principle first derived for the classical case by Gibbs in his study on statistical mechanics. One uses determinants of single-particle wavefunctions rather than products, thereby introducing terms into the Hamiltonian.[†] See also BOGOLUTOV VARIATIONAL PRINCIPLE; HARTREE–FOCK BOGOLYUBOV PRINCIPLE; GIBBS–BOGOLUTOV VARIATIONAL PRINCIPLE; QUANTAL PRINCIPLE; GIBBS THIRD VARIATIONAL PRINCIPLE.

BOGOLUTOV VARIATIONAL PRINCIPLE [measurement]   See BOGOLUTOV PRINCIPLE.

BOHR COMPLEMENTARITY PRINCIPLE [physics] (Niels Bohr, 1885–1962, Danish Physicist) (1) Quantum theory referring to effects such as the wave–particle duality, in which different measurements made on a system reveal it to have either particle-like or wavelike properties.[‡] (2) A single quantum mechanical entity can either behave as a particle or as wave, but never simultaneously as both; a stronger manifestation of the particle nature leads to a weaker manifestation of the wave nature and vice versa.[§] See also HEISENBERG UNCERTAINTY PRINCIPLE.

BOHR PRINCIPLE [chemistry] (Niels Bohr, Danish Physicist, 1885–1962) Describes a nucleus with a positive charge orbited by an electron with a negative charge moving in a circle of radius. The velocity of the electron is equal to the centripetal force of electrostatic attraction.[¶] See BOHR COMPLEMENTARITY PRINCIPLE.

BOHR'S CORRESPONDENCE PRINCIPLE. See BOHR COMPLEMENTARITY PRINCIPLE.

BOHR'S FREQUENCY CORRESPONDENCE PRINCIPLE. See BOHR COMPLEMENTARITY PRINCIPLE.

BOLTZMANN'S PRINCIPLE [engineering] Bohr Principle of Complementarity see Bohr Complementarity Principle (Ludwig Eduard Boltzmann, 1844–1906, Austrian physicist) (1) Probability that a molecule of a gas in thermal equilibrium will have generalized position and momentum coordinates within given infinitesimal rays of values.[‖] (2) Applied to liquid water and vapor, depicts distribution of water molecules between vapor and liquid for pure liquid and for aqueous solutions. Yields the same equation for the osmotic pressure of water in a solution, pI (H_2O) as obtained from a kinetic treatment of Hullett's theory of osmosis. Distribution of water molecules between a liquid for pure liquid and aqueous solutions. When a negative pressure (tension) is applied to liquid water, it lessens (not enhances) the internal tension in the water because the increase in the molar volume of available space lessens the internal tension more than the applied tension encloses it.[**] (3) Entropy is a measure of the number of possible microscopic states (or microstates) of a system in thermodynamic equilibrium, consistent with its macroscopic thermodynamic properties (or macrostate). May be regarded as the foundation of statistical mechanics, which describes thermodynamic systems using the statistical behavior of its constituents.[††]

[*]Bogolubov, N. N., Jr. and Plechko, V. N. (1988), Approximation methods in the Polaron theory, *Riv. Nuovo Cimento* **11**(9).

[†]Logunov, A. A., Bogolubov, N. N., Jr., Kadyshevsky, V. G., and Shumovsky, A. S., eds. (1989), *Proc. 5th Int. Symp. Dedicated to N. N. Bogolubov on His 80th Birthday*, Dubna, Aug. 22–24, 1989, World Scientific Publishing, Teaneck, NJ.

[‡]Clark, S. J. (1988), Bohr's principle of complementarity, *Ann. Int. Med.* **109**(12), 994–995.

[§]Marinescu, I. M. (1970), Formalizing the complementarity principle of Bohr, *Rev. Fiz. Chim. Ser. A* **7**(2), 61–65.

[¶]Daintith, J. (1999), *The Facts on File Dictionary of Chemistry*, 3rd ed., Facts On File, New York, p. 34.

[‖]Herzfeld, K. F. (1914), Observations on Boltzmann's principle, *Silzb. Akad. Wiss. Math. Nat. Klasse* **122**, 1553–1561.

[**]Hammel, H. T. (1995), Boltzmann's principle depicts distribution of water molecules between vapor and liquid for pure liquid and for aqueous solutions, *J. Phys. Chem.* **99**(20), 8392–8400.

[††]Boltzmann, L. (1896), *Vorlesungen über Gastheorie*, 2 vols., Leipzig.

BOLTZMANN'S SUPERPOSITION PRINCIPLE [physics] (Ludwig Eduard Boltzmann, 1844–1906, Austrian Physicist) In a linear viscoelastic material, the accumulated viscoelastic creep strain resulting from a series of stress increments is the superposed sum of the creep response to the individual increments.*

BOLZANO-WEIERSTRASS PRINCIPLE (BWP). [mathematics] (Bernard Placidus Johann Nopomak Bolzano, 1781–1848, Bohemion mathematician) (Karl Theodor Wilhelm Weiestrauss, 1815–1897, German mathematician) Any bounded sequence of real numbers has a convergent subsequence.† See also LIMITED PRINCIPLE OF OMNISCIENCE; BOUNDED SEQUENCE PRINCIPLE; CONSTANT SUBSEQUENCE PRINCIPLE; CONVERGENT SUBSEQUENCE PRINCIPLE; BOLZANO-WEIERSTRASS PRINCIPLE; MONOTONE SEQUENCE PRINCIPLE.

BOND DIRECTIONAL PRINCIPLE [chemistry] (1) The covalent bond, in which bonding electrons are localized between the atoms, is directional; the bond has a specific orientation in space among the bonded atoms. Electrostatic forces in ionic bonds, in contrast, are nondirectional: they have no specific orientation in space. As a result, the specific orientation of electron pairs in covalent molecules imparts a characteristic shape to the molecules.‡ (2) The directions in which the principal tensile, compressive, and shear stresses are located in combined stress analysis. There are three principal directions that are mutually perpendicular.§

*Walker, P. M. B. (1999), *Chambers Dictionary of Science and Technology*, Chambers, New York, p.130.
†Mandelkern, M. (1988) Limited Omniscience and the Bolzano-Weierstrass Principle. Bull London Math Soc v20, p319–320.
‡Wang, J., Clark, B. J., Schmider, H. S., and Vedene, H., Jr. (1996), Topological analysis of electron momentum densities and the bond directional principle: The first-row hydrides, AH, and homonuclear diatomic molecules, A2, *Can. J. Chem.* **74**(6), 1187–1191.
§Tanner, A. C. (1988), The bond directional principle for momentum space wave functions: Comments and cautions, *Chem. Phys.* **123**(2), 241–247.

BONE INDUCTION PRINCIPLE (BIP) [mathematics, physics] See INDUCTION PRINCIPLE.

BOREDOM PRINCIPLE [mathematics] Crick and Watson gave complementary advice to the aspiring scientist based on the insight that to do your best work you need to make your greatest possible effort. Crick made the positive suggestion to work on the subject that most deeply interests you, the thing about which you spontaneously gossip—Crick termed this "the gossip test." Watson made the negative suggestion of avoiding topics and activities that bore you—which I have termed "the boredom principle."¶

BOREL–OKADA PRINCIPLE [mathematics] (Felix Edouard Justin Emile Borel, (1871–1956), French mathematician) (Susumu Okada, Japanese mathematician) If a regular summation method sums Σz^n to $1/(1-z)$ for all z in a subset S of the complex plane, given certain restrictions on S, then the method also gives the analytic continuation of any other function $f(z) = \sum a_n z^n$ on the intersection of S with the Mittag–Leffler star for f.‖

BORIC ACID AFFINITY PRINCIPLE [pharmacology] Water-soluble blue-colored acid derivative and a specific precipitation method for hemoglobin designed as a physician's office test for diabetes.**

BORONIC ACID AFFINITY PRINCIPLE See BORIC ACID AFFINITY PRINCIPLE.

BORWEIN–PREISS PRINCIPLE [mathematics, physics] (Arne Sigvard Eklund, b.6/19/1911, Kiruna, Sweden, nuclear physics, instrumentation, and atomic energy) Approximate solutions of scalar functions. The main idea is to replace the distance function (used in the Ekeland principle) and the power of the

¶Charlton, B. G. (2008), Crick's gossip test and Watson's boredom principle: A pseudo-mathematical analysis of effort in scientific research, *Med. Hypoth.* **70**(1), 1–3.
‖Korevaar, J. (2004), *Tauberian Theory: A Century of Developments*, Springer.
Higa, S. and Kishimoto, S. (1986), Isolation of 2-hydroxycarboxylic acids with a boronate affinity gel, *Anal. Biochem.* **154(1), 71–74.

distance function (used in the Borwein–Preiss principle) by a gauge-type lower semicontinuous function. Geometric methods in variational problems. Assume that (X, d) is a metric space a continuous function $\rho : X \times X \to [0, \infty]$ is a gauge-type function on a complete metric space (X, d), provided that (a) $\rho(x, x) = 0$, for all $x \in X$ and (b) for any $\varepsilon > 0$, there exists $\delta > 0$ such that for all $y, z \in X, \rho(y, z) \le \delta$ implies that $d(y, z) \prec \varepsilon$. The Borwein–Preiss variational principle states that if (X, d) is a complete metric space, and we let $f : X \to R \bigcup \{+\infty\}$ be a first function bounded from below, then, ρ is a gauge-type function and $(\delta_i)_{i=0}^{\infty}$ is a sequence of positive numbers, and if $\varepsilon > 0$ and $z \in X$ satisfy $f(z) \le \inf f + \varepsilon X$, then there exist y and a sequence $\{x_i\} \subset X$ such that (a) $\rho(z, y) \le \varepsilon/\delta_0, \rho(x_i, y) \le \varepsilon/(2^i \delta_0)$; (b) $f(y) + \sum_{i=0}^{\infty} \delta_i \rho(y, x_i) \le f(z)$; and (c) $f(x) + \sum_{i=0}^{\infty} \delta_i \rho(x, x_i) \succ f(y) + \sum_{i=0}^{\infty} \delta_i \rho(y, x_i)$, for all $x \in X\{y\}$.[*] (2) Extension of a complement to Ekeland's variational principle for functionals whose domain is a particular subset of a Banach space. See also PARAMETRIC EKELAND VARIATIONAL PRINCIPLE; EKELAND'S EPSILON VARIATIONAL PRINCIPLE; VECTOR EKELAND VARIATIONAL PRINCIPLE; EKELAND'S PRINCIPLE.

BOUNDARY HARNACK PRINCIPLE [Mathematics, Physics] (Carl Gustau Axel Harnack, 1851–1888, Baltic German mathematician) A Harnack inequality for conditioned Brownian motion.[†] Several closely related theorems about the convergence of sequences of harmonic functions. May be stated as follows:

THEOREM. Let D be a Lipschitz domain and V an open set. For any comp $K \subseteq V$, there exists a constant c_0 such that for all positive harmonic function u and v in D that vanish continuously on $(\partial D) \cap V$ with $u(x) = v(x)$ for so $x \in K \cap D$,

$$c_0^{-1} u(y) < v(y) < c_0 u(y) \quad \text{for all } y \in K \cap D.$$

[*]Hyers, D.H., Isac, G. and Rassias, T.M. (1997) Topics in nonlinear analysis & applications. World Scientific, p. 392.
[†]Bass, R.F. and Burdzy, K. (1989) A probabilistic proof of the boundary harnack principle. University of Washington.

See HARNACK PRINCIPLE; PICARD PRINCIPLE

BOUNDARY OF A BOUNDARY PRINCIPLE [mathematics, physics] For any n-dimensional manifold and any scalar valued $(n-2)$-form α, we have

$$\int_{\partial \partial M} \alpha = 0$$

Thus, we can use Stoke's theorem twice to get

$$\int_M dd\alpha = \int_{\partial M} d\alpha = \int_{\partial \partial M} \alpha = 0$$

We can write (3) at each point $x \in M$ for any neighborhood of x in M, which gives

$$dd\alpha = 0$$

and, since it is true for any form α, this makes relation

$$dd = 0$$

just another expression of the boundary of a boundary principle (1). This fact can be expressed mathematically as follows: operation d on cochains is dual to operation ∂ on chains and in our situation they express the same result.[‡]

The application of the boundary of a boundary principle in electrodymanics is straightforward. The electromagnetic field is assumed to be described by electromagnetic 2-form $F = \frac{1}{2} F_{\mu\nu} dx^\mu \wedge dx^\nu$ (a scalar valued 2-form on spacetime), the form being generated by 4-potential 1-form $A = A_\mu dx^\mu$ via

$$F = dA$$

The boundary of a boundary principle in the 1-2-3 form provides then equations for F:

$$dF = ddA = 0$$

See BOUNDARY PRINCIPLE.

[‡]Kheyfets, Arkady (1986) The boundary of a boundary principle: A unified approach. Foundations of Physics. Springer Netherlands, v16 (5) p483–497.

BOUNDARY-POINT PRINCIPLE [mathematics] (J. M. Bismut, Université Paris-Sud, France) See BISMUT PRINCIPLE.

BOUNDARY PRINCIPLE [mathematics, physics] See also UNIFORM BOUNDEDNESS PRINCIPLE; LOCAL UNIFORM BOUNDEDNESS PRINCIPLE; BOUNDARY PRINCIPLE; BOUNDARY OF A BOUNDARY PRINCIPLE; BOUNDEDNESS PRINCIPLE; NADIRDSHVLLI BOUNDARY PRINCIPLE; BOUNDARY HARNACK PRINCIPLE; BOUNDEDNESS PRINCIPLE; AUSTERITY PRINCIPLE; LOCAL UNIFORM BOUNDEDNESS PRINCIPLE.

BOUNDED SEQUENCE PRINCIPLE (BSP). [mathematics] Any sequence of positive integers is either bounded or unbounded.[*] See also LIMITED PRINCIPLE OF OMNISCIENCE; BOUNDED SEQUENCE PRINCIPLE; CONSTANT SUBSEQUENCE PRINCIPLE; CONVERGENT SUBSEQUENCE PRINCIPLE; BOLZANO-WEIERSTRASS PRINCIPLE; MONOTONE SEQUENCE PRINCIPLE.

BOUNDEDNESS PRINCIPLE [mathematics, physics] See BOUNDARY PRINCIPLE; BANACH–STEINHAUS UNIFORM BOUNDEDNESS PRINCIPLE.

BOYARSKY PRINCIPLE (M. Boyarsky American Mathematician) "If cohomology is parametrized rationally by a character then the Frobenius operation will vary continuously [locally analytically] with the character."

The p-adic gamma function is not only analytic; it is distinguished by a modular property, i.e. $\Gamma_p(x)$ behaves in a simple way under translation of x by elements of Z.

We interpret the Boyarsky principle with this in mind. Let us consider a family of cohomology spaces of fixed finite dimension, $W_{\bar{a}}$, indexed by $\bar{a}$ lying in the image of $U = Q \cap Z_p - Z_p$ in $Q \cap Z_p/Z$. Suppose that each fiber $W_{\bar{a}}$ is provided with bases $X(a)$ described "smoothly" in terms of $a \in U$. Suppose that under the inverse of Frobenius the space $W_{\bar{a}}$ is mapped onto $W_{\bar{b}}$, where $p\bar{b} = \bar{a}$ and let $\gamma(a,b)$ denote the matrix of this transformation relative to the bases $X(a), X(b)$. By

insisting that the representative b of $\bar{b}$ be chosen such that $pb - a = \mu$ lies in a fixed set (say $\mu \in \{0, 1, \ldots, p - 1\}$) we may interpret $\gamma(a,b) = \gamma(a, p^{-1}(a + \mu))$ as a (Boyarsky) function, Γ_B, of one variable, $a \in U$. The principle of Boyarsky asserts that Γ_B may be extended to a locally analytic function on a somewhat larger set, $\underline{S}$, and indeed more explicitly there exist p disks each of radius at least $|p|$ which cover U such that the restriction of Γ_B to each disk is a Krasner analytic function.[†]

BOYCOTT PRINCIPLE [economics] A philosophy bringing about the feud between South African followers and those of the radical United Democratic Front (UDF). Negotiation with the government would achieve more. Appeals to traditional values: discipline, self-improvement through hard work, and the authority of the Zulu chiefs. Does not respect hostility to destructive tactics, including economic sanctions. Violent opposition and foreign economic sanctions were most popular among blacks when a quick victory over apartheid looked plausible. As those hopes faded, so did the willingness to suffer imprisonment or unemployment.[‡]

BOYLE'S PRINCIPLE [geology] (Robert Boyle, 1627–1691, Anglo Irish philosopher, chemist, physicist & inventor) At constant temperatures, the voume of a given mass of gas is inversely proportional to the pressure of the gas.[§]

BRACKET MATCHING PRINCIPLE [electronics, computer science, psychology] See MATCHING PRINCIPLE.

BRAGG–BRENTANO PRINCIPLE [optics] (Sir William Lawrence Bragg, 1890–1971, British physicist, Nobel Prize) Symmetric

[*]Mandelkern, M. (1988) Limited Omniscience and the Bolzano-Weierstrass Principle. Bull London Math Soc v20, p319–320.

[†]Dwork, B. (1983) On the Boyarsky Principle. American Journal of Mathematics, v105, (1), pp. 115–156.

[‡]Economist (1988) South Africa; Time to reflect in the townships. The Economist, p.42.

[§]Stiegeler, S. E. (1977), *A Dictionary of Earth Sciences*, Pica Press, New York (distributed by Universe Books), p.121.

configuration using an X-ray powder diffraction scan.[*]

BRAGG REFLECTION PRINCIPLE [physics] (Sir William Lawrence Bragg, 1890–1971, British physicist, Nobel Prize) (1) Refractive index variations and thickness of the layers are chosen so that partial reflections interfere constructively at a narrow band of wavelengths but not at wavelengths outside the band.[†] (2) The light rays reflected by specular areas are essentially at the same intensity as the supplied light rays and thus diminishes only minimally, the farther the imaging device is located along the observation axis from the object to be observed. The light rays reflected by the diffuse areas, on the other hand, diminish substantially with distance the farther the imaging device is located along the observation axis from the object to be observed. This reflection principle of diffuse areas is commonly referred to as the *Inverse-square law*. Light within the selected narrow band is coupled back into the laser active area, enhancing single-longitudinal-mode operation, or multilongitudinal-mode operation with the modes restricted to lie within the bandwidth of the distributed Bragg reflector's reflection peak.[‡] (3) Any diverging field, including light, decreases as the inverse square of the distance from the source—in this instance the diffuse area.[§] See also REFLECTION PRINCIPLE.

BRAIN'S DIVISION OF FUNCTION PRINCIPLE [psychology] (1) Explores the nature of intuition. Adam Smith's lectures on rhetoric and belles lettres, and their relation to intuition. Smith's early work on grammar not only has relevance to his economics but also used what we now know to be the brain's division of labor function principle. The article also shows that Smith's early writings on ethics and socioeconomic patterns directly related to his economic treatise, the *Wealth of Nations*.[¶] (2) There is strong evidence for an *object shift* in English; that is, just as there is an *endplate potential position* (EPP) high in the sentence where subjects and derived subjects wind up, there is a similar EPP position in the *ventral pallidum* (VP) region, where objects and extracellular matrix (ECM) subjects (among other categories) wind up. A sentence must contain a tensed verbal element and a subject.[‖] See also EXTENDED PROJECTION PRINCIPLE.

BRAUN–LE CHATELIER PRINCIPLE [physics] (Karl Ferdinand Braun, 1850–1918, German inventor, physicist Nobel laureate) (Henry Louis Le Chatelier, 1850–1936, French chemist) aka Le Chatelier-Braun principle (1) Force equilibrium with a magnetic field responds to an abrupt change much less than to a change in the heating power.[**] (2) Equilibrium thermodynamics for two sets of general parameters of the molecular total energy function; when an external force is applied to a system at equilibrium, the system adjusts so as to minimize the effect of the applied force.[††] See PRINCIPLE OF BRAUN–LE CHATELIER.

BRÉZIS–BROWDERS MONOTONE PRINCIPLE (Haim Brézis, 1944- French mathematician) (Felix E. Browder, 1927- American mathematician) See BRÉZIS–BROWDER PRINCIPLE.

[*]Kolb, K. and Macherauch, E. (1964), A back-reflection goniometer based on the Bragg-Grentano principle for the x-ray strain measurement, *Zairyo* **13**(135), 918–919.

[†]Wang, S. (1974), Principles of distributed feedback and distributed Bragg reflector lasers, IEEE J. Quantum *Electron.* **10**(4), 413–427.

[‡]Omori, S., Ishii, H., and Nihei, Y. (1998), Principles of crystallography using Bragg reflection from atomically localized sources, *J. Electron Spectrosc. Related Phenomena* (88–91), 517–522.

[§]Slama, S., von Cube, C., Kohler, M., Zimmermann, C., and Courteille, P. W. (2005), Multiple reflections and diffuse scattering in Bragg scattering at optical lattices, Los Alamos National Laboratory, *Preprint Archives—Quantum Physics*, pp. 1–10.

[¶]Frantz, R. (2000), Intuitive elements in Adam Smith, *J. Socio-Econ.* **29**(1), 1–19.

[‖]Wood, J. D., Alpers, D. H., and Andrews, P. L. (1999), Fundamentals of neurogastroenterology, *Gut* **45**(Suppl. 2), II-6–II-16.

[**]Braun, F. (1910), The so-called LeChatelier-Braun principle, *Ann. Phys.* (Weinheim, Germany), **32**, 1102–1106.

[††]Ehrenfest, P. (1911), The principle of LeChatelier-Braun and the reciprocity laws of thermodynamics, *Z. Phys. Chem.* **77**, 227–244.

BRÉZIS–BROWDER ORDERING PRINCIPLE (Haim Brézis, 1944- French, mathematician) (Felix Browder, 1927- American mathematician)   See BRÉZIS–BROWDER PRINCIPLE.

BRÉZIS–BROWDER PRINCIPLE [mathematics] (Haim Brézis, 1944- French mathematician) (Felix Browder, 1927- American mathematician) (H. Brézis and F. E. Browder) (1) A bridge principle is a method by which one takes two minimal surfaces with boundary, connects their boundaries together with a thin "bridge," and produces a new minimal surface whose boundary is the bridged boundary, and is close to the old minimal surfaces connected by a thin strip. Thus it is a method for constructing new examples of minimal surfaces, with certain properties. Stated mathematically: Stable minimal surfaces in a Riemannian three-manifold $\bar{M}$ may be constructed by means of "bridges." Let M_1 and M_2 be strictly stable compact minimal surfaces with boundary immersed in $\bar{M}^3$, and choose points $x_i \in \partial M_i$, so that x lies on the boundary of the *locally convex hull* $H(M_i)$. Suppose that π is a smooth regular curve in $\bar{M}$, from $x_1 = \pi(0)$ to $x_2 = \pi(1)$, so that $\pi'(0)$ points outward from $H(M_1)$ and $-\pi'(1)$ points outward from $H(M_2)$. Let a short arc (y_i, z_i) containing x_i be removed from ∂M_i, and choose arcs τ from y_1 to y_2 and ρ from z_1 to z_2, each C^2–close to π. Then the Jordan curve γ formed from $\partial M_1 (y_1, z_1), \tau, \partial M_2 (y_2, z_2)$, and ρ is the boundary of a strictly stable immersed minimal surface M, which is C^0–close to the union of M_1 and M_2 with a narrow strip between τ and ρ. (2) A uniform space equipped with a pre-order relation used for proving minimal point theorem; examples include providing an abstract Newton–Kantorovich scheme for solving nonlinear equations and fixed-point theorems for inward set-valued maps, related to Caristi–Kirk's results.[*] See also AXIOM OF CHOICE PRINCIPLE; TURINICI PRINCIPLE.

BRILLOUIN SCATTERING PRINCIPLE [physics] (Leon Brillouin, 1889–1969, French physicist) Scattering of light by the acoustic

modes of vibration in a crystal.[†] (2) Light scattering by acoustic phonons.[‡]

BROENSTED'S PRINCIPLE OF CONGRUENCE [chemistry] (Johannes Nicolaus Brönsted, 1879–1947, Danish chemist)   See BROENSTED'S PRINCIPLE.

BROENSTED'S PRINCIPLE [chemistry] (Johannes Nicolaus Brönsted, 1879–1947, Danish chemist) Interaction between the ions of acid and basic solutions on the exchange of protons during reactions. Acid and basic catalysis of many chemical reactions, recognizing the concept of enthalpy to characterize chemical reactivity.[§] See also PRINCIPLE OF CONGRUENCE.

THE CONTINUITY OF REAL FUNCTIONS The continuity of real functions is an immediate consequence of Brouwer's Continuity Principle, as we showed in Veldman [10] 1982. We now repeat this argument.

We first formulate a generalization of the principle that easily follows from the principle itself.

1. Let X be a subset of $\mathcal{N}$. X will be called a *spread* if and only if the following two conditions are satisfied:

    (i) For every finite sequence $s = \langle s(0), \ldots, s(n-1) \rangle$ of natural numbers one may decide if there exists α in X such that for each $i < n, \alpha(i) = s(i)$.

    (ii) For every α in $\mathcal{N}$, if for each n in $\mathbb{N}$ there exists β in X such that for each $i < n, \alpha(i) = \beta(i)$, then α itself belongs to X.

[†]Walker, P. M. B. (1999), *Chambers Dictionary of Science and Technology*, Chambers, New York, p.145.
[‡]Gammon, R. W. (1968), Crystal Brillouin scattering polarization selection rules, *Proc. Int. Conf. Light Scattering Spectra Solids*, p. 579.
[§]Soerensen, T. S. (1979), Broensted's principle of specific interaction of ions and the mean ionic activity coefficients in aqueous solutions of alkaline earth halides. Calculation of cationic radii from the ASPEV theory, *Acta Chem. Scand. A Phys. Inorg. Chem.* **A33**(8), 583–592.

[*]Brézis, H. and Browder, F. E. (1976), *Adv. Math.* **21**(3), 355–364.

A spread is a closed subset of Baire space $\mathcal{N}$ that satisfies the classically empty condition (i).

2. *Brouwer's Continuity Principle, general formulation:*

For every spread $X \subseteq \mathcal{N}$, for every $R \subseteq X \times \mathbb{N}$,

if $\forall \alpha \in X \, \exists m \, [\alpha R m]$, then

$\forall \alpha \in X \, \exists n \exists m \forall \beta \in X$ [if for every $i < n, \alpha(i) = \beta(i)$, then $\beta R m$].

One may prove 2 from 1 by defining a so-called retraction of $\mathcal{N}$ onto X, that is, a continuous function r from $\mathcal{N}$ onto X such that for every α in $X, r(\alpha) = \alpha$, and then arguing straightforwardly.*

BROUWER'S   PRINCIPLE [mathematics] (Luitzen Egbertus Jan Brouwer, 1881–1966, Dutch mathematician) (1) Concept of the continuum is based on the notion of a choice sequence. In turn, the mathematical treatment of choice sequences is based on two fundamental ideas of Brouwer: the Principle of Continuity and the Principle of Bar Induction (PBI). The intuitive meaning of the first principle is quite clear. It reflects the developing, non-completed nature of choice sequences. As only an initial segment of a choice sequence is available to an observer, only such "finite" information can be used in an intuitionistic assignment of natural numbers to choice sequences. Hence, every such assignment is necessarily continuous.[†] (2) The strongest form of the principle applies to formulas of the form $\forall \alpha \exists \beta A(\alpha, \beta)$, where α and β are function variables. Shows how the principle affords new and simplified proofs of some of the

basic theorems of intuitionistic topology. A continuity axiom for intuitionistic analysis.[‡]

BUCKING MAGNET PRINCIPLE [engineering] Uses opposing magnets to push the magnetic field back. Beside lowering the operating pressure of a magnetron, the bucking magnet also can significantly improve the discharge efficiency and thus the electron density of the discharge of a magnetron even at regular sputtering pressures. The bucking magnet also has the desirable effect of lowering the impedance of the sputtering source and increasing its emissions, due to the increase in electron densities within the discharge brought about through enhanced confinement.[§] See also MAGNET PRINCIPLE; LINEAR MAGNET PRINCIPLE; FRICTION WHEEL DRIVE PRINCIPLE; PLUNGER MAGNET PRINCIPLE; LOCK MAGNET PRINCIPLE; CERAMIC MAGNET PRINCIPLE.

BUILDING BLOCK PRINCIPLE [engineering] Combination of products of a wide range using high- and low-temperature heat transfer applicable to thermal bonding, cooling, heat setting, shrinking, and curing.[¶]

BUOYANCY CAPTURE PRINCIPLE [engineering] Capture efficiency equals the ratio of capture flow rate to total plume flow rate in a confined space.[‖]

BUSBAR PROTECTION PRINCIPLE [engineering] (1) Breaker failure protection. Ensures protection of electric nodes, based on laws such as Kirchhoff, the impedance variation,

*Berger, U., Osswald, H., and Schuster, P. (2001) Reuniting the Antipodes: Constructive and non-standard views of the continuum (Synthese Library) Springer. p 285.

[†]Wiegers, G. A., De Boert, J. L., Meetsma, A. and van Smaalen, S. (2001), Domain structure and refinement of the triclinic superstructure of 1T-TaSe2 by single crystal x-ray diffraction, *Z. Kristallogr.* **216**(1), 45–50.

[‡]Berger, U., Osswald, H. and Schuster, P. (2001) Reuniting the Antipodes: Constructive and Non-standard Views of the Continuum (Synthese Library). Springer, p. 119 and [(Op. Cit. p.285).]

[§]Ray, N. and Waghmare, U. V. (2007), Coupling between magnetic ordering and structural instabilities in perovskite biferroics: A first- principles study, *Condensed Matt.* Phys. Rev B.77, 134112 v. 77, Issue 13.

[¶]Weigelt, O. (1970), Metabolism cage for rats on the building block principle, *Z. Versuch.* **12**(1), 68–76.

[‖]Li, Y. et al. (1997), Residential kitchen range hoods. Buoyancy capture principle and capture efficiency revisited. *Int. J. Indoor Air Qual. Climate* **7**, 151.

and the admittance variation, in order to distinguish internal faults of the electric node from the external faults.* (2) Takes into account the application of the protection and the law of the nodes (the sum of currents entering and leaving a node must equal zero).† Also known as *protection principle; d'Alembert relay principle; distance principle in relay protection; novel shielding principle; principle of protection scale selectivity; principle of reasonableness; principle of structural protection; radiation protection principle; high-impedance principle.*

BUTT-WELDING   PRINCIPLE [engineering] Heat welding depends on three process variables: temperature, pressure, and time. In the process, two parts are loaded into inexpensive holding fixtures and a heated platen is moved between the parts, which are then pressed against opposite sides of the platen until their mating surfaces are plasticated to a predetermined degree.‡ See also WELDING PRINCIPLE; RESISTANCE WELDING PRINCIPLE; TRANSMISSION WELDING PRINCIPLE.

BYGONES PRINCIPLE [mathematics] An economic theory used in business. Economists stress the "extra" or "marginal" costs and benefits of every decision. The idea is to not look backward when making decisions and

stresses the importance of ignoring past costs in future decisionmaking. When making a decision, one should make a hard-headed calculation of the extra costs one will incur and weigh these against its extra advantages. It emphasizes the importance of only taking into account the future costs and benefits when making decisions.§

BYPASS OVERFLOW PRINCIPLE [engineering] Based on fluid pulsation initiative control method, namely, installing the piezoelectric ceramics with main pipeline connections to actuate the servo valve driving shock absorber, producing a secondary pulsating wave and driving shock absorber's overflow to the original pressure. The fluid structure interaction vibration shock counterbalances mutually, causing reduction in fluid pulsation and prevents mechanical fatigue.¶

BYPASS PRINCIPLE [engineering]   (1) An alternating, usually smaller, diversionary flow path in a fluid dynamic a bypass system to avoid some device, fixture or obstruction. (2) Controlling valves to isolate evaporation chamber from environmental motion.‖

BYRNES–MARTIN   INTEGRAL   INVARIANCE PRINCIPLE (C. J. Byrnes, C. F. Martin). See INVARIANCE PRINCIPLE.

*Kasztenny B., Sevov L. and Brunello G. (2001), Digital low-impedance busbar protection, review of principles and approaches, *Proc. 54th Annual Conf. Protective Relay Engineers*, College Station, TX, April 3–5, 2001.

†Hewitson, L., Brown, M. and Ramesh, B. (2004), *Practical Power Systems Protection*, Elsevier, p. 65.

‡Yamaguchi, Y. (1990), Welding by means of ultrasonic processing, *Kagaku Kogyo* **41**(5), 410–419.

§Gottschall, W. C. (1996), Exams as teaching training tools, *Book of Abstracts*, 211th ACS National Meeting, New Orleans, March 24–28,1996.

¶Ouyang, P., Liu, H., and Jiao, Z. (2007), Active control of fluid pulsation based on bypass overflow principle, *Hangkong Xuebao/Acta Aeronaut. Astron. Sin.* **28**(6), 1302–1306.

‖Forsberg, E. (1940), Oil purifying with continuous lubrication, *Gas Oil Power* **35**, 63–66.

C

CACCIOPPOLI–BANACH PRINCIPLE [mathematics] (Renato Caccioppoli, 1904–1959, Italian mathematics) (Stefan Banach, 1892–1945, Polish mathematician) Theorem stating that if a mapping f of a metric space E into itself is a contraction, then there exists a unique element x of E such that $fx = x$.[*] See BANACH PRINCIPLE. Also known as *Banach's fixed-point theorem*.

CALCULATION PRINCIPLE [mathematics] Performs logic and arithmetic digital operations based on numerical data.[†]

CALDERÓN–ZYGMUND PRINCIPLE [mathematics] (Alberto Calderón, 1920–1998, Argentine mathematician) (Antoni Zygmund, 1900–1992, American mathematician) Every singular integral operator is controlled in an appropriate sense by a maximal operator.[‡] See also ORTHOGONAL PROJECTION PRINCIPLE.

CALORIMETRIC MEASUREMENT PRINCIPLE [engineering] Thermochemical characteristics of propellants and explosives; heat of combustion; heat of explosion; heat of formation and heat of reaction.[§]

CAPACITANCE PRINCIPLE [engineering] Leak detection system providing a visual signal and operation of additional warning or shutdown should the system be compromised.[¶] See also CONDUCTANCE PRINCIPLE.

CAPILLARY ELECTROPHORESIS PRINCIPLE [biology] An efficient analytical technique for determination of a wide variety of charged as well of uncharged biomolecules of both low and high molecular weight. High-performance capillary electrophoresis: principles, present possibilities, and future potentialities in studies of low- and high-molecular-weight charged and uncharged biomolecules.[‖] (2) Separation technique performed in capillaries smaller than 100 μm according to several modes for chemical and biochemical analysis. An analytical technique used as a more rapid separation power.[**] See also ION PAIR PRINCIPLE.

CAPTURE PRINCIPLE [engineering, physics] Acquire an additional particle by process by atomic or nuclear process such as substituting a trace element for a lower-valence common element.[††] See also AMINE CAPTURE PRINCIPLE; PHASED CAPTURE PRINCIPLE.

CARATHÉODORY'S PRINCIPLE [physics] (Constantin Carathéodory, 1873–1950, Greek mathematician) (Constantin Carathéodory,

[*]Zabreiko, P. P. and Tarasik, T. V. (2004), The Banach-Caccioppoli principle for operators in K-normal linear spaces, and stochastic differential equations (Russ.), *Dokl. Nats. Akad. Nauk Belarusi* **48**(3), 41–45, 125–126.
[†]Bubeck, B. (1993), Renal clearance determination with one blood sample: improved accuracy and universal applicability by a new calculation principle, *Semin. Nucl. Med.* **23**(1), 73–86.
[‡]Petermichl, S. (2003), Bellman functions and continuous problems, *Proc. 1st Joint Meeting* between the *RSME* and the *AMS*, (American Mathematical Society), Seville, Session, Sevilla, June 18–21, 2003.
[§]Mansson, M. (1988), Basic principles for the determination of calorific values. Chemical abstracts VIII(10), p. 407.

[¶]Hong, S., Woo, J., Shin, H., Jeon, J. U., Pak, Y. E., Colla, E. L., Setter, N., Kim, E., and No, K., (2001), Principle of ferroelectric domain imaging using atomic force microscope, *J. Appl. Phys.* **89**(2), 1377–1386.
[‖]Michaelsen, S. and Sorensen, H. (1994), *Polish J. Food Nutr. Sci.* **3**(1), 5–44.
[**]Robert, F., Bouilloux, J. P., and Denoroy, L. (1991), Capillary electrophoresis: Principle and applications, *Ann. Biol. Clin.* **49**(3), 137–148.
[††]Grubitsch, H. (1966), The area of capture principle with oxygen corrosion in electrolytes, *Werkstoffe Korros.* **17**(8), 679–685.

German mathematician, 1873–1950) (1) Second law of thermodynamics: In the neighborhood of any arbitrary equilibrium state J of a thermally isolated system, there are other equilibrium states J' that are inaccessible from state J by means of a quasistatic process. This axiom allows the existence of an integrating factor for heat transfer in an infinitesimal reversible process for a physical system of any number of degrees of freedom.[*] (2) Light traveling through some substance has a speed that is determined by the substance. The actual path taken by light between any two points, in any combination of substances, is always the path of least time that can be traveled at the required speeds.[†] See also FERMAT'S PRINCIPLE; CONTES PRINCIPLE; STATIONARY TIME PRINCIPLE; THERMODYNAMIC PRINCIPLES.

CARDAN PRINCIPLE [engineering] (Gerolamo Cardano, 1501–1576, Renaissance mathematician)   A heavy body is fixed at one point through an arrangement of rings.[‡]

CARNOT'S PRINCIPLE [physics]   (Nicolas Leonard Sadi Carnot, 1796–1832, French physicist)   No heat engince can be more efficient than a reversible engine working between the same temperatures. The efficiency of a reversible engine is independent of the working substance and depends only on the temperatures between which it is working.[§] Also referred to as *principle of Carnot*.

CAROTHERS' PRINCIPLE [chemistry] (Wallace Hume Carothers, 1896–1937, American chemist)   A comonomer mole ratio of unity results in the highest molecular weights.[¶]

CAROUSEL PRINCIPLE [engineering]   A rotating transport system that transfers and presents workpieces for loading and unloading by a robot or other machine.[∥]

CARTER'S WEAK ANTHROPIC PRINCIPLE (WAP) See WEAK ANTHROPIC PRINCIPLE.

CASTIGLIANO'S PRINCIPLE [engineering] (Carlo Alberto Castigliano, b. 1847–1884, Italian mathematician) 11/9/1847, Asti, Italy).   See LEAST-ACTION PRINCIPLE, MENABREA'S PRINCIPLE; PRINCIPLE OF COMPLEMENTARY ENERGY.

CASUISTIC PRINCIPLE [psychology]   An act with two effects, one right and one wrong, can be performed when fair conditions are met—the affect of activity, the intention, material cause of the act, proportionate reason; also, right or indifferent action, not intrinsically wrong; wrong, although unforeseen, cannot be intended, wrong effect cannot be a means to the right effect; for the wrong effect to occur, there must be proportionate reason.[**] See also PRINCIPLE OF DOUBLE EFFECT; PRINCIPLE OF DO NO HARM; PRINCIPLES OF BIOMEDICAL ETHICS; EXCEPTION-GRANTING PRINCIPLE; JUSTIFYING PRINCIPLE; DOUBLE-EFFECT PRINCIPLE.

CATALYSIS MAXIMIZATION PRINCIPLE [mathematics]   Technique for parameter estimation based on actual expert decisions. Applied to derive a scoring function of Leontief type. The method is applied to the more important problem of deriving a linear scoring function. Hence the results are more directly comparable to linear discriminant analysis.[††] Also

[*]Thewlis, J., ed. (1962), *Encyclopaedic Dictionary of Physics*, Pergamon Press, New York/Oxford/London.

[†]Pogliani, L. and Berberan-Santos, M. N. (2000), Constantin Carathéodory and the axiomatic thermodynamics, *J. Math. Chem.* **28**(1–3), 313–324.

[‡]Cheng, P. L., Nicol, A. C., and Paul, J. P. (2000), Determination of axial rotation angles of limb segments—a new method, *J. Biomech.* **33**(7), 837–843.

[§]Walker, P. M. B. (1999), *Chambers Dictionary of Science and Technology*, Chambers, New York, p. 176.

[¶]Nomura, N., Tsurugi, K., Rajan Babu, T. V., and Kondo, T. (2004), Homogeneous two-component

polycondensation without strict stoichiometric balance via the Tsuji-Trost reaction: Remote control of two reaction sites by catalysis, *J. Am. Chem. Soc.* **126**(17), 5354–5355.

[∥]Heller, Z. H., D'Aquino, M., Alvite, A., Becker, J., Cronin, P., Giegel, J., Hatch, W. P., and Intengan, F. S. (1984), Design principles of the Stratus fluorometric immunoassay instrument, *Biomed. Sci. Instrum.* **20**, 63–72.

[**]McCarrick, P. M. (1995), Principles and theory in bioethics, *Kennedy Inst. Ethics J.* **5**(3), 279–286.

[††]Grinbergs, A. (1998), Equation revealing a new approach to enzyme catalysis maximization principle on symmetry basis, *Latv. Kim. Z.* (2), 89–90.

known as *governing principle*; *principle of mean fitness maximization; maximum decisional efficiency principle.*

CAUCHY'S ARGUMENT PRINCIPLE [mathematics] (Baron Augustin-Louis Cauchy, 1789–1857, French mathematician (1) Used in complex analysis. If a function $f(z)$ is analytic inside and on a contour C, then the integral of $f(z)$ taken around the contour is equal to zero.[*] (2) If a function $f(z)$ is a meromorphic function inside and on some closed contour C, with f having no zeros or poles on C, then a formula holds where N and P denote respectively the number of zeros and poles of $f(z)$ inside the contour C, with each zero and pole counted as many times as its multiplicity and order, respectively. This theorem assumes that the contour C is simple, without self-intersections, and is oriented counterclockwise.[†] Also referred to as *argument principle*.

CAUSALITY PRINCIPLE [psychology, mathematics] (1) An event cannot precede its cause.[‡] (2) Provides a theoretical and practical basis for integrating the activities of various group therapy approaches and for promoting interdisciplinary cooperation. The causality principle serves as the basis of a process- or action-oriented conceptualization of the therapeutic mechanism in psychotherapy. Implications of this principle for the therapist–patient relationship and for group dynamics in psychotherapy groups.[§] (3) Distinct causes have distinct effects (and vice versa); equal causes have equal effects.[¶]

(4) Conservation of the causal order in the time coordinate of any referential frame. A cause always precedes its effect.[‖] (5) Effect of information about one possible cause of an event as inferences regarding another possible cause.[**] (6) Analyzes situations until a minimal set of sufficient causes are identified; then, other possible causes are ignored or dismissed.[††] See also LIGHT PRINCIPLE; PRINCIPLE OF COVARIANCE; PRINCIPLE OF MINIMAL CAUSATION.

CAUTION PRINCIPLE [ecology] New technology must verify that it does not damage, either directly or indirectly, the environment. Burden of proof falls on the perpetrator, not the victim.[‡‡] See also PRECAUTION PRINCIPLE; UNCERTAINTY PRINCIPLE.

CAVALIERI PRINCIPLE [mathematics] [Bonaventura Francesco Cavalieri (Latin, Cavalerius), 1598–1647, Italian mathematician; originally discovered by third-century Chinese mathematician Liu Hui] (1) The volumes of two objects are equal if the areas of their corresponding cross sections are in all cases equal. Two cross sections correspond if they are intersections of the body with planes equidistant from a chosen base plane.[§§] (2) A principle used by Cavalieri in the early development of calculus. If two solids have equal heights and their sections at equal distances from the base have areas that always have a given ratio, then the volumes of the solids are in the same ratio.[¶¶] (3) Solids with the same height and with cross sections of equal area

[*]Walker, P. M. B. (1999), *Chambers Dictionary of Science and Technology*, Chambers, New York, p. 183.
[†]Krantz, S. G. (1999) The Argument Principle. Chapter 5 in Handbook of Complex Variables. Boston, Birkhauser p. 69–78.
[‡]Walker, P. M. B. (1999), *Chambers Dictionary of Science and Technology*, Chambers, New York, p. 183.
[§]Röhrborn, H. (1988), The causality principle of psychotherapy and ancillary methods in group therapy: Considerations for the basis of method combination (in German), *Psychiatr. Neurol. Med. Psychol.* **40**(8), 449–455.
[¶]Op. cit. (Rohrborn)

[‖]Meiman, N. N. (1964), *The Principle of Causality and Asymptotic Behavior of the Scattering Amplitude*. Zh. Eksp. Teor, Fiz., 47, p. 1966–1983.
[**]Bartholdi, E. and Ernst, R. R. (1973), Fourier spectroscopy and the causality principle, *J. Magn. Reson.* **11**(1), 9–19.
[††]Perepelitsa, V. F. (1981), *Causality Principle, Relativity Theory and Faster-than-Light Signals*. Institute of Theoretical and Experimental Physics, ITEP-100 and 165, Moscow.
[‡‡]Pellerin, D. (2004), From the caution principle to the unpredictable medical risk: Which medicine for the future? *Arch. Pediatrie* **11**(3), 197–200.
[§§]Bauer, J. et al. (1995), Estimation of multicellular colon carcinoma spheroids using Cavalieri's principle, *Pathol. Res. Pract.* **191**, 1192.
[¶¶]Bauer, J. (1996), Cavalieri's principle, *Am. J. Dermatopathol.* **6**, 159.

have the same volume, in particular, prisms or cylinders with equal bases and heights have the same volume.* (4) If the lengths of every one-dimensional slice are equal for two regions, then the regions have equal areas.[†] See also STEREOLOGICAL PRINCIPLE; PAPPUS CENTROLD PRINCIPLE.

CAVITATION   PRINCIPLE [engineering] Emulsification produced by disruption of a liquid into a liquid–gas two-phase system when the hydrodynamic pressure in a liquid is reduced to vapor pressure.[‡]

CEILING PRINCIPLE [forensics]   The ceiling principle was a method designed to be conservative in estimating probabilities, providing a frequency that would not overstate the strength of the evidence.[§]

CELL SURFACE RECEPTOR PRINCIPLE [biology]   The epidermal growth factor (EGF) is a potent multiplication-stimulating factor for several types of cultured cells. Molecules bind with a cell surface receptor. The binding event is accompanied by a reduction in the number of EGF receptors. Receptor loss acts to regulate the cellular response to the binding ligand.[¶]

*Michel, R. P. and Cruz-Orive, L. M. (1988), Application of the Cavalieri principle and vertical sections method to lung: Estimation of volume and pleural surface area, *J. Microsc.* **150**(Pt.2), 117–136.

[†]Lukomskii, Yu. Ya., Orlovskii, A. M., and Vasil'ev, V. V. 1973, Use of the Cavalieri principle during the study of metal electrocrystallization, *Izvest. Vysshikh Uchebnykh Zaved. Khim. Khimichesk. Tekhnol.* **16**(8), 1301–1303.

[‡]Yao, M. (2007), Apparatus for rapid wastewater treatment by using ultrasonic cavitation and electromagnetic field, in *Faming Zhuanli Shenqing Gongkai Shuomingshu*.

[§]Cohen, J. E. (1992), The ceiling principle is not always conservative in assigning genotype frequencies for forensic DNA testing, *Am. J. Human Geneti.* **51**(5), 1165–1168.

[¶]Harish, S. Haluk, R., and Steven, W. H. (2007), Cell surface receptors for signal transduction and ligand transport: A design principles study, *PLoS Comput. Biol.* **3**(6), 101.

CENTRIFUGATION PRINCIPLE [physics, biology]   Method applied to determine concentration or precipitate material from a substance by means of the centrifugal force.[∥] The tides can illustrate the complex affect of centrifugal and centripetal forces. There are lunar tides and solar tides. The Earth revolves around the Sun. The Moon revolves around the Earth. The combination of the Moon and the Earth together have a center of mass as they are considered together while revolving around the Sun. This combined mass is referred to as the Earth-Moon system whose center of gravity actually lies about 850 miles (1700 km) below the surface of the Earth. The lunar tides are a result from the interaction not only from the gravitational pull created by revolution of the Earth-Moon system's center of mass but also solar tides around the Earth-Sun center of mass.

Centrifugal forces is the outward directed force action on an object in or on an orbiting object. With the Earth-Moon system, the centripetal force, an inward directed force keeps the bodies from flying off into space as the system orbits the center of mass. On Earth, the tides feel a centrifugal force causing it to move outward.

Newton's principle of gravity, one object exerts a gravitational attraction on another object: the size of this attraction depends on the mass of the objects and on the distance between them. Thus the Earth and the Moon exert a gravitational attraction on each other. Because the revolution around the Earth is not a perfect circle, but rather an elliptical revolution, the Moon exerts more attraction on the near side of the Earth than at Earth's center and less attraction on the far side of the Earth than at Earth's center. Thus, the combinato of centrifugal force and gravitational force produces outward forces on both sides of the Earth. The tides feel both the force of gravity that results form the Moon's attraction and the centrifugal force resulting from the rotation of the Earth-Moon system around its center of mass.**

[∥]Sternbach, H. (1983), Centrifuging the right way, *LaborPraxis* **7**(1–2), 64, 66, 68, 70, 73.

**Marshak, Stephen (2002) Chapter 18, Restless Realm: Oceans and Coasts in Earth: Portrait of a Planet. W. W. Norton and Company.

CENTRIPETAL EXECUTION PRINCIPLE [psychology]   Action grammar governs elementary figure drawings acting like syntactic rule and stabilizes by the age of 8.[*]

CEPHALOCAUDAL PRINCIPLE [psychology] Children develop from head to toe, or cephalocaudally. Initially, the head is disproportionately larger than the other parts of the infant's body. The cephalocaudal theory states that muscular control develops from the head downward: first the neck, then the upper body and the arms, and then the lower trunk and the legs. Motor development from birth to 6 months of age includes initial head and neck control, then hand movements and eye–hand coordination, followed by upper-body control. The subsequent 6 months of life include important stages in learning to control the trunk, arms, and legs for skills such as sitting, crawling, standing, and walking.[†] Also known as PROXIMAL–DISTAL PRINCIPLE; GENERAL-TO-SPECIFIC PRINCIPLE.

CERAMIC MAGNET PRINCIPLE [engineering] The combination of a dielectric permanent magnet and the antenna coil generates energy at the same frequency and modulation as a comparable antenna coil without the permanent magnet, but at a substantially greater amplitude than the conventional antenna.[‡] See also BUCKING MAGNET PRINCIPLE; LINEAR MAGNET PRINCIPLE; FRICTION WHEEL DRIVE PRINCIPLE; PLUNGER MAGNET PRINCIPLE; LOCK MAGNET PRINCIPLE; MAGNET PRINCIPLE.

CERTAINTY EQUIVALENCE PRINCIPLE [physics]   A gravitational field is equivalent to an accelerating reference frame. It is sometimes expressed as the equivalence of gravitational and inertial mass, namely, that objects with the same inertia (resistance to acceleration) experience the same gravitational force.[§]

CERTAINTY FACTOR PROPAGATION PRINCIPLE [mathematics]   See WEAKEST-LINK PRINCIPLE.

CHAINING PRINCIPLE [psychology]   One can train behavior to meet a goal state by successively introducing steps necessary for obtaining the goal state and reinforcement, each time reinforcing a behavioral sequence that includes the newly introduced step as well as all previously introduced steps in appropriate order to achieve the goal state. As behavioral steps become linked to the goal state and reinforcement, each becomes conditioned to reinforce completion of previous steps and to signal continuation to the next step.[¶] See also BACKWARD CHAINING PRINCIPLE; ONE-TRIAL PRINCIPLE; SUCCESSIVE APPROXIMATION PRINCIPLE; RESPONSE-SHAPING PRINCIPLE; TRANSITIONAL STIMULI PRINCIPLE; SCHEDULE OF REINFORCEMENT PRINCIPLE.

CHALMERS PRINCIPLE OF ORGANIZATIONAL INVARIANCE [psychology] (Thomas Chalmers, 1780–1847, Scottish mathematician)   Any two systems with the same fine-grained functional organization will have qualitatively identical experiences.[‖]

CHALMERS PRINCIPLE OF STRUCTURAL COHERENCE (Thomas Chalmers, 1780–1847, Scottish mathematician)   The contents of awareness are those information contents that are accessible to central systems and brought to bear on the control of behavior.

[*]Groc, W. and Pissavy, M. (1972), Identification of precipitates in centripetal-radial immunodiffusion, *Clin. Chim. Acta* **42**(2), 423–431.
[†]Sun, H. and Jensen, R. (1994), Body segment growth during infancy, *J. Biomech.* **27**(3), 265–275.
[‡]Ellingson, W. A., Wong, P. S., Dieckman, S. L., Ackerman, J. L., and Garrido, L. (1989), Magnetic resonance imaging: A new characterization technique for advanced ceramics, *Am. Ceramic Soc. Bull.* **68**(6), 1180–1184, 1186.

[§]Denisov, V. I. and Denisov, M. I. (1999), Verification of Einstein's principle of equivalence using a laser gyroscope in terrestrial conditions, *Phys. Rev. D Particles Fields* **60**(4).
[¶]Seebass, G. (1982), Mediation theory and the problem of psychological discourse on "inner" events: Part 11, *Ratio* **24**(1), 29–43.
[‖]Adamson, A. W. and Williams, R. R., Jr. (1951), *Szilard-Chalmers Reactions. I. Principles of Enrichment*, National Nuclear Energy Series, Manhattan Project Technical Section, Division 4: Plutonium Project, 9 (Radiochem. Studies: The Fission Products, Book 1), p. 176–183.

According to this idea, awareness makes information available for global control of the organism's behavior.[*]

CHAOS SUPPRESSION PRINCIPLE [physics] Parameters of chaotic systems using non-self-resonant incentives chaos theory to achieve strong noise suppression against the background of weak square-wave signal detection.[†]

CHAOTIC PRINCIPLE [physics, biology, psychology]   (1) Uncertainties of position and motion should be no less than the limit set by Planck's constant. A reversible multiparticle system in a stationary state can be regarded as a transitive. Anosov system for the purpose of computing macroscopic properties. As the forcing strength grows, the attractor ceases to be an Anosov system and becomes an axiom A attractor.[‡] (2) Recent developments in the field of chaotic advection in hydrodynamical flows can pertain to the population dynamics of species competing for the same resource in an open aquatic system. If this aquatic environment is homogeneous and well mixed, then classical studies predict competitive exclusion of all but the most perfectly adapted species. (3) Spatial heterogeneity generated by chaotic advection can lead to coexistence. In open flows this imperfect mixing lets the populations accumulate along fractal filaments, where competition is governed by an "advantage of rarity" principle. (4) New algorithms and methods in applied mathematics and statistics, comprising determination of predictive rank, determination of stable solution, determination of measurement times, tailoring models, model choices in nonlinear multivariate regression, and estimation of nonlinear parameters.[§] See also CORRESPONDENCE PRINCIPLE; HEISENBERG PRINCIPLE.

CHARLES PRINCIPLE [physics, chemistry] (1) The volume of a given mass of gas at constant pressure is directly proportional to the absolute thermodynamic temperature.[¶] (2) For a given mass of gas at constant pressure, the volume increase by a constant fraction of the volume at $0°C$ for each Celsius degree rise in temperature.[‖] Also referred to as *Gay–Lussac's law*.

CHARNES COOPER EXTREMAL PRINCIPLE [mathematics] (Abraham Charnes, b. 1917, United States)   Relationships between levels of industrial activity and population group sizes. A reinterpretation of owner–consumer group size as an index of standard of living. Substantial changes in standard of living and other factors can result from relatively minor changes in total resource valuation if these changes are in critical resources.[**]

CHARNOV'S INVARIANCE PRINCIPLE [physics, mathematics, chemistry]   See INVARIANCE PRINCIPLE.

CHEMILUMINESCENCE PRINCIPLE [chemistry] Model gas is analyzed for its content. The difference of the concentration yields the concentration formed by air oxidation.[††] See also FLOW INJECTION CHEMILUMINESCENCE PRINCIPLE.

[*]Jansz, P. V., Wild, G., and Hinckley, S. (2008), Stepped mirrored structures for generating true time delays in stationary optical delay line proof-of- principle experiments for application to optical coherence tomography, *Proc. SPIE* p. 6801 (Photonics: Design, Technology, and Packaging III).

[†]Chen, L. and Wang, D. S. (2007), Detection of weak square wave signals based on the chaos suppression principle with nonresonant parametric drive, *Wuli Xuebao/Acta Phys. Sinica* **56**(9), 5098–5102.

[‡]Gallavotti, G. (1997), Chaotic principle: Some locations to developed turbulence, *J. Stat. Phys.* **86**, 907.

[§]Stribeck, N., Camarillo, A. A., and Bayer, R. K. (2004), Oriented quiescent crystallization of polyethylene studied by USAXS. Part 3: The evolution of crystallite stacking, *Macromol. Chem. Phys.* **205**(11), 1463–1470.

[¶]Walker, P. M. B. (1999), *Chambers Dictionary of Science and Technology*, Chambers, New York, p. 200.

[‖]Daintith, J. (1999), *The Facts on File Dictionary of Chemistry*, 3rd ed., Facts On File, New York, p. 53.

[**]Charnes, A., Littlechild, S., and Rousseau, J. (1974), *On a Simple Resource-Value Transfer Economy*, Texas Univ. at Austin Center for Cybernetic Studies, ADA005906.

[††]Garcia-Campana, A. M. and Baeyens, W. R. G. (2000), Principles and recent analytical applications of chemiluminescence, *Analysis* **28**(8), 686–698.

CHEMIOSMOTIC PRINCIPLE [biology]   (Peter D. Mitchell, 1920–1992, British biochemist) (1) At steady state the phoshorylation potential is in thermodynamic equilibrium with the protonmotive force.* (2) The theory that most adenosine triphosphate (ATP) synthesis in respiring cells comes from the electrochemical gradient across the inner membranes of mitochondria by using the energy of NADH–nicotinamide adenine dinucleotide (NAD$^+$) and nicotinamide adenine dinucleotide phosphate (NADP$^+$), two important cofactors found in cells. NADH is the reduced form of NAD$^+$, and FADH$_2$–flavin adenine dinucleotide (FAD), formed from the breakdown of energy-rich molecules such as glucose.[†] See also MITCHELL'S CHEMIOSMOTIC PRINCIPLES.

CHEMOSTERILANT PRINCIPLE [chemistry, biology]   Compounds that do not affect the universal DNA–RNA genetic mechanism.[‡]

CHETVERIKOV–HARDY–WEINBERG PRINCIPLE (HWP) [biology]   (Igor Vyacheslavovich Chyetverikov, 1909–1987, Soviet engineer) (Godfrey Harold Hardy, 1877–1947, English mathematician) (Wilhelm Weinberg, 1862–1937, German physician) (Godfrey Harold Hardy, British mathematician, 1877–1947) and Wilhelm Weinberg, 1862–1937, German physicist) In the absence of factors causing evolution within propulations (e.g., natural selection, nonrandom meeting, genetic drift, gene flow, mutation pressure), allelic and genotopic frequencies are expected to remain constant over time.[§] See also HARDY–WEINBERG PRINCIPLE.

CHIRAL SEPARATION PRINCIPLES Principle of chiral interaction (three-point rule) postulated in 1952 by Dalgliesh.

CHOICE UNCERTAINTY PRINCIPLE [computers]   It is impossible to make unambiguous choice between near-simultaneous events under a deadline.[¶]

CHURCH–TURING–DEUTSCH   PRINCIPLE (CTD) [computer science] (Alonzo Church, 1903–1995, American mathematician, Alan Mathison Turing, 1912–1954, English mathematician and David Elieser Deutsch, 1953–English physicist)   A universal computing device can simulate every physical process.[‖]

CLAIRVOYANCE PRINCIPLE [computer science]   Object-oriented computer program applying Eiffel programming language to resolve time-consuming memory management.[**]

CLASSICAL MANDREL PRINCIPLE. See SPIRAL MANDREL PRINCIPLE.

CLONAL SELECTION PRINCIPLE [engineering] The clonal selection algorithm (CSA) is a population-based stochastic method, with binary representation of the variables.[††]

COASE PRINCIPLE [mathematics]   (Ronald Coase, b. 12/29/1910, British economist, Nobel Prize)   (1) Relates to the economic efficiency of a government's allocation of property rights. Property rights are clearly defined and tradeable with zero transaction costs, then economic efficiency is independent of the original distribution of property rights. Under these assumptions, the allocation of property rights will have distributional consequences for the parties involved, but

*Kashket, E. R. (1982), Stoichiometry of the proton translocating ATPases of growing and resting aerobic Escherichia coli, *Biochemistry* **21**(22), 5534–5538.
[†]Green, D. E. (1981), A critique of the chemiosmotic model of energy coupling, *Proc. Natl. Acad. Sci. USA*, **78**(4), 2240–2243.
[‡]Borkovec, A. B. (1968), Chemosterilants in entomology and the sterile-male technique, *Proc. Symp. Isotope Radiation Entomology*, 1967, pp. 201–208.
[§]Schaap, T. (1980), The applicability of the Hardy-Weinberg principle in the study of populations, *Ann. Hum. Genet.* **44**(Pt.2), 211–215.

[¶]Denning, P. J. (2007), The choice uncertainty principle, *Commun. ACM* **50**(11), 9–14.
[‖]Church, A., Olszewski, A., Wolenski, J., and Janusz, R. (2006), *Church's Thesis after 70 Years*, Ontos mathematical logic, Volume 1, Ontos verlag.
[**]Howard, R. (1996), *Windows Tech J.* (3), 42–50.
[††]Zhang, Q. B, Wu, T. H., and Liu, B. (2007), Hybrid univariate marginal distribution algorithm based on clonal selection principle, *J. Zhejiang Univ. (Eng. Sci.)* **41**(10), 1715–1718.

will not affect the allocation of resources or the Pareto optimality of a market economy.[*]

CODE DIVISION MULTIPLE ACCESS (CDMA) PRINCIPLE [telecommunication, physics] (1) A spread spectrum technique in which many channels simultaneously occupy the whole of a single wideband channel.[†] (2) Within a radio cell, the base station defines on the "radio interface" to the mobile stations communicating with it the time division multiplex frame and thus the relative timing of the timeslots that are allocated in a link-specific manner to the information interchange in the frequency bands. With regard to the frame clock cycle phase used in operation, the base stations are in principle independent of one another.[‡] (3) Telecommunication apparatuses and second telecommunication apparatuses following the first telecommunication connections, taking into account the items of information in the first timeframes. Wireless telecommunication connections are set up to the second telecommunication apparatuses. Each further telecommunication connection following the first telecommunication connection is set up with priority on the same FDMA frequency. These named telecommunication systems thus also include, for example, the systems often designated *third system generation* in the context of a universal mobile telecommunication, probably based on the frequency division multiple access (FDMA), time division/domain multiple access (TDMA), and code division multiple access (CDMA) principles.[‡] See also FREQUENCY DIVISION MULTIPLE ACCESS (FDMA) PRINCIPLE; TIME DIVISION MULTIPLE ACCESS (TDMA) PRINCIPLE.

[*]Krueger, A. B. (1991), The evolution of unjust-dismissal legislation in the U.S., *Industr. Labor Relations Rev.* ILR Review, Cornell Univ. **44**, 644–660.
[†]Walker, P. M. B. (1999), *Chambers Dictionary of Science and Technology*, Chambers, New York, p. 228.
[‡]Ye, Z., Dai, Y.-T., Jie, S., Zhang, Y.-J., and Xie, S.-Z. (2007), Encoder and decoder of fiber Bragg gratings in optical code division multiple access system based on reconstruction-equivalent-chirp, *Wuli Xuebao* **56**(12), 7034–7038.

COEX PRINCIPLE See STANISLAV GROF'S COEX PRINCIPLE.

COEXTRACTION PRINCIPLE [chemistry] The extraction of a trace constituent with the carrier element in which the trace constituent element would not extract in macro.[§]

COGENERATION PRINCIPLE [engineering] Simultaneous onsite generation of electric energy and process steam or heat from same plant.[¶]

COGNITIVE ECONOMY PRINCIPLE [psychology] Asserting that the task of a category system is to provide maximum information with minimum cognitive effort. Refers to the role played by concepts to reduce the amount of information needed to process each individual experience of a concept instance. Two particularly well known kinds of categorization models implement the cognitive economy principle: classical models and prototype models. In classical models, a category is represented by a set of necessary and sufficient conditions. Typically, these models discard idiosyncratic information about instances in the process of establishing definitions. Likewise, standard prototype models typically discard idiosyncratic information and retain only high-frequency attributes in prototypes. There are numerous other ways to create category.[‖] (2) Underlies both the storing and the retrieval of conceptual information. Accordingly, concepts are defined by the properties and the attributes that establish their identity as well as by their relationships. It is the hierarchical organization of taxonomic relations binding them together that allows people to infer the

[§]Stromberg, N. and Hulth, S. (2001), An ammonium selective fluorosensor based on the principles of coextraction, *Anal. Chim. Acta* **443**(2), 215–225.
[¶]Krause, N. and Loffler, W. (2002), Cogeneration with fuel cells:—*Meteorit. Mchgt. Demonstration Plant. Gas, Wasser, Abwasser* **82**(2), 115–119.
[‖]Viezzer, M. (2006), *Autonomous Concept Formation: An Architecture-Based Analysis*, PhD thesis, Univ. Birmingham, p. 25.

shared properties and attributes that make the conceptual network coherent.[*]

COINDUCTION PRINCIPLE [mathematics, physics]   See INDUCTION PRINCIPLE.

COLLIDER OPERATIONAL PRINCIPLE [physics] Fabrication and displacement characteristics of functionally gradient piezoelectric ceramic actuator.[†] See also PRINCIPLE OF OPERATIONISM; REVISED PRINCIPLE; OPERATIONAL PRINCIPLE.

COLONNETT'S MINIMUM PRINCIPLE [mathematics] (Gustav Colonnetti. b. 11/8/1886, Turin, Italy)   Joint optimization method inspired by the deterministic annealing algorithm for data clustering, which extends previous work on tree-structured vector quantization using informative priors to approximate the unstructured solution while imposing the structural constraint.[‡]

COLOR DOPPLER PRINCIPLE [physics]   See DOPPLER PRINCIPLE.

COMBINATION METHOD PRINCIPLE [mathematics]   Based on a model of generalized contradictory equations and their problem solving that can be expanded as "layer tender" method and "competing for" tender method.[§] See also PANSYSTEMS PRINCIPLE; PERFECTION PRINCIPLE; PANOPTIMIZATION PRINCIPLE; REALITY PRINCIPLE; RELATIVITY PRINCIPLE; COMBINATION PRINCIPLE.

COMBINATION PRINCIPLE [engineering] Repeated combination and complements of various inclinations, superiorities, epitomes, methods, emphases, partialities, and stresses. All the pansystems the transformation method, strengthening method and simplification method can be considered as a referential framework to treat various problems in editorship, including certain concrete work.[¶] See also PERFECTION PRINCIPLE; PANOPTIMIZATION PRINCIPLE; REALITY PRINCIPLE; SIMPLIFICATION PRINCIPLE; STRENGTH PRINCIPLE; RELATIVITY PRINCIPLE;.

COMBINATION–SEPARATION PRINCIPLE (CSP) [engineering] (Walter Ritz, b. 2/22/1878, Sion, Switzerland)   Bergson[‖] Demonstrates the homology of human thought. See also RITZ'S COMBINATION PRINCIPLE.

COMBINATORIAL MATCHING PRINCIPLE [electronics, computer science, psychology]   See MATCHING PRINCIPLE.

COMBINATORIAL PRINCIPLE [mathematics] In proving results in combinatorics use: (1) *Rule of sum*: If we have a ways of doing something and b ways of doing another thing and we cannot do both at the same time, then there are $a + b$ ways to choose one of the actions. (2) *Rule of product*: If we have a ways of doing something and b ways of doing another thing, then there are $a \cdot b$ ways of performing both actions.[**] (3) *Bijective proof principle*: Technique that finds a bijective function $f : A > B$ between two sets A and B and thus proves that both sets have the same number of elements: $|A| = |B|$.[††] (4) *Double-counting principle*: Technique that involves counting the size of a set in two ways in order to show that the two resulting expressions for the size of

[*]Borghi, A. M. and Caramelli, B. (2001), *Taxonomic Relations and Cognitive Economy in Conceptual Organization* (http://laral.istc.cnr.it/borghi/borghicaramelli.pdf).
[†]Xie, J. (1982), Collider operational principle and the Beijing e+e- collider, *Gaoneng Wuli* **4**(1–3), 12.
[‡]de LaCaze-Duthers, F. J. H. (1904), Librairie de C Reinwald
[§]Xuemou, W. and Dinghe, G. (1999), Pansystems cybernetics: Framework, methodology and development, *Kybernetes* (MCB UP Ltd.) **28**(6/7), 679–694.

[¶]Nisio, S. (1966), From Balmer to the combination principle, *Jpn. Stud. Hist. Sci.* (5), 50–74.
[‖]Bergson, B. P. (1995), On the combination-separation principle: A metaphor for evolving system processes, *Technol. Forecast. Social Change* **50**(2), 171–183.
[**]Faigle, U. (1979), The greedy algorithm for partially ordered sets, *Discrete Math.* **28**, 153–159.
[††]Crawley, P. and Dilworth, R. P. (1973), *Algebraic Theory of Lattices*, Prentice-Hall, Englewood Cliffs, NJ.

the set are equal.* (5) *Pigeonhole principle*: Also known as *Dirichlet's principle* or *drawer principle*, states that if n pigeons are put into m pigeonholes, and $n > m$, then at least one pigeonhole must contain more than one pigeon. Another way of stating this would be that m holes can hold at most m objects with one object to a hole; adding another object will force you to reuse one of the holes. More formally, the theorem states that there does not exist an injective function on finite sets whose codomain is smaller than its domain.[†] (6) *Siegel's principle*: The quantitative application to the existence of integer solutions of a system of linear equations in Diophantine approximation.[‡] (7) The inclusion-exclusion principle relates the size of the union of multiple sets the size of each set and the size of each possible intersection of the sets. The smallest example is when there are two sets: the number of elements in the union of A and B is equal to the sum of the number of elements in A and B, minus the number of elements in their intersection. Generally, according to this principle, if $A_1, \ldots, A_n$ are finite sets, then

$$
\left| \cup_{i=1}^{n} A_i \right| = \sum_{i=1}^{n} |A_i| - \sum_{i,j:1 \leq i < j \leq n} |A_i \cap A_j|
$$
$$
+ \sum_{i,j,k:1 \leq i < j < k \leq n} |A_i \cap A_j \cap A_k| - \ldots
$$
$$
+ (-1)^{n-1} |A_1 \cap \ldots \cap A_n|.
$$

See also FUNDAMENTAL PRINCIPLE OF COUNTING; BIJECTION PRINCIPLE; PIGEONHOLE PRINCIPLE; DOUBLE-COUNTING PRINCIPLE; DIRICHLET'S PRINCIPLE; SIEGEL'S PRINCIPLE; INCLUSION–EXCLUSION PRINCIPLE; SIEVE PRINCIPLE.

COMFORT-THROUGH-DISCOMFORT PRINCIPLE. See AARON ANTONOVSKY'S COMFORT-THROUGH-DISCOMFORT PRINCIPLE.

*Edmonds, J. (1970), Submodular functions, matroids and certain polyhedra, *Proc. Int. Conf. Combinatorics*, Gordon & Breach, New York, pp. 69–87.
[†]Faigle, U. (1979), The greedy algorithm for partially ordered sets, *Discrete Math.* **28**, 153–159.
[‡]van Lint, J. H. and Wilson, R. M. (2001), *A Course in Combinatorics*, 2nd ed., Cambridge University Press.

COMMITMENT   PRINCIPLE [psychology] (1) Among the various role factors, the one pertaining to high commitment of the counselor to the helping relationship was seen to be the most facilitative, affecting the broadest range of observer–clients' impressions. In this connection, it was noted that other operational treatments of counselor's behavior differed in emphasis and were not necessarily independent. Thus the factor of commitment was perhaps implicit and accounted also for the several reliable effects obtained with those conditions wherein the counselor was problem-centered or confirmed the client's self-esteem. A further consideration was that the principle could likewise be extended to organize a confused assortment of reports in the literature. For example, the more positive response of clients to eclectic or authoritative or leading types of counseling may have resulted from a more visible display of commitment than in the case of counselors using reflective or nondirective approaches. The overall sense of findings was that normative role expectancy and especially high commitment appear to be crucial factors in the initial facilitation of helping relationships, as is the case in other social transactions. Further, lack of commitment seems to make expert counselors susceptible to downgrading in the impressions formed by others—contrary to the uncritical positive halo ordinarily accorded persons with high status in other types of relationships. Students' perception of counselors with varying statuses and role behaviors in the initial interview.[§] (2) Planning can cover a period over which commitment of resources can be clearly visualized.[¶] See also PRINCIPLE OF PLANNING.

COMMON CLOSURE PRINCIPLE (CCP) [computers] (Robert Cecil Martin, U.S. software consultant)   A package is a binary deliverable

[§]Price, L. Z. and Iverson, M. (1969), *J. Counsel. Psychol.*, **16**(6), 469–475.
[¶]Koontz, H. and O'Donnell, C. (1968), *Principles of Management: An Analysis of Managerial Functions*, 4th ed., McGraw-Hill, New York, 1968 (http://knol.google.com/k/narayana-rao-kvss/principles-of-management-koontz-and/2utb2lsm2k7a/89).

like a. jar file, or a dll as opposed to a namespace like a java package or a C++ namespace. Package cohesion describe what to put inside. (1) Classes that change together are packaged together. (2) Describes principles governing the macro structure of large object-oriented applications. Attempts to gather together in one place all the classes that are likely to change for the same reasons. If two classes are so tightly bound, either physically or conceptually, that they almost always change together, then they belong in the same package. This minimizes the workload related to releasing, revalidating, and redistributing the software. Grouping together into the same package of classes that cannot be closed (protected) against certain types of change. Thus, when a change in requirements comes along, (See also OPEN CLOSED PRINCIPLE) that change has a good chance of being restricted to a minimal number of packages.*

COMMON REUSE PRINCIPLE (CRP) [computers] (Robert Cecil Martin)   A package is a binary deliverable like a. jar file, or a dll as opposed to a namespace like a java package or a C++ namespace. Package cohesion describe what to put inside. (1) Classes that are used together are packaged together. (2) Classes tending to be reused together belong in the same package. Classes are seldom reused in isolation. Generally reusable classes collaborating with other classes are part of the reusable abstraction. The CRP states that these classes belong together in the same package.†

COMMUTATION PRINCIPLE [computer science, engineering, mathematics]   (1) Sampling of various quantities in a specific manner for transmission over a single channel in telemetry. (2) Transfer of current from one channel to another in a gas tube. (3) Does not allow any member of a large category to substitute for any other member. Transitive verbs cannot be commuted with intransitive ones,

nor can count and mass nouns be freely interchanged. The success of the commutation test cannot be a necessary condition, only a sufficient one.‡

COMONOTONIC SURE-THING PRINCIPLE [general]   Restricts Savage's sure-thing principle to comonotonic acts, that is, acts inducing the same ordering on the states of nature in terms of the associated outcomes, and is characterized in full generality by means of a new functional form—cumulative utility—generalizing the Choquet integral. Thus, a common generalization of all existing rank-dependent forms is obtained, including rank-dependent expected utility, Choquet expected utility, and cumulative prospect theory.§ See also SURE-THING PRINCIPLE.

COMPARISON PRINCIPLE [engineering] (1) Operation to match identity, relative magnitude, or sign.¶ (2) Driver circuit with clocked amplifier that generates corresponding pulsewidth-modulated compensating current from linear measuring value produced by the evaluation circuit.‖

COMPENSATION PRINCIPLE [physics]   See BRAUN–LE CHATELIER PRINCIPLE.

COMPETITIVE EXCLUSION PRINCIPLE [biology, mathematics] (G. F. Gause, Soviet biologist, and J. Grinnell, American naturalist)   (1) Ecological law two species competing for exactly the same resources cannot stably coexist over time. One of the two competitors is predicted to be eliminated as a result of the competition.** (2) Theoretical concept following from abstract

*Martin, R. C. (n.d.), *Dependency Inversion Group. Ten Commandments of OO Prorarming* (http://www.butunclebob.com/ArticleS.UncleBob.PrinciplesOfOod).
†Op. cit. (Martin).

‡Alinei, M. and Århammar, N. (1987), *Aspects of Language: Theoretical and Applied Semantics*, Rodopi, p. 249.
§Hong, C.-S. and Wakker, P. (1996), The comonotonic sure-thing principle, *J. Risk Uncertain.* **12**, 5–27.
¶Bitsoris, G. and Gravalou, E. (1995), Comparison principle, positive invariance and constrained regulation of nonlinear systems, *Tomatica* **3**, 217.
‖Slyn'ko, V. I. (2005), On the comparison principle for a hybrid system with delay component, *Dopovidi Natsion. Akad. Nauk Ukrainii* **8**, 62–66.
Hardin, G. (1960), Competitive exclusion principle, *Science* **13, 1292.

mathematical modeling. The conditions under which competitive exclusion must hold are not very well understood; several natural ecosystems are known in which competitive exclusion seems to be violated.* See GAUSE'S PRINCIPLE.

COMPLEMENT   PRINCIPLE [mathematics] Used to count the number of possible outcomes in an experiment. If A is a subset of a universal set U, then $n(A) = n(U) - N$.[†]

COMPLEMENTARITY   PRINCIPLE [physics] (Jacques Babinet, 1794–1872, French physicist)   Diffraction patterns produced by complementary screens are identical; two screens are said to be complementary when the opaque parts of one correspond to transparent part of other. Used to solve diffraction problems in optics and electrodynamics. Sum of the waves diffracted by a certain boundary and by the complement of this boundary is equal to the undistorted wave. Given a diffracting hole of arbitrary shape, the conjugate to the hole produces an identical diffraction pattern.[‡] See also BABINET'S PRINCIPLE.

COMPLEMENTARY   PRINCIPLE [psychology, physics] (Danish physicist Niels Bohr, 1885–1962)   (1) The complementarity approach allows the possibility of accommodating widely divergent human experiences in an underlying harmony, and bringing to light newer prospects and ethical views for the exploration and mitigation of human suffering.[§] (2) It is well known that the electron is a particle. It is equally well known that the electron is also a wave. The wave and particle natures are flagrantly opposite because a certain thing cannot at the same time be a particle (i.e., a substance confined to a very small volume) and a wave (i.e., a field spread out over a large space), but the electron exhibits both, although not simultaneously, as the two natures are mutually exclusive.[¶] (3) The wave–particle duality of the electron presents a most familiar example of complementarity of opposites in the domain of physics.[‖] Also known as *variational principle; dual principle; complementary variational principle; nonconvex variational principle; dual variational principle; principle of complementarity*.

COMPLEMENTARY VARIATIONAL PRINCIPLE See COMPLEMENTARY PRINCIPLE.

COMPLEX KOHN VARIATIONAL PRINCIPLE [chemistry, engineering]   Quantum mechanical calculations or quantum reactive scattering.[**] See also HULTHEN–KOHN VARIATIONAL PRINCIPLE; SCHWINGER VARIATIONAL PRINCIPLE; KOHN VARIATIONAL PRINCIPLE.

COMPLEX MOMENT-PRESERVING PRINCIPLE [mathematics, engineering]   See MOMENT-PRESERVING PRINCIPLE.

COMPLEXITY   PRINCIPLE [mathematics] (Jorma Johannes Rissanen, b. 10/20/1932, Finland)   See RISSANEN'S MINIMUM DESCRIPTION LENGTH PRINCIPLE.

COMPOSING PRINCIPLE [medicine]   (1) Modern dietology, for all internal diseases, involves composition of a balanced diet adapted, above all, to metabolic disorders inherent in a given malady. Strict control

---

*Savile, D. B. O. (1960), Limitations of the competitive exclusion principle, *Science*, **132**, 1761.

[†]Walker, P. M. B. (1999), *Chambers Dictionary of Science and Technology*, Chambers, New York, p. 241.

[‡]Baroody, A. J., Ginsburg, H. P., and Waxman, B. (1983), Children's use of mathematical structure, J. Res. Math. Educ. (Natal. Council Math. Teachers) **14**(3), 156–168.

[§]Strigachev, A. (1985), Interpretation of the complementarity principle, *Godishnik Sofiisk. Univ. Sv. Kliment Okhridski, Fiz. Fakultet* **75**, 7–17.

[¶]Goswami, S. C. (n.d.), *Complementarity Principle: Meeting Ground of Science, Philosophy and Religion. Science, Belief and Conscience–Essential and Peripheral Concepts*, Chemistry Dept., Univ. Delhi, New Delhi, India (http://www.herenow4u.de/eng/complementary_principle_meeti.htm).

[‖]Sergienko, I. V., Gupal, A. M., and Vagis, A. A. (2005), Complementary principles of bases recoding along one chain of DNA, (*T Sitologii a i Genetika*) **39**(6), 71–75.

[**]de Araujo, C. F., Adhikari, S. K., and Tomio, L. (1995), Complex Kohn variational principle for two-nucleon bound-state and scattering with the tensor potential, *J. Comput. Phys.* **118**(2), 200–207.

over the proportions of nutritional components in the diet and limits of permissible increase or restriction of the content of each of them. Substance is given to dietetic approaches to various forms of the disease, including nutritional patterns, caloricity, and qualitative and quantitative contents of the nutrients. A special connective system of peculiar fibrillar structures maintaining dynamic equilibrium in arrangement of collagenous fibers, muscles, vessels, etc.* (2) Partitioning an information or software system into smaller and more manageable and comprehensible parts. Each criterion is derived from a concern or need belonging to a particular area of interest. Many decomposing criteria are based on stakeholders' concerns.[†] See also DECOMPOSITION PRINCIPLE.

COMPOSITION PRINCIPLE [psychology]   The meanings of words or expressions can (or must) be determined prior to, and independently of, the meanings of the propositions in which they occur.[‡] See also CONTEXT PRINCIPLE; FREGE'S PRINCIPLE.

COMPOSITIONALITY PRINCIPLE [psychology] (1) The meaning of a complex expression is determined by the meanings of its constituent expressions and the rules used to combine them. Normally taken to quantify over expressions of some particular language L: (C'). For every complex expression e in L, the meaning of e in L is determined by the structure of e in L and the meanings of the constituents of e in L. (1a) In a meaningful sentence, if the lexical parts are taken out of the sentence, what remains will be the rules of composition. (1b) Every operation of the syntax should be associated with an operation of the semantics that acts on the meanings of the constituents combined by the syntactic operation. (1c) Every construct of the syntax should be associated with a clause of the T schema with an operator in the semantics that specifies how the meaning of the whole expression is built from constituents combined by the syntactic rule.[§] (2) In the tradition of Montague grammar, the interpretation of a language is essentially given by a homomorphism between an algebra of syntactic representations and an algebra of semantic objects.[¶] See also PRINCIPLE OF COMPOSITIONALITY; FREGE'S PRINCIPLE; FREGE'S CONTEXT PRINCPLE; CONTEXT PRINCIPLE.

COMPREHENSION PRINCIPLE [mathematics] (Bertrand Russell, 1872–1970, English mathematician–philosopher)   (1) When two numbers are combined as that the one has always an unit answering to every unit of the other, they are pronounced as equal. Given any condition expressible by a formula $f(x)$, it is possible to form the set of all sets x meeting that condition, denoted $\{x|f(x)\}$. For example, the set of all sets—the universal set—would be $\{x|x = x\}$. (2) (Willard Van Orman Quine, American logician) In his paper "New Foundations for Mathematical Logic," the comprehension principle allows formation of $\{x|f(x)\}$ only for formulas $f(x)$ that can be written in a certain form that excludes the "vicious circle" leading to the paradox. In this approach, there is a universal set.[‖] See also HUME'S PRINCIPLE

COMPRESSIBLE FLOW PRINCIPLE [engineering]   When flow velocity is large, it is necessary to consider the fluid as compressible rather than to assume that it has a constant density.[**]

---

*Wang, X., Zhang, N., and Sun, H. (2008), Comparative analysis of components absorbed into blood after oral administration of Liuwei Dihuang Wan and its related prescriptions, *Zhong. Zhong Yao Za Zhi* **33**(15), 1881–1884.
[†]Parnas, D. L. (1972), On the criteria to be used in decomposing systems into modules, *Commun. ACM* **15**, 12.
[‡]Freiman, T. (1999), Mendelejeff's idea. The first composition principle of periodic system of the elements by mind-maps, *Naturwissen. Unterricht Chem.* **10**(53), 222–227.

[§]Davis, W. (2003), *Meaning, Expression, and Thought*, Cambridge Univ. Press.
[¶]Acero, J. J. (2007), Nonsense and the privacy of sensation, *Sorites*(18), 33–55 (http://www.sorites.org/Issue_18/acero.htm).
[‖]Enderton, H. (2009), Russell's paradox, in *Encyclopædia Britannica* (http://search.eb.com/eb/article-9384405).
[**]Godin, O. A. (1996), Reciprocity principle for waves in a flow of an inhomogeneous compressible fluid, *Dokl. Akad. Nauk* **351**(5), 614–617.

CONCEPT PRINCIPLE [ecology]   Cracking process for recycling waste oils using reactor in the form of an engine operation.*

CONCURRENCY PRINCIPLE [computer science]   (1) Specifies that infrastructure to support a development must be available concurrent with the impact of the development. Developments should not be allowed until financially viable plans to provide infrastructure for them are in place. (2) Computer programming method. (3) Decompose a problem into tasks that can execute simultaneously. Then, you must be able to map these tasks onto units of execution (usually threads or processes) to take advantage of the concurrency in a parallel program.†

CONCURRENT RELATION PRINCIPLE See CONCURRENCY PRINCIPLE.

CONDENSATION PRINCIPLE [mathematics] In set theory, silver machines are devices used for bypassing fine structure theory in proofs of statements holding in *L*. They were invented as a means of proving that the global square theory holds in the constructible universe.‡ See also FINITENESS PRINCIPLE.

CONDENSED-IMAGE PRINCIPLE [engineering] (1) Electromagnetic sources in a chiral medium above the soft/hard boundary. Two consecutive decompositions allow the problem to be reduced to four classical problems. These involve electric and magnetic sources above perfect electric conductor and perfect magnetic conductor boundaries, each involving an isotropic nonchiral medium and possessing a known

solution.§ (2) Time-harmonic problem of TE/TM wave propagation and reflection in a waveguide. Heaviside (Oliver Heaviside, 1850–1925, English physicist) operational calculus and a transmission-line model of the waveguide used to derive the fictitious image generating the reflected field. The image of a pointlike source in front of the waveguide discontinuity is another pointlike source in the mirror-image position and a line source extending from the mirror-image location to infinity.¶ See also COINCIDENT IMAGE PRINCIPLE; MIXED-IMAGE PRINCIPLE; STATIC IMAGE PRINCIPLE.

CONDITIONALITY PRINCIPLE [engineering] See LIKELIHOOD PRINCIPLE.

CONDITIONED-REFLEX PRINCIPLE [psychology]   An organism may respond to a stimulus that was inadequate to elicit the response until paired one or more times with adequate stimulus.‖

CONDUCTANCE PRINCIPLE See CAPACITANCE PRINCIPLE.

CONDUCTION PRINCIPLE [physics]   (1) Passing of electric charge, which can occur by a variety of processes, such as the passage of electrons or ionization. (2) Transmission of energy by a medium that does not involve movement of the medium itself.**

CONFINEMENT SCALE INVARIANCE PRINCPLE [physics]   The confinement times, the ELM, and sawtooth frequencies in the two pulses all scale as expected, suggesting that the

*Reigeluth, C. M., Merrill, M. D., Wilson, B. G., and Spille, R. T. (1980), The elaboration theory of instruction: A model for sequencing and synthesizing instruction, *Instruct. Sci.* (Springer Netherlands) **9**(3), 195–219.

†Hertel, E. and Romer, G. H. (1930), The concurrence of principle and secondary valence, *Ber. Dtsch. Chem. Gesell. (Abteilung) B Abhandlungen* **63B**, 2446–2452.

‡Clar, E. and Sandke, R. (1948), Aromatic hydrocarbons. XLII. The condensation principle, a simple new principle in the structure of aromatic hydrocarbon, *Chem. Ber.* **81**, 52–63.

§Engheta, N. and Mickelson, A. (1982), Transition radiation caused by a chiral plate, *IEEE Trans. Anten. Propag.* **30**(6), 1213–1216.

¶Afande, M. M., Ke, W., Giroux, M., and Bosisio, R. G. (1995), A finite-difference frequency-domain method that introducescondensed nodes and image principle, *IEEE Trans. Microwave Theory Tech.* **43**(4), 838–846.

‖Vartanian, G. A. (1990), Conditioned reflex principle in the system of neurosciences: concepts and prospects, *Vest. Akad. Med. Nauk SSSR* (11), 23–29.

Wussow, S. and Pensky, P. (1972), Procedure for the production of a standard solution for particle counters based on the conduction principle, *Z. Med. Labor.* **13(4), 240–242.

invariance principle is satisfied through the plasma radial extent, despite the differing physical processes taking place in the plasma center, core, and edge regions. Enables one to describe the transport properties in all three regions in terms of the profiles of the basic dimensionless plasma physics parameters $\rho^*(\propto (MT)^{1/2}/aB)$, $\beta(\propto nT/B^2)$, $\nu^*(\propto na/T^2)$ and q $(\propto B\kappa/Rj)$. The thermal diffusivity should have the form

$$\chi \propto \frac{Ba^2}{M} F(\rho^*, \beta, \nu^*, q, \ldots)$$

where the form of the function F will be different in each of the three regions.*

CONSERVATION OF ENERGY PRINCIPLE [chemistry] (Helmholtz, 1847)   In all processes occurring in an isolated system, the energy of the system remains constant.[†]

CONSERVATION OF MASS PRINCIPLE [chemistry] (Lavoisier, 1774)   Matter cannot be created or destroyed. In a chemical reaction, the total mass of the products equals the total mass of the reactants.[‡]

CONSERVATION OF MOMENTUM PRINCIPLE [physics]   (1) The constancy of the sum of the momenta in a closed system.[§] (2) If no net external force acts on a system of particles, the total linear momentum of the system cannot change.[¶] (3) A consequence

*Cordey, J. G., Alper, B., Budny, R., Christiansen, J. P., Coffey, I., Erents, K., Harbour, P., Horton, L. D., Lawson, K., Matthews, G. F., Saibene, G., Sartori, R., Strachan, J., and Thomsen, K. (2000), A demonstration of the "isotope wind tunnel principle" in JET and its use in predicting reactor performance, *JET-P* **99**(56), p. 1.
[†]Daintith, J. (1999), *The Facts on File Dictionary of Chemistry*, 3rd ed., Facts On File, New York, p. 64.
[‡]Walker, P. M. B. (1999), *Chambers Dictionary of Science and Technology*, N.Y.: Chambers, New York, p. 255
[§]Tasch, P. (1952), Conservation of momentum in antiquity: A note on the prehistory of the principle of jet-propulsion, ISIS (an international review devoted to history of science and its cultural influences) **43**(133), 251–252.
[¶]Tolman, R. C. (1934), Suggestions as to the energy-momentum principle in a non-conservative mechanics, *Proc. Natl. Acad. Sci. USA* **20**(7), 437–439.

from Newton's third law of motion. For every action there is an equal and opposite reaction, which is true for bodies free to move as well as for bodies rigidly fixed.[‖]   See also D'ALEMBERT PRINCIPLE; DIFFERENTIAL PRINCIPLE; VARIATIONAL PRINCIPLE; HAMILTON PRINCIPLE; PRINCIPLE OF EQUAL A PRIORI PROBABILITIES; NEWTON'S THIRD PRINCIPLE.

CONSERVATION OF NUMBER PRINCIPLE [mathematics, psychology]   See CONTINUITY PRINCIPLE.

CONSERVATION PRINCIPLE OF MASS [chemistry, physics] (A.-L. Lavoisier, 1743–1794, French chemist)   (1) In energy balancing of any system, it is necessary to include the term mc^2, where m is the sum of the rest masses measured at a zero velocity, of all the bodies forming the system. The rest mass term m_0 derives from the fact the mass of a body depends on its velocity, referred to a particular frame of reference, the frame in which the mass is at rest.** (2) Consists in considering constant and indestructible the quantity of matter existing in nature; that is, in any system and, by extension, in the universe the total quantity of matter is always the same.** (3) Affirms that in any system and, by extension, in the universe, the mass transforms into energy and, vice versa the energy transforms into mass, as in practice it happens in the nuclear reactions and in the ones among subnuclear particles.[††]

CONSTANT-COMPOSITION PRINCIPLE [chemistry]   See CONSTANT-PROPORTIONS PRINCIPLE.

CONSTANT-FLUX PRINCIPLE [chemistry]   When an axis is uncoupled there is no mutual inductance between both axes. Due

[‖]Picker, O., Wietasch, G., Scheeren, T. W., and Arndt, J. O. (2001), Determination of total blood volume by indicator dilution: A comparison of mean transit time and mass conservation principle, *Intens. Care Med.* **27**(4), 767–774.
Corio, P. L. (1971), Mass conservation and multiple stoichiometries, *Trans. Kentucky Acad. Sci.* **32(3–4), 51–56.
[††]de Andrade Martins, R. (1993), Landolt's experiments on conservation of mass, *Quimica Nova* **16**(5), 481–490.

to the flux there cannot be an abrupt change. The projections of main flux over each axis are considered constant in the control loop. Thus main flux and mutual inductance change slowly. From the control point of view it is more important to use a fast solution to compute variables.[*]

CONSTANT-INFUSION PRINCIPLE [medicine] Cardiopulmonary recirculation of dialyzed blood returning to the arterial line.[†]

CONSTANT-PROPORTIONS PRINCIPLE [chemistry] (Proust, 1779)   The proportion of each element in a compound is fixed or constant.[‡] Also referred to as *constant-composition principle; principle of constant composition; principle of definite proportiions.*

CONSTANT SUBSEQUENCE PRINCIPLE (KSP). [mathematics]   Any bounded sequence of positive integers has a constant subsequence. See also LIMITED PRINCIPLE OF OMNISCIENCE; BOUNDED SEQUENCE PRINCIPLE; CONSTANT SUBSEQUENCE PRINCIPLE; CONVERGENT SUBSEQUENCE PRINCIPLE; BOLZANO-WEIERSTRASS PRINCIPLE; MONOTONE SEQUENCE PRINCIPLE.[§]

CONSTRUCTION   PRINCIPLE [engineering] Manner in which something is put together. See also PRINCIPLE OF SIMILARITY.

CONSTRUCTIVE LEAST-UPPER-BOUND PRINCIPLE [mathematics]   Let S be a nonempty subset of R that is bounded above. Then S has a least upper bound if and only if it is *order-located*, in the sense that for all real

numbers α, β with $\alpha < \beta$, either β is an upper bound for S or else there exists $x \in S$ with $x > a$.[¶] See also LIMITED PRINCIPLE OF OMNISCIENCE

CONSUMER   PRINCIPLE [mathematics] (1) Descriptions affecting how nations, industries, or other groups perform and identify protection for interaction between consumers and producers of goods and services.[‖] (2) Survey methods used to determine consumer preferences; for instance, the quota method demands the formulation of a hypothetical model to fit the data, a probabilistic survey does not. With probability sampling, randomization distribution is used to draw conclusions from the sample, and to obtain sampling errors. In a quota sample, comparable estimates of precision cannot be obtained. In general, nonresponse in a quota sample is handled by selection of another respondent fitting the quota. Nonresponse in a probability-based sample can be handled with less complexity in the sample (although this means using some form of modeling). No existing survey exactly matches the ideal picture of probability-based sampling.[**]

CONSUMER PRINCIPLE OF RANDOMIZATION [mathematics]   Offers physicians and patients the option of choosing one of three randomization ratios, such as 30:70 (uncertain or idiosyncratic preference for B, yet willing to be randomized if allocation is weighted in favor of B), 50:50 (absolute uncertainty or complete altruism), or 70:30 (uncertain or idiosyncratic preference for A, yet willing to be randomized if allocation is weighted in favor of A).[††] See also CONSUMER PRINCIPLE.

[*]Plata, E. C., Trad, O., and Ratta, G. (1999), ATP simulation of switching transients in ASD systems including cable modeling and algorithm for damping overvoltage problems. International Conference on Power Systems Transients, Budapest, Hungary, p. 536.
[†]Schneditz, D. et al. (1997), Measurement of access flow (QAC) and cardiopulmonary recirculation (CPR) by constant infusion principle, *JASN Abstracts*, vol. 8, p. 171A.
[‡]Daintith, J. (1999), *The Facts on File Dictionary of Chemistry*, 3rd ed., Facts On File, New York, p. 65.
[§]Mandelkern, M. (1988) Limited Omniscience and the Bolzano-Weierstrass Principle, *Bull. London Math. Soc. 20*, 319–320.

[¶]Bishop, E. and Bridges, D. (1985), Grundlehren der math, *Wissenschaften* **279**, 37.
[‖]Wilting, H. and Kees, V. (2007), Environmental accounting from a producer or a consumer principle: An empirical examination covering the world; paper presented at 16th Intl. Conf. Input-Output Techniques in Istanbul, July 2–7, 2007.
[**]Potter, J. (2007), Consumerism and the public sector: How well does the coat fit? *Public Admin.* **66**(2), 149–164.
[††]Gore, S. M. (1994), The consumer principle of randomisation, *Lancet* **343**(8888), 58.

CONTACT BATTERY PRINCIPLE [engineering] Contact with the anode is made by a metal pin or leaf inserted into the anode mix. The cell has a plastic seal assembly, to keep electrolyte from leaking out of the cell and to keep air from getting in. Contact with the anode collector is made through this seal. The seal contains a fail-safe vent that is activated when the battery internal pressure exceeds a certain level.*

CONTACT ELECTRODE PRINCIPLE [engineering]   Cathode immersed only a few millimeters in the electrolyte.[†]

CONTACTING POWER PRINCIPLE [chemistry, engineering] (Konrad Troxel Semrau, b. 6/5/1919, Chico, CA) . At constant particle size distribution: $Nt = \alpha \propto \gamma^T$. The constants α and γ depend on the physical chemical properties of the system and the particle size distribution. Efficiency of collection is proportional to power expended and that more energy is required to capture finer particles.[‡] See also SEMRAU'S PRINCIPLE; PRINCIPLE OF VIRTUAL POWER.

CONTAGION PRINCIPLE [medicine]   Process whereby disease spreads from one person to another by direct or indirect contact.[§]

CONTAINMENT PRINCIPLE [physics]   There are two kinds of creation: creation of the universe and creation in the universe. On one hand, we have creation (as in cosmogenesis) of the whole universe complete with space and time; on the other, we have creation of things in the space and time of an already existing universe. In the big bang universe, everything including space and time is created; in the steady-state universe, matter is created in the space and time of a universe already created. Failure to distinguish between the two violates the containment principle. The steady-state theory employs creation in the magical sense that at a certain place in space at a certain instant in time there is nothing, and at the same place a moment later there is something. But the creation of the universe does not have this meaning unless we revert to the old belief that time and space are metaphysical and extend beyond the physical universe; in that case, creation of a universe is in principle the same as the creation of a hazelnut. But in fact, uncontained creation (cosmogenesis) is totally unlike contained creation. Cosmogenesis involves the creation of space and time.[¶] See also PHYSICAL PRINCIPLE OF COMPUTATION; PRINCIPLE OF HYPERCOMPUTATION UPPER BOUNDS; GALILEO'S PRINCIPLE OF NATURAL COMPUTATION; GANDALF'S PRINCIPLE OF HIDDEN UNIVERSE COMPUTATION.

CONTEXT-INDEPENDENT   PRINCIPLES [psychology]   (1) Restriction of criminal laws based on the harmful consequences of individual conduct. Translation based on linguistic theories. The translator has to produce in the target language a new text fulfilling all the functions intended by the source text in order to communicate the new text. This requires a good command of the working language and extralinguistic competence.[‖] (2) Only purpose for which power can be rightfully exercised over any member of a civilized community, against that person's will, is to prevent harm to others.[**] See also HARM PRINCIPLE.

*Xie, Y., and Xie, J. (2000), Treatment of chromium-containing electroplating wastewater with contact battery principle, *Guangzhou Huaxue* **25**(4), 40–43.

[†]Pavlov, N. N. and Pavlova, S. N. (1985), Principles of the manufacture of materials for spot contact welding electrodes by powder metallurgy methods, *Trudy Leningra. Politekh. Inst. M. I. Kalinina* **404**, 46–48.

[‡]Dell'Isola, F. and Seppecher, P. (1995), The relationship between edge contact forces, double forces and interstitial working allowed by the principle of virtual power, *C. R. Acad. Sci. Ser. IIb Mecan. Phys. Chim. Astron.*, **321**(8), 303–308.

[§]MacWhirter, P. J. (1997), Shifting paradigms: The hard road to acceptance of the contagion principle in Australia, *Australi. Vet. J.* **75**(7), 515–519.

[¶]Costantino, E. (2007), Pulsation and the creation of life: A review of the science of the life force, *Proc. Radix Conf.*, Albuquerque, NM.

[‖]Omer, H. and Alon, N. (1994), The continuity principle: A unified approach to disaster and trauma, *Am. J. Commun. Psychol.* **22**(2), 273–287.

[**]Popelka, M., Jr. and Harkins, W. D. (1950), The principle of continuity as related to the existence of stable even n-odd p and odd n-even p nuclei, *Phys. Rev.* **77**, 756.

CONTEXT PRINCIPLE [psychology] (Gottlob Frege, 1884–1980) A form of semantic holism holding that a philosopher should "never … ask for the meaning of a word in isolation, but only in the context of a proposition." One of "three fundamental principles" for philosophical analysis, first discussed in the introduction to the *Foundations of Arithmetic* (*Grundlagen der Arithmetik*, 1884). Sometimes called *contextualism*, but should not be confused with the common contemporary use of the term "contextualism" in epistemology or ethics.* See also COMPOSITION PRINCIPLE; FREGE'S PRINCIPLE

CONTINUITY PRINCIPLE [mathematics, psychology] Common practice among social and life scientists to adopt an implied continuity principle when interpreting the results of a statistical analysis. It is often assumed, for example, that data that are observed to deviate only slightly in form from that of the familiar normal curve will only slightly distort the usual estimates of means, standard deviations, correlations, and associated hypothesis tests. With increasing departure from an underlying normal model, the greater this assumption, the greater will be the inaccuracy of the computed statistics. Over the past several decades, research in statistics has demonstrated that a continuity principle of the form described above for normal theory-based statistics is invalid. The classical estimates of means, variances, and correlations have been shown to be highly sensitive to even small departures from an underlying normal model.[†] (2) Unified approach to disaster and trauma; through all stages of disaster, management and treatment should aim at preserving and restoring functional, historical, and interpersonal continuities, at the individual, family, organization, and community levels; misconceptions and errors result when the abnormality bias results in underestimating the probability or extent of expected disruption.[‡] (3) If an analytic identity in any finite number of variables holds for all real values of the variables, then it also holds by analytic continuation for all complex values. Metric properties discovered for a primitive figure remain applicable, without modifications other than changes of signs, to all correlative figures that can be considered to arise from the first.[§] Also known as *conservation of number principle; poncelet's continuity principle; permanence of mathematical relations principle.*

CONTINUITY PRINCIPLE [mathematics, management] If a certain number of solutions are expected, then there will be the same number of solutions in all cases, although some solutions may be imaginary. If any analytic identity in any finite number of variables holds for all real values of the variables, then it also holds by analytic continuation for all complex values.[¶] See also CONSERVATION OF NUMBER PRINCIPLE; PONCELET'S CONTINUITY PRINCIPLE; PERMANENCE OF MATHEMATICAL RELATIONS PRINCIPLE.

CONTINUOUS-FLOW PRINCIPLE [engineering] Used in a variety of fields to measure specific elements or quantities in a large volume. (1) Consists of sucking the suspension to flow through an orifice. As each particle passes through the orifice, it replaces its own volume of electrolyte within the orifice, momentarily changing the resistance value between the

*Forth, H. J. (1967), Cryogenic devices using the continuous flow principle, *Vacuum*, **17**(1), 21–22.
[†]Lind, J. C. and Zumbo, B. D. (1993), The continuity principle in psychological research: An introduction to robust statistics, *Can. Psychol./Psychol. Cana.* **34**(4), 407–414.

[‡]Harsanyi, G. (1980), Flow diagrams of the CONTIFLO chemical analyzer system, *HSI* (*Hungarian Sci. Instrum.*) **50**, 75–78.
[§]Bell, E. T. (1992), *The Development of Mathematics*, Dover Books on Mathematics, Courier Dover Publications, p. 339.
[¶]Bell, E. T. (1945) *The development of Mathematics*, 2nd ed., NY: McGraw-Hill, p. 340.

electrodes.* (2) Data are directly obtained for plotting cumulative particle frequency or concentration versus particle volume (or mass, given constant density).[†] (3) Used in a variety of fields to measure specific elements or quantities in a large volume.[‡]

CONTINUOUS-REINFORCEMENT PRINCIPLE [psychology]   To develop a new behavior that the child has not previously exhibited, arrange for an immediate reward after each correct performance.[§]

CONTINUUM PRINCIPLE [psychology]   The comparison of culturally embedded meanings does not involve the comparison of meanings fixed at different points, but rather involves the comparison of meanings residing within a shared space and differ in their distance and difference from one another.[¶] See also PRINCIPLE OF INTERCULTURAL COMMUNICATION; TRANSACTION PRINCIPLE; PERFORMATIVITY PRINCIPLE; POSITIONALITY PRINCIPLE; PENDULUM PRINCIPLE; SYNERGY PRINCIPLE; SUSTAINABILITY PRINCIPLE.

CONTRACTION PRINCIPLE [mathematics] (Stefan Banach, 1892–1945, Polish mathematician)   See CONTRACTION-MAPPING PRINCIPLE.

CONTRACTION-MAPPING PRINCIPLE [mathematics] (Stefan Banach, 1892–1945, Polish mathematician)   (1) Banach fixed-point theorem guarantees the existence and uniqueness of fixed points of certain self-maps of metric spaces, and provides a constructive method to find those fixed

points.[‖] (2) Large deviations from the law of large numbers for a sequence of independent and identically distributed random variables. For measures corresponding to marginal conditional distributions of diffusion processes.[**] (3) Expression for the rate of convergence of ordinal comparison in exponential.[†] (4) Reduction of noise received by a robot sensor by screening it over a period of time.[**] (5) Device for finding expansion for solutions. Allows for shrinking; a continuos function of a metric space to itself which moves each pair of points closer together.[**] Also known as *averaging principle*; *Gibbs variational principle*; *large deviation principle*.

CONTROLLED CLEARANCE PRINCIPLE [engineering]   (1) External pressure controls the clearance between cylinder wall and piston.[††] (2) Basis of design of a free piston primary pressure standard gauge. The controlled clearance free piston gauge has since become the primary pressure standard at the National Institute of Standards and Technology (NIST).[‡‡]

CONVERGENT SUBSEQUENCE PRINCIPLE (CSP) [mathematics]   Any sequence in N has a subsequence convergent in $N^*\zeta N \cup \{\infty\}$[§§] See also LIMITED PRINCIPLE OF OMNISCIENCE; BOUNDED SEQUENCE

*Imhoff, K. (1948), The continuous-flow principle in sewage treatment, *Sewage Works J.* **20**, 626–628.
[†]Forth, H. J. (1967), Cryogenic devices using the continuous flow principle, *Vacuum* **17**(1), 21–22.
[‡]Harsanyi, G. (1980), Flow diagrams of the CONTIFLO chemical analyzer system, *HSI* (*Hungarian Sci. Instrum.*) **50**, 75–78.
[§]Huitt, W. (1994), Principles for using behavior modification in *Educational Psychology Interactive*, Valdosta State Univ., Valdosta, GA (http://chiron.valdosta.edu/whuitt/col/behsys/behmod.html).
[¶]Cimmino, C. V. (1973), The continuum principle in diagnostic roetgenology, *Am. J. Roentgenol. Radium Ther. Nucl. Med.* **119**(1), 208–209.

[‖]Ho, T. C. (1980), Further development of the uniqueness criteria of the steady state via the contraction mapping principle, *Chem. Eng. Sci.* **35**(4), 867–870; see also Sine, R. C. (1983), *Fixed Points and Nonexpansive Mappings*, AMS Bookstore, p. 121.
[**]Ragusa, C. and Fiorillo, F. (2006), A three-phase single sheet tester with digital control of flux loci based on the contraction mapping principle, *J. Magn. Magn. Mater.* **304**(2), e568–e570.
[††]Newhall, D. H. (1957), New high-pressure technique-controlled clearance principle, *J. Industr. Eng. Chem.* **49**, 1993–1995.
[‡‡]Johnson, D. P. and Newhall, D. H. (1975), The Piston gage as a precise pressure measuring instrument, Paper #52-11RD-2, *Transactions of the American Society of Mechanical Engineers*, 75–301.
[§§]Mandelkern, M. (1988), Limited omniscience and the Bolzano-Weierstrass principle, *Bull. London Math. Soc.* **20**, 319–320.

PRINCIPLE; CONSTANT SUBSEQUENCE PRINCIPLE; CONVERGENT SUBSEQUENCE PRINCIPLE; BOLZANO-WEIERSTRASS PRINCIPLE; MONOTONE SEQUENCE PRINCIPLE

CONVEYING   PRINCIPLE [engineering]
When product is transferred from one conveying element and advanced to the next one, in each case one apparatus is charged with two blank streams. This results in an imbricated stream running to the right and another imbricated stream running to the left, each oriented such that in each case the upper edge of the conveyed blanks is at the front in the conveying direction.* See also DENSE-FLOW CONVEYING PRINCIPLE.

COOPERATION   PRINCIPLE [psychology]
(1) Directs practitioners to join with and utilize the client's ongoing experience as the basis for all communications. Guide to understanding and applying Ericksonian hypnotherapy interventions, with emphasis on psychotherapeutic change and improvement of communication skills in hypnotherapists. (2) Human cognitive processes are geared to achieving the greatest possible cognitive effect for the smallest possible processing effort. (3) Communication is accompanied by a warranty of optimal relevance, which states that the utterance will produce enough effects to make it worth the hearer's processing efforts. (4) Every utterance is interpreted relative to a context that is neither given nor reduced to the situation in which the communication takes place, but that is built for each new utterance and that consists of information in propositional form drawn from the environment, the interpretation of previous utterances, and the hearer's encyclopedic knowledge. (5) Defined in economic terms as an equilibrium between effect and effort that underlies the whole of human cognition.† See also PRINCIPLE OF RELEVANCE.

*Heep, D. and Winkhardt, G. (1998), Simplified application of the vacuum-/pressure conveying principle for pneumatic conveyance, *Bulk Solids Handl.* **18**(2), 245–251.
†Gilligan, S. G. (1987), *Therapeutic Trances: The Cooperation Principle in Ericksonian Hypnotherapy*, Brunner/Mazel, Philadelphia, p. xvii.

COOPERATIVE   PRINCIPLE [psychology]
(1) Proposes a tripartite metaperspective of discourse study; discourse may be studied as utterance, social interaction, or social context. The metaperspective is used to survey uses and critiques of H. P. Grice's cooperative principle (CP) across several fields (e.g., linguistic philosophy, gender studies, teacher research). The scholarship surveyed illustrates the widespread impact of CP on a range of scholarly activity, from work focusing on the narrowest issues of language meaning to work focusing on the broadest issues within the social context of human communication. The survey demonstrates that scholars who study discourse as utterance are most interested in CP maxims and conversational implicature, while they critique the notion of a general cooperative principle; scholars who study discourse as social interaction are more critical of the maxims. Scholars who approach the CP from more than one perspective are likely to find it most useful. The essay concludes that no articulation of how the CP could consistently describe discourse as utterance, social interaction, and social context exists, but such an articulation could address critiques of the CP, open the concept for greater cross-disciplinary use, and provide new vocabulary for describing discourse.‡ (2) The use of different signals for different categories and of the same signal for all members within a category corresponds to the principle of contrast or of mutual exclusivity that children rely on when they assign only one label per category. (3) Children will allow only one lexical entry to occupy a semantic niche. When two words are determined to have similar meanings, one of them is preempted and removed from the lexicon. Different words should have different meanings. (4) Captures facts about the inferences that speakers and addresses make for both conventional and novel words. It accounts for the preemption of novel words by well-established ones, and it holds to the same extent for morphology

‡Lindblom, K. (2001), Cooperating with Grice: A cross-disciplinary metaperspective on uses of Grice's cooperative principle, *J. Pragmatics*, **33**(10), 1601–1623.

as for words and larger expressions. (5) The development of nonlinguistic concepts, the acquisition of language in context, and the use by participants in conversational exchanges to account for those features of language and language acquisition.* See also PRINCIPLE OF UNIQUENESS; PRINCIPLE OF CONVENTIONALITY; PRINCIPLE OF PREEMPTION; PRINCIPLE OF CONTRAST; MUTUAL EXCLUSIVITY PRINCIPLE.

COORDINATION NUMBER PRINCIPLE [physics] (1) Numbers of a nearest neighbors of a position in a space lattice, of an atom or an ion in a solid, or of an anion or cation in a solution.* (2) Comprises oscillator and receive transducers set in recesses thinning thick walls of single straight-bore tubes. Determines oscillation amplitude from corresponding voltage drop. High-strength alloy and long connecting tubes are used to prevent damage from thermal stresses.*

COORDINATIVE   PRINCIPLE [psychology] Age-related changes in the coordination of movement. The power law is robust in the drawing movements of young adults (aged 20–25 years), typically explaining between 65% and 95% of the variance. Nevertheless, in young children a pattern of increasing strength of the power law with advancing age, with an exponent of $1 = 3$ emerging for adults. This suggests that a correlation between velocity and curvature strengthens with maturation of the nervous system and musculature.*

COPERNICAN PRINCIPLE [astronomy] (Nicolaus Copernicus, 1473–1543, Renaissance astronomer.)   (1) Idea that Earth occupies a typical or unexceptional position in the universe.[†] (2) Philosophical statement that no "special" observers should be proposed.

The term refers to the paradigm shift away from the Ptolemaic model of the heavens, which placed Earth at the center of the Solar system.[†] (3) The heavens can be explained without the Earth (or anything else) being in the geometric center of the system, so the assumption we are observing from a special position can be dispensed with.[‡] (4) Immanuel Kant described the effect his critical method would have on contemporary epistemological thinking. The conditions and qualities he ascribed to the subject of knowledge placed humans at the center of all conceptual and empirical experience, and overcame the rationalism–empiricism impasse, characteristic of the seventeenth and eighteenth centuries.[§] (5) Acknowledgment the universe is generally homogeneous and isotropic over large scales. A significant, large-scale deviation from homogeneity and isotropy would be statistically unlikely, and this acknowledgment has been found to be correct in different contexts in prior observations. The universe has heterogeneous structures up to the scale of galactic superclusters, filaments, and great voids, but the universe is essentially homogeneous when considered on scales of at least about 200 million parsecs. The universe is homogeneous over time on noncosmological timescales, but is not isotropic over time beyond the timescales of elementary particle reactions.[¶]

COPERNICUS PRINCIPLE [astronomy] (Nicolaus Copernicus, 1473–1543, Renaissance astronomer)   See COPERNICAN PRINCIPLE.

COPY PROTECTION PRINCIPLE [computer science]   (1) A security precaution to prevent the unauthorized copying of programs and

---

*Tye-Murray, N., Witt, S., and Schum, L. (1995), Effects of talker familiarity on communication breakdown in conversations with adult cochlear-implant users, *Ear Hearing*, 16(5), 459–469.

[†]Lu, X., Xu, X., Wang, N.-Q., and Zhang, Q.-E. (1998), Coordination number principle for cluster modeling of metal oxides. Ab initio cluster modeling of CO chemisorption on ZnO, *Gaodeng Xuexiao Huaxue Xuebao* 19(5), 783–788.

[‡]Saling Lauren, L. and Phillips, J. G. (2002), Age-related changes in the kinematics of curved drawing movements: relationships between tangential velocity and the radius of curvature, *Exp. Aging Res.* **28**(2), 215–229.

[§]Caldwell, R. R. and Stebbins, A. (2008), A test of the copernican principle, *Phys. Rev. Lett.* **100**(19), 191302.

[¶]Uzan, J.-P., Clarkson, C., and Ellis, G. F. R. (2007), *Time Drift of Cosmological Redshifts as a Test of the Copernican Principle*, Los Alamos National Laboratory (1–4), arXiv:0801.0068v1.

files.* (2) Additional hardware or equipment is required for creating the physical irregularities in the storage layer of the disk. Since these irregularities are of a physical nature (i.e., the physical shape or arrangement of the irregular pits is different from that of a normal pit), an infringer or plagiarist may be capable of locating those positions on the disk, in which the copy-protecting information (the irregular pits) are located, whereupon the irregularities thus located may be transferred to, or re-created on, the disk copy.† See also DATA PROTECTION PRINCIPLE; PRIVACY PRINCIPLE.

CORIOLIS MEASUREMENT PRINCIPLE [physics] (Gaspard-Gustove Coriolis, 1792–1843, French mathematician) Found in the solar coronal geometry, modeled as a rectangular domain. The solution bifurcates into two different states when the magnetic helicity integral or the geometric factor, defined as the ratio of the height to the width of the domain, is satisfactorily increased.‡

CORIOLIS PRINCIPLE [physics] (Gaspard-Gustave Coriolis, 1792–1843, French mathematician) (1) In a rotating reference frame, Newton's second law of motion is not valid, but it can be made so if, in addition to the real forces acting on a body, a Coriolis force and a centrifugal force are introduced.§ (2) Acceleration of a particle moving in a relative coordinate system. The total acceleration of the particle, as measured in an inertial coordinate system, may be expressed as the sum of the acceleration within the relative system, the acceleration of the relative system itself, and the Coriolis acceleration.¶ (3) Coriolis acceleration may be considered as coming from the conservation of momentum in a body moving in a direction not parallel to the axis of rotation of the relative system.‖ (4) Mathematically, Coriolis acceleration comes from the differentiation of terms containing the angular velocity in the absolute velocity of the particle.** (5) Calculates the mass flow rate by means of detecting phases in oscillations by electronic, thermal, or other means.†† (6) Acceleration of fluid flowing through straight measuring tube and using zero-point correction method.‡‡

CORNER POINT PRINCIPLE [mathematics] A maximum or minimum value of a linear expression, such as $P = A_x + B_y$, if it exists, will occur at a corner point of the feasible region.‡‡

CORRELATION PRINCIPLE OF THE FOURIER TRANSFORM [chemistry] (1) Efficient implementation of quaternion Fourier transform, convolution, and correlation by two-dimensional (2D) complex fractional Fourier transform (FFT). The correlation performance of a joint fractional Fourier transform correlator (JFRTC) using computer simulation results. (2) A mathematical analysis suggesting use of processing techniques based on a nonlinear transformation and

*Bloom, J. A., Ingemar, J. C., Linnartz, J.-P. M. G., Miller, M. L., and Brendan, C. (1990), Copy protection for DVD video, *Proc. IEEE* **87**(7), 1268.

†Barchan, J. (2001), *Method of Providing an Optical Data Carrier with Identity Information*, US Patent 6,226,770.

‡Sittler, E. C. Jr. and Guhathakurta, M. (1999), Semiempirical two-dimensional magnetohydrodynamic model of the solar corona and interplanetary medium, *Astrophys. J.* **523**(2).

§Walker, P. M. B. (1999), *Chambers Dictionary of Science and Technology*, Chambers, New York, p. 267.

¶Binz, H. (1988), Continuous measurement and dosage of bulk solid flow by the Coriolis principle, *Chem. Ingen. Technik* **60**(11), 894–896.

‖Walker, P. M. B. (1999), *Chambers Dictionary of Science and Technology*, Chambers, New York, p. 267.

Binz, H. (1988), *Chem. Ingen. Technik* **60(11), 894–896.

††Kolahi, K., Gast, Th., and Roeck, H. (1994), Coriolis mass flow measurement of gas under normal conditions, *Flow Meas. Instrum.* **5**(4), 275–283.

‡‡Heinrici, H. (1997), Economical and reliable dispensing of fuels. The Coriolis measurement principle shows its advantages, *Cemento-Hormigon* **68**(777), 1429–1442.

fractional order fractional power fringe-adjusted filter to attain improved performance in terms of discrimination sensitivity and input space–bandwidth utilization.*

CORRESPONDENCE PRINCIPLE [chemistry] (1) The predictions of quantum and classical mechanics must correspond in the limit of very large quantum numbers.[†] (2) Two opposite (conjugate) elements that are intergrated into an elementary system by certain conservation law. Particle and wave characteristics in the same large-scale phenomenon are incompatible rather than complementary. Knowledge of a small-scale phenomenon, however, is essentially incomplete until both aspects are known.[‡] (3) A new law of causality is derived starting from Curie's law of symmetry. For occurring a phenomenon, in a system there should hold three necessary (but not sufficient) conditions: (a) the existence of an ordering or regularity pattern, (b) Curie's dissymmetry condition, and (c) between the system and environment, a matching between geometric or/and energetic characteristics.[§] (4) A phenomenon has some similar or even identical characteristics with the one occurring in a system with a much higher symmetry. It is further shown that this law of causality, tentatively called the Curie–Fulea law, is an expression of Bohr's principle of complementarity.[¶][‖][**] See also BOHR'S CORRESPONDENCE PRINCIPLE, BOHR COMPLEMENTARITY PRINCIPLE; PRINCIPLE OF CORRESPONDENCE.

CORRESPONDING STATES PRINCIPLE [physics, chemistry] (Kenneth Sanborn Pitzer, b. 1/6/1914, Pomona, CA or Ruell Moher Pitzer, b. 5/10/1938, Berkeley, CA—son of Kenneth) (1) Applied to substances when their pressures and temperatures are equal fractions of the critical values.[††] (2) Treatment of polyatomic molecules by quantum mechanics.[‡‡] (3) Use of symmetrry and relativity in the computation of molecular electronic structure and heavy atoms.[§§] (4) Condition when two or more substances are at the same reduced pressures, the same reduced temperatures and the same reduced volumes.[¶¶] *Also known as* PITZER CORRESPONDING STATES PRINCIPLE

COSMIC CENSORSHIP PRINCIPLE [physics] (Sir Roger Penrose, b. Aug. 1931, English mathematical physicist) In general relativity, the cosmic censorship hypothesis (CCH) is a conjecture about the nature of singularities in spacetime. Singularities that arise in the solutions of Einstein's equations

*Davies, L. (2007), *Coriolis Mass Flowmeter with Magnetic Coil Body of Thermal-Conductive Polyphenylene Sulfide*, Eur. Patent Application.

[†]Mita, Y., Kobayashi, D., and Shibata, T. (2003), A convex-corner preservation principle in bulk micromachining and its application to nano- point needles, Transducers '03, *Proc. 12th Int. Conf. Solid-State Sensors, Actuators and Microsystems, Digest of Technical Papers*, Boston, June 8–12, 2003, vol. 2, pp. 1683–1686.

[‡]Ruscic, B. (1986), Fourier transform photoelectron spectroscopy: The correlation function and the harmonic oscillator approximation, *J. Chem. Phys.* **85**(7), 3776–3784.

[§]Walker, P. M. B. (1999), *Chambers Dictionary of Science and Technology*, Chambers, New York, p. 269.

[¶]Anon. (2009), Complementarity principle, in *Encyclopedia Britannica* (http://www.britannica.com/EBchecked/topic/129874/complementarity-principle).

[‖]DeHoop, A. T. (1996), A general correspondence principle for time domain electromagnetic wave and diffusion fields, *Geophys. J. Int.* **127**, 757.

[**]Bohr, N. Rosenfeld, L., and Nielsen, J. R., eds. (1976), *Niels Bohr, Collected Works*, vol. 3, *The Correspondence Principle* (1918–1923), North-Holland, Amsterdam.

[††]Walker, P. M. B. (1999), Chambers Dictionary of Science and Technology, Chambers, New York, p. 269.

[‡‡]Chen, X. Z. and Hou, Y. J. (1997), New shape factor for corresponding states principle, *Chin. J. of Chem. Eng.* **5**, 169.

[§§]Kim, H.-Y., Angela D. Lueking, A. D., Gatica, S. M., Johnson, J. K., and Cole, M. W. (2008) A corresponding states principle for physisorption and deviations for quantum fluids, *Mol. Phys.* **106**(12/13), 1579–1585.

[¶¶]Anon. (n.d.), Corresponding states principle, in *McGraw-Hill Dictionary of Scientific and Technical Terms*, 6th ed., McGraw-Hill, New York.

are typically hidden within event horizons, and therefore cannot be seen from the rest of spacetime. Singularities that are not so hidden are called "naked." The weak cosmic censorship hypothesis conjectures that no naked singularities other than the big bang singularity exist in the universe.* (2) If singularities can be seen from the rest of spacetime, causality may break down, and physics may lose its predictive power. According to the Penrose–Hawking singularity theorems, singularities are inevitable in physically reasonable situations. Still, in the absence of naked singularities, the universe is deterministic—it's possible to predict the entire evolution of the universe, knowing only its condition at a certain moment in time (more precisely, everywhere on a spacelike three-dimensional hypersurface, called the *Cauchy surface*).[†]

COSMOLOGICAL   PRINCIPLE [astronomy] (1) The universe is uniform, homogeneous, and isotropic.[‡] (2) The presence of intelligent life on Earth places limits on the many ways the universe could have developed and could have caused the conditions of temperature that prevails today.[§] (3) Our existence necessarily puts some constraints on the evolution of the universe.[¶] (4) Associated anthropic coincidences support the thesis that God exists and does not support supernaturalism.[∥] See also ANTHROPIC PRINCIPLE.

*Israel, W. (1986), The formation of black holes in nonspherical collapse and cosmic censorship, *Can. J. Phys.* **64**(2), 120–127.

[†]Santiago-German, W. (2003), Surface-gravity inequalities and generic conditions for strong cosmic censorship, *Phys. Rev. D Particles Fields* **68**(8).

[‡]Walker, P. M. B. (1999), *Chambers Dictionary of Science and Technology*, Chambers, New York, p. 270.

[§]Wesson, P. S. (1979), The cosmological principle, *Astronomy*, **7**, 66.

[¶]Budinich, P., Nurowski, P., Raczka, R., and Ramella, M. (1995), On the geometry of large-scale distribution of galaxies. (spontaneous violation of the cosmological principle), *Europhys. Lett.* **30**(6), 373–378.

[∥]Celerier, M. N. (2000), *Supernova Data: Cosmological Constant or Ruling out the Cosmological Principle?* Los Alamos National Laboratory, Preprint Archive, Astrophysics, pp. 1–4.

COST DISTRIBUTION PRINCIPLE [mathematics, engineering]   Optimum tax distribution.**. See also COUNTERCURRENT DISTRIBUTION PRINCIPLE; PRINCIPLE OF CURRENT DISTRIBUTION; CHARGE DISTRIBUTION PRINCIPLE; DISTRIBUTION PRINCIPLE.

COST PRINCIPLE [mathematics] (Josiah Warren,1798–1874, individualist anarchist, inventor, musician, and author, American) (1) It is unethical to charge a higher price for a commodity than the cost of purchasing, producing or acquiring, and bringing it to market.[††] (2) It is a strict interpretation of the labor theory of value, which holds that the value of a commodity is the amount of labor incurred in producing or acquiring it.[‡‡] (3) Josiah Warren, states that "the limit of price is total cost" (including one-time cost, cost of capital and opportunity, and other hidden costs), such that over time in a free, efficient, and competitive market, the price of a good will come arbitrarily close to the total cost in providing it. Striving to achieve with given means a maximum of ends.[§§] Also termed *economic principle*.

COULOMB'S PRINCIPLE [physics]   (Charles Augustin de Coloumb, 1736–1806, French physicist)   The electric force of attraction or repulsion between two point charges is proportional to the product of the charges

Dompere, K. K. (1993), The theory of fuzzy decisions, cost distribution principle in social choice and optimal tax distribution, *Fuzzy Sets Syst. Arch.* **53(3), 253–273.

[††]Machlup, F. (1946), Marginal analysis and empirical research, *Am. Econ. Rev.* (Am. Econ. Assoc.) **36**(4), 519–554.

[‡‡]McKern, B. and Dunning, J. H. (1993), *Transnational Corporations and the Exploitation of Natural Resources*, Transnational Corporations and Management Div., United Nations Dept. Economic and Social Development, Routledge, p. 145.

[§§]Warren, J. (1849), *Equitable Commerce: A New Development of Principles, for the Harmonious* [sic] *Adjustment and Regulation of the Pecuniary, Intellectual, and Moral Intercourse of Mankind: Proposed as Elements of New Society*, 2nd ed., Amos E. Senter, p. 15.

and inversely proportional to the square of the distance between them.*

COULTER ORIFICE PRINCIPLE [biology] (Thomas Coulter, 1793–1843, Irish Botanist) Difference in electrical conductivity between the cells and the medium in which they are suspended is measured by the change in electrical impedance produced as they pass through an orifice.[†]

COULTER PRINCIPLE [biology] (Thomas Coulter, 1793–1843, Irish Botanist) (1) Method of counting in dividual cells by pumping a suspension through an orifice and measuring the change in capacitance as each cell passes through.[‡] (2) Method of counting and measuring microscopic particles such as blood cells immersed in liquid.[§] (3) Electrical sensing zone method of particle size measurement of particle size of powders, suspensions and emulsions.[¶]

COUNT(*P*) PRINCIPLE. See POWDERING PRINCIPLE.

COUNT(*Q*) PRINCIPLE. See POWDERING PRINCIPLE.

COUNTERFLOW PRINCIPLE [engineering] (1) Fluid flow in opposite directions in adjacent parts of an apparatus as in a heat exchanger. (2) Movement in a centrifugal field, improving the selectivity of the classification process and the fineness of the product.[‖]

COUNTING GENERALIZED PRINCIPLE [mathematics] For experiments performed with multiple possible outcomes for each experiment, there are a combination of possible outcomes.[**]

COUPLING PRINCIPLE [physics] Mutual relation between two circuits that permits energy transfer from one to another.[††]

COVARIANCE PRINCIPLE [physics] (1) The laws of physics are the same in every Lorentz reference system.[**] (2) Consisting of two features: (a) the mathematical formulation in terms of Riemannian geometry and (b) the general validity of any Gaussian coordinate system as a spacetime coordinate system in physics.[‡‡] See also PRINCIPLE OF MATERIAL FRAME INDIFFERENCE.

COVARIATION PRINCIPLE [medicine] (1) Attributing illness or death related to a particular illness to the less common of two consistently present causes. (2) Subjects in an experiment related the less common cause as the superior predictor of the event. (3) An individual with less common causal factor was rated as more likely to experience death or disease than subjects with the more common causal factor.[§§]

*Walker, P. M. B. (1999), *Chambers Dictionary of Science and Technology*, Chambers, New York, p. 271.

[†]Ferris, C. D., and Veal, B. L. (1986), Amplifier design considerations for blood cell counter sampling probes, *ISA Trans.* **25**(1), 1–4.

[‡]Walker, P. M. B. (1999), *Chambers Dictionary of Science and Technology*, Chambers, New York, p. 272.

[§]Gutmann, J., Hofmann, G., Ruhenstroth, and Bauer G., (1966), Exact methods for measuring the volume distribution of erythrocytes on the basis of the Coulter principle, *Bibliotheca Haematol.* **24**, 42–53.

[¶]Lallukka, Y., Heikonen, M., Entela, M., and Kiviniemi, L. (1986), Particle size measurement of lactose powder with an electrical sensing zone method (Coulter principle), *Meijeritieteellinen Aikakauskirja*, **44**(2), 63–73.

[‖]Lovasz, S., Nyirady, P., and Romics, I. (2004), A new concept for active ureteric occlusion during percutaneous nephrolithotripsy: The "counterflow" principle, *BJU Int.* **93**(9), 1355–1356.

[**]Ecker, M. W. (1985), The fundamental counting principle; counting without enumerating, *Byte* IO, 425.

[††]Fry, M. and Green, D. E. (1980), Ion-transport chain of cytochrome oxidase: the two chain-direct coupling principle of energy coupling, *Proce. Natl. Acad. Sci. USA* **77**(11), 6391–6395.

[‡‡]Sinanoglu, O. (1984), A principle of linear covariance for quantum mechanics and the electronic structure theory of molecules and other atom clusters, *Theor. Chim. Acta* **65**(4), 233–242.

[§§]Forsterling, F., Buhner, M., and Gall, S. (1998), Attributions of depressed persons: how consistent are they with the covariation principle? *J. Pers. Social Psychol.* **75**(4), 1047–1061.

COVERSET INDUCTION PRINCIPLE [mathematics, physics]   See INDUCTION PRINCIPLE.

CRAPS PRINCIPLE [mathematics]   If the trials are repetitions of a game between two players, and the events are E_1—player 1 wins, E_2—player 2 wins, then the respective conditional probabilities of each player winning a certain repetition, given that someone wins (i.e., given that a draw does not occur) is given. In fact, the result is affected only by the relative marginal probabilities of winning $P[E_1]$ and $P[E_2]$; in particular, the probability of a draw is irrelevant.*

CRITICAL PATH METHOD PRINCIPLE [mathematics]   (1) A procedure used in planning a large program of work.[†] (2) Their main objective is to minimize the waste in needless computations, data transfer, and data storage. The potential of a node is the number of the control pulses or signals (bit-serial words) pass on that node.[‡]

CROSS-CHECK PRINCIPLE [psychology]   Factors influencing results of objective hearing examination in children, i.e., auditory evoked potentials, based on experiences.[§]

CROSS-PRINCIPLE RELATIONSHIPS [geology, philosophy]   (James Hutton, 1726–1797, Scottish geologist, physician, naturalist, chemist, and experimental farmer) (1) A rock or fault is younger than any rock (or fault) through which it cuts.[¶] (2) Can be involved and complex. A matter of not only whether a principle is applied but also the extent to which it is applied. Every principle can be implemented to a certain degree. This can make it challenging to measure the actual level of realization of a principle and the extent to which it does in fact affect others.[||] See also PRINCIPLE OF BENEFICENCE; PRINCIPLE OF ORIGINAL HORIZONTALITY; PRINCIPLE OF LATERAL CONTINUITY; PRINCIPLE OF FAUNAL SUCCESSION.

CRUSHING PRINCIPLE [engineering]   Method used to recycle wire. Example includes crushing while the gyrating movement simultaneously results in a chewing action between plates, thereby crushing the material.** Also known as *pressure crushing principle*.

CRYSTALLOGRAPHIC PRINCIPLES [chemistry]   Based on precipitate morphology with determinants of the connection between the optimum shape and orientation relationship of precipitates in a solid. Precipitate dimensions tend to be inverse to the magnitude of the transformation strain. Precipitates are bounded by unrotated planes (eigenplanes). Interfaces are parallel to the planes of three independent dislocation loop arrays necessary to accommodate the transformation strain completely. These principles are illustrated for different orientation relationships, and it is shown that special features are displayed by invariant-line precipitates.[††]

CRYSTODYNE PRINCIPLE [engineering]   Some contacts, such as crystal and metal or crystal and carbon, generally employed as detectors may produce undamped oscillations of any frequency, exactly as the vacuum tube oscillator. The same contact may also be utilized as an amplifier.[‡‡]

*Fleming, P. E. (2001), A quantum mechanical game of craps: Teaching the superposition principle using a familiar classical analog to a quantum mechanical system, *J. Chem. Educ.* **78**(1), 57–60.

[†]Walker, P. M. B. (1999), *Chambers Dictionary of Science and Technology*, Chambers, New York, p. 280.

[‡]Tsutsumi, J. and Kato, K. (1994), First principles dynamical calculations of atomic diffusion in aluminum, *Trans. Mater. Res. Soc. Jpn.* pp. 205–208.

[§]Jerger, J. F. and Hayes, D. (1976), The cross-check principle in pediatric audiometry, *Arch. Otolaryngol.* **102**(10), 614–620.

[¶]Tarbuck, L. T. (2005), *Earth—an Introduction to Physical Geology*, Pearson Education Canada Inc.

[||]Arneson, R. J. (2004), Moral limits on the demands of beneficence, *Ethics of Assistance*, Chatterjee, D. K., ed., Cambridge Univ. Press.

Nakayama, T. (2007), Crushing principle of beads mill, *Kagaku Sochi*, **49(8), 87–89.

[††]Dahmen, U. (1994), A comparison between three simple crystallographic principles of precipitate morphology, *Metallurg. Mater. Transa. A Phys. Metallurgy Mater. Sci.* **25A**(9), 1857–1863.

[‡‡]Lossev, O. V. (1924), The crystodyne principle, *Radio News* pp. 294–295, 431.

CTD PRINCIPLE See CHURCH–TURING–DEUTSCH PRINCIPLE.

CUBE PRINCIPLE [kinesiology] Intent of system to display or view an object or situation in more than one dimension at a time.

CUE OVERLOAD PRINCIPLE [psychology] The probability of recalling an item declines with the number of items subsumed by its functional retrieval cue. In contrast to a registration interpretation, the cue overload view predicts that if the effects of initial recall and of differential recency are controlled, performance in a delayed test of all items from successive lists will be independent of their presentation order. Buildup of proactive inhibition as a cue overload effect.[*]

CUEING PRINCIPLE [psychology] To teach a child to remember to act at a specific time, arrange for the child to receive a cue for the correct performance just before the action is expected rather than after she or he has performed it incorrectly.[†]

CURIE SYMMETRY PRINCIPLE [mathematics] (Pierre Curie, 1859–1906, French physicist, Nobel Prize) (Applied to nonlinear functional systems) (1) A macroscopic cause never has more elements of symmetry than the effect it produces; for example, a scalar cause cannot produce a vectorial effect.[‡] (2) The symmetry of the effect may occasionally be the same as or higher than that of the causes. However, breaking of this principle occurs in some nonlinear phenomena such as buckling of a cylinder. Sattinger's group-theoretic method mathematically explains how these phenomena, dominated by nonlinear functional equations, break the Curie principle. The result of the group-theoretic method

demonstrates that cases both preserving and breaking the Curie principle occur in the same nonlinear system.[§] (3) Associated with a particular cue, the less effective the cue will be for generating recall.[¶]

CURIE'S PRINCIPLE [physics, mathematics] (Pierre Curie, 1859–1906, French physicist, Nobel Prize) (1) For paramagnetic substances, the madnetic susceptibility is inversely proportional to the absolute temperature.[∥] (2) This principle of symmetry superposition defines the resultant symmetry of a complex structure.[**]

CURRENCY PRINCIPLE [mathematics] Effect on an economy due to control and dependence on a nation's monetary standard such as gold, dollars, floating exchange, or other method.[††]

CURTIN–HAMMETT PRINCIPLE [chemistry] For a reaction that has a pair of reactive intermediates or reactants that interconvert rapidly, each going to a different product, the product ratio will depend only on the difference in the free energy of the transition state going to each product, and not on the equilibrium constant between the intermediates.[‡‡]

[*]Watkins, O. C. and Watkins, M. J. (1975), *J. Exp. Psychol. Human Learn. Memory* **1**(4), 442–452.
[†]Huitt, W. (1994), Principles for using behavior modification, in *Educational Psychology Interactive*, Valdosta State Univ., Valdosta, GA (http://chiron.valdosta.edu/whuitt/col/behsys/behmod.html).
[‡]Sci-Tech Dictionary (n.d.), *McGraw-Hill Dictionary of Scientific and Technical Terms*.

[§]Fulea, A. O. (1988), Chemical shift nonequivalence versus chemical shift equivalence of geminal diastereotopic groups in prochiral assemblies and Curie's principle of symmetry, *Rev. Roumaine Chim.* **33**(1), 39–52.
[¶]Koptsik, V. A. (1983), Symmetry principle in physics, *J. Phy. C Solid State Phys.* **16**(1), 23–34.
[∥]Walker, P. M. B. (1999), *Chambers Dictionary of Science and Technology*, Chambers, New York, p. 291.
[**]Fulea, A. O. (1988), Chemical shift nonequivalence versus chemical shift equivalence of geminal diastereotopic groups in prochiral assemblies and Curie's principle of symmetry, *Rev. Roumaine Chim.* **33**(1), 39–52.
[††]Brown, J. H. and Sibly, R. M. (2006), Life-history evolution under a production constraint, *Proc. Natl. Acad. Sci. USA* **103**(47), 17595–17599.
[‡‡]Alvarez-Ibarra, C., Fernandez-Gonzalez, F., Garcia-Martinez, A.., Perez-Ossorio, R.., and Quiroga, M. L. (1973), Generalized Curtin–Hammett principle and the elucidation of transition state type. Application to stereoselectivity in the lithium aluminum hydride

CURVILINEAR IMPETUS PRINCIPLE [psychology, physics]   If a stimulus directs the motion of an object in a curvilinear trajectory, the object will continue in a curvilinear trajectory even after the stimulus is removed: This is a false belief commonly found in individuals' naive or folk theories of physics.*

CYBERNETIC PRINCIPLE [psychology]   Applies to behaviors of interest to personality and social psychologists. In pursuit of a goal, little is mentioned of what processes are entailed in the pursuit and attainment of a goal. The goal construct provides an entry point into the logic of cybernetic self-regulation.[†]

CYCLIC IDENTITY PRINCIPLE [mathematics]   The sum of any component of the Riemann–Christoffel tensor and two other components obtained from it by cyclic permutation of any three indices, while the fourth is held fixed, is zero.[‡]

CYCLOPENTENE CYCLOADDITION AROMATICATIC HETEROCYCLE INERTIA PRINCIPLE [chemistry]   See INERTIA PRINCIPLE.

CYCLOTRON PRINCIPLE [physics]   Positive ions produced by a beam of electrons flowing along the axis of a uniform magnetic field follow circular trajectories with a radius proportional to momentum and a frequency of rotation inversely proportional to mass.[§]

CYLINDER FORMER PRINCIPLE [engineering]   Suction applied to creation of cylindrical design for packaging or other purposes.[¶]

CYRANO PRINCIPLE [mathematics]   If what seems to be the reason for a change, a decline, or an advance is as obvious as the nose on your face, either it's not the real reason or the movement itself is only a blip.[‖]

CYTOCHEMICAL PRINCIPLE [chemistry]   Direct histocytochemical staining methods on undisrupted tissues, stabilized by chemical fixation, potentially offer perhaps the most reliable approach to the study of enzymes of the cell in relation to the cellis ultrastructure. The atoms, which, for the most part, constitute the biomacromolecules and enzymes of cells and tissues, contribute little to their inherent electron opacity or ability to scatter electrons differentially. The latter property of a substance is responsible for the electron microscopic observation of that substance. Since the introduction of osmiophilic reagents into cytochemistry, the selective deposition of relatively large amounts of polymeric osmium black reaction products at the subcellular sites of insoluble or immobilized enzymes or biomacromole has been observed.[**]

CYTOTOXIC   PRINCIPLE [biochemistry]   Behaves like a true antibody in its specificity and absorption reactions, and in its association with the nondialyzable globulin constituents of serum. Electrolytes may not be necessary for the reaction.[††]

reduction of alkyl aryl ketones, *Tetrahedron Lett.* (29), 2715–2718.

*McCloskey, M. and Kohl, D. (1983), Naive physics: The curvilinear impetus principle and its role in interactions with moving objects, *J. Exp. Psychol.* **9**(1), 146–156.

[†]Markus & Nurius, 1986 Possible Selves. American Psychologist, 417, 954–969.

[‡]*McGraw-Hill Dictionary of Scientific and Technical Terms.* McGraw-Hill (2003) Cyclic Identify, McGraw Hill dictionary of Scientific and technical terms. p. 534

[§]Jernakoff, G. (1958), Mass spectrometer. Institute of Petroleum, ASTM Committee E-14 on Mass Spectrometry. (1976) Advances in Mass Spectrometry, v3, p. 1020.

[¶]Zhao, S. D., Zhang, Z. Y., Zhang, Y., and Yuan, J. H. (2007), The study on forming principle in the process of hydro-mechanical reverse deep drawing with axial pushing force for cylindrical cups, *J. Mater. Process. Technol.* (187–188), 300–303.

[‖]Birinyi, L. J. (1994), Cyrano principle (stock market fluctuations), *Forbes*, **153**, 146.

[**]Hanker, J. S., Seaman, A. R., Weiss, L. P., Ueno, H., Bergman, R. A., and Seligman, A. M. (1964), Osmiophilic reagents: New cytochemical principle for light and electron microscopy, *Science* **146**(3647), 1039–1043.

[††]Kupchan, S. M. (1964), Calotropin, a cytotoxic principle isolated from asclepias curassavica, *Science* **146**, 1685.

D

D'ALEMBERT PRINCIPLE [physics] On a body in motion, the external forces are in equilibrium with the inertial forces.* See DIFFERENTIAL PRINCIPLE; VARIATIONAL PRINCIPLE; HAMILTON PRINCIPLE; PRINCIPLE OF LEAST CONSTRAINT; HAMILTON–LAGRANGE PRINCIPLE; MAUPERTUIS–LAGRANGE PRINCIPLE; LEAST-CONSTRAINT PRINCIPLE; D'ALEMBERT–LAGRANGE PRINCIPLE; GAUSS' PRINCIPLES; JOURDAIN PRINCIPLE; PRINCIPLE OF VIRTUAL WORK; KINETIC PRINCIPLE; MAUPERTUIS–EULER–LAGRANGE PRINCIPLE OF LEAST ACTION; D'ALEMBERT–LAGRANGE VARIATION PRINCIPLE.

D'ALEMBERT–LAGRANGE PRINCIPLE [physics] (Jean le Rond d'Alembert, 1717–1783, French mathematician) (Giuseppe Lodovico Lagrangia, 1736–1813, Italian mathematician) (1) The path of conservative system in configuration space between two configurations is such that the integral of the Lagrangian function over time is a minimum or maximum relative to nearby paths between the same and points and taking the same time.[†] (2) Motion of the underlying system of particles compatible with the collective equilibrium provided that the variations are associated with reversible processes.[‡] See also DIFFERENTIAL PRINCIPLE; VARIATIONAL PRINCIPLE; HAMILTON PRINCIPLE; PRINCIPLE OF LEAST CONSTRAINT; D'ALEMBERT PRINCIPLE; HAMILTON–LAGRANGE PRINCIPLE; MAUPERTUIS–LAGRANGE PRINCIPLE; LEAST-CONSTRAINT PRINCIPLE; GAUSS PRINCIPLE; JOURDAIN PRINCIPLE; PRINCIPLE OF VIRTUAL WORK; KINETIC PRINCIPLE; MAUPERTUIS–EULER–LAGRANGE PRINCIPLE OF LEAST ACTION; D'ALEMBERT–LAGRANGE VARIATION PRINCIPLE; LAGRANGE PRINCIPLE.

D'ALEMBERT–LAGRANGE VARIATION PRINCIPLE See D'ALEMBERT–LAGRANGE PRINCIPLE.

D'ALEMBERT RELAY PRINCIPLE [engineering] See BUSBAR PROTECTION PRINCIPLE.

D'ALEMBERT'S PRINCIPLE [engineering] (Giuseppe Lodovico Lagrangia, 1736–1813, Italian mathematician) A special case of Gauss' principle, restricted by the two conditions that no applied forces and all masses are identical.[§] See also GAUSS' PRINCIPLES; HERTZ'S PRINCIPLE OF LEAST CURVATURE.

DALE'S PRINCIPLE [physics] (Sir Henry Hallett Dale, 1875–1968, English pharmacologist Nobel Prize) Being an aspect of Newton's third law of motion stating that an equal and opposite reaction is true for bodies that are free to move as well as for bodies rigidly fixed.[¶] See also NEWTON'S THIRD PRINCIPLE.

DALTON'S PRINCIPLE OF PARTIAL PRESSURES [chemistry] (John Dalton, 1766–1844, English chemist) (1) The pressure of a gas in a mixture is equal to the pressure that it would exert if it occupied the same volume alone at the same temperature.[‖] (2) The pressure of a mixture of gases is the sum of the partial

---

[*]Walker, P. M. B. (1999), *Chambers Dictionary of Science and Technology*, Chambers, New York, p. 300.
[†]Wang, Y. and Guo, Y.-X. (2005), d'Alembert-Lagrange principle on Riemann-Cartan space, *Wuli Xuebao* **54**(12), 5517–5520.
[‡]Minardi, E. (2005), The magnetic entropy concept, *J. Plasma Phys.* **71**(1), 53–80.

[§]Volosevich, P. Yu (2007), D'Alembert's principle and contemporary concepts of plastic deformation, *Metallofiz. Noveishie Tekhnol.* **29**(10), 1393–1406.
[¶]Sabelli, H. C., Mosnaim, A. D., Vazquez, A. J., Giardina, W. J., Borison, R. L. and Pedemonte, W. A. (1976), Biochemical plasticity of synaptic transmission: A critical review of Dale's principle, *Biol. Psychiatr.* **11**(4), 481–524.
[‖]Walker, P. M. B. (1999), *Chambers Dictionary of Science and Technology*, Chambers, New York, p. 300.

---

pressures of each individual constituent.* See also PRINCIPLE OF MULTIPLE PROPORTIONS.

DANTZIG–WOLFE DECOMPOSITION PRINCIPLE [engineering] (George Bernard Dantzig, 1914–2005, American mathematician) (Phil Wolfe, American mathematician) (G. B. Dantzig and P. Wolfe) Given a real image and given some object models, find the model that under some coordinate transforms, with parameters to be determined, best fits the real image.†

DARWINIAN PRINCIPLE OF NATURAL SELECTION [biology] (Charles Robert Darwin, 1809–1882, English naturalist) (1) Natural selection could not act effectively on individual differences and argued in favor of a saltationist (a very large sudden change) view of evolution through the selection of discontinuous variations.‡ (2) Of survival of the fittest, an analog of the naturally occurring genetic operation of crossover (sexual recombination), and occasional mutation. The crossover operation is designed to create syntactically valid offspring programs (given closure among the set of ingredients).§ (2) Genetic programming combines the expressive high-level symbolic representations of computer programs with the near-optimal efficiency of learning associated with Holland's genetic algorithm.¶ (3) The individuals in the population in intermediate generations of a run of genetic programming (and random subtrees picked from them) differ from the individuals (and their randomly picked subtrees) in the randomly created population of generation of the same run; that is, crossover fragments from intermediate generations of a run of genetic programming are very different from the randomly grown subtree provided by the mutation operation.¶ (4) In the genetic algorithm, the entire population generally improves from generation to generation. The improvement, from generation to generation, in the fitness of the population as a whole is evident by examining the average fitness of the population by generation.‖ See also HALDANE'S PRINCIPLE; PRINCIPLES OF ZOOLOGICAL PHILOSOPHY.

DATA PROTECTION PRINCIPLE [mathematics] Safeguards to protect the integrity, privacy, and security of data.** See COPY PROTECTION PRINCIPLE; PRIVACY PRINCIPLE; FIRST DATA PROTECTION PRINCIPLE; SECOND DATA PROTECTION PRINCIPLE; THIRD DATA PROTECTION PRINCIPLE; FOURTH DATA PROTECTION PRINCIPLE; FIFTH DATA PROTECTION PRINCIPLE; SIXTH DATA PROTECTION PRINCIPLE; SEVENTH DATA PROTECTION PRINCIPLE; EIGHTH DATA PROTECTION PRINCIPLE.

DATA-SENDING PRINCIPLE [computer science] Complex systems are broken down into smaller parts. Each part is then simulated and characterized by a matrix of responses due to resulting equivalent impulses. The solution to the overall system is then computed by subsequently interconnecting the different impulse response matrices with appropriate interfaces.††

DECISION FEEDBACK PRINCIPLE [physics] For a noncyclic channel the elements of the output vector of the decision feedback equalizer can be calculated iteratively. For a cyclic channel, as is the case for block transmission using cyclic extensions, the

*Daintith, J. (1999), *The Facts on File Dictionary of Chemistry*, 3rd ed., Facts On File, New York, p. 73.

†Cheng, R., Fraser Forbes, J. and Yip, W. S. (2008), Dantzig-Wolfe decomposition and plant-wide MPC coordination, *Comput. Chem. Eng.* **32**(7), 1507–1522.

‡Gayon, J. and Cobb, M. (1998), *Darwinism's Struggle for Survival: Heredity and the Hypothesis of Natural Selection*, Cambridge Univ. Press, p. 103.

§Koza, J. R. (2003), *Generic Alrgorithms and Genetic Programming*, Stanford Univ. (http://www.smi.stanford.edu/people/koza).

¶Koza, J. R. (2008), Human-competitive machine invention by means of genetic programming, *Artif. Intell. Eng. Design Anal. Manuf.* **22**, 185–193.

‖Kutschera, U. (2008), Darwin-Wallace principle of natural selection, *Nature* **453**(7191), 27.

**Walker, P. M. B. (1999), *Chambers Dictionary of Science and Technology*, Chambers, New York, p. 303.

††Anon. (2007), Univ. Edinburgh, Records Management Section (http://www.recordsmanagement.ed.ac.uk/infostaff/dpstaff/dp_research/researchannexa.htm).

first elements of the output signal depend on the last ones.* See also QUANTIZED FEEDBACK PRINCIPLE.

DECOMPOSITION PRINCIPLE [physics]   More or less permanent structural breakdown of a molecule into simpler molecules or atoms.†

DECOUPLING   PRINCIPLE [physics] Preventing transfer or feedback of energy from one circuit to another.‡

DECREASING REINFORCEMENT PRINCIPLE [psychology] To encourage a child to continue performing an established behavior with few or no rewards, gradually requires a longer time period or more correct responses before a correct behavior is rewarded.§

DEDUCTIVE CLOSURE PRINCIPLE [mathematics] (Edmund L. Gettier III, b. 1927, Baltimore, MD) For any proposition P, if S is justified in believing P and if P entails Q, and if S deduces Q from P and accepts Q as a result of this deduction, then S is justified in believing Q. (2) If set O is the set of propositions, and operation R is a natural deduction, and provided that p is a member of O, and p deductively entails q, then q is also a member of O.¶ See also PRINCIPLE OF DEDUCIBILITY FOR

*Hogan, W. R. and Wagner, M. M. (1998), Optimal use of communication channels in clinical event monitoring, Proce. AMIA (Am. Medical Informatics Assoc.) *Annual Symp.*, pp. 617–621.
†Frank, T., Klein, A., Costa, E. and Schulz, E. (2005), Low complexity equalization with and without decision feedback and its application to IFDMA, *Proc.* IEEE 16th Intl. Symp. Personal, Indoor and Mobile Radio Communications (PIMRC 2005), Sept. 11–14, 2005, vol. 2, pp. 1219–1223.
‡Xu, Z. B. and Kwong, C. P. (1996), A decomposition principle for complexity reduction of artificial neural networks, *Neural Networks*, **9**, 999.
§Fang, F., Tan, W. and Liu, J. (2008), Pid parameter setting and debugging method based on two-degree-of-freedom control structure and series decoupling principle for coordinated control system of boiler-turbine unit generator set, *Faming Zhuanli Shenqing Gongkai Shuomingshu.*
¶Huitt, W. (1994), Principles for using behavior modification, in *Educational Psychology Interactive*, Valdosta State Univ., Valdosta, GA (http://chiron.valdosta.edu/whuitt/col/behsys/behmod.html).

JUSTIFICATION; STRAIGHT PRINCIPLE; GETTIER'S PRINCIPLE; PRINCIPLE OF EPISTEMIC CLOSURE.

DEEP ECOLOGY'S PRINCIPLES [ecology] The world does not exist as a resource to be freely exploited by humans.‖ See also HANNOVER PRINCIPLES; FOREST STEWARDSHIP COUNCIL (FSC) PRINCIPLES; GLOBAL SULLIVAN PRINCIPLES OF SOCIAL RESPONSIBILITY; MARINE STEWARDSHIP COUNCIL (MSC) PRINCIPLES; PERMACULTURE PRINCIPLES; THE BELLAGIO PRINCIPLES FOR ASSESSMENT; EQUATOR PRINCIPLES; MELBOURNE PRINCIPLES; PRECAUTIONARY PRINCIPLE; SANBORN PRINCIPLES; TODDS' PRINCIPLES OF ECOLOGICAL DESIGN.

DEFAULT   PRINCIPLE [linguistics]   (Yosef Grodzinsky, Canadian linguilt professor, McGill   University.)   Operates   to   compensate for such incomplete grammatic representation in interpretation. A nontheta (non-θ) position in a sentence carries a list of theta roles according to the thematic hierarch, and the noun phrase (NP) that appeared in that position is automatically assigned the thematically highest role as a default value in that position, when the normal assignment fails.**

DEFINING   PRINCIPLES [psychology] (1) Boundaries and meaning represent two essential principles of existence. A boundary is a defining principle, whereas meaning is a supportive liberating principle. Frankl's perception of meaning, becoming conscious of what should be done in a given situation, helps a person better understand the existence of each boundary.†† (2) The core principles are defining characteristics, the necessary conditions for humanitarian response. Organizations such as military forces and for-profit companies may deliver assistance to communities affected

‖Devall, B. and Sessions, G. (1985), Deep Ecology. Gibbs M. Smith. p. 85–88.
**Luper, S. (2001), The epistemic closure principle, in *Stanford Encyclopedia of Philosophy* (http://plato.stanford.edu/entries/closure-epistemic/#CloPr).
††Higawara, H. (1993), The breakdown of Japanese passives and theta-role assignment principle by Broca's aphasics., *Brain Lang.* **45**(3), 323.

by disaster in order to save lives and alleviate suffering, but they are not considered by the humanitarian sector as humanitarian agencies as their response is not based on the core principles.* See also PRINCIPLE OF HUMANITY; PRINCIPLE OF IMPARTIALITY; PRINCIPLE OF INDEPENDENCE; PRINCIPLE OF NEUTRALITY; PRINCIPLE OF PROSELYTISM; EMERGENCY-ACTION PRINCIPLE.

DELAY-LINE PRINCIPLE [physics] A transmission line or an electric network, which, if terminated in its characteristic impedance, will reproduce at its output a waveform applied to its input terminal with little distortion.[†]

DEMICLOSEDNESS PRINCIPLE [psychology] Asymptotic behavior of asymptotically nonexpansive mappings.[‡]

DEMOCRATIC PRINCIPLE [physics] (John Archibald Wheeler, 1911–, American physicist) Definition of inertial motion at some event in spacetime should be physically determined by the relative distribution and motion of all the mass in the universe. "It is not necessary to enter into the mathematics of the theory to state its simple consequence …. Each mass has an 'inertia-contributing' power, a voting power, equal to its mass, there, divided by the distance from there to here."[§] See also MACH'S PRINCIPLE.

DENSE-FLOW CONVEYING PRINCIPLE [engineering] When product is transferred from one conveying element and advanced to the next, in each case one apparatus is charged with two blank streams. This results in an

imbricated stream running to the right and another imbricated stream running to the left, each oriented such that in each case the upper edge of the conveyed blanks is at the front in the conveying direction.[¶] See also CONVEYING PRINCIPLE; PRINCIPLE OF PNEUMATIC CONVEYANCE.

DENSITY MEASURE ANALYST (DMA) PRINCIPLE [engineering] The tube containing the sample is excited to an oscillation. The frequency of this oscillation depends on the mass of the sample cell plus the sample. The nodal points determine the volume of the sample cell. An external excitation force compensates the damping of the oscillator sample. To avoid possible parasitic resonance, the sample cell is mounted on a great countermass.[‖]

DEPENDENCE INJECTION PRINCIPLE [computers] States a high-level modules should not depend on low-level modules. Both should depend on abstractions and abstractions should not depend upon details. Details should depend upon abstractions. Refers to a specific form of decoupling where conventional dependency relationships established from high-level, policy-setting modules to low-level, dependency modules are inverted for the purpose of rendering high-level modules independent of the low-level module implementation detail.[**]

DEPENDENCE INVERSION PRINCIPLE (DIP) [mathematics] (1) Every[††] function that operates on a reference or pointer to a base class should be able to operate on derivatives of that base class without knowing it. This means that the virtual member functions of derived classes must expect no more than the corresponding member functions of the base

*Gould, W. B. (1995), *Int. Forum Logotherapy* **18**(1), 49–52.

[†]Abraham, M. A. (2007), *Sustainability Science and Engineering: Defining Principles*, vol. 1, Elsevier; *Environ. Progress* **26**(1), 11–14.

[‡]Meydan, T. and Elshebani, M. S. M. (1992), Displacement transducers using magnetostrictive delay line principle in amorphous materials, *J. Magn. Magn. Mater.* **112**(1–3), 344–346.

[§]Lin, P. K. et al. (1995), Demiclosedness principle and asymptotic behavior for asymptotically nonexpansive matings, *Nonlinear Anal. Theory Meth. Appl.* **24**, 929.

[¶]Wheeler, J. A. (1990), A journey into gravity and spacetime, Freeman, San Francisco, pp. 232–233.

[‖]Mantéa, C., Yao, A.-F. and Degiovanni, C. (2007), Principal component analysis of measures, with special emphasis on grain-size curves, *Comput. Stat. Data Anal.* **51**(10), 4969–4983.

[**]Martin, R. (1994) OO Design Quality Metrices: an Analysis of Dependencies. C++ Report.

[††]Fowler, M. (2000), *Patterns and Advanced Principles of OOD*, Prentice-Hall, p. 6 (http://www.scribd.com/doc/7214581/Dependency-Injection-Patteren).

class; and should promise no less. Details should depend on abstractions. Abstractions should not depend on details. When high-level modules depend on low-level modules, it becomes very difficult to reuse those high-level modules in different contexts. However, when the high-level modules are independent of the low-level modules, then the high level modules can be reused quite simply.[*] (2) Depend on abstractions, not on concretions. (3) Virtual member functions present in base classes must also be present in the derived classes; and they must do useful work.[†] See also INVERSION OF CONTROL PRINCIPLE; LISKOV SUBSTITUTION PRINCIPLE; OPEN–CLOSED PRINCIPLE.

DERIVATION PRINCIPLE [mathematics]   Process of debugging a formula.[‡]

DESIGN PRINCIPLE [engineering]   Act of conceiving of and planning the structure and parameter values of a system, device, process, or work.

DESORPTION PRINCIPLE [engineering]   Process of removing a sorbed substance by the reverse of absorption or adsorption.[§]

DETAILED BALANCE PRINCIPLE [physics] Hypothesis stating that when a system is in equilibrium, any process occurs with the same frequency as the reverse process.[¶]

DETERMINANT PRINCIPLE [mathematics]   A natural application of the theory of invariants in affine geometry, i.e., projective geometry under adjunction of the plane at infinity.[‖]

DEVILLE–GODEFROY–ZIZLER PRINCIPLE [mathematics] (V. Zizler), (G. Godefroy) Assume that X is a Banach space that admits a smooth Lipschitzian bump function; then, for every lower semicontinuous bounded below function f, there exists a Lipschitzian smooth function g on X such that $f + g$ attains its strong minimum on X, thus extending a result of Borwein and Preiss.[**]

DIAGNOSTIC PRINCIPLE [engineering, psychology, medicine]   This principle addresses the challenges of diagnosis and continuing patient assessment and is embodied in the current International Classification of Diseases (ICD) and DSM methodologies, which are firmly based on the objective assessment of symptoms; it has improved the reliability of schizophrenic diagnosis to levels roughly comparable with those of many other medical conditions of unknown etiology. However, the differences between schizophrenia and the major affective disorders in terms of etiology, pathology, course and outcome, and treatment response now appear less obvious than previously considered, bringing the validity of the current dichotomous classification of "nonorganic" psychoses into

[*]Scott, S. D. (2001), Software Components for Simulation and Optimization, master's thesis, Rice Univ., Houston, TX.
[†]Martin, R. C. (n.d.), *Dependency Inversion Group. Ten Commandments of OO Proramming* (http://www.butunclebob.com/ArticleS.UncleBob. PrinciplesOfOod).
[‡]Kipriyanov, A. A. and Doktorov, A. B. (2005), The analysis of the derivation principles of kinetic equations based on exactly solvable models of the bulk reaction $A + B \boxed{?}$ product, *Chem. Phys.* **320**(1), 21–30.
[§]Fujihara, T., Fujita, S., Takeishi, S., Arinaga, K., Yamaguchi, Y. and Tatsuya, U. (2004), *Target Detection Apparatus and Target Trapping Body, Molecule Adsorption / Desorption Apparatus and Molecule Adsorption / Desoption Method, and Protein Detection Apparatus and Protein Detection Method*, PCT International Patent Application.

[¶]Berg, O. G. (1983), Time-averaged chemical potential of proteins and the detailed-balance principle (an alternative viewpoint), *Proce. Natl. Acad. Sci. USA* **80**(17), 5302–5303.
[‖]Klein, F. (2004), *Elementary Mathematics from an Advanced Standpoint: Geometry*, Courier Dover Publications, p. 150.
[**]Deville, R., Godefroy, G. and Zizler, V. (1993), A smooth variational principle with applications to Hamilton-Jacobi equations in infinite dimensions, *J. Funct. Anal.* **111**(1), 197–212.

question. An agreed-on understanding of clinical material that is essential for scientific investigation was lacking before. Without this, it is unlikely that psychiatry could have made full use of the wide range of new investigative and treatment modalities.[*] [Engineering] Power output of a machine is proportional to the active power input of the motor. The active power curve is computed from the measured control voltage and current over time. The actual active power curve showing the current point status is recorded for each point reversal. It is evaluated automatically and compared with defined limit values. See QUASILOCAL PRINCIPLE.

DIET PRINCIPLE [ecology] (Thomas Doubleday, 1790–1870, English political economist) (Doubleday) Whenever a species or genus is endangered, a corresponding effort is invariably made by nature for its preservation and continuance, by an increase of fecundity or fertility; and that this especially takes place wherever such danger arises from a dimuntion of proper nourishment or food, so that consequently the state of depletion, or the deplethoric state, is favorable to fertility, and that, on the other hand, the plethoric state, or state of repletion, is unfavorable to fertility, in the ratio of the intensity of each state, and this probably throughout nature universally, in the vegetable as well as the animal world. Further, that as applied to humankind this law produces the following consequences, and acts thus: There is in all societies a constant increase going on among that portion of it that is the worst supplied with food; in short, among the poorest. Among those in the state of affluence, and well supplied with food and luxuries, a constant decrease continues. Among those who form the mean or medium between these two opposite states, population is stationary.[†]

DIFFERENT MEASURING PRINCIPLE [mathematics] (1) Measuring system utilized for monitoring and correcting the first measuring system. (2) Measuring the same physical parameters; a "diversitary" design, with the measuring devices operating via different measuring principles, can be particularly advantageous. (3) An optical sensor or a sensor that operates with radar waves.[‡]

DIFFERENTIAL MOBILITY ANALYZER PRINCIPLE [engineering] Eletrically charged particles move in an electric field according to their electrical mobility. The electrical mobility depends mainly on the particle size and electrical charge. The smaller the particle, the higher is the electrical mobility. The higher the electrical charge, the higher is the electrical mobility.[§] Also known as *DMA principle; principle of DMA.*

DIFFERENTIAL PRINCIPLE [physics] See INDETERMINANCY PRINCIPLE.

DIFFERENTIAL PRINCIPLE OF D'ALEMBERT [physics] See D'ALEMBERT–LAGRANGE PRINCIPLE.

DIFFERENTIAL PULSE-CODE MODULATION PRINCIPLE [computer science] (Developed by Alec H. Reeves,1926) A version of pulse-code modulation in which the difference in value between a sample and its predecessor constitutes the transmitted information.[¶] See PRINCIPLE OF PULSE-CODE MODULATION.

DIFFERENTIAL TRANSFORMER PRINCIPLE [engineering] A transformer used to check on two or more sources of signals to a common transmission line.[‖]

[*]Cunningham Owens, D. G. (2000), Principles of practice: A new philosophy of care, *Intl. J. Psychiatr. Clin. Pract.* (special issue: Schizophrenia: Diagnosis and Continuing Treatment) 4(Suppl. 1), S13–S18.

[†]Coontz, S. H. (2003), *Population Theories and the Economic Interpretation*, Routledge p. 42.

[‡]Hayakawa, O., Nakahira, K. and Tsubaki, J. (1995), Evaluation of fine ceramics raw powders with particle size analyzers having different measuring principle and its problem, *J. Ceramic Soc. Jpn.* **103**, 586–592.

[§]Seto, T., Okuyama, K. and Takeuchi, K. (1998), Differential mobility analyzer (DMA) for the measurement and sizing of nano-particles, clusters and ions., *RIKEN Rev.* **17**, 5–6.

[¶]Walker, P. M. B. (1999), *Chambers Dictionary of Science and Technology*, Chambers, New York, p. 325.

[‖]Zamenhof, S., Leidy, G., Greer, S. and Hahn, E. (1957), Differential stabilities of individual heredity determinants in transforming principle, *J. Bacteriol.* **74**(2), 194–199.

DIFFRACTION PRINCIPLE [physics] Any redistribution in space of the intensity of waves resulting from the pressure of an object causing variations of either the amplitude or phase of the waves; found in all types of wave phenomena.*

DIFFUSION TRANSFER PRINCIPLE [optics] Utilizing light and a developer for the instant transfer of an image on white sensitized paper.†

DIGITAL MULTIMETER MEASUREMENT (DMM) PRINCIPLE [engineering] Applying identification techniques allows one to establish the exact relation between the diodes and the frame of an object. Once the objects are located in a common world frame, any calculation can be performed: distance, velocity, acceleration, vibration. The flexible approach of digital multimeter (DMM) allows measurement and analysis of any dynamic six-dimensional (6D) environment.‡

DIP ROLLER PRINCIPLE [engineering] A roll dipping into a liquid container has a "doctor blade" assigned thereto that removes from the circumference of the dip roller that part of the liquid that is not accepted or taken up by the cells of the dip roller.§ See also ROLLER WEDGE PRINCIPLE; THREE-ROLLER PRINCIPLE.

DIRAC PRINCIPLE [physics] (Paul Dirac, 1902–1984, British physicist, Nobel laureate) Theory using the same postulates as the Schrödinger equation, plus the requirement that quantum mechanics conform with the theory of relativity, concluding that an electron must have an inherent angular momentum and magnetic moment.¶

DIRECT ARC PRINCIPLE [engineering] Current must flow through a metal bath in order that the heat developed by the electrical resistance of the metal is added to that radiated from the arcs.‖

DIRECT LIGHT PRINCIPLE [engineering, mathematics] (1) An arrangement for measuring lengths or angles. Radiation produced by a light source is collimated by a condenser and diffracted and reflected by phase grids. The resulting partially diffracted light rays are then directed onto two sets of photodetectors.** (2) Effect of information about one possible cause of an event as inferences regarding another possible cause. Analyzes situations until a minimal set of sufficient causes are identified; then, other possible causes are ignored or dismissed.†† (3) Comprising at least two sensor or feeler elements having different functional or operating principles and a common evaluation circuit for evaluation of the property changes of the sensor elements and for triggering a signal.‡‡ (4) Rest mass of a particle cannot be kept invariant. There must exist a unique preferred inertial frame of reference wherein a particle is absolutely at rest.§§ See also INCIDENT LIGHT PRINCIPLE;

*Lee, S. W. and Lee, S. S. (2008), Application of Huygens-Fresnel diffraction principle for high aspect ratio SU-8 micro-/nanotip array, *Optics Lett.* **33**(1), 40–42.

†Oishi, Y. (1976), *Color Photographic Material Working by the Diffusion Transfer Principle*, US Patent (1976) 3,993 486.

‡Chaves, A. E. V. (2006), Utilization of a digital multimeter for capturing spectral information generated by an atomic emission/absorption spectrometer, *Ingen. Ciencia Quim.* **22**(1), 29–34.

§Gorecki, W., Hladki, T., Rusnak, M., Kowalczyk, A., Selega, R., et al. (2001), Bozena. Role of temper rolling in the manufacture of cold-rolled steel sheets and strips, *Hutnik–Wiadomosci Hutnicze*, **68**(1), 13–18.

¶Walker, P. M. B. (1999), *Chambers Dictionary of Science and Technology*, Chambers, New York, p. 332.

‖Rahm, C. (2005), Twenty-five years of direct purging in electric arc furnaces, *RHI Bull.* (1), 22–27.

**Dieter, M. (1986), Photoelectric position measuring instrument with grids, US Patent 47,663.

††Shaklee, H. and Fischhoff, B. (1978), *Discounting in Multicausal Attribution: The Principle of Minimal Causation*, Decisions and Designs, Inc., accession no. ADA065142.

‡‡Shcherbakov, A. S., Tepichin, R. E. and Aguirre Lopez, A. (2005), A multi-fold Bragg scattering of light by elastic waves with direct transitions between all the light modes, *Proce. SPIE* (Int. Society Optical Engineering).

§§Ohanian, H. C. (1971), Scalar-tensor theories and the principle of equivalence, *Int. J. Theor. Phys.* **4**(4), 273–280.

SCATTERED LIGHT PRINCIPLE; TRANSMITTED LIGHT PRINCIPLE.

DIRECT PRINCIPLE [engineering, mathematics] The objective is to determine distribution satisfying given constraints.* See also ENTROPY OPTIMIZATION PRINCIPLE.

DIRICHLET DRAWER PRINCIPLE See PIGEONHOLE PRINCIPLE.

DIRICHLET'S BOX PRINCIPLE [physics, mathematics, chemistry] (Johann Peter Gustave L. Dirichlet, 1805–1859, German mathematician; principles developed in 1834)   See also DIRICHLET'S BOXING-IN PRINCIPLE; PIGEONHOLE PRINCIPLE.

DIRICHLET'S BOXING-IN PRINCIPLE [measurement, mechanics, probability] (Johann Peter Gustave Lejeune Dirichlet, 1805–1859 German mathematician)   Given a number of boxes and a greater number of objects, at least one box must contain more than one object. There is no perfect matching on an odd number of vertices. In general, if n objects are placed into k boxes, then there exists at least one box containing at least $[n/k]$ objects, where $[n/k]$ is the ceiling function. This principle generalizes the pigeonhole principle, which states that for a fixed bipartition of the vertices, there is no perfect matching between them.[†] See also DIRICHLET'S BOX PRINCIPLE; PIGEONHOLE PRINCIPLE.

DIRICHLET'S PRINCIPLE [mathematics, physics] (dirichlet's box principle) (1) The function being integrated is the square of the gradient of some continuously differentiable function with a particular value on the boundary of the region of integration. The function minimizing the integral is a harmonic function (i.e., the function that solves Laplace's equation with those boundary values).[‡] (2) Dirichlet's principle is mistakenly applied to duality between current and voltage. This duality, long familiar to electricians, is known to mathematicians as *Hodge duality*. Under this duality resistance corresponds to conductance. To a principle yielding upper bounds for resistance there corresponds a principle yielding lower bounds for conductance, and vice versa.[§] (3) There exists a function u that minimizes the function $D[u] = \int \Omega |\bigtriangledown u| 2dv$ (called the *Dirichlet integral*) for $\Omega \subset \Re 2$ or $\Re 3$ among all the functions $u \in C(1)(\Omega) \subset C(0)(\Omega)$ which take on given values on the boundary $\partial\Omega$ of Ω, and that function satisfies $\bigtriangledown 2 = 0$ in Ω, $u|\partial\Omega = f$, $u \in C(2)(\Omega) \cap C(0)(\Omega)$. Weierstrass showed that Dirichlet's argument contained a subtle fallacy. As a result, it can be claimed only if there exists a lower bound to which $D[u]$ comes arbitrarily close without being forced to actually reach it. Kneser, however, obtained a valid proof of Dirichlet's principle. The function minimizing the integral is a harmonic function (i.e., the functions solving Laplace's equation with those bound values).[¶] See also THOMSON PRINCIPLE; DIRICHLET'S BOX PRINCIPLE.

---

*Schmid, A. and Reubi, F. (1951), Comparative determination of cardiac output by means of the Wezler-Boger pulse-wave method and Fick's direct principle, *Cardiologia*, **19**(1), 42–47.
†Rother, F. (1999), Self-repelling polymer chains in finite volume with Dirichlet boundary conditions: crossover from the dilute to the dense phase, *J. Phys. A Math. General* **32**(8), 1439–1459.

‡D'yakonov, E. G. (2001), Some modification of the classical Dirichlet principle, *Doklady Akad. Nauk* **377**(1), 11–16.
§Doyle, P. G. (1998), *Application of Rayleigh's Short-cut Method to Polya's Recurrence Problem*, PhD thesis, Mathematics Dept., Dartmouth College, p. 11.
¶Sario, L. (1963), An integral equation and a general existence theorem for harmonic functions, *J. Comment. Math. Helveti.* **38**(1), 284–292.

DISCRETE MAXIMUM PRINCIPLE [engineering, mathematics] See MAXIMUM PRINCIPLE.

DISCRETE PONTRYAGIN MAXIMUM PRINCIPLE [engineering, mathematics] See MAXIMUM PRINCIPLE.

DISCRETE-TIME PONTRYAGIN'S MINIMUM PRINCIPLE [mathematics] (1) Presents both deterministic and stochastic control problems, in both discrete- and continuous-time, deterministic control systems together with several extensions. Addresses problems with perfect and imperfect state information, as well as minimax control methods (also known as worst-case control problems or games against nature).* (2) To eliminate maintenance errors based on assembly precedence relationship. Assembly sequence planning can be transformed into optimal transition firing sequence(OFS). An OFS must minimize the Hamiltonian function which can be treated as the heuristic information to find the optimal assembly sequence and avoid latent deadlock.[†]

DISCRIMINATION PRINCIPLE [psychology] To teach a child to act in a particular way under one set of circumstances but not in another, help her/him to identify the cues that differentiate the circumstances and reward her/him only when her/his action is appropriate to the cue.[‡]

DISEASE SIMILARITY PRINCIPLE [ecology] See SYNDROME PRINCIPLE.

DISPLACEMENT PRINCIPLE [mathematics, engineering] (Eric Reissner, 1913–1996, German mathematician) (1) Formulation that radiation of an alpha particle reduces the atomic number by 2 and the mass number by 4, and that radiation of a beta particle increases the atomic number by 1, but does not change the mass number.[§] (2) Normally associated with decision theory; it strives to evaluate relative utilities of simple and mixed parameters that can be used to describe outcome. Since management techniques and knowledge and the total environment of managing change constantly, the enterprise that would ensure its managerial competence cannot tolerate managers who are not interested in their continuous development.[¶] (3) To determine the force multiplication coefficient as a function of the load and to correct for the frictional force, which opens up new means of improving the accuracy of machines in transmitting the force unit.[‖] (4) Constraints introduced to the symmetry condition functional with the aid of a Lagrange multiplier.[**] See also VIRTUAL DISPLACEMENT PRINCIPLE; PRINCIPLE OF VIRTUAL WORK; VOLUME DISPLACEMENT PRINCIPLE; REISSNER TWO-FIELD STRESS AND DISPLACEMENT PRINCIPLE; REISSNER TWO-FIELD PRINCIPLE; PRINCIPLE OF MINIMUM POTENTIAL ENERGY; PRINCIPLE OF MAXIMUM COMPLEMENTARY ENERGY; MIXED VARIATIONAL PRINCIPLE; HELLINGER–REISSNER TWO-FIELD DISPLACEMENT AND STRESS MODIFIED VARIATIONAL PRINCIPLE; HELLINGER–REISSNER TWO-FIELD MODIFIED VARIATIONAL PRINCIPLE; FRAEIJS DE VEUBEKE PRINCIPLE; PRINCIPLE OF DISPLACEMENT; ELEMENT DISPLACEMENT PRINCIPLE; ROOTS PRINCIPLE; PRINCIPLE OF DISPLACEMENT.

*Bertsekas, Dimitri P. Dynamic programming and optimal control. Vol. I. Third edition. Athena Scientific, Belmont, MA, 2005.

[†]TANG Xin-min, and ZHONG Shi-sheng 2008 Aero-engine assembly sequence planning based on discrete-time pontryagin's minimum principle. Control and Decision.

[‡]Huitt, W. (1994), Principles for using behavior modification, in *Educational Psychology Interactive*, Valdosta State Univ., Valdosta, GA (retrieved 6/20/2006 from http://chiron.valdosta.edu/whuitt/col/behsys/behmod.html).

[§]Walker, P. M. B. (1999), *Chambers Dictionary of Science and Technology*, Chambers, New York, p. 337.

[¶]Constable, F. H. (1928), Reichinstein's displacement principle, *Proc. Cambridge Phil. Soc.* **24**, 56–64.

[‖]Kathrein, G. (1943), The displacement principle for concrete, *Tonindustrie-Zeitung* (67), 324–326.

[**]Haas, A. (1988), The element displacement principle and its importance for the chemistry of the para-block-elements, *Kontakte* (Darmstadt) **3**, 3–11.

DISSIPATION PRINCIPLE [physics]   (1) Loss or diminution, usually undesirable of power, with the lost power being converted into heat.* (2) The loss of energy from a system caused by the action of forces such as friction.† See also VIRTUAL DISSIPATION PRINCIPLE; D'ALEMBERT'S PRINCIPLE.

DISTANCE PRINCIPLE IN RELAY [engineering] See BUSBAR PROTECTION PRINCIPLE.

DISTANCE PROTECTION PRINCIPLE [engineering]   To be able to selectively detect a short circuit, it is necessary, when the short circuit occurs, to have access to correct values of currents and voltages at the measuring station. With the aid of the measured values at the instance of the fault, the circuit impedance, which corresponds to the positive-sequence impedance from the measuring station to the fault location, can be calculated.‡

DISTANCE-TO-FAULT PRINCIPLE [communications]   In order to display the distance to fault on the site master, several steps internal to the site master have already occurred. First, the site master takes the frequency-domain reflectometry (FDR) measurement, a fancy name for reflected signals versus frequency or commonly known as *return loss* [standing-wave ratio (SWR)] versus frequency. Second, the built-in IFFT (inverse fast Fourier transform) function converts the reflected signals from frequency domain to time domain. With known relative propagation velocity and cable loss in dB/ft (m), the site master transforms the time-domain results into distance and displays the return loss (SWR) versus distance on the liquid crystal display (LCD) screen. Any faults will

appear to have a high return loss associated with a unique distance.§

DISTORTION MINIMIZATION PRINCIPLE [physics]   See ENERGY MINIMIZATION PRINCIPLE; MINIMIZATION PRINCIPLE.

DISTRIBUTED MAXIMUM PRINCIPLE [engineering, mathematics]   See AUTONOMOUS COOPERATION DISTRIBUTED MAXIMUM PRINCIPLE; MAXIMUM PRINCIPLE; MAXIMUM MODULUS PRINCIPLE.

DISTRIBUTION MATCHING PRINCIPLE [electronics, computer science, psychology]   See MATCHING PRINCIPLE.

DISTRIBUTION PRINCIPLE [chemistry, mathematics]   The total energy in a given assembly of molecules is not distributed equally, but the number of molecules having an energy different from the median decreases as the energy difference increases, according to a statistical law.¶ See also COUNTERCURRENT DISTRIBUTION PRINCIPLE; PRINCIPLE OF CURRENT DISTRIBUTION; COST DISTRIBUTION PRINCIPLE. CHARGE DISTRIBUTION PRINCIPLE.

DIVIDE-AND-CONQUER PRINCIPLE [mathematics]   (1) An area of computer display in which units of distance are the same horizontally and vertically so that there is no distortion.‖ (2) Viewing area in which positions are determined by using a Cartesian coordinate system with horizontal and vertical axes.** (3) The corresponding focus on shared novel character states yields a

*Walker, P. M. B. (1999), *Chambers Dictionary of Science and Technology*, Chambers, New York, p. 338.

†Watson, S. J. and Norris, S. A. (2006), Scaling theory and morphometrics for a coarsening multiscale surface, via a principle of maximal dissipation, *Phys. Rev. Lett.* **96**(17).

‡Strom, D. J. (1988), The four principles of external radiation protection: Time, distance, shielding and decay, *Health Phys.* **54**(3), 353–354.

§Lau, Y. (n.d.), *Understanding the Distance-to-Fault Measurement Data* (http://www.rfmarketing.com/rfm/aw/ref/dtf-yl.pdf).

¶Walker, P. M. B. (1999), *Chambers Dictionary of Science and Technology*, Chambers, New York, p. 339.

‖Duque, C. A., Ribeiro, M. V., Ramos, F. R. and Szczupak, J. (2005), Power quality event detection based on the divide and conquer principle and innovation concept, *IEEE Trans. Power Deliv.* **20**(4), 2361–2369.

**Galván-López, E. and Rodríguez-Vázquez, K. (2007), Multiple interactive outputs in a single tree: An empirical investigation, *Book Series Lecture Notes in Computer Science*, Springer, Berlin/Heidelberg, vol. 4445, pp. 341–350.

fast and transparent phylogeny estimation algorithm and heuristic search; shared novelties give evidence of the exclusive common heritage (monophyly) of a subset of the species. They indicate conflict in a split of all species considered, if the split tears them apart. Only the split at the root of the phylogenetic tree cannot have such conflict. Therefore, we can work top–down, from the root to the leaves, by heuristically searching for a minimum-conflict split, and tackling the resulting two subsets in the same way.* See also VIRTUAL WORK PRINCIPLE.

DIVISION PRINCIPLE [psychology, computers] See INFORMATION-PROCESSING PRINCIPLE.

DO-NO-HARM PRINCIPLE [medicine]   Part of the Hippocratic Oath: (1) identify ways in which government policies create incentives and problems that many turn to government to solve; and (2) identify policy changes needed to make government a neutral player in the healthcare system.† See HARM PRINCIPLE.

DONSKER'S INVARIANCE PRINCIPLE [physics, mathematics, chemistry] (Monroe David Donsker, 1925–1991, American mathematician) Relates asymptotic properties for the distribution of random walks to the distribution of Brownian motion.‡ See INVARIANCE PRINCIPLE.

DOPPLER MATCHING PRINCIPLE [electronics, computer science, psychology]   See MATCHING PRINCIPLE.

DOPPLER PRINCIPLE [physics]   (Christian Doppler, 1803–1853, Austrian physicist) (1) The apparent change of frequency because of the relative motion of the source

of radiation and the observer.§ (2) Changes the received frequency of the signal from its transmitted frequency by an amount proportional to the relative velocity of the transmitter and the receiver; change in frequency can be either positive or negative. The return signal from the target is affected in frequency compared to that directed toward the target by motion of the target. The method is used for detecting the difference in wavelength between the incident beam and that reflected from the approaching or receding car or plane.¶ Also known as *microwave Doppler principle; principle of proximal isovelocity surface area; photoelectric barrier principle; principle of the Doppler effect; color Doppler principle; ultrasound Doppler principle; sonic Doppler principle; Doppler radar principle.* (3) Shared novelties with exclusive common heritage of species indicate a conflict by splits in the Phylogenetic tree. Only the root of the tree cannot have such a conflict.

DOPPLER RADAR PRINCIPLE See DOPPLER PRINCIPLE.

DOUBLE-BED   PRINCIPLE [psychology] Remarks concerning acceptance of differing lifestyles.‖

DOUBLE-COUNTING   PRINCIPLE [mathematics]   Technique involving counting the size of a set in two ways in order to show that the two resulting expressions for the size of the set are equal.** See also COMBINATORIAL PRINCIPLE.

DOUBLE-EFFECT   PRINCIPLE [psychology] (Jean Pierre Gury, 1801–1866, Jesuit theologian)   (1) A rule of conduct frequently

---

*Fuellen, G., Wagele, J. W. and Giegerich, R. (2001), Minimum conflict: A divide-and-conquer approach to phylogeny estimation, *Bioinformatics* **17**(12), 1168–1178.

†Goodman, J. C. (2007), Applying the "do no harm" principle to health policy, *J. Legal Med.* **28**(1), 37–52 (http://www.tandf.co.uk/journals/titles/01947648.asp).

‡Watkins, J. C. (1989), Donsker's invariance principle for Lie groups, *Ann. Probab.* **17**(3), 1220–1242.

§Walker, P. M. B. (1999), *Chambers Dictionary of Science and Technology*, Chambers, New York, p. 345.

¶n.a. (2009), Ultrasonics in *Encyclopædia Britannica* (*Encyclopædia Britannica Online*: http://search.eb.com/eb/article-64034).

‖Graff, E. J. (1993), The double bed principle (discrimination against lesbians), *New York Times Mag.* p. 14.

Johannesson, M. (1997), Avoiding double-counting in pharmacoeconomic studies, *PharmacoEconomics* **11(5), p. 385–388.

used in moral theology to determine when a person may lawfully perform an action from which two effects will follow, one bad, and the other good. (2) Four conditions must be verified in order that a person may legitimately perform such an act: The act itself must be morally good or at least indifferent; The agent may not positively will the bad effect but may merely permit it; The good effect must flow from the action at least as immediately (in the order of causality, though not necessarily in the order of time) as the bad effect; and The good effect must be sufficiently desirable to compensate for the allowing of the bad effect. Of these four conditions the first two are general rules of morality. The third and fourth conditions enumerated above pertain specifically to the principle of the double effect.[*][†] See also PRINCIPLE OF DO NO HARM; PRINCIPLES OF BIOMEDICAL ETHICS, EXCEPTION-GRANTING PRINCIPLE; JUSTIFYING PRINCIPLE; CASUISTIC PRINCIPLE.

DOUBLE PRINCIPLE [psychology] Language and cognition (termed the *epical* and *logical dimensions*, respectively) show the most obvious result of the sensory and social deprivation resulting from modern information technology in the development of purely abstract forms of language and cognition lacking authentic reference. Language consisting of logos without epos is an impoverishment and could lead to a totalitarian script that senselessly and endlessly writes itself. En skrift der skriver sig selv—informationsteknologien set som et nyt kapitel i skriftens historie = a script that writes itself—the new information technology seen as a further chapter in the history of script.[‡]

DOUBLE-SURFACTANT-LAYER   PRINCIPLE [biology] Applied to the design of magnetic particles having primary and secondary surfactant layers composed of different fatty acids.[§]

DOUBLE-WIRE-PRESS   PRINCIPLE [engineering] Drying with uniform heat transfer applicable to thermal drying.[¶]

DRILLED-CELL   PRINCIPLE [engineering] See ASSEMBLY SEQUENCE PRINCIPLE.

DRUG DESIGN PRINCIPLES [chemistry] How new drugs are discovered with emphasis on lead identification, lead optimization, classification and kinetics of molecules targeting enzymes and receptors, and prodrug design and applications. Acceptable drug absorption depends on the triad of potency, solubility, and permeability. Poor permeability in orally active peptide-like drugs is usually compensated by very high solubility.[‖]

DRUG PRINCIPLE [biology] Suitable for the destruction of cells in any tumor or other defined class of cells selectively exhibiting a recognizable (surface) entity.[**]

DUAL-FUNCTION   PRINCIPLE [mathematics] Formulated in 1958, an attempt to provide a rationale while at the same time establishing limitations to the expanded role. A principle with multiple applications. Examples include demonstrating how to treat the operations of fuzzy numbers with step form, variables for backorder inventory models, and calculation

[*]Connell, F. J. (1958), Outlines of Moral Theology (2nd ed. Bruce Pub. Co) p. 22–24.

[†]Mangan, J. (1949), An Historical Analysis of the Principle of the Double Effect, ThSt v10 p. 40–61.

[‡]Lauritsen, L. (1984) Affect and Effect: What becomes of Freud's affect in the psychoanalysis of Jacques Lacan? Psyke & Logos textbf5(1), 109–124.

[§]Wooding, A., Kilner, M. and Lambrick, D. B. (1992), "Stripped" magnetic particles. Applications of the double surfactant layer principle in the preparation of water-based magnetic fluids, *J. Colloid Interface Sci.* **149**(1), 98–104.

[¶]Petschauer, F., Brogyanyi, E. and Kappel, J. (1996), New concept for pulp sheet formers based on the double wire press principle, *Proc. 5th Int. Conf. New Available Techniques*, part 2, SPCI, Stockholm, Sweden, pp. 867–874.

[‖]Yan, Bing and Czarnik, A. W., (2002), Optimization of Lead Compounds, Drug Design Principles in Analytical methods in combinatorial chemistry. v6 of Critical Reviews in Combinatorial Chemistry Forces. CRC Press p. 248.

[**]Hoppe, G. (1967), An effective drug principle for prolonging radioiodine therapy of hyperthyroidism, *Radiobiol. Radiother.* **8**(3), 321–324.

of a dropper for infusion apparatuses.* See also FUNCTION PRINCIPLE; FORM-AND-FUNCTION PRINCIPLE; FUNCTIONING PRINCIPLE.

DUAL PRINCIPLE [psychology, physics] (Niels Bohr, 1885–1962, Danish physicist) See COMPLEMENTARY PRINCIPLE.

DUAL PRINCIPLE [engineering, mathematics, politics]   (1) The party of power and dignity of church and state in political affairs.[†] (2) Concerned with minimization of entropy or maximization of cross-entropy.[‡] See also ENTROPY OPTIMIZATION PRINCIPLE.

DUAL VARIATIONAL PRINCIPLE [psychology, physics] (Niels Bohr, 1885–1962, Danish physicist)   See COMPLEMENTARY PRINCIPLE.

DUALITY PRINCIPLE [engineering]   (Joseph Diez Gergonne, 1771–1859 French mathematician)   (1) Dual equations (matrices and vectors replaced by their transposes) of control optimization are the equations for state estimation.[§] (2) All the propositions in projective geometry occur in dual pairs that have the property that, starting from either propositions of a pair, the other can be immediately inferred by interchanging the parts played by the words *point* and *line*.[¶] See also PRINCIPLE OF EFFICIENCY; PRINCIPLE OF DUALITY; CONSERVATION OF NUMBER PRINCIPLE.

DUANE AND HUNT'S PRINCIPLE [physics] The maximum photon energy in an X-ray spectrum is equal to the kinetic energy of the electrons producing the X rays so the maximum frequency, as deduced from quantum mechanics is eV/k, where V is the applied voltage, e is the electronic charge, and k is Planck's constant $\left[\lambda_{min} = \frac{1.2398 \times 10^6}{V_0} \right]$.[‖]

DUHAMMEL'S CONVOLUTION PRINCIPLE [mathematics, engineering]   (Jean-Marie Constant Duhamel, 1797–1872, French mathematician)   (1) Used to derive solutions for time-dependent inner boundary conditions. Can be used to invert a Laplace transform.[**] (2) Spring depletion, where pumping from a well beside a spring also drains water from the spring. A solution for this problem was obtained by assuming that the spring partially penetrates an aquitard (low-permeability bed adjacent to an aquifer) that overlies the pumped aquifer.[††]

DULONG AND PETIT'S PRINCIPLE [chemistry] (Pierre Louis Dulong, 1785–1838, French Physicist) (Alexis Thérèse Petit, 1791–1820, French physicist)   The molar thermal capacity of a solid element is approximately equal to $3R$, where R is a gas constant.[‡‡]

DYNAMIC OPTIMIZATION PRINCIPLE [mathematics]   The aim is to choose the values of the adjustable input variables of a process as a function of time, so as to ensure the best performance over a given period. The most important characteristics of such processes are (1) The nonlinear behavior as a function of adjustable variables such as flow rates, (2) the disturbance patterns acting on the process (given by the heat supply and demand patterns, such as may be caused

---

*Vatikiotis, M. R. J. (1989), *The Military and Democracy in Indonesia* (epress.anu.edu.au/mdap/mobile_devices/ch02.html).

[†]Anon. (2009), Organized labour, in *Encyclopædia Britannica* (http://search.eb.com/eb/article-66944).

[‡]Rassat, A. (2003), Application of the duality principle to chiral icosahedral metal complexes, *Angew. Chem.* **42**(6), 611–613.

[§]Gerver, M. L., Kudryavtsevia, E. A. and Gerver, M. L. (1998), On extremal properties of discrete measures, the universal sequence and the duality principle, *Dokl. Akad. Nauk* **363**(5), 586–589.

[¶]Walker, P. M. B. (1999), *Chambers Dictionary of Science and Technology*, Chambers, New York, p. 360.

[‖]Persic, M. and Rephaeli, Y. (2001), *X-Ray Spectral Components of Starburst Galaxies* (arXiv:astro-ph/0112030v1).

[**]Weisstein, E. W. (2004), Duhamel's convolution principle, in *MathWorld—A Wolfram Web Resource* (http://mathworld.wolfram.com/Duhamels-ConvolutionPrinciple.html).

[††]Hunt, B. (2004), Spring-depletion solution, *J. Hydrol. Eng.* **9**(2), 144–149.

[‡‡]Daintith, John, *The Facts on File Dictionary of Chemistry*, 3rd ed., Facts On File, New York, 1999 p. 84.

by weather variations), and (3) the possibility of different operational modes.* See also OPTIMIZATION PRINCIPLE.

DYNAMIC PRINCIPLE [medicine]   Treatment of fractures of the neck of the femur by means of Pohl's nonblocking fishplate. Consists of an early mobile and strain from splintering of the fractive, a neutralization of shearing and pulling stresses and also of bending strains, with conversion of these forces into a constant pressure on the fractive surfaces.[†] See also DYNAMIC PRINCIPLE OF OSTEOSYNTHESIS.

DYNAMIC PRINCIPLE OF OSTEOSYNTHESIS [medicine]   Treatment of fractures of the neck of the femur by means of Pohl's non-blocking fishplate. Consists of an early mobile and strain-free splinting of the fracture, a neutralization of shearing and pulling stresses, and also of bending strains, with conversion of these forces into a constant pressure on the fracture surfaces.[‡] See also DYNAMIC PRINCIPLE; ILIZAROV PRINCIPLE OF DISTRACTION OSTEOGENESIS.

DYNAMIC TRANSLINEAR PRINCIPLE [mathematics]   Used to realize linear filters and constitutes an interesting approach to the implementation of nonlinear differential equations.[§] See also TRANSLINEAR PRINCIPLE.

DYSONBERG CONFUSION PRINCIPLE [physics] Theory developed by the Cosmic Ray Deflection Society, which contends that cosmic rays have a low level of intelligence, allowing them to be easily confused by actions contrary to what is considered normal. Once confused, cosmic rays tend to retreat on their own to whence they came.[¶]

[‡]Hirschfeld, J. and Krebedunkel, K. (1978), A dynamic principle of osteosynthesis for use in the management of femoral-neck fractures, *Z. Allgemeinmed.* **54**(5), 274–278.

[§]Mulder, J., Van der Woerd, A. C., Serdijn, W. A. and Van Roermund, A. H. M. (1997), An RMS-DC converter based on the dynamic translinear principle, *IEEE J. Solid-State Circuits* **32**(7), 1146–1150.

[¶]Cosmic Ray Deflection Society is a satirical/environmental pseudoscientific organization, based in New Orleans, LA; The Cosmic Ray Deflection Society of North America, Inc. (http://www.geocities.com/SunsetStrip/1483/).

*Von Schalien, R., Toijala, K. and Sourander, M. (1971), Use of proportional operators in dynamic programming of chemical engineering optimization problems. I. Basic principle of the iterative method, *Kemian Teollisuus* **27**(1), 15–22.

[†]Heller, W. (1936), The dynamic principle of thixotropic solidification and its application, *C. R. Acad. Sci.* **202**, 1507–1509.

E

EARLY STREAMER EMISSION (ESE) PRINCIPLE
[physics]   (1) Devices design based on protection, against the traditional Franklin lightning rod. Device stimulates a burst of streamers from its tip before producing streamers induced by the field of a downward leader, approaching from a thundercloud. Lightning protection system to measure high-resolution surface charge density distributions on insulators.*

EARNED VALUE MANAGEMENT (EVM) [project management]   A project management system that combines schedule performance and cost performance to answer the question, "What did we get for the money we spent?" All project steps "earn" value as work is completed. Can be compared to actual costs and planned costs to determine project performance and predict future performance trends. Physical progress is measured in dollars, so schedule performance and cost performance can be analyzed in the same terms.†

ECHO PULSE PRINCIPLE [physics]   See PULSE ECHO PRINCIPLE.

ECHO-SOUNDING   PRINCIPLE [physics]   (1) Use of echoes of pressure waves sent down to the bottom of the sea and reflected, the delay between sending and receiving times giving a measure of the depth.‡ (2) Determination of the depth of water by measuring the time internal between emission of a sonic ultrasonic signal and the return of its echo from the sea bottom.§

ECKMANN–HILTON PRINCIPLE [mathematics]
(Beno Eckmann, 1917–2008, Swiss mathematician) (Peter John Hilton, 1923–, British mathematician)   Monoid structures on a set where one is a homomorphism for the other. Given this, the structures can be shown to coincide, and the resulting monoid demonstrated to be commutative. This can then be used to prove the commutativity of the higher homotopy groups.¶

ECOLOGICAL PRINCIPLE [ecology]   The sum of all environmental factors that act as agents of natural selection. (1) Ecological structure and function are not stable and static. They are often influenced by natural disturbances such as from fires, floods, droughts, storms, outbreaks of disease, or pest infestations.‖ (2) An organism or group of organisms that reflect environmental conditions in a habitat.** (3) Everything that lives transforms other aspects of the energy system into forms that it can use to sustain itself.†† See also PRINCIPLE OF DYNAMIC ECOLOGY.

*Allen, N. L. and Evans, J. C. (2000), New investigations of the early streamer emissions principle, *Science, IEE Proc. Meas. Technol.* **147**(5), 243–248.
†Defense Systems Management College (1997) Earned Value Management Textbook, Chapter 2. Defense Systems Management College, EVM Dept., Fort Belvoir, VA.
‡Baez, J. and Neuchl, M. (n.d.), *Higher-Dimensional Algebra I: Braided Monoidal 2-Categories* (http://math.ucr.edu/home/baez/bm2cat.ps.Z.ts).

§Zdabov, V. M. (1959), Ecological principle in epidemiology, Ceskoslov. *Epidemiol. Mikrobiol. Imunol.* **8**, 356–360.
¶Batanin, M. A. (2008), The Eckmann–Hilton argument and higher operads, *Adv. Math.* **217**(1), 334–385.
‖Alieva, R. M., and Ilyaletdinov, A. N. (1986), Realization of the ecological principle in microbiological industrial wastewater treatment, *Izvest. Akad. Nauk SSSR Ser. Biologich.* **4**, 517–527.
Rumiantsev, S. N. (1972), Ecological principle in the taxonomy of Clostridium genus, *Zh. Mikrobiol. Epidemiol. Immunobiol.* **49(5), 122–126.
††Fan, J.-F. and Li, Z.-W. (2005), Land reclamation in refuse dump of open-pit coal mines and ecological principle, *Liaoning Gongcheng Jishu Daxue Xuebao* **24**(3), 313–315.

ECOLOGOECONOMICAL TERRITORIAL SYNDROME PRINCIPLE [ecology] Compiling an estimate map of the environment state. The territory possesses zones of the ecological–economical syndrome risk and a number of critical zones as well. Maps estimating definite indicators characterizing the state of region during the period to a marker transform are constructed.* See also SYNDROME PRINCIPLE.

ECONOMIC MOTIVATIONAL PRINCIPLES [psychology] See PRINCIPLES OF DETERMINATION.

ECOSYSTEM PRINCIPLE [ecology] Energy flows under transformation encompassing thermodynamics, chemistry, biological energetics, biochemistry, and ecological energetics.[†]

EDGE RAY PRINCIPLE [optics] Any smooth optical surface redistributes light in a continuous manner, which means that adjacent rays from the source end up being adjacent on the target, too. Mathematicians call such redistribution a *topological map*. Therefore, in order to guarantee that all rays from the source are transferred to the target, it is sufficient that rays, which originate at the rim of the source, end up at the rim of the target. This conclusion is a direct transfer of the general mathematical principle topology to optical problems. This powerful tool for the design of nonimaging optics is known as the *edge ray theorem* or *edge ray principle* because traditionally rays intersecting the rim of an object are called *edge rays*. The edge ray principle is the key to tailored optics; for transferring desired optical properties into a mathematical equation, it suffices to consider the edge rays—then all other rays are redirected automatically in the right way.[‡] Also known as *principle of nonimaging*.

EDGEWORTH–PARETO PRINCIPLE [physics] (Francis Ysidro Edgeworth, 1845–1926, Irish philosopher) (Vilfredo Federico Damaso Pareto, 1848–1923, Italian industrialist, economist, philosopher) In general terms of a fuzzy choice function. Its application is justified for a wide class of fuzzy multicriterial choice problems described by certain axioms of "rational" behavior.[§] See also WEAK PARETO PRINCIPLE.

EDINBURGH PRINICPLES [psychology] (1) Adopt an operational philosophy promoting the utmost quality of life of persons with intellectual disabilities affected by dementia and, whenever possible, base services and support practices on a person-centered approach. (2) Affirm individual strengths, capabilities, skills, and wishes; this should be the overriding consideration in any decision-making for and by persons with intellectual disabilities affected by dementia. (3) Involve individuals, their families, and other close supports in all phases of assessment and services planning and provision for the person with an intellectual disability affected with dementia. (4) Ensure that appropriate diagnostic, assessment, and intervention services and resources are available to meet the individual needs, and support healthy aging of persons with intellectual disabilities affected by dementia. (5) Plan and provide support and services optimizing remaining in the chosen home and community of adults with intellectual disabilities affected by dementia. (6) Ensure that persons with intellectual disabilities affected by dementia have the same access to appropriate services and support as afforded to other persons in the general population affected by dementia. (7) Ensure that strategic planning across relevant policy, provider, and advocacy groups involves consideration of the current and future needs of adults with intellectual disabilities affected by dementia.[¶]

*Trofimov, A. M., Shagimardanov, R. A., and Petrova, R. S. (1996), Spatial analysis of the ecological syndrome of the territory, *ERAE* **2**(1).

[†]Zhukinskii, V. N., Oksiyuk, O. P., Oleinik, G. N., and Kosheleva, S. I. (1981), Principles and practice of the ecological classification of land surface water quality, *Gidrobiologich. Zh.* **17**(9), 38–49.

[‡]Ries, H. and Rabl, A. (1994), The edge ray principle of nonimaging optics, *J. Opt. Soc. Am. A* **11**(10), 2627–2632.

[§]Nogin, V. D. (2006), The Edgeworth-Pareto principle in terms of a fuzzy choice function, *Zh. Vychisl. Mat. Mat. Fiz.* **46**(4), 583–592.

[¶]Wilkinson, H. and Janicki, M.P. (2002), The Edinburgh Principles with accompanying guidelines and recommendations. *J. Intell. Disab. Res.* **46**(Pt. 3), 279–284.

EENIE, MEENIE, MINEE, MO PRINCIPLE [psychology] (1) That information to be used in decisionmaking is sometimes condensed in an unsystematic fashion such that the ultimate version includes a somewhat random assortment of details, thereby increasing the possibility that those nuances significant to a particular individual or group will be overlooked.* (2) That when there are conflicting risk elements from various sources and reports, a decisionmaker may look to a professional or expert to avoid making a choice.†

EFFECTIVE STRESS PRINCIPLE [physics] Average normal force per unit area transmitted directly from particle to particle of a rock or soil mass.‡ See also EDUCATIONAL PRINCIPLE.

EHRENFEST ADIABATIC PRINCIPLE [physics] (Paul Ehrenfest, 1880–1933, Austrian physitist) See ADIABATIC PRINCIPLE.

EHRENFEST PRINCIPLE [physics] (Paul Ehrenfest, 1880–1933, Austrian physicist) More commonly known as *Ehrenfest's theorem*, when expectation values are taken, the quantum mechanical equations reduce to the classical form.§

EIGHT PRINCIPLES [pharmacology] One of the basic ways Chinese medicine has to diagnose. It uses the following eight divisions of

symptoms: yin or yang (yin–yang,); superficial or internal (li-biao,); cold or hot (han-re,); deficient or replete (xu-shi,).¶

EIGHTH DATA PROTECTION PRINCIPLE [mathematics] Personal data shall not be transferred to a country or territory outside the European economic area unless that country or territory ensures an adequate level of protection for the rights and freedoms of data subjects in relation to the processing of personal data.‖ See also COPY PROTECTION PRINCIPLE; PRIVACY PRINCIPLE; FIRST DATA PROTECTION PRINCIPLE; SECOND DATA PROTECTION PRINCIPLE; THIRD DATA PROTECTION PRINCIPLE; FOURTH DATA PROTECTION PRINCIPLE; FIFTH DATA PROTECTION PRINCIPLE; SIXTH DATA PROTECTION PRINCIPLE; SEVENTH DATA PROTECTION PRINCIPLE.

80/20 PRINCIPLE [mathematics, industrial engineering] (Vilfredo Frederico Damsco Pareto, 1848–1923, Italian economist; Joseph M. Juran, quality management pioneer) (1) For many phenomena, a small percentage of a population accounts for a large percentage of a particular characteristic of the population. Typically, 80% of consequences stem from 20% (the vital few) of causes; (for example, 20% of accounts produce 80% of turnover; 80% of GDP enriches 20% of the population, and 80% of sales comes from 20% of clients. Used as a rule of thumb, it separates the "vital few" (the 20%) from the "trivial many" (the 80%).** (2) The significant items of a group will normally constitute a relatively small portion of the total.†† (3) Where something is

*Rubin, D. C. (1995), *Memory in Oral Traditions: The Cognitive Psychology of Epic, Ballads and Counting-out Rhymes*, Oxford Univ. Press, New York.

†Beach, L. R. and Mitchell, T. R. (1978), A Contingency Model for the Selection of Decision Strategies. The Academy of Management-Review, v. 3, N. 3; pp 439–449.

‡Khalili, N. (2002), Effective stress in unsaturated soils, *Proc. Workshop on Chemo-Mechanical Coupling in Clays: From Nano-Scale to Engineering Applications*, Maratea, Italy, June 28–30, 2001, pp. 201–209.

§Santillan, A. and Cutanda, H. V. (2008), Analysis of the resonance frequency shift in cylindrical cavities containing a sphere and its prediction based on the Boltzmann-Ehrenfest principle, *J. Acoust. Soc. Am.* **123**(5), 3421.

¶Deng, T., Ergil, K., and Ergil, M. (1999), Practical diagnosis in traditional Chinese medicine, *Elsevier Health Sci.* p. 165.

‖Anon. (2007), Univ. Edinburgh, Records Management Section (http://www.records management.ed.ac.uk/infostaff/dpstaff/dp_res earch/researchannexa.htm).

Koch, R. (2001), *The 80/20 Principle. The Secret of Achieving More with Less*, Nicholas Brealey Publishing, London. *Acad. Manage. Rev.* (Acad. Management) **3(3), 439–449.

††Imundo, L. V. (1993), *Effective Supervisor's Handbook*, AMACOM Div., Am. Mgmt. Assoc., p. 203.

shared among a sufficiently large set of participants, there will always be a number k between 50 and 100 such that $k\%$ is taken by $(100 - k)\%$ of the participants. The number k is not necessarily always 80 but may vary from 50 in the case of equal distribution to nearly 100 in the case of a tiny number of participants taking almost all of the resources.* (4) A law describing the frequency distribution of an empirical relationship fitting the skewed concentration of the variate-values pattern. When the data are plotted graphically, the result is called a *maldistribution curve*.[†] See also PARETO PRINCIPLE; PRINCIPLE OF THE VITAL FEW; PRINCIPLE OF FACTOR SPARSITY.

EINSTEIN'S EQUIVALENCE PRINCIPLE [physics] (1) The free-fall time of bodies in a vacuum, subjected to the gravity force, is independent from the mass.[‡] (2) The acceleration acquired by a body subjected to a gravitational field is physically indistinguishable from the acceleration acquired by the body by means of an accelerated motion of its reference frame.[§] (3) Consists in considering the impossibility of distinguishing, by physical experiments, the acceleration given to a body by the gravity from the one produced by an accelerated motion of the reference frame.[¶] (4) Any acceleration is fully equivalent to an acceleration of the same magnitude and direction due to

gravitational force, in that the physical effects at any instant due to any acceleration are identical to and indistinguishable from the effect that would result from whatever mass distribution would be required at that instant to produce the same acceleration by gravitational force.[‖] See also LEAST-COUPLING PRINCIPLE.

EINTHOVEN'S PRINCIPLE [engineering] (Willem Einthoven, 1860–1927, Dutch physiologist) (Augustus Desire Waller, b.1856–1922, British Scientist; Willem Einthoven, 1860–1927, Dutch physiologist) (1) In electrocardiography, if electrocardiograms are taken with three leads, at any given instant the potential of any wave in a lead is equal to the sum of the potentials in the two other leads.[**] (2) Einthoven's recording is known as the "three lead" electrocardiogram (ECG), with measurements taken from three points on the body (defining the "Einthoven triangle"). The connection establishes a common ground for the body and the recording device (oscilloscope). Establishing the correspondence between the ECG trace and the electrical events in the heart is known as the inverse problem of electrocardiology: solving for the electric sources from the potential generated by those sources on the surface of the body.[††]

EKELAND'S EPSILON VARIATIONAL PRINCIPLE [mathematics, physics] (Arne Sigvard Eklund, b. 6/19/1911, Kiruna, Sweden, nuclear physics, instrumentation, and atomic energy) See EKELAND'S VARIATIONAL PRINCIPLE.

EKELAND'S PRINCIPLE [mathematics, physics] (Arne Sigvard Eklund, b. 6/19/1911, Kiruna, Sweden, nuclear physics, instrumentation, and atomic energy). Extension

*Chen, Y., Chong, P. P., and Tong, Y. (1993), Theoretical foundation of the 80/20 rule, *Scientometrics* (Akademiai Diado, Springer Science) **28**(2), 183–204.

[†]Hart, J. T. (2004), Inverse and positive care laws, *Br. J. Gen Pract.* **54**(509), 890.

[‡]Horvath, J. E., Logiudice, E. A., Riveros, C., and Vucetich, H. (1988), Einstein equivalence principle and theories of gravitation: A gravitationally modified standard model, *Phys. Rev. D Particles Fields* **38**(6), 1754–1760.

[§]Denisov, V. I. and Denisov, M. I. (1999), Verification of Einstein's principle of equivalence using a laser gyroscope in terrestrial conditions, *Phys. Rev. D Particles Fields* **60**(4).

[¶]Lo, C. Y. (1999), Compatibility with Einstein's notion of weak gravity: Einstein's equivalence principle and the absence of dynamic solutions for the 1915 Einstein equation, *Phys. Essays* **12**(3), 508–526.

[‖]Herdegen, A. and Wawrzycki, J. (2002), Is Einstein's equivalence principle valid for a quantum particle? *Phys. Rev. D Particles Fields* **66**(4), 044007/1–044007/5.

[**]Drew, B. J. (2006), Pitfalls and artifacts in electrocardiography, *Cardiol. Clin.* **24**(3), 309–315.

[††]Pico Technology (2008), Electrocardiogram (ECG) project for DrDaq (http://www.picotech.com/applications/ecg.html).

of a complement to Ekeland's variational principle for functionals whose domain is a particular subset of a Banach space. Offers an approximate solution of scalar functions.*

EKELAND'S VARIATIONAL PRINCIPLE [mathematics, measurement, physics] (Arne Sigvard Eklund, b. 6/19/1911, Kiruna, Sweden) If a Gateaux differentiable functional f has a finite lower bound (although it need not attain it), then, for every $\varepsilon > 0$, there exists some point z, such that $\frac{\|f'(z)\| \le}{\varepsilon 1 + h(\|z,\|)}$, where $h : [0, \infty) \to [0, \infty)$ is a continuous function such that $\frac{\int_0^\infty}{11 + h(r) dr = \infty}$. Applications are given to extremum problem and some surjective mappings.[†] See EKELAND'S PRINCIPLE; BORWEIN–PREISS PRINCIPLE.

ELASTIC TUBE PRINCIPLE [mathematics, biology] (1) The esophageal airflow and the interacting viscous–elastic (viscoelastic) tube structure form the aerodynamical and biomechanical.[‡] (2) The wave speed of a one-dimensional longtitudinal mode in an elastic tube depends on the compresibility of the liquid and the elasticity of the tube. A wave traveling along the tube causes liquid displacement in the tube.[§] Also known as ICE TUBE PRINCIPLE.

ELASTICITY PRINCIPLE (NACHGIEBIGKEIT) [psychology] (Sándor Ferenczi, b. Sándor Fraenkel, 1873–1933, Hungarian psychoanalyst) (1) Nachgiebigkeit is the theoretical and clinical idea that the analyst is elastically flexible in functioning with the analysand. The rigid, cool atmospere given

was to tenderness, affection, and empathy.[¶] (2) Important shift away from classical neutrality approach to psychoanalysis. In order to reduce resistance, the analyst should present any interpretations in a tactful, empathetic manner. The analytic work should bend or yield toward the analysand.[‖] See also FRUSTRATION PRINCIPLE; PRINCIPLE OF INDULGENCE; RELAXATION PRINCIPLE.

ELECTRICALLY CONTROLLED BIREFRINGENCE PRINCIPLE [optics, electronics] (1) The spatial resolution for birefringence measurement determined by the width of the laser beam.[**] (2) An optical measurement system adopts a high-speed birefringence measurement method.[††]

ELECTRODYNAMIC PRINCIPLE [engineering] (1) Production of an electromotive force by either (a) motion of a conductor through a magnetic field so as to cut across the magnetic flux or (b) a change in the magnetic flux that treats a conductor.[‡‡] (2) A magnetic system for the contactless guidance of a vehicle moved along a track in which a plurality of magnets are attached to the vehicle and arranged one behind the other in the direction of travel; the magnet system cooperating with nonferromagnetic conductor loops on the track to generate forces.[§§] (3) Energy functional can be transformed into a function

---

[*]Jung, J. S. et al. (1996), Coincidence theorems for set valued mappings and Ekeland's variational principle in fuzzy metric spaces, *Fuzzy Sets Syst.* **79**, 239, 250.

[†]Zhong, C. K. (1997), A generalization of Ekeland's variational principle and a lication to the study of the relation between the weak PS condition and coercivity, *Nonlinear Anal. Theory Meth. Appl.* **29**, 1421.

[‡]Rubinow, S.I. (2003), *Introduction to Mathematical Biology*, Dover Publications. p. 168.

[§]Shin, Y., Chung, J., Kladias, N., Panides, E., Domoto, G. A., and Grigoropoulos, C. P. (2005), Compressible flow of liquid in a standing wave tube, *J. Fluid Mech.* **536**, 321–345.

[¶]Rachman, A. W. and Hutton, L. (2006), Clinical flexibility in the psychoanalytic situation: The elasticity principle, *Psychoanal. Social Work* **13**(1), 21–42.

[‖]Rachman, A. W. and Hutton, L. (2006), *J. Psychoanal. Social Work* (http://haworthpress.com/store/resolve.asp?DOI=10.1300/J032v13n01_02).

[**]Chabicovsky, R. and Stangl, G. (1978) A matrix addressed liquid crystal color display, *Proc. Rec. Development in Biennial Disp. Research Conf.*, 56–58.

[††]Geelhaar, T., Weber, G., Reiffenrath, V., Poetsch, E. (1991) *Liquid-crystal Display device Based on the Principle of Electrically Controlled Birefringence*, Patent WO/1992/013928.

[‡‡]Dobrowolny, M. (1987), The TSS project: Electrodynamics of long metallic tethers in the ionosphere, *Rivista Nuovo Cimento.* (Italian Physical Soc.) **10**(3), 1–83.

[§§]Miericke, J. and Urankar, L. (1976), *Magnet System for use in Electrodynamicly Suspended Vehicles*, US Patent 3,937,150.

of the hot charges on atoms.* (4) Magnetically mounted position stabilized flywheel for motor vehicle generating systems for rotary drive.[†] (5) Winding of transformer charging and discharging capacitor in alternate current directions using electronic changeover switch.[‡] See also ELECTROMAGNETIC PRINCIPLE; ELECTROMAGNETIC INDUCTION PRINCIPLE; PRINCIPLE OF ELECTRONEGATIVITY.

ELECTRODYNAMICAL REPULSION PRINCIPLE [engineering]   See REPULSION PRINCIPLE.

ELECTROMAGNETIC INDUCTION PRINCIPLE [electronics]   (1) Transfer of electrical power from one circuit to another by varying the magnetic linkage.[§] (2) The production of an electromotive force (emf) either by motion of a conductor through a magnetic field in such a manner as to cut across the magnetic flux or by a change in the magnetic flux that threads a conductor. The process by which an emf is induced in one circuit by a change of current in a neighboring circuit is called *mutual induction*. Flux produced by a current in a circuit A threads or links circuit B. When there is a change of current in circuit A, there is a change in the flux linking coil B, and an emf is induced in circuit B while the change is taking place.[¶] See also INDUCTION PRINCIPLE.

ELECTROMAGNETIC PRINCIPLE [engineering] See ELECTROMAGNETIC PRINCIPLE.

ELECTRONEGATIVITY EQUALIZATION PRINCIPLE [mathematics, physics]   The relative ability of an atom to retain or gain electrons.[‖] See ELECTROMAGNETIC PRINCIPLE.

ELECTRONEUTRALITY   PRINCIPLE [physics] In an electrolytic solution the concentration of all the ionic species are such that the solution as a whole is neutral.[**]

ELECTROSTATIC   PRINCIPLE [engineering] Electrostatic loudspeaker with optional circular radiation characteristic.[††] Also known as *foil electric principle*.

ELEMENT DISPLACEMENT PRINCIPLE [mathematics, engineering]   (Eric Reissner, b. 1/5/1913, Aachen, Germany)   See DISPLACEMENT PRINCIPLE.

ELEMENTARY PRINCIPLE ABOUT HUMAN PSYCHOLOGY [psychology]   One's wants and desires have more influence than one's behavior. They influence thinking, as well, and even powers of perception. This is true even with regard to things that would be otherwise intuitively obvious. Psychologists say that when a person is confronted by ideas or facts at odds with preexisting notions, what results is "cognitive dissonance," a sort of static in the human psyche. This "static" has the power to distort or even block perception.[‡‡] See also PRINCIPLE OF COGNITIVE DISSONANCE.

ELISA   PRINCIPLE [biology]   See ENZYME-LINKED IMMUNOSORBENT ASSAY PRINCIPLE.

ELLIOT   WAVE   PRINCIPLE [mathematics]   (Ralph Nelson Elliott, 1871–1948, Marysville, KS, accountant)   (1) Market prices unfold in specific patterns called *waves*. A form of technical analysis that

*Principles of physical science, *Encyclopedia Britannica*.
[†]Harris, C. H. and Piersol, A. G. (1961), *Harris' Shock and Vibration Handbook*, McGraw Hill, p. 442.
[‡]Metzl, K. and Sevcik, F. (1964), Universal device for drop-time control based on the electrodynamic principle, *Chemicke Zvesti* **18** 462–464.
[§]Walker, P. M. B. (1999), *Chambers Dictionary of Science and Technology*, Chambers, New York, p. 385.
[¶]Zhang, L., Yan, Z., and Zhou, D. (2001), Application of electromagnetic induction principle in analyzer, *Huagong Zidonghua Ji Yibiao* **28**(2), 61–62.

[‖]Walker, P. M. B. (1999), *Chambers Dictionary of Science and Technology*, Chambers, New York, p. 386.
[**]Pauling, L. (1964), Electroneutrality principle and the structure of molecules, *Anales Real Soc. Espan. Fis. Quim.* (Madrid) **60**(2–3) (Ser. B), 87–90.
[††]Gillitzer, E., Horn, A., Kalusche, H., and Pohls, M. (1951), *Electroacoustic Transducer Operated Corresponding to the Electrostatic Principle*, Germany DE 825997, 19511227.
[‡‡]Anon. (2003), *Cognitive Dissonance* (http://www.2001principle.net/2003.htm).

investors use to forecast trends in the financial markets and other collective activities.* (2) "Because man is subject to rhythmical procedure, calculations having to do with his activities can be projected far into the future with a justification and certainty heretofore unattainable."[†]

ELLUL PRINCIPLE OF EFFICIENCY [psychology]   See EMANCIPATION PRINCIPLE.

EM INTERACTION PRINCIPLE [mathematics] See ANOKHIN–PARETO INTERACTION PRINCIPLE.

EMANCIPATION PRINCIPLE [psychology] (Karl Heinrich Marx, 1818–1883) "The emancipation of the working classes must be conquered by the working classes themselves." Source of modern theories of social contract and natural rights manifested in the institutionalization of the rights of the individual subjected to the imperatives of total mobilization and tends to be transformed into an obedient soldier or an obedient worker or consumer. Expectation of a recrudescence of the totalitarian tendencies. Imperatives of control; technology represents unrestricted application to all spheres of life, not only technology but also abstract techniques; empowerment of human beings leads instead to their subjugation to the requirements of maximum efficiency; emancipation of individuals from external constraints; source of modern theories of social contract and natural rights manifested in institutionalization of the rights of the individual; despite intended goals of individuals; archetypes lay historical tendency to embody all spheres of historical life.[‡] See also POWER PRINCIPLE; ELLUL PRINCIPLE OF EFFICIENCY; PRINCIPLE OF NEGATIVE FREEDOM; PRINCIPLE OF MAJORITY RULE BASED ON DECISIONMAKING POWER DEMAND; PRINCIPLE OF EMANCIPATION; PRINCIPLE OF FREE DOMAIN; PRINCIPLE OF THE AUTOMOMY OF THE

WILL; PRINCIPLE OF POWER; PRINCIPLE OF EFFICIENCY; PRINCIPLE OF DOMINATION; PRINCIPLE OF SUBJECTIVE FREEDOM; KANT'S PRINCIPLE.

EMERGENCY-ACTION PRINCIPLE [psychology] Core humanitarian principles.[§] See also PRINCIPLE OF HUMANITY; PRINCIPLE OF IMPARTIALITY; PRINCIPLE OF INDEPENDENCE; DEFINING PRINCIPLES; PRINCIPLE OF NEUTRALITY; PRINCIPLE OF PROSELYTISM.

ENANTIODI VERGENT INDUCTION PRINCIPLE [chemistry, engineering, mathematics]   See INDUCTION PRINCIPLE.

END-TO-END PRINCIPLE [psychology, computers] (from 1981 paper, "End-to-end arguments in system design" by Jerome H. Saltzer, David P. Reed, and David D. Clark) (1) Communications protocol operations should be defined to occur at the endpoints of a communications system, or as close as possible to the resource being controlled.[¶] (2) Reliable systems tend to require end-to-end processing to operate correctly, in addition to any processing in the intermediate system. Features in the lowest level of a communications system have costs for all higher-layer clients, even if those clients do not need the features, and are redundant if the clients have to reimplement the features on an end-to-end basis.[‖] (3) Whenever possible, communications protocol operations should be defined to occur at the endpoints of a communications system, or as close as possible to the resource being controlled.[‖]

*Nurock, R. J. (1980), Elliott wave principle—key to stock market profits, *Barron's*, **60**, 18.

[†]Plaut, M. E. (1988), Illusion in medicine: The case for the Elliott wave principle, *J. Med.* **19**(5–6), 269–295.

[‡]Draper, H. (1971), The principle of self-emancipation in Marx and Engels, *Socialist Register* p. 80.

[§]Pictet, J. (1979), *The Fundamental Principles of the Red Cross: Commentary*, International Committee of the Red Cross (http://www.icrc.org/Web/eng/siteeng0.nsf/html/ EA08067453343B76C1256D2600383BC4?Open Document&Style=Custo_Final.3&View=default Body2).

[¶]Borg, E., Wilson, M., and Samuelsson, E. (1998), Towards an ecological audiology: Stereophonic listening chamber and acoustic environmental tests, *Scand. Audiol.* **27**(4), 195–206.

[‖]Saltzer, J., Reed, D., and Clark, D. D. (1984), End-to-end arguments in system design, *Proc. 2nd Int. Conf. Distributed Computing Systems*, pp. 509–512; *ACM Trans. Comput. Syst.* **2**(4), 277–288.

ENERGY BALANCE PRINCIPLE (EBP) [physics] (1) One of the bands of allowed energies separated by forbidden regions arising in a solid when the energy levels of the individual atoms combine.* See also EQUIVALENT BACKGROUND PRINCIPLE.

ENERGY CONSERVATION PRINCIPLE [physics] (Antoine-Laurent Lavoisier, 1743–1794, French chemist) Matter is subjected to physicochemical transformations, but it is not destroyed. Consists in considering constant and indestructible the quantity of matter existing in nature; that is, in any system and, by extension, in the universe the total quantity of matter is always the same.[†] (2) Whatever the form of energy considered (mechanical, thermal, electric, etc.) in any system and, by extension, in the universe, the total energy is constant.[‡] (3) Affirms that in any system and, by extension, in the universe, the mass transforms into energy and, vice versa, the energy transforms into mass, as in practice occurs in nuclear reactions and reactions among subnuclear particles.[§] (5) In energy balancing of any system, it is necessary to include the term mc^2, where m is the sum of the rest masses, that is, the masses, measured at a zero velocity, of all the bodies forming the system. The rest mass term m_0 derives from the fact that the mass of a body depends on its velocity, referred to a particular frame of reference, that is the frame in which the mass is at rest.[¶]

*Walker, P. M. B. (1999), *Chambers Dictionary of Science and Technology*, Chambers, New York, p. 399.
[†]Kangro, H. (1970), Significance of the energy conservation principle for physics of 1920–1932, *Proc. 35th Physikertag., Vorabdrucke Kurzfassungen Fachber.*, pp. 191–195.
[‡]Jinescu, V. V. (1987), Principle of energy conservation and conversion in causal formulation, *Revistade Chimie* (Bucharest, Romania) **38**(6), 469–474.
[§]Radcenco, V. (1991), Energy conservation principle in conservative physical systems based on generalized polytropy, *Revistade Chimie* (Bucharest, Romania) **42**(1–3), 74–82.
[¶]Cao, X., Zhou, R., and Duan, Z. (2006), Energy conservation principle and mathematic model of distillation with heat exchange inside, *Huagong Xuebao* **57**(6), 1351–1356.

ENERGY CONSERVATION PRINCIPLES AND THERMODYNAMICS PRINCIPLES [ecology] (Julius Robert Mayer, 1814–1878, German physician; James Prescott Joules, 1818–1889, English industrialist; Hermann Helmholtz [von], 1821–1894, German physiologist) (1) The energy conservation principle is, together with the fundamental conservation principles of both linear and angular momentum, a fundamental law of nature, verified universally as in the macroscopic physical world as in the microcosm.[‖] (2) An unifying principle that ties all the physical phenomena and furnishes an unitary and integrated vision of all the physical theories, from the celestial mechanics to the quantum mechanics, from the quantum theory of the fields to the standard model.[**]

ENERGY MINIMIZATION PRINCIPLE [physics] See MINIMIZATION PRINCIPLE.

ENERGY MOMENTUM PRINCIPLE [physics] (Albert Einstein, 1879–1955, German-born American physicist) The square of the interval between two near events may be expressed as a quadratic function of four coordinates. When the coefficients are constant, this can be transformed into the sum of four squares, three of which are spatial and one of which is temporal. The Lorentz transformation leaves the form of this expression unchanged, and the "restricted" theory of relativity is developed. The velocity of light is the same for all observers, the change of mass with velocity deduced.[††] See also MOMENTUM PRINCIPLE.

ENERGY PRESERVATION PRINCIPLE [ecology, engineering] See PRINCIPLE OF ZERO WASTE.

[‖]Novella, E. C. and Fernandes, R. M. (1954), Circulation of fluids through conduits. Application of the principle of conservation of matter and the first principle of thermodynamics, *Quim. Industr.* **1**, 7–13.
[**]Davies, K. W. (1982), Application of thermodynamic principles to optimize energy conservation designs, *Proc. 10th Australian Chemical Engineering Conf. Resource Development in the 1980s* (CHEMECA 82), pp. 287–292.
[††]Eddington, A. S. (1924), *Mathematical Theory of Relativity*, Cambridge Univ. Press.

ENERGY PRINCIPLE [mathematics]   See LE CHATELIER–BRAUN PRINCIPLE.

ENERGY STORAGE PRINCIPLE [engineering] (1) In certain railway and mass and/or rapid transit systems, such as in automatic train operations, it is necessary to establish minimal headways or time intervals between trains in order to provide fast and efficient service. When a train enters a block or track section, it is essential that the signal at the exit end not be cleared immediately so that the train may proceed into the next block or track section. Requires that some predetermined time delay be employed in switching the speed signal so that the train can be safely stopped or can proceed within the track section at a safe reduced headway speed that is established by the signal aspect at the exit end of the block.* (2) Whereby a piston element mounted in a pump cylinder of an electromagnetic reciprocating pump displaces quantities of the fuel to be injected during a virtually resistanceless acceleration phase during which the piston element stores kinetic energy, before ejection in the pump area.[†] See also SOLID-STATE ENERGY STORAGE PRINCIPLE.

ENERGY–TIME UNCERTAINTY PRINCIPLE [mathematics, engineering]   See UNCERTAINTY PRINCIPLE.

ENGINE PRINCIPLE [engineering]   (1) (Nicolas Leonard Sadi Carnot, 1796–1832, French physicist) Formulated (in 1824) the concept of the heat engine cycle and the principle of reversibility concerning the limitations on the maximum amount of work that can be obtained from a steam engine operating with a high-temperature heat transfer as its driving force.[‡] (2) (Jonathan Hornblower, 1753–1815, British pioneer of steam

power) A process gas is compressed at low temperatures and subsequently expands at high temperature.[§] See also STIRLING ENGINE PRINCIPLE; INDUCTIVE NUCLEOTHERMAL ENGINE PRINCIPLE.

ENTROPY MAXIMIZATION PRINCIPLE [mathematics] (Claude Elwood Shannon, 1916—2001, American electrical engineer, mathematician)   Method for analyzing the available information in order to determine a unique epistemic probability distribution. Defines a property of a probability distribution, $H(p) = -\sum_\pi \log p$ called *entropy*. The least biased distribution that encodes certain given information is that that maximizes the Shannon entropy $H(p)$ while remaining consistent with the given information.[¶] See also PRINCIPLE OF MAXIMUM ENTROPY; MAXIMUM ENTROPY PRINCIPLE.

ENTROPY MAXIMUM PRINCIPLE [mathematics, physics]   See MAXIMUM ENTROPY PRINCIPLE.

ENTROPY MINIMAX PRINCIPLE [physics] (1) Measure of absence of information about a situation. Transformations between measure spaces, expressing amount of disorder or uncertainty, inherent or produced. Informational macrodynamics: system modeling and simulation methodologies.[‖] (2) Measure of disorder of a system equal to the Boltzmann constant. (3) The natural logarithm action of the number of microscopic states corresponding to the thermodynamic state of the system.[‖] (4) Spectral evaluation of random time series via a variational approach and parallel processing.** (5) Derived from a

---

*Darrow, J. O. G. and Popp, W. R. (1979), *Fail-Safe Timing Circuit*, US Patent 4,150,417.
[†]Casarin, C. and Ibanez, J. G. (1993), Experimental demonstration of the principles of thermal energy storage and chemical heat pumps: Experiments for general, inorganic, or physical chemistry and materials science, *J. Chem. Educ.* **70**(2), 158–162.
[‡]Thermodynamics (2009), *Encyclopædia Britannica*. Encyclopedia online http://search.eb.com/eb/article-9108582.

[§]Hornblower, J. (2009), *Encyclopædia Britannica*. http://search.eb.com/06/article-9041076.
[¶]Bakhareva, I. F. (1973), Principle of the entropy maximum of steady-state systems in the theory of nonequilibrium scaler processes, *Zh. Fiz. Khim.* **47**(11), 2928–2930.
[‖]Lerner, V. S. (1998), Informational macrodynamics: System modelling and simulation methodologies, *Proc. Winter Simulation Conf.* (WSC'98), vol. 1, pp. 563–568.
Choi, K. Y., Yoon, Y. K., and Chang, S. H. (1991), A statistical model for prediction of fuel element failure using the Markov process and entropy minimax principles, *Nuclear Technol.* **93(2), 195–205.

variational approach for deciding between conflicting by patterns on limited data.[*]

ENTROPY OPTIMIZATION PRINCIPLE [engineering, mathematics] Often referenced as contrained optimization, performed through calculus of variation with Lagrange's method of finding the constrained extremum (maximization or minimization) being the preferred method. There are other entropy optimization principles arising from statistical theory for use in a wide assortment of problems and disciplines whose nature require to introduce different types of informational-statistical entropies.[†] See also MAXIMUM ENTROPY PRINCIPLE (MAXENT OR MEP); MINIMUM ENTROPY PRINCIPLE (MINXENT); JAYNES' MAXIMUM ENTROPY PRINCIPLE; KULLBACK'S MINNIMUM CROSS-ENTROPY PRINCIPLE; KULLBACK'S PRINCIPLE OF MINIMUM CROSS-ENTROPY; INVERSE PRINCIPLE; DIRECT PRINCIPLE; DUAL PRINCIPLE; GENERALIZATION OF JAYNES' MAXIMUM ENTROPY PRINCIPLE; GEN MINXENT PRINCIPLE; GENERALIZATION OF KULLBACK'S MINIMUM CROSS-ENTROPY PRINCIPLE; FIRST INVERSE MAXIMUM ENTROPY PRINCIPLE; SECOND INVERSE MAXIMUM ENTROPY PRINCIPLE; THIRD INVERSE MAXIMUM ENTROPY PRINCIPLE; FIRST INVERSE MINIMUM CROSS-ENTROPY PRINCIPLE; SECOND INVERSE MINIMUM CROSS-ENTROPY PRINCIPLE; THIRD INVERSE MINIMUM CROSS-ENTROPY PRINCIPLE; FIRST MINIMUM INTERDEPENDENCE PRINCIPLE; SECOND MINIMUM INTERDEPENDENCE PRINCIPLE; MINIMAX PRINCIPLE; MINIMUM LOSS OF INFORMATION PRINCIPLE; MINIMUM LOSS OF POWER OF DISCRIMINATION PRINCIPLE; MAXIMUM CROSS-ENTROPY PRINCIPLE; EXTENDED MAXIMUM ENTROPY PRINCIPLE; EXTENDED MINIMUM CROSS-ENTROPY PRINCIPLE; SECOND EXTENDED MAXIMUM ENTROPY PRINCIPLE; SECOND EXTENDED MINIMUM CROSS-ENTROPY PRINCIPLE.

[*]Zhu, S.-C. (2003), Statistical modeling and conceptualization of visual patterns, *IEEE Trans. Pattern Anal. Machine Intell.* **25**(6), 691–712.
[†]Corotis, R. B., Schuëller, G. I., and Shinozuka M. (2001), Structural safety and reliability, *Proc. 8th Int. Conf. Structural Safety and Reliability*, ICOSSAR '01, Newport Beach, CA, June 17–22, 2001, Taylor & Francis, p. 51.

ENTROPY PRINCIPLE [mathematics, physics] (1) Measure of absence of information about a situation. Used for estimating probability density function under specified moment constraints.[‡] (2) Expresses amount of disorder or uncertainty, inherent or produced.[§] (3) Measure of disorder of a system equal to the Boltzmann constant.[¶] (4) Famous fold/unfold method to give proof of two properties: associatively of the append operation between lists and idempotence of the reverse operation.[‖] (5) Selection of a probability distribution by the principle of maximum entropy (MAXENT). The relevant probability is the volume fraction of the elementary constituents that belong to a given constituent and undergo a given stimulus. (6) Entropy model for deriving the probability distribution of the equilibrium state.[**] (7) A distribution-free method for estimating the quantile function of a nonnegative random variable using the principle of maximum entropy (MaxEnt) subject to constraints specified in terms of the probability-weighted moments estimated from observed data.[††] (8) Used for estimating probability. (9) An inequality is formed by adding the entropy inequality and a linear combination of the field equations. The factors multiplying the field equations in this linear combination are called *Lagrange multipliers*.[‡‡] (10) Energy functional can be transformed into a function

[‡]Pandey, M. D. (2000), Direct estimation of quantile functions using the maximum entropy principle, *Struct. Safety* **22**(1), 61–79.
[§]Blick, E. F. (1991), *Scientific Analysis of Genesis*, Hearthstone Publishing, chap. 4.
[¶]Panos, C. P. (2001), Universal property of the order parameter in quantum many-body systems, *Phys. Lett. A.* **289**(6), 287–290.
[‖]Yuan-Shun, D., Min, X., Quan, L., and Szu-Hui, N. (2007), Uncertainty analysis in software reliability modelling by Bayesian approach with maximum-entropy principle, *IEEE Trans. Software Eng.* **33**(11), 781–795.
[**]Arminjon, M. and Imbault, D. (2000), Maximum entropy principle and texture formation, *Z. Angew. Math. Mech.* **80**(Suppl. 1), S13–S16.
[††]Pandey, M. D. (2000), Direct estimation of quantile functions using the maximum entropy principle, *Struct. Safety* **22**(1), 61–79.
[‡‡]Liu, I.-S. (1972), Method of Lagrange multipliers for exploitation of the entropy principle, *Arch. Rational Mech. Anal.* **46**(2), 131–148.

of the hot charges on atoms.* See also MAXIMUM ENTROPY PRINCIPLE; PME PRINCIPLE OF MAXIMUM ENTROPY; FUZZY ENTROPY PRINCIPLE; MEP MAXIMUM ENTROPY PRINCIPLE; PRINCIPLE OF MAXIMUM NONADDITIVE ENTROPY; PRINCIPLE OF OPERATIONAL COMPATIBILITY; MERMIN ENTROPY PRINCIPLE; VARIATIONAL MINIMAL PRINCIPLE; MINIMUM CROSS-ENTROPY PRINCIPLE; PRINCIPLE OF MINIMUM CROSS-ENTROPY; GENERALIZED MINIMAL PRINCIPLE; MOMENT-PRESERVING PRINCIPLE.

ENTROPY PRODUCTION PRINCIPLE [mathematics] Maximization of the entropy production during nonequilibrium processes.[†]

ENTROPY VARIATIONAL PRINCIPLE Entropy Variational Principle. [mathematics] Leads to a variety of results, estimates, and reflections of the way complexities emerge and lead to the study of entropy structure, a master invariant for entropy theory. The symbolic extension entropy (sex) function is simply the upper semicontinuous envelope of the entropy function expressed as $h_{sex}(T) = max_p\ h_{sex}(p)$, where $h_{sex}(T) := inf\ \{h_{top}(S) : S$ is a sym ext of $T\}$. See SYMBOLIC EXTENSION ENTROPY (SEX); SEXENTROPY VARIATIONAL PRINCIPLE.

ENZYME-LINKED IMMUNOSORBENT ASSAY PRINCIPLE [biology] An assay method in which antigen or antibody is detected by means of an enzyme chemically coupled either to antibody specific for the antigen or to anti-lg, which, in turn will bind to the specific antibody. Either the antigen or the antibody to be detected is attached to the surface of a small container or to plastic beads, and the specific antibody is allowed

to bind in turn.[‡] (2) The amount of protein adsorbed on the surface can be estimated by an *enzyme-linked immunosorbent assay* (ELISA), which is an immunologic assay modified in order to determine solid-phase bound proteins. The assay is based on the interactions between antigen (protein) and specific antibodies. This method allows the semiquantitive determination of proteins. Tissue culture polystyrene is usually used as a reference material. One of the special advantages of ELISA is that it can be applied to a wide range of proteins, such as fibrinogen, collagens, albumine, fibronectin, and laminine simply by variation of the specific primary antibody.[§] See also ELISA PRINCIPLE.

EÖTVÖS PRINCIPLE [physics] (Vásárosnaményi Báró Eötvös Loránd, better known as Loránd Eötvös or Roland Eotvos, (1848–1919, Hungarian physicist) (1) The molecular surface energy of a substance decreases linearly with temperature, becoming zero about 60°C below the critical point.[¶] (2) Study of the equivalence of gravitational and inertial mass and gravitational gradient on Earth's surface.[‖] (3) Measurements of the gravitational gradient in applied geophysics, such as the location of petroleum deposits. The CGS unit for gravitational gradient is termed *eotvos*.[**]

EPIGENETIC PRINCIPLE [biology] (Erik Erikson, 1902–1944, Danish-German-American development psychologist) Derived from

---

*Raz, T. and Levine, R. D. (1995), On the burning of air, *Chem. Phys. Lett.* **246**(4–5), 405–412.

[†]Pasko, Z., Davor, J., and Srecko, B. (2004), Kirchhoff's loop law and the maximum entropy production principle, *Phys. Rev. E Statist. Nonlinear Soft Matt. Phys.* **70**(5, Pt.2), 056108.

[‡]Walker, P. M. B. (1999), *Chambers Dictionary of Science and Technology*, Chambers, New York, p. 403.

[§]Kobayashi, N. (2006), Immunoassays for environmental chemicals: Principle and development of "hapten ELISA," *Kankyo Gijutsu* **35**(9), 631–638.

[¶]Walker, P. M. B. (1999), *Chambers Dictionary of Science and Technology*, Chambers, New York, p. 404.

[‖]Brans, C. and Dicke, R.H. (1961), Mach's principle and a relativistic theory of gravitation, *Phys. Rev.* **124**, 925—935.

[**]Bod, L., Fishbach, E., Marx, G., and Náray-Ziegler, M. (1991), One hundred years of the Eötvös experiment, *Acta Phys. Hungar.* **69**(3–4), 335–355.

embryologic development.* Every organism is born with a certain purpose, and continues to develop how it was intended to in interrelation with its environment.

EPISTATIC HANDICAP PRINCIPLE [biology]
See HANDICAP PRINCIPLE.

EPISTEMIC CLOSURE PRINCIPLE [philosophy]
Most of us think that we can always enlarge our knowledge base by accepting things that are entailed in (or logically implied by) things we know. The set of things we know is closed under entailment (or under deduction or logical implication), which means roughly that we know anything that follows from what we know. However, some theorists deny that knowledge is, in fact, closed under entailment, and the issue remains controversial. One might speak of two main camps: (1) those who take closure as a firm datum—as obvious enough to rule out any understanding of knowledge that undermines closure and (2) those who want to resolve the controversy by analyzing knowledge and working out the implications for closure.[†]

EPISTEMIC PRINCIPLE [mathematics]   (1) In a great many instances, whatever occurs by a cause that is not free is the natural effect of that cause.[‡] (2) A statement about the reliability of a particular faculty, belief-forming mechanism, or way of forming beliefs. The beliefs formed in a particular way are all true or, alternatively, mostly true. The belief can be undermined and thereby be cast into doubt by casting doubt on the principle behind it.[§]

*Germain, C. B. and Bloom, M. (1999), *Human Behavior in the Social Environment: An Ecological View*, 2nd ed., Columbia Univ. Press, p. 241.
[†]Moussaoui, M. and Seeger, A. (1996), Epsilon maximum principle of pontryagin type and perturbation analysis of convex optimal control problems, *SIAM J. Control Optim.* **34**, 407.
[‡]Luper, S. (2009), The epistemic closure principle, in *Stanford Encyclopedia of Philosophy* (http://plato.stanford.edu/archives/spr2009/entries/closure-epistemic]).
[§]Hintikka, J. (2004), *Analyses of Aristotle*, Springer, p. 187.

EPISTEMOLOGICAL   PRINCIPLE [physics]
Relation between relative and absolute truth in science.[¶]

EPSILON MAXIMUM PRINCIPLE [physics]
Necessary conditions for suboptimality in problems involving terminal control are obtained with the aid of ordinary dynamic systems.[‖]

EQUALIZATION   PRINCIPLE [engineering]
(1) Production of an electromotive force either by motion of a conductor through a magnetic field so as to cut across the magnetic flux or by a change in the magnetic flux that treats a conductor.[**] (2) A magnetic system for the contactless guidance of a vehicle moved along a track in which a plurality of magnets are attached to the vehicle and arranged one behind the other in the direction of travel, the magnet system cooperating with nonferromagnetic conductor loops on the track to generate forces.[††] (3) Energy functional can be transformed into a function of the hot charges on atoms.[‡‡] (4) Effect of all frequency discriminatory means employed in transmitting, recording, amplifying, or signal-handling systems to obtain a desired overall frequency response.[§§] (5) Used to study charge distribution in simple

[¶]DeRose, K. (1992), Descartes, epistemic principles, epistemic circularity, and scientia, *Pacific Philo. Quart.* **73**(33), 220–238.
[‖]Gabasov, R., Kirillova, F. M., and Mordukhovich, B. S. H. (1983), The epsilon-maximum principle for suboptimal control, *Dokl. Akad. Nauk SSSR* **268**(3), 525–529.
[**]Toepel, M. E. (1918), *Automotive Magneto Ignition: Its Principle and Application with Special Reference to Aviation Engines* (http://www.archive.org/stream/automotivemagnet00toeprich/automotivemagnet00toeprich_djvu.txt).
[††]Miericke, J. and Urankar, L. (1976), *Magnet System for Use in Electrodynamicly Suspended* US Patent 3,937,150.
[‡‡]Gilli, P., Pretto, L., and Gilli, G. (2005), Advances in H-Bond Theory in Metodi Teorici e di Calcolo. Societa Chimica Italiani, 34th Congresso Nazionale p. 31.
[§§]Kimura, T. and Yamamoto, Y. (1983), Progress of coherent optical fibre communication systems, *Opt. Quantum Electron.* **15**(1), 1–39.

peptides. * (6) Scheme for calculating the atomic charges in a molecule that gives a new scale of the atomic electronegative and hardness in a certain molecular environment and takes the harmonic mean electronegativity as a reference value of the molecular electronegativity so that the multiple-regression and nonuniform parameters are avoided.† (7) Used to define the Fukui function of the kth atom in a molecule for electrophilic attack and for nucleophilic attack, the softness of an atom in a molecule, and the hardness.‡ Also known as *conventional conscience principle; electronegativity equalization principle; equiprobable principle*.

EQUATOR PRINCIPLES [ecology] Voluntary set of standards for determining, assessing and managing social and environmental risk in project financing. See also HANNOVER PRINCIPLES; FOREST STEWARDSHIP COUNCIL (FSC) PRINCIPLES; GLOBAL SULLIVAN PRINCIPLES OF SOCIAL RESPONSIBILITY; MARINE STEWARDSHIP COUNCIL (MSC) PRINCIPLES; PERMACULTURE PRINCIPLES; THE BELLAGIO PRINCIPLES FOR ASSESSMENT; MELBOURNE PRINCIPLES; PRECAUTIONARY PRINCIPLE; SANBORN PRINCIPLES; TODDS' PRINCIPLES OF ECOLOGICAL DESIGN.§

EQUIAVAILABILITY PRINCIPLE [psychology] (Hanley, G. L. and Levine, M. 1983) Ability of subjects to integrate two independently learned pathways into a composite of one cognitive map.¶ See also SYMMETRY PRINCIPLE.

EQUICONTINUITY PRINCIPLE [mathematics] To prove convergence when, ideally, the number of digits in the arithmetic tends to infinity. Convergence analysis of numerical methods in finite precision requiring the coupling of the arithmetic with all the convergence parameters.‖ Also known as *Banach–Steinam principle; principle of equicontinuity; lax principle of equicontinuity*.

EQUIDISTORTION PRINCIPLE [engineering] A necessary condition for optimal vector quantizers; design principle derived by applying Gersho's theory to multiple disjoint clusters.** See also QUANTIZATION DESIGN PRINCIPLE.

EQUILIBRIUM PRINCIPLE [mathematics, chemistry, psychology] (1) Psychology of decisionmaking. The decisionmaker cannot decide to perform an act that is not an equilibrium of the deliberational process. If about to choose a nonequilibrium act, deliberation carries away from that decision. This sort of equilibrium requirement for individual decision can be seen as a consequence of the expected utility principle.†† (2) *Physics*. The state of a body or physical system at rest or in unaccelerated motion in which the resultant of all forces acting on it is zero and the sum of all torques about any axis is zero.‡‡ (3) *Chemistry*. The state of a chemical

*Itskowitz, P. and Berkowitz, M. L. (1997), Chemical potential equalization principle: Direct approach from density functional theory, *J. Phys. Chem. A* **101**(31), 5687–5691.

†Shen, E.-Z. and Yang, Z.-Z. (1994), Application of electronegativity equalization principle to calculation of atomic charges in a molecule, *Chinese Sci. Bull.* **39**(14), 1195–1199.

‡Gazquez, J. L. and Mendez, F. (1994), The hard and soft acids and bases principle: An atoms in molecules viewpoint, *J. Phys. Chem.* **98**(17), 4591–4593.

§(2006) The "Equator Principles" A financial industry benchmark for determining, assessing and managing social & environmental risk in project financing http://www.equator-principles.com/documents/Equator_Principles.pdf.

¶Brysch, K. A. (1996), Studies in cognitive maps: The equiavailability principle and symmetry, *Environ. Behav.* **28**(2), 183.

‖Schaefer H. H. (1990), A Banach-Steinhaus theorem for weak and order continuous operators, in *Book Series Lecture Notes in Mathematics, Complex Geometry and Analysis*, vol. 1422, pp. 1617–9692.

Ueda, N. and Nakano, R. (1995), Competitive and selective learning method for vector quantizer design equidistortion principle and its algorithm, *Syst. Comput. Jpn.* **26, 34.

††Skyrms, B. (1990), *The Dynamics of Rational Deliberation*, Harvard Univ. Press, Cambridge, MA, p. 28.

‡‡Johansson, S. R. (1974), Equilibrium principles. Mathematical chemistry, *Kemia–Kemi* **1**(11), 727–738.

reaction whose forward and reverse reactions occur at equal rates so that the concentration of the reactants and products does not change with time. Every system between two conditions of matter (systems) is displaced by lowering the temperature, at constant volume, toward that system the formation of which evolves heat.* Also known as *expected utility principle, mobile equilibrium principle; Nash equilibrium principle; Donnan equilibrium principle*; Wardrop equilibrium principle; see also WARDROP'S PRINCIPLE.

EQUIPARTITION PRINCIPLE [physics, chemistry] (Boltzmann's theorem on the equipartition of energy, 1902) (1) When a system is in thermal equilibrium, the kinetic energy is equally divided among the degrees of freedom.[†] (2) The available energy in a closed system eventually distributes itself equally among the degrees of freedom present.[‡] (3) Means *equal division*, as derived from the Latin *equi* from the antecedent, *æquus* (equal or even), and partition from the antecedent, *partitionem* (division, portion). The original concept of equipartition was that the total kinetic energy of a system is shared equally among all of its independent parts, on the average, once the system has reached thermal equilibrium. Equipartition also makes quantitative predictions.[§] See also AUTOMATIC CONTROL PRINCIPLES; PRINCIPLE OF DOMINANT SUBSYSTEMS; PRINCIPLE OF FUNDAMENTAL CHARACTERISTICS; PRINCIPLE OF OPERATION; PRINCIPLE OF CONSTRUCTION AND OPERATION; PRINCIPLE OF SMOOTH FIT.

EQUIPROBABLE PRINCIPLE [mathematics, physics] See ELECTROMAGNETIC PRINCIPLE.

EQUITY OF ACCESS PRINCIPLE [computer science] User-centered, barrier-free, and format-independent by alternate methods, approaches, and access points and minimizing restrictions on access and maximizing dialog.[¶] See also INTELLECTUAL FREEDOM PRINCIPLE; PRIVACY PRINCIPLE; INTELLECTUAL PROPERTY PRINCIPLE; INFRASTRUCTURE PRINCIPLE; UBIQUITY PRINCIPLE; CONTENT PRINCIPLE; PRINCIPLES FOR NETWORKED WORLD.

EQUIVALENCE CHANGE PRINCIPLE [medicine] (Philip Shaffer, 1881–1960, American physician) Ionic oxidation–reduction reactions.[‖]

EQUIVALENCE METHODS PRINCIPLE [mathematics] See TRANSFORMATION PRINCIPLE.

EQUIVALENCE PRINCIPLE [physics, mathematics] (1) A gravitational field is equivalent to an accelerating reference frame. It is sometimes expressed as the equivalence of gravitational and inertial mass, namely, that objects with the same inertia (resistance to acceleration) experience the same gravitational force.[**] (2) An inertial reference frame in a uniform gravitational field is equivalent to a reference frame in the absence of a gravitational field that has a constant acceleration with respect to that inertial frame.[††] (3) The postulation of no more than equivalence of inertial and gravitational mass, can be satisfied by theories that predict a smaller gravitational acceleration for a rotating object than for a nonrotating object. Consequently, only the strong form can be considered acceptable.[‡‡] See also PRINCIPLE OF EQUIVALENCE; WEAK EQUIVALENCE PRINCIPLE; QUANTUM EQUIVALENCE PRINCIPLE; PRINCIPLE OF FLEXIBILITY.

*Quilez, J. (2007), A historical/philosophical foundation for teaching chemical equilibrium, *Proc. 9th Int. History Philosophy & Science Teaching Conf.*, Calgary, Canada, June 24–28, 2007, p. 8.

†Anon. (1989), equipartition, in *Oxford English Dictionary*, 2nd ed., Oxford Univ. Press.

‡Walker, P. M. B. (1999), *Chambers Dictionary of Science and Technology*, Chambers, New York, p. 409.

§Anon. (1989), equipartition, in *Oxford English Dictionary*, 2nd ed., Oxford Univ. Press.

¶Gibbard, A. (1982), The prospective Pareto principle and equity of access to health care, *Health and Society* (Milbank Memorial Fund quarterly). **60**(3), 399–428.

‖Shaffer, P. A. (1933), Reaction velocity and the "equivalence-change principle," *J. Am. Chem. Soc.* **55**, 2169–2170.

Davies, P. C. W. and Fang, J. (1982), Quantum theory and the equivalence principle, *Proc. Roy. Soc. Lond. A Math. Phys. Eng. Sci.* **381(1781), 469–478.

††Parker, L. (1968), Equivalence principle and motion of a gyroscope, *Phys. Rev.* **175**, 1658–1660.

‡‡Frieman, J. A. and Gradwohl, B. A. (1993), Dark matter and the equivalence principle, *Science* **260**, 1441.

EQUIVALENCE PRINCIPLE OF GRAVITATION
[physics] (Albert Einstein) The gravitational force caused by an external object is the same for all objects of the same mass, regardless of their compositions. Applied to several related concepts dealing with gravitation and the uniformity of physical measurements in different frames of reference. They are related to the Copernican idea that the laws of physics should be the same everywhere in the universe, to the equivalence of gravitational and inertial mass, and also to Einstein's assertion that the gravitational "force" as experienced locally while standing on a massive body (such as Earth) is actually the same as the pseudoforce experienced by an observer in a noninertial (accelerated) frame of reference.[*]

EQUIVALENCE PRINCIPLE OF SPECIAL RELATIVITY [physics] The laws of physics look the same in all inertial reference frames.[†]

EQUIVALENT BACKGROUND PRINCIPLE [physics] A means of characterizing the adaptive state of the retina. Relies on the equivalent background principle (EBP) to relate the afterimage from an intense light exposure to a hypothetical "background" of uniform luminance that fades with time. Estimates the recovery of visual sensitivity after optical radiation exposure.[‡]

EQUIVALENT COMPETITION PRINCIPLE [engineering] Competition analysis is more accurate in assessing relative affinities than direct binding curves, particularly with those of lower affinity, where it may be difficult to attain conditions of saturation.[§]

EQUIVALENT ORTHOGONAL PROJECTION PRINCIPLE [computer science] See ORTHOGONAL PROJECTION PRINCIPLE.

EQUIVALENT PROPORTIONS PRINCIPLE [chemistry] When two chemical elements both form compounds with a third element, a compound of the first two elements contains them in the relative proportions that they have in compounds with the third element.[¶] See also PRINCIPLE OF EQUIVALENT PROPORTIONS; also referred to as *reciprocal proportion principle*.

ERROR RECONSTRUCTION PRINCIPLE [psychology] Introduces a self-organizing net that has an architecture of one or more layers and works in two phases: perception and learning. In the perception phase, when a pattern is presented to the bottom layer (input field) the net will pass the signals up and experience a stable dynamic process that converges into an equilibrium. The top−down signal received by the input field is regarded as the reconstruction of the input pattern. In the learning phase, all the weights are modified according to a general principle that the mean-square error between input patterns and their reconstruction is minimized. Potential applications of the net include associative memory, feature extraction, data compression, unsupervised pattern clustering and recognition, and attentional recognition. Future uses might also include interpreting the development of orientation cells in the cortical field, as well as the emergence of mental imagery in the brain. Least-mean-square error reconstruction principle for self-organizing neural nets.[‖] (2) A local learning rule called *least-mean-square error reconstruction* (LMSER) is naturally obtained for training nets consisting of either one or several

[*]Ritter, R. C., Goldblum, C. E., Ni, W. T., Gillies, G. T., and Speake, C. C. (1990), Experimental test of equivalence principle with polarized masses, *Phys. Rev. D Particles Fields* **42**(4), 977−991.
[†]Alvarez, C. and Mann, R. (1997), Testing the equivalence principle in the quantum regime, *Gen. Relativ. and Gravit.* **29**, 245.
[‡]Bowen, R. W. and Hood, D. C. (1983), Improvements in visual performance following a pulsed field of light: A test of the equivalent-background principle, *J. Opt. Soc. Am.* **73**(11), 1551−1556.
[§]Jaffrey, S. R., Haile, D. J., Klausner, R. D., and Harford, J. B. (1993), The interaction between the iron-responsive element binding protein and its cognate RNA is highly dependent upon both RNA sequence and structure, *Nucleic Acids Res.* **21**(19), 4629.
[¶]Daintith, J. (1999), *The Facts on File Dictionary of Chemistry*, 3rd ed., Facts On File, New York, p. 94.
[‖]Qi, J. and Huesman, R. H. (2005), Effect of errors in the system matrix on maximum a posteriori image reconstruction, *Phys. Med. Biol.* **50**(14), 3297−3312.

layers.* See also LEAST-MEAN-SQUARE ERROR RECONSTRUCTION PRINCIPLE; LMSER PRINCIPLE.

ESTIMATION   PRINCIPLE [mathematics] Branch of probability and statistics concerned with deriving information about properties of random variables, stochastic processes, and systems based on observed samples.[†]

EUCLID'S PRINCIPLE [mathematics] (Euclid, 300 BC, Greek mathematician)   States that if p and q are mutually prime, then p^2 and q^2 are also mutually prime. Euclid's second theorem states that the number of primes is infinite.[†] (2) A straight-line segment can be drawn joining any two points. Any straight-line segment can be extended indefinitely in a straight line.[‡] (4) Given any straight-line segement, a circle can be drawn having the segment as radius and an endpoint as center.[§] (5) The whole is greater than the part (i.e., strictly greater than any proper part).[§] (6) If two lines are drawn end to end in such a way that the sum of the inner angles on one side is less than the sum of two right angles, then the two lines in contact must intersect each other on that side if extended far enough.[¶]

EUCLID'S SECOND PRINCIPLE [mathematics] (Euclid, 300 BC, Greek mathematician) The number of primes is infinite. This theorem, also called the *infinitude of primes theorem*, was proved by Euclid in Proposition IX.20 of the *Elements*.[‖]

*Xu, L. (1993), *Neural Networks* **6**(5), 627–648.
[†]Frieden, B. R. (1980), Computational methods of probability and statistics, *Topics Appl. Phys.* **41**(on the computer in optical research), 81–210.
[‡]Ball, W. W. R. and Coxeter, H. S. M. (1987), *Mathematical Recreations and Essays*, 13th ed., Dover, New York, p. 60.
[§]Heath, Sir T. L. (1908), *The Thirteen Books of Euclid's Elements* (reprint), Dover Publications, New York.
[¶]Parker, M. W. (n.d.), Philosophical method and Galileo's paradox of infinity, *Phil. Sci.* **70**, 359–382.
[‖]Shoshi, A. I., Steffen, F. D., Dosch, H. G., and Pirner, H. J. (2002), Confining QCD strings, Casimir scaling, and a Euclidean approach to high-energy scattering, in *High Energy Physics—Phenomenology*, Los Alamos National Laboratory, pp. 1–58.

EULER'S   PRINCIPLE [geology]   (Leonard Euler, 1707–1783, Swiss mathematician) Explains the distribution of conservative plate margins. Plate tectonics requires that all conservative plate margins lie on small circles, the axes of which form the axis of rotation for the relative motion of the plates on either side.[**]

EVAPORATION PRINCIPLE [physics, engineering]   (1) A method of making ice by removing the vapor of an aqueous solution from the space above its surface, causing more liquid to evaporate. The evaporation of the liquid lowers the liquid's temperature, causing it to enter a solid phase when its temperature reaches the solidification point.[††] (2) In order to solidify aqueous solutions by means of the evaporation principle, large amounts of vapor must be removed from the environment. (3) Of the several methods that can be employed to concentrate liquids of the class contemplated, such as centrifuging, creaming, filtration, and evaporation, evaporative methods are still the most preferred.[‡‡] See also SORPTION PRINCIPLE.

EVOLUTIONARY EXTREMAL PRINCIPLE [evolution]   The Perron–Frobenius theorem is used to characterize the equilibrium state of a corresponding abstract symbolic dynamical system by an extremal principle. A thermodynamic formalism for random dynamical systems is developed, and the analyticity of the top Lyapunov exponent is proved under mild assumptions. A fluctuation theory for $A(\theta \{n-1\}\omega)\cdots A(\omega)$ is worked out, by which directionality of mutation and selection

**Stiegeler, S. E. (1977), *A Dictionary of Earth Sciences*, Pica Press, New York (distributed by Universe Books), p.104.
[††]Nuutinen, J., Harvima, I., Lahtinen. M. R., and Lahtinen, T. (2003), Water loss through the lip, nail, eyelid skin, scalp skin and axillary skin measured with a closed-chamber evaporation principle, *Br. J. Dermatol.* **148**(4), 839–841.
[‡‡]Standford, F. C. (2008), Evaporation, in *Kirk-Othmer Separation Technology*, 2nd ed., vol. 1, pp. 1054–1076.

in evolutionary dynamics is proved, using the notion of entropy.[*]

EVOLUTIONARY EXTREMAL PRINCIPLE [evolution]   The Perron Frobenius theorem is used to characterize the equilibrium state of a corresponding abstract symbolic dynamical system by an extremal principle. A thermodynamic formalism for random dynamical systems is developed, and the analyticity of the top Lyapunov exponent is proved under mild assumptions. A fluctuation theory for is worked out, by which directionality of mutation and selection in evolutionary dynamics is proved, using the notion of entropy.

1. The characterization of the equilibrium state of the dynamical system in terms of an extremal principle. A consequence of this principle is the relation

$$\lambda = H + \Phi.$$

2. A fluctuation theory for the dynamical state at the equilibrium. This represents the mathematical description of the mutation event. If $\Delta\lambda$ and ΔH denote the change in growth rate and entropy due to mutation, then we have

$$\Phi > 0 \Rightarrow \Delta\lambda\Delta H < 0,$$
$$\Phi < 0 \Rightarrow \Delta\lambda\Delta H > 0.$$

EVOLUTIONARY   PRINCIPLE [psychology] (1) Views adolescence as a species-typical phenomenon whose biological functions and mechanisms afford a firmer and more generalized basis for understanding. From this ethological standpoint, cognitive development takes a back seat to the analysis of fundamental motivations Evolutionary principles of human adolescence.[†] (2) When a certain species is removed from the habitat in which it evolved, or that habitat changes significantly within a brief period,

the said species will develop aberrant and maladaptive behavior.[‡]

EXCEPTION-GRANTING   PRINCIPLE [psychology]   In moral reasoning, one may justify an act typically associated with on unacceptable wrong effect, as acceptable under a particular set of circumstances.[§] See also PRINCIPLE OF DOUBLE EFFECT; PRINCIPLE OF DO NO HARM; CASUISTIC PRINCIPLE; PRINCIPLES OF BIOMEDICAL ETHICS; JUSTIFYING PRINCIPLE; DOUBLE-EFFECT PRINCIPLE.

EXCEPTION   PRINCIPLE [computer   science] This principle (also known as *management by exception*) is closely related to the parity principle. States managers should concentrate their efforts on matters deviating significantly from the normal and let subordinates handle routine matters.[¶] (2) The more managers concentrate their control on exceptions, the more efficient will be the results of this control.[‖] See also PRINCIPLE OF PLANNING, PARITY PRINCIPLE.

EXCLUSION   PRINCIPLE [biology]   (1)   Two species, A and B, interact when their niches a and B intersect in space and time. The composite element AB may be either more, less, or equally effective in regulation. It is suggested that the first condition exists when the overlap between α and β is small; this leads, on a large scale, to complex biocenotic phenomena in which components of a tremendously diversified biota may coexist. The coupling AB is likely, on the other hand, to be antagonistic to optimization of

[*]Arnold, L., Gundlach, V. M., and Demetrius, L. (1994), Evolutionary formalism for products of positive random matrices, *Ann. Appl. Probab.* **4**(3), 859–901.

[†]Haslam, N. (2003), *Transcult. Psychiatry* 40(4), 607–609.

[‡]Armstrong, J. D., Holm, C. F., Kemp, P. S., and Gilvear, D. J. (2003), Linking models of animal behaviour and habitat management: Atlantic salmon parr and river discharge, *J. Fish Biol.* **63**(Suppl. A), 226–245.

[§]Keenan, J. F. (1993), The function of the principle of double effect, *Theol. Stud.* **54**, p. 294–315.

[¶]Dokker, S. W. A. and Woods, D. D. (1999) To intervene or not to intervene: the dilemma of management by exception. Cognition, Technology and Work, V1, p 86–96.

[‖]Koontz, H. and O'Donnell, C. (1968), *Principles of Management: An Analysis of Managerial Functions*, 4th ed., McGraw-Hill, New York   (http://knol.google.com/k/narayana-rao-kvss/principles-of-management-koontz-and/2utb2l sm2k7a/89).

regulation by the whole community; in this case, it is postulated, one of the components (the less effective regulator acting alone) is purged. This is competitive exclusion, and it develops when A and B intersect greatly. Between these extremes of cooperation and competition lies an area of niche intersection in which AB is not much better or worse than A or B alone. Competitive exclusion.* (2) One of the most general and interesting rules describing intercalative DNA binding by small molecules. It suggests that such binding can occur only at every other base-pair site, reflecting a very large negative cooperativity in the binding process.† (3) Applications that consider probabilities of occurrences from randon or nonrandom events such as molecular reactions, viruses, genetic selection, and plant propagation.‡ (4) One species seeking same ecological environment will survive while the other will expire under a given set of conditions. If there is competition between two species for a common place in a limited microcosm, we can quite naturally extend the premises implied in the logistic equation. The growth rate of each competing species in a mixed population will depend on (a) the potential rate of population increase of a given species and (b) the unutilized opportunity for growth of this species.§ (5) Levy's discovery of the complementarity between random displacements, parameterized by time, and random time, parameterized by position, in the

description of Brownian motion.¶ (6) In philosophy, examination of concepts and codes by which societies operate or define themselves.‖ See also SIZE EXCLUSION PRINCIPLE; COMPETITIVE EXCLUSION PRINCIPLE; MINORITY CYTOTYPE EXCLUSION PRINCIPLE; NEIGHBOR EXCLUSION PRINCIPLE; PAULI EXCLUSION PRINCIPLE; PROBABILISTIC EXCLUSION PRINCIPLE; PRINCIPLE OF COMPLEMENTARITY; GAUSE'S PRINCIPLE.

EXCLUSION–INCLUSION   PRINCIPLE See INCLUSION–EXCLUSION PRINCIPLE.

EXISTENCE PRINCIPLE [biology, psychology] (1) Query about what kind of existence, how much of it, whether we mean existence for you or existence for me, or whether we are asking about some property that it might have.** (2) The methods of survival can be summed under the headings of food, protection (defensive and offensive), and procreation. There are no existing life forms that lack solutions to these problems. Every life form errs, one way or another, by holding a characteristic for too long or developing characteristics that may lead to its extinction. But the developments that bring about success of form are far more striking than their errors.†† See also COMPETITIVE EXCLUSION PRINCIPLE; MINORITY CYTOTYPE EXCLUSION PRINCIPLE; NEIGHBOR EXCLUSION PRINCIPLE; PAULI EXCLUSION PRINCIPLE; PROBABILISTIC EXCLUSION PRINCIPLE; PRINCIPLE OF COMPLEMENTARITY; GAUSE'S PRINCIPLE.

EXISTENCE   PRINCIPLE [philosophy] The principle of the excluded middle is part of the existence principle. Logic's principles, however, are in a very broad sense metaphysical principles; presupposing that there is an

---

*Patten, B. C. (1961), The exclusion principle is recast in the context of a generalized scheme for interspecific interactions, *Science* **134**(3490), 1599–1601.

†Rao, S. N. and Kollman, P. A. (1987), Molecular mechanical simulations on double intercalation of 9 aminoacridine into deoxy-cgcgcg deoxy-GcGcgcg analysis of the physical basis for the neighborexclusion principle, *Proc. Natl. Acad. Sci. USA* **84**(16), 5735–5739.

‡Thompson, D. E. (1988), Violating a not so exclusive exclusion principle, *Sci. News* **133**, 132.

§Ortega, R. and Tineo, A. (1998), An exclusion principle for periodic competitive systems in three dimensions, *Nonlinear Anal. Theory Meth. Appl.* **31**, 883.

¶Brooijmans, N. and Kunt, I. D. (n.d.), Molecular recognition and docking algorthms, *Annu. Rev. Biophy. Biomol. Struct.* **32**, 335–373.

‖Lamont, M. and Fournier, M. (1993), *Cultivating Differences: Symbolic Boundaries and the Making of Inequality*, Univ. Chicago Press, p. 288.

**Gibson, Q. (1998), *The Existence Principle* (Australian Natl. Univ. Dept. Philosophy, Australasian Assoc. Philosophy), Springer, p. 4.

††Williamson, T. (1987–1988), Equivocation and existence, *Proc. Aristotelian Soc.* (new series) **88**, 109–127.

existent world, they say something in very general terms about what it is like.* This means that they say nothing about the concept of existence itself. They are concerned with the nature of the world, not with what is required for there to be a world, treating existence as property to distinguishing kinds of existence.[†]

EXPECTATION MINIMUM (EM) PRINCIPLE [mathematics]   (1) Unit of linear measurement used in priority that is equal to the point size of the type.[‡] (2) New strategy for recovering robust solutions to the ill-posed inverse problem of photothermal science.[§] (3) Average of the results of a large number of measurements of a quantity made on a system in a given state.[¶] (4) In case the measurement disturbs the state, the state is reprepared before each measurement.[∥]

EXPLOSION PRINCIPLE [chemistry]   Chemical reaction or change of state that is effected in an exceedingly short time interval with the generation of a high temperature and large quantity of gas.[**]

EXPOSURE PRINCIPLE [psychology]   People express undue liking for things merely because they are familiar with them. This effect has been nicknamed the "familiarity breeds liking" effect. Conversely, if one has an unduly adverse reaction to something, the intensity of that reaction will fade with increased familiarity. In interpersonal attractiveness research, the term is used to characterize the phenomenon in which the more often a person is seen by someone, the more attractive and intelligent that person appears to be.[††]

EXTENDED MAXIMUM ENTROPY PRINCIPLE [mathematics, physics]   See MAXIMUM ENTROPY PRINCIPLE.

EXTENDED MAXIMUM PRINCIPLE [mathematics]   Extended minimum conditions reduce to the minimum condition of the weak maximum principle. Hence, possibility to set up higher-order approximations of arbitrary order.[‡‡]

EXTENDED MINIMUM CROSS-ENTROPY PRINCIPLE [engineering, mathematics]   The entropy measure is minimized to given constraints in order to obtain the most biased proportional distribution consistent with the given constraints. Every additional consistent constraint may reduce the entropy, but the reduction will not go below the entropy of the most biased distribution for the specified original constraints. The maximum entropy proportional distribution is unique but the minimum entropy proportional distribution may not be unique.[§§] See also ENTROPY OPTIMIZATION PRINCIPLE.

EXTENDED PRINCIPLE [principles]   An extension, prolongation, or expansion of an existing principle.[¶¶] See also EXTENSION PRINCIPLE.

*Gibson, Q. (1998), *The Existence Principle*, Australian Natl. Univ. Dept. Philosophy, Australasian Assoc. Philosophy, Springer, vol. 22, p. 22.
[†]Op. cit. Gibson, p. 176.
[‡]US Dept. Veterans Affairs, Emergency Management Strategic Health Care Group, *Emergency Management (EM) Principles and Practices for Healthcare Systems* (http://www1.va.gov/emshg/page.cfm?pg=122).
[§]Power, J. F. (1997), Expectation minimum A new principle of inverse problem theory in the photothermal sciences: Theoretical characterization of expectation values, *Opt. Eng.* **36**, 487.
[¶]Power, J. F. (2002), Inverse problem theory in the optical depth profilometry of thin films, *Rev. Sci. Instrum.* **73**(12), 4057.
[∥]Margenau, H. (1963), Measurements and quantum states: Part II. Philosophy of science, Annals of Physics, Elsevier **30**(2), 138–157.
Wu, X. and Liu, H. (1995), Studies for explosion principle and prevention of explosion in butadiene systems, *Hunan Daxue Xuebao* **22(2), 58–63.

[††]Westen, D. (2007), The political brain: The role of emotion in deciding the fate of the nation, *PublicAffairs* p. 415.
[‡‡]Christodoulos, A., Floudas, P., and Pardalos, M. (2001), *Encyclopedia of Optimization*, Springer, pp. 440–441.
[§§]Rose, K. (1998), Deterministic annealing for clustering, compression, classification, regression, and related optimization problems, *Proc. IEEE* **86**(11), 2210.
[¶¶]Gaffney, J. (1976), The over-extended principle of totality and some underlying issues, *J. Relig. Ethics* **4**(2), 259–267.

EXTENDED PROJECTION PRINCIPLE (EPP)
[psychology, linguistics] (Avram Noam
Chomsky, b. 12/7/28, American linguist and
philosopher, lectures on government and
binding) (1) Explores the nature of intuition. Adam Smith's lectures on rhetoric and
belle lettres, and their relation to intuition.
(2) There is strong evidence for *object shift* in
English; that is, just as there is an endplate
potential position (EPP) high in the sentence
where subjects and derived subjects wind up,
there is a similar *EPP position* in the ventral
pallidum (VP) region where objects and
extracellular matrix (ECM) subjects (among
other categories) wind up. (3) A sentence
must contain a tensed verbal element and a
subject. (4) Extends the projection principle
with the requirement that clauses have
subjects. (5) Representations at each level
of representation are projections of the
features of lexical items, notably their subcategorization features. (6) If F is a lexical
feature, it is projected at each syntactic level
of representation such as a D structure, S
structure, or other logical form.* See also
BRAIN'S DIVISION OF FUNCTION PRINCIPLE;
PROJECTION PRINCIPLE.

EXTENDED RESOLUTION PRINCIPLE [mathe-
matics] Application to logic programming.[†]

EXTENSION PRINCIPLE An expansion, prolongation, or extended version of an existing
principle. See also EXTENDED PRINCIPLE.

EXTINCTION   PRINCIPLE [engineering]
(1) The fluid pressure in the gas storage
chamber is raised, whereby the fluid at a
temperature low enough to extinguish the
arc is blown against the arc to effect self-
extinction in the process in which current
decreases toward its zero point, while the

low-temperature   arc-extinguishing   fluid
from the puffer chamber is forcibly blown
against the arc to effect forcible extinction in
the process in which the current decreases
toward its zero point.[‡] (2) Comprising
at least two sensor or feeler elements
having different functional or operating
principles and a common evaluation circuit
for evaluation of the property changes of
the sensor elements and for triggering a
signal.[§] (3) Rest mass of a particle cannot be
kept invariant. There must exist a unique
preferred inertial frame of reference wherein
a particle is absolutely at rest.[¶] (4) To stop
a child from acting in a particular way, you
may arrange conditions so that the child
receives no rewards following the undesired
act.[||] See also ARC-EXTINGUISING PRINCIPLE;
SEARCHLIGHT PRINCIPLE.

EXTINCTION   SHIFT   PRINCIPLE [physics]
(1) A pure classical physics look at electromagnetism and gravitation in Euclidean
space. (2) Present a clear, simple, and
intuitive approach to the application of
pure classical physics in Euclidean space
geometry. (3) Step-by-step mathematical
illustrations to the invariance of the wave
equation, the velocity-dependent "effective"
mass as opposed to relativistic mass, the
transverse relative time shift as opposed to
time dilation, the planet Mercury perihelion
drift, the neutron pulsar system PSR1913+16

*Chomsky, N. (1981/1993), *Lectures on Govern-
ment and Binding: The Pisa Lectures*, Mouton de
Gruyter.
[†]Mengin, J. (1995), A theorem prover for default
logic based on prioritized conflict resolution and
an extended resolution principle, p.301. in *Proc.
Eur. Conf. Symbolic and Quantitative Approaches
to Reasoning and Uncertainty*, (ECSQARU'95; Fri-
bourg, Switzerland, July 3–5, 1995), Froidevaux,
C. and Kohlas, J., eds., Lecture Notes in Computer
Science, vol. 946, p. 301.

[‡]Thilo, P. (1985), According to the extinction
principle working smoke detector arrangement and
fire-announce eanlage with such smoke detector
arrangement. Ger. Offen., (n.p.).
[§]Muggli, J. and Labhart, M. (1985), Smoke detector
operating according to the radiation extinction
principle U.S. Patent 7,483,139.
[¶]Belov, P. A. and Simovski, C. R. (2006), Boundary
conditions for interfaces of electromagnetic crystals and the generalized Ewald-Oseen extinction
principle, *Phys. Rev. B Condensed Matt. Mater.
Phys.* **73**(4).
[||]Huitt, W. (1994), Principles for using be-
havior modification, in *Educational Psychology Interactive*, Valdosta State Univ., Valdosta, GA: (http://chiron.valdosta.edu/whuitt/col/
behsys/behmod.html).

perihelion drift, the gravitational redshift, and the solar light-bending effect.*

EXTINGUISHING PRINCIPLE [engineering] The high-temperature fluid near the arc-striking part attributed to the thermal energy emission of the arc is caused to flow back into the gas storage chamber and is mixed with the low-temperature fluid contained therein. The fluid pressure in the gas storage chamber is raised, whereby the fluid at a temperature low enough to extinguish the arc is blown against the arc to effect self-extinction in the process in which current decreases toward its zero point, while the low-temperature arc-extinguishing fluid from the puffer chamber is forcibly blown against the arc to effect forcible extinction in the process in which the current decreases toward its zero point.[†] See also ARC-EXTINGUISHING PRINCIPLE; EXTINCTION PRINCIPLE.

EXTREMAL PRINCIPLE [mathematics] [Pierre de Fermat (1601–1665, French mathematician), c. 1660]   (1) The path taken by a ray of light between two fixed points in an arrangement of mirrors, lenses, and so forth, is the path that takes the least time. The laws of reflection and refraction may be deduced that in a medium of refractive index μ light travels more slowly than in free space by a factor μ. Strictly, the time taken along a true ray path is either less or greater than that for any neighboring path. If all paths in the neighborhood take the same time, the two chosen points are such that light leaving one is focused on the other. The perfect example is exhibited by an elliptical mirror, in which all paths from F_1 to the ellipse and thence to F_2 have the same length. In conventional optical terms, the ellipse has the property that every choice of paths obeys the law of reflection, and every ray from F_1 converges after reflection onto F_2. For a flat reflector the path taken is the shortest of all paths in the vicinity, while for a reflector that is more strongly curved than the ellipse, it is the longest.[‡] (2) *Mechanics* (Pierre-Louis Moreau de Maupertuis, 1698–1759, French mathematician and astronomer). The path taken by a particle between two points, A and B, in a region where the potential $f(r)$ is everywhere defined. Once the total energy E of the particle has been fixed, its kinetic energy T at any point P is the difference between E and the potential energy f at P. If any path between A and B is assumed to be followed, the velocity at each point may be calculated from T, and hence the time t between the moment of departure from A and passage through P. The action for this path is found by evaluating the integral $\grave{o}BA\ (T-f)dt$, and the actual path taken by the particle is that for which the action is minimal.[‡] See also PRINCIPLE OF LEAST ACTION; FERMAT'S PRINCIPLE; FIBRINOLYTIC PRINCIPLE; HAMILTON PRINCIPLE; SCHRÖDINGER PRINCIPLE.

EXTREMAL PRINCIPLE OF INFORMATION (EPI) [mathematics, physics]   See UNCERTAINTY PRINCIPLE.

EXTREMUM PRINCIPLE [engineering, mathematics]   Laws of reflection and refraction; path taken by a ray of light between two fixed points in an arrangement of mirrors, lenses, and other objects is the path that takes the least time.[§] See also MAXIMUM PRINCIPLE, MINIMUM PRINCIPLE; LINDELÖF PRINCIPLE.

EXTREMUM PRINCIPLE OF ENTRANSY DISSIPATION [engineering]   (1) For a fixed-boundary heat flux, the conduction process is optimized when the entransy dissipation is minimized, while for a fixed-boundary temperature the conduction is optimized when the entransy dissipation is maximized (*entransy* is a quantitative parameter of heat transfer capacity). An equivalent thermal resistance for multidimensional conduction problems is defined on the basis of the entransy dissipation, so

*Henault, F. (1968), Computing extinction maps of star nulling interferometers, *Optics Express* **16**(7), 4537–4546.
[†]Akita, K. (1968), Principle of fire extinguishing, *Nenryo Kyokaishi* **47**(492), 238–246.

[‡]Anon. (2009), Physical science, principles of, in *Encyclopædia Britannica* (http://search.eb.com/eb/article-14867).
[§]Taylor, J. E. (1992), *A Global Extremum Principle for the Analysis of Solids Composed of Softening Material*, Danish Center for Applied Mathematics and Mechanics.

that the extremum principle of entransy dissipation can be related to the minimum thermal resistance principle to optimize conduction.[*] (2) Entransy dissipation occurs during heat transfer processes as a measure of the heat transfer irreversibility. For a fixed-boundary heat flux, the conduction process is optimized when the entransy dissipation is minimized, while for a fixed-boundary temperature the conduction is optimized when the entransy dissipation is maximized.[†] See also RESISTANCE PRINCIPLE. MINIMUM THERMAL RESISTANCE PRINCIPLE.

EXTREMUM VARIATIONAL PRINCIPLE [mathematics]   A maximum or minimum value of a function.[‡]

EYE-FOR-AN-EYE   **PRINCIPLE** [psychology] The punishment must be exactly equal to the crime (lex talionis). See also EYE-FOR-AN-EYE PRINCIPLE PRINCIPLE OF EXACT RECIPROCITY; PRINCIPLE OF RETRIBUTIVE JUSTICE; PRINCIPLE OF PROPORTIONATE PUNISHMENT.

EYE PLACEMENT PRINCIPLE [psychology]   In works of art, an eye tends to be set on the center vertical in portraits. One eye as compositionally dominant over the other with the head in a variety of poses. The center of symmetry or the closer eye should be centered at the center of the frame.[‡]

EYE PRINCIPLE [psychology] Multiplicative interaction of vestibular and horizontal conjugate eye position signals in the abducens neurons with disynaptic latency. On the basis of these results, we propose a novel neural mechanism that implements the vestibuloocular reflex (VOR) gain modulation by fixation distance and gaze eccentricity. The new mechanism consists of two principles that are derived from experimental data: (1) the *principle of multiplication*, which states that the effects of stimulating a single labyrinth interact multiplicatively with the position signals of each eye and (2) the *principle of addition*, which states that the effects of stimulating the two labyrinths interact additively.[§]

---

[*]Leonov, A. I. (1988), Extremum principles and exact two-side bounds of potential: Functional and dissipation for slow motions of nonlinear viscoplastic media, *J. Non-Newton. Fluid Mech.* **28**(1), 1–28.

[†]Pleshanov, A. S. (2002), Extremum principles in the theory of thermal conductivity of a solid, *High Temp.* (transl. *Teplofizika Vysokikh Temperatur*) **40**(2), 295–299.

[‡]Spruch, L. (1976), Variational principles, subsidiary extremum principles, and variational bounds, *Proc. 9th (1975) Physics Electronic Atomic Collisions, Invited Lecture, Review Papers, Progress Report*, pp. 685–698.

[§]Tyler, C. W. (1998), *An Eye Placement Principle in 500 Years of Portraits*, Smith Kettlewell Eye Research Inst., San Francisco, CA.

F

FAIL-SAFE PRINCIPLE [psychology]   (1) A failure will result to a "safe" or "inert" mode. (2) Alternate or redundant backup systems will kick in following failure of the primary system. (3) Incorporates the concept that a failure of the system could result in human injury; incorporates design features to protect the human component.[*]

FAJAN'S PRINCIPLE [chemistry]   Rules that deal with variations in the degree of covalent character in ionic compounds in terms of polarization effects.[†]

FALSEHOOD PRINCIPLE [mathematics, logic] Each primitive predicate should be existentially restricted.[‡]

FAN PRINCIPLE [mathematics]   Every detachable bar of a fan is uniform. In its classical contrapositive form, if for every n there exists a path of length n that misses B, then there exists an infinite path that misses B. Also known as *König's lemma*.[§]

FARADAY'S ICE PAIL PRINCIPLE [physics] (Michael Faraday, 1791–1867, English chemist/physicist) Classical experiment that consists of lowering a charged body into a metal pail connected to an electroscope, in order to show that charges reside only on the outside surface of conductors.[¶] See INDUCTIVE PRINCIPLE.

FARADAY'S PRINCIPLE OF ELECTROLYSIS [physics] (Michael Faraday, 1791–1867, English chemist/physicist)   (1) The amounts of different substances liberated or deposited by a given quantity of electricity are proportional to the chemical equivalent weights of those substances.[¶] (2) (a) The amount of chemical change produced is proportional to the electric charge passed. (b) The amount of chemical change produced by a given charge depends on the ion concerned.[||]

FARADAY'S PRINCIPLE OF INDUCTION [physics] (Michael Faraday, 1791–1867, English chemist/physicist)   The emf induced in any circuit is proportional to the rate of change of magnetic flux linked with the circuit.[**]

FAUSTMANN PRINCIPLE [ecology]   (Martin Faustman, German forester)   Incorporates biological and economic parameters to derive a function relating the present value of net revenue to rotation length. Established for use with longer-growing species, to bear on leucaena, a short-rotation, leguminous tree crop. Embodies the application of fundamental economic principles to the choice of management methods and alternative land uses. The price of products is a key input in applications of this principle. Provides new valuation and demand forecasting methods based on macroeconomic growth theory attempting to link the demand for environmental goods, such as forest diversity, to total aggregate consumption.[††]

[*]Zhou, W., Xu, Y., and Simpson, I. (2007), Multiplicative computation in the vestibulo-ocular reflex (VOR), J. Neurophysiol. **97**(4), 2780–2789.
[†]Lee, P. (2002), Ideal principles and characteristics of a fail-safe medication-use system, *Am. J. Health Syst. Pharmacy* 59, n.4, p.369–71.
[‡]Daintith, J. (1999), The Facts on File Dictionary of Chemistry, 3rd ed., Facts On File, New York, p. 98.
[§]Fine, K. (1981), *J. Phil. Logic* **10**(3), 293–307.
[¶]Walker, P. M. B. (1999), *Chambers Dictionary of Science and Technology*, Chambers, New York, p. 431.

[||]Daintith, J. (1999), *The Facts on File Dictionary of Chemistry*, 3rd ed., Facts On File, New York, p. 98.
[**]Walker, P. M. B. (1999), *Chambers Dictionary of Science and Technology*, Chambers, New York, p. 431.
[††]Stone, S. W., Kyle, S. C., and Conrad, J. M. (1993), Application of the Faustmann principle to a short-rotation tree species: an analytical tool for economists, with reference to Kenya and leucaena, *Agroforestry Syst.* **21**(1), 79–90.

FAYOL'S PRINCIPLES OF MANAGEMENT [psychology] (Henri Fayol, 1841–1925, French industrialist)   Business interest must prevail over the interest of employees. Ignorance, ambition, selfishness, laziness, and other human weakness are forever at war with the best interests of the firm. Fayol divided the activities of organizations into six fundamental groups: (1) technical, or production, aspects; (2) commercial aspects (buying, selling, and exchanging goods); (3) financial aspects (the search for, securing of, and efficient use of money); (4) security (protecting the safety of employees and property alike); (5) accounting (including statistics and recordkeeping); and (6) managerial activities (planning, organization, and control).* See also PRINCIPLE OF DIVISION OF WORK; PRINCIPLE OF SPECIALIZATION; PRINCIPLE OF AUTHORITY AND RESPONSIBILITY; PRINCIPLE OF DISCIPLINE; PRINCIPLE OF UNITY OF COMMAND; PRINCIPLE OF UNITY OF DIRECTION; PRINCIPLE OF SUBORDINATION OF INDIVIDUAL TO GENERAL INTEREST; PRINCIPLE OF REMUNERATION; PRINCIPLE OF CENTRALIZATION; PRINCIPLE OF LINES OF COMMAND CHAIN; PRINCIPLE OF SCALAR CHAIN; PRINCIPLE OF ORDER; PRINCIPLE OF EQUITY; PRINCIPLE OF STABILITY OF TENURE; PRINCIPLE OF ESPRIT DE CORPS.

FEAR REDUCTION PRINCIPLE [psychology]   To help a child overcome his/her fear of a particular situation, gradually increase his/her exposure to the feared situation while he/she is otherwise comfortable, relaxed, secure, or rewarded.[†]

FEEDBACK PRINCIPLE [psychology] (Wiener's cybernetics and von Bertalanffy's general system theory)   A system, including a living organism, will modify itself based on the basis of feedback about the state of component parts. This principle is applied broadly, including in learning and behavior, biopsychology, and their intersection: biofeedback.[‡] See also BLACK BOX PRINCIPLE.

FEMININE PRINCIPLE [psychology]   (1) Perspective of a Buddhist feminist. Feminine and masculine principles in Western monotheism, Jungian psychology, and feminist spirituality movements are outlined. It is concluded that masculine and feminine principles quickly become oppressive when it is asserted that women are or should be characteristically more like the feminine principle than the masculine principle or should act according to or manifest the feminine principle more than men do.[§] (2) Certain aspects of the universe, and analogously, of human beings, are associated with traits historically assigned to females, including passivity and receptivity. In traditional Eastern philosophies, including Tibetan Buddhism, it is proposed that the feminine principle function in concert with the masculine principle. According to these philosophies, to become aware, one must surrender expectation and concept (the feminine), to gain wisdom and awareness (the masculine). In analytical psychology, the feminine principle generally characterizes the unconscious, or id, and is represented by the self as "being" mode, which functions according to primary process. The feminine principle operates in conjunction with the masculine principle.[¶] See also MASCULINE PRINCIPLE.

FENG SHUI PRINCIPLES [psychology]   Rules in Chinese philosophy that govern spatial arrangement and orientation in relation to patterns of yin and yang and the flow of energy (qi); the favorable or unfavorable effects

---

*Wood, M. C. (2002), *Henri Fayol: Critical Evaluations in Business and Management*, Taylor & Francis, p. 8.

[†]Huitt, W. (1994), Principles for using behavior modification, in *Educational Psychology Interactive*, Valdosta State Univ., Valdosta, GA (http://chiron.valdosta.edu/whuitt/col/behsys/behmod.html).

[‡]Nicol, David J., and Macfarlane-Dick, Dobra. (2006), Formative assessment and self-regulated learning: A model and seven principles of good feedback practice, *Studies in Higher Education*. V 31(2) p. 199–218.

[§]Gross, R. (1984), The feminine principle in Tibetan Vajrayana Buddhism: Reflections of a Buddhist feminist, *J. Transpers. Psychol.* 16(2), 179–192.

[¶]Weisstub, E. B. (1997), Self as the feminine principle, *J. Anal. Psychol.* 42(3), 425–452; discussion, pp. 453–458.

are taken into consideration in designing and siting buildings and graves and furniture.*

FERENCZI'S RELAXATION PRINCIPLE [psychology] (Sandor Ferenz L, 1873–1933, Hunganan Psycho analyst) (1) Model of psychoanalysis and psychotherapy contains unique dimensions concerning the active and flexible participation of the analyst at both verbal and nonverbal interaction, levels, with emphasis on mutuality between analyst and analysand, an ongoing focus on countertransference analysis, and judicious self-disclosure by the analyst. Ferenczi's relaxation principle and the issue of therapeutic responsiveness.[†] (2) Reparative therapeutic experiences within group psychotherapy. A relaxation term that restores the exact pressure law in the limit of an infinite relaxation parameter.* (3) Occurs when words of opposite meaning are placed in analogous, balanced order.[‡] See also ELASTICITY PRINCIPLE; PRINCIPLE OF INDULGENCE.

FERGUSON PRINCIPLE [physics] (Ferguson, R. G.) A method of loss reserving in its own right which, in every single application, can be specified according to the available sources of information and their degree of credibility and that, in turn, can also be used to check the credibility of these different sources of information. Also used to determine ranges reflecting the uncertainty of the best predictors. Can be used to point out the possible presence of different mechanisms of activity within a series of similar compounds or as a general principle of estimation that can be applied to any development pattern.[§] See also BORNHUETTER–FERGUSON PRINCIPLE.

*American Heritage (2000), Fengshui, *American Heritage Dictionary of the English Language*, 4th ed., Houghton Mifflin, New York.
[†]Rachman, A. W. (1998), Ferenczi's relaxation principle and the issue of therapeutic responsiveness, *Am. J. Psychoanal.* **58**(1), 63–81.
[‡]Phillips, H. (1997), The order of words and patterns of opposition in the Battle of Maldon, *Neophilologus* (Springer, Netherlands), **81**(1), 117–128.
[§]Schmidt, K. D. and Zocher, M. (1985), The Bornhuetter-Ferguson principle, *Casualty Actuarial Soc.* **2**(1), 104.

FERMAT–OSTWALD PRINCIPLE [physics] (Pierre de Fermat, 1601–1665, French mathematician) Fermat's principle of least time and Ostwald's law of intermediate stages are postulates concerning the physical processes that deal with light and heat energy flows, respectively.[¶]

FERMAT'S LEAST-TIME PRINCIPLE [physics] (Pierre de Fermat, 1601–1665, French mathematician) (1) A spatial path that was the same always, because light can be made to shine continuously in a way that makes it look less like a moving object than like a streaming current, so it makes sense to talk of the spatial path as a thing that doesn't change from one time to the next. (2) Electromagnetic wave will take a path that involves the least travel time when propagating between two points.[∥]

FERMAT'S MINIMUM TIME PRINCIPLE [computer science] (Pierre de Fermat 1601–1665, French mathematician)

(1) For computing traveltimes, a reflection horizon is mapped from time to depth by finding the envelope of all possible depths that can be computed with the velocity model. The stress-velocity strain relation for an (isothermal), anisotropic viscoelastic medium can be written as the Boltzmann superposition principle. For a given interface model, the reflection traveltimes are calculated following a four-step procedure:

1. The medium is discretized as a fine grid, with nodes equally spaced along the vertical and horizontal directions. The grid dimension depends on the accuracy required for the traveltime calculations.

2. The first-arrival times from each source and receiver at the nodes of the grid are calculated using the 2-D

[¶]Mosienko, B. A. (2005), Fermat-Ostwald principle, *Z. Physi. Chem.* **219**(12), 1655–1663.
[∥]Drakos, N. (1995), *Relation to the Equations of Motion*, Computer Based Learning Unit, Univ. Leeds (http://www.ph.utexas.edu/~gleeson/httb/section1_3_6_5.html).

Eikonal equation and the finite difference solver.

3. The one-way traveltimes for a source/receiver to each point of the discretized interface are calculated by performing an interpolation among the nearest four grid nodes.

4. For a given source-receiver pair, the reflection location point and the total traveltime are calculated according to the Fermat principle: the reflection point will be the one providing the minimum total traveltime. The general principle is reflected travel times provide long wavelength information about the interface morphology while more refined models can be retrieved by the waveform semblance optimization, using as a starting model the one obtained from traveltime misfit.* See also MAXIMUM DEPTH PRINCIPLE.

(2) The earliest arriving wave has followed a path such that its travel-time is minimum.[†] It can be shown that Snell's law from optics follows from this principle and is given by:

$$\frac{\text{Sin } i_1}{\text{Sin } i_2} = \frac{V_1}{V_2}$$

FERMAT'S PRINCIPLE [mathematics] (Pierre de Fermat, 1601–1665, French mathematician) (1) The path taken by a ray of light between two given points is the one in which the light takes the least time compared with any other possible path. (2) Light traveling between two points seeks path in which the number of waves is equal, in the first approximation, to that in neighboring paths. (3) Path taken by a ray of light in traveling between two points requires either a minimum or a maximum time. (4) Light traveling through some substance has a speed that is determined by the substance. The actual path taken by light between any two points, in any combination of substances, is always the path of least time that can be traveled at the required speeds.[‡] (5) Second law of thermodynamics in the neighborhood of any equilibrium state of a system, there are states that are not accessible by a reversible or irreversible diabolic process.[§] (6) Electromagnetic wave will take a path involving the least traveltime when propagating between two points.[¶] See also CONTES PRINCIPLE; CARATHÉODORY'S PRINCIPLE; OPTIMALITY PRINCIPLE.

FERMAT'S PRINCIPLE [physics] (Pierre de Fermat, 1601–1665, French mathematician) A ray of light traversing one or more media will follow a path that minimizes the time required to pass between two given points.[‖]

FERMAT'S PRINCIPLE OF CONJUNCTIVE PROBABILITY [mathematics] (Pierre de Fermat, 1601–1665, French mathematician) The probability that two events will both occur is hk, where h is the probability that the first event will occur and k is the probability that the second event will occur.[**]

FERMAT'S PRINCIPLE OF LEAST TIME [physics] (Pierre de Fermat, 1601–1665, French mathematician) The path of a ray of light from one point to another will be that taking the least time.[††]

FERMI PRINCIPLE [mathematics] (Enrico Ferml, 1901–1954, Italian physicist) (1) A

*Marzocchi, W. and Zollo, A. (2008), *Conception, Verification and Application of Innovative Techniques to Study Active Volcanoes*, Istituto Nazionale di Geofisica e Vulcanologia, (Italy), p. 341.
[†]Bollinger, G. A. (1980), Blast Vibration Analysis. Southern Illinois University Press, p. 14.
[‡]Forlunato, D. et al. (1995), A Fermat principle for stationary space times and a locations to light rays, *J. Geom. Phys.* **15**, 159.
[§]Giannoni, F. (1997), A timelike extension of Fermat's principle in general relativity. Comparisons with the lightlike case, *Nonlinear Anal. Theory Meth. Appl.* **30**, 759.
[¶]Antonacci, F. and Piccione, P. (1996), A Fermat's principle on Lorentzian manifolds and applications, *Appl. Math. Lett.* **9**(2), 91–95.
[‖]Lide, David R. (2007) Fermates Principle, CRC Handbook of Chemistry and Physics, 88th edition, CRC Press, OH, p. 2–45.
[**]Whittaker, E. T. and Robinson, G. (1967), *The Calculus of Observations: A Treatise on Numerical Mathematics* 4th ed., Dover, New York; p. 317.
[††]Walker, P. M. B. (1999), *Chambers Dictionary of Science and Technology*, Chambers, New York, p. 437.

conflict between an argument of scale and probability, and a lack of evidence. (2) The size and age of the universe suggest that many technologically advanced extraterrestrial civilizations ought to exist. However, this belief seems logically inconsistent with the lack of observational evidence to support it. Either the initial assumption is incorrect and technologically advanced intelligent life is much rarer than believed, current observations are incomplete and human beings have not detected other civilizations yet, or search methodologies are flawed and incorrect indicators are being sought.[*]

FERROFILTER PRINCIPLE [engineering] Magnetic forces for liquid–solid separations are usually applied to replace thickening and filtration rather than for dewatering. Used for removing suspended ferromagnetic impurities from liquids.[†]

FIBRINOLYTIC PRINCIPLE (FP) [mathematics] Focusing by mirrors and lenses finds a natural explanation in the wave theory of light.[‡] See also EXTREMAL PRINCIPLE.

FICK PRINCIPLE [medicine, physics] (Adolf Eugen Fick, 1829–1901, German physiologist) (1) Measurement of cardiac output. Flow or cardiac output is the oxygen uptake of either an organ or the whole body, divided by the oxygen extraction of the tissue being examined.[§] (2) In a real motion, the system acted on by potential forces has a stationary value as compared with near kinetically possible motions, with initial and final positions of the system and times of motion identical with those for real motion.[¶] (3) Direct solution of a flow problem free energy of fluctuations about the equilibrium flow that satisfies the equation of motion.[∥] (4) Equation for determining the heart–minute–volume by O_2 absorption per time and arteriovenous difference.[**] (5) Rate of diffusion of matter across a plane is proportional to the negative of the rate of charge of the concentration of the diffusing substance in the diversion perpendicular to the plane.[††] See also STATIONARY ACTION PRINCIPLE; PRINCIPLE OF LEAST ACTION; LIQUID MOVEMENT PRINCIPLE; GAS MOVEMENT PRINCIPLE; VARIATIONAL PRINCIPLE; STEWART–HAMILTON PRINCIPLE; GUERRA–MORATO VARIATIONAL PRINCIPLE; SADDLE POINT PRINCIPLE; SADDLE POINT ENTROPY PRODUCTION PRINCIPLE; HAMILTON–OSTROGRADSKI PRINCIPLES.

FIELD ORIENTATION PRINCIPLE (FOP) [engineering] (1) Defines the stator current in terms of a rotating coordinate system, where a direct axis is aligned with the instantaneous rotor flux vector and a quadrature axis is perpendicular to the rotor flux vector. Command values are determined using a field orientation principle (FOP), which is known in the art, in conjunction with other known control principles such as a space vector modulation (SVM) and pulsewidth-modulated control (PWM).[‡‡]

FIFO PRINCIPLE See FIRST-IN FIRST-OUT PRINCIPLE.

FIFTEEN-TERM PRINCIPLE See FIRST-IN FIRST-OUT PRINCIPLE.

[*]Frank, I. M.; Frank, A. I., (1978) Applicability of the Fermi principle to optics of ultracold neutrons, *Pis'ma Zh. Eksper. Teoret. Fiz.* **28**(8), 559–560.

[†]Morey, B. (1993), Dewatering, in *Kirk-Othmer Encyclopedia of Chemical Technology* (1SRI International), Wiley.

[‡]Ouyang, C. and Huang, T. F. (1976), Purification and characterization of the fibrinolytic principle of Agkistrodon acutus venom, *Biochim. Biophys. Acta* **439**(1), 146–33.

[§]Walker, P. M. B. (1999), *Chambers Dictionary of Science and Technology*, Chambers, New York, p. 443.

[¶]Kumar, N. (1983), Brownian motion and condensed matter physics classical and quantum diffusion, *Lecture Notes in Physics. Stochastic Processes Formalism and Applications*, Springer, Berlin/Heidelberg, Vol. 184, pp. 166–185.

[∥]Scardovelli, R. and Zaleski, St. (1999), Direct numerical simulation of free-surface and interfacial flow, *Annu. Rev. Fluid Mech.* **31**, 567–603.

[**]Wippermann, C. F. et al. (1996), Continuous measurement of cardiac output by the Fick principle in infants and children: Comparison with the thermodilution method, *Intens. Care Med.* **22**, 467.

[††]Sabyasachi. S. (2007), *Principles of Medical Physiology*, Thieme, p. 254.

[‡‡]Trzynadlowsky, A. (1994), *The Field Orientation Principle in Control of Induction Motors*, Kluwer Academic Publishers.

FIFTH DATA PROTECTION PRINCIPLE [mathematics]   Personal data processed for any purpose or purposes shall not be kept for longer than is necessary for that purpose or those purposes.* See also COPY PROTECTION PRINCIPLE; PROTECTION PRINCIPLE; PRIVACY PRINCIPLE; FIRST DATA PROTECTION PRINCIPLE; SECOND DATA PROTECTION PRINCIPLE; THIRD DATA PROTECTION PRINCIPLE; FOURTH DATA PROTECTION PRINCIPLE; SIXTH DATA PROTECTION PRINCIPLE; SEVENTH DATA PROTECTION PRINCIPLE; EIGHTH DATA PROTECTION PRINCIPLE.

FIFTH PARADOX PRINCIPLE [psychology]   In order to build, you must first tear down.[†]

FIFTH PRINCIPLE OF ENERGETICS [physics] (Howard Thomas Odum, 1924–2002, American ecosystem ecologist) (1) The energy quality factor increases hierarchically. From studies of ecological food chains, Odum proposed energy transformations form a hierarchical series measured by transformity increase.[‡] (2) Flows of energy develop hierarchical webs in which inflowing energies interact and are transformed by work processes into energy forms of higher quality that feed back amplifier actions, helping to maximize the power of the system".[§] See also PRINCIPLES OF ENERGETICS; ZEROTH PRINCIPLE OF ENERGETICS; SECOND PRINCIPLE OF ENERGETICS; THIRD PRINCIPLE OF ENERGETICS; FOURTH PRINCIPLE OF ENERGETICS; MAXIMUM EMPOWER PRINCIPLE; MAXIMUM POWER PRINCIPLE; FIRST PRINCIPLE OF ENERGETICS; SIXTH PRINCIPLE OF ENERGETICS.

50% PRINCIPLE [mathematics] (E. George Schaefer, Dow theorist)   Charles H. Dow's writings applied to major, extended market movements to an average or index indicating a midway (50%) level between the lowest and the peak of a market. During a decline, if the price movement can hold above halfway level of the preceding major advance, it is a very constructive indication. Also during a decline, if the price movement breaks decisively below the midlevel of the previous major advance, it is an extremely bearish indication.[¶]

FILM PRINCIPLE [physics]   Transfer of material or heat across a phase boundary where one or both of the planes are flowing fluids, the main controlling factor is resistance to heat conduction or mass diffusion through a relatively stagnant film of the fluid next to the airface.[‖]

FILTER PRESS PRINCIPLE [engineering] Gradual increase of pressure differential.[**] See also BELT FILTER PRESS PRINCIPLE; PRINCIPLE OF BELT FILTER PRESS.

FINAL ANTHROPIC PRINCIPLE (FAP) [computer science] (John Daud Barrow, 1952–, English Cosmologist Frank Jennings Tipler III, 1947–; mathematical physicist and cosmologist)   Intelligent information processing must come into existence in the universe, and, once it does it will never die out.[††]

FINITE-AREA PRINCIPLE [mathematics] (Lipót Fejér, 1880–1959, Hungarian mathematician)   Asserts that if an image under analytic mapping is of finite area, then the image will converge uniformly on closed arcs of continuity.[‡‡]

*Univ Edinburgh, Records Management Section (2007) (http://www.recordsmanagement.ed.ac.uk/InfoStaff/DPstaff/DP_Research/ResearchAnnex A.htm).

[†]Jones, H. R., Jr. Caplan, L. R., Come, P. C., Swinton, N. W., Jr., and Breslin, D. J., (1983), Cerebral emboli of paradoxical origin, *Ann. Neurol.* **13**(3), 314–319.

[‡]Odum, H. T. (2000), An energy hierarchy law for biogeochemicall cycles, in *Proc Biennial Energy Analysis Research Conf. and Applications of the Emergy Methodology Energy Synthesis: Theory* Brown, M. T., ed., Center for Environmental Policy, Univ. Florida, Gainesville, p. 246.

[§]Odum, H. T. (1994), *Ecological and General Systems: An Introduction to Systems Ecology*, University Press of Colorado p. 251.

[¶]Russel, R. (1988), The 50% principle: A Dow theorist views a critical indicator, *Barrons's* **68**, 69.

[‖]Rosenblad, C. F. (1949), Evaporating liquids on the liquid film principle using indirect heating, particularly strongly foaming liquids. U.S Patent **5** 624–531

**Menth, A., Miller, R., and Stucki, S., (1980) Electrolysis cell for water dissolution. U.S. patent 4, 312, 736.

[††]Hogan, C. J. (2000), Why the universe is just so, *Rev. Modern Phys.* **72**(4), 1149–1161.

[‡‡]Waterman, D. (1966), The local finite-area principle in the half-plane, *Proc. Am. Math. Soc.* **17**, 1012.

FINITENESS PRINCIPLE See CONDENSATION PRINCIPLE.

FIRENNCZI'S RELAXATION PRINCIPLE See RELAXATION PRINCIPLE.

FIRST DATA PROTECTION PRINCIPLE [mathematics]   Personal data shall be processed fairly and lawfully and, in particular, shall not be processed unless: a) The processing is necessary for the purposes of legitimate interests pursued by the data controller or by the third party or parties to whom the data are disclosed, except where the processing is unwarranted in any particular case by reason of prejudice to the rights and freedoms or legitimate interests of the data subject, and b) in the case of sensitive personal data.* *See also* COPY PROTECTION PRINCIPLE; PROTECTION PRINCIPLE; PRIVACY PRINCIPLE; SECOND DATA PROTECTION PRINCIPLE; THIRD DATA PROTECTION PRINCIPLE; FOURTH DATA PROTECTION PRINCIPLE; FIFTH DATA PROTECTION PRINCIPLE; SIXTH DATA PROTECTION PRINCIPLE; SEVENTH DATA PROTECTION PRINCIPLE; EIGHTH DATA PROTECTION PRINCIPLE.

FIRST-IN FIRST-OUT PRINCIPLE [engineering] As the name implies, first in a first out.

FIRST INVERSE MAXIMUM ENTROPY PRINCIPLE [engineering, mathematics]   A set of constraints are determined that will give rise to a given observed or theoretical probability distribution as Jaynes' maximum entropy probability distribution using Shannon entropy as the measure.† *See also* ENTROPY OPTIMIZATION PRINCIPLE.

FIRST INVERSE MINIMUM CROSS-ENTROPY PRINCIPLE [engineering, mathematics] Constraints to obtain a given probability distribution as a result of minimization of a specified generalized measure of cross-entropy from a specified a priori probability distribution is chosen.‡ See also ENTROPY OPTIMIZATION PRINCIPLE.

FIRST MINIMUM INTERDEPENDENCE PRINCIPLE [engineering, mathematics]   The joint probability distribution of variables minimizing the measure of interdependence subject to some constraints being satisfied. The constraints may require specifications of some marginal distributions and some joint moments.§ See also ENTROPY OPTIMIZATION PRINCIPLE; MINIMUM INTERDEPENDENCE PRINCIPLE.

FIRST PARADOX PRINCIPLE [psychology] Positive change requires significant stability.¶

FIRST PRINCIPLE [chemistry]   A powerful tool to study properties of clusters. Derivation of the high-latitude total electron content distribution.‖ See also FIRST-PRINCIPLE REAL-SPACE SIMULATION; FIRST-PRINCIPLE MOLECULAR DYNAMICS; FIRST-PRINCIPLE FORCES.

FIRST-PRINCIPLE MOLECULAR DYNAMICS. See FIRST PRINCIPLE.

FIRST PRINCIPLE OF ENERGETICS [physics] (Howard Thomas Odum, 1924–2002, American ecosystem ecologist)   The increase in internal energy of a system is equal to the amount of energy added to the system by heating, minus the amount lost in the form

---

*Univ. Edinburgh, Records Management Section (2007) (http://www.recordsmanagement.ed. ac.uk/InfoStaff/DPstaff/DP_Research/Research AnnexA.htm).
†Mohammad-Djafari, A. (2006), Maximum entropy and Bayesian inference: Where do we stand and where do we go? *AIP Conf. Proc.* 872 (*Bayesian Inference and Maximum Entropy Methods in Science and Engineering*), pp. 3–14.

‡Srikanth, M., Kesavan, H. K., and Roe, P. H. (2000), Probability density function estimation using the MinMax measure, *IEEE Trans. Syst. Man Cybernet. C Appl. Rev.* **30**(1), 77–83.
§Kapur, J. N. (1984), On minimum interdependence principle, *Indian J. Pure Appl. Math.* **15**(9), 968–977.
¶Price Waterhouse Change Integration Team (1995), *The Paradox Principles: How High Performance Companies Manage Chaos Complexity and Contradiction to Achieve Superior Results*.
‖Lei, Y., Cummins, K., and Lacks, D. J. (2003), First-principles enthalpy landscape analysis of structural recovery in glasses, *J. Polym. Sci. B Polym. Phys.* **41**(19), 2302–2306.

of work done by the system on its surroundings.* See also PRINCIPLES OF ENERGETICS; ZEROTH PRINCIPLE OF ENERGETICS; SECOND PRINCIPLE OF ENERGETICS; THIRD PRINCIPLE OF ENERGETICS; FOURTH PRINCIPLE OF ENERGETICS; MAXIMUM EMPOWERMENT PRINCIPLE; MAXIMUM POWER PRINCIPLE; FIFTH PRINCIPLE OF ENERGETICS; SIXTH PRINCIPLE OF ENERGETICS.

FIRST PRINCIPLE OF THERMODYNAMICS [physics]   Establishes the existence of a property called the *internal energy* of a system. A change in the state of a system can be brought about by a variety of techniques. Work and heat are modes of transferring energy. They are not forms of energy in their own right. *Work* is a mode of transfer that is equivalent (if not the case in actuality) to raising a weight in the surroundings. *Heat* is a mode of transfer arising from a difference in temperature between the system and its surroundings. What is commonly called heat is more correctly called the *thermal motion* of the molecules of a system. (4) The internal energy of an isolated system is conserved; that is, for a system to which no energy can be transferred by the agency of work or of heat, the internal energy remains constant. Implies the equivalence of heat and work for bringing about changes in the internal energy of a system (and heat is foreign to classical mechanics).[†] See also THERMODYNAMIC PRINCIPLES.

FISHER'S PRINCIPLE [mathematics]   (Sir Ronald Aylmer Fisher, 1890-1962, English Statistician)   (1) Explains why the sex ratio of most species is approximately 1:1.[‡] (2) (Dr. Gerhard Fisher, German, Univ. Dresden) "The value of capital is the discounted value of the expected income.... It is found by discounting (or 'capitalising') the value of the income expected from the wealth of property." The behavior of a population of competing firms that is elaborated in terms of Fisher's principle; the rate of change of the moments of this population distribution are functionally related to higher-order moments of the distribution. Different kinds of increasing returns are distinguished, and it is shown how they influence the dynamics of selection.[§] See also KALDOR–VERDOORN PRINCIPLE.

FISHEYE ADAPTATION PRINCIPLE See FITNESS PRINCIPLE.

FITNESS PRINCIPLE [engineering]   Also referred to as the *maximization principle*. Organisms tend to behave so as to maximize their inclusive fitness. This is the principle of sociobiology. *Fitness* refers exclusively to a measure of differential reproductive success.[¶]

FLAP-VALVE PRINCIPLE [engineering]   A hinged flap or disk swinging in only one direction.[‖]

FLEISSNER FLOW THROUGH PRINCIPLE [physics]   See FLOWTHROUGH PRINCIPLE.

FLEXIBILITY PRINCIPLE [organizational psychology]   Building flexibility in planning is beneficial, but cost of building flexibility needs to be evaluated against the benefits. See also PRINCIPLE OF PLANNING.[**]

*Odum, H. T. (2000), An energy hierarchy law for biogeochemical cycles, in *Energy Synthesis: Theory and Applications of the Energy Methodology*, Proc. *1st Biennial Energy Analysis Research Conf.* Brown, M. T., ed., Center for Environmental Policy, Univ. Florida, Gainesville.

[†]Sicard, L. (1967), First principle of thermodynamics, *Bull. Union Physiciens* **61**(493), 264–272.

[‡]Fisher, R. A. (1930), *The Genetical Theory of Natural Selection*, Clarendon Press, Oxford.

[§]Metcalfe, J. S. (1994), Competition, Fisher's principle and increasing returns in the selection process, *J. Evolut. Econo.* 4(4), 327–346.

[¶]Turner, J. H. (2006), *Handbook of Sociological Theory*, Springer, p. 413.

[‖]Lampel, A. et al. (1995), New continence mechanisms for continent urinary diversion based on the flap valve principle. A chronic animal study, *Aktuelle Urologie*, **26**, 103.

[**]Koontz, H. and O'Donnell, C. (1968), *Principles of Management: An Analysis of Managerial Functions*, 4th ed., McGraw-Hill, New York (http://knol.google.com/k/narayana-rao-kvss/principles-of-management-koontz-and/2utb2lsm2k7a/89).

FLOTATION POLISHING PRINCIPLE [engineering]   Employed in the beneficiation of ores. Flotation is based on introducing chemicals and the air into water containing solid particles of different suspended materials, which causes adherence of air to certain suspended solids and makes the particles having air bubbles adhere thereto more lightly than the water. Accordingly, they rise to the top of the water to form a froth, which is skimmed off. Applied in a number of mineral separation processes.* See also PRINCIPLE OF CONSERVATION OF TOTAL VORTICITY; PRINCIPLE OF FLUID ORIFICING; PRINCIPLE OF VIRTUAL WORK; FLOTATION PRINCIPLE.

FLOTATION PRINCIPLE [engineering]   Applied in a number of mineral separation processes, including the selective separation of such minerals as sulfide copper minerals, sulfide lead mineral, sulfide zinc mineral, sulfide molybdenum mineral, and other sulfides from sulfide iron minerals.[†] See also PRINCIPLE OF CONSERVATION OF TOTAL VORTICITY; PRINCIPLE OF FLUID ORIFICING; PRINCIPLE OF VIRTUAL WORK; FLOTATION POLISHING PRINCIPLE.

FLOW-ACTING   PRINCIPLE [engineering] Maps system fluctuations (perturbations acting on an initial state) with dissipations (transitions to different states), thus directing the energy flow along competing reactive and non—reactive pathways.[‡]

FLOW   INJECTION   ANALYZER   PRINCIPLE [chemistry]   When coupled with flowthrough detectors and a plotter, may be used to monitor changes in concentration of dissolved substances in both laboratory and industrial situations.[§]

FLOW INJECTION CHEMILUMINESCENCE PRINCIPLE [chemistry]   Model gas is analyzed for its content. The difference of the concentration yields the concentration formed by air oxidation.[¶] See also CHEMILUMINESCENCE PRINCIPLE.

FLOW   INJECTION   PRINCIPLE [chemistry] Automated   system   for   flow   injection analysis.[‖]

FLOWTHROUGH PRINCIPLE [physics]   Drying with uniform heat transfer applicable to thermal bonding, cooling, heat setting, shrinking, and curing.** See also FLEISSNER FLOWTHROUGH PRINCIPLE.

FLUID   KINETICS   PRINCIPLE [engineering] Branch of mechanics dealing with the properties and behavior of fluids, namely, liquids and gasses. Because of their ability to flow, liquids and gases have properties both in common with and not shared by solids. Studies of fluids in motion, or fluid dynamics, make up a large part of fluid mechanics. Hydrodynamics   and   aerodynamics   are branches of fluid dynamics. A plasma is also a fluid and can be described by many of the principles of fluid mechanics, but the electromagnetic properties must also be taken into account. Plasma in motion is known as *magnetohydrodynamics*.[††]

FLUORESCENCE   POLARIZATION   IMMUNOASSAY PRINCIPLE [physics]   See RADIOIMMUNE ASSAY PRINCIPLE.

*Rogers, J., Sutherland, K. L., Wark, E. E., and Wark, I. W. (1946), Principles of flotation-paraffin-chain salts as flotation reagents, *Am. Inst. Mining Met. Eng. Mining Technol.* **10**(4), 30 (Tech. Publ. 2022).

[†]Stechemesser, H. (1989), New findings concerning flotation principles, *Freiberger Forsch. A.* **790**, 81–100.

[‡]Norris, J. R. and Ribbons, D. W. (1970), *Methods in Microbiology*, Academic Press, p. 236.

[§]Wilson, K. and Walker, J., eds., (2005), *Principles and Techniques of Biochemistry and Molecular Biology*, Cambridge Uni. Press.

[¶]Burguera, J. L. and Burguera, M. (1983), The principles, applications and trends of flow injection analysis for monitoring chemiluminescence reactions, *Acta Cientifica Venezolana*, **34**(2), 79–83.

[‖]Lee, G. P., Chun, H. G., and Lee, G. W. (1995), Principle and application of flow injection analysis, *Anal. Sci. Technol.* **8**(2), 23a–38a.

**Hendricks, R. C. and Sengers, J. V. (1979), Application of the principle of similarity to fluid mechanics, *Proc. 9th. Int. Conf. Properties of Steam, Water Steam: Their Properties and Current Industrial Applications*, 1979, pp. 322–335.

[††]Chen, S., Zhou, N., and Yin, Z. (1996), Fluid kinetics principle and design criteria of atomizer, *Trans. Nonferr. Metals Soc. China* **6**(3), 108–112.

FLUX GATE PRINCIPLE [engineering] (1) Magnetic reproducing head in which magnetic flux is modulated by high-frequency saturating magnetic flux in another part of the magnetic circuit.* (2) Detector that produces an electric signal whose magnitude and phase are proportional to the magnitude and direction of the external magnetic field setting along its axis. Used to indicate the direction of the terrestrial magnetic field.[†]

FLUX INJECTION PRINCIPLE [engineering] (1) Process in which fluxing compounds are introduced into the molten metal by a mechanical device using an inert-gas carrier. (2) At both entrance and outlet, the amounts of tracer in particular flow lines are proportional to their volumetric flow rates. (3) Double-sided nonenwrapping both with and against power-loss testing system.[‡]

FLYWHEEL PRINCIPLE [engineering] Affords the most common and effectual method of equalizing motion. Balanced on its axis, and connected with the machinery so as to turn rapidly around with it and receiving a constant impulse from the moving power, it becomes a magazine or repository of motion.[§]

FO₂ PRINCIPLES [chemistry] Application of dyes such as jig dyeing with textiles.[¶]

FOIL ELECTRIC PRINCIPLE [engineering] See ELECTROSTATIC PRINCIPLE.

*Walker, P. M. B. (1999), *Chambers Dictionary of Science and Technology*, Chambers, New York, p. 463.
†Castellano, M. G. and Chiarello, F. (2002), Superconducting devices for quantum logic gates using magnetic flux states, *Proc. Int. School of Physics, "Enrico Fermi," 148th (Experimental Quantum Computation and Information)*, pp. 493–510.
‡Iranmanesh, H., Beckley, P., and Moses, A. J. (1992), Investigation of a double-sided non-enwrapping "with" and "against" power-loss testing system based on the "flux injection" principle, *J. Magn. Magn. Mater.* **112**(1–3), 74–76.
§Olmsted, D. (1858), *An Introduction to Natural Philosophy: Designed as a Text-book, for the Use the of Students [i]n Yale College* (compiled from various authorities), p. 232.
¶Lindsley, D. H., Speidel, D. H., and Nafziger, R. H., (1968), P-T- fo2 relations for the system iron-oxygen-silicon dioxide, *Am. J. Sci.* **266**(5), 342–360.

FOKKER ACTION PRINCIPLE [physics] (Anton Herman Gerard Fokker, 1890-1939, Dutch American aircraft manufacturer) Application for a system of particles interacting through a linear potential as part of a general relativity theory.[‖]

FORCE BALANCE PRINCIPLE [physics] The inertial force produced deflects the mass from its equilibrium position, and the displacement or velocity of the mass is then converted into an electric signal.[**]

FOREST PRINCIPLE [ecology] Global consensus on the management, conservation, and sustainable development of all types of forest produced at the 1992 United Nations Conference on Environment and Development (UNCED), informally known as the "Earth Summit."[††]

FOREST STEWARDSHIP COUNCIL (FSC) PRINCIPLES. [ecology] An international non-profit, multi-stakeholder organization established in 1993 to promote responsible management of the world's forests. Its main tools for achieving this are standard setting, independent certification and labeling of forest products. This offers customers around the world the ability to choose products from socially and environmentally responsible forestry. See also HANNOVER PRINCIPLES; FOREST STEARDSHIP COUNCIL (FSC) PRINCIPLES; GLOBAL SULLIVAN PRINCIPLES OF SOCIAL RESPONSIBILITY; MARINE STEWARDSHIP COUNCIL (MSC) PRINCIPLES; PERMACULTURE PRINCIPLES; THE BELLAGIO PRINCIPLES FOR ASSESSMENT; EQUATOR PRINCIPLES; MELBOURNE PRINCIPLES; PRECAUTIONARY PRINCIPLE; SANBORN PRINCIPLES;

‖Rivacoba, A. (1984), Fokker- action principle for a system of particles interacting through a linear potential, *Nuovo Cimento Soc. Ital. Fis. B Gen. Phys. Relat. Astron. Plasmas* **84B**(1), 35–42.
Stuart-Watson, D., and Tapson, J. (2004), Simple force balance accelerometer/seismometer based on a tuning fork displacement sensor, *Rev. Sci. Instrum.* **75(9), 3045–3049.
††Anon. (1989), Act No. 37 of 3 April 1989 setting forth the principles for structuring the National Forest Development Plan and creating a Forest Service, *Annu. Rev. Popul. Law* **16**, p. 214.

TODDS' PRINCIPLES OF ECOLOGICAL DESIGN; FOR-REST PRINCIPLE.*

FORM-AND-FUNCTION   PRINCIPLE [psychology, engineering]   (1) The potential use incorporating breakpoint distance following design guidelines to individual patches, based on biogeography. (2) A principle with multiple applications. Examples include information on a study that demonstrates how to treat the operations of fuzzy numbers with step form, variables for backorder inventory models, and basis for calculation of a dropper for infusion apparatuses.[†] See also DUAL-FUNCTION PRINCIPLE; FUNCTION PRINCIPLE; FUNCTIONING PRINCIPLE.

FORMAL PRINCIPLE [psychology]   (1) Sensuous activity accompanies the perception of an object, and the pleasure afforded by this activity is found to be quite different from pleasure in the content as such. The principle of the form impulse (Formtrieb) is to be sought in this perceptive activity. The pleasure felt is the more keen, the more energetic and at the same time unconscious of painful effort the activity is. Aesthetic pleasure on its formal side means, therefore, the natural and harmonious functioning of the perceptive organs forced on us by the structural aspect of an object. There are three sources of formal aesthetic pleasure experience: (1) seeing, (2) hearing, and (3) the activity of what the writer calls our "power of representation through speech". The writer distinguishes life and art. Life has variety; art alone has unity. In art the parts maintain distinction through contrast, tension, and dissonance; these in discreet measure. Out of *unity in variety* arises the demand for clear and comprehensive structure. Art is said to consist of such choice from what is called *Wirklichkeit* (actual life) as allows for design within comprehensible limits. *Adequacy of expression* is the weightiest of all determinations -of the ideal form of the beautiful. In the demand for adequate expression,

form and content become inseparable, two points of view in the same thing. Adequacy of expression does not exclude a certain measure of ugliness, both formal and material. Ugliness is also a source of formal pleasure if, without painful effort, art is able to mold it into harmony with the total. Mr. Meyer concludes that in art that is most beautiful that achieves the desired effect with the least cost. This he calls the *principle of least Kraftmass*.[‡] (2) The *principle of invariance* states that changes in the descriptions of outcomes should not alter one's preference order. Invariance is such an obvious principle that many accounts attempting to formalize the rules of rational decisionmaking use it implicitly, without stating it explicitly. The *principle of dominance* holds that we should choose an option if (a) whatever else happens, that option never turns out to be any worse than any of the other options, and (b) it is possible that that option will turn out better than the other options. The *sunk-cost principle* states that choices should be future-oriented. Because our decisions affect the future but not the past, they should be based on the consequences of the actions, and not on what has already happened. Normative standards of judgment and decisionmaking are often defined in terms of formal, abstract rules such as invariance, dominance, and the sunk-cost principle. These rules depend primarily on the structure of the situation, not on all of the particular substantive details, and they are thought to apply to any situation with the proper structure. Taken as exceptionless principles, they are often used as the standard against which individual behavior should be judged. Even if principles such as dominance or invariance seem completely uncontroversial, there is often a problem in determining when they apply. There are many situations that closely resemble a case in which a normative principle is thought to apply even if it nonetheless seems clear that the principle should not be followed. The *principle* of *"leaky" rationality* indicates that how research on behavioral

*Forest Stewardship Council (1993) http://www.fsc-info.org

[†]Bogaert, J., Salvador-Van Eysenrode, D., Impens, I., and Van Hecke, P. (2001), The interior-to-edge breakpoint distance as a guideline for nature conservation policy, *Environ. Manage.* **27**(4), 493−500.

[‡]Blunt, A. C. (1905), Review of Formprinzip des Schönen (preview), *Psychol. Bull.* **2**(9), 311−312.

decisionmaking challenges normative standards of rationality.* (3) The authority that forms or shapes the doctrinal system of a religion, religious movement, or tradition or a religious body or organization. Tend to be texts or revered leaders of the religion or tradition.[†] See also PRINCIPLE OF THE FORM-IMPULSE, PRINCIPLE OF INVARIANCE; PRINCIPLE OF DOMINANCE; SUNK-COST PRINCIPLE.

FORMULATION PRINCIPLE [chemistry]   Particular mixture of base chemicals and additives required for a product.[‡]

FORTESQUE   PRINCIPLE [engineering] Technique of magnetic stimulation from a pulsed induction coil and the enhanced electromyogram response from hyperthenar muscles obtained when electric and magnetic stimuli are applied simultaneously.[§]

FORWARD CHAINING PRINCIPLE [psychology, computers]   An inference method used in expert systems where the "if" portion of rules are matched against facts to establish new facts.[¶] See BACWARD CHAINING PRINCIPLE; CHAINING PRINCIPLE.

FOUNDER PRINCIPLE [genetics]   The conditional probabilities of the frequencies of a set of genes at any future date depend on the initial composition of the founders of the population and have in general no tendency to revert to the composition of the population from which the founders were themselves derived.[‖]

FOURIER   PRINCIPLE [physics]   (Joseph fourier, 1768-1830, French mathematician) (1) All repeating waveforms can be resolved into sine-wave components consisting of a fundamental and a series of harmonics at multiples of this frequency. It can be extended to prove that non-repeating wveforms occupy a continuous frequency spectrum.[**]

FOURIER TRANSFORM PRINCIPLE [mathematics, physics] (Joseph Fourier, 1768-1830, French mathematician)   A mathematical relation between the energy in a transient and that in a continous energy spectrum of adjacent component frequencies.[*]

FOURIER TRANSFORM INFRARED PHOTOA-COUSTIC SPECTROSCOPY (FTIR) PRINCIPLE [physics] (Joseph Fourier, 1768-1830, French mathematician)   A form of infrared spectroscopy involving interferometric methods to give enhanced resolution; spectra are produced by Fourier transformation of the interferometer output data.[‖]

FOUR-POINT-ONE, EZRIN, RADIXIN, MOESIN (FERM) PRINCIPLE. (Four-point-one, Ezrin, Radixin, Moesin) [biology]   Responsible for membrane binding, such as that involved in the linkage of cytoplasmic proteins to the membrane.[††] See also FERM PRINCIPLE.

FOURTH   DATA   PROTECTION   PRINCIPLE [mathematics]   Personal data shall be accurate and, where necessary, kept up-to-date.[‡‡] See also COPY PROTECTION PRINCIPLE; PRIVACY PRINCIPLE; FIRST DATA PROTECTION PRINCIPLE; PROTECTION PRINCIPLE; SECOND DATA PROTECTION PRINCIPLE; THIRD DATA PROTECTION PRINCIPLE; FIFTH DATA PROTECTION PRINCIPLE; SIXTH DATA PROTECTION PRINCIPLE;

*Keys, D. J. and Schwartz, B. (2007), Perspect. *Psychol. Sci.* **2**(2), 162–180.

[†]Icon Group International (2008), *Teachings: Webster's Quotations, Facts and Phrases*, ICON Group International, Inc., p. 484.

[‡]Soerensen, P. (1980), Formulation principles, *Faerg Lack Scand.* **26**(12), p. 221–223, 230, 232–233.

[§]Fortesque, P., Stark, J., eds. (1992), *Model of a Solar Cell (GaAS, Si, InP), Spacecraft Systems Engineering*, Wiley, New York.

[¶]Moutinho, L., Curry, B., and Rita, P. (1996), *Expert Systems in Tourism Marketing*, Routledge, p. 46.

[‖]Wasserman, M. (1990), Genetics, speciation and the founder principle, *BioScience* **40**, 788.

**Walker, P. M. B. (1999), *Chambers Dictionary of Science and Technology*, Chambers, New York, p. 473.

[††]Li, Q., R., Kulikauskas, R., Nyberg, K., Fehon, R. Karplus, P. A., Bretscher, A., and Tesmer, J. J. G. (2007), Self-masking in an intact ERM-merlin protein: An active role for the central a-helical domain, *J. Mol. Biol.* **365**(5), 1446–1459.

[‡‡]Uni. Edinburgh, Records Management Section (2007) (http://www.recordsmanagement.ed.ac.uk/InfoStaff/DPstaff/DP_Research/ResearchAnnexA.htm).

SEVENTH DATA PROTECTION PRINCIPLE; EIGHTH DATA PROTECTION PRINCIPLE.

FOURTH PARADOX PRINCIPLE [psychology]   True empowerment requires forceful leadership.*

FOURTH PRINCIPLE OF ENERGETICS [biology] (Howard Thomas Odum, 1924–2002, American ecosystem ecologist) Describes the propensities of evolutionary self-organization. The Onsager reciprocal relations are sometimes called the *fourth law of thermodynamics*. As the fourth law of thermodynamics, Onsager reciprocal relations would constitute the fourth principle of energetics. In the field of ecological energetics H. T. Odum considered maximum power, the fourth principle of energetics. Odum also proposed the maximum empowerment principle as a corollary of the maximum power principle, and considered it to describe the propensities of evolutionary self-organization.† See also PRINCIPLES OF ENERGETICS; ZEROTH PRINCIPLE OF ENERGETICS; SECOND PRINCIPLE OF ENERGETICS; THIRD PRINCIPLE OF ENERGETICS; FIRST PRINCIPLE OF ENERGETICS; MAXIMUM EMPOWER PRINCIPLE; MAXIMUM POWER PRINCIPLE; FIFTH PRINCIPLE OF ENERGETICS; SIXTH PRINCIPLE OF ENERGETICS.

FRACTIONATION PRINCIPLE [radiology, chemistry] (1) A system of treatment commonly used in radiotherapy in which doses are given daily or at longer intervals over a period of 3-6 weeks.‡ (2) In treating the mixture to be separated with the inert gas, the gas stream is loaded with a portion of the mixture of substances; the composition of this portion is normally different from that of the starting mixture. Since the compounds of higher volatility are more readily taken up, more of them than the less volatile compounds will move into the inert-gas stream. Thus, the mixture of material entrained by the loaded gas is enriched with components more readily taken up by the gas. If this mixture of substances is released and again contacted with an inert-gas stream under supercritical conditions, the same phenomenon will occur; the loaded gas stream will now be withdrawn again enriched with a still higher concentration of the component of higher volatility as compared with the starting mixture. With repeating loading and releasing, the component of higher volatility becomes increasingly concentrated in the gas stream recoverable in an increasingly pure state.§

FRAEIJS DE VEUBEKE PRINCIPLE [mathematics, engineering] (Eric Reissner, b. 1/5/1913, Aachen, Germany)   Baudouin M. Fraeijs de Veubeke, 1917-1976, French engineer See DISPLACEMENT PRINCIPLE.

FRANCK–CONDON PRINCIPLE [physics, chemistry] (James, Frank, 1882-1964, German-born American physicist; Edward Uhler Condon, 1902-1974, American physicist) (1) An electronic transition takes place so rapidly that a vibrating molecule does not change in internuclear distance appreciably during the transition.¶ (2) Applied quantum mechanics to understand the atom and its nucleus; in ionization, the extreme form of excitation in which an electron is ejected, leaving a positive ion. Minimum energy required for this process is called the *ionization potential*.‖ (3) In any molecular system the transition from one energy state to another is so rapid that the nuclei of the atoms involved can be considered to be

---

*Kudriashov, Yu. B. (2001), Main principles of radiobiology, *Radiat. Biologiia, Radioecol./Rossiiskaia Akad. Nauk*, **40**(5), 531–547.
†Kudryashov, Yu. B. (2004), The main principles of radiation biology, *Effects Low Dose Radiat.* pp. 362–395.
‡Walker, P. M. B. (1999), Chambers Dictionary of Science and Technology, Chambers, New York, p. 473.

§Wallach, D. F. H. Kranz, B. Ferber, E., and Fischer, H. (1972), Affinity density perturbation. New Fractionation principle and its illustration in a membrane separation, *FEBS Lett.*, **21**(1), 29–33.
¶Walker, P. M. B. (1999), *Chambers Dictionary of Science and Technology*, Chambers, New York, p. 476.
‖Turro, N. J. (1991), *Modern Molecular Photochemistry*, University Science Books, p. 32.

stationary during the transition.* (4) Interpretation and analysis of bound free transitions in diatomic molecules. In fast reactions such as electron transfer, there is no time for the geometry to change.[†] (5) The nuclei in a molecule remain essentially stationary while an electronic transition is taking place. The physical interpretation rests on the fact the electrons move much more rapidly than do the nuclei because of their much smaller mass.[‡]

FRANK–STARLING   PRINCIPLE [medicine] (Otto Frank, 1865-1944, German physiologist; Ernest Henry Starling, 1866-1927, British physiologist) (1) A decrease in the ability to empty the ventricle during systole increases the tension on the noninjured parts of the heart during stroke; the ventricle responds to this increase in diastolic tension (preload) by enhancing its contraction. When sarcomeres are stretched to their limits because of progressive ventricular dilatation, increases in preload do not enhance systolic ejection, so the Frank—Starling curve becomes both depressed and flattened. (2) Left ventricular function influenced by aortic pressure and venous filling pressure—stroke work increases to a certain maximum and then decreases. (3) Assessment of the ventricular function based on measurement of cardiac output, which depends on preload, afterload, and contractility. The pressure–volume loop can be constructed by measurement of pressure and volume.[§]

FREE-MARKET PRINCIPLES [chemistry, mathematics]   (1) An array of exchanges that take place in society. Each exchange is undertaken as a voluntary agreement between two people or between groups of people represented by agents. These two individuals (or agents) exchange two economic goods, either tangible commodities or nontangible services. An economic market in which supply and demand are not regulated or are regulated with only minor restrictions. Economic system that allows supply and demand to regulate prices, wages, and similar variables, rather than government policy. Places a premium on innovation and effective marketing.[¶] (2) In ionization, the extreme form of excitation in which that an electron is ejected, leaving a positive ion. Minimum energy required for this process is called the *ionization potential*.[||]

FREGE'S   CONTEXT   PRINCIPLE [psychology] (Friedrich Ludwig Gottlob Free, 1848-1925, German   mathematician) A   word   has meaning only as a constituent of a sentence (alternatively, that a concept contributes a content to a thought only as a constituent or a part of such a thought). Expressions can be identified with the items to which a definite meaning has been given only in a sentence which does make sense.[**] See also COMPOSITIONALITY PRINCIPLE; FREGE'S PRINCIPLE; CONTEXT PRINCPLE.

FREGE'S   PRINCIPLE [mathematics,   psychology] (Friedrich Ludwig Gottlob Frege, 1848–1925,   German   mathematician) (1) An easily implemented system for analyzing a sentence's meaning. The meaning of a complex expression is determined by the meanings of its constituent expressions and the rules used to combine them. In a meaningful sentence, if the lexical parts are taken out of the sentence, what remains will be the rules of composition. Every operation of the syntax should be associated with an operation of the semantics that acts on the

---

*Franck, J. (1926), Elementary processes of photochemical reactions, *Trans. Faraday Soc.* **21**, 536–542.
[†]Ortenberg, F. S. and Antropov, E. T. (1967), Probability of electron-vibrational transitions in diatomic molecules, *Sov. Phys. Usp.* **(9)**, 717–742.
[‡]*CRC Handbook of Chemistry and Physics* (2007), 88th ed., CRC Press, Cleveland, OH.
[§]Smirnov, A. D., Loviagin, E. V., and Fedosenko L. I. (1985), Disorder of the Frank-Starling principle in diseases of the circulatory system, *Fiziologiia cheloveka* **11**(1), 107–112.

[¶]Rothbard, M. N. (2008), Frege market, in *Concise   Encyclopedia   of   Economics* (http://www.econlib.org/library/Enc/FreeMarket.html).
[||]Campbell, E. E. B. and Levine, R. D. (2000), Delayed ionization and fragmentation en route to thermionic emission: Statistics and dynamics *Annu. Rev. Phys. Chem.* **1**(51), 65–98.
[**]Acero, J. J. (2007), Nonsense and the privacy   of   sensation. *Sorites* **18**, 33–55. (http://www.sorites.org/Issue_18/acero.htm).

meanings of the constituents combined by the syntactic operation. As a guideline for constructing semantic theories, this is generally taken to mean that every construct of the syntax should be associated with a clause of the T schema with an operator in the semantics that specifies how the meaning of the whole expression is built from constituents combined by the syntactic rule.* (2) In some general mathematical theories (especially those in the tradition of Montague grammar) the interpretation of a language is essentially given by a homomorphism between an algebra of syntactic representations and an algebra of semantic objects.[†] See also PRINCIPLE OF COMPOSITIONALITY.

FREGEAN FUNDAMENTAL PRINCIPLE [psychology] (Friedrich Ludwig Gottlob Frege, 1848-1925, German mathematician)   The sentences "S believes that a is F" and "S believes that b is F" can differ in truth value even if $a = b$. Crucial premise in the traditional Fregean argument for the existence of semantically relevant senses, individuative elements of beliefs that are sensitive to our varying conceptions of what the beliefs are about. Arguing for Frege's fundamental principle.[‡]

FREQUENCY-DEPENDENT PRINCIPLE [engineering]   The effects of all quasiparticles on the frequency-dependent conductivity originate from phase pinning at random impurities. This principle also has implications for the transport relaxation time, which is relevant for the onset of DW-density wave sliding conduction. It is shown that the transport relaxation time is strongly temperaturé and frequency-dependent and suppresses sliding at low temperatures.[§]

FREQUENCY DISTANCE PRINCIPLE (FDP) [engineering]   Depth-dependent collimator response correction to improve spatial resolutions.[¶]

FREQUENCY DIVISION MULTIPLE ACCESS (FDMA) PRINCIPLE [physics]   (1) Within a radio cell, the base station defines on the "radio interface" to the mobile stations communicating with it the time division multiplex frame and thus the relative timing of the timeslots that are allocated in a link-specific manner to the information interchange in the frequency bands. With regard to the frame clock cycle phase used in operation, the base stations are in principle independent of one another.[‖] (2) Telecommunication apparatus and second telecommunication apparatus following the first telecommunication connections, taking into account the items of information in the first timeframes. Each further telecommunication connection following the first one is set up with priority on the same FDMA frequency. These named telecommunication systems thus also include, for example, the systems often designated third system generation in the context of a universal mobile telecommunication, probably based on FDMA, time division/domain multiple access (TDMA) and code division multiple access (CDMA) principles.[**] See also TIME DIVISION MULTIPLE ACCESS (TDMA) PRINCIPLE; CODE DIVISION MULTIPLE ACCESS (CDMA) PRINCIPLE.

FREQUENCY DIVISION MULTIPLEX [telecommunication, engineering]   (1) A method of multiplex transmission in which individual

*Linton, K. and Bouffard, N. (2004), *Computing Frege's Principle of Compositionality: An Experiment in Computational Semantics*, Carleton Univ. Cognitive Science Technical Report 2004–07 (http://www.carleton.ca/ics/techReports.html).
[†]Szabó, Z. G. (2007) Compositionality. *Stanford Encyclopedia of Philosophy*.
[‡]Frances, B. (1998), *Mind Lang.* **13**(3), 341–346.
[§]Borsdorf, H. and Eiceman, G. A. (2006), Ion mobility spectrometry: Principles and applications, *Appl. Spectrosc. Rev.* **41**(4), 323–375.

[¶]Zeng Gengsheng, L. (2007), Uniform attenuation correction using the frequency-distance principle, *Med. Phys.* **34**(11), 4281–4284.
[‖]Takasago, K., Takekawa, M., Shirakawa, A., and Kannari, F. (2000), Spatial-phase code-division multiple-access system with multiplexed Fourier holography switching for reconfigurable optical interconnection, *Appl. Optics* **39**(14), 2278–2286.
Cosmas, J., Evans, B., Evci, C., Herzig, W., Persson, H., Pettifor, J., Polese, P., Rheinschmitt, R., and Samukic, A. (1995), Overview of the mobile communications programme of RACE II, *Electron. Commun. Eng. J.* **7(4),155–167.

speech or information channels are modulated to separate channels and are then transmitted simultaneously over a cable or microwave link.* (2) Realized using flip-flop circuits. Each binary output of the counter is in a fixed ratio to the frequency of the digital signal, which is transmitted to the counter input. The frequency of the least-significant (rightmost) binary position in the known counters is half as great as the frequency of the digital input signal whose pulses are counted. To increase the counting range, one has to fall back each time on a counter with several binary positions.[†]

FREQUENCY–TEMPERATURE SUPERPOSITION (FTS) PRINCIPLE [physics]   See TIME–TEMPERATURE SUPERPOSITION PRINCIPLE.

FRESNEL–ARAGO PRINCIPLE [optics]   The three laws concerning the condition of interference of polarized light. (1) Two rays of light emanating from the same polarized beam, and polarized in the same plane, interfere in the same was as ordinary light. (2) Two rays of light emanating from the same polarized beam and polarized at right angles to each other will interfere only if they are brought into the same plane of polarization. (3) Two rays of light polarized at right angles and emanating from ordinary light will not interfere if brought into the same plane of polarization.[‡]

FREUDIAN FRUSTRATION PRINCIPLE See FRUSTRATION PRINCIPLE.

FRICTION WHEEL DRIVE PRINCIPLE [engineering]   (1) Friction can be used to convert the mechanical energy of a moving object into heat energy. (2) To achieve very high or very low speed. (3) With linear drives or with friction drives, the motor drives the

cable that connects the elevator car and the counterbalance over a return pulley. The safety device must brake the elevator car from a certain overspeed and smoothly bring the car to a stop.[§] See also BUCKING MAGNET PRINCIPLE; LINEAR MAGNET PRINCIPLE; MAGNET PRINCIPLE; PLUNGER MAGNET PRINCIPLE; LOCK MAGNET PRINCIPLE; CERAMIC MAGNET PRINCIPLE.

FRIEDEL–CRAFTS ACYLATION PRINCIPLE [chemistry] (Charles Friedel, 1832–1899, chemist; James Mason Crafts, 1839–1917, American Chemist)   Acids providing both sulfonating agent and catalyst.[¶]

FRIEDEL–CRAFTS PRINCIPLE [chemistry] Process used for production of high-octane gasoline, synthetic rubber, plastics, and detergents. Any organic reaction brought about by the catalytic action of anhydrous aluminum chloride or related catalysts.[‡]

FRIEDEN–SOFFER PRINCIPLE [physics]   (B. Roy Frieden, professor emeritus American Mathematical Physicist B. H. Soffer z.p.) Ability of translating a statistical problem into a dynamical one with the transformation $f > v$(nonlinear) and $x > t$ (linear). The physical environment can always be characterized by an intrinsic information.[‖]

FRUSTRATION PRINCIPLE [psychology]   All organisms are subjected to various selection pressures at once, and they evolve structures or behavior that inevitably compromise solutions to the conflicting needs. Sometimes there are several such solutions successful enough to give a survival edge.** (2) The process of healthy personality development

*Walker, P. M. B. (1999), *Chambers Dictionary of Science and Technology*, Chambers, New York, p. 479.

†Shen, Z., Zhao, J., and Zhang, X. (2007), Frequency-division multiplexing technique of fiber grating Fabry-Perot sensors, *Guangxue Xuebao*, **27**(7), 1173–1177.

‡*McGraw-Hill Dictionary of Scientific and Technical Terms* (2003), McGraw-Hill, New York.

§Weinberger, K., Silberhorn, G., and Rennetaud, J. M. (2000), Linear Drive for Transportation Equipment, US Patent 6,053,287.

¶Martens, J., Praefcke, K., and Schulze, U. (1976), Organic photochemistry. Organic sulfur compounds. Photo-Friedel-Crafts reactions; a new synthetic principle for heterocyclic compounds, *Synthesis* **8**, 532–533.

‖Hernandex, S. and Clark, J. W. (2001), *Condensed Matter Theories*, Nova Publishers, p. 428.

**Niklas, K. J. (1997), *The Evolutionary Biology of Plants*, Univ. Chicago Press.

is guided by repeated experiences of frustration and gratification. Provides the ideal set of circumstances for ego development. Children who are never appropriately frustrated will be crippled in adulthood when confronted with the inevitable frustrations and disappointments of daily life. On the other hand, children who are exposed to levels of frustration that are beyond their developmental capacities to tolerate, especially when these levels reach the plane of traumatic experiences, are also crippled psychologically in adulthood.[*]

FTIR PRINCIPLE [biology]   See FOURIER TRANSFORM INFRARED PHOTOACOUSTIC SPECTROSCOPY (FTIR) PRINCIPLE.

FTIR-PAS PRINCIPLE [biology]   See FOURIER TRANSFORM INFRARED PHOTOACOUSTIC SPECTROSCOPY (FTIR) PRINCIPLE.

FUBINI PRINCIPLE [mathematics]   (1) If the average number of envelopes per pigeonhole is A, then some pigeonhole will have at least A envelopes.[†] (2) The Fubini Principle is a one-sided version of the Fubini Theorem, which will hold for C^1 fibre spaces. A subset S of a C^1 fibre space B with base space X, fibre Y, and projection $\pi : B => X$, is B-null if and only if for almost every $x \in X$ the set $\pi^{-1}(x)$ is a Y-null set.[‡] See also PIGEONHOLE PRINCIPLE; DIRICHLET'S PRINCIPLE

FUHRER PRINCIPLE [psychology]   Refers to a system with a hierarchy of male leaders that resembles a military structure. Sees each organization as a hierarchy of male leaders, where every leader (Führer, in German) has absolute responsibility in his own area, demands absolute obedience from those below him, and answers only to his superiors.[§]

FUNCTION PRINCIPLE [mathematics]   A principle with multiple applications. Examples include information on a study that demonstrates how to treat the operations of fuzzy numbers with step form, variables for backorder inventory models, and basis for calculation of a dropper for infusion apparatuses.[¶] See also DUAL-FUNCTION PRINCIPLE; FORM-AND-FUNCTION PRINCIPLE; FUNCTIONING PRINCIPLE.

FUNCTIONAL PRINCIPLE [psychology]   Concept of specialization of duties, not authority.[‖]

FUNCTIONING PRINCIPLE [mathematics]   See FUNCTION PRINCIPLE.[**]

FUNDAMENTAL PRINCIPLE [law]   The basis underlying the formulation of jurisprudence. The basis from which other truths can be derived; "first you must learn the fundamentals"; "let's get down to basics" [††]

FUNDAMENTAL PRINCIPLE [mathematics]   (1) Adopted by the UN Economic Commiion for Europe in 1992, following a pioneering proposal by the Conference of European Statisticians. Intended to guide governments and statisticians in establishing and maintaining credible national statistical systems, free from improper political influence. (2) Official statistics provide an indispensable element in the information system of a democratic society, serving the government, the economy, and the public with data about economic, demographic, social, and environmental situations. (3) Need to decide according to strictly professional considerations, including scientific principles

[*]Dyer, F. J. (1999), *Psychological Consultation in Parental Rights Cases*, Guilford Press, p. 156.
[†]Penev, K. (2008) The Fubini principle, Amer. Math. Monthly **115** (3) 245–248.
[‡]Sacksteder, R (1963) A remark on equipartitioning and metric transitivity. Proceedings of the American Mathematical Society, **14** (1) 153–157.
[§]Wilkinson, H. (1999), Editorial: Psychotheraphy, fascism and constitutional history, *Int. J. Psychother.* **4**(2), 117.

[¶]Hsieh, C. H. (2002), Optimization of fuzzy production inventory models, *Inform. Sci.* **146**(1–4), 29–40.
[‖]Popov, V. V. (1967), Functional principle in the physiology of development, *Usp. Sovremennoi Biol.* **64**(2), 294–311.
[**]American Heritage (2000), *American Heritage Dictionary of the English Language*, 4 ed., Houghton Mifflin, New York.
[††]Holt, T. (1998), The fundamental principles and the impact of using statistics for administrative purposes and the national statistical institute, *Statist. J. UN. Econ. Commiss. Eur.* (IOS Press). **15**(3–4), 203–212.

and professional ethics, on the methods and procedures for the collection, processing, storage, and presentation of statistical data. (4) To facilitate a correct interpretation of the data, the statistical agencies are to present information according to scientific standards on the sources, methods, and procedures of the statistics. (5) The statistical agencies are entitled to comment on erroneous interpretation and misuse of statistics. (6) Data for statistical purposes may be drawn from all types of sources, be they statistical surveys or administrative records. Statistical agencies are to choose the source with regard to quality, timeliness, cost, and the burden on respondents. (7) Individual data collected by statistical agencies for statistical compilation, whether they refer to natural or legal persons, are to be strictly confidential and used exclusively for statistical purposes. (8) The laws, regulations, and measures under which the statistical systems operate are to be made public. (9) Coordination among statistical agencies within countries is essential to achieve consistency and efficiency in the statistical system. (10) The use by statistical agencies in each country of international concepts, classifications, and methods promotes the consistency and efficiency of statistical systems at all levels.* See also DUAL-FUNCTION PRINCIPLE; FORM-AND-FUNCTION PRINCIPLE; FUNCTION PRINCIPLE.

FUNDAMENTAL PRINCIPLE OF COUNTING [mathematics]   Suppose that two events, E_1 and E_2, are to be performed in sequence; then, if E1 can be performed in 'm' ways and for each of these ways, E_2 can be performed in n ways, the sequence E_1E_2 can be performed in mn different ways.[†] See also ADDITION PRINCIPLE; MULTIPLICATION PRINCIPLE.

FUNDAMENTAL PRINCIPLES OF CONSERVATION [physics, chemistry] (Antoine Laurent, Lavoisier, 1743–1794, French chemist, financier, and administrator) (1) Consists in considering constant and indestructible the quantity of matter existing in nature; that is, in any system and, by extension, in the universe the total quantity of matter is always the same. (2) Matter is subjected to physic. chemical transformations, but it is not destroyed. (3) Affirms that, whatever is the form of energy that is considered (mechanical, thermal, electric, etc.) in any system and, by extension, in the universe, the total energy is constant. (4) Affirms that in any system and, by extension, in the universe, the mass transforms into energy and, vice versa the energy transforms into mass, as in practice it happens in the nuclear reactions and in the ones among subnuclear particles. (5) The energy balancing of any system, it is necessary to include the term mc^2, where m is the sum of the rest masses, that is, the masses, measured at a zero velocity, of all the bodies forming the system. The rest-mass term m derives from the fact that the mass of a body depends on its velocity, referred to a particular frame of reference, that is, the frame in which the mass is at rest.[‡]

FUNGITOXIC PRINCIPLE [chemistry]   Protectant and eradicant activity of chemicals.[§]

FUNNEL PRINCIPLE [mathematics]   (1) Elements compressed by a force with a varied intensity calculated simultaneously by the number of elements exiting the system.[¶] (2) Primary visual emphasis is given to the first digit of an identification number. Subsequent identification digits are given progressively less importance.[‖] (3) No power can be derived from such a construction as the pressure at the inlet, that is, in the part that

*Moll, F., (1920), *Investigations on Fundamental Principles in Wood Conservation*. (Center such anger fiber GRSptzmdssigkeiten.inder Holkonservierung. Jenz; Gustau Fiscler pp 3-23.

†Mishra, A. K., Dubey, N. K., and Mishra, L. (1993), A fungitoxic principle from the leaves of Prunus persica, *Pharm. Pharmacol. Lett.* **2**(6), 203–206.

‡Schay, G. (2007), *Introduction to Probability with Statistical Applications*, Springer, p. 16.

§Rauch, H. (1966), Slubbing of viscose by application of the funnel principle, *Melliand Textilberichte* (1923–1969), **47**(7), 730–732.

¶Beisswanger, D. A. (1984), *Indexing System for File Folders*, US Patent 4,585,253.

‖Pauly, L. (2001), *Centrifuge Compression Combustion Turbine*, US Patent 6,490,865.

has the largest. The flow of air pressure at the inlet has no power to provide due to its large diameter. (4) As data is written to a storage device, each write is copied to a buffer and then transmitted to a similar device at the recovery site. The local device transmits the next set of data only after receiving confirmation that the current set wa received by and committed to the remote device, so that there is no loss of information integrity. In this way, the remote recovery device maintains state and write order consistency, enabling continuous recovery.* See also INVERTED FUNNEL PRINCIPLE; TIME FUNNEL PRINCIPLE.

FUZZY EMPIRICAL RISK MINIMIZATION (FERM) PRINCIPLE [computer science]   Key theorem of learning theory and bounds on the rate of convergence of learning processes are important theoretical foundations of statistical learning theory. Statistical learning theory based on real-valued random samples has been regarded as a better theory on statistical learning with small sample.[†] See also FERM PRINCIPLE.

FUZZY ENTROPY PRINCIPLE [physics] (1) Used for estimating probability density function under specified moment constraints; measure of absence of information about a situation; uncertainty; transformations between measure spaces; expresses amount of disorder inherent or produce; measure of disorder of a system equal to the Boltzmann constant; the natural log arithmic action of the number of microscopic states corresponding to the thermodynamic state of the system. (2) Entropy model for deriving the probability distribution of the equilibrium state. (3) A distribution free method for estimating the quantile function of a nonnegative random variable using the principle of maximum entropy (MaxEnt) subject to constraints specified in terms of the probability-weighted moments estimated from observed data.[‡§] See also MAXIMUM ENTROPY PRINCIPLE; PME PRINCIPLE OF MAXIMUM ENTROPY; MEP MAXIMUM ENTROPY PRINCIPLE; PRINCIPLE OF MAXIMUM NONADDITIVE ENTROPY; PRINCIPLE OF OPERATIONAL COMPATIBILITY; MERMIN ENTROPY PRINCIPLE; ENTROPY PRINCIPLE.

*McCabe, R. (2008) Do You Need IP-Based Disaster Recovery? http://www.ontrackdatarecovery.com/data-storage-replication-article.

[†]Raz, Y. and Tamer, P. (1996), *System for Currently Updating Database by One Host and Reading the Database by Different Host for the Purpose of Implementing Decision Support Functions*, US Patent 5,852,715.

[‡]Cheng, H. D., Chen, Y. H., and Jiang X. H. (2000), Thresholding using two-dimensional histogram and fuzzy entropy principle, *IEEE Trans. Image Process.* (IEEE Signal Processing Soc. **9**(4), 732–735.

[§]Haa, M.-H. and Tianb, J. (2008), The theoretical foundations of statistical learning theory based on fuzzy number samples, *Inform. Sci.* **178**(16), 3240–3246.

G

GABOR'S HOLOGRAPHIC PRINCIPLE [engineering] See T'HOOFT'S PRINCIPLE; HOLOGRAPHIC PRINCIPLE.

GAGA PRINCIPLE [mathematics] (1) A cornerstone of algebraic geometry is the principle that global analytic objects on a subvariety of projective space are actually algebraic.* (2) (GA for general applications) This is a self contained, reentrant procedure that is suitable for the minimization of many difficult cost functions.[†]

GAIA PRINCIPLE [ecology] (J. E. Lovelock, 1979) Role of biota in maintaining a climatic homeostasis.[‡]

GALILEO EQUIVALENCE PRINCIPLE [physics] (Gălileo Galilei, 1564–1642, Italian physicist) All pointlike particles put at a given point with the same velocity fall with the same acceleration in an external gravitational field.[§] See also NEWTON'S EQUIVALENCE PRINCIPLE; STRONG EQUIVALENCE PRINCIPLE.

GALILEO–NEWTON INERTIA PRINCIPLE [physics] (Gelileo Galilec, 1564–1642, Italian physicist) (Sir Jssoc Neton, 1643–1727, English physicist) If a body is moving without external forces, then it maintains indefinitely its rectilinear and uniform motion (with a constant speed) or, if it is initially at rest, it continues to be at rest.[¶]

GALILEO PRINCIPLE [physics] (Galileo Galitec, 1564–1642, Italian physicist) Science is autonomous and its authority is independent of external authority, whether it is political, religious, or whatever.[‖]

GALILEO RELATIVITY PRINCIPLE [engineering] (Galileo Galiber, 1564–1642, Italian physicist) The direct proportionality between the force applied to a body and its acceleration and the limit case that affirms the persistence of a body in its state of rectilinear and uniform motion, when there is no resultant force acting on it, the rectilinear and uniform motion of the reference frame doesn't modify the laws of mechanics.[**] See also INERTIA PRINCIPLE.

GALILEO'S PRINCIPLE OF NATURAL COMPUTATION [physics] (Galileo Galilec, 1564–1642, Italian physicist) Computers exist in Nature when we abstract the physical entities. (This fact will be referred to as Galileo's *principle of natural computation*.)[‖] See also CONTAINMENT PRINCIPLE; PHYSICAL PRINCIPLE OF COMPUTATION; PRINCIPLE OF HYPER-COMPUTATION UPPER BOUNDS; GANDALF'S PRINCIPLE OF HIDDEN UNIVERSE COMPUTATION.

GANDALF'S PRINCIPLE OF HIDDEN UNIVERSE COMPUTATION [mathematics] All computations of the universe are describable by suitable programs that correspond to the prescription by finite means of

*Topiwala, P. and Rabin, J. M. (1991) The Super Gaga Principle and families of Super Riemann Surfaces, *Proce. Am. Math. Soc.* V. 113(1) 11–20.
[†]Crowcroft, J. (2010) University College London, cs.ucl.ac.uk:/darpa/gaga.shar.
[‡]Accioly, A. and Paszko, R. (2009), Conflict between the classical equivalence principle and quantum mechanics, *Adv. Stud. Theor. Phys.* 3(2), 65–78.
[§]Cucinotta, A. (n.d.), *The Laws of the Physical World* (http://www.peoplephysics.com/physics-laws1.htm).
[¶]Mizushima, M. (2001), Deuterium under gravity (a violation of Galileo's equivalence principle), *Hadronic J.* 24(6), 697–700.

[‖]Makowitz, H. (n.d.), *Galileo's Relativity Principle, the Concept of Pressure, and Complex Characteristics, for the Six-Equation, One-Pressure Model*, Astrophysics Data System (ADS), Smithsonian Astrophysical Observatory under NASA Grant NNX09AB39G (http://adsabs.harvard.edu/abs/1992grpc.rept . . . . M).
[**]Costa, J. F. (2006), *Physics and Computation: Essay on the Unity of Science Through Computation* (http://cmaf.ptmat.fc.ul.pt/preprints/pdf/2006/unity.pdf).

some rational parameters of the system or some computable real numbers.∥ See also CONTAINMENT PRINCIPLE; PHYSICAL PRINCIPLE OF COMPUTATION; PRINCIPLE OF HYPERCOMPUTATION UPPER BOUNDS; GALILEO'S PRINCIPLE OF NATURAL COMPUTATION.

GHANDIAN PRINCIPLES [social work] Social work as an expression of culture is a highly value-laden activity. The emergence of many new ethical issues resulting from technological and scientific advancements suggests a need for greater attention to values and ethics. In this article the authors argue that the thought of Mahatma Gandhi (Mohandas Karamchand Gandhi, 1869–1948), as revealed in his social activism, is relevant to social work ethics and a resource for its ethical enrichment.*

GANDY'S PRINCIPLE [mathematics] (Robin Oliver Gandy, 1919–1995, British Mathematician) (1) Explores computability, tracing its roots, at least in part, to developments in logic and philosophy of mathematics and mechanisms as a model of parallel computation. (2) Set theoretic rank of machine states is bounded; the "omniscient machine" is obviously not bounded in this sense. The hieracrchical structure of actual machines, as reflected by the height of the epsilon trees of their states, is not modified by their operation. The *principle of limitation of hiearchy* is then formulated as follows. If $M = <S, F>$, then the set-theoretic rank of the states of M is bounded.† (3) The embedding of the classes of hereditary finite sets (used to formulate Gandy's principles) into the class of abstract membrane systems is shown. A concept of a hereditary finite multiset is introduced as a special case of hereditary finite sets, and then a representation of finite abstract membrane systems by hereditary finite multisets is presented. In this framework, a counterpart of the idea of reassemblance

of hereditary finite sets is obtained for membrane systems.‡

GARSIA–MILNE INVOLUTION PRINCIPLE [mathematics] (Adriano Mario Garsia, 1928–, American mathematician and Stephen Milne, American mathematician) Euler medal 2007 (1) An equation relating the disjoint union of two finite components and involutions on fixed points denoting a fixed point set. Stipulates components (i.e., outside the fixed point sets), both map each component into the other. Then a cycle of the permutation contains either no fixed points or exactly one element of either.§ (2) A proof for a Rogers–Ramanujan identity bijectively and application toward bijections in a large class of other partition identities.¶ See also TWO-INVOLUTIONS' PRINCIPLE; INVOLUTION PRINCIPLE OF GARSIA–MILNE.

GAS BALANCING PRINCIPLE [organizatonal operations] Designed to establish a robust set of principles in relation to the roles and responsibilities for the industry as it restructures to meet the requirements. Tolerance levels should be designed in a way that reflects the actual technical capabilities of the transmission system. They also point out the particular significance of the extent to which tolerances may be utilized by shippers to offer "balancing gas" or cause balancing costs to be incurred by the Transmission Systems Operator (TSO) that are subsequently socialized.∥

*Walz, T. and Ritchie, H. (2000), Gandhian principles in social work practice: ethics revisited, *Social Work* **45**(3), 213–222.

†Anon. (1978), R. Gandy's principles for mechanisms, *Proc. Kleene Symp. Univ.* Wisconsin, Madison, pp. 123–148.

‡Obtuowicz, Adam (2006), Gandy's principles for mechanisms and membrane computing. *Int. J. Found. Comput. Sci.* **17**(1), 167–181.

§Andrews, G. E. (1986), q-Series and Schur's theorem (sec. 6.2) and Bressoud's proof of Schur's theorem (sec. 6.3) in *q-Series: Their Development and Application in Analysis, Number Theory, Combinatorics, Physics, and Computer Algebra*, American Mathematics Soc., Providence, RI, pp. 53–58.

¶Boulet, C. and Pak, I. (2006), A combinatorial proof of the Rogers-Ramanujan and Schur identities, *J. Combin. Theory Ser. A Archive* (Academic Press), **113**(6), 1019–1030.

∥ERGEG (2005), *Gas Balancing; an ERGEG Discussion Paper for Public Consultation* (http://www.ergeg.org/portal/page/portal/ERGEG_HOME/ERGEG_DOCS/ERGEG_DOCUMENTS_NEW/

GAS MOVEMENT PRINCIPLE [physics] See
FICK PRINCIPLE.

GAS PRINCIPLE [geology] Basic physical
equations relating the pressure, volume,
temperature, and density of a perfect gas.
Includes the equation of state and Boyle's
principle.* See also BOYLE'S PRINCIPLE.

GAS TURBINE PRINCIPLE [engineering]
Waste heat from furnace gases are used to
perate a turbo blower with compressed air
for a blast furnace.[†]

GASKET EFFECT PRINCIPLE [engineering]
Enhanced performance of tape coatings are
ensured when tension is applied during the
wrapping process. Tangential force applied
creates an inward radial force, compressing
layers between the suface.[‡]

GASTROINTESTINAL PRINCIPLE [psychology]
An archaic term used to denote hormones,
such as cholecystokinin, gastrin, and se-
cretin, secreted by mucosal cells of the gastrointestinal tract and absorbed into the
blood.[§]

GAUGE PRINCIPLE [mathematics, physics]
(1) Various types of interaction between
fundamental particles.[¶] (2) The essential
ingredient is a group G, a set of gauge
vector mesons A_i, and a set of matter fields
$\psi(x)$, which transform according to some
representation of G. In this framework, the
internal-group space co-ordinates, which
we shall denote by ξ, play no role. On the
other hand, a natural generalization of the
usual approach would be to assign some
physical meaning to ξ. One may in fact set
up, at each space-time point x_μ, a co-ordinate
system for the internal-symmetry group G.
A local gauge transformation then amounts
to a group space co-ordinate transformation,
which may be different for different points
x_μ.

Mathematically, such a situation calls for
a theory described in terms of fields $\psi(x, \xi)$
instead of the usual $\psi(x)$, which is ξ independent. A gauge transformation would now
not only give $\psi(x, \xi)$ an overall phase factor, but also change ξ to ξ', which is the
result of the group space co-ordinate transformation.[‖] (3) The Lagrangian (of a particle)
should have the same form when a total
time derivative is added to it, and is thus
the same as the principle of form invariance
under a "gauge transformation.**" (4) In electromagnetism the physical fields (E and B in
the case of electromagnetism) are unchanged
by a gauge transformation.[††] (5) Studies on
the renormalization of non-Abelian gauge
therorems, topological phenomena in gauge
field theory and thoughts on the role of black
holes in quantum gravity.[‡‡] See also GAUGE
SYMMETRY PRINCIPLE.

**GAUSE'S COMPETITIVE EXCLUSION PRINCI-
PLE** [ecology, biology] Two species that
compete for the exact same resources cannot
stably coexist. One of the two competitors
will always have an ever so slight advantage
over the other that leads to extinction of
the second competitor in the long run (in a
hypothetical nonevolving system) or (in the
real world) to an evolutionary shift of the

GAS_FOCUS_GROUP/ERGEG_2005-07-2005_
GAS_BALANCING_PUBLIC_CONSULTATION.
pdf).
*Stiegeler, S. E. (1977), *A Dictionary of Earth
Sciences*, Pica Press, New York, p. 121.
†Duffy, M. C. (1999), *The Velox Boiler & Its
Application to Railway Traction*, Newcomen Soc.
vol. 71, p. 71, 229–256.
‡Yurovskii, V. S., Komornitskii-Kuznetsov, V. K.
and Fialka, E. M. (1980), Scientific principles and
experience of designing seals for shafts, *Kauchuk i
Rezina* **4**, 17–19.
§Kaiser, J. H. (1959), Principles of Kneipp therapy
in diseases of the digestive apparatus, *Hippokrates*
30, 481–486.
¶Walker, P. M. B. (1999), *Chambers Dictionary
of Science and Technology*, Chambers, New York,
p. 497.

‖Gavrielides, A., Kuo, T. K. and Lee, S. Y., (1976), A
generalized gauge principle, *Lett. Nuovo Cimento*
2(1/2), p. 55–58.
**Bars, I. and Deliduman, C. (1998), Gauge
principles for multi-superparticles, *Phys. Lett. B*
417(3/4), p. 240–246.
††Tonomura, A. (1997), Gauge principle in view of
electromagnetics. *Suri Kagaku* **404**, 47–55.
‡‡Bleuler, K. (1986), The gauge principle in modern
physics, *NATO ASI Series B Physics* **144** (Fundamental Aspects of Quantum Theory), pp. 279–288.

inferior competitor toward a different ecological niche. As a consequence, competing related species often evolve distinguishing characteristics in areas where they both coexist.* See also COMPETITIVE EXCLUSION PRINCIPLE.

GAUSE'S PRINCIPLE [ecology]   (1) Closely related organisms do not coexist in the same niche, except briefly.[†] (2) One species seeking same ecological environment will survive while the other will expire under a given set of conditions. (3) If there is competition between two species for a common place in a limited microcosm, we can quite naturally extend the premises implied in the logistic equation. The growth rate of each competing species in a mixed population will depend on (a) the potential rate of population increase of a given species (blNl or b2N2) and (b) the unutilized opportunity for growth of this species. (4) Levy's discovery of the complementarity between random displacements, parameterized by time, and random time, parameterized by position, in the description of Brownian motion.[‡] See also COMPETITIVE EXCLUSION PRINCIPLE; PRINCIPLE OF COMPLEMENTARITY.

GAUSS' PRINCIPLES [physics, mechanics] (Carl Friedrich Gauss, 1777–1855, German mathematician, physicist, and astronomer) Also referred to as *Gauss' Law*. (1) Gives the relation between the electric flux flowing out of a closed surface and the charge enclosed in the surface. The net electric flux passing through any closed surface is equal to the total charge enclosed by the surface divided by the permittivity of free space (ϵ_0)". The differential formulation of Gauss's law, div $E = \rho/\epsilon_0$ is one of the four Maxwell equations of electromagnetism. Related is something called "Gauss's law

for magnetism": mathematically the statement is div $B = 0$. In words, it implies the following consequences: (a) magnetic monopoles do not exist; (b) magnetic field lines have no ends—they always form closed loops.[§] (2) With relation to electrostatics and magnetostatics. The surface integral of the normal component of electric displacement over any closed surface in a dielectric is equal to the total electic charge enclosed.[¶] (3) A system of interconnected particles, subjected to a certain set of forces, will have nearly the same motion that the particles would have if disconnected from each other but still subjected to the same forces.[‖] See also DIFFERENTIAL PRINCIPLE; D'ALEMBERT PRINCIPLE; HAMILTON–LAGRANGE PRINCIPLE; MAUPERTUIS–LAGRANGE PRINCIPLE; D'ALEMBERT–LAGRANGE PRINCIPLE; MAUPERTUIS–EULER–LAGRANGE PRINCIPLE OF LEAST ACTION; PRINCIPLE OF LEAST CONSTRAINT, LEAST-CONSTRAINT PRINCIPLE; LAGRANGE PRINCIPLE.

GAUSS' PRINCIPLES OF LEAST CONSTRAINT [physics] (Carl Friedrich Gauss, 1777–1855, German mathematician, physicist, and astronomer) (1) The motion of a system of interconnected material points subjected to any influence is such as to minimize the constraint on the system; here the constraint, during an infinitesimal period of time, is the sum over the points of the product of the mass of the point times the square of its deviation from the position it would have occupied at the end of the time period if it had not been connected to other points.[‖] (2) Motion of the underlying system of particles compatible with the collective equilibrium provided the variations are associated with reversible processes.** (3) The path of conservative system in configuration space between

*Flores, J. C. (1999), A mathematical model for Neanderthal extinction, *J. Theor. Biol.* **191**(3), 295–298.

[†]Walker, P. M. B. (1999), *Chambers Dictionary of Science and Technology*, Chambers, New York, p. 497.

[‡]Darlington, P. J, Jr. (1972), Competition, competitive repulsion, and coexistence, *Proc. Natl. Acad. Sci. USA* **69**(11), 3151–3155.

[§]Pickover, C. A. (2008), *Archimedes to Hawking: And the Great Minds behind Them*, Oxford Univ. Press, p. 283.

[¶]Walker, P. M. B. (1999), *Chambers Dictionary of Science and Technology*, Chambers, New York, p. 497.

[‖]Marion, J. B. (1970), *Classical Dynamics of Particles and Systems*, 2nd ed., Academic Press, New York, p. 198.

**Evans, D. J., Hoover, W. G., Failor, B. H., Moran, B. and Anthony, J. C. (1983), Nonequilibrium

and configuration is such that the integral of the lagrangian function over time is a minimum or maximum relative to nearby paths between the same and points and taking the same time.[*]

GAY-LUSSAC'S PRINCIPLE [chemistry] (Joseph Louis Gay-Lussac, 1778–1850, French chemist/physicist) (1) When gases react, they do so in volumes bearing a simple ratio to one another and to the volumes of the resulting substances in the gaseous state, all volumes being measured at the same temperature and pressure.[†] (2) Gases react in volumes in simple ratios to each other and to the products if they are gases.[‡] See also CHARLES PRINCIPLE.

GEN MINXENT PRINCIPLE [engineering, mathematics] Out of all probability distributions satisfying the given linear constraints, linear or nonlinear, the probability distribution that minimizes the Kullback–Leibler measure of cross-entropy is chosen.[§] See also ENTROPY OPTIMIZATION PRINCIPLE.

GENERAL ADAPTATION SYNDROME (GAS) PRINCIPLE [kinesiology] (Hans Selye, 1907–1982, Canadian endocrinologist) See PRINCIPLES OF TRAINING.

GENERAL MAXENT PRINCIPLE [engineering, mathematics] Use of a generalized measure of entropy or use of any moment constraints, either linear or nonlinear. Out of all probability distributions satisfy that given moment

constraints, linear or nonlinear, the probability distribution that maximizes a given generalized measure of entropy is chosen.[¶] See also JAYNES' MAXIMUM ENTROPY PRINCIPLE; ENTROPY OPTIMIZATION PRINCIPLE.

GENERAL PRINCIPLE OF RELATIVITY [physics] A generalization of Einstein's special relativity theory to accelerating frames of reference that replaces the Newtonian notion of instantaneous action at a distance via the gravitational field with a distortion of space-time due to the presence of mass.[||] See also PRINCIPLE OF GALILEAN RELATIVITY.

GENERAL RELATIVISTIC FERMAT PRINCIPLE [physics] See PRINCIPLE OF LEAST TIME.

GENERAL-TO-SPECIFIC PRINCIPLE [psychology] Children's development is marked by progression from the entire use of the body to the use of specific body parts. This pattern can be best seen through the learned process of grasping. Initially, infants can grossly hold a bottle with both hands at about 4 months of age. After practice and time, 12-month-old infants can hold smaller toys or food in each hand using a pincher grasp. This finger–thumb grasp is more precise than the grasping skill of an infant at 4 months. Just as the child develops a more precise grasp with time and experience, many other motor skills are achieved simultaneously throughout motor development. Each important skill mastered by an infant is considered a motor milestone.[**] See also CEPHALOCAUDAL PRINCIPLE.

GENERALIZATION OF JAYNES' MAXIMUM ENTROPY PRINCIPLE [engineering, mathematics] (Edwin Thompson Jaynes, 1922–1998,

molecular dynamics via Gauss's principle of least constraint, *Phys. Rev. A Atom. Mol. Opt. Phys.* **28**(2), 1016–1021.

[*]Bright, J. N., Evans, D. J. and Searles, D. J. (2005), New observations regarding deterministic, time-reversible thermostats and Gauss's principle of least constraint, *J. Chem. Phys.* **122**(19), 194106.

[†]Walker, P. M. B. (1999), *Chambers Dictionary of Science and Technology*, Chambers, New York, p. 498.

[‡]Daintith, J. (1999), *The Facts on File Dictionary of Chemistry*, 3rd ed., Facts On File, New York, p. 106.

[§]Georgiou, T. T. and Lindquist, A. (2003), Kullback-Leibler approximation of spectral density functions, *IEEE Trans. Inform. Theory* **49**(11), 2910–2917.

[¶]Batle, J., Casas, M., Plastino, A. R. and Plastino, A. (2001), On the "fake" inferred entanglement associated with the maximum entropy inference of quantum states, *J. Phys. A Math. Gen.* **34**(33), 6443–6458.

[||]Walker, P. M. B. (1999), *Chambers Dictionary of Science and Technology*, Chambers, New York, p. 501.

[**]Stanley, J. C. (1984), Use of general and specific aptitude measures in identification: Some principles and certain cautions, *Gifted Child Quart.* **28**(4), 177–180.

American physicist) (1) Use of a generalized measure of entropy or use of any moment constraints, either linear or nonlinear. (2) Out of all probability distributions that satisfy given moment constraints, linear or nonlinear, the probability distribution that maximizes a given generalized measure of entropy is chosen.[*] See also ENTROPY OPTIMIZATION PRINCIPLE.

GENERALIZATION OF KULLBACK'S MINIMUM CROSS-ENTROPY PRINCIPLE [mathematics] (Solomon Kullback, 1907–1994, American mathematician) Fundamental contribution of information theory to statistics to provide a unified framework for dealing with notion of information in a precise and technical sense in various statistical problems. Out of all probability distributions satisfying the given linear constraints, linear or nonlinear, the probability distribution that minimizes the Kullback–Leibler measure of cross-entropy is chosen.[†] See also ENTROPY OPTIMIZATION PRINCIPLE.

GENERALIZED MINIMAL PRINCIPLE [physics] Production of an electromotive force either by motion of a conductor through a magnetic field so as to cut across the magnetic flux or by a change in the magnetic flux that treats a conductor. In a magnetic system for the contactless guidance of a vehicle moved along a track in which a plurality of magnets are attached to the vehicle and arranged one behind the other in the direction of travel, the magnet system cooperating with nonferromagnetic conductor loops on the track to generate forces. The oscillatory movement is thereby maintained by supplying the energy required to maintain the mechanical swinging or oscillation. To accomplish these principles, known circuit arrangements have the common property that the energy needed for sustaining the oscillation is more effectively supplied by a short driving pulse that is generated each time the mechanical system is in the state of its greatest

kinetic energy, and this occurs when the mechanical swinger moves through its central position, which lies between the two extreme positions.[‡] See also ENTROPY PRINCIPLE; VARIATIONAL MINIMAL PRINCIPLE; MAXIMUM ENTROPY PRINCIPLE; MINIMUM CROSS-ENTROPY PRINCIPLE; PRINCIPLE OF MINIMUM CROSS-ENTROPY; MOMENT-PRESERVING PRINCIPLE.

GENERALIZED PRINCIPLE [ecology] One species seeking same ecological environment will survive while the other will expire under a given set of conditions.[§]

GENERALIZED PRINCIPLE OF LEAST ACTION [mathematics] Out of all the paths that could be imagined, the one actually taken can be identified because a special property of paths, called the *action*, takes a minimum there.[¶]

GENERALIZED UNCERTAINTY (GUP) PRINCIPLE [physics] Although the GUP affects the early universe, it does not change the current and future dark-energy-dominated universe significantly.[‖]

GENERALLY ACCEPTED ACCOUNTING PRINCIPLE (GAAP) [mathematics] Consensus of experts and Financial Accounting Standards Board governing accounting practice. GAAP is the standard framework of guidelines for financial accounting. It includes the standards, conventions, and rules that accountants follow in recording and summarizing transactions, and in the preparation of financial statements.

[*]Reiser, B. (1998), Real processing II: Extremal principles of irreversible thermodynamics, relations, generalizations and time dependence, *Phys. A Statist. Theor. Phys.* **253**(1–4), 223–246.
[†]Soofi, E. S. (2000), Principal information theoretic approaches, *J. Am. Statist. Assoc.* **95**.

[‡]Mainland, G. B. and O'Raifeartaigh, L. (1974), Derivation of unified gauge theory from a generalized minimal principle, *Lett. Nuovo Cimento* **10**(733), 7.
[§]Matessi, C. and Jayakar, S. D. (1981), Coevolution of species in competition: A theoretical study, *Proc. Natl. Acad. Sci.* USA, **78**(2), 1081–1084.
[¶]Hoover, W. G. (1995), Temperature, least action, and Lagrangian mechanics, *Phys. Lett. A* **204**(2), 133–135.
[‖]Kim, Y.-W., Lee, H.-W., Myung, Y.-S. and Park, M.-In. (2008), New agegraphic dark energy model with generalized uncertainty principle.

1. *Principle of regularity*. Regularity can be defined as conformity to enforced rules and laws.

2. *Principle of consistency*. The consistency principle requires accountants to apply the same methods and procedures from period to period.

3. *Principle of sincerity*. According to this principle, the accounting unit should reflect in good faith the reality of the company's financial status.

4. *Principle of the permanence of methods*. This principle aims at allowing the coherence and comparison of the financial information published by the company.

5. *Principle of noncompensation*. One should show the full details of the financial information and not seek to compensate a debt with an asset, a revenue with an expense, etc.

6. *Principle of prudence*. This principle aims at showing the reality "as is"; one should not try to make things look prettier than they are. Typically, a revenue should be recorded only when it is certain and a provision should be entered for an expense that is probable.

7. *Principle of continuity*. When stating financial information, one should assume that the business will not be interrupted. This principle mitigates the principle of prudence; assets do not have to be accounted for at their disposable value, but it is accepted that they are at their historical value.

8. *Principle of periodicity*. Each accounting entry should be allocated to a given period, and split accordingly if it covers several periods. If a client prepays a subscription (or lease, etc.), the given revenue should be split to the entire timespan and not counted for entirely on the date of the transaction.

9. *Principle of Full Disclosure/ Materiality*. All information and values pertaining to the financial position of a business must be disclosed in the records.*

*Schmidgall, R. S., Hayes, D. K. and Ninemeier, J. D. (2002), *Restaurant Financial Basics*. John Wiley and Sons, p. 7.

See also PRINCIPLE OF REGULARITY; PRINCIPLE OF CONSISTENCY; PRINCIPLE OF SINCERITY; PRINCIPLE OF THE PERMANENCE OF METHODS; PRINCIPLE OF NONCOMPENSATION; PRINCIPLE OF PRUDENCE; PRINCIPLE OF CONTINUITY; PRINCIPLE OF PERIODICITY; PRINCIPLE OF FULL DISCLOSURE/MATERIALITY.

GENERATIVE PRINCIPLE OF WRITING [psychology] (1) Involves the child's comprehension; writing is created from the repetition of patterns and shapes. When children begin to write, they hypothesize, learn, and understand how the writing system works. Children demonstrate several developmental writing patterns, which are broken down into four principles: (a) *recurring principle*—involves the child's comprehension in that writing is created from the repetition of patterns and shapes; (b) *generative principle*—when children realize that writing is made up of many marks; (c) *sign principle*—when children realize that writing carries a message; (d) *inventory principle*—includes children's use of invented spelling. Once evident, the spelling stages begin to emerge. (2) Adjacent vortices on zero crossings of a phase (orientation) mapping must always alternate in sign. (3) The sign principle correctly predicts the structure of orientation maps measured by optical imaging. (4) States that adjacent singularities must always alternate in sign along any zero-crossing path of the real or the imaginary representation of an orientation map. (5) Generalizes that a tautology must contain a positive clause, and therefore is called a '*generalized sign principle*'.[†] Also known as *Hittorf principle; sign principle; short-path principle; writing principles; autarkness principle*.

GENETIC DECENT MINIMUM PRINCIPLE (GDM) Everyone should have genertic potential which exceeds some minimum threshold.[‡]

[†]Edelman, S. and Flash, T. (1987), A model of handwriting, *Biol. Cyberneti.* **57**(1–2), 25–36.
[‡]Araszkiewicz, Michal (2010) Human Genetic Engineering and the Problems of Distributive Justice. Review in Biological Foundations of Law and Ethics

GENETIC DIFFERENCE PRINCIPLE [genetics] Currency problem and the problem of weight. The most promising of three principles—concerns of just healthcare and distributive justice in general. Given the strains on public funds for other important social programs, the costs of pursuing genetic interventions and the nature of genetic interventions, concludes that a more lax interpretation of the genetic difference principle is appropriate. Genetic inequalities should be arranged to be the greatest reasonable benefit of the least advantaged.* Also termed *genetic equality principle; genetic decent minimum principle*.

GENETIC EQUALITY PRINCIPLE (GE) Everyone should have the same genetic potential.[†]

GEODESIC PRINCIPLE [mathematics] A straight line has the lowest information content of any continuous function.[‡] See also MINIMUM INFORMATION PRINCIPLE.

GEOMETRIC MEAN PRINCIPLE. [economics] aka G Policy aka Sanderson Geometric Mean principle or Sanderson's equalization principle.

1. The advocacy of the G policy is based on postulating a certain subgoal for the investor: 'The subgoal proposed here is the choice of the portfolio that has a greater probability (P') of being as valuable or more valuable than other significantly different portfolio at the end of n years, n being large'.
   If a certain subgoal is desirable for some n, 'n being large', it should also be worth pursuing for the case of $n = 1$.
   1. The G policy maximizes the long-run (geometric) mean rate of return (the growth rate).
   2. The G policy maximizes terminal wealth. And, consequently,

3. The G policy is optimal for any investor, irrespective of his preferences (utility function), as long as he prefers more wealth to less.[§]

Successful genotypes are those that maximize geometric mean fitness across environmental stochasticity, which is contrasted with the maximization of arithmetic mean fitness across demographic stochasticity. The successful genotype maximizes the arithmetic mean of reproductive value in both cases. Effectively, the strategy that maximizes the geometric mean of absolute number of offspring is the same as the strategy that maximizes the arithmetic mean of the relative number of offspring (relative to itself).[¶] Sanderson's equalization principle states that the electronegativity of a molecule is given by the geometric mean of the electronegativities of the isolated atoms (or fragments)

$$\mu_{n_f}^o = \left(\prod_x^{n_f} |\mu_x^o| \right)^{1/n_f}$$

where μ_x^o is the chemical potential of fragment x. Note that the larger the value of n_f, the less accurate the result because of the number of bonds and bonding potentials not being considered in the calculations of $\mu_{n_f}^o$.[‖]

GEOMETRIC PRINCIPLE [geometry] Complementary indents providing guides on adjacent faces.[**] See also IRIS PRINCIPLE.

*Farrelly, C. (2004), The genetic difference principle, *Am. J. Bioethics* **4**(2), W21–W28.
[†](Op. cit., Araszkiewicz, Michal).
[‡]Liern, V. and Olivert, J. (1991), *Extension of Geodesics Principle. Recent Developments in Gravitation*, World Scientific Publishers, River Edge, NJ, 1992, pp. 197–203.

[§]Ophir, Tsvi (1979) The Geometric Mean Principle Revised. J of Banking and Finance v.3, p 301–303.
[¶]Grafen, Alan (1999) Formal Darwinism, the Individual-As-Maximizing-Agent Analogy and Bet-Hedging Proceedings: Biological Sciences, v266, (1421) pp. 799–803.
[‖]Sanderson, R. T. (1976) Chemical Bonds and Bond Energy, 2nd ed.; Academic Press: New York.
[**]Getz, E. H., Getz, M. S. and Getz, E. S. (1988), Application of a geometric principle for locating the mandibular hinge axis through the use of a double recording stylus, *J. Prosthet. Dent.* **60**(5), 553–559.

GESTALT PRINCIPLE OF AREA [psychology] (1) Perceptual organization is the process whereby elementary sensations, unconsciously structured and meaningless when viewed individually, are structured into logical objects.* (2) Good continuation is such a powerful organizing principle that when the figure is restored to its initial state, you may "lose" the new organization you just found as your original organization takes over again. (3) The smaller of two overlapping figures is perceived as a figure, while the larger is regarded as ground. We perceive the smaller square to be a shape on top of the other figure, as opposed to a hole in the larger shape. We can reverse this perception by using shading to get our message across.† See also GESTALT PRINCIPLES.

GESTALT PRINCIPLE OF CLOSURE [psychology] (1) The basic gestalt principles of closure and pragnanz appear to have genuine significance for the process of social change, notably the direction that it assumes. These ideas are crucial where progress—as distinguished from mere change, which may be retrogressive in terms of human satisfactions—is emphasized. They contain the objective foundations for evaluating events so far as field theory is at present able to provide them. Both pragnanz and closure imply that the "good gestalt" is the terminal goal of all organizing forces found in nature. *Closure* (the less controversial of the two terms) refers to the fact that certain segregated but "imperfect" wholes tend toward complete or closed forms. Changing, incomplete systems eventually attain equilibrium. In a hypernationalistic age such as ours, the easiest social illustration is that form of aggressive group action known as imperialism or expansionism. The gestalt

*Kubovy, M. and Pomerantz, J. R. (1981), *Perceptual Organization*, Lawrence Erlbaum, Hillsdale, NJ, p. iii.
†Read, S. J., Vanman, E. J. and Miller, L. C. (1997), Connectionism, parallel constraint satisfaction processes, and gestalt principles: (Re)introducing cognitive dynamics to social psychology, *Person. Social Psychol. Rev.* **1**(1), 26–53.

view of the process of institutional transformation.‡ (2) Perceptual organization is the process whereby we unconsciously structure elementary sensations, which may be quite meaningless when viewed individually, into logical objects.§ (3) Good continuation is such a powerful organizing principle that when the figure is restored to its initial state, you may "lose" the new organization you just found as your original organization takes over again.¶ (4) Tendency to see complete figures and ignore gaps in the border. This helps perception of complete forms even when they are partly occluded by other objects.‖ See also GESTALT PRINCIPLES.

GESTALT PRINCIPLE OF COMMON MOVEMENT When stimulus elements move in the same direction and at the same rate, we tend to see them as part of a single object. This helps us distinguish a moving object from the background.‖

GESTALT PRINCIPLE OF CONTIGUITY [psychology] (1) Perceptual organization is the process whereby we unconsciously structure elementary sensations, which may be quite meaningless when viewed individually, into logical objects.** (2) Good continuation is such a powerful organizing principle that when the figure is restored to its initial state, you may "lose" the new organization you just found as your original organization takes over again.††

‡Hartmann, G. W. (1946), *Psychol. Rev.* **53**(5), 282–289.
§Kubovy, M. and Pomerantz, J. R. (1981), *Perceptual Organization*, Lawrence Erlbaum, Hillsdale, NJ, p. iii.
¶Quinn, P. C. and Bhatt, R. S. (2005), Good continuation affects discrimination of visual pattern information in young infants, *Percept. Psychophys.* **67**(7), 1171–1176.
‖O'Shaughnessy, M. P. and Kayson, W. A. (1982), Effect of presentation time and Gestalt principles of proximity, similarity, and closure on perceptual accuracy, *Percept. Motor Skills* **55**(2), 359–362.
**Kubovy, M. and Pomerantz, J. R. (1981), *Perceptual Organization*, Lawrence Erlbaum, Hillsdale, NJ, p. iii.
††Quinn, P. C. and Bhatt, R. S. (2005), Good continuation affects discrimination of visual pattern information in young infants, *Percept. Psychophys.* **67**(7), 1171–1176.

(3) Forms that touch along a boundary or at a point will be seen as belonging together. This enables organization with different shapes, orientations, sizes, colors, textures, or values into a whole.* See also GESTALT PRINCIPLES.

GESTALT PRINCIPLE OF CONTINUITY (OR GOOD CONTINUATION) [psychology] (1) When lines intersect, we tend to group the line segments in such a way as to form continuous lines with minimal change in direction. This helps us decide which lines belong to which object when two or more objects overlap.[†] (2) Perceptual organization is the process whereby we unconsciously structure elementary sensations, which may be quite meaningless if viewed individually, into logical objects.[‡] (3) Good continuation is such a powerful organizing principle that when the figure is restored to its initial state, you may "lose" the new organization you just found as your original organization takes over again.[§] See also GESTALT PRINCIPLES.

GESTALT PRINCIPLE OF GROUPING [psychology] (1) Examined the nature of the psychological processes that underlie the gestalt principles of grouping by proximity and grouping by similarity. Similarity was defined relative to the principles of grouping by common color and grouping by common shape. Subjects were presented with displays comprising a row of seven colored shapes and asked to rate the degree to which the central target shape grouped with either the right- or the left-flanking shapes. Across the displays the proximal and featural relationships between the target and flankers were varied. Ratings reflected persuasive effects of grouping by proximity and common color; there was only weak evidence for grouping by common shape. Nevertheless, both common color and common shape were shown to override grouping by proximity, under certain conditions. The data also show that to understand how the gestalt principles operate, it appears necessary to consider processes that operate within and between groups of elements initially identified on the basis of proximity.[¶] (2) Set of rules that describe the process whereby we unconsciously structure elementary sensations, which may be quite meaningless when viewed individually, into logical objects.[‖]

GESTALT PRINCIPLE OF ORGANIZATION [psychology] Identifies factors leading to particular forms of perceptual organization.[**]

GESTALT PRINCIPLE OF PROXIMITY [psychology] (1) Tendency to see stimulus elements near each other as part of the same object and those separated as part of different objects. This helps us segregate a large set of elements into a smaller set of objects.[††] (2) Perceptual organization is the process whereby we unconsciously structure elementary sensations, which may be quite meaningless if viewed individually, into logical objects.[‡‡] (3) Good continuation is such a powerful organizing principle that when the

---

*Bennamoun, M. and Boashash, B. (1997), A structural-description-based vision system for automatic object recognition, *IEEE Trans. Syst. Man Cybernet. B Cybernetics* **27**(6), p. 893–906.

[†]Quinn, P. C. and Bhatt, R. S. (2005), Good continuation affects discrimination of visual pattern information in young infants, *Percept. Psychophys.* **67**(7), 1171–1176.

[‡]Kubovy, M. and Pomerantz, J. R. (1981), *Perceptual Organization*, Lawrence Erlbaum, As Hillsdale, NJ, p. iii.

[§]Novick, L. R. and Catley, K. M. (2007), Understanding phylogenies in biology: The influence of a Gestalt perceptual principle, *J. Exp. Psychol. Appl.* **13**(4), 197–223.

[¶]Darsono, A. (1970), From gestalt theory to group theory; evolution of a basic psychologic principle, *Nederlands Tijdschrift Psychol. Grensgebieden* **25**(3), 144–177.

[‖]Han, S., Humphreys, G. W. and Chen, L. (1999), Uniform connectedness and classical Gestalt principles of perceptual grouping, *Percept Psychophys.* **61**(4), 661–674.

[**]American Heritage (2000), *American Heritage Dictionary of the English Language*, 4th ed., Houghton Mifflin, New York.

[††]Gray, P. O. (2001), *Psychology*, Macmillan, p. 301.

[‡‡]Kubovy, M. and Pomerantz, J. R. (1981), *Perceptual Organization*, Lawrence Erlbaum, A Hillsdale, NJ, p. iii.

figure is restored to its initial state, you may "lose" the new organization you just found as your original organization takes over again.* (4) The principle of proximity states that things closer together will be seen as belonging together.† See also GESTALT PRINCIPLES.

GESTALT PRINCIPLE OF SIMILARITY [psychology] (1) Perceptual organization is the process whereby we unconsciously structure elementary sensations, which may be quite meaningless if viewed individually, into logical objects.‡ (2) Good continuation is such a powerful organizing principle that when the figure is restored to its initial state, you may "lose" the new organization you just found as your original organization takes over again.§ (3) Forms sharing visual characteristics such as shape, size, color, texture, value, or orientation will be seen as belonging together. This helps us distinguish between two adjacent or overlapping objects on the basis of a change in their texture elements.¶ See also GESTALT PRINCIPLES.

GESTALT PRINCIPLE OF SYMMETRY (OR GOOD FORM) [psychology] (1) The perceptual system strives to produce elegant, simple, uncluttered, symmetric, regular, and predictable perceptions. Encompasses other gestalt principles, including other ways by which the perceptual system organizes stimuli into their simplest, most easily explained,

arrangement.‖ (2) Perceptual organization is the process whereby we unconsciously structure elementary sensations, which may be quite meaningless when viewed individually, into logical objects.** (3) Good continuation is such a powerful organizing principle that when the figure is restored to its initial state, you may "lose" the new organization you just found as your original organization takes over again.†† (4) Describes the instance where the whole of a figure is perceived rather than the individual parts that make up the figure.‡‡ See also GESTALT PRINCIPLES.

GESTALT PRINCIPLES [psychology] (1) *Gestalt* is a German word meaning *whole*. An approach to perception and other cognitive skills, which stresses the need to understand the underlying organization of these functions, and believes that to dissect experience into its "constituent bits" is to lose its essential meaning.§§ (2) Perceptual organization is the process whereby we unconsciously structure elementary sensations, which may be quite meaningless if viewed individually, into logical objects.¶¶ (3) Good continuation is such a powerful organizing principle that when the figure is restored to its initial state, you may "lose" the new organization you just found as your original organization takes over again.‖‖ (4) Things that are

‖Gray, P. O. (2001), *Psychology*, Macmillan, p. 301.
**Kubovy, M. and Pomerantz, J. R. (1981), *Perceptual Organization*, Lawrence Erlbaum, A Hillsdale, NJ, p. iii.
††Quinn, P. C. and Bhatt, R. S. (2005) Good continuation affects discrimination of visual pattern information in young infants, *Percept. Psychophys.* **67**(7), 1171–1176.
‡‡Koontz, N. A. and Gunderman, R. B. (2008), Gestalt theory: Implications for radiology education, *Am. J. Roentgenol.* **190**(5), 1156–1160.
§§Walker, P. M. B. (1999), *Chambers Dictionary of Science and Technology*, Chambers, New York, p. 473.
¶¶Kubovy, M. and Pomerantz, J. R. (1981), *Perceptual Organization*, Lawrence Erlbaum, Hillsdale, NJ, p. iii.
‖‖Quinn, P. C. and Bhatt, R. S. (2005), Good continuation affects discrimination of visual pattern information in young infants, *Percept. Psychophys.* **67**(7), 1171–1176.

*Quinn, P. C. and Bhatt, R. S. (2005), Good continuation affects discrimination of visual pattern information in young infants, *Percept. Psychophys.* **67**(7), 1171–1176.
†Quinlan, P. T. and Wilton, R. N. (1998), Grouping by proximity or similarity? Competition between the Gestalt principles in vision, *Perception* **27**(4), 417–430.
‡Kubovy, M. and Pomerantz, J. R. (1981), *Perceptual Organization*, Lawrence Erlbaum, A Hillsdale, NJ, p. iii.
§Quinn, P. C. and Bhatt, R. S. (2005), Good continuation affects discrimination of visual pattern information in young infants, *Percept. Psychophys.* **67**(7), 1171–1176.
¶Quinn, P. C., Bhatt, R. S., Brush, D., Grimes, A. and Sharpnack, H. (2002), Development of form similarity as a Gestalt grouping principle in infancy, *Psychol. Sci.* **13**(4), 320–328.

closer together will be seen as belonging to-gether.* (5) The *principle of area* states that the smaller of two overlapping figures is per-ceived as a figure while the larger one is regarded as ground. We perceive the smaller square to be a shape on top of the other figure, as opposed to a hole in the larger shape. We can reverse this perception by using shading to get our message across.† (6) The principle of symmetry describes the instance where the whole of a figure is perceived rather than the individual parts that make up the figure.‡ (7) The principle of closure applies when we tend to see complete figures even when part of the information is missing.§ (8) The principle of similarity states that things that share visual characteristics such as shape, size, color, texture, value, or orientation will be seen as belonging together.¶ See also GESTALT PRINCIPLE OF CONTINUITY; GESTALT PRINCIPLE OF PROXIMITY; GESTALT PRINCIPLE OF CONTIGU-ITY; GESTALT PRINCIPLE OF CLOSURE; GESTALT PRINCIPLE OF AREA; GESTALT PRINCIPLE OF SYM-METRY; GESTALT PRINCIPLE OF SIMILARITY.

GETTIER'S PRINCIPLE [mathematics]   (Ed-mund L. Gettier III, b. 1927, Baltimore, MD)   If person S knows p, and p entails q, then S knows q (this is sometimes called the *straight principle*). A subject may not actually believe q, for example, regardless of whether he or she is justified or warranted. Thus, one might instead might say that

knowledge is closed under known deduction: If, while knowing p, S believes q because S knows that p entails q, then S knows q. An even stronger formulation would be as such: If, while knowing various propositions, S believes p because S knows that they entail p, then S knows p.‖ See also STRAIGHT PRINCIPLE; PRINIIPLE OF EPISTEMIC CLOSURE.

GHILANTEN ANTIMETASTATIC PRINCIPLE [pharmacology]   *Haementeria ghilianii* is a useful antimetastatic agent and is henceforth referred to as *ghilanten*. Leech secretions have been more scientifically studied and have been found to contain a variety of biological products having a wide spectrum of biochemical and pharmacological activi-ties such as anticoagulant, antimetastatic, anaesthetic, antibiotic, and vasodilator.**

GHYBEN–HARZBERG   PRINCIPLE [ecology] Determines the amount of freshwater that can be abstracted from a source that is open to the sea before it becomes contaminated by saltwater.††

GIBBS' ADSORPTION PRINCIPLE [chemistry] (Josiah Willard Gibbs, 1839–1903, American physicist) (1) Gibbs' absorption equation is messier, since one may have more surface species, including charged surfactants, which modify the constant between 1 and 2, depend-ing on the ratio of charged to noncharged surfactants.‡‡ (2) Solutes which lower the surface tension of a solvent tend to be con-centrated at the surface, and conversely.§§

*Bruce, V., Green, P. R. and Georgeson, M. A. (2004), *Visual Perception: Physiology, Psychology, & Ecology*, Psychology Press, p. 123.

†Quinn, P. C. and Bhatt, R. S. (2006), Are some gestalt principles deployed more readily than others during early development? The case of lightness versus form similarity, *J. Exp. Psychol. Human Percept. Perform.* **32**(5), 1221–1230.

‡Han, S., Humphreys, G. W. and Chen L. (1999), Uniform connectedness and classical Gestalt prin-ciples of perceptual grouping, *Percept. Psychophys.* **61**(4), 661–674.

§Coren, S. and Girgus, J. S. (1980), Principles of perceptual organization and spatial distortion: The gestalt illusions, *J. Exp. Psychol. Human Percept. Perform.* **6**(3), 404–412.

¶O'Shaughnessy, M. P. and Kayson, W. A. (1982), Effect of presentation time and Gestalt principles of proximity, similarity, and closure on perceptual accuracy, *Percept. Motor Skills*, **55**(2), 359–362.

‖Klein, P. D. (1983), Real knowledge, *Synthese* (Springer, Netherlands, **55**(2), 143–164.

Blankenship, D. T., Brankamp, R. G., Manley, G. D. and Cardin, A. D. (1990), Amino acid sequence of ghilanten: anticoagulant-antimetastatic princi-ple of the South American leech, Haementeria ghilianii, *Biochem. Biophys. Res. Commun.* **166(3), 1384–1389.

††Walker, P. M. B. (1999), *Chambers Dictionary of Science and Technology*, Chambers, New York, p. 507.

‡‡Jákli, A. and Saupe, A. (2006), Fluids with reduced Dimensionality, in One- and two-dimensional fluids: properties of smectic, lamellar and columnar liquid crystals. CRC Press, p. 50.

§§Walker, P. M. B. (1999), *Chambers Dictionary of Science and Technology*. Chambers, New York, p. 507.

GIBBS–BOGOLUTOV VARIATIONAL PRINCIPLE
[mathematics] (Nikolai Nikolaevich Bogolutov, recognized for his many fundamental
contributions in physics and mathematics)
(1) Works on nonlinear mechanics and the
general theory of dynamical systems. A systematic formulation of the renormalization
program for perturbative computations of the
S matrix.* (2) A form of the fundamental
variational principle first derived for the classical case by Gibbs in his study on statistical
mechanics. One uses determinants of single-
particle wavefunctions rather than products,
thereby introducing terms into the Hamiltonian.[†] See also GIBBS THIRD VARIATIONAL
PRINCIPLE, BOGOLUTOV PRINCIPLE, BOGOLUTOV
VARIATIONAL PRINCIPLE, HARTREE–FOCK VARI-
ATIONAL PRINCIPLE, HFB PRINCIPLE, QUANTAL
PRINCIPLE.

GIBBS–KONOWALOW PRINCIPLE [chemistry]
(Josiah Willard Gibbs, 1839–1903, American physicist) At constant pressure the
equilibrium temperature is a maximum or
minimum when the compositions of the two
phases are identical, and vice versa.[‡]

GIBBS THIRD VARIATIONAL PRINCIPLE
[mathematics] (Josiah Willard Gibbs,
1839–1903, American physicist) (Nikolai
Nikolaevich Bogolutov, recognized for his
many fundamental contributions in physics
and mathematics) (1) A systematic formulation of the renormalization program
for perturbative computations of the S
matrix. In mathematics, among his many
important contributions we cite his work
on nonlinear mechanics and the general
theory of dynamical systems.[§] (2) A form of
the fundamental variational principle first
derived for the classical case by Gibbs in his

---

*Girardeau, M. D. (1964), Variational principle
of Gibbs and Bogolyubov, *J. Chem. Phys.* **41**(9),
2945–2946.
[†]Brown, W. B. (1964), Gibbs' third variational
principle in statistical thermodynamics, *J. Chem.
Phys.* **41**(9), 2945.
[‡]Walker, P. M. B. (1999), *Chambers Dictionary of
Science and Technology*, Chambers, New York, p.
508.
[§]Girardeau, M. D. (1964), Variational principle
of Gibbs and Bogolyubov, *J. Chem. Phys.* **41**(9),
2945–2946.

study on statistical mechanics. One uses determinants of single-particle wavefunctions
rather than products, thereby introducing terms into the Hamiltonian.[¶] Also
known as *Bogolutov principle; Bogolyubov
variational principle; Gibbs–Bogolyubov
variational principle*; See also HARTREE–FOCK
VARIATIONAL PRINCIPLE; QUANTAL PRINCIPLE.

GIBBS VARIATIONAL PRINCIPLE [mathemat-
ics] (Josiah Willard Go'bs, 1839–1903,
American physicist) (Stefan Banach,
1892–1945)

THEOREM.

$$-\beta\phi(\beta) = \lim_{n\to\infty} \frac{1}{n}\log Z_n(\beta)$$

$$= \sup_{P\in,\mathcal{M}_3(\Omega)} \left\{ -\beta\int_\Omega \tfrac{1}{2}\omega_1^2 P(d\omega) - 1_\rho^{(3)}(P) \right\}.$$

*The supremum is attained at the unique
measure* $P = P_{\rho_\beta}$, *which is the infinite
product measure on* $\beta(\Omega)$ *with identical
one-dimensional marginals* $\tilde\rho_\beta$.

We now interpret Theorem III.8.1 physically. The level-1 interpretation involves
the energy observable $U_n(\omega) = \sum_{j=1}^n \tfrac{1}{2}Y_j(\omega)^2$.
Each number $u \in (\tfrac{1}{2}v_1^2, \tfrac{1}{2}v_r^2)$ is a candidate
for the equilibrium value of the energty per
particle $U_n(\omega)/n$. By Theorem III.6.2, U_n/n
tends exponentially to $u_c(\beta)$, and so $u_c(\beta)$
is the actual equilibrium value. Part (a) of
Theorem III.8.1 characterizes this number
as the unique point at which the func-
tion $-\beta u - \bar{I}_\rho^{(1)}(u)$ attains its supremum over
$u \in \mathbb{R}$. The quantity $-\beta^{-1}[\beta u_c(\beta) - I_\rho^{(1)}(u_c(\beta))]$
equals the specific free energy $\phi(\beta)$.

For level-3, each measure $\hat{P} \in \mathcal{M}_3(\hat\Omega)$ defines a measure $\hat{P} \in \mathcal{M}_s(\hat\Omega)$ which is a candidate for the infinite-particle description of
the gas. We define the specific energy in P to
be $u(P) = \int_\Omega \tfrac{1}{2}\omega_1^2 P(d\omega)$, the specific entropy in
P to be $s(P) = -I_\rho^{(3)}(P)$, and the specific free
energy in P to be

$$\phi(\beta : P) = \beta^{-1}(\beta u(P) - s(P))$$

$$= \beta^{-1}\left(\beta\int_\Omega \tfrac{1}{2}\omega_1^2 P(d\omega) + I_\rho^{(3)}(P) \right).$$

[¶]Brown, W. B. (1964), Gibbs' third variational
principle in statistical thermodynamics, *J. Chem.
Phys.* **41**(9), 934, 937.

Any measure $P \in \mathcal{M}_s(\Omega)$ at which $-\beta\phi(\beta; P)$ attains its supremum over $\mathcal{M}_s(\Omega)$ is called a *level-3 equilibrium state*. The same term applies to the corresponding measure $\hat{P} \in \mathcal{M}_s(\hat{\Omega})$. Part (b) of Theorem III.8.1 implies the result, known as the *Gibbs variational principle*.[*]See also CONTRACTION MAPPING PRINCIPLE.

GIFFORD–MCMAHON PRINCIPLE [chemistry] The end of the cold head extends into the gas space of the helium supply vessel and recondenses the helium gas flowing back through the return line. The Gifford–McMahon refrigerator is a modification of the Stirling cycle refrigerator. Thermodynamic principles of refrigeration and liquefaction are identical. However, the analysis and design of the two systems are quite different because of the condition of balanced flow in the refrigerator and unbalanced flow in liquefier systems. Prerequisite refrigeration for gas liquefaction is accomplished in a thermodynamic process when the process gas absorbs heat at temperatures below that of the environment. As mentioned, a process for producing refrigeration at liquefied gas temperatures always involves some equipment at ambient temperature in which the gas is compressed and heat is rejected to a coolant. During the ambient temperature compression process, the enthalpy and entropy, but seldom the temperature of the gas, are decreased. The reduction in gas temperature is usually accomplished by recuperative heat exchange between the cooling and warming gas streams followed by an expansion of the high-pressure stream. This expansion may take place through either a throttling device (isenthalpic expansion) where there is a reduction in temperature only (when the Joule–Thomson coefficient is positive) or in a work-producing device (isentropic expansion) where both temperature and enthalpy are decreased.[†]

GLADSTONE–DALE PRINCIPLE [physics] (J. H. Gladstone) (T. P. Dale) Law relating the refracting index of a gas and its density as it is changed by variations in pressure and temperature.[‡]

GLOBAL EXTREMUM PRINCIPLE [mathematics] (1) An extremum principle is presented covering problems in solid mechanics equilibrium analysis for piecewise linear softening materials. Problems formulated according to this principle are expressed in a mixed "stress and deformation" form. The mechanics interpretation is limited according to linear deformation kinematics. More specialized models, such as an extremum principle in mixed form for linearly elastic materials, an equivalent to the minimum complementary energy principle and a statement of a bound theorem of limit analysis, are identified as special cases within the general formulation. Numerical results are presented for two examples of one-dimensional structures made of inhomogeneous, softening material. The evolution of material degradation is demonstrated via a set of solutions obtained for increasing load. Each solution of the set is produced from a single application of a general-purpose computer program for constrained nonlinear programming problems, operating on a finite-element interpretation of the nonlinear continuum.[§] (2) It is impossible to construct an algorithm that will find a global extremum for an arbitrary function.[¶]

GLOBAL MAXIMUM PRINCIPLE [mathematics] (1) The path of a conservative system in configuration space between and configuration is such that the integral of the lagrangian function over time is a minimum or maximum relative to nearby paths between the same and

[*]Ellis, R. (2005), Entropy, Large Deviations, and Statistical Mechanics (Classics in Mathematics). Springer. p. 123–125.
[†]Flynn, T. M. (n.d.), Cryogenic engineering, in *Refrigeration and Liquefaction*, rev. expanded, 2nd ed., chap. 6.

[‡]Walker, P. M. B. (1999), *Chambers Dictionary of Science and Technology*, Chambers, New York, p. 509.
[§]Taylor, J. E. (1993), A global extremum principle for the analysis of solids composed of softening material, *Int. J. Solids Struct.* **30**(15), 2057–2069.
[¶]Kilian, H. G., Knechtel, W., Heise, B. and Zrinyi, M. (1993), Orientation in network-like polymer systems. The role of extremum principles, *Progress Colloid Polym. Sci.* **92**, 60–80.

points and taking the same time.* (2) Motion of the underlying system of particles compatible with the collective equilibrium provided that the variations are associated with reversible processes.† (3) External counting observables of system indicator histograms, its cycled average equilibrium count rate, and indicator volume of distribution in the body with cycle-averaged cardiac output.‡ (4) The largest overall value of a set, function, or other entity over its entire range. It is impossible to construct an algorithm that will find a global maximum for an arbitrary function.§ See also VARIATIONAL PRINCIPLE; EKELAND'S EPSILON VARIATIONAL PRINCIPLE; EKELAND'S VARIATIONAL PRINCIPLE.

GLOBAL MINIMUM PRINCIPLE [mathematics] The smallest overall value of a set, function, and so on over its entire range. It is impossible to construct an algorithm that will find a global minimum for an arbitrary function.¶ See also GLOBAL EXTREMUM PRINCIPLE.

GLOBAL SEPARATION PRINCIPLE [engineering] (1) If dynamic output control laws are allowed, the global output feedback stabilization problem can be split up into two independent subproblems, which are the nonlinear analogous of the corresponding ones in the linear case: the state feedback stabilization problem and the output injection stabilization problem. (2) Separates the discrete composition and continuous composition of hybrid systems. (3) Design control of continuous processes and supervisory control of discrete processes separately.‖

GLOBAL SULLIVAN PRINCIPLES OF SOCIAL RESONSIBILITY [ecology] (Leon Howard Sullivan, 1922–2001, Baptist minister, civil rights leader) To support economic, social and political justice by companies where they do business; to support human rights and to encourage equal opportunity at all levels of employment, including racial and gender diversity on decision making committees and boards; to train and advance disadvantaged workers for technical, supervisory and management opportunities; and to assist with greater tolerance and understanding among peoples; thereby, helping to improve the quality of life for communities, workers and children with dignity and equality.** See also HANNOVER PRINCIPLES; FOREST STEWARDSHIP COUNCIL (FSC) PRINCIPLES; MARINE STEWARDSHIP COUNCIL (MSC) PRINCIPLES; PERMACULTURE PRINCIPLES; THE BELLAGIO PRINCIPLES FOR ASSESSMENT; EQUATOR PRINCIPLES; MELBOURNE PRINCIPLES; PRECAUTIONARY PRINCIPLE; SANBORN PRINCIPLES; TODDS' PRINCIPLES OF ECOLOGICAL DESIGN.

GOOD LUCK PRINCIPLE [physics] origin of the universe(s) can be put into one of the following three categories: (1) the hypothesis is based on the *design principle* (DP) (i.e., the universe is unique, and it is designed for our benefit; which implies that we are somehow special); (2) the hypothesis is based on the *good luck principle* (GLP) [i.e., the universe is unique, and there is a theme of existence (ToE) that determines the physical constants, and we are just very fortunate that this ToE happens to be compatible with our existence, which implies that we are somehow special]; (3) the hypothesis is based on the *anthropic principle* (AP) (i.e., the universe is not designed and there is no ToE that fixes all physical parameters,

*Bright, J. N., Evans, D. J. and Searles, D. J. (2005), New observations regarding deterministic, time-reversible thermostats and Gauss's principle of least constraint, *J. Chem. Phys.* **122**(19), 194106.
†Evans, D. J., Hoover, W. G., Failor, B. H., Moran, B. and Ladd, A. J. C. (1983), Nonequilibrium molecular dynamics via Gauss's principle of least constraint, *Phys. Rev. A Atom. Mol. Opt. Phys.* **28**(2), p. 1016–1021.
‡Eterovic, D., Tukic, A. and Tocilj, J. (1991), Perfect-mixer retention function by analytical deconvolution of tracer histograms: Aplication to evaluation of left-ventricular contractility and competence, *Phys. Med. Biol.* **36**(12), 1587–1597.
§Kilian, H. G., Knechtel, W., Heise, B. and Zrinyi, M. (1993), Orientation in network-like polymer systems. The role of extremum principles, *Progress Colloid Polym. Sci.* **92**, 60–80.
¶Op. Cit.
‖Bounit, H. and Hammouri, H. (1997), Bounded feedback stabilization and global separation principle of distributed parameter systems, *IEEE Trans. Tomacti Control*, **42**, 414.
**http://www.globalsullivanprinciples.org/principles.htm

but instead there is something akin to a multitude of different worlds of which we happen to inhabit one; we are not special). These three principles (DP, GLP, and AP) are normally intended to be mutually exclusive (i.e., a particular hypothesis can adhere to at most one principle, never to two or all three simultaneously). Depending on the precise details of a particular multiverse hypothesis, the hypothesis could be made to be consistent with either AP or GLP (or even DP). Most multiverse hypotheses are normally written to be consistent with the AP. There is one theme of existence determining the laws of physics in all universes, and our existence in any universe is fortunate.[*]

GOUPILLAUD PRINCIPLE [mathematics] (Pierre Goupillaud, 1917–, American geophysicist) A new impulse inverse formula and it's generalization including arbitrary input, are proposed. The new generalized inverse formula has the advantage of not being necessary to specify the first digital value of the input.[†]

GOVERNING PRINCIPLE [mathematics] See CATALYSIS MAXIMIZATION PRINCIPLE; MAXIMUM DECISIONAL EFFICIENCY PRINCIPLE.

GRAEFFE PRINCIPLE [mathematics] (Eduard Heinrich Graeffe, 1833–1916, Swiss entomologist) Root-finding method among the most popular for finding roots of univariate polynomials.[‡]

GRAHAM'S PRINCIPLE [chemistry] (Thomas Graham, 1805–1869, Scottish Chemist) The velocity of the effusion of a gas is inversely proportional to the square root of the gas' density.[§]

GRAHAM'S PRINCIPLE OF DIFFUSION [chemistry] (Thomas Graham, 1805–1869, Scottlish Chemist) Gases diffuse at a rate that is inversely proportional to the square root of their density. Light molecules diffuse faster than do heavy molecules.[¶]

GRANDFATHER PRINCIPLE [psychology] In large organizations the major decisions have to be approved by the manager's manager.[‖]

GREATER INVOLVEMENT OF PEOPLE LIVING WITH HIV/AIDS (GIPA) PRINCIPLE [pharmacology] See HIV PRINCIPLE.

GREATEST HAPPINESS PRINCIPLE [psychology] (Jeremy Bentham, 1748–1832 English jurist and philosopher) Central tenet of utilitarian moral theory. The correct action in any situation is that that brings the most happiness to the most people. The goodness of an action should not be judged by the decency of its intentions, but by the utility of its consequences.[**]

GRESHAM'S PRINCIPLE [economics] (Sir Thomas Gresham, 1519–1579, English Financier) When two kinds of money having the same denominational value are in circulation, the intrinsically more valuable money will be hoarded and the money of lower intrinsic value will circulate more freely until the intrinsically more valuable money is driven out of circulation; bad money drives out good; principle credited to Sir Thomas Gresham.[††]

GROUND SUPPORT INTERACTION PRINCIPLE [mathematics] See ANOKHIN–PARETO INTERACTION PRINCIPLE.

[*]Moving Finger (2005), *"Bad Science" and the Anthropic Cosmological Principle* (http://www.physicsforums.com/archive/index.php/t-70578.html).
[†]Slob, E. and Ziolkowsk, A. (2006), Aspects of ID seismic modelling using the Goupillaud principle, *Geophys. Prospect.* **41**(2), 135–148.
[‡]Bareiss, E. H. (1960), Resultant procedure and the mechanization of the Graeffe process, *J. Assoc. Comput. Machinery* (JACM) *Arch.* **7**(4), 346–386.
[§]Walker, P. M. B. (1999), *Chambers Dictionary of Science and Technology*, Chambers, New York, p. 519.

[¶]Daintith, J. (1999), The Facts on File Dictionary of Chemistry, 3rd ed., Facts On File, New York, p. 108.
[‖]Nattrass, B. and Altomare, M. (1999), *The Natural Step for Business: Wealth, Ecology and the Evolutionary Corporation*, New Society Publishers, p. 57.
[**]Veenhoven, R. (2004), Happiness as a public policy aim: The greatest happiness principle, in *Positive Psychology in Practice*, Linley, P. A. and Joseph, S. eds., Wiley, Hoboken, NJ, pp. 658–678.
[††]American Heritage (2000), *American Heritage Dictionary of the English Language*, 4th ed., Houghton Mifflin, New York.

GROUP DIVISION PRINCIPLE [psychology, computers] See INFORMATION-PROCESSING PRINCIPLE.

GROUP KNOCKOUT PRINCIPLE [computer science, engineering] (1) To keep the packet losses due to network architecture down to a magnitude of unavoidable packet loss due to nonsystem errors, such as transmission errors or bus faults. If packets are lost, they are dropped in order to maintain the extremely low probability. Retransmission at higher-layer protocol should be facilitated. (2) Applicable to a switch that falls into the category of nonblocking sort (Jpn. *banyan*) architecture. It is based on output buffering by allowing simultaneous arrival of multiple packets to the same output. It is a self-routing, space division, self-modular, packet switch requiring no internal speedup and with the packet loss probabilities below any predetermined design value for loading up to 100%.*

GROWTH-PROMOTING PRINCIPLE [biology, medicine] Formed by the eosinophils; the concept that the principle responsible for the enhancing effects on renal function is formed by the eosinophilsis further supported by the finding that in Cushing's disease, in the absence of renal disease, with basophilic hyalinization, which presumably indicates ineffective secretion of the basophile cells, there is no depression of renal function.† Also termed *ketogenic pituitary principle*.

GUERRA–MORATO VARIATIONAL PRINCIPLE [physics] See FICK PRINCIPLE.

GUIDING PRINCIPLE [psychology] (1) Set of rules to determine or reflect policy intent and an obligation to apply them as agreed to by the organization to comport fully with terms and conditions that will continue perpetually as a method for the organization to consider costs, and improve quality and service to the customer. (2) Committed to maintaining an environment in which all members can work together to further general goals and provide the highest level of care, in any environment.‡

GULDIN'S PRINCIPLE [mathematics] (Paul Guldin, 1577–1643, Swiss mathematician) The surface area and the volume respectively swept out by revolving a plane area about a nonintersecting coplanar axis is the product of the distance moved by its centroid and either its perimeter or its area.§ See PAPPUS' PRINCIPLE; PASCAL'S PRINCIPLE.

GURTIN'S VARIATIONAL PRINCIPLE [engineering] (Morton Edward Gurtin, b. 3/7/34; Carnegie Mellon University, PA) (1) The physical basis of the partial differential equation is the postulate of mass conservation. (2) Tendency to optimally organize itself in response to impelling forces and its ability to store and release energy and takes into account the initial displacement and velocity.¶

GUTENBERG PRINCIPLE [psychology] (Johannes Genstleischzur Laden zum Gutenberg, 1398–1468, German printer) (1) New technology always replicates the old. (2) Fundamentally altering the capacity of one institution or set of powers to claim or maintain complete control over the worldview of others. (3) A certain field of knowledge is conveyed in a subset of language depicted as terminology in written language. Human cognition may enter a new stage. Emphasizing pictorial language in cognition.||

*Kannan, R. et al. (1997), SXmin: A self-routing high-performance ATM switch based on group knockout principle, *IEEE Trans. Commun.* **45**, 710.
†White, H. L., Heinbecker, P. and Rolf, D. (1949), Enhancing effects of growth hormone on renal functino, *Am. J. Physiol.* (Legacy Content) **157**(1), 50.

‡McIntyre, T. (1994), The guiding principle: is hunting with a professional guide "true"hunting? It all depends .... (a hunter's view), *Sports Afield*, **21**, 27.
§Walker, P. M. B. (1999), *Chambers Dictionary of Science and Technology*, Chambers, New York, p. 833.
¶Peng, J. S. et al. (1995), Semi analytic method for dynamic response analysis based on Gurtins variational principle, *Commun. Num. Meth. Engi.* **11**, 297.
||Guo, S.-S. (2005), *Br. J. Educ. Technol.* **36**(5), 911–913.

GYARMATI'S PRINCIPLE [physics]   (1) Where there are threats of serious or irreversible damage, lack of full scientific certainty shall not be used as a reason for postponing cost-effective measures to prevent environmental degradation.* (2) The path of a conservative system in configuration space is such that the integral of the Lagrangian function over time is a minimum or maximum relative to nearby paths between the same endpoints and taking the same time.[†] (3) Motion of the underlying system of particles compatible with the collective equilibrium provided the variations are associated with reversible processes.[‡] See also QUANTUM HAMILTON PRINCIPLE; STEWART–HAMILTON PRINCIPLE; STOCHASTIC HAMILTON VARIATIONAL PRINCIPLE; PRIGOGINE PRINCIPLE; PRINCIPLE OF MINIMUM MECHANICAL ENERGY; PRINCIPLE OF MINIMUM POTENTIAL ENERGY; PRINCIPLE OF MINIMUM KINETIC ENERGY; TOUPIN'S DUAL PRINCIPLE; D'ALEMBERT'S PRINCIPLE; HAMILTON PRINCIPLE.

*Kuokkanen, T. (2002), *International Law and the Environment: Variations on a Theme*, Martinus Nijhoff, p. 265.

[†]Bright, J. N., Evans, D. J. and Searles, D. J. (2005), New observations regarding deterministic, time-reversible thermostats and Gauss's principle of least constraint, *J. Chem. Phys.* v. 122, 194106.

[‡]Evans, D. J., Hoover, W. G., Failor, B. H., Moran, B. Ladd, A. and Anthony, J. C. (1983), Nonequilibrium molecular dynamics via Gauss's principle of least constraint, *Phys. Rev. A Atom. Mol. Opt. Phys.* **28**(2), 1016–1021.

H

H INFINITY MATCHING PRINCIPLE [electronics, psychology]   See MATCHING PRINCIPLE.

H PRINCIPLE [mathematics]   (1) Any differential equation (or relation) can be interpreted as a subset S of an appropriate jet space, and a solution is simply an appropriate function whose jet lies in S. Gromov's strategy for solving differential relations was to first find a section of the jet bundle whose image is in S and then try to show that there is a function whose jet agrees with this section. The first part of this program frequently has an algebraic (or sometime geometric) flavor, while the second part is usually more analytic. A differential equation (or relation) satisfies an h principle if the second part of the abovementioned strategy follows automatically (although not necessarily easily) from the first part. Stated another way, an equation (or relation) satisfies an h principle if its solvability is determined by some algebraic (or geometric) data associated with the problem.* (2) Partial differential relations satisfy an h principle if any section of the jet space whose image lies in P can be approximated by a holonomic section. Thus a partial differential relation satisfying an h principle can then be solved by merely finding a section of P.† (3) Uncertainties of position and motion should be no less than the limit set by Planck's constant.‡ (4) Recent developments in the field of chaotic advection in hydrodynamical flows can pertain to the population dynamics of species competing for the same resource in an open aquatic system. If this aquatic environment is homogeneous and well mixed, then classical studies predict competitive exclusion of all except the most perfectly adapted species.§ (5) Spatial heterogeneity generated by chaotic advection can lead to coexistence. In open flows this imperfect mixing lets the populations accumulate along fractal filaments, where competition is governed by an "advantage of rarity" principle.§ (6) New algorithms and methods in applied mathematics and statistics; comprise determination of predictive rank, determination of stable solution, determination of measurement times, tailoring models, model choices in nonlinear multivariate regression, and estimation of nonlinear parameters.¶ See also HEISENBERG PRINCIPLE; CHAOTIC PRINCIPLE.

H₂ MODEL MATCHING PRINCIPLE [electronics, psychology]   See MATCHING PRINCIPLE.

HADAMARD PRINCIPLE [mathematics]   (Jacques Salomon Hadamard, 1865–1963, French mathematician)   (1) Gives an upper bound for the square of the absolute value of the determinant of a matrix in terms of the squares of the matrix entries.‖ (2) Upper bound is the product over the rows of the matrix of the sum of the squares of the absolute values of the entries in a row; a value used for high-order singularities of integral equations.** (3) Spatial light modulation microdevices used in digital light processing.** (4) Used for various problems

*Gromov, M. L. (1986) *Partial Differential Relations*, Springer, Berlin.
†Eliashberg, Y. M. and Mishachev, N. M. (2002), *Introduction to the h-Principle*, AMS, Providence, RI.
‡Zurek, W. H. (2000), Sub-Planck structure in phase space and its relevance for quantum decoherence., *Nature* **412**, 712–717.

§Karolyi, G., Pentek, A., Scheuring, I., Tamas, T., and Toroczkai, Z. (2000), Chaotic flow: The physics of species coexistence, *Proce. Natl. Acad. Sci. USA* **97**(25), 13661–13665.
¶Hoeskuldsson, A. (1992), The H-principle in modelling with applications to chemometrics, Chemometri. Intell. Lab. Syst. 14(1–3), 139–253.
‖Voutier, P. (1996), An effective lower bound for the height of algebraic numbers, Acta Arithmet. (1), **74**(1), 82.
Hotelling, H. (1943), Some new methods in matrix calculation, *Ann. Math. Stat.*, **14(1), 1–34.

including multicrack, nonuniform loading and both infinite space and half-space problems; requires the points at which the stresses are evaluated be distinct from element nodes.* (5) Used to prove an asymptotic law for the distribution of prime numbers.

HALDANE PRINCIPLE [Research]   [Richard Burdon Haldane, (1856–1928, British philosopher/lawyer),   1904]. Decisions about what to spend research funds on should be made by researchers rather than politicians.†

HALDANE'S PRINCIPLE [biology] (John Burdon Sanderson Haldane, 1892–1964, British geneticist and evolutionary biologist)   Sheer size very often defines what bodily equipment an animal must have: "Insects, being so small, do not have oxygen-carrying bloodstreams. What little oxygen their cells require can be absorbed by simple diffusion of air through their bodies. But being larger means an animal must take on complicated oxygen pumping and distributing systems to reach all the cells."‡ See also PRINCIPLES OF ZOOLOGICAL PHILOSOPHY; DARWINIAN PRINCIPLE OF NATURAL SELECTION.

HALF-CALL PRINCIPLE [physics, computer science]   (1) A call is divided into two halves, and each access in the call is assigned its own half-call for traffic control. The call handling is thus implemented as two half-calls. Minimizes the problem of interaction between supplementary services when the different parties in a call use these services.§ (2) Free electrons in a conductor

have sufficient energy to escape the surface.¶ See also THERMIONIC EMISSIONS PRINCIPLE.

HALL PRINCIPLE [physics]   Magnetic sensor.‖

HAMILTON–LAGRANGE PRINCIPLE [physics] (Sir William Rowan Hamlton, 1805–1866, Inshastronomer, physicical and Mathemetician) (1) The path of conservative system in configuration space is such that the integral of the Lagrangian function over time is a minimum or maximum relative to nearby paths between the same endpoints and taking the same time.** (2) Motion of the underlying system of particles compatible with the collective equilibrium provided that the variations are associated with reversible processes.†† See also DIFFERENTIAL PRINCIPLE; VARIATIONAL PRINCIPLE; HAMILTON PRINCIPLE; PRINCIPLE OF LEAST CONSTRAINT; D'ALEMBERT PRINCIPLE; MAUPERTUIS LAGRANGE PRINCIPLE; PRINCIPLE OF EXTREMAL EFFECTS; LEAST-CONSTRAINT PRINCIPLE; D'ALEMBERT–LAGRANGE PRINCIPLE; GAUSS PRINCIPLE; JOURDAIN PRINCIPLE; PRINCIPLE OF VIRTUAL WORK; KINETIC PRINCIPLE; MAUPERTUIS–EULER–LAGRANGE PRINCIPLE OF LEAST ACTION; D'ALEMBERT LAGRANGE VARIATION PRINCIPLE; LAGRANGE PRINCIPLE.

HAMILTON–OSTROGRADSKI PRINCIPLES [physics]   See FICK PRINCIPLE.

HAMILTON PRINCIPLE [mathematics]   (Sir William Rowan Hamilton, 1805–1865, Irish physicist, astronomer, and mathematician) (1) A fundamental principle in dynamics stating that in a conservative field the motion of a mechanical system can be characterized

---

*Riesenberg, R. and Seifert, T. (1999), Design of spatial light modulator microdevices: Microslit arrays, *Proc. SPIE* **3680**), 406–414.

†Winter, M. W., Kanne, S., Lonschinski, J., Auweter-Kurtz, M., and Fruhauf, H. H. (1999), Proposal for a reentry experiment using a newly developed spectrometer, *Aerothermodynamics for Space Vehicles*, 1998, *Proc. 3rd Eur. Symp.* Eur. Space Agency (special publication) SP, SP–426, European Symposium on pp. 711–716.

‡Manning, J. T. (1984), Males and the advantage of sex, *J. Theor. Biol.* **108**(2), 215–220.

§Haccou, P. and Schneider, M. V. (2004), Modes of reproduction and the accumulation of deleterious mutations with multiplicative fitness effects, *Genetics* **166**(2), 1093–1104.

¶Svennevik, A. C. and Lundberg, S. R. (1995), Patent. 5,710,882. (June 29, 1995, Telefonaktiebolaget LM Ericsson).

‖Kimbler, K. and Bouma, L. G. (1998), *Feature Interactions in Telecommunications and Software Systems*, IOS Press, p. 39.

**Kapur, J. N. (1988), *Mathematical Modeling*, New Age International, p. 202.

††Bright, J. N., Evans, D. J., and Searles, D. J. (2005), New observations regarding deterministic, time-reversible thermostats and Gauss's principle of least constraint, *J. Chem. Phys.* **122**(19), p. 194106.

by requiring that the integral $\int_{t_1}^{t_2} (T - V)dt$ be stationary in an actual motion during the time interval from t_1 to t_2. T and V are the kinetic and potential energies of the system. Methods from the calculus of variations are used to study the circumstances under which the integral is stationary.* (2) The motion of any system between two specified times (t_1 and t_2) follows the path in phase space that makes the action of the system a stationary value where the action integral I is defined as $I = \int_{t_1}^{t_2} L dt$, where L is the Lagrangian of the system (a function of the generalized coordinates of a mechanical system; for a nonrelativistic system it is defined as the difference between the kinetic energy of the system and its potential energy: $L = T - V$).[†] (3) Hamilton's principle states that the true evolution q(t) of a system described by N generalized coordinates q = ($q_1, q_2, \ldots, q_N$) between two specified states $q_1 \stackrel{\text{def}}{=} q(t_1)$ and $q_2 \stackrel{\text{def}}{=} q(t_2)$ at two specified times t_1 and t_2 is an extremum (i.e., a stationary point, a minimum, maximum or saddle point) of the action functional

$$S[q] \stackrel{\text{def}}{=} \int_{t_1}^{t_2} L(q(t), \dot{q}(t), t)dt$$

where $L(q, \dot{q}, t)$ is the Lagrangian function for the system. In other words, any *first-order* perturbation of the true evolution results in (at most) *second-order* changes in S. It should be noted that the action S is a functional, i.e., something that takes as its input a function and returns a single number, a scalar.[‡] In terms of functional analysis. Hamilton's principle states that the true evolution of a physical system is the solution of the functional equation

$$\frac{\delta S}{\delta q(t)} = 0$$

(4) Methods from the calculus of variations are used to study the circumstances under which the integral is stationary.[§] (5) From

---

*Hamilton, W.R. On a General Method in Dynamics., Philosophical Transaction of the Royal Society Part I, p. 247–308;

[†](op. cit)

[‡](op. cit)

[§]The new penguin dictionary of science

which can be derived the equations of motion of a classical dynamical system in which friction or other forms of dissipation of energy do not occur. In the original formulation of Newton's laws of motion, the position of each particle of the system of interest is specified by the Cartesian coordinates of that particle. In many cases, these coordinates are not all independent of each other or do not reflect the structure of the system in a convenient way. It is then advantageous to introduce a system of generalized coordinates that are independent of each other and do reflect any special features of the system such as its symmetry about some center. The number of degrees of freedom of the system is the number of such coordinates required to specify the configuration of the system at any time.[¶] (6) The path of a conservative system in configuration space is such that the integral of the Lagrangian function over time is a minimum or maximum relative to nearby paths between the same endpoints and taking the same time.[‖] (7) Motion of the underlying system of particles compatible with the collective equilibrium provided the variations are associated with reversible processes.[**] (8) An alternative formulation of the differential equations of motion for a physical system as an equivalent integral equation, using the calculus of variations.[††] (9) The motion of any system between two specified times (t_1 and t_2) follows the path in phase space that makes the action of the system a stationary value where the action integral I is defined as $I =$ where L is the Lagrangian of the system (a function of the generalized coordinates of a mechanical system). For a nonrelativistic system it is

---

[¶]Weinstock, R. (1974), *Calculus of Variations: With Applications to Physics and Engineering*, Courier Dover Publications, p. 74.

[‖]Beatty, M. F. (2006), *Dynamics—the Analysis of Motion*, Birkhäuser, p. 496.

[**]Bright, J. N., Evans, D. J., and Searles, D. J. (2005), New observations regarding deterministic, time-reversible thermostats and Gauss's principle of least constraint, *J. Chem. Phys.* **122**(19), 194106.

[††]Evans, D. J., Hoover, W. G., Failor, B. H., Moran, B. Ladd, A. J. C. (1983), Nonequilibrium molecular dynamics via Gauss's principle of least constraint, *Phys. Rev. A* and *Atom. Mol. Opt. Phys.* **28**(2), 1016–1021.

defined as the difference between the kinetic energy of the system and its potential energy: $L = T - V$).* See PRINCIPLE OF LEAST ACTION.

HAMILTON PRINCIPLE AND EXTENDED HAMILTON PRINCIPLE [mathematics] (Sir William Rowan Hamilton, 1805–1865, Irish astronomer, physicist mathematician) Under certain conditions, the actual motion of a dynamical system is found by obtaining the extremal of the Hamiltionian integral function $(T - V)dt$, where T and V are kinetic and potential energies of the system. Optimization models arise as a result of application of one or more of these principles. New and challenging mathematical problems arise in obtaining the results from optimizing models.* See also ENTROPY OPTIMIZATION PRINCIPLE HAMILTON PRINCIPLE.

HAMILTON VARIATIONAL PRINCIPLE [physics] (Sir William Rowan Hamilton, 1805–1865, Irish Physicist, astronomer, mathematician) The path of a conservative system in configuration space is such that the integral of the Lagrangian function over time is a minimum or maximum relative to nearby paths between the same endpoints and taking the same time.[†] See HAMILTON'S PRINCIPLE.

HAMILTONIAN   PRINCIPLE [physics] See HAMILTON PRINCIPLE.

HAMMOND PRINCIPLE [chemistry] (George S. Hammond, 1921–2005, Americn chemist) (1) Deals with the transition state of a chemical reaction. If two states (e.g., a transition state and an unstable intermediate) occur consecutively during a reaction process and have nearly the same energy content, their interconversion will involve only a small reorganization of the molecular structures.[‡]

*Walker, P. M. B. (1999), *Chambers Dictionary of Science and Technology*, Chambers, New York, p. 539.
[†]Evans, D. J., Hoover, W. G., Failor, B. H., Moran, B., and Ladd, A. J. C. (1983), *Phys. Rev. A* **28**(2), 1016-1021.
[‡]Bright, J. N., Evans, D. J., and Searles, D. J. (2005), New observations regarding deterministic, time-reversible thermostats and Gauss's principle of least constraint, *J. Chem. Phys.* **122**(19), 194106.

(2) The hypothesis that, when a transition state leading to an unstable reaction intermediate (or product) has nearly the same energy as that intermediate, the two are interconverted with only a small reorganization of molecular structure.[§]

HANDICAP PRINCIPLE [biology, computer science] Amotz Zahavi, 1928- biologist (1) Females may prefer to mate with conspicuous males because these males, having survived despite their handicaps,[¶] must be extraordinarily fit in other respects.[‖] (2) In the genetic algorithm, the entire population generally improves from generation to generation. The improvement, from generation to generation, in the fitness of the population as a whole is evident by examining the average fitness of the population by generation.[**] (3) Electronic circuits using genetic programming; one was whether any significant number of the randomly created circuits would be simulatable. A second concern was whether the crossover operation would create any significant number of simulatable circuits.[††] (4) Darwinian selection apparently is very effective in quickly steering the population on successive generations into the portion of the

[§]Butler, R. N., Duffy, J. P., Cunningham, D., McArdle, P., and Burke, L. A. (1990), Attempted generation of substituted 1,2,3-triazolium−1-methylene ylides: A new ring expansion to 2,3-dihydro−1,2,4-triazines: Ab initio calculations on 1,2,3-triazolium−1-oxide, −1-imide, and −1-methanide 1,3-dipoles and a striking illustration of the Hammond principle in the cyclization of hetero−1,3,5-trienes, *J. Chem. Soc. Chem. Commun.* **12**, 882–884.
[¶]Ando, T. and Yamataka, H. (1981), Prediction rules of transition-state structures. I. The Leffler-Hammond principle and its development, *Kagaku no Ryoiki* **35**(7), 483–493.
[‖]Bell, G. (1978), The handicap principle in sexual selection, *Evolution*, **32**(4), 872–885.
[**]Eshel, I. (1978), On the handicap principle—a critical defence, *J. Theor. Biol.*, **70**(2), 245–250.
[††]Koza, J. R. , Bennett, F. H., III, Andre, D., and Keane, M. A. (1996), Four problems for which a computer program evolved by genetic programming is competitive with human performance, *Proc. IEEE Int. Conf. Evolutionary Computation*, 1996), pp. 1–10.

search space where parents can beget simulatable offspring by means of the crossover operation.* (5) Including the inhibition of the production of RNA, alteration of the level of synthesis of the RNA, alteration in the size or processing kinetics of the RNA, or alteration in the distribution of RNA production throughout the body.[†] (6) Means of identifying the smallest possible region of DNA that could contain a specific gene. The technique relies on the principle when homologous chromatids pair during meiosis (leading to gamete formation); regions along a chromosome very close together or are tightly linked will recombine very rarely relative to those more distantly spaced.[‡] See also ZAHAVI HANDICAP PRINCIPLE.

HANNOVER PRINCIPLE. [Engineering]   (1) Set of statements about designing buildings and objects with forethought about their environmental impact, their effect on the sustainability of growth, and their overall impact on society. May be summarized as: (a) insist on human rights and sustainability; (b) recognize the interaction of design with the environment; (c) consider the social and "spiritual" aspects of buildings and designed objects; (d) be responsible for the effect of design decisions; (e) ensure that objects have long-term value; (f) eliminate waste and consider the entire life-cycle of designed objects; (g) make use of "natural energy flows" such as solar power and its derivatives; (h) be humble, and use nature as a model for design; (i) share knowledge, strive for continuous improvement, and encourage open communication among stakeholders.[§] (2) Should be seen as a living document committed to the transformation and growth in the understanding of our interdependence with nature, so that they may adapt as our knowledge as the world evolves.[¶] (3) Among the first to comprehensively address the fundamental ideas of sustainability and the built environment, recognizing interdependence with nature and proposing a new relationship that includes responsibilities to protect it. (4) Insists on the rights of humanity and nature to coexist by recognizing their interedenpendence. See also BIOMIMICRY PRINCIPLES; DEEP ECOLOGY'S BASIC PRINCIPLES; FOREST STEWARDSHIP COUNCIL (FSC) (PRRINCIPLES; GLOBAL SULLIVAN PRINCIPLES OF SOCIAL RESPONSIBILITY; MARINE STEWARDSHIP COUNCIL (MSC) PRINCIPLES; NATURAL CAPITALISM PRINCIPLES; MELBOURNE PRINCIPLES; PRECAUTIONARY PRINCIPLE; SANBORN PRINCIPLES; TODDS' PRINCIPLES OF ECOLOGICAL DESIGN.

HARD–SOFT ACID–BASE PRINCIPLE [chemistry]   (1) Emphasizes preferential binding of hard metals by hard donor centers and soft electrophilic reagents (metal centers) by soft nucleophilic atoms or groups (mainly unsaturated, aromatic, or heteroaromatic systems.[‖] (2) Hard acids react strongly with hard bases, and soft acids react strongly with soft bases. (a) *Hard species*: difficult to oxidize (bases) or reduce (acids); low polarizabilities; small radii; higher oxidation states (acids); high pK_a (bases); more positive (acids) or more negative (bases) electronegativities; high charge densities at acceptor (acid) or donor (base) sites. (b) *Soft species*: easy to oxidize (bases) or reduce (acids); high polarizability; large radii; small differences in electronegativities between the acceptor and donor atoms; low charge densities at acceptor and donor sites; often have low-lying empty orbitals (bases); often have a number of d electrons (acids).** (3) Exchange reactions of chemicals from hardest possible species the average

*Johnstone, R. A. (1995), Sexual selection, honest advertisement and the handicap principle: Reviewing the evidence, *Biol. Rev. Cambridge Phil. Soc.* **70**(1), 1–65.

[†]Jackson, D. A., Pombo, A., and Iborra, F. (2000), The balance sheet for transcription: An analysis of nuclear RNA metabolism in mammalian cells, *FASEB J.* **14**, 242–254.

[‡]Siller, S. (1998), The handicap principle: A missing piece of Darwin's puzzle by Amotz Zahavi and Avishag Zahavi, *Nature* **392**(6671), 36.

[§]Townsend, A. K. (1997), *The Smart Office*, Gila Press.

[¶]William McDonough Architects (1992).

[‖]Garnovskii, A. D. et al. (2003), Synthesis of coordination compounds with programmed properties, in *Synthetic Coordination and Organometallic Chemistry*, CRC Press chap. 4.

**Choppin, G. R. (1992), Complexation of metal ions, in *Solvent Extraction Principles and Practice*, 2nd ed., M. Dekker.

hardness becomes greater than that of the reactants.* (4) Acids and bases interact selectively in an exchange reaction of the types AB–CD double reaction AC–BD so as to form bonds of similar to the extent possible, polarity between themselves.[†] (5) Reaction and their feasibility on the basics of change in hardness and change in electronegativity. (6) An exchange reaction proceeds in a direction so as to produce the hardest possible species and the average values of the hardnesses of the product.[‡] Also known as *Pearson's HSAB principle; Penaz principle.*

HARDY–SCHULZE PRINCIPLE [physics]   The observation that the efficiency of an ion used as a coagulating agent is roughly proportional to its state of oxidation.[§]

HARDY–WEINBERG PRINCIPLE [biology] (Godfrey Harold Hardy, 1877–1947, British mathematician; Wilhelm Weinberg, 1862–1937, German physicist) (1) The gene frequencies in a large population remain constant from generation to generation if mating is at random and there is no selection, migration, or mutation. If two alleles A and a are segregating at a locus, and each has a frequency of p and q, respectively, then the frequencies of the genotypes AA, Aa, and aa are p^2, $2pq$, and q^2, respectively.[§] (2) Under certain conditions, after one generation of random mating, the genotype frequencies at a single gene locus will become fixed at a particular equilibrium value. It also specifies that those equilibrium frequencies can be represented as a simple function of the allele frequencies at that locus.[¶] (3) Gene frequencies will remain constant from generation to generation within a population that meets certain assumptions. The population will have the given genotypic frequencies (called *Hardy–Weinberg proportions*) after a single generation of random mating within the population. When violations of this provision occur, the population will not have Hardy–Weinberg proportions. Three such violations are (a) *inbreeding*, which causes an increase in homozygosity for all genes; (b) *assortative mating*, which causes an increase in homozygosity only for those genes involved in the trait that is assortatively mated (and genes in linkage disequilibrium with them); and (c) *small population size*, which causes a random change in genotypic frequencies, particularly if the population is very small. This is due to a sampling effect, and is called *genetic drift*.[‖] See also CHETVERIKOV–HARDY–WEINBERG PRINCIPLE.

HARISH-CHANDRA–SELBERG   PRINCIPLE See SELBERG PRINCIPLE.

HARM PRINCIPLE [psychology] (John Stuart Mill's *On Liberty*)   An appeal to individual actors to internalize a norm of tolerance. Only purpose for which power can be rightfully exercised over any member of a civilized community, against that person's will, is to prevent harm to others. Restriction of criminal laws based on the harmful consequences of individual conduct. Internalized norm of toleration or an external constraint on policy. Sole purpose of law should be to stop people from harming others, and that should people want to participate in victimless crimes (i.e., crimes with no complaining witness, such as gambling or engaging in prostitution), then they should not be restrained in doing so.** (2) "To govern absolutely the dealings of society with the individual in the

*Datta, D. and Singh, S. N. (1991), Pearson's chemical hardness, heterolytic dissociative version of Pauling's bond-energy equation and a novel approach towards understanding Pearson's hard-soft acid-base principle, *J. Chem. Soc. Dalton Trans. Inorg. Chem.* (1972-1999) **6**, 1541–1549.

[†]Gazquez, J. L. (1997), The hard and soft acids and bases principle, *J. Phys. Chem. A* **101**, 4657.

[‡]Ayers, P. W. (2007), The physical basis of the hard/soft acid/base principle, *Faraday Discuss.* **135**, 161–190.

[§]Walker, P. M. B. (1999), *Chambers Dictionary of Science and Technology* Chambers, New York, p. 544.

[¶]Guo, S. and Thompson, E. (1992), Performing the exact test of Hardy-Weinberg proportion for multiple alleles, *Biometrics* **48**(2), 361–372.

[‖]Schaap, T. (1980), The applicability of the Hardy-Weinberg principle in the study of populations, *Ann. Human Genet.* **44**(Pt. 2), 211–215.

Kondo, Y. (2006), The Japanese debate surrounding the doping ban: The application of the harm principle. *Sport in Society*, **9(2), 297–313.

way of compulsion and control, whether the means used be physical force in the form of legal penalties, or the moral coercion of public opinion".[*] See also CONTEXT-INDEPENDENT PRINCIPLES.

HARNACK PRINCIPLE [mathematics, linguistics] (Carl Gurtav Axel Harnack, 1851–1888, German mathematician) A theorem about the behavior of sequences of harmonic functions. Key tool in obtaining many results in classical potential theory. Suppose that D is a smooth domain and u and v are two positive harmonic functions on D that vanish on a subset A. The Harnack principle and the boundary Harnack principle state that u and v tend to zero at the same rate.[†] See also BOUNDARY HARNACK PRINCIPLE.

HARRISON'S PRINCIPLE OF INTERNAL MEDICINE [pharmacology] (Dr. Tinsley R. Harrison, Talladega, AL) A strong basis of clinical medicine interwoven with an understanding of pathophysiology.[‡]

HARTREE–FOCK–BOGOLYUBOV PRINCIPLE (HFB) PRINCIPLE [mathematics] See GIBBS THIRD VARIATIONAL PRINCIPLE.

HARTREE–FOCK VARIATIONAL PRINCIPLE [mathematics] (Nikolai Nikolaevich Bogolutov, physicist and mathematician) A form of the fundamental variational principle first derived for the classical case by Gibbs in his study on statistical mechanics. One uses determinants of single-particle wavefunctions rather than products, thereby introducing terms into the Hamiltonian.[§] See also BOGOLUTOV PRINCIPLE; BOGOLUTOV VARIATIONAL PRINCIPLE; GIBBS–BOGOLUTOV VARIATIONAL PRINCIPLE; QUANTAL PRINCIPLE; GIBBS THIRD VARIATIONAL PRINCIPLE; TIME-DEPENDENT HARTREE–FOCK VARIATIONAL PRINCIPLE; SKYRME–HARTREE–FOCK PRINCIPLE.

HASSE PRINCIPLE. [Mathematics] (Helmut Hasse, 1898–1979, German mathematician) (1) One of the central principles of Diophantine geometry which reduces the problem of the existence of reational points on an algebraic variety over a global field to the analogous problem over local fields.[¶] (2) Let Let k be a global field of characteristic different from 2, $\bar{k}$ its algebraic closure, P the set of places of k and k_ν the completion of k at $\nu \in P$. Let X be a smooth projective geometrically integral curve over $k, K = k(X)$ its field of rational functions and $K_\nu = k_\nu(X)$. Let $W(\square)$ denote the Witt group functor. The classical Hasse principle states that the natural map $W(k) \to \prod_{\nu \in P} W(k_\nu)$ is injective. (3) For an algebraic number field F the mapping $BrF \to \oplus_\nu BrF_\nu$ is ibjective, where ν runs over all places of F.[‖]

Algebraic theory for basic contributions in the bodies of classrooms and the application of padicos numbers. For each prine number p, the set of p-ádicos numbers extends to the usual Arithmetic of the rational numbers in a different way of the extension of the one of the rational numbers for Reals or complexes. The theory of the bodies of local classrooms and diofantina geometry and the local functions zeta.[**]

[*]Diekema, D. S. (2004), Parental refusals of medical treatment: The harm principle as threshold for state intervention, *Theor. Med. Bioethics* **25**(4), p.243–264.

[†]Protter, M., and Weinberger, H. (1976), *Maximum Principles in Differential Equations*, Prentice-Hall, Englewood Cliffs, NJ; see also King, H. (1922), The isolation of muscarine, the potent principle of Amanita muscaria, *J. Chem. Soc. Trans.* **121**, 1743–1753.

[‡]Hogan, D. B. (1999), Did Osler suffer from "paranoia antitherapeuticum baltimorensis"? A comparative content analysis of The Principles and Practice of Medicine and Harrison's Principles of Internal Medicine, 11th edition, *Can. Med. Assoc. J.* **161**(7), 842–845.

[§]Brink, D. M., Giannoni, M. J., and Veneroni, M. (1976), Derivation of an adiabatic time-dependent Hartree-Fock formalism from a variational principle, *Nuclear Phys. A* **A258**(2), 237–256.

[¶]Gurak, S. (1978), On the Hasse norm principle, *J. Reine Angew. Math.* **299/300**, 16–27.

[‖]Hsia, J. S. (1973) On the Hasse Principle for Quadratic Forms Proceedings of the American Mathematical Society, v.39, (3), pp. 468–470

[**]Weisstein, Eric W. "Hasse Principle." From MathWorld–A Wolfram Web Resource. http://mathworld.wolfram.com/HassePrinciple.html (see also LOCAL TO GLOBAL PRINCIPLE) Gouvêa, Fernando Q. p-adic Numbers: An Introduction. 2nd ed.

HASSEL PRINCIPLE [chemistry]   See PRINCIPLE OF HASSEL.

HAUSDORFF MAXIMAL PRINCIPLE [mathematics] (Ernst Friedrich Ferdinand Zermelo, 1871–1953, German mathematician) Every partially ordered set has a linearly ordered subset S that is maximal in the sense that S is not a proper subset of another linearly ordeal subset.[*]

HAUSDORFF'S MAXIMAL PRINCIPLE [mathematics] (Ernst Friedrich Ferdinand Zermelo, 1871–1953, German mathematician) Every partially ordered set contains a Q-maximal chain. Maximal principle called the *principle of finite character*. We define what it means for a property to be of finite character. A nonempty property P is of finite character if a class X has property P iff (iF and only iF) every finite subset of X has property P. Each of the properties $AS, TR, C, AS/C, TR/C, P, L, D$, and R are properties of finite character, while W, F, and T are not.[†]

HCI PRINCIPLE [psychology]   Usability has become the central feature. Models were examined that illustrated the increasing refinement of interactive description that focuses on user's needs.[‡]

HEALTH PRINCIPLE [biology, medicine] Common to all the production processes that large amounts of cell material result at the end and contain, inter alia, mainly natural and homologous nucleic acids (DNA and RNA) but also recombinant plasmid DNA or foreign DNA integrated into the genome of the host cell. It has hitherto been possible only with difficulty to estimate the possible risks on release of this recombinant DNA for the environment. Thus, on the precautionary principle, statutory requirements and regulations have been issued by authorities and institutions in most countries [such as, e.g., in the Federal Republic of Germany, the Central Committee for Biological Safety (ZKBS) of the Federal Board of Health in Berlin], which call for inactivation of the waste material, which is usually liquid, and thus of the amount of nucleic acids present after the production process as protection against a possible risk to the environment.[§] See also PRECAUTIONARY PRINCIPLE; PRINCIPLE OF HABEAS MENTEM.

HEART OF THE RUN PRINCIPLE [engineering] Referred to as a *time cycle distillation system*, it involves three stages: (1) heads are removed, (2) product is removed, and (3) residual distillate remaining in the kettle is removed for subsequent redistillation.[¶]

HEAT CONVERTER PRINCIPLE [engineering] (Based on theory of Mr. Kalisky)   allows using as "fuel" a low-temperature heat. Liquid expansion is transformed to mechanical energy, used for different purposes; it could run various equipments or electric generators.[‖]

HEAT PUMP PRINCIPLE [engineering]   Requires provision to recover the heat normally rejected to cooling water or air in the refrigeration condenser.[**]

HEGELIAN PRINCIPLE [psychology] (Georg Wilhelm Friedrich Hegel, 1770–1831, German philosopher) (1) The explicit adequate

[*]Dudley, R. M. (2002), *Real Analysis and Probability*, Cambridge Univ. Press, p. 19.

[†]Harper, J. M. and Rubin, J. E. (1976), Variations of Zorn's lemma, principles of cofinality, and Hausdorff's maximal principle, *Notre Dame J. Formal Logic* **17**(4), 565–588.

[‡]*Encyclopedia of Human Computer Interaction* (E-Book from Idea Group) (2006), *Improving Dynamic Decision Making through HCI Principles*, Idea Group Reference.

[§]Michaels, D. and Monforton, C. (2005), Manufacturing uncertainty: Contested science and the protection of the public's health and environment, *Am. J. Public Health* **95**(Suppl. 1), S39–S48.

[¶]Nissinen, A., Berrios, X., and Puska, P. (2001), Community-based noncommunicable disease interventions: Lessons from developed countries for developing ones, *Bull. WHO* **79**(10), 963–970.

[‖]Lee, Y.-H. (2007), Current status and prospect of strucural cramics. Korean Ceramic Society, Seramiants vol. 4(3), p. 5–6

[**]Crede, H. (1980), Process for the Recovery of Energy and in Particular for the Recovery of Heat on the Heat Pump Principle, U.S. Patent. 4,192,146

development of the science of knowing will require the use of the mathematical theory of categories: (a) subjective logic is a part of objective logic; (b) thinking itself is a part of being; (c) one of the many aspects of being is (b) itself, which is therefore reflect to manifest itself as (a); and (d) considered as a process within being, (a) is a central feature of thinking which a science of thinking must address.[*] (2) Contradiction and negation have a dynamic quality that at every point in each domain of reality—consciousness, history, philosophy, art, nature, and society—leads to further development until a rational unity is reached that preserves the contradictions as phases and subparts by lifting them up to a higher unity. This whole is mental because it is the mind that can comprehend all of these phases and subparts as steps in its own process of comprehension. It is rational because the same, underlying, logical, developmental order underlies every domain of reality and is ultimately the order of self-conscious rational thought, although only in the later stages of development does it come to full self-consciousness. The rational, self-conscious whole is not a thing or being that lies outside other existing things or minds. Rather, it comes to completion only in the philosophical comprehension of individual existing human minds who, through their own understanding, bring this developmental process to an understanding of itself.[†] (3) Central to Hegel's conception of knowledge and mind (and therefore also of reality) was the notion of identity indifference, that is, that the mind externalizes itself in various forms and objects that stand outside of it or opposed to it, and that, through recognizing itself in them, is "with itself" in these external manifestations, so that they are at one and the same time mind and other-than-mind. This notion of identity in difference, which is intimately bound up with his conception of contradiction and negativity, is a principal feature differentiating Hegel's thought from that of other philosophers.[‡]

HEISENBERG INDETERMINANCY PRINCIPLE [psychology] (Werner Karl Heisenberg, 1901–1976, German physicist) (1) Refers both to common scientific and mathematical concepts of uncertainty and their implications and to another kind of indeterminacy deriving from the nature of definition or meaning. (2) Since all communication between sentient beings must be made in some language (whether that language is human speech or writing, body postures, the action of pheromones, or any of a multitude of other types), all scientific and philosophical hypotheses—and, indeed, all statements in general—are given definition by the "words" (or whatever other units may be appropriate given which type of language is being used) of that language.[§]

HEISENBERG PRINCIPLE [physics, mathematics] (Werner Karl Heisenberg, 1901–1976, German physicist) (1) Uncertainties of position and motion should be no less than the limit set by Planck's constant. (2) Recent developments in the field of chaotic advection in hydrodynamical flows can pertain to the population dynamics of species competing for the same resource in an open aquatic system. (3) Spatial heterogeneity generated by chaotic advection can lead to coexistence. In open flows this imperfect mixing lets the populations accumulate along fractal filaments, where competition is governed by an "advantage of rarity" principle. (4) New algorithms and methods in applied mathematics and statistics, comprising determination of predictive rank, determination of stable solution, determination of measurement times, tailoring models, model choices in nonlinear multivariate regression, and estimation of

[*]Lawvere, F. (1994), *Tools for the Advancement of Objective Logic: Closed Categories and Toposes. The Logical Foundations of cognition*, Oxford Univ. Press, New York, pp. 43–56.
[†]Petronis, A. (2004), The origin of schizophrenia: Genetic thesis, epigenetic antithesis, and resolving synthesis, *Biol. Psychiatry* **55**(10), 965–970.

[‡]Wright, J. M. (1973), Viewpoint. The Hegelian principle and the teaching of English to the now generation, *Eng. J.* **62**, 963.
[§]Pertzoff, V. A. (1933), The possible significance of Heisenberg's principle of indeterminancy to the chemistry of living matter, *J. Urusvati Himalyan Res. Inst.* **3**, 79–81.

nonlinear parameters.* See also H PRINCIPLE; CHAOTIC PRINCIPLE.

HEISENBERG SCATTER PRINCIPLE [physics] (Werner Karl Heisenberg, 1901–1976, German mathematician) (1) An interference pattern; For example, when light strikes, or when electrons strike any object, they scatter. But the scatter is a very well-regulated scatter. (2) The most intuitive approach to the problem is to continue using the Fourier transform in conjunction with some window to provide effective localization, yielding the Gabor or short-time Fourier transform (STFT). While this approach, by sliding this localized window in time, permits a time–frequency or 2D representation, the analysis is constrained by resolution concerns imposed by the Heisenberg uncertainty principle.[†] Also known as *principle of measurement; inaccuracy principle; ultrasonic measurement principle.*

HEISENBERG UNCERTAINTY PRINCIPLE [mathematics, physics] (Werner Karl Heisenberg, 1901–1976, German physicist) (1) No experiments can be performed to furnish uncertainties below the limits defined by the uncertainty relationship.[‡] (2) It is impossible to improve simultaneously the time and frequency resolutions of a signal.[§] (3) Simultaneous precise measurement of certain pairs of quantum characteristics, such as position and momentum, cannot be done, even in principle.[¶] (3) With regard to position and momentum, the more accurately you measure the position of a particle, the less accurately you can measure its momentum, and vice versa.[‖] (4) Certain pairs of physical properties, like position and momentum, cannot both be known in any arbitrary precision. The more precisely one property is known, the less precisely the other can be known.

A mathematical statement of the principle is that every quantum state has the property that the root mean square (RMS) deviation of the position from its mean (the standard deviation of the X-distribution):

$$\Delta X = \sqrt{\langle (X - \langle X \rangle)^2 \rangle}$$

times the RMS deviation of the momentum from its mean (the standard deviation of P):

$$\Delta P = \sqrt{\langle (P - \langle P \rangle)^2 \rangle}$$

can never be smaller than a fixed fraction of Planck's constant:

$$\Delta X \Delta P \geq \frac{\hbar}{2}.$$

Any measurement of the position with accuracy ΔX collapses the quantum state making the standard deviation of the momentum ΔP larger than $\hbar/2\Delta x$.[**] (5) The precept that the accurate measurement of an observable quantity necessarily produces uncertainties in one's knowledge of the values of other observables.[††] (6) There is a fundamental limit on the precision of simultaneous measurements, irrespective of the quality of the measuring equipment used.[‡‡] (7) There is a fundamental limit to the precision with which a position coordinate of a particle and its momentum in that direction can

*Louderback, J. (1997), Any PC can fall victim to the Heisenberg principle (altering PC configurations over the network), *PC Week* **14**, 209.

[†]Sakuma, A. (1999), First principles study on the exchange constants of the 3d transition metals, *J. Phys. Soc. Jpn.* **68**(2), 620–624.

[‡]Lyshevski, S. E. (2002), Molecular computing and processing platforms, in *Handbook of Nanoscience, Engineering and Technology*, CRC Press, 2nd ed., chap. 7.

[§]Niethammer, M., Jianmin, Q., and Jarzynski, J. (2000), Time-frequency representation of Lamb waves using the reassigned spectrogram. L19, *J. Acoust. Soc. Am.* **107**(5, Pt. 1).

[¶]Anon. (2004), The Einstein-Podolsky-Rosen argument in quantum theory, in *Stanford Encyclopedia of Philosophy* (http://plato.stanford.edu/entries/qt-epr).

[‖]Clegg, B. (2006), *The God Effect: Quantum Entanglement, Science's Strangest Phenomenon*, Macmillan, p. 18.

[**]Dirac, P. A. M. (1958), *Principles of Quantum Mechanics*, 4th edn., Clarendon Press, Oxford.

[††]Doherty, A. C., Salman, H., Jacobs, K., Mabuchi, H., and Tan, S. M. (2000) Quantum feedback control and classical control theory, *Phys. Rev. A* **62**(162).

[‡‡]Vaughan, N. and Rajan-Sithamparanadarajah, B. (2005), Meaningful workplace protection factor measurement: Experimental protocols and data treatment, *Ann. Occup. Hygiene* **49**(7), 549–561.

be known simultaneously. Also, there is a fundamental limit to the knowledge of the energy of a particle when it is measured for a finite time.[*] (8) The uncertainty of simultaneous measures of energy and time of the mean of change of a given quantum system.[†] (9) Relation whereby, if one simultaneously measures values of two canonically conjugate variables, such as position and momentum, the product of the uncertainties of their measured values cannot be less than approximately Planck's constant divided by 2π.[‡] (10) Two observable properties of a system that are complementary, in the sense that their quantum mechanical operators do not commute, cannot be specified simultaneously with absolute precision. An example is the position and momentum of a particle; according to this principle, the uncertainties in position Δq and momentum Δp must satisfy the relation $\Delta p\,\Delta q = h/4p$, where h is Planck's constant.[§] (11) [Chemistry] The impossibility of making simultaneous measurements of both the position and the momentum of a subatomic particle with unlimited accuracy. The uncertainty arises because, in order to detect the particle, radiation has to be "bounced" off it, and this process itself disrupts the particle's position.[¶] See also UNCERTAINTY PRINCIPLE; PRINCIPLES OF UNCERTAINTY; QUANTUM PRINCIPLE OF UNCERTAINTY; INFORMATION EXCLUSION PRINCIPLE; INTERDETERMINACY PRINCIPLE.

HEISENBERG'S CORRESPONDENCE PRINCIPLE [physics] (Werner Karl Heisenberg, 1901–1976, German mathematician)

[*]Walker, P. M. B. (1999), *Chambers Dictionary of Science and Technology*, Chambers, New York, p. 1214.
[†]Ogrodnik, B. (2004), The metaphysical dimension of optimizing principles, *Concrescence: the Australas. J. Process Thought* **5**, 2.
[‡]Kuo, C.-D. (2005), The uncertainties in radial position and radial momentum of an electron in the non-relativistic hydrogen-like atom, *Ann. Phys.* **316**(2), 431–439.
[§]*CRC Handbook of Chemistry and Physics* (2007), 88th ed., CRC Press, Cleveland, OH.
[¶]Daintith, J. (1999), *The Facts on File Dictionary of Chemistry* 3rd ed., Facts On File, New York, p. 115.

(1) Quantum mechanics in terms of matrices.[‖] (2) More accurately one determines the position of a particle, the less accurately the motion can be known and vice versa. Relation whereby, if one simultaneously measures values of two canonically conjugate variables, such as position and momentum, the product of the uncertainties of their measured values cannot be less than approximately Planck's constant divided by 2π.[**]

HELLENISM PRINCIPLE [psychology] Ideals associated with classical Greek civilization.[††]

HELLINGER–REISSNER PRINCIPLE [mathematics, engineering] (Ernst Hellinger, 1883–195 nad Eric Reissner, 1913–1996; both German mathematicians) (1) The Hellinger–Reissner generalized variational principle is used to deduce further principles directed especially toward formulation of high-integrity finite elements for plates and curved shells in polynomial parametric representation. The assumed stresses in these principles are derived from displacements and supplemented with stresses derived from stress functions."[‡‡] (2) Theory of elasticity, especially develoment of variational method and behavior of beams, plates, and shells. Independent functions are displacement components and new stress functions.[§§] (3) Improving stiffness of the finite element with linear displacement components and stress components; stationary conditions of its functional may satisfy all its field

[‖]Liu, Q. H. and Hu, B. (2001), The hydrogen atom's quantum-to-classical correspondence in Heisenberg's correspondence principle, *J. Phys. A Math. Gen.* **34**(28), 5713–5719.
[**]McFarlane, S. C. (1992), Angular momentum and Heisenberg's correspondence principle, *J. Phys. B Atom. Mol. Opt. Phys.* **25**(20), 4045–4057.
[††]American Heritage (2000), *American Heritage Dictionary of the English Language*, 4th ed., Houghton Mifflin, New York.
[‡‡]Morley, L. S. D. (1984), Hellinger-Reissner principles for plate and shell finite elements, *Int. J. Num. Meth. Eng.* **20**(4), 773–777.
[§§]Isakhanov, G. V. and Chibiryakov, V. K. (1987), Investigation of the stress-strain state and dynamic behavior of thick plates. Report 1. Method for deriving solving equations, *J. Strength Mater.* **19**(2), 243–251.

equations and boundary conditions if all the variables in the functional are considered as independent variations, but there might exist some kinds of constraints.* (4) Displacements and stresses of the (enhanced assumed strain) (EAS) elements calculated from the strains are identical to those of the corresponding hybrid elements.[†] Also known as *Hu Washizu variational principle; pseudo generalized variational principle*.

HELLINGER-REISSNER TWO-FIELD DISPLACE-MENT AND STRESS MODIFIED VARIATIONAL PRINCIPLE [mathematics, engineering]   See DISPLACEMENT PRINCIPLE.

HELLINGER-REISSNER TWO-FIELD MODIFIED VARIATIONAL PRINCIPLE [mathematics, engineering]   See DISPLACEMENT PRINCIPLE.

HELLY SELECTION PRINCIPLE [mathematics] An infinite bounded family of real functions on the closed interval, which is bounded in variation, contains a pointwise convergent sequence whose limit is a function of bounded variation.[‡]

HELMHOLTZ–KORTEWEG PRINCIPLE [mathematics, physics] (Hermann Ludwig Feerdinand van Holmholts, 1821–1894, German Physician)   (Korteweg-nobro) (1) For a nonvanishing nonconstant complex analytic function defined in a domain, the absolute value of the function cannot attain its minimum at any interior point of the domain.[§] (2) Used for estimating probability density function under specified moment constraints; measure of absence of information about a

situation; uncertainty; transformations between measure spaces; expresses amount of disorder inherent or produce; measure of disorder of a system equal to the Boltzmann constant; the natural log (logarithmic) action of the number of microscopic states corresponding to the thermo dynamic state of the system.[¶] (3) Body force density where ρ and ε are the mass density and the dielectric constant of the liquid, respectively. The net force acting on a volume element dV of the fluid is obtained by a volume integration over equation. Associated with porous, mechanically suitable, hydrophilic materials with a high polar surface energy.[‖] (4) Selection of a probability distribution by the *principle of maximum entropy* (MaXEnt). The material is made of constituents, for example, given crystal orientations. Each constituent is itself made of a large number of elementary constituents. The relevant probability is the volume fraction of the elementary constituents that belong to a given constituent and undergo a given stimulus. Assuming only obvious constraints in MaXEnt means describing a maximally disordered material. This is proved to have the same average stimulus in each constituent.[**] (5) Entropy model for deriving the probability distribution of the equilibrium state.[††] (6) A distribution-free method for estimating the quantile function of a nonnegative random variable using the principle of maximum entropy (MaxEnt) subject to constraints specified in terms of

*Sueoka, T. Kondoh, K., Shirahama, T., and Hanai, M., (1994), A numerical method for improving stiffness of the finite element with linear displacement functions using the Hellinger-Reissner principle, *J. Struct. Construct. Eng. Trans. AIJ* **463**, 51–63.

[†]Georgiadis, H. G. and Grentzelou, C. G. (2006), Energy theorems and the J-integral in dipolar gradient elasticity, *Int. J. Solids Struct.* **43**(18–19), 5690–5712.

[‡]Muldowney, P. (2006), Helly's selection principle and the central limit theorem, *Atti Semin. Mat. Fis. Univ. Modena Reggio Emilia* **54**(1–2), 183–190.

[§]Aldrovandi, R. and Pereira, J. G. (1995), *An Introduction to Geometrical Physics*, World Scientific, p. 454.

[¶]Pezzin, G. (1977), Polymer processing and rheometry: Summaries of the lectures delivered at the joint meeting of the British, Dutch and Italian Societies of Rheology, Pisa, April 13–15, 1977, *J. Rheol. Acta.* **16**(6), 652–668.

[‖]Mugele, F. and Baret, J.-C. (2005), Electrowetting: From basics to applications, *J. Phys. Condens. Matt.* **17**, 710.

**Leuchtag, H. R. (2008), *Voltage-Sensitive Ion Channels: Biophysics of Molecular Excitability*, Springer, p. 48.

[††]Velarde, M. G. (1996), Toward a non-equilibrium non-linear thermodynamics, *Book Series Lecture Notes in Physics. Dynamics of Multiphase Flows across Interfaces*, Springer Berlin/Heidelberg, vol. **467**, pp. 253–267.

the probability-weighted moments estimated from observed data.[*] (7) Production of an electromotive force either by motion of a conductor through a magnetic field so as to cut across the magnetic flux or by a change in the magnetic flux that treats a conductor. In a magnetic system for the contactless guidance of a vehicle moved along a track in which a plurality of magnets are attached to the vehicle and arranged one behind the other in the direction of travel, the magnet system cooperating with nonferromagnetic conductor loops on the track to generate forces. The oscillatory movement is thereby maintained by supplying the energy required for maintaining the mechanical swinging or oscillation. To accomplish these principles, known circuit arrangements have the common property that the energy needed for sustaining the oscillation is more effectively supplied by a short driving pulse that is generated each time the mechanical system is in the state of its greatest kinetic energy; this occurs when the mechanical swinger moves through its central position, which lies between the two extreme positions.[†] (8) Energy functional can be transformed into a function of the hot charges on atoms.[‡] (9) Circuit breaker measures current that flows in current transformer using current measuring unit linked to secondary winding wire of transformer.[§] (10) First applied to the relaxation of magnetized plasma to understand the steady-state profiles of RFP (Reversed Field Pinch) configuration under the constraint of a constant rate of supply and dissipation of helicity. It is also a conjecture that relaxed states could be characterized as the states of minimum dissipation

rather than states of minimum energy.[¶] See also PRINCIPLE OF MINIMUM DISSIPATION; ENTROPY PRINCIPLE; VARIATIONAL MINIMAL PRINCIPLE; MAXIMUM ENTROPY PRINCIPLE; MINIMUM CROSS-ENTROPY PRINCIPLE; GENERALIZED MINIMAL PRINCIPLE; MOMENT-PRESERVING PRINCIPLE. Also known as *Korteweg–Helmholtz variational principle*.

HELMHOLTZ PRINCIPLE [physics, electronics] See THEVENIN'S PRINCIPLE; TRAVELING-WAVE PRINCIPLE.

HELMHOLTZ RESONANCE PRINCIPLE [acoustics, physics] (Hermann Ludwig Ferdinand von Helmholtz, 1821–1894, German physician) An air-filled cavity with an opening. The resonence frequency depends on the stiffness of the cavity and the mass of air that oscillates in the opening.[‖] See TRAVELING-WAVE PRINCIPLE.

HENNING FARRIS AUXILIRY PRINCIPLE [genetics] (Emil Hans Willi Hennig, 1913–1976, German biologist and J.S. Farris, biologist) (1) The presence of apomorphous characters in different species is always reason for suspecting kinship and their origin by convergence should not be assumed a priori.[**] (2) Applied to single character data sets, it can be interpreted as a condition that makes the apomorphic state by necessity mark a true nonophyletic group: the state arose only once and never reverted. That group will be present on any tree that requires only a single origin for that state. Grouping by true synapornorphy would have to behave exactly as parsimony, in the sense that it would lead to preference for the tree(s) on hich no homoplasy is present.[††] (3) Features that on the basis of empirical evidence are deemed sufficiently similar to be called the same at some

[*]Holthuijsen, L. H. (2007), *Waves in Oceanic and Coastal Waters*, Cambridge Uni. Press, p. 218.
[†]Michelsen, P. and Rasmussen, J. J. (1982), *Symposium on Plasma Double Layers*, Risø National Laboratory, DK–4000, Roskilde, Denmark p. 155.
[‡]Young, P. M. and Mohseni, M. (2008), Calculation of DEP and EWOD Forces for Application in Digital Microfluidics. *J. Fluids Eng.* **130**(8), 081603.
[§]Lee, T. H., Wilson, W. R., and Sofianek, J. C. (1957), Current density and temperature of high-current arcs. Power apparatus and systems, Part III, *Trans. Am. Inst. Electric. Eng.* **76**(3), 600–606.

[¶]Lee, S.-H., He, X., Kim, D.-K., Elborai, S.,; Choi, H.,-S., Park, I.-H., and Zahn, M. (2005), Evaluation of the mechanical deformation in incompressible linear and nonlinear magnetic materials using various electromagnetic force density methods, *J. Appl. Phys.* p. 97(10), 10E108–10E108–3
[‖]Walker, P. M. B. (1999), *Chambers Dictionary of Science and Technology*, Chambers, New York, p. 552.
[**](Op.cit. Albert, p. 87)
[††](Op.cit. Albert, p. 87)

level of generality should be treated as putative homologues in phylogenetic analysis.[*]

HENNIG'S AUXILIARY METHODOLOGICAL PRINCIPLE [biology] Emil Hans (Willi Hennig, 1913–1976, German Biologist,) Development of a coherent theory of the investigation and presentation of the relations that exist among species. Similarities should be explained as homologous unless incongruence entails convergence. If convergence were our preferred explanation, similarities would not be taken as evidence of relationships.[†] See also HENNIG'S AUXILIARY PRINCIPLE.

HENNIG'S AUXILIARY PRINCIPLE [biology] (Willi Hennig, German Giulogist, 1913–1976) Development of a coherent theory of the investigation and presentation of the relations that exist among species. A shared derived similarity of a group of organisms must be regarded as their synapomorphy (thus as a homology), unless the paraphyly of the group can be demonstrated with other characters, that are conflicting to the putative synapomorphy but regarded as stronger evidence by quantity or quality. In cases of structural similarity of a sufficient degree, homology must be assumed, while convergence should never been assumed a priori, but only be postulated on the basis of the total evidence of all available characters.[‡]

HENRY'S PRINCIPLE [chemistry] The concentration of a gas in solution is proportional to the partial pressure of that gas in equilibrium. The volume solubility of a gas is independent of pressure.[§] See RAOULT'S PRINCIPLE; HENRY'S PRINCIPLE.

[*]Albert, Victor A. (2006) Parsimony, Phylogeny, and Genomics, Oxford University Press, p. 86.
[†]Schlee, D. (1978), Entomologica Germanica **4**, 377–391.
[‡]Harris, S. R., Gower, D. J., and Wilkinson, M. (2003), Intraorganismal homology, character construction, and the phylogeny of aetosaurian archosaurs (reptilia, diapsida), Syst. Biol. **52**(2), 239–252.
[§]Daintith, J. (1999), The Facts on File Dictionary of Chemistry, 3rd ed, Facts On File, New York, p. 117.

HENSEL'S PRINCIPLE [mathematics] (Keart Wilhelm Sebastian Hensel, 1861-1941), An important result in valuation theory that gives information on finding roots of polynomials.[¶]

HEPATOPROTECTIVE PRINCIPLE [pharmacology] Hepatic markers assessed were lipid peroridation, antiulcer activity.[‖]

HERTZ'S PRINCIPLE OF LEAST CURVATURE [mathematics] (Carl Friedrich Gauss, 1777–1855, German mathematician) Formulation of classical mechanics.[**] See also GAUSS' PRINCIPLES; D'ALEMBERT'S PRINCIPLE.

HESS PRINCIPLE [chemistry] (Germain Henri Hess, 1802–1850, Russian chemist) A derivative of the first law of thermodynamics. The total heat change for a given chemical reaction involving alternative series of steps is independent of the route taken.[††]

HETERODYNE PRINCIPLE [engineering, computer science] In telecommunications and radio astronomy, to heterodyne is to generate new frequencies by mixing two or more signals in a nonlinear device such as a vacuum tube, transistor, diode mixer, Josephson junction, or bolometer.[‡‡]

[¶]Rauen, U., Kerkweg, U., and de Groot, H. (2007), Iron-dependent vs. iron-independent cold-induced injury to cultured rat hepatocytes: A comparative study in physiological media and organ preservation solutions, Cryobiology **54**(1), 77–86.
[‖]Yoshikawa, M., Murakami, T., Inadzuki, M., Hirano, K., Ninomiya, K., Yamahara, J., Matsuda, H. (1997), Hepatoprotective principles from Chinese natural medicine "Bupleuri Radix"-structure-activity relationships and chemical modification of bupleurosides- Proc. 39th Conf. Tennen Yuki Kagobutsu Toronkai Koen Yoshishu, pp. 205–210.
[**]Hertz, H. (2004), The Principles of Mechanics Presented in a New Form, Courier Dover Publications.
[††]Daintith, J. (1999), The Facts on File Dictionary of Chemistry, 3rd ed., Facts On File, New York, p. 117.
[‡‡]West, P. W., Burkhalter, T. S., and Broussard, L. (1952), High-frequency oscillator utilizing heterodyne principle to measure frequency changes

HFB PRINCIPLE. See HARTREE–FOCK–BOGOL-YUBOV PRINCIPLE.

HIGH-IMPEDANCE PRINCIPLE [engineering] Zones of protection are required to adjust their boundaries in accordance with changing busbar configuration calling for switching secondary currents.* See also PROTECTION PRINCIPLE; D'ALEMBERT RELAY PRINCIPLE; DISTANCE PRINCIPLE IN RELAY PROTECTION; NOVEL SHIELDING PRINCIPLE; PRINCIPLE OF PROTECTION-SCALE SELECTIVITY; PRINCIPLE OF REASONABLENESS; PRINCIPLE OF STRUC-TURAL PROTECTION; RADIATION PROTECTION PRINCIPLE; BUSBAR PROTECTION PRINCIPLE.

HITTORF PRINCIPLE [psychology]   See GEN-ERATIVE PRINCIPLE OF WRITING.

HIV PRINCIPLE NEUTRALIZING DETERMINANT [pharmacology]   See PRINCIPLES OF THERAPY OF HIV INFECTION.

HOLDREN'S PRINCIPLE [ecology] (John Holdren, 1944-, American environmentalist, current Science and Technology Director of the White House) "Only one rational path is open to us—simultaneous de-development of the [overdeveloped countries] and semi-development of the underdeveloped countries (UDC's), in order to approach a decent and ecologically sustainable standard of living for all in between. By de-development we mean lower per-capita energy consumption, fewer gadgets, and the abolition of planned obsolescence."[†]

HOLIC PRINCIPLE [mathematics]   There is no real difference between a length and a mass, because everything resolves finally in pure numbers because of an overall quantization. Results that astonishing efficacity of dimensionnal analysis must be related to the

properties of Diophantine equations, specifically with the specific degrees.[‡]

HOLLYWOOD PRINCIPLE [computer science] "Don't call us, we'll call you." It has applications in software engineering.[§]

HOLOGRAPHIC PRINCIPLE [psychology]   The mechanism of consciousness and life energy for human beings is considered to have macroscopic quantum condensationlike characteristics.[¶] (2) discuss psychosomatic health science by constructing a model of human consciousness as information and energy from the viewpoints of quantum theory and holographic cosmology. It is believed that omnipresent information and energy in the whole universe's spacetime would be related to nonlocal information based on the holographic principle, and that its localized state would be recognized as individual consciousness. From light freezing of consciousness information to coherent tunneling photons and its energy transformation in brain cells as described by quantum brain dynamics, a weak brain wave voltage and chemical formation and control of hormones in the brain would occur. As a result, it is believed that the consciousness information would affect the DNA, which would be transferred to the whole human body.[‖] (3) Systemic interaction of individuals with their environments and in systemic populations organization with environments.[**] (4) Analogous characteristics of visual perception appear during lesions of the anterior and posterior sections of the brain's

induced by diverse chemical systems, *Anal. Chem.* **22**, 409–471.
*Kasztenny, B., Sevov, L., and Brunello, G. (2001), Digital low-impedance busbar protection, review of principles and approaches, *Proc. 54th Annual Conf. Protective Relay Engineers*, College Station, TX, April 3-5, 2001.
[†]Holdren, J. and Ehrlich, P. (1971), Introduction, in *Global Ecology*, Holdren J. and Ehrlich, P., eds., p. 3.

[‡]Sanchez, F. M. (2000), *The Extended Dimensional Analysis*, (http://www.grandcosmos.org/english/anal-dim_ang.htm).
[§]Medich, R. (1992), The Hollywood principle, *Premiere* **5**, 16.
[¶]Oku, T. (2004), A study on consciousness and life energy based on quantum theory and holographic principle, *J. Int. Soc. Life Inform. Sci.* **22**(2), 431–437.
[‖]Oku, T. (2007), Consciousness-information-energy medicine—health science based on quantum holographic cosmology, *J. Int. Society Life Inform. Sci.* **25**(1), 140–151.
[**]Li, D. J. and Zhang, S. (2008), *The Holographic Principle and the Language of Genes* (arXiv.org), *e-Print Archive, Physics*, pp. 1–5.

right hemisphere.* (5) Graceful exit mechanism that renders the universe nonsingular by connecting pre–and post–big bang phases smoothly.[†] (6) Topology change and sum over topologies not only take place in storm theory but also are required for consistency with holography.[‡] (7) Density fluctuations of extended inflation type and new inflation type.[§] (8) Interaction of any signal carrier with living material with related physiological and psychophysical events and brain functioning.[¶] (9) Frequency range of visible light where all detectors are phase-blind. Some information about the electromagnetic signal is lost. Holography adds redundant information.[‖] See also T'HOOFT'S PRINCIPLE; GABOR'S HOLOGRAPHIC PRINCIPLE

HOMEODYNAMIC   PRINCIPLE [psychology] (Nathan Ackerman)   In the course of family therapy, families have a basic dynamic, and following an interruption to that dynamic (e.g., an intervention by a psychotherapist), the previous family patterns tend to reemerge even if those patterns are dysfunctional.** See also ROGERS PRINCIPLE.

HOOKE'S   PRINCIPLE [geology,   physics] [Robert Hooke, (1635–1703, English scientist) 1676] (1) A basic statement of linear elasticity representing a constitutive

equation for elastic deformation in tension, and analogous equations apply for other deformation modes.[††] (2) A stressed body deforms to an extent that is proportional to the force applied.[‡‡]

HOOVER PRINCIPLE [physics] (W. H. 'Boss' Hoover,   1849–1932   American   vaccum cleaner manufacturer)   The sack gets full. Named after a popular brand vacuum cleaner. See also BLACK HOLE PRINCIPLE; PRINCIPLE OF BLACK HOLE COMPLEMENTARITY; ANTHROPIC PRINCIPLE; STELLAR PRINCIPLE.

HOPKINS' HOST SELECTION PRINCIPLE [biology]   See HOST SELECTION PRINCIPLE.

HOST   SELECTION   PRINCIPLE [biology] Exploited host is very different from the original host. Chemical experience acquired by the larva of an endopterygote insect can be transferred through the pupal stage to the adult.[§§] Also known as *Hopkin's host selection principle*.

HOT-WIRE AMMETER (ANEMOMETER) PRINCIPLE [engineering]   Measures alternation or direct current by sending it through a fine wire, causing the wire to heat and to expand or sag, deflecting a pointer.[¶¶] (2) Used in research on air turbulence and boundary layers.[¶¶] (3) The resistance of an electrically heated fine wire placed in a gas stream is altered by cooling by an amount that depends on the fluid velocity.[¶¶] (4) Activity from the skin is amplified hydraulically

*Willshaw, D. (1981), *Holography, Associative Memory, and Inductive Generalization. Parallel Models of Associative Memory, Lawrence Erlbaum*.
[†]Bak, D. and Rey, S.-J. C. (2000), Holographic principle and string cosmology, *Quantum Gravit.* **17**(1), 1.1–1.7.
[‡]Ocampo, H. Cardona, A., and Paycha, S. (2003), *Proce. Summer School Geometric and Topological Methods for Quantum Field Theory*, Villa de Leyva, Colombia, July 9–27, 2001, World Scientific, p. 378.
[§]Kalyana, R. S. and Sarkar, T. (1999), Holographic principle during inflation and a lower bound on density fluctuations, *Phys. Lett. B* **450**(1–3), 18, 55–60.
[¶]Nanopoulos, D. (1996), *Theory of Brain Function, Quantum Mechanics and Superstrings*, (arXiv.org http://arxiv.org/abs/hep-ph/9505374).
[‖]Mehta, P. C. and Rampal, V. V. (1993), *Lasers and Holography*, World Scientific, p. 602.
Wilson, L. M. and Fitzpatrick, J. J. (1984), Dialectic thinking as a means of understanding systems-in-development: relevance to Rogers's principles, *Adv. Nurs. Sci.* **6(2), 24–41.

[††]Walker, P. M. B. (1999), *Chambers Dictionary of Science and Technology*, Chambers, New York, p. 568.
[‡‡]Stiegeler, S. E. (1977), *A Dictionary of Earth Sciences*, Pica Press, New York, (distributed by Universe Books), p. 140.
[§§]Wood, D. L. (1963), Studies on host selection by Ips confusus (Leconte), (Coleoptera: Scolytidae), with special reference to Hopkins' host selection principle, *Univ. Calif. Publ. Entomol.* **27**(3), 241–282.
[¶¶]Morgan, A. P. (1914), Wireless telegraphy and telephony simply explained; a practical treatise embracing complete and detailed explanations of the theory and practice of modern radio apparatus and its present day applications, together with a chapter on the possibilities of its future development. 3rd ed., Van Nostrand.

by sealing a plastic cup over the area of interest and mounting sensing elements.* (5) Produce biphasic imbalancing of a Wheatstone bridge for sub-audio pulses.[†]

HOT-WIRE PRINCIPLE [engineering] (1) An indicating instrument depending on the thermal change in resistance of a wire when it carries a current.[‡] (2) The resistance of an electrically heated fine wire placed in a gas stream is altered by cooling by an amount that depends on the fluid velocity.[§]

HOVERCRAFT PRINCIPLE [engineering] Levitates heavy objects for easier moving.[¶]

HU—WASHIZU PRINCIPLE [mathematics] (Hu Haichang, 1928-, Chinese mechanical and astronautic engineer) (1) Methods do not require any intuitive or empirical assumptions, but use only the material properties, tool geometry, and the physical laws of deformation.[‖] (2) Found in the solar coronal geometry, which is modeled as a rectangular domain.[**] (3) The bandlimited function of lowpass type whose energy is minimum among those that have the prescribed sample values on a finite set of sampling points. The minimum energy signal converges uniformly to a bandlimited signal when the number of sampling points is increased infinitely.[††] (4) Free boundary subjected to external magnetic or plasma pressure forces.[‡‡] (5) The path of conservative system in configuration space is such that the integral of the Lagrangian function over time is a minimum or maximum relative to nearby paths between the same endpoints and taking the same time.[§§] See also HELLINGER–REISSNER VARIATIONAL PRINCIPLE; PSEUDO–GENERALIZED VARIATIONAL PRINCIPLE (GVP); MINIMUM TOTAL POTENTIAL ENERGY PRINCIPLE.

HU WASHIZU VARIATIONAL PRINCIPLE [mathematics, engineering] (Eric Reissner) (Hu Haichang, 1928, Chinese mechanical and astronautical engineer) See HELLINGER–REISSNER PRINCIPLE.

HUBBLE CONSTANT PRINCIPLE [physics] (Edwin Powell Hubble, 1889–1953, American astronomer) Indicates the rate at which the universe is expanding. Used to determine the intrinsic brightness and masses of stars in nearby galaxies, examine these properties in more distant galaxies, deduce the amount of dark matter in the universe, obtain the size of galaxy clusters, and serve as a test for theoretical cosmological models.[¶¶]

HOUGH TRANSFORM PRINCIPLE [mathematics] (Pdul V.C. Hough) patented but not credited and (Richard O. Dudd, American professor emeritus electrical engineering)

---

*Jovanovic, J. (1996), Hot-wire anemometry: Principles and signal analysis H. H. Bruun. Measure. Sci. *Technol.* **7**(10), 1540.
[†]Thomas, J. S. G. (1918), Hot wire anemometry—its principles and applications, *J. Soc. Chem. Industry* (London), **37**, 165–701.
[‡]Walker, P. M. B. (1999), *Chambers Dictionary of Science and Technology*, Chambers, New York, p. 571.
[§]Oliveira, A. C., Jr., Doi, I., Diniz, J. A., Swart, J. W., and Simoes, E. W. (2003), Modeling and simulation of static characteristics of a PMOS compatible hot wire principle-based flow microsensor, *Proc. Electrochemical Society*, pp. 437–444.
[¶]Rowan's, (1970), *HOVERCRAFT* (http://www.springhurst.org/students/hovercraft.htm).
[‖]Rifai, S. M., Ferencz, R. M., Wang, W.-P., Spyropoulos, E. T., Lawrence, C., and Melis, M. E. (1998), The role of multiphysics simulation in multidisciplinary analysis, *Proc. 7th AIAA/USAF/NASA/ISSMO Symp. Multidisciplinary Analysis and Optimization*, St. Louis, MO, Sept. 2-4, 1998, *Collection of Technical Papers*, Pt. 2 (A98-39701 10–31).
**Bradley, J. N., Lixin, D., and Fumihito, A. (2008), *Micro/Nanorobots. Springer Handbook of Robotics*, Springer Berlin/Heidelberg, Part Pt. B, pp. 411–450.

[††]Saouma, V. (2006), *Merlin Theory Manual*, Electric Power Research Inst., Univ. Colorado, Dept. Civil Engineering, p. 7-2.
[‡‡]Huang, N.-C. (1965), *On Variational Principles in Finite Elasticity*, NASA Accession No. N65-23214, Report TR-154, p. 24
[§§]Altay, G. and Doekmeci, M. C. (2007), Variational principles for piezoelectric, thermopiezoelectric, and hygrothermopiezoelectric continua revisited, *Mech. Adv. Mater. Struct.* **14**(7), 549–562.
[¶¶]Fahr, H. J. and Zoennchen, J. H. (2006), Cosmological implications of the Machian principle, *Naturwissenschaften* **93**(12), 577–587.

(PeterHort, 1940-, American computer scientist) Method for detecting the shape of object boundaries in binary images.*

HULL'S PRINCIPLE OF SECONDARY REINFORCEMENT [psychology] (Clark Leonard Hull, 1884-1952, American Psychology) Let us state the two principles. *Postulate HI*: "Whenever an effector activity (R) is closely associated with a stimulus afferent impulse or trace (s) and the conjunction is closely associated with the rapid diminution in the motivational stimulus (SD or so), there will result an increment (A) to a tendency for that stimulus to evoke that response." *Corollary ii*: "A neutral receptor impulse which occurs repeatedly and consistently in close conjunction with a reinforcing state of affairs, whether primary or secondary, will itself acquire the power of acting as a reinforcing agent"[†]

HULTHEN–KOHN VARIATIONAL PRINCIPLE [chemistry, engineering] Quantum mechanical calculations; quantum reactive scattering.[‡] See also COMPLEX KOHN VARIATIONAL PRINCIPLE; SCHWINGER VARIATIONAL PRINCIPLE; KOHN VARIATIONAL PRINCIPLE.

HUMAN IMMUNODEFICIENCY VIRUS (HIV) PRINCIPLE [pharmacology] (1) Ongoing HIV replication leads to immune system damage and progression to AIDS. HIV infection is always harmful, and true long-term survival free of clinically significant immune dysfunction is unusual.[§] (2) Plasma HIV RNA levels indicate the magnitude of HIV replication and its associated rate of CD4+ T cell destruction, whereas CD4+ T cell counts indicate the extent of HIV-induced immune damage already suffered. Regular, periodic measurement of plasma HIV RNA levels and CD4+ T cell counts is necessary

to determine the risk for disease progression in an HIV-infected person and to determine when to initiate or modify antiretroviral treatment regimens.[¶] (3) As rates of disease progression differ among HIV-infected persons, treatment decisions should be individualized by level of risk indicated by plasma HIV RNA levels and CD4+ T cell counts.[‖] (4) The most effective means to accomplish durable suppression of HIV replication is the simultaneous initiation of combinations of effective anti-HIV drugs with which the patient has not been previously treated and are not cross-resistant with antiretroviral agents with which the patient has been treated previously.[**] (5) Each antiretroviral drug used in combination therapy regimens should always be used according to optimum schedules and dosages.[††] (6) The same principles of antiretroviral therapy apply to HIV-infected children, adolescents, and adults, although the treatment of HIV-infected children involves unique pharmacologic, virologic, and immunologic considerations.[‡‡] (7) Persons identified during acute primary HIV infection should be treated with combination antiretroviral therapy to suppress virus replication to levels below the limit of detection of sensitive plasma HIV RNA assays.[1104] (8) HIV-infected persons, even those whose viral loads are below detectable limits, should be considered infectious. Therefore, they should be counseled to avoid sexual and drug-use behaviors that are associated with either transmission or acquisition of HIV and other infectious pathogens. (9) Involving

---

*Calleri, M. (1962), Determination of crystal structures by the method of optical transformation, *Periodico di Mineralogia*, **31**, 137–161.

[†]Hull, C. L. (1943), *Principles of Behavior*, Appleton-Century, New York.

[‡]Heiss, P. and Hackenbroich, H. H. (1970), Treatment of nuclear reactions by Kohn's variational principle, *Z. Phys.* **235**(5), 422–430.

[§]Farrell, M. H. and Oski, J. A. (2000), *The Portable Pediatrician*, Elsevier Health Sciences, p. 191.

[¶]Green-Hernandez, C., Singleton, J. K., and Aronzon, D. Z. (2001), *Primary Care Pediatrics*, Lippincott Williams & Wilkins, p. 853.

[‖]Committee on the Public Financing and Delivery of HIV Care (2005), *Public Financing and Delivery of HIV/AIDS Care: Securing the Legacy of Ryan White*, National Academies Press, p. 36

[**]Green-Hernandez, C., Singleton, J. K., and Aronzon, D. Z. (2001), *Primary Care Pediatrics*, Lippincott Williams & Wilkins, p. 853.

[††]Farrell, M. H. and Oski, J. A. (2000), *The Portable Pediatrician*, Elsevier Health Sciences, p. 192.

[‡‡]Anon. (1998), *Report of the NIH Panel to Define Principles of Therapy of HIV Infection. MMWR; Recommendations and Reports* (April 24, 1998), **47**(RR–5), 1–41.

people living with HIV/AIDS (PLWAs) in the AIDS response was enshrined in 1994 when 42 countries prevailed on the Paris AIDS Summit to include the Greater Involvement of People Living with HIV/AIDS Principle (GIPA) in the final declaration. While the GIPA principle was a significant step forward, many AIDS activists felt that it did not go far enough.* See also PRINCIPLES OF THERAPY OF HIV INFECTION; GIPA PRINCIPLE (ANTI-HIV PRINCIPLE)

HUMANIST PRINCIPLE [psychology] (Florian Znaniecki) Principle of developing the individual-in-community is a basic principle of teamwork; one cannot develop the individual without developing the team, and vice versa. Organizational form and people development: The team as a vehicle for developing the individual-in-community.[†] (2) "An observer of cultural life can understand the data observed only if taken with the 'humanistic coefficient,' only if he does not limit his observation to his own direct experience of the data but reconstructs the experience and the data in the social context of the people involved."[‡]

HUME'S PRINCIPLE [mathematics] (George Stephen Boolos, 1940–1996, American philosopher and mathematician) (George Boolos) (1) Hume's principle is the statement that for any properties F and G, the number of F is the number of G if and only if F and G are equinumerous. The *finite Hume principle*: (Finite(F) V finite(G) $[(N_x : F_x = N_x : G_x = Eqz(F_x, G_x)]$. Since this principle says nothing about the circumstances under which infinite properties have the same number, it is strictly weaker than Hume's principle. The author shows that PA_2 can be derived from the finite Hume principle, and that with the same definitions, a theory slightly stronger

than PA_2 is equivalent to the finite Hume principle.[§] (2) The number of F values is equal to the number of G values if there is a one-to-one correspondence (a bijection) between the Fs and the Gs.[¶]

HUND'S PRINCIPLE [chemistry] (Friedrich Hermann Hund, 1896–1997, German physicist) The electronic configuration in degenerate orbitals will have the minimum number of paired electrons.[‖]

HUND'S PRINCIPLE OF MAXIMUM MULTICIPLICY [chemistry] (Friedrich Hermann Hund, 1896–1997, German physicist) Degenerate orbitals are occupied singly before spin pairing occurs.[**] See also AUFBAU PRINCIPLE.

HUYGENS' EQUIVALENCE PRINCIPLE [engineering, mathematics] (Christiaan Huggens, 1629–1695, Dutch physicist) See EQUIVALENCE PRINCIPLE.

HUYGENS' FRESNEL PRINCIPLE [optics] (and Augustin-Jean Fresnel 1788–1827, French physicist) Diffracted waves are expressed by an integral of secondary wavelets over an opening. Diffraction problem on the basis of spatial spectra approach and the most common small-angle approximation. Quantitative description of diffraction started when Fresnel introduced the interference effect to Huygen's principle, where no back wave formed.[††] See also HUYGENS' PRINCIPLE.

HUME'S PRINCIPLE. (George Stephen Boolos, 1940–1996, American philosopher and mathematical logician) *sometimes referenced as* Cantor's Principle.

*Anon. (n.d.), *Principles of Antiretroviral Therapy. Practical Antiretroviral Management Recommendations*, (http://aidsinfo.nih.gov/guidelines).
[†]Dovey, K. (1993), *B. J. Guidance Counsell.* **21**(2), 124–132.
[‡]Von Witzleben, H. D. (1957), Natural sciences & humanism as principles of medical education, *Die Medizinische*(17), 643–646.

[§]Heck, R. G., Jr. (1997), Finitude and Hume's principle, *J. Phil. Logic* **26**(6), 589–617.
[¶]McCullough, L. B. (1999), Hume's influence on John Gregory and the history of medical ethics, *J. Medie. Philo.* **24**(4), 376–395.
[‖]Daintith, J. (1999), *The Facts on File Dictionary of Chemistry*, 3rd ed. Facts On File, New York, p. 119.
[**]Op. cit. Daintith, p. 23.
[††]Thrane, L., Yura, H. T., and Andersen, P. E. (2000), Analysis of optical coherence tomography systems based on the extended Huygens-Fresnel principle, *J. Opt. Society Am. A Optics Image Sci. Vision* **17**(3), 484–490.

Basic Law V can be weakened in other ways. The best-known way is due to George Boolos. A "concept" F is "small" if the objects falling under F cannot be put into one-to-one correspondence with the universe of discourse, that is, if: $\exists R[R$ is 1-to-1 & $\forall x \exists y(xRy\&Fy)]$. Now weaken V to V*: a "concept" F and a "concept" G have the same "extension" if and only if neither F nor G is small or $\forall x(Fx \leftrightarrow Gx)$. V* is consistent if second-order arithmetic is, and suffices to prove the axioms of second-order arithmetic.

Basic Law V can simply be replaced with Hume's Principle, which says that the number of Fs is the same as the number of Gs if and only if the Fs can be put into a one-to-one correspondence with the Gs. This principle, too, is consistent if second-order arithmetic is, and suffices to prove the axioms of second-order arithmetic. This result is termed Frege's Theorem because it was noticed that in developing arithmetic, Frege's use of Basic Law V is restricted to a proof of Hume's Principle; it is from this, in turn, that arithmetical principles are derived. On Hume's Principle and Frege's Theorem, see "Frege's Logic, Theorem, and Foundations for Arithmetic".[1]

HUYGENS' PRINCIPLE [physics] (1) Every element of a wavefront acts as a source of so-called secondary waves.* (1) Each point in a light wavefront may be regarded as a source of secondary waves, the envelope of these secondary waves determining the portions of the wavefront at a later time. (2) Every point on an advancing wavefront can be considered to be a source of secondary waves, or wavelets, and the line or surface tangent to all these wavelets defines a new position of the wavefront.[†]

HUYGENS' PRINCIPLE OF SUPERPOSITION [geology]   See PRINCIPLE OF SUPERPOSITION.

*Walker, P. M. B. (1999), *Chambers Dictionary of Science and Technology*, Chambers, New York, p. 574.

[†]Lindell, L.V. (1996), Huygens' principle in electromagnetics, *IEE Proc. Sci. Meas. Technol.* **143**, 103.

HYBRIDOMA PRINCIPLE [medicine]   Fusion occurring between diploid murine lymphoma cells and splenic plasma cells. The lymphoma cells replicated budding mouse leukemia virus particles. The spleen cells produced specific neutralizing antibodies to the mouse leukemia virus.[‡]

HYDROGENIC PRINCIPLE [physics]   Hydrogen is far more plentiful than life.[§] See also ANTHROPIC PRINCIPLE.

HYDROPHOBIC MATCHING PRINCIPLE [electronics, computer science, psychology]   See MATCHING PRINCIPLE.

HYDROSTATIC PRINCIPLE [physics]   (1) A body immersed wholly or partially in a fluid is buoyed up by a force equal in magnitude to the weight of the volume of fluid that it displaces.[¶] (2) Object immersed in a fluid has an upward force equal to the weight of the fluid displaced by the object.[||] See also ARCHIMEDES' PRINCIPLE; PRINCIPLE OF BUOYANCY.

HYPOCALCEMIC HYPOPHOSPHATEMIC PRINCIPLE [organic chemistry]   A factor that lowers serum calcium and inorganic phosphate in rats has been purified 500-fold from 0.1N HCl extracts of hog thyroid glands. It is distinct from thyroxine and triiodothyronine and appears to be a polypeptide.[**]

[‡]Sinkovics, J. G. (1985), Discovery of the hybridoma principle in 1968-69 immortalization of the specific antibody-producing cell by fusion with a lymphoma cell, *J. Med.* **16**(5–6), 509–524.

[§]Hino, K. (2003), Anomalous electric-field dependence of excitonic Fano-resonance spectra in semiconductor quantum-wells, *Solid State Commun.* **128**(1), 9–13.

[¶]Holan, V., Machackova, J., and Zlosky, P. (1980), New method of leg compression measurement using the hydrostatic principle, *Ceskoslovenska Dermatol.* **55**(3), 172–175.

[||]Bassani, R. (1999), Hydrostatic lubrication: The principle and applications of flow self-regulation, *Appl. Mech. Eng.* **4**(2), 291–310.

[**]Hirsch, P. R. (1964), Thyrocalcitonin: Hypocalcemic hypophosphatemic principle of the thyroid gland, *Science*, **146**, 412.

I

IBM POLLYANNA PRINCIPLE [mathematics] "Machines should work; people should think." Machines should do all the hard work, freeing people to think. Most of the world's major problems result from machines that fail to work, and people who fail to think.[*]

ICE TUBE PRINCIPLE [medicine, physics] See ELASTIC TUBE PRINCIPLE.

ICEBERG PRINCIPLE [computer science] Aggregated data can hide information that is important for the proper evaluation of a situation.[†]

ICHINEN SANZEN PRINCIPLE [psychology] (T'ien-t'ai) Literally, *ichinen sanzen* means "3000 conditions in the single mind." "It cannot be said that 'one mind' precedes all phenomena, and not even that all phenomena precede 'one mind'. None of the two can precede the other. It can be, instead, said that the mind is all phenomena and that all phenomena are the mind."[‡]

IDÉE FIXE PRINCIPLE [psychology] (1) An idea that dominates the mind; a fixed idea; an obsession. For example, the initial step on an escalator not in operation triggers a delay in walking due to an *idée fixe*. (2) An obsession—its incessant return to the same few themes, scenarios, and questions; its meticulous examination and reexamination of banal minutiae for hidden meanings that simply aren't there. (3) Usurps other, more interesting thoughts—is that it is confining, not rebellious, and not fascinating but maddeningly dull.[§]

IDENTITY-MATCHING PRINCIPLE (IMP) [psychology] As a heuristic for understanding and predicting the different effects of nested identifications. According to the IMP, when identifications and relevant behavioral or attitudinal outcomes address the same level of categorization, their relationship will be stronger.[¶]

IDEOMOTOR PRINCIPLE [engineering] Actions represented in sensory format, by which "observing the movements of others influences the quality of one's own performance" and develop two neural models that account for a set of related behavioral studies. Compatibility between observed and executed finger movements: comparing symbolic, spatial, and imitative cues.[‖]

ILIZAROV PRINCIPLE OF DISTRACTION OSTEO-GENESIS. [medicine] (Gaurill Agramovich Ilizarov, 1921–1992, Soviet physician) Induction of new bone between bone surfaces

[*]Matlin, M. W. and Stang, D. J. (1978), The Pollyanna principle: Selectivity in language, memory, and thought. Schenkman Books, Inc.

[†]Bayuga, S., Zeana, C., Sahni, J., Della-Latta, P., el-Sadr, W., and Larson, E. (2002), Prevalence and antimicrobial patterns of Acinetobacter baumannii on hands and nares of hospital personnel and patients: The iceberg phenomenon again, *Heart Lung J. Crit. Care* **31**(5), 382–390.

[‡]Endō, A. (1999), Nichiren Shōnins view of humanity; The final Dharma age and the three thousand realms in one thought-moment, *Jpn. J. Relig. Stud.* **26**(3–4), 240.

[§]McKeever, D. C. (1953), Principles and ideals of intramedullary internal fixation, *Clin. Orthopaed.* 212–219.

[¶]Ullrich, J., Wieseke, J., Christ, O., Schulze, M., and van Dick, R. (2007), The identity-matching principle: Corporate and organizational identification in a franchising system, *Br. J. Manage.* **18**, S29–S44.

[‖]Brass, M., Bekkering, H., Wohlschläger, A., and Prinz, W. (2000), Compatibility between observed and executed finger movements: Comparing symbolic, spatial, and imitative cues. *Brain Cogn.* **44**, 124–143.

that are pulled apart in a gradual, controlled manner.* See also DYNAMIC PRINCIPLE OF OSTEOSYNTHESIS.

IMMERSION HEATER PRINCIPLE [engineering] Electric immersion heaters represent an efficient and economical method of heating process solutions. All heaters operate on the principle of heat exchange via temperature differential. The internal temperature versus the temperature of the fluid to be heated is the differential, or "driving force," that heats the fluid.[†]

IMMERSION PRINCIPLE [chemistry] See PERCOLATION PRINCIPLE.

IMMORTALITY PRINCIPLE [psychology, chemistry] (1) Matter is not limited to the perception of the five senses, energy is indestructible, consciousness can exist independently of the physical body, and there is proof that consciousness does not end with death and much evidence that it continues.[‡] (2) Specific cognitive control processes, located in long-term memory, seem to favor the processing of pleasant information, thus effectively screening out unpleasant information.[§] (3) Positive information is remembered more accurately than negative information.[¶]

*Bignardi, A., Boero, G., Barale, I., Mazzinari, S., and Marro, P. (1982), Our experience in the bloodless treatment of fractures with the DOS (dynamic osteosynthesis) external fixation apparatus based on the Ilizarov principle, *Chirurgia Organi Movimento* **68**(1), 51–68; see also Martin, R. C. (2002), *Agile Software Development: Principles, Patterns, and Practices*, Prentice-Hall.
[†]Ara, K., Wakayama, N., and Kobayashi, K. (1983), Development of an in-vessel water level gage for light water power reactors, US Nuclear Regulatory Commission (report), NUREG/CP, (NUREG/CP-0027-V3, *Proc. Int. Meeting Thermal Nuclear Reactor Safety*, 1982, **3**, pp. 1667–1180.
[‡]Siegel, R. K. (1980), The psychology of life after death, *Am. Psychol.* **35**(10), 911–931.
[§]O'Donohue, W. and Graybar, S. (2008), *Handbook of Contemporary Psychotherapy: Toward an Improved Understanding of Effective Psychotherapy*, Sage, p. 98.
[¶]Ford, L. S. and Suchocki, M. (1977), A Whiteheadian reflection on subjective immortality, *Process Stud.* **7**(1), 1–13.

(4) There are two Taoist approaches to immortality: (a) outer alchemy and (b) inner alchemy. Outer alchemy works from the outside in, while inner alchemy works from the inside out. It is the same as the difference between external and internal martial art. Generally speaking, outer alchemy strengthens the physical body first with moving meditation, and various herbal formulas. Inner alchemy, on the other hand, strengthens the energetic body first with still meditation and cultivating the chi.[ǁ] (5) Principle to immortality is derived from the Taoist "three treasures of the universe." The three treasures of the universe are heaven, Earth, and humankind; and each "treasure" contains its own three treasures. The three treasures of heaven are the sun, moon, and stars. The three treasures of Earth are fire, water, and air. The three treasures of humankind are chi, jing, and shen. In order to receive immortality, we must work on our three treasures to raise our spiritual energy and refine it to the same level, or frequency as the Tao. We live and interact in this physical three-dimensional world to learn and nourish ourselves to achieve realization.[ǁ] Also known as *Pollyanna principle*.

IMMUNOSUPPRESSIVE PRINCIPLE [biology] Cell-mediated immunity, which appears to be the relevant host defense mechanism.[**]

IMP See INTERNAL MODEL PRINCIPLE.

INACCURACY PRINCIPLE [mathematics] Relates to measurement accuracies, that is, properties of the measurement device alone, independent of preparation (density operator).[††] See HEISENBERG SCATTER PRINCIPLE.

INCHWORM PRINCIPLE [engineering] To perform a caterpillarlike motion. Actuators

[ǁ]Bonifonte, S. P. (1999), *Chinese Internal Arts Webring*, Chinese Health Inst. (http://internalart. tripod.com/home/immortality.htm).
[**]Motoki, H., Kamo, I., Kikuchi, M., Ono, Y., and Ishida, N. (1974), Purification of an immunosuppressive principle derived from Ehrlich carcinoma cells. *Gann* **65**(3), p. 269–271.
[††]Martens, H. M. (1990), The inaccuracy principle, *Found. Phys.* **20**(4), 357–380.

are controlled in a cyclic mode. Drive elements for each setting axis contained within monolithic block with three piezoelectric setting elements providing simultaneous clamping and drive functions.*

INCIDENT LIGHT PRINCIPLE [physics, mathematics] The object is illuminated from the side, and the reflected light rays are gathered. The object being examined can be observed against a light field or against a dark field in the case of illumination.[†] See also TRANSMITTED LIGHT PRINCIPLE; DIRECT LIGHT PRINCIPLE.

INCLUSION–EXCLUSION PRINCIPLE [mathematics] (Abraham de Moivre, French mathematician, 1667–1754 also named for Daniel da Silva, Joseph Sylvester, or Henri Poincaré) (1) For any sets A and B, $n(A \cup B) = n(A) + n(B) - n(A \cap B)$. If A and B are finite sets, the number of elements in the union of A and B can be obtained by adding the number of elements in A to the number of elements in B, and then subtracting from this sum the number of elements in the intersection of A and B.[‡] (2) The name comes from the idea that the principle is based on overgenerous inclusion, followed by compensating exclusion. When $n > 2$, the exclusion of the pairwise intersections is (possibly) too severe, and the correct formula is as shown with alternating signs. Where an exact formula (in particular, counting prime numbers using the sieve of Eratosthenes), the formula doesn't offer useful content because the number of terms it contains is excessive. If each term individually can be estimated accurately, the accumulation of errors may imply that the inclusion–exclusion formula isn't directly applicable. In number theory, this difficulty was addressed by others, and a large variety of sieve methods

developed. These, for example, may try to find upper bounds for the "sieved" sets, rather than an exact formula.[§] See also SIEVE PRINCIPLE; COMBINATORIAL PRINCIPLE.

INCOMPATIBLE ALTERNATIVE PRINCIPLE [psychology] To stop a child from acting in a particular way, you may reward an alternative action that is inconsistent with or cannot be performed at the same time as the undesired act.[¶]

INDETERMINACY PRINCIPLE [physics] (1) The more precisely the position is determined, the less precisely the momentum is known in this instant, and vice versa. This principle is often called, more descriptively, the "principle of indeterminacy."[‖] (2) Absolutely precise measurements are impossible, due to interference to the measured quantity, which is inevitably introduced by the measuring instrument. The example used by physicists is the impossibility of observing a subatomic particle. Electron microscopes shower their subjects with electrons, and translate the electron "echoes" into images. Electrons, however, are comparable in mass to their subjects and consequently knock them out of their natural states.[**] See HEISENBERG'S UNCERTAINTY PRINCIPLE; PRINCIPLE OF INDETERMINACY.

INDIFFERENCE PRINCIPLE [mathematics] (John Maynard Keynes, 1883–1946, British economist) (Jacob Bernoulli, 1654–1705, Swiss mathematician) "If there is no known reason for predicating of our subject in terms of one rather than another of several alternatives, then relative to such knowledge the assertions of each of these

*Cusin, P., Sawai, T., and Konishi, S. (2000), Compact and precise positioner based on the Inchworm principle, *J. Micromech. Microeng.* **10**, 516–521.

[†]Gabler, F. and Mitsche, R. (1952), The phase-contrast microscope for incident light and its application in metallography, *Archiv Eisenhuettenwesen* **23**, 145–150.

[‡]O'Connor, L. (1993), The inclusion-exclusion principle and its applications to cryptography, *Cryptologia* **17**(1), 63–79.

[§]Ramaswamy, S. (1997), *Discrete Structures. Inclusion Exclusion Principle* (http://users.csc.tntech.edu/~srini/DM/chapters/review3.6.html).

[¶]Huitt, W. (1994), Principles for using behavior modification, in *Educational Psychology Interactive*, Valdosta state Univ., Valdosta, GA (http://chiron.valdosta.edu/whuitt/col/behsys/behmod.html).

[‖]White, F. E. (1934), Some special cases of the indeterminacy principle, *Proc. Natl. Acad. Sci. USA* **20**(9), 525–529.

[**]Takibaev, Zh. S. and Boos, E. G. (1974), Indeterminacy principle and scaling effect, *Dokl. Akad. Nauk SSSR* **217**(4), 804–807.

alternatives have an equal probability."* See INSUFFICIENT REASON PRINCIPLE.

INDIRECT ARC PRINCIPLE [physics] Arcs pass between electrodes supported above the metal in the furnace, which thus is heated solely by radiation from the arc.[†]

INDUCTION MEASUREMENT PRINCIPLE [mathematics, physics] See INDUCTION PRINCIPLE.

INDUCTION PRINCIPLE [mathematics] (David, Hume, 1711–1776, Scottish philosopher) (1) If P is a property and 0 has P, and whenever a number x has P, then $S(x)$ also has P, and it follows that all numbers have P. One of Peano's Postulates.[‡] (2) A method of reasoning by which a general law or principle is inferred from observed particular instances.[§] (3) An inference from statements of the form "the first term of a series has the property P" and "if the $(n+1)$th term of the series has the property P, so does the nth term" to a statement of the form "all terms of the series have the property P." Complete induction allows the move from "the first term of a series has the property P" and "if all terms of the series before the nth term have the property P, so does the nth term" to "all terms of the series have the property P."[¶] See also APPROXIMATION INDUCTION PRINCIPLE; ASYMMETRIC INDUCTION PRINCIPLE; BONE INDUCTION PRINCIPLE; COINDUCTION PRINCIPLE; COVERSET INDUCTION PRINCIPLE; ENANTIODI VERGENT INDUCTION PRINCIPLE; INFERENCE PRINCIPLE; STATIC INDUCTION PRINCIPLE; LYSOGENIC INDUCTION PRINCIPLE; RECURSIVE INDUCTION PRINCIPLE; MCCARTHY'S RECURSION INDUCTION PRINCIPLE; MAGNETIC INDUCTION PRINCIPLE; OSTEOGENIC INDUCTION

PRINCIPLE; PRINCIPLE OF MATHEMATICAL INDUCTION; PROPORTIONAL INDUCTION PRINCIPLE; STRUCTURAL LINERIZATION PRINCIPLE.

INDUCTIVE NUCLEOTHERMAL ENGINE PRINCIPLE [engineering] Features a compromise (performance vs. technological challenge) that lies between conventional chemical propulsion and other nuclear propulsion concepts. This concept aggregates technologies that have already been mastered separately for different purposes in different working environments.[‖] See also STIRLING ENGINE PRINCIPLE; ENGINE PRINCIPLE.

INDUCTIVE PRINCIPLE [electricity] Repellancy between similarly charged bodies and attraction between oppositely charged bodies and the laws of electrostatic induction.[**] See also FARADAY'S ICE PAIL PRINCIPLE.

INERTIA PRINCIPLE [chemistry] (1) Method in predicting Diels–Alder reactivity of aromatic heterocycles by analysis of mass, momentum entropy, and relations for dissipate.[††] (2) Associated with porous, mechanically suitable, hydrophilic materials with a high polar surface energy.[‡‡] (3) Study of chemical reactivity in terms of the generalized and integral similarity indices on a chemical reaction, such as heat or pressure.[§§] (4) Nature chooses the most economical path

[*]Keynes, J. M. (1921), *Treatise on the Principles of Probability*, Macmillan, London, p. 42.
[†]Sasmal, S. K. and Prasad, S. N. (1986), Melting and casting of steel using prime melter, *Tool Alloy Steels* **20**(1), 25–29.
[‡]Shelah, S. and Spinas, O. (2000), The distributivity numbers of $P(\omega)$/FIN and its square, *Trans. Am. Math. Soc.* **352**(5), 2023–2047.
[§]Bukhdahl, G. (1951), Induction and scientific method, *Mind (New Series)* **60**(237), 16–34.
[¶]Kitcher, P. (1984), *The Nature of Mathematical Knowledge*, Oxford Univ. Press, p. 113.

[‖]Dujarric, C. (2002), The inductive nucleothermal engine principle, ESA D/LAU-F, ESF/PESC Exploratory Workshop Nuclear Fission for Future Energy and Space Propulsion: Thin Film and Other Unconventional Solutions, Rome, Italy.
[**]Goovaerts, H. G., Rompelman, O., and van Geijn, H. P. (1989), A transducer for recording fetal movements and sounds based on an inductive principle, *Clin. Phys. Physiol. Meas.* **10**(Suppl. B), 61–65.
[††]Jursic, B. S. (1999), The inertia principle and implementation in the cycloaddition reaction with aromatic heterocycles performed with AM1 semiempirical and density functional theory study, *THEO CHEM* **459**(1–3), 215–220.
[‡‡]Floch, H. G. and Belleville, P. F. (1994), Damage-resistant sol-gel optical coatings for advanced lasers at CEL-V, *J. Sol-Gel Sci. Technol.* **2**(1–3), 695–705.
[§§]Kaviany, M. (1995), *Principles of Heat Transfer in Porous Media*, Springer, p. 470.

for moving bodies, light rays, chemical reactions, and so on, which allows prediction of the effect will have on a change of conditions.* (6) The path of a conservative system in configuration space is such that the integral of the Lagrangian function over time is a minimum or maximum relative to nearby paths between the same endpoints and taking the same time.[†] See also CYCLOPENTENE CYCLOADDITION AROMATIC HETEROCYCLE INERTIA PRINCIPLE; MALEIMIDE CYCLOADDITION AROMATIC HETEROCYCLE INERTIA PRINCIPLE; ACETYLENE CYCLOADDITION AROMATIC HETEROCYCLE INERTIA PRINCIPLE; LAGRANGE PRINCIPLE; JOURDAIN PRINCIPLE; LAGRANGE PRINCIPLE OF LEAST ACTION; LEAST-MOTION PRINCIPLE; PRINCIPLE OF LEAST MOTION; LEAST-ACTION PRINCIPLE.

INERTIAL SLIDER PRINCIPLE See VERTICAL ROTATING PRINCIPLE.

INFERENCE PRINCIPLE [mathematics, physics] See INDUCTION PRINCIPLE; LINKED INFERENCE PRINCIPLE.

INFIMUM PRINCIPLE [mathematics, physics] (H. P. Geering, engineer) Let X and Y be compact, convex subspaces of two linear spaces. If the continuous functions $gi : X \times Y!R, i = 1, 2, \ldots, n, n+1, \ldots, n + m$, satisfy two rather general conditions [in the paper conditions $(gn1)$ and $(gn2)$], then there exists a point $(a, b)2X \times Y$ such that $gi(a, b) = \inf x2 Xgi(x, b)$ for each $i = 1, 2, \ldots, n$, and $gn + j(a, b) = \inf y2Ygn + j(a, y)$ for each $j = 1, 2, \ldots, m$. The restriction to a nonempty subspace of functions is important. There are upper bounds for the growth rate. Static optimization problems for non-scalar-valued performance criteria. The results are felt to be most useful in problems of static and dynamic optimal estimation.[‡] See also SUPREMUM PRINCIPLE; PONTRYAGIN'S MINIMUM

*Peacock, J. A. (1999), *Cosmological Physics*, Cambridge Univ. Press, p. 5.

[†]Altay, G. and Doekmeci, M. C. (2007), Variational principles for piezoelectric, thermopiezoelectric, and hygrothermopiezoelectric continua revisited, *Mech. Adv. Mater. Struct.* **14**(7), 549–562.

[‡]Kulpa, W ladys law and Szymanski, Andrzej (2004) Infimum Principle. Proceedings of the American Mathematical Society, Vol. **132**, No. 1, pp. 203–210.

PRINCIPLE; VON NEUMANN MINIMAX PRINCIPLE. Also referred to as *Geering's infimum principle*.

INFIMUM PRINCIPLE. [mathematics] In mathematics, particularly set theory, the infimum of a subset of some set is the greatest element that is less than or equal to all elements of the subset. Infima of real numbers are a common special case that is especially important in analysis.

THEOREM. *(The Infimum Principle). Let X and Y be compact and convex subspaces of linear spaces. Let continuous functions $g_i : X \times Y \to R, i = 1, 2, \ldots, n + 1, \ldots, n + m$, satisfy conditions:*

(gn1) The partial function $g_i^y : X \to R$ is quasi-convex for each $y \in Y$ and $i = 1, 2, \ldots, n$. The partial function $g_{ix} : Y \to R$ is quasi-convex for each $x \in X$ and $i = n + 1, n + 2, \ldots, n + m$.

(gn2) The family of functions $\{g_i : i = 1, 2, \ldots, n\}$ is n-compatible with respect to the first variable. The family of functions $\{g_i : i = n + 1, n + 2, \ldots, n + m\}$ is m-compatible with respect to the second variable.

Then there exists a point $(a, b) \in X \times Y$ such that

$$g_i(a, b) = \inf_{x \in X} g_i(x, b) \text{ for each } i$$
$$= 1, 2, \ldots, n, and$$
$$g_{n+j}(a, b) = \inf_{y \in Y} g_{n+j}(a, y) \text{ for each } j$$
$$= 1, 2, \ldots, m.$$

INFORMATION-AND-ENERGY PRINCIPLE [psychology] (1) Describes the use of four principles from N. Wiener's (1950) cybernetics and L. von Bertalanffy's (1968) general system theory in daily psychosomatic medicine clinics. The polygraphic method is used for the diagnosis of psychosomatic disease (black box principle). For the control of psychosomatic symptoms, the biofeedback method (feedback principle) is used.

Systematic desensitization is used to relieve social stresses that cause psychosomatic disease (open—closed system principle). Transactional analysis, which corresponds to the information and energy principle. (2) Transactions between parts of a system, including that of the human ecology, result in the exchange of information and energy, such that changing one part of the system, including through therapeutic intervention, will influence the other parts.[*] See also BLACK BOX PRINCIPLE; FEEDBACK PRINCIPLE; OPEN–CLOSED SYSTEM PRINCIPLE.

INFORMATION DIFUSION PRINCIPLE [computer science, physics] (Rogers) The process of diffusion of molecules (the movement of molecules from a region in which they are highly concentrated to a region in which they are less concentrated until the system reaches a state of equilibrium) may be applied to understand how the spread of information leads to organizational or systemic change.[†]

INFORMATION EXCLUSION PRINCIPLE [engineering] Gain of position information can be maximized only at the expense of momentum information and vice versa.[‡] See also HEISENBERG UNCERTAINTY PRINCIPLE.

INFORMATION MAXIMUM LIKELIHOOD PRINCIPLE [mathematics, physics] See PRINCIPLE OF NETWORK ENTROPY MAXIMIZATION.

INFORMATION-PROCESSING PRINCIPLE [computers, psychology] (1) A single time-divisional synchronizing unit applied to the elastic buffering and the monitoring of the fill rate of the buffer memory performed in synchronizing sequentially arranged disassembly and assembly units. The pointers are processed at one or more processing stages on a time division basis; that is, the processing of at least two signals on the same level of hierarchy is performed over the same physical line.[§] (2) Neural networks constitute one of the research techniques for neural information processing, which is referred to as the *constructive method*, and which aims at clarifying the information-processing principle of a human brain by constructing an appropriate neural circuitry model with full consideration given to the facts known physiologically and results of research, investigating the actions and performance of the model, and comparing the actions and performance of the model with those of the actual human brain. The basic information processing units for this principle are neurosymbols, which are structured to neurosymbolic networks. Within these networks, there exist different types of connections.[¶] See also PROCESSING PRINCIPLE; TIME DIVISION PRINCIPLE; also referred to as *conveyor principle*.

INFRARED PRINCIPLE [engineering] Several molecular bonds absorb near infrared light at well-defined wavelengths. The common bonds are O–H in water, N–H in proteins, and C–H in organics and oils. The absorbance level at specific wavelengths is proportional to the quantity of that constituent in the material.[‖] See also PLANCK RADIATION PRINCIPLE.

INFRASTRUCTURE PRINCIPLE [computer science] (1) User-centered, barrier-free, and format-independent.[**] (2) Alternate methods, approaches, and access

[*]Ishikawa, H. (1979), Psychosomatic medicine and cybernetics, *Psychother. Psychosom.* **31**(1), 361–366.

[†]Anon. (2004), *A Model of Matter: Part 9 Diffusion*, Capital Region Science Education Partnership, National Science Foundation Grant 991 186 (http://www.crsep.org/PerplexingPairs/June1704.ModelofMatterPart9.pdf).

[‡]Hall, M. J. W. (1995), Information exclusion principle for complementary observables, *Phys. Rev. Lett.* **74**(17), 3307–3311.

[§]Oksanen, T., Viitanen, E., Patana, J., and Alatalo, H. (1997), *Method for Disassembling and Assembling Frame Structures Containing Pointers*, US Patent 5,666,351.

[¶]Velik, R. and Bruckner, D. (2008), *Neuro-Symbolic Networks: Introduction to a New Information Processing Principle* (http://publik.tuwien.ac.at/files/PubDat_{166316}.pdf).

[‖]Tesmer, W. H. (1945), The infrared principle: What it is, what it does, *Am. Ceramic Soc. Bull.* **24**, 162–163.

[**]Unger, J. P. and Criel, B. (1995), Principles of health infrastructure planning in less developed countries, *Int. J. Health Plann. Manage.* **10**(2), 113–128.

points.* (3) Minimize restrictions on access and maximize dialog.† See also INTELLECTUAL PROPERTY PRINCIPLE.

INKJET PRINCIPLE [engineering] Relationship between the chambers, passages, and the droplet outlet orifice for a combination ejection of a thin film of stain to make direct application to a printing surface.‡

INLINE PLANT PRINCIPLE [engineering] Fundamental spacing and elevation requirements for equipment arrangement.

INSTANTIATION PRINCIPLE [psychology] (1) If something has a property, then necessarily that "something" must exist. For it not to exist would be a property without an essence, which is impossible. (2) Crucial building block of Descartes' philosophy, which is summed up in the cogito ergo sum argument: I am thinking, therefore I exist.§ (3) The representation of a category includes detailed information about its diverse range of instances. Knowledge about a general category reflects a great deal of detailed information about the diverse range of its instances. Detailed information about instances enters into categorization processes.§ See also COGNITIVE ECONOMY PRINCIPLE.

INSUFFICIENT REASON PRINCIPLE [mathematics] (Johann Bernoulli, 1667–1748, Swiss mathematician) If we are ignorant of the ways in which an event can occur and therefore have no reason to believe that one

way will occur preferentially to another, it will occur equally likely in any way.¶ See also INDIFFERENCE PRINCIPLE.

INTEGRAL INVARIANCE PRINCIPLE [physics, mathematics, chemisty] The principal new ingredients are the use of observation functions and certain integrability conditions, which are particularly well suited for dynamical systems involving control and observations. The integral invariance principle leads to the development of a series of results relating stability, observability, and the converse theorems of Lyapunov theory. Corollaries include apparently diverse stabilizability results for adaptive control, nonlinear control, and passive circuits and systems. An integral-invariance principle for nonlinear systems.‖ See INVARIANCE PRINCIPLE; LASALLE'S INVARIANCE PRINCIPLE.

INTEGRATED OPTICAL LIGHT POINTER PRINCIPLE [engineering] Planar optical sensor platform with transducer and recognition layer for conversion of changes in effective refractive index to measurable variables.**

INTELLECTUAL FREEDOM PRINCIPLE [computer science] See INTELLECTUAL PROPERTY PRINCIPLE.

INTELLECTUAL PROPERTY PRINCIPLE [computer science] (1) User-centered, barrier-free, and format-independent. (2) Alternate methods, approaches, and access points. (3) Minimize restrictions on access and maximize dialog.†† See also INTELLECTUAL FREEDOM PRINCIPLE; PRIVACY PRINCIPLE; INFRASTRUCTURE PRINCIPLE; UBIQUITY PRINCIPLE; EQUITY OF ACCESS PRINCIPLE; CONTENT PRINCIPLE; PRINCIPLES FOR NETWORKED WORLD.

*Bell, C. C. and McKay, M. M. (2004), Constructing a children's mental health infrastructure using community psychiatry principles, *J. Legal Med.* **25**(1), 5–22.
†Rogers, J. C. (2008), Assembling patient-centered medical homes—the promise and price of the infrastructure principles, *Family Med.* **40**(1), p. 11–12.
‡Wehl, W. (2001), *Use of a Print Head that Functions According to the Ink Jet Principle for Producing a Microcomponent and Device for Producing a Microcomponent*, PCT Int. Patent Application, p. 21. German Patent Pub no. wO/2001/002173
§Heit, E. and Barsalou, L. W. (1996), The instantiation principle in natural categories, *Memory* **4**(4), p. 21. 413–51.

¶Freeman, S. R. (2007), Rawls, Routledge, p. 174.
‖Byrnes, C. I. and Martin, C. F. (1995), *IEEE Trans. Autom. Control* **40**(6), 983–994.
**Duveneck, G. L., Dubendorfer, J., Kunz, R. E., and Kraus, G. (1998), *Optical Chemical/Biochemical Sensor*, PCT Patent Application p. 33. German Patent WO/1998/609156
††Cate, F. H. (1996), Intellectual property and networked health information: Issues and principles, *Bull. Med. Library Assoc.* **84**(2), 229–236.

INTELLIGENCE   PRINCIPLE [psychology] (Stephen J. Dick, American Astronomer) A hypothetical central concept of cultural evolution describing a potential binding tendency among all intelligent societies, both terrestrial and extraterrestrial. The maintenance, improvement and perpetuation of knowledge and intelligence is the central driving force of cultural evolution, and that to the extent intelligence can be improved, it will be improved.*

INTENTION-TO-TREAT   PRINCIPLE [psychology] Wiener's cybernetics and von Bertalanffy's general system theory adopted four principles from psychosomatic medicine clinics: (1) the *polygraphic method* for the diagnosis of psychosomatic disease (black box principle), (2) the *biofeedback method* (feedback principle) for the control of psychosomatic symptoms, (3) the *open–closed system principle*, for systematic desensitization to relieve social stresses that cause psychosomatic disease and (4), *transactional analysis*, which corresponds to the information-and-energy principle.† See also ISHIKAWA'S PRINCIPLE; BLACK BOX PRINCIPLE; FEEDBACK PRINCIPLE; OPEN–CLOSED SYSTEM PRINCIPLE.

INTERACTION BALANCE PRINCIPLE [mathermatics] If the interaction variables are let free, then the overall solution is reached when the values they are given independently by the infimal units are consistent.‡ See also MAXIMUM PRINCIPLE.

INTERACTION   EQUIVALENCE   PRINCIPLE [mathematics] See   ARIOKHIN–PARETO INTERACTION PRINCIPLE.

INTERACTION PREDICTION PRINCIPLE [mathermatics]   If the supremal unit predicts the values of the interactions between the sub-processes controlled by the infimal units in order to coordinate their action, then the overall solution is reached when the value of these interactions resulting from the controls suggested by the infimal units is equal to the predicted value.§ See also MAXIMUM PRINCIPLE.

INTERACTION   PRINCIPLE [mathematics] When radiation is incident on a material medium, it is absorbed, scattered, or both, and the radiation emerging relates to the incident on the medium.§ See ARIOKHIN–PARETO INTERACTION PRINCIPLE.

INTERCULTURAL COMMUNICATION PRINCIPLE [psychology] Guides the process of exchanging meaningful and unambiguous information across cultural boundaries, in a way that preserves mutual respect and minimizes antagonism. Culture is a shared system of symbols, beliefs, attitudes, values, expectations, and norms of behavior. It refers to coherent groups of people, whether resident wholly or partly within state territories, or existing without residence in any particular territory.¶

INTERDETERMINACY   PRINCIPLE [engineering, mathematics] See also HEISENBERG UNCERTAINTY   PRINCIPLE,   UNCERTAINTY PRINCIPLE.

INTERFACE   PRINCIPLE [engineering]   See also ASSEMBLY SEQUENCE PRINCIPLE.

INTERFACE SEGREGATION PRINCIPLE (ISP) [computers] (Robert Cecil Martin)   (1) Sometimes class methods have various groupings.‖ (1) Make fine grained interfaces that are

*Dick Steven J. (2003), Cultural Evolution, the Postgiological Universe and SETI. Internaternational Journal of Astrobiology, **v2**, p65–74.

†Lachin, J. M. (2000), Statistical considerations in the intent-to-treat principle, *Controll. Clin. Trials* **21**(3), 167–189.

‡Peraiah, A. (1984), The interaction principle in radiative transfer, *Astrophys. Space Sci.* **105**(1), 209–212.

§Libosvar, C. M. (1988), *Hierarchies in Production Management and Control: A Survey*, Laboratory for Information and Decision Systems, Massachusetts Inst. Technology, LIDS-P-1734, p. 17.

¶Hanssen, I. (2004) From human ability to ethical principle: An intercultural perspective on autonomy, *Medi. Health Care Phil.* **7**(3), 269–279.

‖Boguslawski, P., Szwacki, N., and Gonzalez, B. J. (2006), Interfacial segregation and electrodiffusion of dopants in superlattices, *Phys. Revi. Lett.*, **96**(18), 185 501.

client-specific. (2) Deals with the disadvantages of "fat" interfaces. Classes that have "fat" interfaces are classes whose interfaces are not cohesive. In other words, the interfaces of the class can be broken up into groups of member functions. Each group serves a different set of clients. Thus some clients use one group of member functions, and other clients use the other groups.*

INTERFERENCE OF LIGHT PRINCIPLE [physics] Projection apparatus for creating visible images utilizing light for the construction of its image.[†]

INTERFERENCE PRINCIPLE [physics, quantum mechanics] (1) Whenever two waves occupy the same region of space at the same time they are said to "interfere." The resultant wave is the sum of the two waves.[‡] (2) If the wavefunctions for two particles (a and b) overlap, then the probability density becomes $(a*+b*)(a+b)$ (* denotes complex conjugation). The *square* of the net wavefunction is not simply $a*a + b*b$ (the sum of the squares of the two individual wavefunctions); there is an extra bit equal to $a*b + ab*$ interference terms.[§] (3) The presence of the interference terms in one of the distinguishing characteristics of quantum mechanical behavior of a pair of particles; classical mechanics would predict

a different result.[¶] See also RESOLUTION PRINCIPLE.

INTERFERENCE PRODUCTION PRINCIPLE [ecology] To explain the coexistence of species, ecologists have resorted to a myriad of ecological principle involving productivity, predation pressure, climatic factors, and other mechanisms. Not only is interference probably the most frequently cited mechanism, but it also contains what is likely the most agreed-on principle in ecology. See also COMPETITIVE EXCLUSION PRINCIPLE.[‖]

INTERFEROMETRIC PRINCIPLE [physics] (1) Enables a precision more than 10-fold better than that of ultrasound.** (2) A reflected beam from an object occurs on the beamsplitter, where it is split in two. The split beam is transmitted where the parts combine to give interference fringes. A vibrating beam modulates the frequency of the interferometric signal. Interferometic demodulation approaches allow the extraction of the signal of the vibrating beam.[††] (3) Each receive aperture is split, through electronic beamforming, into "half-beams," and the phase difference for each received signal for each aperture is calculated to provide a measure of the angle of arrival of the echo. The point at which the phase is zero is determined for each aperture and provides an accurate measure of the range. Both amplitude and phase detection are recorded for each aperture, and the system software picks the "best" detection method for each aperture, based on a number of quality

*Martin, R. C., (1995) *Dependency Inversion Group. Ten Commandments of OO Proramming.* (http://www.butunclebob.com/ArticleS.UncleBob. PrinciplesOfOod).

[†]Hoffman, R. and Gross, L. (1970), Reflected-light differential-interference microscopy: Principles, use and image interpretation, *J. Microsc.* 91(3), 149–172.

[‡]Yan, M., Wu, M., Zhang, T., Zhang, Y., and Ruan, T. (1985), Neutron spin interference, principle of spinor superposition and influence of earth gravity on the NSE, *Commun. Theor. Phys.* 4(4), 473–482.

[§]Li, B., Chua, S. J., Fitzgerald, E. A., Chaudhari, B. S., Jiang, S., and Cai, Z. (2004), Intelligent integration of optical power splitter with optically switchable cross-connect based on multimode interference principle in SiGe/Si, *Appl. Phys. Lette.* 85(7), 1119–1121.

[¶]Chen, C. C. and Huang, C. P. (2007), Color filter utilizing film interference principle, *Faming Zhuanli Shenqing Gongkai Shuomingshu*, p. 27

[‖]Vanderineer, J. (1981), The interference production principle: an ecological theory for agriculture, *Bioscience* 3, 361.

**Morariu, Gh. and Mailat, A. (2005), Considerations on oxygen sensing devices based on interferometric principles, *Bull. Transylvania Univ. Brasov, Series A: Mechanics, Electrotechnics and Electronics, Materials Processing, Wood Industry, Silviculture* 12, 397–400.

[††]Kaule, W. (1975), *Interferometric Method and Apparatus for Sensing Surface Deformation of a Workpiece Subjected to Acoustic Energy*, US Patent 4,046,477.

control measurements, and uses this method to calculate depth.[*] See also INTERNAL EFFECT PRINCIPLE; MICHELSON INTERFEROMETRIC PRINCIPLE; RADIATION PRINCIPLE; PRINCIPLE OF LINEAR SUPERPOSITION.

INTERFEROMETRY NEUTRON PRINCIPLE [physics] Neutron interferometry involves exploiting the quantum mechanical nature of neutrons to use them in a manner similar to that using X rays to perform diagnostic measurements.[†] See also RESOLUTION PRINCIPLE; INTERFERENCE PRINCIPLE.

INTERNAL EFFECT PRINCIPLE [physics] (1) Appraisals to derive response relationships between equilibrium concentrations and effects.[‡] (2) General categorization and assessment of effects. (3) Determine the mechanism or mode for developing descriptive and predictive models. See also RADIATION PRINCIPLE; INTERFEROMETRIC PRINCIPLE; MICHELSON INTERFEROMETRIC PRINCIPLE.

INTERNAL MODEL PRINCIPLE [mathematics] (1) Controller properties are necessary for structural stability. This is the converse problem of synthesis for achieving structurally stable controllers in the linear multivariable setting. Necessary structural criteria are obtained for linear multivariable regulators that retain loop stability and output regulation in the presence of small perturbations, of specified types, in system parameters. It is shown that structural stability thus defined requires feedback of the regulated variable, together with a suitably reduplicated model, internal to the feedback loop, of the dynamic structure of the exogenous reference and disturbance signals that the regulator is required to process.

Necessity of these structural features constitutes the *internal model principle*.[§] (2) The repetitive control law is applied to the entire control path.[¶] (3) Mathematical system theory (internal model principle) and the relativistic cybernetics are essentially used to explain how severe head injuries destroy the structural homeostasis of personality by influencing the noncognitive elements via the cognitive elements.[‖] (4) Used to ensure that the expected value of the output approaches zero asymptotically in the presence of persistent deterministic disturbances.[**] (5) The robust H[infinity] control theory, and the regional stability constraints. (6) Conditions under which the online algorithms yield an asymptotic controller. Conditions both for the case where the disturbance input properties are constant but unknown and for the case where they are unknown and time varying are given. An adaptive regulation approach against disturbances consisting of linear combinations of sinusoids with unknown and/or time-varying amplitudes, frequencies, and phases for Single Input Single Output (SISO) Linear Time Invariant (LTI) discrete-time systems is considered. The repetitive control law is applied to the entire control path.[††] (7) In the design of rofust linear feedback rules, equilibrium is prespecified in some way and the model is then built conditionally on that equilibrium specification. Identify the restrictions that determine whether a given

[§]Francis, B. A. and Wonham, W. M. (1975), The internal model principle for linear multivariable regulators, *Appl. Math. Optim.* 2(2), 170–194.

[¶]Tzou, Y.-Y., Jung, S.-L., and Yeh, H.-C. (1999), Adaptive repetitive control of PWM inverters for very low THDAC-voltage regulation with unknown loads, *IEEE Trans. Power Electron.* **14**(5), 973–981.

[‖]George, F. H. (1962), *The Brain as a Computer*, Pergamon Press, Oxford/Addison Wesley, Reading, MA (http://www.archive.org/stream/brainasacomputer007406mbp/brainasacomputer-007406mbp_djvu.txt).

[**]Sparks, A. G. and Bernstein, D. S. (1997), Optimal rejection of stochastic and deterministic disturbances, *J. Dynamic Syst. Meas. Control* **119**(1), 140.

[††]Bodson, M., Sacks, A., and Khosla, P. (1994), Harmonic generation in adaptive feedforward cancellation schemes, *IEEE Trans. Autom. Control* **39**(9), 1939–1944.

[*]Li, Y. (1997), Phase aberration correction using near-field signal redundancy, *IEEE Trans. Ultrasonics Ferroelectrics Freq. Control* **44**(2), 355–371.

[†]Bonse, U. (1979), Principles and methods of neutron interferometry, *Proc. Int. Workshop on Neutron Interferom etry*, 1978, pp. 3–33.

[‡]Golbazi, A. M. and Kefalas, N. D. (1993), Hybrid temperature sensor based upon internal-effect principle, *Proc. SPIE* **1952**, 244.

dynamic system satisfies the conditions for error correction specifications; system parameters such as variable disturbance and design and estimating using the internal model of the disturbance so that the disturbance can be rejected.[*]

INTERNAL NOISE PRINCIPLE [psychology] According to the noisy operator theory, internal noise will often make two identical letters appear to be different but rarely to be identical. As a result, perceived mismatches, but not perceived matches, will be rechecked, and response time (RT) generally will be longer on different trials. Further, since rechecking will often be incomplete, owing to impatience and to the apparent willingness of subjects to accept about a 4% risk of error, some misperceived same pairs will trigger a false-different response. This prediction of more errors on same trials is "good," or benevolent, in that it does not allow the speed advantage for same pairs to be attributed merely to a motor bias toward the same button, as would be true if the prediction were for more errors on different trials (false-same responses).[†] See also PRIMING PRINCIPLE.

INTERNATIONAL SAFE-HARBOR PRIVACY [psychology, computer science]   A set of privacy regulations that store customer data and designed to prevent accidental information disclosure or loss. The principles are outlined as follows: (1) *notice*—a company must provide a data usage statement, to inform the user of how said company will use their data; (2) *choice*—the customer must have the option of "opting out" of any information disclosure; (3) *onward transfer*—if the company chooses to disclose information to another entity, it must adhere to the principles of *choice* and *notice*; (4) *security*—companies must secure their systems to protect against the loss, misuse, disclosure, destruction, and alteration of data; (5) *data integrity*—the data must be processed relevant to the purpose for which they were originally collected; (6) *access*—the customer must have access to his/her data so that he/she can add, edit, or delete data; and (7) *enforcement*—the company must enforce these principles, as well as its own internal policies and procedures, in the aim of preventing accidental or intentional data disclosure or loss.[‡] See also SAFE-HARBOR PRIVACY PRINCIPLE.

INTERPOLATION PRINCIPLE [mathematics] Used to estimate an intermediate value of one (dependent) variable that is a function of a second (independent) variable when values of the dependent variable corresponding to several discrete values of the independent variable are known.[§] Also known as *smallest error computer interpolation principle; speech interpolation principle*.

INTERSECTION PRINCIPLE [physics]   (1) The method for generating a rotating elliptical sensing pattern. (2) Any number of arms intersecting and utilizing phase excitation to generate a rotating magnetic field. This type of pole arrangement is similar to polyphase induction motor stator designs for generating rotating magnetic fields.[¶] See also MAXIMUM PRINCIPLE.

INTERSTITIAL PRINCIPLE [chemistry]   Ratio of the radii in the carbides of Mn, Fe, Ni, and Co, as well as chromium carbide. Hardness values and melting points are lower and stability to mineral acids is not apparent.[‖]

[*]Ohno, H., Ohshima, M., and Hashimoto, I. (1992), Model-based predictive control and internal model principle, *IFAC Symposia Series* (1992) (*Adv. Control Chem. Processes*), **8**, 157–162.

[†]Krueger, L. E. (1983), Probing Proctor's priming principle: The effect of simultaneous and sequential presentation on same-different judgments, *J. Exp. Psychol. Learn. Memory Cogn.* **9**(3), 511–523.

[‡]European Union's comprehensive privacy legislation, the Directive on Data Protection (the Directive) (1999), *International Safe Harbor Privacy Principles* (http://www.ita.doc.gov/td/ecom/shprin.html).

[§]Ma, C., Jin, Z., Tian, F., Yang, N., Yang, S., and Liu, S. (1998), Bandgaps and band offsets in strain-compensated InGaAs/InGaAsP multiple quantum wells, *Proc. SPIE* **3547**, 308–314.

[¶]Sen, P. K., Tsai, M.-T., and Jou, Y.-S. (2007), High-dimension, low-sample size perspectives in constrained statistical inference: The SARSCoV RNA genome in illustration, *J. Am. Statist. Assoc.* **102**(478), p. 686–694.

[‖]Nowotny, H. (1997), Crystal chemistry of complex carbides and related compounds, *Angew. Chem. Int.* **11**(10), p. 906–915.

INTERTIBILITY   PRINCIPLE [mathematics] Linear transformations in algebraic expressions.*

INVARIANCE   PRINCIPLE [mathematics] (1) Any principle that states a physical quantity or physical law possesses invariance under certain transformations.[†] (2) Selection of inference procedures with some permutations or groups of permutations with pa[mathematics] rameters involved in such problems. The persent paper is devoted to a study relating to the characterization of regular cyclically invariant non-singularly estimable full rank problems of linear inference. The recognition of a given linear inferential problem as such an invariant problem may simplify the search for the best experiment when a number of alternative experiments are available. In fact, it has been illustrated that every symmetrical allocation is optimum for inferring about such problems for any convex symmetric criterion.[‡] See also LIKELIHOOD PRINCIPLE, SYMMETRY PRINCIPLE.

INVARIANCE PRINCIPLE [physics] (1) The laws of motion remain unchanged by certain transformations or operations, especially those such as a principle in general relativity theory.[§] (2) States that a physical quantity or physical law possesses invariance under certain transformations.[¶] (3) The first invariance principle implies the law of conservation of linear momentum, while the second implies conservation of angular momentum. The symmetry known as the *homogeneity of time* leads to the invariance principle that the laws of physics remain the same at all times, which, in turn, implies the law of conservation of energy. The symmetries and invariance principles underlying the other conservation laws are more complex, and some are not yet understood.[‖] (4) Those principles dependent on variables in such a way that a specific alteration in the variables does not change the applicable principle.[**] See also AMBARTSUMYAN'S INVARIANCE PRINCIPLE; SCALE INVARIANCE PRINCIPLE; LOCAL INVARIANCE PRINCIPLE; INVARIANCE PRINCIPLE OF GENERAL RELATIVITY; DONSKER'S INVARIANCE PRINCIPLE; PRINCIPLE OF UNDULATORY INVARIANCE; RELAXED INVARIANCE PRINCIPLE; VARIATIONAL PRINCIPLE; PERTURBATION INVARIANCE PRINCIPLE; CHARNOV'S INVARIANCE PRINCIPLE; WEAK INVARIANCE PRINCIPLE; SYMMETRY PRINCIPLE; STRONG INVARIANCE PRINCIPLE; PRINCIPLE OF INVARIANCE; DIRICHLET IS PRINCIPLE; BYRNES–MARTIN INTEGRAL INVARIANCE PRINCIPLE; INTEGRAL INVARIANCE PRINCIPLE.

INVARIANCE PRINCIPLE OF GENERAL RELATIVITY [physics] Laws of motion are the same in all forms of reference, whether accelerated or not.[††] See also INVARIANCE PRINCIPLE.

INVARIANT IMBEDDING PRINCIPLE [physics] See PRINCIPLE OF INVARIANT IMBEDDING.

INVENTORY   PRINCIPLE [psychology] See GENERATIVE PRINCIPLE OF WRITING.

INVERFORM PRINCIPLE [engineering] Development of multiple sheet formation.[‡‡]

*Canuti, M, Valenti, C, and Chiriconi, A. (1989), Considerations on the principles of treatment of aseptic pseudarthrosis of the leg with the Ilizarov method, *Archiv. "Putti" Chirurgia Organi Movimento* **37**(1), 107–120.

[†]He, X. M. and Wang, G. (1995), Law of the iterated logarithm and invariance principle for in estimators, *Proc. Am. Math. Soc.* **123**, 563.

[‡]Villegas, C. (1981) Inner Statistical Inference IIInner Statistical Inference II The Annals of Statistics, **v9**, (4), pp. 768–776

[§]Norton, J. D. (1993), General covariance and the foundations of general relativity: Eight decades of dispute. *Rep. Bog. Phys.* **56**, 791 458. Reports on Proress in Physics **v56** p 791–858

[¶]Bergmann, P. G. (1961), Observables in general relativity, *J. Am. Phys. Soc.* **33**, 510–514.

[‖]Puhalskii, A. (1994), On the invariance principle for the first passage time, *Math. Oper. Res.* **19**, 946.

[**]Yourgrau, W. and Mandelstam, S. (1979), *Variational Principles in Dynamics and Quantum Theory*, Courier Dover Publications, p. 66.

[††]Oldershaw, R. L. (2007), Discrete scale relativity, Los Alamos National Laboratory, *Preprint Archive, Physics*, pp. 1–8.

[‡‡]Robertson, G. L. (1998), *Food Packaging: Principles and Practice*, Marcel Dekker, p. 155.

INVERSE NINJA EFFECTIVENESS PRINCIPLE [psychology] The effectiveness of a group of villains is inversely proportional to the number of villains in the group. While a single enemy is often portrayed as a significant threat to the protagonists, a large group of enemies are significantly less of a threat, and as such are easily defeated. A cliché in works of fiction where minor characters (cannon fodder) are unrealistically ineffective in combat against more important characters (almost always the protagonists "equipped" with character shields). Common in cowboy films, action movies, martial arts films, and comics, generally recognized as bringing a camp appeal to works that employ it because of its use in quickly and effectively heightening a story's dramatic atmosphere.*

INVERSE PRINCIPLE [engineering, mathematics] (1) A probability distribution is given and the objective is to determine the constraints or measure.[†] (2) Determining for a given feedback control law, the performance criterion that is optimal.[‡] See also ENTROPY OPTIMIZATION PRINCIPLE.

INVERSE PROBABILITY PRINCIPLE [mathematics] Also known as *Bayes' theorem*; describes the probability of a hypothesis, given the original data and some new data, is proportional to the probability of the hypothesis, given the original data only, and the probability of the new data, given the original data and the hypothesis.[§]

INVERSE SOLUBILITY PRINCIPLE [chemistry] When metal is quenched in polyquench solution, polymer comes out of the solution and forms a film on the surface of hot metal. This film of polymer controls the cooling rate of metal and helps to achieve desired hardness.[¶] See also FERGUSON PRINCIPLE; LE CHATELIER'S SOLUBILITY PRINCIPLE.

INVERSION OF CONTROL PRINCIPLE (IOC) [mathematics] Used to reduce coupling inherent in object-oriented programming languages.[‖] See also DEPENDENCE INVERSION PRINCIPLE.

INVOLUTION PRINCIPLE OF GARSIA–MILNE [mathematics] See GARSIA–MILNE INVOLUTION PRINCIPLE.

IODOPHOR PRINCIPLE [chemistry] Topical antispetic using a solubilizing agents and carrier for iodine.[**]

ION PAIR PRINCIPLE [biology] Used for the separation and quantification of a wide spectrum of biological components ranging from macromolecules (proteins, lipoproteins, or nucleic acids) to small analytes (amino acids, organic acids, or drugs).[††] Also known as principle of diastereometric ion pair formation; capillary electrophoresis principle.

IRIS DIAPHRAGM PRINCIPLE [mathematics, physics] See IRIS PRINCIPLE.

IRIS PRINCIPLE [mathematics, physics] Conducting plate mounted across a

*Fe (2007) *Dark Sound, the Colour of the Heart* (http://fillinginthepages.blogspot.com/2007/02/excerpt-from-wiki-stormtrooper-effect08.html).

[†]Srikanth, M., Kesavan, H. K., and Roe, P. H. (2000), Probability density function estimation using the MinMax measure., *IEEE Trans. Syst. Man Cybernet. C Appli. Rev.* **30**(1), 77–83.

[‡]Charalambos, D. C. and Hibey, J. L. (1996), Minimum principle for partially observable nonlinear risk-sensitive control problems using measure-valued decompositions, *Stochastics* **57**(3), 247–288.

[§]Bakan, D. (1953), Learning and the principle of inverse probability, *Psychol. Rev.* **60**(6), 360–370.

[¶]Adamska, K. and Voelkel, A. (2005), Inverse gas chromatographic determination of solubility parameters of excipients, *Int. J. Pharm.* **304**(1–2), 11–17.

[‖]Mokry, P., Fukada, E., and Yamamoto, K. (2003), Noise shielding system utilizing a thin piezoelectric membrane and elasticity control, *J. Appl. Phys.* **94**(1), 789–796.

[**]Reske, S. N., Knapp, F.F., Jr., and Winkler, C. (1986), Experimental basis of metabolic imaging of the myocardium with radioiodinated aromatic free fatty acids, *Am. J. Physiol. Imag.* **1**(4), 214–229.

[††]Bjornsdottir, I., Hansen, S. H., and Terabe, (1996) Chiral separation in non-aqueous media by capillary electrophoresis using the ion-pair principle, *J. Chromatogr. A*, **745**(1–2), 37–44.

waveguide to introduce impedance.[*] See also GEOMETRIC PRINCIPLE.

ISAR PRINCIPLE [mathematics, physics] (1) Based on maximizing the output entropy (or information flow) of a neural network with nonlinear outputs. (2) Likelihood and mutual information are, for all practical purposes, equivalent.[†] See also PRINCIPLE OF MAXIMUM DECISION EFFICIENCY; INFORMATION MAXIMUM LIKELIHOOD PRINCIPLE; MAXIMUM LIKELIHOOD ESTIMATION PRINCIPLE; MEMORY COMPENSATION PRINCIPLE; PRINCIPLE OF NETWORK ENTRORPY MAXIMIZATION.

ISHIKAWA'S PRINCIPLE [psychology] See BLACK BOX PRINCIPLE

ISOELECTRONIC PRINCIPLE [chemistry] For small variations in nuclear charge at least, the arrangement of electrons in isoelectronic species is approximately same.[‡]

ISOHYDRIC PRINCIPLE SEE ALSO SOLUBILITY PRINCIPLE [chemistry] All buffer systems which participate in defence of acid-base changes are in equilibrium with each other. There is after all only one value for $[H^+]$ at any moment. An assessment of the concentrations of any one acid-base pair can be utilised to provide a picture of overall acid-base balance in the body.[§]

ISOINVERSION PRINCIPLE [chemistry] (1) Based on the Eyring theory; this is a very general selectivity model that can be applied to a variety of types of selectivity, e.g. stereo-, chemo-, or regioselectivity, as long as two or more selectivity levels are involved. (2) It is a dynamic model taking into account all the selevtivity-relevant reaction components for both planning and optimization of the selectivity steps. Besides being helpful in the evaluation of substrates, catalysts, and auxiliars, the isoinversion principle can also be used for the validation of mechanistic hypotheses. (3) The first dynamic model for monitoring selectivity in chemistry. It is a useful tool for the description and explanation of selection processes in organic chemistry.[¶] See also INTENTION-TO-TREAT PRINCIPLE.

ISOLATION PRINCIPLE [psychology] Men isolate themselves in big ways (by physically going off and exploring the world) and in small ways (by not talking). "Men need to move around," says Stephen Johnson, PhD, founder and director of the Los Angeles Men's Center in Woodland Hills, California. "They need to go out, hunt, bring something back, and then unwind in peace. What a man doesn't need is to relate. There are plenty of times he needs to not relate. Women don't understand this because women think that talking is what siphons off stress. Not so for men. For them, talking after a hard day at the office might even make things worse."[‖]

ISOLOBAL PRINCIPLE [chemistry] (Ronald Hoffmann) Aims to improve the understanding of chemical bonding in molecules by identifying molecules that share a common frontier orbital type, approximate energy, and occupancy just as isoelectronic molecules share the same number of valence electrons and structure.[**]

[*]Borgmann, H. (1972), Basic principles of clinical pupillography. IV. Dependence of pupil width and light reaction on sex, color of iris and refraction, *Albrecht von Graefe's Arch. Clin. Exp. Ophthalmol.* **185**(1), 11–21.

[†]Yang, H. and Soumekh, M. (1993) Blind-velocity SAR/ISAR imaging of a moving target in a stationary background, *IEEE Trans. Image Process.* **2**(1), 80–95.

[‡]Heald, R. R., (1972) Use of the isoelectronic principle in teaching chemistry, *School Sci. Math.* **7**(9), 801–810.

[§]Pitts RF. Mechanisms for stabilizing the alkaline reserves of the body. *Harvey Lect* 1952–1953; **48** 172–209.

[¶]Gypser, A., Kethers, S., and Scharf, H.D., (1995), *The Isoinversion Principle (Part II)* http://www.ch.ic.ac.uk/ectoc/papers/55/ectoc.html Computer Aided Evaluation of Temperature Dependent Selectivity Parameters in Organic Chemistry, Ins. Organische Chemie der RWTH Aachen (http://www—i5.informatik.rwth-aachen.de/neysa/ECTOC).

[‖]Cary, S. (1993), The isolation principle; nobody ever told John Wayne to get in touch with his feelings, *Men's Fitness* **9**, p.66.

[**]DeKock, R., Fehlner, L., and Thomas, P. (1982), On the validity of the isolobal principle: Pentaborane(9) and its ferraborane derivatives, *Polyhedron* **1**(6), 521–523.

ISOMORPHISM   PRINCIPLE [mathematics] One of the results concerning the quotient groups of a particular parent group and the ismorphisms between them.*

ISOPATHIC PRINCIPLE [psychology] Paradoxical rule according to which the cause cures the effect, as when a feeling of guilt is relieved by an exhibition of guilt, namely, hate.[†]

*Walker, P. M. B. (1999), *Chambers Dictionary of Science and Technology*, Chambers, New York, p. 630.

[†]Schmidt, J. M., (1991), The history of tuberculin therapy-its discovery by Robert Koch, its forerunners and further development, *Pneumologie* **45**(10), 776–784.

J

JACKSONIAN PRINCIPLE [mathematics] See FIRST-IN FIRST-OUT PRINCIPLE.

JARMAN–BELL PRINCIPLE [psychology, chemistry] (1) Involves a scaling relationship between metabolism and body size, which suggests that body size is a fundamental tactic in an animal's feeding strategy. Relatively accurate predictions regarding the diets of primates of known body weight follow from this model. In addition, it can be expanded to predict the kinds of adaptations that would appear in animals that deviate from the expected size/diet pattern.* (2) Crisis mentality that exists regardless of whether business is successful. Increasing π-donor ability of solvent should decrease the reaction rate.[†] See also POLANYI–EVANS–BELL PRINCIPLE.

JAYNES' MAXENT PRINCIPLE [engineering, mathematics] (Edwin Thompson Jaynes, 1922-1998, American physicist) Maximizing uncertainty subject to given constraints. Out of all the probability distributions satisfying given constraints, the probability distribution that maximizes the Shannon entropy is chosen.[‡] See also JAYNES' MAXIMUM ENTROPY PRINCIPLE; ENTROPY OPTIMIZATION PRINCIPLE.

JAYNES' MAXIMUM ENTROPY PRINCIPLE [mathematics, physics] (Edwin Thompson Jaynes, 1922-1998, American Physicist) (1) When some information regarding an unknown distribution is available, but not enough to fully define the distribution, then the unique distribution that maximizes the entropy measure, while satisfying the known information about the distribution, should be selected.[‡] (2) Estimates a distribution of a random variable from partial information in the form of a finite number of expectations. Although there are theoretically an infinite number of such distributions, Jaynes argues that there is only one way to select the distribution in a consistent manner on the basis of the information given and no other.[§] (3) Of all the probability distributions that satisfy a given set of constraints, choose the one that maximizes the entropy.[¶] See also PRINCIPLE OF MAXIMUM ENTROPY (PME).

JEWISH PLEASURE PRINCIPLE [psychology] See PLEASURE PRINCIPLE.

JOB-SITE PRIMARY PRINCIPLE [psychology] Most basic or suitable materials and personnel should be allocated to perform specific jobs.[‖]

JOULE'S PRINCIPLE [chemistry] (James Prescott Joule, 1818-1889, British Chemist Physicist) (1) The internal energy of a given mass of gas is a function of temperature alone; it is independent of the pressure and volume of the gas. (2) The molar heat capacity of a solid compound is equal to the sum of the atomic heat capacities of its

*Gaulin, S., (1979), A Jarman/Bell model of primate feeding niches, *Human Ecol.* **7**(1), 1–20.

[†]Nakagawa, N. (2003), Difference in food selection between patas monkeys (Erythrocebus patas) and tantalus monkeys (Cercopithecus aethiops tantalus) in Kala Maloue National Park, Cameroon, in relation to nutrient content, *Primates J. Primatol.* **44**(1), 3–11.

[‡]Plastino, A. and Curado, E. M. F. (2005), Equivalence between maximum entropy principle and enforcing dU = TdS, *Phy. E. Statist. Nonlinear Soft Matt. Phys.* 72.

[§]Wang, S., Schuurmans, D., Peng, F., and Zhao, Y. (2004), Learning mixture models with the regularized latent maximum entropy principle, *IEEE Trans. Neural Networks.* **15**(4), 903–916.

[¶]Rose, K. (1998), Deterministic annealing for clustering, compression, classification, regression, and related optimization problems, *Proc. IEEE* **86**(11), 2213.

[‖]Hanson, S. and Pratt, G. (1991), Job search and the occupational segregation of women, *Ann. Assoc. Am. Geogr.* **81**(2), 229–253.

component elements in the solid state. [Engineering] The heat H liberated by the flow of current I in a conductor with resistance R for a time $t : H = I^2Rt$. This is the basis of all electrical heating, wanted or unwanted. With high-frequency alternating current, R is an effective resistance and I may be confined to a thin skin of the conductor.[*] Also referred to as *Kopp's Principle*.

JOURDAIN PRINCIPLE [physics, chemistry] (1) The path of a conservative system in configuration space is such that the integral of the Lagrangian function over time is a minimum or maximum relative to nearby paths between the same endpoints and taking the same time.[†] (2) Motion of the underlying system of particles compatible with the collective equilibrium provided that the variations are associated with reversible processes.[‡] See also DIFFERENTIAL PRINCIPLE; VARIATIONAL PRINCIPLE; HAMILTON PRINCIPLE; PRINCIPLE OF LEAST CONSTRAINT; D'ALEMBERT PRINCIPLE; HAMILTON–LAGRANGE PRINCIPLE; MAUPERTUIS–LAGRANGE PRINCIPLE; PRINCIPLE OF EXTREMAL EFFECTS; LEAST-CONSTRAINT PRINCIPLE; D'ALEMBERT–LAGRANGE PRINCIPLE; GAUSS' PRINCIPLES; JAYNES' PRINCIPLE; PRINCIPLE OF VIRTUAL WORK; KINETIC PRINCIPLE; MAUPERTUIS–EULER–LAGRANGE PRINCIPLE OF LEAST ACTION; D'ALEMBERT–LAGRANGE VARIATION PRINCIPLE; LAGRANGE PRINCIPLE.

JUDICIAL PRINCIPLE [Law] The basis underlying the formulation of jurisprudence.[§]

JUSTIFYING PRINCIPLE [psychology] An act with two effects, one right and one wrong, can be performed when fair condition are met—the effect of activity, the intention, material cause of the act, proportionate reason; also, right or indifferent action, not intrinsically wrong; wrong, though unforeseen, cannot be intended, wrong effect cannot be means to the right effect; for the wrong effect to occur, there must be proportionate reason.[¶] See also PRINCIPLE OF DOUBLE EFFECT; PRINCIPLE OF DO NO HARM; CASUISTIC PRINCIPLE; PRINCIPLES OF BIOMEDICAL ETHICS; EXCEPTION-GRANTING PRINCIPLE; DOUBLE-EFFECT PRINCIPLE.

JUST-IN-TIME PRINCIPLE [organizational engineering] A process of creative destruction was based on the concept of (*economic order quantity*) (EOQ), which is the generally accepted principle in Western inventory control systems. Prevents a large amount of inventory accumulation.[‖]

[*]Walker, P. M. B. (1999), *Chambers Dictionary of Science and Technology*, Chambers, New York, p. 637.
[†]Altay, G. and Doekmeci, M. C. (2007), Variational principles for piezoelectric, thermopiezoelectric, and hygrothermopiezoelectric continua revisited, *Mech. Adv. Mater. Struct.* **14**(7), 549–562.
[‡]Yi, Z. (2007), Differential variational principles of mechanical systems in the event space, *Wuli Xuebao* **56**(2), 655–660.

[§]American Heritage (2000), *American Heritage Dictionary of the English Language*, 4th ed., Houghton Mifflin, New York.
[¶]Beauchamp, D. E. (1980), Public health and individual liberty, *Annu. Rev. Public Health*, 1, 121–136.
[‖]Urabe, K., Child, J., Kagono, T., and Keiei Gakkai, N. (1988), *Innovation and Management: International Comparisons*, Walter de Gruyter, p. 22.

K

KAHN PRINCIPLE [computer science] (Gilles Kahn, 1946–2006, French computer, scientist) Each node in an asynchronous deterministic network computes a continuous function from input histories to output histories, and the behavior of the network can be characterized as a least fixed point.[*]

KAHN PRINCIPLE FOR INPUT/OUTPUT [engineering] See KAHN'S FIXED-POINT PRINCIPLE

KAHN'S FIXED-POINT PRINCIPLE [engineering] (Glleskahn, 1946–2006, French computer scientist) (1) Defines a simple and general model of networks of concurrently executing, nondeterministic processes that communicate through unidirectional, named ports.[†] (2) A notion of the input/output relation computed by a process is defined, and determinate processes are defined to be processes whose input/output relations are single-valued.[‡] (3) Kahn's principle states that if each process in a dataflow network computes a continuous input/output function, then so does the entire network. Moreover, in that case the function computed by the network is the least fixed point of a continuous functional determined by the structure of the network and the functions computed by the individual processes.[§] Also known as *Kahn principle for input/output.*

KANT'S PRINCIPLE [psychology] (Immanuel Kant, 1724–1804, German philosopher) Kant's (1960) view that evil behavior is possible, albeit difficult to understand, and introduce M. S. Peck's theory (1983) of how evil behavior can manifest itself when a person suffers from malignant narcissism—a complaint that involves acting on principles that are not consciously acknowledged. Ward concludes that Kant's views on evil can be understood with reference to Peck's theory (and vice versa). People who do evil things engage with Kant's principles in incoherent ways. He operates on a principle about interpersonal relating that sees relationships with others largely as fear transactions; either he fears them or they fear him.[¶] See EMANCIPATION PRINCIPLE.

KAPLAN–MEIER PRINCIPLE OF CALCULATION (Kaplon, E. L. nobio) (Carl Alfred Meier, 1905–1995, Swiss psychiateist) [psychology] (1) The idea of mental and psychophysical work, the determination of a unit of work, and the psychophysical homogeneity.[‖] (2) The method of development of the computation theory of graphic procedure for the deduction of mental work, mean duration of the psychophysical and mental operations, deduction of mental work as a particular case

[*]Lynch, N. A. (1989), *A Proof of the Kahn Principle for Input/Output Automata,* M. I. T. Laboratory for Computer Science (http://citeseer.nj.nec.com/lynch89proof.html).
[†]Roncero, A. F. (1956), Comparative diagnostic values of the Karmin complement-fixation test and the standard and presumptive Kahn reactions, *Rev. Asoc. Bioquim. Argentina* **21**, p. 322–327.
[‡]Miravent, J. M., Parodi, A., de Bonomi, E., and Barba, R. (1942), Discussion of results with the Bordet-Wassermann low-temperature fixation reaction and the Kahn reaction in the examination of 23,000 serums, *Rev. Inst. Bacteriol. "Carlos G. Malbran* **10**, 461–466.

[§]Stark, E. W. (1990), *A Simple Generalization of Kahn's Principle to Indeterminate Dataflow Networks,* Semantics for Concurrency, Leicester eprints.kfupm.edu.sa/20988/. Technical Report TR 89-29 Sunyat Stongbrook, Computer Science Department. Semantics for Concurrency, Leicester.
[¶]Steigerwald, Joan (2006), Kant's concept of natural purpose and the reflecting power of judgement, *Stud. Hist. Phil. Biol. Biomed. Sci.* **37**(4), 712–734.
[‖]Chapman, G. B. and Sonnenberg, F. A. (2003), *Decision Making in Health Care: Theory, Psychology, and Applications,* Cambridge Univ. Press, p. 111.

of psychophysical calculation. The characteristic function of mental work (characteristics of mental work, graphs of the psychophysical equivalent of mental work). Also called *product–limit Kaplan–Meier estimator* or *Kaplan–Meier product–limit estimator*.* (3) For each time interval we estimate the probability that those who have survived to the beginning will survive to the end. This is a conditional probability (the probability of being a survivor at the end of the interval on condition the subject was a survivor at the beginning of the interval). Survival to any timepoint is calculated as the product of the conditional probabilities of surviving each time interval.† Also known as *principle of calculation*.

KELVIN–ARNOLD ENERGY PRINCIPLE [mathematics] (William Thomson, 1st Baron Kelvia, 1824–1907, Irish mathmatician) (Vladimir Igorevich Arnold, 1937-; Russian mathematician) (1) Methods do not require any intuitive or empirical assumptions, but use only the material properties, tool geometry, and physical laws of deformation.‡ (2) The bandlimited function of lowpass type whose energy is minimum among those that have the prescribed sample values on a finite set of sampling points. The minimum energy signal converges uniformly to a bandlimited signal when the number of sampling points is increased infinitely.§ (3) Free boundary subjected to external magnetic or plasma pressure forces.¶ (5) The path of a conservative system in configuration space is such that the integral of the Lagrangian function over time is a minimum or maximum relative to nearby paths between the same endpoints and taking the same time.‖ See also ENERGY PRINCIPLE.

KELVIN'S PRINCIPLE [engineering] (William Thomson, 1st Baron Kelvin, 1824–1907, Irish mathematician)   The most economical size of conductor use use for a line is that for which the annual cost of loss is equal to the annual interest and depreciation on that part of the capital cost of the conductor that is proportional to its cross-sectioned area.**

KENNARD PRINCIPLE [organic chemistry, biochemistry] (Kenneth Clayton Kennard, b. 12/18/1926, organic chemist, biochemist) Preparation and properties of light-sensitive materials. Relation between age and outcome for head injury.††

KEPLER'S PRINCPLE [atronomy] (Johannes Kepler, 1571-1630, German mathematician) Includes following three descriptions for planetary motion: (1) the planets describe ellipses with the sun at a focus, (2) the line from the sun to any planet sweeps across equal areas in equal times, and (3) the squares of the periodic times ot the planets are proportional to the cubes of their mean distances from the sun.‡‡

KICK'S PRINCIPLE [physics]   The energy requied for subdivion of a definite amount of material is the same for the same fractional

*Gallin, J. I. and Ognibene, F. P. (2007), *Principles and Practice of Clinical Research*, Academic Press, p. 275.

†Ludbrook, J. and Royse, A. G. (2008), Analysing clinical studies: Principles, practice and pitfalls of Kaplan-Meier plots, *ANZ J. Surg.* **78**(3), 204–210.

‡Miles, John W. (1963), *Principles of Classical Mechanics and Field Theory*, vol. III, Part. 1, *Encyclopedia of Physics*, Flugge, S., ed., Springer, p. 198.

§Davidson, P. A. (1998), On the application of the Kelvin-Arnold energy principle to the stability of forced two-dimensional inviscid flows, *J. Fluid Mech.* **356**, 221–257.

¶Vladimirov, V. A., and Ilin, K. I., (1999) On Arnold's variational principles in fluid mechanics. The Arnold fest: Proceedings of a Conference in Honour of V.I. Arnold for hes sixtieth birthday,

E. Bierstone, B. Knesin, A. Khovanskii and J. E. Marsden eds. Amer. Math. Soc. Providence, R. I.

‖Altay, G. and Doekmeci, M. C. (2007), Variational principles for piezoelectric, thermopiezoelectric, and hygrothermopiezoelectric continua revisited, *Mech. Adv. Mater. Struct.* **14**(7), 549–562.

**Walker, P. M. B. (1999), *Chambers Dictionary of Science and Technology*, Chambers, New York, p. 642.

††Hart, K. and Faust, D. (1988), Prediction of the effects of mild head injury: a message about the Kennard principle, *J. Clin. Psychol.* **44**(5), 780–782.

‡‡Walker, P. M. B. (1999), *Chambers Dictionary of Science and Technology*, Chambers, New York, p. 642.

reduction in average size of the individual particles.*

KICK SIMILARITY PRINCIPLE [engineering] Microhardness of solids.[†]

KINETIC PRINCIPLE [physics]   Newton's third law of motion. For every action there is an equal and opposite reaction; this is true for bodies that are free to move as well as for bodies rigidly fixed.[‡]

KIRCHHOFF'S DIFFRACTION PRINCIPLE [physics]   (Gustav Robert Kirchhoff, 1824–1887, German physicist)   A mathematical description of diffraction based on Huygens' principle.[§]

KIRCHHOFF'S PRINCIPLE [engineering] (Gustav Robert-Kirchholf, 1824–1887, German physicist)   Generalized extensions of Ohm's principle employed in network analysis.[§]

KIRCHHOFF'S VOLTAGE LAW PRINCIPLE [engineering]   See HAMILTON PRINCIPLE.

KITAIGORODSKII'S AUFBAU PRINCIPLE [physics] (Sergei Alexander Kitaigorodskii, physical oceanographer 9/13/1934, Moscow, PhD geophysics, Johns Hopknis Univ.) Physics of air–sea interaction, wave motions in the ocean, and geophysical fluid dynamics. Packing geometry of semiflexible organic molecules in translation monolayer aggregates.[¶]

KNASTER–TARSKI PRINCIPLE [mathematics] A fundamental tool to investigate fixpoint calculus.   (1) Let L be a complete lattice and let $f : L \longrightarrow L$ be an order-preserving function. Then the set of fixed points of f in L is also a complete lattice. The least fixpoint of f is the least element x such that $f(x) = x$. or, equivalently, such that $f(x) \leq x$; the dual holds for the greatest fixpoint, the greatest element x such that $f(x) = x$. (2) Uniform spaces. Every monotone function on a complete lattice has a least fixpoint.[‖] [**] See also BANACH PRINCIPLE.

KOEHLER NEWMAN PRINCIPLE [engineering] Propagation operator should pass linear events without distortion.[††]

KOGAN'S SYMMETRIC ACTION PRINCIPLE [entomology, ecology] (Marcos Kogan, PhD, 6/9/33, Rio de Janeiro, Brazil)   Method explores restricting the performance of the systems subcomponents as if they are experimental control variables and then assessing the benefit of a proposed improvement by extrapolation.[‡‡]

KOHLRAUSCH'S PRINCIPLE [chemistry] (Friedrich Wilhelm Georg Kohlrausch, 1840–1910, German physicist)   The contribution from each ion of an electrolytic solution to the total electrical conductance is independent of the nature of the other ion.[§§]

KOHN VARIATIONAL PRINCIPLE [chemistry, engineering] (Walter Kohn, 1923-, American

---

*Op. cit. Walker, p. 644.

[†]Bernhardt, E. O. (1941), The microhardness of solids in the boundary region of the Kick similarity principle, *Z. Metallkunde* **33**, 135–44.

[‡]Okolov, F. S. (1949), Kinetic principle in determination of ascorbic acid in solution, *Higiene y Salubridad* **14**(3), 25–30.

[§]Walker, P. M. B. (1999), *Chambers Dictionary of Science and Technology*. Chambers, New York, p.646.

[¶]Kitaigorodski, A. I. (1979), *Order and Disorder in the World of Atoms*, MIR/VEB Verl, Moscow/Leipzig.

[‖]Knaster, B. (1928) Un theoreme sur les functions d'ensembles. Ann. Soc. Polon. Math, **6**, p. 133–134.

[**]Jachymski, J. R. (1998), Fixed point theorems in metric and uniform spaces via the Knaster-Tarski principle, *Nonlinear Anal.* **32**(2), p. 225–233.

[††]O'Doherty, R. F. and Taner, M. T. (1995), The Koehler-Newman principle, paper presented at 57th EAEG Meeting, Session: Migration III.

[‡‡]Zhou, X. (1988), Construction of interacting actions of bosonic open strings and closed strings by symmetry principles, *Commun. Theor. Phys.* **10**(2), 199–214.

[§§]Walker, P. M. B. (1999), *Chambers Dictionary of Science and Technology*, Chambers, New York, p. 648.

Quantum mechanical calculations; reactive scattering.* See also COMPLEX KOHN VARIATIONAL PRINCIPLE; HULTHEN–KOHN VARIATIONAL PRINCIPLE; SCHWINGER VARIATIONAL PRINCIPLE.

KOPP'S PRINCIPLE [chemistry]   See JOULE'S PRINCIPLE.

KORENBLUM'S MAXIMUM PRINCIPLE [mathematics] (Boris Korenblum)   Let $A^2(\mathbb{D})$ be the Bergman space over the open unit disk $\mathbb{D}$ in the complex plane. Korenblum's maximum principle states that there is an absolute constant $c \in (0, 1)$, such that whenever $|f(z)| \leq |g(z)|(f, g \in A^2(\mathbb{D}))$ in the annulus $c < |z| < 1$, then $\|f\|_{A^2} \leq \|g\|_{A^2}$.[†]

KRASOVSKII–LASALLE PRINCIPLE [mathematics] (Nikolai Nikolapuich Krasvousky, 1924-, Russian mathematician)   A criterion for the asymptotic stability of a (possibly nonlinear) dynamical system.[‡]

KROGH PRINCIPLE [biology]   Schack August Steen barg Krogh, 1874–1949, Danish zoophysiologist (August Krogh) For any problem in physiology there is an organism in which the question can be most conveniently studied.[§]

KRYLOV MAXIMUM PRINCIPLE [engineering, mathematics] (Nikolay Mitrofanovich Krylou, 1879–1955, Russian mathematician) (Aleksei Nikolaevich Krylov, b. 8/15/1863, Leningrad, Russia)   See MAXIMUM PRINCIPLE.

KULLBACK ENTROPY PRINCIPLE [physics] (Soloman Kullback, 1907–1994, American mathematician) (Edwin Thompson Jaynes, 1922–1998, American mathematician) (Jaynes,1957)   An extension of the principle of insufficient reason of Laplace. Consisting in assigning equal probabilities to two events if there are no reasons to think otherwise. The maximum entropy principle provides an information-theory-based approach to assigning probabilities when incomplete information is given. The uncertainty should be maximized subject to all information given. The minimization of the Kullback–Leibler entropy between the empirical probability and the parametric form is equivalent to the maximization of the corresponding empirical likelihood.[¶]

KULLBACK'S MINIMUM CROSS-ENTROPY PRINCIPLE [engineering, mathematics] (Soloman Kullback, 1907–1994, American mathematician)   (1) From all probability distributions, out of all those that satisfy the given constraints, which is closest to a given a priori probability distribution. Requires choice distribution for which a specified measure of cross-entropy is a minimum for all distributions satisfying the given constraints. (2) Out of all probability distributions satisfying the given linear constraints, the probability distribution that minimizes the Kullback–Leibler measure of cross-entropy is chosen.* See also KULLBACK'S PRINCIPLE OF MINIMUM CROSS-ENTROPY; GEN MINXENT PRINCIPLE; ENTROPY OPTIMIZATION PRINCIPLE.

KULLBACK'S PRINCIPLE OF MINIMUM CROSS-ENTROPY [engineering,   mathematics] See KULLBACK'S MINIMUM CROSS-ENTROPY PRINCIPLE.

KUREPA'S ANTICHAIN PRINCIPLE [mathematics] (Ernst Friedrich Ferdinand Zermelo, 1871–1953, German mathematician)   Every partially ordered set has a maximal antichain.[‖]

*Brown, D. and Light, J. C. (1994), Kohn variational principle for a general finite-range scattering functional, *J. Chem. Phys.* **101**(5), 3723–3728.
[†]Korenblum, B. (1993) A maximum principle for the Bergman space. Trans. Amer. Math. Soc., v337 (2) p. 795–806.
[‡]Krasovskii, N. N. (1959), *Problems of the Theory of Stability of Motion*, Stanford Univ. Press.
[§]Krebs, H. A. (1975), The August Krogh principle: For many problems there is an animal on which it can be most conveniently studied, *J. Exp. Zool.* **194**(1), p.221–226.

[¶]Deco, G. and Schürmann, B. (2000), *Information Dynamics: Foundations and Applications*, Springer, p. 46.
[‖]Harper, J. M. and Rubin, J. E., (1976), Variations of Zorn's lemma, principles of cofinality, and Hausdorff's maximal principle, *Notre Dame J. Formal Logic* **17**(4), 565–588.

L

LAGRANGE FOUR-SQUARE PRINCIPLE [mathematics] (Guiseppe Lodovico Lagrangia, 1736–1813, Italian mathematician) (Joseph Louis Lagrange, b. Giuseppe Lodovico Lagrangia, 1736–1813 Italian mathematician and astronomer) Also known as *Bachet's conjecture*. The theorem appears in the *Arithmetica* of Diophantus, translated into Latin by Bachet in 1621. Every positive integer can be written as the sum of at most four squares.*

LAGRANGE PRINCIPLE [physics, chemistry] (Guiseppe Lodovico Lagrangia, 1736–1813, also known as Joseph Louis Lagrange, Italian mathematician) (1) Nature chooses the most economical path for moving bodies, light rays, chemical reactions, and so on. Study of chemical reactivity in terms of the generalized and integral similarity indices.[†] (2) Allows prediction of the effect that a change of conditions will have on a chemical reaction, such as heat or pressure.[‡] (3) Role in predicting Diels–Alder reactivity of aromatic heterocycles.[§] (3) The path of a conservative system in configuration space is such that the integral of the Lagrangian function over time is a minimum or maximum relative to nearby paths between the same endpoints and taking the same time.[¶] (6) Variational

characterization of the wavefunction ψ; motion of the underlying system of particles compatible with the collective equilibrium provided that the variations are associated with reversible processes.[‖] See also LEAST-MOTION PRINCIPLE; INERTIA PRINCIPLE; PRINCIPLE OF LEAST MOTION; LEAST ACTION PRINCIPLE.

LAGRANGE PRINCIPLE OF LEAST ACTION [chemistry] See LAGRANGE PRINCIPLE.

LAGRANGIAN PRINCIPLE See LAGRANGE PRINCIPLE.

LAMBERT–BEER PRINCIPLE [physics] See BEER–LAMBERT PRINCIPLE.

LAMBERT'S COSINE PRINCIPLE [physics] Johann Heinrich Lambert, 1728–1777, Swiss mathematician The energy emitted from a perfectly diffusing surface in any direction is proportional to the cosine of the angle that that direction makes with the normal.[**]

LAMBERT'S PRINCIPLE [physics] Johann Heinoich Lambert, 1728–1777, Swiss mathematician The illumination of a surface on which the light falls normally from a point source is inversely proportional to the square of the distance of the surface from the source.[**]

LAMONT'S PRINCIPLE [engineering] The permeability of steel, at any flux density, is proportional to the difference betweent that flux density and the saturation value.[††]

*Rouse Ball, W. W. (1908), *Joseph Louis Lagrange (1736–1813), a Short Account of the History of Mathematics*, 4th ed.

[†]Peacock, J. A. (1999), *Cosmological Physics*, Cambridge Univ. Press, p. 5.

[‡]Xu, T. and Pruess, K. (2001), Modeling multiphase non-isothermal fluid flow and reactive geochemical transport in variably saturated fractured rocks: 1. Methodology, *Am. J. Sci.* **301**, 16–33.

[§]Bendikov, M., Wudl, F., and Perepichka, D. F. (2004), Tetrathiafulvalenes, oligoacenenes, and their buckminsterfullerene derivatives: The brick and mortar of organic electronics, *Chem. Rev.* **104**(11), 4891–4946.

[¶]Altay, G. and Doekmeci, M. C. (2007), Variational principles for piezoelectric, thermopiezoelectric, and hygrothermopiezoelectric continua revisited, *Mech. Adv. Mater. Struct.* **14**(7), 549–562.

[‖]Schiller, C. (1990), *Motion Mountain, the Adventure of Physics, Fall, Flow, and Heat* (http://www.docstoc.com/docs/4300415/Motion-Mountain—The-Adventure-of-Physics—part-1-of-6 p.192).

[**]Walker, P. M. B., (1999), *Chambers Dictionary of Science and Technology*, Chambers, New York, p. 652.

[††]Op. cit. Walker, p. 653.

LANDAUER'S PRINCIPLE [computer science] —1927–1999, American physicist (Rolf Landauer, IBM) (1) Any logically irreversible manipulation of information, such as the erasure of a bit or the merging of two computation paths, must be accompanied by a corresponding entropy increase in non-information-bearing degrees of freedom of the information processing apparatus or its environment.* (2) Each bit of lost information will lead to the release of an amount $kT \ln 2$ of heat. On the other hand, if no information is erased, computation is thermodynamically reversible, and requires no release of heat.[†] See also SECOND PRINCIPLE OF THERMODYNAMICS.

LARGE-DEVIATION PRINCIPLE [mathematics] (Stefan Banach, 1892–1945)   (1) Difference between actual value of a controlled variable and the desired value corresponding to the set point. (2) Evolutionary differentiation involving interpolation of new stages in the ancestral pattern of morphogenesis. (3) Angle between incident ray on an object or optical system and the emergent ray.[‡] (4) See CONTRACTION-MAPPING PRINCIPLE. Also known as Principle of Large Deviation (Sathyamangalam Ranga Iyengar Srinivasa Varadhan FRS, 1940-, Indian-American mathematician)[§]

$$p^{\mathrm{HYL}}(\mu) = \sup_{0 \leq x_1 \leq x_0} \left\{ \mu x_0 - f(x_0 - x_1) - \frac{a}{2} \left(2x_0^2 - x_1^2 \right) \right\},$$

The first step in our proof is to bound the interaction term (1.6) above by

$$\frac{a}{2V_1} \left\{ N_1^2 - n_1(1)^2 \right\}$$

and below by

$$\frac{a}{2V_1} \left\{ N_1^2 - \left(\sum_{j=1}^{m_1} n_1(j) \right)^2 \right\}.$$

The principle of large deviations provides a compact way of making rigorous the method of the largest term; applied to the upper bound on the hamiltonian it yields the expression for a lower bound to the prssure $p^{\mathrm{HTL}}(\mu)$. To deal with the lower bound we have first to estimate the entropy involved in grouping together the first m_1 levels; this is achieved by an inequality, the method of large deviations applied to the lower bound then yields the expression for an upper bound to the pressure $p^{\mathrm{HYL}}(\mu)$ and the proof is complete.[¶]

LAROX PRINCIPLE [engineering]   Horizontal filter plates and a moving endless filter cloth for discharge.[‖]

LASALLE'S INVARIANCE PRINCIPLE See KRASOVSKII–LASALLE PRINCIPLE.

LASER SCATTERING PRINCIPLE [mathematics]   See LE CHATELIER–BRAUN PRINCIPLE.

LAW OF CONTRADICTION [mathematics] (1) A principle of logic whereby a proposition cannot be both true and false.** (2) A simultaneous assertion and denial of a proposition; namely, a sentence of the form "A and not A," often symbolized within a formal language as "$A - A$." In formal systems a contradiction is a theorem that is said to be inconsistent. The law of contradiction is the logical principle that a proposition cannot be both asserted and denied.[††]

LAX PRINCIPLE OF EQUICONTINUITY See EQUICONTINUITY PRINCIPLE.

*Smith, E. (2008), Thermodynamics of natural selection III: Landauer's principle in computation and chemistry, *J. Theore. Biol. 252(2), 213–220.*

[†]Daffertshofer, A. and Plastino, A. R., (2005), Landauer's principle and the conservation of information, *Phys. Lett. A* **342**(2), 213–216.

[‡]Greenwood, P. E. and Sun, J. M. (1997), Equivalences of the large deviation principle for Gibbs measures and critical balance in the Ising model, *J. Statist. Phys.* **86**, 149.

[§]van den Berg, M., Lewis, J.T., and Pule, J.V. (1988) The Large Deviation Principle and Some Models of an Interacting Boson Gas. Commun. Math. Phys. 118, p. 61–85.

[¶]Chen, Jinwen (2001) Space-time large deviation lower bounds for Spin particle systems. Physics Letters A, **290**(1–2), 65–71.

[‖]Townsend, I. (2003), Automatic pressure filtration in mining and metallurgy, *Minerals Eng.* **16**(2), 165–173.

**McGraw-Hill Dictionary of Scientific and Technical Terms.*

[††]Weisstein, Eric W. (n.d.) Contradiction Law. Math World—A Wolfram Web Resource. http://mathworld.wolfram.com/ContradictionLaw.html.

LE CHATELIER–BRAUN PRINCIPLE [mathematics] (Henry Louis Le Chatelier, 1850–1936, French chemist) (Karl Ferdinand Brown, 1850–1918, German physicist, Nobel laured) (1) Methods do not require any intuitive or empirical assumptions, but use only the material properties, tool geometry, and the physical laws of deformation.* (2) Found in the solar coronal geometry, which is modeled as a rectangular domain.[†] (3) The bandlimited function of lowpass type whose energy is minimum among those that have the prescribed sample values on a finite set of sampling points. The minimum energy signal converges uniformly to a bandlimited signal when the number of sampling points is increased infinitely.[‡] (4) Free boundary subjected to external magnetic or plasma pressure forces.[§] (5) The path of a conservative system in configuration space is such that the integral of the Lagrangian function over time is a minimum or maximum relative to nearby paths between the same endpoints and taking the same time.[¶] See also LE CHATELIER'S PRINCIPLE; LASER SCATTERING PRINCIPLE; ENERGY PRINCIPLE.

LE CHATELIER MINIMUM PRINCIPLE [chemistry] (Henry Louis Le Chatelier, 1850–1936, French chemist) An excess of reactants on either side of an equation will force the reaction in the opposite direction. If the equilibrium of a system is disturbed by a change in one or more of the determining factors, such as temperature, pressure, or concentration, the system tends to adjust itself to a new equilibrium by attempting to counteract any effects of the change.[‖] See LE CHATELIER'S PRINCIPLE.

LE CHATELIER–SAMUELSON PRINCIPLE [mathematics, economics] (Henry Louis Le Chatelier, 1850–1936, French Chemist) (Paul Anthony Samuelson, 1915–2009, American Economist) Examines the effects of an increase in some final demand on the output levels under the constraint that the production of certain goods is held at its original value. The increase in any output is larger when fewer output levels are kept constant.[**]

LE CHATELIER'S PRINCIPLE [physics, chemistry] Henry Louis Le Chatelier, 1850–1936, French chemist (1) Reaction at equilibrium shifts in response to a change in external conditions in a way that moderates the change. If a stress (temperature increase, in this case) is put on a system at equilibrium, reactions in the system will shift to moderate the stress (reduce the temperature increase).[††] (2) The conversion of an equilibrium-controlled reaction as well as the rate of the forward reaction can be significantly enhanced by selectively removing one of the reaction products from the reaction zone.[‡‡] (3) If a system is at equilibrium

*Balian, R. (2006), *On the Proper Use of Equilibrium Thermodynamics. Theoretical and Mathematical Physics, from Microphysics to Macrophysics*, Springer, Berlin heidelberg pp. 241–306.
[†]Pavon, D. and Wang, B. (2007), Le Chatelier-Braun principle in cosmological physics, Los Alamos National Laboratory, *Preprint Archive, General Relativity and Quantum Cosmology*, pp.1–6.
[‡]Nesis, E. L., and Skibin, Yu. N. (2000), Some special features of the LeChatelier-Braun principle, *J. Eng. Phys. Thermophysics* **73**(4), 859–862.
[§]Planck, M. (1934), The LeChatelier-Braun principle, *Ann. Phys.* **19**, 759–768.
[¶]Altay, G. and Doekmeci, M. C. (2007), Variational principles for piezoelectric, thermopiezoelectric, and hygrothermopiezoelectric continua revisited, *Mech. Adv. Mater. Struct.* **14**(7), p. 549–562.

[‖]Bak, T. (1955), A minimum-principle for nonequilibrium steady states, *J. Phys. Chem.* **59**, 665–668.
[**]Dietzenbacher, E. (1992), The LeChatelier-Samuelson principle revised, *Z. Natl. Okonom* **55**(3), 277–296.
[††]Kaufman, Myron (2002), Principles of Thermodynamics, CRC press, p. 200.
[‡‡]Sunggyu, L. (2005), Encyclopedia of Chemical Processing. Taylor and Francis, p. 35.

and a change is made in the conditions, the equilibrium adjusts so as to oppose the change.* See BRAUN–LE CHATELIER PRINCIPLE.

LEADER PRINCIPLE [psychology]   (1) Refers to a system with a hierarchy of leaders resembling a military structure. Three sources for the Führerprinzip; the first is the Hegelian idea of the *state* ("nothing above the State, nothing against the State, nothing outside the State"). The second source was the *Superman* (Übermensch) by the German philosopher Nietzsche. The final source was the German philosopher Count Hermann Graf Keyserling, who claimed that certain "gifted individuals" were "born to rule," not on the basis of birth or class but of "the laws of nature."† (2) The ideology sees each organization as a hierarchy of leaders, where every leader has absolute responsibility in her/his own area, demands absolute obedience from those below her/him and answers only to her/his superiors.‡ See Also TOTALITARIAN PRINCIPLE.

LEADERSHIP PRINCIPLE See LEADER PRINCIPLE.

LEAST-ACTION   PRINCIPLE [mathematics, physics] (Pierre-Louis Moreau de Maupertuis, 1698—1759, French mathematician) (1) Formulation of the least-action principle contains an analogy of this principle with the method of least squares. Gauss reduced the determination of the motion of a system via the least-action principle to the problem of finding the minimum of a sum of squares under supplementary conditions in the form of linear inequalities.§ (2) A particle moving

between two points under the influence of a force will follow the path along which its total action is least. Action is a quantity related to the average difference between the kinetic energy and the potential energy of the particle along its path. The principle is valid only where no energy is lost from the system, for example, an object moving in free fall in a gravitational field. ¶ (3) For a dynamical system moving under conservative forces, the actual motion of the system from point a to point b takes place in such a way that the action has a stationary value with respect to all other possible paths from a to b with the same kinetic–potential energy.‖ (4) Nature must always operate in the most efficient way possible, so all laws of nature should be described as the minimization of a certain quantity.** (5) In classical mechanics, the motion of a particle along some path always minimizes the difference between its kinetic energy and its potential energy. Mathematically, the motion of a particle always minimizes the Lagragian action functional.†† (6) In general relativity, the motion of a particle on a surface m must be a geodesic of m, so it must minimize the geodesic functional.‡‡ (7) In quantum mechanics, the motion of a subatomic particle from its initial state to its end state must minimize the Feynman path integral, which is a sum over all possible paths a that particle can take from initial state to end state.§§ (8) In quantum

---

*Daintith, J. (1999), *The Facts on File Dictionary of Chemistry*, 3rd, ed., Facts On File, New York, p. 142.
†Anon. (2009), *Führerprinzip*, (http://en.wikipedia.org/wiki/F%C3%BChrerprinzip).
‡Milgram, S. (1963), Behavioral study of obedience, *J. Abnorm. Social Psychol.* **67**, 371–378.
§Ciganova, N. J. and Fradlin, B. N. (1971), The connection between the least action principle and the principle of least squares, *Narisi Istor. Prirodoznav. Tehn* (13), 71–76, 125.

¶Anon. (1995), The Hutchinson Unabridged Encyclopedia Helicon (http://encyclopedia.farlex.com/least+action,+principle+of).
‖DeWitt, B. S. (1957), Dynamical theory in curved spaces. I. A review of the classical and quantum action principles, *Rev. Modern Phys.* **29**, 377–397.
Bejan, A. (2000), From heat transfer principles to shape and structure in nature: Constructal theory, *J. Heat Trans.* **122(3), 430.
††Brenier, Y. (1989), The least action principle and the related concept of generalized flows for incompressible perfect fluids. *J. Am. Math. Soci.* **2**(2), 225–255.
‡‡Taylor, M. E. (1996), *Partial Differential Equations: Basic Theory*, Springer, p. 53.
§§Schweber, S. S. (1994), *QED and the Men Who Made It: Dyson, Feynman, Schwinger, and Tomonaga*, Princeton Univ. Press, p. 39.

gravity, a spectral action principle seeks to apply the least-action principle to mathematical formulations.[*] (9) A free particle moving from (t_1, x_1) to (t_2, x_2) will follow the path $(t, x(t))$ that minimizes the "action integral."[†] (10) Over all curves connecting two points. the problem of minimizing an integral over a set of curves satisfying given boundary conditions is a typical problem in the calculus of variations.[‡] (11) For a system whose total mechanical energy is conserved, the trajectory of the system in configuration space is that path that makes the value of the action stationary relative to nearby paths between the same configurations and for which the energy has the same constant value.[§] (12) Traditional nineteenth-century definition states that nature chooses the most economical path for moving bodies, light rays, chemical reactions, and other processes.[¶] See also FERMAT'S PRINCIPLE OF LEAST TIME; HAMILTON PRINCIPLE; MINIMAL PRINCIPLES; PRINCIPLE OF LEAST ACTION.

LEAST-CONSTRAINT   PRINCIPLE [physics] See LEAST-ACTION PRINCIPLE.

LEAST-COUPLING PRINCIPLE [physics]   See EINSTEIN'S EQUIVALENCE PRINCIPLE.

LEAST-ENERGY PRINCIPLE [physics]   (1) A system is in stable equilibrium only under those conditions for which its potential energy is a minimum.[‖] (2) Potential energy of

a system in stable equilibrium is a minimum relative to that of nearby configurations.[**]

LEAST-MEAN-SQUARE ERROR RECONSTRUCTION PRINCIPLE (LMSER) [mathematics] A local learning rule naturally obtained for training nets consisting of either one or several layers.[††] See also ERROR RECONSTRUCTION PRINCIPLE.

LEAST-MOTION PRINCIPLE [chemistry]   See LAGRANGE PRINCIPLE.

LEAST-NUMBER PRINCIPLE (LNP) [mathematics]   (1) The authors show that IP (open), the induction principle for open (quantifier-free) formulas, proves LNP (open), the least-number principle for open formulas, over *PA*-. Since, as is easily shown, *PA*-proves that LNP (open) implies IP (open), this shows that the two principles are equivalent.[‡‡] (2) If a property or condition holds for some number, then there is a least number satisfying the condition.[§§] Also known as *principle of complete induction*.

LEAST-RESISTANCE   PRINCIPLE [physics] Electricity does not "take the path of least resistance." It takes all paths available— in inverse proportion to the impedance of the paths. Current flows through all available paths. The magnitude of the current

[*]Lanczos, C. (1986), *The Variational Principles of Mechanics*, Courier Dover Publications, p. 132.
[†]Noble, R. J. (1979), Quantum-field transition rates at finite temperatures, *Phys. Rev. D* **20**, 3179–3202.
[‡]Gelfand, I. M., Fomin, S. V., and Silverman, R.A. (2000), *Calculus of Variations*, Courier Dover Publications, p. 159.
[§]Johnson, L. (1992), An ecological approach to biosystem thermodynamics, *J. Biolo. Philo.* **7**(1), 35–60.
[¶]Peacock, J. A. (1999), *Cosmological Physics*, Cambridge Univ. Press, p. 5.
[‖]Walker, P. M. B. (1999), *Chambers Dictionary of Science and Technology*, Chambers, New York, p. 665.

[**]Zhou, L., Jiang, X., Liu, Ji., and Wang, H. (2007), Mathematic model of attrition of quartzite particles as medium material in fluidized bed, *Huagong Xuebao* **58**(11), 2776–2781.
[††]Zhang, B. L., Xu, L., and Fu, M. (1996), Learning multiple causes by competition enhanced least mean square error reconstruction, *Int. J. Neural Syst.* **7**(3), 223–236.
[‡‡]Aoyama, K. and Fukuzaki, Kenji (1997), Equivalence of the induction schema and the least number principle for open formulas, *SUT J. Math.* **33**(2), 149–162.
[§§]Casey, J. B., Evans, W. J., and Powell, W. H. (1981), A descriptor system and principles for numbering closed boron polyhedra with at least one rotational symmetry axis and one symmetry plane, *Inorg. Chem.* **20**(5), 1333–1341.

flowing in each path depends on the voltage and impedance of each path. The lower the impedance of the path (assuming that voltage remains constant), the greater the current. Conversely, the higher the impedance of the path (again assuming that voltage remains constant), the lower the current.* See also PRINCIPLE OF LEAST RESISTANCE.

LEAST-SQUARE   PRINCIPLE [mathematics] (1) Numerical error estimation: (2) Moving average-model: (3) Estimate close to the maximum likelihood estimate.[†]

LEAST-TIME PRINCIPLE [physics]   (1) Light traveling between two points seeks path in which the number of waves is equal, in the first approximation, to neighboring paths. (2) Path taken by a ray of light in traveling between two points requires either a minimum or maximum time. (3) Light traveling through some substance has a speed determined by the substance. The actual path taken by light between any two points, in any combination of substances, is always the path of least time traveled at the required speeds.[‡] (4) Second law of thermodynamics, that in the neighborhood of any equilibrium state of a system, there are states not accessible by a reversible or irreversible diabolic process. (5) Electromagnetic wave will take a path involving the least travel time when propagating between two points.[§] See also CONTES PRINCIPLE; CARATHÉODORY'S PRINCIPLE; FERMAT'S PRINCIPLE; FERMAT'S PRINCIPLE OF LEAST TIME; OPTIMALITY PRINCIPLE.

LEAST-UPPER-BOUND PRINCIPLE [mathematics]   An important property of the real numbers is completeness; every nonempty set of real numbers that is bounded above has a supremum. If, in addition, we define

$\sup(S) = -8$ when S is empty and $\sup(S) = +8$ when S is not bounded above, then every set of real numbers has a supremum. The least-upper-bound property is an example of the aforementioned completeness properties, which is typical for the set of real numbers. If an ordered set S has the property that every nonempty subset of S having an upper bound also has a least upper bound, then S is said to have the least-upper-bound property. As noted above, the set R of all real numbers has the least-upper-bound property. Similarly, the set Z of integers has the least-upper-bound property; if S is a nonempty subset of Z and there is some number n such that every element s of S is less than or equal to n, then there is a least upper bound u for S, an integer that is an upper bound for S and is less than or equal to every other upper bound for S.[¶]

LEFFLER–HAMMOND PRINCIPLE [chemistry] (George S. Hammond, 1921-2005, American chemist) (John E. Leffler, American chemist) Extension of the scope of such correlations to include reactions governed by the Bell–Evans–Polanyi–Leffler–Hammond postulate requires an examination of the forward and reverse central barriers and geometric features of nonidentity reactions for which an extended set of data are available at the 4-31G level. When a transition state has nearly the same energy.[‖] The transition states for reactions involving unstable intermediates can be closely approximated by the intermediates themselves.[**] see HAMMOND PRINCIPLE

LEFSCHETZ   PRINCIPLE [mathematics] (Solomon Lefschetz, 1884-1972, American mathematicis)   If a statement of algebraic geometry is proved for one underlying algebraically closed field, then it is valid for

*Wang, Q. A., (2004) Diffusion laws and least action principle, Los Alamos National Laboratory, *Preprint Archive—Condensed Matter*, pp. 1–9.
[†]Wei, C. Z. (1992), On predictive least squares principles, *Ann. Statist.* **20**(1), 1–42.
[‡]Pmnmny, H. J. (1921), The Einstein spectral line effect, *Phil. Mag.* **41**, 747–749.
[§]Guy, A. T. (1981), An extremum principle for reactions in solid solutions, *Chem. Metall.—Tribute Carl Wagner, Proc. Symp.*, pp.397–403.

[¶]Rudin, W. (1976), *Principle of Mathematical Analysis*, 3rd ed., McGraw-Hill.
[‖]Civcir, P. U. (2008), Prediction rules of transition-state structures, *J. Mol. Structu. (TheoCHEM)*, **848**(1–3), 128–138.
[**]Ando, T. and Yamataka, H. (1981) *Proediction rules of transition-state structures. The Leffler-Hammond principle and its development, Kagaku no Ryoiki*, v35 (7), 483–493.

any algebraic geometry whose underlying field has the same characteristic. The program consists in going through the formal definitions of all concepts of algebraic geometry and in verifying that the principle extends to all statements involving these notions. It is well known that the principle holds for statements in the first-order language of fields.* (2) Use of topological techniques for algebraic geometry over any algebraically closed field K of characteristic 0, by treating K as if it were the complex number field. It roughly asserts that true statements in algebraic geometry over C are true over any algebraically closed field K of characteristic zero. Allows carryover of results obtained using analytic or topological methods for algebraic varieties over C to other algebraically closed-base fields of characteristic 0.[†]

LEGAL PRINCIPLE [law]   The basis underlying the formulation of jurisprudence.[‡]

LEIBNIZ'S PRINCIPLE [physics]   See PRINCIPLE OF INDISCERNIBLES.

LEISURE PRINCIPLE [psychology]   Leisure groups behave in quite different ways from workgroups, and the motivation behind leisure is very different from that behind work. The usual social divisions become less important—and they tend to be very welcoming to newcomers of any class. Individuals can choose those leisure activities that produce the kinds of social behavior and relationships that they enjoy most. There is also a strong link between choice of leisure pursuit and personality. Extroverts tend to be more active and successful in sports as well as in social activities, while violent sports appeal most to "tough-minded" people who don't care how much damage they do to others.[§]

LENGTH–AREA   PRINCIPLE [mathematics] Assume that $f(z)$ is a meromorphic function in the open set Δ, $l(t) = l(t, \Delta)$ is the total length of the level curve $|f(z)| = t$ in Δ, and $A(A \prec +\infty)$ is the area of Δ. Let $p(t) = p(t, \Delta) = (1/2\pi) \int_0^{2\pi} n(\Delta, f = te^{i\theta}) d\theta$.   Then $\int_0^{+\infty} [l(t)^2 / tp(t)] dt \leqslant 2\pi a$, where the integrand is defined to be zero if $p(t) = +\infty$. In particular, for almost all values of t satisfying condition $p(t) \prec +\infty$, it is true that $l(t) \prec +\infty$[¶]

LEONARD PRINCIPLE [physics]   Regulator for asynchronous electric motor and two-channel encoder with torque and shaft position detection.[‖]

LERAY–SCHAUDER PRINCIPLE [mathematics] (Jean Leray, 1906-1998, French mathematician) (Julius Pavel Schauder, 1899-1943, Polish-Ukranian mathematics)   For condensing admissible classes of multifunctions and deductions for some new fixed-point theorems.[**]

LESSER LIMITED PRINCIPLE OF OMNISCIENCE (LLPO) [mathematics]   For each binary sequence $(a_1, a_2, \ldots)$ with at most one term equal to 1, either $a_n = 0$ for all even n or else $a_n = 0$ for all odd n.[††]

*Robinson, A. (1952), *On the Metamathematics of Algebra*, North-Holland, vol. 1, pp.686–694.
[†]Grensing, G. (2002), On ghost fermions, *Eur. Phys. J. C Particles Fields* **23**(2), 377–387.
[‡]American Heritage (2000), *American Heritage Dictionary of the English Language*, 4th ed., Houghton Mifflin, New York.
[§]Argyle, M.ichael (1996), The leisure principle, *Independent News and Media Limited* (Sunday, 7/14/96) (http://www.independent.co.uk/arts-entertainment/the-leisure-principle-1328768.html).

[¶]Kim, W. and Oh, J. J. (2007), Determining the Minimal Length Scale of the Generalized Uncertainty Principle from the Entropy-Area Relationship. Los Alamos National Laboratory, *Preprint Archive, High Energy Physics—Theory*, pp.1–12.
[‖]Soyfer, J. C., Vincent M., and Busson, F. (n.d.), Gilletiodendron glandulosum (Porteres) J. Leonard (Cesalpiniacees). Identification of a neurotoxic principle, *Med. Tropicale: Rev. Corps Sante Colonial* **29**(5), 615–616.
[**]Kosenko, I. I. (2005), Application of the theory of the Leray-Schauder degree for the approximation of oscillations of a satellite on an elliptic orbit, *Dokl. Phys.* **50**(10), 532–534.
[††]Bishop, E. and Bridges, D. (1985), Grundlehren der math, *Wissenschaften*, p. 279.

LEXICAL OPERATING PRINCIPLE [psychology] Provide a means by which children may concentrate on the most likely possibilities for the reference of a particular word. (1) Ability to rapidly map novel symbols revealed whether they possessed the novel name–nameless category (N3C) lexical operating principle. One of the more advanced principles that directs the child's learning is the N3C principle. [A methodological colleague has pointed out that mathematically N3C should be correctly presented as N(sup3)C.] This principle states that when a child hears a novel word in the presence of an unknown object, he or she will immediately map the novel name onto the novel entity. The N3C principle, then, enables young children who are developing typically to map the meanings of new words at a rapid rate and with very little exposure to the new words. Fast mapping has been defined as the child's ability to learn something about a new word without ostensive definition. (2) Increase the likelihood certain possibilities will be applied when a person is determining the referent and/or extension of a word as words applied to language learning and aspects of cognition.* See also PRINCIPLE OF REFERENCE; PRINCIPLE OF EXTENDIBILITY; PRINCIPLE OF OBJECT SCOPE; PRINCIPLE OF CATEGORICAL SCOPE; NOVEL NAME–NAMELESS CATEGORY (N3C) PRINCIPLE; PRINCIPLE OF CONVENTIONALITY.

LIGHT DIFFRACTION PRINCIPLE [engineering, psychology]   (1) When the order of space and time integration's can be exchanged. (2) Stereo vision test card for eye disease diagnosis and cure evaluation has cylindrical lens.[†] See also CONTRACTION PRINCIPLE.

LIGHT PRINCIPLE [physics, mathematics] Referred to as the *first and second postulates of special relativity*. (1) Effect of information about one possible cause of an event as inferences regarding another possible cause. Analyzes situations until a minimal set of sufficient causes are identified, then, other possible causes are ignored or dismissed. (2) Comprising at least two sensor or feeler elements having different functional or operating principles and a common evaluation circuit for evaluation of the property changes in the sensor elements and for triggering a signal.[‡] (3) Illumination is created according to either the transmitted light principle, where the light passes through the transparent or translucent object, or *incident light principle*, where the object is illuminated from the side and the reflected light rays are gathered. The object being examined can be observed against a light field or against a dark field in the case of illumination according to either of the aforesaid principles. In the case of light-field illumination the background appears bright and the object dark, while with dark-field illumination the object appears bright and the background appears dark.[§] (4) An arrangement for measuring lengths or angles using the transillumination or direct light principle.[¶] (5) Projection apparatus for creating visible images that utilizes the interference of light principle for construction of its image.[‖] (6) Searchlight principle, when a source S is placed at the focal point of lens L, an image S' will be located at infinity. Since S subtends an angle. alpha. from L, the image S' will also subtend at an angle. alpha.. Now the illumination at a point on the axis will be determined by the brightness of the image and the solid angle subtended by the image. Accordingly, illumination within

*Eick, O. J., Wittkampf, F. H., Bronneberg, T., and Schumacher, B. (1998) The LETR-principle: A novel method to assess electrode-tissue contact in radiofrequency ablation, J. *Cardiovasc Electrophysiol* **v.9**(11), 1180–1185.

[†]Romski, M. A., et. al., (1996) Mapping the meanings of novel visual symbols by youth with moderate or severe mental retardation, *Am. Jo. Mental Retard.* **100**, 391–402.

[‡]Goetting, H. C. and Schuetze, R. (1989), An optical strain measurement facility for damage mechanics investigations on composites, *Materialwissen. Werkstofftechnik* **20**(10), 344–350.

[§]Le Poidevin, R. (1998) *Questions of Time and Tense*, Oxford Univ. Press, p .129.

[¶]Scheidweiler, A. (1981), *Method of Fire Detection and Fire Detection Installation*, US Patent 4,405,919.

[‖]Spies, A. (1989), *Photoelectric Position-Measuring Arrangement Having a Plurality of Shiftable Grids*, US Patent 5,009,506.

the searchlight cone defined by the lens focal point is constant anywhere within the cone.[*] (7) Rest mass of a particle cannot be kept invariant; must exist a unique preferred inertial frame of reference wherein a particle is absolutely at rest.[†] See also CAUSALITY PRINCIPLE; DIRECT LIGHT PRINCIPLE; INCIDENT LIGHT PRINCIPLE; PRINCIPLE OF COVARIANCE; PRINCIPLE OF MINIMAL CAUSATION; SEARCHLIGHT PRINCIPLE; SCATTERED LIGHT PRINCIPLE; TRANSMITTED LIGHT PRINCIPLE.

LIGHT PROJECTION PRINCIPLE [engineering] (1) Light source reflected as a converging beam or a screen arranged in a beam path or a lens arranged in alignment to a light output.[‡] (2) The projection lens is designed so that flat-state light misses the pupil of the projection lens, allowing very little light to be projected through the lens. But the mirrors are only briefly at the flat state as they make a transition from one landed state to the other. When the mirror is in its OFF state, the reflected light is further removed from the pupil of the projection lens and even less light is collected by the projection lens. When the mirror is in its ON state, the reflected light is directed into the pupil of the projection lens, and nearly all the light is collected by the projection lens and imaged to the projection screen. Because of the large rotation angles of the mirror, the OFF-state light and ON-state light are widely separated, allowing fast projection optics to be used.[§] (3) The main lamp parameter necessary for bright projection systems is luminance.[¶] (4) Application

of three-dimensional measurement for microparts and measurement in a narrow space. Current measurement system is limited to offline inspection demands and unsuitable for online. To overcome this shortcoming, a minimized probe is developed to analyze the phase fringe and reconstruct the point clouds of surface rapidly and accurately.[‖]

LIKE–LIKE PRINCIPLE. [Mathematics]   Two unusual mutations, a double deletion coupled with a point transversion. The lack of disease is no real surprise in part because repeats seem to function adequately in other species and because of transgenic deletions resulting in reduced susceptibility.[**]

LIKELIHOOD PRINCIPLE [engineering, mathematics]   (1) Adjoint distribution of the data and the unobserved variables of inferential interest, considered as a function of the parameters and these inferential variables.[††] (2) With the statistical model (X, P, θ), an inference about is defined as a pair (B, v), where is a family of subsets of and is an extended real-valued function defined on B. If forms a σ-field and is a σ-finite measure over B, is termed an inversion. Let be an inference about given the additional information that $\theta \in F$; the inference is localizable if and $Cv^*(F^*) = v(F^*)$ for an appropriate constant C. Requires inferences drawn about θ, given $x \in A$, depend only on the likelihood function specified up to a multiplicative constant where $A \subset X$.[‡‡] Also known as *Conditionality Principle; Sufficiency Principle.*

[*]Misund, O. A. (1997), Underwater acoustics in marine fisheries and fisheries research, *Rev. Fish Biol. Fisheries.* **7**(1), 1–34.
[†]Renshaw, C. (1997), *The Restoration of Space and Time*, AAAS, Philiadelphia, p. 18.
[‡]Koda, N. J., Bleha, W. P. Jr., Giehll, H. W., Reinsch, S. J., and Robusto, P. F. (1985), *Liquid Crystal Light Value Color Project*, US Patent 4,650,286 (filing date 10/3/85).
[§]Hauser, J. (1995), *Projection Apparatus for Creating Visible Images*, US Patent 5,489,951.
[¶]Hornbeck, L. J. (n.d.), *Digital Light Processing and MEMS: Timely Convergence for a Bright Future* (http://focus.ti.com/download/dlpdmd/115 _Digital_Light_Processing_MEMS_Timely_Convergence.pdf).

[‖]Monch, H., Derra, G., and Fischer, E. (n.d.), *Optimised Light Sources for Projection Displays* (http://sysdoc.doors.ch/PHILIPS/pw4₁₆-4−896.pdf).
[**]Watzenig, D. and Fox, C. (2007), A review of statistical modeling and inference for electrical capacitance tomography, *Meas. Sci. Technol.* **20**, 22.
[††]Huang, Z. (1992), *Innovative Three-Dimensional Structured Light Digital Measurement System and the Probe of the R & D*, National Taipei Unive. Science and Technology, Inst. of Automation Technology.
[‡‡]Bromstad, J. F. (1996), On the generalization of the likelihood function and the likelihood principle, *J. Am. Stat. Assoc.* **91**, 791.

LIMITED PRINCIPLE OF OMNISCIENCE (LPO).
[Mathematics]   For each binary sequence
$(a_1, a_2, \ldots)$ either $a_n = 0$ for all n or else
there exists n such that $a_n = 1$, which is
generally regarded as an essentially nonconstructive principle, for several reasons.
There is a recursive algorithm that, when
applied to any recursively defined binary sequence $(a_1, a_2, \ldots)$, outputs 0 if $a_n = 0$ for all
n, and outputs 1 if $a_n = 1$ for some n, is probably false within recursive function theory.[*]
For any decision sequence $\{a_n\}$, *either $a_n = 0$
or some $a_n = 1$.* See also LIMITED PRINCIPLE
OF OMNISCIENCE; BOUNDED SEQUENCE PRINCI-
PLE; CONSTANT SUBSEQUENCE PRINCIPLE; CON-
VERGENT SUBSEQUENCE PRINCIPLE; BOLZANO-
WEIERSTRASS PRINCIPLE; MONOTONE SEQUENCE
PRINCIPLE[†]

LIMITING ABSORPTION PRINCIPLE [physics,
engineering]   See ABSORPTION PRINCIPLE.

LIMITING AMPLITUDE PRINCIPLE [physics,
engineering, mathematics] (Reflecting bodies)   Every solution of the wave equation
with a harmonic forcing term: in the exterior
of a reflecting body tends to the steady
solution uniformly on bounded sets as tends
to infinity.
Every solution of the wave equation with
a harmonic forcing term:

$$\Box U \equiv \Delta U - U_{tt} = e^{iwt} g(x)$$

in the exterior of a reflecting body tends
to the steady solution $V(x)e^{iwt}$ uniformly
on bounded sets as t tends to infinity.
Here V satisfies the reduced wave equation
$\Delta V + w^2 V = g$, vanishes on the body B and
satisfies Sommerfeld's radiation condition:
$r|V_r + iwV| \to 0$ as $r \to \infty$.[‡] See also ABSORP-
TION PRINCIPLE.

LINDELÖF PRINCIPLE [geometry]   (Ernst
Leonard Lindelöf, 1870–1946, Finnish
topologist)   There is an absolute constant
such that integral mean of a suitably chosen
subharmonic function over a circle is an
increasing function of the radius.[§] Also
known as *Korenblum's Maximum Principle*.
(1) A function holomorphic in a sector and
bounded on the boundary either grows
quite fast or attains its maximum modulus
on the boundary.[¶] (2) An extension of the
Pontryagin maximum principle for Mayer
problems without terminal constraints,
subject to affine control systems. Homogeneous tangent vectors provide a nonlinear,
high-order approximation of the attainable
set where linear approximation proves to be
inadequate. Homogeneous tangent vectors
derive new conditions for optimality effective
for nonlinear optimal control problems
where other high-order tests provide no
conclusive information.[‖] Also known as
Tangent Variation Principle.

LINE-SCANNING PRINCIPLE See   VERTICAL
ROTATING PRINCIPLE.

LINEAR COVARIANCE PRINCIPLE [physics,
mathematics]   (1) The foundation of general
relativity consists of the covariance principle, the equivalence principle, and the field
equation whose source term is subjected to
modification. Assumes any Gaussian system
to be valid as a spacetime coordinate system.
Given the mathematical existence of the co-
moving local Minkowski space along a timelike geodesic in a Lorentz manifold, a crucial

[*]Vercammen, J., Sandra, P., Baltussen, E., Sandra, T., and David, F. (2000), Considerations on
static and dynamic sorptive and adsorptive sampling to monitor volatiles emitted by living plants,
J. *High Resolut. Chromatogr.* **23**(9), 547–553.
[†]Mandelkern, M. (1988) Limited Omniscience and
the Bolzano-Weierstrass Principle. *Bull London
Math Soc* **20**, 319–320.
[‡]Morawetz, C.S. (1962) The limiting amplitude
principle. *Communications on pure and applied
mathematics*, **15**, p349–351.

[§]Qiao, Y.-F., Li, R-J., and Meng, J. (2001),
Lindelof's equations of nonholonomic rotational relativistic systems, *Wuli Xuebao* **50**(9),
1637–1642.
[¶]Vdovicheva, N. K. (1992), Computation of sound
fields in layered medium with plane-parallel flow,
Akusticheskii Zh. **38**(6), 1025–1031.
[‖]Vuorinen, M., Martio, O., and Miklyukov, V.
(1996), Phragmen- Lindelof's principle for quasi-
regular mappings and isoperimetry, *Dokl. Akad.
Nauk* **347**(3), 303–305.

question for a satisfaction of the equivalence principle.[*] (2) The general laws of nature are to be expressed by equations that hold good for all systems of coordinates, that is, are covariant with respect to any substitutions whatsoever (generally covariant). The covariance principle can be considered as consisting of two features: (a) the mathematical formulation in terms of Riemannian geometry and (b) the general validity of any Gaussian coordinate system as a spacetime coordinate system in physics.[†] Also known as *Covariance Principle*; *Principle Of Uniformity*; *Unlearned Principle*.

LINEAR MAGNET PRINCIPLE [engineering] With linear drives or with friction drives, the motor is driving the cable that connects the elevator car and the counterbalance over a return pulley. With the friction wheel drive principle or the linear magnet principle, the cable is connected at the elevator car. All drive types also require independent safety devices in case the drive fails and the elevator car begins to drop. The safety device must brake the elevator car from a certain overspeed and smoothly bring the car to a stop.[‡] See also BUCKING MAGNET PRINCIPLE; MAGNET PRINCIPLE; FRICTION WHEEL DRIVE PRINCIPLE; PLUNGER MAGNET PRINCIPLE; LOCK MAGNET PRINCIPLE; CERAMIC MAGNET PRINCIPLE.

LINEAR TIME INVARIANT (LTI) SEPARATION PRINCIPLE. Aka principle of separation of estimation and control.

Proof of separation principle for LTI systems.

[*]Lo, C. Y. (2001), *The Equivalence Principle, the Covariance Principle and the Question of Self-Consistency in General Relativity*, Applied and Pure Research Inst. (http://www.as158.com/article/show.asp?id=16676).
[†]Sinanoglu, O. (1984), A principle of linear covariance for quantum mechanics and the electronic structure theory of molecules and other atom clusters, *Theoret Chim. Acta*, **65**(4), 233–242.
[‡]Weinberger, K., Silberhorn, G., and Rennetaud, J. M. (1998), *Linear Drive for Transportation Equipment*, US Patent 6,053,287.

Consider the system

$$\dot{x}(t) = Ax(t) + Bu(t)$$
$$y(t) = Cx(t)$$

where
  $u(t)$ represents the input signal,
  $y(t)$ represents the output signal, and
  $x(t)$ represents the internal state of the system

We can design the an observer of the form

$$\hat{x} = (A - LC)\hat{x} + Bu + Lu$$

And state feedback $u(t) = -K\hat{x}$ Define the error e:

$$e = x - \hat{x}$$

Then

$$\dot{e} = (A - LC)^e$$
$$u(t) = -K(x - e)$$

Now we can write the closed-loop dynamics as

$$\begin{bmatrix} x \\ \dot{e} \end{bmatrix} = \begin{bmatrix} A - BK & BK \\ 0 & A - LC \end{bmatrix} \begin{bmatrix} x \\ e \end{bmatrix}$$

Since this is triangular the eigenvalues are just those of $A - BK$ together with those of $A - LC$. Thus the stability of the observer and feedback are independent.[§]

LINGUISTIC RELATIVITY PRINCIPLE [psychology] (Benjamin Lee Whorf, 1897–1941) Sometimes referred to as the Sapir–Whorf hypothesis—relationships between language, mind, and experience. Isolates of experience are universally available to the species by virtue of the nature of the world we live in and its interface with our senses. No one "is free to describe nature with absolute impartiality but is constrained to certain modes of interpretation even when" they think themselves "most free." We cannot be impartial or free of culture when we describe because describing is talking and the words we use are often subtly different in meaning from language to language, even

[§]Brezinski, Claude (2002) Computational Aspects of Linear Control (Numerial Methods and Algorithms). Springer.

when they seem at first thought to be equivalent semantically. Through constant use of habitual patterns of speech, our patterns of attention themselves become linguistically conditioned. All observers are not led by the same physical evidence to the same picture of the universe, unless their linguistic backgrounds are similar, or can in some way be calibrated. What varies relates to which isolates are abstracted for attention and which are ignored. By paying attention to the way our languages work, we can begin to identify the culturally specific patterns involved in drawing isolates from experience and understanding events. Our predilection to segment and organize experience on the basis of unconscious linguistic influences that come to overlay universal perceptual principles during the process of language acquisition is not only a function of what is named in our culture (the specific things, attributes, actions, and events that our speech community labels) but is also, and with much more subtle and elusive implications, a function of logical relationships articulated by grammar.[*] See also PRINCIPLE OF LINGUISTIC RELATIVITY.

LINKED INFERENCE PRINCIPLE [mathematics]   Used to obtain the abstract formulations of various linked inference rules. Included among such rules are linked UR resolution, linked hyperresolution, and linked binary resolution, each of which generalizes the corresponding standard and well-known inference rules.[†] (2) Resolution systems represent automatic theorem proving. They achieve extremely high inference rates and can run continuously for days without running out of storage. They can crack many of the toughest-challenge problems that have been circulated. While they exploit many specialized algorithms, data structures, and optimizations, they rely on unification.[‡]

[*]Whorf, B. L., Carroll, J. B., and Chase, S. (1956), *Language, Thought, and Reality*, Massachusetts Inst. Technology (MIT) Press, p. 28.
[†]Veroff, R. and Wos, L. (1992), The linked inference principle. I. The formal treatment, *J. Automat. Reason* 8(2), 213–274.
[‡]Wos, L., Veroff, R., and Pieper, G. W. (1997), Automated reasoning and its applications: Essays in honor of Larry Wos, MIT Press, p. 23.

LIPOSOLUBLE ACTIVE PRINCIPLE [physics, chemistry]   Fatlike acids, limonene, saline, and silicones and mixtures there of can be used to modify the physical properties of the active principle to encapsulate (with respect to the thermal resistance, solidity to the washing, etc.) without affecting the functional effect of its own active principle. The objective is to have very strong walls during the process of elaboration and storage of the microcapsules, and then diminish the resistance of the wall at a precise moment and by the direct action of the media that varied, and to thus allow the nucleus to escape, slowly, according to the active principle.[§] See also HYDRO-SOLUBLE ACTIVE PRINCIPLES; ACTIVE PRINCIPLE.

LIPPMAN–SCHAW VARIATIONAL PRINCIPLE [engineering]   (1) Electrocapillarity observed and formalized by Lippman in nineteenth century is defined as the change of surface tension of liquid metal (typically mercury) immersed in electrolyte by electric potential (voltage) between the electrolyte and the mercury. The result of the analysis can be expressed by Lippman's equation. (2) While the term *electrocapillarity* refers to a relationship between surface tension change of liquid metal due to applied voltage, *continuous electrowetting* refers to a principle of moving a lump of liquid metal using electric potential.[¶]

LIQUID CRYSTAL PRINCIPLE [engineering] Liquid crystal is prepared in the homeotropic texture and is placed between two polars that cross at a $90°$ degrees angle.[‖]

LIQUID MOVEMENT PRINCIPLE [physics] See FICK PRINCIPLE.

[§]Garcia, R. L. and Esquina, S. V. (2005), *Additive for Domestic Washing Processes* (Richfield, OH), US Patent 11,667,871.
[¶]Kouri, D. J., Huang, Y.; Zhu, W., and Hoffman, D. K. (1994), Variational principles for the time-independent wave-packet-Schroedinger and wave-packet-Lippmann-Schwinger equations, J. of *Chem. Physi.* 100(5), 3662–3671.
[‖]Potier, A., Mehani, K., and Luizy, F. (1980), Plate thermography in post-operative monitoring (author's transl), *Nouv. Presse Med.* 9(38), 2819–2821.

LIQUID RING PRINCIPLE [engineering]   Manipulation of one moving part with no metal-to-metal contact, mechanical pistons, or valves is performed by rotating a ring of liquid compressant. Compressor and condenser located in fan casing and arranged to rotate about common shaft with compressor.*

LISKOV SUBSTITUTION PRINCIPLE (LSP). Bartara Jane (Hulerman) Liskov, 1939-, American engineer) [computers, mathematics] (Bartara Hane Huberman 1939-American engineer marce, American engineer, Barbara Liskov and Jeannette Wing) (Robert Cecil Martin) (1) Every function that operates on a reference or pointer to a base class should be able to operate on derivatives of that base class without knowing it. The virtual member functions of derived classes must expect no more than the corresponding member functions of the base class and should promise no less. (2) Virtual member functions that are present in base classes must also be present in the derived classes; and they must do useful work. The functions operating on pointers or references to base classes will need to check the type of the actual object to make sure that they can operate on it properly.[†] (3) Derived classes must be substitutable for their base classes. (4) Functions using pointers or reference to base classes must be able to use objects of derived classes without knowing it. Foundation for building code that is maintainable and reusable. It states that well-designed code can be extended without modification; that in a well-designed program new features are added by adding new code, rather than by changing old, already working, code. See also DEPENDENCE INVERSION PRINCIPLE; INVERSION OF CONTROL PRINCIPLE; SUBSTITUTION PRINCIPLE. (1) In object-oriented programming, a particular definition of subtype. Let $q(x)$ be a property provable about objects x of type T. Then $q(y)$ should be true for objects y of type S where S is a subtype of T.[‡] (2) Closely related to the design by contract methodology, leading to some restrictions on how contracts can interact with inheritance; preconditions cannot be strengthened in a subclass. You cannot have a subclass that has stronger preconditions than its superclass. Postconditions cannot be weakened in a subclass. You cannot have a subclass that has weaker postconditions than its superclass has.[§] (3) No new exceptions should be thrown by methods of the subclass, except where those exceptions are themselves subtypes of exceptions thrown by the methods of the superclass. A function using a class hierarchy violating the principle uses a reference to a base class, yet must have knowledge of the subclasses. Such a function violates the open/closed principle because it must be modified whenever a new derivative of the base class is created.[¶] See also OPEN–CLOSED PRINCIPLE.

LITTLEWOOD'S SECOND PRINCIPLE [Mathematics]   (John Edensor Littlewood, 1885–1977, British Mathmetician)   (1) Essentials of measure theory in mathematical analysis. A measurable set is almost an open set; a measurable function is almost a continuous function; and a convergent series is almost uniformly convergent.[‖] (2) Given a measurable set T and $\varepsilon > 0$ ther is a set of open intervals with union U such that the symmetric difference of T and U has Lebesgue measure at most ε.[‖] (3) Given a measurable real function f and $\varepsilon > 0$, one can choose an open set V of the real line such

*Mackowiak, J. (1990), Principles of design of columns packed with random and stacked metal Bialecki rings for gas-liquid systems, *Inzynieria Aparatura Chemiczna*, **29**(5–6), 3–8.
[†]Martin, R. C. (2002) *Agile Software Development: Principles, Patterns, and Practices*, Prentice Hall.

[‡]Wos, L., Veroff, R., Pieper, and G. W. (1997), *Automated Reasoning and Its Applications: Essays in Honor of Larry Wos*, MIT Press, p. 23.
[§]Liskov, B. (1987), *Conference on Object Oriented Programming Systems Languages and Applications*; archive addendum to *proc. Object-Oriented Programming Systems, Languages and Applications*, Orlando, FL, ACM Specical Interest Group on Programming Languages, pp.17–34.
[¶]Garcia, R. L. and Esquina, S. V. (2005) Additive for Domestic Washing Process, U. S. Patent application 11/667,871
[‖]Kouri, D.J., Huang, Y., Zhu, W., and Hoffman, D. K. (1994) Variational principles for the time-independent wave-packet-Schroedinger and wave-packet-Lippmann-Schwinger equations. Journal of Chemical Physics, 100(5) 3662–3671.

that f is continuous outside of V, and V has Lebesgue measure at most ε.

LMSER PRINCIPLE [mathematics]   See LEAST-MEAN-SQUARE ERROR RECONSTRUCTION PRINCIPLE.

LOAD DIVISION PRINCIPLE [psychology, computers]   See INFORMATION-PROCESSING PRINCIPLE.

LOAD SEPARATION PRINCIPLE [geometry] [ED: Definitions are similar so keep *2nd* only by *Bernal*   (2) Load separation and using characteristic deformation properties of materials to relate load, displacement, and crack length.[*]

LOCAL–GLOBAL PRINCIPLE [mathematics] The assertion that an equation can be solved over the rational numbers if and only if it can be solved over the real numbers and over the p-adic numbers for every prime p.[†] (2) Holds for quadratic forms over the rational numbers; and more generally over any number field, when one uses all the appropriate local field necessary conditions. (3) Applies to the condition of being a relative norm for a cyclic extension of number fields.[‡] (4) The concept of a "local–global" principle (in the following form) for a predicate P on a set S with some algebraic structure if P is true in S if and only if P is true in S_i for all i, where $S_i \in S$ have the same structure but also additional properties (e.g., S is an integral domain, the S_i are localizations S_m of S at maximal ideals m, and P is the predicate "$a_x = b$ has a solution in $S[S_m]$"). The consideration of localizations

does not suffice for most number-theoretic questions;[§] (5) One of the central principles of Diophantine geometry that reduces the problem of the existence of reational points on an algebraic variety over a global field to the analogous problem over local fields.[¶] See also HASSE NORM PRINCIPLE; HASSE PRINCIPLE.

LOCAL INVARIANCE PRINCIPLE [physics, mathematics, chemistry]   See INVARIANCE PRINCIPLE.

LOCAL SIMULATION PRINCIPLE [geometry, physics]   Relationship between the load and the driving force in the component.[‖]

LOCAL-TO-GLOBAL PRINCIPLE [mathematics]] (Helmut Hasse (1898–1979, German mathematician)   The local to global principle is the method of deriving applications and solutions by specifying \ local" (and deliberately myopic) heuristics, critiques and methods followed by using a powerful general method to \ globalize" this specification into a complete solution. The local to global principle is divided into two parts: local encoding of the problem followed by a globalization step that uses the encoding. The guiding feature of local encodings is that they are usually easy to compute from the data at hand. Any extension that looks like enumeration, search or optimization is best left to the global step. The local step is essentially the translation of your problem into an abstract language that is ready for the globalization step. In constrast globalization methods are often øthe shelf"

[*]Bernal, C. Rink, M., and Frontini, P. (1999), Load separation principle in determination of J-R curve for ductile polymers. Suitability of different material deformation functions used in the normalization method, *Macromol. Symp.* **147**(*Mecha. Behav. Poly. Mater.*), 235–248.
[†]Prestel, A., (1975) A Local-global principle for quadratic froms. *Math.Z.* **142**, p 91–95.
[‡]Nakajima, H. (1996), Instantons and affine Lie algebras, Nuclear Phys. B, Proc. Suppl. 46, 154–161.

[§]Pfister, A. (1996), Various aspects of local-global principles, Seminaire on Number Theory, 1972–1973, Univ. Bordeaux, Talence, Exp. 23, Natl. Research Center, Talence, 1973.
[¶]Nakajima, H. (n.d.), Instantons and affine Lie algebras, *Nuclear Phys. B Proc. Suppl.* **46**, p.154–161.
[‖]Zerbst, U., Heerens J. and Schwalbe, K. H., (1995), Fracture mechanics analysis based on a local simulation principle. *Fatigua and Fracture of Engineering Materials and Structures* **18**(3), 371–376.

in that once you abstract and encode the particulars of your problem you can look for pre-existing useful methods or software to finish your solution. The idea of globalization is to find a best overall or global compromise between competing local criteria. See also HASSE PRINCIPLE. *

LOCAL UNIFORM BOUNDEDNESS PRINCIPLE [physics, mathematics]   See BOUNDARY PRINCIPLE.**

LOCALITY PRINCIPLE [physics]   Distant objects cannot have direct influence on one another; an object is influenced directly only by its immediate surroundings. The relative independence of objects far apart in space (*A* and *B*): external influence on *A* has no direct influence on *B*.[†] See also PRINCIPLE OF LOCALITY; PRINCIPLE OF LOCAL ACTION.

LOCALIZATION   PRINCIPLE [physiology] Specific functions have relatively circumscribed locations in some particular part or organ of the body.[‡]

LOCALIZATION   PRINCIPLE [mathematics] The convergence of the Fourier series of a function at a point depends only on the behavior of the function in some arbitrarily small neighborhood of that point.[§]

LOCARD'S EXCHANGE PRINCIPLE [psychology] (Edmond Locard)   See LOCARD'S PRINCIPLE.

LOCARD'S PRINCIPLE [psychology] (Edmond Locard)   1877-1966, forensic scientist (1) "With contact between two items, there will be an exchange." (2) Applied to crime scenes in which the perpetrators of crimes come into contact with the scene, so they both bring something into the scene and leave with something from the scene. Every contact leaves a trace. (3) "Wherever he steps, whatever he touches, whatever he leaves, even unconsciously, will serve as a silent witness against him. Not only his fingerprints or his footprints, but his hair, the fibers from his clothes, the glass he breaks, the tool mark he leaves, the paint he scratches, the blood or semen he deposits or collects. All of these and more, bear mute witness against him. This is evidence that does not forget. It is not confused by the excitement of the moment. It is not absent because human witnesses are. It is factual evidence. Physical evidence cannot be wrong, it cannot perjure itself, it cannot be wholly absent. Only human failure to find it, study and understand it, can diminish its value."[¶] (4) Criminals leave their presence when they interfere into the surroundings of the crime scene. There are scientific methods to detect the perpetrator's identification, interchange, and recovery of contact trace evidence.[‖] Also known as *Locards Exchange Principle*.

LOCATION PRINCIPLE [mathematics] (1) If *P*(*x*) is a polynomial with real coefficients and *A* and *B* are real numbers such that *P*(*a*) and *P*(*b*) have opposites signs, then between *a* and *b* there is at least one real root *R* of the equation *P*(*x*) = 0. (2) Used in locating the roots of an equation stating that if a continuous function has opposite signs for two values of the independent variable, then it is zero for some value of the variable between these two values.**

LOCK-AND-KEY   PRINCIPLE [pharmacology] (1) Origin of life, biological intersction motifer

*Mount, John. (2009) The Local to Global Principle. http://www.win-vector.com/dfiles/Local ToGlobal.pdf
[†]Goedecker, S. (2001), The locality principle of chemistry and its reflection in O(N) algorithms, *Abstracts of Papers*, 221st ACS Natl. Meeting, San Diego, CA, April 1–5, 2001.
[‡]American Heritage (2000), *American Heritage Dictionary of the English Language*, 4th ed., Houghton Mifflin, New York.
[§]Wyman, D. R. (1997), A source localization principle for linear shift-invariant systems with application to point optical and radioactive sources, *IEEE Trans. Biomed. Eng.* **44**(4), 317–321.

[¶]Locard, E. (1930), The analysis of dust traces. Part III, *Am. J. Police Sci.* (Northwestern Univ.) **1**(5), 496–514.
[‖]Crispino, F. (2008), Nature and place of crime scene management within forensic sciences, *Sci. Justice J. Forens. Sci. Soc.* **48**(1), 24–28.
**Tausworthe, R. C. (2009), *Finding Every Root of a Broad Class of Real Continuous Functions in a Given Interval*, IPN Progress Report, pp. 42–176.

recognition.* (2) Presentation of a single favorable binding conformation.[†] (3) Stringent conformational and orientational demands are likely to reduce substantially the rate of drug—receptor binding.[‡]

LOCK MAGNET PRINCIPLE [engineering] (1) Leakage current circuit breaker of high sensitivity responsive to smooth direct-current leakage currents of both polarities, which is adequate with merely a triggering device.[§] See also BUCKING MAGNET PRINCIPLE; LINEAR MAGNET PRINCIPLE; FRICTION WHEEL DRIVE PRINCIPLE; PLUNGER MAGNET PRINCIPLE; MAGNET PRINCIPLE; CERAMIC MAGNET PRINCIPLE.

LOGIC PRINCIPLE [psychology]   Falls into three clusters in terms of difficulty. The easiest principles were affirming the antecedent, conclusion, and denying the consequent; the second cluster consisted of negative conclusion, minor premise, and class chain; and the most difficult cluster consisted of linear transitivity, nonlinear transitivity, major premise, and conditional chain.[¶] (2) Guides for reasoning within a given field or situation; "economic logic requires it"; "by the logic of war."[‖]

*Eschenmoser, A. (1994), One hundred years of the lock-and-key principle, *Angew. Chem.* **106**(23/24), 2455.
[†]Borovik, A. S. (1997), Perspectives in supramolecular chemistry, in *The Lock and Key Principle. The State of the Art-100 Years on*, Jean-Paul Behr, J.-P., ed. (Univ. L. Pasteur, de Strasbourg); *J. Am. Chem. Soc.* 119.
[‡]Borho, N. and Xu, Y. (2007), Lock-and-key principle on a microscopic scale: The case of the propylene oxide ... ethanol complex, *Angew. Chem. Int. Ed. Engl.* **46**(13), 2276–2279.
[§]Lu, J. Pan, D.-A., and Qiao, L. (2007), The principle of a virtual multi-channel lock-in amplifier and its application to magnetoelectric measurement system. Los Alamos National Laboratory, *Preprint Archive, Physics*, pp. 1–11.
[¶]Shigaki, I. and Wolf, W. (1982) Comparison of class and conditional logic abilities of gifted and normal children, *Child Study J.*, **12**(3), 161–170.
[‖]American Heritage (2000), *American Heritage Dictionary of the English Language*, 4th ed., Houghton Mifflin, New York.

LOSSLESS DIVIDE AND CONQUER PRINCIPLE. See DIVIDE AND CONQUER PRINCIPLE.

LOTTERY PRINCIPLE [biology] (George C. Williams, biology, Princeton Univ.)   Sexual reproduction has many drawbacks, since it requires far more energy than asexual reproduction, and there is some argument about why so many species use it. Sexual reproduction was like purchasing fewer tickets but with a greater variety of numbers and therefore a greater chance of success.**

LOW-ENERGY TEMPERATURE RESPONSE (LETR) PRINCIPLE [medicine]   Stable electrode–tissue contact for radiofrequency ablation of cardiac tachyarrhythmias.[††]

LOW-TURBULENCE DISPLACEMENT FLOW PRINCIPLE [chemistry, physics] (Johann Bernoulli, 1667-1748, Swiss mathematician). Responsible for the beginnings of the calculus of variations. Describes the behavior of a fluid under varying conditions of flow and height. The static pressure (in newtons per square meter) and the fluid density (in kilograms per cubic meter), in relation to the velocity of fluid flow (in meters per second) and the height above a reference surface.[‡‡]

LUCIFER PRINCIPLE [psychology]   A complex of natural rules, each working together to weave a fabric frightenting and appalling. Nature does not abjor evil; she embraces it. She uses it to build. She moves the human world to greater heights or organization, intricacy, and power. Contends that evil is woven into our most basic biological fabric. Takes fresh data from a variety of sciences and shapes them into a perceptual lens, a tool with which to reinterpret the human

Blackorby, C., Bossert, W., and Donaldson D. (1998), Uncertainty and critical-level population principles, *J. Popul. Econom.* **11(1), 1–20.
[††]Eick, O. J. WittKampt, F.H., Bronnoberg T., and Schumacher, B. (1998) The LETR-principle: a novel mathod to assess electrode-tissue contact in radio frequency ablation. *J. Cardiovasc. Electrophysiol*, **9**(11) 1180–1185.
[‡‡]Minasyan, A. S. (1961), Possibilities of the principles of fluidized bed and the through flow in continuous contact catalyst, reactors. Khimiyai Technologya Topliv, Masel, p. 111–123.

experience. Attempts to offer a very different approach to the anatomy of the social organism.*

LUDDITE PRINCIPLE [psychology]   Uses the technological achievements of civilization in a regime that is constructive and not destructive for humans. Methods of work aimed at strengthening theoretical thinking in society and preserving the values of classical rationality can and must find their embodiment in the development of a new type of computer program and a new type of educational broadcast on television. If an action or policy might cause severe or irreversible harm to the public, in the absence of a scientific consensus that harm would not ensue, the burden of proof falls on those who would advocate taking the action. Applied in the context of the impact of human development or new technology on the environment and human health, as both involve complex systems where the consequences of actions may be unpredictable. The concept includes risk prevention, cost effectiveness, ethical responsibilities toward maintaining the integrity of natural systems, and the fallibility of human understanding. Generally applied precaution in daily life (e.g., buying insurance, using seatbelts, or consulting experts before decisions) to larger political arenas. May also be interpreted as the evolution of the ancient medical principle of "first, do no harm" to apply to institutions and institutional decisionmaking processes rather than individuals.†

LUMINESCENCE-QUENCHING   PRINCIPLE [engineering]   See LUQUEN PRINCIPLE.

LUMINESCENCE   PRINCIPLE [engineering] (1) Both the intensity of the luminescence and its temporal activity are modified by the preence of oxygen in the surroundings. If a square-wave singal, of fixed frequency, is passed through a low-pass filter the output signal is shifted in phase (and reduced in amplitude) by amounts wich depend on the cut-off frequency of the filter.‡ (2) Luminescence intensity is given by the convolution product of the sample response function of the system corresponding to the δ-pulse excitation and the instrument response function only in the limit of low excitation efficiency. § (3) The concentration induced luminescence quenching (LUQUEN) in enlarged by radiationless (excited-state) energy transfer within the luminescent material. The integrated optical detection part and the chemo-optical interface, in the concentration induced optical changes, are based on the binding of glucose.¶

LUMPLEY PRINCIPLE [psychology]   (Vincent Baker)   "System (including but not limited to 'the rules') is defined as the means by which the group agrees to imagined events during play." Player contributions are assigned credibility by the other players in the game.‖ Also referred to as the *Baker-care principle*.

LUQUEN PRINCIPLE [Engineering]   (1) Both the intensity of the luminsescence and its temporal activity are modified by the presence of oxygen in the surroundings.**

*Bloom, H. (1997), *The Lucifer Principle: A Scientific Expedition into the Forces of History*, Atlantic Monthly Press, p. 3.

†Pruzhinin, B. I. (2006), New information technologies and the fate of rationality in contemporary culture: A roundtable, Russ. Soc. Sci. Rev. (transl. M. E. Sharpe) **47**(6), 79.

‡Reininger, F., Kolle, C., Trettnak, W., Gruber, W., O'Leary, P., and Binot, R. A. (1996) Preparing for the Future v6(2).

§Barzykin, A.V., and Tachiya, M. (1994) Effect of irradition intensity on luminecence quenching kinetics in micellar systems. *Chemical Physics Letters*, **221** (1–2), 81–85.

¶Lebesgue, H. (1985) Lecons sur les ensembles analytiques et leurs applications. [Lectures on analytic sets and their applications], UMN, 40, 3(243). 9–14.

‖Young, M. J. (2008), Theory 101: System and the shared imagined space—tradition of Gabriele "The Keeper" Pellegrini, *Terra d'IF* **1**(1), 2–3 (http://www.terradif.net/downloads/TDIFNL-001.pdf).

**Reininger, F., Kolle, C., Trettnak, W., Gruber, W., O'Leary, P., and Binot, R. A. (1996), *Optrodes: Stable Oxygen Sensors for Gas and Biological Fluids* (Pt. II). Future, v.6(2). (http://esapub.estin.esa.it/pff/pffv6n2/reiv6n2.htm)

(2) Under given conditions, highly luminescent particles.* (3) Applicability of the Smoluchowski type of a general approach to reaction kinetics in microdisperse systems, such as micelles, and reveals several tacit assumptions of the original model.† (4) Concentration induced quenching is enlarged by radiative energy transfer within the luminescent material; (5) Integrated optical detection with a chemo—optical interface in which the concentration induced optical changes are based on the binding of glucose. Also known as *Luminescence Quenching Principle.*

LUZIN SEPARABILITY PRINCIPLES (Nikola, Nikolaeuich Lazin, 1883–1950, Russion mathematician)   See LUZIN SEPARATION PRINCPLES.

LUZIN SEPARATION PRINCIPLES [mathematics] (Nikola, Nikolaeuich Lazin, 1883–1950, Russion mathematician)   Two theorems in descriptive set theory, proved by N.N. Luzin in 1930 (see [1]). Two sets and without common part, lying in a Euclidean space, are called -separable or Borel separable if there are two Borel sets and without common points containing and, respectively. The first Luzin separation principle states that two disjoint analytic sets (cf. -set; Analytic set) are always - separable. Since there are two disjoint co-analytic sets (cf. -set) that are -inseparable, the following definition makes sense: Two sets and without common points are separable by means of co-analytic sets if there are two disjoint sets and containing and, respectively, each of which is a co-analytic set. Luzin's second separation principle asserts that if from two analytic sets one removes their common part, then the remaining parts are always separable by means of co-analytic sets.‡ See LITTLEWOOD PRINCIPLE.

LYELLIAN PRINCIPLE [geology]   (Charles Lyell, 1797–1875, British geologist)   See PRINCIPLE OF UNIFORMITY.

LYSOGENIC INDUCTION PRINCIPLE [mathematics, physics]   See INDUCTION PRINCIPLE.

---

*Barzykin, A.V., and Tachiya, M. (1998), Unified treatment of luminescence quenching kinetics in micelles with quencher migration on the basis of a generalized smoluchowski approach. (Nat. Inst. Materials and Chemical Research, Tsukuba, Japan), *J. Phys. Chem. B* **102**(7), 1296–1300.
†Hesselink, G. L. J., Kreuwel, H. J. M., Lambeck, P. V., Van de Bovenkamp, H. J., Engbersen, J. F. J., Reinhoudt, D. N., and Popma, T. J. A. (1992), Glucose sensor based on the Luquen principle, *Sensors Actuators A/B (Chemical)* **7**(1–3), 363–366.

‡Luzin, N. N., (1930) Leçons sur les ensembles analytiques et leurs applications, Gauthier-Villars.

M

MACBRIDE PRINCIPLE [psychology] (Seán MacBride, 1904–1988, Nobel Peace Prize, Irish government activist and a founding member of Amnesty International) A corporate code of conduct for US companies doing business in Northern Ireland. They include (1) increasing the representation of individuals from under represented religious groups in the workforce, including managerial, supervisory, administrative, clerical, and technical jobs; (2) adequate security for the protection of minority employees both at the workplace and while traveling to and from work; (3) the banning of provocative religious or political emblems at the workplace; (4) all job openings should be publicly advertised and special recruitment efforts should be made to attract applicants from under-represented religious groups; (5) layoff, recall, and termination procedures should not in practice favor particular religious groupings; (6) the abolition of job reservations, apprenticeship restrictions, and differential employment criteria, which discriminate on the basis of religious or ethnic origin; (7) the development of training programs that will prepare substantial numbers of current minority employees for skilled jobs, including the expansion of existing programs and the creation of new programs to train, upgrade, and improve the skills of minority employees; (8) the establishment of procedures to assess, identify, and actively recruit minority employees with the potential for further advancement; (9) the appointment of a senior management staff member to oversee the company's affirmative action efforts; and (10) in addition to the above, each signatory to the principles is required to report annually to an independent monitoring agency on its progress in the implementation of these principles.*

*Kerrigan, C. and Wheelock, H. (2008), Papers collected by Sean O'Mahony relating to Irish

MACH PRINCIPLE [physics] (Hermann Bondi, 1919–2005, Austrian mathematician and cosmologist and Joseph Samuel) (1) Scientific laws are descriptions of nature, are based on observation and alone can provide deductions that can be tested by experiment and/or observation.[†] (2) "Mass there influences inertia here." (3) There are 11 distinct statements that can be called *Mach principles*, labeled by Mach 0 through Mach 10:

Mach 0: The universe, as represented by the average motion of distant galaxies, does not appear to rotate relative to local inertial frames.

Mach 1: Newton's gravitational constant G is a dynamical field.

Mach 2: An isolated body in otherwise empty space has no inertia.

Mach 3: Local inertial frames are affected by the cosmic motion and distribution of matter.

Mach 4: The universe is spatially closed.

Mach 5: The total energy, angular momentum, and linear momentum of the universe are zero.

Mach 6: Inertial mass is affected by the global distribution of matter.

Mach 7: If you take away all matter, there is no more space.

Mach 8: There is a definite number, of order unity, where ρ is the mean density of matter in the

history and various republican and nationalist movements (1689–2005) with an emphasis on the troubles in Northern Ireland and the contemporary Irish republican movement, 1969–2005, National Library of Ireland, Collection List No. 130, Sean O'Mahony Papers (MSS 44,025–44,310), Accession No. 6,148.

[†]Walker, P. M. B. (1999), *Chambers Dictionary of Science and Technology*, Chambers, New York, p. 700.

universe, and T is the Hubble time.

Mach 9: The theory contains no absolute elements.

Mach 10: Overall rigid rotations and translations of a system are unobservable.*

MACH'S HOLOGRAPHIC PRINCIPLE [physics] Mach's principle is the concept that inertial frames are determined by matter. In addition to bulk matter, one can also add boundary matter. Specification of both boundary and bulk stress tensors uniquely specifies the geometry and thereby the inertial frames.[†]

MACH'S MAGIC PRINCIPLE [physics]   An unspecified demand to get rid of all notions of absolute space.[‡] See also MACH'S PRINCIPLE; PRINCIPLE OF EQUIVALENCE.

MACH'S PRINCIPLE [physics, mathematics] (1) Mach held that "absolute" motion could not, in one's thinking, and should not, in one's terminology, be distinguished from motion relative to the "fixed" stars.[§] (2) An unspecified demand to get rid of all notions of absolute space. It can be developed into different directions.[§] See also PRINCIPLE OF EQUIVALENCE; MACH'S MAGIC PRINCIPLE.

MACHINE-SERVING PRINCIPLE [engineering] Imposed by the production program for which the layout has been designed and is a function of more parameters: number of stations; net production plus part loading–unloading times; buffer capacities; topology; and size scale of the machine layout.[¶]

MACHINE VISION PRINCIPLE [physics, engineering]   An object image might be mapped by means of a sensor, such as a camera, into a 2D digital image. The captured image can be processed to extract image features such as object edges and vertices. Applying 3D coordinate transforms such as translation, rotation, and scaling, one can generate an image by a subsequent perspective transform.[‖] (2) Temperature of an exhaust wind is reduced as a machine outside the room, such as an air conditioner, utilizes the principle of evaporation cooling, the water-circulating-type cold-air device, and the water-flow type-cold-air device.[**] See also PRINCIPLE OF THE HOUGH TRANSFORM.

MAGNET PRINCIPLE [physics, engineering] (1) With linear drives or with friction drives, the motor is driving the cable that connects the elevator car and the counterbalance over a return pulley. With the friction wheel drive principle or the linear magnet principle, the cable is connected at the elevator car. All drive types also require independent safety devices in case the drive fails and the elevator car begins to drop. The safety device must brake the elevator car from a certain overspeed and smoothly bring the car to a stop. (2) The bucking magnet principle uses opposing magnets to push the magnetic field back. Beside lowering the operating pressure of a magnetron, the bucking magnet also can significantly improve the discharge efficiency and thus the electron density of the discharge of a magnetron, even at regular sputtering pressures. The bucking magnet also has the desirable effect of lowering the impedance of the sputtering source and increasing its emissions due to the increase in electron densities within the discharge brought about through enhanced confinement. (3) The tool raising and lowering of devices according to the plunger magnet principle provide operation, independent of the position in which they are being used so that no weight forces occur when electric power is shut off. (4) The lock magnet principle states that a leakage

*Graneau, P. (1995), Mach's magic principle: The unique inertial system, *Phys. Essays* **8**, 376.

[†]Khoury, J. and Parikh, M. (2006), Mach's holographic principle, Los Alamos National Laboratory, *Preprint Archive, High Energy Physics—Theory*, pp. 1–33.

[‡]Pessoa, L. (1996), Mach bands: How many models are possible? Recent experimental findings and modeling attempts, *Vision Res.* **36**(19), 3205–3227.

[§]Bridgman, P. W. (1961), Significance of the Mach principle, *Am. J. Phys.* **29**, 32–36.

[¶]Powell, E. F. and Gough, S. W. (1955), Constant-power principle in abrasion testing, *Rubber World* **132**, 201–210.

[‖]Wahl, F. M. (1986), *Method for Identifying Three-Dimensional Objects Using Two-Dimensional Images*, US Patent: 4,731,860, p. 13.

[**]Otagaki, M. (2005), *Jpn. Kokai Tokkyo Koho* p. 3.

current circuit breaker of high sensitivity will be responsive to smooth direct-current leakage currents of both polarities, and will be adequate with merely a triggering device. (5) The ceramic magnet principle states that the optimum percentages for a given antenna structure have not yet been determined; however, the antenna structures that have been built clearly show that the combination of the dielectric permanent magnet and the antenna coil generates energy at the same frequency and modulation as a comparable antenna coil without the permanent magnet, but at a substantially greater amplitude than the conventional antenna. It is believed that this is caused by the interaction of electromagnetic waves with the ceramic permanent magnet in a manner to utilize both the current and voltage portions of the electromagnetic energy, resulting in a significant measured increase in the power of the antenna. Conventional antennas use only the current part of the electromagnetic energy.* See also BUCKING MAGNET PRINCIPLE; LINEAR MAGNET PRINCIPLE; FRICTION WHEEL DRIVE PRINCIPLE; PLUNGER MAGNET PRINCIPLE; LOCK MAGNET PRINCIPLE; CERAMIC MAGNET PRINCIPLE.

MAGNETIC INDUCTION PRINCIPLE [mathematics, physics] Induced magnetization in magnetic material, by either saturation, coil excitation in a magnetic circuit, or the simple method of stroking with another magnet.[†] See INDUCTION PRINCIPLE.

MAGNETIC MEASUREMENT PRINCIPLE [physics] The bandlimited function of lowpass type whose energy is minimum among those that have the prescribed sample values on a finite set of sampling points. The minimum energy signal converges uniformly to a bandlimited signal when the number of sampling points is increased infinitely.[‡]

*Ireland, J. (1963), *Magnetic Transmission*, US Patent 3,267,310.
[†]Walker, P. M. B. (1999), *Chambers Dictionary of Science and Technology*, Chambers, New York, p. 704.
[‡]Graeffe, Jussi and Nuyan, Seyhan (2005), An online laser caliper measurement for the paper industry. *Proc. of SPIE* **5856**, (Pt.1, *Opt. Meas. Syst. Industr. Inspect. IV*), 318–326.

MAGNETIC RELUCTANCE PRINCIPLE [physics] Measure of the opposition presented to magnetic flux in a magnetic circuit, analogous to resistance in an electric circuit; it is equal to magnetomotive force divided by magnetic flux.[§] See also REPULSION PRINCIPLE.

MAGNETOELASTIC FORCE PRINCIPLE [physics] (1) Inertia results from a relationship of that object to all other matter in the universe: (2) Inertial forces experienced by a body in nonuniform motion are determined by the quantity and distribution of matter in the universe.[¶]

MAGNETOHYDRODYNAMIC PRINCIPLE (MHD) [physics] An electric field is produced in the measured volume that is perpendicular to the magnetic field and perpendicular to the direction of fluid flow.[‖]

MAGNETOOPTICAL DISK PRINCIPLE [physics] (1) Arrangement for modulating a beam of light by passing it through a single crystal of yttrium iron garnet, which provides intensity modulation by using a magnetic field to produce optical reaction; study of the effect of a magnetic field on light passing through a substance in a field.** (2) Employs a laser to read data on the disk, while it needs magnetic field and a laser beam to write data to the magnetooptical (MO) disk. MO disk drive is so designed that an inserted disk will be exposed to a magnet on the label side and to the light (laser beam) on the opposite side.[††] See also MO PRINCIPLE; MAGNETOOPTICAL PRINCIPLE.

[§]Nakano, R. (2008), Non-contacting caliper sensor with sub-micron profile accuracy, *Kami Parupu Gijutsu Taimusu* **51**(2), 11–15.
[¶]Shaw, H. R. (1995), *Craters, Cosmos, and Chronicles: A New Theory of Earth*, Stanford Univ. Press, p. 458.
[‖]Heng, K.-H., Wang, W., Murphy, M. C., and Lian, K. (2000), UV-LIGA microfabrication and test of an a.c.-type micropump based on the magnetohydrodynamic (MHD) principle, *Proc. SPIE* (4177) (Microfluidic Devices and Systems III), p. 174–184.
**Kaneko, M. (2000), Magneto-optical recording, in *Magneto-Optics*, Springer Series in Solid-State Sciences, vol. 128, pp. 271–317.
[††]Roll, K. (2001), Magnetooptical discs, *Schrift. Forschung. Juelich Materie Material* **7**, Institute of Nuclear Problems Minsk, Belarusi (Neue Materialien fuer die Informationstechnik).

MAGNETOOPTICAL EFFECT PRINCIPLE [physics] Employs a laser to read data on the disk, while it needs magnetic field and a laser beam to write data to the magnetic optical disk drive designed to be exposed to a magnet on the label side and to the light (laser beam) on the opposite side.* See also MO PRINCIPLE; MAGNETOOPTICAL PRINCIPLE.

MAGNETOOPTICAL PRINCIPLE [physics] Employs a laser to read data on the disk, while it needs magnetic field and a laser beam to write data to the magnetic optical disk drive designed to be exposed to a magnet on the label side and to the light (laser beam) on the opposite side.† See also MO PRINCIPLE; MAGNETOOPTICAL EFFECT PRINCIPLE.

MAGNETRON PRINCIPLE [engineering] A pulsed microwave radiation source for radar and continuous source for microwave cooking. One of a family of cross-field microwave tubes where electrons generated from a heated cathode move under the combined force of a radial electric field and an axial magnetic field in such a way as to produce microwave radiation in the frequency range 1–40 gigahertz.‡

MAJORITY PRINCIPLE [sociology] Expressed by Grotius: "It is unnatural," he says, "that the majority should submit to the minority— hence the majority naturally counts as the whole, if no compacts or positive law prescribe a different form of procedure."§

MALEIMIDE CYCLOADDITION AROMATIC HETEROCYCLE INERTIA PRINCIPLE [chemistry] See INERTIA PRINCIPLE.

MALTHUSIAN EQUIMARGINAL PRINCIPLE [mathematics] (Thomas Robert Malthus, 1766–1834, British scholar) Under selection pressures the relative numbers of higher- and lower-quality organisms will change until, in equilibrium, not the average but the marginal levels of quality will be equalized. See also ZAHAVI HANDICAP PRINCIPLE (also known as *truthful signaling hypothesis*).¶

MALTHUSIAN PRINCIPLE [geography, mathematics] (Thomas Robert Malthus, 1766–1834, British scholar) (Thomas Robert Malthus, FRS, 1766–1834, English demographer and economist) Only natural causes (e.g., accidents and old age), misery (war, pestilence, and especially famine), moral restraint, and vice could check excessive population growth. Based on the idea that population if unchecked increases at a geometric rate (2, 4, 8, 16, 32, 64, 128, etc.) whereas the food supply grows at an arithmetic rate (1, 2, 3, 4, 5, 6, 7, 8, etc.).‖ See also PRINCIPLE OF POPULATION.

MALUS PRINCIPLE [physics] (Elienne–Louis Malus, 1775–1812, French physicist, mathematician) Expression giving the intensity of a transmitted beam when a plane-polarized beam of light is incident on a polarizer.**

MANAGEMENT PRINCIPLES [geography] Fulfill areas of economic function, corporate structure, health of earnings, fairness to stakeholders, research and development, directions analysis, fiscal policies, production efficiency, sales vigor, and executive evaluation.††

*Sepulveda, B., Calle, A., Lechuga, L. M., and Armelles, G. (2006), Highly sensitive detection of biomolecules with the magneto-optic surface-plasmon-resonance sensor, *Optics Lett.* **31**(8), 1085–1087.
†Miyazawa, H. and Oguchi, T. (1999), First-principles investigation of magneto-optical effect, *Nippon Oyo Jiki Gakkaishi*, **23**(1–2), 355–357.
‡Wirz, P., Przybilla, G., Schuller, K.-H., and Cord, B. (1988), Sputtering cathode on the magnetron principle, U.S. Patent 4,734,183.
§Heinberg, J. G. (1926), History of the majority principle, *Am. Polit. Sci. Rev.* **20**(1), 54.

¶Hausken, K. and Hirshleifer, J. (2008), Truthful signalling, the heritability paradox, and the Malthusian equi-marginal principle, *Theor. Popul. Biol.* **73**(1), 3.
‖Moes, J. E. (1958), A dynamic interpretation of Malthus' principle of Population, *Kyklos*, **11**, 58.
**Walker, P. M. B. (1999), *Chambers Dictionary of Science and Technology*, Chambers, New York, p. 710.
††Iskander, M. R. and Chamlou, N. (2000), *Corporate Governance: A Framework for Implementation, with a Foreword by Sir Adrian Cadbury*, The World Bank Group, The International Bank for Reconstruction and Development.

MANDREL PRINCIPLE [engineering] Mechanism for packaging in paper bags with tubemaking and bottom gluing machine.* See also CLASSICAL MANDREL PRINCIPLE; CLASSICAL PRINCIPLE.

MANLE–SOMMER PRINCIPLE [medicine] Technique applied with use of ultrasonics in surgery.†

MANY-EYE PRINCIPLE [psychology] Parents offer such nondispersers a benefit in that they increase their vigilance while feeding together with retained offspring. In contrast, parents reduce their vigilance while in company of nonrelated flock members according to the "many eyes" principle. Nepotistic vigilance behavior in Siberian jay parents.‡

MARGIN-FOR-ERROR PRINCIPLE [philosophy] See PRINCIPLE KK.

MARGINALITY PRINCIPLE [mathematics, economics] The share of joint output attributable to any single factor of production should depend only on that factor's own contribution to output.§

MARINE STEWARDSHIP COUNCIL (MSC) PRINCIPLES [ecology] An independent non-profit organization with an ecolabel and fishery certification programme. Fisheries that are assessed and meet the standard can use the MSC blue ecolable. The MSC mission is to 'reward sustainable fishing practises'. When fish is bought that has the blue MSC ecolabel, it should indicate that this fishery operates in an environmentally responsible way and does not contribute to the global environmental problem of overfishing.¶ See also

HANNOVER PRINCIPLES; FOREST STEWARDSHIP COUNCIL (FSC) PRINCIPLES; GLOBAL SULLIVAN PRINCIPLES OF SOCIAL RESPONSIBILITY; MARINE STEWARDSHIP COUNCIL (MSC) PRINCIPLES; PERMACULTURE PRINCIPLES; THE BELLAGIO PRINCIPLES FOR ASSESSMENT; EQUATOR PRINCIPLES; MELBOURNE PRINCIPLES; PRECAUTIONARY PRINCIPLE; SANBORN PRINCIPLES; TODDS' PRINCIPLES OF ECOLOGICAL DESIGN.

MARKOV'S PRINCIPLE (MP) [mathematics] (Andrey Andreyevich Markov, Jr., 1903–1979, Russian mathematician) For each binary sequence (a_n), if it is contradictory that all the terms a_n equal 0, then there exists a term equal to 1. This principle is equivalent to a number of simple classical propositions, including the following: (1) for each real number x, if it is contradictory that x equal 0, then $x \neq 0$ (in the sense we mentioned earlier); (2) for each real number x, if it is contradictory that x equal 0, then there exists $y \in R$ such that $xy = 1$; (3) for each one-one continuous mapping $f : [0,1] \to R$, if $x \neq y$, then $f(x) \neq f(y)$.‖

MARSHALL LERNER PRINCIPLE [mathematics] (Alfred Marshall, 1842–1924, English economist and Abba Ptachya Lerner, 1903–1982, American economist) The conditions under which a change in a country's exchange rate will improve its balance of payments. The price elasticity of demand for imports and exports must be greater than unity for improvements to be effected in the balance of payments. As a devaluation of the exchange rate means a reduction in price of exports, demand for these will increase. At the same time, price of imports will rise and their demand will diminish. The net effect on the trade balance will depend on price elasticities. If goods exported are elastic to price, their demand will increase proportionately more than the decrease in price, and total export revenue will increase. Similarly, if goods imported are elastic, total import expenditure will decrease.**

*Gerow, M. R. (1969), Art of Mandrel extrusion of thermoplastics, *Proc. Natl. Tech. Conf., Plastic Packaging*, Soc. Plast. Eng., pp. 155–159.
†Heckman, R. (2001), Ultrasound Use in Cardiothoracic Surgery, *Assoc. of periOperative Registered Nurses Jrnl* **73**, 144–165.
‡Griesser, M. (2003), *Behav. Ecol.* **14**(2), 246–250.
§Langlois, N. E. and Donaldson, C. (1998), Application of the principle of marginal analysis to sampling practice using prostatic chippings as a model, *J. Clin. Pathol.* **51**(2), 104–107.
¶Marine Stewardship Council. http://www.msc.org/

‖Namsrai, K. H. (1985) Nonlocal Quantum Field Theory and Stochastic Quantum Mechanics (Fundamental Theories of Physics). Springer. p. 233.
**Lerner, A. P. (1944), *Economics of Control: Principle of Welfare Economics*. Macmillan Co., NY.

MASCULINE PRINCIPLE [psychology]   (1) Dominance of "yang" or masculine principle and underdevelopment of "yin" or feminine principle in Western society suggests that Western society has forgotten its yin dimension, the positive and creative aspects of tenderness, receptivity, compassion, understanding, surrender, and intuition. The yang dimension, an overdeveloped capacity to dominate, control, and analyze, has been allowed to take over, threatening psychological and spiritual development in the process.* (2) The self as the "being" mode represents the feminine principle and functions according to primary process; the ego represents "doing," the masculine principle and secondary process. Feminine and masculine principles are considered to be of equal significance in both men and women and are not limited to gender. Jung's concept of the self is related to the Hindu metaphysical concepts of Atman and Brahman, whose source was the older Aryan nature-oriented, pagan religion. The prominence of self in analytic psychology and its predominantly feminine symbolism can be understood as Jung's reaction to the psychoanalytic emphasis on ego and to Freud's patriarchal orientation. In the proposed model of the psyche neither ego nor self represents the psychic totality. The interplay of both psychic modes and principles constitutes the psyche and the individuation process.† (3) Certain aspects of the universe, and analogously, of human beings, are associated with traits historically assigned to males, including activity, strength, stability, and generativity. In traditional Eastern philosophies, including Tibetan Buddhism, the masculine principle must function in concert with the feminine principle. According to these philosophies, to become aware, one must surrender expectation and concept (the feminine), to gain wisdom and awareness (the masculine). In analytical psychology, the masculine principle generally characterizes the conscious, or ego, and is represented by the self as "doing" mode, which functions

according to secondary PROCESS. The masculine principle operates in conjunction with the feminine principle.‡ See also FEMININE PRINCIPLE.

MASS-ACTION PRINCIPLE [chemistry, neurology]   (1) At constant temperature, the rate of a chemical reaction is directly proportional to the active mass of the reactants, with the active mass taken as the concentration.§ (2) The cortex of the brain operates as a coordinated system with large masses of neural tissue involved in all complex functioning.¶

MASS CONSERVATION PRINCIPLE [physics] (1789, by the French chemist Lavoisier) (1) The whole amount of matter in nature is constant and indestructible. In any system the total amount of matter is always the same. Matter is subjected to physical or chemical transformations, but it is never destroyed.‖

MASS–ENERGY EQUIVALENCE [physics] A measured quantity of mass is equivalent (according to relativity theory) to a measured quantity of energy.**

MASS PRINCIPLE [acoustics]   Describes the sound transmission through walls. The transmission coefficient is approximately proportional to the inverse of the mass per unit area and to the inverse of the frequency.††

---

‡Op. cit. Weisstub; discussion, pp. 453–458.
§Daintith, J. (1999), *The Facts on File Dictionary of Chemistry*, 3rd ed., Facts On File, New York, p. 150.
¶American Heritage (2000), *American Heritage Dictionary of the English Language*, 4th ed., Houghton Mifflin, New York.
‖Picker, O., Wietasch, G., Scheeren, T. W., and Arndt, J. O. (2001), Determination of total blood volume by indicator dilution: A comparison of mean transit time and mass conservation principle, *Intens. Care Med.* **27**(4), 767–774.
**American Heritage (2000), *American Heritage Dictionary of the English Language*, 4th ed., Houghton Mifflin, New York.
††Walker, P. M. B. (1999), *Chambers Dictionary of Science and Technology*, Chambers, New York, p. 716.

---

*Gaboury, P. (1987), We've been away so long, *Can. J. Counsel.* **21**(2), 142–152.
†Weisstub, E. (1997), Self as the feminine principle, *J. Anal. Psychol.* **42**(3), 425–452.

MASTER–SLAVE PRINCIPLE [sociology]   Optimization measurements; operation of subordinate devices and temporarily storing operation that are not immediately transmitted. Efficiency level that could not be matched by a single unit; transforming the signal to the required strength and reconversion. Unconventional hierarchic system of state quantities regulators from which everyone is controlled by the corresponding component of the vector of the desired state and by the output of the regulator of the governing state quantity.[*]

MATCHING PRINCIPLE [electronics, computer science, psychology]   (1) Approach to color image compression with high compression ratios and good quality of reconstructed images using quantization, thresholding, and edge detection.[†] (2) Guiding principle in the design of training programs, especially in planning for the "process" element in training. Tendency to match and to influence groups trained not to match. In training for professional practice, the mode of training should match the mode of practice.[‡] (3) The established *reflection process* as a framework of possible inputs to a system as a subset of the set of tolerable inputs (those that are judged to produce an acceptable response from the system).[‡] (4) System design having direct relevance to the theory and design of control systems in combination with the method of inequalities.[§] (5) Bridge network of rectangular sections of resistive track arranged with longer sides at right angles to flow.[¶] See also ASYMPTOTIC MATCHING PRINCIPLE; HYDROPHOBIC MATCHING PRINCIPLE; DOPPLER MATCHING PRINCIPLE; PHASE MATCHING PRINCIPLE; BRACKET MATCHING PRINCIPLE; MODEL MATCHING PRINCIPLE; MOMENT MATCHING PRINCIPLE; ZERO-FORCING PRINCIPLE; TIME INDEPENDENCE PRINCIPLE; DISTRIBUTION MATCHING PRINCIPLE; PIDGEONHOLE PRINCIPLE; COMBINATORIAL MATCHING PRINCIPLE; PIGEONHOLE PRINCIPLE; H_2 MODEL MATCHING PRINCIPLE; *H* INFINITY MODEL MATCHING PRINCIPLE; OPTIMAL MODEL MATCHING PRINCIPLE; MINTO PYRAMID PRINCIPLE.

MATRIX MINIMUM PRINCIPLE [physics] (1) Intense surface heat transfer associated with radiation; finite-element solutions display anomalous behaviors.[‖] (2) Continuous differentiability of the functions of the equality constraints and a subdifferential regularity of the function of the inequality constraints.[**] (3) Extremes of a functional dependent on the dynamics of a discrete system with time delay; best sequence is defined as the one that minimizes a performance functional composed of the costs incurred in taking the measurements.[††] (4) Modification of triangle-based adaptive stencils for the solution of scalar hyperbolic conservation laws.[‡‡] See also PONTRYAGIN'S MINIMUM PRINCIPLE.

MATRJOSCHKA PRINCIPLE [physics] (1) Each magnetic pole is surrounded by a magnetic field that is attached to this pole and constant in time. If this magnetic pole is the pole of an electromagnet, that

[*]Braeuniger, B., Armbruster, E., and Schneidereit, M. (1986), *Computer Unit for Automatic Integrated Evaluation of Process Chromatography Data Including Peak Recognition and for Management of Data Collecting Systems, German Patent DD 234, 344, 15p.*

[†]Yang, C.-K. and Tsaib, W.-H. (1998), Color image compression using quantization, thresholding, and edge detection techniques all based on the moment-preserving principle, *Pattern Recogn. Lett.* **19**(2), 205–215.

[‡]Buss, A. H. (1967), Stimulus generalization and the matching principle, *Psychol. Rev.* **74**(1), 40–50.

[§]Smith, E. (1998), Annual meeting of the ACL, *Proc. 36th Annual Meeting Assoc. Computational Linguistics and 17th Int. Conf. Computational Linguistics*, Montreal, Canada, Association for Computational Linguistics, pp. 1499–1501.

[¶]Sandell, N., Jr., Varaiya, P., Athans, M., and Safonov, M. (1978), Survey of decentralized control methods for large scale systems, *IEEE Trans. Autom. Control* **23**(2), 108–128.

[‖]Kay, L. and Wood, M. A. (1988), Application of the Doppler-matching principle to the measurement of decay curves in beam-foil spectroscopy, *J. Phys. E Sci. Instrum.* **21**(6), 591–595.

[**]Athans, M. (1968), The matrix minimum principle, *Inform. Control* **11**, 592–606.

[††]Inst. Electrical Engineers, Inst. Electrical and Electronics Engineers (IEE, IEEE) (1969), *Computer & Control Abstracts*, IEE, vol. 4–32.

[‡‡]Kelly, F. P., Zachary, S., and Ziedins, I. (1996), *Stochastic Networks: Theory and Applications*, Oxford Univ. Press, p. 20.

is, electrically pulsed, a time-dependent magnetic field is created that separates from the pole of the electromagnet and propagates into space (it is therefore called a *magnetic traveling wave*). As a consequence of Maxwell's equations, this magnetic traveling wave is tied to a time-dependent electric field of definite frequencies, which varies significantly over very short distances. This field must come into resonance with the activated complex; thus there will take place an energy transfer from this complex to the magnetic field that can be absorbed by induction. (2) The magnetic traveling wave is formed in principle by an acceleration of the space quantum medium flowing around the axis of the pulsed electromagnet despite the fact that the official physics has another interpretation for this process. The "Matrjoschka"-like geometry of several pulsed electromagnets increases this acceleration.[*] See also PRINCIPLE OF PROPAGATION OF A MAGNETIC TRAVELING WAVE.

MATZA'S PRINCIPLES [criminal justice] Framework for understanding the sentencing decisions of judges who use their sentencing power to reverse waive offenders back to juvenile court.[†]

MAUPERTUIS–EULER–LAGRANGE PRINCIPLE OF LEAST ACTION [physics] (Leonard Euler, 1707–1783, Swiss mathematician) See MAUPERTUIS–LAGRANGE PRINCIPLE.

MAUPERTUIS–JACOBI PRINCIPLE [physics] (Carl Gustav Jacob Jacobi, 1804–1851, Russian mathematician) First enunciated in 1744. The reduced action $S_0 = R_p dq$ is independent of any evolution parameter. Moreover, even its initial parameter t and the corresponding Hamilton function cannot be restored from the reduced-action

problem. Nevertheless, solutions of the corresponding variational problem are the initial trajectories $q_j = r_j(t)$ in the common nonparametric form. All the initial trajectories have one nonparametric form on the surface Q2n—(1) Therefore, for all these trajectories we could introduce common parametric form $q_j = r'_j(et)$ and a new local Hamilton function eH defined on Q2n—(1).[‡]
$$\delta S = \delta \int_{q'}^{q^n} \sqrt{E - V(q)} \sqrt{a_{ij} dq^i dq^j} = 0.[§]$$ See MAUPERTUIS–JACOBI LEAST ACTION PRINCIPLE.

MAUPERTUIS–LAGRANGE PRINCIPLE [physics] (Pierre–Louis Moreau de Maupertuis, 1698–1759, French mathematician) (Giuseppe Lodavico Lagrangia, 1736–1813, Italian mathematician) (1) The path of conservative system in a configuration space is such that the integral of the Lagrangian function over time is a minimum or maximum relative to nearby paths between the same endpoints and taking the same time.[¶] (2) Motion of the underlying system of particles compatible with the collective equilibrium provided that the variations are associated with reversible processes.[‖] See also DIFFERENTIAL PRINCIPLE; VARIATIONAL PRINCIPLE; HAMILTON PRINCIPLE; PRINCIPLE OF LEAST CONSTRAINT; D'ALEMBERT PRINCIPLE; HAMILTON–LAGRANGE PRINCIPLE; LEAST-CONSTRAINT PRINCIPLE; D'ALEMBERT–LAGRANGE PRINCIPLE; GAUSS PRINCIPLE; JOURDAIN PRINCIPLE; JAYNE'S PRINCIPLE; PRINCIPLE OF VIRTUAL WORK; KINETIC PRINCIPLE; MAUPERTUIS–EULER–LAGRANGE PRINCIPLE OF LEAST ACTION; D'ALEMBERT–LAGRANGE VARIATION PRINCIPLE; LAGRANGE PRINCIPLE.

[*]Wang, Z. J., Zhang, L., and Liu, Y. (2004), Spectral (finite) volume method for conservation laws on unstructured grids IV: Extension to two-dimensional systems, *J. Comput. Phys.* **194**(2), 716–741.
[†]Moran-Lopez, J. L., Bennemann, K. H., Cabrera-Trujillo, M., and Dorantes-Davila, J. (1994), Energetics of giant fullerenes and Matrjoschka structures, *Solid State Commun.* **89**(12), 977–981.

[‡]Misner, C., Thorne, K., and Wheeler, A., (1973), *Gravitation*, W. H. Freeman.
[§]Zhu, X., Ghahramani, Z., and Lafferty, J. (2003), Supervised Learning Using Gaussian Fields and Harmonic Functions. Proceedings of the Twentieth International Conference on Machine Learning (ICML-2003), Washington DC.
[¶]Burrow, J. (2008), Examining the influence of Matza's principles of justice and their impact on reverse waiver decisions: Has Kadi-(in)justice survived? *Youth Violence Juv. Justice* **6**(1), 59–82.
[‖]Kanai, A. Tilocca, Selloni, A., and Car, A. (2004), First-principles string molecular dynamics: An efficient approach for finding chemical reaction pathways, *J. Chem. Phys.* **121**(8), 3359.

MAUPERTUIS LEAST-ACTION PRINCIPLE [physics]   See MAUPERTUIS PRINCIPLE.

MAUPERTUIS PRINCIPLE [physics] (Pierre–Louis Moreau de Maupertuis, 1698–1759, French mathematician)   Motion of the underlying system of particles compatible with the collective equilibrium provided that the variations are associated with reversible processes.[*] See also ACTION PRINCIPLE; QUANTUM VARIATONAL PRINCIPLE; HAMILTON'S PRINCIPLE OF LEAST ACTION.

MAXIMAL PRINCIPLE [mathematics]   (1) If f is a holomorphic function, then the modulus $|f|$ cannot exhibit a true local maximum within the domain of f. In other words, either f is a constant function, or, for any point z_0 inside the domain of f there exist other points arbitrarily close to z_0 at which $|f|$ takes larger values. This entails, for example, the consequence that if f is defined and holomorphic in the closed unit disk D, the maximum of $|f|$ (which is certainly defined and attained somewhere because $|f|$ is continuous and the closed disk is a compact space) is attained on the unit circle C. (2) Also referred to as *Zorn's lemma* (every nonempty partially ordered set x in which every linearly ordered subset has an upper bound, has a maximal element), *principles of cofinality* (principle of cofinality is equivalent to a corresponding variation of Zorn's lemma and every linearly ordered set has a well-ordered cofinal subset), *Hausdorff's maximal principle* (every partially ordered set contains a Q-maximal chain), *principle of finite character* (FC.) (for every set x and every property P of finite character, there exists a c-maximal subset of x with the property Z_7), *order extension principle* (OE) (every partial ordering can be extended to a linear ordering), *Kurepa's antichain principle* (every partially ordered set has a maximal antichain), and *principle of finite character* (for every set x and every property P of finite character, there exists a c-maximal subset of x with the property).[†] See

also PRINCIPLE OF COFINALTY; HAUSDORFF MAXIMAL PRINCIPLE; ORDER EXTENSION PRINCPLE; KUREPA'S ANTICHAIN PRINCIPLE.

MAXIMALITY PRINCIPLE [mathematics] (Zorn's lemma) (Ernst Friedrich Ferdinand Zermelo, 1871–1953)   (1) Equivalent to the axiom of choice. Let S be a collection of sets. If, for each chain $C \subset S$, there exists an $X \in S$ such that every element of C is a subset of X, then S contain a maximal element.[‡] (2) Only if the following *maximality principle* holds: the first-order nonlinear differential equation

$$g'(s) = \frac{\sigma^2(g(s))L'(g(s))}{2c(g(s))(L(s) - L(g(s)))}$$

admits a maximal solution $s \mapsto g_x(s)$ which stays strictly below the diagonal in $\mathbb{R}^2$. [In this equation $x \mapsto \sigma(x)$ is the diffusion coefficient and $x \mapsto L(x)$ the scale function of X.] In this case the stopping time

$$\tau_x = \inf\{t > 0 | X_t \le g_x(S_t)\}$$

is proved optimal, and explicit formulas for the payoff are given. The result has a large number of applications and may be viewed as the cornerstone in a general treatment of the maximum process. (3) Any nontrivial lattice with a unit element has a maximal ideal. (4) The product of compact topological spaces is compact. (5) Any set has the same cardinality as x.[‡] See also ZORN–BOURBAKI MAXIMALITY PRINCIPLE.

MAXIMIZATION PRINCIPLE [sociology]   Natural selection results in optimal characteristics for a given environment, because those organisms with characteristics maximizing their fitness will reproduce. The principle is controversial because it does not take into account certain key tenets of evolutionary theory: (1) organisms may not have available to them those characteristics that would maximize their fitness, and (2) the selection pressures may not be such that only those with maximum fitness

[*]Marion, J. B. (1970), *Classical Dynamics of Particles and Systems*, 2nd ed., p. 198.
[†]Biesiada, M. (1995), The power of the Maupertuis-Jacobi principle—dreams and reality, *Chaos, Solitons Fractals* **5**(5), 869–879.

[‡]Peskir, Goran (1988) Optimal Stopping of the Maximum Process: The Maximality Principle. The annals of Probability, v26, (4), pp. 1614–1640.

are able to reproduce. The alternate perspective to maximization would be one of minimum fitness; whatever characteristics allow for reproduction will remain in the gene pool, even if they are not optimal.*

MAXIMUM CONTROL PRINCIPLE [product management] Task control problems existing in manufacturing systems with the concept of state jump system to analyze a manufacturing cell with an unreliable agent. The uncertain factors of the manufacturing cell are addressed in the task control model by utilizing a self-learning method of probability distribution parameters of stochastic events. Given the state jump system, the task control problem is simplified that the optimal task control strategy of the manufactuing cell can be obtained by the combination of the uniform technology and the dynamic programming. The objective function can be stabilized for different initial conditions.[†]

MAXIMIZING PRINCIPLE [mathematics] If an expression functioning as a name can be introduced and explained, then there exists a corresponding object.[‡]

MAXIMUM CROSS-ENTROPY PRINCIPLE [engineering, mathematics] (1) The probabilities that maximize the cross-entropy between classes and minimize the cross-entropy within classes are known subject to given constraints that are chosen.[§] (2) Asserts that the probability distribution with maximum entropy, satisfying the prior knowledge, should be used in the decision problem. Given a specified a prior probability distribution, some constraints, and a measure of cross-entropy,

the probability distribution can be determined with maximum cross-entropy. Every additional constraint consistent with the original constraints can increase the cross-entropy, but whatever the number of additional constraints, the cross-entropy will not exceed the unique maximum value of the cross-entropy.[¶] See also ENTROPY OPTIMIZATION PRINCIPLE.

MAXIMUM DECISIONAL EFFICIENCY (MDE) PRINCIPLE [mathematics] See CATALYSIS MAXIMIZATION PRINCIPLE.

MAXIMUM DEPTH PRINCIPLE [computer science] See FERMAT'S MINIMUM TIME PRINCIPLE.

MAXIMUM EMPOWER PRINCIPLE Density functions of specific distribution are denoted by appropriate names. Thus, if x is a random quantity with a normol distribution of mean μ and standard deviation σ, its probability density function will be denoted $N(X|\mu,\sigma)$. An unknown parameter of a given density function provides a family of possible density functions with some information concerning the population in terms of random sampling. The maximum entropy of the probability distribution subject to whatever information is given should be chosen.

MAXIMUM ENTROPY PRINCIPLE (MAXENT OR MEP) [physics] (1) Used for estimating probability density function under specified moment constraints.[‖] (2) Measure of absence of information about a situation; uncertainty; transformations between measure spaces.[**] (3) Expresses amount of disorder inherent or produced, measure of disorder of a

*Carruthers, P. (1990), *The Metaphysics of the Tractatus*, CUP Archive, p. 173.

[†]Hong-Sen Yan, H.-S., Yang, H.-B., and Dong, H. (2009), Control of Knowledgeable manufacturing cell with an unreliable agent. Journal of Intelligent Manufactuirng. V20(6) p. 671–682.

[‡]Haerten, R. and Kim, J. (1993), Procedures in color Doppler ultrasound—comparison of methods, *Ultraschall Med.* **14**(5), 225–230.

[§]Lorenc, J., Przystawa, J., and Cracknell, A. P., (1980), A comment on the chain subduction criterion, *J. Phys. C Solid State Phys.* **13**(10), 1955–1961.

[¶]Kleemola, J., Teittinen, M., and Karvonen, T. (1996), Modeling crop growth and biomass partitioning to shoots and roots in relation to nitrogen and water availability, using a maximization principle, *Plant* and *Soil* **185**(1), 101–111.

[‖]Haerten, R. and Kim, J. (1993), Procedures in color Doppler ultrasound—comparison of methods, *Ultraschall Med.* **14**(5), 225–230.

[**]Wang X.-L., Yuan, Z.-F., Guo, M.-C., Song, S.-D., Zhang, Q.-Q., and Bao, Z.-M. (2002), Maximum entropy principle and population genetic equilibrium, *Acta Genetica Sinica* **29**(6), 562–564.

system equal to the Boltzmann constant.[*] See also PRINCIPLE OF MAXIMUM ENTROPY (PME); FUZZY ENTROPY PRINCIPLE; PRINCIPLE OF MAXIMUM NONADDITIVE ENTROPY; PRINCIPLE OF OPERATIONAL COMPATIBILITY; MERMIN ENTROPY PRINCIPLE; ENTROPY PRINCIPLE; VARIATIONAL MINIMAL PRINCIPLE; MINIMUM CROSS-ENTROPY PRINCIPLE; PRINCIPLE OF MINIMUM CROSS-ENTROPY; GENERALIZED MINIMAL PRINCIPLE; MOMENT-PRESERVING PRINCIPLE.

MAXIMUM ENTROPY PRODUCTION PRINCIPLE [physics] (1) System with fixed boundary conditions and adequate degrees of freedom will always maximize its production of entropy. They begin by noting that this principle has been demonstrated to apply latitudinally to Earth's real-world climate system by data-driven studies stretching back in time a full quarter-century, and it has more recently been demonstrated to likewise apply in the vertical. The planetary scientists then go on to demonstrate, again by means of actual data, that the maximum entropy production principle additionally applies to the climate systems of Mars and Titan, and suggests that it probably also applies to Venus and may have applicability to planets beyond our solar system as well.[†] (2) Provides the nomological basis for spontaneous order production, for dissolving the postulates of incommensurability between physics and psychology and physics and biology, between thermodynamics and evoluton, as the answer to a question that classical thermodynamics never asked. The classical statement of the second law says that entropy will be maximized, or potentials minimized, but it does not ask or answer the question regarding which out of all available paths a system will take to accomplish this end. The answer to the question is that the system will select the path or assembly of paths out of otherwise available paths that minimizes the potential

or maximizes the entropy at the fastest rate given the constraints. This is a statement of the law of maximum entropy production and the physical selection principle that provides the nomological explanation, as will be seen below, for why the world is in the order production business. Note that the law of maximum entropy production is in addition to the second law. The second law says that only entropy is maximized while the law of maximum entropy production says that it is maximized (potentials minimized) at the fastest rate given the constraints. Like the active nature of the second law, the law of maximum entropy production is intuitively easy to grasp and empirically demonstrated.[‡] See also PRIGOGINE'S PRINCIPLE OF MINIMUM ENTROPY.

MAXIMUM HARDNESS PRINCIPLE. (MHP) [chemistry] (1) Associated with chemical behavior studies; has been scrutinized numerically and shown to be a potent technique for studying molecular electronic structure and gaining a better grasp of different reaction mechanisms. Stability of a chemical species is intimately connected with the corresponding gap between the highest occupied and lowest unoccupied molecular orbitals. It also helps in understanding whether a chemical reaction is favorable. Hardness implies stability; the hardness is found to be a good indicator of the more stable isomer in all cases. (2) The electronic and nuclear repulsion energy changes show good correlation with the relative stability of a species even when the constraint of constant chemical potential is not obeyed. The hardnesses and the energy profiles, as a function of the reaction coordinate, are generally opposite in nature only for the isomerization reactions of O_3H^+ and HSiN, for which there is a negligible variation of the chemical potential.[§]

[*]Canosa, N., Rossignoli, R., and Plastino, A. (1990), Maximum entropy principle for many-body ground states, *Nuclear Phys. A* **A512**(3), 492–508.

[†]La Cour, B. R. and Schieve, W. C. (2000), Tsallis maximum entropy principle and the law of large numbers, *Phys. Rev. E. Statist. Phys. Plasmas Fluids Related Interdisc. Topics* **62**(5B), 7494–7496.

[‡]Ozawa, H., Ohmura, A., Lorenz, R. D., and Pujol, T. (2003), The second law of thermodynamics and the global climate system: A review of the maximum entropy production principle, *Rev. Geophys.* **41**(4), 3.

[§]Swenson, R. (1997), Autocatakinetics, evolution, and the law of maximum entropy production: A principled foundation towards the study of human ecology, *Adv. Human Ecol.* **6**, 1–47.

MAXIMUM IGNORANCE PRINCIPLE [physics] (Edwin Thompson Jaynes, 1922–1998, American physicist) States that we should not use any information that was not given in the problem statement, because then the solution would not be objective anymore.* See also PRINCIPLE OF TRANSFORMATION GROUPS.

MAXIMUM LIKELIHOOD ESTIMATION (MLE) PRINCIPLE [mathematics, physics] See PRINCIPLE OF NETWORK ENTROPY MAXIMIZATION.

MAXIMUM LIKELIHOOD (ML) PRINCIPLE [mathematics] Technique in statistics where the likelihood distribution is so maximized as to produce an estimate to the random variables involved.[†]

MAXIMUM-MINIMUM PRINCIPLE [mathematics] The theorem that provides information concerning the nth eigenvalue of a symmetric operator on an inner product space without necessitating knowledge of the other eigenvalues.[‡] Also known as the *minimax theorem*.

MAXIMUM MODULUS PRINCIPLE [mathematics] States that if f is a holomorphic function, then the modulus $|f|$ cannot exhibit a true local maximum within the domain of f. In other words, either f is a constant function, or, for any point z_0 inside the domain of f there exist other points arbitrarily close to z_0 at which $|f|$ takes larger values.[§] (2) Methodology for optimizing distributed systems by their autonomous activities by means of the cooperative work among them.[¶] See also DISTRIBUTED MAXIMUM PRINCIPLE; MAXIMUM PRINCIPLE; AUTONOMOUS CO-OPERATION DISTRIBUTED MAXIMUM PRINCIPLE; MINIMUM MODULUS PRINCIPLE.

MAXIMUM POWER PRINCIPLE [psychology, engineering] (Alfred J. Lotka) (1) L. R. Vandervert's article discussing parallels between J. Mandler's (1992) theory of conceptual development in infancy, and Vandervert's theory of maximum power principle evolution of the neuroalgorithmic (NA) organization of the brain and cultural-level mental models. A maximum power principle prey–predator scenario depicted the dynamic selective origins of NA underlying Mandler's various image schemas (conceptual primitives; CP). Vandervert proposed that CP were inherited spacetime simulation structures originating in cerebellar state estimating functions that spread to cerebral mapping systems.[‖] (2) During self-organization, system designs develop and prevail that maximize power intake, energy transformation, and those uses that reinforce production and efficiency. (3) High-quality energy maximizes power by matching and amplifying energy. In surviving designs a matching of high-quality energy with larger amounts of low-quality energy is likely to occur. As with electronic circuits, the resultant rate of energy transformation will be at a maximum at an intermediate power efficiency. (4) A potential guide to understanding the patterns and processes of ecosystem development and sustainability. Predicts the selective persistence of ecosystem designs that capture a previously untapped energy source.** See also FOURTH PRINCIPLE OF ENERGETICS.

*Jaynes, E. T. (1973), The Well-Posed Problem, Foundations of Physics **3**, 477–493.

[†]Yu, D. and Chen, Z. (2001), Structure and stability of XeF6 isomers: Density functional theory study and the maximum hardness principle, *THEOCHEM* (540), 29–33.

[‡]Gladstien, K. and Kidd, K. K. (1981), An easy-to-use maximum-likelihood method of estimating the ascertainment probability, *Am. J. Human Genet.* **33**(5), 785–801.

[§]Koshizen, T. and Fulcher, J. (1995), An application of Hamiltonian neurodynamics using Pontryagin's maximum (minimum) principle, *Int. J. Neural Syst.* **6**(4), 425–433.

[¶]Lu, B. L. and Ito, M. (1999), Task decomposition and module combination based on class relations: A modular neural network for pattern classification, *IEEE Trans. Neural Networks* **10**(5), 1244–1256.

[‖]Rossi, H. (1960), The local maximum modulus principle, *Ann. Math.* **72**, 1–11.

Vandervert, L. (1997), Image-schemas as space-time simulation structures: A reply to Fox and Paulin, *New Ideas Psychol.* **15(2), 137–139.

MAXIMUM PRINCIPLE [engineering, mathematics] (J.-M. Bismut, professor at Université Paris-Sud, France)   (1) Intense surface heat transfer associated with radiation, finite-element solutions display anomalous behaviors.* (2) Continuous differentiability of the functions of the equality constraints and a subdifferential regularity of the function of the inequality constraints.† (3) Intranset analysis, small time steps can only improve the accuracy because standard stability theorems limit the maximum time step for a given mesh size. However, in finite-element approximations, small time steps may cause stability problems that lead to physically unreasonable results; a first and universal optimization algorithm; applicable to lumped and spatially distributed systems.‡ (4) Extremes of a functional dependent on the dynamics of a discrete system with time delay; best sequence is defined as the one that minimizes a performance functional composed of the costs incurred in taking the measurements.§ (5) Modification of triangle-based adaptive stencils for the solution of scalar hyperbolic conservation laws.¶ Also known as *Bismut principle; discrete maximum principle; discrete Pontryagin maximum principle; intersection principle; Krylov maximum principle; Pontryagin discrete maximum principle; finite-dimensional maximum principle; Pontryagin principle; Lindelöf principle; global extended maximum*

principle; maximum modulus principle; distributed maximum principle; autonomous cooperation distributed maximum principle.

MAXIMUM PRINCIPLE FOR HARMONIC FUNCTIONS [mathematics] (Johann Peter Gustav Lejeune Dirichlet, 1805–1859, German mathematician)   An important consequence of the mean-value property is the following maximum principle for harmonic functions.

1.8   Maximum Principle: *Suppose Ω is connected, u is real valued and harmonic on Ω, and u has a maximum or a minimum in Ω. Then u is constant.*

PROOF.   Suppose u attains a maximum at $a \in \Omega$. Choose $r > 0$ such that $\bar{B}(a, r) \subset \Omega$. If u were less than $u(a)$ at some point of $B(a, r)$, then the continuity of u would show that the average of u over $B(a, r)$ is less than $u(a)$, contradicting 1.6. Therefore u is constant on $B(a, r)$, proving that the set where u attains its maximum is open in Ω. Because this set is also closed in Ω (again by the continuity of u), it must be all of Ω (by connectivity). Thus u is constant on Ω, as desired.

If u attains a minimum in Ω, we can apply this argument to $-u$.‖

MAXIMUM PRINCIPLES FOR SUBHARMONIC FUNCTIONS [mathematics] (Johann Peter Gustav Lejeune Dirichlet, 1805–1859, ...) The maximum principle for $\mathcal{H}_\infty(\Omega, \Gamma)$ is

$$|F(\zeta)| \leq m(\Omega, \Gamma, \zeta) \limsup_{z \to \theta\Omega} |F(z)|.$$

This inequality follows from the maximum principle for subharmonic functions and the definition of $m(\Omega, \Gamma, \zeta)$.

It is easy to see that if $\mathcal{H}_\infty(\Omega, \Gamma)$ is nonempty then each $m(\Omega, \Gamma, \zeta) > 0$; for any function in $\mathcal{H}_\infty(\Omega, \Gamma)$ may be multiplied by a rational function to produce a function of $\mathcal{H}_\infty(\Omega, \Gamma)$ not vanishing at ζ. One of the main results of the main results of the paper is that all the $\mathcal{H}_\infty(\Omega, \Gamma)$ are nonempty if and

*Cai, T. (2006), *The Maximum Power Principle: An Empirical Investigation, Ecological Modeling,* **190**(3–4), 317–335.

†Lobo, M. and Emery, A. F. (1995), Use of the discrete maximum principle in finite-element analysis of combined conduction and radiation in nonparticipating media, *Num. Heat Transf. B Fundamentals (Intl. J. Comput. Methodol.)* **27**(4), 447–465.

‡Clarke, F. H. (1990), Optimization and nonsmooth analysis, *SIAM* p. 211.

§Armenio, V. and La Rocca, M. (1996), On the analysis of sloshing of water in rectangular containers: Numerical study and experimental validation, *Ocean Eng.* **23**(8), 705–739.

¶Libosvar, C. M. (1988), *Hierarchies in Production Management and Control: A Survey*, Lab. Information and Decision Systems, Massachusetts Inst. Technology (MIT), LIDS-P-1734, p. 17.

‖Axler, S.J., Bourdon, P., and Ramey, W. (2001) Harmonic function theory. Springer, P. 7.

only if $m(\Omega, \zeta)$ is positive. Furthermore, we shall obtain a formula for $m(\Omega, \zeta)$.[*]

MAXIMUM VALUE PRINCIPLE [mathematics] (Peng Shige, 1990) Modern control theory devoted to obtaining the stochastic control system.[†] See also PENG'S PRINCIPLE.

MAXWELL-BOLTZMAN PRINCIPLE See EQUIPARTITION PRINCIPLE.

MCCARTHY'S RECURSION INDUCTION PRINCIPLE [mathematics, physics] Recursion induction principle stated by J. McCarthy by using the famous fold/unfold method elaborated by R. Burstall and J. Darlington. "We thus obtain a very simple and flexible method for proving theorems about LISP functions; we call it the McCarthy method."[‡] See INDUCTION PRINCIPLE.

MCKENZIE   PRINCIPLES [pharmacology] (Robin McKenzie) Stress self-treatment through correct posture and repeated end-range movements performed at a high frequency.[§] See also PRINCIPLE-CENTERED SPINE CARE.

MDE PRINCIPLE See MAXIMUM DECISIONAL EFFICIENCY PRINCIPLE.

MDL PRINCIPLE See MINIMUM DESCRIPTIONLENGTH PRINCIPLE.

MEASURING PRINCIPLE [engineering]   Process of determining the value of some quantity in terms of a standard unit.[¶]

MECHANISM   PRINCIPLE [engineering] Specifications are often borrowed from the optimal mechanism design literature and exclude mechanisms that are natural in a competitive environment; for example, mechanisms that depend on the mechanisms chosen by competitors. An equilibrium for a specific model is robust if and only if it is an equilibrium also for the universal set of mechanisms. A key to the construction is a language for describing mechanisms that is not tied to any preconceived notions of the nature of competition.[‖]

MEDIOCRITY PRINCIPLE [cosmology] There is nothing special about Earth, and by implication the human race. Used either as a heuristic about Earth's position or a philosophical statement about the place of humanity.[**] See also COPERNICAN PRINCIPLE.

MELANOPHORE DILATING PRINCIPLE [biology] (1) Any of the dendritic clear cells of the epidermis that synthesize melanin. (2) A polypeptide hormone produced in the anterior or intermediate lobe of the pituitary gland.[††] (3) Formed by cleavage of the prohormone product of the proopiomelanocortin gene, it induces the formation of the pigment melanin in the melanocytes of the skin and may also have a role in the processes of learning and memory.[‡‡]

MELBOURNE   PRINCIPLE [geography] How cities can become more sustainable. Designed to be read by decisionmakers, and provide a starting point on the journey toward sustainability. The vision promoted is to create environmentally healthy, vibrant, and sustainable cities where people respect

[*]Widom, Harold (1971) maximum principle for multiple-valued analytical functions. Acta Mechanica. Springer, p. 64.
[†]Gardiner, S. J. (1991), Maximum principles for subharmonic functions, *Math. Scand.* **68**(2), 210–220.
[‡]Berthelot, M. (n.d.), Remarks on the principle of maximum work, *Bull. Soc. Chim. France* **43**, 264.
[§]Kott, L. (1982), The McCarthy's recursion induction principle: "Oldy" but "goody," *Calcolo* **19**(1), 59–69.
[¶]Simonsen, R. J. (1998), Principle-centered spine care: McKenzie principles, *Occup. Med.* **13**(1), 167–183.

[‖]Epstein, L. and Peters, M. (1996), A Revelation Principle For Competing Mechanisms, Working Papers peters-96-02, University of Toronto, Department of Economics.
[**]Hine, J. (1977), The principle of least nuclear motion, *Adv. Phys. Org. Chem.* **15**, 1–19.
[††]Kroll, S. S. and Freeman, P. (1989), Striving for excellence in breast reconstruction: The salvage of poor results, *Ann. Plast. Surg.* **22**(1), 58–64.
[‡‡]Cachin, M., Habib, G., and Durlach, J. (1951), Research for a melanophore-dilating principle in the urine in alcoholic cirrhosis, *Bull. Memoires Societe Med. Hopitaux Paris* **67**(23–24), 1012–1017.

one another and nature, to the benefit of all. They provide (1) a holistic approach to making cities sustainable; (2) a framework around which consensus and commitment can be built and strategy developed; (3) a framework in which cities can build their programs and engage their communities; and (4) a framework in which international, regional, and country programs can coalesce strengthen linkages and cooperation.* See also BIOMIMICRY PRINCIPLES; DEEP ECOLOGY'S BASIC PRINCIPLES; FOREST STEWARDSHIP COUNCIL (FSC) PRINCIPLES; GLOBAL SULLIVAN PRINCIPLES OF SOCIAL RESPONSIBILITY; MARINE STEWARDSHIP COUNCIL (MSC) PRINCIPLES; NATURAL CAPITALISM PRINCIPLES; PERMACULTURE PRINCIPLES; THE BELLAGIO PRINCIPLES FOR ASSESSMENT; EQUATOR PRINCIPLES; PRECAUTIONARY PRINCIPLE; SANBORN PRINCIPLES; TODDS' PRINCIPLES OF ECOLOGICAL DESIGN.

MEMORY COMPENSATION PRINCIPLE [mathematics, physics] See PRINCIPLE OF NETWORK ENTROPY MAXIMIZATION.

MENABREA'S PRINCIPLE [physics] [Luigi Federico Menabrea (Count), 1809–1896, French-born Italian politician, prime minister (1867–1869), and scientist] For a system whose total mechanical energy is conserved, the trajectory of the system in configuration space is the path that makes the value of the action stationary relative to nearby paths between the same configurations and for which the energy has the same constant value.[†] See also LEAST-ACTION PRINCIPLE; PRINCIPLE OF LEAST WORK; PRINCIPLE OF ELASTICITY OR MINIMUM WORK.

MENDEL'S PRINCIPLE [biology] (Gregor Johann Mendel, 1822–1884, Augustinian-priest) Dealing with the mechanism of inheritance. (1) *Principle of segregation*—two alleles received one from each parent are segregated in gamete formation, so that each gamete receives one or the other with equal probability. This results in various characteristic ratios in the progeny depending on the parental genotypes and dominance. (2) *Principle of recombination*—two characters determined by two unlinked genes are recombined at random in gamete formation, so that they segregate independently of each other, each according to the principle of segregation.[‡]

MERMIN ENTROPY PRINCIPLE [physics] (Nathonisel David Mermin, American physicist) (1) Used for estimating probability density function under specified moment constraints; measure of absence of information about a situation; uncertainty; transformations between measure spaces; expresses amount of disorder inherent or produce; measure of disorder of a system equal to the Boltzmann constant; the natural log action of the number of microscopic states corresponding to the thermodynamic state of the system.[§] (2) Famous fold/unfold method to give proof of two properties—associatively of the append operation between lists and idempotence of the reverse operation.[¶] (3) A distribution-free method for estimating the quantile function of a nonnegative random variable using the principle of maximum entropy (MaxEnt) subject to constraints specified in terms of the probability-weighted moments estimated from observed data. The relevant probability is the volume fraction of the elementary constituents that belong to a given constituent and undergo

*Collin, R. and Drouet, P. L. (1933), Presence of a melanophore-dilating principle in the tuber cinereum of the guinea pig, *C. R. Seances Societe Biol. Ses Filiales* **112**, 63–65.
[†]Azer, S. A. and Frauman, A. G. (2008), Seeing the wood for the trees: Approaches to teaching and assessing clinical pharmacology and therapeutics in a problem-based learning course, *Anna. Acad. Med. Singapore* **37**(3), 204–206.

[‡]Benvenuto, E. (1984), A brief outline of the scientific debate which preceded the works of Castigliano, *Meccanica* (Springer, Netherlands) **19**(Suppl.) 19–32.
[§]Walker, P. M. B. (1999), *Chambers Dictionary of Science and Technology*, Chambers, New York, pp. 726–727.
[¶]Tachibana, A. (1989), Application of the Mermin entropy principle to the "apparatus" density functional theory, *Int. J. Quantum Chem.* **35**(3), 361–372.

a given stimulus.* (4) Entropy model for deriving the probability distribution of the equilibrium state.[†] (5) Production of an electromotive force either by motion of a conductor through a magnetic field so as to cut across the magnetic flux or by a change in the magnetic flux that treats a conductor. In a magnetic system for the contactless guidance of a vehicle moved along a track in which a plurality of magnets are attached to the vehicle and arranged one behind the other in the direction of travel, the magnet system cooperating with nonferromagnetic conductor loops on the track to generate forces.[‡] See also MAXIMUM ENTROPY PRINCIPLE; FUZZY ENTROPY PRINCIPLE; MAXIMUM ENTROPY PRINCIPLE; PRINCIPLE OF MAXIMUM NONADDITIVE ENTROPY; PRINCIPLE OF OPERATIONAL COMPATIBILITY; ENTROPY PRINCIPLE.

MESH IMPEDANCE PRINCIPLE (MIP) [biology] (1) Protein substance secreted by the intermediate lobe of the pituitary that causes dispersion of pigment granules in the skin.[§] (2) For a fixed starting point the number of iterations of Newton's method to obtain a fixed tolerance is eventually independent. Used to formulate an efficient strategy for the solution of nonlinear equations amounting to the work equivalent of two or three Newton iterates.[¶] (3) The behavior of the discretized process is asymptotically the same as that for the original iteration and consequently, the number of steps required by the two processes to converge to within a given tolerance is essentially the same.[‖]

MESH INDEPENDENCE PRINCIPLE [mathematics, physics] A generalization of the maximum entropy principle (MEP), measurements of a random variable contain errors having some known average value.** See also MUTUAL INFORMATION PRINCIPLE; MAXIMUM ENTROPY PRINCIPLE; MINIMUM CROSS-ENTROPY PRINCIPLE.

METABOLIC PRINCIPLE See COMPOSING PRINCIPLE.

METADATA PRINCIPLE 1 [Archiving] Good metadata conforms to community standards in a way that is appropriate to the materials in the collection, users of the collection, and current and potential future uses of the collection.[††]

METADATA PRINCIPLE 2 [Archiving] Good metadata supports interoperability.[‡‡]

METADATA PRINCIPLE 3 [Archiving] Good metadata uses authority control and content standards to describe objects and collocate related objects.[‖]

METADATA PRINCIPLE 4 [Archiving] Good metadata includes a clear statement of the conditions and terms of use for the digital object. [‖]

METADATA PRINCIPLE 5 [archiving] Good metadata supports the long-term curation and preservation of objects in collections.[‖]

*Kott, L. (1982), The McCarthy's recursion induction principle: "oldy" but "goody," *Calcolo* **19**(1), 59–69.

[†]Anon. (n.d.), *New Engl. Yale Rev.* **37** (http://memory.loc.gov/ndlpcoop/nicmoas/nwng/nwng0037.sgm).

[‡]Siegert, A. J. F. (1949), On the approach to statistical equilibrium, *Phys. Rev.* **76**, 1708–1714.

[§]Ashcroft, N. W. and Mermin, N. D. (2001), *Solid State Physics*, Harcourt Asia Pte Ltd.

[¶]Branham, J. M. (1966), *Motility and Aging of Arbacia Sperm*, vol. 131, Lancaster Press, p. 253.

[‖]Lagoudas, D. C., Ravi-Chandar, K., Sarhc, K., and Popov, P. (2003), Dynamic loading of polycrystalline shape memory alloy rods, *Mech. Materi.* **35**(7), 689–716.

**Cooper, W., Daly, C., Demarteau, M., Fast, J., Hanagaki, K., Johnson, M., Kuykendall, W., Lubatti, H., Matulik, M., Nomerotski, A., Quinn, B., and Wang, J. (2005), Electrical properties of carbon fiber support systems, Los Alamos National Laboratory, *Preprint Archive, High Energy Physics—Experiment*, pp. 1–20.

[††]Rahman, M. M. and Shevade, S. S. (2005), Fluid flow and heat transfer in a composite trapezoidal microchannel, *Proc. ASME Summer Heat Transfer Conf.*, San Francisco, vol. 1, pp. 411–417.

[‡‡]NISO (Natl. Information Standards Org.) (2007), *A Framework of Guidance for Building Good Digital Collections*, 3rd ed., Inst. Museum and Library Services, (http://framework.niso.org/node/5).

METADATA PRINCIPLE 6 [Archiving] Good metadata records are objects themselves and therefore should have the qualities of good objects, including authority, authenticity, archivability, persistence, and unique identification.‖

MICHELSON INTERFEROMETRIC PRINCIPLE [physics] (Albert Abraham Michelson, 1852–1931, American physicist) (1) Study to detect a medium for light to travel, referred to as the *ether*. The Michelson interferometer was a device designed to detect Earth's motion with respect to the ether, consisting of a rotable platform, a beamsplitter, and mirrors. A light beam was sent to the beamsplitter where half of the light followed different paths at right angles to detect a phase indifference between the the parallel and perpependicular paths of the waves. By recombining the two beams, one can determinl the interference. Since the extent of the phase difference depends on the orientation of the device relative to the motion of Earth, the difference will change as the device is rotated. To measure Earth is motion with respect to the ether, one only needs to rotate the platform and observe the size of the resulting phase shift.* (2) Makes use of recombining two waves traveling along different paths in order to measure the phase difference between them via the interference effect.† See also RADIATION PRINCIPLE; INTERNAL EFFECT PRINCIPLE; INTERFEROMETRIC PRINCIPLE.

MICROPUMP PRINCIPLE [engineering] (1) Ratio of the voltage to the current in a mesh when all other meshes are open. (2) Consisting of pump chamber, oscillating membrane, and two truncated pyramid shaped microchannels. A direction-dependent behavior of flow resistance, rectifying an alternating flux.‡

MICROWAVE DOPPLER PRINCIPLE [physics] See DOPPLER PRINCIPLE.

MIDAS TOUCH PRINCIPLE [psychology] (1) Expectations beyond customer expectation. (2) Projects or services initiated exceeding all expectations of achievement.

MILL'S HARM PRINCIPLE [psychology] (John Stuart Mill, 1806–1873, English philosopher and economist) (1) The sole purpose of law should be to stop people from harming others and that, should people want to participate in victimless crimes, crimes with no complaining witness, such as gambling, engaging in prostitution, then they should not be encroached in doing so.§ (2) The sole end for which humankind are warranted, individually or collectively, in interfering with the liberty of action of any of their number, is self-protection. The only purpose for which power can be rightfully exercised over any member of a civilized community, against that person's will, is to prevent harm to others. One's own good, either physical or moral, is not sufficient warrant or justification. One cannot rightfully be compelled to do or forbear because it will be better for one to do so, because it will make one happier, because, in the opinion of others, to do so would be wise, or even right. The only part of the conduct of anyone, for which one is amenable to society, is what concerns others. In the part that merely concerns oneself, one's independence is, of right, absolute. Over oneself, over one's own body and mind, the individual is sovereign.¶ See also HARM PRINCIPLE.

MIMETIC PRINCIPLE [physics] Type of solution for converting voltage into fluid pressure, which uses depolarizing electrodes sealed in an electrolyte and generates through the

*Gerlach, T. and Wurmus, H. (1995), Working principle and performance of the dynamic micropump, *Sensors Actuators A Phys.* **A50**(1–2), 135–140.
†Montilla, I., Pereira, S. F., and Braat, J. J. M. (2005), Michelson wide-field stellar interferometry: Principles and experimental verification, *Appl. Optics* **44**(3), 328–336.
‡NISO (2007), *A Framework of Guidance for Building Good Digital Collections*, 3rd ed., Inst. Museum and Library Services (http://framework.niso.org/node/5).
§Kandpal, H. C., Wasan, A., Vaishya, J. S., and Joshi, K. C. (1996), Space-frequency equivalence principle in a laboratory version of Michelson's stellar interferometer, *Optics Commun.* **132**(5,6), 503–510.
¶Maris, C. (1999), The disasters of war: American repression versus Dutch tolerance in drug policy, *J. Drug Issues* **29**(3), 493–510.

streaming potential effort.* See also PRINCIPLE OF BIOCATALYSIS.

MIN SHENG CHU-I PRINCIPLE [sociology] People's livelihood, socialism, and equalization of land ownership through a just system of taxation.[†] See also THREE PRINCIPLES OF THE PEOPLE.

MINER'S PRINCIPLE [engineering] Method of estimating the fatigue lifetime.[‡]

MINIMAL PRINCIPLES [physics] In the treatment of physical phenomena, it can sometimes be shown, of all the processes or conditions that might occur, that the ones actually occurring are those for which some characteristic physical quantity assumes a minimum value. These processes or conditions are known as *minimal principles*. One simple minimal principle asserts that the state of stable equilibrium of any mechanical system is the state for which the potential energy is a minimum.[§] (2) The global minimum of $R(\phi)$, where the phases are constrained to satisfy all identities among them that are known to exist, attained when the phases are equal to their true values and thus equal to R_T. Replaces the problem of phase determination by finding the global minimum of the function $R(\phi)$ constrained by the identities that the phases must satisfy and suggests strategies for determining the values of the phases in terms of N and the known magnitudes.[¶] See also HAMILTON PRINCIPLE; LEAST-ACTION PRINCIPLE.

*van Bogaert, L.-J. (2006), Rights of and duties to non-consenting patients—informed refusal in the developing world, *Devel. World Bioethics* **6**(1), 13–22.
[†]De Bary, W. T. Chan, W.-T., Watson, B., Tan, C., and Mei, Y. (1964), *Sources of Chinese Tradition*, Columbia Univ. Press, New York.
[‡]Kordonsky, Kh. B. and Gertsbakh, I. (1997), Multiple time scales and the lifetime coefficient of variation: Engineering applications, *Lifetime Data Anal.* **3**(2), 139–156.
[§]Miller, R. et al. (1993), On the application of the minimal principle to solve nk,n structures, *Science* **259**, 1430.
[¶]Hauptman, H. A., and Langs, D. A. (2003), The phase problem in neutron crystallography, *Acta Crystallogr. A, Found. of Crystallogr.* **59**(3), A59, 250–254.

MINIMAX PRINCIPLE [mathematics] This principle consists of two parts. The first is the maximum entropy principle for feature binding (or fusion)—for a given set of observed feature statistics, a distribution can be built to bind feature statistics together by maximizing the entropy over all distributions that reproduce them. The second part is the minimum entropy principle for feature selection—among all plausible sets of feature statistics, the set whose maximum entropy distribution has the minimum entropy. The principal method used heretofore in obtaining the critical point relations is the following. A value b will be called a *critical value* of our functional $f(P)$ if there is a critical point Q of $f(P)$ such that $f(Q)b$. The change in connectivity depends on the type of the critical point P having this critical value. By studying these changes of connectivity, one is able to classify the critical points of $f(P)$ and to obtain the critical point relations.[‖] See also MINIMAX PRINCIPLE FOR EIGENVALUES; MAXIMUM ENTROPY PRINCIPLE; MINIMUM ENTROPY PRINCIPLE.

MINIMAX PRINCIPLE FOR EIGENVALUES [mathematics] This principle consists of two parts. The first is the maximum entropy principle for feature binding (or fusion)—for a given set of observed feature statistics, a distribution can be built to bind feature statistics together by maximizing the entropy over all distributions that reproduce them. The second part is the minimum entropy principle for feature selection—among all plausible sets of feature statistics, the set whose maximum entropy distribution has the minimum entropy.[**] See also MINIMAX PRINCIPLE; MAXIMUM ENTROPY PRINCIPLE.

MINIMIZATION PRINCIPLE [physics] (1) A principle requiring the final state of a system is determined by the attainment of the minimum possible value of a certain

[‖]Walker, P. M. B. (1999), *Chambers Dictionary of Science and Technology*, Chambers, New York, p. 743.
[**]Heisenberg, W. (1949), *The Physical Principles of the Quantum Theory*, Dover Books on Physics and Chemistry, Univ. Chicago Science Series. p. 1.

quantity.* (2) Effective energy dissipation control method is an adaptive correction of an allocation of representative points, by evaluating energy levels simply at a position of each representative point.[†] (3) The potential field setting means provides the potential field in dependence on a function with a tendency responsible for a positional change to cause an increased variation in potential value, as it is at a smaller distance to the contour, and a decreased variation in potential value, as it is at a longer distance from the contour.[‡] (4) Potential energy is set so that a potential variation due to a positional change increases as a distance to a contour line decreases. Accordingly, a representative point nearer to the contour line is drawn in with a stronger pulling force, arriving at a corrected position in a shorter time.[§] (5) Multiphase systems tend to self-lubricate by having the lower-viscosity phase migrate to the shearing surface to minimize resistance to flow. This will provide binding of messages with their source, ensuring the fact that messages from a source not under the wiretapping is not considered valid as is required by the minimization principle in tapping (i.e., listening only to the suspects).[¶] (6) Which one set of data points is translated, rotated, and stretched in several axis directions so as to reduce a distance function between respective points.[†] (7) Analogous to minimizing the sidelobes of the "ambiguity function" in the radar pulse compression technique in the time domain. Therefore, phase codes (relationships) that are desirable

for pulse compression radar generally possess characteristics that can be used to achieve distortion minimization for both in-band and out-of-band distortion products.[‖] Also known as *energy minimization principle; distortion minimization principle*.

MINIMUM COMPLEXITY PRINCIPLE [biology] Efficient in estimating multifurcate branching when the tree is described in the form of rooted one.[**] See also OCCAM'S RAZOR PRINCIPLE.

MINIMUM CROSS-ENTROPY PRINCIPLE [physics] (1) The posterior distribution is the one that minimizes cross-entropy; subject to the new constraint information.[††] (2) Joint optimization method inspired by the deterministic annealing algorithm for data clustering, and tree-structured vector quantization using informative priors to approximate the unstructured solution while imposing the structural constraint.[‡‡] See also ENTROPY PRINCIPLE; PRINCIPLE OF MINIMUM CROSS-ENTROPY; VARIATIONAL MINIMAL PRINCIPLE; GENERALIZED MINIMAL PRINCIPLE; MOMENT-PRESERVING PRINCIPLE.

MINIMUM DESCRIPTION-LENGTH (MDL) PRINCIPLE [mathematics] (1) ISODATA clustering analysis based on the theory of fuzzy sets using the concept of membership degree, and the relation between individuals and their representative classes. (2) Formulates the problem as a minimization of a length of description cost function. In particular, as an attribute selection measure for the induction of decision trees. The best model for encoding data is one that minimizes the sum

*Birkhoff, George D. (1917) Dynamical Systems with Two Degrees of Freedom, *Proc. Natl. Acad. Sci.* V3(4) p. 314–316.
[†]Talman, J. D. (1986), Minimax principle for the Dirac equation, *Phys. Rev. Lett.* **57**(9), 1091–1094.
[‡]von Bertalanffy, L. (1950), An outline of general system theory, *Br. J. Phil. Sci.* **1**(2), 134–165.
[§]Saraydar, C. U., Mandayam, N. B., and Goodman, D. J. (2002), Efficient power control via pricing in wireless data networks, *IEEE Trans. Commun.* **50**(2), 291–303.
[¶]Yokoyama, Y. (1997), *System and Method for Inter-frame Prediction of Picture by Vector-Interpolatory Motion-Compensation Based on Motion Vectors Determined at Representative Points Correctable in Position for Adaptation to Image Contours*, US Patent 5,646,691.

[‖]Collier, J. R., Negulescu, I. I., and Collier, B. J. (2000), *Cellulosic Microfibers*, US Patent 6,511,746.
[**]Ren, F. R. et al. (1995), Construction of molecular evolutionary phylogenetic trees from DNA sequences based on minimum complexity principle, *Comput. Meth. Programs Biomed.* **46**, 121.
[††]Alley, G. D. and Kuo, Y.-L. (1996), *Intermodulation Distortion Reduction Method and Apparatus*, US Patent 5,930,678.
[‡‡]Yee, E. (1991), Reconstruction of the antibody affinity distribution from experimental binding data by a minimum cross-entropy procedure, *J. Theor. Biol.* **153**(2), 205–227.

of the cost of describing the data in terms of the model and the cost of describing the model. Encodes the tree and split tests in an MDL-based code, and determines whether to prune and how to prune each node according to the code length of the node. (3) Best edge configuration is the one that allows the shortest description of the image and its edges. (4) Represents a class of models (hypotheses) by a universal model capable of imitating the behavior of any model in the class and calls for a model class whose representative assigns the largest probability or density to the observed data.[*] Approach relates to apparently rival principles of perception, what types of empirical data the approach can and cannot explain, and how an MDL approach to perception might be augmented to provide an empirically adequate framework for understanding perceptual inference.[†] See also RISSANEN'S MINIMUM DESCRIPTION-LENGTH PRINCIPLE; MAXIMUM LIKELIHOOD PRINCIPLE; PRINCIPLE OF PERCEPTION.

MINIMUM ELECTROPHILICITY PRINCIPLE (MEP) [physics] A tendency in atoms to arrange themselves so that the molecule obtained reaches the minimum electrophilicity. The maximum hardness and minimum polarizability principles (MHP and MPP) cannot predict the major regioisomer of a Diels–Alder reaction; whereas the analysis of global electrophilicities of the products shows that, at least in those cases where diffuse basis sets are used, the major product of the reaction always has the less electrophilicity values.[‡]

*Oliveira, A. L. and Sangiovanni, V. A. (1996), Using the minimum description length principle to infer reduced ordered decision graphs, *Machine Learn.* **25**, 23.
†Chater, N. (2005), A minimum description length principle for perception, in *Advances in Minimum Description Length Theory and Applications*, MIT Press, Cambridge, MA, pp. 385–409.
‡Chaquin, P. (2008), Absolute electronegativity and hardness: An analogy with classical electrostatics suggests an interpretation of the Parr electrophilicity index' as a global energy index' leading to the minimum electrophilicity principle, *Chem. Phys. Lett.* **458**(1–3), 231–234.

MINIMUM ENERGY PRINCIPLE [physics] See HAMILTON PRINCIPLE.

MINIMUM ENTROPY PRINCIPLE (MINXENT) [mathematics, physics]   (1) Out of all probability distributions satisfying the given linear constraints, choose the one that is closest (at minimum distance) to the given a priori distribution. (2) Out of all probability distributions satisfying the given linear constraints, the probability distribution that minimizes the Kullback–Leibler measure of cross entropy is chosen.[§] See also KULLBACK'S MINIMUM CROSS–ENTROPY PRINCIPLE; ENTROPY OPTIMIZATION PRINCIPLE.

MINIMUM INFORMATION PRINCIPLE [mathematics]   (1) Extracting model parameters of a physical process from statistical experimental data. (2) A method for quantifying the information content of continuous functions, and a technique for solving inverse problems where the solution is known to be continuous.[¶] See also PRINCIPLE OF MAXIMUM U UNCERTAINTY.

MINIMUM INTEGRAL PRINCIPLE [mathematics]   For a nonvanishing nonconstant complex analytic function defined in a domain, the absolute value of the function cannot attain its minimum at any interior point of the domain.[‖]

MINIMUM KULLBACK ENTROPY PRINCIPLE [physics]   (Jaynes, 1957) An extension of the Laplacian principle of insufficient reason. Consisting in assigning equal probabilities to two events if there are no reasons to think otherwise. The maximum entropy principle provides an information-theory-based approach to assigning probabilities when

§Sadovsky, M. G. (2003), The method to compare nucleotide sequences based on the minimum entropy principle, *Bull. Math. Biol.* **65**(2), 309–322.
¶Chizhov, A. V. (2000), Extracting model parameters from experimental data using the minimum information principle, *Tech. Phys.* (transl. of *Zh. Tekhnich. Fiz.*) **45**(4), 512–514.
‖Power, J. F. (1997), Expectation minimum—a new principle of inverse problem theory in the photothermal sciences: theoretical characterization of expectation values, *Opt. Eng.* **36**(2), 487–503.

incomplete information is given. The uncertainty should be maximized subject to all information given. The minimization of the Kullback–Leibler entropy between the empirical probability and the parametric form is equivalent to maximization of the corresponding empirical likelihood.[*]

MINIMUM LOSS OF INFORMATION PRINCIPLE [engineering, mathematics] The transformation that leads to (1) a minimum loss of information for a given degree of simplification or (2) a maximum simplification should be chosen. Whenever a complex model is simplified into a less detailed model, by omitting unnecessary details or by aggregating among themselves or with others, there is a loss of information.[†] See also ENTROPY OPTIMIZATION PRINCIPLE.

MINIMUM LOSS OF POWER OF DISCRIMINATION PRINCIPLE [engineering, mathematics] Random vectors transformed into other random vectors will, in general, be less than the discrepancy. The transformation should be as small as possible when random vectors in a dimensional space are transferred into another dimensional space to avoid the loss of power of discrimination.[‡] See also ENTROPY OPTIMIZATION PRINCIPLE.

MINIMUM MODULUS PRINCIPLE [mathematics] (1) States that if f is holomorphic within a bounded domain D, continuous up to the boundary of D, and is nonzero at all points, then the modulus $|f(z)|$ takes its minimum value on the boundary of D. Alternatively, the open mapping theorem, which states that a holomorphic function maps open sets to open sets, can be used to give a proof.[§] See also MAXIMUM MODULUS PRINCIPLE.

MINIMUM NORM PRINCIPLE [physics] Of all possible passive forces for a rigid body at equilibrium subjected to active forces and prescribed loading, the unique force solution compatible with the equilibrium renders a minimum norm on the intensities of all the passive forces. In other words, the passivity of the contact system is completely described as an optimality condition. The minimum norm principle is equivalent to the principle of minimum complementary energy for an elastic contact system.[¶] See also MINIMUM ENERGY PRINCIPLE; PRINCIPLE OF VIBRATION MEASUREMENT.

MINIMUM POLARIZABILITY PRINCIPLE (MPP) [chemistry] "The natural direction of evolution of any system is toward a state of minimum polarizability." It has also been shown that "a system is harder and less polarizable in its ground state than in any of its excited states."[‖]

MINIMUM PRINCIPLE [mathematics] (1) Joint optimization method inspired by the deterministic annealing algorithm for data clustering, extending previous work on tree-structured vector quantization using informative priors to approximate the unstructured solution while imposing the structural constraint.[**] (2) If optimizing only yields short-term solutions, then only a minimal approach can provide the basis for long-term solutions. By taking this approach to satisficing, one has a more powerful tool than merely asking whether an action suffices, namely, whether it is necessary. In fact, optimization becomes minimization when asked whether it is necessary. Assuming the need for future decisions, the minimum principle is a higher-level maximization. Its goal is to maximize resources for both

[*]Deco, G. and Schürmann, B. (2000), *Information Dynamics: Foundations and Applications*, Springer, p. 46.

[†]Anon. (1919), Davis "Revergen" principle of firing furnaces with town's gas, *Gas World* **70**, 189–190.

[‡]Kapur, J. N. (1988), *Mathematical Modeling*, New Age International, p. 203.

[§]Moreno, I., Iemmi, C., Marquez Andres, C. J., and Yzuel, M. J. (2004), Modulation light efficiency of diffractive lenses displayed in a restricted phase-mostly modulation display, *Appl. Opt.* **43**(34), 6278–6284.

[¶]Wang, M. Y. (2005), Passive forces in fixturing and grasping, *Proc. 9th IEEE Conf. Mechatronics and Machine Vision*, pp. 271–285.

[‖]Chattaraj, P. K., Fuentealba, P., Jaque, P., and Toro-Labbe, A. (1999), Validity of the minimum polarizability principle in molecular vibrations and internal rotations: An ab initio SCF study, *J. Phys. Chem. A* **103**, 9307.

[**]Cator, E. A. (1996), Potential theory of Monge Ampere on a Banach space, Minimum principle and Poisson property, *Potential Anal.* **5**, 173.

this decision and those future decisions.* (3) [physics] Any of various principles formulated to account for the observation that certain important quantities tend to be minimized when a physical process takes place.[†] See PRINCIPLE OF THE MINIMUM.

MINIMUM PRINCIPLE FOR HARMONIC FUNCTIONS [mathematics]

(1) States that if f is a harmonic function, then f cannot exhibit a true local maximum within the domain of definition of f. In other words, either f is a constant function, or, for any point inside the domain of f, there exist other points arbitrarily close to at which f takes smaller values.[‡] (2) Holds for the more general subharmonic functions, while superharmonic functions satisfy the minimum principle. The key ingredient for the proof is the fact that, by the definition of a harmonic function, the Laplacian of f is zero. Then, if is a nondegenerate critical point of $f(x)$, we must be seeing a saddle point, since otherwise there is no chance that the sum of the second derivatives of f is zero.[§] See also MINIMUM MODULUS PRINCIPLE; MAXIMUM PRINCIPLE FOR HARMONIC FUNCTIONS.

MINIMUM RESIDUAL (MR) PRINCIPLE [computer science]

See ORTHOGONAL PROJECTION PRINCIPLE.

MINIMUM TOTAL POTENTIAL ENERGY PRINCIPLE [physics, chemistry, biology, engineering]

Asserts that a structure or body shall deform or displace to a position that minimizes the total potential energy, with the lost potential energy being dissipated as heat.

*Charalambous, C. D. and Hibey, J. L. (1996), On the allocation of minimum principle for solving partially observable risk sensitive control problems, *Syst. Control Lett.* **27**, 169.
[†]Galatzer-Levy, R. (1983). Perspective on the regulatory principles of mental function, *Psychoanal. Contemp. Thought* **6**, 255–289.
[‡]Kalugin, P. (2001), A new phasing method based on the principle of minimum charge, *Acta Crystallogr. A Found. Crystallogr.* **A57**(6), 690–699.
[§]Wang, J. H., Sun, W. D. (1999), Online learning vector quantization: a harmonic competition approach based on conservation network, *IEEE Trans. Syst. Man Cybernet. B Cybernetics* **29**(5), 642–653.

Given two possibilities—a low heat content and a high potential energy, or a high heat content and low potential energy, the latter will be the state with the highest entropy, and will therefore be the state towards which the system moves.[¶] See also PRINCIPLE OF MINIMUM ENERGY.

MINTO PYRAMID PRINCIPLE [electronics, computer science, psychology] (Barbara Minto, MBA, Harvard)

(1) Present ideas organized as a pyramid under a single point. Direct toward answering an existing question and the ideas must obey a limited number of logical rules.[∥] (2) Permits people to quite dramatically (a) reduce the time they normally need to produce a first draft, (b) increase its clarity, and (c) decrease its length.[§] See also MATCHING PRINCIPLE; ASYMPTOTIC MATCHING PRINCIPLE; HYDROPHOBIC MATCHING PRINCIPLE; DOPPLER MATCHING PRINCIPLE; PHASE MATCHING PRINCIPLE; BRACKET MATCHING PRINCIPLE; MODEL MATCHING PRINCIPLE; MOMENT MATCHING PRINCIPLE; ZERO—FORCING PRINCIPLE; TIME INDEPENDENCE PRINCIPLE; DISTRIBUTION MATCHING PRINCIPLE; COMBINATORIAL MATCHING PRINCIPLE; PIGEONHOLE PRINCIPLE; H_2 MODEL MATCHING PRINCIPLE; H INFINITY MODEL MATCHING PRINCIPLE; OPTIMAL MODEL MATCHINGPRINCIPLE.

MIN-TSU CHU-I PRINCIPLE [sociology]

Self-determination of the Chinese people as a whole and also for the minority groups within China; Chinese ideology.[**] See also THREE PRINCIPLES OF THE PEOPLE NATIONALISM.

MINXENT PRINCIPLE [engineering, mathematics]

See MINIMUM ENTROPY PRINCIPLE.

MIRANDA-AGMON MAXIMUM PRINCIPLE [mathematics] (Shmuel Agmon, American mathematician)

Contains a detailed analysis of the singularities of the solutions

[¶]Zheng, N., Watson, L. G., Yong-Hing, K. (1997), Biomechanical modelling of the human sacroiliac joint, *Med. Biol. Eng. Comput.* **35**(2), 77–82.
[∥]Minto, B. http://www.barbaraminto.com, Minto Books International.
[**]Bilancia, P. R. (1981), *Dictionary of Chinese Law and Government, Chinese-English*, Stanford Univ. Press, p. 449.

to elliptic boundary value problems near conical points.*

MIRRORING PRINCIPLE [Industrial psychology] Understanding how best to identify and sustain service brands' values. It does this through a focus on the Web design aesthetics used in the Websites of small to large companies, and a comparison of these aesthetics with the preferences of target users. Web design contributes to services branding and a finding of a tendency of the majority of websites to employ what may be termed a "male design aesthetic," and for men and women to have a differential preference as between the male and female design aesthetic, leads to a discussion of the appropriateness of previous service branding models.[†]

MIRRORTRON PRINCIPLE [physics] Where the effect of a traveling magnetic field causes a relative displacement of the plasma electrons relative to the ions, creating a local region of positive potential moving through the plasma.[‡]

MITCHELL'S CHEMIOSMOTIC PRINCIPLE (Peter Dennis Mitchell, 1920–1992, British biochemist) When dinitrophenol (DNP) is ingested, it transfers the H+ions back to the mitochondrion easily and no adenosine triphosphate (ATP) is manufactuered. The energy of the electron separation is dissipated as heat and is not built in as chemical energy in ATP. The loss of this energy-storing compound makes the utilization of food much less efficient, resulting in weight loss.[§] See also CHEMIOSMOTIC PRINCIPLE.

MITROFANOFF PRINCIPLE [physiology, biology] (John Anthony Monti, b. 8/7/1949, Birmingham, AL) (1) Continent urinary diversion is a versatile technique with a predictable success rate applicable to a wide variety of urological conditions. The Mitrofanoff principle expands on a versatile technique.[¶] (2) A catheterizable conduit to the bladder when urinary control is inadequate, creating a submucosal tunnel for implantation of a catheterizable tube. (3) Cosmetic or dermatologic composition with controlled-release photoconvertible carotenoid. (5) Neem tree useful for control of gastrichyper acidity, and ulceration and does not affect pepsinogen activity.[‖] See also MONTI PRINCIPLE.

MITSCHERLICH'S PRINCIPLE [chemistry] (Eilhard Mitscherlich, 1794–1863, German chemist) Substances that crystallize in isomophous forms have similar chemical compositions.[**] See also PRINCIPLE OF ISOMORPHISM.

MITSCHERLICH'S PRINCIPLE OF ISOMORPHISM [chemistry] (Eilhard Mitscherlich, 1794–1863, German chemist) Salts having similar crystalline forms have similar chemical constitutions.[††]

MIXED VARIATIONAL PRINCIPLE [mathematics, engineering] (Eric Reissner) See DISPLACEMENT PRINCIPLE.

MOBILITY PRINCIPLE [psychology] Far from following a linear path. Frequent change in environments, themes, and models as well

*Maz'ya, V. G. and Robmann, J. (1992), On the Agmon-Miranda maximum principle for solutions of strongly elliptic equations in domains of R^n with coninncal points, *Ann. Global Anal. Geom.* **10**, 125–150.

[†]Moss, G., Gunn, R., and Kubacki, K. (2008), Gender and web design: The implications of the mirroring principle for the services branding model, *J. Market. Commun.* **14**(1), 37–57.

[‡]Douglass, S. R. (1993), *The Mirrortron Experiment: a Proof of Principle Test for a Method of Generating High Transient Potentials.* IEEE Intl. Conf. on Plasma Science Oakland, CA. Lawrence Livermore Nat Lab., p. 206

[§]Bettelheim, F. A., Brown, W. H., Campbell, M. K., and Farrell, S. O. (2009), Introduction

to general, organic and biochemistry. Congage Learning. p. 746.

[¶]Woodhouse, C. R. J. and Macneily, A. E. (1994), *Br. J. Urol.* **74**(4), 447–453.

[‖]Tarrado, X., Rodo, J., Sepulveda, J. A., Garcia Aparicio, L., and Morales, L. (2005), Continent urinary diversion: the Mitrofanoff principle, *Cirugia Pediatrica* **18**(1), 32–35.

[**]Daintith, J. (1999), *The Facts on File Dictionary of Chemistry*, 3rd ed., Facts On File, New York, p. 117.

[††]Walker, P. M. B. (1999), *Chambers Dictionary of Science and Technology*, Chambers, New York, p. 747.

sources of inspiration originating from very diverse sources.[*]

MÖBIUS PRINCIPLE See MOEBIUS PRINCIPLE.

MOEBIUS PRINCIPLE [mathematics] (August Ferdinand Möbius, 1790–1868, German mathematician) (1) The elemental unit of process is paradox, the dynamic tension of juxtaposed opposites. But paradox does not simply go unresolved, nor is it merely averted or eliminated by extrinsic means. Instead, an intrinsic resolution of paradox is provided which results in the re-emergence of paradox at another, more complex level. In effect, an idea of change is advanced in which process itself is "processed." When applied to the foundations of biology and physics, a novel concept of "dimensional generation" appears in which space and time undergo continuous internal transformation and organic growth. (2) Expanding cubic capacity as the object itself, and as such could re-create the objects environs. (3) Moebius strip is a two-dimensional surface embedded in three-dimensional space; it can embody the paradoxical union of opposites more concretely than can the lines of the schematic cube, limited as they are to a two-dimensional medium of expression. Points on opposite sides are intimately connected as twisting or dissolving into each other, as being bound up internally. Accordingly, mathematicians define such pairs of points as single points, and the two sides of the Moebius strip as only one side. (4) Essentially similar to the oneness of the perspectively fused Necker cube. There is inside and there is outside. The two are different. Yet they also are one and the same.[†] (5) Common electromagnetic audiowaves are converted into scalar waves, which creates the scalar field; this field is induced by specific information of remedies, organs, emotions, cell organelles, and so on by the audiofiles coming out of the computer.[‡]

MODEL MATCHING PRINCIPLE [electronics, computer science, psychology] See MATCHING PRINCIPLE.

MODELING PRINCIPLE [psychology] To teach a child new ways of behaving, allow the child to observe a prestigeful person performing the desired behavior.[§]

MODULATION PRINCIPLE [physics] Novel active noise control techniques on the basis of feedforward.[¶]

MOE'S PRINCIPLE [mathematics] (K. Moe, Denmark electrical engineer) (1) Used in determining the optimum number of connecting devices relative to the expense involved and the demand for service in the area. The economic principles on which the investigation is based involve the well-known extremum solution and are taken from.[‖] (2) Theory derived by conventional methods of the calculus of variations. Provides productive factors related to random fluctuations in demand. The quantities D and F are given to three decimal places. For most uses of the tables, the values of D are probably sufficiently accurate. But the values of F, especially when close to 1, might well have been carried to more places since the formulas for the average queue length $L = N(1 - F)$ and the average delay W-LT/H involve $(1 - F)$ as a factor. Thus in important ranges of traffic

[*]Huitt, W. (1994). Principles for using behavior modification, in *Educational Psychology Interactive*, Valdosta, State Univ., Valdosta GA (http://chiron.valdosta.edu/whuitt/col/behsys/behmod.html).
[†]Gros, F. (2006), The mobility principle: How I became a molecular biologist, *J. biosci.* **31**(3), 303–308.

[‡]Rosen, S. M. (1994), *Science, Paradox, and the Moebius Principle: The Evolution of a "Transcultural" Approach to Wholeness*, SUNY Press, p. xvii.
[§]Schill, G. and Tafelmair, L. (1971), Study of the synthesis of catenanes and nodules according to the Moebius band principle. Doubly bridged tetraamino-p-benzoquinones, *Synthesis* **10**, 546–548.
[¶]Zalesskii, A.V. (1970), New modulation principle for studying NMR in ferromagnetics, *Pribory i Tekhnika Eksperimenta* **2**, 156–157.
[‖]Samuelson, P. A. (1947), *Foundations of Economic Analysis*, Harvard Univ. Press, Cambridge, MA.

loads, it is impossible to make satisfactory estiamtes of these parameters.*

MOIRE INTERFERENCE PRINCIPLE [physics] See RESOLUTION PRINCIPLE.

MOMENT-MATCHING PRINCIPLE [electronics, computer science, psychology] (1) Approach to color image compression with high compression ratios and good quality of reconstructed images using quantization, thresholding, and edge detection.† (2) Tendency to match and to influence groups trained not to match. In training for professional practice, the mode of training should match the mode of practice. Guiding principle in the design of training programs, especially in planning for the "process" element in training.‡ (3) The established "reflection process" as a framework of possible inputs to a system as a subset of the set of tolerable inputs (those judged to produce an acceptable response from the system).§ See also MATCHING PRINCIPLE; ASYMPTOTIC MATCHING PRINCIPLE; HYDROPHOBIC MATCHING PRINCIPLE; DOPPLER MATCHING PRINCIPLE; PHASE MATCHING PRINCIPLE; BRACKET MATCHING PRINCIPLE; MODEL MATCHING PRINCIPLE; ZERO-FORCING PRINCIPLE; TIME INDEPENDENCE PRINCIPLE; DISTRIBUTION MATCHING PRINCIPLE; COMBINATORIAL MATCHING PRINCIPLE; PIGEONHOLE PRINCIPLE; H_2 MODEL MATCHING PRINCIPLE; H INFINITY MODEL MATCHING PRINCIPLE; OPTIMAL MODEL MATCHING PRINCIPLE; MINTO PYRAMID PRINCIPLE.

*Jensen, A. (1950), *Moe's Principle. An Econometric Investigation Intended as an Aid in Dimensioning and Managing Telephone Plant. Theory and Tables*, The Copenhagen Telephone Co.
†Idris, F. M. and Panchanathan, S. (1997), Spatiotemporal indexing of vector quantized video sequences, *IEEE Trans. Circuits Syst. Video Technol.* **7**(5), 728–740.
‡Strand, P. S. (2000), A modern behavioral perspective on child conduct disorder: Integrating behavioral momentum and matching theory, *Clin. Psychol. Rev.* **20**(5), 593–615.
§Brigo, D., Mercurio, F., Rapisarda, F., and Scotti, R. (2003), Approximated moment-matching dynamics for basket-options pricing, *Quant. Finance* **4**(1), 1–16.

MOMENT-PRESERVING PRINCIPLE [computer science] (1) A novel color block truncation coding (BTC) algorithm for compressing color pixel blocks. (2) Analytical formulas for the quaternion moment block truncation, are derived using quaternion arithmetic and the moment-preserving principle. (3) Approach to color image compression with high compression ratios and good quality of reconstructed images using quantization, thresholding, and edge detection. (4) System design that have direct relevance to the theory and design of control systems in combination with the method of inequalities.¶ Also known as *Complex moment-preserving principle; Quaternion moment-preserving principle.*

MOMENTUM POSITION UNCERTAINTY PRINCIPLE [mathematics, engineering] See UNCERTAINTY PRINCIPLE.

MOMENTUM PRINCIPLE [physics] The mass of the particle multiplied by the velocity of the particle. The total momentum of a system of particles is conserved provided the net external force action on the system is zero. Newton's second law of motion describes what happens if the net external force is not zero: $F = dp/dt$ (where $F = $ mass) is correct only if the mass of the particle is fixed. If the mass can change, then $F = dp/dt$ must be used.‖ See also CANONICAL MOMENTUM PRINCIPLE.

MONOTONE SEQUENCE PRINCIPLE (MSP) [mathematics] Any bounded monotone sequence of real Given a sequence $\{p_n\}$, define, for all m and n,

$$a_n^m \equiv \bigvee_{r=1}^{n} [0 \vee (p_k - m) \wedge 1].$$

numbers converges.** See also LIMITED PRINCIPLE OF OMNISCIENCE; BOUNDED SEQUENCE

¶Pei, S. C. and Cheng, C. M. (1996), A fast two class classifier for 2D data using complex moment preserving principle, *Pattern Recogn.* **29**, 519.
‖Orlando, V. A. and Jennings, F. B. (1954), The momentum principle measures mass rate of flow, *Trans. ASME* **76**, 961–965.
**Mandelkern, M. (1988) Limited Omniscience and the Bolzano-Weierstrass Principle. Bull London Math Soc v20, p. 319–320.

PRINCIPLE; CONSTANT SUBSEQUENCE PRINCI-
PLE; CONVERGENT SUBSEQUENCE PRINCIPLE;
BOLZANO-WEIERSTRASS PRINCIPLE; MONOTONE
SEQUENCE PRINCIPLE.

MONTI PRINCIPLE [physiology, biology] (John
Anthony Monti, b 8/7/1949, Birmingham,
AL) Also known as *Yang–Monti principle;
Monti channel principle;* see MITROFANOFF
PRINCIPLE.

MORAL PRINCIPLE [psychology] (1) In daily
clinical decisions, therapists need a framework by which to define both clinical practices
that are unethical as well as those that maintain therapeutic boundaries in a way that
is both ethically and clinically defensible.
The author examines how the principles of
nonmaleficence, beneficence, fidelity, justice,
universality, and autonomy can be applied to
the decisionmaking process when a therapist
faces an ethical dilemma involving boundary
issues.* (2) A guideline for people's knowing
and acting on the difference between right
and wrong behavior). The first requirement
is that each person should have an equal right
to the most extensive total system of equal
basic liberties compatible with a similar system of liberty for all; this is a guarantee of
fundamental liberties for each person. The
second is that social and economic inequalities are to be arranged so that they are both to
the greatest benefit of the least advantaged
(referred to as the *difference principle*) and
attached to offices and positions open to all
under conditions of fair and equal opportunity (known as the *principle of fair equality of
opportunity*).[†] (3) Refers to the concept of human ethics that pertains to matters of good
and evil, used within three contexts: individual conscience, systems of principles and
judgments, and codes of behavior or conduct.
Personal morality defines and distinguishes

among right and wrong intentions, motivations, or actions, as these have been learned,
engendered, or otherwise developed within
each individual.[‡]

MORAL PRINCIPLE [psychology] Right and
wrong accepted by an individual or a social
group; "the Puritan ethic"; "a person with
old-fashioned values."[§]

MOSELEY'S PRINCIPLE [chemistry] (Henry
Gwyn-Jeffreys Moseley, 1887–1915, English
physicist) Lines in the X-ray spectrum of
elements have frequencies that depend on
the proton number of the element.[¶]

MOUSE BETA AMYRIN ACTIVE PRINCIPLE
[physics, chemistry] Used as immunostimulant, antibacterial, and antiinflammatory.
Copolymer obtainable by polymerization
of monomer mixture containing carboxyl
containing monomer and methacrylic acid
ester.[‖] See also ACTIVE PRINCIPLE; ANTIVI-
RAL ACTIVE PRINCIPLE; POLYURONID ACTIVE
PRINCIPLE.

MUELLER–BRESLAU PRINCIPLE [chemistry]
(1) The ordinate value of an influence
line for any function on any structure is
proportional to the ordinates of the deflected
shape obtained by removing the restraint
corresponding to the function from the
structure and introducing a force causing a
unit displacement in the positive direction. If
a function at a point on a structure, such as
reaction, shear, or moment, is allowed to act
without restraint, the deflected shape of the

[‡]Moore, M. (1996), The moral principle (how weak
people can affect the operation and safety of
nuclear power plants), *Bull. Atom. Sci.* **52**, 2.
[§]American Heritage (2000), *American Heritage
Dictionary of the English Language*, 4th ed.,
Houghton Mifflin, New York.
[¶]Daintith, J. (1999), *The Facts on File Dictionary
of Chemistry*, 3rd ed., Facts On File, New York,
p. 159.
[‖]Subarnas, A., Oshima, Y., Sidik,. and Ohizumi,
Y. (1992), An antidepressant principle of Lobelia
inflata L. (Campanulaceae), *J. Pharm. Sci.* **81**(7),
620–621.

*McGrath, G. (1994), Ethics, boundaries, and
contracts: Applying moral principles, *Trans. Anal.
J.*, 24(1), 6–14.
[†]Small, M. and Dickie, L. (1999), A cinematograph
of moral principles: Critical values for contemporary business and society, *J. Manage. Devel.* **18**(7),
628–638.

structure, to some scale, represents the influence line of the function.* (2) Alternative means to qualitatively develop the influence lines for different functions. The ordinate value of an influence line for any function on any structure is proportional to the ordinates of the deflected shape that is obtained by removing the restraint corresponding to the function from the structure and introducing a force causing a unit displacement in the positive direction.** (3) Influence function as a deflection distribution can be used to analyze the influence functions of complicated structure idealized with various kinds of finite elements by the kinematic method.[†] Also known as *virtual work principle; principle of virtual displacements; principle of virtual forces.*

MULTIBYPASS PRINCIPLE [engineering] Process used to dodge, evade, or circumvent current reaction or process such as circulation to either speed up or slow down.[‡]

MULTICULTURAL PRINCIPLE [psychology] From feminist and multicultural counseling theories that counseling psychologists should consider as they engage in social justice work to include: (a) ongoing self-examination, (b) sharing power, (c) giving voice, (d) facilitating consciousness raising, (e) building on strengths, and (f) leaving clients the tools to work toward social change. Purposeful self-examination to social justice work necessitates that counseling psychologists engage in constant vigilance regarding the assumptions and values underlying their views of the communities they aim to support, the goals they hope to achieve, and the commitments that undergird their work. It also means that counseling psychologists need to be aware of the power dynamics

in their relationships with community members. First, counseling psychologists should approach communities with the belief that communities themselves know what questions or problems they want addressed. Only in this way can our interventions, research questions, and methodologies address the real needs of community members as they themselves characterize those needs. In describing the importance of giving voice to the poor, rural women, for example, are considered in the context of a "social action intervention," and "If we begin our inquiries by trying to understand the subordinate group's experience, the limits of conceptual frameworks based on the dominant group's experience can be revealed." This cannot be achieved without spending real time with community members, as mutual trust is critical. Second, counseling psychologists must find ways to amplify the voices of community members so that others can learn about their needs, wishes, strengths, and vision. This can involve a broad range of activities, including publishing qualitative studies with and about community members, bringing their ideas to policymakers, or working with community members to disseminate their ideas to the media Counselors need to know what kinds of support systems are available in the indigenous culture, such as extended family, community elders, and religious support groups. By fostering the ongoing development of support systems that evolved in the client's own culture, a clinician may ensure that the client can continue to be supported after initial clinical crises and interventions have ended. Translated to the community level, the aim of our collaboration is not to develop a one-sided or hierarchical dependence that may render the community or system helpless once we leave but rather to support strengths that will continue to thrive beyond our explicit involvement.[§]

MULTILEVEL MESH INDEPENDENCE PRINCIPLE [engineering] In solving nonlinear systems arising from elliptic boundary-value

*Fanous, F. (2000), *Influence Lines: Qualitative Influence Lines using the Müller Breslau Principle*, Iowa State Univ., College of Engineering (http://www.public.iastate.edu/~fanous/ce332/ influence/simplecantenvelope.html).

[†]Megson, T. H. G. (2005), *Structural and Stress Analysis*, 2nd ed., Butterworth-Heinemann, p. 647.

[‡]Cai, J. H., Wang, J. J., Zhu, W. X., and Zhou, Y. (1994), Experimental analysis of the multi-bypass principle in pulse tube refrigerators, *Cryogenics* **34**(9), 713–715.

[§]Goodman, L., Liang, B., Helms, J., Latta, R., Sparks, E., and Weintraub, S. (2004), Training counseling psychologists as social justice agents: Feminist and multicultural principles in action, *Counsel. Psychol.* **32**(6), 793–837.

problems, the number of iterations required by Newton's method is often observed to be nearly or completely independent of the underlying meshwidth.*

MULTIPLE-POINT PRINCIPLE (MPP) Fundamental physical parameters assume values that correspond to having a maximal number of different coexisting "phases" for the physically realized vacuum. Originally advanced in connection with theoretical predictions for the values of the three gauge coupling constants. In the original context of predicting the standard model gauge couplings using MPP, the principle asserts that the Planck-scale values of the standard model gauge group couplings coincide with the multiple point, namely, the point that lies in the boundary separating the maximum number of phases in the action parameter space corresponding to the gauge group GAnti—GUT. The idea was developed in the context of lattice gauge theory and the phases usually dismissed as lattice artifacts (e.g., a Higgsed phase, a confined or Coulomb-like phase).[†] Also referred to as *principle of multiple-point criticality*.

MULTIPLE-PULSE   PRINCIPLE [psychology] Method of determining the positions of excitation pulses in a speech frame in a linear predictive speech encoder.[‡]

MULTIPLICATION   PRINCIPLE [mathematics] If one event can occur in m ways and a second can occur independently of the first in n ways, then the two events can occur in mn ways.[§] See also MULTI-TAN H PRINCIPLE;

ADDITION PRINCIPLE; FUNDAMENTAL PRINCIPLE OF COUNTING.

MULTI-TAN H PRINCIPLE [engineering]   (1) If an action can be performed in N_1 ways, and for each of these ways another action can be performed in N_2 ways, then the two events can occur in $(N_1)(N_2)$ ways.[¶] (2) Individually nonlinear (hyperbolic tangent, or tan h) transconductance functions may be separated along the input voltage axis to achieve a much more linear overall function.[‖] See also MULTIPLICATION PRINCIPLE; TRANSLINEAR PRINCIPLE.

MUTUAL   EXCLUSIVITY   PRINCIPLE [psychology]   (1) If a child already knows a label for an object, then a new label for that object should be rejected. Preschool children's use of the mutual exclusivity principle during extension of words applied to familiar and unfamiliar nonsolid substances. (2) The use of different signals for different categories and of the same signal for all members within a category corresponds to the principle of contrast or of mutual exclusivity that children rely on when they assign only one label per category. (3) Children will allow only one lexical entry to occupy a semantic niche. When two words are determined to have similar meanings, one of them is preempted and removed from the lexicon. Different words should have different meanings. (4) Captures facts about the inferences speakers and addresses make for both conventional and novel words. It accounts for the preemption of novel words by well-established ones; and it holds just as much for morphology as it does for words and larger expressions. (5) The development of nonlinguistic concepts, the acquisition of language in context, and the use by participants in conversational exchanges to account for those features of language and language acquisition.[**] See also PRINCIPLE OF

*Layton, W. and Lenferink, H. W. J. (1996), A multilevel mesh independence principle for the Navier-Stokes equations, *SIAM J. Num. Anal.* **33**(1), 17–30.

[†]Bennett, D. L. and Nielsen, H. B. (2004), The multiple point principle: Realized vacuum in nature is maximally degenerate, *Proc. Inst. Math. NAS Ukraine* **50**(2), 629–636.

[‡]Haeberlen, U. (1977), [NMR] line narrowing by multiple pulse techniques. I. Objectives and principles, *NATO Adv. Study Inst. Series B: Physics* **B22** (*Nucl. Magn. Reson. Solids*), 229–237.

[§]Kuhn, W. (1953), The multiplication principle for separating and concentrating processes, *Chem. Ingen. Technik* **25**, 12–18.

[¶]Gilbert, B. (2008), Considering multipliers. Part 1: The wit and wisdom of Dr. Leif—7, *Analog Dialogue* **42**.

[‖]Gilbert, B. (1998), The multi-tanh principle: A tutorial overview, *IEEE J. Solid-State Circuits* **33**(1), 2–17.

[**]Merriman, W. E. and Stevenson, C. M. (1977), Restricting a familiar name in response to learning

UNIQUENESS; PRINCIPLE OF CONVENTIONALITY; PRINCIPLE OF PREEMPTION; PRINCIPLE OF CONTRAST; COOPERATIVE PRINCIPLE.

MUTUAL INFORMATION PRINCIPLE [mathematics] This principle has already been used in various areas, as a generalization of the maximum entropy principle (MEP), in the very common situation where measurements of a random variable contain errors having some known average value. An axiomatic derivation of the MIP is given here, in order to place it in a rigorous mathematical framework with the least possible intuitive arguments. The procedure followed is similar to the one proposed by Shore and Johnson for the minimum cross-entropy principle, and some relationships between the two methods of inductive inference are pointed out.* See also MESH INDEPENDENCE PRINCIPLE; MAXIMUM ENTROPY PRINCIPLE; MINIMUM CROSS-ENTROPY PRINCIPLE.

MUTUALLY EXCLUSIVE AND COLLECTIVELY EXHAUSTIVE (MECE) PRINCIPLE [Computer science] (1) Data should be divided in groups that do not overlap and that cover all the data. (2) If information can be arranged exhaustively and without double-counting in each level of the hierarchy, the way of arrangement is ideal.†

a new one: Evidence for the mutual exclusivity bias in young two-year-olds, *Child Devel.* **68**(2), 211–228.

*Avgeris, T. G. (1983), Axiomatic derivation of the mutual information principle as a method of inductive inference, *Kybernetes* **12**(2), 107–113.

†Burmaster, D. E. and Anderson, P. D. (1994), Principles of good practice for the use of Monte Carlo techniques in human health and ecological risk assessments, *Risk Anal.* **14**(4), 477–481.

N

N3C PRINCIPLE [psychology]   Ability to fast-map involving the ability to learn a new word based on very little input and with incomplete information, the semantic and/or syntactic aspects of the new word will require. (2) A vocabulary spurt included with the insight acquired that all objects have a name.* See also PRINCIPLE OF REFERENCE; PRINCIPLE OF EXTENDIBILITY; PRINCIPLE OF OBJECT SCOPE; PRINCIPLE OF CATEGORICAL SCOPE; NOVEL NAME–NAMELESS CATEGORY (N3C) PRINCIPLE; PRINCIPLE OF CONVENTIONALITY.

NADIRDSHVLLI BOUNDARY PRINCIPLE [physics, mathematics]   See BOUNDARY PRINCIPLE.

NAEGELE'S PRINCIPLE [medicine]   Rule for calculating an expected delivery date; subtract 3 months from the first day of the last menstrual period and add 7 days to that date.[†]

NAMELESS CATEGORY N3C PRINCIPLE [psychology]   (1) Words attached to categories for which the child does not yet have a name. A child who makes an attachment after hearing the new word only once or a few times has been able to fast-map the new word to the corresponding item. (2) Investigates the development of lexical frameworks and the applicability of the specificity hypothesis involving same. (3) Acquisition of lexical application make different inferences about the referent or extension of a particular word depending on timing of introduction.[‡] See also PRINCIPLE OF REFERENCE; PRINCIPLE OF

EXTENDIBILITY; PRINCIPLE OF OBJECT SCOPE; PRINCIPLE OF CATEGORICAL SCOPE; NOVEL NAME–NAMELESS CATEGORY (N3C) PRINCIPLE; PRINCIPLE OF CONVENTIONALITY.

NEBULAR PRINCIPLE [geology] [Pierre-Simon (Marquis de) Laplace (1749–1827), French astronomer/mathematician, 1796]   The material now forming the sun and planets originated as a disk-shaped nebula or gas cloud, which contracted into discrete bodies.[§]

NEBULIZATION PRINCIPLE [physics, chemistry]   Spectrophotometric analysis of high-temperature hydraulic and high-pressure atomic absorption.[¶]

NEGATIVE FEEDBACK PRINCIPLE [biology] (Harold Stephen Black, 1898–1983, American electrical engineer)   (1) Amplification output is fed back into the input, thus producing nearly distortionless and steady amplification.[‖] (2) The maturing cells continuously produce an inhibitor that diffuses down to the basal cell layer, where it inhibits the rate of cell proliferation. The concentration of inhibitor in the basal cell layer is dependent on the number of mature or maturing cells. Thus, when mature cells are lost from the surface, the concentration of inhibitor decrease, allowing the basal cells to divide at a faster rate.[**]

[§]Stiegeler, St. E. (1977), *A Dictionary of Earth Sciences*, Pica Press Universe Books, New York, p. 186.
[¶]Berndt, H. and Yanez, J. (1996), High-temperature hydraulic high-pressure nebulization: A recent nebulization principle for sample introduction (invited lecture), *J. Anal. Atom. Spectrom.* **11**(9), 703–712.
[‖]Black, H. S. (2009), in *Encyclopædia Britannica*. (http://search.eb.com/eb/article-9001793).
[**]Mechkov, C. S. (2005), Investigating, presenting and building electronic circuits with dynamic load by using the heuristic conflict principle, *Proc. 2nd Int. Conf. Computer Science'* 2005, Chalkidiki, Greece, Sept. 30, 2005.

*Mervis, C. B. and Bertrand, J. (1994), Acquisition of the novel name–nameless category (N3C) principle, *Child Devel.* **65**, 1646.
[†]American Heritage (2000), *American Heritage Dictionary of the English Language*, Houghton Mifflin, New York.
[‡]Mervis, C. B. and Bertrand, J. (1994), *Child Devel.* **65**, 1646.

NEGATIVE REINFORCEMENT PRINCIPLE [psychology]   To increase a child's performance in a particular way, you may arrange for him/her to avoid or escape a mild aversive situation by improving his/her behavior or by allowing him to avoid the aversive situation by behaving appropriately.*

NEGLIGENCE PRINCIPLE [psychology]   (1) The existence of real risks associated with product use.† (2) Negligence is the doing of something that a reasonably prudent person would not do, or the failure to do something that a reasonably prudent person would do, under circumstances similar to those shown by the evidence. It is the failure to use ordinary or reasonable care. Ordinary or reasonable care is that care that persons of ordinary prudence would use in order to avoid injury to themselves or others under circumstances similar to those shown by the evidence.‡

NEIGHBORHOOD COHERENCE PRINCIPLE (NCP) [physics, medicine]   A system of interacting cells can produce and maintain a spatial organization by virtue of cell–cell communication.§

NERNST'S DISTRIBUTION PRINCIPLE [chemistry]   (Walther Hermann Nernst, 1864–1941, German chemist)   When a single solute distributes itself between two immiscible solutes, then for each molecular species at a given temperature, there exists a constant ratio of distribution between the two solvents.¶

NERNST'S PRINCIPLE [chemistry]   (Walther Hermann Nernst, 1864–1941, German chemist)   Development of electrode potentials, based on the supposition that an equilibrium is established between tendency of the electode material to pass into solution and that of the ions to be deposited on the electrode.‖

NETWORK PROTOCOL DESIGN PRINCIPLES [mathematics]   (1) Systems engineering applied to create a set of common network protocol designs including effectiveness, reliability, and resiliency.** (2) Effectiveness needs to be specified in such a way that engineers, designers, and in some cases software developers can implement and/or use it. In human–machine systems, its design needs to facilitate routine usage by humans. Protocol layering accomplishes these objectives by dividing the protocol design into a number of smaller parts, each of which performs closely related subtasks, and interacts with other layers of the protocol only in a small number of well-defined ways. Protocol layering allows the parts of a protocol to be designed and tested without a combinatorial explosion of cases, keeping each design relatively simple.†† (3) Reliability ensures that data transmission involves error detection and correction, or some means of requesting retransmission. It is a truism that communication media are always faulty. The conventional measure of quality is the number of failed bits per bits transmitted. This has the useful feature of being a dimensionless figure of merit that can be compared across any speed or type of communication media. (4) Resiliency addresses a form of network failure known as *topological failure*, in which a communications link is cut, or degrades below usable quality.‡

*Huitt, W., (1994) Principles for using behavior modification, in *Educational Psychology Interactive*, Valdosta State Univ., Valdosta, GA (http://chiron.valdosta.edu/whuitt/col/behsys/behmod.html).
†*Brown v. Kendall* (1850), The negligence principle, in *Historical Development of Fault Liability*, p. 29.
‡Jung, David (2007) The Negligence Principle: historical development of fault liability Brown v Kendall (1850) California BAJI § 3.10.
§Phipps, M., Phipps, J., Whitfield, J. F., Ally, A., Somorjai, R. L., and Narang, S. A. (1990), Carcinogenic implications of the neighborhood coherence principle (NCP), *Med. Hypoth.* **31**(4), 289–301.
¶Walker, P. M. B. (1999), *Chambers Dictionary of Science and Technology*, Chambers, New York, p. 778.

‖Walker, P. M. B. (1999) Chambers Dictionary of Science and Technology. New York, N.Y. : Chambers p.778.
**(n.d.) Internetwork protocol http://www.spiritus-temporis.com/internetwork-protocol/network-protocol-design-principles.html.
††Lin, D. (2003), *Method and apparatus for defending against distributed denial of service attacks on TCP servers*. US Patent Application 10/668,952.

NEUMANN'S PRINCIPLE [crystals, physics, chemistry] (Carl Gottfried Neumann, 1832–1925, German mathematician) (1) Physical properties of a crystal are never of lower symmetry than the symmetry of the external form of the crystal. Consequently, tensor properties of a cubic crystal, such as elasticity or conductivity, must have cubic symmetry, and the behavior of the crystal will be isotropic.* (2) One of a pair of atoms or ions in a crystal that are close enough to each other for there to be interaction of significance in the physical problem being studied.†

NEW BLOCKING PRINCIPLE [chemistry, geology] (1) Symmetry elements of the point group of a crystal are included among the symmetry elements of any property of the crystal.‡ (2) Convention swing blocking schemes for distance protection.§

NEW EXCLUSION PRINCIPLE [psychology] A new theory of indirect effects is developed and used to form the basis for valuing interaction types.¶

NEW PRINCIPLE [pharmacology] See BASIC PRINCIPLE.

NEW THERAPEUTIC PRINCIPLE [pharmacology] Considered to be key, novel, essential, or rudimentary to the concept, property, or operation of a new or emerging technology applicable to pharmacy, diagnosis, or medicine.‖

NEWLAND'S PRINCIPLE [chemistry] (David Edward Newland, British Chemist) The observation that when elements are arranged in order of increasing relative atomic mass, there is a similarity beween members eight elements apart.** See also PRINCIPLE OF OCTAVES.

NEWMANN'S PRINCIPLE [mathematics, physics, chemistry] (James Newmann, American mathematician) (1) Taking into account symmetry elements of the system and its properties from Newmann's studies with crystallographics. A method for finding all the symmetries. This approach makes it possible to select symmetries potentially suitable for implementation of different functions.†† (2) Used in the synthesis of devices. Allows to find the physical symmetry of the matrix that may be used for synthesis of a device.‡‡

NEWTON'S ACTION–REACTION PRINCIPLE [physics] (Sir Isaac Newton, 1643–1727, English mathematician) To every action corresponds an equal and opposite reaction. The force with which a body acts on another body is always equal and opposite to the reaction force with which the second body acts on the first one.§§

NEWTON'S FIRST PRINCIPLE [physics] Sir Isaac Newton, 1643–1727, English mathematician Known as *Newton's first law of motion*. An object will move with constant velocity (which may be zero) unless acted on by a net external force.¶¶

*Walker, P. M. B. (1999), *Chambers Dictionary of Science and Technology*, Chambers, New York, p. 779.

†Bhagavantam, S. and Pantulu, P. V. (1967), Generalized symmetry and Neumann's principle, *Proc. Indian Acad. Sci. A* **66**(1), 33–39.

‡Schmidt, G. (1962), Hemicholinium, a new principle for blockade of cholinergic synapses, *Dtsch. Med. Wochenschrift* **87**, 2540–2543.

§Rigano, L., Giammarrusti, G., and Rastrelli, F. (2006), Olive oil: Building block for new active principles, *Cosmet. Technol.* (Milano), **9**(1), 23–29.

¶Russell, H. N. (1924), A new form of the exclusion principle in optical spectra, *Science* **59** 512–513.

‖Ronquist, G., Hugosson, R., Sjölander, U., and Ungerstedt, U. (1992), Treatment of malignant glioma by a new therapeutic principle, *Acta Neurochirurg.* (Springer, Wien) **114**(1–2), 8–11.

**Daintith, J. (1999), *The Facts on File Dictionary of Chemistry*, 3rd ed., Facts On File, New York, p. 162.

††Dmitriyev, V. A. (1997), Application of Newman's principle to electromagnetic N-ports, *IEEE Trans. Circuits Syst. I Fund. Theory Appl.* **44**(6), 513–520.

‡‡Pond, R. C., Bollmann, W. (1979), The symmetry and interfacial structure of bicrystals, *Phil. Trans. Roy. and Soc. Lond. A Math. Phys. Eng. Sci.* **292**(1395), 449–472.

§§Cucinotta, A. (2002), *The Laws of the Physical World* (graduate in physics, Industrial Technical High School "Verona Trento" of Messina) (http://www.peoplephysics.com/physics-laws1.html).

¶¶Munera, H. A. (1993), A quantitative formulation of Newton's first law, *Physics Essays*, v.**6**, n.2, p.173-80.

NEWTON'S PRINCIPLE OF COOLING [physics] Sir Isaac Newton, 1643–1727, English mathematician The rate of cooling of a hot body that is losing heat both by radiation and by natural convection is proportional to the difference in termperature between it and its surroundings.[*]

NEWTON'S PRINCIPLES OF MOTION [physics] Sir Isaac Newton, 1643–1727, English mathematician) The three laws of motion, known as *Newton's laws of motion: first law*—a body at rest will remain at rest and a body in motion will continue in motion with constant speed in a straight line, as long as no unbalanced force acts on it; *second law*—the time rate of change of the momentum of a body is equal in magnitude and direction to the next external force acting on it; *third law*—for every action there is an equal and opposite reaction, or, if body *A* exerts a force on body *B*, then body *B* exerts an equal and opposite force on body *A*.[†] See also PRINCIPLE OF MOMENTUM; PRINCIPLE OF REACTION; PRINCIPLE OF INERTIA.

NEWTON'S SECOND PRINCIPLE [physics] Sir Isaac Newton, 1643–1727, English mathematician Known as *Newton's second law of motion*. The time rate of change of the momentum of a body is equal in magnitude and direction to the next external force acting on it. A body at rest remains at rest, and a body in motion continues to move at a constant velocity unless acted on by an external force. A force acting on a body gives it an acceleration *a* in the direction of the force and of magnitude inversely proportional to the mass *m* of the body: $F = ma$. Whenever a body exerts a force on another body, the latter exerts a force of equal magnitude and opposite direction on the former. This is known as the *weak law of action–reaction*.[‡] See also PRINCIPLE OF INERTIA.

NEWTON'S THIRD PRINCIPLE [physics] Sir Isaac Newton, 1643–1727, English mathematician Known as *Newton's third law of motion*. For every action there is an equal and opposite reaction; this is true for bodies that are free to move as well as for bodies rigidly fixed.[§] See also D'ALEMBERT'S PRINCIPLE; DIFFERENTIAL PRINCIPLE; VARIATIONAL PRINCIPLE; HAMILTON PRINCIPLE; CONSERVATION OF MOMENTUM PRINCIPLE; PRINCIPLE OF EQUAL A PRIORI PROBABILITIES.

NIMBY PRINCIPLE [geology] See NOT IN MY BACKYARD PRINCIPLE.

NIPCO PRINCIPLE [engineering] Component for production of paper and pulp. In every fabrication or conversion process in which material is pressed, stretched, printed, or otherwise guided through rotating rollers pressed against each other, undesired roller deflection occurs. This bending yields a nonuniform distribution of forces in the cross-machine direction or an unparallel gap, which in turn results in nonuniform quality (profile, surface quality, print, etc.) over the width of the end product. The principle comes as close to this theoretical ideal as possible; the forces are generated and introduced directly above the gap through hydrostatic support elements on the inside of the freely rotating roller sleeve.[¶]

NOAH PRINCIPLE [economics] Ethical and philosophical tensions between societal development and preservation of biological diversity.[‖]

NONAGGRESSION PRINCIPLE [psychology] People have the right to act as they choose, as long as they do not infringe on anyone else's right to do the same. A deontological

[*]Walker, P. M. B. (1999) *Chambers Dictionary of Science and Technology*. New York, N.Y. : Chambers p.784.
[†]Cucinotta, A. (2002), *The Laws of the Physical World* (http://www.peoplephysics.com/physics-laws1.html).
[‡]Anon. (n.d.), *Dr. Zemelka's Chiropractic Information Center*, (www.drzemelka.com/thompson2.shtml).

[§]Cornille, P. (1995), The Lorentz force and Newton's third principle, *Can. J. Phys.*, **73**(9/10), 619–625.
[¶]Voith Paper Zürich (n.d.), *The NIPCO System* (http://www.zuerich.voithpaper.com/vp_zuerich_en_technologie.htm.
[‖]Baur, D. C. and Irvin, W. R. (2002), *The Endangered Species Act: Law, Policy, and Perspectives*, American Bar Association, Section of Environment, Energy, and Resources, p. 551.

ethical stance associated with the libertarian movement. It holds that "aggression"—which is defined as the initiation of physical force or the threat of such on persons or their property—is inherently illegitimate. Includes property as a part of the owner; to aggress against someone's property is to aggress against the individual. The only purpose for which power can be rightfully exercised over any member of a civilized community, against that person's will, is to prevent harm to others.*

NONCOERCION PRINCIPLE [psychology] (John Stuart Mill 1869 On Liberty)   "The only purpose for which power can be rightfully exercised over any member of a civilized community, against his will, is to prevent harm to others."† See also HARM PRINCIPLE; NONAGGRESSION PRINCIPLE.

NONCONSTRUCTIVE   PRINCIPLE [mathematics]   Proof by contradiction.‡ See also INDUCTION PRINCIPLE.

NONCONVEX VARIATIONAL PRINCIPLE [psychology, physics] (Niels Bohr, 1885–1962, Danish physicist) See COMPLEMENTARY PRINCIPLE.

NONDUAL PRINCIPLE [communication] (Johannesr, brand S. Wassenaar, Dutchbiochemist) The neuromotor concept aims to overcome the duality of bottom-up and top-down approaches. Since a dual approach does not reflect the nondual nature of living systems, the explanation of modes of action of a living endosystem has to be based on one nondual principle of self-organization that underlies the continuous production of flows of information connecting a systems' inner world with its outer world. Such a principle of self-organization can be seen as

a matter-inherent property, through which all corresponding levels of organization are functionally interconnected within the complete endosystem. The relation between this principle and conditions of the existence of living systems is discussed. Living systems are ruled by a principle of self-organization that is the matter-inherent principle of recurrent causality. The phenomenon of recurrence is consistent with the recurrent architecture of neuroorganic processes that connect endosystems with exosystems. The functional separation of the representations of the two interface borders in the neuropsychic domain leads to the development of interactive consciousness. Self-organizing endomatter, the interactive interface, and the origin of consciousness.§ See also PRINCIPLE OF RECURRENT CAUSALITY.

NONLINEAR   PRINCIPLE [physics] With every action there is always opposed an equal reaction; or, the mutual actions of two bodies on each other are always equal and directed to contrary parts.¶

NONLINEAR PRINCIPLE COMPONENTS ANALYSIS (NLPCA) [mathematics]   Components analysis, which chooses quantifications of the category levels to give the best principal components analysis in some nominated number of dimensions, here, as usual, chosen to be two.‖

NONSPASMOGENIC PRINCIPLE [physics]   A body at rest will remain at rest, and a body in motion will continue in motion with constant speed in a straight line, as long as

---

*Murray, C., Friedman, D., Boaz, D., and Bradford, R.W. (2005), What's right vs. what works, *Liberty* **15**(1).

†Anon. (1998), Principles of harm reduction, *Harm Reduction Coalition, Newsline* (People with AIDS Coalition of New York), pp. 7–8.

‡Bishop, E. and Bridges, D. (1985), Grundlehren der math, *Wissenschaften* **279**.

§Wassenaar, H., Van Roon, W., and ten Hallers, C. (1995), *Commun. Cogn.* **28**(2–3), 187–218.

¶Xue, F.-Z., Wang, J.-Z., Guo, Y.-S., Hu, P., and Wu, X.-S. (2004) Multiple nonlinear statistical method of population genetic structure based on the allelic polymorphism data, *Acta Genet. Sinica* **31**(2), 202–211.

‖Bratina, B., Muškinja, N., and Tovornik, B. (2008), Design of an auto-associative neural network by using design of experiments approach, Book Series Lecture Notes in Computer Science, *Book Knowledge-Based Intelligent Information and Engineering Systems*, Springer, vol. 5177/2008, pp. 25–32.

no unbalanced force acts on it.* See also
PRINCIPLE OF REACTION.

NORTHEY PRINCIPLE [electronics]   Breaking down of complex data structures into flat files.†

NOT IN MY BACKYARD PRINCIPLE [geology] Governed by visions of hazardous waste (HW) dumps of days gone by and of rusty dented drums oozing hazardous liquids into the environment. Prevents the construction of new environmentally sound sites, and forces HW facilities to be built on preexisting, already contaminated sites frequently located in blighted neighborhoods. Often the geology of these locations is less favorable for containment than new sites would be.‡ See also NIMBY PRINCIPLE.

NOTICE AND CHOICE PRINCIPLE [psychology, privacy]   Consisting of the following parts. (1) *Notice.* An organization must inform individuals about the purposes for which it collects and uses information about them, how to contact the organization with any inquiries or complaints, the types of third parties to which it discloses the information, and the choices and means the organization offers individuals for limiting its use and disclosure. (2) *Choice.* An organization must offer individuals the opportunity to choose (opt out) whether their personal information is to be (a) disclosed to a third party or (b) used for a purpose that is incompatible with the purpose(s) for which it was originally collected or subsequently authorized by the individual. (3) *Onward transfer.* To disclose information to a third party, organizations must apply the *notice* and *choice*

principles. Where an organization wishes to transfer information to a third party that is acting as an agent, it may do so if it first either ascertains that the third party subscribes to the principles or is subject to the directive or another adequacy finding or enters into a written agreement with such third party requiring that the third party provide at least the same level of privacy protection as is required. (4) *Security.* Organizations creating, maintaining, using, or disseminating personal information must take reasonable precautions to protect it from loss, misuse, unauthorized access, disclosure, alteration, and destruction. (5) *Data integrity.* (6) *Access.* Individuals must have access to personal information about them that an organization holds and be able to correct, amend, or delete that information where it is inaccurate, except where the burden or expense of providing access would be disproportionate to the risks to the individual's privacy in the case in question, or where the rights of persons other than the individual would be violated. (7) *Enforcement.* Effective privacy protection includes mechanisms for ensuring compliance, recourse for individuals affected by noncompliance, and consequences for the organization when not followed.§ See also SAFE-HARBOR PRIVACY PRINCIPLES.

NOVEL EQUIVALENCE PRINCIPLE [physics] Metric theories of gravitation, including general relativity, postulate the weak (Galilean–Newtonian) equivalence principle (EP): that bodies fall identically regardless of composition and configuration.¶ See also EQUIVALENCE PRINCIPLE; STRONG EQUIVALENCE PRINCIPLE.

NOVEL NAME–NAMELESS CATEGORY (N3C) PRINCIPLE [psychology]   Novel words map to objects for which the child does not yet

*Tomita, J. T. and Feigen, G. A. (1969), Serological identification and physical-chemical properties of the non- spasmogenic principle (NSP) in tetanus toxin, *Immunochemistry* **6**(3), 421–435.

†Berges, H. P. and Aktiengesellschaft, L. (1990), *Method for Operating a Twin Shaft Vacuum Pump According to the Northey Principle and a Twin Shaft Vacuum Pump Suitable for the Implementation of the Method* US Patent 5,049,050.

‡Barbalace, R. C. (2001), *Environmental Justice and the NIMBY Principle*, Environmental Chemistry.com. (http://EnvironmentalChemistry.com/yogi/hazmat/articles/nimby.html).

§Promotion Marketing Assoc., Inc. (2008) *PMA's Comments on the Federal Trade Commission's Proposed Online Behavioral Advertising Self-Regulatory Principles* (http://www.ftc.gov/os/comments/behavioraladprinciples/080411promomktg-assoc.pdf), Filed 4/11/2008.

¶Minakata, H. (1995), Testing the principle of equivalence with neutrinos, *Nuclear Phys. B Proc. Suppl.* **38**(Neutrino, 94), 303–307.

have a name. Provides an implicit input base for the referent rather than an explicit indication (e.g., pointing) of the referent of the new word.[||] See also PRINCIPLE OF REFERENCE; PRINCIPLE OF EXTENDIBILITY; PRINCIPLE OF OBJECT SCOPE; PRINCIPLE OF CATEGORICAL SCOPE; PRINCIPLE OF CONVENTIONALITY.

NOVEL ROUTING PRINCIPLE [computer science] Capable of sensing temporal and spatial changes in network dynamics and effectively balancing the mix of routing approaches best leveraging local conditions.[*]

NOVEL SHIELDING PRINCIPLE [engineering] See BUSBAR PROTECTION PRINCIPLE.

NOVIKOV SELF-CONSISTENCY PRINCIPLE [physics] (Dr. Igor Novikov 1935-, Russian astrophysical) (1) To solve the problem of paradoxes in time travel, asserts that if an event exists that would give rise to a paradox, then the probability of that event is zero. (2) Assumes certain conditions about what sort of time travel is possible. Specifically, it assumes counterfactual definiteness, which is the assertion that there is only one timeline and that multiple alternative timelines do not exist or are not accessible.[†]

NUCLEAR MOTION PRINCIPLE [physics] The hypothesis that, for given reactants, the reactions involving the smallest change in nuclear positions will have the lowest energy of activation.[‡] See also PRINCIPLE OF LEAST NUCLEAR MOTION.

NULL-POINT PRINCIPLE [chemistry] The null-point balance principle is used widely in thermogravimetric analysis (TGA). A sensor on the balance detects the deviation from the null position, and this operates a servomechanism to restore the deviation to the null position.[§]

[||]Mervis, C. B. and Bertrand, J. (1994), Acquisition of the novel name–nameless category (N3C) principle, *Child Devel.* **65**, 1646.

[*]Velasco-Santos, C., Martinez-Hernandez, A. L., and Castano, V. M. (2004), Chemical functionalization on carbon nanotubes: Principles and applications, *Trends Nanotechnol. Res.* 51–78.

[†]Friedman, J., Morris, M., Novikov, I., Echeverria, F., Klinkhammer, G., Thorne, K., and Yurtsever K. (1990), Cauchy problem in spacetimes with closed timelike curves, *Phys. Rev. D.* **42**, 1915.

[‡]Hine, J. (1977), The principle of least nuclear motion, Adv. Phys. Org. Chem. **15**, 1–61.

[§]Alexander, K. S., Riga, A. T., and Haines, P. J. (2004), *Ewing's Analytical Instrumentation Handbook*, 3rd ed., Thermoanalytical Instrumentation and Applications. CRC Press

O

OBJECTS PRINCIPLE 1 [archiving]   A good object exists in a format supporting intended use.*

OBJECTS PRINCIPLE 2 [archiving]   A good object is preservable. The object will not raise unnecessary barriers to remaining accessible over time despite changing technologies.*

OBJECTS PRINCIPLE 3 [archiving]   A good object is meaningful and useful outside its local context. A good digital object should be coherent, meaningful, and usable outside the context in which it was created. Depending on the discipline, objects with these properties may be called "portable," "reusable," or "interoperable."*

OBJECTS PRINCIPLE 4 [archiving]   A good object will be named with a persistent, globally unique identifier that can be resolved to the current address of the object. An *identifier* is a name assigned to an object according to a formal standard, an industry convention, or a local system providing a consistent syntax. Good identifiers will at minimum be locally unique, so that resources within the digital collection or repository can be unambiguously distinguished from each other. Global uniqueness can then be achieved through the addition of a globally unique prefix element, such as a code representing the organization.*

OBJECTS PRINCIPLE 5 [archiving]   A good object can be authenticated. *Authenticity* refers to the degree of confidence that a user can have in the integrity and trustworthiness of an object. Authentication is the act of determining that the object conforms to its documented origin, structure, and history,

and that the object has not been corrupted or changed in an unauthorized way.*

OBJECTS PRINCIPLE 6 [archiving]   A good object has associated metadata. A good object will have descriptive and administrative metadata, and compound objects will have structural metadata to document the relationships between components of the object and ensure proper presentation and use of the components.*

OBLIGATORY CONTOUR PRINCIPLE (OCP) [psychology]   (1) Parameterized in a way that accounts for the stages that children follow in the acquisition of English plural allomorphy; explore some issues concerning the child's acquisition of underspecified phonological representations.† (1) A pivotal constraint in autosegmental phonology. (2) Communication disorders.‡

OCCAM'S RAZOR PRINCIPLE [psychology] [William of Occam (also written Ockham), c. 1285–1349, English philosopher] (1) Assumptions introduced to explain that a thing must not be multiplied beyond necessity, and hence the simplest of several hypotheses is always the best in account for unexplained facts. (2) Requires that ad hoc assumptions be minimized as far as possible in scientific explanations of natural phenomena. (3) There is no other possibility than parsimony to choose between different alternative hypotheses, that explain singular historical happenings, that can only be reconstructed, but not repeated and tested

*NISO (Nat. Information Standards Org.) (2007), *A Framework of Guidance for Building Good Digital Collections*, 3rd ed., Inst. Museum and Library Services (http://framework.niso.org/node/5).

†Ingram, D. (1995), The acquisition of negative constraints, the OCP, and underspecified representations, in *Phonological Acquisition and Phonological Theory*, Lawrence Erlbaum, Hillsdale, NJ, pp. 63–79.
‡Berent, I., Everett, D. L., and Shimron, J. (2001), Do phonological representations specify variables? Evidence from the obligatory contour principle, *Cogn. Psychol.* **42**(1), 1–60.

like scientific experiments.* See also PRINCIPLE OF POLYMERASE CHAIN REACTION; PART–CURRENT PRINCIPLE; PARSIMONY PRINCIPLE; PRINCIPLE OF PARSIMONY.

OCCUPANCY PRINCIPLE [physics]   For any part of a steady-state system through which there is a constant flow of mother substance, the ratio of occupancy to capacity is constant and equal to the reciprocal of the constant flow.[†]

OHM'S PRINCIPLE [physics] (Georg Simon Ohm, 1784–1854, German physicist)   In metallic conductors, at constant temperature and zero magnetic field, the current flowing through a component is proportional to the potential difference between its ends, the constant of proportionality being the conductance of the component.[‡]

OHM'S PRINCIPLE OF HEARING [acoustics] Georg Simon Ohm, 1784–1854, German physicist) A simple harmonic motion of the air is appreciated as a simple tone by the human ear. All other motions of the air are analyzed into their harmonic components which the ear appreciates as such separately.[‡]

OLBI PRINCIPLE [physics]   See OVERLAPPING BIPHASIC IMPULSE PRINCIPLE.

OMNISCIENCE PRINCIPLE [mathematics](1)

DEFINITION.
(Omniscience principles). We define:

- LPO : $\mathbb{N}^{\mathbb{N}} \to \mathbb{N}$,

$$\mathrm{LPO}(p) = \begin{cases} 0 & \text{if } (\exists\, n \in \mathbb{N})\, p(n) = 0 \\ 1 & \text{otherwise,} \end{cases}$$

- LLPO $:\subseteq \mathbb{N}^{\mathbb{N}} \overset{\rightarrow}{\to} \mathbb{N}$,

$$\mathrm{LLPO}(p) \ni \begin{cases} 0 & \text{if } (\forall\, n \in \mathbb{N})\, p(2n) = 0 \\ 1 & \text{if } (\forall\, n \in \mathbb{N})\, p(2n+1) = 0, \end{cases}$$

where dom $(\mathrm{LLPO}) := \{p \in \mathbb{N}^{\mathbb{N}} : p(k) \neq 0$ for at most one $k\}$.

One should notice that the definition of LLPO implies that $\mathrm{LLPO}(0^{\mathbb{N}}) = \{0, 1\}$. The natural numbers $\mathbb{N}$ can be represented by $\delta_{\mathbb{N}}(p) := p(0)$, but for simplicity of notation we will usually work directly with $\mathbb{N}$. Because the various conditions concerning sequences each have several classical forms, which are constructively different, it is necessary to give explicit constructive definitions. For example, a sequence $\{p_n\}_{n=1}^{\infty}$ of positive integers is *bounded* if a suitable bound has been constructed. Classically, *unbounded* may be understood to mean that the existence of a bound is contradictory, but for a constructive study we adopt rather an affirmative meaning: a subsequence $\{p_{n_k}\}_{k=1}^{\infty}$ has been constructed with $p_{n_k} > k$ for all k. Other concepts are similarly interpreted in an affirmative manner. A *decision sequence* is a nondecreasing sequence of 0s and 1s.[§]

See also LIMITED PRINCIPLE OF OMNISCIENCE; BOUNDED SEQUENCE PRINCIPLE; CONSTANT SUBSEQUENCE PRINCIPLE; CONVERGENT SUBSEQUENCE PRINCIPLE; BOLZANO-WEIERSTRASS PRINCIPLE; MONOTONE SEQUENCE PRINCIPLE.[¶]

ONE-STAGE PRINCIPLE [physics]   Process involving one action to effect a change or result.

---

*Lörincz, A., Póczos, B., Szirtes, G., and Takács, B. (2002), Ockham's razor at work: Modeling of the "Homunculus," *Brain Mind* **3**(2), 187–220.
[†]Gillespie (1969) Use of the occupancy principle in studies of calcium metabolism. Calcified Tissue International, Springer v 4(1) p 89–90.
[‡]Walker, P. M. B. (1999), *Chambers Dictionary of Science and Technology*, Chambers, New York, p. 806.

[§]Mandelkern, M. (1988) Limited Omniscience and the Bolzano-Weierstrass Principle. Bull London Math Soc **20**, p. 319–320.
[¶]Brattka, V. and Gerhardi, G. (2009) Weihrauch Degrees, Omniscience Principles and Weak Computability http://arxiv.org/PS_cache/arxiv/pdf/0905/0905.4679v2.pdf

ONE-TRIAL PRINCIPLE [psychology]   (1) Leaving aside for the moment the one-trial principle to which we have referred, there are, as it seems to us, other reasons why in this case we should hesitate to exercise our discretion in favor of admitting the fresh evidence. The appellant's own evidence is, as we have said, not capable of belief. The medical evidence on which she seeks to rely came into existence long after the relevant time; it initially relied heavily on her account, which we have found to be flawed, and it is at variance with the evidence of the other psychiatrists who have been involved in the case. That factor is not, of course, decisive but compare, for example, Binning April 12, 1996 unreported, where the fresh evidence was unchallenged. We recognize, of course, that in Section 23(2)(b) the word used is "may," but we cannot treat this as a case in which had the evidence now sought to be relied on been deployed at trial, the defense of diminished responsibility would have been bound to succeed."* (2) The pairing of a stimulus and response depends on contiguity, and a stimulus pattern gains its full associative strength on the occasion of its first pairing with a response. This rejects the law of frequency as a learning principle (the strength of a learned association depends on number of pairings).[†] See also CHAINING PRINCIPLE; BACKWARD CHAINING PRINCIPLE; SUCCESSIVE APPROXIMATION PRINCIPLE; RESPONSE-SHAPING PRINCIPLE; TRANSITIONAL STIMULI PRINCIPLE; SCHEDULE OF REINFORCEMENT PRINCIPLE.

ONSAGER PRINCIPLE [chemistry]   (Lars Onsager, 1903–1976, American Chemist) (1) For sufficiently slow processes or small deviations from equilibrium, there is a linear dependence between all fluxes and all forces.[‡] (2) An equation based on the Debye–Huckel principle, relating to conductance.[§] See also DEBYE–HUCKEL PRINCIPLE.

OPEN–CLOSED PRINCIPLE [computer science] (Robert Cecil Martin)   (1) Software entities such as classes, modules, and functions should be open for extension but closed for modifications.[¶] (2) New functionality by adding new code, not by editing old code.[‖] (3) You should be able to extend a class' behavior, without modifying it. (a) They are "open for extension"; this means that the behavior of the module can be extended. That we can make the module behave in new and different ways as the requirements of the application change, or to meet the needs of new applications. (b) They are "closed for modification"; the source code of such a module is inviolate. No one is allowed to make source code changes to it. It would seem that these two attributes are at odds with each other. The normal way to extend the behavior of a module is to make changes to that module. A module that cannot be changed is normally thought to have a fixed behavior. How can these two opposing attributes be resolved?**

OPEN–CLOSED SYSTEM PRINCIPLE [psychology]   (1) Describes the use of four principles from N. Wiener's (1950) cybernetics and L. von Bertalanffy's (1968) general system theory in daily psychosomatic medicine clinics. The polygraphic method is used for the diagnosis of psychosomatic disease (black box principle). For the control of psychosomatic symptoms, the biofeedback method (feedback principle) is used. Systematic desensitization is used to relieve social stresses that cause psychosomatic disease (open–closed system

*Ognall, Justice and Lord Chief Justice (1999), *In the Supreme Court of Judicature Queen's Bench Divisional Court, Royal Courts of Justice Strand, London*, WC2A 2LL, May 18, 1999, Criminal Cases Review Commision ex parte Maria Pearson.

[†]Wasserman, E. A. and Miller, R. R. (1997), What's elementary about associative learning? *Ann. Rev. Psychol.* **48**, 573–607.

[‡]Cheryan, M. (2007), Membrane concentration of liquid foods, in *Handbook of Food Engineering*, 2nd ed. CRC Press.

[§]Walker, P. M. B. (1999), *Chambers Dictionary of Science and Technology*, Chambers, New York, p. 810.

[¶]Meyer, B. (1988), *Object-Oriented Software Construction*, Prentice-Hall.

[‖]*The Open-Closed Principle*, C++ Report, Jan. 1996.

**Martin, R. C. (n.d.), *Dependency Inversion Group. Ten Commandments ofOOProgramming* (http://www.butunclebob.com/ArticleS.UncleBob. Principles Of Ood).

principle). Transactional analysis, which corresponds to the information-and-energy principle, is also utilized.* (2) In closed systems, which are considered to be isolated from their environment, the final state is unequivocally determined by the initial conditions, whereas in open systems, which involve inflow and outflow, the same final state may be reached from different initial conditions and in different ways (i.e., equifinality). Human beings are, by nature, open systems, and attempts to understand psychological functioning must take this into account. The types of transactions that take place may be understood in the context of Wiener's cybernetics and von Bertalanffy's general system theory.[†] See BLACK BOX PRINCIPLE.

OPERATING PRINCIPLE [mathematics]   Of two competing theories, the simplest explanation of an entity is to be preferred. Optimal patterns of operation.[‡]

OPERATIONAL PRINCIPLE [physics, psychology]   (1) Interbehavioral operationism, which provides definite criteria for objectivity. (2) The principle of operationism can be expanded and used advantageously in psychological thought by including the control of observational procedure and the development of postulates. The revised principle is that of interbehavioral operationism, which provides definite criteria for objectivity.[§] See also COLLIDER OPERATIONAL PRINCIPLE; PRINCIPLE OF OPERATIONISM; REVISED PRINCIPLE.

OPTICAL BLACK HOLE PRINCIPLE [mathematics]   Estimates of the shielding effectiveness of enclosures having dimensions that are large compared to a wavelength.[¶] See also BLACK HOLE PRINCIPLE; REFLECTIONLESS APERTURE PRINCIPLE.

OPTICAL CONVERGENCY PRINCIPLE [engineering]   (1) Tendency to evolve similar structural or physiological characteristics under similar environmental conditions. (2) Light from a broadband, near-infrared source, and a visible aiming beam is combined and coupled into one branch of a fiber optic Michelson interferometer.[‖] (3) Relative motion solutions and distortion amplitudes determined by the approximate equations converge pointwise toward values by the Schrödinger equation. The converged set of equations defines the effective interaction of two clusters in the elastic energy region; if one cluster is a single particle, this definition overlaps with Feshbach's definition of the optical potential in the elastic energy region.** See also PRINCIPLE OF TIME GATING.

OPTICAL COUPLING PRINCIPLE [engineering]   (1) Enables a continuous recording of the intra ocular pressure and its variations in both human eyes and eyes of even conscious animals.[††] (2) Presence in the sump of a strong electromagnetically driven forced convection, which promotes the production of a fine equipped structure, and the fact that the thickness of the segregation zone tends toward zero. (3) Detection of the movements of single biomolecules, single-chemical reactions, and changes in conformation of single-protein molecules in real time, and the manipulation of single molecules in aqueous solution. (4) New conceptual framework for

*Ishikawa, H. (1979), Psychosomatic medicine and cybernetics, *Psychother, Psychosom.* **31**(1), 361–366.
[†]Bertalanffy, L. v. (1968), *Passages from General System Theory.* In General Systems theory: foundations, development, applications. The International library of systems theory and philosophy.
[‡]Voit, E. O. (2003), Design principles and operating principles: The yin and yang of optimal functioning, *Math. Biosci.* **182**(1), 81–92.
[§]Kantor, J. R. (1938), The operational principle in the physical and psychological sciences, *Psychol. Record* **2**, 3–32.

[¶]Nayakshin, S. (2004), Close stars and accretion in low-luminosity active galactic nuclei, *Month. Notices Roy. Astron. Soc.* **352**(3), 1028–1036.
[‖]Gregorio, P. (1990), *Universal Stereoscopic Viewer Based on a New Principle of Optical Convergency*, US Patent 5,084,781 (filed 4/4/90, Stereovision International S.R.L.).
Bjerrum, J. (1944), A new optical principle for the investigation of step equilibria, *Kgl. Danske Videnskab. Selskab, Math.-fys. Medd.* **21(4).
[††]Marchessault, R. H. and Nedea, M. E. (1989), Optical rotatory dispersion of poly(β-hydroxybutyrate)/poly(β-hydroxyvalerate) for composition determination, *Polym. Commun.* **30**(9), 261–263.

chemomechanical energy transduction in the molecular motor.* Also referred to as *principle of optical isolating; principle of optical coupling; optical isolating principle.*

OPTICAL ISOLATING PRINCIPLE [engineering] See OPTICAL COUPLING PRINCIPLE.

OPTICAL PRINCIPLE [engineering]   (1) Physiological variations in refractive index and radius of the cornea induce an insignificant error in the corneal thickness estimate.[†] (2) When an object is viewed through a very small aperture, a clear image will always be formed because only coherent rays of light are able to pass through, so that the "blur circle" normally formed by an out-of-focus eye is reduced almost to the clear point that would be seen if it were in focus. Provided there is no opacity of the eye or impairment of the retina, the object will appear clear regardless of any refractive error. The image through a single pinhole is very small and dim, but by using a regular array of similar-sized holes it is possible to enlarge the field of vision and improve the overall brightness of the image while still retaining most of the clarity of at least the central area.[‡] See also POPTICAL COUPLING PRINCIPLE.

OPTICAL REVERSIBILITY PRINCIPLE [engineering]   See PREVERSIBILITY PRINCIPLE.

OPTICAL SUPERPOSITION PRINCIPLE [engineering] [J. H. van't Hoff, (1852–1911, Dutch chemist), 1894]   Demonstrates what happens when waves meet and how standing waves are formed by the superposition of identical traveling waves in opposite directions. (1) If a number of independent influences act on the system, the resultant

influence is the sum of the individual influences acting separately. (2) When a situation is a composition of a number of elementary situations, its amplitude is the linear superposition of the amplitudes of the components. (3) The response of a complex system is a product of the responses of those simpler elements that can be considered to constitute the complex system. (4) Light emitted from two sources; the net electric field is the sum of the electric field of each source. (5) Total electric field at a point due to the combined influence of the distribution of point charges is the vector sum of the electric field intensities that the individual point charges would produce at that point if each acted alone. (6) In a linear electrical network, the voltage or current in any element resulting from several sources acting together is the sum of the voltages or currents resulting from each source acting alone. (7) When two or more forces act on a particle at the same time, the resultant force is the vector sum of the two. (8) The sum of any number of solutions to the equations is another solution if characterized by linear homogenoeous differential equations such as optics, acoustics, and quantum theory.[§] See also PQUANTUM MECHANICAL SUPERPOSITION PRINCIPLE; PRINCIPLE OF SUPERPOSITION.

OPTICAL TRIANGULATION PRINCIPLE [engineering]   Optical rotation produced by a compound consisting of two radicals of opposite optical activities is the algebraic sum of the rotations of each radical alone.[¶]

OPTIMAL MODEL MATCHING PRINCIPLE [electronics, computer science, psychology]   See PMATCHING PRINCIPLE.

OPTIMALITY PRINCIPLE [control systems] P. J. Schoemaker's examination of the metaprinciple of optimality. (1) Optimality serves as an epistemological organizing principle that (a) provides a general structure and framework for analysis and (b) enables a

*Vieira, M., Fantoni, A., Koynov, S., Cruz, J., Macarico, A. and Martins, R. (1996), Amorphous and microcrystalline silicon p-i-n optical speed sensors based on the flying spot technique, *J. Non-Cryst. Solids.* **198–200**(Pt. 2), 1193–1197.
†Kwong, C. P. (1987), Development of the Schroedinger equation and Klein-Gordon equation via Bellman's principle of optimality, *Phys. Lett. A* **124**(4–5), 220–222.
‡Heit, E. and Barsalou, L. W. (1996), The instantiation principle in natural categories. Preview, *Memory* **4**(4), 413–451.

§Kapur, J. N. (1988), *Mathematical Modeling*, New Age International, p. 203.
¶Jahne, Bernd, ed. v (2000) Computervision and applications; concise edition Academic Press. p. 197.

meaningful interpretation of measurements. (2) In an optimal system, any portion of the optimal state trajectory is optimal between the states that are joined by it. An optimal sequence of decisions in a multistage decision process problem has the property that whatever the initial state and decisions are, the remaining decisions must constitute an optimal policy with regard to the state resulting from the first decisions.* (3) Find all those economic situations that are jointly optimal, that is, that are such that any deviation from a situation for the benefit of any individual can be only at the cost of some other individuals. Optimization models arise as a result of application of one or more of these principles. New and challenging mathematical problems arise in obtaining the results from optimizing models.[†] (4) Representations that also provide some reduction of information. For example, some connectionist models of categorization essentially learn to form prototypelike summaries of general features. The instantiation principle that we explore here stands in contrast to the cognitive economy principle. (5) For optimal systems, any portion of the optimal state trajectory is optimal between the states that it joins. (4) In an optimal system, any portion of the optimal state trajectory is optimal between the states that are joined by it.[‡] See also PORGANIZING PRINCIPLE; BELLMAN'S PRINCIPLE OF OPTIMALITY; INSTANTIATION PRINCIPLE; ENTROPY OPTIMIZATION PRINCIPLE.

OPTIMIZATION   PRINCIPLE [mathematics] (1) Consists of the following two simple propositions in relation to local and global optimal points: (a) a maximum element x of X is also a maximum element of $A\,X$ if and only if $x\,2\,A$; (b) a maximum element x of $A\,X$ is also a maximum element of X if and only if A contains any maximum element of X. The author proceeds to illustrate that these two statements are a unified basis of Bellman's

*Lin, Y. X. (1985), On a general form of optimization principle, *Qufu Shiyuan Xuebao* (3), 35–42.
[†]Ewens, W. J. (2004), *Mathematical Population Genetics: Theoretical Introduction*, Springer, p. 265.
[‡]Mayr, Ernat (1978), Evolution. Scientific American v. 239, p. 47.

principle in dynamic programming, the branch-and-bound principle, and many other principles in the field of optimization theory. (2) Mechanism that minimizes the total energy required to drive the flow, maintain supply, and support the integrity. (3) Predicts that the time interval can be optimized, verified, and determined more accurately by complete numerical simulations and further verified by means of experimentation.[§] See also PRUIN-AND-RE-CREATE PRINCIPLE; ENTROPY OPTIMIZATION PRINCIPLE; PRINCIPLE OF OPTIMALLY.

OPTIMIZING   PRINCIPLE [mathematics] Large subset of nonoptimal solutions without affecting the search for an optimal solution. The path taken by an object is the path that minimizes the total difference between kinetic energy and potential energy over the path.[¶] See also PKIMURA OPTIMIZING PRINCIPLE.

OPTIMUM PRINCIPLE [mathematics]   For optimal systems, any portion of the optimal state trajectory is optimal between the states that it joins.[‖] See also PORGANIZING PRINCIPLE.

ORBITAL   PHASE   CONTINUITY   PRINCPLE (OPCP) [chemistry]   Based on GI-type electron wavefunctions are those that are simultaneously Eigen-function of the total spin projection operators. In the Woodward–Hoffmann approach one determines the symmetries of the bonding and anti-bonding molecular orbitals and constructs a correlation diagram. In this approach molecular and orbital

[§]Goddard, W. A., III. (1970), Orbital phase continuity principle and selection rules for concerted reactions, *J. Am. Chem. Soc.* **92**(25), 7520–7521.
[¶]Harper, J. M. and Rubin, J. E. (1976), Variations of Zorn's lemma, principles of cofinality, and Hausdorff's maximal principle, *Notre Dame Formal Logic* **17**(4), 565–588.
[‖]Varela, F. (1995), Heinz von Foerster, the scientist, the man prologue to the interview, *Stanford Electronic Humanities Review (SEHR)* **4**(2), (Constructions of the mind), updated July 23, 1995, Heinz von Foerster, the scientist, the man prologue to the interview (http://www.stanford.edu/group/SHR/4-2/text/varela.html).

symmetry are crucial. Bonding GI orbitals are used with OPCP.*

ORDER EXTENSION PRINCIPLE [mathematics] Every partial ordering can be extended to a linear ordering.†

ORDER FROM NOISE PRINCIPLE [psychology] (Heinz von Foerster, 1960)   According to which self-organization can be accelerated by random perturbations. In a paradoxical effect, the role of noise in a complex system might well lead to further organization, now more commonly referred to as *stochastic resonance*.‡

ORGANIC RANKINE CYCLE (ORC) PRINCIPLE [physics]   Use of an organic, high-molecular-mass fluid with a liquid–vapor phase change, or boiling point, occurring at a lower temperature than the water–steam phase change. The fluid allows Rankine cycle heat recovery from lower-temperature sources such as industrial waste heat, geothermal heat, and solar ponds. The low-temperature heat is converted into useful work, which can itself be converted into electricity.§

ORGANIZING   PRINCIPLE [psychology] (1) Organizing principles of interindividual differences underline the importance of variability in social representations. Organizing principles can correspond to systematic variations in the weight individuals or groups give to different dimensions underlying the structure of the field of representation. Organizing principles of involvement in human rights and their social

anchoring in value priorities.¶ (1) Basic in the development of biological systems is the ability to self-organize, a process by which new structures and patterns emerge.‖ (2) Individuals differ to some extent in the way operational factors are organized and function within subsystems and the number of ways in which operating factors in a certain subsystem can be organized in patterns, in order to allow the subsystem to play its functional role in totality.** (3) Orientation selectivity in a model visual system, for the development of biological and synthetic perceptual networks. (4) A theory of considerable integrative, explanatory, and interpretive power for a broadly interdisciplinary social science of human behavior. (5) For optimal systems, any portion of the optional state trajectory is optimal between the states it joins. See also POPTIMALITY PRINCIPLE; SELF-ORGANIZING PRINCIPLE; PRINCIPAL-COMPONENT ANALYSIS; SIMILARITY PRINCIPLE.

ORIFICE PRINCIPLE [engineering]   (1) A differential pressure that varies with flow rate. (2) Difference in electrical conductivity between the cells and the medium in which they are suspended is measured by the change in electrical impedance produced as they pass through an orifice. (3) Substance pushes against an obstacle, past a sensor allowing alterations in flow.†† See also PCOULTER ORIFICE PRINCIPLE; SLOT ORIFICE PRINCIPLE; VARIABLE-ORIFICE PRINCIPLE.

ORTHOGONAL   PRINCIPLE [physics]   (1) Opening or window in a sidewall or endwall of a waveguide or cavity resonator through which energy is transmitted. (2) Computation of an inner metric corresponds to performing a spectral factorization and the

*Goddard, William A., III (1972) Selection rules for chemical reactions using the orbital phase continuity principle. J. Am. Chem Soc **94**(3) p 793–807.

†Felgner, U., and Truss, J. K. (1999) The independence of the Prime Ideal Theoremfromthe Order-Extension Principle. The Journal of Symbolic Logic, **64**(1) 199-215.

‡Joslyn, C. (2003) Cybernetics in Encyclopedia of Computer Science. 4th ed. John Wiley & Sons,

§Quoilin, S. (2007), *Experimental Study and Modeling of a Low Temperature Rankine Cycle for Small Scale Cogeneration*, thesis, Dept. Mechanical Engineering, Univ. Liege.

¶Spini, D. and Doise, W. (1988), *Eur. J. Soc. Psychol.* **28**(4), 603–622.

‖Keren, G. (1993), Optimality as an epistemological organizing principle, *Behav. Brain Sci.* **16**(3), 622.

Aarons R. (1986), The organizing principle, *PC Mag.* **5, 149.

††Hackspacher, F. (2000), *Accumulator for an Air Conditioner Working According to the "Orifice" Principle, in Particular for a Vehicle Air Conditioner*, Eur. Patent Application.

inverse of the outer matrix as a whitening filter.*

ORTHOGONAL PROJECTION PRINCIPLE [computer science]   Data that can no longer reduce the distance at their extremum, define the extremum.† Also known as *minimum residual (MR) principle; equivalent orthogonal projection principle; orthogonality principle; calderón–zygmund principle; birman–schwinger principle.*

ORTHOGONALITY PRINCIPLE See   PDIVIDE-AND-CONQUER   PRINCIPLE;   ORTHOGONAL PROJECTION PRINCIPLE.

OSCILLATING AMATURE PRINCIPLE [engineering]   (1) Stimulates the reservoir using the fluidic oscillating principle. A vortex is formed inside the tool, which produces oscillating pulses on exiting. (2) If the oscillation of the system is started by pulling a weight to any side and released, the weight will oscillate with progressive damping. A switched electromagnet will be positioned near the oscillating weight. The electromagnet will be switched to a current source always at the moment when the iron weight approaches the electromagnet. The electromagnet will be energized with a short current impulse, and the induced field will pull the iron weight toward the electromagnet. This pulling force will act for only a short time and supply to the oscillating weight a small amount of energy, just enough to replace the losses by material and air resistance. The system will oscillate with sustained amplitude forever; supposing the coil to be supplied with electrical energy, this is the basic principle of most electrically excited oscillatory motions.‡

OSTEOGENIC INDUCTION PRINCIPLE [mathematics, physics]   See PINDUCTION PRINCIPLE.

OSTWALD'S DILUTION PRINCIPLE [chemistry]   Breakdown of a molecule into two molecules, atoms, radicals, or ions.§

OVERLAPPING BIPHASIC IMPULSE PRINCIPLE (OLBI) [physics]   Features simultaneous application of two pulses, with the same amplitude and pulsewidth but opposite polarity and with the pacemaker case as indifferent electrode.¶

OVERLOAD PRINCIPLE [kinesiology]   (Dr. Hans Selye)   See PPRINCIPLES OF TRAINING.

OVERSHOOT PRINCIPLE [engineering]   The operator needs to know that something has happened as soon as the knob is turned. The most likely result of violating this principle will be "overshoot." If feedback is not instantaneous, the operator is tempted to keep turning the knob until a change is observed.‖ See also PRINCIPLE OF INSTANT GRATIFICATION; PRINCIPLE OF TOTAL CONTROL.

OXYTOCIC PRINCIPLE [psychology]   An obsolete term for *oxytoxin*. A peptide hormone and neuromodulator with a range of physiological and psychological effects related to reproduction and social behavior.**

§Daintith, J. (1999), *The Facts on File Dictionary of Chemistry*, 3rd ed, Facts On File, New York, p. 82.
¶Çeliker, C., Yazioglu, N., and Ersanli, M. (1997), Clinical experience with single lead DDD stimulation using the overlapping biphasic impulse principle. (XIXth Congr. Eur. Soc. Cardiology, Stockholm, Aug. 24–28, 1997), *Eur Heart J.* **18** (Abstr. Suppl.), 423.
‖Liu, M. Z., Silvern, D. A., Gupte, P. M., Inchiosa, M. A., Jr. and Sanchala, V. (1992), Development of a real-time algorithm for predicting sufentanil plasma levels during cardiopulmonary-bypass surgery using a systems approach, *IEEE Trans. Biomed. Eng.* **39**(6), 658–661.
Gulland, J. M. and Newton, W. H. (1932), Oxytocic principle of the posterior lobe of the pituitary. I. *Biochem. J.* **26, 337–348.

*Shi, J. and Sun, H. H. (1991), Decomposition of nonlinear non-Gaussian process and its application to nonlinear filter and predictor design, *Ann. Biomed. Eng.* **19**(4), 457–472.
†Kukulin, V. I. and Pomerantsev, V. N. (1978), The orthogonal projection method in scattering theory, *Ann. Phys.* **111**(2), 330–363.
‡Ringrose, A., Farkas, R., Nicole, A., and Prost J.-L. (1995), *Automated Analytical Apparatus for Measuring Antigens or Antibodies in Biological Fluids*, US Patent 5,698,450.

P

PACKAGING PRINCIPLE [engineering] (1) A load greater than that the local a device is designed to handle. (2) Amount of sediment that exceeds ability of a stream to transport it and therefore is deposited.[*]

PAIRING PRINCIPLE [organizational psychology] A business process is naturally cyclical because it is based on the transfer of value. For example, a common sale involves the transfer of product in one direction and the transfer of remuneration in the other.[†]

PANOPTIMIZATION PRINCIPLE [mathematics] Simplification with paradoxes and fast sub-optimization with relative simplification and paradoxes, or regret compromises.[‡] See also PERFECTION PRINCIPLE; REALITY PRINCIPLE; (SIMPLIFICATION PRINCIPLE;) (STRENGTH PRINCIPLE;) RELATIVITY PRINCIPLE; COMBINATION PRINCIPLE; PARADOX-TOLERATING PRINCIPLE; EQUIVALENCE PRINCIPLE; SATISFACTION PRINCIPLE.

PANSYSTEMS PRINCIPLE [mathematics] (Wu Xuemou and Guo Dinghe) (1) Mathematical harmony of mechanical systems consisting of 10 principles: perfection, regret, panoptimization, reality, simplification, strength, relativity, combination, paradox seeking, and equivalence.[§] (2) Possess the nature of multilayer networklike and relatively mathematical and technical research with certain relative aesthetic conceptions. (3) Activities or work to develop the prototype materials and manuscripts or semifinished products into certain public promotion materials or products that satisfy certain social demands, such as books, magazines, newspapers, films, TV, electronic publishing materials, or other virtual realities.[¶] (4) The course of the development of humans is always full of contradictions between generalized supplies and demands. The course is also full of paradoxes and regrets to look for solving generalized contradictory equations.[‖] (5) Society is a large-scale dynamical system with supercomplicated nature and shengke, growth and restraint, Wu, X. and Guo, D. published by ideas International, USA 5A special form of panderivative is the so-called panvariation which is a unified form of many concepts of panderivatives in various branches of mathematics, including differential, variation, Frechet variation, Gateau variation, various derivatives, generalized variation and tangent space in functional analysis, etc. The concrete definition is as follows: Let $f,g : G \rightarrow F$, if $[f(x) - g(x)]/(x - x0)$ has definition and is convergent to zero (when $(x - x0) \rightarrow 0$), then f and g are considered to be tangent mutually at $x = x0$. For the given $Q < F \uparrow G = m|m : G \rightarrow F$, if $g \in Q$ and is tangent to f at $x = x0$, then g is called the panvariation or Q-panvariation of f at $x = x0$. It is not unique in general, For the common case, Q is confined to be linear, and g is confined to be unique. Usually we

[*]Wolffe, A. P. (1998), Packaging principle: How DNA methylation and histone acetylation control the transcriptional activity of chromatin, *J. Exp. Zool.* **282**(1–2), 239–244.

[†]Johnston, R. L. and Mingos, D. M. P. (1986), The pairing principle in tensor surface harmonic theory: Definition of a general class of N-atom polar deltahedra with N skeletal electron pairs, *Polyhedron* **5**(12), 2059–2061.

[‡]Xuemou, W. and Dinghe, G. (1999), Pansystems cybernetics: Framework, methodology and development, *Kybernetes* **28**(6/7), 679–694.

[§]Wu, X. and Guo, D. (n.d.), *Pansystems: Methodology and Relativity, from Confucius, Laozi, Descartes to Einstein*, Wuhan Digital Engineering Inst., China (http://www.aideas.com/_pan/00 000 022.htm).

[¶]Vaknin, S. (2000), *E-books and E-publishing*, Project Gutenberg.

[‖]Block, F. and Hirschhorn, L. (1979), New productive forces and the contradictions of contemporary capitalism: A post-industrial perspective, *Theory Society* **7**(3), 363–395.

use the symbol $\delta f(x, x - x)$ to represent $g(x) - f(x)$.* (7) Reveals typical laws of society operations and the leading relations and methodological thoughts of various doctrines or social techniques and skills. See also PANOPTIMIZATION PRINCIPLE; REALITY PRINCIPLE; (SIMPLIFICATION PRINCIPLE;) (STRENGTH PRINCIPLE;) RELATIVITY PRINCIPLE; COMBINATION PRINCIPLE; PARADOX-TOLERATING PRINCIPLE; EQUIVALENCE PRINCIPLE.

PANSYSTEMS RELATIVITY METHOD/PRINCIPLE

[mathematics]   Based on an understanding of a model of generalized contradictory equations and their problem solving that can be expanded as layer-tender method and competing-for tender method.* See also PERFECTION PRINCIPLE; PANOPTIMIZATION PRINCIPLE; REALITY PRINCIPLE; SIMPLIFICATION PRINCIPLE; STRENGTH PRINCIPLE; RELATIVITY PRINCIPLE; COMBINATION PRINCIPLE; PARADOX-TOLERATING PRINCIPLE; EQUIVALENCE PRINCIPLE.

PANTOGRAPH'S PARALLELOGRAM PRINCIPLE

[pharmacology]   Invented in 1603, precursor to the epidiascope (episcope). Uses the parallelogram principle with a grid construction.[†]

PAPERT'S PRINCIPLE

[psychology] (Seymour Papert, 1928–, American mathematician) (1) Phonemic blocks are toys that incorporate Papert's principles of constructionism to help a child gain phonemic awareness. Phonemic blocks are similar to alphabetic blocks, except each block is associated with a sound instead of a letter.[‡] (2) Some of the most crucial steps in mental growth are based not simply on acquiring new skills but also on acquiring new administrative ways to use what one already knows. This is used to explain the results of Jean Piaget's experiments on children's developmental stages.[§]

PAPPUS CENTROLD PRINCIPLE

See CAVALIERI PRINCIPLE.

PAPPUS' PRINCIPLE

[mathematics] (Pappus of Alexandria, 290–350AD, Greek mathematician) (Guldin, seventeenth century) The surface area and the volume respectively swept out by revolving a plane area about a nonintersecting coplanar axis is the product of the distance moved by itscentroid and either its perimeter or its area.[¶] See also PASCAL'S PRINCIPLE.

PARADOX PRINCIPLE

[psychology] (1) Process of physically locating connecting and protecting devices or components.[‖] (2) Understanding of our contradictory nature. (3) Dread of constrictive or expansive polarities promotes dysfunction, extremism, or polarization. (4) Confrontation with or integration of the poles promotes optimal living.[**]

PARADOX-TOLERATING PRINCIPLE

[mathematics]   Based on a sort of understanding of a model of generalized contradictory equations and their problem solving, which can be expanded as layer tender method[††] and competing-for tender method. See also PERFECTION PRINCIPLE; PANOPTIMIZATION PRINCIPLE; REALITY PRINCIPLE; SIMPLIFICATION PRINCIPLE; STRENGTH PRINCIPLE; RELATIVITY PRINCIPLE; COMBINATION PRINCIPLE; PARADOX-TOLERATE PRINCIPLE; EQUIVALENCE PRINCIPLE.

PARALLEL RESOLUTION PRINCIPLE

[psychology]   Five principles related to enhancing an organization or operation, including (1) positive change requires significant stability; (2) to build an enterprise, focus on the individual; (3) focus directly on culture, indirectly on

*Xuemou, W. and Dinghe, G. (1999), Pansystems cybernetics: Framework, methodology and development, *Kybernetes*, **28**(6/7), 679–694.
[†]Breed, C. B. and Hosmer, G. L. (1977), *The Principles and Practice of Surveying*, Wiley, New York.
[‡]Larson, Kevin (2000, August). The phoneme hammer: Using blocks to teach phonemic awareness. University of Texas at Austin.
[§]Minsky, M. (1986), Society of Mind. Whole Earth Review. p. 320.

[¶]Walker, P. M. B. (1999), *Chambers Dictionary of Science and Technology*, Chambers, New York, p. 833.
[‖]Schneider, K. J. (1990), *The Paradoxical Self: Toward an Understanding of Our Contradictory Nature*, Insight Books/Plenum Press, New York.
[**]Hattersley, R. (1987), A party of paradoxical principle. (Socialism a future?) *New Statesman* **113**, 4.
[††]Xuemou, W. and Dinghe, G. (1999), Pansystems cybernetics: Framework, methodology and development, *Kybernetes*, **28**(6/7), 679–694.

values, beliefs, climate, norms, symbols, and philosophy; (4) true empowerment requires forceful leaders; and (5) in order to build, you must first tear down.* See also PARADOX PRINCIPLE.

PARALLELOGRAM PRINCIPLE [engineering] (1) Genetic damage in the inaccessible human germ cells can be estimated by determining the effects on lymphocytes (or other somatic cells) from humans and mice and in germ cells of mice.[†] (2) Estimates can be obtained on the amount of genetic damage that cannot always be assessed directly. (3) Utilized to achieve an efficient selective angular disposition between a pair.[‡]

PARAMETRIC EKELAND VARIATIONAL PRINCIPLE See EKELAND'S PRINCIPLE.

PARAMETERIZED PROOF PRINCIPLE [mathematics] The proof principle by induction states for verifying 8x : P (x) for some property P it suffices to show 8x : [[8y ! x : P (y)] =) P (x)], provided ! is a well-founded partial ordering on the domain of interest. A more general formulation of this proof principle allows for a kind of parameterized partial ordering ! x which naturally arises in some cases. Conditions under which the parameterized proof principle 8x : [[8y ! x x : P (y)] =) P (x)] is sound in the sense that 8x : [[8y ! x x : P (y)] =) P (x)] =) 8x : P (x) holds, and provides counterexamples demonstrating these conditions are essential.[§]

PARAMETERIZED PROOF PRINCIPLE [mathematics, physics] See also INDUCTION PRINCIPLE.

*Nadler, D. A. and Tushman, M. L. (1987), Organizational frame bending: Principles for managing reorientation, *Acad. Manage. Executive*, **3**(3), 194–204.
[†]Sobels, F. H. (1993), Approaches to assessing genetic risks from exposure to chemicals, *Environ. Health Perspect.* 101 (Suppl. 3: *Environ. Mutagen. Human Popul. at Risk*), 327–332.
[‡]Tannehill, J. K., Jr. (n.d.), Weighing accuracy fundamentals for cement production belt feeders, *Asia Cement Magazine*. (Merrick Industries, Inc.).
[§]Gramlich, B. (1995), A note on a parameterized version of the well-founded induction principle. Bulletin of the EATCS 52, pp. 274–278, 1994.

PARETO PRINCIPLE [mathematics, economics] (Vilfredo Federico Damaso Pareto, 1848–1923, Italian sociologist, economist philosopher (1) States that 80% of the problems come from 20% of the causes. In healthcare organizations, for example, it is very likely only 20% (or fewer) of the members use 80% (or more) of the resources. One way to control costs is by analyzing the services provided to those 20% who consume 80% of the resources.[¶] See also 80/20 PRINCIPLE; PRINCIPLE OF FACTOR SPARSITY; EDGEWORTH–PARETO PRINCIPLE.

PARETO PRINCIPLE OF OPTIMAL DESIGN OF EXPERIMENTS [mathematics] (Vilfredo Federico Damaso Pareto, 1848–1923, Italian sociologist, economist, philosopher) Design experiments that give the maximum possible information. Optimization models arise as a result of application of one or more of these principles. New and challenging mathematical problems arise in obtaining the results from optimizing models.[‖] See also ENTROPY OPTIMIZATION PRINCIPLE.

PARIS PRINCIPLE [psychology] Relating to the status and functioning of national institutions for protection and promotion of human rights. A comprehensive series of recommendations on the role, composition, status, and functions of national human rights instruments.[**]

PARITY PRINCIPLE [logic] The parity of F = the parity of G iff F and G differ evenly where two concepts differ evenly if an even number of things fall under one but not the other. As Boolos showed, this principle is satisfiable, but only in finite domains.[††]

[¶]Koch, R. (2001), *The 80/20 Principle: The Secret of Achieving More with Less*, Nicholas Brealey Publishers, London.
[‖]Kapur, J. N. (1988), Mathematical modeling, *New Age Int.* p. 203.
[**]Anon. (1993) (http://www.ihrc.ie/legal_documents/11_paris.asp), Irish Human Rights Commission, IHRC A/RES/48/134, Dec. 20, 1993.
[††]Boolos, G. (2000) Logic, Logic, and Logic. History and Philosophy of Logic, **21**(3), 223–229.

PARR–PEARSON PRINCIPLE [mathematics, chemistry] (Robert Ghormley Parr, 1921–, American Chemist) (Ralph Pearson, American Chemist) (1) Application of mathematics to economic analysis.* (2) Molecules arrange their structure to have the maximum hardness.[†]

PARSIMONY PRINCIPLE [mathematics] Frequently used in system identification on more or less heuristical grounds. Used when estimating the parameters of regression models by the least-squares method.[‡] See PRINCIPLE OF PARSIMONY.

PART–CURRENT PRINCIPLE [engineering] A tube and valve vertical to the center of the measuring tubes, where it travels in both directions to expansion chambers, each with an exit to the atmosphere. A measuring light beam is directed through the measuring tubes and the chambers to diaphragm openings. The use of a $90°$ connection from the exhaust to two measuring tubes and expansion chambers ensures accuracy irrespective of the rate at which the gas is being admitted.[§]

PARTIAL LIKELIHOOD PRINCIPLE Consider two experiments with identical (ψ, θ, ω):

$$E_1 = \{Y_1, (\psi, \theta, \omega), P_1\};$$
$$P_1 = \{f^{(1)}_{\theta, \omega}(y_1, \psi), \theta, \omega \in \Theta\};$$
$$l^{(1)}_{y_1}(\lambda, \theta, \omega) = f^{(1)}_{\theta, \omega}(y_1, \lambda)$$

and

$$E_2 = \{Y_2, (\psi, \theta, \omega), P_2\};$$
$$P_2 = \{f^{(2)}_{\theta, \omega}(y_2, \psi), \theta, \omega \in \Theta\};$$
$$l^{(2)}_{y_2}(\lambda, \theta, \omega) = f^{(2)}_{\theta, \omega}(y_2, \lambda).$$

Assume that (15) holds for both experiments; that is, $l^{(1)}_{y_1}(\lambda, \theta, \omega) = l^{(1)}_1(\lambda, \theta)l^{(1)}_2(\omega)$ and $l^{(2)}_{y_2}(\lambda, \theta, \omega) = l^{(2)}_1(\lambda, \theta)l^{(2)}_2(\omega)$. In addition, assume (a) $\lambda(\psi, y_1) = \lambda(\psi, y_2) = \lambda$, for all ψ, and (b) $l^{(1)}_1(\lambda, \theta) = cl^{(2)}_1(\lambda, \theta)$; c independent of (λ, θ). Then $I_{\lambda, \theta}(E_1, y_1) = I_{\lambda, \theta}(E_2, y_2)$.

In the case where λ does not exist, (a) is superfluous and the PLP says that $I_\theta(E_1, y_1) = I_\theta(E_2, y_2)$ if $l^{(1)}_1(\theta) = cl^{(2)}_1(\theta)$. Next, consider partial sufficiency for (λ, θ), defined as follows.[¶]

PASCAL'S PRINCIPLE [physics] (Blaise Pascal, 1623–1662, French philosopher and mathematician) (1) Also known as *Pascal's law*; pressure exerted at any point on a confined liquid is transmitted undiminished in all directions. (2) For a fluid at rest in a closed container a pressure change in one part is transmitted without loss to every portion of the fluid and to the walls of the container.[‖] See also PAPPUS' PRINCIPLE.

PASCHEN'S PRINCIPLE [physics] (Friedrich Paschen, 1865–1947, German physicist) Observation that the breakdown voltage, at constant temperature, is a function only of the product of the gas pressure and the distance between parallel-plate electrodes.[**]

PATER PRINCIPLE [psychology] Focuses on father–son relationships. Role of fathers in raising of children; effects of paternal love; participation in hands-on care of infants and toddlers.[††]

*Szaleniec, M., Tadeusiewicz, R., and Witko, M. (2008), How to select an optimal neural model of chemical reactivity? *Neurocomput. Arch.* **72**(1-3), 241–256.

[†]Arnett, E. M. and Ludwig, R. T. (1995), On the relevance of the Parr-Pearson principle of absolute hardness to organic chemistry, *J. Am. Chem. Soc.* **117**(24), 6627–6628.

[‡]Stoica, P. and Söderström, T. (1982), On the parsimony principle, *Int. J. Control* **36**(3), 409–418.

[§]Mai, H. (2006), *Device for Measuring the Turbidity of Smoke*, Eur. Patent EP0762113.

[¶]On the Generalization of the Likelihood Function and the Likelihood Principle Jan F. Bjornstad. *Journal of the American Statistical Association*, Vol. 91, No. 434 (Jun., 1996), pp. 791–806. Published by: American Statistical Association p. 801.

[‖]Bloomfield, L. (2006), *How Things Work: The Physics of Everyday Life*, 3rd ed., Wiley, Hoboken, NJ, p. 153.

[**]Walker, P. M. B. (1999), *Chambers Dictionary of Science and Technology*, Chambers, New York, p. 840.

[††]Segell, M. (1995), The pater principle: Fathers and sons face a lifetime of unfinished business, *Esquire*, **123**, 121.

PATERNOSTER'S PRINCIPLE [physics] Pressure applied to a fluid is transmitted undiminished throughout the fluid. Pressure applied to an enclosed fluid is transmitted undiminished to every point in the fluid and to the walls of the container.*

PATH PRINCIPLE [engineering] See also AS-SEMBLY SEQUENCE PRINCIPLE.

PATIENCE PRINCIPLE [mathematics] Patience is always a virtue, but in trading it's a requirement. Despite the fact that traders are often thought to be constantly jumping in and out of stocks, the real money is made by holding on for the big moves. In trading as in life, good things come to those who wait.[†]

PAULI ANTISYMMETRY AND EXCLUSION PRINCIPLE [mathematics] (Wolfgang Ernst Paul, 1900–1958, Austrian physicist) Designed to handle unit loads continuously with a loading–unloading facility at each level.[‡]

PAULI EXCLUSION PRINCIPLE [physics] (Wolfgang Ernst Pauli, 1900–1958, Austrian physicist) (1) Explains the electronic structure of atoms and also the general nature of the periodic table.[§] (2) Applications considering probabilities of occurrences from randon or nonrandom events such as molecular reactions, viruses, genetic selection, and plant propagation.[¶] (3) One species seeking same ecological environment will survive while the other will expire under a given set of conditions. If there is competition between two species for a common place in a limited microcosm, we can quite naturally extend the premises implied in the logistic equation. The growth rate of each competing species in a mixed population will depend on (a) the potential rate of population increase of a given species and (b) the unutilized opportunity for growth of this species.[∥] (4) Electrons are fermions; they cannot occupy the same spatial and energy states simultaneously. Two electrons in an atom cannot have identical quantum numbers; thus, if there are two electrons in the same orbital, their spin quantum numbers must be of opposite sign.[**] See also EXCLUSION PRINCIPLE.

PAULI PRINCIPLE [chemistry] (Wolfgang Ernst Pauli, 1900–1958, Austrian physicist) The Wolfgang Pauli principle of exclusion requiring that, even at a temperature close to absolute zero, all the electrons cannot have zero kinetic energy.[††] (2) Only one electron can occupy each electron orbital, inclusive of spin state. No two electrons in a multielectron atom can have the same set of quantum numbers (n,l,mi,mg); that is, no two electrons can be in the same quantum state.[‡‡] See also U(2) PRINCIPLE.

PAULI-FERM PRINCIPLE [chemistry] (Wolfgang Ernst Pauli, 1900–1958, Austrian physicist) (1) Chemical behavior of atoms depends on the shells of the more loosely bound electrons.[§§] (2) Each electron is required to go into a different orbital. With this procedure, each element acquires a unique structure of electron orbitals that gives it its characteristic atomic and chemical

---

*Klaus, K. (1972), *Multi-Level Parking Apparatus*, US Patent 3,866,766.

[†]Pring, M. J. (1995), *Investment Psychology Explained: Classic Strategies to Beat the Markets*, J Wiley, New York, p. 83.

[‡]Bohm, M. C. and Schutt, J. (1997), On the role of the Pauli antisymmetry principle in pericyclic reactions, *Z. Naturforsch. A* **52**, 727.

[§]Walker, P. M. B. (1999), *Chambers Dictionary of Science and Technology*, Chambers, New York, p. 842.

[¶]Peterson, T. (1990), Testing the Pauli exclusion principle, *Sci. News* **137**, 287.

[∥]Kinoshita, J. (1988), Roll over, Wolfgang? New experiments seek violations of the Pauli exclusion principle, *Sci. Am.* **258**, 25.

[**]*CRC Handbook of Chemistry and Physics* (2007), 88th ed., CRC Press, Cleveland, OH.

[††]Wheeler, A. (2007), Novel Laureates in Physics. Stanford Linear Accelerator Center Library (http://www.slac.stanford.edu/library/nobel).

[‡‡]Aluru, N. R., Leburton, J.-P., McMahon, W., Ravaioli, U., Rotkin, S. V., Staedele, M., Trudy van der Straaten, T., Tuttle, B. R., and Hess, K. (n.d.), *Modeling Electronics at the Nanoscale Handbook of Nanoscience, Engineering and Technology*, 2nd ed. CRC Press.

[§§]Evans, R. D. (1955), *The Atomic Nucleus*, MIT/McGraw Hill (http://www.archive.org/stream/atomicnucleus032805mbp/atomicnucleus032805mbp_djvu.txt).

properties. "A single atomic electron cannot have the same four quantum numbers." This principle rules not only the chemistry but also the physics of the nucleus and of the more fundamental particles, of which it is composed.[*]

PAVLOV'S PRINCIPLE [chemistry]   (Juan Petrovich Pavlov, 1849–1936, Russian psychologist) Each level of a quantized system can include one, two, or more electrons. If there are two electrons, they must have spins in opposite directions.[†]

PEARSON PRINCIPLE [psychology] From behaviorist psychology, when there is a stimulus, even unrelated to the activity, there is a response. The stimulus can be changed and thus, the response can be changed.[‡]

PEARSON–PENAZ   PRINCIPLE [chemistry] See HARD–SOFT ACID–BASE PRINCIPLE.

PEELIAN PRINCIPLE [psychology] (Sir Robert Peel, 1788–1850, Prime minister, United Kingdom) (1) Every police officer should be issued a badge number, to ensure accountability for her/his actions. (2) Whether the police are effective is not measured on the number of arrests, but on the lack of crime. (3) Above all else, an effective authority figure knows that trust and accountability are paramount.[§] See also POLICE PRINCIPLES.

PENAZ PRINCIPLE [chemistry]   See HARD–SOFT ACID–BASE PRINCIPLE.

[*]Demortier, G. (2003), *The Nobel Laureates in Physics (1901–2001). A Century of Nobel Prizes Recipients: Chemistry, Physics and Medicine*, Marcel Dekker, p. 43.
[†]Zhang, B., Dai., Z. G., Mészáros, P., Waxman, E., and Harding, A. K. (2003), High-energy neutrinos from magnetars, *Astrophys. J.* **595**(1), 346–351.
[‡]Jones, L. M., Fontanini, A., Sadacca, B. F., Miller, P., and Katz, D. B. (2007), *Natural Stimuli Evoke Dynamic Sequences of States in Sensory Cortical Ensembles*, Natl. Acad. Science   (http://www.pnas.org/content/104/47/18 772.full?ck=nck).
[§]Lentz, S. A. and Chaires, R. H. (2007), The invention of Peel's principles: A study of policing "textbook" history, *J. Crim. Justice* **35**, 69–79.

PENDULUM   PRINCIPLE [mathematics] (Galileo) Experiments with the pendulum in connection with natural accelerated motion.[¶] See also COMMENSURABILITY PRINCIPLE; CONTINUUM PRINCIPLE; PERFORMATIVITY PRINCIPLE; POSITIONALITY PRINCIPLE; PUNCTUATION PRINCIPLE; UNCERTAINTY PRINCIPLE; SUSTAINABILITY PRINCIPLE; SYNERGY PRINCIPLE.

PENG'S   PRINCIPLE [mathematics]   (Peng Shige, 1947–, Chinese mathematician) Modern control theory devoted to obtaining the stochastic control system., Introduction of "second-order dual" approach to solve the optimal control of stochastic systems. The maximum principle results show the form of random people expected a long period of time is different from the case one more than the uncertainty "second-order items," which shows the random nature of the differences and uncertainties. Introduces backward stochastic differential equations. Compared to the classical stochastic differential equation, backward stochastic differential equations of awareness, understanding and research methods are essentially different.[‖] See also MAXIMUM VALUE PRINCIPLE.

PENNY PRINCIPLE [education] Elementary consumer education program designed for infusion into regular subject matter teaching. Presents goals and starter activities from each project's eight topic areas: money management, income determination, basic economics, advertising, purchasing, energy, insurance, and law.[**]

PERCOLATION PRINCIPLE [chemistry] Percolation extraction is used in oilseed extraction and in the extraction of alkaloids. Immersion extraction is the ability to handle finely ground material. Immersion extraction may be used where the percolation rate of

[¶]Coy, P. (1993), Spain's tilt train is a low tech oldie that keeps on rollin' (train uses pendulum principle and springs to tilt), *Business Week* **3323**, 89.
[‖]Shige, P. (2008), China Industrial and Applied Mathematics Institute of Communications, *CSIAM News* n(8) (http://www.docin.com/p-180321.html).
[**]Farnsworth, B. J. and Dunoskovic, J. H. (1980), The penny principle, *Instructor* **90**(3), 114.

the material to be extracted is too great for effective diffusion.* Also known as *immersion principle*.

PERFECT COSMOLOGICAL PRINCIPLE [physics] (Herman Bondi and Thomas Gold, 1948) A cosmological hypothesis. The universe is on average homogenous and isotropic, as well as constant in time. The universe observed from every point, in every direction, and at every time looks roughly the same. Or using other words; the universe is (roughly) homogeneous in space and time and isotropic in space.[†] Also known as the *strong cosmological principle; generalized Copernican cosmological principle*.

PERFECTION PRINCIPLE [mathematics, psychology] (1) Overall pursuits and considerations of multigoals of the true–good–beautiful–zen. Synthetic panoptimization with overall considerations of the macrocostic and the microcostic, the whole and the part, the far and the near, the vertical and the horizontal, and overall considerations of multiple-index-synthetic synergy, optimization, development. (2) Corresponds to various ideals, idealism, utopias, gentleman-countries, lands of peach blossoms, and fictitious lands of peace away from the turmoil of the world. (3) Presents an ideal limit or tendency that helps people organize a programming or movement and realize a certain relatively rational course.[‡] See also PANSYSTEMS OPERATION EPITOPANSYSTEMS PRINCIPLE; PANOPTIMIZATION PRINCIPLE; REALITY PRINCIPLE; SIMPLIFICATION PRINCIPLE; STRENGTH PRINCIPLE; RELATIVITY PRINCIPLE; COMBINATION PRINCIPLE; PARADOX-TOLERATING PRINCIPLE; EQUIVALENCE PRINCIPLE.

PERFORMATIVITY PRINCIPLE [psychology] Intercultural communication is a reiterative process whereby people from different cultures enact meanings in order to accomplish their tasks.[§] See also PRINCIPLE OF INTERCULTURAL COMMUNICATION; PUNCTUATION PRINCIPLE; POSITIONALITY PRINCIPLE; COMMENSURABILITY PRINCIPLE; CONTINUUM PRINCIPLE; SYNERGY PRINCIPLE; SUSTAINABILITY PRINCIPLE.

PERISTALTIC PRINCIPLE [engineering] Method of pumping fluid.

PERMACULTURE PRINCIPLES [ecology] (Bruce Charles 'Bill' Mollison, 1928-, Australian naturalist and David Holmgren, 1955-, Australian naturalist) An approach to designing human settlements and agricultural systems that mimic the relationships found in natural ecologies. By rapidly training individuals in a core set of design principles, those individuals can design their own environments and build increasingly self-sufficient human settlements—ones that reduce society's reliance on industrial systems of production and distribution.[¶] See also HANNOVER PRINCIPLES; FOREST STEWARDSHIP COUNCIL (FSC) PRINCIPLES; GLOBAL SULLIVAN PRINCIPLES OF SOCIAL RESPONSIBILITY; MARINE STEWARDSHIP COUNCIL (MSC) PRINCIPLES; THE BELLAGIO PRINCIPLES FOR ASSESSMENT; EQUATOR PRINCIPLES; MELBOURNE PRINCIPLES; PRECAUTIONARY PRINCIPLE; SANBORN PRINCIPLES; TODDS' PRINCIPLES OF ECOLOGICAL DESIGN.

PERMANENCE OF MATHEMATICAL RELATIONS PRINCIPLE [mathematics] See CONTINUITY PRINCIPLE.

PERMUTATION PRINCIPLE [mathematics] (Fisher) (1) Essential as an axiom in the original system of quantificational logic.[‖]

*Löser, C., Zehnsdorf, A., Görsch, K. and Seidel, H. (2005), Bioleaching of heavy metal polluted sediment: Kinetics of leaching and microbial sulfur oxidation, *Eng. Life Sci.* **5**(6), 535–549.

[†]Rudnicki, K. (n.d.), The perfect cosmological principle, *Southern Cross Rev.* **52**. (http://southerncrossreview.org/52/rudnicki-5.htm).

[‡]Belson, A. A. (2001), The perfection principle: are you too good for your own good?, *Mademoiselle*, **90**, 310.

[§]McKenzie, J. (2001), *Perform or Else: From Discipline to Performance*, Routledge, p. 164.

[¶]Holmgren, D. (2002), Permaculture: Principles and Pathways Beyond Sustainability. Holmgren Design Services.

[‖]Quine, W., (1940), *Mathematical Logic*, Norton, New York.

(2) Applied test in clinical trials with a binary outcome variable (e.g., success/failure).[*]

PERSONA PRINCIPLE [psychology]   Contains eight major principles and 88 rules to build a successful business by expanding knowledge of image marketing, advertising, and branding.[†]

PERTURBATION INVARIANCE PRINCIPLE [physics, mathematics, chemisty]   See INVARIANCE PRINCIPLE

PETER PRINCIPLE [psychology]   (Laurence Johnston Peter, 1919–1990, Canadian educator) (Raymond Hull, 1919–1985, Canadian playwright)   In a hierarchy every employee tends to rise to her/his level of incompetence.[‡] (2) Rearranges a finite number of symbols; a one-to-one function of a finite set onto itself.

PETERBILT PRINCIPLE [education]   (T. A. Peterman, American logger)   Interactive software that captivates children's minds makes learning a breeze, not a chore.[§]

PHASE ANGLE PRINCIPLE [mathematics, psychology]   Simple and economical technique for measuring voltage phase angles and generator shaft angles.[¶]

PHASE CONTRAST PRINCIPLE [physics] (Frits Zernike, 1888–1966, Univ. of Groningen) Ability to discern materials of different refractive indices despite their transparency. Provides a linear relation between phase retardation and image intensity.[‖]

PHASE MATCHING PRINCIPLE [electronics, computer science, psychology] See MATCHING PRINCIPLE.

PHASE SEPARATION PRINCIPLE Generally described with the following steps: (a) mixing additives and heating to make a hot homogenous solution, (b) extruding the hot solution through a sheet die into a gel-like film, (c) extracting mineral oil and other additives to form microporous structures.[**]

PHASED CAPTURE PRINCIPLE [medicine] Molecules supporting solid-phase capture and molecules mediating the subsequent detection are located on different strands of nucleic acides.[††] See also CAPTURE PRINCIPLE.

PHONETIC PRINCIPLE [physics]   Difference between the phase of a sinusoidally varying quantity and the phase of a second quantity that varies sinusoidally at the same frequency.[‡‡]

PHOTOELECTRIC BARRIER PRINCIPLE [physics]   See DOPPLER PRINCIPLE.

PHRAGMÉN–LINDELÖF PRINCIPLE [mathematics] (Lars Edvard Phragmén, 1863–1937, and Ernst Leonard Lindelöf, 1870–1946) Qualitative properties of solutions; general behavior of solutions of PDE comparison theorems. Every entire function of order zero is unbounded along any ray approaching one.[§§] See also LINDELÖF PRINCIPLE.

[*]Fine, K. (1983), The permutation principle in quantificational logic, *J. Phil. Logic* **12**(1), 33–37.
[†]Armstrong, D. L. and Yu, K. W. (1997), *The Persona Principle: How to Succeed in Business with Image Marketing*, Simon & Schuster, p.12.
[‡]Peter, L. and Hull, R. (1969). *The Peter Principle*, Morrow, Oxford.
[§]Stanton, D. (1988), The Peterbilt principle (interactive software for children), *Compute* **6**, 84.
[¶]Fink, D. G. and Beaty, H. W. (1978), *Standard Handbook for Electrical Engineers*, McGraw-Hill.
[‖]Zernike, F. (1942), Phase-contrast, a new method for microscopic observation of transparent objects. Part I, *Physica* **9**, 686–698.

[**]Zhang, S. S. (R) Li-Ion Battery Separator. Advanced Materials and Methods for Lithium-Ion Batteries. Reserach Signpost. India. v37/661 (2).
[††]Foettingera, A., Leitnera, A., and Lindner, W. (2005), Solid-phase capture and release of arginine peptides by selective tagging and boronate affinity chromatography, *J. Chromatogr. A* **1079**(1-2), 187–196.
[‡‡]Kott, L. (1982), The McCarthy's recursion induction principle: "Oldy" but "goody," *Calcolo* **19**(1), 59–69.
[§§]Phragmén, L. E. and Lindelöf, E. (1908), Sur une extension d'un principe classique de l'analyse et sur quelques propriétés des fonctions monogènes dans le voisinage d'un point singulier, *Acta Math.* **31**(1), 381–406.

PHYSICAL PRINCIPLE OF COMPUTATION [physics] Computers exist in nature by the simple observation of the sky. Moreover, the standard model—the Turing machine—can be described in such a way that it resembles the old astronomical observations in the sky by the ancients.* See also CONTAINMENT PRINCIPLE; PRINCIPLE OF HYPERCOMPUTATION UPPER BOUNDS; GALILEO'S PRINCIPLE OF NATURAL COMPUTATION; GANDALF'S PRINCIPLE OF HIDDEN UNIVERSE COMPUTATION.

PHYSIOCHEMICAL PRINCIPLE [chemistry] Grease and water do not mix. Computational biology can interpret this behavior in terms of capacity of working with large volumes of data embodying a high degree of both similarity and complexity. The complexity of interactions between molecular components of biological systems is such that only a systematic framing of information to underlying principles will enable humans to harness the potential of genomic information. The task facing the scientific community is the systematic integration of biomolecular interactions into a framework capable of making predictions for the purposes of designing biological systems. On balance, this means simply that the deflection or movement can be calibrated to read in terms of mass.† Also known as *physicochemical principle*.

PICARD PRINCIPLE [mathematics] (Charles Emile Picard, 1856–1941, French Mathematician) Key tool in obtaining many results in classical potential theory. Principle of positive singularity, originally obtained by Picard for $m = 2$, is reformulated by Bouligand under the name *Picard principle* as follows. The dimension of the half-module of nonnegative harmonic functions on B with vanishing boundary values zero on $|x| = 1$ is

one. As a result, the Riemann theorem of removable singularities follows, the weak form of which states that the boundedness of $u(x)$ on B yields the existence of $\lim x_o u(x)$. Conversely, the Picard principle can be derived from the Riemann theorem as the original proof of Picard as the Picard principle suggests, and therefore the Picard principle and the Riemann theorem are equivalent in essence. If two nonnegative functions harmonic with respect to a symmetric stable levy process vanish continuously outside a Lipschitz domain, near a part of its boundary, then the ratio of the function is bounded inside the domain, near this part of the boundary.‡

PIGEONHOLE PRINCIPLE [mathematics, engineering] (1) Used for estimating probability density function under specified moment constraints. If m pigeons are put into m pigeonholes, there is an empty hole if and only if there's a hole with more than one pigeon. If $n > m$ pigeons are put into m pigeonholes, there's a hole with more than one pigeon. If n objects are put into p pigeonholes, where $1 = p < n$, then some pigeonhole must contain at least two objects. (2) If a very large set of elements is partitioned into a small number of blocks, then at least one block contains a rather large number of elements. (3) Let $|A|$ denote the number of elements in a finite set A. For two finite sets A and B, there exists a 1–1 (one-to-one) correspondence $f : A-> B$ iff $|A| = |B|$.§ (4) Measure of absence of information, uncertainty, or transformation about a situation or space. Entropy model for deriving the probability distribution of the equilibrium state. Expresses the amount of disorder inherent or produced. Measures the disorder of a system equal to the Boltzmann constant.¶ (5) Famous fold/unfold method to give proof of two properties—associatively of the append operation between lists and

---

*Costa, J. F. (2006), *Physics and Computation: Essay on the Unity of Science through Computation*, Elsevier (http://cmaf.ptmat.fc.ul.pt/preprints/pdf/2006/unity.pdf).
†Yang, Y. (2004), *Physicochemical Principle to Identify Spherical Symmetries in the Genetic Code*, The 35th Meeting of the Division of Atomic, Molecular and Optical Physics, May 25–29, 2004, Tuscon, AZ Depts. Neurochemistry, Physical Organic Chemistry, Nankai Univ., China.

‡Nakai, M. (1976), Picard principle and Riemann theorem, *Tôhoku Math. J.* **2**(28), 277–292.
§Grimaldi, R. P. (1998), *Discrete and Combinatorial Mathematics: An Applied Introduction*, 4th ed., pp. 244–248.
¶Miller, J., Flor, P., Berg, G., and Cabillón, J. G. (2002), Pigeonhole principle, in *Earliest Known Uses of Some of the Words of Mathematics*, Miller, J., ed.

idempotence of the reverse operation.[*] See also INFERENCE PRINCIPLE; PROOF PRINCIPLE; NONCONSTRUCTIVE PRINCIPLE; PARAMETERIZED PROOF PRINCIPLE; STRUCTURAL LINEARIZATION PRINCIPLE; DIRICHLET'S BOXING-IN PRINCIPLE; DIRICHLET'S BOX PRINCIPLE.

PIGEONHOLE PRINCIPLE [mathematics] (aka Dirichlet's Box Principle) A generalized version of this principle states that, if n discrete objects are to be allocated to m containers, then at least one container must hold no fewer than $\lceil n/m \rceil$ objects, where $[\dots]$ denotes the ceiling function.

Dirichlet's Theorem: Let α be an irrational number. Then there are infinitely many integer pairs (h, k) where $k > 0$ such that[†]

$$\left| \alpha - \frac{h}{k} \right| < \frac{1}{k^2}.$$

Also known as *Dirichlet's principle*; *drawer principle*.

PIGOU–DALTON TRANSFER PRINCIPLE [mathematics] (Arthur Cecil Pigou, 1877–1959, English economist) (Edward Hugh Neale Dalton, 1887–1962, British Politician) See TRANSFER PRINCIPLE.

PILATES PRINCIPLE [kinesiology] (Joseph Hubertus Pilates, 1880–1967) (1) Based on a well-constructed philosophical and theoretical foundation (one without mysticism or appeals to unseen forces, divine or otherwise). It is not merely a collection of exercises but a method, developed and refined over more than 80 years' of use and observation.[‡] (2) While Pilates draws from many diverse exercise styles, running the gamut from yoga to Greek ideals to Chinese acrobatics, there are certain inherent ruling principles that bring all these elements together under the Pilates name, namely,

centering, concentration, control, precision, breathing, and flowing movement.[§]

PINCH PRINCIPLE [chemistry] Composite curves Figs. 1–3 determine the target for maximum heat recovery and minimum hot and cold utility. The point of closest approach of the composite curves (where there is DT_{min}) limits their maximum heat recovery potential. This point of closest approach is known as the *heat recovery pinch*. However, not only does the pinch limit the heat recovery potential between the composite curves, it also is the key to achieving targets set by the composite curves in design.[¶]

PIPELINE PRINCIPLE [computer science, mathematics] (1) Without internal registers, the input signals pass operators before a result is available and before a new input signal can be entered. Hence the critical path delay is large resulting in both a low clock rate and maximal achievable sample rate. As a result, pipelining is extremely difficult or even impossible when recursive bottlenecks are present where no additional sample delays can be introduced. As the extent of pipelining continues to be raised, the extra registers will eventually increase the area and power consumption more than what can be motivated by the gain in clock speed and maximal sample rate.[‖] (2) Space can be exchanged for time during the architectural exploration by a number of techniques. This can happen by a sequential treatment of either the bits within a word or the words in an algorithm.[**] (3) The signal-processing

[*]Ajtai, M. (1989), Parity and the pigeonhole principle. Feasible mathematics, *Progr. Comput. Sci. Appl. Logic*, **9** 1–24.

[†]Anon. (n.d.), (http://www.math.wm.edu/~shij/math410-problem-solving/pigeon-hole.pdf).

[‡]Isacowitz, R. (n.d.), Manual completo del método Pilates, *Editorial Paidotribo* (http://www.libreriadeportiva.com/images/portadas/978-84-8019-137-1_indice.pdf).

[§]Marshall, E. (2009), *The Pilates Principles Six Basic Concepts that Guide Classical Contrology Exercises* (http://pilates.suite101.com/article.cfm/the_pilates_principles).

[¶]Sunggyu, L. (n.d.), Pinch design and analysis, Robin Smith, in *Encyclopedia of Chemical Processing*, Smith, R. and Kim, J.-K., eds., **4**, p. 2168. CRC Press.

[‖]Niethammer, W. (1989), The SOR method on parallel computers, *Numerische Mathematik* **56**(2-3), 247–254.

[**]Catthoor, F. (2000), Integrated Processor-level architectures for real-time digital signal processing. Chapter 3, p10–34 in Lessenen literatuer-Lectures and reading. Ingarid Verbauwhede. http://homes.esat.kuleuven.be/n/verbauw.

technique can be used to estimate crital overhead delays of communication, storage, and control costs.

PITZER CORRESPONDING STATES PRINCIPLE [physics, chemistry] (Kenneth Sanborn Pitzer, b. 1/6/1914, Pomona CA or Ruell Moher Pitzer, b. 5/10/1938, Berkeley, CA son of Kenneth) See CORRESPONDING STATES PRINCIPLE.

PIU PRINCIPLE 1 [psychology] A *balance with nature*—emphasizes the distinction between utilizing resources and exploiting them. It focuses on a threshold beyond which deforestation, soil erosion, aquifer depletion, silting, and flooding reinforce one another in urban development, destroying life-support systems. Promotes environmental assessments to identify fragile zones, threatened natural systems, and habitats that can be enhanced through conservation, density control, land use, and open-space planning.* See also PRINCIPLES OF INTELLIGENT URBANISM.

PIU PRINCIPLE 2 [psychology] A *balance with tradition*—integrates plan interventions with existing cultural assets, respecting traditional practices and precedents of style. Intelligent urbanism respects the cultural heritage of a place. It seeks out traditional wisdom in the layout of human settlements, in the order of building plans, in the precedents of style, and in the symbols and signs, which transfer meanings through decorations and motifs. Intelligent urbanism respects the order engendered into building systems through years of adaptation to climate, to social circumstances, to available materials, and to technology. It promotes architectural styles and motifs, which communicate cultural values.* See also PRINCIPLES OF INTELLIGENT URBANISM.

PIU PRINCIPLE 3 [psychology] *Appropriate technology*—promotes building materials, techniques, infrastructural systems, and construction management consistent with people's capacities, geo-climatic conditions,

*Benninger, C. (2001), Principle of intelligent urbanism, *Ekistics* **69**(412), 39–65.

local resources, and suitable capital investments. Accountability and transparency are enhanced by overlaying the physical spread of urban utilities and services on electoral constituent areas, such that people's representatives are interlinked with technical systems. Appropriate technology promotes building materials, techniques, infrastructural systems, and construction management consistent with people's capacities, geoclimatic conditions, local resources, and suitable capital investments.* See also PRINCIPLES OF INTELLIGENT URBANISM.

PIU PRINCIPLE 4 [psychology] *Conviviality*—sponsors social interaction through public domains, in a hierarchy of places, devised for personal solace, companionship, romance, domesticity, neighborliness, community, and civic life. Vibrant societies are interactive, are socially engaging, and offer their members numerous opportunities for gathering.* See also PRINCIPLES OF INTELLIGENT URBANISM.

PIU PRINCIPLE 5 [psychology] *Efficiency*—promotes a balance between the consumption of resources such as energy, time, and finance, with planned achievements in comfort, safety, security, access, tenure, and hygiene. It encourages optimum sharing of land, roads, facilities, services, and infrastructural networks reducing per household costs, while increasing affordability and civic viability.*. See also PRINCIPLES OF INTELLIGENT URBANISM.

PIU PRINCIPLE 6 [psychology] *Human Scale*—encourages ground-level, pedestrian-oriented urban arrangements, based on anthropometric dimensions, as opposed to Amachine-scales.= Walkable, mixed-use urban villages are encouraged, over single-functional blocks, linked by motorways and surrounded by parking lots.* See also PRINCIPLES OF INTELLIGENT URBANISM.

PIU PRINCIPLE 7 [psychology] *Opportunity matrix*—enriches the city as a vehicle for personal, social, and economic development, through access to a range of organizations, services, and facilities, providing a variety of opportunities for education, recreation, employment, business, mobility, shelter, health, safety, and basic needs. The city is an engine of economic growth. Cities are places where individuals can increase their knowledge, skills, and sensitivities. Cities provide access to healthcare and preventive medicine. They provide a great umbrella of services under which the individual can leave aside the struggle for survival, and get on with the finer things of life. The city provides a range of services and facilities, whose realization in villages are the all-consuming functions of rural life. Potable water; sewerage disposal; and energy for cooking, heat, and lighting are all piped and wired in; solid-waste disposal and stormwater drainage are taken for granted. The city offers access through roads, buses, telephones, and the Internet.* See also PRINCIPLES OF INTELLIGENT URBANISM.

PIU PRINCIPLE 8 [psychology] *Regional integration*—envisions the city as an organic part of a larger environmental, socioeconomic, and cultural-geographic system, essential for its sustainability.* See also PRINCIPLES OF INTELLIGENT URBANISM.

PIU PRINCIPLE 9 [psychology] *Balanced movement*—intelligent urbanism sees the city as part of a larger social, economic, and geographic organism—the region. Likewise, it sees the region as an integral part of the city. Planning of the city and its hinterland is a single holistic process. City growth and development is an organic part of a much larger organism. If one does not recognize growth as a regional phenomenon, then development will play a hop-scotch game of moving just a little down the valley, or up the hills, keeping beyond the paths of the city boundary, development regulations, and the urban tax regime. If one does not recognize the wholeness of the city and its region, the city will ruthlessly exploit its surrounds, denuding the hills of trees, quarrying out hillsides for stone, and grassing off the biomass for milk and meat.* See also PRINCIPLES OF INTELLIGENT URBANISM.

PIU PRINCIPLE 10 [psychology] *Institutional integrity*—recognizes that inherent good practices can be realized only through accountable, transparent, competent, and participatory local governance, founded on appropriate databases, due entitlements, civic responsibilities, and duties. The PIU promotes a range of facilitative and promotive urban development management tools to achieve appropriate urban practices, systems, and forms.* See also PRINCIPLES OF INTELLIGENT URBANISM.

PLACE PRINCIPLE [acoustics] Every tone has its own characteristic place or pattern of action in the cochlea and hence its own particular representation by fibers in the auditory nerve.*

PLANCK PRINCIPLE [physics] (Max Planck, 1858–1947, German physicist) Basis of quantum theory, the energy of electromagnetic waves is confined in indivisible packets or quanta, each of which has to be radiated or absorbed as a whole, with the magnitude proportional to frequency. If E is the value of the quantum expressed in energy g (graritational) units and v is the frequency of the radiation, then $E = h(v)$, where h is known as Planck's constant and has dimensions of energy multiploid by x time.[†]

PLANCK RADIATION PRINCIPLE (physics) (Max Planck, 1858–1947, German Physicist) (Wilhelm Weir) (1) An expression for the distribution of energy in the spectrum of a blackbody radiator.[‡] (2) Wavelength of the maximum intensity in the electromagnetic radiation that every solid body emits is inversely proportional to the absolute temperature of the body.[§] See also INFRARED PRINCIPLE.

*E. Wever, and E. G. Lawrence, M. (1952), The place principle in auditory theory, *Proc. Natl. Acad. Sci. USA* **38**(2), 133.
[†]Walker, P. M. B. (1999), *Chambers Dictionary of Science and Technology*, Chambers, New York, p. 875.
[‡]Hornbeck, G. A. (1966), Optical methods of temperature measurement. Applied Optics Information Photonics(IP) **5**(2) p. 179.
[§]Freud, S. (1951), *Formulations Regarding the Two Principles in Mental Functioning. Organization and Pathology of Thought: Selected Sources*, Columbia Univ. Press, New York, pp. 315–328.

PLEASURE PRINCIPLE [psychology, physics] (1) According to Freudian psychology, one has an internal drive (of the id, or unconscious) to seek pleasure and to avoid pain. The instinctual wishes in the unconscious have a preemptory quality—they seek pleasureable discharge and the reduction of unpleasurable tension at all costs. The pressure for such direct and immediate gratification, characteristic of the unconscious system, may arouse conflict during the passage of the wish through the other systems, and as a consequence the instinctual wish becomes subjected to censorship.* (2) Affirmation of existence; binding energy; possessing the quality of a gestalt; an original affirmation of existence, which could correspond to Freud's first meeting of the death instinct.† (3) Seeks continual enjoyment and gratification constantly at odds with society, the world, norms, laws, and so on in its incessant drive for gratification.‡ (4) Obligation to maintain one's health, the avoidance of pain, depression and happiness, the experience of pleasure enjoying life, pleasure of the palate, and sensual pleasure.§ Also referred to as *Beyond The Pleasure Principle*; *Jewish Pleasure Principle*; *Pleasure–Pain Principle*; *Pleasure–Unpleasure Principle*; *Principle Of Delayed Gratification*; *Principle Of Immediate Gratification*.

PLEASURE PRINCIPLE [psychology] (Jewish in herent-from text derived from Yorah, crossing Red sea & Exodus 15:20) The governing principle of the id; the principle that an infant seeks gratification and fails to distinguish fantasy from reality.¶ See also PLEASURE–PAIN PRINCIPLE, PLEASURE–UNPLEASURE PRINCIPLE.

*Ashbery, J. L. (1978) Freud: Beyond the pleasure principle, New York, v.l, n.1, p. 96.
†Cantwell, M. (1988), In New England, a spa with chintz, charm and no nonsense pampering (spas the pleasure principle), *Vogue*, **178**, 238.
‡Volmer, J. (1990), The pleasure principle (psychology of fun), *Health* **5**, 60.
§Flocker, M. (2004), *The Hedonism Handbook: Mastering the Lost Arts of Leisure and Pleasure*, Da Capo Press, p. 15.
¶American Heritage (2000), *American Heritage Dictionary of the English Language*, 4th edi., Houghton Mifflin, New York.

PLEASURE–PAIN PRINCIPLE [psychology] (1) Two principles in mental functioning according to Freud: the pleasure–pain principle (or more simply the pleasure principle) and the reality principle. To this end neuroses, hallucinatory psychoses, and other mental adaptations are discussed as a means of establishing mental equilibrium.‖ (2) The governing principle of the id; the principle that an infant seeks gratification and fails to distinguish fantasy from reality.¹⁷⁵² See also PLEASURE PRINCIPLE; PLEASURE-UNPLEASURE PRINCIPLE;.

PLEASURE–UNPLEASURE PRINCIPLE [psychology] (1) The governing principle of the id; the principle that an infant seeks gratification and fails to distinguish fantasy from reality.¹⁷⁵² (2) The German *Lust* is usually translated as "pleasure" and its opposite (*Unlist*), as "unpleasure". Early translations rendered Unlust as "pain", but this should be reserved for the translation of *Schmerz*, which Freud distinguished from Unlust.** See also PLEASURE PRINCIPLE, PLEASURE–PAIN PRINCIPLE.

PLUNGER MAGNET PRINCIPLE [engineering] The tool raising and lowering of devices provide operation, independent of the position in which they are being used so that no weight forces occur when electric power is shut off.†† See also BUCKING MAGNET PRINCIPLE; LINEAR MAGNET PRINCIPLE; FRICTION WHEEL DRIVE PRINCIPLE; MAGNET PRINCIPLE; LOCK MAGNET PRINCIPLE; CERAMIC MAGNET PRINCIPLE.

PLUS/MINUS INTERACTION PRINCIPLE [mathematics] See ARIOKHIN–PARETO INTERACTION PRINCIPLE.

‖Sandler, J., Holder, A., and Dare, C. (1973) Frames of reference in psychoanalytic psychology: VI. The topographical frame of reference: The unconscious, *Bri. J. Med. Psychol.* **46**(1), 39.
Sandler, J., Holder, A., and Dare, C. (1973), Frames of reference in psychoanalytic psychology: VI. The topographical frame of reference: The unconscious, *Br. J. Med. Psychol.* **46(1), 39.
††Cordes, W. (1980), *Device for Moving Automatic Drawing Machine Tools*, US Patent 4,295,146.

POINCARÉ'S HOLOMORPHIC PRINCIPLE [mathematics]. (Jules Henric Poincare, 1854–1912, French Mathematician) Solutions to holomorphic differential equations are themselves holomorphic functions of time, initial conditions, and parameters.*

POINCARÉ'S PRINCIPLE [mathematics] (Jules Henrl Poincare, 1854–1912, French Mathematician) (1) Concerning the impossibility of detecting absolute motion.[†] (2) Propositions that can be verified by a computation require no proof. (3) On a contractible manifold, all closed forms are exact. While $d^2 = 0$ implies that all exact forms are closed, it is not always true all that closed forms are exact. Used to show that closed forms represent cohomology classes.[‡]

POINT OF PRINCIPLE [Psychology] (1) Allows one to create all probable versions under its subjective theme, as its reality creation/evolvement under that subject is open-ended and unpredictable, relying on choice and option. Rules and laws capture a consensus that certain actions and events will occur under a principle (or a combination of principles).[§] (2) A principled view implies that an individual has a firm understanding of the underlying principle(s) of events and the rules and laws which govern them inherently and according to a consensus.[¶]

POINTER PROCESSING PRINCIPLE [chemistry] Heterocycles have poor solubility, even with "sulfone" chemistries, but they at least form liquid crystals or soluble crystals in strong acids. With relatively low molecular weight, capped, heterocycle oligomers can be processed in an autoclave.[||] See also INFORMATION-PROCESSING PRINCIPLE.

POINTWISE MAXIMUM PRINCIPLE [psychology, mathematics] (1) Instinctive attempt to avoid pain, discomfort, or unpleasant situations; desire to obtain maximum gratification with minimum effort.[**] (2) Activities must be grouped to facilitate the accomplishment of goals; and the manager of each subdivision must have authority to coordinate its activities with the organization as a whole. The more clearly a position or a department defines the of results expected, activities to be undertaken, organization authority delegated, and authority and informational relationships with other positions, the more adequately individuals responsible can contribute toward accomplishing enterprise objectives.[††] See also AUTOMATIC CONTROL PRINCIPLES; PRINCIPLE OF DOMINANT SUBSYSTEMS; PRINCIPLE OF FUNDAMENTAL CHARACTERISTICS; PRINCIPLE OF OPERATION; EQUIPARTITION PRINCIPLE; PRINCIPLE OF CONSTRUCTION AND OPERATION; PRINCIPLE OF SMOOTH FIT.

POLANYI–EVANS–BELL PRINCIPLE [psychology, chemistry] (Michael Polanyi, b. 3/12/1891–1976 Hungarian-British Polymath, Budapest, Hungary) See BELL–EVANS–POLANYI PRINCIPLE; JARMAN–BELL PRINCIPLE.

POLICE PRINCIPLES [psychology] The basic mission for which the police exist is to prevent crime and disorder; The ability of the police to perform their duties is dependent on the public approval of police actions. Police must secure the willing cooperation of the public in voluntary observation of the law to be able to

---

*Weisstein, E. W. (2003), *CRC Concise Encyclopedia of Mathematics*, CRC Press, p. 2271.
[†]Sjödin, T. (1980), A note on Poincaré's Principle and the behavior of moving bodies and clocks, *Z. Naturforsch.* **A35**(10), 997–1000.
[‡]Weisstein, E. W. (2003), *CRC Concise Encyclopedia of Mathematics*, CRC Press, p. 2271.
[§]Bhabha, J., Finch, N., Crock, M., and Schmidt, S., (2007), *Seeking asylum alone: A comparative study of Laws, Policy and Practice in Australia, the U.K. and the U.S.*, Themis Press, p. 60.
[¶]Avineri, S. (1974), *Hegel's Theory of the Modern State*, Cambridge Univ. Press, p. 183.

[||]Das, A. K. (2004), Very high rate chemical reaction: A breakthrough for future of chemistry and chemical engineering, American Chemical Society, Div. Industrial and Engineering Chemistry 227th ACS Natl. Meeting, Anaheim, CA, March 28–April 1, 2004.
[**]Rock, I. (1990), *The Legacy of Solomon Asch: Essays in Cognition and Social Psychology*, Lawrence Erlbaum.
[††]Khosla, D. (2006), *Biologically-Inspired Cognitive Architecture for Integrated Learning, Action and Perception (BICA-LEAP)*, HRL Laboratories, LLC, Contract N00014-05-C-0510.

secure and maintain the respect of the public. The degree of cooperation of the public that can be secured diminishes proportionately to the necessity of the use of physical force. Police seek and preserve public favor not by catering to public opinion, but by constantly demonstrating absolute impartial service to the law. Police use physical force to the extent necessary to secure observance of the law or to restore order only when the exercise of persuasion, advice, and warning is found to be insufficient. Police, at all times, should maintain a relationship with the public that gives reality to the historic tradition that the police are the public and the public are the police; the police are only members of the public who are paid to give full-time attention to duties that are incumbent on every citizen in the interests of community welfare and existence. Police should always direct their action strictly toward their functions, and never appear to usurp the powers of the judiciary. The test of police efficiency is the absence of crime and disorder, not the visible evidence of police action in dealing with it.* See also PEELIAN PRINCIPLE.

POLIQUIN PRINCIPLE [kinesiology] (Charles Poliquin, Canadian Codd)   Contains a basic formatting of training methods and regimens. Intended for the purpose of helping athletes improve their sport and nonathletes to gain muscle mass.[†]

POLISHING PRINCIPLE [mathematics]   (1) A sequence of functions defined on a set S converges pointwise to a function f, of the sequence $f_1(x, f_2)(x_2)$ converges to $f(x)$ for each x in S.[‡] (2) Storage surfaces suspended in endless conveyor device that in vertical direction is carried on closed circular revolving track.[§]

*Davis, E. M. (1972), *Professional Police Principles*, 35 Federal Probation 29.
[†]See   pdfdatabase.com/index.php?q=poliquin+principle.
[‡]Umehara, N., Kirtane, T., Gerlick, R., Jain, V. K., and Komanduri, R. (2006), A new apparatus for finishing large size/large batch silicon nitride (Si3N4) balls for hybrid bearing applications by magnetic float polishing (MFP), *Int. J. Machine Tools Manuf.* **46**(2), 151–169.
[§]Mannhart, J. and Mayer, B. (1998), *Electronic Device*, US Patent 6,111,268.

POLITICAL PRINCIPLE [political science]   (1) In expounding the *Constitution of the United States*, every word must have its due force, and appropriate meaning, for it is evident from the whole instrument, that no word was unnecessarily used, or needlessly added."[¶] (2) The "appropriate meaning" of the Second Amendment text cannot be discovered by isolating it from (a) the political principle on which it was based, (b) its sister articles in the Bill of Rights in which it was placed, and (c) the state constitutional precursors from which it was derived—as this Court is being invited to do by both petitioners and the solicitor general. Rather, the right protected by the Second Amendment, like all of the rights stated in the Bill of Rights, must be examined in its textual, contextual, and historic settings.[‖]

POLLUTER PAYS PRINCIPLE [ecology]   (1) Adopted in 1972 by the member countries of the OECD (Organization for Economic Cooperation and Development) for two reasons: (a) to ensure a better distribution of resources and (b) to avoid producing large deviations in international exchanges and investments. All the agents involved in production, consumption, or leisure are, to a lesser or greater degree, polluters. The costs of pollution control are either those involved in preventive action, or in correcting an unsatisfactory situation from the environmental perspective (curative action), or costs incurred by repairing damage to the environment. (2) Any individual polluting the environment directly or indirectly shall pay the cost of removing the pollution or compensating for it. (3) Socio. political process in which conflicts emerge and must be resolved between competing interests, between people holding different value systems and different principles of judgment, and also between different representations of future states and different visions of the world. (4) Used for internalizing external costs and assigning liability. Its application may seem

[¶]*Holmes v. Jennison*, 39 US (14 Peters) 540, 570–571 (1840).
[‖]Cottrol, R. J. (1990), The Fourteenth Amendment: From political principle to judicial doctrine, *J. Am. Hist.* **76**, 1277.

to entail a need for monetary valuation of damages. This allows environmental impacts and protection questions to be formulated as optimal resource use problems through the extension of traditional cost–benefit analysis techniques.*

POLLYANNA PRINCIPLE [psychology, mathematics]   (1) People give precedence to pleasant over unpleasant events Thus, pleasant items are "spewed" early in list-generating tasks, and the more positive member of an antonym pair is uttered first.[†] (2) Concept in linguistic pragmatics adapted from the "Pollyanna hypothesis" of psychology, and designed to account for the preference on the part of speakers for avoiding or mitigating negative terms and expressions.[‡] See IMMORTALITY PRINCIPLE.

PÓLYA–SZEGÖ   PRINCIPLE [mathematics] (George   Pólya,   1887–1985,   Hungarian mathematics) (Gábor Szegö, 1895–1985, Hugarian mathematician) (1) The norm of the gradient of a scalar function of several variables does not essentially grow under the operation of the symmetric rearrangement. An extension of this principle is known to every rearrangement-invariant norm. (2) Enables the reduction of Sobolev inequalities to considerably more manageable one-dimensional ones involving certain appropriate Hardy integral operators.[§]

POLYMER PRINCIPLE [chemistry]   Stretched chains favor lateral alignment and can lead to crystallization, gelation, or only cluster formation. Crystallization can be prevented by any sort of irregularity in particular when these irregularities heavily perturb the smooth spatial conformation of the chain. The prerequisite for lateral alignment is a stretched conformation. For cellulose this condition is already given by the $h(1,4)$-glycosidic bond. In other cases a transition to a supramolecular structure chain must have occurred with the single chain before bundle formation can become effective. The enormous stiffness of the formed objects is understandable, because in a bundle of aligned chains the segmental motion is seriously restricted.[¶]

POLYURONID   ACTIVE   PRINCIPLE [physics, chemistry]   Copolymer   obtainable   by polymerization of monomer mixture containing carboxyl containing monomer and methacrylic acid ester used as immunostimulant, antibacterial, and antiinflammatory.[‖] See also ACTIVE PRINCIPLE; ANTIVIRAL ACTIVE PRINCIPLE; MOUSE BETA AMYRIN ACTIVE PRINCIPLE.

POME [mathematics]   See PRINCIPLE OF MAXIMUM ENTROPY.

PONCELET'S CONTINUITY PRINCIPLE [mathematics, psychology]   See CONTINUITY PRINCIPLE.

PONTRYAGIN'S   DISCRETE-TIME   MAXIMUM PRINCIPLE [mathematics,   engineering] (DTPMP) First proposed in 1956 by Pontryagin. The objective function formulation is represented as a linear function in terms of the final values of a state vector and a vector of constans.[**] See PONTRYAGIN'S MAXIMUM PRINCIPLE.

PONTRYAGIN'S   DISCRETE-TIME   MINIMUM PRINCIPLE (minimum   principle) First, consider fixed-time optimal control problems with differentiable cost, control constraints, and a given initial state, such as UCP'_c,

*O'Connor, M. (1997), The internalization of environmental costs: Implementing the polluter pays principle in the European Union, *Int. J. Environ. Pollut.* **7**, 450.

[†]Matlin, M. W. and Stang, D. J. (1978), *The Pollyanna Principle: Selectivity in Language, Memory, and Thought*, Schenkman, Cambridge, MA.

[‡]Armstrong, N. and Hogg, C. (2001), The Pollyanna principle in French: A study of variable lexis, *J. Pragmatics* **33**(11), 1757–1785.

[§]Cianchi, A., Esposito, L., Fusco, N., and Trombetti, C. (2008), Aquantitative Pólya-Szegö principle, *J. Reine Angew. Math.*, **614**, 153–189.

[¶]Dumetriu, severian (2004) Polysaecharides: Structured Diversity and Functional Versatility, CRC Press p. 378.

[‖]Atkins, E. D. T., Isaac, D. H., Nieduszynski, I. A., Phelps, C. F., and Sheehan, J. K. (1974), Polyuronides: Their molecular architecture, *Polymer* **15**, 263.

[**]Kapur, J. N. (1988), *Mathematical Modeling*, New Age International, p. 203.

in (4.1.21), the Pontryagin Minimum Principle assumes the following form.

THEOREM. *Suppose that Assumption 4.1.1 is satisfied and that $\hat{u} \in U$ is a local minimizor for the problem UCP'_c Let $\hat{x}(\cdot)$ be the corresponding optimal trajectory, and let $\hat{p}(\cdot)$ be the solution of the corresponding adjoint equation*

$$\tilde{p}(\tau) = -h_x(x^\eta(\tau), u(\tau))^r \hat{p}(\tau).$$
$$= \mathcal{H}_x(\hat{x}(\tau), \hat{u}(\tau), p(\tau))^r$$
$$\tau \in [0.1], \ \hat{p}(1) = \Delta_x F^D(E_0, \tilde{x}(1)). \tag{1a}$$

Where the dot denoted differentiation and the Hamiltonian $\mathcal{H} : R^n \times R^m \times R^n \to R$ is defined by

$$\mathcal{H}(x, u, p) = (p, h(x, u)). \tag{1b}$$

Then for every $v \in U$ and almost all $t \in [0, 1]$.

$$\mathcal{H}_x(\hat{x}(t), \hat{u}(t), \hat{p}(t)) \leq \mathcal{H}(\hat{x}(t), v, \hat{p}(t)).1c$$

$\square$

We will compare this condition to (3f). Condition (48c) states that $\hat{u}(r)$ is a constrained global mininizer for $\mathcal{H}(\hat{x}(t), \cdot, \hat{p}(t))$ for almost all $t \in [0.1]$. Hence, by Exercise. 1.1.4, we must have that, for almost all $t \in [0, 1]$ and all $v \in U$.

$$\mathcal{H}_x(\hat{x}(t), \hat{u}(t) \cdot \hat{p}(t))[v(t) - \hat{u}(t)] \geq 0. \tag{2a}$$

and hence that, for all $v \in U$,

$$\int_0^1 \mathcal{H}_u(\hat{x}(t) \cdot \hat{u}(t) \cdot \hat{p}(t))(v(t) - \hat{u}(t)) dt \geq 0. \tag{2b}$$

Referring to (5.6.13d.e.t), we see that (2b) is identical with (3f), i.e. the Minimum Principle implies (3f). Next, to simplify the mathematics, suppose that $\hat{u} \in U$ is piecewise continuous and that it satisfies (3f). Then (2a) must hold at any $t \in [0, 1]$ at which $\hat{u} = \hat{t}$, say, there exists a $\hat{v} \in U$ such that

$$\mathcal{H}_u(\hat{x})(\hat{t} \cdot \hat{u}(\hat{t}) \cdot \hat{p}(\hat{t}))[\hat{v} - \hat{u}(\hat{t})] < 0 \tag{2c}$$

Then, by continuity of $\mathcal{H}(\ldots)\hat{x}(\cdot), \hat{p}(\cdot)$, and $\hat{u}(\cdot)$ at $\hat{t}$, there must exist an interval $J[\hat{t} - e, \hat{t} - \varepsilon]$ with $\epsilon > 0$, such that (2c) holds with $\hat{t}$ replaced by $t \in I$. If we now define $v^*(t) = \hat{v}$ for all other $t \in [0, 1]$, then we see that $v^*(\cdot)$ violates (2b).*

PONTRYAGIN'S MAXIMUM PRINCIPLE (PMP)

[mathematics] (Lev Semyonovich Pontryagin, 1908—1988, Russian mathematician) (1) A theorem giving a necessary condition for the solution of optimal control problems: let B(T), $T_0 \leq T \leq T$ be a piecewise continuous vector function satisfying certain constraints; in order that the scalar function $S = \Sigma c_i x_j(T)$ be minimum for a process described by the equation $\partial x / \partial T = (\partial H / \partial z_i)[z(T), x(T), \theta(T)]$ with given initial conditions $x(T_0) = x^0$ it is necessary that there exist a nonzero continuous vector function $z(T)$ satisfying $dz/dT = -(\partial H/\partial x_i)$. $[z(T), x(T), \theta(T)], z_i(T) = -c_i$, and that the vector $\theta(T)$ be so chosen that $H[z(T), x(T), \theta(T)]$ is maximum for all $T, T_0 \leq T \leq T$.[†] (2) Topological groups and their character theory, on duality in algebraic topology, and on differential equations with applications to optimal control. (3) Dynamic optimization is particularly concerned with decisionmaking situations or economic growth models and mathematical formulations of problems involving moving objects in which the time variable enters naturally into the optimization problem and therefore also appears in its solution. The objective function takes the form of an integral, while the constraints are described by a system of differential equations. (4) Those control processes are considered that can be described by the system

$$\frac{dx}{dt} = f(x, u), \tag{1}$$

where x is an n-dimensional vector which characterizes the process, u is an r-dimensional control vector, and t is the time.

*Polak, E. (1997) Optimization: Algorithms and Consistent Approximations (Applied Mathematical Sciences). Springer, p. 532–533.
†McGraw Hill (2003) Pontryagin's Maximum Principle, *McGraw Hill Dictionary of Scientific and Technical Terms*.

Corresponding to the system the functional

$$J = \int_{t_0}^{tf_0} (x, u)\, dt \qquad (2)$$

is considered, where f_0 is a known function and the initial and terminal states (but not the initial and terminal times!) of the process are fixed [where it is assumed that there exist controls $u(t)$ that take the process from one state to another].* See also EKE-LAND'S VARIATIONAL PRINCIPLE; PONTRYAGIN'S MINIMUM PRINCIPLE; PONTRYAGIN'S DISCRETE-TIME MINIMUM PRINCIPLE; PONTRYAGIN'S MAXIMUM PRINCIPLE.

PONTRYAGIN'S MINIMUM PRINCIPLE (PMP) [mathematics, physics] (Lev Semyonovich Pontryagin, 1908–1988, Russian mathematician) (1) Depending on the upper-bound value of the control vector, possible driving modes of the states are studied for which particular optimal driving modes are extracted so as to meet the specified constraints and boundary conditions imposed in the problem. (2) Find the control input that minimizes the performance index; this is a so-called variational problem, to minimize, subject to the differential constraint. (3) First the optimal solution is characterized. Such a characterization considerably improves the interpretability of the solution and enhances numerical efficiency. This nominal solution is then coupled with available (generic or specific) measurements to develop a multiloop measurement-based optimization framework. In the outer loop (the optimization loop), the optimal setpoint trajectory for endpoint optimization is calculated. In the inner loop (the control loop), a controller is designed to keep the system on the trajectory specified by the optimizer. This controller is required to be robust to modeling errors, changes in setpoint trajectory, and process disturbances in the batch process.† See also EKELAND IS VARIATIONAL PRINCIPLE; MINIMUM

*Pontryagin, L. S., Boltyanskii, V. G., Gamkrelidze, R. V., and Mishchenko, E. F. (1962), The Mathematical Theory of Optimal Processes. Wiley Interscience.
†Kirk, D. E. (1970), *Optimal Control Theory, an Introduction*, Prentice-Hall.

PRINCIPLE; PONTRYAGIN'S MAXIMUM PRINCIPLE; PONTRYAGIN'S DISCRETE-TIME MINIMUM PRINCIPLE.

POSITIONALITY PRINCIPLE [physics] The dual connection between force and postion: on the one hand, forces are the ultimate causes of observed changes in position; on the other hand, if these forces themselves are subject to change, the latter changes must depend on these same positions. In order to prove centrality, any magnitudes used to characterized the motion of a system be defined only in terms of the relative positions of its mass-points.‡ See also PRINCIPLE OF DECOMPOSITION; PRINCIPLE OF POSITIONAL DETERMINACY.

PONTRYAGIN'S PRINCIPLE [mathematics] Devised for the optimal control of ordinary differential equations, and extended to the optimal control of semilinear elliptic systems and variational inequalities in the case of a distributed control.§

POSITIONING PRINCIPLE [physics, measurement] (1) In the field of geometri seismic, modeling with ray tracing and source and receiver points. Then trace the ray to get the wave's traveltime. Locating a source and then extrapolating the wavefield from the located source to subsurface; after new seismic sources are created by incident wave, extrapolating these new sources, reflect waves, to the positioned geophone. Once the reflected wave reaches the located geophone, a seismic trace is created.¶ See also TWO-POINT PRINCIPLE.

POSITIVE-DISPLACEMENT PRINCIPLE [physics, mathematics] See BOUNDARY PRINCIPLE.

‡Hendricks, V. F. and Hyder, D. J. (2006) Interactions: mathematics, physics and philosophy, 1860–1930, **251**, p. 2
§Bonnans, F. and Casas, E. (1995), An extension of pontryagin's principle for state constrained optimal control of semilinear elliptic equations and variational inequalities, *SIAM J. Control Optim.* **33**, 274.
¶Minami, M. M. H. and Aoyama, T. (2007), Design and implementation of a fully distributed ultrasonic positioning system, *Electron. Commun. Jpn.* (Pt. III: *Fund. Electron. Sci.*) **90**(6), 17–26.

POSNER PRINCIPLE [economics] Observed pattern of government regulation of the economy. Includes the public interest theory and several versions of the interest group or capture theory.*

POTENTIAL ENERGY MINIMIZATION PRINCIPLE [engineering] A fundamental concept used in physics, chemistry, biology, and engineering. It asserts that a structure or body shall deform or displace to a position that minimizes the total potential energy, with the lost potential energy dissipated as heat. The entropy of a system will maximize at equilibrium. Given two possibilities—a low heat content and a high potential energy, or a high heat content and low potential energy—the latter will be the state with the highest entropy, and will therefore be the state toward which the system moves.† See also PRINCIPLE OF MINIMUM POTENTIAL ENERGY. Also referred to as minimum potential energy principle.

POTENTIAL PRINCIPLE [mathematics] Branch of applied mathematics that studies the properties of a potential function without reference to the particular subject in which the function is defined.‡

POWDERING PRINCIPLE [fluids] Confines successive volumes of fluid with a closed space in which the pressure of the fluid is increased as the volume of the closed space is decreased.§ See also COUNT(Q) PRINCIPLE; COUNT(P) PRINCIPLE; DIRICHLET'S PRINCIPLE.

POWER PRINCIPLE [psychology] E. S. Person criticizes modern psychoanalysis for not having integrated the power concept. Self-determination, dominance, and will, as well as powerlessness, but also self-control or autonomy are named as phenomena that need

psychoanalytical illumination. It is, however, not possible to use just parts of Nietzsche's and Adler's notions of power. In that sense psychoanalysis does not need a new power concept because power finds its place beside the sexual and aggression drive within the frame of the ego organization. Although power and its disfigurements are important for the psychoanalytical practice, power is not a relevant concept in the theory of psychoanalysis. Despite the marginalization of the power concept in psychoanalytic theory, considerations regarding power are inseparably linked with the therapy situation itself and our organizations. These thoughts inevitably enter the clinical work, from a therapeutic standpoint, stories, dreams, and fantasies of the patients refer to power; power interactions are an integral part of transference, and countertransference phenomena and power thoughts are almost always involved when therapists commit border violations. Power is defined as an inborn force that allows self-control, self-discipline, and taking hold of the outer world as well as the development of interpersonal power.¶ See EMANCIPATION PRINCIPLE.

PRADO'S PRINCIPLE [Mathematics] See PARETO PRINCIPLE.

PRÄGNANZ PRINCIPLE [psychology] German term meaning "good figure." The law of Prägnanz is sometimes referred to as the "law of good figure" or the "law of simplicity". This law holds that objects in the environment are seen in a way that makes them seem as simple as possible.‖

PRATT'S PRINCIPLE [geology] (John Henry Partt, 1809–1871, British Mathematician) Mechanism of hydrostatic support for Earth's crust.**

*Posner, R. A. (1979), The Homeric version of the minimal state, *Ethics* **90**(1), 27–46.

†Sigmund, O. (2000), A new class of extremal composites, *J. Mech. Phys. Solids* **48**(2), 397–428.

‡Walker, P. M. B. (1999), *Chambers Dictionary of Science and Technology*, Chambers, New York, p. 905.

§Morineau, D. and Alba-Simionesco, C. (2003), Liquids in confined geometry: How to connect changes in the structure factor to modifications of local order, *J. Chem. Phys.*, **118**(20). 9389–9400.

¶Person, E. (2001), Über das Versäumnis, das Machtkonzept in die Theorie zu integrieren, *Z. Individ. psychol.* **26**, n.1, p. 4–23.

‖Van Wagner, K. (2006), *The Laws of Perceptual Organisation*: http://psychology.about.com/od/sensationand[erce[topm//ss/gestaltlaws_3.htm.

**Stiegeler, S. E. (1977), *A Dictionary of Earth Sciences*, Pica Press (distributed by Universe Books), p. 219.

PRECAUTION PRINCIPLE [ecology] New technology must verify that it does not damage, either directly or indirectly, the environment. Burden of proof falls on the perpetrator, not the victim.* Framed in culture and politics and its importance lies in the way in which it captures misgivings about the scientfic method, about the proper relationship between human needs and the health of the natural world, and about the appropriate way to incorporate the rights of future generations into the policy making. To encourage decision makers to consider the likely harmful effects of their activities on the environment before they pursue those activities. See also CAUTION PRINCIPLE; UNCERTAINTY PRINCIPLE.

PRECAUTIONARY PRINCIPLE [biology, medicine] (1) Three categories of uncertainty in relation to risk assessment are defined; uncertainty in effect, uncertainty in cause, and uncertainty in the relationship between a hypothesized cause and effect. The *precautionary principle* (PP) relates to the third type of uncertainty. Three broad descriptions of the PP are set out: (a) uncertainty justifies action, (b) uncertainty requires action, and (c) uncertainty requires a reversal of the burden of proof for risk assessments. Risk analysis under uncertainty, the precautionary principle, and the new EU chemicals strategy.† (2) It is common to all the production processes that large amounts of cell material result at the end and contain, inter alia, mainly natural and homologous nucleic acids (DNA and RNA) but also recombinant plasmid DNA or foreign DNA integrated into the genome of the host cell. It has hitherto been possible only with difficulty to estimate the possible risks on release of this recombinant DNA for the environment. Thus, on the precautionary principle, statutory requirements and regulations have been issued by authorities and institutions in most countries [such as,

e.g., in the Federal Republic of Germany the Central Committee for Biological Safety (ZKBS) of the Federal Board of Health in Berlin], which call for inactivation of the waste material, which is usually liquid, and thus of the amount of nucleic acids present after the production process as protection against a possible risk to the environment.‡ See also PRINCIPLE OF HABEAS MENTEM; HEALTH PRINCIPLE.

PREMACK PRINCIPLE [mathematics] (David Premack, 1925–American Psychologest) (1) Probability differential hypothesis to human resources management; part-time service sector employees' overall quality performance, measured from established organization standards, would be greater.§ (2) Desired and undesired behavior is applied by attacking symptoms that disregard the person; a symptom can be attacked removing defenses and imposing behaviors that are not internalized.¶

PRESSURE CRUSHING PRINCIPLE [engineering]   See CRUSHING PRINCIPLE.

PRESSURE DIFFERENTIAL PRINCIPLE [mathematics, engineering] (Commonly known as a "vacuum box.") Provides an apparatus that does not require manual manipulation during operation and that may be advantageously employed in conjunction with tire changing equipment.‖

*Jordan, A. and O'Riordan, T. (n.d.), *The Precautionary Principle in U.K. Environmental Law and Policy*, CSERGE Working Paper GEC 94-11.
†Rogers, M. D. (2003), *Regul. Toxicol. Pharmacol.* **37**(3), 370–381.

‡VanDyke, J. M. (1996), A lying the precautionary principle to ocean shipments of radioactive materials, *Ocean Devel. Int. Law* **27**, 379.
§Welsh, D. H. B., Bernstein, D. J., and Luthans, F. (1993), Application of the Premack principle of reinforcement to the quality performance of service employees, *J. Org. Behav. Manage.* **13**(1), 9–32.
¶Mazur, J. E. (1975), The matching law and quantifications related to premack's principle, *J. Exp. Psychol.* (*Animal Behav. Process.*), **1**(4) 374–386.
‖Hill, N. L. and Linsenbardt, T. L. (1999), *Method for Manufacturing a Variable Insulated Helically Wound Electrical Coil*, US Patent: 6,138,343.

PRESSURE PRINCIPLE [mathematics, engineering] According to Hagen–Poiseuille's law, the fully developed stationary flow through a microchannel with constant cross section is in linear correspondence with the pressure drop across the channel. In particular, the flow rate at a certain pressure drop depends on the microchannel cross section and length and the viscosity of the fluid, which is usually influenced by temperature.*

PRIEST TANG RUOWANG'S PRINCIPLE [mathematics] (Tang Ruowang, Qing Grand Secretariat, seventeenth century, China) Calculations and completion of a lunar calendar.[†]

PRIGOGINE–GLANSDORFF PRINCIPLE [biology] (Ilya, Viscount Prigogine, 1917–2003, Belgian Chemist) (P. Glansdorff) Any mutation in the system, resulting in the production of a sequence with a higher selective value, is equivalent to a negative variation in the entropy production. (2) A negative fluctuation (which must be randomly significant) must result in a breakdown of the steady state. A comparison between the two "species" (mutant and progenitor), after restoration of the steady state, yields a change in the information content reflected by increased order. The change in entropy per unit time for each "species" is the same if external flows of energy are kept constant and invariable, and if the mutant has the same affinity as its precursor. (3) Provides a link between selection theory and the thermodynamics of irreversible processes. This description, however, does not indicate that almost uniform populations of sequences have to be exchanged to produce more "valued" information and decrease the internal entropy. The exchange becomes complex if different "species" have different free energies and affinities for the rates of formation and decomposition reactions. (4) Where there are threats of serious or irreversible damage, lack of full scientific certainty shall not be used as a reason for postponing cost-effective measures to prevent environmental degradation.[‡] See also QUANTUM HAMILTON PRINCIPLE; STEWART–HAMILTON PRINCIPLE; STOCHASTIC HAMILTON VARIATIONAL PRINCIPLE; GYARMATI'S PRINCIPLE; PRINCIPLE OF MINIMUM MECHANICAL ENERGY; PRINCIPLE OF MINIMUM POTENTIAL ENERGY; PRINCIPLE OF MINIMUM KINETIC ENERGY; TOUPIN'S DUAL PRINCIPLE.

PRIGOGINE PRINCIPLE (Ilya, Viscount-Prigogine, 1917–2003, Belgian Chemist) The global entropy production approaches a minimum as a process becomes stationary. Produces field equations that do not agree with the equations of balance of mass, momentum, and energy.[§]

PRIGOGINE'S PRINCIPLE OF MINIMUM ENTROPY [mathematics] (Edwin Thompson Jaynes, 1922–1998, American physicist) (Ilya, discount-Prigogine, 1917–2003, Bolgianchamst) Deals with the instantaneous direction of evolution of the systems under any nonequilibrium conditions. General differential equations are derived for the time history of a thermodynamic system undergoing irreversible transformations. This is done by using Onsager's principle, and introducing generalized concepts of free energy and thermodynamic potentials. In the configuration space of the state variables q_i, the thermodynamic state of a system is represented by a point of coordinates q_i. When not in equilibrium the system is subject to forces, bot internal and external, which are expressed by

$$\partial D/\partial \dot{q}_i = X_i.$$

---

*Grände, P.-O. and Borgström, P. (2005), An electronic differential pressure flowmeter and a resistance meter for continuous measurement of vascular resistance, *Acta Physiol. Scand.* **102**(2), 224–230.

[†]Yue, G. (1999), *The Mouth that Begs: Hunger, Cannibalism, and the Politics of Eating in Modern China*, Duke Univ. Press, p. 167.

[‡]Niklas, K. J. (1979), Information, entropy, and the evolution of living systems, *Brittonia* **31**(3), 428–430.

[§]Kirkaldy, J. S. (1964), Thermodynamic description of heterogeneous dissipative systems by variational methods. IV. Consequences of a synthesis of the Onsager and Prigogine principles, *Can. J. Phys.* **42**(8), 447–454.

and which is called "dis-equilibrium forces." These forces may be considered proportional to the derivatives of the generalized thermodynamic potential. Considering a system which is not in equilibrium, its instantaneous velocity direction is such that the rate of entropy production is a minimum for all possible velocity vectors satisfying the condition that the power input of the dis-equilibrium forces is constant. A dual minimum entropy production theorem identical with the above can be stated except that the variable Xi and qi are interchanged.[*]

$$\partial D / \partial X_i = \dot{q}_i.$$

PRIMING PRINCIPLE [psychology] A supplementary principle was needed to explain why sequential presentation typically produces a larger speed advantage for same pairs along with no error differential in favor of more false-different responses. The encoding of the first member of a pair biases the subject to encode the second member in the same way. The criterion shift involved produces both a further acceleration of the same response and a tendency to err by pressing the same button, which counteracts the tendency to err by pressing a different button (internal noise principle). By contrast, priming hastens encoding not by a criterion shift but by a sensitivity shift, that is, by increasing the rate at which information is built up.[†] See also INTERNAL NOISE PRINCIPLE; RELATIVE FREQUENCY PRINCIPLE.

PRIMING PRINCIPLE REGIMEN ADMINISTRATION [engineering] Determining appropriate medical treatment with initial dosage used as a primer.[‡]

PRINCETON PRINCIPLES [psychology] (Robert P. George, Princeton Univ.) A summary of the public value of marriage and why society should endorse and support the institution. Marriage is a personal union, intended for the whole of life, of husband and wife; marriage is a profound human good, elevating and perfecting our social and sexual nature. Ordinarily, both men and women who marry are better off as a result. Marriage protects and promotes the wellbeing of children, sustains civil society, and promotes the common good. Marriage is a wealth-creating institution, increasing human and social capital. When marriage weakens, the equality gap widens, as children suffer from the disadvantages of growing up in homes without committed mothers and fathers. A functioning marriage culture serves to protect political liberty and foster limited government. The laws that govern marriage matter significantly. "Civil marriage" and "religious marriage" cannot be rigidly or completely divorced from one another.[§]

PRINCIPAL-COMPONENT ANALYSIS [biology] (1) Name of a multivariate statistical technique basic in the development of biological system's ability to self-organize, a process by which new structures and patterns emerge.[¶] (2) Individuals differ to some extent in the way operational factors are organized and function within subsystems and the number of ways in which operating factors in a certain subsystem can be organized in patterns, in order to allow the subsystem to play its functional role in totality.[‖] (3) Orientation selectivity in a model visual system, for the development of biological

[*]Biot, M. A. (1955) Variational principles in irreversible thermodynamics with application to viscoelasticity, Physical Review, v 97(6), pp. 1463–1469.

[†]Krueger, L. E. (1983), Probing Proctor's priming principle: The effect of simultaneous and sequential presentation on same-different judgments, J. Exp. Psycholo. Learni. Memory Cogn. 9(3), 511–523.

[‡]Kaneko, T., Iwama, H., Tobishima, S., Watanabe, K., Komatsu, T., Takeichi, K., and Tase, C. (1997), Placental transfer of vecuronium administered

with priming principle regimen in patients undergoing cesarean section, Masui. 46(6), 750–754.

[§]Macedo, S. and Robinson, M. (1976), The Princeton Principles on Universal Jurisdiction, Program in Law and Public Affairs, Princeton Univ.

[¶]Strother, S. C., Kanno, L., and Rottenberg, D. A. (1995), Principle component analysis, variance partitioning, and functional connectivity, J. Cerebral Blood Flow Metab. 15(3), 353–360.

[‖]Fingelkurts, A. A. Fingelkurts, A. A. (2001), Operational architectonics of the human brain biopotential field: Towards solving the mind-brain problem, Brain Mind 2(3), 261–296.

and synthetic perceptual networks.* (4) A theory of considerable integrative, explanatory and interpretive power for a broadly interdisciplinary social science of human behavior.† (5) For optimal systems, any portion of the optional state trajectory is optimal between the states that it joins.‡ See also OPTIMALITY PRINCIPLE; SELF-ORGANIZING PRINCIPLE; SIMILARITY PRINCIPLE. (ORGANIZING PRINCIPLE).

PRINCIPLE (1) A *principle* signifies a point (or points) of probability on a subject (i.e., the principle of creativity), which allows for the formation of rule or norm or law by (human) interpration of the phenomena (events) that can be created. The rules, norms, and laws depend on and cocreate a particular context to formulate. A principle is the underlying part (or spirit) of the basis for an evolutionary normative or formative development, which is the object of subjective experience and/or interpretation. For example, the ethics of someone may be seen as a set of principles that the individual obeys in the form of rules, as guidance or law. These principles thus form the basis for such ethics. Reducing a rule to its principle says that, for the purpose at hand, the principle will not/cannot be questioned or further derived (unless you create new rules). This is a convenient way of reducing the complexity of an argumentation. (2) To equip with, establish, or fix in certain rule of conduct, good or ill; beginning, foundation, commencement; a source, or origin; that from which anything proceeds; fundamental substance or energy; primordial substance; ultimate element, or cause. (3) A basic generalization that is accepted as true and that can be used as a basic for reasoning or conduct; a rule or standard, especially of good behavior; a basic truth or law or assumption; a rule of law concerning a natural phenomenon or the function of a mechanical system; rule of personal conduct; an explanation of the working of some device in terms or laws of nature. (4) Elementary principle about human psychology—a person's wants and desires have greater influence than his or her behavior. (5) A general application under a set of circumstances, conditions, processes, or systems.

PRINCIPLE A: BENEFICENCE AND NONMALEFICENCE [linguistics, psychology] A statement in the government and binding theory of syntax that is intended to answer questions.§ (2) *Principle A: beneficence and nonmaleficence.* Psychologists strive to benefit those with whom they work and take care to do no harm. In their professional actions, psychologists seek to safeguard the welfare and rights of those with whom they interact professionally and other affected persons, and the welfare of animal subjects of research. When conflicts occur among psychologists' obligations or concerns, they attempt to resolve these conflicts in a responsible fashion that avoids or minimizes harm. Because psychologists' scientific and professional judgments and actions may affect the lives of others, they are alert to and guard against personal, financial, social, organizational, or political factors that might lead to misuse of their influence. Psychologists strive to be aware of the possible effects of their own physical and mental health on their ability to help those with whom they work.¶

PRINCIPLE B: FIDELITY AND RESPONSIBILITY [linguistics, psychology] A statement in the government and binding theory of syntax that applies to pronouns.‖ (2) Strive to contribute a portion of their professional

*Linsker, R. (1988), Self-organization in a perceptual network, *Computer*, vol. **21**, no. 3, pp. 105–117.

†Liechty, D. (1988), Reaction to mortality: An interdisciplinary organizing principle for the human sciences: Ernest Becker's theory of the denial of death, *Zygon* **33**(1), 45–58.

‡Pearlman, D. A., Caldwell, D. A., Case, J. W., Ross, W. S., Cheatham, T. E. III, DeBolt, S., Ferguson, D., Seibel, G., and Kollman, P. (1995), AMBER, applying molecular mechanics, mode analysis, and energy calculations to simulate the properties of molecules, *Comput. Phys. Commun.* **91**(1–3), 1–41.

§Freidin, R (1992), *Foundations of Generative Syntax*, MIT Press, p. 342.

¶American Psychological Assoc. (2002), *The Ethical Principles of Psychologists and Code of Conduct* (http://www.apa.org/ethics/code2002.html).

‖Freidin, R. (1992), *Foundations of Generative Syntax*, MIT Press, p. 285.

time for little or no compensation or personal reward. *principle B: fidelity and responsibility*. Psychologists establish relationships of trust with those with whom they work. They are aware of their professional and scientific responsibilities to society and to the specific communities in which they work. Psychologists uphold professional standards of conduct, clarify their professional roles and obligations, accept appropriate responsibility for their behavior, and seek to manage conflicts of interest that could lead to exploitation or harm. Psychologists consult with, refer to, or cooperate with other professionals and institutions to the extent needed to serve the best interests of those with whom they work. They are concerned about the ethical compliance of their colleagues' scientific and professional conduct. Psychologists' advantage.[*]

PRINCIPLE C: INTEGRITY [linguistics, psychology] A statement in the government and binding theory of syntax that applies to referring expressions—any noun phrase, not an anaphor or pronoun (which are covered by principles A and B, respectively).[†] (2) *Principle C: integrity*. Psychologists seek to promote accuracy, honesty, and truthfulness in the science, teaching, and practice of psychology. In these activities psychologists do not steal, cheat, or engage in fraud, subterfuge, or intentional misrepresentation of fact. Psychologists strive to keep their promises and to avoid unwise or unclear commitments. In situations in which deception may be ethically justifiable to maximize benefits and minimize harm, psychologists have a serious obligation to consider the need for, the possible consequences of, and their responsibility to correct any resulting mistrust or other harmful effects that arise from the use of such techniques.[‡]

PRINCIPLE D: JUSTICE [psychology] Psychologists recognize that fairness and justice entitle all persons to access to and benefit from the contributions of psychology and to equal quality in the processes, procedures, and services being conducted by psychologists. Psychologists exercise reasonable judgment and take precautions to ensure that their potential biases, the boundaries of their competence, and the limitations of their expertise do not lead to or condone unjust practices.[‡]

PRINCIPLE E: RESPECT FOR PEOPLE'S RIGHTS AND DIGNITY [psychology] Psychologists respect the dignity and worth of all people, and the rights of individuals to privacy, confidentiality, and self-determination. Psychologists are aware that special safeguards may be necessary to protect the rights and welfare of persons or communities whose vulnerabilities impair autonomous decisionmaking. Psychologists are aware of and respect cultural, individual, and role differences, including those based on age, gender, gender identity, race, ethnicity, culture, national origin, religion, sexual orientation, disability, language, and socioeconomic status and consider these factors when working with members of such groups. Psychologists try to eliminate the effect on their work of biases based on those factors, and they do not knowingly participate in or condone activities of others based upon such prejudices.[§]

PRINCIPLES AND PARAMETERS THEORY [linguistics] (Avram Noam Chomsky, 1928, American linguist, philosopher, cognitive scientist, and political activist) A person's syntactic knowledge can be modeled with two formal mechanisms: A finite set of fundamental principles that are common to all languages; e.g., that a sentence must always have a subject, even if it is not overtly pronounced; and a finite set of parametrs that determine syntactic variability amongst languages; e.g., a binary parameter that determines whether or not the subject of a sentence must be overtly pronounced. Within this framework, the goal of linguistics is to identify all of the principles and parameters that are universal to human language. Any

[*]American Psychological Assoc. (2002), *The Ethical Principles of Psychologists and Code of Conduct* (http://www.apa.org/ethics/code2002.html).

[†]Freidin, R. (1992), *Foundations of Generative Syntax*, MIT Press, p. 285.

[‡]American Psychological Assoc. (2002), *The Ethical Principles of Psychologists and Code of Conduct* (http://www.apa.org/ethics/code2002.html).

[§]op.eit. American Psychological Association. (2002)

attempt to explain the syntax of a particular language using a principle or parameter is cross-examined with the evidence available in other languages.*

PRINCIPLE AUTHORITY LEVEL [psychology] Maintenance of authority dolegation requires that decisions within the authority competence of an individual manager be made by him and not be referred upward in the organization.[†]

PRINCIPLE-BASED NEGOTIATING [economics] Negotiating agreements with organized employees. Workers being on one side and management on the other is inconsistent with creating an organizational environment in which everyone is encouraged to cooperate in achieving the organization's goals.[‡]

PRINCIPLE-BASED PARSER [psychology] Hypothesis that all natural languages share a common set of principles. The principles have a set of parameters to allow for different variations. Linguistic knowledge for a particular language consists of the settings of the parameters for the language and a lexicon.[§]

PRINCIPLE-CENTERED LEADERSHIP [economics] Defines the 10 critical characteristics of servant leader as listening, empathy, healing, awareness, persesuasion, conceptualization, foresight, stewardship, commitment to the growth of people, and awareness of the need to build community.[¶]

PRINCIPLE-CENTERED LIVING [psychology] (1) May have more positive feelings about themselves, may be more internally directed in their approach toward wellness, and may be more likely to participate in physical activity. Possess higher levels of self-awareness, self-esteem, and wellbeing of individuals; may have more positive feelings about themselves, both generally and physically, and may be more likely to participate in physical activity.[‖] (2) An approach to life propounded under many names by numerous religions and philosophers of self-improvement. In all its forms, it holds, by adhering to certain essential and timeless principles, rules or habits of behavior and thinking; human beings can navigate the complexity of life with increased clarity and focus to achieve personal and/or professional success and perhaps even spiritual enlightenment and meaning.[**] (3) Learning and performance solutions to assist professionals and organizations to increase their effectiveness in productivity, leadership, communications, and sales.[††]

PRINCIPLE-CENTERED SPINE CARE [pharmacology] Stress self-treatment through correct posture and repeated movements performed at a high frequency.[‡‡] See also MCKENZIE PRINCIPLES.

PRINCIPLE COMMITMENT [psychology] As may be defined in effectiveness in assignment and job performance, is to the department area, and in order, to the department, to the organization, to the profession and the community.[§§]

*Newmeyer, F.J. (2004). Against a parameter-setting approach to language variation. Linguistic Variation Yearbook 4, p. 181–234.

[†]Koonty, H., and O'Donnell, C., Principles of Management: (1972) an analysis of managerial function. McGraw-Hill.

[‡]Grattet, P. (1995), Putting collective back into bargaining (principle based negotiating), *Public Manage.* **77**, 4.

[§]Berwick, R. C., Abney, S. P., and Tenny, C., eds. (1991), *Principle-Based Parsing: Computation and Psycholinguistics* (MIT, Bell Communications Research; Univ. Pittsburgh) Kluwer Academic, Dordrecht, *Studies in Linguistics and Philosophy*, vol. 44, pp. 1–37.

[¶]Lau, E. (2001), *Government of the Future*, Organisation for Economic Co-operation and Development, p. 217.

[‖]Adams, T., Bezner, J., and Steinhardt, M. (1995), Principle-centeredness: A values clarification approach to wellness, *Meas. Eval. Counsel. Develo.* **28**(3), 139–147.

[**]Trachtman, L. (1998), User-centered design: principles to live by, *Assistive Technol.* (by RESNA) **10**(1), 1–2.

[††]Watson, W. (2008), *The Principle Centered Life: Paradox—or Positive Living?* AuthorHouse. p. 4.

[‡‡]Simonsen, R. J. (1998), Principle-centered spine care: McKenzie principles, *Occup. med.* **13**(1), 167–183.

[§§]Cohen, D. J. (2001), In good company: How social capital makes organizations work, *Ubiquity* **1**(42) (ACM, 3).

PRINCIPLE CRITICAL PARAMETERS [psychology]   Required level of operation in order to minimally perform a specific task or set of tasks.*

PRINCIPLE KK [philosophy]   One is always in position to know that what one knows is incompatible with a plausible margin-for-error requirement for inexact knowledge. The requirement is based on the appealing claim that knowledge requires one's belief to be safely true. It is impossible for any creature to reach arbitrary high orders of knowledge with respect to any item of inexact knowledge. Inexact knowledge preserving the safety intuitions while avoiding the implausible consequence no possible creature can satisfy principle KK with respect to any piece of inexact knowledge. The principle provides a more flexible account of higher-order knowledge, allowing principle KK to hold or not between any two orders, and formulated the condition at which it does. Some information carried by knowledge states is necessarily hidden, and subjects cannot rely on direct information about their own states.†

PRINCIPLE MOLECULAR DYNAMICS [chemistry]   First-principles molecular dynamics based on the Car–Parrinello approach has been shown to be successful and reliable in the simulations of many complex chemical systems. In this approach both the atomic and electronic degrees of freedom are treated on an equal footing via a Lagrangian in which the dynamic variables include the coefficients of electronic wavefunction as well as the classical atomic positions and momenta in phase space.‡ See also (FIRST-PRINCIPLE REAL-SPACE SIMULATION;) FIRST-PRINCIPLE MOLECULAR DYNAMICS; (FIRST PRINCIPLE FORCES;) FIRST PRINCIPLE.

*Lognonné, J. L. (1994), 2D-page analysis: A practical guide to principle critical parameters, *Cell Mol. Biol.* **40**(1), 41–55.
†Dutant, J. (2007), *The Margin-for-Error Principle Revised*, Open Sessions of the Mind Association-Aristotelian Society Joint Sessions, Bristol, July 2007, p. 1.
‡Lin, X. and Trout, B. L. (2002), *Chemistry of Sulfur Oxides on Transition Metal Surfaces: Interfacial Applications in Environmental Engineering*.

PRINCIPLE OF A BALL PISTON PUMP [physics]   Equations for pressure distribution, pressure gradient, and shear distribution around the ball further used to determine the equilibrium position of the ball inside the cylinder block and the friction forces acting on the ball.§ See also PRINCIPLE OF COORDINATION.

PRINCIPLE OF ABSOLUTENESS [psychology]   Created world that is the perfect handicraft of a perfect creator; can best be represented by a symmetry and proportion, from outer details to inner center that is a point. Point is representative of creative principles, and path repeats one cosmic principle.¶ See also MANDALA PRINCIPLE.

PRINCIPLE OF ABSOLUTENESS OF RESPONSIBILITY [organizational psychology]   No superior can escape, through delegation, responsibility for the activities of subordinates, for it is she or he who delegated authority and assigned duties. See also PRINCIPLE OF PLANNING§

PRINCIPLE OF ACQUISITION [psychology] (Robert Nozick, 1938–2002)   (1) The appropriation of natural resources no one has ever owned before. For example, from Locke's theory of property, according to which a person (being a self-owner) owns her/his labor, and by "mixing her/his labor" with a previously unowned part of the natural world (e.g., by whittling a stick found in a forest into a spear), thereby comes to own it.§ (2) Under certain assumptions the bounded context parsability condition implies a bounded degree of error. Natural languages are well designed for both parsing and learning.‖ See also (PRINCIPLE OF TRANSFER;) PRINCIPLE OF RECTIFICATION; (NOZICK'S PRINCIPLE;) LOCALITY PRINCIPLE.

PRINCIPLE OF ACQUISITION REFORM [engineering]   Performance improvements made or developed to field a new generation of services or products, support equipment, and

§Jamzadeh, F. (1983), *Study of Losses in a Ball Piston Pump*, PhD thesis, Univ. Wisconsin—Madison.
¶Koontz, H. and O'Donnell, C. (1968), *Principles of Management*, 4th ed., McGraw-Hill.
‖Berwick, R. C. (1985), *The Acquisition of Syntactic Knowledge*, MIT Press, p. 46.

software. These enhancements of existing products or services help maintain plans for growth and sales.[*]

PRINCIPLE OF ACTION [psychology] (1) In political leadership, the brunt of the problem of moral judgment that leads to action is how well the judgment is institutionalized and how effectively the action can be regulated. In particular, public officials must be subjected to some degree of institutional oversight. This follows from two central facts: (a) these officials are in positions to take massive advantage of their roles, whose purpose is public service; and (b) in a complex modern society, individually generated moral principles are not likely to constitute a coherent and consistent body. Oversight must be guided by principles for action that are clear to all.[†] (2) No superiors can escape, through delegation, responsibility for the activities of subordinates, because they have delegated authority and assigned duties. Likewise, the responsibility of the subordinate to one's superior is absolute once one has accepted an assignment and the power to carry it out.[‡] (3) Control is justified only if indicated or experienced deviations from plans are corrected through appropriate planning, organizing, staffing, and directing.[§] See also PRINCIPLE OF PLANNING.

PRINCIPLE OF ACTUALISM (astronomy, geology, paleontology) (James Hutton, 1726–1797, Scottish geologist) Used to interpret past patterns for present-day-processes. It is also known as "the present is the key to the past."[§]

PRINCIPLE OF ALTERNATIVES [organizational psychology] Select the plan that is the most effective and the most efficient to the attainment of a desired goal.[¶] See also PRINCIPLE OF PLANNING.

PRINCIPLE OF AN ICE CREAM MAKER In any dilute solution, the solute molecules can be considered to behave like a gas whose pressure and volume correspond to the osmotic pressure and volume of the solution. Dilution of the solution by adding more solvent causes an "expansion" of the solute "gas," and cooling should result.[||]

PRINCIPLE OF ANGULAR ECHO SPECTROSCOPY [optics] Energy resolution in conventional inelastic neutron scattering spectroscopy allowing exploration of the behavior of condensed matter essentially on the timescale of thermal atomic vibrations. By its unique feature of making resolution and beam monochromatization independent from each other, enables enhancement of energy resolutions, by expansion of the accessible time domain by three or four orders of magnitude.[**]

PRINCIPLE OF ARCHIMEDES [physics] (1) A body immersed in a liquid or gas with the density r is subjected to a force S, the so-called hydrostatic thrust, which is equal to the weight of the liquid volume V removed by the body. If the body is partially immersed in the liquid (if the body floats in the liquid), the hydrostatic thrust equates the weight of the volume of the body part immersed in the liquid. The Archimedes thrust $S = r\ gV$ is generated by the resultant of all the forces that the fluid produces on the surface of the body by means of the hydrostatic or aerostatic pressure.[††] (2) The hydrostatic pressure that a liquid produces because of the gravity force depends on both the density and the height

[*]Higgins, Capt. G. (1997), CAIV—an important principle of acquisition reform, *Program Manager* **26**(1), 44.
[†]Hardin, R. (2006), *Morals for Public Officials. Moral leadership: The Theory and Practice of Power, Judgment and Policy*, Jossey-Bass, San Francisco, pp. 111–125.
[‡]Koontz, H. and O'Donnell, C. (1968), *Principles of Managment*, 4th ed., McGraw-Hill.
[§]Berwick, R. C. (1985), *The Acquisition of Syntactic Knowledge*, MIT Press, p. 46.

[¶]Koontz, H. and O'Donnell, C. (1968), *Principles op Management*, 4th ed., McGraw-Hill.
[||]Flynn, T. M. (2005), Refrigeration and liquefaction, in *Cryogenic Engineering Informa Health Care*, p. 359. 2nd ed., CRC Press, p. 359.
[**]Khasanov, O. K., Fedotova, O. M., and Samartsev, V. V. (2007), Principle of angular echo-spectroscopy, *J. Luminesc.* **127**(1), 55–60.
[††]Zilsel, E. (1942), The genesis of the concept of physical law, *Phil. Rev.* **51**(3), 245–279.

of the liquid inside a container.[*] (3) A hot-air balloon is subjected to an ascensional force F_a, which is given by the difference between the aerostatic thrust S and the weight P: $F_a = S - P = r_c V g - r_h V g$, where r_c and r_h are, respectively, the density of the external cool air and the hot air inside the balloon.[†]

PRINCIPLE OF AREA [psychology]   (1) Area sampling eliminates dependence on the assignment of quotas that may be more or less seriously in error, and does not permit the interviewer discretion in the choice of the individuals to be included in the sample. With appropriate methods of designating areas for coverage in the sample, the probabilities of inclusion of the various elements of the population are known, and consequently, the reliability of results form the sample can be measured and controlled.[‡] (2) States that the smaller of two overlapping figures is perceived as a figure while the larger is regarded as ground. See also GESTALT PRINCIPLE OF CONTINUITY; GESTALT PRINCIPLE OF PROXIMITY; GESTALT PRINCIPLE OF CONTIGUITY; GESTALT PRINCIPLE OF CLOSURE; GESTALT PRINCIPLE OF AREA; GESTALT PRINCIPLE OF SYMMETRY; GESTALT PRINCIPLE OF SIMILARITY.

PRINCIPLE OF ASSUMPTION [psychology] (1) It is possible to influence any life unit with one's thoughts and feelings. One may assume the power and characteristic of any life unit. To assume the feelings of a person or a thing is an outstanding action. To assume the feelings of a plant or tree brings oneness or realization that you are that that

you assumed. Anything can be assumed.[§] See also PRINCIPLE OF EMPATHY.

PRINCIPLE OF ASSURANCE OF OBJECTIVE [organizational psychology]   The task of control is to ensure accomplishment of objectives by detecting potential or actual deviation from plans early enough to permit effective corrective action. See also PRINCIPLE OF PLANNING.[¶]

PRINCIPLE OF ASSURANCE OF OBJECTIVE [psychology]   Control is justified only if indicated or experienced deviations from plans are corrected through appropriate planning, organizing, staff, and directing.[¶] See also PRINCIPLE OF COORDINATION.

PRINCIPLE OF AUTHORITY AND RESPONSIBILITY [psychology] (French industrialist, Henri Fayol)   Authority and responsibility must go together.[‖] See also FAYOL'S PRINCIPLES OF MANAGEMENT.

PRINCIPLE OF AUTHORITY AND RESPONSIBILITY see FAYOL'S PRINCIPLES OF MANAGEMENT

PRINCIPLE OF AVOIDING CONFLICT. [psychology]   If the primary aim of conflict avoidance mechanisms is to ensure that situations involving goal incompatibility do not arise, and if a major source of goal incompatibility is roler resource scarcity, then conflict avoidance mechanisms must initially concentrate upon aoviding a scarcity of desired resources or positions. In this way, relationships between relevant parties will remain cooperative.[**]

PRINCIPLE OF BALANCE [organizational psychology] (Fritz Heider)   (1) Results indicate that (a) subjects were more confident of other's success than their own, (b) the unexpected outcome was more often attributed

[*]Hubbert, M. K. and Rubey, W. W. (1959), Role of fluid pressure in mechanics of overthrust faulting. I. Mechanics of filled porous solids and its application to overthrust faulting, *Geol. Soc. Am. Bull.* **70**(2), 115–166.

[†]Bestaoul, Y. and Hamel, T. (2000), *Dynamic Modelling of Small Autonomous Blimps*, Laboratoire des Systèmes Complexes, CEMIF, Univ. Evry Val d'Essonne, France (http://lsc.univevry.fr/~bestaoui/blimp_modelfin.pdf).

[‡]Hansen, M. and Hauser, P. (1945), Area sampling—some principles of sample design, *Public Opin. Quart.* **9**, 183–193.

[§]Bau, S. (n.d.), Questions (http://dimacs.fzu.edu.cn/people/sbau/quests.html).

[¶]Koontz, H. and O'Donnell, C. (1968), *Principles of Management*, 4th ed., McGraw-Hill.

[‖]Wood, M. C. (2002), *Henri Fayol: Critical Evaluations in Business and Management*, Taylor & Francis, p. 8.

[**]Mitchell, C.R. (1989) The structure of international conflict. Palgrave Mcmillan. p 258.

to luck, (c) other's success was more often attributed to ability and failure to bad luck than self's own success or failure, (d) a positivity bias in recall favored the other, (e) contrast effects occurred for satisfaction ratings, and (f) task performance was a dominant factor influencing confidence and satisfaction ratings. Causes of action, in terms of positivity biases in social perception, and as indicating effects of the social context of performance upon attribution and valence.* (2) The application of principles or techniques must be balanced in the light of the overall effectiveness of the structure in meeting enterprise objectives.† See also PRINCIPLE OF PLANNING.

PRINCIPLE OF BALANCE [psychology]   Task of control is to ensure accomplishment of objectives by detecting potential or actual deviation from plans early enough to permit effective corrective action.‡ See also PRINCIPLE OF COORDINATION.

PRINCIPLE OF BELT FILTER PRESS SEE FILTER PRESS PRINCIPLE Press sludge between the endless and tensioned belts. The pressure increase gradually according the belts run through the rollers with decreasing diameter. With this movement the machine creates a shearing effect which gives benefit to better filtrate drainage.§

PRINCIPLE OF BENEFICENCE [philosophy]   A normative statement of a moral obligation to act for the benefit of others, helping them to further their important and legitimate interests, often by preventing or removing possible harms.¶ See also CROSS-PRINCIPLE RELATIONSHIPS.

PRINCIPLE   OF   BENEFICENCE [psychology] (1) Core duty of help—relationships to nonmaleficence, to altruism, and to paternalism. (2) Aspects of recordkeeping—client access, continuing education requirements, counselor contacts, peer supervision, methods of ensuring beneficial group counseling experiences, etc. (3) Medical use of narcotic drugs continues to be indispensable for the relief of pain and suffering: "While there have been efforts by some governments to ensure the availability of narcotic drugs for medical and scientific purposes, it appears that many others have yet to focus on that obligation."‖

PRINCIPLE OF BIOELECTRICAL IMPEDANCE [medical]   Applied in the study of several physiological and pathological processes. Monitoring the electrical conductivity of the bladder neck has been used as a diagnostic test for urinary urgency and as a method for biofeedback training in patients with urinary incontinence associated with detrusor instability.** See also MESH IMPEDANCE PRINCIPLE.

PRINCIPLE OF BIT/DIGIT-SERIAL DESIGN [engineering]   An important option for hardware sharing is the sequential treatment of the bits in a signal in time. If every bit in a word is processed with individual hardware and communicated over a separate wire, we call this a bit-parallel computation. Then, in every cycle a full 3-bit word is produced. On the other hand, if all the bits are treated in sequence on a single hardware unit.†† See also PRINCIPLE OF BIT-PARALLEL ADDITION.

PRINCIPLE OF BIT-PARALLEL ADDITION [engineering]   Ideally, the area can be smaller

*Feather, N. T. and Simon, J. G. (1971), Attribution of responsibility and valence of outcome in relation to initial confidence and success of self and others, *J. Person. Soc. Psychol.* **18**(2), 173–188.
†Koontz, H. and O'Donnell, C. (1968), *Principles of Management*, 4th ed., McGraw-Hill.
‡Sirgy, M. J. (2002), *The Psychology of Quality of Life*, Springer, p. 201.
§http://www.beltfilterpress.com/beltfilter-press.html.
¶Arneson, R. J. (2004), Moral limits on the demands of beneficence, in *Ethics of Assistance*, Chatterjee, D. K., ed., C. Cambridge Univ. Press.

‖Joranson, D. E. (2004), Improving availability of opioid pain medications testing the principle of balance in Latin America. Preview, *J. Palliative Med.* **7**(1), 105–114.
Rataniyx, R. S., Yazakiz, E., Mawy, A., Piloty, M. A., Rogersy, J., and Williamsy, N. S. (1998), Pelvic bioelectrical impedance measurements to detect rectal filling, *Physiol. Meas.* **19, 528.
††Ingrid, D. I. S. (2000), *Integrated Processor-Level Architectures for Real-Time Digital Signal Processing*, http://www.ee.ucla.edu/ingrid/ee213a/FCchap3p1.pdf. p. 21.

and the clock rate can be faster for the bit-serial case compared to a fully bit-parallel realization. Typically, the maximal clock rate of bit-serial hardware lies between 20 and 50 MHz, which means that the serial sample rate at word level is still lower than what can be achieved in bit-parallel architectures.* See also PRINCIPLE OF BIT/DIGIT-SERIAL DESIGN.

PRINCIPLE OF BIVALENCE [mathematics] For any proposition P, either P is true or P is false. For any proposition P, at a given time, in a given respect, there are three related principles: (1) the principle of bivalence, which states, for any proposition P, that P is either true or false; (2) the principle of the excluded middle, which states, for any proposition P, that P is true or "not-P" is true; and (3) the principle of noncontradiction, which states that for any proposition P, it is not the case that both P is true and "not-P" is true.[†] See also PRINCIPLE OF EXCLUDED MIDDLE; PRINCIPLE OF NONCONTRADICTION.

PRINCIPLE OF BIVALENCE [philosophy] Every proposition or statement is either true or false. It therefore proposes that every statement has a truth value, and there are only two truth values (true or false). Under the standard interpretation of the logical connectives this principle is represented in a formal system as the law of the excluded middle.[‡] See also PRINCIPLE OF EXCLUDED MIDDLE.

PRINCIPLE OF BLACK HOLE COMPLEMENTARITY [physics] (Larus Thorlacius, John Uglum, and Leonard Susskind) Special theory of relativity that although different observers disagree about the lengths of time and space intervals, events take place at definite space-time locations. Elementary particles are made of even smaller constituents. An elementary particle does not resemble a point; rather, it is like a tiny rubber band that can vibrate in many modes. The fundamental mode has the lowest frequency; then there are higher harmonics, which can be superimposed on top of one another. There are an infinite number of such modes, each of which corresponds to a different elementary particle.[§] See also STELLAR PRINCIPLE; BLACK HOLE PRINCIPLE; ANTHROPIC PRINCIPLE.

PRINCIPLE OF BRAUN–LE CHATELIER [physics] (1) Properties of states with minimal entropy production such as to be an adequate generalization; based on the processes that compensate for effects.[¶] (2) Equilibrium thermodynamics for two sets of general parameters of the molecular total energy function; when an external force is applied to a system at equalibrium, the system adjusts as to minimize the effect of the applied force.[‖] See also BRAUN–LE CHATELIER PRINCIPLE; SAMUELSON–LE CHATELIER PRINCIPLE; COMPENSATION PRINCIPLE; LE CHATELIER'S PRINCIPLE.

PRINCIPLE OF BRIQUETTING [engineering] (1) Granulating powder materials by compression and structure of a granulator. (2) Method for producing formed parts by supplying powder materials into pockets provided on surfaces of a pair of rolls that rotate at a constant speed so as to be mold cavities for the formed parts. Process assists with volume reduction, environmental countermeasures, and improvement of chemical reactions.[**]

PRINCIPLE OF BUOYANCY [physics] A body immersed wholly or partially in a fluid that is buoyed up by a force equal in magnitude to the weight of the volume of fluid that it displaces. Object immersed in a fluid has an upward force equal to the weight of the fluid displaced by the

*Op. cit. Ingrid, p. 23.

[†]Bell, J. L. (1993), Hilbert's ϵ=epsilon operator and classical logic, *J. Phil. Logic* **22**(1), 1–18.

[‡]Wright, C. (1986), *Realism, Meaning & Truth*, Blackwell, Oxford.

[§]Susskind, L. (1993), String theory and the principle of black hole complementarity, *Phys. Rev. Lett.* **71**, 2367–2368.

[¶]Fainerman, V. B., Miller, R., and Wüstneck, R. (1996), Adsorption of proteins at liquid/fluid interfaces, *J. Colloid Interface Sci.* **183**(1), 26–34.

[‖]Joos, P. and Serrien, G. (1991), The principle of Braun-LeChatelier at surfaces, *J. Colloid Interface Sci.* **145**(1), 291–294.

[**]Anon. (1910), *The world to-day; a monthly record of human progress*, Hearst's International, International Magazine Co., p. 1038.

object.* See also HYDROSTATIC PRINCIPLE, ARCHIMEDES' PRINCIPLE.

PRINCIPLE OF BURST [optics]   A class of ultrafast imaging techniques with an improved signal-to-noise ratio. Conventional BURST excites a set of equally spaced, narrow strips within an object, and creates an image from a single slice, perpendicular to the direction of the strips.[†]

PRINCIPLE OF CALCULATION [psychology] problem of mental work can be resolved with simple logical elaboration, given (1) the basic principle of calculation (the idea of mental and psychophysical work, the determination of a unit of work, the psychophysical homogeneity of the two series), (2) the method of development of the computation (theory of graphic procedure for the deduction of mental work, mean duration of the psychophysical and mental operations, deduction of mental work as a particular case of psychophysical calculation), and (3) the characteristic function of mental work (characteristics of mental work, graphs of the psychophysical equivalent of mental work). Regarding the application of logic to the study of scientific problems.[‡] See KAPLAN–MEIER PRINCIPLE OF CALCULATION.

PRINCIPLE OF CALORIMETRY [biology] Calorimetry involves the use of a calorimeter. The word *calorimetry* is derived from the Latin word *calor*, meaning heat. Indirect calorimetry calculates heat that living organisms produce from their production of carbon dioxide and nitrogen waste or from their consumption of oxygen. Heat production can be predicted from oxygen consumption this way, using multiple regression.[§]

PRINCIPLE OF CARATHÉODONY [physics] See CARATHÉODORY'S PRINCIPLE.

PRINCIPLE OF CATEGORICAL SCOPE [psychology]   (1) Words can be extended to other objects in the same basic level category. Extension will primarily be based on how close other exemplars are to the prototype or central core of the category, demonstrating the organized structure of children's conceptual categories and word meanings.[¶] (2) A word can be extended to objects in the same basic-level category as the original referent. At the basic level, object kind is defined in terms of perceptual similarity (both overall and in the type of parts something has) and function (particularly in the case of artifacts). In the case of natural kinds (such as gold or chicken), basic-level membership is also a function of deeper, nonvisible commonalities such as similar internal structures. When an out-of-category item is perceptually similar to the original referent and a perceptually similar in-category item is also present, categorical scope predicts category membership—the kind of thing an item is—should win out as the basis for extension.[‖] (3) Primary basis for extension of an object word is basic-level category assignment. See also PRINCIPLE OF REFERENCE; PRINCIPLE OF EXTENDIBILITY; PRINCIPLE OF OBJECT SCOPE; NOVEL NAME–NAMELESS CATEGORY (N3C) PRINCIPLE; PRINCIPLE OF CONVENTIONALITY.

PRINCIPLE OF CAUSALITY [psychology]   (1) In order to assign the sensation of a cause, it is necessary to assign causes for every phenomenon and in order to make this induction universal and necessary, this feeling of need

*Stepanoff, A. J. (1969), *Gravity Flow of Bulk Solids and Transportation of Solids in Suspension*, Wiley, New York.
†Duyn, J. H., van Gelderen, P., Liu, G., and Moonen, C. T. W. (1994), *Fast Volume Scanning with Frequency-Shifted BURST MRI*, Laboratory of Diagnostic Radiology Research, OIR, NIH, Betheseda, MD (http://www.lfmi.ninds.nih.gov/pubpdf/DuynMRM1994b.pdf).
‡Pastore, A. (1932), *Fondementi del Calcolo del Lavoro Mentale. Saggio di Analisi Logica*, Atti della Reale Accademia delle Scienze, Torino, p. 62.

§*The Columbia Electronic Encyclopedia* (2007), Columbia Univ. Press.
¶Barrett, M. D. (1999), *The Development of Language: Love, Money and Daily Routines*, Psychology Press, p. 316.
‖Golinkoff, R., Shuff-Bailey, M., Olguin, R., and Ruan, W. (1995), Young children extend novel words at the basic level: Evidence for the principle of categorical scope, *Devel. Psychol.* **31**(3), 494–507.

must be universal and necessary.* (2) In management, enterprise objectives must be considered when decisions are being made to focus on the over—all effectiveness of a system, process, condition, or service.† See also PRINCIPLE OF CAUSE AND EFFECT; PRINCIPLE OF PURPOSE.

PRINCIPLE OF CAUSE AND EFFECT [psychology] First published in December 1908, *The Kybalion: Hermetic Philosophy*, is by the anonymous "Three Initiates." The seven principles are as follows. *Principle of mentalism* embodies the truth that "all is mind." The "all" as a substantial reality underlying all the outward manifestations and appearances that we know under the terms of "the material universe," the "phenomena of life," "matter," "energy," and in short, all that is apparent to our material senses. It also defines it as "spirit," which is also unknowable and undefinable, but may be considered of as an universal, infinite, living mind; "mental transmutation" means the art of changing and transforming mental states, forms, and conditions, into others. *Principle of correspondence* embodies the truth that there is always a correspondence between the laws of phenomena of the various planes of being and life. There is a harmony, agreement and correspondence between the following "planes": the great physical plane, the great mental plane, and the great spiritual plane. *Principle of vibration* embodies the truth that motion is manifest in everything in the universe, that nothing rests, and that everything moves, vibrates, and circles. The differences between different manifestations of matter, energy, mind, and even spirit, are the result of only different "vibrations." The higher you are on the scale, the higher your rate of vibration will be. Here, "the all" is purported to be at an infinite level of vibration, almost to the point of being at rest. There are said to be millions on millions of varying degrees between "the

all," and the objects of the lowest vibration. *Principle of polarity* embodies the truth that everything is dual, everything has two poles, and everything has its pair of opposites. *Principle of rhythm* embodies the truth that in everything there is manifested a measured motion, a to and a fro; a flow and an inflow; a swing backward and a swing forward, in short, a pendulumlike movement. There is rhythm between every pair of opposites, or poles, and is closely related to the principle of polarity. *Principle of cause and effect* explains that there is a cause for every effect, and an effect for every cause. It also states that there is no such thing as chance, that chance is merely a term indicating cause existing but not recognized or perceived. *Principle of gender* embodies the truth that there is gender manifested in everything. It is stated that the view of the authors does not relate to "sex," but to "beget; to procreate, to generate, to create, or to produce" in this respect. (2) Generally accepted in the physical sciences, but it has only recently been applied to the understanding of human behavior. With the application of this principle, psychology loses its metaphysical element and becomes based on definite, understandable laws rather than on metaphysical guesses. The fundamental principles underlying the science of human behavior are the same as the basic principles underlying the natural sciences. Just as the principle of cause and effect is basic to an understanding of every natural science, so the same principle is indispensable in the science of psychology. The physical sciences deal with the prediction, control, and cause of certain phenomena in nature. Psychology deals with the prediction, control, and cause of human behavior. As the natural sciences are much older than psychology, they have advanced much farther in understanding of the phenomena with which they deal.‡ (3) [Occultism] Explains that there is a cause for every effect, and an effect for every cause. It also states that there is no such thing as chance, that chance is merely a term indicating extant causes not recognized

*Cousin, V. (1838), *Elements of Psychology*, Gould & Newman, Harvard Univ. p.145.
†Meiman, N. N. (1964), *The Principle of Causality and Asymptotic Behavior of the Scattering Amplitude, Akademiya Nauk SSSR Institut Teoreticheskoi 'Eksperimentolnoi Fiski, Moscow IOP of Science*.

‡Moss, F. (1929), Basic principles of behavior, *Appl. Psychol. Houghton Mifflin*, p.3–4.

or perceived.* See also MASCULINE PRINCIPLE; FEMININE PRINCIPLE; PRINCIPLE OF CAUSALITY.

PRINCIPLE OF CELL SURFACE RECEPTOR [physics]   Promotes cell motility, adhesion, and proliferation to assist with morphogenesis, wound repair, inflammation, and metastasis. These processes require massive cell movement and tissue reorganization. Many of the effects are mediated through cell surface receptors directing cell trafficking during physiological and pathological events. When these processes are deregulated, cell behavior becomes uncontrolled, leading to developmental abnormalities, abnormal physiological responses, and tumorigenesis.[†] See also CAVITATION PRINCIPLE.

PRINCIPLE OF CENTRALIZATION [psychology] (Henri Fayol, 1841-1925, French industrialist)   Centralization of authority to be desirable, at least for overall control.[‡] See also FAYOL'S PRINCIPLES OF MANAGEMENT.

PRINCIPLE OF CHANGE [psychology]   Reality is a process that is not static but rather is dynamic and changeable. A thing need not be identical to itself at all because of the fluid nature of reality.[§]

PRINCIPLE OF CHARGE CONSERVATION [physics]   (Benjamin Franklin, 1747) (1) There are two types of charge: positive and negative. Charge may be transferred from one object to another; however, the net charge of a closed system does not change. (2) A conserved quantity cannot be created or destroyed; it can only be transferred.

Therefore, means that the net charge in the universe remains constant.[¶]

PRINCIPLE OF CHARITY [psychology] (Donald Herbert Davidson, 1917-2003, American philosopher)   (1) An approach to understanding a speaker's statements by rendering the best, strongest possible interpretation of an argument's meaning.[||] (2) The goal is to help keep people who are trying to understand or evaluate the truth of an argument from introducing a logical fallacy or other error into an argument that is not inherent to it.[||] (3) We make maximum sense of the words and thoughts of others when we interpret in a way that optimizes agreement.[||] (4) Requires interpreting a speaker's statemetns to be rational and, in the case of any argument, considering its best, strongest possible interpretation. In its narrowest sense, the goal of this methodological principle is to avoid attributing irrationality, logical fallacies or falsehoods to the others' statements, when a coherent, rational interpretation of the statement is available. Simon Blackburn states "it constrians the interpreter to maximize the truth or rationality in the subject's sayings."[**]. See also PRINCIPLE OF HUMANITY.

PRINCIPLE OF CHIRAL INTERACTION [physics] The existence of a molecular magnetic moment can lead to a natural selection of one chirality for certain amino-acids due to coupling with the magnetic field of the earth at the water surface.[††]

PRINCIPLE OF CHOICE OF OPTIMAL DECISIONS [mathematics]   If there is uncertainty, a decision that maximizes expected utility is sought. Optimization

*Three Initiates (2004), *The Kybalion*, Book Tree (the "Three Initiates" who authored *The Kybalion* chose to remain anonymous; as a result, there has been a great deal of speculation about who actually wrote the book.)
[†]Entwistle, J., Hall, C. L., and Turley, E. A. (1998), HA receptors: Regulators of signalling to the cytoskeleton, *J. Cell. Biochem.* **61**(4), 569–577.
[‡]Wood, M. C. (2002), Henri Fayol: Critical Evaluations in Business and Management. Taylor & Francis, p. 8.
[§]Nisbett, R., Peng, K., Choi, I., and Norenzayan, A. (2001), Culture and systems of thought: Holistic versus analytic cognition, *Psychol. Rev.* **108**(2), 291–310.

[¶]*Encyclopedia Britanica*.
[||]Blackburn, S. (1994), Charity, principle of, in *The Oxford Dictionary of Philosophy*, Oxford Univ. Press, p. 62.
[**]Davidson, Donald (1984) [1974]. "Ch. 13: On the Very Idea of a Conceptual Scheme". Inquiries into Truth and Interpreation. Oxford: Clarendon Press.
[††]Gilat, G., and Schulman, L.S., (1985) Chiral interaction, magnitude of the effects and application to natural selection of L-enantiomer. Chemical Physics Letters, v. 121, Issues 1-2, 1, p. 13–16.

models arise as a result of application of one or more of these principles. New and challenging mathematical problems arise in obtaining the results from optimizing models.[*] See also ENTROPY OPTIMIZATION PRINCIPLE.

PRINCIPLE OF CHOOSING OPTIMAL ALGORITHMS [mathematics]   Algorithms which minimize the time taken or maximize the reliability of the results. Optimization models arise as a result of application of one or more of these principles. New and challenging mathematical problems arise in obtaining the results from optimizing models.[*] See also ENTROPY OPTIMIZATION PRINCIPLE.

PRINCIPLE OF CLOSURE [psychology] (K., Koffka, psychologist)   Applies when we tend to see complete figures even when part of the information is missing. "Closed areas are more stable and, therefore, more readily produced than unclosed ones."[†] A general law of perceptual organization. Closure is said to be one of the ways in which dynamic self-distribution in the brain field achieve a better, simpler, more *pregnant* form.[‡] See also PRINCIPLE OF CONTIGUITY; PRINCIPLE OF PROXIMITY; GESTALT PRINCIPLE OF CONTINUITY; GESTALT PRINCIPLE OF PROXIMITY; GESTALT PRINCIPLE OF CONTIGUITY; GESTALT PRINCIPLE OF CLOSURE; GESTALT PRINCIPLE OF AREA; GESTALT PRINCIPLE OF SYMMETRY; GESTALT PRINCIPLE OF SIMILARITY.

PRINCIPLE OF COFINALITY [mathematics] (Ernst Friedrich Ferdinand Zermelo, 1871–1953   German   mathematician) (1) Every linearly ordered set has a well ordered cofinal subset, implies LW: Every linearly ordered set can be well ordered. (2) Equivalent to a corresponding variation of Zorn's lemma. For example, variations of Hausdorff's maximal principle states that

Zorn's Lemma may be as follows: Every nonempty partially ordered set x in which every linearly ordered subset has an upper bound, has a maximal element.[§]

PRINCIPLE OF COGNITIVE DISSONANCE [psychology] (aka Festinger Principle of Cognitive Dissonance) (Leon Festinger, 1919–1989, American social psychologist)   Concept related to the tendency to avoid internal contradictions in certain situations, and as a higher order theory about information processing in the human mind. The question remains: is cognitive dissonance a process intrinsically associated with the way that the mind processes information, or is it caused by such specific contradictions? Doing something unpleasant for a modest payment leads to an internal conflict, which is amended by the reevaluation of the task (attributional bias). The Festinger-Carlsmith experiment became a prototypical framework in the study of cognitive dissonance. Cognitive dissonance has a 'dialectic structure': it relies on the assumption that mental objects associated with conflicting attributional values converge to a economical cognitive output. Cognitive dissonance represents more than just a hypothesis about a specific type of cognitive phenomenon; it is a 'higher order theory' regarding information and behavioral output, based on two axioms: 1. We treat information according to the tendency to diminish contradiction and increase organization, and this can lead to irrational behaviors; 2. This phenomenon takes place outside the conscious sight. Within the field of psychology, cognitive dissonance disavowal the behaviorist assumption that rewards are always associated with the tendency to increase a target-behavior. As revealed in the aforementioned experiment, rewards are inversely correlated with positive evaluations of the rewarded behaviors, thus suggesting that, in the long

---

[*]Kapur, J. N. (1988), *Mathematical Modeling*, New Age International, p. 203.
[†]Koffka, K. (1935), *Principles of Gestalt Psychology*, Hartcourt, Brace, New York, p. 167.
[‡]Postman, L. and Bruner, J. S. (1952), Hypothesis and the principle of closure: The effect of frequency and recency, *J. Psychol. Interdisc. Appl.* **33**, 113.

[§]Harper, J. M., and Rubin, J. E. (1976), Variations of Zorn's lemma, principles of cofinality, and Hausdorff's maximal principle, *Notre Dame J. Formal Logic* **17**(4), 565–588.

run, the former could in fact diminish the occurrence of the latter.*

PRINCIPLE OF COMPARATIVE ADVANTAGE [mathematics] Economic reform did affect trade patterns. China's trade correlated with the dictates of the Heckscher–Ohlin theorem during the early periods of economic reform. These reforms represent a first step in China's move to rationalize its economy and realize the efficiency benefits of trade and production based on comparative advantage.†

PRINCIPLE OF COMPETITIVE EXCLUSION [physics] Charge of an isolated system cannot change.

PRINCIPLE OF COMPETITIVE STRATEGIES [organizational psychology] In a competitive arena, it is important to choose plans in the light of what the competitor will or will not do and navigate according to what competitors are doing or not doing.‡ See also PRINCIPLE OF PLANNING.

PRINCIPLE OF COMPETITIVE STRATEGIES [psychology, mathematics] (1) The development of close mutual relationships between firms and stakeholders who interact with each other may become more pervasive. (2) The management of symbols may become a central principle of competitive strategies designed to perform in markets devoid of physical cues. (3) The emerging array of new technologies can trigger frequent redefinitions of industry boundaries and paradigms. In these environments, many of the actions pioneered by others can serve as templates of "reputation repertoires" that other firms are likely to emulate as they contemplate their sources of competitive advantage in the next millennium.‡

PRINCIPLE OF COMPLEMENTARITY [genetics, biology, physics] (1) One species seeking same ecological environment will survive while the other will expire under a given set of conditions. (2) If there is competition between two species for a common place in a limited microcosm, we can quite naturally extend the premises implied in the logistic equation. The growth rate of each competing species in a mixed population will depend on (a) the potential rate of population increase of a given species and (b) the unutilized opportunity for growth of this species. (3) Levy's discovery of the complementarity between random displacements, parameterized by time, and random time, parameterized by position, in the description of Brownian motion.§ See also COMPETITIVE EXCLUSION PRINCIPLE; GAUSE'S PRINCIPLE.

PRINCIPLE OF COMPLEMENTARY ENERGY [engineering] (Carlo Alberto Castigliano, b. 11/9/1847, Asti, Italy). See LEAST-ACTION PRINCIPLE.

PRINCIPLE OF COMPLETE INDUCTION [mathematics] See LEAST-NUMBER PRINCIPLE; INDUCTION PRINCIPLE.

PRINCIPLE OF COMPOSITION OF THE TRANSLATIONS. [mathematics] Any translation, regardless of how complex, can be considered to be composed of elemental translations, in a series, in parallel or in combinations of both. See also SPATIAL PRINCIPLE; PRINCIPLE OF TRANSLATION QUANTITATIVITY; PRINCIPLE OF THE CENTRE-PERIPHERY; PRINCIPLE OF NUCLEATION; UNIFIED PRINCIPLE OF ACCUMULATED ADVANTAGES.¶

*Dias, Álvaro Machado, Oda, Eduardo, Akiba, Henrique Teruo, Arruda, Leo, Bruder, Luiz Felipe (2009) Is Cognitive Dissonance an Intrinsic Property of the Human Mind? An Experimental Solution to a Half-Century Debate. World Academy fo Science, Engineering and Technology. **54**, p. 784–788.
†Chang, H.-J. (2002), *Kicking Away the Ladder: Development Strategy in Historical Perspective*, Anthem Press.
‡Koontz, H. and O'Donnell, C. (1968), Principles of Management, 4th ed., McGraw-Hill.

§Ruthen, R. (1991), Waves are waves … ; … And particles are particles, and never the twain shall meet. (an apparatus that may test the principle of complementarity), *Sci. Am.* **265**, 21.
¶Ruiz-Banos, R., Bailon-Moreno, R., Jimenez-Contreras, E., and Courtial, J.P. (1999) Structure and dynamics of scientific networks. Part 2: the new Zipf's Law, the cocitations's clusters and the model of the presence of key-words, Scientometrics, 44, p. 235–265.

PRINCIPLE OF COMPOSITIONALITY [psychology] (Donald Davidson, 1917—2003, American philosopher) (1) The meaning of a complex expression is determined by the meanings of its constituent expressions and the rules used to combine them. (2) In a meaningful sentence, if the lexical parts are removed from the sentence, what remains will be the rules of composition. (3) Every operation of the syntax should be associated with an operation of the semantics that acts on the meanings of the constituents combined by the syntactic operation.* (4) Every construct of the syntax should be associated with a clause of the T schema with an operator in the semantics that specifies how the meaning of the whole expression is built from constituents combined by the syntactic rule. In some general mathematical theories this guideline is taken to mean that the interpretation of a language is essentially given by a homomorphism between an algebra of syntactic representations and an algebra of semantic objects.[†] See also FREGE'S PRINCIPLE.

PRINCIPLE OF COMPUTATIONAL EQUIVALENCE [mathematics]   Systems found in the natural world can perform computations up to a maximal level of computational power.[‡]

PRINCIPLE OF COMPUTED TOMOGRAPHY [medicine, mathematics] Combined application of traditional methods of X-ray diagnosis (radiography, tomography) and X-ray computed tomography (CT) allowing increased accuracy of the diagnosis.[§]

PRINCIPLE OF CONCENTRATION [mathematics]   Applied in measure theory, probability

and combinatorics, and has consequences for other fields such as Banach space theory.[¶]

PRINCIPLE OF CONCEPTUAL DESIGN [physics] See also ACTION PRINCIPLE; SCHWINGER QUANTUM ACTION PRINCIPLE.

PRINCIPLE OF CONFERRAL [psychology] The European Union is a union of member states, and all its competences are voluntarily conferred on it by its member states. The EU has no competences by right, and thus any areas of policy not explicitly agreed on in treaties by all member states remain the domain of the member states.[‖]

PRINCIPLE OF CONGREGATION [psychology] (Plato) (1) Complete development through the binding force of love that seeks to achieve the unity of humankind. Cultivation of the right kind can provide a solution to problems that afflict the world.** See also PRINCIPLE OF EROS.

PRINCIPLE OF CONGRUENCE [chemistry] (Johannes Nicolaus Broensted, Danish chemist) (1) Interaction of the ions of acid and basic solutions on the exchange of protons during reactions. (2) Acid and basic catalysis of many chemical reactions, recognizing the concept of enthalpy to characterize chemical reactivity. (3) All acid-based reactions consist simply of the transfer of a proton from one base to another.[††] See also BROENSTED'S PRINCIPLE; BROENSTED'S PRINCIPLE OF CONGRUENCE.

PRINCIPLE OF CONJUGATED CHEMICAL REACTIONS [chemistry]   Means of reducing free energy of a reaction by conjugation of two or more reactions that are thermodynamically nonequivalent. Applying the methods

*Lepore, E. and Smith, B. C. (2006), New York, pp. *The Oxford Handbook of Philosophy of Language*, Clarendon Press/Oxford Univ. Press, 633–666.
[†]Fodor, J. (1995), Comprehending sentence structure, in *Language: An Invitation to Cognitive Science*, 2nd ed., MIT Press, Cambridge, MA, vol. 1, pp. 209–246.
[‡]Wolfram, S. (2002), The principle of computational equivalence, in *A New Kind of Science*, Wolfram Media, Champaign, IL, pp. 5–6, 715–846.
[§]Murry, R. C., Dowdey, J. E., and Christensen, E. E. (1990), *Christensen's Physics of Diagnostic Radiology*, 4th ed., Lippincott Williams & Wilkins, p. 289.

[¶]Fu, Y. (2009), The principle of concentration compactness in Lp(x) spaces and its application, *Nonlinear Anal. Theory Meth. Appl.* v. 71(5-6) p. 1876–1892.
[‖]Gerven, W. (2005), *The European Union: A Polity of States and Peoples*, Stanford Univ. Press, p. 268.
**Hoelzl, M. and Ward, G. (2006), *Religion and Political Thought: Key Readings—Past and Present*, Continuum International Publishing Group, p. 79.
[††]Peters, C. J., et al. (1995), The principle of congruence and its a lication to compressible states, *Fluid Phase Equilib.* **105**, 193.

of stationary kinetics and using the theory of absolute reaction rates, one can obtain a formula for the summation of the free energy of activation of the main reaction and the free energy of activation of the conjugated reaction that precedes it.*

PRINCIPLE OF CONSERVATION OF ENERGY [physics] (Isaac Newton) (1) Under ordinary conditions, matter and energy can be neither created nor destroyed, but only changed from one form to another (matter as a form of energy: $E = mc^2$). Julius Robert Mayer (1814–1878, German physician) used the German word *kraft* (force) for ideas that we now denote with the word *energy*.[†] See also (PRINCIPLE OF CHANGE CONSERVATION).

PRINCIPLE OF CONSERVATION OF MATTER [chemistry] Matter is neither created nor destroyed during any physical or chemical change.[‡]

PRINCIPLE OF CONSERVATION OF MATTER AND ENERGY [physics] (Isaac Newton) Under ordinary conditions, matter and energy can be neither created nor destroyed, but only changed from one form to another. (Julius Robert Mayer (1814–1878) German physician). Used the German word Kraft (force) for ideas we now denote with the word energy.[§] See also PRINCIPLE OF CHANGE CONSERVATION.

PRINCIPLE OF CONSERVATION OF MOMENTUM [physics] If the net external force acting on a system is zero, then the total momentum of the system does not change.[¶]

PRINCIPLE OF CONSERVATION OF TOTAL VORTICITY [physics] (1) Employed in the beneficiation of ores. (2) Flotation is based on introducing chemical and the air into water containing solid particles of different materials suspended therein that causes adherence of air to certain suspended solid and to render the particles having air bubbles thus adhered thereto lighter than the water. Accordingly, they rise to the top of the water to form a froth, which as such is skimmed off.[‖] See also FLOTATION POLISHING PRINCIPLE; PRINCIPLE OF FLUID ORIFICING; PRINCIPLE OF VIRTUAL WORK.

PRINCIPLE OF CONSERVATION OF WORK [engineering] (Carlo Alberto Castigliano, 1847–1884, Italian structural engineer). The coupling of stress and flow is performed in a light fashion. Displacements of a structure are interpolated from stress. The aerodynamic loads, pressure, and shear stresses are transferred in a conservative fashion in such a way that the loads action on the structure are conserved locally and globally over nonmatching unstructured interfaces.[**] See LEAST-ACTION PRINCIPLE.

PRINCIPLE OF CONSTANCY [psychology] A strong tendency to perceive objects as constant in size, shape, color, and other qualities, despite changes that occur in stimulation.[††] See also PRINCIPLE OF CONSTANCY'S OPPOSITE; PRINCIPLE OF PERCEPTUAL CONTEXT; GESTALT PRINCIPLES; also termed *constancy principle*.

PRINCIPLE OF CONSTANCY [psychology] (1) Sometimes called the theory or theorem of constancy. The fundamental theoretical precept of psychoanalysis will be essentially directed to clarifying its relation to associated formulations about (1) the theory of abreactions of accretions of stimuli; (2) two principles of neuronic inertia; (3) the theory

*Polevaya, O. Yu. and Kovalev, I. E. (1978), Principles of synthesis of conjugated antigens, *Pharm. Chem. J.* **12**(2), 155–166.

[†]Hussen, A. M. (2000), *Principles of Environmental Economics: Economics, Ecology and Public Policy*, Routledge, p. 76.

[‡]Walker, P. M. B. (1999), *Chambers Dictionary of Science and Technology*, Chambers, New York, p. 661.

[§]Hussen, A. M. (2000), *Principles of Environmental Economics: Economics, Ecology and Public Policy*, Routledge, p. 76.

[¶]Wolfram, S. (n.d.), *Fundamental Physics*, chap. 9, Wolframscience.com (http://www.physics.unlv. edu/~lenz/PHYS%20180.S09/Chap9.sum.htm).

[‖]Banerjee, P. K. and Morino, L. (1990), *Boundary Element Methods in Nonlinear Fluid Dynamics: Developments in Boundary Element Methods*, Taylor & Francis, p. 260.

[**]Lepage, C. Y. (n.d.), *Conservative Interpolation of Aerodynamic Loads for Aeroelastic Computations*, Concordia Univ., Montreal, Canada.

[††]Needles, W. (1969), The pleasure principle, the constancy principle, and the primary autonomous ego, *J. Am. Psychoanal. Assoc.* **17**, 808–825.

of cathexis; (4) the pleasure and unpleasure principles; (5) Fechner's principle of stability (or of constancy or constant equilibrium); (6) the Nirvana principle; (7) the death and life drives; (7) homeostasis; and (8) the principle of psychic determinism. The intertwining of the principle of constancy with so many associated formulations, put forward at different periods during the development of psychoanalysis, speaks not only for its central position, but also for the different contexts and applications.* (2) Inspired by Sigmund Freud's clinical observations, first with the field of cathartic therapy and then through experiences in the early usage of psychoanalysis. The recognition that memories repressed in the unconscious created increasing tension, and that this was relieved with discharged like phenomena when the unconscious was made conscious. The two principles of "neuronic inertia" are found to offer the key to the ambiguous definition of the principle of constancy. The "original" principles, which sought the complete discharge of energy (or elimination of stimuli), became the foreunner of the death drive; the "extended" principle achieved balances that were relatively constant, but succumbed in the end of complete discharge. This was the predecessor of the life drives. A revision of the principle of constancy was suggested, and it as renamed the Nirvana principle. The former basis for the constancy principle, the extended principle of inertia, became identified with Eros. Freud's later teachings about the Nirvana principle and Eros suggest a continuum of "constancies" embodied in the structural and functional development of the mental apparatus as it evolves from primal unity with the environment (e.g., the mother-child unit) and differentiates in patterns that organize the inner and outer worlds in relation to each other.[†]

PRINCIPLE OF CONSTANCY'S OPPOSITE [psychology]   Sometimes an object or pattern of stimulation will remain constant, but the perceived effect will vary.[1891] See also PRINCIPLE OF CONSTANCY; PRINCIPLE OF PERCEPTUAL CONTEXT; GESTALT PRINCIPLES.

PRINCIPLE OF CONSTANT PROPORTIONS [physics] Every pure substance always contains the same elements combined in the same proportion by weight.[‡] Also referred to as *principle of definite proportions*.

PRINCIPLE OF CONSTRUCTION AND OPERATION [psychology]   A relay monitoring the current; has inverse characteristics with respect to the currents being monitored.[§] See also AUTOMATIC CONTROL PRINCIPLES; PRINCIPLE OF DOMINANT SUBSYSTEMS; PRINCIPLE OF FUNDAMENTAL CHARACTERISTICS; PRINCIPLE OF OPERATION; EQUIPARTITION PRINCIPLE; PRINCIPLE OF SMOOTH FIT.

PRINCIPLE OF CONTIGUITY [psychology] States that things that are closer together will be seen as belonging together.[¶] See also PRINCIPLE OF PROXIMITY; GESTALT PRINCIPLE OF CONTINUITY; GESTALT PRINCIPLE OF PROXIMITY; GESTALT PRINCIPLE OF CONTIGUITY; GESTALT PRINCIPLE OF CLOSURE; GESTALT PRINCIPLE OF AREA; GESTALT PRINCIPLE OF SYMMETRY; GESTALT PRINCIPLE OF SIMILARITY.

PRINCIPLE OF CONTINUITY OF VELOCITY [physics]   Flowing fluid does not experience discontinuous changes at interfaces with other fluids or solids.[‖]

PRINCIPLE OF CONTINUOUS CHOICE [mathematics]   Intuitionistic mathematics diverges from other types of constructive mathematics in its interpretation of the term 'sequence'. There are two basic continuous choices divided into a continuity part and a

*Kanzer M. (1983), The inconstant "principle of constancy". *J. Am. Psychoanal. Assoc.* **31**(4) 843–865.

[†]Kanzer M. (1983), The inconstant "principle of constnacy". *J. Am. Psychoanal. Assoc.* **31**(4) 843–865.

[‡]Walker, P. M. B. (1999), *Chambers Dictionary of Science and Technology*, Chambers, New York, p. 661.

[§]Hewiston, L. G., Ramesh, B., Brown, M. and Balakrishnan, R. (2005), *Practical Power System Protection*, Newnes.

[¶]Canter, D. (1983), The potential of facet theory for applied social psychology, *Qual. Quant.* **17**(1), 35–67.

[‖]Tavoularis, S. (n.d.), Fluid mechanics, in. *Kirk-Othmer Encyclopedia of Chemical Technology*.

choice part. The first part of a continuous choice is denoted by $Cont[N^n, N]$. The second part of the continuous choice is denoted by AC. Any function from N^n to N is continuous. If $P \subset N^n x N$, and for each of $\alpha \in N^N$ there exists $n \in N$ such that $(\alpha, n) \in P$, then there is a function $f : N^N \Rightarrow N$ such that $(\alpha, f(\alpha)) \in P$ for all $\alpha \in N^N$.[*]

PRINCIPLE OF CONTRADICTION [mathematics] (1) A principle of logic whereby a proposition cannot be both true and false. A simultaneous assertion and denial of a proposition; that is, a sentence of the form "A and not B". Formal systems in which a contradiction is a theorem are said to be inconsistent. The law of contradiction is the logical principle that a proposition cannot be both asserted and denied; that is, the theorem of the propositional calculus. For all propositions p, it is impossible for both p and not p to be true.[†] (2) [Psychology] Partly because change is constant, contradiction is constant. Thus old and new, good and bad, exist in the same object or event and indeed depend on one another for their existence.[‡]

PRINCIPLE OF CONTRAST [psychology] (1) The use of different signals for different categories and of the same signal for all members within a category corresponds to the principle of contrast or of mutual exclusivity that children rely on when they assign only one label per category.[§] (2) Children will allow only one lexical entry to occupy a semantic niche. When two words are determined to have similar meanings, one of them is preempted and removed from the lexicon. Different words should have different meanings. (3) Captures facts about the inferences that speakers and addresses

make for both conventional and novel words. It accounts for the preemption of novel words by well established ones; and it holds just as much for morphology as it does for words and larger expressions. (4) The development of nonlinguistic concepts, the acquisition of language in context, and the use by participants in conversational exchanges to account for those features of language and language acquisition.[¶] See also MUTUAL EXCLUSIVITY PRINCIPLE; PRINCIPLE OF UNIQUENESS; PRINCIPLE OF PREEMPTION; PRINCIPLE OF CONVENTIONALITY; COOPERATIVE PRINCIPLE.

PRINCIPLE OF CONTRADICTION [mathematics] (Aristotle) Contradictory statements cannot both at the same time be true; for example, the two propositions "A is B" and "A is not B" are mutually exclusive. A may be B at one time, and not at another; A may be partly B and partly not B at the same time; but it is impossible to predicate of the same thing, at the same time, and in the same sense, the absence and the presence of the same quality.[‖]

PRINCIPLE OF CONTRIBUTION TO OBJECTIVES [psychology] (1) The contribution to mission objectives (calibration/validation). (2) Define the management task in terms of results—or "output." (2) Applies not only to the obvious functions such as sales and production, but to any kind of job and in both the profit and nonprofit sectors. (3) Explains how to construct measurable objectives for each key task and how to ensure that activity and authority are in line.[**]

PRINCIPLE OF CONTROL [psychology] Purpose of every plan and all derivate

[*]Bishop, E. and Bridges, D. (1985), Grundlehren der math, *Wissenschaften Springer-Verlag* p. 279.
[†]Iniguez, J. C. (2003), The second law, metaphysics, and dialectics, *Phys. Chem.* 1–29.
[‡]Nisbett, R., Peng, K., Choi, I., and Norenzayan, A. (2001), Culture and systems of thought: Holistic versus analytic cognition., *Psychol. Rev.* **108**(2), 291–310.
[§]Donaldson, M. C., Grieve, R. B., and Hughes, M. (1990), *Understanding Children: Essays in Honour of Margaret Donaldson*, Blackwell Publishing, p. 12.

[¶]Clark, E. (1987), The principle of contrast: A constraint on language acquisition, in *Mechanisms of Language Acquisition*, MacWhinney, B., ed., Lawrence Erlbaum, Hillsdale, NJ.
[‖]Schlink, B. (1971), On a principle of contradiction in normative logic and jurisprudence, *Theory Decision* **2**(1), 35–48.
[**]Koontz, H. and O'Donnell, C. (1968), *Principles of Management: An Analysis of Managerial Functions*, 4th ed., McGraw-Hill, (http://knol. google.com/k/narayana-rao-kvss/principles-of-management-koontz-and/2utb2lsm2k7a/89).

plans is to contribute positively toward the accomplishment of enterprise objectives.*

PRINCIPLE OF CONTROL RESPONSIBILITY [psychology] (1) The primary responsibility for the exercise of control rests in the manager charged with the execution of plans.* (2) Ensure achievement of objects by detecting deviations from plans and making corrections with maximum efficiency, minimum cost under the responsibility and authority of a designated individual, group, department, division, and so on.* See also PRINCIPLE OF PLANNING.

PRINCIPLE OF CONVENTIONALITY [psychology] (1) A key assumption underlying effective communication is that speakers of a common language tend to use the same words to express certain meanings. According to the principle of contrast, children should reason that if a speaker uses a different name (dax), it is probably because he or she has a different object in mind. This inference should lead children to select the tongs as the referent for dax. The principle of contrast is an assumption about the relation between the meanings—not just the referents—of two linguistic forms. According to the principle of contrast, two names can refer to the same object as long as they have different meanings—a stipulation that renders the principle of contrast compatible with the fact that objects evidently have multiple names (e.g., dog, cat, other animal). In fact, a number of studies have shown that when indications about a contrast in the meanings of two names are provided, even 2-year-olds accept and spontaneously use more than one name as referring to the same object defined in the following way: "For certain meanings, there is a conventional form that speakers expect to be used in the language community, that is, if one does not use the conventional form that might have been expected, it is because one has some other, contrasting meaning in mind." This formulation entails at least three distinct assumptions that speakers and addressees must make: (a) there are conventional forms to express certain meanings, (b) members of

a linguistic community know these forms, and (c) members of a linguistic community expect these forms to be used when speakers intend to express their corresponding meanings.† (2) The use of different signals for different categories and of the same signal for all members within a category corresponds to the principle of contrast or of mutual exclusivity that children rely on when they assign only one label per category. (3) Children will allow only one lexical entry to occupy a semantic niche. When two words are determined to have similar meanings, one of them is preempted and removed from the lexicon. Different words should have different meanings.† (4) For certain meanings, there is a conventional form that speakers expect to use within the language community. Speakers of languages depend on consistency of denotation from one time to the next in the meaning of a word or expression. Captures facts about the inferences that speakers and addresses make for both conventional and novel words. It accounts for the preemption of novel words by well-established ones; and it holds just as much for morphology as it does for words and larger expressions.‡ See also PRINCIPLE OF REFERENCE; PRINCIPLE OF EXTENDIBILITY; PRINCIPLE OF OBJECT SCOPE; PRINCIPLE OF CATEGORICAL SCOPE; NOVEL NAME–NAMELESS CATEGORY (N3C) PRINCIPLE; MUTUAL EXCLUSIVITY PRINCIPLE; PRINCIPLE OF UNIQUENESS; PRINCIPLE OF PREEMPTION; PRINCIPLE OF CONTRAST; COOPERATIVE PRINCIPLE.

PRINCIPLE OF COOPERATION Groups of species of animals, including humans, constitute organismic integrations rather than mere aggregations of individuals. The adoption of this viewpoint has been slow in psychiatry and sociology, although it was emphasized by Tigrant Burrow and associates in 1914. Modern partitive and disordered processes should be replaced by principles and behavior more in line with

*Koontz, H. and O'Donnell, C. (1968).

†Diesendruck, G. (2005), The principles of conventionality and contrast in word learning: An empirical examination, *Devel. Psychol.*, **41**(3), 451–463.

‡Donaldson, M. C., Grieve, R. B., and Hughes, M. (1990), *Understanding Children: Essays in Honour of Margaret Donaldson*, Blackwell Publishing, p. 12

humankind's fundamental phylobiological motivation.* See ROCHDALE PRINCIPLES.

PRINCIPLE OF COORDINATION [psychology, management] (1) Movement coordination must be developed before movement control, which follows from the idea that a person might have coordination without control (e.g., the ability to perform a coordinated movement only in a stereotypical manner, as described earlier), but a person could not have control without coordination. A coordination pattern must have a set of control parameters associated with it; however, movement control in this context is viewed as the ability to successfully perform a particular pattern across a range of parameter values. The second principle states that movement coordination and control should be developed in a hierarchical sequence.† (2) The primary responsibility for the exercise of control rests in the manager charged with the execution of the plans. See also PRINCIPLE OF A BALL PISTON PUMP; PRINCIPLE OF ACTION; PRINCIPLE OF ASSURANCE OF OBJECTIVES; PRINCIPLE OF BALANCE.

PRINCIPLE OF CORRESPONDENCE [psychology] (1) (Agreeableness evokes agreeableness, and hostility evokes hostility.) Perceptions of agency would predict agentic behavior according to the principle of reciprocity (dominance invites submissiveness, and submissiveness invites dominance).‡ (2) Embodies the truth that there is always a correspondence between the laws of phenomena of the various planes of being and life. There is a harmony, agreement, and correspondence between the following "planes": the great physical plane, the great mental plane, and the great spiritual plane.§ (3) Relationships with common and personal goals brings a recognized identity toward agreement on methods and actions, each influencing the others generally along horizontal rather than vertical lines.¶ See also PRINCIPLE OF CAUSE AND EFFECT.

PRINCIPLE OF CORRESPONDING RHEOLOGICAL STATES [mathematics] Viscosity is a function of several variables, including density, the medium, time, and the number and concentration of particles. All variables are expressible in terms of the basic variables of mass, length, and time. The ratio of inertial forces to viscous forces represents a relative viscosity that applies over a concentration range from at least $0 = 0.1$ to 0.5.‖

PRINCIPLE OF CORRESPONDING STATES [mathematics, chemistry] (1) An equation of state written in terms of the reduced properties is a generalized equation that could be applied to any substance. It follows that if two substances are at the same reduced temperature and pressure, then they would have the same reduced volume.** The existence of a universal relation between dimensionless parameters formed using the physical quantities of interest. The existence of such a relation may be established by a dimensional analysis or by use of a mathematical equation, if one exists, connecting the relevant quantities. The interatomic potential u to be of the form

$$u(r) = \varepsilon f(r/\sigma),$$

where r is the inter-atomic distance and the parameters ε and σ have the units of energy and length, respectively. Using the statistical-mechanical expressions for the diffusion coefficient (D) and viscosity (μ) in terms of the ensemble average of the corresponding autocorrelation functions,

*Galt, W. (1940), The principle of cooperation in behavior, *Quart. Rev. Biol.* **15**, 401–410.
†B. A. (1990), Applying principles of coordination in adapted physical education, *Adapt. Phys. Activ. Quart.* **7**(2), 126–142.
‡Foley, J. Elizabeth (2006) Perceived interpersonal climate and interpersonal complementarity. McGII Doctoral Dissertation University.
§Three Initiates (2004), *The Kybalion*, Book Tree (the "Three Initiates" who authored *The Kybalion*

chose to remain anonymous. As a result, there has been a great deal of speculation about who actually wrote the book.)
¶Capria, M. M. (1995), The theory of relativity and the principle of correspondence, *Phys. Essays* **8**, 78.
‖Koontz, H. and O'Donnell, C. (1968), *Principles of Management*, 4th ed., McGraw-Hill.
Simha, R. and Utracki, L. A. (2005), The viscosity of concentrated polymer solutions: Corresponding states principles, *Rheol. Acta.* **12(3), 455–464.

they obtained the following relations: $D^- = D^-(T^-, \rho^-)$ and $\mu^- = \mu^-(T^-, \rho^-)$, where[*]

$$D^- = \frac{D}{\sigma}\sqrt{\frac{m}{\varepsilon}}, \mu^- = \frac{\mu\sigma^2}{\sqrt{m\varepsilon}}, T^- = k_B T/\varepsilon,$$
$$\rho^- = \rho\sigma^3.$$

(3) Similar thermodynamic behavior is possessed by different substances when compared at the same reduced conditions. (3) Refers to conditions where pressure, temperature, and volume are at an equal faction of the critical values. (4) A two-parameter equation of state is a two-parameter corresponding states model. A two-parameter corresponding states model is composed of two-scale factor correlation and a reference fluid equation of state. In a two-parameter equation of states the reference equation of state is the two-parameter equation of state itself. (5) Implies that for polymers, the dominant contribution to the surface tension comes from the cohesive and entropic properties of bulk liquids and is only weakly dependent on the molecular conformations and endgroups.[†] See also CORRESPONDING STATE PRINCIPLE; STRONG PRINCIPLE OF CORRESPONDING STATES.

PRINCIPLE OF COUNTING CONSTANTS [mathematics]   For any kind of intersection theory, we are taking the union of a certain number of constraints. If we have a number N of parameters to adjust (i.e., if we have N degrees of freedom), and a constraint means that we have to "consume" a parameter to satisfy it, then the codimension of the solution set is at most the number of constraints. We do not expect to be able to find a solution if the predicted codimension, that is, the number of independent constraints, exceeds N (in the

linear algebra case, there is always a trivial, null vector solution, which is therefore discounted).[‡]

PRINCIPLE OF COVARIANCE [physics, mathematics]   (1) The laws of physics take the same mathematical form in all inertial reference frames.[§] (2) Effect of information about one possible cause of an event as inferences regarding another possible cause. Analyzes situations until a minimal set of sufficient causes are identified; then, other possible causes are ignored or dismissed.[¶] (3) Comprising at least two sensor or feeler elements having different functional or operating principles and a common evaluation circuit for evaluation of the property changes of the sensor elements and for triggering a signal.[‖] (4) An arrangement for measuring lengths or angles using the transillumination or direct light principle. Used to inspect gemstones or crystals using a magnifying lens or a microscope. The illumination is created either according to the transmitted light principle, the light passes through the transparent or translucent object (incident light principle), or the object is illuminated from the side and the reflected light rays are gathered.[**] (5) Rest mass of a particle cannot be kept invariant; must exist a unique preferred inertial frame of reference wherein a particle is absolutely at rest.[††] See also LIGHT PRINCIPLE; CAUSALITY PRINCIPLE; DIRECT LIGHT PRINCIPLE; EXTINCTION PRINCIPLE; INTERFERENCE OF LIGHT PRINCIPLE; PRINCIPLE OF MINIMAL CAUSATION; SCATTERED LIGHT PRINCIPLE;

[*]Ganesh Prakash, S., Ravia, R., and Chhabrab, R. P. (2004), Corresponding states theory and transport coefficients of liquid metals. Chemical Physics, **302**, 149–159.

[†]Helfand, E. and Rice, S. A. (1960), Principle of corresponding states for transport properties, *Hist. Chem. Phys.* **32**, 1642–1644.

[‡]Dee, G. T. and Sauer, B. B. (1995), The principle of corresponding states for polymer liquid surface tension, *Polymer*, **36**, 1673.

[§]Sommese, A. J. (1978), Submanifolds of Abelian varieties to Rebecca, *Math. Annal.* **233**(3), 229–256.

[¶]Sachs, M. (1993), *Relativity in Our Time: From Physics to Human Relations*, CRC Press, p. 14.

[‖]Friedman, Y. (2004), Physical applications of homogeneous balls, *Progress Math. Phys.* **40**, 1–21.

[**]Post, E. J. (1997), *A History of Physics as an Exercise in Philosophy*, p. 84 (www22.pair.com/csdc/pdf/philos.pdf)

[††]Einstein, A., Lorentz, H. A., Minkowski, H., Weyl, H. (1952), *The Principle of Relativity: A Collection of Original Memoirs on the Special and General Theory of Relativity*, Courier Dover Publications, p. 111.

SEARCHLIGHT PRINCIPLE; TRANSILLUMINATION PRINCIPLE; TRANSLUCENT OBJECT PRINCIPLE; TRANSLUCENT INCIDENT PRRINCIPLE; TRANSMITTED LIGHT PRINCIPLE.

PRINCIPLE OF CRITICAL FLOW [physics] Only the inlet pressure and temperature measurements are needed to determine the flow rate. The flow rate varies linearly with the upstream pressure and is not affected by downstream pressure fluctuation.*

PRINCIPLE OF CRITICAL-POINT CONTROL [organizational psychology] Effective control requires attention to those factors critical to appraising performance against an individual plan. See also PRINCIPLE OF PLANNING.[†]

PRINCIPLE OF CROSS-CUTTING RELATIONSHIPS [geology] A rock or fault is younger than any rock (or fault) through which it cuts.[‡]

PRINCIPLE OF CROSS-ENTROPY [mathematics, physics] (1) Among the many possible distributions satisfying the known constraints, we should choose the one that maximizes the entropy. (2) Select that probability distribution that contains the most uncertainty while still satisfying the original defaults, and this method leads to the unique least committed, or least biased, distribution among all compatible ones. (3) Instead of building separate models for various knowledge sources, the maximum entropy approach will provide a model that meets all these constraints in a unified statistical framework, because they are treated equivalently. (4) Given features and functions, which determine statistics important in modeling a process. A geometric interpretation of this setup is the space of all

(unconditional) probability distributions on *three points*, sometimes called a *simplex*. By imposing no constraints, we find that all probability models are allowable. Imposing one linear constraint is restricted to those that lie on the region. A second linear constraint could determine exactly whether if the two constraints are satisfying. [§] The principle of cross-entropy minimization was first introduced by Kullback. It should be distinguished from the principle of maximum statistical entropy, as expounded by Jaynes. Both of these principles have their origin in Shannon's work; both may be used as general methods of solution in problems requiring *inductive inference*. A general inductive inference problem may be stated as follows. Imagine a physical system (such as a bioelectromagnetic experiment) that has a set of possible states described by the state vector x. Assume x to be distributed according to some unknown probability distribution $q(x)$, and~caltlh is distribution the posterior distribution.[¶] See also POME; PRINCIPLE OF MAXIMUM ENTROPY; MAXIMUM ENTROPY); (MAXENT) MAXIMUM ENTROPY PRINCIPLE; PRINCIPLE OF MINIMUM CROSS-ENTROPY; PRINCIPLE OF CROSS-ENTROPY MINIMIZATION.

PRINCIPLE OF CURRENT DISTRIBUTION [mathematics] The most intuitive network for performing spatial smoothing is the one formed using resistive networks. In such networks a resistive grid receives the input current and each node distributes its current among its neighbors. The output can be taken, for example, by reading the node voltages. We will see that even the most intricate circuits described in this article work based on this simple principle. If all the elements in the network have equivalent impedances, as shown in the figure, one can easily derive the equation relating the output and

*O'Hanian, H. C. and Ruffini, R. (1994), *Gravitation and Spacetime*, 2nd ed., Norton, New York.
[†]Koontz, H. and O'Donnell, C. (1968), *Principles of Management: An Analysis of Managerial Functions*, 4th ed., McGraw-Hill (http://knol.google.com/k/narayana-rao-kvss/principles-of-management-koontz-and/2utb2lsm2k7a/89).
[‡]Tarbuck, E. J., Lutgens, F. K., Tsujita, C. J., and Tasa, Dennis (2004) *Earth—An Introduction to Physical Geology*, Pearson Education Canada.

[§]Shore, I. E. and Johnson, R. W. (1981), Properties of cross-entropy minimization, *IEEE Trans. Inform. Theory* 27(4), 472–482.
[¶]Alavi, F. N., Taylor, J. G., and Ioannides, A. A. (1993), Estimates of current density distributions. I. Applying the principle of cross-entropy minimization to electrographic recordings, *Inverse Problems* 9(6), 623–639.

input currents.* See also COUNTER CURRENT DISTRIBUTION PRINCIPLE; COST DISTRIBUTION PRINCIPLE; CHARGE DISTRIBUTION PRINCIPLE; DISTRIBUTION PRINCIPLE.

PRINCIPLE OF D'ALEMBERT [physics] In any system of bodies mechanically connected in any way, so that their motions may mutually influence one another, if forces equal to the effective forces were applied in direction opposite to their actual directions, these would be in equilibrium with the impressed forces.[†]

PRINCIPLE OF DANDY ROLL [engineering] Developed to allow the production of watermark grades on a continuous sheet.[‡]

PRINCIPLE OF DECOMPOSITION. [physics] The force acting on a single point in a system is the geometric sum of the forces deriving from the other points in the system. The intensity of a force holding between two arbitrary points varies with their positions only. See also POSITION PRINCIPLE; PRINCIPLE OF POSITIONAL DETERMINANCY[§]

PRINCIPLE OF DEDUCIBILITY FOR JUSTIFICATION (PDJ) [mathematics] (Edmund L. Gettier III, b. 1927, Baltimore, MD) For any proposition P, if S is justified in believing P and P entails Q, and S deduces Q from P and accepts Q as a result of this deduction, then S is justified in believing Q (2) If set O is the set of propositions, and operation R is natural deduction, then, provided that p is a member of O, and p deductively entails q, q is

also a member of O.[¶] See also DEDUCTIVE CLOSURE PRINCPLE; STRAIGHT PRINCIPLE; GETTIER'S PRINCIPLE; PRINCIPLE OF EPISTEMIC CLOSURE.

PRINCIPLE OF DELAYED GRATIFICATION [mathematics, psychology] (1) Save up for the items you wish to purchase.[‖] (2) Theistic worldview that life's reward will be granted in the next life as a foundation for belief.[**] See also PRINCIPLE OF IMMEDIATE GRATIFICATION; PLEASURE PRINCIPLE.

PRINCIPLE OF DEFINITE PROPORTIONS [physics] See PRINCIPLE OF CONSTANT PROPORTIONS.

PRINCIPLE OF DELEGATION [psychology] (1) Addresses delegation as a management principle used to obtain desired results through the work of others. Points out the importance of learning and using effective delegations to develop successful leadership and to maintain quality of care. (2) Assign staff members to the level of work that will enable them to utilize their highest abilities over the greatest portion of their workday. This requires that the work be delegated to the experience level at which it can be performed most effectively. See also PRINCIPLE OF PLANNING[††]

PRINCIPLE OF DELEGATION BY RESULTS EXPECTED [organizational psychology] The authority delegated to an individual manager should be adequate to ensure hER/his ability to accomplish the results expected of her/him. See also PRINCIPLE OF PLANNING.[‡‡]

PRINCIPLE OF DIASTEREOMETRIC ION PAIR FORMATION [biology] See ION PAIR PRINCIPLE.

*Moini, A., (1997), *Vision Chips or Seeing Silicon*, Centre for High Performance Integrated Technologies and Systems, Univ. Adelaide, Australia, p. 137.

[†]Verheest, F. and Van den Bergh, N. (1993), D'Alembert's principle of zero virtual power in classical mechanics revisited, *Eur. J. Phys.* **14**, 217–221.

[‡]Holik, H. (2006), *Handbook of Paper and Board*, Wiley-VCH, p. 226.

[§]Hendricks, V. F. and Hyder, D. J. (2006), Interactions: mathematics, physics and philosophy, 1860–1930, **251**, Springer.

[¶]Thalberg, I. (1969), In Defense of Justified True Belief. *J. Philo.* **66**(22), 794.

[‖]Gavosto, E. A., Krantz, S. G. and McCallum, W. G. (1996) *Contemporary Issues in Mathematics Education: Proceedings of a Conference at MSRI*, Cambridge Univ. Press.

[**]Johansson, C. M. (1988), *Musical Discipleship. Discipling Music Ministry: Twenty-first Century Directions* (http://www.ag.org/top/church_workers/wrshp_mus_musical_dscplshp.cfm).

[††]Koontz, H. and O'Donnell, C. (1968), *Principles of Management*, 4th ed., McGraw-Hill.

[‡‡]Koontz, H. and O'Donnell, C. (1968), *Principles of Management*, 4th ed., McGraw-Hill.

PRINCIPLE OF DICHOTOMY See PRINCIPLE OF EXCLUDED MIDDLE.

PRINCIPLE OF DIFFERENTIAL PH MEASURE-MENT [biology]   Enzyme reactions and clinical analysis.*

PRINCIPLE OF DIFFERENTIAL SCANNING CALORIMETRY (DSC) [chemistry]   Two ovens are linearly heated; one oven contains the sample in a pan, while the other contains an empty pan as a reference pan. If no change occurs in the sample during heating, the sample pan and the reference pan are at the same temperature. If a change, such as melting, occurs in the sample, energy is used by the sample and the temperature remains constant in the sample pan, whereas the temperature of the reference pan continues to increase. Therefore, a difference of temperature occurs between the sample pan and reference pan. Heat exchange calorimeters actively exchange heat between the sample and surroundings often during a temperature scanning experiment. The heat flow rate is determined by the temperature difference along the thermal resistance between the sample and the surroundings. Heat-flux DSC uses this principle.†

PRINCIPLE OF DIGIT-SERIAL ADDITION FOR DIGITS OF 2 BITS [computer science]   We can summarize as follows. If a given amount of data has to be processed in a given sample period, the tradeoff will be in favor of digit- or bit- serial addition if the ratio allows an efficient use of the hardware.‡

PRINCIPLE OF DIRECT CONTACT [psychology] Since authority is intended to furnish managers with a tool for managing so as to gain contributions to enterprise objectives, authority delegated to an individual manager should be adequate to ensure her/his ability to accomplish results expected of her/him.§

PRINCIPLE OF DIRECT CONTROL [psychology] (1) Coordination must be achieved through interpersonal, vertical, and horizontal relationships of people in an enterprise.¶ (2) The higher the quality of managers and their subordinates, the less will be the need for indirect controls. (The principle may termed as *principle of reduced controls.* A superior whose department is staffed with higher-quality managers and management subordinates can spend less time in control activities.)¶ See also PRINCIPLE OF PLANNING.

PRINCIPLE OF DIRECT SUPERVISION [psychology]   The higher the quality of managers and their subordinates, the less will be the need for indirect controls.¶

PRINCIPLE   OF   DIRECTING [psychology] Effective direction requires that managers supplement objective methods of supervision with direct personal contact.‖

PRINCIPLE   OF   DISCIPLINE [psychology] (Michel   Foucault,   1926–1984,   French Philosopher and sociologist)   (1) Foucault's management and organization theory, with respect   to   panopticism   "technologies   of self,"** and development of prisons and other   hierarchical   structures.   (2) Clearly defined limits of acceptable behavior are absolutely necessary, so that everyone in an organization   knows   what   can   and   cannot

*Luzzana, M., Perrella, M., and Rossi-Bernardi, L. (1971), An electrometric method for measurement of small pH changes in biological systems, *Anal. Biochem.* **43**(2), 556–563.

†Alexander, K. S., Riga, A. T., and Haines, P. J. (n.d.), *Ewing's Analytical Instrumentation Handbook*, 3rd ed., CRC Press.

‡Verbauwhede, I. (2000), *Integrated Processor-Level   Architectures   for   Real-Time   Digital Signal   Processing*,   Dept.   of   Elec.   Engin., ULCA,   citeseerx.ist.psu.edu/viewdoc/download? doi=10.1.1.132.6477

§Magid, J. Granstedt, A., Dy'rmundsson, O., Kahiluoto, H. and Ruissen, T. eds. (2001), Urban areas—rural areas and recycling—the organic way forward? *Proc. NJF Seminar* 327, Copenhagen, Denmark, Danish Research Centre for Organic Farming, Aug. 20–21, 2001, p. 28.

¶Koontz, H. and O'Donnell, C. (1968), *Principles of Management*, 4th ed., McGraw-Hill.

‖Anbuvelen, K. (1974), *Principles of Management*, Firewall Media, p.196.

**McKinlay, A., and Starkey, K., eds., (1998), Sage, Thousand Oaks, CA, pp. ix, 126–150.

be done.* See also FAYOL'S PRINCIPLES OF MANAGEMENT.

PRINCIPLE OF DISCRIMINATION [Psychology] To discriminate is to behave differently in different situations, and to generalize is to behave similarly in different situations.[†] See also PRINCIPLES OF JUS IN BELLO.

PRINCIPLE OF DISPLACEMENT [mathematics, engineering] (Eric Reissner) b. Describes the neural codes for reproduction of a scene on the basis of information obtained from its separate points. They create the visual picture of the world perceived by the brain.[‡] See DISPLACEMENT PRINCIPLE.

PRINCIPLE OF DISSOLUTION OF FUNCTION [medicine] Lesions in lower centers can inhibit or enhance activity of higher centers.[§] See also DUAL FUNCTION PRINCIPLE; FORM-AND-FUNCTION PRINCIPLE; FUNCTIONING PRINCIPLE.

PRINCIPLE OF DISTAL-TO-PROXIMAL DEVELOPMENT [biology, medicine] The growth of the head/trunk region consistent with cephalocaudal development. However, considerable interindividual differences in the intraindividual growth patterns, in particular for the lower trunk, where regular growth does not start by the age of 15 months.[¶]

PRINCIPLE OF DISTRIBUTIVITY [mathematics] The algebraic distributive law is valid

for classical logic, where both logical conjunction and logical disjunction are distributive over each other. The statement A and (B or C) is equivalent to (A and B) or (A and C).[‖]

PRINCIPLE OF DIVISION OF WORK [Psychology] (Henri Fayol, 1841–1925, French industrialist). (1) It is best to assign workers jobs fairly limited in scope to enable them to develop a high degree of skill.** (2) The better an organization structure reflects a classification of the tasks and activities required for achievement of objectives and assists their coordination through creating a system of interrelated roles; and the more these roles are designed to fit the capabilities and motivations of people available to fill them, the more effective and efficient an organization structure will be.[††] See also PRINCIPLE OF PLANNING; PRINCIPLE OF SPECIALIZATION; FAYOL'S PRINCIPLES OF MANAGEMENT.

PRINCIPLE OF DO NO HARM [Psychology] Above all, a professional of psychology or medicine must do no harm to patients, clients, or research participants. This principle is also known by the Latin phrase *primum non nocere*, or *primum nil nocere*, which means "First, do no harm." It is a reminder- or cautionary-type principle taught to students of psychology and medicine. It is most often mentioned when debating use of an intervention with an obvious chance of harm but a less certain chance of benefit.[‡‡] See also PRINCIPLE OF DOUBLE EFFECT; CASUISTIC PRINCIPLE; PRINCIPLES OF BIOMEDICAL ETHICS; EXCEPTION-GRANTING PRINCIPLE; JUSTIFYING PRINCIPLE; DOUBLE-EFFECT PRINCIPLE.

PRINCIPLE OF DOMINANCE MODIFICATION [evolution] (Sir Ronald Aylmer Fisher, FRS,

*Wood, M. C., (2002), *Henri Fayol: Critical Evaluations in Business and Management*, Taylor & Francis, p. 8.
[†]Hineline, P. (1992), A self-interpretive behavior analysis, *Am. Psychol.* **47**(11), 1274–1286.
[‡]Glezer, V. D. (2008), Meaning of the Weber-Fechner law and the principle of displacement, *Human Physiol.* (MAIK Nauka/Interperiodica, Springer Science–Business Media) **34**(3), 275–281.
[§]Dimitrijevic, D. (1952), A dynamic varient of Jackson's principle of dissolution, *J. Nerv. Mental Disease* **116**, 596–600.
[¶]Van Dam, M., Hallemans, A., and Aerts, P. (2009), Growth of segment parameters and a morphological classification for children between 15 and 36 months, *J. Anat.* **214**(1), 79–90.

[‖]Perkins, D. N. and Simmons, R. (1988), Patterns of misunderstanding: An integrative model for science, math, and programming, Rev. Educa. Res. **58**(3), 303–326.
**Wood, M. C. (2002), *Henri Fayol: Critical Evaluations in Business and Management*, Taylor & Francis, p. 8.
[††]Koontz, H. and O'Donnell, C. (1968), *Principles of Management*, 4th ed., McGraw-Hill.
[‡‡]Baron, J. (1995), Blind justice: Fairness to groups and the do no harm principle, *J. Behav. Decision Making* **8**, 71.

1890–1962, English statistician, evolutionary biologist, eugenicist and geneticist) Recessivity of mutant genes demonstrated the ability of even a minute selection pressure to cause significant evolutionary change. Dominance can be modified by artificial selection.*

PRINCIPLE OF DOMINANT SUBSYSTEMS

[mathematics, chemistry] (1) Sufficient conditions for the existence of local decentralized control laws stabilizing a given large-scale dynamic system with dynamic and parametric uncertainties are derived in terms of controller parameters for incompletely known continuous- and discrete-time systems.[†] (1) Instinctive attempt to avoid pain, discomfort, or unpleasant situations; desire to obtain maximum gratification with minimum effort. (2) Activities must be grouped to facilitate the accomplishment of goals; and the manager of each subdivision must have authority to coordinate its activities with the organization as a whole. The more clearly a position or a department defines the results expected, activities to be undertaken, organization authority delegated, and authority and informational relationships with other positions, the more adequately the individuals responsible can contribute toward accomplishing enterprise objectives. (3) If an activity is designed as a check on the activities of another department, individuals charged with such activity cannot adequately discharge their responsibilities if they report to the department whose activity they are expected to evaluate.[‡] See also AUTOMATIC CONTROL PRINCIPLES; PRINCIPLE OF FUNDAMENTAL CHARACTERISTICS; PRINCIPLE OF OPERATION; EQUIPARTITION PRINCIPLE; PRINCIPLE OF CONSTRUCTION AND OPERATION; PRINCIPLE OF SMOOTH FIT.

---

*Charlesworth, B. (1979) Evidence against Fisher's theory of dominance. Nature **278**, 848–849.
[†]Veselý, V. (1993), Large scale dynamic system stabilization using the principle of dominant subsystems approach, *Kybernetika* **29**(1), 48–61.
[‡]Koontz, H. and O'Donnell, C. (1968), *Principles of Management*, 4th ed., McGraw-Hill.

PRINCIPLE OF DOMINATION [psychology]

(Geórge Herbert Mead, 1863–1931, American Philosopher) Principle of domination for Mead's principle of sociality He sees society as rooted in institutions. The six basic institutions are (1) language, (2) the family, (3) the economy, (4) religion, (5) the polity, and (6) science. All institutions are rooted in social action, and social acts as constituted by any activity require the efforts of two or more persons to be completed. An institution does not merely represent any type of social act, but only a special form of social action. *Institutions* specifically refer to social acts that people carry out with the help of preestablished maxims to satisfy their recurrent impulses. Institutions are the key to both the creation and evolution of human society. Without institutions, human society could have never arisen, and without their alteration, no society could subsequently change. *Domination* refers to the construction of complex social actions through some participants in the social act performing superordinate roles, other participants performing subordinate roles, and everyone assuming the attitudes of "others." Conversely, *sociality* refers to the construction of complex social acts by participants merely assuming each others' attitudes without any special regard to whether they are performing the superordinate or subordinate role.[§] See EMANCIPATION PRINCIPLE.

PRINCIPLE OF DOUBLE EFFECT [psychology]

(Thomas Aquinas, 1224–1274, Italian philosopher) The principle of double effect (PDE) or doctrine of double effect (DDE), sometimes called *double effect* for short, is a thesis in ethics, usually attributed to Thomas Aquinas. Seeks to explain under what circumstances one may act in a way that has both good and bad consequences (a "double effect"). An action having an unintended, harmful effect is defensible on four conditions as follows: (1) an act can produce two immediate effects, one good and one bad—proper applications of the PDE must distinguish between motive and intention, a

---

[§]Athens, L. (2007), Radical interactionism: Going beyond Mead, *J. Theory Soc. Behav.* **37**(2), 37–165.

distinction based on the way the end result is viewed before it actually occurs;* (2) the intention is for the good effect and not the bad; (3) the good effect outweighs the bad effect in a situation sufficiently grave to merit the risk of yielding the bad effect; and (4) the good effect is not achieved through the bad effect.[†] See also PRINCIPLE OF DO NO HARM; CASUISTIC PRINCIPLE; PRINCIPLES OF BIOMEDICAL ETHICS; EXCEPTION-GRANTING PRINCIPLE; JUSTIFYING PRINCIPLE.

PRINCIPLE OF DUALITY [mathematics, geometry] (Jean Victor Poncelet, 1788–1867, French mathematician) (1) A theorem is true if and only if its dual statement is true; that is, if and only if the statement obtained by replacing each object in the original theorem with its dual is true; used extensively in projective geometry, set theory, and Boolean algebra. (2) The connection between lines and points in plane geometry (or between planes and points in solid geometry). A line can be defined by two points, and a point, by the intersection of two lines; in this sense, the line and the point are said to be dual elements in plane geometry. Similarly, connection of points by lines and intersection of lines to give points are dual operations. (3) A statement in which the name of each element is replaced by its dual element and the description of each operation is replaced by its dual operation leads to a dual statement (or dual theorem). (4) In electricity, for any theorem in electric circuit analysis there is a dual theorem in which one replaces quantities with dual quantities; current and voltage, impedance and admittance, and meshes and nodes are examples of dual quantities. (5) In electronics, analogies may be drawn between a transistor circuit and the corresponding vacuum-tube circuit. (6) All the propositions in projective geometry occur in dual pairs that have the property that, starting from either propositions of a pair, the other can be immediately inferred by interchanging the parts played by the words "point" and "line."[‡] (10) Shows the relationship between the principle of duality and the Hough transform. The definition of the Hough transform actually corresponds to an application of the principle of duality. The general mathematical concepts of figures and high-dimensional coordinates in projective space are reexpressed in terms of shapes and mappings used for shape extraction in Euclidean space. Some works have demonstrated the equivalence between the definition of several concepts used in pattern matching and the Hough transform. This definition introduces the formalism of projective geometry to shape extraction and analysis and thus the ideas, properties, and geometric relationships in the projective space can have an interpretation for the development of pattern matching techniques. The notion of a shape in the Hough transform, and how this relationship and the generalization of the principle of duality to high-dimensional spaces, defined by space coordinates, are related to the extensions of the Hough transform. Parametric forms are used to develop a dual analytic expression of general forms. The relationship between the Hough transform and the principle of duality increases our understanding of the dual nature of pattern matching, which can now benefit from established results in projective geometry.[§] See also DUALITY PRINCIPLE; PRINCIPLE OF EFFICIENCY.

PRINCIPLE OF DULONG AND PETIT [chemistry] The atomic heat capacities of solid elements are constant and approximately equal to $25 \text{ mol}^{-1} \text{ K}^{-1}$. Certain elements of low atomic mass and high melting point have,

*Hoffman, R. (1984), Intention, double effect, and single result, *Phil. Phenomenol. Res.* **44**(3), 389–393.

[†]Shineboume, E. A. (1996), Symposium on covert video surveillance Covert video surveillance and the principle of double effect: A response to criticism, *J. Medi. Ethics* **22**, 26.

[‡]Aguado, A. S., Montiel, M. E. and Nixon, M. S. (2000), On the intimate relationship between the principle of duality and the Hough transform. *Proc. Roy. Soc. Lond. A Math. Phys. Eng. Sci.* **456**(1995), 503–526.

[§]Selbourne, D. (1994), Why divorce should be more difficult (duties of parents to their children; extract from David Selboume's *The Principle of Duty*, *The Times* p. 14.

however, much lower atomic heat capacities at ordinary temperatures.[*]

PRINCIPLE OF DYNAMIC ECOLOGY [biology] (1) Principles of dynamic ecology include, surely, food chains and elemental cycles. *Martian biology*: Accumulating evidence favors the theory of life on Mars, but we can expect surprises.[†] (2) Ecological structure and function are not stable and static. They are often influenced by natural disturbances such as from fires, floods, droughts, storms, outbreaks of disease, or pest infestation. (3) An organism or group of organisms that reflect environmental conditions in a habitat. (4) The sum of all environmental factors that act as agents of natural selection. Everything that lives transforms other aspects of the energy system into forms that it can use to sustain itself.[‡] See also ECOLOGICAL PRINCIPLE.

PRINCIPLE OF DYNAMIC OPTIMIZATION [mathematics] Adapts concepts from the economic theory of capital accumulation, which are based on Lagrange multipliers that reflect market prices in the absence of markets. We obtain a sequence of single period decisions, whereby the authority maximizes the current period's return plus the present (or discounted) value of the end-of-period stock. We can now calculate an optimal level of activities by solving for a single cycle $t, t = 0, 1, \ldots$, the optimization problem given by

$$\max_{d_t \geq 0, r_{g_t} \geq 0, k_{t+l} \geq 0, \tilde{k}_t \geq 0} p_t' d_t + \rho \Psi_{t+1}' k_{t+1}$$

subject to the same set of constraints in for cycle t and for given k_t.[§]

PRINCIPLE OF ECONOMY [mathematics] [Guilhelmi Ockam (William of Occam), c. 1285–1349, Franciscan friar, English logician] "Entia non sunt multiplicanda praeter necessitatem" or (entities should not be multiplied unnecessarily).[¶]

PRINCIPLE OF EFFECT [psychology] Thorndike's formulation of the importance of reward in learning, which states that the tendency of a stimulus to evoke a response is strengthened if the response is followed by a satisfactory or pleasant consequence, and is weakened if the response is followed by an annoying or unpleasant consequence.[‖]

PRINCIPLE OF EFFICIENCY [psychology] (1) According to Emerson,[**] an organization or organization structure is efficient if it is structured to make possible accomplishment of enterprise objectives by people with minimum unsought consequences or costs. See also PRINCIPLE OF PLANNING[††] (2) Five of the twelve principles concern the relations between employer and employee, the others deal with methods or institutions and systems occurring in manufacturing concerns. The workings of each principle are explained with the help of many positive and negative illustrations well chosen from actual practices and malpractices. (2) Contrasts the "principle of efficiency" with the "principle of sufficiency." The former is the ideal of the maximum satisfaction of interests; the latter, that of organization, harmony, or unity. These principles conflict if both are taken as standards; but if either of them

[*]Walker, P. M. B. (1999), *Chambers Dictionary of Science and Technology.*, Chambers, New York, p. 661.

[†]Salisbury, F. B. (1962), *Science* **136**(3510), 17–26.

[‡]Jacquemyn, H., Butaye, J., Dumortier, M., Hermy, M., and Lust, N. (2001), Effects of age and distance on the composition of mixed deciduous forest fragments in an agricultural landscape, *J. Veg. Sci.* (Opulus Press, Uppsala, Sweden) **12**, 635–642.

[§]Albersen, P. J. H., Harol, E. D. and Keyzer, M. A. (2003), Pricing a raindrop in a process-based model: general methodology and a case study of the Upper-Zambezi. Physics and Chemistry of the Earth, Parts A/B/C, **28**(4-5), 183–192.

[¶]Ockham's Razor (n.d.), *Encyclopedia Britannica*. (http://www.britannica.com/EBchecked/topic/424706/Ockhams-razor).

[‖]Walker, P. M. B. (1999), *Chambers Dictionary of Science and Technology*, Chambers, New York, p. 661.

[**]Emerson, H. (1912), *Twelve Principles of Efficiency*,

[††]Koontz, H. and O'Donnell, C. (1968), *Principles of Management*, 4th ed., McGraw-Hill.

is chosen, the other is indispensable as a subsidiary.* See EMANCIPATION PRINCIPLE.

PRINCIPLE OF EFFICIENCY OF CONTROLS [psychology] (1) Structured to make possible accomplishment of enterprise objectives by people with the minimum unsought consequences or costs (going beyond the usual thinking of costs entirely in such measurable items as dollars or worker-hours).† (2) The more control approaches and techniques detect and illuminate the causes of potential or actual deviations from plans with the minimum of costs or other unsought consequences, the more efficient these controls will be. See also PRINCIPLE OF PLANNING.‡

PRINCIPLE OF EFFICIENCY OF PLANS [mathematics] The more control approaches and techniques detect and illuminate the causes of potential or actual deviations from plans with the minimum of costs or other unsought consequences, the more efficient are the controls.§ (2) Efficiency is measured by the contribution of the plan to objectives of the enterprise minus the costs and unsought-for consequences in formulating and implementing the plan.¶ See also PRINCIPLE OF CONTRIBUTION OF OBJECTIVES; PRINCIPLE OF PLANNING; PRINCIPLE OF PRIMACY OF PLANNING.

PRINCIPLE OF EFFICIENT COMMUNICATION [psychology] (1) A large part of the variation in natural speech appears along the dimensions of articulatory precision/perceptual distinctiveness. Speaking is considered efficient if the speech sound contains *only* the information needed to understand it. This efficiency is tested by means of a corpus of spontaneous and matched read speech, and syllable and word frequencies as measures of information content (12,007 syllables, 8046 word forms, 1582 intervocalic consonants, and 2540 vowels). It is indeed found that the duration and spectral reduction of consonants and vowels correlate with the frequency of syllables and words. Consonant intelligibility correlates with both the acoustic factors and the syllable and word frequencies. It is concluded that the principle of efficient communication organizes at least some aspects of speech production.‖ (2) Structure and function is affected by the variability of multiple factors as well as by the mean value experienced. While the effects of these two terms can sometimes be independent of each other and separable in space and time, they can also interact; an example might be different effects of the same degree of variation around different mean values of a variable. The hierarchy of spatiotemporal variation produces a cascade of effects across multiple scales.** See also PRINCIPLE OF SOUND ORGANIZING.

PRINCIPLE OF ELASTICITY OR MINIMUM WORK See MENABREA'S PRINCIPLE.

PRINCIPLE OF ELECTROMAGNETIC INDUCTION [physics] Power from battery to charge spark plug.††

PRINCIPLE OF ELECTRONEGATIVITY [mathematics, physics] See ELECTROMAGNETIC PRINCIPLE.

PRINCIPLE OF EMANCIPATION [psychology] See EMANCIPATION PRINCIPLE.

PRINCIPLE OF EMPATHY [physics] (1) Whatever a life unit of whatever kingdom experiences, feels, and thinks, it is possible

*Lee, O. (1945), Value and interest, *J. Phil.* **42**, 141–161.
†Bodik, I., Derco, J., and Hutnan, M. (1991), Biological denitrification—possibilities for improving its efficiency and control, *Vodni Hospodarstvi* (1990–1992), **41**(4), 131–135.
‡Koontz, H. and O'Donnell, C. (1968), *Principles of Management*, 4th ed., McGraw-Hill.
§Devai, I.; Wittner, I.; Bondar, E. (1977), Efficiency of two waste water treating systems based on the activated sludge or the trickling filter principle. II. Results of the examinations between 1974–76, *Acta Biol. Debrecina* **14**, 79–94.
¶Koontz, H. and O'Donnell, C. (1968), *Principles of Management*, 4th ed., McGraw-Hill.

‖van Son, R. J. J. H. and Louis, C. W. (n.d.), *Effects of Stress and Lexical Structure on Speech Efficiency* (http://www.fon.hum.uva.nl/IFA-publications/Eurospeech99/V014/V014.ps).
Györi, G. (2006), Semantic change and cognition, *Cogn. Ling.* **13(2), 123–166.
††Kawagoe, M. (1993), A novel measurement method of gel particle distribution in a bubble column based on electromagnetic induction, *Kenkyu Kiyo—Nara Kogyo Koto Senmon Gakko* **29**, 91–94.

to assume its mental and emotional condition and to experience what it experiences.[*] See also PRINCIPLE OF ASSUMPTION.

PRINCIPLE OF ENERGY CONSERVATION [physics] See PRINCIPLE OF CONSERVATION OF ENERGY.

PRINCIPLE OF ENTROPY [mathematics, physics] A measure of the degree of disorder in a system. (2) Defined in terms of Boltzmann's "$S = k \ln W$," where $S =$ entropy, $k =$ Boltzmann's constant, and $W =$ the number of microstates corresponding to the same macroscopic state.[†] See also PRINCIPLE OF CONSERVATION OF ENERGY.

PRINCIPLE OF ENTROPY MAXIMIZATION [physics] Identifies the interaction network with the highest probability of giving rise to experimentally observed transcript profiles. In its simplest form, the method yields the pairwise interaction network, but it can also be extended to deduce higher-order interactions.[‡]

PRINCIPLE OF EPILEPSY TREATMENT [medicine] Confirm diagnosis of true seizure; evalutate need for treatment initiation; attempt to establish seizure type and syndrome; select treatment on basis of seizure type, spectrum of activity; tolerability, and drug interactions; tailor treatment to individual, and individualize dosage based on response. If seizures persist, either add another drug if the first one is well tolereated and partly effective or switch to another drug if the first drug is poorly tolerated or ineffective.[§]

PRINCIPLE OF EPISTEMIC CLOSURE [mathematics] (Edmund L. Gettier III, b. 1927, Baltimore, MD) If person S knows p, and p entails q, then S knows q (this is sometimes called the "straight principle"). (1) A subject may not actually believe q, for example, regardless of whether he or she is justified or warranted. Thus, one might instead say that knowledge is closed under known deduction: if, while knowing p, S believes q because S knows that p entails q, then S knows q. (2) An even stronger formulation would be as follows: If, while knowing various propositions, S believes p because S knows that they entail p, then S knows p.[¶] See also STRAIGHT PRINCIPLE; GETTIER'S PRINCIPLE.

PRINCIPLE OF EQUAL A PRIORI PROBABILITIES [physics] Normally associated with decision theory to evaluate relative utilities of simple and mixed parameters that can be used to describe outcome.[‖] See also D'ALEMBERT PRINCIPLE; DIFFERENTIAL PRINCIPLE; VARIATIONAL PRINCIPLE; HAMILTON PRINCIPLE; CONSERVATION OF MOMENTUM PRINCIPLE; NEWTON'S THIRD PRINCIPLE.

PRINCIPLE OF EQUICONTINUITY [mathematics] See EQUICONTINUITY PRINCIPLE.

PRINCIPLE OF EQUILIBRIUM [chemistry] See PRINCIPLE OF MASS ACTION.

PRINCIPLE OF EQUIPARTITION OF ENERGY [physics] The equipartition principle relies on the ergodic hypothesis, which basically asserts that a system's path in phase space on a surface of constant energy will, in the long run, visit each region of that surface equally often. To express this more precisely, recall that the dynamical state of a system with n degrees of freedom can be represented by a point in the $2n$-dimensional phase space

[*]Critchfield, K. L. and Benjamin, L. S. (2006), Principles for psychosocial treatment of personality disorder: summary of the APA Division 12 Task Force/NASPR review, *J. Clin. Psychol.* **62**(6), 661–674.

[†]Zanchini, E. (1981), Entropy definition and principle, Termotecnica **35**(5), 254–262.

[‡]Lezon, T. R., Banavar, J. R., Cieplak, M., Maritan, A., and Fedorof, N. V. (2006), Using the principle of entropy maximization to infer genetic interaction networks from gene expression patterns, *Proc. Natl. Acad. Sci. USA* **103**(50), 19033–19038.

[§]Sander, J. W. (2004) The use of antiepileptic drugs—principles and practice, Epilepsia **45**(Suppl. 6), 28–34.

[¶]Vingilis, E. and Burkell, J. (1996), A critique of an evaluation of the impact of hospital bed closures in Winnipeg, Canada: Lessons to be learned from evaluation research methods, *J. Public Health Policy* **17**(4), 409–425.

[‖]Meisels, G. G. (1980), Angular momentum, the principle of equal a priori probabilities, and "intermolecular entropy corrections" in equilibriums of ions and molecules in the gas phase, *J. Am. Chem. Soc.* **102**(20), 6380–6381.

whose axes are the n generalized coordinates q_i and the n generalized momenta p_i of the system. The total energy (kinetic plus potential) of the system can be expressed as a function (called the *Hamiltonian*) of the generalized coordinates and momenta. An isolated system with a given amount of energy is confined to that energy surface in phase space. To each small region on that surface we can assign a weight factor proportional to the volume (of phase space) swept out by that region for an incremental change in energy. Using this weight factor, we can then evaluate the mean values of any variable $X(q,p)$ on the surface. This is called the *phase average of X*, denoted by $\langle X \rangle$. In addition, we can consider the path of an actual system through phase space as a function of time, and assuming that the time average of X converges on a single value, we can denote this value as. The ergodic hypothesis is $x = (x)$.[*]

PRINCIPLE OF EQUIPARTITION OF ENERGY [physics] (1) Each molecular quadratic degree of freedom receives 1/2 kT of energy, a result which had to be modified when quantum mechanics explained certain anomalies, such as discrepancies in the observed specific heats of crystals when the expected thermal energy per degree of freedom is less than the energy necessary to move that degree of freedom up one quantum energy level. (2) Every energy mode has a distribution density function of the same form as,[†]

$$\phi\left(V_x, V_y, V_z\right) = \frac{1}{(2\pi RT)^{3/2}}$$
$$e^{-}(v_x{}^2 + v_y{}^2 + v_z{}^2)/(2RT)$$

from which it follows that if the energy is a quadratic function of either the generalized coordinates or the generalized momenta, then the mean value of that energy mode (in equilibrium) is kT/2.[‡]

[*]Tolman, R. C. (1938), *The Principles of Statistical Mechanics*, Dover Publications, New York, pp. 93–98.
[†]Reif, F. (1965), Statistical Physics. New York: McGraw-Hill Book Company. pp. 246–250.
[‡]n.a. The Ergodic Hypothesis and Equipartition of Energy. http://www.mathpages.com/kmath606/kmath606.htm.

PRINCIPLE OF EQUITY [psychology] (Henri Fayol, 1841–1925, French industrialist) Managers/supervisors elicit loyalty from employees only when they deal with them as individual persons. Employees must be seen as persons, not things to be manipulated.[§] See also FAYOL'S PRINCIPLES OF MANAGEMENT.

PRINCIPLE OF EQUIVALENCE [psychology] (1) The efficiency of a plan is measured by the amount it contributes to objectives offset by the costs and other unsought consequences required formulating and operating it.[¶] (2) Verbal recall and sociometric studies; points to the fact that choice among objects will vary randomly if the objects appear to be of equal value or significance, even when they are widely disparate and unrelated. The fact that much behavior has a probabilistic and not a deterministic nature.[‖]

PRINCIPLE OF EQUIVALENCE [physics] (1) A statement that forms a basic principle in general relativity; observers have no means of distinguishing whether their laboratories are in uniform gravitational fields or accelerated frames of reference.[**] (2) An observer has no way of distinguishing whether his/her laboratory is in a uniform gravitational field or is in an accelerated frame of reference.[††]

PRINCIPLE OF EQUIVALENCE OF MASS AND ENERGY [physics] Einstein's principle stating that a mass is equivalent to an amount of

[§]Wood, M. C. (2002), *Henri Fayol: Critical Evaluations in Business and Management*, Taylor & Francis, p. 8.
[¶]Shelupsky, D. (1996), The principle of equivalence and theories of gravity, *Found. Phys. Lett.* **9**, 475.
[‖]Rozov, A. and Kolominskii, Y. (1965), Principle of Equivalence and the probability of realization of Psychic phenomena Printsip ravnoznachimosti i veroyatnostnoe ponimanie psikhicheskikh yavlenii, *Voprosy Psychologii*. no. 6.
[**]Walker, P. M. B. (1999), *Chambers Dictionary of Science and Technology*, Chambers New York, p. 915.
[††]American Heritage (2000), *American Heritage Dictionary of the English Language*, 4th ed., Houghton Mifflin, New York.

energy and the equation relating these quantities is $E = mc^2$.[*]

PRINCIPLE OF EQUIVALENT PROPORTIONS [chemistry] The proportions in which two elements separately combine with the same weight of a third element are also the proportions in which the first two elements combine together.[†] See also PRINCIPLE OF RECIPROCAL PROPORTIONS; EQUIVALENT PROPORTIONS PRINCIPLE.

PRINCIPLE OF ESPRIT DE CORPS [psychology] (Henri Fayol, 1841–1925, French industrialist). All successful organizations survive only when a feeling of unity pervades the group and that viable organizations cleat? deal with crises as a team.[‡] See also FAYOL'S PRINCIPLES OF MANAGEMENT.

PRINCIPLE OF ETHICAL SUITABILITY [psychology] In the use of means for the "preservation of life," this is an evaluative dynamism that, continuing along the lines set out by classical terminology (ordinary and extraordinary means), tries to apply the contents of moral tradition to the new emerging perspective (proportionate and disproportionate means), underlining the specificity of each term, in a context of ethical systematization able to provide concrete evaluation criteria, at the service of the practical choices of patients and healthcare personnel. The principle of proportionality in therapy: foundations and applications criteria.[§]

PRINCIPLE OF EXACT RECIPROCITY [psychology] See also EYE-FOR-AN-EYE PRINCIPLE; PRINCIPLE OF RETRIBUTIVE JUSTICE; PRINCIPLE OF PROPORTIONATE PUNISHMENT.

PRINCIPLE OF EXCELLENCE [psychology] (According to Peters and Waterman,[¶]) the eight criteria of excellence are (1) a bias for action and the ability to make decisions quickly; (2) Closeness to the customer; (3) autonomy and entrepreneurship; (4) productivity through people; (5) hands-on, value-driven; approach; (6) "stick to the knitting," (7) "simple form, lean staff"; and (8) simultaneous loose–tight properties.[‖]

PRINCIPLE OF EXCLUDED MIDDLE [mathematics] (1) A theory T is said to be inconsistent if it has theorems a formula A and its negation. NA; and it is said to be trivial if every formula of this language is a theorem of the theory. (2) Law in logic, that a statement or proposition is always either true or false, leaving no room for any further alternatives; for instance, it cannot be both true and false. The theorem of the propositional calculus; for any statement, the phrase "or not" is always true. (2) Holds in Q_{eff} for all propositions that are finite truth-functional compounds of elementary propositions. It is argued that if Q_{eff} is restricted by the elimination of commensurability propositions, the resulting system is equivalent to the calculus Q of full quantum logic, which can also be obtained from Q_{eff} by addition of the law of excluded middle. The conclusion reached is that Q_{eff} so restricted is a model for Birkhoff and von Neumann's quantum logical propositional lattice.[**] (3) Law in logic, that a statement or proposition is always either true or false, leaving no room for any further alternatives (e.g., it cannot be both true and false). (4) A proposition is either true or false. As one of the laws of classical logic, it can be symbolically expressed as $P \sim P$. There are two points of view on the matter of "such

[*]Sharma, A. (1998), The generalisation of mass-energy equivalence as $E = mc^2$, *Acta Ciencia Indica Phys.* 24(4), 153–158.
[†]Walker, P. M. B. (1999), *Chambers Dictionary of Science and Technology*, Chambers, New York, p. 661.
[‡]Wood, M. C. (2002), *Henri Fayol: Critical Evaluations in Business and Management*, Taylor & Francis, p. 8.
[§]Calipari, M. (2004), The principle of proportionality in therapy: Foundations and applications criteria. *NeuroRehabilitation* 19(4), 391–397.

[¶]Peters, T. and Waterman, B. (1982), *In Search of Excellence*.
[‖]Bowen, S. A. (2004), Expansion of ethics as the tenth generic principle of public relations excellence: A Kantian theory and model for managing ethical issues, *J. Publ. Relat. Res.* 16(1), 65–92.
[**]Mittelstaedt, P. and Stachow, E. W. (1978), The principle of excluded middle in quantum logic, *J. Phil. Logic* 7(2), 1–208.

and such", and that no third point of view is possible.* See also PRINCIPLE OF BIVALENCE.

PRINCIPLE OF EXTENDIBILITY [psychology] (1) At around the end of the first year, even before children are clear about the basis for extension, they gain the fundamental insight that word labels can be extended at all. This is the principle of extendibility, often presupposed in other word-learning accounts, which permits extension to occur on the basis of a number of types of similarity (e.g., sound, smell, texture, taste), although shape is most frequently used. If shape is not available, children will extend on other criteria in a fixed sequence of preference. Relying on perceptual similarity for extension, and in particular, shape, is a useful albeit not perfect strategy because shape is highly correlated with basic-level category membership.† (2) Words may be used to label similar referents, even if the child has not yet heard someone else label the referent with the specific word. Supports a variety of bases for extension of a word.‡ See also PRINCIPLE OF REFERENCE; PRINCIPLE OF OBJECT SCOPE; PRINCIPLE OF CATEGORICAL SCOPE; NOVEL NAME–NAMELESS CATEGORY (N3C) PRINCIPLE; PRINCIPLE OF CONVENTIONALITY.

PRINCIPLE OF EXTERNAL EFFECTS [economics] The mitigation of negative external effects and the provision of public goods with positive external effects is the sole agenda conceded to the state.§

*Tran-Ba-Huy, P; Pelisse, J. M., Sauvage, J. P., and Pialoux, P. (1976), A proposal of a T.N.M. type of classification of the ear. O.P.A.C. classification of chronic surgical otitis, *Annal. Otolaryngol. Chirurg. Cervico Faciale (Bull. Soc. Otolaryngol.* (Hopitaux Paris) **93**(3), 117–128.
†Burack, J. A., Hodapp, R. M., and Zigler, E. (1998), *Handbook of Mental Retardation and Development*, Cambridge Univ. Press, p. 218.
‡Golinkoff, R. M., Mervis, C. B., and Hirsh-Pasek, K. (1994), Early object labels: The case for a developmental lexical principles framework, *J. Child Lang.* **21**(1), 125–155.
§Caldwell, B. J. and Menger, C. (1990), Carl Menger and his legacy in economics. History of Political Economy Annual Supplement Series. v22 of Annual supplement to History of political economy. Duke University Press, p. 113.

PRINCIPLE OF FACTOR SPARSITY [mathematics] States that for many phenomena, 80% of the consequences stem from 20% of the causes. See also PARETO PRINCIPLE; 80/20 PRINCIPLE.

PRINCIPLE OF FAUNAL SUCCESSION [geology](William Smith, geologist) Based on the observation that sedimentary rock strata contain fossilized flora and fauna, and that fossils succeed each other vertically in a specific, reliable order that can be identified over wide horizontal distances. The fossil content of rocks together with the law of superposition helps determine the time sequence in which sedimentary rocks were laid down.¶ See also PRINCIPLE OF CROSS-CUTTING RELATIONSHIPS; PRINCIPLE OF ORIGINAL HORIZONTALITY; PRINCIPLE OF LATERAL CONTINUITY.

PRINCIPLE OF FAUNAL SUCCESSION [biology] Specific groups of animals have followed, or succeeded, one another in a definite sequence through Earth history.‖

PRINCIPLE OF FILTER BELT PRESS [engineering] Gradual increase of pressure differential.** See also BELT FILTER PRESS PRINCIPLE; FILTER PRESS PRINCIPLE.

PRINCIPLE OF FINITE CHARACTER [mathematics] (Ernst Friedrich Ferdinand Zermelo, 1871–1953, German mathematician) For every set x and every property P of finite character, there exists a c-maximal subset of x with the property Z_7.††

PRINCIPLE OF FLEXIBILITY [organizational psychology] The task of managers is to provide for attaining objectives in the face of

¶Winchester, S. (2001), *The Map that Changed the World*, HarperCollins, New York, pp. 59–91.
‖Holm, L. E. (2004), A common approach for radiological protection of humans and the environment *J. Environ. Radioact.* **72**(1–2), 57–63.
**Ho, Bosco (1992) System for monitoring and/or controlling liquid–solid separation processes. U.S. Patent 5,240,594.
††Harper, J. M. and Rubin, J. E. (1976), Variations of Zorn's lemma, principles of cofinality, and Hausdorff's maximal principle, *Notre Dame J. Formal Logic* **17**(4), 565–588.

changing environments. The more provisions are made for building organization flexibility, the more adequately organization structure can fulfill its purpose.* See also PRINCIPLE OF PLANNING; EQUIVALENCE PRINCIPLE.

PRINCIPLE OF FLEXIBILITY OF CONTROLS [mathematics] The more flexibility can be built into plans, the less the danger of losses incurred through unexpected events, but the cost of flexibility should be weighted against the dangers of future commitments made.[†]

PRINCIPLE OF FLEXIBILITY OF CONTROLS [organizational psychology] If controls are to remain effective despite failure or unforeseen changes in plans, flexibility is required in the design of controls. See also PRINCIPLE OF PLANNING.[‡]

PRINCIPLE OF FLOTATION DEINKING [physics] Process of using air to separate particles by applying chemicals to make particles hydrophobic and attach to air bubbles.[§]

PRINCIPLE OF FLUID ORIFICING [chemistry, physics] (1) Employed in the beneficiation of ores. (2) Flotation is based on the principle that introducing chemical and the air into water containing solid particles of different materials suspended therein, causing adherence of air to certain suspended solid and to causing the particles to have air bubbles thus adhered thereto lighter than the water. Accordingly, they rise to the top of the water to form a froth, which as such is skimmed off. (3) Flotation principle is applied in a number of mineral separation processes, including the selective separation of such minerals as sulfide copper minerals,

sulfide lead mineral, sulfide zinc mineral, sulfide molybdenum mineral, and other sulfides from sulfide iron minerals.[¶] See also PRINCIPLE OF CONSERVATION OF TOTAL VORTICITY; FLOTATION POLISHING PRINCIPLE; PRINCIPLE OF VIRTUAL WORK.

PRINCIPLE OF FOURIER TRANSFORM INFRARED PHOTOACOUSTIC SPECTROSCOPY [biology] Analysis involving enzymatic reactions resulting in variations in pH.[‖] See also FTIR-PAS PRINCIPLE; FTIR PRINCIPLE; FOURIER TRANSFORM INFRARED PHOTOACOUSTIC SPECTROSCOPY PRINCIPLE.

PRINCIPLE OF FREE DOMAIN [psychology] See EMANCIPATION PRINCIPLE; see also WEBER'S PRINCIPLE OF VALUE NEUTRALITY; PRINCIPLE OF NONDOMINATION.

PRINCIPLE OF FRICTIONAL CHARGE GENERATION [physics] Rubbing with a substance of a different composition produces a separation of charge (if one object acquires one unit of postive charge, the other object acquires one unit of negative charge).[**]

PRINCIPLE OF FUNCTIONAL DEFINITION [psychology] (1) The task of managers to provide for attaining objectives in the face of changing environments; the more provisions are made for building in organizational flexibility, the more adequately organization structure can fulfill its purpose.[††] (2) The more clearly a position or a department defines the results expected, activities to be undertaken, organization authority delegated, and authority and informational relationships with

*Koontz, H. and O'Donnell, C. (1968), *Principles of Management, 4th ed., McGraw-Hill.*

†Yiyong, Y., Rencheng, W., Xiaohong, J. and Dewen, J. (2003), Progress in the study on synergetic control principle of human upper extremity and related issues, *J Biomed. Eng. (Shengwu Yixue Gongchengxue Zazhi)*, **20**(4), 738–741.

‡Koontz, H. and O'Donnell, C. (1968), *Principles of Management*, 4th ed., McGraw-Hill.

§Somasundaran, P., Zhang, L., Krishnakumar, S., and Slepetys, R. (1999), Flotation deinking—a review of the principles and techniques, *Progress Paper Recycl* **8**(3), 22–36.

¶Kuipers, J. A. M., Prins, W., and Van Swaaij, W. P. M. (1991) Theoretical and experimental bubble formation at a single orifice in a two-dimensional gas-fluidized bed, *Chem. Eng. Sci.* **46**(11), 2881–2894.

‖Li, L., Lan, S., Liu, R., and Xi, S. (1985), General principle of Fourier transform infrared photoacoustic spectroscopy (FTIR-PAS) and its applications, *Huaxue Tongbao* **10**, 19–21.

Lukac, J. (1984), Application of changes in the surface properties of minerals to their mutual separation in an electrostatic field, *Sbornik Geol. Ved Technol. Geochem.* **19, 197–214.

††Koontz, H. and O'Donnell, C. (1968), *Principles of Management*, 4th ed., Mcbraw-Hill.

other positions, the more adequately individuals responsible can contribute toward accomplishing enterprise objectives.[*] See also PRINCIPLE OF PLANNING.

PRINCIPLE OF FUNCTIONAL DELEGATION [organizational psychology] The more clearly a position or department defines the results expected, activities to be undertaken, organization authority delegated, and informational relationships with other positions, the more adequately individuals responsible can contribute toward accomplishing enterprise objectives.[*] See also PRINCIPLE OF PLANNING.

PRINCIPLE OF FUNDAMENTAL CHARACTERISTICS [medicine, psychology] (1) Instinctive attempt to avoid pain, discomfort, or unpleasant situations; desire to obtain maximum gratification with minimum effort. (2) Activities must be grouped to facilitate the accomplishment of goals; and the manager of each subdivision must have authority to coordinate its activities with the organization as a whole. The more clearly a position or a department defines the results expected, activities to be undertaken, organization authority delegated, and authority and informational relationships with other positions, the more adequately individuals responsible can contribute toward accomplishing enterprise objectives. (3) If an activity is designed as a check on the activities of another department, individuals charged with such activity cannot adequately discharge their responsibilities if they report to the department whose activity they are expected to evaluate.[†] See also AUTOMATIC CONTROL PRINCIPLES; PRINCIPLE OF DOMINANT SUBSYSTEMS; PRINCIPLE OF OPERATION; EQUIPARTITION PRINCIPLE; PRINCIPLE OF CONSTRUCTION AND OPERATION; PRINCIPLE OF SMOOTH FIT.

PRINCIPLE OF FUNDAMENTAL JUSTICE [psychology] Basic procedural rights that are

afforded anyone facing an adjudicative process or procedure that affects fundamental rights.[‡]

PRINCIPLE OF FUNDAMENTALITY On one hand, favors that good that is a necessary precondition for the realization of another one. The principle of dignity, on the other hand, structures the values according to their meaningfulness, and brings the fundamental goods in a moral and ethical context of meaningful priorities. Goods are physical entities existing independently of our individual intentions and are given to us as indispensable factors for our responsible actions. These factors include our physical integrity, mental or physical properties, and in addition, institutional dimensions of life such as matrimony, family, and state. Values, in contrast, are certain stereotyped attitudes or virtues, that can be considered real only inasmuch as they are qualities of human volition; for example, consider the subjective understanding of justice, faith, or solidarity. Consequently, whenever a morally relevant deed is demanded, acting individuals finds themselves in a situation of choice. They must choose between values and goods that are in a permanent constellation of competition among each other. Human beings, thus, always find themselves in a dilemma as to which option they should choose in the constantly conflicting set of alternatives. Viewed negatively, an ethical choice inevitably means choosing the lesser of two evils.[§]

PRINCIPLE OF GAINSHARING [psychology] System whereby the benefits of improved performance are distributed to participating employees.[¶]

[*]Koontz, H. and O'Donnell, C. (1968).
[†]Simoes, W. A. (1976), Fundamental principles and basic characteristics of functional orthopaedic techniques, *Ortodontia* **9**(2), 137–151.

[‡]Sklar, R. (2007), Starson v. Swayze: The Supreme Court speaks out (not all that clearly) on the question of "capacity," *Can. J. Psychiatry* **52**(6), 390–396.
[§]Micewski, E. R. (2006), Terror and terrorism: A history of ideas and philosophical-ethical reflections, *Preview Cultic Stud. Rev.* **5**(2) (special issue on terrorism), 219–243.
[¶]Graham-Moore, B. E., Ross, T. L., and Ross, R. A. (1995), *Gainsharing and Employee Involvement: Plans for Improving Performance*, 2nd ed., reprint, BNA Books, p. 184.

PRINCIPLE OF GALILEAN RELATIVITY [physics] A criterion for judging physical theories, stating that they are inadequate if they do not prescribe the exact same laws of physics in certain similar situations.[*]

PRINCIPLE OF GENDER [psychology] (John William Money, 1921–2006, psychologist, sexologist) (1) John Money brings together for the first time 40 years of research on the subject of gender identity and gender role. In these original clinical studies, Money examines the connection of gender identity/role to genetics, hormones, body morphology, brain chemistry, and social assimilation and learning. For the lay reader, Money introduces each study with both a scientific and historical explanation. Much light is shed on the longstanding debate between nature and nurture, and the prime importance of the critical periods of sexual and gender differentiation. The studies comment on what are and are not politically correct principles of gender, and they show that the Adam principle is, in nature's scheme of things, superimposed on the Eve principle. The development of masculinity is risky business, subject to gender transposition errors. In some rare examples cited in the book, transpositional errors are shown to be associated with paraphilic sexual fixations.[†] (2) Bateman's principle (1948), which states that eggs are more costly than sperm because of larger size and higher nutrient content, is an example of an untested hypothesis that has reached paradigmatic proportions. This difference between eggs and sperm has been used to argue that females invest more in offspring than do males, females will be selective and sexually conservative because they risk the loss of a large investment if they mate with the wrong male, and males will be promiscuous because they have nothing to lose by mating with as many females as possible. The production of millions of tiny sperm has been accepted as evidence that sperm is cheaper than eggs. It could be argued that a female's cost of reproduction is still higher than a male's because the female carries the eggs, gestates, and may have primary responsibility in caring for the young, but this may reflect a mammalian bias. One corollary of the principle of anisogamy is that females have too much to lose by mating with the wrong male.[‡] (3) [occultism] Embodies the idea that gender is manifested in everything. This does not relate explicitly to the commonly understood notion of sex, but rather "to beget; to procreate, to generate, to create, or to produce" in general. Gender is manifested as the masculine and feminine principles, and manifests itself on all planes. Mental gender is described as a hermetic concept that relates to the masculine and feminine principles. It does not refer to the physical gender of someone, nor does it suggest that someone of a certain physical gender necessarily has the same mental gender. Ideally, one wants to have a balanced mental gender. The concept put forth in *The Kybalion* states that gender exists on all planes of existence (physical, mental, and spiritual), and represents different aspects on different planes. It is also stated that everything and everyone contains these two elements or principles. The masculine principle is always in the direction of giving out or expressing, and contents itself with the "will" in its varied phases. The feminine principle is always in the direction of receiving impressions, and has a much more varied field of operation than the masculine does. The feminine conducts the work of generating new thoughts, concepts, and ideas, including the work of the imagination. It is said that there must be a balance in these two forces. Without the feminine, the masculine is apt to act without restraint, order, or reason, resulting in chaos. The feminine alone, on the other hand, is apt to constantly reflect and fail to actually do anything, resulting in stagnation. With both the masculine and feminine working in conjunction, there is thoughtful action that breeds

---

[*]Tykodi, R. J. (1967), Thermodynamics and classical relativity, *Am. J. Phys.* **35**(3), 250–253.

[†]Money, J. (1993), The Adam principle: Genes, genitals, hormones, & gender, *Selected Readings in Sexology*. Prometheus Books, Amherst, NY.

[‡]Tang-Martinez, Z. (2000), Paradigms and primates: Bateman's principle, passive females, and perspectives from other taxa, in *Primate Encounters: Models of Science, Gender, and Society*, Uni. Chicago Press, pp. 261–274.

success. which point out that both the feminine and the masculine fulfill each other.* See also BATEMAN'S PRINCIPLE; MASCULINE PRINCIPLE; FEMININE PRINCIPLE; PRINCIPLE OF CAUSE AND EFFECT.

PRINCIPLE OF GENERAL COVARIANCE [physics] The laws of physics must make the same predictions in all reference frames, not the invariance of physical laws under arbitrary coordinate transformations. This is an extension of the special principle of relativity, which deals only with nonaccelerating frames, and general covariance is a realization of it.[†]

PRINCIPLE OF GENERIC CONSISTENCY (PGC) [psychology] (Alan Gewirth, 1912–2004, American philosopher) Every agent must act in accordance with his or her own and all other agents' generic rights to freedom and wellbeing.[‡] See also EQUIVALENCE PRINCIPLE.

PRINCIPLE OF GLOBAL POSITIONING SYSTEM (GPS) [geology] The Global Positioning System (GPS) has become a widely used tool in geodetic studies of Earth. A global network of dual-frequency GPS receivers continuously tracks the GPS satellites, and the data are distributed to globally accessible online archives. Analysis of these data occurs and the results are made available in the form of satellite ephemerides accurate to a few centimeters and receiver position estimates accurate to a few millimeters. The principle behind GPS is the measurement of distance (or "range") between the receiver and the satellites. Applying a two-dimensional trigonometric measurement to a three-dimensional space.[§]

PRINCIPLE OF GOOD ENOUGH (POGE) [computer science] A rule for software and systems design. It favors quick-and-simple designs over elaborate systems designed by committees. Once the quick-and-simple design is deployed, it can then evolve as needed, driven by user requirements. Ethernet, the Internet protocol, and the World Wide Web are good examples of this kind of design.[¶]

PRINCIPLE OF GOVERNING OF DISSIPATIVE PROCESSES [physics] (Gyarmati) One of the most widely applied variational principles of irreversible thermodynamics. It is formulated to incorporate the whole domain of nonequilibrium thermodynamics. From the original form of the Gyarmati principle, the basic equations of the Onsager's irreversible thermodynamics are derivable, and its formulation extensively exploits the special structure of these equations.[‖] Also termed *governing principle of dissipative processes*.

PRINCIPLE OF GRAEFFE'S METHOD [mathematics] (1) Finding the complex roots of algebraic equations of high degree. (2) Applies a root-squaring method. Given equations are transformed into another whose roots are high powers of those of the original equation; the roots are thus widely separated and are easily found.[**]

PRINCIPLE OF GULDBERG AND WAAGE [chemistry] See PRINCIPLE OF MASS ACTION; PRINCIPLE OF EQUILIBRIUM.

*Three Initiates (2004), *The Kybalion*, Book Tree [the "Three Initiates" who authored *The Kybalion* chose to remain anonymous; as a result, a great deal of speculation has been made about who actually wrote the book.)
[†]O'Hanian, H. C. and Ruffini, R. (1994), *Gravitation and Spacetime*, 2nd edition ed., Norton, New York.
[‡]Hoffman, R. (1991), The dialectical necessity of morality: An analysis and defense of Alan Gewirth's argument to the principle of generic consistency, *Library J.* **116**(21), p. 150.
[§]Filipik, A., Peterlik, I., Hemzal, D., Jan, J., Jirik, R., Zapf, M., and Ruiter, N. (2007), Calibrating an ultrasonic computed tomography system using a time-of-flight based positioning algorithm, *Proc. Annual Int. Conf. IEEE Engineering in Medicine and Biology Society*, 2007, pp. 2146–2149.
[¶]Wilker, S. and Guillaume, P. (1998), International round robin test to determine the stability of DB ball propellants by heat-flow calorimetry, *Proc. 29th Int. Annual Conf. ICT (Energetic Materials)*, pp. 132.1–132.15.
[‖]Van, P. (1996), On the structure of the governing principle of dissipative processes, *J. Nonequlib. Thermodyn.* **21**, 17.
Soli, R. A. and Soonpaa, H. H. (1979), Evaluation of the Shamir-Graeff method for measuring optical constants of thin films, *Appl. Optics* **18(20), 3367–3368.

PRINCIPLE OF HABEAS MENTEM [psychology] (1) The human right to one's own mind. Creative health and the principle of habeas mentem.* (2) Mentem—the right to one's own mind. (a) In our system of justice we have, by building the principle of habeas corpus into precept and precedent, protected the right of a humans to his own bodies. In the coming years, in order to keep our experts from imposing their own ideas and values on the not-so-expert, we may need to weave into all codes of professional conduct the principle of habeas mentem. (b) The term *habeas mentem* was first used by George Kelly.[†] (3) Protecting the health and wellbeing of children is a fundamental value that the chemical industry shares with society. (4) It is common to all the production processes that large amounts of cell material result at the end and contain, inter alia, mainly natural and homologous nucleic acids (DNA and RNA) and also recombinant plasmid DNA or foreign DNA integrated into the genome of the host cell. It has hitherto been possible only with difficulty in estimating the possible risks on release of this recombinant DNA for the environment. Thus, on the precautionary principle, statutory requirements and regulations have been issued by authorities and institutions in most countries [such as, e.g., in the Federal Republic of Germany, the Central Committee for Biological Safety (ZKBS) of the Federal Board of Health in Berlin], which call for inactivation of waste material, which is usually liquid, and thus of the amount of nucleic acids present after the production process as protection against a possible risk to the environment.[‡] See also PRECAUTIONARY PRINCIPLE; HEALTH PRINCIPLE.

PRINCIPLE OF HARMONY OF OBJECTIVES [sociology, psychology] (1) Aspects include

mutual goal setting, a clear statement of objectives, a harmony of objectives, and effort and risk. (2) Ensure that members of ethnic communities are able to exercise their rights and fulfill their responsibilities, and access to and use of government services that are appropriate to their needs and to participate in the public decisionmaking process. (3) Mutual understanding and respect for the individual. (4) Ensure that all beliefs and cultures can prosper in a spirit of tolerance and within the institutional framework. Ensure that the depth of cultural diversity enriches and benefits all sections of society. Enhance the vibrancy of society through support of activities that exhibit cultural diversity.[§] (5) Effective directing depends on the extent to which individual objectives in cooperative activity are harmonized with group objectives.[¶] See also PRINCIPLE OF PLANNING.

PRINCIPLE OF HEAVY MEDIA SEPARATION [engineering] Separation in a sink float process that takes place with the aid of a fluid whose density is between the densities of the components being separated. The particles having a higher density will sink to the bottom while the particles with the lower density will float on the surface. This system permits the separation of non-metallic components, heavy metals and magnesium alloys from aluminum. Separation of different alloys is theoretically possible but difficult due to the variety of densities. The difference of densities between the particles to be separated is a ruling factor of the quality of the separation.[‖]

PRINCIPLE OF HEREDITY [psychology] Fulfillment of human needs that employees will work to satisfy while at the same time contributing to the achievement of enterprise objectives. (2) Fresh, comprehensive, and accurate survey of heredity, supplemented by the most elaborate contribution to its most important recent phase, Mendelism. No better books have been published in recent

*Sanford, F. H. (1956), *Am. J. Public Health* **46**(2), 139–148.

[†]Sanford, F. (1955), George Kelly, in an informal presentation at the 1955 Annual Meeting of the American Psychological Association, Creative health and the principle of habeas mentem; reprinted in *Am. Psychol.* **10**(12), 829–835.

[‡]Sanford, F. H. (1956), Creative health and the principle of habeas mentem, *Am. J. Public Health Nation's Health* **46**(2), 139–148.

[§]Sutton, M. (2003), The father of physical chemistry, *Chem. Britain* **39**(5), 32–34.

[¶]Koontz, H. and O'Donnell, C. (1968), *Principles of Management*, 4th ed., McGraw-Hill.

[‖]Schmitz, Christoph (2006) Handbook of aluminum recycling. Vulkan-Verlag GmbH, p 54.

years for the teacher or student who desires to get in touch with current biological work. Thomson's treatment of heredity is broad enough to include the microscopic study of the germ cells, the experiments in breeding, and the results of the statistical method. The position of the author is indicated by his acceptance of De Vries' origin by mutation without going to the extreme of regarding it as the sole method of origin; his cautious recognition of Medelian principles to explain part, but not all, of inheritance, and his desire to harmonize these results with the conclusions of the statistical school. Mendel's *Principles of Heredity* is an entirely different type of book from Thomson's. It is not a review as much as a fundamental contribution to genetic science and a foresight of the revision of evolutionary conceptions in the light of the new evidence of the factorial analysis of heredity. In the applications to human society, Bateson takes a stand in favor of restraint of the unfit and against proposals for the genetic encouragement of the fit, since "we have little to guide us in estimating the qualities for which society has or may have a use."[*]

PRINCIPLE OF HISTOLOGICAL DETERMINA-TION [pharmacology] Steps developed to perform specific procedures, usually in practice of medicine or other biological sciences such as slide-staining procedures.[†]

PRINCIPLE OF HOLISM [psychology] When two people see the same object, such as a cup, from different perspectives, one person sees some aspects of the cup, and the other sees other aspects. But there must be a god above all individual perspectives who sees the truth about the object. Americans preferred the argument on the basis of noncontradiction in each case, and the Chinese preferred the dialectic one.[‡] See also PRINCIPLE OF RELATIONSHIP; YIN–YANG PRINCIPLE.

PRINCIPLE OF HOMEOPATHY [psychology] (Samuel Hahnemann, 1755–1843, German physician) Homeopathy sought to cure symptoms of disease by use of drugs that induced similar symptoms and restored the patient's "vital force."[§]

PRINCIPLE OF HOMO TYPONS [psychology, biology] Based on the 12 variable loci, typing method. Used to type strains from prolonged epidemic episodes and from genotypic families that are apparently emerging in different parts of the world to analyze the evolution of the pathogenic population in relation to their geographic distribution.[¶]

PRINCIPLE OF HOPE [psychology] (Sigmund Freud) (1) Has two aspects, the "not yet conscious" and the "not yetbecome"; the former is its subjective; the latter, its objective pole. Concerned with daydreams, which are part of everyday life for people of all ages, although the preoccupations of children, adolescents, and adults differ.[||] (2) Death principle is seen to clearly manifest itself in the compulsion of repetition, but the compulsion of repetition is fundamentally ambivalent. Civilization carries with it manifestations of death, for it originated as an answer to it. But its practice is not reduced to the utilization of the "death drive"; on the contrary, its return to the past is preformed under the signum of the "principle of hope," and it is consecrated to the transcendent and progressive quality of eros. The psychoanalysis is based on the compulsion of repetition, but is based on the new quality of this phenomenon: about the

versus analytic cognition, *Psychol. Rev.* **108**(2), 291–310.
[§]Perez, C. and Tomsko, P. (1994), Homeopathy and the treatment of mental illness in the 19th century, *Hosp. Commun. Psychiatry* **45**(10), 1030–1033.
[¶]Troost, F. J., van Baarlen, P., Lindsey, P., Kodde, A., de Vos, W. M., Kleerebezem, M. and Brummer, R.-J. M. (2008), Identification of the transcriptional response of human intestinal mucosa to Lactobacillus plantarum WCFS1, *BMC Genomics* **9**, 374.
[||]Wieseltier, L. (1986), The principle of hope, NY, *in vivo Times Book Review*, p. 44.

[*]Thomson, J. A. (1909), *Heredity*, G. P. Putnam's, New York.
[†]Carey, F. A. (1994), Measurement of nuclear DNA content in histological and cytological specimens: Principles and applications, *J. Pathol.* **172**(4), 307–312.
[‡]Nisbett, R., Peng, K., Choi, I., and Norenzayan, A. (2001), Culture and systems of thought: Holistic

"not yet" of the repetition, about the future and the "utopia" that are inherent to it.*

PRINCIPLE OF HOUGH TRANSFORM REPRESENTATION [mathematics]   A noise-insensitive method of detecting collinear image points, such as straight lines, in images.[†] See also PRINCIPLE OF THE HOUGH TRANSFORM.

PRINCIPLE OF HUMANITARIANISM [psychology]   Fairness in balance in distribution of aid and humanitarian efforts.[‡] See also PRINCIPLE OF MUTUAL HELP.

PRINCIPLE OF HUMANITY [psychology] (Giambattista Vico, 1668–1744)   (1) Humankind shall be treated humanely in all circumstances by saving lives and alleviating suffering, while ensuring respect for the individual. The Code of Conduct for the International Red Cross and Red Crescent Movement and NGOs in Disaster Relief (RC/NGO code) introduces the concept of the humanitarian imperative, which expands to include the right to receive and to give humanitarian assistance. It states that the obligation of the international community is "to provide humanitarian assistance wherever it is needed."[§] (2) Another speaker's beliefs and desires are connected to each other and to reality in some way, and attribute to him or her "the propositional attitudes one supposes one would have oneself in those circumstances."[¶] See also

*Caruso, I. (1968), Aggressivity or death-drive? *Fortschr. Psychoanal. Int. Jahrbuch Weiterentwicklung Psychoanal.* **3**(3), 105–121.
[†]Csenki, L. Alm, E., Torgrip, R. J. O., Aaberg, K. M., Nord, L. I., Schuppe-Koistinen, I., and Lindberg, J. (2007), Proof of principle of a generalized fuzzy Hough transform approach to peak alignment of one-dimensional 1H NMR data, *Anal. Bioanal. Chem.* **389**(3), 875–885.
[‡]Patrnogic, J. (1977), Inter-relationship between general principles of law and fundamental humanitarian principles applicable to the protection of refugees, *Annal. Droit Intl. Med.* **27**, 50–59.
[§]Gauthier, C. C. (1993), Philosophical foundations of respect for autonomy, *Kennedy Inst. Ethics J.*, **3**(1), 21–37.
[¶]Davidson, Donald (1984) [1974]. "Ch. 13: On the Very Idea of a Conceptual Scheme". Inquiries into Truth and Interpretation. Oxford: Clarendon Press.

PRINCIPLE OF CHARITY; PRINCIPLE OF RATIONAL ACCOMMODATION.   See also PRINCIPLE OF IMPARTIALITY; PRINCIPLE OF INDEPENDENCE; DEFINING PRINCIPLES; PRINCIPLE OF NEUTRALITY; PRINCIPLE OF PROSELYTISM; EMERGENCY ACTION PRINCIPLE.

PRINCIPLE OF HYDRAULIC FRACTURING [engineering]   Variations adopted in plants and animals.[‖] See also PRINCIPLE OF INACCESIBILITY.

PRINCIPLE OF HYPERCOMPUTATION UPPER BOUNDS [physics]   Hypercomputation as tool to supersede natural laws and control them is beyond the limits of science (principle of hypercomputation upper bounds).[**] See also CONTAINMENT PRINCIPLE; PHYSICAL PRINCIPLE OF COMPUTATION; GALILEO'S PRINCIPLE OF NATURAL COMPUTATION; GANDALF'S PRINCIPLE OF HIDDEN UNIVERSE COMPUTATION.

PRINCIPLE OF IDENTITY [mathematics]   (1) A "thing" is identical to itself; that is, for all x, $x = x$. (2) If a propositional function F is true of an individual variable x, then F is true of x (symbolically, $\forall x, F(x) \supset F(x)$).[††]

PRINCIPLE OF IMMEDIATE GRATIFICATION [psychology, mathematics]   (1) Buy now and finance the purchase over time. (2) Please instantly and as easily as possible.[‡‡] See also PRINCIPLE OF DELAYED GRATIFICATION; PLEASURE PRINCIPLE.

PRINCIPLE OF IMPARTIALITY [psychology]   (1) Guidance should be client-centered, confidential, open and accessible to all

[‖]Cramer, D. D. (1987), Limited entry extended to massive hydraulic fracturing, *Oil & Gas J.* **85**(50), 40–44.
[**]Costa, J. F. (2006), *Physics and Computation: Essay on the Unity of Science through Computation* (http://cmaf.ptmat.fc.ul.pt/preprints/pdf/2006/unity.pdf).
[††]Lewis, G. N. (1930), The principle of identity and the exclusion of quantum states, *Phys. Rev.* **36**, 1444–1453.
[‡‡]Lazartigues, A., Morales., H., and Planche, P. (2005), Consensus, hedonism: The characteristics of new family and their consequences for the development of children, *L'Encephale* **31**(4 Pt.1), 457–465.

adults, independent in its advice, publicized widely, and capable of contributing to the development of learning opportunities; guidance settings vary greatly, and that impartiality can be assured in services that are not, by definition, independent.[*] (2) Provision of humanitarian assistance must be impartial and not based on nationality, race, religion, or political point of view. It must be based on need alone.[†] See also PRINCIPLE OF HUMANITY; PRINCIPLE OF INDEPENDENCE; DEFINING PRINCIPLES; PRINCIPLE OF NEUTRALITY; PRINCIPLE OF PROSELYTISM; EMERGENCY-ACTION PRINCIPLE.

PRINCIPLE OF INACCESIBILITY [biology] Variations adopted in plants and animals.[‡] See also PRINCIPLE OF HYDRAULIC FRACTURING.

PRINCIPLE OF INCLUSION–EXCLUSION [mathematics] (Abraham de Moivre, 1667–1754, French mathematician)   If $A_1, \ldots, A_n$ are finite sets, then,

$$\left| \bigcup_{i=1}^{n} A_i \right| = \sigma_{i=1}^{n} |A_i| - \sigma_{i,j:1 \geq i < j \geq n} |A_i \cap A_j|$$
$$+ \sigma_{i,j,k:1 \geq i < j < k \geq n} |A_i \cap A_j \cap A_k|$$
$$- \cdots + (-1)^{n-1} |A_1 \cap \cdots \cap A_n|$$

where $|A|$ denotes the cardinality of the set A. For example, taking $n = 2$, we get a special case of double counting; in words, we can count the size of the union of sets A and B by adding $|A|$ and $|B|$ and then subtracting the size of their intersection. The term inclusion-exclusion originates comes from the concept of overgenerous inclusion, followed by compensating exclusion. When $n > 2$, the exclusion of the pairwise intersections is (possibly) too severe, and the correct formula

is as shown with alternating signs.[§] See also SIEVE PRINCIPLE.

PRINCIPLE OF INDEPENDENCE [psychology] (1) Analysis focuses on how Valuing People's four key principles (choice, independence, rights, and inclusion) were drawn on and discussed spontaneously by participants. Each of these four principles has important implications for the provision of services for people with intellectual disabilities and dementia.[¶] (2) Humanitarian agencies must formulate and implement their own policies independently of government policies or actions. See also PRINCIPLE OF HUMANITY; PRINCIPLE OF IMPARTIALITY; DEFINING PRINCIPLES; PRINCIPLE OF NEUTRALITY; PRINCIPLE OF PROSELYTISM; EMERGENCY-ACTION PRINCIPLE.

PRINCIPLE OF INDETERMINACY [physics] See INDETERMINACY PRINCIPLE

PRINCIPLE OF INDIFFERENCE [mathematics] (1) Suffice for the rational assignation of probabilities to possibilities. Bertrand advances a probability paradox, to which the principle is supposed to apply; yet, merely because the problem is ill-posed in a technical sense, applying it leads to a contradiction. Examining an ambiguity in the notion of an ill-posed problem shows that there are precisely two strategies for resolving the paradox: the distinction strategy and the well-posing strategy. The principle applied here was formulated by Jakob Bernoulli as the "principle of insufficient reason," and later by Keynes as "the principle of indifference." "If there is no known reason for predicating of our subject one rather than another of several alternatives, then relatively to such knowledge the assertions of each of these alternatives have an equal probability." The principle is supposed to encapsulate a necessary truth about the relation of possibilities and probabilities: that possibilities of which we have equal ignorance have equal

[*]Payne, J. and Edwards, R. (1997), Impartiality in pre-entry guidance for adults in further education colleges, *Br. J. Guid. Counsel.* **25**(3), 361–375.
[†]Carse, A. L., (1998), Impartial principle and moral context: securing a place for the particular in ethical theory. The *J. Med. Phil.* **23**(2), 153–169.
[‡]Junquera, J., Cohen, M. H., and Rabe, K. M. (2007), Nanoscale smoothing and the analysis of interfacial charge and dipolar densities, Los Alamos National Laboratory, *Preprint Archive—Condensed Matter*, pp. 1–30.

[§]Weisstein, E. W. (n.d.), Inclusion-exclusion principle, *MathWorld—M Wolfram Web Resource* (http://mathworld.wolfram.com/Inclusion-ExclusionPrinciple.html).
[¶]Forbat, L. (2006), An analysis of key principles in Valuing People, *J. Intell. Disabil.* **10**(3), 249–260.

probabilities. Prima facie, the principle is quite unrestricted. It is supposed to apply to any events or sets of events among which we have no reason to discriminate and also supposed to allow equality of ignorance to be sufficient to determine the probabilities.[*] (2) A rule for assigning epistemic probabilities when there are $n > 1$ mutually exclusive and collectively exhaustive possibilities. If the n possibilities are indistinguishable except for their names, then each possibility should be assigned a probability equal to $1/n$.[†]

PRINCIPLE OF INDISCERNIBLES [mathematics] (Gottfried Wilhelm Leibniz, 1646–1716, German mathematician) (1) If there is no way of differentiating between two entities, then they are one and the same entity. In other words, entities x and y are identical if and only if any predicate possessed by x is also possessed by y and vice versa.[‡] (2) No two distinct substances exactly resemble each other.[§] See also LEIBNIZ'S PRINCIPLE; PRINCIPLE OF SUFFICIENT REASON.

PRINCIPLE OF INDIVIDUAL DIFFERENCES [kinesiology] (Dr. Hans Selye) See PRINCIPLE OF TRAINING.

PRINCIPLE OF INDIVIDUALITY OF CONTROLS [physics] (1) Second law of thermodynamics; states that in the neighborhood of any equilibrium state of a system, there are states that are not accessible by a reversible or irreversible diabolic process.[¶] (2) Controls must be consistent with the position, operational responsibility, competence, and needs of the individuals who have to interpret the control measures and exercise control. See also PRINCIPLE OF PLANNING.[‖]

PRINCIPLE OF INDULGENCE [psychology] Infancy is beyond lies, and the tenderness and honesty that children need are denied them by the eroticized and deceptive actions of adults in relation to them.[**].

PRINCIPLE OF INERTIA [physics] A body at rest, will remain at rest, and a body in motion will continue in motion with constant speed in a straight line, as long as no unbalanced force acts on it.[††] See also NEWTON'S PRINCIPLES OF MOTION.

PRINCIPLE OF INERTIAL CONFINEMENT [psychology] Since it is the task of controls to inform people who are expected to act to avoid or correct deviations from plans, effective controls require that they be consistent with the position, operational responsibility, competence, and needs of the individual concerned.[‡‡]

PRINCIPLE OF INFORMATION DIFFUSION [mathematics] Guarantees the existence of reasonable diffusion functions to improve the nondiffusion estimates when the given samples are incomplete. In other words, when X is incomplete, there must exist some approach to pick up fuzzy information of X for more accurately estimating a relationship as function approximation.[§§]

[*]Shackel, N. (2007), Bertrand's paradox and the principle of indifference, *Phil. Sci.* **74**(2), 150–175.
[†]Neapolitan, R. E. (1991), The principle of interval constraints: A generalization of the symmetric Dirichlet distribution, *Math. Biosci.* **103**(1), 33–44.
[‡]Shadmi, Y. (1978), Teaching the exclusion principle with philosophical flavor, *Am. J. Phys.* **46**(8), 844–848.
[§]Troisfontaines, C. (1985), The logical approach to substance and the principle of indiscernibles, *Leibniz Questions Logique*, pp. 94–106.
[¶]Scheibe, R. (1991), Redox-modulation of chloroplast enzymes. A common principle for individual control. *Plant Physiol.* **96**(1), 1–3.
[‖]Koontz, H. and O'Donnell, C. (1968), *Principles of Management*, 4th ed., McGraw-Hill.
[**]Ferenczi, S. (1930), The principle of relaxation and neocatharsis, *Int. J. Psychoanal.* **11**, 428–443.
[††]Smolin, L. (1986), On the nature of quantum fluctuations and their relation to gravitation and the principle of inertia, *Classical Quantum Gravity* **3**(3), 347–359.
[‡‡]Bohachevsky, I. O. and Hafer, J. F. (1977), *Dependence of Sputtering Erosionl of Fuel Pellet Characteristics*. Los Alamos Scientific Lab, Report number LA-6991-MS.
[§§]Chongfu, H. and Yong, S. (2002), *Towards Efficient Fuzzy Information Processing: Using the Principle of Information Diffusion*, Springer, p. 150.

PRINCIPLE OF INHIBITION [kinesiology]   For the proper use of muscles, many of the effects of the stimulus and hence the primary afferent are inhibitory; there is a class of neurons in the spinal cord that are in fact inhibitory interneurons.*

PRINCIPLE OF INSTANT GRATIFICATION [psychology]   If feedback is not instantaneous, the operator is tempted to keep seeking a change.† See also OVERSHOOT PRINCIPLE; PRINCIPLE OF TOTAL CONTROL; PRODUCTION PRINCIPLE.

PRINCIPLE OF INSUFFICIENT REASON [mathematics]   Also known as *principle of indifference*. (John Maynard Keynes, 1883–1946, British economist) Renamed by the economist who was careful to note that it applies only when there is no knowledge indicating unequal probabilities.‡ See also PRINCIPLE OF MAXIMUM ENTROPY.

PRINCIPLE OF INTEGRATING INTERESTS [industrial psychology, physics]   (1) A technique used in selling in which the salesperson, knowing the buyer's personal interests or buying motives, places emphasis on these in the presentation rather than on the features or benefits of the product.§ (2) The connection between the relating of two activities, their interactive influence, and the values thereby created.¶

*Primakoff, H. (1956), Exclusion-principle inhibition of bound hyperon mesonic decay, *Nuovo Cimento* **3**, 1394–1399.
†Bjorklund, E. (2001), Toward a general theory of control knobs, *Proc. Int. Conf. Accelerator and Large Experimental Physics Control Systems* (http://arxiv.org/abs/physics/0111082).
‡Slater, P. B. (2000), High-temperature expansions of Bures and Fisher information priors, *Phys. Rev. E (Statist. Phys., Plasmas, Fluids, Related Interdisc. Topics*, **61**(6), 6087–6090.
§Bradmore, D. (2004), Principle of integrating interests, Monash Univ. Babylon Information Platform, MONASH Marketing Dictionary (http://marketing.businessdictionaries.org/Public-Service-Advertising-Glossary/F/1/Principle_of_Integrating_Interests).
¶Anon. (n.d.), *Experience in the Light of Recent Psychology: Circular Response* (www.follettfoundation.org/CECH3.pdf).

PRINCIPLE OF INTERCULTURAL COMMUNICATION [psychology]   (1) Klyukanov's principles move in a logical progression from features that distinguish separate cultures, to points of commonality, to the tensions inherent in negotiating shared physical and informational space. He concludes with two principles pertaining to the desired end of intercultural communication. Although each principle constitutes a separate chapter, they are not logically isolated. The discussion of commonalities reaffirms the distinctiveness of each culture, while the principles related to tensions demonstrate the dynamic interplay of commonalities.‖ Throughout the book, Klyukanov sacrifices conceptual familiarity for originality, which turns out to be a strength and a weakness at the same time. Clearly, most of the core concepts in the discipline are covered in the text. The difference comes in how they are organized, and how we arrive at our understanding of them.** (2) Consisting of six imperatives: (1) new technologies are creating complex relationships between different cultures; (2) cultural diversity is a fact of life; (3) the ability to communicate with other cultures is a good business; (4) peace stimulates healthy relationships; (5) the better the communication with other cultures, the better understanding is made by ourselves as individuals; and (6) actions and words forces grater consideration toward consequences among cultural understanding.†† See also PUNCTUATION PRINCIPLE; PERFORMATIVITY PRINCIPLE; POSITIONALITY PRINCIPLE; COMMENSURABILITY PRINCIPLE; CONTINUUM PRINCIPLE; SYNERGY PRINCIPLE; SUSTAINABILITY PRINCIPLE.

PRINCIPLE OF INTERFERENCE [psychology]   (1) Pratt's studies, however, have shown that there is a limit to interaction among traces at a psychophysical level. This limit is probably reached within a single sensory

‖Klyukanov, I. E. (2004), *Principles of Intercultural Communication*. Allyn and Bacon.
Smith, L. (2006), Review of principles of intercultural communication, *Int. J. Intercult. Relations* **30(3), 417–420.
††Smith, S. E. (1977), Increasing transcultural awareness: The McMaster-Aga Khan-CIDA Project workshop model, *J. Transcult. Nurs.* **8**(2), 23–31.

modality, and is almost certainly reached when heteromodal stimuli are interpolated. Pratt has also shown that the same kind of changes still occur in these noninteracting traces, and has therefore argued that such changes must be referred to some older principle of fading or disuse. In other words, the principles of interaction and of disintegration are both needed to account for the fate of certain traces. Before it can be argued that assimilation (interference, or interaction) is the sole condition underlying the fate of mnemonic traces, it must be shown that all psychological processes are capable of mutual interference.[*] (2) A technique used in selling in which the salesperson, knowing the buyers' personal interest or buying motives, places emphasis on these in the presentation rather than on the features or benefits of the product.[†]

PRINCIPLE OF INVARIANCE [physics, mathematics, chemistry] Ambartzumyan and Chandrasekhar introduced the principles of invariance of the law of diffuse reflection and the law of darkening and derived the two-dimensional Fredholm integral equations of the second kind for intensity I in the astrophysical problem of the transfer of radiation in a semiinfinite, isotropically scattering atomosphere.[‡] See INVARIANCE PRINCIPLE.

PRINCIPLE OF INVARIANT IMBEDDING [physics] Calculates the wavefields inside and scattered from a strongly (laterally and vertically) heterogenous, anisotropic, inclusion, which may be large but remains compact. The factorization is carried out with respect to direction of average power flow rather than the more conventional factorization with respect to local direction of propagation. Can handle an extreme range of modal wave speeds, and allows continuous as well as discontinuous medium variations on different (wave)length scales. It also, inherently, takes care of critical-angle phenomena. The wavefield solution in the inclusion is coupled to the external field via a boundary element approach.[§] Also known as *invariant imbedding principle*.

PRINCIPLE OF ISOMORPHISM See MITSCHER-LICH'S PRINCIPLE; MITSCHERLICH'S PRINCIPLE OF ISOMORPHISM.

PRINCIPLE OF ISOSTASY [physics]   Measurement of large bright stars can be determined provided the distance is known.[¶]

PRINCIPLE OF JOB DEFINITION [geology, psychology]   (1) Level of Earth's crust is determined by its density; lighter materials rises and heavier material sink. (2) The more precisely the results expected from managers are identified, the more clearly their position can be defined.[‖] (3) Specifications for the job rest on organization requirements and on provision for incentives to induce effective and efficient performance of the tasks involved.[**] See also PRINCIPLE OF PLANNING; (PRINCIPLES OF STAFFING;) STAFFING PRINCIPLES; PRINCIPLE OF MANAGERIAL APPRAISAL; PRINCIPLE OF THE OBJECTIVE OF STAFFING.

PRINCIPLE OF JU [Kinesiology]   Underlies all classical Bujutsu (武術 martial arts) methods and was adopted by the developers of the Budō (武道) martial ways) disciplines. The classical warrior could intercept and momentarily control his enemy's blade when

[*]Burnham, R. W. (1941), Intersensory effects and their relation to memory-theory. *Am. Jrnl. Psych.* 54(4), 473–489.

[†]Mollon, J. D. (2002), The origins of the concept of interference, *Phil. Trans. A Math. Phys. Eng. Sci.* **360**(1794), 807–819.

[‡]Karanjai, S. and Biswas, G. (1990), Solution of integro-differential equations by principle of invariance, *Proc. Natl. Seminar on Mathematics*, Univ. Calcutta, pp. 30–136.

[§]Bellman, R. and Kalaba, R. (1956), On the principle of invariant imbedding and propagation through inhomogeneous media, *Proc. Natl. Acad. Sci. USA* **42**(9), 629–632.

[¶]Soboloev, P. O. and Erinchek, Yu. M. (2002), Implication of the principle of isostasy for study of the crustal structure of Russia, *Dokl. Earth Sci.* **382**(1), 112–116.

[‖]Shimmin, S. (1975), People and Work: Some contemporary issues, *Bri. J. Industr. Med.* **32**(2), p. 93–101.

[**]Koontz, H. and O'Donnell, C. (1968), *Principles of Management*, 4th ed., McGraw-Hill.

attacked, then, in a flash, could counter-attack with a force powerful enough to cleave armour and kill the foe. Terms like "Jūjutsu" (柔術) and "Yawara" (柔) made the principle of Jū the all-pervading one in methods catalogued under these terms. Rooted in the concept of pliancy or flexibility, as understood in both a mental and a physical context. Two aspects are in constant operation, both interchangeable and inseparable. One aspect is that of "yielding", and is manifest in the exponent's actions that accept the enemy's force of attack, rather than oppose him by meeting his force directly with an equal or greater force, when it is advantageous to do so. It is economical in terms of energy to accept the foe's force by intercepting and warding it off without directly opposing it; but the tactic by which the force of the foe is dissipated may be as forcefully made as was the foe's original action. The second aspect makes allowance for situations in which yielding is impossible because it would lead to disaster. In such cases "resistance" is justified. But such opposition to the enemy's actions is only momentary and is quickly followed by an action based on the first aspect of Jū, that of yielding.*

PRINCIPLE OF JUST ENOUGH [engineering] Related to manufacturing costs for materials used and the impact efficiently determining proper quantity needed.[†]

PRINCIPLE OF JUST RETURN [mathematics] (1) Return on investment and benefits of member counting participating in joint pro-grams of space exploration. (2) The need to correlate the financial contributions of the member States to a form of direct remuner-ation is justified by the amount of resources necessary to allow the organisations to carry out their institutional objectives.[‡]

PRINCIPLE OF JUSTICE [Psychology] (1) Just reward function (or micro principle of justice) and the just reward distribution (or macro principle of justice) are linked in a mathematically precise relationship that includes a third element, viz.. the probability distributions of the reward-relevant charac-teristics. b) Points out that the just reward function (a "micro" principle of justice, since it assigns shares of goods to individuals) is not only different from but may also conflict with the just reward distribution (a "macro" principle of justice, since it describes a characteristic or parameter of a reward distribution in a social aggregate). That is, a person may adhere to mutually inconsistent micro and macro principles of justice. Given the probability distributions of the reward-relevant characteristics, the micro and macro principles of justice imply each other.[§] (2) Justice mandates that each social basis for self-respect be equalized (and, as a second priority, maximized).[¶] (3) A guide to ethical practice for provision of counseling services and therapy offering vital benefits, criteria for fair distribution; problems with applying criteria; impact of third parties on the system; problems of accommodat-ing ethnic minorities and impoverished clients.[‖]

PRINCIPLE OF KALMAN FILTER [psychology] Specifications for the job rest on the need for results from plans, the requirements of a clear structure of roles, and the provision

*Bakboot, N. (1969), Shorter notices, *J. Shorter Notices* **5656**(1), 88–91 (Royal Central Asian Soc.) (http://www.informaworld.com/smpp/title~content =t785027059 ~db=all~tab=issueslist~branches= 56, **5656**(1), p. 88–91.
†Dall, W. H. (1876), Botany and zoology structure and movements of the leaves of dionsea music, *Am. J. Sci. Arts* **12**(3), 237.
‡Marini, L. (1999) Brief Observations on the Rela-tionship between the European Space Agency and the European Community in the Light of the Prin-ciple of Just Return. International Organisations and Space Law, Proceedings of the Third ECSL Colloquium, Perugia, Italy, 6-7 May 1999. Harris, R.A., ed. ESA SP-442. ESA/ESTEC, p. 440–447.
§Jasso, Guillermina (1983) Fairness of Individual Rewards and Fairness of the Reward Distribution: Specifying the Inconsistency between the Micro and Macro Principles of Justice. *Soc. Psych. Q.*, **46**(3), 185–199
¶Eyal, N., (2005) Perhaps the most important primary good': Self-respect and Rawls's principles of justice. Politics, Philosophy & Economics, 4(2), 195–219.
‖Smith, S.J. (1985) The principle of justice, Calif. Nurse **81**(5), 3.

for incentives to induce efficient and effective performance.[*]

PRINCIPLE OF KARMA [psychology]   (1) Literally, "accomplished action" and, therefore, the sum of the effects of all causes occurred in the past. (2) The individual responsibility—which excludes any fatalism or intervention of destiny or of a superior will. Provides a theoretical support also for the possibility of changing for better—or for worse—by means of targeted actions.[†]

PRINCIPLE OF KUAN HIS [mathematics]   In general the Chinese are collective, concerned with social harmony over the individual and indirect with their communications.[‡] See also ARIOKHIN–PARETO INTERACTION PRINCIPLE.

PRINCIPLE OF LAND MANAGEMENT [engineering]   Encourage agencies, legislators, property owners, and managers to flow with natural processes.[§] See also USUFRUCT PRINCIPLE.

PRINCIPLE OF LARGE DEVIATIONS [mathematics, physics]

$$p^{\text{HYL}}(\mu) = \sup_{0 \leq x_1 \leq x_0} \left\{ \mu x_0 - f(x_0 - x_1) - \frac{a}{2}(2x_0^2 - x_1^2) \right\},$$

The first step in our proof is to bound the interaction term (1) above by

$$\frac{a}{2V_1} \left\{ N_1^2 - n_1(1)^2 \right\}$$

and below by

$$\frac{a}{2V_1} \left\{ N_1^2 - \left(\sigma_{j=1}^{m_1} n_1(j) \right) \right\}.$$

[*]Zhao, J. and Deng, B. (1997), Direct simultaneous determination of o, m, p-nitronethylbenzene in mixture using spectrophotometry combined with Kalman filter, *Guang Pu* **17**(6), 57–60.

[†]Hanchett E. S. (1992), Concepts from eastern philosophy and Rogers' science of unitary human beings, *Nurs. Sci. Quart.* **5**(4), 164–170.

[‡]Shirokov, M. I. (1961), A general relativistic theory of reactions of the $a + b$ ☐ $c + d + e + \ldots$ type, *Zh. Eksperiment. Teoretich. Fiz.* **40**, 1887–1891.

[§]Dixon-Gough, R. W. and Bloch, P. C. (2008), *The Role of the State and Individual in Sustainable Land Management*, p. 196 (http://www.researchandmarkets.com/reports/357 465).

The principle of large deviations provides a compact way of making rigorous the method of the largest term; applied to the upper bound on the hamiltonian it yields the expression (1) for a lower bound to the pressure $p^{\text{HYL}}(\mu)$. To deal with the lower bound (2) we have first to estimate the entropy involved in grouping together the first m_1 levels; this is achieved by an inequality proved in Sect. 4; the method of large deviations applied to the lower bound (2) then yields the expression (1) for na upper bound to the pressure $p^{\text{HYL}}(\mu)$ and the proof is complete.[¶] See also INVARIANCE PRINCIPLE.

PRINCIPLE OF LASER SPECKLE FLOWGRAPHY [engineering]   Noncontacting strain sensor that is based on tracking laser speckles through a digital correlation technique.[‖]

PRINCIPLE OF LATERAL CONTINUITY [geology]   Layers of sediment initially extend laterally in all directions; in other words, they are laterally continuous. As a result, rocks that are otherwise similar, but are now separated by a valley or other erosional feature, can be assumed to be originally continuous. Layers of sediment do not extend indefinitely; rather, the limits can be recognized and are controlled by the amount and type of sediment available and the size and shape of the sedimentary basin. As long as sediment is transported to an area, it will eventually be deposited. However, as the amount of material lessens away from the source, the layer of that material will become thinner.[**] See also PRINCIPLE OF CROSS-CUTTING RELATIONSHIPS; PRINCIPLE OF ORIGINAL HORIZONTALITY; PRINCIPLE OF FAUNAL SUCCESSION.

PRINCIPLE OF LEADERSHIP FACILITATION [organizational psychology]   The more an

[¶]van den Berg, M., Lewis, J.T., and Pule, J.V. (1988) The Large Deviation Principle and Some Models of an Interacting Boson *Math. Biosci. Gas. Common. Math. Phys.* **118**, p. 61–85.

[‖]Yaoeda, K., Shirakashi, M., Funaki, S., Funaki, H., Nakatsue, T., and Abe, H. (2000), Measurement of microcirculation in the optic nerve head by laser speckle flowgraphy and scanning laser Doppler flowmetry, *Am. J. Ophthalmol.* **129**(6), 734–739.

[**]Tarbuck, L. T. (2005), *Earth—an Introduction to Physical Geology*, Pearson Education Canada Inc.,

organization structure that an authority delegations within it make possible for various managers to design and maintain an environment for performance, the more it will facilitate leadership abilities of managers.* See also PRINCIPLE OF PLANNING.

PRINCIPLE OF LEARNING [psychology] (Wertheimer–Pragnanz principle of learning)   Refers to the fact that the individual's phenomenal field tends to be as simple and as clear as the given conditions allow. How this principle operates to bring about improved adjustment or the better gestalt, is explained as follows. When one grasps a problem situation, its structural features and requirements set up certain strains, stresses, and tensions in the thinker. What happens in real thinking is that these stresses and strains are followed up, yield vectors in the direction of improvement of the situation, and change it accordingly. The changed state is a state of affairs held together by inner forces as a good structure in which there is harmony in the mutual requirements. Also warns of certain dangers such as the seductiveness of shortcut closure. For example, impatient desire tends to produce such a premature closure in the hungry animal separated from its food by a set of bars; the animal focuses only on the near goal and fails to view the situation freely. Such goal blindness is insoluble as long as the focus is on one's own desire rather than on one's own desire as a part of the total situation. Habit blindness and piecemeal attitudes have been demonstrated as other such factors.†

PRINCIPLE OF LEAST ACTION [physics, mathematics] (A. N. Whitehead)   (1) The actual motion of a conserve dynamical system between two points takes place in such a way that the actin has a minimum value with reference to all other paths between the points that correspond to the same energy.‡ (2) The trajectory of the system is the path that makes the value of S stationary relative to nearby paths between the same configurations and for which the energy has the same constant value. The principle is misnamed, as only the stationary property is required. It is a minimum principle for sufficiently short but finite segments of the trajectory.¶ (3) The actual motion is found by finding the extremum of the action integral. Optimization models arise as a result of application of one or more of these principles. New and challenging mathematical problems arise in obtaining the results from optimizing models.§ See also HAMILTON PRINCIPLE, MINIMAL PRINCIPLES; LEAST-ACTION PRINCIPLE; ENTROPY OPTIMIZATION PRINCIPLE. Also referred to as the *principle of cosmic laziness*.

PRINCIPLE OF LEAST ASTONISHMENT [psychology, geology]   When two elements of an interface conflict or are ambiguous, the behavior should be such that which will least surprise the human user or programmer at the time the conflict arises, because the least surprising behavior will usually be the correct one.¶

PRINCIPLE OF LEAST AUTHORITY (POLA) [computer science] (Peter J. Denning, computer scientist)   (1) Every module that can be a process, a user, or a program must be able to see only such information and resources that are immediately necessary. (2) Grant just the least possible amount of privileges to permit a legitimate action, in order to enhance protection of data and functionality from faults (fault tolerance) and malicious behavior (computer security).‖ See also PRINCIPLE OF MINIMAL PRIVILEGE; PRINCIPLE OF LEAST PRIVILEGE.

*Koontz, H. and O'Donnell, C. (1968), *Principles of Management: An Analysis of Managerial Functions*, 4th ed., McGraw-Hill (http://knol.google.com/k/narayana-rao-kvss/principles-of-management-koontz-and/2utb2lsm2k7a/89).
†Torrance, P. (1950), The principle of Pragnanz as a frame of reference for psychotherapy, *J. Consult. Psychol.* **14**(6), 452–457.

‡Walker, P. M. B. (1999), *Chambers Dictionary of Science and Technology*. New York, N.Y. : Chambers p.915.
§Kapur, J. N. (1988), Mathematical modeling, New Age Int. p. 202.
¶Merrill, R. T. (1995), Geomagnetism: Principle of least astonishment, *Nature*, **374**(6524), 674–675.
‖Piyaratn, P. (1982), Doctors' roles in primary health care, *Tropical Doctor*, **12**(4, Pt. 2), 196–202.

PRINCIPLE OF LEAST CONSTRAINT [physics, engineering]   (1) The motions of any number of interconnectd masses under the action of forces deviate as little as possible from the motions of the same masses if disconnected and under the action of the same forces. The motions are such that the constraints are a minimum, the constraint being the sum of the products of each mass and the square of the deviation from the postion that it would occupy if free.* (2) The path of a conservative system in configuration space is such that the integral of the Lagrangian function over time is a minimum or maximum relative to nearby paths between the same endpoints and taking the same time.§ See also D'ALEMBERT'S PRINCIPLE; DIFFERENTIAL PRINCIPLE; VARIATIONAL PRINCIPLE; GAUSS' PRINCIPLE; HAMILTON PRINCIPLE; HAMILTON–LAGRANGE PRINCIPLE; MAUPERTUIS–LAGRANGE PRINCIPLE; PRINCIPLE OF EXTREMAL EFFECTS; LEAST-CONSTRAINT PRINCIPLE; D'ALEMBERT–LAGRANGE PRINCIPLE; JOURDAIN PRINCIPLE; JAYNES' PRINCIPLE; PRINCIPLE OF VIRTUAL WORK; KINETIC PRINCIPLE; MAUPERTUIS–EULER–LAGRANGE PRINCIPLE OF LEAST ACTION; D'ALEMBERT–LAGRANGE VARIATION PRINCIPLE; LAGRANGE PRINCIPLE.

PRINCIPLE OF LEAST EFFORT [mathematics] (1) The maximum entropy principle (MEP) maximizes the entropy subject to the constraint that the effort remains constant. The principle of least effort (PLE) minimizes the effort subject to the constraint that the entropy remains constant. Egghe and Lafouge investigate the relation between these two principles. It is shown that (MEP) is equivalent with the principle "(PLE) or (PME)" where (PME) is (introduced in this paper) the principle of most effort, meaning that the effort is maximized subject to the constraint that the entropy remains constant.† (2) An information-seeking client will tend to use the most convenient search method, in the

least exacting mode available. Information-seeking behavior stops as soon as minimally acceptable results are found. This theory holds true regardless of the user's proficiency as a searcher or level of subject expertise.‡

PRINCIPLE OF LEAST KNOWLEDGE [computer science]   A design guideline for developing software, particularly object-oriented programs. (1) "Talk only to your immediate friends." (2) A given object should assume as little as possible about the structure or properties of anything else, including its subcomponents.§

PRINCIPLE OF LEAST MOTION [chemistry] See LEAST-ACTION PRINCIPLE.

PRINCIPLE OF LEAST NUCLEAR MOTION See NUCLEAR MOTION PRINCIPLE.

PRINCIPLE OF LEAST PRIVILEGE [computer science]   See PRINCIPLE OF LEAST AUTHORITY.

PRINCIPLE OF LEAST RESISTANCE [physics] (1) Investigation on specific problems, such as a mechanism associated with resistance to motion. Performing computer experiments with tools such as molecular dynamics simulations, or create friction and wear machines that probe surface and interface, lubrication, deformation, fracture, and wear behavior. A more global approach searches for general principles describing the path(s) of least resistance using mathematics that chooses the optimal path and aids in understanding a specific problem of resistance to motion, friction, or wear. (2) Electric current flows along all available paths in inverse proportion to the impedance of the paths; that is, the magnitude of the current flowing along a given path is greater for a lower impedance

---

*Walker, P. M. B. (1999), *Chambers Dictionary of Science and Technology*, Chambers, New York, p. 915.
†Egghe, L. and Lafouge, T. (2006), On the relation between the maximum entropy principle and the principle of least effort, *Math. Comput. Model.* **43**(1-2), 1–8.

‡Bierbalm, E. G. (1990), A paradigm for the '90s; in research and practice, library and information science needs a unifying principle; "least effort" is one scholar's suggestion, *Am. Libraries* **21**, p. 18.
§Dahle, L. O., Forsberg, P., Svanberg-Hard, H., Wyon, Y., and Hammar, M. (1997), Problem-based medical education: Development of a theoretical foundation and a science-based professional attitude, *Med. Educa.* **31**(6), 416–424.

and lesser for a higher impedance.* See also LEAST-RESISTANCE PRINCIPLE.

PRINCIPLE OF LEAST SQUARES [mathematics] (1) Unknown quantities in an inconsistent redundant system of linear equations can be estimated according to the condition of minimizing the squares of the residuals.[†] (2) Euler (1778), in his commentary on Daniel Bernoulli's memoir, virtually formulated the principle of least squares.[‡] (3) The best or most probable value obtainable from a set of measurements or observations of equal precision is that value for which the sum of the squares of the errors is a minimum.[§] (4) The best estimates for a,b for fitting the straight line. Optimization models arise as a result of application of one or more of these principles. New and challenging mathematical problems arise in obtaining the results from optimizing models.[¶] See also ENTROPY OPTIMIZATION PRINCIPLE.

PRINCIPLE OF LEAST SURPRISE [mathematics]   When two elements of an interface conflict or are ambiguous, the behavior should be such that will least surprise the human user or programmer when the conflict arises, because the least surprising behavior will usually be the correct one. See also PRINCIPLE OF LEAST ASTONISHMENT.[‖]

PRINCIPLE OF LEAST TIME [physics]   (1) Of all secondary waves (along all possible paths),

the waves with the extrema (stationary) paths contribute most, due to constructive interference. (2) The actual path between two points taken by a beam of light is the one that is traversed in the least time. (3) The light path through the medium is such that the time necessary for its traversal is minimum. (4) At the point where light rays refract, the traveltime is also minimized (i.e., $dt/ds = 0$).[**] See FERMAT'S PRINCIPLE OF LEAST TIME; HUYGENS' PRINCIPLE; EXTREMUM PRINCIPLE OF MECHANICS; VARIATIONAL PRINCIPLE.

PRINCIPLE OF LEAST WAVE CHANGE Fermat's principle of least time has some well-known limitations. It does not, for example, apply to diffraction gratings and holograms, because it does not include the concept of waves. The substitution of least number of waves in flight for least time of flight and the addition of a term that is a function of the grating frequency result in a generalized principle.[††]

PRINCIPLE OF LEAST WORK [physics]   See LEAST-ACTION PRINCIPLE.

PRINCIPLE OF LEX TALIONIS [psychology] (*Code of Hammurabi*; an eye for an eye) (1) May be a universal rule; examines this rule as it applies to the use of both threats and punishments in international and interpersonal interactions.[‡‡] (2) Applies to the broader class of legal systems that specify formulaic penalities for specific crimes, which are thought to be fitting in their severity. Establishment of a body whose purpose was to enact the retaliation and ensure that this was the only punishment. (3) Prescribed "fitting" counterpunishment for an offense. Developed as early civilizations grew and a less well-established system for retribution of wrongs, feuds, and vendettas, threatened

*Schmidt, F. K. (1987), Principle of the least resistance: No longer timely or even false, *Praxis Naturwissen. Chem.* **36**(5), 26–30.
[†]Sheynin, O. B. (1999), The discovery of the principle of least squares, *Historia Sci.* **8**(3), 249–264.
[‡]Sheynin, O. B. (1993), On the history of the principle of least squares, *Arch. Hist. Exact Sci.* **46**(1), 39–54.
[§]Ergun, Sabri (1956) Application of the principle of least squares to families of straight lines, *J. Industr. Eng. Chem.* **48**, 2063–2068.
[¶]Kapur, J. N. (1988), Mathematical modeling, *New Age Int.* p. 202.
[‖]Kang, K., Carlier, D., Reed, J., Arroyo, E. M., Ceder, G., Croguennec, L., and Delmas, C. (2003), Cation disorder and electronic conductivity of layered Li0.9Ni0.45Ti0.55O2 as candidate battery cathode material, *Chem. Mater.* **15**(23), 4503–4507.

**Walker, P. M. B. (1999), *Chambers Dictionary of Science and Technology*, Chambers, New York, p. 915.
[††]Abramson, N. (1989), Principle of least wave change, *J. Opt. Soc. Am. A Optics Image Sci.* **6**(5), 627–629.
[‡‡]Patchen, M. (1993), *Reciprocity of Coercion and Cooperation between Individuals and Nations. Aggression and Violence: Social Interactionist Perspectives*, American Psychological Assoc., Washington, DC, p. 119–144.

the social fabric.* See also PRINCIPLE OF RECIPROCITY.

PRINCIPLE OF LIMITING FACTOR [physics, engineering] (1) Technology relating to the processing machines such as collators and thread-rolling machines. Operation of such machines is, in fact, the rate at which prime materials can be fed into them. (2) The detection of pattern variations, deviations, or defects often requires complex processing equipment and a significant amount of processing time.[†] (3) Consider limiting factor in generating alternatives and selection from alternatives. See also PRINCIPLE OF PLANNING.[‡]

PRINCIPLE OF LINEAR MOMENTUM [physics] Theory of general relativity indicating a relationship. In the systems consisting of two or more bodies the linear momentum is zero, provided by the result of the external forces applied to the system. If F (the resultant of the external forces) $= 0$, there is no variation of the total linear momentum (dP_{total}) of the system per a time unit: $d(P_{total})/dt = F = 0$. Therefore the total linear momentum $P = P^1 + P^2 + \cdots P^n$ of the system maintains a constant value, while the linear momenta of the single bodies can vary, because the pairs of forces inside the system produce equal and opposite variations of the single linear momenta.[§] See also PRINCIPLE OF VIRTUAL POWER; PRINCIPLE OF MATERIAL FRAME INDIFFERENCE, COVARIANCE PRINCIPLE; FUNDAMENTAL PRINCIPLE.

*Capela, A. and Temple, S. (2006), LeX is expressed by principle progenitor cells in the embryonic nervous system, is secreted into their environment and binds Wnt-1, *Devel. Biol.* **291**(2), 300–313.

[†]Kalyuzhnyi, S. V., (2006) Biotechnological hydrogen production: fundamental principles and limiting factors, *Kataliz v. Promyshlennosti* **6**, 33–41.

[‡]Koontz, H. and O'Donnell, C. (1968) *Principles of Management: An Analysis of Managerial Functions*, 4th ed., McGraw-Hill (http://knol.google.com/k/narayana-rao-kvss/principles-of-management-koontz-and/2utb2lsm2k7a/89)

[§]Hertz, H. G. (1982), Momentum conservation determining diffusion in binary molecular and electrolyte solutions, *Z. Phys. Chem. Suppl.* **1**, 7–40.

PRINCIPLE OF LINEAR SUPERPOSITION [physics] See INTERFERENCE PRINCIPLE.

PRINCIPLE OF LINES OF COMMAND CHAIN [psychology] (Henri Fayol, 1841–1925, French industrialist) Organizations need a formalized hierarchy that reflects the flow of authority and responsibility.[¶] See also FAYOL'S PRINCIPLES OF MANAGEMENT; PRINCIPLE OF SCALAR CHAIN.

PRINCIPLE OF LINGUISTIC RELATIVITY [psychology] Sapir–Whorf hypothesis relationships between language, mind, and experience. The language one speaks structures the world and directs one's thought processes; in other words, language structures reality more than it is itself structured by reality. Substantive significance of the linguistic relativity hypothesis when using translations of written personality measures.[∥] See also LINGUISTIC RELATIVITY PRINCIPLE.

PRINCIPLE OF LIQUID DISPLACEMENT [hydrostatics] The volume of a body immersed in a fluid is equal to the volume of the displaced fluid.[**]

PRINCIPLE OF LOCAL ACTION [physics] Action, stating that, in the absence of temperature effects, the stress in a material point is completely determined by the deformation and the deformation history at that point. The actual internal damage parameter D figuring in the constitutive eqn is now assumed to be dependent on the strain (and the strain history) in a limited finite area enclosing the particular material point where the stress has to be evaluated. Consequently local strain peaks will always have a certain transfer to the environment and thus prevent the localization of the damage, which in turn has a suppressing

[¶]Wood, M. C. (2002), *Henri Fayol: Critical Evaluations in Business and Management*, Taylor & Francis, p. 8.

[∥]Harwood, B. T. and Pinxten, R. (1970), Universalism versus relativism in language and thought: Proceedings of a colloquium on the Sapir-Whorf hypotheses, *J. Soc. Psychol.* **81**(1), 3–8.

[**]American Heritage (2000), *American Heritage Dictionary of the English Language*, 4th ed., Houghton Mifflin, New York.

effect on the progressive growth of the deformation.*

PRINCIPLE OF LOCAL STRAIGHTNESS OF CURVES [mathematics, logic] *For any curve C and any point on it, there is a non-degenerate segment of C around the point which is straight.*

This principle allows a very different concept of continuum regarding the real numbers, because in the case of standard (and also non-standard) analysis, this continuum is constructed as a collection of discrete points.

Hence, in an infinitesimal neighborhood of a point, the tangent of a curve in that point exactly coincides with the curve itself: this will allow us a natural application of infinitesimal microscopes[7].

An immediate consequence of the Principle of Local Straightness is the existence of nilpotent infinitesimals. In order to prove this, consider the curve C with equation $y = x^2$. Let Δ be the straight portion of the curve around the origin. So, if Δ is the intersection of the curve with its tangent (the x-axis), it is the set of points x such that $x^2 = 0$. Since Δ has to be non-degenerate, it follows the existence of a nilpotent element not coincident with 0.[†]

PRINCIPLE OF LOCALITY [physics] Distant objects cannot have direct influence on one another; an object is influenced directly only by its immediate surroundings.[‡]

PRINCIPLE OF LOGICAL EFFORT [mathematics] (Ivan Sutherland and Robert Sproull, 1991) A straightforward technique used to estimate delay in a complementary metal oxide semiconductor (CMOS) circuit. Used properly, it can aid in selection of gates for

a given function (including the number of stages necessary) and sizing gates to achieve the minimum delay possible for a circuit. A guideline in very large-scale integrated circuit (VLSI) design to determine the preferred design of several that implement a logical statement—it states that the design with the minimum propagation delay is preferred.[§]

PRINCIPLE OF MACROSCOPIC SEPARABILITY [mathematics] Used for deriving the reciprocal relations in the linear phenomenological coefficients derived from the classical hypothesis of the separability of individual processes. A dissipation function is a minimum with respect to variations in all the unprescribed forces in a steady state. Such a function is found to be the integral of part of the change of the rate of entropy production with respect to the forces. The integrability of such a change is due to the classical principle of separability.[¶]

PRINCIPLES OF MAGNETOHYDRODYNAMICS [physics] Consists of the study of plasmas (i.e., fully ionized gases), containing common fluid flow equations and Maxwell's equations (since the particles making up the ionized gas are electrically charged). Because a plasma consists of charged particles, it conducts electricity and is influenced by (and even generates its own) magnetic fields.[‖]

PRINCIPLE OF MAJORITY RULE BASED ON DECISIONMAKING POWER DEMAND [psychology] See EMANCIPATION PRINCIPLE.

PRINCIPLE OF MANAGEMENT [psychology] (1) In choosing from among alternatives, the more an individual can recognize and solve for those factors that are limiting or critical to the attainment of the desired goal, the more easily that person can select the most

*De Vree, JHP, Brekelmans, WAM., and Van Gils, MAJ. (1995) Comparison of nonlocal approaches in continuum damage mechanics. Computers & Structures, 1995.

[†]Magnani, L. and Li, P. (2007), Model-Based Reasoning in Science, Technology, and Medicine (Studies in Computational Intelligence). Springer, p. 205.

[‡]McWeeny, R. and Amovilli, C. (1999), Locality and nonlocality in quantum mechanics: a two-proton EPR (Einstein-Podolsky-Rosen) experiment, *Int. J. Quantum Chem.* **74**(5), 573–584.

[§]Sutherland, I. E., Sproull, R. F. and Harris, D. F. (1999), *Logical Effort: Designing Fast CMOS Circuits*, Margan Kaufmann.

[¶]Li, J. C. M. (1962), Thermodynamics for nonequilibrium systems. The principle of macroscopic separability and the thermokinetic potential, *J. Appl. Phys.* **33**, 616–624.

[‖]Hutchinson, I. (2005), Principles of magnetohydrodynamics, *Nuclear Technol.* **151**(3), 346.

favorable alternative.* (2) Purpose is to increase efficiency, crystallize the nature of management, improve research, and attain social goals.†

PRINCIPLE OF MANAGEMENT AUDIT [psychology, mathematics] Fulfill areas of economic function, corporate structure, health of earnings, fairness to stakeholders, research and development, directions analysis, fiscal policies, production efficiency, sales vigor, and executive evaluation.‡

PRINCIPLE OF MANAGEMENT DEVELOPMENT [psychology, mathematics] In preparing managers for the uncertainty, threats and opportunities posed by the challenge of frequent and unexpected changes in organizations and markets, an alternative to a traditional Western viewpoint is a holistic approach that embodies balance and harmony, sees more subtle relationships, and avoids the tensions of opposites. The purpose of this article is to report the derivation and application of six holistic principles for management development (quieting the mind, harmony and balance, relinquishing the desire to control, transcending the ego, centeredness, and the power of softness) derived from a non-Western philosophy. (2) Thorough knowledge of officers with an outside board of directors where authority is concentrated and identification of who exercises authority with teamwork with efficiency toward production and conservation of financial resources with ability to foresee changes in the market, and minimize distribution costs with service as a means of achieving profits.§ (3) The objective of

management development is to strengthen existing managers. The most effective means of developing managers is to have the task performed primarily by a manager's superior. See also PRINCIPLE OF PLANNING.¶

PRINCIPLE OF MANAGERIAL APPRAISAL [psychology] (1) The more that management development programs aim at improving the abilities of existing managers in their present positions, as well as enabling them to be promoted, and the more top managers give example and encouragement through participating actively in the leadership and operation of such programs, the more effective such programs will be. (2) Complete appraisal of managers requires appraisal of performance in terms of verifiable objectives and quality of managing. (3) Not only mechanical conservation laws but also rules of frame transformation for constitutive quantities and source quantities are simultaneously derived by changing frame on the assumption that the forms of the first and second laws of thermodynamics are invariant for the changes of frame. (4) Refers to stress–strain relations without inertial forces. (5) The more clearly, that verifiable objectives and required managerial activities are identified, the more precise can be the appraisal of managers against these criteria.‖ (6) Performance must be appraised against the management action required by superiors and against the standard of adherence in practice to managerial principles. See also PRINCIPLE OF PLANNING.** See also STAFFING PRINCIPLES; PRINCIPLES OF JOB DEFINITION; PRINCIPLES OF STAFFING; PRINCIPLE OF THE OBJECTIVE OF STAFFING.

PRINCIPLE OF MASS ACTION [chemistry] Concentrations of reactants and products are related by the equation known as the

*van Fleet, D. D. and Bedeian, A. G. (1977), *A History of the Span of Management*, Academy of Management, p. 356.

†Koontz, H. and O'Donnell, C. (1968), *Principles of Management: An Analysis of Managerial Functions*, 4th ed., McGraw-Hill (http://knol.google.com/k/narayana-rao-kvss/principles-of-management-koontz-and/2utb2lsm2k7a/89).

‡Botha, H. and Boon, J. A. (2003), *The Information Audit: Principles and Guidelines, Libri* (Saur, Germany) **53**, 23–38.

§Shefy, E. and Sadler-Smith, E. (2006), Applying holistic principles in management development. *J. Manage. Devel.* **25**(4), 368–385.

¶Koontz, H. and O'Donnell, C. (1968), *Principles of Management: An Analysis of Managerial Functions*, 4th ed., McGraw-Hill (http://knol.google.com/k/narayana-rao-kvss/principles-of-management-koontz-and/2utb2lsm2k7a/89)

‖Crane, J. S. and Crane, N. K. (2000), A multi-level performance appraisal tool: Transition from the traditional to a CQI approach, *Health Care Manage. Rev.* **25**(2), 64–73.

**Koontz, H. and O'Donnell, C. (1968), *Principles of Management*, 4th ed., McGraw-Hill.

equilibrium constant. The concentration equilibrium constant in also the forward: reverse reaction rate constant ratio.* See MASS-ACTION PRINCIPLE.

PRINCIPLE OF MATERIAL FRAME INDIFFERENCE [physics]   Plays an important role in the development of continuum mechancs by delivering restrictions on the formulation of the constitutive functions of material bodies. Material properties should be independent of observations made by different observers. Since different observers are related by a time-dependent rigid transformation, known as a Euclidean transformation, material frame-indifference is sometimes interpreted as invariance under superposed rigid body motions. Material properties are frame indifferent, or constitutive functions are form invariant relative to change of observers.[†] See also COVARIANCE PRINCIPLE; PRINCIPLE OF VIRTUAL POWER; PRINCIPLE OF LINEAR MOMENTUM; FUNDAMENTAL PRINCIPLE.

PRINCIPLE OF MATERIAL OBJECTIVITY [engineering]   Tensional state is independent (indifferent) of the choice of a coordinate system.[‡]

PRINCIPLE OF MATHEMATICAL INDUCTION [mathematics] (Francesco Maurolico [Maurolycus], 1494–1575, Italian mathematician) (1) From the the following theorem: a subset S of the set N of natural numbers that contains 0 is actually N if either of the following holds: (a) nS implies that $n + 1S$ for all nN, or (b) mS for all $0\ mn$ implies $n + 1S$ for all nN. Proofs by mathematical induction proceed as follows. Let $A(n)$ be a mathematical statement depending on the natural number n. Suppose that one can show that (a) $A(0)$ is true (called the *basis of induction*), and (b) when it is assumed that $A(n)$ is true,

it follows that $A(n+1)$ is also true (called the *inductive step*). Then, by the principle of mathematical induction, $A(n)$ is true for all natural numbers n. (2) Method of mathematical proof typically used to establish a given statement is true of all natural numbers. The method can be extended to prove statements about more general well-founded structures, such as trees known as *structural induction*. (3) The simplest and most common form of mathematical induction proves that a statement holds for all natural numbers n and consists of two steps: (a) the *basis*, showing that the statement holds when $n = 0.2$; (b) the *inductive step*, showing that if the statement holds for $n = m$, then the same statement also holds for $n = m + 1$. The proposition following the word "if" in the inductive step is called the *induction hypothesis* (or *inductive hypothesis*). To perform the inductive step, one assumes the induction hypothesis (that the statement is true for $n = m$) and then uses this assumption to prove the statement for $n = m + 1$.[§] See also INDUCTION PRINCIPLE; WELL-ORDERING PRINCIPLE.

PRINCIPLE OF MAXIMAL DISSIPATION RATE [physics]   Orthogonality condition is equivalent to various extremum principles. Provided that the dissipative force is prescribed, the actual velocity maximizes the dissipation rate subject to the side condition. May also be stated as a principle of maximal rate of entropy production.[¶]

PRINCIPLE OF MAXIMUM BOREDOM [computer science]   In user interface design, programming language design, and ergonomics, the principle (or rule) of least astonishment (or surprise) states that, when two elements of an interface conflict or are ambiguous, the behavior should be such that will least surprise the human user or programmer when the conflict arises, because the least surprising behavior will usually be the correct one.[‖]

*Walker, P. M. B. (1999), *Chambers Dictionary of Science and Technology*, Chambers, New York, p. 661.

[†]Liu, I-S. (2004), On Euclidean objectivity and the principle of material frame-indifference. Continuum Mechanics and Thermodynamics. V16 (1-2) p 177–183.

[‡]Perzyna, P. (1966), Thermodynamics of the rate-type material, *Bull. Acad. Polonaise Sci. Ser. Sci. Techn.* **14**(7), 657–667.

[§]Vacca, G. (1909), Maurolycus, the first discoverer of the principle of mathematical induction, *Bull. Am. Math. Soc.* **16**(2), 70–73.

[¶]Hutchinson, J. W. and Wu, T. Y. (1987), *Advances in Applied Mechanics*, Academic Press, p. 192.

[‖]Guay, M. and Salmoni, A.W. (1987), An examination of self-pacing procedures in human time

PRINCIPLE OF MAXIMUM COMPLEMENTARY ENERGY [mathematics, engineering] (Eric Max Reissner 1913-1996, German mathematician and engineer) See DISPLACEMENT PRINCIPLE.

PRINCIPLE OF MAXIMUM DECISION EFFICIENCY [mathematics, physics] See PRINCIPLE OF NETWORK ENTROPY MAXIMIZATION.

PRINCIPLE OF MAXIMUM ENTROPY (PME)
[mathematics, physics] (Edwin Thompson Jàynes, 1922-1998, American physics) (E. T. Jaynes) (1) Subject to known constraints (called testable information), the probability distribution which best represents the current state of knowledge is the one with largest entropy.* (2) Probability distribution that contains the most uncertainty while still satisfying the original defaults; this method leads to the unique least committed, or least biased, distribution among all compatible ones. (3) Instead of building separate models for various knowledge sources, the maximum entropy approach will provide a model that meets all these constraints in a unified statistical framework, because they are treated equivalently.† (4) When there is only limited information available about the distribution of an asset, it is still possible to price options on the asset by making suitable approximations. Maximum entropy distribution is able to accurately fit a known density, given simulated option prices at different strikes. Bayesian method of statistical inference allows estimates of distribution of a random variable from partial information in the form of a finite number of expectations. Although there are theoretically an infinite number of such distributions, there is only one way to select the distribution in a consistent manner based on the information given and no other. The method will generate a probability function in the discrete case (or density in the continuous case) even if there

is only a single option price. Furthermore, the density is guaranteed to be nonnegative and satisfy the unitary condition on its sum (or integration). The resulting density will be the least prejudiced estimate, compatible with the given price information in the sense that it will be maximally noncommittal with respect to missing or unknown information. The theory presented shows that the Maximum Entropy method depends only on observed option prices, which can be modeled by expectations of a known price functional. The price functional is given by a risk-neutral pricing formula of the form $d_i = D(T)E_Q[c_i(X_T)]$, where $D(T)$ represents the riskless discount rate to time, T, $C_i(X_T)$ denotes the ith option pay-off function at expiry, dependent only on the expiry value of the asset X_T, and d_i is the corresponding option price. Q is the risk-neutral probability measure under which the expectation is calculated.

Let X be a continuous random variable that represents the price or index of some asset or security at some fixed expiry time T in the future. The theory for discrete X closely parallels the continuous case and is, therefore, not presented here. However, we use the discrete results as an approximation to the continuous formulation in the computations discussed in Section V We wish to estimate a probability density p(x) of X on 0 < jc < oo, using only information on prices of certain derivatives of X, which expire at T. The entropy of the distribution $p(x)$ is defined by

$$S(p) = -\int_-^\infty p(x)\log p(x)dx.$$

‡ (5) A probability distribution that has maximum entropy or uncertainty out of all those dittributions that have prescribed moments. Optimization models arise as a result of application of one or more of these principles. New and challenging mathematical problems arise in obtaining the results from

estimation, *Percept. Motor Skills* **64**(3, Pt 2), 1231–1236.
*Jaynes, E. T. (1957). "Information Theory and Statistical Mechanics". Physical Review Series II 106 (4): 620–630.
†Kogan, M. N. (1967), Principle of maximum entropy, *Rarefied Gas Dynam.* **1**, 359–368.

‡Buchen, P. W., and Kelly, M. (1996) The maximum entropy distribution of an asset inferred from option prices Journal of Financial and Quantitative Analysis, v31(1) pp143–159.

optimizing models. * See also MAXENT (MAXI-MUM—ENTROPY); MAXIMUM ENTROPY PRINCIPLE; PRINCIPLE OF MINIMUM CROSS ENTROPY; PRINCI-PLE OF CROSS-ENTROPY; FUZZY ENTROPY PRIN-CIPLE; PRINCIPLE OF MAXIMUM NONADDITIVE ENTROPY; PRINCIPLE OF OPERATIONAL COMPAT-IBILITY; MERMIN ENTROPY PRINCIPLE; ENTROPY PRINCIPLE; ENTROPY OPTIMIZATION PRINCIPLE.

PRINCIPLE OF MAXIMUM ENTROPY ON THE MEAN (PMEM) [mathematics]

The infinite-dimensional optimization problem. The object to be reconstructed x is considered as a random vector. A probability density p on x is then inferred (via the concept of entropy), and finally, as we wish to select a particular object, we shall simply take the expectancy of x under the inferred density.[†]

PRINCIPLE OF MAXIMUM HARDNESS (PMH) [chemistry]

(1) Density functional theory applied to achieve a better understanding of various theoretical tools for describing chemical reactivity.[‡] (2) The energy, the electronic chemical potential, and the molecular hardness together with a similarity index and a thermodynamic index to rationalize the behavior of various intramolecular rearrangement reactions.[§]

PRINCIPLE OF MAXIMUM LIKELIHOOD [mathematics]

The best estimate of a parameter, given a random sample from a population with density function is obtained by maximizing the likelihood functions. Optimization models arise as a result of application of one or more of these principles. New and challenging mathematical problems arise in obtaining the results from optimizing models.[¶] See also ENTROPY OPTIMIZATION PRINCIPLE.

*Kapur, J. N. (1988), *Mathematical Modeling*, New Age International, p. 203.
[†]Maréchal, P. and Lannes, A. (1997), Unification of some deterministic and probabilistic methods for the solution of linear inverse problems via the principle of maximum entropy on the mean, *Inverse Problems* 13(1), 135–151.
[‡]Parr, R. G. and Chattaraj, P. K. (1991), Principle of maximum hardness, *J. Am. Chem. Soc.* 113(5), 1854–1855.
[§]Pearson, R. G. (1993), The principle of maximum hardness, *Acc. Chem. Res.* 26(5), 250–255.
[¶]Kapur, J. N. (1988), *Mathematical Modeling*, New Age International, p. 202.

PRINCIPLE OF MAXIMUM NONADITIVE ENTROPY [physics] (Edwin Thompson Jaynes, 1922-1998, American physicist)

The problem of quantum-state inference and the concept of quantum entanglement are studied using a nonadditive measure in the form of the Tsallis entropy indexed by the positive parameter q.

Philosophy of the Jaynes maximum entropy principle is quite universal. It is actually free from a specific choice of an entropic measure if the measure satisfies some basic properties such as concavity. Additivity of the measure is not absolutely necessary for the principle. A generalization of the Jaynes maximum entropy principle along with an attendant "thermodynamics" sto "nonadditive" cases has been proposed and widely discussed in the area of nonextensive statistical mechanics. This scheme uses the Tsallis entropy:

$$S_q[\hat{\rho}] = \frac{1}{1-q}(Tr\hat{\rho}^q - 1),$$

where q is a positive parameter and describes the degree of nonadditivity. In the limit $q \to 1$, this entropy converges on the von Neumann entropy: $S[\hat{\rho}] = \text{Tr}(\hat{\rho} \ln \hat{\rho})$. The Tsallis quantum entanglement is induced by the principle of maximum Tsallis entropy with the data on the Bell-CHSH observable and have quantified the degree of entanglement by means of the generalized Kullback-Leibler entropy. We have found that strong nonadditivity enhances entanglement in the superadditive regime.[‖]

PRINCIPLE OF MAXIMUM PESSIMISM [economics] (John Templeton)

Once all bad news is completely out and known by everyone, this is the time to begin to invest in the country or company's assets.[**]

PRINCIPLE OF MAXIMUM U UNCERTAINTY [mathematics]

(1) Extracting model parameters of a physical process from statistical

[‖]Abel, S. and Rajagopal, A. K. (1999), Quantum entanglement inferred by the principle of maximum nonadditive entropy. Physical Review A, 60(5), 3461–3466.
[**]Minard, J. (1995), The principle of maximum pessimism (John Templeton's investment principle), *Forbes* 155, 67.

experimental data. (2) A method for quantifying the information content of continuous functions and a technique for solving inverse problems where the solution is known to be continuous.* See also MINIMUM INFORMATION PRINCIPLE; BAYESIAN PRINCIPLE.

PRINCIPLE OF MAXIMUM WORK [chemistry] (Marcellin Pierre Eugène Berthelot, 1827–1907, French chemist) Chemical reactions will tend to evolve in such a manner so as to assemble or dissasemble chemical species in order to yield the maximum amount of chemical energy in the form of work as the reaction progresses.[†] *Also known as* (BERTHELOT PRINCIPLE OF MAXIMUM WORK; BERTHELOT–THOMSON PRINCIPLE.

PRINCIPLE OF MEAN FITNESS MAXIMIZATION [mathematics] See CATALYSIS MAXIMIZATION PRINCIPLE.

PRINCIPLE OF MEASUREMENT [physics] See HEISENBERG SCATTER PRINCIPLE.

PRINCIPLE OF MECHANICS [engineering] (Giuseppe Moletti, 1531—1588) (1) Motion of living organisms is an enigmatic phenomenon. Compared to jet airplanes, the rocket, or the truck, which result from the explosive expansion of high-temperature gases and the motion of sailing ships, waves, or trees blowing in the wind, every living organism represents the successful integration of many biomolecular machines that convert energy from light or raw chemical form into whatever the organism needs—motion, heat, or the construction and disposal of internal structures.[‡] (2) When two or more forces operate on a body in opposite directions, so as to counteract or balance each other, they produce what is called *equilibrium*.

(3) Mathematical foundations of a general mechanics to explain both the action of simple machines and the behavior of heavy bodies in general.[§] (3) Stipates that present conditions and actions no longer precisely decide the future, especially if interactions are assumed Newtonian actions—at a distance. (4) The fabric of space is an all-prevailing continuum. Causality and mechanics will remain valid within the precision of quantum uncertainty.[¶] See also PAULI PRINCIPLE.

PRINCIPLE OF MEDIOCRITY [psychology] (1) There is nothing special about Earth, and by implication the human race.[‖] (2) Used either as a heuristic about Earth's position or as a philosophical statement about the place of humanity. See also MEDIOCRITY PRINCIPLE; ANTHROPIC PRINCIPLE; PLENITUDE PRINCIPLE; UNIFORMITY PRINCIPLE; COPERNICAN PRINCIPLE.

PRINCIPLE OF MENTAL TRANSMUTATION [psychology] The art of changing and transforming mental states, forms, and conditions, into others.[**] See also PRINCIPLE OF CAUSE AND EFFECT.

PRINCIPLE OF MENTALISM [psychology] Embodies the truth that "all is mind." The "all" is a substantial reality underlying all the outward manifestations and appearances that we know under the terms of "the material universe," the "phenomena of life," "matter," "energy," and, in short, all that is apparent to our material senses. It also defines it as "spirit," which is also unknowable and undefinable, but may be considered of as an universal, infinite, living mind.[††] See alsoxs PRINCIPLE OF CAUSE AND EFFECT.

§Tatevskii, V. M. and Spiridonov, V. P. (1968), Superposition principle in quantum mechanics, *Vestnik Moskov. Univ. Ser. 2 Khimiya* **23**(6), 3–14.
¶Zhou, M. (1999), *New Variational Principle in Quantum Mechanics*.
‖Gonzalez, R. (2004), *The Privileged Planet: How Our Place in the Cosmos is Designed for Discovery*, Regnery Publishing.
**Trois initiés (n.d.) *Le Kybalion: Etude sur la Philosophie Hermétique de L'ancienne Egypte et de L'ancienne Grèce*, University of Manne.
††Sivadon, P. (1953), General principles in mental prophylaxis, *Rev. Hygiene Med. Soc.* **1**(4), 322–328.

*Rico-Ramirez, V., Diwekar, Urmila, M., and Morel, B. (2003), Real option theory from finance to batch distillation, *Comput. Chem. Eng.* **27**(12), 1867–1882.
†Solov'ev, Yu. I. and Starosel'skii, P. I. (1962), History of physical chemistry. (principle of maximum work), *Dokl. Akad. Nauk SSSR* **39**, 24–48.
‡Gyarmati, I. (1965), New (local) forms of the principle of least dissipation of energy, *Period. Polytech* **9**(2) 205–207.

PRINCIPLE OF MICROSCOPIC REVERSIBILITY [chemistry] (Richard Chace Tolman 1881–1948, American mathematical physicist) (1) At equilibrium, the rate of every reaction is exactly counterbalanced by the rate of its reverse reaction.* (2) A reversible reaction the mechanism in one direction is exactly the reverse of the mechanism in the other direction. A result of microscopic reversibility is that the series of transition states and intermediates of the forward reaction are mirrored in reverse order in the reverse reaction.[†]

PRINCIPLE OF MICROREVERSIBILITY [physics] (Stephen W. Hawking, 1942-, Univ. Cambridge, British physicist) In most situations, information is scrambled and lost. If the exact details of how the items were scrambled are known, the original order can be reconstructed. Microreversibility, which has always held in classical and quantum physics, is violated by black holes. Because information cannot escape from behind the horizon, black holes are a fundamental new source of irreversibility in nature. Stephen Hawking has introduced evidence contrary to this original principle that he developed.[‡]

PRINCIPLE OF MINIMAL CAUSATION [physics, mathematics] (1) For any given physical system the action on any part or point at a (local) time can depend only on physical systems and points in the past: "We should not depend today on our future." (2) Newton's mechanics is "embryonically noncausal," because one needs knowledge from the future to calculate it; the knowledge of a stone's path is not enough to calculate. Future value must already be known because of the necessity to know. (3) An equation of motion

transporting the information about the system in time has to be of first order in time. (4) Under specific initial conditions and environmental assumptions, processes repeat themselves; there is no difference between time past and time future. (5) Refers to forces, such as gravity, and the matter–energy particles dynamically related to each other on the basis of these forces.[§] See also CAUSALITY PRINCIPLE; Light Principle.

PRINCIPLE OF MINIMAL PRIVILEGE [computer science] See PRINCIPLE OF LEAST AUTHORITY (POLA).

PRINCIPLE OF MINIMUM ASTONISHMENT [mathematics] When two elements of an interface conflict or are ambiguous, the behavior should be that that will least surprise the human user or programmer at the time the conflict arises, because the least surprising behavior will usually be the correct one.[¶] See also PRINCIPLE OF LEAST ASTONISHMENT.

PRINCIPLE OF MINIMUM CHI SQUARE [mathematics] Optimization models arise as a result of application of one or more of these principles. New and challenging mathematical problems arise in obtaining the results from optimizing models.[‖] See also ENTROPY OPTIMIZATION PRINCIPLE.

PRINCIPLE OF MINIMUM CROSS-ENTROPY [physics] (1) Used for estimating probability density function under specified moment constraints, measure of absence of information about a situation, uncertainty, transformations between measure spaces, expresses amount of disorder inherent or produced, measure of disorder of a system

---

*Gregg, B. A. (n.d.), The essential interface: Studies in dye-sensitized solar cells, *Semiconductor Photochemistry and Photophysics*, p. 51 Chap. 2.

[†]Selwyn, M. J. (1993), Application of the principle of microscopic reversibility to the steady-state rate equation for a general mechanism for an enzyme reaction with substrate and modifier, *Biochem. J.* **295**(Pt.3), 897–898.

[‡]Fatu, D. (1981), Considerations on the principle of microreversibility in chemical kinetics, *Bul. Inst. Politehnic Gheorghe Gheorghiu-Dej Bucuresti, Ser. Chim.-Metalurg.* **43**(3), 57–60.

[§]Shaklee, H. and Fischhoff, B. (1978), *Discounting in Multicausal Attribution: The Principle of Minimal Causation*, Decisions and Designs, Inc., McLean, VA.

[¶]Lai, E. C. and Orgogozo, V. (2004), A hidden program in Drosophila peripheral neurogenesis revealed: Fundamental principles underlying sensory organ diversity, *Devel. Biol.* **269**(1), 1–17.

[‖]Kapur, J. N. (1988), *Mathematical Modeling*, New Age International, p. 203.

equal to the Boltzmann constant, the natural logarithm action of the number of microscopic states corresponding to the thermodynamic state of the system.* (2) Provides a probablistic tool to gradually enforce the desired consistency between the leaf layer, where the quantization cost is calculated, and the rest of the tree—thereby imposing the structural constraint on the partition at the limit of zero temperature.[†] See also ENTROPY PRINCIPLE; VARIATIONAL MINIMAL PRINCIPLE; MAXIMUM ENTROPY PRINCIPLE; MINIMUM CROSS-ENTROPY PRINCIPLE; GENERALIZED MINIMAL PRINCIPLE; MOMENT-PRESERVING PRINCIPLE; PRINCIPLE OF MINIMUM DIVERGENCE; PRINCIPLE OF MAXIMUM ENTROPY.

PRINCIPLE OF MINIMUM DISCRIMINATION IN-FORMATION (MDI) [mathematics] (Solomon Kullback, 1903–1994, and Richard Leibler, 1914–2003, American mathematicians and cryptanalysts) Divergence as discrimination information led mathematicians to propose given new facts; a new distribution f should be chosen that is as difficult to discriminate from the original distribution f_0 as possible, so that the new data produces as small an information gain $D_{KL}(f\|f_0)$ as possible.[‡] (2) A distribution that has minimum directed divergence from a given distribution, out of all those that have prescribed moments. Optimization models arise as a result of application of one or more of these principles. New and challenging mathematical problems arise in obtaining the results from optimizing models.[§] See also LAPLACE'S PRINCIPLE OF INSUFFICIENT REASON; PRINCIPLE OF MAXIMUM ENTROPY; PRINCIPLE OF MINIMUM

CROSS-ENTROPY; MINXENT PRINCIPLE; ENTROPY OPTIMIZATION PRINCIPLE.

PRINCIPLE OF MINIMUM DISSIPATION [chemistry, mathematics] See also HELMHOLTZ–KORTEWEG PRINCIPLE; ENTROPY PRINCIPLE; VARIATIONAL MINIMAL PRINCIPLE; MAXIMUM ENTROPY PRINCIPLE; MINIMUM CROSS-ENTROPY PRINCIPLE; GENERALIZED MINIMAL PRINCIPLE; MOMENT-PRESERVING PRINCIPLE.

PRINCIPLE OF MINIMUM DIVERGENCE See PRINCIPLE OF MINIMUM CROSS-ENTROPY.

PRINCIPLE OF MINIMUM ENERGY [physics] For a closed system, with constant external parameters and entropy, the internal energy will decrease and approach a minimum value at equilibrium. *External parameters* generally means the volume, but may include other parameters that are specified externally, such as a constant magnetic field. In contrast, the second law of thermodynamics states that for isolated systems (and fixed external parameters) the entropy will increase to a maximum value at equilibrium. An isolated system has a fixed total energy and mass. A closed system, on the other hand, is a system that is connected to another system, and may exchange energy, but not mass, with the other system. If, rather than an isolated system, we have a closed system, in which the entropy rather than the energy remains constant, then it follows from the first and second laws of thermodynamics that the energy of that system will drop to a minimum value at equilibrium, transferring its energy to the other system.[¶] See also MAXIMUM ENTROPY PRINCIPLE; MINIMUM ENERGY PRINCIPLE.

PRINCIPLE OF MINIMUM ENERGY DISSIPA-TION [mathematics, engineering] See ENTROPY PRINCIPLE; VARIATIONAL MINIMAL PRINCIPLE; MAXIMUM ENTROPY PRINCIPLE; MINIMUM CROSS-ENTROPY PRINCIPLE; GENERALIZED MINIMAL PRINCIPLE; MOMENT-PRESERVING PRINCIPLE; HELMHOLTZ–KORTEWEG PRINCIPLE.

*Yee, E. (1991), Reconstruction of the antibody affinity distribution from experimental binding data by a minimum cross-entropy procedure, *J. Theor. Biol.* **153**(2), 205–227.

[†]Rose, K. (1998), Deterministic annealing for clustering, compression, classification, regression, and related optimization problems, *Proc. IEEE* **86**(11), 2222.

[‡]Liao, S.-C. and Lee, I.-N. (2002), Appropriate medical data categorization for data mining classification techniques, *Med. Informatics Internet in Med.* **27**(1), 59–67.

[§]Kapur, J. N. (1988), *Mathematical Modeling*, New Age International, p. 203.

[¶]Mornev, O. A. (1997), Dynamic principle of minimum energy dissipation for systems with ideal constraints and viscous friction, *Zh. Fizii. Khim.* **71**(12), 2293–2298.

PRINCIPLE OF MINIMUM EXPECTED NUMBER OF OBSERVATIONS IN SEQUENTIAL ANALYSIS

[mathematics] In testing of hypotheses, either keep the error of the first kind and number of observations fixed and try to minimize the error of the second kind, or keep the errors of both kinds at fixed levels and seek to minimize the expected number of observations. Error of first (second) kind arises when a hypothesis that is true or false is rejected or accepted. Optimization models arise as a result of application of one or more of these principles. New and challenging mathematical problems arise in obtaining the results from optimizing models.* See also ENTROPY OPTIMIZATION PRINCIPLE.

PRINCIPLE OF MINIMUM INVENTORY LEVEL

[mathematics, operations research] The optimal ordering time is at the minimum inventory level during the on-sale period.[†]

PRINCIPLE OF MINIMUM KINETIC ENERGY

[physics] (1) Where there are threats of serious or irreversible damage, lack of full scientific certainty shall not be used as a reason for postponing cost-effecitve measures to prevent environmental degradation. (2) The path of a conservative system in configuration space is such that the integral of the Lagrangian function over time is a minimum or maximum relative to nearby paths between the same endpoints and taking the same time. (3) Motion of the underlying system of particles compatible with the collective equilibrium provided that the variations are associated with reversible processes; external counting observables of system indicator histograms, its cycled average equilibrium count rate, and indicator volume of distribution in the body with cycle-averaged cardiac output.[‡] See also QUANTUM HAMILTON PRINCIPLE; STEWART—HAMILTON PRINCIPLE; STOCHASTIC HAMILTON VARIATIONAL PRINCIPLE; PRIGOGINE PRINCIPLE; GYARMATI'S PRINCIPLE; PRINCIPLE OF MINIMUM MECHANICAL ENERGY; PRINCIPLE OF MINIMUM POTENTIAL ENERGY; TOUPIN'S DUAL PRINCIPLE; D'ALEMBERT'S PRINCIPLE; HAMILTON PRINCIPLE.

PRINCIPLE OF MINIMUM MECHANICAL ENERGY

[physics] See QUANTUM HAMILTON PRINCIPLE; STEWART–HAMILTON PRINCIPLE; STOCHASTIC HAMILTON VARIATIONAL PRINCIPLE; PRIGOGINE PRINCIPLE; GYARMATI'S PRINCIPLE; PRINCIPLE OF MINIMUM POTENTIAL ENERGY; PRINCIPLE OF MINIMUM KINETIC ENERGY; TOUPIN'S DUAL PRINCIPLE; D'ALEMBERT'S PRINCIPLE; HAMILTON PRINCIPLE.

PRINCIPLE OF MINIMUM POTENTIAL ENERGY

[mathematics, engineering] (Eric Reissner) In stable equilibrium, the potential energy of a mechanical system is least. Optimization models arise as a result of application of one or more of these principles.[§] See DISPLACEMENT PRINCIPLE; ENTROPY OPTIMIZATION PRINCIPLE.

PRINCIPLE OF MINIMUM SURPRISE

[mathematics] In user interface design, programming language design, and ergonomics, the principle (or rule) of least astonishment (or surprise) states that when two elements of an interface conflict or are ambiguous, the behavior should be such that will least surprise the human user or programmer at the time the conflict arises, because the least surprising behavior will usually be the correct one.[¶]

PRINCIPLE OF MINIMUM TOTAL POTENTIAL ENERGY

[physics] A structure or body shall deform or displace to a position that minimizes the total potential energy, with the lost potential energy being dissipated as heat.[‖]

In this application it is sometimes known as the Ritz method. A form for the deflexion is chosen which satisfies the boundary

*Kapur, J. N. (1988), *Mathematical Modeling*, New Age International, p. 203.
[†]Anon. (1992), Improving contraceptive supply management, *Family Plann. Manager* 1(4), 1–20.
[‡]Lyul'ka, V. A. (2001), On the principle of minimum kinetic energy dissipation in the nonlinear dynamics of viscous fluid, *Tech. Phys.* (transl. of *Zh. Tekhnich. Fiz.*), 46(12), 1501–1503.

[§]Kapur, J. N. (1988), *Mathematical Modeling*, New Age International, p. 202.
[¶]Willems, P. W., Han, K. S., and Hillen, B. (2000), Evaluation by solid vascular casts of arterial geometric optimisation and the influence of ageing, *J. Anat.* 196(Pt.2), 161–171.
[‖]Mansfield, E. H. (1898) The Bending and Stretching of Plates. Cambridge University Press, p. 116.

conditions and which contains a number of disposable parameters. Thus, we may take a linear combination of the form

$$w = \sum_{n=1}^{N} B_n w_n(x,y)$$

or, more generally,

$$w = \sum_{m=1}^{M} \sum_{n=1}^{N} B_{mn} w_{mn}(x,y)$$

where the parameters B_{mn} are determined from the MN equations

$$\frac{\partial \Pi}{\partial B_{mn}} = 0.$$

PRINCIPLE OF MIXTURES [chemistry] Method of determining properties of a mixture by summing for all constituents their value of the corresponding property multiplied by the volume fraction present. Applies generally to the density of any composite, and to such properties as elasticity, tensile strength, thermal and electrical conductivity, and dielectric constant.* Also referred to as *balance principle*.

PRINCIPLE OF MOMENTS [physics]   (1) Under equilibrium conditions, the sum of the forces pulling the body in any direction must be equal to the sum of the forces pulling the body in the opposite direction. (2) The sum of the moments of the forces tending to turn the body in the clockwise direction must be equal to the sum of the moments of the forces tending to turn the body in the counterclockwise direction.† See also PRINCIPLE OF CONDITIONS FOR EQUILIBRIUM.

PRINCIPLE OF MOMENTUM [physics]   If an unbalanced force acts on a body, the body will be accelerated; the magnitude of the acceleration is proportional to the magnitude of the unbalanced force, and the direction

of the acceleration is in the direction of the unbalanced force.‡ See also NEWTON'S SECOND PRINCIPLE; NEWTON'S PRINCIPLES OF MOTION.

PRINCIPLE OF MONOMUTATIONS [biology] (1) Monocytes and macrophages are host defenses in humans. Resistance mutation is observed in patients who had suboptimal single or dual therapies. (2) Mutations may result from reduction in genetic barrier. Reasons for emergence of drug resistance include previous single or dual therapy resulting in alterations in absorption.§

PRINCIPLE OF MULTIPLE CODING [medical] Combination categories for multiple injuries are provided when there is insufficient detail as to the nature of the individual conditions, or for primary tabulation purposes when it is more convenient to record a single code.¶

PRINCIPLE OF MULTIPLE PROPORTIONS [chemistry]   When two elements combine to form more than one compound, the amounts of one of them that combine with a fixed amount of the other exhibit a simple multiple relation.‖ See also DALTON'S PRINCIPLE OF PARTIAL PRESSURES.

PRINCIPLE OF MUTUAL HELP [psychology] Fairness in balance in distribution of aid and humanitarian efforts.** See also PRINCIPLE OF RESPECT FOR NATIONAL SOVEREIGNTY; PRINCIPLE OF HUMANITARIANISM; PRINCIPLE OF AVOIDING CONFLICTS; INTERDEPENDENCE

*Walker, P. M. B. (1999), *Chambers Dictionary of Science and Technology*, Chambers, New York, p. 1006.
†Hu, C.-Y. (1966), Perturbation method based on the principle of moments, *Phys. Rev.* **152**(4), 1116–1119.

‡Tanner, A. C. (1988), The bond directional principle for momentum space wave functions: Comments and cautions, *Chem. Phys.* **123**(2), 241–247.
§Persico, E. (1928), Molecular velocities, states of excitation and the probability of transition into a degenerate gas. I, *Atti della Accademia Nazionale dei Lincei*, **7**, 137–141.
¶Sato, S., Sakai, T., and Okuno, H. (2007), OPT-TWO: Calculation code for two-dimensional MOX fuel models in the optimum concentration distribution, *JAEA-Data/Code* **17**, i–iv, 1–40.
‖Walker, P. M. B. (1999), *Chambers Dictionary of Science and Technology*, Chambers, New York, p. 662.
Stein, C. H., Ward, M., and Cislo, D. A. (1992), The power of a place: Opening the college classroom to people with serious mental illness, *Am. J. Commun. Psychol.* **20(4), 523–547.

PRINCIPLES; MULTIIDENTITY PRINCIPLES; NON INTERFERENCE PRINCIPLE.

PRINCIPLE OF NAPIER'S BONES [mathematics] (John Napier, 1550–1617, Scottish mathematician)   Set rods made of sticks of ivory, spawned tools such as advanced slide rules. The invention served as movable multiplication tables and noted as the first modern step to mechanize calculations.*

PRINCIPLE OF NAVIGATIONAL CHANGE [organizational psychology]   Manager needs to periodically check events of the plan and redraw plans to maintain the move toward a desired goal. See also PRINCIPLE OF PLANNING.[†]

PRINCIPLE OF NEGATIVE FREEDOM [psychology]   See EMANCIPATION PRINCIPLE.

PRINCIPLE OF NEPER'S BONES [mathematics] (John Napier, 1550–1617, Scottish mathematician)   Set rods made of sticks of ivory, spawned tools such as advanced slide rules. The invention served as movable multiplication tables and was noted as the first modern step toward mechanizing calculations.[‡]

PRINCIPLE OF NETWORK ENTROPY MAXIMIZATION [mathematics, physics]   (1) Energy functional can be transformed into a function of the hot charges on atoms. (2) Based on maximizing the output entropy (or information flow) of a neural network with nonlinear outputs.[§]   (3) Likelihood and

mutual information are, for all practical purposes, equivalent.[¶]   Also referred to as *principle of maximum decision efficiency; information maximum likelihood principle; maximum likelihood estimation principle; memory compensation principle.*

PRINCIPLE OF NETWORK MANAGEMENT RESOURCES (NMR) [psychology]   The more planning decisions commit for the future, the more important it is that the manager periodically check on events and expectations and redraw plans to maintain a course toward a desired goal.[‖]

PRINCIPLE OF NEUROLINGUISTIC PROGRAMMING (NLP) [psychology]   Therapeutic or other psychological change may be produced through the conscious manipulation of internal and external states and processes to elicit specific behavioral results. The stem *Neuro* represents understanding that all behavior is the result of neurological processes. *Linguistic* indicates the idea that neural processes are represented, ordered, and sequenced into models and strategies through language and communication systems. *Programming* refers to the process of organizing the components of a system, such as sensory representations, to achieve specific outcomes.**

PRINCIPLE OF NEUTRALITY (NT) [psychology]   (1) Distinguished from the technical tactic of abstinence; the latter is a specific function used to facilitate and foster analytic regression. NT can be defined as it applies to the major subfunctions of the analyst's work ego. Perception of the patient's intrapsychic processes, both empathically and cognitively,

*Caplja, V. (1964), On the question of the origin of the first calculator and application of the principle of Neper's (Napier's) bones in the multiplying mechanism of calculators, *Stroje Zpracování Informací* **10**, 321–325.
[†]Koontz, H. and O'Donnell, C. (1968), *Principles of Management: An Analysis of Managerial Functions*, 4th ed., McGraw-Hill, New York (http://knol. google.com/k/narayana-rao-kvss/principles-of-management-koontz-and/2utb2lsm2k7a/89).
[‡]Caplja, V. (1964), On the question of the origin of the first calculator and application of the principle of Neper's (Napier's) bones in the multiplying mechanism of calculators, *Stroje Zpracování Informací* **10**, 321–325.
[§]Zupanovic, P., Juretic, D., and Botric, S. (2004), Kirchhoff's loop law and the maximum entropy production principle, *Phys. Rev. E Statist. Nonlinear Soft Matt. Phys.* **70**, 5-2. 056108-056108-5

[¶]Lezon, T. R., Banavar, J. R., Cieplak, M., Maritan, A., and Fedoroff, N. V. (2006), Using the principle of entropy maximization to infer genetic interaction networks from gene expression patterns, *Proc. Natl. Acad. Sci. USA* **103**(50), 19033–19038.
[‖]Hobfoll, S. E. and Lilly, R. S. (2006), Resource conservation as a strategy for community psychology, *J. Commun. Psychol.* **21**(2), 128–148.
**Anon. (2000), *History of Neuro Linguistic Programming (NLP)* (http://www.exforsys.com/ tutorials/nlp/history-of-neuro-linguistic-programming.html).

requires an NT of appearance on the analyst's part to minimize the distortion of unfolding transference neurosis. Appropriate interpretive intervention requires NT of action and neutralizing power-related impulses in the service of analytic work.* (2) Not to take sides in hostilities or engage at any time in controversies of a political, racial, religious, or ideological nature. Not only prevent from taking sides in a conflict, but not to "engage at any time in controversies of a political, racial, religious or ideological nature."¶ See also PRINCIPLE OF HUMANITY; PRINCIPLE OF IMPARTIALITY; PRINCIPLE OF INDEPENDENCE; DEFINING PRINCIPLES; PRINCIPLE OF PROSELYTISM; EMERGENCY-ACTION PRINCIPLE.

PRINCIPLE OF NONIMAGING [engineering] See EDGE RAY PRINCIPLE.

PRINCIPLE OF NERNST [physics] (Walther Hermann Nernst, 1864–1941; Nobel Prize, 1920) (aka called the "third law" of thermodynamics) (1) The behavior of the entropy of every system as the absolute zero of the temperature is approached. Particularly, the entropic side of Nernst's theorem (N) states, for every system, if one considers the entropy as a function of the temperature T and of other macroscopic parameters $x; ::::; xn$, the entropy difference $\_TS\_S(T; x1; ::::; xn) - S(T; \_x1; ::::; \_xn)$ goes to zero as $T!0 + \lim T!0 + \_S = 0(1)$ for any choice of $(x1, ; ::::; xn)$ and of $(\_x1; ::::; \_xn)$. This means that the limit $\lim T!0 + S(T; x1; ::::; xn)$ is a constant $S0$ which does not depend on the macroscopic parameters $x1; ::::; xn$. Planck's restatement of (N) is $\lim T!0 + S = 0.$† (2) On the inaccessible state where both entropy and temperature are zero. The definition of statistical entropy in absolute terms makes it clear that S would be zero only in some perfectly well-defined pure quantum state. Other physical definitions of entropy fail to be so specific and leave open the possibility that an arbitrary constant can be added to S. Reconciles the two perspectives by

stating entropy must be zero at zero temperature. In the limited context of classical thermodynamics, the principle of Nernst justifies the existence of the absolute zero, as a lower limit for thermodynamic temperatures.‡

PRINCIPLE OF NONCONTRADICTION [physics] (1) (Parmenides of Elia, Greek philosopher). Either something existed or it didn't.§ (2) (Aristotle) We could not know anything that we do know. Presumably, we could not demarcate the subject matter of any of the special sciences, for example, biology or mathematics, and we would not be able to distinguish between what something is, for example, a human being or a rabbit, and what it is like, for example, pale or white. Aristotle's own distinction between essence and accident would be impossible to draw, and the inability to draw distinctions in general would make rational discussion impossible. Scientific inquiry, reasoning and communication that we cannot do without. Aristotle. Metaphysics IV (Gamma) 3–6, especially 4.¶

PRINCIPLE OF NONMALEFICENCE [psychology] Concentrates on ethical decision-making process and seeks to find a solution to overcoming dilemmas in moral philosophy and answering personal and professional values. (2) Ethical principles governing research in child and adolescent psychiatry; discusses the guidelines for protection of children and adolescents as research subjects. These include the principles of nonmaleficence and beneficence (the risk–benefit ratio), the principle of autonomy (informed consent and confidentiality), and the principle of justice (fair distribution of benefits and

*Poland, W. (1984), On the analyst's neutrality, J. Am. Psychoanal. Assoc. **32**(2), 283–299.
†Landsberg, P.T. (1990) Thermodynamics and Statistical Mechanics, Dover, New York.

‡Michon, G. P. (2005) Final Answers. Classical & Relativistic Thermodynamics, http://www.numericana.com/answer/heat.htm.
§Burt, A. (2000), Perspective: Sex, recombination, and the efficacy of selection—was Weismann right? Evolution; Int. J. Organic Evolut. **54**(2), 337–351.
¶Code, A. (1986), Aristotle's investigation of a basic logical principle: Which science investigates the principle of non contradiction? Can. J. Philo. **16**, 341.

burdens of research).[*] (3) Other ethical principles besides respect for the patient's autonomy assume new importance in this context. In addition to autonomy, two other principles have been important in bioethical decisionmaking; one is really a pair of principles—*beneficence*, which requires us to promote the welfare of other persons, and *nonmaleficence*, which requires us to minimize harm; a further principle is *justice*, which requires the fair apportioning of the benefits and harms of research or clinical care as well as efforts to combat discrimination and prejudice.[†] Ethical decisionmaking processes of family therapists when faced with the dilemma of duty to warn.[‡]

PRINCIPLE OF NONPERFECT SYNCRONIZATION (PNS) [physics] A product-stabilizing factor, the development of which at the transition state lags behind bond changes, increases the intrinsic barrier of a reaction, but lowers it if it develops ahead of bond changes. Because of the generality of this principle, details of transition state structures can be deduced from comparisons of intrinsic barriers within classes of reactions. A major conclusion that has emerged from such comparisons is that reactions that lead to charge delocalization/resonance-stabilized products have transition states in which the development of charge delocalization/resonance stabilization lags behind bond changes and, hence, have relatively high intrinsic barriers. A product-stabilizing factor lagging behind bond changes at the transition state increases the intrinsic barrier, while a product-stabilizing factor that

develops ahead of bond changes lowers the intrinsic barrier.[§]

PRINCIPLE OF NUCLEAR MAGNETIC RESONANCE SPECTROSCOPY [physics] High-resolution nuclear magnetic resonance (HR-NMR) spectroscopy is a powerful tool for both qualitative and quantitative analysis of foods and biological systems. NMR measures the resonant absorption of radiofrequency (RF) waves by the nuclear spins present in a macroscopic sample when the latter is placed in a strong and uniform/constant magnetic field. Most "optical" spectroscopy quantitative analysis methods, including near-infrared (NIR), are based on Lambert−Beer's law.[¶]

PRINCIPLE OF NUCLEATION [mathematics] aka Principle of the Centre-Periphery The translation space is the field that generates a point, which can be called centre or nucleus, which all the actors try to approach in order to improve their strategic advantage.[‖] See also SPATIAL PRINCIPLE; PRINCIPLE OF TRANSLATION QUANTITATIVITY; PRINCIPLE OF COMPOSITION OF THE TRANSLATIONS; PRINCIPLE OF THE CENTRE-PERIPHERY; UNIFIED PRINCIPLE OF ACCUMULATED ADVANTAGES.

PRINCIPLE OF NUMEROLOGY [astrology] The study of numbers inherent in names, irth dates, and other significant objects and events surrounding us, and their meanings and effects. Numbers have properties descending from the structure of our minds and from the structure of the universe, and the

[*]Munir, K., and Earls, F. (1992), Ethical principles governing research in child and adolescent psychiatry, *J. Am. Acad. Child Adolesc. Psychiatry* **31**(3), 408–414.

[†]National Commission for the Protection of Human Subjects of Biomedical and Behavioral Research (1979), *The Belmont Report: Ethical Principles and Guidelines for the Protection of Human Subjects of Research*, Office for the Protection of Subjects from Research Risks (OPRR) Reports, (April 18), pp. 2–8.

[‡]Burkemper, E. M. (1998), *Dissertation Abstr. Int. Sec. A Humanities Soc. Sci.* **58**(8-A), 3319.

[§]Bernasconi, C. F. (2004), The principle of nonperfect synchronization: How does it apply to aromatic systems? *J. Phys. Org. Chem.* **17**(11), 951–956.

[¶]Baianua, I. C., You, T., Costescua, D. M., Lozanoa, P. R., Prisecarua, V., and Nelson, R. L. (n.d.), High-resolution nuclear magnetic resonance and near-infrared determination of soybean oil, protein, and amino acid residues in soybean seeds, in *Oil Extraction and Analysis, Critical Issues and Comparative Studies*, p. 193 chap. 11 American Oil Chemists' Society (AOCS) begining on p86 Principle of Nuclear Magnetic Resonance Spectroscopy [physics].

[‖]Jimenez-Contreras, E. (1992) Las revistas cientificas; el centro y la periferia. Revista Espanola de Documentacion Cientifica, 15, 174–182.

vast majority of religions and cultures attach mystical significance to certain numbers. All of the numbers relating somehow to one's life can be reduced and simplified to one single number, except eleven and twenty-two.[*]

PRINCIPLE OF OBJECT ORIENTED DESIGN In class hierarchies, it should be possible to treat a specialized object as if it were a base class object. All code operating with a pointer or reference to the base class should be completely transparent to the type of the inherited object. It should be possible to substitute an object of one type with another within the same class hierarchy. Inheriting classes should not perform any actions that will invalidate the assumptions made by the base class. Aka Liskov Substitution Principle.[†]

PRINCIPLE OF OBJECT SCOPE [psychology] Words label objects. Words refer to the whole object, rather than to the object's parts or attributes. Develops with the prosodic cues provided by adults when speaking to toddlers.[‡] See also PRINCIPLE OF REFERENCE; PRINCIPLE OF EXTENDIBILITY; PRINCIPLE OF CATEGORICAL SCOPE; NOVEL NAME–NAMELESS CATEGORY (N3C) PRINCIPLE; PRINCIPLE OF CONVENTIONALITY.

PRINCIPLE OF OBSERVATIONAL COMPATIBIL-ITY [mathematics, physics]

PRINCIPLE OF OCCAM'S RAZOR (William of Occam, 1285—1349)   (1) Entities should not be multiplied needlessly; the simplest of two competing theories is to be preferred.[§] (2) The simplest model that accurately represents the data is most desirable. (3) Most techniques for improving generalization in learning are inspired by the well-known principle of Occam's razor, "Causes should not be multiplied beyond necessity."[¶] See also OCCAM'S RAZOR PRINCIPLE, PRINCIPLE OF PARSIMONY.

PRINCIPLE OF OCTAVES [chemistry]   (Newlands, 1863)   Arranges the elements in order of atomic weight and in groups of eight with recurring similarity of properties.[‖] See also; NEWLAND'S PRINCIPLE.

PRINCIPLE OF ONENESS [psychology] (Wayne L. Wang, a Tao philosopher)   (1) Laotzu proclaimed that Tao is manifested as Wu and Yu within a "oneness" state. Wu and Yu transmutate with each other in order to maintain the oneness nature of Tao. "the two manifest simultaneously as different manifestations of the same Tao." (2) All realities in Tao must have oneness in order to ensure a coherent interpretation of the Tao Te Ching. (3) Psychodramatic techniques that promote self-acceptance and improve "here and now" relationships present the method of clinical reasoning that underlies our practice, namely, a process-theory-based approach to therapy that applies beyond psychodrama; in this approach, discontinuity of life story is considered to be a cardinal problem of persons with multiple personality disorder (MPD) and other dissociative disorders. In addition to providing a theoretical model of MPD, process theory provides general guidelines for formulating treatment.[**]

[*]Keller, J. and Keller, J. (2001) The Complete Book of Numerology. Macmillan. p9. ... you will be guided, and often by the numbers! WHAT IS NUMEROLOGY? The principle of numerology is that all of the numbers that relate somewho to your life can be reduced and simplified down to one single number, except for the numbers eleven and twenty-two, ...

[†]Noble, J., Biddle, R., and Tempero, E. (2002) Australian Computer Science Communications archiv. v24, (1) p187–195.

[‡]Golinkoff, R. M., Mervis, C. B., and Hirsh-Pasek, K. (1944), Early object labels: The case for a developmental lexical principles framework, J. Child Lang. **21**(1), 125–155.

[§]American Heritage (2000), *American Heritage Dictionary of the English Language*, 4th ed., Houghton Mifflin, New York.

[¶]Rose, K. (1998), Deterministic annealing for clustering, compression, classification, regression, and related optimization problems, Proc. IEEE, 86(11), 2225.

[‖]Walker, P. M. B. (1999), *Chambers Dictionary of Science and Technology*, Chambers, New York, p. 662.

[**]Kluft, E. S. (1993), *Expressive and Functional Therapies in the Treatment of Multiple Personality Disorder*, Charles C. Thomas, Springfield, IL, p. 169–188.

PRINCIPLE OF OPEN COMPETITION IN PROMO-TION [mathematics] Turning the measurement of a magnetic field into a measurement of frequency.* (2) Managers should be selected from among the best available candidates for the job, whether they are inside or outside the enterprise. See also PRINCIPLE OF PLANNING.¶

PRINCIPLE OF OPERATION [psychology] (1) If an enterprise is to ensure maintenance of the best quality of management, it is necessary to open selection of candidates for promotion to those available both inside and outside the enterprise. (2) "Atomistic" principle of operation of the digital computer, where every computation is broken down into a sequence of simple steps, each of which is computed independently of the problem as a whole.† (3) The theoretical principles are the following: (a) *principle of totality*—the conscious experience must be considered globally (by taking into account all the physical and mental aspects of the individual simultaneously) because the nature of the mind demands that each component be considered as part of a system of dynamic relationships; (b) *principle of psychophysical isomorphism*—a correlation exists between conscious experience and cerebral activity. On the basis of these two principles, the following methodological principles are defined: (a) *phenomenon experimental analysis*—in relation to the *totality principle*, any psychological research should take as a starting point phenomena and not be solely focused on sensory qualities; (b) *biotic experiment*—the school of gestalt established a need to conduct real experiments that sharply contrasted with and opposed classic laboratory experiments. This signified experimenting in natural situations, developed in real conditions, in which it would be possible to reproduce, with

higher fidelity, what would be habitual for a subject.¶ See also PRINCIPLE OF FUNDAMENTAL CHARACTERISTICS.

PRINCIPLE OF OPERATIONAL COMPATIBILITY [mathematics, physics] (1) Direction of knob rotation and direction of the indicator in a linear scale-type display includes Warrick's principle, the scale-side principle, and the clockwise-for-increase principle.‡ (2) The applicable principle depends strongly on control and display positions, as well as, to some extent, the population studied. In this rather specific case of linear displays and rotary controls, all three principles rest on an identity between the pointer and knob directions, but ambiguity arises because opposite sides of the rotary knob move in opposite directions.§ Also referred to as Warrick's principle; scale-side principle; clockwise-for-increase principle.

PRINCIPLE OF OPERATIONISM [physics] Interbehavioral operationism, which provides definite criteria for objectivity.¶ See also COLLIDER OPERATIONAL PRINCIPLE; OPERATIONAL PRINCIPLE.

PRINCIPLE OF OPTICAL BLEACHING [engineering] (Krais, 1929). Applicable in a broad array of fields, industrial use of optical brightening.‖

PRINCIPLE OF OPTICAL COUPLING [engineering] See OPTICAL COUPLING PRINCIPLE.

PRINCIPLE OF OPTICAL ISOLATING [engineering] See OPTICAL COUPLING PRINCIPLE.

*Koontz, H. and O'Donnell, C. (1968), *Principles of Management: An Analysis of Managerial Functions*, 4th ed., McGraw-Hill, New York (http://knol.google.com/k/narayana-rao-kvss/principles-of-management-koontz-and/2utb2lsm2k7a/89).
†Kleine-Horst, L. (2001), *Empiristic Theory of Visual Gestalt Perception (ETVG) Hierarchy and Interactions of Visual Functions*, Köln (http://www.enane.de/cont.htm).

‡Warrick, M. J. (1947), Direction of movement in the use of control knobs to position a visual indicator, in *Psychological Research on Equipment Design*, ed., Fitts, P. M., Research Report 19, Army Air Force, Aviation Psychology Program, Columbus, OH.
§Worringham, C. J., and Beringer, D. B. (1998), Directional stimulus-response compatibility: A test of three alternative principles, *Ergonomics*, **41**(6), 864–880.
¶Bridgman, P. W. (1953), The logic of modern physics, in *Readings of Philosophy of Science*, edited Feigl, H. and Brodbeck, M., eds. Appleton-Century-Crofts.
‖Zheng, C. (1991), Fur optical bleaching, *Zhongguo Pige* **20**(11), 13–15.

PRINCIPLE OF OPTICAL PHASE CONTRAST [engineering] (Frits Zernike, Univ. Groningen, 1888—1966), See PHASE CONTRAST PRINCIPLE.

PRINCIPLE OF OPTIMAL CHOICE OF PLAYERS [mathematics] Each player wants which give maximum lift and minimum drag. Optimization models arise as a result of application of one or more of these principles. New and challenging mathematical problems arise in obtaining the results from optimizing models.¶ See also ENTROPY OPTIMIZATION PRINCIPLE.

PRINCIPLE OF OPTIMAL CHOICE OF PORTFOLIO [mathematics] Portfolios that maximize expected return and minimize variance. Optimization models arise as a result of application of one or more of these principles. New and challenging mathematical problems arise in obtaining the results from optimizing models.* See also ENTROPY OPTIMIZATION PRINCIPLE.

PRINCIPLE OF OPTIMAL FEATURE EXTRACTION IN PATTERN RECOGNITION [mathematics] Features that result in minimum loss of information or in minimum loss of power of discrimination or that lead to minimum variability within classes and maximum variability between classes or to minimum interdependence of components of feature vector. Optimization models arise as a result of application of one or more of these principles. New and challenging mathematical problems arise in obtaining the results from optimizing models.¶ See also ENTROPY OPTIMIZATION PRINCIPLE.

PRINCIPLE OF OPTIMAL RELIABILITY [mathematics] Systems that maximize reliability at given cost or minimize cost for given reliability. Optimization models arise as a result of application of one or more of these principles. New and challenging mathematical problems arise in obtaining the results from optimizing models.¶ See also ENTROPY OPTIMIZATION PRINCIPLE.

*Kapur, J. N. (1988), *Mathematical Modeling*, New Age International, p. 203.

PRINCIPLE OF OPTIMALITY [physics, chemistry] (Richard Ernest Bellman, 1920–1984, American mathematician). (1) The minimum value of a function is a function of the initial state and the initial time. This method is best suited for multistage processes; however, the application of dynamic programming to a continuously operating system leads to a set of nonlinear partial differential equations.† (2) An optimal sequence of decisions in a multistage decision process problem has the property that whatever the initial state and decisions are, the remaining decisions must constitute an optimal policy with regard to the state resulting from the first decisions. (3) For optimal systems, any portion of the optimal state trajectory is optimal between the states it joins.‡ See also OPTIMALITY PRINCIPLE.

PRINCIPLE OF ORDER [psychology] (Henri Fayol, 1841-1925, French industrialist), Categories of gestalt psychology to explain structural and process-related features of phenomena in the domain of artistic work. In particular, we draw on the nexus of the opposing postulates of simplicity and complexity and on the axiom of Gestaltprägnanz in order to explain principles of order of works in the field of arts and music. Gestalt principles are suited for reduction of structural complexity.§ (2) Relationships between various units must be established in a logical, rational manner, to ensure that these units work in harmony.¶ See also FAYOL'S PRINCIPLES OF MANAGEMENT.

†Kim, K and Diwekar, U. (2006) Batch distillation in batch processes, ed. by Korovessi, E., and Linninger, A. A., CRC Press. p. 351–352.
‡Berezovs'kyi, V. Ia. (2005), The principle of optimality in biophysical medicine, *Fiziol. Zh.* (Kiev, Ukraine, 1994) **51**(5), 5–15.
§Reuter, H. and Stadler, M. (2006), Gestaltübergänge in der musik. Vom Wandel der ordnungsprinzipien, *J. Psychol.* **14**(3), 274–301.
¶Wood, M. C. (2002), *Henri Fayol: Critical Evaluations in Business and Management*, Taylor & Francis, p. 8.

PRINCIPLES OF ORGANIC ION RADICAL REACTIVITY [chemistry]   Some ion radicals contain fragment orbitals that suspend an unpaired electron preferentially. Other ion radicals are characterized by delocalization of an unpaired electron along orbitals that are more or less evenly populated with an unpaired electron.*

PRINCIPLE OF ORGANIZATIONAL SUITABILITY [biology, mathematics, psychology]   (1) For optimal systems, any portion of the optimal state trajectory is optimal between the states it joins. (2) Basic in the development of biological systems is the ability to self-organize, a process by which new structures and patterns emerge. (3) Individuals differ to some extent in the way operational factors are organized and function within subsystems and the number of ways in which operating factors in a certain subsystem can be organized in patterns, in order to allow the subsystem to play its functional role in totality. (4) Orientation selectivity in a model visual system, for the development of biological and synthetic perceptual networks. (5) For optimal systems, any portion of the optional state trajectory is optimal between the states it joins.[†] (6) The more controls are designed to reflect the place in the organization structure where responsibility for action lies, the more they will facilitate correction of deviation of events from plans.[‡] See also PRINCIPLE OF PLANNING; OPTIMALITY PRINCIPLE; SELF-ORGANIZING PRINCIPLE; SIMILARITY PRINCIPLE; ORGANIZING PRINCIPLE; PRINCIPAL-COMPONENT ANALYSIS.

PRINCIPLE OF ORGANIZING CELL STRUCTURE [biology]   A self-association of nucleotide-binding proteins used in all cells.[§] See also

PRINCIPLE OF PHYSICAL TIMING; TYPE II CELL STRETCH—ORGANIZING PRINCIPLE.

PRINCIPLE OF ORIGINAL HORIZONTALITY [geology]   (Nicholas Steno, 1638–1686, Danish geologist)   Layers of sediment are originally deposited horizontally. See also PRINCIPLE OF LATERAL CONTINUITY; PRINCIPLE OF CROSS-CUTTING RELATIONSHIPS; PRINCIPLE OF FAUNAL SUCCESSION.[¶]

PRINCIPLE OF ORIGINAL HORIZONTALITY [geology]   Sediments settling out from bodies of water are deposited horizontally or nearly horizontally in layers that lie parallel or nearly parallel to Earth's surface.[‖]

PRINCIPLE OF OSMOSIS [biology]   (1) Movement of solvent through a semipermeable membrane.[**] (2) Used in the preservation of food kept in strong solutions of salt (brine) or sugar (syrup). Any bacteria that gain access to the food become plasmolyzed and are effectively killed by dehydration.

PRINCIPLE OF OVERCOMPENSATION [kinesiology]   (Dr. Hans Selye)   See PRINCIPLE OF TRAINING.

PRINCIPLE OF PARALLELISM [mathematics]   (1) An important claim in the report is that monetary unification and economic integration cannot fruitfully proceed independently of each other. (2) Economic union and monetary union form two integral parts of a single whole and would therefore have to be implemented in parallel. (3) Parallelism, as the guiding principle in constructing monetary union, creates a predisposition for bureaucratic intervention and centralized community decisionmaking.[††]

---

*Todres, Z. V. (n.d.), Basic principles of organic ion radical reactivity, in *Organic Radicals: Chemistry and Applications*, chap. 3.

[†]Vargas, E. A. (2004), The triad of science foundations, instructional technology, and organizational structure, *Span. J. Psychol.* **7**(2), 141–152.

[‡]Koontz, H. and O'Donnell, C. (1968), Principles of Management, 4th ed., McGraw-Hill.

[§]Kostyuk, P. G. (1982), Main principles of the structural organization of ion channels providing depolarization of excitable cell membranes, Dokl. Akad. Nauk SSSR 266(6), 1491–1494.

[¶]Tarbuck, E. J., Lutgens, F. K., and Tsujita, C. Z. (2005) *Earth—an Introduction to Physical Geology*, Pearson Education Canada Inc.

[‖]Yang, X. (1985), Family problem of quarks and leptons in two kinds of composite models, *Gaoneng Wuli Yu Hewuli* **9**(6), 660–668.

[**]Broyer, T. C. (1947), The movement of materials into plants. I. Osmosis and the movement of water into plants, *Bot. Rev.* **13**(1), 1–58.

[††]Reichardt, C. S. (2006), The principle of parallelism in the design of studies to estimate treatment effects, *Psychol. Meth.* **11**(1), 1–18.

PRINCIPLE OF PARITY CONSERVATION [psychology]

The more controls are designed to reflect the place in the organization structure where responsibility for action lies, the more they will facilitate correction of deviation of events from plans.* The quality of space reflection symmetry of subatomic particle interactions. In physics, a property important in the quantum mechanical description of a physical system. In most cases it relates to the symmetry of the wavefunction representing a system of fundamental particles. A parity transformation replaces such a system with a type of mirror image. Stated mathematically, the spatial coordinates describing the system are inverted through the point at the origin; that is, the coordinates x, y, and z are replaced with $-x$, $-y$, and $-z$. In general, if a system is identical to the original system after a parity transformation, the system is said to have *even parity*. If the final formulation is the negative of the original, its parity is odd. For either parity the physical observables, which depend on the square of the wavefunction, are unchanged. A complex system has an overall parity that is the product of the parities of its components.[†] See also PARITY PRINCIPLE. [Physics]

PRINCIPLE OF PARITY OF AUTHORITY [chemistry]

Particle interactions should be indifferent to the direction of time except in rare instances.[‡] See also PARITY PRINCIPLE.

PRINCIPLE OF PARITY OF AUTHORITY AND RESPONSIBILITY [psychology]

(1) Without giving adequate authority, superiors cannot blame their subordinates for not achieving the goals. The authority delegated to subordinates should be equal to their responsibilities. Authority and responsibility go hand in hand. Responsibility without authority is meaningless. Every individual in the organization should be given corresponding authority in to order carry out the assigned task efficiently. There should not be any disparity between the authority granted to and the responsibility imposed on a subordinate.* (2) The responsibility exacted for actions taken under authority delegated cannot be greater than that implied by the authority delegated, nor should it be less.[§] (3) The authority delegated must be consistent with the responsibility assigned to a subordinate.* See also PRINCIPLE OF PARITY; PRINCIPLE OF PLANNING.

PRINCIPLE OF PARSIMONIUS DATA MODELING [psychology, biology]

(1) If you have no evidence to the contrary, assume that your characters are homologous. (2) A shared derived similarity of a group of organisms has to be regarded as their synapomorphy (thus as a homology), unless the paraphyly of the group can be demonstrated with other characters that are conflicting to the putative synapomorphy but regarded as stronger evidence by quantity or quality.[¶] See also PRINCIPLE OF PARSIMONY; AUXILIARY PRINCIPLE.

PRINCIPLE OF PARSIMONY [psychology, biology]

(1) When two theories account for the same facts, the one that is briefer makes fewer assumptions and references to unobservables, and has the greater generality is to be preferred.[‖] (1) Of two meaningful models, the one described by fewer parameters will have the better predictive ability based on new data. (2) William of Ockham (Occam), described assumptions introduced to explain that a thing must not be multiplied beyond necessity, and hence the simplest of several hypotheses is always the best to account for unexplained facts. (3) Requires that ad hoc

*Ziman, J. (1991), *Reliable Knowledge: An Exploration of the Grounds for Belief in Science*, Cambridge Univ. Press, p. 1.

[†]Lee, T.-D. (2009), In *Encyclopædia Britannica* (http://search.eb.com/eb/article-9047597).

[‡]Popov, V. K. and Ivanova, R. S. (1985), Parity principle and kinematic asymmetries in the otolith system, *Kosmichesk. Biolo. Aviakosmichesk. Med.* **19**(3), 53–55.

[§]Koontz, H. and O'Donnell, C. (1968), *Principles of Management*, 4th ed., McGraw-Hill.

[¶]Seasholtz, M. B. and Kowalski, B. (1993), The parsimony principle applied to multivariate calibration, *Anal. Chim. Acta* **277**(2), 165–177.

[‖]Epstein, R. (1984), The principle of parsimony and some applications in psychology, *J. Mind Behav.* **5**(2), 119–130.

assumptions be minimized as far as possible in scientific explanations of natural phenomena. (4) Should be viewed as a tool, not as a claim that evolution always took the most parsimonious path. There is no other possibility than parsimony to choose between different alternative hypotheses that explain singular historical happenings, that can only be reconstructed, but not repeated and tested like scientific experiments.* (5) If you have no evidence to the contrary, assume your characters to be homologous. (6) A shared derived similarity of a group of organisms has to be regarded as their synapomorphy (thus as a homology), unless the paraphyly of the group can be demonstrated with other characters that are conflicting to the putative synapomorphy but regarded as stronger evidence by quantity or quality. (7) Simple explanations are usually better than complicated ones; therefore some aspects of language comprehension might be described and explained more effectively by straightforward variants of discrimination learning principles. Starting with operantly conditioned motor acts and moving up through an increasingly complex network of conditioned relations emerging from and consistent with earlier discrimination learning. Describes the learning abilities required for appropriate responses to a simple artificial language. (8) Preferences for visually presented letters are explained by the visual properties of the letters. (9) When two theories account for the same facts, the one that is briefer, makes fewer assumptions and references to unobservables, and has the greater generality is to be preferred.† Also known as *Occam's (or Occam's razor principle; parsimony principle; auxiliary principle; principle of parsimonius data modelling.*

*Weingarten, S., Bolus, R., Riedinger, M. S., Maldonado, L., Stein, S., and Ellrodt, A. G. (1990), The principle of parsimony Glasgow Coma Scale score predicts mortality as well as the Apache II score for stroke patients, *Stroke* **21**(9), 1280–1282.
†Tarlov, I. M. (1959), The principle of parsimony in medical practice, *NY State J. Med.* **59**(10), 2050–2052.

PRINCIPLE OF PART–WHOLE DETERMINATION PERCEPTUAL   ASSIMILATION [psychology] Variation of the time over which a sequential pattern extends gives rise to a variation in the apparent duration of the periods that compose it. Apparent space and time vary in essentially the same manner when the spatial and temporal stimulus conditions are equivalent.‡

PRINCIPLE   OF   PARTIAL   PRESSURES [chemisry, physics]   See DALTON'S PRINCIPLE OF PARTIAL PRESSURES.

PRINCIPLE OF PARTICLE GROWTH RETARDING [chemistry,   physics]   The   particle growth retarding effect depends on the relative particle size between the raw-material matrix and the particle growth retardant.§

PRINCIPLE OF PASSIVE CONCENTRATION [psychology]   Patients need to be assured that they adhere to the requirements for the correct practice of the exercises; it is only a matter of time before they effectively overcome intrusive thoughts.¶

PRINCIPLE   OF   PERCEPTION [psychology] (1) Perception is to be treated as a part of human behavior, a biological science. This broad definition permits introduction of the complex stimuli of social perception. This is a welcome innovation, and a major contribution. Attention is also given to a number of traditional topics: the idea of threshold, the definition of stimulus, and the problem of isolating perception from other aspects of behavior.‖ (2) First identified by psychologists, who maintain that the smaller the gap between stimuli, the more likely those stimuli are to be seen as belonging together in some sense. The

‡Day, R. (2006), Two principles of perception revealed by geometrical illusions, *Austral. J. Psychol.* **58**(3), 123–129.
§Zsigmondy, R. (1926), The condition in space of colloid particles, *Z. Phys. Chem.* **124**, 145–154.
¶Micah, R. S. (2001), *Autogenic Training: A Mind-Body Approach to the Treatment of Fibromyalgia and Chronic Pain Syndrome*, Haworth Press. p. 103.
‖Bartley, S. H. (1958), *Principles of Perception*, Harper, New York.

gap can be in terms of either space or time. Collective behavior is the result of subjects with similar or related reference signals. Initial exposure to blurred or ambiguous stimuli interferes with accurate perception even after more and better information becomes available. (3) Deviations between observed similarity estimates and values expected.* (4) There are two such principles: whole–part determination of perceived size and space-time reciprocity. The first refers to the determination of the size of intrinsic parts by the whole figure or object and the second, to the modulation of extent by time and vice versa.[†] See also MINIMUM DESCRIPTION-LENGTH PRINCIPLE.

PRINCIPLE OF PERFORMING SIGNAL-PROCESSING OPERATIONS [engineering, physics]   See also PRINCIPLE OF PRESSURE SWING DESORPTION; PRINCIPLES OF DARCY'S LAW.

PRINCIPLE OF PERMANENCE [mathematics] Given any *analytic function* $f(z)$ defined on an *open* (and *connected*) set U of the complex numbers $\mathbb{C}$, and a convergent sequence $\{a_n\}$ that along with its limit L belongs to U, such that $f(a_n) = 0$ for all n, then $f(z)$ is uniformly zero on.[‡]

PRINCIPLE OF PERPENDICULAR MAGNETIZATION [computer science]   See VERTICAL MAGNETIZATION PRINCIPLE.

PRINCIPLE OF PERCEPTUAL CONTEXT [psychology] The role of perceptual context in structuring spatial knowledge. The posterior parietal association areas contribute essentially to the elaboration of exocentric space coordinates based on the perceptual stability - despite body movemnet - of the environmental frame. It is here the perceptual context (principally visual) provides a frame of reference to which the position of objects is referred. Hence, the massive afferentation of this region from segments of the articulated body, together with the information about visual space. The posterior parietal areas as an interface for matching the allocentric description of space with the egocentric one: a function reflected in pathology.[§]

PRINCIPLE OF PERSISTENT IDENTIFIERS [computer science]   (1) References to Web content have been made by using URL hyperlinks. As links are "broken" when content is moved to another location, a reference system based on URLs is inherently unstable and poses risks for continued access to Web resources. To create a more reliable system for referring to published material on the Web, a number of schemes have been developed using namespaces to identify resources, enabling retrieval even if the Web location is unknown. Examples responding to the scheme include Handles, Digital Object Identifiers (DOIs), Archival Resource Keys (ARKs), Persistent Uniform Resource Locators (PURLs), Uniform Resource Names (URNs), National Bibliography Numbers (NBNs), and the Open URL.[¶] (2)The Corporation for National Research Initiatives (CNRI) developed the Handle System (http://www.handle.net/), a resolver application for persistent identifiers called "handles." CNRI maintains a global handle registry as well. Organizations wishing to utilize the Handle System must register a namespace with CNRI. As with the PURL server, organizations have the choice of using the resolver at CNRI together with a local handle application or running their own handle application locally.[‖] See also OBJECTS PRINCIPLE 1; OBJECTS PRINCIPLE 2;

*Kennedy, J. F. and He, M. M. (2000), in *Wine Science: Principle, Practice, Perception*, 2nd ed., by Jackson R. S., ed., Academic Press, London; Carbohydr. Polyms. 59(3), 401.

[†]Day, R. (2006), Two principles of perception revealed by geometrical illusions, *Austral. J. Psychol.* **58**(3), 123–129.

[‡]Wolfram, S. (2002), *A New Kind of Science*, Wolfram Media, Champaign, IL, p. 1168.

[§]Ellen, P. and Thinus-Blanc, C., ed. (1987) Cognitive Processes and Spatial Orientation in Animal and Man: Volume II. Neurophysiology and Developmental Aspects (NATO Science Series D:). Springer, p. 65.

[¶]Werner, H. and Kothe, J. (2006), *Implementing Persistent Identifiers: Overview of Concepts, Guidelines and Recommendations* (http://www.knaw.nl/ecpa/publ/pdf/2732.pdf).

[‖]NISO (Nat. Information Standards Org.(2007), *A Framework of Guidance for Building Good Digital Collections*, 3rd ed., Inst. Museum and Library Services (http://framework.niso.org/node/5).

OBJECTS PRINCIPLE 3; OBJECTS PRINCIPLE 4; OBJECTS PRINCIPLE 5; OBJECTS PRINCIPLE 6.

PRINCIPLE OF PER SURVIVOR PROCESSING [physics] (1) Describe the flow of liquid water and water vapor in the saturated–unsaturated soil below the surface. (2) Necessary for the estimation of unknown parameters, in a per survivor fashion. Introduced by several authors for state complexity reduction in an intersymbol interference, environment, delayed decision feedback sequence estimation, and reduced-state sequence estimation.[*] See also PRINCIPLES OF DARCY'S LAW.

PRINCIPLE OF PHOTOCHEMICAL EQUIVALENCE [chemistry]   Also referred to as *Einstein's Principle of photochemical equivalence.*

PRINCIPLE OF PHOTOTHERMOLYSIS [dermatology]   Selective targeting of an area using a specific wavelength to absorb light into that target area sufficient to damage the target tissue while allowing the surrounding area to remain relatively untouched.[†]

PRINCIPLE OF PHYSICAL TIMING [physics] Moment to attack at the split second when weight, momentum, and strength are gathered for use against themselves. See also TYPE II CELL STRETCH-ORGANIZING PRINCIPLE. [‡]

PRINCIPLE OF PLANETARY ROTATION REVOLUTION [physics]   When photons emitted from the sun enter a planetary magnetic field on the side of the planet facing the sun, photons are deflected by the planetary magnetic field and absorbed at an angle on the planet surface; the absorption of photons will generate attraction force between the planet and the sun, due to interphoton attraction of the radiated photons. The attraction force generated between the sun and the planet will then be resolved at a tangent to the point of the absorbed ion into a rotational force of the planet by trigonometric resolution of the resultant angle of the photon-absorbed ion.[§]

PRINCIPLE   OF   PLANNING [psychology] Should be conceived of, above all, as a coordinative activity. The instruments for this coordinative activity were consciously always of a nonfinancial nature; with one exception, planners never had financial resources of their own. The instruments of the planner were primarily communicative; concepts, plans, and vision documents were to be used to capture the imagination of the various relevant actors, both within the sector departments on the national level [the so-called *horizontal axis* (abscissa) of coordination] and at the other levels of government (the *vertical axis* (ordinate)).[¶] (2) Related to purpose and nature. Consisting of the contribution toward the institutions objectives, efficiency and primacy of plans.[‖]

PRINCIPLE OF PLANNING PREMISES [organizational psychology]   If more people in an organization use common and consistent planning premises, the enterprise planning will be more coordinated.[‡] See also PRINCIPLE OF PLANNING; PRINCIPLE OF CONTRIBUTION OF OBJECTIVES; PRINCIPLE OF EFFICIENCY OF PLANS; PRINCIPLE OF PRIMACY OF PLANNING.

PRINCIPLE OF PLENITUDE [psychology] (Aristotle)   (1) Everything that can happen will

[*]Raheli, R., Polydoros, A., and Tzou, C. K. (1991), The principle of per-survivor processing: a general approach to approximate and adaptive MLSE, *Proc. Global Telecommunications Conf.* (GLOBECOM'91), *Countdown to the New Millennium, Featuring a Mini-Theme on Personal Communications Services*, Dec. 2–5, 1991, pp. 1170–1175.
[†]Bickmore, H. (2005), *Milady's Hair Removal Techniques: A Comprehensive Manual*, Cengage Learning, p. 305.
[‡]Poldrabinek, P. A. and Kazakevich, I. I. (1962), Physical principles of the distribution of erythrocytes in suspensions during the course of time, *Biofizika* **7**, 488–491.

[§]Peter, O. (2008) Principle of planetary rotation revolution. Lulu Enterprises.
[¶]Heftner, E. (1965), The concept of cell metabolism as a principle in planning infusion therapy in accident surgery, *Arch. Orthopad. Unfall-Chirurgie* **57**, 147–155.
[‖]Koontz, H. and O'Donnell, C. (1968), *Principles of Management: An Analysis of Managerial Functions*, 4th ed., McGraw-Hill (http://knol.google.com/k/narayana-rao-kvss/principles-of-management-koontz-and/2utb2lsm2k7a/89).

happen. (2) No possibilities that remain eternally possible will go unrealized.*

PRINCIPLE OF PLURALITY [psychology] (1) Multiple stratification of graphological characteristics, the meanings of which are determined by the whole.[†] (2) Collective action cannot be fully engaged without the acknowledgment of a common objective. The identification of such an objective must not, however, conflict with the diversity of cultural belonging and identities. (3) Applied through legal provisions designed to prevent or repress discrimination based on ethnic belonging, gender, or religion, but it should also be expressed positively through the attribution of greater value to geographic, cultural, and linguistic diversity. See also PRINCIPLE OF SUBSIDIARITY; PRINCIPLE OF RESPONSIBILITY.

PRINCIPLE OF PNEUMATIC CONVEYANCE [engineering]   Conveying is based on the known physical principle that flowing gases under particular conditions are able to carry and transport heavier solids. This principle is utilized technically in a targeted manner in pneumatic conveying.[‡]

PRINCIPLE OF POLARITY [psychology, physics] (1) Embodies the truth that everything is dual, everything has two poles, and everything has its pair of opposites.[§] (2) The principle of polarity has two "roots," namely, the principle of polarity in direction and the principle of polarity in opposition. The first is the principle of originality, meaning that everything that has being is simply itself and no other thing; everything that exists depends for its being on the existence of what it is not; and the third

principle is the principle of sufficient reason, that for everything existing there is a sufficient reason as to why it is what it is, and why it is not what it is not. Both roots are phenomenologically centered in the experience of tension. The concept of tension is employed in an investigation of "process and futurity."[¶] (3) The idea that everything is dual, everything has two poles, and everything has its opposite. All manifested things have two sides, two aspects, or two poles. Everything "is" and "isn't" at the same time, all truths are but half-truths and every truth is half-false; there are two sides to everything, opposites are identical in nature, yet different in degree, extremes meet, and all paradoxes may be reconciled.[‖] See also PRINCIPLE OF CAUSE AND EFFECT.

PRINCIPLE OF POLICY FRAMEWORK [organizational psychology]   If more policies, appropriate to the organization, are expressed in clear terms and form and if managers understand them, the plans of the enterprise will be more consistent. See also PRINCIPLE OF PLANNING.[**]

PRINCIPLE OF POLYMERASE CHAIN REACTION (PCR) [mathematics]   A laboratory technique based on the polymerase chain reaction, which is used to amplify and simultaneously quantify a targeted DNA molecule. It enables both detection and quantification (as absolute number of copies or relative amount when normalized to DNA input or additional normalizing genes) of a specific sequence in a DNA sample.[††] See also OCCAM'S RAZOR PRINCIPLE.

[¶]Dunham, A. (1938), The concept of tension in philosophy, *Psychiatry: J. Study Interpers. Processes* **1**(1), 79–120.
[‖]Three Initiates (2004), *The Kybalion*, Book Tree [the "Three Initiates" who authored *The Kybalion* chose to remain anonymous; as a result, there has been a great deal of speculation about who actually wrote the book.)
[**]Koontz, H. and O'Donnell, C. (1968), *Principles of Managemnet*, 4th ed., McGraw-Hill.
[††]Van Guilder, H. D., Vrana, K. E., and Freeman, W. M. (2008). Twenty-five years of quantitative PCR for gene expression analysis, *Biotechniques* **44**, 619–626.

*Erde, E. L. (1988), Studies in the explanation of issues in biomedical ethics: the example of abortion, *J. Med. Phil.* **13**(4) 329–347.
[†]Pulver, M. (1940), *Symbolik der Handschrift*, Fuessli, Oxford, UK.
[‡]Heep, D. and Winkhardt, G. (1998), Simplified application of the vacuum-/pressure conveying principle for pneumatic conveyance, *Bulk Solids Handl.* **18**(2), 245–251.
[§]Schwarz, R. (1950), The polarity principle in chemistry, *Chemiker-Zeitung* **74**, 13–14.

PRINCIPLE OF POLYMERASE CHAIN REACTION
[psychology] (1) Of two meaningful models, the one described by fewer parameters will have the better predictive ability based on new data. (2) William of Occam described the principle applied in philosophy and science that assumptions introduced to explain a thing must not be multiplied beyond necessity, and hence the simplest of several hypotheses is always the best in account for unexplained facts.* (2) Requires that ad hoc assumptions be minimized as far as possible in scientific explanations of natural phenomena. For phylogenetic systematics, this means that from the millions of theoretically possible cladograms, only those that minimize the number and/or weight of necessary assumptions of nonhomology (homoplasies).[†]

PRINCIPLE OF POPULATION [psychology]
(Thomas Robert Malthus, *Essay on the Principle of Population*) (1) Discusses theories of population and subsistence; the adjustment of households to food supplies and other resources; the changing balance of births and deaths; the environmental impact of population; famine and the prevention of mass mortality; the political economy of health; fertility decline in the course of development; case studies of fertility transition; urbanization, migration, and population, employment; and education. (2) Population is perpetually kept down to the level of the means of subsistence. Thus, among the wandering tribes of America and Asia, the population has so increased as to render necessary the cultivation of Earth. (3) Humans are considered inert, sluggish, and averse from labor. Unless compelled by necessity, with certainty, the world would not have been peopled, but for the superiority of the power of population to the means of subsistence.[‡] See also OCCAM'S RAZOR PRINCIPLE.

*Kato, I. (1990), Principles and applications of polymerase chain reaction (PCR), *Tanpakushitsu Kakusan Koso* **35**(17), 2957–2976.
[†]Fukumaki, Y. (1990), Application of polymerase chain reaction (PCR) to genetic diagnosis of thalassemia, *Igaku no Ayumi* **153**(9), 492–496.
[‡]Short, R. (1998), An essay on the principle of population by Thomas Robert Malthus, *Nature* **395**(6701), 456.

PRINCIPLE OF POSITIONAL DETERMINACY
[physics] (1) Restricts the range of motive concepts that can be applied to systems. A single point in space determines no spatial magnitudes at all, thus it cannot be said to undergo motion, let alone accelerated or forced motion. When two points are given there can be relative motion, but only one direction and only one magnitude are determate. The positional dependence of forces to the elements of a two-point system, derives the two elements of centrality from positional determinacy: 1: Force intensities must be functions of the distance determined by the two-mass-points, would depend on non-observable, and thus esperientially indeterminate properties of the system.[§] 2: The only observable effect of the force acting between two mass-points can be to alter their distance. Finally, all forces in nature must be central forces, and the demand that nature be completely comprehensible entails that all forces are central.[¶] (2) Neither directions nor distances are determinate unless they are referable to empirically given points. Only spatial properties defined with the reference ot the points involved in the system under consideration can be employed to characterize its motion. See also POSITION PRINCIPLE; PRINCIPLE OF DECOMPOSITION *RESEMBLES* TRANSCENDENTAL PRINCIPLES OF NATURAL SCIENCE (EMMANUEL KANT)

PRINCIPLE OF POWER [psychology]   See
EMANCIPATION PRINCIPLE.

PRINCIPLE OF PRÄGNANZ [psychology]
principle of Prägnanz by H. L. Hollingsworth and a statement of the growth principle by Virginia Axline. Hollingsworth explains Prägnanz as follows: "According to the Gestalt law of Pragnanz, each configuration

[§]Hendricks, V. F. and Hyder, D. J. (2006) Interactions: mathematics, physics and philosophy, 1860–1930, v251, Springer.
[¶]Sheppard, E. (2002) The Spaces and Times of Globalization: Place, Scale, Network, and Positionality. *Economic Geography*, **78**(3) 307–330.

strives to be the best possible structure; it moves in such ways as to become as good a Gestalt as it can .... There is in them what might be called a repugnance for, a rejection of, such details as mar their calm and perfection. It is this demand for correct detail, this distaste for an inappropriate item, that is involved in the oughts and musts of the esthetic category."* Note the similarities in the following statement of Miss Axline: "There seems to be a powerful force within the individual which strives continuously for complete self-realization. This force may be characterized as a drive toward maturity, independence, and self-direction. It goes on relentlessly to achieve consummation, but it needs good growing ground to develop a wellbalanced structure."[†]

PRINCIPLE OF PREEMPTION [psychology] (John Quincy Adams) [‡] Any head of a family, widow, or single person over the age of 21 who was a citizen or had declared his or her intention of becoming one, and who was not the owner of 320 acres, could enter the public lands on condition of actual residence and improvement. (1) The use of different signals for different categories and of the same signal for all members within a category corresponds to the principle of contrast or of mutual exclusivity that children rely on when they assign only one label per category. Children will allow only one lexical entry to occupy a semantic niche. When two words are determined to have similar meanings, one of them is preempted and removed from the lexicon. Different words should have different meanings.[§] (3) Captures facts about the inferences speakers and addresses make for both conventional and novel words. It accounts for the preemption of novel

words by well-established ones; and it holds just as much for morphology as it does for words and larger expressions.[‖] (4) The development of nonlinguistic concepts, the acquisition of language in context, and the use by participants in conversational exchanges to account for those features of language and language acquisition. The regular rule applies whenever it is not blocked. When children have an irregular form, such as the preterit verb form *sang*, it blocks application of the regular preempted rule, but when they do not have such a form, they might reasonably produce the incorrect word *singed*. Children often use both the correct and overgeneralized forms at the same time. See also UNIQUENESS PRINCIPLE; MUTUAL EXCLUSIVITY PRINCIPLE; PRINCIPLE OF UNIQUENESS; PRINCIPLE OF CONVENTIONALITY; PRINCIPLE OF CONTRAST; COOPERATIVE PRINCIPLE.

PRINCIPLE OF PRESSURE SWING DESORPTION [physics] Reduction of the pressure of the bed below the pressure used during adsorption. By lowering of the vapor pressure of the absorbed n paraffins, the molecular sieve loading is reduced to maintain equilibrium between the adsorbed and desorbed n paraffins. Low molecular weight. Temperature, pressure, displacement.[¶] See also PRINCIPLE OF PERFORMING SIGNAL-PROCESSING OPERATIONS; PRINCIPLES OF DARCY'S LAW.

PRINCIPLE OF PRIMACY OF PLANNING [psychology] (1) Encompasses contribution of objectives, efficiency of plans, and primacy of planning.[‖] (2) Primary prerequisite for all other functions of management. Every action of the manager follows a planning step. See also PRINCIPLE OF CONTRIBUTION OF OBJECTIVES; PRINCIPLE OF EFFICIENCY OF PLANS; PRINCIPLE OF PLANNING.[**]

*Hollingsworth, H. L. (1949), *Psychology and Ethics*, Ronald Press, New York.
[†]Axline, V. (1950), *Play Therapy*, Houghton Mifflin, Boston, Torrance, P. (1950), The principle of Pragnanz as a frame of reference for psychotherapy, *J. Consult. Psychol.* **14**(6), 452–457.
[‡]Farnam, H. W. and Day, C. (2000), *Chapters in the History of Social Legislation in the United States to 1860*, The Lawbook Exchange, Ltd.
[§]Damon, W. Lerner, R. M., and Eisenberg, N. (2006), *Handbook of Child Psychology*, Wiley, Hoboken, NJ, p. 282.

[¶]Izumi, J. (1992), Application of zeolite adsorbents to pressure swing adsorption, Zeoraito 9(2), 60–68.
[‖]Benshoof, J. (1983), *Ethical Issues in U.S. Family Planning Policy*, Draper Fund Report, no.12, pp. 11–12.
[**]Koontz, H. and O'Donnell, C. (1968), *Principles of Managemnet*, 4th ed., McGraw-Hill.

PRINCIPLE OF PRIORITY [psychology] (1) Causes must precede their effects in time. According to the principle of priority; the cause must be an event that preceded the effect. Other cues are related to the mechanism principle; since the cause must be linked to the effect, it is likely to be an event that occurred in spatial and/or temporal contiguity with that effect. This cue is, of course, weaker than that of priority since causes can be discontiguous with their effects, but they can never succeed those effects in time. Principle of determinism (events are caused) causes precede their effects in time (principle of priority) states that there must be some kind of mechanism linking causes to their effect (principle of mechanism).* (2) The burden is allocated in whole to one community on the basis of selected criteria.† See PRIORITY PRINCIPLE; PRINCIPLE OF PROPORTIONALITY; (PRINCIPLE OF STEPWISE DECISION MAKING).

PRINCIPLE OF PROHIBITION OF REFORMATIO IN PEIUS [law]   (1) Decision from a court of appeal is amended to a worse one. It is in general unfair for the appellate court to impose a more severe sentence when there has been an appeal only on behalf of the defense.‡ (2) One of the primary guidelines behind the awarding of damages in common-law negligence claims. The amount of compensation awarded should put the successful plaintiff in the position he or she would have been had the tortious action not been committed.§ (3) The assertion that a substance known to cause the symptoms of disease will induce a cure of those symptoms if administered in a small amount.¶ See also PRINCIPLE OF RESTITUTIO IN INTEGRUM.

PRINCIPLE OF PROPAGATION OF A MAGNETIC TRAVELING WAVE [physics]   Each pole of a fixed magnet is surrounded by a magnetic field attached to the pole and constant in time. In one of the poles of an electrically pulsed electromagnet, a time-dependent magnetic field is created. This field propagates into space as a traveling magnetic wave. As a consequence of Maxwell's equations, this magnetic traveling wave is tied to an electric traveling wave of the same frequency.‖ See also MATRJOSCHKA PRINCIPLE.

PRINCIPLE OF PROPORTIONALITY [psychology] (1) Moral doctrine on the use of therapeutic means.** (2) Concerns how much force is morally appropriate. (3) Any layer of government should not take any action that exceeds that necessary to achieve the objective of government.† See also PRINCIPLES OF JUS IN BELLO.

PRINCIPLE OF PROPORTIONALITY [industrial psychology]   Burden is distributed in proportion to certain fairness criteria (e.g., responsibility for the burden; existing resources or vulnerability of the host community).†† See also PRINCIPLE OF STEPWISE DECISIONMAKING; PRINCIPLE OF PRIORITY.

PRINCIPLE OF PROPORTIONATE PUNISHMENT [psychology]   (1) "Let the punishment fit the crime," which applies particularly to mirror punishments (which may or may not be proportional). One purposes of the law

---

*Sophian, C. and Huber, A. (1984), Early developments in children's causal judgments, *Child Develo.* **55**(2), 512–526 [see especially Bullock et al. (1982) cited in this paper].
†Nuclear Energy Agency, Organisation for Economic Co-operation and Development (2004), *Stepwise Approach to Decision Making for Long-term Radioactive Waste Management Experience, Issues and Guiding Principles, in Radioactive Waste Management*, OECD, NEA no.4429, p. 36.
‡Council of Europe Committee of Ministers (1992), *Consistency in Sentencing*, Explanatory Memorandum to Recommendation R, vol.92, p. 17.
§*Black's Law Dictionary*.

¶Jonas, B. (2005), *Mosby's Dictionary of Complementary and Alternative Medicine*, Elsevier.
‖Azarenkov, N. A., Olefir, V. P., and Sporov, A. E. (2001), Quadrupole and octopole electromagnetic waves in slightly non-uniform magnetized plasma column, *Physica Scripta* **63**(1), 36–42.
Calipari, M. (2004), The principle of proportionality in therapy: Foundations and applications criteria, *NeuroRehabilitation* **19(4), 391–397.
††Nuclear Energy Agency, Organisation for Economic Co-operation and Development (2004), *Stepwise Approach to Decision Making for Long-term Radioactive Waste Management Experience. Issues and Guiding Principles. Radioactive Waste Management*, OECD. NEA no. 4429, p. 36.

is to provide equitable retaliation for an offended party. It defined and restricted the extent of retaliation.* (2) Applies to the broader class of legal systems that specify formulaic penalities for specific crimes, which are thought to be fitting in their severity. Some propose that this was at least in part intended to prevent excessive punishment at the hands of either an avenging private party or the state.† See also EYE-FOR-AN-EYE PRINCIPLE; PRINCIPLE OF EXACT RECIPROCITY; PRINCIPLE OF RETRIBUTIVE JUSTICE.

PRINCIPLE OF PROSELYTISM [psychology] The provision of aid must not exploit the vulnerability of victims and be used to further political or religious creeds.‡ See also PRINCIPLE OF HUMANITY; PRINCIPLE OF IMPARTIALITY; PRINCIPLE OF INDEPEN-DENCE; DEFINING PRINCIPLES; PRINCIPLE OF NEUTRALITY; EMERGENCY-ACTION PRINCIPLE.

PRINCIPLE OF PROTECTION-SCALE SELEC-TIVITY [engineering] See BUSBAR PROTEC-TION PRINCIPLE.

PRINCIPLE OF PROXIMAL ISOVELOCITY SUR-FACE AREA [physics] See DOPPLER PRINCI-PLE.

PRINCIPLE OF PROXIMITY [psychology] (1) Supposes that (a) the general factor that group formation is due to actual forces of attraction between the members of the group and (b) the more specific factor when the field contains a number of equal parts; those among them that are in greater proximity will be organized into a higher unit. Enables one to predict the actual perception obtained.§ (2) Since managerial operations in organizing, staffing, directing,

and controlling are designed to support the accomplishment of enterprise objectives, planning is the primary requisite of these functions. (3) Things that are closer together will be seen as belonging together.¶ See also PRINCIPLE OF CONTIGUITY; GESTALT PRIN-CIPLE OF CONTINUITY; GESTALT PRINCIPLE OF PROXIMITY; GESTALT PRINCIPLE OF CONTIGUITY; GESTALT PRINCIPLE OF CLOSURE; GESTALT PRINCIPLE OF AREA; GESTALT PRINCIPLE OF SYMMETRY; GESTALT PRINCIPLE OF SIMILARITY.

PRINCIPLE OF PSYCHONEUROIMMUNOLOGY [engineering, psychology, medicine] See QUASILOCAL PRINCIPLE.

PRINCIPLE OF PULSE-CODE MODULATION [computer science] (Developed by Alec H. Reeves, 1926) (1) A procedure of converting an analog signal into a digital signal in which an analog signal is sampled and then the difference between the actual sample value and its predicted value (predicted value is based on previous sample or samples) is quantized and then encoded forming a digital value. (2) DPCM codewords represent differences between samples unlike PCM, where codewords represented a sample value. (3) Basic concept of DPCM—coding a difference is based on the fact that most source signals show significant correlation between successive samples so encoding uses redundancy in sample values, which implies lower bit rate. (4) A part of a digital transmission system or as part of a digital signal processing system.‖ See also ADAP-TIVE DIFFERENTIAL PULSE-CODE MODULATION (ADPCM) PRINCIPLE; DIFFERENTIAL PULSE-CODE MODULATION PRINCIPLE.

PRINCIPLE OF PULSE VARIATION [computer science] (1) The pulse operational neural network and pulse shape analyses are signif-icantly important in biology as well as engi-neering. They can also elucidate the micro-scopic dynamics of neuron signal processing,

---

*Morano, D. V. (1978), Equal protection as a political principle, *J. Value Inquiry* **12**(1), 24–36.
†Ryberg, J. (2004), *The Ethics of Proportionate Punishment: A Critical Investigation*, Springer, p. 1.
‡Gilson, L., Sen, P. D., Mohammed, S., and Mujinja, P. (1994), The potential of health sector non-governmental organizations: Policy options, *Health Policy Plann.* **9**(1), 14–24.
§Krechevsky, I. (1938), An experimental investiga-tion of the principle of proximity in the visual per-ception of the rat, *J. Exp. Psychol.* **22**(6), 497–523.

¶Deutsch, D. (1978), Delayed pitch comparisons and the principle of proximity, *Percept. Psychophys.* **23**(3), p.227–230.
‖Rabbani, M., Ray, L. A., and Sullivan, J. R. (1986), Adaptive predictive coding with applications to radiographs, *Med. Instrum.* **20**(4), 182–191.

including biology-specific phenomena such as the pulse coincidence effect. (2) Quartz crystal microbalances (QCMs) are used in high-technology laboratories and in space applications such as detecting contamination on optical surfaces, measuring outgassing, and determining constituents by means of thermogravimetric analysis. (3) Mind/body model validating the myriad subtle neurotransmitter connections between emotions, mental states, and the immune system. (4) Since the virus can affect the entirety of the person, the entirety of the patient must be considered in making the diagnosis. Any effective therapy must address the entire psychosomatic–spiritual being. (5) A "pulse" is one of a series of regularly recurring, precisely equivalent stimuli. Like the ticks of a metronome or a watch, pulses mark off equal units in the temporal continuum. Although generally established and supported by objective stimuli (sounds), the sense of pulse may exist subjectively. A sense of regular pulses, once established, tends to be continued in the mind and musculature of the listener, even though the sound has stopped. (6) Pulse coincidence can lead to a continuous band of overlapping pulses, and the so-called pulse pileup continuum may occur.* See also STAFF'S DIAGNOSTIC PRINCIPLE; QUASILOCAL PRINCIPLE; PULSE COINCIDENCE PRINCIPLE.

PRINCIPLE OF PURPOSE [psychology] Purpose is derived from a person's or organization's values and beliefs. It is defined in emotional and relational terms, and remains a constant even when the environment changes over time. Going against someone's sense of purpose creates conflict, while action in line with purpose creates harmony and gives people strength and courage. Purpose motivates people in a way that goals are unable to, something shown by the limited effect that increased salary has on the performance of people.† Also known as *principle of causality*.

*Yuan, H.-G. (1981), Principle of pulse variation and its applications, *Proc. Int. Symp. Math. Theory Networks Syst.* **4**, 103—110.
†Oeser, H., Wegener, O. H., and Koeppe, P. (1977), Whole body computer tomography: Principle-purpose-requirements, *Munch. Med. Wochenschrift* **119**(37), 1189–1194.

PRINCIPLE OF QUANTUM ELECTRODYNAMICS [physics] A quantum gauge field theory providing a highly accurate description of the behavior of electrons, positrons, and photons at the subatomic level. In order to describe protons, it is necessary to account for their composite nature and additional interactions—quantum chromodynamics.‡

PRINCIPLE OF QUANTUM REALITY [physics] Scientists describe the natural world in wave possibilities. Only nonmaterial entities capable of turning vibrating energies into actualities are considered.§

PRINCIPLE OF RAID METHOD [physics] Redundant array of inexpensive devices (RAID), mirroring or striping of storage devices, where the storage devices are separated from a controller by a network. Interaction of electrically charged particles with electromagnetic radiation.¶

PRINCIPLE OF RATIONAL ACCOMMODATION [philosophy] (Donald Herbert Davidson, 1917–2003, American philosopher) We make maximum sense of the words and thoughts of others when we interpret in a way that optimises agreement. The principle may be invoked to make sense of a speaker's utterances when one is unsure of their meaning. In particular, the other uses words in the ordinary way; the other makes true statements; the other makes valid arguments; and the other says something interesting.‖ See also PRINCIPLE OF CHARITY; PRINCIPLE OF HUMANITY.

‡Fuerth, R. (1964), Problem of continuity or discontinuity in the fundamental concepts of theoretical physics, *Experientia* **20**(11), 593–598.
§Bartell, L. S. (1985), Perspectives on the uncertainty principle and quantum reality, *J. Chem. Educ.* **62**(3), 192–196.
¶Wang, P. S. S. and Lee, D. C. (2004), *RAID Method and Device with Network Protocol between Controller and Storage Devices*, US Patent 6,834,326 (12/21/04).
‖Davidson, Donald (1984) [1974]. "Ch. 13: On the Very Idea of a Conceptual Scheme". Inquiries into Truth and interpretation. Oxford: Clarendon Press.

PRINCIPLE OF RATIONAL INDICES [crystals] In any natural crystal the indices may be expressed as small whole numbers.*

PRINCIPLE OF RATIONALITY [computer science] (Allen Newell, 1927–1992, computer scientist and cognitive psychologist) "If an agent has knowledge that one of its actions will lead to one of its goals, then the agent will select that action."[†]

PRINCIPLE OF REACTION [physics] Whenever one body exerts a force on another, the second body exerts a force equal in magnitude and opposite in direction on the first body.[‡] See also NEWTON'S PRINCIPLES OF MOTION.

PRINCIPLE OF REASONABLENESS [engineering] See BUSBAR PROTECTION PRINCIPLE.

PRINCIPLE OF RECIPROCAL INHIBITION [kinesiology] Activation of the stretch afferents also produces inhibition of antagonistic muscles; specifically, the flexors of the knee are inhibited through inhibitory interneurons. This type of inhibition is referred to as *reciprocal inhibition*. Typically in a reflex action, there is excitatory activity toward synergist muscles and inhibition toward their antagonists.[§]

PRINCIPLE OF RECIPROCAL PROPORTIONS [chemistry] See PRINCIPLE OF EQUIVALENT PROPORTIONS; EQUIVALENT PROPORTIONS PRINCIPLE.

PRINCIPLE OF RECIPROCITY [mathematics] (1) Formulated in sufficiently general terms to be valid for an arbitrary set of harmonic sources distributed throughout a volume in which also is found any combination of elastic objects (e.g., plates, shells, rods, membranes), on which any system of external forces is acting. The source distribution may be random provided its space correlation function is available. It is shown that certain relations known to exist among the solutions of radiation and diffraction problems are immediate consequences of this principle.[¶] (2) A grouping of units that operate independently but together create an operation that acts to provide performance beyond that available from individual units.[‖]

PRINCIPLE OF RECTIFICATION [psychology] (Robert Nozick, 1938–2002) Governing the proper means of setting right past injustices in acquisition and transfer.** Nozick's entitlement theory is a theory of justice in the distribution of goods (i.e., distributive justice). Consists of three principles of entitlement of holdings; each of these three principles depends on one of three principles of justice in holdings: principles of entitlement of holdings, principle of entitlement of acquisition, principle of entitlement of transfer, principle of exhaustiveness, principles of justice in holdings, principle of justice in acquisition, principle of justice in transfer, and principle of rectification of injustice. See also (PRINCIPLE OF TRANSFER;) PRINCIPLE OF ACQUISITION; (NOZICK'S PRINCIPLE)

*Walker, P. M. B. (1999), *Chambers Dictionary of Science and Technology*, Chambers, New York, p. 662.

[†]Lopotko, A. I. (1971), Problem of rational audiometry and the principle of rationality in audiometric studies, *Vestnik Otorinolaringol.* 34(1), 47–52.

[‡]Spicik, J., Fassati P., Erbenova L., and Fassati, M. (1975), Albumin turbidity reaction in the blood serum. Studies of the principle of reaction by gel filtration and electrophoresis in polyacrylamide gel gradient [author's transl.], *Casopis Lekar Eskych* **114**(9), 270–273.

[§]Ikezuki, M. and Yamagnechi, S. (1996), A study on the mechanism of anxiety reduction through training: a comparison and examination of the stress-model and the distraction-model, *Shinrigaku Kenkyu (Jpn. J. Psychol.)* **67**(1), 9–17.

[¶]Ljamšev, L. M. (1959), A question in connection with the principle of reciprocity in acoustics, *Soviet Phys. Dokl.* **4**, 406–409.

[‖]Eastman, N. (1994), Mental health law: civil liberties and the principle of reciprocity, *Br. Med. J.* **308**(6920), 43–45.

**Wolff, J. (1991), *Robert Nozick Property, Justice and the Minimal State*, Stanford Univ. Press, p. 106. (http://www.westga.edu/~rlane/political/lecture_robertNozick1.html).

PRINCIPLE OF RECURRENT CAUSALITY. [psychology]   The beginning of control of an object can be internal, when it emanates from the organization; external, if the control comes from a foreign object or process; or it could well be an interaction among two or more objects which frequently produces a complementary exaltation or a greater sensibility of the interacting objects supporting the other. The response to this manifests itself as if it were perceived as a signal, to which a message and interpretation is sensed.[*]

PRINCIPLE OF REFERENCE [psychology] Words may be mapped onto the child's representations of objects, actions, events, or attributes in the environment. Acquisition of the notion of nonverbal communicative expression as indicated by comprehension and production of pointing gestures.[†] See also PRINCIPLE OF EXTENDIBILITY; PRINCIPLE OF OBJECT SCOPE; PRINCIPLE OF PRÄGNANZ; PRINCIPLE OF CATEGORICAL SCOPE; NOVEL NAME–NAMELESS CATEGORY (N3C) PRINCIPLE; PRINCIPLE OF CONVENTIONALITY.

PRINCIPLE OF REFLECTION [physics] (1) When a ray of light is reflected at a surface, the reflected ray is found to lie in the plane containing the incident ray and the normal to the surface at the point of incidence. (2) The angle of reflection equals the angle of incidence.[‡]

PRINCIPLE OF REFLECTION OF PLANS [psychology]   This S is a concept that is referred to in the philosophical history and systems of psychology. Human knowledge consists of four classes: (1) things that are knowable simply, (2) things knowable through experience, (3) things knowable that concern one's own actions, and (4) things knowable through human senses.[§] (5) The more controls are designed to deal with and reflect the specific nature and strucuture of plans, the more effective they will serve the interests of the enterprises and their managers.[¶] See also PRINCIPLE OF PLANNING.

PRINCIPLE OF REFORMATIO IN PEIUS [law] (Burrhus Frederic Skinner, 1904–1990, American psychologist) and preventing the appelate court from using the defendant's appeal as an opportunity to modify the judgment of the court below to the detriment of the defendant.[‖] See also PRINCIPLE OF STARE DECISIS.

PRINCIPLE OF REINFORCEMENT [psychology] [B.F. Skinner (David Premack, 1925- American Psychologist) D. Premack, 1959] recognized in the familiar conflict situation in which, for instance, permissive traming is disrupted and alternates with excessive discipline, or the effects of maternal and paternal care operate in different directions. Ideally in society, maternal care and paternal care supplement each other so that the effects of the two are absorbed together. Conflict seems to arise when they function at cross purposes. The principle of reinforcement used here is less a restatement of these mechanisms than a view that includes within the socializing environment the total area of influence to which the child is exposed and at the same tune broadens the definition of nurture so that it is not lumped together

[*]Wassenaar, J.S., Van Roon, W. M. C. and Ten Hallers. C. C. (1995) Self-organizing endomatter, the interactive interface and the origin of consciousness (the principle of recurrent causality, short-and long-looping behavior and psycho-organic phenomena, *Journal of Communication and Cognition*, **28**, (2–3), 145–349.

[†]Markman, E. M., Wasow, J. L., and Hansen, M. B. (2003), Use of the mutual exclusivity assumption by young word learners, *Cogn. Psychol.* **47**(3), 241–275.

[‡]Walker, P. M. B. (1999), *Chambers Dictionary of Science and Technology*, Chambers, New York, p. 662.

[§]Zelenov, V. L. (1980), Physical principles of reflected radiation from bulk materials of biological shielding used at atomic energy plants, *Trudy. Vses. Proekt. Int Teploelektroproekt* **22**, 132–139.

[¶]Koontz, H. and O'Donnell, C. (1968), *Principles of Management: An Analysis of Managerial Functions*, 4th ed., McGraw-Hill (http://knol.google.com/k/narayana-rao-kvss/principles-of-management-koontz-and/2utb2lsm2k7a/89).

[‖]Palumbo, G. (2000) Decision Rules and Optimal Delegation of Information Acquisition. CSEF Working Papers, p. 15. http://www.csef.it/WP/wp42.pdf.

with idiosyncratic elements.* See also PRINCIPLE OF EFFECT; UNIFIED REINFORCEMENT PRINCIPLE.

PRINCIPLE OF RELATIONSHIP [psychology] Because of constant change and contradiction, nothing in either human life or nature is isolated and independent, but instead everything is related. It follows that attempting to isolate elements of some larger whole can only be misleading.[†] See also PRINCIPLE OF HOLISM.

PRINCIPLE OF RELATIVITY [logic, light, mathematics] (1) The laws of mechanics are not affected by a uniform rectilinear motion of the system of coordinates to which they are referred.[‡] (2) Any theorem that expresses various reciprocal relations for the behavior of some physical systems, in which input and output can be interchanged without altering the response of the system to a given excitation.[§] (3) Sensitivity of reversible electroacoustic transducer when used as a microphone divided by the sensitivity when used as a source of sound is independent of the type and construction of the transducer.[¶] See also CAUSALITY PRINCIPLE; CLOCK RETARDATION PRINCIPLE; (EQUAL PASSAGE TIMES PRINCIPLE;) (KINEMATICAL PRINCIPLE;) (KINEMATICAL PRINCIPLE OF EPT;) (KINEMATICAL PRINCIPLE OF EQUAL PASSAGE TIMES;) MACH PRINCIPLE;

*Smith, M. W. (1954), Wild children and the principle of reinforcement, *Child Devel.* **25**(2), 115–123.
[†]Nisbett, R., Peng, K., Choi, I., and Norenzayan, A. (2001), Culture and systems of thought: Holistic versus analytic cognition, *Psychol. Rev.* **108**(2), 291–310.
[‡]Walker, P. M. B. (1999), *Chambers Dictionary of Science and Technology*, Chambers, New York, p. 915.
[§]Lorentz, H. A., Einstein, A., Minkowski, H., and Weyl, H. (n.d.), The principle of relativity, with notes by A. Sommerfeld (Transl. W. Perrett and G. B. Jeffery), *A Collection of Original Memoirs on the Special and General Theory of Relativity*, Dover Publications, New York.
[¶]Bucherer, A. H. (1908), On the principle of relativity and on the electro-magnetic mass of the electron, *Phil. Mag.* (1798–1977), **15**, 316–318.

PRINCIPLE OF COVARIANCE; PRINCIPLE OF EQUIVALENCE; RELATIVITY PRINCIPLE; SPECIAL PRINCIPLE OF RELATIVITY; TELESCOPICAL PRINCIPLE; THIRD PRINCIPLE OF RELATIVITY.

PRINCIPLE OF RELEVANCE [psychology] (1) Helps to explain the phenomenon of narrative creativity that is manifest in the individual creativity a narrator brings to the performance of a traditional tale; explored the process of reminding. (2) Human cognitive processes are geared to achieving the greatest possible cognitive effect for the smallest possible processing effort. (3) Communication is accompanied by a waranty of optimal relevance, which says that the utterance will produce a sufficient number of effects to make it worth the hearer's processing efforts. (4) Every utterance is interpreted relative to a context that is neither given nor reduced to the situation in which the communication takes place, but that is built for each new utterance and contains information in propositional form drawn from the environment, the interpretation of previous utterances, and the hearer's encyclopedic knowledge. (5) Defined in economic terms as an equilibrium between effect and effort that underlies the whole of human cognition.[‖] See also COOPERATION PRINCIPLE.

PRINCIPLE OF RELIABLE INTUITION [psychology] Actual intuitions about counterfactual cases are reliable to the degree to which the counterfactual condition matches experienced actual conditions, and the intuitions about the counterfactual situation match intuitions about the relevant experienced actual conditions.** See also PRINCIPLE OF REPRESENTATION THEORY.

‖Gough, D. (1990), The principle of relevance and the production of discourse: Evidence from Xhosa Folk narrative, Thought and Narrative Thought and Narrative Language. ed by B.K. Brotton and Pellegrini, A.D. Hillsdale, NJ, Erlbaum, p. 199–217.
**Yatsenko, D., McDonnall, D., and Guillory, K. S. (2007), Simultaneous, proportional, multi-axis prosthesis control using multichannel surface EMG, *Conf. Proc. IEEE. Eng. Med. Biol. Soc. 2007*, City, UT, pp. 6134–6137.

PRINCIPLE OF REMUNERATION [psychology] (Henri Fayol, 1841–1925, French industrialist)   (1) Financial arrangements in professional practice are in accord with professional standards that safeguard the best interest of the client and the profession.*

PRINCIPLE OF REPRESENTATION THEORY [psychology]   Actual intuitions about counterfactual cases are reliable to the degree to which the counterfactual condition matches experienced actual conditions, and the intuitions about the counterfactual situation match intuitions about the relevant experienced actual conditions.† See also PRINCIPLE OF RELIABLE INTUITION.

PRINCIPLE OF RESPECT FOR NATIONAL SOVEREIGNTY [psychology]   Fairness in balance in distribution of aid and humanitarian efforts.‡ See also PRINCIPLE OF HUMANITARIANISM; [PRINCIPLE OF AVOIDING CONFLICTS;] [INTERDEPENDENCE PRINCIPLES;] [MULTIIDENTITY PRINCIPLES;] PRINCIPLE OF MUTUAL HELP; (NON INTERFERENCE PRINCIPLE).

PRINCIPLE OF RESPONSIBILITY [psychology] (1) Psychologists respect the rights and reputation of the institute or organization with which they are associated.§ (2) In providing services, psychologists maintain the highest standards of their profession. They accept responsibility for the consequences of their acts and make every effort to ensure that

their services are used appropriately.¶ (3) Responsibilities and duties to the citizens themselves. This leads to underscoring the importance of the educational and teaching function assumed by the various moral and religious traditions. (4) For this reason, the major religions of the world must contribute to the development and multicultural expression of such a charter. (5) Three common principles of governance: responsibility, subsidiarity, and plurality. (6) Reasserts the burden of abiding by rules in times of peace on those acting in war. The issues that arise include the morality of obeying orders (e.g., when one knows those orders to be immoral), as well as the status of ignorance (not knowing of the effects of one's actions). (7) Demands an examination of where responsibility lies in war.‖ 8) The responsibility of subordinates to their superior for authority received by delegation is absolute, and no superiors can escape responsibility for the activities of their subordinates to whom they, in turn, have delegated authority.** See also PRINCIPLES OF JUS IN BELLO; PRINCIPLE OF SUBSIDIARITY; PRINCIPLE OF PLURALITY; PRINCIPLE OF PLANNING.

PRINCIPLE OF RESTITUTIO IN INTEGRUM [law] (Restoration to original condition.)   One of the primary guiding principles behind the awarding of damages in common-law negligence claims. The general rule, as the principle implies, is that the amount of compensation awarded should put the successful plaintiff in a position such that the tortious action had not been committed. Thus the plaintiff should clearly be awarded damages

*Wood, M. C. (2002), *Henri Fayol: Critical Evaluations in Business and Management*, Taylor & Francis, p. 8.

†Cardoso, M. H. and Gomes, R. (2000), Social representations and history: Theoretical and methodological principles for public health, *Cadernos de Saude Publica* **16**(2), 499–506.

‡Faunce, T. A. and Drahos, P. (1998), Trade related aspects of intellectual property rights (TRIPS) and the threat to patients: A plea for doctors to respond internationally, *Med. Law* **17**(3), 299–310.

§American Psychological Association (APA) (1967), *Casebook on Ethical Standards of Psychologists*, Committee on Ethical Standards for Psychology, APA, Washington, DC, p. 63.

¶American Psychological Association (APA) (1987), *Casebook on Ethical Standards of Psychologists*, Committee on Ethical Standards for Psychology, APA, Washington, DC, pp. 5–20.

‖Stepan, J. (1969), Realization of the principle of responsibility in public health as a guaranty of the rights of citizens and health workers. Introductory declaration of the chirman of the permanent commission for medical law, *Ceskoslov. Zdravotnictvi* **17**(1), 9–14.

**Koontz, H. and O'Donnell, C. (1968), *Principles of Management: An Analysis of Managerial Functions*, 4th ed., McGraw-Hill (http://knol.google.com/k/narayana-rao-kvss/principles-of-management-koontz-and/2utb2lsm2k7a/89).

for direct expenses such as medical bills and property repairs and the loss of future earnings attributable to the injury (which often involves difficult speculation about the future career and promotion prospects).* See also PRINCIPLE OF REMORMATIO IN PEIUS; USUFRUCT PRINCIPLE.

PRINCIPLE OF RETRIBUTIVE JUSTICE [psychology] (1) Exodus 21:23–27: "An eye for an eye, a tooth for a tooth." (2) At the root of the nonbiblical form is the belief that one of the purposes of the law is to provide equitable retaliation for an offended party.[†] See also PRINCIPLE OF PROPORTION-ATE PUNISHMENT; PRINCIPLE OF LEX TALIONIS; EYE-FOR-AN-EYE PRINCIPLE; PRINCIPLE OF EXACT RECIPROCITY.

PRINCIPLE OF REVEALED PREFERENCE [psychology] (Paul A. Samuelson, 1915—American economist) (1) a participant's relative preference between two reinforcers is reflected in the participant's consistent choice behavior by the relative numbers of responses emitted to gain units of the alternative reinforcers. A reinforcer who consistently maintains a higher rate of response than an alternative reinforcer, as schedule requirement varies, is said to be shown by the participant's consistent behavior as being preferred. Whereas preference between reinforcers may be measured in terms of either responses or reinforcers, it is conventional in behavioral economics to describe preference using measures of reinforcers. Indicates that differences in both intercepts and slopes of demand functions provide information about relative preferences between reinforcers. Consider differences in intercepts or heights of demand functions. The intercept of the reinforcer demand function with the vertical axis shows the rate of reinforcement that was

obtained when the schedule requirement to be 0. The intercept of the reinforcer demand function is the free-operant measure of preference recommended by Premack, who also pointed out that if one reinforcer is obtained at a higher rate than another when both are freely available, then the more frequently obtained reinforcer is shown as being preferred under these conditions. The slope of a demand function provides information about how a participant adjusts his or her performance when schedule requirements increase. If a participant maintains the rate of one reinforcer more than the rate of an alternative reinforcer when schedule requirements increase, then the reinforcer whose rate was maintained will have a flatter demand function. A reinforcer with a flatter demand function is said to be shown as being preferred, as the participant must emit higher rates of responding to maintain a reinforcer when the schedule requirements increase. Can be used to assess relative preference between reinforcers from the relative shapes of demand functions for the reinforcers. The relative shapes of direct demand functions show how a participant's pattern of exchanging responses for reinforcers changes when schedule requirements change.[‡] (2) A method by which it is possible to discern the best possible option on the basis of consumer behavior. Essentially, this means that the preferences of consumers can be revealed by their purchasing habits.*

PRINCIPLE OF REVERSIBILITY [mathematics, physics] (1) In any ray-tracing procedure the reversal of all the rays must produce the original ray, provided there is no absorption.[§] (2) The position of any point on the screen is a function of the position—and not the history—of the moved point. Any information about the path through which the point was

*Martin, F. F. and Schnably, S. J. (2006), Rights International, Wilson, R., Simon, J., and Tushnet, M. (2006), *International Human Rights and Humanitarian Law: Treaties, Cases, and Analysis*, Cambridge Univ. Press, p. 306.
[†]Dresser, R. (1993), Sanctions for research misconduct: A legal perspective, *Acad. Med.* **68**(9 Suppl.), S39–S43.

[‡]Tustin, D. (2000), Revealed preference between reinforcers used to examine hypotheses about behavioral consistencies, *Behav. Modifi.* **24**(3), 411–424.
[§]Busch, K. W. and Busch, M. A. (1990), *Multielement Detection Systems for Spectrochemical Analysis*, Wiley-Interscience, p. 236.

dragged between its original position and its return to that position.*

PRINCIPLE OF RHYTHM [physics] Embodies the truth that in everything there is manifested a measured motion, a to and a fro; a flow and an inflow; a swing backward and a swing, forward—in short, a pendulumlike movement. There is rhythm between every pair of opposites, or poles, and this is closely related to the principle of polarity.[†] See also PRINCIPLE OF POLARITY; PRINCIPLE OF CAUSE AND EFFECT.

PRINCIPLE OF ROUTE CHOICE [mathematics] (John Glen Wardrop, (1921–1989) English transport analyst) (1) Describes the spreading of trips over alternate routes due to congested conditions.[‡] (2) The journey times in all routes actually used are equal to and less than those that would be experienced by a single vehicle on any unused route. Each user non-cooperatively seeks to minimize the cost of transportation. The traffic flows are usually referred to as *user equilibrium* (UE) flows, since each user chooses the route that is best. Specifically, a user-optimized equilibrium is reached when no user may lower her/his transportation cost through unilateral action.[¶] (3) The stochastic user equilibrium (SUE) wherein no driver can unilaterally change routes to improve his/her perceived travel times.[§] See also WARDROP'S PRINCIPLE.

PRINCIPLE OF SAINT VENANT [physics] See SAINT VENANT'S PRINCIPLE

PRINCIPLE OF SATURATION ANALYSIS [biology] (1) Displacement analysis, or competitive protein binding. (2) Determination of drug and hormone levels in biological fluids.[¶]

PRINCIPLE OF SCALAR CHAIN [psychology] (Henri Fayol, 1841–1925, French industrialist). (1) Organizations need a formalized hierarchy reflecting the flow of authority and responsibility. (2) Responsibility of subordinates to superior for authority received by delegation is absolute, and no superior can escape responsibility for the organization activities of a subordinate.[‖] See also FAYOL'S PRINCIPLE OF MANAGEMENT; PRINCIPLE OF LINES OF COMMAND CHAIN; SCALAR PRINCIPLE.

PRINCIPLE OF SB ELECTRODE [chemistry] Application of pH measurement in oilfield wastewater treatment.[**]

PRINCIPLE OF SEDIMENTATION [geology, physics] States that the speed or velocity with which particles settle out of a liquid medium is dependent on a constant factor and the radius of the particles. (2) The bigger the particle, the faster it will fall out of suspension.[††]

PRINCIPLE OF SELECTION POTENTIAL [physics] As long as there is a positive selection potential at equilibrium, there is the opportunity for new modifiers to invade the population. In this sense, the importance of the selection potential is that unless it is

*Kirgintsev, A. N. (1977), Limitation of the principle of reversibility in systems with solid solutions, *Zh. Fiz. Khim.* **51**(4), 1014.

[†]Trois Initiés (), *Le Kybalion*: Etude sur la Philosophie Hermétique de l'Ancienne Egypte et de l'Ancienne Grèce.

[‡]Wardrop, J. G. (1952), Some theoretical aspects of road traffic research, *Proc. Inst. Civil Eng.* **1**(pt. II), 325–378.

[§]Bobzin, H. (2006), Principles of Network Economics, Birkhäuser, p.169.

[¶]Shaw, W., Smith, J., Spierto, F. W., and Agnese, S. T. (1977), Linearization of data for saturation-type competitive protein binding assay and radioimmunoassay, *Clin. Chim. Acta. Int. Clin. Chem.*, **76**(1), 15–24.

[‖]Wood, M. C. (2002), *Henri Fayol: Critical Evaluations in Business and Management*, Taylor & Francis, p. 8.

[**]Peng, B., Wong, H. F., Liu, Z-P and , W-B (2010) Combined surface enhanced in prepared spectroscopy and first-principles study on electrooxidation of formic acid at sb-modified ptelectrodes. J. Phys. Chem. C. r114(7) pp 3102–3107.

[††]Filitti, W. S. and Galle, C. (1965), Determination of molecular weights of ternary system components by application of the Archibald principle of sedimentation equilibrium, *Bull. la Soc. Chim. Biol.* **47**, 713–718.

reduced, a population can always potentially be invaded by new variants having the appropriate changes in the transformations operating on them.* See also SELECTION POTENTIAL PRINCIPLE.

PRINCIPLE OF SELF-DEFENSE [psychology] Can be extrapolated to anticipate probable acts of aggression, as well as in assisting others against an oppressive government or from another external threat.[†]

PRINCIPLE OF SELF-INTEREST [psychology] (1) Distinguishing self-interest (and components of self-fidelity, self-justice, self-beneficence, self-protection) from selfishness; fear of legal reprisals may lead to actions that substantially impair one's services to clients.[‡] (2) Therapists' reactions to treating traumatized patients. Therapists may need to limit the number of traumatized patients whom they are treating, curtail exposure to vicarious violence in movies and television, and seek support from colleagues. As therapy further progresses, therapists' own needs, the principle of self-interest, may be utilized in the therapeutic relationship.[**]

PRINCIPLE OF SELF-OSCILLATIONS [pharmacology] (Santorio)   A handheld pendulum with which to take a pulse.[§] See also PENDULUM PRINCIPLE.

PRINCIPLE OF SEPARATION [organizational psychology]   If an activity is designed to be a check on the activities of another department, individuals charged with such activity cannot adequately discharge their responsibilities if they report to the department whose activity they are expected to evaluate.[¶] See also PRINCIPLE OF PLANNING.

PRINCIPLE OF SEPARATION [psychology, physics]   (1) (M. R. Mahler's (1968, 1975) attempts to use the separation crisis to maximize growth and self-understanding.[‖] (2) [Literature] (Jean Baudrillard), the early voice, which lasted less than 10 years, and the mature voice, which lasted about 30. The first voice is younger and more conventionally leftist. It was fully embedded in the intellectual debates of the late 1960s. A committed Marxist, the younger Baudrillard wrote on labor and needs, use value, and production. But after this period as a young man, Baudrillard transitioned into a very different thinker in the middle to late 1970s. He developed a whole new theoretical vocabulary that was completely in tune with that decade's historical transformation into digitization, postindustrial economies, immaterial labor, mediation, and simulation. His theories of play and games are at the very heart of this transformation. Through a close reading of several texts, this essay explores Baudrillard's interest in play and games through the concepts of seduction, the fatal strategy, illusion, and what he called the "principle of separation."[**] (3) Chain of direct authority relationships from superior to subordinate throughout the organization.[††] See also PRINCPLE OF SEPARATION–INDIVIDUATION.

*Ratinov, V. B., Enisherlova, S. G., and Zagirova, R. U. (1969), Galvano and potentiostatic methods of investigation and basic principles of selection of admixtures-inhibitors of corrosion of reinforcement of concrete, *Durability* Concrete, 1969, *Preliming Report*, vol. 2, pp. D-135–D–151.
[†]Knecht, T. (2000), Neurobiology of excess self - defense use. Case report with summary of neuroscientific principles, *Archiv Kriminol.* **206**(3–4), 65–72.
[‡]Kinzie, J. D. and Boehnlein, J. K., (1993), Psychotherapy of the victims of massive violence: countertransference and ethical issues, *Am. J. Psychother.* **47**(1), 90–102.
[§]Noszticzius, Z. (1993), Principles of self-organization and self-accelerating reactions in nonequilibrium chemical systems, *Period. Polytech. Phys. Nuclear Sci.* **1**(2), 213–218.

[¶]Koontz, H. and O'Donnell, C. (1968), *Principles of Management*, 4th ed., McGraw-Hill.
[‖]Toomim, M. (1974), *Separation Counseling: A Structured Approach to Marital Crisis. Therapeutic Needs of the Family: Problems, Descriptions and Therapeutic Approaches*, Charles C Thomas, Oxford, UK.
Galloway, A. (2007), Radical illusion (a game against), *Games Cult. J. Interact. Media* **2(4), 376–391.
[††]Schirardin, H. and Bauer, M. (1974), Fluorometric determination of plasma cortisol. Routine method based on the principle of separation of phases, *Feuillets Biol.* **15**(77), 49–54.

PRINCIPLE OF SEPARATION–INDIVIDUATION [psychology] Mahler's principle of separation–individuation, widely recognized as a useful tool for understanding couples difficulties, has been applied traditionally in one-to-one psychotherapy with individual partners. Treatment from a level dominated by symbiotic longings and related ambivalence to one reflective of increased object constancy and empathic mutuality.[*]

PRINCIPLE OF SIMILARITY [psychology, physics] (Wertheimer, 1923) (1) Principle of figure–ground segregation by studying it vis-à-vis the classical gestalt principles of grouping and figure–ground segregation. States, all else being equal, with the most similar elements are grouped together.[†] (2) States that share visual characteristics such as shape, size, color, texture, value, or orientation will be seen as belonging together.[‡] See also PRINCIPLE OF CONTIGUITY; PRINCIPLE OF PROXIMITY; GESTALT PRINCIPLE OF CONTINUITY; GESTALT PRINCIPLE OF PROXIMITY; GESTALT PRINCIPLE OF CONTIGUITY; GESTALT PRINCIPLE OF CLOSURE; GESTALT PRINCIPLE OF AREA; GESTALT PRINCIPLE OF SYMMETRY; GESTALT PRINCIPLE OF SIMILARITY.

PRINCIPLE OF SIMPLICITY [mathematics] (William of Occam) "All things being equal, the simplest solution tends to be the best one."[§] See also OCCAM'S RAZOR PRINCIPLE.

PRINCIPLE OF SOLIDARITY [health care] Contributions depend on the insurants' income and family status because the major goal of the system is to provide equal access to high quality care for all citizens.[¶]

PRINCIPLE OF STEP WISE DECISION MAKING See PRINCIPLE OF PRIORITY AND PRINCIPLE OF PROPORTIONALITY.

PRINCIPLE OF STRONG TRANQUILITY See TRANQUILITY PRINCIPLE

PRINCIPLE OF WEAK TRANQUILITY See TRANQUILITY PRINCIPLE

PRINCIPLE OF SIMPLICITY [mathematics] (William of Occam) "All things being equal, the simplest solution tends to be the best one."[‖] See also OCCAM'S RAZOR PRINCIPLE.

PRINCIPLE OF SMOOTH FIT [psychology, sociology] (1) Instinctive attempt to avoid pain, discomfort, or unpleasant situations; desire to obtain maximum gratification with minimum effort.[**] (2) Activities must be grouped to facilitate the accomplishment of goals; and the manager of each subdivision must have authority to coordinate its activities with the organization as a whole. The more clearly a position or a department defines the of results expected, activities to be undertaken, organization authority delegated, and authority and informational relationships with other positions, the more adequately the individuals responsible can contribute toward accomplishing enterprise objectives.[††] (3) If an activity is designed as a check on the activities of another department, individuals charged with such activity cannot adequately discharge their responsibilities if they report to the department whose activity they are expected to evaluate.[‡‡] See

[*]Applegate, J. S. (1988), Alone together: An application of separation-individuation theory to conjoint marital therapy, *Clin. Soc. Work J.* **16**(4), 418–429.

[†]Pinna, B. (2005), The role of the Gestalt principle of similarity in the watercolor illusion, *Spatial Vision* **18**(2) (special issue in honor of Jacob Beck), 185–207.

[‡]Bellavite, P., Andrioli, G., Lussignoli, S., Signorini, A., Ortolani, R., and Conforti A. (1997), A scientific reappraisal of the "principle of similarity," *Medi Hypoth.* **49**(3), 203–312.

[§]Gendron, G. (1986), Steel man Ken Iverson, (manages Nucor Corp. according to the principle of simplicity)", Inc, **8** 40.

[¶]Sass, H.M. (1992) Introduction the Principle of Solidarity in Health Care Policy. *Journal of Medicine and Philosophy.* **17**(4) 367–370.

[‖]Gendron, G. (1986), Steel man Ken Iverson, (manages Nucor Corp. according to the principle of simplicity)", Inc, **8**, 40.

[**]Sametz, L. (1979), Children, law and child development: The child developmentalist's role in the legal system, *Juv. Family Court J.* **30**, 49.

[††]Bagad V. S. (2008), *Management Information Systems*, Technical Publications, pp. 2–10.

[‡‡]Ma, J. (1992), On the principle of smooth fit for a class of singular stochastic control problems

also AUTOMATIC CONTROL PRINCIPLES; PRINCI-PLE OF DOMINANT SUBSYSTEMS; PRINCIPLE OF FUNDAMENTAL CHARACTERISTICS; PRINCIPLE OF OPERATION; EQUIPARTITION PRINCIPLE; PRINCI-PLE OF CONSTRUCTION AND OPERATION.

PRINCIPLE OF SOMATOMENTAL BALANCE [sociology] (Aaron Antonovsky)   Mutual correspondence of the somatic and mental aspect.* See also SALUTOGENESIS PRINCIPLE; PRINCI-PLE OF THE SALUTOGENETIC TRIAD; COMFORT-THROUGH-DISCOMFORT PRINCIPLE.

PRINCIPLE OF SOUND ORGANIZING [psychology]   (1) A large part of the variation in natural speech appears along the dimensions of articulatory precision/perceptual distinctiveness. Speaking is considered efficient if the speech sound contains *only* the information needed to understand it. This efficiency is tested by means of a corpus of spontaneous and matched read speech, and syllable and word frequencies as measures of information content (12,007 syllables, 8046 word forms, 1582 intervocalic consonants, and 2540 vowels). It is indeed found that the duration and spectral reduction of consonants and vowels correlate with the frequency of syllables and words. Consonant intelligibility correlates with both the acoustic factors and the syllable and word frequencies. It is concluded that the principle of efficient communication organizes at least some aspects of speech production.[†] (2) Structure and function are affected by the variability of multiple factors as well as by the mean value experienced. While the effects of these two terms can sometimes be independent of each other and separable in space and time, they can also interact; effects of the same degree of variation about different mean values of a variable. The hierarchy of spatio temporal variation produces

a cascade of effects across multiple scales.[‡] (3) The relationship between sound management practices and credit lending rates are an important element in building sustainable development. For the financial services sector, change, anticipation, and adaptation to customer needs and market trends is a matter of competitive survival.[§] See also PRIN-CIPLE OF EFFICIENT COMMUNICATION.

PRINCIPLE OF SPACETIME RECIPROCITY [psychology]   If the distance between pairs of stimulated points is different, equal durations of stimulation appear to be different. The greater the distance between stimulated points, the greater the apparent time between stimulating one point and then the other. In these kinds of settings, perceived time and perceived space are reciprocally related. The perceived distance through which we move is a determinant of the time it takes us to do so, and the amount of time it takes to move from one place to another is a determinant of the apparent distance between them. In brief, the principle of space time reciprocity may well extend well beyond the laboratory tasks in the context of which it has been demonstrated. In more speculative terms, the apparent size of the spaces that we occupy may well vary with the time it takes to move through them and vice versa.[¶]

PRINCIPLE OF SPECIAL RELATIVITY [physics] (1) Proposed in 1905 by Albert Einstein in his article "On the electrodynamics of moving bodies." Some three centuries earlier, Galileo's principle of relativity had stated that all uniform motion was relative, and that there was no absolute and well-defined state of rest; a person on the deck of a ship may be at rest in her opinion, but someone

for diffusions, *SIAM J. Control Optim.* **30**(4), 975–999.
*Vuillerme, N., Pinsault, N., Chenu, Ol., Demongeot, J., Payan, Y., and Danilov, Y. (2008), Sensory supplementation system based on electrotactile tongue biofeedback of head position for balance control, *Neurosci. Lett.* **431**(3), 206–210.
†McNellis, M. G. and Blumstein, S. E. (2001), Self-organizing dynamics of lexical access in normals and aphasics, *J. Cogn. Neurosci.* **13**(2) p 151–170.

‡Parker, S. and Wall, T. D. (1998), *Job and Work Design: Organizing Work to Promote Well-Being and Effectiveness*, Sage, p. 121.
§Basel Committee on Banking Supervision (2008), *Principles for Sound Liquidity Risk Management and Supervision* (http://www.bis.org/publ/bcbs144.pdf).
¶Day, R. (2006), Two principles of perception revealed by geometrical illusions, *Austral. J. Psychol.* **58**(3), 123–129.

observing from the shore would say that she was moving. Special relativity overthrows Newtonian notions of absolute space and time by stating that distance and time depend on the observer, and that time and space are perceived differently, depending on the observer.[*] (2) The speed of light is the same for all observers, even if they are in motion relative to one another. Special relativity reveals that c is not just the velocity of a certain phenomenon—light—but rather a fundamental feature of the way space and time are tied together. In particular, special relativity states that it is impossible for any material object to travel as fast as light.[†]

PRINCIPLE OF SPECIALIZATION [psychology] (Henri Fayol, 1841–1925, French industrialist) (1) It is best to assign workers jobs fairly limited in scope to enable them to develop a high degree of skill. (2) Encompassing the purpose, cause, structure, and process of organizing. The attainment of an objective is the purpose of organizing; span of management, the cause; authority, the cement; departmentalized activities, the framework; and effectiveness, the measure in supporting performance. (3) The topography of interaction foci varies for different mental acts. In the cases of perception, the projection cortex is the integration center; in the case of thinking, the associative zones of the cortex are the integrative center. Imaginal thinking is associated primarily with the parietotemporallobe; abstract thinking, with the frontal divisions of the cortex. The two principal functions of consciousness are also spatially separate.[‡] See also PRINCIPLE OF DIVISION OF WORK; FAYOL'S PRINCIPLES OF MANAGEMENT.

PRINCIPLE OF SPINOR SUPERPOSITION [physics] See RESOLUTION PRINCIPLE.

[*]Pierseaux, Y. (2003), The principle of physical identity of units of measure in Einstein's special relativity, *Physica Scripta* **68**(3), C59-C65.
[†]Goldoni, R. (1978), Faster-than-light phenomena in special and in general relativity, *Proc. 1st Tachyons, Monopoles, Related Topics, Interdisciplinary Seminar*, 1 1976, pp.125–140.
[‡]Wood, M. C. (2002), *Henri Fayol: Critical Evaluations in Business and Management*, Taylor & Francis, p. 8.

PRINCIPLE OF STABILITY OF TENURE [psychology] (Henri Fayol, 1841–1925, French industrialist). A high turnover rate is expensive for an organization. Initiative should be encouraged at all levels. and subordinates should be asked to submit plans and new ideas.[¶] See also FAYOL'S PRINCIPLES OF MANAGEMENT.

PRINCIPLE OF STABLE DISEQUILIBRIUM [biology] (1) The approach to stable disequilibrium is slower in larger numbers, and the amount of disequilibrium established is weaker than in a smaller number system. (2) The number of complementary combinations is a function of the number of corresponding matches and of population size. (3) As population size increases, the rate of the approach to stable disequilibrium is slower. (4) There is an optimum selection coefficient that minimizes the transient fixation probability when linkage is present. (5) The absence of linkage disequilibrium is seldom a practical method of testing the hypothesis of balancing selection because it depends strongly on population size in determining linkage disequilibria.[§]

PRINCIPLE OF STAFFING [organizational psychology] The quality of management personnel can be ensured through proper definition of the job and its appraisal in terms of human requirements, evaluation of candidates and incumbents, and appropriate training. Those organizations that have no established job definition, no effective appraisals, and no system for training and development will have to rely on coincidence or sources to fill the positions with able managers.[¶] See also PRINCIPLE OF PLANNING.[†]

PRINCIPLE OF STAFFING [psychology] (1) Division of work to produce more and better work with the same effort. (2) The objective of staffing is to ensure that organization roles are filled by those qualified employees who are able and willing to

[§]Gol'tsman, G. V. (1974), Muscle contraction in the light of the principle of stable disequilibrium, Usp. *Sovremennoi Biol.* **78**(3), 423–433.
[¶]Koontz, H. and O'Donnell, C. (1968), *Principles of Management*, 4th ed., McGraw-Hill.

occupy them. (3) The clearer the definition of organization roles and their human requirement, and the better the technique of manager appraisal and training employed, the higher the managerial quality.* See also STAFFING PRINCIPLES; (PRINCIPLES OF JOB DEFINITION;) PRINCIPLE OF MANAGERIAL APPRAISAL; PRINCIPLE OF THE OBJECTIVE OF STAFFING.

PRINCIPLE OF STAFFING OBJECTIVE [psychology]   The more clearly an organizational role and its translation into human requirements are defined and the better candidates and incumbents are evaluated and trained, the more the quality of personnel can be assured. The objective of managerial staffing is to ensure that organization roles are filled by those qualified personnel who are able and willing to occupy them.[†] See also PRINCIPLE OF STAFFING.

PRINCIPLE OF STAFFING OBJECTIVES [organizational psychology]   The positions provided by the organization structure must be staffed with personnel able and willing to carry out the assigned functions. See also PRINCIPLE OF PLANNING.[‡]

PRINCIPLE OF STANDARDS [organizational psychology]   Effective   control   requires objective, accurate, and suitable controls.[¶] See also PRINCIPLE OF PLANNING.

PRINCIPLE OF STANDARDS [psychology]   The objective of managerial staffing is to ensure that personnel able and willing to occupy them fill organization roles.[¶]

PRINCIPLE OF STATIONARY ACTION [engineering]   (Pierre-Louis   Moreau   de Maupertuis,   1698—1759,   French mathematician)   When   applied   to   the   action of a mechanical system, can be used to obtain the equations of motion for that

system.[§] See also PRINCIPLE OF LEAST ACTION; VARIATIONAL PRINCIPLE; GAUSS' PRINCIPLE OF LEAST CONSTRAINT; HERTZ'S PRINCIPLE OF LEAST CURVATURE.

PRINCIPLE OF STATIONARY PHASE [mathematics]   The asymptotic behaviour of $I(k)$ depends only on the critical points of f. If by choice of g the integral is localized to a region of space where f has no critical point, the resulting integral tends to 0. Consider, for example, the Riemann–Lebesgue lemma. The second statement is that when f is a Morse function, so that the singular points of f are nondegenerate and isolated, then the question can be reduced to the case $n = 1$. In fact, then, choice of g can be made to split the integral into cases with just one critical point P in each. At that point, because the Hessian determinant at P is by assumption not 0, the Morse lemma applies.[¶]

PRINCIPLE OF STATISTICAL MECHANICS [engineering]   The minimum of the free energy determines the distribution at thermal equilibrium.[‖] See also PRINCIPLE OF MINIMAL FREE ENERGY.

PRINCIPLE OF STEAM DISPERSION [engineering]   (1) Steam is injected to add turbulence, thereby preventing hydrocarbons from stagnating, cooling, and ultimately forming coke. Adding colder, possibly wet steam has the opposite effect to that desired—it introduced cold spots that allowed coke to form.** (2) Two basic principles to minimize coking are to avoid dead spots and prevent heat losses: (a) using dome steam or purge steam to sweep

*Wood, J. C. and Wood, M. C. (2002), *Henri Fayol: Critical Evaluations in Business and Management*, Taylor & Francis, p. 203.
[†]Wood, J. C. and Wood, M. C. (2002), *Henri Fayol: Critical Evaluations in Business and Management*, Taylor & Francis, p. 203.
[‡]Koontz, H. and O'Donnell, C. (1968), *Principles of Management*, 4th ed., McGraw-Hill.

[§]Banerjee, A. and Adams, N. P. (1990), Dynamics of classical systems based on the principle of stationary action, *J. Chem. Phys.* **92**(12), 7330–7339.
[¶]Taniguchi, S. (2000), Lévy's stochastic area and the principle of stationary phase (Engl. summary), *J. Funct. Anal.* **172**(1), 165–176.
[‖]Rose, K. (1998), Deterministic annealing for clustering, compression, classification, regression, and related optimization problems, *Proc. IEEE*, **86**(11), 2214.
**McPherson, L. J. (1984) Causes of FCC reactor coke deposits identified, *oil Gas Journal*. 88, 139–143.

out stagnant areas in the disengager system—the purpose of steam is to keep heavy, condensible hydrocarbons out of cooler regions (steam also provides a reduced partial pressure or steam distillation effect on high-boiling-point hydrocarbons, helping them to vaporize at lower temperatures); and (b) using proper insulation on pipes and vessels; cold spots are often caused by heat loss through the walls, in which case increased thermal resistance might help reduce coking. In particular, flanges should be well insulated, especially at the fractionator inlet. Exposed flanges will be a significant source of heat loss. The transfer line, which is a common source of coke deposits, should be as heavily insulated as possible, provided that stress-related problems have been taken into consideration.[*]

PRINCIPLE OF STEPWISE DECISIONMAKING [industrial psychology] Aim of pinpointing current status, to highlight the societal dimension, to analyze its roots in social sciences, and to identify potential guiding principles and issues in implementation. All parties be treated in some sense equally. See also PRINCIPLE OF PARITY; PRINCIPLE OF PRIORTY.[†]

PRINCIPLE OF STEPWISE REVERSIBILITY [industrial operation, industrial psychology] Support continuing a process leading to actions that remain stepwise and reversible.[‡]

PRINCIPLE OF STEREOSCOPIC IMAGE [engineering] (1) Effective control requires objective, accurate, and suitable standards. (2) Displaying two different perspectives (left and right perspectives) alternately. Left and right parallax images are separated in the horizontal direction and observed by the eyes of the observer.[§]

PRINCIPLE OF STEVINO [physics, mathematics] For a liquid that is in static equilibrium in a container, the hydrostatic pressure p at any point of the liquid at a depth h under the free surface of the liquid is given by the formula: $p = p_0 + rgh$, in which r is the density of the liquid, g is the gravity acceleration and p_0 is the atmospheric pressure acting on the free surface of the liquid. The product rgh is the weight of a liquid column with the density r, with base of 1 cm^2 and height h, and is numerically equivalent to the hydrostatic pressure, which depends only on depth h under the free surface of the liquid and produces forces (pressure forces) directed along the perpendicular to the surface that is considered.[¶] Also referred to as *Stevino's principles*.

PRINCIPLE OF STOCHASTIC CONTROL [mathematics] (1) Used to obtain an optimal development policy for the allocation of labor in the labor-surplus economy.[‖] (2) A method for the optimal nonlinear stochastic control of hysteretic systems with parametrically and/or externally random excitations. On the basis of the stochastic dynamical programming principle, a Hamilton–Jacobi–Bellman equation was then established for the given performance index, and this was then solved to give the optimal control force.[**] (3) For optimal control problems of stochastic systems consisting of forward and backward state

[*]Speight, J. G. (n.d.), *The Chemistry and Technology of Petroleum*, 4th ed., CRC Press, p. 542.

[†]Nuclear Energy Agency, Organisation for Economic Co-operation and Development (2004), *Stepwise Approach to Decision Making for Long-term Radioactive Waste Management Experience, Issues and Guiding Principles*, Radioactive Waste Management, OECD 2004, NEA, no. 4429, p. 3.

[‡]Bodansky, D. (2004), *Nuclear Energy: Principles, Practices, and Prospects*, 2nd ed., Springer, p. 360.

[§]Shu, J. (2008), Image reconstruction method adopting x ray volume radiography, *Faming Zhuanli Shenqing Gongkai Shuomingshu*.

[¶]Cucinotta, A. (n.d.), *Archimedes and Stevino's Principles in the Laws of the Physical World*; see People Physics Scientific Culture Website (http://www.peoplephysics.com/physics-laws5.html).

[‖]Lehmann, U. (1977), Stochastic principles in the temporal control of activity behaviour, *Int. J. Chronobiol.* **4**(4), 223–266.

[**]Meyer, K. (1979), Calculation of the spectral density of neutron flux fluctuations due to stochastic vibrations of control elements in a nuclear reactor. Part I. Principles and application to a central control organ, *Kernenergie* **22**(3), 86–90.

variables.* See also STOCHASTIC CONTROL PRINCIPLE; (STOCHASTIC DYNAMICAL PROGRAMMING PRINCIPLE).

PRINCIPLE OF STOCHASTIC MODELING [mathematics]   Method toward the development of quantitative techniques of describing due to effectiveness of heterogeneity and its utility in quantitative study of the uncertainty of the evaluation.[†]

PRINCIPLE OF STRONG INDUCTION. [mathematics]   Let D be a subset of the nonnegative integers Z^* with the properties that (1) the integer 0 is in D and (2) any time that the interval $[O, n]$ is contained in D, one can show that $n + 1$ is also in D. Under these conditions, $D = Z*$. See also PRINCIPLE OF TRANSFINITE INDUCTION, PRINCIPLE OF WEAK INDUCTION[‡]

PRINCIPLE OF STRUCTURAL LAG [psychology] Sociologists Mathilda White Riley and Anne Foner and social psychologist Robert L. Kahn developed description to define policies and practices adapted to the behavior of earlier generations continuing to be enforced, even though the behavior of later generations had changed.[§]

PRINCIPLE OF STRUCTURAL PROTECTION [engineering]   See BUSBAR PROTECTION PRINCIPLE.

*Rico-Ramirez, V. and Diwekar, U. M. (2004), Stochastic maximum principle for optimal control under uncertainty, *Comput. & Chem. Eng.* **28**(12), 2845–2849.

[†]Chilingar, G. V. and Katz, S. (1996), Stochastic modeling and geostatistics-principles, *J. Petrol. Sci. Eng.* **15**(2/4), 396.

[‡]Séroul, R. (2000) Reasoning by Induction." in Programming for Mathematicians. Berlin: Springer-Verlag, pp. 22–25.

[§]Hu, Z. Xiong, K., and Wang, X. (2005), Study on interface failure of shape memory alloy (SMA), reinforced smart structure with damages, *Acta Mech. Sinica* **21**(3), 286–293.

PRINCIPLE OF STRUCTURE–ACTIVITY RELATIONSHIPS (SAR) [chemistry]   Small organic molecules bind to proximal subsites of a protein and are identified, optimized, and linked together to produce high-affinity ligands. Amount of chemical synthesis and time required for the discovery of high-affinity ligands are useful in target-directed drug research. Discovers high-affinity ligands with experimentally derived information. Although the small fragment molecules might bind only in the micromolar to millimolar range to the target protein, the affinity of a linked molecule is, in principle, the product of the binding constants of the individual fragments plus a term that accounts for the changes in binding affinity that are due to linking.[¶]

PRINCIPLE OF SUBJECTIVE FREEDOM [psychology]   See EMANCIPATION PRINCIPLE.

PRINCIPLE OF SUBORDINATION OF INDIVIDUAL TO GENERAL INTEREST [psychology] (Henri Fayol, 1841–1925, French industrialist)   The individual should subordinate self-interest to the general good.[‖] See also FAYOL'S PRINCIPLES OF MANAGEMENT.

PRINCIPLE OF SUBSIDIARITY [psychology] (1) One of the roles of international rules is to ensure as much as possible the cohesion and unity of the whole, by taking into account two conditions: (a) the need to promote our common good or goods; and (b) the need to preserve the greatest possible autonomy for each of the components. Leads to seeking a distribution of jurisdiction among the various levels of authority or responsibility, but it gains to be enhanced by the concept of active subsidiarity in the increasingly frequent situations where jurisdiction is necessarily shared; No major problem, whether related to ensuring security or essential needs, or management of the biosphere, can be taken at only one level of responsibility; this

[¶]Anon. (2000), Optimizing the in vitro activity: SAR by NMR, in *The Principle of SAR by NMR Approach*, CRC Press, p. 239, Fig. 81.

[‖]Wood, M. C. (2002), *Henri Fayol: Critical Evaluations in Business and Management*, Taylor & Francis, p. 8.

therefore requires cooperation among the different levels and leads to the exercise of shared responsibility.* (5) In the practice of active subsidiarity, decisions are taken at the lowest possible level in compliance with the duty to achieve a given result. The definition of this mandatory result is the fruit of collective elaboration. Active subsidiarity thus serves as a foundation for the necessary cooperation of power exercised at different levels.[†] See also PRINCIPLE OF RESPONSIBILITY.

PRINCIPLE OF SUBSIDIARITY [psychology] (*Novarum* of 1891 by Pope Leo XIII) (1) Matters ought to be handled by the smallest (or, the lowest) competent authority.[‡] (2) A central authority should have a subsidiary function, performing only those tasks that cannot be performed effectively at a more immediate or local level; government should undertake only those initiatives that exceed the capacity of individuals or private groups acting independently.[¶] (3) Based on the autonomy and dignity of the human individual, and holds that all other forms of society, from the family to the state and the international order, should be in the service of the human person. Subsidiarity assumes that these human persons are inherently social beings, and emphasizes the importance of small and intermediate-sized communities or institutions, like the family, the church, and voluntary associations, as mediating structures that empower individual action and link the individual to society as a whole.[¶] (4) A middle course between the excesses of laissez-faire capitalism on one hand and the various forms of totalitarianism, which subordinate the individual to the state, on the other hand.[¶]

PRINCIPLE OF SUBSTANCE STABILITY [biology, mathematics] The law of temporal hierarchies of the biological world allows us to pick out of the biomass quasiclosed thermodynamic systems of a specific hierarchy. The use of this law of nature as applied to supramolecular structures of organisms allows us the opportunity of using the methods of equilibrium supramolecular thermodynamics in the examination of open living systems. It has been proved that the second law of thermodynamics in its classic formulation is easy to apply to specific aspects of living systems in order to make calculations, carried out through methods of chemical, supramolecular, and overall hierarchical thermodynamics. "Diets including 'evolutionary young' animal and vegetable foods stimulate longevity and improve the quality of human life. The degree of 'evolutionary' youth of a food product is determined by its chemical composition and supramolecular structure. The chemical composition and supramolecular structure of a product depend, in their turn, on its ontogenetic and phylogenetic ages. An important quantitative measure of the 'gerontological efficiency' of a food product is the Gibbs function of supramolecular structure formation, which characterizes the thermodynamic stability of its supramolecular structure."[§]

PRINCIPLE OF SUFFICIENCY AND NECESSITY [psychology] Argues for assessment of choices for metadata elements and guidelines for values against stated objectives.[¶] See also PRINCIPLE OF SUFFICIENCT REASON; PRINCIPLE OF PARSIMONY; PRINCIPLE OF REPUTATION THEORY-SELF DUALITY.

PRINCIPLE OF SUFFICIENT REASON [psychology] (Gottfried Wilhelm Leibniz, 1 1646—1716, German polymath) In all philosophy there appears to be three fundamental laws: (1) the *principle of originality*, meaning that everything that has being is simply itself and no other thing; (2) the *principle of polarity*, meaning that everything that exists depends for its being on the existence of what it is not; and (3) the *principle of sufficient reason*,

*Carozza, P. G. (2003), Subsidiarity as a structural principle of international human rights law, *Am. J. Int. Law*, **97**(1), 38–79.
[†]Burnett, D. and Blair, C. (2001), Standards for the medical laboratory-harmonization and subsidiarity, *Clin. Chim. Acta* **309**(2), 137–145.
[‡]*Oxford English Dictionary*.

[§]Gladyshev, G. P. (2006), The principle of substance stability is applicable to all levels of organization of living matter, *Int. J. Mol. Sci.* **7**(3), 98–110.
[¶]Svenonius, E. (2000), *The Intellectual Foundation of Information Organization*, MIT Press, p. 75.

meaning that for everything existing there is a sufficient reason for why it is what it is, and why it is not what it is not. The principle of polarity has two "roots": the principle of polarity in direction and the principle of polarity in opposition. Both roots are phenomenologically centered in the experience of tension. The concept of tension is employed in an investigation of "process and futurity." After this the author (Dunham) gives some aspects of his psychology of tension, discussing the opinions of other philosophers.* (1) Anything that happens does so for a definite reason. (2) It denies that contingent events are really so, rather than a description of our ignorance of their detailed causes. It is therefore strongly linked to ideas of determinism. "Thus the sufficient reason, which needs no other reason, must be outside this series of contingent things, and must be found in a substance which is its cause, and which is a necessary being, carrying the reason of its existence with itself. Otherwise, we would not yet have a sufficient reason where one could end the series."† See also PRINCIPLE OF INSUFFICIENT REASON. (3) There is an adequate reason to account for the existence and nature of everything that could conceivably not exist. See also PRINCIPLE OF INDISCERNIBLES; PRINCIPLE OF SUFFICIENCY AND NECESSITY; PRINCIPLE OF PARSIMONY; PRINCIPLE OF REPUTATION THEORY–SELF-DUALITY.

PRINCIPLE OF SUFFICIENT REASON OF ACTING [psychology]   Every human decision is the result of an object that necessarily determines the human's will by functioning as a motive.‡ See also PRINCIPLE OF SUFFICIENT REASON.

PRINCIPLE OF SUFFICIENT REASON OF BECOMING [physics]   If a new state of one or several real objects appears, another state must have preceded it on which the new state follows regularly.¶ See also PRINCIPLE OF SUFFICIENT REASON.

PRINCIPLE OF SUFFICIENT REASON OF BEING [physics]   The position of every object in space and the succession of every object in time is conditioned by another object's position in space and succession in time.¶ See also PRINCIPLE OF SUFFICIENT REASON.

PRINCIPLE OF SUFFICIENT REASON OF KNOWING [mathematics]   If a judgment is to express a piece of knowledge, it must have a sufficient ground. By virtue of this quality, it receives the predicate true. Truth is therefore the reference of a judgment to something different therefrom.¶ See also PRINCIPLE OF SUFFICIENT REASON.

PRINCIPLE OF SUPERPOSITION [mathematics, physics]   (1) The resultant disturbance at a given place and time caused by a number of waves traversing the same space is the vector sum of the disturbances that would have been produced by the individual waves. separately.§ (2) Two or more solutions to a linear equation or set of linear equations can be added together so that their sum is also a solution. (3) When two or more forces act on a particle at the same time, the resultant force is the vector sum of the two. (4) If a number of independent influences act on the system, the resultant influence is the sum of the individual influences acting separately¶ (5) In all theories characterized by linear homogeneous differential equations, such as optics, acoustics, and quantum theory, the principle that the sum of any number of solutions to the equations is another solution.‖ See also SUPERPOSITION PRINCIPLE; also referred to as *Boltzmann–Hopkinson principle of superposition*.

§Walker, P. M. B. (1999), *Chambers Dictionary of Science and Technology*, Chambers, New York, p. 915.

¶Staquet, S. and Espion, B. (2005), Deviations from the principle of superposition and their consequences on structural behavior, *Shrinkage and Creep of Concrete* American Concrete Inst., SP, SP-227, pp. 67–83.

‖Carter, S. E. and Malocha, D. C. (1997), Finite impulse response utilizing the principle of superposition, *IEEE Trans. Ultrason., Ferroelectrics Freq. Control* **44**(2), 386–398.

*Dunham, A. (1938), The concept of tension in philosophy, *Psychiatry J. Study Interpers. Process.* **1**, 79–120.

†Urban, W. (1898), *The History of the Principle of Sufficient Reason. Its Metaphysical and Logical Formulation*, Princeton Contrib. to Philos, Univ. Press.

‡Schopenhauer, A. (1813), *On the Fourfold Root of the Principle of Sufficient Reason*, dissertation.

PRINCIPLE OF SUPERPOSITION [geology] (William Smith) (1) If one set of strata occurs on top of another in a succession, the upper unit was formed later.* (2) The displacement of any point due to the super-position of wave systems is equal to the sum of the displacements of the individual waves at that point; "the principle of superposition is the basis of the wave theory of light."[†] (3) In a series of stratified sedimentary rocks the lowest stratum is the oldest.[¶] Also referred to as *superposition principle; Huygens' principle of superposition*.

PRINCIPLE OF SUPERPOSITION [physics, mathematics, geology] (1) In linear al-gebra, describes how certain physical quantities behave when occurring at the same place and time. (2) In geology, de-scribes the ordering of rock layers. (3) When a situation is a composition of a number of elementary situations, its amplitude is the linear superposition of the amplitudes of the components. (4) Light emitted from two sources; the net electric field is the sum of the electric field of each source. (5) Total electric field at a point due to the combined influence of the distribution of point charges is the vector sum of the electric field intensities that the individual point charges would produce at that point if each acted alone.[‡] (5) In a linear electrical network, the voltage or current in any element resulting from several sources acting together is the sum of the voltages or currents resulting from each source acting alone. (6) When two or more forces act on a particle at the same time, the resultant force is the vector sum of the two. (7) If a number of independent influences act on the system, the resultant influence is the sum of the individual influences acting separately. (8) The sum of any number of solutions to the equations is another solution if characterized by linear homogenoeous differential equations such as optics, acoustics, and quantum theory. (9) Demonstrates what happens when waves meet and how a standing wave is formed by the superposition of identical traveling waves in opposite directions. (10) To determine the stress in a member due to a system of applied forces, the system can be split up into several component forces and their moments and reactions added in order to calculate the total stress.[§] See also BOLTZMANN–HOPKINSON PRINCIPLE OF SUPERPOSITION; QUANTUM MECHANICAL SUPER-POSITION PRINCIPLE; OPTICAL SUPERPOSITION PRINCIPLE.

PRINCIPLE OF SUPERPOSITION OF WAVES [physics] If different sources produce waves that have nonvanishing amplitudes at any point in space, then the amplitude of the re-sultant waves at that point is the linear sum of the amplitudes of each wave.[¶]

PRINCIPLE OF SUPERVISORY TECHNIQUES [organizational psychology] Since people, tasks, and organizational environment vary, techniques of supervision will be most effec-tive if appropriately varied. See also PRINCI-PLE OF PLANNING.[‖]

PRINCIPLE OF SYMMETRIC BRACKET INVARI-ANCE [physics] Used to derive both the quantum operator commutator relation and the time-dependent Schrödinger equation. A c-number dynamical equation is found, which leads to the second quantized field theory of bosons and fermions.[**] See also MOMENTUM PRINCIPLE.

*Stiegeler, S. E. (1977), *A Dictionary of Earth Sciences*, Pica Press/Universe Books, New York, p. 268.

[†]*American Heritage Dictionary of the English Language*, 4th ed., Houghton Mifflin, New York.

[‡]Walker, P. M. B. (1999), *Chambers Dictionary of Science and Technology*, Chambers, New York, p. 915.

[§]Wielandt, E., Danilenko, V. A., and Mikulyak, S. V. (2004), On the correctness of the use of the superposition principle in calculations of the wave fields from an explosion of two charges, *Dopovidi Natl. Akad. Nauk Ukraini.* **1**, 105–110.

[¶]Wielandt, E. (2006), The superposition principle of waves not fulfilled under M. W. Evans' O(3), hypothesis, *Physica Scripta* **74**(5), 539–540.

[‖]Koontz, H. and O'Donnell, C. (1968), *Principles of Management*, 4th ed., McGraw-Hill.

[**]Garavaglia, T. and Kauffmann, S. K. (2002), Principle of symmetric bracket invariance as the origin of first and second quantization, *Int. J. Theore. Phys.* **41**(4), 593–611.

PRINCIPLE OF SYMMETRIC CRITICALITY

[physics]   Given a group action on a space of fields, one can consider the restriction of an action functional S to the group invariant fields to obtain the reduced action $\wedge S$. Palais' PSC asserts that, for any group invariant functional S, critical points of $\wedge S$ within the class of group invariant fields are (group invariant) critical points of S. As Palais emphasized, PSC need be neither well-defined nor valid. Under hypotheses that guarantee PSC makes sense, he goes on to give necessary and sufficient conditions for the validity of PSC in a variety of settings.

DEFINITION. A group action obeys the Principle of Symmetric Criticality (PSC) if about each $x \in M$ there exists a G-invariant open neighborhood U and a G-invariant chain on U such that, for any G-equivariant Lagrangian λ, the reduced field equations are equivalent (in the sense of Appendix B) to the Euler-Lagrange equations of the reduced Lagrangian $\hat{\lambda}$,

$$E(\lambda)(g(\hat{q})) = 0 \Longleftrightarrow E(\hat{\lambda})(\hat{q}) = 0.$$

Also known as Palais' Principle of Symmetric Criticality (PSC).[*]

PRINCIPLE   OF   SYMMETRY [psychology]
Judgments of distance and orientation between the relationship as of learned and inferred points can be made equally. Acquiring summary knowledge of a series of points allows for moments between points that have not been experienced.[†] (2) Describes the instance where the whole of a figure is perceived rather than the individual parts that make up the figure.[‡] See also PRINCIPLE OF CONTIGUITY; PRINCIPLE OF PROXIMITY; GESTALT PRINCIPLE OF CONTINUITY; GESTALT PRINCIPLE OF PROXIMITY; GESTALT

PRINCIPLE OF CONTIGUITY; GESTALT PRINCIPLE OF CLOSURE; GESTALT PRINCIPLE OF AREA; GESTALT PRINCIPLE OF SYMMETRY; GESTALT PRINCIPLE OF SIMILARITY.

PRINCIPLE OF SYNOPTIC ANTAGONISM [Psychology]
Deals with the pharmacology of agonists that activate seven-transmembrane receptors to produce direct bias (activation of selected portions of the possible receptor signaling capability) and indirect bias (ligands that allosterically modify the signaling of the receptor to endogenous agonists to select specific behaviors of the receptor).[§]

PRINCIPLE OF THE ACTIVATED DOUBLE BOND
[chemistry]   (1) A bond in which two atoms share two electron pairs; occurs in unsaturated compounds; represented by an equal sign, as in $H_2C = O$; thus, double bonding. (2) A shift occurring between a pair of valence bonds during a chemical reaction, such as butene 1 ($H_2C = CHCH_2CH_3$), to butene 2 ($H_3C-CH-CH-CH_3$). (3) To cause or accelerate a reaction. (4) To treat charcoal, carbon, or the like to improve their capacity for adsorbing impurities. (5) To induce activity in a system that is static, as in neutron activation of radioactivity. (6) To purify sewage by treating it with air and bacteria. (7) To start the operation of an electrical device, usually by applying an enable signal or power to it. (8) To use liquid as an additive in order to make a cell or battery operational. (9) To apply material to the surface of a cathode to create or increase cathode emissivity. (10) To cause a missile or explosive to be in an active state, ready for firing or explosion. (11) To assign a ship, troop unit, or the like to active service after it has previously been in inactive or reserve status. (12) To bring to active status something that had not previously been in operation.[¶]

[*]Fels, M.E. and Torre, C.G. (2001) The Principle of Symmetric Criticality in General Relativity http://arxiv.org/pdf/gr-qc/0108033
[†]Brysch, K. and Dickinson, J. (1996), Studies in cognitive maps: The equiavailability principle and symmetry, *Environ. Behav.* **28**(2), 183–203.
[‡]Jaeger, F. M. (n.d.), 1918 *Lectures on the Principle of Symmetry and Its Practical Application in all Natural Sciences*. London Chemical News office

[§]Black, J. (1989) Drugs from emasculated hormones: The principle of synoptic antagonism, *Science*, **245**, 486.
[¶]Stetter, H. (1976), New synthetic methods. The catalyzed addition of aldehydes to activated double bonds—a new synthesis principle, Angew. Chem. 88(21), 695–704.

PRINCIPLE OF THE AEROSOL PARTICLE MASS ANALYZER (APM) [physics] Aerosol particles are first passed through a bipolar charger in which bipolar ions are generated by an ion source such as ^{241}Am or ^{83}Kr; the particles are then brought to an equilibrium charge state, and introduced into a thin annular gap between coaxial cylindrical electrodes that rotate at the same angular velocity. In the gap, the particles migrate in the radial direction under the influence of centrifugal and electrostatic forces, and only those particles for which the two forces balance can exit the electrodes. The particle mass that can be classified by this principle ranges roughly from 0.01 to 500 fg (1 fg $= 10^{-15}$ g).*

PRINCIPLE OF THE ARGUMENT [mathematics] If $f(z)$ is a meromorphic function inside and on some closed contour C, with f having no zeros or poles on C, then, where N and P denote, respectively, the number of zeros and poles of $f(z)$ inside the contour C, with each zero and pole counted as many times as its multiplicity and order, respectively. This theorem assumes that the contour C is simple, that is, without self-intersections, and that it is oriented counterclockwise.[†] See also ARGUMENT PRINCIPLE; CAUCHY'S ARGUMENT PRINCIPLE.

PRINCIPLE OF THE ARGUMENT [mathematics] If $f(z)$ is meromorphic in a region R enclosed by a contour γ, let N be the number of complex roots of $f(z)$ in γ, and P be the number poles in γ, with each zero and pole counted as many times as its multiplicity and[‡] order, respectively. Then

$$N - P = \frac{1}{2\pi i} \int_\gamma \frac{f'(z)}{f(z)}$$

Defining $w \equiv f(z)$ and $\sigma \equiv f(\gamma)$ gives

$$N - P = \frac{1}{2\pi i} \int_\sigma \frac{dw}{w}.$$

*Hosokawa, M., Nogi, K., and Naito, M. (2007), *Nanoparticle Technology Handbook*, Elsevier, p.16.
[†]Ahlfors, L. (1979), *Complex Analysis*, McGraw-Hill.
[‡]Duren, P.; Hengartner, W.; and Laugessen, R. S. (1996) The Argument Principle for Harmonic Functions. Amer. Math. Monthly 103, p. 411–415.

PRINCIPLE OF THE CENTRE-PERIPHERY [mathematics] Aka Principle of Nucleation The translation space is the field that generates a point, which can be called centre or nucleus, which all the actors try to approach in ord to improve their strategic advantage.[§] See also SPATIAL PRINCIPLE; PRINCIPLE OF TRANSLATION QUANTITATIVITY; PRINCIPLE OF COMPOSITION OF THE TRANSLATIONS; PRINCIPLE OF NUCLEATION; UNIFIED PRINCIPLE OF ACCUMULATED ADVANTAGES.

PRINCIPLE OF THE AUTONOMY OF THE WILL [psychology] See EMANCIPATION PRINCIPLE.

PRINCIPLE OF THE DEFLECTION BALANCES [chemistry] The deflection or movement can be calibrated to read in terms of mass.[¶]

PRINCIPLE OF THE "DETAINED" ELECTRON [chemistry] Includes an extensive body of phenomena, from the formation of three-electron bonds, to ion pairing, to the distonic stabilization of ion radicals at the expense of separation between their spins and charges. In the donor–acceptor interaction, the acceptor provides its lowest unoccupied molecular orbital (LUMO) and the donor participates at the expense of its highest occupied molecular orbital (HOMO). These orbitals are frontier orbitals. In the corresponding ion radicals, the distribution of an unpaired electron proceeds, naturally, under frontier orbital control.[∥]

PRINCIPLE OF THE DOPPLER EFFECT [physics] See DOPPLER PRINCIPLE.

PRINCIPLE OF THE EQUALITY OF MOMENTS [physics] If any number of pressures act in the same plane, and any point is taken in that plane, and perpendiculars are drawn from it in the directions of all these pressures,

[§]Jimenez-Contreras, E. (1992) Las revistas cientificas: el centro y la periferia. Revista Espanola de Documentacion Cientifica, **15**, 174–182.
[¶]Dollimore, D., (1990), Thermoanalytical instrumentation, *in Analytical Instrumentation Handbook*. Ewing, GW, Ed., Marcel Dekker. New York, p. 905–960.
[∥]Todres, Z. V. (n.d.), 2002 Basic principles of organic ion radical reactivity, in *Organic Ion Radicals: Chemistry and Applications*, chap. 3.

produced if necessary, and if the number of units in each pressure is then multiplied by the number of units in the corresponding perpendicular, then this product is called the *moment* of the pressure *about* the point from which the perpendiculars are drawn, and these moments are said to be measured from this point.[*]

PRINCIPLE OF THE EQUIPARTITION OF ENERGY [chemistry]   The total energy of a molecule in the normal state is divided up equally between in different capacities for holding energy.[†]

PRINCIPLE OF THE EXCLUDED THIRD [psychology]   Since people, tasks, and organizational environment vary, techniques of supervision will be most effective if appropriately varied.[‡]

PRINCIPLE OF THE FORM IMPULSE. (FORMTRIEB) [psychology]   Sensuous activity accompanies the perception of an object, and the pleasure afforded by this activity. Pleasure felt is the more keen, the more energetic and at the same time unconscious of painful effort the activty is. Aesthetic pleasure on its formal side means, therefore, the natural and harmonious functioning of the perceptive organs forced on us by the structural aspect of an object. There are three sources of formal aesthetic pleasure experience of seeing, hearing and speech. Theses represent distinct sense-stresses. Rooted in this special capacities is the one universal category of the beautiful in respect of form namely the ideal form of the beautiful.[§]

PRINCIPLE OF THE HOUGH TRANSFORM [physics]   An object image might be mapped by means of a sensor, such as a camera, into a 2D digital image. The captured image can be processed to extract image features such as object edges and vertices. Applying 3D coordinate transforms such as translation, rotation, and scaling, one can generated an image by a subsequent perspective transform.[¶] See also MACHINE VISION PRINCIPLE.

PRINCIPLE OF THE INTERNATIONAL ECONOMIC ORDER [economics]   Principle of exit, or liquidity, as the basic principle of the international economic order, so as to give your asset holders maximum freedom to move in and out of markets according to short-run profit considerations. A flotilla of international organizations that resemble cooperatives of member states and confer the legitimacy of multilateralism, but that you can control according to the principle of unilateral cooperation. ("We'll cooperate provided we get to set the rules and can veto outcomes we don't like.") In particular, you need some of these organizations to operate a bailout mechanism that gives priority to your creditors and displaces losses from periodic panics on to the citizens of the borrowing country, while at the same time advancing your agenda of worldwide liberalization, privatization and free-capital mobility via conditionality on the bailouts.[‖]

PRINCIPLE OF THE MAXIMUM [mathematics, logic]   (1) For every proposition p, either p or not p, implies p.[**] (2) For a nonconstant complex analytic function defined in a domain, the absolute value of the function cannot attain its maximum at any interior point of the

[*]Ferreira, R. (1963), Principle of electronegativity equalization. I. Bond moments and force constants, *Trans. Faraday Soc.* **59**, 1064–1074.

[†]Walker, P. M. B. (1999), *Chambers Dictionary of Science and Technology*, Chambers, New York, p. 915.

[‡]Koontz, H. and O'Donnell, C. (1968), *Principles of Management*, 4th ed., McGraw-Hill.

[§]Meyer, T. (1904) Formprinzip des Schonen. *Arch. F. Syst. Philos.* **10**, 338–394.

[¶]Lam, S. T., Cho, P. S., Marks, R. J. 2nd, and Narayanan, S. (2004), Three-dimensional seed reconstruction for prostate brachytherapy using Hough trajectories, *Phys. Med. Biolo.* **49**(4), 557–569.

[‖]Vitzthum, W. G. (1949), Economic equality: A principle of the international economic order, Sicht der Bundesrepublik Deutschland. Baden Baden: Nomos Verlagsgesellschaft. (Integration Europas find Ordnung der Weltwirtschaft; Bd. 1), p. 249.

[**]Keynes, J. N. (2008), *Studies and Exercises in Formal Logic*, Read Books, p.580.

domain. (3) A nonconstant analytic function f defined on a domain D cannot assume its maximum modulus on the interior of D. If D is closed, bounded, and simply connected, the maximum modulus theorem states that the maximum value of $|f|$ occurs on the boundary of D.* See also MAXIMUM PRINCIPLE.

PRINCIPLE OF THE MIDDLE WAY [organizational psychology] Inclining to embrace both propositions, finding them each to have merit.[†]

PRINCIPLE OF THE MINIMUM [ecology] The rate at which a plant grows, the size it attains, and its overall health all depend on the amount available to it of the scarcest of its essential nutrients.[‡]

PRINCIPLE OF THE MINIMUM [mathematics, physics] (1) A nonconstant, nonzero analytic function f defined on a domain D cannot assume its minimum modulus on the interior of D. If D is closed, bounded, and simply connected, the maximum modulus theorem states that the maximum value of $1/|f|$, and hence the minimum value of $|f|$, occurs on the boundary of D. (2) Any of various principles formulated to account for the observation that certain important quantities tend to be minimized when a physical process takes place (e.g., per Maupertuis' principle of least action or Gauss' principle of least constraint).[§] See also PRINCIPLE OF LEAST ACTION.

PRINCIPLE OF THE OBJECTIVE OF STAFFING [organizational psychology] The objective of staffing is to ensure that organization roles are filled by those qualified employees who are able and willing to occupy them.[¶] See also (STAFFING PRINCIPLES;) PRINCIPLES OF JOB DEFINITION; PRINCIPLES OF STAFFING; PRINCIPLE OF MANAGERIAL APPRAISAL.

PRINCIPLE OF THE POOR LAW OF 1834 [psychology] Premise that poor people lacked strong or moral character and that public assistance is for all able-bodied citizens except those in public institutions.[‖]

PRINCIPLE OF THE PROPER MODULATING FUNCTION [psychology] (1) Mental operations narrow down a vast array of information available in the environment to a processable number of items to enable information intake, absorption, and digestion. This function can be achieved by preference-selecting (Jpn. *iitoko-dori*) behavior.** (2) The intellectual–physiological modulating function serves as an "automatic alarm device" that changes, stops, and creates objects and means for information behavior through such signals as "excitement–sleep," "shortage–saturation," and "memory–oblivion."[††]

PRINCIPLE OF THE "RELEASED" ELECTRON [chemistry] Deals with ion radicals. All the structural components are coplanar or almost coplanar. In this case, spin density appears to be uniformly or symmetrically distributed over the molecular framework. Spin density distribution has a decisive effect on the thermodynamic stability of ion radicals. In

*Volin, Yu. M. Ostrovskii, G. M., and Slin'ko, M. G. (1963), Use of the principle of the maximum to determine the optimum range for exothermal processes. *Kinetika Kataliz* **4**(5), 760–767.

[†]Nisbett, R., Peng, K., Choi, I., and Norenzayan, A. (2001), Culture and systems of thought: Holistic versus analytic cognition, *Psychol. Rev.* **108**(2), 291–310.

[‡]Walker, P. M. B. (1999), *Chambers Dictionary of Science and Technology*, Chambers, New York, p. 662.

[§]Richardson, I.W. (1969), On the principle of minimum entropy production, *Biophys. J.* **9**(2), 265–267.

[¶]Anon. (2007) Principles of staffing, *MBA—MyNotes* (pgdba.blogspot.com/2007/11/principles-of-staffing.htm).

[‖]Bloy, Marjie (1967) The Poor Law Amendment Act: 14 August 1834 National univ of Singapore (http://www.victorianweb.org/history/poorlaw/plaetextihtml)

**Nagaya, T. (2007), Constraints-programmed macro-model of information selection - limits of information-processing and the basic principle of information behavior, *Electron. Commun. Jpn.* (pt. III: *Fund. Electron. Sci.*) (Wiley Periodicals) 79(9), 1–14.

[††]Salinas, E. and Thier, P. (2000), Gain modulation: A major computational principle of the central nervous system, Neuron **27**(1), 15–21.

general, the stability of ion radicals increases with an enhancement in delocalization.*

PRINCIPLE OF THE SALUTOGENETIC TRIAD [sociology] (Aaron Antonovsky) Balanced diet, exercise, and physical hardening.[†] See also PRINCIPLE OF SOMATOMENTAL BALANCE; SALUTOGENESIS PRINCIPLE; COMFORT-THROUGH-DISCOMFORT PRINCIPLE.

PRINCIPLE OF THE TRANSMISSION ELECTRON MICROSCOPE [engineering] (1) Considerations regarding the choice of conditions for recording images. Choice of accelerating voltage, apertures, specimen stage/holder, magnification, focusing, magnification calibration, resolution tests, image intensifiers/TV displays, microscope maintenance and photography.[‡] (2) To form the final electron microscope image on a transparent transmission phosphor screen mounted beneath the normal fluorescent screen and camera. The first dim image is formed on the transparent screen, and all subsequent operations are performed outside the microscope vacuum. The primary image is coupled optically to a photocathode that emits electrons when excited by light. A lens of very wide aperture or glass fiberoptics are used to make the optical coupling as efficient as possible to minimize the loss of light from the primary image during the coupling process. Electrons emitted by the first photocathode are accelerated by a high potential of several thousand volts and focused onto a second cathode that emits several secondary electrons for each primary electron striking it.[§] (3) Electron

has a wavelength several hundred times shorter than light and the use of electrically charged particles permit focusing as with a light beam.[¶]

PRINCIPLE OF THE UNIQUENESS OF FREE FALL [mathematics] The world line of a freely falling test body is independent of its composition or structure. The uniqueness of free-fall trajectories allows one to regard spacetime as filled with a set of curves, the test body trajectories, which are unique aside from parameterization. When translated into Newtonian language, the uniqueness of free-fall states than any two test bodies must fall with the same acceleration in a given external gravitational field. A short form of the two principles of equivalence is the strong principle of equivalence and the weak principle of equivalence. See EQUIVALENCE PRINCIPLE; STRONG EQUIVALENCE. PRINCIPLE.

PRINCIPLE OF THERMALLY STIMULATED CURRENT (TSC) [chemistry] the objective is to orient polar molecules or pendant polar groups of macromolecules by applying a high voltage field at a high temperature and then quenching the material to a much lower temperature where molecular motion occurs. After this polarization, the material is heated at a constant rate, causing it to depolarize, thereby creating a depolarizing current. This thermally stimulated current can be related directly to molecular mobility, indicating the physical and morphological structure of materials.[‖]

PRINCIPLE OF THIRDS [mathematics] An image can be divided into nine equal parts by two equally spaced horizontal lines and two equally spaced vertical lines. The four points formed by the intersections of these lines can be used to align features in the

*Todres, Z. V. (2002), chap. 3. Basic principles of organic ion radical reactivity, in *Organic Ion Radicals: Chemistry and Applications*, p. 129. CRC Press

[†]Mundt, C. (2002), Psychological perspectives for the development of future diagnostic systems, *Psychopathology* **35**, 145–151.

[‡]Goldstein, J. I. (1978), Principles and practice of STEM (scanning transmission electron microscope), microanalysis, *Proc. Annual Conf. Microbeam Analysis Society*, 13 Paper T3.

[§]Pond, R. C. (1984), Review of the principle contrast effects observed at interphase boundaries using transmission electron microscopy, *J. Microsc.* **135**(3), 213–240.

[¶]Chai, B. F. and Tang, X. M. (1987), Ultrastructural observations on experimental fractures treated with the principle of promoting blood circulation and relieving stasis. A transmission electron microscopy study, *chinese J. Modern Devel. Traditional Med.* **7**(7), 390, 417–419.

[‖]Alexander, K. S., Riga, A. T., and Haines, P. J. (n.d.), chap. 15. Thermoanalytical instrumentation and applications, in *Ewing's Analytical Instrumentation Handbook*, 3rd ed., chap. 15.

photograph. Proponents of this technique claim that aligning a photograph with these points creates more tension, energy, and interest in the photo than simply centering the feature would.*

PRINCIPLE OF TIME INVARIANCE [mathematics] For a nonconstant complex analytic function defined in a domain, the absolute value of the function cannot attain its maximum at any interior point within the domain.[†]

PRINCIPLE OF TIME GATING [optics] By monitoring "time slices" at increasing invervals after a pulse, e.g., every 100 ns, the intensity of the image can be monitored as a function of time. (see also OPTICAL CONVERGENCY PRINCIPLE).[‡]

PRINCIPLE OF TIMING [organizational psychology] If plans are structured to provide a network of derivative plans in sequence, it will be more effective to attain enterprise objectives.[§] See also PRINCIPLE OF PLANNING

PRINCIPLE OF TIME DOMAIN [computer science] See DATA-SENDING PRINCIPLE.

PRINCIPLE OF TISSUE DETERMINATION [biology] (1) Certain tissues of the adult human such as blood, skin, and gut undergo continual turnover and contain a small population of dividing stem cells capable of replenishing the tissue losses as they occur. Other tissues, most notably muscle and liver, contain dormant stem cells that only undergo this process in response to tissue loss or damage while others, such as brain, appear to

lose this regenerative capacity completely. In all cases these cells, known as *somatic stem cells*, exist in exceedingly small numbers and are able to differentiate into only the tissue to which they belong. (2) It was previously believed that when a cell differentiates the characteristics of a particular tissue type (muscle, blood, gut, etc.), it also forfeits the ability to reverse this process and thereby return to its undifferentiated state. (3) In many cases, different cell types must position themselves very precisely so that they can interact appropriately with each other, for example, nerve cells and the muscles that they innervate—or the various components of the eye (lens, retina, optic nerve)[¶]

PRINCIPLE OF TOTAL CONTROL [engineering] The operator must be able to "tweak" a device to the smallest level of precision allowed by the hardware and be able to slew a device throughout its entire range in a "reasonable amount of time." These "tweak" and "slam" rules imply that the "tweak" ideally should have direct access to the raw hardware units of the controlled device.[‖] See also OVERSHOOT PRINCIPLE; PRODUCTION PRINCIPLE; PRINCIPLE OF INSTANT GRATIFICATION.

PRINCIPLE OF TOTAL DIFFERENTIAL [physics] See also D'ALEMBERT'S PRINCIPLE; TOTAL DIFFERENTIAL PRINCIPLE; DIFFERENTIAL PRINCIPLE OF D'ALEMBERT; VARIATIONAL PRINCIPLE; HAMILTON PRINCIPLE.

PRINCIPLE OF TRAINING [kinesiology] (Dr. Hans Selye) (1) Applied to physical stress in three phases: (a) the *alarm phase*—the body will not like the overloaded stress placed on it and will begin to take drastic measures to combat it (b) the *Resistance Phase*—the body will try to resist the stress and (c) the *exhaustion phase*—the body will inevitably

---

*Thomas, J. (2008), Focal point. The principle of thirds: Blame the Greeks, *Essex Photo News* p. 3.
[†]Kuz'menko, V. A. (2003), Physical origin of nonlinear phenomena in optics, Los Alamos National Laboratory, *Preprint Archive—Physics*.
[‡]Stanley W. botchway, S.W., Charnley, M., Haycock, J.W., Parker, A.W., Rochester, D.L., Weinstein, J.A., and Williams, J.A. G. (2008) Time-resolved and two-photon emission imaging microscopy of live cells with inert platinum complexes. *Proc Natl Acad Sci*, **105**(42): 16071–16076.
[§]Koontz, H. and O'Donnell, C. (1968), *Principles of Management*, 4th ed., McGraw-Hill.

[¶]Mikhailov, V. P. (1972), Tissue classification and the phenomenon of metaplasia in light of the principle of tissue determination, *Arkhiv Anat. Gistol. Embriol.* **62**(6), 12–34.
[‖]Jianbo, P. (1999), Study on system optimizing of total quantity control plan of wastewater discharge from nonferrous metals enterprises, Proc. Global Conf. Environmental Control in Mining and Metallurgy (GME '99, Global Metals Environment), Beijing, China, May 24–27, 1999, pp. 362–366.

become exhausted if it doesn't receive rest from the stress. This leads to the belief there must be periods of low intensity between those overloaded stresses taxing the body.[*] (2) The muscles of the body must be forced to work against greater resistance than that to which they are normally accustomed. The resistance may be isometric, isokinetic, or isotonic, depending on the purpose of exercise. This relates to strengthening the heart muscle by causing it to do more work than normal. During exercise, there is an increased return of blood from the veins, which gives the heart resistance to beat against. It is this resistance or loading that causes the heart to develop.[†] (3) There is progression that states that if physical condition is to improve, a person must be exposed to new, higher levels of overload. (4) In order to force "overcompensation," the stress placed on the body must be an overload.[‡] Also known as *principle of individual differences; principle of overcompensation; principle of overload; specific adaptation to imposed demand (SAID) Principle; general adaptation syndrome (GAS) principle; use / disuse principle; specificity principle; three-phase principle*

PRINCIPLE OF TRAINING OF THE MAXIMUM [mathematics] (Rev. Thomas Bayes, 1763). See BAYES PRINCIPLE.

PRINCIPLE OF TRANSFER CODING [automation] Compression noise is a linear tranform of quantization noise, which is usually generated during quantization of tranform coefficients using uniform scalar quantizers. The quantisation noise may bot distribute uniformly as distributions and quantization step sizes vary among tranform coefficients.

Compression noise has a jointly normal distribution, which enables its calculation to have reasonable computation complexity. [§]

PRINCIPLE OF TRANSFINITE INDUCTION [mathematics] (Gerhard, Gentzen. 1909—1945, German mathematician) Over a given decidable relation up_y states, for any formula $M(u)$, $V_u\{V_y[up_y + M(y)] + M(u)\}$) implies $V_uM(u)$. Let $A(x_o, \ldots, x, - \sim)$ denote $k_22 + x_opx_lp, \ldots, oxkP_l$ and let z be the tree given by $A(x)$ as above. For given Q, let $M(u)$ denote $VxQ(x * u)$. Then it is easy to verify that the principle of bar induction over z applied to the formula Q follows from the principle of transfinite induction over p applied to the formula M. Conversely, for given M, let $Q((x_o, \ldots, x, - \sim)$ denote $k > 0 + M(xk-,)$. Then transfinite induction over p for M follows from bar induction over z for Q. See also PRINCIPLE OF [FINITE INDUCTION] [PRINCIPLE OF STRONG INDUCTION] [PRINCIPLE OF WEAK INDUCTION.]

Let A be a well-ordered set and let $P(x)$ be a propostion with domain A. A proof by transfinite induction uses the following steps:

1. Demontrate $P(O)$ is true.
2. Assume $P(b)$ is true for all $b < a$.
3. Prove $P(a)$, using the assumption in (2).
4. The $P(a)$ is true for all a EA.

(1) The use of the principle of mathematical induction in a proof. Induction used in mathematics is often called mathematical induction.[¶] (2) Transfinite induction, like regular induction, is used to show a property P (n) holds for all numbers n. The essential difference is that regular induction is restricted to the natural numbers Z, which are precisely the finite ordinal numbers. The normal inductive step of deriving $P(n + 1)$ from $P(n)$ can fail due to limit ordinals. See also PRINCIPLE OF MATHEMATICAL INDUCTION, PRINCIPLE OF STRONG INDUCTION, PRINCIPLE OF

[*]Glowacki, S. P., Martin S. E., Maurer, A. Baek, W. Green, J. S., and Crouse, S. F. (2004), Effects of resistance, endurance, and concurrent exercise on training outcomes in men, *Med. Sci. Sports Exercise* **36**(12), 2119–2127.
[†]Page, P. and Ellenbecker, T. S., 2003, The scientific and clinical application of elastic resistance, *Human Kinet.* p. 352.
[‡]Hatfield, F. C. (n.d.), *Popular Training Systems: Are They Really "Systems?"* (http://drsquat.com/content/knowledge-base/popular-training-systems-are-they-really-systems).

[§]Li, D. and Loew, M. (2005) Closed-form compression noise in images with known statistics. *Proc. International Society for Optical Engineering (SPIE)*, **5749**, p. 211.
[¶]Buck, R. C. (1963) Mathematical Induction and REcursive Definitions. *Amer. Math. Monthly* **70**, 128–135.

TRANSFINITE INDUCTION, PRINCIPLE OF WEAK INDUCTION.*

PRINCIPLE OF TRANSFORMATION GROUPS (Edwin Thompson Jaynes, 1922–1998, American physicist) One of the logical principles for assigning prior probability distributions. Equivalent states of knowledge should be assigned equivalent epistemic probabilities. Within the classical interpretation of probability theory. Joseph Bertrand introduced the Bertrand paradox in his work Calcul des probabilités (1888) to demonstrate probabilities may not be well defined if the mechanism or method that produces the random variable is not clearly defined. In 1973, Jaynes argued, "random" means that we do not know more about the problem than was stated in the question. This "maximum ignorance" principle is shown to imply that there is really a unique solution, which can be found using transformation invariances. See also MAXIMUM IGNORANCE PRINCIPLE.[†]

PRINCIPLE OF TRANSLATIN QUANTITATIVITY [mathematics] The translation, T, is equal to the variation of the qualities or attributes of the actors, $Q(x)$, as they move in the translation space, x. The translation is therefore the derivate or gradient of the function quality with respect to the coordinates of the translation space. $T(x) = \frac{dQ(x)}{dx}$ See also SPATIAL PRINCIPLE; PRINCIPLE OF COMPOSITION OF THE TRANSLATIONS; PRINCIPLE OF THE CENTRE-PERIPHERY; PRINCIPLE OR NUCLEATION; UNIFIED PRINCIPLE OF ACCUMULATED ADVANTAGES.[‡]

PRINCIPLE OF TRANSITIVITY [mathematics] Subjective conditionals also seem to obey a rule of transitivity, in that if C → A and A → B, then C → B may be inferred. So, for example. 'If this metallic object were heated. It would expand' and 'If it expanded.

It would shatter its container' toether imply that if the object were heated it would shatter its container.[§] See also PRINCIPLE OF CONTRAPOSITION. See also SIMPLIFICATION PRINCIPLE;

PRINCIPLE OF ULTRASONIC PLASTIC WELDING [engineering, physics] Parts to be welded are clamped together under pressure between an ultrasonic horn and fixture. Area of contact between the two parts is reduced to maximize ultrasonic stress. Ultrasonic shear motion breaks up and disperses the oxides and other contaminants at the interface, and the exposed plasticized metal surfaces form a bond under pressure.[¶]

PRINCIPLE OF UNCERTAINTY [physics] (1) The position and momentum of a particle cannot be simultaneously measured with zero uncertainty. The energy and lifetime of an unstable particle cannot be measured with zero uncertainty.[‖] (2) Increasing the precision of a position measurement decreases the precision of the corresponding momentum measurement, and so on.[**] See also HEISENBERG UNCERTAINTY PRINCIPLE.

PRINCIPLE OF UNDULATORY INVARIANCE [physics, mathematics, chemisty] See INVARIANCE PRINCIPLE.

PRINCIPLE OF UNIFORMITARIANISM [geology] The geologic processes taking place in the present operated similarly in the past and can therefore be used to explain past geologic events.[††] See also PRINCIPLE OF UNIFORMITY.

[§]Urbach, P. (1988) What Is a Law of Nature? A Humean Answer. *The British Journal for the Philosophy of Science*, **39**,(2), 193–209.
[¶]Yan, J. Li, D., Dong, Z., and Yang, S. (1998), Analysis and measurement of acoustic power in plastics ultrasonic welding process, *China Welding*, **7**(2), 106–111.
[‖]Montesinos, M. and Torres del Castillo, G. F. (2007), Reply to "Comment on 'symplectic quantization, inequivalent quantum theories, and Heisenberg's principle of uncertainty,' "*Phys. Rev. A Atom. Mole. Opt. Phys.* **75**(6, Pt. B).
[**]Glezer, V. D., Gauzel'man, V. E., and Iakovlev, V. V. (1986), Principle of uncertainty in vision, *Neurophysiology* **18**(3), 307–312.
[††]Belsanova, A. (1967), Actualism and the genesis of ore deposits, *Vest. Ustred. Ustavu Geol.* **42**(2), 81–85.

*Gleason, A. M. (1991) Fundamentals of Abstract Analysis Natick, A. K. Peters, p. 82.
[†]Jaynes, E.T. (1968) Prior Probabilities, IEEE Trans. on Systems *Science and Cybernetics* SSC-4, p.227.
[‡]Ruiz-Banoz, R., Bailon-Moreno, R., Jimenez-Contreras, E., and Courtial, J.P. (1999) Structure and dynamics of scientific networks. Part 1: Fundamentals of the quantitative model of translation. *Scientometrics*, **44**, 217–234.

PRINCIPLE OF UNIFORMITY [mathematics, physics] (1) The periodic orbits in a period interval $[T - 1T; TC_1T]$, although more numerous than the ones in an interval centred at a shorter period T_0, have a compensating smaller importance. The content of this principle is used in the definition of the measure in the Hilbert space of dynamical variables.[*] (2) Assumes any Gaussian system to be valid as a spacetime coordinate system. Given the mathematical existence of the simultaneously moving local Minkowski space along a timelike geodesic in a Lorentz manifold.[†] (3) Quantum mechanics in terms of matrices (Werner Heisenberg) relation whereby, if one simultaneously measures values of two canonically conjugate variables, such as position and momentum, the product of the uncertainties of their measured values cannot be less than approximately Planck's constant divided by 2_{p_i}. (4) Also known as *uniformitarianism*; the concept that the present is the key to the past. (4) The principle that contemporary geologic processes have occurred in the same regular manner and with essentially the same intensity throughout geologic time, and that events of the geologic past can be explained by phenomena observable today.[‡] See also LINEAR COVARIANCE PRINCIPLE; UNLEARNED PRINCIPLE; HEISENBERG UNCERTAINTY PRINCIPLE; PRINCIPLE OF UNITY.

PRINCIPLE OF UNIFORMITY [geology] (Charles, Lyell, 1797–1875, British geologist) (1) The laws of nature were constant: (a) geological processes were the same in the past as at present, (b) the energy of geologic processes was uniform over time, (c) conditions on Earth were approximately constant, and (d) there was no geologic "progress." (2) The contemporary geologic processes have occurred in the same regular manner and with essentially the same intensity throughout geologic time, and events of the geologic past can be explained by phenomena observable today.[**] See also ANTHROPIC PRINCIPLE; LINEAR COVARIANCE PRINCIPLE; LYELLIAN PRINCIPLE; MEDIOCRITY PRINCIPLE; UNLEARNED PRINCIPLE; HEISENBERG UNCERTAINTY PRINCIPLE; PRINCIPLE OF UNITY.

PRINCIPLE OF UNIQUENESS [psychology] (1) Haridas Chaudhuri's triadic principles of uniqueness, relatedness, and transcendence in integral psychology, an extensive review of literature of the concept of self in Western psychology and several Eastern psychospiritual traditions is undertaken to establish universal, cross-cultural support for the experience of self. A parsimonious model for self in integral psychology. According to this model, the process of integral self-realization consists of a harmonious experience of self in all three spheres of consciousness, necessitating a balanced personality. This model stresses the uniqueness of individual constitution and an individualized approach to the process of integral self-realization, a dynamic and evolutionary interpretation of spiritual development and self-realization that advocates healthy ego development and reconciliation of the ego–self dichotomy.[§] (2) The use of different signals for different categories and of the same signal for all members within a category corresponds to the principle of contrast or of mutual exclusivity that children rely on when they assign only one label per category.[¶] (3) Children will allow only one lexical entry to occupy a semantic niche. When two words are determined to have similar meanings, one of them is preempted and removed from the lexicon. Different words should have

---

[*]Hannay, J. H. and Ozorio de Almeida, A. M. (1984), *J. Phys. A Math. Gen.* **17**, 3429.

[†]de Carvalho, T. O. and de Aguiar, M. A. M. (1996), Eigenfunctions of the Liouville operator, periodic orbits and the principle of uniformity, *J. Phys.* **A29**(13), 3597–3615.

[‡]Blatov, V. A., Pol'kin, V. A., and Serezhkin, V. N. (1994), Polymorphism of elementary substances and the principle of uniformity, *Kristallografiya*, **39**(3), 457–463.

[§]Agha-Kazem-Shirazi, B. (1995), Self in integral psychology. Ann Arbor, MI

[¶]Roberts, H. (1950), A note on the principle of uniqueness: A principle that can be used to simplify the stress analysis where some members of a structure are unloaded, *Aircraft Eng. Aerospace Technol.* **22**, 20.

different meanings.* (4) Captures facts about the inferences speakers and addresses make for both conventional and novel words. It accounts for the preemption of novel words by well-established ones; and it holds just as much for morphology as it does for words and larger expressions.† (5) The development of nonlinguistic concepts, the acquisition of language in context, and the use by participants in conversational exchanges to account for those features of language and language acquisition.‡ See also MUTUAL EXCLUSIVITY PRINCIPLE; PRINCIPLE OF CONVENTIONALITY; PRINCIPLE OF PREEMPTION; PRINCIPLE OF CONTRAST; COOPERATIVE PRINCIPLE.

PRINCIPLE OF UNITY [physics]   See LINEAR COVARIANCE PRINCIPLE.

PRINCIPLE OF UNITY OF COMMAND [psychology] (Henri Fayol, 1841–1925, French industrialist)   (1) An employee should receive orders from only one supervisor. Yet, because of a number of interacting variables in any job situation, line and staff as authority become opposed to line and staff as function. (2) The closer is an individual's reporting relationship to a single superior, the less will be the problem of conflict in instructions and the greater the feeling of personal responsibility for results. Encompassing both organization and command structure. (3) A subordinate should report to only one superior. associated with less role conflict and increased employee effectiveness within this bureaucratic setting. Unity of command and job attitudes of managers in a bureaucratic organization.§ See also FAYOL'S PRINCIPLES OF MANAGEMENT; PRINCIPLE OF PLANNING.

*Naigles, L. R. and Lehrer, N. (2002), Language-general and language-specific influences on children's acquisition of argument structure: A comparison of French and English, *J. Child Lang.ge*, v.29, n.3, p.545–566.
†MacWhinney, B. (2004), A multiple process solution to the logical problem of language acquisition, *J. Child Lang.* **31**(4), 883–914.
‡Kalhat, J., (2008), Structural universals and the principle of uniqueness of composition, *Grazer Phil. Studien*, **76**(1), 57–77.
§Wood, M. C. (2002), *Henri Fayol: Critical Evaluations in Business and Management*, Taylor & Francis, p. 8.

PRINCIPLE OF UNITY OF DIRECTION [psychology] (Henri Fayol, 1841–1925, French industrialist)   There should be only one plan, and the person should be responsible for supervising it; all activities that have the same objective should be supervised by one person.‖ See also FAYOL'S PRINCIPLES OF MANAGEMENT.

PRINCIPLE OF UNITY OF OBJECTIVE [psychology] (Henri Fayol)   (1) The closer an individual's reporting relationship to a single superior, the less the problem of conflict in instructions and the greater the feeling of personal responsibility for results. Encompassing both organization and command structure to clarify authority–responsibility relationships.‖ (2) An organization structure is effective if it as a whole, and every part of it, make possible accomplishment of individuals in contributing toward the attainment of enterprise objectives.¶ See also PRINCIPLE OF PLANNING.

PRINCIPLE OF UNIVERSAL DEVELOPMENT [organizational psychology]   The enterprise can tolerate only those managers who are interested in their continuous development. See also PRINCIPLE OF PLANNING.*

PRINCIPLE OF UNIVERSAL GRAVITATION [physics]   Every particle in the universe attracts every other particle with a force that is directly proportional to the product of their masses and inversely and inversely proportional to the square of the distance their centers of gravity.‖

PRINCIPLE OF UNIVERSALITY [psychology] (Ferenczi's three principles: egoism, universality, and guilt)   A universal component has properties that allow it to be used in a large class of systems. If you apply the principle of universality, you can have parts that interact with multiple number of parts. You can make a subpart perform

¶Koontz, H. and O'Donnell, C. (1968), Principles of Management, 4th ed., McGraw-Hill.
‖Sidharth, B. G. (2006), Puzzles of large scale structure and gravitation, *Chaos Solitons Fractals* **30**(2), 312–317.

multiple functions to eliminate other parts.* (2) Reference frames in which the physical systems conserve their state of motion, if they do not interact with other objects, are universal.[†]

PRINCIPLE OF UNIVOCAL DETERMINATION [biology] Postulates that nothing happens but what is related in only one way to the rest of the "given." Formulated with special reference to the origin of diversities of any kind, the principle would demand that any increase with regard to any kind of diversity be referable in only one way to preexisting diversities, corresponding to the increase that is studied; in other words, that every newly originating singularity is referable to a preexisting singularity.[‡]

PRINCIPLE OF VARIANCE MINIMIZATION [physics] Variance minimization provides lower bounds for the lower eigenvalues of some three-particle systems. The procedure does not yield absolute lower bounds, but rather an interval enclosing at least one eigenvalue; on the other hand, it has the advantages of a variational procedure, where basis functions may be arbitrarily chosen.[§] See also VARIATIONAL MINIMAL PRINCIPLE.

PRINCIPLE OF VELOCITY/ACCELERATION COMBINATION [physics] Emphasis on the relative motion bringing out the physical relevance of the terms in the velocity/acceleration equation.[¶]

*Barish, S. and Vida, J. (1998), As far as possible: Discovering our limits and finding ourselves, *Am. J. Psychoanal.* **58**(1), 883–897.
[†]Marinchev, E. (2003), *Universality* (arXiv:physics/0211106v3).
[‡]Driesch, H. A. E. (1908), *The Science and Philosophy of the Organism*, vol. 2, sec. B, The philosophy of the organism; Part I, The indirect justification of entelechy.
[§]Pauli, G. and Kleindienst, H. (1984), A Hartree-Fock-Roothaan analogon using the principle of variance minimization. II. Test of the iteration procedure, *Theor. Chim. Acta* **64**(6), 481–499.
[¶]Thiel, A. Greschner, M. Eurich, C. W., Ammermuller, J., and Kretzberg, J. (2007), Contribution of individual retinal ganglion cell responses to velocity and acceleration encoding, *J. Neurophysiol.* **98**(4), 2285–2296.

PRINCIPLE OF VIBRATION [physics] Embodies the concept that motion is manifest in everything in the universe, that nothing rests, and everything moves, vibrates, and circles. The differences between different manifestations of matter, energy, mind, and even spirit, are the result of only different "vibrations." The higher you are on the scale, the higher your rate of vibration will be. The "All" is purported to be at an infinite level of vibration, almost to the point of being at rest. There are said to be millions on millions of varying degrees between the "all" and the objects of the lowest vibration.[‖]

PRINCIPLE OF VIBRATION MEASUREMENT [physics]   See MINIMUM NORM PRINCIPLE.

PRINCIPLE OF VIRTUAL POWER [physics] Principle of Virtual Forces see Muller-Breslau Principle Muller–Breslau Principle See also PRINCIPLE OF MATERIAL FRAME INDIFFERENCE, COVARIANCE PRINCIPLE; PRINCIPLE OF LINEAR MOMENTUM; FUNDAMENTAL PRINCIPLE.

PRINCIPLE OF VIRTUAL VELOCITY [physics] (1) If there is a system of forces such that their points of application are moved through certain consecutive positions, and those forces are all in such a position in equilibrium that, in respect to any finite motion of the points of application through that series of postions, the aggregate of the work of those forces, which act in directions in which their multiple points of application are made to move, is equal to the aggregate of the work in the opposite direction.[**] The updated Lagrangian rate formulation is employed tio describe the nonlinear problem. The rate type equilibrium equation and the boundary at

[‖]Three Initiates (2004), *The Kybalion*, Book Tree (the "Three Initiates" who authored *The Kybalion* chose to remain anonymous; as a result, there has been a great deal of speculation about who actually wrote the book).
[**]Xing, H.L., Miyamura, T., Makinouchi, A., Homma, T., Kanai, T., Oishi, Y., 2001. Development of high performance finite element software system for simulation of earthquake nucleation and development. In: Moresi, L., Muller, D., Hobbs, B. (Eds.), Exploration Geodynamics. Western Australia, pp. 178–188. nical Report of ACcESS (ES-SCC, UQ), pp. 1–20.

the current configuration are equivalently expressed by a principle of virtual velocity of the form [4–9]

$$\int_v \left\{ (\sigma_{ij}^J - 2\sigma_{ik}D_{kj}) + \delta D_{ij} + \sigma_{jk}L_{ik}\delta L_{ij} \right\} dV$$
$$= \int_{S\Gamma} \dot{F}_i \delta v_i dS + \int_S {}_c 1 \dot{f}_i^1 \delta v_i^1 dS + \int_{s_c^2} \dot{f}_i^2 \delta v_2^2 dS$$

(1) If any number of forces is in equilibrium (being in any way mechanically connected with one another), and if, subject to that connection, their different point of application be made to move, each through any exceedingly small distance, then the aggregate of the work of those forces, whose points of application are made to move toward the directions in which the several forces applied to them act, shall equal the aggregate of the work of those forces, the motions of whose points of application are opposed to the directions of the forces applied to them.* (3) If a machine works under a constant equilibrium of the pressures applied to it, or if it work uniformly, then is the aggregate work of those pressures that tend to acceelarate its motion equal to the aggregate work of those that tend to retard it.†

PRINCIPLE OF VIRTUAL WORK [physics, mathematics] (Johann Bernoulli and Eric Reissner)   (1) The principle of virtual work was stated for the first time in almost modern formulation without proof in 1717 in Johann Bernoulli's letter to Varignon. One is impressed by the fact that Bernoulli explicitly emphasized the validity of the principle for one-, two-, and three-dimensional bodies. As we show, each of these three cases has specific singularities. Let us also note that in Bernoulli's formulation the principle of virtual work is a necessary condition only for equilibrium.‡ (2) The total work done

by all forces acting on a system in static equilibrium is zero for any possible virtual displacement.§ (3) Normally associated with decision theory; it strives to evaluate relative utilities of simple and mixed parameters that can be used to describe outcome. Since management techniques and knowledge and the total environment of managing change constantly, the enterprise that would ensure its managerial competence cannot tolerate managers who are not interested in their continuos development.¶ (4) To determine the force multiplication coefficient as a function of the load and to correct for the frictional force that opens up new means of improving the accuracy of machines in transmitting the force unit.‖ (5) Constraints introduced to the symmetry condition functional with the aid of a Lagrange multiplier.** See also DISPLACEMENT PRINCIPLE.

PRINCIPLE OF VIS VIVA [physics]   (1) If the forces of any system are not in equilibrium with one another, then the difference between the aggregate work of those oriented in the direction of the motions of their multiple points of application, and those oriented in the opposite direction, is equal to one-half the aggregate vis viva of the system. (2) The difference between the aggregate work done on the machine, during any time, by those forces that tend to accelerate the motion, and the aggregate work, during the same time, of those that tend to retard the motion, is equal to the aggregate number of units of work accumulated in the moving parts of the machine during that time if the former aggregate exceeds the latter, and lost from

*Stronge, W. J. and Yu,T. (1995), *Dynamic Models for Structural Plasticity*, Springer, p. 34.

†Moon, F. C. (2008), *Applied Dynamics: With Applications to Multibody and Mechatronic Systems*, Wiley-VCH, p. 64.

‡Chobanov, I. (1976), Aproposition of pure mechanics or proof of a corollary from the definition of equilibrium of systems of forces usually called the principle of virtualwork (Bulgarian summary), *Annuaire Univ. Sofia Fac. Math. Méc.* **71**(2), 229–251.

§Wang, Q. A. (2007), From virtual work principle to least action principle for stochastic dynamics, Inst. Supérieur des Matériaux et Mécaniques Avancés du Mans, France, p. 4.

¶Groenewegen, P. D. (1998), *Alfred Marshall: Critical Responses*, Taylor & Francis, p. 150.

‖Freise, V. (1965), Principle of "virtual work" and the conditions of the thermodynamic equilibrium, *Z. Phys. Chem.* (Muenchen, Germany) **46**(3/4), 199–215.

Sokolov, V. V. Tolmachev, V. V. (1996), Application of generalized virtual work principle in ferrohydrodynamics. Part 1. Magnetic fluid with free magnetization, *Magnitnaya Gidrodinamika* **32(3), 313–317.

them during that time if the former aggregate falls short of the latter.*

PRINCIPLE OF VOLUMES [CHEMISTRY] See GAY-LUSSAC'S PRINCIPLE.

PRINCIPLE OF WATERSTOP FUNCTION [engineering] Waterstops work because of two specific aspects of their design for sealing and resistance.[†] See also VALVE PRINCIPLE; TORTUOUS PATH PRINCIPLE.

PRINCIPLE OF WEAK INDUCTION [mathematics] Let D be a subset of the nonnegative integers Z^* with the properties that (1) the integer 0 is in D and (2) any time that n is in D, one can show that $n + 1$ is also in D. Under these conditions, $D = Z^*$. See also PRINCIPLE OF STRONG INDUCTION, PRINCIPLE OF TRANSFINITE INDUCTION[‡]

PRINCIPLE OF WHOLE–PART DETERMINATION [psychology] (Wertheimer, 1925) "There are wholes, the behavior of which is not determined by that of their individual elements, but where the part processes are themselves determined by the intrinsic nature of the whole." In brief, the whole–part determination in illusory stimulus figures is a special case of a central canon of gestalt psychology, that of relational determination of appearances. Additional instances of it are readily predictable.[§]

PRINCIPLE OF WORK [physics] If no work were lost by friction, the work done by a machine would always be exactly equal to the work put into it.[¶] See also PRINCIPLE OF CONSERVATION OF ENERGY.

PRINCIPLE OF X-RAY COMPUTED TOMOGRAPHY [physics, mathematics] An X-ray generator emits a flux of highly collimated X rays in order to obtain information on the desired section of the object. The X-ray beam goes through the object, and a multidetector placed within the beam axis delivers a current proportional to the number of transmitted photons; the signal is then coded by the associated electronics. A multidirectional analysis of the object allows to reconstitute a matrix that is treated by appropriate processing algorithms. From the tomographic image observed on a screen, precise digital values can be extracted to Facilitate diagnostic and quantitative analysis. ‖

PRINCIPLE OF ZERO WASTE [ecology] (1) Recovery of all discarded items so that nothing is left for disposal. (2) Taking a "whole system" approach to the vast flow of resources and waste through human society by maximizing recycling, minimizing waste, reducing consumption, and ensuring that products are made to be reused, repaired, or recycled back into nature or the marketplace. (3) Redesigns the current, one-way industrial system into a circular system modeled on nature's successful strategies. (4) Challenges poorly designed business systems that "use too many resources to make too few people more productive." (5) Addresses, through job creation and civic participation, increasing wastage of human resources and erosion of democracy. (6) Helps communities achieve a local economy to operate efficiently, maintain good jobs, and provide a measure of self-sufficiency. (7) Eliminate rather than manage waste. (8) Makes recycling a powerful entry point into a critique of excessive consumption, waste, corporate irresponsibility, and

*Hecht, E. (2003), An historico-critical account of potential energy: Is PE really real? *Phys. Teacher* **41**(8), 486–493.

[†]Anon. (2007), *PARCHEM Concrete Repair Flooring Jointing Systems Waterproofing*, Technical Data Sheet, ABN 80 069 961 968 (www.parchem.com.au).

[‡]Séroul, R. (2000) Reasoning by Induction." in Programming for Mathematicians. Berlin: Springer-Verlag, pp. 22–25.

[§]Day, R. (2006), Two principles of perception revealed by geometrical illusions, *Austral. J. Psychol.* **58**(3), 123–129.

[¶]Lazic, Z., Lazic L., Kokovic, M., Pantazis, D., Jurisic, M., Nikolic, S., and Zivanovic, D. (1986) Opening of sports stomatologic centers, planning, organization and principle of work, *Stomatol. Rev.* **15**(3–4), 223–226.

‖Robb, R. A. (1982), X-ray computed tomography: an engineering synthesis of multiscientific principles, *Crit. Rev. Biomed. Eng.* **7**(4), 265–333.

the fundamental causes of environmental destruction.* Also known as *Energy preservation principle*.

PRINCIPLE REMIFENTANIL See PRINCIPLE OF VIRTUAL WORK.

PRINCIPLE THAT THE POLLUTER PAYS. [Ecology, Mathematics] See POLLUTER PAYS PRINCIPLE.

PRINCIPLE TO THE DUAL PRESSURE [engineering]   Linde system discussed earlier. In the dual-pressure Claude cycle, only the gas that is sent through the expansion valve is compressed to high pressure; this reduces the work requirement per unit mass of gas liquefied.[†]

PRINCIPLES FOR NETWORKED WORLD [computer science] (1) User-centered, barrier-free, and format-independent. (2) Alternate methods, approaches, and access points. (3) Minimize restrictions on access and maximize dialog.[‡] See also INTELLECTUAL FREEDOM PRINCIPLE; PRIVACY PRINCIPLE; INTELLECTUAL PROPERTY PRINCIPLE; INFRASTRUCTURE PRINCIPLE; UBIQUITY PRINCIPLE; EQUITY OF ACCESS PRINCIPLE. CONTENT PRINCIPLE.

PRINCIPLES OF ATTENTION STRESS [computer science, psychology]   (1) The demands that various Web applications place on attentional capacities may be measured and compared by employing cognitive psychological knowledge about the type and amount of attention required to perform various tasks. (2) Attention stress is based on many psychological observations; two most important ones are attention shift and selection threshold. Attention shift addresses the issue of "getting

lost," or the experience of a "broken flow." It is usually measured by the number of page refreshes or the amount of hand–eye coordination required to complete a task. Selection threshold deals with the matter of "being overwhelmed." It is observed when users are presented with more than four choices at a time, and their decisions tend to be based on random guess instead of reasoning.[§]

PRINCIPLES OF BIOMEDICAL ETHICS [pharmacology]   An act with two effects, one right and one wrong, can be performed when fair conditions are met—the effect of activity, the intention, material cause of the act, and proportionate reason. Also, right or indifferent action, not intrinsically wrong, wrong, though unforeseen, cannot be intended, wrong effect cannot be a means to the right effect; for the wrong effect to occur, there must be proportionate reason.[¶] See also PRINCIPLE OF DOUBLE EFFECT; PRINCIPLE OF DO NO HARM; CASUISTIC PRINCIPLE; EXCEPTION-GRANTING PRINCIPLE; JUSTIFYING PRINCIPLE; DOUBLE EFFECT PRINCIPLE

PRINCIPLES OF CHEMICAL COMPOSITIONS See CHEMICAL COMPOSITION PRINCIPLES.

PRINCIPLES OF COMPILER DESIGN [computer science]   A classic textbook on compilers for computer programming languages. It is often called the "dragon book" because its cover depicts a knight and a dragon in battle; the dragon is green, and labeled "complexity of compiler construction," while the knight wields a lance labeled "LALR parser generator." The book may be called the "green dragon book"to distinguish it from its successor.[‖]

*Kinsella, S. Knapp, D., et al. (1997), *Zero Waste: Management Principles for the Coming Age of Zero Waste*, GRRN Green Paper.
[†]Flynn, T. M. (n.d.), 2nd ed., Refrigeration and liquefaction, in *Cryogenic Engineering: Revised and Expanded*, Chapter 6.
[‡]Bolt, N. (2002), COSLA 2002 Mid-Winter Meeting, Jan. 18, 2002, Cabildo Complex Arsenal, New Orleans, LA.

[§]Seaward, B. L. (2005), *Managing Stress: Principles and Strategies for Health and Well Being*, Jones & Bartlett Publishers, p. 560.
[¶]Iserson K.V. (1999), Principles of biomedical ethics, *Emerg. Med. Clin. North Am.* **17**(2), 283–306, ix.
[‖]Aho, S. and Ullman's *Compilers: Principles, Techniques and Tools*, (which is the "red dragon book" because the dragon on its cover is red), Addison-Wesley.

PRINCIPLES OF CONDITIONS FOR EQUILIBRIUM [physics]   (1) The sum of the forces pulling the body in any direction must be equal to the sum of the forces pulling the body in the opposite direction. (2) The sum of the moments of the forces tending to turn the body in the clockwise direction must be equal to the sum of the moments of the forces tending to turn the body in the counterclockwise direction.*

PRINCIPLES OF DARCY'S LAW [physics] (1) Describes the flow of liquid water and water vapor in the saturated–unsaturated soil below the surface. (2) Necessary for the estimation of unknown parameters, in a per survivor fashion. Introduced by several authors for state complexity reduction in an intersymbol interference, environment, delayed decision feedback sequence estimation, and reduced state sequence estimation.† See also PRINCIPLE OF PERFORMING SIGNAL-PROCESSING OPERATIONS.

PRINCIPLES OF DETERMINATION [psychology] (1) Role of wishes in motivational systems. (2) The relation between wishes and drive-energic factors. (3) The relation between the putative motion of psychic energy as the potential for activation of psychic structure and its differentiation from drives (libidinal, aggressive, and narcissistic); activity and efficacy of psychic functions requires a concept of energic potential in psychoanalytic metapsychology. (4) Reevalutation of the organization and integration of psychic functions and governing psychic activity. (5) Determining the stimulus conditions governing activation and integration of receptive and response systems and their adaptive integration in assimilative and accommodative patterns of environmental interaction. (6) Reevaluating the regulatory principles governing the organization and integration of psychic functions. Recasting these principles in modified nonenergic terms suggests that regulatory principles play a different role in governing psychic activity. Once divorced from their energic dependence, the regulatory principles can be recast as operational principles determining the stimulus conditions governing activation and integration of receptive and response systems and their adaptive integration in assimilative and accommodative patterns of environmental interaction. Reformulations are suggested for the principles of determination, overdetermination, and multiple function, and for more specific regulatory principles of inertia, constancy, pleasure-unpleasure, reality, repetition, and primary–secondary process.‡ See also REGULATORY PRINCIPLES AND OTHER REGULATORY PRINCIPLES OF INERTIA; REGULATORY PRINCIPLES OF CONSTANCY; REGULATORY PRINCIPLES OF PLEASURE–UNPLEASURE; REGULATORY PRINCIPLES OF REALITY; REGULATORY PRINCIPLES OF REPETITION; REGULATORY PRINCIPLES OF PRIMARY–SECONDARY PROCESS; PRINCIPLES OF OVERDETERMINATION; PRINCIPLES OF MULTIPLE FUNCTION; PRINCIPLES OF ECONOMICS.

PRINCIPLES OF ECONOMICS [mathematics] (Carl Menger, 1840–1921, Austrian economist)   Published in 1871, advances that the marginal utility of goods, rather than the labor inputs that went into making them, is the source of their value. Stresses uncertainty in the making of economic decisions, rather than relying on "homo oeconomicus" or the rational person who was fully informed of all circumstances impinging on her or his decisions. Perfect knowledge never exists, and that therefore all economic activity implies risk. The entrepreneurs' role was to collect and evaluate information and to act on those risks.§

PRINCIPLES OF ENERGETICS [physics]   (Howard Thomas Odum, 1924–2002, American ecosystem ecologist)   A general statement of energy flows under transformation, including laws of

*Pedregal, P. (1998), Equilibrium conditions for young measures, *SIAM J. Control Optim.* (Soc. Industrial and Applied Math.) **36**(3), 797–813.
†Das, A. K. (1997), Generalized Darcy 's law including source effect. Can. J. Petrol. Technol. **36**(6), 57–59.

‡Meissner, W. W. (1995), The economic principle in psychoanalysis: II. Regulatory principles, *Psychoanal. Contemp. Thought* **18**(2), 227–259.
§Menger, C. (1905), *Principles of Economics* (J. Dingwall and B. F. Hoselitz, transls. The Free Press. New York.

thermodynamics that seek a rigorous description. The principles are as Follows:

Zeroth principle of energetics. If two thermodynamic systems A and B are in thermal equilibrium, and B and C are also in thermal equilibrium, then A and C are in thermal equilibrium.

First principle of energetics. The increase in the internal energy of a system is equal to the amount of energy added to the system by heating, minus the amount lost in the form of work done by the system on its surroundings.

Second principle of energetics. The total entropy of any isolated thermodynamic system tends to increase over time, approaching a maximum value.

Third principle of energetics. As a system approaches absolute zero of temperature, all processes cease and the entropy of the system approaches a minimum value or zero for the case of a perfect crystalline substance.

Fourth principle of energetics. Describes the propensities of evolutionary self-organization.

Fifth principle of energetics. The energy quality factor increases hierarchically.

Sixth principle of energetics. Material cycles have hierarchical patterns measured by the emergy/mass ratio that determines its zone and pulse frequency in the energy hierarchy.[*]

PRINCIPLES OF ETHICS [philosophy] (American philosophers Beachump and Childress wrote a book titled *Principles of Biomedical Ethics* in 1979. In this book they formulated some ethical principles, namely, "respect to [For] autonomy," "nonmaleficence," "beneficence," and "justice."[†]

PRINCIPLES OF FLIGHT [physics]   Process by which a heavier-than-air animal or object achieves sustained movement either through the air by aerodynamically generating lift or aerostatically using buoyancy, or movement beyond Earth's atmosphere, in the case of spaceflight.[‡]

PRINCIPLES OF HOUGH TRANSFORM REPRESENTATION [physics]   Used in an automatic technique, which makes use of a parameter space to describe features of interest in images. This method has been widely applied in machine vision for recognition of features in highly structured images.[§]

PRINCIPLES OF HUMAN EFFECTIVENESS Being responsible, planning, executing, cooperating, and listening.[¶]

PRINCIPLES OF INTELLIGENT URBANISM (PIU) [psychology]   (1) A response to rapidly growing low-income cities, with limited resources and inadequate infrastructure levels. (2) Address fragile environments, settings with strong building traditions, issues of soc. economic opportunities and urban management. They are a set of 10 axioms, laying down a value-based framework within which participatory planning can proceed, as follows: *principle 1*—a balance with nature, *principle 2*—a balance with tradition, principle 3—appropriate technology, *principle 4*—conviviality, *principle 5*—efficiency, *principle 6*—human scale, *principle 7*—opportunity matrix, *principle 8*—regional integration, *principle 9*—balanced movement, and *principle 10*—institutional integrity.[‖]

[*]H. T. Odum, (2000), An energy hierarchy law for biogeochemical cycles, in *Energy Synthesis: Theory and Applications of the Emergy Methodology, Proc. Biennial Emergy Analysis Research Conf.* Brown, M. T., ed. Center for Environmental Policy, Univ. Florida, Gainesville.

[†]Golpinarli A: Mevlana'dan sonra Mevlevilik (1207–1253) (Mawlawi order after Mawlana), *Inkilap Yayinlari, Istanbul* (1934), pp. 441–454.

[‡]Project SOAR (1997–1999), *Science in Ohio through Aerospace Resources*, vols. I–III, Natl. Museum of the US Air Force and Air Force Museum Foundation, Inc., Dayton, OH, p. 7

[§]Csenki, L. Alm, Er., Torgrip, R. J. O., Aaberg, K. M., Nord, L. I., Schuppe-Koistinen, I., and Lindberg, J. (2007), Proof of principle of a generalized fuzzy Hough transform approach to peak alignment of one-dimensional 1H NMR data, *Anal. Bioanal. Chem.* **389**(3), 875–885.

[¶]Adams, T., Bezner, J., and Steinhardt, M. (1995), Principle-centeredness: A values clarification approach to wellness, *Meas. Eval. Counsel. Devel.* **28**(3), 139–147.

[‖]Benninger, C. (2001), Principles of intelligent urbanism, *Ekistics*, **69**(412), 39–65.

PRINCIPLES OF ION TRANSFER [chemistry] Reactions can best be discussed for liquid metals as electrodes in contact with a liquid electrolyte containing the respective metal ions. On a liquid interface there are no special sites where atoms have different properties as they do on solid surfaces. In the kinetics of ion discharge or formation at a metal, the ion must pass the electric double layer. The metal ion to be deposited must lose its solvation shell—or, in general, its interaction with the components of the electrolyte—and be incorporated into the metal by the interaction with the metal electrons. There are intermediate stages where the interaction with the electrolyte has already been weakened while the interaction with the metal electrons is still incomplete. The ion has to overcome an energy barrier. This energy barrier is affected by the electric field in the double layer.[*]

PRINCIPLES OF JUSTICE Find their most succinct formulation in Immanuel Kant's definition of (moral) right: "The whole of conditions under which the voluntary actions of any one person can be harmonized in reality with the voluntary actions of every other person, according to the universal law of freedom." "Causa causae est causa effectus." This principle tells us simply that whatever the one who is exposed to injustice undertakes to ward off injustice and to reestablish just conditions, the one who commits the unjust act must ascribe to oneself. What happens to him/her as a result has been triggered by his/her act of injustice and would otherwise not take place.[†]

PRINCIPLES OF JUS IN BELLO [psychology] (1) Requires that the agents of war be held responsible for their actions. With regard to just cause, a policy of war requires a goal, and that goal must be proportional to the other principles of just cause. (2) Rules of just conduct fall under discrimination,

proportionality, and responsibility.[‡] See also PRINCIPLE OF PROPORTIONALITY; PRINCIPLE OF DISCRIMINATION; PRINCIPLE OF RESPONSIBILITY.

PRINCIPLES OF MILITARY STRATEGY [psychology] Many military strategists have attempted to encapsulate successful military strategy. Sun Tzu defined 13 principles in his *The Art of War*, while Napoleon listed 115 maxims. American Civil War General Nathan Bedford Forrest required only one: "Get there firstest with the mostest." The fundamental concepts common to most lists of principles are: (1) the objective, (2) offense, (3) cooperation, (4) concentration (mass), (5) economy, (6) maneuver, (7) surprise, (8) security, and (9) simplicity.[§] See also PRINCIPLE OF WAR; PRINCIPLE OF WARFARE.

PRINCIPLES OF MULTIPLE FUNCTION [psychology] Hierarchic layering of the psychic apparatus, and by maintaining a view of the ego as a dynamic, wishful organization, the existing descriptive, theoretical, and technical categories of psychoanalysis concerned with defense could be transcended. Problem of defense as a special instance of a general problem in psychoanaltic ego psychology, stemming from a lack of systematic development of dynamic propositions concerning the ego system. It is suggested that all defense should be considered to be wishful activity that inherently provides libidinal and aggressive gratification or leads to it, or both, serving counterdynamic purposes at the same time. With this in mind, the therapist would be better able to understand the unconscious status of the defense mechanisms, and the patient's libidinal, as well as aggressive, resistance to the analysis of his/her defenses.[¶] See PRINCIPLES OF DETERMINATION

[*]Gerischer, H. (n.d.), Principles of electrochemistry, in *The CRC Handbook of Solid State Electrochemistry*, Chap. 1.

[†]Micewski, E. R. (2006), Terror and terrorism: A history of ideas and philosophical-ethical reflections, *Cultic Studies Rev.* **5**(2), (special issue on terrorism), 219–243.

[‡]Fichtelberg, A. (2006), Applying the rules of just war theory to engineers in the arms industry, *Sci. Eng. Ethics*, **12**(4), 685–700.

[§]Colonel Charles-Jean-Jacques-Joseph Ardant du Picq (2003), **Battle Studies** [EBook #7294] (http://onlinebooks.library.upenn.edu/webbin/gutbook/lookup?num=7294).

[¶]Schafer, R. (1968), The mechanisms of defense, *Int. J. Psychoanal.* **49**(1), 49–61.

PRINCIPLES OF NEUROLINGUISTIC PROGRAM-MING (NLP) [psychology]   (1) "Multiple viewpoints are useful." Information gathered by considering many different people's points of view will usually be more complete or useful than information provided by just one source. The subjective nature of our experience never fully captures the objective world. Regardless of whether there is an objective absolute "reality," individual people do not in general have access to absolute knowledge of reality but, in fact, only to a set of beliefs that they have built up over time, about reality.* (2) The processes that take place within a human being and between human beings and their environment are complex systems, and they are processes. Our bodies, our societies, and our planet form an ecology of complex systems and subsystems, all of which interact with and mutually influence each other. (3) Whatever people do, they are in fact attempting to fulfill some positive intention (of which they may not be aware). The NLP principles assume that the current behavior exhibited by a person represents the best choice available to them at the time. (4) "Relational quality of a connection between two communicating individuals" is a highly subjective term. Good rapport is characterized by a sense of ease with another, trust, and easy flow of dialog.† (5) There is no failure, only feedback. Not to view communication in terms of success and failure, but rather in terms of competence or lack thereof, or learning and failure to learn. As a field that utilizes trial and error, not all actions are expected to "work"; rather, they are intended to explore, and the results should be utilized as a source of valuable learning and new focus, rather than cause for negativity and despair.‡ (6) Choice is better than no choice (and flexibility is the way one gets

choice). Recognizing "stuckness," and learning how to open it out in accordance with the saying "One choice is no choice, two choices is a dilemma, three [or more] choices is choice." (7) In systems theory the part of the system that can adapt best, be most influential, and has the best chance of achieving its goals, is often not the most forceful part but the part that has most flexibility and least rigidity in its responses. (8) Meaning is in the eye of the recipient. This is an "as if" concept: it may not be true, it may be that the recipient is mistaken, but if you work on the basis that the recipient's understanding of what you say (and not yours) is the important one, it will lead you to communicate in a way that gets the actual message across and heard, even if linguistic gymnastics (i.e., flexibility) are needed to do so. (9) People's resources are their sensory representation systems and the manner in which they are organized. (10) Because of the systemic nature of human lives, a person in a certain situation seldom can see answers that a person standing outside can. So, by moving between different perceptual positions, it is claimed that one can see a problem in new ways, or with less emotional attachment, and thus gather more information and develop new choices of response.§

PRINCIPLES OF NYQUIST'S THEOREM [physics, mathematics]   To accurately reproduce a waveform, one must sample the waveform at two times the highest frequency to be measured.¶

PRINCIPLES OF OVERDETERMINATION [psychology]   See PRINCIPLES OF DETERMINATION.

PRINCIPLES OF PATIENT CARE [psychology] Collection of principles of effective treatment and recovery addressing the context of addictive disorders, the structure of treatment,

*Tosey, P. and Mathison, J. (2006), *Introducing Neuro-Linguistic Programming*, Centre for Management Learning & Development, School of Management, Uni. Surrey.
†O'Connor, J. (2001), *The NLP Workbook—the pillars of NLP*, Thorson.
‡Dilts, R., Grinder, J., Delozier, J., and Bandler, R. (1980), *Neuro-Linguistic Programming*, vol. I: *The Study of the Structure of Subjective Experience*, Meta Publications, Cupertino, CA.

§Thompson, C. K., and McReynolds, L. V. (1986), *Wh* interrogative production in agrammatic aphasia: an experimental analysis of auditory-visual stimulation and direct-production treatment, *J. speech and hearing research*, **29**(2), 193–206.
¶Engert, J., Beyer, J., Drung, D. Kirste, A., and Peters, M. (2007), A noise thermometer for practical thermometry at low temperatures, *Int. J. Thermophys.* **28**(6), 1800–1811.

the process and content of treatment, and the treatment outcome:

1. *Principle 1.* Common dynamics underlie the process of problem resolution occuring in formal treatment, informal care, and "natural" recovery.
2. *Principle 2.* The duration and continuity of care are more closely related to treatment outcome than is the amount or intensity of care.
3. *Principle 3.* Patients treated by substance abuse or mental health specialists experience better outcomes than do patients treated by primary care or nonspecialty providers.
4. *Principle 4.* Treatment settings and counselors who establish a therapeutic alliance, are oriented toward personal growth goals, and are moderately structured tend to promote positive outcomes.
5. *Principle 5.* The common component of effective psychosocial interventions is the focus on helping clients shape and adapt to their life circumstances.
6. *Principle 6.* Among individuals who recognize a problem and are willing or motivated to receive help, formal intervention or treatment leads to better outcomes than does remaining untreated. treated or untreated, an addiction is not an island unto itself.[*]

PRINCIPLES OF PERCEPTION [pharmacology] (1) Strategy for diminishing transfusion requirements in cardiac surgical patients and anesthesia. (2) Within a mental set, perception can be affected by the object being sought and how perception was learned. (3) Perception is related to or classified into categories or events facilitating understanding and comprehension. (4) Perception takes time and is a selective process involving unaccustomed senses that the mind can use only to interpret then analyze the information provided. Perception cannot see everything. (5) Selection is critical since the eye transmits millions of impulses, but the brain can handle only a limited number within a narrow timeframe.[†] See also (GESTALT PRINCIPLES OF PERCEPTION).

PRINCIPLES OF PHILOSOPHY [philosophy] [René Descartes (1596-1650, French mathematician and Philosopher), 1644] Intended to replace Aristotle's philosophy and traditional scholastic philosophy then used in universities and divided into the following four parts: the principles of the human knowledge, the principles of the material things, an objective study of the composition of the universe, and a study of the structure of the land.[‡]

PRINCIPLES OF POLITICAL ECONOMY AND TAXATION [mathematics] (David Ricardo, 1722–1823, English economist) Land rent grows as population increases. It also clearly lays out the theory that all nations can benefit from free trade, even if a nation lacks an absolute advantage in all sectors of its economy.[§]

PRINCIPLES OF PSYCHOLOGY [psychology] William James text describing the four methods in psychology: analysis (i.e., the logical criticism of precursor and contemporary views of the mind), introspection (i.e., the psychologist's study of her/his own states of mind), experiment (e.g., in hypnosis or neurology), and comparison (the use of statistical means to distinguish norms from anomalies).[¶]

PRINCIPLES OF REVOLUTION [psychology] (Michael Bakunin, 1869) Never accept any method other than destruction. The final goal is revolution because evil can be terminated only through violence; and Russian soil will be cleansed only by fire and sword.

---

[*]Moos, R. H. (2003), Principles and puzzles of effective treatment and recovery, *Psychol. Addict. Behav.* **17**(1), 3–12.

[†]Kuppers, E. (1971), Psychophysiological principles of perception, *Archiv Psychiatrie Nervenkrankheiten*, **214**(3), 301–318.

[‡]Descartes, R., and Veitch, J. L. L. D., trans. (2008), *Selections from the Principles of Philosophy*, NuVision Publications. p. 7–15.

[§]Ricardo, D. (1817), *On The Principles of Political Economy and Taxation*, John Murray, London.

[¶]James, W. (1890), *The Principles of Psychology*, Holt.

Furthermore, Bakunin wrote and published a revolutionary catechism in which he gave rules of conduct for terrorists. He coined the phrase "anonymous terrorist," referring to one who was supposed to radically break with society and deny all its laws, customs, and conventions. He advised the terrorists of his time to dispose of the most dangerous enemies (all those who were conspicuous because of their intelligence or particular talents) at the beginning. This was to terrify both the government and the people alike.*

PRINCIPLE OF THE VITAL FEW see PARETO PRINCIPLE

PRINCIPLES OF THEORETICAL LOGIC [mathematics]   David Hilbert's and Wilhelm Ackermann's classic text *Grundzüge der Theoretischen Logik*, on elementary mathematical logic.

PRINCIPLES OF THERAPY OF HIV INFECTION [pharmacology]   The part of an antigen that most reliably induces a protective immune response. The principal neutralizing determinant of HIV is the V3 loop of the envelope glycoprotein gp120. (2) The crystal structure of a complex amino acid peptide from human immunodeficiency virus type 1 (HIV-1) and fragment of a broadly neutralizing antibody and adopts a double-turn conformation, which may be the basis of its conservation in many HIV-1 isolates.[†] See also HIV PRINCIPLE.

PRINCIPLES OF UNCERTAINTY [engineering, mathematics]   (Werner Heisenberg) See UNCERTAINTY PRINCIPLE.

PRINCIPLES OF UNCERTAINTY AND INFORMATION INVARIANCE [psychology]   Uncertainty in a problem situation emerges whenever information pertaining to the situation is deficient in some respect. It may be incomplete, imprecise, fragmentary, not fully reliable, vague, contradictory, or deficient in some other way. Uncertainty can be reduced only by obtaining relevant information. Assume that we can measure the amount of uncertainty associated with a problem situation. Assume further that this amount of uncertainty can be reduced by obtaining relevant information as a result of some action (e.g., observing a new fact, performing an experiment, finding an historical record). Then, the amount of information obtained by the action can be measured by the reduction in uncertainty that results from the action. In this sense, the amount of uncertainty (pertaining to a problem situation) and the amount of information (obtained by a relevant action) are intimately connected. Since this notion of information does not capture the full richness of information in human communicatlon, it is appropriate to refer to it as *uncertainty-based information*.[‡]

PRINCIPLES OF UNITARIAN UNIVERALISM [psychology]   (1) The inherent worth and dignity of every person. (2) Justice, equity, and compassion in human relations. (3) Acceptance of one another and encouragement for spiritual growth in our congregations. (4) A free and responsible search for truth and meaning. (5) The right of conscience and the use of the democratic process within our congregations and in society at large. (6) The goal of world community with peace, liberty, and justice for all. (7) Respect for the interdependent web of all existence of which we are a part.[§] See also U(2) PRINCIPLE.

PRINCIPLES OF UNITY [psychology]   Hakomi therapy principles of unity, mindfulness, nonviolence, organicity, and mind–body holism,

---

*Micewski, E. R. (2006), Terror and terrorism: A history of ideas and philosophical-ethical reflections., *Cultic Studies Rev.* **5**(2), (special issue on terrorism), 219–243.
[†]Anon. (1998), Report of the NIH panel to define principles of therapy of HIV infection (recommendations and reports: morbidity and mortality weekly report; recommendations and reports, CDC, Atlanta), *MMWR* **47**(RR-5), 1–41.

[‡]Klir, G. J. (1992), The role of uncertainty measures and principles in AI, Springer Verlag. v. 617, pp. 245–254.
[§]First Unitarian Church (2008), *The 7 Principles of Unitarian Universalism*, Unitarian Universalist Assoc. Congregations (www.uua.org).

which are- rooted it in a spiritual matrix. The present author's own experience of the idea of "grace-that was contacted in Hakomi workshops is described. The message of grace that Hakomi therapists can offer their clients is unconditional love and forgiveness that allows them to let go of their own pain, mistakes, and transgressions of the past and go on more joyfully with their lives All roads lead home: nourishment, transformation, and grace.* (2) Basis of unity is the unifying or ethics on which all members of an organization or coalition can agree—identifying features of a movement.

PRINCIPLES OF VAPOR–LIQUID EQUILIBRIA [physics] Based on the Mie theory of light scattering with the particles acting as point sources. The dark-field microscopy method consists of depositing a sample on a glass slide and measuring the intensity of light scattered from individual particles viewed in dark-field illumination, and visually accessed or recorded on photographic film for subsequent image analysis. Particles acting as point sources. This method provides an indication of the breadth or narrowness of the size distribution, the presence of outsized and agglomerated particles, and the presence of floc and/or second-generation particles. The method is sensitive to dust and dirt on microscope slides, and special precautions are required to maintain absolute cleanliness.

PRINCIPLES OF WAR [psychology] (1) By means of psychoanalytical concepts, individual sadomasochistic tendencies are projected and translated into terms of group activities during peace and war. The psychological mechanisms of frustration, repression, and projection are evaluated in terms of their relationships to the conscious and unconscious activities promoting war and peace. Certain principles of war prevention are set forth on the basis of individual and group activities. The rationalizations are used as basic reasons for war exposed through analysis of the individual's fundamental learnings toward sadism and/or masochism. A long-term program of research is outlined

to provide civilization with information to combat the practice of war and to protect itself against recurring war. Individual psychological disorders are examined, with the conclusion that a repression of infantile sex impulses results in a confusion between real adult dangers and the dangers attendant to infantile loving. A plea is registered for examining the primitive factors of human nature through an analysis of early human development. The study of war-related neuroses is suggested for beginning a study of war psychology together with an understanding of aim inhibition and sublimation. Until we understand the unconscious causes of war, we cannot prescribe for a lasting peace. War, sadism, and pacifism.† (2) Tenets used by military organizations to focus the thinking of leaders toward successful prosecution of battles and wars. (3) The United States applies nine principles of war in training their officers: (a) *objective*—define a decisive and attainable objective for every military operation; (b) *offensive*—seize, retain, and exploit the initiative; (c) *mass*—apply sufficient force to achieve the objective; (d) *economy of force*—focus the right amount of force on the key objective, without wasting force on secondary objectives; (e) *maneuver*—place the enemy in a position of disadvantage through the flexible application of combat power; (f) *unity of command*—for every objective, there must be a unified effort and one person responsible for command decisions; (g) *security*—never permit the enemy to acquire an unexpected advantage; (h) *surprise*—otherwise known as "audacity"; strike the enemy at a time and/or place the on in a manner for which they are unprepared; and (i) *Simplicity*—prepare clear, uncomplicated plans and clear, concise orders. Officers in the US military sometimes use the acronym "MOUSE MOSS" to remember the first letters of these nine principles. (4) The UK military uses 10 principles of war, as taught to all officers of the Royal Navy, British Army, and Royal Air Force: (a) *selection and maintenance of the aim*—define the end state and ensure that all

---

*Townsend-Simmons, R. (1985), *Hakomi Forum*, no. 3, pp. 46–47.

†von Clausewitz, C. P. G., and Graham, J. J., trans. (1873) On war, N. Trubner, London.

strategy is directed toward achieving it; (n) *concentration of force*—make the best use of military power to achieve the commander's aims by overwhelming the enemy's military capacity; (c) *economy of effort*—make efficient use of forces, conserving energy and materiel to prevent unnecessary depletion; (d) *maintenance of morale*—prevent one's own forces from losing the will to fight; (e) *offensive action*—maintain operational tempo by attacking the enemy; (f) *flexibility*—be able to respond to unexpected changes or attacks and be able to modify one's plans accordingly; (2) *cooperation*—ensure that the maritime, land, and air components work in unison to achieve the end state; (h) *security*—Prevent the enemy from benefiting from lapses in operational security at all stages of the military process; (i) *surprise*—seize the initiative by attacking the enemy when, where, and how they least expect and it; (j) *sustainability*—support, fuel, and guide forces to maintain operational capability.* See also PRINCIPLES OF MILITARY STRATEGY, PRINCIPLES OF WARFARE.

PRINCIPLES OF WARFARE [psychology] (1) Ancient principles (*Book of Deuteronomy*) prescribes how the army was to fight, including dealing with plunder, enslavement of the enemy, protection of women and children, and forbidding the destruction of fruit-bearing trees. (2) Modern principles The Hague and Geneva Conventions. (3) Twenty-first-century issues: Use of private contractors as soldiers or private armies and whether they are mercenaries under international conventions. In addition, several classes of weapons, such as landmines or cluster bombs, have been decried by nongovernmental organizations and some governments as inherently inhumane. Laws of land warfare in such a way as to exclude captives from these organizations from the status of prisoner of war. (4) Descriptive principles of warfare: Sun Tzu's *The Art of War*, written approximately in 400 BC, listed five basic factors for a commander to consider: (a) the *moral law*, or discipline and

*Ardant du Picg, col. C.-J.-J.-J. (2003), *Battle Studies*, [EBook #7294] (http://onlinebooks.library.upenn.edu/webbin/gutbook/lookup?num=7294).

unity of command; (b) *heaven*, or weather factors; (c) *Earth*, or the terrain, (d) the *commander*; and (e) *method and discipline*, which included logistics and supply. (5) Early Western theoreticians: Based on the experiences of the Napoleonic Wars. Jomini's approach was more theoretical than von Clausewitz's. (6) Modern NATO principles of warfare includes objective, offensive action, surprise, concentration, economy of force, security, mobility, and cooperation. (7) The US Army's *Field Manual* (100–5) listed the following basic principles: (a) *objective*—direct every military operation toward a clearly defined, decisive, and attainable objective; "The ultimate military purpose of war is the destruction of the enemy's armed forces and will to fight", (b) *offensive*—seize, retain, and exploit the initiative; even in defense, a military organization is expected to maintain a level of aggressiveness by patrolling and launching limited counter-offensives; (c) *mass*—mass the effects of overwhelming combat power at the decisive place and time; (d) *economy of force*—employ all combat power available in the most effective way possible; allocate minimum essential combat power to secondary efforts; (e) *maneuver*—place the enemy in a position of disadvantage through the flexible application of combat power; (f) *unity of command*—for every objective, seek unity of command and unity of effort; (g) *security*—Never permit the enemy to acquire unexpected advantage; (h) *surprise*—strike the enemy at a time or place or in a manner for which they are unprepared; and (i) *simplicity*—Prepare clear, uncomplicated plans and concise orders to ensure thorough understanding. (8) The British military adds *maintenance of morale and administration* to this list. (9) Russian doctrine is similar, but includes the concept of *annihilation* as well.† See also PRINCIPLES OF WAR; PRINCIPLES OF MILITARY STRATEGY.

PRINCIPLES OF ZOOLOGICAL PHILOSOPHY [biology] (Etienne Geoffroy Saint-Hilaire, 1772–1844, French zoologist) Belief in the "unity of plan"—the idea that all vertebrates, and ultimately all animals, are variations on a single design. Asserts that all vertebrates have fundamentally the same skeletal system, and that each part has a parallel in all other creatures.* See also HALDANE'S PRINCIPLE; DARWINIAN PRINCIPLE OF NATURAL SELECTION.

PRIORITY PRINCIPLE [psychology] (1) The approach is part of a more comprehensive model designed to maximize successful engagement with a minimum amount of professional time and effort: (a) less confrontative, thereby avoiding the reactivity in clients and family members that such confrontational approaches have tended to evoke; (b) takes into account the needs of the chemically dependent person as well as the needs of the larger family and network system; and (c) aimed toward enrolling substance abusers in outpatient (as well as inpatient) treatment, thus placing it more in line with managed care priorities.[†] (2) The more controls are designed to deal with and reflect the specific nature and structure of plans, the more effectively they will serve the interests of the enterprise and its managers. See also PRINCIPLE OF PRIORITY; MCKENZIE PRINCIPLES.

PRIVACY PRINCIPLE [computer Science] (1) User centered, barrier free and format independent. (2) Alternate methods, approaches and access points. (3) Minimize restrictions on access and maximize dialog. (4) Essential component of intellectual freedom and right of all people. Data collected must have well defined limits. Individuals have the right to inspect, correct and remove data. Personal information may be shared only with informed consent.[‡] See also INTELLECTUAL FREEDOM PRINCIPLE; INTELLECTUAL PROPERTY PRINCIPLE; INFRASTRUCTURE PRINCIPLE; [UBIQUITY PRINCIPLE;]

EQUITY OF ACCESS PRINCIPLE; PRINCIPLES FOR NETWORKED WORLD.

PROACTIONARY PRINCIPLE [psychology] Places emphasis on action for the certain consequences of inaction, as opposed to the potential consequences of action.[§] See also PRECAUTIONARY PRINCIPLE.

PROBABLISTIC EXCLUSION PRINCIPLE see EXISTENCE PRINCIPLE An existence principle for boundary value problems for second order functional … An observation density for tracking which solves a problem for identifying identical targets of several independent trackers which may coalesce onto a best fitting group. Exclussion arises naturally from a systematic derivation of the obsrvation density. Describes how occlusion reasoning about solid objects can be incorporated naturally into the framework of a probabilitic distribution association filter (pdaf).

$$p_0(n; z) = b(n)/L^n$$
$$p_1(n; z|v) = b(n-1)\sigma_{k=1}^n g(z_k|v)/nL^{n-1}$$
$$p_2(n; z|v_1, v_2) = b(n-2)\sigma_{i \neq j} \frac{g(z_i|v_1)g(z_j|v_2)}{L^{n-2}n(n-1)}$$

As described so far, the generative model assumes that if a target boundary is present, then the edge detector will detect it. This is unrealistic occasionally the target object's boundary is not detected, because the background and target happen to have similar greyscale values. Hence a final step is added to the generative model. It is assumed that when $c = 1$ there is a small fixed probability q_{01} of the edge detector failing to detect the target boundary, and $q_{11} = 1 - q_{11}$ that it will succeed. This is precisely analogous to the nondetection probabilities used in PDAFs (Bar-Shalom and Fortmann, 1988). Similarly, when $c = 2$, there are fixed probabilities q_{02}, q_{12}, q_{22} that 0, 1, 2 target boundaries are detected successfully. Thus we can define pdfs $\bar{p}$ for the final generative model as

*Count de Buffon's *Histoire Naturelle* (1849–1888; 36 vols.).

[†]Garrett, J., Landau, J., Shea, R., Stanton, M., Baciewicz, G., and Brinkman-Sull, D. (1998), The ARISE intervention: Using family and network links to engage addicted persons in treatment, *J. Substance Abuse Treat.* **15**(4), 333–343.

[‡]Office of the Privacy Commissioner, Australian govt. (1988), *National Privacy Principles*, Private Sector Information Sheet.

[§]More, M. (2004), *Proactionary Principle*, Extropy Institute's Vital Progress Summit I. (http://www.maxmore.com/proactionary.htm).

follows, for[*] the cases $c = 0, 1, 2$:

$$\bar{p}_0(\cdot) = p_0(\cdot)$$
$$\bar{p}_1(\cdot|v) = q_{01}p_0(\cdot) + q_{11}p_1(\cdot|v)$$
$$\bar{p}_2(\cdot|v_1, v_2) = q_{02}p_0(\cdot) + q_{12}(p_1(\cdot|v_1)$$
$$+ p_1(\cdot|v_2))/2 + q_{22}p_2(\cdot|v_1 \cdot v_2)$$

PROCESS PRINCIPLE [psychology, physics] (1) Effectiveness of group treatment, independent of the specific therapeutic orientation used. The authors then briefly review recent findings regarding the relative effectiveness of interventions based on formal theories of change compared to placebo groups, and discuss the implications of these findings for identifying and understanding the mechanisms of change underlying the positive treatment effects observed across a variety of trauma treatment modalities. The authors subsequently focus on two illustrative principles—group development and interpersonal feedback—that show promise for complementing existing trauma treatment protocols. These principles are then concretely applied in the form of a practical case example, illustrating how small-group process principles can be integrated into, and complement, a representative "gold standard" group-based trauma treatment protocol. The authors conclude with a discussion of the incremental benefits that may accompany the integration of group-based knowledge and therapeutic skills.[§] (1) The more competitive are the conditions under which a firm operates (i.e., where others are striving for the same goals), the more important it is that the plans be chosen in the light of what a competitor will or will not do in the same area. (2) Internal finishing process of a nonferromagnetic tubing by applying a linearly traveling magnetic field.[†]

PROCESSING PRINCIPLE [psychology, computers, engineering] (F. I. Craik and R. S. Lockhart) Memory trace is a byproduct of perceptual analysis, and its persistence is a positive function of the depth to which this analysis is performed.[‡] If a test given to a subject deals with meaning, and is conceptually driven, performance will be best if the subject had concentrated primarily on meaning while originally encoding the stimuli. If the test deals primarily with the perceptual or physical aspects, and is data-driven, performance will be best if the subject had concentrated on the physical properties of the stimuli while encoding the material. It is possible to find variables that have differing effects on explicit and implicit tasks not because they are using different memory systems but rather because they are testing different kinds of information.[§] See INFORMATION-PROCESSING PRINCIPLE; also referred to as *encoding specificity principle; principle of transfer-appropriate processing.*

PRODUCER/CONSUMER PRINCIPLE [engineering] Diversion of sediment−water suspension into the sewer system is no longer permitted on grounds of environmental protection. This would produce no cost savings because introduction of the sediment−water suspension into the wastewater network would mean payment of the corresponding fees or taxes to the applicable local authorities for sewage purification; these fees and taxes would be even higher than the expense involved in equipping and maintaining a decantation plant.[¶]

PRODUCT SOLUTION PRINCIPLE [mathematics] Applying computer algebra to develop pressure distribution in reservoirs.[‖]

[‡]Giboin, A. (1979), The depth or levels of processing principle, *L'Année Psychol.* **79**(2), 623–655.
[§]Craik, F. I. M. and Lockhart, R. S. (1972), Levels of procession: A framework for memory research, *J. Verbal Learn. and Verbal Behav.* **11**, 671–684.
[¶]Collet, C., Vargas-Solar, G., and Grazziotin-Ribeiro, H. (1998), Towards a semantic event service for distributed active database applications, *Book Series Lecture Notes in Computer Science*, Springer, Berlin/Heidelberg, vol. 1460, pp. 16–27.
[‖]Chen, H. Y., Poston, S. W., and Raghavan, R. (1991), An application of the product-solution principle for instantaneous source and Green's functions, *Resource Relation SPE Formation Evaluation* (Soc. Petroleum Engineers) **6**(2), 161–168.

[*]Maccormick, J. and Blake, B. (2000) A Probablilistic exclusion principle for tracking multiple objects. *International Journal of Computer Vision* **39**(1) 57–71.
[†]Wicke, E. (1957), A new process principle with the fluidized bed, *Chem. Eng. Sci.* **6**, 160–169.

PRODUCTION PRINCIPLE [engineering] See OVERSHOOT PRINCIPLE; MINIMUM PRODUCTION PRINCIPLE; MAXIMUM PRODUCTION PRINCIPLE.

PROFIT IMPACT OF MARKET STRATEGIES (PIMS) PRINCIPLES [mathematics] (1) Address generally the relationships between market structure, market strategies, and business performance. As an extension of industrial organization economics, the PIMS project was launched in 1972 to explore the causal linkages among the three sets of factors with due emphasis on strategy issues. Identifies a large number of business principles pertaining to relationships between performance and the use of various marketing strategies. The PIMS project has developed a database composed of the market conditions, competitive positions, and financial performance relative to competition of about 3000 US manufacturing businesses.*

PROJECTION PRINCIPLE [psychology] (1) Syntactic theories of scrambling. The extended projection principle (EPP) states that each level of syntactic representation is a uniform projection of the lexical properties of heads. Contrasts cross-linguistic predictions made by a grammar-derived parsing model with those of a top–down model whose functional motivation is nonlinguistic. This latter minimal attachment model significantly predicts difficulty with respect to the processing of languages such as Japanese, which display surface properties different from those in English. This problem is not encountered in a model that recognizes the crucial role of heads in licensing argument structure with respect to processing as well as grammar. Cross-linguistic parsing differences are attributed to the linear and structural positions of licensing heads, which constitute the primary locus of the cross-linguistic variation.† (1) Representations at each level of representation are projections of the features of lexical items,

notably their subcategorization features. (2) If F is a lexical feature, it is projected at each syntactic level of representation (D structure, S structure, logical form). See also EXTENDED PROJECTION PRINCIPLE.

PROOF OF PRINCIPLE [physics] Evidence an application or process will function under specific conditions or circumstances.‡

PROOF PRINCIPLE [mathematics, physics] The initialized invariance rule for invariance and the well-founded liveness rule, are formalized in terms of a "program P leading from one state formula ' to another state formula." This notion, although precisely defined, results in intuitive rules and rather readable proofs.§ See INDUCTION PRINCIPLE; PROOF OF PRINCIPLE.

PROPORTIONAL INDUCTION PRINCIPLE [mathematics, physics] See INDUCTION PRINCIPLE.

PROTECTION PRINCIPLE [evolution] Many species of animals face the continual problem of balancing the trade-off between reducing predation risks and maintaining or increasing their reproductive fitness. The terms of the trade-off are often asyummetric: each separate behavioral decision may lead to only a marginal increase in fitness, but may place the organism's entire future reproduction in jeopardy. Consequently, the orgranism's reproductive value is an important component of most antipredator decision problems. The larger the current reproductive asset, the more important it becomes to protect it. Because reproductive value is usually age and condition dependent, optimal antipredator behavior also often depends on these variables. see also ASSET PROTECTION PRINCIPLE.¶

*Buzzell, R. and Gale, B. (1987), *The PIMS Principles: Linking Strategy to Performance*, Free Press, New York.
†Pritchett, B. (1991), Head position and parsing ambiguity, *J. Psycholing. Res.* **20**(3), 251–270.

‡Hellman, S. (1994), Immunotherapy for metastatic cancer: Establishing a "proof of principle," *JAMA* **271**, 945.
§Manna, Z. and Pnueli, A. (1984), Adequate proof principles for invariance and liveness properties of concurrent programs, *Sci. Comput. Program.* **4**(3), 257–289.
¶Clark, C.W. (1994) Antipredator behavior and the asset-protection principle. *Behavioral Ecology*, **5**, (2) 159–170.

PROXIMAL–DISTAL PRINCIPLE [psychology] See also CEPHALOCAUDAL PRINCIPLE; GENERAL-TO-SPECIFIC PRINCIPLE.

PROXIMATE PRINCIPLE [chemistry] An obsolete term for a substance extracted from its complex form without destroying or altering its chemical properties.*

PROXIMITY COMPATIBILITY PRINCIPLE [computer science] Visually unitary configurations of data values, such as object displays, will support information integration better than will more separable formats. Two mechanisms may underlie object display advantages for integration—one based on relational properties or emergent features and one based on the efficient processing of the lower—level codes themselves.†

PROXIMITY PRINCIPLE [ecology] Waste should be disposed of (or otherwise managed) close to the point at which it is generated, thus aiming to achieve responsible self-sufficiency at a regional or sub regional level.‡

PRUDENT AVOIDANCE PRINCIPLE [psychology] (Prof. Granger Morgan, Carnegie Mellon Univ.) Reasonable efforts to minimize potential risks should be taken when the actual magnitude of the risks is unknown.§

PSEUDO–GENERALIZED VARIATIONAL PRINCIPLE (GVP) [mathematics, engineering] (Eric Reissner) See HELLINGER–REISSNER PRINCIPLE.

PSEUDO-2-PATH PRINCIPLE [physics] Modern wireless communication systems applications for both digital and analog signal processing. Bandpass modulators provide a method for performing analog-to-digital conversion of IF signals.¶

PSYCHOLOGY PRINCIPLES [psychology] Five principles revising the ethical code for psychology. See also PRINCIPLE A: BENEFICENCE AND NONMALEFICENCE; PRINCIPLE B: FIDELITY AND RESPONSIBILITY; PRINCIPLE C: INTEGRITY; PRINCIPLE D: JUSTICE; PRINCIPLE E: RESPECT FOR PEOPLE'S RIGHTS AND DIGNITY‖

PULL PRINCIPLE [engineering] (George J. M. Darrieus) See VERTICAL ROTATING PRINCIPLE.

PULSE COINCIDENCE PRINCIPLE [computer science] See PRINCIPLE OF PULSE VARIATION.

PULSE ECHO PRINCIPLE [physics] (1) Used for detecting the difference in wavelenth between the incident beam and that reflected from the approaching or receding car or plane.** (2) Changes the received frequency of the signal from its transmitted frequency by an amount proportional to the relative velocity of the transmitter and the receiver. (3) Continuous frequency source with a vehicle of known velocity by a means of determining direction, velocity, temperature, and ability to record. (4) Enables one to ensure the velocity of moving phase objects. (5) The change in frequency of all wave motion or signals introduced by relative motion of object to an observer. (6) Intrusion monitoring system in passage opening. A moving reflector will shift the frequency of returning echoes in

*Patil, N. D, and Phadke, M. S. (1989), Proximate principle and energy content of human milk in well-nourished urban mother, *Indian Pediatri.*, **26**(12), 1211–1213.
†Marino, C. J. and Mahan, R. R. (2005), Configural displays can improve nutrition-related decisions: An application of the proximity compatibility principle, *Human Factors* **47**(1), 121–130.
‡Thomson, V. E. (2007), Regionalization of municipal solid waste management in Japan: balancing the proximity principle with economic efficiency, *Environ. Manage.* **40**(1), 12–19.
§Knave, B. (2001), Electromagnetic fields and health outcomes, *Ann. Acad. Med.* (Singapore), **30**(5), 489–493.

¶Tan, S. E., Inoue, T., and Ueno, F. (1993), A design of narrow-band band-pass SCFapos;s using a capacitor-error-free voltage inversion, Int. *Proc. IEEE Symp. Circuits and Systems*, ISCAS, May 3-6, 1993, vol. **2**, pp. 1034–1037.
‖American Psychological Association (2002), *The Ethical Principles of Psychologists and Code of Conduct* (http://www.apa.org/ethics/code2002.html).
Petrov, M. P., Ivanov, A. V., and Paugurt, A.P. (1983), Principles of the treatment of pulse radiofrequency signals by a spin echo method, *Radiospektroskopiya* (Perm), **15, 86–95.

proportion to its velocity.* (7) Short bursts of ultrasound and electronic range gates used to analyze signals from specific depths.† (8) Used for the imaging of soft body tissue; involves the transmitting of short bursts of ultrasonic energy and recording echoes reflected from anatomic structures within the body.‡ Also known as *Doppler Principle; Pulsed Doppler Principle; Echo Pulse Principle*.

PULSE OXIMETRY PRINCIPLE [physics] Pulse oximeters determine arterial blood oxygen saturation by measuring the light absorbance of tissue at two different wavelengths and using the arterial blood pulsation to differentiate between absorbance of arterial blood and other absorbers (ski, bone, venous blood). A good choice of wavelength is where there are large differences in the extinction coefficients of deoxyhaemoglobin and oxyhaemoglobin. Another criterion for the wavelength selection is the relative flatness of the absorption spectra around the chosen wavelength. The two conventional wavelenghts used in pulse oximetry are 660 nm (red) and 940 nm (near infrared).§

PULSE TRANSIT-TIME PRINCIPLE [physics, mathematics] (1) Equals the spacing in time between emission of a transmit pulse and reception of the echo pulse reflected from the surface. (2) Using the round-trip travel time of a pulse reflected from some object to deduce the distance to that object.¶

*Mitchell, M. R., Tarr, R. W., Conturo, T. E., Partain, C. L., and James, A. E., Jr. (1986), Spin echo technique selection: Basic principles for choosing MRI pulse sequence timing intervals, *Radiographics* **6**(2), 245–260.

†Frielinghaus, R. (1989), Echo pulse method. Part I. The principle and fields of application of ultrasound testing of thermoplastics, *Plastverarbeiter*, **39**(11), 50–52, 54, 56.

‡Harada, J. (1997), Principle and clinical application of MRI, *Jpn. J. Clin. Pathol.* **45**(3), 237–241.

§Kyriacou, P. A. (2006) Pulse oximetry in the oesophagus. Institute of Physics Publishing Physiological Measurement. *Physiol. Meas.* **27**, r1–r35.

¶Foo, J. Y. A. Wilson, S. J. Williams, G., Harris, M.-A., and Cooper, D. (2005), Pulse transit time as a derived noninvasive mean to monitor arterial distensibility changes in children, *J. Human Hypertension*, **19**(9), 723–729.

PULSED DOPPLER PRINCIPLE [physics]   See PULSE ECHO PRINCIPLE.

PUNCTUATION PRINCIPLE [psychology]   See TRANSACTION PRINCIPLE.

PUNGENT PRINCIPLE [chemistry] (1) Feeding deterrence, repellency, toxicity, sterility, and growth-disruptive activities. Azadirachtin, the major bioactive principle of *Azadirachta indica* and azadirachtin-based formulations, shows a wide array of pest control properties also induce insect repellent, antifeedant, and growth inhibitory effects in various insects isolated from either the whole plant or its parts such as leaf, flower, seed, stem, bark, root, and rhizomes.‖ (2) Olfactory repellents include cinnamic aldehyde, methyl nonyl ketone, and essence of red pepper, otherwise known as *capsicum* and *quinine*.** (3) Bind to a receptor at the nerve ending and cause a release of a variety of neuropeptides and results in afferent nerve conduction initially. There are two main phases of action: first an excitation and then a desensitization of the nerve to stimulation. The excitation results in the "hot" of hot pepper or the burning, tingling sensation. Used as a counterirritant in various medicinal preparations to soothe irritated skin and diminish pain of a variety of inflammatory conditions, including but not restricted to arthritis, osteoarthritis, myalgias, neuralgias, lumbago, and low-back pain. Ability to deplete and prevent reaccumulation of substance from local sensory nerve terminals.†† Also called *bioactive principle*.

PUNISHMENT PRINCIPLE [psychology]   To stop a child from acting in a certain way,

‖Parmar, B. S. and Walia, S. (2001), Prospects and problems of phytochemical biopesticides, in *Phytochemical Biopesticides*, Koul, O. and Dhaliwal, G. S., (eds.). Harwood Academic Publishers, Amsterdam, chap. 9.

Matsuoka, H. et al. (1997), Formation of thioxopyrrolidines and dithiocarbamates from 4methylthio 3 butenyl isothiocyanates, the pungent principle of radish, in aqueous media, *Biosci. Biotechnol. Biochem.* **61, 2109.

††Garnett, H. and Grier, J. (1908), The pungent principle of ginger, *Pharma. J.* **79**, 118–120, 156.

deliver an aversive stimuli immediately after the action occurs. Since punishment results in increased hostility and aggression, it should be used only infrequently and in conjunction with reinforcement.[*]

PUSH PRINCIPLE [engineering] (George J. M. Darrieus)   See VERTICAL ROTATING PRINCIPLE.

PYTHAGORAS PRINCIPLE [physics] (Pythagoras of Samos, 570 BC - 495 BC, Greek Philosopher)   (1) The square of the length of the hypotenuse of a right triangle is equal to the sum of the squares of the lengths of its two legs. (2) The power and esoteric side of numbers remains as yet unknown but were intensively studied by Pythagoras. Pythagoras referred to numbers in the following terms: monad—1, duad—2, triad—3, tetrad—4, pentad—5, hexad—6, heptad—7, ogdoad—8, ennead—9, and decad—10.[†] See also PRINCIPLE OF NUMEROLOGY; PRINCIPLE OF ASSUMPTION.

[*]Huitt, W. (1994), Principles for using behavior modification, *Educational Psychology Interactive*, Valdosta State Univ., Valdosta, GA (http://chiron.valdosta.edu/whuitt/col/behsys/behmod.html).

[†]Euclid, (1956), *The Elements*, 2nd ed. (transl. with introduction and commentary by Sir Thomas L. Heath), Dover (3 vols.).

Q

QCC PRINCIPLE See QUALITY CONTROL CHART PRINCIPLE.

QUADRATIC EQUIVALENCE PRINCIPLE [mathematics]   A nondegenerate linear programming problem is equivalent to a separable quadratic one, obtained by adding a special penalty term to the original problem.[*]

QUALITY CONTROL CHART PRINCIPLE (QCC) [mathematics]   Determination of low concentration of pesticides in potable water; enabling improved statistical quality control.[†]

QUANTAL PRINCIPLE [mathematics]   See GIBBS THIRD VARIATIONAL PRINCIPLE.

QUANTIZATION COMMUTES WITH REDUCTION PRINCIPLE [mathematics]   Describes the relationship of the geometric quantization of a Hamiltonian manifold with the quantization of its symplectic quotient.[‡]

QUANTIZATION DESIGN PRINCIPLE [physics]   A necessary condition for optimal vector quantizers; design principle derived by applying Gersho's theory to multiple disjoint clusters.[§] See also EQUIDISTORTION PRINCIPLE.

QUANTIZED FEEDBACK PRINCIPLE [physics]   Using in the proposed clock generator a new synchronization technique, which even in pulse transmission systems has a strong bandwidth limitation, one can generate clock signals from the received pulse signals.[¶] See also DECISION FEEDBACK PRINCIPLE.

QUANTUM ACTION PRINCIPLE [engineering]   (1) Penetration of a substance into the body of another. (2) Reduction in intensity.[‖] See also ACTION PRINCIPLE; PRINCIPLE OF CONCEPTUAL DESIGN; SCHWINGER' QUANTUM ACTION PRINCIPLE.

QUANTUM EQUIVALENCE PRINCIPLE (QEP) [physics, mathematics]   (1) A natural mapping of paths in a curved space onto the paths in the corresponding (tangent) flat space may be used to reduce the curved-spacetime path integral to the flat-spacetime path integral. The dynamics of the particle in a curved spacetime is then expressed in terms of an integral over paths in the flat (Minkowski) spacetime.[**] (2) The relation of internal symmetry quantum numbers to the geometry of the spacetime manifold.[††] See EQUIVALENCE PRINCIPLE.

QUANTUM HAMILTON PRINCIPLE [physics]   Obtained by the combination of two

[*]Lin, J. and Wan, B. (1986), Quadratic equivalence principle and point convexifying technique in optimization and control of linear steady state systems, *Large Scale Syst.* **11**(1), 59–68.

[†]Vogelgesang, J. (1991), The quality control chart principle: Application to the routine determination of pesticide residues in water, *Fresenius' J. Anal. Chem.* **340**(6), 384–388.

[‡]Guillemin, V. and Sternberg, S. (1982), The quantization of the symplectic quotient. *Invent. Math.* **67**(3), 515–538.

[§]Vilaprinyo, E., Alves, R., and Sorribas, A. (2006), Use of physiological constraints to identify quantitative design principles for gene expression in yeast adaptation to heat shock *BMC Bioinformatics* p. 7.

[¶]Ahn, C., Doherty, A. C., and Landahl, A. J. (2002), Continuous quantum error correction via quantum feedback control, *Phys. Rev. A Atom. Mol. Opt. Phys.* **65**(4-A).

[‖]Thoss, M. and Englert, B. G. (1996), A quantum action principle for open systems, *Lett. Math. Phys.* **37**, 293.

[**]Mensky, M. B. (1996), Classical and quantum equivalence principle in terms of the path group, *J. Relativistes* **96**(P. I) (Ascona, 1996) *Helv. Phys. Acta* **69**(3), 301–304.

[††]Osipova, A. V. (1984), (2-SCS) Quantum equivalence principle, *Proc. Sir Arthur Eddington Centenary Symp.*, Nagpur, India, 1984, vol. 3, pp. 236–244.

variational principles, a variational characterization of the logarithm of the wavefunction. The role of background noise has the intuitive meaning of attempting to contrast the more classical mechanical features of the system by trying to maximize the action in the first principle and by trying to increase the entropy in the second. The first variational principle includes as special cases both the Guerra–Morato variational principle and Schrödinger original variational derivation of the time-independent equation.* See also STEWART–HAMILTON PRINCIPLE; STOCHASTIC HAMILTON VARIATIONAL PRINCIPLE; PRIGOGINE PRINCIPLE; GYARMATI'S PRINCIPLE; PRINCIPLE OF MINIMUM MECHANICAL ENERGY; PRINCIPLE OF MINIMUM POTENTIAL ENERGY; PRINCIPLE OF MINIMUM KINETIC ENERGY; TOUPIN'S DUAL PRINCIPLE; D'ALEMBERT'S PRINCIPLE; HAMILTON PRINCIPLE.

QUANTUM LINEARITY PRINCIPLE [geometry] The sum of the squares of the lengths of the diagonals of a parallelogram is equal to twice the sum of the squares of the length of its sides.[†]

QUANTUM MECHANICAL SUPERPOSITION PRINCIPLE [physics] (1) When a situation is a composition of a number of elementary situations, its amplitude is the linear superposition of the amplitudes of the components.[‡] (2) When a quantum mechanical transistion can take place via more than one path, then the probability of that transistion is calculated as the square of the sum (superposition) of the probabilities for each path.[§] See also PRINCIPLE OF SUPERPOSITION; OPTICAL SUPERPOSITION PRINCIPLE.

QUANTUM MECHANICS PRINCIPLE [physics] (1) Description of motion and interaction of particles at the small scales where the discrete nature of the physical world becomes important. (2) The very idea of making an observation implies that what is observed is totally distinct from the person observing it. This paradox is avoided by taking note of the fact that all real observations are, in their last stages, classically describable. (3) Small particles do not possess a position until you observe them. Once observed, there is one-and-only-one position (up to measurement uncertainties) of the particle. (4) All energy exists in packets, called quanta, and that the quanta of electromagnetic radiation, called photons, act like particles.[¶]

QUANTUM PRINCIPLE OF UNCERTAINTY [engineering, physics] (Werner Heisenberg, 1901–1976, German physicist) (1) Relation whereby, if one simultaneously measures values of two canonically conjugate variables, such as position and momentum, the product of the uncertainties of their measured values cannot be less than approximately Planck's constant divided by $2p_i$.[‖] (2) The more accurately one determines the position of a particle, the less accurately the motion can be known and vice versa.[**] See also UNCERTAINTY PRINCIPLE.

QUARTZ MICROBALANCE PRINCIPLE [physics] Gravimetric sensor that can be used in the dynamic regime to determine a mass/potential transfer function. The principle is equivalent to classical electrochemical impedance measurements; the only difference is the determination of mass changes given by the quartz crystal microbalance rather than current

*Pavon, Michele. (1995), Hamilton's principle in stochastic mechanics, *J. Math. Phys.* v. 36, p.6774.
†Rylov, Yu. A. (1991), The problem of pair production and a statistical principle, *Phys. Essays* **4**(3), 300–314.
‡Roos, M. (1966), Principle of superposition in quantum mechanics, *Commentaliones Physico Math.* **33**(1), 1–17.
§Odagiri, M. (1957), Further studies on quantum conditions (commutation relations) and Heisenberg's principle of uncertainty from the viewpoint of chemical superposition of states, *Sci. Rep. Soc. Res. Theoret. Chem.* (Osaka) **3**, 13–21.

¶Nikulov, A. (2007), Fundamental differences between application of basic principles of quantum mechanics on atomic and higher levels, Los Alamos National Laboratory, *Preprint Archive—Condensed Matter*.
‖Ion, D. B. and Ion, M. L. D. (2007), Principle of minimum distance in space of states as new principle in quantum physics, *Romanian Rep. Phys.* **59**(4), 1045–1080.
Seife, C. (2005), Do deeper principles underlie quantum uncertainty and nonlocality? *Science* **309(5731), 98.

changes following sine-wave modulations of the applied potential.*

QUASILOCAL   PRINCIPLE [pharmacology] (1) Quartz crystal microbalances (QCMs) are used in high-technology laboratories and in space applications such as detecting contamination on optical surfaces, measuring outgassing, and determining constituents by

means of thermogravimetric analysis.[†] See also STAFF'S DIAGNOSTIC PRINCIPLE; PRINCIPLE OF PSYCHONEUROIMMUNOLOGY; DIAGNOSTIC PRINCIPLE.

QUATERNION MOMENT-PRESERVING PRINCIPLE [mathematics]   See MOMENT-PRESERVING PRINCIPLE.

---

*Atige, J. et al. (1995), Versatile microcontrolled gas sensor array system using the quartz microbalance principle and pattern recognition methods, *Sensors Actuators B Chem.* **26**, 181.

†Oks, E. A. and Gavrilenko, V. P. (1983), Quasilocal principle of measurements of electric fields in plasma according to satellites of helium forbidden lines, *Pis'ma Zh. Tekhnich. Fiz.* **9**(5), 257–260.

R

RADAR PRINCIPLE [physics] (1) Based on the property of electromagnetic waves to propagate at a constant speed within a homogeneous nonconducting medium, and to reflect a part of the energy at the interface of different media. The distance to an irradiated interface may be determined from the measured time difference that elapses between the transmitted and received waves of corresponding wavelengths. The electromagnetic waves have to be modulated in order to measure the transit time between the transmission and return of the measuring signal at the transmission point.* (2) Using a voltage-controlled oscillator determining the radar frequency, and a voltage source controlling the voltage-controlled oscillator, in which case the radar frequencies are modulated according to the frequency modulated continuous wave (FMCW) process by a corresponding control of the voltage source. The frequency of the voltage-controlled oscillator is measured at least regularly during the frequency sweep, and a nonlinear frequency characteristic is corrected during the frequency sweep through a corresponding control of the voltage source.* (3) Microwave energy is transmitted in the direction of the liquid level and the bottom of the tank from an antenna located above the liquid level. A measuring signal reflected from the liquid level is received by an antenna, and the liquid level may be determined from the transit time of the measuring signal. Signals other than the measuring signals (i.e., spurious signals, in particular a signal reflected on the bottom, and which is usually relatively strong), are filtered out or otherwise accounted for by the measuring technique. The actual distance between the antenna and the bottom of the tank (the actual bottom distance) are known, and the permittivity and permeability values for the liquid in the tank are at least approximately known. This leads to reliable results in a method that is simpler than the analytic technique, because the transit time is determined for the bottom signal and an approximate bottom distance is determined from this transit time. Then the ratio of the approximate bottom distance to the known actual bottom distance, the ratio of apparent to actual liquid level, and therefore the actual liquid level, are determined.* (4) In accordance with which microwaves of a transmission frequency are sent out by the vehicle, a part thereof is reflected back and mixed with the microwave signal at transmission frequency. In this way Doppler signals are produced which, after pulse formation, are evaluated as Doppler pulse signals by a frequency measurement in the time range plus direction-of-travel evaluation, so as to form digital Doppler signal values.[†] See also DOPPLER PRINCIPLE.

RADIATION PRINCIPLE [physics] (1) *Radiation* refers to the transfer of energy (heat) by means of electromagnetic waves. As such, heat transfer by radiation can occur across a vacuum (prime example: the sun warms the Earth even though about 93 million miles of empty space exists between the sun and Earth),[‡] (2) All objects are composed of continually vibrating atoms, with higher-energy atoms vibrating more frequently. The vibration of all charged particles, including these atoms, generates electromagnetic waves. The higher the temperature of an object, the faster the vibration, and thus the higher the spectral radiant energy. As a result, all objects are continually emitting radiation at a

[†]Klysz, G., Balayssac, J. P., and Laurens, S. (2004), Spectral analysis of radar surface waves for nondestructive evaluation of cover concrete, *NDT & E Int.* **37**(3), 221–227.
[‡]Pu, R., and King, M. D., (n.d.) *The Fundamentals of Electromagnetic Radiation Principles*, Dept. Atmospheric and Oceanic Science, Univ. Maryland.

*Van Der Pol, R. (1995), *Process for Measuring the Level of Fluid in a Tank According to the Radar Principle*, US Patent 5,438,867.

rate with a wavelength distribution that depends on the temperature of the object and its spectral emissivity.[*] See also INTERFEROMETRIC PRINCIPLE; INTERNAL EFFECT PRINCIPLE; MICHELSON INTERFEROMETRIC PRINCIPLE.

RADIATION PROTECTION PRINCIPLE [engineering]   See BUSBAR PROTECTION PRINCIPLE.

RADIOIMMUNOASSAY PRINCIPLE [physics] (1) Method of detection permitting exact and specific determinations of specifically bindable substances.[†] (2) Method depends on the fact that a conjugate of a fluorescing substance and a small molecule, for example a hapten, moves very quickly in solution so that polarized light that is passed through the solution and excites the fluorescing substance is depolarized in the case of the emission, due to the rotation of the conjugate taking place between impingement and emission. The degree of depolarization is thereby inversely proportional to the rate of rotation of the molecule.[‡] See also FLUORESCENCE POLARIZATION IMMUNOASSAY PRINCIPLE.

RADIOISOTOPE FIELD SUPPORT PRINCIPLE [engineering]   Sensing device for measuring accelerations and angular displacements relative to inertial space.[§]

RADISH PUNGENT PRINCIPLE [chemistry] (1) Formation of thioxopyrrolidines and dithiocarbamates from 4-methylthio-3-butenyl isothiocyanates in aqueous media.[¶]

[*]Tanaka, T. and Shintake, T., (eds.,) (n.d.), *Conceptupal Design Report*, SCSS XFEL, R&D Group, RIKEN Harima Inst./SPring-8, Hyogo, Japan.

[†]Rodbard, D. (1988), Radioimmunoassays and 2-site immunoradiometric "sandwich" assays: Basic principles, *Radioisotopes* **37**(10), 590–594.

[‡]Addison, G. M., Hales, C. N., and Miles, L. E. M. (1971), Principles and modifications of radioimmunological assays of protein hormones, *Proc. Symp. Immunologic Methods in Endocrinology*, 1970, pp. 22–25.

[§]Pay, R. (1967), High performance gyros seen from radioisotope techniques; utilizing the radioisotope field support principle, *Technol. Week* **20**, 40.

[¶]Matsuoka, H., Uda, Y., Mitani, K., Yoneyama, K., nad Maeda, Y. (1996), Easy preparation

RANKING PRINCIPLE REF...RANKING PRINCIPLE PROGRAMMING BY RECALL LEVEL [psychology]   (1) Emergence and transmission of organizational culture in small, growing organizations founded by teams was examined using data from a laboratory study of 40 simulated organizations and a field study of 4 actual organizations. The dimension of culture examined was the nature of human relations, which was measured using instruments based on Fiske's theory of elementary relational forms. Results indicate that the sex composition of the founding team affected both preferred and perceived norms for social interaction in the groups. In the laboratory, simulated organizations founded by women emphasized communal sharing and equality-matching principles more, and authority-ranking principles less, than did organizations founded by men, but the differences were small. The two naturally occurring organizations founded by all-female teams seemed to develop more consistent, unified cultures than did the two organizations founded by mixed-sex teams, and both female-founded organizations strongly resisted attempts to change the culture. Disagreement about relational norms contributed to the breakup of one mixed-sex founding team, and to the firing of an early hire in the other organization founded by a mixed-sex team. In the laboratory, females in groups founded by women reached consensus about relational norms more quickly than did males in groups founded by men. Female newcomers also perceived the relational culture of their new organizations more accurately than male newcomers did, when accuracy was measured as agreement with established group members. Female newcomers in previously all-male groups judged themselves to adjust to their new groups better than did male newcomers joining previously all-female groups. When newcomers arrived alone, they initially agreed more closely with the founders about relational culture than did newcomers arriving in a cohort.

method for 2-thioxopyrrolidine derivatives including 3-hydroxymethylene-2-thioxopyrrolidine, an antimicrobial degradation product of radish pungent principle, via (E,Z)-4-methoxy-3-butenyl isothiocyanate, *Biosci. Biotechnol. Biochem.* **60**(5), 914–915.

This tendency was strongest for newcomers who both arrived alone and were the only representative of their gender in the organization. The pattern of results is explained by a multicultural analysis of diversity in groups.* (2) Physiological influences on the framework for thinking and information behavior that are governed by memory and recollection standards. (3) Memories are accumulated physiologically in some part of the brain throughout one's life. But in our daily lives, the volume of memories recalled is like only the tip of an iceberg. (4) The number of items that hover around the point of easy recollection and can actually be recalled is comparatively small, although many memories are kept in the depths of one's consciousness. (5) People can understand things only by ranking them—as "celebrities," known and unknown—due to the constraints of memory and the limitations of the information intake frame. While the total human information space is vast, it is divided into many layers as a result of physiological specifications. These layers are the recall level, the recognition level (known–foggy–unknown), and the nonrecognition level.[†] See also AUTHORITY-RANKING PRINCIPLE.

RAOULT'S PRINCIPLE [chemistry] (Francois-Mane Raoult, 1830–1901, French chemist) (1) The vapor pressure of an ideal solution at any temperature is the sum of the vapor pressures of all components.[‡] (2) A relationship between the pressure exerted by the vapor of a solution and the presence of a solute. The principle states that the partial vapor pressure of a solvent above a solution is proportional to the mole fraction of the solvent in the solution and that the proportionality constant is the vapor pressure of pure solvent,

at the given temperature.[§] See also BABO'S PRINCIPLE; HENRY'S PRINCIPLE.

RAYLEIGH–JEANS   PRINCIPLE [physics] (John William Strut, 3rd Baron Rayleigh, 1842–1919, English physicist) (Sir James Hopwood Jeans, 1877–1946, English physicist, astronomer, mathematician) An expression for the distribution of energy in the spectrum of a blackbody radiator.[¶] See also PLANCK RADIATION PRINCIPLE.

RAYLEIGH   PRINCIPLE [physics]   (John William Strut, 3rd Baron Rayleigh, 1842–1919, English physicist) (1) Used to address problems related to guided waves. Can calculate partial derivatives of (a) the phase velocity with respect to either an elatic modulus or a density at either constant frequency or constant wavenumber, (b) group velocity, (c) changes in the phase velocity due to slightly anisotropic formation, and (d) changes due to a slightly irregular borehole. Avoids numerical differential, which can sometimes yield inaccurate results. (2) Also known as *Rayleigh scattering*, used to explain why the sky appears blue. Any photon coming from the sky from a location other than directly from the sun did not travel in a straight line to reach your eye. Instead, the photon must have been scattered from one or more air molecules en route. *Rayleigh scattering* refers to the probability of a photon scattering from an air molecule as proportional to $1/(\lambda_4)$. Short-wavelength photons (like blue) are much more likely to be scattered by air molecules than are long-wavelength photons (like red). This also explains the orange/red tint to the sun near sunrise or sunset (especially on hazy days). (3) Rayleigh's criterion is also used for determining when two images are discernable as two separate objects. Since light is a wave, it undergoes diffraction when passing through an aperture (such as the pupil of the eye or the lens of a telescope or

*Arrow, H. (1997), Standing out and fitting in: Culture, gender, and socialization in growing organizations. Univ. of Illinois.
[†]Nagaya, T. (n.d.), Constraints-programmed macro-model of information selection—limits of information-processing and the basic principle of information behavior, *Electron. Commun. Jpn.* (Pt. III: *Fund. Electron Sci.*) **79**(9), 1–14.
[‡]Walker, P. M. B. (1999), *Chambers Dictionary of Science and Technology*, Chambers, New York, p. 960.

[§]Daintith, J. (1999), *The Facts on File Dictionary of Chemistry*, 3rd ed., Facts On File, New York, p. 206.
[¶]Walker, P. M. B. (1999), *Chambers Dictionary of Science and Technology*, Chambers, New York, p. 962.

microscope). Rayleigh's criterion states that to separate two closely spaced objects, their angular separation must be at least large enough that the first diffraction minimum surrounding one object is no closer to the second object than its central diffraction maximum.[*]

RAYLEIGH–RITZ MINIMUM PRINCIPLE [mathematics, physics] (John William Strutt, 3rd Baron Rayleigh, 1842–1919, English physicist) (Walter Ritz, 1878–1909, Swiss-born German physicist)   Let A be a $n \times n$ Hermitian matrix. As with many other variational results on eigenvalues, one considers the Reyleigh–Ritz quotient $R:C^n \to R$ defined by

$$R(x) = \frac{(Ax, x)}{\|x\|^2}$$

where $(\cdot, \cdot)$ denotes the Euclidean inner product on $\mathbf{C}^n$. Equivalently, the Rayleigh–Ritz quotient can be replaced by

$$f(x) = (Ax, x), \|x\| = 1.$$

For Hermitian matrices, the range of the continuous function $R(x)$, or $f(x)$, is a compact subset $[a, b]$ of the real line. The maximum b and the minimum a are the largest and smallest eigenvalue of A, respectively, The min-max theorem is a refinement of this fact.

LEMMA. Let S_k be a k dimensional subspace.

1. If the eigenvalues of A are listed in increasing order $\lambda_1 \leq \ldots \leq \lambda_k \leq \ldots \leq \lambda_n$, then there exists $x \in S_k, \|x\| = 1$ such that $(Ax, x) \geq \lambda_k$.
2. Similarly, if the eigenvalues of A are listed in decreasing order $\lambda_1 \geq \ldots \geq \lambda_k \geq \ldots \geq \lambda_n$, then there exists $y \in S_k, \|y\| = 1$ such that $(Ay, y) \leq \lambda_k$.[†]

See also RITZ'S COMBINATION PRINICIPLE.

[*]Haber, F. and Lowe, F. (1910), An interferometer for chemists according to Rayleigh's principle, *Angew. Chemie.* **23**, 1393–1398.
[†]Hall, R. L. (1979), Complement to the Rayleigh-Ritz principle for the energy spectrum of a system composed of identical particles, *Phys. Rev. C* **20**(3), 1155–1160.

REACTION PRINCIPLE [biology]   Makes it possible to produce symmetric and unsymmetric polyhydroxy ethers wherein setting a sequence called *primer* as an extension point, after synthesizing a double-strand extension from the primer using single-strand nucleic acid as a template, the finished double-strand is denatured into a single strand, and then repeats double-strand synthesis.[‡]

REACTIVITY/SELECTIVITY PRINCIPLE [geology]   Concept of a reaction series for the principal rock-forming mineral.[§]

REALITY PRINCIPLE [chemistry, psychology, political science]   (1) Capacity of an atom or molecule to combine chemically with another.[¶] (2) In psychoanalysis, the idea the striving for narcissistic pleasure can never be absolute, but instead must be balanced against competing demands placed on the self by other persons and situations.[‖] See also PERFECTION PRINCIPLE; REGRET PRINCIPLE; SIMPLIFICATION PRINCIPLE; STRENGTH PRINCIPLE; RELATIVITY PRINCIPLE; COMBINATION PRINCIPLE; PARADOX-TOLERATING PRINCIPLE; EQUIVALENCE PRINCIPLE.

REALITY PRINCIPLE [psychology]   (1) Freud has distinguished between two principles of mental functioning: the pleasure principle and the reality principle. (2) Under the sway of the pleasure principle, wishes give rise to vivid images of the objects that would satisfy them. The wished-for objects may be in actuality difficult, dangerous, or impossible to

[‡]Bowen, N. L. (1922), The reaction principle in petrogenesis, *J. Geol.* **30**, 177–198; See also Bogachev, A. I. (1959), The Bowen reaction principle, *Izvest. Karel'sk. Kol'sk. Filialov Akad. Nauk SSSR*, (1), 67–70.
[§]Roth, M. and Mayr, H. (1995), The coexistence of the reactivity selectivity principle and of linear free energy relationships: A diffusion clock for determining carbocation reactivities, *Angew. Chem. Int. Ed. Engl.* **34**, 2250.
[¶]Rutsch, W. and Cech, M. A. (2007), Effects to improve the quality of life: Color, performance and protection from Ciba Specialty Chemicals, *Chem. Gesellschaft Chimia* **61**, 33–41.
[‖]Anon. (2005), Reality prinicple, *Webster's New World College Dictionary*, Wiley, P., Cleveland, OH.

attain. While the pleasure principle remains dominant, there is complete disregard for these limitations of fact and logic. The wish proceeds directly to the imagined possession of its object. This wish-fulfillment fantasy brings momentary delight. But it also brings eventual dissatisfaction. The physical needs of the organism require physical objects to satisfy them. Under the necessity of attaining physical objects, the reality principle is gradually instituted. The imperious demand of the wish is subjected to criticism from the point perspective of possibility and prudence, and is also implemented with knowledge of ways and means. The sway of the pleasure principle becomes restricted. The mere image of the wished-for thing is devaluated; it pales in comparison with the physical object. Wish-fulfillment fantasies are criticized as idle dreaming and cease to be the major preoccupation of waking consciousness.* (1) The mental activity leading to instinctual gratification by accommodating to the demands of the real world; it is acquired during development.† (2) The fundamental component of the ego; as a child grows, he/she becomes aware of the real environment and the need to accommodate to it.‡ See also PLEASURE PRINCIPLE.

RECAPITULATION PRINCIPLE [biology] (1) Haeckel's biogenetic rule.§ (2) Stages in the evolution of the species are reproduced during the developmental stages of the individual.¶ Also referred to as *biogenetic principle; Haeckel's principle.*

RECENCY PRINCIPLE [psychology] (1) The last item in a series of items carries unique temporal information that is rapidly lost over time.‖ (2) Used by police interrogators to have subjects incriminate themselves without realizing it, and more importantly, without speaking further about the topic. It exploits a known phenomenon where people focus on the most recent topic. By quickly shifting to an unrelated topic after something incriminating was said, the interrogator can avoid having to elaborate on the statement, and avoid protest from the subject.**

RECIPROCITY PRINCIPLE [physics] (1) The interchange of radiation source and detector will not change the level of radiation at the latter, whatever the shilding arrangement between them.†† (2) Demands of the external environment and that the individual adjusts to these inescapable requirements so that the ultimately secures satisfaction of her/his instinctual wishes.‡‡ (3) Occurs in nonestablished heat propagation in a solid; specifically if the heat source in one location undergoes a temperature change, then there will also be a same temperature change at the second location. (4) The antenna can function as either a transmit antenna or a receive antenna, or both simultaneously, unless a nonreciprocal device or material is used in the construction.§§

RECIRCULATING PRINCIPLE [engineering] Water from an upper steam–water disengaging drum is carried down to the lower end of banks of water-filled tubes in which stream is generated as the mixture rises to the steam separating drum under the

*Wolfenstein, M. (1944), *Character & Personality* **13**, 135–151.
†Walker, P. M. B. (1999), *Chambers Dictionary of Science and Technology*, Chambers, New York, p. 965.
‡American Heritage (2000), *American Heritage Dictionary of the English Language*, 4th ed., Houghton Mifflin, New York.
§Hilbig, R., Roesner, H., and Rahmann, H. (1981), Phylogenetic recapitulation of brain ganglioside composition during ontogenetic development, *Compar. Biochem. Physiol. B* **68**(2), 301–306.
¶Walker, P. M. B. (1999), *Chambers Dictionary of Science and Technology*, Chambers, New York, p. 965.

‖Underwood, B., Lund, A., and Malmi, R. (1978), The recency principle in the temporal coding of memories, *Am. J. Psychol.* **91**(4), 563–573.
Carlsson, G. (1997), Memory for words and drawings in children with hemiplegic cerebral palsy, *Scand. J. Psychol.* **38(4), 265–273.
††Walker, P. M. B. (1999), *Chambers Dictionary of Science and Technology*, Chambers, New York, p. 966.
‡‡Geller, E. S. (2001), *The Psychology of Safety Handbook*, CRC Press, p. 374.
§§Kreis, R., Slotboom, J., Pietz, J., Jung, B., and Boesch, C. (2001), Quantitation of localized (31)P magnetic resonance spectra based on the reciprocity principle, *J. Magn. Reson.* **149**(2), 245–250.

influence of the heat absorbed in the furnace of the boiler.[*]

RECURRING PRINCIPLE [psychology]   (1) See GENERATIVE PRINCIPLE OF WRITING.

RECURSIVE INDUCTION PRINCIPLE [mathematics, physics]   See INDUCTION PRINCIPLE.

RED QUEEN'S PRINCIPLE [biology] (L. M. Van Valen)   (1) An explanation for an unexpected pattern of extinction; namely, except for mass extinction, extinction is independent of the age of the taxon but is selective in other ways. The probability of extinction in an adaptively unified group is stationary over geologic time, varying around a constant mean. The sum of the momentary absolute fitnesses of interacting species in a biota is constant. (2) In *Through the Looking Glass*, Lewis Carroll (pseudonym of Charles Lutwidge Dodgson) states: "Now here, you see, it takes all the running you can do, to keep in the same place."[†]

REFLECTED BEAM PRINCIPLE [physics] (1) In an acoustic system consisting of a fluid medium with boundary surfaces and subject to no impressed body forces.[‡] (2) Sensitivity of reversible electroacoustic transducer when used as a microphone divided by the sensitivity when used as a source of sound is independent of the type and construction of the transducer.[§] (3) Any theorem that expresses various reciprocal relations for the behavior of some physical systems, in which input and output can be interchanged

without altering the response of the system to a given excitation.[¶]

REFLECTION PRINCIPLE [physics]   (1) Two laws concerning wave propagation: (a) incident beam, reflected beam, and normal to surface are coplanar; and (b) the beams make equal angles with the normal.[‖] (2) Refractive index variations and thickness of the layers are chosen so that partial reflections interfere constructively at a narrow band of wavelengths but not at wavelengths outside the band.[**] (3) Light within the selected narrow band is coupled back into the laser active area enhancing single longitudinal mode operation, or multilongitudinal mode operation with the modes restricted to lie within the bandwidth of the distributed Bragg reflector reflection peak.[††] (4) Stated simply, the light rays reflected by specular areas are essentially at the same intensity as the supplied light rays and thus only diminish minimally the farther the imaging device is located along the observation axis from the object to be observed. The light rays reflected by the diffuse areas, on the other hand, diminish substantially with distance the farther the imaging device is located along the observation axis from the object to be observed. This reflection principle of the diffuse areas is commonly referred to as the *inverse square law*.[‡‡] (5) Any diverging field, including light, decreases as the inverse square of the distance from the source—in this instance the diffuse area.[§§] See also BRAGG REFLECTION PRINCIPLE;

[*]Hoffmann-Schiffner, R. (1957), The cooling-water operation cycle, its operation as to practicability and economy, *Wasserwirtschaft Wassertechnik*, **7**, 436–437.

[†]Tong, D. A. (1996), Healthcare must be wary of Red Queen Principle, *Modern Healthcare* **26**(46), 88.

[‡]Zhang, W. and Sanders, B. C. (1994), Atomic beam splitter: Reflection and transmission by a laser beam, *J. Phys. B Atom. Mol. Opt. Phys.* **27**(4), 795–808.

[§]Luo, Z.-Y., Yang, L.-F., and Chen, Y.-C. (2005), Phase-shift algorithm research based on multiple-beam interference principle, *Wuli Xuebao* **54**(7), 3051–3057.

[¶]Suksin, V. S. (1974), Principles of the energy resolution of a beam of charged particles by quadrupole lenses, *Zh. Tekhnich. Fiz.* **44**(3), 481–490.

[‖]Walker, P. M. B. (1999), *Chambers Dictionary of Science and Technology*, Chambers, New York, p. 972.

[**]Asselmeyer, F. (1979), *Refractometer with Respect to the Fresnel Reflection Principle*. Chemical Abstracts Service.

[††]Bagemihl, F. (1964), Analytic continuation and the Schwarz reflection principle, *Proc. Natl. Acad. Sci. USA* **51**(3), 378–380.

[‡‡]Bart, A. G., Ditiatev, A. E., and Kozhanov, V. M. (1989), Analysis of signal transmission in interneuronal synapses based on the reflection principle, *Dokl. Akad. Nauk SSSR*, **306**(6), 1503–1507.

[§§]Sukhov, A. B. (1995), On multidimensional reflection principle, *Dokl. Akad. Nauk* **344**(5), 599–600.

FRESNEL REFLECTION PRINCIPLE; SCHWARZ RE-
FLECTION PRINCIPLE.

REFLECTIONLESS APERTURE PRINCIPLE
[mathematics] Obtain estimates (within
10 dB) of the shielding effectiveness of
enclosures having dimensions that are
large compared to a wavelength.* See also
BLACK HOLE PRINCIPLE; OPTICAL BLACK HOLE
PRINCIPLE.

REGENERATIVE BRAKING PRINCIPLE [phys-
ics] Efficient means to ensure substantially
100% onboard receptivity of the regenerated
energy.[†]

REGENERATIVE HEATING PRINCIPLE [engi-
neering] (1) Increased thermal efficiency.
(2) A continuous flow path for extraction
steam condensate to return to the feedwater
system.[‡]

REGENERATIVE PRINCIPLE [physics]
(1) Heat exchange provided by a halo-
gen atmosphere and that would also
concentrate the light rays into a concentrated beam in a more efficient manner
without the use of a separate outer envelope
or bulb. (2) Heating gases escaping from the
upper portion of combustion chamber are
reversed and led into a grating consisting of
heat storage bricks. The heating gases leave
the grating at the lower end of the hot-blast
stove. The heating phase is followed by the
cold-blasting phase, wherein the cold blast
is led from below in an upward direction
and warmed up thereby. The resultant hot
blast is discharged from the hot-blast stove

through a hot-blast outlet arranged in the
combustion chamber.[§]

REGRET PRINCIPLE [psychology] (1) Minimum decisional regret (MDR) estimation
principle Bowman's assumption was that
managers make reasonably good decisions on
average. More precisely, his assumption was
that the biases in their decisions are expected
to be small, but their variance may be imac-
ceptably large and costly. This can be viewed
as suggesting that managerial and perhaps
other expert decisionmaking is generally expected to be close to optimal in some sense.
Therefore, it is reasonable to think in terms
of how far from being optimal such decisions may have been and to construct related
measures of distance and validation tests
for the assumption of approximate optimality. Minimizing the average or total distance
from optimal plans is consistent with Bowman's assumption. Thus, this method seeks
the cost estimates for which the associated
model-optimal plans are as near to their data
analogs as possible, on average.[¶] (2) With no-
regret principles, the main idea is to develop
a contingency strategy for the area at risk in
combination with a wide-scale improvement
of the quality of life in the metropolitan area.
The integration of different policy-analytical
methods can support decisionmakers in this
complex decisionmaking process by developing a scientific framework involving problem structuring, designing alternatives, and
evaluation and selection of choice options.
Deploy the scenario methodology in combination with multicriteria analysis and fuzzy
set theory, as a useful learning tool for the
governance of complex dynamic systems. In
current debates on policymakers' possible reactions to uncertainty, the use of scenarios
in combination with proper decision support
tools is an indispensable tool for assisting
policymakers in complex planning strategies, where decision processes and actions

*Kentzinger, E., Dohmen, L., Alefeld, B., Ruecker,
U., Stellbrink, J., Ioffe, A., Richter, D., and
Brueckel, Th. (2004), KWS-3, the new focusing-
mirror ultra small-angle neutron scattering in-
strument and reflectometer at Juelich, *Phys. B
Condensed Matt.* **350**(1-3, Suppl. 1).
[†]Lee, T. H., Hong, M. H., and Seo, H. S. (2005),
*Method for Sensing Amount of Laundry of Washing
Machine.* Eur. Patent Application.
[‡]Skiepko, T. (1993), Some essential principles for
adjustment of seal clearances in rotary regenera-
tors, *Heat Transfer Eng.* **14**(2), 27–43.

[§]Edmundson, J. T. (1989), Regenerative firing of
low calorific value gas for high temperature processes, *Revue Metallurgie/Cahiers Inform. Tech.*
86(1), 55–62.
[¶]Troutt, M., Pang, W., and Hou, S. (2006), Behavioral estimation of mathematical programming
objective function coefficients, *Manage. Sci.* **52**(3),
422–434.

are characterized by deep uncertainty due to real-word complexity and a rapidly changing reality, imperfect information on decision alternatives, and the multiplicity of actors involved.* (3) Concerning the leading role of generalized supplies and demands and their contradictions in social things and ideals. (4) Relatively commanding observocontrol roles of generalized supplies and demands. (5) Describes the grim situations for the human history, reality, and the future. It is the foundation to analyze and treat the social problems, including the problems concerning ethics, politics, economics, military, psychology, ecology, sociology, history, humanities, religions, pedagogy, and so on concerning editorship as a theory and practice with social characteristics. (6) One can minimize the proportion of maximum expected loss over minimum expected loss. See also PERFECTION PRINCIPLE; REALITY PRINCIPLE; SIMPLIFICATION PRINCIPLE; STRENGTH PRINCIPLE; RELATIVITY PRINCIPLE; COMBINATION PRINCIPLE; PARADOX-TOLERATING PRINCIPLE; EQUIVALENCE PRINCIPLE.

REGULATORY PRINCIPLES AND OTHER REGULATORY PRINCIPLES OF INERTIA [psychology] Calls for reevaluating the regulatory principles governing the organization and integration of psychic functions. Recasting these principles in modified nonenergic terms suggests that regulatory principles play a different role in governing psychic activity. Once divorced from their energic dependence, the regulatory principles can be recast as operational principles determining the stimulus conditions governing activation and integration of receptive and response systems and their adaptive integration in assimilative and accommodative patterns of environmental interaction. Reformulations are suggested for the principles of determination, overdetermination, and multiple function, and for more specific regulatory principles of inertia, constancy, pleasure–unpleasure, reality,

*Torrieri, F., Concilio, G., and Nijkamp, P. (2002), Decision support tools for urban contingency policy. A scenario approach to risk management of the Vesuvio area in Naples, Italy, *J. Conting. Crisis Manage.* **10**(2), 95–112.

repetition, and primary–secondary process.[†] See PRINCIPLES OF DETERMINATION.

REGULATORY PRINCIPLES OF CONSTANCY [psychology]   See PRINCIPLES OF DETERMINATION.

REGULATORY PRINCIPLES OF PLEASURE–UNPLEASURE [psychology]   See PRINCIPLES OF DETERMINATION.

REGULATORY PRINCIPLES OF PRIMARY–SECONDARY PROCESS [psychology] See PRINCIPLES OF DETERMINATION.

REGULATORY PRINCIPLES OF REALITY [psychology]   See PRINCIPLES OF DETERMINATION.

REGULATORY PRINCIPLES OF REPETITION [psychology]   See PRINCIPLES OF DETERMINATION.

REICHENBACH'S COMMON CAUSE PRINCIPLE [mathematics] (George Sheridan Reichenbach, b. 5/25/1929, Waterbury, CT)   (1) Probabilistic correlations formulated in terms of relativistic quantum field theory.[‡] (2) If two events are correlated, then either there is a causal connection between the correlated events that is responsible for the correlation or there is a third event, a so-called (Reichenbachian) common cause, which brings about the correlation.[§] (3) Simultaneous correlated events have a prior common cause that screens off the correlation. (4) Simultaneous correlated events must have prior common causes. (5) A correlation between events A and B indicates either that A causes B, or that B causes A, or that A and B have a common cause. It also seems that causes always occur

[†]Meissner, W. (1995), The economic principle in psychoanalysis: II. Regulatory principles, *Psychoanal. Contemp. Thought*, **18**(2), 227–259.
[‡]Rédei, M. (1997), Reichenbach's common cause principle and quantum field theory, *Found. Phys.* **27**(10), 1309–1321.
[§]Rédei, M. (2001), Reichenbach's common cause principle and quantum correlations. Non-locality and modality (Cracow, 2001), *NATO Sci. Ser. II Math. Phys. Chem.*, (Kluwer Acad. Publ.) **64**, 259–270.

before their effects and, thus, that common causes always occur before the correlated events. When $\Pr(A + B) > \Pr(A) \times \Pr(B)$ for simultaneous events A and B, there exists an earlier common cause C of A and B, such that $\Pr(A/C) > \Pr(A/\sim C)$, $\Pr(B/C) > \Pr(B/\sim C)$, $\Pr(A + B/C) = \Pr(A/C) \times \Pr(B/C)$, and $\Pr(A + B/\sim C) = \Pr(A/\sim C) \times \Pr(B/\sim C)$. C is said to "screen off" the correlation between A and B when A and B are uncorrelated conditional on C.*

REINTERPRETATION PRINCIPLE [physics] (E. C. G. Stuckelberg and R. P. Feynman) A particle traveling backward in time can be interpreted as its antiparticle traveling forward in time.†

REISSNER TWO-FIELD PRINCIPLE [mathematics, engineering] (Eric Reissner, 1913–1996, German mathematician) See DISPLACEMENT PRINCIPLE.

REISSNER TWO-FIELD STRESS AND DISPLACEMENT PRINCIPLE [mathematics, engineering] (Eric Reissner, 1913–1996, German mathematician) See DISPLACEMENT PRINCIPLE.

REISSNER VARIATIONAL PRINCIPLE [mathematics, engineering] (Eric Reissner, 1913–1996, German mathematician) (1) There are two variational principles in elastostatics: Green's principle and Castigliano's principle. These are special cases of Reissner's variational principle. Green's principle and Castigliano's principle have been generalized in a symmetric manner to include elastodynamic problems.‡ (3) Theory of elasticity, especially development of variational method and behavior of beams, plates, and shells. Simultaneous independent variations in deflection and moments.§ See

*Janzing, D. (2007), On causally asymmetric versions of Occam's Razor and their relation to thermodynamics, Los Alamos National Laboratory, *Preprint Archive—Condensed Matter*, pp. 1–53.
†Kamoi, K. and Kamefuchi, S. (1971), Quantum field theory of tachyons, *Progress Theor. Phys.* **45**(5), 1646–1661.
‡Karnopp, B. H. (1968), Reissner's variational principle and elastodynamics, *ActaMech.* **6**, 158–164.
§Wu, C.-I. and Vinson, J. R. (1969), Nonlinear oscillations of plates composed of composite materials, *J. Compos. Mater.* 3, 548–561.

also HAMILTON–OSTROGRADSKĬI–REISSNER VARIATIONAL PRINCIPLE.

RELATIVE FREQUENCY PRINCIPLE [mathematics] (1) Probabilities are assigned to events that indicate likelihood that the event will occur. They should agree with the concept of relative frequency. Probabilities can be expressed mathematically but cannot be expected since, after all, they are merely probabilities. Properties of relative frequency alone does not provide a rigorous self-consistent probability theory.¶ (2) [psychology] When more than two letters are used to form pairs, the number of unique different pairs will exceed the number of unique same pairs, and a particular different pair will occur less frequently than will a particular same pair. It is conceivable, then, that the slower processing of different pairs is due to their lower relative frequency of occurrence.‖ See also INTERNAL NOISE PRINCIPLE; PRIMING PRINCIPLE.

RELATIVITY PRINCIPLE [mathematics, physics] See PRINCIPLE OF RELATIVITY.

RELAXATION PRINCIPLE [psychology] (Sándor Ferenczi) The analyst purposely creates an easy atmosphere by adjusting her or his own attitude to the unconscious mood of the patient. The significance of emotional attitudes in the psychoanalytical situation experiments in the 1930s expanded the analytic method so that difficult cases where regression predominated could be treated. Model of psychoanalysis and psychotherapy contains unique dimensions concerning the active and flexible participation of the analyst at both verbal and nonverbal interaction, levels, with emphasis on mutuality between analyst and analysand, an ongoing focus on counter transference analysis, and

¶Shafer, G. and Vovk, V. (2006), The sources of Kolmogorov's Grundbegriffe, Statist. Sci. **21**(1), 70–98.
‖Krueger, L. E. (1983), Probing Proctor's priming principle: The effect of simultaneous and sequential presentation on same-different judgments, *J. Exp. Psychol. Learn. Memory Cogn.* **9**(3), 511–523.

judicious self-disclosure by the analyst.* See FERENCZI'S RELAXATION PRINCIPLE.

RELAXED INVARIANCE PRINCIPLE [physics, mathematics, chemistry] See INVARIANCE PRINCIPLE.

RELEASE REUSE EQUIVALENCY PRINCIPLE (REP) [computers] (Robert Cecil Martin) A package is a binary deliverable like a. jar file, or a dll as opposed to a namespace like a java package or a C++ namespace. Package cohesion describe what to put inside. (1) The granule of reuse is the granule of release. (2) The granule of reuse can be no smaller than the granule of release. Anything that we reuse must also be released. Clearly, a packages is a candidate for a releasable entity. It might be possible to release and track classes, but there are so many classes in a typical application that this would almost certainly overwhelm the release tracking system.[†]

RELUCTANCE PRINCIPLE [engineering] System of dynamic braking in which the electric drive motors are used as generators and return the kinetic energy of the motor armature and load to the electric supply system.[‡]

REPULSION   PRINCIPLE [engineering] (1) Where a rotor formed of an electrically conductive material rotates relative to a stator, comprising magnets, which give rise to a rotationally symmetric magnetic field concentric with the rotation axis of the rotor.[§] (2) Measure of the opposition presented to magnetic flux in a magnetic circuit, analogous to resistance in an electric circuit; it is equal to magnetomotive force divided

by magnetic flux.[¶] Also known as *magnetic reluctance principle; electrodynamical repulsion principle*.

RESISTANCE PRINCIPLE [mathematics, physics] Sucking the suspension to flow through an orifice. As each particle passes through the orifice, it replaces its own volume of electrolyte within the orifice, momentarily changing the resistance value between the electrodes. This change produces a voltage pulse of short duration having a magnitude proportional to particle volume, and the resultant series of pulses is electronically processed.[‖]

RESISTANCE WELDING PRINCIPLE [engineering] After the joint has been brought to melting heat, an upset motion follows. The burr resulting from upsetting is removed by a grinder or a file.[**] See also BUTT-WELDING PRINCIPLE; TRANSMISSION WELDING PRINCIPLE.

RESOLUTION PRINCIPLE [physics] (1) Force which tends to increase the distance between two bodies having like electric charges, or the force between atoms or molecules at very short distances which keeps them apart.[††] (2) Quantifiers are constrained to range over acts of ground tensions specified by a class of context-free grammars called *T grammars*; special relative positions and the polarization directions.[‡‡] (3) Alignment accuracy or difference existing between refractive and

*Alexander, F. (1933), *Am. J. Orthopsychiatry* **3**, 35–43.

[†]Martin, R. C. and Dependency Inversion Group (2005), *Ten Commandments of OO Proramming* (http://www.butunclebob.com/ArticleS.UncleBob. PrinciplesOfOod).

[‡]Rahman, M. A. (1993), Modern electric motors in electronic world, *Proc. Int. Conf. Industrial Electronics, Control, and Instrumentation*, IECON '93, Nov. 15–19, 1993, vol. 2, pp. 644–648.

[§]Lembke, T. (2000), *Magnetic Bearings*, US Patent 6,118,199.

[¶]Willeke, G., Bitnar, B., Wendl, M., Kloc, C., Bucher, E., and Vallêra, A. (1998), Electromagnetic ribbon: proposal of a novel method for silicon sheet generation, *Semiconduct. Sci. Technol.* **13**, 440–443.

[‖]Gear, A. R. L. (1978), *Continuous-Flow, Resistive-Particle Counting Apparatus*, US Patent 4,103,229.

[**]Ritter, G., Ritter, K., Scherr, R., and Jursche, K. (1992), *Grid Welding Machine Operating According to the Electrical Resistance Welding Principle*, US Patent 5,134,269.

[††]Poort, E. R. and de With, P. H. N. (2004), Resolving requirement conflicts through non-functional decomposition, *4th Working IEEE/IFIP Conference on Software Architecture* (WICSA'04), p. 145.

[‡‡]Gips, J. and Stiny, G. (1980), Production systems and grammars: A uniform characterization, *Environ. Plann. B* **7**(4), 399–408.

diffractive elements.* Also known as *principle of spinor superposition; Moire interference principle; interference principle*.

RESONANCE PRINCIPLE [physics]   (1) Smallest increment in distance that can be distinguished and acted on by an automatic control system. (2) The maximum number of lines that can be discerned on the screen at a distance equal to tube height. (3) In radar, the minimum separation between two targets, in angle or range at which they can be distinguished on a radar screen. (4) Process of separating a racemic mixture into the two component optical isomers.[†] (5) For a measurement of energy or momentum of a collection of particles, the difference between the highest and lowest energies at which the response of an instrument to a beam of monoenergetic particles is at least half its maximum value, divided by the energy of the particles. (6) The procedure of breaking up a vectorial quantity into its components.[‡]

RESOURCE   CONSUMPTION   PRINCIPLE [physics, biology]   (1) Within a local volume of tissue, the resource is consumable by a subset of synaptic elements and is therefore in limited supply for the population of synapses residing in the volume. Over short timescales, this hypothesis suggests that the shift of resources into intrasynaptic and intradendritic compartments permits a local tissue volume to select a set of functioning synapses at the expense of other possible sets of functioning connections. . . . atoms and molecules from extracellular to intrasynaptic compartments represents the consumption of a shared, limited resource available to local volumes of neural tissue. Such consumption results in a dramatic competition among synapses for resources necessary for their function. In this paper, Montague[§] explores a theory in which this resource consumption plays a critical role in the way local volumes of neural tissue operate. On short timescales, this principle of resource consumption permits a tissue volume to choose those synapses that function in a particular context and thereby helps to integrate the many neural signals that impinge on a tissue volume at any given moment. On longer timescales, the same principle aids in the stable storage and recall of information. The theory provides one framework for understanding how cerebral cortical tissue volumes integrate, attend to, store, and recall information. (2) Phenomenon exhibited by an alternating current circuit in which there are relatively large currents near certain frequencies, and a relatively unimpeded oscillation of energy from a potential to a kinetic form; a special case of the physics definition. (3) A physical system acted on by an external periodic driving force in which the resulting amplitude of oscillation of the system becomes large when the driving-force frequency approaches a natural free oscillation frequency of the system. In general, any phenomenon that is greatly enhanced at frequencies or energies that are at or very close to a given characteristic value.[§]

RESPECTIVE INSPECTING PRINCIPLE [engineering]   A method of inspecting pattern defects of an object to be inspected comprising the steps of preparing a plurality of inspecting units for the object, each said inspecting unit having a proper inspecting principle, storing groups of domain signals in a memory device corresponding to addresses of the object, each said group of domain signals comprising inspection signals and

*Chao, S. D., Hayashi, M., Lin, S. H., and Schlag, E. W. (1999), The I-mixing dynamics of high rydberg states and the high resolution principle of ZEKE spectroscopy, *Book of Abstracts*, 217th ACS Natl. Meeting, Anaheim, CA, March 21–25, 1999.
[†]Bjorkman, A. (1952), Shaking of high-pressure vessels according to the resonance principle, *J. Industr. Eng Chemi.* **44**, 2459–2463.
[‡]Wolf, F. (1937), The limits of validity of the resonance principle in charge transfer, *Annal. Phy.* **30**, 313–332.

[§]Montague, P. R. (1996), The resource consumption principle: Attention and memory in volumes of neural tissue, *Proc. Natl. Acad. Sci. USA* **93**(8), 3619–3623.

inhibition signals, each said group of domain signals also having a relation with the respective prepared inspecting unit, scanning and binarizing an image of the object, reading out the domain signals of every group stored in the memory device according to an address of an inspecting position on the object, controlling the operation of each inspecting unit according to each kind of the respective domain signal read out from the memory device at the step, and forming an inspection decision according to an output signal of the inspecting units.*

RESPONSE-SHAPING PRINCIPLE [psychology] One may shape behavioral responses to meet a desired state by rewarding successively closer approximations of the desired behavioral state.[†] See also CHAINING PRINCIPLE; BACKWARD CHAINING PRINCIPLE; ONE-TRIAL PRINCIPLE; SUCCESSIVE APPROXIMATION PRINCIPLE; TRANSITIONAL STIMULI PRINCIPLE; SCHEDULE OF REINFORCEMENT PRINCIPLE.

REVELATION   PRINCIPLE [mathematics] (1) The guiding principle for the theory of implementation and mechanism design under imperfect information. It states that the range of implementable outcomes is simply the set of outcomes that give no incentive to agents to misrepresent their type.[‡] (1) With one agent, payoffs holds, and with more than one agent, the result fails and direct mechanisms may by suboptimal. (2) Truth-telling, direct revelation mechanisms can generally be designed to achieve the Nash equilibrium outcome of other mechanisms. (3) For every indirect

mechanism there exists a direct mechanism that produces the same outcome.[§]

REVENUE RECOGNITION PRINCIPLE [mathematics] One of the four main principles in the generally accepted accounting principles (GAAPs). It is also the main difference between cash-basis accounting and accrual basis accounting. In cash-basis accounting revenues are simply recognized when cash is received no matter when and how the services were performed or goods delivered. In accrual-basis accounting revenues are recognized when they are (1) realized or realizable and (2) earned no matter when cash is received.[¶] See also MATCHING PRINCIPLE.

REVERSE-HEAT-ENGINE   PRINCIPLE [engineering] A heat engine converts heat from a high-temperature area to work. Coolant substance changes from gas to liquid as it goes from higher to lower temperature. This change from gas to liquid is a phase transition and releases energy.[‖]

REVERSED   FICK   PRINCIPLE [mathematics, engineering] (Fick = cardiac index × (arterial − mixed venous oxygen content) Automated measurements of respiratory gas exchange for the determination of oxygen uptake.[**]

REVERSIBILITY PRINCIPLE [physics] (Jamais Cascio) (1) If a ray of light travels from one point to another through an optical system along a particular path, a ray can also proceed in the reverse direction along the same

---

*Hironori Endo, N. (2000) Printing method and printer that effect dot dropout inspection and recording medium prerecorded with program therefore. U.S. Patent 6,478,400.

[†]Bacsik, Z., Mink, J., and Keresztury, G. (2004), FTIR Spectroscopy of the atmosphere. I. Principles and methods, *Appl. Spectrosc. Rev.* **39**(3), 295–363.

[‡]Bester, H. and Strausz, R. (2001), Contracting with imperfect commitment and the revelation principle: The single agent case, *Econometrica*, **69**(4), 1077–1098 (http://www.jstor.org/stable/2692255).

[§]Li, S. H. (1995), A unified framework for implementation and the revelation principle, *Econ. Lett.* **49**, 335.

[¶]Anon. (2009) *Preliminary FASB-IASB Views on Revenue Recognition. Defining Issues*, 09-2, p. 2.

[‖]Traverso, M., (2004), *Reverse Heat Engines*, Washington Univ. St. Louis, MO (http://www.chemistry.wustl.edu/~courses/genchem/Tutorials/Fridge/reverse.htm).

[**]Weyland, A., Weyland, W., Sydow, M., Weyland, C., and Kettler, D. (1994), Measurement of oxygen uptake by the reversed Fick principle and respiratory gas monitoring. Does intrapulmonary oxygen uptake account for systematic differences between methods? *Anaesthesist* **43**(10), 658–666.

path.* (2) When considering the development or deployment of beneficial technologies with uncertain, but potentially significant, negative results, any decision should be made with a strong bias toward the ability to step back and reverse the decision should harmful outcomes become more likely. The determination of possible harmful results must be grounded in science but recognize the potential for people to use the technology in unintended ways, must include a consideration of benefits lost by choosing not to move forward with the technology, and must address the possibility of serious problems resulting from the interaction of the new technology with existing systems and conditions. This consideration of reversibility should not cease on the initial decision to go forward or hold back, but should be revisited as additional relevant information emerges.[†] Also referred to as *optical reversibility principle*.

REVISED PRINCIPLE [physics] Interbehavioral operationism, which provides definite criteria for objectivity.[‡] See also COLLIDER OPERATIONAL PRINCIPLE; PRINCIPLE OF OPERATIONISM; OPERATIONAL PRINCIPLE.

REVISION PRINCIPLE [physics] If a beam of light is reflected back on itself, it will traverse the same path or paths as it did before reversal.[§]

RF MARKETING PRINCIPLE [engineering] Radiofrequency products for digital cellular, personal communication services and cordless phones that will perform each of the major RF functions within wireless handsets: receiver, synthesizer, modulator, and power amplifier.[¶] See also DISTANCE-TO-FAULT PRINCIPLE.

RICOEUR'S PRINCIPLE OF DISTANCIATION [pharmacology] (Paul Ricouer, b. 1913–2005, French philosopher) (1) In the world of action, there is a general need for making what is alien and making it understandable. This is the principle of struggle between the "otherness" transforming all spatio temporal distance into cultural estrangement and the "oneness" by which all understanding aims at the extension of self-understanding. Reading is the remedy by which the meaning of the test is rescued from the estrangement of distanciation and put in a new proximity, which suppresses and preserves the cultural distance and includes the otherness within the oneness.[‖] (2) Understand the "work" of nursing as a hermeneutic process itself where healing unfolds in the everyday discourse of life, enhancing understanding of aging and retirement as aspects of loss and the relationship to personal identity.[**]

RIECKE'S PRINCIPLE [chemistry] (Edgar Eric K. Riecke, b. 1944-, Spencer, IA) Explains recrystallization in metamorphic rocks. Stressed grains in a rock will dissolve more readily than will unstressed grains in the same rock, and material may be transported between the two.[††]

RISSANEN'S MINIMUM DESCRIPTION-LENGTH PRINCIPLE [mathematics] (Jorma J. Rissanen, 1932-, Finnish Information Theorist) (1) Describes the complexity of the molecular phylogenetic tree by three terms related to tree topology, the sum of the branch lengths, and the difference between the model

*Thewlis, J., ed., (1962), *Encyclopaedic Dictionary of Physics*, New York Oxford London. Pergamon Pross.

[†]Cascio, J. (2006), *Reversibility Principle* (http://www.sentientdevelopments.com/2006/03/cascios-reversibility-principle.html).

[‡]Dutant, J. (2007), The margin-for-error principle revised, *Proc. Open Sessions of the Mind Assoc.—Aristotelian Society Joint Sessions*, Bristol, July 2007, p. 1.

[§]Ding, L. and Mukaidono, M. (2007), A proposal on approximate reasoning based on revision principle and fixed value law, *Syst. Comput. Jpn.* (Wiley Periodicals) **22**(13), 20–27.

[¶]Nicholson, P. F. (1990), Personal communication systems: Integration opportunities, *Proc. IEE Colloquium on CT2 / CAI and DECT Cordless Telecommunications*, Nov. 27, 1990, pp. 1/1-1/9.

[‖]Ricœur, P. and Valdés, M. J. (1991), *A Ricoeur Reader: Reflection and Imagination*, Univ. Toronto Press.

[**]Geanellos, R. (2000), Exploring Ricoeur's hermeneutic theory of interpretation as a method of analysing research texts, *Nurs. Inquiry* **7**(2), 112–119.

[††]Griggs, D. (1939), Creep of rocks, *J. Geol.* **47**, 225–251.

and the data measured by logarithmic likelihood.[*] (2) The best model for encoding data minimizes the sum of the cost of describing the data in terms of the model and the cost of describing the model.[†] (3) Encodes the tree and split tests in a minimum description-length (MDL)-based code, and determines whether to prune and how to prune each node on the basis of the code length of the node.[‡] (4) Electric power-line transmission-line fault detection. (5) Energy spectra of electrons emitted from laser-irradiated low density.[§] Also known as *complexity principle; maximum likelihood principle; minimum description-length principle; Turner's complexity principle*.

RITTINGER'S PRINCIPLE [chemistry] The energy required in a crushing operation is directly proportional to the area of fresh surface produced.[¶]

RITZ'S COMBINATION PRINCIPLE [physics] (Walter Ritz, 1878–1909, Swiss-born German physicist) (1) Spectra of all atoms contain combinations of two lines whose frequencies sum up to the frequency of a third line. There is a simple mathematical scheme for distribution of spectral lines of atoms.[‖]

[*]Ren, F., Tanaka, H., and Gojobori, T. (1995), Construction of molecular evolutionary phylogenetic trees from DNA sequences based on minimum complexity principle, *Comput. Meth. Programs Biomed.* **46**(2), 121–130.

[†]Verbeek, J. J. (2000), *An Information Theoretic Approach to Finding Word Groups for Text Classification*, master's thesis, University of Amsterdam, Inst. Language, Logic and Computation.

[‡]Marinucci, M. (2008), *Automatic Prediction and Model Selection*, doctoral dissertation, Philosophy, Univ. Complutense de Madrid.

[§]Ren, F., Tanaka, H., and Gojobori, T. (1995), Construction of molecular evolutionary phylogenetic trees from DNA sequences based on minimum complexity principle, *Comput. Meth. Programs Biomed.* **46**(2), 121–130.

[¶]Walker, P. M. B. (1999), *Chambers Dictionary of Science and Technology*, Chambers, New York, p. 993.

[‖]Van-Veld, R. D. and Meissner, K. W. (1956), Interferrometric wave-length measurements of germanium lines of a hollow-cathode discharge, *J. Opt. Soc. Am.* **46**, 598–604.

See also COMBINATION–SEPARATION PRINCIPLE; RAYLEIGH–RITZ MINIMUM PRINCIPLE.

ROCHDALE PRINCIPLE [economics] A set of ideals for the operation of cooperatives. They were first set out by the Rochdale Society of Equitable Pioneers in Rochdale, England, in 1844, and have formed the basis for the principles on which co-operatives around the world operate to this day. The implications of the Rochdale Principles are a focus of study in co-operative economics. The original Rochdale Principles were officially adopted by the International Co-operative Alliance (ICA) in 1937 as the Rochdale Principles of Co-operation.[**]

ROCKET PRINCIPLE [physics] The greater the mass of the escaping charge, or the higher its velocity, the greater the thrust it produces on the chamber.[††] See also COMBINATION PRINCIPLE, COMBINATION–SEPARATION PRINCIPLE; RITZ'S COMBINATION PRINCIPLE.

ROGERS' PRINCIPLES [psychology] (Carl Rogers, 1902–1987, American psychologist) Necessary and sufficient conditions of therapeutic personality change.[‡‡]

ROLLER WEDGE PRINCIPLE [physics] See also DIP ROLLER PRINCIPLE.

ROOTS PRINCIPLE [mathematics, engineering] (Eric Reissner) (1) Design of fractional order controller based on the method from poles distribution of the characteristic equation in the complex plane. Values of dominant roots are designed for the quality requirement of the control circuit.[§§] (2) Pump

[**]Conover, Milton (1959) The Rochdale Principles in American Co-Operative Associations. The *Western Political Quarterly*, v12, (1, Part 1), pp. 111–122.

[††]Cramp, L. G. (1997), *UFOs & Anti-Gravity—Piece for a Jig Saw* (http://www.scribd.com/doc/6348097/Cramp-UFOs-AntiGravity-Piece-for-a-Jig-Saw-1997#document_metadata).

[‡‡]Wilson, L. M. and Fitzpatrick, J. J. (1984), Dialectic thinking as a means of understanding systems-in-development: relevance to Rogers's principles, *Adv. Nurs. Sci.* **6**(2), 24–41.

[§§]Petras, I. (1999), The fractional-order controllers: Methods for their synthesis and application, *J. Electrical Eng.* **50**(9-10), 284–288.

used to increase the pumping speed at low pressures, thus extending the operating pressure range.* See DISPLACEMENT PRINCIPLE.

ROTARY VANE PUMP PRINCIPLE [engineering] (Wolfhart Willimczik, German physicist) In contrast to the classical rotary vane pump, where several loose vanes rotate inside a cylinder guided in radial slots, rotate one single vane axial on the flat end surface of a stationary ring shaped piston.[††] Also known as *system Willimczik*.

ROTATIONAL   ACCELERATION   PRINCIPLE [physics]   Deceleration of the rotating mass is required after the transfer of the torsional impulse in order to start a new operating cycle.[†]

RUIN-AND-RE-CREATE PRINCIPLE [mathematics]   (1) Solutions of problems are partly, but significantly, ruined and rebuilt or recreated afterward. Performing this type of change frequently, one can obtain astounding results for classical optimization problems.[‡] (2) Learning and teaching method by means of building and then destroying and rebuilding for purpose of mastery and improvements.[§]

RULE   CHAINING   PRINCIPLE [computers] Adapting concepts of the compiler theory and knowledge engineering, for acquisition and representation of knowledge about the program. Redundant and conflicting knowledge about the program under study is recognized and solved by means of an embedded truth maintenance system. As a result of fault diagnosis, rules for fault classification are used.[¶] See also CHAINING PRINCIPLE.

RULE OF PRODUCT PRINCIPLE [mathematics] aka Principle of Choice. If a procedure can be broken down into first and second stages, and if there are m possible outcomes for the first stage and if for each of these outcomes, there are n possible outcomes for the second stage, then the total procedure can be carried out in the designated order, in (n × m) ways.[‖] See also INCLUSION-EXCLUSION PRINCIPLE; RULE OF SUM PRINCIPLE.

RULE OF SUM PRINCIPLE [mathematics]   A basic counting principle. If we have a ways of doing something and b ways of doing another thing and we can not do both at the same time, then there are $a + b$ ways to choose one of the actions. More formally, the rule of sum is a fact about set theory. It states that sum of the sizes of a finite collection of pairwise disjoint sets is the size of the union of these

$$|S_1| + |S_2| + \cdots + |S_n| = |S_1 \cup S_2 \cup \cdots \cup S_n|$$

sets.[**] See also INCLUSION-EXCLUSION PRINCIPLE; RULE OF PRODUCT PRINCIPLE.

RYDBERG–RITZ   COMBINATION   PRINCIPLE [physics] (Walter Ritz, 1878–1909, Swiss theoretical   physicist)   (Johannes   Robert Rydberg, 1854–1919, Swedish physicist) (1) Explains relationship of the spectral lines for all atoms. The spectral lines of any

*Din, J.-P., Liu, X.-B., Wang, H.-L., and Lu, J.-G. (2004), Maintenance and improvement of Edwards rotary vacuum pump, *Zhipu Xuebao* **25**(3), 189–192.

[†]Gerstenberger, F. and Jaedicke, L. (Jan. 25, 1983), *Recoilless Air Weapon*, US Patent: 4,369,759.

[‡]Schrimpf, G., Schneider, J., Stamm-Wilbrandt, H., and Dueck, G. (2000), Record breaking optimization results using the ruin and recreate principle, *J. Comput. Phys.* **159**(2), 139–171.

[§]Boire, R. (2003), *A Healing Approach to Teaching: A Case Study*, thesis, Dept. Educational Administration, Univ. Saskatchewan, Saskatoon.

[¶]Belli, F. and Crisan, R. (1996), Towards automation of checklist-based code-reviews. (Dept. Electrical/Electronic Engineering, Paderborn Univ.), *Proc. 7th Int. Symp. Software Reliability Engineering*, pp. 24–33.

[‖](n.a.) (2001) Lecture Notes. Simon Fraser University   http://web.viu.ca/math/SFUstat270/lecturenote.htm

**(n.a.) (2001) Lecture Notes. Simon Fraser University   http://web.viu.ca/math/SFUstat270/lecturenote.htm

element include frequencies that are either the sum or the difference of the frequencies of two other lines.* (2) An atom can be excited to higher energy either spontaneously or via absorption of a photon. However, according to the principles of quantum mechanics, these excitations can only occur at certain energy intervals.†

*Ibrahimi, M., Babay, A., Lemoine, B., and Rohart, F. (1999), Pressure-induced frequency lineshifts in the 2 band of ammonia: An experimental test of the Rydberg-Ritz principle, *J. Mol. Spectrosc.* **193**(2), 277–284.

†Belli, S., Buffa, G., and Tarrini, O. (1997), On the extension of the Rydberg-Ritz principle to the collisional relaxation of molecular and atomic lines, *Chem. Phys. Lett.* **271**(4–6), 349–354.

S

SADDLE POINT ENTROPY PRODUCTION Principle [physics] See FICK PRINCIPLE.

SADDLE POINT PRINCIPLE [meathematics, physics] (1) A *saddle point* of a surface is a point reminiscent of the inner part of a horse's saddle or of a geographic pass between two mountains.* (2) Leads to asymptotic estimates or even full asymptotic expansions depending on the nature of the problem. Its principle is to use a saddle point crossing path, then estimate the integrand locally near this saddle point (at which point the integrand achieves its maximum), and deduce by termwise integration an asymptotic expansion of the integral itself.[†] See also STATIONARY ACTION PRINCIPLE; PRINCIPLE OF LEAST ACTION; STEWART–HAMILTON PRINCIPLE; FICK PRINCIPLE.

SAFE-HARBOR PRIVACY PRINCIPLES [psychology, computer science] A set of privacy regulations that store customer data and designed to prevent accidental information disclosure or loss. The list of principles are outlined as follows: (1) *Notice*—a company must provide a data usage statement, to inform the user of how said company will use their data; (2) *choice*—the customer must have the option of "opting out" of any information disclosure; (3) *onward transfer*—if the company chooses to disclose information to another entity, it must adhere to the principles of choice and notice; (4) *security*—companies must secure their systems to protect against the loss, misuse, disclosure, destruction, and alteration of data; (5) *data integrity*—the data must be processed relevant to the purpose for which they were originally collected; (6) *access*—the customer must have access to his/her data so

that he/she can add, edit, or delete data; and (7) *enforcement*—the company must enforce these principles, as well as its own internal policies and procedures, in the aim of preventing accidental or intentional data disclosure or loss.[‡] See also INTERNATIONAL SAFE-HARBOR PRIVACY.

SAFE PRINCIPLE [general science] Provide a solution that may avoid other issues, whether known or unknown, to other beneficial existing conditions.[§]

SAINT VENANT'S PRINCIPLE [physics] (Adhémar Jean Claude Barré deSaint-Venant, 1797–1886, French mechanician and mathematician) (1) Point where all the first partial derivatives of a function vanish but that is not a local maximum or minimum.[¶] (2) Study of differentiable functions and their derivatives from the perspective of saddle points, especially applicable to the calculus of variations.[‖] (3) In its original form where bounded bodies are considered, it is shown by means of a counterexample that in order to ensure the validity of this principle, some restriction must be imposed on the shape of the body.[**] (4) If two sets of loadings

*Flajolet, P. and Sedgewick, R. (2008), *Analytic Combinatorics*, Cambridge Univ. Press, p. 541.
[†]Pomraning, I.G.C. and Clark, M., Jr. (1963), The variational method applied to the monoenergetic Boltzmann equation, *Nuclear Sci. Eng.* **16**, 147–154.

[‡]Warrington, T. (2003), The E-privacy imperative: Protect your customers' Internet privacy and ensure your company's survival in the electronic age, *J. Consumer Market.* **20**(3), 269.
[§]Eastwood, M.A., (1999) Interaction of dietary antioxidants in vivo: how fruit and vegetables prevent disease? Association of Physicians J Med, v.92, p.527–530.
[¶]Horgan, C.O., and Payne, L.E., (1997) Saint Venant's principle in linear isotropic elasticity for incompressible or nearly incompressible materials, *J. Elasticity*, **46**, 43.
[‖]Borrelli, A. and Patria, M. C. (1997), Decay and other estimates for a semi infinite magnetoelastic cylinder: Saint Venant's principle, *Int. J. Nonlinear Mech.* **32**, 1087.
[**]Knops, R. J., and Payne, L. E., (1996) The effect of a variation in the elastic moduli on Saint Venant's principle for a half cylinder, *J. Elasticity* **44**, 161.

are statically equivalent at each end of a cylinder with free lateral surface, then the difference in strain fields is negligible, except possibly near the ends.* See also PRINCIPLE OF SAINT VENANT.

SALUTOGENESIS PRINCIPLE [psychology] (Aaron Antonovsky) Optimal focus is one of exploring how health is achieved and maintained (salutogenesis) rather than how disease is caused (pathogenesis). It is a systems view, in which "information" can originate anywhere in the natural hierarchy of a human being—from atoms, molecules, organs, whole bodies, families, and entire societies—and can be transmitted anywhere, making a difference in the overall functioning of the organism. It also includes the principle of comfort through discomfort.† See also AARON ANTONOVSKY'S COMFORT-THROUGH-DISCOMFORT PRINCIPLE; PRINCIPLE OF SO-MATOMENTAL BALANCE; PRINCIPLE OF THE SALUTOGENETIC TRIAD; COMFORT-THROUGH-DISCOMFORT PRINCIPLE.

SAMPLING PRINCIPLE [mathematics] (1) Survey methods used to determine consumer preferences; for instance, the quota method demands the formulation of a hypothetical model to fit the data, whereas a probabilistic survey does not. With probability sampling, randomization distribution is used to draw conclusions from the sample, and to obtain sampling errors. In a quota sample, comparable estimates of precision cannot be obtained. In general, nonresponse in a quota sample is handled by selection of another respondent fitting the quota. Nonresponse in a probability-based sample can be handled with less complexity in the sample (although this entails the use of some form of modeling). No existing survey exactly matches the ideal picture of probability-based sampling.‡ (2) A method and an arrangement

for qualitative potential measurement at surface—wave filters with an electron beam measuring device.§ (3) A method for measuring electrical signals with the assistance of an electron probe, whereby the signal is respectively repeatedly sampled in succession at different phase points, enables rapid registration of an electrical signal at a circuit node of an integrated circuit and simultaneously permits registration when at least one noise signal appears between two clock edges of a basic pulse rate of the integrated circuit.¶ See also CONSUMER PRINCIPLE.

SAMUELSON–LE CHATELIER PRINCIPLE [economics] (Paul Samuelson, 1915–2009, American Economist) (Henry Louis Le Chatelier, 1850–1936, French chemist) Long-run demand is "more elastic" than short-run demand. Formally, the statement applies to smooth demand functions for sufficiently small price changes.‖

SANBORN PRINCIPLES [ecology] Develops ecological responsiveness for optimized efficiency in residential and commercial buildings, preparing school and public buildings for community education, geothermal power, water filtration, transportation accessibility and minimization of waste. See also BIOMIMICRY PRINCIPLES; DEEP ECOLOGY'S BASIC PRINCIPLES; FOREST STEWARDSHIP COUNCIL (FSC) PRINCIPLES; GLOBAL SULLIVAN PRINCIPLES OF SOCIAL RESPONSIBILITY; MARINE STEWARDSHIP COUNCIL (MSC) PRINCIPLES; NATURAL CAPITALISM PRINCIPLES; PERMACULTURE PRINCIPLES; THE BELLAGIO PRINCIPLES FOR ASSESSMENT; EQUATOR PRINCIPLES; MELBOURNE PRINCIPLES; PRECAUTIONARY PRINCIPLE; TODDS' PRINCIPLES OF ECOLOGICAL DESIGN.**

*Chirita, S. and Quintanilla, R. (1996), On Saint Venant's principle in linear elastodynamics, *J. Elasticity* **42**, 201.

†Schaefer, J., Nierhaus, K. H., Lohff, B., Peters, T., Schaefer, T., and Vos, R. (1998), Mechanisms of autoprotection and the role of stress-proteins in natural defenses, autoprotection, and salutogenesis, *Med. Hypotheses* **51**(2), 153–163.

‡ERSA, (2007), ESRA Conf., June 8, 2007, The European Survey Research Association.

§Feuerbaum, H. P. and Tobolka, G. (1983), *Method and Arrangement for Quantitative Potential*, US Patent 4,412,191.

¶Lewis, M. (1984), The surface acoustic wave oscillator—a natural and timely development of the quartz crystal oscillator, *Proc. 28th Annual Symp. Frequency Control*, pp. 304–314.

‖Murata, Y., Nossa, T., and Morishima, M. (1972), *The Working of Econometric Models*, Cambridge Univ. Press.

**Harwood, B. (1994) The First Sanborn Sustainability Conference. *Proceedings of the National Passive Solar Conference 21*, p. 193–199.

SANDERSON'S ELECTRONEGATIVITY PRINCIPLE [physics] Equalization principle states that "all the constituent atoms in a molecule have the same electronegativity value given by the geometric mean of the electronegativities of the pertinent isolated atoms." There are two hardness-related principles: the hard–soft acid–base (HSAB) principle and the maximum hardness principle (MHP). While the former states "hard likes hard and soft likes soft," the statement of the latter is "there seems to be a rule of nature that molecules arrange themselves so as to be as hard as possible."[*]

SANGAMON'S PRINCIPLE [pharmacology] (Neal Town Stephenson, b. 10/31/1959, author)   Simple molecules make better recreational drugs than do complex ones, because you never know what side effects more complicated compounds will have.[†]

SATELLITE TEST OF THE EQUIVALENCE PRINCIPLE (STEP) [space]   STEP was conceived as a joint NASA/ESA mission with equal participation by both agencies. ESA's contribution to the program would be the spacecraft; NASA would provide the launcher and half of the instrument, while the other half of the instrument would be provided by various European agencies. STEP was in competition with three other programs, INTEGRAL, PRISMA, and MARSNET.[‡]

SATIATION PRINCIPLE [psychology] To stop a child from acting in a particular way, you may allow her/him to continue (or insist that she/he continue) performing the undesired act until she/he tires of it.[§]

SATISFACTION PRINCIPLE [psychology] (1) (S. Juillerat, 1995.) Investigating a two-dimensional approach to the assessment of student satisfaction: validation of the student satisfaction inventory (SSI). Researchers in the consumer field frequently define satisfaction in relation to meeting and/or exceeding customers' expectations. The SSI also produces two other scores, the importance score and the satisfaction score, which represent the strength of a student's expectation and the level of satisfaction that the expectation has been met, respectively.[¶] (2) Simplification with paradoxes and fast suboptimization with relative simplification and paradoxes or regret compromises. See also PERFECTION PRINCIPLE; REGRET PRINCIPLE; REALITY PRINCIPLE; [SIMPLIFICATION PRINCIPLE;] [STRENGTH PRINCIPLE;] RELATIVITY PRINCIPLE; COMBINATION PRINCIPLE; PARADOX-TOLERATING PRINCIPLE; EQUIVALENCE PRINCIPLE.

SATURATION PRINCIPLE [mathematics] A small number of randomly chosen units (the sample) of a total population (the universe) will end to have the same characteristics, and the same proportion, as the population as a whole.[‖]

SAVAGE PRINCIPLE [mathematics] Also known as the *regret criterion*, it is a technique used in decision theory; a criterion is used to construct a regret matrix in which each outcome entry represents a regret defined as the difference between best possible outcome and the given outcome; the matrix is then used as in decisionmaking under risk with expected regret as the decision-determining quality.[**]

[*]Chattaraj, P. K., Fuentealba, P., Jaque, P., and Toro-Labbe, A. (1999), Validity of the minimum polarizability principle in molecular vibrations and internal rotations: An ab Initio SCF study, *J. Phys. Chem. A* **103**, 9307.
[†]Stephenson, N. T. (1988), *Zodiac: An Eco-Thriller*, Atlantic Monthly Press.
[‡]PW Worden, P.W., Jr (1993) *Satellite test of the equivalence principle*. Final Technical Report, Stanford Univ., CA.
[§]Huitt, W. (1994), Principles for using behavior modification, in *Educational Psychology Interactive*, Valdosta State Univ., Valdosta, GA (http://chiron.valdosta.edu/whuitt/col/behsys/behmod.html).

[¶]Read, S., V., E. and Miller, L. (1997), Connectionism, parallel constraint satisfaction processes, and Gestalt principles: (Re)introducing cognitive dynamics to social psychology, *Pers. Soc. Psychol. Rev.* **1**(1), 26–53.
[‖]Mason, M. F., Hawley, G., and Smith, A. (1947), *Application of the Saturation Principle to the Estimation of Functional Hepatic Mass in Normal Dogs* (ajplegacy.physiology.org/cgi/reprint/152/1/42.pdf).
[**]Anon. (2003), Savage principle. *McGraw-Hill Dictionary of Scientific and Technical Terms*.

SAVAGE PRINCIPLE [physics]   (1) Condition that occurs when a transistor is driven so that it becomes biased in the forward direction.* (2) Decay rate of a given radionuclide is equal to its rate of production in an induced nuclear reaction.[†] (3) Voltage applied to an ionization chamber is high enough to collect all the ions formed by radiation but not high enough to produce ionization by collision; condition in which the partial pressure of any fluid constituent is equal to its maximum possible partial pressure under the existing environmental conditions. Any increase in the amount of constituent will initiate within it a change to a more condensed state.[‡]

SCALAR PRINCIPLE [mathematics, industrial psychology]   (Henry   Fayol,   1841–1925, French industrialist)   (1) Technique used in decision theory; a criterion is used to construct a regret matrix in which each outcome entry represents a regret defined as the difference between the best possible outcome and the given outcome; the matrix is then used as in decisionmaking under risk with expected regret as the decision-determining quality.[§] (2) The clearer the line of authority from the ultimate authority for management in an enterprise (CEO) to every subordinate position, the more effective will be decision-making and organization communication at various levels in the organization.[¶] See also PRINCIPLE OF PLANNING; PRINCIPLE OF SCALAR CHAIN.

SCALE INVARIANCE PRINCIPLE [mathematics] (Derivations of Benford's law)   If there is any universal law for significant digits, then it should be independent of the units and scales as well as the base system employed by society. Units or scales are merely arbitrary measuring rods compared against other entities. Measurement is some very specific entity for measuring, such as heights of people, heights of horses, heights of buildings, lengths of rivers, distances between global cities, ages of people, or lifespan of a single bacterium (these examples constitute seven different measurements in all). The assumption measurements are not just numerous, but enormous. Changes in scale will have such varying and independent effects on measurements in terms of leading digits will all sum up to nothing. In other words, even though a change in scale would revolutionize digital leadership for almost each and every measurement (individually) on that dimension, yet, the changes for almost all measurements do take totally different turns, leaving the net results on digital leadership unaffected.[‖] See also INVARIANCE PRINCIPLE.

SCALING PRINCIPLE [mathematics, physics] (1) If you take any shape and "scale it up" so that every line in it is f times bigger, then the area will be larger by f_2. The volume of a 3D shape scales like $f_3(d_3)$.[**] (2) The area of a figure is proportional to the square of any of its linear dimensions. So, if any line in the figure has a length $= d$, then the area will be $r * d_2$, and the r will stay the same if you change d, as long as you change the whole figure proportionately.[††] See also CHARGE SCALING PRINCIPLE; VOLTAGE SCALING PRINCIPLE.

*Sexton, F. W. (2003), Destructive single-event effects in semiconductor devices and ICs, *IEEE Trans. Nuclear Sci.* **50**(3), 603–621.

[†]Nico, J. S. and Snow, W. M. (2005), Fundamental neurtron physics, *Annu. Rev. Nuclear Particle Sci.* **55**, 27–69.

[‡]Vandeweert, F., Vervaecke, S., Wyczawska, P. L., and Silverans, R. E. (2005), Particle-induced desorption from self-assembled monolayers: Signatures of an intricate balance between inter- and intramolecular interactions resolved by laser-ionization mass spectrometry, *Proc. 15th Int. Conf. Secondary Ion Mass Spectrometry*, Sept. 12–16, 2005, Manchester, UK, p. 222.

[§]Wood, Michael C., (2002) *Henri Fayol: Critical Evaluations in Business and Management*, Taylor & Francis, p. 8.

[¶]Koontz, H. and O'Donnell, C. (1968), *Principles of Management: An Analysis of Managerial Functions*, 4th ed., McGraw-Hill, New York (http://knol.google.com/k/narayana-rao-kvss/principles-of-management-koontz-and/2utb2lsm2k7a/89).

[‖]Benford, F. (1938), The law of anomalous numbers, Proc. Am. Phil. Soc. 78(4), 551–572.

[**]Young, T. and Anderson, H. (2005), Tests of an approximate scaling principle for dynamics of classical fluids, *J. Phys. Chem. B* **10**(7) 2985–2944.

[††]Livingston, E. H. and Lee, S. (2001), Body surface area prediction in normal-weight and obese patients, *Am. J. Physiol. Endocrinol. Metab.* **281**(3), E586–E591.

SCATTERED LIGHT PRINCIPLE [mathematics] (1) Effect of information about one possible cause of an event as inferences regarding another possible cause.* (2) Analyzes situations until a minimal set of sufficient causes are identified, then, other possible causes are ignored or dismissed. (3) Comprising at least two sensor or feeler elements having different functional or operating principles and a common evaluation circuit for evaluation of the property changes of the sensor elements and for triggering a signal.† (4) When a source is placed at the focal point of lens, an image will be located at infinity. The illumination at a point on the axis will be determined by the brightness of the image and the solid angle subtended by the image. The size of the solid angle subtended by the source of illumination will be limited by the lens diameter. This solid angle will be equal to the area of the lens divided by d_2 (d = distance from the lens) and the illumination beyond the distance to the lens. Accordingly, illumination within the searchlight cone defined by the lens focal point is constant anywhere within the cone.‡ (5) Rest mass of a particle cannot be kept invariant. (11) Must exist a unique preferred inertial frame of reference wherein a particle is absolutely at rest. See also EXTINCTION PRINCIPLE; INCIDENT LIGHT PRINCIPLE; TRANSMITTED LIGHT PRINCIPLE; SEARCHLIGHT PRINCIPLE; DIRECT LIGHT PRINCIPLE.

SCATTERED WAVE VARIATIONAL PRINCIPLE (SWVP) [physics] (1) Examines what happens when a light wave or a photon is incident on an atom. All the processes of transmission, reflection, and refraction are macroscopic manifestations of scattering effects on the atomic and subatomic levels.§

(2) When a photon (or light wave) encounters an atom, there are two possibilities: (1) the atom may scatter the light, redirecting it without changing its frequency or energy; or (2) it may absorb the light, using the energy to make a quantum jump into an excited energy state (more precisely, one of its electrons makes the jump). With absorption it is likely that the excitation energy will rapidly be transferred to atomic motion, via collisions, thus producing thermal energy before the atom decays back to the lower energy state reemitting a photon. The electric and magnetic fields of the light drive the electron cloud of the atom into oscillation, causing it to reradiate in all directions at that same frequency.¶ (3) Air tends to scatter more blue light than red light. For sunlight coming laterally through the atmosphere, more blue light than red light is scattered toward the ground, so the sky appears blue! When the sun is low on the horizon, light passes through a greater thickness of air; the blue is scattered out and we see the leftover red light coming directly along the line of sight to the sun, causing red sunsets.‖ (4) In dense, homogeneous media, where the spacing of atoms is much less than the wavelength of the light, it turns out that very little light gets scattered in the backward direction or any direction perpendicular to the propagation of the wave, but most is propagated in the forward direction. This allows light to propagate through dielectrics.** See also WAVE VARIATIONAL PRINCIPLE.

SCHEDULE OF REINFORCEMENT PRINCIPLE [psychology] (1) There are three central assumptions of this theory: the activation of

*Callan, R. and Larder, B. (2002), The development and demonstration of a probabilistic diagnostic and prognostic system (ProDAPS) for gas turbines, *IEEE Aerospace Conf. Proc.*, vol. 6, pp. 6-3083–6-3094.

†Scheidweiler, A. (1983), *Method of Fire Detection and Fire Detection Installation*, US Patent 4,405,919.

‡Bukowski, R. W. (1979), Smoke measurements in large-and small-scale fire testing—Part II, *J. Fire Technolo.* **15**(4), 271–281.

§Sun, Y. and Kouri, D. J. (n.d.), Scattered wave variational principle for atom—diatom reactive

scattering: Hybrid basis set calculations, *Chem. Phys. Lett.* **179**(1–2), 142–146.

¶Sun, Y., Kouri, D. J. and Truhlar, D. G. (1990), A comparative analysis of variational methods for inelastic and reactive scattering, *Nuclear Phys. A* **508**, 41–61.

‖Anon. (n.d.), *Scattering: The Colours of the Sky* (http://www.itp.uni-hannover.de/~zawischa/ITP/scattering.html).

**Kunzig, R. (2001), Trapping light: This is the future, and it moves at 186,000 miles per second, *Discover* (published online at http://discovermagazine.com/2001/apr/feattrap).

behavior by incentives, constraints on responding, and memory as the mediator of reinforcement. An important ancillary assumption is the continual indexing of memory by responses, including observing and consummatory responses, which will displace the memory of target responses by the memory of incentives (or their consummation).* (2) The schedule of providing reinforcement for a behavior displayed under particular circumstances will determine characteristics of learning, including the probability that behavior will be displayed again in those circumstances, and the number of times that the behavior will continue to be displayed in those circumstances after the reinforcement is no longer provided.[†] See SUCCESSIVE APPROXIMATION PRINCIPLE; CHAINING PRINCIPLE; BACKWARD CHAINING PRINCIPLE; ONE-TRIAL PRINCIPLE; RESPONSE-SHAPING PRINCIPLE; TRANSITIONAL STIMULI PRINCIPLE; SKINNER PRINCIPLE.

SCHEIMPFLUG PRINCIPLE [mathematics] (Theodor Scheimpflug, Austrian 1865–1911 Austria mathematician) Deals with the change of the focus plane when tilting the front standard of a view camera. When an oblique tangent is extended from the film plane, and another is extended from the lens plane, they meet at a point. To obtain the correct focus on a subject, a plane formed by the desired area of critical focus within the subject(s) should, if extended, intersect with the previously explained point. If this is achieved, then the focus should be as good as possible from a particular location of the camera.[‡]

SCHEINER PRINCIPLE [biology] (Donald M. Scheiner, b. 3/12/1932, American Chemist) (1) Yields an objective assessment of the refractive state of the photoreceptor image plane.[§] (2) Positions where the same amino acid substitution has occurred in various lineages, and that same substitution will also occur within a single species as a polymorphism if a sufficient number of individuals are examined.[¶]

SCHRÖDINGER VARIATIONAL PRINCIPLE [physics] (Erwin Rudolf Josef Alexander Schrödinger, 1887–1961, Austrian physicist) Among all possible configurations or histories of a physical system, the system realizes the one minimizing some specified quantity. Variational methods are used in physics both for theory construction and for calculational purposes. For any field theory, only the Lagrange density needs to be given; the field equations are then derivable as the corresponding Euler–Lagrange equations. A similar technique makes it possible to derive the Schrödinger equation and the Dirac equation in quantum mechanics from specific Lagrange density functions. This method has great procedural advantages. For example, it facilitates a check of whether the theory satisfies certain invariance principles (such as relativistic invariance or rotational invariance) by simply ascertaining whether the Lagrange density satisfies them. The variational method also plays an important role in quantum mechanical calculations. For the computation of needed quantities in terms of functions that result from the solution of differential equations, it is always of great advantage to use formulas having the special form required to make them stationary with respect to small variations of the input functions in the vicinity of the unknown, exact solutions.[‖] See also Extremal Principle; FERMAT'S PRINCIPLE; HAMILTON' PRINCIPLE; PRINCIPLE OF LEAST ACTION; LEAST-ACTION PRINCIPLE; MINIMAL PRINCIPLES.

*Killeen, P. R. (1994), Mathematical principles of reinforcement, *Behav. Brain Sci.* **17**(1), 105–172.

[†]Bezzina, G., Body, S., Cheung, T. H. C., Hampson, C. L., Deakin, J. F. W., Anderson, I. M., Szabadi, E., and Bradshaw, C. M. (2008), Effect of quinolinic acid-induced lesions of the nucleus accumbens core on performance on a progressive ratio schedule of reinforcement: implications for inter-temporal choice, *Psychopharmacology* **197**(2), 339–350.

[‡]Larmore, L. (1965), *Introduction to Photographic Principles*, Dover Publications, New York.

[§]Hodos, W., Fitzke, F. W., Hayes, B. P., and Holden, A. L. (1985), Experimental myopia in chicks ocular refraction by electroretiongraphy, *Invest. Ophthalmol. Visual Sci.* **26**(10), 1423–1430.

[¶]Jiang, B. C. (1988), Steady state of accommodation during observation of a Scheiner image, *Am. J. Optometry Physiol. Opt.* **65**(10), 809–813.

[‖]Moore, W. J. (1992), *Schrödinger: Life and Thought*, Cambridge Univ. Press, p. 392.

SCHWARZ REFLECTION PRINCIPLE [mathematics] (Karl Hermann Amandus Schwarz; 1843–1921, German mathematician) (1) To obtain the analytic continuation of a given function $f(z)$ analytic in a region R, whose boundary contains a segment of the real axis, into a region reflected from R through this segment, one takes the complex conjugate function $f(\bar{z})$.* (2) Let $f(z)$ be analytic in a region R of the complex plane, where the boundary of R contains a segment of the real axis on which $f(z)$ assumes real values. Then the analytic continuation of $f(z)$ into the region obtained by reflecting R across the real axis is $[f(\bar{z})]^*$, where $z\bar{z}$ and * indicate complex conjugate. (3) A physical quantity or physical law posses invariance under certain transformations. (4) Laws of motion are the same in all forms of reference whether accelerated or not. Symmetric points are preserved under a Mobius transformation.

> THEOREM. *Let* $\Omega = \{x : |x| \leq a\}, T > 0$, *and* u *a solution of the heat equation (1.5) in* $\Omega \times (0, T)$ *such that* u *is continuous in* $\bar{\Omega} \times (0, T)$ *and vanishes on* $\partial\Omega \times (0, T)$. *Then* u *can be uniquely contined as a solution of the heat equation into all of* $R^n \times (0, T)$ *where the domain of dependence of* u *at the point* (r, θ, t) *for* $r > a$ *is the line segment joining* $(a^2/r, \theta, t)$ *to* (a, θ, t).[†]

See also LINDELÖF PRINCIPLE; MAXIMUM PRINCIPLE; SYMMETRY PRINCIPLE; INVARIANCE PRINCIPLE.

SCHWINGER PRINCIPLE [pharmacology] (Jullian Seymonr Schwinger, 1918–1994, American physicist) (1) The high-carbohydrate, low-fat, moderate-protein diet turns people into diabetics, makes them age faster and contract degenerative diseases, and keeps them fat and unhealthy. (2) A diet low in carbohydrates and high in proteins will bring patients' blood sugar levels down and improve their lipid levels. The body will either gain or lose weight as to achieve its optimum weight.[‡] See also VARIATIONAL PRINCIPLE.

SCHWINGER QUANTUM ACTION PRINCIPLE [physics] See PRINCIPLE OF CONCEPTUAL DESIGN; QUANTUM ACTION PRINCIPLE.

SCHWINGER VARIATIONAL PRINCIPLE [physics] (Julian Seymonr Schwinger, 1918–1994, American physicist) A method used in electromagnetic theory, or similar disciplines, to calculate an approximate value or a linear of quadratic functional, such as a scattering amplitude or reflection coefficient, when the function for which the functional is evaluated is the solution of an integral equation.[§] See also COMPLEX KOHN VARIATIONAL PRINCIPLE; HAMILTON PRINCIPLE; HULTHEN–KOHN VARIATIONAL PRINCIPLE; KOHN VARIATIONAL PRINCIPLE.

SCIENTIFIC PRINCIPLE [mathematics, physics] (Aristotle) Generally accepted in the relevant scientific community. Discovery crossing the line between the experimental and demonstrable stage. The evidential force of the principle must be recognized; the deduction is made to have gained general acceptance in the particular field in which it belongs.

SCOTCH YOKE PRINCIPLE [mathematics, engineering] Antifriction bearings and mountings provide simple and effective means of maintaining positive alignment even at great temperature extremes and provide a substantially friction-free environment to work more efficiently and smoothly while resisting asymmetric drag and reversing torque. Rotational motion is imparted to a reciprocating component. An advantage of the scotch yoke layout is that there is little or no side thrust. The component may be driven by one or more pairs of working chambers mounted on either one or all sides of it.[¶]

*Bagemihl, F. (1964), Analytic continuation and the Schwarz reflection principle, *Proc. Natl. Acad. Sci. USA* **51**(3): 378–380.

[†]Colton, D. S. (1981), Reflection principles for solutions of parabolic equations, *Proc. Am. Math. Soc.* **82**(1), 87–94.

[‡]Kar, S. and Mandal, P. (1997), Correlated basis functions for studies on positron collisions using Schwinger's principle, *J. Phys. B Atom. Mol. Opt. Phys.* **30**, L627.

[§]Maleki, N., and Macek, J. (1980), Schwinger's variational principle for electron-ion scattering, *Phys. Rev. A Atom. Mol. Opt. Phys.* **21**(5), 1403–1411.

[¶]Hinderks, M. V. (2008), *Improved Reciprocating Devices*, World Patent WO/2008/045036.

SEARCHLIGHT PRINCIPLE [mathematics] See also EXTINCTION PRINCIPLE; INCIDENT LIGHT PRINCIPLE; SCATTERED LIGHT PRINCIPLE; TRANSMITTED LIGHT PRINCIPLE; DIRECT LIGHT PRINCIPLE.

SECOND DATA PROTECTION PRINCIPLE [mathematics] Personal data shall be obtained only for one or more specified and lawful purposes, and shall not be further processed in any manner incompatible with that purpose or those purposes.* An act to make new provision for the regulation of the processing of information relating to individuals, including the obtaining, holding, use, or disclosure of such information.[†] See also COPY PROTECTION PRINCIPLE; PRIVACY PRINCIPLE; FIRST DATA PROTECTION PRINCIPLE; THIRD DATA PROTECTION PRINCIPLE; FOURTH DATA PROTECTION PRINCIPLE; FIFTH DATA PROTECTION PRINCIPLE; SIXTH DATA PROTECTION PRINCIPLE; SEVENTH DATA PROTECTION PRINCIPLE; EIGHTH DATA PROTECTION PRINCIPLE.

SECOND EXTENDED MAXIMUM ENTROPY PRINCIPLE [engineering, mathematics] An unknown parameter of a given density function of a known population's functional form that provides a family of possible density functions with some information concerning the population in terms of random sampling; the maximum entropy of the probability distribution subject to whatever information is given should be chosen.[‡] See also ENTROPY OPTIMIZATION PRINCIPLE.

SECOND EXTENDED MINIMUM CROSS-ENTROPY PRINCIPLE [engineering, mathematics] A density function so that the given distribution proportion is as close to the distribution as possible to minimize the entropy of a discrete distribution as relative to the distribution proportion should be chosen.[‡] See also ENTROPY OPTIMIZATION PRINCIPLE.

SECOND INVERSE MAXIMUM ENTROPY PRINCIPLE [engineering, mathematics] A set of constraints are determined that will give rise to a given observed or theoretical probability distribution as Jaynes' maximum entropy probability distribution using generalized measure of entropy.[‡] See also ENTROPY OPTIMIZATION PRINCIPLE.

SECOND INVERSE MINIMUM CROSS-ENTROPY PRINCIPLE [engineering, mathematics] Measure of cross-entropy to obtain a given probability distribution when the cross-entropy for a given constraint and a given a priori probability distribution is chosen.[‡] See also ENTROPY OPTIMIZATION PRINCIPLE.

SECOND MINIMUM INTERDEPENDENCE PRINCIPLE [engineering, mathematics] The random vector with known probability distribution is transformed into another random vector by means of a transformation that is chosen to minimize the interdepence.[§] See also ENTROPY OPTIMIZATION PRINCIPLE.

SECOND PRINCIPLE OF ENERGETICS [physics] (Howard Thomas Odum, 1924–2002, American ecosystem ecologist) The total entropy of any isolated thermodynamic system tends to increase over time, approaching a maximum value.[¶] See also PRINCIPLES OF ENERGETICS; ZEROTH PRINCIPLE OF ENERGETICS; FIRST PRINCIPLE OF ENERGETICS; THIRD PRINCIPLE OF ENERGETICS; FOURTH PRINCIPLE OF ENERGETICS; MAXIMUM EMPOWER PRINCIPLE; MAXIMUM POWER PRINCIPLE; FIFTH PRINCIPLE OF ENERGETICS; SIXTH PRINCIPLE OF ENERGETICS.

SECOND PRINCIPLE OF THERMODYNAMICS [physics] (William Thomson, 1st Baron Kelvin (Lord Kelvin) 1824–1907, Irish mathematical physicist (Rudolf Juliup Emanvel Clausices, 1822–1888, German physicist and mathematician) (Lord Kelvin and R. Clausius) (1) The distinction between

---

*Data Protection Act (July 16, 1998), Chapter 29.
[†]Univ. Edinburgh, Records Management Section (2007) (http://www.recordsmanagement.ed.ac.uk/InfoStaff/DPstaff/DP_Research/Research AnnexA.htm).
[‡]Kesavan, H. K. and Kapur, J. N. (1989), The generalized maximum entropy principle, *IEEE Trans. Syst. Man Cybernet.* **19**(5), 1042–1052.

[§]Kapur, J. N. (1984), On minimum interdependence principle, *Indian J. Pure Appl. Math.*, **15**(9), 968–977.
[¶]Helm, G. (1898), *Die Energetic*, Verlag, Leipzig.

spontaneous and nonspontaneous processes. A process is spontaneous if it occurs without needing to be driven. In other words, spontaneous changes are natural changes, like the cooling of hot metal and the free expansion of a gas. No cyclic engine operates without a heat sink, and heat does not transfer spontaneously from a cool to a hotter body, respectively. The property of entropy is introduced to formulate the law quantitatively in exactly the same way the properties of temperature and internal energy are introduced to render the zeroth and first laws quantitative and precise. The entropy S of a system is a measure of the quality of the energy it stores. When a given quantity of energy is transferred as heat, the change in entropy is large if the transfer occurs at a low temperature and small if the temperature is high. Thus, energy and matter tend to disperse in disorder, and this dispersal is the driving force of spontaneous change.* (2) The entropy of a closed system cannot decrease—together with the definition of thermodynamic temperature. For, if the number of possible logical states of a computation were to decrease as the computation proceeded forward (logical irreversibility), this would constitute a forbidden decrease of entropy, unless the number of possible physical states corresponding to each logical state were to simultaneously increase by at least a compensating amount, so that the total number of possible physical states was no smaller than originally (total entropy has not decreased).[†] See also LANDAUER'S PRINCIPLE; THERMODYNAMIC PRINCIPLES.

SELBERG PRINCIPLE [mathematics] (Atle Selberg, 1917–2007, Norwegian mathematician) (Harish-Chandra 1923–1983, Indian mathematician) (1) Let be a simple algebraic group over a non-Archimedean local field F. An element $h \in G$ is noncompact if it is contained in no compact subgroup of G. (2) If is a conjugacy class of noncompact, regular, semisimple elements in G, then the orbital integral over of any coefficient of supercuspidal representation of G is zero.(2) Harish-Chandra has shown how to evaluate the integral of f with respect to the G-invariant measure on any regular semisimple conjugacy class. In fact suppose that h is a regular semisimple element of G. The Cartan subgroup T which centralizes h may be assumed to be stable with respect to a fixed Cartan involution θ. In other words, there is a θ-stable decomposition

$$T = T_I T_{\mathbb{R}},$$

where T_I is compact and $T_{\mathbb{R}}$ is a vector group. Then according to Harish-Chandra,

$$\int_{T_{\mathbb{R}} \backslash G} f(x^{-1}hx)dx = \varepsilon(T)\Theta_\omega(f)\Theta_\omega(h),$$

where Θ_ω is the character of ω and $\varepsilon(T)$ equals 1 if T is compact and is 0 otherwise. Implicit in this formula is the absolute convergence of the integral on the left. The vanishing statement (the case that T is noncompact) is sometimes known as the Selberg principle. The purpose of this paper to establish a formula which generalizes.[‡] See also HARISH-CHANDRA–SELBERG PRINCIPLE.

SELECTION POTENTIAL PRINCIPLE [biology] As long as there is a positive selection potential at equilibrium, there is the opportunity for new modifiers to invade the population. Unless it is reduced, a population can always potentially be invaded by new variants having the appropriate changes in the transformations operating on them.[§] See also PRINCIPLE OF SELECTION POTENTIAL.

*Paglietti, A. (1976), Some remarks on the local form of the second principle of thermodynamics, *Lettere Al Nuovo Cimento* (Italian Physical Society) **16**(15), 475–478.
†Iribarne, J. V. and Godson, W. L. (1981), *Atmospheric Thermodynamics*, 2nd rev. ed, Springer Science & Business, p. 35.

‡Arthur, J. (1976) The characters of discrete series as orbital integrals. Inventiones Mathematicae. Springer Berlin/Heidelberg, **32**(3) 205–261.
§Ratinov, V. B., Enisherlova, S. G., and Zagirova, R. U. (1969), Galvano- and potentiostatic methods of investigation and basic principles of selection of admixtures-inhibitors of corrosion of reinforcement of concrete, *Proc. Inst. Symp. Durability Concrete*, preliminary report, vol. 2, pp. D-135–D-151.

SELF-ORGANIZING PRINCIPLE. [biology] Allows groups of individuals to converge on the same selection among multiple possible alternatives. The principle is important for distributed coordination, since it allows groups, such as schools of fish and flocks of birds, to behave as if their members were parts of a single organism or system without the need for a centralized decision maker.[*] See ORGANIZING PRINCIPLE; also termed *self-organization principle.*

SELF-SENSING PRINCIPLE [engineering] A system able to affect or determine its own internal structure. A function of large-scale control systems modifying modes of control action or the structure of the control system in response to changes in system objectives, contingency events, and so forth.[†]

SEMRAU'S PRINCIPLE [engineering] (Konrad Troxel Semrau, b. 6/5/1919 Chemical engineer) (1) At constant particle size distribution: $N_t = \alpha P^\Gamma / T$. The constants α and γ depend on the physical chemical properties of the system and the particle size distribution. (2) The efficiency of collection is proportional to power expended, and more energy is required to capture finer particles.[‡] Furthermore, the correlation is not general because different parameters are obtained for differing emissions being controlled by different devices. Semrau's power law principle can be applied, especially for interpolating or extrapolating data

$$N_t = \alpha P^\gamma$$

for an existing scrubber: where: Nt = number of contacting units P = total contacting power α, γ = constants that depend on gas, particle size distribution, and device. See also CONTACTING POWER PRINCIPLE.

[*]Baldassarre, G., Parisi, D., and Nolfi, S. (2002), Distributed coordination of simulated robots based on self-organization. *Artificial Life* **12**(3).

[†]Ren, S., Bian, C., and Liu, J. (2007), Study on estimation of rotor position of self-sensing active magnetic bearings, *Int. Symp. Nonlinear Dynamics, J. Phys. Conf. Series* 96, IOP (Inst. physics).

[‡]Cooper, C.D. (2007) Air Pollution Control methods in Kirk-Othmer Encyclopedia of Chemical Technology. John Wiley & Sons, Inc.

SEPARATION DETECTION PRINCIPLE [physics] Gas chromatography interfaced with mass spectrometry to produce the "hyphenated" method GC-MS.[§]

SEPARATION PRINCIPLE [ecology] (1) (The classical separation principle.) Disjoint sets in a Polish space can be separated by an alternated union of closed sets. This can be generalized to the separation of sets by alternated unions of sets.[¶] (2) In the theory of stochastic control for linear models, when the objective is to minimize a quadratic form in the state and control variables, the main result is the so-called separation principle. An optimal control can be found by (a) finding a formula for an optimal control, assuming that the state of the system is known; and (b) in this formula, substituting for the known state the best estimate for the state of the system from the observed output.[‖] (3) For division into partial components that are conducted to zones of different states or components. (4) Mixture can be separated into desirable and undesirable components by methods known from fractional distillation.[**]

SERENDIPITY PRINCIPLE [mathematics] States that the solution to the linear quadratic Gaussian problem separates into the optimal deterministic controller in which the state used is obtained as the output to an optimal state estimator.[††]

[§]Girard, J. (2005), *Principles of Environmental Chemistry*, Jones & Bartlett Publishers, p. 271.

[¶]Makarov, V. M., Veselov, I. N., Vandishev, A. B., and Timofeev, N. I. (1996), Scientific and technical aspects of the problem of hydrogen purification based on the membrane separation principle, *Proc. 11th Conf. Hydrogen Energy Progress World Hydrogen Energy*, Stuttgart, June 23–28, 1996, vol. 1, pp. 879–884.

[‖]Kirchgessner, M. and Muller, H. L. (1996), Effect of Hay's separation principle on the energy metabolism, *Ann. Nutr. Metab.* **40**(6), 336–342.

[**]Terabe, S., (1992) Capillary electrophoresis: separation principle and application. Seikagaku, *J. Jpn. Biochem. Soc.* **64**(2), 111–115.

[††]Christoffersen, E. E. (2004), *The Principle of Serendipity*, Aarhus.

SERENITY PRINCIPLE [psychology] Applies principles of mental health to an understanding of the nature of addiction and offers for counselors and clients serenity as an immediate goal rather than a distant dream.*

SEVEN LAWS OF TRAINING [kinesiology] (Fredevice CottatField, [aka DrSquat], 1942- American, Philosophy and sport psychology) (1) *Law of individual differences.* We all have different abilities, bodies, and weaknesses, and we all respond differently (to a degree) to any given system of training. These differences should be taken into consideration when designing your training program. (2) *The overcompensation principle.* Mother nature overcompensates for training stress by giving you bigger and stronger muscles. (3) *The overload principle.* To make mother nature overcompensate, you must stress your muscles beyond what they're already used to. (4) *The SAID principle.* The acronym for *specific adaptation to imposed demands.* Each organ and organelle responds to a different form of stress. (5) *The use / disuse principle.* "Use it or lose it" means that your muscles hypertrophy with use and atrophy with disuse. (6) *The GAS principle.* The acronym for *general adaptation syndrome*; this law states that there must be a period of low-intensity training or complete rest following periods of high-intensity training. (7) *The specificity principle.* You'll get stronger at squats by doing squats as opposed to leg presses, and you'll achieve greater endurance for the marathon by running long distances than you will by (say) cycling long distances.†

SEVEN-TERM PRINCIPLE [mathematics, engineering] Calibrates a network analyzer having two test ports and at least four measuring locations by successive measurement of the transmission and reflection parameters at three calibration standards successively connected in arbitrary sequence between the two test ports. Entailing the following steps: (1) providing an electrical line whose characteristic impedance is known and whose electrical propagation constant is unknown and complex; (2) using in a first calibration measurement said electrical line to provide first measured value; (3) using in second and third calibration measurements, respectively, at least one 2-port connected between the two test ports, the 2-port being formed by concentrated components, to provide second and third measured values, respectively; and (4) calculating the electrical propagation constant of said electrical line from said first, second, and third measured values.‡ See also (BLOOD SEPARATION PRINCIPLE;) (OPERATING SEPARATION) (PRINCIPLE; PHASE SEPARATION PRINCIPLE.)

SEVENTH DATA PROTECTION PRINCIPLE [mathematics] Appropriate technical and organizational measures shall be taken against unauthorized or unlawful processing of personal data and against accidental loss or destruction of, or damage to, personal data.§ An act to make new provision for the regulation of the processing of information relating to individuals, including the obtaining, holding, use, or disclosure of such information. See also COPY PROTECTION PRINCIPLE; PRIVACY PRINCIPLE; FIRST DATA PROTECTION PRINCIPLE; SECOND DATA PROTECTION PRINCIPLE; THIRD DATA PROTECTION PRINCIPLE; FOURTH DATA PROTECTION PRINCIPLE; FIFTH DATA PROTECTION PRINCIPLE; SIXTH DATA PROTECTION PRINCIPLE; EIGHTH DATA PROTECTION PRINCIPLE.

SEVENTH JOINT PRINCIPLE [engineering] See also ASSEMBLY SEQUENCE PRINCIPLE.

SHMUSHKEVICH PRINCIPLE [mathematics, physics] (Vladimer Naumvovich Gribov, 1930–1997, Russian Physicist) The connection between the static bootstrap and the current commutator algebra. The bootstrap condition may be interpreted as follows. Suppose that there is a system of

---

*Bailey J. (1990), Living without stress (excerpts from *The Serenity Principle: Finding Inner Peace in Recovery*), *Minneapolis St. Paul Mag.* **18**, 56.
†Hatfield, F. C. (1987), *Seven "Laws" of Training* (http://drsquat.com/content/knowledge-base/fresh-look-strength).
‡Schiek, B. and Eul-Hermann, J. (1991), *Method of Calibrating a Network Analyzer*, US Patent: 4,982,164.
§Data Protection Act (1998), Chapter 29.

particles, any one of which is a composite particle of the others, that is, that all Z factors vanish. This condition leads to the vertex symmetry.[*]

SEX ENTROPY VARIATIONAL PRINCIPLE [mathematics]   hsex(T) = maxμ hsex(μ).[†]

SHOOTER PRINCIPLE [engineering]   (1) Allowing one to change terrain in real time, effectively getting the drop on enemies, blocking them off and creating cover strategically to get the upper hand.[‡] (2) Ink printer modules operating according to the edge shooter principle equipped with piezoelectric actuators is provided for use in printers for postage meter machines and correspondingly has a large number of nozzles arranged in columns.[§] See also EDGE SHOOTER PRINCIPLE; FACE SHOOTER PRINCIPLE.

SIEGEL'S PRINCIPLE [mathematics]   (Garldwing Siegel, 1896–1981, German mathematician)   The quantitative application to the existence of integer solutions of a system of linear equations in Diophantine approximation.[¶] See also COMBINATORIAL PRINCIPLE.

SIEVE PRINCIPLE [mathematics]   (Abraham de Moivre, 1667–1754, French mathematician), (Eratosthenes of Cyrene, 276 BC–195 BC, Greek mathematician)   Count the size of the union of sets A and B by adding $|A|$ and $|B|$ and then subtracting the size of their intersection. The name is based on overgenerous inclusion, followed by compensating exclusion. When $n > 2$ the exclusion of the pairwise intersections is (possibly) too severe, and the correct formula is as shown with alternating signs.[‖] The *sieve of Eratosthenes* is the number-theoretic version of the principle of inclusion-exclusion. In the typical application of the sieve of Eratosthenes one is concerned with estimating the number of elements of a set A that are not divisible by any of the primes in the set $\mathcal{P}$.

Möabius inversion formula can be written as

$$\sum_{\substack{d|n \\ d|P}} \mu(d) = \sum_{d| \gcd(n,P)} \mu(d) = \begin{cases} 1, & \text{if } \gcd(n,P) = 1, \\ 0, & \text{if } \gcd(n,P) > 1. \end{cases}$$

If we set $P = \prod_{p \in \mathcal{P}} p$ then (1) is the characteristic function of the set of numbers that are not divisible by any of the primes in $\mathcal{P}$.[**] If we set A_d to be the number of elements of A that are divisible by d, then the number of elements of A that are divisible by d, then the number of elements of A that are not divisible by any primes in $\mathcal{P}$ can be expressed as

$$\sum_{n \in A} \sum_{\substack{d|n \\ d|P}} \mu(d) = \sum_{d|P} \mu(d) \sum_{n \in A d|n} 1 = \sum_{d|P} \mu(d) A_d.$$

See also PRINCIPLE OF INCLUSION-EXCLUSION

SIGN PRINCIPLE [psychology, physics]   A linear relationship is shown to exist between the width of the experimental point spread function of the optical imaging method and the resulting measurement error in vortex location. A recent theorem derived in the context of random optical wavefields (the *sign principle*) is shown to predict alternation of vortex signs in near-neighbor vortices along zero-crossing curves of the complex-valued orientation map function. This

[*]Watanabe, K., (1966), Shmushkevich Principle and Current Algebra. Il Nuovo Cimento, v.XLII A, n.3, p.1832.

[†]Boyle, M. (n.d.) Entropy on shrinking scales, and the entropy theory of symbolic extensions. University of Maryland.

[‡]Day 1 Studios (2008), *LucasArts Entertainment* (http://xbox360.gamezone.com/gzreviews/p32458_03.htm).

[§]Wehl, W. R., (1989), Ink-jet printing: The present state of the art, *Proc. Conf. VLSI and Computer Peripherals, VLSI and Microelectronic Application's in Intelligent Peripherals and their Interconnection Networks*, CompEuro '89, Siemens AG, Berlin, May 8–12, 1989, pp. 2/46–2/52.

[¶]Struppeck, T. and Vaaler, J. (1989), Inequalities for heights of algebraic subspaces and the Thue-Siegel principle. Analytic number theory, *Progress Math.* **85**, 493–528.

[‖]Zhong, S. N. and Feng, S. (1993), You A sieve principle of the best compromise solution and application. *J. Syst. Eng.* **8**(1), 91–101.

[**]El'natanov, B.A. (1983). A brief outline of the history of the development of the sieve of Eratosthenes. Istor.-Mat. Issled. 27, p. 238–259.

theorem leads to a novel visualization of cortical orientation maps that is applied to experimental optical recording data from monkey visual cortex provided by the laboratory of G. Blasdel, showing a perfect alternating arrangement of singularities, as predicted by the sign principle.* See GENERATIVE PRINCIPLE OF WRITING.

SIMILARITY PRINCIPLE [biology, giology, mathematics]   (1) A function defined on an open subset of the complex plane is said to be approximately analytic (a term coined by L. Bers in the mid-1950s) if on for some constant. An approximately analytic function has a local representation, where is holomorphic and is continuous. Thus the zero sets of and are identical, in particular, has only isolated zeros on which it vanishes to finite order. (2) *Planar vector fields*. Let be an open subset of the complex plane and consider the equation (1) Every continuous solution w of has the form

$$w = e^g h,$$

for some holomorphic function h and Hölder continuous g. Thus w and h are "similar" in the sense that both $\frac{w}{h}$ and $\frac{h}{w}$ are bounded away from zero on compact sets.

The Similarity Principle holds for any elliptic vector field L, since in appropriate coordinates, L becomes a multiple of $\frac{\partial}{\partial z}$.[†] The similarity principle states that, if and are Hölder continuous, then solutions of (1) have the form, where F is a nonvanishing Hölder continuous function and is holomorphic. In particular, the zeros of are isolated and of finite order, thus quantitatively similar to zeros of holomorphic functions. (2) A theory of considerable integrative, explanatory, and interpretive power for a broadly interdisciplinary social science of human behavior. Individuals differ to some extent in the way operational factors are organized and function within subsystems and the number of ways in which operating factors in a certain subsystem can be organized in patterns, to enable the subsystem to play its functional role in totality.[‡] (3) Orientation selectivity in a model visual system, for the development of biological and synthetic perceptual networks. (4) Basic in the development of biological systems is the ability to self-organize, a process by which new structures and patterns emerge. Formulated according to which the statistical pattern of the pelagic population is identical in all scales sufficiently large in comparison with the molecular one. From this principle, a power law is obtained analytically for the pelagic animal biomass distribution over the animal sizes. A hypothesis is presented according to which, under fixed external conditions, the O_2 exchange intensity of an animal is governed only by its mass and density and by the specific absorbing capacity of the animal's respiratory organ. From this hypothesis a power law is obtained by the method of dimensional analysis for the exchange intensity mass dependence. The known empirical values of the exponent of this power law are interpreted as an indication that the O_2-absorbing organs of the animals can be represented as so-called fractal surfaces. The biological principle of the decrease in specific exchange intensity with increase in animal mass is discussed.[§] See also PRINCIPLE OF SIMILARITY; KICK SIMILARITY PRINCIPLE; OPTIMALITY PRINCIPLE; SELF-ORGANIZING PRINCIPLE; PRINCIPLE COMPONENT ANALYSIS; ORGANIZING PRINCIPLE.

SIMILIA PRINCIPLE [pharmacology]   (1) A substance that causes certain symptoms in a healthy organism may be applied in a curative sense to a diseased organism with

*Freund, I. and Shvartsman, N. (1994), Wave-field phase singularities: The sign principle, *Phys. Rev. A Atom. Mol. Opt. Phys.* **50**(6), 5164–5172.

†Berhanu, S., Hounie, J., and Santiago, P. (2000) A similarity principle for complex vector fields and applications. *Transactions of the American Mathematical Society*, **353**,(4) 1661–1675.

‡Harvey, J. H., Ickes, W. J., and Kidd, R. F. (1976), *New Directions in Attribution Research*, Wiley Monographs in Applied Econometrics Series, Lawrence Erlbaum, vol. 1, p. 172.

§Barenblatt, G. I. and Monin, A. S. (1983), Similarity principles for the biology of pelagic animals, *Proc. Natl. Acad. Sci. USA* **80**(11), 3540–3542.

identical symptoms.* (2) Classical homeopath tries to find a substance that fits the patient's complaints as closely as possible. Unusual symptoms that do not fit the symptom complexes recognized by conventional medicine may be considered even more important than the regular symptoms. (3) Similar principle states that to cure a sick person, one should look for something that provokes similar symptoms in healthy people.[†]

Suppose that $AvB \to_s C$ implies $A \to_s C$ (call this the 'Simplification Principle'). Presumably. $A \to_s C$ is equivalent to $((A \& B) v (A \& \sim B)) \to_s C$ which, according to this principle, entails $A\&B \to_s C$; hence. $A \to_s C$ entails $A\&B \to_s C$. 'If A or B occurred, then C would happen' implies 'If A occurred, then C would happen'. Hence, they must deny that subjunctive conditionals with disjunctive antecedents are of the form $AvB \to_s C$. Both tentatively suggest that $A \to_s C \& B \to_s C$ would better express such subjunctive conditionals, though as Ellis conceded, there is little more than the need to accommodate the simplification principle itself to justify such a re-interpretation.[‡] See also PRINCIPLE OF CONTRAPOSITION. See also SIMPLIFICATION PRINCIPLE;

The principle of contraposition seems essential to scientific reasoning, and it explains why a well-established theory asserting that $P \to_s Q$ is often used as the rationale for a diagnostic test of the falsity of P, it being argued that if Q were false, then P would be false

too.[§] See also PRINCIPLE OF CONTRAPOSITION. See also SIMPLIFICATION PRINCIPLE;

SIMILIA PRINCIPLE OF HOMEOPATHY [pharmacology] One might treat a disease according to its cause. In psychotherapy, for example, one might treat a patient experiencing distress after a trauma by exposing the patient to an artificial trauma similar to the one that originally caused the distress.[¶]

SINGLE-RESPONSIBILITY   PRINCIPLE [computers] (Robert Cecil Martin) (1) A class should have one, and only one, reason to change.[‖] Every object should have a single responsibility, and that all its services should be narrowly aligned with that responsibility. The principle forms part of cohesion, a concept widely used in software development. Included among the principles of class design.[**] (2) There should never be more than one reason for a class to change.[††] See also PRINCIPLES OF OBJECT-ORIENTED DESIGN; OPEN–CLOSED PRINCIPLE; LISKOV SUBSTITUTION PRINCIPLE; DEPENDENCE INVERSION PRINCIPLE; INTERFACE SEGREGATION PRINCIPLE.

SINK/FLOAT PRINCIPLE [physics], Type of beneficiation process among four major categories: (1) the economic recovery of additional distinctly separate products, such as dolomite limestone at the Mascot plant; (2) elimination of a large portion of run-of-the-mine ore prior to grinding and flotation, to allow increased throughput with low capital cost; (3) processing of a sedimentary deposit, in which the valuable mineral is contained in the matrix

*Van Wijk, R. and Wiegant, F. A. (1997), The similiar principle as a therapeutic strategy: A research program on stimulation of self-defense in disordered mammalian cells, *Altern. Ther. Health Medi.* **3**(2), 33–38.

[†]Menges, G. T. (1996), Cultured mammalian cells in homeopathy research. The similia principle in self recovery, *Tijdschrift Diergeneesk.* **121**(22), 660–661.

[‡]Urbach, P. (1988) What Is a Law of Nature? A Humean Answer. *The British Journal for the Philosophy of Scince*, **39**,(2), 193–209.

[§]Urbach, P. (1988) What Is a Law of Nature? A Humean Answer. *The British Journal for the Philosophy of Science*, **39**,(2), 193–209.

[¶]Opmeer, R. J. (1997), How the Similia principle of homeopathy resolved an emergency. Case history of ivermectin poisoning in a collie, *Tijdschrift Diergeneesk.* **122**(2), 36–39.

[‖](Martin, R. C. (2002), *Agile Software Development: Principles, Patterns, and Practices*, Robert C. Martin, Prentice-Hall.

[**]Martin, R. C. *Dependency Inversion Group. Ten Commandments of OO Proramming* (http://www.butunclebob.com/ArticleS.UncleBob.PrinciplesOfOod).

[††]Page-Jones, M. (1988), *The Practical Guide to Structured Systems Design*, 2d. ed., Yourdon Press Computing Series, chap. 6, p. 82.

separating the boulders; and (4) simplification of "around" support programs in mines by allowing greater freedom of action in the extractive pattern.[*] See also PRINCIPLES OF HEAVY MEDIA SEPARATION; ARCHIMEDES PRINCIPLE.

SIPHON PRINCIPLE [pharmacology] Health program suggesting that "like cures like"; drugs or treatments for some diseases can be applied to similar ailments.[†]

SIPPING PRINCIPLE [physics] Feed enters the bottom of the heating tubes and as heats, steam begins to form. The ascending force of this steam produced during the boiling, liquid and vapors flow upward in parallel flow. At the same time the production of vapor increases and the product is pressed as a thin film on the walls of the tubes, and the liquid rises upward. This concurrent upward movement has the beneficial effect of creating a high degree of turbulence in the liquid.[‡]

SIXTH DATA PROTECTION PRINCIPLE [mathematics] Personal data shall be processed in accordance with the rights of data subjects under this Act. The rights of data subjects are to (1) request a copy of personal data (subject access), (2) prevent processing likely to cause substantial damage or distress, (3) prevent processing for direct marketing purposes, and (4) not be subject to automated decision taking.[§] See also COPY PROTECTION PRINCIPLE; PROTECTION PRINCIPLE; PRIVACY PRINCIPLE; FIRST DATA PROTECTION PRINCIPLE; SECOND DATA PROTECTION PRINCIPLE; THIRD DATA PROTECTION PRINCIPLE; FOURTH DATA PROTECTION PRINCIPLE; FIFTH DATA PROTECTION PRINCIPLE; SEVENTH DATA PROTECTION PRINCIPLE; EIGHTH DATA PROTECTION PRINCIPLE.

SIXTH PRINCIPLE OF ENERGETICS [physics] (Howard Thomas Odum, 1924–2002, American ecosystem ecologist) Material cycles have hierarchical patterns measured by the emergy/mass ratio that determines its zone and pulse frequency in the energy hierarchy.[¶] See also PRINCIPLES OF ENERGETICS; ZEROTH PRINCIPLE OF ENERGETICS; SECOND PRINCIPLE OF ENERGETICS; THIRD PRINCIPLE OF ENERGETICS; FOURTH PRINCIPLE OF ENERGETICS; MAXIMUM POWER PRINCIPLE; FIFTH PRINCIPLE OF ENERGETICS; FIRST PRINCIPLE OF ENERGETICS.

SIZE PRINCIPLE [psychology, engineering] (Henneman) Motor units are recruited in order of increasing size. This serves two important purposes: (1) it minimizes the development of fatigue by using the most fatigue-resistant muscle fibers most often (holding more fatigue-prone fibers in reserve until needed to achieve higher forces), and (2) it permits equally fine control of force at all levels of force output (e.g., using smaller motor units when only small, refined amounts of force are required).[‖] See also OPTIMALITY PRINCIPLE; SELF-ORGANIZING PRINCIPLE; PRINCIPAL-COMPONENT ANALYSIS; ORGANIZING PRINCIPLE; BODY SIZE PRINCIPLE.

SKINNER'S PRINCIPLES [psychology] Burrhus Frederic Skinner (B.F. Skinner), 1904–1990, American psychologist Skinner's principles of progressive disambiguation, and his classification of autoclitics into their different types can be fit into the more conventional account of syntax. Demonstrating behavior change and the variables responsible for that change was fundamental to the Skinnerian view: "If reasonable order was not discoverable, we could scarcely be effective in dealing with human affairs. The methods of science are

[*]Doyle, E. N. (1970), The sink-float process in lead-zinc concentration, *Proc. World Symp. Mining and Metallurgy of Lead & Zinc*, vol. 1, p. 38.
[†]Akito, T., Teruyoshi, A., Nobufumi, U., Naoki, K., Noriaki, U., and Koji, I. (2006), Application of siphon principle to fluid drainage in transurethral surgery, *Int. J Urol.* (Jpn. Urol. Assoc.) **13**(8), 1156–1157.
[‡]Fritz, G., Hoffmann, H., and Seidelmann, K. (1988), Wet-sipping process with recirculation principle. Experience with an improved fuel element leak test at Muelheim-Kaerlich nuclear power plant, *VGB Kraftwerkstechnik* **68**(1), 9–12.
[§]Univ. Edinburgh, Records Management Section (7/2007), (http://www.recordsmanagement.ed.ac.uk/InfoStaff/DPstaff/DP_Research/ResearchAnnexA.htm).

[¶]Odum, H. T. and Pinkerton, R. T. (1955), Time's speed regulator, *Am. Sci.* **43**(2), 331.
[‖]Ertas, M. et al. (1995), Can the size principle be detected in conventional EMG recordings? *Muscle Nerve*, **18**, 435.

designed to clarify these uniformities and make them explicit"* See SUCCESSIVE APPROXIMATION PRINCIPLE; CHAINING PRINCIPLE; BACKWARD CHAINING PRINCIPLE; ONE-TRIAL PRINCIPLE; RESPONSE-SHAPING PRINCIPLE; TRANSITIONAL STIMULI PRINCIPLE; SCHEDULE OF REINFORCEMENT PRINCIPLE.

SKYRME HARTREE-FOCK PRINCIPLE [physics] (Tony Hilton Royle Skyrme, 1922–1987 British Physicist) (Donlas Rayner Hortree, 1897–1958, English mathematician) (Vladimir Alerk Sandrouich Fock, 1898–1974, Soviet physicist) (1) Two-body correlation to be included not only in the ground-state energy but also in the constraints.† (2) Widths of resonant states have an important effect on the pairing properties of nuclei close to the drop line.‡ (3) Energies of excitations are determined essentially by the coordination number of the excited atom.§ (4) Statistical exchange potential can be given in a universal form and then applied in wave mechanical calculations (the sum of the exchange potential and the electrostatic potential).¶ See also TIME-DEPENDENT HARTREE–FOCK VARIATIONAL PRINCIPLE; HARTREE–FOCK VARIATIONAL PRINCIPLE.

SLOT ORIFICE PRINCIPLE [engineering]   See ORIFICE PRINCIPLE.

SLOTTED BEAM CONNECTION PRINCIPLE [engineering]   Individual wires are designed to be screwed down at their separable ends and

to which conductors are permanently joined at the back end.‖ See also CONTACT PRINCIPLE.

SMALLEST ERROR COMPUTER INTERPOLATION PRINCIPLE [mathematics]   See INTERPOLATION PRINCIPLE.

SNELL'S PRINCIPLE [physics] (Wille brord Snellius, 1580–1626, Dutchastronomer) Relating angles of incidence and refraction of a ray of light at the boundary between media of different refractive index. The two rays and the normal at the point of incidence on the boundary lie in the same plane.**

SOCIAL SUSTAINABILTY PRINCIPLE [gerontology]   the rise in demand for services and features linked to the ageing process, the increase in dependence, and the crisis of informal support. Implies reformulation of the regulatory, care, economic, administrative, cultural, and axiological framework enabling a response to the needs of long-term care without compromising the welfare of future generations.††

SODIUM HEAT ENGINE PRINCIPLE [geology] Heat conduction across a temperature gradient.‡‡

SOLID-STATE ENERGY STORAGE PRINCIPLE [engineering] (1) Fuel injection devices whose electrically driven reciprocating pumps work according to the so-called solid-state energy storage principle. In certain railway and mass and/or rapid transit systems, such as in automatic train

---

*Skinner, B. F. (1953), *Science and Human Behavior*, New York, Macmillan, p. 16.

†Stoitsov, M. V., Dobaczewski, J., Nazarewicz, W., and Ring, P. (2005), *Comput. Phys. Commun.* **167** (43).

‡Grasso, M., Sandulescu, N., Giai, N. V., and Liotta, R. J. (2001), Pairing and continuum effects in nuclei close to the drip line, *Phys. Rev. C* **64**(6).

§Yamagami, M., Shimizu, Y. R., and Nakatsukasa, T., (2009), Optimal pair density functional for description of nuclei with large neutron excess. *Phys. Rev. C.* 064301.

¶Dobaczewskia, J. and Dudek, J. (1997), Solution of the Skyrme-Hartree-Fock equations in the Cartesian deformed harmonic oscillator basis I. The method, *Comput. Phys. Commun.* **102**(1–3), 166–182.

‖Antler, M. (2000), *Electrical connectors, Kirk-Othmer Encyclopedia of Chemical Technology*, Wiley.

**Walker, P. M. B. (1999), *Chambers Dictionary of Science and Technology*. Chambers New York, p. 1067.

††Garcés, J., Ródenas, F., and Sanjosé, V. (2003). Towards a new welfare state: The social sustainability principle and health care strategies, **65**(3), 201–215.

‡‡Betz, B. H., Sungu, S., and Vu, H. V. (1994), Sodium heat engine system: Space application. (Int. Conf. Thermoelectrics, Kansas City, MO), *AIP Conf. Proc.* **316**(1), 13.

operations, it is necessary to establish minimal headways or time intervals between trains in order to provide fast and efficient service. When a train enters a block or track section, it is essential that the signal at the exit end not be cleared immediately so that the train may proceed into the next block or track section. Requires that some predetermined time delay be employed in switching the speed signal so that the train can be safely stopped or can proceed within the track section at a safe reduced headway speed that is established by the signal aspect at the exit end of the block.* (2) Whereby a piston element mounted in a pump cylinder of an electromagnetic reciprocating pump displaces quantities of the fuel to be injected during a virtually resistanceless acceleration phase during which the piston element stores kinetic energy, before ejection in the pump area.[†] See also ENERGY STORAGE PRINCIPLE.

SOLUBILITY PRINCIPLE [chemistry]   (1) Like dissolves like. A polar solvent will dissolve polar compounds but will not dissolve nonpolar compounds.[‡] (2) Nernst[2] was the first to advance the theory that at a given temperature the solubility of a difficultly soluble electrolyte in water or in aqueous solutions of other electrolytes is dependent on a constant called the solubility product, which is proportional to the concentrations of the ions of the salt, each raised to the power corresponding to the number resulting from one molecule. The constant is an important one in the theory of precipitation and solution and particularly useful in calculations of the solubility of a precipitate in mixtures that are not too concentrated. In Nernst's text-book

on physical chemistry.[3] the relation for a difficultly soluble binary salt—such as silver acetate—in water and in solutions containing a salt with a common ion, is developed as follows: calling the total concentrations of the difficultly soluble salt m_0 and m in the saturated water solution and in the salt solution respectively,[§] See also LE CHATELIER'S INVERSE SOLUBILITY PRINCIPLE; DRUG DESIGN PRINCIPLES; PRINCIPLE OF NERNST.

SONIC DOPPLER PRINCIPLE [physics]   See DOPPLER PRINCIPLE.

SORPTION PRINCIPLE [physics]   (1) Reduce the pressure of a gas in an atmosphere. The gas is absorbed on a granular sorbent material such as a molecular sieve in a metal container; when this sorbent-filled container is immersed in liquid nitrogen, the gas is sorbed.[¶] (2) These methods for making ice remove vapor of an aqueous solution from the space above the surface of the solution, which causes more liquid to evaporate. The evaporation of the liquid lowers the liquid's temperature, causing it to enter a solid phase when its temperature reaches the solidification point temperature.[‖] (3) In order to solidify aqueous solutions by means of the evaporation principle, large amounts of vapor must be removed from the environment.[**] See also EVAPORATION PRINCIPLE.

SOVEREIGNTY PRINCIPLE [political science] Military force should be seen as the last resort. Using it, even to maintain international law and order, necessarily means a setback for the new thinking, a reversal of the trend toward establishing a new type of international relations. Objective is to harmonize

*Heimberg, W., Hellmich, W., Kogl, F., and Malatinszky, P. (1995), *Fuel Injection Device According to the Solid-State Energy Storage Principle for Internal Combustion Engines*, US Patent 5,469,828.
[†]Larson, C. O., Jr., Smith, J. S., Chapman, J. H., Slimon, S. A., Trahan, J. D., Brozek, R. J., Franco, A., McGarvey, J. J., Rosen, M. E., and Pasque, M. K. (1998), *Reciprocating Pump with Imperforate Piston*, US Patent 5,758,666.
[‡]Hugus, Z. Z., Jr. and Hentz, F. C., Jr. (1985), A different look at the solubility-product principle, *J. Chem. Educ.* **62**(8), 645–647.

[§]Henderson, L. A. (1908) A diagrammatic representation of equilibria between acids and bases in solution. *J. Am. Chem. Soc.*, **30**(6), 954–960.
[¶]Lee, H. and Stahl, D. E. US (1972), Patent 3,788,036.
[‖]Peter, M.-L. and Relner, E., (1993), *Ice Making System and Method Utilizing the Sorption Principle*, US Patent 5,207,073.
[**]Maier-Laxhuber, Peter, Engelhardt, Reiner, Worz, Reiner and Becky, Andreas, (1995), *Cooling System Having a Vacuum Tight Steam Operating Manifold*, US Patent 5,415,012.

national, regional, and global interests, to assert a single universal scale of democratic values providing for the freedom of choice, a variety of forms of social development, economic and political pluralism, the supremacy of international law, and human rights.[*]

SPACETIME UNCERTAINTY PRINCIPLE [chemistry, physics]   (1) (Werner Heisenberg) Relation whereby, if one simultaneously measures values of two canonically conjugate variables, such as position and momentum, the product of the uncertainties of their measured values cannot be less than approximately Planck's constant divided by $2p_i$. (2) The spacetime uncertainty principle would produce an interesting physical picture in string theory, maybe also in M theory. Spacetime in itself is quantized at the short distance and the concept of spacetime as a continuum manifold cannot be extrapolated beyond the fundamental string scale ls. It is also important to point out this principle seems to be consistent with the recent nonperturbative formulations of M theory and IIB superstring where this principle is realized implicitly in the form of the noncommutative geometry. Moreover, in terms of the *conformal constraint* coming from the Schild action.[†] M-theory, from theoretical physics, is an extension of string theory for eleven dimensions. These dimensionality's exceed the dimensionality of five superstring theories in ten dimensions. The 11-dimensional theory unites all string theoies (and supersedes them). Though a full description of the theoy is not yet known, the low-entropy dynamics are known to be supergravity interacting with 2- and 5-dimensional membranes. M-theory is the unique supersymmetric theory in eleven dimensions, with its low-entropy matter content and interactions fully determined, and can be obtained as the strong coupling limit of type IIA string theory because a new dimension of space emerges as the coupling constant increases.

[*]Pankin, B. D. (1991), The dangers of nationalism: The development of the sovereignty principle in international law (Boris D. Pankin speech), *Vital Speeches* **58**, 5.
[†]Oda, Ichiro, (1998), Supersymmetric IIB matrix models from space-time uncertainty principle and topological symmetry, *Nuclear Phys.* **B516** 167.

SPAN OF MANAGEMENT PRINCIPLE [organizational psychology]   There is a limit at each managerial position on the number of persons an individual can effectively manage. But this number is not a fixed number and will vary in accordance with underlying variables of the situation. See also PRINCIPLE OF PLANNING.[‡]

SPATIAL PRINCIPLE [mathematics]   Existence of a space with temporal and geometric components of the Husdorff-Besicovitch type for which the spatial dimensions are fractionary.[§] See also PRINCIPLE OF TRANSLATION QUANTITATIVITY; PRINCIPLE OF COMPOSITION OF THE TRANSLATIONS; PRINCIPLE OF THE CENTRE-PERIPHERY; PRINCIPLE OF NUCLEATION; UNIFIED PRINCIPLE OF ACCUMULATED ADVANTAGES.

SPARSITY OF EFFECTS PRINCIPLE [mathematics]   (1) Only a few effects in a factorial experiment will be statistically significant.[¶] (2) States that a system is usually dominated by main effects and low-order interactions. Thus it is most likely that main (single factor) effects and two-factor interactions are the most significant responses. Higher-order interactions such as three-factor interactions are very rare.[‖] See also HIERARCHICAL ORDERING PRINCIPLE.

SPECIAL PRINCIPLE OF RELATIVITY [mathematics, physics] (Albert Einstein, 1908) (1) A system of mechanics applicable at high velocities in the absence of gravitation; a generalization of Newtonian mechanics. The speed of light c is the same for all observers, no matter how they are moving; that the laws of physics are the same in all inertial frames;

[‡]Koontz, H. and O'Donnell, C. (1968), *Principles of Management*, A 4th ed., McGraw-Hill.
[§]Mandelbrot, B.B. (1977) The Fractal Geometry of Nature, Freeman, NY.
[¶]Wu, C. F. J. and Hamada, M. (2000), *Experiments: Planning, Analysis, and Parameter Design Optimization*, Wiley, New York.
[‖]Simpson, T. W., Peplinski, J. D., Koch, P. N., and Allen, J. K. (1997), On the use of statistics in design and the implications for deterministic computer experiments, *Proc. DETC'97 1997 ASME Design Engineering Technical Conf.*, Sacramento, Cal., ASME, Sept. 14–17, 1997, p. 3.

and that all such frames are equivalent. No object may have a velocity in excess of the speed of light, and two events that appear simultaneous to one observer need not be so for another.* (2) Physical laws should be the same in all inertial reference frames, but that they may vary across noninertial ones.? Physical laws are the same in any vehicle moving at constant velocity as they are in a vehicle at rest. A consequence is that observers in a noninertial reference frame cannot determine an absolute speed or direction of their travel in space; they may only speak of their travel relative to some other object.[†] (3) A criterion for judging physical theories, stating that they are inadequate if they do not prescribe the exact same laws of physics in certain similar situations. These types of principles have been successfully applied throughout science, whether implicitly (as in Newtonian mechanics) or explicitly (as in Albert Einstein's special relativity and general relativity theories).[‡] See PRINCIPLE OF RELATIVITY.

SPECIFIC ADAPTATION TO IMPOSED DEMANDS (SAID) PRINCIPLE [kinesiology] (Dr. Hans Selye)   See PRINCIPLE OF TRAINING.

SPECIFICITY PRINCIPLE [kinesiology] (Dr. Hans Selye)   See PRINCIPLE OF TRAINING.

SPECTROSCOPIC PRINCIPLES [physics] of Near-infrared (NIR) absorption spectra occur because chemical bonds within molecules can vibrate and many molecular groups can rotate, thus generating series of different energy levels between which rapid, IR-(or NIR)-induced transitions can occur. Principle on which both principle-component regression (PCR) and partial lease-squares (PLS) are based; stems from the observation that although many different variations usually make up a spectrum (e.g., interconstituent interactions, instrument variations, or differences in sample handling), after proper data pretreatments [(such as baseline corrections, light-scattering corrections, e.g., mass spectrometry calorimetry (MSC)], the largest variations remaining in the calibration set would be due only to the chemical composition variations of the standard samples. The main purpose of both PCR and PLS model is then to calculate a set of "variation spectra," representing only the variations caused by composition. Such calculated variation spectra are sometimes called *loading vectors, principal components,* or more frequently, *factors*. The calculation of such spectra usually involves an iterative process that manipulates n samples of proper numerical values called *eigenvectors*; for this reason, PCR and PLS algorithms are also called *eigenvector methods*. Once the factors are calculated, they are utilized instead of the raw spectra for building the calibration model; therefore, the possibility of overfitting can be minimized by choosing the correct number of factors. Although the concepts of PLS and PCR are similar, the approaches to the calculation of the factors (loading vectors) are quite different. The PCR algorithm calculates the factors independently of the concentration information, whereas the PLS algorithm utilizes *both* the concentration and spectral information of the calibration set to calculate the factors.[§]

SPEECH INTERPOLATION PRINCIPLE [mathematics]   See INTERPOLATION PRINCIPLE.

SPIJKER'S PRINCIPLE [mathematics]   The image on the Riemann sphere of any circle under a complex rational mapping with numerator and denominator having degrees no greater than n and length no greater than $2n$ n.[¶]

SPIPRO PRINCIPLE [engineering, horticulture] (Werner Heisenberg)   (1) Organic food

*Walker, P. M. B. (1999), *Chambers Dictionary of Science and Technology*, Chambers, New York p. 1080.
[†]Anderson, J. L. (1967), *Principles of Relativity Physics*, Academic Press, New York.
[‡]Silberstein L. (1914) *The Theory of Relativity.* MacMillan & Co.

[§]Luthria, D. L. (2004) Oil Extraction and Analysis: AOCS Publishry Pg. 193. Critical issues and competitive studies.
[¶]Edelman, A. and Kostlan, E. (1995), How many zeros of a random polynomial are real? Bull. Am. Math. Soc. 32, 1–37.

production, with emphasis on the spiral combining propulsion (SPIPRO) principle.[*] (2) Aspiral procest driven by synergice effect of critied factors and can be reflected by improvement of informationization level. [†] See also HEISENBERG UNCERTAINTY PRINCIPLE.

SPIRAL MANDREL PRINCIPLE See CLASSICAL MANDREL PRINCIPLE.

SPIRAL PRINCIPLE [mathematics, engineering] Presenting respective axially interengaging spiral-shaped walls, and drive means connected between the elements to produce a relative translatory circular movement. The displacement is effected by two displacement elements or units, each of which essentially consists of a baseplate with a spiral wall formed thereon defining a spiral recess. The spiral wall of each displacement element axially interpenetrates the spiral recess of the other element.[‡] See also HEISENBERG UNCERTAINTY PRINCIPLE.

SQUASH-AND-STRETCH   PRINCIPLE [animation] One of 12 principles of animation. Provides a sense of weight and flexibility to drawn objects and can be applied to simple objects, like a bouncing ball, or more complex constructions, like the musculature of a human face. Taken to an extreme point, a figure stretched or squashed to an exaggerated degree can have a comical effect. In realistic animation, however, the most important aspect of this principle is the fact that an object's volume does not change when squashed or stretched. If the length of a ball is stretched vertically, its width (in three dimensions; also its depth) needs to contract correspondingly horizontally.[§] Also termed *assimilate–stretch principle*.

STABLE   ABSTRACTIONS   PRINCIPLE   (SAP) [computers] (Robert Cecil Martin) (1) Abstractness increases with stability. (2) Sets up a relationship between stability and abstractness. A stable package should also be abstract so that its stability does not prevent it from being extended. On the other hand, principle states that an instable package should be concrete since its instability allows the concrete code within it to be easily changed.[¶]

STABLE   DEPENDENCES   PRINCIPLE   (SDP) [computers] (Robert Cecil Martin) (1) Depend on the direction of stability. (2) The *I* metric of a package should be larger than the *I* metrics of the packages it depends on: *I* metrics should decrease in the direction of dependence.[¶]

STACKABILITY   PRINCIPLE [engineering] See also ASSEMBLY SEQUENCE PRINCIPLE.

STAFFING PRINCIPLES See PRINCIPLE OF THE OBJECTIVE OF STAFFING.

STAFF'S   DIAGNOSTIC   PRINCIPLE [pharmacology] (1) Quartz crystal microbalances (QCMs) are used in high-technology laboratories and in space for applications such as detecting contamination on optical surfaces, measuring outgassing, and determining constituents by means of thermogravimetric analysis.[‖] (2) Mind/body model validating the myriad subtle neurotransmitter connections between emotions, mental states, and the immune system; this mind/body model has validated the myriad subtle neurotransmitter connections between emotions, mental states, and the immune system. (3) Since the virus can affect the entirety of the person, the entirety of the patient must be considered

[*]JianYing, S. and YiTong, L. (n.d.), SPIPRO principles on green food development, *J. Shanghai Jiaotong Univ. Agric. Sci.* p 428–431.

[†]Chen, D (2006) SPIPRO mode research from catching-up to leapfrogging. *Studies in Science of Science* **24**(S) 67–73.

[‡]Fischer, B., Kabelitz, H. Pfaff, H., and Schmitz, A., (1981), *Positive Displacement Machine with Elastic Suspension*, US Patent 4,300,875.

[§]Thomas, F. and Johnston, O. (1981), *The Illusion of Life: Disney Animation*, Hyperion, pp. 47–69.

[¶]Martin, R. C. (2003) *Dependency Inversion Group. Ten Commandments of OO Proramming* (http://www.butunclebob.com/ArticleS.UncleBob.PrinciplesOfOod).

[‖]Caruso, F., Rodda, E., Furlong, D. N., Kenichi, N., and Okahata, Y. (1997), Quartz crystal microbalance study of DNA immobilization and hybridization for nucleic acid sensor development, *Anal. Chem.* **69**(11), 2043–2049.

in making the diagnosis.* See also PRINCIPLE OF PSYCHONEUROIMMUNOLOGY; QUASILOCAL PRINCIPLE; DIAGNOSTIC PRINCIPLE.

STAM'S PRINCIPLE [physics]   The synthesis and analysis of single-particle position and momentum densities of many-body systems such as atoms and molecules.[†]

STANFORD   EQUIVALENCE   PRINCIPLE [physics] (Paul W. Worden, Jr., 1983) A program intended to test the uniqueness of free fall to the ultimate possible accuracy. The program is conducted in two phases: first, a ground-based version of the experiment, which should have a sensitivity to differences in rate of fall of one part in 10^{12}, followed by an orbital experiment with a sensitivity of one part in 10^{17} or better.[‡]

STANISLAV GROF'S COEX PRINCIPLE [psychology] (Stanislav Grof, b. 1931, Prague, Czechoslovakia, transpersonal psychology) A system of condensed experience (COEX) includes a specific constellation of memories (and related fantasies) from different life periods of the individual that have a similar basic theme or contain similar elements and are associated with a strong emotional charge of the same quality.[§]

STATIC IMAGE PRINCIPLE [geology]   Can be generalized to anisotropic media in the static case applicable to geologic media exhibiting electrical conductivity caused by fractures and fissures in the rock.[¶]

*Martin, P. (1999), *The Healing Mind: The Vital Links Between Brain and Behavior, Immunity and Disease*, Macmillan, p. 89.

[†]Romera, E. (2002), Stam's principle D-dimensional uncertainty-like relationships and some atomic properties, *Mol. Phys.* **100**(20), 3325–3329.

[‡]Worden, P. W., Jr., Everitt, C. W. Francis, and Bye, M. (1994), The Stanford equivalence principle program, *Relativistic Gravitational Experiments in Space*, NASA, pp. 137–140.

[§]Grof, S. (1986), *Beyond the Brain: Birth, Death, and Transcendence in Psychotherapy* (Suny Series, Transpersonal & Humanistic Psychology), State Univ. New York Press, p. 97.

[¶]Ermutlu, M. E. (1994), Static image principle for anisotropic layered medium using transmission line analogy, *Radio Sci.* **29**(4), 739–749.

STATIC INDUCTION PRINCIPLE [mathematics, physics]   See INDUCTION PRINCIPLE.

STATIONARY ACTION PRINCIPLE [physics] (Sir William Rowan Hamilton, 1805–1865) (1) The dynamics of a physical system is determined by a variational problem for a functional based on a single function, the Lagrangian, which contains all physical information concerning the system and the forces acting on it. The variational problem is equivalent to and allows for the derivation of the differential equations of motion of the physical system.[‖] (2) In a real motion, the system acted on by potential forces has a stationary value as compared with near kinetically possible motions, with initial and final positions of the system and times of motion identical with those for real motion. Richard Feynman's path integral formulation of quantum mechanics is based on a stationary action principle, using path integrals. Maxwell's equations can be derived as conditions of stationary action.[**] See also PRINCIPLE OF LEAST ACTION; STEWART–HAMILTON PRINCIPLE; GUERRA–MORATO VARIATIONAL PRINCIPLE; HAMILTON–OSTROGRADSKI PRINCIPLES.

STATIONARY PHASE PRINCIPLE [mathematics, physics]   In the computation of oscillatory integrals, the contributions of nonstationary points of the phase are smaller than any power of, for[††] For short wavelengths, at a point P far away, the electrical quantities are, approximately, only dependent on the electrical state of the points on the wave surface near the foot of the normal passing through P.

Taking into account the surface equation in the vicinity of N and extending the integral

[‖]Hamilton, W. R., (1834), On a general method in dynamics, *Phil. Trans. Roy. Soc. Part I* 247–308.

[**]Arnold, V. I. (1989), *Mathematical Methods of Classical Mechanics*, 2nd ed., Springer-Verlag, pp. 59–61.

[††]Drabowitch, S., Papiernik, A., Griffiths, H., Encinas, J., and Smith, B.L. (2005) Modern Antennas. Springer p. 181.

to infinity, which the stationary phase principle permits us to do, we find

$$E'(\text{P}) = E_N \sqrt{\frac{R_1 R_2}{(R_1 + p)(R_2 + p)}} \exp\left(-j\frac{2\pi}{\lambda}p\right)$$

where R_1, R_2 are the principal radii of curvature.

The stationary phase principle establishes a correspondence from one point to another point of two successive wavefronts.

STATIONARY TIME PRINCIPLE [physics] (1) Light traveling between two points seeks path that the number of waves is equal, in the first approximation, to that in neighboring paths. (See FERMAT'S PRINCIPLE.) (2) Path taken by a ray of light in traveling between two points requires either a minimum or maximum time.* (3) Light traveling through some substance has a speed that is determined by the substance. The actual path taken by light between any two points, in any combination of substances, is always the path of least time that can be traveled at the required speeds. (4) According to the second law of thermodynamics, in the neighborhood of any equilibrium state of a system, certain states are not accessible by a reversible or irreversible diabolic process.[†] See also CONTES PRINCIPLE; LEAST-TIME PRINCIPLE; CARATHÉODORY'S PRINCIPLE; FERMAT'S PRINCIPLE; OPTIMALITY PRINCIPLE.

STATISTICAL EXCLUSION PRINCIPLE [optics] Electromagnetic wave will take a path involving the least travel time when propagating between two points.[‡] See also PAULI EXCLUSION PRINCIPLE.

STATUTORY ACCOUNTING PRINCIPLE [insurance] A set of accounting rules for insurance companies set forth by the National Association of Insurance Commissioners. Used to prepare the statutory financial statements of insurance companies.[§]

STAUNTON'S SINGLE-MOVE PRINCIPLE [mathematics] (Haward Stauton, 1810–1874, English chesomaster   Chess; limiting time allowed to make a move during a game of chess.[¶] See also SUDDEN-DEATH PRINCIPLE; TASSILO VON HEYDEBRAND UND DER LASA PRINCIPLE; STENO'S PRINCIPLE.

STEENBECK MINIMUM PRINCIPLE [physics] (Wilhelm Steen back, German Film editing) Related to the power dissipated in the cathode root, which decreases when the glow changes to an arc.[‖]

STELLAR PRINCIPLE [physics]   Black holes originate from stars. See also BLACK HOLE PRINCIPLE; PRINCIPLE OF BLACK HOLE COMPLEMENTARITY; ANTHROPIC PRINCIPLE.

STENO'S PRINCIPLE OF LATERAL CONTINUITY [geology] (Nicholas Steno, 1638–1686) Danish qeologist)   Disassociated but similar strata can be considered from the same depositional period because strata do not simply end. If a layer in a dig contains a type of soil and artifics similar to those of another layer located nearby, but not in direct association, so long as the similarities are sufficient (nature of the soil, nature of the artifacts), the two layers are considered to be from the same time period.[**]

*Veselago, V. G. (2002), Formulating Fermat's principle for light traveling in negative refraction materials, *Phys. Usp.* **45**(10), 1097–1099.
[†]Thomson, C. J. (1997), Complex rays and wave packets for decaying signals in inhomogeneous, anisotropic and anelastic media, *Stud. Geophys. Geodaet.* (Springer) **41**(4), 345–381.
[‡]Byczuk, K. and Spalek, J. (1995), Universality classes, statistical exclusion principle and properties of interacting fermions, *Phys. Rev. B* **51**, 7934.

[§]Connolly, J. (1997), Codification action deferred until 1998, *Codification of Statutory Accounting Principle Guidelines*, National Underwriter Property & Casualty-Risk & Benefits Management.
[¶]*Encyclopedia Britannica.*
[‖]Yahya, A. A. and Harry, J. E. (1999), Factors affecting the glow-to-arc transition at the cathode of an electric discharge at atmospheric pressure, *Int. J. Electron.* **86**(6), 755–762.
[**]Bevel, T. and Gardner, R. M. (2008), *Bloodstain Pattern Analysis with an Introduction to Crime Scene Reconstruction*, 3rd ed., CRC Press, p. 331.

STENO'S PRINCIPLE OF ORIGINAL HOR-IZONTALITY [geology] (Nicholas Steno, 1638–1686, Danish geologist) (1) Rock layers form in the horizontal position, and any deviations from this position are due to the rocks being disturbed later. (2) If a solid body is enclosed on all sides by another solid body, then of the two bodies, the one that first became hard is the body that, in the mutual contact, expresses on its own surface the properties of the other surface.[**]

STENO'S PRINCIPLE OF SUPERPOSITION [geology] (Nicholas Steno, 1638–1686, Danish geologist) Geologic strata are deposited in a time order, oldest to youngest, unless otherwise disturbed. If three layers of strata and artifacts are uncovered, the one at the bottom is the oldest, and the one in the middle is younger than the bottom layer, but older than the layer above it. Each layer of strata is followed in age by each subsequent overlying stratum.[*]

STEREOLOGICAL PRINCIPLE See CAVALIERI PRINCIPLE.

STEVINO'S PRINCIPLES [physics, mathematics] See PRINCIPLE OF STEVINO.

STENO'S PRINCIPLE [geology] (Nicolas Steno 1638–1686, Danish geologist) (1) Earth's crust contains a chronological history of geologic events and that the history may be deciphered by careful study of the strata and fossils.[†] (2) A physical system seems to have a single-minded purpose to search for, find, and follow the path that minimizes the action between the initial and final states.[‡] See also HAMILTON PRINCIPLE; PRINCIPLE OF LEAST ACTION.

[*]Op. cit. Bevel and Gardner, p. 440.
[†]Steno, N. (2001), *Encyclopædia Britannica Online* (http://www.britannica.com/EBchecked/topic/565278/Nicolaus-Steno).
[‡]Simon, R., Herrmann, G., and Amende, I. (1990), Comparison of three different principles in the assessment of coronary flow reserve from digital angiograms, *Int. J. Cardiac Imag.* **5**(2–3), 203–212.

STEWART-HAMILTON PRINCIPLE [cardiology] aka Hamilton Stewart Principle (G. N. Stewart, 1897 and W.F. Hamilton, 1932) (1) Cardiac output (Q) is the volume of blood being pumped by the heart, in particular by a ventricle in a minute. This is measured in dm^3/min (1 dm^3 equals 1000 cm^3 or 1 litre). An average cardiac output would be 5 L/min for a human male and 4.5 L/min for a female.[§]

$$\text{Cardiac output} = \frac{\text{Quantity of indicator}}{\int_0^\infty \text{Concentration of indicator}_{dt}}$$

(2) A formula for measuring cardiac output following the rapid intravenous injection of an indicator dye: $F = i/ct$, in whch F represents the blood flow in liters per minute; i, the injected substance in milligrams; $c >$, the average dye concentration of the primary curve; and t, the duration of the primary curve in seconds, i.e., the time from appearance to disappearance of the dye at a fixed site if there were no recirculation of the dye.[¶]

STIRLING ENGINE PRINCIPLE [engineering] (Robert Stirling, 1790–1878, Scoltish clergyman/inventor) (1) A process gas is compressed at low temperatures and subsequently expands at high temperature. (2) Cylinder having spacially separated external means in heat conductive relationship with the interior of the cylinder and establishing separated hot and cold zones therein. (3) Piston operating as displacement as well as regenerator piston movable in said cylinder.[‖] See also ENGINE PRINCIPLE.

STOCHASTIC CONTROL PRINCIPLE [mathematics] A method for the optimal nonlinear stochastic control of hysteretic systems with

[§]Fronek, M.D. and Ganzi, V. (1960) Measurement of flow in single blood vessels including cardiac output by local thermodilution. *Circulation Research*, **8**, p175–182.
[¶]formula. (2007). Dorland's Illustrated Medical Dictionary. Philadelphia, W. B. Saunders, http://www.credoreference.com/entry/ehsdorland/formula
[‖]Ehrig, D. and Jacobs, H. (1987), *Hot Gas Engine Operating in Accordance with the Stirling Principle*, US Patent 4,691,515.

parametrically and/or externally random excitations. When control is needed, stochastically select a class of workpieces to process.* (2) Consisting of forward and backward state variables, under the assumption that the diffusion coefficient does not contain the control variable, but the control domain need not be convex.† See also MAXIMUM CONTROL PRINCIPLE; PRINCIPLE OF STOCHASTIC CONTROL.

STOCHASTIC HAMILTON VARIATIONAL PRINCIPLE [physics] (William Rowan Hamilton, 1805–1865, Jrish Mathematician) The path of a conservative system in configuration space between and configuration is such the integral of the lagrangian function over time is a minimum or maximum relative to nearby paths between the same endpoints and taking the same time. ‡ See also QUANTUM HAMILTON PRINCIPLE; STEWART–HAMILTON PRINCIPLE; STOCHASTIC HAMILTON VARIATIONAL PRINCIPLE; PRIGOGINE PRINCIPLE; GYARMATI'S PRINCIPLE; PRINCIPLE OF MINIMUM MECHANICAL ENERGY; PRINCIPLE OF MINIMUM POTENTIAL ENERGY; PRINCIPLE OF MINIMUM KINETIC ENERGY; TOUPIN'S DUAL PRINCIPLE; D'ALEMBERT'S PRINCIPLE; HAMILTON PRINCIPLE.

STOCHASTIC HAMILTONS PRINCIPLE [engineering] (William Rowan Hamilton, 1805–1865, Jrish mathematician) Predicting and minimizing the magnitude and limits of random deviations of a control system through optimizing the design of the controller.§

STOCHASTIC MAXIMUM PRINCIPLE [mathematics] (1) The admissible controls are square-integrable; not only a necessary but also a sufficient condition for optimality is obtained. (2) *Stochastic optimization* refers to the minimization (or maximization) of a function in the presence of randomness in the optimization process. The randomness may be present as either noise in measurements or Monte Carlo randomness in the search procedure, or both.¶ See also STOCHASTIC OPTIMUM PRINCIPLE.

STOCHASTIC OPTIMUM PRINCIPLE [mathematics] *Stochastic optimization* refers to the minimization (or maximization) of a function in the presence of randomness in the optimization process. The randomness may be present as either noise in measurements or Monte Carlo randomness in the search procedure, or both.‖ See also STOCHASTIC MAXIMUM PRINCIPLE.

STOKES PRINCIPLE [physics] (George Gabncle Stokes, 1819–1903, British Mathematician) (1) Expression for the resisting force offered by a fluid of dynamic viscosity to a sphere of radius moving through it at a steady velocity to indicate the terminal velocity of a sphere of a specific density falling under gravitational acceleration through a fluid of density.** (2) In vector calculus, used to convert the integral of the curl of a vector function over some region into a line integral of the same function along the boundary of the region.††

STOMACHIC PRINCIPLE [pharmacology] Describing any medicine promoting the functional activity of the stomach. An agent that stimulates the appetite and gastric secretion. An agent, such as a medicine, that

*Yan, H.-S., Yang, H.-B., and Dong, H., (2008), Control of knowledgeable manufacturing cell with an unreliable agent, *J. Intell. Manuf.* v20 (6) pp 671–682.
†Xu, W. (1996), Stochastic maximum principle for optimal control problem of forward and backward system, *J. Austral. Math. Soc. B* **37**, 172–185.
‡LuFeng Yang, A. Y. T., Leung, L. Y., and Wong, C. W. Y. (2005), Stochastic spline Ritz method based on stochastic variational principle, *Eng. Struct.* **27**(3), 455–462.
§Jumarie, G. (1995), Stochastic Hamilton's principle and noisy sliding surfaces in the tracking control of manipulators, *Robotica* **13**, 209.

¶Cadenillas, A. K. (1995), The stochastic maximum principle for linear, convex optimal control with random coefficients, *SIAM Control Optim.* **33**, 590.
‖Weisstein, E. W. (2006), Stochastic optimization, *MathWorld—a Wolfram Web Resource* (http://mathworld.wolfram.com/StochasticOptimization.html).
**Walker, P. M. B. (1999), *Chambers Dictionary of Science and Technology*, Chambers, New York, p. 1108.
††Yeh, P. and Gu, C. (1995), *Landmark Papers on Photorefractive Nonlinear Optics*, World Scientific.

strengthens or stimulates the stomach.* See also STOCHASTIC OPTIMUM PRINCIPLE.

STRAIGHT PRINCIPLE [mathematics] (Edmund L. Gettier III, b. 1927, Baltimore, MD) If person S knows p, and p entails q, then S knows q (this is sometimes called the "straight principle"). (1) A subject may not actually believe q, for example, regardless of whether he or she is justified or warranted. Thus, one might instead say that knowledge is closed under known deduction: if, while knowing p, S believes q because S knows that p entails q, then S knows q. (2) An even stronger formulation would be as follows: If, while knowing various propositions, S believes p because S knows that they entail p, then S knows p.† See also GETTIER'S PRINCIPLE; PRINCIPLE OF EPISTEMIC CLOSURE.

STRONG ANTHROPIC PRINCIPLE (SAP) [astronomy, genetics] (Brandon Carter, b. 1942, Australian theoretical physicist) (1) The universe (and hence the fundamental parameters on which it depends) must be such as to admit the creation of observers within it at some stage. To paraphrase Descartes, "cogito ergo mundus talis est." The Latin tag ['I think, therefore the world is such" (as it is)] makes it clear that "must" indicates a deduction from the fact of our existence; the statement is thus a truism.‡ (2) The universe must have those properties that allow life to develop within it at some stage in its history. The "must" is an imperative. There exists one possible universe "designed" with the goal of generating and sustaining "observers." This implies the purpose of the universe is to give rise to intelligent life, with the laws of nature and their fundamental constants set to

ensure that life as we know it will emerge and evolve.§ (3) The universe must be constructed in such a way that intelligent life will come into existence. The weak anthropic principle states that any observed universe must be constructed in such a way that intelligent life might evolve, no matter how improbable. The *negative anthropic principle* (NAP), states that a universe could (must?) exist where it is impossible for intelligent life to evolve. Our universe's existence would not disprove the possibility of a NAP universe, just as the weak anthropic principle does not require that a universe with intelligent life exist, just that it must be possible. So WAP and SAP are not logically equivalent. The strong stellar principle states stars will form. A "weak stellar principle" would say that *Maybe* stars would form and maybe they never would.¶

STRONG-AXIS PRINCIPLE [engineering] A part should be designed to compensate for inaccuracies of construction; all bevels and guides must be oriented in the strong axis of assembly. Howe derived a set of design principles for "kit of parts" and robotic construction system: (1) *strong-axis principle*—a part should be designed to compensate for inaccuracies of construction; all bevels and guides must be oriented in the strong axis of assembly; (2) *seventh joint principle*—it is advantageous to have a mounting mechanism built into the part to facilitate assembly or disassembly of the part; (3) *assembly sequence principle*—the construction sequence should be planned to take into account the access space requirements of the robotic systems. Parts that will be hidden should be installed by the robot first); (4) *robot and component interface principle*—the grasp points on the components should be designed in parallel

*Stephenson, J., Churchill, J. M., and Burnett, G. T. (1834), *Medical Botany, Or, Illustrations and Descriptions of the Medicinal Plants of the London, Edinburgh, and Dublin Pharmacopœias: Comprising a Popular and Scientific Account of Poisonous Vegetables Indigenous to Great Britain.*
†Luper, S. (2001), The epistemic closure principle, *Stanford Encyclopedia of Philosophy*.
‡Carter, B. (1974), Large number coincidences and the anthropic principle in cosmology, *Proc. 63rd IAU Symp. Confrontation of Cosmological Theories with Observational Data*, Reidel, Dordrecht, pp. 291–298.

§Barrow J. D., and Tipler, F. J. (1986), *The Anthropic Cosmological Principle*, Oxford Univ. Press.
¶Dean (2005), *Anthropic Principle. Dean's World Defending the Liberal Tradition in History, Science, and Philosophy* (see http://74.125.47.132/search?q=cache:T9aK0CzLDuQJ:www.deanesmay.com/posts/1130850322.shtml+%22STELLAR+PRINCIPLE%22&hl=en&ct=clnk&cd=3&gl=us).

with the robot's grab (factors to be considered include ease of manufacture, appearance, transportability, and balancing of heavy components. Also suggested two additional principles to improve buildability of the kit-of-parts system; (5) *stackability principle*—for the purpose of compact transportation and accessibility, parts should be designed for easy stacking and storage; (6) *path principle*—the movement trajectory of the parts, from storage, through assembly to final installed position, should be considered to prevent collision of parts or objects. A well-designed path could also improve the speed of assembly. If the structures in each bay are separated, automated construction principles can be facilitated because the building can be constructed one bay at a time, optimizing the area that can be covered by the robotic work cell.* See also ASSEMBLY SEQUENCE PRINCIPLE.

STRONG COMPARISON PRINCIPLE [mechanics, physics] (1) The existence and the uniqueness of a continuous solution.[†] (2) Individuals with a strong comparison orientation in particular tended to engage in and to respond to social comparison.[‡] (3) Solution operators for quasilinear, second-order parabolic equations that are autonomous or time-periodic, and also in delay equations.[§] (4) The "generalized" Dirichlet problem (in the sense of viscosity solutions) for quasilinear elliptic and parabolic equations in the

case when losses of boundary conditions can actually occur.[¶]

STRONG CORRESPONDENCE PRINCIPLE [mechanics, physics] (1) Whenever a new scientific theory or model is proposed, the new theory must overlap with all of the verified predictions of the old theory. Particle and wave characteristics in the same large-scale phenomenon are incompatible rather than complementary. Knowledge of a small-scale phenomenon, however, is essentially incomplete until both aspects are known.[‖] (2) Quantum mechanics has a classical limit in which it is equivalent to classical mechanics.[**]

STRONG EQUIVALENCE PRINCIPLE [physics, astronomy, mathematics] (1) A gravitational field is equivalent to an accelerating reference frame. It is sometimes expressed as the equivalence of gravitational and inertial mass, namely, that objects with the same inertia (resistance to acceleration) experience the same gravitational force. (2) An inertial reference frame in a uniform gravitational field is equivalent to a reference frame in the absence of a gravitational field that has a constant acceleration with respect to that inertial frame. (3) Gravitational masses and inertial masses are one and the same thing. The m in Newton's second law, $F = ma$, and the m in the equation describing the gravitational force on an object $F = mg$, are identical.[††] (2) An ideal observer in a gravitational field can choose a reference frame in which gravitation goes unnoticed.[‡‡]

*Howe, A. S. (1997), Designing for automated construction, *Proc. 14th Int. Symp. Automation and Robotics in Construction* (ISARC14), June 1997, p. 49.

[†]Da Lio, F. (2002), Comparison results for quasilinear equations in QAnnular domains and applications. *Commun. Partial Differ. Eq.* **27**(1,2), 283–323.

[‡]Van der Zee, K., Oldersma, F., Buunk, B. P., and Bos, D. (1998), Social comparison preferences among cancer patients as related to neuroticism and social comparison orientation, *J. Pers Soc. Psychol.* **75**(3), 801–810.

[§]Tai-Ping, L. (1997), Regularization of solution operators for quasilinear hyperbolic-parabolic partial differential equations, *J. Partial Differ. Eq.* **10**(4), 347–354.

[¶]Roselli, P., and Sciunzi, B. (2007), A strong comparison principle for the Laplacian, *Proc. Am. Math. Soc.* **135**, 3217–3224.

[‖]Radder, H. (1991), Heuristics and the generalized correspondence principle, *Br. Phil. Sci.* **42**, 195–226.

[**]Bohr, N. (1921), Atomic structure. *Nature* **108**, 208–209.

[††]Roscoe, D. F. (1995), Gravitation, global four momentum conservation and the strong equivalence principle, *Astrophys. Space Sci.* **227**, 119.

[‡‡]Accioly, A. and Paszko, R. (2009), Conflict between the classical equivalence principle and quantum mechanics, *Adv. Stud. Theor. Phys.* **3**(2), 65–78.

STRONG INVARIANCE PRINCIPLE [physics] See INVARIANCE PRINCIPLE.

STRONG PRINCIPLE [physics, mathematics, chemisty] See INVARIANCE PRINCIPLE.

STRONG PRINCIPLE OF CORRESPONDING STATES [chemistry, physics] See PRINCIPLE OF CORRESPONDING STATES; CORRESPONDING STATES PRINCIPLE.

STRUCTURAL LINEARIZATION PRINCIPLE [mathematics, physics] An induction principle for processes was given that allows one to apply model-checking techniques to parameterized families of processes. A limitation of the induction principle is that it does not apply to the case in which one process depends directly on a parameterized number of processes, which grows without bound. This would seem to preclude its application to families of N processes interconnected in a star topology. Nonetheless, we show that if the dependence can be computed incrementally, then the direct dependence on the parameterized number of processes may be reexpressed recursively in terms of a linear cascade of processes, yielding in effect a 'linearization' of the interprocess dependences and allowing the induction principle to apply.[*] See INDUCTION PRINCIPLE.

INDUCTION PRINCIPLE *Given two sets of-proccesses,* $\{P^1, \ldots, P^N\}$ *and* $\{q1, \ldots, q^N\}$, *and an integer* $1 \leq m < N$, *such that for all* $m \leq i < N, p^{i+1} = \phi(p^i)$ *and* $q^{i+1} = \phi(q^i)$: *if*

$$\otimes_i^m = {}_1p^i \leq q^m$$

and

$$q_m \otimes p^{m+1} \leq q^{m+1}$$

then

$$\otimes_i^N = {}_1pi \leq q^N$$

Thus, by proving two propositions about arrays of fixed size m_i we may draw a conclusion about an array of arbitrary size N. In particular given a task $T_i q^N \leq T$ implies $\otimes_i^N = {}_1p^i \leq T$ that is: in order to verify that the system $\otimes_i^N = {}_1pi$ performs task T it is

[*]Kurshan, R. P., Orda, A., and Sachs, S.R. (1994) *A structural Linearization Principle for Processes.* Kluwer Academic Publisher. pp. 3

sufficient to verify that q^N performs T q^i is called the process *invariant*.

We now apply the Induction Principle to the linearized doorway example. Denote $p^0 = D_1 \otimes D_2, p^i = E^i$ for $1 \leq i \leq N$. The linearized doorway system is given by. $S(N) = p^0 \otimes (\otimes_{i=1}^N p^i)$.

We choose the following invariant, valid for $i > 3$:

$$q^i = p^0 \otimes (\otimes_{k=i}^{i-2} p^k) \otimes p^{-1-3}$$

where p^i is a modified version of an arbitrary p^i, as explained in the following. The variables of p^i are defined exactly as in p^i. A process p^i non-deterministically choses values for all of its variables. However, the values of the propagation variables are constrained as follows:

$$S_p^i = 1 \iff S^i = 1$$
$$R_l^i = 1 \iff (\hat{p}^i, reqiest = true)$$
$$K^i = 1 \iff (\hat{p}^i, reques = false)$$
$$* ((X_1^i = win + lose) + (X_2^i = win + lose))$$
$$Z^i = 1 \iff (\hat{p}^i state = initial) * (X_2^i = win)$$

SUBRANGING PRINCIPLE [physics] Part of a special issue on analog and signal processing circuits. Ensures accuracy and wide dynamic range. Subranging architectures are more efficient with regard to power dissipation than are flash architectures.[†]

SUBSPACE DECOMPOSITION PRINCIPLE [mathematics] A subset of the range of values that a function may assume. (2) The subspace decomposition scheme and the Gafni–Bertsekas dual-cone decomposition scheme are shown to produce identical iterates near nondegenerate stationary points (and in particular near nonsingular minimizers) in polyhedra, and a new local superlinear convergence theorem is proved

[†]Werner, G., Andrey, G., and Rainer, M. L. (1995), *Scanning Analog-to-Digital Converter Operating under the Subranging Principle*, US Patent 5,442,575.

for the subspace decomposition principle in nonpolyhedral sets.*

SUBSTITUTION PRINCIPLE [Physics] (1) Related to the design by contract methodology, leading to some restrictions on how contracts can interact with inheritance: Preconditions cannot be strengthened in a subclass and postconditions cannot be weakened in a subclass. In addition, the principle implies no new exceptions should be thrown by methods of the subclass, except where those exceptions are themselves subtypes of exceptions thrown by the methods of the superclass.[†] (2) [Psychology] To change reinforcers when a previously effective reward is no longer controlling behavior, present it just before (or as soon as possible after) the time you present the new, hopefully more effective reward.[‡] See also LISKOV SUBSTITUTION PRINCIPLE.

SUCCESSIVE APPROXIMATION PRINCIPLE [psychology] To teach children to act in a manner in which they have seldom or never before behaved, and reward successive steps to the final behavior.[‡] See RESPONSE-SHAPING PRINCIPLE.

SUNSET PRINCIPLE [economics] Compromise solutions to contentious social policies should carry a time limit.[§]

SUPERGAGA PRINCIPLE [genetics]
(1) GAGA factor is an important chromosomal protein involved in establishing specific nucleosome arrays and in regulating gene transcription in *Drosophila melanogaster*. GAGA protein required for correct chromatin architecture at diverse chromosomal sites.[¶] (2) Acts as a transacting factor binding to the cis-acting regulatory element in the hAT(1) promoter, which is necessary for the basal and growth factor(s)–mediated transcriptional activation of the hAT(1) gene.[‖] (3) Component of at least some types of PcG complexes and may participate in the assembly of PcG complexes at PREs.[**] (4) Binds to multiple sites as a large oligomer and induces bending of the promoter DNA.[††]

SUPERPOSITION PRINCIPLE [Geology]   See PRINCIPLE OF SUPERPOSITION

SUPERPOSITION PRINCIPLE [geology, physics] (1) Strata that overlie other strata are always younger, except in strongly folded areas.[‡‡] (2) Each stress is accompanied by the same strains whether it acts alone or in conjunction with others; it is true so long as the total stress does not exceed the limit of proportionality. In vibrations and wave motion, one set of vibrations or waves is unaffected by the presence of another set. For example, two sets of ripples on water will pass through one another without mutual interaction, so, at a particular instant, the resultant disturbance at any point traversed by both sets of waves is the sum of the two

*Dunn, J. C. (1991), A subspace decomposition principle for scaled gradient projection methods: Global theory, *SIAM J. Control Optim.* **29**, 1150–1175.

[†]Liskov, B., (1987), *Proc. Conf. Object Oriented Programming Systems, Languages and Applications*, Orlando, Fl., ACM, New York, p.17–34.

[‡]Huitt, W. (1994), Principles for using behavior modification, *Educational Psychology Interactive* Valdosta State Univ., Valdosta, GA (http://chiron.valdosta.edu/ whuitt / col/ behsys/ behmod. html).

[§]Boston, T. D. (2002), *Leading Issues in Black Political Economy*, Transaction Publishers, p. 516.

[¶]Granok, H., Leibovitch, B. A., and Elgin, S. C. R., (2001), A heat-shock-activated cDNA encoding GAGA factor rescues some lethal mutations in the Drosophila melanogaster Trithorax-like gene, *Genet. Res.* **78**(1), 13–21.

[‖]Hughes, J. D., Estep, P. W., Tavazoie, S., and Church, G. M. (2000), Computational identification of Cis-regulatory elements associated with groups of functionally related genes in Saccharomyces cerevisiae, J. Mol. Biol. **296**(5), 1205–1214.

[**]Pirrotta, V. (1998), Polycombing the genome, PcG, trxG, and chromatin silencing, *Cell.* **93**, 333–336.

[††]Katsani, K. R., Nasser Hajibagheri, M. A., and Verrijzer, C. P. (1999), Co-operative DNA binding by GAGA transcription factor requires the conserved BTB/POZ domain and reorganizes promoter topology, *EMBO J.* **18**, 698–708.

[‡‡]Walker, P. M. B., (1999), *Chambers Dictionary of Science and Technology*, Chambers, New York, p.1130.

component disturbances. The superposition of two vibrations, y_1 and y_2, both of frequency f, produces a resultant vibration of the same frequency; its amplitude and phase are functions of the component amplitudes and phases. Thus if $y_1 = a_1 \sin(2ft + _1) y_2 = a_2 \sin(2ft + _2)$, then the resultant vibration, y, is given by $y_1 + y_2 = A \sin(2ft +)$, where amplitude A and phase are both functions of a_1, a_2, $_1$, and $_2$. (3) If a number of independent influences act on the system, the resultant influence is the sum of the individual influences acting separately. (4) Perception of objects already understood, that is, determined and related to other objects. The response of a complex system is a result of the responses of those simpler elements from which the complex system can be considered to be made. As long as superposition can be considered as holding, that is, as long as the assumption of a linear system is valid, the responses of very complex systems are obtained easily.[*] See also QUANTUM MECHANICAL SUPERPOSITION PRINCIPLE; PRINCIPLE OF SUPERPOSITION; TUNING FORK PRINCIPLE.

SUPREMUM PRINCIPLE [mathematics] Given a subset S of a partially ordered set T, the supremum of S, if it exists, is the least element of T greater than or equal to each element of S. If the supremum exists, it may or may not belong to S. If the supremum exists, it is unique. Suprema are often considered for subsets of real numbers, rational numbers, or any other well-known mathematical structure for an element to be "greater than or equal-to" another element. The definition generalizes to more abstract setting of order theory for arbitrary partially ordered sets.[†] See also INFIMUM PRINCIPLE.

SURE-THING PRINCIPLE [mathematics, decision theory] (L. J. Savage, 1954 States if the utility functions associated with two

alternatives agree on a subset of the set of all relevant states of the world, then this subset can be ignored in comparing the utilities of the two alternatives. L. J. Savage used this principle to resolve the so-called Allais paradox, assuming the equivalence of certain simple and compound gambles.[‡] Directly violated by the Allais paradox in the context of risk, and by the Ellsberg paradox in the context of uncertainty.[§] See also COMONOTONIC SURE-THING PRINCIPLE.

SURVIVOR   PRINCIPLE [economics] The most efficient production method is the onesed by the firm(s) able to survive the competition in the long run.[¶]

SUSPENSION PRINCIPLE [engineering] Great weights used to provice support over large spans.[‖]

SUSTAINABILITY   PRINCIPLE [psychology] See TRANSACTION PRINCIPLE.

SUSTAINED   YIELD   PRINCIPLE [ecology] (1) Preservation and conservation of natural resources program.[**] (2) Guiding principle for a global society entering the new millennium, superseding almost all others within the environment and development communities.[††]

[‡]Wakker, P. (1996), The sure thing principle and the comonotonic sure thing principle: An axiomatic analysis, *J. Math. Econ.* **25**, 213.
[§]Savage, L. J. (1972), *The Foundations of Statistics*, 2nd ed., Wiley, New York.
[¶]Peterson, R. D. (1982), The survivor principle and small firm entry decisions, *Jo. Small Business Manage.* **20**, 13.
[‖]Zhu, H. and Fang, L. (2006), Suspension principle and digital control for bearingless permanent magnet slice motors, *Proc. 5th Int. Power Electronics and Motion Control Conf.*, IPEMC '06, Aug. 2006, CES/IEEE, 5 vol. 2, pp. 1–4.
[**]US Bureau and Land Management, Public Lands, Title 43, US *Code of Federal Regulations*. (http://ecfr.gpoaccess.gov/cgi/t/text/text-idx?c=ecfr&sid=d8b010fac79d911d6f62b0d0f4e2b70b&rgn=div5&view=text&node=43:2.1.1.5.98&idno=43).
[††]Turner II, B. L., (1997), The sustainability principle in global agendas: Implications for understanding Land-use/cover change, *Geogr. J.* **163**.

[*]Creenberger, D. M. etal. (1995), Multiparticle interferometry and the superposition principle, *Physics Today* **46**, 22.
[†]Rudin, W., (1976), *Principles of Mathematical Analysis*, 3rd ed., McGraw-Hill.

SWITCHING PRINCIPLE [physics] (1) Of dopaminergic systems, modulates the timing, timesharing and initiation of responses (program—control).* (2) In higher-nervous-system activity is to be distinguished from complex conditioned reflexes. In the former, the same stimulus elicits different kinds of activity in response to the situation. In the latter, different activities are elicited in response to different stimuli.† (3) Circuits made up of ideal digital devices including theories of circuits and networks for telephone switching, digital computing, digital control, and data processing.‡ See also ADIABATIC SWITCHING PRINCIPLES; BUSH–MOSTELLER FREQUENT CUE SWITCHING PRINCIPLE; CURRENT-SWITCHING PRINCIPLE; FREEZE/THAW SWITCHING PRINCIPLE; OPTICAL SWITCHING PRINCIPLE.

SYMMETRY PRINCIPLE [physics, mathematics] (1) The centroid of a geometric figure (line, area, or volume) is at a point on a line or plane of symmetry of the figure.§ (2) Any principle that states that a physical quantity or physical law possesses invariance under certain transformations.¶ (3) Any principle stating a physical quantity or physical law possesses invariance under certain transformations.‖ See also ANTISYMMETRY PRINCIPLE; INVARIANCE PRINCIPLE; SCHWARZ REFLECTION PRINCIPLE; SIMILARITY PRINCIPLE; LINDELÖF PRINCIPLE.

SYNCHRONOUS SAMPLING PRINCIPLE [mathematics] The vibration is very irregular, the observation of vibratory movement is almost impossible. These technical difficulties hav erestricted a large-scale data acquisition for studying the patterns of vocal fold vibration in various modes of phonation.** See SAMPLING PRINCIPLE.

SYNDROME PRINCIPLE [geology] The ecologoeconomical territorial syndrome principle is of primary importance while compiling the estimate map of the environment state. The territory possesses zones of the ecological–economical syndrome risk and a number of critical zones as well. Maps estimating definite indicators characterizing the state of a region during the period to a marker transform are constructed.†† See also ECOLOGOECONOMICAL TERRITORIAL SYNDROME PRINCIPLE.

SYNERGY PRINCIPLE [psychology] Gregory Bateson "Connects all the living creatures" and "defines the vast generalization," as Gregory Bateson (1979)‡‡ states. It is a pattern that "all existences follow" in their evolution of consciousness, as the ancient I-ching philosophy states. Specifically, it refers to a life-producing pattern—the two inseparable processes of differentiation and integration—which exists in matter, life, and mind.§§ See TRANSACTION PRINCIPLE.

SYSTEM GUIDING PRINCIPLE [pharmacology] Group of signs and symptoms that together characterize a disease. "The system must ensure choice of physician and preserve patient/physician relationships. The system must focus on providing care that

*Oades, R. D. (1985), The role of noradrenaline in tuning and dopamine in switching between signals in the CNS, *Neurosci. Biobehav. Rev.* **9**(2), 261–282.

†Vartanyan, G. A., Pirogov, A. A., and Shabaev, V. V. (1987) Informational switching hypothesis of conditioned-reflex activity, *J. Neurosc., Behav. Physiolo.* **17**(5), 363.

‡Boylestad, R. L. and Nashelsky, L. (2005), *Electronic Devices and Circuit Theory*, 9th ed., Prentice-Hall.

§Anon. (1994), *McGraw-Hill Dictionary of Scientific and Technical Terms*.

¶Lin, S.-K. (1999), Symmetry principle and similarity principle, *Book of Abstracts, 218th ACS Natl. Meeting*, New Orleans.

‖Nazareth, J. L. (1997), Deriving potential functions via a symmetry principle for nonlinear equations, *Operations Res. Lett.* **21**, 147.

**Ijonda, K., Kiritani, H., Imagawa H., Hajime Hirose, H., and Hashimoto, K. (1985), High-speed digital recording of vocal fold vibrationusing a solid-state image sensor, *Ann. Bull. RILP*(19), 47–53.

††Trofimov, A. M., Shagimardanov, R. A., and Petrova, R. S. (1996), Spatial analysis of the ecological syndrome of the territory, ERAE: **2**(1).

‡‡(Gregory Bateson 1904–1980 British anthropologist) Bateson, G. (2000) Steps to an Ecology of Mind. Univ. of Chicago Press

§§Tang, Y. (2000), The synergy principle, human action and evolution of consciousness, 1st *Int. Electronic Seminar on Wholeness*.

is safe, timely, efficient, effective, patient-centered and equitable," and fragment quality, driving instead toward systems that enable value-driven, patient-centered care.[*]

SYVA EMIT PRINCIPLE [pharmacology] (Leo Sailard, 1898–1964, Hungarian physicest) (Chalmers Ashoy Johnson, 1931–, American Political)   Method of determining presence of opiates in urine.[†]

SZILARD–CHALMERS   PRINCIPLE [physics] Process in which a nuclear transformation occurs with no change of atomic number, but with breakdown of chemical bond. This leads to formation of free active radicals from which material of high specific activity can be separated chemically.[‡]

[*]Ross, K. (Nov./Dec. 2006), Physicians' Congress The next chapter in health care, *Colorado Med.* p. 367.

[†]Colbert, D. L. and Gooch, J. C. (1992), An in-house opiate enzymoimmunoassay based on the Syva EMIT principle, *Clin. Chem.* **38**, 1483–1485.

[‡]Walker, P. M. B. (1999), *Chambers Dictionary of Science and Technology*, Chambers, New York, 1142.

T

TAKAHASHI'S MINIMIZATION PRINCIPLE [mathematics] (1) Conditions that are applicable to sets with finite number of elements, to corroborate or falsify the hypothesis that the elements of that set are samples of the Pareto set. These conditions lead to several generic criteria that can be employed in the evaluation of algorithms as multiobjective optimization mechanisms.* (2) The minimality of a submanifold in a hypersphere (centered at the origin) is equivalent to the immersion vector being an eigenvector of the Laplacian.[†] (3) Minimal immersions of a Riemannian manifold into the sphere are just those immersions whose coordinate functions in the ambient Euclidean space are eigenfunctions of the Laplace–Beltrami operator in the induced metric.[‡] (4) When the position vector of a submanifold in the Euclidean space is an eigenvector of the Beltrami–Laplace operator is minimal.[§]

TANGENT VARIATION PRINCIPLE [mathematics] Gives an upper bound for an arbitrary meromorphic function and for a large class of curves.[¶] See LINDEL'S PRINCIPLE.

*Oettli, W. and Thera, M. (1993), Equivalents of Ekeland's principle, *Bull. Austral. Math. Soc.* **48**, 385–392.

[†]Xingxiao, L. (1996), Fullness and Scalar Curvature of the Totally Real Submanifolds in $S^6(1)^1$, Note de Matematica 16(1), 105–115

[‡]do Carmo, M. P. and Wallach, N. R. (1971), Minimal immersions of spheres into spheres, *Ann. Math.* **93**, 43–62.

[§]Garay, O. J. (1990), An extension of Takahashi's theorem, *Geometriae Dedicata.* (Springer) **34**(2), 105–112.

[¶]Sukiasyan, G. A. (2004), On level sets of quasiconformal mappings, in *Value Distribution Theory and Related Topics*, Barsegian, G. et al., eds., Kluwer Academic Publishers, Boston, see *Adv. Complex Anal. Appl.* **3**, 75–92.

TAO PRINCIPLE The ultimate principle of the universe.[‖]

TAPER PRINCIPLE [engineering] Tapers are a standard alignment principle used in the machine tool industry. The taper principle has a downside with respect to spear tip alignment, however—the longer the taper, the better the alignment, but past a critical angle the taper through friction is also a drive attachment mechanism (the whole power of the drill press is transmitted through only the friction of the tapered quill). A tip designed with only the taper principle at the adapter interface will tend to jam (if too long) or wobble (if too short) or both (as the ice pick tip exhibits).**

TASSILO VON HEYDEBRAND UND DER LASA PRINCIPLE [telegraphy] (Tassilo Von Heydebrand un der Lasa, nineteenth-century German author and player) (1) The third, and most popular, principle for time controls was a flexible system proposed by Tassilo von Heydebrand und der Lasa, who proposed that each player be allowed a bank of time in which to play a predetermined number of moves, such as 2 hours for 30 moves. This principle, adopted for the vast majority of competitions from 1861 on, permits each player to budget time, playing some moves quickly and taking as much as an hour or more on others. In addition, a player who made the prescribed number of moves, such as 30 in the example above, would get an additional time budget, such as one hour for the next 15 moves. (2) Principle of single-move time limits was abandoned in all except

[‖]American Heritage (2000), *American Heritage Dictionary of the English Language*, 4th ed., Houghton Mifflin, New York.
**Boothroyd, G. and Knight, W. A. (2006), *Fundamentals of Machining and Machine Tools*, CRC Press.

postal games (in which players had a preset number of days to respond to a move) and some forms of quick or speed chess, such as games in which players must move every 5 or 10 seconds. (3) Sometimes called "sudden death," was also considered—and abandoned—in the early days of competitive chess. With a sudden-death format a set amount of time is allowed for all a player's moves in a game.[*]

TEAKETTLE   PRINCIPLE [mathematics] (1) Refers to the practice of reducing a given problem to one that has been solved previously. (2) Refers to the practice, common among mathematicians, of reducing a given problem to one that has been solved previously. The name itself arises from a long-running joke in the mathematical community involving a mathematician and an engineer. The mathematician and the engineer arrive in the kitchen to make tea; both fill a pot with water, put the pot on the stove, and boil the water. Not a big problem to solve. The next day, they go to make tea again, but find that the pot is already full of water. The engineer will put the pot on the stove; the mathematician will throw out the water—"reducing the problem to a previously solved problem".[†]

TELECENTRIC   LENS   PRINCIPLE [physics] (1) For the telecentric lens, your eye focuses at infinity, and the entrance pupil of the lens remains in focus with no further refocusing of your eye as you move the lens closer or farther away. (2) Three types of telecentric lenses are appropriate for metrology vision systems. In an *object-sided telecentric lens*, the lens and aperture are configured so that the principal ray from the object to the lens runs parallel to the lens' optical axis. A small change in the distance from the object to the object-sided telecentric lens does not change the magnification of the resulting image. In a conventional lens, the principal ray is not parallel to the optical axis, so the lens will

magnify points on an object based on their distances from the lens. (3) The image from an *image-sided telecentric lens* is insensitive to small changes in the position of the image plane. In a camera, small differences in the distance between the lens and an image detector do not affect the size of the image. Thus, these small changes do not affect the accuracy of measurements made on the resulting images. (4) A *bilateral telecentric lens* combines the advantages of both object- and image-sided telecentric lenses into one form, thus providing the highest degree of measurement accuracy for objects with different heights. A bilateral telecentric lens accurately reproduces dimensional relationships within its telecentric depth, and it isn't susceptible to small differences in the distance between the lens and the camera's sensor. In general, these lenses provide a means of accurately imaging 3D objects when critical measurements are necessary. (5) Rays generated by a light point are parallel in infinite distance from that point. A lens that is positioned perpendicularly to these parallel rays bundles them in the focal point. Thus the focal point is a mapping of the infinitely remote light point. (6) Avoid the problem of perspective distortion. Aperture is positioned directly at its focal point. Hence, only parallel (or almost parallel) rays are able to pass through this aperture; therefore, as the reflecting object seems to be infinitely remote, there can be no perspective distortion.[‡] See also TELECENTRIC PRINCIPLE.

TELECENTRIC PRINCIPLE [optics, mathematics] Utilize the diffraction effects so that those fringes of good contrast may be formed, enabling more accurate recordings of grids and thus better accuracy of strain measurements.[§] See also TELECENTRIC LENS PRINCIPLE.

TELESCOPICAL   PRINCIPLE [mathematics, physics]   Mach's principle of the relativity

---

[*]Anon. (2001), Chess, *Encyclopædia Britannica Online* (http://search.eb.com/eb/article-80444).
[†]Lawler, E. (2001), *Combinatorial Optimization: Networks and Matroids*, Dover Publications, p. 13 (originally publ. 1976).

[‡]Petrozzo, R. A. and Stuart, W. (2001), Telecentric lenses simplify noncontact metrology. The proper lens lets inspection systems accurately dimension 3-D parts, *Test Measure. World* p. 6.
[§]Luxmoore, A. (1974), A method to shift moiré fringes using gap effect, *Exp. Mech.* **14**(12), 507–508.

of inertia and of the Mach–Einstein doctrine on the determination of inertia by gravitation. These principles are neither philosophical nor epistemological postulates but well-defined physical axioms with exactly analytical expressions. The fundamental principle is the Galilean reciprocity of motions. According to this generalized Galilean invariance, the principal functions of analytical dynamics (Lagrangian L and Hamiltonian H) depending on the differences AB of the coordinate vectors A and B of the velocity differences $AB = A - B$ only. The Galilean reciprocity of motions means that neither the vectors A and B nor the accelerations A of one particle have physical significance.[*] See PRINCIPLE OF RELATIVITY.

TENSION–RELAXATION PRINCIPLE [psychology] Interval shifting between tension and relaxation.[†] See FERENCZI'S RELAXATION PRINCIPLE; RELAXATION PRINCIPLE.

TESLA PRINCIPLE [engineering]   (Nikola Tesla, 1856–1943, Croatian-American electrical engineer) Used to describe certain reversible processes. A reversible process in engineering is a process or operation of a system or device such that a net reverse in operation will accomplish the converse of the original function. Some systems could be reversed and operated in a complementary manner.[‡]

THALWEG PRINCIPLE [geography] Signifying the line of greatest slope along the bottom of a valley. It thus marks the natural direction of a watercourse. Determines national boundaries at the thalweg of a river separating two states. Defines the border between two states separated by a watercourse as lying along the thalweg.[§]

[*]Treder, H. J. (1979), Telescopical principles in the theory of gravitation (Mach's principle, relativity of inertia according to Mach and einstein and Hertz mechanics), *Ann. Phys.* (Germany) **36**, 4–19.
[†]Bühler, C., (n.d.), Maturation and motivation, *Dialectica* **5**(3-4), 312–361.
[‡]Anon. (1911), Tesla's new monarch of machines, *New York Herald Tribune* (Oct. 15, 1911).
[§]*Oxford English Dictionary.*

THERAPEUTIC PRINCIPLE [pharmacology] (1) Effect of treatment on medical problem. (2) Describes focus of condition and correction to be ameliorated.[¶] (3) Any treatment that will focus on treatment of organs, circulation, or property toward healing, curative, restorative, recuperative, rehabilitative, or remedy.[‖]

THERMAL COMPENSATION PRINCIPLE [engineering] Calculates and adjusts for the effects of thermal growth on large structures and long-axis travel to enable controlled environment accuracies from machines running in the harsh atmosphere of plants without air conditioning.[**]

THERMAL SIMILARITY PRINCIPLE [biology] Biological similarities can be extended to the problems of thermal similarity by introducing quantity of heat (H) and temperature (t) into the fundamental system ($M = $ mass, $L = $ length, $T = $ time). From the general equation of thermal similarity it is possible to calculate the allometric exponents for several variables related to heat transfer processes. In order to establish a correlation between the metabolic rate of homeothermic animals and the thermal similarity criterion, the range of biological similarities can be adjusted in accordance with the empirical allometric equation for metabolic rate versus body weight. Finally, the dimensional analysis of 12 thermal variable leads to five independent similarity criteria, expressed as π [long dash] numbers, which are both invariant and dimensionless. Through the principle of thermal simlarity it is possible to analyze theoretically different species of homeothermy.[††]

[¶]Jørgensen, T. B., Sørensen, A. M., and Jansen, E. C. (2008), Iatrogenic systemic air embolism treated with hyperbaric oxygen therapy, *Acta Anaesthesiol. Scand.* **52**(4), 566–568.
[‖]Patel, C. (1984), Yogic therapy, in *Principle and Practice of Stress Management*, Woolfolk, R. L. and Leherer, P. M., eds., Guilford Press, New York, p. 207.
[**]Chen, J. S., Yuan, J., and Ni, J. (1996), Thermal error modelling for real-time error compensation, *int. J. Adv. Manuf. Technol.* **12**(4), 266–275.
[††]De La Barra, B. L. and Gunther, B. (1965), Thermal similarity principle, dimensional analysis

THERMAL-TIME-OF-FLIGHT   PRINCIPLE [mathematics, physics]   (1) Performs three-dimensional distribution measurements.* (2) When the lasers that influence the vertical motion of the atoms are "detuned" in relation to each other for a short time and in a defined way, a well-aimed upward "push" can be given to the cooled and trapped atoms; they fly up at an initial speed of a few meters per second, rise until the gravitational force has consumed their kinetic energy and fall down again on the same path in a scenario reminiscent of a fountain.[†] (3) In a similar way as in a conventional atomic clock the energy state of the atoms is manipulated and checked. At the beginning the atoms are prepared in a single energy state; During the up/down movement they fly, however, through the same microwave field.[‡] (4) Neutron energy is measured by time of flight.[§]

THERMIONIC EMISSIONS PRINCIPLE [mathematics, physics]   Any of a class of devices that convert heat directly into electricity using thermionic emission rather than first changing it to some other form of energy.[¶] See also HALF-CALL PRINCIPLE.

THERMODYNAMIC   PRINCIPLES [physics] (1) Laws governing the transformation of energy. Useful for assessing the efficiencies of devices transforming heat into work and

and homeothermy, *Acta Physiol. Latino Am.* **15**(4), 378–385.
*Crescini, D., Marioli, D., and Taroni, A., (2001), Thick film flow sensor based on the thermal time-of-flight principle, paper presented at 17th IEEE Instrumentation and Measurement Technology Conf., May 1–4, 2000, Baltimore, MD.
[†]Riis, E. (1995), Motion of laser-cooled atoms in an optical potential, *Phys. Rev. A* **52**(2), 52.
[‡]Tsai, C. C., Freeland, R. S., Vogels, J. M., Boesten, H. M. J. M., Verhaar, B. J., and Heinzen, D. J. (1997), Two-color photoassociation spectroscopy of ground state Rb2, *Phys. Rev. Lett.* **79**(7), 1245–1248.
[§]Lou, T. P. (2003), *Compact D-D / D-T Neutron Generators and Their Applications*, Lawrence Berkeley Natl. Lab., (Univ. California), Paper LBNL, pp. 74–75.
[¶]Thermionic power (2001), in *Encyclopædia Britannica Online* (http://www.britannica.com/EBchecked/topic/591517/thermionic-power).

devices using external sources of work to transfer heat from a hot system to cooler sinks, and for discussing the spontaneity of chemical reactions and the work that they can be used to generate. (2) Classical thermodynamics consists of a collection of mathematical relations between observables, and as such is independent of any underlying model of matter (in terms, for instance, of atoms). However, interpretations in terms of the statistical behavior of large assemblies of particles greatly enriches the understanding of the relations established by thermodynamics. (3) First principle of thermodynamics establishes the existence of a property called the *internal energy* of a system. A change in the state of a system can be brought about by a variety of techniques (J. P. Joule). Shows same change of state is brought about by a given quantity of work regardless of the manner in which the work is done. The difference between the work of adiabatic change and the work of nonadiabatic change is called *heat*. (4) Work and heat are modes of transferring energy. They are not forms of energy in their own right. Work is a mode of transfer that is equivalent (if not the case in actuality) to raising a weight in the surroundings. Heat is a mode of transfer arising from a difference in temperature between the system and its surroundings. What is commonly called heat is more correctly called the *thermal motion* of the molecules of a system. (5) The internal energy of an isolated system is conserved; that is, for a system to which no energy can be transferred by the agency of work or of heat, the internal energy remains constant. Implies the equivalence of heat and work for bringing about changes in the internal energy of a system (and heat is foreign to classical mechanics). (6) The second principle of thermodynamics describes the distinction between spontaneous and nonspontaneous processes. A process is spontaneous if it occurs without needing to be driven. In other words, spontaneous changes are natural changes, such as the cooling of hot metal and the free expansion of a gas (Lord Kelvin and R. Clausius). No cyclic engine operates without a heat sink, and heat does not transfer spontaneously from a cool body to a hotter body, respectively. The property of entropy is introduced to formulate the law quantitatively in exactly the same way in which the properties of temperature and internal energy are introduced to render the zeroth and first laws quantitative and precise. The entropy S of a system is a measure of the quality of the energy it stores. When a given quantity of energy is transferred as heat, the change in entropy is large if the transfer occurs at a low temperature and small if the temperature is high. Thus, energy and matter tend to disperse in disorder, and this dispersal is the driving force of spontaneous change. (7) The third principle of thermodynamics states that, in the neighborhood of an equilibrium state of a thermodynamic system, there exist states that are not accessible by either reversible or irreversible adiabatic processes. (8) The practical significance of the second law is that it limits the extent to which the internal energy may be extracted from a system as work. In order for a process to generate work, it must be spontaneous. For the process to be spontaneous, it is necessary to discard some energy as heat in a sink of lower temperature. In other words, nature in effect exacts a tax on the extraction of energy as work. There is, therefore, a fundamental limit on the efficiency of engines that convert heat into work. The quantitative limit on the efficiency, which is defined as the work produced divided by the heat absorbed from the hot source, was first derived by the entropy S (Carnot cycle). (9) Asserts that absolute zero is unattainable in a finite number of steps for any process. Therefore, heat can never be completely converted into work in a heat engine. Entropy change accompanying any process approaches zero as the temperature approaches zero. All perfect crystalline substances have zero entropy at absolute zero.* (10) First law entails a formulation of the principle of energy conservation. The increase in the internal energy of a system equals the heat absorbed by the system from its surroundings minus the work done by the system on its surroundings. The second law of thermodynamics relates to the availability of energy in a system for conversion to useful work. In order for a

system to perform work, it must have the capacity for spontaneous change toward equilibrium.[†] See also CARATHÉODORY'S PRINCIPLE; FIRST PRINCIPLE OF THERMODYNAMICS; SECOND PRINCIPLE OF THERMODYNAMICS; THIRD PRINCIPLE OF THERMODYNAMICS.

THERMOSIPHON PRINCIPLE [physics] The tendency of heated liquids to rise. Also termed *thermosyphon principle*.

THETA ROLE ASSIGNMENT PRINCIPLE [linguistics, psychology] (Noam Chomsky) Theory of universal grammar. Consists of various separate modules such as case theory, theta theory, binding theory with parameters, each taking only a limited range of values, accounting for variations among languages.[‡] The head parameter, the directionality parameter of the theta assignment, and the distinction between the internal argument and external argument are retained in their grammar.[§] See also INFORMATION-PROCESSING PRINCIPLE; PRINCIPLES AND PARAMETERS THEORY.

THEVENIN'S PRINCIPLE [physics, electronics] (Leon Charles Thevenin, 1857–1926, French telegraph engineer) The source behind two accessible terminals may be regarded as a constant voltage generator in series with a source impedance. The value of the voltage is that appearing with terminals open-circuited, and the impedance is that measured at the terminal with all voltage sources open-circuited.[¶] See TRAVELING-WAVE PRINCIPLE; NORTON'S PRINCIPLE.

THIRD DATA PROTECTION PRINCIPLE [mathematics] Personal data shall be adequate, relevant, and not excessive in relation to the purpose or purposes for which they are

---

*Anon. (2005), *McGraw-Hill Encyclopedia of Science and Technology*, McGraw-Hill, New York.

[†]Khoury, F. M. (2004), *Thermodynamics and Phase Equilibria Multistage Separation Processes*, 3rd ed., chap. 01. CRC Press.
[‡]Higawara, H. (1993), The breakdown of Japanese passives and theta-role assignment principle by Broca's aphasics, *Brain Lang.* **45**(3), 319.
[§]Op. cit. Higawara, p. 337.
[¶]Walker, P. M. B. (1999), *Chambers Dictionary of Science and Technology*, Chambers, New York, p. 1164.

processed.* See also COPY PROTECTION PRIN-
CIPLE; PROTECTION PRINCIPLE; PRIVACY PRINCI-
PLE; FIRST DATA PROTECTION PRINCIPLE; SEC-
OND DATA PROTECTION PRINCIPLE; FOURTH DATA
PROTECTION PRINCIPLE; FIFTH DATA PROTEC-
TION PRINCIPLE; SIXTH DATA PROTECTION PRIN-
CIPLE; SEVENTH DATA PROTECTION PRINCIPLE;
EIGHTH DATA PROTECTION PRINCIPLE.

**THIRD INVERSE MAXIMUM ENTROPY PRINCI-
PLE** [mathematics] Leads to a determination of a measure of entropy so that the given
probability distribution is obtained by maximizing this measure of entropy subject to
given constraints.† See also ENTROPY OPTI-
MIZATION PRINCIPLE.

**THIRD INVERSE MINIMUM CROSS-ENTROPY
PRINCIPLE** [mathematics] (Edwin Thompson
Jaynes, 1922–1998, American physicist)
Identifies an a priori probability distribution
so that a given probability distribution
is obtained when a specified measure of
cross-entropy is minimized from the given
contstraints.‡ See also ENTROPY OPTIMIZATION
PRINCIPLE.

THIRD PARADOX PRINCIPLE [mathematics,
physics] Network problems allowing calculation of the performance of a device
from its thermal properties only.§ See also
HELMHOLTZ PRINCIPLE.

THIRD PRINCIPLE OF ENERGETICS [physics]
(Howard Thomas Odum, 1924–2002, American ecosystem ecologist) As a system approaches absolute zero of temperature, all
processes cease and the entropy of the system
approaches a minimum value or zero for the

case of a perfect crystalline substance.¶ See
also PRINCIPLES OF ENERGETICS; ZEROTH PRIN-
CIPLE OF ENERGETICS; SECOND PRINCIPLE OF
ENERGETICS; FIRST PRINCIPLE OF ENERGETICS;
FOURTH PRINCIPLE OF ENERGETICS; MAXIMUM
EMPOWER PRINCIPLE; MAXIMUM POWER PRIN-
CIPLE; FIFTH PRINCIPLE OF ENERGETICS; SIXTH
PRINCIPLE OF ENERGETICS.

THIRD PRINCIPLE OF RELATIVITY [mathemat-
ics, physics] See PRINCIPLE OF RELATIVITY;
PRINCIPLE OF SOLIDARITY.

THIRD PRINCIPLE OF THERMODYNAMICS
[physics] (Walther Hermann Nernst,
1864–1941, German physicist/chemist)
(1) States that in the neighborhood of an
equilibrium state of a thermodynamic system, there exist states that are not accessible
by either reversible or irreversible adiabatic
processes. (2) The practical significance of
the second law is that it limits the extent to
which the internal energy may be extracted
from a system as work. In order for a process
to generate work, it must be spontaneous.
For the process to be spontaneous, it is
necessary to discard some energy as heat in
a sink of lower temperature. In other words,
nature in effect exacts a tax on the extraction
of energy as work. There is therefore a
fundamental limit on the efficiency of
engines that convert heat into work. The
quantitative limit on the efficiency is defined
as the work produced divided by the heat
absorbed from the hot source. (3) Asserts
that absolute zero is unattainable in a finite
number of steps for any process. Therefore,
heat can never be completely converted
into work in a heat engine. Entropy change
accompanying any process approaches zero
as the temperature approaches zero. All
perfect crystalline substances have zero
entropy at absolute zero. (4) Requires that
the entropy vanishes at zero temperature.
The entropy of a thermodynamical system is
determined by its energy up to an arbitrary

*Anon. (2007), Univ. Edinburgh, Records Manage-
ment Section (http://www.recordsmanagement.
ed.ac.uk/InfoStaff/DPstaff/DP_Research/Research
AnnexA.htm).
†Kesavan, H. K. and Kapur, J. N. (1989), The
generalized maximum entropy principle. *IEEE
Trans. Syst. Man Cybernet.* **19**(5), 1042–1052.
‡Barford, L. (2004), Sequential Bayesian bit error
rate measurement, *IEEE Trans. Instrum. Measure.*
53(4), 947–954.
§Simon, I., Bârsan, N., Bauer, M., and Weimar,
U. (2001), Micromachined metal oxide gas sen-
sors: opportunities to improve sensor performance,
Sensors Actuators B Chem. **73**(1), 1–26.

¶Odum, H. T. (2000), An energy hierarchy law
for biogeochemical cycles, *Proc. 1st Biennial Energy Analysis Research Conf. Energy Synthesis:
Theory and Applications of the Energy Methodology*, Brown, M. T., ed., Center for Environmental
Policy, Univ. Florida, Gainesville.

constant. In our case, the computation of the entropy from the energy of a shell outside the horizon is however a subtle matter. The derivation of the entropy for generic T must be performed before removing the brick wall, since the removal is possible only for $T = TH$. Any attempt to determine at this stage the arbitrary constant by the third principle, leads to divergences when later removing the brick wall.[*] See also THERMODYNAMIC PRINCIPLES; NERNST'S PRINCIPLE; FIRST PRINCIPLE OF THERMODYNAMICS; SECOND PRINCIPLE OF THERMODYNAMICS.

THOMPSON'S PRINCIPLE [mathematics] (James Thomson, 1822–1892, British engineer and physicist whose younger brother William Thomson (Lord Kelvin).) The dual variational principle (Thompson's principle) also determines a bound for the effective tensor σ_*. The diagonal coefficient β_*^{11} of the inverse tensor $\beta = \sigma^{-1}$ is

$$\beta_*^{11} = \min_{j \in \mathcal{J}} \langle j \cdot \sigma^{-1} j \rangle,$$

where

$$\mathcal{J} = \{j : \nabla \cdot j = 0, \quad \langle j \rangle = i_1, \quad j \text{ is 1-periodic}\}.$$

Thompson's principle leads to upper estimates of the coefficients of the inverse tensor σ_*^{-1} (which are the *lower* estimates of the tensor σ_*). Again, using the constant trial function, one obtains the inequality

$$\beta_*^{11} \geq \langle i_1 \cdot \sigma^{-1} i_1 \rangle,$$

which leads to

$$\sigma_*^{-1} \geq \langle \sigma^{-1} \rangle = \sum_{i=1}^{N} m_i \sigma_i^{-1} = \sigma_h^{-1},$$

where

$$\sigma_h = \left(\sum_{i=1}^{N} m_i \sigma_i^{-1} \right)^{-1}$$

[*]Battaglia, F. and George, T. F. (1998), *Fundamentals in Chemical Physics*, Springer, p. 96.

denotes the harmonic mean. This bound is called the *Voigt bound* (Voigt, 1928) or the *harmonic mean bound.*[†]

THOMSON PRINCIPLE [mathematics] (William Thomson, 1st Baron Kelvin, 1824–1907, Irish mathematician)[‡] (1) Let G=(V, E) be a connected graph, and $W_e \in 2E$, (strictly positive) conductances. Let $s, t \in V, s \neq t$. Amongst all unit flows through G from s to t, the flow that satisfies the Kirchhoff laws is the unique s/t-flow i that minimizes the dissipated energy. That is, $E(i) = inf\{E(j) : j$ a unit flow from s to $t\}$.[§] (2) The unique function minimizing the integral of the sum of squares among all differentiable functions having given boundary values and the unique function is the solution of Laplace's equation with those boundary values. (3) Dirichlet's principle is mistakenly applied to duality between current and voltage. This duality, long familiar to electricians, is known to mathematicians as *Hodge duality*. Under this duality resistance corresponds to conductance (i.e., inverse resistance), so to a principle yielding upper bounds for resistance, there is a corresponding principle yielding upper bounds for conductance, that is, lower bounds for resistance.[¶] See also DIRICHLET'S PRINCIPLE.

THOMSON–BERTHELOT PRINCIPLE [chemistry] (William Thomson, 1st Baron Kelvin, 1824–1907, Irish mathematician) (Mercellin Pierre Eugene Berthelot, 1827–1907, French chemist) Heat released in a chemical reaction is directly related to the chemical affinity. In the absence of the application of external energy, chemical reactions that

[†]Cherkaev, Andrej (2000) Variational Methods for Structural Optimization (Applied Mathematical Sciences Vol. 140). Springer, p. 62.
[‡]Stoddart A. W. J. (1971), An extension of the Thomson principle, Proc. of the Amer. Math. Soc. 29(3) 561–5.
[§]Grimmett, G. R. (2009) Probability on Graphs; Lecture Notes on Stochastic Processes on Graphs and Lattices. Centre for Mathematical Sciences, University of Cambridge p. 9.
[¶]Milanesi, G., and Mintchev, M. (2005) Aspects of Finite Temperature Quantum Field Theory in a Black Hole Background. arXiv:hep-th/0509080.

release the greatest heat are favored over others.[*]

T'HOOFT'S PRINCIPLE [physics] (Gerardus t'Hooft, b. 1946, Nobel Prize Dutch theoretical physicist) Quantum gravity can be postulated as a dissipative deterministic system, where quantum states at the "atomic scale" can be understood as equivalence classes of primordial states governed by a dissipative deterministic dynamics law at the "Planck scale." Theoretical techniques for making quantitative predictions concerning electroweak interactions.[†] See also GABOR'S HOLOGRAPHIC PRINCIPLE; HOLOGRAPHIC PRINCIPLE.

3-DAY PRINCIPLE [computer science, medicine] Used in clinical event monitoring.[‡] See DATA-SENDING PRINCIPLE.

THREE-DAY PRINCIPLE [sociology] Since leaders convert dreams into reality, this approach emphasizes the interrelationships between individuals, teams, and the organization as a whole. By linking individual with organizational vision and values, a spirit of partnership develops, creating an empowered organization. (2) A bill that cannot survive a 3-day scrutiny of its provisions if it is a bill that should not be enacted.

THREE-PHASE PRINCIPLE [kinesiology] (Dr. Hans Selye) Leads to the belief that there must be periods of low or no intensity between those overloaded stresses that tax the body.[§] See also PRINCIPLE OF TRAINING.

[*]Anon. (2003), *McGraw-Hill Dictionary of Scientific and Technical Terms*, McGraw-Hill.
[†]Vermillon, Stey (2003) Dustoff. No Compromise! No Rationdigation! No Hestation! Fly the mission 1. wild N Woolly Publishing.
[‡]Willmen, H. R. et al. (1996), Indications for a endectomy according to the 3 day principle; ten years' experience, *Chirug* **67**, 826.
[§]Panteleev, V. A., and Urman, P. N., (1979), Principles of diffusion in a three-phase system, *Izvest. Vysshikh Uchebnykh Zavedenii, Fiz.* **22**(11), 78–82.

THREE-POINT LIGHTING PRINCIPLE [physics] (1) Employs a KEY light (performing the job of the sun), the BACK KEY (as a counterpoint to the key) and, finally, a soft or FILL light. (2) Using only a BACK KEY light and a couple of foreground reflectors (a Lastolite reflector or a sheet of polystyrene foam reflect the BACK light into the foreground to be used as a foreground fill. Provides two or more balancing light sources from a single lamp. For a harder reflection, place some kitchen foil on the surface of the polystyrene sheet. Using the same reflector material, treat the sun's light as a background key (with the subject in front of, but at an angle to, the sun), and then bounce the sun's rays back into the foreground.[¶]

THREE PRINCIPLES OF APPEAL [psychology] (Aristotle) These principles are ethos, pathos, and logos. *Ethos* is how well the speaker convinces the audience they are qualified to speak on the particular subject. Persuasion is achieved by the speaker's personal character when the speech is so spoken as to make us think her/him credible. *Pathos* is an appeal to the audience's emotions. It can be in the form of metaphor, simile, a passionate delivery, or even a simple claim that a matter is unjust. Persuasion may come through the hearers, when the speech stirs their emotions. *Logos* is logical appeal used to describe facts and figures that support the speaker's topic. Since data are difficult to manipulate, especially if from a trusted source, logos may sway cynical listeners. Having a logos appeal also enhances ethos (see above) because information makes the speaker seem knowledgeable and prepared to his audience. Persuasion is effected through the speech itself when we have proved a truth or an apparent truth by means of the persuasive arguments suitable to the case in question.[‖]

[¶]Kitayama, K. and Murata, M. (2002), Optical packet pulling dropping insertion method and that device in optical network. *Jpn. Kokai Tokkyo Koho* Japanese patent.
[‖]Hollingworth, H. L. (2001), Advertising and elling: principles of appeal and response. NY, D. Appleton and Co., http://www. archive. org/ stream/ advertisingselli00holluoft,.

THREE-ROLLER PRINCIPLE [physics] Tension measurement where the outer two reference rollers are fixed to create a known angle of wrap over the middle sensing roller. The middle roller is part of a precision strain gauge sensing system that measures the resulting force on the roller. The strain gauge measures a change in resistance and converts this to a highly accurate voltage signal.* See also DIP ROLLER PRINCIPLE; WEDGE PRINCIPLE; ROLLER WEDGE PRINCIPLE.

TIAN–CALVET PRINCIPLE [mathematics, physics] (1) To apply the Tian equation correction to the initial power difference–temperature data to determine whether this correction makes the differentiel scanning calorimeter (DSC) quantities independent of scan rate. (2) Controller measuring principle heat flux after Tian and Calvet: $dW/dt = K^1 E + K^2(dE/dt) - t =$ time; $T =$ temperature difference; $E = gT$, where g is a calorimeter constant; $dW/dt =$ rate of heat production; $K^1 = p/g = $ a calorimeter constant; $K^2 = u/g = $ a calorimeter constant.[†]

TIAN–CALVET PRINCIPLE [measurement, physics] (1) Applying the Tian equation correction to the initial power difference–temperature data to determine whether this correction renders the DSC quantities independent of scan rate. A similar dynamic correction has been developed (2) Controller measuring principle heat flux DSC after Tian and Calvet: $dW/dt = K^1 E + K^2(dE/dt) - t =$ time; $T =$ temperature difference; $E = gT$, where g is a calorimeter constant; $dW/dt =$ rate of heat production; $K^1 = p/g = $ a calorimeter constant; $K^2 = u/g = $ a calorimeter constant.

TILTED LARGE DEVIATION PRINCIPLE [mathematics] Allows one to generate a new large deviation principle from an old one by "tilting," that is, integration against an exponential functional. It can be seen as an alternative formulation of Varadhan's lemma.[‡]

TIME-DEPENDENT HARTREE–FOCK (TDHF) VARIATIONAL PRINCIPLE [physics] (1) Two-body correlation to be included not only in the ground-state energy but also in the constraints. (2) Widths of resonant states have an important effect on the pairing properties of nuclei close to the drop line. (3) Energies of excitations are determined essentially by the coordination number of the excited atom rather than the native of the local environment. (4) Statistical exchange potential can thus be given in a universal form and then applied in wave mechanical calculations (the sum of the exchange potential and the electrostatic potential).[§] See also SKYRME–HARTREE–FOCK PRINCIPLE; HARTREE–FOCK VARIATIONAL PRINCIPLE.

TIME DIVISION MULTIPLE ACCESS (TDMA) PRINCIPLE [mathematics] (1) Within a radio cell, the base station defines on the "radio interface" to the mobile stations communicating with it the time division multiplex frame and thus the relative timing of the timeslots that are allocated in a link-specific manner to the information interchange in the frequency bands. With regard to the frame clock cycle phase used in operation, the base stations are in principle independent of one another. (2) Telecommunication apparatuses and second telecommunication apparatuses following the first telecommunication connections, taking into account the items of information in the first timeframes. (3) Each further telecommunication connection following the first one is set up with priority on

*Masson, C., Parniere, P., Penelle, R., and Pernot, M. (1973), Rolling recrystallization textures of thin sheet metal. Three-dimensional representation. Principles and application, *Memoires Sci. Rev. Metallurgie* **70**(4), 271–280.

[†]Naumann, R. (1991), New possibilities in the application of the Tian-Calvet principle in thermal analysis, *Chemie Labor Biotechnik* **42**(11), 613–617.

[‡]den Hollander, F. (2000), *Large Deviations*, Fields Inst. Monograph 14, Am. Math. Soc., Providence, RI, p. 143.

[§]Lichtner, P. C., Griffin, J. J., Schultheis, H., Schultheis, R., and Volkov, A. B. (1979), Time-dependent Hartree-Fock phase: unique implication of variational principle, *Phys. Lett. B* **88B**(3-4), 221–225.

the same Frequency division multiple access (FDMA) frequency.*

TIME DIVISION PRINCIPLE [psychology, computers]   A single time-divisional synchronizing unit applied to the elastic buffering and the monitoring of the fill rate of the buffer memory performed in synchronizing sequentially arranged disassembly and assembly units. The pointers are processed at one or more processing stages on a time division basis, that is, the processing of at least two signals on the same level of hierarchy is performed over the same physical line.[†] See also INFORMATION-PROCESSING PRINCIPLE.

TIME FUNNEL PRINCIPLE [mathematics] Changes to critical data are continually copied to the mirror system with a defined time delay. In case a system crash occurs, all data are still available on the mirror system until the point before the error occurred. Production is simply switched, and work continues there until it can be resumed.[‡] See also FUNNEL PRINCIPLE.

TIME INDEPENDENCE PRINCIPLE [mathematics]   (1) Approach to color image compression with high compression ratios and good quality of reconstructed images using quantization, thresholding, and edge detection. (2) In training for professional practice the mode of training should match the mode of practice. (3) Guiding principle in the design of training programs, especially in planning for the "process" element in training. (4) The established "reflection process" as a framework of possible inputs to a system as a subset of the set of tolerable inputs (those that are judged to produce an acceptable

response from the system). (5) System design that has direct relevance to the theory and design of control systems in combination with the method of inequalities. (6) Tendency to match and to influence groups trained not to match. (7) Bridge network of rectangular sections of resistive track arranged with longer sides at right angles to flow.[§] See also MATCHING PRINCIPLE; ASYMPTOTIC MATCHING PRINCIPLE; HYDROPHOBIC MATCHING PRINCIPLE; DOPPLER MATCHING PRINCIPLE; PHASE MATCHING PRINCIPLE; BRACKET MATCHING PRINCIPLE; MODEL MATCHING PRINCIPLE; MOMENT-MATCHING PRINCIPLE; ZERO-FORCING PRINCIPLE; DISTRIBUTION MATCHING PRINCIPLE; COMBINATORIAL MATCHING PRINCIPLE; PIGEON-HOLE PRINCIPLE; H_2 MODEL MATCHING PRINCIPLE; H INFINITY MATCHING PRINCIPLE; OPTIMAL MODEL MATCHING PRINCIPLE; MINTO PYRAMID PRINCIPLE.

TIME REVERSAL INVARIANCE PRINCIPLE [physics]   Particle interactions should be indifferent to the direction of time (kaon).[¶] Kaons were essential in establishing the fact particle interactions should be indifferent to the direction of time as served also for the fundations of the Standard Model of partile physics, such as the quark model of hadrons and the theory of quark mixing. In particle physics, a kaon, also called a K-meson and denoted K[nb 1]) is any one of a group of four mesons destinguished by the fact they carry a quantum number called strangeness.

TIME–TEMPERATURE SUPERPOSITION (TTS) PRINCIPLE [physics]   Allows for the estimation of the temperature shift factor for temperatures other than those for wich the material was tested and can be applied to other temperatures. Also used to determine the material behavior temperature dependence or high-frequency, temperature-independent, to expand the time or frequency at a given temperature. Also known as time (frequency) temperature

*Xu, D. (2001), A unified microscale-parameter approach to solidification-transport phenomena-based macrosegregation modeling for dendritic solidification: Part I. Mixture average-based analysis, *Metallurg. Mater. Trans. B Process Metallurgy Mater. Process. Sci.* **32B**(6), 1129–1141.

[†]Molotkov, S. N. (2004), Integration of quantum cryptography into fiber-optic telecommunication systems, *JETP Lett.* **79**(11), 559–570.

[‡]Kumar, S., Ma, B., Tsai, C. J., Wolfson, H., and Nussinov, R. (1999), Folding funnels and conformational transitions via hinge-bending motions, *Cell Biochem. Biophys.* **31**(2), 141–164.

[§]Murphy, G. L. (1977), A unified Einstein-Yukawa theory, *Progress Theor. Phys.* **58**(5), 1622–1626.

[¶]Frazer, W. R. (1966), Time-reversal invariance in S-matrix theory of strong and electromagnetic interactions of hadrons, *Phys. Rev.* **150**(4), 1244–1248.

superposition (FTS) principle[*] See also; TIME TEMPERATURE SHIFT PRINCIPLE.

TIMELESS PRINCIPLES Timeless principles are both universal and constant, and that building a life on a foundation of positive, character-based principles results in lasting internal peace.[†]

TIMING PRINCIPLE [physics] Fundamental feature of quantum theory: when a situation is a composition of a number of elementary situations, its amplitude is the linear superposition of the amplitudes of the components.[‡] See also INTERNAL NOISE PRINCIPLE; PRIMING PRINCIPLE.

TINKERBELL PRINCIPLE [logic] A single and powerful concept stating that something works because everyone believes that it will.[§]

TIRF PRINCIPLE [physics] See TOTAL INTERNAL REFLECTION FLUORESCENCE PRINCIPLE.

TODDS' PRINCIPLES OF ECOLOGICAL DESIGN [ecology] Places nature at the center of the desing network process for biological equity.[¶] See also BIOMIMICRY PRINCIPLES; DEEP ECOLOGY'S BASIC PRINCIPLES; FOREST STEWARDSHIP COUNCIL (FSC) PRINCIPLES; GLOBAL SULLIVAN PRINCIPLES OF SOCIAL RESPONSIBILITY;

MARINE STEWARDSHIP COUNCIL (MSC) PRINCIPLES; NATURAL CAPITALISM PRINCIPLES; PERMACULTURE PRINCIPLES; THE BELLAGIO PRINCIPLES FOR ASSESSMET; EQUATOR PRINCIPLES; MELBOURNE PRINCIPLES; PRECAUTIONARY PRINCIPLE; SANBORN PRINCIPLES.

TOGETHER/SEPARATE PRINCIPLE [communications] Creates distance where it does not physically exist and provides the opportunity to examine the 'together/separate' principle. Imagine face-to-Face with two participants standing in spaces A and B, facing in the direction of the barrier. The barrier separates them visually, but they are acoustically together and can hand an object to each other with minimal movement through space. This is a space that has been articulated for a non-technology-mediated together/separate experience. A house can be designed-its rooms, corridors and the spaces between them-for a series of such together separate experiences. The same holds true for communications technology: when video equipment is set-up, it enables the participants to be visually together as well, but they are joined by electronic space, and they remain physically separate. These variations in perceptual levels for seeing, hearing and touching permit the two participants to be simultaneously together and separate, whether with or without technology. Thisability to be together and separate at the same time has a quality of complexity that is different from the all-or-nothing by which a binary unit of information (bit) is represented.[‖]

TOPOLOGICAL TRANSVERSALITY PRINCIPLE [mathematics] (S. Bernstein, 1912) Applications of the topological transversality theorem to boundary-value problems for second-order ordinary differential equations. In order to prove existence of solutions, the topological transversality theorem is

[*]Kobayashi, M. et al. (1995), Application of the time temperature superposition principle to the shear and uniaxial elongational plow for polymer inorganic particle composite melt, *Kobunshi Ronbunshu* **52**, 478.

[†]Adams, T., Bezner, J., and Steinhardt, M. (1995), Principle-centeredness: A values clarification approach to wellness, *Measur. Eval. Counsel. Devel.* **28**(3), 139–147.

[‡]Koyama, K., Kakoi, H., Miyao, H., Kawasaki, J., and Kawazoe, T. (1993), The circulatory responses to tracheal intubation using the timing principle, *Jpn. J. Anesthesiol.* **42**(5), 690–693.

[§]Center for Science, Mathematics, and Engineering Education, US Natl. Research Council. Computer Science and Telecommunications Board (1998), *Developing a Digital National Library for Undergraduate Science, Mathematics, Engineering, and Technology Education: Report of a Workshop*, National Academies Press, p. 84.

[¶]Edwards, A. R. (2005) The sustainability revolution: portrait of a paradigm shift. New Society Publishers, p. 104.

[‖]Mitropoulos, M. (1991), A sequence of video-to-video installations illustrating the together/separate principle, with reference to two-way interactive cable TV systems, *Leonardo (Connectivity: Art Interact. Telecommun.)* **24**(2), 207–211.

combined with the technique of a priori bounds on solutions.[*]

TORTUOUS PATH PRINCIPLE [engineering] Profiles with more complex cross sections have a much greater surface area. They present a much greater resistance and more difficult path for water to seep around the section.[†] See also PRINCIPLE OF WATERSTOP FUNCTION; VALVE PRINCIPLE.

TORRICELLI'S PRINCIPLE [physics] (Evangelista Torricelli, 1608–1647, Italian mahtematician/physicist) A fluid flows from a hole in an open-topped container with a speed equal to the speed that any object would acquire were it to be dropped through a height equal to the head of water above the opening.[‡]

TORSION BALANCE PRINCIPLE [physics] (1) The gravitation force between two masses is equal to $G_m M/r^2$ according to the Newtonian law of gravity. (2) A torsion balance consists of a lever attached to a piece of wire. When a force is applied to the lever, it exerts a torque on the wire, causing the wire to rotate. Analagously to the manner in which a spring will spring back when you exert a force on it, the wire exerts a restoring torque, proportional to the angle through which it has been twisted. As a result, the size of the force applied to the lever can be deduced by measuring the angle through which the wire has been twisted.[§] See TOTAL VIRTUAL ACTION PRINCIPLE.

TORTUOUS PATH PRINCIPLE [physics] See PRINCIPLE OF WATERSTOP FUNCTION.

[*]Mitra, A. N. (2000), Markov-Yukawa transversality principle and 3D-4D interlinkage of Bethe-Salpeter amplitudes, Los Alamos National Laboratory, *Preprint Archive-High Energy Physics, Phenomenology*, pp. 1–38.
[†]Anon. (Sept. 2007), *PARCHEM Concrete Repair Flooring Jointing Systems Waterproofing*, Technical Data Sheet, ABN 80 069 961 968 (www.parchem.com.au).
[‡]Walker, P. M. B. (1999), *Chambers Dictionary of Science and Technology*, Chambers, New York, p. 1179.
[§]Kalle, K. (1957), Small areometer for the precise determination of salinity and density in sea-water, *Dtsch. Hydrograph. Z.* **10**, 99–106.

TOTAL DIFFERENTIAL PRINCIPLE [physics] If a function $f(p,q)$ and p, q both depend on t, then the total differential of f involves not only dp and dq but dt as well via the chain rule.[¶] See also D'ALEMBERT'S PRINCIPLE; PRINCIPLE OF TOTAL DIFFERENTIAL; DIFFERENTIAL PRINCIPLE OF D'ALEMBERT; VARIATIONAL PRINCIPLE; HAMILTON PRINCIPLE.

TOTAL INTERNAL REFLECTION FLUORESCENCE PRINCIPLE [physics] A light beam penetrates an interface between two phases of different refractive index and is totally reflected in the higher-refractive-index medium if the angle of incidence exceeds a critical angle.[‖]

TOTAL INTERNAL REFLECTION FLUORESCENCE MICROSCOPY (TIRFM) PRINCIPLE [biology] Makes possible to visualize single fluorophores in living cells. Used to visualize molecules both in vitro and in vivo. Observes cell substrate adhesions. An excitation light beam illuminates the meniscus of two media obliquely from a high to a low diffractive index with an incident angle greater than the critical angle of total internal reflection. Under these conditions, an electromagnetic field clled "the evanescent field" rises from the interface into the medium with a lower diffractive index. Unlike normal beam of light that travels in a straight line infinitely, the evanescent field propagates parallel to the interface vanishing exponentially with the distance from the interface. Therefore, using the evanescent field the excitation depth can be limited to a very narrow range that allows the exclusive observation of the cell-substrate adhesion. The decay length of the evanescent field along the depth of the field depends on the incident angle wavelength of the excitation beam

[¶]Otsuki, N., Hisada, M., Nagataki, Sh., and Kamada, T. (1995), A proposal of evaluation method for concrete admixture with the principle of total differential, *Zairyo*, **44**(500), 643–648.
[‖]Lehr, H. P., Reimann, M., Brandenburg, A., Sulz, G., and Klapproth, H. (2003), Real-time detection of nucleic acid interactions by total internal reflection fluorescence, *Anal. Chem.* **75**(10), 2414–2420.

and diffractive indices of both media.* (2) Optical technique utilized to observe single-molecule fluorescence at surfaces and interfaces. The technique is commonly employed to investigate the interaction of molecules with surfaces, an area that is of fundamental importance to a wide spectrum of disciplines in cell and molecular biology.[†]

TOTAL SIMILARITY PRINCIPLE [biology] Frequency judgments for a stimulus will be increased, rather than decreased, by the presentations of similar stimuli. For two different stimuli, A_1 and A_2, which have been presented an equal number of times, the frequency judgment for A_1 will be greater than the judgment for A_2 if other stimuli B similar to A_1 have also been presented. The models make this prediction because the total similarity of A_1 to what is in memory will include similarities to A_1 traces as well as similarities to B traces. Another possibility is that in some cases, the judgment for A_2 will be greater than that for A_1. A frequency judgment for a stimulus A will be increased by presentations of a similar stimulus B regardless of the number of A presentations. An evaluation of the total similarity principle: effects of similarity on frequency judgments.[‡]

TOTAL VIRTUAL ACTION PRINCIPLE [physics] Used to measure minute gravitational, electrostatic, or magnetic forces.[§]

TOTALITARIAN PRINCIPLE [physics] (Murray Gell-Mann, physicist) "Everything which is not forbidden, is compulsory." Any interaction that is not forbidden by a small number of simple conservation laws is not only allowed but also must be included in the sum over all "paths" that contribute to the outcome of the interaction.[¶]

TRACEABILITY MEASUREMENT PRINCIPLE [physics] (1) Methods do not require any intuitive or empirical assumptions, but use only the material properties, tool geometry, and the physical laws of deformation. (2) Found in the solar coronal geometry, which is modeled as a rectangular domain. It is shown that the solution bifurcates into two different states when the magnetic helicity integral or the geometric factor, defined as the of the domain height–width ratio, is satisfactorily increased. (3) The bandlimited function of lowpass type whose energy is minimum among those that have the prescribed sample values on a finite set of sampling points. The minimum energy signal converges uniformly to a bandlimited signal when the number of sampling points is increased infinitely. (4) Free boundary subjected to external magnetic or plasma pressure forces. (5) The path of a conservative system in configuration space is such that the integral of the Lagrangian function over time is a minimum or maximum relative to nearby paths between the same endpoints and taking the same time.[∥] Also known as *minimum norm principle; principle of vibration measurement; coriolis measurement principle; magnetic measurement principle; minimum energy principle.*

TRAFFIC MEASUREMENT PRINCIPLE [mathematics] Key problem for modern network designers is to characterize and/or model the "bursty" traffic arising in broadband networks with a view on predicting and guaranteeing performance.[**]

*Sako, Y., and Uyemura, T. (2000), Total internal reflection fluorescence *microscopy for single-molecule imaging in living cells.* Cell Struct. Funct. **27**(5), 357–365.

[†]Axelrod, D., Hellen, E. H., and Fulbright, R. M. (1992), Total internal reflection fluorescence, *Topics Fluoresc. Spectrosc.*, **3**, 289–343.

[‡]Jones, C. M. and Heit, E. (1993), An Evaluation of the Total Similarity Principle: Effects of Similarity on Frequency Judgments. J. Exp. Psychol. Learning, memory, and cognition, **19**(4), 799–812.

[§]Dvoracek, Z. and Horak, Z. (1967), Atomic Z expansions and electronic structure of molecules: Second-order one-center calculation of the hydrogen-molecule ground state, *J. Chem. Phys.* **47**(4), 1211–1217.

[¶]Buttner, S. (1993), Endangerment of medicine by totalitarian basic principles, *Diskussionsforum Medizinische Ethik* **3**, IX–XI.

[∥]De Bievre, P., Kaarls, R., Peiser, H. S., Rasberry, S. D., and Reed, W. P. (1996), Measurement principles for traceability in chemical analysis, *Accredit. Qual. Assur.* **1**(1), 3–13.

**Montgomery, M., and De Veciana, G. (1996), On the relevance of time scales in performance

TRANQUILITY   PRINCIPLE [mathematics]
(1) The classification of a subject or object does not change while it is being referenced. There are two forms to the tranquility principle: (1) principle of strong tranquility states security levels, which do not change during the lifetime of the system; and (2) principle of weak tranquility states security levels, which do not change in a way that violates the rules of a given security policy. (2) an object's security level/label will not change during an operation (such as read or write); the weak tranquility principle means that an object's security level/label may change in a way that does not violate the security policy during an operation.[*] See also PRINCIPLE OF STRONG TRANQUILITY; PRINCIPLE OF WEAK TRANQUILITY.

TRANSACTION   PRINCIPLE [psychology]
Inter cultural communication is a process whereby people from different cultures move within a negotiation zone from positions to interests, in search of an acceptable resolution.[†] Also known as *principle of intercultural communication; punctuation principle; performativity principle; positionality principle; commensurability principle; continuum principle; pendulum principle; synergy principle; sustainability principle.*

TRANSCENDENT   PRINCIPLE [mathematics]
Dictates the choice of a theoretical framework, and according to a plot that orients the way the analysis unfolds.[‡]

TRANSFER   PRINCIPLE [mathematics, economics] (1) A notion of simple property of theories is introduced and it is shown that if P is a simple property of theories, A countable admissible, and M a structure in A, then $\text{Th}A(M)$ has property P iff $\text{Th}\infty\omega(M)$

has property P. Given a theorem which tells when a 2-definable theory on a countable set has some property P, the "transfer principle" yields the theorem: Let A be countable admissible and let M be a model in A. Then $\text{Th}A(M)$ has property P.[§] (2) For polynomials in several variables irreducible over the rational field, there always exist rational specialzations of some of the variables which preserve irreducibility.[¶]. (3) This axiom requires the inequality measure to rise (or at least not fall) in response to a mean-preserving spread; an income transfer from a poorer person to a richer person should register as a rise (or at least not as a fall) in inequality, and an income transfer from a richer to a poorer person should register as a fall (or at least not as an increase) in inequality. (5) Redistributing consumption from a high-consumption individual to an otherwise identical individual with lower consumption leads to a socially preferred allocation.[‖] See also PIGOU–DALTON TRANSFER PRINCIPLE; DIFFUSION TRANSFER PRINCIPLE.

TRANSFORMATION PRINCIPLE [mathematics]
(1) Rules for transforming equations involving an incircle to equations about excircles. (2) A confidence interval for a parameter may be used to construct an interval for any monotonic transformation of the parameter. (3) Based on a model of generalized contradictory equations and their problem-solving, which can be expanded as a layer-tender method and a competing-for tender method.[**] See also EQUIVALENCE METHOD PRINCIPLE; PANSYSTEMS OPERATION EPITOPANSYSTEMS PRINCIPLE; PANOPTIMIZATION PRINCIPLE.

[§]Nadel, M. E. (1976) A Transfer principle for Simple properties of theories. *Proc. Amer. Math. Soc.* **59**(2) 353–357.
[¶]Zannier, U. (2009) On the Hilbert Irreducibility Theorem. *Rend. Sem. Mat. Univ. Pol.* torino **67**(1), 1–14.
[‖]Ravallion, M. (1998), Does aggregation hide the harmful effects of inequality on growth? *Econ. Lett.* **61**(1), 73–77.
[**]Fried, M. D. and Volklein, H. (1992) The Embedding Probem Over a Hilbertian PAC-Field. The Annals of Mathematics, Second Series, **135**(3), pp. 469–481.

oriented traffic characterizations, *IEEE 15th INFOCOM96*, National Science Foundation Grant NCR-9409722.
[*]Landwehr, C. E. (1981), Formal models for Computer Security, *ACM Comput. Surv. (CSUR) archive.* **13**(3) pp. 247–278.
[†]Berne, E. (1964), Principles of transactional analysis, *Current Psychiatr. Thera.* **25**, 35–45.
[‡]Anthony, W. A. (2004), The principle of personhood: The Field, transcendent principle, *Psychiatr. Rehab. J.* **27**(3), 205.

TRANSFORMING PRINCIPLE [genetics] (Frederick Griffith, 1928) An agent causing genetic transformation in bacteria later determined to be DNA.* Also known as *Griffith transforming principle*.

TRANSILLUMINATION PRINCIPLE [physics] An arrangement for measuring lengths or angles. There are two main types of oximeters, a so-called transillumination oximeter based on light passing through the tissue, and an oximeter based on light reflected from the tissue. In practice, the principle is the same in both measurements, and normal oximeters are able to perform both measurements and their intermediate forms reasonably well. An objective in the measurements is generally to use the transillumination principle, because then the obtained signal amplitude is about fivefold compared to that of the reflection principle.†

TRANSITIONAL STIMULI PRINCIPLE [psychology] See CHAINING PRINCIPLE.

TRANSLINEAR PRINCIPLE (TLP) [communications] Made the analysis of circuits possible in a way that previous views of BJTs as linear current amplifiers did not allow. TLP was later extended to include other elements that obey an exponential current–voltage relationship. In a closed loop containing an even number of translinear elements with an equal number arranged clockwise and counterclockwise, the product of the currents through the clockwise translinear elements equals the product of the currents through the counterclockwise translinear elements. [mathematics] (Barrie Gilbert) (1) Used to realize linear filters and constitutes an interesting approach to the implementation of nonlinear differential equations. (2) Linear filters and an approach to the implementation of nonlinear differential equations.‡

TRANSMISSION WELDING PRINCIPLE [engineering] Welding concept essentially works on the contour welding principle, whereby the laser spot follows a contour and the component is sequentially welded.§ See BUTT-WELDING PRINCIPLE; RESISTANCE WELDING PRINCIPLE.

TRANSMITTED LIGHT PRINCIPLE [physics, mathematics] The light passes through the transparent or translucent object. The object being examined can be observed against a light field or against a dark field in the case of illumination.¶ See also INCIDENT LIGHT PRINCIPLE.

TRANSPONDER PRINCIPLE [mathematics, physics] (1) Voltage levels modulated by an oscillator using a simple mixer or logic gauge. Resulting signal is connected to an impedance switcher. This impedance switching changes the reflection coeeficient, thus modulating the incoming signal. (2) Ratio in decibels of the stattering parameters between matched and mismatched impedance, thus increasing the working range of the system.‖

TRANSPORTING PRINCIPLE [mathematics] (1) Transmitter–receiver capable of accepting the challenge of an interrogator and automatically transmitting an appropriate reply. (2) Advances in electronic navigation and digital charging technology. It also covers traditional navigation such as celestial, plotting, and dead reckoning.**

TRANSVERSE FLUX PRINCIPLE [physics] The path of a magnetic flux is perpendicular to

*Luria, S. E. (1986), The transforming principle: Discovering that genes are made of DNA, *Sci. Am.* **254**, 24.

†Donn, S. M. and Faix, R. G. (1985), Transillumination in neonatal diagnosis, *Clin. Perinatol.* **12**(1), 3–20.

‡Gilbert, B. (1975), Translinear circuits: A proposed classification, *Electron. Lett.* **11**(1), 14–16.

§Zybko, J. (n.d.), *Three-Dimensional Laser Welding*, LEISTER Technologies, LLC and J. W. Chen, LEISTER Process Technologies (http://www.plasticsdecorating.com/articlesdisplay.asp?ID=64).

¶Hetzer, U. (2001), *Configuration for Detecting Mail Items*, US Patent Appl. US 2001/0041041A1.

‖Claes, W., Sansen, W., and Puers, R. (2005), *Design of Wireless Autonomous Datalogger IC's*, Springer.

**Bowditch, N. (2002), *The American Practical Navigator—Bowditch: An Epitome of Navigation*, Paradise Cay Publications.

the direction of the rotor rotation.* See also
MAXIMUM PRINCIPLE.

TRAPEZOIDAL   PRINCIPLE [mathematics]
Estimation of the area of an irregular figure.
The area is equal to the common width of
the strips multiplied by the sum of half the
first and half the last ordinates plus all the
others in between.[†]

TRAVELTIME   PRINCIPLE [physics] Pulses
containing a definite number of periodic
waves propagate with sound velocity toward
a measured object. The pulses are then reflected from the measured object and picked
up by the receiver with a lag time equal
to the elapsed time between the emission
and reception of pulses. The received pulses
of ultrasonic waves are transformed to the
electrical signal by means of the piexolectric
effect.[‡] See also FERMAT'S MINIMUM TIME
PRINCIPLE.

TRAVELING-WAVE   PRINCIPLE [physics] At
any given frequency the current flow in
any impedance, connected to two terminals
of a linear bilateral network containing
generators of the same frequency, is equal
to the current flow in the same impedance
when it is connected to a voltage generator
whose generated voltage is the voltage
at the terminals in question with the
impedance removed, and whose series
impedance is the impedance of the network
looking back from the terminals into the
network with all generators replaced by
their internal impedances.[§] Also known as
*bekesy traveling-wave principle; Helmholtz
resonance principle; Thevenin's principle*.

**TRIADIC PRINCIPLE OF GENETIC PSYCHO-
ANALYSIS** [psychology] Method consisting
of the patient's seeing a male therapist and
a female therapist in regularly alternating
interviews. The two members of the *dual*
team do not act as separate therapists
but consider themselves, as the patient
soon learns to see them, participants in
a single treatment process. This method
is based on the fact that the constellation
into which the child is born and is later
exposed to personality-shaping influences is
preeminently a triadic one: father, mother,
and child.[¶]

TRIANGLE TRANSFORMATION PRINCIPLE [ge-
ometry] Provides rules for transforming
equations involving an incircle to equations
about excircles.[‖]

TRIANGULAR NORM EXTENSION PRINCIPLE
[mathematics] Let T be a continuous
Archimedean t-norm with strictly convex,
twice differentiable additive generator. Then
the T-sum of any linear fuzzy numbers is a
linear fuzzy number if and only if the t-norm
T is a Yager's t-norm T_s^Y for some $s > 1$.[**]

**TRIANGULAR NORM EXTENSION PRINCIPLE (T-
NORM EXTENSION PRINCIPLE)** [mathematics]
(1) There is an associative binary operation
on [0;1] that is commutative, nondecreasing
in each place, and such that $T(a_1 = a$ for all
$a_2)$ [0;1]. (2) The use of fuzzy numbers for calculation with imprecisely known quantities.
Fuzzy numbers are combined using extended
arithmetic operations developed using the extension principle. When a general t norm is
used for the intersection operator, a general
t-norm extension principle is obtained. The
result is that convergence to a crisp set is
obtained for all Archimedean t-norm intersection operators. Proved by deriving it from
a similar result for the law of large numbers
for dependence bounds. Dependence bounds
arise in probabilistic arithmetic when nothing is known of the joint distribution of two

*Masberg, Ul. (2005), *Multi-Axle Direct Drive*,
German Patent Application, p. 8.
[†]Walker, P. M. B. (1999), *Chambers Dictionary
of Science and Technology*, Chambers, New York,
p. 1191.
[‡]Ripka, P., and Tipek, A. (2007) Modern Sensors
Handbook. Wiley ISIE, p. 334.
[§]Tonndorf, J. (1979), A rational approach to the
traveling wave phenomenon, *Am. J. Otolaryngol.*
1(1), 83–93.

[¶]Flescher, J. (1966), *Father, Mother and Child*, D.
T. R. B. Editions, Oxford, UK.
[‖]Ziff, R. M. (2006), Generalized cell-dual-cell trans-
formation and exact thresholds for percolation,
Phys. Rev. E Statist. Nonlinear Soft Matt. Phys.
73(1 Pt. 2), 016134.
**Kolesarova, A. (1995) Triangular Norm-based
addition of linear fuzzy numbers, *Tatra Mountains
Math. Publ.* **6**, 75–81.

random variables apart from the marginal distributions. The bridge used to connect dependence bounds with fuzzy number operations gives a probabilistic interpretation of fuzzy number combination.[*]

TRICHROMATIC PRINCIPLE [physics] Colors are recorded through three color-sensitive channels: red, green, and blue. (2) Color stimuli can be matched by additive mixtures of three fixed primary stimuli. The principle comes from the experimental laws of color matching, and has been confirmed by the presence of three different color-sensitive receptor cells in the human eye. Its consequence is that all color stimuli can be represented as linear combinations of three primary stimuli, which can be seen as coordinates that span a 3D tristimulus space.[†]

TTS PRINCIPLE [physics] See TIME–TEMPERATURE SUPERPOSITION PRINCIPLE.

TUMOR-INDUCING PRINCIPLE (TIP) [genetics, medicine] (1) Consists of four phases: (a) hazard identification, (b) exposure assessment, (c) hazard assessment or dose–response assessment, and (d) risk characterization. (2) Suggests that some DNA sequences of bacterial origin are included in tumorous cells of tissue cultures. (3) Two sets of genes and controlling opine synthesis are transferred from *Agrobacterium tumefaciens* to the plants.[‡]

TUNING FORK PRINCIPLE [physics] Response of an object or a system that vibrates in step (synchrony) with an externally applied vibration. Acoustical resonance is the vibration induced in a string of a given pitch when a note of the same pitch is produced nearby, in the sound box of an instrument such as a guitar, or in the mouth or nasal cavity when speaking. Mechanical resonance, such as that produced in a bridge by wind or by marching soldiers, can eventually produce swings wide enough to cause the bridge's destruction. Resonance in frequency-sensitive electric circuits makes it possible for certain communication devices to accept signals of some frequencies while rejecting others. Magnetic resonance occurs when electrons or atomic nuclei respond to the application of magnetic fields by emitting or absorbing electromagnetic radiation.[§] See also SUPERPOSITION PRINCIPLE.

TUNING PRINCIPLE [pharmacology] In noradrenergic systems is particularly important for the formation of associations and neural plasticity (interference control).[¶]

TURINICI PRINCIPLE [mathematics] (Mihai Turinici) Monotone principle of fixed points.[‖] See also AXIOM OF CHOICE PRINCIPLE; BRÉZIS–BROWDER PRINCIPLE.

TURNER'S COMPLEXITY PRINCIPLE [mathematics] (Jorma Johannes Rissanen, b. 10/20/1932, Finland, information theorist) See RISSANEN'S MINIMUM DESCRIPTION-LENGTH PRINCIPLE.

20/80 PRINCIPLE [mathematics] (Vilfredo Pareto, Italian economist) See 80/20 PRINCIPLE.

TWO-INVOLUTION PRINCIPLE [mathematics] If $M = M^+ \dot\cup M^-$ is a disjoint decomposition of the finite set M, and $\sigma, \tau \in S_M$ are sign-reversing involutions such that $M\sigma, M_\tau \subseteq M^+$, then each orbit of $G := \langle \sigma, \tau \rangle$ either consists of a fixed point of G alone, or it contains exactly one fixed point of σ and exactly one fixed point of τ, or it

---

[*]Williamson, R. C. (1991), The law of large numbers for fuzzy variables under a general triangular norm extension principle, *Fuzzy Sets Syst.* **41**(1), 55–81.

[†]Perry, J. W. (1928), Principles of colorimetry and practical color measurement, *Eidgenoss. Materialprufungsanstalt E. T. H. Zurich, Ber.*, no. 38 (Diskussionsber. no. 15), pp. 49–56.

[‡]Braun, A. C. (1951), Recovery of tumor cells from effects of the tumor-inducing principle in crown gall, *Science* **113**(2945), 651–653.

[§]Crowell, B. (2000), *Vibrations and Waves*. Light and Matter.

[¶]Oades, R. D. (1985), The role of noradrenaline in tuning and dopamine in switching between signals in the CNS, *Neurosci. Biobehav. Rev.* **9**(2), 261–282.

[‖]Maday, Y. and Turinici, G. (2003), Parallel in time algorithms for quantum control: Parareal time discretization scheme, *Int. J. Quantum Chem.* **93**(3), 223–228.

contains a fixed point of neither σ nor τ. This means in particular that there is a canonical bijection between $M\sigma$ and M_τ, namely, the γ that maps $m \in M\sigma$ onto the fixed point of τ that lies in the orbit $G(m)$ of m.[*] See GARSIA–MILNE INVOLUTION PRINCIPLE.

TWO-POINT CONTROL PRINCIPLE [physics] (1) Two control electrodes and one earth electrode. (2) The measuring electrode and container wall (or ground probe) form a capacitor with the material as the dielectric. A change in material level in the container causes a change in the capacitance.[†]

TWO-POINT PRINCIPLE [physics, psychology] (1) The force of attraction or repulsion between two-point electric charges is directly proportional to the product of the charges and inversely proportional to the square of the distance between them. The law also holds for magnetic poles.[‡] (2) The minimum distance on the skin between two point stimuli at which they are perceived as separate stimulus points rather than as a single stimulus point.[§] See also POSITIONING PRINCIPLE.

TWO-POLE COMPLEMENTARY PRINCIPLE [physics] (Niels Bohr, 1885–1962, Danish physicist) (1) Allows the possibility of accommodating widely divergent human experiences in an underlying harmony, and bringing to light newer prospects and ethical views for the exploration and mitigation of human suffering. (2) It is well known that the electron is a particle. It is equally well known that the electron is also a wave. The wave and particle natures are flagrantly opposite because a certain thing cannot be a particle (i.e., substance confined to a very small volume) and a wave (i.e., a field spread out over a large space) at the same time, but the electron exhibits both, although not simultaneously, as the two natures are mutually exclusive. The wave–particle duality of the electron presents a most familiar example of complementarity of opposites in the domain of physics.[¶]

TYPE II CELL STRETCH-ORGANIZING PRINCIPLE [biology]   The main types of collagen in connective tissues are Types II, III, V, and IX. Type II is the collagen of skin and bone and, by far, the most abundant in the body. Type II is found in the cartilage. Type III is found in skin, blood vessels, and internal organs. Type V is found in bone, skin, tendons, ligaments, and cornea. Types IV and VIII are network-forming collagens that polymerize to form the sheetlike network basal laminae and anchoring fibril beneath stratified squamous epithelia.[‖] See also PRINCIPLE OF ORGANIZING CELL STRUCTURE; PRINCIPLE OF PHYSICAL TIMING.

TYPED RESOLUTION PRINCIPLE [mathematics] (1) A typed language as an ordinary many-sorted first-order logic expanded with a set of special constructs. It is motivated by the observation that in real-world problems reasoning pertains to two kinds of problems: (a) reasoning about types and the type structure of the problem domain and (b) general reasoning associated with the kernel of the problem. (2) One primitive service might be to implement an identity (or sanity) check on

[*]Kerber, Adalbert (2005) Applied Finite Group Actions (Algorithms and Combinatorics) Springer p. 77.

[†]Sapcon Intruments Pvt. Ltd. (n.d.), *Capacitance Continuance Level Indicator* (http://www.sapconinstruments.com/pdf/Products3_ContinuousLevelIndicator(ILC_Series).pdf).

[‡]Deverell, M. H. and Whimster, W. F. (1989), A method of image registration for three-dimensional reconstruction of microscopic structures using an IBAS 2000 image analysis system, *Pathol. Res. Pract.* **185**(5), 602–605.

[§]Jiao, Y., Stillinger, F. H., and Torquato, S. (2007), Modeling heterogeneous materials via two-point correlation functions: basic principles, *Phys. Rev. E Statist. Nonlinear Soft Matter Phys.* **76**(3 Pt.1), 031110.

[¶]Yang, P. and Guo, M. (1989), Interaction of some non-platinum metal anticancer complexes with nucleotides and DNA and the two-pole complementary principle (TPCP) arising therefrom, *Metal-Based Drugs* **5**(1), 41–48.

[‖]Koichi, N. (2006), Mechanotransduction and cellular response—a challenge toward development of mechano-pharmacology, *J. Pharm. Soc. Jpn.* **126**(8), 565–577.

a given resolution data element.* (3) Permits a separation of type reasoning from general reasoning even when the background typing theory shares the same language constructs with the kernel formulation. Such a typing theory is required for an accurate formulation of the type structure of a computer program that contains partial functions and predicate subtypes; is also useful for efficiently proving certain theorems from mathematics and logic by typed (sorted) formulation and deduction. (4) Consists of two special forms of binary resolution. One is the binary resolution of two typed clauses on their kernels, called *kernel resolution*. The other is the binary resolution on the type restriction of a typed clause and the head of a typing rule, called *TP resolution*. Kernel resolution is used to deduce consequences from the general relations contained in the kernel formulation. TP resolution is used to carry out type reasoning: type checking and TP deduction. Type checking detects those consequences of kernel resolution whose type restrictions cannot be satisfied in the given typing theory.[†]

*Hammond, T. (1999), *DOI and Identity* (http://www.doi.org/mail-archive/ref-link/msg00058.html).

[†]Wang, T. C. (1995), A typed resolution principle for deduction with conditional typing theory, *Artif. Intell.* **75**(2), 161–194.

U

U(2) PRINCIPLE [physics] (1) The mathematical groups U(2) and SU(2) are related to each other. This would be an example of how group theory is applied to particle and nuclear physics: the group-theoretic structure associated with U(2) [or SU(2), as appropriate] implies relationships among transistion amplitudes, which, in turn, implies "selection rules" indicating which transitions between different states (resonances) are forbidden and which transitions are allowed. (2) Selection of rules to predict candidates for nuclear molecular resonance.* See also PAULI PRINCIPLE.

ULTRAFILTER PRINCIPLE [mathematics] (1) Every filter on a nonempty set is included in some ultrafilter.[†] (2) A prime ideal theorem guarantees the existence of certain types of subsets in a given abstract algebra. A common example is the Boolean prime ideal theorem, which states that ideals in a Boolean algebra can be extended to prime ideals. A filter on a set X is a collection of nonempty subsets of X that is closed under finite intersection and under superset. An ultrafilter is a maximal filter. Every filter on a set X is a subset of some ultrafilter on X (a maximal filter of nonempty subsets of X).[‡]

ULTRASONIC MEASUREMENT PRINCIPLE [physics] See HEISENBERG SCATTER PRINCIPLE; DOPPLER PRINCIPLE.

ULTRASOUND DOPPLER PRINCIPLE [physics] See DOPPLER PRINCIPLE.

UNCERTAINTY PRINCIPLE [mathematics, physics] (Werner Heisenberg) (1) There is a fundamental limit to the precision with which a position coordinate of a particle and its momentum in that direction can be known simultaneously. Also, there is a fundamental limit to knowledge of the energy of a particle when it is measured for a finite time. In both statements, the product of the uncertainties in the measurements of the two quantities involved must be greater than $h/2p_i$, where h is Planck's constant.[§] (2) Quantum mechanics in terms of matrices relation whereby, if one simultaneously measures values of two canonically conjugate variables, such as position and momentum, the product of the uncertainties of their measured values cannot be less than approximately Planck's constant divided by $2p_i$.[¶] (3) Selection of a probability distribution by the principle of maximum entropy (MaXent). The material is made of constituents, for instance, given crystal orientations. Each constituent itself consists of a large number of elementary constituents. The relevant probability is the volume fraction of the elementary constituents that belong to a given constituent and undergo a given stimulus.[‖] (4) A distribution-free method for estimating the quantile function of a nonnegative random variable using the principle of maximum entropy (MaXent) subject to constraints specified in terms of the probability-weighted

*Donth, E. J. (1991), U(2) program. Part 2. First order calculation of fine structure constant and Weinberg angle, *Wissenschaf. Z. Tech. Hochschule Carl Schorlemmer Leuna-Merseburg* **33**(2), 265–279.
[†]Rav, Y. (1976), The ultrafilter principle implies that the projective limit of compact Hausdorff spaces is nonempty, *Bull. Acad. Polon. Sci. S'er. Sci. Math. Astronom. Phys.* **24**(8), 559–562.
[‡]Bergstrom, J. (1978), Ultrafiltration without dialysis for removal of fluid and solutes in uremia, *Clin. Nephrol.* **9**(4), 156-164.

[§]Walker, P. M. B. (1999), *Chambers Dictionary of Science and Technology*, Chambers, New York, p.1214.
[¶]Szabo, R. J. (2003), Quantum field theory on noncommutative spaces, *Phys. Rept.* **378**, 207–299.
[‖]Arminjon, M. and Imbault, D. (2000), Maximum entropy principle and texture formation, *Z. Angew. Math. Mech.* **80** (Suppl. 1° 1), 13–16.

moments estimated from observed data.* (4) Used for estimating probability density function under specified moment constraints. (5) Measure of absence of information about a situation; uncertainty; transformations between measure spaces; expresses amount of disorder inherent or produced. (6) Measure of disorder of a system equal to the Boltzmann constant; the natural log action of the number of microscopic states corresponding to the thermodynamic state of the system.† (7) Intercultural communication is a process whereby people from different cultures constantly search for knowledge of how to interact with one another against the background of uncertainty.‡ (8) New technology must verify that it does not damage, either directly or indirectly, the environment. Burden of proof falls on the perpetrator not the victim. (9) Continuous-time versus reaction events number in stochastic kinetics.§ See HEISENBERG UNCERTAINTY PRINCIPLE; Also referred to as *Maximum Information Entropy Principle; Extended Maximum Entropy Principle; Modification of the Extended Maximum Entropy Principle; Principle of Maximum Entropy; Principle of Maximum nonadditive Entropy; Principle of Operational Compatibility; Frieden Soffer Principle; Jaynes' Maximum Entropy Principle; Extremal Principle of Information; Minimax Entropy Principle; Minimum Entropy Principle; Entropy Maximum Principle; Inference Principle; Pprinciple of Observational Compatibility; Gibbs Principle; Kullback's Minimum Cross-entropy Principle; Minimum Relative Entropy Principle; Measurement Principle; Interdeterminacy Principle; Jaynes' Principle; Momentum Position Uncertainty Principle; energy–time Uncertainty Principle; Precaution Principle;*

Caution Principle; Principles of Uncertainty; Quantum Principle of Uncertainty; Principle of Intercultural Communication; Punctuation Principle; Performativity Principle; Positionality Principle; Commensurability Principle; Continuum Principle; Pendulum Principle; Transaction Principle; Synergy Principle; Sustainability Principle.

UNCERTAINTY THRESHOLD PRINCIPLE [economics]   If the parameter uncertainties exceed a threshold, then optimal long-term stabilization policies do not exist. If the model uncertainly, as quantified by an algebraic threshold function dependent upon the means and standard deviations of the white parameters, increases over a certain value, then the optimal control problem does not have a solutio as the terminal time goes to infinity.¶ See also UNCERTAINTY PRINCIPLE.

UNIFIED PRINCIPLE [engineering, physics, mathematics]   (Werner Heisenberg) Relation whereby, if one simultaneously measures values of two canonically conjugate variables, such as position and momentum, the product of the uncertainties of their measured values cannot be less than approximately Planck's constant divided by $2p_i$.‖ See also HEISENBERG UNCERTAINTY PRINCIPLE.

UNIFIED PRINCIPLE OF ACCUMULATED ADVANTAGES [mathematics]   The translation, T, is directly proportional to the strategic advantage, s (function of spatial, temporal, or geometric coordinates), and the intrinsic advantage, q (function of the qualities or attributes of the actor or actors). Mathematically, it is expressed by the so-called Fundamental Equation of the Unified Scientometric Model

*Pande, M. D. (2000), Direct estimation of quantile functions using the maximum entropy principle, *Struct. Safety* **22**(1), 61–79.

†Wirt, A. and Patterson, B. D. (1993), The measure of order and disorder in the distribution of species in fragmented habitat, *Oecologia* **96**(3), 373–382.

‡Klyukanov, I. E. (n.d.), *Principles of Intercultural Communication* (http://inlawsvsoutlaws. weebly.com/intercultural-communication.html).

§Vlad, M. O. (1993), Continuous time versus reaction events number in stochastic kinetics. A chemical uncertainty principle, *Reaction Kinet. Catal. Lett.* **49**(1), 161–166.

¶Athans, M., Ku, R., and Gershwin, S.B. (1976). The Uncertainty Threshold Principle, Proc. IEEE Conference on Decision and Control, Clearwater, Fla., Dec. 1976.

‖Bailon-Moreno, R., Jurado-Alameda, E., Ruiz-Banos, R., and Courtial, J. P. (2005), The unified scientometric model. Fractality and transfractality, *Scientometrics* **63**(2), 231–257.

$T = ksq$.* See also SPATIAL PRINCIPLE; PRIN-CIPLE OF TRANSLATION QUANTITATIVITY; PRIN-CIPLE OF COMPOSITION OF THE TRANSLATIONS; PRINCIPLE OF THE CENTRE-PERIPHERY; PRINCI-PLE OF NUCLEATION.

UNIFIED REINFORCEMENT PRINCIPLE [psychology] Provides a unified account of conditioning in that learning in both the clas-sical (Pavlovian) and operant (instrumental) procedures is interpreted as the outcome of a common set of biobehavioral processes.[†] See also PRINCIPLE OF REINFORCEMENT.

UNIFIED VARIATIONAL PRINCIPLE [engi-neering] (1) Fundamental equations of the medium; remaining equations are con-straints. The constraints are incorporated into the variational principle, and a unified variational principle is derived.[‡]

UNIFORM ACCESS PRINCIPLE [mathematics] (Bertrand Meyer, b. 1950, France, creator of Eiffel programming language) (1) "All services offered by a module should be avail-able through a uniform notation, which does not betray whether they are implemented through storage or through computation." (2) Basic rules of object-oriented program-ming: that the basic dynamic structure is the object, represented statically by its generat-ing class, and that all accesses to the object should be through one of the features of that class.[§]

UNIFORM BOUNDEDNESS PRINCIPLE [space, mathematics, linguistics] (1) A family of pointwise bounded, real-valued continuous functions on a complete metric space X is uni-formly bounded on some open subset of X.[¶] (2) A uniform boundedness principle (UBP) is a result that asserts that a pointwise bounded sequence or family of functions defined on a common domain is uniformly bounded on some family of subsets of the domain space. For example, the UBP for normed linear spaces (NLSs) states that if a sequence of continuous linear operators from a Banach space into an NLS is pointwise bounded, then the sequence is uniformly bounded on the unit ball of the domain space and, hence, is uniformly bounded on the family of bounded subsets of the domain space.[‖] See also BOUND-ARY PRINCIPLE.

UNIFORMITY PRINCIPLE [mathematics] (Ed-win Powell Hubble, 1889–1953, American astronomer) The laws of physics are not dif-ferent from the laws of physics anywhere else in the Universe.[**]

Let X be closed convex symmetric set in R^v with center $x^0 = 0$. Then it follows that

$$\min_{JEF} \text{Prod}\ \{X_f \in X\} = \text{Prod}\ \{X_u \in X\}$$

UNLEARNED PRINCIPLE [physics] Linear re-lation between elbow and shoulder dynamic torque in natural pointing movements in the sagittal plane. The present study investigates if the process of learning to reach involves dis-covering this linearity principle.[††] See LINEAR COVARIANCE PRINCIPLE.

*Bailon-Moreno, R. (2003) Ingeniera del conocimiento y vigilancia tecnologica aplicada a la investigaction en el campo de los tensioactivos. Desarrollo de un modelo ciencimetrico unificado. Ph.D. Thesis. Universidad de Granada.

[†]Donahoe, J. W. and Palmer, D. C. (1989), The interpretation of complex behavior: Some reactions to parallel distributed processing, *J. Exp. Anal. Behav.* **51**, 399–416.

[‡]Fu, Y. H. (1997), Stream function wave derived by unified variational principle of water gravity wave, *China Ocean Eng.* **1**, 187.

[§]Meyer, B. (2006), Business plus pleasure, in *Eiffel: Analysis, Design and Programming Language*, 2nd ed., Standard ECMA-367.

[¶]Dickmeis, W. and Nessel, R. J. (1981), A uniform boundedness principle with rates and an applica-tion to linear processes. Functional analysis and approximation (Oberwolfach, 1980), *Int. Ser. Nu-mer. Math.* **60**, 311–322.

[‖]Staszewski, R. (1990), Murphy's law of limiting dilution cloning, *Statist. Med.* **9**(4), 457–461.

[**]Barmish, B. R., and Lagoa, C. M. (1996) The Uniformity Principle: A New Tool for Proba-bilistic Robustness Analysis. Scientific Commons. http://en.scientificcommons.org/43161454

[††]Zaal, F. T. J. M., Daigle, K. Gottlieb, G. L., and Thelen, E. (1999), An unlearned principle for controlling natural movements, *J. Neurophysiol.* **82**(1), 255–259.

UNREALITY PRINCIPLE [psychology] Why and how learners might respond positively to language practice based on television advertisements of the "fantasy" type. It considers how the form and content that reflect the advertisers' original purpose help to make "fantasy" ads appropriate and accessible as input to language practice with learners at a wide range of levels of English proficiency.[*]

UNSUPPORTED AREA PRINCIPLE [engineering] (P. W. Bridgman, Harvard Uni., Nobel Laureate) The pressure in the vessel is transmitted through the plunger to the steel surfaces, which initially are nearly line contacts. Thus, the pressure in the seal surface greatly exceeds the pressure in the vessel because most of the area of the plunger is unsupported.[†] (2) Self-sealing types of ring in which the stress across the jointing faces is automatically maintained at a higher value than the pressure to be sealed are used almost exclusively for chemical process vessels designed for pressures above about 20 MPa (2900 psi). These rings are sometimes known as *unsupported area* or *Bridgman seals*, following Bridgman's development of the unsupported area principle in the 1920s to seal pressures of 1200 MPa

(174,000 psi) in laboratory equipment.[‡] Basis of design of a free piston primary pressure standard gauge. The controlled clearance free piston gauge has since become the primary pressure standards at the National Institute of Standards and Technology.[§] See also CONTROLLED CLEARANCE PRINCIPLE.

UPPER BOUNDEDNESS PRINCIPLE [mathematics] Let $\Omega = R^n$, and let $\Phi(P, \mu)$ be the kernel of the Newtonian potential. According to the upper boundedness principle, if $\Phi(P, \mu)$ is bounded from above by S_{mu}, then it is bounded on Ω also, where μ is such that $S_{mu} \subset K$, with K a compact set, $K \subset \Omega$.[¶] $K \subset \Omega$

USE/DISUSE PRINCIPLE [kinesiology] (Dr. Hans Selye) See PRINCIPLE OF TRAINING.

USUFRUCT PRINCIPLE [geography] (Thomas Jefferson) (1) The right to utilize and enjoy the profits and advantages of something belonging to another as long as the property is not damaged or altered in any way. (2) The concept can be traced as far back as ancient Roman law. The right to make full use of and to profit from a thing without injuring the substance of the thing itself.[‖]

[*]Lynch, A. J. (1985), The "unreality principle": One use of television commercials, *ELT J.* 39(2), 115–120.

[†]Armington, Alton F. (1997), Silica, synthetic quartz crystals, in *Kirk-Othmer Encyclopedia of Chemical Technology*, Wiley, New York.

[‡]Faupel, J. H. and Fisher, F. E. (1982), *Engineering Design*, McGraw Hill, New York, p. 285.

[§]Johnson, D. P. and Newhall, D. H. (n.d.), The piston gage as a precise pressure measuring instrument, Paper 52-11RD-2, *Trans. Am. Soci. Mechanical Engineers*, pp. 75–301.

[¶]Ito, Kiyosi (1993) Encyclopedic dictionary of mathematics. Cambridge (Mass.); London : The MIT Press, p. 1249.

[‖]Laing, A. (1976), Subject comes into my head: Jefferson's usufruct principle, *Nation* **223**, 7.

V

VACUUM PRINCIPLE [physics] (1) Ideally, a region of space completely devoid of any form of matter: (2) A totally empty space. (3) An enclosed region of space in which the pressure has been reduced (below normal atmospheric pressure) sufficiently so that processes occurring within the region are unaffected by the residual matter. (4) The space housing an infinite sea of electrons having negative energy, which is observed normally as empty space. (5) Freeing up the energy attached to all belongings, relationships, habits, and obligations and allowing room and energy.[*]

VALID ANALYTICAL MEASUREMENTS PRINCIPLES [mathematics] (1) Analytical measurements should be made using methods and equipment treated to ensure that they are fit for their purpose. (2) The specificity, detection limit, quantitation limit, working range, accuracy, precision, recovery, and ruggedness are listed as typical parameters used in method validation.[†] (3) Simple waterstop profiles based on dumbbells are cast into the edges of adjacent concrete panels that act as baffles. In the event of joints opening as drying shrinkage or other movement occurs, the edge bulbs of the profile act as anchors. These induce tensions across the waterstop, resulting in a sealing effect at the inner faces of the edge bulbs.[‡] See also TORTUOUS PATH PRINCIPLE; PRINCIPLE OF WATERSTOP FUNCTION.

VAN'T HOFF'S PRINCIPLE [chemistry] (Jacobus Henricus ran't Hoff, 1852–1911, Dutch chemist) The osmotic pressure of a dilute solution is equal to the pressure that the dissolved substance would exert if it were in the gaseous state and occupied the same volume as the solution at the same temperature.[§]

VARADHAN'S CONTRACTION PRINCIPLE [mathematics] (Sathyamangalam Ranga Iyengar Srinivasa Varadhan FRS, 1940-, Indian-American mathematician) (1) If the first class of measures obeys the large deviation principle then the same is true for the second one.[¶] (2) A sequence of approximation of the measurable Itô map by continuous maps uniformly converging on the level sets of the initial good rate function and that the resulting sequence of family of laws are exponentially good approximations of the laws of he solutions.[‖]

VARIABLE-ORIFICE PRINCIPLE [engineering] Substance pushes against an obstacle, past a sensor, allowing alterations in flow.[**] See also ORIFICE PRINCIPLE.

VARIABLE REINFORCEMENT PRINCIPLE [psychology] To improve or increase a

[*]Matsuyama, Y. (2006), Industrial instrumentation: Principle and application. 8. Pressure measurement, *Kagaku Kogaku* **70**(6), 303–306.

[†]Plesch, R. (1973), Examination of Poisson-distributed measuring values. *Fresenius' Zeitschrift fuer Analytische Chemie*, **265**(2), v.265 n.2, p. 114–121.

[‡]Anon. (2007), *PARCHEM Concrete Repair Flooring Jointing Systems Waterproofing*, Technical Data Sheet ABN 80 069 961 968 (www. parchem.com.au).

[§]Walker, P. M. B. (1999), *Chambers Dictionary of Science and Technology*, Chambers, New York, p. 1228.

[¶]Varadhan, S.R.S. (1966) Asymptotic probabilities and differential equations. Commun. Pure Appl. Math. 19, 261–286

[‖]Gautier, E. (2004) Uniform large deviations for the nonlinear Schroedinger equation with multiplicative noise. http://arxiv.org/pdf/math.AP/0412319.

[**]Fox, L. S. and Westaway, S. F. (1988), Performance of a venturi eductor as a feeder in a pneumatic conveying system, *Powder Bulk Eng.* **2**(3), 33–36.

child's performance of a certain activity, provide the child with an intermittent reward.[*]

VARIABLE-TIME POSITION CHANNEL ASSIGNMENT PRINCIPLE [communications] See INFORMATION-PROCESSING PRINCIPLE.

VARIATIONAL MINIMAL PRINCIPLE [physics] Methods based on the principle that, among all possible configurations or histories of a physical system, the system realizes the one that minimizes some specified quantity.[†] See also ENTROPY PRINCIPLE; MAXIMUM ENTROPY PRINCIPLE; MINIMUM CROSS-ENTROPY PRINCIPLE; PRINCIPLE OF MINIMUM CROSS-ENTROPY; GENERALIZED MINIMAL PRINCIPLE; MOMENT-PRESERVING PRINCIPLE; PRINCIPLE OF VARIANCE MINIMIZATION.

VARIATIONAL PRINCIPLE [mathematics, mechanics] (1) A technique for solving boundary-value problems that is applicable when the given problem can be rephrased as a minimization problem.[‡] (2) A way of expressing fundamental laws of physics, particularly mechanics, in terms of the form of function that minimizes or maximizes an integral rather than in terms of differential equations.[§] (3) Any of various principles formulated to account for the observation that when a physical process takes place, certain important quantities tend to have little variation from a minimum or maximum value.[¶] See also HAMILTONS PRINCIPLE.

VARIATIONAL PRINCIPLE FOR TIME-DEPENDENT HARTREE–FOCK EQUATIONS [physics] (1) Two-body correlation should be included not only in the ground-state energy but also in the constraints. (2) Widths of resonant states have an important effect on the pairing properties of nuclei close to the drop line. (3) Energies of excitations are determined essentially by the coordination number of the excited atom rather than the native of the local environment. (4) Statistical exchange potential can thus be given in a universal form and then applied in wave mechanical calculations (the sum of the exchange potential and the electrostatic potential).[‖]

VECTOR EKELAND VARIATIONAL PRINCIPLE See EKELAND'S PRINCIPLE.

VECTOR QUANTIZER DESIGN EQUIDISTORTION PRINCIPLE [mathematics] (1) Approximate solutions of partial differential equations can often be obtained by utilizing the fact the function makes a certain integral an extremum satisfies a partial differential equation. (2) Technique for solving boundary-value problems that is applicable when the given problem can be rephrased as a minimization problem. (3) Pressure pump and refrigeraton unit in moistureser line downstream of pump operating on jet pump.[**]

VENTURI PRINCIPLE [physics] (Giovanni B. Venturi, 1746—1822, Italian physicist) (1) Specially designed constriction in a pipe causing a pressure drop when fluid flows through it. (2) A pipe containing a fluid whose level will change as a function of the speed of fluid flow past its opening in accord with Bernoulli's law.[††]

[*]Huitt, W. (1994), Principles for using behavior modification, *Educational Psychology Interactive*, Valdosta State Univ., Valdosta, GA (http://chiron.valdosta.edu/whuitt/col/behsys/behmod.html).

[†]Miller R., DeTitta, G. T., Jones R., Langs, D. A., Weeks, C. M., and Hauptman H. A. (1993), On the application of the minimal principle to solve unknown structures, *Science* **259**(5100), 1430–1433.

[‡]Nohara, A. (n.d.), Variational principle for high temperature plasticity in alloy type solid solutions, *J. Phys. Soc. Jpn.* **65**, 472.

[§]Chen, W. J. et al. (1996), Variational principle with nonlinear complementarity for three dimensional contact problems and its numerical method, *Sci. China Ser. a Math. Phys. Astron.* **39**, 528.

[¶]Blouquin, R. and Joulin, G. (1996), On a variational principle for reaction/radiation/conduction equilibria, *Combun Sci. Technol.* **1**, 375.

[‖]Kull, H.-J. and Pfirsch, D. (2000), Generalized variational principle for the time-dependent Hartree-Fock equations for a Slater determinant, *Phys. Rev. E Statist. Phys. Plasmas Fluids Related Interdisc. Topics* **61**(5-B), 5940–5947.

[**]Jia, L. and Zha Hongyuan, Z. (2002), Simultaneous classification and feature clustering using discriminant vector quantization with applications to microarray data analysis, *Proc. IEEE Computer Society Bioinformatics Conf.* pp. 246–255.

[††]Schafer, R., Jungmann, M., and Werthschutzky, R. (2000), Self diagnosis in flow measurement by combining venturi and vortex-principle, *Technisches Messen* **67**(9), 361–366.

VERGENT INDUCTION PRINCIPLE [mathematics]   See INDUCTION PRINCIPLE.

VERIFICATION   PRINCIPLE [mathematics] The statement is literally meaningful if and only if it is either analytically or empirically verifiable.*

VERTICAL   MAGNETIZATION   PRINCIPLE [mathematics]   Special recording media in the form of rigid magnetic storage disks, flexible individual disks (floppy disks), or magnetic tapes. The information density in the recording medium can be increased advantageously.† Also known as *principle of perpendicular magnetization*.

VERTICAL   ROTATING   PRINCIPLE [engineering] (George J. M. Darrieus)   (1) Originally consisting of two or more vertical sails or paddles that are blown around their vertical axis by the wind. A Persian vertical axis design dating from about 200 BC channeled the wind with walls toward the paddles of the machine, which were then pushed around by the wind. (2) Consists of two or more flexible airfoil blades, which are attached to both the top and bottom of a rotating vertical shaft, giving the machine the appearance of a giant egg whisk. The wind blowing over the airfoil contours of the blade create aerodynamic lift, which actually pulls along the blades.‡ See also INERTIAL SLIDER PRINCIPLE; LINE-SCANNING PRINCIPLE.

VICIOUS   CIRCLE   PRINCIPLE. [philosophy, mathematics] (Bertrand Russell, 1872–1970, philosopher, logician, mathematician, historian, socialist, pacifist and social critic) (1) 'Whatever involves all of a collection must not be one of that collection'; or 'If, provided a certain collection had a total, it would have members only definable in terms

of that total, then the said collection has no total.'§ (2) It is a matter of the set E of all numbers that can be defined using a finite number of words, and of a certain number N defined from E using a "diagonal procedure." By its very definition, N is distinct from each element of E and does not therefore belong to E. And yet, it does belong to E because this definition has itself been formulated using a finite number of words: hence the contradiction. E is defined as the set of all those numbers which one can define using a finite number of words without introducing the notion of the set E itself. Without this, the definition of E would contain a vicious circle.¶ (3) Paradox reference to troublesome sets (such as the set of all sets that are not members of themselves) could be avoided by arranging all sentences into a hierarchy (beginning with sentences about individuals at the lowest level, sentences about sets of individuals at the next lowest level, sentences about sets of sets of individuals at the next lowest level, etc.). Explains why the unrestricted comprehension axiom fails: propositional functions, such as the function "x is a set," should not be applied to themselves since self-application would involve a vicious circle. It is possible to refer to a collection of objects for which a given condition (or predicate) holds only if they are all at the same level or of the same "type."‖

VIRTUAL DISPLACEMENT PRINCIPLE [mathematics] (Eric Reissner)   The integral form among mixed forms can be formed in combined expressions according to need. Using

---

*Ayer, A. J. (1946), *Language, Truth and Logic*, 2nd ed., Dover, New York.
†Lu, M., Liang, Y., Wang, L., and Li, L. (1993), A new production method for vertical magnetic recording dielectric thin film, *Cailiao Kexue Jinzhan* **7**(1), 38–45.
‡Drerup, B. (1984), Principles of measurement of vertebral rotation from frontal projections of the pedicles, *J. Biomech.* **17**(12), 923–935.

§Whitehead, A. N. and Russell, B. A. W., (1910) Principia Mathematica, v1, p.1
¶De Rouilhan, P. (1992) Philippe Russell and the Vicious Circle Principle. Philosophical Studies: An International Journal for Philosophy in the Analytic Tradition, v65, (1/2) American Philosophical Association Pacific Division Meetting 1991, pp. 169–182.
‖Paukov V. S., Ugriumov A. I., and Khritinin D. F. (1983), Morphology of the brain and heart in alcoholic psychoses, *Zh. Nevropatologii i Psikhiatrii Imeni S. S. Korsakova* (Moscow, 1952) **83**(7), 1061–1066.

by combined expressions, virtual work principles and all kinds of variation principles can be proved.* See DISPLACEMENT PRINCIPLE.

VIRTUAL DISSIPATION PRINCIPLE [physics] See DISSIPATION PRINCIPLE.

VIRTUAL FORCE PRINCIPLE [mathematics]

VIRTUAL VELOCITY INCREMENT PRINCIPLE [mathematics] The incremental finite-element equations for geometric nonlinear analysis of piezoelectric smart structures are developed using a total Lagrange approach by emplaying virtual velocity incremental variational principles. Geometrically non-linear transient vibration response and control of plates with piezoelectric patches subjected to pulse loads with active damping on the plates by coupling a self-sensing and negative velocity feedback algorithm in a closed control loop. The numerical results show that piezoelectric actuators can introduce significant damping and suppress transient vibration effectively.[†]

VIRTUAL VELOCITY PRINCIPLE [mathematics] See VIRTUAL VELOCITY INCREMENT PRINCIPLE.

VIRTUAL WORK PRINCIPLE [mathematics, labor] (1) Strives to evaluate relative utilities of simple and mixed parameters that can be used to describe outcome. If a system in static equilibrium undergoes an infinitesimal displacement consistent with the constraints on the system, then the total work done on the system is zero. The total work done by all forces acting on a system in static equilibrium is zero for any infinitesimal displacement from equilibrium that is consistent with the constraints of the system. (2) Concept of work; for example, "in the determination of the equilibrium

state of a static system, potential energy was introduced instead of the workfunction . . . ." The latter represents the input of work to the system by the applied forces. Clearly, in practice this is a far more difficult measure to establish than the output of work from the system specified in terms of a well-defined change of the system configuration.[‡] (3) The correlation between element forces and corresponding deformations is derived from the weighted integral of the constitutive force—deformation relation. The constitutive force–deformation relation of the control sections of the beam and of the end rotational springs has the form of a differential relation that is derived by extending the simple standard solid model according to the endochronic theory.[§] (4) The force distribution within the element is based on interpolation functions that satisfy equilibrium. The relation between element forces and corresponding deformations is derived from the weighted integral of the constitutive force–deformation relation.[¶] See also DIVIDE-AND-CONQUER PRINCIPLE.

VITAL PRINCIPLE [pharmacology] Bodily functions are due to a vitality existing in all living creatures going back to at least 384–322 BC. While vitalist ideas have been commonplace in traditional medicine, attempts to construct workable scientific models date from the 1600s, when it was argued that matter existed in two radically different forms, observable by their behavior with regard to heat. These two forms of matter were termed *organic* and *inorganic*. Inorganic matter could be melted, but could also be restored to its former condition by removing the heat. Organic compounds "cooked" when heated, transforming into

*Luo, J.-H, Liu, G.-D., and Shang, S.-P. (2003), Research on a systematic methodology for theory of elasticity, *Appl. Math. Mech.* **24**(7), (Shanghai Univ.).

[†]Gao, J. X., and Shen, Y. P. (2003), Active control of geometrically nonlinear transient vibration of composite plates with piezoelectric actuators, *J. Sound Vibr.* **264**(4), 911–928.

[‡]Franksen, O. I. (1980), The virtual work principle—a unifying systems concept, in *Structures and Operations in Engineering and Management Systems*, p. 64. Academic Press.

[§]Kamegaya, H. and Asano, N. (1982), A finite element analysis of steady viscous flow based on a hybrid type of virtual work principle, *Bull. JSME* **25**(203), 774–781.

[¶]Wang, Q. A. (2007), From virtual work principle to maximum entropy for nonequilibrium system, Los Alamos National Laboratory, *Preprint Archive, Condensed Matter*.

new forms that could not be restored to the original. It was argued that the essential difference between the two forms of matter was the "vital force," present only in organic material.[*]

VOLLEY PRINCIPLE [pharmacology] Attempts to account for the maximum theoretical limit for the neuronal firing of action potentials and the small timescales over which sound discrimination must occur. In this theory, the organ of cortices in the cochlear that transduces the sound into action potentials must combine multiple stimuli along the cochlear nerve within a volley in order to encode high-frequency auditory stimuli.[†]

VOLUME DISPLACEMENT PRINCIPLE [mathematics] (Eric Reissner)   See DISPLACEMENT PRINCIPLE; LE CHATELIER–BRAUN PRINCIPLE; ARCHIMEDE'S PRINCIPLE; CAVALIERI PRINCIPLE.

VONNEUMANN MINIMAX PRINCIPLE [Mathematics] (J. von Neumann, 1903–1957, Hungarian American Mathematician)

THE MINIMAX THEOREM. *For every m ×
n matrix A there are stochastic vectors x**
*and y** such that*

$$\max_x \min_y \{xAy\} = \min_y \max_x \{xAy\} = x^*Ay^*$$

*where the minimum is taken over stochastic
n-vectors y and the maximum over stochastic m-vectors x. See also HASSE PRINCIPLE;
INFIMUM PRINCIPLE*[‡]

VON STERNBERG PRINCIPLE [filmography] (Jonas Sternberg, 1894–1969, Austrian-American Filmographer).   Make a likeness memorable by planning his films not around dialogue but around the performers' "dramatic encounter with light," in insisting that the "dead space" between the camera and subject be filled and enlivened, and above all in seeing every story in terms of "spiritual power" rather than star quality.[§]

VOPĚNKA'S PRINCIPLE [Mathematics] (Petr Vopěnka, May 16, 1935, Prague) (1) For every proper class of binary relations (with set-sized domain), there is one elementarily embeddable into another. Equivalently, for every predicate P and proper class S, there is a nontrivial elementary embedding $j:(V\kappa, \in, P \to (V\lambda, \in, P)$ for some κ and λ in S. A cardinal κ is Vopěnka if and only if Vopěnka's principle holds in $V\kappa$ (allowing arbitrary $S \in V\kappa$ as proper classes).[¶] (2) Vopenka's principle is a compactness principle. Compactness allows the passage from small (finite) to big (infinite). If we now replace "finite" by "being a set" and "infinite" by "being a proper class," we have another interpretation of small and big. Every proper class of L structures contains two structures, A and B, such that A is elementarily embeddable in B. It is shown that Vopěnka's principle is equivalent to each of the following: Every finitely generated logic has a compact cardinal; Every finitely generated logic has a global Hanf number; A logic is finitely generated if it is obtained from first-order logic by the adjunction of finitely many Lindström quantifiers; A logic is M-compact cardinal just in case there is a cardinal K such that the logic is [K, K]-compact.[‖]

VORTEX PRINCIPLE [mathematics]   (1) The motion of a dynamical system in the configuration space at any instant in the possible

[*]Haigh, E. L. (1977), The vital principle of Paul Joseph Barthez: The clash between monism and dualism, *Med. His.* **21**(1), 1–14.

[†]Hanekom, J. J. (2000), What do cochlear implants teach us about the encoding of frequency in the auditory system? *South Afr. J. Commun. Disorders* **47**, 49–56.

[‡]Chandru, Vijay and Rao, M.R., (1999) Part 4. Minimax and Cake. In 175 Years of Linear Programming. Resonance. Springer India, in co-publication with Indian Academy of Sciences v.4, n. 7, p.4–13.

[§]Von Sternberg, J. (1963), Von Sternberg principle, *Proc. Metallurgical Society Conf. On the Background and Beginning of Metallography in Germany*, vol. 27, pp. 163–169.

[¶]Friedman, H. M. (2005) Embedding Axioms http://www.cs.nyu.edu/pipermail/fom/2005-August/009023.html

[‖]Makowsky, J. A. (1985), Vopěnka's principle and compact logics. J. Symbolic Logic 50 no.1, p.42–48.

(compatible with the constraints) variation of the associated kinetic energy is to be distributed into two components: the longitudinal component (which represents the work done by the exterior forces and the local inertia forces in an arbitrary possible displacement), and the transversal component (the flow of the vortex of impulses per unit of time through an element of area, made by the actual and possible displacements.[*] (2) Vortices interact with their environment by stimulating or overpowering other swirls on the verge of birth and development. In the course of vortex generation, numerous forces are exerted on the shedder surface.[†] (3) The aeroelastic phenomenon is known as *control surface reversal* when the lift or rolling rate vanishes at a sufficiently large ratio of flow dynamic pressure to airfoil or wing stiffness.[‡] (4) A leading-edge control surface can be used to counteract control surface reversal, and, a leading-edge control surface can entirely cancel the tendency of the trailing-edge control surface to undergo reversal.[§] See also VORTEX SHEDDING PRINCIPLE.

VORTEX SHEDDING PRINCIPLE [physics] Fluid passing around a "bluff body" will produce downstream vortices, the frequency of which is proportional to the fluid velocity.[¶]

[*]Aržanyh, I. S. (1949), The vortex principle in analytical dynamics, *Dokl. Akad. Nauk SSSR* **65**, 613–616.

[†]Schafer, R., Jungmann, M., and Werthschutzky, R. (2000), Self diagnosis in flow measurement by combining venturi and vortex principle, *Technisches Messen* **67**(9), 361–366.

[‡]Perry, H., Corey, R. C., and Elliott, M. A. (1950), Continuous gasification of pulverized coal with oxygen and steam by the vortex principle, *Trans. ASME* **72** 599–610.

[§]Rafferty, E. H., Kletschka, H. D., Wynyard, M., Larkin, J. T., Smith, L. V., and Cheathem, B. (1968), Artificial heart. I. Application of nonpulsatile force-vortex principle, *Minnesota Med.* **51**(1), 11–16.

[¶]Zhang, H., Huang, Y., and Sun, Z. (2006), A study of mass flow rate measurement based on the vortex shedding principle, *Flow Measure. Instrum.* **17**(1), 29–38.

W

WARD–LEONARD PRINCIPLE [engineering] (Hanry Ward Leonard; 1861–1915, American electrical engineer) (1) Large variable-speed direct current drives with motor generator sets. (2) Extensive maintenance of commutator and brush maintenance.[*]

WARDROP'S PRINCIPLE [mathematics] (John Glen Wardrop 1942, English transport analyst) (1) Concepts related to the idea of Nash equilibrium in game theory. However, in transportation networks, there are many players, making the analysis more difficult than in games with small numbers of players.[†] (2) Network equilibrium models used for the prediction of traffic patterns in transportation networks that are subject to congestion. Wardrop stated two principles that formalize this notion of equilibrium and introduced the alternative behavior postulate of the minimization of the total travel costs.[‡] (3) The journey times in all routes actually used are equal to and less than those that would be experienced by a single vehicle on any unused route. Each user noncooperatively seeks to minimize her/his cost of transportation. The traffic flows that satisfy this principle are usually referred to as *user equilibrium* (UE) flows, since each user chooses the route that is best. Specifically, a user-optimized equilibrium is reached when no user may lower her/his transportation cost

through unilateral action.[§] (4) At equilibrium the average journey time is minimum. This implies that each user behaves cooperatively in choosing her/his own route to ensure the most efficient use of the whole system. Traffic flows satisfying Wardrop's second principle are generally deemed *system-optimal* (SO). Economists argue that this can be achieved with marginal-cost road pricing.[¶] See also PRINCIPLE OF ROUTE CHOICE.

WAVE VARIATIONAL PRINCIPLE [physics] Light propagating in free space. Also concerns what happens inside matter. All the processes of transmission, reflection, and refraction are macroscopic manifestations of scattering effects on the atomic and subatomic levels. (3) When a photon (or light wave) encounters an atom, there are two possibilities: (1) the atom may scatter the light, redirecting it without changing its frequency or energy; or (2) it may absorb the light, using the energy to make a quantum jump into an excited energy state (more precisely, one of its electrons makes the jump). With absorption it is likely that the excitation energy will rapidly be transferred to atomic motion, via collisions, thus producing thermal energy before the atom decays back to the lower energy state reemitting a photon.[‖] (4) Random scattering occurs only for media such as gases where the atoms are far apart and randomly placed at distances far greater than the wavelength

---

[*]Anon. (1915), Home-made charging panel for small batteries, *Electrical World* **66**.

[†]Maugeri, A., Oettli, W., and Schläger, D., (1997), A flexible form of Wardrop's principle for traffic equilibria with side constraints. Equilibrium problems with side constraints. Lagrangian theory and duality, II (Scilla, 1996), *Rend. Circ. Mat. Palermo* (48, Suppl, 2), 185–193.

[‡]Florian, M. and Hearn, D. (1995), Network equilibrium models and algorithms, *Handbook in OR & MS*, Ball, M. O. et al., eds., chap. 6, vol. 8, pp. 485–550.

[§]Inst. Operations Research and the Management Sciences (1999), *OR/MS Today*, Lionheart Publishing. Marietta, GA.

[¶]Namatame, A. (2004), The design of desired collectives with agent-based simulation, *J. Soc. Comple.* **2**(1), 8.

[‖]Sun, Y. and Kouri, D. J. (1991), Scattered wave variational principle for atom-diatom reactive scattering: Hybrid basis set calculations, *Chem. Phys. Lett.* **179**(1–2), 142–146.

---

of the light. In dense, homogeneous media, where the spacing of atoms is much less than the wavelength of the light, it turns out that very little light gets scattered in the backward direction or any direction perpendicular to the propagation of the wave, but most is propagated in the forward direction. This allows light to propagate through dielectrics.* See also SCATTERED WAVE VARIATIONAL PRINCIPLE.

WAŻEWSKI PRINCIPLE [mathematics] (Tadeusz Ważewski, 1896–1972, Polish mathematician) (1) In consideration to problems of eigenvalues and eigenfunctions for certain class of fourth order ordinary differential equation $x = Df(x)^1 w.$[†] (2) A significant tool in the study of asymptotic behavior of solutions of ordinary differential equations (ODEs). A direct extension of this principle to retarded functional differential equations (RFDEs) can be made in a natural way.[‡]

WEAK-AND-STRONG EINSTEINIAN EQUIVALIENCE PRINCIPLE [mathematics] See EQUIVALIENCE PRINCIPLE.

WEAK ANTHROPIC PRINCIPLE (WAP) [astronomy, genetics] (Brandon Carter, b. 1942, Australian theoretical physicist) (1) We must be prepared to realize that our location in the universe is necessarily privileged to the extent of being compatible with our existence as observers. Location is a spacetime position.[§]

*Mielke, S. L., Truhlar, D. G., and Schwenke, D. W. (1991), Improved techniques for outgoing wave variational principle calculations of converged state-to-state transition probabilities for chemical reactions, *J. Chem. Phys.* **95**(8), 5930–5939.
[†]Olech, C. (1998) On the Ważewski equation. Proceedings of the Conference "Topological Methods in Differential Equations and Dynamical Systems" (Kraków-Przegorza□y, 1996). Univ. Iagel. Acta Math. No. 36, 55–64.
[‡]Rybakowski, Krzysztof P., (1980), Ważewski's principle for retarded functional differential equations. J. Differential Equations, v.36, n.1, p. 117–138.
[§]Carter, B. (1974), Large number coincidences and the anthropic principle in cosmology. *Proc. IAU Symp. 63, Confrontation of Cosmological Theories with Observational Data*, Reidel, Dordrech, pp. 291–298.

(2) The observed values of all physical and cosmological quantities are not equally probable, but they assume values restricted by the requirement that there exist sites where carbon-based life can evolve and by the requirements that the universe be old enough for it to have already done so. This definition is restricted to carbon-based life and fundamental physical constants, such as the fine-structure constant, the number of dimensions in the universe, and the cosmological constant.[¶] See also ANTHROPIC PRINCIPLE.

WEAK EQUIVALENCE PRINCIPLE (WEP) [mathematics] All the laws of motion for freely falling particles are the same as in an unaccelerated reference frame.[‖] See EQUIVALENCE PRINCIPLE; STRONG EQUIVALENCE PRINCIPLE; PRINCIPLE OF THE UNIQUENESS OF FREE FALL.

WEAK GALILEAN–NEWTONIAN EQUIVALENCE PRINCIPLE [mathematics] The weak equivalence principle extends to the strong (Einsteinian) equivalence principle: Nonrotating free fall is locally indistinguishable from uniform motion absent gravitation. Linear acceleration relative to an inertial frame in special relativity is locally identical to being at rest in a gravitational field. A local reference frame always exists in which gravitation vanishes. Local Lorentz invariance (absolute velocity does not exist) and position invariance. All local free-fall frames are equivalent. The strong equivalence principle embraces all laws of nature; all reference frames accelerated or not, in a gravitational field or not, rotating or not, anywhere at any time (frame covariance; global diffeomorphism invariance aside from the big bang). The equivalence principle demands that gravitational fields contain a local Minkowski (flat) spacetime reference frame (free fall). Gravitational fields cannot have a stress-energy tensor if free fall exists. If no reference frame makes gravitation locally vanish, spacetime curvature is

[¶]Barrow, J. D. and Tipler, F. J. (1986), *The Anthropic Cosmological Principle*, Oxford Univ. Press.
[‖]Mureika, J. R. (1997), An investigation of equivalence principle violations using solar neutrino oscillations in a constant gravitational potential (Univ. Southern Calif.) *Phys. Rev. D.* 56, (4 p. 2408–2418)

counter-demonstrated as a violated equivalence principle.* See EQUIVALENCE PRINCIPLE; STRONG; EQUIVALENCE PRINCIPLE; PRINCIPLE OF THE UNIQUENESS OF FREE FALL.

WEAK INVARIANCE PRINCIPLE [physics mathematics, chemistry] A gravitational field is equivalent to an accelerating reference frame. It is sometimes expressed as the equivalence of gravitational and inertial mass; specifically, the objects with the same inertia (resistance to acceleration) experience the same gravitational force.[†] See also DONSKER'S INVARIANCE PRINCIPLE; EQUIVALENCE PRINCIPLE; STRONG EQUIVALENCE PRINCIPLE.

WEAKEST-LINK   PRINCIPLE [mathematics] (1) Materials may be regarded as only as strong as their weakest portion.[‡] (2) No security solution is stronger than the weakest component of the solution. The strength of an argument is determined by its weakest link. (3) The strength of an argument is determined by the strength of its immediate subarguments, times the strength of its top rule. See also CERTAINTY FACTOR PROPAGATION PRINCIPLE.

WEDDLE'S RULE [mathematics] The foundation of the process of interpolation on which the process of approximated quadrature is based. The theoretical validity of interpolation formuls, and hence of all approximate quadrature formulas derived from them, rests on the famous theorem of Weierstrass, which states that any continuous function can be approximated over an interval, to any desired degree of accuracy, by a polynomial. The accuracy of the approximation in any given problem depends on (1) the nature of the function under consideration and (2) the magnitude of the interval h between the equidistant values of the independent variable.[§]

WEDGE   PRINCIPLE [physics] A wedge, chisel, and screwdriver are examples of tools using the wedge principle. Variations in angles will provide greater or lesser mechanical advantage. In the case of a chisel, the angle plus the sharpness of the edge are key determinants. A screwdriver applies force to a screw that is the actual wedge.[¶] See also DIP ROLLER PRINCIPLE.

WEIGHTED MAXIMUM PRINCIPLE [mathematics] Classical maximum principle works only for second-order equations. Recent developments of maximum principles for higher-order equations predict an extension of the results for other geometric shapes and physical systems, in particular, piezoelectric ones. Singularities at corner points crucially affect the maximum principle for fourth-order equations, including methods to deduce a weighted maximum principle.[‖]

WELDING PRINCIPLE [engineering] A laser beam is transmitted through a glass wafer and absorbed by the silicon at the silicon–glass interface. After the joint has been brought to melting heat, an upset motion follows. The burr resulting from upsetting is removed by a grinder or a file.[**] See also BUTT-WELDING PRINCIPLE; RESISTANCE WELDING PRINCIPLE; TRANSMISSION WELDING PRINCIPLE.

WELL-ORDERING   PRINCIPLE [mathematics] (1) The proposition that every set can be endowed with an order so that it becomes

*Halprin, A., Kim, and Hang, B. (1979), Mapping Lorentz invariance violations into equivalence principle violations, *Ann. Phys.* **118**, 156.

[†]Marshak, R. E. (1963), *Theory of Weak Interactions of Elementary Particles*, US Atomic Energy Commission NP-13877, p. 177.

[‡]Wolstenholme, L. C., (1995), A nonparametric test of the weakest-link principle (Engl. summary), *Technometrics*, **37**(2), 169–175.

[§]Scarborough, J. B. (1927), Questions and discussions: Discussions: On the relative accuracy of Simpson's rules and Weddle's rule., *Am. Math. Monthly* **34**(3), 370.

[¶]Fisher, L. (2003), *How to Dunk a Doughnut: The Science of Everyday Life*, Arcade Publishing, p. 66.

[‖]Sweers, G. H. (n.d.) (2008), *Singulariteiten en Maximum Principe in de Breukmechanica van Dunne Vaste Structuren*, NARCIS—Natl. Academic Research and Collaborations Information System.

[**]Novozhilov, N. M. and Standen, N. (trans) (1979), Fundamental physics of gas shielded arc Weldig. *Gordone Breach Science Publ, ltd*, p. 215.

a well-ordered set. (2) The assumption that every set can be given an order so that it becomes a well-ordered set (a set A is a well-ordered set if every subset of A has a first or minimal element). This principle is logically equivalent to the axiom of choice, Tukey's lemma, Zorn's lemma, and Hausdorff maximality principle.[*] Also known as *Zorn's principle; Hausdorff maximality principle*; see also AXIOM OF CHOICE PRINCIPLE.

WHEATSTONE BRIDGE PRINCIPLE [engineering]   (1) Apparatus for measuring electrical resistance using a null indicator, comprising two parallel resistance branches, each branch consisting of two resistances in series.[†] (2) Consolidation of sediments by pressure. (3) Water is squeezed out and cohering particles are brought within the limits of mutual molecular attraction.[‡] (4) Joining two metals by applying heat to melt and fuse them with or without filler metal.[§]

WHITE-BEAR PRINCIPLE [psychology]   (with Fyodor Dostoevsky)   Proves the impossibility of suppressing an image—by trying not to think of something, we find we continually think of it.[¶]

WIEDEMANN–FRANZ   PRINCIPLE [physics] (Gustav Wiedemann, 1826-1899, German physicist) (Rudolph, Franz, German physicist)   The ratio of thermal to electrical conductivity of any metal equals the absolute temperature multiplied by a formula containing the Boltzmann's constant.[‖]

WIEN'S PRINCIPLE [physics]   (Wilholm Carl Werner Otto Fritz Frans Wien, 1864-1928, German physicist)   Relating to radiation from a blackbody. (a) the displacement principle, (b) the emissive power, and (c) the distribution principle.[**]

WISE-USE   PRINCIPLE [electronics]   Measures resistance by comparing the current flowing through one part of a bridge with a known current flowing through another part.[††]

WOLFE'S DECOMPOSITION PRINCIPLE [engineering]   (Phil Wolfe, mathematician)   See DANTZIG–WOLFE DECOMPOSITION PRINCIPLE.

WOLTJER MINIMUM ENERGY PRINCIPLE [mathematics]   (Lodewijk Wolter)   See HAMILTON PRINCIPLE.

WOLTJER–TAYLOR MINIMUM ENERGY PRINCIPLE [mathematics]   (Lodewijk Wolter)   See HAMILTON PRINCIPLE.

WORKFUNCTION   PRINCIPLE [engineering]   Sensing is dedicated for operation as signal contributions for multiple detection.[‡‡]

WORKING PRINCIPLE [engineering]   Contribution to novel operations.

WRITING PRINCIPLES [psychology]   Learning how print signifies language requires learning both a general principle of symbolic representation and the specific correspondence rules for a particular language. All writing systems work by

[*]Swanson, L. G. and Hansen, R. T. (1988), The equivalence of the multiplication, pigeonhole, induction, and well ordering principles, *Int. J. Math. Educ. Sci. Technol.* **19**(1), 129–131.
[†]Walker, P. M. B. (1999), *Chambers Dictionary of Science and Technology*, Chambers, New York, p.1258.
[‡]Howe, D. A., Craig, J. R., Harris, R. L., Jr., and Hadidiacos, C. G. (1969), A low-cost temperature control unit for experimental petrologic studies, *J. Geol. Educ.* **17**(2), 35–37.
[§]Moerbe, K. and Doerfel, C. (1973), Measurements of electric resistance used in corrosion control, *Chemische Technik* **25**(2), 92–94.
[¶]MacGowan, C. E. and Schuberth, C. (2006), Melding chemistry with geology: A cross-disciplinary program of study for middle/secondary grade educators, *Abstracts of Papers, 231st ACS Natl. CMF., Atlanta, GA, March 26-30, 2006.*

[‖]Walker, P. M. B. (1999), *Chambers Dictionary of Science and Technology*, Chambers, New York, p. 1260.
[**]E. Rüchardt (1936). Zur Entdeckung der Kanalstrahlen vor fünfzig Jahren. Naturwissenschaften **24**(30), 57–62.
[††]Farup, I., Gatta, C., and Rizzi, A. (2007), A multiscale framework for spatial gamut mapping, *IEEE Trans. Image Process.* **16**(10), 2423–2435.
[‡‡]Doll, T. and Eisele, I. (1998), Low power gas sensors based on the work function principle, *Proc. 43rd Int. Wissenschaftliches Kolloquium-Technische Univ. Ilmenau*, vol. 2, pp.243–248.

using conventional forms as a written representation of spoken language, so the general insight is the same for children learning all written languages. It is the specific correspondence rules that demand different kinds of insights—the symbolic form designates the word in a particular writing system. In alphabetic languages, there is a rough mapping between the phonetic and graphic systems; the more sounds that are in the word, the more letters that will be required to record it. This concept, sometimes called the *alphabetic principle*, is difficult for preschool children to acquire. Young children appeal to semantic properties of the words rather than to their sounds to make decisions about how the word should be written; for instance, claiming train, a big thing, needs more letters than caterpillar, a small thing. The word size task thus requires detailed knowledge relating letters to sounds. Children can recognize that print represents a meaning (moving word task) without understanding the specific mapping principle governing such representation (word size task). This ability to develop more detailed representations of abstract knowledge has been called *analysis of representational structures.** See ALPHABETIC PRINCIPLE; GENERATIVE PRINCIPLE OF WRITING.

*Bialystok, E., Shenfield, T., and Codd, J. (2000), Languages, scripts, and the environment: Factors in developing concepts of print, *Devel. Psychol.* **36**(1), 66–76.

X

X PRINCIPLE [genetics, psychology] (1) Serves as a focal point for research on the cyclic activation and inactivation of gene loci. (2) Good leadership is based on a combination of talent, knowledge, and skills. At a higher level, extreme personal leadership requires passion, along with absolute confidence, a willingness to learn and grow, and a burning desire to excel. A set of eleven principles for extreme personal leadership and outlines a clear-cut strategy for both aspiring and seasoned leaders.*[†]

The actual mathematical relation is known as Fick's first law of diffusion and its one dimensional representation reads

$$J = -D\frac{\partial c}{\partial x} \quad [mol \text{ m}^{-2}\text{s}^{-1}]$$

where J [mol m^{-2}s^{-1}] is the flux, D [m^{2}s^{-1}] is the coefficient of diffusion, c [mol L^{-1} or mol m^{-3}] is the concentration of molecules and x [mm or m] is the thickness in the direction of diffusion or mass flow. In unsteady-state diffusion where the boundary conditions in terms of concentration and flux are continuously changing, the process of diffusion is more complicated. It is expressed in terms of rate of change of concentration and takes up the form of Ficks' second law of diffusion that reads

$$\frac{\partial c}{\partial t} = D \cdot \frac{\partial^2 c}{\partial x^2} [mol \text{ m}^{-3}\text{s}^{-1}]$$

where t [s] is time.

---

*VandeBerg, J. L. (2005) Developmental aspects of X chromosome inactivation in eutherian and matatherian mammals. Journal of Experimental Zoology, v228 (2) pp 271–286.
[†]Crawford, C. J. (2005) Corporate Rise: The X Principles of Extreme Personal Leadership. Xceo, p.5

Y

YAGI PRINCIPLE [engineering] Optimum performance requires separate adjustments in a number of parameters, such as the array length and height, diameter, and spacing of the directors and reflectors. By introducing the notion of a surface wave traveling along the array, it is possible to demonstrate experimentally the interrelationship between these parameters. With this, the gain then depends only on the phase velocity of the surface wave (which is a function of the height, diameter, and spacing of the directors) and on the choice of the reflector.[*]

YANG PRINCIPLE [psychology] (1) The active, male cosmic principle in Chinese dualistic philosophy, "yin and yang together produce everything that comes into existence."[†] (2) Fear and desire are regarded as two sides of the same coin. They coexist with the same intensity while maintaining a balance. Thus, the fear of death and failures in social adaptation and self-actualization is interpreted as a reflection of the underlying desire for life in Morita therapy.[‡] See also MASCULINE PRINCIPLE; FEMININE PRINCIPLE.

YANG–MONTI PRINCIPLE [physiology, biology, chemistry] See MITROFANOFF PRINCIPLE.

YANGI PRINCIPLE [management] "You ain't gonna need it" (YANGI) highlights the value of delaying an investment decision in the face of uncertainty about the return on the investment. The option to delay implementation

may add a great deal of value, because the future is likely to resolve the uncertainty.[§]

YAO PRINCIPLE [medicine, mathematics] (Andrew Chi-Chih Yao, b. 12/24/46, Chinese computer scientist) Proposed to establish a lower bound on the expected worst-case cost of a randomized algorithm, and evaluate the minimum cost of a deterministic algorithm on a probability distribution over the problem instances.[¶]

YIN PRINCIPLE [psychology] The passive, female cosmic principle in Chinese dualistic philosophy. The interaction of yin and yang maintains the harmony of the universe."[‖] See also MASCULINE PRINCIPLE; FEMININE PRINCIPLE.

YIN–YANG PRINCIPLE [psychology, medicine] (1) The compensatory and dialectal thinking method combining positive analysis and negative analysis. (2) A universal methodology for developing a general theory to handle

[*]Ehrenspeck, H. and Poehler, H. (n.d.), A new method for obtaining maximum gain from yagi antennas, *IRE Trans. Antenn. Propag.* **7**(4), 379–386.

[†]Pickett, J. P., Exec. editor (2000) Yang. *American Heritage Dictionary of the English Language*, 4th ed., Houghton Mifflin, NY. p. 1992.

[‡]Ishiyama, F. (1986), Morita therapy: Its basic features and cognitive intervention for anxiety treatment, *Psychother. Theory Res. Pract. Train.* **23**(3), 375–381.

[§]Erdogmus, H. and Favaro, J. (2002), Keep your options open: Extreme programming and the economics of flexibility, in *Extreme Programming Perspectives*, Marchesi, M., Succi, G., Wells, D., and Williams, L., eds., Addison-Wesley, pp. 503–552.

[¶]Takahisa, U., Toshimitsu, H., Kuniko, M., Yoshiki, Y., Atsushi, I., and Minoru, U. (2006), Effects of switching to wen-jing-tang (unkei-to) from preceding herbal preparations selected by eight-principle pattern identification on endocrinological status and ovulatory induction in women with polycystic ovary syndrome, *Am. J. Chinese Medici.* **34**(2), 177–187.

[‖]Pickett, J. P., Exec. editor (2000) Yang. *American Heritage Dictionary of the English Language*, 4th ed., Houghton Mifflin, NY. p. 1992.

the complex problems of uncertainty and find concrete approaches to them.* See also PRINCIPLE OF CHANGE, PRINCIPLE OF CONTRA-DICTION, PRINCIPLE OF RELATIONSHIP, PRINCIPLE OF HOLISM, PRINCIPLE OF THE MIDDLE WAY.

*Nisbett, R., Peng, K., Choi, I., and Norenzayan, A. (2001), Culture and systems of thought: Holistic versus analytic cognition, *Psychol. Rev.* **108**(2), 291–310.

Z

ZADEH'S EXTENSION PRINCIPLE [engineering] (Lotd Asker-Zadeh, 1921-, American mathematician) (Lotfi Asker Zadeh, b. 2/04/21, Iranian-American mathematician; Founder of Fuzzy logic, math, and set theory) (1) Provides guidance for the design process to result in functionally reliable engineering systems satisfying the requirements of some safety specification. (2) A tool for performing arithmetic and logical operations in the fuzzy environment is the t norm. Only one t norm, the min operator, is congruent with results of engineering experiments.[*]

ZAHAVI HANDICAP PRINCIPLE [mathematics] Also known as *truthful signaling hypothesis.* See also MALTHUSIAN EQUIMARGINAL PRINCIPLE.

ZERMELO'S AXIOM OF CHOICE See AXIOM OF CHOICE PRINCIPLE.

SERO AGGRESSION PRINCIPLE See ANTICOERKION PRINCIPLE.

ZERO-BIAS PRINCIPLE [physics] Consists in replacing "bare" estimators associated with an observable by an improved or "renormalized" estimator to obtain more accurate averages. Improved estimators obey a zero-variance zero-bias (ZVZB) property similar to the usual zero-variance zero-bias property of the energy with the local energy as improved estimator.[†] Also known as a *ZVZB principle.*

ZERO-FORCING PRINCIPLE [electronics, computer science] (1) The established "reflection process" as a framework of possible inputs to a system as a subset of the set of tolerable inputs (those judged to produce an acceptable response from the system).[‡] (2) Constrains the signal component at the output of the equalizer to be free of intersymbol interference (ISI). More explicitly, this implies that the product of the transfer functions of the dispersive and hence frequency-selective channel and the channel equalizer results in a "frequency-flat" constant, implying that the concatenated equalizer restores the perfect allpass channel transfer function.[§] See also ASYMMETRICAL MATCHING PRINCIPLE; ASYMPTOTIC MATCHING PRINCIPLE; BRACKET MATCHING PRINCIPLE; COMBINATORIAL MATCHING PRINCIPLE; DISTRIBUTION MATCHING PRINCIPLE; DOPPLER MATCHING PRINCIPLE; H INFINITY MODEL MATCHING PRINCIPLE; H_2 MODEL MATCHING PRINCIPLE; HYDROPHOBIC MATCHING PRINCIPLE; MATCHING PRINCIPLE; MINTO PYRAMID PRINCIPLE; MODEL MATCHING PRINCIPLE; MOMENT MATCHING PRINCIPLE; OPTIMAL MODEL MATCHING PRINCIPLE; PHASE MATCHING PRINCIPLE; PIGEONHOLE PRINCIPLE; TIME INDEPENDENCE PRINCIPLE.

ZERO-VARIANCE ZERO-BIAS PRINCIPLE [physics] See ZERO BIAS PRINCIPLE.

ZEROTH PRINCIPLE OF ENERGETICS [physics] (Howard Thomas Odum, American ecologist) If two thermodynamic systems A and B are in thermal equilibrium, and B and C are also in thermal equilibrium, then A and C

---

[*]Gazdik, L. (1996), Zadeh's extension principle in design reliability, *Fuzzy Sets Syst.* **83**, 169.
[†]Assaraf, R. and Caffarel, M. (2003), Zero-variance zero-bias principle for observables in quantum Monte Carlo: Application to forces, *J. Chem. Phys.* **119**(20), 10536.

[‡]Hiroyoshi, O. and Sato, Y. (1994), Blind equalization based on time-independency principle, *Proc. Int. Symp. Information Theory & Its Applications,* ISITA '94: Barton, ACT, Inst. Engineers, Australia, pp. 1289–1294.
[§]Hanzo, L., Wong, C. H., and Yee, M. S. (n.d.), *Adaptive Wireless Transceivers: Turbo-Coded, Turbo-Equalised and Space-Time Coded TDMA, CDMA, MC-CDMA and OFDM Systems,* Dept. Electronics and Computer Science, Univ. Southampton, UK, p. 488.

---

are in thermal equilibrium.* See also FIRST PRINCIPLE OF ENERGETICS; SECOND PRINCIPLE OF ENERGETICS; THIRD ENERGETICS PRINCIPLE OF ENERGETICS; FOURTH PRINCIPLE OF ENERGETICS; FIFTH PRINCIPLE OF ENERGETICS; SIXTH PRINCIPLE OF ENERGETICS; MAXIMUM POWER PRINCIPLE.

ZEROTH PRINCIPLE OF THERMODYNAMICS [physics] Establishes the existence of a property called *temperature*. This law is based on the observation that if a system A is in thermal equilibrium with a system B, and if system B is in thermal equilibrium with a system C, then it A invariably will be found to be in equilibrium with C if the two systems are placed in mutual contact. This law suggests that a numerical scale can be established for the common property, and if A, B, and C have the same numerical values of this property, then they will be in mutual thermal equilibrium if they were placed in contact. This property is now called the *temperature*.[†] See also THERMODYNAMIC PRINCIPLES.

ZINTL–KLEMM PRINCIPLE [chemistry] (Eduard Zintl, 1898-1941, German chemist) (Wilhelm Klemm, German chemist) Generally, compounds obeying a modified octet rule, correlating the number bonds per atom with the number of valence electrons per electron acceptor, and as long as they are natural numbers, two electrons per covalent bond between two anions are available. Explains the mechanism leading to polyanionic clustering. In addition the chemical stability of the clusters is strongly influenced by the volume of the alkali metal separating these clusters. The larger the donor atoms, the more stable the Zintl ions. This requires the formal charge transfer of all the valence electrons from the electropositive to the electronegative element. Crystalline intermetallic compounds in the K–Te system.[‡]

ZORN'S MAXIMALITY PRINCIPLE [mathematics] (Max August Zorn, 1906-1993, American mathematics) (Max August Zorn, b. 1906, Germany) See AXIOM OF CHOICE PRINCIPLE.

ZORN'S MAXIMUM PRINCIPLE [mathematics] (Max August Zorn, 1906-1993, American mathematics) (Max August Zorn, b. 1906, Germany) Commonly known as *Zorn's lemma*. Every nonempty directed set x, in which every antichain has a least upper bound, has a maximal element.[§] See AXIOM OF CHOICE PRINCIPLE.

*Odum, H. T. and Pinkerton, R. T. (1955), Time's speed regulator, *Am. Sci.* **43**(2), 331.

[†]Reif, F. (1965), Statistical thermodynamics, in *Fundamentals of Statistical and Thermal Physics*, McGraw-Hill, New York, chap. 3, p. 102.

[‡]Seifert-Lorenz, K. and Hafner, J. (n.d.), *The Zintl-Klemm Principle*, Institut für Materialphysik and Center for Computational Material Science, Univ. Wien, Sensengasse 8, A-1090.

[§]Harper, J. M. and Rubin, J. E. (1976), Variations of Zorn's lemma, principles of cofinality, and Hausdorff's maximal principle, *Notre Dame J. Formal Logic* **17**(4), 565–588.

PRINCIPLES-APPLICATIONS

A

ABDOMINAL DISEASE Mitrofanoff Principle

ABDUCTION IN LEARNING DIAGNOSTIC KNOWL-EDGE Extended Resolution Principle

ABELIAN GROUP Action Principle; Schwinger's Quantum Action Principle; Variational Principle

ABSORPTION Active Principle; Correspondence Principle; Principle of Measurement

ACCELERATION Equivalence Principle, Inference Principle; Jaynes' Principle; Kullback's Minimum Cross-entropy Principle; Maximum Information Entropy Principle; Measurement Principle; Measuring Principle; Minimax Entropy Principle; Minimum Energy Principle; Minimum Norm Principle; Minimum Relative Entropy Principle; Modification of The Extended Maximum Entropy Principle; Principle of Maximum Entropy; Principle of Maximum Nonadditive Entropy; Principle of Observational Compatibility; Principle of Operational Compatibility; Uncertainty Principle, Entropy Maximum Principle; Extended Maximum Entropy Principle; Extremal Principle of Information; Frieden-Soffer Principle; Gibbs Principle; Inference Principle; Jaynes' Maximum Entropy Principle; Coriolis Principle

ACIDITY. Active Principle; Azimuth Measurement Inference Principle; Jaynes' Principle; Kullback's Minimum Cross-entropy Principle; Maximum Information Entropy Principle; Measurement Principle; Measuring Principle; Minimax Entropy Principle; Minimum Energy Principle; Minimum Norm Principle; Minimum Relative Entropy Principle; Modification of The Extended Maximum Entropy Principle; Principle of Maximum Entropy; Principle of Maximum Nonadditive Entropy; Principle of Observational Compatibility; Principle of Operational Compatibility; Uncertainty Principle

ACIDS AND BASES Hard-soft acid-base Principle

ACOUSTICS Matching Principle, Guiding Principle,

ACTION FUNCTIONAL Coriolis Principle Variational Principle

ACTION MECHANICAL QUANTITY Spacetime Uncertainty Principle

ACTIVE CONTROL Internal Model Principle; Matching Principle

ACTUATOR Comparison Principle, Maximum Principle; Micropump Principle

AD CONVERSION Division Principle; Load Division Principle; Time Division Principle; Variable Time Position Channel Assignment Principle

ADAPTIVE CONTROL Ambartsumian Invariance Principle; Charnov's Invariance Principle; Donsker's Invariance Principle; Internal Model Principle; Invariance Principle; Invariance Principle of General Relativity; Local Invariance Principle; Maruyama's Invariance Principle; Perturbation Invariance Principle; Principle of Dominant Subsystems; Principle of Invariance; Principle of Undulatory Invariance; Relaxed Invariance Principle; Scale Invariance Principle; Separation Principle; Strong Principle; Variational Principle; Weak Invariance Principle

ADAPTIVE MEDIAN FILTER Entropy Maximum Principle; Extended Maximum Entropy Principle; Extremal Principle of Information; Frieden-Soffer Principle; Gibbs Principle; Inference Principle; Jaynes' Maximum Entropy Principle; Jaynes' Principle; Kullback's Minimum Cross-entropy Principle; Maximum Entropy Principle; Maximum Information

Dictionary of Scientific Principles, by Stephen Marvin
Copyright © 2011 John Wiley & Sons, Inc.

Entropy Principle; Measurement Principle; Minimax Entropy Principle; Minimum Relative Entropy Principle; Modification of The Extended Maximum Entropy Principle; Principle of Maximum Entropy; Principle of Maximum Nonadditive Entropy; Principle of Observational Compatibility; Principle of Operational Compatibility; Uncertainty Principle

ADAPTIVE SYSTEM Equivalent Background Principle; Internal Model Principle

ADDITIVE FUNCTIONAL Ambartsumian Invariance Principle; Charnov's Invariance Principle; Donsker's Invariance Principle; Invariance Principle; Invariance Principle of General Relativity; Local Invariance Principle; Maruyama's Invariance Principle; Perturbation Invariance Principle; Principle of Invariance; Principle of Undulatory Invariance; Relaxed Invariance Principle; Scale Invariance Principle; Strong Principle; Variational Principle; Weak Invariance Principle

ADIABATIC APPROXIMATION Capture Principle; Variational Principle

ADJOINT PROCESSES Maximum Principle; Mueller-Breslau Principle; Maximum Principle

ADRENERGIC AGENTS Fick Principle, Priming Principle

ADSORPTION OF METALS Hard-soft acid-base Principle

AEROSPACE Detailed Balance Principle, Flotation Polishing Principle; Half-call Principle

AESTHETIC JUDGEMENT Eye Placement Principle

AGE COMPRESSION Minimum Description Length Principle

AGNETOSTATICS Projection Principle

AGRICULTURE Active Principle, Polluter Pays Principle

AGROECOSYSTEM MANAGEMENT Optimality Principle; Organizing Principle; Principle of component analysis; Self-organizing Principle; Similarity Principle

AGRONOMY Active Principle; Maximization Principle

AIR Derivation Principle; Principle of Corresponding States

AIR TIGHTNESS Principle of Kalman Filter

ALCURONIUM Priming Principle

ALFENTANIL Priming Principle; Principle of Remifentanil

ALGEBRA Boundary Principle; Coriolis Principle; Ekeland's variational Principle; Ekeland's Epsilon Variational Principle; Global Maximum Principle; Matching Principle; Maximum Principle; Principle of Causality; T-norm Extension Principle; Variational Principle; Orthogonality Principle; Powdering Principle

ALGEBRAIC STRUCTURE Principle of Representation theoretic Self dualtiy

ALGORITHMS Inertia Principle; Lagrange Principle of Least Action; Least Action Principle; Least Motion Principle; Minimum Principle; Moment Matching Principle; Newmann's Principle; Permutation Principle; Principle of molecular dynamics; Principle of Least Motion; Sufficiency Principle; Variational Principle; Internal Model Principle; August Krogh Principle; Inference Principle; Jaynes' Principle; Kullback's Minimum Cross-entropy Principle; Matching Principle; Maximum Information Entropy Principle; Maximum Likelihood Principle; Measurement Principle; Measuring Principle; Minimal Principle; Minimax Entropy Principle; Minimum Description Length Principle; Minimum Energy Principle; Minimum Norm Principle; Minimum Relative Entropy Principle; Modification of The Extended Maximum Entropy Principle; Parallel Resolution Principle; Principle of Maximum Entropy; Principle of Maximum Nonadditive Entropy; Principle of Observational Compatibility;

Principle of Operational Compatibility; Quaternion Moment Preserving Principle; Uncertainty Principle

ALKANES Principle of Congruence, Reactivity/Selectivity Principle

ALLOY PHASE DIAGRAM First Principle

ALMOST SURE CONVERGENCE Ambartsumian Invariance Principle; Charnov's Invariance Principle; Donsker's Invariance Principle; Invariance Principle; Invariance Principle of General Relativity; Local Invariance Principle; Maruyama's Invariance Principle; Perturbation Invariance Principle; Principle of Invariance; Principle of Undulatory Invariance; Relaxed Invariance Principle; Scale Invariance Principle; Strong Principle; Variational Principle; Weak Invariance Principle

ALTERNATIVES IN COMPLEX SYSTEMS Kogan's Symmetric Action Principle

ALUMINUM Principle of molecular dynamics, Entropy Maximum Principle; Extended Maximum Entropy Principle; Extremal Principle of Information; Frieden-Soffer Principle; Gibbs Principle; Inference Principle; Jaynes' Maximum Entropy Principle; Jaynes' Principle; Kullback's Minimum Cross-entropy Principle; Maximum Entropy Principle; Maximum Information Entropy Principle; Measurement Principle; Minimax Entropy Principle; Minimum Relative Entropy Principle; Modification of The Extended Maximum Entropy Principle; Principle of Maximum Entropy; Principle of Maximum Nonadditive Entropy; Principle of Observational Compatibility; Principle of Operational Compatibility; Uncertainty Principle

AMBIGUITY Principle of Information Processing

AMINO ACIDS Minimum Description Length Principle, Hiv Principle

AMINO ACTIVE PRINCIPLE

AMNESIA Averaging Principle; Contraction Principle; Gibbs Variational Principle; Large Deviation Principle

AMNIOTIC FLUID VOLUME Inverse Problem Theory Principle

AMORPHOUS MATERIALS Decoupling Principle; Principle of molecular dynamics

AMOXICILLIN Entropy Maximum Principle; Extended Maximum Entropy Principle; Extremal Principle of Information; Frieden-Soffer Principle; Gibbs Principle; Inference Principle; Jaynes' Maximum Entropy Principle; Jaynes' Principle; Kullback's Minimum Cross-entropy Principle; Maximum Entropy Principle; Maximum Information Entropy Principle; Measurement Principle; Minimax Entropy Principle; Minimum Relative Entropy Principle; Modification of The Extended Maximum Entropy Principle; Principle of Maximum Entropy; Principle of Maximum Nonadditive Entropy; Principle of Observational Compatibility; Principle of Operational Compatibility; Uncertainty Principle

AMPLIFIER FOR SUPER SHORT PULSE Boundedness Principle; Principle of Causality

AMPLITUDES Absorption Principle; Correspondence Principle; Pauli Principle, Principle of Causality

ANAEROBIC THERMOPHILIC BACTERIA Minimum Description Length Principle

ANAL INCONTINENCE Mitrofanoff Principle; Monti Principle

ANALGESICS Color Doppler Principle; Doppler Principle; Doppler Radar Principle; Microwave Doppler Principle; Photoelectric Barrier Principle; Principle of Double Effect; Principle of Proximal Isovelocity Surface Area; Principle of The Doppler Effect; Sonic Doppler Principle; Ultrasound Doppler Principle

ANALOG DIGITAL CONVERSION Inference Principle; Jaynes' Principle; Kullback's Minimum Cross-entropy Principle; Maximum Information Entropy Principle; Measurement Principle; Measuring Principle; Minimax Entropy Principle; Minimum Energy Principle; Minimum Norm Principle; Minimum Relative Entropy Principle;

Modification of The Extended Maximum Entropy Principle; Principle of Maximum Entropy; Principle of Maximum Nonadditive Entropy; Principle of Observational Compatibility; Principle of Operational Compatibility; Uncertainty Principle

ANALOGUE CIRCUITS Matching Principle; New Therapeutic Principle

ANALOGUES Extremal Principle; Korenblum's Maximum Principle; Maximum Principle; Multi Tanh Principle; Phragmen-Lindel of Principle

ANALYSIS Coriolis Principle; Entropy Maximum Principle; Extended Maximum Entropy Principle; Extremal Principle of Information; Frieden-Soffer Principle; Gibbs Principle; Heisenberg's Correspondence Principle; Inference Principle; Jaynes' Maximum Entropy Principle; Jaynes' Principle; Kullback's Minimum Cross-entropy Principle; Maximum Entropy Principle; Maximum Information Entropy Principle; Measurement Principle;Mmemory Compensation Principle; Minimax Entropy Principle; Minimum Description Length Principle; Minimum Principle; Minimum Relative Entropy Principle; Mitrofanoff Principle; Modification of The Extended Maximum Entropy Principle; Monti Principle; Principle of Differential Ph Measurement; Principle of Maximum Entropy; Principle of Maximum Nonadditive Entropy; Principle of Observational Compatibility; Principle of Operational Compatibility; Uncertainty Principle, Maximum Principle, Newmann's Principle

ANALYSIS ON MANIFOLDS Extremum Principle; Hamilton's Principle; Maupertuis-Jacobi Principle; Maximum Principle; Saint-venant's Principle; Variational Principle

ANALYTICAL MECHANICS D'Alembert Principle; D'Alembert Lagrange Variation Principle; Differential Principle; Hamilton Principle; Inertia Principle; Lagrange Principle of Least Action; Least Action Principle; Least Motion Principle; Principle of Least Motion; Total Virtual Action Principle; Variational Principle; Virtual Force Principle

ANALYTICAL METHOD Action Principle; Active Principle; Hiv Principle; Inference Principle; Jaynes' Principle; Kullback's Minimum Cross-entropy Principle; Maximum Information Entropy Principle; Measurement Principle; Measuring Principle; Minimax Entropy Principle; Minimum Energy Principle; Minimum Norm Principle; Minimum Relative Entropy Principle; Modification of The Extended Maximum Entropy Principle;Principle of critical parameters; Principle of Maximum Entropy; Principle of Maximum Nonadditive Entropy; Principle of Observational Compatibility; Principle of Operational Compatibility; Schwinger's Quantum Action Principle; Uncertainty Principle

ANATOMY Entropy Maximum Principle; Extended Maximum Entropy Principle; Extremal Principle of Information; Frieden-Soffer Principle; Gibbs Principle; Inference Principle; Jaynes' Maximum Entropy Principle; Jaynes' Principle; Kullback's Minimum Cross-entropy Principle; Maupertuis Principle; Maximum Entropy Principle; Maximum Information Entropy Principle; Mckenzie Principles; Measurement Principle; Minimax Entropy Principle; Minimum Entropy Principle; Minimum Relative Entropy Principle; Mitrofanoff Principle; Modification of The Extended Maximum Entropy Principle; Monti Principle; Principle of Maximum Entropy; Principle of Maximum Nonadditive Entropy; Principle of Observational Compatibility; Principle of Operational Compatibility; Uncertainty Principle; Elastic Tube Principle; Functioning Principle; Ice Tube Principle; Mckenzie Principles; Mitrofanoff Principle; Monti Principle; Priming Principle Regimen Administration; Principle-centered Spine Care; Three-point Principle

ANEMIA Antipernicious Anemia Principle

ANESTHESIA Fick Principle; Priming Principle; Priming Principle Regimen Administration; Principle of Remifentanil; Timing Principle; Type II Cell Stretch-organizing Principle

ANGULAR MOMENTUM Conservation of Momentum Principle; D'Alembert Principle; Differential Principle; Hamilton Principle;

Heisenberg's Correspondence Principle; Newton's Third Principle; Principle of Correspondence; Principle of Equal A Priori Probabilities; Projection Principle; Variational Principle

ANIMAL BEHAVIOR Epistatic Handicap Principle; Formulation Principle; Handicap Principle; Mouse Beta Amyrin Active Principle; Maximizing Principle

ANIMAL HOST VIRUSES Antisymmetry Principle; Drug Principle; Ekeland's variational Principle; Ekeland's Epsilon Variational Principle; Ekeland's Variational Principle; Global Maximum Principle; Hiv Principle; Variational Principle

ANIMALS Inference Principle; Jaynes' Principle; Kullback's Minimum Cross-entropy Principle; Maximum Information Entropy Principle; Measurement Principle; Measuring Principle; Minimax Entropy Principle; Minimum Energy Principle; Minimum Norm Principle; Minimum Relative Entropy Principle; Modification of The Extended Maximum Entropy Principle; Optimality Principle; Organizing Principle; Principle of component analysis; Principle of Maximum Entropy; Principle of Maximum Nonadditive Entropy; Principle of Observational Compatibility; Principle of Operational Compatibility; Self-organizing Principle; Similarity Principle; Uncertainty Principle

ANISOTROPHIES Boltzmann's Principle; Caratheodory's Principle; Conte's Principle; Fermat's Principle; Least Time Principle; Micropump Principle; Principle of Caratheodony; Saint-Venant's Principle; Stationary Time Principle

ANISOTROPIC   SCATTERING Ambartsumian Invariance Principle; Charnov's Invariance Principle; Donsker's Invariance Principle; Invariance Principle; Invariance Principle of General Relativity; Local Invariance Principle; Maruyama's Invariance Principle; Perturbation Invariance Principle; Principle of Invariance; Principle of Undulatory Invariance; Relaxed Invariance Principle; Scale Invariance Principle; Strong Principle; Variational Principle; Weak Invariance

Principle; Anisotropy; Saint Venant's Principle; Anomalous Diffusion; Correspondence Principle; Anosov Systems; Saint-Venant's Principle;

ANTENNA ANALYZER RF Marketing Principle

ANTHROPOLOGY Optimization   Principle; Run and Recreate Principle

ANTI NEOPLASTICS Color Doppler Principle; Doppler Principle; Doppler Radar Principle; Microwave Doppler Principle; Photoelectric Barrier Principle; Principle of Proximal Isovelocity Surface Area; Principle of The Doppler Effect; Sonic Doppler Principle; Ultrasound Doppler Principle

ANTIARRHYTHMIC   EFFECTS Antiarrhythmic Principle; Anthropic Cosmological Principle; Anthropic Principle; Cosmological Principle

ANTIBODIES Affinity Principle; Principle of Measurement; Proof of Principle, Deviation Principle

ANTIDEPRESSANT   DRUG Therapeutic   Principle

ANTIINFLAMMATORY   ACTION Antiinflammation Principle; Antiphlogistic Principle

ANTISEPTICS Iodophor Principle

AORTA Drug Principle; Principle of optical phase contrast; Optimality Principle; Stewart-Hamilton Principle

APPENDIX Mitrofanoff Principle; Mitrofanoff Principle; Monti Principle; Three Day Principle

APPLIED MATHEMATICS Averaging Principle; Contraction Principle; Gibbs Variational Principle; Large Deviation Principle; Principle of Causality

APPROXIMATION Working Principle; Principle of per-survivor processing, Absorption Principle; Limiting Absorption Principle; Limiting Amplitude Principle; Maximum Principle; Revision Principle, Decision

Feedback Principle; Entropy Maximum Principle; Equivalent Background Principle; Extended Maximum Entropy Principle; Extremal Principle of Information; Frieden-Soffer Principle; Gibbs Principle; Inference Principle; Jaynes' Maximum Entropy Principle; Jaynes' Principle; Kullback's Minimum Cross-entropy Principle; Maximum Entropy Principle; Maximum Hardness Principle; Maximum Information Entropy Principle; Maximum Principle; Measurement Principle; Minimax Entropy Principle; Minimum Description Length Principle; Minimum Entropy Principle; Minimum Relative Entropy Principle; Modification of The Extended Maximum Entropy Principle; Minimum Principle; Variational Principle; Principle of Maximum Entropy; Principle of Maximum Nonadditive Entropy; Principle of molecular dynamics; Principle of Observational Compatibility; Principle of Operational Compatibility; Uncertainty Principle

APW CALCULATIONS Principle of molecular dynamics

AQUEOUS SOLUBILITY Corresponding States Principle; Hard-soft acid-base Principle; Pitzer Corresponding States Principle; Principle of Correspondence

ARCHIMEDEAN RIESZ SPACES Entropy Maximum Principle; Extended Maximum Entropy Principle; Extremal Principle of Information; Frieden-Soffer Principle; Gibbs Principle; Inference Principle; Jaynes' Maximum Entropy Principle; Jaynes' Principle; Kullback's Minimum Cross-entropy Principle; Maximum Entropy Principle; Maximum Information Entropy Principle; Measurement Principle; Minimax Entropy Principle; Minimum Entropy Principle; Minimum Relative Entropy Principle; Modification of The Extended Maximum Entropy Principle; Principle of Maximum Entropy; Principle of Maximum Nonadditive Entropy; Principle of Observational Compatibility; Principle of Operational Compatibility; Uncertainty Principle

ARGON Principle of Corresponding States

ARITHMETIC CODING Conventional Conscience Principle; Electronegativity Equalization Principle; Equalization Principle; Equiprobable Principle

AROMATICITY Maximum Hardness Principle; Pauli Antisymmetry; Exclusion Principle

ARTERY Hard-soft acid-base Principle; Penaz Principle

ARTIFICIAL INTELLIGENCE First Principle; Minimum Cross-entropy (Me) Principle; Minimum Description Length Principle; Orthogonality Principle; Principle of Information Processing, Film Principle

ASSOCIATED STRUCTURES AND FUNCTIONS Critical Path Method Principle; Eye Placement Principle; Functioning Principle; Gestalt Principle; Hopkins' Host Selection Principle; Host Selection Principle; Superposition Principle

ASSOCIATION OF INFINITELY DIVISIBLE RANDOM VECTORS Ambartsumian Invariance Principle; Charnov's Invariance Principle; Donsker's Invariance Principle; Invariance Principle; Invariance Principle of General Relativity; Local Invariance Principle; Maruyama's Invariance Principle; Perturbation Invariance Principle; Principle of Invariance; Principle of Undulatory Invariance; Relaxed Invariance Principle; Scale Invariance Principle; Strong Principle; Variational Principle; Weak Invariance Principle

ASTROPHYSICAL PLASMA Hamilton's Principle; Kirchhoff's Voltage Law Principle; Minimum Energy Principle; Variational Principle; Woltjer Minimum Energy Principle; Woltjer-Taylor Minimum Energy Principle

ASYMMETRY Principle of Maximum Entropy, Minimum Principle, Immunosuppresive Principle

ASYMPTOTIC EVALUATION Ambartsumian Invariance Principle; Averaging Principle; Charnov's Invariance Principle; Contraction Principle; Donsker's Invariance Principle; Entropy Maximum Principle; Extended

Maximum Entropy Principle; Extremal Principle of Information; Frieden-Soffer Principle; Gibbs Principle; Gibbs Variational Principle; Inference Principle; Invariance Principle; Invariance Principle of General Relativity; Jaynes' Maximum Entropy Principle; Jaynes' Principle; Kullback's Minimum Cross-entropy Principle; Large Deviation Principle; Local Invariance Principle; Maruyama's Invariance Principle; Maximum Entropy Principle; Maximum Information Entropy Principle; Measurement Principle; Minimum Entropy Principle; Minimum Relative Entropy Principle; Modification of The Extended Maximum Entropy Principle; Parsimony Principle; Principle of Parsimonius Data Modeling Perturbation Invariance Principle; Principle of Invariance; Principle of Maximum Entropy; Principle of Maximum Nonadditive Entropy; Principle of Observational Compatibility; Principle of Operational Compatibility; Principle of Undulatory Invariance; Relaxed Invariance Principle; Scale Invariance Principle; Strong Principle; Uncertainty Principle; Variational Principle; Weak Invariance Principle

ASYMPTOTIC NORMALITY Entropy Maximum Principle; Extended Maximum Entropy Principle; Extremal Principle of Information; Frieden-Soffer Principle; Gibbs Principle; Inference Principle; Jaynes' Maximum Entropy Principle; Jaynes' Principle; Kullback's Minimum Cross-entropy Principle; Maximum Entropy Principle; Maximum Information Entropy Principle; Maximum Likelihood Principle; Measurement Principle; Minimax Entropy Principle; Minimum Relative Entropy Principle; Modification of The Extended Maximum Entropy Principle; Principle of Maximum Entropy; Principle of Maximum Nonadditive Entropy; Principle of Observational Compatibility; Principle of Operational Compatibility; Uncertainty Principle

ASYNCHRONOUS INTERFACE CIRCUITS Minimum Description Length Principle; Resource Consumption Principle

ATM NETWORK Data Sending Principle; Principle of Timedomain Diakoptics; Traffic Measurement Principle

ATOM MOLECULE COLLISION Action Principle; Ckvp; Complex Kohn Variational Principle; HULTHÉN–KOHN Variational Principle; Kohn Variational Principle; Schwinger Variational Principle; Schwinger's Quantum Action Principle; Stationary Action Principle; Variational Principle

ATOMS Boltzmann's Principle; Contraction Mapping Principle; Contraction Principle; Maximum Hardness Principle; Variational Principle, Aufbau Principle; Bond Directional Principle; Maximum Hardness Principle, Priming Principle

ATRIAL STIMULATION Overlapping Biphasic Impulse Principle

ATTENUATION Saint Venant's Principle

ATTRACTORS Certainty Equivalence Principle; Equivalence Principle

ATTRIBUTE Minimum Description Length Principle, Minimum Description Length Principle

AUDIOLOGY Cross Check Principle, Inverse Problem Theory Principle

AUTOCORRELATION FUNCTION Principle of molecular dynamics

AUTOMATED SEGMENTATION Entropy Maximum Principle; Extended Maximum Entropy Principle; Extremal Principle of Information; Frieden-Soffer Principle; Gibbs Principle; Inference Principle; Jaynes' Maximum Entropy Principle; Jaynes' Principle; Kullback's Minimum Cross-entropy Principle; Maximum Entropy Principle; Maximum Information Entropy Principle; Measurement Principle; Minimax Entropy Principle; Minimum Relative Entropy Principle; Modification of The Extended Maximum Entropy Principle; Principle of Maximum Entropy; Principle of Maximum Nonadditive Entropy; Principle of Observational Compatibility; Principle of Operational Compatibility; Uncertainty Principle

AUTOMATIC ANALYSER Inference Principle; Jaynes' Principle; Kullback's Minimum Cross-entropy Principle; Maximum Information Entropy Principle; Measurement Principle; Measuring Principle; Minimax Entropy Principle; Minimum Energy Principle; Minimum Norm Principle; Minimum Relative Entropy Principle; Modification of The Extended Maximum Entropy Principle; Principle of Maximum Entropy; Principle of Maximum Nonadditive Entropy; Principle of Observational Compatibility; Principle of Operational Compatibility; Uncertainty Principle

AUTOMATIC CONTROLS Principle of Dominant Subsystems; Principle of Fundamental Characteristics; Principle of Smooth Fit; Equipartition Principle; Principle of Construction and Operation; Maximum Principle; Minimum Principle

AUTOMATIC KNOT SELECTION Minimum Description Length Principle

AUTOMATIC SWING HYDRAULIC DECODER D'Alembert's Principle; Differential Principle; Differential Principle of D'Alembert; Hamilton Principle

AUTOMATION Minimum Information Principle

AUTOMATIZED ELECTRIC DRIVES Optimum Principle; Stochastic Optimum Principle

AUTOMOBILE HEADLAMP Projection Principle

AUTONOMOUS DELAY DIFFERENCE SYSTEMS Ambartsumian Invariance Principle; Charnov's Invariance Principle; Donsker's Invariance Principle; Invariance Principle; Invariance Principle of General Relativity; Local Invariance Principle; Maruyama's Invariance Principle; Perturbation Invariance Principle; Principle of Invariance; Principle of Undulatory Invariance; Relaxed Invariance Principle; Scale Invariance Principle; Strong Principle; Variational Principle; Weak Invariance Principle

AUTONOMOUS SYSTEM Autonomous Cooperation Distributed Maximum Principle; Distributed Maximum Principle; Maximum Principle; Maximum Principle

AUTOREGRESSION Ambartsumian Invariance Principle; Charnov's Invariance Principle; Donsker's Invariance Principle; Invariance Principle; Invariance Principle of General Relativity; Local Invariance Principle; Maruyama's Invariance Principle; Perturbation Invariance Principle; Principle of Invariance; Principle of Undulatory Invariance; Relaxed Invariance Principle; Scale Invariance Principle; Strong Principle; Variational Principle; Weak Invariance Principle

AXIOM OF CHOICE Aapproximation Induction Principle; Asymmetric Induction Principle; Bbone Induction Principle; Coinduction Principle; Coverset Induction Principle; Enantiodi Vergient Induction Principle; Induction Principle; Inference Principle; Lysogenic Induction Principle; Magnetic Induction Principle; Mccarthy's Recursion Induction Principle; Non Constructive Principle; Osteogenic Induction Principle; Parameterized Proof Principle; Pigeon Hole Principle; Proof Principle; Proportional Induction Principle; Recursive Induction Principle; Static Induction Principle; Structural linearization Principle

AZIMUTH MEASUREMENT Inference Principle; Jaynes' Principle; Kullback's Minimum Cross-entropy Principle; Maximum Information Entropy Principle; Measurement Principle; Measuring Principle; Minimax Entropy Principle; Minimum Energy Principle; Minimum Norm Principle; Minimum Relative Entropy Principle; Modification of The Extended Maximum Entropy Principle; Principle of Maximum Entropy; Principle of Maximum Nonadditive Entropy; Principle of Observational Compatibility; Principle of Operational Compatibility; Uncertainty Principle

B

BAHADUR REPRESENTATION Ambartsumian Invariance Principle; Charnov's Invariance Principle; Donsker's Invariance Principle; Invariance Principle; Invariance Principle of General Relativity; Local Invariance Principle; Maruyama's Invariance Principle; Perturbation Invariance Principle; Principle of Invariance; Principle of Undulatory Invariance; Relaxed Invariance Principle; Scale Invariance Principle; Strong Principle; Variational Principle; Weak Invariance Principle

BANACH SPACE Ambartsumian Invariance Principle; Charnov's Invariance Principle; Donsker's Invariance Principle; Invariance Principle; Invariance Principle of General Relativity; Local Invariance Principle; Maruyama's Invariance Principle; Mesh Independence Principle; Minimum Principle; Perturbation Invariance Principle; Principle of Invariance; Principle of Large Deviations; Principle of Undulatory Invariance; Relaxed Invariance Principle; Scale Invariance Principle; Strong Principle; Variational Principle; Weak Invariance Principle; Delphin Vane Principle; Maximum Principle

BAND LIMITED FUNCTION Hamilton's Principle; Kirchhoff's Voltage Law Principle; Minimum Energy Principle; Variational Principle; Woltjer Minimum Energy Principle; Woltjer-Taylor Minimum Energy Principle

BARDEEN-COOPER-SCHRIEFFER MODEL Blossoming Principle; Hartree Fock Variational Principle; Skyrme Hartree Fock Principle; Time Dependent Hardtree Fock Variational Principle; Variational Principle

BAYES ESTIMATION Entropy Maximum Principle; Extended Maximum Entropy Principle; Extremal Principle of Information; Frieden-Soffer Principle; Gibbs Principle; Inference Principle; Jaynes' Maximum Entropy Principle; Jaynes' Principle; Kullback's Minimum Cross-entropy Principle; Maximum Entropy Principle; Maximum Information Entropy Principle; Measurement Principle; Minimax Entropy Principle; Minimum Entropy Principle; Minimum Relative Entropy Principle; Modification of The Extended Maximum Entropy Principle; Principle of Kalman Filter; Principle of Maximum Entropy; Principle of Maximum Nonadditive Entropy; Principle of Observational Compatibility; Principle of Operational Compatibility; Uncertainty Principle

BAYES FACTOR Minimum Description Length Principle

BAYESIAN NETWORKS Entropy Maximum Principle; Extended Maximum Entropy Principle; Extremal Principle of Information; Frieden-Soffer Principle; Gibbs Principle; Inference Principle; Jaynes' Maximum Entropy Principle; Jaynes' Principle; Kullback's Minimum Cross-entropy Principle; Maximum Entropy Principle; Maximum Information Entropy Principle; Measurement Principle; Minimax Entropy Principle; Minimum Description Length Principle; Minimum Entropy Principle; Minimum Relative Entropy Principle; Modification of The Extended Maximum Entropy Principle; Principle of Maximum Entropy; Principle of Maximum Nonadditive Entropy; Principle of Observational Compatibility; Principle of Operational Compatibility; Uncertainty Principle

BEAM FOIL SPECTROSCOPY DECAY CURVE Matching Principle

BEAM MECHANICS D'Alembert's Principle; Energy Principle; Gyarmati's Principle; Hamilton Principle; Mueller-Breslau Principle; Prigogine Principle; Principle of Minimum Kinetic Energy; Principle of Minimum Mechanical Energy; Principle of Minimum Potential Energy; Quantum

Hamilton Principle; Stewart-Hamilton Principle; Stochastic Hamilton Variational Principle; Toupin's Dual Principle; Variational Principle; Virtual Force Principle; Woltjer Minimum Energy Principle; Total Virtual Action Principle

BEHAVIOR Babinet's Principle; Boundary Harnack Principle; Complementarity Principle; Frequency-dependent Principle; Gestalt Principle; Harnack Principle; Maximization Principle; Obligatory Contour Principle; Picard Principle; Principle of Congruence; Salutogenesis Principle; Superposition Principle; Unified Principle; Variational Principle; Epistatic Handicap Principle; Handicap Principle

BEHAVIORAL BIOLOGY Epistatic Handicap Principle; Formulation Principle; Handicap Principle; Mouse Beta Amyrin Active Principle; Nameless Category N3C Principle; Antidepressant Principle; Mouse Beta Amyrin Active Principle

BELL THEOREM Action Principle; Principle of Maximum Nonadditive Entropy; Schwinger's Quantum Action Principle

BENCHMARK TEST Inference Principle; Jaynes' Principle; Kullback's Minimum Cross-entropy Principle; Maximum Information Entropy Principle; Measurement Principle; Measuring Principle; Minimax Entropy Principle; Minimum Energy Principle; Minimum Norm Principle; Minimum Relative Entropy Principle; Modification of The Extended Maximum Entropy Principle; Principle of Maximum Entropy; Principle of Maximum Nonadditive Entropy; Principle of Observational Compatibility; Principle of Operational Compatibility; Uncertainty Principle

BENEFICIAL EFFECT Color Doppler Principle; Doppler Principle; Doppler Radar Principle; Microwave Doppler Principle; Photoelectric Barrier Principle; Principle of Double Effect; Principle of Proximal Isovelocity Surface Area; Principle of The Doppler Effect; Sonic Doppler Principle; Ultrasound Doppler Principle

BIAS ROBUST ESTIMATION Ambartsumian Invariance Principle; Charnov's Invariance Principle; Donsker's Invariance Principle; Invariance Principle; Invariance Principle of General Relativity; Local Invariance Principle; Maruyama's Invariance Principle; Perturbation Invariance Principle; Principle of Invariance; Principle of Undulatory Invariance; Relaxed Invariance Principle; Scale Invariance Principle; Strong Principle; Variational Principle; Weak Invariance Principle

BIMODAL SIZE DISTRIBUTION Entropy Maximum Principle; Extended Maximum Entropy Principle; Extremal Principle of Information; Frieden-Soffer Principle; Gibbs Principle; Inference Principle; Jaynes' Maximum Entropy Principle; Jaynes' Principle; Kullback's Minimum Cross-entropy Principle; Maximum Entropy Principle; Maximum Information Entropy Principle; Measurement Principle; Minimax Entropy Principle; Minimum Relative Entropy Principle; Modification of The Extended Maximum Entropy Principle; Principle of Maximum Entropy; Principle of Maximum Nonadditive Entropy; Principle of Observational Compatibility; Principle of Operational Compatibility; Uncertainty Principle

BINARY SYSTEM Corresponding States Principle; Permutation Principle; Pitzer Corresponding States Principle

BIOBUSINESS Antiviral Active Principle; Optimality Principle; Organizing Principle; Principle of component analysis; Self-organizing Principle; Similarity Principle

BIOCHEMICAL MECHANISMS August Krogh Principle

BIOCHEMICAL METHODS Active Principle; Antiulcer Principle; Boric Acid Affinity Principle; Boronic Acid Affinity Principle;Principle of critical parameters; Dale's Principle

BIOCHEMISTRY Active Principle; Adaptive Principle; Antidepressant Principle; Antiulcer Principle; Anthropic Cosmological

Principle; Anthropic Principle; Antiarrhythmic Principle; Antiviral Active Principle; Asset Protection Principle; August Krogh Principle; Bond Directional Principle; Dale's Principle; Cosmological Principle; Fick Principle; Hiv Principle; Hopkins' Host Selection Principle; Host Selection Principle; Maximum Likelihood Principle; Mimetic Principle; Priming Principle; Priming Principle Regimen Administration Mouse Beta Amyrin Active Principle; Optimality Principle; Organizing Principle; Principle of component analysis; Principle of critical parameters; Self-organizing Principle; Similarity Principle; SimiliaPrinciple; Sweet Principle; Type II Cell Stretch-organizing Principle; Unlearned Principle

BIOCYBERNETICS Entropy Maximum Principle; Extended Maximum Entropy Principle; Extremal Principle of Information; Extremum Principle; Frieden-Soffer Principle; Gibbs Principle; Inference Principle; Jaynes' Maximum Entropy Principle; Jaynes' Principle; Kullback's Minimum Cross-entropy Principle; Maximum Entropy Principle; Maximum Information Entropy Principle; Maximum Likelihood Principle; Measurement Principle; Minimax Entropy Principle; Minimum Description Length Principle; Minimum Entropy Principle; Minimum Relative Entropy Principle; Modification of The Extended Maximum Entropy Principle; Neighborhood Coherence Principle; Optimizing Principle; Principle of Maximum Entropy; Principle of Maximum Nonadditive Entropy; Principle of Observational Compatibility; Principle of Operational Compatibility; Uncertainty Principle

BIODEGRADABLE POLYMER Active Principle

BIODIVERSITY Optimality Principle; Organizing Principle; Principle of component analysis; Self-organizing Principle; Similarity Principle

BIOENGINEERING Constant Infusion Principle; Entropy Maximum Principle; Extended Maximum Entropy Principle; Extremal Principle of Information; Frieden-Soffer Principle; Gibbs Principle; Inference Principle; Jaynes' Maximum Entropy Principle;

Jaynes' Principle; Kullback's Minimum Cross-entropy Principle; Maximum Entropy Principle; Maximum Information Entropy Principle; Measurement Principle; Minimax Entropy Principle; Minimum Entropy Principle; Minimum Relative Entropy Principle; Modification of The Extended Maximum Entropy Principle; Principle of Maximum Entropy; Principle of Maximum Nonadditive Entropy; Principle of Observational Compatibility; Principle of Operational Compatibility; Uncertainty Principle; Orthogonality Principle

BIOLOGICAL SAMPLE Capillary Electrophoresis Principle; Ion Pair Principle; Principle of Diastereomeric Ion Pair Formation

BIOLOGICAL SIGNAL PROCESSING Hamilton's Principle; Kirchhoff's Voltage Law Principle; Minimum Energy Principle; Variational Principle; Woltjer Minimum Energy Principle; Woltjer-Taylor Minimum Energy Principle

BIOLOGY Antiarrhythmic Principle; Casuistic Principle; Double Effect Principle; Ecological Principle; Elastic Tube Principle; Exception Granting Principle; Eye Placement Principle; Justifying Principle; Optimality Principle; Organizing Principle; Principle of component analysis; Principle of Do No Harm; Principle of Double Effect; Principle of Dynamic Ecology; Principles of Biomedical Ethics; Self-organizing Principle; Similarity Principle; Wheatstone Bridge Principle, Archimedes' Principle; D'Alembert Principle; D'Alembert Lagrange Variation Principle; Differential Principle; Hamilton Principle; Hiv Principle; Hydrostatic Principle; Maximum Likelihood Principle; Minimum Description Length Principle; Principle of Relevance; Timing Principle; Unlearned Principle; Variational Principle; Resource Consumption Principle; Precautionary Principle, Ice Tube Principle

BIOMEDICAL ENGINEERING Flow Acting Principle, Hamilton's Principle; Kirchhoff's Voltage Law Principle; Minimum Energy Principle; Variational Principle; Woltjer Minimum Energy Principle; Woltjer-Taylor Minimum Energy Principle

BIOPHYSICS Action Principle; Active Principle; Anthropic Cosmological Principle; Anthropic Principle; Antiviral Active Principle; August Krogh Principle; Constant Infusion Principle; Cosmological Principle; D'Alembert Principle; D'Alembert Lagrange Variation Principle; Drilled Cell Principle; Differential Principle; Entropy Maximum Principle; Extended Maximum Entropy Principle; Extremal Principle of Information; Extremum Principle; Frieden-Soffer Principle; Functioning Principle; Gibbs Principle; Hamilton Principle; Hiv Principle; Inference Principle; Jaynes' Maximum Entropy Principle; Jaynes' Principle; Kullback's Minimum Cross-entropy Principle; Maximum Entropy Principle; Maximum Information Entropy Principle; Maximum Likelihood Principle; Measurement Principle; Minimax Entropy Principle; Minimum Description Length Principle; Minimum Entropy Principle; Minimum Relative Entropy Principle; Modification of The Extended Maximum Entropy Principle; Neighborhood Coherence Principle; Optimizing Principle; Overlapping Biphasic Impulse Principle; Principle of Comparative Advantage; Principle of Maximum Entropy; Principle of Maximum Nonadditive Entropy; Principle of Observational Compatibility; Principle of Operational Compatibility; Schwinger's Quantum Action Principle; Uncertainty Principle; Variational Principle; Working Principle

BIOT SLOW WAVE Three-day Principle; Educational Principle; Ehrenfest Principle

BIOTECHNOLOGY Proximity Compatibility Principle; Orthogonality Principle

BIPOLAR INTEGRATED CIRCUITS Duality Principle; Multi Tanh Principle; Proof of Principle

BIT MAP Entropy Principle; Generalized Minimal Principle; Maximum Entropy Principle; Minimum Cross-entropy Principle; Moment Preserving Principle; Principle of Minimum Cross-entropy; Variational Minimal Principle

BIVARIATE CUMULATIVE PROBIT REGRESSION MODEL Ambartsumian Invariance Principle;

Charnov's Invariance Principle; Donsker's Invariance Principle; Invariance Principle; Invariance Principle of General Relativity; Local Invariance Principle; Maruyama's Invariance Principle; Perturbation Invariance Principle; Principle of Invariance; Principle of Undulatory Invariance; Relaxed Invariance Principle; Scale Invariance Principle; Strong Principle; Variational Principle; Weak Invariance Principle

BLADDER Mitrofanoff Principle; Monti Principle; Principle of Measurement

BLIND EQUALIZATION Matching Principle; Time independency Principle

BLOCK TRUNCATION CODING Entropy Principle; Generalized Minimal Principle; Maximum Entropy Principle; Minimum Cross-entropy Principle; Moment Preserving Principle; Principle of Minimum Cross-entropy; Quaternion Moment Preserving Principle; Variational Minimal Principle

BLOOD CIRCULATION Inference Principle; Jaynes' Principle; Kullback's Minimum Cross-entropy Principle; Maximum Information Entropy Principle; Measurement Principle; Measuring Principle; Minimax Entropy Principle; Minimum Energy Principle; Minimum Norm Principle; Minimum Relative Entropy Principle; Modification of The Extended Maximum Entropy Principle; Principle of Maximum Entropy; Principle of Laser Speckle Flowgraphy; Principle of Maximum Nonadditive Entropy; Principle of Observational Compatibility; Principle of Operational Compatibility; Uncertainty Principle, Principle of Laser Speckle Flowgraphy, Antithrombotic Principle; Fick Principle, Constant Infusion Principle

BODY FORCES Saint-Venant's Principle; Principle of Corresponding States

BOGOMOLNYI PRASAD SOMMER FIELD ENTROPY Correspondence Principle

BOLTZMANN EQUATION Principle of Material Frame Indifference; Variational Principle

BOND FORMATION Electromagnetic Induction Principle; Maximum Hardness Principle

BONES Antiviral Active Principle; Mckenzie Principles; Principle-centered Spine Care Syndrome Principle; Unlearned Principle, Archimedes' Principle; Hydrostatic Principle

BOOTSTRAP   APPROXIMATION Parsimony Principle; Principle of Parsimonius Data Modeling

BOSONIZATION Action Principle; Schwinger's Quantum Action Principle; Variational Principle

BOUNDARY CONDITION Maximum Principle; Spacetime Uncertainty Principle; Hamilton's Principle; Kirchhoff's Voltage Law Principle; Minimum Energy Principle; Variational Principle; Woltjer Minimum Energy Principle; Woltjer-Taylor Minimum Energy Principle

BOUNDARY ELEMENT METHOD Correspondence Principle; Huygen's's Principle; Variational Principle

BOUNDARY VALUE PROBLEM Maximum Principle; Total Virtual Action Principle; Variational Principle; Virtual Work Principle; Mesh Independence Principle; Minimum Principle; Existence Principle; Principle of Correspondence; Saint-Venant's Principle

BOUNDS Averaging Principle; Contraction Principle; Gibbs Variational Principle; Large Deviation Principle

BRAIN Entropy Maximum Principle; Extended Maximum Entropy Principle; Extremal Principle of Information; Frieden-Soffer Principle; Gibbs Principle; Inference Principle; Jaynes' Maximum Entropy Principle; Jaynes' Principle; Kullback's Minimum Cross-entropy Principle; Maximum Entropy Principle; Maximum Information Entropy Principle; Measurement Principle; Minimax Entropy Principle; Minimum Entropy Principle; Minimum Relative Entropy Principle; Modification of The Extended Maximum Entropy Principle; Principle of Information Processing; Principle of Maximum Entropy; Principle of Maximum Nonadditive Entropy; Principle of Observational Compatibility; Principle of Operational Compatibility; Superposition Principle; Uncertainty Principle

BRAIN MAPPING Entropy Maximum Principle; Extended Maximum Entropy Principle; Extremal Principle of Information. Frieden-Soffer Principle; Gibbs Principle; Inference Principle; Jaynes' Maximum Entropy Principle; Jaynes' Principle; Kullback's Minimum Cross-entropy Principle; Maximum Entropy Principle; Maximum Information Entropy Principle; Measurement Principle; Minimax Entropy Principle; Minimum Entropy Principle; Minimum Relative Entropy Principle; Modification of The Extended Maximum Entropy Principle; Principle of Maximum Entropy; Principle of Maximum Nonadditive Entropy; Principle of Observational Compatibility; Principle of Operational Compatibility; Uncertainty Principle

BRAND Entropy Maximum Principle; Extended Maximum Entropy Principle; Extremal Principle of Information; Frieden-Soffer Principle; Gibbs Principle; Inference Principle; Jaynes' Maximum Entropy Principle; Jaynes' Principle; Kullback's Minimum Cross-entropy Principle; Maximum Entropy Principle; Maximum Information Entropy Principle; Measurement Principle; Minimax Entropy Principle; Minimum Entropy Principle; Minimum Relative Entropy Principle; Modification of The Extended Maximum Entropy Principle; Principle of Maximum Entropy; Principle of Maximum Nonadditive Entropy; Principle of Observational Compatibility; Principle of Operational Compatibility; Uncertainty Principle

BRANSDICKE THEORY OF GRAVITY Anthropic Cosmological Principle; Anthropic Principle; Cosmological Principle

BREAKWATER Entropy Maximum Principle; Extended Maximum Entropy Principle; Extremal Principle of Information; Frieden-Soffer Principle; Gibbs Principle; Inference Principle; Jaynes' Maximum Entropy Principle; Jaynes' Principle; Kullback's Minimum Cross-entropy Principle; Maximum Entropy Principle; Maximum Information Entropy Principle; Measurement Principle; Minimax Entropy Principle; Minimum Entropy Principle; Minimum Relative Entropy Principle; Modification of The Extended Maximum Entropy Principle; Principle of Maximum

Entropy; Principle of Maximum Nonadditive Entropy; Principle of Observational Compatibility; Principle of Operational Compatibility; Uncertainty Principle

BROAD BAND EDGE TURBULENCE Mesh Independence Principle; Multilevel Mesh Independence Principle

BROAD CAST SPAWNING Bateman's Principle; Broensted's Principle of Congruence; Principle of Congruence

BROWNIAN MOTION Ambartsumian Invariance Principle; Charnov's Invariance Principle; Competitive Exclusion Principle; Correspondence Principle; Donsker's Invariance Principle; Invariance Principle; Invariance Principle of General Relativity; Local Invariance Principle; Maruyama's Invariance Principle; Perturbation Invariance Principle; Principle of Complementarity; Principle of Invariance; Principle of Maximum Entropy; Principle of Undulatory Invariance; Relaxed Invariance Principle; Scale Invariance Principle; Strong Principle; Variational Principle; Weak Invariance Principle; Principle of Large Deviations; Separation Principle

BROWNIAN SNAKE Averaging Principle; Contraction Principle; Gibbs Variational Principle; Large Deviation Principle

BROWSE Maximizing Principle

BSPLINES Complementary Principle; Complementary Variational Principle; Dual Principle; Dual Variational Principle; Nonconvex Variational Principle; Variational Principle

BUCKLING Energy Preservation Principle; Hamilton's Principle; Variational Principle

BULK EFFECT DEVICES Inference Principle; Jaynes' Principle; Kullback's Minimum Cross-entropy Principle; Maximum Information Entropy Principle; Measurement Principle; Measuring Principle; Minimax Entropy Principle; Minimum Energy Principle; Minimum Norm Principle; Minimum Relative Entropy Principle; Modification of The Extended Maximum Entropy Principle; Principle of Maximum Entropy; Principle of Maximum Nonadditive Entropy; Principle of Observational Compatibility; Principle of Operational Compatibility; Uncertainty Principle

C

CALCIUM Antiviral Active Principle; Archimedes' Principle; Hydrostatic Principle, Action Principle; Schwinger's Quantum Action Principle; Stationary Action Principle

CALCULATION Minimum Description Length Principle; Protection Principle, Variational Principle

CALCULUS OF VARIATIONS Decoupling Principle; Saint-Venant's Principle

CALIBRATION Principle of Measurement; Seven term Principle, Fifteen Term Principle

CANTILEVER D'Alembert's Principle; Gyarmati's Principle; Hamilton Principle; Prigogine Principle; Principle of Minimum Kinetic Energy; Principle of Minimum Mechanical Energy; Principle of Minimum Potential Energy; Quantum Hamilton Principle; Stewart-Hamilton Principle; Stochastic Hamilton Variational Principle; Toupin's Dual Principle; Variational Principle

CAPTURE EFFICIENCY Capillary Electrophoresis Principle; Ion Pair Principle; Principle of Diastereomeric Ion Pair Formation

CARBOHYDRATES Boric Acid Affinity Principle; Boronic Acid Affinity Principle; Hiv Principle, Deviation Principle

CARBON DIOXIDE Corresponding States Principle; Pitzer Corresponding States Principle;, Proof of Principle

CARCINOMA OF THE ORAL CAVITY Caution Principle; Precaution Principle; Uncertainty Principle

CARDIAC SYSTEM Anthropic Cosmological Principle; Anthropic Principle; Cosmological Principle

CARDIOVASCULAR SYSTEM Active Principle; Adaptive Principle; Constant Infusion Principle; Consumer Principle; Antiarrhythmic Principle; Antithrombotic Principle; Drilled Cell Principle; Drug Principle; Overlapping Biphasic Impulse Principle, Antithrombotic Principle; Drug Principle; Hard-soft acid-base Principle; Penaz Principle, Overlapping Biphasic Impulse Principle, Priming Principle

CARISTI FIXED POINT THEOREM Educational Principle; Ehrenfest Principle; Ekeland's Principle

CASCADE CONNECTION Principle of Conservation of Energy; Processing Principle

CASTIGLIANO VARIATIONAL THEOREM Least Action Principle; Principle of Least Work

CATALYSIS Working Principle; Variational Principle; Principle of molecular dynamics

CATALYTIC REACTION Maximum Principle; Variational Principle

CATEGORY Principle of Representationtheoretic Selfdualtiy

CATHETER Hard-soft acid-base Principle; Penaz Principle, Mitrofanoff Principle

CAUTIOUS MONOTONY Entropy Maximum Principle; Extended Maximum Entropy Principle; Extremal Principle of Information; Frieden-Soffer Principle; Gibbs Principle; Inference Principle; Jaynes' Maximum Entropy Principle; Jaynes' Principle; Kullback's Minimum Cross-entropy Principle; Maximum Entropy Principle; Maximum Information Entropy Principle; Measurement Principle; Minimax Entropy Principle; Minimum Entropy Principle; Minimum Relative Entropy Principle; Modification of The Extended Maximum Entropy Principle;

Dictionary of Scientific Principles, by Stephen Marvin
Copyright © 2011 John Wiley & Sons, Inc.

Principle of Maximum Entropy; Principle of Maximum Nonadditive Entropy; Principle of Observational Compatibility; Principle of Operational Compatibility; Uncertainty Principle

CELESTIAL DYNAMICS Principle of Relativity, Maupertuis-Jacobi Principle; Equivalence Principle

CELL BIOLOGY Action Principle; Antiarrhythmic Principle; August Krogh Principle; Caution Principle; Dale's Principle; Drug Principle; Neighborhood Coherence Principle; Precaution Principle; Uncertainty Principle; Resource Consumption Principle; Schwinger's Quantum Action Principle; SimiliaPrinciple; Uncertainty Principle

CELLS Action Principle; Antipernicious Anemia Principle; Cavalieri Principle; Matching Principle; Mechanism Principle, Neighborhood Coherence Principle; Schwinger's Quantum Action Principle; Stereological Principle; Proof of Principle; Working Principle

CENTRAL EXTENSION Hasse Norm Principle, Hasse Principle; Functioning Principle

CENTRAL LIMIT THEOREM Ambartsumian Invariance Principle; Charnov's Invariance Principle; Donsker's Invariance Principle; Invariance Principle; Invariance Principle of General Relativity; Local Invariance Principle; Maruyama's Invariance Principle; Perturbation Invariance Principle; Principle of Invariance; Principle of Undulatory Invariance; Relaxed Invariance Principle; Scale Invariance Principle; Strong Principle; Variational Principle; Weak Invariance Principle; Principle of Large Deviations

CENTRAL NERVOUS SYSTEM Competitive Exclusion Principle; Principle of Complementarity

CHANNEL IDENTIFICATION Correspondence Principle; Principle of per-survivor processing; Principles of Darcy's Law

CHAOS THEORY Certainty Equivalence Principle; Correspondence Principle; Entropy Maximum Principle; Extended Maximum Entropy Principle; Extremal Principle of Information; Frieden-Soffer Principle; Gibbs Principle; Inference Principle; Jaynes' Maximum Entropy Principle; Jaynes' Principle; Kullback's Minimum Cross-entropy Principle; Maximum Entropy Principle; Maximum Information Entropy Principle; Measurement Principle; Minimax Entropy Principle; Minimum Entropy Principle; Minimum Relative Entropy Principle; Modification of The Extended Maximum Entropy Principle; Principle of Maximum Entropy; Principle of Maximum Nonadditive Entropy; Principle of Observational Compatibility; Principle of Operational Compatibility; Uncertainty Principle; Variational Principle

CHAPMAN KOLMOGOROV EQUATION Entropy Maximum Principle; Extended Maximum Entropy Principle; Extremal Principle of Information; Frieden-Soffer Principle; Gibbs Principle; Inference Principle; Jaynes' Maximum Entropy Principle; Jaynes' Principle; Kullback's Minimum Cross-entropy Principle; Maximum Entropy Principle; Maximum Information Entropy Principle; Measurement Principle; Minimax Entropy Principle; Minimum Entropy Principle; Minimum Relative Entropy Principle; Modification of The Extended Maximum Entropy Principle; Principle of Maximum Entropy; Principle of Maximum Nonadditive Entropy; Principle of Observational Compatibility; Principle of Operational Compatibility; Uncertainty Principle

CHARGE DENSITY Principle of molecular dynamics; Projection Principle

CHARGE DISTRIBUTION Electromagnetic Induction Principle, Equipartition Principle; Principle of molecular dynamics; Working Principle

CHEMICAL CORRESPONDENCE Principle; Equipartition Principle; Inertia Principle; Lagrange Principle of Least Action; Least Action Principle; Least Motion Principle; Mouse Beta Amyrin Active Principle; Neighborhood Coherence Principle; Optimality Principle; Organizing Principle; Principle of component analysis; Principle of Corresponding States; Principle of Least Motion; Quartz Microbalance Principle; Self-organizing Principle; Similarity Principle

CHEMICAL ENGINEERING Asymptotic Matching Principle; Equipartition Principle; Flow Acting Principle; Principle of Corresponding Rheological States

CHEMICAL PLANT Displacement Principle; Element Displacement Principle; Maximum Principle; Principle of Displacement; Roots Principle; Virtual Displacement Principle, Principle of Corresponding States; Pulseecho Principle

CHEMICAL POTENTIAL Chemical Reaction. Maximum Hardness Principle, Maximum Principle; Variational Principle; Working Principle

CHEMICALS AND BIOCHEMICALS Antithrombotic Principle; Priming Principle Regimen Administration; Principle of molecular dynamics; Similia Principle of Homeopathy

CHEMISTRY Aufbau Principle; Maximum Hardness Principle; Sweet Principle; Reactivity/Selectivity Principle, Pauli Antisymmetry and Exclusion Principle, Coordination Number Principle

CHEMOMETRICS Parsimony Principle; Principle of Parsimonius Data Modeling

CHEMOTHERAPY Similia Principle of Homeopathy, Antiviral Active Principle

CHERN SIMONS VORTICES Mach's Principle

CHI SQUARE Entropy Maximum Principle; Extended Maximum Entropy Principle; Extremal Principle of Information; Frieden-Soffer Principle; Gibbs Principle; Inference Principle; Jaynes' Maximum Entropy Principle; Jaynes' Principle; Kullback's Minimum Cross-entropy Principle; Maximum Entropy Principle; Maximum Information Entropy Principle; Measurement Principle; Minimax Entropy Principle; Minimum Entropy Principle; Minimum Relative Entropy Principle; Modification of The Extended Maximum

Entropy Principle; Principle of Maximum Entropy; Principle of Maximum Nonadditive Entropy; Principle of Observational Compatibility; Principle of Operational Compatibility; Uncertainty Principle

CHIN IDEOLOGY Min Sheng Chui Principle; Minch'Uan Chui Principle; Mintsu Chui Principle

CHIRAL COMPOUND Action Principle; Schwinger's Quantum Action Principle; Superposition Principle, Hamilton Variational Principle; Variational Principle

CHOICE Epistatic Handicap Principle; Handicap Principle, Extended Resolution Principle; Principle of Uncertainty; Tnorm Extension Principle

CHORDATA Inference Principle; Jaynes' Principle; Kullback's Minimum Cross-entropy Principle; Maupertuis Principle; Maximizing Principle; Maximum Information Entropy Principle; Mckenzie Principles; Measurement Principle; Measuring Principle; Minimax Entropy Principle; Minimum Energy Principle; Minimum Norm Principle; Minimum Relative Entropy Principle; Modification of The Extended Maximum Entropy Principle; Precautionary Principle; Principle of Maximum Entropy; Principle of Maximum Nonadditive Entropy; Principle of Observational Compatibility; Principle of Operational Compatibility; Three-point Principle; Uncertainty Principle

CHRONOLOGY PROTECTION Caratheodory's Principle; Conte's Principle; Fermat's Principle; Least Time Principle; Principle of Caratheodony; Stationary Time Principle

CIRCUIT ANALYSIS Matching Principle; New Therapeutic Principle; Variational Principle

CIRCUIT PLACEMENT Selforganization Principle; Circuit Switching Network. Dale's Principle; Circuit Theory. Hamilton's Principle; Kirchhoff's Voltage Law Principle; Minimum Energy Principle; Newmann's Principle; Principle of Conservation of Energy; Processing Principle; Variational Principle; Woltjer Minimum Energy Principle; Woltjer-Taylor Minimum Energy Principle

CIVIL ENGINEERING Flow Acting Principle; Principles of Darcy's Law, Decision Feedback Principle; Wise Use Principle

CLAMPED PLATE D'Alembert's Principle; Gyarmati's Principle; Hamilton Principle; Prigogine Principle; Principle of Minimum Kinetic Energy; Principle of Minimum Mechanical Energy; Principle of Minimum Potential Energy; Quantum Hamilton Principle; Stewart-Hamilton Principle; Stochastic Hamilton Variational Principle; Toupin's Dual Principle

CLASS CLASSIFIER Complex Moment Preserving Principle; Entropy Principle; Generalized Minimal Principle; Maximum Entropy Principle; Minimum Cross-entropy Principle; Moment Preserving Principle; Principle of Minimum Cross-entropy; Variational Minimal Principle

CLIMATOLOGY ENVIRONMENTAL SCIENCES Optimization Principle; Run and Recreate Principle

CLINICAL PHARMACOLOGY Casuistic Principle; Double Effect Principle; Exception Granting Principle; Justifying Principle; Principle of Do No Harm; Principle of Double Effect; Principles of Biomedical Ethics

CLINICAL TRIAL Permutation Principle; Priming Principle

CLINICAL VALIDATION Fick Principle; Clock Retardation Principle; Principle of Relativity

CLUSTERING Complex Moment Preserving Principle; Entropy Principle; Generalized Minimal Principle; Maximum Entropy Principle; Minimum Cross-entropy Principle; Moment Preserving Principle; Principle of Minimum Cross-entropy; Variational Minimal Principle, Conventional Conscience Principle; Electronegativity Equalization Principle; Equalization Principle; Equiprobable Principle

COERCIVITY RESEARCH FRONTS Ekeland's variational Principle; Ekeland's Epsilon Variational Principle; Ekeland's Variational Principle; Global Maximum Principle; Variational Principle

COILS Inference Principle; Jaynes' Principle; Kullback's Minimum Cross-entropy Principle; Maximum Information Entropy Principle; Measurement Principle; Measuring Principle; Minimax Entropy Principle; Minimum Energy Principle; Minimum Norm Principle; Minimum Relative Entropy Principle; Modification of The Extended Maximum Entropy Principle; Principle of Maximum Entropy; Principle of Maximum Nonadditive Entropy; Principle of Observational Compatibility; Principle of Operational Compatibility; Uncertainty Principle

COLLINEAR COLLISION Action Principle; Principle of Corresponding States; Schwinger's Quantum Action Principle; Stationary Action Principle

COLOR Color Projection Principle; Entropy Principle; Generalized Minimal Principle; Maximum Entropy Principle; Minimum Cross-entropy Principle; Moment Preserving Principle; Principle of Laser Speckle Flowgraphy; Principle of Minimum Cross-entropy; Projection Principle; Quaternion Moment Preserving Principle; Variational Minimal Principle

COLOR QUANTIZATION Entropy Principle; Generalized Minimal Principle; Maximum Entropy Principle; Minimum Cross-entropy Principle; Moment Preserving Principle; Principle of Minimum Cross-entropy; Variational Minimal Principle, Derivation Principle

COMBINATORIAL MATHEMATICS Averaging Principle; Contraction Principle; Gibbs Variational Principle; Large Deviation Principle; Matching Principle; Maximum Principle; Minimum Description Length Principle; Principle of Relativity; Pontryagin's Minimum Principle

COMMUNTATION RELATION Uncertainty Principle; D'Alembert Lagrange Principle; D'Alembert Lagrange Variation Principle; D'Alembert Principle; Gauss Principle; Hamilton Lagrange Principle; Jayne's Principle; Jourdain Principle; Jourdain's Principle; Kinetic Principle; LagrangePrinciple; Least Constraint Principle; Maupertius

Lagrange Principle; Maupertuis Euler Lagrange; Principle of Least Action; Principle of Extremal Effects; Principle of Least Constraint; Principle of Maximal Effect; Principle of Virtual Work

COMPARATIVE AND EXPERIMENTAL MORPHOLOGY Ecological Principle; Fisher's Principle; Formulation Principle; Hopkins' Host Selection Principle; Host Selection Principle; Principle of Dynamic Ecology

COMPARATIVE STUDY Hard-soft acid-base Principle; Minimum Cross-entropy (Me) Principle; Penaz Principle

COMPETITIVE LEARNING Equidistortion Principle; Quantization Design Principle

COMPLEMENTARITY Competitive Exclusion Principle; Principle of Complementarity

COMPLEMENTARY OBSERVABLES Heisenberg Uncertainty Principle; Information Exclusion Principle

COMPLETE THEORY Bogolyubov Principle; Bogolyubov Variational Principle; Compensation Principle; Gibbs-Bogolyubov Variational Principle; Gibbs Third Variational Principle; Hartree Fock Bogolyubov Principle; Hfb Principle; Quantal Principle

COMPLEX MOMENTS Complexity Principle; Entropy Principle; Generalized Minimal Principle; Maximum Entropy Principle; Minimum Cross-entropy Principle; Moment Preserving Principle; Principle of Minimum Cross-entropy; Variational Minimal Principle

COMPONENT REGRESSION Parsimony Principle; Principle of Parsimonius Data Modeling

COMPOSITE MATERIAL Minimum Cross-entropy (Me) Principle; Minimum Principle; Saint-Venant's Principle; Superposition Principle; Orthogonality Principle

COMPRESSION Minimum Description Length Principle, Principle of Congruence, Multi Tanh Principle, Principle of Measurement

COMPUTATION THEORY Minimum Description Length Principle, Orthogonality Principle

COMPUTATIONAL BIOLOGY Extremum Principle; Fisheye Adaptation Principle; Maupertuis Principle; Neighborhood Coherence Principle; Optimization Principle; Optimizing Principle; Principle of Comparative Advantage; Run and Recreate Principle

COMPUTATIONAL COMPLEXITY Derivation Principle; Entropy Maximum Principle; Extended Maximum Entropy Principle; Extremal Principle of Information; Frieden-Soffer Principle; Gibbs Principle; Inference Principle; Jaynes' Maximum Entropy Principle; Jaynes' Principle; Kullback's Minimum Cross-entropy Principle; Matching Principle; Maximum Entropy Principle; Maximum Information Entropy Principle; Measurement Principle; Minimax Entropy Principle; Minimum Description Length Principle; Minimum Entropy Principle; Minimum Relative Entropy Principle; Modification of The Extended Maximum Entropy Principle; Powdering Principle; Principle of Maximum Entropy; Principle of Maximum Nonadditive Entropy; Principle of Observational Compatibility; Principle of Operational Compatibility; Resolution Principle; Uncertainty Principle; Powdering Principle; Minimum Description Length Principle

COMPUTATIONAL LINGUISTICS Entropy Maximum Principle; Extended Maximum Entropy Principle; Extremal Principle of Information; Frieden-Soffer Principle; Gibbs Principle; Inference Principle; Jaynes' Maximum Entropy Principle; Jaynes' Principle; Kullback's Minimum Cross-entropy Principle; Maximum Entropy Principle; Maximum Information Entropy Principle; Measurement Principle; Minimax Entropy Principle; Minimum Entropy Principle; Minimum Relative Entropy Principle; Modification of The Extended Maximum Entropy Principle; Principle of Maximum Entropy; Principle of Maximum Nonadditive Entropy; Principle of Observational Compatibility; Principle of Operational Compatibility; Uncertainty Principle

COMPUTATIONS Projection Principle, Maximum Likelihood Principle, Correspondence Principle

COMPUTER ARCHITECTURE Minimum Description Length Principle; Parallel Resolution Principle

COMPUTER HARDWARE Parallel Resolution Principle; Proximity Compatibility Principle; Orthogonality Principle

COMPUTER NETWORK Displacement Principle; Element Displacement Principle; Maximum Principle; Principle of Displacement; Roots Principle; Virtual Displacement Principle

COMPUTER PROGRAM D'Alembert's Principle; Gyarmati's Principle; Hamilton Principle; Minimum Cross-entropy (Me) Principle; Mueller-Breslau Principle; Orthogonality Principle; Prigogine Principle; Principle of Minimum Kinetic Energy; Principle of Minimum Mechanical Energy; Principle of Minimum Potential Energy; Quantum Hamilton Principle; Stewart-Hamilton Principle; Stochastic Hamilton Variational Principle; Toupin's Dual Principle

COMPUTER SIMULATION Entropy Maximum Principle; Extended Maximum Entropy Principle; Extremal Principle of Information; First Principle; Frieden-Soffer Principle; Gibbs Principle; Inference Principle; Jaynes' Maximum Entropy Principle; Jaynes' Principle; Kullback's Minimum Cross-entropy Principle; Maximum Entropy Principle; Maximum Information Entropy Principle; Maximum Likelihood Principle; Measurement Principle; Minimax Entropy Principle; Minimum Description Length Principle; Minimum Entropy Principle; Minimum Relative Entropy Principle; Modification of The Extended Maximum Entropy Principle; Principle of molecular dynamics; Principle of Maximum Entropy; Principle of Maximum Nonadditive Entropy; Principle of Observational Compatibility; Principle of Operational Compatibility; Uncertainty Principle, Principle of Corresponding States

COMPUTER SOFTWARE Maximum Likelihood Principle; Orthogonality Principle; Parallel Resolution Principle; Proximity Compatibility Principle; D'Alembert Lagrange Principle; D'Alembert Lagrange Variation Principle; D'Alembert Principle; Gauss Principle; Hamilton Lagrange Principle; Jaynes' Principle; Jourdain Principle; Kinetic Principle; Lagrange Principle; Least Constraint Principle; Maupertius Lagrange Principle; Maupertuis Euler Lagrange Principle of Least Action; Principle of Extremal Effects; Principle of Least Constraint; Principle of Maximal Effect; Principle of Maximum Entropy; Principle of Virtual Work; Uncertainty Principle

COMPUTER VISION Entropy Maximum Principle; Extended Maximum Entropy Principle; Extremal Principle of Information; Frieden-Soffer Principle; Gibbs Principle; Inference Principle; Jaynes' Maximum Entropy Principle; Jaynes' Principle; Kullback's Minimum Cross-entropy Principle; Maximum Entropy Principle; Maximum Information Entropy Principle; Measurement Principle; Minimax Entropy Principle; Minimum Description Length Principle; Minimum Entropy Principle; Minimum Relative Entropy Principle; Modification of The Extended Maximum Entropy Principle; Principle of Maximum Entropy; Principle of Maximum Nonadditive Entropy; Principle of Observational Compatibility; Principle of Operational Compatibility; Uncertainty Principle

COMPUTERS Archimedes' Principle; Clairvoyance Principle; D'Alembert Lagrange Principle; D'Alembert Lagrange Variation Principle; D'Alembert Principle; Gauss Principle; Hamilton Lagrange Principle; Hydrostatic Principle; Jaynes' Principle; Jourdain Principle; Kinetic Principle; Lagrange Principle; Least Constraint Principle; Maupertius Lagrange Principle; Maupertuis Euler Lagrange Principle of Least Action; Principle of Extremal Effects; Principle of Least Constraint; Principle of Maximal Effect; Principle of Virtual Work; Uncertainty Principle

CONDENSED DATA Entropy Maximum Principle; Extended Maximum Entropy Principle; Extremal Principle of Information;

Frieden-Soffer Principle; Gibbs Principle; Inference Principle; Jaynes' Maximum Entropy Principle; Jaynes' Principle; Kullback's Minimum Cross-entropy Principle; Maximum Entropy Principle; Maximum Information Entropy Principle; Measurement Principle; Minimax Entropy Principle; Minimum Entropy Principle; Minimum Relative Entropy Principle; Modification of The Extended Maximum Entropy Principle; Principle of Maximum Entropy; Principle of Maximum Nonadditive Entropy; Principle of Observational Compatibility; Principle of Operational Compatibility; Uncertainty Principle

CONDITIONAL DEDUCTION Entropy Maximum Principle; Extended Maximum Entropy Principle; Extremal Principle of Information; Frieden-Soffer Principle; Gibbs Principle; Inference Principle; Jaynes' Maximum Entropy Principle; Jaynes' Principle; Kullback's Minimum Cross-entropy Principle; Maximum Entropy Principle; Maximum Information Entropy Principle; Measurement Principle; Minimax Entropy Principle; Minimum Entropy Principle; Minimum Relative Entropy Principle; Modification of The Extended Maximum Entropy Principle; Principle of Maximum Entropy; Principle of Maximum Nonadditive Entropy; Principle of Observational Compatibility; Principle of Operational Compatibility; Uncertainty Principle

CONDITIONAL LIMIT THEOREM Entropy Maximum Principle; Extended Maximum Entropy Principle; Extremal Principle of Information; Frieden-Soffer Principle; Gibbs Principle; Inference Principle; Jaynes' Maximum Entropy Principle; Jaynes' Principle; Kullback's Minimum Cross-entropy Principle; Maximum Entropy Principle; Maximum Information Entropy Principle; Maximum Principle; Measurement Principle; Minimax Entropy Principle; Minimum Entropy Principle; Minimum Relative Entropy Principle; Modification of The Extended Maximum Entropy Principle; Principle of Maximum Entropy; Principle of Maximum Nonadditive Entropy; Principle of Observational Compatibility; Principle of Operational Compatibility; Uncertainty Principle

CONDITIONAL LOGIC Entropy Maximum Principle; Extended Maximum Entropy Principle; Extremal Principle of Information; Frieden-Soffer Principle; Gibbs Principle; Inference Principle; Jaynes' Maximum Entropy Principle; Jaynes' Principle; Kullback's Minimum Cross-entropy Principle; Maximum Entropy Principle; Maximum Information Entropy Principle; Measurement Principle; Minimax Entropy Principle; Minimum Entropy Principle; Minimum Relative Entropy Principle; Modification of The Extended Maximum Entropy Principle; Principle of Maximum Entropy; Principle of Maximum Nonadditive Entropy; Principle of Observational Compatibility; Principle of Operational Compatibility; Uncertainty Principle

CONDITIONAL TRANSITION PROBABILITIES Correspondence Principle; Conditionality Principle; Sufficiency Principle

CONDUCTIVITY OF COMPOSITES Minimum Cross-entropy (Me) Principle, Maximum Likelihood Principle; Minimum Description Length Principle; Quartz Microbalance Principle

CONFOCAL LASER SCANNING MICROSCOPY Cavalieri Principle; Inverse Problem Theory Principle; Stereological Principle

CONJUGATE VARIABLE Competitive Exclusion Principle; Principle of Complementarity

CONJUNCTIVE DEDUCTION Entropy Maximum Principle; Extended Maximum Entropy Principle; Extremal Principle of Information; Frieden-Soffer Principle; Gibbs Principle; Inference Principle; Jaynes' Maximum Entropy Principle; Jaynes' Principle; Kullback's Minimum Cross-entropy Principle; Maximum Entropy Principle; Maximum Information Entropy Principle; Measurement Principle; Minimax Entropy Principle; Minimum Entropy Principle; Minimum Relative Entropy Principle; Modification of The Extended Maximum Entropy Principle; Principle of Maximum Entropy; Principle of Maximum Nonadditive Entropy; Principle of Observational Compatibility; Principle of Operational Compatibility; Uncertainty Principle

CONNECTEDNESS Ambartsumian Invariance Principle; Charnov's Invariance Principle; Donsker's Invariance Principle; Invariance Principle; Invariance Principle of General Relativity; Local Invariance Principle; Maruyama's Invariance Principle; Perturbation Invariance Principle; Principle of Invariance; Principle of Undulatory Invariance; Relaxed Invariance Principle; Scale Invariance Principle; Strong Principle; Variational Principle; Weak Invariance Principle

CONNECTION QUALITY Air Distribution Principle; Charge Distribution Principle; Cost Distribution Principle; Counter Current Distribution Principle; Distribution Principle; Principle of Current Distribution

CONNECTIVE TISSUE Archimedes' Principle; Hydrostatic Principle; Mckenzie Principles; Principle-centered Spine Care; Unlearned Principle

CONSERVATION Echo Sounding Principle; Ecological Principle; Optimality Principle; Organizing Principle; Precautionary Principle; Principle of component analysis; Principle of Dynamic Ecology; Protection Principle; Self-organizing Principle; Similarity Principle; Wheatstone Bridge Principle, Mach Principle; Principle of Equivalence, Minimum Cross-entropy (Me) Principle

CONSERVATION OF FLUX Hamilton's Principle; Kirchhoff's Voltage Law Principle; Minimum Energy Principle; Variational Principle; Woltjer Minimum Energy Principle; Woltjer-Taylor Minimum Energy Principle

CONSTANT ALPHA Hamilton's Principle; Kirchhoff's Voltage Law Principle; Minimum Energy Principle; Variational Principle; Woltjer Minimum Energy Principle; Woltjer-Taylor Minimum Energy Principle

CONSTITUENTS Active Principle; Antithrombotic Principle; Stomachic Principles

CONSTITUTIVE EQUATION Correspondence Principle; Minimum Cross-entropy (Me) Principle; Principle of Material Frame Indifference; Principle of Minimum Dissipation; Saint-Venant's Principle

CONSTRAINT THEORY Comparison Principle; Entropy Maximum Principle; Extended Maximum Entropy Principle; Extremal Principle of Information; Frieden-Soffer Principle; Gibbs Principle; Inference Principle; Jaynes' Maximum Entropy Principle; Jaynes' Principle; Kullback's Minimum Cross-entropy Principle; Maximum Entropy Principle; Maximum Information Entropy Principle; Maximum Principle; Measurement Principle; Minimax Entropy Principle; Minimum Entropy Principle; Minimum Relative Entropy Principle; Modification of The Extended Maximum Entropy Principle; Principle of Maximum Entropy; Principle of Maximum Nonadditive Entropy; Principle of Observational Compatibility; Principle of Operational Compatibility; Uncertainty Principle, Maximum Principle

CONSTRUCTION Capture Principle; Construction Principle, Constant Infusion Principle, Saint-Venant's Principle, Decomposition Principle; Wise Use Principle

CONSUMER BEHAVIOR Entropy Maximum Principle; Extended Maximum Entropy Principle; Extremal Principle of Information; Frieden-Soffer Principle; Gibbs Principle; Inference Principle; Jaynes' Maximum Entropy Principle; Jaynes' Principle; Kullback's Minimum Cross-entropy Principle; Maximum Entropy Principle; Maximum Information Entropy Principle; Measurement Principle; Minimax Entropy Principle; Minimum Entropy Principle; Minimum Relative Entropy Principle; Modification of The Extended Maximum Entropy Principle; Principle of Maximum Entropy; Principle of Maximum Nonadditive Entropy; Principle of Observational Compatibility; Principle of Operational Compatibility; Uncertainty Principle, Maximum Principle

CONTINENCE Fitness Principle; Mitrofanoff Principle; Monti Principle

CONTINENT EFFERENT LIMB Elastic Tube Principle; Ice Tube Principle

CONTINUOUS SYSTEM Separation Principle; Matching Principle; New Therapeutic Principle, Maximum Likelihood Principle;

Maximum Principle; Minimum Principle; Pontryagin's Maximum Principle; Principle of Dominant Subsystems

CONTINUUM MECHANICS D'Alembert's Principle; Energy Preservation Principle; Gyarmati's Principle; Hamilton Principle; Hamilton's Principle; Minimum Cross-entropy (Me) Principle; Prigogine Principle; Principle of Material Frame Indifference; Principle of Minimum Kinetic Energy; Principle of Minimum Mechanical Energy; Principle of Minimum Potential Energy; Quantum Hamilton Principle; Stewart-Hamilton Principle; Stochastic Hamilton Variational Principle; Toupin's Dual Principle

CONTRACTION Kitaigorodski's Autbau Principle; Principle of Relativity; Contraction Principle; Contagion Principle; Gibbs Variational Principle

CONTROL ENGINEERING Equipartition Principle; Principle of Dominant Subsystems; Principle of Construction and Operation

CONTROL SWITCH Master-slave Principle; Ambartsumian Invariance Principle; Charnov's Invariance Principle; Donsker's Invariance Principle; Internal Model Principle; Invariance Principle; Invariance Principle of General Relativity; Local Invariance Principle; Maruyama's Invariance Principle; Master-slave Principle; Maximum Principle; Perturbation Invariance Principle; Principle of Invariance; Principle of Undulatory Invariance; Relaxed Invariance Principle; Scale Invariance Principle; Strong Principle; Variational Principle; Weak Invariance Principle

CONTROL SYSTEMS Matching Principle; Maximum Principle; Minimum Information Principle; Minimum Principle; Pontryagin's Principle, Anokhinpareto Interaction Principle. Principle of Smooth Fit

CONTROL THEORY Internal Model Principle; Matching Principle; Principle of Dominant Subsystems

CONTROLLED SYSTEMS Ekeland's variational Principle; Ekeland's Epsilon Variational Principle; Ekeland's Variational Principle; Global Maximum Principle; Maximum Principle; Minimum Cross-entropy (Me) Principle; Variational Principle

CONVERGENCE Additional Principle; Ambartsumian Invariance Principle; Charnov's Invariance Principle; Delphin Vane Principle; Donsker's Invariance Principle; Internal Model Principle; Invariance Principle; Invariance Principle of General Relativity; Local Invariance Principle; Maruyama's Invariance Principle; Maximum Principle; Minimum Cross-entropy (Me) Principle; Perturbation Invariance Principle; Principle of Invariance; Principle of Undulatory Invariance; Relaxed Invariance Principle; Scale Invariance Principle; Strong Principle; Variational Principle; Weak Invariance Principle

CONVERGENCE THEOREM Hamilton's Principle; Kirchhoff's Voltage Law Principle; Minimum Energy Principle; Variational Principle; Woltjer Minimum Energy Principle; Woltjer-Taylor Minimum Energy Principle

COOPER PAIR Bogolyubov Principle; Bogolyubov Variational Principle; Gibbs-Bogolyubov Variational Principle; Gibbs Third Variational Principle; Hartree Fock Bogolyubov Principle; Hfb Principle; Quantal Principle; Hardtreefock Variational Principle; Hartreefock Variational Principle; Minimum Cross-entropy (Me) Principle; Skyrme Hartree Fock Principle; Time Dependent Hardtree Fock Variational Principle

COOPERATIVE CONTROL Displacement Principle; Element Displacement Principle; Maximum Principle; Principle of Displacement; Roots Principle; Virtual Displacement Principle; Antidepressant Principle

COORDINATE ANALYSIS Principle of Gps, Fundamental Principle, Correspondence Principle

CORONA Hamilton's Principle; Kirchhoff's Voltage Law Principle; Minimum Energy Principle; Variational Principle; Woltjer Minimum Energy Principle; Woltjer-Taylor Minimum Energy Principle

CORRELATION Corresponding States Principle; Pitzer Corresponding States Principle;, Principle of Corresponding States; Principle of Measurement, Frequency Distance Principle, Relative Frequency Principle

CORRESPONDENCE RELATIONS IN ELASTICITY Duality Principle

CORRESPONDING ELASTIC PROBLEM Principle of Correspondence; Corresponding State Principle; Principle of Corresponding States

CORRESPONDING STATES Strong Principle

CORROSION Hamilton's Principle; Kirchhoff's Voltage Law Principle; Minimum Energy Principle; Variational Principle; Woltjer Minimum Energy Principle; Woltjer-Taylor Minimum Energy Principle

CORTICOTROPIN RELEASING HORMONE NEURONS Cavalieri Principle; Stereological Principle

COSMIC BACKGROUND EXPLORER Strong Equivalence Principle, Principle of Relativity

COSMIC NO HAIR CONJECTURE Caratheodory's Principle; Conte's Principle; Fermat's Principle; Fermats Principle; Least Time Principle; Principle of Caratheodony; Stationary Time Principle

COSMOLOGICAL DENSITY PERTURBATIONS Strong Equivalence Principle

COSMOLOGY Anthropic Cosmological Principle; Anthropic Principle; Corresponding States Principle; Cosmological Principle; Pitzer Corresponding States Principle; Principle of Relativity

COST FAVORABLE Process Displacement Principle; Element Displacement Principle; Principle of Displacement; Roots Principle; Virtual Displacement Principle

COVARIANCE Principle of Relativity; Covariance Principle; Rinciple of Material Frame Indifference

CRITICAL POINT Minimal Principle; Minimum Cross-entropy (Me) Principle, Complementary Principle; Complementary Variational Principle; Dual Principle; Dual Variational Principle; Nonconvex Variational Principle; Variational Principle

CROSS VALIDATION Parsimony Principle; Principle of Parsimonius Data Modeling

CRUSHING TRANSITION Correspondence Principle

CRYSTAL STRUCTURE Correspondence Principle; Minimal Principle; Entropy Maximum Principle; Extended Maximum Entropy Principle; Extremal Principle of Information; Frieden-Soffer Principle; Gibbs Principle; Inference Principle; Jaynes' Maximum Entropy Principle; Jaynes' Principle; Kullback's Minimum Cross-entropy Principle; Maximum Entropy Principle; Maximum Information Entropy Principle; Measurement Principle; Minimax Entropy Principle; Minimum Entropy Principle; Minimum Relative Entropy Principle; Modification of The Extended Maximum Entropy Principle; Principle of Maximum Entropy; Principle of Maximum Nonadditive Entropy; Principle of Observational Compatibility; Principle of Operational Compatibility; Uncertainty Principle

CRYSTALS Magnetron Principle; Principle of molecular dynamics, Huygen's Principle; Pitzer Corresponding States Principle; Equivalence Principle

CUBIC EQUATION OF STATE Corresponding States Principle; Pitzer Corresponding States Principle; Principle of Correspondence

CURRENT MEASURING DEVICE Electrodynamic Principle; Electromagnetic Induction Principle; Electromagnetic Principle; Principle of electronegativity

CYLINDRICAL FLUX TUBE Hamilton's Principle; Kirchhoff's Voltage Law Principle; Minimum Energy Principle; Variational Principle; Woltjer Minimum Energy Principle; Woltjer-Taylor Minimum Energy Principle

CYTOCHEMISTRY Asset Protection Principle; Entropy Maximum Principle; Extended Maximum Entropy Principle; Extremal Principle of Information; Frieden-Soffer Principle; Gibbs Principle; Inference Principle; Jaynes' Maximum Entropy Principle; Jaynes' Principle; Kullback's Minimum Cross-entropy Principle; Maximum Entropy Principle; Maximum Information Entropy Principle; Measurement Principle; Minimax Entropy Principle; Minimum Entropy Principle; Minimum Relative Entropy Principle; Modification of The Extended Maximum Entropy Principle; Optimality Principle; Organizing Principle; Principle of component analysis; Principle of Maximum Entropy; Principle of Maximum Nonadditive Entropy; Principle of Observational Compatibility; Principle of Operational Compatibility; Self-organizing Principle; Similarity Principle; Type II Cell Stretch-organizing Principle; Uncertainty Principle

CYTOGENETICS August Krogh Principle; Extremum Principle; Optimizing Principle

CYTOLOGY Asset Protection Principle; August Krogh Principle; Entropy Maximum Principle; Extended Maximum Entropy Principle; Extremal Principle of Information; Frieden-Soffer Principle; Gibbs Principle; Inference Principle; Jaynes' Maximum Entropy Principle; Jaynes' Principle; Kullback's Minimum Cross-entropy Principle; Maximum Entropy Principle; Maximum Information Entropy Principle; Measurement Principle; Minimax Entropy Principle; Minimum Entropy Principle; Minimum Relative Entropy Principle; Modification of The Extended Maximum Entropy Principle; Principle of Maximum Entropy; Principle of Maximum Nonadditive Entropy; Principle of Observational Compatibility; Principle of Operational Compatibility; Type II Cell Stretch-organizing Principle; Uncertainty Principle

CYTOLOGY AND CYTOCHEMISTRY Antiarrhythmic Principle; Dale's Principle; Neighborhood Coherence Principle; Organizing Principle; Principle of component analysis; Self-organizing Principle; Similarity Principle; Reflection Principle; Timing Principle; D'Alembert Relay Principle; Rotection Principle

D

DAMPING Equivalence Principle; Quartz Microbalance Principle, Orthogonality Principle, Maximum Principle

DATA ANALYSIS Entropy Maximum Principle; Extended Maximum Entropy Principle; Extremal Principle of Information; Frieden-Soffer Principle; Gibbs Principle; Inference Principle; Jaynes' Maximum Entropy Principle; Jaynes' Principle; Kullback's Minimum Cross-entropy Principle; Maximum Entropy Principle; Maximum Information Entropy Principle; Measurement Principle; Minimax Entropy Principle; Minimum Entropy Principle; Minimum Relative Entropy Principle; Modification of The Extended Maximum Entropy Principle; Principle of Maximum Entropy; Principle of Maximum Nonadditive Entropy; Principle of Observational Compatibility; Principle of Operational Compatibility; Uncertainty Principle

DATA COMPRESSION Conventional Conscience Principle; Electronegativity Equalization Principle; Equalization Principle; Equiprobable Principle; Minimum Description Length Principle; Data Erasing. Magnetooptical Disk Principle; Mo Principle; Data Processing. D'Alembert Lagrange Principle; D'Alembert Lagrange Variation Principle; D'Alembert Principle; Gauss Principle; Hamilton Lagrange Principle; Jaynes' Principle; Jourdain Principle; Kinetic Principle; Lagrange Principle; Least Constraint Principle; Maupertius Lagrange Principle; Maupertuis Euler Lagrange Principle of Least Action; Principle of Extremal Effects; Principle of Least Constraint; Principle of Maximal Effect; Principle of Virtual Work; Proximity Compatibility Principle; Uncertainty Principle

DCAC CONVERSION Master-slave Principle; Electrodynamic Principle; Electromagnetic Induction Principle; Electromagnetic Principle; Principle of electronegativity

DECISION Entropy Maximum Principle; Extended Maximum Entropy Principle; Extremal Principle of Information; Frieden-Soffer Principle; Gibbs Principle; Inference Principle; Jaynes' Maximum Entropy Principle; Jaynes' Principle; Kullback's Minimum Cross-entropy Principle; Matching Principle; Maximum Entropy Principle; Maximum Information Entropy Principle; Measurement Principle; Minimax Entropy Principle; Minimum Entropy Principle; Minimum Relative Entropy Principle; Modification of The Extended Maximum Entropy Principle; Principle of Maximum Entropy; Principle of Maximum Nonadditive Entropy; Principle of Observational Compatibility; Principle of Operational Compatibility; Uncertainty Principle

DECISION FEEDBACK EQUALIZATION Memory Compensation Principle; Mitrofanoff Principle; Monti Principle

DECISION MAKING Conduction Principle; Displacement Principle; Element Displacement Principle; Maximization Principle; Maximum Principle; Precautionary Principle; Principle of Displacement; Relative Frequency Principle; Roots Principle; Virtual Displacement Principle, Additional Principle

DECISION THEORY Maximization Principle; Maximum Principle; Mde Principle, Optimality Principle

DECISION TREES Active Principle; Minimum Description Length Principle, Complementary Principle; Complementary Variational Principle; Dual Principle; Dual Variational Principle; Nonconvex Variational Principle; Variational Principle

DEFLECTION Deviation Principle; Minimum Cross-entropy (Me) Principle, Mueller-Breslau Principle

DEFORMABLE POROUS MEDIA Educational Principle; Ehrenfest Principle

DEFORMATION Least Action Principle; Minimal Principle; Principle of Least Work, Orthogonality Principle

DEFORMATION EFFECTS Entropy Maximum Principle; Extended Maximum Entropy Principle; Extremal Principle of Information; Frieden-Soffer Principle; Gibbs Principle; Inference Principle; Jaynes' Maximum Entropy Principle; Jaynes' Principle; Kullback's Minimum Cross-entropy Principle; Maximum Entropy Principle; Maximum Information Entropy Principle; Measurement Principle; Minimax Entropy Principle; Minimum Entropy Principle; Minimum Relative Entropy Principle; Modification of The Extended Maximum Entropy Principle; Principle of Maximum Entropy; Principle of Maximum Nonadditive Entropy; Principle of Observational Compatibility; Principle of Operational Compatibility; Uncertainty Principle

DEFORMATION MECHANISMS Entropy Maximum Principle; Extended Maximum Entropy Principle; Extremal Principle of Information; Frieden-Soffer Principle; Gibbs Principle; Inference Principle; Jaynes' Maximum Entropy Principle; Jaynes' Principle; Kullback's Minimum Cross-entropy Principle; Maximum Entropy Principle; Maximum Information Entropy Principle; Measurement Principle; Minimax Entropy Principle; Minimum Entropy Principle; Minimum Relative Entropy Principle; Modification of The Extended Maximum Entropy Principle; Principle of Maximum Entropy; Principle of Maximum Nonadditive Entropy; Principle of Observational Compatibility; Principle of Operational Compatibility; Uncertainty Principle, Saint-Venant's Principle

DELAY SYSTEMS Ekeland's variational Principle; Ekeland's Epsilon Variational Principle; Ekeland's Variational Principle; Global Maximum Principle; Maximum Principle; Minimum Cross-entropy (Me) Principle; Variational Principle

DELAYED DECISION FEEDBACK SEQUENCE ESTIMATION Principle of per-survivor processing; Principles of Darcy's Law

DELAYED PLUMAGE MATURATION Epistatic Handicap Principle; Handicap Principle

DELAYED SPONTANEOUS RUPTURE Mitrofanoff Principle

DELIVERY Fick Principle; Reversed Fick Principle, Extremal Principle; Fibrinolytic Principle

DENSITY Active Principle; Allee's Principle; Corresponding States Principle; Pitzer Corresponding States Principle; Principle of Corresponding States; Dma Principle, Maximum Hardness Principle

DENSITY FUNCTIONAL METHOD Conventional Conscience Principle; Electronegativity Equalization Principle; Equalization Principle; Equiprobable Principle; First Principle; Principle of molecular dynamics; Geometric Mean Principle; Hard-soft acid-base Principle; Maximum Hardness Principle; Principle of electronegativity

DEPTH PERCEPTION Adjacency Principle; Entropy Maximum Principle; Extended Maximum Entropy Principle; Extremal Principle of Information; Frieden-Soffer Principle; Gibbs Principle; Inference Principle; Jaynes' Maximum Entropy Principle; Jaynes' Principle; Kullback's Minimum Cross-entropy Principle; Maximum Entropy Principle; Maximum Information Entropy Principle; Measurement Principle; Minimax Entropy Principle; Minimum Entropy Principle; Minimum Relative Entropy Principle; Modification of The Extended Maximum Entropy Principle; Principle of Maximum Entropy; Principle of Maximum Nonadditive Entropy; Principle of Observational Compatibility; Principle of Operational Compatibility; Uncertainty Principle

DESIGN Equipartition Principle; Minimum Cross-entropy (Me) Principle, Reflection Principle, Composing Principle, Proximity Compatibility Principle

DETERMINANT Fluorescence Polarization Immunoassay Principle; Inertia Principle; Lagrange Principle of Least Action; Least Action Principle; Least Motion Principle; Maximum Principle; Minimal Principle; Principle of Least Motion

DETERMINATION Superposition Principle; Directional Waves Spectrum Estimation; Entropy Maximum Principle; Extended Maximum Entropy Principle; Extremal Principle of Information; Frieden-Soffer Principle; Gibbs Principle; Inference Principle; Jaynes' Maximum Entropy Principle; Jaynes' Principle; Kullback's Minimum Cross-entropy Principle; Maximum Entropy Principle; Maximum Information Entropy Principle; Measurement Principle; Minimax Entropy Principle; Minimum Entropy Principle; Minimum Relative Entropy Principle; Modification of The Extended Maximum Entropy Principle; Principle of Maximum Entropy; Principle of Maximum Nonadditive Entropy; Principle of Observational Compatibility; Principle of Operational Compatibility; Uncertainty Principle; Dirichlet Principle; Ambartsumian Invariance Principle; Charnov's Invariance Principle; Donsker's Invariance Principle; Invariance Principle; Invariance Principle of General Relativity; Local Invariance Principle; Maruyama's Invariance Principle; Perturbation Invariance Principle; Principle of Invariance; Principle of Undulatory Invariance; Relaxed Invariance Principle; Scale Invariance Principle; Strong Principle; Variational Principle; Weak Invariance Principle; Maximum Principle; Saint-Venant's Principle; Strong Comparison Principle

DEUTERIUM ATOMS Ckvp; Complex Kohn Variational Principle; Hulthen-Kohn Variational Principle; Kohn Variational Principle; Minimum Cross-entropy (Me) Principle; Schwinger Variational Principle

DEVELOPED COUNTRIES Maximizing Principle

DEVELOPMENT Action Principle; August Krogh Principle; Investment Principle; John Templeton Principle; Minimum Cross-entropy (Me) Principle; Nameless Category N3C Principle; Priming Principle Regimen Administration; Principle of Maximum Pessimism; Resource Consumption Principle; Schwinger's Quantum Action Principle; Serendipity Principle; Type II Cell Stretch-organizing Principle

DEVELOPMENT STABILITY Epistatic Handicap Principle; Handicap Principle

DEVELOPMENTAL BIOLOGY Maupertuis Principle; Nameless Category N3C Principle; Neighborhood Coherence Principle; Optimality Principle; Organizing Principle; Priming Principle Regimen Administration; Principle of component analysis; Self-organizing Principle; Similarity Principle; Type II Cell Stretch-organizing Principle

DIAGNOSIS Active Principle; Cavalieri Principle; Principle of Measurement; Stereological Principle; Three Day Principle

DIAGNOSTIC Boric Acid Affinity Principle; Boronic Acid Affinity Principle; Constant Infusion Principle, Occam's Razor Principle

DIATOMIC MOLECULE Action Principle; Ckvp; Complex Kohn Variational Principle; HULTHÉN–KOHNVariational Principle; Kohn Variational Principle; Minimum Cross-entropy (Me) Principle; Schwinger Variational Principle; Schwinger's Quantum Action Principle; Stationary Action Principle

DIELECTRIC MATERIALS Minimum Cross-entropy (Me) Principle; Superposition Principle

DIFFERENTIAL CROSS SECTION Ckvp; Complex Kohn Variational Principle; Ulthenkohn Variational Principle; Kohn Variational Principle; Minimum Cross-entropy (Me) Principle; Schwinger Variational Principle

DIFFERENTIAL EQUATION Principle of Maximum Entropy; Dirichlet Principle; Duality Principle; Ekeland's variational Principle; Ekeland's Epsilon Variational Principle; Ekeland's Variational Principle; Global Maximum Principle; Maximum Principle; Minimum Cross-entropy (Me) Principle; Principle of Material Frame Indifference; Principle of Minimum Dissipation; Variational Principle

DIFFERENTIAL INCLUSION Ambartsumian Invariance Principle; Charnov's Invariance Principle; Donsker's Invariance Principle; Invariance Principle; Invariance Principle of General Relativity; Local Invariance

Principle; Maruyama's Invariance Principle; Perturbation Invariance Principle; Principle of Invariance; Principle of Undulatory Invariance; Relaxed Invariance Principle; Scale Invariance Principle; Strong Principle; Variational Principle; Weak Invariance Principle; Maximum Principle

DIFFERENTIAL PRESSURE SENSOR D'Alembert's Principle; Differential Principle; Differential Principle of D'Alembert; Hamilton Principle; Principle of Total Differential; Total Differential Principle; Variational Principle

DIFFUSE REFLECTION Ambartsumian Invariance Principle; Charnov's Invariance Principle; Donsker's Invariance Principle; Invariance Principle; Invariance Principle of General Relativity; Local Invariance Principle; Maruyama's Invariance Principle; Perturbation Invariance Principle; Principle of Invariance; Principle of Undulatory Invariance; Relaxed Invariance Principle; Scale Invariance Principle; Strong Principle; Variational Principle; Weak Invariance Principle

DIFFUSION Boric Acid Affinity Principle; Boronic Acid Affinity Principle; Harnack Principle; Maximum Principle; Minimum Cross-entropy (Me) Principle; Minimum Description Length Principle; Superposition Principle. Auxiliary Principle, Principle of Parsimony

DIFFUSION COEFFICIENT Maximality Principle; Maximum Principle; Principle of Corresponding States

DIGESTIVE SYSTEM Fick Principle; Ice Tube Principle; Mitrofanoff Principle; Monti Principle, Extremal Principle; Fibrinolytic Principle, Elastic Tube Principle

DIGITAL MEASURING EQUIPMENT Division Principle; Load Division Principle; Time Division Principle; Variable Time Position Channel Assignment Principle

DIGITAL SIMULATION Matching Principle; New Therapeutic Principle; Superposition Principle; Water Wave. Entropy Maximum

Principle; Extended Maximum Entropy Principle; Extremal Principle of Information; Frieden-Soffer Principle; Gibbs Principle; Inference Principle; Jaynes' Maximum Entropy Principle; Jaynes' Principle; Kullback's Minimum Cross-entropy Principle; Maximum Entropy Principle; Maximum Information Entropy Principle; Measurement Principle; Minimax Entropy Principle; Minimum Entropy Principle; Minimum Relative Entropy Principle; Modification of The Extended Maximum Entropy Principle; Principle of Maximum Entropy; Principle of Maximum Nonadditive Entropy; Principle of Observational Compatibility; Principle of Operational Compatibility; Uncertainty Principle

DIGITAL TECHNIQUES D'Alembert Lagrange Principle; D'Alembert Lagrange Variation Principle; D'Alembert Principle; Gauss Principle; Hamilton Lagrange Principle; Jayne's Principle; Jaynes' Principle; Jourdain Principle; Kinetic Principle; Lagrange Principle; Least Constraint Principle; Maupertius Lagrange Principle; Maupertuis Euler Lagrange Principle of Least Action; Principle of Extremal Effects; Principle of Least Constraint; Principle of Least Constraint; Principle of Maximal Effect; Principle of Virtual Work; Uncertainty Principle

DIMENSIONS Minimum Cross-entropy (Me) Principle, Fundamental Principle, Entropy Maximum Principle; Extended Maximum Entropy Principle; Extremal Principle of Information; Frieden-Soffer Principle; Gibbs Principle; Inference Principle; Jaynes' Maximum Entropy Principle; Jaynes' Principle; Kullback's Minimum Cross-entropy Principle; Maximum Entropy Principle; Maximum Information Entropy Principle; Measurement Principle; Minimax Entropy Principle; Minimum Entropy Principle; Minimum Relative Entropy Principle; Modification of The Extended Maximum Entropy Principle; Principle of Maximum Entropy; Principle of Maximum Nonadditive Entropy; Principle of Observational Compatibility; Principle of Operational Compatibility; Uncertainty Principle

DIRECT BULK SYNCHRONOUS PARALLEL ALGRORITHMS Contraction Principle; Gibbs Variational Principle; Large Deviation Principle

DIRECT CURRENT CONVERTERS Dynamic Translinear Principle; Translinear Principle

DIRECT VOLTAGE Division Principle; Load Division Principle; Time Division Principle; Variable Time Position Channel Assignment Principle

DIRECTION Entropy Maximum Principle; Extended Maximum Entropy Principle; Extremal Principle of Information; Frieden-Soffer Principle; Gibbs Principle; Inference Principle; Jaynes' Maximum Entropy Principle; Jaynes' Principle; Kullback's Minimum Cross-entropy Principle; Maximum Entropy Principle; Maximum Information Entropy Principle; Measurement Principle; Minimax Entropy Principle; Minimum Entropy Principle; Minimum Relative Entropy Principle; Modification of The Extended Maximum Entropy Principle; Principle of Maximum Entropy; Principle of Maximum Nonadditive Entropy; Principle of Observational Compatibility; Principle of Operational Compatibility; Uncertainty Principle

DISC INFILTROMETER Educational Principle; Ehrenfest Principle

DISCHARGE ELEMENT Discrete Maximum Principle; Discrete Pontryagin Maximum Principle; Dmp; Krylov Maximum Principle; Maximum Principle; Pontryagin Discrete Maximum Principle

DISCRETE TIME SYSTEMS Principle of Dominant Subsystems, Maximum Principle, Minimum Cross-entropy (Me) Principle; Minimum Principle; Commutation Principle; Maximum Principle; Pontryagin's Maximum Principle

DISCRETED BOLTZMANN EQUATION Entropy Maximum Principle; Extended Maximum Entropy Principle; Extremal Principle of Information; Frieden-Soffer Principle; Gibbs Principle; Inference Principle; Jaynes' Maximum Entropy Principle; Jaynes'

Principle; Kullback's Minimum Cross-entropy Principle; Maximum Entropy Principle; Maximum Information Entropy Principle; Measurement Principle; Minimax Entropy Principle; Minimum Entropy Principle; Minimum Relative Entropy Principle; Modification of The Extended Maximum Entropy Principle; Principle of Maximum Entropy; Principle of Maximum Nonadditive Entropy; Principle of Observational Compatibility; Principle of Operational Compatibility; Uncertainty Principle

DISCRIMINATORS AND MIXERS Multi Tanh Principle; Disease Similarity Principle; Syndrome Principle

DISPERSE SYSTEM Autonomous Cooperation Distributed Maximum Principle; Distributed Maximum Principle; Maximum Principle, Superposition Principle, Saint-Venant's Principle

DISPLAY DESIGN Proximity Compatibility Principle; Principle of Stereoscopic Image

DISTANCE MEASUREMENT Proximity Compatibility Principle; Distance Measuring Instrument Operating; Echo Sounding Principle; Triangulation Principle; Distance Principle In Relay Protection; Protection Principle

DISTRIBUTED CONSTANT CIRCUIT Principle of Conservation of Energy; Processing Principle

DISTRIBUTED PARAMETER CONTROL SYSTEMS Ekeland's Epsilon Variational Principle; Ekeland's Variational Principle; Global Maximum Principle; Maximum Principle; Minimum Cross-entropy (Me) Principle; Pointwise Maximum Principle; Variational Principle

DISTRIBUTION Active Principle; Detailed Balance Principle; Minimum Cross-entropy (Me) Principle, Decomposition Principle; Wise Use Principle, Powdering Principle; Principle of Measurement

DISTURBANCE Ecological Principle; Internal Model Principle; Principle of Dynamic Ecology

DIVERGENCE Contraction Principle; Gibbs Variational Principle; Large Deviation Principle

DNA Occam's Razor Principle; Principle of Pcr; Two-pole complementary Principle, Active Principle; Mdl Principle, Epistatic Handicap Principle; Handicap Principle; Uncertainty Principle

DOMAIN OF PROGRAMMING Active Principle; Dominant Subsystem Principle; Principle of Dominant Subsystems

DOSAGE Priming Principle Regimen Administration, Protection Principle, Priming Principle, Working Principle

DRAGGING Inertia Principle; Lagrange Principle of Least Action; Least Action Principle; Least Motion Principle; Principle of Least Motion

DROPS Entropy Maximum Principle; Extended Maximum Entropy Principle; Extremal Principle of Information; Frieden-Soffer Principle; Gibbs Principle; Inference Principle; Jaynes' Maximum Entropy Principle; Jaynes' Principle; Kullback's Minimum Cross-entropy Principle; Maximum Entropy Principle; Maximum Information Entropy Principle; Measurement Principle; Minimax Entropy Principle; Minimum Entropy Principle; Minimum Relative Entropy Principle; Modification of The Extended Maximum Entropy Principle; Principle of Maximum Entropy; Principle of Maximum Nonadditive Entropy; Principle of Observational Compatibility; Principle of Operational Compatibility; Uncertainty Principle

DRUG ABSORPTION Working Principle, Occam's Razor Principle; Principle of Pcr, Antiarrhythmic Principle

DUAL MODE Entropy Maximum Principle; Extended Maximum Entropy Principle; Extremal Principle of Information; Frieden-Soffer Principle; Gibbs Principle; Inference Principle; Jaynes' Maximum Entropy Principle; Jaynes' Principle; Kullback's Minimum Cross-entropy Principle; Maximum Entropy Principle; Maximum Information Entropy

Principle; Measurement Principle; Minimax Entropy Principle; Minimum Entropy Principle; Minimum Relative Entropy Principle; Modification of The Extended Maximum Entropy Principle; Principle of Maximum Entropy; Principle of Maximum Nonadditive Entropy; Principle of Observational Compatibility; Principle of Operational Compatibility; Uncertainty Principle

DUAL SURFACES Interference Principle; Interferometry Neutron Principle; Moire Interference Principle; Principle of Spinor Superposition; Resolution Principle

DUALITY OF LIGHT Bogolyubov Principle; Bogolyubov Variational Principle; Compensation Principle; Gibbs-Bogolyubov Variational Principle; Gibbs Third Variational Principle; Hartree Fock Bogolyubov Principle; Hfb Principle; Quantal Principle

DUST PARTICLE SIZE MEASURING APPARATUS Energy Principle; Laser Scattering Principle; Le Chatelierbraun Principle

DYAR'S RULE Investment Principle; John Templeton Principle; Principle of Maximum Pessimism

DYNAMIC PROGRAMMING Caratheodory's Principle; Conte's Principle; Fermat's Principle; Fermats Principle; Least Time Principle; Maximum Principle; Minimum Cross-entropy (Me) Principle; Minimum Principle; Mueller-Breslau Principle; Optimality Principle; Principle of Caratheodony; Stationary Time Principle

DYNAMICAL SYSTEM Ambartsumian Invariance Principle; Charnov's Invariance Principle; Deviation Principle; Dirichlet Principle; Donsker's Invariance Principle; Invariance Principle; Invariance Principle of General Relativity; Local Invariance Principle; Maruyama's Invariance Principle; Perturbation Invariance Principle; Principle of Invariance; Principle of Undulatory Invariance; Relaxed Invariance Principle; Scale Invariance Principle; Strong Principle; Variational Principle; Weak Invariance Principle; Commutation Principle; Hamilton's Principle; Maupertuis-Jacobi Principle;

Minimum Cross-entropy (Me) Principle; Saint-Venant's Principle; Treatment of Pde; Saint-Venant's Principle

DYNAMICS D'Alembert's Principle; Gyarmati's Principle; Hamilton Principle; Pontryagin's Maximum Principle; Prigogine Principle; Principle of Minimum Kinetic Energy; Principle of Minimum Mechanical Energy; Principle of Minimum Potential Energy; Quantum Hamilton Principle; Stewart-Hamilton Principle; Stochastic Hamilton Variational Principle; Toupin's Dual Principle, Maupertuis-Jacobi Principle

E

EARTHWORK Wise Use Principle, Decomposition Principle; Wise Use Principle

ECHELLE Proof of Principle

ECOLOGY Echo Sounding Principle; Extremal Principle; Frequency-dependent Principle; Investment Principle; John Templeton Principle; Optimality Principle; Organizing Principle; Precautionary Principle; Principle of Maximum Pessimism, Active Principle; Competitive Exclusion Principle; Ecological Principle; Extinguishing Principle; Optimization Principle; Principle of Comparative Advantage; Principle of Dynamic Ecology; Run and Recreate Principle; Wheatstone Bridge Principle; Optimality Principle; Organizing Principle; Principle of component analysis; Self-organizing Principle; Similarity Principle

EFFECTIVE ACTION FIELD Action Principle; Minimum Cross-entropy (Me) Principle; Schwinger's Quantum Action Principle

EFFECTIVE MEDIUM MODEL Saint-Venant's Principle, Dirichlet Principle; Ambartsumian Invariance Principle; Charnov's Invariance Principle; Donsker's Invariance Principle; Invariance Principle; Invariance Principle of General Relativity; Local Invariance Principle; Maruyama's Invariance Principle; Perturbation Invariance Principle; Principle of Invariance; Principle of Undulatory Invariance; Relaxed Invariance Principle; Scale Invariance Principle; Strong Principle; Variational Principle; Weak Invariance Principle

EIGEN RAY TRACING Caratheodory's Principle; Conte's Principle; Fermat's Principle; Fermats Principle; Least Time Principle; Principle of Caratheodony; Stationary Time Principle

EIGENVALUE DOMAINS Boric Acid Affinity Principle; Boronic Acid Affinity Principle; Harnack Principle; Minimum Cross-entropy (Me) Principle; Minimax Principle; Momentum Principle; Saint Venant's Principle

80/20 RULE Pareto Principle

EIKONAL EQUATION Caratheodory's Principle; Conte's Principle; Fermat's Principle; Fermats Principle; Least Time Principle; Principle of Caratheodony; Stationary Time Principle

EINSTEIN SMOLUCHOWSKI PROMEASURE Averaging Principle; Correspondence Principle; Entropy Maximum Principle; Extended Maximum Entropy Principle; Extremal Principle of Information; Frieden-Soffer Principle; Gibbs Principle; Inference Principle; Jaynes' Maximum Entropy Principle; Jaynes' Principle; Kullback's Minimum Cross-entropy Principle; Maximum Information Entropy Principle; Measurement Principle; Minimax Entropy Principle; Minimum Entropy Principle; Minimum Relative Entropy Principle; Modification of The Extended Maximum Entropy Principle; Principle of Maximum Entropy; Principle of Maximum Nonadditive Entropy; Principle of Observational Compatibility; Principle of Operational Compatibility; Uncertainty Principle

EINSTEIN DOLSKY ROSEN PARADOX Action Principle; Schwinger's Quantum Action Principle; Principle of Maximum Entropy

EKELAND Minimum Cross-entropy (Me) Principle; Ekeland's variational Principle; Pontryagin Maximum Principle

ELASTIC STABILITY D'Alembert's Principle; Gyarmati's Principle; Hamilton Principle; Prigogine Principle; Principle of Minimum Kinetic Energy; Principle of

Minimum Mechanical Energy; Principle of Minimum Potential Energy; Quantum Hamilton Principle; Stewart-Hamilton Principle; Stochastic Hamilton Variational Principle; Toupin's Dual Principle

ELASTICITY Energy Principle; Hamilton's Principle; Least Action Principle; Mueller-Breslau Principle; Orthogonality Principle; Principle of Correspondence; Principle of Least Work; Saint-Venant's Principle; Woltjer Minimum Energy Principle

ELASTODYNAMICS Caratheodory's Principle; Conte's Principle; Fermat's Principle; Fermats Principle; Least Time Principle; Principle of Caratheodony; Stationary Time Principle

ELASTOMER D'Alembert Lagrange Principle; D'Alembert Lagrange Variation Principle; D'Alembert Principle; Gauss Principle; Hamilton Lagrange Principle; Jayne's Principle; Jourdain Principle; Kinetic Principle; Lagrange Principle; Least Constraint Principle; Maupertius Lagrange Principle; Maupertuis Euler Lagrange Principle of Least Action; Principle of Extremal Effects; Principle of Least Constraint; Principle of Maximal Effect; Principle of Virtual Work

ELECTRIC CIRCUITS Hamilton's Principle; Kirchhoff's Voltage Law Principle; Minimum Energy Principle; Variational Principle; Woltjer Minimum Energy Principle; Woltjer-Taylor Minimum Energy Principle

ELECTRIC COMPONENTS Principle of Fundamental Characteristics

ELECTRIC CURRENT Minimal Principle; Monti Principle, Inference Principle; Jaynes' Principle; Kullback's Minimum Cross-entropy Principle; Maximum Information Entropy Principle; Measurement Principle; Easuring Principle; Minimax Entropy Principle; Minimum Energy Principle; Minimum Norm Principle; Minimum Relative Entropy Principle; Modification of The Extended Maximum Entropy Principle; Principle of Maximum Entropy; Principle of Maximum Nonadditive Entropy; Principle of Observational Compatibility; Principle

of Operational Compatibility; Uncertainty Principle

ELECTRIC NETWORK Duality Principle, Multi Tanh Principle; Principle of Fundamental Characteristics

ELECTRIC RESISTANCE Division Principle; Load Division Principle; Minimal Principle; Principle of Fundamental Characteristics; Time Division Principle; Variable Time Position Channel Assignment Principle

ELECTRIC SENSING DEVICES Inference Principle; Jaynes' Principle; Kullback's Minimum Cross-entropy Principle; Maximum Information Entropy Principle; Measurement Principle; Measuring Principle; Minimax Entropy Principle; Minimum Energy Principle; Minimum Norm Principle; Minimum Relative Entropy Principle; Modification of The Extended Maximum Entropy Principle; Principle of Maximum Entropy; Principle of Maximum Nonadditive Entropy; Principle of Observational Compatibility; Principle of Operational Compatibility; Uncertainty Principle

ELECTRICAL ENGINEERING Black hole Principle; General Relativistic Fermat Principle; Optical Black Hole Principle; Processing Principle; Reflectionless Aperture Principle, Minimum Information Principle

ELECTRICITY Asymptotic Matching Principle; Gas Turbine Principle; General Relativistic Fermat Principle; General Relativistic Fermat Principle; Principle of Fundamental Characteristics; Black hole Principle; Optical Black Hole Principle; Reflectionless Aperture Principle, Engine Principle

ELECTROCHEMISTRY Derivation Principle; New Principle, Quartz Microbalance Principle

ELECTRODYNAMICS Babinet's Principle; Complementarity Principle; Minimum Description Length Principle; Repulsion Principle, Principle of Correspondence, Correspondence Principle

ELECTROENCEPHALOGRAPHY Entropy Maximum Principle; Correspondence Principle; Extended Maximum Entropy Principle; Extremal Principle of Information; Frieden-Soffer Principle; Gibbs Principle; Inference Principle; Jaynes' Maximum Entropy Principle; Jaynes' Principle; Kullback's Minimum Cross-entropy Principle; Maximum Information Entropy Principle; Measurement Principle; Minimax Entropy Principle; Minimum Entropy Principle; Minimum Relative Entropy Principle; Modification of the Extended Maximum Entropy Principle; Principle of Maximum Entropy; Principle of Maximum Nonadditive Entropy; Principle of Observational Compatibility; Principle of Operational Compatibility; Principle of Remifentanil; Uncertainty Principle

ELECTROHYDRODYNAMICS Hamilton's Principle; Kirchhoff's Voltage Law Principle; Minimum Energy Principle; Variational Principle; Woltjer Minimum Energy Principle; Woltjer-Taylor Minimum Energy Principle

ELECTROMAGNETIC FIELDS Black hole Principle; Optical Black Hole Principle; Equivalence Principle; Huygen's Principle; Precautionary Principle; Principle of Fundamental Characteristics; Projection Principle; Reflectionless Aperture Principle

ELECTROMAGNETIC RADIATION Equivalent Background Principle; Huygen's Principle; Minimum Description Length Principle

ELECTROMAGNETIC SHIELDING Black hole Principle; Equivalence Principle; Optical Black Hole Principle; Reflectionless Aperture Principle

ELECTROMAGNETIC WAVES Black hole Principle; Equivalence Principle; Huygen's Principle; Maximum Likelihood Principle; Minimum Description Length Principle; Optical Black Hole Principle; Principle of Relativity; Projection Principle; Reflectionless Aperture Principle

ELECTROMAGNETISM Principle of Relativity; Strong Equivalence Principle, Equivalence Principle; Magnetron Principle

ELECTRON STRUCTURE Equivalence Principle; Mach Principle; Principle of Equivalence

ELECTRONEGATIVITY Electronegativity Equalization Principle; Electromagnetic Induction Principle, Equalization Principle; Geometric Mean Principle; Hard-soft acid-base Principle; Maximum Hardness Principle; Maximum Hardness Principle; Principle of electronegativity

ELECTRONIC CIRCUITS Matching Principle; New Therapeutic Principle; Pseudo-2-path Principle

ELECTRONIC COMPONENTS Memory Compensation Principle; Mitrofanoff Principle; Monti Principle; Pseudo-2-path Principle

ELECTRONIC EQUIPMENT Black hole Principle; Optical Black Hole Principle; Reflectionless Aperture Principle

ELECTRONICS Approximation Induction Principle; Archimedes' Principle; Asymmetric Induction Principle; Asymptotic Matching Principle; Biot's Variational Principle; Bbone Induction Principle; Coverset Induction Principle; Data Sending Principle; Deming Principle; Dynamic Optimization Principle; Enantiodi Vergient Induction Principle; Flow Acting Principle; Half-call Principle; Hydrostatic Principle; Induction Measurement Principle; Lysogenic Induction Principle; Magnetic Induction Principle; Maximum Likelihood Principle; Osteogenic Induction Principle; Principle of Timedomain Diakoptics; Proportional Induction Principle; Structural linearization Principle; Upside Down Deming Principle

ELECTRONICS AND COMMUNICATIONS Black hole Principle; Optical Black Hole Principle; Reflectionless Aperture Principle

ELECTRONICS AND COMPONENTS Memory Compensation Principle; Mitrofanoff Principle; Monti Principle

ELECTROPHORESIS Capillary Electrophoresis Principle; Ion Pair Principle; Principle of Diastereomeric Ion Pair Formation

ELECTROPHYSIOLOGY Hamilton's Principle; Kirchhoff's Voltage Law Principle; Minimum Energy Principle; Variational Principle; Woltjer Minimum Energy Principle; Woltjer-Taylor Minimum Energy Principle

ELECTROSTATIC LOUD SPEAKER Electrostatic Principle; Foil Electric Principle; Electrostatic Projection Principle; Projection Principle

ELEMENTARY PARTICLE Huygen's Principle; Pauli Principle, Minimum Description Length Principle, Minimal Principle, Equivalence Principle

ELLIPTIC OPERATOR Boric Acid Affinity Principle; Boronic Acid Affinity Principle; Harnack Principle

ELONGATION Principle of Correspondence, Timetemperature Superposition Principle

EM SHIELDING EFFECTIVENESS Black hole Principle; Optical Black Hole Principle; Reflectionless Aperture Principle

EMBRYOLOGY Maupertuis Principle; Nameless Category N3C Principle; Neighborhood Coherence Principle; Optimality Principle; Organizing Principle; Principle of component analysis; Self-organizing Principle; Similarity Principle; Type II Cell Stretch-organizing Principle

EMPIRICAL DISTRIBUTION Correspondence Principle; Entropy Maximum Principle; Extended Maximum Entropy Principle; Extremal Principle of Information; Frieden-Soffer Principle; Gibbs Principle; Inference Principle; Jaynes' Maximum Entropy Principle; Jaynes' Principle; Kullback's Minimum Cross-entropy Principle; Maximum Information Entropy Principle; Measurement Principle; Minimax Entropy Principle; Minimum Entropy Principle; Minimum Relative Entropy Principle; Modification of The Extended Maximum Entropy Principle; Principle of Maximum Entropy; Principle of Maximum Nonadditive Entropy; Principle of Observational Compatibility; Principle of Operational Compatibility; Uncertainty Principle

ENDOCRINE SYSTEM Boric Acid Affinity Principle; Boronic Acid Affinity Principle; Optimality Principle; Organizing Principle; Principle of component analysis; Self-organizing Principle; Similarity Principle; Therapeutic Principle; Type II Cell Stretch-organizing Principle; Principle of Measurement, Timing Principle; Type II Cell Stretch-organizing Principle

ENERGY Aufbau Principle; Equivalence Principle; Minimum Description Length Principle; Protection Principle, Extremum Principle

ENERGY CONSERVATION Crushing Principle; Mach Principle; Principle of Equivalence; Principle of Corresponding States, Principle of molecular dynamics; Equivalence Principle; Principle of Conservation of Energy; Principle of Material Frame Indifference; Processing Principle

ENGINE TEST STAND Inference Principle; Jaynes' Principle; Kullback's Minimum Cross-entropy Principle; Maximum Information Entropy Principle; Measurement Principle; Measuring Principle; Minimax Entropy Principle; Minimum Energy Principle; Minimum Norm Principle; Minimum Relative Entropy Principle; Modification of The Extended Maximum Entropy Principle; Principle of Maximum Entropy; Principle of Maximum Nonadditive Entropy; Principle of Observational Compatibility; Principle of Operational Compatibility; Uncertainty Principle

ENGINEERING MANAGEMENT Consumer Principle; General Relativistic Fermat Principle; Polluter Pays Principle

ENGINEERING Approximation Induction Principle; Argument Fetching Principle; Asymmetric Induction Principle; Asymptotic Matching Principle; Biot's Variational Principle; Bbone Induction Principle; Capture Principle; Coverset Induction Principle; Dale's Principle; Deming Principle; Dynamic Optimization Principle; Enantiodi Vergient Induction Principle; Flotation Polishing Principle; Half-call Principle; Induction Measurement Principle; Lysogenic

Induction Principle; Magnetic Induction Principle; Maximum Likelihood Principle; Osteogenic Induction Principle; Principle of A Ball Piston Pump; Proportional Induction Principle; Sodium Heat Engine Principle; Structural Linearization Principle, Unified Variational Principle; Upside Down Deming Principle; Orthogonality Principle, Polluter Pays Principle

ENGINEERING MATHEMATICS Adaptive Principle; Contraction Principle; Gibbs Variational Principle; Large Deviation Principle; Memory Compensation Principle; Mitrofanoff Principle; Monti Principle; Principle of Causality

ENTROPY Active Principle; Bateman's Principle; Competitive Exclusion Principle; Correspondence Principle; Engine Principle; Entropy Maximum Principle; Entropy Minimax Principle; Extended Maximum Entropy Principle; Extremal Principle of Information; First Principle; Frieden-Soffer Principle; Gibbs Principle; Inference Principle; Jaynes' Maximum Entropy Principle; Jaynes' Principle; Kullback's Minimum Cross-entropy Principle; Maximum Hardness Principle; Maximum Information Entropy Principle; Maximum Principle; Measurement Principle; Minimax Entropy Principle; Minimum Description Length Principle; Minimum Entropy Principle; Minimum Relative Entropy Principle; Modification of The Extended Maximum Entropy Principle; Principle of Complementarity; Principle of Material Frame Indifference; Principle of Maximum Entropy; Principle of Maximum Nonadditive Entropy; Principle of Observational Compatibility; Principle of Operational Compatibility; Principle of Uncertainty; Uncertainty Principle

ENTROPY CORRECTION INTERMOL POLEMIC Conservation of Momentum Principle; D'Alembert Principle; Differential Principle; Hamilton Principle; Newton's Third Principle; Principle of Equal A Priori Probabilities; Variational Principle

ENTROPY OPTIMIZATION Correspondence Principle; Entropy Maximum Principle;

Extended Maximum Entropy Principle; Extremal Principle of Information; Frieden-Soffer Principle; Gibbs Principle; Inference Principle; Jaynes' Maximum Entropy Principle; Jaynes' Principle; Kullback's Minimum Cross-entropy Principle; Maximum Information Entropy Principle; Measurement Principle; Minimax Entropy Principle; Minimum Entropy Principle; Minimum Relative Entropy Principle; Modification of The Extended Maximum Entropy Principle; Principle of Maximum Entropy; Principle of Maximum Nonadditive Entropy; Principle of Observational Compatibility; Principle of Operational Compatibility; Uncertainty Principle

ENUMERATIVE CODING Principle of Measurement

ENVIRONMENT Active Principle; Hard-soft acid-base Principle; China Ninth Principle

ENVIRONMENTAL AND INDUSTRIAL TOXICOLOGY Optimization Principle; Run and Recreate Principle

ENVIRONMENTAL BIOLOGY Ecological Principle; Extremal Principle; Frequency-dependent Principle; Optimality Principle; Optimization Principle; Organizing Principle; Precautionary Principle; Principle of component analysis; Principle of Comparative Advantage; Principle of Dynamic Ecology; Run and Recreate Principle; Self-organizing Principle; Similarity Principle; Wheatstone Bridge Principle

ENVIRONMENTAL HEALTH Minimal Principle; Optimization Principle; Run and Recreate Principle; Working Principle

ENVIRONMENTAL INTERACTION Economic Principle; Principles of Determination; Principles of Multiple Function; Principles of Overdetermination; Regulatory Principles and Other Regulatory Principles of Inertia; Regulatory Principles of Constancy; Regulatory Principles of Pleasureunpleasure; Regulatory Principles of Primarysecondary Process; Regulatory Principles of Reality; Regulatory Principles of Repetition

ENVIRONMENTAL MANAGEMENT Precautionary Principle; Principle of Briquetting; Protection Principle

ENVIRONMENTAL POLICY Polluter Pays Principle; Precautionary Principle

ENVIRONMENTAL PROTECTION Carousel Principle; Precautionary Principle; Polluter Pays Principle; Protection Principle

ENVIRONMENTAL SEX DETERMINATION Epistatic Handicap Principle; Handicap Principle

ENZYMES Principle of Measurement, Maximizing Principle, ELISA Principle, Mimetic Principle, Affinity Principle; Principle of Differential Ph Measurement

EOMETRIC QUANTITY Correspondence Principle; Entropy Maximum Principle; Extended Maximum Entropy Principle; Extremal Principle of Information. Frieden-Soffer Principle; Gibbs Principle; Inference Principle; Jaynes' Maximum Entropy Principle; Jaynes' Principle; Kullback's Minimum Cross-entropy Principle; Maximum Information Entropy Principle; Measurement Principle; Minimax Entropy Principle; Minimum Entropy Principle; Minimum Relative Entropy Principle; Modification of The Extended Maximum Entropy Principle; Principle of Maximum Entropy; Principle of Maximum Nonadditive Entropy; Principle of Observational Compatibility; Principle of Operational Compatibility; Uncertainty Principle

EPR PARADOX Minimum Description Length Principle; Principle of Maximum Entropy; Equal Passage Times Principle; Principle of Relativity

EQUATION OF MOTION D'Alembert Lagrange Principle; D'Alembert Lagrange Variation Principle; D'Alembert's Principle; Gauss Principle; Gyarmati's Principle; Hamilton Lagrange Principle; Hamilton Principle; Internal Model Principle; Jayne's Principle; Jourdain Principle; Kinetic Principle; Lagrange Principle; Least Constraint Principle; Maupertius Lagrange Principle; Maupertuis Euler Lagrange Principle of Least Action;

Prigogine Principle; Principle of Extremal Effects; Principle of Least Constraint; Principle of Maximal Effect; Principle of Minimum Kinetic Energy; Principle of Minimum Mechanical Energy; Principle of Minimum Potential Energy; Principle of Virtual Work; Quantum Hamilton Principle; Stewart-Hamilton Principle; Stochastic Hamilton Variational Principle; Toupin's Dual Principle

EQUATION OF STATE Minimum Principle; Principle of Correspondence, Minimum Description Length Principle

EQUATION RESOLUTION Maximum Principle

EQUATIONAL THEOREM PROVER Extension Principle

EQUATIONS Drug Principle, Corresponding States Principle; Pitzer Corresponding States Principle;, Saint-Venant's Principle, Pontryagins Principle

EQUATIONS OF MOTION Gauss' Principle; Optical Principle; Principle of Material Frame Indifference

EQUATIONS OF STATE Corresponding States Principle; Pitzer Corresponding States Principle; Principle of Congruence; Principle of Corresponding States

EQUIDAE Maximizing Principle; Equilibrium Analysis With Elastic/Stiffer Energy; Energy Principle; Global Extremum Principle; Woltjer Minimum Energy Principle

EQUILIBRIUM STATE Correspondence Principle; Entropy Maximum Principle; Extended Maximum Entropy Principle; Extremal Principle of Information; Frieden-Soffer Principle; Gibbs Principle; Inference Principle; Jaynes' Maximum Entropy Principle; Jaynes' Principle; Kullback's Minimum Cross-entropy Principle; Maximum Information Entropy Principle; Measurement Principle; Minimax Entropy Principle; Minimum Entropy Principle; Minimum Relative Entropy Principle; Modification of The Extended Maximum Entropy Principle; Principle of Maximum Entropy; Principle of

Maximum Nonadditive Entropy; Principle of Observational Compatibility; Principle of Operational Compatibility; Uncertainty Principle; Equilibrium Structures; Principle of molecular dynamics

EQUIPMENT DESIGN Inference Principle; Jaynes' Principle; Kullback's Minimum Cross-entropy Principle; Maximum Information Entropy Principle; Measurement Principle; Measuring Principle; Minimax Entropy Principle; Minimum Energy Principle; Minimum Norm Principle; Minimum Relative Entropy Principle; Modification of The Extended Maximum Entropy Principle; Principle of Maximum Entropy; Principle of Maximum Nonadditive Entropy; Principle of Observational Compatibility; Principle of Operational Compatibility; Uncertainty Principle; Equivalence. Minimum Description Length Principle; Equivalence of Ensembles. Correspondence Principle; Entropy Maximum Principle; Extended Maximum Entropy Principle; Extremal Principle of Information; Frieden-Soffer Principle; Gibbs Principle; Inference Principle; Jaynes' Maximum Entropy Principle; Jaynes' Principle; Kullback's Minimum Cross-entropy Principle; Maximum Information Entropy Principle; Measurement Principle; Minimax Entropy Principle; Minimum Entropy Principle; Minimum Relative Entropy Principle; Modification of The Extended Maximum Entropy Principle; Principle of Maximum Entropy; Principle of Maximum Nonadditive Entropy; Principle of Observational Compatibility; Principle of Operational Compatibility; Uncertainty Principle

ERGODIC TRANSFORMATIONS Contagion Principle; Contraction Principle; Gibbs Variational Principle; Large Deviation Principle

ERGODICITY Ambartsumian Invariance Principle; Charnov's Invariance Principle; Dirichlet Principle; Donsker's Invariance Principle; Invariance Principle; Invariance Principle of General Relativity; Local Invariance Principle; Maruyama's Invariance Principle; Perturbation Invariance Principle; Principle of Invariance; Principle of Undulatory Invariance; Relaxed Invariance

Principle; Scale Invariance Principle; Strong Principle; Variational Principle; Weak Invariance Principle

ERROR RATE Contraction Principle; Gibbs Variational Principle; Large Deviation Principle

ERRORS Principle of Material Frame Indifference; Erythrocytes. Inference Principle; Jaynes' Principle; Kullback's Minimum Cross-entropy Principle; Maximum Information Entropy Principle; Measurement Principle; Measuring Principle; Minimax Entropy Principle; Minimum Energy Principle; Minimum Norm Principle; Minimum Relative Entropy Principle; Modification of The Extended Maximum Entropy Principle; Principle of Maximum Entropy; Principle of Maximum Nonadditive Entropy; Principle of Observational Compatibility; Principle of Operational Compatibility; Uncertainty Principle

ESTIMATION Active Principle; Correspondence Principle; Entropy Maximum Principle; Extended Maximum Entropy Principle; Extremal Principle of Information; Frieden-Soffer Principle; Gibbs Principle; Inference Principle; Jaynes' Maximum Entropy Principle; Jaynes' Principle; Kullback's Minimum Cross-entropy Principle; Maximum Entropy Principle; Maximum Information Entropy Principle; Measurement Principle; Minimax Entropy Principle; Minimum Entropy Principle; Minimum Relative Entropy Principle; Modification of The Extended Maximum Entropy Principle; Principle of Maximum Entropy; Principle of Maximum Nonadditive Entropy; Principle of Observational Compatibility; Principle of Operational Compatibility; Uncertainty Principle

ESTIMATOR Ambartsumian Invariance Principle; Charnov's Invariance Principle; Donsker's Invariance Principle; Invariance Principle; Invariance Principle of General Relativity; Local Invariance Principle; Maruyama's Invariance Principle; Perturbation Invariance Principle; Principle of Invariance; Principle of Undulatory Invariance; Relaxed Invariance Principle; Scale Invariance Principle; Strong Principle; Variational Principle; Weak Invariance Principle

ETHICS Casuistic Principle; Double Effect Principle; Exception Granting Principle; Justifying Principle; Principle of Do No Harm; Principle of Double Effect; Principles of Biomedical Ethics

EULER LAGRANGE EQUATION Least Action Principle; Principle of Least Work; Maximum Principle; Pontryagin Maximum Principle

EVOKED POTENTIAL MONITORING Hamilton's Principle; Kirchhoff's Voltage Law Principle; Minimum Energy Principle; Variational Principle; Woltjer Minimum Energy Principle; Woltjer-Taylor Minimum Energy Principle

EVOLUTION AND ADAPTATION Entropy Principle; Extremum Principle; Fitness Principle; Frequency-dependent Principle; Fuzzy Entropy Principle; Maxent Principle; Maximum Entropy Principle; Mep Maximum Entropy Principle; Mermin Entropy Principle; Optimizing Principle; Pme Principle of Maximum Entropy; Principle of Maximum Nonadditive Entropy; Principle of Operational Compatibility

EVOLUTION Active Principle; Epistatic Handicap Principle; Extremum Principle; Fisher's Principle; Fisheye Adaptation Principle; Fitness Principle; Frequency-dependent Principle; Handicap Principle; Maximizing Principle; Optimizing Principle; Uncertainty Principle

EVOLUTION EQUATION Equivalence Principle; Maximum Principle; Saint-Venant's Principle

EVOLUTION OF MATERNAL CHARACTERS Epistatic Handicap Principle; Handicap Principle

EVOLUTIONARY LARGE SCALE OPTIMIZATION Correspondence Principle; Entropy Maximum Principle; Extended Maximum Entropy Principle; Extremal Principle of Information; Frieden-Soffer Principle; Gibbs Principle; Inference Principle; Jaynes' Maximum Entropy Principle; Jaynes' Principle; Kullback's Minimum Cross-entropy Principle; Maximum Information Entropy Principle; Measurement

Principle; Minimax Entropy Principle; Minimum Entropy Principle; Minimum Relative Entropy Principle; Modification of The Extended Maximum Entropy Principle; Principle of Maximum Entropy; Principle of Maximum Nonadditive Entropy; Principle of Observational Compatibility; Principle of Operational Compatibility; Uncertainty Principle

EVOLVING SYSTEM PROCESSES Combination-separation Principle; Ritz's Combination Principle

EXCESS ENTHALPY PREDICTION Corresponding States Principle; Pitzer Corresponding States Principle; Principle of Correspondence

EXCESS MOLAR ENTHALPIES Corresponding States Principle; Pitzer Corresponding States Principle; Principle of Correspondence

EXCITED STATE First Principle; Maximum Principle; Minimum Description Length Principle

EXPECTED UTILITY Extension Principle; Maximization Principle; Principle of Uncertainty; Tnorm Extension Principle

EXPERIMENTAL DESIGN Correspondence Principle; Entropy Maximum Principle; Extended Maximum Entropy Principle; Extremal Principle of Information; Frieden-Soffer Principle; Gibbs Principle; Inference Principle; Jaynes' Maximum Entropy Principle; Jaynes' Principle; Kullback's Minimum Cross-entropy Principle; Maximum Information Entropy Principle; Measurement Principle; Minimax Entropy Principle; Minimum Entropy Principle; Minimum Relative Entropy Principle; Modification of The Extended Maximum Entropy Principle; Principle of Information Processing; Principle of Maximum Entropy; Principle of Maximum Nonadditive Entropy; Principle of Observational Compatibility; Principle of Operational Compatibility; Uncertainty Principle

EXPERT SYSTEMS Correspondence Principle; Entropy Maximum Principle; Extended Maximum Entropy Principle; Extremal

Principle of Information; Frieden-Soffer Principle; Gibbs Principle; Inference Principle; Jaynes' Maximum Entropy Principle; Jaynes' Principle; Kullback's Minimum Cross-entropy Principle; Maximum Information Entropy Principle; Measurement Principle; Minimax Entropy Principle; Minimum Entropy Principle; Minimum Relative Entropy Principle; Modification of The Extended Maximum Entropy Principle; Principle of Maximum Entropy; Principle of Maximum Nonadditive Entropy; Principle of Observational Compatibility; Principle of Operational Compatibility; Uncertainty Principle, Principle of Information Processing, Principle of Uncertainty

EXTERNAL MAGNETIC PRESSURE FORCES
Hamilton's Principle; Kirchhoff's Voltage Law Principle; Minimum Energy Principle; Variational Principle; Woltjer Minimum Energy Principle; Woltjer-Taylor Minimum Energy Principle

F

FAINT BLUE GALAXIES Caratheodory's Principle; Conte's Principle; Fermat's Principle; Fermats Principle; Least Time Principle; Principle of Caratheodony; Stationary Time Principle

FASCIAE Archimedes' Principle; Hydrostatic Principle; Mckenzie Principles; Principle-centered Spine Care; Syndrome Principle; Unlearned Principle

FAST COMPUTATION Principle of molecular dynamics

FAST STOCHASTIC GLOBAL OPTIMIZATION Correspondence Principle; Entropy Maximum Principle; Extended Maximum Entropy Principle; Extremal Principle of Information; Frieden-Soffer Principle; Gibbs Principle; Inference Principle; Jaynes' Maximum Entropy Principle; Jaynes' Principle; Kullback's Minimum Cross-entropy Principle; Maximum Information Entropy Principle; Measurement Principle; Minimax Entropy Principle; Minimum Entropy Principle; Minimum Relative Entropy Principle; Modification of The Extended Maximum Entropy Principle; Principle of Maximum Entropy; Principle of Maximum Nonadditive Entropy; Principle of Observational Compatibility; Principle of Operational Compatibility; Uncertainty Principle

FAST TWO DIMENSIONAL ENTROPIC THRESHOLDING ALGORITHM Entropy Principle; Generalized Minimal Principle; Maximum Entropy Principle; Minimum Cross-entropy Principle; Moment Preserving Principle; Principle of Minimum Cross-entropy; Variational Minimal Principle; Complexity Principle

FAULT DIAGNOSIS Active Principle; First Principle; Maximum Entropy Principle; New Therapeutic Principle

FAULT LOCATION Maximum Entropy Principle; New Therapeutic Principle; Protection Principle

FEATURE MODEL Action Principle; Schwinger's Quantum Action Principle

FECAL INCONTINENCE Mitrofanoff Principle; Mitrofanoff Principle; Monti Principle

FEED COMPOSITION Maximizing Principle

FEED FORWARD Maximum Entropy Principle, Parsimony Principle; Pontryagin's Maximum Principle, Principle of Parsimonius Data Modeling

FEEDBACK Desoption Principle; Dirichlet Principle; Maximum Principle

FEEDBACK CONTROL Commutation Principle; Neighborhood Coherence Principle

FERMION Action Principle; Minimal Principle; Pauli Principle; Schwinger's Quantum Action Principle

FERROUS ALLOYS Correspondence Principle; Entropy Maximum Principle; Extended Maximum Entropy Principle; Extremal Principle of Information; Frieden-Soffer Principle; Gibbs Principle; Inference Principle; Jaynes' Maximum Entropy Principle; Jaynes' Principle; Kullback's Minimum Cross-entropy Principle; Maximum Information Entropy Principle; Measurement Principle; Minimax Entropy Principle; Minimum Entropy Principle; Minimum Relative Entropy Principle; Modification of The Extended Maximum Entropy Principle; Principle of Maximum Entropy; Principle of Maximum Nonadditive Entropy; Principle of Observational Compatibility; Principle of Operational Compatibility; Uncertainty Principle

FIELD EQUATION Corresponding States Principle; Huygen's Principle; Pitzer

Corresponding States Principle; D'Alemberts Principle; Dissipation Principle; Virtual Dissipation Principle

FILTER CIRCUITS (TRANSMISSION LINE SEGMENTS) Reciprocity Principle; Filter Performance. Correspondence Principle; Entropy Maximum Principle; Extended Maximum Entropy Principle; Extremal Principle of Information; Frieden-Soffer Principle; Gibbs Principle; Inference Principle; Jaynes' Maximum Entropy Principle; Jaynes' Principle; Kullback's Minimum Cross-entropy Principle; Maximum Information Entropy Principle; Measurement Principle; Minimax Entropy Principle; Minimum Entropy Principle; Minimum Relative Entropy Principle; Modification of The Extended Maximum Entropy Principle; Principle of Maximum Entropy; Principle of Maximum Nonadditive Entropy; Principle of Observational Compatibility; Principle of Operational Compatibility; Uncertainty Principle

FILTERING METHOD Correspondence Principle; Entropy Maximum Principle; Extended Maximum Entropy Principle; Extremal Principle of Information; Frieden-Soffer Principle; Gibbs Principle; Inference Principle; Jaynes' Maximum Entropy Principle; Jaynes' Principle; Kullback's Minimum Cross-entropy Principle; Maximum Information Entropy Principle; Measurement Principle; Minimax Entropy Principle; Minimum Entropy Principle; Minimum Relative Entropy Principle; Modification of The Extended Maximum Entropy Principle; Principle of Maximum Entropy; Principle of Maximum Nonadditive Entropy; Principle of Observational Compatibility; Principle of Operational Compatibility; Uncertainty Principle

FINITE DEFORMATION ELASTOPLASTICITY Educational Principle; Ehrenfest Principle; Principle of Correspondence

FINITE DOMAIN Correspondence Principle; Entropy Maximum Principle; Extended Maximum Entropy Principle; Extremal Principle of Information; Frieden-Soffer Principle; Gibbs Principle; Inference Principle; Jaynes' Maximum Entropy Principle;

Jaynes' Principle; Kullback's Minimum Cross-entropy Principle; Maximum Information Entropy Principle; Measurement Principle; Minimax Entropy Principle; Minimum Entropy Principle; Minimum Relative Entropy Principle; Modification of The Extended Maximum Entropy Principle; Principle of Maximum Entropy; Principle of Maximum Nonadditive Entropy; Principle of Observational Compatibility; Principle of Operational Compatibility; Uncertainty Principle

FINITE ELEMENT ANALYSIS Maximum Principle; Projection Principle, Mesh Independence Principle, Hamilton's Principle

FINITE ELEMENT METHOD D'Alembert Lagrange Principle; D'Alembert Lagrange Variation Principle; D'Alembert Principle; Energy Principle; Gauss Principle; Gyarmati's Principle; Hamilton Lagrange Principle; Hamilton Principle; Hellinger-Reissner Principle; Hu Washizu Variational Principle; Huygen's Principle; Jayne's Principle; Jourdain Principle; Kinetic Principle; Lagrange Principle; Least Constraint Principle; Least Square Principle; Maupertius Lagrange Principle; Maupertuis Euler Lagrange Principle of Least Action; Mueller-Breslau Principle; Prigogine Principle; Principle of Extremal Effects; Principle of Least Constraint; Principle of Maximal Effect; Principle of Minimum Kinetic Energy; Principle of Minimum Mechanical Energy; Principle of Minimum Potential Energy; Principle of Virtual Work; Pseudogeneralized Variational Principle; Quantum Hamilton Principle; Stewart-Hamilton Principle; Stochastic Hamilton Variational Principle; Toupin's Dual Principle; Virtual Force Principle; Woltjer Minimum Energy Principle; Working Principle

FINITE ELEMENT THERMAL RADIATION Discrete Maximum Principle; Discrete Pontryagin Maximum Principle; Dmp; Krylov Maximum Principle; Pontryagin Discrete Maximum Principle

FINITE IMPULSE RESPONSE FILTERS Complementary Principle; Complementary Variational Principle; Dual Principle; Dual Variational Principle; Nonconvex Variational Principle; Variational Principle

FINITE RANGE SCATTERING FUNCTION Complex Kohn Variational Principle; HULTHÉN–KOHNVariational Principle; Kohn Variational Principle; Schwinger Variational Principle

FIRST PASSAGE TIME Ambartsumian Invariance Principle; Charnov's Invariance Principle; Dirichlet Principle; Donsker's Invariance Principle; Invariance Principle; Invariance Principle of General Relativity; Local Invariance Principle; Maruyama's Invariance Principle; Perturbation Invariance Principle; Principle of Invariance; Principle of Undulatory Invariance; Relaxed Invariance Principle; Scale Invariance Principle; Strong Principle; Variational Principle; Weak Invariance Principle; First Principle Real-space Simulation

FIRST ORDER POLES Minimum Description Length Principle; First-Principle Forces; First Principle; Principle of molecular dynamics

FLASH BLINDNESS Equidistortion Principle; Quantization Design Principle

FLAT OPEN FLOAT CHAMBER Bell Principle; Jarman Bell Principle; Polamyi Evans Bell Principle

FLOATING PERMANENT MAGNETS Inference Principle; Jaynes' Principle; Kullback's Minimum Cross-entropy Principle; Maximum Information Entropy Principle; Measurement Principle; Measuring Principle; Minimax Entropy Principle; Minimum Energy Principle; Minimum Norm Principle; Minimum Relative Entropy Principle; Modification of The Extended Maximum Entropy Principle; Principle of Maximum Entropy; Principle of Maximum Nonadditive Entropy; Principle of Observational Compatibility; Principle of Operational Compatibility; Uncertainty Principle

FLOW Three-day Principle; Correspondence Principle; Educational Principle; Ehrenfest Principle; Entropy Maximum Principle; Extended Maximum Entropy Principle; Extremal Principle of Information; Frieden-Soffer Principle; Gibbs Principle; Inference Principle; Jaynes' Maximum Entropy Principle; Jaynes' Principle; Kullback's Minimum Cross-entropy Principle; Maximum Information Entropy Principle; Measurement Principle; Minimax Entropy Principle; Minimum Entropy Principle; Minimum Relative Entropy Principle; Modification of The Extended Maximum Entropy Principle; Principle of Maximum Entropy; Principle of Maximum Nonadditive Entropy; Principle of Observational Compatibility; Principle of Operational Compatibility; Uncertainty Principle

FLOW CYTOMETRY Inference Principle; Jaynes' Principle; Kullback's Minimum Cross-entropy Principle; Maximum Information Entropy Principle; Measurement Principle; Measuring Principle; Minimax Entropy Principle; Minimum Energy Principle; Minimum Norm Principle; Minimum Relative Entropy Principle; Modification of The Extended Maximum Entropy Principle; Principle of Maximum Entropy; Principle of Maximum Nonadditive Entropy; Principle of Observational Compatibility; Principle of Operational Compatibility; Uncertainty Principle

FLOW MEASUREMENT Inference Principle; Jaynes' Principle; Kullback's Minimum Cross-entropy Principle; Maximum Information Entropy Principle; Measurement Principle; Measuring Principle; Minimax Entropy Principle; Minimum Energy Principle; Minimum Norm Principle; Minimum Relative Entropy Principle; Modification of The Extended Maximum Entropy Principle; Principle of Maximum Entropy; Principle of Maximum Nonadditive Entropy; Principle of Observational Compatibility; Principle of Operational Compatibility; Uncertainty Principle; Coriolis Principle

FLOW RATE CONTROLLER FOR AIR DUCTS Geometric Principle; Iris Diaphragm Principle; Iris Principle

FLOW VELOCITY Compressible Flow Principle; Correspondence Principle; Entropy Maximum Principle; Extended Maximum Entropy Principle; Extremal Principle of Information; Frieden-Soffer Principle; Gibbs Principle; Inference Principle; Jaynes'

Maximum Entropy Principle; Jaynes' Principle; Kullback's Minimum Cross-entropy Principle; Maximum Information Entropy Principle; Measurement Principle; Measuring Principle; Minimax Entropy Principle; Minimum Energy Principle; Minimum Entropy Principle; Minimum Norm Principle; Minimum Relative Entropy Principle; Modification of The Extended Maximum Entropy Principle; Principle of Maximum Entropy; Principle of Maximum Nonadditive Entropy; Principle of Observational Compatibility; Principle of Operational Compatibility; Uncertainty Principle, Principle of Laser Speckle Flowgraphy, Principle of Measurement

FLUCTUATION PHENOMENA Minimum Description Length Principle; Principle of Maximum Entropy; Competitive Exclusion Principle; Principle of Complementarity

FLUCTUATION Action Principle; Epistatic Handicap Principle; Handicap Principle; Schwinger's Quantum Action Principle

FLUID Absorption Principle; Compressible Flow Principle; Hard-soft acid-base Principle; Hsab Principle, Flow Acting Principle

FLUID DYNAMICS Conservation of Momentum Principle; D'Alembert Principle; Differential Principle; Hamilton Principle; Newton's Third Principle; Principle of Equal A Priori Probabilities; Principle of Minimum Dissipation; Variational Principle; Least Action Principle; Orthogonality Principle; Principle of Least Work, Product Solution Principle

FLUID MEDIUM Discrete Maximum Principle; Discrete Pontryagin Maximum Principle; Dmp; Krylov Maximum Principle; Maximum Principle; Pontryagin Discrete Maximum Principle; Working Principle

FLUID WAVE Correspondence Principle; Entropy Maximum Principle; Extended Maximum Entropy Principle; Extremal Principle of Information; Frieden-Soffer Principle; Gibbs Principle; Inference Principle; Jaynes' Maximum Entropy Principle; Jaynes' Principle; Kullback's Minimum Cross-entropy Principle; Maximum Information Entropy

Principle; Measurement Principle; Minimax Entropy Principle; Minimum Entropy Principle; Minimum Relative Entropy Principle; Modification of The Extended Maximum Entropy Principle; Principle of Maximum Entropy; Principle of Maximum Nonadditive Entropy; Principle of Observational Compatibility; Principle of Operational Compatibility; Uncertainty Principle

FLUIDIC GAS FLOWMETER Inference Principle; Jaynes' Principle; Kullback's Minimum Cross-entropy Principle; Maximum Information Entropy Principle; Measurement Principle; Measuring Principle; Minimax Entropy Principle; Minimum Energy Principle; Minimum Norm Principle; Minimum Relative Entropy Principle; Modification of The Extended Maximum Entropy Principle; Principle of Maximum Entropy; Principle of Maximum Nonadditive Entropy; Principle of Observational Compatibility; Principle of Operational Compatibility; Uncertainty Principle

FLUORESCENCE Caratheodory's Principle; Conte's Principle; Fermat's Principle; Fermats Principle; Least Time Principle; Principle of Caratheodony; Stationary Time Principle

FLUORESCENCE PROBE Caratheodory's Principle; Conte's Principle; Fermat's Principle; Fermats Principle; Least Time Principle; Principle of Caratheodony; Stationary Time Principle

FLUORESCENCE SPECTRUM Caratheodory's Principle; Conte's Principle; Fermat's Principle; Fermats Principle; Least Time Principle; Principle of Caratheodony; Stationary Time Principle; Fluorimetric Detection. Monti Principle; Flux Gate Magnetometer. Inference Principle; Jaynes' Principle; Kullback's Minimum Cross-entropy Principle; Maximum Information Entropy Principle; Measurement Principle; Measuring Principle; Minimax Entropy Principle; Minimum Energy Principle; Minimum Norm Principle; Minimum Relative Entropy Principle; Modification of The Extended Maximum Entropy Principle; Principle of Maximum Entropy; Principle of Maximum Nonadditive Entropy;

Principle of Observational Compatibility; Principle of Operational Compatibility; Uncertainty Principle

FOLLOWER FORCE D'Alembert's Principle; Gyarmati's Principle; Hamilton Principle; Prigogine Principle; Principle of Minimum Kinetic Energy; Principle of Minimum Mechanical Energy; Principle of Minimum Potential Energy; Quantum Hamilton Principle; Stewart-Hamilton Principle; Stochastic Hamilton Variational Principle; Toupin's Dual Principle

FORAGING Maximizing Principle, Area Ratio Principle; Epistatic Handicap Principle, Handicap Principle

FORCE FREE FIELD Hamilton's Principle; Kirchhoff's Voltage Law Principle; Minimum Energy Principle; Variational Principle; Woltjer Minimum Energy Principle; Woltjer-Taylor Minimum Energy Principle

FOREIGN TECHNOLOGY Minimum Information Principle; Orthogonality Principle; Principle of molecular dynamics; Principle of Relativity; Projection Principle

FORGETTING Contraction Principle; Contraction Principle; Gibbs Variational Principle; Large Deviation Principle

FORMAL CONSTANTS Hard-soft acid-base Principle

FORMAL LANGUAGES Correspondence Principle; Entropy Maximum Principle; Extended Maximum Entropy Principle; Extremal Principle of Information; Frieden-Soffer Principle; Gibbs Principle; Inference Principle; Jaynes' Maximum Entropy Principle; Jaynes' Principle; Kullback's Minimum Cross-entropy Principle; Maximum Information Entropy Principle; Measurement Principle; Minimax Entropy Principle; Minimum Entropy Principle; Minimum Relative Entropy Principle; Modification of The Extended Maximum Entropy Principle; Principle of Maximum Entropy; Principle of Maximum Nonadditive Entropy; Principle of Observational Compatibility; Principle of Operational Compatibility; Uncertainty Principle

FORMATION IMAGE Caratheodory's Principle; Conte's Principle; Fermat's Principle; Fermats Principle; Least Time Principle; Principle of Caratheodony; Stationary Time Principle

FORMULATION D'Alembert's Principle; Gyarmati's Principle; Hamilton Principle; Maximum Hardness Principle; Prigogine Principle; Principle of Minimum Kinetic Energy; Principle of Minimum Mechanical Energy; Principle of Minimum Potential Energy; Quantum Hamilton Principle; Stewart-Hamilton Principle; Stochastic Hamilton Variational Principle; Toupin's Dual Principle

FORMULATION MATHEMATICS Action Principle; Schwinger's Quantum Action Principle

FOUNDATION ENGINEERING Hamilton's Principle; Kirchhoff's Voltage Law Principle; Minimum Energy Principle; Variational Principle; Woltjer Minimum Energy Principle; Woltjer-Taylor Minimum Energy Principle

FOUNDATIONS Minimum Description Length Principle; Principle of Maximum Entropy; Principle of Measurement

FOURIER SERIES Bernoulli Principle, Uncertainty Principle, Equivalence Principle

FOURIER TRANSFORMS Frequency-dependent Principle; Principle of Maximum Entropy; Cyclotron Principle; Magnetron Principle

FREE BOUNDARY Comparison Principle; Hamilton's Principle; Kirchhoff's Voltage Law Principle; Minimum Energy Principle; Variational Principle; Woltjer Minimum Energy Principle; Woltjer-Taylor Minimum Energy Principle, Principle of Smooth Fit

FREE ENERGY Saint-Venant's Principle

FREE FALLING Equivalent Background Principle; Principle of Causality

FREE MAGNETIC FIELDS Hamilton's Principle; Kirchhoff's Voltage Law Principle; Minimum Energy Principle; Variational

Principle; Woltjer Minimum Energy Principle; Woltjer-Taylor Minimum Energy Principle

FREQUENCY RESPONSE Master-slave Principle; Maximum Entropy Principle; Minimum Information Principle

FREUDIAN THEORY Beyond The Pleasure Principle; Jewish Pleasure Principle; Pleasure Pain Principle; Pleasure Principle; Pleasurepain Principle

FRICTION Minimum Principle, Least Action Principle; Principle of Least Work

FRONTIER MOLECULAR ORBITAL Hard-soft acid-base Principle

FUEL INJECTION DEVICE Energy Storage Principle; Solidstate Energy Storage Principle

FUNCTION EVALUATION Contraction Principle; Gibbs Variational Principle; Large Deviation Principle

FUNCTION FEATURE MODEL Action Principle; Schwinger's Quantum Action Principle

FUNCTION MATHEMATICS Hellinger-Reissner Principle; Hu Washizu Variational Principle; Pseudogeneralized Variational Principle

FUNCTION THEORY Heisenberg's Correspondence Principle; Minimum Description Length Principle; Minimum Principle; Principle of Maximum Entropy; Principle of Correspondence

FUNCTIONAL ANALYSIS Korenblum's Maximum Principle; Minimal Principle, Heisenberg's Correspondence Principle; Minimum Principle; Minimum Description Length Principle

FUNCTIONAL INTEGRAL Contraction Principle; Gibbs Variational Principle; Large Deviation Principle

FUNCTIONAL LAW OF THE ITERATED LOGARITHM Ambartsumian Invariance Principle; Charnov's Invariance Principle; Dirichlet

Principle; Donsker's Invariance Principle; Invariance Principle; Invariance Principle of General Relativity; Local Invariance Principle; Maruyama's Invariance Principle; Perturbation Invariance Principle; Principle of Invariance; Principle of Undulatory Invariance; Relaxed Invariance Principle; Scale Invariance Principle; Strong Principle; Variational Principle; Weak Invariance Principle

FUNCTIONS Active Principle; Concept Principle; Correspondence Principle; Ekeland's variational Principle; Ekeland's Epsilon Variational Principle; Ekeland's Principle; Ekeland's Variational Principle; Entropy Maximum Principle; Equivalent Background Principle; Extended Maximum Entropy Principle; Extremal Principle of Information; Frieden-Soffer Principle; Gibbs Principle; Global Maximum Principle; Inference Principle; Jaynes' Maximum Entropy Principle; Jaynes' Principle; Kullback's Minimum Cross-entropy Principle; Maximum Information Entropy Principle; Measurement Principle; Minimax Entropy Principle; Minimum Entropy Principle; Minimum Relative Entropy Principle; Modification of The Extended Maximum Entropy Principle; Multi Tanh Principle; Principle of Maximum Entropy; Principle of Maximum Nonadditive Entropy; Principle of Observational Compatibility; Principle of Operational Compatibility; Uncertainty Principle; Variational Principle

FUNCTIONS OF A COMPLEX VARIABLE Extremal Principle; Korenblum's Maximum Principle; Maximum Principle; Schwarz Reflection Principle, Korenblum's Maximum Principle; Maximum Principle; Phragmen-Lindelof Principle; Fundamental Principle; Principle of Material Frame Indifference

FUZZY CONTROLLER Combination-separation Principle; Ritz's Combination Principle; Working Principle; Correspondence Principle; Entropy Maximum Principle; Extended Maximum Entropy Principle; Extremal Principle of Information; Frieden-Soffer Principle; Gibbs Principle; Inference Principle; Jaynes' Maximum Entropy Principle; Jaynes' Principle; Kullback's Minimum Cross-entropy Principle; Maximum Information Entropy

Principle; Measurement Principle; Minimax Entropy Principle; Minimum Entropy Principle; Minimum Relative Entropy Principle; Modification of The Extended Maximum Entropy Principle; Principle of Maximum Entropy; Principle of Maximum Nonadditive Entropy; Principle of Observational Compatibility; Principle of Operational Compatibility; Uncertainty Principle

FUZZY LOGIC Extension Principle; Flares. Hamilton's Principle; Kirchhoff's Voltage Law Principle; Minimum Energy Principle; Variational Principle; Woltjer Minimum Energy Principle; Woltjer-Taylor Minimum Energy Principle

FUZZY MAPPING Ekeland's variational Principle; Ekeland's Epsilon Variational Principle; Ekeland's Variational Principle; Global Maximum Principle; Variational Principle

FUZZY PARTICLE LATTICES Extension Principle; Principle of Uncertainty; Tnorm Extension Principle

FUZZY SETS Active Principle; Minimum Information Principle; Principle of Information Processing

G

GALAXIES EVOLUTION Correspondence Principle; Entropy Maximum Principle; Extended Maximum Entropy Principle; Extremal Principle of Information; Frieden-Soffer Principle; Gibbs Principle; Inference Principle; Jaynes' Maximum Entropy Principle; Jaynes' Principle; Kullback's Minimum Cross-entropy Principle; Maximum Information Entropy Principle; Measurement Principle; Minimax Entropy Principle; Minimum Entropy Principle; Minimum Relative Entropy Principle; Modification of The Extended Maximum Entropy Principle; Principle of Maximum Entropy; Principle of Maximum Nonadditive Entropy; Principle of Observational Compatibility; Principle of Operational Compatibility; Uncertainty Principle

GALAXIES KINEMATICS AND DYNAMICS Correspondence Principle; Entropy Maximum Principle; Extended Maximum Entropy Principle; Extremal Principle of Information; Frieden-Soffer Principle; Gibbs Principle; Inference Principle; Jaynes' Maximum Entropy Principle; Jaynes' Principle; Kullback's Minimum Cross-entropy Principle; Maximum Information Entropy Principle; Measurement Principle; Minimax Entropy Principle; Minimum Entropy Principle; Minimum Relative Entropy Principle; Modification of The Extended Maximum Entropy Principle; Principle of Maximum Entropy; Principle of Maximum Nonadditive Entropy; Principle of Observational Compatibility; Principle of Operational Compatibility; Uncertainty Principle

GALAXY NUCLEI Caratheodory's Principle; Conte's Principle; Fermat's Principle; Fermats Principle; Least Time Principle; Principle of Caratheodony; Stationary Time Principle

GALILEAN INVARIANCE Principle of Correspondence, Principle of Relativity

GAME THEORY Maximization Principle; Maximum Principle; Mde Principle; Epistatic Handicap Principle

GAS Principle of Corresponding States

GAS DYNAMICS Inference Principle; Jaynes' Principle; Kullback's Minimum Cross-entropy Principle; Maximum Information Entropy Principle; Measurement Principle; Measuring Principle; Minimax Entropy Principle; Minimum Description Length Principle; Minimum Energy Principle; Minimum Norm Principle; Minimum Relative Entropy Principle; Modification of The Extended Maximum Entropy Principle; Principle of Maximum Entropy; Principle of Maximum Nonadditive Entropy; Principle of Observational Compatibility; Principle of Operational Compatibility; Uncertainty Principle

GAS SENSORS Quartz Microbalance Principle, Wolfe's Decomposition Principle

GASTRICULCERATION Active Principle; Azimuth Measurement. Inference Principle; Jaynes' Principle; Kullback's Minimum Cross-entropy Principle; Maximum Information Entropy Principle; Measurement Principle; Measuring Principle; Minimax Entropy Principle; Minimum Energy Principle; Minimum Norm Principle; Minimum Relative Entropy Principle; Modification of The Extended Maximum Entropy Principle; Principle of Maximum Entropy; Principle of Maximum Nonadditive Entropy; Principle of Observational Compatibility; Principle of Operational Compatibility; Uncertainty Principle

GAUGE THEORY Gauge Principle; Inference Principle; Jaynes' Principle; Kullback's Minimum Cross-entropy Principle; Maximum Information Entropy Principle; Measurement Principle; Measuring Principle; Minimax

Dictionary of Scientific Principles, by Stephen Marvin
Copyright © 2011 John Wiley & Sons, Inc.

Entropy Principle; Minimum Energy Principle; Minimum Norm Principle; Minimum Relative Entropy Principle; Modification of The Extended Maximum Entropy Principle; Principle of Maximum Entropy; Principle of Maximum Nonadditive Entropy; Principle of Observational Compatibility; Principle of Operational Compatibility; Uncertainty Principle

GEDANKEN EXPERIMENTS Minimum Description Length Principle; Gene-active Principle; Mdl Principle

GENETIC ALGORITHMS Active Principle; Correspondence Principle; Entropy Maximum Principle; Extended Maximum Entropy Principle; Extremal Principle of Information; Frieden-Soffer Principle; Gibbs Principle; Inference Principle; Jaynes' Maximum Entropy Principle; Jaynes' Principle; Kullback's Minimum Cross-entropy Principle; Maximum Information Entropy Principle; Measurement Principle; Minimax Entropy Principle; Minimum Entropy Principle; Minimum Relative Entropy Principle; Modification of The Extended Maximum Entropy Principle; Principle of Maximum Entropy; Principle of Maximum Nonadditive Entropy; Principle of Observational Compatibility; Principle of Operational Compatibility; Uncertainty Principle

GENETICS Active Principle; August Krogh Principle; Extremum Principle; First Principle; Founder Principle; Frequency-dependent Principle; Matching Principle; Optimizing Principle; Serendipity Principle

GENOTYPE Uncertainty Principle, Epistatic Handicap Principle; Handicap Principle

GEODESIC Caratheodory's Principle; Conte's Principle; Fermat's Principle; Fermats Principle; Huygen's Principle; Least Time Principle; Principle of Caratheodony; Stationary Time Principle, Maupertuis-Jacobi Principle

GEODESIC EQUATION Caratheodory's Principle; Conte's Principle; Fermat's Principle; Fermats Principle; Least Time Principle; Principle of Caratheodony; Stationary Time Principle

GEOMETRIC FUNCTION THEORY Extremum Principle; Korenblum's Maximum Principle; Maximum Principle; Phragmen-Lindelof Principle; Schwarz Reflection Principle

GEOMETRIC INTEGRATION THEORY Hamilton's Principle; Maupertuis-Jacobi Principle; Saint-Venant's Principle

GEOMETRICAL OPTICS Caratheodory's Principle; Conte's Principle; Fermat's Principle; Fermats Principle; Least Time Principle; Principle of Caratheodony; Stationary Time Principle

GEOMETRY Austerity Principle; Boundary Harnack Principle; Boundary of A Boundary Principle; Boundary Principle; Boundedness Principle; Ekeland's variational Principle; Ekeland's Epsilon Variational Principle; Ekeland's Principle; Ekeland's Variational Principle; Equivalence Principle; Global Maximum Principle; Local Uniform Boundedness Principle; Minimum Description Length Principle; Nadirdshvlli Boundary Principle; Uniform Boundedness Principle; Variational Principle; Principle of Causality

GIBBS MEASURE Correspondence Principle; Entropy Maximum Principle; Extended Maximum Entropy Principle; Extremal Principle of Information; Frieden-Soffer Principle; Gibbs Principle; Inference Principle; Jaynes' Maximum Entropy Principle; Jaynes' Principle; Kullback's Minimum Cross-entropy Principle; Maximum Information Entropy Principle; Measurement Principle; Minimax Entropy Principle; Minimum Entropy Principle; Minimum Relative Entropy Principle; Modification of The Extended Maximum Entropy Principle; Principle of Maximum Entropy; Principle of Maximum Nonadditive Entropy; Principle of Observational Compatibility; Principle of Operational Compatibility; Uncertainty Principle; Contraction Principle; Contraction Principle; Gibbs Variational Principle; Large Deviation Principle

GIBBS SAMPLING Conventional Conscience Principle; Electronegativity Equalization Principle; Equalization Principle; Equiprobable Principle

GLOBAL ANALYSIS Hamilton's Principle; Maupertuis-Jacobi Principle; Saint Venant's Principle

GNC ALGORITHM Correspondence Principle; Entropy Maximum Principle; Extended Maximum Entropy Principle; Extremal Principle of Information; Frieden-Soffer Principle; Gibbs Principle; Inference Principle; Jaynes' Maximum Entropy Principle; Jaynes' Principle; Kullback's Minimum Cross-entropy Principle; Maximum Information Entropy Principle; Measurement Principle; Minimax Entropy Principle; Minimum Entropy Principle; Minimum Relative Entropy Principle; Modification of The Extended Maximum Entropy Principle; Principle of Maximum Entropy; Principle of Maximum Nonadditive Entropy; Principle of Observational Compatibility; Principle of Operational Compatibility; Uncertainty Principle

GOODNESS OF FIT TEST Correspondence Principle; Entropy Maximum Principle; Extended Maximum Entropy Principle; Extremal Principle of Information; Frieden-Soffer Principle; Gibbs Principle; Inference Principle; Jaynes' Maximum Entropy Principle; Jaynes' Principle; Kullback's Minimum Cross-entropy Principle; Maximum Information Entropy Principle; Measurement Principle; Minimax Entropy Principle; Minimum Entropy Principle; Minimum Relative Entropy Principle; Modification of The Extended Maximum Entropy Principle; Principle of Maximum Entropy; Principle of Maximum Nonadditive Entropy; Principle of Observational Compatibility; Principle of Operational Compatibility; Uncertainty Principle; Strong Invariance Principle

GRADED INDEX FIBER Caratheodory's Principle; Conte's Principle; Fermat's Principle; Fermats Principle; Least Time Principle; Principle of Caratheodony; Stationary Time Principle

GRAVITATION Antenna Principle; Corresponding States Principle; Equivalence Principle; Mach Principle; Pitzer Corresponding States Principle; Principle of Equivalence

GRAVITATION ALLENS EFFECT Caratheodory's Principle; Conte's Principle; Fermat's Principle; Fermats Principle; Least Time Principle; Principle of Caratheodony; Stationary Time Principle

GRAVITATIONAL FIELDS Caratheodory's Principle; Conte's Principle; Equivalence Principle, Fermat's Principle; Fermats Principle; Least Time Principle; Principle of molecular dynamics; Principle of Caratheodony; Stationary Time Principle

GRAVITATIONAL INTERACTIONS Principle of molecular dynamics, Minimum Description Length Principle

GRAVITATIONAL LENSES Caratheodory's Principle; Conte's Principle; Fermat's Principle; Fermats Principle; Least Time Principle; Principle of Caratheodony; Stationary Time Principle

GRAVITY Principle of molecular dynamics; Principle of Equivalence; Principle of Representationtheoretic Selfdualtiy; Quantum Equivalence Principle; Weak Equivalence Principle, Principle of molecular dynamics

GREEDY SEARCH TECHNIQUES Active Principle; Mdl Principle, Correspondence Principle; Entropy Maximum Principle; Extended Maximum Entropy Principle; Extremal Principle of Information; Frieden-Soffer Principle; Gibbs Principle; Inference Principle; Jaynes' Maximum Entropy Principle; Jaynes' Principle; Kullback's Minimum Cross-entropy Principle; Maximum Information Entropy Principle; Measurement Principle; Minimax Entropy Principle; Minimum Entropy Principle; Minimum Relative Entropy Principle; Modification of The Extended Maximum Entropy Principle; Principle of Maximum Entropy; Principle of Maximum Nonadditive Entropy; Principle of Observational Compatibility; Principle of Operational Compatibility; Uncertainty Principle

GREEN FUNCTION Product solution Principle; Positive Displacement Principle, Pauli Antisymmetry and Exclusion Principle

GROUP WARE Autonomous Cooperation Distributed Maximum Principle; Distributed Maximum Principle; Maximum Principle

GROWTH Contraction Mapping Principle; Contraction Principle; Investment Principle; John Templeton Principle; Principle of Maximum Pessimism

H

HALL EFFECT DEVICES Inference Principle; Jaynes' Principle; Kullback's Minimum Cross-entropy Principle; Maximum Information Entropy Principle; Measurement Principle; Measuring Principle; Minimax Entropy Principle; Minimum Energy Principle; Minimum Norm Principle; Minimum Relative Entropy Principle; Modification of The Extended Maximum Entropy Principle; Principle of Maximum Entropy; Principle of Maximum Nonadditive Entropy; Principle of Observational Compatibility; Principle of Operational Compatibility; Uncertainty Principle

HALL OUTPUT Inference Principle; Jaynes' Principle; Kullback's Minimum Cross-entropy Principle; Maximum Information Entropy Principle; Measurement Principle; Measuring Principle; Minimax Entropy Principle; Minimum Energy Principle; Minimum Norm Principle; Minimum Relative Entropy Principle; Modification of The Extended Maximum Entropy Principle; Principle of Maximum Entropy; Principle of Maximum Nonadditive Entropy; Principle of Observational Compatibility; Principle of Operational Compatibility; Uncertainty Principle

HALOGALACTIQUE Caratheodory's Principle; Conte's Principle; Fermat's Principle; Fermats Principle; Least Time Principle; Principle of Caratheodony; Stationary Time Principle

HAMILTON EQUATION Caratheodory's Principle; Conte's Principle; Fermat's Principle; Fermats Principle; Least Time Principle; Principle of Caratheodony; Stationary Time Principle

HAMILTONIAN Action Principle; Maximum Principle; Minimum Description Length Principle; Schwinger's Quantum Action Principle; Wolfe's Decomposition Principle, Principle of Correspondence, Hamilton's Principle; Maupertuis-Jacobi Principle; Wolfe's Decomposition Principle, Minimum Principle; Least Action Principle; Principle of Least Work, Energy Preservation Principle; Saint-Venant's Principle, Pontryagin's Maximum Principle

HAMILTONJACOBI CONTROL SYSTEMS Maximum Principle; Pontryagin Maximum Principle

HARMFUL EFFECT Casuistic Principle; Double Effect Principle; Exception Granting Principle; Justifying Principle; Principle of Do No Harm; Principle of Double Effect; Principles of Biomedical Ethics

HARMONIC Extremum Principle, Minimum Description Length Principle; Wolfe's Decomposition Principle

HARMONIC SUPPRESSION Principle of Fundamental Characteristics

HARTREEFOCK METHOD Action Principle; Schwinger's Quantum Action Principle; Wolfe's Decomposition Principle; Hard-soft acid-base Principle; Hsab Principle

HEAD ACCELERATION Active Principle

HEALTH CARE Health Principle; Precautionary Principle; Principle of Habeas Mentem; Salutogenesis Principle; Proximity Compatibility Principle

HEALTH SERVICES AND MEDICAL CARE Contagion Principle; Health Principle; Precautionary Principle; Principle of Habeas Mentem

HEART MUSCLE Working Principle; Heart Pathology. Drilled Cell Principle; Overlapping Biphasic Impulse Principle; Priming Principle

Dictionary of Scientific Principles, by Stephen Marvin
Copyright © 2011 John Wiley & Sons, Inc.

HEAT Consumer Principle; Double Inlet Principle; Equipartition Principle; Principle of Construction and Operation

HEAT SHOCK PROTEIN Similia Principle Transfer; Entropy Minimax Principle; Equipartition Principle; Saint-Venant's Principle

HEIGHT Correspondence Principle; Entropy Maximum Principle; Extended Maximum Entropy Principle; Extremal Principle of Information; Frieden-Soffer Principle; Gibbs Principle; Inference Principle; Jaynes' Maximum Entropy Principle; Jaynes' Principle; Kullback's Minimum Cross-entropy Principle; Maximum Information Entropy Principle; Measurement Principle; Minimax Entropy Principle; Minimum Entropy Principle; Minimum Relative Entropy Principle; Modification of The Extended Maximum Entropy Principle; Principle of Maximum Entropy; Principle of Maximum Nonadditive Entropy; Principle of Observational Compatibility; Principle of Operational Compatibility; Uncertainty Principle

HEISENBERG EQUATION Uncertainty Principle; Heisenberg Scatter Principle; Principle of Measurement

HELICITY INJECTION Hamilton's Principle; Kirchhoff's Voltage Law Principle; Minimum Energy Principle; Variational Principle; Woltjer Minimum Energy Principle; Woltjer-Taylor Minimum Energy Principle

HELMHOLTZ EQUATION Saint-Venant's Principle; Wolfe's Decomposition Principle; Helmholtzkorteweg Principle; Principle of Minimum Dissipation

HEMIVARIATIONAL INEQUALITIES Ekeland's variational Principle; Ekeland's Epsilon Variational Principle; Ekeland's Variational Principle; Global Maximum Principle; Global Maximum Principle; Maximum Principle; Variational Principle

HEMODIALYSISTHERAPEUTIC METHOD Constant Infusion Principle

HETEROGENEOUS MEDIUM Ambartsumian Invariance Principle; Charnov's Invariance Principle; Dirichlet Principle; Donsker's Invariance Principle; Invariance Principle; Invariance Principle of General Relativity; Local Invariance Principle; Maruyama's Invariance Principle; Perturbation Invariance Principle; Principle of Invariance; Principle of Undulatory Invariance; Relaxed Invariance Principle; Scale Invariance Principle; Strong Principle; Variational Principle; Weak Invariance Principle

HIERARCHICAL BAYES MODELS Equidistortion Principle; Quantization Design Principle

HIERARCHICAL STRUCTURE Minimum Description Length Principle, Master-slave Principle; Maximum Principle

HIGH FREQUENCY Contraction Principle; Gibbs Variational Principle; Large Deviation Principle; Superposition Principle, Protection Principle

HIGH-PRESSURE VAPOR LIQUID EQUILIBRIA Corresponding States Principle; Pitzer Corresponding States Principle

HIGH STRAIN D'Alembert Lagrange Principle; D'Alembert Lagrange Variation Principle; D'Alembert Principle; Gauss Principle; Hamilton Lagrange Principle; Jayne's Principle; Jourdain Principle; Kinetic Principle; Lagrange Principle; Least Constraint Principle; Maupertius Lagrange Principle; Maupertuis Euler Lagrange Principle of Least Action; Principle of Extremal Effects; Principle of Least Constraint; Principle of Maximal Effect; Principle of Virtual Work

HILBERT SPACE Inference Principle; Jaynes' Principle; Korenblum's Maximum Principle; Kullback's Minimum Cross-entropy Principle; Maximum Information Entropy Principle; Maximum Principle; Measurement Principle; Measuring Principle; Mesh Independence Principle; Minimax Entropy Principle; Minimum Energy Principle; Minimum Norm Principle; Minimum Relative Entropy Principle; Modification of The Extended Maximum Entropy Principle; Principle of Maximum Entropy; Principle of Maximum Nonadditive Entropy; Principle of Observational Compatibility; Principle of Operational Compatibility; Uncertainty Principle; Wolfe's Decomposition Principle

HISTOLOGY Correspondence Principle; Entropy Maximum Principle; Extended Maximum Entropy Principle; Extremal Principle of Information; Frieden-Soffer Principle; Gibbs Principle; Inference Principle; Jaynes' Maximum Entropy Principle; Jaynes' Principle; Kullback's Minimum Cross-entropy Principle; Maupertuis Principle; Maupertuis Principle; Maximum Information Entropy Principle; Maximum Likelihood Principle; Mckenzie Principles; Measurement Principle; Minimax Entropy Principle; Minimum Entropy Principle; Minimum Relative Entropy Principle; Modification of The Extended Maximum Entropy Principle; Principle of Maximum Entropy; Principle of Maximum Nonadditive Entropy; Principle of Observational Compatibility; Principle of Operational Compatibility; Uncertainty Principle

HISTORY Contagion Principle; Health Principle; Precautionary Principle; Principle of Habeas Mentem; Serendipity Principle

HITTING TIME Boric Acid Affinity Principle; Boronic Acid Affinity Principle; Harnack Principle

HOLISTIC RESOURCE MANAGEMENT Optimality Principle; Organizing Principle; Principle of component analysis; Self-organizing Principle; Similarity Principle

HOLOGRAPHIC HOMOLOGY Gabor's Holographic Principle; Holographic Principle; T Hooft Principle

HOMEOSTASIS Optimality Principle; Organizing Principle; Principle of component analysis; Self-organizing Principle; Similarity Principle

HOMOGENEOUS DIRICHLET BOUNDARY CONDITIONS Pontryagin Minimum Principle

HOMOGENEOUS MEDIUM Ambartsumian Invariance Principle; Charnov's Invariance Principle; Dirichlet Principle; Donsker's Invariance Principle; Invariance Principle; Invariance Principle of General Relativity; Local Invariance Principle; Maruyama's Invariance Principle; Perturbation Invariance Principle; Principle of Invariance; Principle of Undulatory Invariance; Relaxed Invariance Principle; Scale Invariance Principle; Strong Principle; Variational Principle; Weak Invariance Principle

HUMAN Hard-soft acid-base Principle; Penaz Principle; Priming Principle; Salutogenesis Principle

HUMAN MEDICINE Construction Principle; Health Principle; Mckenzie Principles; Precautionary Principle; Principle of Habeas Mentem; Principle-centered Spine Care

HUMAN PRIMATES Cross Check Principle; Occam's Razor Principle; Principle of Laser Speckle Flowgraphy; Wolfe's Decomposition Principle

HUMIDITY PROFILE RETRIEVAL Correspondence Principle; Entropy Maximum Principle; Extended Maximum Entropy Principle; Extremal Principle of Information; Frieden-Soffer Principle; Gibbs Principle; Inference Principle; Jaynes' Maximum Entropy Principle; Jaynes' Principle; Kullback's Minimum Cross-entropy Principle; Maximum Information Entropy Principle; Measurement Principle; Minimax Entropy Principle; Minimum Entropy Principle; Minimum Relative Entropy Principle; Modification of The Extended Maximum Entropy Principle; Principle of Maximum Entropy; Principle of Maximum Nonadditive Entropy; Principle of Observational Compatibility; Principle of Operational Compatibility; Uncertainty Principle

HYBRID SYSTEM Ambartsumian Invariance Principle; Charnov's Invariance Principle; Dirichlet Principle; Donsker's Invariance Principle; Invariance Principle; Invariance Principle of General Relativity; Local Invariance Principle; Maruyama's Invariance Principle; Perturbation Invariance Principle; Principle of Invariance; Principle of Undulatory Invariance; Relaxed Invariance Principle; Scale Invariance Principle; Strong Principle; Variational Principle; Weak Invariance Principle

HYBRID TEMPERATURE SENSOR BASED Interferometric Principle; Internaleffect Principle; Radiation Principle

HYDROCARBON Corresponding States Principle; Pitzer Corresponding States Principle; Principle of Congruence

HYDRODYNAMIC CELLULAR AUTOMATA Averaging Principle; Correspondence Principle; Entropy Maximum Principle; Extended Maximum Entropy Principle; Extremal Principle of Information; Frieden-Soffer Principle; Gibbs Principle; Inference Principle; Jaynes' Maximum Entropy Principle; Jaynes' Principle; Kullback's Minimum Cross-entropy Principle; Maximum Information Entropy Principle; Measurement Principle; Minimax Entropy Principle; Minimum Entropy Principle; Minimum Relative Entropy Principle; Modification of The Extended Maximum Entropy Principle; Principle of Maximum Entropy; Principle of Maximum Nonadditive Entropy; Principle of Observational Compatibility; Principle of Operational Compatibility; Uncertainty Principle

HYDRODYNAMICS Consumer Principle; D'Alembert's Principle; Gyarmati's Principle; Hamilton Principle; Orthogonality Principle; Prigogine Principle; Principle of Minimum Kinetic Energy; Principle of Minimum Mechanical Energy; Principle of Minimum Potential Energy; Quantum Hamilton Principle; Stewart-Hamilton Principle; Stochastic Hamilton Variational Principle; Toupin's Dual Principle, Least Action Principle; Principle of Least Work

HYDROGEN ATOMS Action Principle; Minimum Description Length Principle; Schwinger Principle; Schwinger's Quantum Action Principle; Stationary Action Principle; Wolfe's Decomposition Principle

HYDROGEN FLUORIDES Action Principle; Schwinger's Quantum Action Principle; Stationary Action Principle

HYDROGEN MOLECULES Action Principle; Ckvp; Complex Kohn Variational Principle; Hulthén–Kohn Variational Principle; Kohn Variational Principle; Schwinger Variational Principle; Schwinger's Quantum Action Principle; Stationary Action Principle; Wolfe's Decomposition Principle

HYDROLOGY ANALYSIS Principle of Maximum Entropy; Hydromagnetics and Plasmas. Hamilton's Principle; Kirchhoff's Voltage Law Principle; Minimum Energy Principle; Variational Principle; Woltjer Minimum Energy Principle; Woltjer-Taylor Minimum Energy Principle

HYPERBOLIC EQUATIONS Correspondence Principle; Entropy Maximum Principle; Extended Maximum Entropy Principle; Extremal Principle of Information; Frieden-Soffer Principle; Gibbs Principle; Inference Principle; Jaynes' Maximum Entropy Principle; Jaynes' Principle; Kullback's Minimum Cross-entropy Principle; Maximum Information Entropy Principle; Measurement Principle; Minimax Entropy Principle; Minimum Entropy Principle; Minimum Relative Entropy Principle; Modification of The Extended Maximum Entropy Principle; Principle of Maximum Entropy; Principle of Maximum Nonadditive Entropy; Principle of Observational Compatibility; Principle of Operational Compatibility; Uncertainty Principle, Maximum Principle; Wolfe's Decomposition Principle

HYPOTHESIS Maximum Principle; Contraction Principle; Gibbs Variational Principle; Large Deviation Principle

I

IDENTIFICATION Maximum Entropy Principle; Quartz Microbalance Principle; Radish Pungent PPrinciple

IMAGE FORMING Caratheodory's Principle; Conte's Principle; Fermat's Principle; Fermats Principle; Least Time Principle; Principle of Caratheodony; Stationary Time Principle

IMAGE PROCESSING Active Principle; Correspondence Principle; Entropy Maximum Principle; Extended Maximum Entropy Principle; Extremal Principle of Information; Frieden-Soffer Principle; Gibbs Principle; Hamilton's Principle; Inference Principle; Jaynes' Maximum Entropy Principle; Jaynes' Principle; Kirchhoff's Voltage Law Principle; Kullback's Minimum Cross-entropy Principle; Maximum Information Entropy Principle; Measurement Principle; Minimax Entropy Principle; Minimum Energy Principle; Minimum Entropy Principle; Minimum Principle; Minimum Relative Entropy Principle; Modification of The Extended Maximum Entropy Principle; Principle of Maximum Entropy; Principle of Maximum Nonadditive Entropy; Principle of Observational Compatibility; Principle of Operational Compatibility; Uncertainty Principle; Variational Principle; Woltjer Minimum Energy Principle; Woltjer-Taylor Minimum Energy Principle

IMAGE QUALITY Correspondence Principle; Entropy Maximum Principle; Extended Maximum Entropy Principle; Extremal Principle of Information; Frieden-Soffer Principle; Gibbs Principle; Inference Principle; Jaynes' Maximum Entropy Principle; Jaynes' Principle; Kullback's Minimum Cross-entropy Principle; Maximum Information Entropy Principle; Measurement Principle; Minimax Entropy Principle; Minimum Entropy Principle; Minimum Relative Entropy Principle; Modification of The Extended Maximum Entropy Principle; Principle of Maximum

Entropy; Principle of Maximum Nonadditive Entropy; Principle of Observational Compatibility; Principle of Operational Compatibility; Uncertainty Principle

IMAGE RECOGNITION Active Principle; Correspondence Principle; Entropy Maximum Principle; Extended Maximum Entropy Principle; Extremal Principle of Information; Frieden-Soffer Principle; Gibbs Principle; Inference Principle; Jaynes' Maximum Entropy Principle; Jaynes' Principle; Kullback's Minimum Cross-entropy Principle; Maximum Information Entropy Principle; Measurement Principle; Minimax Entropy Principle; Minimum Entropy Principle; Minimum Relative Entropy Principle; Modification of The Extended Maximum Entropy Principle; Principle of Maximum Entropy; Principle of Maximum Nonadditive Entropy; Principle of Observational Compatibility; Principle of Operational Compatibility; Uncertainty Principle

IMAGE SEGMENTATION Active Principle; Complexity Principle; Entropy Principle; Generalized Minimal Principle; Maximum Entropy Principle; Minimum Cross-entropy Principle; Moment Preserving Principle; Principle of Minimum Cross-entropy; Variational Minimal Principle

IMAGE SHAPE ANALYSIS Hamilton's Principle; Kirchhoff's Voltage Law Principle; Minimum Energy Principle; Variational Principle; Woltjer Minimum Energy Principle; Woltjer-Taylor Minimum Energy Principle

IMAGE SIGNAL PROCESSING Correspondence Principle; Entropy Maximum Principle; Extended Maximum Entropy Principle; Extremal Principle of Information; Frieden-Soffer Principle; Gibbs Principle; Inference Principle; Jaynes' Maximum Entropy Principle; Jaynes' Principle; Kullback's Minimum Cross-entropy Principle; Maximum

Information Entropy Principle; Measurement Principle; Minimax Entropy Principle; Minimum Entropy Principle; Minimum Relative Entropy Principle; Modification of The Extended Maximum Entropy Principle; Principle of Maximum Entropy; Principle of Maximum Nonadditive Entropy; Principle of Observational Compatibility; Principle of Operational Compatibility; Uncertainty Principle

IMAGES Entropy Principle; Generalized Minimal Principle; Maximum Entropy Principle; Minimum Cross-entropy Principle; Moment Preserving Principle; Principle of Minimum Cross-entropy; Variational Minimal Principle, Equidistortion Principle; Quantization Design Principle, Ink Jet Principle

IMAGING Maximum Likelihood Principle, Function Principle, Active Principle; Working Principle

IMMERSED BODY Wolfe's Decomposition Principle

IMMUNE SYSTEM Inference Principle; Jaynes' Principle; Kullback's Minimum Cross-entropy Principle; Maximum Information Entropy Principle; Measurement Principle; Measuring Principle; Minimax Entropy Principle; Minimum Energy Principle; Minimum Norm Principle; Minimum Relative Entropy Principle; Modification of The Extended Maximum Entropy Principle; Principle of Maximum Entropy; Principle of Maximum Nonadditive Entropy; Principle of Observational Compatibility; Principle of Operational Compatibility; Uncertainty Principle, Drug Principle

IMMUNOCONJUGATES Diagnostic Principle; Principle of Psychoneuroimunology; Quasilocal Principle; Staff's Diagnostic Principle

IMMUNOHISTOCHEMICAL STAINING TECHNIQUE Action Principle; Schwinger's Quantum Action Principle

IMMUNOLOGY Immunosuppresive Principle, Inference Principle; Jaynes' Principle; Kullback's Minimum Cross-entropy Principle;

Maximum Information Entropy Principle; Measurement Principle; Measuring Principle; Minimax Entropy Principle; Minimum Energy Principle; Minimum Norm Principle; Minimum Relative Entropy Principle; Modification of The Extended Maximum Entropy Principle; Principle of Maximum Entropy; Principle of Maximum Nonadditive Entropy; Principle of Observational Compatibility; Principle of Operational Compatibility; Uncertainty Principle

IMPEDANCE Hamilton's Principle; Kirchhoff's Voltage Law Principle; Minimum Energy Principle; Variational Principle; Wolfe's Decomposition Principle; Woltjer Minimum Energy Principle; Woltjer-Taylor Minimum Energy Principle, Synchronous Sampling Principle

IN VITRO STUDIES Action Principle; Schwinger's Quantum Action Principle; Inaccuracy Principle; Principle of Measurement

INCOMPRESSIBLE MATERIAL D'Alembert Lagrange Principle; D'Alembert Lagrange Variation Principle; D'Alembert Principle; Gauss Principle; Hamilton Lagrange Principle; Jayne's Principle; Jourdain Principle; Kinetic Principle; Lagrange Principle; Least Constraint Principle; Maupertius Lagrange Principle; Maupertuis Euler Lagrange Principle of Least Action; Principle of Extremal Effects; Principle of Least Constraint; Principle of Maximal Effect; Principle of Virtual Work

INCONTINENCE Mitrofanoff Principle; Variational Principle

INDENTATION CONTACT Energy Principle; Woltjer Minimum Energy Principle, Energy Principle; Woltjer Minimum Energy Principle

INDEPENDENT RANDOM VECTORS Contraction Principle; Gibbs Variational Principle; Large Deviation Principle

INDETERMINANCY Compensation Principle; Minimum Description Length Principle; Principle of Measurement; Indeterminate Structure; Least Action Principle; Principle of Least Work

INDOOR AIR QUALITY Capillary Electrophoresis Principle; Ion Pair Principle; Principle of Diastereomeric Ion Pair Formation

INERTIA Covariance Principle; Machs Magic Principle; Machs Principle; Principle of Material Frame Indifference

INERTIAL MASS Principle of Molecular Dynamics; Principle of Equivalence; Quantum Equivalence Principle; Weak Equivalence Principle

INFERENCE Active Principle; Correspondence Principle; Entropy Maximum Principle; Extended Maximum Entropy Principle; Extremal Principle of Information; Frieden-Soffer Principle; Gibbs Principle; Inference Principle; Jaynes' Maximum Entropy Principle; Jaynes' Principle; Kullback's Minimum Cross-entropy Principle; Maximum Information Entropy Principle; Measurement Principle; Minimax Entropy Principle; Minimum Entropy Principle; Minimum Relative Entropy Principle; Modification of The Extended Maximum Entropy Principle; Principle of Maximum Entropy; Principle of Maximum Nonadditive Entropy; Principle of Observational Compatibility; Principle of Operational Compatibility; Uncertainty Principle

INFINITE DILUTION ACTIVITY COEFFICIENTS Principle of Correspondence; Corresponding States Principle; Pitzer Corresponding States Principle;

INFORMATION Active Principle; Causality Principle; Maximum Likelihood Principle; Minimal Principle; Principle of Minimal Causation; Principle of Uncertainty; Sufficiency Principle

INFORMATION PROCESSING Active Principle; Principle of Information Processing; Relative Frequency Principle, Entropy Minimax Principle; Matching Principle

INFORMATION SYSTEMS Correspondence Principle; Entropy Maximum Principle; Extended Maximum Entropy Principle; Extremal Principle of Information; Frieden-Soffer Principle; Gibbs Principle; Inference

Principle; Jaynes' Maximum Entropy Principle; Jaynes' Principle; Kullback's Minimum Cross-entropy Principle; Maximum Information Entropy Principle; Measurement Principle; Minimax Entropy Principle; Minimum Entropy Principle; Minimum Relative Entropy Principle; Modification of The Extended Maximum Entropy Principle; Principle of Maximum Entropy; Principle of Maximum Nonadditive Entropy; Principle of Observational Compatibility; Principle of Operational Compatibility; Uncertainty Principle

INFORMATION THEORY Principle of Maximum Entropy, Pareto Principle, Active Principle; Contraction Principle; Contraction Principle; Correspondence Principle; Entropy Maximum Principle; Extended Maximum Entropy Principle; Extremal Principle of Information; Frieden-Soffer Principle; Gibbs Principle; Gibbs Variational Principle; Inference Principle; Jaynes' Maximum Entropy Principle; Jaynes' Principle; Kullback's Minimum Cross-entropy Principle; Large Deviation Principle; Maximum Information Entropy Principle; Maximum Likelihood Principle; Maximum Principle; Measurement Principle; Minimal Principle; Minimax Entropy Principle; Minimum Cross-entropy (Me) Principle; Minimum Entropy Principle; Minimum Relative Entropy Principle; Modification of The Extended Maximum Entropy Principle; Principle of Maximum Entropy; Principle of Maximum Entropy; Principle of Maximum Nonadditive Entropy; Principle of Minimum Cross-entropy; Principle of Observational Compatibility; Principle of Operational Compatibility; Uncertainty Principle; Principle of Per-survivor Processing; Principles of Darcy's Law

INFRARED Minimum Description Length Principle, Work Function Principle, Minimum Description Length Principle

INFRARED REFLECTANCE SPECTROSCOPY Parsimony Principle; Principle of Parsimonius Data Modeling

INHIBITION OF PROSTAGLANDIN BIOSYNTHESIS Antiinflammation Principle, Antimaximum Principle

INITIAL VALUE PROBLEM Dirichlet Principle; Total Virtual Action Principle; Virtual Force Principle; Maximum Principle

INORGANIC COMPOUND Action Principle; Ckvp; Complex Kohn Variational Principle; Hulthén–Kohn Variational Principle; Kohn Variational Principle; Schwinger Variational Principle; Schwinger's Quantum Action Principle; Stationary Action Principle

INPUT OUTPUT MEASUREMENTS Maximum Entropy Principle; New Therapeutic Principle

INSPECTION FIXTURE Inference Principle; Jaynes' Principle; Kullback's Minimum Cross-entropy Principle; Maximum Information Entropy Principle; Measurement Principle; Measuring Principle; Minimax Entropy Principle; Minimum Energy Principle; Minimum Norm Principle; Minimum Relative Entropy Principle; Modification of The Extended Maximum Entropy Principle; Principle of Maximum Entropy; Principle of Maximum Nonadditive Entropy; Principle of Observational Compatibility; Principle of Operational Compatibility; Uncertainty Principle

INSTANTON Action Principle; Schwinger's Quantum Action Principle; Spacetime Uncertainty Principle

INSTINCT Beyond The Pleasure Principle; Jewish Pleasure Principle; Pleasure Pain Principle; Pleasure Principle; Pleasure Pain Principle

INSTITUTIONALIZATION Ellul Principle of Efficiency; Emancipation Principle; Kant's Principle; Power Principle; Principle of Domination; Principle of Efficiency; Principle of Emancipation; Principle of Emancipation; Principle of Free Domain; Principle of Majority Rule Based On Decision Making Power Demand; Principle of Power; Principle of Subjective Freedom; Principle of The Authomomy of The Will

INSTRUMENTS AND MEASUREMENT Principle of Causality; Principle of Construction and Operation; Optical Principle

INSULATION Hamilton's Principle; Kirchhoff's Voltage Law Principle; Minimum Energy Principle; Variational Principle; Woltjer Minimum Energy Principle; Woltjer-Taylor Minimum Energy Principle

INTAKE SYSTEMS Casuistic Principle; Double Effect Principle; Exception Granting Principle; Justifying Principle; Principle of Do No Harm; Principle of Double Effect; Principles of Biomedical Ethics

INTEGRAL EQUATION Ambartsumian Invariance Principle; Charnov's Invariance Principle; Dirichlet Principle; Donsker's Invariance Principle; Invariance Principle; Invariance Principle of General Relativity; Local Invariance Principle; Maruyama's Invariance Principle; Perturbation Invariance Principle; Principle of Invariance; Principle of Undulatory Invariance; Relaxed Invariance Principle; Scale Invariance Principle; Strong Principle; Variational Principle; Weak Invariance Principle; Huygen's Principle

INTEGRAL Caratheodory's Principle; Conte's Principle; Fermat's Principle; Fermats Principle; Least Time Principle; Maximum Principle; Principle of Caratheodony; Stationary Time Principle

INTEGUMENTARY SYSTEMS Investment Principle; John Templeton Principle; Principle of Maximum Pessimism

INTELLIGENCE Correspondence Principle; Entropy Maximum Principle; Extended Maximum Entropy Principle; Extremal Principle of Information; Frieden-Soffer Principle; Gibbs Principle; Inference Principle; Jaynes' Maximum Entropy Principle; Jaynes' Principle; Kullback's Minimum Cross-entropy Principle; Maximum Information Entropy Principle; Measurement Principle; Minimax Entropy Principle; Minimum Entropy Principle; Minimum Relative Entropy Principle; Modification of The Extended Maximum Entropy Principle; Principle of Maximum Entropy; Principle of Maximum Nonadditive Entropy; Principle of Observational Compatibility; Principle of Operational Compatibility; Uncertainty Principle

INTERACTION PARAMETER Corresponding States Principle; Pitzer Corresponding States Principle

INTERFERENCE Babinet's Principle; Bohr Principle of Complementarity; Bohrs Complementarity Principle; Bohr's Correspondence Principle; Compensation Principle; Complementarity Principle; Correspondence Principle; Minimum Description Length Principle, Pulseecho Principle

INTERMITTENT CURRENT Inference Principle; Jaynes' Principle; Kullback's Minimum Cross-entropy Principle; Maximum Information Entropy Principle; Measurement Principle; Measuring Principle; Minimax Entropy Principle; Minimum Energy Principle; Minimum Norm Principle; Minimum Relative Entropy Principle; Modification of The Extended Maximum Entropy Principle; Principle of Maximum Entropy; Principle of Maximum Nonadditive Entropy; Principle of Observational Compatibility; Principle of Operational Compatibility; Uncertainty Principle

INTERNAL WAVES Conservation of Momentum Principle; D'Alembert Principle; Differential Principle; Hamilton Principle; Newton's Third Principle; Principle of Equal A Priori Probabilities; Variational Principle

INTERSYMBOL INTERFERENCE Principle of per-survivor processing; Principles of Darcy's Law

INTESTINAL ABSORPTION Correspondence Principle; Entropy Maximum Principle; Extended Maximum Entropy Principle; Extremal Principle of Information; Frieden-Soffer Principle; Gibbs Principle; Inference Principle; Jaynes' Maximum Entropy Principle; Jaynes' Principle; Kullback's Minimum Cross-entropy Principle; Maximum Information Entropy Principle; Measurement Principle; Minimax Entropy Principle; Minimum Entropy Principle; Minimum Relative Entropy Principle; Modification of The Extended Maximum Entropy Principle; Principle of Maximum Entropy; Principle of Maximum Nonadditive Entropy; Principle of Observational Compatibility; Principle

of Operational Compatibility; Uncertainty Principle; Working Principle

INTRAPULMONARY OXYGEN CONSUMPTION Reversed Fick Principle, Extremal Principle; Fibrinolytic Principle

INVARIANCE Comparison Principle; Huygen's Principle; Pauli Principle; Principle of Relativity; Substitution Principle

INVERSE PROBLEM Active Principle; Minimum Description Length Principle; Principle of Corresponding States; Equivalence Principle

INVERSE PROPORTIONAL HAZARDS MODEL Parsimony Principle; Principle of Parsimonius Data Modeling

INVERTEBRATA Ecological Principle; First Principle; Frequency-dependent Principle; Hopkins' Host Selection Principle; Host Selection Principle; Principle of Dynamic Ecology

ION MOL GAS POLEMIC Conservation of Momentum Principle; D'Alembert Principle; Differential Principle; Hamilton Principle; Newton's Third Principle; Principle of Equal A Priori Probabilities; Variational Principle

IONS Bond Directional Principle; Cyclotron Principle; Maximum Hardness Principle

IONTOPHORETIC DEVICE OR TRANSDERMAL DELIVERY OF MEDICINAL Monti Principle

IRREDUCIBLE SCIENTIFIC UNCERTAINTY Precautionary Principle.

IRREVERSIBLE Orthogonality Principle; Irreversible Thermodynamics. Equipartition Principle; Principle of Maximum Entropy

ISDN Memory Compensation Principle; Mitrofanoff Principle; Monti Principle

ISI CHANNELS Principle of Per-survivor Processing; Principles of Darcy's Law

ISOBARIC VAPOR LIQUID EQUILIBRIA Corresponding States Principle; Pitzer Corresponding States Principle;, Principle of Correspondence

ISOTACHOPHORESIS Calorimetric Measurement Principle; Hard-soft acid-base Principle

ISOTOPE TECHNIQUES Maximum Likelihood Principle; Optimization Principle; Run and Recreate Principle

ISOTROPIC MEDIUM Ambartsumian Invariance Principle; Charnov's Invariance Principle; Donsker's Invariance Principle; Invariance Principle; Invariance Principle of General Relativity; Local Invariance Principle; Maruyama's Invariance Principle; Perturbation Invariance Principle; Principle of Invariance; Principle of Undulatory Invariance; Relaxed Invariance Principle; Scale Invariance Principle; Strong Principle; Variational Principle; Weak Invariance Principle; Principle of Relativity; Active Principle; August Krogh Principle

J

JOINTS Archimedes' Principle; Hydrostatic Principle; Mckenzie Principles; Principle-centered Spine Care; Syndrome Principle; Unlearned Principle

K

KAISER BESSEL WINDOW FUNCTIONS Complementary Principle; Complementary Variational Principle; Dual Principle; Dual Variational Principle; Nonconvex Variational Principle; Variational Principle

KALMAN FILTER Principle of Kalman Filter, Maximum Likelihood Principle

KINEMATIC THEORY Mueller-Breslau Principle; Kinematical Principle; Principle of Relativity; Principle of Relativity; Kinematical Principle of Equal Passage Times; Principle of Relativity

KINETIC ENERGY Principle of Molecular Dynamics; Principle of Minimum Dissipation

KINETIC THEORY OF GASES Principle of Material Frame Indifference

KINETICS Maximum Principle; Principle of molecular dynamics; Reactivity/Selectivity Principle, Value Principle

KIRCHHOFF'S VOLTAGE LAW Hamilton's Principle; Kirchhoff's Voltage Law Principle; Minimum Energy Principle; Variational Principle; Woltjer Minimum Energy Principle; Woltjer-Taylor Minimum Energy Principle

KITCHEN VENTILATION Capillary Electrophoresis Principle; Ion Pair Principle; Principle of Diastereomeric Ion Pair Formation

KNOWLEDGE ENGINEERING TOOLS Correspondence Principle; Entropy Maximum Principle; Extended Maximum Entropy Principle; Extremal Principle of Information; Frieden-Soffer Principle; Gibbs Principle; Inference Principle; Jaynes' Maximum Entropy Principle; Jaynes' Principle; Kullback's Minimum Cross-entropy Principle; Maximum Information Entropy Principle; Measurement Principle; Minimax Entropy Principle; Minimum Entropy Principle; Minimum Relative Entropy Principle; Modification of The Extended Maximum Entropy Principle; Principle of Maximum Entropy; Principle of Maximum Nonadditive Entropy; Principle of Observational Compatibility; Principle of Operational Compatibility; Uncertainty Principle

LAGRANGE EQUATION Action Principle; Caratheodory's Principle; Conte's Principle; D'Alemberts Principle; Dissipation Principle; Virtual Dissipation Principle; Fermat's Principle; Fermats Principle; Handicap Principle; Least Time Principle; Principle of Caratheodony; Schwinger's Quantum Action Principle, Stationary Time Principle

LANGUAGE Correspondence Principle; Entropy Maximum Principle; Extended Maximum Entropy Principle; Extremal Principle of Information; Frieden-Soffer Principle; Gibbs Principle; Inference Principle; Jaynes' Maximum Entropy Principle; Jaynes' Principle; Kullback's Minimum Cross-entropy Principle; Maximum Information Entropy Principle; Measurement Principle; Minimax Entropy Principle; Minimum Entropy Principle; Minimum Relative Entropy Principle; Modification of The Extended Maximum Entropy Principle; Principle of Maximum Entropy; Principle of Information Processing; Principle of Maximum Nonadditive Entropy; Principle of Observational Compatibility; Principle of Operational Compatibility; Uncertainty Principle

LAPLACE APPROXIMATIONS Contraction Principle; Gibbs Variational Principle; Large Deviation Principle

LAPLACE OPERATORS Absorption Principle; Ambartsumian Invariance Principle; Charnov's Invariance Principle; Donsker's Invariance Principle; Invariance Principle; Invariance Principle of General Relativity; Local Invariance Principle; Maruyama's Invariance Principle; Minimum Description Length Principle; Perturbation Invariance Principle; Principle of Invariance; Principle of Undulatory Invariance; Relaxed Invariance Principle; Scale Invariance Principle; Strong Principle; Variational Principle; Weak Invariance Principle; Maximum Hardness Principle, Maximum Principle

LARGE DEVIATION Contraction Mapping Principle; Contraction Principle; Averaging Principle; Contraction Principle; Contraction Principle; Correspondence Principle; Entropy Maximum Principle; Extended Maximum Entropy Principle; Extremal Principle of Information; Frieden-Soffer Principle; Gibbs Principle; Gibbs Variational Principle; Inference Principle; Jaynes' Maximum Entropy Principle; Jaynes' Principle; Kullback's Minimum Cross-entropy Principle; Large Deviation Principle; Maximum Information Entropy Principle; Maximum Principle; Measurement Principle; Minimax Entropy Principle; Minimum Entropy Principle; Minimum Relative Entropy Principle; Modification of The Extended Maximum Entropy Principle; Principle of Maximum Entropy; Principle of Maximum Nonadditive Entropy; Principle of Observational Compatibility; Principle of Operational Compatibility; Uncertainty Principle; Principle of Large Deviations

LARGE SCALE STRUCTURE Corresponding States Principle; Cosmological Principle; Pitzer Corresponding States Principle; Strong Equivalence Principle

LARGE SCALE SYSTEM Autonomous Cooperation Distributed Maximum Principle; Distributed Maximum Principle; Maximum Principle, Principle of Dominant Subsystems, Energy Storage Principle

LARGE TIME Contraction Principle; Contraction Principle; Correspondence Principle; Entropy Maximum Principle; Extended Maximum Entropy Principle; Extremal Principle of Information; Frieden-Soffer Principle; Gibbs Principle; Gibbs Variational Principle; Inference Principle; Invariance Principle; Invariance Principle; Jaynes' Maximum Entropy Principle; Jaynes' Principle; Kullback's Minimum Cross-entropy Principle; Large Deviation Principle; Maximum Information

Entropy Principle; Measurement Principle; Minimax Entropy Principle; Minimum Entropy Principle; Minimum Relative Entropy Principle; Modification of The Extended Maximum Entropy Principle; Principle of Maximum Entropy; Principle of Maximum Nonadditive Entropy; Principle of Observational Compatibility; Principle of Operational Compatibility; Uncertainty Principle

LATTICE GASES Averaging Principle; Correspondence Principle; Entropy Maximum Principle; Extended Maximum Entropy Principle; Extremal Principle of Information; Frieden-Soffer Principle; Gibbs Principle; Inference Principle; Jaynes' Maximum Entropy Principle; Jaynes' Principle; Kullback's Minimum Cross-entropy Principle; Maximum Information Entropy Principle; Measurement Principle; Minimax Entropy Principle; Minimum Entropy Principle; Minimum Relative Entropy Principle; Modification of The Extended Maximum Entropy Principle; Principle of Maximum Entropy; Principle of Maximum Nonadditive Entropy; Principle of Observational Compatibility; Principle of Operational Compatibility; Uncertainty Principle

LATTICE HOMOMORPHISMS Correspondence Principle; Entropy Maximum Principle; Extended Maximum Entropy Principle; Extremal Principle of Information; Frieden-Soffer Principle; Gibbs Principle; Inference Principle; Jaynes' Maximum Entropy Principle; Jaynes' Principle; Kullback's Minimum Cross-entropy Principle; Maximum Information Entropy Principle; Measurement Principle; Minimax Entropy Principle; Minimum Entropy Principle; Minimum Relative Entropy Principle; Modification of The Extended Maximum Entropy Principle; Principle of Maximum Entropy; Principle of Maximum Nonadditive Entropy; Principle of Observational Compatibility; Principle of Operational Compatibility; Uncertainty Principle

LATTICE LORENTZ GAS Ambartsumian Invariance Principle; Charnov's Invariance Principle; Charnov's Invariance Principle; Donsker's Invariance Principle; Invariance Principle; Invariance Principle of General Relativity; Local Invariance Principle; Maruyama's Invariance Principle; Perturbation Invariance Principle; Principle of Invariance; Principle of Undulatory Invariance; Relaxed Invariance Principle; Scale Invariance Principle; Strong Principle; Variational Principle; Weak Invariance Principle

LAW D'Alembert's Principle; Gyarmati's Principle; Hamilton Principle; Majority Principle; Prigogine Principle; Principle of Correspondence; Principle of Minimum Kinetic Energy; Principle of Minimum Mechanical Energy; Principle of Minimum Potential Energy; Quantum Hamilton Principle; Stewart-Hamilton Principle; Stochastic Hamilton Variational Principle; Toupin's Dual Principle

LAW OF THE ITERATED LOGARITHM Invariance Principle; Ambartsumian Invariance Principle; Charnov's Invariance Principle; Donsker's Invariance Principle; Invariance Principle; Invariance Principle of General Relativity; Local Invariance Principle; Maruyama's Invariance Principle; Perturbation Invariance Principle; Principle of Invariance; Principle of Undulatory Invariance; Relaxed Invariance Principle; Scale Invariance Principle; Strong Principle; Variational Principle; Weak Invariance Principle

LEARNING Active Principle; Composing Principle; Inertia Principle; Lagrange Principle of Least Action; Least Action Principle; Least Motion Principle; Principle of Least Motion

LEARNING ALGORITHM Equidistortion Principle; Quantization Design Principle; Correspondence Principle; Entropy Maximum Principle; Extended Maximum Entropy Principle; Extremal Principle of Information; Frieden-Soffer Principle; Gibbs Principle; Inference Principle; Jaynes' Maximum Entropy Principle; Jaynes' Principle; Kullback's Minimum Cross-entropy Principle; Maximum Information Entropy Principle; Measurement Principle; Minimax Entropy Principle; Minimum Entropy Principle; Minimum Relative Entropy Principle; Modification of The Extended Maximum Entropy Principle;

Principle of Maximum Entropy; Principle of Maximum Nonadditive Entropy; Principle of Observational Compatibility; Principle of Operational Compatibility; Uncertainty Principle

LEARNING MONOTONE DNF FORMULAS Contraction Principle; Gibbs Variational Principle; Large Deviation Principle

LEAST MEDIAN SQUARES Ambartsumian Invariance Principle; Charnov's Invariance Principle; Donsker's Invariance Principle; Invariance Principle; Invariance Principle of General Relativity; Local Invariance Principle; Maruyama's Invariance Principle; Perturbation Invariance Principle; Principle of Invariance; Principle of Undulatory Invariance; Relaxed Invariance Principle; Scale Invariance Principle; Strong Principle; Variational Principle; Weak Invariance Principle

LEAST SQUARES METHOD Least Square Principle; Minimum Description Length Principle; Principle of Kalman Filter; Virtual Work Principle

LEGISLATION Caution Principle; Precaution Principle; Uncertainty Principle

LENGTH Correspondence Principle; Entropy Maximum Principle; Extended Maximum Entropy Principle; Extremal Principle of Information; Frieden-Soffer Principle; Gibbs Principle; Inference Principle; Jaynes' Maximum Entropy Principle; Jaynes' Principle; Kullback's Minimum Cross-entropy Principle; Maximum Entropy Principle; Maximum Information Entropy Principle; Measurement Principle; Minimax Entropy Principle; Minimum Entropy Principle; Minimum Relative Entropy Principle; Modification of The Extended Maximum Entropy Principle; Principle of Maximum Entropy; Principle of Maximum Nonadditive Entropy; Principle of Observational Compatibility; Principle of Operational Compatibility; Uncertainty Principle; Unified Principle

LEVEE Correspondence Principle; Entropy Maximum Principle; Extended Maximum Entropy Principle; Extremal Principle

of Information; Frieden-Soffer Principle; Gibbs Principle; Inference Principle; Jaynes' Maximum Entropy Principle; Jaynes' Principle; Kullback's Minimum Cross-entropy Principle; Maximum Information Entropy Principle; Measurement Principle; Minimax Entropy Principle; Minimum Entropy Principle; Minimum Relative Entropy Principle; Modification of The Extended Maximum Entropy Principle; Principle of Maximum Entropy; Principle of Maximum Nonadditive Entropy; Principle of Observational Compatibility; Principle of Operational Compatibility; Uncertainty Principle

LEVEL 1 ENTROPY FUNCTION Contagion Principle; Contraction Principle; Gibbs Variational Principle; Large Deviation Principle

LEVEL 2 ENTROPY FUNCTION Contagion Principle; Contraction Principle; Gibbs Variational Principle; Large Deviation Principle

LEVICIVITA CONNECTION Caratheodory's Principle; Conte's Principle; Fermat's Principle; Fermats Principle; Least Time Principle; Principle of Caratheodony; Stationary Time Principle

LEVINSON'S CLOSURE THEOREMS Hamilton's Principle; Kirchhoff's Voltage Law Principle; Minimum Energy Principle; Variational Principle; Woltjer Minimum Energy Principle; Woltjer-Taylor Minimum Energy Principle

LEVY DISTRIBUTIONS Least Action Principle; Competitive Exclusion Principle; Principle of Complementarity; Principle of Least Work

LIGHT ABSORPTION Inference Principle; Jaynes' Principle; Kullback's Minimum Cross-entropy Principle; Maximum Information Entropy Principle; Measurement Principle; Measuring Principle; Minimax Entropy Principle; Minimum Energy Principle; Minimum Norm Principle; Minimum Relative Entropy Principle; Modification of The Extended Maximum Entropy Principle; Principle of Maximum Entropy; Principle of Maximum Nonadditive Entropy; Principle of Observational Compatibility; Principle of Operational Compatibility; Uncertainty Principle; Principle of Causality

LIGHT LIKE GEODESICS Caratheodory's Principle; Conte's Principle; Fermat's Principle; Fermat; Least Time Principle; Principle of Caratheodony; Stationary Time Principle

LIGHT POLARIZATION Hamilton Variational Principle; Light Projection Principle; Projection Principle

LIGHT PROPAGATION Caratheodory's Principle; Conte's Principle; Fermat's Principle; Fermat; Least Time Principle; Principle of Caratheodony; Stationary Time Principle

LIGHT RAYS Caratheodory's Principle; Conte's Principle; Fermat's Principle; Fermat; Least Time Principle; Principle of Caratheodony; Stationary Time Principle

LIGHT SCATTERING Caratheodory's Principle; Conte's Principle; Fermat's Principle; Fermat; Inference Principle; Jaynes' Principle; Kullback's Minimum Cross-entropy Principle; Least Time Principle; Maximum Information Entropy Principle; Measurement Principle; Measuring Principle; Minimax Entropy Principle; Minimum Energy Principle; Minimum Norm Principle; Minimum Relative Entropy Principle; Modification of The Extended Maximum Entropy Principle; Principle of Caratheodony; Principle of Maximum Entropy; Principle of Maximum Nonadditive Entropy; Principle of Observational Compatibility; Principle of Operational Compatibility; Stationary Time Principle; Uncertainty Principle

LIMIT THEOREMS Correspondence Principle; Entropy Maximum Principle; Extended Maximum Entropy Principle; Extremal Principle of Information; Frieden-Soffer Principle; Gibbs Principle; Inference Principle; Jaynes' Maximum Entropy Principle; Jaynes' Principle; Kullback's Minimum Cross-entropy Principle; Maximum Information Entropy Principle; Measurement Principle; Minimax Entropy Principle; Minimum Entropy Principle; Minimum Relative Entropy Principle; Modification of The Extended Maximum Entropy Principle; Principle of Maximum Entropy; Principle of Maximum Nonadditive Entropy; Principle of Observational Compatibility; Principle of Operational Compatibility; Uncertainty Principle

LIMITS Principle of molecular dynamics; Lindel of Principle; Phragmen-Lindelof Principle; Schwarz Reflection Principle; Lindel'of Principle; Extremum Principle; Korenblum's Maximum Principle

LINE FORMATION Hamilton's Principle; Kirchhoff's Voltage Law Principle; Minimum Energy Principle; Variational Principle; Woltjer Minimum Energy Principle; Woltjer-Taylor Minimum Energy Principle

LINE INTEGRAL Equivalence Principle

LINE SHAPES Competitive Exclusion Principle; Principle of Complementarity

LINE VOLTAGE Hamilton's Principle; Kirchhoff's Voltage Law Principle; Minimum Energy Principle; Variational Principle; Woltjer Minimum Energy Principle; Woltjer-Taylor Minimum Energy Principle

LINEAR ALGEBRA Contraction Principle; Gibbs Variational Principle; Large Deviation Principle, Stochastic Maximum Principle

LINEAR CCD (CHARGE COUPLED DEVICE) ALTITUDE SYSTEM Collider Operational Principle; Operational Principle; Principle of Operationism; Revised Principle

LINEAR CONNECTIONS Principle of Equivalence; Linear Covariance Principle; Unlearned Principle

LINEAR NON EQUITHERMODYNAMICS Energy Principle; Laser Scattering Principle; Le Chatelierbraun Principle

LINEAR QUADRATIC CONTROL Ekeland's variational Principle; Ekeland's Epsilon Variational Principle; Ekeland's Variational Principle; Equivalence Principle; Global Maximum Principle; Maximum Principle; Variational Principle

LINEAR SYSTEM Separation Principle; Comparison Principle; Internal Model Principle; Maximum Principle, Fundamental Principle

LIPSCHITZ FUNCTION Maximum Principle, Ekeland's Epsilon Variational Principle; Ekeland's Principle; Ekeland's Variational Principle; Global Maximum Principle; Maximum Principle; Variational Principle

LIQUID JETS Correspondence Principle; Entropy Maximum Principle; Extended Maximum Entropy Principle; Extremal Principle of Information; Frieden-Soffer Principle; Gibbs Principle; Inference Principle; Jaynes' Maximum Entropy Principle; Jaynes' Principle; Kullback's Minimum Cross-entropy Principle; Maximum Information Entropy Principle; Measurement Principle; Minimax Entropy Principle; Minimum Entropy Principle; Minimum Relative Entropy Principle; Modification of The Extended Maximum Entropy Principle; Principle of Maximum Entropy; Principle of Maximum Nonadditive Entropy; Principle of Observational Compatibility; Principle of Operational Compatibility; Uncertainty Principle

LIQUID QUANTITY DETECTOR OF TANK Bekesy Travelling Wave Principle; Helmholtz Resonance Principle; Thevenin's Principle; Traveling Wave Principle

LIQUID VAPOR EQUILIBRIUM Corresponding States Principle; Pitzer Corresponding States Principle; Principle of Congruence

LIQUIDS Corresponding States Principle; Pitzer Corresponding States Principle;

LITERATURE Bruegelian Principle; Foucaldian Principle of Exclusion; Principle of Reversal

LOCAL IMAGE PIXEL VARIANCE Correspondence Principle; Entropy Maximum Principle; Extended Maximum Entropy Principle; Extremal Principle of Information; Frieden-Soffer Principle; Gibbs Principle; Inference Principle; Jaynes' Maximum Entropy Principle; Jaynes' Principle; Kullback's Minimum Cross-entropy Principle; Maximum Information Entropy Principle; Measurement Principle; Minimax Entropy Principle; Minimum Entropy Principle; Minimum Relative Entropy Principle; Modification of The Extended Maximum Entropy Principle; Principle of Maximum Entropy; Principle of Maximum Nonadditive Entropy; Principle of Observational Compatibility; Principle of Operational Compatibility; Uncertainty Principle

LOCAL TIME Ambartsumian Invariance Principle; Charnov's Invariance Principle; Charnov's Invariance Principle; Donsker's Invariance Principle; Invariance Principle; Invariance Principle of General Relativity; Local Invariance Principle; Maruyama's Invariance Principle; Perturbation Invariance Principle; Principle of Invariance; Principle of Undulatory Invariance; Relaxed Invariance Principle; Scale Invariance Principle; Strong Principle; Variational Principle; Weak Invariance Principle

LOCATION SCATTER Invariance Principle; Invariance Principle; Ambartsumian Invariance Principle; Charnov's Invariance Principle; Charnov's Invariance Principle; Donsker's Invariance Principle; Invariance Principle; Invariance Principle of General Relativity; Local Invariance Principle; Maruyama's Invariance Principle; Perturbation Invariance Principle; Principle of Invariance; Principle of Undulatory Invariance; Relaxed Invariance Principle; Scale Invariance Principle; Strong Principle; Variational Principle; Weak Invariance Principle

LOGIC PROGRAMMING Active Principle; Parallel Resolution Principle; Extension Principle

LONGTIME TAIL OF VELOCITY Invariance Principle, Ambartsumian Invariance Principle; Charnov's Invariance Principle; Donsker's Invariance Principle; Invariance Principle; Invariance Principle of General Relativity; Local Invariance Principle; Maruyama's Invariance Principle; Perturbation Invariance Principle; Principle of Invariance; Principle of Undulatory Invariance; Relaxed Invariance Principle; Scale Invariance Principle; Strong Principle; Variational Principle; Weak Invariance Principle

LORENTZ FORCE Conservation of Momentum Principle; D'Alembert Principle; Differential Principle; Hamilton Principle; Newton's Third Principle; Principle of Equal A Priori Probabilities; Variational Principle

LORENTZ INVARIANCE OF LOCALIZATION (LIL) Ambartsumian Invariance Principle; Charnov's Invariance Principle; Donsker's

Invariance Principle; Invariance Principle; Invariance Principle of General Relativity; Local Invariance Principle; Maruyama's Invariance Principle; Perturbation Invariance Principle; Principle of Invariance; Principle of Undulatory Invariance; Relaxed Invariance Principle; Scale Invariance Principle; Schwinger's Quantum Action Principle; Strong Principle; Variational Principle; Weak Invariance Principle

LORENTZ MANIFOLD Caratheodory's Principle; Fermat's Principle; Fermat; Least Time Principle; Principle of Caratheodony; Stationary Time Principle; Conte's Principle; Fermat's Principle; Fermats Principle

LORENTZ TRANSFORMATION Minimum Description Length Principle; Principle of Relativity

LORENTZIAN GEOMETRY Caratheodory's Principle; Conte's Principle; Fermat's Principle; Fermat; Least Time Principle; Principle of Caratheodony; Stationary Time Principle

LORENTZIAN SPACE TIMES Caratheodory's Principle; Conte's Principle; Fermat's Principle; Fermat; Least Time Principle; Principle of Caratheodony; Stationary Time Principle

LOSS RATE D'Alembert's Principle; Gyarmati's Principle; Hamilton Principle; Prigogine Principle; Principle of Minimum Kinetic Energy; Principle of Minimum Mechanical Energy; Principle of Minimum Potential Energy; Quantum Hamilton Principle; Stewart-Hamilton Principle; Stochastic Hamilton Variational Principle; Toupin's Dual Principle

LOWER SEMICONTINUITY Ekeland's Epsilon Variational Principle; Ekeland's Variational Principle; Global Maximum Principle; Variational Principle

LQTYPE PROBLEM Maximum Principle; Ekeland's Epsilon Variational Principle; Ekeland's Variational Principle; Global Maximum Principle; Variational Principle

LUBRICATING OIL SUPPLY TO GEAR WHEEL TRANSMISSION MONITORING SENSOR Calculation Principle

LUMINOSITY FUNCTION OF THE CFARED SHIFT SURVEY Caratheodory's Principle; Conte's Principle; Fermat's Principle; Fermat; Least Time Principle; Principle of Caratheodony; Stationary Time Principle

LUMO MOLECULAR ORBITAL Saint-Venant's Principle

LYAPUNOV FUNCTION Ambartsumian Invariance Principle; Charnov's Invariance Principle; Donsker's Invariance Principle; Extremum Principle; Invariance Principle; Invariance Principle of General Relativity; Local Invariance Principle; Maruyama's Invariance Principle; Perturbation Invariance Principle; Principle of Invariance; Principle of Undulatory Invariance; Relaxed Invariance Principle; Scale Invariance Principle; Strong Principle; Variational Principle; Weak Invariance Principle

M

MACH EINSTEIN UNIVERSE Maximum Hardness Principle; Mach's Principle; Maximum Hardness Principle

MACHINING SPEED Maximum Principle, Machine Serving Principle, Minimum Information Principle

MAGNET POSITION Inference Principle; Jaynes' Principle; Kullback's Minimum Cross-entropy Principle; Maximum Information Entropy Principle; Measurement Principle; Measuring Principle; Minimax Entropy Principle; Minimum Energy Principle; Minimum Norm Principle; Minimum Relative Entropy Principle; Modification of The Extended Maximum Entropy Principle; Principle of Maximum Entropy; Principle of Maximum Nonadditive Entropy; Principle of Observational Compatibility; Principle of Operational Compatibility; Uncertainty Principle

MAGNETIC AND ELECTRIC FIELDS Hamilton's Principle; Kirchhoff's Voltage Law Principle; Minimum Energy Principle; Variational Principle; Woltjer Minimum Energy Principle; Woltjer-Taylor Minimum Energy Principle

MAGNETIC ENERGY MINIMISATION Hamilton's Principle; Kirchhoff's Voltage Law Principle; Minimum Energy Principle; Variational Principle; Woltjer Minimum Energy Principle; Woltjer-Taylor Minimum Energy Principle

MAGNETIC FIELDS Hamilton's Principle; Kelvinarnol'D Energy Principle; Kirchhoff's Voltage Law Principle; Minimum Energy Principle; Variational Principle; Woltjer Minimum Energy Principle; Woltjer-Taylor Minimum Energy Principle, Principle of molecular dynamics

MAGNETIC FLUX INTENSITY Inference Principle. Jaynes' Principle; Kullback's Minimum Cross-entropy Principle; Maximum Information Entropy Principle; Measurement Principle; Measuring Principle; Minimax Entropy Principle; Minimum Energy Principle; Minimum Norm Principle; Minimum Relative Entropy Principle; Modification of The Extended Maximum Entropy Principle; Principle of Maximum Entropy; Principle of Maximum Nonadditive Entropy; Principle of Observational Compatibility; Principle of Operational Compatibility; Uncertainty Principle

MAGNETIC FLUX TUBES Hamilton's Principle; Kirchhoff's Voltage Law Principle; Minimum Energy Principle; Variational Principle; Woltjer Minimum Energy Principle; Woltjer-Taylor Minimum Energy Principle

MAGNETIC INSTRUMENTS AND TECHNIQUES Inference Principle; Jaynes' Principle; Kullback's Minimum Cross-entropy Principle; Maximum Information Entropy Principle; Measurement Principle; Measuring Principle; Minimax Entropy Principle; Minimum Energy Principle; Minimum Norm Principle; Minimum Relative Entropy Principle; Modification of The Extended Maximum Entropy Principle; Principle of Maximum Entropy; Principle of Maximum Nonadditive Entropy; Principle of Observational Compatibility; Principle of Operational Compatibility; Uncertainty Principle

MAGNETIC INSULATION THEORY Hamilton's Principle; Kirchhoff's Voltage Law Principle; Minimum Energy Principle; Variational Principle; Woltjer Minimum Energy Principle; Woltjer-Taylor Minimum Energy Principle

MAGNETIC RESONANCE IMAGING Correspondence Principle; Entropy Maximum Principle; Extended Maximum Entropy Principle; Extremal Principle of Information; Frieden-Soffer Principle; Gibbs Principle; Inference Principle; Jaynes' Maximum Entropy

Principle; Jaynes' Principle; Kullback's Minimum Cross-entropy Principle; Maximum Information Entropy Principle; Measurement Principle; Minimax Entropy Principle; Minimum Entropy Principle; Minimum Relative Entropy Principle; Modification of The Extended Maximum Entropy Principle; Principle of Maximum Entropy; Principle of Maximum Nonadditive Entropy; Principle of Observational Compatibility; Principle of Operational Compatibility; Uncertainty Principle

MAGNETIC SENSORS Inference Principle; Jaynes' Principle; Kullback's Minimum Cross-entropy Principle; Maximum Information Entropy Principle; Measurement Principle; Measuring Principle; Minimax Entropy Principle; Minimum Energy Principle; Minimum Norm Principle; Minimum Relative Entropy Principle; Modification of The Extended Maximum Entropy Principle; Principle of Maximum Entropy; Principle of Maximum Nonadditive Entropy; Principle of Observational Compatibility; Principle of Operational Compatibility; Uncertainty Principle

MAGNETIC STRUCTURE Hamilton's Principle; Kirchhoff's Voltage Law Principle; Minimum Energy Principle; Variational Principle; Woltjer Minimum Energy Principle; Woltjer-Taylor Minimum Energy Principle

MAGNETO HYDRODYNAMICS RESEARCH (MHD) Hamilton's Principle; Kelvin Arnol'D Energy Principle; Kirchhoff's Voltage Law Principle; Minimum Energy Principle; Variational Principle; Woltjer Minimum Energy Principle; Woltjer-Taylor Minimum Energy Principle

MAGNETO STATIC LOUD SPEAKER Electronegativity Equalization Principle; Geometric Mean Principle; Principle of electronegativity

MANEUVERS D'Alembert's Principle; Gyarmati's Principle; Hamilton Principle; Prigogine Principle; Principle of Minimum Kinetic Energy; Principle of Minimum Mechanical Energy; Principle of Minimum Potential Energy; Quantum Hamilton Principle; Stewart-Hamilton Principle; Stochastic Hamilton Variational Principle; Toupin's Dual Principle

MANNERS Correspondence Principle; Entropy Maximum Principle; Extended Maximum Entropy Principle; Extremal Principle of Information; Frieden-Soffer Principle; Gibbs Principle; Inference Principle; Jaynes' Maximum Entropy Principle; Jaynes' Principle; Kullback's Minimum Cross-entropy Principle; Maximum Information Entropy Principle; Measurement Principle; Minimax Entropy Principle; Minimum Entropy Principle; Minimum Relative Entropy Principle; Modification of The Extended Maximum Entropy Principle; Principle of Maximum Entropy; Principle of Maximum Nonadditive Entropy; Principle of Observational Compatibility; Principle of Operational Compatibility; Uncertainty Principle

MAPPING Austerity Principle; Boundary Harnack Principle; Boundary of A Boundary Principle; Boundary Principle; Boundedness Principle; Local Uniform Boundedness Principle; Nadirdshvlli Boundary Principle; Uniform Boundedness Principle

MARKOV ANALYSIS Correspondence Principle; Entropy Maximum Principle; Extended Maximum Entropy Principle; Extremal Principle of Information; Frieden-Soffer Principle; Gibbs Principle; Inference Principle; Jaynes' Maximum Entropy Principle; Jaynes' Principle; Kullback's Minimum Cross-entropy Principle; Maximum Information Entropy Principle; Measurement Principle; Minimax Entropy Principle; Minimum Entropy Principle; Minimum Relative Entropy Principle; Modification of The Extended Maximum Entropy Principle; Principle of Maximum Entropy; Principle of Maximum Nonadditive Entropy; Principle of Observational Compatibility; Principle of Operational Compatibility; Uncertainty Principle

MARKOV CHAINS Ambartsumian Invariance Principle; Charnov's Invariance Principle; Donsker's Invariance Principle; Invariance Principle; Invariance Principle of General Relativity; Local Invariance Principle; Maruyama's Invariance Principle; Perturbation Invariance Principle; Principle of Invariance; Principle of Undulatory Invariance; Relaxed Invariance Principle;

Scale Invariance Principle; Strong Principle; Variational Principle; Weak Invariance Principle

MARKOV PROCESS Action Principle; Contraction Mapping Principle; Contraction Principle; Gibbs Variational Principle; Invariance Principle; Invariance Principle; Large Deviation Principle; Ambartsumian Invariance Principle; Charnov's Invariance Principle; Correspondence Principle; Donsker's Invariance Principle; Entropy Maximum Principle; Extended Maximum Entropy Principle; Extremal Principle of Information; Frieden-Soffer Principle; Gibbs Principle; Inference Principle; Invariance Principle; Invariance Principle of General Relativity; Jaynes' Maximum Entropy Principle; Jaynes' Principle; Kullback's Minimum Cross-entropy Principle; Local Invariance Principle; Maruyama's Invariance Principle; Maximum Information Entropy Principle; Measurement Principle; Minimax Entropy Principle; Minimum Entropy Principle; Minimum Relative Entropy Principle; Modification of The Extended Maximum Entropy Principle; Perturbation Invariance Principle; Principle of Invariance; Principle of Maximum Entropy; Principle of Maximum Nonadditive Entropy; Principle of Observational Compatibility; Principle of Operational Compatibility; Principle of Undulatory Invariance; Relaxed Invariance Principle; Scale Invariance Principle; Strong Principle; Uncertainty Principle; Variational Principle; Weak Invariance Principle; Active Principle; Maximum Principle; Mdl Principle

MARKOV RANDOM FIELDS Entropy Minimax Principle; Equidistortion Principle; Quantization Design Principle

MARTINGALE Ambartsumian Invariance Principle; Charnov's Invariance Principle; Charnov's Invariance Principle; Donsker's Invariance Principle; Invariance Principle; Invariance Principle of General Relativity; Local Invariance Principle; Maruyama's Invariance Principle; Perturbation Invariance Principle; Principle of Invariance; Principle of Undulatory Invariance; Relaxed Invariance Principle; Scale Invariance Principle;

Separation Principle; Strong Principle; Variational Principle; Weak Invariance Principle

MARTINGALES CENTRAL LIMIT THEOREM Ambartsumian Invariance Principle; Charnov's Invariance Principle; Donsker's Invariance Principle; Invariance Principle; Invariance Principle of General Relativity; Local Invariance Principle; Maruyama's Invariance Principle; Perturbation Invariance Principle; Principle of Invariance; Principle of Undulatory Invariance; Relaxed Invariance Principle; Scale Invariance Principle; Strong Principle; Variational Principle; Weak Invariance Principle

MASS Cyclotron Principle; Minimum Description Length Principle; Principle of molecular dynamics

MASS TRANSFER Double Inlet Principle; Equipartition Principle; Maximizing Principle

MASS VELOCITY DEPENDENCE Conservation of Momentum Principle; D'Alembert Principle; Differential Principle; Hamilton Principle; Newton's Third Principle; Principle of Equal A Priori Probabilities; Variational Principle

MASSIVE CURRENT SHEET Hamilton's Principle; Kirchhoff's Voltage Law Principle; Minimum Energy Principle; Variational Principle; Woltjer Minimum Energy Principle; Woltjer-Taylor Minimum Energy Principle

MATCHING Working Principle, Cavalieri Principle; Stereological Principle, Conservation of Momentum Principle; D'Alembert Principle; Differential Principle; Hamilton Principle; Newton's Third Principle; Principle of Equal A Priori Probabilities; Variational Principle, Matching Principle, Absorption Principle; Matching Principle; Principle of Fundamental Characteristics, Principle of Causality, Optical Principle; Pulse echo Principle, Quartz Microbalance Principle

MATCHING Maximum Entropy Principle, Minimum Description Length Principle

MATERIAL DEDUCTION Correspondence Principle; Entropy Maximum Principle; Extended Maximum Entropy Principle; Extremal Principle of Information; Frieden-Soffer Principle; Gibbs Principle; Inference Principle; Jaynes' Maximum Entropy Principle; Jaynes' Principle; Kullback's Minimum Cross-entropy Principle; Maximum Information Entropy Principle; Measurement Principle; Minimax Entropy Principle; Minimum Entropy Principle; Minimum Relative Entropy Principle; Modification of The Extended Maximum Entropy Principle; Principle of Maximum Entropy; Principle of Maximum Nonadditive Entropy; Principle of Observational Compatibility; Principle of Operational Compatibility; Uncertainty Principle

MATERIALS Minimum Information Principle, Principle of Critical Energy, Pulse Echo Principle, Telecentric Principle

MATHEMATICAL ANALYSIS Contraction Principle; Gibbs Variational Principle; Large Deviation Principle, Ekeland's variational Principle; Ekeland's Epsilon Variational Principle; Ekeland's Variational Principle; Ekeland's Variational Principle; Equivalence Principle; Global Maximum Principle; Global Maximum Principle; Maximum Likelihood Principle; Maximum Principle; Pontryagin Maximum Principle; Variational Principle

MATHEMATICAL BIOLOGY Active Principle; Extremum Principle; Fisheye Adaptation Principle; Maupertuis Principle; Maximum Likelihood Principle; Neighborhood Coherence Principle; Maupertuis Principle; Optimization Principle; Optimizing Principle; Run and Recreate Principle, Extremal Principle; Fitness Principle; Maupertuis Principle; Principle of Comparative Advantage

MATHEMATICAL FORMULA Ambartsumian Invariance Principle; Charnov's Invariance Principle; Donsker's Invariance Principle; Invariance Principle; Invariance Principle of General Relativity; Local Invariance Principle; Maruyama's Invariance Principle; Perturbation Invariance Principle; Principle of Invariance; Principle of Undulatory Invariance; Relaxed Invariance Principle; Scale Invariance Principle; Strong Principle; Variational Principle; Weak Invariance Principle

MATHEMATICAL METHOD Huygen's's Principle; Conservation of Momentum Principle; D'Alembert Principle; Differential Principle; Hamilton Principle; Newton's Third Principle; Principle of Equal A Priori Probabilities; Variational Principle; Mathematical Model. Inertia Principle; Lagrange Principle of Least Action; Least Action Principle; Least Motion Principle; Maximizing Principle; Minimal Principle; Neighborhood Coherence Principle; Principle of Kalman Filter; Principle of Least Motion; Uncertainty Principle; Active Principle; August Krogh Principle; Correspondence Principle; Entropy Maximum Principle; Extended Maximum Entropy Principle; Extremal Principle of Information; Fitness Principle; Frieden-Soffer Principle; Gibbs Principle; Inference Principle; Jaynes' Maximum Entropy Principle; Jaynes' Principle; Kullback's Minimum Cross-entropy Principle; Maximum Entropy Principle; Maximum Information Entropy Principle; Maximum Likelihood Principle; Measurement Principle; Minimax Entropy Principle; Minimum Entropy Principle; Minimum Relative Entropy Principle; Modification of The Extended Maximum Entropy Principle; Mueller-Breslau Principle; Powdering Principle; Principle of Critical Energy; Principle of Maximum Entropy; Principle of Maximum Nonadditive Entropy; Principle of Observational Compatibility; Principle of Operational Compatibility; Uncertainty Principle

MATHEMATICS Chaotic Principle; Composing Principle; Conte's Principle; Correspondence Principle; Dirichlet Principle; Ekeland's variational Principle; Ekeland's Epsilon Variational Principle; Ekeland's Principle; Ekeland's Variational Principle; Entropy Maximum Principle; Equivalence Principle; Extended Maximum Entropy Principle; Extremal Principle of Information; Fermat's Principle; Fermats Principle; Frieden-Soffer Principle; Gibbs Principle; Global Maximum Principle; Hamilton's Principle; Inference Principle; Invariance Principle; Jaynes'

Maximum Entropy Principle; Jaynes' Principle; Kullback's Minimum Cross-entropy Principle; Least Time Principle; Maximum Information Entropy Principle; Maximum Principle; Measurement Principle; Minimal Principle; Minimax Entropy Principle; Minimum Entropy Principle; Minimum Relative Entropy Principle; Modification of The Extended Maximum Entropy Principle; Principle of Caratheodony; Principle of Maximum Entropy; Principle of Maximum Nonadditive Entropy; Principle of Observational Compatibility; Principle of Operational Compatibility; Principle of The Maximum; Stationary Time Principle; Uncertainty Principle; Variational Principle; Weakest Link Principle

MATRIX ALGEBRA Ekeland's variational Principle; Ekeland's Epsilon Variational Principle; Ekeland's Variational Principle; Global Maximum Principle; Maximum Principle; Minimum Description Length Principle; Variational Principle; Powdering Principle

MATRIX LIQUID CRYSTAL DISPLAY Birefringe Principle; Birefringence Principle; Electrically Controlled Birefringence Principle

MAXIMUM EFFICIENCY Ellul Principle of Efficiency; Emancipation Principle; Kant's Principle; Power Principle; Principle of Domination; Principle of Efficiency; Principle of Emancipation; Principle of Emancipation; Principle of Free Domain; Principle of Majority Rule Based On Decision Making Power Demand; Principle of Power; Principle of Subjective Freedom; Principle of The Autonomy of The Will

MAXIMUM ENTROPY (POME) Air Distribution Principle; Charge Distribution Principle; Cost Distribution Principle; Counter Current Distribution Principle; Distribution Principle; Principle of Current Distribution; Principle of Uncertainty; Maximum Entropy Method (Mem). Correspondence Principle; Entropy Maximum Principle; Extended Maximum Entropy Principle; Extremal Principle of Information; Frieden-Soffer Principle; Gibbs Principle; Inference Principle; Jaynes'

Maximum Entropy Principle; Jaynes' Principle; Kullback's Minimum Cross-entropy Principle; Maximum Information Entropy Principle; Measurement Principle; Minimax Entropy Principle; Minimum Entropy Principle; Minimum Information Principle; Minimum Relative Entropy Principle; Modification of The Extended Maximum Entropy Principle; Principle of Maximum Entropy; Principle of Maximum Nonadditive Entropy; Principle of Observational Compatibility; Principle of Operational Compatibility; Uncertainty Principle; D'Alembert Lagrange Principle; D'Alembert Lagrange Variation Principle; D'Alembert Principle; Gauss Principle; Hamilton Lagrange Principle; Jaynes' Principle; Jourdain Principle; Kinetic Principle; Lagrange Principle; Least Constraint Principle; Maupertius Lagrange Principle; Maupertuis Euler Lagrange Principle of Least Action; Principle of Extremal Effects; Principle of Least Constraint; Principle of Maximal Effect; Principle of Virtual Work; Uncertainty Principle; Saint-Venant's Principle

MAXIMUM ENTROPY SPECTRAL ESTIMATION D'Alembert Lagrange Principle; D'Alembert Lagrange Variation Principle; D'Alembert Principle; Gauss Principle; Hamilton Lagrange Principle; Jayne's Principle; Jourdain Principle; Kinetic Principle; Lagrange Principle; Least Constraint Principle; Maupertius Lagrange Principle; Maupertuis Euler Lagrange Principle of Least Action; Principle of Extremal Effects; Principle of Least Constraint; Principle of Maximal Effect; Principle of Virtual Work; Uncertainty Principle

MAXIMUM LIKELIHOOD ESTIMATION Mde Principle; Principle of Per-survivor Processing; Principles of Darcy's Law

MEASUREMENT ERRORS Inference Principle; Jaynes' Principle; Kullback's Minimum Cross-entropy Principle; Maximum Information Entropy Principle; Measurement Principle; Measuring Principle; Minimax Entropy Principle; Minimum Energy Principle; Minimum Norm Principle; Minimum Relative Entropy Principle; Modification of The Extended Maximum Entropy Principle;

Principle of Maximum Entropy; Principle of Maximum Nonadditive Entropy; Principle of Observational Compatibility; Principle of Operational Compatibility; Uncertainty Principle

MEASUREMENT THEORY Adaptive Principle; Maximum Hardness Principle; Minimum Description Length Principle; Principle of Measurement

MEASURING MASS Coriolis Principle; Measuring Principle; Minimax Entropy Principle; Minimum Energy Principle; Minimum Relative Entropy Principle; Modification of The Extended Maximum Entropy Principle; Principle of Maximum Entropy; Principle of Maximum Nonadditive Entropy; Principle of Observational Compatibility; Principle of Operational Compatibility; Schwinger's Quantum Action Principle; Uncertainty Principle, Coulter Principle

MECHANICAL SYSTEM D'Alembert's Principle; Gyarmati's Principle; Hamilton Principle; Micropump Principle; Prigogine Principle; Principle of Minimum Kinetic Energy; Principle of Minimum Mechanical Energy; Principle of Minimum Potential Energy; Quantum Hamilton Principle; Stewart-Hamilton Principle; Stochastic Hamilton Variational Principle; Toupin's Dual Principle

MECHANICS Correspondence Principle; Least Action Principle; Maximum Principle; Principle of Least Work; Principle of Material Frame Indifference; Orthogonality Principle, Heisenberg's Correspondence Principle, D'Alembert's Principle; Gyarmati's Principle; Hamilton's Principle; Maupertuis Principle; Prigogine Principle; Principle of Corresponding States; Principle of Minimum Potential Energy; Quantum Hamilton Principle; Principle of Minimum Mechanical Energy; Stewart-Hamilton Principle; Stochastic Hamilton Variational Principle; Toupin's Dual Principle

MECHANISM Uncertainty Principle; Mechanism of Polymerbased Separations. Hard-soft acid-base Principle; Principle of Minimum Kinetic Energy

MEDIAN FILTERS Parsimony Principle; Principle of Parsimonius Data Modeling

MEDICAL EQUIPMENT Cross Check Principle; Inference Principle; Jaynes' Principle; Kullback's Minimum Cross-entropy Principle; Maximum Information Entropy Principle; Measurement Principle; Measuring Principle; Minimax Entropy Principle; Minimum Energy Principle; Minimum Norm Principle; Minimum Relative Entropy Principle; Modification of The Extended Maximum Entropy Principle; Overlapping Biphasic Impulse Principle; Principle of Maximum Entropy; Principle of Maximum Nonadditive Entropy; Principle of Observational Compatibility; Principle of Operational Compatibility; Uncertainty Principle

MEDICAL PSYCHOLOGY AND SOCIOLOGY Babinet's Principle; Complementarity Principle; Gestalt Principle; Obligatory Contour Principle

MEDICAL SCIENCE Mitrofanoff Principle; Monti Principle; Critical Path Method Principle; Fick Principle; Health Principle; Mckenzie Principles; Nameless Category N3C Principle; Optimization Principle; Overlapping Biphasic Impulse Principle; Precautionary Principle; Principle of Habeas Mentem; Principle-centered Spine Care; Run and Recreate Principle; Sufficiency Principle, Pareto Principle

MEDICINE Health Principle; Precautionary Principle; Principle of Habeas Mentem; Salutogenesis Principle

MEMBERSHIP Principle of Uncertainty, Principle of Information Processing; Zadehs Extension Principle

MEMBRANES CELL BIOLOGY Antiarrhythmic Principle; Dale's Principle; Drug Principle; HIV Principle

METABOLIC DISORDERS Boric Acid Affinity Principle; Boronic Acid Affinity Principle

METABOLIC PATHWAYS Principle of Remifentanil

METABOLIC STIMULATORS Antiarrhythmic Principle

METABOLISM Antimaximum Principle; Boric Acid Affinity Principle; Boronic Acid Affinity Principle; Dale's Principle; Principle of Remifentanil

METEOROLOGICAL ELEMENT Correspondence Principle; Entropy Maximum Principle; Extended Maximum Entropy Principle; Extremal Principle of Information; Frieden-Soffer Principle; Gibbs Principle; Inference Principle; Jaynes' Maximum Entropy Principle; Jaynes' Principle; Kullback's Minimum Cross-entropy Principle; Maximum Information Entropy Principle; Measurement Principle; Minimax Entropy Principle; Minimum Entropy Principle; Minimum Relative Entropy Principle; Modification of The Extended Maximum Entropy Principle; Principle of Maximum Entropy; Principle of Maximum Nonadditive Entropy; Principle of Observational Compatibility; Principle of Operational Compatibility; Uncertainty Principle

METEOROLOGY General Relativistic Fermat Principle; Principle of Construction and Operation

METHODOLOGY Action Principle; Autonomous Cooperation Distributed Maximum Principle; Distributed Maximum Principle; Inference Principle; Jaynes' Principle; Kullback's Minimum Cross-entropy Principle; Maximum Information Entropy Principle; Maximum Modulus Principle; Maximum Principle; Measurement Principle; Measuring Principle; Minimal Principle; Minimax Entropy Principle; Minimum Energy Principle; Minimum Norm Principle; Minimum Relative Entropy Principle; Modification of The Extended Maximum Entropy Principle; Principle of critical parameters; Principle of Maximum Entropy; Principle of Maximum Nonadditive Entropy; Principle of Measurement; Principle of Observational Compatibility; Principle of Operational Compatibility; Schwinger's Quantum Action Principle; Uncertainty Principle

METHODS Action Principle; Bond Directional Principle; Dale's Principle; Ecological Principle; Elastic Tube Principle; Frequency-dependent Principle; Gestalt Principle; Ice Tube Principle; Investment Principle; John Templeton Principle; Principle of Dynamic Ecology; Principle of Maximum Pessimism; Schwinger's Quantum Action Principle

METHYLTERT BUTYLETHER SYNTHESIS KINETICS Corresponding States Principle; Pitzer Corresponding States Principle

METRIC REGULARITY Maximum Principle; Pontryagins Principle, Maximum Hardness Principle

METRIC SPACE Ekeland's Epsilon Variational Principle; Ekeland's Variational Principle; Global Maximum Principle; Variational Principle; Contraction Principle; Gibbs Variational Principle; Large Deviation Principle

METRICS Minimum Description Length Principle

MICOCANONICAL DISTRIBUTION Correspondence Principle; Entropy Maximum Principle; Extended Maximum Entropy Principle; Extremal Principle of Information; Frieden-Soffer Principle; Gibbs Principle; Inference Principle; Jaynes' Maximum Entropy Principle; Jaynes' Principle; Kullback's Minimum Cross-entropy Principle; Maximum Information Entropy Principle; Measurement Principle; Minimax Entropy Principle; Minimum Entropy Principle; Minimum Relative Entropy Principle; Modification of The Extended Maximum Entropy Principle; Principle of Maximum Entropy; Principle of Maximum Nonadditive Entropy; Principle of Observational Compatibility; Principle of Operational Compatibility; Uncertainty Principle

MICROELECTRODES Inference Principle; Jaynes' Principle; Kullback's Minimum Cross-entropy Principle; Maximum Information Entropy Principle; Measurement Principle; Measuring Principle; Minimax Entropy Principle; Minimum Energy Principle; Minimum Norm Principle; Minimum Relative Entropy Principle; Modification of The Extended

Maximum Entropy Principle; Principle of Maximum Entropy; Principle of Maximum Nonadditive Entropy; Principle of Observational Compatibility; Principle of Operational Compatibility; Uncertainty Principle

MICROORGANISM Principle of Maximum Entropy, Occam's Razor Principle; Principle of Polymerase Chain Reaction (PCR)

MICROSCOPIC   CRITERIA Electronegativity Equalization Principle; Geometric Mean Principle; Principle of electronegativity

MICROSCOPY Maximum Likelihood Principle; Minimum Description Length Principle, Action Principle; Schwinger's Quantum Action Principle

MICROSTRUCTURAL   EFFECTS Correspondence Principle; Entropy Maximum Principle; Extended Maximum Entropy Principle; Extremal Principle of Information; Frieden-Soffer Principle; Gibbs Principle; Inference Principle; Jaynes' Maximum Entropy Principle; Jaynes' Principle; Kullback's Minimum Cross-entropy Principle; Maximum Information Entropy Principle; Measurement Principle; Minimax Entropy Principle; Minimum Entropy Principle; Minimum Relative Entropy Principle; Modification of The Extended Maximum Entropy Principle; Principle of Maximum Entropy; Principle of Maximum Nonadditive Entropy; Principle of Observational Compatibility; Principle of Operational Compatibility; Uncertainty Principle, Minimal Principle; Minimum Principle

MINIMAX   TECHNIQUES Maximum Decisional Efficiency (MDE) Principle; Minimax Theorem. Ekeland's variational Principle; Ekeland's Epsilon Variational Principle; Ekeland's Variational Principle; Global Maximum Principle; Variational Principle

MINIMISATION Active Principle; Minimum Description Length Principle

MINIMUM DESCRIPTION LENGTH Active Principle; Minimum Description Length Principle; Active Principle; Mdl Principle

MINIMUM ENERGY WAVEFORMS Hamilton's Principle; Kirchhoff's Voltage Law Principle; Minimum Energy Principle; Variational Principle; Woltjer Minimum Energy Principle; Woltjer-Taylor Minimum Energy Principle

MINIMUM   ENTROPY   METHODS Correspondence   Principle;   Entropy   Maximum Principle; Extended Maximum Entropy Principle; Extremal Principle of Information; Frieden-Soffer Principle; Gibbs Principle; Inference Principle; Jaynes' Maximum Entropy Principle; Jaynes' Principle; Kullback's Minimum Cross-entropy Principle; Maximum Information Entropy Principle; Measurement Principle; Minimax Entropy Principle; Minimum Cross-entropy (Me) Principle; Minimum Entropy Principle; Minimum Relative Entropy Principle; Modification of The Extended Maximum Entropy Principle; Principle of Maximum Entropy; Principle of Maximum Nonadditive Entropy; Principle of Observational Compatibility; Principle of Operational Compatibility; Uncertainty Principle

MINKOWSKI SPACE Conte's Principle; Fermats Principle; Least Time Principle; Maximum Hardness Principle; Principle of Caratheodony; Stationary Time Principle, Huygen's Principle

MIXED   METHOD D'Alembert   Lagrange Principle; D'Alembert Lagrange Variation Principle; D'Alembert Principle; Gauss Principle; Hamilton Lagrange Principle; Jayne's Principle; Jourdain Principle; Kinetic Principle; Lagrange Principle; Least Constraint Principle; Maupertius Lagrange Principle; Maupertuis Euler Lagrange Principle of Least Action; Principle of Extremal Effects; Principle of Least Constraint; Principle of Maximal Effect; Principle of Virtual Work

MIXING RULE Corresponding States Principle; Pitzer Corresponding States Principle;

MIXING   SEQUENCE Ambartsumian   Invariance Principle; Charnov's Invariance Principle; Donsker's Invariance Principle; Invariance Principle; Invariance Principle of General Relativity; Local Invariance

Principle; Maruyama's Invariance Principle; Perturbation Invariance Principle; Principle of Invariance; Principle of Undulatory Invariance; Relaxed Invariance Principle; Scale Invariance Principle; Strong Principle; Variational Principle; Weak Invariance Principle; Invariance Principle

MODEL BASED VISION APPROACH Correspondence Principle; Entropy Maximum Principle; Extended Maximum Entropy Principle; Extremal Principle of Information; Frieden-Soffer Principle; Gibbs Principle; Inference Principle; Jaynes' Maximum Entropy Principle; Jaynes' Principle; Kullback's Minimum Cross-entropy Principle; Maximum Information Entropy Principle; Measurement Principle; Minimax Entropy Principle; Minimum Entropy Principle; Minimum Relative Entropy Principle; Modification of The Extended Maximum Entropy Principle; Principle of Maximum Entropy; Principle of Maximum Nonadditive Entropy; Principle of Observational Compatibility; Principle of Operational Compatibility; Uncertainty Principle

MODEL COLOSSAL MAGNETORESISTANCE Correspondence Principle; Entropy Maximum Principle; Extended Maximum Entropy Principle; Extremal Principle of Information; Frieden-Soffer Principle; Gibbs Principle; Inference Principle; Jaynes' Maximum Entropy Principle; Jaynes' Principle; Kullback's Minimum Cross-entropy Principle; Maximum Information Entropy Principle; Measurement Principle; Minimax Entropy Principle; Minimum Entropy Principle; Minimum Relative Entropy Principle; Modification of The Extended Maximum Entropy Principle; Principle of Maximum Entropy; Principle of Maximum Nonadditive Entropy; Principle of Observational Compatibility; Principle of Operational Compatibility; Uncertainty Principle

MODEL POROELASTIC HALF SPACE Three-day Principle; Educational Principle; Ehrenfest Principle

MODEL PREDICTION Corresponding States Principle; Pitzer Corresponding States Principle;

MODEL SELECTION Active Principle; Maximum Likelihood Principle; Parsimony Principle; Principle of Parsimonius Data Modeling; Quartz Microbalance Principle

MODELING Decomposition Principle; Equivalence Principle; Mde Principle; Maximum Entropy Principle; Maximum Principle; Minimum Description Length Principle; Minimum Principle; New Therapeutic Principle; Principle of Information Processing; Wise Use Principle, Balance Principle

MODELS Hard-soft acid-base Principle; Maximum Likelihood Principle; Penaz Principle; Principle of Maximum Entropy, Computational Biology. Epistatic Handicap Principle; Extremum Principle; Gestalt Principle; Maximization Principle; Neighborhood Coherence Principle; Principle of Comparative Advantage; Optimizing Principle

MOISTURE MEASURING INSTRUMENTS Principle of Construction and Operation; Multi Bypass Principle

MOLECULAR CHEMICAL POTENTIALS Hard-soft acid-base Principle; Saint-Venant's Principle

MOLECULAR DYNAMICS First Principle; Maximizing Principle; Working Principle

MOLECULAR EVOLUTION Active Principle; Handicap Principle; Uncertainty Principle, Complexity Principle

MOLECULAR STRUCTURE Active Principle; Antiviral Active Principle; Maximizing Principle; Minimal Principle; Saint-Venant's Principle; Weak Invariance Principle

MOLTING STRATEGY Investment Principle; John Templeton Principle; Principle of Maximum Pessimism

MOMENTUM Cyclotron Principle; Principle of Measurement; Schwinger Principle

MONITORING Hard-soft acid-base Principle; Minimum Information Principle; Penaz Principle; Principle of Kalman Filter

MONTE CARLO METHOD Protection Principle; Principle of Maximum Entropy; Maximum Principle

MORPHOGENESIS Maupertuis Principle; Nameless Category N3C Principle; Neighborhood Coherence Principle; Optimality Principle; Organizing Principle; Principle of component analysis; Self-organizing Principle; Similarity Principle; Type II Cell Stretch-organizing Principle

MORPHOLOGICAL FILTERING Correspondence Principle; Entropy Maximum Principle; Extended Maximum Entropy Principle; Extremal Principle of Information; Frieden-Soffer Principle; Gibbs Principle; Inference Principle; Jaynes' Maximum Entropy Principle; Jaynes' Principle; Kullback's Minimum Cross-entropy Principle; Maximum Information Entropy Principle; Measurement Principle; Minimax Entropy Principle; Minimum Entropy Principle; Minimum Relative Entropy Principle; Modification of The Extended Maximum Entropy Principle; Principle of Maximum Entropy; Principle of Maximum Nonadditive Entropy; Principle of Observational Compatibility; Principle of Operational Compatibility; Uncertainty Principle

MORPHOLOGY Additional Principle; Ekeland's variational Principle; Ekeland's Epsilon Variational Principle; Ekeland's Variational Principle; Global Maximum Principle; Unlearned Principle; Variational Principle; Maupertuis Principle

MORPHOMETRY Cavalieri Principle; Stereological Principle

MORSE POTENTIAL Minimum Description Length Principle

MORSE THEORY Conte's Principle; Fermat's Principle; Fermats Principle; Least Time Principle; Principle of Caratheodony; Stationary Time Principle

MOTION Curvilinear Impetus Principle; D'Alembert's Principle; Gyarmati's Principle; Hamilton Principle; Inertia Principle; Lagrange Principle of Least Action; Least

Action Principle; Least Motion Principle; Prigogine Principle; Principle of Least Motion; Principle of Minimum Kinetic Energy; Principle of Minimum Mechanical Energy; Principle of Minimum Potential Energy; Quantum Hamilton Principle; Stewart-Hamilton Principle; Stochastic Hamilton Variational Principle; Toupin's Dual Principle, Active Principle; Minimum Description Length Principle

MULTI COMPARISON ANALOG TO DIGITAL CONVERTERS Interpolation Principle; Smallest Error Computer Interpolation Principle; Speech Interpolation Principle; Interpolation Principle

MULTIDIMENSIONAL SYSTEM Fundamental Principle, Pointwise Maximum Principle, Principle of Maximum Entropy, Maximum Principle

MULTIDISCIPLINARY SCIENCES Generalized Principle of Least Action; Weak Equivalence Principle; Absorption Principle; Limiting Absorption Principle; Limiting Amplitude Principle

MULTIPLE EXPERTS Maximum Decisional Efficiency (MDE) Principle; Multiple Imagescorrespondence Principle; Entropy Maximum Principle; Extended Maximum Entropy Principle; Extremal Principle of Information; Frieden-Soffer Principle; Gibbs Principle; Inference Principle; Jaynes' Maximum Entropy Principle; Jaynes' Principle; Kullback's Minimum Cross-entropy Principle; Maximum Information Entropy Principle; Measurement Principle; Minimax Entropy Principle; Minimum Entropy Principle; Minimum Relative Entropy Principle; Modification of The Extended Maximum Entropy Principle; Principle of Maximum Entropy; Principle of Maximum Nonadditive Entropy; Principle of Observational Compatibility; Principle of Operational Compatibility; Uncertainty Principle

MULTIPLE OUTLIERS Ambartsumian Invariance Principle; Charnov's Invariance Principle; Donsker's Invariance Principle; Invariance Principle; Invariance Principle of General Relativity; Local Invariance

Principle; Maruyama's Invariance Principle; Perturbation Invariance Principle; Principle of Invariance; Principle of Relativity; Principle of Undulatory Invariance; Relaxed Invariance Principle; Scale Invariance Principle; Strong Principle; Variational Principle

MULTIVARIABLE SYSTEMS Master-slave Principle; Principle of Dominant Subsystems, Parsimony Principle; Principle of Parsimonius Data Modeling

MULTIVARIATE   CALIBRATION Parsimony Principle; Principle of Parsimonius Data Modeling

MUSHROOM SHAPED FLOAT Inference Principle; Jaynes' Principle; Kullback's Minimum Cross-entropy Principle; Maximum Information Entropy Principle; Measurement

Principle; Measuring Principle; Minimax Entropy Principle; Minimum Energy Principle; Minimum Norm Principle; Minimum Relative Entropy Principle; Modification of The Extended Maximum Entropy Principle; Principle of Maximum Entropy; Principle of Maximum Nonadditive Entropy; Principle of Observational Compatibility; Principle of Operational Compatibility; Uncertainty Principle

MUTATION BIAS Entropy Principle; Fuzzy Entropy Principle; Maxent Principle; Maximum Entropy Principle; Mep Maximum Entropy Principle; Mermin Entropy Principle; Pme Principle of Maximum Entropy; Principle of Maximum Nonadditive Entropy; Principle of Operational Compatibility

MYRICETIN3OBETAD   GLUCURONIDEARACHI-DONICACID Antimaximum Principle

N

N ATOMIC POSITION VECTORS Minimal Principle

NARCISSISM Beyond The Pleasure Principle; Jewish Pleasure Principle; Pleasure Pain Principle; Pleasure Principle; Pleasurepain Principle

NATURAL SELECTION Competitive Exclusion Principle; Handicap Principle; Uncertainty Principle

NAVIER STOKES EQUATIONS Maximum Principle; Mesh Independence Principle; Multilevel Mesh Independence Principle; Multilevel Mesh Independence Principle; Principle of Minimum Dissipation; Reflection Principle

NAVIGATION Inference Principle; Jaynes' Principle; Kullback's Minimum Cross-entropy Principle; Maximum Information Entropy Principle; Maximum Principle; Measurement Principle; Measuring Principle; Minimax Entropy Principle; Minimum Energy Principle; Minimum Norm Principle; Minimum Relative Entropy Principle; Modification of The Extended Maximum Entropy Principle; Principle of Maximum Entropy; Principle of Maximum Nonadditive Entropy; Principle of Observational Compatibility; Principle of Operational Compatibility; Uncertainty Principle

NEAR INFRARED COUNTS Conte's Principle; Fermat's Principle; Fermats Principle; Least Time Principle; Principle of Caratheodony; Stationary Time Principle

NEAREST NEIGHBOR DISTRIBUTION Competitive Exclusion Principle; Principle of Complementarity

NECESSARY AND SUFFICIENT CONDITION Maximum Principle, Huygen's's Principle; Pointwise Maximum Principle

NERVOUS SYSTEM D'Alembert Principle; D'Alembert Lagrange Variation Principle; Dale's Principle; Differential Principle; Early Streamer Emission Principle; Priming Principle; Resource Consumption Principle; Reflection Principle; Fick Principle; Mitrofanoff Principle; Gestalt Principle; Hamilton Principle; Mckenzie Principles; Mouse Beta Amyrin Active Principle; Mouse Beta Amyrin Active Principle; Nameless Category N3C Principle; Principle of Remifentanil; Principle-centered Spine Care; Therapeutic Principle; Size Principle; Superposition Principle; Variational Principle, Antidepressant Principle; Network Models Extension Principle; Principle of Uncertainty

NEURAL NETWORKS Active Principle; Correspondence Principle; Decomposition Principle; Entropy Maximum Principle; Equidistortion Principle; Extended Maximum Entropy Principle; Extremal Principle of Information; Frieden-Soffer Principle; Gibbs Principle; Inference Principle; Jaynes' Maximum Entropy Principle; Jaynes' Principle; Kullback's Minimum Cross-entropy Principle; Maximum Information Entropy Principle; Measurement Principle; Minimax Entropy Principle; Minimum Entropy Principle; Minimum Relative Entropy Principle; Modification of The Extended Maximum Entropy Principle; Parsimony Principle; Principle of Maximum Entropy; Principle of Maximum Nonadditive Entropy; Principle of Observational Compatibility; Principle of Operational Compatibility; Principle of Parsimonius Data Modeling; Quantization Design Principle; Size Principle; Sufficiency Principle; Uncertainty Principle; Zadehs Extension Principle, Inertia Principle; Lagrange Principle of Least Action; Least Action Principle; Least Motion Principle; Principle of Least Motion, Equidistortion Principle; Quantization Design Principle

Dictionary of Scientific Principles, by Stephen Marvin
Copyright © 2011 John Wiley & Sons, Inc.

NEURAL NETWORKS Minimum Description Length Principle; Parsimony Principle; Principle of Parsimonius Data Modeling; Successive Principle

NEUROLOGICAL DISORDER Mitrofanoff Principle; Correspondence Principle; Entropy Maximum Principle; Extended Maximum Entropy Principle; Extremal Principle of Information; Frieden-Soffer Principle; Gibbs Principle; Inference Principle; Jaynes' Maximum Entropy Principle; Jaynes' Principle; Kullback's Minimum Cross-entropy Principle; Maximum Information Entropy Principle; Measurement Principle; Minimax Entropy Principle; Minimum Entropy Principle; Minimum Relative Entropy Principle; Modification of The Extended Maximum Entropy Principle; Principle of Maximum Entropy; Principle of Maximum Nonadditive Entropy; Principle of Observational Compatibility; Principle of Operational Compatibility; Uncertainty Principle

NEUROLOGICAL MODELS Correspondence Principle; Entropy Maximum Principle; Extended Maximum Entropy Principle; Extremal Principle of Information; Frieden-Soffer Principle; Gibbs Principle; Inference Principle; Jaynes' Maximum Entropy Principle; Jaynes' Principle; Kullback's Minimum Cross-entropy Principle; Maximum Information Entropy Principle; Measurement Principle; Minimax Entropy Principle; Minimum Entropy Principle; Minimum Relative Entropy Principle; Modification of The Extended Maximum Entropy Principle; Principle of Maximum Entropy; Principle of Maximum Nonadditive Entropy; Principle of Observational Compatibility; Principle of Operational Compatibility; Uncertainty Principle

NEUROLOGY HUMAN MEDICINE Fick Principle; Mckenzie Principles; Nameless Category N3C Principle; Principle-centered Spine Care

NEURONS Parsimony Principle; Principle of Information Processing; Principle of Parsimonius Data Modeling

NEWTON METHOD Mesh Independence Principle; Minimum Description Length Principle

NOISE MEASUREMENT (POLLUTION) Principle of Fundamental Characteristics, Maximum Entropy Principle

NONCOMMUTATIVE GEOMETRY Unified Principle; Nonconstructive Model Theory. Maximum Entropy Principle

NONCONVEX ENERGY PROBLEMS Ekeland's variational Principle; Ekeland's Epsilon Variational Principle; Ekeland's Variational Principle; Global Maximum Principle; Variational Principle

NONEQUILIBRIUM MOLECULAR DYNAMICS Chaotic Principle, Principle of Maximum Entropy

NON INVASIVE METHOD Hard-soft acid-base Principle; Penaz Principle

NONLINEAR CONTROL SYSTEMS Equivalence Principle; Pontryagin Minimum Principle; Principle of Dominant Subsystems

NON LINEAR EFFECT D'Alembert Lagrange Principle; D'Alembert Lagrange Variation Principle; D'Alembert Principle; Gauss Principle; Hamilton Lagrange Principle; Jayne's Principle; Jourdain Principle; Kinetic Principle; Lagrange Principle; Least Constraint Principle; Maupertius Lagrange Principle; Maupertuis Euler Lagrange Principle of Least Action; Principle of Extremal Effects; Principle of Least Constraint; Principle of Maximal Effect; Principle of Virtual Work; Virtual Force Principle

NON LINEAR EQUATION Maximality Principle; Mesh Independence Principle

NON LINEAR REGRESSION Ambartsumian Invariance Principle; Charnov's Invariance Principle; Donsker's Invariance Principle; Invariance Principle; Invariance Principle of General Relativity; Local Invariance Principle; Maruyama's Invariance Principle; Perturbation Invariance Principle; Principle of Invariance; Principle of Relativity; Principle of Undulatory Invariance; Relaxed Invariance Principle; Scale Invariance Principle; Strong Principle; Variational Principle

NONLINEAR RANDOM WAVE EQUATIONS Contraction Principle; Gibbs Variational Principle; Large Deviation Principle

NONLINEAR STOCHASTIC OSCILLATORS Correspondence Principle; Entropy Maximum Principle; Extended Maximum Entropy Principle; Extremal Principle of Information; Frieden-Soffer Principle; Gibbs Principle; Inference Principle; Jaynes' Maximum Entropy Principle; Jaynes' Principle; Kullback's Minimum Cross-entropy Principle; Maximum Information Entropy Principle; Measurement Principle; Minimax Entropy Principle; Minimum Entropy Principle; Minimum Relative Entropy Principle; Modification of The Extended Maximum Entropy Principle; Principle of Maximum Entropy; Principle of Maximum Nonadditive Entropy; Principle of Observational Compatibility; Principle of Operational Compatibility; Uncertainty Principle

NON LINEAR SYSTEM Ambartsumian Invariance Principle; Charnov's Invariance Principle; Donsker's Invariance Principle; Invariance Principle; Invariance Principle of General Relativity; Kelvinarnol'D Energy Principle; Local Invariance Principle; Maruyama's Invariance Principle; Perturbation Invariance Principle; Principle of Invariance; Principle of Relativity; Principle of Undulatory Invariance; Relaxed Invariance Principle; Scale Invariance Principle; Strong Principle; Variational Principle

NONLINEAR SYSTEMS Byrnesmartin Interval Invariance Principle; Comparison Principle; Integral Invariance Principle; Mesh Independence Principle

NONPARAMETRIC REGRESSION Inverse Problem Theory Principle; Parsimony Principle; Principle of Parsimonius Data Modeling

NON PARAMETRIC TEST Ambartsumian Invariance Principle; Charnov's Invariance Principle; Donsker's Invariance Principle; Invariance Principle; Invariance Principle of General Relativity; Local Invariance Principle; Maruyama's Invariance Principle; Perturbation Invariance Principle; Principle of

Invariance; Principle of Relativity; Principle of Undulatory Invariance; Relaxed Invariance Principle; Scale Invariance Principle; Strong Principle; Variational Principle

NONREVERSIBILITY Ambartsumian Invariance Principle; Charnov's Invariance Principle; Donsker's Invariance Principle; Invariance Principle; Invariance Principle of General Relativity; Local Invariance Principle; Maruyama's Invariance Principle; Perturbation Invariance Principle; Principle of Invariance; Principle of Relativity; Principle of Undulatory Invariance; Relaxed Invariance Principle; Scale Invariance Principle; Strong Principle; Variational Principle

NONSMOOTH OPTIMIZATION Ekeland's variational Principle; Ekeland's Epsilon Variational Principle; Ekeland's Variational Principle; Global Maximum Principle; Variational Principle

NON SYMMETRIC INTERACTION Ambartsumian Invariance Principle; Charnov's Invariance Principle; Donsker's Invariance Principle; Invariance Principle; Invariance Principle of General Relativity; Local Invariance Principle; Maruyama's Invariance Principle; Perturbation Invariance Principle; Principle of Invariance; Principle of Relativity; Principle of Undulatory Invariance; Relaxed Invariance Principle; Scale Invariance Principle; Strong Principle; Variational Principle

NUCLEAR FUSION REACTORS FUELS AND PLASMAS Hamilton's Principle; Kirchhoff's Voltage Law Principle; Minimum Energy Principle; Variational Principle; Woltjer Minimum Energy Principle; Woltjer-Taylor Minimum Energy Principle

NUCLEON BOND SCATTERING WITH TENSOR POTENTIAL Ckvp; Complex Kohn Variational Principle; Hulthén–Kohn Variational Principle; Kohn Variational Principle; Schwinger Variational Principle

NUMERICAL CALCULATION Active Principle; Principle of Maximum Entropy; Principle of Measurement; Saint-Venant's Principle

NUMERICAL METHOD Minimum Description Length Principle; Minimum Principle; Correspondence Principle; Entropy Maximum Principle; Extended Maximum Entropy Principle; Extremal Principle of Information; Frieden-Soffer Principle; Gibbs Principle; Inference Principle; Jaynes' Maximum Entropy Principle; Jaynes' Principle; Kullback's Minimum Cross-entropy Principle; Maximum Information Entropy Principle; Measurement Principle; Minimax Entropy Principle; Minimum Entropy Principle; Minimum Relative Entropy Principle; Modification of The Extended Maximum Entropy Principle; Principle of Critical Energy; Principle of Maximum Entropy; Principle of Maximum Nonadditive Entropy; Principle of Observational Compatibility; Principle of Operational Compatibility; Uncertainty Principle; Virtual Force Principle; Contraction Principle; Contraction Principle; Gibbs Variational Principle; Large Deviation Principle

NUTRITION Archimedes' Principle; Health Principle; Hydrostatic Principle; Investment Principle; John Templeton Principle; Monti Principle; Precautionary Principle; Precautionary Principle; Principle of Habeas Mentem; Principle of Maximum Pessimism

NUTRITIONAL STATUS AND METHODS Archimedes' Principle; Health Principle; Hydrostatic Principle; Investment Principle; John Templeton Principle; Precautionary Principle; Principle of Habeas Mentem; Principle of Maximum Pessimism

NYQUIST INTERVAL Hamilton's Principle; Kirchhoff's Voltage Law Principle; Minimum Energy Principle; Variational Principle; Woltjer Minimum Energy Principle; Woltjer-Taylor Minimum Energy Principle

O

OCEAN SURFACE Conservation of Momentum Principle; D'Alembert Principle; Differential Principle; Hamilton Principle; Newton's Third Principle; Principle of Equal A Priori Probabilities; Variational Principle

OCEAN WAVE Correspondence Principle; Entropy Maximum Principle; Extended Maximum Entropy Principle; Extremal Principle of Information; Frieden-Soffer Principle; Gibbs Principle; Inference Principle; Jaynes' Maximum Entropy Principle; Jaynes' Principle; Kullback's Minimum Cross-entropy Principle; Maximum Information Entropy Principle; Measurement Principle; Minimax Entropy Principle; Minimum Entropy Principle; Minimum Relative Entropy Principle; Modification of The Extended Maximum Entropy Principle; Principle of Maximum Entropy; Principle of Maximum Nonadditive Entropy; Principle of Observational Compatibility; Principle of Operational Compatibility; Uncertainty Principle

OPACITY Hamilton's Principle; Kirchhoff's Voltage Law Principle; Minimum Energy Principle; Variational Principle; Woltjer Minimum Energy Principle; Woltjer-Taylor Minimum Energy Principle

OPEN CHANNEL SURGES AND ROLL WAVES Conservation of Momentum Principle; D'Alembert Principle; Differential Principle; Hamilton Principle; Newton's Third Principle; Principle of Equal A Priori Probabilities; Variational Principle

OPHTHALMOLOGY Function Principle; Superposition Principle

OPTICAL FILTERS Frequency Distance Principle; New Principle; Projection Principle

OPTICAL HOMOGENEITY Interference Principle; Interferometry Neutron Principle; Moire Interference Principle; Principle of Spinor Superposition; Resolution Principle

OPTICAL METRIC Conte's Principle; Fermat's Principle; Fermat; Least Time Principle; Principle of Caratheodony; Principle of Causality; Projection Principle; Stationary Time Principle

OPTICAL PROPERTIES Equidistortion Principle; Quantization Design Principle

OPTICAL TECHNOLOGY Adaptive Principle; Optical Principle. Principle of Causality; Proximity Compatibility Principle

OPTICS Action Principle; Schwinger's Quantum Action Principle, Adaptive Principle; Babinet's Principle; Beam Splitter Principle; Caratheodory's Principle; Complementarity Principle; Conte's Principle; Correspondence Principle; Entropy Maximum Principle; Extended Maximum Entropy Principle; Extremal Principle of Information; Fermat Principle; Frieden-Soffer Principle; Gibbs Principle; Inference Principle; Inference Principle; Inverse Problem Theory Principle; Jaynes' Maximum Entropy Principle; Jaynes' Principle; Kullback's Minimum Cross-entropy Principle; Kullback's Minimum Cross-entropy Principle; Least Time Principle; Maximum Information Entropy Principle; Maximum Information Entropy Principle; Measurement Principle; Measuring Principle; Minimax Entropy Principle; Minimum Energy Principle; Minimum Entropy Principle; Minimum Norm Principle; Minimum Relative Entropy Principle; Modification of The Extended Maximum Entropy Principle; Modification of The Extended Maximum Entropy Principle; Principle of Causality; Principle of Caratheodony; Principle of Maximum Entropy; Principle of Maximum Nonadditive Entropy; Principle of Maximum Nonadditive Entropy; Principle of Operational Compatibility; Stationary Time Principle; Uncertainty Principle; Optical Principle, Equivalent Background Principle

Dictionary of Scientific Principles, by Stephen Marvin
Copyright © 2011 John Wiley & Sons, Inc.

OPTIMAL CONTROL Ekeland's variational Principle; Ekeland's Epsilon Variational Principle; Ekeland's Variational Principle; Equivalence Principle; Global Maximum Principle; Investment Principle; John Templeton Principle; Principle of Maximum Pessimism Maximum Entropy Principle; Minimum Description Length Principle; Minimum Information Principle; Pontryagin's Principle, Pontryagin Minimum Principle; Pontryagin's Maximum Principle; Separation Principle; Variational Principle; Weak Invariance Principle; D'Alembert's Principle; Gyarmati's Principle; Hamilton Principle; Prigogine Principle; Principle of molecular dynamics; Principle of Minimum Kinetic Energy; Principle of Minimum Mechanical Energy; Principle of Minimum Potential Energy; Quantum Hamilton Principle; Stewart-Hamilton Principle; Stochastic Hamilton Variational Principle; Toupin's Dual Principle, Pointwise Maximum Principle

OPTIMAL ESTIMATION Maximum Hardness Principle; Optimal Model Matching Principle. Maximum Entropy Principle

OPTIMAL PRESTRESS Ekeland's variational Principle; Ekeland's Epsilon Variational Principle; Ekeland's Principle; Ekeland's Variational Principle; Global Maximum Principle; Variational Principle

OPTIMIZATION Equivalence Principle; Hamilton's Principle; Kirchhoff's Voltage Law Principle; Maximum Hardness Principle; Minimum Energy Principle; Pointwise Maximum Principle; Variational Principle; Weak Invariance Principle; Woltjer Minimum Energy Principle; Woltjer-Taylor Minimum Energy Principle

OPTIMIZATION PROBLEM Weak Invariance Principle; Autonomous Cooperation Distributed Maximum Principle; Distributed Maximum Principle; Equivalence Principle; Maximum Principle; Weak Invariance Principle

OPTIMIZATION TECHNIQUES Autonomous Cooperation Distributed Maximum Principle; Decomposition Principle; Distributed

Maximum Principle; Maximum Principle; Wise Use Principle; Ekeland's variational Principle; Ekeland's Epsilon Variational Principle; Ekeland's Variational Principle; Global Maximum Principle; Mde Principle; Pointwise Maximum Principle; Pontryagin Maximum Principle; Variational Principle; Equipartition Principle; Minimum Principle; Mueller-Breslau Principle; Saint-Venant's Principle; Weak Invariance Principle; Protection Principle

ORDER STATISTICS Correspondence Principle; Entropy Maximum Principle; Extended Maximum Entropy Principle; Extremal Principle of Information; Frieden-Soffer Principle; Gibbs Principle; Inference Principle; Jaynes' Maximum Entropy Principle; Jaynes' Principle; Kullback's Minimum Cross-entropy Principle; Maximum Information Entropy Principle; Measurement Principle; Minimax Entropy Principle; Minimum Entropy Principle; Minimum Relative Entropy Principle; Modification of The Extended Maximum Entropy Principle; Principle of Maximum Entropy; Principle of Maximum Nonadditive Entropy; Principle of Observational Compatibility; Principle of Operational Compatibility; Uncertainty Principle

ORDERED DATA Correspondence Principle; Entropy Maximum Principle; Extended Maximum Entropy Principle; Extremal Principle of Information; Frieden-Soffer Principle; Gibbs Principle; Inference Principle; Jaynes' Maximum Entropy Principle; Jaynes' Principle; Kullback's Minimum Cross-entropy Principle; Maximum Information Entropy Principle; Measurement Principle; Minimax Entropy Principle; Minimum Entropy Principle; Minimum Relative Entropy Principle; Modification of The Extended Maximum Entropy Principle; Principle of Maximum Entropy; Principle of Maximum Nonadditive Entropy; Principle of Observational Compatibility; Principle of Operational Compatibility; Uncertainty Principle

ORDERING KINETICS Corresponding States Principle; Pitzer Corresponding States Principle; Principle of Correspondence

ORDINARY DIFFERENTIAL EQUATIONS ON MANIFOLDS Hamilton's Principle; Maupertuis-Jacobi Principle; Minimum Principle; Weak Invariance Principle

ORGANIZATIONS Economic Principle; Principles of Determination; Principles of Multiple Function; Principles of Overdetermination; Regulatory Principles and Other Regulatory Principles of Inertia; Regulatory Principles of Constancy; Regulatory Principles of Pleasureunpleasure; Regulatory Principles of Primarysecondary Process; Regulatory Principles of Reality; Regulatory Principles of Repetition

ORGANOTIN COMPLEXES Electronegativity Equalization Principle; Geometric Mean Principle; Principle of electronegativity

ORTHOPEDICS Mckenzie Principles; Principle-centered Spine Care; Three-point Principle

OVERLAPPING CURVED SHAPED Geometric Principle; Iris Diaphragm Principle; Iris Principle

P

PALAIS-SMALE CONDITION Conte's Principle; Fermat's Principle; Least Time Principle; Principle of Caratheodony; Stationary Time Principle

PANCREAS Boric Acid Affinity Principle; Boronic Acid Affinity Principle

PARABOLIC EQUATION Comparison Principle; Weak Invariance Principle; Pontryagin Maximum Principle

PARAMAGNETIC COMPONENTS Torsion Balance Principle; Active Principle; Correspondence Principle; Entropy Maximum Principle; Equivalence Principle; Extended Maximum Entropy Principle; Extremal Principle of Information; Frieden-Soffer Principle; Gibbs Principle; Inference Principle; Internal Model Principle; Jaynes' Maximum Entropy Principle; Jaynes' Principle; Kullback's Minimum Cross-entropy Principle; Maximum Entropy Principle; Maximum Hardness Principle; Maximum Information Entropy Principle; Measurement Principle; Minimax Entropy Principle; Minimum Entropy Principle; Minimum Relative Entropy Principle; Modification of The Extended Maximum Entropy Principle; Principle of Dominant Subsystems; Principle of Maximum Entropy; Principle of Maximum Nonadditive Entropy; Principle of Observational Compatibility; Principle of Operational Compatibility; Principle of per-survivor processing; Principles of Darcy's Law; Separation Principle; Uncertainty Principle

PARAMETER ESTIMATION FOR 3 PARAMETER Autonomous Cooperation Distributed Maximum Principle; Distributed Maximum Principle; Maximum Principle

PARAMETERS Maximum Likelihood Principle; Minimum Description Length Principle

PARAMETRIC DOWN CONVERSION Action Principle; Schwinger's Quantum Action Principle

PARASITOLOGY Hopkins' Host Selection Principle; Host Selection Principle; Similia Principle of Homeopathy

PARATHYROID HORMONE RELATED PROTEIN Type II Cell Stretch-organizing Principle; Timing Principle

PARETO OPTIMIZATION Educational Principle; Ehrenfest Principle; Ekeland's Principle

PARTIAL AUTOCORRELATIONS Parsimony Principle; Principle of Parsimonius Data Modeling

PARTIAL DIFFERENTIAL EQUATIONS Equivalence Principle; Minimum Principle; Weak Invariance Principle, Schwarz Reflection Principle

PARTIAL LEAST SQUARES Parsimony Principle; Principle of Parsimonius Data Modeling

PARTIAL RESIDUAL PLOTS Ambartsumian Invariance Principle; Charnov's Invariance Principle; Donsker's Invariance Principle; Invariance Principle; Invariance Principle of General Relativity; Local Invariance Principle; Maruyama's Invariance Principle; Perturbation Invariance Principle; Principle of Invariance; Principle of Relativity; Principle of Undulatory Invariance; Relaxed Invariance Principle; Scale Invariance Principle; Strong Principle; Variational Principle

PARTIAL SUM Ambartsumian Invariance Principle; Charnov's Invariance Principle; Donsker's Invariance Principle; Invariance Principle; Invariance Principle of General Relativity; Local Invariance Principle; Maruyama's Invariance Principle; Perturbation Invariance Principle; Principle of Invariance; Principle of Relativity; Principle of Undulatory Invariance; Relaxed Invariance Principle; Scale Invariance Principle; Strong Principle; Variational Principle

PARTICLE INTERACTION Ambartsumian Invariance Principle; Charnov's Invariance Principle; Donsker's Invariance Principle; Invariance Principle; Invariance Principle of General Relativity; Local Invariance Principle; Maruyama's Invariance Principle; Perturbation Invariance Principle; Principle of Invariance; Principle of Relativity; Principle of Undulatory Invariance; Relaxed Invariance Principle; Scale Invariance Principle; Strong Principle; Variational Principle

PARTICLE LIKE AND WAVE LIKE PROPERTIES Babinet's Principle; Complementarity Principle, Bohr Principle of Complementarity; Bohrs Complementarity Principle; Bohr's Correspondence Principle; Correspondence Principle

PARTICLE SIZE Coulter Principle; Principle of Maximum Entropy

PARTICLE SYSTEM Ambartsumian Invariance Principle; Charnov's Invariance Principle; Donsker's Invariance Principle; Invariance Principle; Invariance Principle of General Relativity; Local Invariance Principle; Maruyama's Invariance Principle; Perturbation Invariance Principle; Principle of Invariance; Principle of Relativity; Principle of Undulatory Invariance; Relaxed Invariance Principle; Scale Invariance Principle; Strong Principle; Variational Principle

PARTICLES IN ELECTROMAGNETIC FIELDS Minimum Description Length Principle; Principle of Correspondence; Unified Principle

PARTITION FUNCTION Spacetime Uncertainty Principle; Weak Invariance Principle

PATH INTEGRAL Action Principle; Schwinger's Quantum Action Principle; Weak Invariance Principle

PATHOGEN Occam's Razor Principle; Principle of Pcr; Polluter Pays Principle

PATHOLOGICAL RESEARCH Cavalieri Principle; Stereological Principle

PATHOLOGY Action Principle; Boric Acid Affinity Principle; Boronic Acid Affinity Principle; Casuistic Principle; Constant Infusion Principle; Constant Infusion Principle; Double Effect Principle; Exception Granting Principle; Hopkins' Host Selection Principle; Host Selection Principle; Justifying Principle; Maximum Likelihood Principle; Mckenzie Principles; Mitrofanoff Principle; Monti Principle; Monti Principle; Nameless Category N3C Principle; Precautionary Principle; Principle of Do No Harm; Principle of Double Effect; Principle-centered Spine Care; Principles of Biomedical Ethics; Similia Principle of Homeopathy; SimiliaPrinciple; Syndrome Principle; Three-point Principle

PATIENT Action Principle; Functioning Principle; Health Principle; Maximum Likelihood Principle; Mitrofanoff Principle; Monti Principle; Overlapping Biphasic Impulse Principle; Precautionary Principle; Principle of Habeas Mentem; Salutogenesis Principle; Three-point Principle; Working Principle

PATTERN RECOGNITION Active Principle; Correspondence Principle; Entropy Maximum Principle; Extended Maximum Entropy Principle; Extremal Principle of Information; Frieden-Soffer Principle; Gibbs Principle; Inference Principle; Jaynes' Maximum Entropy Principle; Jaynes' Principle; Kullback's Minimum Cross-entropy Principle; Maximum Entropy Principle; Maximum Information Entropy Principle; Measurement Principle; Minimax Entropy Principle; Minimum Entropy Principle; Minimum Relative Entropy Principle; Modification of The Extended Maximum Entropy Principle; Principle of Maximum Entropy; Principle of Maximum Nonadditive Entropy; Principle of Observational Compatibility; Principle of Operational Compatibility; Uncertainty Principle, Quartz Microbalance Principle, Working Principle

PATTERNS Complexity Principle; Entropy Principle; Generalized Minimal Principle; Maximum Entropy Principle; Minimum Cross-entropy Principle; Moment Preserving Principle; Principle of Minimum Cross-entropy; Variational Minimal Principle

PD DOPED SNO2 THIN FILMS Active Principle; Principle of Corresponding States; Quartz Microbalance Principle

PEDIATRICS Cross Check Principle; Nameless Category N3C Principle; Syndrome Principle; Weak Invariance Principle, Cpm Principle; Critical Path Method Principle; Nameless Category N3C Principle

PENCILS Minimax Principle; Momentum Principle; Weak Invariance Principle

PENETRATION OF CHARGE Discrete Maximum Principle; Discrete Pontryagin Maximum Principle; Dmp; Krylov Maximum Principle; Maximum Principle; Pontryagin Discrete Maximum Principle

PENG ROBINSON INTERACTION PARAMETERS Corresponding States Principle; Pitzer Corresponding States Principle; Principle of Correspondence

PENICILLINS Correspondence Principle; Entropy Maximum Principle; Extended Maximum Entropy Principle; Extremal Principle of Information; Frieden-Soffer Principle; Gibbs Principle; Inference Principle; Jaynes' Maximum Entropy Principle; Jaynes' Principle; Kullback's Minimum Cross-entropy Principle; Maximum Information Entropy Principle; Measurement Principle; Minimax Entropy Principle; Minimum Entropy Principle; Minimum Relative Entropy Principle; Modification of The Extended Maximum Entropy Principle; Principle of Maximum Entropy; Principle of Maximum Nonadditive Entropy; Principle of Observational Compatibility; Principle of Operational Compatibility; Uncertainty Principle

PEPTIDES AND AMINO ACIDS Antiarrhythmic Principle; Boric Acid Affinity Principle; Boronic Acid Affinity Principle; Hiv Principle; Monti Principle

PERFORMANCE ANALYSIS Calorimetric Measurement Principle; Magnetron Principle

PERIODIC POTENTIALS Minimax Principle; Momentum Principle; Weak Invariance Principle

PERMANENT DISPLACEMENT OF LIQUEFIED GROUND Hamilton's Principle; Kirchhoff's Voltage Law Principle; Minimum Energy Principle; Variational Principle; Woltjer Minimum Energy Principle; Woltjer-Taylor Minimum Energy Principle

PERMANENT MAGNETS Minimum Energy Principle; Minimum Norm Principle; Processing Principle; Transverse Flux Principle

PERRON FROBENIUM EQUATION Correspondence Principle; Entropy Maximum Principle; Extended Maximum Entropy Principle; Extremal Principle of Information; Frieden-Soffer Principle; Gibbs Principle; Inference Principle; Jaynes' Maximum Entropy Principle; Jaynes' Principle; Kullback's Minimum Cross-entropy Principle; Maximum Information Entropy Principle; Measurement Principle; Minimax Entropy Principle; Minimum Entropy Principle; Minimum Relative Entropy Principle; Modification of The Extended Maximum Entropy Principle; Principle of Maximum Entropy; Principle of Maximum Nonadditive Entropy; Principle of Observational Compatibility; Principle of Operational Compatibility; Uncertainty Principle

PERSONALITY DEVELOPMENT Beyond The Pleasure Principle; Jewish Pleasure Principle; Maximization Principle; Pleasure Pain Principle; Pleasure Principle; Pleasurepain Principle

PERTURBATION Ambartsumian Invariance Principle; Charnov's Invariance Principle; Donsker's Invariance Principle; Internal Model Principle; Invariance Principle; Invariance Principle of General Relativity; Local Invariance Principle; Maruyama's Invariance Principle; Minimum Description Length Principle; Perturbation Invariance Principle; Pontryagin's Principle; Principle of molecular dynamics; Principle of Invariance;

Principle of Relativity; Principle of Undulatory Invariance; Relaxed Invariance Principle; Scale Invariance Principle; Strong Principle; Variational Principle

PETROV TYPE N Huygen's Principle; Pfa Resin. Principle of Measurement; Ph Electrodes. Inference Principle; Jaynes' Principle; Kullback's Minimum Cross-entropy Principle; Maximum Information Entropy Principle; Measurement Principle; Measuring Principle; Minimax Entropy Principle; Minimum Relative Entropy Principle; Modification of The Extended Maximum Entropy Principle; Principle of Maximum Entropy; Principle of Maximum Nonadditive Entropy; Principle of Observational Compatibility; Principle of Operational Compatibility; Uncertainty Principle

PHARMACEUTICALS Antiviral Active Principle; Percolation Principle

PHARMACODYNAMICS Monti Principle; Principle of Remifentanil; Working Principle; Antiarrhythmic Principle

PHARMACOGNOSY Antiviral Active Principle; Monti Principle; Mouse Beta Amyrin Active Principle; Sweet Principle

PHARMACOGNOSY AND PHARMACEUTICAL BOTANY Adaptive Principle; Antidepressant Principle; Antiulcer Principle; Monti Principle

PHARMACOKINETICS Correspondence Principle; Entropy Maximum Principle; Extended Maximum Entropy Principle; Extremal Principle of Information; Frieden-Soffer Principle; Gibbs Principle; Inference Principle; Jaynes' Maximum Entropy Principle; Jaynes' Principle; Kullback's Minimum Cross-entropy Principle; Maximum Information Entropy Principle; Measurement Principle; Minimax Entropy Principle; Minimum Entropy Principle; Minimum Relative Entropy Principle; Modification of The Extended Maximum Entropy Principle; Principle of Maximum Entropy; Principle of Maximum Nonadditive Entropy; Principle of Observational Compatibility; Principle of Operational Compatibility; Principle of Remifentanil; Uncertainty Principle; Weak Invariance Principle

PHARMACOLOGIC   TREATMENTS Casuistic Principle; Double Effect Principle; Exception Granting Principle; Justifying Principle; Principle of Do No Harm; Principle of Double Effect; Principles of Biomedical Ethics

PHARMACOLOGY Antiarrhythmic Principle; Antithrombotic Principle; Antiviral Active Principle; Casuistic Principle; Double Effect Principle; Drilled Cell Principle; Drug Principle; Exception Granting Principle; Immunosuppresive Principle; Justifying Principle; Mimetic Principle; Monti Principle; Mouse Beta Amyrin Active Principle; Optimality Principle; Organizing Principle; Priming Principle Regimen Administration; Principle of component analysis; Principle of Do No Harm; Principle of Double Effect; Principle of Double Effect; Principle of Remifentanil; Principles of Biomedical Ethics; Self-organizing Principle; Similarity Principle; Similia Principle of Homeopathy; Sweet Principle; Syndrome Principle; Therapeutic Principle; Type II Cell Stretch-organizing Principle; Weak Invariance Principle; Working Principle, Adaptive Principle, Monti Principle

PHASE COEXISTENCE REGION Contraction Principle; Gibbs Variational Principle; Large Deviation Principle

PHASE DIAGRAM Weak Invariance Principle; Braun Lechatelier Principle; Compensation Principle; Le Chatelier's Principle; Principle of Braunlechatelier; Samuelson-Le Chatelier Principle

PHASE EQUILIBRIUM Corresponding States Principle; Pitzer Corresponding States Principle;

PHASE PROBLEM Multiple Pulse Principle; Principle of Corresponding States; Principle of molecular dynamics; Weak Invariance Principle

PHASE SEPARATION DYNAMICS Corresponding States Principle; Pitzer Corresponding States Principle;, Principle of Correspondence

PHASE SHIFT Principle of Corresponding States; Schwinger Principle

PHASE SPACE Minimum Description Length Principle; Spacetime Uncertainty Principle; Uncertainty Principle; Weak Invariance Principle

PHILOSOPHY Aristotle's Theory of Principles; Casuistic Principle; Double Effect Principle; Exception Granting Principle; Justifying Principle; Principle of Do No Harm; Principle of Double Effect; Color Doppler Principle; Doppler Principle; Doppler Radar Principle; Double Effect Principle; Exception Granting Principle; Justifying Principle; Microwave Doppler Principle; Photoelectric Barrier Principle; Principle of Proximal Isovelocity Surface Area; Principle of The Doppler Effect; Principles of Biomedical Ethics; Salutogenesis Principle; Sonic Doppler Principle; Ultrasound Doppler Principle

PHOTOMETRY Inference Principle; Jaynes' Principle; Kullback's Minimum Cross-entropy Principle; Maximum Information Entropy Principle; Measurement Principle; Measuring Principle; Minimax Entropy Principle; Minimum Energy Principle; Minimum Norm Principle; Minimum Relative Entropy Principle; Modification of The Extended Maximum Entropy Principle; Principle of Maximum Entropy; Principle of Maximum Nonadditive Entropy; Principle of Observational Compatibility; Principle of Operational Compatibility; Uncertainty Principle

PHOTON Babinet's Principle; Bogolyubov Principle; Bogolyubov Variational Principle; Complementarity Principle; Gibbs-Bogolyubov Variational Principle; Gibbs Third Variational Principle; Hartree Fock Bogolyubov Principle; Pauli Principle; Principle of molecular dynamics; Quantal Principle, Minimum Description Length Principle

PHOTOSYNTHESIS August Krogh Principle, Maximum Entropy Principle

PHYLOGENETIC COMPONENT Active Principle; Mdl Principle, Phylogenyactive Principle; Mdl Principle

PHYSICAL CONSTANT Anthropic Cosmological Principle; Anthropic Principle; Cosmological Principle; Physical Covariance; Principle of molecular dynamics; Principle of Causality

PHYSICAL LAW Maximum Hardness Principle; Physical Organic Chemistry; Nuclear Motion Principle

PHYSICAL PROPERTIES OF GASES; LIQUIDS AND SOLIDS Consumer Principle; Priming Principle; Principle of Fundamental Characteristics

PHYSICS Chaotic Principle; Flotation Polishing Principle; Pauli Antisymmetry and Exclusion Principle; Principle of Equivalence; Weak Invariance Principle; Equivalent Background Principle; Mach's Principle

PHYSIOLOGY August Krogh Principle; Babinet's Principle; Complementarity Principle; Correspondence Principle; Drug Principle; Drug Principle; Echo Sounding Principle; Entropy Maximum Principle; Extended Maximum Entropy Principle; Extremal Principle of Information; Frequency-dependent Principle; Frieden-Soffer Principle; Gibbs Principle; Hopkins' Host Selection Principle; Host Selection Principle; Inference Principle; Jaynes' Maximum Entropy Principle; Jaynes' Principle; Kullback's Minimum Cross-entropy Principle; Maximum Information Entropy Principle; Mckenzie Principles; Measurement Principle; Measuring Principle; Minimax Entropy Principle; Minimum Description Length Principle; Minimum Energy Principle; Minimum Entropy Principle; Minimum Norm Principle; Minimum Relative Entropy Principle; Modification of The Extended Maximum Entropy Principle; Mouse Beta Amyrin Active Principle; Principle of Maximum Entropy; Principle of Maximum Nonadditive Entropy; Principle of Observational Compatibility; Principle of Operational Compatibility; Principle-centered Spine Care; Schwinger's Quantum Action Principle; Superposition Principle; Three-point Principle; Type II Cell Stretch-organizing Principle; Uncertainty Principle; Unlearned Principle; Optimality Principle; Organizing Principle; Principle of Component Analysis; Self-organizing Principle; Similarity Principle

PHYSIOLOGY AND BIOCHEMISTRY D'Alembert Principle; D'Alembert Lagrange Variation Principle; Differential Principle; Drug Principle; Hamilton Principle; Variational Principle

PHYSIOLOGY AND PATHOLOGY Fisher's Principle; Frequency-dependent Principle; Hopkins' Host Selection Principle; Host Selection Principle, Ecological Principle; Principle of Dynamic Ecology

PISCES Inference Principle; Jaynes' Principle; Kullback's Minimum Cross-entropy Principle; Maximum Information Entropy Principle; Measurement Principle; Measuring Principle; Minimax Entropy Principle; Minimum Energy Principle; Minimum Relative Entropy Principle; Modification of The Extended Maximum Entropy Principle; Principle of Maximum Entropy; Principle of Maximum Nonadditive Entropy; Principle of Observational Compatibility; Principle of Operational Compatibility; Schwinger's Quantum Action Principle; Uncertainty Principle

PITUITARY Optimality Principle; Organizing Principle; Principle of component analysis; Self-organizing Principle; Similarity Principle; Type II Cell Stretch-organizing Principle

PLANE WAVE PROPAGATION Weak Invariance Principle, Hamilton Variational Principle

PLANT PHYSIOLOGY Antiarrhythmic Principle; Antiulcer Principle; Asset Protection Principle; August Krogh Principle; Monti Principle; Mouse Beta Amyrin Active Principle; Sweet Principle

PLANT SPECIES Optimality Principle; Organizing Principle; Principle of Component Analysis; Self-organizing Principle; Similarity Principle

PLANTS Asset Protection Principle; Matching Principle; Percolation Principle

PLAPLACIAN Antidepressant Principle, Optimality Principle; Organizing Principle; Principle of Component Analysis; Self-organizing Principle; Similarity Principle

PLASMA COATING PROCESS CONTROL PYROMETRY Inference Principle; Jaynes' Principle; Kullback's Minimum Cross-entropy Principle; Maximum Information Entropy Principle; Measurement Principle; Measuring Principle; Minimax Entropy Principle;

Minimum Energy Principle; Minimum Relative Entropy Principle; Modification of The Extended Maximum Entropy Principle; Principle of Maximum Entropy; Principle of Maximum Nonadditive Entropy; Principle of Observational Compatibility; Principle of Operational Compatibility; Schwinger's Quantum Action Principle; Uncertainty Principle

PLASMA FLOW Hamilton's Principle; Kirchhoff's Voltage Law Principle; Minimum Energy Principle; Variational Principle; Woltjer Minimum Energy Principle; Woltjer-Taylor Minimum Energy Principle; Plasma Instability; Weak Invariance Principle

PLASMA MAGNETOHYDRODYNAMICS Hamilton's Principle; Kirchhoff's Voltage Law Principle; Minimum Energy Principle; Variational Principle; Woltjer Minimum Energy Principle; Woltjer-Taylor Minimum Energy Principle

PLASMA PRESSURE FORCES Hamilton's Principle; Kirchhoff's Voltage Law Principle; Minimum Energy Principle; Minimum Norm Principle; Variational Principle; Woltjer Minimum Energy Principle

PLASMAS Correspondence Principle; Entropy Maximum Principle; Extended Maximum Entropy Principle; Extremal Principle of Information; Frieden-Soffer Principle; Gibbs Principle; Inference Principle; Jaynes' Maximum Entropy Principle; Jaynes' Principle; Kullback's Minimum Cross-entropy Principle; Maximum Information Entropy Principle; Measurement Principle; Minimax Entropy Principle; Minimum Entropy Principle; Minimum Relative Entropy Principle; Modification of The Extended Maximum Entropy Principle; Principle of Maximum Entropy; Principle of Maximum Nonadditive Entropy; Principle of Observational Compatibility; Principle of Operational Compatibility; Uncertainty Principle

PLASTIC Principle of Briquetting; Correspondence Principle; Entropy Maximum Principle; Extended Maximum Entropy Principle; Extremal Principle of Information; Frieden-Soffer Principle; Gibbs Principle; Inference Principle; Jaynes' Maximum Entropy

Principle; Jaynes' Principle; Kullback's Minimum Cross-entropy Principle; Maximum Information Entropy Principle; Measurement Principle; Minimax Entropy Principle; Minimum Entropy Principle; Minimum Principle; Minimum Relative Entropy Principle; Modification of The Extended Maximum Entropy Principle; Principle of Material Frame Indifference; Principle of Maximum Entropy; Principle of Maximum Nonadditive Entropy; Principle of Minimum Dissipation; Principle of Observational Compatibility; Principle of Operational Compatibility; Uncertainty Principle; Weak Invariance Principle

PLASTIC SPIN Principle of Material Frame Indifference; Principle of Minimum Dissipation

PLASTICITY Principle of Critical Energy; Principle of Material Frame Indifference; Principle of Minimum Dissipation. Weak Invariance Principle

PLATE D'Alembert's Principle; Gyarmati's Principle; Hamilton Principle; Prigogine Principle; Principle of Minimum Kinetic Energy; Principle of Minimum Mechanical Energy; Principle of Minimum Potential Energy; Quantum Hamilton Principle; Stewart-Hamilton Principle; Stochastic Hamilton Variational Principle; Toupin's Dual Principle; Weak Invariance Principle

POINT ESTIMATION Correspondence Principle; Entropy Maximum Principle; Extended Maximum Entropy Principle; Extremal Principle of Information; Frieden-Soffer Principle; Gibbs Principle; Inference Principle; Jaynes' Maximum Entropy Principle; Jaynes' Principle; Kullback's Minimum Cross-entropy Principle; Maximum Information Entropy Principle; Measurement Principle; Minimax Entropy Principle; Minimum Entropy Principle; Minimum Relative Entropy Principle; Modification of The Extended Maximum Entropy Principle; Principle of Maximum Entropy; Principle of Maximum Nonadditive Entropy; Principle of Observational Compatibility; Principle of Operational Compatibility; Uncertainty Principle

POLLUTION Polluter Pays Principle, Optimization Principle; Run and Recreate Principle

POLYCRYSTALLINE MATERIALS Principle of Material Frame Indifference; Principle of Minimum Dissipation

POLYMER MELTS Priming Principle, Corresponding States Principle; Pitzer Corresponding States Principle; Principle of Correspondence

POLYNOMIALS Weak Invariance Principle, Fundamental Principle, Gaga Principle; Super Gaga Principle

POLYSTYRENE Time Temperature Superposition Principle; Principle of Maximum Entropy

POPULATION DYNAMICS Active Principle; Minimum Description Length Principle, Hamilton Variational Principle; Weak Invariance Principle; Population Genetics; Extremum Principle; Frequency-dependent Principle; Optimizing Principle

POPULATION STUDIES Extremum Principle; Frequency-dependent Principle; Optimizing Principle

POSITION MEASUREMENT Inference Principle; Jaynes' Principle; Kullback's Minimum Cross-entropy Principle; Maximum Information Entropy Principle; Measurement Principle; Measuring Principle; Minimax Entropy Principle; Minimum Energy Principle; Minimum Relative Entropy Principle; Modification of The Extended Maximum Entropy Principle; Principle of Maximum Entropy; Principle of Maximum Nonadditive Entropy; Principle of Measurement; Principle of Observational Compatibility; Principle of Operational Compatibility; Schwinger's Quantum Action Principle; Uncertainty Principle; Weak Invariance Principle; Position Selectiv Speed Measurement; Digital Multimeter Measurement Principle

POSITIVE OPERATORS Correspondence Principle; Entropy Maximum Principle; Extended Maximum Entropy Principle; Extremal Principle of Information; Frieden-Soffer Principle; Gibbs Principle; Inference Principle; Jaynes' Maximum Entropy Principle; Jaynes' Principle; Kullback's Minimum Cross-entropy

Principle; Maximum Information Entropy Principle; Measurement Principle; Minimax Entropy Principle; Minimum Entropy Principle; Minimum Relative Entropy Principle; Modification of The Extended Maximum Entropy Principle; Principle of Maximum Entropy; Principle of Maximum Nonadditive Entropy; Principle of Observational Compatibility; Principle of Operational Compatibility; Uncertainty Principle

POSITRON Schwinger Principle; Possibilistic Uncertainty; Zadehs Extension Principle

POSSIBILITY THEORY Maximum Hardness Principle; Minimum Information Principle; Principle of Uncertainty

POSTOPERATIVE COMPLICATION Woltjer-Taylor Minimum Energy Principle; Mitrofanoff Principle; Monti Principle

POTENTIAL ENERGY Equivalence Principle; Maximizing Principle; Priming Principle; Weak Invariance Principle, Principle of Minimum

POTENTIAL ENERGY SURFACE Maximizing Principle; Maximizing Principle; Saint-Venant's Principle

POTENTIAL SURFACE Action Principle; Maximum Likelihood Principle; Stationary Action Principle

POTENTIALS Contraction Mapping Principle; Contraction Principle, Weak Invariance Principle

POWER ELECTRONIC APPLICATIONS Inference Principle; Jaynes' Principle; Kullback's Minimum Cross-entropy Principle; Maximum Information Entropy Principle; Measurement Principle; Measuring Principle; Minimax Entropy Principle; Minimum Energy Principle; Minimum Relative Entropy Principle; Modification of The Extended Maximum Entropy Principle; Principle of Maximum Entropy; Principle of Maximum Nonadditive Entropy; Principle of Observational Compatibility; Principle of Operational Compatibility; Schwinger's Quantum Action Principle; Uncertainty Principle

POWER ELECTRONICS Inference Principle; Jaynes' Principle; Kullback's Minimum Cross-entropy Principle; Master-slave Principle; Maximum Information Entropy Principle; Measurement Principle; Measuring Principle; Minimax Entropy Principle; Minimum Energy Principle; Minimum Relative Entropy Principle; Modification of The Extended Maximum Entropy Principle; Principle of Maximum Entropy; Principle of Maximum Nonadditive Entropy; Principle of Observational Compatibility; Principle of Operational Compatibility; Schwinger's Quantum Action Principle; Uncertainty Principle

PREDICTION OF DROPLET SIZE AND VELOCITY DISTRIBUTIONS IN SPRAYS Correspondence Principle; Entropy Maximum Principle; Extended Maximum Entropy Principle; Extremal Principle of Information; Frieden-Soffer Principle; Gibbs Principle; Inference Principle; Jaynes' Maximum Entropy Principle; Jaynes' Principle; Kullback's Minimum Cross-entropy Principle; Maximum Information Entropy Principle; Measurement Principle; Minimax Entropy Principle; Minimum Entropy Principle; Minimum Relative Entropy Principle; Modification of The Extended Maximum Entropy Principle; Principle of Maximum Entropy; Principle of Maximum Nonadditive Entropy; Principle of Observational Compatibility; Principle of Operational Compatibility; Uncertainty Principle

PRESSURE Belt Filter Press Principle; Casuistic Principle; Double Effect Principle; Dynamic Principle; Dynamic Principle of Osteosynthesis; Exception Granting Principle; Justifying Principle; Principle of Do No Harm; Principle of Double Effect; Principle of Filter Belt Press; Principles of Biomedical Ethics; Superposition Principle

PRESTRESSED PLATE D'Alembert's Principle; Gyarmati's Principle; Hamilton Principle; Prigogine Principle; Principle of Minimum Kinetic Energy; Principle of Minimum Mechanical Energy; Principle of Minimum Potential Energy; Quantum Hamilton Principle; Stewart-Hamilton Principle; Stochastic Hamilton Variational Principle; Toupin's Dual Principle

PRESURVIVOR PROCESSING Principle of Persurvivor Processing; Principles of Darcy's Law

PREVENTION Carousel Principle; Monti Principle; Uncertainty Principle, Coordination Number Principle, Explosion Principle

PREVENTION OF NONLINEAR SIGNAL LOSS Ambartsumian Invariance Principle; Charnov's Invariance Principle; Donsker's Invariance Principle; Invariance Principle; Invariance Principle of General Relativity; Local Invariance Principle; Maruyama's Invariance Principle; Perturbation Invariance Principle; Principle of Invariance; Principle of Relativity; Principle of Undulatory Invariance; Relaxed Invariance Principle; Scale Invariance Principle; Strong Principle; Variational Principle

PRICE Autonomous Cooperation Distributed Maximum Principle; Distributed Maximum Principle; Maximum Principle

PRINCIPAL COMPONENT ANAYLSIS Parsimony Principle; Principle of Parsimonius Data Modeling

PRINCIPAL COMPONENTS OF NATURAL IMAGES Equidistortion Principle; Quantization Design Principle

PROBABILISTIC APPROACH D'Alembert's Principle; Gyarmati's Principle; Hamilton Principle; Prigogine Principle; Principle of Corresponding States; Principle of Minimum Kinetic Energy; Principle of Minimum Mechanical Energy; Principle of Minimum Potential Energy; Quantum Hamilton Principle; Stewart-Hamilton Principle; Stochastic Hamilton Variational Principle; Toupin's Dual Principle; Weak Invariance Principle

PROBABILISTIC LOGIC Correspondence Principle; Entropy Maximum Principle; Extended Maximum Entropy Principle; Extremal Principle of Information; Frieden-Soffer Principle; Gibbs Principle; Inference Principle; Jaynes' Maximum Entropy Principle; Jaynes' Principle; Kullback's Minimum Cross-entropy Principle; Maximum Information Entropy Principle; Measurement Principle; Minimax Entropy Principle; Minimum Entropy Principle; Minimum Information Principle; Minimum Relative Entropy Principle; Modification of The Extended Maximum Entropy Principle; Principle of Maximum Entropy; Principle of Maximum Nonadditive Entropy; Principle of Observational Compatibility; Principle of Operational Compatibility; Uncertainty Principle

PROBABILISTIC MODUS PONENS Correspondence Principle; Entropy Maximum Principle; Extended Maximum Entropy Principle; Extremal Principle of Information; Frieden-Soffer Principle; Gibbs Principle; Inference Principle; Jaynes' Maximum Entropy Principle; Jaynes' Principle; Kullback's Minimum Cross-entropy Principle; Maximum Information Entropy Principle; Measurement Principle; Minimax Entropy Principle; Minimum Entropy Principle; Minimum Relative Entropy Principle; Modification of The Extended Maximum Entropy Principle; Principle of Maximum Entropy; Principle of Maximum Nonadditive Entropy; Principle of Observational Compatibility; Principle of Operational Compatibility; Uncertainty Principle

PROBABILISTIC SYSTEMS Correspondence Principle; Entropy Maximum Principle; Extended Maximum Entropy Principle; Extremal Principle of Information; Frieden-Soffer Principle; Gibbs Principle; Inference Principle; Jaynes' Maximum Entropy Principle; Jaynes' Principle; Kullback's Minimum Cross-entropy Principle; Maximum Information Entropy Principle; Measurement Principle; Minimax Entropy Principle; Minimum Entropy Principle; Minimum Relative Entropy Principle; Modification of The Extended Maximum Entropy Principle; Principle of Maximum Entropy; Principle of Maximum Nonadditive Entropy; Principle of Observational Compatibility; Principle of Operational Compatibility; Uncertainty Principle

PROBABILITY Active Principle; Conservation of Momentum Principle; Contraction Principle; Correspondence Principle; D'Alembert Principle; Differential Principle; Entropy

Maximum Principle; Extended Maximum Entropy Principle; Extremal Principle of Information; Frieden-Soffer Principle; Gibbs Principle; Gibbs Variational Principle; Hamilton Principle; Inference Principle; Jaynes' Maximum Entropy Principle; Jaynes' Principle; Kullback's Minimum Cross-entropy Principle; Large Deviation Principle; Likelihood Principle; Maximum Information Entropy Principle; Measurement Principle; Minimax Entropy Principle; Minimum Cross-entropy (Me) Principle; Minimum Description Length Principle; Minimum Entropy Principle; Minimum Relative Entropy Principle; Modification of The Extended Maximum Entropy Principle; Newton's Third Principle; Principle of Equal A Priori Probabilities; Principle of Maximum Entropy; Principle of Maximum Nonadditive Entropy; Principle of Observational Compatibility; Principle of Operational Compatibility; Sufficiency Principle; Uncertainty Principle; Variational Principle; Powdering Principle, Momentum Principle

PROBABILITY DISTRIBUTION Active Principle; Correspondence Principle; Entropy Maximum Principle; Extended Maximum Entropy Principle; Extremal Principle of Information; Frieden-Soffer Principle; Gibbs Principle; Inference Principle; Jaynes' Maximum Entropy Principle; Jaynes' Principle; Kullback's Minimum Cross-entropy Principle; Maximum Information Entropy Principle; Measurement Principle; Minimax Entropy Principle; Minimum Entropy Principle; [Minimum Relative Entropy Principle; Modification of The Extended Maximum Entropy Principle; Principle of Maximum Entropy; Principle of Maximum Nonadditive Entropy; Principle of Observational Compatibility; Principle of Operational Compatibility; Uncertainty Principle; Ambartsumian Invariance Principle; Charnov's Invariance Principle; Detailed Balance Principle; Donsker's Invariance Principle; Invariance Principle; Invariance Principle of General Relativity; Local Invariance Principle; Maruyama's Invariance Principle; Perturbation Invariance Principle; Principle of Invariance; Principle of Relativity; Principle of Undulatory Invariance; Relaxed Invariance Principle; Scale Invariance

Principle; Strong Principle; Variational Principle; Weak Invariance Principle

PROBABILITY DISTRIBUTION FUNCTIONS Minimum Information Principle, Discrete Time Systems. Principle of Corresponding States

PROBABILITY THEORY Contraction Principle; Gibbs Variational Principle; Large Deviation Principle; Maupertuis Principle; Powdering Principle; Principle of Uncertainty

PROBABILITY WEIGHTED MOMENT Correspondence Principle; Entropy Maximum Principle; Extended Maximum Entropy Principle; Extremal Principle of Information; Frieden-Soffer Principle; Gibbs Principle; Inference Principle; Jaynes' Maximum Entropy Principle; Jaynes' Principle; Kullback's Minimum Cross-entropy Principle; Maximum Information Entropy Principle; Measurement Principle; Minimax Entropy Principle; Minimum Entropy Principle; Minimum Relative Entropy Principle; Modification of The Extended Maximum Entropy Principle; Principle of Maximum Entropy; Principle of Maximum Nonadditive Entropy; Principle of Observational Compatibility; Principle of Operational Compatibility; Uncertainty Principle

PROBES Inference Principle; Jaynes' Principle; Kullback's Minimum Cross-entropy Principle; Maximum Information Entropy Principle; Measurement Principle; Measuring Principle; Minimax Entropy Principle; Minimum Energy Principle; Minimum Relative Entropy Principle; Modification of The Extended Maximum Entropy Principle; Principle of Maximum Entropy; Principle of Maximum Nonadditive Entropy; Principle of Observational Compatibility; Principle of Operational Compatibility; Schwinger's Quantum Action Principle; Uncertainty Principle

PROBLEM SOLVING Ekeland's variational Principle; Ekeland's Epsilon Variational Principle; Ekeland's Variational Principle; Global Maximum Principle; Minimum Information Principle; Variational Principle

PROGRAMMING Relative Frequency Principle, Maximum Entropy Principle; Powdering Principle, Clairvoyance Principle

PROJECTION Overshoot Principle; Projection Principle, Parsimony Principle; Principle of Parsimonius Data Modeling

PROPOFOL Principle of Double Effect; Propositional Variables; Correspondence Principle; Energy Preservation Principle; Entropy Maximum Principle; Extended Maximum Entropy Principle; Extinguishing Principle; Extremal Principle of Information; Frieden-Soffer Principle; Gibbs Principle; Global Extremum Principle; Inference Principle; Jaynes' Maximum Entropy Principle; Jaynes' Principle; Kullback's Minimum Cross-entropy Principle; Maximum Information Entropy Principle; Measurement Principle; Minimax Entropy Principle; Minimum Entropy Principle; Minimum Relative Entropy Principle; Modification of The Extended Maximum Entropy Principle; Principle of Maximum Entropy; Principle of Maximum Nonadditive Entropy; Principle of Observational Compatibility; Uncertainty Principle

PROTEINS Affinity Principle; Antipernicious Anemia Principle; Boric Acid Affinity Principle; Boronic Acid Affinity Principle; HIV Principle; Maximum Entropy Principle; Weak Invariance Principle

PROXIMATE ORDER Uncertainty Principle; Proximity Compatibility; Proximity Compatibility Principle

PSYCHIATRY Babinet's Principle; Complementarity Principle; Gestalt Principle; Nameless Category N3C Principle; Obligatory Contour Principle; Principle of Relevance; Salutogenesis Principle, Therapeutic Principle

PSYCHIC ACTIVITY Economic Principle; Principles of Determination; Principles of Multiple Function; Principles of Overdetermination; Regulatory Principles and Other Regulatory Principles of Inertia; Regulatory Principles of Constancy; Regulatory Principles of Pleasureunpleasure; Regulatory Principles of Primary Secondary Process; Regulatory Principles of Reality; Regulatory Principles of Repetition

PSYCHOANALYTIC THEORY Beyond The Pleasure Principle; Jewish Pleasure Principle; Pleasure Pain Principle; Pleasure Principle

PSYCHOLOGY Economic Principle; Maximization Principle; Principles of Determination; Principles of Multiple Function; Principles of Overdetermination; Regulatory Principles and Other Regulatory Principles of Inertia; Regulatory Principles of Constancy; Regulatory Principles of Pleasureunpleasure; Regulatory Principles of Primarysecondary Process; Regulatory Principles of Reality; Regulatory Principles of Repetition; Relative Frequency Principle; Psychology and SociologyPrinciple of Relevance

PSYCHOPHARMACOLOGY Babinet's Principle; Complementarity Principle; Mouse Beta Amyrin Active Principle

PSYCHOPHYSICAL FUNCTIONS Extension Principle; Principle of Uncertainty; Psychophysics of Vision; Principle of Information Processing; Principle of Per-survivor Processing

PSYCHOSOMATICS Competitive Exclusion Principle; Principle of Complementarity

PUBLIC HEALTH Caution Principle; Health Principle; Minimal Principle; Optimization Principle; Precaution Principle; Precautionary Principle; Principle of Habeas Mentem; Run and Recreate Principle; Uncertainty Principle; Salutogenesis Principle, Contagion Principle; Health Principle; Precautionary Principle; Principle of Habeas Mentem

PULSATING CURRENT Inference Principle; Jaynes' Principle; Kullback's Minimum Cross-entropy Principle; Maximum Information Entropy Principle; Measurement Principle; Measuring Principle; Minimax Entropy Principle; Minimum Energy Principle; Minimum Relative Entropy Principle; Modification of The Extended Maximum Entropy Principle; Principle of Maximum Entropy; Principle of Maximum Nonadditive Entropy; Principle of Observational Compatibility; Principle of Operational Compatibility; Schwinger's Quantum Action Principle; Uncertainty Principle

PULSE TUBE REFRIGERATOR Multi Bypass Principle; Casuistic Principle; Consumer Principle; Double Effect Principle; Double Inlet Principle; Exception Granting Principle; Justifying Principle; Principle of Do No Harm; Principle of Double Effect; Principles of Biomedical Ethics

PYROMETER PLASMA SPRAYING CONTROL Inference Principle; Jaynes' Principle; Kullback's Minimum Cross-entropy Principle; Maximum Information Entropy Principle; Measurement Principle

PYROMETERS Inference Principle; Jaynes' Principle; Kullback's Minimum Cross-entropy Principle; Maximum Information Entropy Principle; Measurement Principle; Measuring Principle; Minimax Entropy Principle; Minimum Energy Principle; Minimum Relative Entropy Principle; Modification of The Extended Maximum Entropy Principle; Principle of Maximum Entropy; Principle of Maximum Nonadditive Entropy; Principle of Observational Compatibility; Principle of Operational Compatibility; Schwinger's Quantum Action Principle; Uncertainty Principle

Q

QED Anthropic Cosmological Principle; Anthropic Principle; Cosmological Principle, Minimum Description Length Principle

QUANTILE Correspondence Principle; Energy Preservation Principle; Entropy Maximum Principle; Extended Maximum Entropy Principle; Extinguishing Principle; Extremal Principle of Information; Frieden-Soffer Principle; Gibbs Principle; Global Extremum Principle; Inference Principle; Jaynes' Maximum Entropy Principle; Jaynes' Principle; Kullback's Minimum Cross-entropy Principle; Maximum Information Entropy Principle; Measurement Principle; Minimax Entropy Principle; Minimum Entropy Principle; Minimum Relative Entropy Principle; Modification of The Extended Maximum Entropy Principle; Principle of Maximum Entropy; Principle of Maximum Nonadditive Entropy; Principle of Observational Compatibility; Uncertainty Principle

QUANTITATIVE ANALYSIS (ANALYTICAL CHEMISTRY) Principle of Measurement; Capillary Electrophoresis Principle; Ion Pair Principle; Principle of Diastereomeric Ion Pair Formation

QUANTITATIVE GENETICS Entropy Principle; Fuzzy Entropy Principle; Maxent Principle; Maximum Entropy Principle; Mermin Entropy Principle; Principle of Maximum Entropy; Principle of Maximum Nonadditive Entropy; Principle of Operational Compatibility, Handicap Principle

QUANTITATIVE SOLUTIONS Three-day Principle; Educational Principle; Ehrenfest Principle

QUANTIZATION Action Principle; Maximum Likelihood Principle; Weak Invariance Principle

QUANTUM BEHAVIOR Minimum Description Length Principle; Quantum Cosmology; Caratheodory's Principle; Conte's Principle; Fermat's Principle; Fermat; Least Time Principle; Principle of Caratheodony; Stationary Time Principle

QUANTUM ELECTRODYNAMICS Action Principle; Maximum Likelihood Principle; Weak Invariance Principle

QUANTUM FIELD THEORY Principle of Causality; Quantum Linearity Principle; Spacetime Uncertainty Principle; Weak Invariance Principle, Galileo Principle; Principle of Molecular Dynamics, Minimal Principle

QUANTUM MECHANICS Action Principle; Babinet's Principle; Bogolyubov Principle; Bogolyubov Variational Principle; Bohr Principle of Complementarity; Bohrs Complementarity Principle; Bohr's Correspondence Principle; Complementarity Principle; Correspondence Principle; Gibbs-Bogolyubov Variational Principle; Gibbs Third Variational Principle; Hartree Fock Bogolyubov Principle; Heisenberg Uncertainty Principle; Inference Principle; Jaynes' Principle; Kullback's Minimum Cross-entropy Principle; Maupertuis Principle; Maximum Hardness Principle; Maximum Information Entropy Principle; Maximum Likelihood Principle; Measurement Principle; Measuring Principle; Minimax Entropy Principle; Minimum Energy Principle; Minimum Relative Entropy Principle; Modification of The Extended Maximum Entropy Principle; Principle of Maximum Entropy; Principle of Maximum Nonadditive Entropy; Principle of Observational Compatibility; Principle of Operational Compatibility; Principle of Representationtheoretic Selfdualtiy; Principles of Uncertainty; Quantal Principle; Quantum Principle of Uncertainty;

Schwinger's Quantum Action Principle; Uncertainty Principle; Unified Principle; Weak Invariance Principle; Minimum Description Length Principle

QUANTUM OPTICS Babinet's Principle; Complementarity Principle; Principle of Causality

QUANTUM STATE INFERENCE Principle of Maximum Entropy; Quantum Statistical Mechanics; Competitive Exclusion Principle; Principle of Complementarity

QUANTUM THEORY Ckvp; Complex Kohn Variational Principle; Fermats Principle; Hulthén–Kohn Variational Principle; Kohn Variational Principle; Maupertuis Principle; Maximum Hardness Principle; Minimum Description Length Principle; Principle of Molecular Dynamics; Quantum Linearity Principle; Schwinger Variational Principle; Weak Invariance Principle

QUASARS Caratheodory's Principle; Conte's Principle; Fermat's Principle; Least Time Principle; Mitrofanoff Principle; Principle of Caratheodory; Stationary Time Principle

QUASISTABLE LAW Competitive Exclusion Principle; Principle of Complementarity

QUATERNION MOMENT PRESERVING Entropy Principle; Generalized Minimal Principle; Maximum Entropy Principle; Minimum Cross-entropy Principle; Moment Preserving Principle; Principle of Minimum Cross-entropy; Variational Minimal Principle

QUERY Entropy Principle; Fuzzy Entropy Principle; Maxent Principle; Maximum Entropy Principle; Mep Maximum Entropy Principle; Mermin Entropy Principle; Pme Principle of Maximum Entropy; Principle of Maximum Nonadditive Entropy; Principle of Operational Compatibility

QUEUING Averaging Principle, Correspondence Principle; Energy Preservation Principle; Entropy Maximum Principle; Extended Maximum Entropy Principle; Extinguishing Principle; Extremal Principle of Information; Frieden-Soffer Principle; Gibbs Principle; Global Extremum Principle; Inference Principle; Jaynes' Maximum Entropy Principle; Jaynes' Principle; Kullback's Minimum Cross-entropy Principle; Maximum Information Entropy Principle; Measurement Principle; Minimax Entropy Principle; Minimum Entropy Principle; Minimum Relative Entropy Principle; Modification of The Extended Maximum Entropy Principle; Principle of Maximum Entropy; Principle of Maximum Nonadditive Entropy; Principle of Observational Compatibility; Uncertainty Principle

QUIESCENT PROMINENCES Hamilton's Principle; Kirchhoff's Voltage Law Principle; Minimum Energy Principle; Minimum Norm Principle; Variational Principle; Woltjer Minimum Energy Principle

R

RADIATION HAZARDS Equidistortion Principle; Quantization Design Principle; Radiation Health Optimization Principle; Run and Recreate Principle; Protection Principle; Radiation Protection Principle

RADIATION RESISTANCE Equivalent Background Principle; Radiation Shielding; Principle of Maximum Entropy Radiation Antiviral Active Principle; Optimization Principle; Principle of Corresponding States; Run and Recreate Principle, Functioning Principle, Absorption Principle, Equivalent Background Principle

RADIATIVE TRANSFER Hamilton's Principle; Kirchhoff's Voltage Law Principle; Minimum Energy Principle; Minimum Norm Principle; Variational Principle; Woltjer Minimum Energy Principle; Ambartsumian Invariance Principle; Charnov's Invariance Principle; Donsker's Invariance Principle; Invariance Principle; Invariance Principle of General Relativity; Local Invariance Principle; Maruyama's Invariance Principle; Perturbation Invariance Principle; Principle of Invariance; Principle of Relativity; Principle of Undulatory Invariance; Relaxed Invariance Principle; Scale Invariance Principle; Strong Principle

RADIOLOGY Correspondence Principle; Energy Preservation Principle; Entropy Maximum Principle; Extended Maximum Entropy Principle; Extinguishing Principle; Extremal Principle of Information; Frieden-Soffer Principle; Functioning Principle; Gibbs Principle; Global Extremum Principle; Inference Principle; Jaynes' Maximum Entropy Principle; Jaynes' Principle; Kullback's Minimum Cross-entropy Principle; Maximum Information Entropy Principle; Measurement Principle; Minimax Entropy Principle; Minimum Entropy Principle; Minimum Relative Entropy Principle; Modification of The Extended Maximum Entropy Principle;

Principle of Maximum Entropy; Principle of Maximum Nonadditive Entropy; Principle of Observational Compatibility; Uncertainty Principle; Optimization Principle; Run and Recreate Principle

RADIONUCLIDE IMAGING Correspondence Principle; Energy Preservation Principle; Entropy Maximum Principle; Extended Maximum Entropy Principle; Extinguishing Principle; Extremal Principle of Information; Frieden-Soffer Principle; Gibbs Principle; Global Extremum Principle; Inference Principle; Jaynes' Maximum Entropy Principle; Jaynes' Principle; Kullback's Minimum Cross-entropy Principle; Maximum Information Entropy Principle; Measurement Principle; Minimax Entropy Principle; Minimum Entropy Principle; Minimum Relative Entropy Principle; Modification of The Extended Maximum Entropy Principle; Principle of Maximum Entropy; Principle of Maximum Nonadditive Entropy; Principle of Observational Compatibility; Uncertainty Principle

RANDOM EFFECTS Conditionality Principle, Competitive Exclusion Principle; Principle of Complementarity, Hamilton Variational Principle; Weak Invariance Principle

RANDOM FIELD Ambartsumian Invariance Principle; Charnov's Invariance Principle; Donsker's Invariance Principle; Invariance Principle; Invariance Principle of General Relativity; Local Invariance Principle; Maruyama's Invariance Principle; Perturbation Invariance Principle; Principle of Invariance; Principle of Relativity; Principle of Undulatory Invariance; Relaxed Invariance Principle; Scale Invariance Principle; Strong Principle; Variational Principle

RANDOM GRAPHS Contraction Principle; Gibbs Variational Principle; Large Deviation Principle

RANDOM MEASURE Ambartsumian Invariance Principle; Charnov's Invariance Principle; Donsker's Invariance Principle; Invariance Principle; Invariance Principle of General Relativity; Local Invariance Principle; Maruyama's Invariance Principle; Perturbation Invariance Principle; Principle of Invariance; Principle of Relativity; Principle of Undulatory Invariance; Relaxed Invariance Principle; Scale Invariance Principle; Strong Principle; Variational Principle

RANDOM PROCESSES Competitive Exclusion Principle; Equivalence Principle; Minimum Description Length Principle; Principle of Complementarity; Contraction Principle; Contraction Principle; Gibbs Variational Principle; Large Deviation Principle

RANDOM VARIABLE SEQUENCE Ambartsumian Invariance Principle; Charnov's Invariance Principle; Donsker's Invariance Principle; Invariance Principle; Invariance Principle of General Relativity; Local Invariance Principle; Maruyama's Invariance Principle; Perturbation Invariance Principle; Principle of Invariance; Principle of Relativity; Principle of Undulatory Invariance; Relaxed Invariance Principle; Scale Invariance Principle; Strong Principle; Variational Principle

RANDOM VARIABLES Correspondence Principle; Energy Preservation Principle; Entropy Maximum Principle; Extended Maximum Entropy Principle; Extinguishing Principle; Extremal Principle of Information; Frieden-Soffer Principle; Gibbs Principle; Global Extremum Principle; Inference Principle; Invariance Principle; Jaynes' Maximum Entropy Principle; Jaynes' Principle; Kullback's Minimum Cross-entropy Principle; Maximum Information Entropy Principle; Measurement Principle; Minimax Entropy Principle; Minimum Entropy Principle; Minimum Relative Entropy Principle; Modification of The Extended Maximum Entropy Principle; Permutation Principle; Principle of Maximum Entropy; Principle of Maximum Nonadditive Entropy; Principle of Observational Compatibility; Uncertainty Principle; Powdering Principle, Principle of Corresponding States

RANDOM WALK Ambartsumian Invariance Principle; Charnov's Invariance Principle; Donsker's Invariance Principle; Invariance Principle; Invariance Principle of General Relativity; Local Invariance Principle; Maruyama's Invariance Principle; Perturbation Invariance Principle; Principle of Invariance; Principle of Relativity; Principle of Undulatory Invariance; Relaxed Invariance Principle; Scale Invariance Principle; Strong Principle; Variational Principle

RANDOM WAVE Correspondence Principle; Energy Preservation Principle; Entropy Maximum Principle; Extended Maximum Entropy Principle; Extinguishing Principle; Extremal Principle of Information; Frieden-Soffer Principle; Gibbs Principle; Global Extremum Principle; Inference Principle; Jaynes' Maximum Entropy Principle; Jaynes' Principle; Kullback's Minimum Cross-entropy Principle; Maximum Information Entropy Principle; Measurement Principle; Minimax Entropy Principle; Minimum Entropy Principle; Minimum Relative Entropy Principle; Modification of The Extended Maximum Entropy Principle; Principle of Maximum Entropy; Principle of Maximum Nonadditive Entropy; Pinciple of Observational Compatibility; Uncertainty Principle

RANDOMIZATION Conduction Principle; Conditionality Principle, Consumer Principle; Working Principle

RATE FUNCTION Contraction Principle; Gibbs Variational Principle; Large Deviation Principle

RATIONAL APPROACH Three-day Principle; Educational Principle; Ehrenfest Principle

RATIONAL FUNCTION Action Principle; Maximum Likelihood Principle; Stationary Action Principle

RATIOS Area Principle, Correspondence Principle; Energy Preservation Principle; Entropy Maximum Principle; Extended Maximum Entropy Principle; Extinguishing Principle; Extremal Principle of Information; Frieden-Soffer Principle; Gibbs Principle; Global

Extremum Principle; Inference Principle; Jaynes' Maximum Entropy Principle; Jaynes' Principle; Kullback's Minimum Cross-entropy Principle; Maximum Information Entropy Principle; Measurement Principle; Minimax Entropy Principle; Minimum Entropy Principle; Minimum Relative Entropy Principle; Modification of The Extended Maximum Entropy Principle; Principle of Maximum Entropy; Principle of Maximum Nonadditive Entropy; Principle of Observational Compatibility; Uncertainty Principle

RAY THEORY Caratheodory's Principle; Conte's Principle; Fermat's Principle; Least Time Principle; Mitrofanoff Principle; Principle of Caratheodony; Stationary Time Principle

RAYLEIGH DISTRIBUTION D'Alembert's Principle; Gyarmati's Principle; Hamilton Principle; Prigogine Principle; Principle of Minimum Kinetic Energy; Principle of Minimum Mechanical Energy; Principle of Minimum Potential Energy; Quantum Hamilton Principle; Stewart-Hamilton Principle; Stochastic Hamilton Variational Principle; Toupin's Dual Principle

RAYLEIGH FLOW Correspondence Principle; Energy Preservation Principle; Entropy Maximum Principle; Extended Maximum Entropy Principle; Extinguishing Principle; Extremal Principle of Information; Frieden-Soffer Principle; Gibbs Principle; Global Extremum Principle; Inference Principle; Jaynes' Maximum Entropy Principle; Jaynes' Principle; Kullback's Minimum Cross-entropy Principle; Maximum Information Entropy Principle; Measurement Principle; Minimax Entropy Principle; Minimum Entropy Principle; Minimum Relative Entropy Principle; Modification of The Extended Maximum Entropy Principle; Principle of Maximum Entropy; Principle of Maximum Nonadditive Entropy; Principle of Observational Compatibility; Uncertainty Principle

REACTION MECHANISM First Principle; Maximizing Principle; Nuclear Motion Principle

REACTIONS Principle of Differential Ph Measurement; Reactivity/Selectivity Principle

REACTIVE COLLISION Action Principle; Ckvp; Complex Kohn Variational Principle; Hulthén–Kohn Variational Principle; Kohn Variational Principle; Maximum Likelihood Principle; Schwinger Variational Principle; Stationary Action Principle

REACTIVITY Hard-soft acid-base Principle; Inertia Principle; Lagrange Principle of Least Action; Least Action Principle; Least Motion Principle; Principle of Least Motion; Saint-Venant's Principle

REAL HYPER SURFACES Conte's Principle; Fermat's Principle; Least Time Principle; Mitrofanoff Principle; Stationary Time Principle

REASONING Curvilinear Impetus Principle; Minimum Information Principle

RECEPTIVE FIELDS Principle of Information Processing; Principle of Per-survivor Processing

RECIPROCITY Desoption Principle, Maximum Hardness Principle, Micropump Principle, Fermats Principle; Minimal Principle

RECOGNITION Active Principle; Maximum Entropy Principle; Minimum Description Length Principle

RECOMBINATIVE HYDROGEN DESORPTION Electronegativity Equalization Principle; Geometric Mean Principle; Principle of electronegativity

RECONSTRUCTION Duality Principle; Projection Principle; Woltjer-Taylor Minimum Energy Principle

RECORD AND REPRODUCTION CHARACTERISTIC Magnetooptical Disk Principle; Mo Principle

RECTUM Mitrofanoff Principle; Variational Principle, Proof of Principle

REDUCED ROTATION Heisenberg's Correspondence Principle; Reduced State Sequence Estimation; Principle of per-survivor processing; Principles of Darcy's Law

REFLECTED WAVE Correspondence Principle; Energy Preservation Principle; Entropy Maximum Principle; Extended Maximum Entropy Principle; Extinguishing Principle; Extremal Principle of Information; Frieden-Soffer Principle; Gibbs Principle; Global Extremum Principle; Inference Principle; Jaynes' Maximum Entropy Principle; Jaynes' Principle; Kullback's Minimum Cross-entropy Principle; Maximum Information Entropy Principle; Measurement Principle; Minimax Entropy Principle; Minimum Entropy Principle; Minimum Relative Entropy Principle; Modification of The Extended Maximum Entropy Principle; Principle of Maximum Entropy; Principle of Maximum Nonadditive Entropy; Principle of Observational Compatibility; Uncertainty Principle

REFLECTIONLESS APERTURE Black hole Principle; Optical Black Hole Principle; Reflectionless Aperture Principle

REFLECTIVITY Correspondence Principle; Energy Preservation Principle; Entropy Maximum Principle; Extended Maximum Entropy Principle; Extinguishing Principle; Extremal Principle of Information; Frieden-Soffer Principle; Gibbs Principle; Global Extremum Principle; Inference Principle; Jaynes' Maximum Entropy Principle; Jaynes' Principle; Kullback's Minimum Cross-entropy Principle; Maximum Information Entropy Principle; Measurement Principle; Minimax Entropy Principle; Minimum Entropy Principle; Minimum Relative Entropy Principle; Modification of The Extended Maximum Entropy Principle; Principle of Maximum Entropy; Principle of Maximum Nonadditive Entropy; Principle of Observational Compatibility; Uncertainty Principle

REFRIGERATION Consumer Principle; Double Inlet Principle; Multi Bypass Principle, Principle of Construction and Operation

REFRIGERATORS Casuistic Principle; Double Effect Principle; Exception Granting Principle; Justifying Principle; Multi Bypass Principle; Principle of Do No Harm; Principle of Double Effect; Principles of Biomedical Ethics

REGRESSION DIAGNOSTICS Ambartsumian Invariance Principle; Charnov's Invariance Principle; Donsker's Invariance Principle; Invariance Principle; Invariance Principle of General Relativity; Local Invariance Principle; Maruyama's Invariance Principle; Perturbation Invariance Principle; Principle of Invariance; Principle of Relativity; Principle of Undulatory Invariance; Relaxed Invariance Principle; Scale Invariance Principle; Strong Principle; Variational Principle

REGRESSION PREDICTION Parsimony Principle; Principle of Parsimonius Data Modeling

REGULAR MEDIAN FILTER Correspondence Principle; Energy Preservation Principle; Entropy Maximum Principle; Extended Maximum Entropy Principle; Extinguishing Principle; Extremal Principle of Information; Frieden-Soffer Principle; Gibbs Principle; Global Extremum Principle; Inference Principle; Jaynes' Maximum Entropy Principle; Jaynes' Principle; Kullback's Minimum Cross-entropy Principle; Maximum Information Entropy Principle; Measurement Principle; Minimax Entropy Principle; Minimum Entropy Principle; Minimum Relative Entropy Principle; Modification of The Extended Maximum Entropy Principle; Principle of Maximum Entropy; Principle of Maximum Nonadditive Entropy; Principle of Observational Compatibility; Uncertainty Principle

RELATIVISTIC EFFECTS Minimum Description Length Principle, Maximum Hardness Principle

RELATIVISTIC MECHANICS Hamilton's Principle; Principle of Material Frame Indifference

RELATIVITY Caratheodory's Principle; Conte's Principle; Fermat's Principle; Galileo Principle; Hamilton's Principle; Huygen's Principle; Least Time Principle; Machs Principle; Principle of Caratheodory; Stationary Time Principle; Equivalence Principle, Covariance Principle; Principle of Equivalence; Principle of Material Frame Indifference; Principle of Correspondence; Principle of Material Frame Indifference; Strong

Equivalence Principle, Maximum Hardness Principle, Hamilton's Principle, Maximum Hardness Principle, Principle of Molecular Dynamics

RELAXATION Hamilton's Principle; Kirchhoff's Voltage Law Principle; Minimum Energy Principle; Minimum Norm Principle; Variational Principle; Woltjer Minimum Energy Principle, Kelvin Arnold Energy Principle, Fermats Principle; Pome

RELEVANCE Entropy Principle; Fuzzy Entropy Principle; Maxent Principle; Maximum Entropy Principle; Mep Maximum Entropy Principle; Mermin Entropy Principle; Pme Principle of Maximum Entropy; Principle of Maximum Nonadditive Entropy; Principle of Operational Compatibility

REPRESENTATION   INVARIANCE Minimum Cross-entropy (Me) Principle, Principle of Representationtheoretic Selfdualtiy

REPRINTS Relative Frequency Principle

REPRODUCIBILITY Capillary Electrophoresis Principle; Ion Pair Principle; Principle of Diastereomeric Ion Pair Formation

REPRODUCTIVE   SYSTEM First   Principle; Frequency-dependent   Principle;   Mitrofanoff Principle; Monti Principle; Priming Principle

RESEARCH Competitive Exclusion Principle; Principle of Complementarity, Fermats Principle; Scientific Principle

RESERVOIRS Monti Principle, Productsolution Principle, Woltjer-Taylor Minimum Energy Principle

RESIDUAL STRESS Correspondence Principle; Energy Preservation Principle; Entropy Maximum Principle; Extended Maximum Entropy Principle; Extinguishing Principle; Extremal Principle of Information; Frieden-Soffer Principle; Gibbs Principle; Global Extremum Principle; Inference Principle; Jaynes' Maximum Entropy Principle; Jaynes' Principle; Kullback's Minimum Cross-entropy Principle; Maximum

Information Entropy Principle; Measurement Principle; Minimax Entropy Principle; Minimum Entropy Principle; Minimum Principle; Minimum Relative Entropy Principle; Modification of The Extended Maximum Entropy Principle; Principle of Maximum Entropy; Principle of Maximum Nonadditive Entropy; Principle of Observational Compatibility; Uncertainty Principle

RESONANCE FREQUENCY D'Alembert's Principle; Gyarmati's Principle; Hamilton Principle; Prigogine Principle; Principle of Minimum Kinetic Energy; Principle of Minimum Mechanical Energy; Principle of Minimum Potential Energy; Quantum Hamilton Principle; Stewart-Hamilton Principle; Stochastic Hamilton Variational Principle; Toupin's Dual Principle

RESONANT DOMAIN Optical Black Hole Principle; Reflectionless Aperture Principle

RESOURCE MANAGEMENT Ecological Principle; Optimality Principle; Organizing Principle; Precautionary Principle; Principle of Component Analysis; Principle of Dynamic Ecology; Self-organizing Principle; Similarity Principle; Wheatstone Bridge Principle

RESPIRATION August Krogh Principle; Optimality Principle; Organizing Principle; Principle of component analysis; Self-organizing Principle; Similarity Principle; Type II Cell Stretch-organizing Principle

RESPIRATORY SYSTEM Drilled Cell Principle; Drug Principle; Optimality Principle; Organizing Principle; Principle of Component Analysis; Self-organizing Principle; Similarity Principle; Type II Cell Stretch-organizing Principle, Timing Principle

RESPONSE SELECTIVITIES Principle of Information Processing; Principle of Per-survivor Processing

RETINA Equidistortion Principle; Quantization Design Principle

RETRIEVAL Archimedes' Principle; Hydrostatic Principle; Principle of Corresponding States

REUSE Decomposition Principle; Wise Use Principle

RHYTHMICAL ACTIVITIES Gabor's Holographic Principle; T Hooft Principle

RICCATI TYPE DIFFERENTIAL EQUATION Ekeland's variational Principle; Ekeland's Epsilon Variational Principle; Ekeland's Variational Principle; Global Maximum Principle; Variational Principle

RIEMANNIAN GEOMETRY Caratheodory's Principle; Conte's Principle; Fermat's Principle; Least Time Principle; Mitrofanoff Principle; Principle of Caratheodory; Stationary Time Principle

RIEMANNIAN MANIFOLD Fermat's Principle; Mitrofanoff Principle; Principle of Caratheodory

RIGIDITY Hellinger-Reissner Principle; Hu Washizu Variational Principle; Pseudogeneralized Variational Principle

RISK ANALYSIS Fermats Principle; Precautionary Principle, Maximization Principle

RISK Caution Principle; Negligence Principle; Precaution Principle; Uncertainty Principle

ROBUSTNESS Internal Model Principle; Invariance Principle; Invariance Principle; Maximum Entropy Principle; Ambartsumian Invariance Principle; Charnov's Invariance Principle; Donsker's Invariance Principle; Invariance Principle; Invariance Principle of General Relativity; Local Invariance Principle; Maruyama's Invariance Principle; Perturbation Invariance Principle; Principle of Invariance; Principle of Relativity; Principle of Undulatory Invariance; Relaxed Invariance Principle; Scale Invariance Principle; Strong Principle; Variational Principle

ROCURONIUM Optimality Principle; Organizing Principle; Principle of component analysis; Self-organizing Principle; Similarity Principle; Timing Principle; Type II Cell Stretch-organizing Principle; Working Principle

ROTATIONS Heisenberg's Correspondence Principle; Principle of Material Frame Indifference; Principle of Minimum Dissipation, Minimum Description Length Principle

RUBBER D'Alembert Lagrange Principle; D'Alembert Lagrange Variation Principle; D'Alembert Principle; Gauss Principle; Hamilton Lagrange Principle; Jayne's Principle; Jourdain Principle; Kinetic Principle; Lagrange Principle; Least Constraint Principle; Maupertius Lagrange Principle; Maupertuis Euler Lagrange Principle of Least Action; Principle of Extremal Effects; Principle of Least Constraint; Principle of Maximal Effect; Principle of Virtual Work

S

S MATRIX Ckvp; Complex Kohn Variational Principle; Hulthén–Kohn Variational Principle; Kohn Variational Principle; Schwinger Variational Principle

SAMPLING Conditionality Principle, Maximum Entropy Principle; New Therapeutic Principle, Hamilton's Principle; Kirchhoff's Voltage Law Principle; Minimum Energy Principle; Minimum Norm Principle; Variational Principle; Woltjer Minimum Energy Principle

SCANNING TUNNELING MICROSCOPE Minimum Description Length Principle; Maximizing Principle

SCATTER CORRECTION Parsimony Principle; Principle of Parsimonius Data Modeling

SCATTERER TIME SCALE Invariance Principle; Ambartsumian Invariance Principle; Charnov's Invariance Principle; Donsker's Invariance Principle; Invariance Principle; Invariance Principle of General Relativity; Local Invariance Principle; Maruyama's Invariance Principle; Perturbation Invariance Principle; Principle of Invariance; Principle of Relativity; Principle of Undulatory Invariance; Relaxed Invariance Principle; Scale Invariance Principle; Strong Principle; Variational Principle

SCATTERING Equivalence Principle; Pauli Principle, Minimum Description Length Principle, Principle of Causality, Schwinger's Variational Principle, Maximum Entropy Principle; Projection Principle; Unified Principle

SCHRODINGER EQUATION Blossoming Principle; Equivalence Principle; Minimum Description Length Principle; Positive Displacement Principle; Hardtree Fock Variational Principle; Skyrme Hartree Fock Principle; Time Dependent Hardtree Fock Variational Principle, Uncertainty Principle

SCHWARZS CHILD Minimum Description Length Principle, Principle of Molecular Dynamics; Principle of Causality

SCHWARZ'S LEMMA Korenblum's Maximum Principle; Phragmenlindel of Principle; Schwarz Reflection Principle; Extremum Principle, Tangent Variation Principle

SECOND LAW OF THERMODYNAMIC Capture Principle; Minimum Description Length Principle

SEDATION Principle of Double Effect; Casuistic Principle; Double Effect Principle; Exception Granting Principle; Justifying Principle; Principle of Do No Harm; Principle of Double Effect; Principles of Biomedical Ethics

SEEDED REGION GROWING Complex Moment Preserving Principle; Entropy Principle; Generalized Minimal Principle; Maximum Entropy Principle; Minimum Cross-entropy Principle; Moment Preserving Principle; Principle of Minimum Cross-entropy; Variational Minimal Principle

SEISMIC VELOCITY Three-day Principle; Educational Principle; Ehrenfest Principle

SELF ADJOINT OPERATOR Absorption Principle, Minimax Principle; Momentum Principle

SELF ORGANIZATION OF NEURONS Correspondence Principle; Energy Preservation Principle; Entropy Maximum Principle; Extended Maximum Entropy Principle; Extinguishing Principle; Extremal Principle of Information; Frieden-Soffer Principle; Gibbs Principle; Global Extremum Principle; Inference Principle; Jaynes' Maximum Entropy Principle; Jaynes' Principle; Kullback's Minimum Cross-entropy Principle; Maximum Information Entropy Principle; Measurement Principle; Minimax Entropy Principle; Minimum Entropy Principle; Minimum Relative

Dictionary of Scientific Principles, by Stephen Marvin
Copyright © 2011 John Wiley & Sons, Inc.

Entropy Principle; Modification of The Extended Maximum Entropy Principle; Principle of Maximum Entropy; Principle of Maximum Nonadditive Entropy; Principle of Observational Compatibility; Uncertainty Principle

SELF-ORGANIZING NEURALNETS Error Reconstruction Principle; Least Mean Square Error Reconstruction Principle

SELF-ORGANIZING SYSTEMS Correspondence Principle; Energy Preservation Principle; Entropy Maximum Principle; Extended Maximum Entropy Principle; Extinguishing Principle; Extremal Principle of Information; Frieden-Soffer Principle; Gibbs Principle; Global Extremum Principle; Inference Principle; Jaynes' Maximum Entropy Principle; Jaynes' Principle; Kullback's Minimum Cross-entropy Principle; Maximum Information Entropy Principle; Measurement Principle; Minimax Entropy Principle; Minimum Entropy Principle; Minimum Relative Entropy Principle; Modification of The Extended Maximum Entropy Principle; Principle of Maximum Entropy; Principle of Maximum Nonadditive Entropy; Principle of Observational Compatibility; Uncertainty Principle

SELF SENSING ACTIVE MAGNETIC BEARINGS Approximation Induction Principle; Asymmetric Induction Principle; Bbone Induction Principle; Coverset Induction Principle; Enantiodi Vergient Induction Principle; Induction Measurement Principle; Lysogenic Induction Principle; Magnetic Induction Principle; Osteogenic Induction Principle; Proportional Induction Principle; Structural linearization Principle

SEMIMARTINGALE Ambartsumian Invariance Principle; Charnov's Invariance Principle; Donsker's Invariance Principle; Invariance Principle; Invariance Principle of General Relativity; Local Invariance Principle; Maruyama's Invariance Principle; Perturbation Invariance Principle; Principle of Invariance; Principle of Relativity; Principle of Undulatory Invariance; Relaxed Invariance Principle; Scale Invariance Principle; Strong Principle; Variational Principle

SENSE ORGANS Critical Path Method Principle; Early Streamer Emission Principle; Sweet Principle; Eye Placement Principle; Gestalt Principle; Hopkins' Host Selection Principle; Host Selection Principle; Superposition Principle; Sweet Principle

SENSING DEVICES AND TRANSDUCERS Inference Principle; Jaynes' Principle; Kullback's Minimum Cross-entropy Principle; Maximum Information Entropy Principle; Measurement Principle; Measuring Principle; Minimax Entropy Principle; Minimum Energy Principle; Minimum Relative Entropy Principle; Modification of The Extended Maximum Entropy Principle; Principle of Maximum Entropy; Principle of Maximum Nonadditive Entropy; Principle of Observational Compatibility; Principle of Operational Compatibility; Schwinger's Quantum Action Principle; Uncertainty Principle

SENSITIVITY Equivalent Background Principle, Mueller-Breslau Principle; Pontryagins Principle

SENSORY RECEPTION Eye Placement Principle; Hopkins' Host Selection Principle; Host Selection Principle

SEQUENCES Relative Frequency Principle, Active Principle; Minimum Description Level Principle, New Therapeutic Principle; Maximum Entropy Principle

SERVOMECHANISM Ambartsumian Invariance Principle; Charnov's Invariance Principle; Donsker's Invariance Principle; Invariance Principle; Invariance Principle of General Relativity; Local Invariance Principle; Maruyama's Invariance Principle; Perturbation Invariance Principle; Principle of Invariance; Principle of Relativity; Principle of Undulatory Invariance; Relaxed Invariance Principle; Scale Invariance Principle; Strong Principle; Variational Principle; Minimum Information Principle

SESTIMATOR MULTIVARIATE LOCATION Ambartsumian Invariance Principle; Charnov's Invariance Principle; Donsker's Invariance Principle; Invariance Principle; Invariance Principle of General Relativity; Local

Invariance Principle; Maruyama's Invariance Principle; Perturbation Invariance Principle; Principle of Invariance; Principle of Relativity; Principle of Undulatory Invariance; Relaxed Invariance Principle; Scale Invariance Principle; Strong Principle; Variational Principle

SETS Comparison Principle; Tnorm Extension Principle, Contraction Principle; Gibbs Variational Principle; Large Deviation Principle; Maximum Entropy Principle; Minimum Description Length Principle; Transfert Principle

SEXUAL SELECTION Bateman's Principle; Entropy Principle; Fuzzy Entropy Principle; Handicap Principle; Maxent Principle; Maximum Entropy Principle; Mep Maximum Entropy Principle; Mermin Entropy Principle; Principle of Maximum Entropy; Principle of Maximum Nonadditive Entropy; Principle of Operational Compatibility

SHANNON INFORMATION THEORY Correspondence Principle; Energy Preservation Principle; Entropy Maximum Principle; Extended Maximum Entropy Principle; Extinguishing Principle; Extremal Principle of Information; Frieden-Soffer Principle; Gibbs Principle; Global Extremum Principle; Inference Principle; Jaynes' Maximum Entropy Principle; Jaynes' Principle; Kullback's Minimum Cross-entropy Principle; Maximum Information Entropy Principle; Measurement Principle; Minimax Entropy Principle; Minimum Entropy Principle; Minimum Relative Entropy Principle; Modification of The Extended Maximum Entropy Principle; Principle of Maximum Entropy; Principle of Maximum Nonadditive Entropy; Principle of Observational Compatibility; Uncertainty Principle

SHAPE AND POSITION ERROR Inference Principle; Jaynes' Principle; Kullback's Minimum Cross-entropy Principle; Maximum Information Entropy Principle; Measurement Principle; Measuring Principle; Minimax Entropy Principle; Minimum Energy Principle; Minimum Relative Entropy Principle; Modification of The Extended Maximum Entropy Principle; Principle of Maximum

Entropy; Principle of Maximum Nonadditive Entropy; Principle of Observational Compatibility; Principle of Operational Compatibility; Schwinger's Quantum Action Principle; Uncertainty Principle

SHAPE FACTOR Priming Principle

SHAPING MECHANISM Caratheodory's Principle; Conte's Principle; Least Time Principle; Principle of Caratheodony; Stationary Time Principle

SHIELDED ENCLOSURE Black Hole Principle; Optical Black Hole Principle; Reflectionless Aperture Principle

SIGNAL PROCESSING Active Principle; Correspondence Principle; Energy Preservation Principle; Entropy Maximum Principle; Extended Maximum Entropy Principle; Extinguishing Principle; Extremal Principle of Information; Frieden-Soffer Principle; Gibbs Principle; Global Extremum Principle; Hamilton's Principle; Inference Principle; Jaynes' Maximum Entropy Principle; Jaynes' Principle; Kirchhoff's Voltage Law Principle; Kullback's Minimum Cross-entropy Principle; Maximum Entropy Principle; Maximum Information Entropy Principle; Maximum Information Entropy Principle; Measurement Principle; Measuring Principle; Minimax Entropy Principle; Minimum Energy Principle; Minimum Entropy Principle; Minimum Norm Principle; Minimum Relative Entropy Principle; Modification of The Extended Maximum Entropy Principle; Principle of Laser Speckle Flowgraphy; Principle of Maximum Entropy; Principle of Maximum Entropy; Principle of Maximum Nonadditive Entropy; Principle of Maximum Nonadditive Entropy; Principle of Observational Compatibility; Principle of Operational Compatibility; Principle of Persurvivor Processing; Principles of Darcy's Law; Schwinger's Quantum Action Principle; Uncertainty Principle; Variational Principle; Woltjer Minimum Energy Principle; Uncertainty Principle

SIGNAL PROCESSING EQUIPMENT Inference Principle; Jaynes' Principle; Kullback's Minimum Cross-entropy Principle; Maximum

Information Entropy Principle; Measurement Principle; Measuring Principle; Minimax Entropy Principle; Minimum Energy Principle; Minimum Relative Entropy Principle; Modification of The Extended Maximum Entropy Principle; Principle of Maximum Entropy; Principle of Maximum Nonadditive Entropy; Principle of Observational Compatibility; Principle of Operational Compatibility; Schwinger's Quantum Action Principle; Uncertainty Principle

SIMULATED ANNEALING Correspondence Principle; Energy Preservation Principle; Entropy Maximum Principle; Extended Maximum Entropy Principle; Extinguishing Principle; Extremal Principle of Information; Frieden-Soffer Principle; Gibbs Principle; Global Extremum Principle; Inference Principle; Jaynes' Maximum Entropy Principle; Jaynes' Principle; Kullback's Minimum Cross-entropy Principle; Maximum Information Entropy Principle; Measurement Principle; Minimax Entropy Principle; Minimum Entropy Principle; Minimum Relative Entropy Principle; Modification of The Extended Maximum Entropy Principle; Orthogonality Principle; Principle of Maximum Entropy; Principle of Maximum Nonadditive Entropy; Principle of Observational Compatibility; Uncertainty Principle

SIMULATING FLUIDS Correspondence Principle; Energy Preservation Principle; Entropy Maximum Principle; Extended Maximum Entropy Principle; Extinguishing Principle; Extremal Principle of Information; Frieden-Soffer Principle; Gibbs Principle; Global Extremum Principle; Inference Principle; Jaynes' Maximum Entropy Principle; Jaynes' Principle; Kullback's Minimum Cross-entropy Principle; Maximum Information Entropy Principle; Measurement Principle; Minimax Entropy Principle; Minimum Entropy Principle; Minimum Relative Entropy Principle; Modification of The Extended Maximum Entropy Principle; Principle of Maximum Entropy; Principle of Maximum Nonadditive Entropy; Principle of Observational Compatibility; Uncertainty Principle

SIMULATION MODEL Correspondence Principle; Energy Preservation Principle; Entropy

Maximum Principle; Extended Maximum Entropy Principle; Extinguishing Principle; Extremal Principle of Information; Frieden-Soffer Principle; Gibbs Principle; Global Extremum Principle; Inference Principle; Jaynes' Maximum Entropy Principle; Jaynes' Principle; Kullback's Minimum Cross-entropy Principle; Maximum Information Entropy Principle; Measurement Principle; Minimax Entropy Principle; Minimum Entropy Principle; Minimum Relative Entropy Principle; Modification of The Extended Maximum Entropy Principle; Principle of Maximum Entropy; Principle of Maximum Nonadditive Entropy; Principle of Observational Compatibility; Uncertainty Principle; Virtual Work Principle

SINGLE PHOTON EMISSION (SPECT) Frequency-dependent Principle; Correspondence Principle; Energy Preservation Principle; Entropy Maximum Principle; Extended Maximum Entropy Principle; Extinguishing Principle; Extremal Principle of Information; Frieden-Soffer Principle; Gibbs Principle; Global Extremum Principle; Inference Principle; Jaynes' Maximum Entropy Principle; Jaynes' Principle; Kullback's Minimum Cross-entropy Principle; Maximum Information Entropy Principle; Measurement Principle; Minimax Entropy Principle; Minimum Entropy Principle; Minimum Relative Entropy Principle; Modification of The Extended Maximum Entropy Principle; Principle of Maximum Entropy; Principle of Maximum Nonadditive Entropy; Principle of Observational Compatibility; Uncertainty Principle

SIX MEMBERED COMPOUND First Principle; Sizecavalieri Principle; Correspondence Principle; Handicap Principle; Stereological Principle; Unified Principle

SIZE DEPENDENCE Investment Principle; John Templeton Principle; Principle of Maximum Pessimism

SIZES Correspondence Principle; Energy Preservation Principle; Entropy Maximum Principle; Extended Maximum Entropy Principle; Extinguishing Principle; Extremal Principle of Information; Frieden-Soffer Principle; Gibbs Principle; Global Extremum

Principle; Inference Principle; Jaynes' Maximum Entropy Principle; Jaynes' Principle; Kullback's Minimum Cross-entropy Principle; Maximum Information Entropy Principle; Measurement Principle; Minimax Entropy Principle; Minimum Entropy Principle; Minimum Relative Entropy Principle; Modification of The Extended Maximum Entropy Principle; Principle of Maximum Entropy; Principle of Maximum Nonadditive Entropy; Principle of Observational Compatibility; Uncertainty Principle

SKELETAL SYSTEM Three-point Principle, Archimedes' Principle; Hydrostatic Principle; Unlearned Principle

SLIGHTLY ROUGH FACET MODEL Conservation of Momentum Principle; D'Alembert Principle; Differential Principle; Hamilton Principle; Newton's Third Principle; Principle of Equal A Priori Probabilities; Variational Principle

SLOPE Principle of Maximum Entropy, Priming Principle; Small Angle Light Scattering; Superposition Principle

SMALL ANGLE SCATTERING Caratheodory's Principle; Conte's Principle; Fermat's Principle; Least Time Principle; Mitrofanoff Principle; Proof of Principle; Stationary Time Principle

SOBOLEV SPACES Contraction Principle; Contraction Principle; Gibbs Variational Principle; Large Deviation Principle, Conte's Principle; Fermat's Principle; Least Time Principle; Mitrofanoff Principle; Proof of Principle; Stationary Time Principle

SOCIAL CONTRACTS Ellul Principle of Efficiency; Emancipation Principle; Kant's Principle; Power Principle; Principle of Domination; Principle of Efficiency; Principle of Emancipation; Principle of Free Domain; Principle of Majority Rule Based On Decision Making Power Demand; Principle of Power; Principle of Subjective Freedom; Principle of The Authomomy of The Will

SOIL MECHANICS Correspondence Principle; Hamilton's Principle; Kirchhoff's Voltage Law Principle; Minimum Energy Principle; Minimum Norm Principle; Variational Principle; Woltjer Minimum Energy Principle

SOLAR CORONA Hamilton's Principle; Kirchhoff's Voltage Law Principle; Minimum Energy Principle; Minimum Norm Principle; Variational Principle; Woltjer Minimum Energy Principle

SOLAR MAGNETIC FIELD Hamilton's Principle; Kirchhoff's Voltage Law Principle; Minimum Energy Principle; Minimum Norm Principle; Variational Principle; Woltjer Minimum Energy Principle

SOLIDS Bond Directional Principle; Maximizing Principle, Global Extremum Principle

SOLUTIONS OF WAVE EQUATIONS Competitive Exclusion Principle; Principle of Complementarity; Correspondence Principle

SPACE Cosmic Principle; Maximum Hardness Principle, Principle of Relativity

SPACE TIME Correspondence Principle; Galileo Principle; Hamilton's Principle; Principle of molecular dynamics; Spacetime Uncertainty Principle; Huygen's Principle; Unified Principle; Principle of Causality

SPACECRAFT DYNAMICS D'Alembert's Principle; Gyarmati's Principle; Hamilton Principle; Prigogine Principle; Principle of Minimum Kinetic Energy; Principle of Minimum Mechanical Energy; Principle of Minimum Potential Energy; Quantum Hamilton Principle; Stewart-Hamilton Principle; Stochastic Hamilton Variational Principle; Toupin's Dual Principle

SPACES Founder Principle; Fundamental Principle, Korenblum's Maximum Principle, Minimum Principle

SPATIAL SMOOTHING Active Principle; Principle of Corresponding States; Quartz Microbalance Principle

SPATIAL VARIABLES MEASUREMENT Inference Principle; Jaynes' Principle; Kullback's Minimum Cross-entropy Principle; Maximum Information Entropy Principle; Measurement Principle; Measuring Principle; Minimax Entropy Principle; Minimum Energy Principle; Minimum Relative

Principle; Inference Principle; Jaynes' Maximum Entropy Principle; Jaynes' Principle; Kullback's Minimum Cross-entropy Principle; Maximum Information Entropy Principle; Measurement Principle; Minimax Entropy Principle; Minimum Entropy Principle; Minimum Relative Entropy Principle; Modification of The Extended Maximum Entropy Principle; Principle of Maximum Entropy; Principle of Maximum Nonadditive Entropy; Principle of Observational Compatibility; Uncertainty Principle

SPIRITUAL RELIEF Diagnostic Principle; Principle of Psychoneuroimmunology; Quasilocal Principle; Staff's Diagnostic Principle

SPONTANEOUS LOCALIZATIONS Correspondence Principle, Unlearned Principle, Stationary Time Principle

SPRAY NOZZLES Correspondence Principle; Energy Preservation Principle; Entropy Maximum Principle; Extended Maximum Entropy Principle; Extinguishing Principle; Extremal Principle of Information; Frieden-Soffer Principle; Gibbs Principle; Global Extremum Principle; Inference Principle; Jaynes' Maximum Entropy Principle; Jaynes' Principle; Kullback's Minimum Cross-entropy Principle; Maximum Information Entropy Principle; Measurement Principle; Minimax Entropy Principle; Minimum Entropy Principle; Minimum Relative Entropy Principle; Modification of The Extended Maximum Entropy Principle; Principle of Maximum Entropy; Principle of Maximum Nonadditive Entropy; Principle of Observational Compatibility; Uncertainty Principle

SPURIOUS SIGNAL NOISE Frequency Distance Principle; Multi Tanh Principle; Action Principle; Maximum Likelihood Principle

STABILITY Comparison Principle; Internal Model Principle; Saint-Venant's Principle, Principle of Material Frame Indifference; Principle of Minimum Dissipation

STABILITY OF VARIATION Ekeland's variational Principle; Ekeland's Epsilon Variational Principle; Ekeland's Variational Principle; Global Maximum Principle; Variational Principle

STABILIZATION Ambartsumian Invariance Principle; Charnov's Invariance Principle; Comparison Principle; Donsker's Invariance Principle; Invariance Principle; Invariance Principle of General Relativity; Local Invariance Principle; Maruyama's Invariance Principle; Perturbation Invariance Principle; Principle of Dominant Subsystems; Principle of Invariance; Principle of Relativity; Principle of Undulatory Invariance; Relaxed Invariance Principle; Scale Invariance Principle; Strong Principle; Variational Principle, Handicap Principle

STABLE PROCESS Ambartsumian Invariance Principle; Charnov's Invariance Principle; Donsker's Invariance Principle; Invariance Principle; Invariance Principle of General Relativity; Local Invariance Principle; Maruyama's Invariance Principle; Perturbation Invariance Principle; Principle of Invariance; Principle of Relativity; Principle of Undulatory Invariance; Relaxed Invariance Principle; Scale Invariance Principle; Strong Principle; Variational Principle

STATE ESTIMATION Equivalence Principle; Maximum Entropy Principle; New Therapeutic Principle

STATIONARY ERGODIC AND REVERSIBLE MEASURES Invariance Principle; Ambartsumian Invariance Principle; Charnov's Invariance Principle; Donsker's Invariance Principle; Invariance Principle; Invariance Principle of General Relativity; Local Invariance Principle; Maruyama's Invariance Principle; Perturbation Invariance Principle; Principle of Invariance; Principle of Relativity; Principle of Undulatory Invariance; Relaxed Invariance Principle; Scale Invariance Principle; Strong Principle; Variational Principle

STATIONARY EULER FLOWS Hamilton's Principle; Kirchhoff's Voltage Law Principle; Minimum Energy Principle; Minimum Norm Principle; Variational Principle; Woltjer Minimum Energy Principle

STATIONARY SEQUENCE Ambartsumian Invariance Principle; Charnov's Invariance Principle; Donsker's Invariance Principle; Invariance Principle; Invariance Principle of General Relativity; Local Invariance Principle; Maruyama's Invariance Principle; Perturbation Invariance Principle; Principle of Invariance; Principle of Relativity; Principle of Undulatory Invariance; Relaxed Invariance Principle; Scale Invariance Principle; Strong Principle; Variational Principle

STATISTICAL DECISION Correspondence Principle; Energy Preservation Principle; Entropy Maximum Principle; Extended Maximum Entropy Principle; Extinguishing Principle; Extremal Principle of Information; Frieden-Soffer Principle; Gibbs Principle; Global Extremum Principle; Inference Principle; Jaynes' Maximum Entropy Principle; Jaynes' Principle; Kullback's Minimum Cross-entropy Principle; Maximum Entropy Principle; Maximum Information Entropy Principle; Measurement Principle; Minimax Entropy Principle; Minimum Entropy Principle; Minimum Relative Entropy Principle; Modification of The Extended Maximum Entropy Principle; Principle of Maximum Entropy; Principle of Maximum Nonadditive Entropy; Principle of Observational Compatibility; Uncertainty Principle

STATISTICAL DISTRIBUTION Competitive Exclusion Principle; Principle of Complementarity; Powdering Principle

STATISTICAL ENSEMBLE Quantum Linearity Principle; Spacetime Uncertainty Principle

STATISTICAL ESTIMATION Correspondence Principle; Energy Preservation Principle; Entropy Maximum Principle; Extended Maximum Entropy Principle; Extinguishing Principle; Extremal Principle of Information; Frieden-Soffer Principle; Gibbs Principle; Global Extremum Principle; Inference Principle; Jaynes' Maximum Entropy Principle; Jaynes' Principle; Kullback's Minimum Cross-entropy Principle; Maximum Entropy Principle; Maximum Information Entropy Principle; Measurement Principle; Minimax Entropy Principle; Minimum Entropy Principle; Minimum Relative Entropy Principle; Modification of The Extended Maximum Entropy Principle; Principle of Maximum

Entropy; Principle of Maximum Nonadditive Entropy; Principle of Observational Compatibility; Uncertainty Principle

STATISTICAL INFERENCE ENGINE Correspondence Principle; Energy Preservation Principle; Entropy Maximum Principle; Extended Maximum Entropy Principle; Extinguishing Principle; Extremal Principle of Information; Frieden-Soffer Principle; Gibbs Principle; Global Extremum Principle; Inference Principle; Jaynes' Maximum Entropy Principle; Jaynes' Principle; Kullback's Minimum Cross-entropy Principle; Maximum Information Entropy Principle; Measurement Principle; Minimax Entropy Principle; Minimum Entropy Principle; Minimum Relative Entropy Principle; Modification of The Extended Maximum Entropy Principle; Principle of Maximum Entropy; Principle of Maximum Nonadditive Entropy; Principle of Observational Compatibility; Uncertainty Principle

STATISTICAL MECHANICS Action Principle; Bogolyubov Principle; Bogolyubov Variational Principle; Correspondence Principle; Gibbs-Bogolyubov Variational Principle; Gibbs Third Variational Principle; Hamilton's Principle; Hartree Fock Bogolyubov Principle; Maupertuis Principle; Maupertuis-Jacobi Principle; Maximum Likelihood Principle; Priming Principle; Quantal Principle

STATISTICAL METHOD Maximum Entropy Principle; Adaptive Principle; Correspondence Principle; Energy Preservation Principle; Entropy Maximum Principle; Extended Maximum Entropy Principle; Extinguishing Principle; Extremal Principle of Information; Frieden-Soffer Principle; Gibbs Principle; Global Extremum Principle; Inference Principle; Jaynes' Maximum Entropy Principle; Jaynes' Principle; Kullback's Minimum Cross-entropy Principle; Maupertuis Principle; Maximum Information Entropy Principle; Measurement Principle; Minimax Entropy Principle; Minimum Entropy Principle; Minimum Relative Entropy Principle; Modification of The Extended Maximum Entropy Principle; Neighborhood Coherence

Principle; Optimization Principle; Optimizing Principle; Principle of Corresponding States; Principle of Maximum Entropy; Principle of Maximum Nonadditive Entropy; Principle of Observational Compatibility; Run and Recreate Principle; Stationary Time Principle; Uncertainty Principle; Contraction Principle; Contraction Principle; Gibbs Variational Principle; Large Deviation Principle; Memory Compensation Principle; Mitrofanoff Principle; Monti Principle, Reversed Fick Principle

STATISTICAL MODEL Active Principle; Correspondence Principle; Energy Preservation Principle; Entropy Maximum Principle; Extended Maximum Entropy Principle; Extinguishing Principle; Extremal Principle of Information; Frieden-Soffer Principle; Gibbs Principle; Global Extremum Principle; Inference Principle; Jaynes' Maximum Entropy Principle; Jaynes' Principle; Kullback's Minimum Cross-entropy Principle; Maximum Information Entropy Principle; Measurement Principle; Minimax Entropy Principle; Minimum Entropy Principle; Minimum Relative Entropy Principle; Modification of The Extended Maximum Entropy Principle; Principle of Maximum Entropy; Principle of Maximum Nonadditive Entropy; Principle of Observational Compatibility; Uncertainty Principle

STATISTICAL THERMODYNAMICS Corresponding States Principle; Minimal Principle; Pitzer Coresponding States Principle;

STATISTICS AND PROBABILITY Contraction Principle; Gibbs Variational Principle; Large Deviation Principle; Ambartsumian Invariance Principle; Averaging Principle; Charnov's Invariance Principle; Charnov's Invariance Principle; Correspondence Principle; Likelihood Principle; Maximum Entropy Principle; Sufficiency Principle; Donsker's Invariance Principle; Invariance Principle; Invariance Principle of General Relativity; Local Invariance Principle; Maruyama's Invariance Principle; Perturbation Invariance Principle; Principle of Correspondence; Principle of Invariance; Principle of Relativity; Principle of Undulatory Invariance; Relaxed Invariance Principle; Scale Invariance Principle; Strong Principle; Variational Principle

STEADYSTATE ELECTROMAGNETIC FIELDS Projection Principle; Maximum Hardness Principle

STELLAR ATMOSPHERE Ambartsumian Invariance Principle; Charnov's Invariance Principle; Donsker's Invariance Principle; Invariance Principle of General Relativity; Local Invariance Principle; Maruyama's Invariance Principle; Perturbation Invariance Principle; Principle of Correspondence; Principle of Invariance; Principle of Relativity; Principle of Undulatory Invariance; Relaxed Invariance Principle; Scale Invariance Principle; Strong Principle; Variational Principle; Hamilton's Principle; Kirchhoff's Voltage Law Principle; Minimum Energy Principle; Minimum Norm Principle; Variational Principle; Woltjer Minimum Energy Principle

STELLAR CORONAE Hamilton's Principle; Kirchhoff's Voltage Law Principle; Minimum Energy Principle; Minimum Norm Principle; Variational Principle; Woltjer Minimum Energy Principle

STELLAR MAGNETIC FIELDS Hamilton's Principle; Kirchhoff's Voltage Law Principle; Minimum Energy Principle; Minimum Norm Principle; Variational Principle; Woltjer Minimum Energy Principle

STELLAR MAGNETISM Hamilton's Principle; Kirchhoff's Voltage Law Principle; Minimum Energy Principle; Minimum Norm Principle; Variational Principle; Woltjer Minimum Energy Principle

STELLARATOR Protection Principle; Step; Satellite Test of The Equivalence Principle; Principle of Molecular Dynamics

STEREOLOGY Cavalieri Principle; Stereological Principle, Principle of Stereoscopic Image

STIMULUS Economic Principle; Principles of Determination; Principles of Multiple Function; Principles of Overdetermination; Regulatory Principles and Other Regulatory Principles of Inertia; Regulatory Principles of Constancy; Regulatory Principles of Pleasure Unpleasure; Regulatory Principles of Primarysecondary Process; Regulatory Principles of Reality; Regulatory Principles of Repetition, Maximum Entropy Principle

STOCHASTIC CALCULUS Maupertuis Principle, Active Principle; Minimum Description Level Principle, Principle of Large Deviations, Equivalence Principle, Precautionary Principle.

STOCHASTIC INTEGRAL Maximality Principle; Ambartsumian Invariance Principle; Charnov's Invariance Principle; Donsker's Invariance Principle; Invariance Principle; Invariance Principle of General Relativity; Local Invariance Principle; Maruyama's Invariance Principle; Perturbation Invariance Principle; Principle of Correspondence; Principle of Invariance; Principle of Relativity; Principle of Undulatory Invariance; Relaxed Invariance Principle; Scale Invariance Principle; Strong Principle; Variational Principle

STOCHASTIC MODEL Correspondence Principle; Energy Preservation Principle; Entropy Maximum Principle; Extended Maximum Entropy Principle; Extinguishing Principle; Extremal Principle of Information; Frieden-Soffer Principle; Gibbs Principle; Global Extremum Principle; Inference Principle; Jaynes' Maximum Entropy Principle; Jaynes' Principle; Kullback's Minimum Cross-entropy Principle; Maximum Information Entropy Principle; Measurement Principle; Minimax Entropy Principle; Minimum Entropy Principle; Minimum Relative Entropy Principle; Modification of The Extended Maximum Entropy Principle; Principle of Maximum Entropy; Principle of Maximum Nonadditive Entropy; Principle of Observational Compatibility; Uncertainty Principle; Memory Compensation Principle; Mitrofanoff Principle; Monti Principle

STOCHASTIC PROCESS Ambartsumian Invariance Principle; Charnov's Invariance Principle; Contraction Principle; Donsker's Invariance Principle; Gibbs Variational Principle; Invariance Principle; Invariance Principle of General Relativity; Large Deviation Principle; Local Invariance Principle; Maruyama's Invariance Principle; Maximality Principle; Perturbation Invariance

Principle; Principle of Correspondence; Principle of Invariance; Principle of Relativity; Principle of Undulatory Invariance; Relaxed Invariance Principle; Scale Invariance Principle; Strong Principle; Variational Principle; Correspondence Principle; Energy Preservation Principle; Entropy Maximum Principle; Extended Maximum Entropy Principle; Extinguishing Principle; Extremal Principle of Information; Frieden-Soffer Principle; Gibbs Principle; Global Extremum Principle; Inference Principle; Jaynes' Maximum Entropy Principle; Jaynes' Principle; Kullback's Minimum Cross-entropy Principle; Maximum Information Entropy Principle; Measurement Principle; Minimax Entropy Principle; Minimum Entropy Principle; Minimum Relative Entropy Principle; Modification of The Extended Maximum Entropy Principle; Principle of Maximum Entropy; Principle of Maximum Nonadditive Entropy; Principle of Observational Compatibility; Uncertainty Principle

STRAIN Dynamic Principle; Dynamic Principle of Osteosynthesis, Conditionality Principle

STRAIN RATE Principle of Material Frame Indifference; Principle of Minimum Dissipation

STRATIFIED MATERIAL D'Alembert's Principle; Equivalence Principle; Gyarmati's Principle; Hamilton Principle; Prigogine Principle; Principle of Minimum Kinetic Energy; Principle of Minimum Mechanical Energy; Principle of Minimum Potential Energy; Quantum Hamilton Principle; Stewart-Hamilton Principle; Stochastic Hamilton Variational Principle; Toupin's Dual Principle

STREAMS Conservation of Momentum Principle; D'Alembert Principle; Differential Principle; Hamilton Principle; Newton's Third Principle; Principle of Equal A Priori Probabilities; Variational Principle

STRESS ANALYSIS Equivalence Principle; Least Action Principle; Least Square Principle; Minimum Principle; Mueller-Breslau Principle; Principle of Least Work; Virtual Work Principle, Principle of Caratheodory

STRESS FUNCTION Hellinger-Reissner Principle; Hu Washizu Variational Principle; Pseudogeneralized Variational Principle

STRESS REDUCING Displacement Principle; Element Displacement Principle; Principle of Displacement; Roots Principle; Virtual Displacement Principle

STRESS STRAIN RELATIONSHIPS Three-day Principle; Educational Principle; Ehrenfest Principle; Principle of Material Frame Indifference

STRONG LASER PLASMA Caratheodory's Principle; Conte's Principle; Fermat's Principle; Least Time Principle; Mitrofanoff Principle; Principle of Measurement; Proof of Principle

STRONG RESONANCE Drug Principle

STRUCTURAL ANALYSIS Least Action Principle; Mueller-Breslau Principle; Principle of Least Work, Fermats Principle; Orthogonality Principle

STRUCTURE Active Principle; Correspondence Principle; Energy Preservation Principle; Entropy Maximum Principle; Extended Maximum Entropy Principle; Extinguishing Principle; Extremal Principle of Information; Frieden-Soffer Principle; Gibbs Principle; Global Extremum Principle; Inference Principle; Jaynes' Maximum Entropy Principle; Jaynes' Principle; Kullback's Minimum Cross-entropy Principle; Maximum Information Entropy Principle; Measurement Principle; Minimax Entropy Principle; Minimum Entropy Principle; Minimum Relative Entropy Principle; Modification of The Extended Maximum Entropy Principle; Principle of Maximum Entropy; Principle of Maximum Nonadditive Entropy; Principle of Observational Compatibility; Uncertainty Principle

SUBJUGATION Ellul Principle of Efficiency; Emancipation Principle; Kant's Principle; Power Principle; Principle of Domination; Principle of Efficiency; Principle of Emancipation; Principle of Free Domain; Principle of Majority Rule Based On Decision Making Power Demand; Principle of Power; Principle of Subjective Freedom; Principle of The Authomomy of The Will

SUBORDINATION Extremum Principle; Phragmen-Lindelof Principle; Schwarz Reflection Principle; Korenblum's Maximum Principle

SUBSTITUENT CONSTANTS Electronegativity Equalization Principle; Geometric Mean Principle; Principle of Electronegativity

SUBTRACTION TECHNIQUE Correspondence Principle; Energy Preservation Principle; Entropy Maximum Principle; Extended Maximum Entropy Principle; Extinguishing Principle; Extremal Principle of Information; Frieden-Soffer Principle; Gibbs Principle; Global Extremum Principle; Inference Principle; Jaynes' Maximum Entropy Principle; Jaynes' Principle; Kullback's Minimum Cross-entropy Principle; Maximum Information Entropy Principle; Measurement Principle; Minimax Entropy Principle; Minimum Entropy Principle; Minimum Relative Entropy Principle; Modification of The Extended Maximum Entropy Principle; Principle of Maximum Entropy; Principle of Maximum Nonadditive Entropy; Principle of Observational Compatibility; Uncertainty Principle

SUCCINYL CHOLINE Working Principle; Type II Cell Stretch-organizing Principle, Optimality Principle; Organizing Principle; Rinciple Component Analysis. Self-organizing Principle; Similarity Principle; Timing Principle

SUN Antenna Principle; Hamilton's Principle; Kirchhoff's Voltage Law Principle; Kirchhoff's Voltage Law Principle; Minimum Energy Principle; Minimum Energy Principle; Minimum Norm Principle; Variational Principle; Woltjer Minimum Energy Principle

SUPER CRITICAL CARBON DIOXIDE Corresponding States Principle; Pitzer Corresponding States Principle;

SURFACE STRUCTURE Correspondence Principle; Dirichlet Principle; Maximizing Principle; Surface Tensioncorrespondence Principle; Priming Principle; Minimal Principle

T

TAGGED PARTICLE Ambartsumian Invariance Principle; Charnov's Invariance Principle; Donsker's Invariance Principle; Invariance Principle; Invariance Principle of General Relativity; Local Invariance Principle; Maruyama's Invariance Principle; Perturbation Invariance Principle; Principle of Correspondence; Principle of Invariance; Principle of Relativity; Principle of Undulatory Invariance; Relaxed Invariance Principle; Scale Invariance Principle; Strong Principle; Variational Principle

TEACHING Correspondence Principle; Indeterminacy Principle; Maximum Hardness Principle; Principle Molecular Dynamics

TECHNOLOGY Ellul Principle of Efficiency; Emancipation Principle; Kant's Principle; Power Principle; Principle of Domination; Principle of Efficiency; Principle of Emancipation; Principle of Free Domain; Principle of Majority Rule Based On Decision Making Power Demand; Principle of Power; Principle of Subjective Freedom; Principle of The Authomomy of The Will, Correspondence Principle

TELECOMMUNICATION CHANNEL Faustmann Principle, Principle of Per-survivor Processing; Principles of Darcy's Law

TELEOLOGY Anthropic Cosmological Principle; Anthropic Principle; Cosmological Principle; Telescopical Principles; Maximum Hardness Principle

TEMPERATURE Analytical Principle; Priming Principle, Correspondence Principle; First Principle; Projection Principle, Multi Bypass Principle; Principle of Caratheodony, Phase Angle Principle

TENSORS Principle of Molecular Dynamics; Principle of Material Frame Indifference; Principle of Minimum Dissipation

TERNARY LOGIC Correspondence Principle; Energy Preservation Principle; Entropy Maximum Principle; Extended Maximum Entropy Principle; Extinguishing Principle; Extremal Principle of Information; Frieden-Soffer Principle; Gibbs Principle; Global Extremum Principle; Inference Principle; Jaynes' Maximum Entropy Principle; Jaynes' Principle; Kullback's Minimum Cross-entropy Principle; Maximum Information Entropy Principle; Measurement Principle; Minimax Entropy Principle; Minimum Entropy Principle; Minimum Relative Entropy Principle; Modification of The Extended Maximum Entropy Principle; Principle of Maximum Entropy; Principle of Maximum Nonadditive Entropy; Principle of Observational Compatibility; Uncertainty Principle

TEST PROVISION Inference Principle; Jaynes' Principle; Kullback's Minimum Cross-entropy Principle; Maximum Information Entropy Principle; Measurement Principle; Measuring Principle; Minimax Entropy Principle; Minimum Energy Principle; Minimum Relative Entropy Principle; Modification of The Extended Maximum Entropy Principle; Principle of Maximum Entropy; Principle of Maximum Nonadditive Entropy; Principle of Observational Compatibility; Principle of Operational Compatibility; Schwinger's Quantum Action Principle; Uncertainty Principle

TEST STATISTIC Ambartsumian Invariance Principle; Charnov's Invariance Principle; Donsker's Invariance Principle; Invariance Principle; Invariance Principle of General Relativity; Local Invariance Principle; Maruyama's Invariance Principle; Perturbation Invariance Principle; Principle of Correspondence; Principle of Invariance; Principle of Relativity; Principle of Undulatory Invariance; Relaxed Invariance Principle; Scale Invariance Principle; Strong Principle; Variational Principle

TESTING METHOD Inference Principle; Jaynes' Principle; Kullback's Minimum Cross-entropy Principle; Maximum Information Entropy Principle; Measurement Principle; Measuring Principle; Minimax Entropy Principle; Minimum Energy Principle; Minimum Relative Entropy Principle; Modification of The Extended Maximum Entropy Principle; Principle of Maximum Entropy; Principle of Maximum Nonadditive Entropy; Principle of Observational Compatibility; Principle of Operational Compatibility; Schwinger's Quantum Action Principle; Uncertainty Principle

THEOREM PROVING Active Principle; Contraction Principle; Denseflow Conveying Principle; Ekeland's Principle; Extended Resolution Principle; Gibbs Variational Principle; Large Deviation Principle; Resolution Principle

THEOREMS Deviation Principle; Uncertainty Principle, Tnorm Extension Principle

THEORETICAL STUDY Action Principle; Caratheodory's Principle; Ckvp; Competitive Exclusion Principle; Complex Kohn Variational Principle; Fermat's Principle; Maximum Hardness Principle; Hulthén–Kohn Variational Principle; Kohn Variational Principle; Least Time Principle; Maximizing Principle; Maximum Likelihood Principle; Minimum Principle; Mitrofanoff Principle; Principle of Caratheodony; Principle of Causality; Principle of Complementarity; Principle of Dominant Subsystems; Principle of Material Frame Indifference; Principles of Darcy's Law; Principle of Molecular Dynamics; Principle of Congruence; Principle of Measurement; Proof of Principle; Schwinger Principle; Schwinger Variational Principle; Stationary Action Principle

THEORETICAL STUDY GALACTIC HALOS Caratheodory's Principle; Conte's Principle; Fermat's Principle; Least Time Principle; Mitrofanoff Principle; Principle of Measurement; Proof of Principle

THEORY AND MODELS OF CHAOTIC SYSTEMS Correspondence Principle; Quaternion Moment Preserving Principle

THEORY OF CONTROLLED SYSTEMS Anistropic Principle; Ariokhin Pareto Interaction Principle; E M Interaction Principle; Ground Support Interaction Principle; Interaction Equivalence Principle; Interaction Principle; Plus Minus Interaction Principle; Principle of Kuan His

THERAPEUTIC METHOD Mitrofanoff Principle; Monti Principle; Overlapping Biphasic Impulse Principle; Type II Cell Stretch-organizing Principle; Monti Principle

THERAPY Casuistic Principle; Constant Infusion Principle; Double Effect Principle; Exception Granting Principle; Justifying Principle; Mitrofanoff Principle; Monti Principle; Principle of Do No Harm; Principle of Double Effect; Principles of Biomedical Ethics; Mckenzie Principles; Principle-Centered Spine Care

THERMAL PROPERTIES Correspondence Principle; Energy Preservation Principle; Entropy Maximum Principle; Extended Maximum Entropy Principle; Extinguishing Principle; Extremal Principle of Information; Frieden-Soffer Principle; Gibbs Principle; Global Extremum Principle; Inference Principle; Jaynes' Maximum Entropy Principle; Jaynes' Principle; Kullback's Minimum Cross-entropy Principle; Maximum Information Entropy Principle; Measurement Principle; Minimax Entropy Principle; Minimum Entropy Principle; Minimum Relative Entropy Principle; Modification of The Extended Maximum Entropy Principle; Principle of Maximum Entropy; Principle of Maximum Nonadditive Entropy; Principle of Observational Compatibility; Uncertainty Principle

THERMODYNAMIC FUNCTIONS Competitive Exclusion Principle; Correspondence Principle; Principle of Complementarity; Fermats Principle

THERMODYNAMIC MODEL Corresponding States Principle; Pitzer Corresponding States Principle;

THERMODYNAMIC PROPERTIES Equipartition Principle; Principle of Congruence, Extended Principle

THERMODYNAMICS Consumer Principle; D'Alemberts Principle; Dissipation Principle; Equipartition Principle; Maximum Entropy Principle; Minimum Principle; Principle of Material Frame Indifference; Thermodynamic Principle; Virtual Dissipation Principle; Priming Principle; Minimum Principle

THICKNESS Cavalieri Principle; Maximum Entropy Principle; Stereological Principle, Functioning Principle

THREE DIMENSIONAL IMAGES Huygen's Equivalence Principle; Inverse Problem Theory Principle; Majority Principle; Minimum Principle; Orthogonality Principle; Triangulation Principle; Variational Principle

THREE DIMENSIONAL MODEL Energy Principle; Hamilton's Principle; Minimum Principle; Woltjer Minimum Energy Principle

THREE VALUED LOGIC Correspondence Principle; Energy Preservation Principle; Entropy Maximum Principle; Extended Maximum Entropy Principle; Extinguishing Principle; Extremal Principle of Information; Frieden-Soffer Principle; Gibbs Principle; Global Extremum Principle; Inference Principle; Jaynes' Maximum Entropy Principle; Jaynes' Principle; Kullback's Minimum Cross-entropy Principle; Maximum Information Entropy Principle; Measurement Principle; Minimax Entropy Principle; Minimum Entropy Principle; Minimum Relative Entropy Principle; Modification of The Extended Maximum Entropy Principle; Principle of Maximum Entropy; Principle of Maximum Nonadditive Entropy; Principle of Observational Compatibility; Uncertainty Principle

TIME EVOLUTION Principle of molecular dynamics, Correspondence Principle; Fermats Principle; Pome

TIME FACTORS Correspondence Principle; Energy Preservation Principle; Entropy Maximum Principle; Extended Maximum Entropy Principle; Extinguishing Principle; Extremal Principle of Information; Frieden-Soffer Principle; Gibbs Principle; Global Extremum Principle; Inference Principle;

Jaynes' Maximum Entropy Principle; Jaynes' Principle; Kullback's Minimum Cross-entropy Principle; Maximum Information Entropy Principle; Measurement Principle; Minimax Entropy Principle; Minimum Entropy Principle; Minimum Relative Entropy Principle; Modification of The Extended Maximum Entropy Principle; Principle of Maximum Entropy; Principle of Maximum Nonadditive Entropy; Principle of Observational Compatibility; Uncertainty Principle

TIME RESOLVED FLUOROIMMUNOASSAY Inference Principle; Jaynes' Principle; Kullback's Minimum Cross-entropy Principle; Maximum Information Entropy Principle; Measurement Principle; Measuring Principle; Minimax Entropy Principle; Minimum Energy Principle; Minimum Relative Entropy Principle; Modification of The Extended Maximum Entropy Principle; Principle of Maximum Entropy; Principle of Maximum Nonadditive Entropy; Principle of Observational Compatibility; Principle of Operational Compatibility; Schwinger's Quantum Action Principle; Uncertainty Principle

TIMEDELAY SYSTEM Ekeland's variational Principle; Ekeland's Epsilon Variational Principl; Ekeland's Variational Principle; Global Maximum Principle; Variational Principle

TIMESERIES Parsimony Principle; Principle of Parsimonius Data Modeling, Active Principle; Timing Principle. Internalnoise Principle; Working Principle

TIMOSHENKO BEAM D'Alembert's Principle; Gyarmati's Principle; Hamilton Principle; Prigogine Principle; Principle of Minimum Kinetic Energy; Principle of Minimum Mechanical Energy; Principle of Minimum Potential Energy; Quantum Hamilton Principle; Stewart-Hamilton Principle; Stochastic Hamilton Variational Principle; Toupin's Dual Principle

TISSUE Inference Principle; Jaynes' Principle; Kullback's Minimum Cross-entropy Principle; Maximum Information Entropy Principle; Mckenzie Principles; Measurement Principle; Measuring Principle;

Minimax Entropy Principle; Minimum Energy Principle; Minimum Relative Entropy Principle; Modification of The Extended Maximum Entropy Principle; Principle of Corresponding States; Principle of Maximum Entropy; Principle of Maximum Nonadditive Entropy; Principle of Observational Compatibility; Principle of Operational Compatibility; Schwinger's Quantum Action Principle; Uncertainty Principle

TOLERABILITY Therapeutic Principle, Alarp Principle; As Low As Reasonably Practical Principle

TOMOGRAPHY Correspondence Principle; Functioning Principle; Weak Invariance Principle, Minimum Cross-entropy (Me) Principle; Principle of Computed Tomograph

TOPOLOGY Contraction Principle; Gibbs Variational Principle; Inertia Principle; Lagrange Principle of Least Action; Large Deviation Principle; Least Action Principle; Least Motion Principle; Principle of Least Motion

TOTAL CROSS SECTION Ckvp; Complex Kohn Variational Principle; HULTHÉN–KOHNVariational Principle; Kohn Variational Principle; Schwinger Variational Principle

TOTAL LINE CURRENT Hamilton's Principle; Kirchhoff's Voltage Law Principle; Minimum Energy Principle; Minimum Norm Principle; Variational Principle; Woltjer Minimum Energy Principle

TOXICOLOGY Immunosuppresive Principle; Optimization Principle; Run and Recreate Principle; Similia Principle of Homeopathy; Precautionary Principle.

TRACER DIFFUSION Ambartsumian Invariance Principle; Charnov's Invariance Principle; Donsker's Invariance Principle; Invariance Principle; Invariance Principle of General Relativity; Local Invariance Principle; Maruyama's Invariance Principle; Perturbation Invariance Principle; Principle of Correspondence; Principle of Invariance; Principle of Relativity; Principle of Undulatory Invariance; Relaxed Invariance Principle; Scale Invariance Principle; Strong Principle; Variational Principle

TRACHEA Drug Principle; Timing Principle; Type II Cell Stretch-organizing Principle; Working Principle

TRAINING Asymmetrical Matching Principle; Principle of Information Processing; Principle of Per-survivor Processing

TRAJECTORIES Ferm Principle; Inertia Principle; Lagrange Principle of Least Action; Least Action Principle; Least Motion Principle; Principle of Least Motion

TRANSFERABLE NET ATOMIC CHARGES Electronegativity Equalization Principle; Geometric Mean Principle; Principle of Electronegativity

TRANSFERFUNCTION MODELS Parsimony Principle; Principle of Parsimonius Data Modeling

TRANSFORMATIONS Competitive Exclusion Principle; Principle of Complementarity, Principle of Uncertainty, Correspondence Principle

TRANSFORMS Correspondence Principle; Minimal Principle; Transient. Machs Principle, Principle of Maximum Entropy

TRANSIENTS Inference Principle; Jaynes' Principle; Kullback's Minimum Cross-entropy Principle; Maximum Information Entropy Principle; Measurement Principle; Measuring Principle; Minimax Entropy Principle; Minimum Energy Principle; Minimum Relative Entropy Principle; Modification of The Extended Maximum Entropy Principle; Principle of Maximum Entropy; Principle of Maximum Nonadditive Entropy; Principle of Observational Compatibility; Principle of Operational Compatibility; Schwinger's Quantum Action Principle; Uncertainty Principle

TRANSITIVE CHAINING Correspondence Principle; Energy Preservation Principle; Entropy Maximum Principle; Extended Maximum Entropy Principle; Extinguishing Principle; Extremal Principle of Information; Frieden-Soffer Principle; Gibbs Principle; Global Extremum Principle; Inference Principle;

Jaynes' Maximum Entropy Principle; Jaynes' Principle; Kullback's Minimum Cross-entropy Principle; Maximum Information Entropy Principle; Measurement Principle; Minimax Entropy Principle; Minimum Entropy Principle; Minimum Relative Entropy Principle; Modification of The Extended Maximum Entropy Principle; Principle of Maximum Entropy; Principle of Maximum Nonadditive Entropy; Principle of Observational Compatibility; Uncertainty Principle

TRANSITIVITY Extension Principle; Majority Principle; Maximum Hardness Principle

TRANSMISSION LINES Hamilton's Principle; Kirchhoff's Voltage Law Principle; Minimum Energy Principle; Minimum Norm Principle; Variational Principle; Woltjer Minimum Energy Principle

TRANSMISSION SCALAR THEORY Caratheodory's Principle; Conte's Principle; Fermat's Principle; Least Time Principle; Mitrofanoff Principle; Principle of Maximum Entropy; Variational Principle; Woltjer Minimum Energy Principle; Principle of Measurement; Proof of Principle

TRANSPORT COEFFICIENTS Fermats Principle; Pome, Action Principle; Maximum Likelihood Principle, Correspondence Principle

TRANSVERSALITY CONDITION Pontryagins Principle, Hamiltonian Principle, Woltjer-Taylor Minimum Energy Principle

TURBIDITY Caratheodory's Principle; Conte's Principle; Fermat's Principle; Least Time Principle; Mitrofanoff Principle; Principle of Measurement; Proof of Principle

TWO DIMENSION Action Principle; Maximum Likelihood Principle, Nonlinear Principle, Kelvin Arnold Energy Principle, Minimum Principle, Extremum Principle, Principle of critical parameters

220 VOLT GRID Master-slave Principle

TWO POINT BOUNDARY VALUE Ekeland's variational Principle; Ekeland's Epsilon Variational Principle; Ekeland's Variational Principle; Global Maximum Principle; Variational Principle, Pontryagin Minimum Principle

U

ULTRASOFT XRAY SPECTRUM RECOVERY Principle of Corresponding States

ULTRASONIC APPLICATIONS Pulseecho Principle; Ultrasonic Measurement Principle; Principle of Caratheodony

ULTRASONIC SENSORS Inference Principle; Jaynes' Principle; Kullback's Minimum Cross-entropy Principle; Maximum Information Entropy Principle; Measurement Principle; Measuring Principle; Minimax Entropy Principle; Minimum Energy Principle; Minimum Relative Entropy Principle; Modification of The Extended Maximum Entropy Principle; Principle of Maximum Entropy; Principle of Maximum Nonadditive Entropy; Principle of Observational Compatibility; Principle of Operational Compatibility; Schwinger's Quantum Action Principle; Uncertainty Principle, Principle of Caratheodory

ULTRASONICS Capacitance Principle; Conductance Principle, Cross Check Principle, Principle of Caratheodony, Pulse Echo Principle

UNCERTAIN EVIDENCE Correspondence Principle; Energy Preservation Principle; Entropy Maximum Principle; Extended Maximum Entropy Principle; Extinguishing Principle; Extremal Principle of Information; Frieden-Soffer Principle; Gibbs Principle; Global Extremum Principle; Inference Principle; Jaynes' Maximum Entropy Principle; Jaynes' Principle; Kullback's Minimum Cross-entropy Principle; Maximum Information Entropy Principle; Measurement Principle; Minimax Entropy Principle; Minimum Entropy Principle; Minimum Relative Entropy Principle; Modification of The Extended Maximum Entropy Principle; Principle of Maximum Entropy; Principle of Maximum Nonadditive Entropy; Principle of Observational Compatibility. Uncertainty Principle

UNCERTAINTY Maximization Principle; Precautionary Principle; Principle of Uncertainty; Uncertainty Principle, Fermats Principle; Minimum Cross-entropy (Me) Principle

UNIFIED CONCEPTUAL FRAMEWORK Correspondence Principle; Entropy Maximum Principle; Extended Maximum Entropy Principle; Extremal Principle of Information; Frieden-Soffer Principle; Gibbs Principle; Global Extremum Principle; Inference Principle; Jaynes' Maximum Entropy Principle; Jaynes' Principle; Kullback's Minimum Cross-entropy Principle; Maximum Information Entropy Principle; Measurement Principle; Minimax Entropy Principle; Minimum Entropy Principle; Minimum Relative Entropy Principle; Modification of The Extended Maximum Entropy Principle; Principle of Maximum Entropy; Principle of Maximum Nonadditive Entropy; Principle of Observational Compatibility; Uncertainty Principle

UNIFORM CENTRAL LIMIT THEOREM Ambartsumian Invariance Principle; Charnov's Invariance Principle; Donsker's Invariance Principle; Invariance Principle; Invariance Principle of General Relativity; Local Invariance Principle; Maruyama's Invariance Principle; Perturbation Invariance Principle; Principle of Correspondence; Principle of Invariance; Principle of Relativity; Principle of Undulatory Invariance; Relaxed Invariance Principle; Scale Invariance Principle; Strong Principle; Variational Principle

UNIFORM CONVERGENCE Ambartsumian Invariance Principle; Charnov's Invariance Principle; Donsker's Invariance Principle; Invariance Principle; Invariance Principle of General Relativity; Local Invariance Principle; Maruyama's Invariance Principle; Perturbation Invariance Principle; Principle of Correspondence; Principle of Invariance; Principle of Relativity; Principle

Dictionary of Scientific Principles, by Stephen Marvin
Copyright © 2011 John Wiley & Sons, Inc.

of Undulatory Invariance; Relaxed Invariance Principle; Scale Invariance Principle; Strong Principle; Variational Principle

UNIFORM SCALE DISTRIBUTIONS Ambartsumian Invariance Principle; Charnov's Invariance Principle; Donsker's Invariance Principle; Invariance Principle; Invariance Principle of General Relativity; Local Invariance Principle; Maruyama's Invariance Principle; Perturbation Invariance Principle; Principle of Correspondence; Principle of Invariance; Principle of Relativity; Principle of Undulatory Invariance; Relaxed Invariance Principle; Scale Invariance Principle; Strong Principle; Variational Principle

UNIQUENESS Principle of Uncertainty; Strong Comparison Principle; Uncertainty Principle, Reflection Principle

UNSTEADY FLOW Conservation of Momentum Principle; D'Alembert Principle; Differential Principle; Hamilton Principle; Newton's Third Principle; Principle of Equal A Priori Probabilities; Variational Principle

UNSUPERVISED LEARNING Active Principle; Measurement Principle, Correspondence Principle; Energy Preservation Principle; Entropy Maximum Principle; Extended Maximum Entropy Principle; Extinguishing Principle; Extremal Principle of Information; Frieden-Soffer Principle; Gibbs Principle; Global Extremum Principle; Inference Principle; Jaynes' Maximum Entropy Principle; Jaynes' Principle; Kullback's Minimum Cross-entropy Principle; Maximum Information Entropy Principle; Measurement Principle; Minimax Entropy Principle; Minimum Entropy Principle; Minimum Relative Entropy Principle; Modification of The Extended Maximum Entropy Principle; Principle of Maximum Entropy; Principle of Maximum Nonadditive Entropy; Principle of Observational Compatibility; Uncertainty Principle

UNSUPERVISED MULTISTAGE SEGMENTATION Entropy Principle; Fuzzy Entropy Principle; Maxent Principle; Maximum Entropy Principle; Maximum Entropy Principle; Mermin

Entropy Principle; Pme Principle of Maximum Entropy; Principle of Maximum Nonadditive Entropy; Principle of Operational Compatibility

URETER Mitrofanoff Principle; Monti Principle; Woltjer-Taylor Minimum Energy Principle

URETHRA Mitrofanoff Principle; Monti Principle; Woltjer-Taylor Minimum Energy Principle

URINARY DIVERSION Elastic Tube Principle; Ice Tube Principle; Mitrofanoff Principle; Monti Principle; Variational Principle; Woltjer-Taylor Minimum Energy Principle

URINARY INCONTINENCE Mitrofanoff Principle; Monti Principle; Variational Principle; Woltjer-Taylor Minimum Energy Principle

URINARY SYSTEM Monti Principle; Projection Principle, Consumer Principle; Ekeland's variational Principle; Ekeland's Epsilon Variational Principle; Ekeland's Variational Principle; Global Maximum Principle; Monti Principle; Projection Principle; Variational Principle, Elastic Tube Principle; Ice Tube Principle

URINE ANALYSIS Inference Principle; Jaynes' Principle; Kullback's Minimum Cross-entropy Principle; Maximum Information Entropy Principle; Measurement Principle; Measuring Principle; Minimax Entropy Principle; Minimum Energy Principle; Minimum Entropy Principle; Minimum Relative Entropy Principle; Modification of The Extended Maximum Entropy Principle; Principle of Maximum Entropy; Principle of Maximum Nonadditive Entropy; Principle of Observational Compatibility; Principle of Operational Compatibility; Schwinger's Quantum Action Principle; Uncertainty Principle

UROLOGY Woltjer-Taylor Minimum Energy Principle; Monti Principle; Projection Principle

USTATISTICS Ambartsumian Invariance Principle; Charnov's Invariance Principle; Donsker's Invariance Principle; Invariance Principle; Invariance Principle of General Relativity; Local Invariance Principle; Maruyama's Invariance Principle; Perturbation Invariance Principle; Principle of Correspondence; Principle of Invariance; Principle of Relativity; Principle of Undulatory Invariance; Relaxed Invariance Principle; Scale Invariance Principle; Strong Principle; Variational Principle

V

VACUUM Magnetron Principle, Principle of Caratheodony, Modulation Principle, Principle of Molecular Dynamics

VARIATIONAL CALCULUS D'Alembert Lagrange Principle; D'Alembert Lagrange Variation Principle; D'Alembert Principle; D'Alembert's Principle; Energy Principle; Equivalence Principle; Gauss Principle; Gyarmati's Principle; Hamilton Lagrange Principle; Hamilton Principle; Hamilton's Principle; Inertia Principle; Jayne's Principle; Jourdain Principle; Jourdain's Principle; Kinetic Principle; Lagrange Principle; Lagrange Principle of Least Action; Least Action Principle; Least Constraint Principle; Least Motion Principle; Least Square Principle; Maupertius Lagrange Principle; Maupertuis Euler Lagrange Principle of Least Action; Minimum Principle; Mueller-Breslau Principle; Prigogine Principle; Principle of Extremal Effects; Principle of Least Constraint; Principle of Least Motion; Principle of Maximal Effect; Principle of Minimum Kinetic Energy; Principle of Minimum Mechanical Energy; Principle of Minimum Potential Energy; Principle of Virtual Work; Quantum Hamilton Principle; Stewart-Hamilton Principle; Stochastic Hamilton Variational Principle; Total Virtual Action Principle; Toupin's Dual Principle; Virtual Force Principle; Virtual Work Principle; Woltjer Minimum Energy Principle

VARIATIONAL METHOD Equivalence Principle; Variational Principle; Schwinger Principle

VARIATIONAL TECHNIQUES August Krogh Principle; Ekeland's variational Principle; Ekeland's Epsilon Variational Principle; Ekeland's Variational Principle; Equivalence Principle; Global Maximum Principle; Pontryagin Maximum Principle; Variational Principle

VARYING ACTION D'Alembert's Principle; Gyarmati's Principle; Hamilton Principle; Prigogine Principle; Principle of Minimum Kinetic Energy; Principle of Minimum Mechanical Energy; Principle of Minimum Potential Energy; Quantum Hamilton Principle; Stewart-Hamilton Principle; Stochastic Hamilton Variational Principle; Toupin's Dual Principle

VASCULAR DISEASE Antiarrhythmic Principle; Antithrombotic Principle; Monti Principle, Drug Principle

VECTOR Correspondence Principle; Energy Preservation Principle; Entropy Maximum Principle; Extended Maximum Entropy Principle; Extinguishing Principle; Extremal Principle of Information; Frieden-Soffer Principle; Gibbs Principle; Inference Principle; Jaynes' Maximum Entropy Principle; Jaynes' Principle; Kullback's Minimum Cross-entropy Principle; Maximum Information Entropy Principle; Measurement Principle; Minimax Entropy Principle; Minimum Entropy Principle; Minimum Relative Entropy Principle; Modification of The Extended Maximum Entropy Principle; Principle of Maximum Entropy; Principle of Maximum Nonadditive Entropy; Principle of Observational Compatibility; Uncertainty Principle, Stationary Time Principle, Equidistortion Principle; Quantization Design Principle; Orthogonality Principle

VEGETATION Principle of Maximum Entropy, Active Principle; Measurement Principle, Matching Principle

VELOCITY Correspondence Principle; Entropy Maximum Principle; Extended Maximum Entropy Principle; Extinguishing Principle; Extremal Principle of Information; Frieden-Soffer Principle; Gibbs Principle; Global Extremum Principle; Inference

Kirchhoff's Voltage Law Principle; Minimum Energy Principle; Variational Principle; Woltjer Minimum Energy Principle; Woltjer-Taylor Minimum Energy Principle, Equivalent Background Principle, Rinciple of Information Processing; Principle of per-survivor processing, Optical Principle, Light Diffraction Principle

VOLTAGE Principle of Caratheodory, Pseudo-2-path Principle, Minimum Information Principle, Principle of Corresponding States

VOLUME Cavalieri Principle; Stationary Time Principle; Stereological Principle, Priming Principle, Huygen's Equivalence Principle, Principle of Briquetting

VON MISES STATISTICS Contraction Principle; Gibbs Variational Principle; Large Deviation Principle

W

WATER Three-day Principle; Educational Principle; Ehrenfest Principle; Principle of Construction and Operation, Optimization Principle; Run and Recreate Principle

WATER DEPTH Correspondence Principle; Entropy Maximum Principle; Extended Maximum Entropy Principle; Extinguishing Principle; Extremal Principle of Information; Frieden-Soffer Principle; Gibbs Principle; Inference Principle; Jaynes' Maximum Entropy Principle; Jaynes' Principle; Kullback's Minimum Cross-entropy Principle; Maximum Information Entropy Principle; Measurement Principle; Minimax Entropy Principle; Minimum Entropy Principle; Minimum Relative Entropy Principle; Mitrofanoff Principle; Modification of The Extended Maximum Entropy Principle; Principle of Maximum Entropy; Principle of Maximum Nonadditive Entropy; Principle of Observational Compatibility; Uncertainty Principle

WATER WAVE Correspondence Principle; Entropy Maximum Principle; Extended Maximum Entropy Principle; Extinguishing Principle; Extremal Principle of Information; Frieden-Soffer Principle; Gibbs Principle; Inference Principle; Jaynes' Maximum Entropy Principle; Jaynes' Principle; Kullback's Minimum Cross-entropy Principle; Maximum Information Entropy Principle; Measurement Principle; Minimax Entropy Principle; Minimum Entropy Principle; Minimum Relative Entropy Principle; Mitrofanoff Principle; Modification of The Extended Maximum Entropy Principle; Principle of Maximum Entropy; Principle of Maximum Nonadditive Entropy; Principle of Observational Compatibility; Uncertainty Principle; Hamilton's Principle

WAVE HEIGHT Correspondence Principle; Entropy Maximum Principle; Extended Maximum Entropy Principle; Extinguishing Principle; Extremal Principle of Information;

Frieden-Soffer Principle; Gibbs Principle; Inference Principle; Jaynes' Maximum Entropy Principle; Jaynes' Principle; Kullback's Minimum Cross-entropy Principle; Maximum Information Entropy Principle; Measurement Principle; Minimax Entropy Principle; Minimum Entropy Principle; Minimum Relative Entropy Principle; Mitrofanoff Principle; Modification of The Extended Maximum Entropy Principle; Principle of Maximum Entropy; Principle of Maximum Nonadditive Entropy; Principle of Observational Compatibility; Uncertainty Principle

WAVE MOTION Correspondence Principle; Entropy Maximum Principle; Extended Maximum Entropy Principle; Extinguishing Principle; Extremal Principle of Information; Frieden-Soffer Principle; Gibbs Principle; Inference Principle; Jaynes' Maximum Entropy Principle; Jaynes' Principle; Kullback's Minimum Cross-entropy Principle; Maximum Information Entropy Principle; Measurement Principle; Minimax Entropy Principle; Minimum Entropy Principle; Minimum Relative Entropy Principle; Mitrofanoff Principle; Modification of The Extended Maximum Entropy Principle; Principle of Maximum Entropy; Principle of Maximum Nonadditive Entropy; Principle of Observational Compatibility; Uncertainty Principle

WAVE NUMBER D'Alembert's Principle; Gyarmati's Principle; Hamilton Principle; Prigogine Principle; Principle of Minimum Kinetic Energy; Principle of Minimum Mechanical Energy; Principle of Minimum Potential Energy; Quantum Hamilton Principle; Stewart-Hamilton Principle; Stochastic Hamilton Variational Principle; Toupin's Dual Principle

WAVE PROPAGATION Anemometer Principle; Caratheodory's Principle; Conte's Principle; Fermat's Principle; Huygen's Principle; Least Time Principle; Maximum Entropy

Dictionary of Scientific Principles, by Stephen Marvin
Copyright © 2011 John Wiley & Sons, Inc.

Principle; Mitrofanoff Principle; Principle of Conservation of Energy; Principle of Measurement; Proof of Principle; Processing Principle

WAVEFORM Hamilton's Principle; Invariance Principle; Kirchhoff's Voltage Law Principle; Magnetron Principle; Minimum Energy Principle; Minimum Norm Principle; Variational Principle; Woltjer Minimum Energy Principle; Principle of Fundamental Characteristics

WAVEFRONT Conte's Principle; Fermat's Principle; Least Time Principle; Mitrofanoff Principle; Principle of Measurement; Proof of Principle

WAVELENGTH DEPENDENCES Minimum Information Principle; Protection Principle

WEAK CONVERGENCE Correspondence Principle; Entropy Maximum Principle; Extended Maximum Entropy Principle; Extinguishing Principle; Extremal Principle of Information; Frieden-Soffer Principle; Gibbs Principle; Inference Principle; Jaynes' Maximum Entropy Principle; Jaynes' Principle; Kullback's Minimum Cross-entropy Principle; Maximum Information Entropy Principle; Measurement Principle; Minimax Entropy Principle; Minimum Entropy Principle; Minimum Relative Entropy Principle; Mitrofanoff Principle; Modification of The Extended Maximum Entropy Principle; Principle of Maximum Entropy; Principle of Maximum Nonadditive Entropy; Principle of Observational Compatibility; Uncertainty Principle; Ambartsumian Invariance Principle; Charnov's Invariance Principle; Donsker's Invariance Principle; Invariance Principle; Invariance Principle of General Relativity; Local Invariance Principle; Maruyama's Invariance Principle; Perturbation Invariance Principle; Principle of Correspondence; Principle of Invariance; Principle of Relativity; Principle of Undulatory Invariance; Relaxed Invariance Principle; Scale Invariance Principle; Strong Principle; Variational Principle

WEAK GRAVITATIONAL FIELD Principle of Molecular Dynamics, Conte's Principle;

Fermat's Principle; Least Time Principle; Mitrofanoff Principle; Principle of Measurement; Proof of Principle

WEAK INVARIANCE Ambartsumian Invariance Principle; Charnov's Invariance Principle; Donsker's Invariance Principle; Invariance Principle; Invariance Principle of General Relativity; Local Invariance Principle; Maruyama's Invariance Principle; Perturbation Invariance Principle; Principle of Correspondence; Principle of Invariance; Principle of Relativity; Principle of Undulatory Invariance; Relaxed Invariance Principle; Scale Invariance Principle; Strong Principle; Variational Principle

WEAK PHASE OBJECTS Babinet's Principle; Complementarity Principle

WEAK PS CONDITION Ekeland's variational Principle; Ekeland's Epsilon Variational Principle; Ekeland's Variational Principle; Global Maximum Principle; Variational Principle

WEIGHT DISTRIBUTION Correspondence Principle; Entropy Maximum Principle; Extended Maximum Entropy Principle; Extinguishing Principle; Extremal Principle of Information; Frieden-Soffer Principle; Gibbs Principle; Inference Principle; Jaynes' Maximum Entropy Principle; Jaynes' Principle; Kullback's Minimum Cross-entropy Principle; Maximum Information Entropy Principle; Measurement Principle; Minimax Entropy Principle; Minimum Entropy Principle; Minimum Relative Entropy Principle; Mitrofanoff Principle; Modification of The Extended Maximum Entropy Principle; Principle of Maximum Entropy; Principle of Maximum Nonadditive Entropy; Principle of Observational Compatibility; Uncertainty Principle

WIENER PROCESS Ambartsumian Invariance Principle; Charnov's Invariance Principle; Donsker's Invariance Principle; Invariance Principle; Invariance Principle of General Relativity; Local Invariance Principle; Maruyama's Invariance Principle; Perturbation Invariance Principle; Principle of Correspondence; Principle of Invariance; Principle of Relativity; Principle of Undulatory Invariance; Relaxed Invariance

X

XRAYS Fermats Principle; Principle of Corresponding States; Weak Invariance Principle

Z

ZERO POINT Caratheodory's Principle; Conte's Principle; Least Time Principle; Principle of Caratheodony; Stationary Time Principle

ZONE ELECTROPHORESIS Capillary Electrophoresis Principle; Ion Pair Principle; Principle of Diastereomeric Ion Pair Formation

ZOOLOGY Inference Principle; Jaynes' Principle; Kullback's Minimum Cross-entropy Principle; Maximum Information Entropy Principle; Measurement Principle; Measuring Principle; Minimax Entropy Principle; Minimum Energy Principle; Minimum Relative Entropy Principle; Modification of The Extended Maximum Entropy Principle; Precautionary Principle; Principle of Maximum Entropy; Principle of Maximum Nonadditive Entropy; Principle of Observational Compatibility; Principle of Operational Compatibility; Schwinger's Quantum Action Principle; Uncertainty Principle

three-place common logarithms

N	0	1	2	3	4	5	6	7	8	9
0		000	301	477	602	699	778	845	903	954
1	000	041	079	114	146	176	204	230	255	279
2	301	322	342	362	380	398	415	431	447	462
3	477	491	505	519	531	544	556	568	580	591
4	602	613	623	633	643	653	663	672	681	690
5	699	708	716	724	732	740	748	756	763	771
6	778	785	792	799	806	813	820	826	833	839
7	845	851	857	863	869	875	881	886	892	898
8	903	908	914	919	924	929	934	940	944	949
9	954	959	964	968	973	978	982	987	991	996
10	000	004	009	013	017	021	025	029	033	037
11	041	045	049	053	057	061	064	068	072	076
12	079	083	086	090	093	097	100	104	107	111
13	114	117	121	124	127	130	134	137	140	143
14	146	149	152	155	158	161	164	167	170	173
15	176	179	182	185	188	190	193	196	199	201
16	204	207	210	212	215	217	220	223	225	228
17	230	233	236	238	241	243	246	248	250	253
18	255	258	260	262	265	267	270	272	274	276
19	279	281	283	286	288	290	292	294	297	299
20	301	303	305	308	310	312	314	316	318	320
21	322	324	326	328	330	332	334	336	338	340
22	342	344	346	348	350	352	354	356	358	360
23	362	364	365	367	369	371	373	375	377	378
24	380	382	384	386	387	389	391	393	394	396
25	398	400	401	403	405	407	408	410	412	413
26	415	417	418	420	422	423	425	427	428	430
27	431	433	435	436	438	439	441	442	444	446
28	447	449	450	452	453	455	456	458	459	461
29	462	464	465	467	468	470	471	473	474	476
30	477	479	480	481	483	484	486	487	489	490
31	491	493	494	496	497	498	500	501	502	504
32	505	507	508	509	511	512	513	515	516	517
33	519	520	521	522	524	525	526	528	529	530
34	531	533	534	535	537	538	539	540	542	543
35	544	545	547	548	549	550	551	553	554	555
36	556	558	559	560	561	562	563	565	566	567
37	568	569	571	572	573	574	575	576	577	579
38	580	581	582	583	584	585	587	588	589	590
39	591	592	593	594	596	597	598	599	600	601
40	602	603	604	605	606	607	609	610	611	612
41	613	614	615	616	617	618	619	620	621	622
42	623	624	625	626	627	628	629	630	631	632
43	633	634	635	636	637	638	639	640	641	642
44	643	644	645	646	647	648	649	650	651	652
45	653	654	655	656	657	658	659	660	661	662
46	663	664	665	666	667	667	668	669	670	671
47	672	673	674	675	676	677	678	679	679	680
48	681	682	683	684	685	686	687	688	688	689
49	690	691	692	693	694	695	695	696	697	698
N	0	1	2	3	4	5	6	7	8	9

N	0	1	2	3	4	5	6	7	8	9
50	699	700	701	702	702	703	704	705	706	707
51	708	708	709	710	711	712	713	713	714	715
52	716	717	718	718	719	720	721	722	723	723
53	724	725	726	727	728	728	729	730	731	732
54	732	733	734	735	736	736	737	738	739	740
55	740	741	742	743	744	744	745	746	747	747
56	748	749	750	751	751	752	753	754	754	755
57	756	757	757	758	759	760	760	761	762	763
58	763	764	765	766	766	767	768	769	769	770
59	771	772	772	773	774	775	775	776	777	777
60	778	779	780	780	781	782	782	783	784	785
61	785	786	787	787	788	789	790	790	791	792
62	792	793	794	794	795	796	797	797	798	799
63	799	800	801	801	802	803	803	804	805	806
64	806	807	808	808	809	810	810	811	812	812
65	813	814	814	815	816	816	817	818	818	819
66	820	820	821	822	822	823	823	824	825	825
67	826	827	827	828	829	829	830	831	831	832
68	833	833	834	834	835	836	836	837	838	838
69	839	839	840	841	841	842	843	843	844	844
70	845	846	846	847	848	848	849	849	850	851
71	851	852	852	853	854	854	855	856	856	857
72	857	858	859	859	860	860	861	862	862	863
73	863	864	865	865	866	866	867	867	868	869
74	869	870	870	871	872	872	873	873	874	874
75	875	876	876	877	877	878	879	879	880	880
76	881	881	882	883	883	884	884	885	885	886
77	886	887	888	888	889	889	890	890	891	892
78	892	893	893	894	894	895	895	896	897	897
79	898	898	899	899	900	900	901	901	902	903
80	903	904	904	905	905	906	906	907	907	908
81	908	909	910	910	911	911	912	912	913	913
82	914	914	915	915	916	916	917	918	918	919
83	919	920	920	921	921	922	922	923	923	924
84	924	925	925	926	926	927	927	928	928	929
85	929	930	930	931	931	932	932	933	934	934
86	934	935	936	936	937	937	938	938	939	939
87	940	940	941	941	942	942	942	943	943	944
88	944	945	945	946	946	947	947	948	948	949
89	949	950	950	951	951	952	952	953	953	954
90	954	955	955	956	956	957	957	958	958	959
91	959	960	960	961	961	962	962	963	963	964
92	964	964	965	965	966	966	967	967	968	968
93	968	969	969	970	970	971	971	972	972	973
94	973	974	974	975	975	975	976	976	977	977
95	978	978	979	979	980	980	980	981	981	982
96	982	983	983	984	984	985	985	985	986	986
97	987	987	988	988	989	989	989	990	990	991
98	991	992	992	993	993	993	994	994	995	995
99	996	996	997	997	997	998	998	999	999	000
N	0	1	2	3	4	5	6	7	8	9

Basic
Mathematics
For Electronics
Fifth
Edition

Nelson M. Cooke
Late President
Cooke Engineering Company

Herbert F. R. Adams
Former Chief Instructor
Electronics Division
Pacific Vocational Institute
Burnaby, British Columbia

Peter B. Dell
Chief Instructor
Electronics Division
Pacific Vocational Institute
Burnaby, British Columbia

Gregg Division
McGraw-Hill Book Company
New York   Atlanta   Dallas
St. Louis   San Francisco   Auckland   Bogotá   Guatemala
Hamburg   Johannesburg   Lisbon   London   Madrid
Mexico   Montreal   New Delhi   Panama   Paris   San Juan
São Paulo   Singapore   Sydney   Tokyo   Toronto

Sponsoring Editor: George Z. Kuredjian
Editing Supervisor: Mitsy Kovacs
Design and Art Supervisor: Caryl Valerie Spinka
Production Supervisor: Priscilla Taguer

Text Designer: Sharkey Design
Cover Designer: Sharkey Design

Library of Congress Cataloging in Publication Data

Cooke, Nelson Magor.
  Basic mathematics for electronics.

  Includes index.
  1. Electric engineering—Mathematics.
2. Electronics—Mathematics.   I. Adams, Herbert F. R.
II. Dell, Peter B.   III. Title.
TK153.C63 1982   510′.246213 82-226
ISBN 0-07-012514-7   AACR2

  34567890 DODO 89876543

ISBN 0-07-012514-7

Contents

Preface

This is the fifth edition of the textbook originally entitled *Mathematics for Electricians and Radiomen* by the late Nelson M. Cooke. This new edition stands as a continuing monument to Mr. Cooke's memory. The authors sincerely hope that the electronics technicians of the metric age will find this enlarged and updated volume to be as useful to them as the preceding editions were to their predecessors.

As its title suggests, the book has been designed to provide the basic mathematics needed for a study of electronics. This was the objective of all the preceding editions, and the overall objective still is to provide learners with the basic mathematical tools to solve the sometimes complex problems that arise in the study of electronics.

Two major groups of learners are considered in the pursuit of this objective—those without any previous electronics background, and those who have some previous knowledge of electronics but who have been away from formal academic studies for two years or more. As a consequence, this text can be used by both the recent high-school graduate and the more mature or adult student/learner. Whichever classification the learner falls into, this text will provide the basic mathematics required for electronic studies. For the learner who wishes to continue his or her studies beyond the original intent of this text, we have provided a background for studies in digital techniques.

We should stress that, while this text will apply electronic principles to illustrate the mathematical applications of a concept, it is *not* intended to be a textbook in electronic theory. It is designed to be used in conjunction with theoretical and practical studies in electronics. The chapters which represent applications of the mathematical development will fit the various courses of study, and they may be stressed, omitted, repeated, or adjusted to accommodate the requirements of the electronics course being followed by individuals or classes. Our objective, once again, is to provide the basis for a clearer mathematical understanding of the electrical and electronic principles encountered in your studies. The material is offered in ''block'' form: algebra, trigonometry, logarithms, and computer mathematics.

Teachers will find that studies in one main subject may be interrupted to fit suitably into another, or the topics may be interleaved, so that, after the initial chapters, studies in algebra and trigonometry may proceed together. Everything possible has been done to provide logical subdivisions within the book and still promote flexibility. Some sections dealing with practical applications may be delayed until the appropriate theory or laboratory work has been covered. In addition, the block form will be found helpful by the self-directed learner.

Another consideration is the necessity for justifying and developing the mathematical procedures. Too often students are expected to accept on trust many empirical statements simply because the statements appear in print.

The authors believe that a sound understanding of the mathematical principles will result in greater ease in problem solving.

The authors expect that students will come to the study of this book with a good working knowledge of arithmetic. Students may find it helpful to complete the pretests in the answer booklet available through their teachers. Teachers will find the pretests helpful in determining the extent of preliminary work that has not been covered. Students who have been away from formal schooling for some years will find help in a companion volume, *Arithmetic Review for Electronics,* also published by McGraw-Hill.

To acknowledge the changes which are characteristic of electronics, many application problems have been replaced with more modern circuitry. The two computer mathematics chapters, Numbering Systems for Computers and Boolean Algebra, have been enlarged to include hexadecimal conversions in order to reflect the increasing importance of these applications.

In this second metric edition, we have found it continues to be desirable to distinguish between the ''symbol'' of pure metric notation for the units of measurement (A for ampere) and the circuit diagram and algebraic ''symbol'' (I for electric current). We hope that the two symbols will be used correctly by all concerned. The units of measurement in the electronic field have been metric (now SI) for many years, but the *language* and symbolic usage are still new to many practitioners. Teachers and students alike must learn to think and solve problems by using the metric language. For that essential reason, very little is offered in this edition by way of conversion from the traditional measurement system to SI or vice versa, although a number of exercises involving adjustments in formulas have been retained.

Many people have shared in the continuing effort to make this a useful and valuable text. In addition to the friends and colleagues acknowledged in preceding editions, special thanks must be offered to those users who kindly wrote to the authors or to McGraw-Hill with comments and suggestions. Proposals and criticisms are always welcome. It is requested that comments be addressed to us in care of the publisher.

Herbert F. R. Adams
Peter B. Dell

Algebra— General Numbers

2

In general, arithmetic consists of the operations of addition, subtraction, multiplication, and division of a type of numbers represented by the digits 0, 1, 2, 3, . . . , 9. By using the above operations or combinations of them, we are able to solve many problems. However, a knowledge of mathematics limited to arithmetic is inadequate and a severe handicap to anyone interested in acquiring an understanding of electric circuits. Proficiency in performing even the most simple operations of algebra will enable you to solve problems and determine relations that would be impossible with arithmetic alone.

2-1 THE GENERAL NUMBER

Algebra may be thought of as a continuation of arithmetic in which letters and symbols are used to represent definite quantities whose actual values may or may not be known. For example, in electrical and electronics texts, it is customary to represent current by the letter I or i, voltage by V or v, resistance by R or r, etc. The base of a triangle is often represented by b, and the altitude may be specified as a. Such letters or symbols used for representing quantities in a general way are known as *general numbers* or *literal numbers*.

The importance of the general-number idea cannot be overemphasized. Although the various laws and facts concerning electricity can be expressed in English, they are more concisely and compactly expressed in mathematical form in terms of general numbers. As an example, Ohm's law states, in part, that the current in a certain part of a circuit is proportional to the potential difference (voltage) across that part of the circuit and inversely proportional to the resistance of that part. This same statement, in mathematical terms, is

$$I = \frac{V}{R}$$

where I represents the current, V is the potential difference, and R is the resistance. Such an expression is known as a *formula*.

Although expressing various laws and relationships of science as formulas gives us a more compact form of notation, that is not the real

value to many decimal places, or to many "significant figures," when the circuit is to be constructed with a standard off-the-shelf resistor made to, say, ±10% accuracy. (Obviously a shunt to be made by hand may well be accurate to $\frac{1}{2}\%$, and then this argument would not apply.)

An electronic calculator can be relied upon to give a satisfactory answer (three significant figures) to most of the problems at the level of study in this text. Answers computed by logarithms or by long multiplication or division will disagree with calculator answers and with each other—if they are taken to enough decimal places.

There are occasions, of course, when three significant figures may not be sufficient: accountants and auditors will want your financial calculations to be correct to the nearest cent, even when thousands of dollars are involved; the FCC will not be satisfied with a carrier frequency correct to only three significant figures; logarithms and trigonometric functions are given to four places, or five, or ten, and the answers achieved will reflect the accuracy of the tables used; angles greater than 90° must be converted into equivalents less than 90° for purposes of calculations, and they should not be rounded off prior to conversion. All the answers in this text reflect these notions, and you are accordingly encouraged to start using a good calculator early in your career. (See the section in the Appendix on calculators before purchasing one.)

1-12 METRICATION

Since the publication of the third edition of this book, both Canada and the United States have taken lengthy strides toward the adoption of the metric system of units. The metric system has undergone many refinements during its lifetime, from cgs through mks to mksa. Now, in the most logical refinement of all, the International System of Metric Units, known universally as SI (for Système International d'Unités), is being phased into the North American measurement scene. Since the SI metric units of electricity are those with which most of the users of this book are already familiar (amperes, volts, ohms, hertz, etc.), it only remains for us to learn to think in terms of meters for length and kilograms for weight (mass) to become proficient in the basic electronics requirements of metric units.

The SI units which you must know in order to study this textbook are introduced in Chap. 7, which was completely rewritten for the fourth edition.

wherever you open the book, these numbers show the section or sections covered on the pages in view. For example, Sec. 4-10 is easily found by leafing through the book while noting the inclusive numbers.

1-10 ABBREVIATIONS

Every profession and every technology has its own jargon—the particular words and phrases which describe the phenomena with which it deals. Electronics is particularly noteworthy in this respect, with inductance, capacitance, resistance, impedance, and frequency leading a host of others. Each phenomenon must be measured and described in understandable units so that other workers in the field will be able to understand exactly what is involved. After establishing such a vocabulary and list of units, the next logical development is a system of abbreviations—shorthand symbols which everyone will recognize as standing for the units and dimensions of the technology. For many years there was no single agreed-upon list of electronics abbreviations, and most of us had to be able to recognize several variations as acceptable abbreviations of the same term. For instance, A, a, amp, Amp, amps, and Amps were all used to represent *amperes,* depending upon the teacher, the author, and the publisher involved.

Even today, the exhaustive list of standard abbreviations recommended by the Institute of Electrical and Electronics Engineers is not wholly acceptable to all branches of the industry, and local variations and established forms continue to be used. Some publishers are still reluctant to use the single-letter abbreviations for fear of introducing ambiguities. Some of us were reluctant to adopt hertz (Hz) in place of cycles per second (c/s or c/sec or cps), not so much that we did not honor Hertz as that the name does not make obvious the "per time" relationship involved in frequency. However, the name is now being used widely, and Hz is used in this edition as a reflection of what you may expect when you step out of school and into industry.

One drawback to all this is that although we have, at the publisher's request, attempted to be uniform in the matter of abbreviations, you will nevertheless meet, and must be able to deal with, several variations for many years to come. However, in an attempt to keep before you the dimensional aspect of units, we have used rev/min rather than rpm, Ω/m rather than Ω/M, and so on. You should study the tables of symbols and abbreviations carefully and repeatedly so that you achieve an early and complete mastery of abbreviations. It has been our aim to supply the full name the first time the term is used, follow it immediately with the abbreviation in parentheses, and then use the abbreviation at every opportunity thereafter.

1-11 SIGNIFICANT FIGURES

The resistors, capacitors, and other devices used in electronic circuitry are often manufactured to convenient tolerances; 5%, 10% and 20% are the most common. Accordingly, it is meaningless to calculate a resistance

intended to convey. Proficiency in mathematics depends upon thorough understanding of each step as it is encountered so that it can be used to master the step which follows.

1-6 IMPORTANCE OF PROBLEMS

Full advantage should be taken of the many problems distributed through the text. There is no approach to a full and complete understanding of any branch of mathematics other than the solution of numerous problems. Application of what has been learned from the text to practical problems in which you are primarily interested will not only help with the subject matter of the problem but also serve the purpose of fixing in mind the mathematical principles involved.

In general, the arrangement of problems is such that the most difficult problems appear at the end of each group. It is apparent that the working of the simpler problems first will tend to make the more difficult ones easier to solve. The home-study student is, therefore, urged to work all problems in the order given. At times, this may appear to be useless, and you may have the desire to proceed to more interesting things; but time spent in working problems will amply repay you in giving you a depth of understanding to be obtained in no other manner. This does not mean that progress should cease if a particular problem appears to be impossible to solve. Return to such a problem when your mind is fresh, or mark it for solution during a review period.

1-7 ILLUSTRATIVE EXAMPLES

Each of the illustrative examples in this book is intended to make clear some important principle or method of solution. The subject matter of the examples will be more thoroughly assimilated if, after careful analysis of the problem set forth, you make an independent solution and compare the method and results with the illustrative example.

1-8 REVIEW

Too much stress cannot be placed upon the necessity for frequent and thorough review. Points that have been missed in the original study of the text will often stand out clearly upon careful review. A review of each chapter before proceeding with the next is recommended.

1-9 SECTION REFERENCES

Throughout this book you will be referred to earlier sections for review or to bring to attention similar material pertaining to the subject under discussion. For convenience and ready reference, two sets of numbers are printed at the top of each right-hand page. The first number refers to the *first* text or problem section on the left-hand page, and the second number refers to the *last* text or problem section on the right-hand page. Thus,

for most of our basic ideas of electrical phenomena are based upon mathematical reasoning and stated in mathematical terms. This is a fortunate circumstance, for it enables us to build a structure of electrical knowledge with precision, assembling and expressing the components in clear and concise mathematical terms and arranging the whole in logical order. Without mathematical assistance, technicians must be content with the long and painful process of accumulating bits of information, details of experience, etc., and they may never achieve a thorough understanding of the field in which they live and work.

1-3 MATHEMATICS—A TEACHER

In addition to laying a foundation for technical knowledge and assisting in the practical application of knowledge already possessed, mathematics offers unlimited advantages in respect to mental training. The solution of a problem, no matter how simple, demands logical thinking for it to be possible to state the facts of the problem in mathematical terms and then proceed with the solution. Continued study in this orderly manner will increase your mental capacity and enable you to solve more difficult problems, understand more complicated engineering principles, and cope more successfully with the everyday problems of life.

1-4 METHODS OF STUDY

Before beginning detailed study of this text, you should carefully analyze it, in its entirety, in order to form a mental outline of its content, scope, and arrangement. You should make another preliminary survey of each individual chapter before attempting detailed study of the subject matter. After the detailed study, you should, before proceeding to new material, work problems until all principles are fixed firmly in your mind.

In working problems, the same general procedure is recommended. First, analyze a problem in order to determine the best method of solution. Then state the problem in mathematical terms by utilizing the principles that are applicable. If you make but little progress, it is probable you have not completely mastered the principles explained in the text, and a review is in order.

The authors are firm believers in the use of a workbook, preferably in the form of a loose-leaf notebook, which contains all the problems you have worked, together with the numerous notes made while studying the text. Such a book is an invaluable aid for purposes of review. The habit of jotting down notes while reading or studying should be cultivated. Such notes in your own words will provide a better understanding of a concept.

1-5 RATE OF PROGRESS

Home-study students should guard against too rapid progress. There is a tendency, especially in studying a chapter whose contents are familiar or easy to comprehend, to hurry on to the next chapter. Hasty reading may cause the loss of the meaning that a particular section or paragraph is

Introduction

1

In the legions of textbooks on the subject of mathematics, all the basic principles contained here have been expounded in admirable fashion. However, students of electricity, radio, and electronics have need for a course in mathematics that is directly concerned with application to electric and electronic circuits. This book is intended to provide those students with a sound mathematical background as well as further their understanding of basic circuitry.

1-1 MATHEMATICS—A LANGUAGE

The study of mathematics may be likened to the study of a language. In fact, mathematics *is* a language, the language of number and size. Just as the rules of grammar must be studied in order to master English, so must certain concepts, definitions, rules, terms, and words be learned in the pursuit of mathematical knowledge. These form the vocabulary or structure of the language. The more a language is studied and used, the greater becomes the vocabulary; the more mathematics is studied and applied, the greater its usefulness becomes.

There is one marked difference, however, between the study of a language and the study of mathematics. A language is based on words, phrases, expressions, and usages that have been brought together through the ages in more or less haphazard fashion according to the customs of the times. Mathematics is built upon the firm foundation of sound logic and orderly reasoning and progresses smoothly, step by step, from the simplest numerical processes to the most complicated and advanced applications, each step along the way resting squarely upon the steps which have been taken before. This makes mathematics the fascinating subject that it is.

1-2 MATHEMATICS—A TOOL

As the builder works with a square and compasses, so the engineer employs mathematics. A thorough grounding in mathematics is essential to proficiency in any of the numerous branches of engineering. In no other branch is this more apparent than in the study of electrical and electronic subjects,

value of formulas. As you attain proficiency in algebra, the value of general formulas will become more apparent. Our studies of algebra will consist mainly of learning how to add, subtract, multiply, divide, and solve general algebraic expressions, or formulas, in order to attain a better understanding of the fundamentals of electricity and related fields.

2-2 SIGNS OF OPERATION

In algebra the signs of operation $+$, $-$, $\times$, and $\div$ have the same meanings as in arithmetic. The sign $\times$ is generally omitted between literal numbers. For example, $I \times R$ is written IR and means that I is to be multiplied by R. Similarly, $2\pi fL$ means 2 times π times f times L. There are times when the symbol $\cdot$ is used to denote multiplication. Thus $I \times R$, $I \cdot R$, and IR all mean I times R.

2-3 THE ORDER OF SIGNS OF OPERATION

In performing a series of different operations, we will follow convention and perform first the multiplications, next the divisions, and then the additions and subtractions. Thus,

$$16 \div 4 + 8 + 4 \times 5 - 3 = 4 + 8 + 20 - 3 = 29$$

2-4 ALGEBRAIC EXPRESSIONS

An *algebraic expression* is one that expresses or represents a number by the signs and symbols of algebra. A *numerical algebraic expression* is one consisting entirely of signs and numerals. A *literal algebraic expression* is one containing general numbers or letters. For example, $8 - (6 + 2)$ is a numerical algebraic expression and I^2R is a literal algebraic expression.

2-5 THE PRODUCT

As in arithmetic, a *product* is the result obtained by multiplying two or more numbers. Thus, 12 is the product of 6×2.

2-6 THE FACTOR

If two or more numbers are multiplied together, each of them or the product of any combination of them is called a *factor* of the product. For example, in the product $2xy$, 2, x, y, $2x$, $2y$, and xy are all factors of $2xy$.

2-7 COEFFICIENTS

Any factor of a product is known as the *coefficient* of the product of the remaining factors. In the foregoing example, 2 is the coefficient of xy, x is the coefficient of $2y$, y is the coefficient of $2x$, etc. It is common practice to speak of the numerical part of an expression as the *coefficient* or as the *numerical coefficient*. If an expression contains no numerical coefficient, 1 is understood to be the numerical coefficient. Thus, $1abc$ is the same as abc.

2-8 PRIMES AND SUBSCRIPTS

When, for example, two resistances are being compared in a formula or it is desirable to make a distinction between them, they may be represented by R_1 and R_2 or R_a and R_b. The small numbers or letters written at the right of and below the R's are called *subscripts*. They are generally used to denote different values of the same units.

R_1 and R_2 are read "R sub one" and "R sub two" or simply "R one" and "R two."

Care must be used in distinguishing between subscripts and exponents. Thus V^2 is an indicated operation that means $V \cdot V$, whereas V_2 is used to distinguish one quantity from another of the same kind.

Primes and *seconds,* instead of subscripts, are often used to denote quantities. Thus one current might be denoted by I' and another by I''. The first is read "I prime" and the latter is read "I second." I' resembles I^1 (I to the first power), but in general this causes little confusion.

2-9 EVALUATION

To *evaluate* an algebraic expression is to find its numerical value. In Sec. 2-1, it was stated that in algebra certain signs and symbols are used to represent definite quantities. Also, in Sec. 2-4, an algebraic expression was defined as one that represents a number by the signs and symbols of algebra. We can find the numerical, or definite, value of an algebraic expression only when we know the values of the letters in the expression.

Example 1
Find the value of $2ir$ if $i = 5$ and $r = 11$.

Solution
$$2ir = 2 \times 5 \times 11 = 110$$

Example 2
Evaluate the expression $23V - 3ir$ if $V = 10$, $i = 3$, and $r = 22$.

Solution
$$23V - 3ir = 23 \times 10 - 3 \times 3 \times 22$$
$$= 230 - 198 = 32$$

Example 3
Find the value of $\dfrac{V}{R} - 3I$ if $V = 230$, $R = 5$, and $I = 8$.

Solution
$$\frac{V}{R} - 3I = \frac{230}{5} - 3 \times 8 = 46 - 24 = 22$$

PROBLEMS 2-1

Note The accuracy of answers to numerical computations is, in general, that obtained with an electronic calculator.

1. (a) What does the expression $(25)(R)$ mean?
    (b) What is the meaning of $6 \cdot r$?
    (c) What does $0.25I$ mean?
2. What is the value of:
    (a) $5i$ when $i = 7$ amperes (A)?
    (b) $4Z$ when $Z = 16$ ohms (Ω)?
    (c) $16V$ when $V = 110$ volts (V)?
3. One electrolytic capacitor costs $2.75.
    (a) What will one gross of capacitors cost?
    (b) What will n capacitors cost?
4. One dozen resistors cost a total of $2.04.
    (a) What is the cost of each resistor?
    (b) What is the cost of p resistors?
5. The current in a certain circuit is $25I$ A. What is the current if it is reduced to one-half its original value?
6. There are three resistances, of which the second is twice the first and one-sixth the third. If R represents the first resistance, what expressions describe the other two?
7. There are four capacitances, of which the second is two-thirds the first, the third is six times the second, and the fourth is twelve times the third. If C represents the first capacitance, in picofarads (pF), what expressions describe the other three?
8. If $P = 3$, $X = 5$, and $\psi = 12$, evaluate:

    (a) $P + \psi$ (b) $\psi + X - P$ (c) $\dfrac{\psi}{X}$

    (d) $\dfrac{X - P}{\psi}$ (e) $\dfrac{P + \psi}{X}$

9. Write the expression which will represent each of the following:
    (a) A resistance which is R Ω greater than 16 Ω.
    (b) A voltage which is 220 V more than v V.
    (c) A current which is I A less than i A.
10. A circuit has a resistance of 125 Ω. Express a resistance which is R Ω less than six times this resistance.
11. An inductance L_1 exceeds another inductance L_2 by 125 millihenrys (mH). Express the inductance L_2 in terms of L_1.
12. When two capacitors C_1 and C_2 are connected in series, the resultant capacitance C_s of the combination is expressed by the formula

$$C_s = \frac{C_1 C_2}{C_1 + C_2}$$

    What is the resultant capacitance if:
    (a) 5 pF is connected in series with 15 pF?
    (b) 150 pF is connected in series with 475 pF?

13. The current in any part of a circuit is given by $I = \dfrac{V}{R}$, in which I is the current in amperes through that part, V is the electromotive force (emf) in volts across that part, and R is the resistance in ohms of that part. What will be the current through a circuit with:
(a) An emf of 220 V and a resistance of 5 Ω?
(b) An emf of 50 V and a resistance of 200 Ω?

14. The time interval between the transmission of a radar pulse and the reception of the pulse's echo off a target is $t = \dfrac{2R}{c}$ seconds (s), where t is the time interval in seconds, R is the range in kilometers (km), and c is the speed of light, at which radio waves travel. [c = 300 000 kilometers per second (km/s).] What is the time between the transmission of a pulse and the reception of its echo from a target at a distance of 124 km?

15. The relation $t = \dfrac{2R}{c}$ in Prob. 14 is applicable to the transmission of sound in air and water. Owing to lower speeds of transmission, R is usually expressed in meters (m), and c is usually expressed in meters per second (m/s). (In air, $c \simeq 335$ m/s, and in salt water, $c \simeq 1460$ m/s. The sign $\simeq$ means "is approximately equal to.")
(a) What is the time between the transmission of a short pulse of sound through air and the reception of its echo at a distance of 500 m?
(b) What time will elapse if the sound pulse is transmitted through seawater at the same distance?

16. The relationship between the wavelength λ of a wave, the frequency f in hertz (Hz, or cycles per second), and the speed c at which the wave is propagated is $\lambda = \dfrac{c}{f}$. If λ is expressed in meters, then c must be expressed in meters per second; that is, λ and c must be expressed in the same units of length.
(a) What is the wavelength in kilometers of a radio wave having a frequency of 980 kilohertz (kHz)? (980 kHz = 980 000 Hz; c = 300 000 km/s = 300 000 000 m/s.)
(b) What is the wavelength in meters of a radio wave having a frequency of 121.5 megahertz (MHz)? (121.5 MHz = 121 500 000 Hz; c = 300 000 km/s = 300 000 000 m/s.)

17. The distance between a dipole antenna and its reflector is usually one-fifth of a wavelength. What will be this spacing for a signal at 205 MHz in meters?

2-10 EXPONENTS

To express "x is to be taken as a factor four times," we could write $xxxx$, but the general agreement is to write x^4 instead.

An *exponent*, or *power*, is a number written at the right of and above a second number to indicate how many times the second number is to be taken as a factor. The number to be multiplied by itself is called the *base*.

Thus, I^2 is read "I square" or "I second power" and means that I is to be taken twice as a factor; e^3 is read "e cube" or "e third power" and means that e is to be taken as a factor three times. Likewise, 5^4 is read "5 fourth power" and means that 5 is to be taken as a factor four times; thus,

$$5^4 = 5 \times 5 \times 5 \times 5 = 625$$

When no exponent, or power, is indicated, the exponent is understood to be 1. Thus, x is the same as x^1.

2-11 THE RADICAL SIGN

The radical sign $\sqrt{}$ has the same meaning in algebra as in arithmetic; $\sqrt{e}$ means the square root of e, $\sqrt[3]{x}$ means the cube root of x, $\sqrt[4]{i}$ means the fourth root of i, etc. The small number in the angle of a radical sign, like the 4 in $\sqrt[4]{i}$, is known as the *index* of the root.

2-12 TERMS

A *term* is an expression containing literal and/or numerical parts which are not separated by plus or minus signs. Terms may be parts of larger expressions in which the terms are separated by plus or minus signs. $3V^2$, IR, and $-2v$ are all terms of the expression $3V^2 + IR - 2v$.

Although the value of a term depends upon the values of the literal factors of the term, it is customary to refer to a term whose sign is plus as a *positive term*. Likewise, we refer to a term whose sign is minus as a *negative term*.

Terms having the same literal parts are called *like terms* or *similar terms*. $2a^2bx$, $-a^2bx$, $18a^2bx$, and $-4a^2bx$ are like terms.

Terms that are not alike in their literal parts are called *unlike terms* or *dissimilar terms*. $5xy$, $6ac$, $9I^2R$, and VI are *unlike terms*.

An algebraic expression consisting of but one term is known as a *monomial*.

A *polynomial*, or *multinomial*, is an algebraic expression consisting of two or more terms.

A *binomial* is a polynomial of two terms. Some examples of binomials are: $v + ir$, $a - 2b$, and $2x^2y + xyz^2$ are binomials.

A *trinomial* is a polynomial consisting of three terms. For example, $2a + 3b - c$, $IR + 3v - V^2$, and $8ab^3c + 3d + 2xy$ are trinomials.

PROBLEMS 2-2

1. If $a = 3$, $b = 6$, and $c = 2$, evaluate the following:
   (a) $2abc$
   (b) $5a^2b + 3c$
   (c) $a^2b^2c^2$
   (d) $12ac^2 - 2b^2$
   (e) $\sqrt{4a^2b^2}$
   (f) $5\sqrt{9b^2c^2} + 3a^2$

2. If $V = 110$, $I = 6$, and $R = 25$, evaluate the following:

(a) $5VI$

(b) VI^2R

(c) $I^2R + \dfrac{12V^2}{R}$

(d) $\dfrac{25I^3R^2}{6IR} - \sqrt{\dfrac{100V^2}{R}}$

(e) $\dfrac{36V^2IR}{I^3R} - 3R^2$

3. State which of the following are monomials, binomials, and trinomials:

(a) $\dfrac{V}{I}$

(b) I^2R

(c) $2\pi fL$

(d) $a + jb$

(e) $\Phi + \theta + 90°$

(f) $V_s - V_g$

(g) $I + \sqrt{\dfrac{P}{R}} + \dfrac{V}{R}$

(h) $a^2 + 2ab + b^2$

(i) $\dfrac{5(\mu V_g)^2}{16r_p}$

(j) $\dfrac{1}{R_1} + \dfrac{1}{R_2} + \dfrac{1}{R_3}$

4. In Probs. 1, 2, and 3, state which expressions are polynomials.

5. Write the following statements in algebraic symbols:

(a) I is equal to V divided by R.

(b) V is equal to I times R.

(c) P is equal to R times the square of I.

(d) R_1 is equal to the sum of R_2 and R_3.

(e) K is equal to M divided by the square root of the product of L_1 and L_2.

(f) R_p is equal to the product of R_1 and R_2 divided by their sum.

(g) The meter multiplier N is equal to the meter resistance R_m divided by the shunt resistance R_s all plus 1.

6. The approximate inductance of a single-layer air-core coil, such as used in the tuning circuits of radio receivers, can be calculated by the formula

$$L = \frac{2.54r^2n^2}{9r + 10l} \qquad \text{microhenrys } (\mu\text{H})$$

where L = inductance, μH

r = radius of winding, centimeters (cm)

n = number of turns of wire in winding

l = length of coil, cm

What is the inductance of a coil that is 3 cm in diameter and 10 cm long and has 150 turns of wire?

7. The winding in Prob. 6 is removed from the coil form and smaller wire is substituted, so that, in the same length of coil, the number of turns is tripled. What is the inductance?

8. The power in any part of an electric circuit is given by the formula

$$P = I^2R \qquad \text{watts (W)}$$

where P = power, W

I = current, A

R = resistance, Ω

Find the power expended when:

(a) The current is 0.25 A and the resistance is 10 000 Ω.

(b) The current is 30 A and the resistance is 0.5 Ω.

9. In Prob. 8, if the resistance is kept constant, what happens to the power if the current is (a) doubled, (b) tripled, (c) halved?

10. The power in any part of an electric circuit is also given by the formula

$$P = \frac{V^2}{R} \quad \text{W}$$

where P = power expended, W
V = electromotive force, V
R = resistance, Ω
What happens to the power if:
(a) The voltage is doubled and the resistance is unchanged?
(b) The voltage is halved and the resistance is unchanged?
(c) The resistance is doubled and the voltage is unchanged?
(d) The resistance is halved and the voltage is unchanged?

Algebra— Addition and Subtraction

3

The problems of arithmetic deal with positive numbers only. A *positive number* may be defined as any number greater than zero. Accepting this definition, we know that when such numbers are added, multiplied, and divided, the results are always positive. Such is the case in subtraction if a number is subtracted from a larger one. However, if we attempt to subtract a number from a smaller one, arithmetic furnishes us with neither a rule for carrying out this operation nor a meaning for the result.

3-1 NEGATIVE NUMBERS

Limiting our knowledge of mathematics to positive numbers would place us under a severe handicap, for there are many instances when it becomes necessary to deal with numbers that are called negative. Often, a negative number is defined as a number less than zero. Numerous examples of the uses of negative numbers could be cited. For example, zero degrees on the Celsius thermometer has been chosen as the temperature of melting ice—commonly referred to as freezing temperature. Now, everyone knows that in some climates it gets much colder than "freezing." Such temperatures are referred to as so many "degrees below zero." How shall we state, in the language of mathematics, a temperature of "10 degrees below zero"? Ten degrees above zero would be written 10°. Because 0° is the reference point, it is logical to assume that 10° below zero would be written as $-10°$, which, for our purposes, makes it a negative number.

Therefore, we see that a definition making a negative number less than zero is not completely correct. A negative number is some quantity away from a reference point in one direction (the defined negative direction), whereas the same positive quantity is simply the same quantity in the opposite direction (the defined positive direction).

Negative numbers are prefixed with the minus sign. Thus, negative 2 is written -2, negative $3ac$ is written $-3ac$, etc. If no sign precedes it, a number is assumed to be positive.

3-2 PRACTICAL NEED FOR NEGATIVE NUMBERS

The need for negative numbers often arises in the consideration of voltages or currents in electric and electronic circuits. It is common practice to select the ground, or earth, as a point of zero potential. This does not mean, however, that there can be no potentials below ground, or zero, potential. Consider the case of the three wire feeders connected as shown in Fig. 3-1.

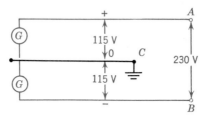

Fig. 3-1 Two 115-V generators connected in series with neutral wire grounded.

The generators G, which maintain a voltage of 115 V each, are connected in series so that their voltages add to give a voltage of 230 V across points A and B, and the neutral wire is grounded at C. Since C is at ground, or zero, potential, point A is 115 V positive with respect to C and point B is 115 V negative with respect to C. Therefore, the voltage at A with respect to ground, or zero, potential could be denoted as 115 V and the voltage at B with respect to ground could be denoted as -115 V.

Similar conditions exist in semiconductor circuits, as illustrated by the schematic diagram symbol of a 2N5457 N-channel junction field-effect transistor (NJFET) in Fig. 3-2. The drain current indicated by the arrow flows through the source resistor R and creates a potential of 4 V across it. Since the source is N-type material and the gate is P-type material, a *back-biased* PN junction is produced. The result is that the gate voltage, with respect to the source terminal, is negative. When measured with respect to ground or common, the source voltage is $+4$ V and the gate is at 0 V. The gate voltage, when measured with respect to the source terminal, is -4 V. This is usually written as $V_{GS} = -4$ V.

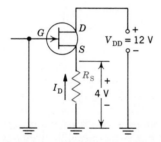

Fig. 3-2 The gate G is negative with respect to source S.

3-3 THE MATHEMATICAL NEED FOR NEGATIVE NUMBERS

From a purely mathematical viewpoint the need for negative numbers can be seen from the following series of operations in which we subtract successively larger numbers from 5:

5	5	5	5	5
0	1	2	3	4
5	4	3	2	1

5	5	5	5
5	6	7	8
0	−1	−2	−3

The above subtractions result in the remainders becoming less until zero is reached. When the remainder becomes less than zero, the fact is indicated by placing the negative sign before the remainder. This is one reason for defining a negative number as a number less than zero. Mathematically, the definition is correct if we consider only the signs that precede the numbers.

You must not lose sight of the fact, however, that as far as magnitude, or size, is concerned, a negative number may represent a larger absolute value than some positive number. *The positive and negative signs simply denote reference from zero.* For example, if some point in an electric circuit is 1000 V negative with respect to ground, you can say so by writing −1000 V. But if you make good contact with your body between that point and ground, your chances of being electrocuted are just as good as if that point were positive 1000 V with respect to ground—and you wrote it +1000 V! In this case, *how much* is far more important than a matter of sign preceding the number. Similarly, −1000 V is greater than +500 V, but of different polarity. −$10 000 is greater than +$6000, except that it is owed, rather than owned.

3-4 THE ABSOLUTE VALUE OF A NUMBER

The numerical, or absolute, value of a number is the value of the number without regard to sign. Thus, the absolute values of numbers such as −1, +4, −6, and +3 are 1, 4, 6, and 3, respectively. Note that different numbers, such as −9 and +9, may have the same absolute value. To specify the absolute value of a number, such as Z, we write $|Z|$. This is often referred to as "the modulus of Z," or simply, "mod Z."

3-5 ADDITION OF POSITIVE AND NEGATIVE NUMBERS

Positive and negative numbers can be represented graphically as in Fig. 3-3. Positive numbers are shown as being directed toward the right of zero, which is the reference point, whereas negative numbers are directed toward the left.

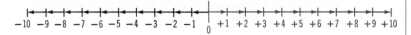

Fig. 3-3 Graphical representation of numbers from −10 to +10.

Such a scale of numbers can be used to illustrate both addition and subtraction as performed in arithmetic. Thus, in adding 3 to 4, we can begin at 3 and count 4 units to the right to obtain the sum 7. Or, because these are positive numbers directed toward the right, we could draw them to scale, place them end to end, and measure their total length to obtain a length of 7 units in the positive direction. This is illustrated in Fig. 3-4.

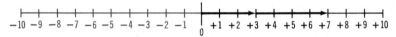

Fig. 3-4 Graphical addition of 3 and 4 to obtain 7.

In like manner, −2 and −3 can be added to obtain −5 as shown in Fig. 3-5.

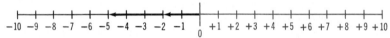

Fig. 3-5 Illustrating the addition of −2 and −3. The result is −5.

Note that adding −3 and −2 is the same as adding −2 and −3 as in the foregoing example. The sum −5 is obtained, as shown in Fig. 3-6.

Fig. 3-6 Adding −3 and −2 is the same as adding −2 and −3. Each result is −5.

Suppose we want to add +6 and −10. We could accomplish this on the scale by first counting 6 units to the right and from *that* point counting 10 units to the left. In so doing, we would end up at −4, which is the sum of +6 and −10. Similarly, we could have started by first counting 10 units to the left, from zero, and from that point counting 6 units to the right for the +6. Again we would have arrived at −4.

Adding +6 and −10 can be accomplished graphically as in Fig. 3-7. The +6 is drawn to scale, and then the tail of the −10 is joined with the head of +6. The head of the −10 is then on −4. As would be expected, the same result is obtained by first drawing in the −10 and then the +6.

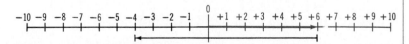

Fig. 3-7 Graphical addition of +6 and −10.

The following examples can be checked graphically in order to verify their correctness:

$$
\begin{array}{r} +8 \\ +4 \\ \hline +12 \end{array} \qquad
\begin{array}{r} +9 \\ -3 \\ \hline +6 \end{array} \qquad
\begin{array}{r} +6 \\ -9 \\ \hline -3 \end{array} \qquad
\begin{array}{r} -5 \\ +2 \\ \hline -3 \end{array} \qquad
\begin{array}{r} -7 \\ +9 \\ \hline +2 \end{array} \qquad
\begin{array}{r} -17 \\ -14 \\ \hline -31 \end{array}
$$

Consideration of the above examples enables us to establish the following rule:

RULE

1. To add two or more numbers with like signs, find the sum of their absolute values and prefix this sum with the common sign.
2. To add a positive number to a negative number, find the difference of their absolute values and prefix to the result the sign of the number that has the greater absolute value.

When three or more algebraic numbers that differ in signs are to be added, find the sum of the positive numbers and then the sum of the negative numbers. Add these sums algebraically, and use Rule 2 to obtain the total algebraic sum.

The *algebraic sum* of two or more numbers is the result obtained by adding the numbers according to the preceding rules. Hereafter, the word "add" will mean "find the algebraic sum."

PROBLEMS 3-1

Add:

1.
$$
\begin{array}{r} 28 \\ 43 \\ \hline \end{array}
$$
2.
$$
\begin{array}{r} 36 \\ -18 \\ \hline \end{array}
$$
3.
$$
\begin{array}{r} -82 \\ 36 \\ \hline \end{array}
$$
4.
$$
\begin{array}{r} -18 \\ -47 \\ \hline \end{array}
$$

5.
$$
\begin{array}{r} 124 \\ -96 \\ \hline \end{array}
$$
6.
$$
\begin{array}{r} 165 \\ -572 \\ \hline \end{array}
$$
7.
$$
\begin{array}{r} -286 \\ -795 \\ \hline \end{array}
$$
8.
$$
\begin{array}{r} 0.0007 \\ -0.0052 \\ \hline \end{array}
$$

9.
$$
\begin{array}{r} 175.03 \\ -2.75 \\ 36.28 \\ \hline \end{array}
$$
10.
$$
\begin{array}{r} -97.63 \\ 5.74 \\ -26.32 \\ \hline \end{array}
$$
11.
$$
\begin{array}{r} 7\frac{1}{2} \\ -3\frac{1}{2} \\ \hline \end{array}
$$
12.
$$
\begin{array}{r} -6\frac{1}{4} \\ 2\frac{1}{8} \\ \hline \end{array}
$$

13.
$$
\begin{array}{r} -3\frac{1}{32} \\ -7\frac{3}{16} \\ \hline \end{array}
$$
14.
$$
\begin{array}{r} -\frac{5}{8} \\ 3\frac{1}{4} \\ \hline \end{array}
$$
15.
$$
\begin{array}{r} \frac{1}{3} \\ -\frac{1}{5} \\ \hline \end{array}
$$

3-6 THE SUBTRACTION OF POSITIVE AND NEGATIVE NUMBERS

We may think of subtraction as the process of determining what number must be added to a given number in order to produce another given number.

Thus, when we subtract 5 from 9 and get 4, we have found that 4 must be added to 5 in order to obtain 9. From this it is seen that subtraction is the inverse of addition.

Example 1
$(+5) - (+2) = ?$

Solution
In this example the question is asked, "What number added to $+2$ will give $+5$?" Using the scale of Fig. 3-8, start at $+2$ and count to the right (positive direction), until you reach $+5$. This requires three units. Then, the difference is $+3$, or $(+5) - (+2) = +3$.

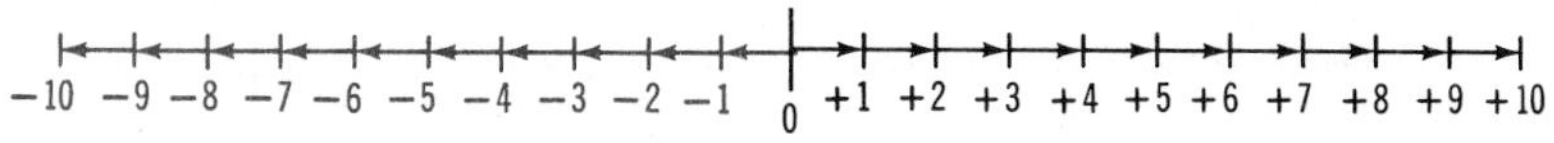

$$\begin{array}{ccccccccccccccccccccc} & -10 & -9 & -8 & -7 & -6 & -5 & -4 & -3 & -2 & -1 & 0 & +1 & +2 & +3 & +4 & +5 & +6 & +7 & +8 & +9 & +10 \end{array}$$

Fig. 3-8 Scale for graphical subtraction of positive and negative numbers.

Example 2
$(+5) - (-2) = ?$

Solution
In this example the question is asked, "What number added to -2 will give $+5$?" Using the scale, start at -2 and count the number of units to $+5$. This requires seven units, and because it was necessary to count in the positive direction, the difference is now shown as $+7$, or $(+5) - (-2) = +7$.

Example 3
$(-5) - (+2) = ?$

Solution
In this example the question is, "What number added to $+2$ will give -5?" Again using the scale, we start at $+2$ and count the number of units to -5. This requires seven units, but because it was necessary to count in the negative direction, the difference is -7, or now shown as $(-5) - (+2) = -7$.

Example 4
$(-5) - (-2) = ?$

Solution
Here the question is, "What number added to -2 will give -5?" Using the scale, we start at -2 and count the number of units to -5. This requires three units in the negative direction. This yields $(-5) - (-2) = -3$.

Summing up Examples 1 to 4, we have the following subtractions:

$$\begin{array}{cccc} +5 & +5 & -5 & -5 \\ +2 & -2 & +2 & -2 \\ \hline +3 & +7 & -7 & -3 \end{array}$$

A study of the foregoing subtractions illustrates the following principles:

1. Subtracting a positive number is equivalent to adding a negative number of the same absolute value.
2. Subtracting a negative number is equivalent to adding a positive number of the same absolute value.

These principles can be used for the purpose of establishing the following rule:

RULE

To subtract one number from another, change the sign of the subtrahend and add algebraically.

As in arithmetic, the number to be subtracted is called the *subtrahend*. The number from which the subtrahend is subtracted is called the *minuend*. The result is called the *remainder* or *difference*.

$$\begin{aligned}
\text{Minuend} &= -642 \\
\text{Subtrahend} &= \underline{\quad 403} \\
\text{Remainder} &= -1045
\end{aligned}$$

PROBLEMS 3-2

Subtract the second line from the first:

1. $\begin{array}{r} 87 \\ \underline{26} \end{array}$ 2. $\begin{array}{r} 25 \\ \underline{-96} \end{array}$ 3. $\begin{array}{r} -362 \\ \underline{-575} \end{array}$ 4. $\begin{array}{r} -125 \\ \underline{252} \end{array}$

5. $\begin{array}{r} 596 \\ \underline{-398} \end{array}$ 6. $\begin{array}{r} 0.009\ 25 \\ \underline{0.072\ 54} \end{array}$ 7. $\begin{array}{r} -3.08 \\ \underline{-6.92} \end{array}$ 8. $\begin{array}{r} 5\frac{2}{3} \\ \underline{-2\frac{3}{4}} \end{array}$

9. $\begin{array}{r} -12\frac{7}{16} \\ \underline{-2\frac{3}{8}} \end{array}$ 10. $\begin{array}{r} -\frac{1}{2} \\ \underline{\frac{5}{8}} \end{array}$

11. How many degrees must the temperature rise to change from (*a*) $+6°$ to $+73°$, (*b*) $-12°$ to $+14°$, and (*c*) $-273°$ to $-114°$?
12. How many degrees must the temperature fall to change from (*a*) $+212°$ to $+32°$, (*b*) $+55°$ to $-16°$, and (*c*) $-6°$ to $-42°$?
13. What amount of money is required to change an account from a debit of $124.50 to a credit of $240.30?
14. A certain point in a circuit is 570 V negative with respect to ground. Another point in the same circuit is 115 V positive with respect to ground. What is the potential difference between the two points?
15. In Fig. 3-2 what is the potential difference between the drain D and source S?

3-7 ADDITION AND SUBTRACTION OF LIKE TERMS

In arithmetic, it is never possible to add unlike quantities. For example, we should not add inches and gallons and expect to obtain a sensible answer. Neither should we attempt to add volts and amperes, kilohertz and microfarads, ohms and watts, etc. So it goes on through algebra—we can never add quantities unless they are expressed in the same units.

The addition of two like terms such as $6VI + 12VI = 18VI$ can be checked by substituting numbers for the literal factors. Thus, if $V = 1$ and $I = 2$,

$$
\begin{aligned}
6VI &= 6 \times 1 \times 2 = 6 \times 2 = 12 \\
12VI &= 12 \times 1 \times 2 = 12 \times 2 = 24 \\
\hline
18VI &= 18 \times 1 \times 2 = 18 \times 2 = 36
\end{aligned}
$$

From the foregoing, it is apparent that like terms may be added or subtracted by adding or subtracting their coefficients.

The addition or subtraction of unlike terms cannot be carried out but can only be indicated, because the unlike literal factors may stand for entirely different quantities.

Example 5
Addition of like terms:

$$
\begin{array}{ccc}
-3i^2r & -16IR & 13jIX \\
8i^2r & 14IR & -20jIX \\
\hline
5i^2r & -3IR & -32jIX \\
& \hline & \hline \\
& -5IR & -39jIX
\end{array}
$$

Example 6
Subtraction of like terms:

$$
\begin{array}{ccc}
-8e_1 & 6iZ & -28L^2R \\
3e_1 & -13iZ & -29L^2R \\
\hline
-11e_1 & 19iZ & L^2R
\end{array}
$$

Example 7
Addition of unlike terms:

$$
\begin{array}{cc}
3v & -3r \\
-3IX & 4R \\
4V & -16R_t \\
\hline
3v - 3IX + 4V & 4R - 3r - 16R_t
\end{array}
$$

$$
\begin{array}{c}
3VI \\
10I^2R \\
-46W \\
\hline
3VI + 10I^2R - 46W
\end{array}
$$

3-8 ADDITION AND SUBTRACTION OF POLYNOMIALS

Polynomials are added or subtracted by arranging like terms in the same column and then combining terms in each column, as with monomials.

Example 8
Addition of polynomials

$$
\begin{array}{rrr}
-3ab & + 6cd & + \ x^2y \\
14ab & & - 5x^2y \\
\underline{ab} & \underline{- 3cd} & \\
12ab & + 3cd & - 4x^2y
\end{array}
\qquad
\begin{array}{rrr}
6V & + 3RI & - 8IZ \\
& RI & - 2IZ \\
\underline{-7V} & & \underline{+ 3IZ} \\
-V & + 4RI & - 7IZ
\end{array}
$$

Example 9
Subtraction of polynomials:

$$
\begin{array}{rrr}
3mn & + 16pq & - \ xy^2 \\
\underline{-9mn} & & \underline{+ 7xy^2} \\
12mn & + 16pq & - 8xy^2
\end{array}
\qquad
\begin{array}{rrr}
11R & + 4x & \\
\underline{15R} & & \underline{- 18Z} \\
-4R & + 4x & + 18Z
\end{array}
$$

PROBLEMS 3-3

Add (Problems 1 through 16):

1. $2i, \ 6i, \ -5i, \ 8i$
2. $5i^2r, \ 10i^2r, \ -26i^2r, \ 3i^2r$
3. $27IZ, \ 165IZ, \ -64IZ, \ -32IZ, \ 16IZ$
4. $65IR, \ -8.7IR, \ IR, \ -16.6IR, \ 15.2IR$
5. $3i + 16I, \ -8i - 12I$
6. $8jX, \ 26jX, \ -30jX, \ 18R, \ -5jX, \ 12R$
7. $25IR, \ + 3V, \ -4IR - 2V, \ - 18IR + 12V$
8. $12\Omega, \ 2\omega, \ -16\omega, \ 4\Omega$
9. $25\phi + 41\theta, \ 36\theta - 82\phi, \ -53\phi + 51\theta$
10. $5L, \ 4R, \ -27L, \ -5Z, \ 36L, \ 7R - 2Z$
11.
$$
\begin{array}{rrrr}
6i^2r & + 8W & - 6vi & + 32w \\
-3i^2r & + 3W & + 8vi & + 18w \\
\underline{24i^2r} & \underline{- W} & \underline{- 5vi} & \underline{- w}
\end{array}
$$

12.
$$
\begin{array}{rrr}
25IX & - 16IZ & + 3IR \\
14IZ & + 2IX & - IR \\
\underline{8IR} & + 4IX & - 3IZ
\end{array}
$$

13.
$$
\begin{array}{rrr}
1.65vI & + 3.07W & - 1.46I^2r \\
0.025W & - 1.11vI & - 0.85I^2r \\
\underline{3.06I^2r} & + 0.92vI & + 0.725W}
\end{array}
$$

14. $2.15vi + 1.64 \dfrac{v^2}{r} - 3.82i^2r, \ 0.57 \dfrac{v^2}{r} + 1.94i^2r$

15. $\frac{1}{4}\pi ft$, $-3\pi Z$, $-\frac{2}{3}\pi ft$, $\frac{3}{16}\pi ft$, $\frac{7}{8}\pi Z$

16. $47IR + 3IZ$ to $-15IR - 4IZ$.

17. From $25\phi + 3\theta$ subtract $15\phi - 7\theta$.

18. From $17.2\omega L + 5X_C - 13.2Z$ subtract $4.5\omega L - 3.2X_C + 5.6Z$.

19. From the sum of $26.2\,\dfrac{V^2}{R} + 14.6VI - 3I^2R$ and

    $6.2I^2R - 3.8VI + 19.6\,\dfrac{V^2}{R}$ subtract $27.2VI - 2.6I^2R - 1.8\,\dfrac{V^2}{R}$.

20. Subtract $9.5X_C + \dfrac{3.26}{\omega C}$ from the sum of $-8.7X_C + \dfrac{2.46}{\omega C}$ and

    $-4.6X_C - \dfrac{1.98}{\omega C}$.

21. Take $1.25IR + 0.64IX - 2.81IZ$ from $-0.06IR + 0.23IX + 1.09IZ$.

22. How much more than $5V_g - 2iR$ is $3V_g + 6iR$?

23. What must be added to $3\psi + 2.8\lambda$ to obtain $9.64\psi - 4.3\lambda$?

24. What must be subtracted from $16.2\gamma - 3.3\alpha + 2.8\beta$ to obtain $8.1\alpha + 1.7\gamma - 2.6\beta$?

3-9 SIGNS OF GROUPING

Often it is necessary to express or group together quantities that are to be affected by the same operation. Also, it is desirable to be able to represent that two or more terms are to be considered as one quantity.

In order to meet the above requirements, signs of grouping have been adopted. These signs are the *parentheses* (   ), the *brackets* [   ], the *braces* {   }, and the *vinculum* _____. The first three are placed around the terms to be grouped as in $(V - IR)$, $[a + 3b]$, and $\{x^2 + 4y\}$. All have the same meaning: that the enclosed terms are to be considered as one quantity.

Thus, $16 - (12 - 5)$ means that the quantity $(12 - 5)$ is to be subtracted from 16. That is, 5 is to be subtracted from 12, and then the remainder 7 is to be subtracted from 16 to give a final remainder of 9. In like manner, $V - (IR + v)$ means that the sum of $(IR + v)$ is to be subtracted from V.

Carefully note that the sign preceding a sign of grouping, as the minus sign between V and $(IR + v)$ above, is a sign of *operation* and does not denote that $(IR + v)$ is a negative quantity.

The vinculum is used mainly with radical signs and fractions, as in

$$\sqrt{7245} \qquad \text{and} \qquad \frac{a + b}{x - y}$$

In the latter case the vinculum denotes the division of $a + b$ by $x - y$, in addition to grouping the terms in the numerator and denominator. When studying later chapters, you will avoid many mistakes by remembering that *the vinculum is a sign of grouping*.

In working problems involving signs of grouping, the operations within the signs of grouping should be performed first.

Example 10

$a + (b + c) = ?$

Solution

This means, "What result will be obtained when the sum of $b + c$ is added to a?" Because both b and c are denoted as positive, it follows that we can write

$$a + (b + c) = a + b + c$$

because it makes no difference in which order we add.

Example 11

$a + (b - c) = ?$

Solution

This means, "What result will be obtained when the difference of $b - c$ is added to a?" Again, because it makes no difference in which order we add, we can write

$$a + (b - c) = a + b - c$$

Example 12

$a - (b + c) = ?$

Solution

Here the sum of $b + c$ is to be subtracted from a. This is the same as if we first subtract b from a and from the remainder subtract c. Therefore,

$$a - (b + c) = a - b - c$$

Because this is subtraction, we could change the signs and add algebraically, remembering that b and c are denoted as positive, as shown below:

$$\begin{array}{r} a \\ b + c \\ \hline a - b - c \end{array}$$

Example 13

$a - (-b - c) = ?$

Solution

This means that the quantity $-b - c$ is to be subtracted from a. Performing this subtraction, we obtain

$$\begin{array}{r} a \\ -b - c \\ \hline a + b + c \end{array}$$

Therefore, $a - (-b - c) = a + b + c$

A study of Examples 10 to 13 enables us to state the following:

RULES

1. Parentheses or other signs of grouping preceded by a plus sign can be removed without any other change.
2. To remove parentheses or other signs of grouping preceded by a minus sign, change the sign of every term within the sign of grouping.

Although not apparent in the examples, another rule can be added as follows:

3. If parentheses or other signs of grouping occur one within another, remove the inner grouping first.

Examples

$$(x + y) + (2x - 3y) = x + y + 2x - 3y = 3x - 2y$$

$$3a - (2b + c) - a = 3a - 2b - c - a = 2a - 2b - c$$

$$10x - (-3x - 4y) + 2y = 10x + 3x + 4y + 2y = 13x + 6y$$

$$x - [2x + 3y - (3x - y) - 4x] = x - [2x + 3y - 3x + y - 4x]$$
$$= x - 2x - 3y + 3x - y + 4x$$
$$= 6x - 4y$$

PROBLEMS 3-4

Simplify by removing the signs of grouping and combining the similar terms:

1. $(x - 3y - 4) - (x + 4y - 7)$
2. $(5\lambda + 3\theta) - (-4\lambda + 5\theta + 6)$
3. $(4R + 5Z + 6X) - (9X - 6R + 5Z - 3)$
4. $6I^2R + [-5VI + (-I^2R - 3VI) - 7VI] + 5$
5. $8\dfrac{V^2}{R} - \left[-6I^2R - \left(5\dfrac{V^2}{R} - \overline{6I^2R + 3\dfrac{V^2}{R} + 3VI} \right) \right]$
6. $X_L - \{3L - [2R - (X_L + 5L)]\}$
7. $5\alpha - \{\alpha + \beta - [\gamma + \alpha + \beta - (\alpha + \beta + \gamma) - 3\alpha] - 3\beta\}$
8. $-\{-\theta - [\phi + \omega - 2\phi - (\omega + \phi) - \phi] - 2\theta - 3\omega\}$
9. $4a - [-5a - (-6b + 3c) - (8a - \overline{4b - 3c})]$
10. $5.4R - 2.6Z - \overline{1.5IX - 7R} - [4.6Z - (3X_C - 5.7IX) - 4.32R + 27]$

3-10 INSERTING SIGNS OF GROUPING

To enclose terms within signs of grouping preceded by a plus sign, rewrite the terms without changing their signs.

Example 14

$$a + b - c + d = a + (b - c + d)$$

To enclose terms within signs of grouping preceded by a minus sign, rewrite the terms and change the signs of the terms enclosed.

Example 15

$$a + b - c + d = a + b - (c - d)$$

No difficulty need be encountered when inserting signs of grouping because, by removing the signs of grouping from the result, the original expression should be obtained.

Example 16

$$x - 3y + z = x - (3y - z)$$
$$= x - 3y + z$$

PROBLEMS 3-5

1. Enclose the last three terms of each of the following expressions in parentheses preceded by a plus sign:

   (a) $3X + X_C - X_L + Z$ (b) $\alpha + 6\beta - 3\phi + \lambda$

   (c) $5W + 6I^2R - 3VI + 7I^2Z$ (d) $\dfrac{V^2}{R} - 3I^2R + 7I^2Z - 4VI$

   (e) $8\lambda + 3\mu - 7\theta - 3\phi + 6\alpha$

2. Enclose the last three terms of each of the following expressions in parentheses preceded by a minus sign:

   (a) $8VI + 5I^2R - 6W + 4\dfrac{V^2}{R}$ (b) $5a + 3b - 6c + 4d + 5e$

   (c) $8\omega + 13\phi - 3\lambda + 2.7r$ (d) $4.6° - 3\theta + 3.8^r - 0.52\phi$

   (e) $\dfrac{V^2}{R} + W + I^2Z - 6VI$

3. Write the amount by which N is less than $(X^2 + R^2)$.

4. The sum of two currents is 526 milliamperes (mA). The larger of the two currents is i mA. What is the smaller?

5. The difference between two voltages is 16.8 V. The smaller voltage is v V. What is the greater?

6. Write the amount by which X_L exceeds $\dfrac{1}{2\pi fC}$.

7. What is the larger part of Z if $\sqrt{r^2 + x^2}$ is the smaller part?

8. Write the amount by which V is greater than $v - IR$.

9. Write the amount by which P exceeds $I^2R + \dfrac{V^2}{R}$.

10. The difference between two numbers is 19.6. If the larger number is β, what is the smaller?

11. Write the smaller part of X_C if $\dfrac{1}{2\pi fC_1}$ is the larger part.

12. The difference between two numbers is X^2 and the larger of the two is Z^2. Write the relationship which describes R^2, which is the smaller of the two.

Algebra— Multiplication and Division

4

Multiplication is often defined as the *process of repeated addition*. Thus, 2×3 may be thought of as adding 2 three times, or $2 + 2 + 2 = 6$.

Considering multiplication as a shortened form of addition is not satisfactory, however, when the multiplier is a fraction. For example, it would not be sensible to say that $5 \times \frac{2}{7}$ was adding 5 two-sevenths of a time. This problem could be rewritten as $\frac{2}{7} \times 5$, which would be the same as adding $\frac{2}{7}$ five times. But this is only a temporary help, for if two fractions are to be multiplied together, as $\frac{3}{4} \times \frac{5}{6}$, the original definition of multiplication will not apply. However, the definition has been extended to include such cases, and the product of $5 \times \frac{2}{7}$ is taken to mean 5 multiplied by 2 and this product divided by 7; that is, by $5 \times \frac{2}{7}$ is meant $\frac{5 \times 2}{7}$. Also,

$$\frac{3}{4} \times \frac{5}{6} = \frac{3 \times 5}{4 \times 6} = \frac{15}{24}$$

4-1 MULTIPLICATION OF POSITIVE AND NEGATIVE NUMBERS

Because we are now dealing with both positive and negative numbers, it becomes necessary to determine what sign the product will have when combinations of these numbers are multiplied.

When only two numbers are to be multiplied, there can be but four possible combinations of signs, as follows:

(1) $\qquad (+2) \times (+3) = ?$
(2) $\qquad (-2) \times (+3) = ?$
(3) $\qquad (+2) \times (-3) = ?$
(4) $\qquad (-2) \times (-3) = ?$

Combination (1) means that $+2$ is to be added three times:

$$(+2) + (+2) + (+2) = +6$$
or $\qquad (+2) \times (+3) = +6$

In the same manner, combination (2) means that -2 is to be added three times:

$$(-2) + (-2) + (-2) = -6$$

or

$$(-2) \times (+3) = -6$$

Combination (3) means that $+2$ is to be subtracted three times:

$$-(+2) - (+2) - (+2) = -6$$

or

$$(+2) \times (-3) = -6$$

Note that this is the same as subtracting 6 once, -6 being thus obtained.

Combination (4) means that -2 is to be subtracted three times:

$$-(-2) - (-2) - (-2) = +6$$

or

$$(-2) \times (-3) = +6$$

This may be considered to be the same as subtracting -6 once; and because subtracting -6 once is the same as adding $+6$, we obtain $+6$ as above.

From the foregoing we have these rules:

RULES

1. The product of two numbers having like signs is positive.
2. The product of two numbers having unlike signs is negative.
3. If more than two factors are multiplied, Rules 1 and 2 are to be used successively.
4. The product of an even number of negative factors is positive. The product of an odd number of negative factors is negative.

These rules can be summarized in general terms as follows:

- Rule 1 $(+a)(+b) = +ab$
- Rule 1 $(-a)(-b) = +ab$
- Rule 2 $(+a)(-b) = -ab$
- Rule 2 $(-a)(+b) = -ab$
- Rule 3 $(-a)(+b)(-c) = +abc$
- Rule 4 $(-a)(-b)(-c)(-d) = +abcd$
- Rule 4 $(-a)(-b)(-c) = -abc$

PROBLEMS 4-1

Find the products of the following factors:

1. $3, 4$
2. $6, -5$
3. $-9.1, -1.5$
4. $-1.7, 6.5, -7.3$
5. $\dfrac{3}{16}, -\dfrac{5}{8}, \dfrac{1}{2}$
6. $\dfrac{2}{3}, -\dfrac{3}{4}, -\dfrac{7}{8}$
7. $-0.025, -0.0005, -2.5, -0.03$
8. $3000, -0.06, 250, -0.002$
9. $-e, -i, t$
10. $q, -r, -s, t$
11. $2\pi f, L_1, L_2$
12. $\theta^2, \phi^2, \lambda^2$

13. $\dfrac{1}{2}, \dfrac{1}{\pi}, \dfrac{1}{f}, \dfrac{1}{C_p}$ **14.** $\dfrac{1}{a}, -\dfrac{1}{b}, -\dfrac{1}{c}, -\dfrac{1}{d}$

15. $\psi, \dfrac{1}{\theta}, -\dfrac{1}{\phi}, \mu$

4-2 GRAPHICAL REPRESENTATION

Our system of representing numbers is a graphical one, as previously illustrated in Fig. 3-3. It might be well at this time to consider certain facts regarding multiplication.

When a number is multiplied by any other number except 1, we can think of the operation as having changed the absolute value of the multiplicand. Thus, 3 in $\times$ 4 becomes 12 in, 6 A $\times$ 3 becomes 18 A, etc. Such multiplications could be represented graphically by simply extending the multiplicand the proper amount, as shown in Fig. 4-1.

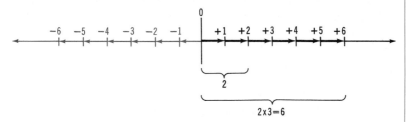

Fig. 4-1 Representation of the multiplication 2 $\times$ 3 = 6.

The multiplication of a negative number by a positive number is shown in Fig. 4-2.

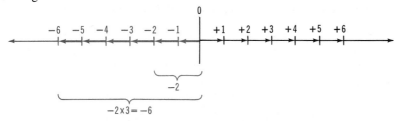

Fig. 4-2 Representation of the multiplication $-2 \times 3 = -6$.

From these examples, it is evident that a positive multiplier simply changes the absolute value, or magnitude, of the number being multiplied. What happens if the multiplier is negative? As an example, consider $2 \times (-3) = -6$. How will this be represented graphically?

Now, $2 \times (-3) = -6$ is the same as

$$2 \times (+3) \times (-1) = -6$$

Therefore, let us first multiply 2×3 to obtain $+6$ and represent it as shown in Fig. 4-1. We must multiply by -1 to complete the problem and in so doing should obtain -6, but -6 must be represented as a number six units in length and directed toward the left, as illustrated in Fig. 4-2.

We therefore agree that multiplication by −1 causes counterclockwise rotation of a number in a direction that will be exactly opposite from its original direction. This is illustrated in Fig. 4-3.

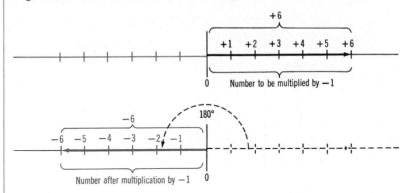

Fig. 4-3 Multiplication by −1 rotates multiplicand counterclockwise through 180°.

If both multiplicand and multiplier are negative, as in

$$(-2) \times (-3) = +6$$

the representation is as illustrated in Fig. 4-4. Again,

$$(-2) \times (-3) = +6$$

is the same as

$$(-2) \times (+3) \times (-1) = +6$$

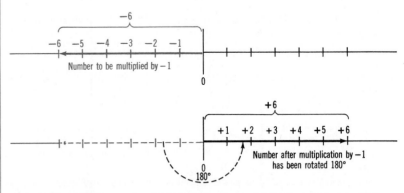

Fig. 4-4 Illustration of −6 rotated counterclockwise through 180° to become +6 owing to multiplication by −1.

The product has an absolute value of 6, and at the same time there has been rotation to +6 because of multiplication by −1.

The foregoing representations are applicable to division also, since the law of signs is the same as in multiplication.

The important thing to bear in mind is that multiplication or division by −1 causes counterclockwise rotation of a number to a direction exactly opposite the original direction. The number −1, when used as a multiplier or divisor, should be considered as an *operator* for the purpose of rotation. You should clearly understand this concept; you will encounter it later.

4-3 LAW OF EXPONENTS IN MULTIPLICATION

As explained in Sec. 2-10, an exponent indicates how many times a number is to be taken as a factor. Thus $x^4 = x \cdot x \cdot x \cdot x$, $a^3 = a \cdot a \cdot a$, etc.

Because $\qquad x^4 = x \cdot x \cdot x \cdot x$

and $\qquad x^3 = x \cdot x \cdot x$

then $\qquad x^4 \cdot x^3 = x \cdot x \cdot x \cdot x \cdot x \cdot x \cdot x = x^7$

or $\qquad x^4 \cdot x^3 = x^{4+3} = x^7$

Thus, we have the rule:

RULE

To find the product of two or more powers having the same base, add the exponents.

Examples

$$a^3 \cdot a^2 = a^{3+2} = a^5$$
$$x^4 \cdot x^4 = x^{4+4} = x^8$$
$$6^2 \cdot 6^3 \cdot 6^5 = 6^{2+3+5} = 6^{10}$$
$$a^2 \cdot b^3 \cdot b^3 \cdot a^5 = a^{2+5} \cdot b^{3+3} = a^7 b^6$$
$$e \cdot e^3 = e^{1+3} = e^4$$
$$3^2 \cdot 3^4 = 3^{2+4} = 3^6$$
$$e^a \cdot e^b = e^{a+b}$$

From the foregoing examples, it is seen that the law of exponents can be expressed in the well-known general form

$$a^m \cdot a^n = a^{m+n}$$

where $a \neq 0$ and m and n are literal numbers and may represent any number of factors.

4-4 MULTIPLICATION OF MONOMIALS

RULES

1. Find the product of the numerical coefficients and give it the proper sign, plus or minus, according to the rules for multiplication (Sec. 4-1).

2. Multiply this numerical product by the product of the literal factors. Use the law of exponents as applicable.

Example 1

Multiply $3a^2b$ by $4ab^3$.

Solution

$$(3a^2b)(4ab^3) = +(3 \cdot 4) \cdot a^{2+1} \cdot b^{1+3}$$
$$= 12a^3b^4$$

Example 2

Multiply $-6x^3y^2$ by $3xy^2$.

Solution

$$(-6x^3y^2)(3xy^2) = -(6 \cdot 3) \cdot x^{3+1} \cdot y^{2+2}$$
$$= -18x^4y^4$$

Example 3

Multiply $-5e^2x^4y$ by $-3e^2x^2p$.

Solution

$$(-5e^2x^4y)(-3e^2x^2p) = +(5 \cdot 3)e^{2+2} \cdot p \cdot x^{4+2} \cdot y$$
$$= 15e^4px^6y$$

PROBLEMS 4-2

Multiply:

1. $x^3 \cdot x^2$
2. $-b^3 \cdot b^5$
3. $e^2 \cdot e^3 \cdot -e^5$
4. $-\lambda \cdot \lambda^2 \cdot -\theta^3$
5. $(2m^2)(3m^2)$
6. $(6\alpha)(-3\beta^3)$
7. $(4x)(5m^3)(-3x^2m)$
8. $(-5\mu)^2$
9. $(am^n)(bm^p)$
10. $(13b^x)(-2b^{a+y})$
11. $(2p)^3$
12. $(-5\lambda^2)^3$
13. $(-3a^2b^3cd^2)(-2abc^2d^5)$
14. $\left(\frac{1}{4}a^3\right)\left(-\frac{2}{3}ab^2\right)$
15. $\left(\frac{3}{16}X_L\right)\left(\frac{2}{3}M\right)(-2\pi)$
16. $(14a^2b^3cd)\left(-\frac{2}{7}ab^2de\right)$
17. $(0.5v^2i)(3i^2r)(-0.05vi)(w)$
18. $\left(\frac{5}{16}\theta\phi\right)\left(-\frac{3}{25}\mu\theta\right)\left(-\frac{24}{27}\theta^2\omega\right)$
19. $(a^3)^2$
20. $(3p^q)^r$

4-5 MULTIPLICATION OF POLYNOMIALS BY MONOMIALS

Another method of graphically representing the product of two numbers is as shown in Fig. 4-5. The product $5 \times 6 = 30$ is shown as a rectangle whose sides are 5 and 6 units in length; therefore, the rectangle contains 30 square units.

Similarly, the product of $5(6 + 9)$ can be represented as illustrated in Fig. 4-6.

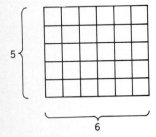

Fig. 4-5 Graphical representation of the multiplication $5 \times 6 = 30$.

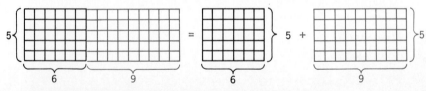

Fig. 4-6 Grahical representation of the multiplication $5(6 + 9) = 5 \times 6 + 5 \times 9 = 75$.

Thus, $5(6 + 9) = 5 \times 15$
                                            $= 75$

Also, $5(6 + 9) = (5 \times 6) + (5 \times 9)$
                                    $= 30 + 45$
                                    $= 75$

In like manner the product

$$a(c + d) = ac + ad$$

can be illustrated as in Fig. 4-7.

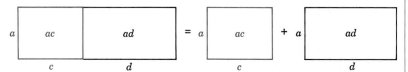

Fig. 4-7 Illustration of the product $a(c + d) = ac + ad$.

From the foregoing, you can show that

$$3(4 + 2) = 3 \times 4 + 3 \times 2 = 12 + 6 = 18$$
$$4(5 + 3 + 4) = 4 \times 5 + 4 \times 3 + 4 \times 4$$
$$= 20 + 12 + 16 = 48$$
$$x(y + z) = xy + xz$$
$$p(q + r + s) = pq + pr + ps$$

Note that, in all cases, each term of the polynomial (the terms enclosed in parentheses) is multiplied by the monomial. From these examples, we develop the following rule:

RULE

To multiply a polynomial by a monomial, multiply each term of the polynomial by the monomial and write in succession the resulting terms with their proper signs.

Example 4

$3x(3x^2y - 4xy^2 + 6y^3) = ?$

Solution

Multiplicand $= 3x^2y - 4xy^2 + 6y^3$
Multiplier $= 3x$
Product $= 9x^3y - 12x^2y^2 + 18xy^3$

Example 5

$-2ac(-10a^3 + 4a^2b - 5ab^2c + 7bc^2) = ?$

Solution

Multiplicand $= -10a^3 + 4a^2b - 5ab^2c + 7bc^2$
Multiplier $= -2ac$
Product $= 20a^4c - 8a^3bc + 10a^2b^2c^2 - 14abc^3$

Example 6

Simplify $5(2e - 3) - 3(e + 4)$.

Solution

First multiply $5(2e - 3)$ and $3(e + 4)$, and then subtract the second result from the first, thus:

$$5(2e - 3) - 3(e + 4) = (10e - 15) - (3e + 12)$$
$$= 10e - 15 - 3e - 12$$
$$= 7e - 27$$

PROBLEMS 4-3

Multiply:

1. $3a + 5b$ by 6
2. $2a + 3$ by $3a$
3. $2R_1 + 4R_2$ by $2I^2$
4. $5.8a - jb$ by b^2
5. $\lambda^2 + 2\theta - 3\mu$ by 4.7ϕ
6. $2\alpha^3 - 3\alpha^2 + 4\alpha$ by -5α
7. $4\alpha^3\beta + 3\alpha^2\beta^2 - 5\alpha\beta^3$ by $0.5\alpha\beta$
8. $2\theta^2\phi - 5\alpha\theta^2 - 4\alpha\beta + 3$ by $3\alpha\phi$
9. $-5a^2r_1 - 2ar_1^2 + 6r_1^3$ by $-3ar_2$
10. $3\omega^2L_1^2 - 5\omega^2M + 7\omega^2L_2^2$ by $-2\omega L_1L_2$
11. $\frac{1}{2}I^2R - \frac{1}{4}I^2R^2 - \frac{1}{3}iZ$ by $\frac{2}{3}iIZ$
12. $8ab + 4ab^2 + 4$ by $-\frac{1}{4}ab^2$
13. $\frac{I^2R}{4} - \frac{i^2r}{2} + \frac{P}{6}$ by $12IP$
14. $5\mu^2k^2 + 3\eta k - 2\mu\eta^2$ by $-3\theta\omega$
15. $0.025E^3Z^2 + 0.05EZ^4 - 1.67Z^5$ by $6.28IZ$

Simplify:

16. $3ars(-4ar^2 + 2rs - 6as^2)$
17. $3(6\phi - 5\theta) - 3(\phi + 2\theta)$
18. $\mu(\alpha - j\beta) + \mu(\alpha + j\beta)$
19. $\theta(\theta^2 + \phi) - \phi(\theta + \phi^2)$
20. $3Z(2I^2 - i^2) - Z(6I^2 - 5i^2)$
21. $0.5\omega(6\pi + 5\eta\omega - \pi\omega^2) - 3\pi(0.7\omega - \eta\pi + 2\omega^3)$
22. $\frac{1}{2}\gamma\beta(4\gamma^2\beta - 2\gamma\beta^2 - 10\gamma^3 + 5\beta^3)$
23. $8\lambda\left(\frac{E^2}{3} + \frac{Ee}{2} - \frac{e^2}{16}\right)$
24. $5\theta(2\theta^2 + 3\theta\phi - 6\phi^2) - 3(6\theta^3 - 2\theta^2\phi - 7\theta\phi^2)$
25. $0.25I\left(\frac{R}{5} - \frac{R_1}{10} + \frac{3R_2}{5}\right) + 1.5(0.05IR - 0.375IR_2)$
26. $\frac{1}{2}\lambda\mu(6\lambda^2\mu - 5\lambda\mu^2 + 12\lambda - 4\mu)$
27. $4\theta(3\theta^2 + 2\theta\phi - 3\phi^2) - 2\theta(6\theta^2 + 4\theta\phi - 6\phi^2)$
28. $5\gamma^2\left(\frac{\gamma\lambda}{3} - \frac{\beta\lambda^2}{5} - \frac{\theta\beta^2}{10}\right) + 6\beta^2\left(\frac{\beta\lambda}{2} + \frac{\gamma\lambda}{5} - \frac{\gamma^2\theta}{12}\right)$
29. $6\left(\frac{s}{3} - \frac{s}{2} + \frac{2s}{3}\right) + 8\left(\frac{s}{4} - \frac{s}{2} + \frac{3s}{4}\right)$
30. $0.5I^2(2R_1 + 3R_2 - 5R_3) - 0.8I^2(-0.5R_1 - 2R_2 - 0.25R_3)$

4-6 MULTIPLICATION OF A POLYNOMIAL BY A POLYNOMIAL

It is apparent that

$$(3 + 4)(6 - 3) = 7 \times 3 = 21$$

The above multiplication can also be accomplished in the following manner:

$$
\begin{aligned}
(3 + 4)(6 - 3) &= 3(6 - 3) + 4(6 - 3) \\
&= (18 - 9) + (24 - 12) \\
&= 9 + 12 \\
&= 21
\end{aligned}
$$

Similarly,

$$
\begin{aligned}
(2a - 3b)(a + 5b) &= 2a(a + 5b) - 3b(a + 5b) \\
&= (2a^2 + 10ab) - (3ab + 15b^2) \\
&= 2a^2 + 10ab - 3ab - 15b^2 \\
&= 2a^2 + 7ab - 15b^2
\end{aligned}
$$

From the foregoing, we have the following:

RULE

To multiply polynomials, multiply every term of the multiplicand by each term of the multiplier and add the partial products.

Example 7

Multiply $2i - 3$ by $i + 2$.

Solution

$$
\begin{aligned}
\text{Multiplicand} &= 2i - 3 \\
\text{Multiplier} &= \underline{i + 2} \\
i \text{ times } (2i - 3) &= 2i^2 - 3i \\
2 \text{ times } (2i - 3) &= \underline{\qquad 4i - 6} \\
\text{Adding, product} &= 2i^2 + \quad i - 6
\end{aligned}
$$

Example 8

Multiply $a^2 - 3ab + 2b^2$ by $2a^2 - 3b^2$.

Solution

$$
\begin{aligned}
\text{Multiplicand} &= a^2 - 3ab + 2b^2 \\
\text{Multiplier} &= \underline{2a^2 - 3b^2} \\
2a^2 \text{ times} & \\
(a^2 - 3ab + 2b^2) &= 2a^4 - 6a^3b + 4a^2b^2 \\
-3b^2 \text{ times} & \\
(a^2 - 3ab + 2b^2) &= \underline{\qquad\qquad - 3a^2b^2 + 9ab^3 - 6b^4} \\
\text{Adding, product} &= 2a^4 - 6a^3b + \quad a^2b^2 + 9ab^3 - 6b^4
\end{aligned}
$$

Products obtained by multiplication can be tested by substituting any convenient numerical values for the literal numbers. It is not good practice to substitute the numbers 1 and 2. If there are exponents, then the use of

1 will not be a proof of correct work, for 1 to any power is still 1. Similarly, if addition should be involved, the use of 2 could give an incorrect indication, because $2 + 2 = 4$ and $2 \times 2 = 4$.

Example 9

Multiply $a^2 - 4ab - b^2$ by $a + b$, and test by letting $a = 3$ and $b = 4$.

Solution

$$
\begin{array}{lll}
a^2 - 4ab - b^2 & = 9 - 48 - 16 = & -55 \\
\underline{a + b} & = 3 + 4 = & \underline{7} \\
a^3 - 4a^2b - ab^2 & & -385 \\
\underline{ a^2b - 4ab^2 - b^3} & & \\
a^3 - 3a^2b - 5ab^2 - b^3 & = 27 - 108 - 240 - 64 & = -385
\end{array}
$$

PROBLEMS 4-4

Multiply

1. $\alpha + 1$ by $\alpha + 1$ 2. $\alpha + 1$ by $\alpha - 1$
3. $\alpha - 1$ by $\alpha - 1$ 4. $\beta + 2$ by $\beta + 2$
5. $\beta + 3$ by $\beta - 3$ 6. $\beta - 3$ by $\beta - 3$
7. $p + 3$ by $p + 5$ 8. $X_C - 6$ by $X_C - 4$
9. $r - 11$ by $r + 3$ 10. $j + 2$ by $j - 2$

Note Parentheses or other signs of grouping are often used to indicate a product. Thus, $(ir + v)(2ir - 3v)$ means $ir + v$ multiplied by $2ir - 3v$. Perform the indicated multiplications:

11. $(m + 4)(m + 2)$ 12. $(4C + L)(3C + L)$
13. $(\alpha + 7\beta)(3\alpha - 6\beta)$ 14. $(ax + bx)(cx + dx)$
15. $(2\theta + \lambda)(3\theta - 5\lambda)$ 16. $(17VI - 2I^2R)(2VI - 6I^2R)$
17. $(3m + 2n)(2m - 3n)$ 18. $(1.5\psi + 0.5\phi)(2\psi + 1.75\phi)$

19. $(R - 3Z)(5R - 2Z)$ 20. $\left(\dfrac{1}{3}m - \dfrac{1}{2}q\right)\left(\dfrac{3}{4}m + \dfrac{5}{6}q\right)$

21. $(2a^2 + 5a - 1)(3a + 1)$ 22. $(3\theta^2 - 4\theta - 7)(\theta + 3)$
23. $(R + r)(2R^2 - 4Rr + 2r^2)$ 24. $(x + y)(x + y)(x + y)$
25. $(a + b)(a - b)(a - b)$ 26. $(p - q)(p - q)(p - q)$
27. $(\theta - \phi)(\theta + \phi)(\theta - \phi)$ 28. $(IR + P)(I^2R^2 - 2IRP + P^2)$
29. $(a^2 + 2ab + b^2)(a + b)$ 30. $(x + 1)^2$
31. $(x + y)^2$ 32. $(x - y)^2$
33. $(M - N)^2$ 34. $(2\theta\phi + \psi + 1)^2$
35. $(2\alpha + 2w)^3$
36. $(3\alpha + 7)(\alpha - 5) + (2\alpha - 3)(4\alpha - 1)$
37. $2(3I^2R + 1)(4I^2R - 5) - 4(2I^2R - 2)(I^2R + 3)$
38. $4(3\theta - 2\phi + \lambda)(2\theta + 2\phi - \lambda) - 6(\theta + 2\phi + \lambda)(2\theta - \phi - 2\lambda)$
39. $3a(2a + b - 1)^2 + 2a(a + 2b + 1)^2$
40. $3\theta^2(5\omega - \lambda + \theta)^2 - \theta^2(\omega + 7\lambda - 2\theta)^2$

4-7 DIVISION

The division of algebraic expressions requires the development of certain rules and new methods in connection with operations involving exponents. However, if you have mastered the processes of the preceding sections, algebraic division will be an easy subject.

For the purpose of review the following definitions are given:

1. The *dividend* is a number, or quantity, that is to be divided.
2. The *divisor* is a number by which a number, or quantity, is to be divided.
3. The *quotient* is the result obtained by division. That is,

$$\frac{\text{Dividend}}{\text{Divisor}} = \text{Quotient}$$

4-8 DIVISION OF POSITIVE AND NEGATIVE NUMBERS

Because division is the inverse of multiplication, the methods of the latter will serve as an aid in developing methods for division. For example,

because $6 \times 4 = 24$
then $24 \div 6 = 4$
and $24 \div 4 = 6$

These relations can be used in applying the rules for multiplication to division.

All the possible cases can be represented as follows:

$$(+24) \div (+6) = ?$$
$$(-24) \div (+6) = ?$$
$$(+24) \div (-6) = ?$$
$$(-24) \div (-6) = ?$$

Because division is the inverse of multiplication, we apply the rules for multiplication of positive and negative numbers and obtain the following:

$(+24) \div (+6) = +4$ because $(+4) \times (+6) = +24$
$(-24) \div (+6) = -4$ because $(-4) \times (+6) = -24$
$(+24) \div (-6) = -4$ because $(-4) \times (-6) = +24$
$(-24) \div (-6) = +4$ because $(+4) \times (-6) = -24$

Therefore, we have the following:

RULE

To divide positive and negative numbers,

1. If dividend and divisor have like signs, the quotient is positive.
2. If dividend and divisor have unlike signs, the quotient is negative.

PROBLEMS 4-5

Divide the first number by the second in Probs. 1 to 10:

1. 25, 5
2. $-16, 4$
3. $-30, -6$
4. $-6.4, -800$
5. $-\frac{2}{3}, \frac{1}{2}$
6. $\frac{21}{64}, \frac{7}{16}$
7. $2\pi fC, -1$
8. $R, V - v$
9. $V \times 10^8, L_v$
10. $\omega L, Q$

Supply the missing divisors:

11. $\frac{-24}{?} = 4$
12. $\frac{16}{?} = -2$
13. $\frac{75}{?} = -\frac{1}{3}$

14. $-\frac{27}{?} = -\frac{1}{3}$
15. $\frac{-\frac{15}{16}}{?} = \frac{3}{8}$

4-9 THE LAW OF EXPONENTS IN DIVISION

By previous definition of an exponent (Sec. 2-10),

$$x^6 = x \cdot x \cdot x \cdot x \cdot x \cdot x$$

and

$$x^3 = x \cdot x \cdot x$$

Then

$$x^6 \div x^3 = \frac{x^6}{x^3} = \frac{\cancel{x} \cdot \cancel{x} \cdot \cancel{x} \cdot x \cdot x \cdot x}{\cancel{x} \cdot \cancel{x} \cdot \cancel{x}} = x^3$$

This result is obtained by canceling common factors in numerator and denominator. The above could be expressed as

$$x^6 \div x^3 = \frac{x^6}{x^3} = x^{6-3} = x^3$$

In like manner,

$$\frac{a^7}{a^3} = a^{7-3} = a^4$$

From the foregoing, it is seen that the law of exponents can be expressed in the general form

$$a^m \div a^n = \frac{a^m}{a^n} = a^{m-n}$$

where $a \neq 0$ and m and n are general numbers.

4-10 THE ZERO EXPONENT

Any number, except zero, divided by itself results in a quotient of 1. Thus,

$$\frac{6}{6} = 1$$

Also,

$$\frac{a^3}{a^3} = 1$$

Therefore,

$$\frac{a^3}{a^3} = a^{3-3} = a^0 = 1$$

Then, in the general form,

$$\frac{a^m}{a^n} = a^{m-n}$$

If $\qquad\qquad m = n$

then $\qquad\qquad m - n = 0$

and $\qquad\qquad \frac{a^m}{a^n} = a^{m-n} = a^0 = 1$

The foregoing leads to the definition that

Any base, except zero, affected by zero exponent is equal to 1.

Thus, a^0, x^0, y^0, 3^0, 4^0, etc., all equal 1.

4-11 THE NEGATIVE EXPONENT

If the law of exponents in division is to apply to all cases, it must apply when n is greater than m. Thus,

$$\frac{a^2}{a^5} = \frac{\cancel{a} \cdot \cancel{a}}{\cancel{a} \cdot \cancel{a} \cdot a \cdot a \cdot a} = \frac{1}{a^3}$$

or $\qquad\qquad \frac{a^2}{a^5} = a^{2-5} = a^{-3}$

Therefore, $\qquad a^{-3} = \frac{1}{a^3}$

Also, $\qquad\qquad a^{-n} = \frac{1}{a^n}$

This leads to the definition that

Any base affected by a negative exponent is the same as 1 divided by that same base but affected by a positive exponent of the same absolute value as the negative exponent.

Examples

$$x^{-4} = \frac{1}{x^4}$$

$$2^{-2} = \frac{1}{2^2} = \frac{1}{4}$$

$$3^{-3} = \frac{1}{3^3} = \frac{1}{27}$$

$$\frac{4^3}{4^5} = \frac{4 \times 4 \times 4}{4 \times 4 \times 4 \times 4 \times 4}$$

$$= \frac{1}{4 \times 4} = \frac{1}{4^2} = 4^{-2}$$

or $\qquad\qquad \frac{4^3}{4^5} = 4^{3-5} = 4^{-2}$

It follows, from the consideration of negative exponents, that

Any *factor* of an algebraic term may be transferred from numerator to denominator, or vice versa, by changing the sign of the exponent of the *factor*.

Example 10

$$3a^2x^3 = \frac{3a^2}{x^{-3}} = \frac{3}{a^{-2}x^{-3}} = \frac{3x^3}{a^{-2}}$$

4-12 DIVISION OF ONE MONOMIAL BY ANOTHER

RULE

To divide one monomial by another,

1. Find the quotient of the absolute values of the numerical coefficients and affix the proper sign according to the rules for division of positive and negative numbers (Sec. 4-8).
2. Determine the literal coefficients with their proper exponents, and write them after the numerical coefficient found in 1 above.

Example 11

Divide $- 12a^3x^4y$ by $4a^2x^2y$.

Solution

$$\frac{- 12a^3x^4y}{4a^2x^2y} = - 3ax^2$$

Example 12

Divide $-7a^2b^4c$ by $-14ab^2c^3$. Express the quotient with positive exponents.

Solution

$$\frac{- 7a^2b^4c}{- 14ab^2c^3} = \frac{ab^2}{2c^2}$$

Example 13

Divide $15a^{-2}b^2c^3d^{-4}$ by $-5a^2bc^{-1}d^{-2}$. Express the quotient with positive exponents.

Solution

$$\frac{15a^{-2}b^2c^3d^{-4}}{- 5a^2bc^{-1}d^{-2}} = - \frac{3bc^4}{a^4d^2}$$

Division can be checked by substituting convenient numerical values for the literal factors or by multiplying the divisor by the quotient, the product of which should result in the dividend.

PROBLEMS 4-6

Divide:

1. $20x^4y^6$ by $5x^2y^2$	2. $- 32x^8y^4z^6$ by $- 8x^4yz^5$
3. $180\theta^4\phi^3\psi^5$ by $-9\theta^3\phi\psi^2$	4. $-96\omega^4L^8M^2$ by $-24\omega^2L^6M^2$
5. $108X_C^4Z^5$ by $-81X_C^3Z^3$	6. $-10^3a^{12}b^9c^{14}d^7$ by $10a^5b^5c^4d^4$

7. $-33\eta^6\lambda^4\pi^{10}$ by $-11\eta^2\lambda\pi^9$

8. $13k^4\Delta^3\varepsilon^5$ into $-39k^8\Delta^4\varepsilon^5$

9. $-\dfrac{7}{16}m^4n^5p^2$ into $-\dfrac{21}{64}m^5n^7p^3$

10. $-\dfrac{5}{8}x^4y^{12}z^8$ into $-\dfrac{25}{48}x^3y^{14}z^6$

11. $\dfrac{18c^9d^2e^3}{-2c^8d^2e^3}$

12. $\dfrac{33\theta^3\phi^2\alpha}{3\theta^6\phi\alpha^2}$

13. $\dfrac{108\lambda^5\psi^6Q^2}{-27\lambda^2\psi^2Q^2}$

14. $\dfrac{35i^4r^5p^3w^5}{-0.7ir^5p^4w^3}$

15. $\dfrac{-9a^{-3}b^4c^{-4}d^2}{-27a^{-2}b^{-3}c^2d^{-2}}$

16. $\dfrac{-21rs^2t^{-4}u^6}{-63r^2s^{-2}t^{-3}u^2}$

17. $\dfrac{13\phi^2\theta^{-6}\psi\Omega^{-1}}{-52\phi^{-4}\theta^6\psi^2\Omega^2}$

18. $\dfrac{\dfrac{7}{16}x^3y^4\alpha^{-3}}{\dfrac{3}{8}x^{-2}y^3\alpha^{-1}}$

19. $\dfrac{360\alpha^4\beta^3\gamma^{-7}}{0.004\alpha^2\beta^{-2}\gamma^{-5}}$

20. $\dfrac{-0.000\ 256I^4R^3Z}{0.016I^{-2}R^{-1}Z^3}$

4-13 DIVISION OF A POLYNOMIAL BY A MONOMIAL

Because $\qquad\qquad 2 \times 8 = 16$

then $\qquad\qquad\qquad \dfrac{16}{2} = 8$

Also, because $\qquad 3(a + 4) = 3a + 12$

then $\qquad\qquad\qquad \dfrac{3a + 12}{3} = a + 4$

Similarly, because $\qquad 3I(2R + 3r) = 6IR + 9Ir$

then $\qquad\qquad\qquad \dfrac{6IR + 9Ir}{3I} = 2R + 3r$

From the foregoing we have the following:

RULE
To divide a polynomial by a monomial,

1. Divide each term of the dividend by the divisor.
2. Unite the results with the proper signs obtained by the division.

Example 14
Divide $8a^2b^3c - 12a^3b^2c^2 + 4a^2b^2c$ by $4a^2b^2c$.

Solution

$$\frac{8a^2b^3c - 12a^3b^2c^2 + 4a^2b^2c}{4a^2b^2c} = 2b - 3ac + 1$$

Example 15
Divide $-27x^3y^2z^5 + 3x^4y^2z^4 - 9x^4y^3z^5$ by $-3x^3y^2z^4$.

Solution

$$\frac{-27x^3y^2z^5 + 3x^4y^2z^4 - 9x^4y^3z^5}{-3x^3y^2z^4} = 9z - x + 3xyz$$

PROBLEMS 4-7

Divide:

1. $8x + 10y$ by 2
2. $12\theta - 6\phi$ by 3
3. $108\alpha^2 - 81\beta^2$ by 9
4. $16\phi^6 - 8\phi^4 + 24\phi^2$ by $4\phi^2$
5. $24R_1 + 48R_2 - 32R_3$ by 8
6. $X_C^6 - 12X_C^4 - 18X_C^2$ by $3X_C$
7. $0.025\mu^4\pi^2 + 50\mu^2\pi^4$ by $5\mu\pi^3$
8. $8.1\alpha^3\beta^2\gamma + 7.2\alpha^2\beta\gamma^3 - 3.6\alpha\beta\gamma$ by $0.09\alpha^2\beta^2\gamma^2$

9. $\dfrac{1}{2}m^2$ into $\dfrac{3}{20}m^5 - \dfrac{7}{10}m^3 - \dfrac{3}{5}m$

10. $-\dfrac{3}{4}I^2R$ into $\dfrac{5}{16}I^4R^2 - \dfrac{3}{8}I^2R + \dfrac{3}{10}I^{-2}R^{-1} + \dfrac{3}{4}I^{-4}R^{-2}$

11. $\dfrac{102xyz + 170x^2yz^2 - 85x^3yz^3 - 51x^5y^5z}{17xyz}$

12. $R^2(I + i) + r^2(I + i)$ by $I + i$

13. $8(\theta + \phi)^2 - 16(\theta + \phi)^4 + 12(\theta + \phi)^6$ by $4(\theta + \phi)$

14. $\lambda(\alpha^2 + \beta^2)^2 - \pi(\alpha^2 + \beta^2)^2$ by $-(\alpha^2 + \beta^2)^2$

15. $8\pi(VI + P)^4 - 32\pi(VI + P)^2 + 96\pi(VI + P)$ by $16\pi(VI + P)^2$

16. $\dfrac{6I^2(R + r)(R - r) + 10I^4(R + r)^2(R - r)^2 - 12I^8(R + r)^4(R - r)^4}{-2I(R + r)(R - r)}$

17. $\dfrac{5I\left(\omega L - \dfrac{1}{\omega C}\right) - 10I^3\left(\omega L - \dfrac{1}{\omega C}\right)^3 - 25I^5\left(\omega L - \dfrac{1}{\omega C}\right)^5}{5I^2\left(\omega L - \dfrac{1}{\omega C}\right)^2}$

18. $\dfrac{(2\beta + 7)(\beta + 1) - (3\beta + 2)(\beta + 1)^2}{\beta + 1}$

19. $\dfrac{-36\omega(\theta + \phi)(\theta - \phi) + 12\omega(\theta + \phi)^2(\theta - \phi)^2 - 24\omega(\theta + \phi)^3(\theta - \phi)^3}{4\omega(\theta + \phi)^2}$

20. $\dfrac{54V^2(R + R_1)(r + r_1) + 36V^4(R + R_1)^2(r + r_1)^2 - 108V^6(R + R_1)^3(r + r_1)^3}{9V(R + R_1)(r + r_1)}$

4-14 DIVISION OF ONE POLYNOMIAL BY ANOTHER

RULE

To divide one polynomial by another,

1. Arrange the dividend and divisor in ascending or descending powers of some common literal factor.

2. Divide the first term of the dividend by the first term of the divisor, and write the result as the first term of the quotient.
3. Multiply the entire divisor by the first term of the quotient; write the product under the proper terms of the dividend; and subtract the product from the dividend.
4. Consider the remainder a new dividend, and repeat 1, 2, and 3 until there is no remainder or until there is a remainder that cannot be divided by the divisor.

Example 16

Divide $x^2 + 5x + 6$ by $x + 2$.

Solution

Write the divisor and dividend in the usual positions for long division and eliminate the terms of the dividend, one by one:

$$
\begin{array}{r}
x + 3 \\
x + 2 \overline{)x^2 + 5x + 6} \\
\underline{x^2 + 2x} \\
3x + 6 \\
\underline{3x + 6}
\end{array}
$$

x, the first term of the divisor, divides into x^2, the first term of the dividend, x times. Therefore, x is written as the first term of the quotient. The product of the first term of the quotient and the divisor $x^2 + 2x$ is then written under like terms in the dividend and subtracted. The first term of the remainder then serves as a new dividend, and the process of division is continued.

This result can be checked by multiplying the divisor by the quotient.

$$
\begin{array}{rl}
\text{Divisor} = & x + 2 \\
\text{Quotient} = & \underline{x + 3} \\
& x^2 + 2x \\
& \underline{ 3x + 6} \\
\text{Dividend} = & x^2 + 5x + 6
\end{array}
$$

Example 17

Divide $a^2b^2 + a^4 + b^4$ by $-ab + b^2 + a^2$.

Solution

First arrange the dividend and divisor according to step 1 of the rule. Because there are no a^3b or ab^3 terms, allowance is made by supplying 0 terms. Thus,

$$
\begin{array}{r}
a^2 + ab + b^2 \\
a^2 - ab + b^2 \overline{)a^4 + 0 + a^2b^2 + 0 + b^4} \\
\underline{a^4 - a^3b + a^2b^2} \\
a^3b \\
\underline{a^3b - a^2b^2 + ab^3} \\
a^2b^2 - ab^3 + b^4 \\
\underline{a^2b^2 - ab^3 + b^4}
\end{array}
$$

Example 18

Divide $4 + x^4 + 3x^2$ by $x^2 - 2$.

Solution

$$x^2 - 2 \overline{\smash{)}\, x^4 + 3x^2 + 4} \qquad \begin{array}{l} x^2 + 5 \end{array}$$

$$\underline{x^4 - 2x^2}$$
$$5x^2 + 4$$
$$\underline{5x^2 - 10}$$
$$14 = \text{remainder}$$

The result is written $\quad x^2 + 5 + \dfrac{14}{x^2 - 2}$

which is as it would be written in an arithmetical division that did not divide out evenly.

PROBLEMS 4-8

Divide:

1. $x^2 + 2x + 1$ by $x + 1$
2. $9p^2 + 9p - 40$ by $3p - 5$
3. $\theta^2 + 7\theta + 12$ by $\theta + 4$
4. $12\omega^2 + 29\omega + 14$ by $4\omega + 7$
5. $6E^2 - 22E + 12$ by $3E - 2$
6. $6\phi^2 - 13\phi\theta + 6\theta^2$ by $2\phi - 3\theta$
7. $3R^3 + 9R^2 - 7R - 4RZ - 12Z - 21$ by $R + 3$
8. $\phi^3 + 3\phi^2\omega + 3\phi\omega^2 + \omega^3$ by $\phi + \omega$
9. $K^3 + 6K^2 + 7K - 8$ by $K - 1$
10. $12\lambda^2 - 36\phi^2 - 11\lambda\phi$ by $4\lambda - 9\phi$
11. $E^2 - e^2$ by $E - e$ 12. $E^3 - e^3$ by $E - e$
13. $E^4 - e^4$ by $E - e$ 14. $V^3 + I^3R^3$ by $V + IR$
15. $V^4 - I^4R^4$ by $V^2 - I^2R^2$ 16. $\theta^5 + \phi^5$ by $\theta + \phi$
17. $X^6 - Y^6$ by $X + Y$ 18. $X^6 + Y^6$ by $X^2 + Y^2$
19. $\theta^3 + 3\theta^2\phi + 3\theta\phi^2 + \phi^3$ by $\theta + \phi$
20. $L_1^4 - L_2^4$ by $L_1 + L_2$
21. $6R_2^3 - R_2^2 - 14R_2 + 3$ by $3R_2^2 + 4R_2 - 1$
22. $1 + 2m^4 + 4m^2 - m^3 + 7m$ by $3 + m^2 - m$
23. $30E^4 + 3 - 82E^2 - 5E + 11E^3$ by $3E^2 - 4 + 2E$
24. $\dfrac{1}{8}\theta^3 - \dfrac{9}{4}\theta^2\phi + \dfrac{27}{2}\theta\phi^2 - 27\phi^3$ by $\dfrac{1}{2}\theta - 3\phi$
25. $6R^2 - \dfrac{5}{6}R - \dfrac{1}{6}$ by $2R - \dfrac{1}{2}$
26. $n^3 - \dfrac{9}{5}n^2 - \dfrac{9}{25}n - \dfrac{27}{125}$ by $n - \dfrac{3}{5}$
27. $36x^2 + \dfrac{1}{9}y^2 + \dfrac{1}{4} - 4xy - 6x + \dfrac{1}{3}y$ by $6x - \dfrac{1}{3}y - \dfrac{1}{2}$
28. $\dfrac{1}{27}K^3 - \dfrac{1}{12}K^2 + \dfrac{1}{16}K - \dfrac{1}{64}$ by $\dfrac{1}{3}K - \dfrac{1}{4}$
29. $\dfrac{3}{2}L_1^2 - L_1 - \dfrac{8}{3}$ into $\dfrac{9}{16}L_1^4 - \dfrac{3}{4}L_1^3 - \dfrac{7}{4}L_1^2 + \dfrac{4}{3}L_1 + \dfrac{16}{9}$
30. $R_1^7 + \left(\dfrac{V}{I}\right)^7$ by $R_1 + \dfrac{V}{I}$

Equations

5

In the preceding chapters, considerable time has been spent in the study of the fundamental operations of algebra. These fundamentals will be of little value unless they can be put to practical use in the solution of problems. This is accomplished by use of the equation, the most valuable tool in mathematics.

5-1 DEFINITIONS

An *equation* is a mathematical statement that two numbers, or quantities, are equal. The *equality sign* (=) is used to separate the two equal quantities. The terms to the left of the equality sign are known as the *left member* of the equation, and the terms to the right are known as the *right member* of the equation. For example, in the equation

$$3E + 4 = 2E + 6$$

$3E + 4$ is the left member and is equal to $2E + 6$, which is the right member.

An *identical equation,* or *identity,* is an equation whose members are equal for all values of the literal numbers contained in the equation. The equation

$$4I(r + R) = 4Ir + 4IR$$

is an identity because if $I = 2$, $r = 3$, and $R = 1$, then

$$4I(r + R) = 4 \cdot 2(3 + 1) = 32$$

Also,
$$4Ir + 4IR = 4 \cdot 2 \cdot 3 + 4 \cdot 2 \cdot 1$$
$$= 24 + 8 = 32$$

Any other values of I, r, and R substituted in the equation will produce equal numerical results in the two members of the equation.

An equation is said to be *satisfied* if, when numerical values are substituted for the literal numbers, the equation becomes an identity. Thus, the equation

$$ir - iR = 3r - 3R$$

is satisfied by $i = 3$, because when we substitute this value in the equation, we obtain

$$3r - 3R = 3r - 3R$$

which is an identity.

A *conditional equation* is one consisting of one or more literal numbers that is not satisfied by all values of the literal numbers. Thus, the equation

$$e + 3 = 7$$

is not satisfied by any value of e except $e = 4$.

To *solve* an equation is to find the value or values of the unknown number that will satisfy the equation. This value is called the *root* of the equation. Thus, if

$$i + 6 = 14$$

the equation becomes an identity only when i is 8, and therefore 8 is the root of the equation.

5-2 AXIOMS

An *axiom* is a truth, or fact, that is self-evident and needs no formal proof. The various methods of solving equations are derived from the following axioms:

1. If equal numbers are added to equal numbers, the sums are equal.

    ### Example 1

If		$x = x$
then		$x + 2 = x + 2$
because, if	$x = 4,$	$4 + 2 = 4 + 2$
or		$6 = 6$

    Therefore, *the same number can be added to both members of an equation without destroying the equality.*

2. If equal numbers are subtracted from equal numbers, the remainders are equal.

    ### Example 2

If		$x = x$
then		$x - 2 = x - 2$
because, if	$x = 4,$	$4 - 2 = 4 - 2$
or		$2 = 2$

    Therefore, *the same number can be subtracted from both members of an equation without destroying the equality.*

3. If equal numbers are multiplied by equal numbers, their products are equal.

Example 3

If $$x = x$$
then $$3x = 3x$$
because, if $$x = 4, \quad 3 \cdot 4 = 3 \cdot 4$$
or $$12 = 12$$

Therefore, *both members of an equation can be multiplied by the same number without destroying the equality.*

4. If equal numbers are divided by equal numbers, their quotients are equal.

Example 4

If $$x = x$$

then $$\frac{x}{2} = \frac{x}{2}$$

because, if $$x = 4, \quad \frac{4}{2} = \frac{4}{2}$$

or $$2 = 2$$

Therefore, *both members of an equation can be divided by the same number without destroying the equality, except that division by zero is not allowed.*

5. Numbers that are equal to the same number or equal numbers are equal to each other.

Example 5

If $$a = x \quad \text{and} \quad b = x$$
then $$a = b$$
because, if $$x = 4, \quad a = 4$$
and $$b = 4$$

Therefore, *an equal quantity can be substituted for any term of an equation without destroying the equality.*

6. Like powers of equal numbers are equal.

Example 6

If $$x = x$$
then $$x^3 = x^3$$
because, if $$x = 4, \quad 4^3 = 4^3$$
or $$64 = 64$$

Therefore, *both members of an equation can be raised to the same power without destroying the equality.*

7. Like roots of equal numbers are equal.

Example 7

If $x = x$

then $\sqrt{x} = \sqrt{x}$

because, if $x = 4,$ $\sqrt{4} = \sqrt{4}$

or $2 = 2$

Therefore, *like roots of both members of an equation can be extracted without destroying the equality*.

5-3 NOTATION

In order to shorten the *explanations* of the solutions of various equations, we shall employ the letters **A, S, M,** and **D** for "add," "subtract," "multiply," and "divide," respectively. Thus,

- **A:** 6 will mean "add 6 to both members of the equation."
- **S:** $-6x$ will mean "subtract $-6x$ from both members of the equation."
- **M:** $-3a$ will mean "multiply both members of the equation by the factor $-3a$."
- **D:** 2 will mean "divide both members of the equation by 2."

5-4 THE SOLUTION OF EQUATIONS

A considerable amount of time and drill must be spent in order to become proficient in the solution of equations. It is in this branch of mathematics that you will find you must be familiar with the more elementary parts of algebra.

Some of the methods used in the solutions are very easy—so easy, in fact, that there is a tendency to employ them mechanically. This is all very well, but you should not become so mechanical that you forget the reason for performing certain operations.

We shall begin the solution of equations with very easy cases and attempt to build up general methods of procedure for all equations as we proceed to the more difficult problems.

If you are studying equations for the first time, you are urged to study the following examples carefully until you thoroughly understand the methods and the reasons behind them.

Example 8

Find the value of x if $x - 3 = 2$.

Solution

In this equation, it is seen by inspection that x must be equal to 5. However, to make the solution by the methods of algebra, proceed as follows:

Given $x - 3 = 2$
A: 3, $x = 2 + 3$ (Axiom 1)
Collecting terms, $x = 5$

Example 9

Solve for e if $e + 4 = 12$.

Solution

Given	$e + 4 = 12$	
S: 4,	$e = 12 - 4$	(Axiom 2)
Collecting terms,	$e = 8$	

Example 10

Solve for i if $3i + 5 = 20$.

Solution

Given	$3i + 5 = 20$	
S: 5,	$3i = 20 - 5$	(Axiom 2)
Collecting terms,	$3i = 15$	
D: 3,	$i = 5$	(Axiom 4)

Example 11

Solve for r if $40r - 10 = 15r + 90$.

Solution

Given	$40r - 10 = 15r + 90$	
S: 15r,	$40r - 10 - 15r = 90$	(Axiom 2)
A: 10,	$40r - 15r = 90 + 10$	(Axiom 1)
Collecting terms,	$25r = 100$	
D: 25	$r = 4$	(Axiom 4)

From the foregoing examples, it will be noted that adding or subtracting a term from both members of an equation is equivalent to *transposing* that number from one member to the other and changing its sign. This fact leads to the following rule:

RULE

A *term* can be transposed from one member of an equation to the other provided that its sign is changed.

By transposing all terms containing the unknown to the left member and all others to the right member and then collecting terms and dividing both members by the numerical coefficient of the unknown, the equation has been solved for the value of the unknown.

5-5 CANCELING TERMS IN AN EQUATION

Example 12

Solve for x if $x + y = z + y$.

Solution

Given	$x + y = z + y$	
S: y,	$x = z$	(Axiom 2)

The term y in both members of the given equation does not appear in the next equation as the result of subtraction. The result is the same as if the term were dropped from both members. This fact leads to the following rule:

RULE

If the same *term* preceded by the same sign occurs in both members of an equation, it can be canceled.

5-6 CHANGING SIGNS IN AN EQUATION

Example 13

Solve for x if $8 - x = 3$.

Solution

Given	$8 - x = 3$	
S: 8,	$-x = 3 - 8$	(Axiom 2)
M: -1,	$x = -3 + 8$	(Axiom 3)
Collecting terms,	$x = 5$	

Note that multiplication by -1 has the effect of changing the signs of all terms. This gives the following rule:

RULE

The signs of all the *terms* of an equation can be changed without destroying the equality.

Although the foregoing rules involving mechanical methods are valuable, you should not lose sight of the fact that they are all derived from fundamentals, or axioms, as outlined in Sec. 5-2.

5-7 CHECKING THE SOLUTION

If there is any doubt that the value of the unknown is correct, the solution can be checked by substituting the value of the unknown in the original equation. If the two members reduce to an identity, the value of the unknown is correct.

Example 14

Solve and test $3i + 14 + 2i = i + 26$.

Solution

Given	$3i + 14 + 2i = i + 26$
Transposing,	$3i + 2i - i = 26 - 14$
Collecting terms,	$4i = 12$
D: 4,	$i = 3$

Test by substituting $i = 3$ in given equation.

Check

$$(3 \cdot 3) + 14 + (2 \cdot 3) = 3 + 26$$
$$9 + 14 + 6 = 3 + 26$$
$$29 = 29$$

PROBLEMS 5-1

Solve for the unknown in the following equations:

1. $3x - 6 = 6$
2. $4\theta - 1 = 3\theta + 3$
3. $k - 10 = 5 + 4k$
4. $I - 9I = -6I - 2$
5. $6p + 3 - 2p = 27$
6. $16 - 9\mu = 5\mu - 12$
7. $11\pi - 22 = 4\pi + 13$
8. $5M + 2 = 3 + 4M$
9. $21 - 15IR = -8IR - 7$
10. $27Q + 22 = 30 + 17Q$
11. $8\alpha - 5(4\alpha + 3) = -3 - 4(2\alpha - 7)$
12. $3(\lambda - 2) - 10(\lambda - 6) = 5$
13. $4 + 3(E - 7) = 16 + 2(5E + 1)$
14. $4(K - 5) - 3(K - 2) = 2(K - 1)$
15. $0 = 18 - 4Q + 27 + 9Q - 3 + 16Q$
16. $25R_1 - 19 - [3 - (4R_1 - 15)] - 3R_1 + (6R_1 + 21) = 0$
17. $19 - 5I(4I + 1) = 40 - 10I(2I - 1)$
18. $(\phi + 5)(\phi - 4) + 4\phi^2 = (5\phi + 3)(\phi - 4) + 2(\phi - 4) + 64$
19. $6(\beta - 1)(\beta - 2) - 4(\beta + 2)(\beta + 1) = 2(\beta + 1)(\beta - 1) - 24$
20. $18 - 3Z(2Z + 1) - [3 - 2(Z + 2)(Z - 3)] = 18 - 6Z - 4(Z - 5)(Z + 2)$

5-8 FORMING AND SOLVING EQUATIONS

As previously stated, we are continually trying to express certain laws and relations in the language of mathematics.

Examples

Words	Algebraic Symbols
The sum of the voltages V and v	$V + v$
The difference between resistances R and R_1	$R - R_1$
The excess of current I_1 over current I_2	$I_1 - I_2$
The number of centimeters in l meters	$100l$
The number of cents in d dollars	$100d$
The voltage V is equal to the product of the current I and the resistance R	$V = IR$

The solution of most problems consists in writing an equation that connects various observed data with known facts. This, then, is nothing more than translating from ordinary English, or words, into the symbolic language of mathematics. In relatively simple problems the translation can be made directly, almost word by word, into algebraic symbols (Examples 15 and 16).

It is almost impossible to lay down a set of rules for the solution of general problems, for they could not be made applicable to all cases.

However, no rules will be needed if you thoroughly understand what is to be translated into the language of mathematics from the wording or facts of the problem at hand. The following outline will serve as a guide:

1. Read the problem carefully so that you understand every fact in it and recognize the relationships between the facts.
2. Determine what is to be found (the unknown quantity), and denote it by some letter. If there are two unknowns, try to represent one of them in terms of the other. If there are more than two unknowns, try to represent all but one of them in terms of that one.
3. Find two expressions which, according to the facts of the problem, represent the same quantity, and set them equal to each other. You can then solve the resulting equation for the unknown.

Example 15

Five times a certain voltage diminished by 3,

$\quad$ 5 $\quad \times \quad$ V $\quad - \quad$ 3

gives the same result as the voltage increased by 125.

$\quad \quad = \quad \quad$ V $\quad + \quad$ 125

That is $\quad \quad \quad \quad 5V - 3 = V + 125$

or $\quad \quad \quad \quad \quad \quad V = 32 \text{ V}$

Example 16

What number increased by 42 is equal to 110?

$\quad \quad$ x $\quad + \quad$ 42 $\quad = \quad$ 110

That is, $\quad \quad \quad \quad x + 42 = 110$

or $\quad \quad \quad \quad \quad \quad x = 68$

Check

$68 + 42 = 110$

PROBLEMS 5-2

1. The sum of two voltages is V V. One voltage is 75 V. What is the other?
2. The difference between two resistances is 10.5 Ω. One resistance is R Ω. What is the other?
3. How great a distance d will you travel in t hours (h) at r kilometers per hour (km/h)?
4. What is the fraction f whose numerator n is 3 less than its denominator?
5. An electric timer has a guarantee of y years. We have been using it for t years. For how many years longer will the guarantee apply?
6. An oscilloscope is guaranteed for q years, and it has been in service for m months. How much longer is it covered by the guarantee?
7. At what speed must a missile be traveling to cover Z km in t min?
8. From what number must 8 be subtracted in order that the remainder may be 27?

9. If a certain voltage is doubled and the result is diminished by 15, the remainder is 205 V. What is the voltage?

10. The volume of a parts container is v cubic centimeters (cm³). Express the height in centimeters if the width is w cm and the length is l cm.

11. Write algebraically that the current is equal to the voltage divided by the resistance.

12. Write algebraically that the power dissipated by a resistor is equal to the square of the current multiplied by the resistance.

13. A stockroom is twice as long as it is wide, and its perimeter is 36 m. Find its length and width.

14. A multimeter and an oscilloscope together cost $574. The oscilloscope costs $356 more than the meter. Find the cost of (a) the oscilloscope and (b) the multimeter.

15. Find the three sides of a triangle whose perimeter is 23.5 m if one side is 6.5 m shorter than the second side and one-half the third side.

16. The sum of the three angles in any triangle is 180°. The smallest angle in a given triangle is one-half the second angle and 52° smaller than the largest angle. How many degrees does each angle contain?

17. Write algebraically that the square on the hypotenuse h of a right triangle is equal to the sum of the squares on the other two sides, which are identified as a and b.

18. The sum of two consecutive numbers is 31. What are the numbers?

19. The sum of three consecutive numbers is 192. What are the numbers?

20. Write algebraically that the product of the impressed emf V and the resultant current I in a circuit is equivalent to the square of the emf divided by R, the resistance in the circuit.

5-9 LITERAL EQUATIONS—FORMULAS

A *formula* is a rule, or law, generally pertaining to some scientific relationship expressed as an equation by means of letters, symbols, and constant terms.

Example 17
The area A of a rectangle is equal to the product of its base b and its altitude h. This statement written as a formula is

$$A = bh$$

Example 18
The power P expended in an electric circuit is equal to its current I squared times the resistance R of the circuit. Stated as a formula

$$P = I^2R$$

The ability to handle formulas is of the utmost importance. The usual formula expresses one quantity in terms of other quantities, and it is often desirable to solve for *any* quantity contained in a formula. This is readily accomplished by using the knowledge gained in solving equations.

Example 19

The voltage V across a part of a circuit is given by the current I through that part of the circuit times the resistance R of that part. That is,

$$V = IR$$

Suppose V and I are given but it is desired to find R.

Given	$V = IR$	
D: I,	$\dfrac{V}{I} = R$	(Axiom 4)
or	$R = \dfrac{V}{I}$	

Similarly, if we wanted to solve for I,

Given	$V = IR$	
D: R,	$\dfrac{V}{R} = I$	(Axiom 4)
or	$I = \dfrac{V}{R}$	

Example 20

Solve for I if $v = V - IR$.

Solution

Given	$v = V - IR$	
Transposing,	$IR = V - v$	
D: R,	$I = \dfrac{V - v}{R}$	(Axiom 4)

Example 21

Solve for C if $X_C = \dfrac{1}{2\pi f C}$.

Solution

Given	$X_C = \dfrac{1}{2\pi f C}$	
D: X_C,	$1 = \dfrac{1}{2\pi f C X_C}$	(Axiom 4)
M: C,	$C = \dfrac{1}{2\pi f X_C}$	(Axiom 3)

It will be noted from the foregoing examples that if the numerator of a member of an equation contains but one term, any *factor* of that term may be transferred to the denominator of the other member as a *factor*. In like manner if the denominator of a member of an equation contains but one term, any *factor* of that term may be transferred to the numerator of the other member as a *factor*. These mechanical transformations simply make use of Axioms 3 and 4, and you should not lose sight of the real reasons behind them.

PROBLEMS 5-3

Given: Solve for:

1. $Q = CV$ C and V

2. $I = \dfrac{V}{Z}$ V and Z

3. $R^2 = Z^2 - X^2$ Z^2 and X^2

4. $R = \dfrac{P}{I^2}$ P and I^2

5. $L = \dfrac{Rm}{K}$ R, K, and m

6. $R_2 = R_t - R_1 - R_3$ R_t, R_1, and R_3

7. $f = \dfrac{\upsilon}{\lambda}$ λ and υ

8. $C = 2\pi r$ r and π

9. $R = \dfrac{\omega L}{Q}$ L, Q, and ω

10. $L = \dfrac{X_L}{2\pi f}$ X_L and f

11. $C = \dfrac{1}{2\pi f X_C}$ X_C and f

12. $S = 2\pi rh$ r and h

13. $H = \dfrac{\phi}{A}$ ϕ and A

14. $N_s = \dfrac{V_s N_p}{V_p}$ N_p, V_s, and V_p

15. $B = \dfrac{V10^8}{Lv}$ V, L, and v

16. $T = ph + 2A$ h and A

17. $V_s I_s = V_p I_p$ V_s

18. $L = \dfrac{F}{Hi}$ F, H, and i

19. $R = \dfrac{V - v}{I}$ I, V, and v

20. $\mu = g_m r_p$ g_m

21. $t = \dfrac{\theta}{\omega}$ θ and ω

22. $h = \dfrac{V^2}{2g}$ g and V^2

23. $V_0 = 2V - V_t$ V and V_t

24. $n = \dfrac{\omega}{2\pi}$ ω

25. $A = \dfrac{4}{3}\pi r^3$ r^3

26. $\mu = \dfrac{B^2 Al}{8\omega}$ l, ω, and A

27. $C = \dfrac{F(R - r)}{Z_t}$ Z_t, F, R, and r

28. $r = \dfrac{F}{4\pi^2 n^2 m}$ m and F

29. $R_L = \dfrac{V_b - v_b}{i}$ $\qquad\qquad$ V_b, v_b, and i

30. $t = \dfrac{T(C - F)}{C}$ $\qquad\qquad$ T and F

31. $R = \dfrac{\rho l}{d^2}$ $\qquad\qquad$ l, ρ, and d^2

32. $PF = \dfrac{R}{X}$ $\qquad\qquad$ R and X

33. $C = \dfrac{0.0884KA(n - 1)}{d}$ $\qquad\qquad$ A and n

34. $M = k\sqrt{L_1 L_2}$ $\qquad\qquad$ k

35. $Z_r = \dfrac{L}{RC}$ $\qquad\qquad$ L, C, and R

36. $F = \dfrac{vI}{2kT_g}$ $\qquad\qquad$ T_g

37. $\omega = \dfrac{\eta\beta}{\gamma\alpha}$ $\qquad\qquad$ β

38. $\dfrac{P_{so}}{P_{no}} = \dfrac{P_{si}}{P_{ni}}$ $\qquad\qquad$ P_{no}

39. $\rho = \dfrac{Qe}{hv}$ $\qquad\qquad$ Q

40. $V_b = \dfrac{V_B + V_{pt}}{W}$ $\qquad\qquad$ V_{pt}

41. $V_2 = (1 - \omega^2 LC_2)V_3$ $\qquad\qquad$ C_2

42. $4a = \dfrac{h + 2b}{v}$ $\qquad\qquad$ b

43. $Q = I_p p' + I_n n$ $\qquad\qquad$ I_n

44. $G_o = G + \dfrac{g_m}{1 + n}$ $\qquad\qquad$ g_m

45. $\omega_{01} = \dfrac{1}{C(R_1 + R_2)}$ $\qquad\qquad$ R_1

Note When solving numerical problems which involve the solution of formulas, always solve the formula algebraically for the wanted factor before substituting the numerical values. This procedure permits you to check your work more easily. The reason is that the letters retain their identity through the various algebraic procedures, whereas once numbers are added, multiplied, etc., their identity is lost and your audit becomes more difficult.

46. The power P in any part of an electric circuit is given by

$$P = \frac{V^2}{R}\,\text{W}$$

in which V is the emf applied to that part of the circuit and R is the resistance of that part. What is the resistance of a circuit in which 1.21 W is expended at an emf of 110 V?

47. The voltage drop V across any part of a circuit can be computed by the formula $V = IZ$ V, where I is the current in amperes through

that part of the circuit and Z is the impedance in ohms of that part. Give the impedance of a circuit in which a voltage drop of 460 V is produced by a current flow of 0.115 A?

48. To find the frequency f of an alternator in hertz (Hz), that is, cycles per second, the number of pairs of poles P is multiplied by the speed of the armature S in revolutions per second (rev/s), $f = PS$. A tachometer connected to the rotor of a 60-Hz alternator reads 3600 revolutions per minute (rev/min). How many poles has the alternator?

49. For radio waves, the relationship between frequency f in megahertz and wavelength λ in meters is expressed by the formula

$$f = \frac{3 \times 10^2}{\lambda} \text{ MHz}$$

What is the wavelength of a radio wave at 60 MHz?

50. The length of a broadband dipole L_{fD} used for television reception can be computed by the formula $L_{fD} = \frac{141.3}{f}$ m, where f is the frequency in megahertz. The folded dipole shown in Fig. 5-1 is 0.7976 m long. For what frequency was it constructed?

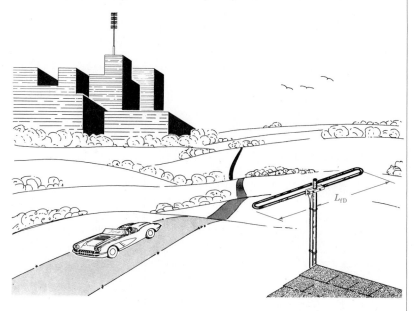

Fig. 5-1 Folded dipole of Prob. 50.

5-10 RATIO AND PROPORTION

Because proportions are special forms of equations, it is expedient to look now at the twin subjects ratio and proportion.

A ratio is a comparison of two things expressed in one of two ways: first, the "old-fashioned" method, $a:b$, pronounced "a is to b"; and second, as found in newer books, $\frac{a}{b}$. If the ratio of x to y is 1 to 4, or $\frac{1}{4}$, then x is one-quarter of y. Alternatively, y is four times as great as x.

Example 22
Write the ratio of 25 cents (¢) to $3.00.

Solution
25¢ to $3.00 may be written simply as 25¢:$3.00, but this does not tell us much. It is more helpful to convert both quantities to the same units:

$$\frac{25¢}{\$3.00} = \frac{25¢}{300¢} = \frac{1}{12}$$

Note that the two parts being compared are given the same units, in this case cents. Therefore, when the simplification is performed, not only the numbers but also the units are canceled. Thus a *true ratio* is a "pure," or dimensionless, number. Note also that a ratio may be an integer, that is, a fraction whose denominator is 1.

PROBLEMS 5-4
Write as a fraction the ratio of:

1. 3 cm to 12 cm
2. 12 square meters (m²) to 18m²
3. 15 000 Ω to 12 000 Ω
4. $5.00 to 25¢
5. Write two different sets of numbers in the ratio 2:3.
6. Write two different sets of numbers in the ratio 0.125:1.
7. A recipe for ceramic insulators calls for 8 parts of type A clay to 24 parts of type B. What is the ratio of type A to type B?
8. In Prob. 7, what is the ratio of the weight of type A to the weight of the total mixture?
9. The mechanical advantage (MA) of any machine is the ratio of load moved to effort applied. What is the MA of a system in which 24 kilograms (kg) of effort just starts motion of a 768-kg load?
10. In a certain alloy, 55% of the material is copper and 22% is zinc. What is the ratio of zinc to copper?

Just as ratios compare two things, so proportions are equalities of pairs of ratios.

When we draw a map to scale, the proportions on the map should equal those on the ground. If the scale is 1 cm to 10 km, then a trip which is 3 cm on the map must be 30 km on the ground. The proportion here is $\frac{1\,cm}{3\,cm} = \frac{10\,km}{30\,km}$ and, since the units cancel, our true proportion is an equality of two pure numbers. We could also write this proportion as $\frac{1\,cm}{10\,km} = \frac{3\,cm}{30\,km}$. Note that it is essential that the units on one side of the proportion be equal to those on the other side. This provides one good way of checking your work. If you perform a wrong operation such as multiplying instead of dividing, you will find that your units will reveal an error. The solution may read "cm/km = cm · km" and such an imbalance of units is a sure indication that you have made an error.

The usual purpose of proportions is to solve one part when the other three parts are known.

Example 23

Given the proportion $\frac{18}{a} = \frac{6}{5}$, solve for a.

Solution

Obeying the usual rules of equations,

$$18 \times 5 = 6 \times a$$
$$a = 15$$

In the older form of writing ratios and proportions, $\frac{a}{b} = \frac{c}{d}$ would be written $a:b::c:d$ and pronounced "a is to b as c is to d." The elements on the outsides of the proportion were called the "extremes," and those in the middle the "means." Based on these definitions, you can prove the old law of proportions:

RULE

In a proportion, the product of the means equals the product of the extremes.

PROBLEMS 5-5

Find the missing term in each of the following proportions:

1. $\frac{5}{8} = \frac{?}{16}$ 2. $\frac{3}{7} = \frac{?}{56}$ 3. $\frac{?}{15} = \frac{40}{3}$

4. $\frac{80}{?} = \frac{60}{12}$ 5. $\frac{X}{90} = \frac{60}{360}$ 6. $\frac{5}{i} = \frac{0.2}{36}$

7. $\frac{6}{IR} = \frac{9}{12}$ 8. $\frac{0.6}{1.2} = \frac{0.4}{d}$ 9. $\frac{0.007}{0.200} = \frac{Q}{0.04}$

10. $\frac{16}{Z} = \frac{Z}{4}$

5-11 VARIATION AND PROPORTIONALITY

Often, in the study of electronics, you will hear such expressions as "the current is proportional to the voltage and inversely proportional to the resistance" and "the force is jointly proportional to the charges and inversely proportional to the square of the distance between them."

Sometimes an equivalent expression is used: "the current varies directly as the voltage," etc.

Two forms may be used to express mathematically the words "the current varies as the voltage." The first uses the symbol of proportionality:

$I \propto V$. The second substitutes for the symbol $\propto$ the equivalent "$= k$," where k is the "konstant" of proportionality: $I = kV$. Other symbols such as b, c, n, etc., also are used as constants.

Similarly, the expression "the current is inversely proportional to the resistance" may be written $I \propto \frac{1}{R}$ or $I = k \frac{1}{R}$ or simply $I = \frac{k}{R}$.

"Jointly proportional" means "proportional to the product," so that "the force is jointly proportional to the masses" may be written $F \propto m_1 m_2$ or $F = k m_1 m_2$.

Often, past experience, tables, and measurements, as well as calculations, may reveal the value of the constant of proportionality. For example, we know that the circumference of a circle is proportional to the radius. We may write this $C \propto R$ or $C = kR$. However, from previous knowledge, we can replace the general constant k by the known constant of proportionality 2π, and we can write $C = 2\pi R$.

Example 24

If a varies directly as ρ and if $a = 8$ when $\rho = 4$, what will be the value of a when $\rho = 7$?

Solution

$a \propto \rho = k\rho$. We know that $8 = k4$, from which $k = 2$. Substitute this value of k into the second condition:

$$a = k \times 7 = 2 \times 7 = 14$$

PROBLEMS 5-6

Write the following expressions in "proportionality" form and in "equation" form:

1. The distance D varies directly as the rate R.
2. The cost C varies directly as the weight W.
3. The capacitance C varies directly as the area A.
4. The reactance X_L varies jointly with the frequency f and the inductance L.
5. The capacitive reactance X_C varies inversely as the capacitance C.
6. The resistance varies directly as the length l and inversely as the cross-sectional area A.
7. The period T of vibration of a reed is directly proportional to the square root of the length l.
8. The volume of a sphere V is proportional to the cube of the radius r.
9. The volume of a gas V varies inversely as the pressure P.
10. The ratio of the similar areas A_1 and A_2 is proportional to the square of the ratio of corresponding lengths l_1 and l_2.
11. The illumination L of an object varies inversely as the square of the distance d from the source of light.
12. If the current I varies directly as the voltage V and if $I = 0.5$ A when $V = 30$ V, what will be the value of I when $V = 75$ V?
13. In a certain varistor the current is proportional to the square of the voltage. If $I = 0.006$ A when $V = 110$ V, what voltage will produce a current of 1.5 A?

14. The resistance R of a wire varies directly as the wire length l and inversely as the square of the wire diameter d. If $R = 3.277 \ \Omega$ when $l = 1$ km and if $d = 2.588$ mm, what will be the resistance of a 500-m length of wire 1.45 mm in diameter?

15. The load that a beam of given thickness can carry safely is directly proportional to the beam's width and inversely proportional to its length. If a beam 10 m long and 50 mm wide can support 9000 kg, what is the load that could be supported by a beam of identical thickness 25 m long and 75 mm wide?

Powers of 10

6

Very few people enjoy performing numerical computations simply for the joy of "figuring." The practical person wants concrete answers; therefore, such a person should use whatever tools or devices are available to help arrive at those answers with a minimum expenditure of time and effort. The engineer would normally simplify his calculations, especially when he is dealing with large numbers. Such a simplification device is the use of powers of ten.

6-1 SIGNIFICANT FIGURES

In mathematics, a number is generally considered as being exact. For example, 220 would mean 220.0000, etc., for as many added zeros as desired. A meter reading, however, is always an *approximation*. We might read 220 V on a certain switchboard type of voltmeter, whereas a precision instrument might show that voltage to be 220.3 V and a series of precise measurements might show it to be 220.36 V. It should be noted that the position of the decimal point does not determine the accuracy of a number. For example, 115 V, 0.115 kV, and 115 000 mV are of identical value and are equally accurate.

Any number representing a measurement, or the amount of some quantity, expresses the accuracy of the measurement. The figures required are known as *significant figures*.

The *significant figures* of any number are the figures 1, 2, 3, 4, . . . , 9, in addition to such ciphers, or zeros, as may occur between them or to their right or as may have been retained in properly rounding them off. Thus, the number of significant figures is an indication of *precision*, not *accuracy*.

Examples

0.002 36	is correct to *three* significant figures.
3.141 59	is correct to *six* significant figures.
980 000.0	is correct to *seven* significant figures.
24.	is correct to *two* significant figures.
24.0	is correct to *three* significant figures.
0.025 00	is correct to *four* significant figures.

PROBLEMS 6-1

To how many significant figures have the following numbers been expressed?

1. 2.718 28
2. 0.000 003 14
3. 300 000
4. 23.0055
5. 1.00
6. 1
7. 0.000 01
8. 6.28
9. 0.000 025 38
10. 2 726.375

6-2 ROUNDED NUMBERS

A number is *rounded off* by dropping one or more figures at its right. If the last figure dropped is 6 or more, we increase the last figure retained by 1. Thus 3867 would be rounded off to 3870, 3900, or 4000. If the last figure dropped is 4 or less, we leave the last figure retained as it is. Thus 5134 would be rounded off to 5130, 5100, or 5000. If the last figure dropped is 5, add 1 if it will make the last figure retained *even*; otherwise, do not. Thus, $55.75 = 55.8$, but $67.65 = 67.6$.

6-3 DECIMALS

An important consideration arises in making computations involving decimals. Electrical engineers and particularly electronics engineers are, unfortunately, required to handle cumbersome numbers ranging from extremely small fractions of electrical units to very large numbers, as represented by radio frequencies. The fact that these wide limits of numbers are encountered in the same problem does not simplify matters. This situation is becoming more complicated owing to the trend to the higher radio frequencies with attendant smaller fractions of units represented by circuit components.

The problem of properly placing the decimal point and thus reducing unnecessary work presents little difficulty to the person who has a working knowledge of the powers of 10.

Of course, using a calculator will place the decimal point accurately, leaving no room for confusion. More on calculators may be found in the Appendix.

6-4 POWERS OF 10

The powers of 10 are sometimes termed the "engineer's shorthand" or "scientist's shorthand." A thorough knowledge of the powers of 10 and the ability to apply the theory of exponents will greatly assist you in determining an approximation.

There was a time when the terms engineering notation and scientific notation were used interchangeably. Recently, however, a distinction has been made:

■ *Scientific notation* describes the rewriting of any number as a number between 1 and 10 times the appropriate power of 10:

$$12\ 345 = 1.234\ 5 \times 10^4$$

■ *Engineering notation* describes the more specialized technique of rewriting any number so as to use the third powers of 10, for which there are SI prefixes. This technique calls for the use of numbers between 0.1 and 1000 times the appropriate third power of 10. For example,

$$12\ 345 = 12.345 \times 10^3$$

If a unit of measure were involved, the 10^3 would be replaced by the prefix *kilo*.

Some calculators will permit you to FIX a readout to either SCI or ENG notation, and a choice of decimal places will be displayed.

TABLE 6-1

Number	Power of 10	Expressed in English
$0.000\ 001 =$	$10^{-6} =$	ten to the negative *sixth* power
$0.000\ 01 =$	$10^{-5} =$	ten to the negative *fifth* power
$0.000\ 1 =$	$10^{-4} =$	ten to the negative *fourth* power
$0.001 =$	$10^{-3} =$	ten to the negative *third* power
$0.01 =$	$10^{-2} =$	ten to the negative *second* power
$0.1 =$	$10^{-1} =$	ten to the negative *first* power
$1 =$	$10^0 =$	ten to the *zero* power
$10 =$	$10^1 =$	ten to the *first* power
$100 =$	$10^2 =$	ten to the *second* power
$1\ 000 =$	$10^3 =$	ten to the *third* power
$10\ 000 =$	$10^4 =$	ten to the *fourth* power
$100\ 000 =$	$10^5 =$	ten to the *fifth* power
$1\ 000\ 000 =$	$10^6 =$	ten to the *sixth* power

If a calculator is not used for computation, the powers of 10 enable one to work all problems by using convenient whole numbers. Either method offers a convenient way to obtain a final answer with the decimal point in its proper place.

Some of the multiples of 10 may be represented as shown in Table 6-1. From the table it is seen that any decimal fraction may be written as a whole number times some negative power of 10. This may be expressed by the following rule.

RULE

To express a decimal fraction as a whole number times a power of 10, move the decimal point to the right and count the number of places to the original point. The number of places counted is the proper negative power of 10.

Examples

$$0.006\ 87 = 6.87 \times 10^{-3}$$
$$0.000\ 048\ 2 = 4.82 \times 10^{-5}$$
$$0.346 = 34.6 \times 10^{-2}$$
$$0.086\ 43 = 86.43 \times 10^{-3}$$

Also, it is seen that any large number can be expressed as some smaller number times the proper power of 10. This can be expressed by the following rule.

RULE

To express a large number as a smaller number times a power of 10, move the decimal point to the left and count the number of places to the original decimal point. The number of places counted will give the proper positive power of 10.

Examples

$$435 = 4.35 \times 10^2$$
$$964\ 000 = 96.4 \times 10^4$$
$$6\ 835.2 = 6.835\ 2 \times 10^3$$
$$5723 = 5.723 \times 10^3$$

PROBLEMS 6-2

Express the following numbers to three significant figures and write them as numbers between 1 and 10 times the proper power of 10:

1. 643 000	**2.** 13.6	**3.** 6534
4. 0.0963	**5.** 0.000 000 009 435	**6.** 8 743 000
7. 0.367	**8.** 59 235	**9.** 250×10^{-3}
10. $0.000\ 086 \times 10^6$	**11.** $0.000\ 399 \times 10^8$	
12. $0.000\ 399\ 5 \times 10^8$	**13.** 259×10^{-4}	
14. 0.031 415 9	**15.** 276 492.536 24	
16. $1\ 254\ 325 \times 10^{-12}$	**17.** 0.000 000 107 52	
18. $0.000\ 008\ 145\ 73 \times 10^{12}$	**19.** 3.000 725	
20. $0.000\ 055\ 55 \times 10^{-3}$		

6-5 MULTIPLICATION WITH POWERS OF 10

In Sec. 4-3 the law of exponents in multiplication was expressed in the general form

$$a^m \cdot a^n = a^{m+n} \qquad \text{(where } a \neq 0)$$

This law is directly applicable to the powers of 10.

Example 1

Multiply 1000 by 100 000.

Solution

$$1000 = 10^3$$
and $$100\ 000 = 10^5$$
then
$$1000 \times 100\ 000 = 10^3 \times 10^5$$
$$= 10^{3+5}$$
$$= 10^8$$

Example 2

Multiply 0.000 001 by 0.001.

Solution

$$0.000\ 001 = 10^{-6}$$
and
$$0.001 = 10^{-3}$$
then
$$0.000\ 001 \times 0.001 = 10^{-6} \times 10^{-3}$$
$$= 10^{-6+(-3)}$$
$$= 10^{-6-3} = 10^{-9}$$

Example 3

Multiply 23 000 by 7000.

Solution

$$23\ 000 = 2.3 \times 10^4$$
and
$$7000 = 7 \times 10^3$$
then
$$23\ 000 \times 7000 = 2.3 \times 10^4 \times 7 \times 10^3$$
$$= 2.3 \times 7 \times 10^7$$
$$= 16.1 \times 10^7, \text{ or } 161\ 000\ 000$$

Example 4

Multiply 0.000 037 by 600.

Solution

$$0.000\ 037 \times 600 = 3.7 \times 10^{-5} \times 6 \times 10^2$$
$$= 3.7 \times 6 \times 10^{-3}$$
$$= 22.2 \times 10^{-3}, \text{ or } 0.0222$$

Example 5

Multiply 72 000 × 0.000 025 × 4600.

Solution

$$72\ 000 \times 0.000\ 025 \times 4600$$
$$= 7.2 \times 10^4 \times 2.5 \times 10^{-5} \times 4.6 \times 10^3$$
$$= 7.2 \times 2.5 \times 4.6 \times 10^2$$
$$= 82.8 \times 10^2, \text{ or } 8280$$

You will find that by expressing all numbers as numbers between 1 and 10 times the proper power of 10, the determination of the proper place for the decimal point will become a matter of inspection.

PROBLEMS 6-3

Multiply the following. Although not all factors are expressed to three significant figures, express answers to three significant figures as numbers between 1 and 10 times the proper power of 10.

1. $10\ 000 \times 0.01 \times 0.0001$
2. $0.000\ 01 \times 10^5 \times 100$
3. 0.0004×980
4. $0.000\ 25 \times 16 \times 10^{-4} \times 20 \times 10^5$
5. $0.000\ 008\ 4 \times 0.005 \times 0.000\ 17$
6. $35\ 000\ 000 \times 680 \times 10^{-9} \times 5.5 \times 10^{-5}$
7. $9.34 \times 10^{12} \times 628\ 000 \times 0.000\ 053 \times 10^{-3}$

8. $500 \times 10^{-6} \times 782 \times 10^4 \times 0.000\ 037 \times 10^{-8}$
9. $5\ 960\ 000 \times 0.000\ 888 \times 604 \times 10^{-5}$
10. $2.846 \times 10^3 \times 0.009\ 438 \times 10^6 \times 0.6848 \times 10^4$

The alternating-current (ac) inductive reactance of a circuit or an inductor is given by

$$X_L = 2\pi f L \qquad \Omega$$

where X_L = inductive reactance, Ω
$\quad\ f$ = frequency of alternating current, Hz
$\quad\ L$ = inductance of circuit, or inductor, henrys (H)
Compute the inductive reactance when:

11. $f = 60$ Hz, $L = 0.015$ H
12. $f = 1000$ Hz, $L = 0.015$ H
13. $f = 1\ 000\ 000$ Hz, $L = 0.015$ H
14. $f = 60$ Hz, $L = 1.5$ H
15. $f = 10\ 000$ Hz, $L = 0.000\ 003\ 5$ H

6-6 DIVISION WITH POWERS OF 10

The law of exponents in division (Secs. 4-9 to 4-11) can be summed up in the following general form:

$$\frac{a^m}{a^n} = a^{m-n} \qquad (\text{where } a \neq 0)$$

Example 6

$$\frac{10^5}{10^3} = 10^{5-3} = 10^2$$

or

$$\frac{10^5}{10^3} = 10^5 \times 10^{-3} = 10^2$$

Example 7

$$\frac{72\ 000}{0.0008} = \frac{72 \times 10^3}{8 \times 10^{-4}}$$

$$= \frac{72}{8} \times 10^{3+4}$$

$$= 9 \times 10^7$$

or

$$\frac{72\ 000}{0.0008} = \frac{72 \times 10^3}{8 \times 10^{-4}}$$

$$\frac{72\ 000}{0.0008} = \frac{72}{8} \times 10^3 \times 10^4$$

$$= 9 \times 10^7$$

Example 8

$$\frac{169 \times 10^5}{13 \times 10^5} = \frac{169}{13} \times 10^{5-5}$$

$$= 13 \times 10^0$$

$$= 13 \times 1 = 13$$

or

$$\frac{169 \times 10^5}{13 \times 10^5} = 13$$

It is apparent that powers of 10 which are factors that have the same exponents in numerator and denominator can be canceled. Also, you will note that powers of 10 which are factors can be transferred at will from denominator to numerator, or vice versa, if the sign of the exponent is changed when the transfer is made (Sec. 4-11).

6-7 APPROXIMATIONS

Multiplying 37 by 26 is very close to multiplying 40 by 25. The approximation 1000 is ''within the order'' of the actual product, 962. Usually, approximations which are within reason may be arrived at, and they serve as a guide to what the actual answer should be.

Such approximations should be made quickly before undertaking the exact calculations. The ''order'' of the calculated answer should be of the ''order'' of the approximation. If you expect an answer of the order of 1000 and you actually come up with 940 or 1050, the answer is probably correct. If, however, you arrive at an answer of 9.62, you should suspect that you have lost a factor of 10^2 somewhere, and you should check out your calculations. Although approximations will not guarantee the correctness of the calculated answer, they will reveal possible errors.

6-8 COMBINED MULTIPLICATION AND DIVISION

Combined multiplication and division is most conveniently accomplished by alternately multiplying and dividing until the problem is completed.

Example 9

Simplify $\dfrac{0.000\ 644 \times 96\ 000 \times 3300}{161\ 000 \times 0.000\ 001\ 20}$

Solution

First convert all numbers in the problem to numbers between 1 and 10 times their proper power of 10, thus:

$$\frac{6.44 \times 10^{-4} \times 9.6 \times 10^4 \times 3.3 \times 10^3}{1.61 \times 10^5 \times 1.2 \times 10^{-6}} = \frac{6.44 \times 9.6 \times 3.3 \times 10^4}{1.61 \times 1.2}$$

The problem as now written consists of multiplication and division of simple numbers. The answer approximates to

$$\frac{6 \times 10 \times 3 \times 10^4}{2 \times 1} = 90 \times 10^4$$

If the remainder of the problem is computed by calculator, then the answer 1056 can easily be adjusted to read 105.6×10^4, or 1.056×10^6. If the problem is solved without the aid of a calculator, there are no small decimals and no cumbersome large numbers to handle.

To solve this problem manually, first simplify the equation.

$$\frac{6.44 \times 9.6 \times 3.3 \times 10^4}{1.61 \times 1.2} = \frac{6.44 \times 8 \times 3.3 \times 10^4}{1.61}$$

Then find the products of the numerator and the denominator, respectively.

$$6.44 \times 8 \times 3.3 \times 10^4 = 170.016 \times 10^4$$
$$1.61 = 1.61$$

Finally, divide the numerator by the denominator.

$$\frac{170.016 \times 10^4}{1.61} = 105.6 \times 10^4$$

If we desire to express the answer in powers of 10, we would write 1.056×10^6, but written out without the power of 10, it would be 1 056 000.

6-9 RECIPROCALS

In electrical and electronics problems, many of the formulas that are used involve reciprocals. Some examples are

$$\frac{1}{R_t} = \frac{1}{R_1} + \frac{1}{R_2}$$
$$X_c = \frac{1}{2\pi f C}$$
$$f = \frac{1}{2\pi \sqrt{LC}}$$

The *reciprocal* of a number is 1 divided by that number. Finding a reciprocal presents no difficulty if the powers of 10 are used properly. Also, many calculators offer the convenience of a $\frac{1}{x}$ key.

Example 10

Simplify $\dfrac{1}{40\ 000 \times 0.000\ 25 \times 125 \times 10^{-6}}$.

Solution

First convert all numbers in the denominator to numbers between 1 and 10 times their proper powers of 10, thus:

$$\frac{1}{4 \times 10^4 \times 2.5 \times 10^{-4} \times 1.25 \times 10^{-4}} = \frac{10^4}{4 \times 2.5 \times 1.25}$$

Multiplying the factors of the denominator results in

$$\frac{10^4}{12.5}$$

Instead of writing out the numerator as 10 000 and then dividing by 12.5, we could write the numerator as two factors in order better to divide mentally. That is, we can write the problem as

$$\frac{10^2 \times 10^2}{12.5} \quad \text{or} \quad \frac{100}{12.5} \times 10^2 = 8 \times 10^2$$

This method is of particular advantage because of the ease of estimating the final result.

If the final result is a decimal fraction, rewriting the numerator into two factors allows fixing the decimal point with the least effort.

Example 11

Simplify $\dfrac{1}{625 \times 10^4 \times 2000 \times 64\ 000}$.

Solution

First convert all numbers in the denominator to numbers between 1 and 10 times their proper powers of 10, thus:

$$\frac{1}{6.25 \times 10^6 \times 2 \times 10^3 \times 6.4 \times 10^4} = \frac{10^{-13}}{6.25 \times 2 \times 6.4}$$

Multiplying the factors in the denominator results in

$$\frac{10^{-13}}{80}$$

Instead of writing out the numerator as 0.000 000 000 000 1 and dividing it by 80, we write the numerator as two factors in order better to divide mentally:

$$\frac{10^2 \times 10^{-15}}{80} \quad \text{or} \quad \frac{100}{80} \times 10^{-15} = 1.25 \times 10^{-15}$$

If the value of the denominator product were over 100 and less than 1000, we would break up the numerator so that one of the factors would be 10^3, or 1000, and so on. This method will always result in a final quotient of a number between 1 and 10 times the proper power of 10.

PROBLEMS 6-4

Perform the indicated operations. Round off the figures in the results, if necessary, and express answers to three significant figures as a number between 1 and 10 times the proper power of 10:

1. $\dfrac{0.000\ 25}{500}$

2. $\dfrac{10}{0.000\ 125 \times 80\ 000}$

3. $\dfrac{0.6043}{5763}$

4. $\dfrac{420 \times 0.036}{0.0090}$

5. $\dfrac{0.256 \times 338 \times 10^{-9}}{865\ 000}$

6. $\dfrac{1}{6.28 \times 452\ 000 \times 0.000\ 155}$

7. $\dfrac{2804 \times 74.23}{0.000\ 900\ 6 \times 0.008\ 040}$

8. $\dfrac{1000}{248\ 000 \times 5630 \times 10^{-3} \times 0.000\ 090\ 3 \times 10^2}$

9. $\dfrac{1 \times 10^6}{6.28 \times 10^3 \times 2500 \times 10^3 \times 0.25 \times 10^{-6}}$

10. $\dfrac{1}{6.28 \times 400 \times 10^6 \times 50 \times 10^{-12}}$

11. $\dfrac{150 \times 216 \times 1.78}{4.77 \times 10^2 \times 1.23 \times 6.03 \times 10^4}$

12. $\dfrac{65.3 \times 10^{-6} \times 504 \times 10^6 \times 12\ 700}{312 \times 10^6 \times 0.007 \times 6.82}$

The ac capacitive reactance of a circuit, or capacitor, is given by the formula

$$X_C = \frac{1}{2\pi f C} \qquad \Omega$$

where X_C = capacitive reactance, Ω
f = frequency of the alternating current, Hz
C = capacitance of the circuit, or capacitor, farads (F)
Compute the capacitive reactances when:

13. $f = 60$ Hz, $C = 0.000\ 004$ F
14. $f = 28\ 000\ 000$ Hz, $C = 0.000\ 000\ 000\ 025$ F
15. $f = 225\ 000\ 000\ 000\ 000$ Hz, $C = 0.000\ 000\ 000\ 563$ F

6-10 THE POWER OF A POWER

It becomes necessary, in order to work a variety of problems utilizing the powers of 10, to consider a few new definitions concerning the laws of exponents before we study them in algebra. This, however, should present no difficulty.

In finding the power of a power the exponents are multiplied. That is, in general,

$$(a^m)^n = a^{mn} \qquad \text{(where } a \neq 0)$$

Example 12

$$100^3 = 100 \times 100 \times 100 = 1\ 000\ 000 = 10^6$$
or $\qquad 100^3 = 10^2 \times 10^2 \times 10^2 = 10^6$
then $\qquad 100^3 = (10^2)^3 = 10^{2 \times 3} = 10^6$

Numbers can be factored when raised to a power in order to reduce the labor in obtaining the correct number of significant figures, or properly fixing the decimal point.

Example 13

$$19\ 000^3 = (1.9 \times 10^4)^3$$
$$= 1.9^3 \times 10^{4 \times 3} = 6.859 \times 10^{12}$$

Example 14

$$0.000\ 007\ 5^2 = (7.5 \times 10^{-6})^2$$
$$= 7.5^2 \times 10^{(-6) \times 2}$$
$$= 56.25 \times 10^{-12}$$
$$= 5.625 \times 10^{-11}$$

In Example 13, 19 000 was factored into 1.9×10^4 in order to allow an easy mental check. Because 1.9 is nearly 2 and $2^3 = 8$, it is apparent that the result of cubing 1.9 must be 6.859, not 0.6859 or 68.59.

In Example 14, the 0.000 007 5 was factored for the same reason. We know that $7^2 = 49$; therefore, the result of squaring 7.5 must be 56.25, not 0.5625 or 5.625.

If your calculator offers x^y or y^x, review your instruction manual and practice raising to powers on your calculator.

6-11 THE POWER OF A PRODUCT

The power of a product is the same as the product of the powers of the factors. That is, in general,

$$(abc)^m = a^m b^m c^m$$

Example 15

$$(10^5 \times 10^3)^3 = 10^{5 \times 3} \times 10^{3 \times 3}$$
$$= 10^{15} \times 10^9 = 10^{24}$$

or

$$(10^5 \times 10^3)^3 = (10^8)^3 = 10^{8 \times 3} = 10^{24}$$

6-12 THE POWER OF A FRACTION

The power of a fraction equals the power of the numerator divided by the power of the denominator. That is,

$$\left(\frac{a}{b}\right)^m = \frac{a^m}{b^m}$$

Example 16

$$\left(\frac{10^5}{10^3}\right)^2 = \frac{10^{5 \times 2}}{10^{3 \times 2}} = \frac{10^{10}}{10^6} = 10^4$$

The above can be solved by first clearing the exponents inside the parentheses and then raising to the required power. Thus,

$$\left(\frac{10^5}{10^3}\right)^2 = (10^{5-3})^2 = (10^2)^2 = 10^4$$

6-13 THE ROOT OF A POWER

The root of a power in exponents is given by

$$\sqrt[n]{a^m} = a^{m \div n} \qquad \text{(where } a \text{ and } n \neq 0\text{)}$$

Example 17

$$\sqrt{25 \times 10^8} = \sqrt{25} \times \sqrt{10^8} = 5 \times 10^{8 \div 2} = 5 \times 10^4$$

Example 18

$$\sqrt[3]{125 \times 10^6} = \sqrt[3]{125} \times \sqrt[3]{10^6} = 5 \times 10^{6 \div 3} = 5 \times 10^2$$

In the general case when m is evenly divisible by n, the process of extracting roots is comparatively simple. When m is not evenly divisible by n, the result obtained by extracting the root is a fractional power.

Example 19

$$\sqrt{10^5} = 10^{5 \div 2} = 10^{\frac{5}{2}}, \text{ or } 10^{2.5}$$

Such fractional exponents are encountered in various phases of engineering mathematics and are conveniently solved by the use of logarithms. However, in using the powers of 10, the fractional exponent is cumbersome for obtaining a final answer. It becomes necessary, therefore, to devise some means of extracting a root whereby an integer can be obtained as an exponent in the final result. The means found is to express the number, the root of which is desired, as some number times a power of 10 that is evenly divisible by the index of the required root. As an example, suppose it is desired to extract the square root of 400 000. Though it is true that

$$\sqrt{400\ 000} = \sqrt{4 \times 10^5}$$
$$= \sqrt{4} \times \sqrt{10^5}$$
$$= 2 \times 10^{2.5}$$

we have a fractional exponent that is not readily reduced to actual figures. However, if we express the number differently, we obtain an integer as an exponent. Thus,

$$\sqrt{400\ 000} = \sqrt{40 \times 10^4}$$
$$= \sqrt{40} \times \sqrt{10^4}$$
$$= 6.32 \times 10^2$$

It will be noted that there are a number of ways of expressing the above square root, such as

$$\sqrt{400\ 000} = \sqrt{0.4 \times 10^6}$$

or
$$\sqrt{4000 \times 10^2}$$

or
$$\sqrt{0.004 \times 10^8}$$

All are equally correct, but you should try to write the problem in a form that will allow a rough mental approximation in order that the decimal may be properly placed with respect to the significant figures.

PROBLEMS 6-5

Perform the indicated operations. When answers do not come out in round numbers, express them to three significant figures.

1. $(10^3)^4$
2. $(10^{-4})^3$
3. $(10^2 \times 10^3)^4$
4. $(4 \times 10^{-4})^2$
5. $(5 \times 10^3)^4$
6. $(3 \times 10^{-2})^3$

7. $(2 \times 10^4 \times 8 \times 10^{-5})^2$
8. $\left(\dfrac{32 \times 10^3}{8 \times 10^4}\right)^2$

9. $\sqrt{0.0625 \times 0.0004}$
10. $\sqrt{0.000\ 36 \times 0.009}$

11. $\sqrt{36 \times 10^2 \times 25 \times 10^{-2}}$
12. $\sqrt[3]{27 \times 10^{-3} \times 8 \times 10^{12}}$

13. $\dfrac{1}{6.28\sqrt{250 \times 10^{-3} \times 10^{-9}}}$
14. $\left(\dfrac{63 \times 10^6 \times 460 \times 10^{-12}}{5.1 \times 10^{-6}}\right)^2$

The resonant frequency of a circuit is given by the formula

$$f = \frac{1}{2\pi\sqrt{LC}} \quad \text{Hz}$$

where f = resonant frequency, Hz
 L = inductance of circuit, H
 C = capacitance of circuit, F

Compute the resonant frequencies when:

15. $L = 0.000\ 045$ H, $C = 0.000\ 000\ 000\ 250$ F
16. $L = 0.000\ 018$ H, $C = 100 \times 10^{-12}$ F
17. $L = 8 \times 10^{-6}$ H, $C = 56.3 \times 10^{-12}$ F
18. $L = 0.000\ 23$ H, $C = 0.000\ 000\ 000\ 5$ F
19. $L = 70.4 \times 10^{-6}$ H, $C = 250 \times 10^{-12}$ F
20. $L = 40$ H, $C = 7 \times 10^{-6}$ F

6-14 ADDITION AND SUBTRACTION WITH POWERS OF 10

Sometimes it becomes necessary, when making calculations, to perform additions and subtractions with powers of 10. These operations present no difficulties if you remember that you are dealing with the addition and subtraction of terms as described in Sec. 3-7. For example, you would not write $3x^2 + 5x^3 = 8x^5$, because $3x^2$ and $5x^3$ are unlike quantities. Similarly, you would not write

$$3 \times 10^2 + 5 \times 10^3 = 8 \times 10^5$$

because 3×10^2 and 5×10^3 also are unlike quantities.

The foregoing addition of

$$3 \times 10^2 + 5 \times 10^3$$

can be performed by either of two methods. You can convert the numbers so that no powers of 10 are involved and write $300 + 5000 = 5300$. Also, you can rewrite the terms to be added so that like powers of 10 are added, such as

$$3 \times 10^2 + 50 \times 10^2 = 53 \times 10^2$$

or $\quad\quad 0.3 \times 10^3 + 5 \times 10^3 = 5.3 \times 10^3$

This is the same as adding like terms.

Example 20
Add 8.3×10^4 and 3.6×10^2.

Solution

$$
\begin{aligned}
8.3 \times 10^4 &= 83\ 000 &=& 830 \quad\ \times 10^2 \\
3.6 \times 10^2 &= \quad\quad 360 &=& \underline{\quad 3.6 \times 10^2} \\
&\ 83\ 360 &=& 833.6 \times 10^2 \\
& &=& 8.336 \times 10^4
\end{aligned}
$$

PROBLEMS 6-6

Perform the indicated operations. Express all answers (*a*) in ordinary form and (*b*) to three significant figures as numbers between 1 and 10 times the proper power of 10.

1. $3 \times 10^3 + 1 \times 10^2$
2. $25 \times 10^6 + 3.4 \times 10^3$
3. $1.73 \times 10^{12} + 2.46 \times 10^{12}$
4. $2 \times 10^3 + 4 \times 10^{-1}$
5. $6.28 \times 10^6 - 159 \times 10^{-3}$

Units and Dimensions

7

As previously stated, the solution of every practical problem, when a concrete answer is desired, eventually reduces to an arithmetical computation; that is, the answer reduces to some *number*. In order for this answer, or number, to have a concrete meaning, it must be expressed in some *unit*. For example, if you were told that the resistance of a circuit is 16, the information would have no meaning unless you knew to what unit the 16 referred.

From the foregoing it is apparent that the expression for the magnitude of any physical quantity must consist of two parts. The first part, which is a number, specifies "how much"; the second part specifies the unit of measurement, or "what," as, for example, in 16 Ω, 20 A, or 100 m, the Ω, A, or m.

It is necessary, therefore, before beginning the study of circuits, to define a few of the more common electrical and dimensional units used in electrical and electronics engineering.

7-1 SYSTEMS OF MEASUREMENT

Over the years, the systems by which we have made measurements have changed considerably. We do not often now deal with grains of corn or the length of a man's forearm. Occasionally the civil engineer surveying an antenna site will talk about "chains" when we would have said "hundreds of meters." We in electronics are primarily concerned with three specific fields of measurement: distance-mass, time, and electric charge. All the electrical quantities are fundamentally related to these measurements, as you will discover if you study "higher mathematics."

Generally speaking, North Americans have used two main systems for measuring some quantities, whereas the units of other quantities are the same in both systems. One of these systems is the so-called English fps (foot-pound-second) or "traditional" system, which was widely, almost exclusively, used by engineers in English-speaking countries until very recently. The other is the metric mks (meter-kilogram-second) system, which has grown in importance over the last century. The most modern refinement of the mks system is called SI, for Système International d'Unités. It is used by well over 90% of the world's population, and it has become even more widely used in the last few years with the conversion of Great Britain, India, Australia, Canada, and, as metric legislation is passed, a growing number of industries in the United States to the SI metric system. You should note that in both the English system and the SI metric system several of the units are the same: seconds, volts, ohms, and amperes, especially.*

7-2 THE ENGLISH SYSTEM

The English system, developed over many centuries, contains many quite arbitrary relationships between the units and no systematic correlations. It is being superseded by the very logical SI metric units (Sec. 7-3). The small list given here shows only a very few of the many conversions which have been developed over the years.

$$12 \text{ inches (in)} = 1 \text{ foot (ft)}$$
$$3 \text{ feet (ft)} = 1 \text{ yard (yd)}$$
$$5280 \text{ feet (ft)} = 1 \text{ statute mile (mi)}$$
$$16 \text{ ounces (oz)} = 1 \text{ pound (lb)}$$
$$2000 \text{ pounds (lb)} = 1 \text{ ton}$$

7-3 THE SI METRIC SYSTEM

The metric system is a relatively newer, more orderly system related originally to the measurement of the earth itself. It uses decimal relationships throughout, rather than the arbitrary, hard-to-memorize conversions of the English system. The metric system started out as the centimeter-gram-second (cgs) system. Later, the cgs units were modified; the basic defined units were then the meter, the kilogram, and the second, and the name of the system was changed to mks. Later still, the ampere was added in order to elevate electrical units to the "physical" ones, and the name changed again to mksa.

The most recent development, the SI, is the result of many years of concentrated international cooperation and agreement, and has been published by the General Conference on Weights and Measures and the International Organization for Standardization (ISO). This system defines seven base measuring units, but it goes much farther in including other specific details. Altogether, SI involves:

- Seven base measuring units
- Two "supplementary" units
- An added collection of related units which can be defined in terms of the seven base units
- An orderly use of decimal calculations, with powers of 10 notation, and special word prefixes representing numbers
- An international system of symbols and notation
- A comprehensive system of national and international standards

*A quite thorough introduction to the SI metric system was written by Mr. Adams and published by McGraw-Hill in a revised edition in June 1974. It is entitled *SI Metric Units: An Introduction*. Students and teachers with no background in the metric system at all may find it useful.

Some of the SI units have little direct meaning to us in electronics and are described below only for the sake of providing you with a complete list. Others are daily necessities, and you will find them used repeatedly throughout this book.

7-4 THE SEVEN BASE SI UNITS

Length/meter   The meter was originally defined to be 1×10^{-7} of the length of the line of longitude passing through Paris from the equator to the North Pole. The present-day definition is that the standard meter is the length of 1 650 763.73 wavelengths in a vacuum of the radiation corresponding to the unperturbed transition between the energy levels $2p_{10}$ and $5d_5$ of the krypton 86 atom. This orange-red line has a wavelength of $6\ 057.802 \times 10^{-10}$ m.

$$1 \text{ meter (m)} = 39.370\ 079 \text{ inches (in)}$$

Mass/kilogram   The kilogram is simply defined to be the mass of a special cylinder of platinum-iridium alloy which is in the safekeeping of the International Bureau of Weights and Measures. This cylinder is called the *International Prototype Kilogram*.

$$1 \text{ kilogram (kg)} = 2.204\ 622\ 6 \text{ pounds (lb)}$$

Time/second   The second is specifically defined as the duration of 9 192 631 870 periods of the radiation corresponding to the transition between the two hyperfine levels of the ground state of the atom of cesium 133. There are other special definitions of time based on the sun, stars, and moon, but the definition given here is one which can be duplicated in laboratories of Bureaus of Standards anywhere.

Electric current/ampere   That intensity of electric current known as an ampere is defined as the constant current which, if maintained in two straight parallel conductors of infinite length and of negligible cross section in a vacuum exactly one meter apart, will produce a force between the conductors of 2×10^{-7} newton (N) per meter length of wire. The circuit symbol* for current is I, and the unit symbol for amperes is A.

Temperature/kelvin   The defined SI unit of temperature is the kelvin. The freezing point of pure water is 273.15 K (*not* °K). Ordinary temperature readings will be made on the Celsius scale, on which the freezing point of pure water is 0°C and on which the boiling point is 100°C (= 373.15 K).

Luminous intensity/candela   The standard SI unit of luminous intensity is the candela (cd). This is the amount of luminosity which will produce a luminous flux of 1 lumen (l) within a solid angle of 1 steradian (sr). (We will not concern ourselves with illumination in this book.)

---

*Circuit symbols appear in circuit or schematic diagrams, and they are used as "quantity" symbols in algebraic formulas. They are always printed in *italic* type. Unit symbols (not abbreviations) are used for units of measure. They are always printed in roman type (not slanting).

Molecular substance/mole   The mole is the standard SI unit which gives the gram molecular weight of a substance. (We will not concern ourselves with this more or less pure science unit in this book.)

Angles/radians and steradians   In addition to the seven base units, there are two supplementary units for the measurement of angles: the radian for the measurement of plane angles and the steradian for the measurement of solid angles. We will study plane angles in Chap. 23.

7-5 THE ADDITIONAL DEFINED SI UNITS

In addition to the base or standard SI units, there are 16 other units which are used so often that they have been given special names. Many of them are important to us in electronics, and they are listed first in the descriptions which follow. Again, all 16 units are listed for the benefit of users of this book who are studying beyond the book's limitations.

Electric charge/coulomb   The coulomb is defined as the ampere-second. A reverse definition is that one ampere is the current intensity when one coulomb (C) flows in a circuit for one second. The coulomb is also defined as $6.241\,96 \times 10^{18}$ electronic charges. The circuit symbol for charge is Q.

Electric potential/volt   The volt is the practical unit of electromotive force (emf), or potential difference. It is defined as the watt per ampere. (See watt below.) A more common understanding of the volt is that it is the potential difference which will drive a current of one ampere through a resistance of one ohm. The circuit symbols for voltage are V and v, and the unit symbol for volt is V.

Electric resistance/ohm   The ohm is the practical unit of resistance. It is defined as the volt per ampere; that is, the ohm is the amount of resistance which limits the current flow to one ampere when the applied electromotive force (potential) is one volt. The circuit symbol for resistance is R, and the unit symbol for ohms is Ω.

Electric conductance/siemens   The siemens is the practical unit of conductance. It is the reciprocal of the ohm, since the conductance is the reciprocal of the resistance. The relationship between ohms and siemens is given by

$$G = \frac{1}{R} \text{ siemens}$$

If resistance is thought of as representing the difficulty with which an electric current is forced to flow through a circuit, conductivity may be thought of as the ease with which a current will pass through the same circuit.

A conductance of one siemens will permit a current flow of one ampere under an electrical pressure of one volt. The siemens is a new unit honoring a pioneer in electricity. Formerly, the unit of conductance was called the mho. The circuit symbol for conductance is *G,* and the unit symbol for siemens is S.

Electric capacitance/farad   The farad is the unit of capacitance. It is an ampere-second per volt. A circuit, or capacitor, is said to have a capacitance of one farad when a change of one volt per second across it produces a current of one ampere. The circuit symbol for capacitance is *C,* and the unit symbol for farad is F. Capacitance will be further discussed in Chap. 32.

Electric inductance/henry   The henry is the unit of inductance. It is a volt-second per ampere. A circuit, or inductor, is said to have a self-inductance of one henry when a counterelectromotive force of one volt is generated within it by a rate of change of current of one ampere per second. The circuit symbol for inductance is *L,* and the unit symbol for henry is H. Inductance will be further discussed in Chap. 32.

Frequency/hertz   The SI unit of frequency is the hertz, which was formerly called the cycle per second. Since *cycle* is not a unit of measure as such, it is sufficient to decribe the hertz as the reciprocal of time.

Magnetic flux/weber   Magnetic flux is fully described as the volt-second.

Magnetic flux density/tesla   Tesla is the special name given to the "density" relationship of webers per square meter.

Luminous flux/lumen   This SI unit relates the amount of radiant energy in terms of candelas of luminous intensity multiplied by the solid angle in steradians from which the radiant flux "flows."

Illumination/lux   This unit describes the lumens per square meter relationship of luminous flux.

Energy/joule   The SI unit for energy of all forms—mechanical work, electric energy, heat quantity, etc.—is the joule. The joule (J) may be expressed in terms of newtons of force multiplied by the distance in meters through which the force moves in the direction of its application.

Force/newton   The newton is the SI unit describing joules per meter.

Pressure/pascal   The pascal (Pa) is defined as newtons per square meter. It is a very small unit of measurement.

Power/watt   The watt is the SI unit for power of all forms—electric, mechanical, and so on. It is defined as the energy in joules expended per unit of time in seconds. The circuit symbol for power is *P,* and the unit symbol for watt is W.

In direct-current circuits the power in watts is the product of the voltage and the current, or

$$P = VI \qquad W$$

The watthour is the unit of electric energy, and its abbreviation is Wh. It is the amount of energy delivered by a power of one watt over a period of one hour.

Customary temperature/degree Celsius   Ordinary (nonscientific) temperature measurements will be made on the Celsius scale, which is related to the Kelvin scale by $°C = K - 273.15$.

7-6 SOME SI METRIC INTERRELATIONSHIPS

The fundamental relationships between metric units are decimal. The following equations show some of the simpler multiples and submultiples. You will be involved in many such conversions as you continue your studies in electronics.

$$1 \text{ millimeter (mm)} = \frac{1}{1000} \text{ meter} = 10^{-3} \text{ m}$$

$$1 \text{ centimeter (cm)} = \frac{1}{100} \text{ meter} = 10^{-2} \text{ m}$$

$$1 \text{ kilometer (km)} = 1000 \text{ meters} = 10^{3} \text{ m}$$

$$1 \text{ gram (g)} = \frac{1}{1000} \text{ kilogram (kg)}$$

$$= 10^{-3} \text{ kg}$$

7-7 RELATIONS BETWEEN THE SYSTEMS

Since the metric system is based on a decimal plan and the English system is not, there is no one numerical factor or constant which can be used for the conversion of one system to the other. Table 6 in the Appendix contains some conversion factors, and a few approximate equivalents are given here for your convenience:

$$1 \text{ inch (in)} = 2.540 \text{ centimeters (cm) (exactly)}$$
$$1 \text{ foot (ft)} = 0.3048 \text{ meter (m) (exactly)}$$
$$1 \text{ meter (m)} = 39.37 \text{ inches (in)}$$
$$1 \text{ mile (mi)} = 1.609 \text{ kilometers (km)}$$
$$1 \text{ kilometer (km)} = 0.6214 \text{ mile (mi)}$$
$$1 \text{ kilogram (kg)} = 2.205 \text{ pounds (lb)}$$
$$1 \text{ pound (lb)} = 0.4536 \text{ kilogram (kg)}$$

If you are unfamiliar with the metric system, try to visualize these relationships for future convenience. What is the weight in kilograms of a loaf of bread in your community? What is the distance in kilometers from your home to your work? What is your height in centimeters?

The units of time (seconds) and of electricity are identical in the two systems, and we will now deal with them in more detail.

Since the SI units are becoming increasingly important as Canada and the United States convert to the metric system, you should make the habit of thinking in metric units. Do not keep translating metric quantities into the old English units.

7-8 FREQUENCY

A current which reverses itself at intervals is called an *alternating current*. When this current rises from zero value to maximum value and returns to zero and then increases to maximum value in the opposite direction and again returns to zero, it is said to have completed *one cycle*. The number of times this cycle is repeated in one second is known as the *frequency* of the alternating current. The frequency of the average house current is 60 cycles per second (cps), and that of radio waves may be as high as several hundred million cycles per second. Note that frequency involves our other main unit, time, by measuring the number of events per second. In both the English and SI systems,

$$60 \text{ seconds (s)} = 1 \text{ minute (min)}$$
$$60 \text{ minutes (min)} = 1 \text{ hour (h)}$$
$$24 \text{ hours (h)} = 1 \text{ day (d)}$$

The International Electrotechnical Commission (IEC), the International Organization for Standardization (ISO), and the Conférence Générale des Poids et Mesures (CGPM) have adopted the name *hertz* (Hz) as the unit of frequency.

$$1 \text{ hertz} = 1 \text{ cycle per second}$$

7-9 RANGES OF UNITS

As stated in Sec. 6-7, the fields of communication and electrical engineering embrace extremely wide ranges in values of the foregoing units. For example, at the input of a radio receiver, we deal in millionths of a volt, whereas the output circuit of a transmitter may develop hundreds of thousands of volts. An electric clock might consume a fraction of a watt, whereas the powerhouse furnishing this power probably has a capability of millions of watts.

Furthermore, two of these units, the henry and the farad, are very large units, especially the latter. The average radio receiver employs inductances ranging from a few millionths of a henry, as represented by tuning inductance, to several henrys for power filters. The farad is so large that even the largest capacitors are rated in millionths of a farad. Smaller capacitors used in radio circuits are often rated in terms of so many millionths of one-millionth of a farad.

The use of some power of 10 is very convenient in converting to larger multiples or smaller fractions of the basic units, called *practical units*.

7-10 DECIMAL MULTIPLIERS

Some of the more common multipliers and their unit names are explained below, and all of the multipliers are listed in Table 7-1.

MILLIUNITS The milliunit is one-thousandth of a unit. Thus, it takes 1000 millivolts to equal 1 volt, 500 milliamperes to equal 0.5 ampere,

TABLE 7-1 DECIMAL MULTIPLIERS

Number		Power of 10		Expressed in English		Prefix	Abbreviation
0.000 000 000 000 000 001	=	10^{-18}	=	ten to the negative *eighteenth* power	=	atto	a
0.000 000 000 000 001	=	10^{-15}	=	ten to the negative *fifteenth* power	=	femto	f
0.000 000 000 001	=	10^{-12}	=	ten to the negative *twelfth* power	=	pico	p
0.000 000 001	=	10^{-9}	=	ten to the negative *ninth* power	=	nano	n
0.000 001	=	10^{-6}	=	ten to the negative *sixth* power	=	micro	μ
0.001	=	10^{-3}	=	ten to the negative *third* power	=	milli	m
1	=	10^{0}	=	ten to the *zero* power	=	unit	
1 000	=	10^{3}	=	ten to the *third* power	=	kilo	k
1 000 000	=	10^{6}	=	ten to the *sixth* power	=	mega	M
1 000 000 000	=	10^{9}	=	ten to the *ninth* power	=	giga	G
1 000 000 000 000	=	10^{12}	=	ten to the *twelfth* power	=	tera	T
1 000 000 000 000 000	=	10^{15}	=	ten to the *fifteenth* power	=	peta	P
1 000 000 000 000 000 000	=	10^{18}	=	ten to the *eighteenth* power	=	exa	E

etc. This unit is commonly used with volts, amperes, henrys, and watts. It is abbreviated m; thus, 10 mH = 10 millihenrys.* Mathematically, milli = 10^{-3}, and 1 mW = 10^{-3} W.

MICROUNITS The microunit is one-millionth of a unit. That is, it takes 1 000 000 microamperes to make 1 ampere, 2 000 000 microfarads to equal to 2 farads, etc. This unit, abbreviated μ (greek letter mu), is commonly used with volts, amperes, ohms, siemens, henrys, and farads. Thus,

$$5 \ \mu F = 5 \text{ microfarads}$$

Mathematically, micro = 10^{-6}, and 1 μs = 10^{-6} s.

PICOUNITS The picounit is one-millionth of one-millionth of a unit; that is, 1 farad is equivalent to 1 000 000 000 000, or 10^{12}, picofarads. This unit is seldom used for anything other than farads. It is represented by p; thus, 250 pF = 250 picofarads. Mathematically, pico = 10^{-12}. Older texts use the micromicrounit, abbreviated μμ. Thus,

$$2 \ \mu\mu F = 2 \text{ micromicrofarads} = 2 \text{ pF}$$

KILOUNITS The kilounit is 1000 basic units. Thus, a kilovolt is equivalent to 1000 volts. This unit is commonly used with cycles, volts, amperes, ohms, watts, and volt-amperes. It is abbreviated k. Thus, 35 kW means 35 kilowatts; 2000 hertz (Hz) = 2 kilohertz (kHz). Mathematically, kilo = 10^{3}.

MEGAUNITS The megaunit is 1 000 000 basic units. Thus, 1 megohm is equal to 1 000 000 ohms. This unit is used mainly with ohms

---

*See Table 3 in the Appendix for abbreviations and unit symbols.

and hertz. It is abbreviated M. Thus, 3 MHz = 3 megahertz. Mathematically, mega = 10^6.

7-11 PREFERRED DECIMAL MULTIPLIERS

Table 7-1 gives the prefix names and abbreviations for the *third powers of 10*. These are the preferred powers, and, therefore, the preferred prefixes. Almost every calculation in electronics will result in a quantity involving one of the third powers of 10 prefixes: *milli*watts, *micro*amperes, *mega*hertz.

In a few cases, prefixes are also used for other powers of 10. Table 7-2 lists these denigrated, or nonstandard, powers of 10.

TABLE 7-2   DENIGRATED POWERS OF TEN

Number	Power of 10		Expressed in English		Prefix	Abbreviation
100	= 10^2	=	ten squared	=	hecto	h
10	= 10^1	=	ten	=	deca	da
0.1	= 10^{-1}	=	ten to the negative *first* power	=	deci	d
0.01	= 10^{-2}	=	ten to the negative *second* power	=	centi	c

In order to use the preferred third powers of 10, quantities will normally be expressed as numbers between 0.1 and 1000.

In Examples 1 to 3, express numbers by using preferred third powers of 10 prefixes.

Example 1
0.01 A

Solution

$$0.01 \text{ A} = 1 \times 10^{-2} \text{ A}$$

Rewriting to a third power,

$$1 \times 10^{-2} = 10 \times 10^{-3}$$

Therefore,
$$0.01 \text{ A} = 10 \times 10^{-3} \text{ A} = 10 \text{ mA}$$

Example 2
1320 kHZ

Solution

$$1320 \text{ kHz} = 1.32 \times 10^3 \text{ kHz}$$
$$1320 \text{ kHz} = 1.32 \times 10^3 \times 10^3 \text{ Hz}$$
$$= 1.32 \times 10^6 \text{ Hz}$$
$$= 1.32 \text{ MHz}$$

Example 3
0.872 H

Solution

This is a perfectly good number and need not be changed. However, some people may prefer to rewrite it as 872 mH, which is equally good.

TABLE 7-3   CONVERSION FACTORS

Multiply	By	To Obtain
Picounits	10^{-6}	Microunits
Picounits	10^{-9}	Milliunits
Picounits	10^{-12}	Units
Microunits	10^6	Picounits
Microunits	10^{-3}	Milliunits
Microunits	10^{-6}	Units
Milliunits	10^9	Picounits
Milliunits	10^3	Microunits
Milliunits	10^{-3}	Units
Units	10^{12}	Picounits
Units	10^6	Microunits
Units	10^3	Milliunits
Units	10^{-3}	Kilounits
Units	10^{-6}	Megaunits
Kilounits	10^3	Units
Kilounits	10^{-3}	Megaunits
Megaunits	10^6	Units
Megaunits	10^3	Kilounits

7-12 DECIMAL CONVERSION FACTORS

Often it becomes necessary to convert microamperes to milliamperes, gigahertz to kilohertz, megawatts to watts, and so on. The more common conversions in simplified form are listed in Table 7-3.

Example 4
Convert 8 μF to farads.

Solution
$$8 \ \mu F = 8 \times 10^{-6} \ F$$

Example 5
Convert 250 mA to amperes.

Solution
$$250 \ mA = 250 \times 10^{-3} \ A$$
$$= 2.50 \times 10^{-1} \ A$$
or
$$= 0.250 \ A$$

Example 6
Convert 1500 W to kilowatts.

Solution
$$1500 \ W = 1500 \times 10^{-3} \ kW$$
or
$$= 1.5 \ kW$$

Example 7
Convert 200 000 Ω to megohms.

Solution
$$200 \ 000 \ \Omega = 200 \ 000 \times 10^{-6} \ M\Omega = 0.2 \ M\Omega$$

Example 8
Convert 2500 kHz to meghertz.

Solution

$$2500 \text{ kHz} = 2500 \times 10^{-3} \text{ MHz} = 2.500 \text{ MHz}$$

Example 9
Convert 0.000 450 S to microsiemens.

Solution

$$0.000\ 450 \text{ S} = 0.000\ 450 \times 10^6 \text{ } \mu\text{S}$$
or $$= 450\mu\text{S}$$

Example 10
Convert 5 μs to seconds.

Solution

$$5 \text{ } \mu\text{s} = 5 \times 10^{-6} \text{ s}$$

PROBLEMS 7-1
Express answers to three significant figures as numbers between 1 and 10 times the proper power of 10:

1.	4300 V	= (a) ____mV	= (b) ____μV	= (c) ____kV	
2.	6.85 A	= (a) ____mA	= (b) ____μA		
3.	1.35 V	= (a) ____kV	= (b) ____μV	= (c) ____mV	
4.	125 mA	= (a) ____μA	= (b) ____A		
5.	3300 Ω	= (a) ____kΩ	= (b) ____MΩ	= (c) ____S	
6.	50 μF	= (a) ____F	= (b) ____pF		
7.	20 000 pF	= (a) ____F	= (b) ____μF		
8.	16.5 mH	= (a) ____H	= (b) ____μH		
9.	347 W	= (a) ____kW	= (b) ____mW	= (c) ____μW	
10.	25.3 s	= (a) ____ms	= (b) ____μs		
11.	1320 kHz	= (a) ____MHz	= (b) ____Hz		
12.	47 kΩ	= (a) ____Ω	= (b) ____MΩ	= (c) ____S	
13.	400 mW	= (a) ____W	= (b) ____kW		
14.	220 μH	= (a) ____mH	= (b) ____H		
15.	15 kHz	= (a) ____MHz	= (b) ____Hz		
16.	8 μs	= (a) ____ms	= (b) ____s	= (c) ____ns	
17.	0.055 A	= (a) ____μA	= (b) ____mA		
18.	325 kV	= (a) ____V	= (b) ____MV		
19.	2.7 MΩ	= (a) ____Ω	= (b) ____kΩ		
20.	3.7 kWh	= (a) ____Wh	= (b) ____mWh		
21.	3350 mH	= (a) ____μH	= (b) ____H		
22.	506 MHz	= (a) ____kHz	= (b) ____Hz		
23.	0.000 50 μF	= (a) ____pF	= (b) ____F		
24.	1500 ms	= (a) ____μs	= (b) ____s	= (c) ____ns	
25.	2.5 S	= (a) ____μS	= (b) ____Ω		
26.	5000 μS	= (a) ____S	= (b) ____Ω		
27.	2350 μA	= (a) ____mA	= (b) ____A		
28.	0.15 kV	= (a) ____V	= (b) ____mV		

29. 150 MW   = (a) ____W   = (b) ____kW
30. 980 000 Hz = (a) ____kHz = (b) ____MHz

7-13 INTERSYSTEM CONVERSIONS

In the early sections of this chapter we briefly reviewed the two systems with which we most often deal, and we listed some common conversion factors. Some books of tables give hundreds of such interrelationships, and you will meet them as you continue your studies.

You must realize that, without the units, your calculations are incomplete. When measurements are added, subtracted, multiplied, or divided, then the units pertaining to those measurements must also take part in the calculations.

Example 11

Add 6 V and 12 V.

Solution

$$6 \text{ V} + 12 \text{ V} = 18 \text{ V}$$

Example 12

Add 3 m and 75 cm.

Solution

In the metric system, the values of the prefixes represent decimal multipliers:

(a) 75 cm $= \dfrac{75}{100}$ m $= 0.75$ m. Adding:

$$
\begin{array}{rcl}
3 \text{ m} & = & 3 \quad\ \text{m} \\
+75 \text{ cm} & = & +0.75 \text{ m} \\
\hline
 & & 3.75 \text{ m}
\end{array}
$$

(b) Alternatively, 3 m = 300 cm. Adding:

$$
\begin{array}{rcl}
3 \text{ m} & = & 300 \text{ cm} \\
+75 \text{ cm} & = & +\ 75 \text{ cm} \\
\hline
 & & 375 \text{ cm}
\end{array}
$$

Either answer is correct, but one form may be more acceptable than the other under some circumstances. The same person might properly describe the height of a child as 112 cm and later refer to a folded dipole antenna as 1.12 m long.

Example 13

What is the speed of an object that traverses 30 m in 2 s?

Solution

Speed is given in units of distance per unit of time. In this case, the speed is

$$\frac{30 \text{ m}}{2 \text{ s}} = 15 \frac{\text{m}}{\text{s}} \qquad \text{(usually written m/s*)}$$

*m/s (*a shilling fraction*) has exactly the same meaning as $\frac{\text{m}}{\text{s}}$ (*a built-up fraction*); the *only* difference is in the manner of printing.

Example 14

What is the area of a room 12 m long and 18 m wide?

Solution

Areas are given in square measure:

$$(12 \text{ m})(18 \text{ m}) = 216 \text{ square meters (m}^2)*$$

Example 15

$$3\,\Omega + 6\,\Omega = 9\,\Omega$$
$$230\text{ V} - 115\text{ V} = 115\text{ V}$$

Example 16

$$2\text{ m} \times 4\text{ m} = 2 \times 4 \times \text{m} \times \text{m} = 8\text{ m}^2$$
$$0.7\text{ m} \times 1.6\text{ m} = 0.7 \times 1.6 \times \text{m} \times \text{m} = 1.12\text{ m}^2$$
$$20\text{ cm} \times 1.2\text{ m} = 0.2\text{ m} \times 1.2\text{ m} = 0.24\text{ m}^2$$
$$3\text{ m} \times 5\text{ m} \times 2\text{ m} = 3 \times 5 \times 2 \times \text{m} \times \text{m} \times \text{m}$$
$$= 30\text{ m}^3$$
$$6\text{ m} \times 10\text{ m} = 6 \times 10\text{ meters} \times \text{meters}$$
$$= 60\text{ meters}^2 = 60\text{m}^2$$

In a ratio between identical units, such as $\frac{60\text{ m}}{12\text{ m}}$, the unit symbols cancel and the result of the division is a "pure" number with no dimension.

Example 17

$$\frac{60\text{ m}}{12\text{ m}} = \frac{60\,\cancel{\text{m}}}{12\,\cancel{\text{m}}} = 5$$

When quantities having different units are multiplied or divided, the result must express the operation.

Example 18

$$4\text{ m} \times 5\text{ kg} = 4 \times 5 \times \text{m} \times \text{kg}$$
$$= 20\text{ kg} \cdot \text{m}$$

Example 19

$$\frac{30\text{ m}}{10\text{ s}} = \frac{30}{10}\frac{\text{m}}{\text{s}} = 3\frac{\text{m}}{\text{s}}$$
$$= 3\text{ m/s} = 3\text{ m} \cdot \text{s}^{-1}$$

Example 20

$$\frac{45\,\Omega}{15\text{ m}} = \frac{45}{15}\frac{\Omega}{\text{m}} = 3\,\Omega/\text{m}$$

In Example 19 note that m/s is read as "meters per second," and in Example 20 note that Ω/m is read as "ohms per meter." Per means *divided by*.

*It is generally preferred that areas be written in the form 8 m² rather than 8 sq m.

Thus some of the equivalent lengths stated in Sec. 7-7 can be expressed as follows:

- There are 2.540 cm/in.
- There is 0.3048 m/ft.
- There are 1.609 km/mi.
- There are 39.37 in/m.
- There is 0.6214 mi/km.

Using relations in forms such that units are treated mathematically as literal factors facilitates conversions and assures that results will be obtained with correct units.

Example 21
Convert 3 in to centimeters.

Solution

$$3 \text{ in} \times 2.54 \frac{\text{cm}}{\text{in}} = 3 \times 2.54 \cdot \cancel{\text{in}} \cdot \frac{\text{cm}}{\cancel{\text{in}}}$$
$$= 7.62 \text{ cm}$$

Example 22
How many meters are there in 236 ft?

Solution

$$236 \text{ ft} \times 0.3048 \frac{\text{m}}{\text{ft}} = 236 \times 0.3048 \cdot \cancel{\text{ft}} \cdot \frac{\text{m}}{\cancel{\text{ft}}}$$
$$= 71.93 \text{ m}$$

Example 23
A certain resistance wire has a resistance of 3 Ω/m. What is the resistance of 6 m of this wire?

Solution

$$3 \frac{\Omega}{\text{m}} \times 6 \text{ m} = 3 \times 6 \cdot \frac{\Omega}{\cancel{\text{m}}} \cdot \cancel{\text{m}} = 18 \ \Omega$$

Example 24
Convert 1500 kHz to hertz.

Solution

There are 10^3 Hz per kilohertz, that is, $10^3 \frac{\text{Hz}}{\text{kHz}}$. Then

$$1500 \text{ kHz} = 1500 \frac{\text{kcycles}}{\text{s}} \times 10^3 \frac{\text{cycles}}{\text{kcycle}}$$
$$= 1500 \times 10^3 \frac{\cancel{\text{kcycles}}}{\text{s}} \cdot \frac{\text{cycles}}{\cancel{\text{kcycles}}}$$
$$= 1.5 \times 10^6 \text{ cycles/s}$$
$$= 1.5 \times 10^6 \text{ Hz}$$

Example 25

The wavelength λ of a radio wave in meters, the frequency f of the wave in hertz, and the velocity of propagation c in meters per second are related to one another by the formula

$$\lambda = \frac{c}{f}$$

or

$$\lambda = \frac{3 \times 10^8}{f} \quad m$$

Derive a formula for wavelength expressed in feet.

Solution

Since there are 3.28 ft/m, this factor must be applied to express λ in feet. Thus,

$$\lambda = \frac{3 \times 10^8}{f} \, m \times 3.28 \frac{ft}{m}$$

$$= \frac{3 \times 3.28 \times 10^8}{f} \cdot m \cdot \frac{ft}{m}$$

$$= \frac{9.84 \times 10^8}{f} \, ft*$$

Example 26

By using the formula $\lambda = (3 \times 10^8)/f$ m, derive a formula for wavelength in meters when the frequency is expressed in megahertz.

Solution

In the above formula f is expressed in hertz and it is desired to express the frequency in megahertz. Since

$$MHz = Hz \times 10^6,$$

this is substituted for f in the formula. Thus,

$$\lambda = \frac{3 \times 10^8}{f \times 10^6} \, m = \frac{3 \times 10^2}{f} \, m = \frac{300}{f} \, m$$

PROBLEMS 7-2

1. 9 ft   = (a) _____in = (b) _____cm = (c) _____mm
2. 3500 mm = (a) _____km = (b) _____ft = (c) _____yd
3. 2.05 m   = (a) _____in = (b) _____cm = (c) _____yd
4. 15 840 ft = (a) _____km = (b) _____mi = (c) _____cm
5. 5064 yd   = (a) _____mi = (b) _____m = (c) _____km
6. An automobile is traveling at a rate of 90 mi/h. What is its speed in meters per second?

*Some users of this book will be interested in the 1972 report issued by the National Bureau of Standards in Boulder, Colorado, which gives the value of c as 299 792.4562 km/s ± 1.1 m/s. 3×10^8 is a sufficiently accurate approximation for the correct solution to all the problems in this book, and, indeed, for most of the problems ever to be solved by the majority of electronics technicians anywhere.

7. The radius of No. 14 wire is 32 thousandths of an inch. What is its diameter in millimeters?

8. Radio waves are often referred to by wavelength instead of frequency. The wavelength of waves at a frequency of 3000 MHz is 10 cm. What is that wavelength in millimeters?

9. A power transmission line 120 km long is found to have a total inductance of 0.4488 H. What is the inductance per kilometer?

10. The capacitance of a power line is measured at 4.98 nF/km. What is the capacitance per meter?

11. A transmission line 250 m long is found to have an attenuation loss of 0.15 decibel (dB). What is the attenuation in decibels per hundred meters?

12. A twisted-pair transmission line 200 m long has a loss of 42 dB. What is the loss in decibels per centimeter?

13. The high-frequency resistance of No. 10 copper wire was measured some years ago by using a 6-ft length of wire. At 100 MHz, the resistance was found to be 0.588 Ω. What was the resistance in ohms per centimeter at the same frequency?

14. The speed of free electrons in random motion is approximately 100 000 m/s. What is this speed in miles per hour?

15. The speed of electrons ''drifting'' in an electric current flow is about 0.2 cm/s. What is this speed in inches per minute?

7-14 PRACTICAL CONSIDERATIONS

In Secs. 6-7 and 7-9 and in several instances through the use of examples and problems, attempts have been made to emphasize that extremely wide ranges in values of units are encountered in electrical and electronics computations. This has been done in order to impress you with the necessity of exercising care in making computations if you are to obtain accurate results. For example, in computing inductive reactances, the frequency may be in megahertz and the inductance in microhenrys. In radar and other applications we are concerned with the velocity of propagation of radio waves (3×10^8 m/s) and with time intervals in microseconds. This is equally true in television reception, particularly as it relates to the production of duplicate images. As an example, Fig. 7-1 illustrates how a television receiver can receive a picture signal from a transmitting station by different paths. The direct wave is received from the transmitter along one path, while the other signal arrives at the receiving antenna via a path 1 km longer than the direct path as a result of being reflected. Because the velocity of radio waves is 3×10^8 m/s, the reflected signal arrives at the receiver $1/(3 \times 10^5)$ s, or about 3.3 μs, later than the signal received via the direct path between transmitter and receiver. Since the electron scanning beam will scan one horizontal line in approximately 55 μs, on a picture 50 cm wide the beam scans about 1 cm in 1.1 μs. Therefore, the reflected signal arriving 3.3 μs late will produce a second picture 3 cm to the right in the direction of scanning as shown in Fig. 7-2. This duplicate image produced by the reflected wave is called a *ghost*.

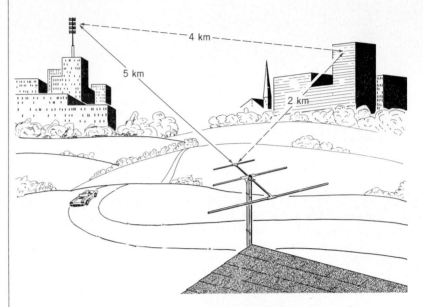

Fig. 7-1 Antenna receiving picture signal via two paths.

Fig. 7-2 Television ghost. (*Courtesy of Radio Corporation of America*)

7-15 SIGNIFICANT FIGURES

The subjects of accuracy and significant figures were discussed in Secs. 6-4 and 6-5. Now that we have some idea of the various units used in electrical and electronics problems, two questions arise:

1. To how many significant figures should an answer be expressed?
2. How can we definitely show that an answer is correct to just so many significant figures?

The answer to the first question is comparatively easy. No answer can be more accurate than the figures, or data, used in the problem. It is safe to assume that the values of the average circuit components and calibrations of meters that we use in our everyday work are not known beyond three significant figures. Therefore, in the future we will round off long answers and express them to three significant figures. The exception will be when it is necessary to carry figures out in order to demonstrate or obey some fact or law, carefully.

The second question brings up some interesting points. As an example, suppose we have a resistance of 500 000 Ω and we want to write this value so that it will be apparent to anyone that the figure 500 000 is correct to three significant figures. We can do so by writing

$$500 \times 10^3 \, \Omega \quad \text{or} \quad 500 \, \text{k}\Omega$$
$$50.0 \times 10^4 \, \Omega$$
$$5.00 \times 10^5 \, \Omega \quad \text{or} \quad 0.500 \, \text{M}\Omega, \text{etc.}$$

Any one of these expressions definitely shows that the resistance is correct to three significant figures. Similarly, suppose we had measured the capacitance of a capacitor to be 3500 pF. How can we specify that the figure 3500 is correct to three significant figures? Again we can do so by writing

$$350 \times 10^1 \, \text{pF}$$
$$35.0 \times 10^2 \, \text{pF}$$
$$3.50 \times 10^3 \, \text{pF} \quad \text{or} \quad 3.50 \, \text{nF}, \text{etc.}$$

As in the preceding example, there are definitely three figures in the first factor that show the degree of accuracy.

7-16 CALCULATIONS WITH UNITS

In Sec. 7-14 we emphasized the necessity of keeping track of the units involved when performing calculations. The necessity becomes even more apparent when decimal multipliers of basic units are involved or when you are unsure how to proceed with a solution involving units of different measurements such as decibels and meters, ohms and meters, and hours and kilometers.

As long as your calculations are made in basic units, which are directly related, you will have no difficulty. For example, you know that

$$\text{Ohms} = \frac{\text{volts}}{\text{amperes}}$$

and

$$\text{Ohms} \neq \frac{\text{volts}}{\text{milliamperes}}$$

The milliamperes must be converted to amperes in order to keep the basic relationship in units. Therefore,

$$\text{Ohms} = \text{the number of} \frac{\text{volts}}{\text{milliamperes} \times 10^{-3}}$$

Of course, you could make up your own formulas for special cases and write, for example,

$$\text{Ohms} = \text{the number of} \frac{\text{volts} \times 10^3}{\text{milliamperes}}$$

but the task would be endless. Some frequently used formulas are derived for convenience, and you will derive some of them in Problems 7-3. However, when performing calculations, you will never go wrong if you first convert to basic units.

Example 27

The voltage across a circuit is 250 V, and the current is 5 mA. What is the resistance of the circuit?

Solution

Since ohms $= \dfrac{\text{volts}}{\text{amperes}}$, it is necessary to convert the current of 5 mA into amperes before calculating:

$$R = \frac{V}{I} = \frac{250}{5 \times 10^{-3}}$$
$$= 50 \times 10^3 \ \Omega$$
$$= 50 \ k\Omega$$

Example 28

A current of 150 μA flows through a resistance of 30 kΩ. What is the voltage across the resistance?

Solution

Since the current is in microunits and the resistance is in kilounits, both must be converted into basic units (amperes and ohms) before calculating:

$$\text{Volts} = \text{amperes} \times \text{ohms}$$
or
$$V = I \times R$$
$$= (150 \times 10^{-6})(30 \times 10^3)$$
$$= 4.5 \ V$$

You will encounter cases in which you may be unsure how to proceed, particularly when you deal with units of differing measurements such as Ω/m, μF/km, m/s, kg/m², and dB/100 m. Keeping track of your units and handling them as literal numbers will ensure a correct numerical answer expressed in the proper units.

Example 29

How long will it take to travel 225 km at an average speed of 45 km/h?

Solution

Here we have kilometers and kilometers per hour, and we know the answer must be expressed in hours. Also, we know that

$$\text{Distance} = \text{speed} \times \text{time}$$
or
$$\text{Time} = \frac{\text{distance}}{\text{speed}}$$
That is
$$h = \frac{km}{\dfrac{km}{h}}$$
$$= km \cdot \frac{h}{km} = h$$

Knowing that the answer will be expressed in the proper unit, we can complete the calculation:

$$\text{Time} = \frac{225 \text{ km}}{45 \frac{\text{km}}{\text{h}}}$$

$$= \frac{225}{45} \text{ km} \cdot \frac{\text{h}}{\text{km}}$$

$$= 5 \text{ h}$$

Example 30

A 3-km roll of No. 10 wire is measured and found to have a resistance of 9.81 Ω. What is the resistance of 100 m of this wire?

Solution

The resistance must be expressed in ohms. Since the measurement was $\frac{9.81 \ \Omega}{300 \text{ m}}$,

$$\frac{9.81 \ \Omega}{3000 \text{ m}} = 3.27 \times 10^{-3} \frac{\Omega}{\text{m}}$$

Then the resistance of 100 m of this wire is

$$3.27 \times 10^{-3} \frac{\Omega}{\cancel{\text{m}}} \times 100 \ \cancel{\text{m}} = 0.327 \ \Omega$$

This could be written as 0.327 Ω/100 m.

In the problems which follow, you will be asked to make conversions to accommodate readings in units which do not exactly fit the formulas relating the dimensions, as in Example 27, in which 5 mA had to be converted into amperes before proceeding. You will also be asked to convert the basic or classic formulas to adjust for units other than the basic ones. When both of these are asked for in one problem, follow this rule:

RULE

Adjust the units in which the measurements were made so that they will agree with the units for which the formula was developed. Then convert to other units as required.

PROBLEMS 7-3

1. The capacitive reactance of a circuit, or a capacitor, is given by the formula

$$X_C = \frac{1}{2\pi f C} \qquad \Omega$$

where X_C = capacitive reactance, Ω
$\quad f$ = frequency, Hz
$\quad C$ = capacitance of circuit, or capacitor, F

Show that $\qquad X_C = \frac{159 \times 10^3}{fC} \qquad \Omega$

when f = frequency, MHz
$\quad C$ = capacitance, pF

2. Referring to Prob. 1, what is the capacitive reactance of a capacitor of 0.000 50 µF at a frequency of 4000 MHz?

3. The inductive reactance of a circuit, or an inductor, is given by the formula

$$X_L = 2 \pi f L \qquad \Omega$$

where X_L = inductive reactance, Ω
 f = frequency, Hz
 L = inductance of circuit, or inductor, H
Derive a formula for X_L
when f = frequency, MHz
 L = inductance, µH

4. Referring to Prob. 3, an amplifier coil has an inductance of 27 µH. What is its inductive reactance at 6 MHz?

5. The resonant frequency of any circuit is given by the formula

$$f = \frac{1}{2\pi \sqrt{LC}} \qquad \text{Hz}$$

where f = frequency, Hz
 L = inductance of circuit, H
 C = capacitance of circuit, F
Derive a formula expressing f in megahertz
when L = inductance, µ H
 C = capacitance, pF

6. Referring to Prob. 5, what is the resonant frequency of a circuit with an inductance of 0.25 µH and a capacitance of 16 pF?

7. In copper conductors used in transmission lines, the depth of penetration of high-frequency currents is given by the formula

$$\delta = \frac{2.61 \times 10^{-3}}{\sqrt{f}} \qquad \text{in}$$

where f = frequency, Hz
Derive a formula for current penetration depth in centimeters when f is the frequency in megahertz.

8. Referring to Prob. 7, to what depth in millimeters will a current at 3.75 GHz penetrate a copper conductor?

9. The high-frequency resistance of a round copper wire or of round copper tubing was found in an old handbook to be

$$R_{ac} = 9.98 \times 10^{-4} \frac{\sqrt{f}}{d} \qquad \Omega/\text{ft}$$

where R_{ac} = high-frequency resistance, Ω/ft
 f = frequency, MHz
 d = outside diameter of conductor, in
Derive a formula for R_{ac} in ohms per centimeter when f is given in megahertz and d is given in centimeters.

10. Referring to Prob. 9, No. 36 wire has a diameter of 0.127 mm. What is the resistance per centimeter of the wire at a frequency of 85 MHz?

11. Use the formula in Example 25 to show that $\lambda = \dfrac{3 \times 10^4}{f}$ cm when f is given in megahertz.

12. Use the formula in Example 25 to derive a formula for wavelength (λ) in centimeters when f is given in kilohertz.

13. The midfrequency of television channel 4 is 69 MHz. Using the formula derived in Prob. 12, what is the length of one wavelength in centimeters?

14. The great majority of television receiving antennas consist of various combinations of dipoles. A dipole antenna is one that is approximately one-half wavelength long (0.5λ), as illustrated in Fig. 7-3.

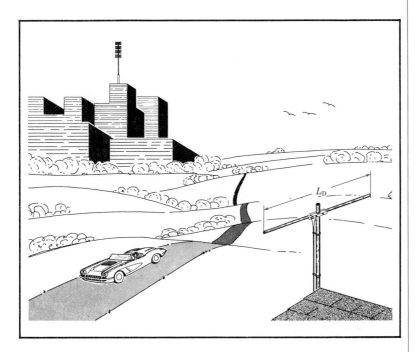

Fig. 7-3 Dipole antenna.

The actual length is slightly less than a half wave owing to "end effect" caused by the capacitance of the antenna, and it has been determined that dipoles used for television reception should be approximately 6% shorter than one-half wavelength. Use the formula derived in Prob. 12 to derive a formula for the length of a dipole antenna in centimeters when the frequency is in megahertz.

15. The midfrequency of television channel 13 is 213 MHz. Using the formula derived in Prob. 14, what length would you make a receiving antenna for this channel?

16. If a wire approximately one-half wavelength long is placed behind a dipole antenna, it acts as a reflector and increases the directivity of the antenna. This results in the reception of stronger signals when the dipole and the reflector are pointed at the transmitting station as illustrated in Fig. 7-4. For best results, the reflector should be 5% longer than the dipole. Referring to the formula for the length of a dipole derived in Prob. 14, derive a formula for the length of a reflector in centimeters when f is in megahertz.

Fig. 7-4 Dipole antenna with reflector.

17. The distance between a dipole and its reflector should be approximately one-fifth of one wavelength (0.2λ) as shown in Fig. 7-4. Referring to previously derived formulas, compute the following dimensions for the midfrequency of television channel 10, which is 195 MHz: (*a*) length of dipole, (*b*) length of reflector, and (*c*) spacing between dipole and reflector.

18. The directivity of a dipole-reflector combination, as shown in Fig. 7-4, can be increased by the addition of a conductor in front of the dipole as illustrated in Fig. 7-5. This wire or tube, which is known as a director, is usually placed one-tenth wavelength (0.1λ) from the dipole, and it should be about 5% shorter than the dipole. Derive a formula for the length of a director in centimeters when f is in megahertz.

Fig. 7-5 Dipole antenna with reflector and director.

19. Referring to Fig. 7-5, compute the following dimensions for the midfrequency of television channel 10, which is 195 MHz: (*a*) length of dipole, (*b*) length of reflector, (*c*) length of director, (*d*) spacing between dipole and reflector, and (*e*) spacing between dipole and director.

20. Ohm's law may be stated in the form $V = IR$, where V is measured in volts, I in amperes, and R in ohms. What voltage will appear across a resistor measuring 680 MΩ when a current of 0.250 μA flows through the resistor?

Ohm's Law—
Series Circuits

8

Ohm's law for the electric circuit is the foundation of electric circuit analysis and is therefore of fundamental importance. The various relations of Ohm's law are easily learned and are readily applied to practical circuits. A thorough knowledge of these relations and their applications is essential for understanding electric circuits.

This chapter is concerned with the study of Ohm's law in dc series circuits as applied to *parts* of a circuit. For this reason, the internal resistance of a source of voltage, such as a generator or a battery, and the resistance of the wires connecting the parts of a circuit are not discussed in this chapter.

8-1 THE ELECTRIC CIRCUIT

An electric circuit consists of a source of voltage connected by conductors to the apparatus that is to use the electric energy.

An electric current will flow between two points in a conductor when a difference of potential exists across those points. The most generally accepted concept of an electric current is that it consists of a motion, or flow, of electrons from the negative toward a more positive point in a circuit. The force that causes the motion of electrons is called an *electromotive force*, a *potential difference*, or a *voltage*, and the opposition to the motion is called *resistance*.

The basic theories of electrical phenomena and the methods of producing currents are not within the scope of this book. You will find them adequately treated in the great majority of textbooks on the subject.

8-2 OHM'S LAW

Ohm's law for the electric circuit, reduced to plain terms, states the relation that exists among voltage, current, and resistance. One way of stating this relation is as follows: The voltage across any *part* of a circuit is proportional to the product of the current through that *part* of the circuit and the resistance of that *part* of the circuit. Stated as a formula, the foregoing is expressed as

$$V = IR \qquad \text{V} \tag{1}$$

where V = voltage, or potential difference, V
I = current, A
R = resistance, Ω

If any two factors are known, the third can be found by solving Eq. (1). Thus,

$$I = \frac{V}{R} \quad \text{A} \tag{2}$$

and

$$R = \frac{V}{I} \quad \Omega \tag{3}$$

8-3 METHODS OF SOLUTION

The general outline for working problems given in Sec. 5-8 is applicable to the solution of circuit problems. In addition, a neat, simplified diagram of the circuit should be drawn for each problem. The diagram should be labeled with all the known values of the circuit such as voltage, current, and resistances. In this manner the circuit and problem can be visualized and understood. Solving a problem by making purely mechanical substitutions in the proper formulas is not conducive to gaining a complete understanding of any problem.

Example 1

How much current will flow through a resistance of 150 Ω if the applied voltage across the resistance is 117 V?

Solution

The circuit is represented in Figs. 8-1 and 8-2.

$$\text{Given} \qquad V = 117 \text{ V} \quad R = 150 \text{ }\Omega$$
$$I = ?$$
$$I = \frac{V}{R} = \frac{117}{150} = 0.780 \text{ A}$$

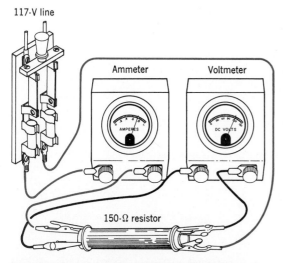

Fig. 8-1 Sketch of the circuit of Example 1 showing how the parts are connected to form the circuit.

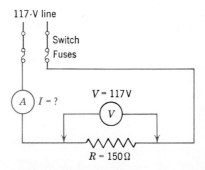

Fig. 8-2 Schematic circuit diagram of Example 1.

Example 2

A voltmeter connected across a resistance reads 22 V, and an ammeter connected in series with the resistance reads 2.60 A. What is the value of the resistance?

Solution

The circuit is represented in Fig. 8-3.

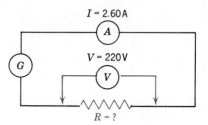

Fig. 8-3 Circuit of Example 2.

Given

$$V = 220 \text{ V} \qquad I = 2.60 \text{ A}$$
$$R = ?$$
$$R = \frac{V}{I} = \frac{220}{2.60} = 84.6 \; \Omega$$

Example 3

A current of 1.40 A flows through a resistance of 450 Ω. What should be the reading of a voltmeter connected across the resistance?

Solution

The diagram of the circuit is shown in Fig. 8-4.

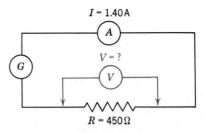

Fig. 8-4 Circuit of Example 3.

Given

$$I = 1.40 \text{ A} \qquad R = 450 \; \Omega$$
$$V = ?$$
$$V = IR = 1.40 \times 450 = 630 \text{ V}$$

Example 4

A measurement shows a potential difference of 63.0 μV across a resistance of 300 Ω. How much current is flowing through the resistance?

Solution

The circuit is represented in Fig. 8-5.

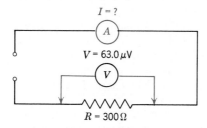

Fig. 8-5 Circuit of Example 4.

Given

$$V = 63.0 \ \mu V = 6.3 \times 10^{-5} \ V \qquad R = 300 \ \Omega$$
$$I = ?$$
$$I = \frac{V}{R} = \frac{6.3 \times 10^{-5}}{300} = \frac{6.3 \times 10^{-7}}{3.00}$$
$$= 2.1 \times 10^{-7} \ A$$

or
$$I = 0.21 \ \mu A$$

Example 5

A current of 8.60 mA flows through a resistance of 500 Ω. What voltage exists across the resistance?

Solution

The circuit is represented in Fig. 8-6.

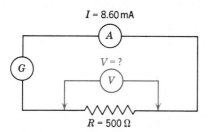

Fig. 8-6 Circuit of Example 5.

Given

$$I = 8.60 \ mA = 8.60 \times 10^{-3} \ A \qquad R = 500 \ \Omega$$
$$V = ?$$
$$V = IR = 8.60 \times 10^{-3} \times 500$$
$$= 8.60 \times 10^{-3} \times 5 \times 10^{2}$$
$$= 8.60 \times 5 \times 10^{-1} = 4.30 \ V$$

Carefully note, as illustrated in Examples 4 and 5, that the equations expressing Ohm's law are in units, that is, volts, amperes, and ohms.

PROBLEMS 8-1

1. How much current will flow through a 50.0 Ω resistor if a potential of 220 V is applied across the resistor?

2. A certain soldering iron draws 1.35 A from a 120-V line. What is the resistance of the heating unit of the soldering iron?

3. What current will flow when an emf of 440 V is impressed across a 71.0-Ω resistor?

4. A milliammeter that is connected in series with a 10-kΩ resistor reads 8.0 mA. What is the voltage across the resistor?

5. A microvoltmeter connected across a 500-Ω resistor reads 40 μV. What current is flowing through the resistor?

6. What voltage is required to cause a current flow of 6.2 mA through a resistance of 7.1 kΩ?

7. A certain milliammeter, with a scale of 0 to 1.0 mA, has a resistance of 32 Ω. If this milliammeter is connected directly across a 120-V line, how much current will flow through the meter? What conclusion do you draw?

8. The current flowing through a 3.3-kΩ resistor is 4.3 mA. What should a voltmeter read when it is connected across the resistor?

9. The cold resistance of a carbon filament lamp is 210 Ω, and the hot resistance is 189 Ω. What is the current flow (a) the instant the lamp is switched across a 120-V line and (b) when constant operating temperature is reached?

10. A type 1N4455 semiconductor PN diode drops 0.7 V across its anode to cathode when connected in series with a 1-kΩ resistor and a 10-V direct-current source.

(a) Determine the current passed by the diode.

(b) Determine the resistance of the diode while it is conducting.

(c) What would be the resistance of the diode if the source voltage were increased to 17.5 V?

(d) If the answers to parts (b) and (c) are not the same, what reasons can you give?

8-4 POWER

In specifying the rating of electrical equipment, it is customary to state not only the voltage at which the equipment was designed to operate but also the rate at which the equipment produces or consumes electric energy.

The rate of producing or consuming energy is called *power*, and electric energy is measured in watts or kilowatts. Thus, your study lamp may be rated 100 W at 120 V and a generator may be rated 2000 kW at 440 V.

Electric motors were formerly rated in terms of the mechanical energy output, measured in *horsepower*, which they could develop. The conversion from electric energy to this older unit of mechanical energy is given by the relation

$$746 \text{ W} = 1 \text{ horsepower (hp)}$$

With the advent of metrication, the SI unit of power will be used more and more, and the *watt*, or *joule per second*, will replace the horsepower rating.

$$1 \text{ watt (W)} = 1 \text{ J/s}$$

Because users of this book will undoubtedly be called upon to handle older motors rated in horsepower, a number of problems involving this older unit have been retained in this edition.

8-5 THE WATT

Energy is expended at a rate of one wattsecond (Ws or W · s)* (Joule, J) every second when one volt causes a current of one ampere to flow. In this case, we say that the power represented when one volt causes one ampere to flow is one watt. This relation is expressed as

$$P = VI \quad \text{W} \tag{4}$$

This is a useful equation when the voltage and current are known.

Because, by Ohm's law, $V = IR$, this value of V can be substituted in Eq. (4). Thus,

$$P = (IR)I$$
or $$P = I^2R \quad \text{W} \tag{5}$$

This is a useful equation when the current and resistance are known.

By substituting the value of I of Eq. (2) in Eq. (4),

$$P = V\frac{V}{R}$$
or $$P = \frac{V^2}{R} \quad \text{W} \tag{6}$$

This is a useful equation when the voltage and resistance are known.

WATTHOURS—KILOWATTHOURS The consumer of electric energy pays for the amount of energy used by his electrical equipment. This is measured by instruments known as *watthour* or *kilowatthour meters*. These meters record the amount of energy taken by the consumer.

Electric energy is sold at so much per kilowatthour (kWh). One watthour of energy is consumed when one watt of power continues in action for one hour. Similarly, 1 kWh is consumed when the power is 1000 W and the action continues for 1 h or when a 100-W rate persists for 10 h, etc. Thus, the amount of energy consumed is the product of the power and the time. Perhaps in time kilowatthour meters will be replaced by megajoule meters.

8-6 LOSSES

The study of the various forms in which energy may occur and the transformation of one kind of energy into another has led to the important principle known as the principle of the *conservation of energy*. Briefly, this states that energy can never be created or destroyed. It can be transformed from one form to another, but the total amount remains unchanged. Thus, an electric motor converts electric energy into mechanical

*The use of the center dot in the symbols for wattsecond and watthour is preferred in general physics relationships. However, it is customarily omitted in electricity and electronics usage.

energy, the incandescent lamp changes electric energy into heat energy, the loudspeaker converts electric energy into sound energy, the generator converts mechanical energy into electric energy, etc. In each instance the transformation from one type of energy to another is not accomplished with 100% efficiency because some energy is converted into heat and does no useful work as far as that particular conversion is concerned.

Resistance in a circuit may serve a number of useful purposes, but unless it has been specifically designed for heating or dissipation purposes, the energy transformed in the resistance generally serves no useful purpose.

8-7 EFFICIENCY

Because all electrical equipment contains resistance, some heat always develops when current flows. Unless the equipment is to be used for producing heat, the heat due to the resistance of the equipment represents wasted energy. No electrical equipment or other machine is capable of converting energy received into useful work without some loss.

The power that is furnished a machine is called the machine's *input*, and the power received from a machine is called its *output*. The efficiency of a machine is equal to the ratio of the output to the input. That is,

$$\text{Efficiency} = \frac{\text{output}}{\text{input}} \qquad (7)$$

It is evident that the efficiency, as given in Eq. (7), is always a decimal, that is, a number less than 1. Naturally, in Eq. (7), the output and input must be expressed in the same units. Hence, if the output is expressed in kilowatts, then the input must be expressed in kilowatts; if the output is expressed in horsepower, then the input must be expressed in horsepower; etc.

Example 6
A voltage of 110 V across a resistor causes a current of 5 A to flow through the resistor. How much power is expended in the resistor?

Solution 1
The circuit is represented in Fig. 8-7.

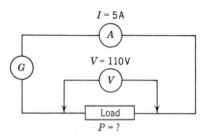

Fig. 8-7 Circuit of Example 6.

Given $V = 110$ V $I = 5$ A $P = ?$

Using Eq. (4), $P = VI = 110 \times 5 = 550$ W

Solution 2

Find the value of the resistance and use it to solve for P. Thus, using Eq. (3),

$$R = \frac{V}{I} = \frac{110}{5} = 22 \ \Omega$$

Using Eq. (5), $P = I^2R = 5^2 \times 22 = 5 \times 5 \times 22$
$$= 550 \ \text{W}$$

Solution 3

Using Eq. (6),

$$P = \frac{V^2}{R} = \frac{110^2}{22} = \frac{110 \times 110}{22} = 550 \ \text{W}$$

Solving a problem by two methods serves as an excellent check on the results, for there is then little chance of making the same error twice, as happens too often when the same method of solution is repeated.

Example 7

A current of 2.5 A flows through a resistance of 40 Ω.
(a) How much power is expended in the resistor?
(b) What is the potential difference across the resistor?

Solution 1

The circuit is represented in Fig. 8-8.

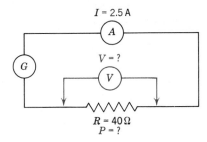

Fig. 8-8 Circuit of Example 7.

Given $I = 2.5$ A $R = 40 \ \Omega$ $P = ?$ $V = ?$

(a) $P = I^2R = 2.5^2 \times 40$
            $= 2.5 \times 2.5 \times 40 = 250$ W

(b) $V = IR = 2.5 \times 40 = 100$ V

Solution 2

(a) Find V, as above, and use it to solve for P. Thus,

$$P = \frac{V^2}{R} = \frac{100^2}{40} = \frac{100 \times 100}{40} = 250 \ \text{W}$$

or $P = VI = 100 \times 2.5 = 250$ W

Example 8

A voltage of 1.732 V is applied across a 500-Ω resistor.
(*a*) How much power is expended in the resistor?
(*b*) How much current flows through the resistor?

Solution

A diagram of the circuit is shown in Fig. 8-9.

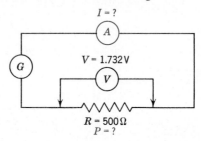

Fig. 8-9 Circuit of Example 8.

Given $\qquad\qquad V = 1.732 \text{ V} \qquad R = 500 \text{ }\Omega$
$$P = ? \qquad I = ?$$

(*a*) $\qquad\qquad P = \dfrac{V^2}{R} = \dfrac{1.732^2}{500} = \dfrac{1.732^2}{5 \times 10^2}$

$$= \dfrac{1.732^2}{5} \times 10^{-2} = 0.006 \text{ W}$$

or $\qquad\qquad P = 6 \text{ mW}$

(*b*) $\qquad\qquad I = \dfrac{V}{R} = \dfrac{1.732}{500} = \dfrac{1.732}{5} \times 10^{-2}$

$$= 0.346 \times 10^{-2} \text{ A}$$

or $\qquad\qquad I = 3.46 \text{ mA}$

Check the foregoing solution for power by using an alternative method.

Example 9

(*a*) What is the hot resistance of a 100-W 110-V lamp?
(*b*) How much current does the lamp take?
(*c*) At 4¢/kWh, how much does it cost to operate this lamp for 24 h?

Solution 1

The circuit is represented in Fig. 8-10.

Given $\qquad\qquad P = 100 \text{ W} \qquad V = 110 \text{ V}$

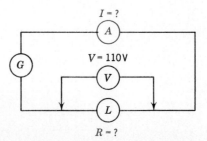

Fig. 8-10 Circuit of Example 9.

(*a*) Because the power and voltage are known and the resistance is unknown, an equation that contains these three must be used. Thus,

$$P = \frac{V^2}{R}$$

hence

$$R = \frac{V^2}{P} = \frac{110^2}{100} = 121 \ \Omega$$

(*b*)

$$I = \frac{V}{R} = \frac{110}{121} = 0.909 \text{ A}$$

(*c*) If the lamp is lighted for 24 h, it will consume

$$100 \times 24 = 2400 \text{ Wh} = 2.40 \text{ kWH}$$

At 4¢/kWh the cost will be

$$2.4 \times 4 = 9.6¢$$

Solution 2

The current may be found first by making use of the relation

$$P = VI$$

which results in $I = \frac{P}{V} = \frac{100}{110} = 0.909$ A

The resistance can now be determined by

$$R = \frac{V}{I} = \frac{110}{0.909} = 121 \ \Omega$$

The solution can be checked by

$$P = I^2R = 0.909^2 \times 121 = 100 \text{ W}$$

which is the power rating of the lamp as given in the example. The cost is computed as before.

Example 10

A motor delivering 6.50 mechanical horsepower is drawing 26.5 A from a 220-V line.
(*a*) How much electric power is the motor taking from the line?
(*b*) What is the efficiency of the motor?
(*c*) If power costs 3¢/kWh, how much does it cost to run the motor for 8 h?

Solution

A diagram of the circuit is shown in Fig. 8-11.

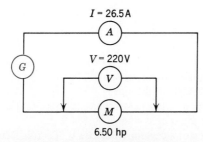

Fig. 8-11 Circuit of Example 10.

Given $\qquad$ $V = 220$ V $\qquad$ $I = 26.5$ A

and mechanical horsepower

$$P = 6.5\,hp = 6.5 \times 746$$
$$= 4850\,W = 4.85\,kW$$

(a) The power taken by the motor is

$$P = VI = 220 \times 26.5 = 5830\,W$$
$$= 5.83\,kW$$

(b) Efficiency $= \dfrac{output}{input} = \dfrac{4.85}{5.83} = 0.832$
$$= 83.2\%$$

(c) Because the motor consumes 5.83 kW, in 8 h it will take

$$5.83 \times 8 = 46.6\,kWh$$

At 3¢/kWh, the cost will be $\qquad$ $46.6 \times 0.03 = \$1.40$

Note The cost was computed in two steps for the purpose of illustrating the solution. When you have become familiar with the method, the cost should be computed in one step. Thus,

$$Cost = 5.83 \times 8 \times 0.03 = \$1.40$$

From the foregoing examples, it will be noted that computations involving power consist mainly in the applications of Ohm's law. Little trouble will be encountered if each problem is given careful thought and the systematic procedure previously outlined is followed in finding the solution.

PROBLEMS 8-2

1. 7.5 hp = (a) ＿＿＿W = (b) ＿＿＿kW
2. 29.84 kW = (a) ＿＿＿W = (b) ＿＿＿hp
3. What current is drawn by a 100-W soldering iron that is connected to a 120-V line?
4. How much power is expended in a 120-Ω resistor through which a current of 15 A flows?
5. What is the electric horsepower of a generator which delivers a current of 50.9 A at 220 V?
6. A voltmeter connected across a 2.2-kΩ resistor reads 120 V. How much power is being expended in the resistor?
7. A diesel engine is rated at 1500 hp. What is its electrical rating in kilowatts?
8. An ammeter is connected in the circuit of a 440-V motor. When the motor is running, the ammeter reads 2.27 A. How much power is being absorbed from the line?
9. The resistance of a certain ammeter is 0.012 Ω. Determine the power expended in the meter when it reads 3 A.
10. The resistance of a certain voltmeter is 300 kΩ. Determine the power expended in the voltmeter when it is connected across a 220-V line.

11. A type 2N5458 JFET used as an audio amplifier is self-biased by a 47-kΩ source resistor. A voltmeter connected across the resistor indicates 4.7 V dc.
   (a) What is the current through the resistor?
   (b) What is the continuous power radiated by the resistor while operating under these conditions?

12. A type 2N5459 JFET is operating with a 2.2 kΩ source bias resistor through which flows a current of 9 mA.
   (a) How much power is being expended by the resistor?
   (b) What is the voltage across the resistor?
   (c) What is the voltage at the gate with respect to the source terminal?

13. An emf of 90 μV is applied across a 390-Ω resistor.
   (a) How much power is expended in the resistor?
   (b) How much current will flow through the resistor?

14. A 1-kΩ resistor in the emitter circuit of a 2N1414 transistor produces a voltage drop of 6 V between collector and emitter.
   (a) What is the emitter current?
   (b) What is the power loss in this bias resistor?

15. A radar antenna motor is delivering 10 hp. A kilowattmeter that measures the power taken by the motor reads 8.24 kW.
   (a) What is the efficiency of the motor?
   (b) At 2.5¢/kWh, how much would it cost to run the motor continuously for 5 days?

16. A 440-V 10-hp forced-draft fan motor has an efficiency of 80%.
   (a) How many kilowatts does it consume?
   (b) How much current does it draw from the line?
   (c) At 2.5¢/kWh, how much would it cost to run this motor continuously for 1 week?

17. A generator which is 80% efficient delivers 50 A at 220 V. What must be the output of the diesel engine which drives the generator?

18. 23.9 kW is required to operate a 25-hp forced-draft fan motor.
   (a) What is its efficiency?
   (b) How much power is lost in the motor?

19. A generator delivers 80 A at 220 V with an efficiency of 88%. How much power is lost in the generator?

20. A 230-V 7½-hp motor, which has an efficiency of 85%, is driving a radio transmitter 2-kV generator which has an efficiency of 80%. The motor is running fully loaded.
   (a) How much power does the motor take from the line?
   (b) How much current does the motor draw?
   (c) How much power will the generator deliver?
   (d) How much current will the generator deliver?
   (e) What is the overall efficiency; that is, what is the efficiency from motor input to generator output?

8-8 RESISTANCES IN SERIES

So far, our studies of the electric circuit have taken into consideration but one electrical component in the circuit, excluding the source of voltage. This is all very well for the purpose of becoming familiar with simple

Ohm's law for power relations. However, practical circuits consist of more than one piece of equipment as far as circuit computations are concerned.

In a *series circuit* the various components comprising the circuit are so connected that the current, starting from the voltage source, must flow through each circuit component, in turn, before returning to the other side of the source.

There are three important facts concerning series circuits that must be borne in mind in order to understand thoroughly the action of such circuits and to facilitate their solution. In a series circuit:

1. The total voltage is equal to the sum of the voltages across the different parts of the circuit.
2. The current in any part of the circuit is the same.
3. The total resistance of the circuit is equal to the sum of the resistances of the different parts.

Point 1 is practically self-evident. If the sum of all the potential differences (voltage drops) around the circuit were not equal to the applied voltage, there would be some voltage left over which would cause an increase in current. This increase in current would continue until it caused enough voltage drop across some resistance just to balance the applied voltage. Hence,

$$V_t = V_1 + V_2 + V_3 + \cdots \tag{8}$$

Point 2 is self-evident, for the circuit components are so connected that the current must flow through each part in turn and there are no other paths back to the source.

To some, point 3 might not be self-evident. However, because it is agreed that the current I in Figs. 8-12 and 8-13 flows through all resistors,

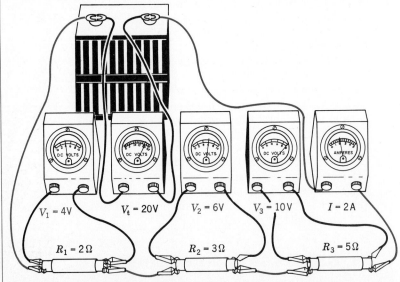

Fig. 8-12 Three resistors connected in series with a voltmeter connected across each resistor. The sum of the voltages across the resistors is equal to the battery voltage.

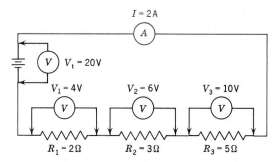

Fig. 8-13 Schematic diagram of the circuit represented in Fig. 8-12.

Eq. (8) can be used to demonstrate the truth of point 3. Thus, by dividing each member of Eq. (8) by I, we have

$$\frac{V_t}{I} = \frac{V_1 + V_2 + V_3}{I} + \cdots$$

or

$$\frac{V_t}{I} = \frac{V_1}{I} + \frac{V_2}{I} + \frac{V_3}{I} + \cdots$$

and by substituting R for $\frac{V}{I}$, we have

$$R_t = R_1 + R_2 + R_3 + \cdots \qquad (9)$$

Note V_t and R_t are used to denote total voltage and total resistance, respectively.

Example 11

Three resistors $R_1 = 30\ \Omega$, $R_2 = 160\ \Omega$, and $R_3 = 40\ \Omega$ are connected in series across a generator. A voltmeter connected across R_2 reads 80 V. What is the voltage of the generator?

Solution

Figure 8-14 is a diagram of the circuit.

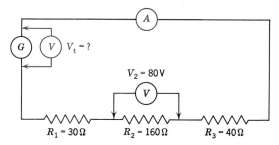

Fig. 8-14 Circuit of Example 11.

$$I = \frac{V_2}{R_2} = \frac{80}{160} = 0.5\ \text{A}$$
$$R_t = R_1 + R_2 + R_3$$
$$= 30 + 160 + 40 = 230\ \Omega$$
$$V_t = IR_t = 0.5 \times 230 = 115\ \text{V}$$

Example 12

A 300-Ω relay must be operated from a 120-V line. How much resistance must be added in series with the relay coil to limit the current through it to 250 mA?

Solution 1

The circuit is represented in Fig. 8-15. For a current of 250 mA to flow in a 120-V circuit, the total resistance must be

$$R_t = \frac{V}{I} = \frac{120}{0.250} = 480 \ \Omega$$

Because the relay coil has a resistance of 300 Ω, the resistance to be added is

$$R_x = R_t - R_c = 480 - 300 = 180 \ \Omega$$

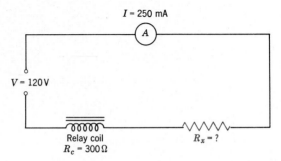

Fig. 8-15 Circuit of Example 12.

Solution 2

For 0.250 A to flow through the relay coil, the voltage across the coil must be

$$V_c = IR_c = 0.250 \times 300 = 75 \ V$$

Because the line voltage is 120 V, the voltage across the added resistance must be

$$V_x = V_t - V_c = 120 - 75 = 45 \ V$$

Then the value of resistance to be added is

$$R_x = \frac{V_x}{I} = \frac{45}{0.250} = 180 \ \Omega$$

Example 13

Three resistors $R_1 = 20 \ \Omega$, $R_2 = 50 \ \Omega$, and $R_3 = 30 \ \Omega$ are connected in series across a generator. The current through the circuit is 2.5 A.
(a) What is the generator voltage?
(b) What is the voltage across each resistor?
(c) How much power is expended in each resistor?
(d) What is the total power expended?

Solution

The circuit is represented in Fig. 8-16.

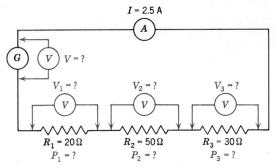

Fig. 8-16 Circuit of Example 13.

(a)
$$R_t = R_1 + R_2 + R_2 = 20 + 50 + 30$$
$$= 100 \; \Omega$$

$$V = IR_t = 2.5 \times 100 = 250 \text{ V}$$

(b)
$$V_1 = IR_1 = 2.5 \times 20 = 50 \text{ V}$$
$$V_2 = IR_2 = 2.5 \times 50 = 125 \text{ V}$$
$$V_3 = IR_3 = 2.5 \times 30 = 75 \text{ V}$$

Check

$$V = V_1 + V_2 + V_3$$
$$= 50 + 125 + 75 = 250 \text{ V}$$

(c) Power in R_1, $P_1 = V_1 I = 50 \times 2.5 = 125 \text{ W}$

Check

$$P_1 = I^2 R_1 = 2.5^2 \times 20 = 125 \text{ W}$$

Power in R_2, $P_2 = V_2 I = 125 \times 2.5 = 312.5 \text{ W}$

Check

$$P_2 = I^2 R_2 = 2.5^2 \times 50 = 312.5 \text{ W}$$

Power in R_3, $P_3 = V_3 I = 75 \times 2.5 = 187.5 \text{ W}$

Check

$$P_3 = I^2 R_3 = 2.5^2 \times 30 = 187.5 \text{ W}$$

(d) Total power, $P_t = P_1 + P_2 + P_3$
$$= 125 + 312.5 + 187.5 = 625 \text{ W}$$

Check

$$P_t = I^2 R_t = 2.5^2 \times 100 \times 625 \text{ W}$$

or
$$P_t = \frac{V^2}{R_t} = \frac{250^2}{100} = 625 \text{ W}$$

PROBLEMS 8-3

1. Three resistors, $R_1 = 330\ \Omega$, $R_2 = 680\ \Omega$, and $R_3 = 570\ \Omega$, are connected in series across 110 V.
   (a) How much current flows in the circuit?
   (b) What is the voltage drop across R_2?
   (c) How much power is expended in R_1?

2. Three resistors, $R_1 = 2.2\ k\Omega$, $R_2 = 5.7\ k\Omega$, and $R_3 = 1.5\ k\Omega$, are connected in series across 450 V.
   (a) How much current flows through the circuit?
   (b) What is the voltage drop across each resistor?

3. A 115-V soldering iron which is rated at 100 W is to be used on a 220-V line.
   (a) How much resistance must be connected in series with the iron to limit the current to rated value?
   (b) If a standard resistor of 150 Ω is used in place of the calculated value, what minimum power rating must be specified for it?
   (c) If the standard resistor of (b) is used, what actual power will be delivered to the soldering iron?

4. Four identical 100-W lamps are connected in series across a 440-V line. The hot resistance of each lamp is 121 Ω.
   (a) What is the current through the lamps?
   (b) What is the voltage drop across each lamp?
   (c) What is the power dissipated by each lamp?

5. Three identical lamps are connected in series across a 440-V line. If the current through the lamps is 820 mA, what is the hot resistance of each lamp?

6. Three resistors, R_1, R_2, and R_3, are connected in series across a 470-V power supply. A voltmeter connected across R_1 reads 76 V. When connected across R_2, the voltmeter reads 51 V. R_3 is 150 kΩ.
   (a) What is the current flowing through the circuit?
   (b) What is the value of R_1?
   (c) What is the value of R_2?
   (d) What is the wattage dissipated by each resistor?

7. Three resistors of 12, 18, and 47 Ω are connected in series across a 12-V source. If the current through the circuit is 153 mA, what is the resistance of the connecting wires and connections?

8. The visual readout system of a CPU terminal uses eight cathode-ray tubes. To conserve power drawn from the central control unit (CCU) power supply, the video monitors are connected in series directly across the 115-V line.

   Six of the monitors require 12.6-V filament voltage, and the remaining two require only 6.3 V. The filament current of all the monitors, to maintain *correct* operating temperature, is 210 mA. What value of series ballast resistor must be used to allow operation from the 115-V line?

9. Three resistors, $R_1 = 1.2\ \Omega$, R_2, and R_3, are connected in series across a 125-V generator, which delivers a current of 27.8 A. The voltage drop across R_3 is 50 V.
   (a) What is the value of R_3?
   (b) What is the value of R_2?

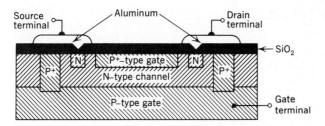

Fig. 8-17 (a) Block model of an N-channel junction FET.

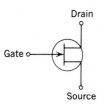

(b) Usual schematic symbol.

(c) How much power is expended in the circuit?

10. Four resistors, $R_1 = 820\ \Omega$, $R_2 = 270\ \Omega$, $R_3 = 1.5\ \text{k}\Omega$, and $R_4 = 390\ \Omega$, are connected in series across a generator. The voltage appearing across R_3 is 504 V.

(a) What is the generator voltage?

(b) What is the power being dissipated by each resistor?

8-9 BIAS RESISTORS—FIELD-EFFECT TRANSISTORS

The field-effect transistor (FET) (Fig. 8-17) was introduced as a more suitable replacement for vacuum tubes than the bipolar junction transistor (BJT). The major reason was that the FET, like the vacuum tube, is a voltage-controlled device whereas the BJT is a current-operated device. Other reasons include the fact that the FET is normally ON, because it has only one major current carrier, whereas the BJT must have current flowing through all of the electrodes to set up the no-signal biasing conditions. When all the electrodes are passing current, the BJT has both electrons and holes (positive charges) passing, and recombining, among all three electrodes. The FET is controlled by the external circuitry, but the BJT's "fixed" parameters determine what external bias circuitry must be used (R_1 and R_2, Fig. 8-18, and R_B, Fig. 8-19). In other words, parameters of the FET do not control the external circuitry design to the extent that those of the BJT do.

To clarify FET biasing, consider the schematic diagram of Fig. 8-20. The FET shown in the diagram is an N-channel junction FET or, simply,

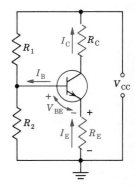

Fig. 8-18 Universal dc circuit.

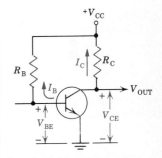

Fig. 8-19 Base bias of a common emitter BJT.

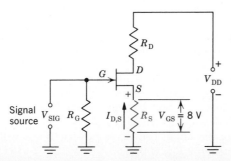

Fig. 8-20 Gate G is biased −8V with respect to source S.

an NJFET. The N channel is indicated by the direction of the arrow of the gate terminal, which is similar to the arrow on the emitter electrode of the BJT. It indicates the type, NPN or PNP, of base material used. In Fig. 8-20 the arrow pointing in indicates that the material from source to drain is N-type semiconductor material (usually silicon).

The FET of Fig. 8-20 indicates an N channel and a P-type gate; it is similar to any PN junction diode. The resistor R_S connected to the source terminal and ground passes a current I_{DS} from source to drain. This current will develop a voltage drop across the resistor R_S with polarity as shown. With a positive potential at the source terminal and, by connection, a negative potential at the gate terminal with respect to the source terminal, there now exists a reverse bias condition between the gate and source. Current flow through the junction is thereby prevented.

A further examination of Fig. 8-20 shows that there is a positive potential at the drain terminal which is much larger than the potential at the source terminal. Because of the difference, electrons will flow from the source to the drain and back to the supply V_{DD}. The gate terminal, while not passing current, will have a potential of -4 V when measured with respect to the source. We can say that the FET has a gate bias voltage V_{GS} of -4 V.

The bias voltage just discussed is typical of a voltage drop across R_S. When the value of R_S is increased, the voltage at the gate will become more negative with respect to the source. It is usual to describe this type of biasing as either *source bias* or *self-bias*.

With the drain supply voltage V_{DD} maintaining the drain positive with respect to the source, electrons will flow from source to drain and introduce the current I_{DS}.

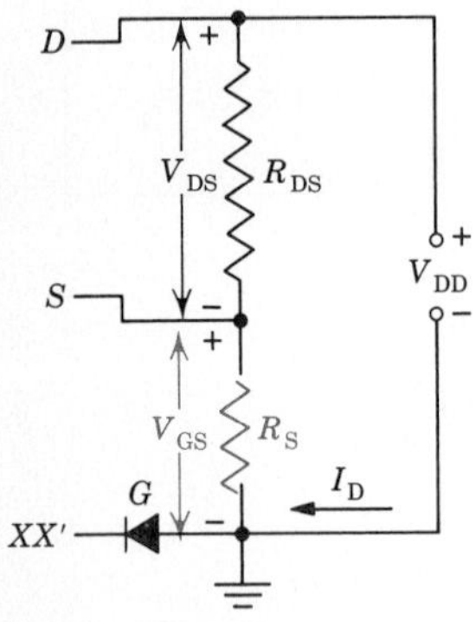

The schematic of Fig. 8-20 may be replaced by its equivalent dc circuit, as shown in Fig. 8-21. The purpose of the equivalent circuit is to illustrate that the N-type bar, when considered on its own, is nothing more than a piece of semiconductor material having a resistance r_{DS} which is made variable by the introduction of the P-type gate. By eliminating the external resistor R_D between the drain and the V_{DD} supply, we may show that the sum of the voltages V_{DS} and V_{RS} must equal the supply V_{DD}.

$$V_{DS} + V_{RS} = V_{DD}$$

You should notice that this formula does not take into account the voltage V_{GS}. That should not be too alarming, since V_{GS} provides bias, and the gate draws no current.

Fig. 8-21 Equivalent circuit of Fig. 8-20. Point *XX'* is connected into N-type bar between *D* and *S*.

Example 14

A 2N5459 JFET is to be operated as a class A audio amplifier. In that class of operation, the drain current I_D is 6 mA when the drain voltage V_{DS} is $+8$ V with respect to the source and the gate voltage V_{GS} is -4 V with respect to the source (achieved by voltage drop V_{RS}).

(*a*) What value of source-biasing resistor R_S is required?

(*b*) What power is dissipated by R_S?

(*c*) Ignoring the drain load resistor, what is the value of the drain voltage supply V_{DD}?

(*d*) What is the total power taken from the drain voltage supply?

Solution
The circuit is shown schematically in Fig. 8-22.

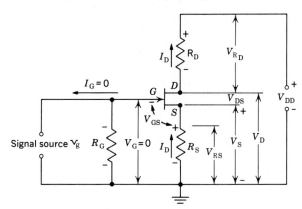

Fig. 8-22 Circuit of Example 14. NJFET using self/source bias.
$V_S = V_{RS} = I_D R_S.$ $V_G = 0$ V. $I_G = 0$ A.

(a)
$$R_S = \frac{V_S}{I_D} = \frac{4}{0.006} = \frac{4}{6 \times 10^{-3}}$$
$$= \frac{4}{6} \times 10^3 = 667 \ \Omega$$

(b)
$$P_{RS} = \frac{V_S{}^2}{R_S} = \frac{4^2}{667} = 24 \ \text{mW}$$

Check
$$P_{RS} = I_D{}^2 R_S = (6 \times 10^{-3})^2 \times 667 = 24 \ \text{mW}$$

(c) $V_{DD} = V_{DS} + V_{RS} = 8 + 4 = 12$ V

(d) $P = I_D \times V_{DD} = 6 \times 10^{-3} \times 12 = 72$ mW

Example 15
The FET of Example 14 is to work into a dc load resistance of $R_D = 2.2$ kΩ. What drain supply voltage V_{DD} will be required?

Solution
The circuit is illustrated in Fig. 8-23. The voltage across the load resistor R_D is

$$V_{RD} = I_D R_D = 0.006 \times 2200 = 13.2 \ \text{V}$$
$$V_{DD} = I_D R_D + V_{DS} + V_{GS}$$
$$= 13.2 + 8 + 4$$
$$V_{DD} = 25.2 \ \text{V}$$

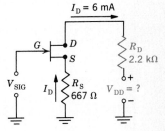

Fig. 8-23 Circuit of Example 15.

Example 16
The FET type 2N5458 has the following characteristics:

$$V_{DD} = 25 \ \text{V}, \ V_S = 6 \ \text{V}, \ \text{and} \ V_{DS} = 15 \ \text{V when} \ I_G = 0.$$

Use the circuit shown in Fig. 8-24 and determine values for (a) I_D, (b) R_S, (c) V_{GS}, and (d) the total current I_T.

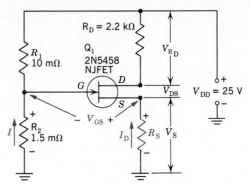

Fig. 8-24 Circuit of Example 16.

Solution

$$V_D = V_{DD} - I_D R_D \quad \text{or} \quad V_D = V_{DS} + V_S$$
$$= 15 + 6$$
$$V_D = 21 \text{ V}$$

From this value for V_D we can determine that the voltage $I_D R_D$ is:

$$V_{DD} - V_D \quad \text{or} \quad 25 - 21 = 4 \text{ V}$$

From that voltage the value of I_D can easily be calculated:
(a) $I_D R_D = V_D$; therefore,

$$I_D = \frac{4}{2.2 \times 10^3} = 1.8 \text{ mA}$$

(b) With $I_D = 1.8$ mA, $I_D R_S = V_S$; therefore

$$R_S = \frac{V_S}{I_D} = \frac{6}{1.8} \times 10^3 = 3.3 \text{ k}\Omega$$

(c) To determine the value of V_{GS}, we must first establish the value of the voltage drop across R_2. We can do so in either of two ways:

$$V_{DD} = I(R_1 + R_2)$$

Solve for current I necessary to produce IR_2. Alternatively,

$$V_{R_2} = \frac{R_2}{R_1 + R_2} \times V_{DD} = \frac{1.5 \times 10}{11.5 \times 10^6} \times 25$$

$$= 3.261 \text{ V}$$

The voltage V_{GS} is the difference between V_{R_2} and V_S; Therefore,

$$V_{GS} = 3.261 - 6 = -2.739 \text{ V}$$

(d)
$$I_{R_2} = \frac{3.261}{1.5 \times 10^6} = 0.002 \text{ mA}$$

$$I_T = I_D + I_{R_2}$$
$$= 1.8 + 0.02 = 1.802 \text{ mA}$$

8-10 BIAS RESISTORS—TRANSISTORS

Proper operation of a transistor circuit requires that the emitter-base junction of the transistor be forward-biased and that the collector-base junction be reverse-biased, as shown in diagram form in Fig. 8-25. The evolution of late forms of discrete transistors is illustrated in Fig. 8-26. But whether discrete or in integrated circuits, all transistors require biasing and biasing is generally achieved by means of resistors.

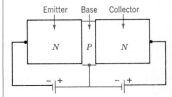

Fig. 8-25 NPN transistor biased for proper operation. The N-type emitter is forward-biased for low effective resistance, and the N-type collector is reverse-biased for high effective resistance.

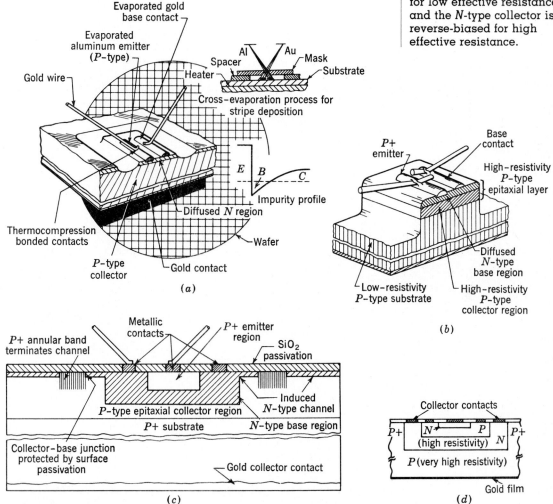

Fig. 8-26 Evolution of transistors. (a) Diffused-base mesa, (b) epitaxial mesa, (c) annular, (d) basic integrated. (*Courtesy of Lothar Stern, "Fundamentals of Integrated Circuits," Hayden Book Co., Inc., 1968*)

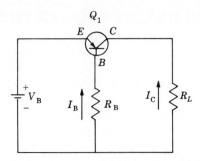

Fig. 8-27 Simple single-battery transistor biasing circuit for *PNP* transistor Q_1.

Sometimes the use of two different batteries is avoided by utilizing bias resistors, as in tube circuits. In addition, resistor values are chosen to limit current flows to acceptable levels. Figure 8-27 shows a simple circuit in which transistor Q_1 is supplied by a single battery V_B. The resistor in the base circuit R_B is chosen to regulate the base-emitter current I_B, and the output signal is taken across the load resistor R_L as the collector current I_C flows through it.

Example 17
In Fig. 8-27, assuming that the voltage drop across the emitter-base junction is negligible, what must be the value of R_B if the base current must be limited to 80 μA? $V_B = 6$ V.

Solution

$$R_B = \frac{V_B}{I_B} = \frac{6}{80 \times 10^{-6}} = 75 \text{ k}\Omega$$

When two batteries are used, as in Fig. 8-28, an analysis based upon constant-emitter-current bias reveals that

$$R_E = \frac{V_{EE}}{I_E}$$
$$I_C = \alpha I_E + I_{CO}$$
$$I_B = (1 - \alpha)I_E - I_{CO}$$

where I_{CO} = the very small leakage current in the collector circuit at room temperature
α = the current amplification factor under certain circuit arrangements; its value is usually slightly less than 1

Example 18
In Fig. 8-28, the applied emf $V_{EE} = 12$ V and the specifications for transistor Q_1 indicate that the emitter current I_E should be limited to 10 mA. What value of resistor R_E should be chosen?

Solution

$$R_E = \frac{V_{EE}}{I_E} = \frac{12}{10 \times 10^{-3}} = 1.2 \text{ k}\Omega$$

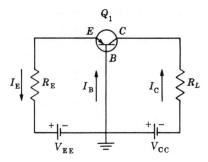

Fig. 8-28 *PNP* transistor Q_1 biased by means of two batteries, V_{EE} and V_{CC}.

Example 19

For the circuit of Fig. 8-28, $V_{EE} = 12$ V, $I_E = 8$ mA, $\alpha = 0.95$, and $I_{CO} = 50$ μA. Find (*a*) R_E, (*b*) I_C, and (*c*) I_B.

Solution

(*a*)
$$R_E = \frac{V_{EE}}{I_E} = \frac{12}{0.008} = 1.5 \text{ k}\Omega$$

(*b*)
$$I_C = \alpha I_E + I_{CO}$$
$$= (0.95)(0.008) + 0.000\,050$$
$$= 7.65 \text{ mA}$$

(*c*)
$$I_B = (1 - 0.95)(0.008) - 0.000\,050$$
$$= 350 \text{ μA}$$

PROBLEMS 8-4

1. The 2N5459 JFET is to be used as an audio amplifier operating class A. When the drain voltage V_D with respect to the source is 15 V, the drain current is 9 mA and the gate-to-source bias is -4.5 V.
   (*a*) What value of source resistor is required for this bias?
   (*b*) Disregarding any drain load resistance, what is the drain supply voltage V_{DD}?

2. The NPN transistor shown schematically in Fig. 8-18 is to be biased for operation as a common-emitter class A audio amplifier. The data book shows the following characteristics: $I_B = 0.1$ mA; forward current gain $\beta = 50$; $V_{BE} = 0.2$ V; $R_C = 1.5$ kΩ; $R_1 = 9.8$ kΩ. If the voltage at the emitter measured with respect to ground is 1 V,
   (*a*) Determine the value of I_C and I_E when $V_{CC} = 12$ V.
   (*b*) Determine the values of R_2 and R_E.
   (*c*) What will be the value of the current through R_2?
   (*d*) What voltage will be measured at the collector with respect to ground?

3. The FET circuit shown in Fig. 8-24 is biased so that $V_S = 6$ V. $V_{DS} = 15$ V. With all other components and voltage supplies remaining the same,
   (*a*) Determine the value of R_S.
   (*b*) Determine the value of I_D.

    (*c*) What will be the total current drawn from the 25-V source?

    (*d*) How much power is drawn from the supply?

4.   For the circuit used in Prob. 3 and the data given,

    (*a*) Determine the voltage drop across R_1.

    (*b*) Determine the voltage drop across R_2.

    (*c*) What will be the value of V_{GS} when $V_S = 6$ V?

    (*d*) Which resistor in this circuit will dissipate the most power?

5.   A microphone preamplifier is wired as shown in Fig. 8-19, with $R_C = 10$ kΩ, $I_B = 0.01$ mA, $V_{BE} = 0.3$ V, and $V_{CC} = 20$ V. If the transistor has a forward current gain $\beta = 50$, determine (*a*) the value of I_C and I_E, (*b*) the value of R_B, (*c*) the magnitude and polarity of the voltage V_{CE}, and (*d*) the power dissipated by the collector of the transistor.

6.   In Fig. 8-27, assuming that the voltage drop across the emitter-base junction is negligible, what must be the value of R_B if the base current must be limited to 90 μA? $V_B = 6$ V.

7.   It is desired to operate a transistor in grounded-base connection (Fig. 8-28) with a fixed bias of 6 V. The maximum current in the base circuit is 100 μA.

    (*a*) What is the value of the resistor which will provide this voltage?

    (*b*) What is the power which this resistor must radiate?

8.   In the circuit of Fig. 8-28, emf $V_{EE} = 6$ V and the emitter current I_E should be limited to 8 mA. What value of resistor R_E should be chosen?

9.   In the circuit of Fig. 8-28, what value should R_E be if $V_{EE} = 30V$ and I_E must be kept to 12 mA or less?

10.   In the circuit of Fig. 8-28, $V_{EE} = 12$ V and $V_C = 15$ V. $I_E = 10$ mA, $\alpha = 0.98$, and $I_{CO} = 75$ μA. Find (*a*) R_E, (*b*) I_C, and (*c*) I_B.

Resistance— Wire Sizes

9

The effects of resistance in series circuits were discussed in the preceding chapter. However, in order to prevent confusion while the more simple relations of Ohm's law were being discussed, the nature of resistance and the resistance of wires used for connecting sources of voltage with their respective loads were not mentioned.

In the consideration of practical circuits two important features must be taken into account: the resistance of the wires between the source of power and the electronic equipment that is to be furnished with power and the current-carrying capacity of these wires for a given temperature rise.

9-1 RESISTANCE

There is a wide variation in the ease (conductance) of current flow through different materials. No material is a perfect conductor, and the amount of opposition (resistance) to current flow within it is governed by the specific resistance and the length, cross-sectional area, and temperature of the material. Thus, for the same material and cross-sectional area, a long conductor will have a greater resistance than a shorter one. That is, *the resistance of a conductor of uniform cross-sectional area is directly proportional to the length of the conductor*. This is conveniently expressed as

$$\frac{R_1}{R_2} = \frac{L_1}{L_2} \tag{1}$$

where R_1 and R_2 are the resistances of conductors with lengths L_1 and L_2, respectively.

Example 1
The resistance of No. 8 copper wire is 2.06 Ω/km. What is the resistance of 175 m of the wire?

Solution
Given $R_1 = 2.06\ \Omega$, $L_1 = 1000$ m, and $L_2 = 175$ m, $R_2 = ?$ Upon solving Eq. (1) for R_2, we have

$$R_2 = \frac{R_1 L_2}{L_1} \frac{\Omega \cdot m}{m} = \frac{2.06 \times 175}{1000} \frac{\Omega \cdot \cancel{m}}{\cancel{m}} = 0.3605\ \Omega$$

For the same material and length, one conductor will have more resistance than another with a larger cross-sectional area. That is, *the resistance of a conductor is inversely proportional to the cross-sectional area of the conductor.* Expressed as an equation,

$$\frac{R_1}{R_2} = \frac{A_2}{A_1} \qquad (2)$$

where R_1 and R_2 are the resistances of conductors with cross-sectional areas A_1 and A_2, respectively.

Because most wires are drawn round, Eq. (2) can be rearranged into a more convenient form. For example, let A_1 and A_2 represent the cross-sectional areas of two equal lengths of round wires with diameters d_1 and d_2, respectively. Because the area A of a circle of a diameter d is given by

$$A = \frac{\pi d^2}{4}$$

then $\qquad A_1 = \dfrac{\pi d_1^2}{4} \qquad$ and $\qquad A_2$ is $= \dfrac{\pi d_2^2}{4}$

Substituting in Eq. (2)

$$\frac{R_1}{R_2} = \frac{\dfrac{\pi d_2^2}{4}}{\dfrac{\pi d_1^2}{4}}$$

or $\qquad\qquad \dfrac{R_1}{R_2} = \dfrac{d_2^2}{d_1^2} \qquad\qquad (3)$

Hence, the resistance of a round conductor varies inversely as the square of the diameter.

Example 2

A rectangular conductor with a cross-sectional area of 1.04 square millimeters (mm²) has a resistance of 0.075 Ω. What would be its resistance if its cross-sectional area were 2.08 mm²?

Solution

Given $R_1 = 0.075$ Ω, $A_1 = 1.04$ mm², and $A_2 = 2.08$ mm², $R_2 = ?$
Solving Eq. (2) for R_2,

$$R_2 = \frac{R_1 A_1}{A_2} \frac{\Omega \cdot mm^2}{mm^2}$$
$$= \frac{0.075 \times 1.04}{2.08} \frac{\Omega \cdot \cancel{mm^2}}{\cancel{mm^2}} = 0.0375 \ \Omega$$

Example 3

A round conductor with a diameter of 0.25 mm has a resistance of 8 Ω. What would be its resistance if its diameter were 0.5 mm?

Solution

Given $d_1 = 0.25$ mm, $R_1 = 8$ Ω, and $d_2 = 0.5$ mm, $R_2 = ?$
Solving Eq. (3) for R_2,

$$R_2 = \frac{R_1 d_1^2}{d_2^2} \frac{\Omega \cdot mm^2}{mm^2} = \frac{8 \times 0.25^2}{0.5^2} \frac{\Omega \cdot \cancel{mm^2}}{\cancel{mm^2}} = 2 \ \Omega$$

Hence, if the diameter is doubled, the cross-sectional area is increased four times and the resistance is reduced to one-quarter of its original value.

PROBLEMS 9-1

1. Number 14 copper wire has a resistance of 8.28 Ω/km.
   (a) What is the resistance of 500 m of this wire?
   (b) What is the resistance of 20 m of this wire?
2. Number 30 copper wire has a resistance of 340 Ω/km.
   (a) What is the resistance of 800 m of this wire?
   (b) What is the resistance of 1 m?
3. Using the information of Prob. 2, what is the resistance of a coil that has a mean diameter of 40 mm and is wound with 6280 turns of No. 30 copper wire?
4. The resistance of a 1-km run of No. 10 copper wire telephone line is measured and found to be 3.277 Ω.
   (a) What is the resistance per meter?
   (b) What is the resistance of a 720-m line?
   (c) What is the resistance of 20 m?
5. The telephone line of Prob. 4 is replaced with No. 8 copper wire, which has a resistance of 2.061 Ω/km. What is the resistance of the 720-m section?
6. A length of square conductor that is 0.5 cm on a side has a resistance of 0.0756 Ω. What will be the resistance of a similar length of 1.5 cm square conductor?
7. One kilometer of No. 6 wire, which has a diameter of 4.115 mm, has a resistance of 1.297 Ω. What is the resistance of 1 km of No. 2 wire whose diameter is 6.543 mm?
8. The resistance of 30 m of a specially drawn wire is found to be 32.1 Ω. A coil wound with identical wire has a measured resistance of 702 Ω. What is the length of wire in the coil?
9. It is desired to wind a milliammeter shunt having a resistance of 4.62 Ω, and No. 40 enameled copper wire with a resistance of 3.54 kΩ/km is available. What length of wire is required?
10. It is desired to wind a microammeter shunt having a resistance of 0.280 Ω, and No. 36 enameled copper wire with a resistance of 1.36 kΩ/km is available. What length of wire is required?

9-2 MICROHM-METER

Equations (1) and (2) from Sec. 9-1 can be combined to form the compound proportion

$$\frac{R_1}{R_2} = \frac{L_1}{L_2} \cdot \frac{A_2}{A_1}$$

Such ratios are extremely helpful in solving problems when sufficient information is available. However, as a statement regarding a single conductor, we fall back on the simple proportionality (Sec. 5-11)

$$R \propto \frac{L}{A}$$

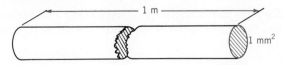

Fig. 9-1 Representation of 1 microhm-meter conductor.

When this proportionality is written as an equation with a constant of proportionality, it becomes

$$R = \rho \frac{L}{A} \quad \Omega \tag{4}$$

where R = resistance of wire

ρ = specific resistance of material of which wire is made

L = length of wire

A = cross-sectional area of wire

From an algebraic rearrangement of Eq. (4), you can see that the units of ρ, the specific resistance, must relate to the units of L and A:

$$\rho = \frac{RA}{L}$$

In the older, or traditional, system of units, L was measured in feet and A was measured in circular mils (the circular-mil area was defined to be equal to the square of the diameter when the diameter was given in mils, or thousandths of an inch). This meant ρ was expressed in Ω-cmils/ft, commonly pronounced Ω/cmil-ft. (See Secs. 9-2 and 9-3 of the third edition of this book if you require more information about this superseded set of units.)

In the SI metric system, the unit of length is the meter and the unit of area is the square meter. Using those units, ρ would be expressed as

$$\rho = \frac{RA}{L} = \frac{\Omega \cdot m^2}{m} = \Omega \cdot m$$

Sometimes you will see tables of specific resistance giving values of ρ in ohm-meters. However, since more realistic sizes of wire will be given in square millimeters (see Appendix, Table 5), more often than not you will see practical values of ρ as:

$$\rho = \frac{RA}{1} = \frac{\Omega \cdot mm^2}{m} = \frac{\Omega \cdot m^2 \times 10^{-6}}{m}$$
$$= \Omega \cdot m \times 10^{-6} = \text{microhm-meters} (\mu\Omega \cdot m)$$

Table 9-1 gives specific resistances of common conductive materials in microhm-meters.

Eq. (4) and its definition block may now be adjusted to read

$$R = \rho \frac{L}{A} \quad \Omega \tag{5}$$

where R = resistance of wire, Ω

ρ = specific resistance of wire, $\mu\Omega \cdot m$

L = length of wire, m

A = cross-sectional area of wire, mm²

TABLE 9-1 SPECIFIC RESISTANCE AT 20°C

Material	$\mu\Omega \cdot$ m
Aluminum	0.028 24
Brass	0.070 0
Constantan	0.490 0
Copper, hard drawn	0.017 71
Gold	0.024 4
Iron	0.100 0
Lead	0.220 0
Mercury	0.957 8
Nickel	0.078 0
Silver	0.015 9
Tin	0.115 0
Zinc	0.058 0

Example 4

What is the resistance at 20°C of a copper wire that is 250 m long and 1.63 mm in diameter?

Solution

Given $L = 250$ m, $d = 1.63$ mm ($r = 0.815$ mm), and also, from Table 9-1, $\rho = 0.017\ 71\ \mu\Omega \cdot$ m, $R = ?$ Substituting in Eq. (4),

$$R = \frac{0.017\ 71 \times 250}{\pi(0.815)^2} = 2.122\ \Omega$$

Example 5

The resistance of a conductor 1 km long and 2.05 mm in diameter is found to be 4.82 Ω at 20°C. What is the specific resistance of the wire?

Solution

Given $L = 1000$ m, $d = 2.05$ mm ($r = 1.025$ mm), and $R = 4.82\ \Omega$, $\rho = ?$ Solving Eq. (5) for ρ,

$$\rho = \frac{RA}{L} = \frac{4.82\ \pi(1.025)^2}{10^3} \qquad \mu\Omega \cdot \text{m}$$
$$= 15.91 \times 10^{-3}$$
$$= 0.015\ 91\ \mu\Omega \cdot \text{m}$$

Example 6

A roll of copper wire is found to have a resistance of 2.54 Ω at 20°C. The diameter of the wire is measured as 1.63 mm. How long is the wire?

Solution

Solving Eq. (5) for L,

$$L = \frac{RA}{\rho} \qquad \text{m}$$

Substituting known values,

$$L = \frac{2.54\pi(0.815)^2}{0.017\ 71}\ \text{m}$$
$$= 299\ \text{m}$$

PROBLEMS 9-2

Note In the following problems, consider that all the wire temperatures are 20°C.

1. What is the resistance of a copper wire 75 m long and 0.361 mm in diameter?
2. With reference to Prob. 1, what is the resistance of an otherwise identical wire of aluminum?
3. With reference to Prob. 1, what is the resistance of an otherwise identical wire of iron?
4. What is the resistance of 200 m of copper wire with a diameter of 0.254 mm?
5. A special alloy wire 10 m long and 0.079 mm in diameter has a resistance of 78 Ω. What is the specific resistance of the alloy?
6. A constantan wire that has a specific resistance of 0.49 $\mu\Omega \cdot$m has a diameter of 0.511 mm and a length of 1.1 m. What is its resistance?
7. How many kilometers of copper wire 3.264 mm in diameter will it take to make 5.00 Ω of resistance?
8. What is the resistance of the wire in Prob. 7 in ohms per kilometer?
9. A coil of copper wire has a resistance of 2.38 Ω. If the diameter of the wire is 2.05 mm, find the length of the wire.
10. What is the resistance of 2 km of the wire in Prob. 9?

9-3 TEMPERATURE EFFECTS

In the preceding section the specific resistance of certain materials was given at a temperature of 20°C. The reason for stating the temperature is that the resistance of all pure metals increases with a rise in temperature. The results of experiments show that over ordinary temperature ranges this variation in resistance is directly proportional to the temperature. Hence, for each degree rise in temperature above some reference value, each ohm of resistance is increased by a constant amount α, called the *temperature coefficient of resistance*. The relation between temperature and resistance can be expressed by the equation

$$R_t = R_0(1 + \alpha t) \qquad \Omega \qquad (6)$$

where R_t = Resistance at a temperature of t°C*
$\quad R_0$ = resistance at 0°C
$\quad \alpha$ = temperature coefficient of resistance at 0°C

The temperature coefficient for copper is 0.004 27. That is, if a copper wire has a resistance of 1 Ω at 0°C, it will have a resistance of 1 + 0.004 27 = 1.004 27 Ω at 1°C. The value of the temperature coefficient for copper is essentially the same as that for most of the unalloyed metals such as gold, silver, aluminum, and lead.

The value of α varies with the temperature at which it is determined: 0.004 27 is valid only when the reference temperature is 0°C. If 20°C is taken as the reference temperature, the value of α changes to 0.003 93. Thus it is important to always relate to the reference temperature for which

---

*°C stands for degrees Celsius.

the value of α is specified. If the resistance is known at a certain temperature and you are required to determine the resistance corresponding to some other temperature, you *must* first calculate, as an intermediate step, the resistance for the reference temperature.

Example 7

The resistance of a coil of copper wire is 34 Ω at 15°C. What is the resistance at 70°C?

Solution 1

Using $\alpha = 0.004\ 27$ and relating to 0°C, $R_t = 34\ \Omega$ and $t = 15$ (15°C − 0°C).

$$R_t = R_0(1 + \alpha t)$$

Solving for R_0, $\quad R_0 = \dfrac{R_t}{1 + \alpha t} = \dfrac{34}{1 + (0.004\ 27)(15)}$

$$= \dfrac{34}{1.064\ 05}$$

$$= 31.95\ \Omega$$

Using R_0, solve for R_{70}:

$$R_{70} = 31.95(1 + 0.004\ 27 \times 70)$$
$$= 31.95(1.2989)$$
$$= 41.5\ \Omega$$

Solution 2

Using $\alpha = 0.003\ 93$ and relating to 20°C, $R_t = 34\ \Omega$ and

$$t = -5\ (15°C - 20°C)$$

Solving for R_{20}, $\quad R_{20} = \dfrac{34}{1 + (0.003\ 93)(-5)}$

$$= \dfrac{34}{1 - 0.019\ 65}$$

$$= \dfrac{34}{0.980\ 35}$$

$$= 34.68\ \Omega$$

Using R_{20}, solve for R_{70}:

$$R_{70} = 34.68(1 + 0.003\ 93 \times 50)$$
$$= 34.68(1.1965)$$
$$= 41.5\ \Omega$$

A more convenient relation is derived by assuming that the proportionality between resistance and temperature extends linearly to the point where copper has a resistance of 0 Ω at a temperature of −234.5°C. This results in the ratio

$$\frac{R_2}{R_1} = \frac{234.5 + t_2}{234.5 + t_1} \tag{7}$$

where R_1 = resistance of copper in ohms at a temperature of t_1°C
$\quad\quad\ R_2$ = resistance of copper in ohms at a temperature of t_2°C

Example 8

The resistance of a coil of copper wire is 34 Ω at 15°C. What is the coil's resistance at 70°C?

Solution

Given $R_1 = 34$ Ω, $t_1 = 15$°C, and $t_2 = 70$°C. $R_2 = ?$ Solving Eq. (7) for R_2,

$$R_2 = \frac{234.5 + t_2}{234.5 + t_1} R_1$$

Substituting the known values,

$$R_2 = \frac{234.5 + 70}{234.5 + 15} \times 34 = 41.5 \ \Omega$$

The specifications for electric machines generally include a provision that the temperature of the coils, etc., when the machines are operating under a specified load for a specified time, must not rise more than a certain number of degrees. Temperature rise can be computed by measuring the resistance of the coils at room temperature and again at the end of the test.

Example 9

The field coils of a shunt motor have a resistance of 90 Ω at 20°C. After the motor was run for 3 h, the resistance of the field coils was 146 Ω. What was the temperature of the coils?

Solution

Given $R_1 = 90$ Ω, $t_1 = 20$°C, $R_2 = 146$ Ω. $t_2 = ?$ Solving Eq. (7) for t_2,

$$t_2 = \frac{234.5 + t_1}{R_1} R_2 - 234.5$$

Substituting the known values,

$$t_2 = \frac{234.5 + 20}{90} \times 146 - 234.5$$
$$= 413 - 234.5 = 178.5°$$

The actual temperature rise is

$$t_2 - t_1 = 178.5° - 20° = 158.5°.$$

PROBLEMS 9-3

1. The resistance of a coil of copper wire at 40°C is 5.38 Ω. What will be the resistance at 0°C?

2. If the resistance of a copper coil is 3.07 Ω at 0°C, what will it be at 20°C?

3. The dc resistance of an inductor is 19.5 Ω at 80°C. What will be the resistance when the inductor is operated at an ambient temperature of 20°C?

4. The resistance of the primary winding of a transformer was 2.95 Ω at 20°C. After operation for 3 h, the resistance increased to 3.28 Ω. What was the final operating temperature?

5. The specifications for a high-power transformer included a provision that the transformer was to operate continuously under full load with the winding temperature not to exceed 55°C. The resistance of the primary coil was measured before the transformer was put on test, at 22°C, and found to be 52.7 Ω. After a day's test at rated load, the resistance was again measured and was found to be 60.0 Ω. Did the transformer meet the specifications?

9-4 WIRE MEASURE

Wire sizes are designated by numbers in a system known as the American wire gage (formerly Brown and Sharpe gage). These gage numbers, ranging from 0000, the largest size, to 40, the smallest size, are based on a constant ratio between successive gage numbers. The wire sizes and other pertinent data are listed in Table 5 in the Appendix.

Inspection of the wire table will reveal the progression formed by the wire sizes. As the sizes become smaller, every third gage number results in one-half the cross-sectional area and, therefore, double the resistance.

Example 10

Number 10 wire has a cross-sectional area of 5.261 mm² and a resistance of 3.277 Ω/km. Three sizes smaller, No. 13 wire has an area of 2.63 mm² (almost exactly one-half of 5.261), and a resistance of 6.56 Ω (almost double 3.277). Similarly, half the resistance of No. 10 wire is provided by No. 7: 1.634 Ω/km, with an area of 10.55 mm² (double 5.261).

9-5 FACTORS GOVERNING WIRE SIZE IN PRACTICE

From an electrical viewpoint, three factors govern the selection of the size of wire to be used for transmitting current:

1. The safe current-carrying capacity of the wire
2. The power lost in the wire
3. The allowable voltage variation, or the voltage drop, in the wire

It must be remembered that the length of wire, for the purpose of computing wire resistance and its effects, is always twice the distance from the source of power to the load (outgoing and return leads).

Example 11

A motor receives its power through No. 4 wire from a generator located at a distance of 1 km. The voltage across the motor is 220 V, and the current taken by the motor is 19.8 A. What is the terminal voltage of the generator?

Solution

The circuit is represented in Fig. 9-2. Note that it consists of a simple series circuit which can be simplified to that of Fig. 9-3. The resistance

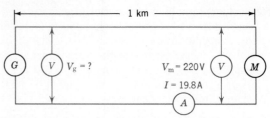

Fig. 9-2 Generator G supplying power to motor M at a distance of 1 km.

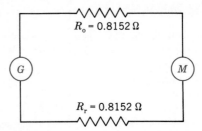

Fig. 9-3 Simplified form of circuit shown in Fig. 9-2.

of the 1 km of No. 4 wire from the generator to the motor is represented by R_0; reference to Table 5 shows it to be 0.8152 Ω. Similarly, the resistance from the motor back to the generator, which is represented by R_r, also is 0.8152 Ω. The voltage drop in each wire is

$$V = IR_0 = IR_r = 19.8 \times 0.8152 = 16.14 \text{ V}$$

Since the applied voltage must equal the sum of all the voltage drops around the circuit (Sec. 8-8), the terminal voltage of the generator is

$$V_g = 220 + 16.14 + 16.14 = 252.28 \text{ or } 252 \text{ V}$$

Since the resistance out R_0 is equal to the return resistance R_r, the foregoing solution is simplified by taking twice the actual wire distance for the length of wire that constitutes the resistance of the feeders. Therefore, the length of No. 4 wire between generator and motor is 2 km, which results in a line resistance R_L of

$$2 \times 0.8152 = 1.6304 \text{ Ω}$$

The circuit can be further simplified as shown in Fig. 9-4. Thus, the generator terminal voltage is

$$V_g = 220 + IR_L$$
$$= 220 + (19.8 \times 1.6304) = 252 \text{ V}$$

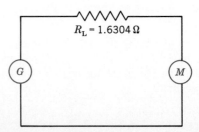

Fig. 9-4 Equivalent circuit of circuits shown in Figs. 9-2 and 9-3.

The power lost in the line is

$$P_L = I^2 R_L$$
$$= 19.8^2 \times 1.6304 = 639 \text{ W}$$

The power taken by the motor is

$$P_M = V_M I$$
$$= 220 \times 19.8$$
$$= 4356 \text{ W} = 4.356 \text{ kW}$$

The power delivered by the generator is

$$P_G = P_L + P_M$$
$$= 639 + 4356 = 4995 \text{ W}$$

$$\text{Efficiency of transmission} = \frac{\text{power delivered to load}}{\text{power delivered by generator}}$$

$$= \frac{4356}{4995} = 0.872 = 87.2\% \qquad (8)$$

The efficiency of transmission is obtainable in terms of the generator terminal voltage V_G and the voltage across the load V_L. Because

$$\text{Power delivered to load} = V_L I$$

and

$$\text{Power delivered by generator} = V_G I$$

substituting in Eq. (8) gives us

$$\text{Efficiency of transmission} = \frac{V_L I}{V_G I}$$

$$= \frac{V_L}{V_G} \qquad (9)$$

and substituting the voltages in Eq. (9) gives us

$$\text{Efficiency of transmission} = \frac{220}{252}$$

$$= 0.873 = 87.3\%$$

PROBLEMS 9-4

Note All wires in the following problems are of copper with characteristics as listed in Table 5 of the Appendix.

1. (a) What is the resistance of 2500 m of No. 00 wire?
   (b) What is its weight?
2. (a) What is the resistance of 1.8 m of No. 8 wire?
   (b) What is its weight?
3. (a) What is the length of a 120-kg coil of No. 12 wire?
   (b) What is its resistance?
4. (a) What is the length of a 100-kg coil of No. 16 wire?
   (b) What is its resistance?

5. A telephone cable consisting of several pairs of No. 19 wire connects two cities 40 km apart. If a pair is short-circuited at one end, what will be the resistance of the loop thus formed?

6. A relay is to be wound with 1500 turns of No. 22 wire. The average diameter of a turn is 46 mm.
    (*a*) What will be the resistance?
    (*b*) What will be the weight of the coil?

7. Fifteen kilowatts of power is to be transmitted 200 m from a generator that maintains a constant terminal voltage of 240 V. If not over 5% line drop is allowed, what size wire must be used?

8. A generator with a constant brush potential of 230 V is feeding a motor 100 m away. The feeders are No. 6 wire, and the motor current is 27.7 A.
    (*a*) What would a voltmeter read if connected across the motor brushes?
    (*b*) What is the efficiency of transmission?

9. A motor requiring 34 A at 230 V is located 125 m from a generator that maintains a constant terminal voltage of 240 V.
    (*a*) What size wire must be used between generator and motor in order to supply the motor with rated current and voltage?
    (*b*) What will be the efficiency of transmission?

10. A 25-hp 230-V motor is to be installed 120 m from a generator that maintains a constant potential of 240 V.
    (*a*) If the motor is 84% efficient, what size wire should be used between motor and generator?
    (*b*) If the wire specified in (*a*) is used, what will be the motor voltage under rated load condition?

Special Products and Factoring

10

In the study of arithmetic, it is necessary to memorize the multiplication tables as an aid to rapid computation. Similarly, in the study of algebra, certain forms of expressions occur so frequently that it is essential to be able to multiply, divide, or factor them by inspection.

10-1 FACTORING

To *factor* an algebraic expression means to find two or more expressions that, when multiplied, will result in the original expression.

Example 1
$2 \times 3 \times 4 = 24$. Thus, 2, 3, and 4 are some of the factors of 24.

Example 2
$b(x + y) = bx + by$. b and $(x + y)$ are the factors of $bx + by$.

Example 3
$(x + 4)(x - 3) = x^2 + x - 12$. The quantities $(x + 4)$ and $(x - 3)$ are the factors of $x^2 + x - 12$.

10-2 PRIME NUMBERS

A number that has no factor other than itself and unity is known as a *prime number*. Thus, 3, 5, 13, x, and $(a + b)$ are prime numbers.

10-3 SQUARE OF A MONOMIAL

At this point you should review the law of exponents for multiplication in Sec. 4-3.

Example 4
$$(2ab^2)^2 = (2ab^2)(2ab^2) = 4a^2b^4$$

Example 5
$$(-3x^2y^3)^2 = (-3x^2y^3)(-3x^2y^3) = 9x^4y^6$$

By application of the rules for the multiplication of numbers having like signs and the law of exponents, we have the following rule:

RULE

To square a monomial, square the numerical coefficient, multiply this product by the literal factors of the monomial, and multiply the exponent of each letter by 2.

10-4 CUBE OF A MONOMIAL

Example 6

$$(3a^2b)^3 = (3a^2b)(3a^2b)(3a^2b) = 27a^6b^3$$

Example 7

$$(-2xy^3)^3 = (-2xy^3)(-2xy^3)(-2xy^3)$$
$$= -8x^3y^9$$

Note that the cube of a *positive* number is always *positive* and that the cube of a *negative* number is always *negative*. Again, by application of the rules for the multiplication of positive and negative numbers and the law of exponents, we have the following rule:

RULE

To cube a monomial, cube the numerical coefficient, multiply this product by the literal factors of the monomial, multiply the exponent of each letter by 3, and affix the same sign as the monomial.

PROBLEMS 10-1

Find the values of the following indicated powers:

1. $(xy)^2$

2. $(\theta\lambda)^3$

3. $(vi^2Z)^3$

4. $\pi\left(\dfrac{D}{2}\right)^2$

5. $(-4\pi\phi)^2$

6. $(3\alpha^2\omega^3)^2$

7. $(-2IR)^3$

8. $\left(3\dfrac{v}{i}\right)^2$

9. $2\pi(X_L)^2$

10. $\left(-3\dfrac{ir}{v}\right)^3$

11. $-\left(\dfrac{1}{2\pi fC}\right)^2$

12. $-(-13x^3y)^2$

13. $\left(-\dfrac{5P^2}{VI}\right)^3$

14. $\dfrac{(V_sN_p)^2}{V_p{}^3}$

15. $-\left(\dfrac{V^2}{2g}\right)^3$

16. $\left(\dfrac{120f}{N}\right)^2$

17. $\left(\dfrac{B^2Al}{8\omega}\right)^3$

18. $-(2\pi fL)^3$

19. $-\left(\dfrac{4}{3}\pi R^3\right)^2$

20. $\left(\dfrac{5}{8}u^2v^3wx^4y^5\right)^3$

21. $\left(\dfrac{x^4y^6}{p^5}\right)^3$

10-5 SQUARE ROOT OF A MONOMIAL

The *square root* of an expression is one of two equal factors of the expression.

Example 8

$\sqrt{3}$ is a number such that

$$\sqrt{3} \cdot \sqrt{3} = 3$$

Example 9

$\sqrt{n}$ is a number such that

$$\sqrt{n} \cdot \sqrt{n} = n$$

Because $\qquad (+2)(+2) = +4$
and $\qquad (-2)(-2) = +4$

it is apparent that 4 has two square roots, $+2$ and -2. Similarly, 16 has two square roots, $+4$ and -4.

In general, every number has two square roots equal in magnitude, one positive and one negative. The positive root is known as the *principal root;* if no sign precedes the radical, the positive root is understood. Thus, in practical numerical computations, the following is understood:

$$\sqrt{4} = +2$$

and $\qquad -\sqrt{4} = -2$

In dealing with literal numbers, the values of the various factors often are unknown. Therefore, when we extract a square root, we affix the double sign $\pm$ to denote "plus or minus."

Example 10

Since $a^4 \cdot a^4 = a^8$ and $(-a^4)(-a^4) = a^8$,

then $\qquad \sqrt{a^8} = \pm a^4$

Example 11

Since $x^2y^3 \cdot x^2y^3 = x^4y^6$ and $(-x^2y^3)(-x^2y^3) = x^4y^6$,

then $\qquad \sqrt{x^4y^6} = \pm x^2y^3$

From the foregoing examples, we formulate the following:

RULE

To extract the square root of a monomial, extract the square root of the numerical coefficient, divide the exponents of the letters by 2, and affix the $\pm$ sign.

Example 12

$$\sqrt{4a^4b^2} = \pm 2a^2b$$

Example 13

$$\sqrt{\tfrac{1}{9}x^2y^6z^4} = \pm \tfrac{1}{3}xy^3z^2$$

Note

A perfect monomial square is one that is positive and has a perfect square numerical coefficient and has only even numbers as exponents.

10-6 CUBE ROOT OF A MONOMIAL

The *cube root* of a monomial is one of the three equal factors of the monomial.

Because $$(+2)(+2)(+2) = 8$$

then $$\sqrt[3]{8} = 2$$

Similarly, $$(-2)(-2)(-2) = -8$$

and $$\sqrt[3]{-8} = -2$$

From this it is evident that the cube root of a monomial has the same sign as the monomial itself.

Because $$x^2y^3 \cdot x^2y^3 \cdot x^2y^3 = x^6y^9$$

then $$\sqrt[3]{x^6y^9} = x^2y^3$$

The above results can be stated as follows:

RULE

To extract the cube root of a monomial, extract the cube root of the numerical coefficient, divide the exponents of the letters by 3, and affix the same sign as the monomial.

Example 14
$$\sqrt[3]{8x^6y^3z^{12}} = 2x^2yz^4$$

Example 15
$$\sqrt[3]{-27a^3b^9c^6} = -3ab^3c^2$$

Note

A perfect cube monomial has a positive or negative perfect cube numerical coefficient and exponents that are exactly divisible by 3.

PROBLEMS 10-2

Find the value of the following:

1. $\sqrt{a^2}$ 2. $\sqrt{\omega^4}$ 3. $\sqrt{9i^2}$

4. $\sqrt{6^2}$ 5. $\sqrt{(-\omega)^2}$ 6. $\sqrt{100m^2n^{12}}$

7. $\sqrt{25\lambda^4\Omega^6}$ 8. $5\sqrt{64\phi^4}$ 9. $\sqrt[3]{27x^6}$

10. $\sqrt[3]{-64\theta^3}$ 11. $\sqrt[3]{(-2)^6}$ 12. $\sqrt{4\pi^2f^2L^2 \times 10^2}$

13. $\sqrt{169m^4n^2p^6}$ 14. $\sqrt[5]{32\lambda^5\psi^{10}}$ 15. $\sqrt[3]{270^6\phi^{12}\omega^3}$

16. $\sqrt{121x^{10}y^{12}z^6}$ 17. $\sqrt{\dfrac{256\pi^2r^2x^4}{289z^6\phi^4}}$ 18. $\sqrt{\dfrac{25m^4n^2p^8}{64a^4b^2c^6}}$

19. $-\sqrt{\dfrac{625r^6s^4t^8}{16x^6z^{10}}}$ 20. $\sqrt[3]{\dfrac{-8\pi^3X_L^3}{27Z^6X_C^{12}}}$ 21. $-\sqrt[3]{\dfrac{-64a^3\omega^6}{125x^6z^{12}}}$

22. $\sqrt{\dfrac{196h^2n^4p^6}{121a^2b^4c^2}}$ 23. $\sqrt{\dfrac{25v^2t^2}{256a^8b^2x^2}}$ 24. $\sqrt[3]{-\dfrac{1}{64}a^3b^{12}c^{15}}$

10-7 POLYNOMIALS WITH A COMMON MONOMIAL FACTOR

Type: $a(b + c + d) = ab + ac + ad$

RULE

To factor a polynomial whose terms contain a common monomial factor:

1. Determine by inspection the greatest common factor of its terms.
2. Divide the polynomial by this factor.
3. Write the quotient in parentheses preceded by the monomial factor.

Example 16

Factor $3x^2 - 9xy^2$.

Solution

The common monomial factor of both terms is $3x$.

$$\therefore 3x^2 - 9xy^2 = 3x(x - 3y^2)$$

Example 17

Factor $2a - 6a^2b + 4ax - 10ay^3$.

Solution

Each term contains the factor $2a$.

$$\therefore 2a - 6a^2b + 4ax - 10ay^3 = 2a(1 - 3ab + 2x - 5y^3)$$

Example 18

Factor $14x^2yz^3 - 7xy^2z^2 + 35xz^5$.

Solution

Each term contains the factor $7xz^2$.

$$\therefore 14x^2yz^3 - 7xy^2z^2 + 35xz^5 = 7xz^2(2xyz - y^2 + 5z^3)$$

PROBLEMS 10-3

Factor:

1. $2a + 6$ 2. $\dfrac{1}{3}m + \dfrac{q}{3}$

3. $3\theta + \theta\phi + 4\theta\omega$ 4. $\dfrac{1}{2}ab^3 - \dfrac{1}{6}a^2b^2 + \dfrac{1}{8}a^3b$

5. $20ir - 10iz$

6. $480\theta^3\phi - 1440\theta^2\phi^2 + 1080\theta\phi^3$

7. $\dfrac{a^2y^2}{9} + \dfrac{a^3y}{3} - \dfrac{ay^3}{12}$

8. $4\omega^4 X_L^4 - 12\omega X_L + 28\omega^2 X_L^2$

9. $2a^3b^2c + 8a^2bc^3 + 12a^2b^2c^2$

10. $\dfrac{1}{4}I^2R^2Z^2 - \dfrac{1}{12}IRZ^2 + \dfrac{1}{16}I^2RZ$

11. $36\alpha^4\beta^3\omega^2 - 72\alpha^2\beta^2\omega^5 + 180\alpha^2\beta^5\omega^2$

12. $540\theta^3\lambda^2\phi + 810\theta^2\lambda\phi^3 - 1080\lambda^3\phi^2$

13. $\dfrac{1}{64}I^2i^3 + \dfrac{1}{76}Ii^4 - \dfrac{1}{48}I^3i^2$

14. $\dfrac{1}{36}X_L^3X_C^2 - \dfrac{1}{18}X_L^2X_C^3 + \dfrac{1}{72}X_L^2X_C^2$

15. $720\eta^4\theta^2\phi^3\omega + 1080\eta^2\theta^4\phi\omega^2 + 600\eta^3\theta^3\phi\omega^2 - 480\eta\theta^6\phi\omega$

10-8 SQUARE OF A BINOMIAL

Type: $\qquad (a + b)^2 = a^2 + 2ab + b^2$

The multiplication

$$
\begin{array}{r}
a + b \\
a + b \\
\hline
a^2 + ab \\
\quad + ab + b^2 \\
\hline
a^2 + 2ab + b^2
\end{array}
$$

results in the formula

$$(a + b)^2 = a^2 + 2ab + b^2$$

which can be expressed by the following rule:

RULE

To square the sum of two terms, square the first term, add twice the product of the two terms, and add the square of the second term.

Example 19

Square $2b + 4cd$.

Solution

$$(2b + 4cd)^2 = (2b)^2 + 2(2b)(4cd) + (4cd)^2$$
$$= 4b^2 + 16bcd + 16c^2d^2$$

Example 20

Let x and y be represented by lengths. Then

$$(x + y)^2 = x^2 + 2xy + y^2$$

can be illustrated graphically as shown in Fig. 10-1.

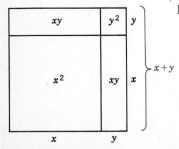

Fig. 10-1 Graphical illustration of $(x + y)^2 = x^2 + xy + y^2$.

The multiplication

$$a - b$$
$$a - b$$
$$a^2 - ab$$
$$- ab + b^2$$
$$a^2 - 2ab + b^2$$

results in the formula

$$(a - b)^2 = a^2 - 2ab + b^2$$

which can be expressed as follows:

RULE

To square the difference of two terms, square the first term, subtract twice the product of the two terms, and add the square of the second term.

Example 21

Square $3a^2 - 5xy$.

Solution

$$(3a^2 - 5xy)^2 = (3a^2)^2 - 2(3a^2)(5xy) + (5xy)^2$$
$$= 9a^4 - 30a^2xy + 25x^2y^2$$

Example 22

Let x and y be represented by lengths. Then

$$(x - y)^2 = x^2 - 2xy + y^2$$

can be illustrated graphically as shown in Fig. 10-2. x^2 is the large square. The figure shows that the two rectangles taken from x^2 leave $(x - y)^2$. Since an amount y^2 is a part of one xy that has been subtracted from x^2 and is outside x^2, we must add it. Hence, we obtain

$$(x - y)^2 = x^2 - 2xy + y^2$$

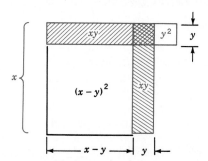

Fig. 10-2 Graphical illustration of $(x - y)^2 = x^2 - 2xy + y^2$.

Mentally, practice squaring sums and differences of binomials by following the foregoing rules. Proficiency in these and later methods will greatly reduce the labor in performing multiplications.

PROBLEMS 10-4

Mentally, square the following:

1.	$\theta + 3$	2.	$a + 6$	3.	$m - R$
4.	$I - 5$	5.	$\alpha + 16$	6.	$p - 4$
7.	$3X - R$	8.	$2r + 3R$	9.	$F - f$
10.	$2\alpha - 3\beta$	11.	$5\theta + 4\phi$	12.	$2\lambda - 5\mu$
13.	$9r_1 - 3r_2$	14.	$m^2 + 6$	15.	$1 + X_L^2$
16.	$2\theta^2 - 13\phi$	17.	$6v^2 - 2t^3$	18.	$20 + 2$
19.	$30 - 3$	20.	$30 + 5$	21.	$6\pi R^2 - 2\pi r^2$

22. $2\pi f L_1 - Z$ 23. $1.5\theta^2 - 0.5\alpha$ 24. $\dfrac{1}{2}R_1 + \dfrac{1}{4}R_2$

25. $\dfrac{3}{4}X^2 - \dfrac{1}{2}Z^2$ 26. $\dfrac{1}{3}\phi^3\lambda + \dfrac{1}{2}\alpha^2$ 27. $6\phi^2\omega - \dfrac{1}{4}\lambda^2$

Expand:

28. $(a + 5)^2$ 29. $\left(x + \dfrac{1}{2}\right)^2$ 30. $\left(\alpha + \dfrac{1}{3}\right)^2$

31. $\left(\dfrac{1}{2} - E\right)^2$ 32. $\left(\mu - \dfrac{1}{12}\right)^2$ 33. $(1 + e^3)^2$

34. $\left(X_1^2 + \dfrac{2}{3}\right)^2$ 35. $\left(L^2 - \dfrac{7}{8}P\right)^2$ 36. $\left(\dfrac{X}{2} + Y\right)^2$

37. $\left(\dfrac{b}{3} + \dfrac{m}{2}\right)^2$ 38. $(3 + 2ab)^2$ 39. $\left(R_1 - \dfrac{5}{8}R\right)^2$

40. $\left(2\phi + \dfrac{3}{4}\theta^2\right)^2$

41. Develop a graphical illustration of $(x + y)(x - y)$.

10-9 SQUARE ROOT OF A TRINOMIAL

In the preceding section, it was shown that

$$(a + b)^2 = a^2 + 2ab + b^2$$

and
$$(a - b)^2 = a^2 - 2ab + b^2$$

From these and other binomials that have been squared, it is evident that a trinomial is a perfect square if

1. Two terms are squares of monomials and are positive.
2. The other term is twice the product of the monomials and has affixed either a plus or a minus sign.

Example 23

$x^2 + 2xy + y^2$ is a perfect trinomial square because x^2 and y^2 are the squares of the monomials x and y, respectively, and $2xy$ is twice the product of the monomials. Therefore,

$$x^2 + 2xy + y^2 = (x + y)^2$$

Example 24

$4a^2 - 12ab + 9b^2$ is a perfect trinomial square because $4a^2$ and $9b^2$ are the squares of $2a$ and $3b$, respectively, and the other term is $-2(2a)(3b)$. Therefore,

$$4a^2 - 12ab + 9b^2 = (2a - 3b)^2$$

RULE

To extract the square root of a perfect trinomial square, extract the square root of the two perfect square monomials and connect them with the sign of the remaining term.

Example 25

Supply the missing term in $x^4 + ? + 16$ so that the three terms will form a perfect trinomial square.

Solution

The missing term is twice the product of the monomials whose squares result in the two known terms; that is, $2(x^2)(4) = 8x^2$. Hence,

$$x^4 + 8x^2 + 16 = (x^2 + 4)^2$$

Example 26

Supply the missing term in $25a^2 + 30ab + ?$ so that the three terms will form a perfect trinomial square.

Solution

The square root of the first term is $5a$. The missing term is the square of some number N such that $2(5a)(N) = 30ab$. Then by multiplying, we obtain $10aN = 30ab$, or $N = 3b$. Therefore,

$$25a^2 + 30ab + 9b^2 = (5a + 3b)^2$$

PROBLEMS 10-5

Supply the missing terms so that the three terms form perfect trinomial squares:

1. $e^2 + ? + 9$
2. $I^2 + ? + 4$
3. $\lambda^2 - ? + 4$
4. $F^2 - ? + f^2$
5. $25x^2 - ? + y^2$
6. $25X_c^2 - ? + 4$
7. $49\omega^2 + ? + \pi^2$
8. $100L_1^2 + ? + 16M^2$
9. $4m^2 + ? + 9p^2$
10. $r^2 - 2rs + ?$
11. $V^2 + 2VI + ?$
12. $? - 90xy + 25y^2$
13. $? + 80pq + 100q^2$
14. $Z^2 + 12XZ + ?$
15. $\frac{1}{9}\theta^2\phi^2 - ? + \frac{1}{4}\omega^2$
16. $\frac{1}{16}\eta^4 - \frac{1}{4}\eta^2\theta + ?$
17. $? - \frac{\pi\phi}{3} + \frac{1}{4}\phi^2$
18. $\frac{R_1^2}{64} - \frac{1}{24}XR_1 + ?$

Extract the square roots of the following:

19. $M^2 + 2M + 1$
20. $a^2 - 10ab + 25b^2$
21. $16q_1^2 + 8q_1q_2 + q_2^2$
22. $V^2 + 12VI + 36I^2$
23. $9\alpha^4\beta^2 + 54\alpha^2\beta\gamma + 81\gamma^2$
24. $64\omega^2\lambda^2 + 16\omega\lambda\Omega^2 + \Omega^4$

25. $\dfrac{9}{25}\pi^2R^4 + \dfrac{4}{5}\pi R^2 + \dfrac{4}{9}$

26. $9R_1{}^2 + \dfrac{4}{25}r^2 - \dfrac{12}{5}R_1r$

27. $\dfrac{10\phi\lambda}{21} + \dfrac{25\phi^2}{36} + \dfrac{4\lambda^2}{49}$

28. $-\dfrac{4Z^4M^2}{27} + \dfrac{4Z^8}{81} + \dfrac{M^4}{9}$

10-10 PRIME FACTORS OF AN EXPRESSION

In factoring a number, all of the number's prime factors should be obtained. After an expression is factored once, it may be possible to factor it again.

Example 27

Find the prime factors of $12i^2r + 12iIr + 3I^2r$.

Solution

$$12i^2r + 12iIr + 3I^2r = 3r(4i^2 + 4iI + I^2)$$
$$= 3r(2i + I)(2i + I)$$
$$= 3r(2i + I)^2$$

PROBLEMS 10-6

Find the prime factors of the following:

1. $3ac + 6bc$

2. $15qrx + 35rtx$

3. $2\lambda\theta^2 + 4\theta\lambda\phi + 2\phi^2\lambda$

4. $5V^2i^2 - 10VIi^2R + 5I^2i^2R^2$

5. $24\alpha^4 + 120\alpha^3\beta + 150\alpha^2\beta^2$

6. $\dfrac{2V^4}{3r} + \dfrac{4V^2v^2}{r} + \dfrac{6v^4}{r}$

7. $\dfrac{20f_0\omega_1{}^2}{\omega} - \dfrac{40f_0\omega_1\omega_2}{\omega} + \dfrac{20f_0\omega_2{}^2}{\omega}$

8. $\dfrac{3X_L{}^2}{4X_C} + \dfrac{3KMX_L}{4X_C} + \dfrac{3K^2M^2}{16X_C}$

9. $\dfrac{5r\lambda^2}{16e} - \dfrac{5f^2r\lambda}{2e} + \dfrac{5f^4r}{e}$

10. $\dfrac{200Ff^2}{C} + \dfrac{480Ffx}{C} + \dfrac{288Fx^2}{C}$

10-11 PRODUCT OF THE SUM AND DIFFERENCE OF TWO NUMBERS

Type: $\qquad (a + b)(a - b) = a^2 - b^2$

The multiplication of the sum and difference of two general numbers, such as

$$\begin{array}{r} a + b \\ a - b \\ \hline a^2 + ab \\ -\,ab - b^2 \\ \hline a^2 \qquad - b^2 \end{array}$$

results in the formula

$$(a + b)(a - b) = a^2 - b^2$$

which can be expressed by the following:

RULE
The product of the sum and difference of two numbers is equal to the difference of their squares.

Example 28

$$(3x + 4y)(3x - 4y) = 9x^2 - 16y^2$$

Example 29

$$(6ab^2 + 7c^3d)(6ab^2 - 7c^3d) = 36a^2b^4 - 49c^6d^2$$

PROBLEMS 10-7

Multiply by inspection:

1. $(\theta + 2)(\theta - 2)$

2. $(\phi - 4)(\phi + 4)$

3. $(I + i)(I - i)$

4. $(I + Z)(I - Z)$

5. $(3Q - 2L)(3Q + 2L)$

6. $(2\pi R_1 + 2\pi R_2)(2\pi R_1 - 2\pi R_2)$

7. $\left(\frac{2}{3}VI + P\right)\left(\frac{2}{3}VI - P\right)$

8. $\left(\frac{3}{4}\omega^2 - \frac{2}{5}\lambda\right)\left(\frac{3}{4}\omega^2 + \frac{2}{5}\lambda\right)$

9. $\left(\frac{2V^2}{R} + \frac{3I^2R}{P}\right)\left(\frac{2V^2}{R} - \frac{3I^2R}{P}\right)$

10. $\left(\frac{\theta^2}{2\phi} + \frac{3\alpha}{2\beta}\right)\left(\frac{\theta^2}{2\phi} - \frac{3\alpha}{2\beta}\right)$

10-12 FACTORING THE DIFFERENCE OF TWO SQUARES

RULE
To factor the difference of two squares, extract the square root of the two squares, add the roots for one factor, and subtract the second root from the first for the other factor.

Example 30

$$x^2 - y^2 = (x + y)(x - y)$$

Example 31

$$9a^2c^4 - 36d^6 = (3ac^2 + 6d^3)(3ac^2 - 6d^3)$$

PROBLEMS 10-8

Factor:

1. $a^2 - b^2$
2. $I_1^2 - I_2^2$
3. $4\theta^2 - 16\phi^2$
4. $4I^2 - 9r^2$
5. $\dfrac{1}{4} - \theta^2$
6. $\dfrac{\alpha^2}{\beta^2} - \dfrac{4\gamma^2}{9}$
7. $1 - 225\omega^2$
8. $\dfrac{1}{E_1^2} - \dfrac{1}{e^2}$
9. $81\theta^2\mu^2 - 1$
10. $\dfrac{1}{X_C^2} - \dfrac{V^2}{Q^2}$
11. $9c^2 - a^2 + 2ab - b^2$

Solution: $\quad 9c^2 - a^2 + 2ab - b^2 = 9c^2 - (a^2 - 2ab + b^2)$
$$= [3c + (a - b)][3c - (a - b)]$$
$$= (3c + a - b)(3c - a + b)$$

12. $(\theta^2 + 4\theta\phi + 4\phi^2) - \omega^2$
13. $36m^2 - 81p^2q^2 + 9a^2b^2 - 36abm$
14. $16I^2 - V^2 + \dfrac{14V}{X} - \dfrac{49}{X^2}$
15. $100acl - 144l^2 + 25a^2 + 100c^2l^2$

10-13 PRODUCT OF TWO BINOMIALS HAVING A COMMON TERM

Type: $\qquad (x + a)(x + b) = x^2 + (a + b)x + ab$

The multiplication

$$
\begin{array}{r}
x + a \\
x + b \\
\hline
x^2 + ax \\
+ bx + ab \\
\hline
x^2 + ax + bx + ab
\end{array}
$$

when factored, results in $x^2 + (a + b)x + ab$.

This type of formula can be expressed as follows:

RULE

To obtain the product of two binomials having a common term, square the common term, multiply the common term by the algebraic sum of the second terms of the binomials, find the product of the second terms, and add the results.

Example 32

Find the product of $x - 7$ and $x + 5$.

Solution

$$(x - 7)(x + 5) = x^2 + (-7 + 5)x + (-7)(+5)$$
$$= x^2 - 2x - 35$$

Example 33

$$(ir + 3)(ir - 6) = i^2r^2 + (+3 - 6)ir + (+3)(-6)$$
$$= i^2r^2 - 3ir - 18$$

Although the preceding examples have been written out in order to illustrate the method, the actual multiplication should be performed mentally. In Example 33, write the i^2r^2 term first. Then glance at the $+3$ and -6, note that their sum is -3 and their product is -18, and write down the complete product.

PROBLEMS 10-9

Mentally, mutliply the following:

1. $(\theta + 4)(\theta + 3)$

2. $(Q + 1)(Q + 2)$

3. $(R + 1)(R - 2)$

4. $(\phi - 3)(\phi - 2)$

5. $(\theta + 6)(\theta + 3)$

6. $(2r + 3)(2r + 2)$

7. $(3\theta - 2)(3\theta + 1)$

8. $(4x + 2)(4x - 4)$

9. $(I - 3)(I - 4)$

10. $\left(\frac{1}{2}P + 2\right)\left(\frac{1}{2}P + 6\right)$

11. $(\alpha - 1)\left(\alpha - \frac{1}{4}\right)$

12. $(\lambda + 6)\left(\lambda - \frac{1}{3}\right)$

13. $\left(IR + \frac{1}{2}\right)\left(IR - \frac{1}{3}\right)$

14. $(2f + 12)\left(2f - \frac{1}{4}\right)$

15. $\left(\alpha + \frac{2}{3}\right)\left(\alpha + \frac{1}{3}\right)$

16. $\left(\frac{1}{X} + 5\right)\left(\frac{1}{X} - 2\right)$

17. $\left(\frac{1}{\sqrt{LC}} - f\right)\left(\frac{1}{\sqrt{LC}} - 3f\right)$

18. $\left(vt - \frac{1}{12}\right)\left(vt + \frac{1}{2}\right)$

19. $\left(\alpha\beta^2 + \frac{1}{10}\right)\left(\alpha\beta^2 + \frac{1}{5}\right)$

20. $\left(I + \frac{6V}{R}\right)\left(I - \frac{4V}{R}\right)$

10-14 FACTORING TRINOMIALS OF THE FORM $a^2 + ba + c$

A trinomial of the form $a^2 + ba + c$ can be factored if it is the product of two binomials having a common term.

RULE

To factor a trinomial of the form $a^2 + ba + c$, find two numbers whose sum is b and whose product is c. Add each of them to the square root of the first term for the factors.

Example 34

Factor $a^2 + 7a + 12$.

Solution

If this expression will factor, it will take the form

$$a^2 + 7a + 12 = (a + \quad)(a + \quad)$$

where the two blanks represent numbers whose product is 12 and whose sum is 7. The factors of 12 are

$$1 \times 12$$
$$2 \times 6$$
$$3 \times 4$$

The first two pairs will not do because the sum of neither pair is 7. The third pair gives the correct sum.

$$\therefore a^2 + 7a + 12 = (a + 3)(a + 4)$$

Example 35

Factor $x^2 - 15x + 36$.

Solution

Since the 36 is positive, its factors must bear the same sign; also, since -15 is negative, it follows that both factors must be negative. The factors of 36 are

$$1 \times 36$$
$$2 \times 18$$
$$3 \times 12$$
$$4 \times 9$$
$$6 \times 6$$

Inspection of these factors shows that 3 and 12 are the required numbers.

$$\therefore x^2 - 15x + 36 = (x - 3)(x - 12)$$

Example 36

Factor $e^2 - e - 56$.

Solution

Since we have -56, the two factors must have unlike signs. The sum of the factors must equal -1; therefore, the negative factor of -56 must have the greater absolute value. The factors of 56 are

$$1 \times 56$$
$$2 \times 28$$
$$4 \times 14$$
$$7 \times 8$$

Since the factors 7 and 8 differ in value by 1, we have

$$e^2 - e - 56 = (e + 7)(e - 8)$$

PROBLEMS 10-10

Factor:

1. $a^2 + 3a + 2$

2. $\theta^2 + 8\theta + 15$

3. $R^2 + 8R + 12$

4. $\alpha^2 - 9\alpha + 14$

5. $\beta^2 + 2\beta - 24$

6. $R^2 - 6R - 72$

7. $\theta^2 + 10\theta + 24$

8. $\omega^2 - 14\omega + 24$

9. $t^2 + 9t - 22$

10. $\alpha^2 t^2 - 13\alpha t + 36$

11. $Z^4 + 8Z^2 - 20$

12. $f^2 - 17f - 60$

13. $\pi^2 + \pi - 56$

14. $I^2 R^2 + 4VIR + 3V^2$

15. $\omega^2 - \omega f - 6f^2$

16. $\theta^4 - 4\theta^2\phi - 12\phi^2$

17. $\theta^2 - \dfrac{5}{6}\theta + \dfrac{1}{6}$

18. $q^4 - \dfrac{1}{4}q^2 - \dfrac{1}{8}$

19. $\phi^4 + \dfrac{\phi^2}{10} - \dfrac{1}{50}$

20. $v^4 i^4 + 2v^2 i^2 P - 3P^2$

10-15 PRODUCT OF ANY TWO BINOMIALS

Type: $(ax + b)(cx + d)$

Up to the present, if it was desired to multiply $5x - 2$ by $3x + 6$, we multiplied in the following manner:

$$\begin{array}{r} 5x - 2 \\ 3x + 6 \\ \hline 15x^2 - 6x \\ + 30x - 12 \\ \hline 15x^2 + 24x - 12 \end{array}$$

Note that $15x^2$ is the product of the first terms of the binomials and the last term is the product of the last terms of the binomials. Also, the middle term is the sum of the products obtained by multiplying the first term of each binomial by the second term of the other binomial.

The preceding example can be written in the following manner:

$$\begin{array}{c} 5x - 2 \\ 3x + 6 \\ \hline 15x^2 + 24x - 12 \end{array}$$

The middle term $(+24x)$ is the sum of *cross products* $(5x)(+6)$ and $(3x)(-2)$, which is obtained by multiplying the first term of each binomial by the second term of the other.

The usual method of obtaining this product is indicated by the following solution:

$$(5x - 2)(3x + 6) = 15x^2 + 24x - 12$$

RULE
For finding the product of any two binomials,

1. The first term of the product is the product of the first terms of the binomials.
2. The second term is the algebraic sum of the product of the two outer terms and the product of the two inner terms.
3. The third term is the product of the last terms of the binomials.

Example 37

Find the product of $(4e + 7j)(2e - 3j)$.

Solution
The only difficulty encountered in obtaining such products mentally is that of finding the second term.

$$(4e)(-3j) = -12ej$$
$$(7j)(2e) = 14ej$$
$$(-12ej) + (14ej) = +2ej$$
$$\therefore (4e + 7j)(2e - 3j) = 8e^2 + 2ej - 21j^2$$

Example 38
Find the product $(7r^2 + 8Z)(8r^2 - 9Z)$.

Solution

1. The first term of the product is $(7r^2)(8r^2) = 56r^4$.
2. Since $(7r^2)(-9Z) = -63r^2Z$ and $(8Z)(8r^2) = 64r^2Z$, the second term is $(-63r^2Z) + (64r^2Z) = +r^2Z$.
3. The third term is $(8Z)(-9Z) = -72Z^2$.

$$\therefore (7r^2 + 8Z)(8r^2 - 9Z) = 56r^4 + r^2Z - 72Z^2$$

By repeated drills you should acquire skill enough that you can readily obtain such products mentally. This type of product is frequently encountered in algebra, and the ability to multiply rapidly will save you much time.

PROBLEMS 10-11
Multiply:

1. $(x + 2)(x - 5)$
2. $(IR - 4)(IR + 3)$
3. $(3\phi + 1)(2\phi + 3)$
4. $(2R + 6)(3R + 5)$
5. $(3j - 2)(4j + 2)$
6. $(7\lambda + 5)(2\lambda - 3)$
7. $(2\omega + 5)(3\omega - 1)$
8. $(7\theta + 3)(3\theta + 7)$
9. $\left(\frac{1}{2}\omega + 8\right)\left(\frac{1}{2}\omega - 4\right)$
10. $\left(\frac{3}{\theta} - 6\right)\left(\frac{2}{\theta} - 12\right)$
11. $(2Z + IR)(3Z + 5IR)$
12. $(I - 18)(I + 6)$

13. $(3X - 20)(5X + 2)$

14. $(12M - 3)(3M - 12)$

15. $(15\theta - 2)(\theta - 5)$

16. $(5 + 4p)(4 - 5p)$

17. $(5 - 3\pi)(7 - 2\pi)$

18. $(3\phi + 4)(5\phi + 3)$

19. $(3\alpha + 5\beta)(2\alpha + 7\beta)$

20. $(3x + 7)(4x - 5)$

21. $(2a - 7t)(2a - 5t)$

22. $(a + 0.5)(a - 0.3)$

23. $(\omega + 0.7f)(\omega - 0.2f)$

24. $(IR - 0.9)(IR + 1)$

25. $\left(\dfrac{x}{8} + \dfrac{\lambda}{4}\right)(2x - 16\lambda)$

26. $\left(8\delta - \dfrac{2}{3\eta}\right)\left(9\delta - \dfrac{1}{2\eta}\right)$

27. $\left(6Z + \dfrac{1}{2IR}\right)\left(4Z + \dfrac{1}{3IR}\right)$

28. $\left(12\pi L - \dfrac{2}{3\pi C}\right)\left(10\pi L - \dfrac{2}{3\pi C}\right)$

29. $(0.2p - 0.7q)(0.8p - 0.3q)$

30. $\left(4m + \dfrac{3r}{p}\right)\left(6m - \dfrac{2r}{3p}\right)$

10-16 FACTORING TRINOMIALS OF THE TYPE $ax^2 + bx + c$

The method of factoring trinomials of the type $ax^2 + bx + c$ is best illustrated by examples.

Example 39
Factor $3a^2 + 5a + 2$.

Solution
It is apparent that the two factors are binomials and the product of the end terms must be $3a^2$ and 2. Therefore, the binomials to choose from are

$$(3a + 1)(a + 2)$$
and
$$(3a + 2)(a + 1)$$

However, the first factors when multiplied result in a product of $7a$ for the middle term. The second pair of factors when multiplied give a middle term of $5a$. Therefore,

$$3a^2 + 5a + 2 = (3a + 2)(a + 1)$$

Example 40
Factor $6e^2 + 7e + 2$.

Solution
Again, the end terms of the binomial factors must be so chosen that their products result in $6e^2$ and 2. Both last terms of the factors are of like signs, for the last term of the trinomial is positive. Also, both last terms of the factors must be positive, for the second term of the

trinomial is positive. One of the several methods of arranging the work is as shown below. The tentative factors are arranged as if for multiplication:

Trial Factors	Products	
$(6e + 1)(e + 2) = 6e^2 + 13e + 2$		Wrong
$(6e + 2)(e + 1) = 6e^2 + 8e + 2$		Wrong
$(3e + 1)(2e + 2) = 6e^2 + 8e + 2$		Wrong
$(3e + 2)(2e + 1) = 6e^2 + 7e + 2$		Right

It is seen that any combination of the trial factors when multiplied results in the correct first and last term.

$$\therefore 6e^2 + 7e + 2 = (3e + 2)(2e + 1)$$

Note

This may seem to be a long process, but with practice, most of the factor trials can be tested mentally.

Example 41

Factor $12i^2 - 17i + 6$.

Solution

The third term of this trinomial is $+6$; therefore, its factors must have like signs. Since the second term is negative, the cross products must be negative. Then it follows that both factors of 6 must be negative. Some of the combinations are as follows:

Trial Factors	Products	
$(2i - 3)(6i - 2) = 12i^2 - 22i + 6$		Wrong
$(2i - 2)(6i - 3) = 12i^2 - 18i + 6$		Wrong
$(3i - 3)(4i - 2) = 12i^2 - 18i + 6$		Wrong
$(3i - 2)(4i - 3) = 12i^2 - 17i + 6$		Right

$$\therefore 12i^2 - 17i + 6 = (3i - 2)(4i - 3)$$

Example 42

Factor $8r^2 - 14r - 15$.

Solution

The factors of -15 must have unlike signs. The signs of these factors must be so arranged that the cross product of greater absolute value is minus, because the middle term of the trinomial is negative.

Trial Factors	Products	
$(8r + 3)(r - 5) = 8r^2 - 37r - 15$		Wrong
$(4r + 5)(2r - 3) = 8r^2 - 2r - 15$		Wrong
$(4r + 3)(2r - 5) = 8r^2 - 14r - 15$		Right

Example 43

Factor $6R^2 - 7R - 20$.

Note

Many students prefer the following method to the trial-and-error method of the foregoing examples.

Solution

Multiply and divide the entire expression by the coefficient of R^2. The result is

$$\frac{36R^2 - 42R - 120}{6}$$

Take the square root of the first term, which is $6R$, and let that be some other letter such as x. Then, if

$$6R = x$$

by substituting the value of $6R$ in the above expression, we obtain

$$\frac{x^2 - 7x - 120}{6}$$

This results in an expression with a numerator easy to factor. Thus,

$$\frac{x^2 - 7x - 120}{6} = \frac{(x + 8)(x - 15)}{6}$$

Substituting $6R$ for x in the last expression, we obtain

$$\frac{(6R + 8)(6R - 15)}{6}$$

Factoring the numerator,

$$\frac{2(3R + 4)3(2R - 5)}{6}$$

Canceling, $\quad 6R^2 - 7R - 20 = (3R + 4)(2R - 5)$

Note

The denominator will *always* cancel out.

Example 44

Factor $4V^2 - 8VI - 21I^2$.

Solution

Multiplying and dividing by the coefficient of V^2,

$$\frac{16V^2 - 32VI - 84I^2}{4}$$

Let the square root of the first term $4V = x$.

Then $\quad \dfrac{x^2 - 8Ix - 84I^2}{4} = \dfrac{(x + 6I)(x - 14I)}{4}$

Substituting for x, $\quad \dfrac{(4V + 6I)(4V - 14I)}{4}$

Factoring, $\quad \dfrac{2(2V + 3I)2(2V - 7I)}{4}$

Canceling, $\quad 4V^2 - 8VI - 21I^2 = (2V + 3I)(2V - 7I)$

PROBLEMS 10-12

Factor:

1. $\omega^2 - 3\omega - 10$
2. $6\theta^2 + 7\theta + 2$
3. $8m^2 - 2m - 15$
4. $3I^2 - 14I + 8$
5. $6x^2 + 11x - 10$
6. $3\alpha^2 - 10\alpha + 7$
7. $9\phi^2 + 18\phi + 8$
8. $18L_1^2 + 31L_1 + 6$
9. $2\alpha^2 - \alpha\beta - 21\beta^2$
10. $10P^2 - 17PW + 3W^2$
11. $40m^2 + 2m - 21$
12. $20\lambda^2 - 22\lambda\phi + 6\phi^2$
13. $80I^2 + 14lw - 6w^2$
14. $18q^2 + 57qr + 35r^2$
15. $24\beta^4 - 30\beta^2\gamma + 9\gamma^2$
16. $42y^2z^2 + 11wyz - 20w^2$
17. $27l^2m^2 + 15lmw - 2w^2$
18. $2\mu^2 - 18\pi^2$
19. $6\psi^2 - 24\Omega^2$
20. $24m^4 - 43m^2p + 18p^2$
21. $15x^2 - 7\Delta x - 2\Delta^2$
22. $\alpha^2 - \dfrac{5\alpha}{6} + \dfrac{1}{6}$
23. $48\theta^2 + 5\theta + \dfrac{1}{8}$
24. $10Z^2 - \dfrac{3Z}{2} + \dfrac{1}{20}$
25. $0.18\theta^2 - 2$

10-17 SUMMARY

In this chapter, various cases of products and factoring have been treated separately in the different sections. Frequently, however, it becomes necessary to apply the principles underlying two or more cases to a single problem. It is very important, therefore, that you recognize the standard form for various types of algebraic expressions in order that you can apply the method of solution as needed. The standard forms are summarized in Table 10-1.

TABLE 10-1

General Type	Factors	Section
$ab + ac + ad$	$a(b + c + d)$	10-7
$a^2 + 2ab + b^2$	$(a + b)^2$	10-8
$a^2 - 2ab + b^2$	$(a - b)^2$	10-8
$a^2 - b^2$	$(a + b)(a - b)$	10-12
$a^2 + (b + c)a + bc$	$(a + b)(a + c)$	10-13
$acx^2 + (bc + ad)x + bd$	$(ax + b)(cx + d)$	10-15

Problems 10-13 are included as a review of the entire chapter. If you can work all of them, you thoroughly understand the contents of this chapter. If not, a review of the doubtful parts is suggested, for a good working knowledge of special products and factoring makes it possible to do the following:

1. Multiply, divide, and factor very quickly in your head (mentally).
2. Find the solutions to problems which can be solved by (quick mental) factoring.

PROBLEMS 10-13

Find the value of the following:

1. $(-4\omega L)^2$

2. $(-3\lambda^2\phi^3\omega)^3$

3. $\left(\dfrac{a^3b^3cd^2}{p^2q^3r}\right)^4$

4. $-\sqrt{64a^2x^4y^2z^6}$

5. $\sqrt{\dfrac{144I^2R^2}{169F^2X_C^4}}$

6. $\sqrt[3]{-64\alpha^6\beta^3\gamma^9}$

7. $-\sqrt[3]{\dfrac{125l^3m^6}{27x^{12}y^{15}z^3}}$

8. $\sqrt{\dfrac{625I^4R^2P^2}{64V^2W^6}}$

9. $-\sqrt[3]{-216\theta^3\phi^6\omega^6}$

10. $125a\sqrt[3]{\dfrac{a^5x^6z^8}{a^2x^3z^2}}$

Factor:

11. $IR^2 - Ir^2$

12. $I^2R_1 + I^2R_2 + I^2R_3$

13. $\dfrac{3v^2}{8r_1} + \dfrac{5v^2}{8r_2} - \dfrac{7v^2}{8r_3}$

14. $4.8\omega L_1 - 0.24\omega L_2 + 1.2\omega L_3$

15. $\dfrac{7}{16}xk - \dfrac{3}{16}xl - \dfrac{9}{16}xm$

16. $\dfrac{2g_m r_p}{3} - 6g_m R_p + \dfrac{16g_m R_p'}{3}$

Mentally, find the products:

17. $(R + 12)^2$

18. $(2\theta - 3\phi^2)^2$

19. $\left(12I^2 + \dfrac{2}{3}\right)^2$

20. $(0.5\omega M - 0.3\omega L)^2$

21. $\left(\dfrac{5}{9}\beta - 3\lambda\right)^2$

22. $(0.4X_C + 0.8X_L)^2$

Supply the missing term so that the three terms form a trinomial square:

23. $r^2 + ? + 9$

24. $9\alpha^2 - ? + 4\beta^2$

25. $? + 28Qr + 4r^2$

26. $64Z^6 - 32Z^3 + ?$

27. $\mu^2 + \dfrac{\mu\lambda}{2} + ?$

28. $? + \dfrac{3}{2}L^2M + \dfrac{9}{64}M^2$

Extract the square roots of the following:

29. $m^2 + 10m + 25$

30. $\theta^2 + 14\theta\phi + 49\phi^2$

31. $16\alpha^2 + 80\alpha\beta + 100\beta^2$

32. $F^2 - \dfrac{2Ff}{3} + \dfrac{f^2}{9}$

33. $\dfrac{\phi^2}{36} - \dfrac{\phi\lambda}{6} + \dfrac{\lambda^2}{4}$

34. $\dfrac{4V^2}{25} - \dfrac{12VX}{5} + 9X^2$

Factor:

35. $24iR + 42IR$

36. $6pq - 27pr$

37. $3ir^2 + 18ir + 27i$

38. $16\pi^3Dr^2 + 48\pi^2CDr + 36\pi C^2D$

39. $768\theta^2\omega - 1920\theta\phi\omega + 12\phi^2\omega$

40. $\dfrac{12P^2}{VI} - \dfrac{144PW}{VI} + \dfrac{432W^2}{VI}$

Find the products:

41. $(\alpha + 2\beta)(\alpha - 2\beta)$

42. $(2IR - 3V)(2IR + 3V)$

43. $(Z - 12)(Z + 12)$

44. $(8\theta + 7\phi)(8\theta - 7\phi)$

45. $\left(\dfrac{24V}{IR} + 2P\right)\left(\dfrac{24V}{IR} - 2P\right)$

46. $(0.8\lambda + 0.3\Omega)(0.8\lambda - 0.3\Omega)$

Factor:

47. $Q^2 - 1$

48. $25 - f_o^2$

49. $4\omega^2L^2 - \dfrac{1}{16\omega^2C^2}$

50. $\dfrac{4}{25}\alpha^2\beta^4 - \dfrac{9}{16}\lambda^2$

51. $0.0025\psi^2 - 0.36\mu^2$

52. $0.01X_L^2 - 0.81X_C^2$

Find the quotients:

53. $(\lambda^2 - 4) \div (\lambda + 2)$

54. $(4L^2 - 9C^2) \div (2L + 3C)$

55. $\left(\dfrac{1}{9}\alpha^2 - \dfrac{4}{49}\beta^2\right) \div \left(\dfrac{1}{3}\alpha - \dfrac{2}{7}\beta\right)$

56. $(0.25\theta^2 - 0.04\delta^2) \div (0.5\theta - 0.2\delta)$

57. $\left(\dfrac{9}{25}e^2 - \dfrac{16}{81}i^2r^2\right) \div \left(\dfrac{3}{5}e + \dfrac{4}{9}ir\right)$

58. $\dfrac{L^2 + 2LM + M^2 - 25}{L + M + 5}$

Find the products:

59. $(\kappa + 2)(\kappa - 4)$

60. $(3 - Q)(4 - 2Q)$

61. $(0.2X_C - 3)(X_C + 0.5)$

62. $(0.1Z + 0.6R)(0.3Z - R)$

63. $\left(A - \dfrac{1}{3}\right)\left(A + \dfrac{1}{5}\right)$ **64.** $(5\lambda + 12)\left(\dfrac{1}{4}\lambda - \dfrac{1}{5}\right)$

65. $(8\mu + 6g_m)(3\mu - 2g_m)$ **66.** $(2\pi fL + X_C)(2\pi fL - 3X_C)$

67. $(2R - r)(0.3R + 0.2r)$ **68.** $(0.5\alpha + 8\beta)(0.2\alpha + 0.4\beta)$

69. $\left(4\phi + \dfrac{2\theta}{3}\right)\left(6\phi - \dfrac{\theta}{2}\right).$ **70.** $\left(\dfrac{2v}{3} - \dfrac{4s}{t}\right)\left(\dfrac{v}{2} - \dfrac{6s}{t}\right)$

Factor (remove any common factors first):

71. $6z^2 + 11z + 3$ **72.** $12I_1^2 - 2I_1 - 4$

73. $\lambda^2 - 8\lambda + 15$ **74.** $e^2 - 0.2e - 0.03$

75. $x^2 - 2.6x + 1.2$ **76.** $A^2 - \dfrac{3A}{40} - \dfrac{1}{40}$

77. $12R^2 + 8RX - 15X^2$ **78.** $3\alpha^2\beta^2\gamma^2 + \alpha\beta\gamma\Omega - 10\Omega^2$

79. $2V^2 + 0.1VIR - 0.15I^2R^2$ **80.** $l^2 - 0.3lq - 0.4q^2$

81. $\dfrac{X_C^2}{9} + \dfrac{2X_C Z}{3} + Z^2$ **82.** $a^2 + \dfrac{2a}{b} + \dfrac{1}{b^2}$

83. $16\pi^2 - 2\pi f - 5f^2$ **84.** $5\theta^2\omega - 5\phi^2\omega$

85. $3x^2 - 12$ **86.** $288f^2\lambda - \dfrac{2\lambda}{9}$

87. $\dfrac{3V^2}{2i} - \dfrac{18Vv}{2i} + \dfrac{54v^2}{2i}$ **88.** $\dfrac{x^3}{9Z} - \dfrac{x^2y}{6Z} + \dfrac{xy^2}{16Z}$

89. $\dfrac{a^2c}{2d} - \dfrac{145abc}{144d} + \dfrac{b^2c}{2d}$

90. $\dfrac{2VR_1^2}{3I} - \dfrac{1898VR_1R_2}{1350I} + \dfrac{2VR_2^2}{3I}$

Algebraic Fractions

11

Algebraic fractions play an important role in mathematics, especially in equations for electric and electronic circuits.

At this time, if you feel you have not thoroughly mastered arithmetical fractions, you are urged to review them. Despite the current emphasis on metrication and the increasing use of decimal fractions in measurement, a good foundation in arithmetical fractions is essential, for every rule and operation pertaining to them is applicable to algebraic fractions. It is a fact that anyone who really knows arithmetical fractions rarely has trouble with algebraic fractions.

11-1 THE DEGREE OF A MONOMIAL

The degree of a monomial is determined by the number of literal factors the monomial has.

Thus, $6ab^2$ is a monomial of the third degree because $ab^2 = a \cdot b \cdot b$; $3mn$ is a monomial of the second degree. From these examples, it is seen that the degree of a monomial is the sum of the exponents of the literal factors (letters).

In such an expression as $5X^2Y^2Z$, we speak of the whole term as being of the fifth degree, X and Y as being of the second degree, and Z as being of the first degree.

The above definition for the degree of a monomial does not apply to letters in a denominator.

11-2 THE DEGREE OF A POLYNOMIAL

The degree of a polynomial is taken as the degree of the term of highest degree. Thus, $3ab^2 - 4cd - d$ is a polynomial of the *third degree* and $6x^2y + 5xy^2 + x^2y^2$ is a polynomial of the *fourth degree*.

11-3 HIGHEST COMMON FACTOR

A factor of each of two or more expressions is a *common factor* of those expressions. For example, 2 is a common factor of 4 and 6; a^2 is a common factor of a^3, $(a^2 - a^2b)$, and $(a^2x^2 - a^2y)$.

The product of all the factors common to two or more numbers, or expressions, is called the *highest common factor*. That is, the highest common factor is the expression of highest degree that will divide each of the numbers, or expressions, without a remainder. It is commonly abbreviated HCF.

Example 1
Find the HCF of

$$6a^2b^3(c + 1)(c + 3)^2$$

and
$$30a^3b^2(c - 2)(c + 3)$$

Solution

6 is the greatest integer that will divide both expressions. The highest power of a that will divide both is a^2. The highest power of b that will divide both is b^2. The highest power of $(c + 3)$ that will divide both is $(c + 3)$. $(c + 1)$ and $(c - 2)$ will not divide both expressions.

$$\therefore 6a^2b^2(c + 3) = \text{HCF}$$

RULE
To determine the HCF:

1. Determine all the prime factors of each expression.
2. Take the common factors of all the expressions and give to each the lowest exponent it has in any of the expressions.
3. The HCF is the product of all the common factors as obtained in the second step.

Example 2
Find the HCF of

$$50a^2b^3c(x + y)^3(x - y)^4$$

and
$$75a^2bc^2(x + y)^2(x - y)$$

Solution

$$50a^2b^3c(x + y)^3(x - y)^4 = 2 \cdot 5 \cdot 5a^2b^3c(x + y)^3(x - y)^4$$

$$75a^2bc^2(x + y)^2(x - y) = 3 \cdot 5 \cdot 5a^2bc^2(x + y)^2(x - y)$$

$$\therefore \text{HCF} = 5^2a^2bc(x + y)^2(x - y) = 25a^2bc(x + y)^2(x - y)$$

Example 3
Find the HCF of

$$v^2 + vr \qquad v^2 + 2vr + r^2 \qquad \text{and} \qquad v^2 - r^2$$

Solution

$$v^2 + 2vr = v(v + r)$$

$$v^2 + 2vr + r^2 = (v + r)^2$$

$$v^2 - r^2 = (v + r)(v - r)$$

$$\therefore \text{HCF} = v + r$$

PROBLEMS 11-1

Find the HCF of:

1. 24, 40
2. 50, 125, 625
3. $4\theta^2\phi$, $12\theta\phi\omega$, $36\theta\phi^2\omega$
4. $16\lambda^2\omega$, $48\lambda^2\theta$, $36\lambda^2\phi$
5. $0.5a^3b^2c$, $0.25a^2b^2c^2$, $0.1a^2bc^3$
6. $39x^4y^2z^3$, $78x^3y^3z^3$, $156x^2y^4z^3$
7. $39I^2R$, $195I^2R^2$, $36IR$
8. $18\alpha\beta^2\gamma^3$, $162\alpha^2\beta^3\gamma$, $220\alpha\beta^3\gamma^2$
9. $X_L^2 - X_C^2$, $X_L^2 + X_LX_C$
10. $m^2 + 2mn + n^2$, $m^2 - n^2$
11. $E^2 - 2E + 1$, $E^2 - 1$, $3E^2 - 3E$
12. $12\pi + 4\phi$, $9\pi^2 + 6\pi\phi + \phi^2$, $9\pi^2 - \phi^2$
13. $L_1L_2 + 2\sqrt{L_1L_2}M + M^2$, $5\sqrt{L_1L_2} + 5M$
14. $9I^2R^2 - 24VIR + 16V^2$, $3I^2R^2 - 10VIR + 8V^2$, $15I^2R^2 - 17VIR - 4V^2$
15. $10I^2 + 25\dfrac{VI}{R} + 15\dfrac{V^2}{R^2}$, $30I + 45\dfrac{V}{R}$, $40I^2 + 40\dfrac{VI}{R} - 30\dfrac{V^2}{R^2}$

11-4 MULTIPLE

A number is a *multiple* of any one of its factors. For example, some of the multiples of 4 are 8, 16, 20, and 24. Similarly, some of the multiples of $a + b$ are $3(a + b)$, $a^2 + 2ab + b^2$, and $a^2 - b^2$. A *common multiple* of two or more numbers is a multiple of each of them. Thus, 45 is a common multiple of 1, 3, 5, 9, and 15.

11-5 LOWEST COMMON MULTIPLE

The smallest number that will contain each one of a set of factors is called the *lowest common multiple* of the factors. Thus, 48, 60, and 72 are all common multiples of 4 and 6, but the lowest common multiple of 4 and 6 is 12.

The lowest common multiple is abbreviated LCM.

Example 4

Find the LCM of $6x^2y$, $9xy^2z$, and $30x^3y^3$.

Solution

$$6x^2y = 2 \cdot 3 \cdot x^2y$$
$$9xy^2z = 3^2 \cdot xy^2z$$
$$30x^3y^3 = 2 \cdot 3 \cdot 5 \cdot x^3y^3$$

Because the LCM must contain *each* of the expressions, it must have 2, 3^2, and 5 as factors. Also, it must contain the literal factors of highest degree, or x^3y^3z.

$$\therefore \text{LCM} = 2 \cdot 3^2 \cdot 5 \cdot x^3y^3z = 90x^3y^3z$$

RULE

To determine the LCM of two or more expressions, determine all the prime factors of each expression. Find the product of all the different prime factors, taking each factor the greatest number of times it occurs in any one expression.

Example 5

Find the LCM of

$$3a^3 + 6a^2b + 3ab^2$$
$$6a^4 - 12a^3b + 6a^2b^2$$
$$9a^3b - 9ab^3$$

Solution

$$3a^3 + 6a^2b + 3ab^2 = 3a(a + b)^2$$
$$6a^4 - 12a^3b + 6a^2b^2 = 2 \cdot 3 \cdot a^2(a - b)^2$$
$$9a^3b - 9ab^3 = 3^2 \cdot ab(a + b)(a - b)$$
$$\therefore \text{LCM} = 2 \cdot 3^2 \cdot a^2b(a + b)^2(a - b)^2$$
$$= 18a^2b(a + b)^2(a - b)^2$$

PROBLEMS 11-2

Find the LCM of the following:

1. 12, 70, 210

2. 22, 154, 231

3. 40, 72, 180

4. $a^3b^2c,\ a^2bc^3$

5. $\theta^4\phi^3\lambda^2\omega,\ \theta^2\phi^2\lambda^3\mu\omega$

6. $2\alpha^4\beta^2,\ 10\alpha^2\beta^2\gamma^3,\ 15\alpha^3\beta^3\gamma$

7. $5m^3n^2p^2,\ 20m^2np,\ 45mnp^4$

8. $I^2,\ 3IR,\ 17I^2R^2$

9. $t - 3,\ t^2 - 5t + 6$

10. $X^2 - 11X + 30,\ X^2 - 9X + 20$

11. $\mu^2 + 3\mu,\ \mu^2 + 5\mu,\ \mu^2 + 8\mu + 15$

12. $6 + 4\psi,\ 3 - 2\psi,\ 9 - 12\psi + 4\psi^2$

13. $6\theta^2 + 7\theta - 3,\ 44\theta^2 + 88\theta + 33,\ 66\theta^2 + 11\theta - 11$

14. $4X_L^2 + 12X_LX_C + 8X_C^2,\ 2X_L^2 + 10X_LX_C + 12X_C^2,$
    $X_L^2 + 4X_LX_C + 3X_C^2$

15. $8Q^2 - 38\dfrac{\omega LQ}{R} + 35\dfrac{\omega^2L^2}{R^2},\ Q^2 - \dfrac{\omega^2L^2}{R^2},\ 2Q^2 - 9\dfrac{\omega LQ}{R} + 7\dfrac{\omega^2L^2}{R^2}$

11-6 DEFINITIONS

A fraction is an indicated division. Thus, we indicate 4 divided by 5 as $\frac{4}{5}$ (read four-fifths). Similarly, *X divided by Y* is written $\frac{X}{Y}$ (read *X* divided by *Y* or *X* over *Y*).

The quantity above the horizontal line is called the *numerator* and that below the line is called the *denominator* of the fraction. The numerator and denominator are often called the *terms* of the fraction.

11-7 OPERATIONS ON NUMERATOR AND DENOMINATOR

As in arithmetic, when fractions are to be simplified or affected by one of the four fundamental operations, we find it necessary to make frequent use of the following important principles:

1. The numerator and the denominator of a fraction can be multiplied by the same number or expression, except zero, without changing the value of the fraction.

2. The numerator and the denominator can be divided by the same number or expression, except zero, without changing the value of the fraction.

Example 6

$$\frac{2}{3} = \frac{2 \times 3}{3 \times 3} = \frac{6}{9} = \frac{2}{3}$$

Also,

$$\frac{6}{9} = \frac{6 \div 3}{9 \div 3} = \frac{2}{3} = \frac{6}{9}$$

Example 7

$$\frac{x}{y} = \frac{x \cdot a}{y \cdot a} = \frac{ax}{ay} = \frac{x}{y}$$

Also,

$$\frac{ax \div a}{ay \div a} = \frac{x}{y} \quad \text{(where } a \neq 0\text{)}$$

No new principles are involved in performing these operations, for multiplying or dividing both numerator and denominator by the same number, except zero, is equivalent to multiplying or dividing the fraction by 1 in any form convenient for our use, such as

$$\frac{2}{2}, \frac{4}{4}, \frac{10}{10}$$

or

$$\frac{-1}{-1}$$

It will be noted that, in the foregoing principles, multiplication and division of numerator and denominator by zero are excluded. When any expression is multiplied by zero, the product is zero.

For example, $6 \times 0 = 0$. Therefore, if we multiplied both numerator and denominator of some fraction by zero, the result would be meaningless. Thus,

$$\frac{5}{6} \neq \frac{5 \times 0}{6 \times 0}$$

because

$$\frac{5 \times 0}{6 \times 0} = \frac{0}{0}$$

Division by zero is meaningless. Some people say that any number divided by zero results in a quotient of infinity, denoted by ∞. If we accept this, we immediately impose a severe restriction on operations with even simple equations. For example, let us assume for the moment that any number divided by zero *does* result in infinity. Then if

$$\frac{4}{0} = \infty$$

by following Axiom 3, we should be able to multiply both sides of this equation by 0. If so, we obtain

$$4 = \infty \cdot 0$$

which we know is not sensible. Obviously, there is a fallacy here; therefore, we shall simply say at this time that *division by zero is not a permissible operation.*

11-8 EQUIVALENT FRACTIONS

Examples 6 and 7 show that when a numerator and a denominator are multiplied or divided by the same number, except zero, we change the *form* but not the value of the given fraction. Therefore, two fractions having the same value but not the same form are called *equivalent fractions*.

PROBLEMS 11-3

Supply the missing terms:

1. $\dfrac{3}{7} = \dfrac{?}{42}$

2. $\dfrac{7}{16} = \dfrac{?}{144}$

3. $\dfrac{1}{x} = \dfrac{?}{x^2 y}$

4. $\dfrac{2\theta}{7\phi} = \dfrac{?}{35\phi\omega}$

5. $\dfrac{3ab}{25c} = \dfrac{?}{75cd}$

6. $\dfrac{\alpha}{\alpha + 3} = \dfrac{?}{(\alpha + 3)(\alpha - 2)}$

7. $\dfrac{t - 1}{t - 3} = \dfrac{?}{t^2 - 4t + 3}$

8. $\dfrac{7 + \theta}{\theta - 1} = \dfrac{?}{\theta^2 - 1}$

9. $\dfrac{i + \alpha}{\alpha - 3\beta} = \dfrac{?}{6\alpha - 18\beta}$

10. $\dfrac{x - 2y}{2x + y} = \dfrac{?}{2x^2 + 5xy + 2y^2}$

11. Change the fraction $\dfrac{3}{16}$ into an equivalent fraction whose denominator is 64.

12. Change the fraction $\dfrac{7}{25}$ into an equivalent fraction whose denominator is 150.

13. Change the fraction $\dfrac{\omega L}{R}$ into an equivalent fraction whose denominator is $R^3 - RX^2$.

14. Change the fraction $\dfrac{L + 2}{L - 2}$ into an equivalent fraction whose denominator is $L^2 - 4$.

15. Change the fraction $\dfrac{Q}{VC + 1}$ into an equivalent fraction whose denominator is $2V^2C^2 - VC - 3$.

11-9 REDUCTION OF FRACTIONS TO THEIR LOWEST TERMS

If the numerator and denominator of a fraction have no common factor other than 1, the fraction is said to be in its lowest terms. Thus, the fractions $\frac{2}{3}, \frac{3}{5}, \frac{x}{y}$, and $\frac{x+y}{x-y}$ are in their lowest terms, for the numerator and denominator of each fraction have no common factor except 1.

The fractions $\frac{4}{6}$ and $\frac{3x}{9x^2}$ are not in their lowest terms, for $\frac{4}{6}$ can be reduced to $\frac{2}{3}$ if both numerator and denominator are divided by 2. Similarly, $\frac{3x}{9x^2}$ can be reduced to $\frac{1}{3x}$ by dividing both numerator and denominator by $3x$.

RULE

To reduce a fraction to its lowest terms, factor the numerator and denominator into prime factors and cancel the factors common to both.

Cancellation as used in the rule really means that we actually *divide* both terms of the fraction by the *common factors*. Then, to reduce a fraction to its lowest terms, it is only necessary to divide both numerator and denominator by the highest common factor, which leaves an equivalent fraction.

Example 8

Reduce $\frac{27}{108}$ to lowest terms.

Solution

$$\frac{27}{108} = \frac{\not3 \cdot \not3 \cdot \not3}{2 \cdot 2 \cdot 3 \cdot \not3 \cdot \not3} = \frac{1}{4}$$

Example 9

Reduce $\frac{24x^2yz^3}{42x^2yz^2}$ to lowest terms.

Solution

$$\frac{24x^2yz^3}{42x^2yz^2} = \frac{2 \cdot 2 \cdot 2 \cdot \not3 \cdot x^2\,yz^3}{\not2 \cdot \not3 \cdot 7 \cdot x^2yz^2} = \frac{4z}{7}$$

Actually, the solution to Example 9 need not have been written out, for it can be seen by inspection that the HCF of both terms of the fraction is $6x^2yz^2$, which we divide into both terms in order to obtain the equivalent fraction $\frac{4z}{7}$.

Also, in reducing fractions, we may resort to direct cancellation as in arithmetic.

Example 10

Reduce $\frac{x^2 - y^2}{x^3 - y^3}$ to lowest terms.

Solution

$$\frac{x^2 - y^2}{x^3 - y^3} = \frac{(x + y)\cancel{(x - y)}}{\cancel{(x - y)}(x^2 + xy + y^2)}$$

$$= \frac{x + y}{x^2 + xy + y^2}$$

Example 11

Reduce to lowest terms

$$\frac{r^2 - R^2}{r^2 + 3rR + 2R^2}$$

Solution

$$\frac{r^2 - R^2}{r^2 + 3rR + 2R^2} = \frac{\cancel{(r + R)}(r - R)}{(r + 2R)\cancel{(r + R)}} = \frac{r - R}{r + 2R}$$

PROBLEMS 11-4

Reduce to lowest terms:

1. $\dfrac{36}{48}$ 2. $\dfrac{72}{729}$ 3. $\dfrac{12}{156}$

4. $\dfrac{15}{240}$ 5. $\dfrac{a^3b^2}{a^4b^5}$ 6. $\dfrac{3\theta^2\phi}{12\theta\phi^3}$

7. $\dfrac{125I^2R}{25IR^2}$ 8. $\dfrac{320\theta\lambda^3\mu\phi^2}{800\theta^2\lambda\mu\phi^3}$ 9. $\dfrac{x^2}{x^3 + xy^2}$

10. $\dfrac{7.5p + 0.5q}{2.5pq}$ 11. $\dfrac{a^2 + 2ab + b^2}{a^2 - b^2}$

12. $\dfrac{4m - 4n}{m^2 - n^2}$ 13. $\dfrac{2x^2 + 5xy + 3y^2}{6x + 9y}$

14. $\dfrac{\alpha^2 + 3\alpha\beta - 10\beta^2}{2\alpha^2 + 11\alpha\beta + 5\beta^2}$ 15. $\dfrac{\pi^2\omega^2 - 9\lambda^2\omega^2}{3\pi^2\omega - 8\pi\lambda\omega - 3\lambda^2\omega}$

11-10 SIGNS OF FRACTIONS

As stated in Sec. 11-6, a fraction is an indicated division or an indicated quotient. Heretofore, all our fractions have been positive, but now we must take into account three signs in working with an algebraic fraction: the sign of the numerator, the sign of the denominator, and the sign preceding the fraction. By the law of signs in division, we have

$$+ \frac{+12}{+6} = + \frac{-12}{-6} = - \frac{+12}{-6} = - \frac{-12}{+6} = +2$$

or, in general,

$$+ \frac{+a}{+b} = + \frac{-a}{-b} = - \frac{+a}{-b} = - \frac{-a}{+b}$$

Careful study of the above examples will show the truths of the following important principles:

1.   The sign before either term of a fraction can be changed if the sign before the fraction is changed.

2. If the signs of both terms are changed, the sign before the fraction must not be changed.

That is, we can change *any two* of the three signs of a fraction without changing the value of the fraction.

It must be remembered that, when a term of a fraction is a polynomial, changing the sign of the term involves changing the sign of *each term of the polynomial*.

Changing the signs of both numerator and denominator, as mentioned in the second principle above, can be explained by considering both terms as multiplied or divided by −1, which, as previously explained, does not change the value of the fraction.

Multiplying (or dividing) a quantity by −1 twice does not change the value of the quantity. Hence, multiplying each of the two factors of a product by −1 does not change the value of the product. Thus,

$$(a - 4)(a - 8) = (-a + 4)(-a + 8)$$
$$= (4 - a)(8 - a)$$

Also,

$$(a - b)(c - d)(e - f) = (b - a)(d - c)(e - f)$$

The validity of these illustrations should be checked by multiplication.

Example 12

Change $-\dfrac{a}{b}$ to three quivalent fractions having different signs.

Solution

$$-\frac{a}{b} = \frac{-a}{b} = \frac{a}{-b} = -\frac{-a}{-b}$$

Example 13

Change $\dfrac{a - b}{c - d}$ to three equivalent fractions having different signs.

Solution

$$\frac{a - b}{c - d} = \frac{-a + b}{-c + d} = -\frac{-a + b}{c - d} = -\frac{a - b}{-c + d}$$

Example 14

Change $\dfrac{a - b}{c - d}$ to a fraction whose denominator is $d - c$.

Solution

$$\frac{a - b}{c - d} = \frac{-a + b}{-c + d} = \frac{b - a}{d - c}$$

PROBLEMS 11-5

Express as fractions with positive numerators:

1. $-\dfrac{-a}{x}$

2. $\dfrac{-IR}{V_1 - v}$

3. $\dfrac{-2\pi fL}{X_L - X_C}$

4. $-\dfrac{\sqrt{L_1L_2}}{\omega L}$ 5. $\dfrac{-\omega L}{R_1 - R_2}$ 6. $\dfrac{-\pi - \omega}{\alpha - \beta}$

Express as fractions with positive denominators:

7. $\dfrac{IR}{-V - v}$ 8. $\dfrac{\mu V_g}{-(R_p + R_L)}$

9. $\dfrac{\pi R^2}{-(A_1 - A_2)}$ 10. $-\dfrac{\theta + \phi}{-2\lambda^2}$

Reduce to lowest terms:

11. $\dfrac{a - b}{b - a}$ 12. $\dfrac{I - i}{-(i^2 - I^2)}$

13. $\dfrac{\theta - \phi}{\phi^2 - \theta^2}$ 14. $\dfrac{x^2 - 2xy + y^2}{y^2 - 2yx + x^2}$

15. $\dfrac{\pi^2 - 8\pi + 16}{20 - \pi - \pi^2}$ 16. $\dfrac{-4s^2t^2 + 3stv - v^2}{2s^2t^2 + stv - v^2}$

11-11 COMMON ERRORS IN WORKING WITH FRACTIONS

It has been demonstrated that a fraction may be reduced to lower terms by dividing both numerator and denominator by the same number (Sec. 11-9). Mistakes are often made by canceling parts of numerator and denominator that are not factors. For example,

$$\frac{5 + 2}{7 + 2} = \frac{7}{9}$$

Here is a case in which both terms of the fraction are polynomials and the terms, even if alike, can never be canceled. Thus,

$$\frac{5 + 2}{7 + 2} \neq \frac{5}{7}$$

because canceling terms has changed the value of the fraction. Similarly, it would be incorrect to cancel the x's in the fraction $\dfrac{6a - x}{6b - x}$, for the x's are not factors. At the same time, it is incorrect to cancel the 6's because, although they are factors of terms in the fraction, they are not factors of the complete numerator and denominator. Therefore, it is apparent that $\dfrac{6a - x}{6b - x}$ cannot be reduced to lower terms, for neither term (numerator or denominator) can be factored.

It is permissible to cancel x's in the fraction $\dfrac{6x}{ax + 5x}$, because each term of the denominator contains the common factor x. The denominator may be factored to give $\dfrac{6x}{x(a + 5)}$, the result being that x is a factor in both terms of the fraction. Note, however, that the single x in the numerator cancels both x's in the denominator.

Thus, we cannot remove, or cancel, like *terms* from the numerator and denominator of a fraction. Only like *factors* can be removed, or canceled.

Another important fact to be remembered is that adding the same number to or subtracting the same number from both numerator and denominator changes the value of the fraction. That is,

$$\frac{3}{4} \neq \frac{3+2}{4+2} \qquad \text{because the latter equals } \frac{5}{6}$$

Likewise,

$$\frac{3}{4} \neq \frac{3-2}{4-2} \qquad \text{because the latter equals } \frac{1}{2}$$

Similarly, squaring or extracting the same root of numerator and denominator results in a different value. For example,

$$\frac{3}{4} \neq \frac{3^2}{4^2} \qquad \text{because the latter equals } \frac{9}{16}$$

Likewise,

$$\frac{16}{25} \neq \frac{\sqrt{16}}{\sqrt{25}} \qquad \text{because the latter equals } \frac{4}{5}$$

Students sometimes thoughtlessly make the error of writing 0 (zero) as the result of the cancellation of all factors. For example,

$$\frac{4x^2y(a+b)}{4x^2y(a+b)} = 1, \, not \, 0$$

Another serious, although common, mistake is forgetting that the fraction bar, or vinculum, is a sign of grouping, so that $-\dfrac{x-y}{x}$ really means $-\left(\dfrac{x-y}{x}\right)$, or $-\left(\dfrac{x}{x}-\dfrac{y}{x}\right)$, or $-\left(1-\dfrac{y}{x}\right)$, and it does not reduce to $-(1-y)$.

Note that the *vinculum* is a sign of grouping and, when a minus sign precedes a fraction having a polynomial numerator, all the signs of the numerator must be changed in order to complete the process of subtraction.

Thus, $-\dfrac{x-y}{x}$ simplifies to $\dfrac{y}{x} - 1$.

11-12 CHANGING MIXED EXPRESSIONS TO FRACTIONS

In arithmetic, an expression such as $3\frac{1}{3}$ is called a *mixed number;* $3\frac{1}{3}$ means $3 + \frac{1}{3}$. Similarly, in algebra, an expression such as $x + \dfrac{y}{z}$ is called a *mixed expression.* Because

$$4\frac{2}{3} = 4 + \frac{2}{3} = \frac{4}{1} + \frac{2}{3} = \frac{12}{3} + \frac{2}{3} = \frac{14}{3}$$

then, $\qquad x + \dfrac{y}{z} = \dfrac{x}{1} + \dfrac{y}{z} = \dfrac{xz}{z} + \dfrac{y}{z} = \dfrac{xz + y}{z}$

Also, $\quad 3x^2 - 4x + \dfrac{3}{x^2 - 1} = \dfrac{3x^2}{1} - \dfrac{4x}{1} + \dfrac{3}{x^2 - 1}$

$$= \dfrac{3x^2(x^2 - 1)}{x^2 - 1} - \dfrac{4x(x^2 - 1)}{x^2 - 1} + \dfrac{3}{x^2 - 1}$$

$$= \dfrac{3x^4 - 3x^2 - 4x^3 + 4x + 3}{x^2 - 1}$$

11-13 REDUCTION OF A FRACTION TO A MIXED EXPRESSION

As would be expected, reducing a fraction to a mixed expression is the reverse of changing a mixed expression to a fraction. That is, a fraction can be changed to a mixed expression by dividing the numerator by the denominator and adding to the quotient thus obtained the remainder, which is written as a fraction.

Example 15

Change $\dfrac{12x^3 + 16x^2 - 8x - 3}{4x}$ to a mixed expression.

Solution

Divide each term of the numerator by the denominator. Thus,

$$\dfrac{12x^3 + 16x^2 - 8x - 3}{4x} = 3x^2 + 4x - 2 - \dfrac{3}{4x}$$

Example 16

Change $\dfrac{a^2 + 1}{a - 2}$ to a mixed expression.

Solution

By division,

$$
\begin{array}{r}
a^2 + 1 \,\big|\, \underline{a - 2} \\
 a + 2 \\
\underline{a^2 - 2a} \\
2a + 1 \\
\underline{2a - 4} \\
5
\end{array}
$$

$$\therefore \dfrac{a^2 + 1}{a - 2} = a + 2 + \dfrac{5}{a - 2}$$

PROBLEMS 11-6

Change the following mixed expressions to fractions:

1. $2\dfrac{1}{8}$

2. $5\dfrac{3}{16}$

3. $a + \dfrac{b}{c}$

4. $R - \dfrac{V}{I}$

5. $4 - \dfrac{5}{F}$

6. $\dfrac{3}{Q} - \dfrac{5}{Q^2}$

7. $4 + \dfrac{2}{\pi + 1}$

8. $\theta + \dfrac{\theta}{2\pi}$

9. $R - 1 - \dfrac{V}{I}$

10. $5 + \dfrac{5x - 30}{x^2 - 2x}$ 11. $\dfrac{9}{x^2} - \dfrac{14}{2x} - 2$ 12. $4 - \dfrac{4}{c} - \dfrac{8}{c^2}$

13. $1 + \dfrac{6}{R} - \dfrac{7}{R^2}$ 14. $\dfrac{a + b}{4} - \dfrac{a - b}{8}$

15. $1 - \dfrac{4\lambda + 1}{9\lambda^2 - 1}$ 16. $\dfrac{x - 1}{2x} - \dfrac{x^2 - 1}{3x^2}$

17. $\dfrac{45}{\theta^2} + \dfrac{14}{\theta} - \dfrac{\theta + 1}{\theta - 1}$ 18. $2 - \dfrac{12Q - 2}{Q^2 - 1}$

19. $2\alpha^2 - 1 - \dfrac{4}{\alpha^2 - 3}$ 20. $1 - \dfrac{50\omega\pi - 30\pi^2}{(5\omega - 3\pi)(3\omega + 5\pi)}$

Reduce the following fractions to mixed expressions:

21. $\dfrac{83}{16}$ 22. $\dfrac{231}{32}$

23. $\dfrac{x^2 + y^2}{x^2}$ 24. $\dfrac{32\alpha^2 - 16\alpha + 4}{4\alpha}$

25. $\dfrac{R^3 + 6R^2 + 7R - 8}{R - 1}$ 26. $\dfrac{x^2 + 5x + 6}{x - 1}$

27. $\dfrac{E^4 - e^4 - 1}{E + e}$

28. $\dfrac{6\phi^5 - \phi^4 + 4\phi^3 - 5\phi^2 - \phi + 20}{2\phi^2 - \phi + 3}$

29. $\dfrac{2x^3 + 2x^2 + x + 2}{x^2 + 1}$

30. $\dfrac{2\alpha^3 + \alpha\beta^2}{\alpha + \beta}$

11-14 REDUCTION TO THE LOWEST COMMON DENOMINATOR

The *lowest common denominator* (LCD) of two or more fractions is the lowest common multiple of their denominators.

Example 17

Reduce $\frac{1}{3}$ and $\frac{3}{5}$ to their LCD.

Solution

The LCM of 3 and 5 is 15. To change the denominator of $\frac{1}{3}$ to 15, we must multiply the 3 by 5 (15 ÷ 3). So that the value of the fraction will not be changed, we must also multiply the numerator by 5. Hence,

$$\frac{1}{3} = \frac{1}{3} \times \frac{5}{5} = \frac{5}{15}$$

For the second fraction, we must multiply the denominator by 3 in order to obtain a new denominator of 15 (15 ÷ 5). Again we must also multiply the numerator by 3 to maintain the original value of the fraction. Hence,

$$\frac{3}{5} = \frac{3}{5} \times \frac{3}{3} = \frac{9}{15}$$

Example 18

Reduce $\dfrac{4a^2b}{3x^2y}$ and $\dfrac{6cd^2}{4xy^2}$ to their LCD.

Solution

The LCM of the two denominators is $12x^2y^2$. This is the LCD.

For the first fraction the LCD is divided by the denominator. That is,

$$12x^2y^2 \div 3x^2y = 4y$$

Multiplying both numerator and denominator by $4y$, we have

$$\frac{4a^2b}{3x^2y} = \frac{4a^2b}{3x^2y} \cdot \frac{4y}{4y} = \frac{16a^2by}{12x^2y^2}$$

For the second fraction we follow the same procedure.

$$12x^2y^2 \div 4xy^2 = 3x$$

Multiplying both numerator and denominator by $3x$, we have

$$\frac{6cd^2}{4xy^2} = \frac{6cd^2}{4xy^2} \cdot \frac{3x}{3x} = \frac{18cd^2x}{12x^2y^2}$$

RULE

To reduce fractions to their LCD:

1. Factor each denominator into its prime factors and find the LCM of the denominators. This is the LCD.
2. For each fraction, divide the LCD by the denominator and multiply both numerator and denominator by the quotient thus obtained.

Example 19

Reduce $\dfrac{3x}{x^2 - y^2}$ and $\dfrac{4y}{x^2 - xy - 2y^2}$ to their LCD.

Solution

$$\frac{3x}{x^2 - y^2} = \frac{3x}{(x + y)(x - y)}$$
$$\frac{4y}{x^2 - xy - 2y^2} = \frac{4y}{(x + y)(x - 2y)}$$

The LCM of the two denominators, and therefore the LCD, is $(x + y)(x - y)(x - 2y)$.

For the first fraction, the LCD divided by the denominator is:

$$(x + y)(x - y)(x - 2y) \div (x + y)(x - y) = x - 2y.$$

$$\therefore \frac{3x}{(x + y)(x - y)} = \frac{3x(x - 2y)}{(x + y)(x - y)(x - 2y)}$$

For the second fraction, the LCD divided by the denominator is:

$$(x + y)(x - y)(x - 2y) \div (x + y)(x - 2y) = x - y.$$

$$\therefore \frac{4y}{(x + y)(x - 2y)} = \frac{4y(x - y)}{(x + y)(x - 2y)(x - y)}$$

To check the solution, the fractions having the LCD can be changed into the original fractions by cancellation.

PROBLEMS 11-7

Convert the following sets of fractions to equivalent sets having their LCD:

1. $\dfrac{1}{2}, \dfrac{3}{7}, \dfrac{2}{5}$

2. $\dfrac{3}{16}, \dfrac{5}{8}, \dfrac{7}{32}$

3. $\dfrac{3}{4}, \dfrac{7}{16}, \dfrac{5}{12}$

4. $\dfrac{1}{x}, \dfrac{1}{y}$

5. $\dfrac{\theta}{\phi}, \dfrac{\lambda}{\omega}$

6. $\dfrac{1}{ir}, \dfrac{1}{\omega}, \dfrac{i}{\omega}$

7. $\dfrac{e}{r}, \dfrac{1}{ir}, ei$

8. $\dfrac{Q}{L_1}, \dfrac{1}{L_2}, \dfrac{\sqrt{L_1 L_2}}{M}$

9. $\dfrac{1}{a-b}, \dfrac{1}{a+b}$

10. $\dfrac{x}{y}, \dfrac{2x+y}{x-y}$

11. $\dfrac{3}{\phi-\pi}, \dfrac{4}{\phi+\pi}$

12. $\dfrac{3\phi}{1-\phi^2}, \dfrac{2}{\phi+1}, \dfrac{2}{1-\phi}$

13. $\dfrac{a}{c+d}, \dfrac{b}{c-d}, \dfrac{a-b}{d-c}$

14. $\dfrac{1}{2M+2}, \dfrac{5}{3M-3}, \dfrac{3M-1}{1-M^2}$

15. $\dfrac{\pi^2-\phi^2}{\pi\phi}, \dfrac{\pi\phi-\phi^2}{\pi\phi-\pi^2}$

16. $\dfrac{R+3Z}{4R^2+12RZ+8Z^2}, \dfrac{R+Z}{4R^2+20RZ+24Z^2}, \dfrac{R+2Z}{R^2+4RZ+3Z^2}$

11-15 ADDITION AND SUBTRACTION OF FRACTIONS

The sum of two or more fractions having the same denominator is obtained by adding the numerators and writing the result over the common denominator.

Example 20

$$\frac{2}{7}+\frac{1}{7}+\frac{5}{7} = \frac{2+1+5}{7}$$
$$= \frac{8}{7}$$

Example 21

$$\frac{3v}{R+r}+\frac{v}{R+r}+\frac{5v}{R+r} = \frac{3v+v+5v}{R+r}$$
$$= \frac{9v}{R+r}$$

To subtract two fractions having the same denominator, subtract the numerator of the subtrahend from the numerator of the minuend and write the result over their common denominator.

Example 22

$$\frac{4}{5}-\frac{3}{5} = \frac{4-3}{5} = \frac{1}{5}$$

Example 23

$$\frac{a}{x} - \frac{b}{x} = \frac{a - b}{x}$$

Example 24

$$\frac{a}{x} - \frac{b - c}{x} = \frac{a - b + c}{x}$$

Note that *the vinculum is a sign of grouping* and that, when a minus sign precedes a fraction having a polynomial numerator, all the signs in the numerator must be changed in order to complete the process of subtraction.

We thus have the following rule:

RULE
To add or subtract fractions having unlike denominators:

1. Reduce them to equivalent fractions having their LCD.
2. Combine the numerators of these equivalent fractions, in parentheses, and give each the sign of the fraction. This is the numerator of the result.
3. The denominator of the result is the LCD.
4. Simplify the numerator by removing parentheses and combining terms.
5. Reduce the fraction to the lowest terms.

Example 25

Simplify $\dfrac{a - 5}{6x} - \dfrac{2a - 5}{16x}$.

Solution

$$\frac{a - 5}{6x} - \frac{2a - 5}{16x} = \frac{8(a - 5)}{48x} - \frac{3(2a - 5)}{48x}$$

$$= \frac{8(a - 5) - 3(2a - 5)}{48x}$$

$$= \frac{8a - 40 - 6a + 15}{48x}$$

$$= \frac{2a - 25}{48x}$$

Check

Let $a = 6$, $x = 1$.

$$\frac{a - 5}{6x} = \frac{1}{6} \qquad \frac{2a - 5}{16} = \frac{7}{16}$$

$$\frac{1}{6} - \frac{7}{16} = \frac{8 - 21}{48} = -\frac{13}{48}$$

Also, $$\frac{2a - 25}{48} = \frac{12 - 25}{48} = -\frac{13}{48}$$

Solution is correct.

Example 26

Simplify $x^2 - xy + y^2 - \dfrac{2y^3}{x + y}$.

Solution

$$x^2 - xy + y^2 - \frac{2y^3}{x + y}$$

$$= \frac{(x + y)x^2}{x + y} - \frac{(x + y)xy}{x + y} + \frac{(x + y)y^2}{x + y} - \frac{2y^3}{x + y}$$

$$= \frac{x^3 + x^2y - x^2y - xy^2 + xy^2 + y^3 - 2y^3}{x + y}$$

$$= \frac{x^3 - y^3}{x + y}$$

PROBLEMS 11-8

Perform the following indicated additions and subtractions:

1. $\dfrac{1}{2} + \dfrac{2}{5} - \dfrac{3}{7}$

2. $\dfrac{7}{32} - \dfrac{5}{8} + \dfrac{3}{16}$

3. $\dfrac{3}{4} - \dfrac{7}{16} - \dfrac{5}{12}$

4. $\dfrac{5a}{4} - \dfrac{a}{5} + \dfrac{7a}{3}$

5. $\dfrac{7IR}{8} + \dfrac{2IR}{3} - \dfrac{3IR}{16}$

6. $\dfrac{1}{I} + \dfrac{1}{i}$

7. $\dfrac{\alpha}{\beta} - \dfrac{\gamma}{\delta}$

8. $\dfrac{3p}{4q} - \dfrac{p}{6q} - \dfrac{5p}{30q}$

9. $\dfrac{a}{x} - \dfrac{b}{y} - \dfrac{c}{z}$

10. $\dfrac{3}{r_1} + \dfrac{2}{r_2} + \dfrac{5}{r_1 r_2}$

11. $\dfrac{10}{I^2} - \dfrac{3}{R} + \dfrac{4}{I^2 R}$

12. $\dfrac{3\alpha}{\phi\lambda} + \dfrac{2\phi}{\alpha\lambda} + \dfrac{6\lambda}{\alpha\phi}$

13. $\dfrac{3I - i}{2} + \dfrac{5I + 2i}{3}$

14. $\dfrac{a + 4}{7} - \dfrac{a - 1}{3}$

15. $\dfrac{2}{\alpha - \beta} + \dfrac{1}{\alpha + \beta}$

16. $\dfrac{3}{2e + 4} + \dfrac{5}{e + 2}$

17. $\dfrac{5}{L_1 - 2} - \dfrac{2}{L_1 + 6}$

18. $\dfrac{a}{c + d} + \dfrac{b}{c - d} - \dfrac{a - b}{d - c}$

19. $\dfrac{1}{2\theta + 2} - \dfrac{5}{3\theta - 3} + \dfrac{3\theta - 1}{1 - \theta^2}$

20. $\dfrac{8}{\alpha^2 - 9} - \dfrac{2}{\alpha^2 - 5\alpha + 6}$

21. $\dfrac{2}{I^2 + 7I} - \dfrac{3}{I} + \dfrac{3}{I - 7}$

22. $\dfrac{11R_1 - 2}{3R_1^2 - 3} - \dfrac{5R_1 + 1}{2R_1^2 - 2}$

23. $\dfrac{21}{14 - \pi} - \dfrac{35 - 2\pi^2}{\pi^2 - 11\pi - 42}$

24. $\dfrac{2L - 4M}{2L - 2M} - \dfrac{3M^2 - 3LM}{L^2 - 2LM + M^2}$

25. $\dfrac{\theta + \phi}{\theta - \phi} - \dfrac{\theta - \phi}{\theta + \phi} + \dfrac{4\theta\phi}{\theta^2 - \phi^2}$

26. $\dfrac{2X_C}{2X_C + 3X_L} - \dfrac{3X_L}{2X_C - 3X_L} + \dfrac{8X_L^2}{4X_C^2 - 9X_L^2}$

27. $\dfrac{E - 1}{E^2 - 9E + 20} - \dfrac{E + 1}{E^2 - 11E + 30}$

28. $a + b - \dfrac{a^2 - b^2}{a - b} + 1$

29. $\dfrac{\omega^2 + 3\omega + 9}{\omega^2 - 3\omega + 9} - \dfrac{54}{\omega^3 + 27} - \dfrac{\omega - 3}{\omega + 3}$

30. $\dfrac{\theta + 3\pi}{4\theta^2 + 12\theta\pi + 8\pi^2} + \dfrac{\theta + 2\pi}{\theta^2 + 4\theta\pi + 3\pi^2} - \dfrac{\theta + \pi}{4\theta^2 + 200\theta\pi + 24\pi^2}$

11-16 MULTIPLICATION OF FRACTIONS

The methods of multiplication of fractions in algebra are identical with those in arithmetic.

The product of two or more fractions is the product of the numerators of the fractions divided by the product of the denominators.

Example 27

$$\frac{2}{3} \times \frac{3}{5} = \frac{6}{15}$$

Example 28

$$\frac{a}{b} \cdot \frac{x}{y} = \frac{ax}{by}$$

When a factor occurs one or more times in *any* numerator and in *any* denominator of the product of two or more fractions, it can be canceled the same number of times from both. This process results in the product of the given fractions in lower terms.

Example 29

Multiply $\dfrac{6x^2y}{7b}$ by $\dfrac{21b^2c}{24xy^2}$.

Solution

$$\frac{6x^2y}{7b} \cdot \frac{21b^2c}{24xy^2} = \frac{3bcx}{4y}$$

Example 30

Simplify

$$\frac{2a^2 - ab - b^2}{a^2 + 2ab + b^2} \cdot \frac{a^2 - b^2}{4a^2 + 4ab + b^2}$$

Solution

$$\frac{2a^2 - ab - b^2}{a^2 + 2ab + b^2} \cdot \frac{a^2 - b^2}{4a^2 + 4ab + b^2}$$

$$= \frac{(2a + b)(a - b)}{(a + b)(a + b)} \cdot \frac{(a + b)(a - b)}{(2a + b)(2a + b)}$$

$$= \frac{(a - b)(a - b)}{(a + b)(2a + b)} = \frac{a^2 - 2ab + b^2}{2a^2 + 3ab + b^2}$$

It is very important that you understand clearly what we are allowed to cancel in the numerators and the denominators. The *whole* of an expression is always canceled, *never one term*. For example, in the expression $\frac{8a}{a-5}$, it is not permissible to cancel the a's and obtain $\frac{8}{-5}$. It must be remembered that the denominator $a - 5$ denotes *one quantity*. Because of the parentheses, we would not cancel the a's if the expression were written $\frac{8a}{(a-5)}$. However, the parentheses are not needed; for the *vinculum, which is also a sign of grouping, serves the same purpose.* We will consider this again in the next chapter.

11-17 DIVISION OF FRACTIONS

As with multiplication, the methods of division of fractions in algebra are identical with those of arithmetic. Therefore, to divide by a fraction, invert the divisor fraction and proceed as in the multiplication of fractions.

Example 31

$$\frac{5}{2} \div \frac{2}{3} = \frac{5}{2} \cdot \frac{3}{2} = \frac{15}{4}$$

Example 32

$$\frac{ab^2}{xy} \div \frac{a^2b}{xy^2} = \frac{ab^2}{xy} \cdot \frac{xy^2}{a^2b}$$
$$= \frac{by}{a}$$

Example 33

$$\frac{x}{y} \div \left(a + \frac{b}{c} \right) = \frac{x}{y} \div \frac{ac+b}{c}$$
$$= \frac{x}{y} \cdot \frac{c}{ac+b} = \frac{cx}{y(ac+b)} = \frac{cx}{acy+by}$$

Students often ask why we must invert the divisor and multiply by the dividend in dividing fractions. As an example, suppose we have $\frac{a}{b} \div \frac{x}{y}$. The dividend is $\frac{a}{b}$, and the divisor is $\frac{x}{y}$. Now

$$\text{Quotient} \times \text{divisor} = \text{dividend}$$

Therefore, the quotient must be a number such that, when multiplied by $\frac{x}{y}$, it will give $\frac{a}{b}$ as a product. Then,

$$\left(\frac{a}{b} \cdot \frac{y}{x} \right) \cdot \frac{x}{y} = \frac{a}{b}$$

Hence, the quotient is $\frac{a}{b} \cdot \frac{y}{x}$, which is the dividend multiplied by the inverted divisor.

PROBLEMS 11-9

Simplify:

1. $\dfrac{2}{3} \times \dfrac{5}{7} \times \dfrac{21}{40}$

2. $\dfrac{12}{35} \times \dfrac{5}{24} \times \dfrac{42}{21}$

3. $\dfrac{5}{16} \times \dfrac{6}{25} \times \left(\dfrac{-4}{15}\right)$

4. $\dfrac{5}{9} \div \dfrac{15}{18}$

5. $\dfrac{7}{8} \div \dfrac{7}{32}$

6. $-\dfrac{2}{3}\left(-\dfrac{5}{16} \div \dfrac{15}{64}\right)$

7. $\dfrac{4x^3}{5y} \times \dfrac{15y^4}{3x^2}$

8. $3p\left(\dfrac{5r}{6p^2} \times \dfrac{7pr}{15}\right)$

9. $\dfrac{400\theta\phi^2\omega}{21\theta^2\phi^3\omega^2} \div \dfrac{100\theta^3\phi\omega^2}{21\theta\phi^2\omega}$

10. $\dfrac{\pi r^2 h}{3} \div 2\pi r$

11. $\dfrac{\omega L}{R} \div 2\pi f L$

12. $\left(\dfrac{m^2 + 4m}{m}\right)\left(\dfrac{m^2}{m^3 + 4m^2}\right)$

13. $\dfrac{4}{x - y} \div \dfrac{x^2 - y^2}{x^2 + 2xy + y^2}$

14. $\dfrac{4\theta^2 - 1}{\theta^3 - 16\theta} \div \dfrac{2\theta - 1}{\theta - 4}$

15. $\dfrac{25x^2 - y^2}{9x^2z - 4z} \div \dfrac{5x - y}{3xz - 2z}$

16. $\dfrac{I^2 - 4i^2}{Ii + 2i^2} \cdot \dfrac{2i}{I - 2i}$

17. $\dfrac{\lambda^2 - 2\lambda\mu + \mu^2}{4\lambda - 4\mu} \cdot \dfrac{4\lambda + 4\mu}{\phi^3 - 3\phi^2 + 2\phi} \cdot \dfrac{\phi^3 - \phi^2}{\phi\lambda^2 - \phi\mu^2}$

18. $\dfrac{F^2 + 2F + 1}{P^3 - PZ^2} \cdot \dfrac{P^2 - Z^2}{5F^3 + 10F^2 + 5F} \cdot \dfrac{F^2P - 10FP + 25P}{F^2 - 110F + 525}$

19. $\dfrac{\theta^2 - 2\theta - 3}{-6\phi^2} \cdot \dfrac{5\theta\phi^6 + 25\phi^6}{\theta^2\phi - 8\theta + \theta\phi - 8} \cdot \dfrac{48\phi^2 - 60\phi^3}{5\theta^2\phi^3 + 100\phi^3 - 75\phi^3}$

20. $\dfrac{R^2 - r^2}{r^2 + Rr} \cdot \dfrac{R(R - r)}{(R - r)^2} \div \dfrac{R^2 - 3Rr + 2r^2}{Rr - 2r^2}$

21. $\dfrac{\alpha^2 - 6\alpha + 8}{\alpha^2} \cdot \dfrac{7\alpha^4 + 7\alpha^3}{\alpha^2 - 11\alpha + 28} \div \dfrac{\alpha^2 - \alpha - 2}{2\alpha^2 - 14\alpha}$

22. $\dfrac{16I^4R^2 - 9}{4\left(I^2R + \dfrac{3}{4}\right)} \cdot \dfrac{I^4R^2 - 3I^2R - 28}{2I^4R^2 - 32} \div \dfrac{8I^4R^2 - 62I^2R + 42}{8I^2R - 32}$

23. $\left(4 - \dfrac{4}{c} - \dfrac{8}{c^2}\right)\left(\dfrac{3c^4 - 6c^3}{2c^2 - 2c - 4}\right)\left(\dfrac{2c^2 + 8c}{3c^3 + 6c^2 - 24c}\right)$

24. $\left(m - \dfrac{m^2}{m}\right)\left(\dfrac{m^2 - n^2}{m^2 + mn}\right)\left(\dfrac{m + n}{m^2 + mn}\right)$ 0

25. $\left(\dfrac{5\phi^5 - 5\phi^4}{\phi^2 - \phi - 20}\right)\left(\dfrac{\phi^2 + 11\phi + 28}{5\phi - 5}\right) \div \left(\dfrac{\phi^4 + 9\phi^3 + 14\phi^2}{\phi^2 - 3\phi - 10}\right)$ ϕ^2

26. $\left(\dfrac{I^2 + I - 6}{I^4 - 9I^2}\right)\left(I^2 + 4I + \dfrac{12I}{I - 3}\right) \div \left(\dfrac{I^2 - I - 2}{I^2 - 6I + 9}\right)$ $($

27. $\left(\dfrac{\omega L + R}{2} + \dfrac{\omega L - R}{4}\right)\left(\dfrac{4}{9\omega^2 L^2 + 6\omega LR + R^2}\right)$ $\dfrac{1}{3\omega\lambda + R}$

28. $\left(\dfrac{45}{\theta^2} + \dfrac{14}{\theta} + 1\right)\left(\dfrac{3\theta^3 + 60^2}{\theta^2 + 18\theta + 81}\right)\left(\dfrac{\theta^2 + 13\theta + 36}{\theta^2 + 9\theta + 20}\right)\left(\dfrac{1}{3\theta + 3}\right)$

29. $\left(\dfrac{6m^2 - 2m}{-9m^2 + 4m + 2}\right)\left(\dfrac{2}{m^2} + \dfrac{10}{m} + 12\right)\left(\dfrac{4m + 1}{9m^2 - 1} - 1\right)\left(\dfrac{m^3}{4m^2 + 2m}\right)$ $2m$

30. $\left(\dfrac{1}{f^2} + \dfrac{2}{f} + 1\right)\left(\dfrac{f^3 - f^2}{f^2 - 5f - 6}\right)\left(2 - \dfrac{12f - 2}{f^2 - 1}\right)$ $= 2f$

11-18 COMPLEX FRACTIONS

A *complex fraction* is one with one or more fractions in its numerator, denominator, or both. The name is an unfortunate one. There is nothing complex or intricate about such compounded fractions, as we shall see.

RULE
To simplify a complex fraction, reduce both numerator and denominator to simple fractions; then perform the indicated division.

Example 34
Simplify $\dfrac{\dfrac{1}{3} + \dfrac{1}{5}}{4 - \dfrac{1}{5}}$.

Solution

$$\frac{\dfrac{1}{3} + \dfrac{1}{5}}{4 - \dfrac{1}{5}} = \frac{\dfrac{5 + 3}{15}}{\dfrac{20 - 1}{5}}$$

$$= \frac{\dfrac{8}{15}}{\dfrac{19}{5}} = \frac{8}{15} \times \frac{5}{19} = \frac{8}{57}$$

Example 35

Simplify $\dfrac{5 - \dfrac{1}{a + 1}}{3 + \dfrac{2}{a + 1}}$.

Solution

$$\frac{5 - \dfrac{1}{a + 1}}{3 + \dfrac{2}{a + 1}} = \frac{\dfrac{5(a + 1) - 1}{a + 1}}{\dfrac{3(a + 1) + 2}{a + 1}}$$

$$= \frac{\dfrac{5a + 4}{a + 1}}{\dfrac{3a + 5}{a + 1}}$$

$$= \frac{5a + 4}{a + 1} \cdot \frac{a + 1}{3a + 5}$$

$$= \frac{5a + 4}{3a + 5}$$

Note

It is evident that if the same factor occurs in both numerators of a complex fraction, the factors can be canceled. Also, if a factor occurs in both denominators, it can be canceled. Thus, $(a + 1)$ could have been canceled in Example 35 after the numerators and denominators were reduced from mixed expressions to simple fractions.

Example 36

Simplify $\dfrac{\dfrac{a}{b} + \dfrac{a + b}{a - b}}{\dfrac{a}{b} - \dfrac{a - b}{a + b}}$.

Solution

$$\frac{\dfrac{a}{b} + \dfrac{a + b}{a - b}}{\dfrac{a}{b} - \dfrac{a - b}{a + b}} = \frac{\dfrac{a(a - b) + b(a + b)}{b(a - b)}}{\dfrac{a(a + b) - b(a - b)}{b(a + b)}}$$

$$= \frac{\dfrac{a^2 - ab + ab + b^2}{b(a - b)}}{\dfrac{a^2 + ab - ab + b^2}{b(a + b)}}$$

$$= \frac{\dfrac{a^2 + b^2}{b(a - b)}}{\dfrac{a^2 + b^2}{b(a + b)}} = \frac{a + b}{a - b}$$

PROBLEMS 11-10

Simplify:

1. $\dfrac{2 + \dfrac{1}{3}}{\dfrac{1}{3} - 3}$

2. $\dfrac{2}{\dfrac{1}{7} + \dfrac{1}{2}}$

3. $\dfrac{\left(\dfrac{5}{8}\right)^2 - \dfrac{16}{25}}{\dfrac{5}{8} + \dfrac{4}{5}}$

4. $\dfrac{\theta + \dfrac{1}{\phi}}{\theta - \dfrac{1}{\phi}}$

5. $\dfrac{Q}{\dfrac{1}{\omega L_1} + \dfrac{1}{\omega L_2}}$

6. $\dfrac{\dfrac{i^2}{8} - 8}{1 + \dfrac{i}{8}}$

7. $\dfrac{I}{I - \dfrac{V}{r}}$

8. $\dfrac{5\theta + \dfrac{2\lambda}{5\phi}}{\dfrac{2\lambda}{5\theta} + 5\phi}$

9. $\dfrac{\dfrac{E^2}{e^2} - 1}{\dfrac{E^2 + e^2}{2Ee} + 1}$

10. $\dfrac{\dfrac{\lambda + \pi}{\lambda^2 + \pi^2} - \dfrac{1}{\lambda + \pi}}{\dfrac{1}{\lambda + \pi} - \dfrac{\lambda}{\lambda^2 + \pi^2}}$

11. $\dfrac{\dfrac{l}{l + w}}{1 + \dfrac{w}{l - w}}$

12. $\dfrac{\omega + 2 - \dfrac{15}{\omega}}{1 - \dfrac{8}{\omega} + \dfrac{15}{\omega^2}}$

13. $\dfrac{1 - \dfrac{a - b}{a + b}}{1 + \dfrac{a - b}{a + b}}$

14. $\dfrac{\dfrac{\theta}{\theta + \phi} - \dfrac{\theta}{\theta - \phi}}{\dfrac{\theta}{\theta + \phi} + \dfrac{\theta}{\theta - \phi}}$

15. $\dfrac{\dfrac{I - i}{I + i} + \dfrac{I + i}{I - i}}{\dfrac{I - i}{I + i} - \dfrac{I + i}{I - i}}$

16. $\dfrac{L_1}{Q - \dfrac{1}{Q + \dfrac{1}{Q}}} - \dfrac{L_1}{Q + \dfrac{1}{Q - \dfrac{1}{Q}}}$

Fractional Equations

An equation containing a fraction in which the unknown occurs in a denominator is called a *fractional equation*. Equations of this type are encountered in many problems involving electric and electronic circuits. Simple fractional equations, wherein the unknown appeared only as a factor, were studied in earlier chapters.

12-1 FRACTIONAL COEFFICIENTS

A number of problems lead to equations containing *fractional coefficients*. This type of equation is included in this chapter because the methods of solution apply to fractional equations also.

Example 1

$$\frac{3x}{4} + \frac{3}{2} = \frac{5x}{8} \quad \text{and} \quad \frac{x}{2} + \frac{x}{3} = 5$$

are equations having fractional coefficients.

Example 2

$$\frac{60}{x} - 3 = \frac{60}{4x} \quad \text{and} \quad \frac{x-2}{x} = \frac{4}{5}$$

are fractional equations.

You are familiar with the methods of solving simple equations that do not contain fractions. An equation involving fractions can be changed to an equation containing no fractions by canceling the denominators and then solved as heretofore. To accomplish this we have the following rule:

RULE

To solve an equation containing fractions:

1. First clear the equation of fractions by multiplying every term by the LCD of the whole equation. (This will permit canceling all denominators.)
2. Solve the resulting equation.

Example 3

Given $\frac{5x}{12} - 13 = \frac{x}{18}$. Solve for x.

Solution

Given

$$\frac{5x}{12} - 13 = \frac{x}{18}$$

M: 36, the LCD,

$$\frac{36 \cdot 5x}{12} - 36 \cdot 13 = \frac{36x}{18}$$

Canceling,

$$\frac{\overset{3}{\cancel{36}} \cdot 5x}{\cancel{12}} - 36 \cdot 13 = \frac{\overset{2}{\cancel{36}}x}{\cancel{18}}$$

Simplifying, $\quad\quad\quad\quad 15x - 468 = 2x$

Collecting terms, $\quad\quad\quad\quad\quad 13x = 468$

D: 13, $\quad\quad\quad\quad\quad\quad\quad\quad x = 36$

Check

Substitute 36 for x in the original equation:

$$\frac{5 \cdot 36}{12} - 13 = \frac{36}{18}$$

Clearing fractions, $\quad\quad 15 - 13 = 2$

$$2 = 2$$

Example 4

Given $\frac{e - 4}{9} = \frac{e}{10}$. Solve for e.

Solution

Given

$$\frac{e - 4}{9} = \frac{e}{10}$$

M: 90, the LCD,

$$\frac{90(e - 4)}{9} = \frac{90e}{10}$$

Canceling,

$$\frac{\overset{10}{\cancel{90}}(e - 4)}{\cancel{9}} = \frac{\overset{9}{\cancel{90}}e}{\cancel{10}}$$

Simplifying, $\quad\quad 10(e - 4) = 9e$

or $\quad\quad\quad\quad 10e - 40 = 9e$

Collecting terms, $\quad 10e - 9e = 40$

or $\quad\quad\quad\quad\quad\quad\quad e = 40$

Check

Substitute 40 for e in the original equation:

$$\frac{40 - 4}{9} = \frac{40}{10}$$

Clearing fractions, $\quad\quad\quad\quad 4 = 4$

Note that when the fractions were cleared and the equation was written in simplified form in the above solution, the resulting equation was

$$10(e - 4) = 9e$$

which is equivalent to multiplying each member by the denominator of the other member and expressing the resulting equation with no denominators. This is called *cross multiplication*. You will see the justification of this if each member is expressed as a fraction having the LCD. Although the method is convenient, it must be remembered that *cross multiplication is permissible only when each term of an equation has the same denominator*.

Problems 12-1
Solve the following equations:

1. $\dfrac{\phi}{2} - \dfrac{\phi}{4} = 2$

2. $\dfrac{x}{3} = \dfrac{x}{6} + 4$

3. $\dfrac{3\alpha}{2} + \dfrac{\alpha}{4} = 10 + \dfrac{\alpha}{2}$

4. $I - \dfrac{1}{4} = \dfrac{2I}{5} - \dfrac{1}{16}$

5. $\omega - \dfrac{4\omega}{7} = 2\omega - \dfrac{11}{16}$

6. $\dfrac{1}{3} + \dfrac{Z}{5} = \dfrac{Z}{3}$

7. $\dfrac{6 + 3\phi}{4} + \dfrac{12 - 2\phi}{15} = \dfrac{6\phi}{5} - \dfrac{37}{60}$

8. $\dfrac{F}{6} + \dfrac{F - 3}{18} = \dfrac{3 + 3F}{12}$

Note If a fraction is negative, the sign of each term of the numerator must be changed after removing the denominator. (See Sec. 11-10.) Remember that *the vinculum is a sign of grouping*.

9. $3 - \dfrac{1 + \lambda}{2} = \dfrac{2\lambda - 3}{3}$

10. $\dfrac{4I + 3}{5} - \dfrac{I - 5}{10} = \dfrac{I}{3}$

11. $\dfrac{\omega + 2}{2} - \dfrac{\omega - 3}{3} = 0$

12. $x - \dfrac{3 + 4x}{5} + \dfrac{2x - 3}{6} - \dfrac{5x - 4}{15} = 0$

13. $\dfrac{1}{16}(3\theta - 10) - \dfrac{1}{8}(5\theta - 6) = \dfrac{1}{2}(7\theta + 16)$

Note $\dfrac{1}{16}(3\theta - 10) = \dfrac{3\theta - 10}{16}$

14. $\dfrac{2}{3}(z + 1) - \dfrac{3}{4}(z + 2) = \dfrac{1}{6}(z + 1)$

15. $\dfrac{1}{2}\left(\dfrac{5}{16} + \dfrac{1}{4}m\right) + 3 = \dfrac{1}{8}\left(3m - \dfrac{1}{3}\right)$

12-2 EQUATIONS CONTAINING DECIMALS

An equation containing decimals is readily solved by first clearing the equation of the decimals. This is accomplished by multiplying both members by a power of 10 that corresponds to the largest number of decimal places appearing in any term.

Example 5
Solve $0.75 - 0.7a = 0.26$.

Solution

Given	$0.75 - 0.7a = 0.26$
M: 100,	$75 - 70a = 26$
Collecting terms,	$70a = 49$
D: 70,	$a = 0.7$

Check
Substitute 0.7 for a in the original equation:

$$0.75 - 0.7 \cdot 0.7 = 0.26$$
$$0.75 - 0.49 = 0.26$$
$$0.26 = 0.26$$

If decimals occur in any denominator, multiply both numerator and denominator of the fraction by a power of 10 that will reduce the decimals to integers.

Example 6
Solve $\dfrac{5m - 1.33}{0.02} - \dfrac{m}{0.05} = 1083.5$.

Solution

Given $\dfrac{5m - 1.33}{0.02} - \dfrac{m}{0.05} = 1083.5$. Multiplying numerator and denominator of each fraction by 100,

$$\frac{500m - 133}{2} - \frac{100m}{5} = 1083.5$$

The equation is then solved and checked by the usual methods.

PROBLEMS 12-2
Solve the following equations:

1. $0.4Q = 16$
2. $0.05e = 0.20$
3. $0.8\theta = 1.6 + 0.4\theta$
4. $0.125x - 0.02 = 0.035x + 0.025$
5. $0.3r + 4 = 0.7r - 8$
6. $\phi + 2.6 - 0.2\phi = 1.4 + 0.3\phi$
7. $16.5 - 1.5(2R - 0.5) - 15.6 + 2.1(R + 0.3) = 0.03$
8. $0.2 - 0.5(V - 2) - V = 18.7 + 0.8(V + 4)$

9. $\dfrac{0.5b}{6} - \dfrac{0.2b - 0.5}{30} = \dfrac{0.3b + 0.3}{15}$

10. $\dfrac{0.5(\theta - 5)}{3.75} = \dfrac{0.3(\theta + 5)}{7.5} - \dfrac{0.2(3\theta - 2)}{5}$

11. $\dfrac{1.3a - 1.5}{30} = \dfrac{0.4a + 0.3}{5}$

12. $\dfrac{0.8r_i - 0.1}{3} - \dfrac{0.2r_i - 0.5}{5} + \dfrac{0.6r_i + 1.5}{15} - 0.25r_i = 2.75$

13. $\dfrac{\lambda - 2}{0.05} - 70 = \dfrac{\lambda - 4}{0.08}$

14. $\dfrac{0.2(\omega - 1)}{0.5(\omega + 5)} - \dfrac{0.3(1 - \omega)}{0.7(\omega + 5)} - \dfrac{29}{140} = 0$

15. $(0.7\alpha - 0.7)(0.2 + \alpha) = (1 - 1.4\alpha)(0.1 - 0.5\alpha)$

12-3 FRACTIONAL EQUATIONS

Fractional equations are solved in the same manner as equations containing fractional coefficients (Sec. 12-1). That is, every term of the equation must be multiplied by the LCD.

Example 7

Solve $\dfrac{x + 2}{3x} - \dfrac{2x^2 + 3}{6x^2} = \dfrac{1}{2x}$.

Solution

Given
$$\frac{x + 2}{3x} - \frac{2x^2 + 3}{6x^2} = \frac{1}{2x}$$

M: $6x^2$, the LCD,
$$\frac{6x^2(x + 2)}{3x} - \frac{6x^2(2x^2 + 3)}{6x^2} = \frac{6x^2}{2x}$$

Canceling,
$$\frac{\overset{2x}{\cancel{6x^2}(x + 2)}}{\cancel{3x}} - \frac{\cancel{6x^2}(2x^2 + 3)}{\cancel{6x^2}} = \frac{\overset{3x}{\cancel{6x^2}}}{\cancel{2x}}$$

Rewriting,
$$2x(x + 2) - (2x^2 + 3) = 3x$$

Simplifying,
$$2x^2 + 4x - 2x^2 - 3 = 3x$$

Collecting terms,
$$4x - 3x = 3$$

or
$$x = 3$$

Check

Substituting 3 for x in the original equation,
$$\frac{3 + 2}{9} - \frac{18 + 3}{54} = \frac{1}{6}$$

That is,
$$\frac{30}{54} - \frac{21}{54} = \frac{9}{54}$$

Example 8

Solve $\dfrac{8a + 2}{a - 2} - \dfrac{2a - 1}{3a - 6} + \dfrac{3a + 2}{5a - 10} + 5 = 15$.

Solution

Given $\qquad \dfrac{8a + 2}{a - 2} - \dfrac{2a - 1}{3a - 6} + \dfrac{3a + 2}{5a - 10} + 5 = 15$

Factoring denominators,

$$\dfrac{8a + 2}{a - 2} - \dfrac{2a - 1}{3(a - 2)} + \dfrac{3a + 2}{5(a - 2)} + 5 = 15$$

M: $15(a - 2)$, the LCD,

$$\dfrac{15(a - 2)(8a + 2)}{a - 2} - \dfrac{15(a - 2)(2a - 1)}{3(a - 2)} + \dfrac{15(a - 2)(3a + 2)}{5(a - 2)} + 15(a - 2)(5) = 15(a - 2)(15)$$

Canceling,

$$\dfrac{15(a - 2)(8a + 2)}{a - 2} - \dfrac{\overset{5}{15}(a - 2)(2a - 1)}{3(a - 2)} + \dfrac{\overset{3}{15}(a - 2)(3a + 2)}{5(a - 2)} + 15(a - 2)(5) = 15(a - 2)(15)$$

Rewriting,

$$15(8a + 2) - 5(2a - 1) + 3(3a + 2) + 15(a - 2)(5) = 15(a - 2)(15)$$

Simplifying,

$$120a + 30 - 10a + 5 + 9a + 6 + 75a - 150 = 225a - 450$$

Collecting terms,

$$120a - 10a + 9a + 75a - 225a = -30 - 5 - 6 + 150 - 450$$
$$-31a = -341$$
$$a = 11$$

Check the solution by the usual method.

PROBLEMS 12-3 *(do 1-20)*

Solve the following equations:

1. $\dfrac{3}{I} + \dfrac{5}{I} = 4$

2. $2 - \dfrac{2}{V} = \dfrac{10}{V}$

3. $\dfrac{16}{q} - 5 = \dfrac{3}{q} - \dfrac{2}{q}$

4. $\dfrac{3}{5R} - \dfrac{1}{15} + \dfrac{7}{5R} + \dfrac{2}{5} = 1$

5. $\dfrac{1}{\phi} - 1 - \dfrac{3}{2\phi} = 1 - \dfrac{1}{\phi}$

6. $\dfrac{5}{3x} + \dfrac{13}{12} + \dfrac{2}{x} = 2$

7. $\dfrac{12 - \omega}{\omega} - \dfrac{4}{\omega} = \dfrac{6}{\omega}$

8. $\dfrac{4}{8 + 2L} = \dfrac{3}{20 - 2L}$

9. $\dfrac{40 - \pi}{24\pi} + \dfrac{5}{6} - \dfrac{40 + \pi}{8\pi} = 0$

10. $\dfrac{10}{W} - 3 = \dfrac{2 - W}{W}$

11. $\dfrac{40 + v_o}{8v_o} - \dfrac{5}{6} - \dfrac{40 - v_o}{24v_o} - 0$

12. $\dfrac{6m - 17}{3m + 3} - \dfrac{2m - 5}{9 + m} = 0$

13. $\dfrac{6}{x - 1} - \dfrac{5}{1 - x} - \dfrac{8}{x - 1} + \dfrac{x}{1 - x} = 0$

14. $\dfrac{3}{5 + R} + \dfrac{R}{R + 2} = \dfrac{R + 4}{R + 5}$

15. $\dfrac{27 - \alpha}{\alpha + 1} + \alpha = 1 + \alpha$

16. $\dfrac{5 + R}{5 - R} - \dfrac{16R}{25 - R^2} + \dfrac{5 - R}{5 + R} + 2 = 0$

17. $\dfrac{\omega + 3}{\omega - 8} - \dfrac{5 - \omega}{\omega + 1} = \dfrac{2\omega^2 - 2}{\omega^2 - 7\omega - 8}$

18. $\dfrac{2\phi + 7}{6\phi - 4} - \dfrac{17\phi + 7}{9\phi^2 - 4} - \dfrac{3\phi - 5}{9\phi + 6} = 0$

19. $\dfrac{9\alpha + 17}{\alpha^2 - 2\alpha - 48} - \dfrac{2\alpha + 1}{2\alpha - 16} + \dfrac{2\alpha - 1}{2\alpha + 12} = 0$

20. $\dfrac{a - 7}{a + 2} - \dfrac{6}{a + 3} = \dfrac{a^2 - a - 42}{a^2 + 5a + 6}$

21. A can do a piece of work in 8 h, and B can do it in 6 h; how long will it take them to do it together?

Solution: Let n = number of hours it will take them to do it together.

Now A does $\frac{1}{8}$ of the job in 1 h; therefore, A will do $\frac{n}{8}$ in n h. Also,

B does $\frac{1}{6}$ of the job in 1 h; therefore, B will do $\frac{n}{6}$ in n h. Then they

will do $\frac{n}{8} + \frac{n}{6}$ in n h.

The entire job will be completed in n h, which we may represent by $\frac{8}{8}$ or $\frac{6}{6}$ of itself, which is 1.

$$\therefore \frac{n}{8} + \frac{n}{6} = 1$$

M: 24, the LCD, $3n + 4n = 24$

$$7n = 24$$

$$n = 3\tfrac{3}{7} \text{ h}$$

22. A technician can install a television transmission line in 5 h, and the technician's helper can do it in 8 h. In how many hours should they be able to do it if they work together?

23. A water tank can be filled in 1 h and 10 min if one pipe is used. If a different pipe is used, it takes 1 h and 45 min to fill the tank. How long will it take to fill the tank if both pipes are used?

24. A can do a piece of work in a days, and B can do it in b days. Derive a general formula for the number of days it would take both together to do the work.

Solution: Let x = number of days it will take both together.

Now A will do $\frac{x}{a}$ of the job in x days. Also, B will do $\frac{x}{b}$ of the job in x days.

Then $\dfrac{x}{a} + \dfrac{x}{b} = 1$

M: ab, $bx + ax = ab$

Factoring, $x(a + b) = ab$

D: $(a + b)$, $x = \dfrac{ab}{a + b}$

Alternate solution: Let x = number of days it will take both together.

Then $\dfrac{1}{x}$ = part that both together can do in 1 day; $\dfrac{1}{a}$ = part that A alone can do in 1 day; and $\dfrac{1}{b}$ = part that B can do in 1 day.

Now, $$\dfrac{1}{a} + \dfrac{1}{b} = \dfrac{1}{x}$$

M: $abx,$ $\qquad\qquad\qquad\qquad bx + ax = ab$

Factoring, $\qquad\qquad\qquad x(b + a) = ab$

D: $(a + b),$ $\qquad\qquad\qquad x = \dfrac{ab}{a + b}$

25. A can do a piece of work in a days, B in b days, and C in c days. Derive a general formula for the number of days it would take them to do the work together.

26. A tank can be filled by one of two pipes in 3 h and by the other of the two in 5 h. It can be emptied by the drain pipe in 6 h. If all three pipes are open, how long will it take to fill the tank?

27. Three circuits are connected to a storage battery. Circuit 1 completely discharges the battery in 20 h, circuit 2 in 15 h, and circuit 3 in 12 h. All circuits are connected to the battery in parallel. In how many hours will the battery be discharged?

28. A tank can be filled by one of two pipes in x h and by the other of the two in y h; it can be emptied by a drain pipe in z h. Derive a general formula for the number of hours required to fill the tank with all pipes open.

29. A bottle contains 1 liter (L) of a mixture of equal parts of acid and water. How much water must be added to make a mixture that will be one-tenth acid?

Solution: Let n = number of liters of water to be added.

$$1\ L = \text{amount of original mixture}$$

and $\qquad\quad 0.5\ L$ = amount of acid

Hence $\qquad\quad n + 1$ = amount of new one-tenth acid mixture

Now, $\qquad\quad \dfrac{1}{10} = \dfrac{\text{amount of acid}}{\text{total mixture}}$

Then, $\qquad\quad \dfrac{1}{10} = \dfrac{0.5}{n + 1}$

$$n + 1 = 5$$
$$n = 4\ \text{L of water to be added}$$

30. How much metal containing 25% copper must be added to 20 kg of pure copper to obtain an alloy having 50% copper?

Solution: Let x = desired amount of metal containing 25% copper.

Then $\qquad\quad 0.25x$ = amount of copper in this metal

$\qquad\quad 20 + 0.25x$ = amount of copper in mixture

$$x + 20 = \text{total weight of mixture}$$
$$0.5(x + 20) = \text{amount of copper in mixture}$$
$$0.5x + 10 = 20 + 0.25x$$
$$x = 40 \text{ kg}$$

31. How much 10% nickel alloy must be added to 10 kg of 30% nickel alloy to form a 12% nickel alloy?

32. A full radiator contains 50 L of a 30% mixture of antifreeze. How much antifreeze is required to obtain a 45% mixture?

   Solution: The radiator now contains 50 L of 30% antifreeze = 15 L. We want it to contain 50 L of 45% antifreeze = 22.5 L. But to get the mixture we want, we must drain off some quantity of 30% mixture and replace it with 100% antifreeze. Let the volume replaced be x L:

$$15 - 0.3x + x = 22.5$$
$$x = 10.7 \text{ L}$$

33. A diesel engine driving a 100-kW generator for an isolated communications center has a 200-L cooling system which, during the summer, contains a 20% antifreeze solution. At 50¢ per liter, what is the total cost of increasing the cold-weather protection by making the coolant 55% antifreeze?

34. A fighter plane traveling at 900 km/h leaves its base at 9:00 A.M. in order to overtake a bomber which departed from the same base at 7:00 A.M. and is traveling at 475 km/h. How much time is required for the fighter to overtake the bomber?

35. The sum of two numbers is 625. When the larger is divided by the smaller, the quotient is 24. Find the numbers.

36. The numerator of a fraction is 54 greater than the denominator. When 9 is subtracted from each term, the quotient is 4. What is the value of the fraction?

37. The sum of three consecutive numbers is $4\frac{1}{2}$. Find the numbers.

38. A certain number, plus 23, is divided by the same number plus 12. The quotient is $\frac{4}{3}$. What is the number?

39. The perimeter of a stock room is 20 m. The room is four times as long as it is wide. What are its dimensions?

40. A screened room is two-thirds as wide as it is long. If it had been 3 m wider and 3 m shorter, its area would have been 3 m² larger. What are its dimensions?

12-4 LITERAL EQUATIONS

Equations in which some or all of the numbers are replaced by letters are called *literal equations;* they were studied in Chap. 5. Having attained more knowledge of algebra, such as factoring and fractions, we are now ready to proceed with the solution of more difficult literal equations, or formulas. No new methods are involved in the actual solutions—we are prepared to solve a more complicated equation simply because we have available more tools with which to work. Again, we point out that the ability to solve formulas is of utmost importance.

Example 9

Given $$I = \frac{V}{R + r}, \text{ solve for } r.$$

Solution

Given $$I = \frac{V}{R + r}$$

M: $(R + r)$, $\qquad I(R + r) = V$

Removing parentheses, $\qquad IR + Ir = V$

S: IR, $\qquad Ir = V - IR$

D: I, $\qquad r = \frac{V - IR}{I}$

Example 10

Given $$S = \frac{RL - a}{R - 1}, \text{ solve for } L.$$

Solution

Given: $$\frac{RL - a}{R - 1} = S$$

M: $(R - 1)$, $\qquad RL - a = S(R - 1)$

A: a, $\qquad RL = S(R - 1) + a$

D: R, $\qquad L = \frac{S(R - 1) + a}{R}$

Example 11

Given $$\frac{a}{x - b} = \frac{2a}{x + b}, \text{ solve for } x.$$

Solution

Given $$\frac{a}{x - b} = \frac{2a}{x + b}$$

M: $(x^2 - b^2)$, the LCD, $\qquad \dfrac{(x^2 - b^2)a}{x - b} = \dfrac{(x^2 - b^2)2a}{x + b}$

Canceling, $\qquad \dfrac{\overset{x + b}{\cancel{(x^2 - b^2)}}a}{\cancel{x - b}} = \dfrac{\overset{x - b}{\cancel{(x^2 - b^2)}}2a}{\cancel{x + b}}$

Rewriting, $\qquad (x + b)a = (x - b)2a$

Removing parentheses, $\qquad ax + ab = 2ax - 2ab$

Collecting terms, $\qquad ax - 2ax = -2ab - ab$

or $\qquad -ax = -3ab$

M: -1, $\qquad ax = 3ab$

D: a, $\qquad x = 3b$

Note

The last two steps can be combined into a single step by dividing $-ax = -3ab$ by $-a$ to obtain $x = 3b$.

Check

Substitute $3b$ for x in the given equation:

$$\frac{a}{3b - b} = \frac{2a}{3b + b}$$

Simplifying,

$$\frac{a}{2b} = \frac{2a}{4b}$$

or

$$\frac{a}{2b} = \frac{a}{2b}$$

PROBLEMS 12-4

Given: Solve for:

1. $Y_d = \dfrac{LbV_d}{2aV_0}$ $\qquad\qquad\qquad$ $V_0,\ V_d$

2. $\mathbf{V}_{max} = \dfrac{V + V_{pt}}{\omega}$ $\qquad\qquad$ $V_{pt},\ \omega,\ \mathbf{V}_{max}$

3. $I = \dfrac{V_b - v}{R}$ $\qquad\qquad\qquad$ $V_b,\ v$

4. $C = \dfrac{\omega_{01}}{R_1 + R_2}$ $\qquad\qquad\qquad$ $R_1,\ \omega_{01}$

5. $C_2 = \dfrac{V_3 - V_2}{\omega^2 L V_3}$ $\qquad\qquad\qquad$ $V_2,\ L$

6. $I_1 = \dfrac{V_1 - I_2(R + s)}{R}$ $\qquad\quad$ $V_1,\ s,\ R$

7. $\alpha = \dfrac{R_t - R_0}{R_0 t}$ $\qquad\qquad\qquad$ $R_t,\ t$

8. $I_{\lambda_2} = \dfrac{V_{e_2} + V_\lambda - V_2}{R_b}$ $\qquad\quad$ $V_\lambda,\ V_2$

9. $v = \dfrac{Vr}{R + r}$ $\qquad\qquad\qquad$ $r,\ R$

10. $\mu = \dfrac{g_m}{g_m{}' - g_m}$ $\qquad\qquad\quad$ $g_m{}',\ g_m$

11. $\omega^2 C_1 C_2 R_3 = \dfrac{1}{R_1 + R_2}$ $\qquad$ R_1

12. $\mu = \dfrac{2G_L + g_p - 2G_2}{G_2 - G_L}$ $\qquad$ $G_2,\ G_L,\ g_p$

13. $\dfrac{V_0}{I_0} = \dfrac{R_0}{1 - \mu\beta}$ $\qquad\qquad\quad$ β

14. $\beta_m = \dfrac{m\pi a}{a + b}$ $\qquad\qquad\qquad$ $a,\ b$

15. $C_0 = \dfrac{a - b}{a + b}$ $\qquad\qquad\qquad$ $a,\ b$

16. $\gamma = \dfrac{I_n}{I_n + I_p}$ $\qquad\qquad$ I_p, I_n

17. $\dfrac{E}{I} = \dfrac{Z_1 Z_2 + Z_2 Z_3 + Z_3 Z_1}{Z_3}$ $\qquad$ Z_1, Z_2, Z_3

18. $Z_0 = \dfrac{R_a R}{(\mu + 1)R + R_a}$ $\qquad$ R, R_a, μ

19. $\dfrac{V}{V_1} = \dfrac{AR_y}{(A + 1)R_x + R_y}$ $\qquad$ R_x, R_y, A

20. $B_c = \dfrac{\pi\sqrt{2}DFf_b}{\sqrt{2}D + F}$ $\qquad$ D, F

21. $Z_{ab}{}^2 = \dfrac{X_s{}^2 R}{X_p - X_s - R}$ $\qquad$ X_p, R

22. $Z_1 = \dfrac{(\mu + 1)R_1 R + R_a(R_1 + R)}{R_a + R}$ $\qquad$ R_a, R

23. $I_2 R - V_n = \left(\dfrac{R_1}{R_1 + R_2}\right) V_n$ $\qquad$ V_n, R_1

24. $2C_2 R_3 = \sqrt{2} - C_1 R_1 \left(\dfrac{R_2}{R_3} - 1\right)$ $\qquad$ R_1, R_2, C_1

25. $F = \dfrac{9}{5}C + 32$ $\qquad$ C

26. $r = \dfrac{\mu V_g - PR_p}{P}$ $\qquad$ P, μ

27. $V_1 = \dfrac{BI_0 R_0}{R + R_0}$ $\qquad$ R, R_0

28. $f_{out} = \dfrac{C_1 f_{in}}{C_1 + C_2}$ $\qquad$ f_{in}, C_1, C_2

29. $K = \dfrac{\mu m N}{g(R_H + r)}$ $\qquad$ R_H, r

30. $r_p = \dfrac{GR_{pg}}{g_m R_{pg} - G}$ $\qquad$ G, g_m

31. $\alpha = \dfrac{Z_1 + Z_2 - R}{Z_1(1 - k) + Z_2}$ $\qquad$ R, Z_1, k

32. $X = \dfrac{K}{(f_1 - f_2) - (f_0 - f_2)}$ $\qquad$ f_1, K

33. $F_{12} = \left(\dfrac{2f}{\alpha}\right)\left(1 + \dfrac{F_s}{F_2}\right)$ $\qquad$ F_2, f

34. $\mu\beta = \dfrac{2N}{2L + N}$ $\qquad\qquad$ L, N

35. $H_2S = \left(\dfrac{1}{R_1}\right)\left(\dfrac{S}{S + \alpha}\right)$ $\qquad$ α, S

36. $\dfrac{n_2 - n_1}{n_1} = \dfrac{-h\nu}{kT}$ $\qquad\qquad$ n_1

37. $v_1 = \mu v_g\left(\dfrac{R_p}{R_p + Z_1}\right)$ $\qquad$ Z_1, R_p

38. $F = 1 + 2\left(\dfrac{T_s}{T_a}\right)\left(\dfrac{1}{X}\right)$ $\qquad$ T_a, X, T_s

39. $X = \left(1 - \dfrac{C_v}{C_0}\right)\left(\dfrac{f_c}{f}\right)$ $\qquad$ C_v, C_0

40. $\dfrac{C_3}{C_1 + C_2} = \dfrac{R_3}{\dfrac{1}{R_1} + \dfrac{1}{R_2}}$ $\qquad$ C_2, R_1

41. $\mu = \dfrac{\omega s}{2}\left(\dfrac{1}{v_0{}'} - \dfrac{1}{v_m{}'}\right)$ $\qquad$ $v_0{}', v_m{}'$

42. $\alpha = 1 + \dfrac{1}{\mu_0}\left(1 + 1.5\dfrac{d_2}{d_1}\right)$ $\qquad$ d_1

43. $\dfrac{V - v_0}{v_0} = \dfrac{R_2}{R_1}\left(\dfrac{i_1 + i_2}{i_1}\right)$ $\qquad$ v_0, i_1

44. $Z_{am}{}^2 = R\dfrac{(X_p - X_s)Z_{ab}{}^2}{Z_{ab}{}^2 + X_s{}^2}$ $\qquad$ $Z_{ab}{}^2, X_p$

45. $\mu_1 = \dfrac{G(\mu_2\beta_2 - 1)}{G\beta_1(\mu_2\beta_2 - 1) - \mu_2}$ $\qquad$ G, β_1

46. $C_g = C_{gf} + C_{gp}\left(1 + \dfrac{\mu R_b}{r_p + r_b}\right)$ $\qquad$ R_b, C_{gp}

47. $\sigma_0 = 2\pi\lambda^2\left(\dfrac{\gamma_1}{\gamma_1 + \gamma_f}\right)\left(\dfrac{2I_f + 1}{2I_1 + 1}\right)$ $\qquad$ I_1, I_f

48. $K_\varepsilon{}^2\left(1 + \dfrac{\tan^2 K_a}{\varepsilon_p{}^2}\right) = -a^2$ $\qquad$ $\varepsilon_p{}^2$

49. $n' = \dfrac{\lambda}{\pi d_0}\left(\dfrac{1 - d_1}{d_1 - d_0}\right)$ $\qquad$ λ, d_1

50. $I_2 = \dfrac{VR_0}{R_1R_0 + R_1R_2 + R_2R_0}$ $\qquad$ VR_0, R_1

51. $\quad R = -\left(\dfrac{1}{k}\right)\left(\dfrac{Z_1 R_2}{Z_2 \alpha} + \dfrac{R_2}{\alpha} + \dfrac{Z(1-\alpha)}{\alpha}\right)$ $\qquad R_2,\, Z,\, \alpha$

52. $\quad \dfrac{V_b - V_c}{\mu} = V_c + V_s\left(\dfrac{R_p}{R_1 + R_p}\right)$ $\qquad R_1,\, V_s$

53. $\quad \dfrac{r_1}{r_1 + r_2} = \dfrac{r_3}{r_3 + r_4}$ $\qquad r_1,\, r_3,\, r_4$

54. $\quad \dfrac{S^2}{N^2} = \dfrac{\alpha F}{2f\left(1 + \dfrac{F_s}{F_2}\right)}$ $\qquad F_s,\, F_2$

55. $\quad V_{out} = \dfrac{Q}{C_f}\ \dfrac{1}{1 + \dfrac{1}{G} + \dfrac{C_d}{C_{fg}}}$ $\qquad G$

56. $\quad T_m = \dfrac{T}{\dfrac{\omega_{32} k v_{12} T_m}{\omega_{21} h v_{12}} - 1}$ $\qquad T,\, h$

57. $\quad \dfrac{P_L}{2p} = \dfrac{\omega \varepsilon_2 p_2 (\tan \delta)}{2CN(p_1 + p_2)}$ $\qquad p_2$

58. $\quad a_2 = \dfrac{FC}{(\Omega_1 - B)(\Omega_2 - B) + c^2}$ $\qquad \Omega_1$

59. $\quad \dfrac{1}{R_p} = \dfrac{1}{R_1} + \dfrac{1}{R_2}$ $\qquad R_p,\, R_1,\, R_2$

60. $\quad i_s = \dfrac{v}{L\left(S_s + \dfrac{R}{L}\right)}$ $\qquad L,\, R$

61. $\quad \dfrac{V_0}{V} = \dfrac{\mu}{\mu + 1 + \dfrac{R_a}{R_3}}$ $\qquad R_3,\, R_a,\, \mu$

62. $\quad \dfrac{\omega_{01} L}{\dfrac{R_1 R_2}{R_1 + R_2}} = 1$ $\qquad L,\, R_1$

63. $\quad M = \dfrac{k}{1 + \dfrac{N}{4\pi} k} H_0$ $\qquad \pi,\, k$

64. $\quad HS = \dfrac{\dfrac{1}{C}}{S + \dfrac{1}{R_c C}}$ $\qquad C,\, R_c$

65. $d = b + \dfrac{2b}{\dfrac{X}{X'} + \dfrac{X'}{X}}$ $\qquad$ b

66. $(G_2)(p) = \dfrac{A(p + \omega_1)}{p + \omega - \dfrac{AC_2}{C_1 + C_2}p}$ $\qquad$ C_2

67. $\dfrac{V_0}{V} = \dfrac{\mu R_1 + R_a}{\mu R_1 + R_a + (R_s + R_1)\left(1 + \dfrac{R_a}{R_3}\right)}.$ $\qquad$ R_a, R_s, μ

68. $\dfrac{V_0}{V} = \dfrac{h_{fe} + 1 + \dfrac{h_{ie}}{R_B}}{h_{fe} + 1 + h_{ie}\left(\dfrac{1}{R_B} + \dfrac{1}{R_E}\right)}$ $\qquad$ V, R_B

69. $R_0 = \left(\dfrac{1}{1 + \mu\dfrac{R_1}{R_a}}\right)\left(\dfrac{1}{\dfrac{1}{R_s + R_1} + \dfrac{1}{R_2}}\right)$ $\qquad$ R_a, R_2, μ

70. $R_{in} = R_E\left[\dfrac{h_{fe} + 1 + h_{ie}\left(\dfrac{1}{R_E} + \dfrac{1}{R_B}\right)}{1 + \dfrac{h_{ie}}{R_B}}\right]$ $\qquad$ R_B, R_E

71. $R_i = R_1\dfrac{\mu + R_a\left(\dfrac{1}{R_1} + \dfrac{1}{R_2} + \dfrac{1}{R_3}\right)}{1 + R_a\left(\dfrac{1}{R_1} + \dfrac{1}{R_2} + \dfrac{1}{R_3}\right)}$ $\qquad$ R_a

72. $MH = \dfrac{4\pi r^2}{T^2\left(1 + \dfrac{\alpha}{\dfrac{1}{2}\pi - \alpha}\right)}$ $\qquad$ α, π

73. $\dfrac{\alpha - \dfrac{\pi}{\alpha - \beta}}{\alpha + \dfrac{\pi}{\alpha - \beta}} - 1 = \dfrac{\alpha}{\beta}$ $\qquad$ π, β

74. The force between two magnetic poles of strength S_1 and S_2 at a separation of d cm is

$$F = \dfrac{10S_1S_2}{d^2} \qquad \text{micronewtons } (\mu N)$$

When the poles are separated by a distance of 60 cm, a force of 15 μN exists between them. $S_2 = 90$ units. What is the value of S_1?

75. The force acting to close the air gap of a simple electromagnetic relay is

$$F = \frac{B^2 A}{2\mu} \quad \text{N}$$

What will be the value of A, the cross-sectional area of the gap, in square meters, which will permit a flux density B of 64×10^3 webers per square meter (Wb/m²) to exert a force F of 96 N? μ, the permeability of air, is $4\pi \times 10^{-7}$ SI units.

76. When two impedances Z_1 and Z_2 are connected in parallel, the resultant joint impedance Z_p is

$$Z_p = \frac{Z_1 Z_2}{Z_1 + Z_2}$$

Solve for Z_2.

77. Using the formula given in Prob. 76, what is the value of Z_2 when $Z_p = 3\ \Omega$ and $Z_1 = 6\ \Omega$?

78. $\dfrac{N_p}{N_s} = \dfrac{V_p}{V_s}$; $V_p = 100$, $V_s = 20$, $N_p = 400$. Find the value of N_s?

79. $\dfrac{V_1}{V_2} = \dfrac{R_1}{R_2}$; $V_1 = 16.2$ V, $V_2 = 34$ V, $R_1 = 47.7\ \Omega$. What is R_2?

80. Corresponding temperature readings in Fahrenheit degrees (°F) can be obtained from a Celsius thermometer by the use of the formula $F = \dfrac{9}{5} C + 32$, where C is the temperature in degrees Celsius.

When the temperature is 77°F, what is the Celsius temperature?

81. Use the formula given in Prob. 80 to find the temperature at which the Fahrenheit and Celsius temperatures are equal, that is, at which $F = C$.

82. $L_t = L_0 + L_0 \alpha t$. If $L_t = 15$, $\alpha = 8.33 \times 10^{-2}$, and $t = 6$, what is the value of L_0?

83. $R_t = R_0 (1 + 0.0042t)\ \Omega$. What is the resistance R_0 at 0°C if, at a temperature $t = 59.5$°C, the resistance $R_t = 40\ \Omega$?

84. $P = \dfrac{LI^2}{2}$. The energy P stored in a circuit is 1250 joules (J). If the current $I = 2.5$ A, find the value of the coefficient of self-induction L.

85. When two capacitors C_1 and C_2 are connected in series, the resultant total capacitance can be computed by means of the equation

$$\frac{1}{C_s} = \frac{1}{C_1} + \frac{1}{C_2}$$

If $C_s = 2$ pF and $C_2 = 6$ pF, what is the value of C_1?

86. The joint conductance $\dfrac{1}{R_p}$ siemens of three resistances connected in parallel is expressed by

$$\frac{1}{R_p} = \frac{1}{R_1} + \frac{1}{R_2} + \frac{1}{R_3}$$

Solve for R_p.

87. A lens formula is $\frac{1}{f} = \frac{1}{p} + \frac{1}{q}$. What is the value of p when $q = 80$ and $f = 50$?

88. Use the lens equation that was given in Prob. 87 to find the image distance q when the focal length $f = 10$ cm and the object distance $p = 40$ cm.

89. $P = \frac{V^2}{R}$. (a) How is the value of P changed when V is doubled? (b) How is the value of P changed when R is doubled?

90. A source of emf consists of n cells in parallel, and each cell has an emf of V V and an internal resistance of r Ω. The current that flows through a load of R Ω is given by the relation

$$I = \frac{V}{R + \dfrac{r}{n}} \quad \text{A}$$

Solve for r and R.

91. Use the formula stated in Prob. 90 to find the value of R when $V = 2.1$ V, $r = 0.6$ Ω, $I = 2$ A, and $n = 4$ cells.

92. Use the formula stated in Prob. 90 to find n in terms of I and V when $R = 32$ Ω and $r = 0.1$ Ω.

93. A source of emf consists of n cells in series, and each cell has an emf of V V and an internal resistance of r Ω. The current flowing through a load of R Ω is given by the relation

$$I = \frac{nV}{R + nr} \quad \text{A}$$

Solve for R and n.

94. Use the formula stated in Prob. 93 to find the number of identical cells of internal resistance $r = 0.6$ Ω each, if they provide an emf of $V = 2.1$ V each, when they drive a current $I = 2$ A through a load $R = 4.5$ Ω.

95. A signal voltage v_g is supplied to the gate of an FET which has a voltage gain expressed as the ratio of drain voltage v_d to gate signal voltage v_g. The gain can be expressed as follows:

$$\text{Gain} = \frac{i_d r_d}{r_s + \dfrac{1}{g_m}} = \frac{v_d}{v_g}$$

where g_m is the transconductance in siemens and r_s is the ac source resistance in ohms. Solve for r_s and g_m.

96. Use the formula in Prob. 95 to find the value of i_d if $r_s = 400$ Ω, $r_d = 5$ kΩ, and $g_m = 0.005$ S, producing a gain of 8.33.

97. Does $\frac{IR + V}{R} = I + V$? Explain your answer.

98. If $I = \frac{V}{R_1 + R_2 + R_3}$, does $R_2 = \frac{V}{R_1 + R_2 + I}$? Explain your answer.

99. $S = V_0 t + \frac{1}{2}gt^2$. What is the value of the initial velocity V_0 in terms of S, g, and t?

100. Using the formula stated in Prob. 99, what is the acceleration due to gravity g if the initial velocity $V_0 = 3$ m/s, $S = 520.5$ m, and $t = 10$ s?

101. A radiosonde is dropped from an airplane and falls freely until its parachute opens. Ten seconds after its parachute opens, it has fallen an additional 1 km. What was its velocity when the parachute opened?

102. If

$$\frac{a}{b} = \frac{a - \dfrac{x}{a - b}}{a + \dfrac{x}{a - b}} - 1$$

what is the value of x when $b = 4.62$ and $a = 3$?

103. The transconductance g_m of a JFET can be found by dividing the incremental changes in signal drain current i_d by the signal gate voltage v_{gs}. That is,

$$g_m = \frac{\Delta i_d}{v_{gs}}$$

Solve for v_{gs} and write an expression to explain Δi_d.

104. Use the formula in Prob. 103 to determine the value of v_{gs} when $\Delta i_d = 7.5$ μA and $g_m = 7500$ μS.

105. $I_C = \dfrac{V_{CC} - V_C}{R_C}$. What will be the value of V_{CC} when $I_C = 4.9$ mA and the collector voltage V_C measured with respect to ground is -10 V? R_C has a value of 5 kΩ. With respect to the data given and the calculated values, is this transistor type PNP or NPN?

106. $V = L\dfrac{I_1 - I_2}{t}$. What is the change in current when a voltage $V = 1$ kV is induced in an inductance $L = 5$ H in time $t = 0.5$ s?

107. $I = C\dfrac{V_1 - V_2}{t}$. What is the change of voltage which will produce a current flow of $I = 0.05$ A during the discharge of a 15-μF capacitor in 0.0294 s?

108. $R_a = \dfrac{R_1 R_3}{R_1 + R_2 + R_3}$. Three resistances $R_1 = ?$, $R_2 = 3$ Ω, and $R_3 = 2.14$ Ω are connected in delta to produce an equivalent Y-circuit branch $R_a = 0.6$ Ω. Find R_1.

109. In transistor parameters, $\beta = \dfrac{\alpha}{1 - \alpha}$. Solve for α in terms of β.

110. Using the formula stated in Prob. 109, what is α when $\beta = 284.7$?

Ohm's Law— Parallel Circuits

Most of the systems employed for the distribution of electric energy consist of parallel circuits; that is, a source of emf is connected to a pair of conductors, known as *feeders*, and various types of load are connected across the feeders. A simple distribution circuit consisting of a motor and a bank of five lamps is represented schematically in Fig. 13-1 and pictorially in Fig. 13-2. The motor and the lamps are said to be in *parallel*, and it is evident that the current supplied by the generator divides between the motor and the lamps.

In this chapter you will analyze parallel circuits and solve parallel circuit problems. The solution of a parallel circuit generally consists in reducing the entire circuit to a single equivalent resistance that could replace the original circuit without any change in the supply voltage or current.

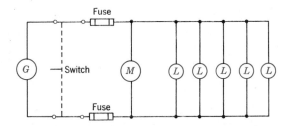

Fig. 13-1 Schematic diagram of a generator G connected to a motor M in parallel with a bank of five lamps L.

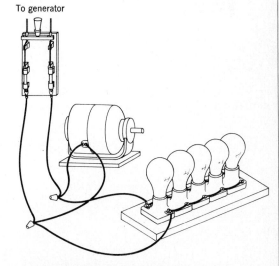

Fig. 13-2 Illustration of circuit shown schematically in Fig. 13-1.

13-1 TWO RESISTANCES IN PARALLEL

The schematic diagram of Fig. 13-3 and the accompanying circuit shown in Fig. 13-4 represent two resistors R_1 and R_2 connected in parallel across a source of voltage V. An examination of the circuit arrangement brings out two important facts:

1. The same voltage exists across the two resistors.
2. The total current I_t delivered by the generator enters the paralleled resistors at junction a, divides between the resistors, and leaves the parallel circuit at junction b. Thus, the sum of the currents I_1 and I_2, which flow through R_1 and R_2, respectively, is equal to the total current I_t.

By making use of these facts and applying Ohm's law, it is easy to derive equations that show how paralleled resistances combine. From 1 above,

$$I_1 = \frac{V}{R_1} \qquad I_2 = \frac{V}{R_2} \qquad \text{and} \qquad I_t = \frac{V}{R_p}$$

where R_p is the joint resistance of R_1 and R_2, or the equivalent resistance of the parallel combination. From 2 above,

$$I_t = I_1 + I_2 \tag{1}$$

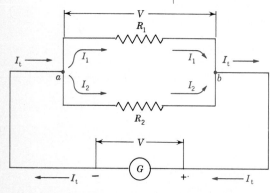

Fig. 13-3 Resistors R_1 and R_2 connected in parallel across generator G, which maintains a potential of V V.

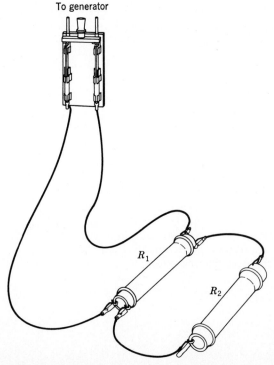

Fig. 13-4 Illustration of schematic circuit shown in Fig. 13-3.

Substituting in Eq. (1) the value of the currents,

$$\frac{V}{R_p} = \frac{V}{R_1} + \frac{V}{R_2}$$

D: V,

$$\frac{1}{R_p} = \frac{1}{R_1} + \frac{1}{R_2} \qquad (2)$$

Equation (2) states that the total conductance (Sec. 7-5) of the circuit is equal to the sum of the parallel conductances of R_1 and R_2; that is,

$$G_t = G_1 + G_2 \qquad (3)$$

It is evident, therefore, that, when resistances are connected in parallel, each additional resistance represents another path (conductance) through which current will flow. Hence, increasing the number of resistances in parallel increases the total conductance of the circuit and thus decreases the equivalent resistance of the circuit.

Example 1
What is the joint resistance of the circuit of Fig. 13-3 if $R_1 = 6\ \Omega$ and $R_2 = 12\ \Omega$?

Solution 1
Given $R_1 = 6\ \Omega$ and $R_2 = 12\ \Omega$, $R_p = ?$

Substituting the known values in Eq. (2),

$$\frac{1}{R_p} = \frac{1}{6} + \frac{1}{12} = 0.1667 + 0.0833$$

or

$$\frac{1}{R_p} = 0.250$$

Solving for R_P, $R_p = \dfrac{1}{0.250} = 4.0\ \Omega$

Solution 2
A more convenient formula for the joint resistance of two parallel resistances is obtained by solving Eq. (2) for R_p. Thus,

$$R_p = \frac{R_1 R_2}{R_1 + R_2} \qquad (4)$$

Hence, the joint resistance of two resistances in parallel is equal to the product of the resistances divided by the sum.
Substituting the values of R_1 and R_2 in Eq. (4),

$$R_p = \frac{6 \times 12}{6 + 12} = \frac{72}{18} = 4.0\ \Omega$$

Thus, the paralleled resistors R_1 and R_2 are equivalent to a single resistance of 4.0 Ω. Note that the joint resistance is *less* than either of the resistances in parallel.

Example 2

(*a*) Give the joint resistance of the circuit of Fig. 13-3 if $R_1 = 21\ \Omega$ and $R_2 = 15\ \Omega$? (*b*) If the generator supplies 12 V across points *a* and *b*, what is the generator (line) current?

Solution 1

(*a*)
$$R_p = \frac{R_1 R_2}{R_1 + R_2} = \frac{21 \times 15}{21 + 15} = 8.75\ \Omega$$

(*b*)
$$I_t = \frac{V}{R_t} = \frac{12}{8.75} = 1.371\ \text{A}$$

Solution 2

Since 12 V exists across both resistors, the current through each can be found and the two currents can be added to obtain the total current. Thus,

Current through R_1,
$$I_1 = \frac{V}{R_1} = \frac{12}{21} = 0.571\ \text{A}$$

Current through R_2,
$$I_2 = \frac{V}{R_2} = \frac{12}{15} = 0.8\ \text{A}$$

Total current,
$$I_t = I_1 + I_2 = 0.571 + 0.8 = 1.371\ \text{A}$$

Hence,
$$R_p = \frac{V}{I_t} = \frac{12}{1.371} = 8.75\ \Omega$$

From the foregoing, it is evident that R_1 and R_2 could be replaced by a single resistor of 8.75 Ω, connected between *a* and *b*, and the generator would be working under the same load conditions. Also, it is apparent that when a current enters a junction of resistors connected in parallel, the current divides between the branches in inverse proportion to the resistances; that is, the greatest current flows through the least resistance.

Example 3

In the circuit shown in Fig. 13-3, $R_1 = 25\ \Omega$, $V = 220$ V, and $I_t = 14.3$ A. What is the resistance of R_2?

Solution 1

The current through R_1 is

$$I_1 = \frac{V}{R_1} = \frac{220}{25} = 8.8\ \text{A}$$

Since
$$I_t = I_1 + I_2$$

the current through R_2 is

$$I_2 = I_t - I_1 = 14.3 - 8.8 = 5.5\ \text{A}$$

Then
$$R_2 = \frac{V}{I_2} = \frac{220}{5.5} = 40\ \Omega$$

Solution 2

$$R_p = \frac{V}{I_t} = \frac{220}{14.3} = 15.4\ \Omega$$

Solving Eq. (2) or (4) for R_2,

$$R_2 = \frac{R_1 R_p}{R_1 - R_p} = \frac{25 \times 15.4}{25 - 15.4} = 40\ \Omega$$

PROBLEMS 13-1

1. Two 330-Ω resistors are connected in parallel. What is the equivalent resistance?
2. Two resistors, one of 1500 Ω and the other of 4700 Ω, are connected in parallel. What is the equivalent resistance of the combination?
3. What is the joint resistance of 68 kΩ in parallel with 82 kΩ?
4. What is the equivalent resistance of 27 kΩ in parallel with 1.5 kΩ?
5. What is the equivalent resistance of:
   (a) Two 100-Ω resistors in parallel?
   (b) Two 680-kΩ resistors in parallel?
   (c) Two 3.9-kΩ resistors in parallel?
6. State a general formula for the total resistance R_p of two equal resistances of R Ω connected in parallel.
7. In the circuit of Fig. 13-3, how much generator voltage would be required to deliver a total current of 3.63 A through a parallel combination of $R_1 = 220\ \Omega$ and $R_2 = 270\ \Omega$?
8. How much power would be absorbed by the 270-Ω resistor of Prob. 7?
9. In the circuit shown in Fig. 13-3, $I_t = 20.3$ mA, $V = 220$ V, and $R_1 = 12$ kΩ. What is the resistance of R_2?
10. How much power is dissipated by R_1 of Prob. 9?
11. How much total power is drawn from the generator of Prob. 9?
12. In the circuit of Fig. 13-3, $R_1 = 18$ kΩ and the current through R_2 is 14.71 mA. A total current $I_t = 70.27$ mA flows through the parallel combination. What is the resistance of R_2?
13. How much power is expended in R_2 of Prob. 12?
14. How much power is drawn from the generator of Prob. 12?
15. What is the generated voltage of Prob. 12?

13-2 THREE OR MORE RESISTANCES IN PARALLEL

The procedure for deriving a general equation for the joint resistance of three or more resistances in parallel is the same as that of the preceding section. For example, Fig. 13-5 represents three resistors R_1, R_2, and R_3

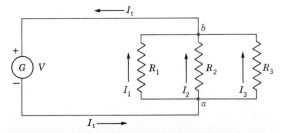

Fig. 13-5 Resistors R_1, R_2, and R_3 connected in parallel.

connected in parallel across a source of voltage V. The total line current I_t splits at junction a into currents I_1, I_2, and I_3, which flow through R_1, R_2, and R_3, respectively. Then

$$I_1 = \frac{V}{R_1} \qquad I_2 = \frac{V}{R_2}$$

$$I_3 = \frac{V}{R_3} \qquad I_t = \frac{V}{R_p}$$

where R_p is the joint resistance of the parallel combination.

Since $\qquad\qquad\qquad I = I_1 + I_2 + I_3$

by substituting, $\qquad\qquad \dfrac{V}{R_p} = \dfrac{V}{R_1} + \dfrac{V}{R_2} + \dfrac{V}{R_3}$

D: V, $\qquad\qquad\qquad \dfrac{1}{R_p} = \dfrac{1}{R_1} + \dfrac{1}{R_2} + \dfrac{1}{R_3}$ (5)

From Eq. (5), it is evident that the total conductance of the circuit is equal to the sum of the paralleled conductances of R_1, R_2, and R_3; that is,

$$G_p = G_1 + G_2 + G_3$$

In like manner, it can be demonstrated that the joint resistance R_p of any number of resistances connected in parallel is

$$\frac{1}{R_p} = \frac{1}{R_1} + \frac{1}{R_2} + \frac{1}{R_3} + \frac{1}{R_4} + \frac{1}{R_5} + \cdots$$

Or, in terms of conductances,

$$G_p = G_1 + G_2 + G_3 + G_4 + G_5 + \cdots$$

Example 4

What is the joint resistance of the circuit of Fig. 13-5 if $R_1 = 5\ \Omega$, $R_2 = 10\ \Omega$, and $R_3 = 12.5\ \Omega$?

Solution

Substituting the known values in Eq. (5),

$$\frac{1}{R_p} = \frac{1}{5} + \frac{1}{10} + \frac{1}{12.5}$$

$$= 0.2 + 0.1 + 0.08$$

or $\qquad\qquad \dfrac{1}{R_p} = 0.38$

Solving for R_p, $\qquad R_p = \dfrac{1}{0.38} = 2.63\ \Omega$

If Eq. (5) is solved for R_p, the result is

$$R_p = \frac{R_1 R_2 R_3}{R_1 R_2 + R_1 R_3 + R_2 R_3}$$ (6)

It is seen that Eq. (6) is somewhat cumbersome for computing the joint resistance of three resistances connected in parallel. However, you should recognize such expressions for three or more resistances in parallel, for you will encounter them in the analysis of networks.

Finding the joint resistance of any number of resistors in parallel is facilitated by arbitrarily assuming a voltage to exist across the parallel combination. The currents through the individual branches that *would* flow if the assumed voltage were actually impressed are added to obtain the total line current. The assumed voltage divided by the total current results in the joint resistance of the combination.

In order to avoid decimal quantities, that is, currents of less than 1 A, the assumed voltage should be numerically greater than the highest resistance of any parallel branch.

Example 5

Three resistances $R_1 = 10\ \Omega$, $R_2 = 15\ \Omega$, and $R_3 = 45\ \Omega$ are connected in parallel. Find their joint resistance.

Solution

Assume $V_a = 100$ V to exist across the combination.

Current through R_1, $I_1 = \dfrac{V_a}{R_1} = \dfrac{100}{10} = 10$ A

Current through R_2, $I_2 = \dfrac{V_a}{R_2} = \dfrac{100}{15} = 6.67$ A

Current through R_3, $I_3 = \dfrac{V_a}{R_3} = \dfrac{100}{45} = 2.22$ A

Total current, $I_t = 18.89$ A

Joint resistance, $R_p = \dfrac{V_a}{I_t} = \dfrac{100}{18.89} = 5.3\ \Omega$

PROBLEMS 13-2

1. What is the equivalent resistance of 10 Ω, 15 Ω, and 30 Ω connected in parallel?
2. What is the joint resistance of 150 Ω, 470 Ω, and 470 Ω connected in parallel?
3. Three resistors of 12 Ω, 330 Ω, and 8.2 Ω are connected in parallel. What is their joint resistance?
4. Three resistors of 10 Ω, 100 Ω, and 1000 Ω are connected in parallel. Find the joint resistance of the combination.
5. What is the equivalent resistance of 22 Ω, 15 Ω, 33 Ω, and 47 Ω connected in parallel?
6. Four resistors of 8.2 Ω, 1.5 Ω, 2.7 Ω, and 3.3 Ω are connected in parallel. What is the equivalent resistance of the combination?
7. What is the joint resistance of:
   (a) Three 6.3-kΩ resistors in parallel?
   (b) Four 68-kΩ resistors in parallel?
8. What is the joint resistance of:
   (a) Three 100-kΩ resistors connected in parallel?
   (b) Four 100-kΩ resistors connected in parallel?
   (c) Five 100-kΩ resistors connected in parallel?
9. State a general formula for the resistance R_p of n equal resistances of $R\ \Omega$ connected in parallel.
10. In the circuit of Fig. 13-5, the total current $I_t = 18.03$ A, $R_1 = 100\ \Omega$, $R_2 = 150\ \Omega$, and $V = 475$ V. What is the resistance of R_3?
11. If the values of Prob. 10 are used, what is the power delivered to the 150-Ω resistor?

12. What would be the resistance in Prob. 10 if the 150-Ω resistor were shorted out?

13. In the circuit of Fig. 13-5, $R_1 = 12\ \Omega$, $R_2 = 18\ \Omega$, $I_3 = 4.545$ A, and $V = 100$ V. Find (*a*) the value of R_3 to two significant figures and (*b*) the total power delivered to the circuit.

14. In the circuit of Fig. 13-5, $R_2 = 510\ \Omega$, $R_3 = 270\ \Omega$, $I_t = 4.38$ A, and $I_1 = 1.52$ A. Find the value of R_1 to two significant figures.

15. In the circuit of Fig. 13-5, $R_1 = R_2 = 5$ kΩ, and R_3 is disconnected. $I_t = 0.40$ A. What must be the value of R_3 connected into the circuit to result in a total current of 0.50 A?

16. A 10-kΩ 100-W resistor, a 15-kΩ 50-W resistor, and a 100-kΩ 10-W resistor are connected in parallel.
  (*a*) What is the maximum voltage which may be applied without exceeding the rating of any resistor?
  (*b*) What is the total current drawn by the combination when the voltage of part (*a*) is applied?

13-3 COMPOUND CIRCUITS

The solution of circuits containing combinations of series and parallel branches generally consists in reducing the parallel branches to equivalent series circuits and combining them with the series branches. No set rules can be formulated for the solution of all types of such circuits, but from the examples that follow you will be able to build up your own methods of attack.

Example 6
Find the total resistance of the circuit represented in Fig. 13-6.

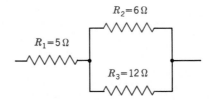

Fig. 13-6 Series-parallel circuit of Example 6.

Solution
Note that the parallel branch of Fig. 13-6 is the circuit of Example 1. Since the equivalent series resistance of the parallel branch is

$$\frac{R_2 R_3}{R_2 + R_3}$$

the circuit reduces to two resistances in series, the total resistance of which is

$$R_t = R_1 + \frac{R_2 R_3}{R_2 + R_3} = 5 + \frac{6 \times 12}{6 + 12} = 9.0\ \Omega$$

Example 7

Find the total resistance of the circuit represented in Fig. 13-7.

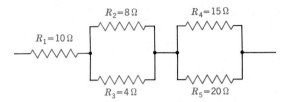

Fig. 13-7 Circuit of Example 7 consisting of one resistance in series with two parallel branches.

Solution

The circuit of Fig. 13-7 is similar to that shown in Fig. 13-6, but with an additional parallel branch. By utilizing the expression for the joint resistance of two resistances in parallel, the entire circuit reduces to three resistances in series, the total resistance of which is

$$R_t = R_1 + \frac{R_2 R_3}{R_2 + R_3} + \frac{R_4 R_5}{R_4 + R_5}$$

$$= 10 + \frac{8 \times 4}{8 + 4} + \frac{15 \times 20}{15 + 20}$$

$$= 21.2 \ \Omega$$

Example 8

Find the total resistance between points a and b in Fig. 13-8.

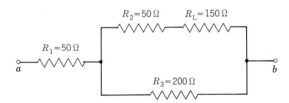

Fig. 13-8 Circuit of Example 8.

Solution

Since R_2 and R_L are in series, they must be added before being combined with R_3. Again, by utilizing the expression for the joint resistance of two resistances in parallel, the entire circuit reduces to two resistances in series. Thus, the total resistance is

$$R_t = R_1 + \frac{R_3(R_2 + R_L)}{R_2 + (R_2 + R_L)}$$

$$= 50 + \frac{200(50 + 150)}{200 + 50 + 150}$$

$$= 150 \ \Omega$$

Note that the circuit of Fig. 13-8 is identical with that of Fig. 13-9. The latter is the customary method for representing T networks, often

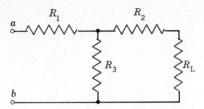

Fig. 13-9 Circuit of Example 8 illustrated in T-network form.

encountered in communication circuits, where R_L is the load or receiving resistance.

Example 9

Find the resistance between points a and b in Fig. 13-10.

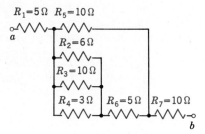

Fig. 13-10 Circuit of Example 9.

Solution

In many instances a circuit diagram that *appears* to be complicated can be better understood and analyzed by redrawing it in a form which is more simplified. For example, Fig. 13-11 represents the circuit of Fig. 13-10.

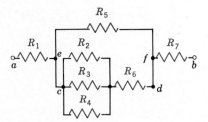

Fig. 13-11 Simplified circuit of Example 9.

First, find the equivalent series resistance of the parallel group formed by R_2, R_3, and R_4 and add it to R_6, which will result in the resistance R_{cd} between points c and d. Now, combine R_{cd} with R_5, which is in parallel, to give an equivalent series resistance R_{ef} between points e and f. The circuit is now reduced to an equivalence of R_1,

R_{ef}, and R_7 in series, which are added to obtain the total resistance R_{ab} between points a and b. The joint resistance of R_2, R_3, and R_4 is 1.67 Ω, which, when added to R_6, results in a resistance $R_{cd} = 6.67$ Ω between c and d. The equivalent series resistance R_{ef} between points e and f, formed by R_{cd} and R_5 in parallel, is 4.0 Ω. Therefore, the resistance R_{ab} between points a and b is

$$R_{ab} = R_1 + R_{ef} + R_7 = 19\ \Omega$$

PROBLEMS 13-3

1. In the circuit of Fig. 13-12, $R_1 = 510$ Ω, $R_2 = 300$ Ω, $R_3 = 470$ Ω, and $V_G = 230$ V. What is the total current I_t of the circuit?

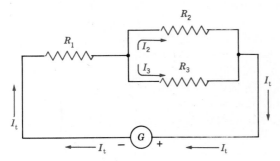

Fig. 13-12 R_1 connected in series with R_2 and R_3 in parallel.

2. In Prob. 1, how much power is expended in R_3?
3. In Prob. 1, if R_1 is short-circuited, how much power is expended in R_2?
4. In Prob. 1, what will be the total current I_t if R_2 is open-circuited?
5. In the circuit of Fig. 13-12, $R_1 = 62$ kΩ, $R_2 = 15$ kΩ, and $I_t = 3.26$ mA and the voltage across R_3 is 27.9 V. Find (a) V_G, (b) R_3, (c) R_t, (d) I_2, (e) I_3.
6. In Prob. 5, how much current will the generator supply if R_3 is short-circuited?
7. In the circuit of Fig. 13-12, $R_t = 5.562$ kΩ, $R_1 = 3.9$ kΩ, $V_G = 1000$ V, and $I_2 = 135.4$ mA. Find (a) voltage across R_1, (b) voltage across R_2, (c) resistance of R_2 to two significant figures, (d) resistance of R_3 to two significant figures, (e) total current I_t, (f) current through R_3, (g) total power expended in the circuit.
8. In Prob. 7, if R_1 is short-circuited, (a) how much power will be expended in R_2 and (b) how much current will flow through R_3?
9. In the circuit of Fig. 13-9, R_1, R_2, and R_3 are 200-Ω resistors and $R_L = 470$ Ω. What is the effective resistance between points a and b?
10. In the circuit of Fig. 13-9, $R_1 = R_2 = R_3 = 300$ Ω and $R_L = 600$ Ω. What is the resistance between points a and b?
11. In the circuit of Fig. 13-9, $R_1 = R_2 = R_L = 300$ Ω and $R_3 = 600$ Ω. What is the resistance between points a and b?

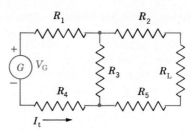

Fig. 13-13 Circuit of Prob. 12.

12. In the circuit of Fig. 13-13, $R_1 = R_2 = R_4 = R_5 = 10 \ \Omega$ and $R_3 = R_L = 600 \ \Omega$. If a voltage of 30 V exists across R_L, what is the total current I_t?

13. In the circuit of Fig. 13-14, the generator voltage $V_G = 3500$ V, $R_4 = 1.5 \ k\Omega$, $R_2 = 6.8 \ k\Omega$, $I_2 = 52.9$ mA, $R_3 = 2.7 \ k\Omega$, and $I_t = 273$ mA. Find to two significant figures (*a*) resistance of R_1, (*b*) resistance of R_5, and (*c*) power expended in R_3.

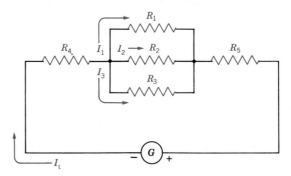

Fig. 13-14 Series-parallel circuit of Prob. 13.

14. In the circuit represented in Fig. 13-15, find the total current I_t.

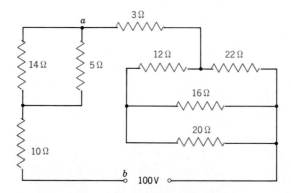

Fig. 13-15 Circuit of Prob. 14.

15. If, in Fig. 13-15, points *a* and *b* are short-circuited, find the total power expended.

16. What is the total curent I_t in the circuit shown in Fig. 13-16?

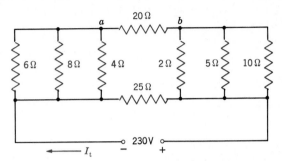

Fig. 13-16 Circuit of Prob. 16.

17. In the circuit of Prob. 16, what is the current flow through the 5-Ω resistor?

18. What would be the power expended in the circuit of Fig. 13-16 if points a and b were short-circuited?

Meter Circuits

14

Chapters 8 and 13 dealt with the study of Ohm's law as applied to series and parallel circuits, and in Chap. 9 consideration was given to the effects of resistance in current-carrying conductors. The principles and methods learned therein are applied in the present chapter to circuits relating to *dc instruments* used for servicing electrical, radio, and other electronic equipment.

14-1 DIRECT-CURRENT INSTRUMENTS— BASIC METER MOVEMENT

The most common measuring instruments used with electric and electronic circuits are the *voltmeter* and the *ammeter*. As the names imply, a voltmeter is an instrument used to measure voltage and an ammeter is a current-measuring instrument.

The great majority of meters used with direct currents employ the D'Arsonval movement illustrated in Fig. 14-1. This movement utilizes a

Fig. 14-1 D'Arsonval meter movement. (*Courtesy of Western Electrical Instrument Corporation*)

Fig. 14-2 0–1 milliammeter. (*Courtesy of Triplett Electrical Instrument Company*)

coil of wire mounted on jeweled bearings between the pole pieces of a permanent magnet. When direct current flows through the coil, a magnetic field is set up around the coil, thereby producing a force which, in conjunction with the magnetic field of the permanent magnet, causes the coil to rotate from the no-current position. Since the arc of rotation is proportional to the amount of current passing through the coil, a pointer can be attached to the coil and the deflection of the pointer over a calibrated scale can be used to indicate values of current.

The *sensitivity* of a current-indicating meter is the amount of current necessary to cause full-scale deflection of the pointer. For example, an instrument of wide usage is the 0–1 milliammeter illustrated in Fig. 14-2. This meter has a sensitivity of 1 mA because, when a current of 1 mA flows through the meter, the pointer indicates full-scale deflection. This particular meter has an internal resistance of 55 Ω. Other meter movements have different sensitivities with various values of internal resistance.

14-2 MULTIRANGE CURRENT METERS

Instead of utilizing a number of meters to make various current measurements, it is common practice to select a meter movement with sufficient sensitivity and, with the aid of one or more shunts, extend the range and therefore the usefulness of the meter. A shunt, in this application, is a resistor that is shunted (connected in parallel) across the meter coil as shown in Fig. 14-3.

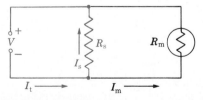

Fig. 14-3 Total current I_t consists of current I_s, which flows through shunt resistor R_s, and the meter current I_m, which flows through the coil of the meter. That is, $I_t = I_s + I_m$.

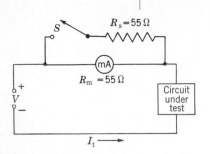

Fig. 14-4 Total current I_t flows through the milliammeter, which indicates a full-scale deflection of 1 mA.

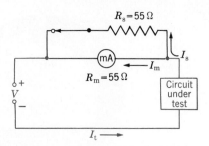

Fig. 14-5 Total current I_t divides equally between meter resistance R_m and shunt resistance R_s. $I_t = I_m + I_s = 1$ mA and $I_s = I_m = 0.5$ mA.

A meter such as illustrated in Fig. 14-2, with a resistance of 55 Ω, is connected to measure the circuit current of Fig. 14-4. In this condition the switch S is open and the meter indicates a full-scale deflection of 1 mA. In Fig. 14-5 the switch S is closed, thereby shunting the 55-Ω resistor R_s across the meter. Since the meter resistance and shunt resistance are equal, the circuit current I_t divides equally between them and the meter reads 0.5 mA.

In Fig. 14-4, with the switch open, the meter would indicate actual values of current. In Fig. 14-5, with the switch closed, circuit current would be obtained by multiplying the meter readings by a factor of 2 or by re-marking the scale as shown in Fig. 14-6.

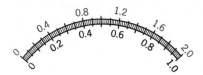

Fig. 14-6 Multirange meter scale.

Example 1

A 0–1 milliammeter has an internal resistance of 70 Ω. Design a circuit that will allow this meter to be used as a multirange meter having the ranges 0–1, 0–10, and 0–100 mA and 0–1 A.

Solution

The circuit is shown in Fig. 14-7. The switch S is used for range selection by switching in the proper shunt resistor. In its present position no shunt resistor is used and therefore the meter is connected to measure within its basic range of 0–1 mA. At full-scale deflection the voltage across the meter will be

$$V_m = I_t R_m = 0.001 \times 70 = 7 \times 10^{-2} \text{ V}$$

Since whatever shunt resistor is in use will be in parallel with the resistance of the meter R_m, the same voltage will appear across the shunt resistance. That is,

$$V_m = V_s = 7 \times 10^{-2} \text{ V}$$

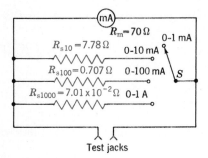

Fig. 14-7 Circuit for extending range of 0–1 milliammeter. Test leads from jacks are connected in series with circuit in which current is to be measured.

When the 0- to 10-mA range is used, the switch S will connect R_{s10} in parallel with the meter and therefore its internal resistance R_m. For full-scale deflection, 1 mA must flow through the meter coil, which leaves 9 mA to flow through R_{s10}. For this condition the value of R_{s10} must be

$$R_{s10} = \frac{V_s}{I_s} = \frac{7 \times 10^{-2}}{9 \times 10^{-3}} = 7.78 \; \Omega$$

Similarly, when the 0- to 100-mA range is placed in operation by switching to shunt resistor R_{s100}, full-scale deflection 1 mA still must flow through the meter coil, leaving 99 mA to flow through R_{s100}. Then,

$$R_{s100} = \frac{V_s}{I_s} = \frac{7 \times 10^{-2}}{99 \times 10^{-3}} = 0.707 \; \Omega$$

Likewise, when the 0- to 1-A (0- to 1000-mA) range is used, 999 mA must flow through the shunt resistor for full-scale deflection.

$$\therefore R_{s1000} = \frac{V_s}{I_s} = \frac{7 \times 10^{-2}}{999 \times 10^{-3}} = 0.0701 \; \Omega$$

It will be noted that only basic Ohm's law was used in Example 1. This was done to emphasize the usefulness of the law. Also, special seldom-used formulas are difficult to remember and handbooks for ready reference are not always available on the job. Actually, you can find the resistance of a meter shunt by using your knowledge of current distribution in parallel circuits. For the 0- to 10-mA range of Example 1, the 70-Ω meter movement must carry 1 mA and the shunt resistor must carry 9 mA. Since the shunt carries nine times the meter current, the shunt resistance must be one-ninth the resistance of the meter, or $\frac{1}{9} \times 70 = 7.78 \; \Omega$.

Similarly, for the range of 0 to 100 mA, the meter movement still must carry 1 mA, leaving 99 mA to flow through the shunt. Therefore, the resistance of the shunt will be one ninety-ninth of the resistance of the meter movement, or $\frac{1}{99} \times 70 = 0.707 \; \Omega$.

Now that the principles of meter shunts are understood, it is left as an exercise for you to show that

$$R_s = \frac{R_m}{N-1} \quad \Omega \tag{1}$$

where R_s = shunt resistance, Ω
R_m = meter resistance, Ω
N = ratio obtained by dividing new full-scale reading by basic full-scale reading, both readings in same units

The ratio N is known as the *multiplying power* of the shunt resistor, that is, the factor by which the basic meter scale is multiplied when the shunt resistor R_s is connected in parallel with the meter resistance R_m. From Eq. (1),

$$N = \frac{R_m}{R_s} + 1$$

Example 2

By what factor must the scale readings be multiplied when a resistance of 100 Ω is connected across a meter movement of 400 Ω?

Solution

$$N = \frac{R_m}{R_s} + 1 = \frac{400}{100} + 1 = 5$$

14-3 SHUNTING METHODS

Although mechanical details are not shown in Fig. 14-7, it is necessary to use a shorting switch in this type of circuit to avoid damage to the meter movement. When operation requires switching from one shunt to another, the new shunt must be connected before contact with the shunt in use is broken. If it is not, the entire circuit current will flow through the meter movement while the switch is moving from one contact to another.

By another method of switching, illustrated in Fig. 14-8, shunts are connected into the circuit by the two-pole rotary switch which makes

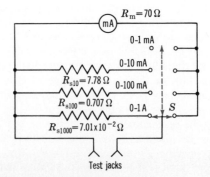

Fig. 14-8 Method of switching shunts.

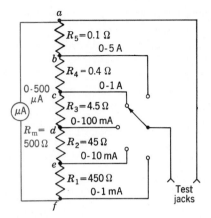

Fig. 14-9 Multicurrent test meter using universal shunt.

connections between two sets of contacts. With this arrangement, the meter movement is protected by an open circuit when the operator switches from one shunt to another.

Still another method of employing shunts is shown in Fig. 14-9. This is known as the *Ayrton,* or *universal,* shunt. In addition to other advantages, it provides a safe and convenient method of switching from one range to another. The total shunt resistance, which is permanently connected across the meter, generally has the same resistance as the meter movement. The value of the resistance for each range shunt can be computed by dividing the total circuit resistance $R_{a-f} + R_m$ by the multiplying power N. This is demonstrated by the development which follows:

$$R_{a-e}(I_t - I_m) = (R_{e-f} + R_m)I_m$$

$$= (R_{a-f} - R_{a-e} + R_m)I_m$$

$$R_{a-e}I_t - R_{a-e}I_m = R_{a-f}I_m - R_{a-e}I_m + R_mI_m$$

$$R_{a-e}I_t = R_{a-f}I_m + R_mI_m$$

$$R_{a-e} = \frac{I_m}{I_t}(R_{a-f} + R_m)$$

$$R_{a-e} = \frac{1}{N}(R_{a-f} + R_m) \tag{2}$$

where R_{a-e} = portion of Ayrton shunt which is connected in shunt for the meter connection at point e. In Fig. 14-9, this is $R_{a-e} = R_2 + R_3 + R_4 + R_5$.

R_{a-f} = total Ayrton shunt (In Fig. 14-9, this is illustrated as $R_1 + R_2 + R_3 + R_4 + R_5$.)

R_m = meter movement resistance

N = multiplier for switch setting (In Fig. 14-9, $N = 2$ for setting at f, $N = 20$ for setting at e, and so on.)

For example, the 0–500 microammeter movement has a resistance R_m of 500 Ω and total shunt resistance R_{a-f} connected across the meter is

500 Ω. When the switch is on the 0- to 1-mA position, the multiplying power N is 2.

For the 0- to 10-mA range, N would be 20 because 10 mA is 20 times the original full scale of 0.5 mA. Therefore, the required shunt for this range is

$$R_{a-e} = \frac{R_{a-f} + R_m}{N} = \frac{500 + 500}{20} = 50 \ \Omega$$

Since the entire shunt resistance is 500 Ω

$$R_1 = R_{a-f} - R_{a-e} = 500 - 50 = 450 \ \Omega$$

When the switch is connected to the 0- to 100-mA range, N becomes 200 and R_1 and R_2 in series (R_{d-f}) form the shunt. That is,

$$R_{a-d} = \frac{R_{a-f} + R_m}{N} = \frac{500 + 500}{200} = 5 \ \Omega$$

Note $\qquad R_{a-d} = \dfrac{2R_m}{N} \qquad$ when $R_{a-f} = R_m$

Since $\qquad\qquad R_1 = 450 \ \Omega \qquad$ and $\qquad R_{a-d} = 5 \ \Omega$

then $\qquad\qquad\qquad R_2 = R_{a-f} - (R_1 + R_{a-d})$
$$= 500 - (450 + 5)$$
$$= 45 \ \Omega$$

The values of the remaining shunts are computed in the same manner.

PROBLEMS 14-1

1. A 0–1 milliammeter has an internal resistance of 53 Ω. What shunt resistance is required to extend the meter range to 0–50 mA?

2. A meter movement with a sensitivity of 100 μA has an internal resistance of 1250 Ω. How much shunt resistance is required to result in a 0- to 10-mA range?

3. The meter in Prob. 1 is being used as a multicurrent instrument. The shunt for the 0- to 50-mA range is burned out, but a spool of No. 30 enamel-covered copper wire is on hand. How much of this wire is needed to wind a substitute shunt?

4. A 0–1 milliammeter has an internal resistance of 46 Ω. If this meter is shunted with a 0.939-Ω resistor, by what must the meter readings be multiplied to obtain the correct values of current?

5. It is desired to use the milliammeter illustrated in Fig. 14-2 as a multicurrent meter. What values of shunts are required for the following ranges: (a) 0–10 mA, (b) 0–100 mA, (c) 0–1 A, (d) 0–10 A?

6. In the circuit of Fig. 14-10, the total shunt resistance is equal to the resistance of the meter movement. Find the values of R_1, R_2, R_3, R_4, and R_5?

7. A 0–1 milliammeter is available. Design an Ayrton shunt to permit it to be used for the following ranges: (a) 0–10 mA, (b) 0–100 mA, (c) 0–1 A, (d) 0–10 A. The meter resistance is 1500 Ω.

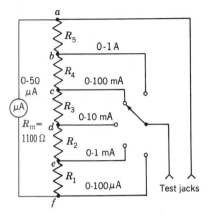

Fig. 14-10 Multicurrent meter circuit of Prob. 6.

14-4 VOLTMETERS

In Fig. 14-11, a voltage of 1 V is impressed across a circuit consisting of a 0–1 milliammeter in series with a variable resistor. The resistor is so adjusted that the circuit is limited to 1 mA; therefore, the meter indicates a full-scale deflection, or a reading of 1 mA. If the resistor is unchanged and the voltage is reduced to 0.5 V, then the circuit current will be reduced to one-half its original value and the meter will read 0.5 mA. Even though the meter deflection is the result of current flow, actually the meter can be used as a 0–1 voltmeter, indicating 1 V in the first instance and 0.5 V when the voltage is reduced.

Similarly, if the resistor is adjusted to a higher safe value so that the application of 150 V causes full-scale deflection, the instrument can be used as a 0–150 voltmeter. In that case voltage values will be obtained by multiplying the basic scale readings by a factor of 150 or by substituting a new scale as shown in Fig. 14-12.

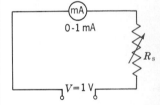

Fig. 14-11 Basic circuit of milliammeter used to indicate voltage.

Fig. 14-12 Panel voltmeter. (*Courtesy of Western Electrical Instrument Corporation*)

Example 3

It is desired to use the milliammeter of Fig. 14-2 as a 0–10 voltmeter. What resistance R_{mp} must be connected in series with the instrument to accomplish this?

Solution

The additional series resistance is called a *multiplier* resistance, and its value must be such that, when it is added to the resistance of the meter movement, the total resistance will limit the current through the instrument to 1 mA when 10 V is applied. The circuit is shown in Fig. 14-13. R_{mp} is the multiplier resistance, and $R_m = 55\ \Omega$ is the resistance of the meter movement.

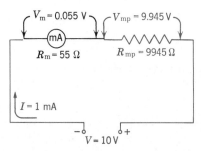

Fig. 14-13 Voltmeter circuit of Example 3.

If 10 V is to be applied across the two series resistances as shown in Fig. 14-13, in order to limit the current to 1 mA, 0.055 V must appear across the meter because

$$V_m = IR_m = 10^{-3} \times 55 = 0.055\ \text{V}$$

The remaining voltage, which is $10 - 0.055 = 9.945$ V, must appear across R_{mp}. Accordingly,

$$R_{mp} = \frac{V_{mp}}{I} = \frac{9.945}{10^{-3}} = 9945\ \Omega$$

If a $10\,000$-Ω resistor is used as a multiplier, with 10 V applied to the jacks, and if an observer could discern the difference, the voltage reading would be in error by only 0.05 V. (What percent error does this represent?)

Example 4

A 0–50 microammeter, with a resistance of 1140 Ω, is to be used as a 0–100 voltmeter. What value of multiplier resistance is needed?

Solution

For full-scale deflection the voltage across the meter must be limited to

$$V_m = IR_m = 50 \times 10^{-6} \times 1140$$
$$= 0.057\ \text{V}$$

The voltage that remains across the multiplier is $100 - 0.057 = 99.943$ V which results in

$$R_{mp} = \frac{V_{mp}}{I} = \frac{99.943}{50 \times 10^{-6}} = 1\,998\,860 \; \Omega$$

Naturally, a 2-MΩ resistor would be used.

14-5 VOLTMETER SENSITIVITY

The *sensitivity* of a voltmeter is expressed in the number of ohms in the multiplier for each volt of range. For example, the voltmeter of Example 3 has a range of 10 V and a multiplier of 10 000 Ω, which results in a sensitivity of 1000 Ω/V. The voltmeter of Example 4 has a sensitivity of 20 000 Ω/V.

14-6 VOLTMETER LOADING EFFECTS

The sensitivity of a voltmeter is a good indication of the meter's accuracy. This is particularly true when the voltages in the low-current circuits often encountered in electronic equipment are measured. For example, a 0–150 voltmeter with a sensitivity of 200 Ω/V would give excellent service, say as a power switchboard meter, at a reasonable cost. However, it would not be satisfactory for some other applications. In Fig. 14-14, two 60-kΩ resistors are connected in series across 120 V. In this condition, 60 V will appear across each resistor. If the voltmeter is connected across R_2 as shown in Fig. 14-15, the joint resistance R_p of R_2 and R_{mp} becomes

$$R_p = \frac{R_2 R_{mp}}{R_2 + R_{mp}} = 20\,000 \; \Omega$$

The total resistance of the circuit is now

$$R_t = R_1 + R_p = 60\,000 + 20\,000$$
$$= 80\,000 \; \Omega = 80 \; k\Omega$$

This results in a circuit current of

$$I_t = \frac{V}{R_t} = \frac{120}{80\,000} = 1.5 \times 10^{-3} \; A$$

Therefore, the voltage existing across R_2 due to the shunting effect of the voltmeter is

$$V_p = I_t R_p = 1.5 \times 10^{-3} \times 20\,000$$
$$= 30 \; V$$

It is left as an exercise to show that if the voltmeter of Example 4 is used to measure the voltage across R_2, the reading will be 59.1 V.

14-7 MULTIRANGE VOLTMETERS

Using a single multiplier provides only one voltmeter range. Similar to the usage of current-measuring instruments, it has become practice to increase the usefulness of an instrument by selecting a meter movement

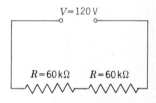

Fig. 14-14 The current through the resistors is 1 mA, and the voltage across each resistor is 60 V.

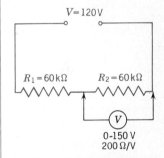

Fig. 14-15 A 30 000-Ω voltmeter connected across R_2. Total circuit current is now 1.5 mA, and the voltage across R_2 is 30 V.

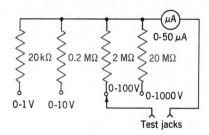

Fig. 14-16 A 0–50 microammeter used with multipliers for multirange voltmeter.

of sufficient sensitivity and, with the use of several multipliers, use the instrument as a multirange voltmeter. Such an arrangement is shown in Fig. 14-16.

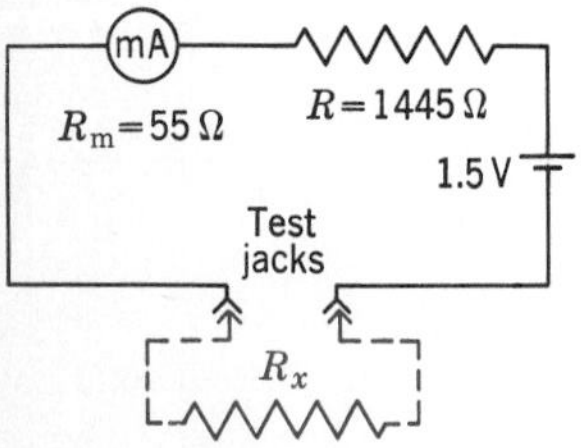

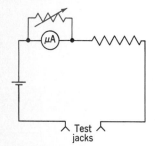

Fig. 14-17 A 0–1 milliammeter used in ohmmeter circuit.

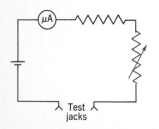

Fig. 14-18 Ohmmeter circuit with variable shunt resistance.

14-8 OHMMETERS

Owing to the fact that a change in the resistance of a circuit will cause a change in the current in that circuit, a current-measuring instrument can be calibrated to indicate values of resistance required for a given change in current. Such a calibrated instrument is called an *ohmmeter*.

In the schematic diagram shown in Fig. 14-17, the 0–1 milliammeter of Fig. 14-2 is connected in series with a 1.5-V battery and a resistance of 1445 Ω. Since the total resistance of the circuit is 1500 Ω, if the test jacks are short-circuited, the meter will read full scale. If the short circuit is removed and a resistance R_x of 1500 Ω is connected across the jacks, the meter will indicate half-scale deflection because now the total circuit resistance is 3000 Ω. Therefore, at full-scale deflection the meter scale could be marked 0 Ω of external circuit resistance, and at half scale it could be marked 1500 Ω. Similarly, other values of known resistance could be used to calibrate the scale throughout its range. Also, unknown resistances can be used to calibrate the scale by making use of the relation

$$R_x = R_c \frac{I_1 - I_2}{I_2} \qquad \Omega \qquad (3)$$

where R_x = unknown resistance, Ω

R_c = circuit resistance when test jacks are short-circuited, Ω

I_1 = current when test jacks are short-circuited, A

I_2 = current when R_x is connected in circuit, A

Use your knowledge of Ohm's law and Axiom 5 (Sec. 5-2) to derive Eq. (3).

As a provision for compensating for battery aging and maintaining calibration, variable resistors controlled from the instrument panel are connected in ohmmeter circuits by either of the methods as illustrated in Figs. 14-18 and 14-19. In each case the test leads are short-circuited and the resistor control is adjusted until the meter reads full scale, or 0 Ω. An

Fig. 14-19 Ohmmeter circuit with variable series resistance.

Fig. 14-20 Multimeter; see the arrangement of shunts and multipliers on the selector switch. (*Courtesy of Triplett Electrical Instrument Company*)

example of such a control is the "Ω ADJ" on the instrument as shown in Fig. 14-20.

Since zero resistance between the test jacks results in maximum current and larger values of resistance result in less current, certain types of ohmmeter scales are marked with numbers increasing from right to left as illustrated on the ohms scale in Fig. 14-20.

In practice, the use of the ordinary ohmmeter should be limited from about one-tenth of to ten times the center-scale resistances reading because of the small deflection changes at the ends of the scale. For this reason multirange ohmmeters are employed for changing midscale values, and the ranges generally are designed to multiply the basic scale by some power of 10.

14-9 MULTIMETERS

For the purposes of convenience and economy, meters combining the functions and desired ranges of ammeters, voltmeters, and ohmmeters are incorporated into one instrument called a multimeter, one type of which is illustrated in Fig. 14-20. If the test leads are plugged into the proper pin jacks and the rotary switch is switched to the proper function and range, the instrument can be utilized for several functions.

PROBLEMS 14-2

1. In the circuit of Fig. 14-21: (*a*) What voltages are across R_1 and R_2? (*b*) A 0–100 voltmeter with a sensitivty of 1000 Ω/V is connected across R_1. What is the reading of the voltmeter?

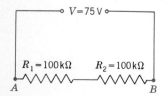

$V=75\,\text{V}$

$R_1=100\,\text{k}\Omega$ $R_2=100\,\text{k}\Omega$

A B

Fig. 14-21 Circuit of Probs. 1 and 2.

2. In the circuit of Fig. 14-21:
   (*a*) A 0–100 voltmeter with a sensitivity of 20 000 Ω/V is connected across R_1. What is the voltmeter reading?
   (*b*) What will the voltmeter read if connected across points *A* and *B*?
   (*c*) When the voltmeter is connected across points *A* and *B*, what current flows through R_2?

3. What are the values of the multiplier resistors R_1, R_2, R_3, and R_4 in Fig. 14-22?

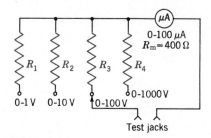

0-100 μA
$R_m=400\,\Omega$

R_1 R_2 R_3 R_4

0-1 V   0-10 V   0-100 V   0-1000 V

Test jacks

Fig. 14-22 Multirange voltmeter circuit of Prob. 3.

4. Refer to Eq. (1). Did you show that $R_s = \dfrac{R_m}{N-1}\ \Omega$?

5. Refer to the end of Sec. 14-6. Did you show that the voltmeter reading will be 59.1 V?

6. Refer to Eq. (3). Did you show that $R_x = R_c\dfrac{I_1 - I_2}{I_2}$?

14-10 DIGITAL METERS

The D'Arsonval meter movements discussed in this chapter have been in service for at least three generations, and they will continue to find many applications for years to come. These analog devices have long service lives (or their successors have, after their users progress past the beginner stage), and they are superior to digital meters when it comes to monitoring varying values. However, when accuracy of reading, ease of reading, and many accurate repetitions of a reading for the same value of parameter are required, digital displays are preferable.

The various methods employed to convert analog signals to digital readouts involve mathematical concepts beyond the scope of this chapter. They require the use of graphs (Chap. 16) and time constants (Chap. 34), and they rely heavily on operational amplifiers and a variety of readout devices.

With a profusion of LSI and VLSI devices now on the market, digital meters which may be accurate to ±0.01%, ±1 digit are available. (You may find it interesting to investigate the meanings of the various ways in which meter accuracy may be stated.)

Divider Circuits and Wheatstone Bridges

15

In this chapter consideration is given to voltage and current divider circuits. Computations involving voltages and currents in these circuits are simply applications of Ohm's law to series and parallel circuits.

The source of power for radio and television receivers, amplifiers, and similar electronic equipment generally consists of a filtered direct voltage which has been obtained from a rectified alternating voltage. For reasons of economy and design considerations, rectifier power supplies are usually so designed that only the highest voltage desired is available at the output. In most applications, however, other voltages are needed. For example, cathode ray tubes require higher voltages than other circuit elements require. Screen grids require yet other voltages. Also, bias voltages are required. These voltages can be made available from single sources of voltage by the use of *voltage dividers*.

15-1 VOLTAGE DIVIDERS

That several values of voltage can exist around a circuit was first demonstrated in Sec. 8-8 and Figs. 8-12 and 8-13. A similar situation exists when tapped resistors, or resistors in series, are connected across the output of a power supply as illustrated in Fig. 15-1. This represents a simple *voltage divider*.

Since the resistors are of equal value, one-third of the 300-V output voltage will appear across each one. Therefore, since terminal D is at zero or ground potential, terminal C will be $+100$ V with respect to D, terminal B will be $+200$ V, and terminal A will be $+300$ V.

In addition to serving as a voltage divider, the total resistance connected across the output of a power supply generally serves as a *load resistor* and as a *bleeder*. The latter serves to "bleed off" the charge of the filter capacitors after the rectifier is turned off. As a compromise between output voltage regulation and efficiency of operation, the total value of the voltage divider resistance is so designed that the bleeder current will be about 10% of the full-load current. The bleeder current in Fig. 15-1 with no loads connected to the various voltage divider terminals is

$$I = \frac{V}{R_1 + R_2 + R_3} = \frac{300}{75\,000}$$
$$= 4.00 \text{ mA}$$

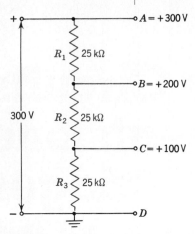

Fig. 15-1 Voltage divider consisting of three 25-kΩ resistors connected across 300-V power supply.

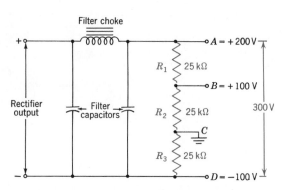

Fig. 15-2 Voltage divider grounded at C.

The grounded point of a voltage divider is generally used as the reference point for circuit voltages supplied by the voltage divider. In Fig. 15-1, this is at grounded terminal D.

If the power supply output voltage is grounded at no other point, the voltage divider can be grounded at an intermediate point so as to obtain both positive and negative voltages. For example, if the voltage divider resistors of Fig. 15-1 are grounded as shown in Fig. 15-2, the voltage relations change. Terminal D is now -100 V with respect to ground, B is $+100$ V, and A is $+200$ V.

15-2 VOLTAGE DIVIDERS WITH LOADS

The voltage dividers of Figs. 15-1 and 15-2 have no loads connected to them; only the bleeder current of 4 mA flows through the voltage divider resistors. When loads are connected to the various terminals, the resulting additional currents must be taken into consideration because they affect the operating voltages. For example, assume a load of $R_4 = 50\,000$ Ω connected between terminals C and D of Fig. 15-1. Under these conditions, the resistance between terminals C and D is

$$R_{CD} = \frac{R_3 R_4}{R_3 + R_4} = \frac{25\,000 \times 50\,000}{25\,000 + 50\,000}$$
$$= 16\,700 \ \Omega = 16.7 \text{ k}\Omega$$

The total resistance between terminals A and D is

$$R_{AD} = R_1 + R_2 + R_{CD} = 66\,700 \ \Omega$$

resulting in a total current of

$$I_t = \frac{V}{R_{AD}} = 4.50 \text{ mA}$$

The voltage across terminals B and D is

$$V_{BD} = I_t R_{BD}$$
$$= 188 \text{ V} \qquad \text{(instead of 200 V)}$$

and across terminals C and D it is

$$V_{CD} = I_t R_{CD}$$
$$= 75 \text{ V} \qquad \text{(instead of 100 V)}$$

The circuit is shown in Fig. 15-3.

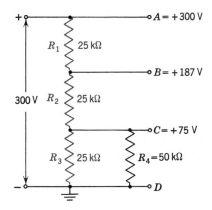

Fig. 15-3 Load of 50 kΩ connected across terminals C and D.

Show that, if an additional load of $R_5 = 50$ kΩ is connected across terminals B and D, the terminal voltages would be as illustrated in Fig. 15-4.

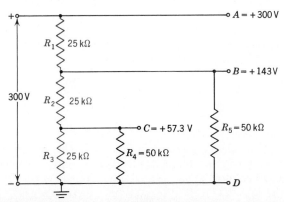

Fig. 15-4 Loads $R_4 = R_5 = 50$ kΩ connected to voltage divider.

Example 1

Design a voltage divider circuit for a 250-V power supply. The connected loads are 60 mA at 250 V and 40 mA at 150 V. Allow a 10% bleeder current.

Solution

The circuit is shown in Fig. 15-5. The total load current is 100 mA; therefore, the bleeder current, which flows through R_2, is 10 mA. Since the voltage across R_2 is 150 V,

$$R_2 = \frac{150}{10 \times 10^{-3}} = 15\ 000\ \Omega = 15\ \text{k}\Omega$$

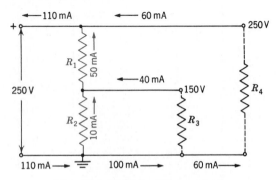

Fig. 15-5 Circuit of Example 1.

The current flowing through R_1 is $40 + 10 = 50$ mA, and the voltage across R_1 is $250 - 150 = 100$ V. Then

$$R_1 = \frac{100}{50 \times 10^{-3}} = 2000\ \Omega$$

Example 2

What are the values of the voltage divider resistors in Fig. 15-6 if the bleeder current is 10% of the total load current?

Solution

The total load current I_L is

$$I_L = 50 + 40 + 30 = 120\ \text{mA}$$

The bleeder current is

$$I_B = 0.1 \times 120 = 12\ \text{mA}$$

The complete circuit is shown in Fig. 15-7. The voltage across R_3 is 150 V, and only the bleeder current of 12 mA flows through this resistor. Therefore,

$$R_3 = \frac{150}{12 \times 10^{-3}} = 12.5\ \text{k}\Omega$$

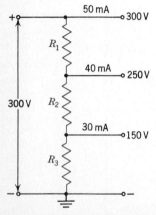

Fig. 15-6 Voltage divider of Example 2.

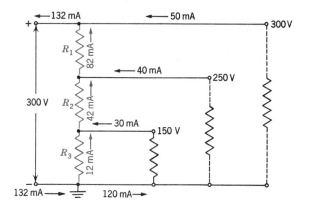

Fig. 15-7 Complete circuit of Example 2.

The 30-mA load current of the 150-V load terminal combines with the bleeder current of 12 mA for a total of 42 mA through R_2, across which is 100 V. Therefore,

$$R_2 = \frac{100}{42 \times 10^{-3}} = 2.38 \text{ k}\Omega$$

Similarly, 82 mA flows through R_1, across which is 50 V. Then

$$R_1 = \frac{50}{82 \times 10^{-3}} = 610 \ \Omega$$

Note
Resistors $R_1 = 610 \ \Omega$, $R_2 = 2380 \ \Omega$, and $R_3 = 12\ 500 \ \Omega$ are not readily available commercially. Try substituting standard preferred values of $R_1 = 560 \ \Omega$, $R_2 = 2.4 \text{ k}\Omega$, and $R_3 = 12 \text{ k}\Omega$ for the computed values, and determine how this would affect the loads.

Example 3
Find the values of the voltage divider resistors of Fig. 15-8. The -50-V bias terminal draws no current, and the bleeder current is 10% of the total load current.

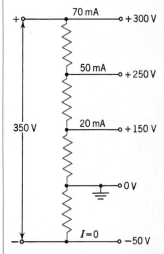

Fig. 15-8 Voltage divider of Example 3.

Solution
The total load current I_L is

$$I_L = 70 + 50 + 20 = 140 \text{ mA}$$

The bleeder current is

$$I_B = 0.1 I_L = 0.1 \times 140 = 14 \text{ mA}$$

The complete circuit is illustrated in Fig. 15-9. There is a voltage of 50 V across R_4, and the total current of 154 mA flows through this resistor. Therefore

$$R_4 = \frac{50}{154 \times 10^{-3}} = 325 \ \Omega$$

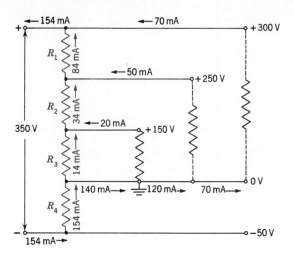

Fig. 15-9 Complete circuit of Example 3.

Since R_3 carries only the bleeder current and the voltage across this resistor is 150 V,

$$R_3 = \frac{150}{14 \times 10^{-3}} = 10.7 \text{ k}\Omega$$

In like manner,

$$R_3 = \frac{100}{34 \times 10^{-3}} = 2.94 \text{ k}\Omega$$

and

$$R_1 = \frac{50}{84 \times 10^{-3}} = 595 \text{ }\Omega$$

Note

As a problem, substitute the commercially available preferred values of $R_1 = 620 \text{ }\Omega$, $R_2 = 3 \text{ k}\Omega$, $R_3 = 11 \text{ k}\Omega$, and $R_4 = 300 \text{ }\Omega$ for the computed values, and determine how the loads would be affected.

PROBLEMS 15-1

1. The vertical attenuator of an oscilloscope is illustrated in Fig. 15-10. With an input voltage of 60 V, what voltages appear between the switch positions and the input to the vertical amplifier?
   Note No current flows from the circuit.

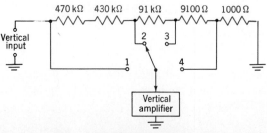

Fig. 15-10 Circuit of Prob. 1.

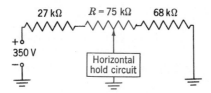

Fig. 15-11 Circuit of Prob. 2.

2. The horizontal hold control of a television receiver is illustrated in Fig. 15-11. What range of control voltage is available from the potentiometer to the horizontal hold control?
    Note The horizontal hold draws no current from the circuit.
3. What is the power dissipated by each of the resistors and the potentiometer of Prob. 2?
4. Determine the values of the voltage divider resistors of Fig. 15-12 if a total of 180 mA is drawn from the power supply.

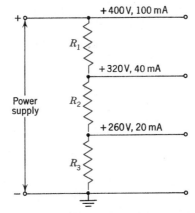

Fig. 15-12 Circuit of Probs. 4 and 5.

5. What is the total power expended in the voltage divider of Prob. 4, and what power is dissipated by each of the resistors?
6. What are the values of the voltage divider resistors of Fig. 15-13 if the bleeder current is 10% of the total load current?

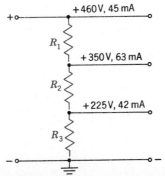

Fig. 15-13 Circuit of Probs. 6, 7, and 8.

7. What is the power dissipated by each of the resistors in Prob. 6?
8. What is the total power delivered by the voltage source in Prob. 6?
9. What are the values of the voltage divider resistors of Fig. 15-14 if the bleeder current is 10 mA?
10. What wattage ratings should be used for the resistors in Prob. 9?

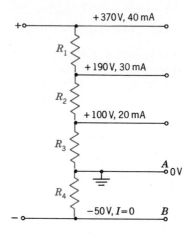

Fig. 15-14 Circuit of Probs. 9, 10, and 11.

11. What is the total power delivered by the voltage source of Prob. 9?
12. If the biasing resistor R_4 of Fig. 15-14 became open-circuited, what would be the voltage between terminals A and B?
13. Referring to Sec. 15-2, did you show that Fig. 15-4 is the result when Fig. 15-3 is changed by the addition of a 50-kΩ load?
14. Referring to Example 2, did you try substituting standard 5% preferred values into the voltage divider of Fig. 15-7?
15. Referring to Example 3, did you try substituting standard 5% preferred values into the voltage divider of Fig. 15-9?

15-3 CURRENT DIVIDERS

We have seen that voltage dividers are employed to develop voltage drops across series resistors. Each voltage drop is proportional to the resistance value related to the total series resistance. When resistors are connected in parallel, the voltage is the same across each, but the current is divided in *inverse* proportion.

Example 4

In Fig. 15-15, since

$$V = \text{voltage drop across } R_1 = I_1 R_1$$

and
$$V = \text{voltage drop across } R_2 = I_2 R_2$$

and
$$I_t = I_1 + I_2$$

Therefore,
$$I_1 R_1 = I_2 R_2$$

and
$$I_1 R_1 = (I_t - I_1)R_2 = I_t R_2 - I_1 R_2$$

Collecting like terms, $I_1 R_1 + I_1 R_2 = I_t R_2$
$$I_1(R_1 + R_2) = I_t R_2$$

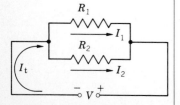

Fig. 15-15 Current divider circuit of Example 4.

so that
$$I_1 = I_t \left(\frac{R_2}{R_1 + R_2} \right) \qquad (1)$$

Note carefully that the numerator in Eq. (1) is R_2 and not R_1. You should now prove to your own satisfaction that

$$I_2 = I_t \left(\frac{R_1}{R_1 + R_2} \right) \qquad (2)$$

PROBLEMS 15-2

1. In Fig. 15-15, $R_1 = 100\ \Omega$ and $R_2 = 300\ \Omega$. $V = 25$ V. Find (a) I_1; (b) I_2.
2. In Fig. 15-15, $R_1 = 2.2\ \text{k}\Omega$ and $R_2 = 4.7\ \text{k}\Omega$. $V = 150$ V. Find (a) I_1; (b) I_2.
3. In Fig. 15-15, $I_t = 150$ mA. $R_1 = 680\ \Omega$ and $R_2 = 1.5\ \text{k}\Omega$. Find (a) I_1; (b) I_2.
4. In Fig. 15-15, it is required that $I_1 = 25$ mA and $I_2 = 75$ mA. A resistance of 200 Ω has been established for R_1.
    (a) What value must be used for R_2?
    (b) What emf must be applied to achieve the required output?
5. In Fig. 15-15, the source of emf is replaced with a constant current source capable of delivering 5 A under all conditions. $R_1 = 100\ \text{k}\Omega$ is a shunt resistor permanently connected across the constant current source. $R_2 = 1.2\ \text{k}\Omega$ is a load driven by the shunted constant current source. Find I_2.

15-4 WHEATSTONE BRIDGE CIRCUITS

The accuracy of resistance measurements by the voltmeter-ammeter method is limited, mainly because of errors in the meters and the difficulty of reading the meters precisely. Probably the most widely used device for precise resistance measurement is the Wheatstone bridge, the circuit diagram of which is shown in Fig. 15-16.

Resistors R_1, R_2, and R_3 are known values, and R_x is the resistance to be measured. In most bridges, R_1 and R_2 are adjustable in ratios of 1:1, 10:1, 100:1, etc., and R_3 is adjustable in small steps. In measuring a resistance, R_3 is adjusted until the galvanometer reads zero, and in this

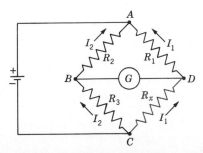

Fig. 15-16 Schematic diagram of Wheatstone bridge.

condition the bridge is said to be "balanced." Since the galvanometer reads zero, it is evident that the points B and D are exactly at the same potential; that is, the voltage drop from A to B is the same as from A to D. Expressed as an equation,

$$V_{AD} = V_{AB}$$

or
$$I_1R_1 = I_2R_2 \tag{3}$$

Similarly, the voltage drop across R_x must be equal to that across R_3; hence,

$$I_1R_x = I_2R_3 \tag{4}$$

Dividing Eq. (4) by Eq. (3),

$$\frac{I_1R_x}{I_1R_1} = \frac{I_2R_3}{I_2R_2}$$

$$\therefore \frac{R_x}{R_1} = \frac{R_3}{R_2} \tag{5}$$

Equation (5) is the fundamental equation of the Wheatstone bridge. By solving it for the only unknown, R_x, the value of the resistance under measurement can be computed.

Since the balance conditions of the bridge are not directly related to the voltage of the energy source, you may be tempted to think that the value of the voltage is immaterial. Obviously, for mere mathematical analysis, any convenient voltage may be assumed. In practice, however, the resistors used in Wheatstone bridges are very precise, and they are usually delicate. You must be sure that their power-dissipating capabilities are not exceeded by applying too high a voltage.

Example 5

In the circuit illustrated in Fig. 15-16, $R_1 = 10\ \Omega$, $R_2 = 100\ \Omega$, and $R_3 = 13.9\ \Omega$. If the bridge is balanced, what is the value of the unknown resistance?

Solution

Solving Eq. (5) for R_x,

$$R_x = \frac{R_1R_3}{R_2}$$

Substituting the known values,

$$R_x = \frac{10 \times 13.9}{100} = 1.39\ \Omega$$

Locating the point at which a telephone cable or a long control line is grounded is simplified by the use of two circuits that are modifications of the Wheatstone bridge. These are the Murray loop and the Varley loop.

Figure 15-17 represents the method of locating the grounded point in a cable by using a Murray loop. A spare ungrounded cable is connected to the grounded cable at a convenient location beyond the grounded point G. This forms a loop of length L, one part of which is the distance x from the point of measurement to the grounded point G. The other part of the

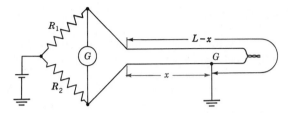

Fig. 15-17 Murray loop.

loop is then $L - x$. These two parts of the loop form a bridge with R_1 and R_2, which are adjusted until the galvanometer shows no deflection. Because this results in a balanced bridge circuit,

$$\frac{R_2}{R_1} = \frac{x}{L - x} \tag{6}$$

Solving for x,

$$x = \frac{R_2}{R_1 + R_2} L \tag{7}$$

Example 6

A Murray loop is connected as in Fig. 15-17 to locate a ground in a cable between two cities 60 km apart. The lines forming the loop are identical. With the bridge balanced, $R_1 = 645\ \Omega$ and $R_2 = 476\ \Omega$. How far is the grounded point from the test end?

Solution

Substituting the known values in Eq. (7),

$$x = \frac{476}{645 + 476} \times 2 \times 60 = 50.95 \text{ km}$$

If the two cables forming the loop are not the same size, the relations of Eq. (7) can be used to compute the resistance R_x of the grounded cable from the point of measurement to the grounded point. Then if R_L is the resistance of the entire loop.

$$R_x = \frac{R_2}{R_1 + R_2} R_L \tag{8}$$

Example 7

A Murray loop is connected as shown in Fig. 15-17. The grounded cable is No. 19 wire, and wire of a different size is used to complete the loop. The resistance of the entire loop is 126 Ω, and when the bridge is balanced, $R_1 = 342\ \Omega$ and $R_2 = 217\ \Omega$. How far is the ground from the test end?

Solution

Substituting the known values in Eq. (8),

$$R_x = \frac{217}{342 + 217} \times 126 = 48.9\ \Omega$$

Since No. 19 wire has a resistance of 24.6 Ω/km, 48.9 Ω represents 1.852 km of wire between the test end and the grounded point.

PROBLEMS 15-3

1. In the Wheatstone bridge of Fig. 15-16, $R_1 = 0.001 \ \Omega$, $R_2 = 1 \ \Omega$, $R_3 = 52.4 \ \Omega$. What is the value of the unknown resistance?

2. In the Wheatstone bridge of Fig. 15-16, the ratio of R_2:R_3 is 100:1. R_1 is 6.28 Ω. What is the unknown resistor?

3. In the Wheatstone bridge, the ratio of R_1:R_2 is 1000 and R_x is believed to be 22.6 Ω. At what setting of R_3 may a balance be expected?

4. A ground exists on one conductor of a lead-covered No. 19 pair. A Murray loop is used to locate the fault by connecting the pair together at the far end (Fig. 15-17). When the bridge circuit is balanced, $R_1 = 33.3 \ \Omega$ and $R_2 = 21.7 \ \Omega$. If the cable is 2.2 km long, how far from the test end is the cable grounded?

5. Several No. 8 wires run between two cities located 65 km apart. One wire becomes grounded, and a Murray loop is used in one city to locate the fault by connecting two of the wires in the other city. When the bridge is balanced, $R_1 = 716 \ \Omega$ and $R_2 = 273 \ \Omega$. How far from the test end is the wire grounded?

6. A No. 6 wire, which is known to be grounded, is made into a loop by connecting a wire of different size at its far end. The resistance of the loop thus formed is 5.62 Ω. When a Murray loop is connected and balanced, the value of R_1 is 16.8 Ω and that of R_2 is 36.2 Ω. How far from the test end does the ground exist?

7. As a research project, discover the details of the Varley loop and develop its equation, which is similar to that for the Murray loop.

Graphs

16

A graph is a pictorial representation of the relationship between two or more quantities. Everyone is familiar with various types of graphs or graphic charts. They are used extensively in magazines, newspapers, annual reports, and trade journals published for engineers, manufacturers, and others concerned with relative values. It is difficult to conceive how engineers could dispense with them.

We have already used simple graphic representations in Chap. 3, and here we will develop a few of the uses of straight-line graphs. In later chapters we will use graphs in working out the solutions of problems and in quickly presenting information in varied forms.

The notions presented here are fundamental to the use of all graph forms, and we are paving the way for some important and interesting topics which will follow in later chapters.

16-1 LOCATING POINTS ON A GRAPH

The accurate location of points is vital, and the manner of marking points can help or seriously hinder in arriving at a correct solution to a problem. One of the most common methods of locating a point is by using a large dot (Fig. 16-1). But this is the poorest form of location, and Fig. 16-1 illustrates why. Do you draw the line through the center, through the top, or through the bottom of a large dot? Can you be sure where the center is? The possibility of introducing errors is great, and you should study the variations of error illustrated in the various parts of Fig. 16-1.

A more acceptable way to mark a point is to use an X, with the intersection marking the spot, or else a circled dot, ⊙, with the tiny point marking the spot and the circle attracting your attention to it. These correct methods are illustrated in Fig. 16-2, and they should be used in all your graph-drawing practice.

A second important item to watch always is the placing of the points. If there is a choice, the points should be far apart, so that the line joining them spans the most important area of the graph. Thus, any error in locating the points themselves is minimized. If the points are located close together and an error is made in locating either one point or both points, then other useful locations ''outside'' the points plotted will be subject to

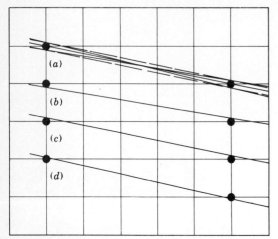

Fig. 16-1 Illustration of errors introduced by the use of large dots to locate points. (*a*) Instead of a single fine line, a broad range of possibilities is presented. (*b*) Shall we join the outside edges of the dots? (*c*) Should we join from top to top (or bottom to bottom)? (*d*) Should we just pick a line that somehow touches both dots somewhere?

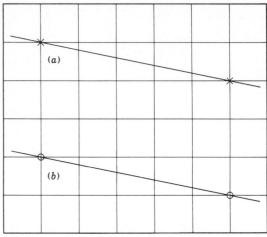

Fig. 16-2 Illustration of the correct method of locating points: After the line is drawn, the small point locations are still indicated, but only a single line can be drawn between the points.

greater error. This fault is illustrated in Fig. 16-3, in which the two circled dots have been plotted slightly off their desired locations. The line joining them comes some distance away from the X points, which should lie on the line. In Fig. 16-4, the two circled dots are again plotted slightly off their desired locations, but since they are widely separated, the amount of error of intermediate points is less.

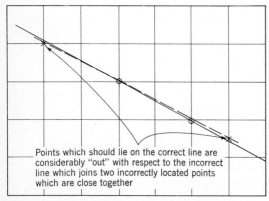

Fig. 16-3 Illustration of the error introduced when points are plotted close together. If the points are slightly incorrect, then useful points "outside" the plotted are are even further off, and the error is enlarged.

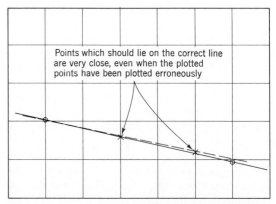

Fig. 16-4 Illustration of the reduction in error when incorrectly plotted points are far apart, so that the line joining them spans the working area of the graph. The error in locating each circled dot is the same as the error in Fig. 16-3, but the × locations are closer to the incorrect line which joins the plotted points.

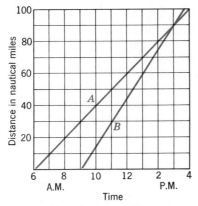

Fig. 16-5 Graph of Example 1.

16-2 SOLVING PROBLEMS BY MEANS OF GRAPHS

In many instances there arise problems involving relationships that, though readily solved by usual arithmetical or algebraic methods, are more clearly understood when solved graphically. It is also true that there are many problems which can be solved graphically with less labor than is required for the purely mathematical solutions. The following illustrative examples will show how some problems can be worked graphically.

Example 1

Steamship A sails from New York at 6 A.M., steaming at an average speed of 10 knots (kn). (A knot is a measure of speed and is one nautical mile per hour.) The same day, at 9 A.M., steamship B sails from New York, steering the same course as A but steaming at 15 kn. (a) How long will it take B to overtake A? (b) What will be the distance from New York at that time?

Solution

Choose convenient scales on graph paper, and plot the distance in nautical miles (nmi) covered by each vessel against the time in hours, as shown in Fig. 16-5. This is conveniently accomplished by making a table like Table 16-1.

TABLE 16-1

Time, o'clock	Distance Covered by A, nmi	Distance Covered by B, nmi
6 A.M.	0	0
8	20	0
10	40	15
12	60	45
2 P.M.	80	75
4	100	105

It will be noted that the graphs of the two distances intersect at 90 nmi, or at 3 P.M. This means the two ships will be 90 nmi from New York at 3 P.M. Because both are steering the same course, *B* will overtake *A* at this time and distance.

The graphic solution furnishes us with other information. For example, by measuring the vertical distance between the graphs, we can determine how far apart the ships will be at any time. Thus, at 11 A.M. the ships will be 20 nmi apart, at 1 P.M. they will be 10 nmi apart, etc.

Example 2

Ship *A* is 200 nmi at sea, and ship *B* is in port. At 8 A.M., *A* starts toward the port, making a speed of 20 kn. At the same time, *B* leaves port at a speed of 30 kn to intercept *A*. After traveling 2 h, *B* is delayed for 1 h and 40 min at the lightship. *B* then continues on its course to intercept *A*. (*a*) At what time will the two ships meet? (*b*) How far will they be from port at that time?

Solution

Figure 16-6 is a graph showing the conditions of the problem. The graph is constructed as in Example 1. A table of distances against time is made up; a convenient scale is chosen; and the points are plotted and joined with a straight line.

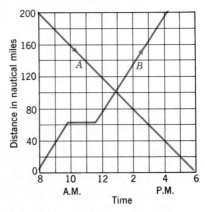

Fig. 16-6 Graph of Example 2.

The intersection of the graphs illustrates that the ships will meet 100 nmi from port at 1 P.M. Why is there a horizontal portion in the graph of *B*'s distance from port? If *A* and *B* continue their speeds and courses, at what time will *A* reach port? At what time will *B* arrive at *A*'s 10 A.M. position? What will be the distance between the ships at that time?

PROBLEMS 16-1

1. A circuit consists of a 10-Ω resistor R_c connected across a variable emf V_v. Plot the current I through the resistor against the voltage V across the resistor as V_v is varied in 10-V steps from 0 to 100 V. What conclusion do you draw from this graph?

2. A circuit consists of a 50-Ω resistor R_L connected across the variable emf V_v of Prob. 1. On the same graph sheet as your solution to Prob. 1, plot the current I through R_L against the voltage V as V_v is varied between 0 and 100 V. What conclusion do you draw from the pair of graphs?

3. The distance s covered by a moving object is equal to the product of the object's velocity v and the time t during which the object is moving; that is, $s = vt$. Plot the distance in kilometers traveled by an automobile averaging 55 km/h against time for every hour from 9 A.M. to 6 P.M. What conclusions do you draw from the graph?

4. A variable resistor R_v is connected across a generator which maintains a constant voltage V_c of 120 V. Plot the current I through the resistor as the resistance is varied in 5-Ω steps between 5 and 50 Ω. What conclusions do you draw from this graph?

Solve these problems graphically:

5. Train A leaves a city at 8 A.M. traveling at the rate of 80 km/h. Two hours later train B leaves the same city, on the same track, traveling at the rate of 120 km/h.
   (*a*) At what time does train B overtake A?
   (*b*) How far from the starting point will the trains be at the time of part (*a*)?
   (*c*) How far apart will the trains be 2 h after B starts?

6. Two people start toward each other from points 144 km apart, the first traveling at 96 km/h and the second at 64 km/h.
   (*a*) How long will it be before they meet?
   (*b*) How far will each have traveled when they meet?
   (*c*) How far apart will they be after 30 min of travel?

7. A owns a motor that consumes 10 kWh per day, and B owns a motor that consumes 30 kWh per day. Beginning on the first day of a 30-day month, A's motor runs continuously. B's motor runs for 1 day, is idle for 4 days, then runs for 2 days, is idle for 6 days, and then runs continuously for the rest of the month. On what days of the month will A's and B's power bills be the same?

8. The owner of a radio store decides to pay the salespeople according to either of two plans. The first plan provides for a fixed salary of $25 per week plus a commission of $3 for each radio sold. According to the second plan, a salesperson may take a straight commission of $4 for each radio set sold. Determine at which point the second plan becomes more attractive for an energetic salesperson.

16-3 COORDINATE NOTATION

Let us suppose you are standing on a street corner and a stranger asks you for directions to some prominent building. You tell the stranger to go four blocks east and five blocks north. By these directions, you have automatically made the street intersection a *point of reference*, or *origin*, from which distances are measured. From this point you could count distances to any point in the city, using the blocks as a unit of distance and pairs of directions (east, north, west, or south) for locating the various points.

To draw a graph, we had to use two lines of reference, or *axes*. These correspond to the streets meeting at right angles. Also, in fixing a point on a graph, it was necessary to locate that point by pairs of numbers. For example, when we plot distance against time, we need one number to represent the time and another number to represent the distance covered in that time.

So far, only positive numbers have been used for graphs. To restrict graphs to positive values would impose just as severe a handicap as if we were to restrict algebra to positive numbers. Accordingly, a system must be established for plotting pairs of numbers either or both of which may be positive or negative. In such a system, a sheet of squared paper is divided into four sections, or quadrants, by drawing two intersecting axes at right angles to each other. The point *O*, at the intersection of the axes, is called the *origin*. The horizontal axis is generally known as the *x axis*, and the vertical axis is called the *y axis*.

There is nothing new about measuring distances along the *x axis*; it is the basic system described in Sec. 3-5 and shown in Fig. 3-3. That is, we agree to regard distances along the *x axis* to the *right* of the origin as *positive* and those to the *left* as *negative*. Also, we consider distances along the *y axis* as *positive* if *above* the origin and *negative* if *below* the origin. In effect, we have simply added to our method of graphical representation as originally outlined in Fig. 3-3.

With this system of representation, which is called a system of *rectangular coordinates*, we are able to locate any pair of numbers regardless of the signs. Because this system was developed by the French mathematician René Descartes, you will often hear it referred to as the system of *Cartesian coordinates*.

Example 3
Referring to Fig. 16-7,

- Point *A* is in the first quadrant. Its *x* value is $+3$, and its *y* value is $+4$.
- Point *B* is in the second quadrant. Its *x* value is -4, and its *y* value is $+5$.
- Point *C* is in the third quadrant. Its *x* value is -5, and its *y* value is -2.
- Point *D* is in the fourth quadrant. Its *x* value is $+5$, and its *y* value is -3.

Thus, every point on the surface of the paper corresponds to a pair of coordinate numbers that completely describe the point.

The two signed numbers that locate a point are called the *coordinates* of that point. The *x* value is called the *abscissa* of the point, and the *y* value is called the *ordinate* of the point.

In describing a point in terms of its coordinates, the abscissa is always stated first. Thus, to locate point *A* in Fig. 16-7, we write $A = (3, 4)$, meaning that, to locate point *A*, we count three divisions to the right of

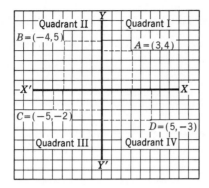

Fig. 16-7 System of rectangular coordinates.

the origin along the x axis and up four divisions along the y axis. In like manner, we completely describe point B by writing $B = (-4, 5)$. Also,

$$C = (-5, -2) \quad \text{and} \quad D = (5, -3)$$

PROBLEMS 16-2

1. On a map, which lines correspond to the x axis, latitude or longitude?
2. Plot the following points: $(2, 3)$, $(-6, -1)$, $(3, -7)$, $(0, -6)$, $(0, 0)$, $(-8, 0)$.
3. Plot the following points: $(-1.5, 10)$, $(-6.5, -7.5)$, $(3.6, -4)$, $(0, 2.5)$, $(6.5, 8.5)$, $(3.5, 0)$.
4. Using Fig. 16-8, give the coordinates of the points A, B, C, D, E, F, G, H, I, J, K, L, M, and N.

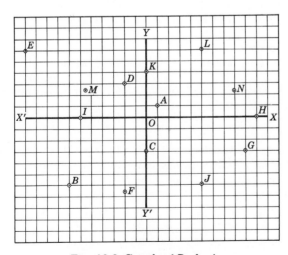

Fig. 16-8 Graph of Prob. 4.

5. Plot the following points: $A = (-1, -2)$, $B = (5, -2)$, $C = (5, 4)$, $D = (-1, 4)$. Connect these points in succession. What kind of figure is $ABCD$? Draw the diagonals DB and CA. What are the coordinates of the point of intersection of the diagonals?

16-4 GRAPHS OF LINEAR EQUATIONS

A relation between a pair of numbers, not necessarily connected with physical quantities such as those in foregoing exercises, can be expressed by a graph.

Consider the following problem: The sum of two numbers is equal to 5. What are the numbers? Immediately it is evident there is more than one pair of numbers that will fulfill the requirements of the problem. For example, if only positive numbers are considered, we have, by addition,

$$
\begin{array}{cccccc}
0 & 1 & 2 & 3 & 4 & 5 \\
\underline{5} & \underline{4} & \underline{3} & \underline{2} & \underline{1} & \underline{0} \\
5 & 5 & 5 & 5 & 5 & 5
\end{array}
$$

Similarly, if negative numbers are included, we can write

$$
\begin{array}{cccccc}
-1 & -2 & -3 & -4 & -5 & -6 \\
\underline{+6} & \underline{+7} & \underline{+8} & \underline{+9} & \underline{+10} & \underline{+11} \\
5 & 5 & 5 & 5 & 5 & 5
\end{array}
$$

and so on, indefinitely.

Also, if fractions or decimals are considered, we have

$$
\begin{array}{cccc}
1.5 & -3.75 & -1.63 & -8.36 \\
\underline{3.5} & \underline{+8.75} & \underline{+6.63} & \underline{+13.36} \\
5 & 5 & 5 & 5
\end{array}
$$

and so on, indefinitely.

It follows that there are an infinite number of pairs of numbers whose sum is 5.

Let x represent any possible value of one of these numbers, and let y represent the corresponding value of the second number. Then

$$x + y = 5$$

For any value assigned to x, we can solve for the corresponding value of y. Thus, if $x = 1$, $y = 4$. Also, if $x = 2$, $y = 3$. Likewise, if $x = -4$, $y = 9$, because, by substituting -4 for x in the equation, we obtain

$$-4 + y = 5$$

or

$$y = 9$$

In this manner, there may be obtained an unlimited number of values for x and y that satisfy the equations, some of which are listed below.

If $x =$	-6	-4	-2	0	2	4	6	8	10
Then $y =$	11	9	7	5	3	1	-1	-3	-5
Coordinates of	A	B	C	D	E	F	G	H	I

With the tabulated pairs of numbers as coordinates, the points are plotted and connected in succession, as shown in Fig. 16-9. The line drawn through these points is called the *graph of the equation* $x + y = 5$.

Regardless of what pairs of numbers (coordinates) are chosen from the graph, it will be found that each pair satisfies the equation. For example, the point P has coordinates $(15, -10)$; that is, $x = 15$ and $y = -10$.

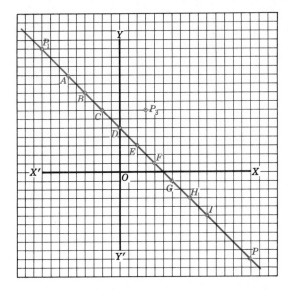

Fig. 16-9 Graph of the equation $x + y = 5$.

These numbers satisfy the equation because $15 - 10 = 5$. Likewise, the point P_1 has coordinates $(-9, 14)$ that also satisfy the equation because $-9 + 14 = 5$. The point P_3 has coordinates $(3, 7)$. This point is not on the line, nor do its coordinates satisfy the equation; for $3 + 7 \neq 5$. The straight line, or graph, can be extended in either direction, always passing through points whose coordinates satisfy the conditions of the equation. This is as would be expected, for there is an infinite number of pairs of numbers called *solutions* that, when added, are equal to 5.

PROBLEMS 16-3

1. Graph the equation $x - y = 8$ by tabulating and plotting five pairs of values for x and y that satisfy the equation. Can a straight line be drawn through these points? Plot the point $(4, 4)$. Is it on the graph of the equation? Do the coordinates of this point satisfy the equation? From the graph, when $x = 0$, what is the value of y? When $y = 0$, what is the value of x? Do these pairs of values satisfy the equation?

2. Graph the equation $2x + 3y = 6$ by tabulating and plotting at least five pairs of values for x and y that satisfy the equation. Can a straight line be drawn through these points? Plot the point $(-15, 12)$. Is this point on the graph of the equation? Do the coordinates satisfy the equation? Plot the point $(10, -5)$. Is this point on the graph of the equation? Do the coordinates satisfy the equation? From the graph, when $x = 0$, what is the value of y? When $y = 0$, what is the value of x? Do these pairs of values satisfy the equation?

16-5 VARIABLES

When two variables, such as x and y, are so related that a change in x causes a change in y, then y is said to be a *function* of x. By assigning values to x and then solving for the value of y, we make x the *independent variable* and y the *dependent variable*.

The above definitions are applicable to all types of equations and physical relations. For example, in Fig. 16-5, distance is plotted against time. The distance covered by a body moving at a constant velocity is given by

$$s = vt$$

where s = distance
v = velocity
t = time

In this equation and therefore in the resulting graph, the distance is the dependent variable because it depends upon the amount of time. The time is the independent variable, and the velocity is a constant.

Similarly, in Prob. 1 of Problems 16-1, the formula $I = \dfrac{V}{R}$ is used to obtain values for plotting the graph. Here the resistance R is the constant, the voltage V is the independent variable, and the current I is the dependent variable.

In Prob. 4 of Problems 16-1, the same formula $I = \dfrac{V}{R}$ is used to obtain coordinates for the graph. Here the voltage V is a constant, the resistance R is the independent variable, and the current I is the dependent variable.

From these and other examples, it is evident, as will be illustrated in Sec. 16-6, that the graph of an equation having variables of the first degree is a straight line. This fact does not apply to variables in the denominator of a fraction as in the case above where R is a variable. However, $I = \dfrac{V}{R}$ is not an equation of the first degree as far as R is concerned because, by the law of exponents, $I = VR^{-1}$.

It is general practice to plot the independent variable along the horizontal, or x axis, and the dependent variable along the vertical, or y axis.

In plotting the graph of an equation, it is convenient to solve the equation for the dependent variable first. Values are then assigned to the independent variable in order to find the corresponding values of the dependent variable.

If an equation or formula contains more than two variables, we must, after choosing the dependent variable, decide which is to be the independent variable for each separate investigation, or graphing. For example, consider the formula

$$X_L = 2\pi fL$$

where X_L = inductive reactance of an inductor, Ω
f = frequency, Hz
L = inductance, H
2π = 6.28 . . .

In this case, we can vary either the frequency f or the inductance L in order to determine the effect upon the inductive reactance X_L, but we must not vary both at the same time. Either f must be fixed at some constant value and L varied or L must be fixed. A little thought will show the difficulty of plotting, on a plane, the variations X_L if f and L are varied simultaneously.

16-6 THE GRAPH-EQUATION RELATIONSHIPS

Each one of the equations that have been plotted is of the *first degree* (Sec. 11-1) and contains *two unknowns*. From the graphs the following important facts are obtained:

1. The graph of an equation of the first degree is a straight line.
2. The coordinates of every point on the graph satisfy the conditions of the equation.
3. The coordinates of every point not on the graph do not satisfy the conditions of the equation.

Because the graph of every equation of the first degree results in a straight line, as stated under 1 above, first-degree equations are called *linear equations*. Also, because such equations have an infinite number of solutions, they are called *indeterminate equations*.

As x changes in value in such an equation, the value of y also changes. Hence, x and y are called *variables*.

Now consider Fig. 16-9, the graph of $x + y = 5$. This equation may be written in the form $y = -x + 5$. Here y is called the dependent variable, because its value depends upon the value of x; and x is called the independent variable, because we may assign to it any value we choose.

Notice in the graph first of all that the y intercept, the point where the curve cuts the y axis, is at the point $x = 0$, $y = 5$, and this value is revealed in the equation $y = -x + 5$ because, at the y axis, $x = 0$ and y then equals 5.

Second, note the slope of the line. For every step in the x direction (positive to the right), there is a downward (negative) step in the y direction. By definition, the slope of a line is the ratio of the change in the y values between two points to the corresponding change in x values between the same two points:

$$\text{Slope} = \frac{\Delta y}{\Delta x}$$

where the symbol Δ (Greek letter delta) means "the change in."

Figure 16-9 has been redrawn in Fig. 16-10 to show the changes in x and y between two arbitrarily selected points B and H. The slope of the graph equals

$$\frac{\Delta y}{\Delta x} = \frac{-12}{+12} = -1$$

Now see in the equation $y = -x + 5$ that the slope, -1, is indicated in the coefficient of x.

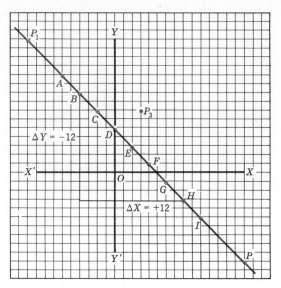

Fig. 16-10 Fig. 16-9 redrawn to show $m = \dfrac{\Delta y}{\Delta x}$.

Therefore, when we write the original equation $x + y = 5$ in standard form

$$y = -x + 5$$

the slope of the line is the coefficient of the x term and the y intercept is the constant term.

The general form of equation for a straight line is

$$y = mx + b$$

where y = dependent variable
 x = independent variable
 m = slope of the curve (straight line)
 b = value of the y intercept

16-7 METHODS OF PLOTTING

To graph a linear equation of two variables,

1. Convert the equation to the standard form $y = mx + b$ to indicate quickly the values of the slope m and the y intercept b.
2. Choose a suitable value for x, substitute it into the standard form equation, and solve for the corresponding value of y. This results in one solution, or one set of coordinates.
3. Choose another value for x, and again solve for y. This second x value should be reasonably well spaced from the first (see Figs. 16-3 and 16-4).
4. Plot the two points whose coordinates were calculated in steps 2 and 3. Connect them with a fine straight line.

5.   Check the resulting graph by solving for and plotting a third point. This third point must lie on the same straight line or its extension.

Example 4

Graph the equation $2x - 5y = 10$.

Solution

1.   Rewrite the equation in the standard form: $y = \frac{2}{5}x - 2$.
2.   Always plot first the value of y when $x = 0$. This value is immediately obtained from the "-2" of the equation, which shows the y intercept. This inspection results in a point, which we shall call A, whose coordinates are $(0, -2)$.
3.   Now choose some value of x. Any value will serve, but one which cancels the denominator of the fractional coefficient will be the best choice. Let $x = 5$ and, by solving the equation, obtain $y = 0$. This gives the second point, B, at $(5, 0)$. (Sometimes it may be more convenient to choose, as the second point, the value of $y = 0$ and solve for x.)
4.   Choose another value of x in order to solve for the third (check) point. Let $x = -10$. Then $y = -6$, and this gives point C at $(-10, -6)$.
5.   Draw the line of the equation by joining the three points. The points and the finished graph are shown in Fig. 16-11.

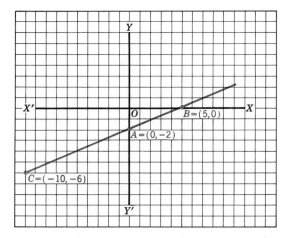

Fig. 16-11 Graph of the equation $2x - 5y = 10$.

When x was set equal to zero, the resulting point A had coordinates that located the point where the graph crossed the y axis. This point is called the *y intercept*. Likewise, when y was set equal to zero, the resulting point B had coordinates that located the point where the graph crossed the x axis. This point is called the *x intercept*. Not only are these easy methods of locating two points with which to graph the equation but also these two points give us the exact location of the intercepts. The intercepts are important, as will be shown later.

The x intercept is often referred to as the *root* or *zero* of the equation.

An alternative method of plotting straight-line graphs is to use the information obtained from the standard form $y = mx + b$. If we locate the y intercept b immediately and then step over and up (or down) in accordance with the slope m, we can locate additional points. If, for example, $y = 2x + 9$, then the y intercept is at $+9$ and the slope is $+2:1$.

Follow the development of the graph in Fig. 16-12. First plot the y intercept, $+9$. Then step one unit in the positive x direction and two units in the positive y direction and plot the first point. Next, since $+\frac{2}{1} = \frac{-2}{-1}$, again starting at the y intercept, step one unit in the negative x direction and two units in the negative y direction and plot the second point. If these two points are too close together to be reliable, space them better by moving greater distances in the x and y directions while keeping the ratio $\frac{\Delta y}{\Delta x}$ equal to 2:1 ($= m$). Finally, join the two points so located with a straight line which passes through the third, or test, point, the y intercept.

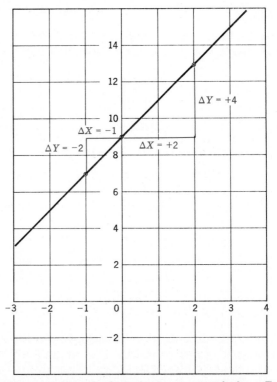

Fig. 16-12 Alternative method of plotting a straight line: First, locate y intercept, given by the constant in the standard form equation.

Second, step off Δx and Δy so that $\frac{\Delta y}{\Delta x} = m$, or slope, also given in the standard form of the equation, first in the $+x$ Direction and then in the $-x$ Direction.

PROBLEMS 16-4

Graph the following equations and determine the x and y intercepts:

1. $5x + 4y = 12$ **2.** $2x - y = 8$
3. $x - 3y = 3$ **4.** $2x + y = 9$
5. Plot the following equations on the same sheet of graph paper (same axes), and carefully study the results: (a) $x - y = -8$; (b) $x - y = -5$; (c) $x - y = 0$; (d) $x - y = 4$; (e) $x - y = 8$. Are the graphs parallel? Note that all left members of the given equations are identical. Solve each of these equations for y and write them in a column, thus:

$$(a)\ y = x + 8$$
$$(b)\ y = x + 5$$
$$(c)\ y = x + 0$$
$$(d)\ y = x - 4$$
$$(e)\ y = x - 8$$

In each equation, does the last term of the right member represent the y intercept?

When the equations are solved for y, as above, each coefficient of x is $+1$. All the graphs slant to the right because the coefficient of each x is positive. Each time an x increases one unit, note that the corresponding y increases one unit. That is because the coefficient of x in each equation is 1.

6. Plot the following equations on the same sheet of graph paper (same axes), and carefully study the results.
(a) $4x - 2y = -30$; (b) $4x - 2y = -16$; (c) $4x - 2y = 0$; (d) $4x - 2y = 12$; (e) $4x - 2y = 30$; (f) $8x - 4y = 60$.
Are all the graphs parallel? Again note that all left members are identical. Does the graph of Eq. (f) fall on that of Eq. (e)? Note that (e) and (f) are *identical equations*. Why?
Solve each of these equations, except (f), for y and write them in a column, thus:

$$(a)\ y = 2x + 15$$
$$(b)\ y = 2x + 8$$
$$(c)\ y = 2x + 0$$
$$(d)\ y = 2x - 6$$
$$(e)\ y = 2x - 15$$

In each equation, does the last term of the right member represent the y intercept? When linear equations are written in this form, this last term is known as the *constant term*.

Are all the coefficients of the x's positive? That is why all the graphs slant upward to the right. Lines slanting in this manner are said to have *positive slopes*.

Each time an x increases or decreases one unit, note that y respectively increases or decreases two units. That is because the coefficient of each x is 2. If a graph has a *positive slope*, an increase or decrease in *x* always results in a corresponding increase or decrease in *y*. In these equations, each line has a slope of $+2$, the coefficient of each *x*.

7. Plot the following equations on the same set of axes: (*a*) $x + 2y = 18$; (*b*) $x + 2y = 10$; (*c*) $x + 2y = 0$; (*d*) $x + 2y = -14$; (*e*) $x + 2y = -22$; (*f*) $3x + 6y = -66$.

Are all the graphs parallel? How should you have known they would be parallel without plotting them?

Does the graph of (*f*) fall on that of (*e*)? How should you have known (*e*) and (*f*) would plot the same graph without actually plotting them? Solve each equation for *y* as in Probs. 5 and 6. Does the constant term denote the *y* intercept in each case? Is the coefficient of each *x* equal to $\frac{1}{2}$? The minus sign means that each graph has a *negative slope;* that is, the lines slant downward to the right. Thus, when *x* increases, *y* decreases, and vice versa. The $\frac{1}{2}$ slope means that, when *x* varies one unit, *y* is changed $\frac{1}{2}$ unit. Therefore, the variations of *x* and *y* are completely described by saying the slope is $-\frac{1}{2}$.

8. Plot the following equations on the same set of axes: (*a*) $x - 4y = 0$; (*b*) $x - 2y = 0$; (*c*) $x - y = 0$; (*d*) $2x - y = 0$; (*e*) $4x - y = 0$; (*f*) $4x + y = 0$; (g) $2x + y = 0$; (*h*) $x + y = 0$; (*i*) $x + 2y = 0$; (*j*) $x + 4y = 0$. Solve the equations for *y*, as before, and carefully analyze your results.

16-8 EQUATIONS DERIVED FROM GRAPHS

Often we obtain a set of readings relating two variables and want to know whether there is any definite relationship between the variables. This investigation makes use of both the graph showing the relationship and our understanding of the standard form of a straight-line equation.

$$y = mx + b$$

1. Plot the observed values carefully on a graph. If a straight-line relationship is indicated, draw it.
2. Sometimes one or more points appear to be off the trend. There may or may not be errors in these readings. For the present, we will *assume* that they are errors.
3. If the trend is a straight line, but some points are off, try to draw the line so that there is an equal number of floating points above and below the line. (Use a transparent straightedge.)
4. The straight-line result must now obey the law $y = mx + b$.

Example 5

Given the following set of readings, draw the graph and determine the law relating the variables:

x	-2	2	4	6
y	-7	5	11	17

Solution

First, plot the points as they have been given, and try them with a straightedge for a straight-line relationship. Since, in Fig. 16-13 a straight line is indicated, draw the line that joins the points. The y intercept is seen to be -1. This gives the value of b in the standard form. Then, to determine the slope m, choose any two convenient points, reasonably spaced, say $(2, 5)$ and $(6, 17)$. The difference between the points in the y direction is $17 - 5 = 12$. The difference between the points in the x direction is $6 - 2 = 4$.

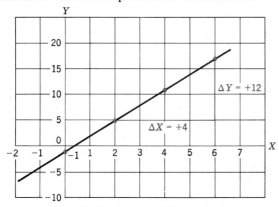

Fig. 16-13 Graph of Example 5.

$$\text{Then the slope } m = \frac{\Delta y}{\Delta x} = \frac{17 - 5}{6 - 2} = \frac{+12}{+4}$$
$$= +3$$

and the relationship is $y = 3x - 1$.

Example 6

Given the readings relating P and V, determine the law relating them:

P	-4	-2	2	6	10
V	17	11	-5	-22	-38

Solution

Plot the points and test for a straight-line relationship. Because some of the points are not quite on the line, draw the straight line which will balance the floating points (Fig. 16-14). Now the V intercept is seen to be $+2$, and the equation relating P and V will be of the form $V = mP + 2$. To evaluate the slope m, choose any two convenient points on the line, and arrive at m:

$$m = \frac{\Delta V}{\Delta P} = \frac{-38 - (-22)}{10 - 6}$$
$$= \frac{-16}{+4} = -4$$

and the relationship is seen to be

$$V = -4P + 2$$

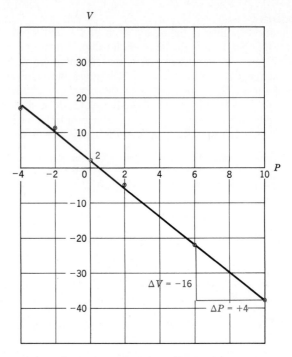

Fig. 16-14 Graph of Example 6.

Referring to Examples 5 and 6, see how m may be found algebraically by realizing that $\Delta y = y_2 - y_1$, the difference of the values of y when going from point 1 to point 2, and $\Delta x = x_2 - x_1$, the difference of the values of x going from point 1 to point 2. Then

$$m = \frac{\Delta y}{\Delta x} = \frac{y_2 - y_1}{x_2 - x_1}$$

Always call your starting point 1 and your finishing point 2. That will yield the correct *sign* as well as the correct *value* of the slope.

PROBLEMS 16-5

1. What is the $y = mx + b$ form equation for the graph of Fig. 16-11?
2. A series of readings shows values of y for predetermined values of x:

x	5	10	15	20	25	30
y	100	200	300	400	500	600

Plot values of y against values of x and determine the values of the constants m and b which connect x and y in the form $y = mx + b$.

3. A laboratory experiment relates x and y as follows:

x	10	20	30	40	50	60
y	2.35	3.5	4.6	5.75	6.9	8.0

What is the equation, in the form $y = \alpha x + \theta$, which relates x and y?

4. The following is a series of readings relating s and t:

t	50	125	210	250	360	435
s	0.36	0.30	0.23	0.20	0.11	0.05

Plot s against t and, assuming s and t are connected by a law of the form $s = u + qt$, find u and q.

5. The following is a set of laboratory readings relating R and T:

T	30	75	150	210	270	300	360	390	425	450
R	0.38	0.35	0.31	0.26	0.22	0.195	0.16	0.13	0.12	0.10

Plot the graph of R versus T and determine the formula which relates them.

6. A comparison of Celsius (C) and Fahrenheit (F) temperatures is given in the following table:

°C	0	10	38	60	100
°F	32	50	100	140	212

Plot °F against °C.
(a) Determine from the graph the relationship between the two temperature scales in the form, $F = \theta C + \phi$.
(b) From the graph, what is the Fahrenheit equivalent of 25°C?
(c) From the graph, what is the Celsius equivalent of 165°F?

7. The readings of current flow I through a certain resistor as the emf V is changed are given in the following table:

V	10	20	30	40	50	60	70	80	90	100	V
I	0.2125	0.4255	0.638	0.851	1.062	1.278	1.49	1.702	1.915	2.125	A

(a) From the graph, what is the ratio $\dfrac{\text{change in voltage}}{\text{change in current}}$ $\left(\dfrac{\Delta V}{\Delta I}\right)$?

(b) What is the ratio $\dfrac{\text{change in current}}{\text{change in voltage}}$ $\left(\dfrac{\Delta I}{\Delta V}\right)$?

(c) From Ohm's law, what is the resistance of the resistor?
(d) What conclusions do you draw from your answers to questions (a), (b), and (c)?

8. The following is a series of readings of the avalanche breakdown of a zener diode:

V	−14.2	−14.4	−14.6	−14.7	−14.8	−14.9	−15	−15.1	−15.2	−15.3	−15.4	−15.5	V
I	0	0	0	−10	−18.9	−28.2	−37.4	−46.8	−56	−65.2	−74.6	−83.9	mA

Plot the graph of I versus V and determine:

(a) What $\dfrac{\Delta I}{\Delta V}$ is after the voltage goes more negative than 14.6 V.

(b) What the ratio is for voltages less negative than 14.6 V.

9. When the base current I_B of an NPN transistor is 0.01 mA, the readings of collector-to-emitter voltage V_{CE} produce the following set of readings for the collector current I_C. Plot I_C against V_{CE}.

V_{CE}	12.5	14.3	16.5	18.5	20.0	21.5	23.2	24.4	V
I_C	0.9	1.1	1.2	1.4	1.6	1.62	1.65	1.69	mA

(a) Over what range of voltages is I_C considered constant?

(b) What would be your interpretation of I_C if I_B were doubled?

(c) How do you interpret the current gain of the transistor over the tabulated voltages? Explain your observations.

(d) What is the value of beta $\left(\beta = \dfrac{I_C}{I_B} \right)$ over the tabulated spread of current? How do you explain the differences?

10. The readings of current versus applied voltage for a tunnel diode are as follows:

V_1	0.002	0.008	0.011	0.016	0.02	0.023	0.027	0.03	0.04	0.07	0.095	0.105	0.115
I_1	0.1	0.2	0.3	0.4	0.5	0.6	0.7	0.8	0.9	1.0	0.9	0.8	0.7

0.125	0.135	0.145	0.160	0.20	0.32	0.39	0.42	0.43	0.45	0.46	0.47	0.48	V
0.6	0.5	0.4	0.3	0.2	0.1	0.2	0.3	0.4	0.5	0.7	0.8	0.9	mA

(a) Draw the graph of I_1 versus V_1.

(b) Note specifically the range of voltages which makes the tunnel diode act like a negative resistance.

(c) Note the ranges of voltages which make the tunnel diode act like a positive resistance.

Simultaneous Equations

17

Many times in electronics we find several circuit conditions applying at the same time and therefore requiring interlocking solutions. Accordingly, the study of simultaneous equations and their most common methods of solution is a vital one for electronics technicians.

The subject of simultaneous equations also provides us with an excellent application of the linear graphs discussed in Chap. 16. This chapter leans heavily on the notions presented there, although, once the meaning of simultaneous solutions is understood, we can quickly move on to various algebraic methods of solution.

17-1 GRAPHICAL SOLUTION OF SIMULTANEOUS LINEAR EQUATIONS

The graphs of the equations

$$x + 2y = 12$$

and

$$3x - y = 1$$

are shown in Fig. 17-1. The point of intersection of the lines has the coordinates (2, 5); that is, the x value is 2 and the y value is 5. Now this

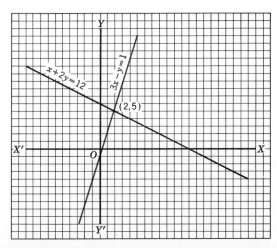

Fig. 17-1 Graph of the equations $x + 2y = 12$ and $3x - y = 1$.

point is on both of the graphs; it follows, therefore, that the x and y values should satisfy both equations. Substituting 2 for x and 5 for y in each equation results in the identities

$$2 + 10 = 12$$

and
$$6 - 5 = 1$$

From this it is observed that, if the graphs of two linear equations intersect, they have one common set of values for the variables, or one common solution. Such equations are called *simultaneous linear equations*.

Because two straight lines can intersect at only one point, there can be only one common set of values or one common solution that satisfies both equations.

Two equations, each with two variables, are called *inconsistent equations* when their plotted lines are parallel to each other. Because parallel lines do not intersect, there is no common solution for two or more inconsistent equations.

Considerable care must be used in graphing equations, for a deviation in the graph of either equation will cause the intersection to be in the wrong place and hence will lead to an incorrect solution.

PROBLEMS 17-1

Solve the following pairs of equations graphically, and check your solutions by substituting them into each of the original equations:

1. $x + 4y = 14$
    $x - 4y = -2$

2. $6x - y = 15$
    $2x + 5y = 21$

3. $x + y = 8$
    $x - y = 2$

4. $x + 2y = 26$
    $4x - y = 32$

5. $9E + 2I = 34$
    $6E + 5I = -14$

6. $l - 8m = 0$
    $l + m = 45$

7. $7\alpha + 3\beta = -23$
    $5\beta + 4\alpha = -23$

8. $8F - f = 0$
    $3f + 4F = 14$

9. $3I_1 + 7i = 50$
    $5I_1 - 2i = 15$

10. $2Z_2 + 6Z_1 = 7$
    $4Z_2 - 3Z_1 = 9$

17-2 SOLUTION OF SIMULTANEOUS LINEAR EQUATIONS BY ADDITION AND SUBTRACTION

It has been shown in preceding sections that an unlimited number of pairs of values of variables satisfy one linear equation. Also, it can be determined graphically whether there is one pair of values, or solution, that will satisfy two given linear equations. The solution of two simultaneous linear equations can also be found by algebraic methods, as illustrated in the following examples:

Example 1
Solve the equations $x + y = 6$ and $x - y = 2$.

Solution

Given

$$x + y = 6 \qquad (a)$$
$$x - y = 2 \qquad (b)$$

Add (a) and (b),

$$2x = 8 \qquad (c)$$

D: 2 in (c),

$$x = 4$$

Substitute this value of x in (a),

$$4 + y = 6$$

Collect terms,

$$y = 2$$

The common solution for (a) and (b) is

$$x = 4 \qquad y = 2$$

Check

Substitute in (a), $4 + 2 = 6$
Substitute in (b), $4 - 2 = 2$

In Example 1 the coefficients of y in Eqs. (a) and (b) are the same except for sign. That being so, y can be *eliminated* by adding these equations, and the resulting sum is an equation in one unknown. This method of solution is called *elimination by addition*.

Because the coefficients of x are the same in Eqs. (a) and (b) of Example 1, x could have been eliminated by subtracting either equation from the other, and an equation containing only y as a variable would have been the result. This method of solution is called *elimination by subtraction*. The remaining variable x would have been solved for in the usual manner by substituting the value of y in either equation.

Example 2

Solve the equations $3x - 4y = 13$ and $5x + 6y = 9$.

Solution

Given

$$3x - 4y = 13 \qquad (a)$$
$$5x + 6y = 9 \qquad (b)$$

M: 3 in (a),

$$9x - 12y = 39 \qquad (c)$$

M: 2 in (b),

$$10x + 12y = 18 \qquad (d)$$

Add (c) and (d)

$$19x = 57 \qquad (e)$$

D: 19 in (e),

$$x = 3 \qquad (f)$$

Substitute this value of x in (a),

$$9 - 4y = 13 \qquad (g)$$

Collect terms,

$$-4y = 4 \qquad (h)$$

D: -4 in (h),

$$y = -1$$

The common solution for (a) and (b) is

$$x = 3 \qquad y = -1$$

Check

Substitute in (a), $9 + 4 = 13$
Substitute in (b), $15 - 6 = 9$

In Example 2 the coefficients of x and y in Eqs. (a) and (b) are not the same. The coefficients of y were made the same absolute value in the Eqs. (c) and (d) in order to eliminate y by the method of addition.

Example 3
Solve the equations $4a - 3b = 27$ and $7a - 2b = 31$.

Solution

Given	$4a - 3b = 27$	(a)
	$7a - 2b = 31$	(b)
M: 7 in (a)	$28a - 21b = 189$	(c)
M: 4 in (b),	$28a - 8b = 124$	(d)
Subtract (d) from (c),	$-13b = 65$	(e)
D: -13 in (e),	$b = -5$	(f)
Substitute this value of b in (a),		
	$4a + 15 = 27$	(g)
Collect terms,	$4a = 12$	(h)
D: 4 in (h),	$a = 3$	(i)

The common solution for (a) and (b) is

$$a = 3 \qquad b = -5$$

Check
Substitute the values of the variables (a) and (b) as usual.

In Example 3 the coefficients of a and b in Eqs. (a) and (b) are not the same. The coefficients of a were made the same absolute value in the Eqs. (c) and (d) in order to eliminate a by the method of subtraction.

RULE
To solve two simultaneous linear equations having two variables by the method of elimination by addition or subtraction:

1. If necessary, multiply each equation by a number that will make the coefficients of one of the variables of equal absolute value.
2. If these coefficients of equal absolute value have like signs, subtract one equation from the other; if they have unlike signs, add the equations.
3. Solve the resulting equation.
4. Substitute the value of the variable found in step 3 in one of the original equations, and then solve the resulting equation for the remaining variable.
5. Check the solution by substituting in both the original equations.

PROBLEMS 17-2
Solve for the unknowns by the method of addition and subtraction:

1. $2a + b = 9$
   $4a - b = 6$

2. $V - 4I = 9$
   $2V - 2I = 6$

3. $5Z + 2R = 16$
   $3Z - R = 3$

4. $4V + 3I = -1$
   $5V + I = 7$

5. $R_1 - 3R_2 = -8$
   $3R_1 + R_2 = 6$

6. $5\theta + 4\phi = 12$
   $\theta - 2\phi = 8$

7. $s + t = 0$
   $3s + 7t = 8$

8. $2\alpha - \beta = 3$
   $4\beta + 3\alpha = 10$

9. $5M + L = 11$
   $3M + 2L = 8$

10. $4p - 3q = 5$
    $9p - 8q = 0$

11. $3I_1 - 4I_2 = 17$
    $I_1 + 3I_2 = -3$

12. $3Z_1 + Z_2 = 14$
    $Z_1 + 2Z_2 = 13$

13. $V + 3v = 11$
    $4V + 7v = 29$

14. $I + 3i = 25$
    $4i + I = 31$

15. $3\lambda - 7\pi - 19 = 0$
    $2\lambda - \pi = 9$

16. $5\alpha + 1 = -3\beta$
    $7\beta + 3\alpha - 15 = 0$

17. $0.3V + 0.2v = -0.9$
    $0.5V = -1.9 - 0.3v$

18. $0.9X_L + 0.04X_C = 9.4$
    $0.05X_L + 2.5 = 0.3X_C$

19. $0.03I - 0.54 = -0.02i$
    $21 - i = I$

20. $0.4L + 1.6 = 0.9X$
    $0.7X + 0.2 = 0.6L$

21. Solve the problems of Problems 17-1 by the method of addition and subtraction, and confirm the answers obtained by the graphical method.

17-3 SOLUTION BY SUBSTITUTION

Another common method of solution is called *elimination by substitution*.

Example 4

Solve the equations $16x - 3y = 10$ and $8x + 5y = 18$.

Solution

Given

$$16x - 3y = 10 \qquad (a)$$
$$8x + 5y = 18 \qquad (b)$$

Solve (a) for x in terms of y,

$$x = \frac{10 + 3y}{16} \qquad (c)$$

Substitute this value of x in (b),

$$8\left(\frac{10 + 3y}{16}\right) + 5y = 18 \qquad (d)$$

M: 16 in (d), $8(10 + 3y) + 80y = 288 \qquad (e)$

Expand (e), $80 + 24y + 80y = 288 \qquad (f)$

Collect terms in (f), $104y = 208 \qquad (g)$

D: 104 in (g), $y = 2 \qquad (h)$

Substitute value of y in (a),

$$16x - 6 = 10 \qquad (i)$$

Collect terms in (i), $16x = 16 \qquad (j)$

D: 16 in (j), $x = 1 \qquad (k)$

Check

Usual method.

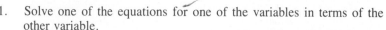

Not only is the method of substitution a very useful one; it also serves to emphasize that the values of the variables are the same in both equations. The method of solving by substitution can be stated as follows:

RULE

To solve by substitution:

1. Solve one of the equations for one of the variables in terms of the other variable.
2. Substitute the resultant value of the variable, found in step 1, in the remaining equation.
3. Solve the equation obtained in step 2 for the second variable.
4. In the simpler of the original equations, substitute the value of the variable found in step 3 and solve the resulting equation for the remaining unknown variable.

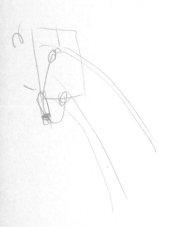

PROBLEMS 17-3

Solve by the method of substitution:

1. $2V - I = 4$
   $2V + 3I = 12$

2. $a + 2b = 6$
   $3a - 10 = 2b$

3. $4I = -2 - 2i$
   $3I + 12 = 2i$

4. $\pi - 8\omega = 0$
   $\pi + \omega = 45$

5. $5\alpha - 8\beta = 0$
   $8\alpha - 13\beta = -1$

6. $5I_1 + 7I_2 = 74$
   $5I_2 - 7I_1 = 0$

7. $3 + 4V = 15v$
   $2 - 9v = -2V$

8. $3X_L + 20 = 8X_C$
   $3X_C - 44 = -8X_L$

9. $4\theta - 164 = 10\phi$
   $3\theta - 2\phi = 68$

10. $3\lambda_1 + 11 = 4\lambda_2$
    $3\lambda_2 = 9 + 2\lambda_1$

11. $3f + 5F = -9$
    $17 - 4f = -3F$

12. $18 - 6I_1 = 8I_2$
    $5I_1 + 4I_2 - 22 = 0$

13. $16 - 2\gamma = 3\delta$
    $\delta - 52 = -4\gamma$

14. $2\pi - 8 = \omega$
    $2\omega + 3\pi = 5$

15. $5 + \varepsilon = 2\psi$
    $3\varepsilon + 4\psi = 20$

16. $4X_L - 9X_C = -16$
    $7X_C + 2 = 6X_L$

17. $0.6\theta + 1.7\phi = 3.5$
    $1.4\theta - 3.9 = 0.3\phi$

18. $0.6I + 0.8i = 2.6$
    $7.0 - 0.5I = -0.3i$

19. $1.2a - 2b = 1$
    $1.4a - 1.5b = 1.5$

20. $0.6L + 0.2M = 2040$
    $0.5L + 0.3M = 1860$

21. Solve Probs. 1 to 20 graphically, and confirm the answers obtained algebraically.

17-4 SOLUTION BY COMPARISON

In the method of solution by comparison we solve for the value of the same variable in each equation in terms of the other variable and place these values equal to each other. The result is an equation having only one unknown.

Example 5

Solve the equations $x - 4y = 14$ and $4x + y = 5$.

Solution

Given

$$x - 4y = 14 \qquad (a)$$
$$4x + y = 5 \qquad (b)$$

Solve (a) for x in terms of y,

$$x = 14 + 4y \qquad (c)$$

Solve (b) for x in terms of y,

$$x = \frac{5 - y}{4} \qquad (d)$$

Equate values of x in (c) and (d),

$$14 + 4y = \frac{5 - y}{4} \qquad (e)$$

M: 4 in (e),

$$56 + 16y = 5 - y \qquad (f)$$

Collect terms in (f)

$$17y = -51 \qquad (g)$$

D: 17 in (g),

$$y = -3$$

Substitute the value of y in (a),

$$x + 12 = 14$$

Collect terms,

$$x = 2$$

Check

Usual method.

PROBLEMS 17-4

Solve by the method of comparison:

1. $3I + 2i = 5$
    $I + i = 2$

2. $3Z - 2R = 7$
    $Z + 2R = 5$

3. $\lambda + 2\pi = -2$
    $15\lambda - 106 = 4\pi$

4. $4E + 3e = 15$
    $2E + 11e = 36$

5. $4x + 2y = 20$
    $1 + 2y = 3x$

6. $5L_1 + 24 = 6L_2$
    $9L_2 - 22 = 4L_1$

7. $2\alpha - 5\beta - 7 = 0$
    $7\alpha - 2\beta - 40 = 0$

8. $2M - 24Q = 0$
    $3M - 20Q = 16$

9. $0.7p - 0.6q = 6.3$
    $0.9p - 1.3 = -0.2q$

10. $2.8I - 2.7i = 19.9$
    $6 + 5i = 2.1I$

Solve Probs. 1 to 10 graphically and by the other algebraic methods.

17-5 FRACTIONAL FORM

Simultaneous linear equations having fractions with numerical denominators are readily solved by first clearing the fractions from the equations and then by solving by means of any method which is considered to be most convenient.

Example 6

Solve the equations $\frac{x}{4} + \frac{y}{3} = \frac{7}{12}$ and $\frac{x}{2} - \frac{y}{4} = \frac{1}{4}$.

Solution

Given

$$\frac{x}{4} + \frac{y}{3} = \frac{7}{12} \qquad (a)$$

$$\frac{x}{2} - \frac{y}{4} = \frac{1}{4} \qquad (b)$$

M: 12, the LCD, in (a), $\qquad 3x + 4y = 7$
M: 4, the LCD, in (b), $\qquad 2x - y = 1$

The resulting equations contain no fractions. Inspection of them shows that solution by addition is most convenient. The solution is

$$x = 1 \qquad y = 1$$

PROBLEMS 17-5

Solve the following sets of equations:

1. $\frac{a}{7} + \frac{4b}{7} = 2$

    $\frac{a}{4} - b = -\frac{1}{2}$

2. $\frac{A}{3} + \frac{B}{5} = -\frac{3}{5}$

    $\frac{3A}{34} - \frac{2B}{17} = -\frac{1}{2}$

3. $\frac{6\theta}{13} + \frac{8\phi}{13} = 2$

    $\frac{\theta}{7} - \frac{3\phi}{35} = 2$

4. $2E - \frac{15e}{26} = 3\frac{1}{13}$

    $\frac{13E}{33} - \frac{8e}{99} = 1$

5. $\varepsilon + \eta = 45$

    $\frac{\varepsilon}{8} - \eta = 0$

6. $\frac{I}{3} + \frac{i}{5} = -\frac{1}{15}$

    $\frac{7i}{30} + \frac{I}{10} = \frac{1}{2}$

7. $\frac{X_L}{4} - \frac{X_C}{8} = \frac{1}{2} = \frac{X_L}{12} + \frac{X_C}{8}$

8. $\frac{Z_1 + 2Z_2}{24} - \frac{Z_2 - 5}{4} = \frac{Z_1 + Z_2 + 1}{36}$

    $\frac{Z_1 - 2}{12} = \frac{5 + Z_2}{3} - \frac{2Z_2 + 6}{6}$

9. $\frac{\lambda - \theta}{3} + \frac{5\lambda}{6} = \frac{5 - \theta}{6} - \frac{1 + \lambda}{4}$

    $\frac{\lambda + 2}{5} = \theta - \frac{1}{4}$

10. $\frac{4I - i}{15} + \frac{1}{8} = 2i - \frac{12I}{5}$

    $i - I = \frac{3}{16}$

17-6 FRACTIONAL EQUATIONS

When variables occur in denominators, it is generally easier to solve without clearing the equations of fractions.

Example 7

Solve the equations $\dfrac{5}{x} - \dfrac{6}{y} = -\dfrac{1}{2}$ and $\dfrac{2}{x} - \dfrac{3}{y} = -1$.

Solution

Given

$$\frac{5}{x} - \frac{6}{y} = -\frac{1}{2} \qquad (a)$$

$$\frac{2}{x} - \frac{3}{y} = -1 \qquad (b)$$

M: 2 in (a),

$$\frac{10}{x} - \frac{12}{y} = -1 \qquad (c)$$

M: 5 in (b),

$$\frac{10}{x} - \frac{15}{y} = -5 \qquad (d)$$

Subtract (d) from (c),

$$\frac{3}{y} = 4$$

$$y = \tfrac{3}{4}$$

Substitute $\tfrac{3}{4}$ for y in (b),

$$\frac{2}{x} - 4 = -1$$

Collect terms,

$$\frac{2}{x} = 3$$

$$\therefore x = \tfrac{2}{3}$$

Check

Usual method.

PROBLEMS 17-6

Solve the following sets of equations:

1. $\dfrac{1}{R} + \dfrac{1}{Z} = \dfrac{7}{12}$

   $\dfrac{1}{R} - \dfrac{1}{Z} = \dfrac{1}{12}$

2. $\dfrac{2}{E_1} + \dfrac{3}{E_2} = \dfrac{13}{6}$

   $\dfrac{1}{E_1} + \dfrac{1}{E_2} = \dfrac{1}{6}$

3. $\dfrac{2}{X_L} - \dfrac{3}{X_C} = \dfrac{7}{55}$

   $\dfrac{1}{X_L} + \dfrac{1}{X_C} = \dfrac{27}{55} - \dfrac{1}{X_L}$

4. $\dfrac{6}{p} = \dfrac{1}{3}$

   $\dfrac{5}{q} - \dfrac{3}{p} = \dfrac{1}{4}$

5. $\dfrac{7}{\theta} + \dfrac{1}{\phi} = \dfrac{51}{80}$

   $\dfrac{4}{\phi} - \dfrac{4}{\theta} = \dfrac{11}{20}$

6. $\dfrac{4}{a-1} = \dfrac{3}{1-b}$

   $\dfrac{7}{2a-39} = \dfrac{5}{2b-5}$

7. $\dfrac{G-5}{5} - \dfrac{Y+3}{3} = -1$

   $\dfrac{18}{Y-1} = \dfrac{27}{G-12}$

8. $\dfrac{\lambda + 3\pi}{7} + 1 = \pi$

   $\dfrac{2}{\lambda} - \dfrac{4}{\pi} = 0$

9. $\dfrac{1}{M} + \dfrac{1}{L_1} = \dfrac{4}{15}$

$\dfrac{19}{15} - \dfrac{3}{M} = \dfrac{2}{L_1}$

10. $\dfrac{1}{\pi} + \dfrac{1}{\lambda} = 3\dfrac{31}{35}$

$\dfrac{1}{2\pi} + \dfrac{1}{4\lambda} = 1\dfrac{19}{35}$

17-7 LITERAL EQUATIONS IN TWO UNKNOWNS

The solution of literal simultaneous equations involves no new methods of solution. In general, it will be found that the addition or subtraction method will suffice for most cases.

Example 8

Solve the equations $ax + by = c$ and $mx + ny = d$.

Solution

Given

$$ax + by = c \qquad (a)$$
$$mx + ny = d \qquad (b)$$

First eliminate x.

M: m in (a), $amx + bmy = cm$ (c)
M: a in (b), $amx + any = ad$ (d)
Subtract (d) from (c), $bmy - any = cm - ad$ (e)
Factor (e), $y(bm - an) = cm - ad$ (f)

D: $(bm - an)$ in (f). $y = \dfrac{cm - ad}{bm - an}$

Now go back to (a) and (b), and eliminate y.

M: n in (a), $anx + bny = cn$ (g)
M: b in (b), $bmx + bny = bd$ (h)
Subtract (h) from (g), $anx - bmx = cn - bd$ (i)
Factor (i), $x(an - bm) = cn - bd$ (j)

D: $(an - bm)$ in (j), $x = \dfrac{cn - bd}{an - bm}$

$$= \dfrac{bd - cn}{bm - an}$$

Example 9

Solve the equations

$$\dfrac{a}{x} + \dfrac{b}{y} = \dfrac{1}{xy} \qquad \text{and} \qquad \dfrac{c}{x} + \dfrac{d}{y} = \dfrac{1}{xy}$$

Solution

Given

$$\dfrac{a}{x} + \dfrac{b}{y} = \dfrac{1}{xy} \qquad (a)$$
$$\dfrac{c}{x} + \dfrac{d}{y} = \dfrac{1}{xy} \qquad (b)$$

First eliminate y, although it makes no difference which variable is eliminated first.

M: xy, the LCD, in (a), $ay + bx = 1$ (c)

M: xy, the LCD, in (b), $\qquad cy + dx = 1 \qquad (d)$
M: c in (c), $\qquad acy + bcx = c \qquad (e)$
M: a in (d), $\qquad acy + adx = a \qquad (f)$
Subtract (f) from (e), $\qquad bcx - adx = c - a \qquad (g)$
Factor (g), $\qquad x(bc - ad) = c - a \qquad (h)$

D: $(bc - ad)$ in (h), $\qquad x = \dfrac{c - a}{bc - ad}$

Now go back to (a) and (b) to eliminate x, and find

$$y = \frac{b - d}{bc - ad}$$

PROBLEMS 17-7

Given: Solve for

1. $4\alpha - \beta = P$ $\qquad\qquad$ α and β
    $\beta + 2\alpha = Q$

2. $3\pi + 2\lambda = x$ $\qquad\qquad$ π and λ
    $2\pi - \lambda = y$

3. $V + IR = a$ $\qquad\qquad$ V and IR
    $3V + 7IR = b$

4. $4L_1 + 3L_2 = C$ $\qquad\qquad$ L_1 and L_2
    $3L_1 - 2L_2 = C$

5. $6\theta + 5\phi = \alpha$ $\qquad\qquad$ θ and ϕ
    $3\phi - 4\theta = \beta$

6. $5r + 3R = Z_1$ $\qquad\qquad$ R and r
    $3r + 7R = Z_2$

7. $0.04X_C + 0.3X_L = Z_1$ $\qquad\qquad$ X_C and X_L
    $0.02X_C + 0.3X_L = Z_2$

8. $\dfrac{R_L}{4} + \dfrac{R_p}{3} = R_T$ $\qquad\qquad$ R_L and R_p

    $\dfrac{R_L}{2} - \dfrac{R_p}{4} = R_1$

9. $\dfrac{1}{R_1} + \dfrac{1}{R_2} = \dfrac{1}{R_p}$ $\qquad\qquad$ R_1 and R_2

    $\dfrac{3}{R_1} + \dfrac{2}{R_2} = \dfrac{1}{R_t}$

10. $\dfrac{1}{3}(Z_1 - Z_2) = Z_1 - Z_2 - X_C$ $\qquad\qquad$ Z_1 and Z_2

    $\dfrac{2}{5}Z_1 - Z_2 = 0$

17-8 EQUATIONS CONTAINING THREE UNKNOWNS

In the preceding examples and problems, two equations were necessary to solve for two unknown variables. For problems involving three variables, three equations are necessary. The same methods of solution apply for problems involving three variables.

Example 10
Solve the equations

$$2x + 3y + 5z = 0 \qquad (a)$$
$$6x - 2y - 3z = 3 \qquad (b)$$
$$8x - 5y - 6z = 1 \qquad (c)$$

Solution
Choose a variable to be eliminated. Let it be x.

M: 3 in (a),
$$6x + 9y + 15z = 0 \qquad (d)$$
$$6x - 2y - 3z = 3 \qquad (b)$$

Subtract (b) from (d),
$$11y + 18z = -3 \qquad (e)$$

M: 4 in (a),
$$8x + 12y + 20z = 0 \qquad (f)$$
$$8x - 5y - 6z = 1 \qquad (c)$$

Subtract (c) from (f),
$$17y + 26z = -1 \qquad (g)$$

This gives Eqs. (e) and (g) in two variables y and z. Solving them, we obtain $y = 3$, $z = -2$.
Substitute these values into (a),

$$2x + 9 - 10 = 0 \qquad (h)$$

Collect terms,
$$2x = 1 \qquad (i)$$

D: 2 in (i),
$$x = \tfrac{1}{2}$$

Check
Substitute the values of the variables in the equations.

PROBLEMS 17-8
Solve:

1.
$$\theta + 3\phi + 4\pi = 14$$
$$\theta + 2\phi + \pi = 7$$
$$2\theta + \phi + 2\pi = 2$$

2.
$$X_L - X_C + R = 2$$
$$X_C + R + X_L = 6$$
$$X_L - R + X_C = 0$$

3.
$$R_1 + 2R_2 + R_3 = 9$$
$$R_2 + R_3 + 2R_1 = 16$$
$$2R_3 + R_1 + R_2 = 3$$

4.
$$a - 2b + c = 3$$
$$a + b + 2c = 1$$
$$2a - b + c = 2$$

5.
$$\frac{1}{R_L} - \frac{1}{R_p} - \frac{1}{R_1} = \frac{1}{120}$$
$$\frac{1}{R_L} + \frac{1}{R_p} - \frac{1}{R_1} = \frac{49}{120}$$
$$\frac{1}{R_p} - \frac{1}{R_1} - \frac{1}{R_L} = \frac{-31}{120}$$

6.
$$\frac{1}{a} - \frac{1}{b} - \frac{1}{c} = 1$$
$$\frac{1}{b} - \frac{1}{a} - \frac{1}{c} = 1$$
$$\frac{1}{c} - \frac{1}{a} - \frac{1}{b} = 1$$

7.
$$0.1r - 0.1R + 0.6R_L = 4.1$$
$$2r + 3R + 6R_L = 70$$
$$\tfrac{3}{40}r + \tfrac{1}{20}R - \tfrac{1}{40}R_L = \tfrac{1}{2}$$

8.
$$E_1 - E_2 - E_3 = \alpha$$
$$E_3 - E_1 - E_2 = \beta$$
$$E_2 - E_3 - E_1 = \gamma$$

9. $s - t = 8$
    $2v - 6 = s - 2$
    $3v - 12 = 3t$

10. $a + 5 = c$
    $7b = 3c - 1$
    $2b - a = c - 9$

17-9 METHODS OF SOLUTION OF PROBLEMS

It is convenient to solve a problem involving more than one unknown by setting up a system of simultaneous equations according to the statements of the problem.

Example 11

When a certain number is increased by one-third of another number, the result is 23. When the second number is increased by one-half of the first number, the result is 29. What are the numbers?

Solution

Let x = first number and y = second number.

Then
$$x + \frac{1}{3}y = 23 \qquad (a)$$

Also,
$$y + \frac{1}{2}x = 29 \qquad (b)$$

Solving the equations, we obtain $x = 16$, $y = 21$.

Check

When 16, the first number, is increased by one third of 21, we have
$$16 + 7 = 23$$

When 21, the second number, is increased by one-half of 16, we have
$$21 + 8 = 29$$

Example 12

Two airplanes start from Omaha at the same time. The plane traveling west has a speed 130 km/h faster than that of the plane traveling east. At the end of 4 h they are 2600 km apart. What is the speed of each plane?

Solution

Let x = rate of plane flying west and y = rate of plane flying east.

Then $\qquad x - y = 130 \qquad\qquad\qquad (a)$
Since   Rate × time = distance
then $\qquad\qquad 4x$ = distance traveled by plane flying west
and $\qquad\qquad 4y$ = distance traveled by plane flying east
hence, $\qquad 4x + 4y = 2600 \qquad\qquad (b)$

Solving Eqs. (a) and (b), we obtain

$$x = 390 \text{ km/h}$$
$$y = 260 \text{ km/h}$$

Check

Substitute these values of x and y into the statements of the example.

Often it is possible to derive a formula from known data and thereby eliminate terms which are not desired or cannot be used conveniently in some investigation.

Example 13

The effective voltage V of an alternating voltage is equal to 0.707 times its maximum value V_{max}. That is,

$$V = 0.707V_{max} \tag{1}$$

Also, the average value V_{av} is equal to 0.637 times the maximum value. That is,

$$V_{av} = 0.637V_{max} \tag{2}$$

It is desired to express the effective value V in terms of the average value V_{av}.

Solution

V_{max} must be eliminated.

Solving Eq. (1) for V_{max}, $\qquad\qquad V_{max} = \dfrac{V}{0.707}$

Solving Eq. (2) for V_{max}, $\qquad\qquad V_{max} = \dfrac{V_{av}}{0.637}$

By Axiom 5, $\qquad\qquad\qquad\quad \dfrac{V}{0.707} = \dfrac{V_{av}}{0.637}$

Solving for V, $\qquad\qquad\qquad V = 1.11V_{av} \tag{3}$

Equation (3) shows that the effective value of an alternating voltage is 1.11 times the average value of the voltage.

Example 14

You know that in a dc circuit $P = VI$ and also that $P = I^2R$. Derive a formula for V in terms of I and R.

Solution

It is evident that P must be eliminated. Because both equations are equal to P, we can equate them (Axiom 5) and obtain

$$VI = I^2R$$

D: I,

$$V = IR \tag{4}$$

Example 15

The quantity of electricity Q, in coulombs, in a capacitor is equal to the product of the capacitance C and the applied voltage V. That is,

$$Q = CV \tag{5}$$

The total voltage across capacitors C_a and C_b connected in series $V = V_a + V_b$. Find C in terms of C_a and C_b.

Solution

Solve for V, V_a, and V_b. Thus

$$V = \frac{Q}{C}$$

$$V_a = \frac{Q}{C_a}$$

and

$$V_b = \frac{Q}{C_b}$$

Then, since

$$V = V_a + V_b$$

By substitution

$$\frac{Q}{C} = \frac{Q}{C_a} + \frac{Q}{C_b}$$

D: Q,

$$\frac{1}{C} = \frac{1}{C_a} + \frac{1}{C_b}$$

M: CC_aC_b, the LCD

$$C_aC_b = CC_b + CC_a$$

Transposing,

$$CC_a + CC_b = C_aC_b$$

D: $(C_a + C_b)$

$$C = \frac{C_aC_b}{C_a + C_b} \tag{6}$$

This is the formula for the resultant capacitance C of two capacitors C_a and C_b connected in series.

PROBLEMS 17-9

1. The sum of two currents is I_t A, and their difference is I_d A. What are the currents?
2. Find two numbers whose sum is 19 and whose difference is 5.
3. If 1 is added to each term of a fraction, the value of the fraction becomes 0.75; and if 1 is subtracted from each term, the value of the fraction becomes 0.5. What is the fraction?
4. In a right triangle, the acute angles are complementary (that is, they add up to 90°). What are the angles if their difference is 40°?
5. The difference between the two acute angles of a right triangle is $\alpha°$. Find the angles.
6. The sum of the three angles of any triangle is 180°. Find the three angles of a particular triangle if the smallest angle is one-third the middle angle and the largest is 5° larger than the middle one.
7. A TV repair technician goes to the parts dealer for an assortment of common resistors and capacitors. The sales clerk replies: "We have two such assortments: 20 resistors and 8 capacitors for $3.60, or 60 resistors and 40 capacitors for $14.00. Both assortments come under the same discount schedule." "I'll take the larger selection," says the technician, "if you'll figure out the price of one resistor and one capacitor."
   Help the sales clerk.
8. A takes 1 h longer than B to walk 30 km, but if A's pace were doubled, then A would take 2 h less than B. Find their rates of walking.

9. In 3 h, L drives 40 km farther than Q does in 2 h. In 6 h, Q drives 180 km more than L does in 4 h. Find their average rates of driving.

10. $v = gt$ and $s = \frac{1}{2}gt^2$. Solve for v in terms of s and t.

11. $C = \dfrac{Q}{V}$ and $W = \dfrac{QV}{2}$. Solve for W in terms of C and Q.

12. $I = \dfrac{V}{R}$ and $P = I^2R$. If $P = 2.7$ kW and $V = 180$ V, find the current I and the resistance R.

13. $v = u + at$ and $s = \frac{1}{2}(u + v)t$. Find the distance s in terms of initial velocity u and acceleration a and time t.

14. Use the information of Prob. 13 to show that $v^2 = u^2 + 2as$.

15. $\text{Gain} = \dfrac{i_d r_d}{v_{gs}}$ and $\text{gain} = \dfrac{v_d}{i_d\left(r_s + \dfrac{1}{g_m}\right)}$.

Solve for gain in terms of r_d and g_m when $r_s = 0$.

16. $R = 2D_L fL$ and $Q = \dfrac{2\pi fL}{R}$. Solve for D_L in terms of π and Q.

17. $R = \omega LQ$, $Q = \dfrac{\omega L}{r}$, and $\omega^2 = \dfrac{1}{LC}$. Solve for R in terms of L, C, and r.

18. $I = \dfrac{V}{R}$ and $I_1 = \dfrac{V}{R + R_1}$. Solve for R in terms of R_1, I, and I_1.

19. $Q = It$ coulombs (C), and $I = \dfrac{CV}{t}$ A. Solve for Q in terms of C and V.

20. Given $P = VI$ W, $I = \dfrac{V}{R}$ A, and $H = 0.24I^2Rt$. Solve for H in terms of P and t.

21. Use the data of Prob. 20 to find H when $V = 30$ V over a time $t = 10$ s if the heater resistance $R = 300$ Ω.

22. Given $I_aR_a = I_bR_b$, $\dfrac{Q_a}{Q_b} = \dfrac{C_a}{C_b}$, $I_a = \dfrac{Q_a}{t}$, and $I_b = \dfrac{Q_b}{t}$, show that $R_aC_a = R_bC_b$.

23. $I_p = \dfrac{\mu V_g}{R + R_p}$ and $V_p = I_pR$. Solve for R in terms of R_p, μ, V_p, and V_g.

24. Use the data of Prob. 23 to find V_p when $\mu = 50$, $V_g = 5$ V, $I_p = 12.5$ mA, and $R_p = 10$ kΩ.

25. The three-Varley method of cable fault location yields the following relationships:

$$R_AV_1 + R_AR_Y = R_AV_2 - R_YR_B$$
$$R_AV_2 + R_AR_X = R_AV_3 - R_XR_B$$
$$R_AV_1 + R_AR_T = R_AV_3 - R_TR_B$$

Solve these three equations for R_Y, R_X, and R_T, and show that $R_T = R_Y + R_X$.

26. Given $V = I_x(R + R_x)$, $V = I_a(R + R_a)$, and $V = IR$. Show that

$$R_x = R_a \times \dfrac{\dfrac{I - I_x}{I_x}}{\dfrac{I - I_a}{I_a}}$$

27. Given three star-delta transformation equations:

$$R_a = \frac{R_1 R_3}{R_1 + R_2 + R_3}$$

$$R_b = \frac{R_1 R_2}{R_1 + R_2 + R_3}$$

$$R_c = \frac{R_2 R_3}{R_1 + R_2 + R_3}$$

Solve for R_1, R_2, and R_3 in terms of R_a, R_b, and R_c.

28. If three resistors R_1, R_2, and R_3 are connected in parallel so that the total circuit current I_t is divided into I_1, I_2, and I_3, respectively, determine relationships similar to Eq. (1) in Chap. 15 for I_1, I_2, and I_3 in terms of R_1, R_2, R_3, and I_t.

Determinants

18

In Chap. 17 we learned four methods of solving simultaneous equations of the second order, and we used some of those methods to solve equation sets of the third order. Indeed, some of the methods we learned are limited to solving simultaneous equations of the second order, while others may be used to solve third-, fourth-, fifth-, or even higher-order systems.

However, after about the third order, the method of repeated addition and subtraction, with its attendant multiplication, becomes tedious. In this chapter we shall investigate a "mechanical" method of solving simultaneous equations. This method, known as the method of determinants, is usually not introduced until students are well along in advanced mathematics, so we are not going to study all the fascinating developments which the whole subject of determinants may involve. (That would take a separate book of its own.) Instead, we are going to see how determinants may be put to work for us in order to simplify our solutions to simultaneous equations.

18-1 SECOND-ORDER DETERMINANTS

In sec. 17-2 we learned how to solve pairs of simultaneous equations by the method of addition and subtraction. Let us apply this method to a pair of *general equations:*

$$a_1x + b_1y = c_1 \tag{1}$$

$$a_2x + b_2y = c_2$$

where a_1, a_2, b_1, b_2, c_1, and c_2 represent any numbers, positive or negative, integers or fractions, or zero, Let us solve these general equations for x:

$$a_1x + b_1y = c_1 \tag{a}$$

$$a_2x + b_2y = c_2 \tag{b}$$

M: b_2 in (a), $\qquad a_1b_2x + b_1b_2y = b_2c_1 \tag{c}$

M: b_1 in (b), $\qquad a_2b_1x + b_1b_2y = b_1c_2 \tag{d}$

Subtract (d) from (c),

$$(a_1b_2 - a_2b_1)x = b_2c_1 - b_1c_2 \qquad (e)$$

Solve for x,

$$x = \frac{b_2c_1 - b_1c_2}{a_1b_2 - a_2b_1} \qquad (2)$$

It is left as an exercise for you to prove similarly that

$$y = \frac{a_1c_2 - a_2c_1}{a_1b_2 - a_2b_1} \qquad (3)$$

Observe that we have kept the literal factors in alphabetical order for convenience in checking.

Note several interesting facts about these two solutions:

1. Their denominators are identical, and they contain only the coefficients of x and y.
2. The numerator for the solution of y contains no y coefficients.
3. The numerator for the solution of x contains no x coefficients.

For a few minutes, let us consider just the denominator: $a_1b_2 - a_2b_1$. We are going to define a new, alternative method of writing this expression.

$$a_1b_2 - a_2b_1 = \begin{vmatrix} a_1 & b_1 \\ a_2 & b_2 \end{vmatrix}$$

This arrangement is called the *determinant* of the denominator. It is a mechanical statement made up of two horizontal *rows* and two vertical *columns* of two elements each, and it is a *second-order determinant*. Whenever this form appears, it is understood to mean $a_1b_2 - a_2b_1$. To obtain this evaluation of the determinant, we perform diagonal multiplication, first of all downward to the right to obtain

$$\begin{vmatrix} a_1 & b_1 \\ a_2 & b_2 \end{vmatrix} \; a_1b_2 \qquad \text{(this is, by definition, positive multiplication)}$$

and, second, we multiply upward to the right to obtain

$$\begin{vmatrix} a_1 & b_1 \\ a_2 & b_2 \end{vmatrix} - a_2b_1 \qquad \text{(this is, by definition, negative multiplication)}$$

RULE

1. The diagonal multiplication in determinants derives its sign from the direction of the multiplication, and not primarily from any algebraic signs of the elements being multiplied.
2. After the individual steps of multiplication, with the appropriate sign of the multiplication affixed, the products are added algebraically to form the evaluation of the determinants.

Example 1

Evaluate the determinant

$$\begin{vmatrix} -3 & 2 \\ 5 & 1 \end{vmatrix}$$

Solution

Perform the signed diagonal multiplication:

$$+(-3)(1) - (5)(2) = -3 - 10 = -13$$

PROBLEMS 18-1

Evaluate the following determinants:

1. $\begin{vmatrix} 4 & 1 \\ 2 & 1 \end{vmatrix}$ 2. $\begin{vmatrix} 1 & 8 \\ 3 & -12 \end{vmatrix}$ 3. $\begin{vmatrix} 2 & -8 \\ 3 & 5 \end{vmatrix}$

4. $\begin{vmatrix} -1 & 4 \\ 3 & 12 \end{vmatrix}$ 5. $\begin{vmatrix} 3 & -4 \\ 3 & -4 \end{vmatrix}$ 6. $\begin{vmatrix} 3 & -2 \\ 1 & 2 \end{vmatrix}$

7. $\begin{vmatrix} 9 & 4 \\ 15 & -6 \end{vmatrix}$ 8. $\begin{vmatrix} -3 & -7 \\ -4 & -2 \end{vmatrix}$ 9. $\begin{vmatrix} 0.8 & 0.2 \\ 0.5 & 0.1 \end{vmatrix}$

10. $\begin{vmatrix} -0.06 & 0.02 \\ 0.05 & -1.6 \end{vmatrix}$ 11. $\begin{vmatrix} a & b \\ a & b \end{vmatrix}$ 12. $\begin{vmatrix} a & b \\ x & y \end{vmatrix}$

13. $\begin{vmatrix} b & a \\ y & x \end{vmatrix}$ 14. $\begin{vmatrix} a & x \\ b & y \end{vmatrix}$ 15. $\begin{vmatrix} b & y \\ a & x \end{vmatrix}$

16. $\begin{vmatrix} y & x \\ b & a \end{vmatrix}$

18-2 SOLUTION OF EQUATIONS

Consider Eqs. (2) and (3) to be the solutions for x and y in the general equations (1):

$$x = \frac{b_2 c_1 - b_1 c_2}{a_1 b_2 - a_2 b_1} \qquad y = \frac{a_1 c_2 - a_2 c_1}{a_1 b_2 - a_2 b_1}$$

or, in determinant form:

$$x = \frac{\begin{vmatrix} b_2 & c_2 \\ b_1 & c_1 \end{vmatrix}}{\begin{vmatrix} a_1 & b_1 \\ a_2 & b_2 \end{vmatrix}} \qquad (4)$$

$$y = \frac{\begin{vmatrix} a_1 & c_1 \\ a_2 & c_2 \end{vmatrix}}{\begin{vmatrix} a_1 & b_1 \\ a_2 & b_2 \end{vmatrix}} \qquad (5)$$

Let us see how the determinant form may be developed directly from the original equations without performing the intervening additions and subtractions. Given the original equations:

$$a_1x + b_1y = c_1 \qquad (1)$$
$$a_2x + b_2y = c_2$$

First, produce the determinant of the denominator by setting, in order, the coefficients of the unknowns:

$$\begin{vmatrix} a_1 & b_1 \\ a_2 & b_2 \end{vmatrix}$$

Second, by using the denominator determinant as a base, develop the determinant of the numerator of the solution for x by replacing the column of x coefficients by the corresponding column of constants (the right-hand sides of the equations). Then complete the new determinant by putting in the column of the y coefficients in its original position:

$$\begin{vmatrix} c_1 & b_1 \\ c_2 & b_2 \end{vmatrix}$$

Confirm that this determinant is identical in value with

$$\begin{vmatrix} b_2 & c_2 \\ b_1 & c_1 \end{vmatrix}$$

given as Eq. (4), but easier to develop automatically.

Third, still using the denominator determinant as a starting place, develop the determinant of the numerator of y by replacing the column of y coefficients by the column of constants and leaving the column of x coefficients in its original position:

$$\begin{vmatrix} a_1 & c_1 \\ a_2 & c_2 \end{vmatrix}$$

Last, put these three determinants together to form the full solution statements:

$$x = \frac{\begin{vmatrix} c_1 & b_1 \\ c_2 & b_2 \end{vmatrix}}{\begin{vmatrix} a_1 & b_1 \\ a_2 & b_2 \end{vmatrix}} \qquad (4)$$

$$y = \frac{\begin{vmatrix} a_1 & c_1 \\ a_2 & c_2 \end{vmatrix}}{\begin{vmatrix} a_1 & b_1 \\ a_2 & b_2 \end{vmatrix}} \qquad (5)$$

RULE

To solve two simultaneous equations having two variables by the method of determinants:

1. Form the denominator determinant by using the coefficients of the unknowns in their correct rows and columns.
2. Form the x numerator determinant by replacing the column of x coefficients in the denominator determinant by the column of constants.
3. Form the y numerator determinant by replacing the column of y coefficients in the denominator by the column of constants.
4. Combine the three determinants so formed to produce the pair of solution equations.

Example 2

Solve the simultaneous equations

$$3p + 2q = 8$$
$$5p + q = 11$$

Solution

The denominator determinant is

$$\begin{vmatrix} 3 & 2 \\ 5 & 1 \end{vmatrix}$$

Using this determinant as a base, the determinant for the numerator of p must be $\begin{vmatrix} 8 & 2 \\ 11 & 1 \end{vmatrix}$ and the determinant for the numerator of q must

be $\begin{vmatrix} 3 & 8 \\ 5 & 11 \end{vmatrix}$. Thus,

$$p = \frac{\begin{vmatrix} 8 & 2 \\ 11 & 1 \end{vmatrix}}{\begin{vmatrix} 3 & 2 \\ 5 & 1 \end{vmatrix}} \quad \text{and} \quad q = \frac{\begin{vmatrix} 3 & 8 \\ 5 & 11 \end{vmatrix}}{\begin{vmatrix} 3 & 2 \\ 5 & 1 \end{vmatrix}}$$

When evaluating determinants, *always* evaluate the denominator first. (The reason will be explained soon.) The value of the denominator is

$$+(3)(1) - (5)(2) = -7$$

The numerator of p has the value

$$+(8)(1) - (11)(2) = -14$$
$$p = \frac{-14}{-7} = 2$$

The numerator of q has the value $+(3)(11) - (5)(8) = -7$, and

$$q = \frac{-7}{-7} = 1$$

18-3 CONSISTENCY OF EQUATIONS

For solving systems of second-order simultaneous equations, there are three main possibilities:

1. The equations may represent straight lines which intersect. These are said to be *independent equations*. They are in no way related to each other except that the unknowns have similar symbols, A, b, x, θ, etc., and one pair of values constitutes the whole solution.
2. The equations may represent superimposed lines. These are said to be *dependent equations*. They are related to each other, and every solution of the one is also a solution of the other. There is an endless number of solutions.
3. The equations may represent parallel lines. These are said to be *inconsistent equations*. They differ only in the constant terms (the y intercepts), and there is no solution for one equation which satisfies the other.

The values of the denominator and the numerators quickly show us into which classification any system of simultaneous equations falls:

1. To be independent, the denominators may not equal zero.
2. To be dependent, the denominator is zero and the numerators equal zero.
3. To be inconsistent, the denominator is zero and at least one of the numerators does not equal zero.

This is why we evaluate the denominator first. If it is zero, there is no single set of values which will constitute the entire solution, and, in electronics problems, there is no use investigating further.

PROBLEMS 18-2

Solve these systems of simultaneous equations by using determinants:

1. $4a - 3b = 10$
    $3a + b = 14$

2. $4x + y = 15$
    $2x + 3y = 15$

3. $2\theta + \pi = 22$
    $3\theta - 5\pi = 20$

4. $R_1 + 3R_2 = 23$
    $R_1 - 3R_2 = 5$

5. $I + 4i = -5$
    $2I + i = 4$

6. $3V + 2V_g = 1$
    $V_g + V = -2$

7. $4r_p + 3r_L = 3$
    $6r_p - 9r_L = 0$

8. $4X_C + 3X_L = 2.9$
    $30X_L = 17 - 8X_C$

9. $0.5R_1 + 0.2R_2 = 315$
    $0.6R_1 - 54 = 0.03R_2$

10. $Z_1 = 9300 - Z_2$
    $192 + 0.06Z_2 = 0.04Z_1$

18-4 THIRD-ORDER DETERMINANTS

When solving sets of three simultaneous equations, naturally, we arrive at third-order determinants consisting of three columns and three rows of three elements each, such as

$$\begin{vmatrix} 3 & 1 & 2 \\ 2 & 6 & 5 \\ 4 & 8 & 1 \end{vmatrix} \qquad \begin{vmatrix} a_1 & b_1 & c_1 \\ a_2 & b_2 & c_2 \\ a_3 & b_3 & c_3 \end{vmatrix}$$

Now, when we multiply on the diagonal, we find a slight complication. Multiplying the main diagonal is simple:

$$\begin{vmatrix} 3 & 1 & 2 \\ 2 & 6 & 5 \\ 4 & 8 & 1 \end{vmatrix} = +(3)(6)(1) = +18$$

but the next diagonal gets complicated:

$$\begin{vmatrix} 3 & 1 & 2 \\ 2 & 6 & 5 \\ 4 & 8 & 1 \end{vmatrix} = +(1)(5)(4) = +20$$

and also the next:

$$\begin{vmatrix} 3 & 1 & 2 \\ 2 & 6 & 5 \\ 4 & 8 & 1 \end{vmatrix} = +(2)(8)(2) = +32$$

And you can see that the negative diagonals will be just as complicated. So we devise a method of notation which gets around this complication and enables us to perform straight-line multiplication. First, we set down the determinant in its usual form, with three columns and three rows. Then, to the right of this determinant, we repeat the first two columns. This process straightens out the diagonals

$$\begin{vmatrix} 3 & 1 & 2 \\ 2 & 6 & 5 \\ 4 & 8 & 1 \end{vmatrix} \begin{matrix} 3 & 1 \\ 2 & 6 \\ 4 & 8 \end{matrix}$$

$$= -(1)(2)(1) = -2$$
$$= +(3)(6)(1) = +18$$

and we obtain, with a complete program of diagonal multiplication, the value of the determinant $= -100$.

Example 3
Evaluate the determinant

$$\begin{vmatrix} 2 & -1 & 4 \\ 1 & 6 & 5 \\ 7 & -3 & -2 \end{vmatrix}$$

Solution

Rewrite the determinant and repeat the first two columns outside to the right:

$$\begin{vmatrix} 2 & -1 & 4 \\ 1 & 6 & 5 \\ 7 & -3 & -2 \end{vmatrix} \begin{matrix} 2 & -1 \\ 1 & 6 \\ 7 & -3 \end{matrix}$$

Then perform the diagonal multiplication, signed, as for second-order determinants and obtain

$$-24 - 35 - 12 - 168 + 30 - 2 = -211$$

Example 4

Solve the third-order set of simultaneous equations:

$$\begin{aligned} a + 2b + c &= 7 \\ 2a + b + 2c &= 2 \\ a + 3b + 4c &= 14 \end{aligned}$$

Solution

First, write and evaluate the denominator determinant:

$$\begin{vmatrix} 1 & 2 & 1 \\ 2 & 1 & 2 \\ 1 & 3 & 4 \end{vmatrix} \begin{matrix} 1 & 2 \\ 2 & 1 \\ 1 & 3 \end{matrix} = -9$$

Second, develop the determinant for the numerator of a, replacing the column of a coefficients by the column of constants, and evaluate it:

$$\begin{vmatrix} 7 & 2 & 1 \\ 2 & 1 & 2 \\ 14 & 3 & 4 \end{vmatrix} \begin{matrix} 7 & 2 \\ 2 & 1 \\ 14 & 3 \end{matrix} = 18$$

Third, combine the denominator and numerator to evaluate a:

$$a = \frac{18}{-9} = -2$$

You should immediately prove that $b = 4$ and $c = 1$.

PROBLEMS 18-3

Evaluate these third-order determinants:

1. $\begin{vmatrix} 1 & 1 & 1 \\ 2 & -1 & -1 \\ 3 & 2 & -5 \end{vmatrix}$

2. $\begin{vmatrix} 1 & 3 & 1 \\ 5 & 40 & 6 \\ -2 & -25 & -3 \end{vmatrix}$

3. $\begin{vmatrix} 2 & 3 & 32 \\ 5 & -2 & 0 \\ 4 & -8 & -41 \end{vmatrix}$

4. $\begin{vmatrix} -3 & -2 & 3 \\ 0 & -7 & 2 \\ 0 & 7 & -4 \end{vmatrix}$

5. $\begin{vmatrix} 4 & 6 & 8 \\ -10 & -3 & 4 \\ 2 & 12 & -20 \end{vmatrix}$

6. $\begin{vmatrix} 3 & 8 & 3.2 \\ 12 & 20 & 16.5 \\ -16 & -12 & -7.8 \end{vmatrix}$

Solve these simultaneous equations by using determinants:

7. $x + y + z = 15$
$2x - y - z = 0$
$3x + 2y - 5z = 14$

8. $R_1 + R_2 + R_3 = 3$
$5R_1 - 2R_2 + 6R_3 = 40$
$-2R_1 + 3R_2 - 3R_3 = -25$

9. $2\alpha + 3\beta + 2\gamma = 32$
$5\alpha - \gamma = 2\beta$
$4\alpha - 8\beta = 3\gamma - 41$

10. $3r + 5p - 2q = -3$
$p + q = 4r$
$3p - 7q + 2r = -42$

11. $4V + 6v + 8(IR) = 6$
$4(IR) - 10V - 3v = -5$
$12v - 20(IR) + 12V = 5$

12. $12I_1 + 20I_2 + 10I_3 = 16.5$
$8I_2 - 6I_3 + 3I_1 = 3.2$
$20I_3 - 16I_1 - 12I_2 = -7.8$

18-5 MINORS

The method of diagonal multiplication works perfectly for both second- and third-order determinants. Unfortunately, it will not work for higher-order systems. Thus, if we are required to evaluate by determinants a fourth- or fifth-order set of equations such as might arise from the solution of a complicated circuit (see Chap. 22), we must work out another useful system. Since we can do this without difficulty, we will not try to prove the statement above. (Even many "higher mathematics" texts say simply: Do not use diagonal multiplication for fourth-order determinants or higher.)

This is how minors come about: Let us evaluate the general third-order determinant:

$$\begin{vmatrix} a_1 & b_1 & c_1 \\ a_2 & b_2 & c_2 \\ a_3 & b_3 & c_3 \end{vmatrix}\begin{matrix} a_1 & b_1 \\ a_2 & b_2 \\ a_3 & b_3 \end{matrix}$$
$$= a_1b_2c_3 + a_3b_1c_2 + a_2b_3c_1 - a_3b_2c_1 - a_1b_3c_2 - a_2b_1c_3 \quad (6)$$

Consider the terms which involve the value a_1. These may be collected to yield $a_1(b_2c_3 - b_3c_2)$, which in turn could be written

$$a_1\begin{vmatrix} b_2 & c_2 \\ b_3 & c_3 \end{vmatrix}$$

where the new second-order determinant is called the *minor of the element a_1*.

We can develop this minor from the original third-order determinant by selecting the element a_1, crossing out the other elements of the row and column which contain a_1, and writing the minor with the elements remaining.

$$\begin{vmatrix} \cancel{a_1} & \cancel{b_1} & \cancel{c_1} \\ \cancel{a_2} & b_2 & c_2 \\ \cancel{a_3} & b_3 & c_3 \end{vmatrix}$$

yields
$$\begin{vmatrix} b_2 & c_2 \\ b_3 & c_3 \end{vmatrix}$$

RULE

To find the *minor* of any element in a determinant, select the element, cross out the row and column containing that element, and write the lower-order determinant which contains all the other elements that remain.

Thus, in the third-order determinant of Eq. (6), the minor of the element b_3 is

$$\begin{vmatrix} a_1 & c_1 \\ a_2 & c_2 \end{vmatrix}$$

Example 5

Evaluate the minor of 2 in the determinant

$$\begin{vmatrix} 1 & 4 & 0 \\ 3 & 1 & 5 \\ 5 & 6 & 2 \end{vmatrix}$$

Solution

Striking out the elements in the row and column containing the 2 yields

$$\begin{vmatrix} 1 & 4 \\ 3 & 1 \end{vmatrix} = +1 - 12 = -11$$

PROBLEMS 18-4

Write and evaluate the *minors* of the indicated elements:

1. $\begin{vmatrix} 2 & 3 & 2 \\ -4 & -1 & 3 \\ 5 & 2 & ⑥ \end{vmatrix}$

2. $\begin{vmatrix} 3 & 1 & -1 \\ ⑧ & -2 & 2 \\ -13 & -3 & -1 \end{vmatrix}$

3. $\begin{vmatrix} 3 & 2 & 5 \\ ⊖2 & -3 & 4 \\ 6 & 5 & 0 \end{vmatrix}$

4. $\begin{vmatrix} -3 & -7 & 16 \\ -8 & 2 & 84 \\ 2 & 3 & ⊖26 \end{vmatrix}$

5. $\begin{vmatrix} 2 & 3 & -5 \\ 3 & 0 & 4 \\ 0 & ⊖2 & 7 \end{vmatrix}$

6. $\begin{vmatrix} -8 & ⊖13 & 10 \\ 0 & 2 & 5 \\ 2 & 10 & -20 \end{vmatrix}$

7. $\begin{vmatrix} 0 & 0 & 4 \\ 0 & 2 & 4 \\ -4 & ⑩ & 0 \end{vmatrix}$

8.* $\begin{vmatrix} 3 & 6 & -3 & 2 \\ 2 & -2 & 2 & -1 \\ 5 & ㉕ & 0 & 3 \\ 0 & 5 & -5 & 1 \end{vmatrix}$

*Hint The minor of any element of a fourth-order determinant will be a third-order determinant which may itself be evaluated by the diagonal method or by second-step cofactors, which are discussed in the following section.

18-6 COFACTORS

A simple step converts the *minor* into a *cofactor*. When evaluating a complete determinant by the method of cofactors, we first find the minors of all the elements in any given row or column. Then we convert these minors into cofactors by assigning them algebraic signs according to this simple rule:

RULE

Each element of a determinant, regardless of its actual algebraic value, has a cofactor sign according to its place in the determinant. The signs are found by a checkerboard arrangement:

$$\begin{vmatrix} + & - & + \\ - & + & - \\ + & - & + \end{vmatrix}$$

The only thing to remember is to always start the upper left-hand corner (the element in row 1 and column 1) with a + sign. All the rest follows automatically, regardless of the number of elements in the determinant.

Example 6

Evaluate the following determinant by means of cofactors:

$$\begin{vmatrix} 1 & 4 & 0 \\ -3 & 1 & 5 \\ 5 & 6 & -2 \end{vmatrix}$$

Solution

Choose any convenient row or column, and, one after the other, set down the individual elements of that row or column, together with their minors:

$$4\begin{vmatrix} -3 & 5 \\ 5 & -2 \end{vmatrix} \quad 1\begin{vmatrix} 1 & 0 \\ 5 & -2 \end{vmatrix} \quad 6\begin{vmatrix} 1 & 0 \\ -3 & 5 \end{vmatrix}$$

Then assign the cofactor signs according to the checkerboard plan:

$$-4\begin{vmatrix} -3 & 5 \\ 5 & -2 \end{vmatrix} \quad +1\begin{vmatrix} 1 & 0 \\ 5 & -2 \end{vmatrix} \quad -6\begin{vmatrix} 1 & 0 \\ -3 & 5 \end{vmatrix}$$

Evaluate each minor, multiply its value by the element of which it is the minor, and add algebraically according to the cofactor signs and the actual algebraic sign of the multiplications:

$$-4(6 - 25) + 1(-2 - 0) - 6(5 - 0) = 76 - 2 - 30 = 44$$

You should immediately evaluate the same third-order determinant by the cofactors of the elements of each other row and column in turn. The answer must always be 44.

Example 7

Solve this set of simultaneous equations by means of cofactors:

$$\begin{aligned} 2p + 10q + 5r &= 9 \\ -3p + 9q + 4r &= -3 \\ 7p - 6q - r &= 17 \end{aligned}$$

Solution

Using the information now at hand, we may immediately set up the determinant form of solution:

$$p = \frac{\begin{vmatrix} 9 & 10 & 5 \\ -3 & 9 & 4 \\ 17 & -6 & -1 \end{vmatrix}}{\begin{vmatrix} 2 & 10 & 5 \\ -3 & 9 & 4 \\ 7 & -6 & -1 \end{vmatrix}} \qquad q = \frac{\begin{vmatrix} 2 & 9 & 5 \\ -3 & -3 & 4 \\ 7 & 17 & -1 \end{vmatrix}}{\begin{vmatrix} 2 & 10 & 5 \\ -3 & 9 & 4 \\ 7 & -6 & -1 \end{vmatrix}}$$

$$r = \frac{\begin{vmatrix} 2 & 10 & 9 \\ -3 & 9 & -3 \\ 7 & -6 & 17 \end{vmatrix}}{\begin{vmatrix} 2 & 10 & 5 \\ -3 & 9 & 4 \\ 7 & -6 & -1 \end{vmatrix}}$$

Always evaluate the denominator first. To solve by means of cofactors, we choose any row or column in the denominator determinant, evaluate the minors, and multiply by the elements, adding algebraically and using the checkerboard signs.

$$\begin{vmatrix} 2 & 10 & 5 \\ -3 & 9 & 4 \\ 7 & -6 & -1 \end{vmatrix}$$

$$= -(-3)\begin{vmatrix} 10 & 5 \\ -6 & -1 \end{vmatrix} + (9)\begin{vmatrix} 2 & 5 \\ 7 & -1 \end{vmatrix} - (4)\begin{vmatrix} 2 & 10 \\ 7 & -6 \end{vmatrix}$$

$$= 3(-10 + 30) + 9(-2 - 35) - 4(-12 - 70)$$

$$= 60 - 333 + 328$$

$$= 55$$

Since the denominator is not zero, we should evaluate the numerators, in turn, of *p, q,* and *r.* The numerator of

$$p = \begin{vmatrix} 9 & 10 & 5 \\ -3 & 9 & 4 \\ 17 & -6 & -1 \end{vmatrix}$$

$$= +(17)\begin{vmatrix} 10 & 5 \\ 9 & 4 \end{vmatrix} - (-6)\begin{vmatrix} 9 & 5 \\ -3 & 4 \end{vmatrix} + (-1)\begin{vmatrix} 9 & 10 \\ -3 & 9 \end{vmatrix}$$

$$= +110$$

Therefore, $p = \dfrac{110}{55} = 2$. Now prove that $q = -1$ and $r = 3$.

18-7 USEFUL PROPERTIES OF DETERMINANTS

The evaluation of determinants by the methods of diagonal multiplication or cofactors will yield the correct answers if you keep close watch on your arithmetic and the algebraic signs of positive and negative diagonals or of the checkerboard cofactor signs. There are, however, a few very useful properties of determinants which will simplify the process of evaluation. These properties are described briefly below, and it is left to you to perform the diagonal multiplication or cofactor evaluation methods to confirm them immediately when you meet them.

1. When all the elements of any row (or column) are zero, the value of the determinant is zero:

$$\begin{vmatrix} a_1 & b_1 & 0 \\ a_2 & b_2 & 0 \\ a_3 & b_3 & 0 \end{vmatrix} = 0$$

Example 8
Evaluate the determinant:

$$\begin{vmatrix} 2 & 4 & -3 \\ 0 & 0 & 0 \\ -4 & 6 & 1 \end{vmatrix}$$

Solution

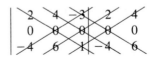

Each diagonal multiplication introduces a factor of zero. Therefore, each diagonal product is zero, and the value of the determinant is zero.

2. When all the elements to the right (or left) of the principal diagonal are zero, the value of the determinant is the product of the elements of the principal diagonal:

$$\begin{vmatrix} a_1 & 0 & 0 \\ a_2 & b_2 & 0 \\ a_2 & b_3 & c_3 \end{vmatrix} = a_1 b_2 c_3$$

(It is left to you to prove that this is true for fourth-order determinants also.)

Example 9
Evaluate the determinant:

$$\begin{vmatrix} 3 & 8 & 5 \\ 0 & -2 & 7 \\ 0 & 0 & -5 \end{vmatrix}$$

Solution

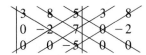

All of the diagonal multiplications except the first one, through the principal diagonal, are zero. Therefore, the value of the determinant is $(3)(-2)(-5) = 30$.

3. Interchanging all the rows and columns gives the identical result, both absolute value and algebraic sign:

$$\begin{vmatrix} a_1 & a_2 & a_3 \\ b_1 & b_2 & b_3 \\ c_1 & c_2 & c_3 \end{vmatrix} = \begin{vmatrix} a_1 & b_1 & c_1 \\ a_2 & b_2 & c_2 \\ a_3 & b_3 & c_3 \end{vmatrix}$$

4. Interchanging two rows (or columns) gives the same absolute value but the opposite algebraic sign:

$$\begin{vmatrix} c_1 & b_1 & a_1 \\ c_2 & b_2 & a_2 \\ c_3 & b_3 & a_3 \end{vmatrix} = - \begin{vmatrix} a_1 & b_1 & c_1 \\ a_2 & b_2 & c_2 \\ a_3 & b_3 & c_3 \end{vmatrix}$$

5. When the corresponding elements of any two rows (or columns) are identical or proportional, the value of the determinant is zero:

$$\begin{vmatrix} a_1 & ka_1 & c_1 \\ a_2 & ka_2 & c_2 \\ a_3 & ka_3 & c_3 \end{vmatrix} = 0 \qquad (k \text{ may } = 1)$$

Example 10
Evaluate the determinant:

$$\begin{vmatrix} 3 & 5 & 6 \\ 2 & -1 & 4 \\ 7 & 4 & 14 \end{vmatrix}$$

Solution

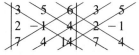

Diagonal multiplication yields are zero value. Observation of the first and third columns shows that col 3 = 2 × col 1.

6. A common factor of any row (or column) may be factored out as a common factor of the whole determinant:

$$\begin{vmatrix} a_1 & b_1 & kc_1 \\ a_2 & b_2 & kc_2 \\ a_3 & b_3 & kc_3 \end{vmatrix} = k \begin{vmatrix} a_1 & b_1 & c_1 \\ a_2 & b_2 & c_2 \\ a_3 & b_3 & c_3 \end{vmatrix}$$

Example 11

Evaluate the determinant

$$\begin{vmatrix} 3 & 6 & 2 \\ -2 & 8 & 5 \\ 40 & 30 & -70 \end{vmatrix}$$

Solution

$$\begin{vmatrix} 3 & 6 & 2 \\ -2 & 8 & 5 \\ 40 & 30 & -70 \end{vmatrix} = 10 \begin{vmatrix} 3 & 6 & 2 \\ -2 & 8 & 5 \\ 4 & 3 & -7 \end{vmatrix} \begin{matrix} 3 & 6 \\ -2 & 8 \\ 4 & 3 \end{matrix}$$

$$= 10(-253) = -2530$$

7. When the elements of any row (or column) are increased by a constant times the corresponding elements of any other row (or column), the value of the determinant is unchanged. (k may equal 1, -1, or any other positive or negative integer or fraction):

$$\begin{vmatrix} a_1 & b_1 & ka_1 + c_1 \\ a_2 & b_2 & ka_2 + c_2 \\ a_3 & b_3 & ka_3 + c_3 \end{vmatrix} = \begin{vmatrix} a_1 & b_1 & c_1 \\ a_2 & b_2 & c_2 \\ a_3 & b_3 & c_3 \end{vmatrix}$$

Example 12

Evaluate the determinant

$$\begin{vmatrix} 2 & 8 & 3 \\ 3 & 7 & -6 \\ -1 & 2 & 1 \end{vmatrix}$$

Solution

If the spaces filled by the elements 8, 3, and -6 can be converted to zeros, the evaluation of the determinant will be the product of the elements of the principal axis. Or if any two spaces in any row or column can be adjusted to zero, the evaluation becomes a single element times its cofactor.

Using the principle introduced above, let us attempt to eliminate the element 3. We will multiply each element of the third row by -3 and add the result to the corresponding elements of the first row:

$$\begin{vmatrix} 2 & 8 & 3 \\ 3 & 7 & -6 \\ -1 & 2 & 1 \end{vmatrix} = \begin{vmatrix} 2 + (-3)(-1) & 8 + (-3)(2) & 3 + (-3)(1) \\ 3 & 7 & -6 \\ -1 & 2 & 1 \end{vmatrix}$$

$$= \begin{vmatrix} 5 & 2 & 0 \\ 3 & 7 & -6 \\ -1 & 2 & 1 \end{vmatrix}$$

Then, to eliminate the -6, we will multiply the third row by 6 and add the results to the second row:

$$\begin{vmatrix} 5 & 2 & 0 \\ 3 & 7 & -6 \\ -1 & 2 & 1 \end{vmatrix} = \begin{vmatrix} 5 & 2 & 0 \\ 3 + (6)(-1) & 7 + (6)(2) & -6 + (6)(1) \\ -1 & 2 & 1 \end{vmatrix}$$

$$= \begin{vmatrix} 5 & 2 & 0 \\ -3 & 19 & 0 \\ -1 & 2 & 1 \end{vmatrix}$$

This determinant may be evaluated by the product of the element 1 and its cofactor:

$$\begin{vmatrix} 5 & 2 & 0 \\ -3 & 19 & 0 \\ -1 & 2 & 1 \end{vmatrix} = +1 \begin{vmatrix} 5 & 2 \\ -3 & 19 \end{vmatrix}$$

$$= 95 + 6 = 101$$

You should test this solution by the diagonal multiplication of the original determinant. Alternatively, the simplification may continue by removal of the element 2 in the first row. If we add to the first row the product of $-\frac{2}{19}$ (second row),

$$\begin{vmatrix} 5 + (-\frac{2}{19})(-3) & 2 + (-\frac{2}{19})(19) & 0 + (-\frac{2}{19})(0) \\ -3 & 19 & 0 \\ -1 & 2 & 1 \end{vmatrix}$$

$$= \begin{vmatrix} 5\frac{6}{19} & 0 & 0 \\ -3 & 19 & 0 \\ -1 & 2 & 1 \end{vmatrix}$$

Evaluation by the principal diagonal yields

$$(5\tfrac{6}{19})(19)(1) = 101$$

With practice, the addition of a fraction in the form $-\dfrac{a_x}{a_y} a_y$ will reveal itself as a valuable tool.

8. When the elements of any row (or column) may be written as sums, the determinant may be written as the sum of two determinants with the rows (or columns) of the sum elements in their corresponding places:

$$\begin{vmatrix} a_1 & b_1 & p_1 + q_1 \\ a_2 & b_2 & p_2 + q_2 \\ a_3 & b_3 & p_3 + q_3 \end{vmatrix} = \begin{vmatrix} a_1 & b_1 & p_1 \\ a_2 & b_2 & p_2 \\ a_3 & b_3 & p_3 \end{vmatrix} + \begin{vmatrix} a_1 & b_1 & q_1 \\ a_2 & b_2 & q_2 \\ a_3 & b_3 & q_3 \end{vmatrix}$$

Now apply these fundamental properties of determinants in the solution of the following problems and problems like them in later chapters.

PROBLEMS 18-5

Evaluate the following determinants by means of the *cofactors* of the indicated rows or columns:

1.
$$\begin{vmatrix} 5 & 41 & 6 \\ 2 & 1 & -2 \\ -4 & 8 & 3 \end{vmatrix} \quad \text{Row 1}$$

2.
$$\begin{vmatrix} 2 & 1 & -1 \\ 4 & -3 & -1 \\ 3 & -2 & 2 \end{vmatrix} \quad \text{Col 2}$$

3.
$$\begin{vmatrix} 5 & 2 & 3 \\ 4 & -3 & 12 \\ 0 & 5 & -8 \end{vmatrix} \quad \text{Row 3}$$

4.
$$\begin{vmatrix} -26 & 3 & 2 \\ 84 & 2 & -10 \\ 16 & -7 & 4 \end{vmatrix} \quad \text{Col 1}$$

5.
$$\begin{vmatrix} 3 & 0 & 21.7 \\ 2 & 3 & 15.3 \\ 0 & -2 & 1.9 \end{vmatrix} \quad \text{Col 3}$$

6.
$$\begin{vmatrix} 2 & 4 & 10 \\ -8 & -16 & -13 \\ 0 & 16 & 2 \end{vmatrix} \quad \text{Row 2}$$

7.
$$\begin{vmatrix} 3 & 2 & -3 & 2 \\ 2 & -3 & 2 & -1 \\ 5 & 4 & 0 & 3 \\ 0 & 8 & -5 & 1 \end{vmatrix} \quad \text{Col 2}$$

Hint The cofactors of elements in a fourth-order determinant will themselves be third-order determinants which may be evaluated by diagonals or by cofactors.

8.
$$\begin{vmatrix} 2 & 16 & 12 & -10 & -2 \\ 5 & 2 & 2 & 3 & -9 \\ 11 & 0 & 0 & 5 & 4 \\ 5 & 0 & 2 & 15 & 4 \\ 0 & -4 & 10 & -8 & 0 \end{vmatrix} \quad \text{Row 3}$$

Solve by using determinants and cofactors:

9.
$$\begin{aligned} 5I_1 + 2I_2 + 6I_3 &= 41 \\ 2I_1 + 3I_2 - 2I_3 &= 1 \\ -4I_1 - I_2 + 3I_3 &= 8 \end{aligned}$$

10.
$$\begin{aligned} 2\theta + \phi - \lambda &= 3 \\ 3\theta - 2\phi + 2\lambda &= 8 \\ 4\theta - 3\phi - \lambda &= -13 \end{aligned}$$

11.
$$\begin{aligned} 3\alpha + 2\beta + 3\gamma &= 5 \\ -2\alpha - 3\beta + 12\gamma &= 4 \\ 6\alpha + 5\beta - 8\gamma &= 0 \end{aligned}$$

12.
$$\begin{aligned} 2I_1 + 3I_2 + 2I_3 &= -26 \\ -8I_1 + 2I_2 - 10I_3 &= 84 \\ -3I_1 - 7I_2 + 4I_3 &= 16 \end{aligned}$$

13.
$$\begin{aligned} 3R_1 + 4R_3 &= 21.7 \\ 2R_1 + 3R_2 - 5R_3 &= 15.3 \\ 7R_3 - 2R_2 &= 1.9 \end{aligned}$$

14.
$$\begin{aligned} 2x + 4y + 10z &= 10 \\ -8x - 16y + 5z &= -13 \\ 16y - 20z &= 2 \end{aligned}$$

15.
$$\begin{aligned} 3\varepsilon + 2\eta - 3\kappa + 2\lambda &= 6 \\ 2\varepsilon - 3\eta + 2\kappa - \lambda &= -2 \\ 5\varepsilon + 4\eta + 3\lambda &= 25 \\ 8\eta - 5\kappa + \lambda &= 5 \end{aligned}$$

16. $2I_1 + 16I_2 + 12I_3 - 10I_4 - 2I_5 = 100$
      $5I_1 + 2I_2 + 2I_3 + 3I_4 - 9I_5 = 0$
      $11I_1 + 5I_4 + 4I_5 = 100$
      $5I_1 + 2I_3 + 15I_4 + 4I_5 = 100$
      $-4I_2 + 10I_3 - 8I_4 = 0$

17.   Solve selected problems from Chap. 17 by means of determinants.

18.   Use determinants for the solution of appropriate problems throughout the remainder of this book.

19.   The vidicon tubes of a color TV camera receive light information that is transformed into color and luminance signals. The output of the three vidicon tubes is measured in the following proportions. Solve for the resulting red, green, and blue percentages:

$$R - 0.7G + 0.3B = -0.08$$
$$0.6R + 0.2G + 0.2B = 0.32$$
$$0.5R + 0.5G - B = 0.335$$

If a white picture is Y, where Y is the *luminance signal* made up of the red, green, and blue components, show that:

$$Y = 0.3R + 0.59G + 0.11B$$

Batteries

19

In preceding discussions of electric circuits, to avoid confusion, all sources of electromotive force have been considered to be sources of constant potential, and nothing has been said of their internal resistances. At the same time, no mention has been made of the actual sources of the emf. In this chapter we will consider both of these factors. First of all, electrical devices that produce electric energy, as well as those that consume energy, have a certain amount of internal resistance which materially affects their operation. The application of Ohm's law to the internal resistance of batteries is the feature topic of this chapter. And despite the prevalence of utility power supply, batteries are still useful, indeed necessary, sources of portable power. For this reason, the electronics technician should be aware of the problems which arise in the use of batteries.

19-1 ELECTROMOTIVE FORCE

A battery is a device which converts chemical energy into electric energy. Essentially, it consists of a cell, or several cells connected in series or parallel, conveniently packaged. The emf of the battery is the total voltage developed by the chemical action. However, not all of this total voltage is available for doing useful work in an external circuit, because some of it is needed to overcome the internal resistance of the battery itself. The voltage which is supplied to the external circuit is known as the terminal voltage; that is,

$$\text{Terminal voltage} = \text{emf} - \text{internal voltage drop}$$

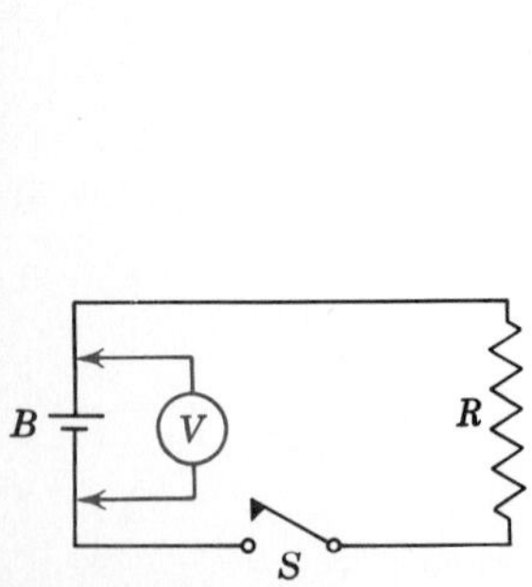

19-2 BATTERIES

The word *battery* is taken to mean two or more *cells* connected to each other, although a single cell is often referred to as a battery.

Figure 19-1 represents a circuit by which the voltage existing across the cell can be read with the resistance connected across the battery or with the resistance disconnected from the circuit.

Fig. 19-1 High-resistance voltmeter used for measuring electromotive force of a cell.

The emf of a cell is the total amount of voltage developed by the cell. For all practical purposes the emf of a cell can be read with a high-resistance voltmeter connected across the cell when the cell is not supplying current to any other circuit, as when the switch S, Fig. 19-1, is open.

When a cell supplies current to an external circuit, as when the switch in Fig. 19-1 is closed, it will be found that the voltmeter no longer reads the open-circuit voltage (emf) of the cell. The reason is that part of the emf is used in forcing current through the resistance of the cell and the remainder is used in forcing current through the external circuit. Expressed as an equation,

$$V = V_t + Ir \qquad (1)$$

where V is the emf of the cell or group of cells and V_t is the voltage measured across the terminals while forcing a current I through the internal resistance r. Since I also flows through the external circuit of resistance R, Eq. (1) can be written

$$V = IR + Ir$$

or
$$V = I(R + r) \qquad (2)$$

Example 1

A cell whose internal resistance is 0.15 Ω delivers 0.50 A to a resistance of 2.85 Ω. What is the emf of the cell?

Solution

Given $r = 0.15$ Ω, $R = 2.85$ Ω, and $I = 0.50$ A.

From Eq. (2), $V = 0.50(2.85 + 0.15) = 1.5$ V

Example 2

Figure 19-2 represents a cell with an emf of 1.2 V and an internal resistance r of 0.2 Ω connected to a resistance R of 5.8 Ω. How much current flows in the circuit?

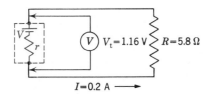

Fig. 19-2 Circuit of Example 2.

Solution

Solving Eq. (2) for the current,

$$I = \frac{V}{R + r} \qquad (3)$$

$$= \frac{1.2}{5.8 + 0.2} = 0.2 \text{ A}$$

Note the significance of Eq. (3). It says that the current which flows in a circuit is proportional to the emf of the circuit and inversely proportional to the *total* resistance of the circuit. This is Ohm's law for the *complete circuit*.

Example 3

A cell with an emf of 1.6 V delivers a current of 2 A to a circuit of 0.62 Ω. What is the internal resistance of the cell?

Solution

Solving Eq. (2) for the internal resistance,

$$r = \frac{V - IR}{I} \tag{4}$$

$$= \frac{1.6 - 2 \times 0.62}{2} = 0.18 \, \Omega$$

Therefore, the significance of Eq. (4) is that a voltage equal to $V - IR$ is sending the current I through the internal resistance r.

Since Eq. (4) can be rearranged to

$$r = \frac{V}{I} - R$$

and

$$\frac{V}{I} = R_t$$

Eq. (4) can be written

$$r = R_t - R$$

or

$$R_t = R + r \tag{5}$$

Equation (5) states simply that the resistance of the entire circuit is equal to the resistance of the external circuit plus the internal resistance of the source of the emf.

PROBLEMS 19-1

1. A battery taken off the shelf gives a voltmeter reading of 9 V. When connected across a 24-Ω circuit, it drives a current of 360 mA. What is its internal resistance?

2. A 24-cell battery measures 38.4 V on open circuit. If the total internal resistance is 7.2 Ω, how much current will flow through a 430-Ω circuit?

3. A 6-V battery drives a current of 1 A through a 5.6-Ω load. What is the internal resistance of the battery?

4. With the circuit of Prob. 3, how much power is absorbed by the internal resistance of the battery?

5. With the circuit of Prob. 3, (*a*) how much power is delivered to the load and (*b*) what is the efficiency of the circuit?

19-3 CELLS IN SERIES

If *n* identical cells are connected in series, the emf of the combination will be *n* times the emf of each cell. Similarly, the total internal resistance of the circuit will be *n* times the internal resistance of each cell. By modifying

Eq. (2), the expression for the current through an external resistance of $R\ \Omega$ is

$$I = \frac{nV}{R + nr} \qquad (6)$$

Example 4

Six cells, each having an emf of 2.1 V and an internal resistance of 0.1 Ω, are connected in series, and a resistance of 3.6 Ω is connected across the combination.
(a) How much current flows in the circuit?
(b) What is the terminal voltage of the group?

Solution

Figure 19-3 is a diagram of the circuit. The resistance nr represents the total internal resistance of all cells in series.

(a) $I = \dfrac{nV}{R + nr} = \dfrac{6 \times 2.1}{3.6 + 6 \times 0.1} = 3.0$ A

(b) The terminal voltage of the group is equal to the total emf minus the voltage drop across the internal resistance. From Eq. (1),

$$\begin{aligned} V_t &= nV - Inr \\ &= 6 \times 2.1 - 3 \times 6 \times 0.1 \\ &= 10.8 \text{ V} \end{aligned}$$

Since the terminal voltage exists across the external circuit, a more simple relation is

$$V_t = IR = 3 \times 3.6 = 10.8 \text{ V}$$

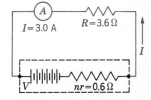

Fig. 19-3 Circuit of Example 4.

19-4 CELLS IN PARALLEL

If n identical cells are connected in parallel, the emf of the group will be the same as the emf of one cell and the internal resistance of the group will be equal to the internal resistance of one cell divided by the number of cells in parallel, that is, to $\dfrac{r}{n}$. By modifying Eq. (2), the expression for the current through an external resistance of $R\ \Omega$ is

$$I = \frac{V}{R + \dfrac{r}{n}} \qquad (7)$$

Example 5

Three cells, each with an emf of 1.4 V and an internal resistance of 0.15 Ω, are connected in parallel, and a resistance of 1.35 Ω is connected across the group.
(a) How much current flows in the circuit?
(b) What is the terminal voltage of the group?

Solution

Figure 19-4 is a diagram of the circuit. The resistance $\dfrac{r}{n}$ represents the internal resistance of the group.

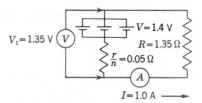

Fig. 19-4 Circuit of Example 5.

(a) $I = \dfrac{V}{R + \dfrac{r}{n}} = \dfrac{1.4}{1.35 + \dfrac{0.15}{3}} = 1.0\ A$

(b) $V_t = IR = 1.0 \times 1.35 = 1.35\ V$

PROBLEMS 19-2

1. The emf of a cell is 1.5 V, and the internal resistance of the cell is 0.15 Ω. When the cell supplies current to a load, the voltage drop across the internal resistance is 0.2 V.
   (a) What is the terminal voltage?
   (b) What is the current flow?
   (c) What is the connected load?

2. A cell whose emf is 1.4 V is supplying 1.5 A to a 0.733-Ω circuit.
   (a) What is the internal resistance of the cell?
   (b) How much power is lost in the cell?

3. A cell of emf 1.6 V develops a terminal voltage of 1.48 V when delivering 250 mA to an external circuit.
   (a) What is the internal resistance of the cell?
   (b) How much power is expended in the cell?
   (c) What is the resistance of the external circuit?
   (d) How much power is absorbed by the load circuit?
   (e) What is the efficiency of the power transfer?

4. A high-resistance voltmeter reads 2 V when connected across the terminals of an open-circuit cell. What will the meter read when a 5-A current is delivered to a 0.22-Ω load if the internal resistance of the cell is 0.18 Ω?

5. Using the data and results of Prob. 4, how much current would flow if the cell itself were short-circuited?

6. A cell with an emf of 2 V and an internal resistance of 0.1 Ω is connected to a load consisting of a variable resistor.
   (a) Plot the power delivered to the load as the load resistance is varied in 0.01-Ω steps from 0.05 to 0.15 Ω. What conclusion do you draw from this graph?
   (b) Plot the efficiency of power transfer over the same resistance range. What conclusion do you draw?

7. Six identical cells, each of emf 1.5 V and internal resistance 0.1 Ω, are connected in series across a load resistor, and they deliver a circuit current of 1.0 A.
   (a) What is the resistance of the load?
   (b) How much power is absorbed by the battery?
   (c) How much current would flow if the battery were short-circuited?

8. If the cells in Prob. 7 are connected in parallel, how much power will be delivered to the load?

9. Ten cells of emf 1.5 V and internal resistance 0.6 Ω each are connected in series across a load of 33 Ω.
   (a) How much current will flow in the circuit?
   (b) What will be the terminal voltage of the battery?
   (c) How much power will be delivered to the load?

10. If the cells of Prob. 9 are connected in parallel across the same load, how much current will flow?

11. Twelve identical cells are hooked up so that four groups of three cells each in series are connected in parallel as shown in Fig. 19-5. The emf of each cell is 1.6 V, and each cell has an internal resistance of 0.2 Ω. If the load R is 0.85 Ω and the measured current flow through R is 4.8 A:
   (a) What is the terminal voltage of the battery?
   (b) What is the emf of each cell?
   (c) How much power is expended in each cell?

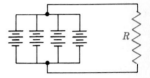

Fig. 19-5 Circuit of Prob. 11.

12. The cells of Prob. 11 are so arranged that there are two-cells-per-series groups (six groups in parallel).
   (a) How much power is dissipated in R?
   (b) How much current flows through each cell?

13. Each cell of a six-cell storage battery has an emf of 2.0 V and an internal resistance of 0.01 Ω. The battery is to be charged from a 14-V line.
   (a) How much resistance must be connected in series with the battery to limit the charging current to 15 A?
   (b) What current would flow if the battery were disconnected from the charging circuit and short-circuited?

14. Sixteen storage batteries of three cells each are to be charged in series from a 115-V line. Each cell has an emf of 2.1 V and an internal resistance of 0.02 Ω.
   (a) How much resistance must be connected in series with the battery to limit the charging current to 10 A?
   (b) How much power is dissipated in the entire circuit?
   (c) How much power is dissipated in the series charging resistance?
   (d) What current would flow if the batteries were disconnected from the charging circuit and short-circuited?

15. Six identical cells that are connected in series deliver 4 A to a circuit of 2.7 Ω. When two of the same cells are connected in parallel, they deliver 5 A to an external resistance of 0.375 Ω. What are the emf and internal resistance of each cell?

   Solution:
   Let V = emf of each cell
   r = internal resistance of each cell
   I = current in external circuit
   R = resistance of external circuit
   For the series connection,
   $6V$ = emf of six cells in series
   and
   $6r$ = internal resistance of six cells in series

Substituting in Eq. (2),

$$6V = 4(2.7 + 6r) = 10.8 + 24r \qquad (a)$$

For the parallel connection,

$V =$ emf of cells in parallel

and

$\dfrac{r}{2} =$ internal resistance of two cells in parallel

Substituting in Eq. (2),

$$V = 5\left(0.375 + \frac{r}{2}\right)$$

or $\qquad\qquad 2V = 3.75 + 5r \qquad (b)$

Solve Eqs. (a) and (b) simultaneously to obtain

$$V = 2.0 \text{ V}$$
and $\qquad\qquad r = 0.05 \ \Omega$

16. Ten identical cells connected in series send a current of 3 A through a 1-Ω circuit. When three of the cells are connected in parallel, they send a current of 6 A through an external resistance of 0.1 Ω. What are the emf and internal resistance of each cell?

17. Five cells connected in series send a current of 5 A through a resistance of 0.4 Ω. When four of the cells are connected in parallel, they send 1 A through 1.35 Ω. What are the emf and internal resistance of each cell?

18. Twelve cells in series, each with an emf of 2.0 V, send a certain current through a 2.4-Ω circuit. The same current flows through a 0.24-Ω circuit when five of the cells are connected in parallel. What is the value of the current and what is the internal resistance of each cell?

19. A cell with an internal resistance of 0.035 Ω sends a 3-A current through an external circuit. Another cell, having the same emf but with an internal resistance of 0.385 Ω, sends a current of 2 A through the external circuit when substituted for the first cell. What is the emf of the cells and what is the resistance of the external circuit?

20. A cell sends a 20-A current through an external circuit of 0.04 Ω. When the resistance of the external circuit is increased to 3.96 Ω, the current is 0.4 A. What is the emf and what is the internal resistance of the cell?

Exponents and Radicals

20

In earlier chapters, examples and problems have been limited to those containing exponents and roots that consisted of integers. In this chapter the study of exponents and radicals is extended to include new operations that will enable you to solve electrical formulas and equations of a type hitherto omitted. In addition, new ideas that will be of fundamental importance in your study of alternating currents are introduced.

20-1 FUNDAMENTAL LAWS OF EXPONENTS

As previously explained, if n is a positive integer, a^n means that a is to be taken as a factor n times. Thus, a^4 is defined as being a shortened form of notation for the product $a \cdot a \cdot a \cdot a$. The number a is called the *base,* and the number n is called the *exponent.*

For the purpose of review, the fundamental laws for the use of *positive-integer exponents* are listed below:

$$a^m \cdot a^n = a^{m+n} \qquad \text{(Sec. 4-3)} \quad (1)$$

$$a^m \div a^n = a^{m-n} \quad \text{(when } n < m\text{) (Sec. 4-9)} \quad (2)$$

$$= \frac{1}{a^{n-m}} \quad \text{(when } n > m\text{)}$$

$$(a^m)^n = a^{mn} \qquad \text{(Sec. 6-11)} \quad (3)$$

$$(ab)^m = a^m b^m \qquad \text{(Sec. 6-12)} \quad (4)$$

$$\left(\frac{a}{b}\right)^m = \frac{a^m}{b^m} \qquad (b \neq 0) \qquad (5)$$

20-2 ZERO EXPONENT

If a^0 is to obey the law of exponents for multiplication as was stated under Eq. (1) of the preceding section, then

$$a^m \cdot a^0 = a^{m+0} = a^m$$

Also, if a^0 is to obey the law of exponents for division, then

$$\frac{a^m}{a^0} = a^{m-0} = a^m$$

Therefore, the zero power of any number, except zero, is defined as being equal to 1, for 1 is the only number that, when used to multiply another number, does not change the value of the multiplicand.

20-3 NEGATIVE EXPONENTS

If a^{-n} is to obey the multiplication law, then

$$\frac{a^n}{a^n} = a^{n-n} = a^0 = 1$$

In Sec. 4-11, it was shown that a *factor* can be transferred from one term of a fraction to the other if the sign of its exponent is changed, that is, from numerator to denominator, or vice versa.

PROBLEMS 20-1

By making use of the five fundamental laws of exponents, write the results of the indicated operations:

1.	$a^4 \cdot a^3$	**2.**	$\pi^2 \cdot \pi^5$	**3.**	$x^2 \cdot x$
4.	$\theta^3 \cdot \theta^7$	**5.**	$p^q p^r$	**6.**	$\lambda^{2x} \cdot \lambda^{5x}$
7.	$I^\alpha \cdot I^\beta$	**8.**	$m^{x+y} \cdot m^{x-y}$	**9.**	$x^8 \div x^5$
10.	$\alpha^{5.3} \div \alpha^{2.7}$	**11.**	$X^{5y} \div X^2$	**12.**	$e^{\pi+2} \div e^3$
13.	$\theta^{\alpha+\beta} \div \theta^{\alpha-\beta}$	**14.**	$\psi^{\alpha+\beta} \div \psi^{\alpha-\gamma}$	**15.**	$(I^3)^3$
16.	$(f^2)^5$	**17.**	$(x^2y^3)^3$	**18.**	$(IR^2t)^3$
19.	$(a^x)^4$	**20.**	$(a^4)^x$	**21.**	$(-x^l y^m z^p)^4$

22. $(-a^\pi b^\lambda)^3$ **23.** $\left(\dfrac{V}{R}\right)^2$ **24.** $\left(\dfrac{R_1 R_2}{R_3}\right)^3$

25. $\left(\dfrac{\omega^3}{2\pi f^2}\right)^6$ **26.** $\left(\dfrac{Z_1^2}{Z_3 Z_4}\right)^2$ **27.** $\left(\dfrac{-X_C^2}{X_L}\right)^3$

28. $\left(\dfrac{\pi D^2}{4}\right)^4$ **29.** $\left(\dfrac{\alpha^{3x}}{\alpha^{x+2}}\right)^2$ **30.** $\left(\dfrac{a^{3\pi}}{a^{5\lambda}}\right)^{4\gamma}$

Express with all positive exponents:

31. $I^2 R^{-1}$ **32.** $x^{-3} y^{-2}$ **33.** $y^{-\pi} z^{3\lambda}$

34. $16 L_1^{-2} L_2^{-2}$ **35.** $\theta^4 \phi^{-3} \lambda^{-2x}$ **36.** $(\pi R^2)^{-2i}$

37. $\dfrac{a^{-3} b}{c^{-1}}$ **38.** $\left(\dfrac{Z_1 Z_2}{Z_4}\right)^{-3}$ **39.** $\dfrac{3I^3 R^{-2}}{12 I^2 r^{-3}}$

40. $\dfrac{\alpha^3}{2(4\beta\gamma)^{-2}}$

20-4 FRACTIONAL EXPONENTS

The meaning of a base affected by a fractional exponent is established by methods similar to those employed in determining meanings for zero or negative exponents. If we assume that Eq. (1) of Sec. 20-1 holds for fractional exponents, we should obtain, for example,

$$a^{\frac{1}{2}} \cdot a^{\frac{1}{2}} = a^{\frac{1}{2}+\frac{1}{2}} = a^1 = a$$

Also,
$$a^{\frac{1}{3}} \cdot a^{\frac{1}{3}} \cdot a^{\frac{1}{3}} = a^{\frac{1}{3}+\frac{1}{3}+\frac{1}{3}} = a^1 = a$$

That is, $a^{\frac{1}{2}}$ is one of two equal factors of a, and $a^{\frac{1}{3}}$ is one of three equal factors of a. Therefore, $a^{\frac{1}{2}}$ is the square root of a, and $a^{\frac{1}{3}}$ is the cube root of a. Hence,

$$a^{\frac{1}{2}} = \sqrt{a}$$
and
$$a^{\frac{1}{3}} = \sqrt[3]{a}$$

Likewise,
$$a^{\frac{2}{3}} \cdot a^{\frac{2}{3}} \cdot a^{\frac{2}{3}} = a^{\frac{2}{3}+\frac{2}{3}+\frac{2}{3}}$$
$$= a^{\frac{6}{3}} = a^2$$

Hence,
$$(a^{\frac{2}{3}})^3 = a^2$$
or
$$a^{\frac{2}{3}} = \sqrt[3]{a^2}$$

In a fractional exponent, the denominator denotes the root and the numerator denotes the power of the base.

In general,
$$a^{\frac{m}{n}} = \sqrt[n]{a^m}$$

Example 1
$$a^{\frac{3}{5}} = \sqrt[5]{a^3}$$

Example 2
$$(-8)^{\frac{1}{3}} = \sqrt[3]{-8} = -2$$

PROBLEMS 20-2

Find the value of:

1. $16^{\frac{1}{2}}$

2. $(-27)^{\frac{1}{3}}$

3. $16^{\frac{1}{4}}$

4. $-(-32)^{\frac{1}{5}}$

5. $(-64a^6b^3c^{12})^{\frac{1}{3}}$

6. $(L_1{}^4L_2{}^4)^{\frac{1}{2}}$

7. $(I^4R^2)^{\frac{3}{2}}$

8. $(\theta^3\pi^6)^{\frac{2}{3}}$

9. $\left(\dfrac{27\lambda^9}{\omega^{12}}\right)^{\frac{2}{3}}$

10. $\left(\dfrac{r^{12}R^8}{16V^4}\right)^{\frac{3}{4}}$

Express with radical signs:

11. $9^{\frac{1}{2}}$

12. $8\alpha^{\frac{1}{3}}$

13. $(8\alpha)^{\frac{1}{3}}$

14. $6^{\frac{2}{3}}$

15. $\theta^{\frac{3}{4}}\lambda^{\frac{3}{4}}$

16. $x^{\frac{2}{3}}y^{\frac{3}{2}}$

Express with fractional exponents:

17. $\sqrt{a^3}$

18. $\sqrt[3]{x^2}$

19. $\sqrt[3]{16E}$

20. $\sqrt[3]{a^2b^4c^6}$

21. $\sqrt[3]{\theta^2\omega^4}$

22. $\alpha\sqrt[5]{\beta^2}$

23. $\sqrt[5]{\alpha^2\beta^2}$ **24.** $4L\sqrt{\omega^3}$ **25.** $2\pi\sqrt[3]{16f^3}$

26. $5\alpha^2\sqrt[5]{-32\alpha^3\beta^7}$

20-5 RADICAND

The meaning of the radical sign was explained in Sec. 2-11. The number under the radical sign is called the *radicand*.

20-6 SIMPLIFICATION OF RADICALS

The form in which a radical expression is written can be changed without altering the numerical value of the expression. Such a change is desirable for many reasons. For example, addition of several fractions containing different radicals in the denominators would be more difficult than addition with the radicals removed from the denominators. Similarly, it will be shown later that

$$\frac{1}{\sqrt{3}} = \frac{\sqrt{3}}{3}$$

It is apparent that the value to several decimal places could be computed more easily from the second fraction than from the first.

Because we are chiefly concerned with radicals involving a square root, only that type will be considered.

20-7 REMOVING A FACTOR FROM THE RADICAND

Since, in general, $\sqrt{ab} = \sqrt{a} \cdot \sqrt{b}$, the following is evident:

RULE

A radicand can be separated into two factors one of which is the greatest perfect square it contains. The square root of this factor can then be written as the coefficient of a radical the other factor of which is the radicand.

Example 3

$$\begin{aligned}
\sqrt{27} &= \sqrt{9 \cdot 3} \\
&= \sqrt{9} \cdot \sqrt{3} \\
&= \pm 3\sqrt{3}
\end{aligned}$$

Example 4

$$\begin{aligned}
\sqrt{8} &= \sqrt{4 \cdot 2} \\
&= \sqrt{4} \cdot \sqrt{2} \\
&= \pm 2\sqrt{2}
\end{aligned}$$

Example 5

$$\sqrt{75} = \sqrt{25 \cdot 3}$$
$$= \sqrt{25} \cdot \sqrt{3}$$
$$= \pm 5\sqrt{3}$$

Example 6

$$\sqrt{200a^5b^3c^2d} = \sqrt{100a^4b^2c^2} \cdot \sqrt{2abd}$$
$$= \pm 10a^2bc\sqrt{2abd}$$

Simplify by removing factors from the radicand:

1. $\sqrt{8}$

2. $\sqrt{32}$

3. $\sqrt{18}$

4. $\sqrt{24}$

5. $\sqrt{50}$

6. $\sqrt{20}$

7. $\sqrt{80}$

8. $\sqrt{28}$

9. $\sqrt{720}$

10. $\sqrt{27x^4}$

11. $\sqrt{12\theta^2\phi^4}$

12. $\sqrt{99A^3D}$

13. $5\sqrt{96I^2R}$

14. $3\pi\sqrt{72r^3z^5\pi^3}$

15. $6\omega\sqrt{63f^4F^3T^5}$

16. $7x\sqrt{147xy^2z^3D^3}$

17. $3\alpha^2\sqrt{242\alpha^5\beta^7\gamma^8}$

18. $8\sqrt{567X_L^2Z_1^4}$

19. $2r^3\sqrt{588\pi^4L^4X_L^2}$

20. $50\sqrt{289\theta^5\lambda^7}$

20-8 SIMPLIFYING RADICALS CONTAINING FRACTIONS

Since

$$\sqrt{\frac{4}{9}} = \pm\frac{2}{3}$$

and

$$\frac{\sqrt{4}}{\sqrt{9}} = \pm\frac{2}{3}$$

then

$$\sqrt{\frac{4}{9}} = \pm\frac{\sqrt{4}}{\sqrt{9}}$$

Also,

$$\sqrt{\frac{16}{25}} = \pm\frac{4}{5}$$

and

$$\frac{\sqrt{16}}{\sqrt{25}} = \pm\frac{4}{5}$$

then

$$\sqrt{\frac{16}{25}} = \frac{\sqrt{16}}{\sqrt{25}}$$

Or, in general terms,

$$\sqrt{\frac{a}{b}} = \frac{\sqrt{a}}{\sqrt{b}}$$

The above relation permits simplification of radicals containing fractions by removing the radical from the denominator. This process, by which the denominator is made a rational number, is called *rationalizing the denominator*.

RULE
To rationalize the denominator:

1. Multiply both numerator and denominator by a number that will make the resulting denominator a perfect square.
2. Simplify the resulting radical by removing factors from the radicands.

Example 7

$$\sqrt{\frac{2}{5}} = \sqrt{\frac{2}{5} \cdot \frac{5}{5}}$$
$$= \sqrt{\frac{10}{25}}$$
$$= \frac{\sqrt{10}}{\sqrt{25}}$$
$$= \pm \frac{\sqrt{10}}{5}$$

Example 8

$$\sqrt{\frac{1}{2}} = \sqrt{\frac{1}{2} \cdot \frac{2}{2}}$$
$$= \sqrt{\frac{2}{4}}$$
$$= \frac{\sqrt{2}}{\sqrt{4}}$$
$$= \pm \frac{\sqrt{2}}{2}$$

Example 9

$$\frac{3}{\sqrt{6}} = \frac{3}{\sqrt{6}} \cdot \frac{\sqrt{6}}{\sqrt{6}}$$
$$= \pm \frac{1}{2}\sqrt{6}$$

Example 10

$$\sqrt{\frac{3a}{5x}} = \sqrt{\frac{3a}{5x} \cdot \frac{5x}{5x}}$$
$$= \sqrt{\frac{15ax}{25x^2}}$$
$$= \frac{\sqrt{15ax}}{\sqrt{25x^2}}$$
$$= \pm \frac{1}{5x}\sqrt{15ax}$$

PROBLEMS 20-4

Simplify the following:

1. $\sqrt{\dfrac{1}{3}}$ 2. $\sqrt{\dfrac{1}{7}}$ 3. $\sqrt{\dfrac{2}{5}}$

4. $\sqrt{\dfrac{4}{7}}$ 5. $\sqrt{\dfrac{3}{4}}$ 6. $\sqrt{\dfrac{7}{15}}$

7. $\dfrac{8}{\sqrt{2}}$ 8. $\dfrac{9}{\sqrt{3}}$ 9. $\dfrac{1}{\sqrt{\lambda}}$

10. $\dfrac{21\sqrt{35}}{\sqrt{7}}$ 11. $\sqrt{\dfrac{9}{16\theta}}$ 12. $\sqrt{\dfrac{Q}{R}}$

13. $\theta\sqrt{\dfrac{\lambda}{\theta}}$ 14. $\pi\sqrt{\dfrac{X_L}{2\pi fL}}$ 15. $\sqrt{\dfrac{\alpha^2}{\gamma}}$

16. $\dfrac{2F}{f_0}\sqrt{\dfrac{f_0}{F}}$ 17. $\dfrac{\pi R^2}{A}\sqrt{\dfrac{A}{\pi}}$ 18. $\sqrt{\dfrac{E-e}{E+e}}$

19. $\sqrt{X_L^2 - \left(\dfrac{X_L}{4}\right)^2}$ 20. $\sqrt{R^2 + \left(\dfrac{R}{3}\right)^2}$

21. $\sqrt{Q^4 - \left(\dfrac{Q}{3}\right)^4}$

20-9 ADDITION AND SUBTRACTION OF RADICALS

Terms that are the same except in respect to their coefficients are called *similar terms*. Likewise, *similar radicals* are defined as radicals that have the same index and the same radicand and differ only in their coefficients. For example, $-2\sqrt{5}$, $3\sqrt{5}$, and $\sqrt{5}$ are similar radicals.

Similar radicals can be added or subtracted in the same way that similar terms are added or subtracted.

Example 11

$$3\sqrt{6} - 4\sqrt{6} - \sqrt{6} + 8\sqrt{6} = 6\sqrt{6}$$

Example 12

$$\sqrt{12} + \sqrt{27} = 2\sqrt{3} + 3\sqrt{3}$$
$$= 5\sqrt{3}$$

Note that, in the simplification of radicals, the positive root is assumed.

Example 13

$$\sqrt{48x} + \sqrt{\dfrac{x}{3}} + \sqrt{3x} = 4\sqrt{3x} + \dfrac{1}{3}\sqrt{3x} + \sqrt{3x} = \dfrac{16}{3}\sqrt{3x}$$

If the radicands are alike, then factors removed are assumed to be positive roots. If the radicands are not alike and cannot be reduced to a common radicand, then the radicals are dissimilar terms and addition and subtraction can only be indicated. Thus the following statement can be made:

RULE
To add or subtract radicals:

1. Reduce the radicals to their simplest form.
2. Combine similar radicals, and assume positive square roots of factors removed from the radicands.
3. Indicate addition or subtraction of dissimilar radicals.

PROBLEMS 20-5
Simplify:

1. $5\sqrt{3} - 2\sqrt{3}$
2. $3\sqrt{5} + 2\sqrt{20}$
3. $5\sqrt{5} - \sqrt{80}$
4. $\sqrt{63} - \sqrt{28}$
5. $m\sqrt{3} - p\sqrt{3} + q\sqrt{3}$
6. $\alpha\sqrt{2} + \beta\sqrt{8} - \gamma\sqrt{50}$
7. $5\sqrt{48} + 2\sqrt{108} - \sqrt{12}$
8. $2\sqrt{\frac{1}{3}} + \sqrt{\frac{1}{3}}$
9. $7\sqrt{5} - \frac{15}{\sqrt{5}} - 16\sqrt{\frac{5}{16}}$
10. $6\sqrt{27} + 5\sqrt{32}$
11. $4\sqrt{\frac{1}{8}} + 6\sqrt{\frac{1}{2}} + 2\sqrt{2}$
12. $\frac{R_1}{3} + \sqrt{\frac{16R_1^2}{3}}$
13. $\sqrt{\frac{4}{5}} - \sqrt{\frac{9}{15}}$
14. $\sqrt{\frac{\varepsilon + \eta}{\varepsilon - \eta}} + \sqrt{\frac{\varepsilon - \eta}{\varepsilon + \eta}}$
15. $\sqrt{\frac{\pi}{8}} - \sqrt{\frac{\pi}{32}}$
16. $\sqrt{\frac{7R^2}{16V}} + \sqrt{\frac{M^2V}{28}} - 4\sqrt{\frac{63}{16V}}$

20-10 MULTIPLICATION OF RADICALS

Obtaining the product of radicals is the inverse of removing a factor, as will be shown in the following examples:

Example 14
$$3\sqrt{3} \cdot 5\sqrt{4} = 15\sqrt{3 \cdot 4}$$
$$= 15 \cdot 2\sqrt{3}$$
$$= 30\sqrt{3}$$

Example 15
$$4\sqrt{3a} \cdot 2\sqrt{6a} = 8\sqrt{3a \cdot 6a}$$
$$= 8\sqrt{18a^2}$$
$$= 8\sqrt{9 \cdot 2a^2}$$
$$= 24a\sqrt{2}$$

Example 16

Multiply $3\sqrt{2} + 2\sqrt{3}$ by $4\sqrt{2} - 3\sqrt{3}$.

Solution

$$
\begin{array}{l}
3\sqrt{2} + 2\sqrt{3} \\
\underline{4\sqrt{2} - 3\sqrt{3}} \\
24 \quad + 8\sqrt{6} \\
\underline{\quad\quad - 9\sqrt{6} - 18} \\
24 \quad - \sqrt{6} - 18 = 6 - \sqrt{6}
\end{array}
$$

PROBLEMS 20-6

Perform the indicated operations:

1. $\sqrt{2} \cdot \sqrt{3}$
2. $\sqrt{12} \cdot \sqrt{3}$
3. $2\sqrt{10} \cdot \sqrt{2}$
4. $8\sqrt{5} \cdot 4\sqrt{15}$
5. $2\sqrt{8} \cdot 3\sqrt{5}$
6. $\sqrt{6} \cdot \sqrt{24}$
7. $\sqrt[3]{2} \cdot \sqrt[3]{4}$
8. $\sqrt{\frac{7}{16}} \cdot \sqrt{\frac{21}{3}}$
9. $(\sqrt{A - D})^2$
10. $(\varepsilon + \sqrt{3})(\varepsilon - \sqrt{3})$
11. $(\sqrt{\alpha} - \sqrt{\alpha - 7})^2$
12. $(3 + \sqrt{5})^2$
13. $\sqrt{(\theta - \phi)^2}$
14. $(2\sqrt{5} + 3\sqrt{2})(\sqrt{5} + 5\sqrt{2})$
15. $\sqrt{6\pi} \cdot \sqrt{12\alpha^2\pi}$
16. $\sqrt{2(x^2 - 4x + 4)} \cdot \sqrt{\dfrac{8}{4x^2 + 16x + 16}}$
17. $(-1 - \sqrt{3})(3 - 3\sqrt{3})$
18. $(4 + 2\sqrt{3})(2 - \sqrt{3})$
19. $\left(\dfrac{36 - 9\sqrt{5}}{2}\right)\left(\dfrac{2\sqrt{5} + 8}{11}\right)$
20. $\dfrac{(\sqrt{\alpha} - \sqrt{\beta})(\alpha + 2\sqrt{\alpha\beta} + \beta)}{\alpha - \beta}$

20-11 DIVISION

An indicated root whose value is irrational but whose radicand is rational is called a *surd*. Thus, $\sqrt[3]{3}$, $\sqrt{2}$, $\sqrt[4]{5}$, $\sqrt{3}$, etc., are surds. If the indicated root is the square root, then the surd is called a *quadratic surd*. For example, $\sqrt{2}$, $\sqrt{5}$, $\sqrt{6}$, $\sqrt{15}$ are quadratic surds. Then, by extending the definition, such expressions as $3 + \sqrt{2}$ and $\sqrt{3} - 6$ are called *binomial quadratic surds*.

It is important that you become proficient in the multiplication and division of binomial quadratic surds. One method of solving ac circuits, which will be discussed later, makes wide use of these particular operations. Multiplication of such expressions was covered in the preceding section. However, a new method is necessary for division.

Consider the two expressions $a - \sqrt{b}$ and $a + \sqrt{b}$. They differ only in the sign between the terms. These expressions are *conjugates;* that is, $a - \sqrt{b}$ is called the conjugate of $a + \sqrt{b}$, and $a + \sqrt{b}$ is called the conjugate of $a - \sqrt{b}$. Remember this meaning of "conjugate," for it is the same with reference to certain circuit characteristics.

To divide a number by a binomial quadratic surd, rationalize the divisor (denominator) by multiplying both dividend (numerator) and divisor by the conjugate of the divisor.

Example 17

$$\frac{1}{3 + \sqrt{2}} = \frac{3 - \sqrt{2}}{(3 + \sqrt{2})(3 - \sqrt{2})}$$

$$= \frac{3 - \sqrt{2}}{7}$$

Example 18

$$\frac{1}{3\sqrt{3} - 1} = \frac{3\sqrt{3} + 1}{(3\sqrt{3} - 1)(3\sqrt{3} + 1)}$$

$$= \frac{3\sqrt{3} + 1}{26}$$

Example 19

$$\frac{3 - \sqrt{2}}{4 + \sqrt{2}} = \frac{(3 - \sqrt{2})(4 - \sqrt{2})}{(4 + \sqrt{2})(4 - \sqrt{2})}$$

$$= \frac{14 - 7\sqrt{2}}{14}$$

$$= \frac{2 - \sqrt{2}}{2}$$

Note

In each of the foregoing examples the resulting denominator is a rational number. In general, the product of two conjugate surd expressions is a rational number. This important fact is widely used in the solution of ac problems.

PROBLEMS 20-7

Perform the indicated division:

1. $\dfrac{2\sqrt{10}}{\sqrt{8}}$

2. $\dfrac{3}{3 - \sqrt{2}}$

3. $\dfrac{8}{3 + \sqrt{7}}$

4. $\dfrac{7}{2\sqrt{3} - 2}$

5. $\dfrac{9}{3 - 3\sqrt{3}}$

6. $\dfrac{x + \sqrt{y}}{x - \sqrt{y}}$

7. $\dfrac{x - \sqrt{y}}{x + \sqrt{y}}$ 8. $\dfrac{3 - \sqrt{5}}{2 + \sqrt{5}}$ 9. $\dfrac{3 + 2\sqrt{3}}{2 + 2\sqrt{3}}$

10. $\dfrac{\sqrt{R} + \sqrt{Z}}{\sqrt{R} - \sqrt{Z}}$ 11. $\dfrac{\sqrt{2} + 3}{\sqrt{3} + 2}$ 12. $\dfrac{50 + j35}{8 + j5}$

Hint   Maintain order j35, j5, etc. and treat terms containing algebraic symbol j as if they were radicals.

20-12 THE OPERATOR j

In our studies so far, we have met with several mathematical symbols which actually indicate *commands*; $+$, $-$, $\times$, $\div$, and $\sqrt{}$ are all symbols which actually tell us to perform some specific operation. In Sec. 3-5, for instance, we saw that the minus sign is equivalent to a rotation of a quantity through 180°, and, by definition, this rotation is in the positive, or counterclockwise, direction.

Now we must meet the operator j, which also provides a rotation, not of 180°, but of 90°. You have noticed that all the algebraic symbols used so far in this book are printed in *italic* (slanting) type. The operator j, however, is printed in roman (regular) type to distinguish it as an operator and to constantly remind the student that it is not just another algebraic symbol. The operator j is extremely useful in the solution of electronic circuits, and although the idea is simple and straightforward—*just rotate through 90° in a counterclockwise (ccw) direction*—it is essential that we understand exactly how to operate with it. In Fig. 20-1, the line *OA*, which lies on the *x* axis and is *a* units long, can be operated on by the operator j to become *ja*, a line of the same length as before but now rotated ccw through 90° to lie on the *y* axis.

Note how the rotated quantity is described: first is given the symbol of the operator j, and then the quantity which has been operated upon, *a*.

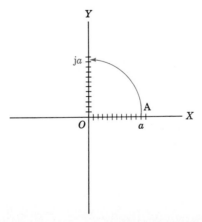

Fig. 20-1 Representation of a quantity affected by the operator j.

Thus, when a is "j'd", it becomes ja. This practice of placing the operator first draws attention to the fact that we are not dealing with some quantity j multiplied by some other quantity a, but that the j operator is operating on the quantity a. The algebraic symbol ja represents for us the geometric symbol of a line rotated through 90° in a counterclockwise direction.

Any quantity operated upon by j will rotate through 90° in a counterclockwise direction; similarly, any quantity operated upon by $-$j will rotate through 90° in a clockwise direction. (See Fig. 20-2.)

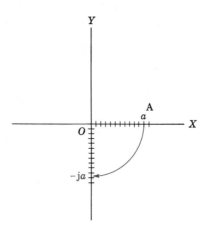

Fig. 20-2 Representation of a quantity affected by the operator $-$j.

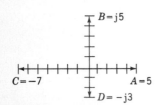

Fig. 20-3 Comparison of quantities $A = +5$, $B = +$j5, $C = -7$, and $D = -$j3.

Figure 20-3 relates four different quantities for purposes of review: $A = 5$, $B =$ j5, $C = -7$, and $D = -$j3.

A quantity may be j-operated more than once. If we start with a quantity ja, as in Fig. 20-1, and j it again, we cause it to rotate through an additional 90° ccw, as shown in Fig. 20-4.

j(ja) may be written jj$a$, or, more simply, j^2a. Similarly, j^3 indicates that a quantity has been operated on three times in succession; that is, it

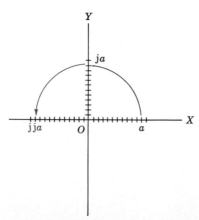

Fig. 20-4 Representation of repeated operation by j.

has been rotated through 90° ccw three times in succession. Figures 20-5 and 20-6 indicate repeated rotations resulting from repeated operations by j and $-$j.

Note, in passing, a very interesting point about j^2a: j-ing a twice in succession brings it to the same point as a single operation with a minus sign. From this graphic illustration, you can see that

$$j^2 = -1$$

and

$$j = \sqrt{-1}$$

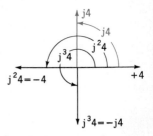

Fig. 20-5 Repeated rotation of numbers in counterclockwise direction.

This added relationship, $j = \sqrt{-1}$, is an extremely interesting one, because so far, in the removal of factors from radicands, all the radicands have been positive numbers. In Sec. 20-13, we will use the important relationship $j = \sqrt{-1}$ to factor negative radicands and to determine (or, at least represent) the square roots of negative numbers.

First, however, let us continue with the fascinating relationships exhibited by repeated operations with j. Since $j^2 = -1$, then j^3 must equal $j(-1)$, or $-j$, and j^4 must equal j^2j^2, that is, $(-1)(-1) = +1$. The truth of these statements can be justified by the following considerations:

$$\sqrt{-1} \cdot \sqrt{-1} = -1$$

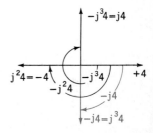

Fig. 20-6 Repeated rotation of numbers in clockwise direction.

That is,

$$j \cdot j = -1$$
$$\therefore j^2 = -1$$

Also,

$$\sqrt{-1} \cdot \sqrt{-1} \cdot \sqrt{-1} = -1 \cdot \sqrt{-1} = -j$$

That is,

$$j \cdot j \cdot j = j^3$$
$$\therefore j^3 = -j$$

Also,

$$\sqrt{-1} \cdot \sqrt{-1} \cdot \sqrt{-1} \cdot \sqrt{-1} = (\sqrt{-1} \cdot \sqrt{-1})(\sqrt{-1} \cdot \sqrt{-1})$$
$$= (-1)(-1) = 1$$

That is,

$$j \cdot j \cdot j \cdot j = j^4$$
$$\therefore j^4 = 1$$

Similarly, it can be shown that successive multiplication by each $+$j rotates the number 90° in a counterclockwise direction.

If we consider successive multiplication by $-$j, we have

$$(-\sqrt{-1})(-\sqrt{-1}) = -1$$

That is,

$$(-j)(-j) = j^2$$
$$\therefore (-j)^2 = -1$$

Also,

$$(-\sqrt{-1})(-\sqrt{-1})(-\sqrt{-1}) = (-1)(-\sqrt{-1}) = \sqrt{-1}$$

That is,

$$(-j)(-j)(-j) = (j^2)(-j)$$
$$= (-1)(-j) = j$$
$$\therefore (-j)^3 = j$$

To demonstrate that $(-j)^4 = 1$ and $\dfrac{1}{j} = -j$ is left as an exercise for you.

Note the convenience of the graphic method of representation of the j operations, Figs. 20-5 and 20-6. This method is an advantageous one because, if we can *visualize* a graph or diagram when we come up against certain types of numbers and equations, we often have a better understanding of the manner in which the quantities vary or are related.

One special note must be drawn to your attention: Long before the operator j was found to have practical application in electrical and electronics calculations, mathematicians used the symbol *i* to represent $\sqrt{-1}$. When electrical theory adopted the symbol *i* for instantaneous current flow in a circuit, we switched the mathematicians' *i* to j for our symbol of rotation through 90° ccw. Sometimes in your reading you will meet *i* instead of j, but you will know what it really means: "Rotate the quantity operated upon by 90° in a counterclockwise direction."

As a mathematical definition, j is sometimes referred to as the "complex operator," but, as we have seen, there is nothing particularly complex about j.

20-13 INDICATED SQUARE ROOTS OF NEGATIVE NUMBERS

So far, in the removal of factors from radicands, all the radicands have been positive numbers. Also, we have extracted the square roots of positive numbers only. How shall we proceed to factor negative radicands, and what is the meaning of the square root of a negative number?

According to our laws for multiplication, no number multiplied by itself or raised to any even power will produce a negative result. For example, what does $\sqrt{-25}$ mean when we know of no number that, when multiplied by itself, will produce -25?

The indicated square root of a negative number is known as an *imaginary number*. It is probable that this name was assigned before mathematicians could visualize such a number and that the word "imaginary" was originally used to distinguish such numbers from the so-called "real numbers" previously studied. In any event, calling such a number imaginary might be considered unfortunate, because in working with circuits such numbers become very real in the physical sense. If you accidentaly touch a large capacitor that is highly charged, you are likely to be killed by some of those "imaginary" volts. This will be discussed later.

To avoid the difficulty of operations with the indicated square roots of negative numbers, or imaginary numbers, it becomes necessary to introduce a new type of number. That is, we agree that every imaginary number can be expressed as the product of a positive number and $\sqrt{-1}$.

Example 20

$$\begin{aligned}\sqrt{-25} &= \sqrt{(-1)25} \\ &= \sqrt{-1}\sqrt{25} \\ &= \sqrt{-1}\cdot 5\end{aligned}$$

As we saw in Sec. 20-12, $\sqrt{-1}$ may be represented by the operator j, and we may now rewrite $\sqrt{-1} \cdot 5$ as j5.

Example 21

$$\sqrt{-16} = \sqrt{(-1)16}$$
$$= \sqrt{-1}\sqrt{16}$$
$$= \sqrt{-1} \cdot 4 = j4$$

Example 22

$$\sqrt{-X^2} = \sqrt{(-1)X^2}$$
$$= \sqrt{-1}\sqrt{X^2}$$
$$= \sqrt{-1} \cdot X = jX$$

Example 23

$$-\sqrt{-4X^2} = -\sqrt{(-1)4X^2} = -\sqrt{-1}\sqrt{4X^2}$$
$$= -\sqrt{-1} \cdot 2X = -j2X$$

PROBLEMS 20-8

Express the following by using the operator j:

1. $\sqrt{-36}$ 2. $\sqrt{-64}$ 3. $\sqrt{-144}$

4. $\sqrt{-\theta^2}$ 5. $-\sqrt{-z^2}$ 6. $-\sqrt{-49\omega^2}$

7. $\sqrt{-I^4X^2}$ 8. $\sqrt{\dfrac{-Q^4}{\omega^2 L^2}}$ 9. $-5\sqrt{-49}$

10. $2\sqrt{-48}$ 11. $\sqrt{\dfrac{-16}{121}}$ 12. $-\sqrt{\dfrac{169}{-\alpha^2}}$

13. $\sqrt{\dfrac{-32}{75}}$ 14. $-\sqrt{-\lambda^2\pi}$ 15. $-\sqrt{\dfrac{-V^2}{P}}$

16. Did you demonstrate that $(-j)^4 = 1$?

17. Did you demonstrate that $\dfrac{1}{j} = -j$?

20-14 COMPLEX NUMBERS

If a "real" number is united to an "imaginary" number by a plus or a minus sign, the expression thus obtained is called a *complex number*. Thus, $3 - j4$, $a + jb$, $R + jX$, etc., are complex numbers. At this time, we shall consider, not their graphical representation, but simply how to perform the four fundamental operations algebraically. Figure 20-7 shows the representation of the complex number $a + jb$.

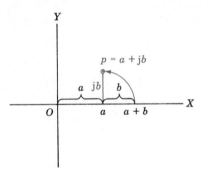

Fig. 20-7 Representation of a complex number $a + jb$. a lies in OX; b is rotated through 90° counterclockwise; the point p represents the "sum" of a and jb.

20-15 ADDITION AND SUBTRACTION OF COMPLEX NUMBERS

Combining a real number with an imaginary number cannot be accomplished by the usual methods of addition and subtraction; these processes can only be expressed. As an example—if we have the complex number $5 + j6$, this is as far as we can simplify it at this time. We should not attempt to add 5 and j6 arithmetically, for these two numbers are at right angles to each other, and such an operation would be meaningless. However, we *can* add and subtract complex numbers by treating them as ordinary binomials.

Example 24
Add $3 + j7$ and $4 - j5$.

Solution

$$\begin{array}{r} 3 + j7 \\ 4 - j5 \\ \hline 7 + j2 \end{array}$$

Example 25
Subtract $-15 - j6$ from $-5 + j8$.

Solution

$$\begin{array}{r} -5 + j8 \\ -15 - j6 \\ \hline 10 + j14 \end{array}$$

PROBLEMS 20-9
Find the indicated sums:

1.	$3 + j12$	**2.**	$14 + j3$	$25 + j8$	$96 - j22$
	$2 + j8$		$12 + j3$	$16 - j10$	$32 - j5$

5. 47 − j3   **6.** 32   **7.** 20 + j3   **8.** 26 − j6
       125 + j8   5 + j6   − j5   31

9 to 16.   Subtract the lower complex number from the upper in each of the above problems.

20-16 MULTIPLICATION OF COMPLEX NUMBERS

As in addition and subtraction, complex numbers are treated as ordinary binomials when multiplied. However, when writing the result, we must not forget that $j^2 = -1$.

Example 26
Multiply $4 - j7$ by $8 + j2$.

Solution

$$
\begin{array}{r}
4 - j7 \\
8 + j2 \\
\hline
32 - j56 \\
+ j8 - j^2 14 \\
\hline
32 - j48 - j^2 14
\end{array}
$$

Since $j^2 = -1$, the product is

$$32 - j48 - (-1)(14) = 32 - j48 + 14$$
$$= 46 - j48$$

Example 27
Multiply $7 + j3$ by $6 + j2$.

Solution

$$
\begin{array}{r}
7 + j3 \\
6 + j2 \\
\hline
42 + j18 \\
+ j14 + j^2 6 \\
\hline
42 + j32 + j^2 6 = 36 + j32
\end{array}
$$

20-17 DIVISION OF COMPLEX NUMBERS

As in the division of binomial quadratic surds, we simplify an indicated division by rationalizing the denominator in order to obtain a "real" number as divisor (Sec. 20-11). We do this by multiplying by the conjugate in the usual manner.

Example 28

$$\frac{10}{1 + j2} = \frac{10(1 - j2)}{(1 + j2)(1 - j2)} = \frac{10(1 - j2)}{1 - j^2 4}$$

$$= \frac{10(1 - j2)}{5}$$

$$= 2(1 - j2)$$

Example 29

$$\frac{5 + j6}{3 - j4} = \frac{(5 + j6)(3 + j4)}{(3 - j4)(3 + j4)}$$

$$= \frac{15 + j38 + j^2 24}{9 - j^2 16}$$

$$= \frac{-9 + j38}{25}$$

Example 30

$$\frac{a + jb}{a - jb} = \frac{(a + jb)(a + jb)}{(a - jb)(a + jb)}$$

$$= \frac{a^2 + j2ab + j^2 b^2}{a^2 - j^2 b^2}$$

$$= \frac{a^2 + j2ab - b^2}{a^2 + b^2}$$

PROBLEMS 20-10

Find the indicated products:

1. $(3)(1 - j3)$
2. $(6 + j2)(2 + j3)$
3. $(8 - j9)(6 + j3)$
4. $(3 - j5)(6 - j7)$
5. $(\theta + j\phi)(\theta + j\phi)$
6. $(R - jX_C)(R + jX_L)$

Find the quotients:

7. $\dfrac{1}{1 + j1}$
8. $\dfrac{10}{1 - j3}$
9. $\dfrac{1 + j1}{1 - j1}$

10. $\dfrac{1 - j1}{1 + j1}$
11. $\dfrac{8}{8 + j8}$
12. $\dfrac{3 + j2}{6 - j5}$

13. $\dfrac{6}{6 - jx}$
14. $\dfrac{\theta + j\phi}{\theta - j\phi}$
15. $\dfrac{R + j\omega X}{R - j\omega X}$

16. $\dfrac{j3}{2 + j3}$
17. $\dfrac{j\phi}{\theta - j\phi}$
18. $\dfrac{1 + j\dfrac{\omega}{\omega_o}}{1 - j\dfrac{\omega}{\omega_o}}$

19. $\dfrac{R}{\dfrac{1}{j\omega C} + R + j\omega L}$

20. Write in the form $a + jb$: $\dfrac{(1 + j\omega\tau_1)(1 + j\omega\tau_2)}{\mu_o - \beta}$

20-18 RADICAL EQUATIONS

An equation in which the unknown occurs in a radicand is called an *irrational* or *radical equation*. To solve such an equation, so arrange it that the radical is the only term in one member of the equation. Then eliminate the radical by squaring both members of the equation.

Example 31
Given $\sqrt{3x} = 6$; solve for x.

Solution
$$\sqrt{3x} = 6$$
Squaring, $\qquad 3x = 36$
D: 3, $\qquad x = 12$

Check
Substituting 12 for x in the given equation,
$$\sqrt{3 \cdot 12} = 6$$
$$\sqrt{36} = 6$$
$$6 = 6$$

Example 32
Given $\sqrt{2x + 3} = 7$; solve for x.

Solution
$$\sqrt{2x + 3} = 7$$
Squaring, $\qquad 2x + 3 = 49$
S: 3, $\qquad 2x = 46$
D: 2, $\qquad x = 23$

Check
$$\sqrt{2 \cdot 23 + 3} = 7$$
$$\sqrt{49} = 7$$
$$7 = 7$$

Example 33
The time for one complete swing of a simple pendulum is given by

$$t = 2\pi \sqrt{\frac{L}{g}}$$

where t = time, s
L = length of pendulum
g = acceleration due to gravity

Solution

Given $\qquad\qquad\qquad t = 2\pi \sqrt{\dfrac{L}{g}}$ $\qquad\qquad$ (a)

Squaring (a), $\qquad\qquad t^2 = 4\pi^2 \dfrac{L}{g}$ $\qquad\qquad$ (b)

M: g in (b), $\qquad\qquad gt^2 = 4\pi^2 L$ $\qquad\qquad$ (c)

D: t^2 in (c), $\qquad\qquad g = \dfrac{4\pi^2 L}{t^2}$ $\qquad\qquad$ (d)

Rewrite (c), $\qquad\qquad 4\pi^2 L = gt^2$ $\qquad\qquad$ (e)

D: $4\pi^2$ in (e), $\qquad\qquad L = \dfrac{gt^2}{4\pi^2}$

Example 34

Given $V = I_pZ_p + j\omega MI_s$ and $I_sZ_s = -j\omega MI_p$. Show that

$$V = I_p\left[Z_p + \frac{(\omega M)^2}{Z_s}\right]$$

Solution

Since I_s does not appear in the final equation, it must be eliminated. Solving the given equations for I_s,

$$I_s = \frac{V - I_pZ_p}{j\omega M} \qquad (a)$$

$$I_s = \frac{-j\omega MI_p}{Z_s} \qquad (b)$$

Equating the right members of (a) and (b),

$$\frac{V - I_pZ_p}{j\omega M} = \frac{-j\omega MI_p}{Z_s}$$

M: $j\omega M$ $\qquad V - I_pZ_p = \dfrac{-j^2\omega^2M^2I_p}{Z_s}$

Substituting -1 for j^2 in the right member,

$$V - I_pZ_p = \frac{\omega^2M^2I_p}{Z_s}$$

A: $I_pZ_p,$ $\qquad V = I_pZ_p + \dfrac{(\omega M)^2I_p}{Z_p}$

Factoring the right member,

$$V = I_p\left[Z_p + \frac{(\omega M)^2}{Z_s}\right]$$

PROBLEMS 20-11

Solve the following equations:

1. $\sqrt{x} = 2$
2. $\sqrt{R} = 6$
3. $\sqrt{\gamma} = 3$
4. $\sqrt{i} + 1 = 4$
5. $\sqrt{Z} - 5 = 20$
6. $\sqrt{\theta + 3} = 7$
7. $\sqrt{M - 3} = 8$
8. $2\sqrt{\theta - 2} = 6$
9. $4\sqrt{\lambda + 3} - 2 = 6$
10. $\sqrt{\dfrac{7K + 4}{2}} = 4$
11. $3\sqrt{\phi + 3} = 2\sqrt{3\phi - 12}$

Given: $\qquad\qquad\qquad\qquad\qquad\qquad\qquad$ Solve for:

12. $E = \sqrt{\dfrac{\eta\phi}{\omega^2\theta}}$ $\qquad\qquad\qquad\qquad\qquad\qquad$ ϕ

13. $i_s = \&\sqrt{2P_rP_s}$ $\qquad\qquad\qquad\qquad\qquad\qquad$ P_r

14. $\dfrac{i_s}{i_n} = \sqrt{\dfrac{\rho P_s}{e(\Delta f)}}$ $\qquad\qquad P_s$

15. $\dfrac{S}{N} = \alpha\sqrt{\eta\tau}$ $\qquad\qquad \eta$

16. $\lambda = \dfrac{4\pi}{\gamma Q}\sqrt{\dfrac{KFTS(\Delta f)}{NP_0}}$ $\qquad\qquad \dfrac{S}{N},\ P_0$

17. $\dfrac{V}{C} = \sqrt{\dfrac{1}{\dfrac{w-a}{w} + \dfrac{\varepsilon_1 a}{w}}}$ $\qquad\qquad w$

18. $\gamma = \sqrt{\dfrac{1 - \mu_x\eta E}{\omega X}}$ $\qquad\qquad \mu_x$

19. $Y_n = G\sqrt{\left(\dfrac{n^2-1}{n}\right)^2 Q_2 + 1}$ $\qquad\qquad Q_2$

20. $Z_t = R\sqrt{1 + \left(\dfrac{f}{f_0}\right)^4}$ $\qquad\qquad f_0$

21. $G_a = \sqrt{G_1 + \dfrac{G_1}{R_{eq} + \dfrac{G_L}{g_m{}^2}}}$ $\qquad\qquad g_m{}^2$

22. At a resonant frequency of f Hz, the inductive reactance X_L of a circuit of L H is $X_L = \omega L$ Ω and the capacitive reactance X_C of a circuit with a capacitance of C F is $X_C = \dfrac{1}{\omega C}$ Ω. $\omega = 2\pi f$. At the resonant frequency, with both inductance and capacitance in the circuit, $X_L = X_C$. Solve for the resonant frequency f in terms of π, L, and C.

23. Use the formula for the resonant frequency derived in Prob. 22 to find the value of C in picofarads in a case when $f = 1.4$ MHz and $L = 51.7$ μH.

24. Use the formula that was derived in Prob. 22 to find the value of f when $C = 47$ nF and $L = 15$ nH.

25. $f = \dfrac{1}{\sqrt{2\pi\,\dfrac{LC_aC_b}{C_a + C_b}}}$. Solve for C_a.

26. In a conductor through which current I flows, energy E_m existing in the magnetic field about the line is $\dfrac{LI^2}{2}$ J, where L is the inductance of the line per unit length. An equal energy E_c exists in the electrostatic field of the line, equal to $\dfrac{CV^2}{2}$ J, where C is the capacitance of the line per unit length. If the surge impedance Z_o of the line is $\dfrac{V}{I}$ Ω, show that $Z_o = \sqrt{\dfrac{L}{C}}$.

27. Given $\Delta = \dfrac{4}{\pi} \sqrt{1 + \left(\dfrac{\pi\tau\omega}{4}\right)^2}$, show that

$$\omega = \pm \frac{1}{\pi\tau} \sqrt{(\pi\Delta + 4)(\pi\Delta - 4)}$$

28. Given $\sqrt{\dfrac{1}{\tau_1\tau_2} - \dfrac{1}{4\tau_2{}^2}} = 786$ and $\dfrac{1}{2\tau_2} = 78.6$, solve for τ_1.

29. Show that $KV_p{}^{\frac{3}{2}} = KV_p \sqrt{V_p}$.

30. A West Coast semiconductor products manufacturer, in a design for a 100-W 10-MHz power amplifier, equates the actual output circuit to its equivalent:

$$\frac{R_L \left(\dfrac{1}{j\omega C_7}\right)}{R_L + \dfrac{1}{j\omega C_7}} = R'_L + \frac{1}{j\omega C'_7}$$

(a) Show that

$$C_7 = \frac{1}{\omega R_L} \sqrt{\frac{R_L}{R'_L} - 1}$$

and

$$C'_7 = C_7 \left[1 + \left(\frac{1}{\omega C_7 R_L}\right)^2 \right]$$

(b) If $R_L = 50\ \Omega$, $R'_L = 12.5\ \Omega$, and $\omega = 2\pi \times 10^7$, show that $C_7 = 551$ pF and $C'_7 = 735$ pF.

Quadratic Equations

In preceding chapters the study of equations has been limited mainly to equations which contain the unknown quantity in the first degree. This chapter is concerned with equations of the second degree, which are called quadratic equations.

21-1 DEFINITIONS

In common with polynomials (Sec. 11-2), the degree of an equation is defined as the degree of the term of highest degree in it. Thus, if an equation contains the square of the unknown quantity and no higher degree, it is an equation of the second degree, or a *quadratic equation*.

A quadratic equation that contains terms of the second degree only of the unknown is called a *pure quadratic equation*. For example,

$$x^2 = 25 \qquad R^2 - 49 = 0 \qquad 3x^2 = 12 \qquad ax^2 + c = 0$$

are pure quadratic equations.

A quadratic equation that contains terms of *both* the first and the second degree of the unknown is called an *affected* or a *complete quadratic equation;* $x^2 + 3x + 2 = 0$, $3x^2 + 11x = -2$, $ax^2 + bx + c = 0$, etc., are affected, or complete, quadratic equations.

When a quadratic equation is solved, values of the unknown that will satisfy the conditions of the equation are found.

A value of the unknown that will satisfy the equation is called a *solution* or a *root* of the equation.

21-2 SOLUTION OF PURE QUADRATIC EQUATIONS

As stated in Sec. 10-5, every number has two square roots that are equal in magnitude but opposite in sign. Hence, all quadratic equations have two roots. In pure quadratic equations, the absolute values of the roots are equal but of opposite sign.

Example 1

Solve the equation $x^2 - 16 = 0$.

Solution

Given $\qquad\qquad\qquad x^2 - 16 = 0$

A: 16, $\qquad\qquad\qquad\quad\ x^2 = 16$

$\sqrt{\ }$ (see note below), $\qquad\ x = \pm 4$

Check

Substituting in the equation either $+4$ or -4 for the value of x, because either squared results in $+16$, we have

$$(\pm 4)^2 - 16 = 0$$

or $\qquad\qquad\qquad 16 - 16 = 0$

Note

Hereafter, the radical sign will mean "take the square root of both members of the preceding or designated equation."

Example 2

Solve the equation $5R^2 - 89 = 91$.

Solution

Given $\qquad\qquad\qquad 5R^2 - 89 = 91$

A: 89, $\qquad\qquad\qquad\quad 5R^2 = 180$

D: 5, $\qquad\qquad\qquad\qquad R^2 = 36$

$\sqrt{\ }$, $\qquad\qquad\qquad\qquad\ R = \pm 6$

Check

$$5(\pm 6)^2 - 89 = 91$$
$$5 \times 36 - 89 = 91$$
$$180 - 89 = 91$$
$$91 = 91$$

Example 3

Solve the equation

$$\frac{I + 4}{I - 4} + \frac{I - 4}{I + 4} = \frac{10}{3}$$

Solution

Given $\qquad\qquad \dfrac{I + 4}{I - 4} + \dfrac{I - 4}{I + 4} = \dfrac{10}{3}$

Clearing fractions,

$$3(I + 4)(I + 4) + 3(I - 4)(I - 4) = 10(I - 4)(I + 4)$$

Expanding,

$$3I^2 + 24I + 48 + 3I^2 - 24I + 48 = 10I^2 - 160$$

Collecting terms, $\qquad -4I^2 = -256$

D: -4, $\qquad\qquad\qquad\ I^2 = 64$

$\sqrt{\ }$, $\qquad\qquad\qquad\qquad I = \pm 8$

Check

By the usual method.

PROBLEMS 21-1

Solve the following:

1. $E^2 - 25 = 0$
2. $s^2 - 49 = 0$
3. $i^2 + 36 = 225$
4. $\theta^2 - 0.25 = 0$
5. $5\omega^2 - 180 = 0$
6. $\phi^2 - 0.0004 = 0.0012$
7. $\lambda^2 - \dfrac{9}{121} = 0$
8. $49I^2 - 144 = 0$
9. $5\mu^2 = 3\dfrac{1}{5}$
10. $5x^2 - 0.0308 = 0.0817$
11. $2(m + 1) - m(m - 3) - 5m = 0$
12. $\dfrac{28}{R^2 - 9} = \dfrac{R + 3}{R - 3} - 1 + \dfrac{R - 3}{R + 3}$
13. $\dfrac{3\lambda - 18}{6} + \dfrac{90 + 9\lambda - 4\lambda^2}{3\lambda} = 0$
14. $6a(4a - 3) + 3(6a - 16) = 0$
15. $X_C = \dfrac{24 - X_C + (X_C - 1)^3}{2 + X_C{}^2} - 2$

21-3 COMPLETE QUADRATIC EQUATIONS—SOLUTION BY FACTORING

As an example, let it be assumed that all that is known about two expressions x and y is that $xy = 0$. We know that it is impossible to find the value of either unless the value of the other is known. However, we do know that, if $xy = 0$, *either $x = 0$ or $y = 0$*; for the product of two numbers can be zero if, and only if, one of the numbers is zero.

Example 4

Solve the equation $x(5x - 2) = 0$.

Solution

Here we have the product of two numbers x and $(5x - 2)$, equal to zero; and in order for the equation to be satisfied, one of the numbers must be equal to zero. Therefore, $x = 0$, or $5x - 2 = 0$. Solving the latter equation, we have $x = \frac{2}{5}$. Hence,

$$x = 0 \qquad \text{or} \qquad x = \tfrac{2}{5}$$

Check

If $x = 0$,

$$x(5x - 2) = 0(5 \cdot 0 - 2) = 0(-2) = 0$$

If $x = \frac{2}{5}$,

$$x(5x - 2) = \tfrac{2}{5}(5 \cdot \tfrac{2}{5} - 2) = \tfrac{2}{5}(2 - 2) = 0$$

It is evident that the roots of a complete quadratic may be of unequal absolute value and may or may not have the same signs.

It is incorrect to say $x = 0$ *and* $x = \frac{2}{5}$, for actually x cannot be equal to both 0 and $\frac{2}{5}$ at the same time. This will be more apparent in the following examples.

Example 5

Solve the equation $(x - 5)(x + 3) = 0$.

Solution

Again we have the product of two numbers, $(x - 5)$ and $(x + 3)$, equal to zero. Hence, either

$$x - 5 = 0 \quad \text{or} \quad x + 3 = 0$$
$$\therefore x = 5 \quad \text{or} \quad x = -3$$

Check

If $x = 5$,

$$(x - 5)(x + 3) = (5 - 5)(5 + 3)$$
$$= 0(8) = 0$$

If $x = -3$,

$$(x - 5)(x + 3) = (-3 - 5)(-3 + 3)$$
$$= (-8)0 = 0$$

Example 6

Solve the equation $x^2 - x - 6 = 0$.

Solution

Given	$x^2 - x - 6 = 0$
Factoring,	$(x - 3)(x + 2) = 0$
Then, if $x - 3 = 0$,	$x = 3$
Also, if $x + 2 = 0$,	$x = -2$

$\therefore x = 3$ or -2

Check

If $x = 3$,

$$x^2 - x - 6 = 3^2 - 3 - 6 = 9 - 3 - 6$$
$$= 0$$

If $x = -2$,

$$x^2 - x - 6 = (-2)^2 - (-2) - 6$$
$$= 4 + 2 - 6 = 0$$

Example 7

Solve the equation $(V - 3)(V + 2) = 14$.

Solution

Given $(V - 3)(V + 2) = 14$
Expanding, $V^2 - V - 6 = 14$
S: 14, $V^2 - V - 20 = 0$
Factoring, $(V - 5)(V + 4) = 0$
Then, if $V - 5 = 0$, $V = 5$
Also, if $V + 4 = 0$, $V = -4$
$\therefore V = 5$ or -4

Check

If $V = 5, (V - 3)(V + 2) = (5 - 3)(5 + 2)$
$= (2)(7) = 14$

If $V = -4, (V - 3)(V + 2) = (-4 - 3)(-4 + 2)$
$= (-7)(-2) = 14$

PROBLEMS 21-2

Solve by factoring:

1. $\alpha^2 + 5\alpha + 4 = 0$

2. $e^2 + 2e - 15 = 0$

3. $R^2 + 14 = 9R$

4. $x^2 = 5x - 6$

5. $\lambda^2 = 2 - \lambda$

6. $\psi^2 = 17\psi - 60$

7. $E^2 + 40 = 22E$

8. $26 + 11L - L^2 = 0$

9. $\dfrac{2Q - 13}{Q - 5} = \dfrac{7Q - 5}{5Q - 7}$

10. $\dfrac{8}{\kappa} + \kappa + 2 = \dfrac{2}{\kappa} - 3$

11. $\alpha + 32 + \dfrac{20}{\alpha} = 5 - \dfrac{30}{\alpha}$

12. $\dfrac{160}{I^2} = \dfrac{26}{I} - 1$

13. $\dfrac{1}{Z - 4} - 1 = \dfrac{-2}{Z - 2}$

14. $\dfrac{2F - 6}{17 - F} = 1 - \dfrac{2}{F - 2}$

15. $\dfrac{4}{2i + 2} + \dfrac{i}{3i + 7} - \dfrac{11}{4i + 4} = 0$

21-4 SOLUTION BY COMPLETING THE SQUARE

Some quadratic equations are not readily solved by factoring, but frequently such quadratic equations are readily solved by another method known as *completing the square*.

In Problems 10-5, missing terms were supplied in order to form a perfect trinomial square. This is the basis for the method of completing the square. For example, in order to make a perfect square of the expression $x^2 + 10x$, 25 must be added as a term to obtain $x^2 + 10x + 25$, which is the square of the quantity $x + 5$.

Example 8

Solve the equation $x^2 - 10x - 20 = 0$.

Solution

Inspection shows that the given equation cannot be factored with integral numbers. Therefore, the solution will be accomplished by the method of completing the square.

Given $\qquad\qquad\qquad\qquad x^2 - 10x - 20 = 0$

A: 20, $\qquad\qquad\qquad\qquad\quad x^2 - 10x = 20$

Squaring one-half the coefficient of x and adding to both members,

$$x^2 - 10x + 25 = 20 + 25$$

Collecting terms, $\qquad x^2 - 10x + 25 = 45$

Factoring, $\qquad\qquad\qquad (x - 5)^2 = 45$

$\sqrt{}$, $\qquad\qquad\qquad\qquad x - 5 = \pm 6.71$

A: 5, $\qquad\qquad\qquad\qquad\qquad x = 5 \pm 6.71$

or $\qquad\qquad\qquad\qquad\qquad x = 11.71 \text{ or } -1.71$

The above answers are correct to three significant figures. The values of x are more precisely stated by maintaining the radical sign in the final roots. That is, if

$$(x - 5)^2 = 45$$

$\sqrt{}$, $\qquad\qquad x - 5 = \pm\sqrt{45}$

or $\qquad\qquad x - 5 = \pm 3\sqrt{5}$

A: 5, $\qquad\qquad\qquad x = 5 \pm 3\sqrt{5}$

That is, $\qquad\qquad x = 5 + 3\sqrt{5} \qquad$ or $\qquad 5 - 3\sqrt{5}$

Example 9

Solve the equation $3x^2 - x - 1 = 0$.

Solution

Given $\qquad\qquad\qquad 3x^2 - x - 1 = 0$

D: 3 (because the coefficient of x^2 must be 1),

$$x^2 - \tfrac{1}{3}x - \tfrac{1}{3} = 0$$

Transposing the constant term,

$$x^2 - \tfrac{1}{3}x = \tfrac{1}{3}$$

Squaring one-half the coefficient of x and adding to both members,

$$x^2 - \tfrac{1}{3}x + \tfrac{1}{36} = \tfrac{1}{3} + \tfrac{1}{36}$$

Collecting terms, $\quad x^2 - \tfrac{1}{3}x + \tfrac{1}{36} = \tfrac{13}{36}$

Factoring,

$$(x - \tfrac{1}{6})^2 = \tfrac{13}{36}$$

$\sqrt{}$, $\qquad\qquad\qquad x - \dfrac{1}{6} = \pm \dfrac{\sqrt{13}}{6}$

$$\therefore x = \dfrac{1 + \sqrt{13}}{6} \quad \text{or} \quad \dfrac{1 - \sqrt{13}}{6}$$

To summarize the method, we have the following:

RULE

To solve by completing the square:

1. If the coefficient of the square of the unknown is not 1, divide both members of the equation by the coefficient.
2. Transpose the constant terms (those not containing the unknown) to the right member.
3. Find one-half the coefficient of the unknown of the first degree, square the result, and add this square to both members of the equation. This makes the left member a perfect trinomial square.
4. Take the square root of both members of the equation and write the $\pm$ sign before the square root of the right member.
5. Solve the resulting simple equation.

PROBLEMS 21-3

Solve by completing the square:

1. $x^2 - 8x + 12 = 0$
2. $\alpha^2 - 4\alpha - 45 = 0$
3. $E^2 - 15E + 54 = 0$
4. $\Omega^2 + 5\Omega + 6 = 0$
5. $i^2 - 27i = -50$
6. $63 - a^2 = 2a$
7. $\theta^2 + 2 = 3\theta$
8. $e^2 - 6 = e$
9. $M^2 = 22M + 48$
10. $24E^2 = 2E + 1$
11. $3 + \theta = \theta^2 - 3$
12. $17I - 42 = I^2 + 2I - 16$
13. $\phi = \dfrac{60}{\phi} + 4$
14. $1 + \dfrac{12}{f} + \dfrac{35}{f^2} = 0$
15. $\dfrac{7(R - 4)}{R - 3} - (R - 2) = \dfrac{R - 4}{2}$
16. $\dfrac{Z - 1}{Z + 1} = \dfrac{Z - 2}{Z + 2} - 6$

21-5 STANDARD FORM

Any quadratic equation can be written in the general form

$$ax^2 + bx + c = 0$$

This is called the *standard form* of the quadratic equation. When it is written in this way, a represents the coefficient of the term containing x^2, b represents the coefficient of the term containing x, and c represents the constant term. Note that all terms of the equation, when written in standard form, are in the left member of the equation.

Example 10

Given $2x^2 + 5x - 3 = 0$. In this equation, $a = 2$, $b = 5$, and $c = -3$.

Example 11

Given $R^2 - 5R - 6 = 0$. In this equation, $a = 1$, $b = -5$, and $c = -6$.

Example 12

Given $9E^2 - 25 = 0$. In this equation, $a = 9$, $b = 0$, and $c = -25$.

21-6 THE QUADRATIC FORMULA

Because the standard form

$$ax^2 + bx + c = 0$$

represents *any* quadratic equation, it follows therefore that the roots of $ax^2 + bx + c = 0$ represent the roots of *any* quadratic equation. Therefore, if the standard quadratic equation can be solved for the unknown, the values, or roots, thereby obtained will serve as a formula for finding the roots of *any* quadratic equation.

This formula is derived by solving the standard form by the method of completing the square as follows:

Given $\qquad\qquad ax^2 + bx + c = 0$

Divide by a (Rule 1): $\qquad x^2 + \dfrac{bx}{a} + \dfrac{c}{a} = 0$

Transpose the constant term (Rule 2):

$$x^2 + \frac{bx}{a} = -\frac{c}{a}$$

Add the square of one-half the coefficient of x to both members (Rule 3):

$$x^2 + \frac{bx}{a} + \frac{b^2}{4a^2} = \frac{b^2}{4a^2} - \frac{c}{a}$$

Factor the left member, and add terms in the right member:

$$\left(x + \frac{b}{2a}\right)^2 = \frac{b^2 - 4ac}{4a^2}$$

Take the square root of both members:

$$x + \frac{b}{2a} = \pm \frac{\sqrt{b^2 - 4ac}}{2a}$$

Subtract $\dfrac{b}{2a}$: $\qquad\qquad x = -\dfrac{b}{2a} \pm \dfrac{\sqrt{b^2 - 4ac}}{2a}$

Collect terms of the right member:

$$x = \frac{-b \pm \sqrt{b^2 - 4ac}}{2a}$$

This equation is known as the quadratic formula.

Instead of attempting to solve a quadratic equation by factoring or by completing the square, we now make use of the quadratic formula. Upon becoming proficient in the use of the formula, you will find this method a convenience.

Example 13

Solve the equation $5x^2 + 2x - 3 = 0$.

Solution

Comparing this equation with the standard form

$$ax^2 + bx + c = 0$$

we have $a = 5$, $b = 2$, and $c = -3$. Substituting in the quadratic formula,

$$x = \frac{-b \pm \sqrt{b^2 - 4ac}}{2a}$$

$$= \frac{-2 \pm \sqrt{2^2 - 4 \cdot 5 \cdot (-3)}}{2 \cdot 5}$$

Hence,

$$x = \frac{-2 \pm \sqrt{64}}{10}$$

$$= \frac{-2 \pm 8}{10}$$

$$= \frac{-2 + 8}{10} \quad \text{or} \quad \frac{-2 - 8}{10}$$

$$\therefore x = \tfrac{3}{5} \text{ or } -1$$

Check

Substitute the values of x in the given equation.

Note

It must be remembered that the expression $\sqrt{b^2 - 4ac}$ is the square root of the *quantity* $(b^2 - 4ac)$ *taken as a whole.*

Example 14

Solve the equation $\dfrac{3}{5 - R} = 2R$.

Solution

Clearing the fractions results in $2R^2 - 10R + 3 = 0$. Comparing this equation with the standard form

$$ax^2 + bx + c = 0$$

we have $a = 2$, $b = -10$, and $c = 3$. Substituting in the quadratic formula,

$$x = \frac{-b \pm \sqrt{b^2 - 4ac}}{2a}$$

$$R = \frac{-(-10) \pm \sqrt{(-10)^2 - 4 \cdot 2 \cdot 3}}{2 \cdot 2}$$

Hence,

$$R = \frac{10 \pm \sqrt{76}}{4}$$

Factoring the radicand,

$$R = \frac{10 \pm 2\sqrt{19}}{4}$$

Dividing both terms of the fraction by 2,

$$R = \frac{5 \pm \sqrt{19}}{2}$$

$$= \frac{5 + \sqrt{19}}{2} \quad \text{or} \quad \frac{5 - \sqrt{19}}{2}$$

$$\therefore R = 4.68 \text{ or } 0.320$$

These final answers are correct to three significant figures. Check the solution by the usual method.

21-7 TESTING SOLUTIONS

Now that we can obtain solutions to quadratic equations by means of the quadratic formula, there will be two answers that are possible so long as $b^2 - 4ac \neq 0$. One of these answers we may call α:

$$\alpha = \frac{-b + \sqrt{b^2 - 4ac}}{2a}$$

and the other we may call β:

$$\beta = \frac{-b - \sqrt{b^2 - 4ac}}{2a}$$

By suitable combinations of α and β, we can achieve two useful relationships, the proof of which we leave to you as an exercise:

$$\alpha + \beta = \frac{-b}{a} \tag{1}$$

$$\alpha \cdot \beta = \frac{c}{a} \tag{2}$$

Whenever you obtain answers to quadratic equations by means of the formula (or any other means), you may quickly test them for accuracy. The sum of the two answers must equal $-\frac{b}{a}$, and the product of the two must equal $\frac{c}{a}$.

Example 15

Solve the equation $6x^2 - 2x - 4 = 0$, and test the answers.

Solution

Using the quadratic formula, $x = 1$ or $x = -\frac{2}{3}$. Applying the tests:

$$\alpha + \beta = 1 - \frac{2}{3} = +\frac{1}{3}$$

$$-\frac{b}{a} = -\frac{-2}{6} = +\frac{1}{3}$$

and

$$\alpha \cdot \beta = (1)\left(-\frac{2}{3}\right) = -\frac{2}{3}$$

$$\frac{c}{a} = \frac{-4}{6} = -\frac{2}{3}$$

The tests show that the solutions obtained are correct. You should make a habit of applying the tests to every solution to quadratic equations that you obtain.

PROBLEMS 21-4

Solve the following equations by using the quadratic formula, and apply the tests of Eqs. (1) and (2):

1. $\theta^2 = 4 - 3\theta$

2. $\lambda^2 + 7\lambda = 18$

3. $2I + 35 = I^2$

4. $\alpha^2 - 4\alpha + 3 = 0$

5. $15 - 14q = 8q^2$

6. $3I^2 - 7I + 2 = 0$

7. $5 = 6Z^2 - 3Z$

8. $5(R + 2) = 2R(R - 1)$

9. $24 - \dfrac{2}{m} - \dfrac{1}{m^2} = 0$

10. $\dfrac{2}{I_1} + \dfrac{3}{I_1} = \dfrac{1}{I_1^2} - 14$

11. $\dfrac{4 - R_1}{1 - R_1} - \dfrac{12}{3 - R_1} = 0$

12. $\dfrac{2}{\lambda + 3} = \dfrac{3}{\lambda - 2} - 1$

13. $\dfrac{7}{\beta - 3} - \dfrac{1}{2} = \dfrac{\beta - 2}{\beta - 4}$

14. $\dfrac{36}{(I + 3)^2} - \dfrac{I + 2}{I + 3} = 1$

15. $7i + 5 = \dfrac{21i^3 - 16}{3i^2 - 4}$

16. $4 - E - \dfrac{1}{2E} = -\dfrac{E^2 + 25}{7E}$

21-8 THE GRAPH OF A QUADRATIC EQUATION—THE PARABOLA

In Chap. 16 we spent some time on the drawing of graphs, especially graphs of unity-power (first-degree) equations, or linear graphs. Graphs of quadratic equations also may be drawn, and in this section we will investigate the common methods of producing such graphs and also a method of predicting the shape of graphs just from the equation itself in the same way that we learned to use the standard form $y = mx + b$ to predict the slope and y intercept of linear graphs.

All the quadratic equations we have studied so far have contained only one unknown, but that is because we looked at special cases. In the algebraic solution of quadratics, the standard form $ax^2 + bx + c = 0$ is sufficient, because we want to know the values of x which will satisfy this standard form equation. However, to draw a graph requires two variables, an independent one x and a dependent one y, so we rewrite the standard equation:

$$y = ax^2 + bx + c$$

Then, by plotting values of y for given values of x, we can draw the complete graph. Note that the algebraic solutions so far in this chapter have simply let $y = 0$, that is, the algebraic solutions have given us the x intercepts for the equation of the general form

$$y = ax^2 + bx + c$$

Example 16

Graph the equation $x^2 - 10x + 16 = 0$.

Solution

Set the equation equal to y:

$$y = x^2 - 10x + 16$$

Make a table of the values of y corresponding to assigned values of x. (See Eq. Fig. 21-1)

EQ. FIG. 21-1

If $x =$	0	1	2	3	4	5	6	7	8	9	10
Then $x^2 =$	0	1	4	9	16	25	36	49	64	81	100
$10x =$	0	10	20	30	40	50	60	70	80	90	100
$x^2 - 10x =$	0	-9	-16	-21	-24	-25	-24	-21	-16	-9	0
$\therefore y = x^2 - 10x + 16 =$	16	7	0	-5	-8	-9	-8	-5	0	7	16

Plotting the corresponding values of x and y as pairs of coordinates and drawing a smooth curve through the points results in the graph shown in Fig. 21-1.

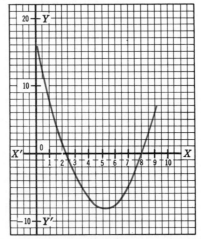

Fig. 21-1 Graph of the equation $y = x^2 - 10x + 16$.

From Fig. 21-1 it is apparent that the graph has two x intercepts at $x = 2$ and $x = 8$. That is, when $y = 0$, the graph crosses the x axis at $x = 2$ and $x = 8$. This is to be expected; for when $y = 0$, the given equation

$$x^2 - 10x + 16 = 0$$

can be solved algebraically to obtain $x = 2$ or 8. Hence, it is evident that the points at which the graph crosses the x axis denote the values of x when $y = 0$, which are the roots of the equation.

Another interesting fact regarding this graph is that the curve goes through a *minimum value*. Suppose it is desired to solve for the coordinates of the point of minimum value. First, if the equation is changed to standard form, we obtain $a = 1$, $b = -10$, and $c = 16$. If the value of $\dfrac{-b}{2a}$ is computed, the result is the x value, or abscissa, of the minimum point on the curve. That is,

$$x = -\frac{b}{2a} = -\frac{-10}{2 \times 1} = \frac{10}{2} = 5$$

Substituting this value of x in the original equation,

$$y = x^2 - 10x + 16$$
$$y = 5^2 - 10 \times 5 + 16 = -9$$

Thus, the point $(5, -9)$ is where the curve passes through a minimum value. That is, the dependent variable y is a minimum and equal to -9 when x, the independent variable, is equal to 5.

A third point of interest is that the parabola, as the graph of the quadratic is called, is symmetrical about its turning point, which lies midway between the two intercepts. Indeed, this can be seen from a revision of the quadratic formula:

$$x = \frac{-b \pm \sqrt{b^2 - 4ac}}{2a}$$

which may appropriately be written as

$$x = \frac{-b}{2a} \pm \frac{\sqrt{b^2 - 4ac}}{2a}$$

from which we can see that, with the turning point at $\frac{-b}{2a}$, the values of the x intercepts, or roots, of the graph will be offset from the x value of the turning point by amounts equal to $\pm \frac{\sqrt{b^2 - 4ac}}{2a}$.

Look now at some of the main possibilities concerning the appearance of parabolas:

1. They may open upward or downward (Fig. 21-2).

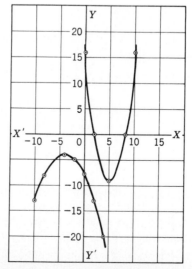

Fig. 21-2 Quadratic graph may open upward or downward.

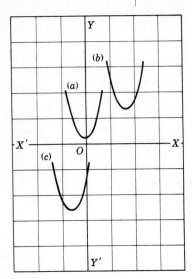

Fig. 21-3 Quadratic graphs may be symmetrical about the y axis or about a line parallel with the y axis.

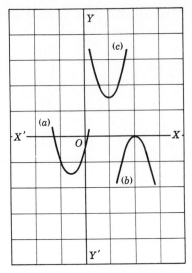

Fig. 21-4 Quadratic graphs may cut the x axis in two places, in one place, or not at all.

2. They may be symmetrical about the y axis or about some line parallel to the y axis (Fig. 21-3).

3. They may (a) cut the x axis in two places, (b) touch the x axis (cut it in one place), or (c) not touch the x axis at all (Fig. 21-4).

It is possible to decide many of these possibilities from the values of a particular quadratic equation. This general equation $y = ax^2 + bx + c$ offers many possibilities and a restriction:

1. a, the coefficient of the square term, may be any number, positive or negative, but *not* zero. (Why?)

2. b, the coefficient of the unity-power term, may be any number, positive or negative, *or* zero.

3. c, the constant term, may be any number, positive or negative, *or* zero.

Now, what is the effect of these algebraic possibilities on the graph? You should, at this point, arm yourself with graph paper and confirm the following statements:

RULE

The effect of a on the graph of the quadratic equation:

The value of a in the quadratic equation governs the steepness of the parabola. When a is large, the parabola is very steep, approaching a needlelike shape. When a is small, the parabola is shallow, approaching a dished shape.

Let $b = c = 0$ and plot the comparison graph $y = x^2$, in which the value of a is 1. Then plot various graphs of $y = ax^2$, letting a equal, in succession, 2, 5, 10, $\frac{1}{2}$, and $\frac{1}{4}$. If all these are plotted on the same graph sheet, with different colors or dashed lines, etc., then the effect of the value of a will be impressed on your mind forever.

RULE
The effect of a on the appearance of the parabola:

The algebraic sign of a will determine the opening of the parabola. $+a$ causes the curve to open upward, and the turning point is the *minimum* value. $-a$ causes the parabola to open downward, and the turning point is the *maximum* value.

You have already plotted a number of graphs with $+a$. Now plot a few graphs of $y = -ax^2$, letting $a = 2, 5, 10, \frac{1}{2}$, and $\frac{1}{4}$.

RULE
The effect of c on the appearance of the parabola:

The constant c in the quadratic equation determines the y intercept, and therefore the amount of vertical shift of the parabola. When c is positive, the curve is raised to cut the y axis above the x axis. When c is negative, the curve cuts the y axis below the x axis.

Let $a = 1$ and $b = 0$ and vary the value of c in the equation $y = x^2 + c$. Draw the curves when $c = +5$ and -5, and compare with the standard parabola $y = x^2$.

RULE
The effect of b on the appearance of the parabola:

The factor b in the quadratic equation determines the rotational shift of the turning point of the graph. When b is positive, the turning point shifts in a positive (ccw) direction about its "original" position, and when b is negative, the turning point shifts in a negative (cw) direction about its original position.

Let $a = 1$ and $c = 0$, and vary the value of b in the equation $y = x^2 + bx$. Draw the curves when $b = +2, +5, F10, -2, -5$, and -10. Next, repeat these curves with $a = -1$, and then draw curves for $y = -x^2 - bx$.

Example 17
Plot the curve $y = 27 - 3x - 4x^2$.

Solution
Predict, first of all, what effect the various coefficients will have on the graph:

1. The value of a is 4, so that the curve will be reasonably steep.
2. The algebraic sign of a is minus, so that the curve will open downward.
3. The constant term is $+27$, so that the curve cuts the y axis at $+27$, well above the x axis. Since the curve opens downward and the y intercept is above the x axis, the curve will cut the x axis in two places. The special equation $27 - 3x - 4x^2 = 0$ will have two definite solutions.

4. The value of b is -3, so that the turning point will be shifted from the "ideal" value of $x = 0$, $y = 27$ in the clockwise direction. The turning point will then be at a value of y greater than 27 and at some value of x to the left, or minus, side of the y axis.

With these predictions, together with a sketch of the probable appearance of the curve (Fig. 21-5), you may assign values to x and calculate the corresponding values of y:

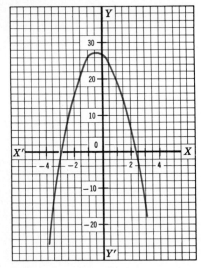

Fig. 21-5 Graph of the equation $y = 27 - 3x - 4x^2$.

If $x =$		-4	-3	-2	-1	0	1	2	3
Then $3x =$		-12	-9	-6	-3	0	3	6	9
$27 - 3x =$		39	36	33	30	27	24	21	18
$x^2 =$		16	9	4	1	0	1	4	9
$4x^2 =$		64	36	16	4	0	4	16	36
$\therefore y = 27 - 3x - 4x^2 =$	-25	0	17	26	27	20	5	-18

Plotting the corresponding values of x and y as pairs of coordinates and drawing a smooth curve through them results in the graph shown in Fig. 21-5.

From the graph of the equation $y = 27 - 3x - 4x^2$, Fig. 21-5, it is observed:

1. The roots (solution) of the equation are denoted by the x intercepts. These are $x = -3$ and $x = 2.25$. They can be checked algebraically to obtain

$$27 - 3x - 4x^2 = 0$$
Factoring, $$(3 + x)(9 - 4x) = 0$$
$$x = -3 \text{ or } 2.25$$

2. The parabola opens *downward* because the coefficient of x^2 is negative ($a = -4$.)
3. Because the parabola opens downward, the graph goes through a *maximum* value. The point of maximum value is found in the same manner as the minimum point of Example 16. That is,

$$x = \frac{-b}{2a} = \frac{-(-3)}{2(-4)} = -\frac{3}{8}$$

Substituting $-\frac{3}{8}$ for x in the original equation,

$$y = 27 - 3(-\tfrac{3}{8}) - 4(-\tfrac{3}{8})^2 = 27.6$$

Thus, the dependent variable y is a maximum and equal to 27.6 when x, the independent variable, is equal to $-\frac{3}{8}$.

Example 18
Graph the equations

$$y = x^2 - 8x + 12 \qquad\qquad (a)$$
$$y = x^2 - 8x + 16 \qquad\qquad (b)$$
$$y = x^2 - 8x + 20 \qquad\qquad (c)$$

Solution
1. Based on an analysis of the values of a, b, and c, predict the probable appearance of each curve.
2. As before, and for each equation, make up a table of values of y corresponding to chosen values of x. Using these x and y values as pairs of coordinates, plot the graphs of the equations. These graphs are shown in Fig. 21-6.
   The coefficients of the equations are the same except for the values of the constant term c.

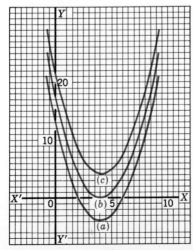

Fig. 21-6 Graphs of the equations of Example 18.

From the graph of the equations of Example 18, it is observed that:

1. The curve of (*a*) intercepts the *x* axis at $x = 2$ and $x = 6$, and the roots of the equation are thus denoted as $x = 2$ or 6.
2. The curve of (*b*) just touches the *x* axis at $x = 4$. Solving (*b*) algebraically shows that the roots are *equal*, both roots being 4.
3. The curve of (*c*) does not intersect or touch the *x* axis. Solving (*c*) algebraically results in the imaginary roots $x = 4 \pm j2$.
4. All curves pass through minimum values at points having equal *x* values. This is as expected, for the *x* value of a maximum or a minimum is given by $x = \dfrac{-b}{2a}$, and these values are equal in each of the given equations.
5. Checking the *y* values of the minima, it is seen that they must be affected by the constant terms, for, as previously mentioned, the other coefficients of the equations are the same.

21-9 GRAPHICAL SOLUTIONS

From the foregoing, it must now be obvious that quadratic equations can be solved graphically by letting the equation $ax^2 + bx + c = 0$ take the more general form $ax^2 + bx + c = y$ or, as seen more often, $y = ax^2 + bx + c$. Then the two *x* intercepts of the graph will give the roots of the original equation. It is for this reason that you will often hear the solutions to a quadratic equation referred to as the *zeros* of the equation—they occur when $y = 0$.

PROBLEMS 21-5

Select problems from Problems 21-1, 21-2, 21-3, and 21-4 and solve them graphically to confirm the algebraic solutions. Predict what the graphs will look like before plotting calculated values of *x* and *y*.

21-10 THE DISCRIMINANT

The quantity $b^2 - 4ac$ under the radical in the quadratic formula is called the *discriminant* of the quadratic equation. The two roots of the equation are

$$x = \frac{-b + \sqrt{b^2 - 4ac}}{2a}$$

and

$$x = \frac{-b - \sqrt{b^2 - 4ac}}{2a}$$

Now, if $b^2 - 4ac = 0$, it is apparent that the two roots are equal. Also, if $b^2 - 4ac$ is *positive*, each of the roots is a *real* number. But if $b^2 - 4ac$ is *negative*, the roots are *imaginary*. Therefore, there is a direct relationship between the value of the discriminant and the roots, and hence the graph, of a quadratic equation.

For example, the discriminants of the equations of Example 18 in the preceding section are

$$b^2 - 4ac = (-8)^2 - 4 \cdot 1 \cdot 12 = 16$$
$$b^2 - 4ac = (-8)^2 - 4 \cdot 1 \cdot 16 = 0$$
$$b^2 - 4ac = (-8)^2 - 4 \cdot 1 \cdot 20 = -16$$

Upon checking these values with the curves of Fig. 21-6 and also checking the values of the discriminants found in the preceding exercises with their respective curves, it is evident that the roots of a quadratic equation are:

1. Real and unequal if and only if $b^2 - 4ac$ is positive.
2. Real and equal if and only if $b^2 - 4ac = 0$.
3. Imaginary and unequal if and only if $b^2 - 4ac$ is negative.
4. Rational if and only if $b^2 - 4ac$ is a perfect square.

21-11 MAXIMA AND MINIMA

As earlier stated, in the general quadratic equation $ax^2 + bx + c = 0$ the relation $x = \dfrac{-b}{2a}$ gives the value of the independent variable x at which the dependent variable y will be maximum or minimum. Then by substituting this value of x, the independent variable, in the equation, the corresponding value of y can be obtained. Also, it has been shown that the function will be maximum if a, the coefficient of x^2, is negative because the curve opens downward. Similarly, if the coefficient of x^2 is positive, the curve will pass through a minimum because the curve opens upward.

This knowledge facilitates the solution of many problems that heretofore would have involved considerable labor.

Example 19

A source of emf V, with an internal resistance r, is connected to a load of variable resistance R. What will be the value of R with respect to r when maximum power is being delivered to the load?

Solution

The circuit can be represented as shown in Fig. 21-7. By Ohm's law, the current flowing through the circuit is

$$I = \frac{V}{r + R} \qquad (a)$$

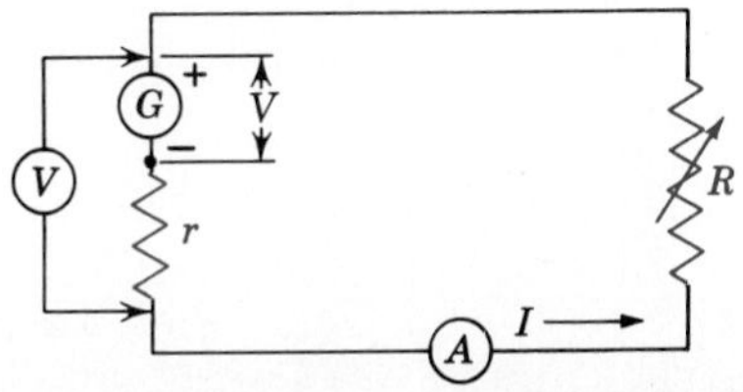

Fig. 21-7 Circuit of Example 19.

The power delivered to the external circuit is

$$P = V_t I = I^2 R \qquad (b)$$

where V_t is the terminal voltage of the source and

$$V_t = V - Ir \qquad (c)$$

Now the terminal voltage V_t will decrease as the current I increases. Therefore, the power P supplied to the load is a function of the two variables V_t and I. Substituting Eq. (c) in Eq. (b),

$$P = (V - Ir)I = VI - I^2 r$$

that is,

$$P = -rI^2 + VI \qquad (d)$$

Equation (d) is a quadratic in I, where

$$a = -r \quad \text{and} \quad b = V$$

Then, since, for maximum conditions, $I = \dfrac{-b}{2a}$,

$$I = \frac{-b}{2a} = \frac{-V}{2(-r)} = \frac{V}{2r} \qquad (e)$$

which is the value of the current through the circuit when maximum power is being delivered to the load. Substituting Eq. (e) in Eq. (a),

$$\frac{V}{2r} = \frac{V}{r + R} \qquad (f)$$

Solving Eq. (f) for R,

$$R = r \qquad (g)$$

Equation (g) shows that maximum power will be delivered to any load when the resistance of that load is equal to the internal resistance of the source of emf. This is one of the important concepts in electronics engineering. For example, we are concerned with obtaining maximum power output from several types of power amplifier. We obtain it when the amplifier load resistance matches the output resistance of the associated components. Also, maximum power is delivered to an antenna circuit when the impedance of the antenna is made to match that of the transmission line that feeds the antenna.

In Fig. 21-8, power delivered to the load is plotted against values of

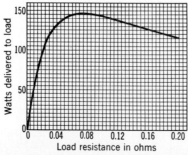

Fig. 21-8 Power delivered to load plotted against load resistance.

the load resistance R_L when a storage battery with an emf V of 6.6 V and an internal resistance $r = 0.075 \; \Omega$ is used. See the circuit in Fig. 21-9.

It is apparent that, when the battery or any other source of emf is delivering maximum power, half the power is lost within the battery. Under these conditions, therefore, the efficiency is 50%.

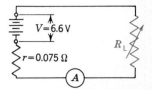

Fig. 21-9 Load resistance R_L is varied to obtain power values plotted in Fig. 21-8.

PROBLEMS 21-6

1. Graph the following equations all on the same sheet with the same axes:

   (a) $x^2 - 6x - 16 = 0$ (b) $x^2 - 6x - 7 = 0$
   (c) $x^2 - 6x = 0$ (d) $x^2 - 6x + 5 = 0$
   (e) $x^2 - 6x + 9 = 0$ (f) $x^2 - 6x + 12 = 0$
   (g) $x^2 - 6x + 15 = 0$

   Does changing the constant term change only the vertical positions of the graphs and the solutions of the equations? Do all graphs pass through minimum values at the same value of x?

2. Solve the equations of Prob. 1 algebraically. Do these solutions check the graphs of the equations? Test your solutions by means of the quadratic tests.

3. Compute the discriminant for each equation of Prob. 1. Do you see any connection between the value and the graph?

4. Compute the minimum value of the dependent variable y for each equation of Prob. 1. Does the value check with the graph?

5. What do you see from the graphs of Prob. 1 when x is equal to zero?

21-12 SUMMARY

Several methods are available for solving quadratic equations. All quadratic equations can be solved by factoring, by completing the square, by use of the quadratic formula, or by graphical methods. However, some of these methods involve unnecessary work for certain forms or types of quadratic equations; therefore, one tries to choose the most convenient method for a particular equation. For example, a pure quadratic equation is readily solved merely by reducing the equation to its simplest form and extracting the square root of both members of the equation in order to obtain the two roots, which are equal in absolute value but of opposite sign (Sec. 21-2).

In practical problems involving complete quadratic equations the numerical coefficients are such that you will seldom be able to solve the equation readily by factoring. Also, solution by completing the square sometimes can become a chore. Probably the most widely used method is solution by use of the quadratic formula, which, if you forget it, can be found in most handbooks and put to use whenever needed.

Solution by graphical methods allows you to visualize the variation of quantities and serves to check computations. In any event, through solving many problems, you will develop your own methods of attack.

In solving problems involving quadratic equations, care must be used because two answers (roots) are obtained. In all cases both roots will satisfy the mathematics of the equation, but in some cases only one root will satisfy the conditions of the problem. Therefore, we reject the obviously impossible or the impractical answer and retain the one that is consistent with the physical conditions of the problem.

Example 20

The square of a certain number plus four times the number is 12. Find the number.

Solution

Let	x = the number
Then	x^2 = the square of the number
and	$4x$ = four times the number
From the problem	$x^2 + 4x = 12$
S: 12,	$x^2 + 4x - 12 = 0$
Factoring,	$(x + 6)(x - 2) = 0$
Then	$x = -6$ or 2

Both roots satisfy the equation and the condition of the problem; therefore, both answers are correct.

Example 21

Find the dimensions of a right triangle if the hypotenuse of the triangle is 40 m and the base exceeds the altitude by 8 m.

Solution

In any right triangle, Fig. 21-10, $c^2 = a^2 + b^2$.

Since	$c = 40$
and	$a = b - 8$
then	$1600 = (b - 8)^2 + b^2$

Are both roots of this equation consistent with the physical conditions of the problem?

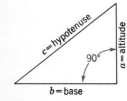

Fig. 21-10 In any right triangle, $c^2 = a^2 + b^2$.

Example 22

A storage battery has an emf of 6.3 V and an internal resistance of 0.015 Ω. The battery is used to drive a dynamotor requiring 300 W. What current then will the battery deliver to the dynamotor, and what will be the voltage reading across the battery terminals while this current is supplied?

Solution

The circuit is represented in Fig. 21-11.

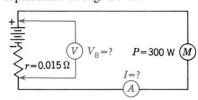

Fig. 21-11 Circuit of Example 22.

Let P = power consumed by dynamotor
  = 300 W
  V_B = voltage across battery terminals when dynamotor is delivering 300 W

Since $\qquad I = \dfrac{P}{V_B}$

then $\qquad I = \dfrac{300}{V_B}$

Now $\qquad V_B = 6.3 - rI$

Substituting for r, $\qquad V_B = 6.3 - 0.015I$

Substituting for I, $\qquad V_B = 6.3 - 0.015 \times \dfrac{300}{V_B}$

Multiplying, $\qquad V_B = 6.3 - \dfrac{4.5}{V_B}$

Clearing fractions, $\qquad V_B{}^2 = 6.3V_B - 4.5$

Transposing, $\qquad V_B{}^2 - 6.3V_B + 4.5 = 0$

This equation is a quadratic in V_B; hence, $a = 1$, $b = -6.3$, and $c = 4.5$. Substituting these values in the quadratic formula,

$$V_B = \frac{-(-6.3) \pm \sqrt{(-6.3)^2 - 4 \cdot 1 \cdot 4.5}}{2 \cdot 1}$$

or $\qquad V_B = \dfrac{6.3 \pm \sqrt{21.7}}{2}$

$$\therefore V_B = 5.48 \text{ V or } 0.82 \text{ V}$$

$$I = \frac{300}{V_B} = \frac{300}{5.48} = 54.7 \text{ A}$$

Why was 5.48 V chosen instead of 0.82 V in the above solution?

PROBLEMS 21-7

1. Compute the discriminant, and tell what it shows, in each of these equations:
   (a) $x^2 - 8x + 12 = 0$
   (b) $9x^2 - 42x + 49 = 0$
   (c) $4x^2 - 20x + 30 = 0$

2. Find two positive consecutive even numbers whose product is 288.
3. Find two positive consecutive odd numbers whose product is 483.
4. Can the sides of a right triangle ever be consecutive integers? If so, find the integers.
5. Find the dimensions of a rectangular parking lot whose area is 22 400 m^2 and whose perimeter is 600 m.
6. Separate 156 into two parts such that one part is the square of the other.
7. One number is 20 less than another, and the difference of their squares is 9200. What are the numbers?
8. $F = \dfrac{Wv^2}{32r}$
   (a) Solve for v.
   (b) If W is doubled and r is halved, what happens to F?
   (c) What is W if $F = 12$, $r = 1\frac{1}{3}$, and $v = 16$?
9. Given $P = \dfrac{kV^2}{nR}$. Solve for V. If k and n are doubled and P and R are held constant, what happens to V?

10. $R_t = \dfrac{r}{\left(\dfrac{d_o}{d_i}\right)^2} - 1$. Solve for $\dfrac{d_o}{d_i}$.

11. $P = \dfrac{R(r^2 + x^2)}{r(Rr + Xx)}$. Solve for r and x.

12. The following relations exist in the Wien bridge:

$$\omega^2 = \frac{1}{R_1 R_2 c_1 c_2} \quad \text{and} \quad \frac{c_1}{c_2} = \frac{R_b - R_2}{R_a R_1}$$

Solve for c_1 and c_2 in terms of resistance components and ω.

13. Kinetic energy (KE) is equal to one-half the product of mass m in kilograms and the square of velocity v in meters per second; that is, KE $= \frac{1}{2}mv^2$ joules. Find the value of v when KE $= 1.1 \times 10^6$ J and $m = 2.2$ kg.

14. A ball rolls down a slope and travels a distance $d = 6t + \frac{1}{2}t^2$ meters in t s. Solve for t.

15. The distance through which an object will fall in t s is $s = \frac{1}{2}gt^2$ meters, where $g = 9.81$ m/s². The velocity v attained after t s is $v = gt$ m/s. Solve for the velocity in terms of g and s.

16. If an object is thrown straight up with a velocity of v m/s, its height t s later is given by $h = vt - 4.9t^2$ meters. If a rocket were fired upward with a velocity of 1176 m/s, neglecting air resistance:
(a) At what time would its height be 15 km on the way up?
(b) At what time would its height be 15 km on the way down?
(c) At what time would it attain its maximum height?
(d) What maximum height would it attain?
Attempt these solutions both graphically and algebraically.

17. Use the formula for height in Prob. 16 to derive a formula for maximum height attained for any initial velocity v.

18. In an ac series circuit containing resistance R in ohms and inductance L in henrys, the current I may be computed from the formula

$$I = \frac{V}{\sqrt{R^2 + \omega^2 L^2}} \quad \text{A}$$

where V is the emf in volts applied across the circuit. Find the value of R to three significant figures if $V = 282$ V, $I = 2$ A, $\omega = 2\pi f$, $f = 60$ Hz, and $L = 0.264$ H.

19. In an ac circuit containing R Ω resistance and X_C Ω reactance, the impedance is

$$Z = \sqrt{R^2 + X_C^2} \quad \Omega$$

Find the value of R if $Z = 130$ Ω and $X_C = 120$ Ω.

20. The susceptance of an ac circuit that contains R Ω resistance and X Ω reactance is

$$B = \frac{X}{R^2 + X^2} \quad \text{siemens (S)}$$

Find the value of R to three significant figures when $B = 0.008$ S and $X = 100$ Ω.

21. The equivalent noise resistance R_N of a bipolar NPN transistor depends upon collector current I_C, emitter current I_E, leakage current I_{CO}, and base-emitter internal junction resistance r_e expressed in terms of the mutual transconductance $g_m \cong \dfrac{1}{r_e}$. The empirical formula relating these parameters is

$$R_N = \frac{2.5I_C^2}{g_m I_E^2} + \frac{20I_C I_{CO}}{g_m^2 I_E}$$

Show that the ratio of collector current to emitter current is

$$\frac{I_C}{I_E} = \frac{-20I_{CO}}{5g_m} + \frac{1}{5g_m} \sqrt{400I_{CO}^2 + 10R_N g_m^3}$$

22. Did you prove that $\alpha + \beta = \dfrac{-b}{a}$?

23. Did you prove that $\alpha \cdot \beta = \dfrac{c}{a}$?

24. Find the two combinations of resistance of R_2 and R_3 that will satisfy the circuit conditions of Fig. 21-12.

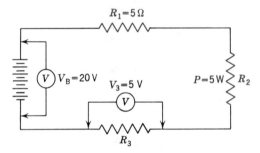

Fig. 21-12 Circuit of Prob. 24.

25. The circuit conditions as shown in Fig. 21-13 existed when the generator G was supplying current to the circuit. When the generator was disconnected, an ohmmeter connected between points A and B read 60 Ω.
(a) What was the circuit current?
(b) What was the generator voltage?
(c) What is the value of each resistor?

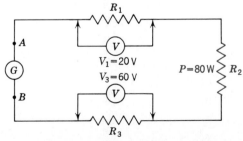

Fig. 21-13 Circuit of Prob. 25.

26.   In the circuit of Fig. 21-14, the resistor *ABC* represents a potentiometer with total resistance (*A* to *C*) of 25 000 Ω. $R_1 = 5000\ \Omega$,
across which is 60 V.
(*a*) What is the resistance from *A* to *B*?
(*b*) How much current flows from *B* to *C*?

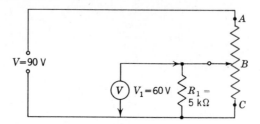

Fig. 21-14 Circuit of Prob. 26.

27.   What are the meter readings in the circuit of Fig. 21-15?

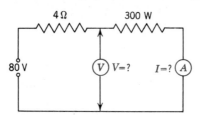

Fig. 21-15 Circuit of Prob. 27.

28.   When two capacitors C_1 and C_2 are connected in series, the total
capacitance C_t of the combination is always less than either of the
two capacitors. That is,

$$C_t = \frac{C_1 C_2}{C_1 + C_2}$$

Suppose we have a tuning capacitor that varies from 200 to 300 pF;
that is, it has a *change* in capacitance of 100 pF. What value of
fixed capacitor should be connected in series with the tuning capacitor
to limit the total *change* of circuit capacitance to 50 pF?

Network Simplification

An understanding of Kirchhoff's laws, plus the ability to apply the laws in analyzing circuit conditions, will give you a better insight into the behavior of circuits. Furthermore, you will be able to solve circuit problems that, with only a knowledge of Ohm's law, would be very difficult in some cases and impossible in others.

22-1 DIRECTION OF CURRENT FLOW

As stated in Sec. 8-1, the most generally accepted concept of an electric current is that the current consists of a motion of electrons from a negative toward a more positive point in a circuit. That is, a positively charged body is taken to be one that is deficient in electrons, whereas a negatively charged body carries an excess of electrons. When the two bodies are joined by a conductor, electrons flow from the negatively charged body to the positively charged one. Hence, if two such points in a circuit are *maintained* at a difference of potential, a *continuous* flow of electrons, or current, will take place from negative to positive. Therefore, in the consideration of Kirchhoff's laws, current will be thought of as flowing from the negative terminal of a source of emf, through the external circuit, and back to the positive terminal of the source. Thus, in Fig. 22-1, the current flows away from the negative terminal of the battery, through R_1 and R_2, and back to the positive terminal of the battery. Note that point b is positive with respect to point a and that point d is positive with respect to point c.

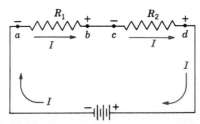

Fig. 22-1 Current I flowing from − to + through the connected circuit.

22-2 STATEMENT OF KIRCHHOFF'S LAWS

In 1847, G. R. Kirchhoff extended Ohm's law by two important statements which have become known as Kirchhoff's laws. These laws can be stated as follows:

1. The algebraic sum of the currents at any junction of conductors is zero.

That is, at any point in a circuit, there is as much current flowing away from the point as there is flowing toward it.

2. The algebraic sum of the emf's and voltage drops around any closed circuit is zero.

That is, in any closed circuit, the applied emf is equal to the voltage drops around the circuit.

These laws are straightforward and need no proof here; for the first is self-evident from the study of parallel circuits, and the second was stated in different words in Sec. 8-8. When properly applied, they enable us to set up equations for any circuit and solve for the unknown circuit components, voltages, or currents as required.

22-3 APPLICATION OF SECOND LAW TO SERIES CIRCUITS

The second law is considered first because of its applications to problems with which you are already familiar.

Figure 22-2 represents a 20-V generator connected to three series resistors. The validity of Kirchhoff's second law was shown in Sec. 8-8; that is, in any closed circuit the applied emf will be equal to the sum of the voltage drops around the circuit. Thus, neglecting the internal resistance of the generator and the resistance of the connecting wires in Fig. 22-2,

$$V = IR_1 + IR_2 + IR_3 \tag{1}$$

or
$$20 = 2I + 3I + 5I$$

Hence,
$$I = 2 \text{ A}$$

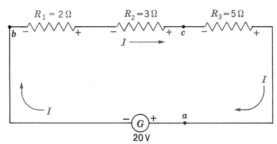

Fig. 22-2 The sum of the voltage drops across the resistors is equal to the applied emf.

Equation (1) is satisfactory for a circuit containing one source of emf. By considering the circuit from a different viewpoint, however, the voltage relations around the circuit become more understandable. For example, by starting at any point in the circuit, such as point a, we proceed completely around the circuit in the direction of current flow, remembering that, when current passes through a resistance, there is a voltage drop that represents a loss and therefore is subtractive. Also, in going around the circuit, sources of emf represent a gain in voltage if they tend to aid current flow and therefore are additive. By this method, according to the second law, the algebraic sum of all emf's and voltage drops around the circuit is zero.

For example, in starting at point a in Fig. 22-2 and proceeding around the circuit in the direction of current flow, the first thing encountered is the positive terminal of a source of emf of 20 V. Because this causes current to flow in the direction we are going, it is written $+20$. This is easily remembered, for the positive terminal was the first one encountered; therefore, write it plus. Next comes R_1, which is responsible for a drop in voltage due to the current I passing through it. Hence, this voltage drop is written $-IR_1$ or $-2I$, for R_1 is known to be 2 Ω. R_2 and R_3 are treated in a similar manner because both represent voltage *drops*. This completes the trip around the circuit, and by equating the algebraic sum of the emf and voltage drops to zero,

$$20 - 2I - 3I - 5I = 0 \qquad (2)$$

or
$$I = 2 \text{ A}$$

Note that Eq. (2) is simply a different form of Eq. (1). If the polarities of the sources of emf are marked, they will serve as an aid in remembering whether to add or subtract. In going around the circuit, if the first terminal of a source of emf is positive, the emf is added; if negative, the emf is subtracted.

The point at which to start around the circuit is purely a matter of choice, for the algebraic sum of all voltages around the circuit is equal to zero. For example, starting at point b,

$$-2I - 3I - 5I + 20 = 0$$
$$I = 2 \text{ A}$$

Starting at point c,

$$-5I + 20 - 2I - 3I = 0$$
$$I = 2 \text{ A}$$

Example 1
Find the amount of current flowing in the circuit that is represented in Fig. 22-3 if the internal resistance of battery V_1 is 0.3 Ω, that of V_2 is 0.2 Ω, and that of V_3 is 0.5 Ω.

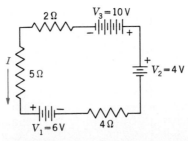

Fig. 22-3 Circuit of Example 1.

Solution

Figure 22-4 is a diagram of the circuit in which the internal resistances are represented in color as an aid in setting up the circuit equation. Beginning at point a and going around the circuit in the direction of current flow,

$$6 - 0.3I - 4I - 0.2I - 4 + 10 - 0.5I - 2I - 5I = 0$$

Hence, $\qquad I = 1 \text{ A}$

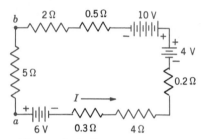

Fig. 22-4 Circuit of Example 1 illustrating internal resistances of the batteries.

In more complicated circuits the direction of the current is often in doubt. However, this need cause no confusion, for the direction of current flow can be *assumed* and the circuit equation can then be written in the usual manner. If the current results in a negative value when the equation is solved, the negative sign denotes that the assumed direction was wrong. As an example, let it be assumed that the current in the circuit of Fig. 22-4 flows in the direction from a to b. Then, starting at point a and going around the circuit in the assumed direction,

$$-5I - 2I - 0.5I - 10 + 4 - 0.2I - 4I - 0.3I - 6 = 0$$
$$\therefore I = -1 \text{ A}$$

As stated above, the minus sign shows that the assumed direction of the current was wrong; therefore, the current flows in the direction from b to a.

PROBLEMS 22-1

1. Three resistors, $R_1 = 22 \text{ k}\Omega$, $R_2 = 39 \text{ k}\Omega$, and $R_3 = 33 \text{ k}\Omega$, are connected in parallel across a 12-V power supply whose internal resistance is 1.8 kΩ. How much current is drawn from the source?
2. The resistors in Prob. 1 are replaced by new values $R_1 = 2.2 \text{ k}\Omega$, $R_2 = 3.9 \text{ k}\Omega$, and $R_3 = 3.3 \text{ k}\Omega$. How much current will be drawn from the source?
3. Three resistors, $R_1 = 68 \text{ k}\Omega$, $R_2 = 22 \text{ k}\Omega$, and $R_3 = 18 \text{ k}\Omega$, are connected in series across a signal generator whose internal resistance is 600 Ω. If 0.500 mA flows through the circuit, what is the terminal voltage of the generator?
4. What is the value of R_4 in Fig. 22-5?

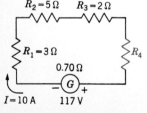

Fig. 22-5 Circuit of Prob. 4.

5. A motor that draws 16 A at 234 V is connected to a generator through two No. 8 copper feeders each of which is 500 m long. What is the generator terminal voltage?

6. A generator with a terminal voltage of 117 V is supplying 63 A to a load through two feeders that are each 1500 m long. If the feeders are No. 0 copper, what is the voltage across the load?

7. (a) How much current flows in the circuit of Fig. 22-6?
   (b) What is the terminal voltage of the 12-V battery?

8. (a) How much current flows in the circuit of Fig. 22-7?
   (b) What is the terminal voltage of the generator?

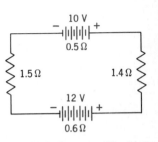

Fig. 22-6 Circuit of Prob. 7.

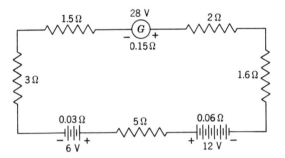

Fig. 22-7 Circuit of Prob. 8.

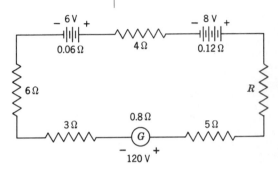

Fig. 22-8 Circuit of Prob. 9.

9. A current of 5 A flows through the circuit of Fig. 22-8. What is the value of R?

10. How much current flows in the circuit of Fig. 22-9?

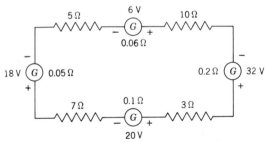

Fig. 22-9 Circuit of Prob. 10.

22-4 SIMPLE APPLICATIONS OF BOTH LAWS

Although the circuits of the following examples can be solved by Ohm's law, they are included here because you are familiar with such circuits. You will have no trouble in solving circuits that appear to be complicated if you understand the applications of Kirchhoff's laws to simple circuits, for all circuits are combinations of the fundamental series and parallel circuits.

Example 2

A generator supplies 7 A to two resistances of 40 and 30 Ω connected in parallel. Neglecting the internal resistance of the generator and the resistance of the connecting wires, find the generator voltage and the current through each resistance.

Solution

Figure 22-10 is a diagram of the circuit. From our knowledge of parallel circuits, it is evident that the line current I divides at junction c into the branch currents I_1 and I_2. Similarly, I_1 and I_2 combine at junction f to form the line current I. Therefore,

$$I = I_1 + I_2$$

which is the same as

$$I - I_1 - I_2 = 0 \tag{3}$$

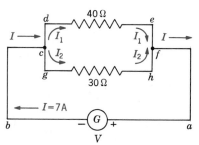

Fig. 22-10 Circuit of Example 2.

These are algebraic expressions for Kirchhoff's first law. When they are used in conjunction with the second law, they facilitate the solution of circuits.

If we start at point a and go around the circuit in the direction of current flow, the equation for the voltages around path $abcdefa$ is

$$V - 40I_1 = 0$$
$$I_1 = \frac{V}{40} \tag{4}$$

The equation for the voltages around path $abcghfa$ is

$$V - 30I_2 = 0$$
$$I_2 = \frac{V}{30} \tag{5}$$

Substituting the known values in Eq. (3),

$$7 - \frac{V}{40} - \frac{V}{30} = 0$$
$$V = 120 \text{ V}$$

$I_1 = 3$ A and $I_2 = 4$ A are found from Eqs. (4) and (5), respectively.

Example 3

Two 6-V batteries, each with an internal resistance of 0.05 Ω, are connected in parallel to a load resistance of 9.0 Ω. How much current flows through the load resistance?

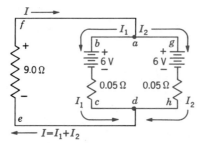

Fig. 22-11 Circuit of Example 3.

Solution

Figure 22-11 is a diagram of the circuit. In the circuit, two identical sources of emf are connected in parallel to supply the line current I to the load resistance. Again,

$$I = I_1 + I_2$$

or $$I - I_1 - I_2 = 0$$

Starting at junction a, the equation for the voltages around path $abcdefa$ is

$$6 - 0.05I_1 - 9I = 0$$

Solving for I_1, $$I_1 = 120 - 180I \qquad (6)$$

Starting at junction a, the equation for the voltages around path $aghdefa$ is

$$6 - 0.05I_2 - 9I = 0$$

Solving for I_2, $$I_2 = 120 - 180I \qquad (7)$$

As would be expected, I_1 and I_2 are equal. Substituting the values of I_1 and I_2 in Eq. (3),

$$I - (120 - 180I) - (120 - 180I) = 0$$

Hence, $$I = 0.6648 \text{ A}$$

The foregoing solution assumes three unknowns I, I_1, and I_2. However, in writing the equations for the voltages around any path, only two unknowns can be used, for $I = I_1 + I_2$. Thus, around path $abcdefa$,

$$6 - 0.05I_1 - 9(I_1 + I_2) = 0$$

Collecting terms, $$9.05I_1 + 9I_2 = 6 \qquad (8)$$

Voltages around path $aghdefa$,

$$6 - 0.05I_2 - 9(I_1 + I_2) = 0$$

Collecting terms, $$9I_1 + 9.05I_2 = 6 \qquad (9)$$

Since Eqs. (8) and (9) are simultaneous equations, they can be solved for I_1 and I_2. Hence,

$$I_1 = 0.3324 \text{ A}$$
$$I_2 = 0.3324 \text{ A}$$

and $$I = I_1 + I_2 = 0.6648 \text{ A}$$

PROBLEMS 22-2

1. A power supply supplies a total of 1.46 A to two resistors of 75 and 43 Ω connected in parallel. What is the terminal voltage of the power supply?

2. A battery supplies 5.53 A to three resistors of 2, 2.7, and 3 Ω connected in parallel. What is the terminal voltage of the battery?

3. A generator with an internal resistance of 0.05 Ω supplies 15.2 A to three resistors of 8, 4, and 10 Ω connected in parallel. What is the generator terminal voltage?

4. A battery supplies 9.7 A to four resistors that are 110, 50, 100, and 200 Ω connected in parallel. What is the voltage across the resistors?

5. (a) What is the value of the current in the circuit of Fig. 22-12?
   (b) How much power is expended in each of the batteries?

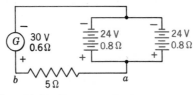

Fig. 22-12 Circuit of Probs. 5 and 6.

6. How much power would be expended in each battery in the circuit of Fig. 22-12 if the load resistance were changed from 10 to 0.5 Ω?

7. (a) What is the generator current in the circuit of Fig. 22-13?
   (b) In what direction does the current flow?

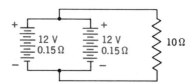

Fig. 22-13 Circuit of Probs. 7 and 8.

8. (a) What is the value of the generator current in the circuit shown in Fig. 22-13 if the generator emf voltage is decreased to 12 V?
   (b) In what direction does the current flow?

22-5 FURTHER APPLICATIONS OF KIRCHHOFF'S LAWS

In preceding examples and problems if two sources of emf have been connected to the same circuit, the values of emf and internal resistance have been equal. However, there are many types of circuits that contain more than one source of power, each with a different emf and different internal resistance.

Example 4

Figure 22-14 represents two batteries connected in parallel and supplying current to a resistance of 2 Ω. One battery has an emf of 6 V

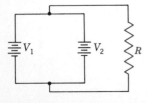

Fig. 22-14 Circuit of Example 4.

and an internal resistance of 0.15 Ω, and the other battery has an emf of 5 V and an internal resistance of 0.05 Ω. Determine the current through the batteries and the current in the external circuit. Neglect the resistance of the connecting wires.

Solution

Draw a diagram of the circuit representing the internal resistance of the batteries, and label the circuit with all the known values as shown in Fig. 22-15. Label the unknown currents, and mark the direction in which each current is assumed to flow. There are three currents of unknown value in the circuit: I_1, I_2, and the current I which flows through the external circuit. However, because $I = I_1 + I_2$, the unknown currents can be reduced to two unknowns by considering a current of $I_1 + I_2$ A flowing through the external circuit.

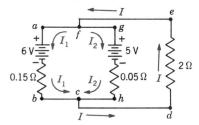

Fig. 22-15 Circuit of Example 4 labeled with known values.

For the path $abcdefa$, $6 - 0.15I_1 - 2(I_1 + I_2) = 0$
Collecting terms, $2.15I_1 + 2I_2 = 6$ (10)
For the path $ghcdefg$, $5 - 0.05I_2 - 2(I_1 + I_2) = 0$
Collecting terms, $2I_1 + 2.05I_2 = 5$ (11)

Equations (10) and (11) are simultaneous equations that, when solved, result in

$$I_1 = 5.64 \text{ A}$$
and
$$I_2 = -3.07 \text{ A}$$

The negative sign of the current I_2 denotes that this current is flowing in a direction opposite to that assumed. The value of the line current is

$$I = I_1 + I_2 = 5.64 + (-3.07)$$
$$= 2.57 \text{ A}$$

Now check this solution by changing the direction of I_2 in Fig. 22-15 and rewriting the voltage equations accordingly, while remembering that now, at junction f, for example, $I + I_2 - I_1 = 0$. This will demonstrate that it is immaterial which way the arrows point, for the signs preceding the current values, when found, determine whether or not the assumed directions are correct. As previously mentioned, however, it must be remembered that going through a resistance in a direction opposite to the current arrow represents a voltage (rise) which must be added, whereas going through a resistance in the direction of the current arrow represents a voltage (drop) which must be subtracted.

Example 5

Figure 22-16 represents a network containing three unequal sources of emf. Find the current flowing in each branch.

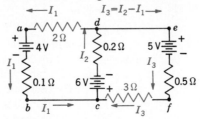

Fig. 22-16 Circuit of Example 5.

Solution

Assume directions for I_1, I_2, and I_3, and label as shown in Fig. 22-16.

Although three unknown currents are involved, they can be reduced to two unknowns by expressing one current in terms of the other two. This is accomplished by applying Kirchhoff's first law to some junction such as c. By considering current flow toward a junction as positive and that flowing away from a junction as negative,

$$I_1 + I_3 - I_2 = 0$$
$$I_3 = I_2 - I_1 \tag{12}$$

Since there are now only two unknown currents I_1 and I_2, Kirchhoff's second law may be applied to any two different closed loops in the network.

For path *abcda*,

$$4 - 0.1I_1 + 6 - 0.2I_2 - 2I_1 = 0$$

Collecting terms,
$$2.1I_1 + 0.2I_2 = 10 \tag{13}$$

For path *efcde*,
$$5 - 0.5(I_2 - I_1) - 3(I_2 - I_1) + 6 - 0.2I_2 = 0$$

Collecting terms,
$$3.5I_1 - 3.7I_2 = -11 \tag{14}$$

Equations (13) and (14) are simultaneous equations that, when solved, result in

$$I_1 = 4.109 \text{ A}$$
and
$$I_2 = 6.860 \text{ A}$$

Substituting in Eq. (12),

$$I_3 = 6.860 - 4.109 = 2.751 \text{ A}$$

The assumed directions of current flow are correct because all values are positive.

The solution can be checked by applying Kirchhoff's second law to a path not previously used. When the current values are substituted in the equation for this path, an identity should result. Thus, for path *adefcba*,

$$2I_1 + 5 - 0.5(I_2 - I_1) - 3(I_2 - I_1) + 0.1I_1 - 4 = 0$$

Collecting terms,

$$5.6I_1 - 3.5I_2 = -1 \qquad (15)$$

The substitution of the numerical values of I_1 and I_2 in Eq. (15) verifies the solution within reasonable limits of accuracy.

22-6 OUTLINE FOR SOLVING NETWORKS

In common with all other problems, the solution of a circuit or a network should not be started until the conditions are analyzed and it is clearly understood what is to be found. Then a definite procedure should be adopted and followed until the solution is completed.

To facilitate solutions of networks by means of Kirchhoff's laws, the following procedure is suggested:

1. Draw a large, neat diagram of the network, and arrange the circuits so that they appear in their simplest form.
2. Letter the diagram with all the known values such as sources of emf, currents, and resistances. Carefully mark the polarities of the known emf's.
3. Assign a symbol to each unknown quantity.
4. Indicate with arrows the assumed direction of current flow in each branch of the network. The number of unknown currents can be reduced by assigning a direction to all but one of the unknown currents at a junction. Then, by Kirchhoff's first law, the remaining current can be expressed in terms of the others.
5. Using Kirchhoff's second law, set up as many equations as there are unknowns to be determined. So that each equation will contain some relation that has not been expressed in another equation, each circuit path followed should cover some part of the circuit not used for other paths.
6. Solve the resulting simultaneous equations for the values of the unknown quantities.
7. Check the values obtained by substituting them in a voltage equation that has been obtained by following a circuit path not previously used.

PROBLEMS 22-3

1. In the circuit of Fig. 22-17, (a) how much current flows through R_3 and (b) how much power is expended in R_2?

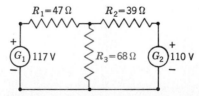

Fig. 22-17 Circuit of Probs. 1 and 2.

2. In the circuit of Fig. 22-17, R_3 becomes short-circuited. (*a*) How much current flows through the short circuit? (*b*) How much power is supplied by generator G_1?

3. In the circuit of Fig. 22-18, (*a*) how much current flows through R and (*b*) how much current flows, and in what direction, through the batteries when R is open-circuited?

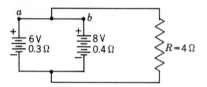

Fig. 22-18 Circuit of Probs. 3 and 4.

4. In the original circuit of Fig. 22-18, R is shunted by a resistor of 1 Ω.
   (*a*) How much power is expended in the shunting resistor?
   (*b*) What is the terminal voltage of the 6-V battery?

5. In the circuit of Fig. 22-19, if the internal resistance of the generator is neglected, (*a*) how much power is being supplied by the generator and (*b*) what is the voltage across R?

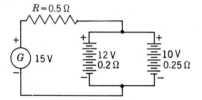

Fig. 22-19 Circuit of Probs. 5 and 6.

6. In the circuit of Fig. 22-19, the generator has an internal resistance of 0.15 Ω. If the connections of the generator are reversed, (*a*) how much power will be dissipated in R and (*b*) what will be the terminal voltage of the 10-V battery?

7. In the circuit of Fig. 22-20, (*a*) how much power is dissipated in R_4 and (*b*) what is the voltage across R_1?

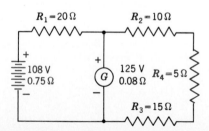

Fig. 22-20 Circuit of Probs. 7 and 8.

8. If R_1 in the circuit of Fig. 22-20 is short-circuited, (*a*) what is the voltage across R_4 and (*b*) how much power is dissipated in the battery?

9. In the circuit of Fig. 22-21, battery *A* has an emf of 114 V and an internal resistance of 1.5 Ω. Battery *B* has an emf of 108 V and an internal resistance of 1 Ω. Each generator has an emf of 122 V and an internal resistance of 0.05 Ω. The resistance of each feeder is 0.02 Ω.

    (*a*) How much current flows through battery *A*?

    (*b*) How much power is expended in battery *B*?

10. In the circuit of Fig. 22-22, (*a*) how much power is expended in R_5 and (*b*) how much power is expended in generator G_2?

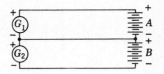

Fig. 22-21 Circuit of Prob. 9.

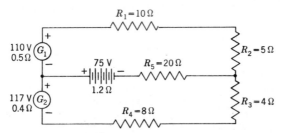

Fig. 22-22 Circuit of Probs. 10 and 11.

11. If the connections of the battery in the circuit of Fig. 22-22 are reversed, (*a*) what is the voltage across R_5 and (*b*) how much power is expended in the entire circuit?

12. Figure 22-23 represents a bank of batteries supplying power to loads R_a and R_b, with R_1, R_2, and R_3 representing the lumped line resistance. R_b is disconnected, and R_a draws 50 A. Neglecting the internal resistance of the generator and batteries, (*a*) what is the voltage across R_2 and (*b*) how much current is flowing in the batteries and in what direction?

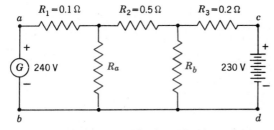

Fig. 22-23 Circuit of Probs. 12, 13, and 14.

13. R_b is connected in the circuit of Fig. 22-23 and draws 75 A. If R_a draws 50 A, (*a*) what is the voltage across R_b and (*b*) how much power is expended in R_2?

14. In the circuit that is shown in Fig. 22-23 the loads are adjusted until R_a draws 150 A and R_b draws 25 A. How much power is lost in R_2?

22-7 EQUIVALENT STAR AND DELTA CIRCUITS

Example 6
Determine the currents through the branches of the network shown in Fig. 22-24 and find the equivalent resistance between points a and c.

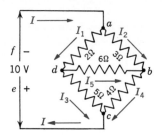

Fig. 22-24 Circuit of Example 6.

Solution
Assume directions for all the currents, and label them on the figure. By Kirchhoff's second law.

Path *efabce*, $\qquad 3I - 3I_1 + 4I - 4I_2 = 10 \qquad$ (16)

Path *efadce*, $\qquad 2I_1 + 5I_2 = 10 \qquad$ (17)

Path *adba*, $\qquad 2I_1 + 6I_1 - 6I_2 - 3I + 3I_1 = 0 \qquad$ (18)

Collecting like terms,

Equation (16) becomes $\qquad 7I - 3I_1 - 4I_2 = 10 \qquad$ (19)

Equation (17) becomes $\qquad 2I_1 + 5I_2 = 10 \qquad$ (20)

Equation (18) becomes $\qquad -3I + 11I_1 - 6I_2 = 0 \qquad$ (21)

Equations (19), (20), and (21) permit us to write

$$I = \frac{\begin{vmatrix} 10 & -3 & -4 \\ 10 & 2 & 5 \\ 0 & 11 & -6 \end{vmatrix}}{\begin{vmatrix} 7 & -3 & -4 \\ 0 & 2 & 5 \\ -3 & 11 & -6 \end{vmatrix}} = 2.879 \text{ A}$$

The equivalent resistance between points a and c is

$$\frac{V}{I} = \frac{10}{2.879} = 3.47 \ \Omega$$

By expressing the branch currents in terms of other currents and labeling the circuit accordingly, this problem can be solved with a smaller number of equations. This is left as an exercise for you.

You will note, from the solution of Example 5, that the solution by Kirchhoff's laws of networks containing such configurations can become complicated. There are many cases, however, in which such networks can be replaced with more convenient equivalent circuits.

The three resistors R_1, R_2, and R_3 in Fig. 22-25a are said to be connected in *delta* (Greek letter Δ). R_a, R_b, and R_c in Fig. 22-25b are connected in *star,* or Y.

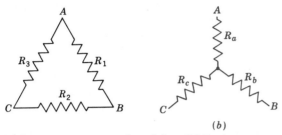

(b)

Fig. 22-25 (a) Resistors connected in delta. (b) Resistors connected in star or Y.

If these two circuits are to be made equivalent, then the resistance between terminals A and B, B and C, and A and C must be the same in each circuit. Hence, in Fig. 22-25a the resistance from A to B is

$$R_{AB} = \frac{R_1(R_2 + R_3)}{R_1 + R_2 + R_3} \tag{22}$$

In Fig. 22-25b the resistance from A to B is

$$R_{AB} = R_a + R_b \tag{23}$$

Equating Eqs. (22) and (23),

$$R_a + R_b = \frac{R_1R_2 + R_1R_3}{R_1 + R_2 + R_3} \tag{24}$$

Similarly,

$$R_b + R_c = \frac{R_1R_2 + R_2R_3}{R_1 + R_2 + R_3} \tag{25}$$

and

$$R_a + R_c = \frac{R_1R_3 + R_2R_3}{R_1 + R_2 + R_3} \tag{26}$$

Equations (24), (25), and (26) are simultaneous and, when solved, result in

$$R_a = \frac{R_1R_3}{R_1 + R_2 + R_3} = \frac{R_1R_3}{\Sigma R_\Delta} \tag{27}$$

$$R_b = \frac{R_1R_2}{R_1 + R_2 + R_3} = \frac{R_1R_2}{\Sigma R_\Delta} \tag{28}$$

and

$$R_c = \frac{R_2R_3}{R_1 + R_2 + R_3} = \frac{R_2R_3}{\Sigma R_\Delta} \tag{29}$$

Since Σ (Greek letter sigma) is used to denote "the summation of,"

$$\Sigma R_\Delta = R_1 + R_2 + R_3$$

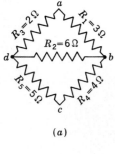

(a)

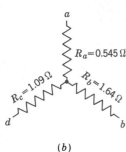

(b)

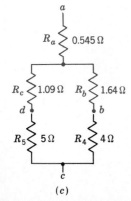

(c)

Fig. 22-26 Circuits of Example 8.

Example 7

In Fig. 22-25a, $R_1 = 2\ \Omega$, $R_2 = 3\ \Omega$, and $R_3 = 5\ \Omega$. What are the values of the resistances in the equivalent Y circuit of Fig. 22-25b?

Solution

$$\Sigma R_\Delta = 2 + 3 + 5 = 10\ \Omega$$

Substituting in Eq. (27), $\quad R_a = \dfrac{2 \times 5}{10} = 1\ \Omega$

Substituting in Eq. (28), $\quad R_b = \dfrac{2 \times 3}{10} = 0.6\ \Omega$

Substituting in Eq. (29), $\quad R_c = \dfrac{3 \times 5}{10} = 1.5\ \Omega$

Example 8

Determine the equivalent resistance between points a and c in the circuit of Fig. 22-26a.

Solution

Convert one of the delta circuits of Fig. 22-26a to its equivalent Y circuit. Thus, for the delta abd,

$$\Sigma R\Delta = 3 + 6 + 2 = 11\ \Omega$$

The equivalent Y resistances, which are shown in Fig. 22-26b, are

$$R_a = \frac{3 \times 2}{11} = 0.545\ \Omega$$

$$R_b = \frac{3 \times 6}{11} = 1.64\ \Omega$$

and $\qquad R_c = \dfrac{2 \times 6}{11} = 1.09\ \Omega$

The equivalent Y circuit is connected to the remainder of the network as shown in Fig. 22-26c and is solved as an ordinary series-parallel combination. Thus,

$$R_{ac} = R_a + \frac{(R_c + R_5)(R_b + R_4)}{R_c + R_5 + R_b + R_4}$$

$$= 0.545 + \frac{(1.09 + 5)(1.64 + 4)}{1.09 + 5 + 1.64 + 4}$$

$$= 3.47\ \Omega$$

Note that the values of Fig. 22-26 are the same as those of Fig. 22-24.

The equations for converting a Y circuit to its equivalent delta circuit are obtained by solving Eqs. (27), (28), and (29) simultaneously. This results in

$$R_1 = \frac{\Sigma R_Y}{R_c} \tag{30}$$

$$R_2 = \frac{\Sigma R_Y}{R_a} \tag{31}$$

$$R_3 = \frac{\Sigma R_Y}{R_b} \tag{32}$$

where $\qquad \Sigma R_Y = R_a R_b + R_b R_c + R_a R_c$

A convenient method for remembering the Δ to Y and Y to Δ conversions is illustrated in Fig. 22-27.

In converting from Δ to Y, each equivalent Y resistance is equal to the product of the two *adjacent* Δ resistances divided by the sum of the Δ resistances. For example, R_1 and R_3 are adjacent to R_a; therefore,

$$R_a = \frac{R_1 R_3}{\Sigma R_\Delta}$$

In converting from Y to Δ, each equivalent Δ resistance is found by dividing ΣR_Y by the *opposite* Y resistance. For example, R_1 is opposite R_c; therefore,

$$R_1 = \frac{\Sigma R_Y}{R_c}$$

Note

A comparison of the Y network of Fig. 22-25b with the network formed by R_1, R_2, and R_3 of Fig. 13-9 will show the common interchangeability of the names T and Y and π (pi) and Δ (delta) in electronics circuitry.

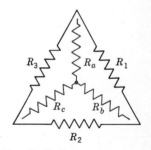

Fig. 22-27 Resistance equivalents.

PROBLEMS 22-4

1. In the Δ circuit of Fig. 22-25a, $R_1 = 12\ \Omega$, $R_2 = 15\ \Omega$, and $R_3 = 18\ \Omega$. Determine the resistances of the equivalent Y circuit.
2. In the Δ circuit of Fig. 22-25a, $R_1 = 120\ \Omega$, $R_2 = 240\ \Omega$, and $R_3 = 300\ \Omega$. Determine the resistances of the equivalent Y circuit.
3. In the π circuit of Fig. 22-25a, $R_1 = R_2 = R_3 = 500\ \Omega$. Determine the resistances of the equivalent T circuit.
4. In the Y circuit of Fig. 22-25b, $R_a = 8\ \Omega$, $R_b = 16\ \Omega$, and $R_c = 40\ \Omega$. Determine the resistances of the equivalent Δ circuit.
5. In the T circuit of Fig. 22-25b, $R_a = 4.7\ k\Omega$, $R_b = 3.3\ k\Omega$, and $R_c = 1.8\ k\Omega$. Determine the resistances of the equivalent π circuit.
6. In the T circuit of Fig. 22-25b, $R_a = R_b = R_c = 1.5\ k\Omega$. Determine the resistances of the equivalent π circuit.

In Probs. 7 to 17, solve the circuits by both the Δ to Y conversion and Kirchhoff's laws:

7. In the circuit of Fig. 22-28, $R_1 = 20\ \Omega$, $R_2 = 10\ \Omega$, $R_3 = 45\ \Omega$, $R_4 = 12\ \Omega$, $R_5 = 15\ \Omega$, and $V = 1.5$ V. What is the value of I?
8. How much current flows through R_5 of Prob. 7?
9. How much current is flowing through R_2 of Prob. 7?
10. In the circuit of Fig. 22-28, $R_1 = 25\ \Omega$, $R_2 = 10\ \Omega$, $R_3 = 15\ \Omega$, $R_4 = 50\ \Omega$, $R_5 = 30\ \Omega$, and $V = 50$ V. What is the value of I?
11. How much current is flowing through R_2 of Prob. 10?
12. In the circuit of Fig. 22-28, $R_1 = ?$, $R_2 = 10\ \Omega$, $R_3 = 15\ \Omega$, $R_4 = 12\ \Omega$, $R_5 = 8\ \Omega$, $V = 32$ V, and $I = 2.39$ A. What is the resistance of R_1?

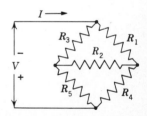

Fig. 22-28 Circuit of Probs. 7 to 12.

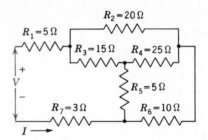

Fig. 22-29 Circuit of Probs. 13, 14, and 15.

13. Determine the value of the current I in Fig. 22-29 if $V = 100$ V.
14. How much current flows through R_4 of Prob. 13?
15. How much current flows through R_5 of Prob. 13?
16. How much current flows through the load resistance R_L in the Fig. 22-30?

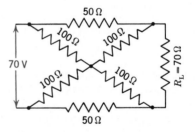

Fig. 22-30 Circuit of Prob. 16.

17. How much current does the signal generator G supply to the circuit of Fig. 22-31?

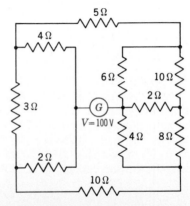

Fig. 22-31 Circuit of Prob. 17.

22-8 THEVENIN AND NORTON EQUIVALENTS

Often a knowledge of the actual components inside a power supply circuit is immaterial so long as we can measure the open-circuit output voltage and the short-circuit output current. From these easy measurements, we can picture a model of the circuit which will behave in exactly the same way as the original so far as any external connected circuit is concerned. Figure 22-32 illustrates this idea.

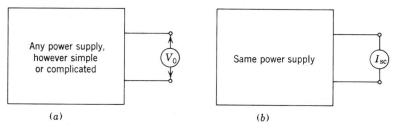

(a)   (b)

Fig. 22-32 (a) Any power supply will deliver a particular open-circuit (no-load) emf V_0, which may be measured by an infinite-resistance voltmeter. (b) Any power supply will deliver a short-circuit (maximum load) current I_{sc}, which may be measured by a zero-resistance ammeter.

A voltmeter with extremely high resistance can make a reasonable measurement of the open-circuit emf which the power supply generates. And an ammeter with very low resistance can make a reasonable measurement of the short-circuit current which the power supply can deliver. Two such models of power supplies are available to us:

Thevenin's theorem suggests that the power supply of Fig. 22-32 can be pictured as consisting of a simple equivalent source of constant emf V_{Th} in series with an equivalent resistance R_{Th}. Figure 22-33 shows the Thevenin equivalent of the circuit of Fig. 22-32. Obviously,

$$V_{Th} = V_0 \tag{33}$$

$$R_{Th} = \frac{V_0}{I_{sc}} \tag{34}$$

$$I_L = \frac{V_{Th}}{R_{Th} + R_L} \tag{35}$$

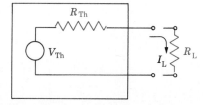

Fig. 22-33 Thevenin's equivalent of power supply of Fig. 22-32; that is, a source of constant emf V_{Th} in series with internal resistance R_{Th}.

Norton's theorem suggests that the power supply of Fig. 22-32 can be pictured as consisting of a simple equivalent source of constant current I_N in parallel with an equivalent resistance R_N. Figure 22-34 shows the Norton equivalent of the circuit of Fig. 22-32. You can see that

$$R_N = R_{Th} \tag{36}$$

$$I_N = \frac{V_{Th}}{R_{Th}} \tag{37}$$

$$I_L = \frac{R_N}{R_N + R_L} \cdot I_N \tag{38}$$

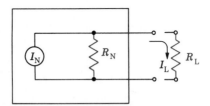

Fig. 22-34 Norton's equivalent of power supply of Fig. 22-32; that is, a source of constant current I_N in parallel with internal resistance R_N.

You should apply your knowledge of parallel resistances in order to prove Eq. (38).

The solution to network problems may sometimes be simplified by applying one or the other of these two theorems.

Example 9

Use Thevenin's theorem to solve the current flow through the load resistor R in the circuit of Fig. 22-14. $V_1 = 6$ V with an internal resistance of 0.15 Ω and $V_2 = 5$ V with an internal resistance of 0.05 Ω; and $R = 2$ Ω.

Solution

Redraw the circuit to show R as the load to be connected and the rest of the circuit as a power supply (Fig. 22-35).

Determine the open-circuit voltage which would appear across the terminals ab of Fig. 22-35. A high-resistance voltmeter would measure

$$V_{Th} = V_0 = V_2 + I_c r_2$$

The circulating current I_c is found by applying Ohm's law to the internal circuit:

$$I_c = \frac{6 - 5}{0.15 + 0.05} = \frac{1}{0.20} = 5 \text{ A}$$

and $\qquad V_{Th} = 5 + 5(0.05) = 5 + 0.25$

$$= 5.25 \text{ V}$$

Determine the circuit resistance which would be seen by an ohmmeter connected to terminals ab with the sources of emf shorted and rep-

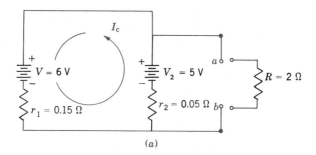

(a)

(b)   *(c)*

Fig. 22-35 (*a*) Redrawn from Fig. 22-14 for Thevenin's solution.
(*b*) Redrawn from Fig. 22-14 for Thevenin's equivalent circuit.
(*c*) Redrawn from Fig. 22-14 for Norton's equivalent circuit.

resented by their internal resistances. Under such circumstances, an ohmmeter would see r_1 and r_2 in parallel:

$$R_{\text{Th}} = \frac{0.15 \times 0.05}{0.15 + 0.05} = 0.0375 \ \Omega$$

Thus, the Thevenin equivalent circuit (Fig. 22-35*b*) is a constant source of 5.25 V in series with 0.0375 Ω. Then the current through the 2-Ω "load" is

$$I_R = \frac{5.25}{2.0375} = 2.58 \text{ A}$$

(Compare with Example 4, Sec. 22-5.)

Example 10
Solve Example 9 by using Norton's theorem.

Solution
As before, determine the equivalent internal resistance of the "power supply" as seen by a connected load:

$$R_N = R_{\text{Th}} = 0.0375 \ \Omega$$

Then determine the current which the "power supply" would drive through a short circuit across terminals *ab*. This may be done by using a Thevenin open-circuit approach and finding

$$I_N = \frac{V_{\text{Th}}}{R_{\text{Th}}} = \frac{5.25}{0.0375} = 140 \text{ A}$$

from which

$$I_R = \frac{0.0375}{2.0375} \times 140 = 2.58 \text{ A}$$

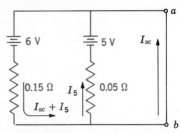

Fig. 22-36 Determination of $I_N = I_{sc}$ for Fig. 22-35.

Alternatively, determine from first principles what the short-circuit current through *ab* would be. Figure 22-36 shows this approach. Using Kirchhoff's laws,

$$0.15(I_{sc} + I_5) + 0.05I_5 = 1$$
$$0.015(I_{sc} + I_5) = 6$$

from which $$I_{sc} = 140 \text{ A}$$
and $$I_R = 2.58 \text{ A}$$

22-9 OUTLINE FOR THEVENIN AND NORTON SOLUTIONS

The following systematic procedure will simplify the utilization of these two circuit simplification theorems:

1. Determine the *leg* of a circuit through which the current flow is to be determined and redraw the circuit, omitting that part.
2. Consider the balance of the circuit to be a power supply whose terminals are eventually to deliver current to the part omitted. Often it is helpful to letter all connecting points in the original circuit to make sure that the equivalent has been drawn correctly.
3. Determine the voltage which would be indicated by a voltmeter connected across the open-circuit terminals of the "power supply." This is V_{Th}.
4. Short-circuit all the internal sources of emf, leaving them represented by their internal resistances, and determine the resistance which would be indicated by an ohmmeter connected across the open-circuit terminals of the power supply. This is $R_{Th} = R_N$.
5. Determine $I_N = \dfrac{V_{Th}}{R_{Th}}$, or
6. Determine the value of I_N as the current which the power supply would drive through an ammeter connected across its terminals.
7. Use Eq. (35) or (38) to determine the current flow through the reconnected "load."

Example 11

Determine the current I_5 through the 6-Ω resistor of Fig. 22-24.

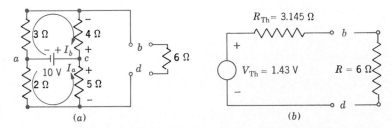

Fig. 22-37 (a) Redrawn from Fig. 22-24 for Thevenin's solution of current I_s through resistor across points bd. (b) Thevenin's equivalent circuit for (a).

Solution

Redraw the circuit. Omit the 6-Ω bridging resistor and let the balance of the circuit be a power supply which will later serve the 6-Ω load (Fig. 22-37).

When terminals bd are open-circuited, the 10-V source will drive currents I_a and I_b through the power supply internal circuitry, thereby producing voltage drops across the 4- and 5-Ω resistors with the polarities indicated:

$$I_a = \frac{10}{7} = 1.43 \text{ A}$$

$$I_b = \frac{10}{7} = 1.43 \text{ A}$$

$$V_4 = 1.43 \times 4 = 5.72 \text{ V}$$

$$V_5 = 1.43 \times 5 = 7.15 \text{ V}$$

A voltmeter across terminals bd will measure

$$V_{Th} = 7.15 - 5.72 = 1.43 \text{ V}$$

When the 10-V internal source is shorted, its internal resistance being zero, an ohmmeter across terminals bd will measure

$$R_{Th} = \frac{3 \times 4}{3 + 4} + \frac{2 \times 5}{2 + 5}$$
$$= 1.715 + 1.43$$
$$= 3.145 \text{ }\Omega$$

Then the Thevenin equivalent to the power supply is a constant 1.43 V in series with 3.145 Ω.

$$I_s = \frac{1.43}{6 + 3.145} = 156 \text{ mA}$$

(Compare with Example 6, Sec. 22-7.)

PROBLEMS 22-5

1. A power supply delivers an open-circuit emf of 120 V. An ammeter that is connected across its terminals measures a short-circuit current of 150 A.
   (a) What is the Thevenin circuit equivalent to the power supply so far as any connected load is concerned?
   (b) What is the Norton equivalent to the power supply?

2. A power supply delivers an open-circuit emf of 6 V. An ammeter that is connected across its terminals measures a short-circuit current of 220 mA.
   (a) What is the Thevenin equivalent circuit?
   (b) What is the Norton equivalent circuit?

3. What is the Thevenin equivalent circuit of the "power supply" portion of the circuit of Fig. 22-28 for the solution of the current through R_2 if $R_1 = 20 \ \Omega$, $R_2 = 10 \ \Omega$, $R_3 = 45 \ \Omega$, $R_4 = 12 \ \Omega$, $R_5 = 15 \ \Omega$, and $V = 1.5$ V? What is the current flow through R_2?

4. What is the Thevenin equivalent circuit of the power supply portion of the circuit of Fig. 22-28 for the solution of the current through R_2 if $R_1 = 25 \ \Omega$, $R_2 = 10 \ \Omega$, $R_3 = 15 \ \Omega$, $R_4 = 50 \ \Omega$, $R_5 = 30 \ \Omega$, and $V = 50$ V? What is the current through R_3?

5. What is the Norton equivalent circuit of the power supply portion of the circuit of Prob. 3 for the solution of the current through R_5? What is the current through R_5?

6. What is the Thevenin equivalent circuit of the power supply portion of the circuit of Prob. 4 for the solution of the current through R_1? What is the current through R_1?

Angles

23

This chapter deals with the study of angles as an introduction to the branch of mathematics called *trigonometry*. The word "trigonometry" is derived from two Greek words meaning "measurement" or "solution" of triangles.

Trigonometry is both algebraic and geometric in nature. It is not confined to the solution of triangles but forms a basis for more advanced subjects in mathematics. A knowledge of the subject paves the way for a clear understanding of ac and related circuits.

23-1 ANGLES

In trigonometry, we are concerned primarily with the many relations that exist among the sides and angles of triangles. In order to understand the meaning and measurement of angles, it is essential that you thoroughly understand these relationships.

An angle is formed whenever two straight lines meet at a point. In Fig. 23-1a, lines *OA* and *OX* meet at the point *O* to form the angle *AOX*. Too, in Fig. 23-1b, the angle *BOX* is formed by lines *OB* and *OX* meeting at the point *O*. This point is called the *vertex* of the angle, and the two lines are called the *sides* of the angle. The size, or magnitude, of an angle is a measure of the difference in directions of the sides. Thus, in Fig. 23-1, angle *BOX* is a larger angle than *AOX*. The lengths of the sides of an angle have no bearing on the size of the angle.

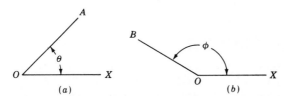

Fig. 23-1 Formation of angles.

In geometry it is customary to denote an angle by the symbol ∠. If this notation is used, "angle *AOX*" would be written ∠*AOX*.

An angle is also denoted by the letter at the vertex or by a supplementary letter placed inside the angle. Thus, angle *AOX* is correctly denoted by ∠*AOX*, ∠*O*, or ∠θ. Also, *BOX* could be written ∠*BOX*, ∠*O*, of ∠φ.

If equal angles are formed when one straight line intersects another, the angles are called *right angles*. In Fig. 23-2, angles *XOY*, φ, *X'OY'*, and α are all right angles.

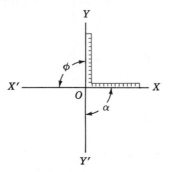

Fig. 23-2 Right angles.

An *acute angle* is an angle that is less than a right angle. For example, in Fig. 23-3a, ∠α is an acute angle.

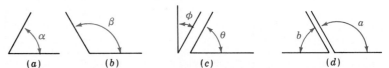

(a) (b) (c) (d)

Fig. 23-3 (a) Acute angle; (b) obtuse angle; (c) complementary angles; (d) supplementary angles.

An *obtuse angle* is an angle that is greater than a right angle. For example, in Fig. 23-3b, ∠β is an obtuse angle.

Two angles whose sum is one right angle are called *complementary angles*. Either one is said to be the *complement* of the other. For example, in Fig. 23-3c, angles φ and θ are complementary angles; φ is the complement of θ, and θ is the complement of φ.

Two angles whose sum is two right angles (a straight line) are called *supplementary angles*. Either one is said to be the *supplement* of the other. Thus, in Fig. 23-3d, angles *b* and *a* are supplementary angles; *b* is the supplement of *a*, and *a* is the supplement of *b*.

23-2 GENERATION OF ANGLES

In the study of trigonometry, it becomes necessary to extend our concept of angles beyond the geometric definitions stated in Sec. 23-1. An angle should be thought of as being generated by a line (line segment or half ray) that starts in a certain initial position and rotates about a point called the *vertex* of the angle until it stops at its final position. The original position of the rotating line is called the *initial side* of the angle, and the final position is called the *terminal side* of the angle.

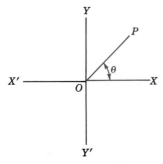

Fig. 23-4 Angle θ in standard position.

An angle is said to be in *standard position* when its vertex is at the origin of a system of rectangular coordinates and its initial side extends in the positive direction along the *x* axis. Thus, in Fig. 23-4, the angle θ is in standard position. The vertex is at the origin, and the initial side is on the positive *x* axis. The angle has been generated by the line *OP* revolving, or sweeping, from *OX* to its final position.

An angle is called a *positive angle* if it is generated by a line revolving counterclockwise. If the generating line revolves clockwise, the angle is called a *negative angle*. In Fig. 23-5, all angles are in standard position. ∠θ is a positive angle that was generated by the line *OM* revolving counterclockwise from *OX*. φ is also a positive angle whose terminal side is *OP*. α is a negative angle that was generated by the line *OQ* revolving in a clockwise direction from the initial side *OX*. β is also a negative angle whose terminal side is *ON*.

If the terminal side of an angle that is in standard position lies in the first quadrant, then that angle is said to be *an angle in the first quadrant*, etc. Thus, θ in Fig. 23-4 and θ in Fig. 23-5 are in the first quadrant. Similarly, in Fig. 23-5, β is in the second quadrant, φ is in the third quadrant, and α is in the fourth quadrant.

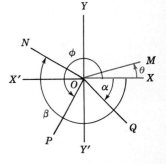

Fig. 23-5 Generation of angles.

23-3 THE SEXAGESIMAL SYSTEM

There are several systems of angular measurement. The three most commonly used are the right angle, the circular (or natural) system, and the sexagesimal system. The right angle is almost always used as a unit of angular measure in plane geometry and is constantly used by builders, surveyors, etc. However, for the purposes of trigonometry, it is an inconvenient unit because of its large size.

The unit most commonly used in trigonometry is the *degree*, which is one-ninetieth of a right angle. The degree is defined as the angle formed by one three hundred sixtieth part of a revolution of the angle-generating line. The degree is divided into 60 equal parts called *minutes,* and the minute into 60 equal parts called *seconds*. The word "sexagesimal" is derived from a Latin word pertaining to the number 60.

Instead of dividing the degrees into minutes and seconds, we shall divide them decimally for convenience. For example, instead of expressing an angle of 43 degrees 36 minutes as 43°36′, we write 43.6°.

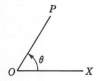

Fig.23-6 Angle to be measured.

The actual measurement of an angle consists in finding how many degrees and a decimal part of a degree there are in the angle. This can be accomplished with a fair degree of accuracy by means of a *protractor,* which is an instrument for measuring or constructing angles.

To measure an angle *XOP,* as in Fig. 23-6, place the center of the protractor indicated by *O* at the vertex of the angle with, say, the line *OX* coinciding with one edge of the protractor as shown in Fig. 23-7. The magnitude of the angle, which is 60°, is indicated where the line *OP* crosses the graduated scale.

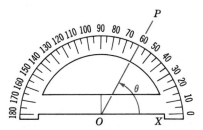

Fig. 23-7 Using protractor to measure angle *XOP* of Fig. 23-6.

To construct an angle, say 30° from a given line *OX,* place the center of the protractor on the vertex *O.* Pivot the protractor about this point until *OX* is on a line with the 0° mark on the scale. In this position, 30° on the scale now marks the terminal side *OP* as shown in Figs. 23-8 and 23-9.

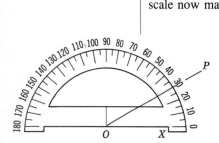

Fig. 23-8 Using protractor to construct angle.

Fig. 23-9 30° angle constructed by protractor in Fig. 23-8.

23-4 ANGLES OF ANY MAGNITUDE

In the study of trigonometry, it will be necessary to extend our concept of angles in order to include angles greater than 360°, either positive or negative. Thinking of an angle being generated, as was explained in Sec. 23-2, permits consideration of angles of any size; for the generating line can rotate from its initial position in a positive or negative direction to produce any size angle, even one greater than 360°. Fig. 23-10 illustrates how an angle of +750° is generated. However, for the purpose of ordinary computation, we consider such an angle to be in the same quadrant as its terminal side, with a magnitude equal to the remainder after the largest multiple of 360° it will contain has been subtracted from

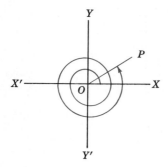

Fig. 23-10 Generation of 750° angle.

it. Thus, in Fig. 23-10, the angle is in the first quadrant and, geometrically, is equal to $750° - 720° = 30°$.

PROBLEMS 23-1

1. What is the complement of (a) 68°, (b) 23°, (c) 41°, (d) 170°, (e) 255°, (f) −10°?
2. What is the supplement of (a) 75°, (b) 153°, (c) 258°, (d) 270°, (e) 350°, (f) −150°?
3. Construct two complementary angles each in standard position on the same pair of axes.
4. Construct two supplementary angles each in standard position on the same pair of axes.
5. By using a protractor, construct the following angles and place them in standard position on rectangular coordinates. Indicate by arrows the direction and amount of rotation necessary to generate these angles: (a) 45°, (b) 160°, (c) 220°, (d) 315°, (e) 405°, (f) −60°, (g) −315°, (h) −300°, (i) −390°, (j) −850°.
6. Through how many degrees does the minute hand of a clock turn in (a) 20 min, (b) 40 min?
7. Through how many right angles does the minute hand of a clock turn from 10:30 A.M. to 5:00 P.M. of the same day?
8. Through how many degrees per minute do (a) the second hand, (b) the minute hand, (c) the hour hand of a clock rotate?
9. A motor armature has a speed of 3600 rev/min. What is the angular velocity (speed) in degrees per second?
10. The shaft of the motor armature in Prob. 9 is directly connected to a pulley 300 mm in diameter. What is the pulley rim speed in meters per second?

23-5 THE CIRCULAR, OR NATURAL, SYSTEM

The circular, or natural, system of angular measurement is sometimes called *radian measure* or *π measure*. The unit of measure is the *radian*. [In this book the abbreviation for *radian* is "rad" when used with units (0.55 rad/s); but an angle of 0.55 radian is written symbolically with a

Roman superscript "r" (0.55ʳ) in order to parallel the use of the degree symbol (288°).]

A radian is an angle that, when placed with its vertex at the center of a circle, intercepts an arc equal in length to the radius of the circle. Thus, in Fig. 23-11, if the length of the arc *AP* equals the radius of the circle, then angle *AOP* is equal to one radian. Figure 23-12 shows a circle divided into radians.

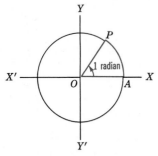

Fig. 23-11 Angle *AOP* = 1ʳ.

Fig. 23-12 Circle divided into 2πʳ.

The circular system of measure is used extensively in electrical and electronics formulas and is almost universally used in the higher branches of mathematics.

From geometry, it is known that the circumference of a circle is given by the relation

$$C = 2\pi r \qquad (1)$$

where *r* is the radius of the circle. Dividing both sides of Eq. (1), by *r*, we have

$$\frac{C}{r} = 2\pi \qquad (2)$$

Now Eq. (2) says simply that the ratio of the circumference to the radius is 2π; that is, the circumference is 2π times longer than the radius. Therefore, a circle must contain 2π radians (2πʳ). Also, since the circumference subtends 360°, it follows that

$$2\pi^r = 360°$$
$$\pi^r = 180°$$

or
$$1^r = \frac{180°}{\pi} = 57.2959° \cong 57.3° \qquad (3)$$

From Eq. (3), the following is evident:

■ To reduce radians to degrees, multiply the number of radians by $\frac{180°}{\pi^r}$ (≅57.3).

■ To reduce degrees to radians, multiply the number of degrees by $\frac{\pi^r}{180°}$ (≅0.017 45).

Example 1

Reduce 1.7^r to degrees.

Solution

$$1^r = 57.3°$$

Hence, $1.7^r = 1.7 \times 57.3 = 97.4°$

Example 2

Convert $15.6°$ to radians.

Solution

$$1° = 0.01745^r$$

Hence, $15.6° = 15.6 \times 0.017\ 45 = 0.272^r$

PROBLEMS 23-2

1. Express the following angles in radians, first in terms of π and second as decimals: (a) $60°$, (b) $120°$, (c) $165°$, (d) $225°$, (e) $285°$, (f) $5°$.

2. Express the following angles in degrees: (a) 2^r, (b) 0.6^r, (c) $\frac{1^r}{\pi}$, (d) $\frac{2\pi^r}{3}$, (e) $\frac{5\pi^r}{6}$, (f) $0.610\ 87^r$.

3. Through how many radians does the second hand of a clock turn between 6:35 A.M. and 9:20 A.M. of the same day?

4. Through how many radians does the hour hand of a clock turn in 40 min?

5. Through how many radians does the minute hand of a clock turn in 1 h 15 min?

6. What is the angular velocity in radians per second of (a) the second hand, (b) the minute hand, (c) the hour hand of a clock?

7. The speed of a rotating switch is 400 rev/min. What is the angular velocity of the switch in radians per second (rad/s)?

8. A radar antenna rotates at 6 rev/min. What is its angular velocity in radians per second?

9. A radar antenna has an angular velocity of π rad/s. What is its speed of rotation in revolutions per minute?

10. What is the approximate angular velocity of the earth in radians per minute (rad/min)?

23-6 GONS AND GRADS

In some parts of Europe, as a stage in furthering decimalization, the right angle is divided into one hundred equal parts known as *grads* (from the German) or, internationally, as *gons* (from the Greek). Each gon (grad) may be subdivided into 100 centigons (centigrads). Sometimes angles measured in this system will be written 20^g, sometimes 20 gon or 20 grad (no *s* for plural). The manner of notation will be a matter for future international agreement. Some calculators now available to electronics technicians offer alternative angle calculations in gons (grads).

$$1 \text{ right angle} = 90° = \frac{\pi^r}{2} = 100^g \tag{4}$$

From Eq. (4), the following is evident:

■ To reduce gons to degrees, multiply the number of gons by $\frac{90°}{100^g}$ (= 0.9).

■ To reduce degrees to gons, multiply the number of degrees by $\frac{100^g}{90°}$ (= 1.111).

■ To reduce gons to radians, multiply the number of gons by $\frac{\pi/2^r}{100^g}$ ($\cong$ 0.0157).

■ To reduce radians to gons, multiply the number of radians by $\frac{100^g}{\pi/2^r}$ ($\cong$ 63.7).

PROBLEMS 23-3

1. Express the following angles in gons: (*a*) 45°, (*b*) 30°, (*c*) 60°, (*d*) 120°, (*e*) 225°, (*f*) 315°.

2. Express the following angles in degrees; (*a*) 50^g, (*b*) 20^g, (*c*) 75^g, (*d*) 150^g, (*e*) 200^g, (*f*) 400^g.

3. Express the following angles in radians: (*a*) 50^g, (*b*) 20^g, (*c*) 75^g, (*d*) 150^g, (*e*) 200^g, (*f*) 400^g.

4. Express the following angles in gons: (*a*) $\frac{\pi^r}{4}$, (*b*) $\frac{\pi^r}{6}$, (*c*) $\frac{2\pi^r}{5}$, (*d*) $\frac{3\pi^r}{2}$, (*e*) $\frac{5\pi^r}{6}$, (*f*) 1.5708^r.

If you have an electronic calculator which gives angles in gons (grads), you may convert angles less than 90° as follows:

■ Enter degrees, 45°; call for sin and read 0.707 11; convert to gons and call for arcsin to read 50^g.

■ Enter gons, 50^g; call for sin and read 0.707 11; convert to rad and call for arcsin to read 0.785 40^r.

23-7 SIMILAR TRIANGLES

Two triangles are said to be *similar* when their corresponding angles are equal. That is, similar triangles are identical in shape but may not be the same in size. The important characteristic of similar triangles is that a direct proportionality exists between corresponding sides. The three triangles of Fig. 23-13 have been so constructed that their corresponding

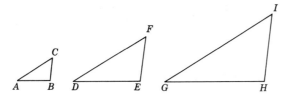

Fig. 23-13 Similar triangles.

angles are equal. Therefore, the three triangles are similar, and their corresponding sides are proportional. This leads to the proportions

$$\frac{AB}{AC} = \frac{DE}{DF} = \frac{GH}{GI}$$

$$\frac{BC}{AB} = \frac{EF}{DE} = \frac{HI}{GH} \qquad \text{etc.}$$

As an example, if $AB = 0.5$ cm, $DE = 1$ cm, and $GH = 1.5$ cm, then DF is twice as long as AC and GI is three times as long as AC. Similarly, HI is three times as long as BC, and EF is twice as long as BC.

The properties of similar triangles are used extensively in measuring distance, such as the distances across bodies of water or other obstructions and the heights of various objects. In addition, the relationship between similar triangles forms the very basis of trigonometry.

Since the sum of the three angles of any triangle is 180°, it follows that if two angles of a triangle are equal to two angles of another triangle, the third angle of one must also be equal to the third angle of the other. Therefore, two triangles are similar if two angles of one are equal to two angles of the other.

If the numerical values of the necessary parts of a triangle are known, the triangle can be drawn to scale by using compasses, protractor, and ruler. The completed figure can then be measured with protractor and ruler to obtain the numerical values of the unknown parts. This is conveniently accomplished on squared paper.

Note

In the following problems the sides and angles of all triangles will be as represented in Fig. 23-14. That is, the angles will be represented by the capital letters A, B, and C and the sides opposite the angles will be correspondingly lettered a, b, and c.

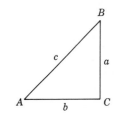

Fig. 23-14 Triangle for Probs. 3 to 10.

PROBLEMS 23-4

1. The sides of a triangular plot are 8, 12, and 16 m. The shortest side of a scale triangle is 3 m. How long are the other two sides of the smaller triangle?
2. Two triangles are similar. The sides of the first are 18, 30, and 36 in. The longest side of the second is 20 mm. How long are the other two sides of the second triangle?

Solve the following triangles by graphical methods:

3. $b = 3, A = 53.1°, C = 90°$

4. $a = 15, b = 20, c = 25$

5. $b = 4, A = 80°, C = 80°$

6. $b = 5, c = 4.75, A = 110°$

7. $a = 10, B = 100°, C = 46.2°$

8. $a = 4.95, c = 7, B = 45°$

9. $a = 15.4, b = 20, C = 29.3°$

10. $a = 35, c = 35, A = 60°$

23-8 THE RIGHT TRIANGLE

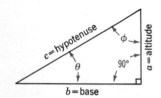

Fig. 23-15 Right triangle.

If one of the angles of a triangle is a right angle, the triangle is called a *right triangle*. Then, since the sum of the angles of any triangle is 180°, a right triangle contains one right angle and two acute angles. Also, the sum of the acute angles must be 90°. This relation enables us to find one acute angle when the other is given. For example, in the right triangle shown in Fig. 23-15, if $\theta = 30°$, then $\phi = 60°$.

Since all right angles are equal, if an acute angle of one right triangle is equal to an acute angle of another right triangle, the two triangles are similar.

The side of a right triangle opposite the right angle is called the *hypotenuse*. Thus, in Fig. 23-15, the side c is the hypotenuse. When a right triangle is in standard position as in Fig. 23-15, the side a is called the *altitude* and the side b is called the *base*.

Another very important property of a right triangle is that the square of the hypotenuse is equal to the sum of the squares of the other two sides. That is,

$$c^2 = a^2 + b^2$$

This relationship provides a means of computing any one of the three sides if two sides are given.

Example 3

A chimney is 40 m high. What is the length of its shadow at a time when a vertical post 2 m high casts a shadow that is 2.1 m long?

Solution

BC in Fig. 23-16 represents the post, and EF represents the stack. Because the rays of the sun strike both chimney and post at the same angle, right triangles ABC and DEF are similar. Then, since

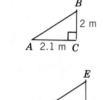

Fig. 23-16 Similar right triangles of Example 3.

$$\frac{DF}{AC} = \frac{EF}{BC}$$

by substituting,

$$\frac{DF}{2.1} = \frac{40}{2}$$

or

$$DF = 42 \text{ m}$$

Example 4

What is the length of a in the triangle of Fig. 23-17?

Solution

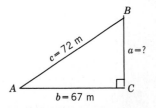

Given	$c^2 = a^2 + b^2$
Transposing,	$a^2 = c^2 - b^2$
$\sqrt{}$,	$a = \sqrt{c^2 - b^2}$
Substituting,	$a = \sqrt{72^2 - 67^2} = \sqrt{695}$
	$a = 26.4$ m

Fig. 23-17 Right triangle of Example 4.

PROBLEMS 23-5

In the following right triangles, solve for the indicated elements:

1. $a = 56$, $b = 15$, $A = 75°$. Find c and B.
2. $a = 24$, $c = 30$, $A = 53.1°$. Find b and B.
3. $b = 78$, $c = 80$, $B = 77°$. Find a and A.
4. An instrument plane flies north at the rate of 650 kn, and a hurricane hunter flies east at 1100 kn. If both planes start from the same place at the same time, how far apart will they be in 2 h?
5. In Fig. 23-18, if $AC = 18$ m, $BC = 24$ m, and $AE = 9$ m, find the length of DE.

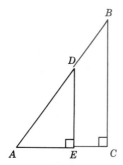

Fig. 23-18 Similar right triangles of Probs. 5, 6, and 7.

6. In Fig. 23-18, if $AD = 30$ cm, $DB = 20$ cm, and $BC = 40$ cm, what is the length of DE?
7. In Fig. 23-18, $AE = 12$ m, $EC = 12$ m, and $AB = 46.5$ m. What is the length of DE?
8. The top of an antenna tower is 40 m above the ground. The tower is to be guyed at a point 6 m below its top to a point on the ground 18 m from the base of the tower. What is the length of the guy?
9. A transmitter antenna tower casts a shadow 270 m long at a time when a meterstick held upright with one end touching the ground casts a shadow 1.8 m long. What is the height of the tower?
10. The tower in Prob. 9 is to be guyed from its top with a 230-m guy wire. How far out from the base of the tower may the guy be anchored?

Trigonometric Functions

24

In the preceding chapter it was shown that plane geometry furnishes two important properties of right triangles. These are

$$A + B = 90°$$

and

$$a^2 + b^2 = c^2$$

The first relation makes it possible to find one acute angle when the other is known. By means of the second relation, any one side can be computed if the other two sides are known. These relations, however, furnish no methods for computing the magnitude of an acute angle when two sides are given. Also, we cannot, by using these relations, compute two sides of a right triangle if the other side and one acute angle are given. If we had only that amount of knowledge, we would be forced to resort to actual measurement by *graphical methods*.

The results obtained by graphical methods are unsuitable for use in many problems, for even with the greatest care and large-scale drawings the degree of accuracy is definitely limited. There is, then, an evident need for certain other relations in which the sides of a right triangle and the angles are united. Such relations form the foundation of trigonometry.

24-1 TRIGONOMETRIC FUNCTIONS ARE RATIOS

In Sec. 23-7 we saw that triangles may be similar regardless of their respective sizes. For example, in Fig. 24-1, the two triangles ABC and DEF are similar, and

$$\frac{AB}{AC} = \frac{DE}{DF} \qquad \frac{BC}{AC} = \frac{EF}{DF} \qquad \text{etc.}$$

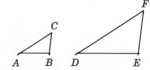

Fig. 24-1 Similar triangles.

Even if one of the pair of similar triangles is tilted (Fig. 24-2), the ratios still hold, since the triangles themselves have not changed in any of their

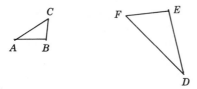

Fig. 24-2 Similar triangles of Fig. 24-1, except triangle *DEF* has been rotated.

dimensions. We may, however, have to look a little harder to see that this is so.

Consider the 30°–60°–90° triangle developed by bisecting an equilateral triangle (Fig. 24-3). First of all, you should confirm that, if the hypotenuse is 2 units long, then the base *AC* will be 1 unit long and the altitude *CB* will be $\sqrt{3}$ units long. Then consider the truth of the following statement:

In the 30°–60°–90° triangle, regardless of its size, the ratio of the base to the hypotenuse will always be 0.5000.

You should draw several 30°–60°–90° triangles of different sizes and prove to your complete satisfaction that this statement *must* always be true.

If the triangle were now rotated so that the side *CB* were the base and *AC* the altitude, the above statement would have to be adjusted. Therefore, we should rename the parts of the triangle so that there can be no possibility of misunderstanding a statement about it. The most convenient way to refer to a side of a triangle is to relate the side to the angles in the triangle. For instance, the hypotenuse is always the longest side, it is always opposite the right angle, and it is always adjacent to (forms) each of the other two angles. We can always refer to it as simply the hypotenuse without introducing any possibility of being misunderstood.

In the 30°–60°–90° triangle with which we are dealing, the side *AC* is always the side *opposite* the 30° angle, and it is always the side *adjacent* to the 60° angle, regardless of the letter designation given it or the orientation of the triangle.

Similarly, the side *CB* is always opposite the 60° angle, and it is always adjacent to the 30° angle, regardless of the symbols used to identify the side or how the triangle is tilted. These side-angle relationships are illustrated in Fig. 24-4. They must be memorized, because they will be used continuously henceforth.

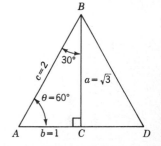

Fig. 24-3 Equilateral triangle divided into two equal 30°–60°–90° right triangles.

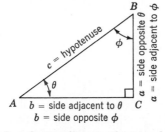

Fig. 24-4 Side-angle relationships in the standard right triangle.

For the rest of this chapter and the next, we shall be dealing only with right triangles. The hypotenuse is always the longest side and is opposite the right angle. The other two sides will be designated according to their relationships to the acute angles.

You should immediately confirm, while using sketches as required, the truth of the following statements relating to the sides of the 30°–60°–90° triangle, first as they apply to the 30° angle and then as they apply to the 60° angle:

1. In the 30°–60°–90° triangle, regardless of its size or orientation, the ratio of the side opposite the 30° angle to the hypotenuse will always be 0.5000.
2. In the 30°–60°–90° triangle, regardless of its size or orientation, the ratio of the side adjacent to the 30° angle to the hypotenuse will always be 0.866.
3. In the 30°–60°–90° triangle, regardless of its size or orientation, the ratio of the side opposite the 30° angle to the side adjacent to the 30° angle will always be 0.577.
4. In the 30°–60°–90° triangle, regardless of its size or orientation, the ratio of the side opposite the 60° angle to the hypotenuse will always be 0.866.
5. In the 30°–60°–90° triangle, regardless of its size or orientation, the ratio of the side adjacent to the 60° angle to the hypotenuse will always be 0.5000.
6. In the 30°–60°–90° triangle, regardless of its size or orientation, the ratio of the side opposite the 60° angle to the side adjacent to the 60° angle will always be 1.732.

It is left as an exercise for you to develop the three similar statements for the 45°–45°–90° triangle. (Why only three statements?)

Now, student, stop and look at these statements. See what they really mean. Make sure that their message is plain. When you fully understand the import of the relationships between sides of triangles, you will have trigonometry in the palm of your hand forever. We do not say that all of trigonometry is simple. But to grasp quickly the fact that the trigonometric functions are merely ratios of sides of triangles is to resolve most of the difficulties which stand in the way of students who have never properly understood how simple the functions of trigonometry really are.

The word "trigonometry" just means "measurement of triangles," and one of the most useful tools in the measurement of triangles is the ratios of sides.

"In the triangle, regardless of its size or orientation" means that, so long as the angles made by the sides are specified, the triangle itself may be formed by:

1. Three lines on a piece of paper
2. A ladder, the ground, and the wall of a house
3. An antenna mast, its shadow on the ground, and the line of sight from the end of the shadow to the top of the mast
4. The lines of sight between two surveyors and a distant landmark
5. A mast, a guy wire, and the ground between the foot of the mast and the guy anchor

6. Any other system which uses three straight lines to form three enclosed angles

The entire statement, "In the . . . triangle . . . will always be . . ." is quite a mouthful, far too lengthy for convenience, and it is often abbreviated. For instance, statement 1 above becomes

$$\frac{\text{opp } 30°}{\text{hyp}} = 0.500$$

or

$$\frac{\text{opp}}{\text{hyp}} \, 30° = 0.500$$

and all the other parts of the statement are understood to apply. Statement 2 becomes

$$\frac{\text{adj}}{\text{hyp}} \, 30° = 0.866$$

and statement 3 becomes

$$\frac{\text{opp}}{\text{adj}} \, 30° = 0.577$$

You should now write similar abbreviations for statements 4, 5, and 6 and check your work for the 45°–45°–90° triangle to show your own statements 7, 8, and 9 may be written

$$\frac{\text{opp}}{\text{hyp}} \, 45° = 0.7071$$

$$\frac{\text{adj}}{\text{hyp}} \, 45° = 0.7071$$

$$\frac{\text{opp}}{\text{adj}} \, 45° = 1.000$$

Example 1

A triangular piece of farm land is to be used as an "antenna farm." It is in the shape of a 30°–60°–90° triangle the shortest side of which is 600 m long (Fig. 24-5). What are the dimensions of the other two sides?

Solution

By using the ratios which have been discovered above and drawing a sketch of the triangle to show the relationships between the sides and angles, we find that the 600-m side must be adjacent to the 60° angle. Then we have

$$\frac{600}{\text{hyp}} \, 60° = 0.500$$

from which

$$\text{hyp} = \frac{600}{0.5} = 1200 \text{ m}$$

and

$$\frac{600}{\text{adj}} \, 30° = 0.577$$

from which

$$\text{adj } 30° = \frac{600}{0.577} = 1040 \text{ m}$$

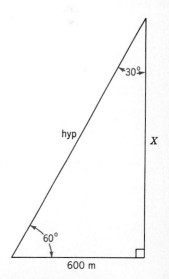

Fig. 24-5 Triangle of Example 1.

Even these abbreviations are more than we require for everyday use, and we now introduce the proper *trigonometric names* for the different ratios (*functions*). θ is the "general angle," just as x is the "general number."

1. The ratio $\dfrac{\text{opp}}{\text{hyp}}$ θ is properly called sine θ, abbreviated to sin θ.

2. The ratio $\dfrac{\text{adj}}{\text{hyp}}$ θ is properly called cosine θ, abbreviated to cos θ.

3. The ratio $\dfrac{\text{opp}}{\text{adj}}$ θ is properly called tangent θ, abbreviated to tan θ.

It must be clearly understood that the names sine, cosine, and tangent are meaningless in themselves; you must relate them to angles of triangles. To say simple "cosine" means nothing. But "cos 60°" means, very specifically, the ratio of the side adjacent to the 60° angle of a 30°–60°–90° triangle to the hypotenuse of the same triangle.

In the general triangle, Fig. 24-6, it will be seen that there exist *six* possible trigonometric functions. Three of them we have already discovered, and the others are reciprocals of those three.

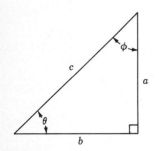

Fig. 24-6 Standard right triangle, as used in electronics problems.

$$\frac{\text{opp}}{\text{hyp}} \theta = \sin \theta = \frac{a}{c}$$

$$\frac{\text{adj}}{\text{hyp}} \theta = \cos \theta = \frac{b}{c}$$

$$\frac{\text{opp}}{\text{adj}} \theta = \tan \theta = \frac{a}{b}$$

$$\frac{\text{hyp}}{\text{opp}} \theta = \text{cosecant } \theta \ = \csc \theta = \frac{c}{a}$$

$$\frac{\text{hyp}}{\text{adj}} \theta = \text{secant } \theta \quad = \sec \theta = \frac{c}{b}$$

$$\frac{\text{adj}}{\text{opp}} \theta = \text{cotangent } \theta = \cot \theta = \frac{b}{a}$$

The cosecant, secant, and cotangent should always be thought of as the reciprocals of the sine, cosine, and tangent, respectively. This is shown easily by considering the reciprocal of sin θ:

$$\frac{1}{\sin \theta} = \frac{1}{\dfrac{a}{c}} = \csc \theta$$

You should confirm the other two reciprocal functions.

These definitions should be memorized so thoroughly that you can tell instantly any ratio of either acute angle of a right triangle, regardless of its position.

The sine, cosine, and tangent are the ratios most frequently used in practical work. If they are carefully learned, the others are easily remembered because they are reciprocals.

The fact that the numerical value of any one of the trigonometric functions (ratios) depends only upon the magnitude of the angle θ is of fundamental importance. This is established by considering Fig. 24-7. There, the angle θ is generated by the line *AD* revolving about the point *A*. From the points *B*, *B'*, and *B''*, perpendiculars are let fall to the initial line, or adjacent side, *AX*. These form similar triangles *ABC*, *AB'C'*, and *AB''C''* because all are right triangles that have a common acute angle θ (Sec. 23-8). Hence,

$$\frac{BC}{AB} = \frac{B'C'}{AB'} = \frac{B''C''}{AB''}$$

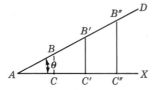

Fig. 24-7 The values of the functions depend only on the size of the angle.

Each of these ratios defines the sine of θ. Similarly, it can be shown that this property is true for each of the other functions. Therefore, the size of the right triangle is immaterial, for only the *relative* lengths of the sides are of importance.

Each one of the six ratios will change in value whenever the angle changes in magnitude. Thus, it is evident that the ratios are really functions of the angle under consideration. If the angle is considered to be the independent variable, then the six functions (ratios) and the relative lengths of the sides of the triangles are dependent variables.

Example 2

Calculate the functions of the angle θ in the right triangle shown in Fig. 24-6 if $a = 6$ mm and $c = 10$ mm.

Solution

Since $c^2 = a^2 + b^2$,

then
$$b = \sqrt{c^2 - a^2} = \sqrt{100 - 36}$$
$$= \sqrt{64} = 8 \text{ mm}$$

Applying the definitions of the six functions,

$$\sin\theta = \tfrac{6}{10} = \tfrac{3}{5} \qquad \cos\theta = \tfrac{8}{10} = \tfrac{4}{5}$$

$$\tan\theta = \tfrac{6}{8} = \tfrac{3}{4} \qquad \cot\theta = \tfrac{8}{6} = \tfrac{4}{3}$$

$$\sec\theta = \tfrac{10}{8} = \tfrac{5}{4} \qquad \csc\theta = \tfrac{10}{6} = \tfrac{5}{3}$$

What would be the values of the above functions if $a = 6$ m, $b = 8$ m, and $c = 10$ m?

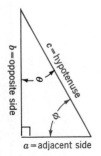

Fig. 24-8 Right triangle for determining functions of angle φ.

24-2 FUNCTIONS OF COMPLEMENTARY ANGLES

By applying the definitions of the six functions to the angle φ given in Fig. 24-8 and noting the positions of the adjacent and opposite sides for this angle, we obtain

$$\sin \phi = \frac{\text{opp}}{\text{hyp}} = \frac{b}{c} \qquad \csc \phi = \frac{\text{hyp}}{\text{opp}} = \frac{c}{b}$$

$$\cos \phi = \frac{\text{adj}}{\text{hyp}} = \frac{a}{c} \qquad \sec \phi = \frac{\text{hyp}}{\text{adj}} = \frac{c}{a}$$

$$\tan \phi = \frac{\text{opp}}{\text{adj}} = \frac{b}{a} \qquad \cot \phi = \frac{\text{adj}}{\text{opp}} = \frac{a}{b}$$

Upon comparing these with the original definitions given for the triangle of Fig. 24-2, we find the following relations:

$$\sin \phi = \cos \theta \qquad \cos \phi = \sin \theta$$
$$\tan \phi = \cot \theta \qquad \cot \phi = \tan \theta$$
$$\sec \phi = \csc \theta \qquad \csc \phi = \sec \theta$$

Since $\phi = 90° - \theta$, the above relations can be written

$$\sin (90° - \theta) = \cos \theta \qquad \cos (90° - \theta) = \sin \theta$$
$$\tan (90° - \theta) = \cot \theta \qquad \cot (90° - \theta) = \tan \theta$$
$$\sec (90° - \theta) = \csc \theta \qquad \csc (90° - \theta) = \sec \theta$$

The above can be stated in words as follows: *A function of an acute angle is equal to the cofunction of its complementary angle.* This enables us to find the function of every acute angle greater than 45° if we know the functions of all angles less than 45°. For example, sin 56° = cos 34°, tan 63° = cot 27°, cos 70° = sin 20°, etc.

24-3 CONSTRUCTION OF AN ANGLE WHEN ONE FUNCTION IS GIVEN

When the trigonometric function of an acute angle is given, the angle can be constructed geometrically by using the definition for the given function. Also, the magnitude of the resulting angle can be measured by the use of a protractor.

Example 3

Construct the acute angle whose tangent is $\frac{9}{10}$.

Solution

Erect perpendicular lines AC and BC, preferably on cross-sectional paper. Measure off 10 units along AC and 9 units along BC. Join A and B and thus form the right triangle ABC. Since tan $A = \frac{9}{10}$; A is the required angle. Measuring A with a protractor shows it to be an angle of approximately 42°. See the construction in Fig. 24-9.

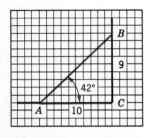

Fig. 24-9 Construction of acute angle whose tangent is $\frac{9}{10}$.

Example 4

Find by construction the acute angle whose cosine is $\frac{3}{4}$.

Solution

Erect perpendicular lines AC and BC. Measure off three units along AC. (Let three divisions of the cross-sectional paper be equal to one unit for greater accuracy.) With A as a center and with a radius of 4 units, draw an arc to intersect the perpendicular at B. Connect A and B. $\cos A = \frac{3}{4}$; therefore A is the required angle. Measuring A with a protractor shows it to be an angle of approximately $41.4°$. The construction is shown in Fig. 24-10.

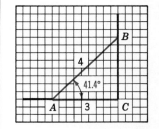

Fig. 24-10 Construction of acute angle whose cosine is $\frac{3}{4}$.

PROBLEMS 24-1

1. In Fig. 24-11, what are the values of the trigonometric functions for the angles θ and ϕ in terms of ratios of the sides, a, b, and c?

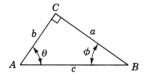

Fig. 24-11 Right triangle of Prob. 1.

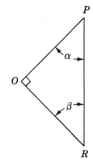

Fig. 24-12 Right triangle of Probs. 2 and 3.

2. As shown in Fig. 24-12, (a) $\sin \alpha = $? (b) $\sin \beta = $? (c) $\cot \beta = $? (d) $\sec \alpha = $? (e) $\tan \alpha = $?

3. In Fig. 24-12, (a) $\dfrac{OP}{OR} = $ tan? (b) $\dfrac{PR}{PO} = $ sec? (c) $\dfrac{OR}{PR} = $ cos? (d) $\dfrac{OP}{RP} = $ sin? (e) $\dfrac{PR}{RO} = $ csc?

4. The three sides of a right traingle are 5, 12, and 13. Let α be the acute angle opposite the side 5 and let β be the other acute angle. Write the six functions of α and β.

5. In Fig. 24-13, if $X = R$, find the six functions of θ.

6. In Fig. 24-13, if $R = \frac{1}{2}Z$, find the sine, cosine, and tangent of θ.

7. In Fig. 24-13, if $X = 2R$, find the sine, cosine, and tangent of θ.

8. (a) $\sin \theta = \frac{2}{3}$, $\csc \theta = $? (b) $\sec \alpha = 2$, $\cos \alpha = $? (c) $\cot \beta = \frac{7}{8}$, $\tan \beta = $? (d) $\cos \phi = \frac{5}{16}$, $\sec \phi = $? (e) $\tan \phi = 12$, $\cot \phi = $? (f) $\csc \alpha = 4$, $\sin \alpha = $?

9. The three sides of a right triangle are 6, 8, and 10. Write the six functions of the largest acute angle.

10. Write the other functions of an acute angle whose cosine is $\frac{4}{5}$.

11. In a right triangle, $c = 5$ cm and $\cos A = \frac{4}{5}$. Construct the triangle, and write the functions of the angle B.

12. State which of these is greater if $\theta \neq 0°$ and is less than $90°$: (a) $\sin \theta$ or $\tan \theta$, (b) $\cos \theta$ or $\cot \theta$, (c) $\sec \theta$ or $\tan \theta$, (d) $\csc \theta$ or $\cot \theta$.

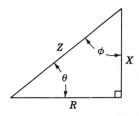

Fig. 24-13 Right triangle of Probs. 5, 6, and 7.

24-4 FUNCTIONS OF ANY ANGLE

The notion of trigonometric functions has been introduced from the point of view of right triangles because this allows for an easy introduction which most students can follow with assurance. However, the total concept applies to far more than just right triangles and to far more than angles between 0° and 90°. In Chap. 27 we shall investigate a few interesting and useful relationships in nonright triangles. For the moment, we will concentrate on the trigonometric functions of any angle.

In Chap. 23 we found the concepts of angles were extended to include angles in any quadrant and both positive and negative angles. In the Fig. 24-14 the line *r* revolves about the origin of the rectangular coordinate system in a counterclockwise (positive) direction. This line, which generates the angle θ, is known as the *radius vector*. The initial side of θ is the positive *x* axis, and the terminal side is the radius vector. If a perpendicular is let fall from any point *P* along the radius vector, in any of the quadrants, a right triangle *xyr* will be formed with *r* as a hypotenuse of constant unit length and with *x* and *y* having lengths equal to the respective coordinates of *P*.

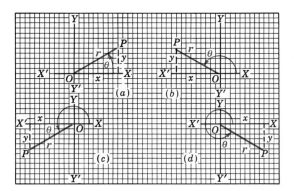

Fig. 24-14 Radius vector **r** generating angles.

We then define the trigonometric functions of θ as follows:

$$\sin \theta = \frac{y}{r} = \frac{\text{ordinate}}{\text{radius}} \qquad \csc \theta = \frac{r}{y} = \frac{\text{radius}}{\text{ordinate}}$$

$$\cos \theta = \frac{x}{r} = \frac{\text{abscissa}}{\text{radius}} \qquad \sec \theta = \frac{r}{x} = \frac{\text{radius}}{\text{abscissa}}$$

$$\tan \theta = \frac{y}{x} = \frac{\text{ordinate}}{\text{abscissa}} \qquad \cot \theta = \frac{x}{y} = \frac{\text{abscissa}}{\text{ordinate}}$$

Since the values of the six trigonometric functions are entirely independent of the position of the point *P* along the radius vector, it follows that they depend only upon the position of the radius vector, or the size of the angle. Therefore, for every angle there is one, and only one, value of each function.

24-5 SIGNS OF THE FUNCTIONS

The signs of the functions of angles in various quadrants are very important. If you remember the signs of the abscissas (x values) and the ordinates (y values) in the four quadrants, you will encounter no trouble.

For angles in the first quadrant, as shown in Fig. 24-14a, the x and y values are positive. Since the length of the radius vector r is always considered positive, it is evident that all functions of angles in the first quadrant are positive. For angles in the second quadrant, as in Fig. 24-14b, the x values are negative values and the y values are positive. Therefore, the sine and its reciprocal are positive and the other four functions are negative. Similarly, the signs of all the functions can be checked from their definitions as given in the preceding section. You should verify each part of Table 24-1.

TABLE 24-1

Quadrant	$\sin \theta$	$\cos \theta$	$\tan \theta$	$\cot \theta$	$\sec \theta$	$\csc \theta$
I	+	+	+	+	+	+
II	+	−	−	−	−	+
III	−	−	+	+	−	−
IV	−	+	−	−	+	−

If the proper signs for the sine and cosine are fixed in mind, the other signs will be remembered because of an important relationship.

$$\frac{\sin \theta}{\cos \theta} = \frac{\dfrac{y}{r}}{\dfrac{x}{r}} = \frac{y}{x}$$

Since
$$\tan \theta = \frac{y}{x}$$

then
$$\frac{\sin \theta}{\cos \theta} = \tan \theta$$

If the sine and the cosine have like signs, the tangent is positive, and if they have unlike signs, the tangent is negative. Because the signs of the sine, cosine, and tangent always agree with signs of the respective reciprocals, the cosecant, secant, and cotangent, the signs for the latter are obtainable from the signs of the sine and cosine as outlined above. Fig. 24-15 will serve as an aid in remembering the signs.

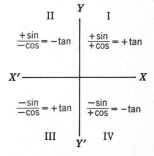

Fig. 24-15 Signs of functions in quadrants.

PROBLEMS 24-2

In what quadrant or quadrants is θ for each of the following conditions?

1. $\sin \theta$ is positive.
2. $\cos \theta$ is positive.
3. $\sin \theta$ is negative.
4. $\tan \theta$ is negative.
5. $\cos \theta$ is negative
6. $\sin \theta$ is positive, $\cos \theta$ negative.
7. $\tan \theta$ and $\sin \theta$ both positive.
8. $\cot \theta$ negative, $\cos \theta$ negative.

9. tan θ negative, cos θ positive.
10. All functions of θ are positive.
11. tan θ = 6
12. cos θ = $-\frac{3}{4}$
13. Is there an angle whose cosine is negative and whose secant is positive?
14. When tan θ = $\frac{3}{4}$, find the value of

$$\frac{\sin \theta - \csc \theta}{\cot \theta - \sec \theta}$$

Give the signs of the sine, cosine, and tangent of each of the following angles:

15. 32°
16. 210°
17. 98°
18. 350°
19. −175°
20. $\frac{\pi^r}{3}$
21. $\frac{-3\pi^r}{4}$
22. −72°
23. 780°

Find the value of the radius vector *r* for each of the following positions of *P*, and then find the trigonometric functions of the angle θ (∠*XOP*). Keep answers in fractional form.

24. (−9, 12)

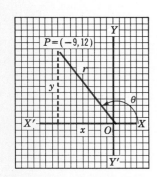

Fig. 24-16 Diagram of Prob. 24.

Solution: Draw the radius vector *r* from *O* to *P* = (−9, 12) as shown in Fig. 24-16. Hence, θ is an angle in the second quadrant with a side adjacent that has an *x* value of −9 and a side opposite that has a *y* value of 12.

Then

$$r = \sqrt{x^2 + y^2} = \sqrt{(-9)^2 + (12)^2} = 15$$

Hence, by definition,

$$\sin \theta = \frac{y}{r} = \frac{12}{15} = \frac{4}{5} \qquad \csc \theta = \frac{r}{y} = \frac{15}{12} = \frac{5}{4}$$

$$\cos \theta = \frac{x}{r} = \frac{-9}{15} = -\frac{3}{5} \qquad \sec \theta = \frac{r}{x} = \frac{15}{-9} = -\frac{5}{3}$$

$$\tan \theta = \frac{y}{x} = \frac{12}{-9} = -\frac{4}{3} \qquad \cot \theta = \frac{x}{y} = \frac{-9}{12} = -\frac{3}{4}$$

25. (12, −5)

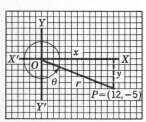

Fig. 24-17 Diagram of Prob. 25.

Solution: Draw the radius vector *r* from *O* to *P* as in Fig. 24-17. θ is an angle in the fourth quadrant with a side adjacent that has an *x* value of 12 and a side opposite that has a *y* value of −5.

Then

$$r = \sqrt{x^2 + y^2} = \sqrt{12^2 + (-5)^2} = 13$$

Hence, by definition,

$$\sin \theta = \frac{y}{r} = -\frac{5}{13} \qquad \csc \theta = \frac{r}{y} = -\frac{13}{5}$$

$$\cos \theta = \frac{x}{r} = \frac{12}{13} \qquad \sec \theta = \frac{r}{x} = \frac{13}{12}$$

$$\tan \theta = \frac{y}{x} = -\frac{5}{12} \qquad \cot \theta = \frac{x}{y} = -\frac{12}{5}$$

26. (3, 4)   27. (12, 5)   28. (−3, 4)

29. (−4, −5)   30. (3, 3)   31. (4, −3)

32. (−8, 6)   33. (−5, −3)   34. (8, 8)

24-6 COMPUTATION OF THE FUNCTIONS

In Sec. 24-1 we developed the functions of 30°, 45°, and 60° by merely using simple notions about right triangles. These angles are very important and will be used often, so they and their trigonometric functions are worthy of the time you spend in this development. At the same time, their use will make it easy for some students to quickly relearn trigonometry a few years hence if their work has been such that they haven't required it immediately. In Chap. 25 we will extend our notions of trigonometric functions and develop and use the tables prepared by expert mathematicians for our use and convenience.

24-7 FUNCTIONS OF 0°

For an angle of 0°, both the initial and terminal sides are on OX. At any distance a from O, choose the point P as shown in Fig. 24-18. Then the coordinates of P are $(a, 0)$. that is, the x value is equal to a units, and the y value is zero. Since the radius vector r is equal to a, by definition,

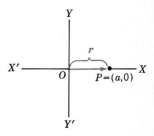

Fig. 24-18 $\theta = 0°$, $x = a$, and $y = 0$.

$$\sin 0° = \frac{y}{r} = \frac{0}{r} = 0 \qquad \csc 0° = \frac{r}{y} = \frac{a}{0} = \infty$$

$$\cos 0° = \frac{x}{r} = \frac{a}{a} = 1 \qquad \sec 0° = \frac{r}{x} = \frac{a}{a} = 1$$

$$\tan 0° = \frac{y}{x} = \frac{0}{a} = 0 \qquad \cot 0° = \frac{x}{y} = \frac{a}{0} = \infty$$

By $\frac{a}{0} = \infty$ is meant the value of $\frac{a}{y}$ as y approaches zero without limit.

Thus, as y gets nearer and nearer to zero, $\frac{a}{y}$ gets larger and larger. Therefore,

$\frac{a}{y}$ is said to *approach* infinity as y approaches zero. However, $\frac{a}{0}$ does not actually result in a quotient of infinity, for division by zero is meaningless.

Determining the functions of 90°, 180°, and 270° is accomplished by the same method as that used for 0°. This is left as an exercise for you.

24-8 THE RANGES OF THE FUNCTIONS

As the radius vector **r** starts from *OX* and revolves about the origin in a positive (counterclockwise) direction, the angle θ is generated and varies in magnitude continuously from 0° to 360° through the four quadrants. Figure 24-19 illustrates the manner in which the sine, cosine, and tangent vary as the angle θ changes in value.

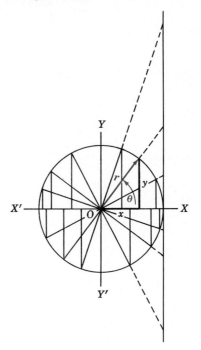

Fig. 24-19 Lengths of lines showing the ranges of sin θ, cos θ, and tan θ.

Quadrant I. As θ increases from 0° to 90°,

■ *x* is positive and decreases from **r** to 0.
■ *y* is positive and increases from 0 to **r**.

Therefore,

$$\sin \theta = \frac{y}{r} \text{ is } \textit{positive} \text{ and increases from 0 to 1.}$$

$$\cos \theta = \frac{x}{r} \text{ is } \textit{positive} \text{ and decreases from 1 to 0.}$$

$$\tan \theta = \frac{y}{x} \text{ is } \textit{positive} \text{ and increases from 0 to } \infty.$$

Quadrant II. As θ increases from 90° to 180°,

■ x is negative and increases from 0 to $-r$.
■ y is positive and decreases from r to 0.

Therefore,

$\sin \theta = \dfrac{y}{r}$ is *positive* and decreases from 1 to 0.

$\cos \theta = \dfrac{x}{r}$ is *negative* and increases from 0 to -1.

$\tan \theta = \dfrac{y}{x}$ is *negative* and decreases from $-\infty$ to 0.

Quadrant III. As θ increases from 180° to 270°,

■ x is negative and decreases from $-r$ to 0.
■ y is negative and increases from 0 to $-r$.

Therefore,

$\sin \theta = \dfrac{y}{r}$ is *negative* and increases from 0 to -1.

$\cos \theta = \dfrac{x}{r}$ is *negative* and decreases from -1 to 0.

$\tan \theta = \dfrac{y}{x}$ is *positive* and increases from 0 to ∞.

Quadrant IV. As θ increases from 270° to 360°,

■ x is positive and increases from 0 to r.
■ y is negative and decreases from $-r$ to 0.

Therefore,

$\sin \theta = \dfrac{y}{r}$ is *negative* and decreases from -1 to 0.

$\cos \theta = \dfrac{x}{r}$ is *positive* and increases from 0 to 1.

$\tan \theta = \dfrac{y}{x}$ is *negative* and decreases from $-\infty$ to 0.

Students often become confused in comparing the variations of the functions, when represented as lines, with the actual numerical values of the functions. For example, in quadrant II as the angle θ increases from 90 to 180°, we say that cos θ increases from 0 to $-r$. Actually, the abscissa representing the cosine is getting *longer*; confusion results from not remembering that a negative number is always greater than zero in the defined negative direction. The *lengths* of the lines representing the functions, when compared with the radius vector, indicate only the *magnitude* of the function. The positions of the lines, with respect to the *x* or *y* axis, specify the signs of the functions.

24-9 LINE REPRESENTATION OF THE FUNCTIONS

By representing the functions as lengths of lines, we are able to obtain a mental picture of the manner in which the functions vary as the radius vector *r* revolves and generates angles. Since we are primarily concerned with the sine, cosine, and tangent, only these functions will be represented graphically.

In Fig. 24-20 the radius vector *r*, with a length of one unit, is revolving about the origin and generating the angle θ. Then, in each of the four quadrants,

$$\sin \theta = \frac{BC}{r} = \frac{BC}{1} = BC$$

and

$$\cos \theta = \frac{OC}{r} = \frac{OC}{1} = OC$$

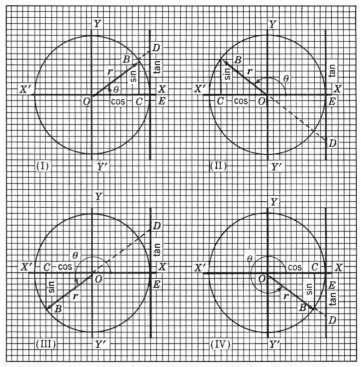

Fig. 24-20 Line representation of functions.

It is evident *that the sine of an angle can be represented by the ordinate (y value) of any point where the end of the radius vector coincides with the circumference of the circle.* Hence, the length BC represents sin θ in all quadrants, as shown in Fig. 24-20. Note that the ordinate gives both the sign and the magnitude of the sine in any quadrant. Thus, in quadrants I and II, sin θ = +0.6; in quadrants III and IV, sin θ = −0.6. That is, when the radius vector is above the x axis, the ordinate and therefore the sine are positive. When the radius vector is below the x axis, the ordinate and therefore the sine are negative.

Similarly, *the cosine of an angle can be represented by the abscissa (x value) of any point where the end of the radius vector coincides with the circumference of the circle.* Hence, the length OC represents cos θ in all quadrants, as shown in Fig. 24-20. The abscissa gives both the sign and the magnitude of the cosine in any quadrant. Thus, in quadrants I and IV, cos θ = +0.8; in quadrants II and III, cos θ = −0.8. That is, when the radius vector is to the right of the y axis, the abscissa and therefore the cosine are positive. When the radius vector is to the left of the y axis, the abscissa, and therefore the cosine, are negative.

In Fig. 24-20, the radius vector has been extended to intersect the line DE, which has been drawn tangent to the circle at the positive x axis. Since by construction, DE is perpendicular to OX, OBC and ODE are similar right triangles, for they have a common acute angle BOC. From the similar triangles,

$$\frac{BC}{OC} = \frac{DE}{OE}$$

Then, in each of the four quadrants,

$$\tan \theta = \frac{BC}{OC} = \frac{DE}{OE} = \frac{DE}{1} = DE$$

From the above, it is evident that *the tangent of an angle can be represented by the ordinate (y value) of any point where the extended radius vector intersects the tangent line.* The ordinate gives both the sign and the magnitude of the tangent in any quadrant. Thus, in quadrants I and III, tan θ = +0.75; in quadrants II and IV, tan θ = −0.75.

PROBLEMS 24-3

1. What is the least value sin θ may have?
2. What is the least value cos θ may have?
3. What is the greatest value csc θ may have in the first quadrant?
4. What is the greatest value sec θ may have in the fourth quadrant?
5. Can the secant and cosecant have values between −1 and +1?
6. What is the greatest value sin θ may have in the (*a*) first quadrant, (*b*) second quadrant, (*c*) third quadrant, and (*d*) fourth quadrant?
7. What is the greatest value cos θ may have in the (*a*) first quadrant, (*b*) second quadrant, (*c*) third quadrant, and (*d*) fourth quadrant?

Trigonometric Tables

25

For the purpose of making computations, it is evident that a table of trigonometric functions would be helpful. Such a table could be made by computing the functions of all angles by graphical methods. However, that would be laborious and the resulting functions would not be accurate.

Fortunately, mathematicians have calculated the values of the trigonometric functions by the use of advanced mathematics and have tabulated the results. These tables are known as *tables of natural functions* to distinguish them from *tables of the logarithms of the functions*.

25-1 GIVEN AN ANGLE—TO FIND THE DESIRED FUNCTION

How to use the calculator for natural functions is best illustrated by examples. *Tables* are available in a number of printed forms, in addition to being programmed into a variety of "scientific" calculators. You should check with your instructors to determine whether or not printed tables are available for your use in the event of any emergency. It is left as an exercise for you to become acquainted with printed tables. Trigonometric values in this book are taken from a scientific hand-held calculator rounded off to five decimal places. You should carefully check the instruction manual for your own calculator as you follow the examples below. At the same time, refer to the convenient three-place tables inside the front covers, and compare their values with those on your calculator.

Example 1
Find the sine of 36.7°.

Solution
Key   36.7
Key   SIN
Read   0.597 63

$$\sin 36.7° = 0.597\,63$$

Example 2
Find the cosine of 7.9°.

Solution
Key   7.9
Key   COS
Read   0.990 51

$$\cos 7.9° = 0.990\ 51$$

Example 3
Find the tangent of 79.1°.

Solution
Key   79.1
Key   TAN
Read   5.192 93

$$\tan 79.1° = 5.192\ 93$$

Example 4
Find the sine of 26.42°.

Solution
Key   26.42
Key   SIN
Read   0.444 95

$$\sin 26.42° = 0.444\ 95$$

Example 5
Find the cosine of 53.77°.

Solution

$$\cos 53.77° = 0.591\ 03$$

Example 6
Find the tangent of 48.13°.

Solution

$$\tan 48.13° = 1.115\ 69$$

PROBLEMS 25-1

1. Find the sine, cosine, and tangent of (a) 18°, (b) 68°, (c) 9.3°, (d) 52.5°, (e) 2.6°.
2. Find the sine, cosine, and tangent of (a) 12°, (b) 88.7°, (c) 70.2°, (d) 0.8°, (e) 20.1°.
3. Find the sine, cosine, and tangent of (a) 1.94°, (b) 57.36°, (c) 38.91°, (d) 40.28°, (e) 55.37°.
4. Find the sine, cosine, and tangent of (a) 7.39°, (b) 12.18°, (c) 32.65°, (d) 41.55°, (e) 3.17°.

25-2 INVERSE TRIGONOMETRIC FUNCTIONS

Frequently some form of notation is needed in order to express an angle in terms of one of its functions. For example, in Sec. 24-3 Example 3 dealt with an angle whose tangent was $\frac{9}{10}$. Similarly, in Example 4 of the same section, we considered an angle whose cosine was $\frac{3}{4}$.

If $\sin \theta = x$, then θ is an angle whose sine is x. It has been agreed to express such a relation by the notation

$$\theta = \sin^{-1} x \qquad \text{or} \qquad \theta = \arcsin x$$

Both are read "θ is equal to the angle whose sine is x" or "the inverse sine of x." For example, the tangent of 32.7° is 0.759 04. Stated as an inverse function, this would be written

$$37.2° = \arctan 0.759\ 04$$

Similarly, in the case of a right triangle labeled as in Fig. 24-8, we should write $\theta = \arctan \frac{a}{b}$, $\theta = \arccos \frac{b}{c}$, etc. In this book, we shall not use the notation "$\theta = \sin^{-1} x$" (although it appears on some calculators), for we prefer not to use an exponent when no exponent is intended. Although this form of notation is used in a number of texts, you will find that nearly all recent mathematics and engineering texts are using the "$\theta = \arcsin x$" form of notation. Because more advanced mathematics employs trigonometric functions affected by exponents, it is evident that confusion would eventually result from utilizing the other notation for specifying the inverse functions.

25-3 GIVEN A FUNCTION—TO FIND THE CORRESPONDING ANGLE

As in Sec. 25-1, the use of the calculator is best illustrated by examples. The results in degrees are rounded off to two decimal places.

Example 7
Find the angle whose sine is 0.235 14.

Solution
Key   0.235 14
Key   INV SIN
Read   13.60

$$\arcsin 0.235\ 14 = 13.6°$$

Example 8
Find θ if $\cos \theta = 0.033\ 16$

Solution
Key   0.033 16
Key   INV COS
Read   88.09 97

$$\arccos 0.033\ 16 = 88.10°$$

Example 9

Find θ if θ = arctan 1.142 29.

Solution

Key   1.142 29
Key   INV TAN
Read 48.79

$$\text{arctan } 1.142\ 29 = 48.79°$$

Example 10

Find θ if θ = arcsin 0.445 26.

Solution

$$\text{arcsin } 0.445\ 26 = 26.44°$$

Example 11

Find θ if cos θ = 0.373 15.

Solution

$$θ = 68.09°$$

Example 12

Find θ if θ = arctan 0.591 87.

Solution

$$θ = 30.62°$$

25-4 ACCURACY

In our considerations of ac circuits, we shall confine our accuracy of component values to three significant figures and angles to the nearest tenth of a degree. This, except for isolated cases, will more than meet all practical requirements.

Inside the front cover of this book is a three-place table of sines, cosines, and tangents for each degree from 0° to 90°. With the confidence gained from working with the components that form all but the most precise circuits, you will find that this table will serve many of your needs.

You should study the tables at this point, and compare them with your calculator to satisfy yourself that, for angles up to about 6°, the values of sin θ and tan θ are within 0.55% of each other and, at 10°, the difference is only 1.52%.

Compare your own calculator with other brands. You will find many different readings in the seventh and eighth decimal places, but identical values to five places.

PROBLEMS 25-2

1. Find the angles having the following values as sines:
    (a) 0.453 99, (b) 0.116 67, (c) 0.878 82, (d) 0.644 12, (e) 0.037 34.

2. Find the angles having the following values as cosines:
 (*a*) 0.965 93, (*b*) 0.190 81, (*c*) 0.998 72, (*d*) 0.866 90, (*e*) 0.343 17.
3. Find the angles whose tangents are (*a*) 12.429, (*b*) 0.008 73, (*c*) 0.842 08, (*d*) 1.651 20, (*e*) 0.482 34.

4. Find θ if:
 (*a*) θ = arctan 1.356 37
 (*b*) θ = arccos 0.486 34
 (*c*) θ = arcsin 0.273 96
 (*d*) θ = arccos 0.048 85
 (*e*) θ = arcsin 0.518 03

5. Find θ if:
 (*a*) θ = arccos 0.973 74
 (*b*) θ = arctan 0.009 25
 (*c*) θ = arcsin 0.963 06
 (*d*) θ = arctan 0.893 15
 (*e*) θ = arcsin 0.732 66

25-5 FUNCTIONS OF ANGLES GREATER THAN 90°

You will note that the trigonometric functions inside the front cover have been tabulated only for angles of 0° to 90°. The signs and magnitudes for angles in all quadrants were considered in the preceding chapter, and it is evident that any table of functions must be combined with a method of expressing any angle in terms of an angle of the first quadrant in order to make use of the table of functions. Similarly, all calculators provide the ability to deal with angles greater than 90°.

25-6 TO FIND THE FUNCTIONS OF AN ANGLE IN THE SECOND QUADRANT

In Fig. 25-1, let θ represent any angle in the second quadrant. From any point P on the radius vector r, draw the perpendicular y to the horizontal axis. The acute angle that r makes with the horizontal axis is designated by ϕ. Consequently, because $\theta + \phi = 180°$, θ and ϕ are supplementary angles.

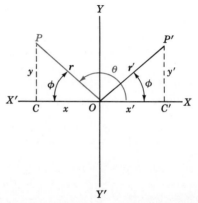

Fig. 25-1 θ and ϕ are supplementary angles; $\theta + \phi = 180°$.

Hence,

$$\phi = 180° - \theta$$

Now construct the angle XOP' in the first quadrant equal to ϕ, make r' equal to r, and draw y' perpendicular to OX. Since the right triangles OPC and $OP'C'$ are equal, $x = -x'$ and $y = y'$. Then

$$\sin(180° - \theta) = \frac{y}{r} = \frac{y'}{r'} = \sin\phi$$

$$\cos(180° - \theta) = \frac{x}{r} = \frac{-x'}{r'} = -\cos\phi$$

$$\tan(180° - \theta) = \frac{y}{x} = \frac{y'}{-x'} = -\tan\phi$$

These relationships show that, in all respects, the function of an angle has the same absolute value as the same function of its supplement. That is, if two angles are supplementary, their sines are equal in all respects and their cosines and tangents are equal in magnitude but opposite in sign.

Example 13

$$\sin 140° = \sin(180° - 140°)$$
$$= \sin 40° = 0.642\ 79$$

$$\cos 100° = -\cos(180° - 100°)$$
$$= -\cos 80° = -0.173\ 65$$

$$\tan 175° = -\tan(180° - 175°)$$
$$= -\tan 5° = -0.087\ 49$$

Note that your calculator gives the correct algebraic sign as well as the numerical value.

25-7 TO FIND THE FUNCTION OF AN ANGLE IN THE THIRD QUADRANT

In the triangles shown in Fig. 25-2, let θ represent any angle in the third quadrant and let ϕ be the acute angle that the radius vector r makes with the horizontal axis.
Then

$$\phi = \theta - 180°$$

Now construct the angle XOP' in the first quadrant equal to ϕ, make r' equal to r, and draw y and y' perpendicular to the horizontal axis. Since the right triangles OPC and $OP'C'$ are equal, $x = -x'$ and $y = -y'$
Then

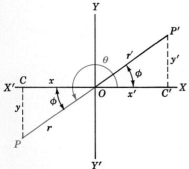

Fig. 25-2 θ is in the third quadrant; φ = θ − 180°.

$$\sin(\theta - 180°) = \frac{y}{r} = \frac{-y'}{r'} = -\sin\phi$$

$$\cos(\theta - 180°) = \frac{x}{r} = \frac{-x'}{r'} = -\cos\phi$$

$$\tan(\theta - 180°) = \frac{y}{x} = \frac{-y'}{-x'} = \tan\phi$$

These relationships show that the function of an angle in the third quadrant has the same absolute value as the same function of the acute angle between the radius vector and the horizontal axis. The signs of the functions are the same as for any angle in the third quadrant, as discussed in Sec. 24-5.

Example 14

$$\sin 200° = -\sin(200° - 180°)$$
$$= -\sin 20° = -0.342\ 02$$
$$\cos 260° = -\cos(260° - 180°)$$
$$= -\cos 80° = -0.173\ 65$$
$$\tan 234° = \tan(234° - 180°)$$
$$= \tan 54° = 1.376\ 38$$

25-8 TO FIND THE FUNCTIONS OF AN ANGLE IN THE FOURTH QUADRANT

In Fig. 25-3, let θ represent any angle in the fourth quadrant and let φ be the acute angle that the radius vector *r* makes with the horizontal axis.

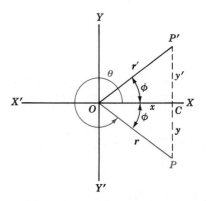

Fig. 25-3 θ is in the fourth quadrant; φ = 360° − θ.

Then

$$\phi = 360° - \theta$$

Now construct the angle XOP' in the first quadrant equal to ϕ, make r' equal to r, and draw y and y' perpendicular to the horizontal axis. Since the right triangles OPC and $OP'C$ are equal, $y = -y'$. Then

$$\sin(360° - \theta) = \frac{y}{r} = \frac{-y'}{r'} = -\sin\phi$$

$$\cos(360° - \theta) = \frac{x}{r} = \frac{x}{r'} = \cos\phi$$

$$\tan(360° - \theta) = \frac{y}{x} = \frac{-y'}{x} = -\tan\phi$$

These relationships show that the functions of an angle in the fourth quadrant have the same absolute value as the same functions of an acute angle in the first quadrant equal to $360° - \theta$. The signs of the functions, however, are those for an angle in the fourth quadrant, as discussed in Sec. 24-5.

Example 15

$$\sin 300° = -\sin(360° - 300°)$$
$$= -\sin 60° = -0.866\ 03$$
$$\cos 285° = \cos(360° - 285°)$$
$$= \cos 75° = 0.258\ 82$$
$$\tan 316° = -\tan(360° - 316°)$$
$$= -\tan 44° = -0.965\ 69$$

25-9 TO FIND THE FUNCTIONS OF AN ANGLE GREATER THAN 360°

Any angle θ greater than $360°$ has the same trigonometric functions as θ minus an integral multiple of $360°$. That is, a function of an angle larger than $360°$ is found by dividing the angle by $360°$ and finding the required function of the remainder. Thus θ in Fig. 25-4 is a positive angle of $955°$. To find any function of $955°$, divide $955°$ by $360°$, which gives 2 with a remainder of $235°$. Hence,

$$\sin 955° = \sin 235° = -\sin(235° - 180°)$$
$$= -\sin 55° = -0.819\ 15$$
$$\cos 955° = \cos 235° = -\cos(235° - 180°)$$
$$= -\cos 55° = -0.573\ 58$$
$$\tan 955° = \tan 235° = \tan(235° - 180°)$$
$$= \tan 55° = 1.428\ 15$$

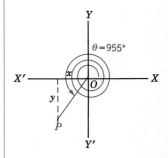

Fig. 25-4 $\theta = 955°$.

When using a calculator it is not necessary to go through the above divisions and subtractions. Simply enter 955 and enter the required functions. The calculator will give the correct sign and result.

25-10 TO FIND THE FUNCTIONS OF A NEGATIVE ANGLE

In Fig. 25-5, let $-\theta$ represent a negative angle in the fourth quadrant made by the radius vector r and the horizontal axis. Construct the angle θ in the first quadrant equal to $-\theta$, make r' equal to r, and draw y and

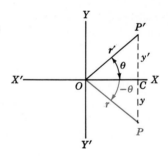

Fig. 25-5 $-\theta$ Generated by clockwise rotation.

y' perpendicular to the horizontal axis. Since the right triangles OPC and $OP'C$ are equal, $y = -y'$. Then

$$\sin(-\theta) = \frac{y}{r} = \frac{-y'}{r'} = -\sin\theta$$

$$\cos(-\theta) = \frac{x}{r} = \frac{x}{r'} = \cos\theta$$

$$\tan(-\theta) = \frac{y}{x} = \frac{-y'}{x'} = -\tan\theta$$

These relationships are true for any values of $-\theta$ regardless of the quadrant or the magnitude of the angle.

Example 16

$$\sin(-65°) = -\sin 65° = -0.906\ 31$$
$$\cos(-150°) = -\cos 150° = -\cos(180° - 150°)$$
$$= -\cos 30° = -0.866\ 03$$
$$\tan(-287°) = -\tan 287° = -\tan(360° - 287°)$$
$$= -(-\tan 73°) = 3.270\ 85$$

Test your calculator, as in Sec. 25-10:
Key   65
Key $+/-$
Key   sin
Read $-0.906\ 31$

25-11 TO REDUCE THE FUNCTIONS OF ANY ANGLE TO THE FUNCTIONS OF AN ACUTE ANGLE

It has been shown in the preceding sections that all angles can be reduced to terms of $(180° - \theta)$, $(\theta - 180°)$, $(360° - \theta)$, or θ. These results can be summarized as follows:

RULE

To find any function of any angle θ, take the same function of the acute angle formed by the terminal side (radius vector) and the *horizontal* axis and prefix the proper algebraic sign for that quadrant.

When finding the functions of angles, you should make a sketch showing the approximate location of the angle. This procedure will clarify the trigonometric relationships, and many errors will be avoided by using it.

Example 17
Find the functions of 143°.

Solution
Construct the angle 143°, and mark the signs of the radius vector, abscissa, and ordinate, as shown in Fig. 25-6. (The radius vector is always positive.) Since $180° - 143° = 37°$ the acute angle for the functions is 37°. Hence,

$$\sin 143° = \sin 37° = 0.601\ 82$$
$$\cos 143° = -\cos 37° = -0.798\ 64$$
$$\tan 143° = -\tan 37° = -0.753\ 55$$

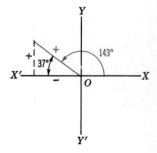

Fig. 25-6 $180° - 143° = 37°$.

   In this and the following examples, confirm that your calculator gives both the values and the signs of the functions.

Example 18
Find the functions of 245°.

Solution
Construct the angle 245° as shown in Fig. 25-7.

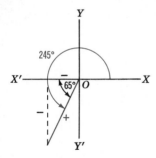

Fig. 25-7 $245° - 180° = 65°$.

Since $245° - 180° = 65°$, the acute angle for the functions is 65°. Hence,

$$\sin 245° = -\sin 65° = -0.906\ 31$$
$$\cos 245° = -\cos 65° = -0.422\ 62$$
$$\tan 245° = \tan 65° = 2.144\ 51$$

Example 19
Find the functions of 312°.

Solution
Construct the angle 312° as shown in Fig. 25-8.
Since $360° - 312° = 48°$, the acute angle for the functions is 48°. Hence,

$$\sin 312° = -\sin 48° = -0.743\ 14$$
$$\cos 312° = \cos 48° = 0.669\ 13$$
$$\tan 312° = -\tan 48° = -1.110\ 61$$

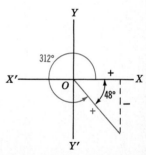

Fig. 25-8 $360° - 312° = 48°$.

Example 20

Find the functions of 845°.

Solution

845° ÷ 360° = 2 + 125°. Therefore, the functions of 125° will be identical with those of 845°. The construction is shown in Fig. 25-9.

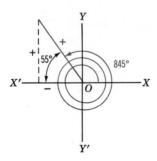

Fig. 25-9 Functions of 845° are the same as those of 125°.

Since 180° − 125° = 55°, the acute angle for the functions is 55°. Hence,

$$\sin 845° = \sin 55° = 0.819\ 15$$
$$\cos 845° = -\cos 55° = -0.573\ 58$$
$$\tan 845° = -\tan 55° = -1.428\ 15$$

Example 21

Find the functions of −511°.

Solution

−511° ÷ 360° = −(1 + 151°). Therefore, the functions of −151° will be identical with those of −511°. The construction is shown in Fig. 25-10. Since 180° − 151° = 29°, the acute angle for the functions is 29°. Hence,

$$\sin(-151°) = -\sin 29° = -0.484\ 81$$
$$\cos(-151°) = -\cos 29° = -0.874\ 62$$
$$\tan(-151°) = \tan 29° = 0.554\ 31$$

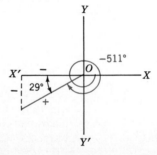

Fig. 25-10 Functions for −511° are the same as those of −151°.

25-12 ANGLES CORRESPONDING TO INVERSE FUNCTIONS

Now that we are able to express all angles as acute angles in order to use the table of functions from 0° to 90°, it has probably occurred to you that an important distinction exists between the direct trigonometric functions and the inverse trigonometric functions. Each trigonometric function of any given angle has only one value, whereas a given function corresponds to an infinite number of angles. For example, an angle of 30° has only one sine value, which is 0.500 00, but an angle whose sine is 0.500 00 (arcsin 0.500 00) may be taken as 30°, 150°, 390°, 510°, etc.

To avoid confusion, it has been agreed that the values of arcsin θ and arctan θ which lie between +90° and −90°, in the first and fourth quadrants, are to be known as the *principal values* of arcsin θ and arctan θ. The principal value is often denoted by using a capital letter, as Arcsin θ. Thus,

	Arcsin	0.575 01	=	35.1°
		Key	0.575 01	
		Key	INV SIN	
		Read	35.1	
and	Arcsin (−0.998 03)	=	−86.4°	
		Key	0.998 03	
		Key	+/−	
		Key	INV SIN	
		Read	−86.4	
Also	Arctan	1.482 56	=	56°
and	Arctan (−0.069 93)	=	−4°	

The principal values of arccos θ are taken as the values between 0° and 180° and are denoted by Arccos θ. Thus,

	Arccos 0.173 65	=	80°
and	Arccos (−0.981 63)	=	169°

Take special note of the principal angles because functions of angles greater than 90° may lead to errors of interpretation.

Example 22
Key   120
Key   SIN
Read   0.866 03
Key   INV SIN
Read   60

The calculator does not "remember" that it was originally working from a second quadrant angle.

Example 23
Key   240
Key   COS
Read   −0.500 00
Key   INV COS
Read   120

Example 24
Key 300
Key COS
Read 0.500 00
Key INV COS
Read 60

Example 25
Key 240
Key TAN
Read 1.732 05
Key INV TAN
Read 60

PROBLEMS 25-3

1. Find the sine, cosine, and tangent of (*a*) 107°, (*b*) 160°, (*c*) 130.1°, (*d*) 147.5°, (*e*) 176.2°.

2. Find the sine, cosine, and tangent of (*a*) 183°, (*b*) 235°, (*c*) 217.8°, (*d*) 180.9°, (*e*) 268.1°.

3. Find the sine, cosine, and tangent of (*a*) 280°, (*b*) 318°, (*c*) 349.9°, (*d*) 300.1°, (*e*) 359.5°.

4. Find the sine, cosine, and tangent of (*a*) 461°, (*b*), 510°, (*c*) 480.5°, (*d*) 523.2°, (*e*) 539.3°.

5. Find the sine, cosine, and tangent of (*a*) 905°, (*b*) −17.1°, (*c*) 940.7°, (*d*) −362.6°, (*e*) 1260.2°.

6. Find θ if:
   (*a*) θ = Arccos 0.969 02   (*b*) θ = Arcsin 0.582 12
   (*c*) θ = Arccos (−0.455 55)   (*d*) θ = Arctan (−3.510 53)
   (*e*) θ = Arcsin (−0.377 84)

7. Find φ if:
   (*a*) ϕ = Arctan (−1.076 13)   (*b*) ϕ = Arccos (−0.027 92)
   (*c*) ϕ = Arcsin 0.780 43   (*d*) ϕ = Arccos (−0.976 30)
   (*e*) ϕ = Arctan (−2.732 63)

8. The illumination on a surface that is not perpendicular to the rays of light from a light source is given by the formula

$$E = \frac{F \cos \theta}{d^2} \qquad \text{lux (lx)}$$

where E = illumination at a point on the surface, lx
   F = intensity of light output of source, lumens (lm)
   d = distance of source of light to surface, m
   θ = angle between incident light ray and a line perpendicular to the surface

Solve for F, d, and θ.

9. In the formula of Prob. 8, find the value of d if F = 900 lm, θ = 48°, and E = 300 lx.

10. A 100-W lamp has a total light output of 1700 lm. Disregarding reflection, compute the illumination at a point on a surface 3 m from the lamp if the plane of the surface is at an angle of 30° to the incident rays.

11. In the formula of Prob. 8, at what angle of the plane of the surface to the incident ray will the illumination be the greatest?

12. The illumination on a horizontal surface from a source of light at a given vertical distance from the surface is given by the formula

$$E_h = \frac{F}{h^2} \cos^3 \theta \quad \text{lx}$$

where E_h = illumination at a point on horizontal surface, lx
F = intensity of light output from source of light, lm
h = vertical distance from horizontal surface to source of light, m
θ = angle between incident ray and vertical line, as shown in Fig. 25-11.

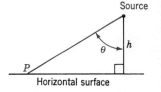

Fig. 25-11 Illumination at P from source.

Note $\cos^3 \theta$ means $(\cos \theta)$ raised to the third power.

Solve for F, h, and θ.

13. Use the formula of Prob. 12 to solve for E_h if $F = 3260$ lm, $h = 4$ m, and $\theta = 18°$.

14. Use the formula of Prob. 12 to solve for F if $E_h = 330$ lx, $h = 4.5$ m, and $\theta = 50°$.

15. According to illumination experts, 1000 to 1500 lx of illumination on a printed page should be provided for study purposes. A 60-W, 850-lm lamp is suspended 2 m above a reading table. The reflector used projects 70% of the light downward. Does this produce a satisfactory amount of illumination on a book directly below the lamp?

16. To produce 1250 lx on the book in Prob. 15, what lumen-rating lamp should be used?

17. Snell's law states that, when a wave of electromagnetic energy passes from one dielectric material to another, the ratio of the sines of the angles of incidence θ_1 and refraction θ_2 is inversely proportional to the square root of the ratio of the dielectric relative permittivities (Fig. 25-12). That is,

$$\frac{\sin \theta_1}{\sin \theta_2} = \sqrt{\frac{\varepsilon_2}{\varepsilon_1}}$$

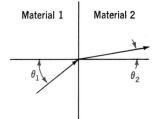

Fig. 25-12 Diagram for Prob. 17.

If the angle of incidence $\theta_1 = 70°$, material 1 is lucite, $\varepsilon_1 = 2.6$, and material 2 is mica, $\varepsilon_2 = 5.4$, what is the angle of refraction θ_2?

Solution of Right Triangles

26

One of the most important applications of trigonometry is the solution of triangles, both right and oblique. This chapter is concerned with the former. The right triangle is probably the most universally used geometric figure; with the aid of trigonometry, it is applied to numerous problems in measurement that otherwise might be impossible to solve.

A large percentage of the problems relating to the analysis of ac circuits and networks involves the solution of the right triangle in one form or another.

26-1 FACTS CONCERNING RIGHT TRIANGLES

Before we proceed with the actual solutions of right triangles, we will review the following useful facts regarding the properties of the right triangle:

1. The square of the hypotenuse is equal to the sum of the squares of the other two sides ($c^2 = a^2 + b^2$).
2. The acute angles are complements of each other; that is, the sum of the two acute angles is 90° ($A + B = 90°$).
3. The hypotenuse is greater than either of the other two sides and is less than their sum.
4. The greater angle is opposite the greater side, and the greater side is opposite the greater angle.

These facts will often be a material aid in checking computations made by trigonometric methods.

26-2 PROCEDURE FOR SOLUTION OF RIGHT TRIANGLES

Every triangle has three sides and three angles, and these are called the six *elements* of the triangle. To *solve* a triangle is to find the values of the unknown elements.

A triangle can be solved by two methods:

1. By constructing the triangle accurately from known elements with scale, protractor, and compasses. The unknown elements can then be measured with the scale and the protractor.
2. By computing the unknown elements from those that are known.

The first method has been used to some extent in preceding chapters. However, as previously discussed, the graphical method is cumbersome and has a limited degree of accuracy.

Trigonometry, combined with simple algebraic processes, furnishes a powerful tool for solving triangles by the second method listed above. Moreover, the degree of accuracy is limited only by the number of significant figures to which the elements have been measured and the number of significant figures available in the table of functions or the calculator used for the solution.

As pointed out in earlier chapters, every type of problem should be approached and solved in a planned and systematic manner. Only in this way are the habits of clear and ordered thinking developed, the principles of the problem understood, and the possibility of errors reduced to a minimum. With the foregoing in mind, we list the following suggestions for solving right triangles as a guide:

1. Make a reasonable sketch of the triangle, and mark the known (given) elements. This shows the relation of the elements, helps you choose the functions needed, and will serve as a check for the solution. List what is to be found.
2. To find an unknown element, select a formula that contains two known elements and the required unknown element. Substitute the known elements in the formula, and solve for the unknown.
3. As a rough check on the solution, compare the results with the drawing. To check the values accurately, note whether they satisfy relationships different from those already employed for the solution of the values being checked. A convenient check for the sides of a right triangle is the relation

$$a^2 = c^2 - b^2 = (c + b)(c - b)$$

4. In the computations, round off the numbers representing the lengths of sides to three significant figures and all angles to the nearest tenth of a degree. This means that the values of the functions employed in computations are to be used to only three significant figures. As previously stated, such accuracy is sufficient for ordinary *practical circuit* computations.

Heretofore, the right triangles used in figures for illustrative examples have been lettered in the conventional manner, as shown in Figs. 24-4, 24-11, etc. At this point the notation for the various elements will be changed to that of Fig. 26-1. In no way does this change of lettering have any effect on the fundamental relations existing among the elements of a right triangle, nor are any new ideas involved in connection with the trigonometric

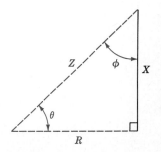

Fig. 26-1 Lettering of "standard" electrical right triangle.

functions. Because certain ac problems will employ this form of notation, this is a convenient place to introduce it in order that you may become accustomed to solving right triangles lettered in this manner.

The following sections illustrate all the possible conditions encountered in the solution of right triangles.

26-3 GIVEN AN ACUTE ANGLE AND A SIDE NOT THE HYPOTENUSE

Example 1

Given $R = 30.0$ and $\theta = 25.0°$. Solve for Z, X, and ϕ.

Solution

The construction is shown in Fig. 26-2.

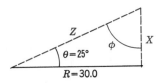

Fig. 26-2 Construction for solution of Example 1.

$$\phi = 90° - \theta = 90° - 25° = 65°$$

An equation containing the two known elements and one unknown is

$$\tan \theta = \frac{X}{R}$$

Solving for X, $\qquad X = R \tan \theta$

Substituting the values of R and $\tan \theta$,

$$X = 30 \times 0.466 = 14.0$$

Also, since $\qquad \sin \theta = \frac{X}{Z}$

Solving for Z, $\qquad Z = \frac{X}{\sin \theta}$

Substituting the values of X and $\sin \theta$,

$$Z = \frac{14.0}{0.423} = 33.1$$

This solution can be checked by using some relation other than the relations already used. Thus, substituting values in

$$X^2 = (Z + R)(Z - R)$$

results in
$$14.0^2 = (33.1 + 30.0)(33.1 - 30.0)$$
$$196 = 63.1 \times 3.10 = 196$$

Since all results were rounded off to three significant figures, the check shows the solution to be correct for this degree of accuracy.

The value of Z can be checked by employing a function not used in the solution. Thus, since

$$R = Z \cos \theta$$

by substituting the values,

$$30 = 33.1 \times 0.906$$

Still another check could be made by use of an inverse function employing two of the elements found in the solution. For example,

$$\phi = \arccos \frac{X}{Z} = \arccos \frac{14.0}{33.1}$$

$$= \arccos 0.423 = 65°$$

Example 2

Given $X = 106$ and $\theta = 36.4°$. Solve for Z, R, and ϕ.

Solution

The construction is shown in Fig. 26-3.

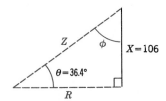

Fig. 26-3 Triangle of Example 2.

$$\phi = 90° - \theta = 90° - 36.4° = 53.6°$$

An equation containing two known elements and one unknown is

$$\sin \theta = \frac{X}{Z}$$

Solving for Z,

$$Z = \frac{X}{\sin \theta}$$

Substituting the values of X and $\sin \theta$,

$$Z = \frac{106}{0.593} = 179$$

Also, since

$$\cos \theta = \frac{R}{Z}$$

solving for R,

$$R = Z \cos \theta$$

Substituting the values of Z and $\cos \theta$,

$$R = 179 \times 0.805 = 144$$

Check the solution by one of the methods previously explained.

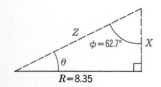

Fig. 26-4 Triangle of Example 3.

Example 3

Given $R = 8.35$ and $\phi = 62.7°$. Find Z, X, and θ.

Solution

The construction is shown in Fig. 26-4.

$$\theta = 90° - \phi = 90° - 62.7° = 27.3°$$

When θ is found, the methods to be used in the solution of this example become identical with those of Example 1. Hence,

$$X = R \tan \theta = 8.35 \tan 27.3°$$

$$= 8.35 \times 0.516 = 4.31$$

$$Z = \frac{X}{\sin \theta} = \frac{4.31}{\sin 27.3°}$$

$$= \frac{4.31}{0.459} = 9.39$$

Check the solution by the method considered most convenient.

Example 4

Given $X = 1290$ and $\phi = 41.9°$. Find Z, R, and θ.

Solution

The construction is shown in Fig. 26-5.

$$\theta = 90° - \phi = 90° - 41.9° = 48.1°$$

When θ is found, the methods to be used in the solution of this example become identical with those of Example 2. Hence,

$$Z = \frac{X}{\sin \theta} = \frac{1290}{\sin 48.1°} = \frac{1290}{0.744} = 1730$$

$$R = Z \cos \theta = 1730 \cos 48.1°$$

$$= 1730 \times 0.688 = 1190$$

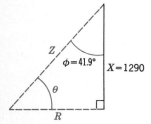

Fig. 26-5 $X = 1290$, $\phi = 41.9°$.

Check the solution by the method considered most convenient.

With the exception of finding the unknown acute angle, which involves subtraction, any of the foregoing examples and the following problems may be solved using previously described calculator keystrokes.

PROBLEMS 26-1

Solve the following right triangles for the unknown elements. Check each by making a construction and by substituting into a formula not used in the solution:

1. $R = 22.0$, $\theta = 34.7°$
2. $X = 4.39$, $\phi = 86.5°$
3. $X = 424$, $\phi = 45°$
4. $R = 8.10$, $\phi = 21°$
5. $R = 63.5$, $\theta = 24.9°$
6. $X = 1530$, $\theta = 73.5°$
7. $R = 8.85 \times 10^5$, $\theta = 27.7°$
8. $R = 222$, $\phi = 26.3°$

9. $X = 867, \theta = 57.3°$

10. $R = 0.230, \theta = 77°$

11. $X = 124, \theta = 51.1°$

12. $X = 0.0929, \theta = 6.4°$

13. $R = 0.105, \theta = 63.9°$

14. $r = \frac{2}{3}, \theta = 51.9°$

15. $X = \frac{3}{8}, \theta = 82.4°$

16. $R = \frac{1}{\sqrt{2}}, \theta = 45°$

26-4 GIVEN AN ACUTE ANGLE AND THE HYPOTENUSE

Example 5

Given $Z = 45.3$ and $\theta = 20.3°$. Find R, X, and ϕ.

Solution

The construction is shown in Fig. 26-6.

$$\phi = 90° - \theta = 90° - 20.3° = 69.7°$$

An equation containing two known elements and one unknown is

$$\cos \theta = \frac{R}{Z}$$

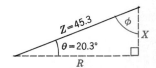

Fig. 26-6 $Z = 45.3$, $\theta = 20.3°$.

Solving for R,

$$R = Z \cos \theta$$

Substituting the values of Z and $\cos \theta$,

$$R = 45.3 \times 0.938 = 42.5$$

Another convenient equation is

$$\sin \theta = \frac{X}{Z}$$

Solving for X,

$$X = Z \sin \theta$$

Substituting the values of Z and $\sin \theta$,

$$X = 45.3 \times 0.347 = 15.7$$

The solution can be checked by any of the usual methods.

Example 6

Given $Z = 265$ and $\phi = 22.4°$. Find R, X, and θ.

Solution

The construction is shown in Fig. 26-7.

$$\theta = 90° - \phi = 90° - 22.4° = 67.6°$$

When θ is found, this triangle is solved by the methods used in Example 1. Hence,

$$R = Z \cos \theta = 265 \cos 67.6° = 265 \times 0.381 = 101$$
$$X = Z \sin \theta = 265 \sin 67.6° = 265 \times 0.924 = 245$$

Check the solution by one of the several methods.

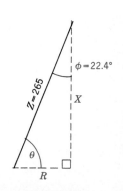

Fig. 26-7 $Z = 265$, $\phi = 22.4°$.

PROBLEMS 26-2

Solve the following right triangles for the unknown elements. Check each by construction and by substituting into an equation not used in the solution.

1. $Z = 76.2$, $\phi = 75°$
2. $Z = 464$, $\theta = 23.6°$
3. $Z = 47.6$, $\theta = 69.1°$
4. $Z = 179$, $\phi = 77.7°$
5. $Z = 1 \times 10^4$, $\phi = 51.6°$
6. $Z = 60$, $\theta = 48.2°$
7. $Z = 0.948$, $\phi = 79.6°$
8. $Z = 610$, $\phi = 79.7°$
9. $Z = 5.10$, $\theta = 52.3°$
10. $Z = 0.342$, $\phi = 73.2°$

26-5 GIVEN THE HYPOTENUSE AND ONE OTHER SIDE

Example 7

Given $Z = 38.3$ and $R = 23.1$. Find X, θ, and ϕ.

Solution

The construction is shown in Fig. 26-8.

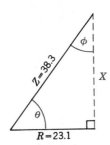

Fig. 26-8 Triangle of Example 7.

An equation containing two known elements and one unknown is

$$\cos \theta = \frac{R}{Z}$$

Substituting the values of R and Z,

$$\cos \theta = \frac{23.1}{38.3} = 0.603$$
$$\therefore \theta = 52.9°$$
$$\phi = 90° - \theta$$
$$= 90° - 52.9°$$
$$= 37.1°$$

Then, since $\qquad \sin \theta = \frac{X}{Z}$

Solving for X, $\qquad X = Z \sin \theta$

Substituting the values of Z and $\sin \theta$,

$$X = 38.3 \times 0.798 = 30.6$$

Example 8

Given $Z = 10.7$ and $X = 8.10$. Find R, θ, and ϕ.

Solution

The construction is shown in Fig. 26-9.

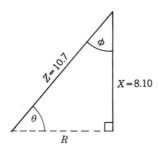

Fig. 26-9 Triangle of Example 8.

An equation containing two known elements and one unknown is

$$\sin \theta = \frac{X}{Z}$$

Substituting the values of X and Z,

$$\sin \theta = \frac{8.10}{10.7} = 0.757$$

$$\therefore \theta = 49.2°$$

$$\phi = 90° - \theta$$

$$= 90° - 49.2°$$

$$= 40.8°$$

Then, since $\qquad \cos \theta = \dfrac{R}{Z}$

Solving for R, $\qquad R = Z \cos \theta$

Substituting the values of Z and $\cos \theta$,

$$R = 10.7 \times 0.653 = 6.99$$

PROBLEMS 26-3

Solve the following right triangles and check each graphically and algebraically as in the preceding problems:

1. $Z = 229, X = 200$
2. $Z = 2160, R = 1200$
3. $Z = 47.6, R = 17$
4. $Z = 3100, R = 3060$
5. $Z = 0.742, R = 0.734$
6. $Z = 407, X = 57.0$
7. $Z = 1 \times 10^4, X = 6210$
8. $Z = 39.7, R = 11.4$
9. $Z = 1.08, R = 0.667$
10. $Z = 0.342, R = 0.327$

26-6 GIVEN TWO SIDES NOT THE HYPOTENUSE

Example 9
Given $R = 76.0$ and $X = 37.4$. Find Z, θ, and ϕ.

Solution
The construction is shown in Fig. 26-10.

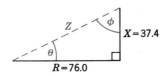

Fig. 26-10 Triangle of Example 9.

An equation containing two known elements and one unknown is

$$\tan \theta = \frac{X}{R}$$

Substituting the values of X and R,

$$\tan \theta = \frac{37.4}{76.0} = 0.492$$
$$\therefore \theta = 26.2°$$
$$\phi = 90° - \theta = 90° - 26.2° = 63.8°$$

$Z = 84.7$ can be found by one of the methods explained in the preceding sections.

PROBLEMS 26-4
Solve the following right triangles and check as in the preceding problems:

1. $R = 35.5$, $X = 6.19$
2. $R = 11.5$, $X = 6.94$
3. $X = 5.30$, $R = 4.79$
4. $R = 76.3$, $X = 277$
5. $X = 20.3$, $R = 430$
6. $X = 50.6$, $R = 10.3$
7. $R = 5.43$, $X = 48.4$
8. $R = \dfrac{\sqrt{3}}{2}$, $X = \dfrac{1}{2}$
9. $X = 0.290$, $R = 0.280$
10. $X = 4.01$, $R = 5.25$

26-7 TERMS RELATING TO MISCELLANEOUS TRIGONOMETRIC PROBLEMS

If an object is higher than an observer's eye, the *angle of elevation* of the object is the angle between the horizontal and the line of sight to the object. This is illustrated in Fig. 26-11.

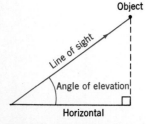

Fig. 26-11 Angle of elevation.

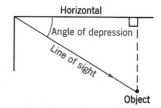

Fig. 26-12 Angle of depression.

If an object is lower than an observer's eye, the *angle of depression* of the object is the angle between the horizontal and the line of sight to the object. This is illustrated in Fig. 26-12.

The *horizonal distance* between two points is the distance from one of the two points to a vertical line that is drawn through the other. Thus, in Fig. 26-13, the line *AC* is a vertical line through the point *A* and *CB* is a horizontal line through the point *B*. Then the horizontal distance from *A* to *B* is the distance between *C* and *B*.

Fig. 26-13 Vertical and horizontal distances.

The *vertical distance* between two points is the distance from one of the two points to the horizontal line drawn through the other. Thus, the vertical distance from *A* to *B*, in Fig. 26-13, is the distance between *A* and *C*.

Calculations of distance in the vertical plane are made by means of right triangles having horizontal and vertical sides. The horizontal side is usually called the *run*, and the vertical side is called the *rise* or *fall*, as the case may be.

The *slope* or *grade* of a line is the rise or fall divided by the run. Thus, if a road rises 5 m in a run of 100 m, the grade of the road is

$$5 \div 100 = 0.05 = 5\%$$

PROBLEMS 26-5

1. What is the angle of inclination of a stairway with the floor if the steps have a tread of 27 cm and a rise of 18 cm?
2. What angle does an A-frame rafter make with the horizontal if it has a rise of 3.75 m in a run of 1.5 m?
3. A transmission line rises 2.65 m in a run of 36 m. What is the angle of elevation of the line with the horizontal?
4. A radio tower casts a shadow 174 m long, and at the same time the angle of elevation of the sun is 41.7°. What is the height of the tower?
5. An antenna mast 120 m tall casts a shadow 62 m long. What is the angle of elevation of the sun?

6. At a horizontal distance of 85 m from the foot of a radio tower, the angle of elevation of the top is found to be 31°. How high is the tower?

7. A telephone pole 12.5 m high is to be guyed from its middle, and the guy is to make an angle of 45° with the ground. Allowing 1 m extra for splicing, how long must the guy wire be?

8. An extension ladder 15 m long rests against a vertical wall with its foot 4 m from the wall. (Do not use Pythagoras' theorem to solve.)
   (a) What angle does the ladder make with the ground?
   (b) How far up the wall does the ladder reach?

9. A ladder 15 m long can be so placed that it will reach a point on a wall 12 m above the ground. By tipping it back without moving its foot, it can be made to reach a point on another wall 10 m above the ground. What is the horizontal distance between the walls?

10. From the top of a cliff 58 m high, the angle of depression of a boat is 28.6°. How far out is the boat?

11. In order to find the width *BC* of a river, a distance *AB* was laid off along the bank, the point *B* being directly opposite a tree *C* on the opposite side, as shown in Fig. 26-14. If the angle *BAC* was observed to be 62.9° and *AB* was measured at 50 m, find the width of the river.

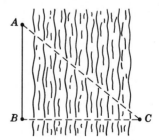

Fig. 26-14 Measuring across a river.

12. In order to measure the distance *AC* across a swamp, a surveyor lays off a line *AB* such that the angle *BAC* = 90°, as shown in Fig. 26-15. At point *B*, 240 m from *A*, the surveyor observes that angle *ABC* = 59.1°. Find the distance *AC*.

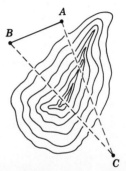

Fig. 26-15 Measuring across a pond or swamp.

26-8 THE AREA OF TRIANGLES

A convenient use of trigonometry is the calculation of the area of a triangle. In Fig. 26-16, the area of the triangle *ABC*, from previous knowledge, is known to be

$$A = \tfrac{1}{2}ab$$

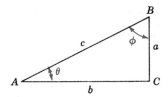

Fig. 26-16 Area of a right triangle.

But $b = c \sin \phi$ and $a = c \sin \theta$, from which we can write

$$A = \tfrac{1}{2}ac \sin \phi$$
or
$$A = \tfrac{1}{2}bc \sin \theta$$

Either of these expressions may be stated:
The area of a triangle is one-half the product of any two sides times the sine of the angle between them.

You should prove that the formula holds for the more general case of the triangle of Fig. 26-17.

Hint

Draw an altitude perpendicular to the base.

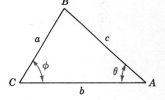

Fig. 26-17 Area of an oblique triangle.

PROBLEMS 26-6

1. In the right triangle of Fig. 26-16, $a = 15$ m and $c = 40$ m.
   (*a*) What is the angle ϕ?
   (*b*) What is the area of the triangle by the sine formula?
   (*c*) What is the length of b?
   (*d*) What is the area by the formula $A = \tfrac{1}{2}$(base)(altitude)?

2. In the triangle of Fig. 26-17, $a = 77$ mm, $b = 96.4$ mm, $c = 72$ mm, $\phi = 47.5°$, and $\theta = 52°$.
   (*a*) What is the angle opposite side b?
   (*b*) What is the area of the triangle by the sine formula?
   (*c*) What is the length of altitude h?
   (*d*) What is the area by the formula $A = \tfrac{1}{2}$(base)(altitude)?

3. In the triangle of Fig. 26-17, $a = 100$ mm, $c = 86.5$ mm, and angle *CBA* $= 107°$. What is the area of the triangle?

Trigonometric Identities and Equations

27

So far, our studies in trigonometry have been confined to the solution of *right triangles,* but there are times when other types of problems must be considered. In this chapter, we shall develop some useful relationships between the trigonometric functions and also solve oblique triangles.

27-1 SIMPLE IDENTITIES

Consider the right triangle *ABC* (Fig. 27-1). From our studies in trigonometry we know that:

$$\sin \theta = \frac{X}{Z}$$

and
$$\cos \theta = \frac{R}{Z}$$

The ratio of these two functions is

$$\frac{\sin \theta}{\cos \theta} = \frac{\dfrac{X}{Z}}{\dfrac{R}{Z}} = \frac{X}{R} = \tan \theta \qquad (1)$$

This interesting and useful relationship is the simplest of a group of trigonometric interrelationships called *identities.* We shall develop a few of the simpler identities and then tabulate them for convenience.

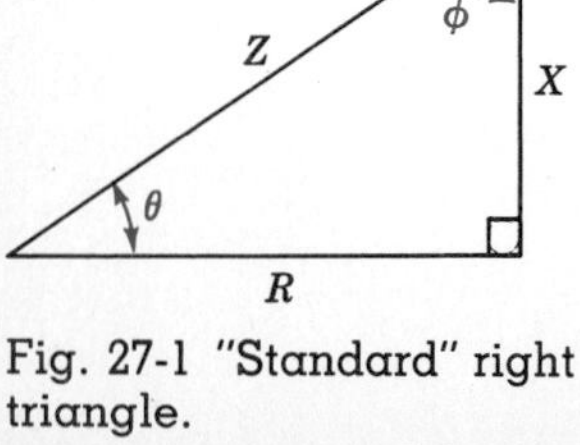

27-2 THE PYTHAGOREAN IDENTITIES

In the triangle of Fig. 27-1, we can readily see that

$$X^2 + R^2 = Z^2$$

the statement of Pythagoras' theorem. Dividing the entire equation by Z^2:

$$\frac{X^2}{Z^2} + \frac{R^2}{Z^2} = \frac{Z^2}{Z^2}$$

Fig. 27-1 "Standard" right triangle.

from which we can see that

$$(\sin \theta)^2 + (\cos \theta)^2 = 1$$

which is usually written (as in 12 of Problems 25-3)

$$\sin^2 \theta + \cos^2 \theta = 1 \qquad (2)$$

This is the first of the interrelationships known as the *Pythagorean identities* because they are derived from Pythagoras' theorem. You should now repeat the process twice, dividing first by X^2 and then by R^2 to develop the other two Pythagorean identities:

$$1 + \cot^2 \theta = \csc^2 \theta \qquad (3)$$
$$\tan^2 \theta + 1 = \sec^2 \theta \qquad (4)$$

These relationships will prove quite useful in the advanced study of electronics because many of the mathematical descriptions of electrical and electronics phenomena are described by rather complicated combinations of trigonometric functions, which may often be simplified by the use of identities. Here we shall confine ourselves to achieving some practice in manipulation of identities.

No set rule may be established about simplifying or proving identities. Usually, one side of the identity is manipulated until it is shown to be equal to the other side. Sometimes, each side is developed into the same equivalent in order to arrive at an obvious equality.

Example 1

Show that $\dfrac{\tan^2 \theta}{\sec^2 \theta} + \dfrac{\cot^2 \theta}{\csc^2 \theta} = 1$.

Solution

(*a*) One possible method of solution uses the fundamental relationships between the trigonometric functions:

$$\frac{\tan^2 \theta}{\sec^2 \theta} + \frac{\cot^2 \theta}{\csc^2 \theta} = \frac{\left(\dfrac{\sin \theta}{\cos \theta}\right)^2}{\left(\dfrac{1}{\cos \theta}\right)^2} + \frac{\left(\dfrac{1}{\tan \theta}\right)^2}{\left(\dfrac{1}{\sin \theta}\right)^2}$$

$$= \sin^2 \theta + \sin^2 \theta \left(\frac{\cos^2 \theta}{\sin^2 \theta}\right)$$

$$= 1$$

(*b*) An alternative solution is to start with the Pythagorean identities, which suggests itself from the square relationships in the problem:

$$\frac{\tan^2 \theta}{\sec^2 \theta} + \frac{\cot^2 \theta}{\csc^2 \theta} = \frac{\sec^2 \theta - 1}{\sec^2 \theta} + \frac{\csc^2 \theta - 1}{\csc^2 \theta}$$

$$= 1 - \frac{1}{\sec^2 \theta} + 1 - \frac{1}{\csc^2 \theta}$$

$$= 2 - (\cos^2 \theta + \sin^2 \theta)$$

$$= 2 - 1 = 1$$

PROBLEMS 27-1

Prove that the following equations are identities:

1. $\cos \theta \tan \theta = \sin \theta$

2. $(\sec \phi + \tan \phi)(\sec \phi - \tan \phi) = 1$

3. $\cos^2 \lambda - \sin^2 \lambda = 1 - 2 \sin^2 \lambda$

4. $\sin^4 \alpha - \cos^4 \alpha = \sin^2 \alpha - \cos^2 \alpha$

5. $\dfrac{2 \tan \phi}{1 + \tan^2 \phi} = 2 \sin \phi \cos \phi$

6. $\dfrac{\cos^2 \phi}{1 - \sin \phi} = 1 + \sin \phi$

7. $(1 + \tan^2 \beta) \cos^2 \beta = 1$

8. $\tan \theta + \cot \theta = \sec \theta \csc \theta$

9. $(\sin \theta + \cos \theta)^2 + (\sin \theta - \cos \theta)^2 = 2$

10. $1 - 2 \sin^2 \omega = 2 \cos^2 \omega - 1$

11. $\tan^2 \psi - \sin^2 \psi = \tan^2 \psi \sin^2 \psi$

12. $\dfrac{1 - 2 \cos^2 \alpha}{\sin \alpha \cos \alpha} = \dfrac{\sin^2 \alpha - \cos^2 \alpha}{\sin \alpha \cos \alpha}$

13. $\dfrac{1 - \tan^2 \theta}{1 + \tan^2 \theta} = \cos^2 \theta - \sin^2 \theta$

14. $\sec \phi - \cos \phi = \sqrt{(\tan \phi + \sin \phi)(\tan \phi - \sin \phi)}$

15. $\cot \theta \cos \theta = \csc \theta - \sin \theta$

16. $\dfrac{\sin \theta + \tan \theta}{1 + \tan^2 \theta} = \sin \theta \tan \theta$

17. $\tan \lambda + \cot \lambda = \dfrac{\csc^2 \lambda + \sec^2 \lambda}{\csc \lambda \sec \lambda}$

18. $(\tan \alpha - \sin \alpha)^2 + (1 - \cos \alpha)^2 = (1 - \sec \alpha)^2$

19. $\dfrac{1 - \sin \omega}{1 + \sin \omega} = (\sec \omega - \tan \omega)^2$

20. $\dfrac{\tan \alpha + \tan \beta}{\cot \alpha + \cot \beta} = \tan \alpha \tan \beta$

27-3 LAW OF SINES

Consider the triangle ABC (Fig. 27-2). This is not a right triangle, and therefore we have no relationships which we can use to solve it, that is, to relate the various sides and angles in order to find the unknown dimensions in an actual numerical problem. But if we were to develop within it our own right triangles, we might derive some useful relationships.

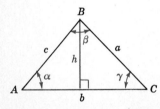

Fig. 27-2 Nonright triangle cannot be solved by simple trigonometric relationships.

First of all, we redraw the triangle, Fig. 27-3, and from the vertex B we drop the altitude h perpendicular to the base b. This yields two right triangles, from which we develop the relationships:

$$h = c \sin \alpha \quad \text{and} \quad h = a \sin \gamma$$

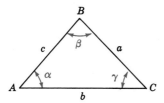

Fig. 27-3 Redrawn from Fig. 27-2 with altitude h perpendicular to base b.

Then, equating things equal to the same thing (Axiom 5, Sec. 5-2):

$$c \sin \alpha = a \sin \gamma$$

We rewrite this equation in the simple easy-to-remember form

$$\frac{a}{\sin \alpha} = \frac{c}{\sin \gamma} \tag{5}$$

You should immediately prove the more general statement:

$$\frac{a}{\sin \alpha} = \frac{b}{\sin \beta} = \frac{c}{\sin \gamma} \tag{6}$$

Example 2
Given the triangle *MPL*, Fig. 27-4, find the values of m and p.

Solution
First of all, solve for

$$\lambda = 180° - (80° + 30°) = 70°$$

Then, using the law of sines,

$$\frac{10}{\sin 70°} = \frac{p}{\sin 80°} = \frac{m}{\sin 30°}$$

so that

$$p = \frac{10 \sin 80°}{\sin 70°} = 10.5$$

and

$$m = \frac{10 \sin 30°}{\sin 70°} = 5.32$$

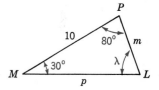

Fig. 27-4 Triangle of Example 2.

Note
To be able to use the law of sines, it is necessary for us to know certain specific data: two sides and the angle opposite one of them or two angles and the side opposite one of them.

PROBLEMS 27-2

Referring to Fig. 27-2, solve the following triangles:

1. $a = 8.04$, $\alpha = 57°$, $\beta = 53°$
2. $a = 19$, $\beta = 80°$, $\gamma = 88°$
3. $b = 16.3$, $\alpha = 44°$, $\beta = 61°$
4. $c = 760$, $\alpha = 68°$, $\beta = 42°$
5. $b = 76$, $\alpha = 20°$, $\beta = 52°$
6. $b = 3.26$, $\alpha = 25°$, $\beta = 41°$
7. $c = 7.6$, $\beta = 60°$, $\gamma = 112°$
8. $a = 600$, $\beta = 17.6°$, $\gamma = 105.9°$
9. $b = 58$, $\alpha = 9.2°$, $\gamma = 115.3°$
10. $c = 635$, $\alpha = 15.5°$, $\beta = 26°$
11. Two observers who are 1500 m apart on a horizontal plane observe a radiosonde balloon in the same vertical plane as themselves and between themselves. The angles of elevation are 72° and 75°. Find the height of the balloon.
12. A 50-m antenna mast stands on the edge of the roof of the studio building. From a point on the ground at some distance from the base of the building, the angles of elevation of the top and bottom of the mast are respectively 76.5° and 54.5°. How high is the building?

27-4 LAW OF COSINES

Sometimes we are not given data suitable for solving a triangle by means of the law of sines, but another useful relationship can be readily developed. Using the triangle ABC of Fig. 27-2, copied as Fig. 27-5 and adjusted with an altitude h perpendicular to the base and rising to the vertex, so that the base is divided into parts x and y,

$$h^2 = c^2 - x^2 = a^2 - y^2$$

from which

$$\begin{aligned}
a^2 &= c^2 - x^2 + y^2 \\
&= c^2 - x^2 + (b - x)^2 \\
&= c^2 - x^2 + b^2 - 2bx + x^2 \\
&= b^2 + c^2 - 2bx
\end{aligned}$$

but

$$x = c \cos \alpha$$

and

$$a^2 = b^2 + c^2 - 2bc \cos \alpha \qquad (7)$$

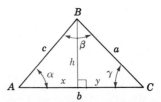

Fig. 27-5 Redrawn from Fig. 27-2. Altitude h divides base b into parts x and y.

See how straightforward this statement may be: "In any triangle, the square of any one side is equal to the sum of the squares of the other two sides minus twice their product times the cosine of the angle between them." You should prove that this statement holds true for right triangles, to become Pythagoras' theorem.

Like the law of sines, the law of cosines has a rhythm that makes it easy to memorize one part and simply rotate the other parts into duplicate statements. However, besides merely memorizing the result, you should prove that all parts of the full statement of the law of cosines are true:

$$a^2 = b^2 + c^2 - 2bc\,\cos\alpha$$

$$b^2 = a^2 + c^2 - 2ac\,\cos\beta \tag{8}$$

$$c^2 = a^2 + b^2 - 2ab\,\cos\gamma \tag{9}$$

The careful use of these three equations, together with what we have learned about the *signs* of the cosine, will enable us to prepare any triangle so that we may complete its solution by means of the law of sines.

Example 3
Acute triangle. Solve the triangle of Fig. 27-6.

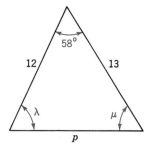

Fig. 27-6 Triangle of Example 3.

Solution
Using the law of cosines:

$$p^2 = 12^2 + 13^2 - 2 \times 12 \times 13 \times \cos 58°$$
$$= 144 + 169 - 312 \cos 58°$$
$$= 147.9$$
$$p = 12.2$$

Now, having at least two sides and the angle opposite one of them, we may, if we wish, complete the solution by means of the law of sines instead of repeating the cosine solution.

$$\frac{12.2}{\sin 58°} = \frac{12}{\sin \mu}$$

from which $\quad\quad\quad \mu = 65°$

Similarly $\quad\quad\quad \lambda = \arcsin \dfrac{12 \sin 58°}{12.2} = 65°$

Test
$$58° + 56.8° + 65° = 179.8°$$

Example 4

Oblique triangle. Solve the triangle of Fig. 27-7.

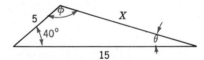

Fig. 27-7 Triangle of Example 4.

Solution

Since the information given is not sufficient to use the law of sines, check to see if the law of cosines may be applied. Knowing two sides and the angle between them is sufficient:

$$X^2 = 5^2 + 15^2 - 2 \times 5 \times 15 \times \cos 40°$$
$$= 135.1$$
$$X = 11.6$$

Then, using the law of sines,

$$\theta = \arcsin \frac{5 \sin 40°}{11.6} = 16.1°$$

and

$$\phi = \arcsin \frac{15 \sin 40°}{11.6} = 56.1°$$

Test

$$40° + 16.1° + 56.1° = 112.2° \qquad \text{Oh.}$$

From Fig. 27-7, the side of length 15, being the longest side, *must* be opposite the largest angle, which we have calculated as 56.1°. Since this must be the largest angle, since it *could* be obtuse (greater than 90°, an angle in the second quadrant), and since all that our calculations guarantee is that $\phi = \arcsin 0.831$, then perhaps ϕ is $180° - 56.1° = 123.9°$.

Testing this possibility,

$$40° + 16.1° + 123.9° = 180°$$

We have arrived at the correct solution.

Be sure to test your solutions.

Example 5

The three sides of a triangle are given, and it is required to solve the angles. (Note that if just three angles are given, there is an infinite number of solutions.) Solve the triangle of Fig. 27-8.

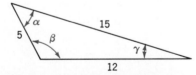

Fig. 27-8 Triangle of Example 5.

Solution

If any angle in the triangle can be obtuse, it will be angle β. (Why?) We will defer solving for β for now. Consider the angle α. It is related, by the law of cosines, as follows:

$$12^2 = 5^2 + 15^2 - 2 \times 5 \times 15 \times \cos \alpha$$

from which $\qquad \alpha = \arccos \dfrac{5^2 + 15^2 - 12^2}{2 \times 5 \times 15} = 45.1°$

You should confirm that

$$\gamma = \arccos \dfrac{12^2 + 15^2 - 5^2}{2 \times 12 \times 15} = 17.2°$$

Then $\qquad \beta = 180° - (45.1° + 17.2°) = 117.7°$

Alternatively, starting the solution for β,

$$15^2 = 5^2 + 12^2 - 2 \times 5 \times 12 \times \cos \beta$$

from which

$$\beta = \arccos(-0.466)$$

This negative cosine indicates immediately that β must be an angle between 90° and 180°, and we find it to be

$$180° - 62.3° = 117.7°$$

PROBLEMS 27-3

Referring to Fig. 27-2, solve the following triangles:

1. $b = 5.2, c = 8, \alpha = 63°$
2. $a = 544, b = 805, \gamma = 80°$
3. $a = 0.17, b = 0.785, \gamma = 132°$
4. $a = 2.6, c = 8.45, \beta = 48.8°$
5. $a = 1600, b = 3260, \gamma = 147.7°$
6. $b = 0.0945, c = 0.0980, \alpha = 5°$
7. $a = 3, b = 5, c = 7$
8. $a = 2000, b = 4000, c = 6000$
9. $a = 1280, b = 3260, c = 3935$
10. $a = 25, b = 30, c = 50$
11. The diagonals of a parallelogram are 130 mm and 180 mm, and they intersect at an angle of 38°. What are the sides of the parallelogram?
12. Using the data of Prob. 11, but *not* your results, what is the area of the parallelogram? (After obtaining a solution, check it by means of a different computational method.)

27-5 THE SUM IDENTITIES

Often in the solution of antenna and modulation problems we come upon various combinations such as sin (θ + φ) and cos (θ − φ). It is often convenient to resolve these forms into the products of simple trigonometric functions.

Consider triangle PQR, Fig. 27-9, with the altitude h dividing the angle RPQ into the angles, α and β. Since the area of the whole triangle must be equal to the sum of the areas of the two component triangles,

$$\tfrac{1}{2}qr \sin(\alpha + \beta) = \tfrac{1}{2}qh \sin \alpha + \tfrac{1}{2}rh \sin \beta$$

from which

$$\sin(\alpha + \beta) = \frac{h}{r} \sin \alpha + \frac{h}{q} \sin \beta$$

which yields

$$\sin(\alpha + \beta) = \sin \alpha \cos \beta + \cos \alpha \sin \beta \quad (10)$$

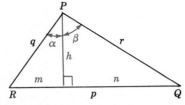

Fig. 27-9 Triangle adjusted for development of the sum identities.

Again, using the same triangle, Fig. 27-9, and the law of cosines,

$$(m + n)^2 = q^2 + r^2 - 2qr \cos(\alpha + \beta)$$

from which

$$\cos(\alpha + \beta) = \frac{q^2 + r^2 - m^2 - n^2 - 2mn}{2qr}$$

$$= \frac{q^2 - m^2}{2qr} + \frac{r^2 - n^2}{2qr} - \frac{2mn}{2qr}$$

$$= \frac{2h^2}{2qr} - \frac{2mn}{2qr}$$

$$= \frac{h}{q} \cdot \frac{h}{r} - \frac{m}{q} \cdot \frac{n}{r}$$

which converts to $\quad \cos(\alpha + \beta) = \cos \alpha \cos \beta - \sin \alpha \sin \beta \quad (11)$

27-6 THE DIFFERENCE IDENTITIES

Sometimes, instead of functions of the sum of two angles, it is necessary to deal with the differences of two angles: In triangle PQR, Fig. 27-10,

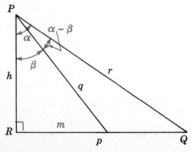

Fig. 27-10 Triangle adjusted for development of the difference identities.

the line q divides the vertex into two angles, β and $\alpha - \beta$. As in the sum identity, the area of the whole triangle is equal to the sum of the parts:

$$\tfrac{1}{2}hr \sin \alpha = \tfrac{1}{2}hq \sin \beta + \tfrac{1}{2}qr \sin (\alpha - \beta)$$

from which

$$\sin (\alpha - \beta) = \frac{hr \sin \alpha - hq \sin \beta}{qr}$$

$$= \frac{h}{q} \sin \alpha - \frac{h}{r} \sin \beta$$

which yields

$$\sin (\alpha - \beta) = \sin \alpha \cos \beta - \cos \alpha \sin \beta \qquad (12)$$

And, as before, using the law of cosines:

$$(p - m)^2 = q^2 + r^2 - 2qr \cos (\alpha - \beta)$$

from which

$$\cos (\alpha - \beta) = \frac{q^2 + r^2 - p^2 - m^2 + 2mp}{2qr}$$

$$= \frac{q^2 - m^2}{2qr} + \frac{r^2 - p^2}{2qr} + \frac{2mp}{2qr}$$

$$= \frac{h}{q} \cdot \frac{h}{r} + \frac{m}{q} \cdot \frac{p}{r}$$

which yields

$$\cos (\alpha - \beta) = \cos \alpha \cos \beta + \sin \alpha \sin \beta \qquad (13)$$

Example 6
Simplify the expression

$$\sin (\theta + 45°) + \cos (\theta + 45°)$$

Solution
Using Eqs. (10) and (11) and substituting the equivalent product expressions:

$$\sin (\theta + 45°) + \cos (\theta + 45°)$$
$$= \sin \theta \cos 45° + \cos \theta \sin 45° + \cos \theta \sin 45° - \sin \theta \sin 45°$$
$$= 0.7071 \sin \theta + 0.7071 \cos \theta + 0.7071 \cos \theta - 0.7071 \sin \theta$$
$$= 1.4142 \cos \theta$$

TABLE 27.1   TRIGONOMETRIC IDENTITIES AND USEFUL RELATIONSHIPS

$$\tan \theta = \frac{\sin \theta}{\cos \theta} \qquad \cot \theta = \frac{\cos \theta}{\sin \theta}$$

$$\sin^2 \theta + \cos^2 \theta = 1$$
$$1 + \tan^2 \theta = \sec^2 \theta$$
$$1 + \cot^2 \theta = \csc^2 \theta$$

$$\frac{a}{\sin \alpha} = \frac{b}{\sin \beta} = \frac{c}{\sin \gamma}$$

$$a^2 = b^2 + c^2 - 2bc \cos \alpha$$

$$\sin (\theta + \phi) = \sin \theta \cos \phi + \cos \theta \sin \phi$$
$$\cos (\theta + \phi) = \cos \theta \cos \phi - \sin \theta \sin \phi$$
$$\sin (\theta - \phi) = \sin \theta \cos \phi - \cos \theta \sin \phi$$
$$\cos (\theta - \phi) = \cos \theta \cos \phi + \sin \theta \sin \phi$$

PROBLEMS 27-4

Using the sum and difference relationships, simplify:

1. $\sin(\theta + 30°) + \cos(\theta + 30°)$
2. $\sin(45° - \theta) - \cos(45° + \theta)$
3. $\sin(\theta - 60°) + \cos(\theta + 60°)$
4. $\sin(\theta - 30°) - \cos(\theta - 45°)$

Given $\sin\theta = \frac{3}{5}$ and $\sin\phi = \frac{5}{12}$, evaluate:

5. $\cos(\theta + \phi)$
6. $\sin(\theta - \phi) - \cos(\theta - \phi)$
7. Use Eq. (10) to show that $\sin 2\theta = 2\sin\theta\cos\theta$.
8. Use Eq. (11) to show that $\cos 2\theta = \cos^2\theta - \sin^2\theta$.
9. When a VHF direction-finding array is fed in modulation phase quadrature, the two fields about the antennas are

$$V_1 = K\cos\theta\cos pt\cos\omega t$$
$$V_2 = K\sin\theta\sin pt\cos\omega t$$

Show that the total field $V_t = V_1 + V_2 = K\cos\omega t\cos(pt - \theta)$.

10. Use Eqs. (11) and (13) to show that

$$\tfrac{1}{2}\cos(\omega t - pt) - \tfrac{1}{2}\cos(\omega t + pt) = \sin pt\sin\omega t$$

It is based on this relationship that an amplitude-modulated carrier wave is shown to consist of a fundamental and two sidebands. The equation of the modulated carrier wave is

$$v = V\sin\omega t + mV\sin\omega t\sin pt$$

where m is the depth of modulation, and your work in this problem shows the correctness of the substitution:

$$v = V\sin\omega t + \tfrac{1}{2}mV\cos(\omega t - pt) - \tfrac{1}{2}mV\cos(\omega t + pt)$$

where $V\sin\omega t$ represents the original carrier and the other two parts represent the difference and sum sideband frequencies whose amplitudes are each one-half that of the carrier voltage when $m = 1$.

Elementary Plane Vectors

28

Many physical quantities can be expressed by specifying a certain number of units. For example, the volume of a tank may be expressed as so many cubic meters, the temperature of a room as a certain number of degrees, and the speed of a moving object as a number of linear units per unit of time such as kilometers per hour or meters per second. Such quantities are *scalar quantities,* and the numbers that represent them are called *scalars*. A scalar quantity is one that has only magnitude; that is, it is a quantity fully described by a number, but it does not involve any concept of direction.

28-1 VECTORS

Many other types of physical quantities need to be expressed more definitely than is possible by specifying magnitude alone. For example, the velocity of a moving object has a direction as well as a magnitude. Also, a force due to a push or a pull is not completely described unless the direction as well as the magnitude of the force is given. In addition, electric circuit analysis is built up around the idea of expressing the directions and magnitudes of voltages and currents. Quantities which have both magnitude and direction are called *vector quantities*. A vector quantity is conveniently represented by a directional straight-line segment called a magnitude and whose head points in the direction of the vector quantity.

Example 1

If a vessel steams northeast at a speed of 15 kn, its speed can be represented by a line whose length represents 15 kn, to some convenient scale, as shown in Fig. 28-1. The direction of the line represents the direction in which the vessel is traveling. Thus the line **OA** is a vector that completely describes the velocity of the vessel.

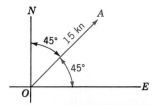

Fig. 28-1 Vector **OA** of Example 1.

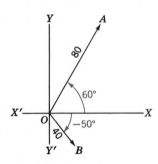

Fig. 28-2 Vector Diagram of Example 2.

Example 2

In Fig. 28-2, the vector **OA** represents a force of 80 N pulling on a body at O in a direction of 60°. The vector **OB** represents a force of 40 N acting on the same body in a direction of 310° or $-50°$.

Two vectors are equal if they have the same magnitude and direction. Thus, in Fig. 28-3, vectors **A, B,** and **C** are equal.

Fig. 28-3 Vectors **A, B,** and **C** are equal.

28-2 NOTATION

As you progress in the study of vectors, you will find that vectors and scalars satisfy different algebraic laws. For example, a scalar when reduced to its simplest terms is simply a number and as such obeys all the laws of ordinary algebraic operations. Since a vector involves direction, in addition to magnitude, it does not obey the usual algebraic laws and therefore has an analysis peculiar to itself.

From the foregoing, it is apparent that it is desirable to have a notation that indicates clearly which quantities are scalars and which are vectors. Several methods of notation are used, but you will find little cause for confusion; for most authors specify and explain their particular system of notation.

A vector can be denoted by two letters, the first indicating the origin, or initial point, and the second indicating the head, or terminal point. This form of notation was used in Examples 1 and 2 of the preceding section. Sometimes a small arrow is placed over these letters to emphasize that the quantity considered is a vector. Thus, $\overrightarrow{OA}$ could be used to represent the vector from O to A as in Fig. 28-2. In most texts, vectors are indicated by boldface type; thus, **A** denotes the vector A. Other common forms of specifying a vector quantity, as, for example, the vector A, are $\bar{A}$, $\dot{A}$, $\underset{\cdot}{A}$, and $\mathbf{A}$.

28-3 ADDITION OF VECTORS

Scalar quantities are added algebraically.

Thus   20 cents + 8 cents = 28 cents
and   16 insulators − 7 insulators = 9 insulators

Since vector quantities involve direction as well as magnitude, they cannot be added algebraically unless their directions are parallel. Figure 28-4 illustrates vectors **OA** and **AB**. Vector **OA** can be considered as a motion from O to A, and vector **AB** as a motion from A to B. Then the sum of the vectors represents the sum of the motions from O to A and from A to B, which is the motion from O to B. This sum is the vector **OB;** that is, the vector sum of **OA** and **AB** is **OB**. Therefore, the sum of two vectors is the vector joining the initial point of the first to the terminal point of the second if the initial point of the second vector is joined to the terminal point of the first vector as shown in Fig. 28-4.

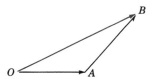

Fig. 28-4 Vector **OB** is the vector sum of **OA** and **AB**.

In Fig. 28-5, vectors **OC** and **OD** are equal to vectors **OA** and **AB**, respectively, of Fig. 28-4. In Fig. 28-5, however, the vectors start from the same origin. That their sum can be represented by the diagonal of a parallelogram of which the vectors are adjacent sides is evident by comparing Figs. 28-4 and 28-5. This is known as the *parallelogram law* for the composition of forces, and it holds for the composition or addition of all vector quantities.

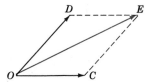

Fig. 28-5 Resultant vector **OE** is the vector sum of **OC** and **OD**.

The addition of vectors that are not at right angles to each other will be considered in Sec. 28-7. At this time, it is sufficient to know that two forces acting simultaneously on a point, or an object, can be replaced by a single force called the *resultant*. That is, the resultant force will produce the same effect on the object as the joint action of the two forces. Thus, in Fig. 28-4 the vector **OB** is the resultant of vectors **OA** and **AB**. Similarly, in Fig. 28-5, the vector **OE** is the resultant of the vectors **OC** and **OD**. Note that **OB** = **OE**.

Example 3

Three forces, A, B, and C, are acting on point O, as is shown in Fig. 28-6. Force A exerts 150 N at an angle of 60°; B exerts 100 N at an angle of 135°; and C exerts 150 N at an angle of 260°. What is the resultant force on point O?

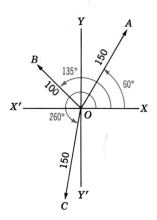

Fig. 28-6 Vector diagram of Example 3.

Solution

The resultant of vectors A, B, and C can be found graphically by either of two methods.

(*a*) First draw the vectors to scale. Find the resultant of any two vectors, such as OA and OC, by constructing a parallelogram with OA and OC as adjacent sides. Then the resultant of OA and OC will be the diagonal OD of the parallelogram $OADC$ as was illustrated in Fig. 28-7. In effect, there are now two forces, OB and OD, acting on point O. The resultant of these two forces is found as before by constructing a parallelogram with OB and OD as adjacent sides. The resultant force on point O is then the diagonal OE of the parallelogram $OBED$. By measurement with scale and protractor, OE is found to be 57 N acting at an angle of 112°.

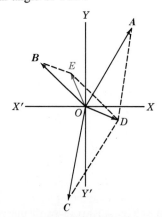

Fig. 28-7 **OE** is the vector sum of vectors **A, B,** and **C.**

(*b*) Draw the vectors to scale as shown in Fig. 28-8, joining the initial point of *C* to the terminal point of *B*. The vector drawn from the point *O* to the terminal point of *C* is the resultant force, and measurements show it to be the same as that found by the method that is illustrated in Fig. 28-7.

A figure such as *OABCO*, in Fig. 28-8, is called a *polygon of forces*. The vectors can be joined in any order as long as the initial point of one vector joins the terminal point of another vector and the vectors are drawn with the proper magnitude and direction. The length and direction of the line that is necessary to close the polygon, that is, the line from the original initial point to the terminal point of the last vector drawn, constitute a vector that represents the magnitude and the direction of the resultant.

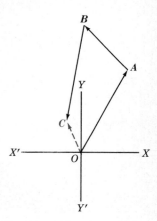

Fig. 28-8 **OC** is the vector sum of vectors **A**, **B**, and **C**.

PROBLEMS 28-1

1 to 4. Find the magnitude and direction, with respect to the positive *x* axis, of the vectors shown in Figs. 28-9 to 28-12.

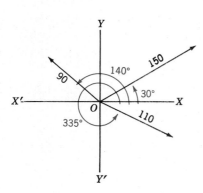

Fig. 28-9 Vector diagram of Prob. 1.

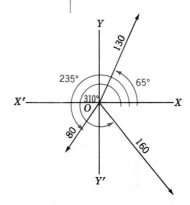

Fig. 28-10 Vector diagram of Prob. 2.

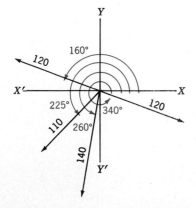

Fig. 28-11 Vector diagram of Prob. 3.

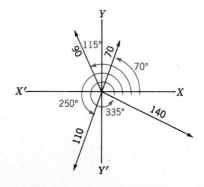

Fig. 28-12 Vector diagram of Prob. 4.

28-4 COMPONENTS OF A VECTOR

From what has been considered regarding combining or adding vectors, it follows that a vector can be resolved into components along any two specified directions. For example, in Fig. 28-4, the vectors *OA* and *AB* are components of the vector *OB*. If the directions of the components are so chosen that they are right angles to each other, the components are called *rectangular components*.

By placing the initial point of a vector at the origin of the *x* and *y* axes, the rectangular components are readily obtained either graphically or mathematically.

Example 4

A vector with a magnitude of 10 makes an angle of 53.1° with the horizontal. What are the vertical and horizontal components?

Solution

The vector is illustrated in Fig. 28-13 as the directed line segment *OA*. Its length drawn to scale represents the magnitude of 10, and it makes an angle of 53.1° with the *x* axis.

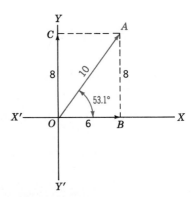

Fig. 28-13 Vertical and horizontal components of vector.

The *horizontal component* of *OA* is the horizontal distance (see Sec. 26-7) from *O* to *A* and is found graphically by projecting the vector *OA* upon the *x* axis. Thus the vector *OB* is the horizontal component of *OA*.

The *vertical component* of *OA* is the vertical distance from *O* TO *A* and is found graphically by projecting the vector *OA* upon the *y* axis. Similarly, the vector *OC* is the vertical component of *OA*. Finding the horizontal and vertical components of *OA* by mathematical methods is simply a problem in solving a right triangle as was outlined in Sec. 26-4. Hence,

$$OB = 10 \cos 53.1° = 6$$

and $$OC = BA = 10 \sin 53.1° = 8$$

Check

$$0 = \arctan \tfrac{8}{6} = \arctan 1.33 = 53.1°$$
$$10^2 = 6^2 + 8^2 = 36 + 64 = 100$$

The foregoing can be summarized as follows:

RULE

1. The horizontal component of a vector is the projection of the vector upon a horizontal line, and it equals the magnitude of the vector multiplied by the cosine of the angle made by the vector with the horizontal.
2. The vertical component of a vector is the projection of the vector upon a vertical line, and it equals the magnitude of the vector multiplied by the sine of the angle made by the vector with the horizontal.

Example 5

An airplane is flying on a course of 40° at a speed of 400 km/h. How many kilometers per hour is the plane advancing in a due eastward direction? In a direction due north?

Solution

Draw the vector diagram as shown in Fig. 28-14. (Courses are measured from the north.) The vector *OB*, which is the horizontal component of *OA*, represents the velocity of the airplane in an eastward direction. The vector, *OC*, which is the vertical component of *OA*, represents the velocity of the airplane in a northward direction.

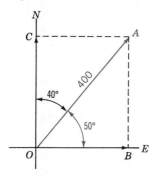

Fig. 28-14 Vector diagram of Example 5.

Again, the process of finding the magnitude of *OB* and *OC* resolves into a problem of solving the right triangle *OBA*. Hence,

$$OB = 400 \cos 50° = 257 \text{ km/h eastward}$$
$$OC = BA = 400 \sin 50° = 306.5 \text{ km/h northward}$$

If the vector diagram has been drawn to scale, an approximate check can be made by measuring the lengths of *OB* and *OC*. Such a check will disclose any large errors in the mathematical solution.

Example 6

A radius vector of unit length is rotating about a point with a velocity of $2\pi^r$/s. What are its horizontal and vertical components (*a*) at the end of 0.15 s, (*b*) at the end of 0.35 s, (*c*) at the end of 0.75 s?

Solution

(*a*) At the end of 0.15 s the rotating vector will have generated $2\pi \times 0.15 = 0.942^r$, or $0.942 \times 57.3° = 54°$ as shown in Fig. 28-15.

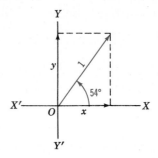

Fig. 28-15 When $t = 0.15$ s, angle $\theta = 54°$.

The horizontal component, measured along the x axis, is

$$x = 1 \cos 54° = 0.588$$

The vertical component, measured along the y axis, is

$$y = 1 \sin 54° = 0.809$$

Check the solution by measurement or any other method considered convenient.

(*b*) At the end of 0.35 s the rotating vector will have generated an angle of $2\pi \times 0.35 = 2.20^r$, or $2.20 \times 57.3° = 126°$ as shown in Fig. 28-16. The horizontal component, measured along the x axis, is

$$x = 1 \cos 126° = 1(-\cos 54°)$$
$$= -0.588 \qquad \text{(Sec. 25-11)}$$

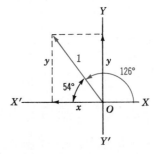

Fig. 28-16 When $t = 0.35$ s, angle $\theta = 126°$.

The vertical component, measured along the y axis, is

$$y = 1 \sin 126° = 1 \sin 54° = 0.809 \qquad \text{(Sec. 25-11)}$$

Check by some convenient method.

(*c*) At the end of 0.75 s the rotating vector will have generated $2\pi \times$ 0.75 = 4.71^r, or 4.71 × 57.3° = 270° as shown in Fig. 28-17.

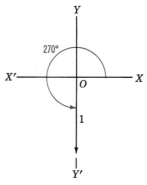

Fig. 28-17 When *t* = 0.75 s, angle **θ** = 270°.

The horizontal component is

$$x = 1 \cos 270° = 0$$

The vertical component is

$$y = 1 \sin 270° = -1$$

PROBLEMS 28-2

Find the horizontal and vertical components, denoted by *x* and *y*, respectively, of the following vectors. Check the mathematical solution of each by drawing a vector diagram to scale.

1. 30 at 65.5° (This is commonly written 30 $\underline{/65.5°}$)

2. 99 $\underline{/22.8°}$ 3. 0.865 $\underline{/87.2°}$ 4. 1800 $\underline{/120°}$

5. 46.3 $\underline{/180°}$ 6. 0.987 $\underline{/295.5°}$ 7. 185.5 $\underline{/252.2°}$

8. 27.8275 $\underline{/90°}$ 9. 30.8 $\underline{/157.3°}$ 10. 1600 $\underline{/270°}$

11. The resultant of two forces acting at right angles is a force of 765 N which makes an angle of 17.8° with one of the forces. Find the component forces.

12. A test missile was fired at an angle of 82° from the horizontal. At a particular instant its velocity was 1950 km/h. Find its horizontal velocity at that instant in meters per second.

13. A jet fighter leaves its base and flies 1200 m southeast. How far east does it go?

14. Resolve a force of 250 N into two rectangular components one of which is 155 N.

15. The resultant of two forces acting at right angles is 199 N. One of the forces is 150 N. What is the other?

28-5 PHASORS

Early in this chapter we discovered the difference between scalar quantities, which involve magnitude only, and vectors, which involve both magnitude and direction. When electrical units are shown on paper, with the length of the line indicating the magnitude and the direction of the line indicating the phase relationship, they may be thought of as *vectors*. However, when an emf is impressed across a circuit, its *polarity* is not *direction* in the sense of vector definition. The paper representation as vectors serves a valuable purpose in our circuit calculations, but the electrical quantities are not true vectors. Since the angular separation of electrical units always represents *time* revealed as a *phase* relationship, scientists and engineers prefer to use the term *phasors* when discussing electrical "vectors."

On paper (in a "uniplanar" representation) there is no difference between phasors and vectors. The operations of conversion between rectangular and polar forms are the same. The summation of perpendicular components is the same. But since our purpose is to study the mathematics of electronics in an electronics environment and our communication is with electronics and scientific people, we will use the expressions *phasor* and *phasor summation* throughout the remaining chapters of this book.

28-6 PHASOR SUMMATION OF RECTANGULAR COMPONENTS

If two forces that are at right angles to each other are acting on a body, their resultant can be found by the usual methods of phasor summation as outlined in Sec. 28-3. However, the resultant can be obtained by geometric or trigonometric methods; for the problem is that of solving for the hypotenuse of a right triangle when the other two sides are given, as outlined in Sec. 26-6.

Example 7

Two phasors are acting at a point. One with a magnitude of 6 is directed along the horizontal to the right of the point, and the other with a magnitude of 8 is directed vertically above the point. Find their resultant.

Solution 1

In Fig. 28-18 the horizontal phasor, with a magnitude of 6, is shown as **OB**. The vertical phasor, with a magnitude of 8, is shown as **OC**. The resultant of these two phasors can be obtained graphically by completing the parallelogram of forces *OCAB*, as was outlined in Sec. 28-3. Thus, the magnitude of the resultant will be represented by the length of **OA** in Fig. 28-18. The angle, or direction of the resultant, can be measured with the protractor.

Graphical methods have a limited degree of accuracy, as pointed out in earlier sections. They should be used as an approximate check for more precise mathematical methods.

Solution 2

Since **BA** = **OC** in Fig. 28-18, *OBA* is a right triangle the hypotenuse of which is the resultant **OA**. Therefore, the magnitude of the resultant is

$$OA = \sqrt{OB^2 + BA^2} = \sqrt{6^2 + 8^2}$$
$$= 10$$

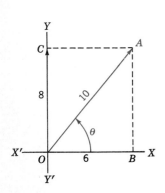

Fig. 28-18 Addition of rectangular components

The angle, or direction of the resultant, is

$$\theta = \arctan \frac{BA}{OB} = \arctan \frac{8}{6} = \arctan 1.33$$
$$= 53.1°$$

Although the method of Solution 2 is accurate and mathematically correct, there are several operations involved. For example, in finding the magnitude, 6 and 8 must be squared, these squares must be added, and then the square root of this sum must be extracted. This involves four operations.

Solution 3

Since OBA is a right triangle for which OB and BA are given, the hypotenuse (resultant) can be computed as explained in Sec. 26-6. Hence,

$$\tan \theta = \frac{BA}{OB} = 1.33$$

$$\therefore \theta = 53.1°$$

Then
$$OA = \frac{OB}{\cos 53.1°} = \frac{6}{0.6} = 10$$

or
$$OA = \frac{BA}{\sin 53.1°} = \frac{8}{0.8} = 10$$

The method of Solution 3 is to be preferred, owing to the minimum number of operations involved.

It should be noted that Example 4 of Sec. 28-4 involves the same quantities as those that were used in the example of this section and that Figs. 28-13 and 28-18 are alike. In the earlier example a vector that is resolved into its rectangular components is given. In the example of this section, the same components are given as vectors which are added vectorially to obtain the vector of the first example. From this it is apparent that resolving a vector into its rectangular components and adding vectors that are separated by 90° are inverse operations. Basically, either problem resolves itself into the solution of a right triangle.

PROBLEMS 28-3

Find the resultants of the following sets of phasors.

1. $64.3 \angle 0°$ and $415 \angle 90°$
2. $10.6 \angle 0°$ and $2.04 \angle 90°$
3. $1.23 \angle 90°$ and $1.47 \angle 0°$
4. $45.4 \angle 0°$ and $153 \angle 90°$
5. $351 \angle 0°$ and $94.8 \angle 90°$
6. $459 \angle 0°$ and $405 \angle 0°$
7. $307 \angle 0°$ and $124 \angle 180°$
8. $5.27 \angle 180°$ and $6.0 \angle 90°$
9. $310 \angle 270°$ and $185 \angle 90°$
10. $323 \angle 270°$ and $323 \angle 0°$
11. $2.34 \angle 180°$ and $7.30 \angle 270°$
12. $84.2 \angle 0°$, $34.4 \angle 90°$, and $37 \angle 90°$
13. $23.5 \angle 270°$, $32 \angle 90°$, $26.5 \angle 0°$, and $51 \angle 180°$

14. $167\underline{/270°}$, $252\underline{/0°}$, $143.8\underline{/180°}$, and $81.3\underline{/90°}$

15. $12.1\underline{/0°}$, $72.3\underline{/270°}$, $51.9\underline{/90°}$, $2.7\underline{/270°}$, $8.6\underline{/90°}$, and $31.6\underline{/180°}$

16. Check your calculated answers graphically.

28-7 PHASOR SUMMATION OF NONRECTANGULAR COMPONENTS

Often we are called upon to resolve into a resultant a set of phasors which are not themselves perpendicular (Fig. 28-19). The best analytical method of arriving at a solution is to apply the methods already developed in this chapter.

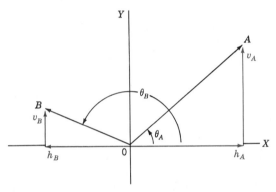

Fig. 28-19 Summation of nonrectangular phasors by resolution into rectangular components.

The first step is to find the perpendicular components of each of the phasors to be added and determine their magnitudes and directions. These are shown in Fig. 28-19 as h_A and v_A, the components of phasor A, and h_B and v_B, the components of phasor B.

Second, these components are added algebraically. The horizontal components are added to obtain the resultant horizontal phasor, and then the vertical components are added to obtain the resultant vertical phasor:

$$h_R = h_A + h_B$$

and

$$v_R = v_A + v_B$$

taking into consideration the signs as well as the magnitudes of the components.

Finally, the resultant is the phasor summation of the new perpendicular components:

$$R = \sqrt{h_R{}^2 + v_R{}^2}$$

$$\theta_R = \arctan\frac{v_R}{h_R}$$

$$R = \frac{h_R}{\cos\theta} = \frac{v_R}{\sin\theta}$$

Example 8
Find the resultant of two phasors $500\,\underline{/\,36.9°}$ and $142\,\underline{/\,135°}$.

Solution
Sketch the two phasors in the standard position (Fig. 28-20), and then resolve each phasor into its perpendicular components:

$$h_{500} = 500 \cos 36.9° = 400$$
$$h_{142} = 142 \cos 135° = -142 \cos 45° = -100$$
$$v_{500} = 500 \sin 36.9° = 300$$
$$v_{142} = 142 \sin 135° = 142 \sin 45° = 100$$

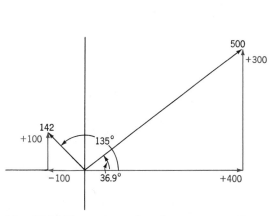

Fig. 28-20 Nonrectangular phasor summation of Example 8.

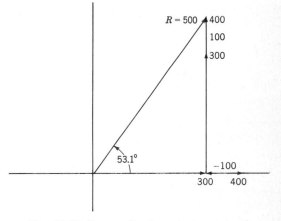

Fig. 28-21 Perpendicular components of Fig. 28-20 resolved into resultant R.

Add these components algebraically to obtain the new horizontal and vertical resultants:

$$h_R = +400 - 100 = 300$$
$$v_R = +300 + 100 = 400 \qquad \text{(Fig. 28-21)}$$

The angle θ_R, which R makes with the x axis, is

$$\theta = \arctan \frac{400}{300} = 53.1°$$

and the resultant R of the two resultant perpendicular components is

$$R = \frac{300}{\cos 53.1°}$$

or

$$R = \frac{400}{\sin 53.1°} = 500$$

This process of analysis of phasors into their components and synthesis of resultant components into a final phasor resultant may be applied to any number of phasors.

PROBLEMS 28-4

Find the resultants of the following sets of phasors. Check your solutions graphically:

1. $217 \underline{/63.8°}$ and $110 \underline{/40.3°}$ 2. $799 \underline{/48.7°}$ and $233 \underline{/120.2°}$

3. $100 \underline{/40.3°}$ and $39.6 \underline{/315°}$ 4. $7.65 \underline{/17.8°}$ and $4.34 \underline{/137.5°}$

5. $10.7 \underline{/32.8°}$, $42.0 \underline{/81.2°}$, and $61.2 \underline{/221.4°}$

If your electronic calculator is equipped with a polar-rectangular key (usually designated P → R), it is time to consult your instruction manual and determine the steps for *automatic* conversion from $5 \underline{/36.9°}$ to $4 + j3$, and vice versa. Attempt each of the conversions in this chapter by using the P → R key. Then you may use the key regularly in the following chapters.

Periodic Functions

In Sec. 24-9, it was shown that the trigonometric functions could be represented by the ratios of lengths of certain lines to the unit radius vector. Also, in Sec. 24-8, the variation of the functions was represented by lines.

The complete variation of the functions is more clearly illustrated and better understood by plotting the continuous values of the functions on rectangular coordinates.

29-1 THE GRAPH OF THE SINE CURVE $y = \sin x$

The equation $y = \sin x$ can be plotted just as the graphs of algebraic equations are plotted, that is, by assigning values to the angle x (the independent variable), computing the corresponding value of y (the dependent variable), plotting the points whose coordinates are thus obtained, and drawing a smooth curve through the points. This is the same procedure as used for plotting linear equations in Chap. 16 and for plotting quadratic equations in Chap. 21.

The first questions that come to mind in preparing to graph this equation are, "What values will be assigned to x? Will they be in radians or degrees?" Either might be used, but it is more reasonable to use radians. In Sec. 23-5, it was shown that an angle measured in radians can be represented by the arc intercepted by the angle on the circumference of a circle of unit radius. Since, as previously mentioned, the functions of an angle can be represented by suitable lengths of lines, it follows that if an angle is expressed in radian measure, both the angle and its functions can be expressed in terms of a common unit of length. Therefore, we shall select a suitable unit of length and plot both x and y values in terms of that unit. Then, to graph the equation $y = \sin x$, the procedure is as follows:

1. Assign values to x.
2. From the calculator or a table, determine the corresponding values of y (Table 29-1).
3. Take each pair of values of x and y as coordinates of a point, and plot the point.
4. Draw a smooth curve through the points.

TABLE 29-1

x, degrees	x, radians (π measure)	x, radians (unit measure)	y (sin x)	Point
0	0	0	0	$P_0 = (0, 0)$
30	$\frac{\pi}{6}$	0.52	0.50	$P_1 = (0.52, 0.50)$
60	$\frac{\pi}{3}$	1.05	0.87	$P_2 = (1.05, 0.87)$
90	$\frac{\pi}{2}$	1.57	1.00	$P_3 = (1.57, 1.00)$
120	$\frac{2\pi}{3}$	2.09	0.87	$P_4 = (2.09, 0.87)$
150	$\frac{5\pi}{6}$	2.62	0.50	$P_5 = (2.62, 0.50)$
180	π	3.14	0	$P_6 = (3.14, 0)$

It is not necessary to tabulate values of sin x between π and 2π radians (180 to 360°), for these values are negative but equal in magnitude to the sines of the angles between 0 and π radians (0 to 180°). To plot the curve, the angle and the function should have the same unit or scale; that is, one unit on the y axis should be the same length as that representing 1 radian on the x axis. When the curve is so plotted, it is called a *proper sine curve*, as shown in Fig. 29-1. This wave-shaped curve is called the *sine curve* or *sinusoid*.

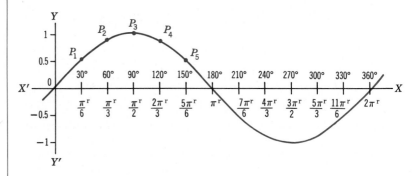

Fig. 29-1 Graph of the equation $y = \sin x$.

If additional values of x are chosen, both positive and negative, the curve continues indefinitely in both directions while repeating in value. Note that, as x increases from 0 to $\frac{\pi}{2}$ $\left(\text{or } \frac{1}{2}\pi\right)$, sin x increases from 0 to 1; as x increases from $\frac{1}{2}\pi$ to π, sin x decreases from 1 to 0; as x increases from π to $\frac{3\pi}{2}$, sin x increases from 0 to -1; and as x increases from

$\frac{3\pi}{2}$ to 2π, sin x decreases from -1 to 0. Thus the curve repeats itself for every multiple of 2π radians.

29-2 THE GRAPH OF THE COSINE CURVE $y = \cos x$

By following the procedure for plotting the sine curve, you can easily verify that the graph of $y = \cos x$ appears as shown in Fig. 29-2.

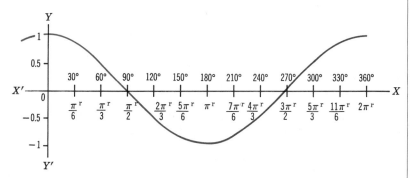

Fig. 29-2 Graph of the equation $y = \cos x$.

Note that, as x increases from 0 to $\frac{1}{2}\pi$, cos x decreases from 1 to 0; as x increases from $\frac{1}{2}\pi$ to π, cos x increases from 0 to -1; as x increases from π to $\frac{3\pi}{2}$, cos x decreases from -1 to 0; and as x increases from $\frac{3\pi}{2}$ to 2π, cos x increases from 0 to 1. If additional values of x are chosen, both positive and negative, the curve will repeat itself indefinitely in both directions. The cosine curve is identical in shape with the sine curve except that there is a difference of 90° between corresponding points on the two curves. Another similarity between the curves is that both curves repeat their values for every multiple of 2π radians ($2\pi^r$).

29-3 THE GRAPH OF THE TANGENT CURVE $y = \tan x$

The graph of the equation $y = \tan x$, shown in Fig. 29-3, has characteristics different from those of the sine or cosine curve. The curve slopes upward and to the right. At points where x is an odd multiple of $\frac{1}{2}\pi$, the curve is discontinuous. This is to be expected from the discussion of the tangent function in Sec. 24-8.

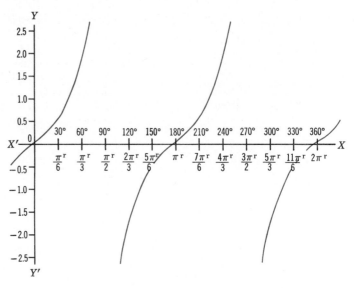

Fig. 29-3 Graph of the equation $y = \tan x$.

The tangent curve repeats itself at intervals of π radians (π^r), and it is thus seen to be a series of separate curves, or branches, rather than a continuous curve.

PROBLEMS 29-1

1. Plot the equation $y = \sin x$ from -2π to $2\pi^r$.
2. Plot the equation $y = \cos x$ from -2π to $2\pi^r$.
3. Plot the equation $y = \cot x$ from 0 to $2\pi^r$.
4. Plot the equation $y = \sec x$ from 0 to $2\pi^r$.
5. Plot the equation $y = \csc x$ from 0 to $2\pi^r$.
6. Plot the equations $y = \sin^2 x$ and $y = \cos^2 x$ on the same coordinates and to the same scale. In computing points, remember that when a negative number is squared, the result is positive. Add the respective ordinates of the curves for several different values of angle, and plot the results. What conclusion do you draw from these results?

29-4 PERIODICITY

From the graphs plotted in the preceding figures and from earlier considerations of the trigonometric functions, it is evident that each trigonometric function repeats itself exactly in the same order and at regular intervals. A function that repeats itself periodically is called a *periodic function*. From the definition, it is apparent that the trigonometric functions are periodic functions.

Owing to the fact that many natural phenomena are periodic in character, the sine and cosine curves lend themselves ideally to graphical representation and mathematical analysis of these recurrent motions. For example, the rise and fall of tides, motions of certain machines, the vibrations of a pendulum, the rhythm of our bodily life, sound waves, and water waves are familiar happenings that can be represented and analyzed by the use of sine and cosine curves. An alternating current follows these variations, as will be shown in Chap. 30, and it is because of this fact that you must have a good grounding in trigonometry. It is essential that you understand the mathematical expressions for various periodic functions and especially their applications to ac circuits.

The tangent, cotangent, secant, and cosecant curves are not used to represent recurrent happenings, for although these curves are periodic, they are discontinuous for certain values of angles.

29-5 ANGULAR MOTION

The *linear velocity* of a point or object moving in a particular direction is the rate at which distance is traveled by the point or object. The unit of velocity is the distance traveled in unit time when the motion of the point or object is uniform, such as kilometers per hour, meters per second, or centimeters per second.

The same concept is used to measure and define *angular velocity*. In Fig. 29-4 the radius vector r is turning about the origin in a counterclockwise direction to generate the angle θ. The *angular velocity* of such a rotating line is the rate at which an angle is generated by rotation. When the rotation is uniform, the unit of angular velocity is the angle generated per unit of time. Thus, angular velocity is measured in degrees per second or radians per second, the latter being the more widely used.

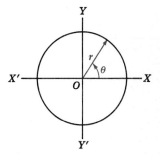

Fig. 29-4 Radius vector **r** generates angle **θ**.

Angular velocity may be expressed as revolutions per minute or revolutions per second. For example, if f is the number of revolutions per second of the vector of Fig. 29-4, then $2\pi f$ is the number of radians generated per second. The angular velocity in radians per second is denoted by ω (Greek letter omega). Thus, if the radius vector is rotating f revolutions per second,

$$\omega = 2\pi f \quad \text{rad/s}$$

If the armature of a generator is rotating at 1800 rev/min, which is 30 rev/s, it has an angular velocity of

$$\omega = 2\pi f = 2\pi \times 30 = 188.4^{r}/s$$

where we have again used ^{r} as the symbol for radians.

The total angle θ generated by a rotating line in t s at an angular velocity of ω^{r}/s is

$$\theta = \omega t \quad \text{rad}$$

Thus the angle generated by the armature in 0.01 s is

$$\theta = \omega t = 188.4 \times 0.01 = 1.884^{r}$$

or

$$\theta = 1.884 \times \frac{180}{\pi} = 108°$$

Example 1

A flywheel has a velocity of 300 rev/min. (*a*) What is its angular velocity? (*b*) What angle will be generated in 0.2 s? (*c*) How much time is required for the wheel to generate 628^r?

Solution

(*a*) $$f = \frac{300 \text{ rev/min}}{60} = 5 \text{ rev/s}$$

Then $$\omega = 2\pi f = 2\pi \times 5 = 10\pi \text{ or } 31.4^r/\text{s}$$

(*b*) $$\theta = \omega t = 10\pi \times 0.2 = 2\pi^r$$
$$\theta = 360°$$

(*c*) Since $$\theta = \omega t$$

then $$t = \frac{\theta}{\omega} = \frac{628}{10\pi} = 20 \text{ s}$$

PROBLEMS 29-2

1. What is the angular velocity, in terms of π^r/s, of (*a*) the hour hand of a clock, (*b*) the minute hand of a clock, and (*c*) the second hand of a clock?

2. Express the angular velocity of 1800 rev/min in (*a*) radians per second and (*b*) degrees per second.

3. If a satellite circles the earth in 80 min, what is its average angular velocity in (*a*) degrees per minute and (*b*) radians per second?

4. A revolution counter on an armature shaft recorded 900 rev in 30 s. What was the value of the shaft's angular velocity in (*a*) radians per minute and (*b*) degrees per minute?

5. The radius vector *r* of Fig. 29-4 is rotating at the rate of 3600 rev/s. What is the value of θ in radians at the end of:
   (*a*) 0.01 s, (*b*) 0.001 s, and (*c*) 0.5 ms?

6. If the radius vector *r* of Fig. 29-4 is rotating at the rate of 1 rev/s, what is the value of sin ωt at the end of:
   (*a*) 0.001 s, (*b*) 0.1 s, (*c*) 0.5 s, and (*d*) 0.95 s?

29-6 PROJECTION OF A POINT HAVING UNIFORM CIRCULAR MOTION

In Fig. 29-5 the radius vector *r* rotates about a point in a counterclockwise direction with a uniform angular velocity of 1 rev/s. Then every point on the radius vector, such as the end point *P*, rotates with uniform angular velocity. If the radius vector starts from 0°, at the end of $\frac{1}{12}$ s it will have rotated 30° or 0.5236^r, to P_1, at the end of $\frac{1}{6}$ s, it will have rotated to P_2 and generated an angle of 60°, or 1.047^r, etc.

The projection of the end point of the radius vector, that is, its ordinate value at any time, can be plotted as a curve. The plotting is accomplished by extending the horizontal diameter of the circle to the right for use as an *x* axis along which time is to be plotted. Choose a convenient length

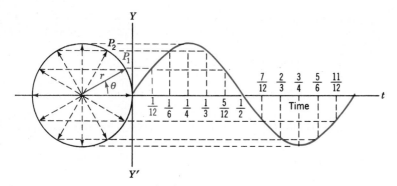

Fig. 29-5 Radius vector generating sine curve.

along the x axis and divide it into as many intervals as there are angle values to be plotted. In Fig. 29-5, projections have been made every 30°, starting from 0°. Therefore, the x axis is divided into 12 divisions; and since one complete revolution takes place in 1 s, each division on the time axis will represent $\frac{1}{12}$ s, or 30° rotation.

Through the points of division on the time axis (x axis), construct vertical lines, and through the corresponding points (made by the end point of the radius vector at that particular time) draw lines parallel to the time axis. Draw a smooth curve through the points of intersection. Thus the resulting sine curve traces the ordinate of the end point of the radius vector for any time t, and from it we could obtain the sine value for any angle generated by the radius vector.

As the vector continues to rotate, successive revolutions will generate repeating, or periodic, curves.

Since the y value of the curve is proportional to the sine of the generated angle and the length of the radius vector, we have

$$y = r \sin \theta$$

Then, since the radius vector rotates through $2\pi^r$ in 1 s, the y value at any time t is

$$y = r \sin 2\pi t$$
or $$y = r \sin 6.28t$$

which is the equation of the sine curve of Fig. 29-5.

From the foregoing considerations, it is apparent that if a straight line of length r rotates about a point with a uniform angular velocity of ω^r per unit time, starting from a horizontal position when the time $t = 0$, the projection y of the end point upon a vertical straight line will have a motion that can be represented by the relationship

$$y = r \sin \omega t \qquad (1)$$

This equation is of fundamental importance in describing the motion of any object or quantity that varies periodically, or with *simple harmonic motion*. Thus the value of an alternating emf at any instant can be completely described in terms of such an equation; that is, if the motion or variation can be represented by a sine curve, it is said to be *sinusoidal* or to vary *sinusoidally*.

Example 2

A crank 150 mm long, starting from 0°, turns in a counterclockwise direction at the rate of 1 rev in 10 s.

(a) What is the equation for the projection of the crank handle upon a vertical line at any instant? That is, what is the vertical distance from the crankshaft at any time?

(b) What is the vertical distance from the handle to the shaft at the end of 3 s?

(c) At the end of 8 s?

Solution

(a) The general equation for the projection of the end point on a vertical line is

$$y = r \sin \omega t \tag{1}$$

where r = length of rotating object
ω = angular velocity, rad/s
t = time at any instant, s

Then, since the crank makes 1 rev, or $2\pi^r$, in 10 s, the angular velocity is

$$\omega = \frac{2\pi}{10} = \frac{\pi}{5}, \text{ or } 0.628^r/s$$

Substituting the values of r and ω in Eq. (1),

$$y = 150 \sin 0.628t \text{ mm}$$

(b) At the end of 3 s, the crank will have turned through

$$0.628 \times 3 = 1.88^r$$

which is $1.88 \times \dfrac{180}{\pi} = 108°$. Substituting this value for $0.628t$ in Eq. (1) results in

$$y = 150 \sin 108° = 150 \times 0.951 = 142.6 \text{ mm}$$

which is the vertical distance of the handle from the shaft at the end of 3 s.

(c) At the end of 8 s the crank will have turned through

$$0.628 \times 8 = 5.02^r$$

which is $5.02 \times \dfrac{180}{\pi} = 288°$. Substituting this value for $0.628t$ in the above equation results in

$$y = 150 \sin 288° = 150 \times (-0.951) = -142.6 \text{ mm}$$

which is the vertical distance of the handle from the shaft at the end of 8 s. The negative sign denotes that the handle is *below* the shaft; that is, the distance is measured downward, whereas the distance in (b) above was taken as positive, or *above* the shaft.

If it is desired to express the projection of the end point of the radius vector upon the horizontal, the relation is

$$y = r \cos \omega t \tag{2}$$

which, when plotted, results in a cosine curve. Thus, in the foregoing example, the horizontal distance (Sec. 26-7) between the handle and shaft at the end of 8 s will be

$$y = 150 \cos 288° = 150 \times 0.309$$
$$= 46.4 \text{ mm}$$

29-7 AMPLITUDE

The graphs of Figs 29-1, 29-2, and 29-5 have an equal amplitude of 1, that is, an equal vertical displacement from the horizontal axis. The value of the radius vector r determines the amplitude of a general curve, and for this reason the factor r in the general equation

$$y = r \sin \omega t$$

is called the *amplitude factor*. Thus the amplitude of a periodic curve is taken as the maximum displacement, or value, of the curve. It is apparent that, if the length of the radius vector which generates a sine wave is varied, the amplitude of the sine wave will be varied accordingly. This is illustrated in Fig. 29-6.

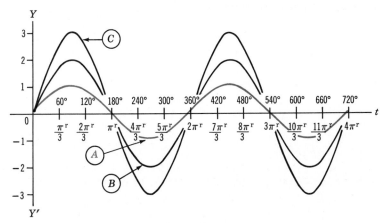

Fig. 29-6 A: $y = \sin \theta$, B: $y = 2 \sin \theta$, C: $y = 3 \sin \theta$.

29-8 FREQUENCY

When the radius vector makes one complete revolution, regardless of its starting point, it has generated one complete sine wave; hence, we say the sine wave has gone through one complete *cycle*. Thus the number of cycles occurring in a periodic curve in a unit of time is called the *frequency* of the curve. For example, if the radius vector rotated 5 rev/s, the curve describing its motion would go through 5 cycles in 1 s of time. The frequency f in hertz is obtained by dividing the angular velocity ω by 360° when the latter is measured in degrees or by 2π when measured in radians. That is,

$$f = \frac{\omega}{2\pi} \quad \text{Hz} \tag{3}$$

Curves for different frequencies are shown in Fig. 29-7.

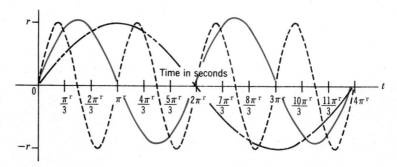

Fig. 29-7 $y = r \sin t$ ———, $y = r \sin 2t$ –, $y = r \sin \frac{1}{2}t$ ——— — ———

In the equation $y = r \sin \frac{1}{2}t$, since $\omega t = \frac{1}{2}t$, the angular velocity ω is 0.5 rad/s. That is, at the end of 2π, or 6.28 s, the curve has gone through one-half cycle, or 3.14^r of an angle, as shown in Fig. 29-7.

In the equation $y = r \sin t$, since $\omega t = t$, the angular velocity ω is 1 rad/s. Thus at the end of 2π s the curve has gone through one complete cycle, or $2\pi^r$ of angle.

Similarly, in the equation $y = r \sin 2t$, the angular velocity ω is 2 rad/s. Then at the end of 2π s the curve has completed two cycles, or $4\pi^r$ of angle.

29-9 PERIOD

The time T required for a periodic function, or curve, to complete one cycle is called the *period*. Hence, if the frequency f is given by

$$f = \frac{\omega}{2\pi} \quad \text{Hz}$$

it follows that

$$T = \frac{2\pi}{\omega} = f^{-1} \quad \text{s} \tag{4}$$

For example, if a curve repeats itself 60 times in 1 s, it has a frequency of 60 Hz and a period of

$$T = \frac{1}{60} = 0.0167 \text{ s}$$

Similarly, in Fig. 29-7, the curve represented by $y = r \sin \frac{1}{2}t$ has a frequency of

$$\frac{\omega}{2\pi} = \frac{0.5}{2\pi} = 0.0796 \text{ Hz}$$

and a period of 12.6 s. The curve of $y = r \sin t$ has a frequency of

$$\frac{\omega}{2\pi} = \frac{1}{2\pi} = 0.159 \text{ Hz}$$

and a period of 6.28 s. The curve of $y = r \sin 2t$ has a frequency of 0.318 Hz and a period of 3.14 s.

29-10 PHASE

In Fig. 29-8, two radius vectors are rotating about a point with equal angular velocities of ω and separated by the constant angle θ. That is, if r starts from the horizontal axis, then r_1 starts ahead of r by the angle θ and maintains this angular difference.

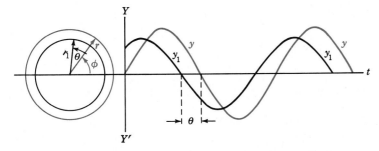

Fig. 29-8 $y = r \sin \omega t$, $y_1 = r_1 \sin (\omega t + \theta)$.

When $t = 0$, r starts from the horizontal axis to generate the curve $y = r \sin \omega t$. At the same time, r_1 is ahead of r by an angle θ; hence, r_1 generates the curve

$$y_1 = r_1 \sin (\omega t + \theta)$$

It will be noted that this *displaces* the y_1 curve along the horizontal by an angle θ as shown in the figure.

The angular difference θ between the two curves is called the *phase angle,* and since y_1 is *ahead* of y, we say that y_1 leads y. Thus, in the equation $y_1 = r_1 \sin (\omega t + \theta)$, θ is called the *angle of lead*. In the Fig. 29-8, y_1 leads y by 30°; therefore, the equation for y_1 becomes

$$y_1 = r_1 \sin (\omega t + 30°)$$

In Fig. 29-9, the radius vectors r and r_1 are rotating about a point with equal angular velocities of ω, except that now r_1 is *behind* r by a constant angle θ. The phase angle between the two curves is θ, but in this case, y_1 lags y. Hence the equation for the curve generated by r_1 is

$$y_1 = r_1 \sin (\omega t - \theta)$$

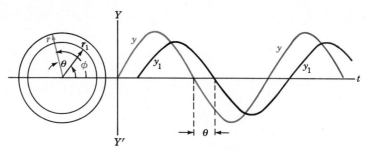

Fig. 29-9 $y = \mathbf{r} \sin \omega t$, $y_1 = \mathbf{r}_1 \sin (\omega t - \theta)$.

In Fig. 29-9, the *angle of lag* is $\theta = 60°$; therefore, the equation for y_1 becomes

$$y_1 = \mathbf{r}_1 \sin (\omega t - 60°)$$

29-11 SUMMARY

The general equation

$$y = r \sin (\omega t \pm \theta) \tag{5}$$

describes a periodic event, and its graph results in a periodic curve. By choosing the proper values for the three arbitrary constants r, ω, and θ, you can describe or plot any periodic sequence of events because a change in any one of them will change the curve accordingly. Hence,

> If r is changed, the *amplitude* of the curve will be changed proportionally. For this reason, r is called the *amplitude factor*.
> If ω is changed, the *frequency,* or period, of the curve will be changed. Thus, ω is called the *frequency factor*.
> If θ is changed, the curve is moved along the time axis with no other change. Thus, if θ is made larger, the curve is displaced to the left and results in a leading phase angle. If θ is made smaller, the curve is moved to the right and results in a lagging phase angle. Hence the angle θ in the general equation is called the *phase angle* or *the angle of lead or lag*.

Example 3

Discuss the equation

$$y = 147 \sin (377t + 30°)$$

Solution

Given $y = 147 \sin (377t + 30°)$.

Comparing the given equation with the general equation, it is seen that $r = 147$, $\omega = 377$ rad/s, and $\theta = 30°$. Therefore, the curve represented by this equation is a sine curve with an amplitude of 147. The angular velocity is 377; hence, the frequency is

$$f = \frac{\omega}{2\pi} = \frac{377}{2\pi} = 60 \text{ Hz}$$

and the period is

$$T = f^{-1} = \frac{1}{60} = 0.0167 \text{ s}$$

The curve has been displaced to the left 30°; that is, it leads the curve $y = r \sin 377t$ by a phase angle of 30°. Therefore, when $t = 0$, the curve begins at an angle of 30° with a value of

$$y = 147 \sin (\omega t + 30°)$$

$$= 147 \sin (0° + 30°)$$

$$= 147 \times 0.5 = 73.5$$

PROBLEMS 29-3

In the following equations of periodic curves, specify (a) amplitude, (b) angular velocity (c) frequency, (d) period, and (e) angle of lead or lag with respect to a curve of the same frequency but having no displacement angle.

1. $y = 100 \sin (2\pi t + 40°)$ 2. $y = 157 \sin (377t - 12°)$

3. $i = 0.750 \sin (628t + 3°)$ 4. $i = I_{max} \sin (31.4t - 20°)$

5. $v = V_{max} \sin (157t - 17°)$ 6. $i_c = I_{c_{max}} \sin (1000\pi t + 37°)$

Plot the curves that represent the following motions:

7. $y = \sin 2\pi t$ 8. $y = 10 \sin 10t$
9. $v = 141 \sin 120t$ 10. $i = 0.5 \sin (120t + 30°)$
11. $i = 1.3 \sin (120t - 20°)$ 12. $y = 16 \sin (377t + 10°)$
13. A radar antenna that is 60 cm long rotates in a horizontal plane at 20 rev/s in a counterclockwise direction, starting from east.
    (a) Plot the curve that shows the projection of the antenna on a north-south centerline at any time.
    (b) Write the equation for the curve.
    (c) What is the distance of the end of the antenna from the east-west line at the end of 0.08 s?
    (d) What is the distance of the end of the antenna from the north-south line at the end of 0.1 s?
    (e) Through how many radians will the antenna turn in 0.25 s?
14. A radar scope scanning line rotates on the face of the oscilloscope just as a spoke on a wheel rotates with the wheel. If a scan line that is 175 mm long rotates in a positive direction at 12 sweeps/s, starting from a position 40° below the horizontal:
    (a) Plot the curve that shows the projection of the line upon a vertical reference line at any time.
    (b) Write the equation of the curve.
    (c) Find the vertical projection of the line at the end of 0.0375 s.
    (d) Find the horizontal projection of the line at the end of 0.833 s.
    (e) Through how many radians will the line sweep in 2.5 s?

Alternating Currents— Fundamental Ideas

30

Thus far we have considered direct voltages and direct currents, that is, voltages that do not change in polarity and currents that do not change in their directions of flow.

In this chapter, you will begin the study of mathematics as applied to alternating currents. An *alternating current* is one that alternates, or changes its direction, periodically.

The fact that over 90% of the electric energy produced is generated in the form of alternating current makes this subject very important, for the operation of all radio and communication circuits is based on ac phenomena. The first requisite in the study of electronics engineering is a solid foundation in the principles of alternating currents.

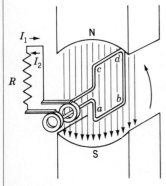

Fig. 30-1 Representation of elementary alternator.

30-1 GENERATION OF AN ALTERNATING ELECTROMOTIVE FORCE

A coil of wire that has its ends connected to slip rings and is rotating in a counterclockwise direction in a uniform magnetic field is illustrated in Fig. 30-1. That an alternating emf will be generated in the coil is apparent from a consideration of generated currents. For example, when the side of the coil ab moves from its present position away from the S pole, the emf generated in it will be directed from b to a; that is, a will be positive with respect to b. At the same time, the side of the coil cd is moving away from the N pole, thus cutting magnetic lines of force with a motion opposite that of ab. Then the emf generated in cd will be directed from c to d and will add to the emf from b to a to send a current I_1 through the resistance R.

When the coil has rotated 90° from the position shown in Fig. 30-1, the plane of the coil is perpendicular to the magnetic field, and at this instant the sides of the coil are moving parallel to the magnetic field, thus cutting no lines of force. There is no emf generated at this instant.

As the side of the coil ab begins to move up toward the N pole, the emf generated in it will now be directed from a to b. Similarly, because the side of the coil cd is now moving down toward the S pole, the emf in cd will be directed from d to c. This reversal of the direction of generated emf is due to a change of direction of motion with respect to the direction of the lines of force. Therefore, the flow of current I_2 through R will be in the direction indicated by the arrow.

When the coil rotates so that the plane of the coil is again perpendicular to the lines of force (270° from the position shown in Fig. 30-1), no emf will be generated at that instant. Rotation beyond this position, however, causes the generation of an emf such that current flows in the original direction I_1. Such an emf, which periodically reverses its direction, is known as an *alternating electromotive force,* and the resulting current is known as an *alternating current.*

In some engineering textbooks the generation of an emf is explained as due to the change of magnetic flux through the rotating coil. In the final analysis, the results are the same. Here we are interested mainly in the behavior of the circuits connected to sources of alternating currents.

30-2 VARIATION OF AN ALTERNATING ELECTROMOTIVE FORCE

The first questions that come to mind are, "In what manner does an alternating emf vary? How can we represent that variation graphically?"

Figure 30-2 illustrates a cross section of the elementary alternator of Fig. 30-1. The circles represent either side of the rotating coil at successive instants during the rotation.

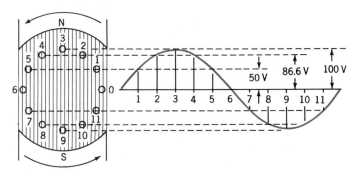

Fig. 30-2 Generation of voltage sine wave.

When a conductor passes through a magnetic field, there must be a component of its velocity at right angles to the lines of force in order to generate an emf. For example, a conductor must actually *cut* lines in order to develop an emf the amount of which will be proportional to the number of lines cut and the rate of cutting.

From studies of rotation and a consideration of Fig. 30-2, it is evident that the component of horizontal velocity of the rotating conductor is proportional to the sine of the angle of rotation. Because the horizontal velocity is perpendicular to the magnetic field, it is this component that develops an emf. For example, at position 0, where the angle of rotation is zero, the conductor is moving parallel to the field; hence, no voltage is generated. As the conductor rotates toward 90°, the component of horizontal velocity becomes greater, thus generating a higher voltage.

Therefore, the sine curve of Fig. 30-2 is a graphical representation of the induced emf in a conductor rotating in a uniform magnetic field. The voltage starts from zero, increases in a positive direction to a maximum value (100 V in the figure) at 90°, decreases to zero at 180°, increases in the opposite or negative direction until it attains maximum negative value at 270°, and finally decreases to zero value again at 360°. It follows, then, that the induced emf can be completely described by the relation

$$v = V_{max} \sin \theta \qquad V \tag{1}$$

where v = instantaneous value of emf at any angle θ, V
V_{max} = maximum value of emf, V
θ = angular position of coil

30-3 VECTOR REPRESENTATION

Since the sine wave of emf is a periodic function, a simpler method of representing the relation of the emf induced in a coil to the angle of rotation is available. The rotating conductor can be replaced by a rotating radius vector whose length represents the magnitude of the maximum generated voltage V_{max}. Then the instantaneous value for any position of the conductor can be represented by the vertical component of the vector (Sec. 28-4).

In Fig. 30-3, which is the vector diagram for the conductor at position *0* in Fig. 30-2, the vector V_{max} is at 0° position and therefore has no vertical component. Thus the value of the emf in this position is zero. Or, since

$$v = V_{max} \sin \theta$$

by substituting the values of V_{max} and θ,

$$v = 100 \sin 0° = 0$$

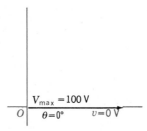

Fig. 30-3 $v = \sin 0° = V$.

In Fig. 30-4, which is the vector diagram for the conductor at position 2 in Fig. 30-2, the coil has moved 60° from the zero position. The vector V_{max} is therefore at an angle of 60° from the reference axis, and the instantaneous value of the induced emf is represented by the vertical component of V_{max}. Then, since

$$v = V_{max} \sin \theta$$

by substituting the values of V_{max} and θ,

$$v = 100 \sin 60° = 86.6 \text{ V}$$

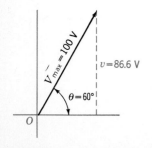

Fig. 30-4 $v = 100 \sin 60° = 86.6$ V.

Example 1

What is the instantaneous value of an alternating emf that has reached 58° of its cycle? The maximum value is 500 V.

Solution

Draw the vector diagram to scale as shown in Fig. 30-5. The instantaneous value is the vertical component of the vector V_{max}. Then, since

$$v = V_{max} \sin \theta$$

by substituting the values of V_{max} and θ,

$$v = 500 \sin 58° = 424 \text{ V}$$

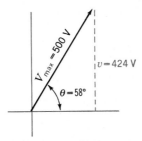

Fig. 30-5 $v = 500 \sin 58° = 424$ V.

Example 2

What is the instantaneous value of an alternating emf when the emf has reached 216° of its cycle? The maximum value is 163 V.

Solution

Draw the vector diagram to scale as shown in Fig. 30-6. The instantaneous value is the vertical component of the vector V_{max}. Then, since

$$v = V_{max} \sin \theta$$

by substituting the values of V_{max} and θ,

$$v = 163 \sin 216° = 163[-\sin(216° - 180°)]$$
$$= 163(-\sin 36°) = -95.8 \text{ V}$$

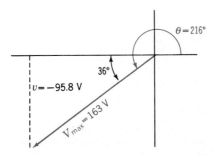

Fig. 30-6 $v = 163 \sin(-36°) = -95.8$ V.

A vector diagram drawn to scale should be made for every ac problem. It will give you a better insight into the functioning of alternating currents and at the same time serve as a good check on the mathematical solution.

Since the current in a circuit is proportional to the applied voltage, it follows that an alternating emf which varies periodically will produce a current of similar variation. Hence, the instantaneous current of a sine wave of alternating current is given by

$$i = I_{max} \sin \theta \quad \text{A} \quad\quad (2)$$

where i = instantaneous value of current, A
I_{max} = maximum value of current, A
θ = angular position of coil

PROBLEMS 30-1

1. An alternating current has a maximum value of 165 A. What are the instantaneous values of this current at the following points in its cycle:
(a) 18°, (b) 67°, (c) 136°, (d) 242°, (e) 326°?

2. The instantaneous value of an alternating emf at 17° is 34.2 V. What is its maximum value?

3. The instantaneous value of an alternating emf at 334.4° is − 190 V. What is its maximum value?

4. An alternating current has a maximum value of 750 mA. What are the instantaneous values of the current at the following points in its cycle:
(a) 26°, (b) 341°, (c) 210°, (d) 297°, (e) 162°?

5. The instantaneous value of an alternating emf is 110 V at 71°. What will the value be at 232°?

6. The instantaneous value of an alternating emf at 289° is − 22 V. What will the value be at 142°?

7. The instantaneous value of an alternating current at 99.9° is 3.2 A. What will the value be at 199.9°?

8. An alternating current has a maximum value of 365 mA. At what angles will it be 80% of its positive maximum value?

9. At what angles are the instantaneous values of an alternating current equal to 50% of the maximum negative value?

10. What is the instantaneous value of an alternating emf 110° after its maximum positive value of 165 V?

30-4 CYCLES, FREQUENCY, AND POLES

Each revolution of the coil in Fig. 30-1 results in one complete *cycle* which consists of one positive and one negative loop of the sine wave (Sec. 29-8). The number of cycles generated in 1 s is called the *frequency* of the alternating emf, and the *period* is the time required to complete one cycle. One half cycle is called an *alternation*. Thus, by a 60-Hz alternating current is meant that the current passes through 60 cycles per second, which results in a period of 0.0167 s. Also, a 60-Hz current completes 120 alternations per second.

Figure 30-7 represents a coil rotating in a four-pole machine. When one side of the coil has rotated from position 0 to position 4, it has passed under the influence of an N and an S pole, thus generating one complete

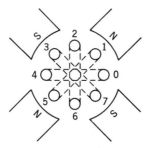

Fig. 30-7 Elementary four-pole alternator.

sine wave, or electrical cycle. This corresponds to 2π electrical radians, or 360 electrical degrees, although the coil has rotated only 180 space degrees. Therefore, in one complete revolution the coil will generate two complete cycles, or 720 electrical degrees, so that for every *space degree* there result two *electrical time degrees*.

In any alternator the armature, or field, must move an angular distance equal to the angle formed by two consecutive like poles in order to complete one cycle. It is evident, then, that a two-pole machine must rotate at twice the speed of a four-pole machine to produce the same frequency. Therefore, to find the frequency of an alternator in hertz (cycles per second), *the number of pairs of poles is multiplied by the speed of the armature in revolutions per second*. That is,

$$f = \frac{PS}{60} \qquad \text{Hz} \tag{3}$$

where f = frequency, Hz
P = number of pairs of poles
S = rotational speed of armature, or field, rev/min

Example 3
What is the frequency of an alternator that has four poles and rotates at a speed of 1800 rev/min?

Solution

$$f = \frac{2 \times 1800}{60} = 60 \text{ Hz}$$

30-5 EQUATIONS OF VOLTAGES AND CURRENTS

Since each cycle consists of 360 electrical degrees, or 2π electrical radians, the variation of an alternating emf can be expressed in terms of time. Thus, a frequency of f Hz results in $2\pi f$ r/s, which is denoted by ω (Sec. 29-5). Hence, the instantaneous emf at any time t is given by the relation

$$v = V_{\text{max}} \sin \omega t \qquad \text{V} \tag{4}$$

The instantaneous current is

$$i = I_{\text{max}} \sin \omega t \qquad \text{A} \tag{5}$$

You should review Secs. 29-6 to 29-10 to ensure a complete understanding of the relations between the general equation for a periodic function and Eqs. (4) and (5). Thus, V_{max} and I_{max} are the amplitude factors of their respective equations, and ω is the frequency factor.

Example 4

Write the equation of a 60-Hz alternating voltage that has a maximum value of 156 V.

Solution

The angular velocity ω is 2π times the frequency, or

$$2\pi \times 60 = 377^r/\text{s}$$

Substituting 156 V for V_{max} and 377 for ω in Eq. (4),

$$v = 156 \sin 377t \qquad \text{V}$$

Example 5

Write the equation of a radio-frequency (RF) current of 700 kHz that has a maximum value of 21.2 A.

Solution

$I_{max} = 21.2$ A and $f = 700$ kHz $= 7 \times 10^5$ Hz. Then

$$\omega = 2\pi f = 2\pi \times 7 \times 10^5$$
$$= 4.4 \times 10^6$$

Substituting these values in Eq. (5),

$$i = 21.2 \sin (4.4 \times 10^6)t \qquad \text{A}$$

Example 6

If the time $t = 0$ when the voltage of Example 4 is zero and increasing in a positive direction, what is the instantaneous value of the voltage at the end of 0.002 s?

Solution

Substituting 0.002 for t in the equation for the voltage,

$$v = 156 \sin (377 \times 0.002)$$
$$= 156 \sin 0.754^r \qquad \text{V}$$

where 0.754 is the time angle in *radians*. Then, since $1^r \cong 57.3°$,

$$v = 156 \sin (0.754 \times 57.3°)$$
$$= 156 \sin 43.2°$$

Hence, $\qquad v = 107$ V

PROBLEMS 30-2

1. An alternator with 40 poles has a speed of 1200 rev/min and develops a maximum emf of 314 V.
   (a) What is the frequency of the alternating emf?
   (b) What is the period of the alternating emf?
   (c) Write the equation for the instantaneous emf at any time t.

2. An alternator with 8 poles has a speed of 3600 rev/min and develops a maximum voltage of 120 V.
    (a) What is the frequency of the alternating emf?
    (b) Write the equation for the instantaneous value of the emf at any time t.

3. A 400-Hz generator which develops a maximum emf of 250 V has a speed of 1200 rev/min.
    (a) How many poles has it? (b) Write the equation of the voltage.
    (c) What is the value of the voltage when the time $t = 2$ ms?

4. An 800-Hz alternator generates a maximum of 163 V at the rate of 4000 rev/min.
    (a) How many poles has it?
    (b) Write the equation for the voltage.
    (c) What is the value of the emf when time $t = 500$ μs?

5. At what speed must a 12-pole 60-Hz alternator be driven in order to develop its rated frequency?

6. The equation for a certain alternating current is

    $$i = 84.6 \sin 377t \text{ mA}$$

    What is the frequency of the current?

7. The equation for an alternating emf is

    $$v = 0.05 \sin (3.14 \times 10^9)\, t \text{ V}$$

    What is the frequency of the emf?

8. The equation for an alternating current is

    $$i = (2.75 \times 10^{-2}) \sin (2.7 \times 10^7)\, t \quad\quad \text{A}$$

    (a) What is the maximum instantaneous current?
    (b) What is the frequency?

9. A 500-MHz current has a maximum instantaneous value of 30 μA. Write the equation describing the current.

10. A broadcasting station operating at 1430 kHz develops a maximum potential of 0.362 mV across a listener's antenna. Write the equation for this emf.

30-6 AVERAGE VALUE OF CURRENT OR VOLTAGE

Since an alternating current or voltage is of sine-wave form, it follows that the average current or voltage of one cycle is zero owing to the reversal of direction each half-cycle. The term *average value* is usually understood to mean the average value of one alternation without regard to positive or negative values. The average value of a sine wave, such as that shown in Fig. 30-2, can be computed to a fair degree of accuracy by taking the average of many instantaneous values between two consecutive zero points of the curve, the values chosen being separated by equal values of angle. Thus, the average value is equal to the average height of any voltage or current loop. The exact average value is $2 \div \pi \cong 0.637$ times the maximum value. Thus, if I_{av} and V_{av} denote the average values of alternating current and voltage, respectively, we obtain

$$I_{av} = \frac{2}{\pi}I_{max} \cong 0.637I_{max} \qquad A \qquad (6)$$

and

$$V_{av} = \frac{2}{\pi}V_{max} \cong 0.637V_{max} \qquad V \qquad (7)$$

Example 7

The maximum value of an alternating voltage is 622 V. What is the average value?

Solution

$$V_{av} = 0.637V_{max} = 0.637 \times 622$$
$$= 396 \text{ V}$$

30-7 EFFECTIVE VALUE OF CURRENT OR VOLTAGE

If a direct current of I A is caused to flow through a resistance of R Ω, the resulting energy converted into heat equals I^2R W. We should not expect an alternating current with a maximum value of 1 A to produce as much heat as a direct current of 1 A, for the former does not maintain a constant value. Thus, the above ac ampere is not as effective as the dc ampere. The *effective value* of an alternating current is rated in terms of direct current; that is, an alternating current has an effective value of 1 A if, when it flows through a given resistance, it produces heat at the same rate as a dc ampere would.

The effective value of a sine wave of current can be computed to a fair degree of accuracy by taking equally spaced instantaneous values and extracting the square root of their average, or mean, squared values. For this reason, the effective value is often called the *root-mean-square* (rms) value. The exact effective value of an alternating current or voltage is $1/\sqrt{2} \cong 0.707$ times the maximum value. Thus, if I and V denote the effective values of current and voltage, respectively, we obtain

$$I = \frac{I_{max}}{\sqrt{2}} \cong 0.707I_{max} \qquad A \qquad (8)$$

and

$$V = \frac{V_{max}}{\sqrt{2}} = 0.707V_{max} \qquad V \qquad (9)$$

It should be noted that all meters, unless marked to the contrary, read effective values of current and voltage.

Example 8

The maximum value of an alternating voltage is 311 V. What is the effective value?

Solution

$$V = 0.707V_{max} = 0.707 \times 311$$
$$= 220 \text{ V}$$

Example 9

An ac ammeter reads 15 A. What is the maximum value of the current?

Solution 1

Since
$$I = 0.707I_{max}$$

then
$$I_{max} = \frac{I}{0.707}$$

Substituting 15 A for I,
$$I_{max} = \frac{15}{0.707} = 21.2 \text{ A}$$

Solution 2

Since
$$I = \frac{I_{max}}{\sqrt{2}}$$

then
$$I_{max} = I\sqrt{2} = 1.41I$$

Substituting for I,
$$I_{max} = 1.41 \times 15 = 21.2 \text{ A}$$

Hence the maximum value of an alternating current or voltage is equal to 1.41 times the effective value.

PROBLEMS 30-3

1. What is the average value of an alternating emf whose maximum value is 77 V?
2. What is the maximum value of an alternating current whose average value is 56 mA?
3. The average value of an alternating emf is 10.5 V. What is the maximum value?
4. The maximum value of an alternating current is 173 μA. What is the average value?
5. The maximum value of an alternating emf is 180 V. What is the effective value?
6. An rms voltmeter indicates 117 V of alternating emf. What is the maximum value of the emf?
7. What is the effective value of an alternating current which has a maximum value of 30 A?
8. What is the effective value of an alternating emf which has an average value of 125 V?
9. What is the average value of an alternating current which has an effective value of 258 mA?
10. An rms ammeter indicates an alternating current reading of 33.8 A. What is the average value of the current?

30-8 PHASE RELATIONS—PHASE ANGLES

Nearly all ac circuits contain elements, or components, that cause the voltage and current to pass through their corresponding zero values at different times. The effects of such conditions are given detailed consideration in the next chapter.

If an alternating voltage and the resulting alternating current of the same frequency pass through corresponding zero values at the same instant, they are said to be *in phase*.

If the current passes through a zero value before the corresponding zero value of the voltage, the current and voltage are *out of phase* and the current is said to *lead* the voltage.

Figure 30-8 illustrates a phasor diagram and the corresponding sine waves for a current of i A leading a voltage of v V by a *phase angle* of θ (Sec. 29-10). Hence, if the voltage is taken as reference, the general equation of the voltage is

$$v = V_{max} \sin \omega t \quad \text{V} \tag{10}$$

and the current is given by

$$i = I_{max} \sin (\omega t + \theta) \quad \text{A} \tag{11}$$

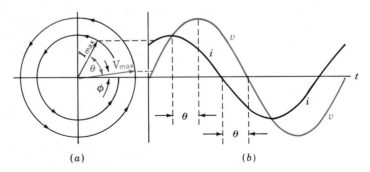

Fig. 30-8 Current i leads voltage v by phase angle θ.

The instantaneous values of the voltage and current for any angle ϕ of the voltage are

$$v = V_{max} \sin \phi \quad \text{V} \tag{12}$$

and

$$i = I_{max} \sin (\phi + \theta) \quad \text{A} \tag{13}$$

Example 10

In Fig. 30-8, the maximum values of the voltage and the current are 156 V and 113 A, respectively. The frequency is 60 Hz, and the current leads the voltage by 40°.

(*a*) Write the equation for the voltage at any time t.

(*b*) Write the equation for the current at any time t.

(*c*) What is the instantaneous value of the current when the voltage has reached 10° of its cycle?

Solution

Given

$$\text{Maximum voltage} = V_{max} = 156 \text{ V}$$
$$\text{Maximum current} = I_{max} = 113 \text{ A}$$
$$\text{Frequency} = f = 60 \text{ Hz}$$
$$\text{Phase angle} = \theta = 40° \text{ lead}$$
$$\text{Voltage angle} = \phi = 10°$$

Draw a vector diagram as shown in Fig. 30-8a. (The circles are not necessary; they simply denote rotation of the vectors.)

(a) Substituting given values in Eq. (10),

$$v = 156 \sin 2\pi \times 60t$$

or
$$v = 156 \sin 377t \text{ V}$$

(b) Substituting given values in Eq. (11),

$$i = 113 \sin (377t + 40°) \text{ A}$$

Note

The quantity 377t is in *radians*.

(c) Substituting given values in Eq. (13),

$$i = 113 \sin (10° + 40°)$$

or
$$i = 113 \sin 50° = 86.6 \text{ A}$$

Figure 30-9 illustrates a vector diagram and the corresponding sine waves for a current of i A lagging a voltage of v V by a *phase angle* of θ. Therefore, if the voltage is taken as reference, the general equation of the voltage will be as given by Eq. (10) and the current will be

$$i = I_{max} \sin (\omega t - \theta) \qquad \text{A}$$

The instantaneous value of the current for any angle ϕ of the voltage is

$$i = I_{max} \sin (\phi - \theta) \qquad \text{A}$$

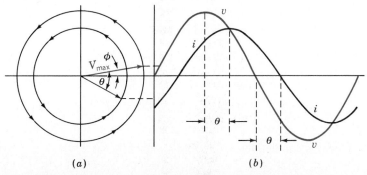

(a)　　　　　　　　　　(b)

Fig. 30-9 Current i lags voltage v by phase angle θ.

Example 11

In Fig. 30-9, the maximum values of the voltage and the current are 170 V and 14.1 A, respectively. The frequency is 800 Hz, and the current lags the voltage by 40°.

(a) Write the equation for the voltage at any time t.
(b) Write the equation for the current at any time t.
(c) What is the instantaneous value of the current when the voltage has reached 10° of its cycle?

Solution

Given

$$\text{Maximum voltage} = V_{max} = 170V$$
$$\text{Maximum current} = I_{max} = 14.1\,A$$
$$\text{Frequency} = f = 800\,Hz$$
$$\text{Phase angle} = \theta = 40° \text{ lag}$$
$$\text{Voltage angle} = \phi = 10°$$

Draw a vector diagram as shown in Fig. 30-9a.

(a) Substituting given values in Eq. (10),

$$v = 170 \sin 2\pi \times 800t$$

or

$$v = 170 \sin 5030t \text{ V}$$

(b) Substituting given values in Eq. (14),

$$i = 14.1 \sin (5030t - 40°) \text{ A}$$

(c) Substituting given values in Eq. (15),

$$i = 14.1 \sin (10° - 40°)$$

or

$$i = 14.1 \sin (-30°) = -7.05 \text{ A}$$

Example 12

In a certain ac circuit a current of 14 A lags a voltage of 220 V by an angle of 60°. What is the instantaneous value of the voltage when the current has completed 245° of its cycle?

Note

Unless otherwise specified, all voltages and currents are to be considered *effective* values.

Solution

Draw the vector diagram as shown in Fig. 30-10.

$$V_{max} = \sqrt{2}V = \sqrt{2} \times 220 = 311 \text{ V}$$
$$\phi = 245° + \theta = 245° + 60°$$
$$= 305° = -55°$$

Then, substituting the values of V_{max} and θ in Eq. (12),

$$v = 311 \sin (-55°) = -255 \text{ V}$$

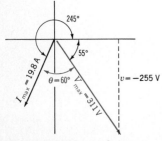

Fig. 30-10 Phasor diagram of Example 12.

PROBLEMS 30-4

1. A 60-Hz alternator generates a maximum emf of 165 V and delivers a maximum current of 6.5 A. The current leads the voltage by an angle of 36°.
    (a) Write the equation for the current at any time t.
    (b) What is the instantaneous value of the current when the emf has completed 60° of its cycle?

2. A 25-Hz alternator generates 6.6 kV at 700A. The current leads the voltage by an angle of 22°.
    (a) Write the equation for the current at any time t.
    (b) How much of the voltage cycle will have been completed the first time that the instantaneous current rises to 465 A?

3. In the alternator of Prob. 1, what will be the instantaneous value of the current when the voltage has completed 200° of its cycle?

4. In the alternator of Prob. 2, what will be the instantaneous value of the current when the voltage has completed 350° of its cycle?

5. A 50-Hz alternator generates 2.3 kV with a current of 200 A. The phase angle is 25° lagging.
    (a) Write the equation for the current at any time t.
    (b) What is the instantaneous value of the current when the voltage has completed 192° of its cycle?

6. In the alternator of Prob. 5, what is the instantaneous value of the current when the voltage has completed 17° of its cycle?

7. A 60-Hz alternator generates a maximum of 170 V and delivers a maximum current of 42.4 A. If the instantaneous value of the current is 22.5 A when the instantaneous value of the emf is 112 V, what is the phase angle between the current and the emf?

8. In Prob. 7, what will be the instantaneous value of the emf when the instantaneous value of the current is -39.3 A for the first time?

9. A 400-Hz alternator develops 30 A at 230 V. If the instantaneous value of the emf is -85.8 V when the instantaneous value of the current is 23.5 A, what is the phase angle between current and emf?

10. (a) Write the equation for the current in Prob. 9.
    (b) In Prob. 9, what will be the instantaneous value of the current when the emf has reached its maximum value negatively?

Phasor Algebra

31

In Sec. 1-2 we commented briefly on the use of mathematics as a tool in electronics. One of the most valuable of all the mathematical tools, certainly the most valuable in the solution of ac circuits, is the operator j together with complex numbers. The complex number operations to be developed in this chapter are so important in electronics that some of the more expensive electronic calculators have special function keys to simplify even further the mathematical processes involved.

31-1 PHASOR DEFINITIONS

Real and imaginary numbers Complex numbers were introduced in Secs. 20-14 to 20-18, and, in keeping with traditional methods of notation, Secs. 20-13 and 20-14 used the expressions "real number" and "imaginary number." In the complex number $a + jb$, theoretical mathematicians refer to a as the *real part*, and to b as the *imaginary part*, of the complex number.

Phase Since our development of the operator j did not depend upon any imaginary features, we now abandon the traditional definitions in favor of the more realistic expressions of electrical engineering. We will refer to a as the *in-phase* portion of the complex number and to b as the *out-of-phase* portion. These expressions are more in keeping with the correct understanding of the phase relationships of ac circuits which will be developed further in Chap. 32.

Rectangular form The form $a + jb$ is referred to as the *rectangular form* of the complex number. Later, in circuit analyses, we shall use the form $R + jX$.

Polar form When the rectangular components a and b of a complex number are resolved into a single magnitude r rotated through an angle θ from a reference axis, the resultant form $r \underline{/\theta}$ is referred to as the *polar form*. Later, in circuit analyses, we shall use the form $Z \underline{/\theta}$.

Any complex number is fully described by either its rectangular form or its polar form.

You must be wholly satisfied with the following relationships, which are described in Fig. 31-1, before continuing the study of this chapter.

If $\qquad\qquad a + jb = r\underline{/\theta}$

then $\qquad\qquad a = r\cos\theta$

and $\qquad\qquad b = r\sin\theta$

and $\qquad\qquad \theta = \arctan\dfrac{b}{a}$

and $\qquad\qquad r = \dfrac{a}{\cos\theta} = \dfrac{b}{\sin\theta}$

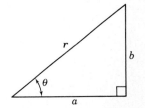

Fig. 31-1 Interrelations of rectangular and phasor forms of complex numbers: $a = r\cos\theta$ and $b = r\sin$ $\theta = \arctan\dfrac{b}{a}$ and $r = \dfrac{a}{\cos\theta} = \dfrac{b}{\sin\theta}$

In Chap. 28 we developed the idea of *vectors* and introduced the *phasor* as the two-dimensional electrical refinement of a vector to describe the polar or phasor relationships. In Sec. 28-6 we developed *phasor summation,* and you can see, by comparing that section with Sec. 20-15, that a phasor relationship may be simply described by a complex number. *Phasor algebra* is merely the systematic analytical form of right triangle calculations developed in Chaps. 26 and 28.

31-2 ADDITION AND SUBTRACTION OF PHASORS IN RECTANGULAR FORM

As stated in Sec. 20-15, complex numbers, or phasors in rectangular form, can be added or subtracted by treating them as ordinary binomials.

Example 1
Add $4.60 + j2.82$ and $2.11 - j8.10$.

Solution

$$
\begin{array}{r}
4.60 + j2.82 \\
2.11 - j8.10 \\
\hline
6.71 - j5.28
\end{array}
$$

Express the sum in polar form,

$$6.71 - j5.28 = 8.54\underline{/-38.2°}$$

Example 2
Subtract $3.7 + j4.62$ from $14.6 - j8.84$.

Solution

$$
\begin{array}{r}
14.6 - j8.84 \\
3.7 + j4.62 \\
\hline
10.9 - j13.46
\end{array}
$$

Express the result in polar form,

$$10.9 - j13.46 = 17.3\underline{/-51°}$$

PROBLEMS 31-1

Perform the indicated operations and express the answers in both rectangular and polar forms. Check your results by graphical methods:

1. $(8.3 - j11.3) + (12.4 + j22.6)$
2. $(18.4 + j25) + (81.2 - j110)$
3. $(400 + j298) + (700 + j102)$
4. $(16.95 - j17.8) + (-11.33 - j22.2)$
5. $(115 + j925) + (-557 - j184)$
6. $(-488 - j603) + (172 + j168)$
7. $(23.8 - j4^\wedge.5) - (12.6 - j8.1)$
8. $(8.37 - j3.4) - (-6.53 + j10.2)$
9. $(1100 - j200) - (-1400 - j600)$
10. $(75.3 - j38.7) - (137.4 + j47.1)$
11. $(32.6 + j3.4) - (22.6 - j5.6)$
12. $(-16.5 - j13.7) - (-16.5 + j86.3)$

31-3 MULTIPLICATION OF PHASORS IN RECTANGULAR FORM

Multiplication of complex numbers was explained in Sec. 20-16, where it was shown that phasors expressed in terms of their retangular components are multiplied by treating them as ordinary binomials.

Example 3

Multiply $8 + j5$ by $10 + j9$.

Solution

$$
\begin{array}{r}
8 + j5 \\
10 + j9 \\
\hline
80 + j50 \\
+ j72 + j^2 45 \\
\hline
80 + j122 + j^2 45
\end{array}
$$

Since $j^2 = -1$, the product is

$$80 + j122 + (-1)45 = 80 + j122 - 45 = 35 + j122$$

Expressing the product in polar form,

$$35 + j122 = 127 \underline{/74°}$$

Example 4

Multiply $80 + j39$ by $35 - j50$.

Solution

$$
\begin{array}{r}
80 + j39 \\
35 - j50 \\
\hline
2800 + j1365 \\
- j4000 - j^2 1950 \\
\hline
2800 - j2635 - j^2 1950
\end{array}
$$

Since $j^2 = -1$, the product is

$$2800 - j2635 - (-1)1950 = 2800 - j2635 + 1950$$
$$= 4750 - j2635$$

Expressing the product in polar form,

$$4750 - j2635 = 5430 \,\underline{/-29°}$$

31-4 DIVISION OF PHASORS IN RECTANGULAR FORM

As explained in Sec. 20-17, division of complex numbers, or phasors in rectangular form, is accomplished by rationalizing the denominator in order to obtain an in-phase number for a divisor. Multiplying a complex number by its conjugate always results in a product that is a simple number not affected by the operator j.

Example 5

Find the quotient of $\dfrac{50 + j35}{8 + j5}$.

Solution

Multiply both dividend and divisor (numerator and denominator) by the conjugate of the divisor, which is $8 - j5$. Thus,

$$\frac{50 + j35}{8 + j5} \cdot \frac{8 - j5}{8 - j5} = \frac{400 + j30 - j^2 175}{64 - j^2 25}$$
$$= \frac{575 + j30}{89}$$

That is,

$$\frac{575 + j30}{89} = \frac{575}{89} + j\frac{30}{89}$$
$$= 6.46 + j0.337$$

Express the quotient in polar form,

$$6.46 + j0.337 \cong 6.46 \,\underline{/3.0°}$$

Example 6

Simplify $\dfrac{10}{3 + j4}$.

Solution

Multiply both numerator and denominator by the conjugate of the denominator, which is $3 - j4$. Thus,

$$\frac{10}{3 + j4} \cdot \frac{3 - j4}{3 - j4} = \frac{10(3 - j4)}{9 - j^2 16}$$
$$= \frac{30 - j40}{25} = 1.2 - j1.6$$

Express the quotient in polar form,

$$1.2 - j1.6 = 2.0 \,\underline{/-53.1°}$$

Problems 31-2

Perform the indicated operations and express the answers in both rectangular and polar form:

1. $(5 + j4)(2 - j6)$
2. $(12 + j14)(22 + j17)$
3. $(2.5 + j7.6)(3.8 - j1.5)$
4. $(470 - j35.0)(330 + j0.621)$
5. $(6.8 - j4.6)(5.6 - j7.2)$
6. $(2.7 - j9)(12 - j8)$
7. $(4 - j2) \div (3 + j5)$
8. $(7 - j5) \div (10 - j14)$
9. $(20 - j16) \div (3 + j5)$
10. $1 \div (12 - j9)$

31-5 ADDITION AND SUBTRACTION OF POLAR PHASORS

As explained in preceding sections, phasors expressed in polar form can be added or subtracted by graphical methods only if their directions are parallel.

In order to add or subtract them algebraically, phasors must be expressed in terms of their rectangular components.

Example 7

Add $5.40 \underline{/31.5°}$ and $8.37 \underline{/-75.4°}$.

Solution

Converting the phasors into their rectangular components,

$$5.40 \underline{/31.5°} = 5.40(\cos 31.5° + j\sin 31.5°) = 4.60 + j2.82$$
$$8.37 \underline{/-75.4°} = 8.37(\cos 75.4° - j\sin 75.4°) = \underline{2.11 - j8.10}$$

Adding, $\qquad\qquad\qquad\qquad\qquad$ Sum $= 6.71 - j5.28$

Expressing the sum in polar form,

$$6.71 - j5.28 = 8.54 \underline{/-38.2°}$$

Note that the phasors of this example are the same as those of Example 1 of Sec. 31-2.

Example 8

Subtract $5.92 \underline{/51.3°}$ from $17.1 \underline{/-31.2°}$.

Solution

Converting the phasors into their rectangular components,

$$17.1 \underline{/-31.2°} = 17.1(\cos 31.2° - j\sin 31.2°) = 14.6 - j8.86$$
$$5.92 \underline{/51.3°} = 5.92(\cos 51.3° + j\sin 51.3°) = \underline{3.7 + j4.62}$$

Subtracting, $\qquad\qquad\qquad\qquad\qquad$ Result $= 10.9 - j13.48$

Expressing the result in polar form,

$$10.9 - j13.48 = 17.3 \underline{/-51°}$$

Note that the phasors of this example are the same as those of Example 2 of Sec. 31-2.

PROBLEMS 31-3

Perform the indicated operations and express the results in both polar and rectangular form. Check results graphically.

1. $14 \angle -53.7° + 25.8 \angle 61.2°$
2. $31 \angle 53.7° + 137 \angle -53.7°$
3. $500 \angle 36.6° + 710 \angle 8.27°$
4. $24.6 \angle -46.4° + 24.9 \angle 242.8°$
5. $933 \angle 82.9° + 590 \angle 198.3°$
6. $777 \angle -129° + 241 \angle 44.3°$
7. $50.6 \angle -61.9° - 15 \angle -32.7°$
8. $9 \angle -22.2° - 12.1 \angle 57.4°$
9. $1110 \angle 10.3° - 1510 \angle 203.4°$
10. $85 \angle -27.15° - 145 \angle 18.91°$
11. $1000 \angle -53.1° - 1500 \angle -53.1°$
12. $10.64 \angle -53° - 22.35 \angle 62.5°$

31-6 MULTIPLICATION OF POLAR PHASORS

In Example 3 of Sec. 31-3, it was shown that

$$(8 + j5)(10 + j9) = 127 \angle 74°$$

Now $\qquad 8 + j5 = 9.44 \angle 32°$

and $\qquad 10 + j9 = 13.45 \angle 42°$

Multiplying the magnitudes and adding the angles,

$$(9.44 \times 13.45) \angle 32° + 42° = 127 \angle 74°$$

which is the same product as that obtained by multiplying the phasors when expressed in terms of their rectangular components.

Similarly, in Example 4 of Sec. 31-3, it was shown that

$$(80 + j39)(35 - j50) = 5430 \angle -29°$$

Now $\qquad 80 + j39 = 89.0 \angle 26°$

and $\qquad 35 - j50 = 61.0 \angle -55°$

Multiplying the magnitudes and adding the angles,

$$(89 \times 61.0) \angle 26° + (-55°) = 5430 \angle -29°$$

which is the same product as that obtained by multiplying the phasors when the phasors are expressed in terms of their rectangular components.

From the foregoing, it is evident that the product of two polar phasors is found by multiplying the magnitudes and adding the angles of the phasors algebraically.

31-7 DIVISION OF POLAR PHASORS

In Example 5 of Sec. 31-4, it was shown that

$$\frac{50 + j35}{8 + j5} = 6.46 \angle 3.0°$$

Now
$$50 + j35 = 61.0 \angle 35°$$
and
$$8 + j5 = 9.44 \angle 32°$$

Dividing the magnitudes and subtracting the angle of the divisor from the angle of the dividend,

$$\frac{61.0 \angle 35°}{9.44 \angle 32°} = \frac{61.0}{9.44} \angle 35° - 32° = 6.46 \angle 3.0°$$

which is the same quotient as that obtained by dividing the phasors when expressed in terms of their rectangular components.

Similarly, in Example 6 of Sec. 31-4, it was shown that

$$\frac{10}{3 + j4} = 2.0 \angle -53.1°$$

Since 10 is a positive number, it is plotted on the 0° axis (Sec. 3-5) and expressed as

$$10 \angle 0°$$
Now
$$3 + j4 = 5 \angle 53.1°$$

Dividing the magnitudes and subtracting the angle of the divisor from the angle of the dividend,

$$\frac{10 \angle 0°}{5 \angle 53.1°} = \frac{10}{5} \angle 0° - 53.1°$$
$$= 2.0 \angle -53.1°$$

which is the same quotient as that obtained by dividing the phasors when expressed in terms of their rectangular components.

From the foregoing, it is evident that the quotient of two polar phasors is found by dividing the magnitudes of the phasors and subtracting the angle of the divisor from the angle of the dividend.

31-8 EXPONENTIAL FORM

In the preceding two sections it has been demonstrated that angles are added when phasors are multiplied and they are subtracted when one phasor is divided by another. These operations can be further justified from a consideration of the sine and cosine when expanded in series form.

By Maclaurin's theorem, a treatment of which is beyond the scope of this book, $\cos \theta$ and $\sin \theta$ can be expanded into series form as follows:

$$\cos \theta = 1 - \frac{\theta^2}{2!} + \frac{\theta^4}{4!} - \frac{\theta^6}{6!} + \cdots \tag{1}$$

$$\sin \theta = \theta - \frac{\theta^3}{3!} + \frac{\theta^5}{5!} - \frac{\theta^7}{7!} + \cdots \tag{2}$$

The symbol $n!$ denotes the product of 1, 2, 3, 4, . . ., n and is read "factorial n." Thus, 5! (factorial five) is $1 \times 2 \times 3 \times 4 \times 5$. Similarly, it can be shown that

$$\varepsilon^{j\theta} = 1 + j\theta - \frac{\theta^2}{2!} - j\frac{\theta^3}{3!} + \frac{\theta^4}{4!} + j\frac{\theta^5}{5!} - \frac{\theta^6}{6!} - j\frac{\theta^7}{7!} + \cdots \qquad (3)$$

where ε is the base of the natural system of logarithms $\cong 2.718$. By collecting and factoring j terms, Eq. (3) can be written

$$\varepsilon^{j\theta} = \left(1 - \frac{\theta^2}{2!} + \frac{\theta^4}{4!} - \frac{\theta^6}{6!} + \cdots\right) + j\left(\theta - \frac{\theta^3}{3!} + \frac{\theta^5}{5!} - \frac{\theta^7}{7!} + \cdots\right) \qquad (4)$$

Note that the first term of the right member of Eq. (4) is $\cos \theta$ as given in Eq. (1) and that the second term in the right member of Eq. (4) is $j \sin \theta$. Therefore,

$$\varepsilon^{j\theta} = \cos \theta + j \sin \theta \qquad (5)$$

This expression, $\cos \theta + j \sin \theta$, is often referred to as cis θ, and some texts will actually refer to the *cis function*. You should bear in mind that *cis* is simply an abbreviation for $\cos + j \sin$.

Since a phasor, such as $Z\angle\theta$, can be expressed in terms of its rectangular components by the relation

$$Z\angle\theta = Z(\cos \theta + j \sin \theta) \qquad (6)$$

it follows from Eqs. (5) and (6) that

$$Z\angle\theta = Z\varepsilon^{j\theta} \qquad (7)$$

Similarly, it can be shown that

$$Z\angle{-\theta} = Z\varepsilon^{-j\theta} \qquad (8)$$

Equations (7) and (8) show that the angles of phasors can be treated as exponents.

Two vectors $Z_1\angle\theta$ and $Z_2\angle\phi$ are multiplied by multiplying the magnitudes and adding the angles of the phasors algebraically. That is,

$$(Z_1\angle\theta)(Z_2\angle\phi) = Z_1 Z_2 \angle{\theta + \phi}$$

Also,

$$\frac{Z_1\angle\theta}{Z_2\angle\phi} = \frac{Z_1}{Z_2}\angle{\theta - \phi}$$

and

$$\frac{Z_a\angle\theta}{Z_b\angle{-\phi}} = \frac{Z_a}{Z_b}\angle{\theta + \phi}$$

Example 9

Multiply $Z_1 = 8.4\angle{15°}$ by $Z_2 = 10.5\angle{20°}$.

Solution

$$Z_1 Z_2 = 8.4 \times 10.5 \angle{15° + 20°}$$
$$= 88.2\angle{35°}$$

Example 10

Multiply $Z_a = 164 \underline{/-39°}$ by $Z_b = 2.2 \underline{/-26°}$)

Solution

$$Z_a Z_b = 164 \times 2.2 \underline{/-39° + (-26°)}$$
$$= 361 \underline{/-65°}$$

Example 11

Divide $Z_1 = 54.2 \underline{/47°}$ by $Z_2 = 18 \underline{/16°}$.

Solution

$$\frac{Z_1}{Z_2} = \frac{54.2}{18} \underline{/47° - 16°} = 3.01 \underline{/31°}$$

Example 12

Divide $Z_a = 886 \underline{/18°}$ by $Z_b = 31.2 \underline{/-50°}$

Solution

$$\frac{Z_a}{Z_b} = \frac{886}{31.2} \underline{/18° - (-50°)}$$
$$= 28.4 \underline{/68°}$$

31-9 POWERS AND ROOTS OF POLAR PHASORS

In addition to following the laws of exponents for multiplication and division, phasor angles can be used as any other exponents are used when powers or roots of phasors are desired. For example, to square a phasor, the magnitude is squared and the angle is multiplied by 2. Similarly, the root of a phasor is found by extracting the root of the magnitude and dividing the angle by the index of the root.

Example 13

Find the square of $Z_1 = 14 \underline{/18°}$.

Solution

$$Z_1^2 = (14 \underline{/18°})^2 = 14^2 \underline{/18° \times 2}$$
$$= 196 \underline{/36°}$$

Example 14

Find the square root of $Z_a = 625 \underline{/60°}$.

Solution

$$\sqrt{Z_a} = \sqrt{625 \underline{/60°}}$$
$$= \sqrt{625} \underline{/60° \div 2}$$
$$= \pm 25 \underline{/30°}$$

Our treatment of this subject at this time is necessarily limited to the features which are of immediate use to us in our present studies. You will find in advanced work in mathematics that DeMoivre's theorem proves that there are as many answers to a root problem as there are roots to be taken: the third root of a phasor has three answers, each of the same magnitude but at a different angle. For our immediate purposes, however, Examples 13 and 14 show the basic operations.

PROBLEMS 31-4

Perform the indicated operations and express the results in both polar and rectangular form:

1. $5\underline{/53.1°} \times 6.7\underline{/-63.4°}$

2. $21.4\underline{/52.6°} \times 25.5\underline{/25.6°}$

3. $(9.9\underline{/69.9°})(8.8\underline{/82.2°})$

4. $(183.3\underline{/-11°})(3.26\underline{/11°})$

5. $(8.24\underline{/-34°})(9.07\underline{/-52.6°})$

6. $(9.5\underline{/-71.6°})(8.26\underline{/-7.6°})$

7. $10\underline{/53.2°} \div 5\underline{/36.8°}$

8.

9. $92.3\underline{/-12.5°} \div 81\underline{/-64.6°}$

10. $3.86\underline{/-79.57°} \div 13.9\underline{/69°}$

$1\underline{/0°} \div 20\underline{/-36.8°}$

11. $\dfrac{66.8\underline{/13°}}{4.73\underline{/24°}}$

12. $\dfrac{1.87\underline{/-180°}}{3.54\underline{/-180°}}$

Perform the indicated operations:

13. $\sqrt{144\underline{/30°}}$ 14. $\sqrt{1024\underline{/-17°}}$ 15. $(1.7\underline{/22°})^2$

16. $(0.31\underline{/-60°})^2$ 17. $\sqrt[3]{64\underline{/270°}}$ 18. $\sqrt[3]{1728\underline{/-21.9°}}$

19. $(3\underline{/11°})^3$ 20. $(2\underline{/-16°})^5$

At the end of Chapter 28, we drew your attention to the polar-rectangular conversion key available on some calculators. Now it is time for you to check on some other calculator feaures which may be available to you. If your calculator offers $\Sigma+$ and $\Sigma-$ features, it may also offer an opportunity to add or subtract rectangular components which have been computed and are stored in scratchpad memory. Check your instruction manual for this added service. You may wish to practice it on the examples and problems of this chapter and the following chapters.

Alternating Currents— Series Circuits

32

Because of the phenomena that occur in them, ac circuits make a very interesting subject for study. In several ways, they are unlike circuits that carry direct currents. The product of the voltage and current is seldom equal to the reading of a wattmeter connected in the circuit; the current may lag or lead the voltage; and the potential difference across an inductance or capacitance may be several times the supply voltage. This chapter deals with the computation of such effects in series circuits.

32-1 DEFINITIONS

In Chap. 9 we investigated *resistance* and defined it as the amount of opposition to current flow within a conductor. It may be helpful to think of it as the electrical phenomenon which always tends to oppose the flow of electric current and which always converts some of the energy of the current electricity into heat energy. This heat energy is dissipated, usually by radiation, and is *lost* so far as the circuit is concerned. In some cases, of course, the purpose of the circuit is to provide a conversion of electric energy into heat energy. This heat energy is then radiated away from the circuit, and it represents lost energy so far as the circuit is concerned.

In this chapter, we will also investigate relationships which are involved when alternating current flows under the influence of alternating emf's because, when inductance and/or capacitance is involved in the circuit, we must abandon Ohm's law as a specific method of computation.

Inductance is the electrical phenomenon which always tends to oppose a change in electric current and which always converts some of the energy of current electricity into stored electromagnetic energy. This electromagnetic energy is stored by the inductance when the circuit increases, and is released into the circuit when the current decreases. It is found that the current flow through an inductance lags the applied emf by 90 electrical degrees.

Capacitance is the electrical phenomenon which always tends to oppose a change in voltage and which converts some of the energy of current electricity into stored electrostatic energy. This electrostatic energy is stored by the capacitance as an electric charge on the plates of a capacitor when the applied emf is increasing, and it is released into the circuit when the applied emf is decreasing. It is found that the voltage across a capacitor lags the current flow ''through'' the capacitor by 90 electrical degrees.

It is the 90° phase angles between voltage and current in *ac* circuits containing inductance and capacitance, together with their associated resistances, that really bring the trigonometric functions into play. You should make a special effort to resolve any difficulties which may still exist in your ability to solve right triangles by trigonometry (Chap. 26) and the j operator (Chaps. 20 and 31). Also, you should be confident in the use of the trigonometric operations on your calculator.

32-2 THE RESISTIVE CIRCUIT

Figure 32-1 represents a 60-Hz alternator supplying 220 V to two resistances connected in series.

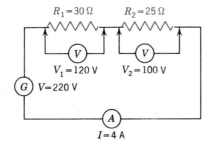

Fig. 32-1 Alternator supplying resistive circuit.

This circuit contains resistance only; therefore, Ohm's law applies in every respect. The internal resistance of the alternator and the resistance of the connecting wires being neglected, the current through the circuit is given by the familiar relation

$$I = \frac{V}{R_t} = \frac{V}{R_1 + R_2} = \frac{220}{30 + 25} = \frac{220}{55}$$
$$= 4 \, A$$

Again, as with direct currents, the voltage drops, or potential differences, across the resistances are

$$V_1 = IR_1 = 4 \times 30 = 120 \, V$$
$$V_2 = IR_2 = 4 \times 25 = \underline{100 \, V}$$
$$\text{Applied voltage} = 220 \, V$$

In an ac circuit containing only resistance, the voltage and current are in phase; that is, the voltage and current pass through corresponding parts of their cycles at the same instant.

From the above it follows that if

$$v = V_{max} \sin \omega t = 311 \sin 377t \, V$$

is the equation for the alternator voltage of Fig. 32-1, then the current through the circuit is

$$i = I_{max} \sin (\omega t + \theta) = I_{max} \sin (\omega t + 0°)$$
$$= 5.66 \sin 377t \, A$$

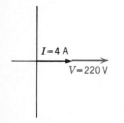

Fig. 32-2 Phasor diagram for circuit of Fig. 32-1.

Figure 32-2 is the phasor diagram for the circuit of Fig. 32-1. It will be noted that the voltage phasor and the current phasor coincide. This is what the equations for the voltage and current would make you expect, for they differ only in amplitude factors. The frequency factors are equal, and the phase angle is 0° (Secs. 29-7 to 29-9).

It is evident that Ohm's law says nothing about maximum, average, or effective values of current and voltage. Any of these values can be used; that is, maximum voltage can be used to find maximum current, average voltage can be used to find average current, etc. Naturally, maximum voltage is not used to find effective current unless the proper conversion constant is introduced into the equation. As previously stated, all voltage and current values here are to be considered as effective values unless otherwise specified (Sec. 30-7).

32-3 POWER IN THE RESISTIVE CIRCUIT

In dc circuits the power is equal to the product of the voltage and the current (Sec. 8-5). This is true of ac circuits for *instantaneous values* of voltage and current. That is, the *instantaneous power* is

$$p = vi \quad \text{VA} \tag{1}$$

and is measured in *voltamperes* or *kilovoltamperes*, abbreviated VA and kVA (or sometimes V · A and kV · A), respectively.*

When a sine wave of voltage is impressed across a resistance, the relations among voltage, current, and power are as shown in Fig. 32-3. The voltage existing across the resistance is in phase with the current flowing through the resistance. The power delivered to the resistance at any instant is represented by the height of the power curve, which is the product of the instantaneous values of voltage and current at that instant. The shaded area under the power curve represents the total power delivered to the circuit during one complete cycle of voltage. It will be noted that the power curve is of sine-wave voltage. Also, the power curve lies entirely above the *x* axis; there are no negative values of power.

The maximum height of the power curve is the product of the maximum values of voltage and current. Stated as an equation

$$P_{max} = V_{max}I_{max} \tag{2}$$

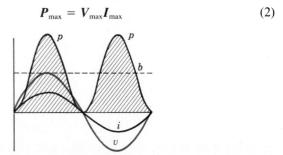

Fig. 32-3 Power curves for circuit containing only resistance.

* The dot is preferred in general physics relationships, but is is customarily omitted in electricity and electronics use.

The average power delivered to a resistance load is represented by the height of the line *ab* in Fig. 32-3, which is half the maximum height of the power curve, or its average height. Then, since

$$\text{Average power} = P = \tfrac{1}{2}P_{\max}$$

by dividing both members of Eq. (2) by 2 we obtain

$$\tfrac{1}{2}P_{\max} = \tfrac{1}{2}V_{\max}I_{\max}$$

Substituting for the value of $\tfrac{1}{2}P_{\max}$ and factoring the denominator of the right member,

$$P = \frac{V_{\max}I_{\max}}{\sqrt{2}\sqrt{2}}$$

Substituting for the values in the right member (Sec. 30-7),

$$P = VI \qquad \text{W} \tag{3}$$

Hence, the alternating power consumed by a resistance load is equal to the product of the effective values of voltage and current. As in dc circuits, alternating power is measured in watts and kilowatts.

Example 1

What is the power expended in the resistances of Fig. 32-1?

Solution

$$\text{Voltage across } R_1 = V_1 = 120 \text{ V}$$
$$\text{Voltage across } R_2 = V_2 = 100 \text{ V}$$
$$\text{Current through circuit} = I \ = 4 \text{ V}$$
$$\text{Power expended in } R_1 = P_1 = V_1I = 120 \times 4 = 480 \text{ W}$$
$$\text{Power expended in } R_2 = P_2 = V_2I = 100 \times 4 = \underline{400 \text{ W}}$$
$$\text{Total} = 880 \text{ W}$$

Also, the total power is

$$P_t = VI = 220 \times 4 = 880 \text{ W}$$

Because $P = VI$, the usual Ohm's law relations hold for resistances in ac circuits. Hence,

$$P = I^2R \qquad \text{W} \tag{4}$$

and

$$P = \frac{V^2}{R} \qquad \text{W} \tag{5}$$

Thus, the power consumed by R_1 of Fig. 32-1 can be computed by using Eq. (4) or (5). Hence,

$$P_1 = I^2R_1 = 4^2 \times 30 = 480 \text{ W}$$

or

$$P_1 = \frac{V_1^2}{R_1} = \frac{120^2}{30} = 480 \text{ W}$$

PROBLEMS 32-1

1. A 400-Hz alternator supplies 88 V across a combination of three series resistors of 150, 67, and 22 Ω.
    (a) How much current flows in the circuit?
    (b) Write the equation for the alternator voltage at any time t.
    (c) Write the equation for the circuit current at any time t.
    (d) What is the voltage measured across the 67-Ω resistor?
    (e) How much power is dissipated by the 22-Ω resistor?
    (f) What is the instantaneous value of the current when the instantaneous emf is 26 V?
2. Given the circuit of Fig. 32-4:
    (a) Write the equation for the emf of the alternator at any time t.
    (b) Write the equation for the total current of the circuit.
    (c) What is the voltage across R_3?
    (d) How much power is dissipated in R_2?
    (e) How much current flows through R_1?
    (f) What is the instantaneous value of the total current when the instantaneous alternator emf is 36.5 V?

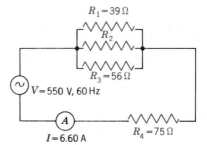

Fig. 32-4 Circuit of Probs. 2 and 3.

3. In Fig. 32-4, what is the instantaneous value of the voltage across R_2 when the instantaneous current through R_4 is 2.75 A?
4. A 10-kHz signal generator is connected to a 600-Ω resistive load. A milliwattmeter indicates that the resistor is dissipating 800 mW. What is the maximum instantaneous voltage developed at the generator terminals?
5. What is the equation of the current in Prob. 4?

32-4 THE INDUCTIVE CIRCUIT

A circuit, or an inductance coil, has the property of inductance when there is set up in it an emf due to a *change* of current through it. Thus, a circuit has an inductance of 1 H when a change of current of 1 A/s induces an emf of 1 V. Expressed as an equation,

$$V_{av} = L\frac{I}{t} \quad V \tag{6}$$

where V_{av} is the average voltage induced in a circuit of L H by a *change* of current of I A in t s.

An alternating current of I_{max} A makes *four changes* during each cycle. These are

1. From zero to maximum positive value
2. From maximum positive value to zero
3. From zero to maximum negative value
4. From maximum negative value to zero

The time required for one complete cycle of alternating currents is $T = f^{-1}$ s (Sec. 29-9), and each of the above changes occurs in one-quarter of the time required for the completion of one cycle. Then the time for each change is $(4f)^{-1}$ s. Substituting this value of t, and I_{max} for I, in Eq. (6), we have

$$V_{av} = L\frac{I_{max}}{(4f)^{-1}} = 4fLI_{max} \qquad (7)$$

Equation (7) is cumbersome if used in its present form, for it contains an average-voltage term and a maximum-current term. The equation can be expressed in terms of the relation between average and maximum values as given in Sec. 30-6:

$$V_{av} = \frac{2}{\pi}V_{max}$$

Substituting in Eq. (7) for this value of V_{av}, we have

$$\frac{2}{\pi}V_{max} = 4fLI_{max}$$

which becomes

$$V_{max} = 2\pi fLI_{max} \qquad (8)$$

Because both voltage and current in Eq. (8) are now in terms of maximum values, effective values can be used. Thus,

$$V = 2\pi fLI \qquad V \qquad (9)$$

The factors $2\pi fL$ in Eqs. (8) and (9) represent a reaction due to the frequency of the alternating current and the amount of inductance contained in the circuit. Hence, the alternating voltage of V required to cause a current of I A with a frequency of f Hz to flow through an inductance of L H is given by Eq. (9). That is, the voltage must overcome the reaction $2\pi fL$, which is called the *inductive reactance*. From Eq. (9) the inductive reactance, which is denoted by X_L and expressed in ohms, is given by

$$\frac{V}{I} = 2\pi fL$$

or
$$X_L = 2\pi fL = \omega L \qquad \Omega \qquad (10)$$

where f = frequency, Hz
  L = inductance, H

Note the similarity of the relations between voltage and current for inductive reactance and resistance. Both inductive reactance and resistance offer an opposition to a flow of alternating current; both are expressed in ohms; and both are equal to the voltage divided by the current. Here the similarity ends; there is no inductive reactance to steady-state direct currents because there is no *change* in current, and, as explained later, inductive reactances consume no alternating power.

Figure 32-5 represents a 60-Hz alternator delivering 220 V to a coil having an inductance of 0.165 H. The opposition, or inductive reactance, to the flow of current is

$$X_L = 2\pi fL = 2\pi \times 60 \times 0.165$$
$$= 62.2 \; \Omega$$

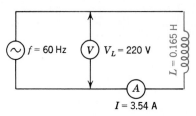

$$I = 3.54 \; A$$

Fig. 32-5 V_L = 220 V, L = 0.165 H.

It is impossible to construct an inductance containing no resistance; but in order to simplify basic considerations, we shall consider the coil of Fig. 32-5 as being an inductance with negligible resistance. (The effects of inductance and resistance acting together are discussed in Sec. 32-8.) The current in the circuit due to the action of voltage and inductive reactance is

$$I = \frac{V_L}{X_L} = \frac{220}{62.2} = 3.54 \; A$$

Example 2

What is the inductive reactance of an inductance of 17 μH at a frequency of 2500 kHz?

Solution

$$f = 2500 \; kHz = 2.5 \times 10^6 \; Hz$$
$$L = 17 \; \mu H = 1.7 \times 10^{-5} \; H$$
$$X_L = 2\pi fL = 2\pi \times 2.5 \times 10^6 \times 1.7 \times 10^{-5}$$
$$= 2\pi \times 1.7 \times 2.5 \times 10 = 267 \; \Omega$$

Example 3

An inductor is connected to 115 V, 60 Hz. An ammeter connected in series with the coil reads 0.714 A. On the assumption that the coil contains negligible resistance, what is its inductance?

Solution

$$V_L = 115 \; V$$
$$f = 60 \; Hz$$
$$I = 0.714 \; A$$
$$X_L = \frac{V_L}{I} = \frac{115}{0.714} = 161 \; \Omega$$

Since $\quad X_L = 2\pi fL$

then $\quad L = \dfrac{X_L}{2\pi f} = \dfrac{161}{2\pi \times 60} = 0.427 \; H$

In a circuit containing inductance, a change of current induces an emf of such polarity that it always opposes the change of current. Because an alternating current is constantly changing, in an inductive circuit there is always present a reaction that opposes this change. The net effect, in a *purely inductive circuit,* is to cause the *current to lag the voltage by* 90°. This is illustrated by the phasor diagram of Fig. 32-6, which shows the voltage of the circuit of Fig. 32-5 to be at maximum positive value when the current is passing through zero.

The instantaneous voltage across the inductance is given by

$$v = V_{max} \sin \omega t \quad \text{V}$$

or

$$v = 311 \sin 377t \quad \text{V}$$

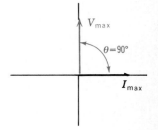

Fig. 32-6 Current lags voltage by 90°.

Since the current lags the voltage by a phase angle θ of 90°, the equation for the current through the inductance is

$$i = I_{max} \sin (\omega t - \theta) \quad \text{A} \qquad (11)$$

or

$$i = 5 \sin (377t - 90°) \quad \text{A} \qquad (12)$$

If the voltage has completed $\phi°$ of its cycle, the instantaneous current is

$$i = 5 \sin (\phi - 90°) \quad \text{A} \qquad (13)$$

Example 4

What is the instantaneous value of the current in Fig. 32-5 when the voltage has completed 120° of its cycle?

Solution

Draw a phasor diagram of the current and voltage relations as shown in Fig. 32-7. The instantaneous value of the current is found from Eq. (13) and is

$$
\begin{aligned}
i &= I_{max} \sin (\phi - 90°) \\
&= 5 \sin (120° - 90°) \\
&= 5 \sin 30° \\
&= 2.5 \text{ A}
\end{aligned}
$$

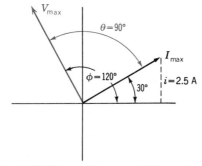

Fig. 32-7 Phasor diagram of Example 4.

PROBLEMS 32-2

1. What is the reactance of a 15-mH coil at 60 Hz?
2. What is the reactance of a 15-mH coil at 1 kHz?

3. What is the reactance of a 15-mH coil at 1 MHz?
4. What is the inductance of a coil that exhibits a reactance of 754 Ω at a frequency of 400 Hz?
5. A tuning coil in a radio transmitter has an inductance of 270 μH. What is its reactance at a frequency of 1.5 MHz?
6. At what frequency will a television set coil with an inductance of 3.25 μH offer a reactance of 3740 Ω?
7. Assuming negligible resistance, what would be the current flow through an inductance of 0.067 H at a voltage of 100 V, 800 Hz?
8. What would be the equation of the current in Prob. 7?
9. A current of 379 μA at 2.5 V flows through a 5.25-μH coil. Assuming negligible resistance, what is the frequency of the applied emf?
10. An emf described by the equation $v = 311 \sin 314t$ V is applied to an inductor of 1.65 H. What is the equation of the current flow, assuming negligible resistance?
11. What is the instantaneous value of the current in Prob. 10 when the emf has completed 45° of its cycle?
12. What is the instantaneous value of the applied voltage in Prob. 10 when the current has completed 210° of its cycle?
13. What happens to the inductive reactance of a circuit when the inductance is fixed but the frequency of the applied emf is (*a*) doubled, (*b*) tripled, (*c*) halved?
14. What happens to the inductive reactance of a circuit when the frequency of the applied emf is held constant and the inductance is varied?

32-5 THE CAPACITIVE CIRCUIT

A capacitance is formed between two conductors when there is an insulating material between them. A circuit, or a capacitor, is said to have a capacitance of one farad when a *change* of one volt per second produces a current of one ampere. Expressed as an equation,

$$I_{av} = C\frac{V}{t} \quad A \tag{14}$$

where I_{av} is the average current in amperes that is caused to flow through a capacitance of C by a *change* of V V in t s.

In all probability the above definition does not clearly indicate to you *how much* electricity, or charge, a given capacitor will contain. Perhaps a more understandable definition is that a circuit, or a capacitor, has a capacitance of one farad when a difference of potential of one volt will produce on it one coulomb of charge. Expressed as an equation,

$$Q = CV \quad C \tag{15}$$

where Q is the charge in coulombs placed on a capacitor of C F by a difference of potential of V V across the capacitor.

It was shown in Sec. 32-4 that the time t required for one change of an alternating emf was $(4f)^{-1}$ s. Thus, if an alternating emf of V_{max} V at a frequency of f Hz is impressed across a capacitor of C F, by substituting the above value of t and V_{max} for V in Eq. (14),

$$I_{av} = C\frac{V_{max}}{(4f)^{-1}} = 4fCV_{max} \quad \text{A} \qquad (16)$$

Again, as in Eq. (7), the above equation contains an average term and a maximum term. As given in Sec. 30-6,

$$I_{av} = \frac{2}{\pi}I_{max} \quad \text{A}$$

Substituting in Eq. (16) for this value of I_{av}, we have

$$\frac{2}{\pi}I_{max} = 4fCV_{max}$$

which becomes

$$I_{max} = 2\pi fCV_{max} \quad \text{A} \qquad (17)$$

Because both voltage and current in Eq. (17) are now in terms of maximum values, effective values can be used. Thus,

$$I = 2\pi fCV \quad \text{A} \qquad (18)$$

The factors $2\pi fC$ represent a reaction due to the frequency of the alternating emf and the amount of capacitance; hence, it is evident that the amount of current in a purely capacitive circuit depends upon these factors. As in the case of resistive circuits and inductive circuits, the opposition to the flow of current is obtained by dividing the voltage by the current. Then, from Eq. (18),

$$\frac{V}{I} = \frac{1}{2\pi fC} \quad \Omega \qquad (19)$$

The right member of Eq. (19), which represents the opposition to a flow of alternating current in a purely capacitive circuit, is called the *capacitive reactance*. It is denoted by X_C and expressed in ohms. Thus,

$$X_C = \frac{1}{2\pi fC} = \frac{1}{\omega C} \quad \Omega \qquad (20)$$

where f = frequency, Hz
  C = capacitance, F

Figure 32-8 represents a 60-Hz alternator delivering 220 V to a capacitor having a capacitance of 14.5 μF. The opposition, or capacitive reactance, to the flow of current is

$$X_C = \frac{1}{2\pi fC} = \frac{1}{2\pi \times 60 \times 14.5 \times 10^{-6}}$$

$$= \frac{10^4}{2\pi \times 6 \times 1.45} = 183 \ \Omega$$

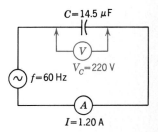

Fig. 32-8 V_C = 220 V, C = 14.5 μF.

Neglecting the resistance of the connecting leads and the extremely small losses at low frequencies in a well-constructed capacitor, the current in the circuit due to the action of the voltage and capacitive reactance is

$$I = \frac{V_C}{X_C} = \frac{220}{183} = 1.20 \ \text{A}$$

Example 5

What is the capacitive reactance of a 350-pF capacitor at a frequency of 1200 kHz?

Solution

$$f = 1200 \text{ kHz} = 1.2 \times 10^6 \text{ Hz}$$
$$C = 350 \text{ pF} = 3.5 \times 10^{-10} \text{ F}$$
$$X_C = \frac{1}{2\pi fC}$$
$$= \frac{1}{2\pi \times 1.2 \times 10^6 \times 3.5 \times 10^{-10}}$$
$$= \frac{10^4}{2\pi \times 1.2 \times 3.5} = 379 \ \Omega$$

Example 6

A capacitor is connected across 110 V, 60 Hz. A milliammeter connected in series with the capacitor reads 350 mA. What is the capacitance of the capacitor?

Solution

$$V_C = 110 \text{ V}$$
$$f = 60 \text{ Hz}$$
$$I = 350 \text{ mA} = 0.350 \text{ A}$$
$$X_C = \frac{V_C}{I} = \frac{110}{0.35} = 314 \ \Omega$$

since

$$X_C = \frac{1}{2\pi fC}$$

often

$$C = \frac{1}{2\pi fX_C} = \frac{1}{2\pi \times 60 \times 314}$$
$$= \frac{10^{-3}}{2\pi \times 6 \times 3.14}$$
$$= 8.44 \times 10^{-6} \text{ F} = 8.44 \ \mu\text{F}$$

Because current flows in a capacitor only when the voltage across the capacitor is changing, it is evident that, when an alternating voltage is impressed, current is flowing at all times because the potential difference across the capacitor is constantly changing. Furthermore, the greatest amount of current will flow when the voltage is changing most rapidly, and this occurs when the voltage passes through zero value. This property, in conjunction with the effects of the counter emf, *causes the current to lead the voltage by 90° in a purely capacitive circuit*. This is illustrated by the vector diagram of Fig. 32-9, which shows the current through the circuit of Fig. 32-8 to be at maximum positive value when the voltage is passing through zero.

The instantaneous voltage across the capacitor is given by

$$v = V_{max} \sin \omega t \qquad \text{V} \qquad (21)$$

or

$$v = 311 \sin 377t \text{ V} \qquad (22)$$

Therefore, the equation for the current is

$$i = I_{max} \sin (377t + \theta) \qquad \text{A} \qquad (23)$$

or

$$i = 1.70 \sin (377t + 90°) \text{ A} \qquad (24)$$

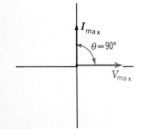

Fig. 32-9 Current leads voltage by 90°.

If the voltage has completed $\phi°$ of its cycle, the instantaneous current is

$$i = I_{max} \sin(\phi + 90°) \quad A \tag{25}$$

Example 7

What is the instantaneous value of the current in the circuit shown in Fig. 32-8 when the voltage has completed 35° of its cycle?

Solution

Draw a phasor diagram of the current and voltage relations as shown in Fig. 32-10. The instantaneous value of the current is found from Eq. (25) and is

$$i = I_{max} \sin(\phi + 90°)$$
$$= 1.70 \sin(35° + 90°)$$
$$= 1.70 \sin 125° = 1.39 \, A$$

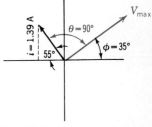

Fig. 32-10 Phasor diagram for Example 7.

32-6 CAPACITORS IN SERIES

Figure 32-11 represents two capacitors C_1 and C_2 connected in series with a voltage V across the combination. Because the capacitors are in series, the same quantity of electricity must be sent into each of them. Then, if V_1 and V_2 represent the potential differences across C_1 and C_2, respectively, Q represents the quantity of electricity in each capacitor and C_t is the capacitance of the combination. Hence,

$$V = \frac{Q}{C_t}$$

$$V_1 = \frac{Q}{C_1}$$

and

$$V_2 = \frac{Q}{C_2}$$

Since

$$V = V_1 + V_2 \tag{26}$$

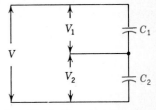

Fig. 32-11 Capacitors C_1 and C_2 connected in series.

by substituting the values for all voltages into Eq. (26),

$$\frac{Q}{C_t} = \frac{Q}{C_1} + \frac{Q}{C_2}$$

or

$$\frac{1}{C_t} = \frac{1}{C_1} + \frac{1}{C_2} \tag{27}$$

Equation (27) resolves into

$$C_t = \frac{C_1 C_2}{C_1 + C_2} \tag{28}$$

The above illustrates the fact that capacitors in series combine like resistances in parallel; that is, the reciprocal of the combined capacitance of capacitors in series is equal to the sum of the reciprocals of the capacitances of the individual capacitors.

Example 8

What is the capacitance of a 6-μF capacitor in series with a capacitor of 4μF?

Solution

$$C_t = \frac{6 \times 4}{6 + 4} = 2.4 \; \mu F$$

PROBLEMS 32-3

1. What is the capacitive reactance of a 22-μF capacitor at a frequency of 400 Hz?
2. What is the capacitive reactance of a 22-μF capacitor at a frequency of 1 kHz?
3. What is the capacitive reactance of a 22-μF capacitor at a frequency of 100 kHz?
4. What is the reactance of a 50-pF capacitor at a frequency of 12 GHz?
5. The filter capacitance in a radio receiver is 0.0016 μF. What is its reactance at a frequency of 720 kHz?
6. What is the reactance of the capacitor of Prob. 5 if the frequency is increased to 1320 kHz?
7. How much current will flow in a capacitor of 6.3 pF when 475 V at 1 kHz is impressed across the capacitor, neglecting resistance?
8. What will be the current in the capacitor of Prob. 7 if the frequency is increased to 12 kHz?
9. When a 120-V 800-Hz emf is impressed across a capacitor, the current flow is 2.41 A. What is the capacitance?
10. A current of 452 mA flows through a 5-μF capacitor when the frequency of the applied emf is 60 Hz. What is the voltage?
11. What is the equation for the current in Prob. 10?
12. What is the instantaneous value of the current in Prob. 10 when the emf has completed 230° of its cycle?
13. What is the resulting capacitance when a 220-pF capacitor is connected in series with a 500-pF capacitor?
14. Two capacitors, 20 and 200 pF, are connected in series. What is the resultant capacitance?
15. If an emf of 80 V at 15 kHz is impressed across the series circuit of Prob. 14, what will be the resultant current flow, neglecting resistance?
16. What happens to the capacitive reactance of a circuit when the capacitance is fixed but when the frequency of the applied emf is (a) doubled, (b) tripled, (c) halved?
17. What happens to the capacitive reactance of a circuit when the frequency of the applied emf is held constant and the capacitance is varied?
18. Neglecting the resistance of the connecting wires in Fig. 32-12:
    (a) Write the equation for the emf of the alternator.
    (b) Write the equation for the circuit current.
    (c) What is the voltage across C_1?
    (d) What is the voltage across C_2?

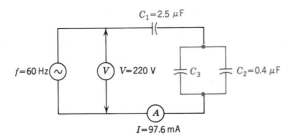

Fig. 32-12 Circuit of Prob. 18.

32-7 POWER IN CIRCUITS CONTAINING ONLY INDUCTANCE OR CAPACITANCE

Figure 32-13 illustrates the voltage, current, and power relations when a sine wave of emf is impressed across an inductor whose resistance is negligible.

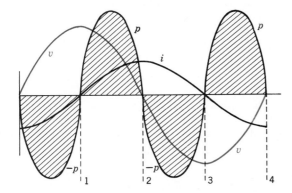

Fig. 32-13 Voltage, current, and power in an inductive circuit.

When the current is increasing from zero to maximum positive value, during the time interval from 1 to 2, power is being taken from the source of emf and is being stored in the magnetic field about the coil. As the current through the inductor decreases from maximum positive value to zero, during the time from 2 to 3, the magnetic field is collapsing, thus returning its power to the circuit. Thus during the intervals from 1 to 2 and from 3 to 4, the inductor is taking power from the source that is represented by the *positive* power in the figure. During the intervals from 0 to 1 and 2 to 3, the inductor is returning power to the source that is represented by the *negative* power in the figure. As previously stated, the instantaneous power is equal to the product of the voltage and current, it is positive when the voltage and current are of like sign and negative when they are of unlike sign. Note that between points 3 and 4, although both the voltage and the current are negative, the power is positive.

When an alternating emf is impressed across a capacitor, power is taken from the source and stored in the capacitor as the voltage increases from zero to maximum positive value. As the voltage decreases from maximum positive value to zero, the capacitor discharges and returns power to the source. As in the case of the inductor, half of the power loops are positive and half are negative; therefore, no power is expended in either circuit, for the power alternately flows to and from the source. This power is called *reactive* or *apparent power* and is given by the relation

$$P = VI \qquad \text{VA}$$

32-8 RESISTANCE AND INDUCTANCE IN SERIES

It has been explained that in a circuit containing only resistance the voltage applied across the resistance and the current through the resistance are in phase and that in a circuit containing only reactance the voltage and current are 90° out of phase. However, circuits encountered in practice contain both resistance and reactance. Such a condition is shown in Fig. 32-14, where an alternating emf of 100 V is impressed across a combination of 6 Ω resistance in series with 8 Ω inductive reactance.

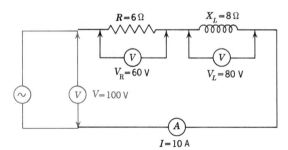

Fig. 32-14 Series circuit containing resistance and inductance.

As with dc circuits, the sum of the voltage drops around the circuit comprising the load must equal the applied emf. In the consideration of resistance and reactance, however, we are dealing with voltages that can no longer be added or substracted arithmetically. That is because the voltage drop across the resistance is in phase with the current and the voltage drop across the inductive reactance is 90° ahead of the current.

Because the current is the same in all parts of a series circuit, we can use it as a reference and plot the voltage across the resistance and that across the inductive reactance, as shown in Fig. 32-15. The resultant of these two voltages, which can be treated as rectangular components (see Sec. 28-4), must be equal to the applied emf. Hence, if IR and IX_L are the potential differences across the resistance and inductive reactance, respectively,

$$V = \sqrt{(IR)^2 + (IX_L)^2} \qquad \text{V} \tag{29}$$

or $\qquad V = \sqrt{60^2 + 80^2} = 100 \text{ V}$

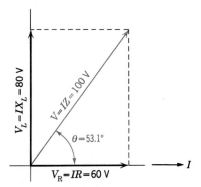

Fig. 32-15 Phasor diagram for circuit of Fig. 32-14.

The phase angle θ between voltage and current can be found by using any of the trigonometric functions. For example,

$$\tan \theta = \frac{IX_L}{IR} = \frac{80}{60} = 1.33$$
$$\therefore \theta = 53.1°$$

and it is apparent from the phasor diagram that the current through the circuit lags the applied voltage by this amount.

Although the foregoing demonstrates that the *phasor summation* of the voltage across the resistance is equal to the applied emf, no relation between applied voltage and circuit current has been given as yet.

Since $\qquad V = \sqrt{(IR)^2 + (IX_L)^2}$

then $\qquad V = \sqrt{I^2R^2 + I^2X_L^2}$

Factoring, $\qquad V = \sqrt{I^2(R^2 + X_L^2)}$

Hence, $\qquad V = I\sqrt{R^2 + X_L^2} \qquad$ V $\qquad$ (30)

As previously stated, the applied voltage divided by the current results in a quotient that represents the opposition offered to the flow of current. Hence, from Eq. (30),

$$\frac{V}{I} = \sqrt{R^2 + X_L^2} \qquad (31)$$

The expression $\sqrt{R^2 + X_L^2}$ is called the *impedance* of the circuit. It is denoted by Z and measured in ohms. Therefore

$$Z = \sqrt{R^2 + X_L^2} \qquad \Omega \qquad (32)$$

Applying Eq. (32) to the circuit of Fig. 32-14,

$$Z = \sqrt{6^2 + 8^2} = 10 \ \Omega$$

and $\qquad I = \frac{V}{Z} = 10 \text{ A}$

From Eq. (31), Eq. (32) can be written

$$V = IZ = I\sqrt{R^2 + X_L^2}$$

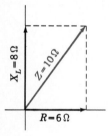

Fig. 32-16 Z can be plotted as phasor sum of R and X_L.

The foregoing illustrates that the factor I is common to all expressions, which is the same as saying that the current is the same in all parts of the circuit. Because this condition exists, it is permissible to plot the resistance and reactance as rectangular components as shown in Fig. 32-16; hence, the impedance of a series circuit is simply the phasor sum of the resistance and reactance. The various methods used in solving for the impedance are the same as those given for phasor summation of rectangular components in Example 4 of Sec. 28-4. Note that the values are identical.

Example 9

A circuit consisting of 120 Ω resistance in series with an inductance of 0.35 H is connected across a 440-V 60-Hz alternator. Determine (a) the phase angle between voltage and current, (b) the impedance of the circuit, and (c) the current through the circuit.

Solution

(a) Drawing and labeling the circuit is left to you. The inductive reactance is

$$X_L = 2\pi f L = 2\pi \times 60 \times 0.35$$
$$= 132 \ \Omega$$

Draw the phasor impedance diagram as shown in Fig. 32-17. Then, since

$$\tan \theta = \frac{X_L}{R} = \frac{132}{120} = 1.10$$
$$\therefore \theta = 47.7°$$

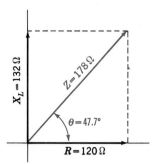

Fig. 32-17 Impedance phasor diagram for circuit of Example 9.

Note that the phase angle denotes the position of the applied voltage with respect to the current, which is taken as a reference. Thus an inductive series circuit always has a "lagging" phase angle which is a *positive angle* when resistance, reactance, and impedance are plotted vectorially.

(b)
$$Z = \frac{R}{\cos \theta} = \frac{120}{\cos 47.7°} = 178 \ \Omega$$

or
$$Z = \frac{X_L}{\sin \theta} = \frac{132}{\sin 47.7°} = 178 \ \Omega$$

(c)
$$I = \frac{V}{Z} = \frac{440}{178} = 247 \ A$$

32-9 RESISTANCE AND CAPACITANCE IN SERIES

Figure 32-18 represents a circuit in which an alternating emf of 100 V is applied across a combination of 6 Ω resistance in series with 8 Ω capacitive reactance. Note the similarity between the circuits shown in Figs. 32-14 and 32-18. Both have the same values of resistance and absolute values of reactance. However, in the circuit of Fig. 32-18 the voltage drop across the capacitance reactance is 90° behind the current. Again using the current as a reference, because it is the same in all parts of the circuit, the voltage across the resistance and the voltage across the capacitive reactance are plotted as shown in Fig. 32-19 and treated as rectangular components of the applied emf. The impedance of the circuit is found in the same manner as that of the inductive circuit, that is, by phasor summation of the rectangular components. The phase angle is found by the same method.

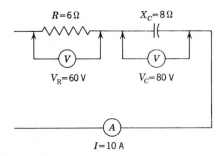

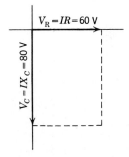

Fig. 32-18 Series circuit consisting of resistance and capacitance.

Fig. 32-19 Phasor diagram for circuit of Fig. 32-18.

$$\tan \theta = \frac{X_C}{R} = \frac{8}{6} = 1.33$$
$$\therefore \theta = -53.1°$$

In the capacitive circuit the current leads the voltage, and we prefix the impedance angle with a minus sign because of its position (Sec. 23-2).

Example 10
A circuit consisting of 175 Ω resistance in series with a capacitor of 5.0 μF is connected across a source of 150 V, 120 Hz. Determine (a) the phase angle between voltage and current, (b) the impedance of the circuit, and (c) the current through the circuit.

Solution
(a) Drawing and labeling the circuit is left to you. The capacitive reactance is

$$X_C = \frac{1}{2\pi fC}$$
$$= \frac{1}{2\pi \times 120 \times 5 \times 10^{-6}}$$
$$= \frac{10^4}{2\pi \times 1.2 \times 5} = 265 \ \Omega$$

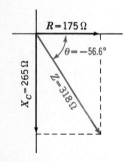

Fig. 32-20 Impedance phasor diagram for Example 10.

Draw the impedance diagram as shown in Fig. 32-20. Then, since

$$\tan\theta = \frac{X_C}{R} = \frac{265}{175} = 1.51$$

$$\therefore \theta = -56.6°$$

Thus the current is leading the voltage by 56.6°, as shown by the impedance phasor diagram.

(b)　　　　$Z = \dfrac{R}{\cos\theta} = \dfrac{175}{\cos 56.6°} = 318\ \Omega$

or　　　　$Z = \dfrac{X_C}{\sin\theta} = \dfrac{265}{\sin 56.6°} = 318\ \Omega$

(c)　　　　$I = \dfrac{V}{Z} = \dfrac{150}{318} = 0.472\ A$

PROBLEMS 32-4

1. A series circuit consists of a 1.5-H inductor which has a resistance of 35 Ω. It is supplied with 220 V, 60 Hz. Find
   (a) The inductive reactance.
   (b) The impedance of the coil.
   (c) The current flowing through the coil.
   (d) The equation of the current.
   (e) The voltage across the resistance of the coil.
   (f) The voltage across the inductance of the coil.
   (g) Why (e) + (f) does not equal 220 V.

2. A 500-V 8-MHz source is connected to a series circuit consisting of a 3.3-kΩ resistor and a 500-μH inductor of negligible resistance. Find
   (a) The inductive reactance of the inductor.
   (b) The impedance of the circuit.
   (c) The current flowing through the circuit.
   (d) The phase angle of the current.
   (e) The voltage across the resistor.
   (f) The voltage across the inductor.

3. In the circuit of Prob. 2, the applied emf is held constant while the frequency is decreased.
   (a) Why will this cause the current to rise?
   (b) When the current is twice that found in Prob. 2, find the impedance, the frequency, and the phase angle.

4. A 25-mH choke has a measured resistance of 40 Ω at 400 Hz. This choke is connected across 48 V at 400 Hz. Find (a) the impedance of the choke and (b) the current flow.

5. A 120-V 60-Hz source energizes a series circuit consisting of a 330-Ω resistor and a 22-μF capacitor. Find
   (a) The capacitive reactance.
   (b) The impedance of the circuit.
   (c) The current flow through the circuit.
   (d) The voltage across the resistor.
   (e) The voltage across the capacitor.

6. If the frequency of the 120-V source in Prob. 5 is doubled, what will be the current flow through the circuit?
7. What will be the impedance of the circuit of Prob. 5 if a 150-μF capacitor is connected in series with the original circuit?
8. What will be the current flow in the circuit of Prob. 5 if a 6.7-kΩ resistor is connected in series with the original circuit?
9. A series circuit consisting of a 1-kΩ resistor and a 150-pF capacitor is connected across 600 V at 4.3 MHz. Find
   (a) The impedance of the circuit.
   (b) The current flowing through the circuit.
   (c) The voltage across the resistor.
   (d) The voltage across the capacitor.
10. In the circuit of Prob. 9, a 50-pF capacitor is connected in series with the original capacitor. Find
    (a) The current flow through the new circuit.
    (b) The voltage across the resistor.
    (c) The voltage across the 150-pF capacitor.
    (d) The voltage across the 50-pF capacitor.

32-10 RESISTANCE, INDUCTANCE, AND CAPACITANCE IN SERIES

It has been shown that inductive reactance causes the current to lag the voltage and that capacitance reactance causes the current to lead the voltage; hence, these two reactions are exactly opposite in effect. Figure 32-21 represents a series circuit consisting of resistance, inductance, and capacitance connected across an alternator that supplies 220 V, 60 Hz. Now

$$\omega = 2\pi f = 2\pi \times 60 = 377$$
$$\therefore X_L = \omega L = 377 \times 0.35 = 132 \ \Omega$$

and

$$X_C = \frac{1}{\omega C} = \frac{1}{377 \times 13 \times 10^{-6}}$$

$$= \frac{10^3}{3.77 \times 1.3} = 203 \ \Omega$$

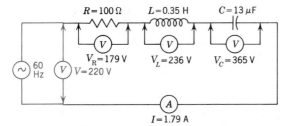

Fig. 32-21 Series circuit consisting of R, L, and C.

Figure 32-22 is an impedance phasor diagram of the conditions existing in the circuit. Since X_L and X_C are oppositely directed phasors, it is evident that the resultant reactance will have a magnitude equal to their algebraic sum and will be in the direction of the greater. Therefore, the net reactance of the circuit is a capacitive reactance of 72 Ω as illustrated in Fig. 32-22. Thus the entire circuit could be replaced by an equivalent series circuit consisting of 100 Ω resistance and 72 Ω capacitive reactance, provided that the frequency of the alternator remained constant.

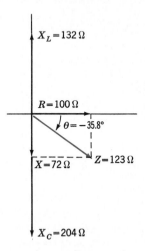

Fig. 32-22 Impedance phasor diagram for circuit of Fig. 32-21.

The impedance, current, and potential differences are found by the usual methods.

$$\tan \theta = \frac{X_C}{R} = \frac{72}{100} = 0.72$$
$$\therefore \theta = -35.8°$$
$$Z = \frac{X}{\sin \theta} = \frac{72}{\sin 35.8°} = 123 \ \Omega$$
$$I = \frac{V}{Z} = \frac{220}{123} = 1.79 \ A$$
$$V_R = IR = 1.79 \times 100 = 179 \ V$$
$$V_L = IX_L = 1.79 \times 132 = 236 \ V$$
$$V_C = IX_C = 1.79 \times 204 = 365 \ V$$

Note that the potential difference across the reactances is greater than the emf impressed across the entire circuit. This is reasonable, for the applied emf is across the impedance of the circuit, which is a smaller value, in ohms, than the reactances. Because the current is common to all circuit components, it follows that the greatest potential difference will exist across the component offering the greatest opposition.

32-11 POWER IN A SERIES CIRCUIT

It has been shown that, in a circuit consisting of resistance only, no power is returned to the source of emf. Also, it has been shown that a circuit containing reactance alone consumes no power; that is, a reactance alternately receives and returns all power to the source. It is evident, therefore, that in a circuit containing both resistance and reactance there must be some power expended in the resistance and also some returned to the source by the reactance. Figure 32-23 represents the relation among voltage, current, and power in the circuit of Fig. 32-21.

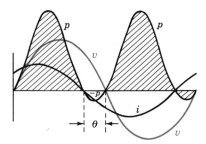

Fig. 32-23 Voltage, current, and power relations for circuit shown in Fig. 32-21.

As previously stated, the instantaneous power in the circuit is equal to the product of the applied voltage and the current through the circuit. When the voltage and current are of the same sign, they are acting together and taking power from the source. When their signs are unlike, they are operating in opposite directions and power is returned to the source. The *apparent power* is

$$P_a = VI \qquad \text{VA} \qquad (33)$$

and the actual power taken by the circuit, which is called the the *true power* or *active power,* is

$$P = I^2R \qquad \text{W} \qquad (34)$$

or

$$P = V_R I \qquad \text{W} \qquad (35)$$

where V_R is the potential difference across the resistance of the circuit.

The *power factor* (PF) of a circuit is the ratio of the true power to the apparent power. That is,

$$\text{PF} = \frac{P}{P_a} \qquad (36)$$

Substituting the value of P from Eq. (34) and that of P_a in Eq. (33),

$$\text{PF} = \frac{I^2R}{VI} = \frac{IR}{V}$$

Then, since

$$V = IZ$$

$$\text{PF} = \frac{IR}{IZ}$$

or

$$\text{PF} = \frac{R}{Z} \qquad (37)$$

Hence, the power factor of a series circuit can be obtained by dividing the resistance of a circuit by its impedance. The power factor is often expressed in terms of the angle of lead or lag. From preceding vector diagrams, it is evident that

$$\frac{R}{Z} = \cos\theta$$

$$\therefore PF = \cos\theta \tag{38}$$

From Eq. (36), $P = (P_a)(PF)$

Substituting for P_a, $P = (VI)(PF)$

Substituting for the PF, $P = VI\cos\theta \tag{39}$

From the foregoing it is seen that the power expended in a circuit can be obtained by utilizing different relations. For example, in the circuit of Fig. 32-21.

$$P = I^2R = 1.79^2 \times 100 = 320\text{ W}$$
$$P = V_RI = 179 \times 1.79 = 320\text{ W}$$

and

$$P = VI\cos\theta$$
$$= 220 \times 1.79 \times \cos 35.8°$$
$$= 320\text{ W}$$

The power factor of a circuit can be expressed as a decimal or as a percent. Thus the power factor of this circuit is

$$\cos\theta = \cos 35.8° = 0.812$$

Expressed as percent,

$$PF = 100\cos 35.8° = 81.2\%$$

32-12 NOTATION FOR SERIES CIRCUITS

In Sec. 3-5, it was shown that positive and negative "real" numbers could be represented graphically by plotting them along a horizontal line. The positive numbers were plotted to the right of zero, and the negative numbers were plotted to the left. This idea was expanded in Sec. 16-3, where the original horizontal line was made the x axis of a system of rectangular coordinates.

In Sec. 20-12 the system of representation was extended to include what we referred to as "imaginary" numbers by agreeing to plot the numbers along the y axis, the letter j being used as a symbol of 90° operation. Thus, when some number is prefixed with j, it means that the vector which the number represents is to be rotated through an angle of 90°. The rotation is positive, or in a counterclockwise direction, when the sign of j is positive and negative, or in a clockwise direction, when the sign of j is negative. In Sec. 31-1 we saw that "real" and "imaginary" are more properly called "in phase" and "out of phase."

From the foregoing, it is evident that resistance, when plotted on an impedance phasor diagram, is considered as an in-phase number because it is plotted along the x axis. In this instance the term *real* may well define resistance, for only the opposition to the flow of current consumes power.

Since reactances are displaced 90° from resistance in an impedance phasor diagram, it follows that inductive reactance can be prefixed with a plus j and capacitive reactance with a minus j. Thus, an inductive reactance of 75 Ω would be written j75 Ω and plotted on the positive y axis and a capacitance reactance of 86 Ω would be written $-$j86 Ω and plotted on the negative y axis.

In Sec. 31-1 we saw that a vector can be completely described in terms of either its polar or its rectangular components. For example, the circuit of Fig. 32-14 can be described as consisting of an impedance of 10 Ω at an angle of 53.1°, which would be written

$$Z = 10 \underline{/53.1°} \ \Omega$$

where the angle sign is included for emphasis and the number of degrees denotes the angle that the vector makes with the positive x axis. This is known as *polar form*. Since this impedance is made up of 6 Ω of resistance and 8 Ω of inductive reactance, we can write

$$Z = R + jX_L = 6 + j8 \ \Omega$$

This is known as *rectangular form*.

In converting from rectangular to polar form, use the usual methods of solution of right triangles that were developed in Chap. 26 and shown in Fig. 32-24.

Circuit	Impedance phasor	Z Rectangular form	Z Polar form
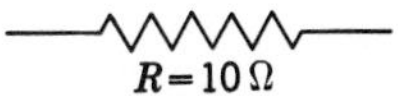 $R=10\,\Omega$	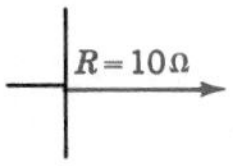 $R=10\Omega$	$Z=10+j0 \ \Omega$	$Z=10\underline{/0°} \ \Omega$
 $X_L=7\,\Omega$	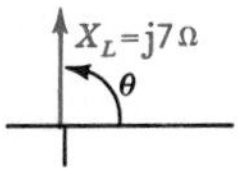 $X_L=j7\Omega$	$Z=0+j7 \ \Omega$	$Z=7\underline{/90°} \ \Omega$
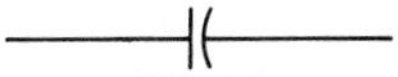 $X_C=6\,\Omega$	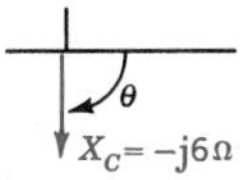 $X_C=-j6\Omega$	$Z=0-j6 \ \Omega$	$Z=6\underline{/-90°} \ \Omega$
 $R=4\,\Omega$ $X_L=3\,\Omega$	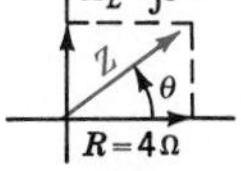 $X_L=j3\Omega$ $R=4\Omega$	$Z=4+j3 \ \Omega$	$Z=5\underline{/36.9°} \ \Omega$
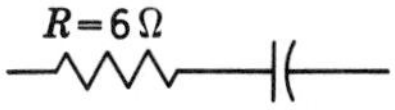 $R=6\,\Omega$ $X_C=8\,\Omega$	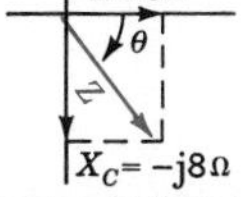 $R=6\Omega$ $X_C=-j8\Omega$	$Z=6-j8 \ \Omega$	$Z=10\underline{/-53.1°} \ \Omega$
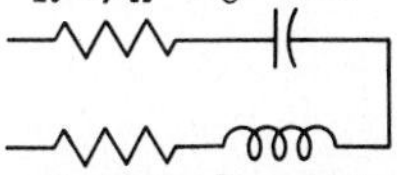 $R=7\,\Omega$ $X_C=40\,\Omega$ $R=13\,\Omega$ $X_L=20\,\Omega$	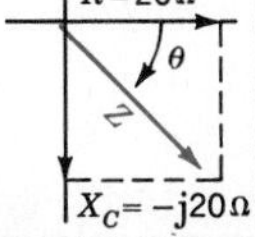 $R=20\Omega$ $X_C=-j20\Omega$	$Z=20-j20 \ \Omega$	$Z=28.2\underline{/-45°} \ \Omega$

Fig. 32-24 Phasor notation for series circuits.

The rectangular form is a very convenient method of notation. For example, instead of writing, "A series circuit of 4 Ω resistance and 3 Ω capacitive reactance," we can write, "A series circuit of 4 − j3 Ω." Figure 32-24 shows the various types of series circuits with their proper impedance phasor diagram and corresponding notation.

Note that the sign of the phase angle is the same as that of j in the rectangular form. 4 − j3 converts to a polar form with a negative angle 5 $/-36.9°$. It must be understood that neither the rectangular form nor the polar form is a method for solving series circuits. The two forms are simply convenient kinds of notation that completely describe circuit conditions from both electrical and mathematical viewpoints.

Example 11

Find the phasor impedance of the following series circuit:

$$250 - j100 \ \Omega$$

Solution

Given

$$Z = R - jX = 250 - j100 \ \Omega$$

$$\tan \theta = \frac{X}{R} = \frac{100}{250} = 0.400$$

$$\therefore \theta = -21.8°$$

$$Z = \frac{X}{\sin \theta} = \frac{100}{\sin 21.8°} = 269 \ \Omega$$

or

$$Z = \frac{R}{\cos \theta} = \frac{250}{\cos 21.8°} = 269 \ \Omega$$

Hence,

$$Z = 269 \ \underline{/-21.8°} \ \Omega$$

Converting from polar form, in which the magnitude and angle are given, to rectangular form is simplified by making use of the trigonometric functions. Since

$$R = Z \cos \theta \qquad \Omega$$

$$X = Z \sin \theta \qquad \Omega$$

and

$$Z = R \pm jX \tag{40}$$

by substitution,

$$Z = Z \cos \theta + jZ \sin \theta \tag{41}$$

Factoring,

$$Z = Z(\cos \theta + j \sin \theta) \qquad \Omega \tag{42}$$

The $\pm$ sign is omitted in Eqs. (41) and (42) because, if the proper angles are used (positive or negative), the respective sine values will determine the proper sign of the reactance component.

Example 12

A series circuit has an impedance of 269 Ω with a leading power factor of 0.928. What are the reactance and resistance of the circuit?

Solution

Given $Z = 269$ Ω and PF $= 0.928$. The power factor, when expressed as a decimal, is equal to the cosine of the phase angle. Hence,

if $\qquad\qquad 0.928 = \cos\theta$

then $\qquad\qquad\qquad \theta = -21.8°$

The angle was given the minus sign because a "leading power factor" means the current leads the voltage. Therefore,

$$Z = 269\underline{/-21.8°}\ \Omega$$

Substituting these values in Eq. (41),

$$Z = 269\cos 21.8° - j269\sin 21.8°$$
$$= 250 - j100\ \Omega$$

32-13 THE GENERAL SERIES CIRCUIT

In a series circuit consisting of several resistances and reactances, the total resistance of the circuit is the sum of all the series resistances and the total reactance is the algebraic sum of the series reactances. That is, the total resistance is

$$R_t = R_1 + R_2 + R_3 + \cdots$$

and the reactance of the circuit is

$$X = j(\omega L_1 + \omega L_2 + \omega L_3 + \cdots) - j\left(\frac{1}{\omega C_1} + \frac{1}{\omega C_2} + \frac{1}{\omega C_3} + \cdots\right)$$

Hence, the impedance is

$$Z = R_t \pm jX \qquad \Omega$$

As an alternate method, such a circuit can always be reduced to an equivalent series circuit by combining inductances and capacitances before computing reactances. Thus, the total inductance is

$$L_t = L_1 + L_2 + L_3 + \cdots$$

and the capacitance of the circuit is obtained from

$$\frac{1}{C_t} = \frac{1}{C_1} + \frac{1}{C_2} + \frac{1}{C_3} + \cdots$$

However, when voltage drops across individual reactances are desired, it is best to find the equivalent circuit by combining reactances.

Example 13

Given the circuit of Fig. 32-25, which is supplied by 220 V, 60 Hz. Find the (*a*) equivalent series circuit, (*b*) impedance of the circuit, (*c*) current, (*d*) power factor, (*e*) power expended in the circuit, (*f*) apparent power, (*g*) voltage drop across C_1, and (*h*) power expended in R_2.

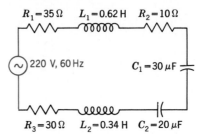

$R_1 = 35\,\Omega$ $L_1 = 0.62\,H$ $R_2 = 10\,\Omega$

220 V, 60 Hz $C_1 = 30\,\mu F$

$R_3 = 30\,\Omega$ $L_2 = 0.34\,H$ $C_2 = 20\,\mu F$

Fig. 32-25 Series circuit of Example 13.

Solution

(*a*)
$$R_t = R_1 + R_2 + R_3 = 35 + 10 + 30$$
$$= 75\ \Omega$$
$$\omega = 2\pi f = 2\pi \times 60 = 377$$
$$L_t = L_1 + L_2 = 0.62 + 0.34 = 0.96\ H$$
$$X_L = \omega L = 377 \times 0.96 = 362\ \Omega$$
$$X_{C_1} = \frac{1}{\omega C_1} = \frac{1}{377 \times 30 \times 10^{-6}}$$
$$= \frac{10^3}{3.77 \times 3} = 88.4\ \Omega$$
$$X_{C_2} = \frac{1}{\omega C_2} = \frac{1}{377 \times 20 \times 10^{-6}}$$
$$= \frac{10^3}{3.77 \times 2} = 132.6\ \Omega$$
$$X_C = 88.4 + 132.6 = 221\ \Omega$$
$$X = X_L - X_C = 362 - 221 = 141\ \Omega$$

The equivalent series circuit consists of a resistance of 75 Ω and an inductive reactance of 141 Ω. That is,

$$Z = 75 + j141\ \Omega$$

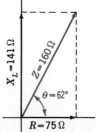

$X_L = 141\,\Omega$ $Z = 160\,\Omega$ $\theta = 62°$ $R = 75\,\Omega$

Fig. 32-26 Impedance phasor diagram for circuit of Fig. 32-25.

The impedance phasor diagram for the equivalent circuit is shown in Fig. 32-26.

(b) $\tan \theta = \dfrac{X}{R_t} = \dfrac{141}{75} = 1.88$

$\therefore \theta = 62°$

$Z = \dfrac{R}{\cos \theta} = \dfrac{75}{\cos 62°} = 160 \ \Omega$

Hence, $Z = 160 \angle 62° \ \Omega$

(c) $I = \dfrac{V}{Z} = \dfrac{220}{160} = 1.38 \ \text{A}$

(d) $\text{PF} = \cos \theta = \cos 62° = 0.470$

Expressed as a percent, PF = 47.0%

(e) $P = VI \cos \theta = 220 \times 1.38 \times \cos 62° = 143 \ \text{W}$

or $P = I^2R = 1.38^2 \times 75 = 143 \ \text{W}$

(f) $P_a = VI = 220 \times 1.38 = 304 \ \text{VA}$

(g) $V_{C_1} = IX_{C_1} = 1.38 \times 88.4 = 122 \ \text{V}$

(h) $P_{R_2} = I^2R_2 = 1.38^2 \times 10 = 19 \ \text{W}$

You will find it convenient to compute the value of the angular velocity $\omega = 2\pi f$ for all ac problems, for this factor is common to all reactance equations.

As with all electric circuit problems, a neat diagram of the circuit, with all known circuit components, voltages, and currents clearly marked, should be made. In addition, a phasor or impedance diagram should be drawn to scale in order to check the mathematical solution.

PROBLEMS 32-5

Given the circuit of Fig. 32-27, with values as listed in Table 32-1. Draw an impedance phasor diagram for each circuit and find (a) the impedance of the circuit, (b) the current flowing through the circuit, (c) the equation of the current, (d) the PF of the circuit, and (e) the power expended in the circuit.

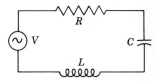

Fig. 32-27 Circuit for Probs. 1 to 10.

TABLE 32-1   PROBLEMS 1 TO 10

Problems	V, V	f	R	L	C
1	220	60 Hz	200 Ω	2 H	10 μF
2	450	1 kHz	67 Ω	5 mH	50 μF
3	110	50 Hz	2 kΩ	5.6 H	2.2 μF
4	850	400 Hz	500 Ω	2.5 H	100 μF
5	1200	5 MHz	220 Ω	67 μH	20 pF
6	1000	8 GHz	330 Ω	0.08 μH	0.005 pF
7	117	60 Hz	15 Ω	4.5 mH	2500 μF
8	2	10 kHz	27 Ω	3.5 μH	1.5 μF
9	1760	2.5 MH	500 Ω	12.5 μH	850 pF
10	110	60 Hz	50 Ω	300 mH	22.0 μF

11.   A choke coil, when it is connected across a 230-V dc source, draws 1.15 A. When it is connected across 230 V, 60 Hz, the current is 665 mA.
(*a*) What is the resistance of the coil?
(*b*) What is the inductive reactance of the coil?
(*c*) What is the inductance of the coil?

12.   Assuming that the resistance of the coil in Prob. 11 is unchanged, how much power would the coil draw when it is connected across 230 V, 400 Hz?

13.   The following 60-Hz impedances are connected in series:

$$Z_1 = 30 - j40 \ \Omega \qquad Z_2 = 5 + j12 \ \Omega$$
$$Z_3 = 8 - j6 \ \Omega \qquad Z_4 = 4 + j4 \ \Omega$$

(*a*) What is the resultant impedance of the circuit?
(*b*) What value of pure reactance must be added in series to make the PF of the circuit 80% leading?

14.   The meters represented in Fig. 32-28 are connected such a short distance from an inductive load that line drop from meters to load is negligible. What is the equivalent series circuit of the load?

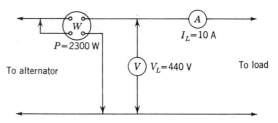

Fig. 32-28 Circuit of Prob. 14.

15.   A single-phase induction motor, with 440 V across its input terminals, delivers 10.8 mechanical horsepower at an efficiency of 90% and a PF of 86.6%.
(*a*) What is the line current?
(*b*) How much power is taken by the motor?

16.   Given any series circuit, for example, 110 V at 60 Hz applied across $3 + j4 \ \Omega$. On the same set of axes and to the same scale, plot instantaneous values of the applied emf v, the potential difference across the resistance R, and the potential difference across the reactance X. What is your conclusion?

32-14 SERIES RESONANCE

It has been shown that the inductive reactance of a circuit varies directly as the frequency and that the capacitive reactance varies inversely as the frequency. That is, the inductive reactance will increase and the capacitive reactance will decrease as the frequency is increased, and vice versa. Then, for any value of inductance and capacitance in a circuit, there is a frequency at which the inductive reactance and the capacitive reactance are equal. This is called the *resonant frequency* of the circuit. Since, in a series circuit,

$$Z = R + j\left(\omega L - \frac{1}{\omega C}\right) \quad \Omega$$

at resonance,

$$\omega L = \frac{1}{\omega C} \qquad (43)$$

Hence,

$$Z = R$$

Therefore, at the resonant frequency of a series circuit, the resistance is the only circuit component that limits the flow of current, for the net reactance of the circuit is zero. Thus the current is in phase with the applied voltage, which results in a circuit power factor of 100%.

Example 14

There is impressed 10 V at a frequency of 1 MHz across a circuit consisting of a coil of 92.2 μH in series with a capacitance of 275 pF. The effective resistance of the coil at this frequency is 10 Ω, and both the resistance of the connecting wires and the capacitance are negligible.

(*a*) What is the impedance of the circuit?
(*b*) How much current flows through the circuit?
(*c*) What are the voltages across the reactances?

Solution

The resistance of the coil is treated as being in series with its inductive reactance.

(*a*)
$$\omega = 2\pi f = 6.28 \times 10^6$$
$$X_L = \omega L$$
$$= 6.28 \times 10^6 \times 92.2 \times 10^{-6}$$
$$= 6.28 \times 92.2 = 579 \ \Omega$$
$$X_C = \frac{1}{\omega C}$$
$$= \frac{1}{6.28 \times 10^6 \times 275 \times 10^{-12}}$$
$$= \frac{10^4}{6.28 \times 2.75} = 579 \ \Omega$$

Since
$$X_L = X_C$$
then
$$Z = R = 10 \ \Omega$$
(*b*)
$$I = \frac{V}{Z} = \frac{10}{10} = 1 \ A$$
(*c*)
$$V_C = IX_C = 1 \times 579 = 579 \ V$$
$$V_L = IX_L = 1 \times 579 = 579 \ V$$

Note that the voltages across the inductance and capacitance are much greater than the applied voltage.

An inductance has a *quality* or *merit*, denoted by Q, that is defined as the ratio of its inductive reactance to its resistance at a given frequency. Thus,

$$Q = \frac{\omega L}{R} \tag{44}$$

Then, at resonance,

$$V_C = V_L = I\omega L \quad \text{V}$$

Substituting for I,

$$V_C = V_L = \frac{V\omega L}{R} \quad \text{V}$$

Substituting for $\frac{\omega L}{R}$,

$$V_C = V_L = VQ \quad \text{V} \tag{45}$$

Because the average radio circuit has purposely been designed for high Q values, it is seen that very high voltages can be developed in resonant series circuits.

32-15 RESONANT FREQUENCY

The resonant frequency of a circuit can be determined by rewriting Eq. (43). Thus,

$$2\pi f L = \frac{1}{2\pi f C}$$

$$\therefore f = \frac{1}{2\pi\sqrt{LC}} \quad \text{Hz} \tag{46}$$

where f, L, and C are in the usual units, hertz, henrys, and farads, respectively when $Q > 10$.

Example 15

A series circuit consists of an inductance of 500 μH and a capacitor of 400 pF. What is the resonant frequency of the circuit?

Solution

$$L = 500\mu\text{H} = 5 \times 10^{-4}\,\text{H}$$

$$C = 400\,\text{pF} = 4 \times 10^{-10}\,\text{F}$$

$$f = \frac{1}{2\pi\sqrt{LC}}$$

$$= \frac{1}{2\pi\sqrt{5 \times 10^{-4} \times 4 \times 10^{-10}}}$$

$$= \frac{10^7}{2\pi\sqrt{20}}$$

$$= 356\,000\,\text{Hz}$$

or

$$f = 356\,\text{kHz}$$

From Eq. (46) it is evident that the resonant frequency of a series circuit depends *only* upon the LC product. This means there is an infinite number of combinations of L and C that will resonate to a particular frequency.

Example 16

How much capacitance is required to obtain resonance at 1500 kHz with an inductance of $45\mu H$?

Solution

$$f = 1500 \text{ kHz} = 1.5 \times 10^6 \text{ Hz}$$
$$L = 45 \text{ }\mu H = 4.5 \times 10^{-5} \text{ H}$$
$$\omega = 2\pi f = 2\pi \times 1.5 \times 10^6 = 9.42 \times 10^6$$

From Eq. (46), $\qquad C = \dfrac{1}{(2\pi f)^2 L} = \dfrac{1}{\omega^2 L}$

$$\therefore C = \frac{1}{(9.42 \times 10^6)^2 \times 4.5 \times 10^{-5}}$$

$$= 250 \text{ pF}$$

PROBLEMS 32-6

1. 100 V 10 kHz is impressed across a series circuit consisting of a 220-pF capacitor of negligible resistance and an 800-mH coil with effective resistance of 125 Ω.
   (a) How much current flows through the circuit?
   (b) How much power does the circuit absorb from the source?
   (c) What are the voltages across the capacitor and the coil?
2. What is the Q of the coil in Prob. 1?
3. At what frequency would the circuit of Prob. 1 be resonant?
4. What type and value of "pure reactance" must be added to the circuit of Prob. 1 to make the circuit resonant at 10 kHz?
5. A tuning capacitor has a range of capacitance between 20 pF and 350 pF.
   (a) What inductance must be connected in series with it to provide a lowest resonant frequency of 550 kHz?
   (b) What will then be the highest resonant frequency?
6. What is the equivalent circuit of a series circuit when operating at (a) resonant frequency, (b) at a frequency less than resonant frequency, and (c) at a frequency higher than resonant frequency?

Alternating Currents— Parallel Circuits

33

Parallel circuits are the kind of circuit most commonly encountered. The average distribution circuit has many types of loads all connected in parallel with each other: lighting circuits, motors, transformers for various uses, etc. The same is true of electronic circuits, which range from the most simple parallel circuits to complex networks.

This chapter deals with the solution of parallel circuits. Such a solution may reduce a parallel circuit to an equivalent series circuit that, when connected to the same source of emf as the given parallel circuit, would result in the same line current and phase angle; that is, the alternator would "see" the same load.

33-1 RESISTANCES IN PARALLEL

It was explained in Secs. 32-1 and 32-2 that, in an ac circuit containing resistance only, the voltage, current, and power relations are the same as in dc circuits. However, in order to build a foundation from which all parallel circuits can be analyzed, the case of paralleled resistances must be considered from a phasor viewpoint.

Figure 33-1 represents a 60-Hz 220-V alternator connected to three resistances in parallel.

Neglecting the internal resistance of the alternator and the resistance of the connecting wires, the emf of the alternator is impressed across each

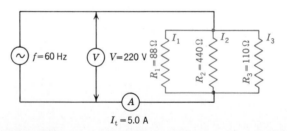

Fig. 33-1 Alternator connected to three resistors in parallel.

of the three resistances. If I_1, I_2, and I_3 represent the currents flowing through R_1, R_2, and R_3, respectively, then by Ohm's law,

$$I_1 = 2.5 \text{ A}$$
$$I_2 = 0.5 \text{ A}$$
$$I_3 = 2.0 \text{ A}$$

Since all currents are in phase, the total current flowing in the line, or external circuit, will be equal to the sum of the branch currents or 5.0 A. The phasor diagram for the three currents is illustrated in Fig. 33-2. All currents are plotted in phase with the applied emf, which is used as a reference phasor because the voltage is common to all resistances. Then, using rectangular phasor notation,

$$I_1 = 2.5 + j0 \text{ A}$$
$$I_2 = 0.5 + j0 \text{ A}$$
$$\underline{I_3 = 2.0 + j0 \text{ A}}$$
$$I_t = 5.0 + j0 \text{ A} = 5.0\underline{/0°} \text{ A}$$

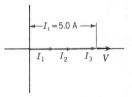

Fig. 33-2 Phasor diagram for the circuit of Fig. 33-1.

As with all other circuits, the equivalent series impedance, which in this case is a pure resistance, is found by dividing the voltage across the circuit by the total current. That is,

$$Z = \frac{V}{I_t} = \frac{220}{5} = 44 \text{ }\Omega = 44\underline{/0°} \text{ }\Omega$$

33-2 CAPACITORS IN PARALLEL

Figure 33-3 represents two capacitors C_1 and C_2 connected in parallel across a voltage V. The quantity of charge in capacitor C_1 will be

$$Q_1 = C_1 V \qquad (1)$$

and that in capacitor C_2 will be

$$Q_2 = C_2 V \qquad (2)$$

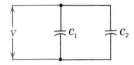

Fig. 33-3 Capacitors C_1 and C_2 connected in parallel.

Since the total quantity in both capacitors is $Q_1 + Q_2$, then

$$Q_1 + Q_2 = C_p V \qquad (3)$$

where C_p is the total capacitance of the combination.

Then adding Eqs. (1) and (2),

$$Q_1 + Q_2 = C_1 V + C_2 V$$

or

$$Q_1 + Q_2 = (C_1 + C_2)V$$

Substituting the value of $Q_1 + Q_2$ from Eq. (3),

$$C_p V = (C_1 + C_2)V$$

which results in

$$C_p = C_1 + C_2 \qquad (4)$$

From the foregoing, it is apparent that capacitors in parallel combine like resistances in series; that is, the capacitance of paralleled capacitors is equal to the sum of the individual capacitances.

Example 1

What is the capacitance of a 6-μF capacitor in parallel with a 4-μF capacitor?

Solution

$$C_p = 6 + 4 = 10 \ \mu F$$

33-3 INDUCTANCE AND CAPACITANCE IN PARALLEL

When a purely inductive reactance and a capacitive reactance are connected in parallel, as shown in Fig. 33-4, the currents flowing through these reactances differ in phase by 180°.

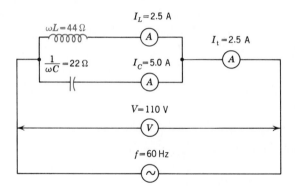

Fig. 33-4 X_L and X_C connected in parallel.

The current flowing through the inductor is

$$I_L = \frac{V}{X_L} = \frac{V}{\omega L} = \frac{110}{44} = 2.5 \text{ A}$$

and that through the capacitor is

$$I_C = \frac{V}{X_C} = \omega C V = \frac{110}{22} = 5.0 \text{ A}$$

In series circuits, the current was used as the reference phasor because the current is the same in all parts of the circuit. In parallel circuits there are different values of currents in various parts of a circuit; therefore, the current cannot be used as the reference phasor.

Since the same voltage exists across two or more parallel branches, the applied voltage can be used as the reference phasor as illustrated in Fig. 33-5.

Note that the current I_L through the inductor is plotted as *lagging* the alternator voltage by 90° and the current I_C through the capacitor is *leading*

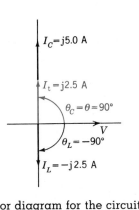

Fig. 33-5 Phasor diagram for the circuit of Fig. 33-4.

the voltage by 90°. The total line current I_t, which is the phasor sum of the branch currents, is leading the applied voltage by 90°. That is, using rectangular phasor notation,

$$
\begin{aligned}
I_L &= 0 - j2.5 \text{ A} \\
I_C &= 0 + j5.0 \text{ A} \\
\hline
I_t &= 0 + j2.5 \text{ A} = 2.5 \underline{/90°} \text{ A}
\end{aligned}
$$

Since the line current leads the alternator voltage by 90°, the equivalent series circuit consists of a capacitive reactance of

$$\frac{V}{I_t} = \frac{110}{2.5} = 44 \; \Omega$$

That is, the parallel circuit could be replaced by a 60.3-μF capacitor which would result in a current of 2.5 A leading the voltage by 90°; in other words, the alternator would not sense the difference.

Note the difference between reactances in series and reactances in parallel. In a series circuit the *greatest* reactance of the circuit results in the equivalent series circuit containing the same kind of reactance. For this reason, it is said that reactances, or voltages across reactances, are the controlling factors of series circuits. In a parallel circuit the *least* reactance of the circuit, which passes the greatest current, results in the equivalent series circuit containing the same kind of reactance. For this reason, it is said that currents are the controlling factors of parallel circuits.

33-4 ASSUMED VOLTAGES

The solutions of the great majority of parallel circuits are facilitated by assuming a voltage to exist across a parallel combination. The currents through each branch, due to the assumed voltage, are then added vectorially to obtain the total current. The assumed voltage is then divided by the total current, the quotient being the joint impedance of the parallel branches.

To avoid small decimal quantities, the assumed voltage should be greater than the largest impedance of any parallel branch.

Example 2

Given the circuit of Fig. 33-6. What are the impedance and the power factor of the circuit at a frequency of 2.5 MHz?

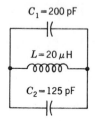

$C_1 = 200\,\text{pF}$

$L = 20\,\mu\text{H}$

$C_2 = 125\,\text{pF}$

Fig. 33-6 Circuit of Example 2.

Solution

C_1 and C_2 are in parallel; hence, the total capacitance is

$$C_p = C_1 + C_2 = 200 + 125 = 325 \text{ pF}$$

This simplifies the circuit to a capacitor C of 325 pF in parallel with an inductance L of 20 μH.

$$\omega = 2\pi f = 2\pi \times 2.5 \times 10^6 = 1.57 \times 10^7$$
$$X_L = \omega L = 1.57 \times 10^7 \times 2 \times 10^{-5} = 314\,\Omega$$
$$X_C = \frac{1}{\omega C} = \frac{1}{1.57 \times 10^7 \times 325 \times 10^{-12}}$$
$$= \frac{10^3}{1.57 \times 3.25} = 196\,\Omega$$

Assume 1000 V across the parallel branch. Then the current through the capacitors is

$$I_C = \frac{V_a}{X_C} = \frac{1000}{196} = 5.10 \text{ A}$$

and the current through the inductance is

$$I_L = \frac{V_a}{X_L} = \frac{1000}{314} = 3.18 \text{ A}$$

Since I_C leads the assumed voltage by 90° and I_L lags the assumed voltage by 90°, they are plotted with the assumed voltage as reference phasor as shown in Fig. 33-7. Then the total current I_t that would flow because of assumed voltage would be the phasor summation of I_C and I_L. Performing phasor summation:

$$I_C = 0 + j5.10 \text{ A}$$
$$I_L = 0 - j3.18 \text{ A}$$
$$I_t = 0 + j1.92 \text{ A} = 1.92 \underline{/90°} \text{ A}$$

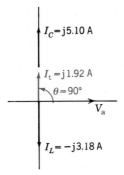

Fig. 33-7 Phasor diagram for circuit of Example 2.

Again, since the total current leads the voltage by 90°, the equivalent series circuit consists of a capacitor whose capacitive reactance is

$$\frac{V_a}{I_t} = \frac{1000}{1.92} = 521 \ \Omega$$

Since $\theta = 90°$,   PF $= \cos \theta = 0$

You should solve the circuit of Fig. 33-6 with different values of assumed voltages.

33-5 RESISTANCE AND INDUCTANCE IN PARALLEL

When a resistance and an inductive reactance are connected in parallel, as represented in Fig. 33-8, the currents that flow differ in phase by 90°. The current flowing through the resistance is

$$I_R = \frac{V}{R} = \frac{120}{20} = 6.0 \ A$$

and that through the inductance is

$$I_L = \frac{V}{\omega L} = \frac{120}{15} = 8.0 \ A$$

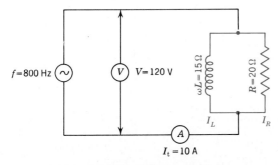

Fig. 33-8 R and X_L in parallel.

Since the current through the resistance is in phase with the applied voltage and the current through the inductance lags the applied voltage by

$90°$, I_R and I_L are plotted with the applied emf as reference phasor as shown in Fig. 33-9. Then the total current I_t, or line current, is the phasor sum of I_R and I_L. Performing phasor summation,

$$
\begin{aligned}
I_R &= 6.0 + j0 \quad A \\
\underline{I_L} &= \underline{0 \quad - j8.0\ A} \\
I_t &= 6.0 - j8.0\ A
\end{aligned}
$$

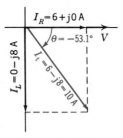

Fig. 33-9 Phasor diagram for circuit of Fig. 33-8.

Hence, the total current, which consists of an in-phase component of 6.0 A and a 90° lagging component of 8.0 A, is expressed in terms of its rectangular components. The magnitude and phase angle are then found by the usual trigonometric methods. Thus,

$$I_t = 10 \underline{/-53.1°}\ A$$

The power factor of the circuit is

$$
\begin{aligned}
PF &= \cos\theta = \cos(-53.1°) \\
&= 0.60 \text{ lagging}
\end{aligned}
$$

The true power expended in the circuit is

$$
\begin{aligned}
P &= VI\cos\theta = 120 \times 10 \times 0.60 \\
&= 720\ W
\end{aligned}
$$

or

$$P = I_R^2 R = 6^2 \times 20 = 720\ W$$

The equivalent impedance, or total impedance, of the circuit is

$$Z_t = \frac{V}{I_t} = \frac{120}{10} = 12\ \Omega$$

Since the entire circuit has a lagging PF of 0.60, it follows that the equivalent series circuit consists of a resistance and an inductive reactance in series, the phasor sum of which is 12 Ω at a phase angle θ such that $\cos\theta = 0.60$. Therefore, $\theta = 53.1°$, and

$$
\begin{aligned}
Z_t &= 12 \underline{/53.1°}\ \Omega \\
&= 12(\cos 53.1° + j\sin 53.1°) \\
&= 7.2 + j9.6\ \Omega
\end{aligned}
$$

From the above, it is evident that the parallel circuit of Fig. 33-8 could be replaced by a series circuit of 7.2 Ω resistance and 9.6 Ω inductive reactance and that the alternator would be working under exactly the same load conditions as before.

In order to justify such solutions, solve for the equivalent impedance of the circuit of Fig. 33-8 by using an assumed voltage and then using the *actual* voltage to obtain the power.

33-6 RESISTANCE AND CAPACITANCE IN PARALLEL

When resistance and capacitive reactance are connected in parallel, as represented in Fig. 33-10, the current through the resistance is in phase with the voltage across the parallel combination, and the current through the capacitive reactance leads this voltage by 90°.

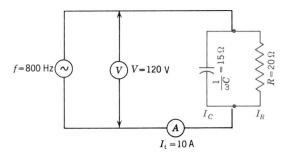

Fig. 33-10 R and X_C in parallel.

The circuit of Fig. 33-10 is similar to that of Fig. 33-8 except that Fig. 33-10 contains a capacitive reactance of 15 Ω in place of the inductive reactance of 15 Ω. The phasor diagram of currents is illustrated in Fig. 33-11, and it is evident that the total current is

$$I_t = 6.0 + j8.0 \text{ A} = 10\underline{/53.1°} \text{ A}$$

The power factor of the circuit is

$$\text{PF} = \cos \theta = \cos 53.1°$$
$$= 0.60 \text{ leading}$$

Similarly, the total impedance of the circuit is 12 Ω; and since the circuit has a leading PF of 0.60, it follows that the equivalent series circuit consists of resistance and capacitive reactance in series the phasor sum of which is 12 Ω at a phase angle θ such that cos θ = 0.60. Therefore,

$$\theta = -53.1°$$
and
$$Z_t = 12\underline{/-53.1°} \text{ Ω}$$
$$= 7.2 - j9.6 \text{ Ω}$$

If the parallel circuit of Fig. 33-10 were replaced by a series circuit of 7.2 Ω resistance and 9.6 Ω capacitive reactance, the alternator would be working under exactly the same load conditions as before.

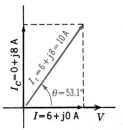

Fig. 33-11 Phasor diagram for circuit of Fig. 33-10.

33-7 RESISTANCE, INDUCTANCE, AND CAPACITANCE IN PARALLEL

When resistance, inductive reactance, and capacitive reactance are connected in parallel, as represented in Fig. 33-12, the line current is the phasor sum of the several currents.

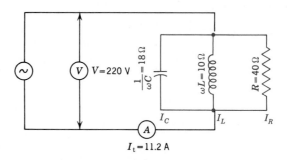

Fig. 33-12 *L*, *C*, and *R* in parallel.

The currents through the branches are

$$I_R = \frac{220}{40} = 5.5 \text{ A}$$

$$I_L = \frac{220}{10} = 22 \text{ A}$$

$$I_C = \frac{220}{18} = 12.2 \text{ A}$$

Performing phasor summation of these currents as shown in Fig. 33-13,

$$
\begin{aligned}
I_R &= 5.5 + j0 && \text{A} \\
I_L &= 0 \;\;- j22 && \text{A} \\
\underline{I_C} &= \underline{0 \;\;+ j12.2} && \underline{\text{A}} \\
I_t &= 5.5 - j9.8 \text{ A} = 11.2 \underline{/-60.7°} \text{ A} \\
\text{PF} &= \cos(-60.7°) = 0.489 \text{ lagging}
\end{aligned}
$$

The total impedance is

$$Z_t = \frac{V}{I_t} = \frac{220}{11.2} = 19.6 \ \Omega$$

Since the circuit has a lagging PF of 0.489, the equivalent series circuit consists of a resistance and an inductive reactance. The phasor sum of these must be 19.6 Ω at a phase angle θ such that cos θ = 0.489. Therefore, θ = 60.7° and

$$Z_t = 19.6 \underline{/60.7°} \ \Omega = 9.59 + j17.1 \ \Omega$$

which are the values which constitute the equivalent series circuit.

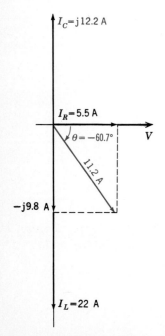

Fig. 33-13 Phasor diagram for circuit of Fig. 33-12.

Example 3

Given the circuit represented in Fig. 33-14. Solve for the equivalent series circuit at a frequency of 5 MHz.

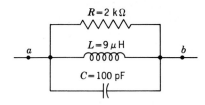

Fig. 33-14 Circuit of Example 3.

Solution

$$f = 5\,\text{MHz} = 5 \times 10^6\,\text{Hz}$$
$$L = 9\,\mu\text{H} = 9 \times 10^{-6}\,\text{H}$$
$$C = 100\,\text{pF} = 10^{-10}\,\text{F}$$
$$\omega = 2\pi f = 2\pi \times 5 \times 10^6 = 3.14 \times 10^7$$
$$X_L = \omega L = 3.14 \times 10^7 \times 9 \times 10^{-6}$$
$$= 283\,\Omega$$

$$X_C = \frac{1}{\omega C} = \frac{1}{3.14 \times 10^7 \times 10^{-10}} = \frac{10^3}{3.14}$$
$$= 318\,\Omega$$

Assume $V_a = 1000$ V applied between a and b.

$$I_R = \frac{V_a}{R} = \frac{1000}{2000} = 0.50\,\text{A}$$

$$I_L = \frac{V_a}{X_L} = \frac{1000}{283} = 3.54\,\text{A}$$

$$I_C = \frac{V_a}{X_C} = \frac{1000}{318} = 3.14\,\text{A}$$

The total current I_t is the phasor sum of the three branch currents as represented in the phasor diagram of Fig. 33-15. Adding vectorially,

$$I_R = 0.50 + j0$$
$$I_L = 0 \quad - j3.54\,\text{A}$$
$$\underline{I_C = 0 \quad + j3.14\,\text{A}}$$
$$I_t = 0.50 - j0.40\,\text{A} = 0.640\,\underline{/-38.7°}\,\text{A}$$
$$\text{PF} = \cos(-38.7°) = 0.78\,\text{lagging}$$

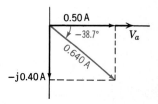

Fig. 33-15 Phasor diagram of circuit of Fig. 33-14.

The total impedance Z_t, which is the impedance between points a and b, is

$$Z_t = Z_{ab} = \frac{V_a}{I_t} = \frac{1000}{0.64} = 1560\,\Omega$$

Since the current is lagging the voltage, the equivalent series circuit consists of a resistance and an inductive reactance. The phasor sum of these is

1560 Ω at a phase angle θ such that $\cos \theta = 0.78$. Therefore, $\theta = 38.7°$ and

$$Z_t = 1560 \underline{/38.7°}\ \Omega = 1220 + j976\ \Omega$$

That is, the equivalent series circuit is a resistance of $R = 1220\ \Omega$ and an inductive reactance of $\omega L = 976\ \Omega$. Since

$$\omega L = 976\ \Omega$$

then
$$L = \frac{976}{\omega} = \frac{976}{3.14 \times 10^7} = 31.1\ \mu H$$

which results in the equivalent circuit as represented in Fig. 33-16 with the impedance phasor diagram of Fig. 33-17.

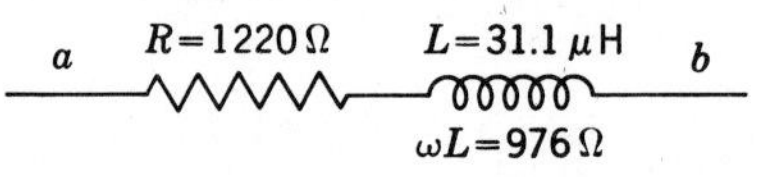

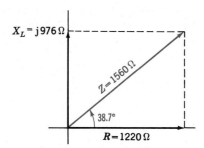

$$\begin{array}{ccc} & R=1220\,\Omega & L=31.1\,\mu H \\ a & \mathrm{\wedge\wedge\wedge} & \mathrm{\infty\!\infty\!\infty} \quad b \\ & & \omega L=976\,\Omega \end{array}$$

Fig. 33-16 Equivalent series circuit of Example 3.

Fig. 33-17 Impedance phasor diagram for equivalent series circuit.

PROBLEMS 33-1

1. What is the resulting capacitance when a 500-pF capacitor is connected in parallel with a 220-pF capacitor?
2. Two capacitors, 50 and 500 pF, are connected in parallel. A current of 200 mA, 2.7 GHz, flows through the 500-pF capacitor. How much current flows through the 50-pF capacitor?
3. Neglecting the resistance of the connecting wires in Fig. 32-12,
    (a) Write the equation for the emf of the alternator.
    (b) Write the equation for the circuit current.
    (c) What is the voltage across C_1?
    (d) What is the capacitance of C_3?
    (e) How much current flows through C_2?
4. In Fig. 33-18, $R = 200\ \Omega$, $L = 2$ H, $C = 5\ \mu F$, $V = 220$ V, and $f = 60$ Hz.
    (a) What is the ammeter reading?
    (b) How much power is expended in the circuit?
    (c) What is the equivalent series circuit?
    (d) What is the power factor?
    (e) What is the equation of the current flowing through the ammeter?

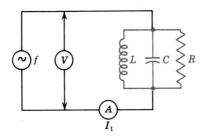

Fig. 33-18 Circuit of Probs. 4 to 6.

5.   Using the other values of Prob. 4, what must be the value of the inductance in the circuit in order to obtain a PF of (a) 0.8 lagging and (b) 1.0?

6.   In Fig. 33-18, $R = 500\ \Omega$, $L = 6$ nH, $C = 0.02$ pF, $V = 1$ kV, and $f = 8$ GHz.
(a) What is the reading of the ammeter?
(b) What parallel capacitance must be added to the circuit in order to achieve unity PF?

33-8 PHASOR IMPEDANCES IN PARALLEL

It was shown in Sec. 13-2 that the reciprocal of the equivalent resistance R_p of several resistances in parallel is expressed by the relation

$$\frac{1}{R_p} = \frac{1}{R_1} + \frac{1}{R_2} + \frac{1}{R_3} + \frac{1}{R_4} + \cdots$$

and that when two resistances R_1 and R_2 are connected in parallel, the equivalent resistance is

$$R_p = \frac{R_1 R_2}{R_1 + R_2}$$

An analogous condition exists when two or more impedances are connected in parallel. By following the line of reasoning used for resistances in parallel, the reciprocal of the equivalent impedance of several impedances in parallel is found to be

$$\frac{1}{Z_p} = \frac{1}{Z_1} + \frac{1}{Z_2} + \frac{1}{Z_3} + \frac{1}{Z_4} + \cdots \tag{5}$$

Similarly, the equivalent impedance Z_p of two impedances Z_1 and Z_2 connected in parallel is

$$Z_p = \frac{Z_1 Z_2}{Z_1 + Z_2} \tag{6}$$

Note that the impedances of Eqs. (5) and (6) are in polar form.

Example 4

Find the equivalent impedance of the circuit of Fig. 33-19.

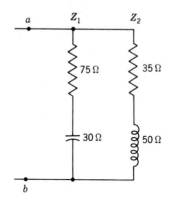

Fig. 33-19 Circuit of Example 4.

Solution

First express the given impedances in both rectangular and polar forms.

$$Z_1 = 75 - j30 = 80.8\underline{/-21.8°}\ \Omega$$
$$Z_2 = 35 + j50 = 61.0\underline{/55°}\ \Omega$$

As pointed out in Sec. 31-5, phasors in polar form cannot be added algebraically; they must be added in terms of their rectangular components. Therefore, when the given impedance values are substituted in Eq. (6), the impedances in the denominator must be in rectangular form so that the indicated addition can be carried out. Substituting,

$$Z_p = \frac{(80.8\underline{/-21.8°})(61.0\underline{/55°})}{(75 - j30) + (35 + j50)}$$
$$= \frac{4930\underline{/33.2°}}{110 + j20}$$

Because the denominator is in rectangular form and the numerator is in polar form, the denominator must be converted to polar form so the indicated division can be completed. Thus, performing phasor summation of the terms of the denominator,

$$Z_p = \frac{4930\underline{/33.2°}}{112\underline{/10.3°}}$$
$$= \frac{4930}{112}\underline{/33.2° - 10.3°}$$
$$= 44\underline{/22.9°}\ \Omega$$

Example 5

Find the equivalent impedance of the circuit of Fig. 33-20.

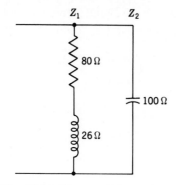

Fig. 33-20 Circuit of Example 5.

Solution

Expressing the impedance in rectangular and polar form,

$$Z_1 = 80 + j26 = 84.1 \underline{/18°} \ \Omega$$

$$Z_2 = 0 - j100 = 100 \underline{/-90°} \ \Omega$$

Substituting these values in Eq. (6),

$$Z_p = \frac{(84.1 \underline{/18°})(100 \underline{/-90°})}{(80 + j26) + (0 - j100)}$$

$$= \frac{8410 \underline{/-72°}}{80 - j74}$$

Performing the phasor summation in the denominator,

$$Z_p = \frac{8410 \underline{/-72°}}{109 \underline{/-42.8°}}$$

$$\therefore Z_p = 77.2 \underline{/-29.2°} \ \Omega$$

The equivalent series circuit is found by the usual method of converting from rectangular form to polar form, namely,

$$77.2 \underline{/-29.2°} = 77.2(\cos 29.2° - j \sin 29.2°)$$

$$= 77.2 \cos 29.2° - j77.2 \sin 29.2°$$

$$= 67.4 - j37.7 \ \Omega$$

33-9 SERIES-PARALLEL CIRCUITS

An equation for the equivalent impedance of a series-parallel circuit is obtained in the same manner as the equation for the equivalent resistance of a combination of resistances in series and parallel as was outlined in

Sec. 13-3. For example, in the circuit represented in Fig. 33-21, the total impedance is

$$Z_t = Z_s + \frac{Z_1 Z_2}{Z_1 + Z_2} \tag{7}$$

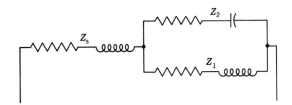

Fig. 33-21 Series-parallel circuit of Example 6.

Example 6

In the circuit which is shown in Fig. 33-21, $Z_s = 12.4 + j25.6\ \Omega$, $Z_1 = 45 + j12.9\ \Omega$, and $Z_2 = 35 - j75\ \Omega$. Determine the equivalent impedance of the circuit.

Solution

Since Z_1 and Z_2 must be multiplied, it is necessary to express them in polar form.

$$Z_1 = 45 + j12.9 = 46.8 \underline{/16°}\ \Omega$$

and
$$Z_2 = 35 - j75 = 82.8 \underline{/-65°}\ \Omega$$

Substituting the values in Eq. (7),

$$Z_t = (12.4 + j25.6) + \frac{(46.8 \underline{/16°})(82.8 \underline{/-65°})}{(45 + j12.9) + (35 - j75)}\ \Omega$$

The solution is completed in the usual manner and results in

$$Z_t = 53.2 \underline{/20°}\ \Omega$$

From the foregoing examples, it is evident that an equation for the impedance of a network is expressed exactly as in direct-current problems, impedances in polar form being substituted for the resistances.

PROBLEMS 33-2

1. What is the equivalent impedance of two impedances $Z_1 = 151 \underline{/4.07°}\ \Omega$ and $Z_2 = 50 \underline{/53.1°}\ \Omega$ connected in parallel?

2. What is the equivalent impedance of two impedances $Z_a = 148.5 \underline{/42.2°}\ \Omega$ and $Z_b = 145 \underline{/-12.7°}\ \Omega$ connected in parallel?

3. What is the equivalent impedance of two impedances $Z_1 = 73.8 - j34.4\ \Omega$ and $Z_2 = 30 + j40\ \Omega$ connected in parallel?

4. What is the equivalent impedance of two impedances $Z_a = 276 - j180$ Ω and $Z_b = 117 - j18.6$ Ω connected in parallel?

5. What is the equivalent impedance of two impedances $Z_x = 60.5\underline{/20°}$ Ω and $Z_y = 100 + j0$ Ω connected in parallel?

6. What is the equivalent impedance of two impedances $Z_1 = 355\underline{/12°}$ Ω and $Z_2 = 0 - j100$ Ω connected in parallel?

7. What is the equivalent impedance of two impedances $Z_z = 251\underline{/-3°}$ Ω and $Z_L = 0 + j70$ Ω connected in parallel?

8. The joint impedance of two parallel impedances is $53.5\underline{/-42.4°}$ Ω. One of the impedances is $168\underline{/27°}$ Ω. What is the other?

9. What impedance must be connected in parallel with $64.9 + j45.4$ Ω to produce $43.7 + j155.5$ Ω?

10. In Fig. 33-22, $Z_s = 9.4 + j6.6$ Ω, $Z_1 = 78.5 - j35$ Ω, and $Z_2 = 33.6 + j48$ Ω. What is the single equivalent impedance Z_t?

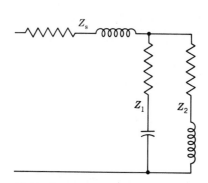

Fig. 33-22 Circuit for Probs. 10, 11, and 12.

11. In Fig. 33-22, $Z_s = 111.5\underline{/21°}$ Ω, $Z_1 = 27.7 - j50$ Ω, and $Z_2 = 150 + j76.2$ Ω. What is Z_t?

12. In Fig. 33-22, $Z_s = 5 + j3.9$ Ω, $Z_1 = 57.2\underline{/-61°}$ Ω, and $Z_2 = 168\underline{/27°}$ Ω. What is Z_t?

13. The primary current I_p of a coupled circuit is expressed by the equation

$$I_p = \frac{V}{Z_p + \dfrac{(\omega M)^2}{Z_s}} \quad \text{A}$$

Compute the value of I_p when $V = 110\underline{/0°}$ V, $Z_p = 12 + j40$ Ω, $Z_s = 18 + j50$ Ω, and $\omega M (= 2\pi f \times$ mutual inductance$) = 15$.

14. The secondary current I_s of a coupled circuit is expressed by the equation

$$I_s = \frac{-j\omega M V}{Z_p Z_s + (\omega M)^2} \quad \text{A}$$

Compute the value of I_s if $\omega M = 15$, $V = 20$ V, $Z_p = 6 + j8$ Ω, and $Z_s = 20 + j12$ Ω.

33-10 PARALLEL RESONANCE

Communication circuits and electronic networks contain resonant parallel circuits. Figure 33-23 represents a typical parallel circuit consisting of an inductor and capacitor in parallel. The resistance of the capacitor, which is very small, can be neglected, and the resistance R represents the effective resistance of the inductor.

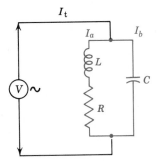

Fig. 33-23 Parallel LC circuit. R represents effective resistance of L.

At low frequencies the inductive reactance is a low value whereas the capacitive reactance is high. Hence, a large current flows through the inductive branch and a small current flows through the capacitive branch. The phasor sum of these currents causes a large lagging line current which, in effect, results in an equivalent series circuit of low impedance consisting of resistance and inductive reactance. At high frequencies the inductive reactance is large and the capacitive reactance is small. This results in a large leading line current with an attendant equivalent series circuit of low impedance consisting of resistance and capacitive reactance.

There is one frequency, between those mentioned above, at which the lagging component of current through the inductive branch is equal to the leading current through the capacitive branch. This condition results in a small line current that is in phase with the voltage across the parallel circuit and therefore an impedance that is equivalent to a very high resistance.

The resonant frequency of a parallel circuit is often a source of confusion to the student studying parallel resonance for the first time. The reason is that different definitions for the resonant frequency are encountered in various texts. Thus, the resonant frequency of a parallel circuit can be defined by any one of the following as:

1. The frequency at which the parallel circuit acts as a pure resistance.
2. The frequency at which the line current becomes minimum.

3. The frequency at which the inductive reactance equals the capacitive reactance. This is the same definition as that for the resonant frequency of a series circuit. That is,

$$\omega L = \frac{1}{\omega C}$$

or

$$f_r = \frac{1}{2\pi\sqrt{LC}} \tag{8}$$

A little consideration of these definitions will convince you that, in high-Q circuits, the three resonant frequencies differ by an amount so small as to be negligible.

In the circuit of Fig. 33-23,

$$I_b = \frac{V}{\dfrac{1}{\omega C}} = \omega C V$$

Also,

$$I_a = \frac{V}{R + j\omega L}$$

Rationalizing (Sec. 20-17)

$$I_a = \frac{V}{R + j\omega L} \cdot \frac{R - j\omega L}{R - j\omega L} = \frac{V(R - j\omega L)}{R^2 + (\omega L)^2}$$

$$= \frac{VR}{R^2 + (\omega L)^2} - j\frac{\omega L V}{R^2 + (\omega L)^2}$$

In order to satisfy the first definition for resonant frequency, the line current must be in phase with the applied voltage; that is, the out-of-phase, or quadrature, component of the current through the inductive branch must be equal to the current through the capacitive branch. Thus,

$$\frac{\omega L V}{R^2 + (\omega L)^2} = \omega C V$$

D: ωV,

$$\frac{L}{R^2 + (\omega L)^2} = C$$

M: $[R^2 + (\omega L)^2]$,

$$L = [R^2 + (\omega L)^2]C \tag{9}$$

or

$$\frac{L}{C} - R^2 = (\omega L)^2$$

Hence,

$$\omega = \frac{\sqrt{\dfrac{L}{C} - R^2}}{L}$$

$$= \sqrt{\frac{1}{LC} - \frac{R^2}{L^2}}$$

Substituting $2\pi f$ for ω,

$$2\pi f = \sqrt{\frac{1}{LC} - \frac{R^2}{L^2}}$$

Thus, the resonant frequency is

$$f = \frac{1}{2\pi}\sqrt{\frac{1}{LC} - \frac{R^2}{L^2}} \tag{10}$$

If the Q of the inductance is at all large, then $\omega L \gg R$, which, for all practical purposes, makes the term $\dfrac{R^2}{L^2}$ in Eq. (10) of such low value that it can be neglected, and Eq. (10) is thus reduced to Eq. (8).

Work out several examples with different circuit values, and compare the resonant frequencies obtained from the formulas. In this connection, it is left to you as an exercise to show that in a parallel-resonant circuit, as represented in Fig. 33-23, the line current and applied voltage will be in phase (unity power factor) when

$$R^2 = X_L(X_C - X_L) \qquad (11)$$

33-11 IMPEDANCE OF PARALLEL-RESONANT CIRCUITS

When a parallel circuit is operating at the frequency at which it acts as a pure resistance, it has unity PF, and the line current I_t (Fig. 33-23) consists of the in-phase component of I_a. That is,

$$I_t = \frac{VR}{R^2 + (\omega L)^2} \quad \text{A} \qquad (12)$$

Then, since

$$Z_t = \frac{V}{I_t} \quad \Omega$$

substituting in Eq. (12) for I_t,

$$\frac{V}{Z_t} = \frac{VR}{R^2 + (\omega L)^2}$$

Hence,

$$Z_t = \frac{R^2 + (\omega L)^2}{R} \quad \Omega \qquad (13)$$

From Eq. (9),

$$R^2 + (\omega L)^2 = \frac{L}{C}$$

Substituting this value in Eq. (13),

$$Z_t = \frac{L}{CR} \quad \Omega \qquad (14)$$

Example 7

In the circuit of Fig. 33-23, let $L = 203$ μH, $C = 500$ pF, and $R = 6.7\ \Omega$.

(a) What is the resonant frequency of the circuit?

(b) What is the impedance of the circuit at resonance?

Solution

(a)

$$f = \frac{1}{2\pi\sqrt{LC}}$$

$$= \frac{1}{2\pi\sqrt{2.03 \times 10^{-4} \times 5 \times 10^{-10}}}$$

$$= 500\ \text{kHz}$$

(b)

$$Z_t = \frac{L}{CR}$$

$$= \frac{203 \times 10^{-6}}{500 \times 10^{-12} \times 6.7}$$

$$= \frac{203}{5 \times 6.7} \times 10^4$$

$$= 60.6\ \text{k}\Omega$$

If the value of C is unknown, Eq. (14) can be used in different form. Thus, by multiplying both numerator and denominator by ω,

$$Z_t = \frac{\omega L}{\omega CR} = \frac{1}{\omega C} \frac{\omega L}{R}$$

Since at resonance,

$$\omega L = \frac{1}{\omega C}$$

then

$$Z_t = \frac{(\omega L)^2}{R} \quad \Omega \qquad (15)$$

Moreover, since

$$Q = \frac{\omega L}{R}$$

substituting in Eq. (15),

$$Z_t = \omega L Q \quad \Omega \qquad (16)$$

Example 8

In the circuit of Fig. 33-23, let $L = 70.4$ μH and $R = 5.31$ Ω. If the resonant frequency of the circuit is 1.2 MHz, determine (a) the impedance of the circuit at resonance and (b) the capacitance of the capacitor.

Solution

$$f = 1.2\,\text{MHz} = 1.2 \times 10^6\,\text{Hz}$$
$$\omega = 2\pi f = 2\pi \times 1.2 \times 10^6 = 7.54 \times 10^6$$

(a)
$$Z_t = \frac{(\omega L)^2}{R} = \frac{(7.54 \times 10^6 \times 70.4 \times 10^{-6})^2}{5.31}$$
$$= 53.1\,\text{k}\Omega$$

(b) Since, at resonance, $\omega L = \frac{1}{\omega C}$ and $\omega L = 531$ Ω,

then
$$\frac{1}{\omega C} = 531\,\Omega$$

Hence,
$$C = \frac{1}{531\omega} = 250\,\text{pF}$$

What is the Q of this circuit?

PROBLEMS 33-3

1. An inductor of 16 μH and a capacitor of 50 pF are connected in parallel as shown in Fig. 32-23. If the effective resistance of the coil is 22 Ω, find:
   (a) The resonant frequency of the circuit according to definition 1 (Sec. 33-10).
   (b) The resonant frequency according to definition 3.
   (c) The Q of the coil by using the frequency of part (b).
2. Repeat Prob. 1 for an effective resistance of the coil of 44 Ω.

3. An inductor of 10 mH with a Q of 800 is connected in parallel with a 200-pF capacitor.

   (a) What is the resonant frequency of the circuit?

   (b) What is the impedance of the circuit at resonance?

   (c) What is the effective resistance of the inductor?

4. If the circuit of Prob. 3 is energized with 600 V at the resonant frequency, how much power will it absorb?

5. A coil with a Q of 71.6 is connected in parallel with a capacitor, and this circuit resonates at 356 kHz. The impedance at resonance is found to be 64 kΩ. What is the value of the capacitor?

6. An inductor is connected in parallel with a 254-pF capacitor, and the circuit is found to resonate at 999 kHz. A circuit magnification meter indicates that the Q of the inductor is 90.

   (a) What is the value of the inductance?

   (b) What is the effective resistance of the inductor?

   (c) What is the impedance of the circuit at resonance?

7. If the circuit of Prob. 6 is connected to 20 V at the resonant frequency, how much power will it absorb?

8. If the circuit of Prob. 6 is connected to a 20-V source at 499 kHz, (a) how much power will it absorb and (b) what will be the PF of the circuit?

9. If the circuit of Prob. 6 is connected to a 20-V source at 1499 kHz, what will be the PF?

10. An inductor with a measured Q of 100 resonates with a capacitor at 7.496 MHz with an impedance of 65.9 kΩ. What is the value of the inductance?

11. What is the capacitance of the test capacitor in Prob. 10?

12. 18.9 mA is the total current drain when a capacitor is in resonance with an inductor at 1.5 MHz and the parallel circuit is energized with a 1-kV source. The Q of the inductor is measured at 99.7. What is the value of the capacitor?

33-12 EQUIVALENT Y AND Δ CIRCUITS

When networks contain complex impedances, the equations for converting from a Δ network to an equivalent Y network, or vice versa, are derived by methods identical with those of Sec. 22-7. Thus, in Fig. 33-24, each equivalent Y impedance is equal to the product of the two *adjacent* Δ impedances divided by the summation of the Δ impedances, or

$$Z_a = \frac{Z_1 Z_3}{\Sigma Z_\Delta} \tag{17}$$

$$Z_b = \frac{Z_1 Z_2}{\Sigma Z_\Delta} \tag{18}$$

and

$$Z_c = \frac{Z_2 Z_3}{\Sigma Z_\Delta} \tag{19}$$

where

$$\Sigma Z_\Delta = Z_1 + Z_2 + Z_3$$

and all impedances are expressed in polar form.

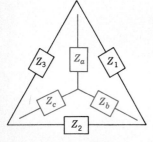

Fig. 33-24 Equivalent Y and Δ impedances.

Similarly, each equivalent Δ impedance is equal to the summation of the Y impedances divided by the *opposite* Y impedance. Thus,

$$Z_1 = \frac{\Sigma Z_Y}{Z_c} \tag{20}$$

$$Z_2 = \frac{\Sigma Z_Y}{Z_a} \tag{21}$$

and

$$Z_3 = \frac{\Sigma Z_Y}{Z_b} \tag{22}$$

where

$$\Sigma Z_Y = Z_a Z_b + Z_b Z_c + Z_a Z_c$$

and all impedances are expressed in polar form.

Example 9

In Fig. 33-24,

$$Z_1 = 7.07 + j7.07 \ \Omega$$
$$Z_2 = 4 + j3 \ \Omega$$

and

$$Z_3 = 6 - j8 \ \Omega$$

What are the values of the equivalent Y circuit?

Solution

Express all impedances in both rectangular and polar forms.

$$Z_1 = 7.07 + j7.07 = 10 \underline{/45°} \ \Omega$$
$$Z_2 = 4 + j3 = 5 \underline{/36.9°} \ \Omega$$
$$Z_3 = 6 - j8 = 10 \underline{/-53.1°} \ \Omega$$
$$\Sigma Z_\Delta = (7.07 + j7.07) + (4 + j3) + (6 - j8)$$
$$= 17.2 \underline{/6.91°} \ \Omega$$

Substituting in Eq. (17),
$$Z_a = \frac{(10 \underline{/45°})(10 \underline{/-53.1°})}{17.2 \underline{/6.91°}}$$
$$= 5.62 - j1.51 \ \Omega$$

Substituting in Eq. (18),
$$Z_b = \frac{(10 \underline{/45°})(5 \underline{/36.9°})}{17.2 \underline{/6.91°}}$$
$$= 0.752 + j2.81 \ \Omega$$

Substituting in Eq. (19),
$$Z_c = \frac{(5 \underline{/36.9°})(10 \underline{/-53.1°})}{17.2 \underline{/6.91°}}$$
$$= 2.67 - j1.14 \ \Omega$$

The solution can be checked by converting the above Y-network equivalents back to the original Δ by using Eqs. (20), (21), and (22).

Example 10

Determine the equivalent impedance between points a and c shown in Fig. 33-25.

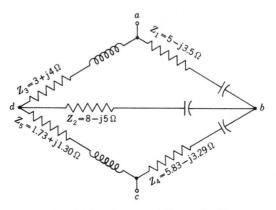

Fig. 33-25 Circuit of Example 19.

Solution

Convert one of the Δ circuits of Fig. 33-25 to its equivalent Y circuit. Thus, for the delta abd,

$$Z_1 = 5 - j3.5 = 6.1 \underline{/-35°}\ \Omega$$

$$Z_2 = 8 - j5 = 9.44 \underline{/-32°}\ \Omega$$

$$Z_3 = 3 + j4 = 5 \underline{/53.1°}\ \Omega$$

$$\Sigma Z_\Delta = (5 - j3.5) + (8 - j5) + (3 + j4)$$

$$= 16.6 \underline{/-15.7°}\ \Omega$$

Substituting in Eq. (17),
$$Z_a = \frac{(6.1\underline{/-35°})(5\underline{/53.1°})}{16.6\underline{/-15.7°}}$$

$$= 1.84 \underline{/33.8°}$$

$$= 1.53 + j1.02\ \Omega$$

Substituting in Eq. (18),
$$Z_b = \frac{(6.1\underline{/-35°})(9.44\underline{/-32°}}{16.6\underline{/-15.7°}}$$

$$= 3.47 \underline{/-51.3°}$$

$$= 2.17 - j2.71\ \Omega$$

Substituting in Eq. (19),
$$Z_c = \frac{(9.44\underline{/-32°})(5\underline{/53.1°})}{16.6\underline{/-15.7°}}$$

$$= 2.84 \underline{/36.8°}$$

$$= 2.27 + j1.70\ \Omega$$

The equivalent Y impedances are shown in Fig. 33-26.

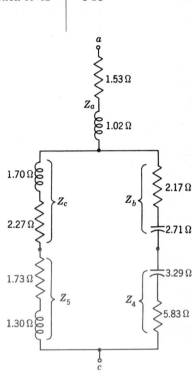

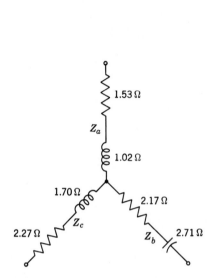

Fig. 33-26 Equivalent Y impedances for circuit of Fig. 33-25.

Fig. 33-27 Equivalent Y impedances connected to remainder of circuit of Fig. 33-25.

The equivalent Y impedances are connected to the remainder of the circuit as shown in Fig. 33-27 and solved as an ordinary series-parallel circuit. See the following equations for Fig. 33-27.

$$Z_{ac} = Z_a + \frac{(Z_c + Z_5)(Z_b + Z_4)}{Z_c + Z_5 + Z_b + Z_4}$$

$$= 1.53 + j1.02 + \frac{[(2.27 + j1.70) + (1.73 + j1.30)][(2.17 - j2.71) + (5.83 - j3.29)]}{(2.27 + j1.70) + (1.73 + j1.30) + (2.17 - j2.71) + (5.83 - j3.29)}$$

$$= 5.45 + j2.0 \ \Omega$$

As we saw in Sec. 22-7, the Δ network is more generally referred to in electronics as a π network and the Y or star network is often known as the T network. In the problems which follow, the two sets of expressions are used interchangeably.

PROBLEMS 33-4

1. In the circuit of Fig. 33-24, $Z_1 = 20 + j30 \ \Omega$, $Z_2 = 25 + j50 \ \Omega$, $Z_3 = 30 - j10 \ \Omega$. Find the impedances of the equivalent Y circuit.
2. In the circuit of Fig. 33-24, $Z_1 = 3 + j4 \ \Omega$, $Z_2 = 12 + j5 \ \Omega$, $Z_3 = 8 - j6 \ \Omega$. Find the equivalent Y-circuit values.

3. In the circuit of Fig. 33-24,

$$Z_a = 46.4 \underline{/75.55°}\ \Omega$$
$$Z_b = 43.8 \underline{/-45.45°}\ \Omega$$
$$Z_c = 56.4 \underline{/-37.45°}\ \Omega$$

Find the impedances of the equivalent π circuit.

4. In the circuit of Fig. 33-24, $Z_a = 50.9 \underline{/86.8°}\ \Omega$, $Z_b = 62.7 \underline{/-20.2°}\ \Omega$, and $Z_c = 44.5 \underline{/8.8°}\ \Omega$. Find the equivalent Δ-circuit values.

5. In the circuit of Fig. 33-28, $Z_1 = 78 \underline{/22.6°}\ \Omega$, $Z_2 = 80 \underline{/-53.1°}\ \Omega$, $Z_3 = 50 \underline{/45°}\ \Omega$, $Z_4 = 39 \underline{/-67.4°}\ \Omega$, and $Z_5 = 100 \underline{/36.9°}\ \Omega$. Find Z_{ab}.

6. In Prob. 5, if $V = 100 \underline{/0°}$ V, find the current flow through impedance Z_4.

7. In the circuit of Fig. 33-28, $Z_1 = 102 + j190\ \Omega$, $Z_2 = 134 - j33\ \Omega$, $Z_3 = 380 - j210\ \Omega$, $Z_4 = 30 - j40\ \Omega$, and $Z_5 = 80 - j60\ \Omega$. What is the equivalent impedance Z_{ab}?

8. In Prob. 7, if $V = 440$ V, how much current flows through Z_5?

9. In Prob. 7, if $V = 200$ V, how much power is expended in Z_4?

10. In Prob. 7, if $V = 200$ V, how much current flows through Z_2?

11. In Fig. 33-28, $Z_1 = 90 - j120\ \Omega$, $Z_2 = 115 - j18\ \Omega$, $Z_3 = 168 - j58\ \Omega$, $Z_4 = 50 + j0\ \Omega$, and $Z_5 = 0 + j25\ \Omega$. Determine the equivalent impedance Z_{ab}.

12. In Prob. 11, if $V = 100$ V, how much current flows through Z_5?

13. In Prob. 11, if $V = 100$ V, how much power is expended in Z_1?

14. In Fig. 33-29, $Z_1 = 3 + j4\ \Omega$, $Z_2 = 37 \underline{/77.5°}\ \Omega$, $Z_3 = 40 \underline{/-80°}\ \Omega$, $Z_4 = 64 - j50\ \Omega$, $Z_5 = 15 + j85\ \Omega$, $Z_6 = 40 - j36\ \Omega$, $Z_7 = 10 \underline{/-53.1°}\ \Omega$, and $V = 120$ V. How much current flows through Z_7?

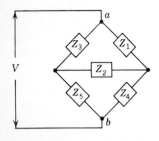

Fig. 33-28 Circuit for Probs. 5 to 13.

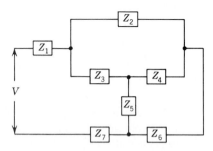

Fig. 33-29 Circuit for Prob. 14.

15. In Fig. 33-30, $Z_1 = 254 \underline{/88.6°}\ \Omega$, $Z_2 = 306 \underline{/86.1°}\ \Omega$, $Z_3 = 437 \underline{/-73.6°}\ \Omega$, $Z_4 = 177 \underline{/-87°}\ \Omega$, $Z_5 = 288 \underline{/87.5°}\ \Omega$, $Z_6 = 250 \underline{/89.1°}\ \Omega$, and $Z_L = 680 \underline{/0°}\ \Omega$. Determine the equivalent impedance Z_{ab}.

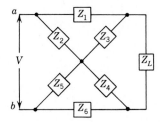

Fig. 33-30 Circuit for Probs. 15 to 18.

16. In Prob. 15, if $V = 475$ V, how much current flow through the load impedance Z_L?

17. In Fig. 33-30, $Z_1 = 63 + j5 \ \Omega$, $Z_2 = 12 + j60 \ \Omega$, $Z_3 = 20 + j90 \ \Omega$, $Z_4 = 18 + j86 \ \Omega$, $Z_5 = 8 + j52 \ \Omega$, $Z_6 = 47 + j2 \ \Omega$, $Z_L = 600 + j0 \ \Omega$. Determine the equivalent impedance Z_{ab}.

18. In Prob. 17, if $V = 135$ V, how much power is dissipated in the load impedance Z_L?

Logarithms

34

In problems pertaining to engineering, there often occurs the need for numerical computations involving multiplication, division, powers, or roots. Some of these problems can be solved more readily by the use of logarithms than by ordinary arithmetical processes.

Credit for the invention of logarithms is chiefly due to John Napier, whose tables appeared in 1614. This was an extremely important event in the development of mathematics; for by the use of logarithms:

1. Multiplication is reduced to addition.
2. Division is reduced to subtraction.
3. Raising to a power is reduced to one multiplication.
4. Extracting a root is reduced to one division.

In some phases of engineering, computation by logarithms used to be utilized to a great extent because of the high degree of accuracy desired and the amount of labor that was thereby saved. Because the electronic calculator is convenient and because calculator results meet the ordinary demands for accuracy in problems relating to electronics, it is not necessary to make wide use of logarithms for computations in the general field. However, it is essential that people in the electrical engineering field and, more particularly, the electronics engineer and technician have a thorough understanding of logarithmic processes.

34-1 DEFINITION

The *logarithm* of a quantity is the exponent of the power to which a given number, called the *base,* must be raised in order to equal the quantity.

Example 1
Since $10^3 = 1000$, then $3 = $ logarithm of 1000 to the base 10.

Example 2
Since $2^3 = 8$, then $3 = $ logarithm of 8 to the base 2.

Example 3
Since $a^x = b$, then $x = $ logarithm of b to the base a.

34-2 NOTATION

If
$$b^x = N \qquad (1)$$

then x is the logarithm of N to the base b. It may be helpful to mentally translate this expression to "x is the power to which b must be raised to obtain N." This statement is abbreviated by writing

$$x = \log_b N \qquad (2)$$

It is evident that Eqs. (1) and (2) mean the same thing and are simply different methods of expressing the same relation among b, x, and N. Equation (1) is called the *exponential form*, and Eq. (2) is called the *logarithmic form*.

As an aid in remembering that a *logarithm is an exponent*, Eq. (1) can be written in the form

$$(\text{Base})^{\log} = \text{number}$$

The following example illustrates relations between exponential and logarithmic forms.

Example 4

Exponential Notation	Logarithmic Notation
$2^4 = 16$	$4 = \log_2 16$
$3^5 = 243$	$5 = \log_3 243$
$25^{0.5} = 5$	$0.5 = \log_{25} 5$
$10^2 = 100$	$2 = \log_{10} 100$
$10^4 = 10\,000$	$4 = \log_{10} 10\,000$
$a^b = c$	$b = \log_a c$
$\varepsilon^x = y$	$x = \log_\varepsilon y$

From the foregoing examples, it is apparent that any positive number, other than 1, can be selected as a base for a system of logarithms. Because 1 raised to any power is 1, it cannot be used as a base.

Based on the definitions in Eqs. (1) and (2), you should satisfy yourself with the correctness of the following statement:

$$\log_a a^b = b$$

PROBLEMS 34-1

Express the following equations in logarithmic form:

1. $10^2 = 100$ 2. $10^3 = 1000$ 3. $7^2 = 49$
4. $4^3 = 64$ 5. $4^{0.5} = 2$ 6. $\varepsilon^1 = \varepsilon$
7. $a^1 = a$ 8. $10^1 = 10$ 9. $a^0 = 1$
10. $1 = 10^0$

Express the following equations in exponential form:

11. $3 = \log_{10} 1000$ 12. $5 = \log_{10} 100\,000$
13. $2 = \log_5 25$ 14. $3 = \log_4 64$

15. $0 = \log_6 1$
16. $0 = \log_a 1$
17. $4 = \log_5 625$
18. $0.5 = \log_9 3$
19. $s = \log_r t$
20. $2x = \log_3 M$

Find the value of x:

21. $3^x = 9$
22. $2^x = 16$
23. $10^x = 1\ 000\ 000$
24. $x = \log_2 32$
25. $4^x = 2$
26. $\log_8 x = 3$
27. Show that $\log_{10} 100 = \log_{10} 100\ 000 - \log_{10} 1000$.
28. Show that $\log_p p = 1$.
29. What are the logarithms to the base 2 of 2, 4, 8, 16, 32, 64, 128, 256, and 512?
30. What are the logarithms to the base 3 of 3, 9, 27, 81, 243, 729, and 2187?

34-3 LOGARITHM OF A PRODUCT

The logarithm of a product is equal to the sum of the logarithms of the factors.

Consider the two factors M and N, and let x and y be their respective logarithms to the base a; then,

$$x = \log_a M \qquad (3)$$

and
$$y = \log_a N \qquad (4)$$

Writing Eq. (3) in exponential form,

$$a^x = M \qquad (5)$$

Writing Eq. (4) in exponential form,

$$a^y = N \qquad (6)$$

Then
$$M \cdot N = a^x \cdot a^y = a^{x+y}$$
$$\therefore \log_a (M \cdot N) = x + y = \log_a M + \log_a N$$

Example 5

$$2 = \log_{10} 100 \qquad \text{or} \qquad 10^2 = 100$$
$$4 = \log_{10} 10\ 000 \qquad \text{or} \qquad 10^4 = 10\ 000$$

Then
$$100 \times 10\ 000 = 10^2 \cdot 10^4$$
$$= 10^{2+4} = 10^6$$
$$\therefore \log_{10} (100 \times 10\ 000) = 2 + 4$$
$$= \log_{10} 100 + \log_{10} 10\ 000$$

The above proposition is also true for the product of more than two factors. Thus, by successive applications of the proof, it can be shown that

$$\log_a (A \cdot B \cdot C \cdot D) = \log_a A + \log_a B + \log_a C + \log_a D$$

34-4 LOGARITHM OF A QUOTIENT

The logarithm of the quotient of two numbers is equal to the logarithm of the dividend minus the logarithm of the divisor.

As in Sec. 34-3, let

$$x = \log_a M \tag{3}$$

and
$$y = \log_a N \tag{4}$$

Writing Eq. (3) in exponential form,

$$a^x = M \tag{5}$$

Writing Eq. (4) in exponential form,

$$a^y = N \tag{6}$$

Dividing Eq. (5) by Eq. (6),

$$\frac{a^x}{a^y} = \frac{M}{N}$$

That is,
$$a^{x-y} = \frac{M}{N} \tag{7}$$

Writing Eq. (7) in logarithmic form,

$$x - y = \log_a \frac{M}{N} \tag{8}$$

Substituting in Eq. (8) for the values of x and y,

$$\log_a M - \log_a N = \log_a \frac{M}{N}$$

Example 6

$$2 = \log_{10} 100 \qquad \text{or} \qquad 10^2 = 100$$
$$4 = \log_{10} 10\ 000 \qquad \text{or} \qquad 10^4 = 10\ 000$$

Then
$$\frac{10\ 000}{100} = \frac{10^4}{10^2} = 10^{4-2} = 10^2$$

$$\therefore \log_{10} \frac{10\ 000}{100} = 4 - 2 = \log_{10} 10\ 000 - \log_{10} 100$$

34-5 LOGARITHM OF A POWER

The logarithm of a power of a number equals the logarithm of the number multiplied by the exponent of the power.

Again, let
$$x = \log_a M \tag{3}$$

Then
$$M = a^x \tag{9}$$

Raising both sides of Eq. (9) to the nth power,

$$M^n = a^{nx} \tag{10}$$

Writing Eq. (10) in logarithmic form,

$$\log_a M^n = nx \tag{11}$$

Substituting in Eq. (11) for the value of x,

$$\log_a M^n = n \log_a M$$

Example 7

$$2 = \log_{10} 100 \quad \text{or} \quad 100 = 10^2$$

Since
$$(10^2)^2 = 10^{2 \cdot 2} = 10^4 = 10\ 000$$

then
$$\log_{10} 10\ 000 = 4$$

$$\therefore \log_{10} 100^2 = 2 \log_{10} 100 = 2 \cdot 2 = 4$$

34-6 LOGARITHM OF A ROOT

The logarithm of a root of a number is equal to the logarithm of the number divided by the index of the root.

Again, let
$$x = \log_a M \tag{3}$$

Then
$$M = a^x \tag{9}$$

Extracting the nth root of both sides of Eq. (9),

$$M^{1/n} = a^{x/n} \tag{12}$$

Writing Eq. (12) in logarithmic form,

$$\log_a M^{1/n} = \frac{x}{n} \tag{13}$$

Substituting in Eq. (13) for the value of x,

$$\log_a M^{1/n} = \frac{\log_a M}{n}$$

Example 8

$$4 = \log_{10} 10\ 000 \quad \text{or} \quad 10\ 000 = 10^4$$

Since
$$\sqrt{10\ 000} = \sqrt{10^4} = 10^{4/2} = 10^2 = 100$$

then
$$\log_{10} \sqrt{10\ 000} = \frac{\log_{10} 10\ 000}{2} = \frac{4}{2} = 2$$

34-7 SUMMARY

It is evident that if the logarithms of numbers instead of the numbers themselves are used for computations, then *multiplication, division, raising to powers,* and *extracting roots* are replaced by *addition, subtraction, multiplication,* and *division,* respectively. Because you are familiar with the laws of exponents, especially as applied to the powers of 10, the foregoing operations with logarithms involve no new ideas. The sole idea behind logarithms is that every positive number can be expressed as a power of some base. That is,

$$\text{Any positive number} = (\text{base})^{\log}$$

34-8 THE COMMON SYSTEM OF LOGARITHMS

Since 10 is the base of our number systems, both integral and decimal, the base 10 has been chosen for a system of logarithms. This system is called the *common system* or *Briggs's system*. The natural system, of which the base to five decimal places is 2.718 28, will be discussed later.

Hereafter, when no other base is stated, the base will be 10. For example, $\log_{10} 625$ will be written log 625, the base 10 being understood.

34-9 THE NATURAL SYSTEM OF LOGARITHMS

In the number system there exist certain special numbers whose value is not absolutely determined, but which are themselves extremely valuable to us. You are already familiar with π, which has a value which is approximately $\frac{22}{7}$.

Another useful number is ε, which has a value of approximately 2.718 28. This unusual number turns out to be extremely valuable when used as a base for logarithms. Because it can be shown to be related to *natural* events, like the decay of charge on a capacitor which is discharged through a resistor or the decay of current when the magnetic field about an inductance collapses, it is called the base of the *natural logarithms*. Tables of natural logarithms, or logarithms to the base ε, are to be found in many published books of tables. A good "scientific" calculator will deliver powers of ε and logarithms to the base ε ($\ln x$). In Sec. 34-25 we will see how to change logarithms to the base 10 into logarithms to the base ε or to other bases.

The notation for logarithms to the base ε is shown variously as $\log_\varepsilon$ or ln (pronounced "lon").

34-10 DEVELOPING A TABLE OF LOGARITHMS

Table 34-1 illustrates the connection between the power of 10 and the logarithms of certain numbers.

TABLE 34-1

Exponential Form	Logarithmic Form
$10^4 = 10\ 000$	$\log 10\ 000 = 4$
$10^3 = 1000$	$\log 1000 = 3$
$10^2 = 100$	$\log 100 = 2$
$10^1 = 10$	$\log 10 = 1$
$10^0 = 1$	$\log 1 = 0$
$10^{-1} = 0.1$	$\log 0.1 = -1$
$10^{-2} = 0.01$	$\log 0.01 = -2$
$10^{-3} = 0.001$	$\log 0.001 = -3$
$10^{-4} = 0.0001$	$\log 0.0001 = -4$

Inspection of Table 34-1 shows that only powers of 10 have integers for logarithms. Also, it is evident that the logarithm of any number between

10 and 100, for example, is between 1 and 2; that is, it is 1 plus a decimal. Similarly, the logarithm of any number between 100 and 1000 is between 2 and 3, and so on. Therefore, to represent all numbers, it is necessary for us to develop the fractional powers which represent numbers between 1 and 10. Then, by using powers of 10 to convert any number to a number between 1 and 10 times the appropriate power of 10 (Chap. 6), we may use our new fractional powers of 10 instead of just integral powers of 10 to find the logarithm of any number.

In Sec. 20-4 we saw that $a^{1/2} = \sqrt{a}$. Accordingly, we can see that

$$10^{0.5} = 10^{1/2} = \sqrt{10} = 3.162\ 277\ 66$$

which gives us the first intermediate step in our table of logarithms between 1 and 10:

$$\log_{10} 3.162\ 277\ 66 = 0.5$$

Similarly,

$$10^{0.25} = (10^{0.5})^{0.5} = \sqrt{3.162\ 277\ 66}$$
$$= 1.778\ 279$$

or
$$\log_{10} 1.778\ 279 = 0.25$$

By repeating the square root operation time after time, we can obtain

$$\log_{10} 1.334 = 0.125 \qquad \text{etc.}$$

Then, by applying the laws of exponents developed in Sec. 4-3 and summarized in Sec. 20-1, we can determine that

$$3.162\ 277\ 66 \times 1.778\ 279 = 10^{0.5} \times 10^{0.25}$$
$$= 10^{0.75} = 5.623\ 41$$

or
$$\log 5.623\ 41 = 0.75$$

Repeated applications of this method give us such additional logarithms as

$$\log 4.2173 = 0.625$$

and
$$\log 2.37 = 0.375$$

You should use the values now developed to prove that $10^{0.75} \times 10^{0.25} = 10$, as a check on our method.

These various values can be plotted on a graph, as in Fig. 34-1, and the more convenient logarithms can be picked off the curve, or other more

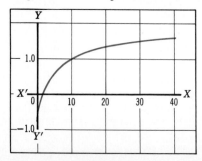

Fig. 34-1 Graph of the equation $y = \log_{10} x$.

sophisticated methods of higher mathematics may be applied to yield Table 34-2 of logarithms of numbers between 1 and 10:

TABLE 34-2   LOGARITHMS OF NUMBERS BETWEEN 1 AND 10

Number	Logarithm	Number	Logarithm
1	0.000 00	6	0.778 15
2	0.301 03	7	0.845 10
3	0.477 12	8	0.903 09
4	0.602 06	9	0.954 24
5	0.698 97	10	1.000 00

Since we convert every number to its equivalent number between 1 and 10 times the appropriate power of 10, every logarithm we will ever look up will be a decimal fraction. Because of this universality of decimals as logarithms, almost every table of logarithms published omits the decimal point: log 2 will appear as simply 301 03 instead of 0.301 03.

From the foregoing discussion it will be evident that every logarithm has two parts: a decimal part which we read from the table of logarithms and an integer which we must provide each time from our knowledge of powers of 10.

Example 9

Determine the logarithm of 200.

Solution

First, rewrite the number in standard form:

$$200 = 2.00 \times 10^2$$

Since log 2.00 = 0.301 03, this number could be written

$$2.00 \times 10^2 = 10^{0.301\ 03} \times 10^2 = 10^{2.301\ 03}$$

This power to which 10 is raised to be equal to 200 is the logarithm of 200. In other words,

$$\log 200 = 2.301\ 03$$

This logarithm is made up of two parts: the decimal part from the table and the integer part which we developed from the "power of 10." Similarly, log 2000 = 3.301 03.

In the same manner, referring to Table 34-1, it follows that the logarithm of a number between 0.1 and 0.01 will be -2 followed by a decimal number from the table and the logarithm of a number between 0.001 and 0.0001 will be -4 followed by a decimal. Note carefully that the decimal number taken from the table is always positive, so that the logarithm of a number between 0.001 and 0.0001, which we have seen will be -4 followed by a *positive* decimal, may be written as -3 followed by a *negative* decimal.

The integral part of the logarithm, which we provide by ourselves, is called the *characteristic,* and it may be positive, negative, or zero. The fractional part, which is taken from the table, is called the *mantissa*, and, when taken from the table, is always *positive*. This is not the case when using a calculator. See Example 11, Solution 2.

34-11 THE CHARACTERISTIC

The use of the base 10 makes it possible to simplify computation by means of logarithms and to express logarithms in a compact tabular form. Any number, however large or small, may be expressed as a number between 1 and 10 multiplied by the appropriate whole-number power of 10. We developed, in Sec. 34-10, a rudimentary table of logarithms of numbers between 1 and 10 and saw that these powers of 10 (logarithms) must be numbers between 0 and 1, usually expressed as decimal fractions, called *mantissas*.

Let us now turn our attention to logarithms of numbers greater than 10. In Example 9 we converted 200 into 2.00×10^2 and saw that the logarithm of 200 is 2.301 03. The mantissa 0.301 03 is taken from the table to represent the 2, a quantity between 1 and 10. The whole number portion, 2, the *characteristic*, is simply the whole-number power of 10 obtained when 200 is written 2.00×10^2.

Example 10
What is the logarithm of 50 000?

Solution 1

$$50\ 000 = 5.00 \times 10^4$$
$$= 10^{0.698\ 97} \times 10^4$$
$$= 10^{4.698\ 97}$$
$$\log 50\ 000 = 4.698\ 97$$

Solution 2
Without actually rewriting 50 000 as 5×10^4, mentally determine the power of 10 necessary to do so. Move the decimal point four places to the right, and record 4 (with the decimal point) as the characteristic. Then determine the mantissa from the table: 0.698 97.

$$\log 50\ 000 = 4.698\ 97$$

When you are required to determine the log of a number less than 1, follow the same procedure exactly to find the *negative* whole-number power of 10.

Example 11
What is the logarithm of 0.005 00?

Solution 1

$$0.005\ 00 = 5.00 \times 10^{-3}$$
$$= 10^{0.698\ 97} \times 10^{-3}$$

There are several alternative methods of showing the logarithm of a fraction, and each is useful for a different class of problem.

Solution 2

Directly combine the two parts of the logarithm

$$10^{0.698\ 97} \times 10^{-3} = 10^{-2.301\ 03}$$

here, $-3.000\ 00 + 0.698\ 97 = -2.301\ 03$, a totally negative number.

$$\log 0.005\ 00 = -2.301\ 03^*$$

This *negative logarithm* is the form provided by scientific pocket calculators. It cannot be taken directly from most of the commonly available tables.

Solution 3

Combine the two parts, but clearly identify the *negative characteristic* and the *positive mantissa*. This is done by rewriting the -3 characteristic as $\bar{3}$. Pronounced "bar three," it signifies that the characteristic alone is negative when it is combined with the positive mantissa:

$$\log 0.005\ 00 = \bar{3}.698\ 97$$

Solution 4

Present the two parts in a form that keeps them separately identifiable:

$$\log 0.005\ 00 = 0.698\ 97 - 3$$
$$or \qquad = 7.698\ 97 - 10$$
$$or \qquad = 3.698\ 97 - 6$$

or any other combination of numbers that has a net positive mantissa 0.698 97 and a net negative characteristic equal to -3 ($7 - 10 = -3$, etc.).

RULE

When any given number is expressed as a number between 1 and 10 multiplied by the appropriate whole-number power of 10, the whole-number power is the characteristic of the logarithm of the given number. The foregoing is illustrated in Table 34-3.

TABLE 34-3

Number	Standard Notation	Characteristic
682	6.82×10^2	2
3765	3.765×10^3	3
14	1.4×10^1	1
1	1×10^0	0
0.004 25	4.25×10^{-3}	-3 or $7 - 10$
0.1	1×10^{-1}	-1 or $9 - 10$
0.000 072	7.2×10^{-5}	-5 or $5 - 10$

*Pocket calculator displays ("readouts") *cannot* show a simultaneous mixture of negative characteristic with positive mantissa.

34-12 THE MANTISSA

Note that all numbers whose logarithms are given below have the same significant figures. These logarithms were obtained by first finding log 2.207 from a table, as will be discussed later. The remaining logarithms were then obtained by applying the properties of logarithms as stated in Secs. 34-3 and 34-4.

$$\log 2207 = \log 1000(2.207) = \log 1000 + \log 2.207 = \quad 3 + 0.343\ 80$$
$$\log 220.7 = \log 100(2.207) = \log 100 + \log 2.207 = \quad 2 + 0.343\ 80$$
$$\log 22.07 = \log 10(2.207) = \log 10 + \log 2.207 = \quad 1 + 0.343\ 80$$
$$\log 2.207 = \log 1(2.207) = \log 1 + \log 2.207 = \quad 0 + 0.343\ 80$$
$$\log 0.2207 = \log \frac{2.207}{10} = \log 2.207 - \log 10 = -1 + 0.343\ 80$$
$$\log 0.022\ 07 = \log \frac{2.207}{100} = \log 2.207 - \log 100 = -2 + 0.343\ 80$$

From the above examples, it is apparent that the mantissa is not affected by a shift of the decimal point. That is, *the mantissa of the logarithm of a number depends only on the sequence of the significant figures in the number*. Because of this, 10 is ideally suited as a base for a system of logarithms to be used for computation.

PROBLEMS 34-2

Write the characteristics of the logarithms of the following numbers:

1.	37	**2.**	226	**3.**	688
4.	20.6	**5.**	7.27	**6.**	72.7
7.	727	**8.**	0.727	**9.**	0.000 727
10.	958 16	**11.**	95.816	**12.**	0.095 816
13.	1002	**14.**	10.02	**15.**	0.000 100 2
16.	1 002 000	**17.**	0.004	**18.**	2.65×10^6
19.	3.3×10^3	**20.**	8×10^{-12}		

Find the value of each of the following expressions:

21. $\log 100 + \log 0.001$ **22.** $\log \sqrt{100}$ **23.** $\log \sqrt{\dfrac{1000}{10}}$

24. $\log \sqrt{1000} - \log \sqrt{100}$ **25.** $\log \sqrt{0.001}$

Write the following expressions in expanded form:

26. $\log \dfrac{278 \times 9.36}{81.1}$

Solution: $\log \dfrac{278 \times 9.36}{81.1} = \log 278 + \log 9.36 - \log 81.1$

27. $\log \dfrac{6792 \times 20.9}{176}$ **28.** $\log \dfrac{3.66 \times (4.71 \times 10^2)}{3.42 \times 7280}$

29. $\log \sqrt{\dfrac{512 \times 0.36}{2\pi \times 177}}$ 30. $\log \sqrt[5]{32\ 000 \times 286 \times 159}$

31. $\log \left(\dfrac{159 \times 0.837}{82.2}\right)^3$ 32. $\log \dfrac{pq^2r}{wy}$

Given log 27.36 = 1.437 12, write the logarithms of the following numbers:

33.	2.736	34.	2736	35.	0.027 36
36.	0.000 273 6	37.	27 360	38.	2736×10^{-4}
39.	27.36×10^6	40.	$0.002\ 736 \times 10^{-3}$	41.	27.36×10^{-12}

Given log 7.57 = 0.879 10, find the numbers that correspond to the following logarithms:

42.	1.879 10	43.	3.879 10	44.	5.879 10 − 10
45.	6.879 10	46.	9.879 10 − 10	47.	3.879 10 − 10
48.	2.879 10	49.	10.879 10	50.	2.879 10 − 10

34-13 TABLES OF LOGARITHMS

Because the characteristic of the logarithm of any number is obtainable by inspection, it is necessary to tabulate only the mantissas of the logarithms of numbers. Though mantissas can be computed by use of advanced mathematics, for convenience the mantissas of the logarithms to a number of significant figures have been computed and arranged in tables. You should check with your instructor as to the requirements of your particular course so far as printed tables are concerned. Inside the front cover of this book is a three-place table of mantissas. You will find that this table will serve most of your needs when working with logarithms related to electronic applications.

34-14 TO FIND THE LOGARITHM OF A GIVEN NUMBER

Table 34-4 is a portion of a typical table of logarithms. You should work through the following examples by using your calculator as well as the table.

TABLE 34-4

N	0	1	2	3	4	5	6	7	8	9
40	60206	60314	60423	60531	60638	60746	60853	60959	61066	61172
41	61278	61384	61490	61595	61700	61805	61909	62014	62118	62221
42	62325	62428	62531	62634	62737	62839	62941	63043	63144	63246
43	63347	63448	63548	63649	63749	63849	63949	64048	64147	64246

Examination of the table shows that the first column has N at the top. N is an abbreviation for "number." The other columns are labled 0, 1, 2, 3, 4, . . ., 9. Therefore, any number consisting of three significant figures has its first two figures in the N column and its third figure in another column. This will be illustrated in the following examples. When finding the logarithm of a number, always write the characteristic at once, before looking for the mantissa.

Example 12
Find log 40.

Solution
$40 = 4 \times 10^1$; therefore, the characteristic is 1.
Since 40 has no third significant figure other than zero, the mantissa of 40 is found at the right of 40 in the N column, in the column headed 0. It is 0.602 06.

$$\therefore \log 40 = 1.602\ 06$$

Example 13
Find log 416.

Solution
$416 = 4.16 \times 10^2$; therefore, the characteristic is 2.
The first two digits of 416 are found in the N column, and the third digit is found in the column headed 6. Then the mantissa is read in the row containing 41 and in the column headed 6. It is 0.619 09.

$$\therefore \log 416 = 2.619\ 09$$

Similarly,
$$\log 4.16 = 0.619\ 09$$
$$\log 41.6 = 1.619\ 09$$
$$\log 4160 = 3.619\ 09$$
$$\log 0.004\ 16 = 7.619\ 09 - 10, \text{ etc.}$$

That is, the mantissa of any number having 416 as significant figures is 0.619 09.

Example 14
Find log 4347.

Solution
$4347 = 4.437 \times 10^3$; therefore, the characteristic is 3.
Since 4347 is between 4340 and 4350, its mantissa must be between the mantissas of 4340 and 4350.

$$\text{Mantissa of } 4350 = 0.638\ 49$$
$$\text{Mantissa of } 4340 = \underline{0.637\ 49}$$
$$\text{Difference} = 0.001\ 00$$

The *tabulator difference* between these mantissas is 0.001 00, and it is apparent that an *increase* of 10 in the number causes the mantissa to *increase* by 0.001 00. Therefore, an increase of 7 in the number will increase the mantissa 0.7 as much. Hence the increase in the mantissa will be $0.001\ 00 \times 0.7 = 0.000\ 70$, and the mantissa of 4347 will be

$$0.637\ 49 + 0.000\ 70 = 0.638\ 19$$
$$\therefore \log 4347 = 3.638\ 19$$

Similarly,
$$\log 43.47 = 1.638\ 19$$
$$\log 4.347 = 0.638\ 19$$
$$\log 434\ 700 = 5.638\ 19$$
$$\log 0.000\ 434\ 7 = 6.638\ 19 - 10, \text{etc.}$$

That is, the mantissa of any number having 4347 as significant figures is 0.638 19.

The foregoing process of finding the mantissa, called *interpolation*, is based on the assumption that the increase in the logarithm is proportional to the increase in the number.

Example 15

Find log 0.000 042 735.

Solution

$$0.000\ 042\ 735 = 4.2735 \times 10^{-5}$$
$$\log 0.000\ 042\ 735 = 5.63079 - 10$$

Summarizing, we have the following rules.

RULE

To find the logarithm of a number containing three significant figures:

1. Determine the characteristic.
2. Locate the first two significant figures in the column headed N.
3. In the same row and in the column headed by the third significant figure, find the required mantissa.

RULE

To find the logarithm of a number containing more than three significant figures:

1. Determine the characteristic.
2. Find the mantissa for the first three significant figures of the number.
3. Find the next higher mantissa, and take the tabular difference of the two mantissas.
4. Add to the lesser mantissa the product of the tabular difference and the remaining figures of the number considered as a decimal.

Note

To students who use engineering-type calculators:

Following the steps of Example 13, we see that the logarithm of 0.004 27 consists of two parts, 7.6304 and -10. The single number combination of these two parts is $-10.000\ 00 + 7.6304 = -2.3696$.

Alternatively, log 0.004 27 may be written 3.6304. Again, the logarithm consists of two parts, -3 and $+0.6304$. These also combine to yield -2.3696.

Keying 0.004 27 into a scientific calculator gives a logarithm reading of $-2.369\ 572\ 1$, which rounds off to the value above.

These negative logarithms are perfectly legitimate. But if further work is to be done using the tables, it may be necessary to convert them into logarithms with positive mantissas with negative (barred) characteristics.

PROBLEMS 34-3
Find the logarithms of the following numbers:

1.	7	2.	700	3.	70
4.	263	5.	721	6.	438
7.	103	8.	400	9.	382 000
10.	0.000 028 8	11.	9264	12.	5 989 000
13.	0.1101	14.	281 300	15.	252.66
16.	989 900	17.	3.142×10^{-6}	18.	202.8×10^7
19.	6.28	20.	3.1416	21.	2.7183
22.	159.1	23.	0.000 471	24.	864 000
25.	69 990	26.	2 003 000	27.	2.003×10^6
28.	0.000 03	29.	5×10^{-12}	30.	84.37×10^{-5}

34-15 TO FIND THE NUMBER CORRESPONDING TO A GIVEN LOGARITHM

The number corresponding to a given logarithm is called the *antilogarithm* and is written "antilog." For example, if log 692 = 2.840 11, then the number corresponding to the logarithm 2.840 11 is 692. That is,

$$\text{antilog } 2.840\ 11 = 692$$

To find the antilog of a given logarithm, we reverse the process of finding the logarithm when the number is given.

Example 16
Find the number whose logarithm is 3.910 09.

Solution 1
The characteristic tells us only the position of the decimal point. Therefore, to find the significant figures of the number (antilog), the mantissa must be found in a suitable table. To the left of the mantissa 0.910 09, in column N, find the first two significant figures of the number, which are 81; and at the head of the column of the mantissa, find the third significant figure, which is 3. Hence, the number has the significant figures 813. The position of the decimal point is fixed by the characteristic; and because the characteristic is 3, there must be four figures to the left of the decimal point.

Thus,   antilog 3.910 09 = 8130
Similarly,   antilog 0.910 09 = 8.13
                antilog 7.910 09 $-$ 10 = 0.008 13
                antilog 6.910 09 = 8.13×10^6, etc.

A change in the characteristic changes only the position of the decimal.

Note how an electronic calculator performing this operation in "scientific or engineering notation" reads out the antilogarithm as a number between 1 and 10 multiplied by the appropriate power of 10.

PROBLEMS 34-4

Find the antilogarithms of the following logarithms:

1.	0.477 12	2.	2.477 12	3.	1.477 12
4.	2.551 45	5.	2.807 54	6.	2.873 32
7.	2.004 32	8.	2.698 97	9.	5.383 82
10.	6.698 10 − 10	11.	3.925 57	12.	6.669 50
13.	9.990 87 − 10	14.	5.151 37	15.	2.534 699
16.	5.391 464	17.	5.174 060 − 10	18.	9.847 14
19.	0.797 96	20.	0.497 206	21.	0.434 249
22.	2.576 226	23.	6.9921 − 10	24.	5.872 74
25.	4.902 997 7	26.	6.751 664	27.	3.237 544
28.	5.778 15 − 10	29.	8.903 09 − 20	30.	6.539 45 − 10

34-16 ADDITION AND SUBTRACTION OF LOGARITHMS

Since the mantissa of a logarithm is always positive, care must be exercised in adding or subtracting logarithms.

Adding logarithms with positive characteristics is the same as adding arithmetical numbers.

Example 17

Add the logarithms 2.764 21 and 4.304 64.

Solution

$$
\begin{array}{r}
2.764\ 21 \\
\underline{4.304\ 64} \\
7.068\ 85
\end{array}
$$

When adding logarithms with negative characteristics, you must bear in mind that the mantissas are always positive.

Example 18

Add the logarithms $\bar{4}.326\ 52$ and $6.284\ 37$.

Solution

The mantissas are added as positive numbers, and the characteristics are added algebraically:

$$
\begin{array}{r}
\bar{4}.326\ 52 \\
\underline{6.284\ 37} \\
\text{Sum} = 2.610\ 89
\end{array}
$$

Example 19

Add the logarithms $\bar{4}.328\ 30$, $\bar{3}.764\ 22$, and $\bar{1}.104\ 82$.

Solution

$$
\begin{array}{r}
\bar{4}.328\ 30 \\
\bar{3}.764\ 22 \\
\bar{1}.104\ 82 \\
\hline
\text{Sum} = \bar{7}.197\ 34
\end{array}
$$

In Example 19 the sum of the *mantissas* is 1.197 34, and the 1 must be carried over for addition with the characteristics. Since the 1 from the mantissa sum is positive and the characteristics are negative, the two are added algebraically to obtain -7.

Example 20

Subtract the logarithm 6.986 02 from the logarithm 4.107 37.

Solution

$$
\begin{array}{r}
4.107\ 37 \\
6.986\ 02 \\
\hline
\text{Remainder} = \bar{3}.121\ 35
\end{array}
$$

Example 21

Subtract the logarithm $\bar{5}.785\ 67$ from the logarithm $\bar{2}.672\ 58$.

Solution

$$
\begin{array}{r}
\bar{2}.672\ 58 \\
\bar{5}.785\ 67 \\
\hline
\text{Remainder} = 2.886\ 91
\end{array}
$$

In Example 21, in order to subtract the mantissas, it was necessary to subtract 7 from 6. The 6 was made to be 16 (1.6). To compensate for this *increase* in the value of the mantissa, the characteristic is changed from -2 to -3. In effect, *borrowing* 1 from the -2 characteristic makes it -3.

Another method of handling logarithms whose characteristics are negative is to express them as logarithms with a positive characteristic and write the proper multiple of negative 10 after the mantissa.

Example 22

Add the logarithms $\bar{4}.326\ 52$ and 6.284 37.

Solution

$$
\bar{4}.326\ 52 = 6.326\ 52\ -\ 10
$$

$$
\begin{array}{r}
6.326\ 52\ -\ 10 \\
6.284\ 37 \\
\hline
\text{Sum} = 12.610\ 89\ -\ 10 = 2.610\ 89
\end{array}
$$

Note that this is the same as Example 18.

If -10, -20, -30, -40, etc., appear in the sum after the mantissa and the characteristic is greater than 9, subtract from both characteristic and mantissa a multiple of 10 that will make the characteristic less than 10.

Example 23

Add the logarithms $\bar{4}.328\ 30$, $\bar{3}.764\ 22$, and $\bar{1}.104\ 82$.

Solution

$$
\begin{array}{r}
6.328\ 30\ -\ 10 \\
7.764\ 22\ -\ 10 \\
\underline{9.104\ 82\ -\ 10} \\
23.197\ 34\ -\ 30 \\
\text{Sum}\ =\quad 3.197\ 34\ -\ 10
\end{array}
$$

Note that this is the same as Example 19.

When a larger logarithm is subtracted from a smaller, the characteristic of the smaller should be increased by 10 and -10 should be written after the mantissa to preserve equality.

Example 24

Subtract the logarithm $6.986\ 02$ from the logarithm $4.107\ 37$.

Solution

$$
\begin{array}{r}
4.107\ 37\ =\ 14.107\ 37\ -\ 10 \\
\underline{6.986\ 02} \\
\text{Remainder}\ =\quad 7.121\ 35\ -\ 10
\end{array}
$$

Also, when a negative logarithm is subtracted from a positive logarithm, the characteristic of the minuend should be made positive by adding to it the proper multiple of 10 and writing that multiple negative after the mantissa in order to preserve equality.

Example 25

Subtract the logarithm $5.785\ 63\ -\ 10$ from the logarithm $1.672\ 57$.

Solution

Adding 10 to the characteristic,

$$
\begin{array}{r}
1.672\ 57\ =\ 11.672\ 57\ -\ 10 \\
\underline{5.785\ 63\ -\ 10} \\
\text{Remainder}\ =\quad 5.886\ 94
\end{array}
$$

Example 26

Subtract the logarithm $8.675\ 43\ -\ 20$ from the logarithm $2.462\ 58$.

Solution

Adding 20 to the characteristic,

$$
\begin{array}{r}
2.462\ 58\ =\ 22.462\ 58\ -\ 20 \\
\underline{8.675\ 43\ -\ 20} \\
\text{Remainder}\ =\quad 13.787\ 15
\end{array}
$$

PROBLEMS 34-5

Add the following logarithms:

1. 2.824 12 + 3.127 37
2. 6.203 81 + 1.536 90
3. $\bar{6}$.232 86 + 4.170 33
4. 8.203 65 − 10 + 1.927 32
5. $\bar{3}$.464 80 + $\bar{2}$.808 86
6. 9.352 82 − 10 + 5.865 33 − 10

Perform the indicated subtractions:

7. 3.258 79 − 0.699 01
8. 0.434 36 − $\bar{3}$.572 82
9. $\bar{2}$.628 58 − $\bar{4}$.280 71
10. $\bar{4}$.392 64 − 2.610 24
11. 3.293 78 − (9.437 87 − 10)
12. 9.538 65 − 10 − (9.749 32 − 10)

34-17 MULTIPLICATION WITH LOGARITHMS

It was shown in Sec. 34-3 that the logarithm of a product is equal to the sum of the logarithms of the factors. This property, with the aid of the tables, is of value in multiplication.

Example 27

Find the product of 2.79 × 684

Solution

Let p = the desired product; then

$$p = 2.79 \times 684 \qquad (14)$$

Taking the logarithms of both members of Eq. (14),

$$\log p = \log 2.79 + \log 684$$

Looking up the logarithms, tabulating them, and adding them,

$$\log 2.79 = 0.445\ 60$$
$$\log 684 = \underline{2.835\ 06}$$
$$\log p = 3.280\ 66$$

Interpolating to find the value of p,

Antilog 280	= 1.90546|
Add for 6	= 266|
Add for 6	= 26|6
Antilog 3.280 66	= 1.90838|6 × 10^3

There is no need to express the result of the above interpolation beyond five significant figures. It is correct to report $p = 1.9084 \times 10^3$.

Example 28

Given $X_L = 2\pi fL$. Find the value of X_L when $f = 10\ 600\ 000$ and $L = 0.000\ 025\ 1$. Use $2\pi = 6.28$.

Solution

$$X_L = 6.28 \times 10\ 600\ 000 \times 0.000\ 025\ 1$$

Taking logarithms,

$$\log X_L = \log 6.28 + \log 10\ 600\ 000 + \log 0.000\ 025\ 1$$

Tabulating,

$$\begin{aligned}
\log 6.28 &= 0.797\ 96 \\
\log 10\ 600\ 000 &= 7.025\ 31 \\
\log 0.000\ 025\ 1 &= \underline{5.399\ 67 - 10} \\
\log X_L &= 13.222\ 94 - 10 = 3.222\ 94
\end{aligned}$$

By interpolation,

$$X_L = 1.6709 \times 10^3$$

In using logarithms, a form should be written out for all the work before beginning any computations. The form should provide places for all logarithms as taken from the table and for other work necessary to complete the problem.

34-18 COMPUTATION WITH NEGATIVE NUMBERS

Because a negative number has an imaginary logarithm, the logarithms of negative numbers cannot be used in computation. However, the numerical results of multiplications and divisions are the same regardless of the algebraic signs of the factors. Therefore, to make computations involving negative numbers, first determine whether the final result will be positive or negative. Then find the numerical value of the expression by logarithms. Consider all numbers as positive, and affix the proper sign to the result.

PROBLEMS 34-6

Compute by logarithms:

1. 8×32
2. 47×5
3. 5×50
4. 0.6×24
5. $3 \times 18 \times 0.7$
6. $12 \times (-16)$
7. $(-95) \times 2.6$
8. $0.007 \times (-22)$
9. 296×8.02
10. $0.425 \times (-0.0036)$
11. 37.7×266
12. $3250 \times (-2.03)$
13. $5.243 \times (-0.1872)$
14. $3 \times 6 \times 47$
15. $2.84 \times 72.4 \times 369$
16. $6.01 \times 444 \times 0.009\ 13$
17. $(-0.003\ 96) \times 500 \times 681$
18. $14.83 \times (-2.222) \times 0.1123$
19. $242.6 \times 471.8 \times 0.000\ 082\ 17$
20. $(-4627) \times 9126 \times (-7336)$

34-19 DIVISION BY LOGARITHMS

It was shown in Sec. 34-4 that the logarithm of the quotient of two numbers is equal to the logarithm of the dividend minus the logarithm of the divisor. This property allows division by the use of logarithms.

Example 29

Find the value of $\dfrac{948}{237}$ by using logarithms.

Solution

Let q = quotient.

Then
$$q = \frac{948}{237}$$

Taking logarithms, $\qquad \log q = \log 948 - \log 237$

Tabulating, $\qquad \log 948 = 2.976\ 81$
$$\log 237 = \underline{2.374\ 75}$$

Subtracting, $\qquad \log q = 0.602\ 06$
Taking antilogs, $\qquad q = 4$

Example 30

Find the value of $\dfrac{-24.68}{682\ 700}$ by using logarithms.

Solution

By inspection the quotient will be negative. Let

$$q = \text{quotient}$$

Then
$$q = \frac{-24.68}{682\ 700}$$

Taking logarithms, $\qquad \log q = \log 24.68 - \log 682\ 700$

Interpolating and tabulating, $\qquad \log 24.68 = 11.392\ 35 - 10$
$$\log 682\ 700 = \underline{\ \ 5.834\ 23\ }$$

Subtracting, $\qquad \log q = \ \ 5.558\ 12 - 10$

Taking antilogs, and inserting a minus sign,

$$q = -3.615 \times 10^{-5}$$

Note

$\log 24.68 = 1.392\ 35$, but 10 was added to the characteristic and subtracted after the mantissa in order to facilitate the subtraction of a larger logarithm, as explained in Sec. 34-16.

PROBLEMS 34-7

Compute by logarithms:

1. $\dfrac{12}{4}$ 2. $\dfrac{81}{9}$ 3. $\dfrac{340}{17}$

4. $\dfrac{1920}{-6.4}$

5. $\dfrac{0.245}{-0.000\ 35}$

6. $\dfrac{426}{-1137}$

7. $\dfrac{-2325}{4.023}$

8. $\dfrac{0.000\ 517\ 9}{-3.648}$

9. $\dfrac{3906}{0.000\ 800\ 2}$

10. $\dfrac{-25.83}{-0.003\ 142}$

34-20 COMBINED MULTIPLICATION AND DIVISION

When it is necessary to perform several steps of multiplication and division in one problem, make up a skeleton form indicating the logarithmic operations required. Insert the characteristics as you make up the skeleton. Then, turning to the tables, insert the logarithms of all the factors, first of the numerator, and then of the denominator. The systematic tabulation of the operations will make it easier to keep in touch with the steps to be followed.

Example 31

Evaluate, by means of logarithms,

$$N = \frac{(14.63)^2}{0.003\ 62 \times 8767}$$

Solution

First, prepare the skeleton form of the logarithmic operations:

$$\begin{aligned}
\log 14.63 &= 1. \\
&\ \underline{\times 2} \\
\log \text{numerator} &= = \\
\log 0.003\ 62 &= \overline{3}. \\
\log 8767 &= 3.\underline{} \qquad + \\
\log \text{denominator} &= = \underline{} \quad - \\
\log N &= \\
N &=
\end{aligned}$$

Then complete the tabulation, referring to the tables, and perform the calculations:

$$\begin{aligned}
\log 14.63 &= 1.165\ 23 \\
&\ \underline{\times 2} \\
\log \text{numerator} &= 2.330\ 46 = 2.330\ 46 \\
\log 0.003\ 62 &= \overline{3}.558\ 71 \\
\log 8767 &= \underline{3.942\ 85} \\
\log \text{denominator} &= 1.501\ 56 = \underline{1.501\ 56} \quad - \\
\log N &= 0.828\ 90 \\
N &= 6.7436
\end{aligned}$$

Example 32

Evaluate, by means of logarithms

$$\phi = \frac{(64.28)(0.009\ 73)}{(4006)(0.051\ 34)(0.002\ 085)}$$

Solution

Make up the skeleton of the operations:

$$\log 64.28 = 1.$$
$$\log 0.009\ 73 = \bar{3}. \underline{\qquad} \quad +$$
$$\log \text{numerator} = \qquad\qquad =$$
$$\log 4006 = 3.$$
$$\log 0.051\ 34 = \bar{2}.$$
$$\log 0.002\ 085 = \bar{3}. \underline{\qquad} \quad +$$
$$\log \text{denominator} = \qquad = \underline{\qquad} \quad -$$
$$\log \phi =$$
$$\phi =$$

Then, from the tables, complete the tabulation and perform the calculations:

$$\log 64.28 = 1.808\ 08$$
$$\log 0.009\ 73 = \bar{3}.988\ 11 \quad +$$
$$\log \text{numerator} = \bar{1}.796\ 19 = \bar{1}.796\ 19$$
$$\log 4006 = 3.602\ 70$$
$$\log 0.051\ 34 = \bar{2}.710\ 46$$
$$\log 0.002\ 085 = \bar{3}.319\ 16 \quad +$$
$$\log \text{denominator} = \bar{1}.632\ 32 = \bar{1}.632\ 32 \quad -$$
$$\log \phi = 0.163\ 87$$
$$\phi = 1.458\ 4$$

PROBLEMS 34-8

Use logarithms to compute the results of the following:

1. $\dfrac{2.4 \times 3.5}{1.7}$

2. $\dfrac{5.6 \times 8.9}{4.7 \times 9.3}$

3. $\dfrac{22.1 \times 1.08}{12.65 \times 0.78}$

4. $\dfrac{86.3 \times 0.0297}{0.0379}$

5. $\dfrac{-0.536}{734.4 \times 0.005\ 83}$

6. $\dfrac{2.006}{3.142 \times 0.833}$

7. $\dfrac{0.000\ 009\ 207}{4.98 \times 0.000\ 000\ 707}$

8. $\dfrac{1}{6.28 \times 427\ 000\ 000 \times 0.000\ 050}$

9. $\dfrac{1}{4.73 \times 5222 \times 0.000\ 680\ 7}$

10. $\dfrac{6.28 \times 0.000\ 159 \times 326}{0.003\ 68 \times 436 \times 0.0278}$

34-21 RAISING TO A POWER BY LOGARITHMS

It was shown in Sec. 34-5 that the logarithm of a power of a number is equal to the logarithm of the number multiplied by the exponent of the power.

Example 33

Find by logarithms the value of 12^3.

Solution

$$\log 12^3 = 3 \log 12$$
$$\log 12 = 1.079\ 18$$

M: 3

$$\frac{3}{3.237\ 54} = \log 1728$$
$$\therefore 12^3 = 1728$$

Example 34

Find by logarithms the value of $0.056\ 3^5$.

Solution

$$\log 0.056\ 3^5 = 5 \log 0.056\ 3$$
$$\log 0.056\ 3 = 8.750\ 51 - 10$$

M: 5

$$5 \log 0.0563 = \frac{5}{43.752\ 55 - 50}$$
$$= 3.752\ 55 - 10$$
$$\text{antilog } 3.7525 - 10 = 5.6566 \times 10^{-7}$$
$$\therefore 0.0563^5 = 5.6566 \times 10^{-7}$$

Example 35

Find by logarithms the value of 5^{-3}.

Solution

By the laws of exponents,

Then

$$5^{-3} = \frac{1}{5^3}$$

$$\log 5^{-3} = \log 1 - \log 5^3$$
$$= \log 1 - 3 \log 5$$

Multiplying,

$$\log 5 = 0.698\ 97$$

$$3 \log 5 = \frac{3}{2.096\ 91}$$

$$\log 1 = 10.000\ 00 - 10$$
$$3 \log 5 = 2.096\ 91$$
$$\log 5^{-3} = \frac{}{7.903\ 09 - 10}$$
$$\text{antilog } 7.903\ 09 - 10 = 0.008$$
$$\therefore 5^{-3} = 0.008$$

34-22 EXTRACTING ROOTS BY LOGARITHMS

It was shown in Sec. 34-6 that the logarithm of a root of a number is equal to the logarithm of the number divided by the index of the root.

Example 36

Find by logarithms the value of $\sqrt[3]{815}$.

Solution

By the laws of exponents,

$$\sqrt[3]{815} = 815^{\frac{1}{3}}$$

Then

$$\log 815^{\frac{1}{3}} = \frac{1}{3} \log 815$$

$$\log 815 = 2.911\ 15$$

$$\frac{1}{3} \log 815 = \frac{2.911\ 15}{3} = 0.970\ 38$$

$$\text{antilog } 0.970\ 38 = 9.3406$$

$$\therefore \sqrt[3]{815} = 9.34 \text{ to three significant figures}$$

Example 37

Find by logarithms the value of $\sqrt[4]{0.009\ 55}$.

Solution

$$\sqrt[4]{0.009\ 55} = 0.009\ 55^{\frac{1}{4}}$$

Then

$$\log 0.009\ 55^{\frac{1}{4}} = \frac{1}{4} \log 0.009\ 55$$

$$\log 0.009\ 55 = 7.980\ 00 - 10$$

$$\frac{1}{4} \log 0.009\ 55 = 1.995\ 00 - 2.5$$

This result, though correct, is not in the standard form for a negative characteristic. This inconvenience can be obviated by writing the logarithm in such a manner that the negative part when divided results in a quotient of -10. Thus,

$$\log 0.009\ 55 = 7.980\ 00 - 10$$

would be written

$$\log 0.009\ 55 = 37.980\ 00 - 40$$

Since it is necessary to divide the logarithm by 4 in order to obtain the fourth root, 30 was subtracted from the negative part to make it exactly divisible by 4. Therefore, to preserve equality, it was necessary to add 30 to the positive part. Then

$$\log \sqrt[4]{0.009\ 55} = \frac{37.980\ 00 - 40}{4}$$

$$= 9.495\ 00 - 10$$

$$\text{antilog } 9.495\ 00 - 10 = 0.312\ 61$$

$$\therefore \sqrt[4]{0.009\ 55} = 0.312\ 61$$

34-23 FRACTIONAL EXPONENTS

Computations involving fractional exponents are made by combining the operations of raising to powers and extracting roots.

Example 38

Find by logarithms the value of $\sqrt[4]{0.0542^3}$.

Solution

$$\sqrt[4]{0.0542^3} = 0.0542^{\frac{3}{4}}$$

Then
$$\log 0.0542^{\frac{3}{4}} = \frac{3}{4} \log 0.0542$$

$$\log 0.0542 = 8.734\ 00 - 10$$
$$3 \log 0.0542 = 26.202\ 00 - 30$$

Adding 10 to the characteristic and subtracting 10 from the negative part in order to make it evenly divisible by 4,

$$3 \log 0.0542 = 36.202\ 00 - 40$$
$$\frac{3}{4} \log 0.0542 = \frac{36.202\ 00 - 40}{4}$$
$$= 9.050\ 50 - 10$$
$$\text{antilog } 9.050\ 50 - 10 = 0.112\ 33$$
$$\therefore \sqrt[4]{0.0542^3} = 0.112$$

Instead of adding 10 to the characteristic, as above, it would also have been correct to subtract 10 from the characteristic and add 10 to the mantissa, and thus obtain $16.202\ 00 - 20$. It is immaterial what numbers are added and subtracted as long as the resulting negative characteristic portion will yield an integral quotient.

PROBLEMS 34-9

Use logarithms to compute the results of the following:

1. 12.8^2 2. 82.3^5 3. 0.0176^4 4. 0.463^6

5. $\sqrt[3]{180}$ 6. $\sqrt[4]{782}$ 7. $1237^{\frac{1}{4}}$ 8. $0.643^{\frac{1}{5}}$

9. $0.862^{\frac{1}{2}}$ 10. $\sqrt[6]{4258}$ 11. $127^{\frac{2}{3}}$ 12. $\sqrt[3]{2.61^4}$

13. $30.6^{\frac{3}{2}}$ 14. $164^{\frac{2}{3}}$

15. $\sqrt{\dfrac{196 \times 0.083}{12.1}}$ 16. $\sqrt[3]{\dfrac{(-0.436) \times 30.8}{0.0287}}$

17. $\left(\dfrac{224}{363}\right)^{\frac{3}{2}}$ 18. $\left(\dfrac{9764}{238.3}\right)^{1.5}$

19. $\sqrt[6]{0.000\ 028\ 6} \times \sqrt[4]{629}$ 20. $\left(\dfrac{0.000\ 000\ 587}{0.000\ 001\ 72}\right)^{\frac{5}{2}}$

34-24 PRECAUTIONS TO BE OBSERVED

We have now investigated the common operations involving the use of logarithms in performing mathematical computations. We know that, to use logarithms to perform multiplications, we add logarithms; to perform division, we subtract logarithms; to raise to a power, we multiply the logarithm by the power; and to extract a root, we divide the logarithm by the root.

There are times, however, when a problem introduces the *use of logarithms*, apart from the employment of logarithms in computing an arithmetical solution. Consider carefully the following examples:

Example 39
Compute, by means of logarithms, 125×13.6

Solution
This is a standard multiplication problem of the type which we successfully mastered in Problems 34-6. We find the logarithm of each number, add the logarithms, and take the antilogarithm of the sum to determine the value:

$$\begin{array}{r} \log 125 = 2.096\ 91 \\ \log 13.6 = \underline{1.133\ 54} \\ \log \text{answer} = 3.230\ 45 \\ \text{Answer} = \text{antilog } 3.230\ 45 = 1.7 \times 10^3 \end{array}$$

Example 40
Compute $(\log 125)(\log 13.6)$.

Solution
This problem calls for us to multiply the logarithm of 125, whatever that may be, by the logarithm of 13.6, whatever that may be. We can determine what these logarithms are and rewrite the problem:

$$(\log 125)(\log 13.6) = (2.096\ 91)(1.133\ 54)$$

In other words, we have replaced the log expressions in the problems with the numbers which *are* the logarithms as called for. Then, having made this substitution, we perform the actual required operation, that is, multiply 2.096 91 by 1.133 54, to obtain 2.38.

Note carefully that this problem did not introduce the addition of logarithms and the taking of antilogarithms in order to arrive at an answer. It *may* have suited our convenience to perform the necessary multiplication by means of logarithms, but that would introduce an additional problem.

Example 41
Compute, by means of logarithms,

$$(\log 125)(\log 13.6)$$

Solution

As in Example 40, first rewrite the problem:

$$(\log 125)(\log 13.6) = (2.096\ 91)(1.133\ 54)$$

To perform this multiplication operation by means of logarithms, we follow the usual procedures of interpolation, addition of logarithms, and subsequent finding of the antilogarithm:

$$
\begin{aligned}
\log 2.096\ 91 &= 0.321\ 633\ 1 \\
\log 1.133\ 54 &= \underline{0.054\ 448\ 4} \\
\log \text{answer} &= 0.376\ 081\ 5 \\
\text{Answer} &= \text{antilog } 0.376\ 081\ 5 \\
&= 2.377\ 284\ 13 = 2.38
\end{aligned}
$$

In Example 41, because the problem called for a logarithmic performance of arithmetic, we performed logarithmic calculations. In Example 40 we arrived at the same value by other methods, despite the fact that logarithms *appeared* in the problem.

It is essential that you be aware at all times of the difference between performing operations by means of logarithms and performing operations which somehow involve the logarithms of numbers. This difference will appear in several of the problems of Chap. 35. Problems 34-10 are included at this point to give you practice in recognizing the different types of problems which may arise.

PROBLEMS 34-10

Evaluate the following:

1. $\log 37.2 + \log 9.83$ 2. $\log 16.3 - \log 7.03$

3. $\log 3.68 - \log 5.66$ 4. $(\log 87.2)(\log 15.7)$

5. $\dfrac{\log 265}{\log 17.6}$ 6. $\dfrac{\log 20.3}{\log 65.2}$

7. $(\log 3.97)\left(\dfrac{\log 16.3}{\log 8.6}\right)$ 8. $(\log 224)^2$

9. $\log 224^2$ 10. $\dfrac{\log 0.987}{\log 3.5}$

11 to 17. Evaluate Probs. 4 to 10 by using logarithms for all calculations.

34-25 CHANGE OF BASE

In Problems 34-1 we found logarithms of numbers to many bases besides 10, and it is often convenient for us to be able to find the logarithms of numbers to certain bases other than 10 without developing a set of tables for other bases. An interesting development shows us how this may be achieved.

$$N = a^x \qquad (15)$$

which we may rewrite

$$x = \log_a N \qquad (16)$$

Taking logarithms of both sides of Eq. (15) to the base b:

$$\log_b N = \log_b a^x \qquad (17)$$

Substituting Eq. (11) into Eq. (17),

$$\log_b N = x \log_b a \qquad (18)$$

Substituting Eq. (16) into Eq. (18),

$$\log_b N = \log_a N \cdot \log_b a \qquad (19)$$

Since it can be shown that

$$\log_b a = \frac{1}{\log_a b} \qquad (20)$$

Equation (19) may be written in the form

$$\log_b N = \frac{\log_a N}{\log_a b} \qquad (21)$$

If, then, we have a table of logarithms to the base 10 and find it necessary to produce the logarithm of any number to any other base b, we simply divide the logarithm to the base 10 of the given number by the logarithm to the base 10 of the other base number b:

$$\log_b N = \frac{\log_{10} N}{\log_{10} b} \qquad (22)$$

We are primarily concerned with the natural system of logarithms, which has for its base the number $\varepsilon = 2.718\ 28 \ldots$ (Sec. 34-9). Many relationships in electronics as well as other branches of science involve logarithms to this base.

Although we will be developing and using logarithms to the base ε in the Sec. 34-26, you will often find that only tables of logarithms to the base 10 are immediately available. Using the relationship expressed in formulas (19) and (21), you will be able to perform the necessary operations.

$$\log_\varepsilon N = 2.302\ 59 \log_{10} N \qquad (23)$$
$$\log_{10} N = 0.434\ 29 \log_\varepsilon N \qquad (24)$$

Example 42

$$\log_\varepsilon 1000 = 2.302\ 59 \log_{10} 1000$$
$$= 2.305\ 59 \times 3$$
$$= 6.907\ 77$$

Example 43

$$\log_{10} 100 = 0.434\ 29 \log_\varepsilon 100$$
$$= 0.434\ 29 \times 4.6052$$
$$= 2.0000$$

Example 44

Given $x = \log_\varepsilon 48$. Solve for x.

Solution

$$\log_\varepsilon 48 = 2.302\ 59 \log_{10} 48$$
$$= 2.302\ 59 \times 1.681\ 24$$
$$x = 3.8712$$

34-26 NATURAL LOGARITHMS

Because so many calculations in electronics do involve logarithms to the base ε, you will use your calculator keys LN or lnx often. If you do not have these keys, keep Eqs. (23) and (24) close at hand. A few special notes will simplify your use of these natural logarithms:

1. The laws of logarithms [Eqs. (6), (8), (11), and (13)] apply to any logarithmic system, regardless of the base used. Therefore, natural logarithms may be used instead of common logarithms, if you prefer, for any problem involving multiplication, division, raising to powers, or extracting roots.
2. For convenience, we often replace the notation $\log_\varepsilon$ with the special symbol ln, pronounced lon.
3. Since $\ln \varepsilon = 1$, the characteristics in natural logarithms do not represent powers of 10. (They represent powers of ε.) Accordingly, your caculator gives the *entire* natural logarithm of a number, and not just its mantissa.

Example 45

$$\ln 2.70 = 0.993\ 25$$
$$\ln 2.72 = 1.000\ 63$$
$$\ln 5.05 = 1.619\ 39$$
$$\ln 7.38 = 1.998\ 77$$
$$\ln 7.39 = 2.000\ 13$$

4. Because the characteristic represents a power of ε, it is necessary to build up natural logarithms of very large and very small numbers, using Eq. (6) for the purpose.

Example 46

Using your calculator, find the natural logarithm of 127.4

Solution

Key   LN or lnx
Key   127.4
Read   4.847 33

Example 47
Find ln 0.001 274.

Solution
Key   LN or lnx
Key   .001 274
Read $-6.665\ 59$

34-27 GRAPH OF $y = \log_{10} x$

The graph of $y = \log_{10} x$ is shown in Fig. 34-1. A study of the graph shows the following:

1. A negative number has no real logarithm.
2. The logarithm of a positive number less than 1 (a decimal between 0 and 1) is negative.
3. The logarithm of 1 is zero.
4. The logarithm of a positive number greater than 1 is positive.
5. As the number approaches zero, its logarithm decreases without limit.
6. As the number increases indefinitely, its logarithm increases without limit.

Is the method of interpolation that treats a short distance on the logarithmic curve as a straight line sufficiently accurate for computation?

34-28 LOGARITHMIC EQUATIONS

An equation in which there appears the logarithm of some expression involving the unknown quantity is called a *logarithmic equation*.

Logarithmic equations have wide application in electric circuit analysis. In addition, the communications engineer uses them in computations involving decibels and transmission line characteristics.

Example 48
Solve the equation $4 \log x + 3.796\ 00 = 4.699\ 09 + \log x$.

Solution

Given	$4 \log x + 3.796\ 00 = 4.699\ 09 + \log x$
Transposing,	$4 \log x - \log x = 4.699\ 09 - 3.796\ 00$
Collecting terms,	$3 \log x = 0.903\ 09$
D: 3,	$\log x = 0.301\ 03$

Using a calculator,
Key   0.301 03
Key   INV LOG *or* 10^x
Read   2.00

In solving logarithmic equations, the logarithm of the unknown, as $\log x$ in Example 48, is considered as any other literal *coefficient*. That is, in general, the rules for solving ordinary algebraic equations apply in logarithmic equations.

A common error made by students in solving logarithmic equations is confusing coefficients of logarithms with coefficients of the unknown. For example,

$$3 \log x \neq \log 3x$$

because the left member denotes the product of 3 times the logarithm of x, whereas the right member denotes the logarithm of the quantity 3 times x, that is, log $(3x)$.

Example 49

Given $500 = 276 \log \dfrac{d}{0.05}$. Solve for d.

Solution 1

Given	$500 = 276 \log \dfrac{d}{0.05}$
Then	$500 = 276 (\log d - \log 0.05)$
D: 276,	$1.81 = \log d - \log 0.05$
Transposing,	$\log d = 1.81 + \log 0.05$

Substituting $8.698\ 97 - 10$ for log 0.05,

$$\log d = 1.81 + 8.698\ 97 - 10$$

Collecting terms, $\log d = 0.508\ 97$

Using a calculator, Key 10^x, read $d = 3.23$

Solution 2

Given	$500 = 276 \log \dfrac{d}{0.05}$
D: 276,	$1.81 = \log \dfrac{d}{0.05}$
Taking antilogs of both members,	$64.57 = \dfrac{d}{0.05}$
Solving for d,	$d = 3.23$

34-29 EXPONENTIAL EQUATIONS

An equation in which the unknown appears in an exponent is called an *exponential equation*. In the equation

$$x^3 = 125$$

it is necessary to find some value of x that, when cubed, will equal 125. In this equation *the exponent is a constant*.

In the *exponential equation*

$$5^x = 125$$

the situation is different. The *unknown appears as an exponent*, and it is now necessary to find what power 5 must be raised to obtain 125.

Some exponential equations can be solved by inspection. For example, the value of x in the foregoing equation is 3. In general, taking the logarithms of both sides of an exponential equation will result in a logarithmic equation that can be solved by the usual methods.

Example 50

Given $4^x = 256$. Solve for x.

Solution

Given $$4^x = 256$$

Taking the logarithms of both members,

$$\log 4^x = \log 256$$

or $$x \log 4 = \log 256$$

D: log 4, $$x = \frac{\log 256}{\log 4}$$

From the log tables, $$x = \frac{2.408}{0.602} = 4$$

Using a calculator,
Key   256
Key   LOG
Read   2.40824

Key   4
Key   LOG
Read   0.602 06

$$x = \frac{2.408\ 24}{0.602\ 06} = 4$$

Check

$$4^4 = 256$$

Example 51

Given $5^{x-3} = 52$. Solve for x.

Solution

Given $$5^{x-3} = 52$$

Taking the logarithms of both members,

$$\log 5^{x-3} = \log 52$$

or $$(x - 3) \log 5 = \log 52$$

D: log 5, $$x - 3 = \frac{\log 52}{\log 5}$$

Using a calculator, $$x - 3 = \frac{1.716\ 00}{0.698\ 97}$$

A: 3, $$x = \frac{1.716\ 00}{0.698\ 97} + 3$$

or $$x = 5.455$$

How would you check this solution?

PROBLEMS 34-11

Solve the following equations:

1. $x = \log_\varepsilon 226$
2. $x = \log_\varepsilon 4.38$
3. $\log x + 3 \log x = 6$
4. $\log x + \log 6x = 8.5$

(**Hint** $\log 6x = \log 6 + \log x$)

5. $\log 5x + 2 \log x = 6.88$
6. $\log \dfrac{P}{3} = 0.573$
7. $\log \dfrac{P_1}{14} = 2.86$
8. $\log \dfrac{12}{V} = 3$
9. $\log x^2 - \log x = 6.75$
10. $x^4 = 462$
11. $4^x = 167$
12. $5^x = 37.3$
13. $2^m = 0.88$
14. $3^{q-3} = 14$
15. $4^{3x} = 14$
16. $M^{2.3} = 25$
17. $x = \log_6 1296$
18. $x = \log_3 2187$
19. If $10 \log L_2 = \frac{3}{2}(10 \log L_1)$, solve for L_1.
20. If $20 \log \dfrac{2Z_1}{2Z_1 - Z_a} = 20 \log \dfrac{-Z_b}{-Z_b + \dfrac{Z_1}{2}}$,

    solve for Z_1 in terms of Z_a and Z_b.

21. If $V_g = \dfrac{2.3T}{11\ 600} \log \dfrac{I_0}{I_g}$, solve for I_0.

22. If $i = \dfrac{V}{L}\, t\varepsilon^{S_c t}$, solve for S_c.

23. If $i_c = \dfrac{V}{R}\, \varepsilon^{-(t/RC)}$, solve for (a) V, (b) C, (c) t.

24. If $I_k = AT^2\varepsilon^{-(B/T)}$, solve for (a) A, (b) B.

25. If $i_L = \dfrac{V}{R}(1 - \varepsilon^{-Rt/L})$, solve for (a) V, (b) L, (c) t.

26. If $q = CV(1 - \varepsilon^{-t/RC})$, solve for (a) V, (b) R, (c) t.

27. If $I_p + I_g = K\left(V + \dfrac{V_p}{\mu}\right)^{\frac{3}{2}}$, solve for (a) V, (b) V_p, (c) μ.

28. In an inductive circuit, the equation for the growth of current is given by

$$i = \frac{V}{R}(1 - \varepsilon^{-Rt/L}) \qquad \text{A} \qquad (25)$$

where i = current, A

t = any elapsed time after switch is closed, s
V = constant impressed voltage, V
L = inductance of the circuit, H
R = circuit resistance, Ω
ε = base of natural system of logarithms

A circuit of 0.75-H inductance and 15-Ω resistance is connected across a 12-V battery. What is the value of the current at the end of 0.06 s after the circuit is closed?

Solution: The circuit is shown in Fig. 34-2.

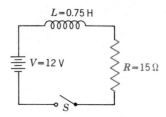

Fig. 34-2 Circuit of Probs. 28 to 31.

Given

$$i = \frac{V}{R}(1 - \varepsilon^{-(Rt/L)})$$

Substituting the known values,

$$i = \frac{12}{15}(1 - \varepsilon^{-(15 \times 0.06/0.75)})$$

$$i = 0.8(1 - \varepsilon^{-1.2})$$

Multiplying,

$$i = 0.8 - 0.8\varepsilon^{-1.2}$$

or

$$i = 0.8 - \frac{0.8}{\varepsilon^{1.2}} \qquad (26)$$

Now evaluate $\varepsilon^{1.2}$,

$$\log_{10}\varepsilon^{1.2} = 1.2 \log_{10}\varepsilon = 1.2 \times 0.434\ 29$$
$$= 0.521\ 15$$

Taking antilogs, $\quad \varepsilon^{1.2} = 3.32$

Substituting the value of $\varepsilon^{1.2}$ in Eq. (26),

$$i = 0.8 - \frac{0.8}{3.32} = 0.559 \text{ A}$$

The growth of the current in the circuit of Fig. 34-2 is shown graphically in Fig. 34-3.

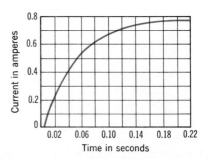

Fig. 34-3 Graph of current in *RL* circuit of Prob. 28.

29. The inductance of the circuit in Fig. 34-2 is halved and the resistance is thus reduced to 0.71 times its original value. If other circuit values remain the same, what will be the value of the current 0.08 s after the switch is closed?

30. Using the circuit values for the circuit of Fig. 34-2, what will be the value of the current (a) 0.005 s after the switch is closed and (b) 0.5 s after the switch is closed?

31. In the circuit of Fig. 34-2, after the switch is closed, how long will it take the current to reach 50% of its maximum value?

32. If $\frac{L}{R}$ is substituted for t in the equation

$$i = \frac{V}{R}(1 - \varepsilon^{-(Rt/L)})$$

show that the value of the current will be 63.2% of its steady-state value. The numerical value of L/R in seconds is known as the *time constant* of the inductive circuit. It is useful in determining the rapidity with which current rises or falls in one inductive circuit in comparison with others.

33. A 220-V generator shunt field has an inductance of 12 H and a resistance of 80 Ω. How long after the line voltage is applied does it take for the current to reach 75% of its maximum value?

34. A relay of 1.2-H inductance and 500-Ω resistance is to be used for keying a radio transmitter. The relay is to be operated from a 110-V line, and 0.175 A is required to close the contacts. How many words per minute will the relay carry if each word is considered as five letters of five impulses per letter? The time of opening of the contacts is the same as the time required to close them.

Hint $0.175 = \frac{110}{500}(1 - \varepsilon^{-(500t/1.2)})$. t is the time required to close the relay.

35. How many words per minute would the relay of Prob. 34 carry if 50 Ω resistance were connected in series with it? The line voltage remains at 110 V.

36. In a capacitive circuit the equation for the current is given by

$$i = \frac{V}{R}\varepsilon^{-(t/RC)} \qquad A \qquad\qquad (27)$$

where i = current, A
  t = any elapsed time after switch is closed, s
  V = impressed voltage, V
  C = capacitance of the circuit, F
  R = circuit resistance, Ω
  ε = base of natural system of logarithms

A capacitance of 500 μF in series with 1 kΩ is connected across a 50-V generator.
(a) What is the value of the current at the instant the switch is closed?

Hint $t = 0$.

(b) What is the value of the current 0.02 s after the switch is closed? The circuit is shown in Fig. 34-4.

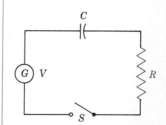

Fig. 34-4 Circuit of Probs. 36 and 37.

37. In the circuit of Fig. 34-3, how long after the switch is closed will the current have decayed to 30% of its initial value if $V = 110$ V, $R = 500$ Ω, $C = 20$ μF, and $i = \dfrac{0.3\,V}{R}.$ $t = ?$

Solution: $i = \dfrac{0.3V}{R} = \dfrac{0.3 \times 110}{500} = 0.066$ A

Substituting in Eq. (27), $0.066 = \dfrac{110}{500}\varepsilon^{-(t/500 \times 20 \times 10^{-6})}$

Simplifying, $0.066 = 0.22\varepsilon^{-(t/10^{-2})}$

or $0.066 = 0.22\varepsilon^{-100t}$

D: 0.22, $0.3 = \varepsilon^{-100t}$

By the law of exponents, $0.3 = \dfrac{1}{\varepsilon^{100t}}$

M: ε^{100t} $0.3\varepsilon^{100t} = 1$

D: 0.3, $\varepsilon^{100t} = 3.33$

Taking logarithms, $\log_{10} \varepsilon^{100t} = \log_{10} 3.33$

That is, $100t \log_{10} \varepsilon = \log_{10} 3.33$

Then $100t \times 0.4343 = 0.5224$

or $43.43t = 0.5224$

$\therefore t = 0.012$ s

The decay of the current in the circuit of Fig. 34-4 is shown graphically in Fig. 34-5.

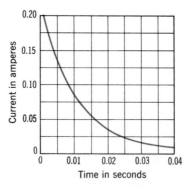

Fig. 34-5 Graph of current in *RC* circuit of Prob. 37.

38. A 20-μF capacitor in series with a resistance of 680 Ω is connected across a 110-V source.
(*a*) What is the initial value of the current?
(*b*) How long after the switch is closed will the current have decayed to 36.8% of its initial value?
(*c*) Is the time obtained in (*b*) equal to *CR* s? The product of *CR*, in seconds, is the time constant of a capacitive circuit.

39. The quantity of charge on a capacitor is given by

$$q = CV(1 - \varepsilon^{-(t/CR)}) \qquad C \qquad (28)$$

where q is the quantity of electricity in coulombs.

(a) Calculate the charge q in coulombs on a capacitor of 50 μF in series with a resistance of 3.3 kΩ, 0.008 s after being connected across a 70-V source.

(b) What is the voltage across the capacitor at the end of 0.02 s?

40. A key-click filter consisting of a 2-μF capacitor in series with a resistance is connected across the keying contacts of a transmitter. If the average time of impulse is 0.004 s, calculate the value of the series resistance required in order that the capacitor can discharge 90% in this time.

Hint Under steady-state conditions, $q = CV$. Then

$$0.9CV = CV(1 - \varepsilon^{-(t/RC)})$$

41. The emission current in amperes of a heated filament is given by

$$I = AT^2 \varepsilon^{-(B/T)} \qquad A \qquad (29)$$

For a tungsten filament, $A = 60$ and $B = 52\ 400$. Find the current of such a filament at a temperature $T = 2500$ K.

42. An important triode formula is

$$I_p + I_g = K\left(V_g + \frac{V_p}{\mu}\right)^{\frac{3}{2}} \qquad A \qquad (30)$$

where I_p = plate current, A
  I_g = grid current, A
  V_g = grid voltage, V
  V_p = plate voltage, V
  μ = amplification factor

Calculate $I_p + I_g$ if $K = 0.0005$, $V_g = 6$ V, $V_p = 270$ V, and $\mu = 15$.

43. The diameter of No. 0000 wires is 11.68 mm, and that of No. 36 is 0.127 mm. There are 38 wire sizes that are between No. 0000 and No. 36; therefore, the ratio between cross-sectional areas of successive sizes is the thirty-ninth root of the ratio of the area of No. 0000 wire to that of No. 36 wire, or $\sqrt[39]{\dfrac{11.68^2}{0.127^2}}$. Compute the value of this ratio. Because this ratio is nearly equal to $\sqrt[3]{2}$, we can use the approximation that the cross-sectional area of a wire doubles for every decrease of three sizes, as explained in Sec. 9-4. Calculate the percent error introduced by using $\sqrt[3]{2}$.

Applications of Logarithms

35

We have seen that logarithms can be extremely useful in the performance of arithmetic operations. Multiplication, division, raising to powers, and extracting roots are important applications of logarithms which will be explored further in this chapter.

Similarly, proficiency in the use of logarithmic equations is an essential part of the electronics technician's mathematical toolbox. The broad application of these equations to computers, power measurement, amplification, attenuators, and transmission lines all testify to the equations' importance.

In this chapter, we will see how logarithmic calculations are applied to the fields mentioned above and we will investigate briefly two extremely important applications of logarithms to our everyday work in electronics—preferred values and decibels.

35-1 PREFERRED VALUES

In the determination of the values of resistors, capacitors, and inductors which may be required in a circuit, such as those calculated in Sec. 15-2, we often find that the values available off the shelf are not identical with our calculated values. We may desire to have a 620-Ω resistor, and the lab assistant says, "Use a 560- or a 680-Ω. Either will be close enough." How can the lab assistant say so, unhesitatingly? How does the lab assistant know? In other words, how do we arrive at *preferred values*?

Under the prompting of the industry as a whole, the Electrical Industries Association has established lists of suggested figures for the guidance of manufacturers and technicians. Several series of values are normally listed, depending on the quality of service required. Most commonly used are the $R6$ and $R12$ series, which list the 6 and 12 values that cover all the requirements for 20% and 10% tolerances, respectively. Becoming more and more called upon is the $R24$ series, which gives values for 5% tolerance. Naturally, the price of the more exact values is considerably higher than the price of the other values, and the $R6$ and $R12$ values meet the demands of ordinary service quite satisfactorily.

Each of the series is developed from a logarithmic progression based on an appropriate root of 10. To develop the $R6$ series, we take the $\frac{1}{6}$, $\frac{2}{6}$,

$\frac{3}{6}$, $\frac{4}{6}$, $\frac{5}{6}$, and $\frac{6}{6}$ roots of 10, in order. Table 35-1 shows the development of the $R6$ series of preferred values.

TABLE 35-1   *R36* SERIES OF PREFERRED VALUES

x	$10^{\frac{x}{6}}$	Preferred Value	Difference	Percent Difference	Max % Error
0	1.000	1.0	0.5	50	± 20
1	1.468	1.5	0.7	46	18.9
2	2.155	2.2	1.1	50	20
3	3.162	3.3	1.4	42.5	17.5
4	4.642	4.7	2.1	44.6	18.3
5	6.813	6.8	3.2	47	19.1
6	10.	10	5.0	50	20
		15			

You should confirm, by using logarithms, that $10^{\frac{4}{6}} \cong 4.642$. Now, the calculated values may be rounded off to easy-to-remember two-significant-figure numbers in order to arrive at the preferred values. Naturally, all these values may be multiplied by any power of 10, so that memorizing six numbers is all that is needed to cover the entire range of 20% values. The maximum error of $\pm 20\%$ has been arrived at by choosing desired values midway between the two preferred values and determining the percentage error. If we required a 4-kΩ resistor, choosing either 3.3-kΩ or 4.7-kΩ would not introduce more than a 20% error. Obviously, then, any value closer to a preferred value than one midway between the two must be closer than 20% tolerance. The advantages to manufacturers, sales agencies, and technicians will be obvious at once.

When greater accuracy (less tolerance) is required, we may use the $R12$ series for $\pm 10\%$ or even the $R24$ series for $\pm 5\%$ values. Naturally, the 5% shows all the values in the 10% and 20% series plus intermediate values to round out the series.

PROBLEMS 35-1

1. Using successive twelfth roots of 10, by logarithms, list the preferred values and the maximum percentage errors for the $R12$ series of preferred values.
2. Using successive twenty-fourth roots of 10, by logarithms, list the preferred values and the maximum percentage errors for the $R24$ series of preferred values.
3. The standard published values of capacitors made by a prominent manufacturer follow the $R10$ series. By using successive tenth roots of 10, determine the nominal value of electrolytic capacitors available from this manufacturer between 100 and 1000 pF. What will be the probable published tolerance?
4. The permeability ratings of a popular line of potentiometer cores follows the $R5$ series. By using successive fifth roots of 10, develop the nominal values between 1 and 100 mH. What will be the probable published tolerance?

35-2 POWER RATIOS—THE DECIBEL

The Weber-Fechner law states that "the minimum change in stimulus necessary to produce a perceptible change in response is proportional to the stimulus already existing." With respect to our sense of hearing, this means that the ear considers as equal changes of sound intensity those changes which are in the same *ratio*.

The above is more easily understood from a consideration of sound intensities. Any volume of sound must be changed approximately 25% before the ear notes a change in volume. If the volume is increased by this amount, in order for the ear to detect another increase in volume, the new value must be increased by an additional 25%. For example, the output of an amplifier delivering 16 W would have to be increased to a new output of 20 W in order for the ear to discern the increase in volume. Then, in order for the ear to detect an additional increase in volume, the output would have to be increased 25% of 20 W to a new output of 25 W.

From the foregoing it is apparent that a *change* of volume, for example, from 10 to 20 mW (a 10-mW change), would seem the same as the *change* from 100 to 200 mW (a 100-mW change) because $\frac{20}{10} = \frac{200}{100}$. Since these changes in hearing response are equally spaced on a logarithmic scale, it follows that the ear responds logarithmically to variations in sound intensity. Therefore, any unit used for expressing power gains or losses in communication circuits must, in order to be practical, vary logarithmically.

One of the earliest of such units was the international transmission unit, the *bel* (B), so called to honor the inventor of the telephone, Alexander Graham Bell. The definition of the bel is

$$\text{bel} = \log_{10} \frac{P_2}{P_1}$$

where P_1 is the initial, or reference, power and P_2 is the final, or referred, power.

In normal practice, the number of bels is quite small and is invariably a decimal number; and so a derived unit, the *decibel*, is used as the practical indicator of power ratio. The abbreviation for decibel is dB. A difference of 1 dB between two sound intensities is just discernible to the ear. Since deci means one-tenth, a decibel is one-tenth the size of a bel, and

$$\text{Number of decibels} = \text{dB} = 10 \log \frac{P_2}{P_1} \tag{1}$$

You should refer to the second paragraph of this section and prove that the difference between two discernible sound intensities is 0.969 dB.

Example 1
A power of 10 mW is required to drive an AF amplifier. The output of the amplifier is 120 mW. What is the gain, expressed in decibels?

Solution
$P_1 = 10$ mW, and $P_2 = 120$ mW. dB $= ?$ Substituting in Eq. (1),

$$\text{dB} = 10 \log \frac{120}{10} = 10 \log 12$$
$$= 10.8 \text{ dB gain}$$

Example 2

A network has a loss of 16 dB. What power ratio corresponds to this loss?

Solution

Given

$$dB = 10 \log \frac{P_2}{P_1} \qquad (1)$$

Substituting 16 for dB,

$$16 = 10 \log \frac{P_2}{P_1}$$

D: 10,

$$1.6 = \log \frac{P_2}{P_1}$$

Taking antilogs of both members,

$$39.8 = \frac{P_2}{P_1}$$

Thus, a loss of 16 dB corresponds to a power ratio of 39.8:1.

Because dB is 10 times the log of the power ratio, it is evident that power ratios of 10 = 10 dB, 100 = 20 dB, 1000 = 30 dB, etc. Therefore, it could have been determined by inspection that the 16-dB loss in the preceding example represented a power ratio somewhere between 10 and 100. This is evident by the figure 1 of 16 dB. The second digit 6 of 16 dB is ten times the logarithm of 3.98; hence, 16 dB represents a power ratio of 39.8.

A loss in decibels is customarily denoted by the minus sign. Thus, a loss of 16 dB is written -16 dB.

Expressing the gain or loss of various circuits or apparatus in decibels obviates the necessity of computing gains or losses by multiplication and division. Because the decibel is a logarithmic unit, the total gain of a circuit is found by adding the individual decibel gains and losses of the various circuit components.

Example 3

A dynamic microphone with an output of -85 dB is connected to a preamplifier with a gain of 60 dB. The output of the preamplifier is connected through an attenuation pad with a loss of 10 dB to a final amplifier with a gain of 90 dB. What is the total gain?

Solution

In this example, all decibel values have been taken from a common reference level. Because the microphone is 85 dB below reference level, the preamplifier brings the level up to $-85 + 60 = -25$ dB. The attenuation pad then reduces the level to $-25 - 10 = -35$ dB, and the final amplifier causes a net gain of $-35 + 90 = 55$ dB. Hence, it is apparent that the overall gain in any system is simply the algebraic sum of the decibel gains or losses of the associated circuit components. Thus, $-85 + 60 - 10 + 90 = 55$ dB gain.

35-3 POWER REFERENCE LEVELS

It is essential that you remember that the decibel is not an absolute quantity; it merely represents a change in power relative to the level at some different time or place. It is meaningless to say that a given amplifier has an output of so many dB unless that output is referred to a specific power level. If we know what the output power is, then the *ratio* of that output power to the specific input power may be expressed in dB.

Several reference levels ("zero-reference" or "zero-dB") have been developed within the industry. Some of these have already been dropped generally; some are used in isolated communities or within individual companies; others are in general use throughout the entire electronics industry. Some of the more common levels are discussed below.

dBm The most common reference level used in the telephone industry is one milliwatt. And since many radio and television programs are carried between studio and transmitter by telephone systems, we should be able to understand telephone transmission engineers when they talk about relative powers. The rather widespread use of the expression "decibels above or below one milliwatt" is usually abbreviated $\pm$ dBm. Signal power in communications systems is almost always being amplified (multiplication) or attenuated (division). It is far more convenient to add or subtract dB than to calculate the power in milliwatts or watts by long processes of multiplication or division. Thus, when a telephone engineer speaks of a power level of 25 dBm, the listeners can readily understand that, if $P_1 = 1$ mW, P_2 is 25 dB higher.

Example 4

What is the output power represented by a level of 25 dBm?

Solution

dBm means "decibels referred to a reference power level of 1 mW"; that is, $P_1 = 1$ mW. Then, an amplification of 25 dB means:

$$25 = 10 \log_{10} \frac{P_2}{1 \text{ mW}}$$
$$\log P_2 = 2.5$$
$$P_2 = 316.23 \text{ mW}$$

Because circuits do not amplify or attenutate all frequencies by the same amount, the industry often reserves the term dBm for an input signal of a single-frequency (pure) sine wave (often 400 Hz or 1 kHz). However, dBm is often applied to more complex waveforms because of the convenience of calculations.

6 mW Several radio receiver and audio amplifier manufacturers use 0.006 W (6 mW) as their reference, or zero-dB, level.

Example 5

How much power is represented by a gain of 23 dB if zero level is 6 mW?

Solution 1

Substituting 23 for dB and 6 for P_1 in Eq. (1),

$$23 = 10 \log \frac{P_2}{6}$$

D: 10,
$$2.3 = \log \frac{P_2}{6}$$

Taking antilogs of both members,

$$199.5 = \frac{P_2}{6}$$
$$\therefore P_2 = 1197 \text{ mW}$$

Check

$$23 = 10 \log \frac{1197}{6}$$
$$23 = 10 \log 199.5$$
$$23 = 10 \times 2.3$$

Solution 2

$$2.3 = \log \frac{P_2}{6}$$

or $\qquad 2.3 = \log P_2 - \log 6$

Transposing, $\qquad \log P_2 = 2.3 + \log 6$

Substituting the value of log 6, $\quad \log P_2 = 2.3 + 0.778$

$$\log P_2 = 3.078$$

Taking antilogs, $\qquad P_2 = 1197 \text{ mW}$

Example 6

How much power is represented by -64 dB if zero level is 6 mW?

Solution 1

Substituting -64 for dB and 6 for P_1 in Eq. (1),

$$-64 = 10 \log \frac{P_2}{6}$$

D: 10, $\qquad -6.4 = \log \frac{P_2}{6}$

The left member of the above equation is a logarithm with a negative mantissa because the entire number 6.4 is negative. Hence, to express this logarithm with a positive mantissa, the equation is written

$$3.6 - 10 = \log \frac{P_2}{6}$$

Taking antilogs of both members,

$$3.98 \times 10^{-7} = \frac{P_2}{6}$$
$$\therefore P_2 = 2.39 \times 10^{-6} \text{ mW}$$

Check

$$-64 = 10 \log \frac{2.39 \times 10^{-6}}{6}$$
$$= 10 \log 3.98 \times 10^{-7}$$
$$-64 = 10(3.6 - 10)$$
$$-64 = -64$$

Solution 2

$$-6.4 = \log \frac{P_2}{6}$$

Then
$$-6.4 = \log P_2 - \log 6$$

Transposing,
$$\log P_2 = \log 6 - 6.4$$

Substituting the value of log 6,
$$\log P_2 = 0.778 - 6.4$$
$$= (10.78 - 10) - 6.4$$
$$\therefore P_2 = 2.39 \times 10^{-6}\, \text{mW}$$

If the larger power is always placed in the numerator of the power ratio, the quotient will always be greater than 1; therefore, the characteristic of the logarithm of the ratio will always be zero or a positive value. In this manner the use of a negative characteristic is avoided. As an illustration, from Example 6,

$$-6.4 = \log \frac{P_2}{6}$$

which is the same as
$$6.4 = \log 6 - \log P_2$$

Hence,
$$6.4 = \log \frac{6}{P_2}$$

It is always apparent whether there is a gain or a loss in decibels; therefore, the proper sign can be affixed after working the problem.

VU   The volume unit, abbreviated VU, is used in broadcasting, and it is based on the amplitude of the program frequencies throughout the system. The standard volume indicator (VU meter) is calibrated in decibels with zero level corresponding to 1 mW of power in a 600-Ω line under steady-state conditions, usually at a frequency between 35 Hz and 10 kHz. Owing to the ballistic characteristics of the instrument, the scale markings are referred to as volume units and correspond to dBm only in the case of steady-state sine-wave signals.

dBRN AND dBA   The signal-to-noise ratio is very important in most electronic amplifiers and communications circuits. When engineers establish a reference noise level, then the signal power may be expressed as being so many dB above this arbitrary reference level. The expression ''decibels referred to an arbitrary reference noise level'' is abbreviated dBRN. Often this reference noise level is set at -90 dBm. You should confirm that this represents 1 pW of power.

Then, when an original established reference noise level is adjusted to some new level, as it sometimes is in the telephone industry, the abbreviation dBA indicates ''decibels referred to some adjusted reference noise level.''

dBRAP   A sound may be heard by ''the average human ear'' (whatever that is) if it has a power of 10^{-16} W or more. This minimum power represents the threshold of hearing, and it is called reference acoustical power. Any noise or signal of any kind must be above the minimum power to be heard, and it may then be compared with the minimum power. Thus, dBRAP means a power ratio in dB when $P_1 = 10^{-16}$ W. Sound engineers often call the number of dBRAP by the name *phons*.

OTHER SPECIALIZED TERMS   Other reference levels, used in more specialized fields are:

- ■ dBW   dB referred to 1 W as zero-dB reference level.
- ■ dBk   dB referred to 1 kW as reference level.
- ■ dBV   dB referred to 1 V as zero reference signal level.

These, and many other zero reference levels, need introduce no great problem to you. It is only necessary to remember that dB represents a power ratio which must be referred to some original or arbitrary reference level.

35-4 CURRENT AND VOLTAGE RATIOS

Fundamentally, the decibel is a measure of the ratio of two powers. However, voltage ratios and current ratios can be utilized for computing the decibel gain or loss provided that the input and output impedances are taken into account.

In the following derivations, P_1 and P_2 will represent the power input and power output, respectively, and R_1 and R_2 will represent the input and output impedances, respectively. Then

$$P_1 = \frac{V_1^2}{R_1} \quad \text{and} \quad P_2 = \frac{V_2^2}{R_2}$$

Since

$$dB = 10 \log \frac{P_2}{P_1}$$

substituting for P_1 and P_2,

$$dB = 10 \log \frac{\dfrac{V_2^2}{R_2}}{\dfrac{V_1^2}{R_1}}$$

$$\therefore dB = 10 \log \frac{V_2^2 R_1}{V_1^2 R_2}$$

$$= 10 \log \left(\frac{V_2}{V_1}\right)^2 \frac{R_1}{R_2}$$

$$= 10 \log \left(\frac{V_2}{V_1}\right)^2 + 10 \log \frac{R_1}{R_2}$$

$$= 20 \log \frac{V_2}{V_1} + 10 \log \frac{R_1}{R_2} \tag{2}$$

$$= 20 \log \frac{V_2 \sqrt{R_1}}{V_1 \sqrt{R_2}} \tag{3}$$

Similarly, $\qquad P_1 = I_1^2 R_1 \qquad$ and $\qquad P_2 = I_2^2 R_2$

Then, since $\qquad\qquad\qquad dB = 10 \log \dfrac{P_2}{P_1}$

by substituting for P_1 and P_2,

$$dB = 10 \log \frac{I_2^2 R_2}{I_1^2 R_1}$$

$$= 20 \log \frac{I_2}{I_1} + 10 \log \frac{R_2}{R_1} \qquad (4)$$

$$= 20 \log \frac{I_2 \sqrt{R_2}}{I_1 \sqrt{R_1}} \qquad (5)$$

If, in both the above cases, the impedances R_1 and R_2 are *equal*, they will cancel and the following formulas will result:

$$\text{Number of dB} = 20 \log \frac{V_2}{V_1} \qquad (6)$$

and

$$\text{Number of dB} = 20 \log \frac{I_2}{I_1} \qquad (7)$$

It is evident that voltage or current ratios can be translated into decibels *only* when the impedances across which the voltages exist or into which the currents flow are taken into account.

Example 7

An amplifier has an input resistance of 200 Ω and an output resistance of 6400 Ω. When 0.5 V is applied across the input, a voltage of 400 V appears across the output. (*a*) What is the power output of the amplifier? (*b*) What is the gain in decibels?

Solution

(*a*) $\qquad$ Power output $= P_o = \dfrac{V_o^2}{R_o}$

$$= \frac{400^2}{6400}$$

$$= 25 \text{ W}$$

(*b*) $\qquad$ Power input $= P_i = \dfrac{V_i^2}{R_i}$

$$= \frac{0.5^2}{200}$$

$$= 1.25 \times 10^{-3} \text{ W}$$

$\qquad$ Power gain $= 10 \log \dfrac{P_o}{P_i}$

$$= 10 \log \frac{25}{1.25 \times 10^{-3}}$$

$$= 43 \text{ dB}$$

Check the solution by substituting the values of the voltages and resistances in Eq. (3).

$$dB = 20 \log \frac{V_o}{V_i} \sqrt{\frac{R_i}{R_o}}$$

$$= 20 \log \frac{400}{0.5} \sqrt{\frac{200}{6400}}$$

$$= 43$$

35-5 THE MERIT, OR GAIN, OF AN ANTENNA

The merit of an antenna, especially one designed for directive transmission or reception, is usually expressed in terms of antenna *gain*. The gain is generally taken as the ratio of the power that must be supplied some standard-comparison antenna to the power that must be supplied the antenna under test in order to produce the same field strengths in the desired direction at the receiving antenna. Similarly, the gain of one antenna over another could be taken as the ratio of the respective radiated fields.

The "effective radiated power" of an antenna is the product of the antenna power and the antenna power gain.

Example 8

One kilowatt is supplied to a rhombic antenna, which results in a field strength of 20 μV/m at the receiving station. In order to produce the same field strength at the receiving station, a half-wave antenna, properly oriented and located near the rhombic, must be supplied with 16.6 kW. What is the gain of the rhombic?

Solution

Because the same antenna is used for reception, both transmitting antennas deliver the same power to the receiver. Hence,

$$dB = 10 \log \frac{P_2}{P_1} = 10 \log \frac{16.6}{1} = 12.2$$

PROBLEMS 35-2

1. How many decibels correspond to a power ratio of (*a*) 20, (*b*) 25, (*c*) 62.5, (*d*) $\frac{1}{177}$?

2. Referred to equal impedances, how many decibels correspond to a voltage ratio of (*a*) 42, (*b*) 100, (*c*) $\frac{1}{130}$, (*d*) $\frac{7}{180}$?

3. If 0 dB is taken as 6 mW, how much voltage across a 90-Ω load does it represent?

4. If 0 dB is taken as 6 mW, how much voltage across a 600-Ω load does it represent?

5. What is the voltage across a 600-Ω line at zero dBm?

6. What is the voltage across a 600-Ω line at 10 dBm?

7. If reference level is taken as 12.5 mW, how much voltage across a 300-Ω load does it represent?

8. If reference level is taken as 12.5 mW, how much voltage across a 600-Ω load does it represent? How much current flows through the load?

9. If 0 dB is 6 mW, compute the power in milliwatts and the voltage across a 600-Ω load for the following output power meter readings: (a) 3 dB, (b) 10 dB, (c) -10 dB, (d) -80 dB.

10. If 0 dB is 1 mW, compute the power in milliwatts and the voltage across a 600-Ω load for the following output meter readings: (a) 5 dB, (b) 10 dB, (c) 20 dB, (d) -10 dB.

11. An amplifier is rated as having a 90-dB gain. What power ratio does this represent?

12. The amplifier of Prob. 11 has equal input and output impedances. What is the ratio of the output current to the input current?

13. An amplifier has a gain of 60 dBm. If the input power is 1 mW, what is the output power?

14. If a high-selectivity tuned circuit has a very high Q, spurious signals which are 10% lower or higher in frequency will be attenuated at least 50 dB. What power ratio is represented by this level?

15. The manufacturer of a high-fidelity 100-W power amplifier claims that hum and noise in the amplifier is 90 dB below full power output. How much hum and noise power does this represent?

16. In the amplifier of Prob. 14, what will be the dB level of noise to signal when the amplifier is producing 3 W of output power?

17. A network has a loss of 80 dB. What power ratio corresponds to this loss?

18. If the network in Prob. 17 has equal input and output impedances, what is the ratio of the output voltage to the input voltage?

19. In single-sideband operation, the signals appearing in the unwanted set of sidebands should be attenuated by at least 30 dB. What is the ratio of output powers of the desired signal to the unwanted signal?

20. The noise level of a certain telephone line used for wired music programs is 60 dB down from the program level of 12.5 mW. How much noise power is represented by this level?

21. A certain crystal microphone is rated at -80 dB. There is on hand a final AF amplifier rated at 60 dB. How much gain must be provided by a preamplifier in order to drive the final amplifier to full output if an attenuator pad between the microphone and preamplifier has a loss of 20 dB? (All dB ratings are taken from the same reference.)

22. The output of a 200-Ω dynamic microphone is rated at -81.5 dB from a reference level of 6 mW. This microphone is to be used with an amplifier which is to have a power output of 25 W. What gain must be provided between the microphone and the amplifier output?

23. If the amplifier of Prob. 21 has an output impedance of 2.7 kΩ, what is the overall voltage ratio from microphone output to amplifier output?

24. What is the equivalent power amplification in the amplifier discussed in Prob. 23?

25. It is desired to use the amplifier of Prob. 21 with a phonograph pickup which is rated at -20 dBm. To keep from overloading the amplifier, how much loss must be introduced between pickup and input?

26. An amplifier has a normal output of 30 W. A selector switch is arranged to reduce the output in 5-dB steps. What power outputs correspond to reduction in output of 5, 10, 15, 20, 25, and 30 dB?

27. An amplifier is operating at 37 dBm with a gain of 50 dB. The input resistance of the amplifier is 22 kΩ. What is the input voltage to the amplifier?

28. A type 2N45 transistor has the following ratings when used as a class A power amplifier:

  ■ Collector voltage V, -20
  ■ Emitter current, mA   5
  ■ Input impedance, Ω 10
  ■ Source impedance, Ω 50
  ■ Load impedance, Ω 4500
  ■ Power output, mW   45
  ■ Power gain, dB   23

  What is the power input?

29. An amplifier has an input impedance of 600 Ω and an output impedance of 6000 Ω. The power output is 30 W when 1.9 V is applied across the input.
  (a) What is the voltage gain of the amplifier?
  (b) What is the power gain in decibels?
  (c) What is the power input?

30. An amplifier has an input impedance of 500 Ω and an output impedance of 4500 Ω. When 0.10 V is applied across the input, a voltage of 350 V appears across the output.
  (a) What is the power output of the amplifier?
  (b) What is the power gain in decibels?
  (c) What is the voltage gain of the amplifier?

31. A dynamic microphone with an output level of -72 dB is connected to a speech amplifier consisting of three voltage amplifier stages. The first voltage amplifier stage has a voltage gain of 100, and the second has a voltage gain of 9. The interstage transformer between the second and third voltage amplifier stages has a step-up ratio of 3:1, and the third stage has a voltage gain of 8. The driver stage and modulator have a gain of 23 dB. If zero power level is 6mW, what is the output power of the modulator?
  (**Hint**   Transformers do not introduce *power changes.*)

Microphone	Amplifier	Amplifier	Transformer	Amplifier	Modulator
-72 dB	100	9	3	8	23 dB

32. How many decibels gain is necessary to produce a 60-μW signal in 600-Ω telephones if the received signal supplies 9 μV to the 80-Ω line that feeds the receiver?

33. In the receiver of Prob. 32, if the overall gain is increased to 96 dB, what received signal will produce the 60-μW signal in the telephones?

34. The voltage across the 600-Ω telephones is adjusted to 1.73 V. When the AF filter is cut in, the voltage is reduced to 1.44 V. What is the "insertion loss" of the filter?

35. The input power to a 50-km line is 10 mW, and 40 μW is delivered at the end of the line. What is the attenuation in decibels per kilometer?

36. It is desired to raise the power level at the end of the line discussed in Prob. 35 to that of the original output. What is the voltage gain of the required amplifier?

37. In Prob. 35, what is the ratio of input power to output power?

38. One of the original attenuation units was the *neper*, which is given by

$$\text{Number of nepers} = \log_\varepsilon \frac{I_1}{I_2}$$

Since

$$\text{Number of dB} = 20 \log_{10} \frac{I_1}{I_2}$$

what is the relation between nepers and decibels for equal impedances?

Hint $\log_\varepsilon \dfrac{I_1}{I_2} = 2.30 \log_{10} \dfrac{I_1}{I_2}$

39. A television transmitting antenna has a power gain of 8.6 dB. If the power input to the antenna is 15 kW, what is the effective radiated power?

40. When 500 W is supplied to a directive antenna, the result is a field strength of 5 μV/m at a receiving station. In order to produce the same field strength at the same receiving station, the standard-comparison antenna must be supplied with 8 kW. What is the decibel gain of the directive antenna?

41. A rhombic transmitting antenna produces a field strength of 98 μV/m at a receiving test station. The standard-comparison antenna delivers a field strength of 5 μV/m. What is the decibel gain of the rhombic antenna?

42. A broadcasting station is rated at 1 kW. If the received signals vary as the square root of the radiated power, how much gain in decibels would be apparent to a nearby listener if the broadcasting station doubled its power?

35-6 TRANSMISSION LINES

A transmission line is a device that consists of one or more electric conductors and is designed for the purpose of transferring electric energy from one point to another. It has a wide variety of uses: in one form it can carry electric power to a city several miles distant from the power plant; in another form it can be used for carrying chain broadcast programs from one studio to several broadcast stations; and in still another form it can carry RF energy from a radio transmitter to an antenna or from an antenna to a radio receiver.

The most common types of transmission lines are:

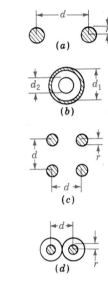

1. The two-wire open-air line as shown in Fig. 35-1a. This line consists of two parallel conductors whose spacing is carefully held constant.
2. The concentric-conductor line, as illustrated in Fig. 35-1b, which consists of tubular conductors one inside the other.
3. The four-wire open-air line as shown in Fig. 35-1c. In this type of line the diagonally opposite wires are connected to each other for effecting an electrical balance.
4. The twisted-pair line, as shown in Fig. 35-1d, which may consist of lamp cord, a telephone line, or other insulated conductors.

Any conductor has a definite amount of self-inductance, capacitance, and resistance per unit length. These properties account for the behavior of transmission lines in their various forms and uses.

The derivations of the transmission line equations that follow can be found in advanced engineering texts.

Fig. 35-1 Types of transmission lines.

35-7 THE INDUCTANCE OF A LINE

The inductance of a two-wire open-air line is given by the equation

$$L = 0.621\, l\left(0.161 + 1.48 \log_{10} \frac{d}{r}\right) \times 10^{-3}\,\text{H} \qquad (8)$$

where L = inductance of line and return, H
l = length of line, km
d = distance between conductor centers
r = radius of each wire, same units as d

Example 9

What is the inductance of a line 145 km long consisting of No. 0000 copper wire spaced 1.5 m apart?

Solution

Diameter of No. 0000 copper wire = 11.68 mm; thus radius = 5.84 mm.

$$\frac{d}{r} = \frac{1500}{5.84} = 256.85$$

$$\log_{10} 256.85 = 2.409\,68,\ \text{say } 2.41$$

Then
$$L = 0.621 \times 145(0.161 + 1.48 \times 2.41)\,\text{mH}$$
$$= 336\,\text{mH}$$

For radio frequencies, more accurate results are obtained by the approximate relation

$$L \cong 9.21 \times 10^{-9} \log_{10} \frac{d}{r} \qquad \text{H/cm} \qquad (9)$$

Where L is the inductance in henrys per centimeter and d and r have the same values as in Eq. (8).

35-8 THE CAPACITANCE OF A LINE

The capacitance of a two-wire open-air line is

$$C = \frac{0.0121\,l}{\log\dfrac{d}{r}} \quad \mu F \tag{10}$$

where C = capacitance of line, μF
l = length of line, km
d = distance between wire centers
r = radius of wire, in same units as d

Example 10

What is the capacitance per kilometer of a line consisting of No. 00 copper wire spaced 1.2 m apart?

Solution

Diameter of No. 00 copper wire = 9.266 mm; thus radius = 4.633 mm.

$$\frac{d}{r} = \frac{1200}{4.633} = 259$$

$$\log_{10} 259 = 2.413\ 30 \qquad \text{say } 2.413$$

Then

$$C = \frac{0.0121}{2.413}\ \mu F/km$$

$$= 5.01\ nF/km$$

For radio frequencies, more accurate results are obtained by the equation

$$C \cong \frac{1}{9.21 \times 10^{-9}\,c^2 \log_{10} \dfrac{d}{r}} \quad F/cm \tag{11}$$

where C is the capacitance in farads per centimeter, c is the velocity of light (3×10^{10} cm/s), and d and r have the same values as in Eq. (10).

The capacitance of submarine cables and of cables laid in metal sheaths is given by

$$C = \frac{0.0241\,Kl}{\log_{10} \dfrac{d_1}{d_2}} \quad \mu F \tag{12}$$

where C = capacitance of line, μF
K = relative dielectric constant of insultation
l = length of line, km
d_1 = inside diameter of outer conductor
d_2 = outside diameter of inner conductor

Example 11

A No. 14 copper wire is lead-sheathed. The wire is insulated with 3 mm gutta percha ($K = 4.1$). What is the capacitance of 1 km of this cable?

Solution

$$d_2 = \text{diameter of No. 14}$$
$$= 1.63 \text{ mm}$$
$$d_1 = 1.63 + 2(3) = 7.63 \text{ mm}$$
$$\log \frac{d_1}{d_2} = \log \frac{7.63}{1.63} = \log 4.687$$
$$= 0.6703$$
$$l = 1 \text{ km}$$

Then
$$C = \frac{0.0241Kl}{\log \dfrac{d_1}{d_2}}$$

$$= \frac{0.0241 \times 4.1 \times 1}{0.6703} = 0.147 \ \mu\text{F}$$

PROBLEMS 35-3

1. What is the inductance of a 120-km line consisting of two No. 00 wires spaced 1 m between centers?

2. What is the inductance of a 30-km line consisting of two No. 6 copper wires spaced 60 cm between centers?

3. A 3.7-km transmission line consists of two No. 0 solid copper wires spaced 40 cm between centers. Determine (*a*) the inductance and (*b*) the capacitance of the line.

4. If the spacing of the line of Prob. 3 were 1 m between centers, what would be (*a*) the inductance and (*b*) the capacitance?

5. A 40-km long two-wire line is to be constructed of No. 0 solid copper wire. What must be the minimum spacing between centers to keep the capacitance below 0.250 μF?

6. A 22-km two-wire line consisting of No. 00 solid copper wire is spaced 180 mm between wire centers. What is the capacitance of the line in microfarads per kilometer?

7. A lead-sheathed underground cable is to be constructed with solid copper wire covered with 12.5 mm of rubber insulation ($K = 4.3$) If then the maximum capacitance per kilometer must be limited to 0.15 μF, ±10%, what size conductor should be used?

8. A lead-sheathed cable which consists of No. 0 copper wire with 12.5 mm of rubber insulation ($K = 4.3$) is broken. A capacitance bridge measures 0.26 μF between the conductor and the sheath. How far out is the open circuit?

9. What is the capacitance per kilometer of the cable of Prob. 8?

10. The cable of Prob. 7 becomes open-circuited 5 km out. What reading will be given on a capacitance bridge?

11. The value of the current in a line at a point l km from the source of power is given by

$$i = I_0 \varepsilon^{-\kappa l}$$

where I_0 is the current at the source and κ is the attenuation constant. In a certain line, with $\kappa = 0.02$ dB/km, find the length of line where i is 10% of the original current I_0.

12. If the attenuation of a line is 0.012 dB/km, how far out from the power source will the current have decreased to 70.7% of its original value?

13. A two-wire open-air transmission line is used to couple a receiving antenna to the receiver. The line is 155 m long, and it consists of No. 10 wire spaced 15 cm between centers. Using Eqs. (9) and (11), find:
    (a) Inductance per centimeter of line
    (b) Capacitance per centimeter of line
    (c) Inductance of the entire line
    (d) Capacitance of the entire line

14. A two-wire open-air transmission line is used to couple a radio transmitter to an antenna. The line is 250 m long, and it consists of No. 14 wire spaced 14 cm between centers. Using Eqs. (9) and (11), find the (a) inductance of the line and (b) capacitance of the line.

35-9 CHARACTERISTIC IMPEDANCES OF RF TRANSMISSION LINES

The most important characteristic of a transmission line is the *characteristic impedance*, denoted by Z_0 and expressed in ohms. This impedance is often called *surge impedance, surge resistance,* or *iterative impedance*.

The value of the characteristic impedance is determined by the construction of the line, that is, by the size of the conductors and their spacing. At radio frequencies, the characteristic impedance can be considered to be a resistance the value of which is given by

$$Z_0 = \sqrt{\frac{L}{C}} \quad \Omega \tag{13}$$

where L and C are the inductance and capacitance, respectively, per unit length of line as given in Eqs. (9) and (11). The unit of length selected for L and C is immaterial as long as the *same* unit is used for both.

Substituting the values of L and C for a two-wire open-air transmission line in Eq. (13) results in

$$Z_0 = 276 \log_{10} \frac{d}{r} \quad \Omega \tag{14}$$

where d is the spacing between wire centers and r is the radius of the conductors *in the same units as d*. Note that the characteristic impedance is *not* a function of the length of the line.

Equation (14) is valid when you can conveniently neglect the capacitance of each line to ground. If you cannot, replace $\frac{d}{r}$ with $\frac{d}{D}$, where D is the diameter of the conductors.

Example 12

A transmission line is made of No. 10 wire spaced 30 cm between centers. What is the characteristic impedance of the line?

Solution

$d = 30$ cm $= 300$ mm. Diameter of No. 10 wire $= 2.588$ mm; therefore $r = 1.294$ mm.

$$Z = 276 \log \frac{d}{r} = 276 \log \frac{300}{1.294}$$

$$= 276 \log 231.8 = 276 \times 2.365$$

$$= 653 \ \Omega$$

The characteristic impedance of a concentric line is given by

$$Z_0 = 138 \log_{10} \frac{d_1}{d_2} \qquad \Omega \qquad (15)$$

where d_1 is the inside diameter of the outer conductor and d_2 is the outside diameter of the inner conductor.

Example 13

The outer conductor of a concentric transmission line consists of copper tubing 1.6 mm thick with an outside diameter of 25 mm. The copper tubing which forms the inner conductor is 0.8 mm thick with an outside diameter of 6 mm. What is the characteristic impedance of the line?

Solution

$$d_1 = 25 - (2 \times 1.6) = 21.8 \text{ mm}$$
$$d_2 = 6 \text{ mm}$$
$$Z_0 = 138 \log \frac{d_1}{d_2} = 138 \log \frac{21.8}{6}$$
$$= 138 \log 3.633$$
$$= 138 \times 0.5603 = 77.3 \ \Omega$$

PROBLEMS 35-4

1. What is the characteristic impedance of a two-line open-air transmission line consisting of No. 10 wire spaced 150 mm between centers?

2. It is desired to use No. 14 wire to provide a transmission line with a characteristic impedance of approximately 500 Ω. What logical spacing between centers should be used?

3. If a 50-mm spacing is used for the line of Prob. 2, what percentage of error is introduced by assuming that the line does have a characteristic impedance of 500 Ω?

4. It is necessary to construct a 600-Ω transmission line to couple a radio transmitter to its antenna, and No. 10 wire is readily available. What should be the spacing between wire centers?

5. The impedance at the center of a half-wave antenna is approximately 74 Ω. For maximum power transfer between transmission line and antenna, the impedance of the line must match that of the antenna. Is it physically possible to construct an *open-wire* line with a characteristic impedance as low as 74 Ω?

6. Plot a graph of the characteristic impedance in ohms against the ratio $\frac{d}{r}$ for two-wire open-air transmission lines. Use values of $\frac{d}{r}$ between 1 and 150.

7. It is desired to construct a 600-Ω two-wire line at a certain radio station. In the stock room there are on hand a large number of 30-cm spreader insulators. That is, these spreaders will space the *wires* 30 cm. What size wire should be ordered to obtain as nearly as possible the desired impedance if the 30-cm spreaders are used?

Hint $d = 30 + 2r$.

8. What outside-diameter tubing should be used to construct a quarter-wave matching stub having an impedance of approximately 300 Ω if spreaders 40 mm long are used?

9. The outer conductor of a concentric transmission line is a copper pipe 5 mm thick with an outside diameter of 70 mm. The inner conductor is a copper rod 6 mm in diameter. What is the characteristic impedance of the line?

10. The inside diameter of the outer conductor of a coaxial line is 10 mm. The surge impedance is 90 Ω. What is the diameter of the inner conductor?

11. Plot a graph of the characteristic impedance in ohms against the ratio $\frac{d_1}{d_2}$ for concentric transmission lines. Use values of $\frac{d_1}{d_2}$ between 2 and 10.

12. A particular grade of twisted-pair transmission line, which has a surge impedance of 72 Ω, has a loss of 0.2 dB/m. For a 30-m length of line, determine (*a*) the total loss in decibels and (*b*) the efficiency of transmission.

Hint % efficiency $= \dfrac{\text{power output}}{\text{power input}} \times 100$

13. The twisted-pair line of Prob. 12 is replaced by a coaxial cable that has a loss of 0.01 dB/m. What is the new efficiency of transmission?

14. For a two-wire transmission line, the attenuation in decibels *per meter of wire* is given by the equation

$$\alpha = \frac{0.0157\,R_{ac}}{\log_{10}\dfrac{d}{r}} \qquad \text{dB/m} \qquad (16)$$

where R_{ac} is the ac resistance of one meter of *wire*. One kilowatt of power, at a frequency of 16 MHz, is delivered to a 460-m line consisting of No. 8 wire spaced 30 cm between centers. The RF resistance of No. 8 wire is 49 times the dc resistance. (*a*) What is the line loss in decibels and (*b*) what is the efficiency of transmission?

15. If the spacing of the line in Prob. 14 should be changed to 20 cm between centers, what will be (*a*) the line loss in decibels and (*b*) the efficiency of transmission?

16. For a concentric transmission line, the attenuation in decibels per meter of *line* is expressed by the relation

$$\alpha = \frac{38.33\,\sqrt{f}(d_1 + d_2)10^{-6}}{d_1 d_2 \log_{10}\dfrac{d_1}{d_2}} \qquad \text{dB/m} \qquad (17)$$

where d_1 and d_2 are in centimeters and have the same meaning as in Eq. (15), and f is the frequency in megahertz. A concentric line 400 m long consists of an outer conductor with an inside diameter of 3.2 cm and an inner conductor that is 0.8 cm in diameter. At a frequency of 27.8 MHz, what is (*a*) the line loss in dB and (*b*) the efficiency of transmission.

17. The capacitance of a vertical antenna which is shorter than one-quarter wavelength at its operating frequency can be computed by the equation

$$C_a = \frac{55.77l}{\left(\log\varepsilon\,\dfrac{200l}{d} - 1\right)\left[1 - \left(\dfrac{fl}{75}\right)^2\right]} \qquad \text{pF} \qquad (18)$$

where C_a = capacitance of antenna, pF
l = height of antenna, m
d = diameter of antenna conductor, cm
f = operating frequency, MHz

Determine the capacitance of a vertical antenna that is 85 m high and consists of 1.5-cm wire. The antenna is being operated at a frequency of 214 kHz.

18. The RF resistance of a copper concentric transmission line can be computed by

$$R_{ac} = 8.33\,\sqrt{f}\left(\frac{1}{d_1} + \frac{1}{d_2}\right) \times 10^{-3} \qquad \Omega/\text{m} \qquad (19)$$

where f = frequency, MHz
d_1 = inside diameter of outer conductor, cm
d_2 = outside diameter of inner conductor, cm

What is the resistance of a concentric line 76 m long operating at 132 MHz if d_1 = 3.8 cm and d_2 = 0.48 cm?

19. If an antenna is matched to a coaxial transmission line, the percent efficiency is given by

$$\eta = \frac{100R_T}{Z_0 + R_T} \qquad \% \qquad (20)$$

where Z_0 = characteristic impedance of the concentric line
R_T = effective resistance of the line due to attenuation, obtainable from the line constants

$$R_T = Z_0 \left(\varepsilon^{\left(\frac{rl}{Z_0}\right)} - 1 \right) \qquad \Omega \qquad (21)$$

where r = RF resistance per meter of line as found in Eq. (19)
l = length of line, m

Find the efficiency of transmission of a matched concentric transmission line with a characteristic impedance of 300 Ω. The line is 24 m long, and it has an RF resistance of 0.72 Ω/m.

20. What is the efficiency of transmission of a matched concentric transmission line with a characteristic impedance of 90 Ω if the line is 335 m long and has an RF resistance of 0.33 Ω/m?

35-10 GRAPHS INVOLVING LOGARITHMIC VALUES

In your further studies in mathematics for electronics, you will spend considerable time investigating graphic displays involving logarithmic values. In Chap. 16 we used only graph paper with equally spaced lines, so the blank sheets consisted of a multitude of little squares. However, in some cases we mathematically changed the *values* of the equal-sized steps by changing the scales. For instance, we plotted milliamperes against volts, not against millivolts. We did not squeeze our graph into a tiny corner of the sheet in order to plot very small values in amperes; we ''stretched'' the scale of current by a factor of 1000 to best use the graph sheet.

Sometimes it happens that one set of values to be graphed covers a very large range. For instance, it is quite common to hear of amplifiers capable of handling frequencies between 20 and 20 000 Hz without appreciable distortion. (Some manufacturers will actually specify the capabilities of their audio equipment to 200 000 Hz!) Normally, there is very little, if any, change over large ranges of frequency, and the changes with which we are normally concerned occur at the very low and very high ends of the graph. So rather than plot values on equally spaced graph paper, we compress the available spaces by plotting the frequency logarithmically. Fig. 35-2 illustrates a typical frequency response curve for an audio amplifier. From this figure you can see that the ''roll-off'' for the amplifier starts at about 5 kHz and continues to roll off at approximately 3 dB per octave.

You will encounter various other types of semilog and logarithmic graph papers as you continue your studies. Do not downgrade the value of logarithms. You will find logarithms to be useful in almost any field of electronics application.

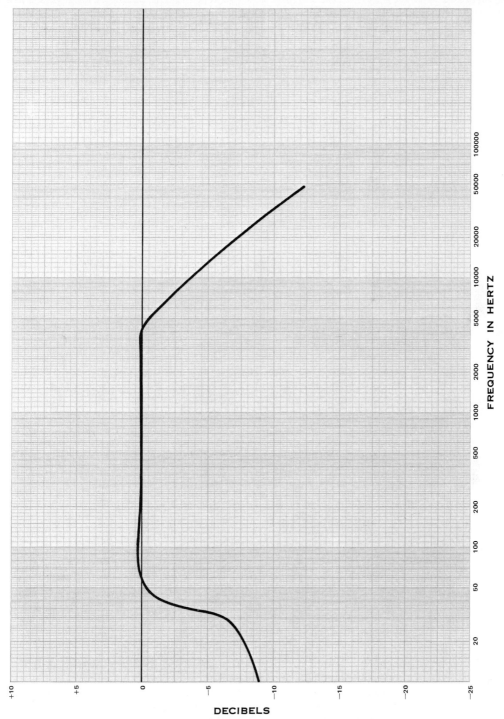

Fig. 35-2 Typical record-playback response curve. Frequency plotted logarithmically; Decibels (a logarithmic value) plotted linearly. Commercially published graphs may show horizontal scale (and vertical lines) of 1, 2, 5 values only. Courtesy of Keuffel & Esser Company, Morristown, New Jersey.

Number Systems for Computers

36

Have you ever wondered about our numbering systems—the seldom-discussed "philosophy" of how we count? In this chapter, we shall explore the background of counting systems and apply the knowledge gained to the electronic computing field.

36-1 NUMBERS IN GENERAL

Recall from Sec. 6-17 how we referred to the problem of adding 5×10^3 to 3×10^2:

$$5 \times 10^3 = 5000$$
$$3 \times 10^2 = \underline{300}$$
$$5 \times 10^3 + 3 \times 10^2 = 5300 = 5.3 \times 10^3$$

In other words, a number like 5300 may be thought of as being made up of two separate parts, 5×10^3 and 3×10^2. Similarly, all the numbers in our decimal system may be broken down into different factors multiplied by suitable powers of 10. For example, 5328 may be thought of as—indeed, it really is:

5000	or	5×10^3
300		3×10^2
20		2×10^1
8		8×10^0

and we could write 5328 in the form

$$5 \times 10^3 + 3 \times 10^2 + 2 \times 10^1 + 8 \times 10^0$$

In fact, the very way we place the digits in their appropriate places carries out the sense of powers of 10. In many elementary schools, students learn what the "place names" of the digits are in a long number like the one that follows:

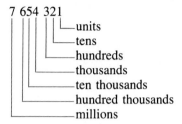

7 654 321
- units
- tens
- hundreds
- thousands
- ten thousands
- hundred thousands
- millions

and so on, and we would pronounce the whole number by using most of those place names: "seven million, six hundred fifty-four thousand, three hundred twenty-one."

36-2 BINARY NUMBERS

When we talk about decimal numbers, or the decimal number system, we mean we are counting in units of 10. That is, our numbering system has a *radix* of 10.

In the *binary* system, which is used extensively in digital computers, the radix is 2 and every number in the system represents an appropriate factor times the suitable power of 2:

$$
\begin{array}{rclcl}
& & & & \textit{or, in binary} \\
0 & = & 0 \times 2^0 & = & 0_2 \\
1 & = & 1 \times 2^0 & = & 1_2 \\
2 & = & 1 \times 2^1 + 0 \times 2^0 & = & 10_2 \\
3 & = & 1 \times 2^1 + 1 \times 2^0 & = & 11_2 \\
4 & = & 1 \times 2^2 + 0 \times 2^1 + 0 \times 2^0 & = & 100_2
\end{array}
$$

Stop and be sure. 100_2, that is, what at first appears to be one hundred in the binary numbering system, means: from the position of the digits,

$$1 \times 2^2 + 0 \times 2^1 + 0 \times 2^0$$

or

$$4 + 0 + 0 = 4$$

If you are sure, go on. If you are not sure, go back to the introduction and start the chapter again. When you are sure of the notion that a power of 2 must be connected with each digit in the binary number and the particular power depends upon the location of the digit in the number, go on and prove the following extension of the binary table:

$$
\begin{array}{rclcrcl}
5 & = & 101_2 & \qquad & 11 & = & 1011_2 \\
6 & = & 110_2 & & 12 & = & 1100_2 \\
7 & = & 111_2 & & 13 & = & 1101_2 \\
8 & = & 1000_2 & & 14 & = & 1110_2 \\
9 & = & 1001_2 & & 15 & = & 1111_2 \\
10 & = & 1010_2 & & 16 & = & 10000_2
\end{array}
$$

Example 1

Write the decimal equivalent of the number 10011001_2.

Solution

Taking our cue from the position of the digits in the number and keeping track of the appropriate powers of 2, we convert each digit to its decimal equivalent, evaluating from the right:

$$
\begin{array}{rcl}
1 \times 2^0 &=& 1_{10} \\
0 \times 2^1 &=& 0_{10} \\
0 \times 2^2 &=& 0_{10} \\
1 \times 2^3 &=& 8_{10} \\
1 \times 2^4 &=& 16_{10} \\
0 \times 2^5 &=& 0_{10} \\
0 \times 2^6 &=& 0_{10} \\
1 \times 2^7 &=& \underline{128_{10}} \\
10011001_2 &=& 153_{10}
\end{array}
$$

You should use the subscripts to designate the *system* in which you are counting until you are satisfied with your confidence in intersystem conversions.

PROBLEMS 36-1

Write the following binary numbers in decimal form:

1.	000101	**2.**	001010	**3.**	000001	**4.**	001011
5.	000111	**6.**	100111	**7.**	101010	**8.**	110001
9.	100011	**10.**	111101				

Now let us consider the reverse operation: converting a decimal number into its binary equivalent. Again we are looking for factors (either 1 or 0) times suitable powers of 2. The number 153, for instance, contains 128, which is 2^7. The remainder, $153 - 128 = 25$, contains 16, which is 2^4. The next remainder, $25 - 16 = 9$, contains 8, which is 2^3, and the last remainder, $9 - 8 = 1$, is 2^0. However, to write the complete binary equivalent, we must show the factors (zero) of 2^6, 2^5, 2^2, and 2^1.

$$153_{10} = 10011001_2$$

Obviously, it would be a tremendous help to know the whole 2^x table, and students anticipating advanced studies in computer designing, programming, or servicing will make these conversions by memorizing the 2^x table, say, to $2^{10} = 1024$. However, a ready mechanical method of arriving at the same binary number, without forgetting the missing powers of 2, is to convert the multiplication process into one of repeated division:

Radix Divisor	Decimal Number to Be Converted	Remainder
2	) 153	1
2	) 76	0
2	) 38	0
2	) 19	1
2	) 9	1
2	) 4	0
2	) 2	0
	1	

Read Up

Writing the quotient and the remainders in order "backwards," we arrive at

$$153_{10} = 10011001_2$$

PROBLEMS 36-2

Convert the following decimal numbers into binary form:

1.	6	**2.**	12	**3.**	18	**4.**	23	**5.**	31
6.	88	**7.**	97	**8.**	126	**9.**	177	**10.**	361

36-3 OCTAL NUMBERS

Modern computers speak to us in binary numbers, but their internal workings are often in octal numbers; the computers translate their octal results into binary readouts. Octal numbers are based on a counting system whose radix is 8:

$$
\begin{aligned}
&&&Or\\
0_{10} &= 0 \times 8^0 && = 0_8\\
1 &= 1 \times 8^0 && = 1_8\\
2 &= 2 \times 8^0 && = 2_8\\
3 &= 3 \times 8^0 && = 3_8\\
4 &= 4 \times 8^0 && = 4_8\\
5 &= 5 \times 8^0 && = 5_8\\
6 &= 6 \times 8^0 && = 6_8\\
7 &= 7 \times 8^0 && = 7_8\\
8 &= 1 \times 8^1 + 0 \times 8^0 &&= 10_8\\
9 &= 1 \times 8^1 + 1 \times 8^0 &&= 11_8\\
10 &= 1 \times 8^1 + 2 \times 8^0 &&= 12_8\\
11 &= 1 \times 8^1 + 3 \times 8^0 &&= 13_8\\
12 &= 1 \times 8^1 + 4 \times 8^0 &&= 14_8\\
13 &= 1 \times 8^1 + 5 \times 8^0 &&= 15_8\\
14 &= 1 \times 8^1 + 6 \times 8^0 &&= 16_8\\
15 &= 1 \times 8^1 + 7 \times 8^0 &&= 17_8\\
16 &= 2 \times 8^1 + 0 \times 8^0 &&= 20_8
\end{aligned}
$$

Just as the binary system uses digits up to, but not including, 2, so the octal system uses only digits below its radix, 8.

Write the following number in decimal form:

$$2731_8$$

As in the binary, we take our cue from the position of the digits in the number and introduce the appropriate powers of 8, reading from the right:

$$
\begin{aligned}
1 \times 8^0 &= 1 \\
3 \times 8^1 &= 24 \\
7 \times 8^2 &= 448 \\
\underline{2 \times 8^3} &= \underline{1024} \\
2731_8 &= 1497_{10}
\end{aligned}
$$

PROBLEMS 36-3

Convert the following octal numbers into their decimal equivalents:

1.	00002	**2.**	00017	**3.**	00063	**4.**	00102
5.	00077	**6.**	00100	**7.**	01124	**8.**	01035
9.	06270	**10.**	22453				

The conversion of decimal numbers to octal equivalents is achieved in the same fashion as in the binary, except that the divisor is the radix 8 instead of 2:

Example 2

Radix Divisor	Decimal Number to Be Converted	Remainder
8	) 1497	1
8	) 187	3
8	) 23	7
	2	

Read up

$$1497_{10} = 2731_8$$

PROBLEMS 36-4

Convert the following decimal numbers to their octal equivalents:

1.	25	**2.**	37	**3.**	84	**4.**	127	**5.**	165
6.	477	**7.**	823	**8.**	1062	**9.**	3928	**10.**	5000

36-4 SYSTEMS WITH ANY RADIX

Just as we have developed binary numbers with radix 2 or octal numbers with radix 8, so we may develop any number system. Consider, for example, quinary numbers: the digits in a quinary number will consist of appropriate factors times suitable powers of 5. The factors may be 0, 1, 2, 3, and 4, but not 5 or higher.

$$22_5 = 2 \times 5^1 + 2 \times 5^0 = 12_{10}$$

PROBLEMS 36-5

Write the following decimal numbers in the systems of the indicated radices:

	Number:	Radix:		Number:	Radix:
1.	9	3	**2.**	12	4
3.	27	5	**4.**	256	16
5.	256	4	**6.**	565	3
7.	1728	12	**8.**	1728	7
9.	5280	6	**10.**	672	5

Write the decimal equivalents of the numbers given:

11.	224_5	**12.**	163_7	**13.**	323_4	**14.**	201_3
15.	003_{12}	**16.**	0725_9	**17.**	0106_8	**18.**	2388_9
19.	51402_6	**20.**	73006_8				

36-5 CONVERSION BETWEEN SYSTEMS

We have already seen how to convert from any numbering system to decimal and from decimal to any other system. Thus, if we should be required to convert a number with any given radix a into a system with some other radix b, we could do so in two steps: (1) convert the given number into its decimal equivalent and (2) convert the decimal equivalent into the new system.

Example 3

Convert 5134_6 into its binary equivalent.

Solution

In the first step, convert 5134_6 to 1138_{10}. In the second step, convert 1138_{10} to 10001110010_2.

Actually, most of the conversions which concern us are between the binary and the octal systems.

Example 4

Convert 1772_8 into its binary equivalent.

Solution

$$1772_8 = 1018_{10} = 1111111010_2 = 1\ 111\ 111\ 010_2$$

In the various numbering systems, no change of value is introduced if we add zeros to the *left* of a number, so that we may change the appearance of $1\ 111\ 111\ 010_2$ to $001\ 111\ 111\ 010_2$ without introducing any value change but yielding a number which consists of a quantity of groups of three binary digits. (BInary digiTS are often referred to as *bits*.)

By good advance planning, (1) the octal numbering system uses ordinary arabic numerals up to 7 and (2) the largest binary number consisting of three digits is 7 ($= 111_2$). If we evaluate each digit in the octal number into its three-bit binary equivalent, we arrive at

$$\begin{array}{cccc} 1 & 7 & 7 & 2_8 \\ 001 & 111 & 111 & 010_2 \end{array}$$

Thus, $1772_8 = 001\ 111\ 111\ 010_2$.

Example 5
Convert 5317_8 into its binary equivalent.

Solution
Replace each octal digit in turn with its binary three-bit equivalent:

$$5317_8 = 101\ 011\ 001\ 111_2$$

Example 6
Convert 10110011001_2 to its octal equivalent.

Solution
From the right, mark off the given binary number into groups of three bits:

$$010\ 110\ 011\ 001$$

Replace each three-bit group with its regular decimal equivalent to arrive at the octal equivalent of the number:

$$010\ 110\ 011\ 001_2 = 2631_8$$

PROBLEMS 36-6
Convert the following octal numbers to their binary equivalents:

1.	361_8	**2.**	277_8	**3.**	532_8	**4.**	465_8
5.	106_8	**6.**	737_8	**7.**	5266_8	**8.**	4137_8
9.	7777_8	**10.**	1000_8				

Convert the following binary numbers to their octal equivalents:

11.	000101_2	**12.**	011001_2	**13.**	11101_2
14.	0011_2	**15.**	$110\ 111\ 101_2$	**16.**	100100100_2
17.	110011010_2	**18.**	$001\ 001\ 111_2$	**19.**	10101010_2
20.	10111010_2				

36-6 BINARY ADDITION

The addition of two quantities $a + b$, may, in binary devices, have only four possible values:

$$\begin{array}{cc} 0 + 0 = 0 & 0 + 1 = 1 \\ 1 + 0 = 1 & 1 + 1 = 10 \end{array}$$

because of the dichotomous (two-state, on-off, open-closed, flipped-flopped, 1-0) nature of switching devices, and therefore the sum S of the addition $a + b$ will be limited to the four possible answers shown above. The first three forms present no difficulty, and we can add binary numbers which involve them very easily:

$$
\begin{array}{c}
11001 \\
\underline{00100} \\
11101
\end{array}
\quad \text{or} \quad
\begin{array}{c}
25 \\
\underline{4} \\
29
\end{array}
$$

But the addition of $1 + 1$ involves us in a two-part answer: 10. The 0 part of this answer is the *sum,* and the 1 part is the *carry.* This is similar to ordinary arithmetic. When the addition of two numbers requires it, say, $9 + 5$, we "put down 4 and carry 1."

Example 7
Add 100110 and 110101.

Solution
Set the two numbers down in traditional addition form, one above the other. Addition of $0 + 0$, $0 + 1$, and $1 + 0$ involves nothing new. When adding $1 + 1$, put down 0 and carry 1 over to the next stage of addition:

$$
\begin{array}{c}
1 \\
100110 \\
\underline{110101} \\
1011011
\end{array}
\quad \text{or} \quad
\begin{array}{c}
38 \\
\underline{53} \\
91
\end{array}
$$

PROBLEMS 36-7
Add the following binary numbers:

1. $\begin{array}{c} 010001 \\ \underline{101000} \end{array}$
2. $\begin{array}{c} 100101 \\ \underline{010101} \end{array}$
3. $\begin{array}{c} 1001101 \\ \underline{0100011} \end{array}$
4. $\begin{array}{c} 0110110 \\ \underline{0100111} \end{array}$

5. $\begin{array}{c} 100111 \\ \underline{010101} \end{array}$
6. $\begin{array}{c} 101111 \\ \underline{010111} \end{array}$
7. $\begin{array}{c} 100011 \\ \underline{011110} \end{array}$
8. $\begin{array}{c} 110010 \\ \underline{011010} \end{array}$

9. $\begin{array}{c} 011010 \\ \underline{011010} \end{array}$
10. $\begin{array}{c} 100101 \\ \underline{111011} \end{array}$

11 to 20.   Prove each of your answers by converting the individual parts into their decimal equivalents.

36-7 SUBTRACTION OF BINARY NUMBERS

Similarly to binary addition, binary subtraction is limited to four possibilities:

$$
\begin{array}{c}
0 - 0 = 0 \\
1 - 0 = 1 \\
1 - 1 = 0 \\
0 - 1 = 1
\end{array}
$$

and carry 1 (or "borrow" 1)

When we are subtracting one ordinary number from another and come upon a step involving $5 - 8$, we borrow 1 from the digit to the left of the 5, subtract 8 from 15, and obtain 7. Binary subtraction is no different.

Example 8
Subtract 0110 from 1011.

Solution
Set the numbers in column form, the subtrahend below the minuend. When we must subtract 1 from 0, we borrow 1 from the number to the left of the 0 to make it 10. Then, $10 - 1 = 1$:

$$
\begin{array}{r}
\overset{\frown}{1}011 \\
-\,0110 \\
\hline
0101
\end{array}
\qquad
\begin{array}{r}
11 \\
-\;6 \\
\hline
5
\end{array}
$$

Example 9

$$
\begin{array}{r}
\overset{\frown}{1}000101 \\
-\,0110011 \\
\hline
0010010
\end{array}
$$

You should convert these two binary numbers into their equivalent decimal numbers and test the solution.

PROBLEMS 36-8
Perform the following binary subtractions:

1. $\begin{array}{r} 010011 \\ -\,001010 \end{array}$
2. $\begin{array}{r} 011011 \\ -\,010111 \end{array}$
3. $\begin{array}{r} 001101 \\ -\,000100 \end{array}$

4. $\begin{array}{r} 110111 \\ -\,011101 \end{array}$
5. $\begin{array}{r} 111000 \\ -\,010001 \end{array}$
6. $\begin{array}{r} 110100 \\ -\,101111 \end{array}$

7. $\begin{array}{r} 110110 \\ -\,011111 \end{array}$
8. $\begin{array}{r} 100111 \\ -\,100011 \end{array}$
9. $\begin{array}{r} 111111 \\ -\,111010 \end{array}$

10. $\begin{array}{r} 100110 \\ -\,100101 \end{array}$

11 to 20.   Prove each answer by converting all parts of each problem into their equivalent decimal forms.

36-8 SUBTRACTION BY ADDING COMPLEMENTS

One of the oldest rules in subtraction is "change the sign and add." This policy makes binary subtraction extremely simple. Changing the sign of a binary number is like changing the condition of a switch. *On* becomes *off,* and *open* becomes *closed*. *Flipped* becomes *flopped,* 1 becomes 0, and 0 becomes 1.

Example 10

Subtract 01101 from 11001 by means of complementation.

Solution

Rewrite the problem, changing the subtrahend to its 1's complement; then add:

$$
\begin{array}{ccc}
11001 & 11001 & 25 \\
-01101 \text{ becomes} & +10010 \text{ that is,} & -13 \\
\hline
& 101011 & 43
\end{array}
$$

43?! Well, when the answer to such a process comes out with one more digit than the number of digits we had to start with, we transfer this extra digit as an "end-carry"* and add it back in:

$$
\begin{array}{l}
11001 \\
10010 \\
\hline
101011 \\
+ \quad \longrightarrow 1 \\
\hline
01100 \qquad \text{which is } 12_{10}
\end{array}
$$

Example 11

Perform the subtraction 11101101 − 01001011 by means of 1's complement.

Solution

Rewrite the subtrahend into its 1's complement and add. Bring down the extra 1, if any, and add it as an end-carry:

$$
\begin{array}{ll}
11111 & \\
11101101 & 237 \\
+10110100 & -75 \\
\hline
110100001 & \\
\quad \longrightarrow 1 & \\
\hline
10100010 & 162
\end{array}
$$

PROBLEMS 36-9

Perform the following subtractions by means of complementation:

1.	110010 − 100111	**2.**	101101 − 010010
3.	011001 − 001101	**4.**	001101 − 000110
5.	010101 − 001001	**6.**	101011 − 001010
7.	111101 − 110010	**8.**	101111 − 001100
9.	110010 − 001101	**10.**	001110 − 001001

11 to 20. Prove each answer by converting all the parts into their decimal equivalents.

* This function is built into the calculator/computer circuitry.

36-9 BINARY MULTIPLICATION

Since 1 times anything is the thing itself and 0 times anything is 0, binary multiplication is very easy.

Example 12
Multiply 1101 by 100.

Solution
Set down the numbers as for ordinary multiplication and multiply in the usual way. Add the partial answer rows in binary form:

$$
\begin{array}{rr}
1101 & 13 \\
\times\ 100 & \times\ 4 \\
\hline
0000 & \\
0000 & \\
\underline{1101} & \\
110100 & 52 \\
\end{array}
$$

Example 13
Multiply 10011 by 101.

Solution
As before, multiply by long multiplication methods. There is no need to write a complete line of 0's–just set down the right-hand 0 and shift the line for the following multiplier one step to the left:

$$
\begin{array}{rr}
10011 & 19 \\
\times\ 101 & \times\ 5 \\
\hline
10011 & \\
\underline{100110} & \\
1011111 & 95 \\
\end{array}
$$

PROBLEMS 36-10

Multiply:

1.	101111 by 10	2.	110011 by 11
3.	100101 by 101	4.	010111 by 100
5.	101001 by 111	6.	110011 by 110
7.	100111001 by 1001	8.	11001110 by 1101
9.	101001101 by 1001		111001111 by 1011

11 to 20.   Prove each solution to Probs. 1 to 10 by converting all parts into their equivalent decimal forms.

36-10 BINARY DIVISION

Dividing by binary numbers is as easy as multiplying. Either the divisor is smaller than the dividend and the quotient is 1 or the divisor is larger than the dividend and the quotient is 0.

Example 14

Divide 1000001 by 101.

Solution

Write the numbers as for ordinary long division. Will the three-bit divisor go into the first three bits of the dividend or not? If it will, put down a 1 as the first item in the quotient, and carry on. If it will not, bring down the next digit in the dividend, and put down a 0 as the first item of the quotient:

```
          01101                    13
    101)1000001                 5)65
        000
        1000
         101
         0110
          101
          0010
          0000
          0101
           101
            xx
```

PROBLEMS 36-11

Perform the following divisions:

1. 010101 by 111
2. 011110 by 101
3. 011011 by 011
4. 010100 by 100
5. 110000 by 1000
6. 001001011 by 1111
7. 001101100 by 1001
8. 101000100 by 10010
9. 101111100 by 10011
   1101010011 by 10111

11 to 20. Prove each answer by converting all parts into their decimal equivalents.

36-11 FRACTIONAL BINARY NUMBERS

To this point, as in preceding editions, any binary arithmetic operations have involved whole-number quantities. But computers and pocket calculators would be seriously limited in their operation if they were able to function only with whole numbers. A computer may receive its *commands* in analog form or in any one of a host of binary codes, but internally it performs the desired arithmetic operations by using only the binary digits 1 and 0.

Obviously, in its internal workings, the computer must be able to translate both whole and fractional numbers in the denary system into their binary equivalents to permit proper operation of the arithmetic unit within the device.

From this point on, we shall use the word "denary" for a number to radix ten, in order to avoid any confusion between decimal numbers and decimal fractions, which are often referred to simply as "decimals."

Example 15

Add $1.00 + $4.00 + $0.25 and express the answer in binary form.

Solution

$$1 + 4 + 0.25 = \$5.25$$

Which may be written as

$$1 + 4 + \tfrac{1}{4} = \$5\tfrac{1}{4}$$

But in binary form,

$$\tfrac{1}{4} = \tfrac{1}{2^2} = 2^{-2} \qquad (\text{and } 2^2 = 4,\ 2^0 = 1)$$

Therefore, $5.25 in binary form is shown as

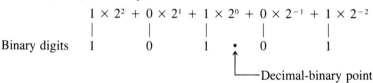

$$1 \times 2^2 + 0 \times 2^1 + 1 \times 2^0 + 0 \times 2^{-1} + 1 \times 2^{-2}$$

Binary digits   1   0   1   .   0   1

└─ Decimal-binary point

The binary solution is $5.25_{10} = 101.01_2$. Also, whether it be dollars, meters, or any unit with the denary number 5.25, the binary equivalent will always be 101.01. As with most numbering systems, the position of the integers is the most important point to note.

In the preceding example, note the use of the negative exponents; they are the key to expressing fractional denary numbers as fractional binary numbers by using the binary digits 1 and/or 0. The method of conversion presupposes that the fraction to be converted *has* a denominator with a base number of 2.

Examples

$$0.5 \quad = \tfrac{1}{2} = 1 \times 2^{-1} = 2^{-1} = 0.1_2$$

$$0.25 \quad = \tfrac{1}{4} = 1 \times 2^{-2} = 2^{-2} = 0.01_2$$

$$0.125 = \tfrac{1}{8} = 1 \times 2^{-3} = 2^{-3} = 0.001_2$$

$$0.875 = \tfrac{7}{8} = 7 \times 2^{-3} \qquad = 0.111_2$$

Therefore, $\tfrac{1}{2} + \tfrac{1}{4} + \tfrac{1}{8} \qquad = \tfrac{7}{8} = 0.111_2$

The positions of the binary digits provide the solution for the sum of $\tfrac{1}{2} + \tfrac{1}{4} + \tfrac{1}{8}$. However, there are many fractions with denominators that are not convenient numbers when converted to 2. The conversion of such fractions can be accomplished by successive multiplication by the desired base number, in this case 2, as illustrated below.

Example 16
Convert 0.25 denary into its binary equivalent.

Solution

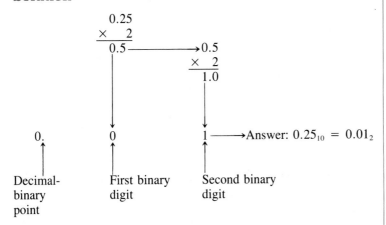

$$
\begin{array}{r}
0.25 \\
\times\ \ 2 \\
\hline
0.5 \\
\end{array}
\longrightarrow 0.5
$$

$$
\begin{array}{r}
\times\ \ 2 \\
\hline
1.0 \\
\end{array}
$$

0.   0   1 ——→Answer: $0.25_{10} = 0.01_2$

Decimal- First binary Second binary
binary   digit   digit
point

Example 17
Convert $\frac{1}{5}$ denary into its binary equivalent.

Solution

$\frac{1}{5} = 0.2$

$$
\begin{array}{r}
0.2 \\
\times\ \ 2 \\
\hline
0.4 \\
\end{array}
\longrightarrow 0.4
$$

$$
\begin{array}{r}
\times\ \ 2 \\
\hline
0.8 \\
\end{array}
\longrightarrow 0.8
$$

$$
\begin{array}{r}
\times\ \ 2 \\
\hline
1.6 \\
\end{array}
\longrightarrow 0.6
$$

$$
\begin{array}{r}
\times\ \ 2 \\
\hline
1.2 \\
\end{array}
\longrightarrow 0.2
$$

$$
\begin{array}{r}
\times\ \ 2 \\
\hline
0.4 \\
\end{array}
\rightarrow \text{etc.}
$$

0.0 0 1 1 0 Binary digits

Decimal-binary
point
  $0.2 = 0.00110 \ldots$

Note that it is only the fractional part that is moved and then multiplied by 2. The whole-number portion is brought down to form part of the new binary number, whether it be 1 or 0.

The binary equivalent of 0.2 appears to be 0.00110, but the reconversion of the binary fraction shows:

$$0 \times 2^{-1} = 0$$
$$0 \times 2^{-2} = 0$$
$$1 \times 2^{-3} = 2^{-3} = 0.125$$
$$1 \times 2^{-4} = 2^{-4} = \underline{0.0625}$$
$$0.1875 \qquad \text{Which is } not \text{ 0.2.}$$

This difference does provide some question about accuracy, since 0.1875 is 93.75% of 0.2, or -6.25%. In most pocket calculators each binary digit occupies at least one circuit, and 0.2 denary will use four circuits to provide an answer that is only 93.75% accurate. To provide greater accuracy, more binary digits must be used. To be within 1% accurate of 0.2, at least eight digits must be used; and that entails more circuits. If 0.2 denary is shown as 0.00110011 binary, the error is only -0.39%, but at the cost of using eight or nine circuits.

By now it should be obvious that 0.2 denary is a recurring binary number, and many calculators and most computers are programmed to recognize the binary sequence 0.001100110011 as denary 0.2. The 12-digit binary number could present a size problem when discrete components are used. The point is worth mentioning because 0.2 will appear often, even in other denary fractions. The *recognition circuit* in the calculator is only one solution. Another solution is to use a different *input code* to the calculator or computer, such as the octal or hexadecimal numbering system, for example.

The point being made is that less expensive calculators will probably have no more accuracy than a "good" slide rule would once have had, especially if the operator has been rounding off throughout the problem and then expressing the final answer to only three decimal places. The user of the calculator should be aware that any result given to the *n*th decimal place should not be assumed more accurate than three significant figures from a slide rule in the *final* answer. It is not our intent to discredit the calculator; we want only to make the operator aware that if the value of any constant, say π for example, is rounded off to three significant digits, the error introduced will be compounded throughout the complete calculation.

To prove the point, and the frequency with which 0.2 occurs in most calculations done with electronic calculators, complete the following problems. Express the binary answer to within 1% of the original denary fraction and reconvert the answer obtained to prove its accuracy.

PROBLEMS 36-12

Express the following fractions as binary numbers to within $\pm 1\%$.

1. $\frac{9}{10}$ 2. $\frac{4}{5}$ 3. $\frac{7}{10}$ 4. $\frac{3}{5}$ 5. $\frac{5}{6}$

6. $\frac{1}{2}$ 7. $\frac{3}{10}$ 8. $\frac{9}{16}$ 9. $\frac{1}{10}$ 10. $\frac{5}{11}$

36-12 ARITHMETIC OPERATIONS WITH FRACTIONAL BINARY NUMBERS

ADDITION AND SUBTRACTION   The rules for binary addition were outlined in Sec. 36-6; they are

$$0 + 0 = 0 \qquad 0 + 1 = 1$$
$$1 + 0 = 1 \qquad 1 + 1 = 10$$

The rules continue to apply; and as in addition with denary fractions and mixed numbers, the decimal-binary point is aligned.

Example 18

Add 10001.01 binary to 111.111 binary and express the answer in both binary and denary form.

Solution

$$
\begin{array}{l}
10001.010 \\
+\ 111.111 \\
\hline
11001.001_2
\end{array}
$$

$$
\begin{aligned}
11001.001_2 = 1 \times 2^4 &= 16 \\
1 \times 2^3 &= 8 \\
0 \times 2^2 &= 0 \\
0 \times 2^1 &= 0 \\
1 \times 2^0 &= 1 \\
0 \times 2^{-1} &= 0 \\
0 \times 2^{-2} &= 0 \\
1 \times 2^{-3} &= \underline{0.125}
\end{aligned}
$$

Therefore, $11001.001_2 = 16 + 8 + 1 + \frac{1}{8} = 25.125_{10}$

The rules for binary subtraction outlined in Sec. 36-7 are applied to direct subtraction. Again, remember to align the binary point.

Example 19

$$
\begin{array}{lrl}
1111.1110 &=& 15.875 \\
-\ \underline{111.1111} &=& -\ \underline{7.9375} \\
111.1111_2 &=& 7.9375_{10}
\end{array}
$$

The same result could have been found by using complement addition. First find the subtrahend's complement (111.1111) by making its number of digits the same as for the minuend (1111.111) without changing the value of either number.

Example 20

Using the data from Example 19,

$$
\begin{array}{l}
1111.1110 \longleftarrow \text{Zero added} \\
-\,0111.1111 \\
\uparrow \\
\text{Zero added}
\end{array}
$$

$$
\begin{array}{ll}
\text{Complement of } 0111.1111 &= 1000.0000 \\
 & \ 1111.1110 \\
\text{Add complement,} & \underline{+\,1000.0000} \\
 & 10111.1110 \\
 & \raisebox{0pt}{$\llcorner\!\!\longrightarrow 1$} \\
 & \ \overline{0111.1111_2}
\end{array}
$$

This agrees with the answer found by using *direct* subtraction. The point to note is that the end-carry is carried to the very last digit, whether it be part of the fraction, as in this case, or part of the whole number, as in Sec. 36-8.

PROBLEMS 36-13

Perform the indicated operations on the following fractions, and express the binary answer to within 1% accuracy:

1. $\dfrac{5}{16} + 0.101_2$

2. $\dfrac{11}{32} + \dfrac{4}{9}$

3. $\dfrac{117}{128} + \dfrac{19}{64}$

4. $\dfrac{11}{132} + \dfrac{17}{24}$

5. $\dfrac{33}{36} + \dfrac{43}{48}$

6. $\dfrac{19}{32} - \dfrac{23}{64}$

7. $1.010101_2 - 0.625_{10}$

8. $\dfrac{59}{256} - \dfrac{987}{1024}$

9. $5.375_{10} - 101.1111_2$

10. $37.0875 - 27.125$

11 to 15. Perform the subtractions of Probs. 6 to 10 by using complementary addition. Express the answers in both denary and binary forms.

MULTIPLICATION AND DIVISION The rules of binary arithmetic (Sec. 36-9) for multiplication still apply, along with the rules for denary arithmetic to place the decimal-binary point. That is,

$$1 \times 1 = 1$$

and all other combinations equal zero.

Example 21
Multiply 1.101 by 1.111

Solution

```
      1.101
      1.111
      1 101
     11 01
    110 1
  1 101
 11.000 011
```

The binary point is placed just as it would be in denary multiplication: the number of places after the point is the sum of the places in the multiplicand and the multiplier. Therefore, with six places of binary digits after the two binary points, $1.101 \times 1.111 = 11.000\ 011$. This can be easily proved:

$$
\begin{array}{rl}
1.101 & = 1.625 \\
1.111 & = 1.875 \\
\hline
11.000\ 011 & = 3.046\ 875
\end{array}
$$

Or

$$1.101 = 1\tfrac{5}{8}$$
$$1.111 = 1\tfrac{7}{8}$$

Which gives

$$\tfrac{13}{8} \times \tfrac{15}{8} = \tfrac{195}{64}$$
$$= 3 + (1 \times 2^{-5}) + (1 \times 2^{-6})$$
$$= 11 + 0.00001 + 0.000001$$
$$= 11.000011_2$$

By inspection, multiplication of whole binary numbers is a series of additions and shifts to the left and multiplication of fractional binary numbers is a series of additions and shifts to the right. In other words, in binary fractional multiplication the number gets smaller, as it does in denary fractional multiplication.

PROBLEMS 36-14

Perform the indicated multiplications. Show all binary working, and express the answers in both binary and denary form.

1. $(11.011)(1.011)$
2. $(111)(0.101)$
3. $(11111)(1.001)$
4. $(1000000)(0.000001)$
5. $(1100)(0.11)$
6. $(0.0101)(0.101)$
7. $(0.0001)(11000)$
8. $(111)(0.111)$
9. $(11111)(0.1001)$
10. $(0.1111)(0.011)$

DIVISION   When in Sec. 36-10 if we had divided 1111_2, by 10_2, we would have produced the following result

$$
\begin{array}{r}
0111 \\
10\overline{)1111} \\
10 \\
\hline
11 \\
10 \\
\hline
11 \\
10 \\
\hline
1
\end{array}
$$

The accepted answer was 111 with $\tfrac{1}{10}$ remainder; the denary answer was 7.5. The denary equivalent of 7.5 can be shown from

$$1111_2 = 15 \quad \text{and} \quad 10_2 = 2$$
$$\frac{15}{2} = 7.5$$

The binary equivalent of 7.5 can be shown to be

$$7.5 = (7 \times 2^0) + (1 \times 2^{-1})$$
$$= 111 + 0.1$$
$$= 111.1_2$$

It is *not* necessary to convert the binary fractions into their denary equivalents before writing them as fractional binary numbers. Division with binary fractional numbers is no more complicated than it is with denary fractional numbers—add zeros where necessary without changing the value of the number.

Example 22
Divide 1101_2 by 10_2.

Solution

```
        110.1
  10)1101
     10
     10
     10
      010 ← Added zero places
       10    the binary point.
       00
```

The final answer is 110.1_2.

In Example 22 the division was accomplished in the same manner that division by a denary fraction would be. Division by a denary fraction is achieved by moving the decimal points in both dividend and divisor to convert the divisor into a whole number:

$$\frac{6.25}{0.5} = \frac{62.5}{5}$$

Similarly, division by a binary fraction is achieved by moving the binary points, as shown in the following example.

Example 23
Divide 0.01_2 by 0.11_2.

Solution

$$\frac{0.01}{0.11} = \frac{1.0}{11.0}$$

```
         0.0101
  11)1.0000
       11
      100
       11
        1   More zeros would be
            considered at this point.
```

The preceding examples have served their purpose, which is to illustrate that binary division is no more involved than denary division is.

Division of whole binary numbers is no more than a series of subtractions and shifts of the binary point to the right. Fractional division is a series of subtractions; but depending upon the magnitude of the divisor, the result may be a shift of the binary point to the left or right.

If the divisor is of lesser magnitude than the dividend, the shift will be to the left, indicating the resulting binary number is greater in magnitude. A shift to the right would indicate that the quotient had a divisor greater than the dividend, and the result would be a smaller binary number, as in Example 23.

Example 24
Divide 1.01_2 by 0.1_2.

Solution

$$0.1)\overline{1.01} \quad \text{(quotient } 10.1\text{)} \quad \text{Or } 101_2 \text{ divided by } 10_2$$

The answer, 10.1_2, is a shift to the right; the quotient is larger than the divisor.

Note that the binary numbers have been *treated* in most respects in the same way as denary numbers. That is to say, if the binary point in the divisor was moved two places to the right, the corresponding move was made in the dividend.

PROBLEMS 36-15
Perform the indicated divisions, and express answers in both denary and binary forms to within 1% accuracy:

1. $111111_2 \div 101_2$
2. $101.01_2 \div 0.011_2$
3. $100000_2 \div 0.0011_2$
4. $100000_2 \div 11.11_2$
5. $1.111_2 \div 0.1111_2$
6. $10101_2 \div 0.011101_2$
7. $0.0101_2 \div 0.00101_2$
8. $0.110001_2 \div 101_2$
9. $0.0001_2 \div 0.0000011_2$
10. $1000000_2 \div 11.111_2$

36-13 FRACTIONAL OCTAL NUMBERS

In earlier sections in this chapter we have shown how a denary number could be converted into a binary or octal number or, in fact, a number to any radix. Throughout these conversions only whole numbers were considered. The three-bit code was developed to illustrate how to convert from binary to octal systems, and vice versa, without having to return through the denary system. This code uses a series of short steps, and it also illustrates the influence of octal numbers and their fractional parts compared to binary fractional numbers.

Certain binary fractions encountered have presented an accuracy problem unless a considerable number of digits were used. The main problem seemed to come from a denary fraction of 0.2; to present that number in its binary equivalent to within 1% accuracy, 12 or more binary digits were required. The same denary fraction can be represented by its octal equivalent by using only four octal digits, and the accuracy is better than 1%.

Example 25

Convert 0.2 denary into its binary and octal equivalents to within 1% accuracy.

Solution

From Ex. 17, 0.2_{10} = 0.001100110011_2

2^0 0 = 0   = 0.00000
2^{-1} 0 = 0   = 0.00000
2^{-2} 0 = 0   = 0.00000
2^{-3} 1 = 1 × 2^{-3} = 0.12500
2^{-4} 1 = 1 × 2^{-4} = 0.06250
2^{-5} 0 = 0   = 0.00000
2^{-6} 0 = 0   = 0.00000
2^{-7} 1 = 1 × 2^{-7} = 0.007813
2^{-8} 1 = 1 × 2^{-8} = 0.003906
2^{-9} 0 = 0   = 0.00000
2^{-10} 0 = 0   = 0.00000
2^{-11} 1 = 1 × 2^{-11} = 0.000488
2^{-12} 1 = 1 × 2^{-12} = 0.000244
                            0.199951   Which is 99.9756% of 0.2

To convert denary 0.2 to its octal equivalent, either of two methods may be used: (1) repeated multiplication by 8 or (2) applying the three-bit code. By the first method,

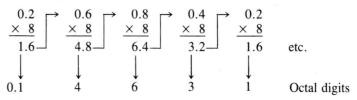

$$
\begin{array}{ccccc}
0.2 & 0.6 & 0.8 & 0.4 & 0.2 \\
\times\ 8 & \times\ 8 & \times\ 8 & \times\ 8 & \times\ 8 \\
1.6 & 4.8 & 6.4 & 3.2 & 1.6 \quad \text{etc.} \\
\downarrow & \downarrow & \downarrow & \downarrow & \downarrow \\
0.1 & 4 & 6 & 3 & 1 \quad \text{Octal digits}
\end{array}
$$

This shows that 0.2_{10} equals 0.14631_8. The proof follows.

8^0 0   = 0.0000
8^{-1} 1 = 1 × 8^{-1} = 0.12500
8^{-2} 4 = 4 × 8^{-2} = 0.06250
8^{-3} 6 = 6 × 8^{-3} = 0.011719
8^{-4} 3 = 3 × 8^{-4} = 0.00073
8^{-5} 1 = 1 × 8^{-5} = 0.00003
                            0.199979   Which is 99.989% of 0.2.

The second method, using the three-bit code, saves considerable time, especially as we already have the binary equivalent. That is,

$$
\begin{array}{ccccc}
0.2 = 0. & 001 & 100 & 110 & 011 \\
& \downarrow & \downarrow & \downarrow & \downarrow \\
0. & 1 & 4 & 6 & 3
\end{array}
$$

These are the first four digits obtained by using method 1, and 0.1463_8 is within the required 1% limit, with an accuracy of 99.99975%.

Obviously, since each octal digit represents three binary bits, the octal system must give greater accuracy for fewer digit places. Even greater accuracy can be obtained by using base 16, or the hexadecimal system; see Sec. 36-14. The point to note is that octal-binary conversions are made relatively simple by use of the three-bit code.

RULE

To use the three-bit code to convert fractional binary numbers into their octal equivalents, group the binary digits into groups of three, starting at the binary point and working to the right.

Example 26

Convert 0.100001 to its octal equivalent.

Solution

$$0.\ \underset{\downarrow}{100}\ \underset{\downarrow}{001} = \frac{1}{2} + \frac{1}{64} = \frac{33}{64} = 0.516\ (0.515625)\ \text{(binary solution)}$$

$$0.\ \ \ 4\ \ \ \ 1\ \ = \frac{4}{8} + \frac{1}{64} = \frac{33}{64} = 0.516\ (0.515625)\ \text{(octal solution)}$$

Check

Convert denary to octal and then to binary.

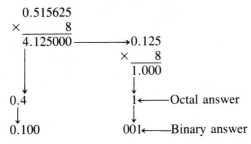

Therefore, $0.100001_2 = 0.41_8 = 0.515625_{10}$, which was the original number.

PROBLEMS 36-16

Use the three-bit code to perform the binary to octal conversions of the following:

1.	1010.0101_2	**2.**	110.0011_2	**3.**	1001.1111_2
4.	1011.1_2	**5.**	111.10101_2	**6.**	1001.011_2
7.	1111.1111_2	**8.**	1.001101_2	**9.**	0.010101_2
10.	1000.101101_2				

36-14 HEXADECIMAL NUMBERING SYSTEM

In the preceding sections of this chapter, denary numbers have been converted to numbers with any radix, although particular attention was given to base 2 (binary) and base 8 (octal) systems. Now that striking advances have been made in computers, microprocessors, and word processors, one other system should be given more than a cursory treatment. It is the hexadecimal system, which has base 16.

The hexadecimal system is an *alphanumeric code* used to solve the data entry problems of microprocessor units, but it is related to the binary numbering system needed by the computer sections that perform the arithmetic functions. The system requires that each of the numbers from 0 through 16 in the denary system be expressed in no more than four binary bits, generally termed a *byte*.

This presents one minor problem: if each denary integer is to be represented by four bits, any number greater than 9 will require eight or more bits. The system is already used in what is known as a weighted 8421 binary-coded decimal (BCD). That is, $9 = 1001$ in BCD, and $12 = 0001\ 0010$ in BCD. However, in their binary equivalents:

$$9_{10} = 1001_2 \qquad 13_{10} = 1101_2$$
$$10_{10} = 1010_2 \qquad 14_{10} = 1110_2$$
$$11_{10} = 1011_2 \qquad 15_{10} = 1111_2$$
$$12_{10} = 1100_2 \qquad 16_{10} = 10000_2$$

Obviously, the binary equivalent of denary 16 requires more than four bits. But as in all numbering systems, the greatest allowable number is less than the base number by one. For that reason 16 cannot appear in hexadecimal; the largest number is 15.

To differentiate between existing codes using four binary bits and the hexadecimal system, the following convention was established: 10 is represented by A, 11 by B, 12 by C, 13 by D, 14 by E, and 15 by F. Note that 16 is *not* represented by a letter for the reason given in the preceding paragraph. It follows that there must be some rule for numbers that are 16 or greater.

RULE
(for denary numbers greater than 16)

$$1_{10} = 0001_{16}$$
$$5_{10} = 0005_{16}$$
$$15_{10} = 000F_{16}$$
$$16_{10} = 0010_{16}$$
$$17_{10} = 0011_{16}$$
$$27_{10} = 001B_{16}$$

Note that the value of 27 denary is the sum of 16_{10} and 11_{10}. But since 11_{10} is represented by B and 16_{10} by 0010_{16},

$$27_{10} = 000B_{16} + 0010_{16}$$
$$= 001B_{16}$$

Hexadecimal equivalents of denary numbers greater than 16 can be derived in the same way, but for the conversion of very large numbers hexadecimal arithmetic must be used. It is not discussed in this text; instead it is left to more advanced studies in computer systems.

The hexadecimal system is related to the binary system; Table 36-1 shows how. The table also includes octal numbers; they are supplied for use with one of the methods for converting denary numbers into hexadecimal numbers. The method for deriving binary and octal equivalents of denary numbers is repeated division by the new base number. It can be applied to get hexadecimal equivalents also, but there are certain rules that must be observed when dealing with remainders.

TABLE 36-1 RELATIONS AMONG FOUR NUMBER SYSTEMS

Denary	Hexadecimal	Binary	Octal
0	00	0000	0
1	01	0001	1
2	02	0010	2
3	03	0011	3
4	04	0100	4
5	05	0101	5
6	06	0110	6
7	07	0111	7
8	08	1000	10
9	09	1001	11
10	0A	1010	12
11	0B	1011	13
12	0C	1100	14
13	0D	1101	15
14	0E	1110	16
15	0F	1111	17

Example 27

Convert the denary number 934 into its hexadecimal equivalent.

Solution

Divide 934 by 16 and obtain 58 with remainder 6,
Divide 58 by 16 and obtain 3 with remainder 10,
3 is not divisible by 16, and the answer is $934_{10} = $
$3\ 10\ 06_{16}$.

$$16\overline{)934(\ 6}$$
$$16\underline{)\ 58(10}$$
$$3$$

But 10_{10} is represented by A in base 16; therefore, the correct answer is $934_{10} = 3A6_{16}$

Example 16 illustrates that care must be taken with remainders, especially 10, 11, 12, 13, 14, and 15. Those remainders must be written into final answers by substituting the alphabetic designations for them. That may appear to be clumsy and open to error, but the operator of a word- or microprocessor can supply information to the main terminal with an ordinary typewriter keyboard. The letters A, B, C, D, E, and F can be interpreted as base 10 integers, or 10, 11, 12, 13, 14, and 15 can be entered in their equivalent alphabetic forms. Usually the latter course is taken. But most minicomputer terminals and pocket calculators, with the exception of some liquid-crystal devices, do not have the capability for an alphanumeric display. The operator of such a machine must recognize that 031006 on an LED display is 3A6 in the hexadecimal system.

As an exercise, show that $187_{10} = 0BB_{16}$.

Did you show that

$$16\underline{)187(11}$$
$$16\underline{)11(11}$$
$$0$$

Reading up, $\qquad\qquad 187_{10} = 0\ 11\ 11_{16}$

But $11_{10} = B_{16}$; therefore, $187_{10} = 0BB_{16}$.

The division method can become cumbersome for very large, very small, or fractional numbers. An alternative method is to use octal numbers coupled with the three-bit-code conversion into binary number equivalents.

Example 28

Convert 187_{10} into its binary, octal, and hexadecimal equivalents.

Solution

Conversion to octal,

$$8\underline{)187(3}$$
$$8\underline{)23(7}$$
$$2 \qquad 187_{10} = 273_8$$

Using the three-bit-code,

$$273_8 = 010\ 111\ 011_2 \qquad \text{and} \qquad 187_{10} = 010111011_2$$

The next step is to divide the binary number 010111011_2 into four-bit bytes starting at the least significant bit (LSB): $1011\ 1011_2$. From Table 36-1,

$$1011_2 = 11_{10} \qquad \text{and} \qquad 11_{10} = B_{16}$$

Therefore, the hexadecimal answer is

$$1011\ 1011_2 = BB_{16} = 187_{10}$$

This answer confirms the solution by the division method, which is what we set out to do.

The major advantage of using the three-bit code is in the reconversion from hexadecimal to denary.

Example 29

Convert BB_{16} to its denary equivalent.

Solution

$B_{16} = 1011_2$; therefore, $BB_{16} = 10111011_2$. Using the three-bit code on 10111011_2

$$010\ 111\ 011_2$$
$$|\qquad|\qquad|$$
$$2\quad 7\quad 3_8 \ = 2_{10} \times 8^2 + 7_{10} \times 8^1 + 3_{10} \times 8^0$$
$$= 128_{10} + 56_{10} + 3_{10}$$
$$= 187_{10}$$

Alternatively, the conversion could have been made from the hexadecimal number without the intermediate steps:

$$BB_{16} = (B \times 16^1) + (B \times 16^0) = (11 \times 16) + (11 \times 1)$$
$$= 176_{10} + 11_{10}$$
$$187_{10}$$

By either method, however, conversion from hexadecimal to denary numbers, although straightforward, is time-consuming. For this reason tables have been constructed to provide fast and easy conversion in either direction.

Since the principal use of the hexadecimal system is in connection with byte-organized machines, computer operators usually become quite adept with either or both the octal and hexadecimal systems. We suggest that you learn to recognize the binary, octal, and hexadecimal equivalents for the denary integers from 0 through 15. You will find the ability useful not only for study of this chapter but also for the study of any other BCD systems such as 8421 and excess-3 weighted codes.

Before we examine the conversion tables, a summary of denary to hexadecimal conversion will help you to speed up the process without using the division method.

1. Convert denary numbers greater than 16 to their octal equivalents (easy mental arithmetic).
2. Convert octal to binary three-bit code; one digit octal = three digits binary.
3. Rewrite the binary triads into one binary "word." Start at LSB and group into binary tetrads (four-bit bytes). Maximum denary value per tetrad = 15.
4. Convert the binary tetrad into hexadecimal equivalents. Remember

10	11	12	13	14	15
A	B	C	D	E	F

Example 30

Convert 144 into its hexadecimal equivalent.

Solution

$144_{10} = 220_8$; then

$$2 \quad 2 \quad 0_8$$
$$| \quad | \quad |$$
$$010 \ 010 \ 000_2 = 010010000_2$$

Regrouping,

$$1001 \ 0000_2 \qquad \text{From Table 36-1, } 1001_2 = 9_{16}$$
$$| \qquad |$$
$$9 \quad 0 \qquad = 90_{16}$$

Check

$$(9 \times 16^1) + (0 \times 16^0) = 144_{10} = 90_{16}$$

Do *not* drop the zero; it places the integer 9 in the correct position.

Having once mastered this technique, you will find it a very simple task to convert denary to hexadecimal—and, by reversing the process, hexadecimal to denary—provided that only whole-number integers are involved. Fractional hex numbers will be considered later.

PROBLEMS 36-17

Use the octal-triad-binary to hexadecimal-tetrad-binary method to convert the following denary numbers to hexadecimal numbers:

1. 261_{10} **2.** 396_{10} **3.** 1341_{10} **4.** 512_{10} **5.** 28_{10}

Reverse the triad-tetrad-binary process to convert the following hexadecimal numbers into denary numbers.

6. $1C_{16}$ **7.** 108_{16} **8.** $53E_{16}$ **9.** 190_{16} **10.** 400_{16}

The usefulness of Table 36-2 is best seen in the conversion of hexadecimal integers into denary integers. Given, say, 700_{16}, we first locate 7_{16} in the position 3 of the table and read across for the denary equivalent, which is 1792_{10}. Since both position 1 and position 2 are equal to 0, $700_{16} = 1792_{10}$.

The application of Table 36-2 for hexadecimal to denary conversion is as follows:

TABLE 36-2   DENARY-HEXADECIMAL INTERCONVERSION

Hex	Denary	Hex	Denary	Hex	Denary	Hex	Denary
0	0	0	0	0	0	0	0
1	4096	1	256	1	16	1	1
2	8192	2	512	2	32	2	2
3	12288	3	768	3	48	3	3
4	16384	4	1024	4	64	4	4
5	20480	5	1280	5	80	5	5
6	24576	6	1536	6	96	6	6
7	28672	7	1792	7	112	7	7
8	32768	8	2048	8	128	8	8
9	36864	9	2304	9	144	9	9
A	40960	A	2560	A	160	A	10
B	45056	B	2816	B	176	B	11
C	49152	C	3072	C	192	C	12
D	53248	D	3328	D	208	D	13
E	57344	E	3584	E	224	E	14
F	61440	F	3840	F	240	F	15
16^3		16^2		16^1		16^0	
4		3		2		1	
↑		↑		↑		↑	

Hexadecimal positions

1. Identify the position of each hexadecimal integer.
2. Look up the denary equivalent for the hexadecimal integer position located in step 1.
3. Repeat the procedure in step 2 for *all* the integers for the four hexadecimal positions. Read from left to right for the complete hexadecimal number.
4. Add all the denary values found from the four positions. The sum is the solution to the problem

Example 31

Use Table 36-2 to convert 2234_{16} to its denary equivalent.

Solution 1

In the fourth position is a 2 corresponding to	8192_{10}
In the third position is a 2 corresponding to	512_{10}
In the second position is a 3 corresponding to	48_{10}
In the first position is a 4 corresponding to	4_{10}
Adding the four positions of the denary values,	8756_{10}

Solution 2

2234_{16}
$$= 2_{10} \times 16_{10}{}^3 + 2_{10} \times 16_{10}{}^2 + 3_{10} \times 16_{10}{}^1 + 4_{10} \times 16_{10}{}^0 = 8756_{10}$$
$$= \quad 8192_{10} \quad + \quad 512_{10} \quad + \quad 48_{10} \quad + \quad 4_{10} \quad = 8756_{10}$$

Solution 2 of Example 20 confirms Solution 1, which was obtained by using Table 36-2. It does, however, involve raising 16_{10} to several different powers, which the table does for you. Once the hexadecimal positions have been located, the arithmetic is reduced to just addition. Now what remains is to show that Table 36-2 can be used for denary to hexadecimal conversions also.

Example 32

Convert 5003_{10} into its hexadecimal equivalent by using Table 36-2.

Solution

In the table, locate the denary number closest to but not greater than 5003_{10}. It is found in position 4. The hex 1 value is 4096_{10}, which is 907_{10} below 5003_{10}.

Write down in position 4 hexadecimal integer 1_{16}. Now locate the number closest to but not greater than 907_{10} and record the hexadecimal integer. It is 768_{10} in hex position 3. Record the value, 3_{16}. The difference between 907_{10} and 768_{10} is 139_{10}. The number closest to but not greater than 139_{10} is in hex position 2. The value is 128_{10}, and the difference is 11_{10}. Record the position 2 hexadecimal integer 8_{16}.

The first hexadecimal position is very straightforward, since $11_{10} = B_{16}$. We now have the complete set of hexadecimal integers for the denary number: $5003_{10} = 138B_{16}$.

Proof of the hexadecimal equivalent can be given in the same manner as in Solution 2 of Example 20. It is left for you as an exercise.

Did you show that

$$5003_{10} = 4096_{10} + 768_{10} + 128_{10} + 11_{10}$$

Or that

$$138B_{16} = (1_{10} \times 16_{10}{}^3) + (3_{10} \times 16_{10}{}^2) + (8_{10} \times 16_{10}{}^1) + (11_{10} \times 16_{10}{}^0)$$

A quick check of the calculations will show that the values are equal and that denary to hexadecimal conversion with the aid of Table 36-2 is possible.

Once you identify the position and hex location, simple addition and subtraction yield the denary number equivalent to the given hexadecimal number or, if the denary is given, the equivalent hexadecimal number. Table 36-2 is only part of a composite table which includes powers of 16 greater than the third power and fractional conversion tables out to at least 12 decimal places. For this study, however, we shall confine Table 36-2 to an upper limit of the third power. The fractional conversions are dealt with in Sec. 36-15.

PROBLEMS 36-18

Use Table 36-2 to express the following as denary numbers.

1. $1C4_{16}$ 2. 801_{16} 3. $35F_{16}$ 4. $AC1_{16}$ 5. 109_{16}

Use the table to express the following as hexadecimal numbers.

6. 288_{10} 7. 792_{10} 8. 560_{10} 9. 2684_{10} 10. 1056_{10}

36-15 FRACTIONAL HEXADECIMAL NUMBERS

In the preceding section, Table 36-2 was used for interconversion of hexadecimal and denary numbers. The construction of a table for fractional conversions requires more careful treatment, especially when the conversion is from hexadecimal to denary, than that of a table for whole-number conversions.

We have already discussed the accuracy aspect, the number of binary bits required for $\pm 1\%$ representation of the equivalent denary fraction. Each hexadecimal integer represents four binary bits, so a four-integer hexadecimal requires the accuracy of 16 binary bits. The one denary fraction that produced some problems when converted to binary was the denary quantity 0.2_{10}; in binary it was 0.001100110011_2 for an accuracy of better than 99%. By using the four-bit-byte conversion from binary to hexadecimal, we can show that $0.2_{10} = 0.333_{16}$ for an accuracy of 99.976% (rounded off to three decimal places). This is an error of only -0.024%.

Although accuracy is a major point, the space required to enter 0.333_{16} is obviously smaller than that for writing 0.001100110011_2 with less accuracy. That may appear to be splitting hairs, but the point to be stressed is that a calculator, computer, or microprocessor with an accuracy as good as -0.024% using binary numbers is not very common. Unless in a laboratory situation, most of the more common calculating devices are no better than $\pm 1\%$.

One last word on the denary quantity 0.2_{10}: If your calculator is able to convert different base numbers into one another and it does not have an alphanumeric display, it will represent the value 0.2_{10} in digital form as 0.030303_{16}. That is true of the Hewlett-Packard model 29C. Using the HP-29C FIXED to four decimal places, $0.2_{10} = 0.0303_{16}$. Note that 0.0303 displayed on the eight-segment LED readout of the HP-29C corresponds to 0.33_{16}; See Table 36-1. Converting 0.33_{16} back to denary in the conventional manner,

$$
\begin{aligned}
0.33_{16} &= 3_{10} \times 16_{10}^{-1} + 3_{10} \times 16_{10}^{-2} \\
&= 3_{10}(0.0625_{10}) + 3_{10}(0.003906_{10}) \\
&= 0.1875_{10} + 0.011719_{10} \\
&= 0.199219_{10} \qquad 99.6\% \text{ accurate}
\end{aligned}
$$

In Sec. 36-14 we demonstrated different ways to convert denary whole numbers into their hexadecimal equivalents. Earlier we showed how to convert the fractional portion of a denary number to either its binary or octal equivalent. The method, successive multiplication by the desired base number, is valid, although time-consuming, for fractional hexadecimal numbers also. The following examples illustrate fractional denary and hexadecimal number conversion by using the values set by Table 36-3.

Example 22 shows that the multiplication method has validity for hexadecimal numbers. It also shows, by using several denary fractions as examples, how the method was used to construct the table. One denary fraction is that used earlier, 0.2_{10}.

TABLE 36-3   FRACTIONAL HEXADECIMAL CONVERSIONS

x	Denary $x \times 16^{-1}$	Hex Pos 1	Denary $x \times 16^{-2}$	Hex Pos 2	Denary $x \times 16^{-3}$	Hex Pos 3
1	0.0625	0.1	0.003 906 250	0.01	0.000 244 141	0.001
2	0.1250	0.2	0.007 812 500	0.02	0.000 488 281	0.002
3	0.1875	0.3	0.011 718 750	0.03	0.000 732 422	0.003
4	0.2500	0.4	0.015 625 000	0.04	0.000 976 563	0.004
5	0.3125	0.5	0.019 531 250	0.05	0.001 220 703	0.005
6	0.3750	0.6	0.023 437 500	0.06	0.001 464 844	0.006
7	0.4375	0.7	0.027 343 750	0.07	0.001 708 984	0.007
8	0.5000	0.8	0.031 250 000	0.08	0.001 953 125	0.008
9	0.5625	0.9	0.035 156 250	0.09	0.002 197 266	0.009
A	0.6250	0.A	0.039 062 500	0.0A	0.002 441 406	0.00A
B	0.6875	0.B	0.042 968 750	0.0B	0.002 685 547	0.00B
C	0.7500	0.C	0.046 875 000	0.0C	0.002 929 688	0.00C
D	0.8125	0.D	0.050 781 250	0.0D	0.003 173 828	0.00D
E	0.8750	0.E	0.054 687 500	0.0E	0.003 417 969	0.00E
F	0.9375	0.F	0.058 593 750	0.0F	0.003 662 109	0.00F

Table 36-3 has been rounded off to nine places of decimals for convenience of table size, and most calculators have only nine-place readouts.

Example 33

Convert the following denary fractions to their hexadecimal equivalents: (a) 0.2_{10} and (b) 0.071044922_{10}.

Solution

(a) Multiply 0.2_{10} by 16_{10}.

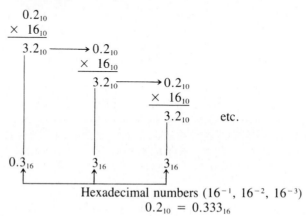

Hexadecimal numbers $(16^{-1}, 16^{-2}, 16^{-3})$
$$0.2_{10} = 0.333_{16}$$

(b) Multiply 0.071044922_{10} by 16_{10}.

$$
\begin{array}{r}
0.071044922_{10} \\
\times \qquad 16_{10} \\
\hline
0.710449220 \\
0.426269532 \\
\hline
1.136718752
\end{array}
$$
$\longrightarrow 0.136718752_{10}$

$$
\begin{array}{r}
0.136718752_{10} \\
\times \qquad 16_{10} \\
\hline
1.367187520 \\
0.820312512 \\
\hline
2.187500032
\end{array}
$$
$\longrightarrow 0.187500032_{10}$

$$
\begin{array}{r}
0.187500032_{10} \\
\times \qquad 16_{10} \\
\hline
1.875000320 \\
1.125000192 \\
\hline
3.000000412
\end{array}
$$

$0.1_{16} \qquad 2_{16} \qquad 3_{16}$

Hexadecimal number, 0.123_{16}

Example 22 illustrates that the multiplication *method* is quite lengthy and therefore time-consuming. The reconversion is somewhat more compact although still cumbersome.

Example 34
Convert 0.123_{16} into its denary equivalent.

Solution
As with all numbering systems, the *position* of the integers is all-important:

$$0.123_{16} = (1_{10} \times 16_{10}^{-1}) + (2_{10} \times 16_{10}^{-2}) + (3_{10} \times 16_{10}^{-3})$$
$$= 0.06250_{10} + 0.0078125_{10} + 0.000732422_{10}$$
$$= 0.071044922_{10}$$

Because of the sometimes involved fractional multiplications, Table 36-3 was constructed. Examples 22 and 23 illustrate the need for a better method, especially for dealing with fractional quantities. Table 36-3 can, with practice, provide a quick conversion from hexadecimal to denary fractions and, with a little more care, from denary fractions to their hexadecimal equivalents. For reason of their simplicity, Tables 36-2 and 36-3 should be used to interconvert from hexadecimal to denary and denary to hexadecimal. Particular attention should be paid, as always, to the position of the integers in either system.

Example 35
Positions to the right of the decimal point are shown as

$$\text{Position } 1_{16} = 16_{10}^{-1} \quad \text{or} \quad 0.06250_{10}$$
$$\text{Position } 2_{16} = 16_{10}^{-2} \quad \text{or} \quad 0.003\ 906\ 250_{10}$$
$$\text{Position } 3_{16} = 16_{10}^{-3} \quad \text{or} \quad 0.000\ 244\ 141_{10}$$

Succeeding positions to the right will be the successive negative powers of 16. This progression, along with the increasing powers for the whole numbers in Table 36-2, lets computer programmers convert readily in either direction from the decimal point. Although "commercial" conversion tables would be expansions of Tables 36-2 and 36-3, in this text we will not go beyond 16^3 to the left, and 16^{-3} to the right of the decimal point.

Table 36-3 was constructed by using the multiplication method as follows:

$$1_{10} \times 16_{10}^{-1} = 0.06250_{10}$$

$$
\begin{array}{r}
0.06250_{10} \\
\times \quad 16_{10} \\
\hline
0.62500 \\
.37500 \\
\hline
1.00000_{10} \\
\downarrow \\
0.1_{16}
\end{array}
$$

It is then reasonable to say that $1_{10} \times 16_{10}^{-1} = 0.06250_{10} = 0.1_{16}$. By using the same procedure, it can be shown that the following statements are correct. It is left to the student as an exercise to validate the statements:

$$1_{10} \times 16_{10}^{-2} = 0.003\ 906\ 250_{10} = 0.01_{16}$$
$$1_{10} \times 16_{10}^{-3} = 0.000\ 244\ 141_{10} = 0.001_{16}$$

The general formula for constructing Table 36-3 can be stated as

$$x \times 16^{-1}, \qquad x \times 16^{-2}, \qquad x \times 16^{-3},$$

where x takes on any integral value from 1 to 15 denary. To reduce the size of the table, zero has been omitted. The table headings, reading from left to right, provide the value of x, the corresponding denary value, and the hexadecimal equivalent and its respective position.

Example 36

What is the denary equivalent of 0.07_{16}?

Solution

Locate $x = 7$; then read across to column hex position 2 = 0.07_{16}. To the left of hex position 2, read the denary number $0.027\ 343\ 750_{10}$.

Example 37

Use Table 36-3 to perform the following conversions:
(a) $0.35B_{16}$ to denary and (b) $0.431\ 396\ 484_{10}$ to hexadecimal.

Solution

(a) Locate 0.3_{16}, hex position 1, and read
  denary number $\qquad\qquad\qquad\qquad 0.1875_{10}$
  Locate 0.05_{16}, hex position 2, and read
  denary number $\qquad\qquad\qquad\qquad 0.019\ 531\ 250_{10}$
  Locate $0.00B_{16}$, hex position 3, and read
  denary number $\qquad\qquad\qquad\qquad \underline{0.002\ 685\ 547_{10}}$
  Add the three denary numbers $\qquad\quad 0.219\ 716\ 797_{10}$

This sum is the solution: $0.35B_{16} = 0.219\ 716\ 797_{10}$.

Therefore, the hexadecimal to denary conversion is achieved by locating the denary equivalents for the corresponding hex positions and then adding them. Without the table, the solution could have involved some nine-place decimal multiplication.

(b) Converting $0.431\ 396\ 484_{10}$ to hexadecimal will require more careful thought. The first step is to locate the denary number closest to but not higher in value than the given number. If the denary number were an *exact* equivalent of a hexadecimal number, the problem would be solved, but examination of the table shows this *not* to be the case.

By inspection, the first digit to the right of the decimal point of the denary number is 4. The denary number closest to but not greater than 0.4 is located where x is equal to 6. It has the value 0.3750, which makes 6_{16} the first hexadecimal digit to the right of the hexadecimal point.

Now 0.3750_{10} is subtracted from the original denary number to provide a difference of $0.056\ 396\ 484_{10}$. It too is not an exact hexadecimal equivalent. In position 2, where x is E, we find the hex value closest to but not greater than 0.056. It is $0.054\ 687\ 500_{10}$, and, subtracted from the preceding remainder, it yields $0.001\ 708\ 984_{10}$. Therefore, we record E_{16} as the second hexadecimal number and look in the table for a denary value that is closest to but not greater than $0.001\ 708\ 984_{10}$. This time we find the exact number; it is the denary equivalent corresponding to x equal to 7. In the third position, therefore, we put 7_{16}.

With this last value our hexadecimal equivalent is complete, and $0.431\ 396\ 484_{10} = 0.6E7_{16}$. It is left as an exercise for you to confirm this answer by using the table or any other conversion method.

From the examples and the following problems you will see that conversion from fractional hexadecimal numbers to the denary equivalents is easier than conversion from the fractional denary number to the equivalent hexadecimal fraction. Practice and use of the conversion tables will remove this difficulty, and it will soon be apparent that the tables are speedier and more accurate to at least nine places of decimals.

However, if the tables are not available or you have the time to perform the successive multiplications, fractional denary hexadecimal conversions are, in principle, no more complex than fractional binary or octal conversions.

PROBLEMS 36-19

Convert the following hexadecimal numbers to their denary equivalents by using the values from Table 36-2 or 36-3.

1. $138B_{16}$ 2. $0.B7C_{16}$ 3. $0.34F_{16}$
4. $0.56B_{16}$ 5. $AA.1A_{16}$

Convert the following denary numbers into their hexadecimal equivalents by using Tables 36-2 and 36-3.

6. 5002_{10} 7. $0.717\ 773\ 438_{10}$ 8. 1342_{10}
9. 639_{10} 10. $15.796\ 142\ 578_{10}$

Boolean Algebra

37

More and more, electronic devices are being put to work in computing machines and controlling machines. First, electronic tubes superseded relays, and then transistors took the place of tubes. Now, even more exotic devices are being added to the list of computer and control components.

And with these applications of electronic devices, there is a growing need for technologists to know at least something about the logic operations of computers. The subject, generally, is known as *Boolean algebra* in honor of George Boole (1815–1864), who developed the work upon which the subject is now based. It is also often referred to as propositional calculus, mathematical logic, and truth-functional logic.

Here we are going to explore the basic ideas of Boolean algebra to see how we can put *logic* to work for us in two ways: (1) to describe circuits mathematically, after they have been designed or assembled and (2) to design circuits mathematically before they are assembled. We are not going to do any work in the *philosophical* field, where logic and its algebra are extremely useful. Several excellent books have been written from that point of view, whereas there has been little introductory work from the point of view of switching or logic circuits.

37-1 THE SYMBOLS OF LOGIC CIRCUITRY

Different associations, different manufacturers, different authors, and different publishers have their own ideas as to what symbols should be used in logic circuits. Table 37-1 shows the ANSI Y32.14 standard symbols which will be used in this book. However, you must be prepared to recognize others in other technical publications.

TABLE 37-1

37-2 THE SYMBOLS OF MATHEMATICAL LOGIC

Just as the symbol for resistance appearing in circuit diagrams is replaced in the electronics mathematics by the symbol R, so the symbols of logic circuitry shown in Table 37-1 are replaced in the logic mathematics by their own special mathematical symbols which are shown in Table 37-2. Let us look further into the meanings of the circuit symbols and see what mathematical expressions are required.

TABLE 37-2   LOGICAL MATHEMATICAL SYMBOLS

	AND	OR	NOT
Symbols used in this text	$\cdot$ juxtaposition	$+$	$-$
Other symbols sometimes used	&	V	

AND   The AND symbol means that an output signal will be produced by the particular device, regardless of the total amount of circuitry involved, only when both the a and b input signals are applied. Our mathematical counterpart must carry this meaning of AND.

OR   The OR symbol means that an output will be produced by the device when either the a input or the b input signal is applied or when *both* are applied. Our mathematical replacement must give this meaning of "either . . . OR . . ., or both."

NOT   The NOT (inverter) symbol means that either (1) there will *not* be an output when the input signal *is* applied or (2) there *is* an output when the input signal *is not* applied. Our mathematical symbol must carry the meaning of "not" or "reversed."

Now we must develop mathematical operators, sometimes referred to as *truth functors*, which will simply and effectively describe these circuit requirements. Table 37-2 shows the variety of symbols used in the literature, and, again, the symbol at the head of each column is the one to be used throughout this book. As well as know the appearance of the symbols and their general purpose, we must take particular pains to be able to pronounce the symbols.

AND $a \cdot b$ May be pronounced:
a and b
both a and b
the logical product of a and b
a conjunct b
the conjunction of a and b
a in series with b
if, and only if, a as well as b

OR $a + b$ May be pronounced:

> a or b or both
> either a or b (or both)
> the inclusive OR of a and b
> the disjunction of a and b
> the alternation of a and b
> the logical sum of a and b
> a in parallel with b
> at least one of a and b
> if, and only if, a or b or both
> true if, and only if, a or b or both

NOT $\overline{a}$ May be pronounced:

> not a
> the complement of a
> the inverse of a
> the negation of a
> the rejection of a
> it is false that a
> a is not assertable
> "not a" is true
> the valence of a is false

These pronunciations are the ones often met with in dealing with logic statements. Those appearing at the end of each group are the ones more usually found in philosophical statements, and they are included as a general-interest addition to our main study. At the same time, special symbols are often used for the *exclusive* OR operator, when we want to say "either a OR b, but not both together." Note that our definition of OR does not suit this requirement. However, we will say this in symbol form later *without* using any other special symbol.

AGGREGATE SYMBOLS: ( ), [ ]   In addition to the operator symbols are the symbols of aggregation, already met with in Sec. 3-9. Everything inside an aggregate symbol is subject to the operator symbol which may be applied to the aggregate: $\overline{(a + b)}$ means "when input signal a or input signal b or both are applied, there will be no output signal." (Can you see that this could be said, "not a and not b"?)

TRUTH SYMBOLS; 1, 0   In addition to the operators and aggregates, we require "truth symbols" to say whether a signal is *true* or *false,* whether there is a signal or there is not a signal, whether a switch is closed or open. Sometimes the letters T and F are used for these designations, but more frequently 1 and 0 are used. (See how these *two possible states* lead us into applications of *binary* arithmetic.)

Thus, if switch . . . a . . . is closed, its value is 1. When switch . . . c . . . is open, its value is 0.

Example 1

Express in logical mathematical symbols the statement, "It is raining and the wind is blowing."

Solution

First of all, select identification symbols to stand for the two propositions which make up the statement, say r for "it is raining" and b for "the wind is blowing." Second, since these two propositions are connected, we must choose the operational symbol which will represent AND, using the $\cdot$ or mere juxtaposition of the identification symbols.

$$\text{"It is raining and the wind is blowing"} = r \cdot b$$
$$or = rb$$

Example 2

Express in logical symbols the statement, "Either switch p is open when switch q is closed or switch p is closed when switch q is open."

Solution

Select identification symbols:

$$p = \text{switch } p \text{ closed}$$
$$\bar{p} = \text{switch } p \text{ open}$$
$$q = \text{switch } q \text{ closed}$$
$$\bar{q} = \text{switch } q \text{ open}$$

Then select the operational symbols to represent the conditions:

1. The requirements of "either . . . or . . ." are met by the use of $+$ = OR.
2. The requirements of "when" = "at the same time" = AND is met with $\cdot$ or juxtaposition

"Either switch p is open when switch q is closed or switch p is closed when switch q is open" $= \bar{p}q + p\bar{q}$.

PROBLEMS 37-1

By using s to represent "We are going to school" and l to represent "We are learning something new," write in symbolic form the following statements:

1. We are going to school, and we are learning something new.
2. We are going to school, but we are not learning something new.
3. Either we are going to school or we are learning something new, or both.
4. We are not going to school, but we are learning something new.
5. When we are going to school, then we are learning something new.
6. We are not going to school; therefore, we are not learning anything new.
7. Either we are not going to school or we are learning something new, or both.
8. We are neither going to school nor learning something new.

9. We are (*a*) both going to school and learning something new or else (*b*) we are not going to school and we are not learning something new.

10. Either we are going to school or we are learning something new, but not both.

37-3 THE AXIOMATIC TAUTOLOGIES

In Sec. 5-2 we have already learned that an axiom is a statement which is so self-evident that it need not be formally proved. And a tautology is nothing more than a statement or equation which shows two different ways of saying the same thing. This is a specific mathematician's version of the dictionary definition. For example, $\sin \theta = \dfrac{\text{opp}}{\text{hyp}} \theta$ is a tautology. Sometimes it is convenient to use one relationship; sometimes the other.

While philosophical logic introduces many tautologies and develops them with great care, the following brief introduction will serve the purposes of most students working in this text. Some, who go on to computer or control engineering, will want to study further to broaden their scope in the subject.

T.1 $$a \cdot a = a$$

This is the *redundancy law of multiplication*. It means that whenever a circuit design calls for a contact on relay *a* to be closed and later calls for another contact on the same relay *a* to be closed in series with the first, we really need only a single contact on relay *a*.

T.2 $$a + a = a$$

This is the *redundancy law of addition*. It means that when a circuit calls for a contact on relay *a* to be closed and later for another contact on the same relay to be closed in parallel with the first, we need only a single contact on relay *a*.

These first two tautologies, or laws, really say, "Saying the same thing over and over again does not make it any more true."

T.3 $$a \cdot b = b \cdot a$$

This is the *commutative law of multiplication*. In the mathematics of logic, as in many other systems (but not all), it does not matter what the order of the multiplication is or, in switching algebra, what the physical order of the switches in series is.

T.4 $$a + b = b + a$$

This is the *addition law of commutation*. It does not matter whether *a* is in parallel with *b* or *b* is in parallel with *a*.

T.5 $$(a \cdot b)c = a \cdot (b \cdot c)$$

This is the *associative law of multiplication* and means, again, that the order of switches in series or the order of factors in multiplication does not matter.

T.6
$$(a + b) + c = a + (b + c)$$

The *associative law of addition,* which is applied in the same way as T.5 and in ordinary algebra.

T.7
$$\bar{\bar{a}} = a$$

This is the *law of double complementation,* and it means that an inverted inversion has the same effect as the original proposition. (A switch, which can only be open or closed, if changed in position twice, is back in its original position.)

Note

Ordinary English grammar does not follow this definition because we do not always understand that two negatives make a positive in an ordinary English statement.

T.8
$$a + \bar{a} = 1$$

This is the *first law of complementation.* Since the circuit will always give an output signal if one contact is normally closed and the other, in parallel, is normally open, a *true* indication will always appear.

T.9
$$a \cdot \bar{a} = 0$$

This is the *second law of complementation.* It is impossible to achieve an output signal with one contact open in series with another that is closed.

T.10
$$a(b + \bar{b}) = a$$

This tautology says that a contact a in series with a circuit that is always operating (T.8) will have the same effect as if that contact were alone.

T.11
$$a + (b \cdot \bar{b}) = a$$

Any contact a in parallel with a permanent open circuit (T.9) will have the same effect as a alone.

T.12
$$\overline{a \cdot b} = \bar{a} + \bar{b}$$

This is the first of De Morgan's *laws of negation.* Some serious thought, coupled with the work which will follow, will prove the truth of this and the next tautology.

T.13
$$\overline{a + b} = \bar{a} \cdot \bar{b}$$

The second of De Morgan's laws of negation.

Some additional tautologies will be found on page 649, and they will be referred to in the text below.

37-4 TRUTH TABLES

Analysis of circuits by mathematical logic may be carried out by purely algebraic means, using the tautologies, and this method will be investigated shortly. But another useful method of analyzing circuits is the method of *truth tables.* These are a fairly systematic mechanical method of examining

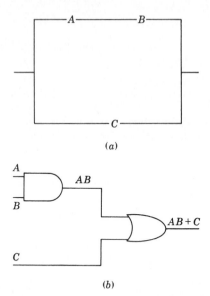

Fig. 37-1 Switching circuit for $a \cdot b + c$ in Table 37-3.

the possible combinations of truths (or circuit conditions) existing in a particular problem. For instance, consider the circuit in Fig. 37-1, which may be described mathematically as $a \cdot b + c$. We may set up the truth table for this circuit to determine which combinations of closed (1) or open (0) conditions of the switches will produce an output signal.

The first step is to list the three possible contacts, a, b, and c, and the possibilities appearing in the formula. This step gives us the row of headings across the top of Table 37-3. Under these headings there will appear eight rows of data and calculations: 2^3, where the 2 represents the two possible states 1 and 0 and the 3 represents the three different switches, or contacts, a, b, and c. Note the mechanical method of establishing the possible combinations: each of the contacts will be open for half of the possibilities, and each will be closed for half. By making the first half of the eight possibilities for a 1 and the second half 0, half of a's 1 conditions will see b 1 and half will see b 0, and so on.

TABLE 37-3   TRUTH TABLE FOR $ab + c$

Combination	a	b	c	$a \cdot b$	$a \cdot b + c$
1	1	1	1	1	1
2	1	1	0	1	1
3	1	0	1	0	1
4	1	0	0	0	0
5	0	1	1	0	1
6	0	1	0	0	0
7	0	0	1	0	1
8	0	0	0	0	0

Now, referring to the circuit of Fig. 37-1 and Table 37-3, check the circuit for each row of combinations:

■ Combination 1. When switches a, b, and c are closed (1), there is a complete circuit through the series leg (ab) and a complete circuit through the parallel switch c. Then the two closed parallel circuits will give a true (1) result, and there will be an output signal.

■ Combination 2. When both a and b are closed, then even with c open, there will be an output signal and again the last, or *total circuit*, column reads 1.

■ Combination 3. Here a and c are closed and b is open. Hence, even when the series leg is an open circuit (0), the closed switch c in parallel yields an output signal.

■ Combination 4. When a is the only closed switch, open b prevents a signal getting through the series leg and open c in parallel means that there will be *no* output signal from the circuit. The final column reads 0.

■ Combination 5. In combinations 5 through 8, since a is open, the condition of b has no effect, since the series leg is of necessity open. (See column ab.) Switch c, in parallel with this open circuit, determines that there will be an output signal when c is closed and no output signal when c is open.

You must satisfy yourself that there are no other possible switch combinations and that there will be a complete circuit, or an output signal, only for combinations 1, 2, 3, 5, and 7 and no output signal for combinations 4, 6, and 8. The formula for the circuit, $ab + c$, is sometimes said to be a tautology for the five closed combinations, although this is a loose use of the word.

PROBLEMS 37-2

Prepare the truth tables for the following expressions.

1. $ac + bc$
2. $c(a + b)$
3. $\overline{a + b + c}$
4. $\bar{a} \cdot \bar{b} \cdot \bar{c}$
5. $a + \bar{a} = 1$
6. $a(b + c) = ab + ac$
7. $a(a + b) = a$
8. $a + ab = a$
9. $p + \bar{p}q = p + q$
10. $\overline{a \cdot b} = \bar{a} + \bar{b}$
11. $\overline{a + b} = \bar{a} \cdot \bar{b}$
12. $a + bc = (a + b)(a + c)$
13. $\overline{x + y} + \overline{x + z} = \overline{x + yz}$
14. $(p + q)(\bar{q} + r)(q + 1) = p\bar{q} + qr$
15. $(a + b)(\bar{a} + c)(b + c) = \bar{a}b + ac$
16. $(a + c)(a + d)(b + c)(b + d) = ab + cd$

37-5 PROPOSITIONAL INVESTIGATIONS

Sometimes it happens that a proposed circuit is described in Boolean algebra in a rather complicated manner and it is possible to use the tautologies to simplify it.

Example 3

A designer asks for a circuit which will perform the following switching function:

$$\overline{a + b} + \overline{a + c}$$

Can we simplify the circuit requirements before drawing modules from stock and putting them together as requisitioned?

Solution

Choosing the appropriate tautologies (and here practice is the only cure), we alter the appearance of the original problem formula and see what might be done. (In the example, each step below has been identified with the number of tautology applied. (Refer to the simple Boolean relationships section that follows.)

- ▪ Given $\overline{a + b} + \overline{a + c}$
- ▪ T.13 $\overline{a + b}$ may be written $\overline{a} \cdot \overline{b}$
- ▪ T.13 $\overline{a + c}$ may be written $\overline{a} \cdot \overline{c}$

and the formula becomes $\overline{a} \cdot \overline{b} + \overline{a} \cdot \overline{c}$

- ▪ T.14 $\overline{a}(\overline{b} + \overline{c})$
- ▪ T.12 $\overline{a}(\overline{b \cdot c})$
- ▪ T.13 $\overline{a + bc}$

Compare the original circuit, as requested, with the simplified version (Fig. 37-2a versus b). You should prepare a truth table for the two circuits and prove that the two forms are tautological, that is, when one set of switches is true, then the other also is true for all possible identical combinations. Check also to satisfy yourself that there are no combinations other than 2^3.

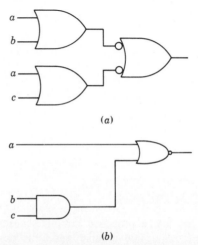

(a)

(b)

Fig. 37-2 Equivalent switching combinations of Example 3.

SIMPLE BOOLEAN RELATIONSHIPS

$a \cdot a = a$

$a + a = a$

$a \cdot b = b \cdot a$ $\qquad\qquad a + 0 = a$

$a + b = b + a$ $\qquad\qquad a + 1 = 1$

$(a \cdot b) \cdot c = a \cdot (b \cdot c)$ $\qquad\quad a \cdot 0 = 0$

$(a + b) + c = a + (b + c)$ $\qquad a \cdot 1 = a$

$(\overline{\overline{a}}) = a$

$a + \overline{a} = 1$

$a \cdot \overline{a} = 0$

$a(b + \overline{b}) = a$

$a + (b \cdot \overline{b}) = a$

$\overline{(a \cdot b)} = \overline{a} + \overline{b}$

$\overline{(a + b)} = \overline{a} \cdot \overline{b}$

$a(b + c) = ab + ac$

$a + bc = (a + b)(a + c)$

$a + \overline{a}b = a + b$

$a(a + b) = a$

$a + ab = a$

PROBLEMS 37-3

Use truth tables to prove the following statements:

1. $\overline{a}b(a + b) = \overline{a}b$

2. $(a + b)(\overline{a} + c)(b + c) = \overline{a}b + ac$

3. $(\overline{a}b + a)(\overline{a}b + c) = (a + b)(\overline{a} + c)(b + c)$

4. $a(\overline{a} + b)(\overline{a} + b + c) = ab$

5. $abc(a + b + c) = abc(ab + bc + ac) + abc(abc + ab)$

6. $\overline{q}t + qt + \overline{q} \cdot \overline{t} = \overline{q}(qt) + \overline{q}(q \cdot t)$

7. $st + vw = (s + v)(s + w)(t + v)(t + w)$

8. $ABC + A\overline{B}C + AB\overline{C} + A\overline{B}\overline{C} + \overline{A}BC + \overline{A}\overline{B}C + \overline{A}B\overline{C} = A + B + C$

9. $(\alpha + \beta)(\alpha + \gamma) = \alpha + \beta\gamma$

10. $\overline{(a \cdot b + bc + ac)} = \overline{a} \cdot \overline{b} + \overline{b} \cdot \overline{c} + \overline{a} \cdot \overline{c}$

37-6 SWITCHING NETWORKS

Although actual switches may be so adjusted that some contacts *make* before others *break,* or vice versa, or some close or open in a special sequence, in general, every individual switch is either open or closed, off or on, flipped or flopped. This two-state condition lends itself to binary operation (1 or 0), and to Boolean analysis. When a switch is closed, it provides, theoretically, perfect permittance to a current flow, and when it is open, perfect hindrance. It is convenient to define Y_{pq} as the permittance of a circuit between the points p and q and Z_{pq} as the hindrance of the circuit between the same points. Obviously, $Y_{pq} = \bar{Z}_{pq}$.

Example 4

Write the expressions for the permittance and the hindrance of the circuit of Fig. 37-3.

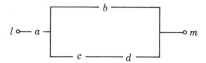

Fig. 37-3 Switching circuit of Example 4.

Solution

To write the expression for the permittance of the circuit Y_{lm}, we agree that

$$Y_{lm} = Y_a(Y_b + Y_cY_d)$$

where Y_a is the permittance of switch a, and so on. We may write this simply as

$$Y_{lm} = a(b + cd)$$

and we understand that the letter designation for a switch without an overbar indicates that the switch is closed, that is, offers perfect permittance. Studying the circuit, you can see that when contact a is closed and then either b or c and d in series is closed, the circuit will offer permittance—there will be an output signal.

Similarly, the hindrance of a contact, that is, an open switch, is indicated by the letter designation with an overbar, so tht Z_{lm} must be written:

$$Z_{lm} = \bar{a} + (\bar{b})(\bar{c} + \bar{d})$$

When contact a is open, or else when both b is open and either c or d is open, then there will be no output signal—or perfect hindrance. You should prepare a set of truth tables to show that $Y_{lm} = \bar{Z}_{lm}$.

PROBLEMS 37-4

1. Write the expressions for (*a*) the hindrance and (*b*) the permittance of the circuit of Fig. 37-4.
2. Write the expressions for (*a*) the hindrance and (*b*) the permittance of the circuit of Fig. 37-5.

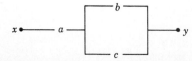

Fig. 37-4 Switching circuit for Prob. 1.

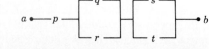

Fig. 37-5 Switching circuit for Prob. 2.

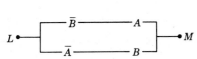

Fig. 37-6 Switching circuit for Prob. 3.

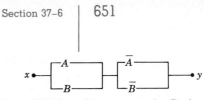

Fig. 37-7 Switching circuit for Prob. 4.

3. Write the expressions for (a) the hindrance and (b) the permittance of the circuit of Fig. 37-6.
4. Write the expressions for (a) the hindrance and (b) the permittance of the circuit of Fig. 37-7.
5. Write the expressions for (a) the hindrance and (b) the permittance of the circuit of Fig. 37-8.

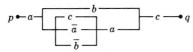

Fig. 37-8 Switching circuit for Prob. 5.

Draw the circuits for the following expressions:

6. $Y_{pq} = a(b + c)(ad)$

7. $Y_{lm} = xy(\bar{y}z + \bar{x})a$

8. $Y_{ab} = [\alpha(\beta + \bar{\gamma}) + \beta]\gamma$
9. $Z_{cd} = A[BC + C(\bar{A} + B)] + \bar{B} \cdot \bar{C}$
10. $Y_{pq} = \bar{A} \cdot \bar{B}(C + D)\bar{B} + \bar{D}$

Equivalent switching networks may be developed mathematically by using the tautologies of Boolean algebra, whereby somewhat complicated circuits may be reduced to circuits which will perform identical services with less hardware or, alternatively, to circuits which will perform identical services with readily available, although not simpler, hardware.

Example 5
Given the switching network of Fig. 37-9, develop a simpler circuit which will provide an identical switching service.

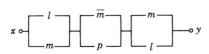

Fig. 37-9 Switching circuit of Example 5.

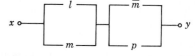

Fig. 37-10 Simpler circuit equivalent of Fig. 37-9.

Solution
Write either the permittance or hindrance function of the circuit:

$$Y_{xy} = (l + m)(\bar{m} + p)(m + l)$$
$$\text{T.4: } (l + m)(\bar{m} + p)(l + m)$$
$$\text{T.2: } (l + m)(\bar{m} + p)$$

That is, the network shown in Fig. 37-9 may be replaced by that shown in Fig. 37-10. You should prepare a truth table to prove that the two circuits are tautological.

PROBLEMS 37-5

1. By using the appropriate tautologies, develop a simpler circuit to replace that of Fig. 37-11.

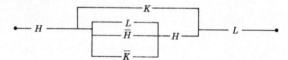

Fig. 37-11 Switching circuit for Prob. 1.

2. Develop a simpler circuit to replace that of Fig. 37-12.

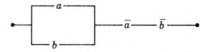

Fig. 37-12 Switching circuit for Prob. 2.

3. Develop a simpler circuit to replace that of Fig. 37-13.

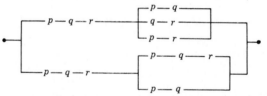

Fig. 37-13 Switching circuit for Prob. 3.

4. Develop a simpler circuit to replace that of Fig. 37-14.

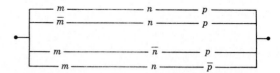

Fig. 37-14 Switching circuit for Prob. 4.

5. Develop a simpler circuit to replace that of Fig. 37-15.

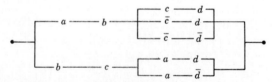

Fig. 37-15 Switching circuit for Prob. 5.

6 to 10. Check each of your solutions above by means of truth tables.

37-7 COMPUTER GATING APPLICATIONS

The standard computer gating symbols are shown in Table 37-1. These simple symbols (and the circuits for which they stand) may be combined into *adders* or *half-adders* or other more complex components. Let us look at a few of the simple tautologies as they would appear in gating configurations.

Example 6

Tautology T.14 states that $a(b + c) = ab + ac$. The two circuit configurations are shown in Fig. 37-16.

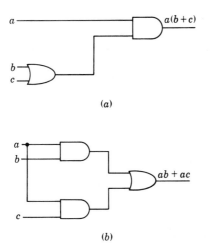

(a)

(b)

Fig. 37-16 Equivalent circuits of inclusive OR.

Investigation

You should check the two parts of Fig. 37-16 and satisfy yourself that the two circuits do perform the same functions. Then, by preparing a truth table for the two statements, you will see that when $a(b + c)$ is 1, so also is $ab + ac$, and when $a(b + c)$ is 0, so also is $ab + ac$. Then, since the two forms have been proved by tracing and by truth table to be tautological, the end results of using one will be identical with those of using the other. There may be times when availability of circuit wiring boards or parts may make it more desirable to use one circuit rather than the other, but the results will be the same regardless of the circuit configuration chosen.

You can see, then, that it may often be convenient to spend time exploring the possibilities mathematically, before even breadboarding a circuit, in order to reduce the total number of components or the number of different components required.

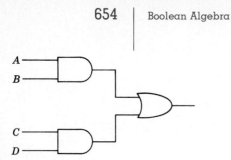

Fig. 37-17 Switching circuit for Prob. 1.

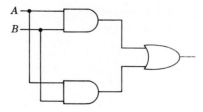

Fig. 37-18 Switching circuit for Prob. 2.

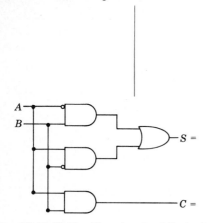

Fig. 37-19 Half-adder circuit of Prob. 3.

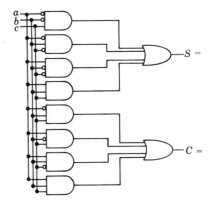

Fig. 37-20 Full-adder circuit of Prob. 4.

PROBLEMS 37-6

1. Write the output expression for the circuit of Fig. 37-17 and develop an alternate circuit. Test your answer by means of a truth table.

2. Write the output expression for the circuit of Fig. 37-18 and develop an alternate circuit. Test your answer by means of a truth table.

3. The *half-adder* circuit produces two outputs, a sum S and a carry C. The circuit is shown in Fig. 37-19. Show that the same result can be achieved by using three AND gates, one OR gate, and one INVERTER.

4. The classic *full adder*, shown in Fig. 37-20, involves the two quantities to be added (a and b) by a digital computer, plus the carry from the preceding step (c_p). The circuit requires eight AND gates, two OR gates, and nine INVERTERS. Show that the carry portion of the output may be simplified with a saving of one AND gate and three INVERTERS.

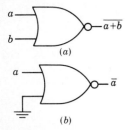

Fig. 37-21(a) The NOR gate delivers a negation of the OR function: $\overline{a + b} = \overline{a} \cdot \overline{b}$. (b) When an input is Grounded, it represents a 0. $\overline{a + 0} = \overline{a} \cdot 1 = \overline{a}$, and we have a NOT gate.

37-8 NOR COMBINATIONS

Over the last few years, many manufacturers have found it convenient for a number of reasons to build their logic circuits as multiples of a single type of gate. Often NOR gates (Fig. 37-21) are used because of the simplicity of circuit elements and design. In Problems 37-7 you will be asked to determine the gating equivalents of various combinations of NOR gates.

PROBLEMS 37-7

1. Write the output expression for the gating circuit of Fig. 37-22, and determine the simplest equivalent function.
2. Write the output expression for the gating circuit of Fig. 37-23, and determine the simplest equivalent function.
3. Write the output expression for the gating circuit of Fig. 37-24, and determine the simplest equivalent function.
4. Write the output expression for the gating circuit of Fig. 37-25, and determine the simplest equivalent function.

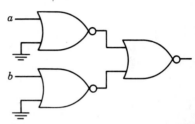

Fig. 37-23 NOR gate combination for Prob. 2.

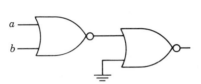

Fig. 37-22 NOR gate combination for Prob. 1.

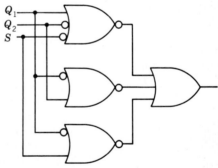

Fig. 37-25 NOR gate combination for Prob. 4.

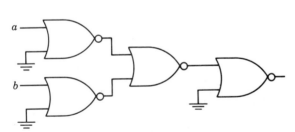

Fig. 37-24 NOR gate combination for Prob. 3.

Appendixes

A-1 THE CALCULATOR

Because of the rapid proliferation of hand-held electronic calculators, many students and, unfortunately, some teachers have decided that all their mathematical work is at an end—just punch a few keys, and the calculator will give them the answer. They have failed to realize that the calculator is able to perform the donkey work of calculation very quickly, but with no guarantees. All calculator work is subject to the law of GIGO: garbage in, garbage out. If you do not understand the principles involved in the mathematical processes, you will not be able to instruct your calculator correctly or interpret the results which it presents to you. It is still your mathematical skill that counts. And when—Heaven forbid!—your calculator batteries run down or your calculator itself is out of reach, all calculation depends upon your personal skill. Study these notes carefully. They will give you a solid foundation for estimating and calculating—and for instructing your calculator.

POCKET CALCULATORS

Advances in integrated-circuit technology have enabled manufacturers to produce reasonably priced hand-held calculators capable of performing extremely complicated calculations very rapidly. Of course, the cheaper calculators are designed to perform only basic arithmetic, with perhaps square roots or percentages thrown in. Naturally, calculators equipped to deal with the additional complications with which people in the various fields of electronics have to deal are considerably more expensive.

It is assumed that you have or soon will have a suitable calculator and, further, that you will refer as often as necessary to the instruction manual or manuals that came with it. Users of pocket calculators should not overlook the importance of the *ideas* of powers of 10, even if they do not immediately feel the urgency to use the techniques extensively. Most pocket calculators give only six to ten places in their readouts, and that is insufficient for dealing with the wide range of units—picowatts to megohms, at least—with which we are involved.

SELECTING YOUR CALCULATOR

Do not be in a rush to buy a low-priced calculator simply because you cannot afford a better one now. On the other hand, do not buy the most expensive model because it has "everything." Take time to discuss with your teachers, and with senior students if possible, which features are essential and which others may be worth serious consideration. For your greatest satisfaction, try more than one *brand,* more than one *type,* and more than one *model* before you buy. Look over the manuals and handbooks that are provided by the manufacturers; the quality of the software may be a guide to the quality of the product.

In the absence of other advice, and to achieve the greatest satisfaction in solving electronics circuit problems, you should choose a calculator which provides, as a minimum, the following features:

- Arithmetic. $+$, $-$, $\times$, $\div$, $\sqrt{x}$, y^x or x^y
- Logarithms. $\log x$, $\ln x$, 10^x, ε^x

- ■ Trigonometry. sin x, cos x, tan x, arcsin x, arccos x, arctan x (or $\sin^{-1} x$, $\cos^{-1} x$, $\tan^{-1} x$)
- ■ Angle selection. Degrees and radians, if not also gons (grads)
- ■ Decimal point. Fixed and floating, preferably selectable and easy to convert to engineering notation
- ■ Memory. One or more (preferably more) memory registers, and preferably with an automatic memory stack (or "scratchpad memory")
- ■ Special keys. π to a suitable number of places
- ■ Special functions. P $\leftrightarrow$ R, rectangular to polar conversions

In addition to those basic requirements, you should consider whether you will also find it worthwhile to pay more to obtain some useful extras:

- ■ Metric conversions. Check these against tables; do they give Imperial gallons, U.S. dry gallons, or U.S. wet gallons?
- ■ Choice of notation. FIX SCI or ENG
- ■ Statistical calculations. $\bar{x}$, s, $\Sigma +$, $\Sigma -$, %, $\triangle$%

There are as many other extras available as manufacturers think you can be persuaded to pay for.

- ■ Programming. When you are solving the same *form* of problem, such as evaluating determinants, only the numbers of which change, it can be very convenient to be able to enter a program to direct the calculator's mathematical processes. Should you consider a *keystroke* programming feature or a system of insertable *magnetic cards?* Or should you save money and key each step of the problem each time?
- ■ Printout. Very often it is sufficient to report the *answer* to a problem. At other times, in order to check your work or justify an answer, it may be desirable to show *every step* of an involved calculation. Would it be worth the extra cost to you to have a calculator which can print out the entire process as well as deliver an answer? (How much is a roll of printout paper for the calculator you are considering?)

Another matter for serious consideration, *before* you buy any calculator, is the *power supply*. Does the calculator which appeals to you operate on batteries alone, or can it also operate, through an ac adapter, on 115 V? Are the batteries rechargeable? In an emergency, can you fit in ordinary cells? What is the manufacturer's estimate of the operating life of a new battery? If the batteries are rechargeable, does recharging take place if you are operating the calculator off the line? Should you consider buying a spare battery pack to be left charging while you are at school or work?

Ask about service, especially warranty service. Are parts and service available locally? Is there a minimum guaranteed turnaround time for service at some distant depot? A maximum time? Will the manufacturer still be in business when you require parts or service?

Check the readout. Can you read it in bright sunlight? Will you ever want to? Is an LED readout easier to see than LCD? Is blue easier to see than red? Is the angle of readout suitable in your normal working position?

Give serious thought to the method of notation which will best satisfy all your needs. Some calculators use an input system, called "direct algebraic notation," which uses this keying program:

KEY   2
KEY   +
KEY   3
KEY   =
Read   5

Others use a system called "reverse Polish notation (RPN)":

KEY   2
KEY   ENTER
KEY   3
KEY   +
Read   5

Do not commit yourself to either system until you have tried both. The authors suggest that RPN has advantages which may well be worth the extra few minutes spent in becoming familiar with it.

MATHEMATICAL TABLES

Since this edition does not contain several pages of tables of logarithms and trigonometric functions, it is essential that your calculator be able to provide at least the features listed above. In addition, it is up to you to know the condition of the power supply so your calculator will not let you down. For emergencies, you will find three-place tables inside the front covers. From time to time you should attempt solutions by using those tables so you will be able to use them when the need arises.

ACCURACY OF CALCULATORS

Different brands of calculators are programmed in different ways. Some manufacturers indicate the accuracy* of their readouts; others will make no mention of how correct their readouts are, regardless of the precision† they indicate. Consider, for example, π:

$$\pi = 3.141\ 592\ 65 \qquad \text{to } nine \text{ significant figures}$$

Many teachers have mistakenly taught their students that $\pi = \frac{22}{7}$, although that is merely the simplest common fraction whose value is closely approximate to that of π. $\frac{22}{7} = 3.142\ 9$; that is, it is good enough to *three* significant figures. However, consider $\frac{355}{113}$ as a more accurate approximation of π. Check your own calculator. Does it have a selector to display π automatically? (You will be using π often.) If not, to what accuracy should you be working? Will you be using 20% components? 10%? 5%? 1%? In the past, electronics students learned that a 10-inch (in) slide rule could be read to 0.1%, which would guarantee three significant figures—adequate for all practical purposes. (Standard switchboard meters, for example, are seldom correct to within 3%.)

* Accuracy is defined as the conformity of an indicated value to an accepted standard or true value. The difference between an observed value and the true value is called the inaccuracy, or error. The smaller the error, the less the inaccuracy and the greater the accuracy.

† Precision is defined as the quality of being sharply defined or stated. A measure of the precision of a quantity is the number of distinguishable alternatives from which the quality was selected, which is sometimes indicated by the number of significant digits the quantity contains. For example, A quartz watch can indicate very *precisely* the elapsed time in seconds, but it may indicate very *inaccurately* the time of day because of a faulty setting.

The answers to problems in this book should be taken to three significant figures (see Sec. 6-1) unless otherwise specified. That will eliminate almost all arguments between users of different brands and models of calculator. We suggest that you round off a display of five or more figures to three. That will meet almost all circuit requirements unless it is obvious from the problem that greater precision is required. There will certainly be times when three figures are inadequate; you must, for example, keep the frequency of your transmitter to a closer precision than three figures.

In your later studies in this book you will study binary fractions. 0.2 denary becomes 0.001100110 binary—accurate to about 2%. To achieve better accuracy for such number system conversions could require 12 to 15 digits!

A-2 TABLES

TABLE 1 MATHEMATICAL SYMBOLS

$\times$ or $\cdot$	multiplied by	$\cong$	is approximately equal to	$\therefore$	therefore
$\div$ or $/$	divided by	$\neq$	does not equal	$\angle$	angle
$+$	positive, plus, add, OR	$>$	is greater than	$\perp$	perpendicular to
$-$	negative, minus, subtract	$\gg$	is much greater than	$\parallel$	parallel to
$\pm$	positive or negative, plus or minus	$<$	is less than	$|n|$	absolute value of n
$\mp$	negative or positive, minus or plus	$\ll$	is much less than	$\triangle$	increment of
$=$ or $::$	equals	$\geq$	greater than or equal to	$\%$	percent
$\equiv$	identity	$\leq$	less than or equal to	$\propto$	is proportional to

TABLE 2 LETTER SYMBOLS

Term	Symbol	Term	Symbol	Term	Symbol	Term	Symbol
Altitude	a	Drain	D	Number of turns	n	Speed of light	c
Area	A	Electromotive force	V, v	Ohm	Ω	Temperature	t
Base	B	Emitter	E	Period (of time)	T	Time	t
Capacitance	C	Frequency	f	Power	P	Transistor	Q
Cathode	K	Gate	G	Reactance	X	Tube (valve)	V
Collector	C	Impedance	Z	Resistance	R, r	Voltage	V, v
Current	I, i	Inductance	L	Resonant frequency	f_r	Wavelength	λ
Diode	D	Length	l	Rise time	t_r	Width	ω

TABLE 3 ABBREVIATIONS AND UNIT SYMBOLS

Term	Abbreviation	Term	Abbreviation	Term	Abbreviation
Alternating current	ac	Circular	cir	Counter electromotive force	cemf
Ampere	A	Clockwise	cw		
Ampere-hour	Ah	Cologarithm	colog	Cubic	. . . ^{3}
Amplitude modulation	AM	Continuous wave	CW	Cubic centimeter	cm^3
Antilogarithm	antilog	Cosecant	csc	Cubic foot	ft^3
Audio frequency	AF	Cosine	cos	Cubic inch	in^3
Bel	B	Cotangent	cot	Cubic meter	m^3
Candela	cd	Coulomb	C	Cubic yard	yd^3
Centimeter	cm	Counterclockwise	ccw	Cycles per second	Hz

TABLE 3 ABBREVIATIONS AND UNIT SYMBOLS (*Continued*)

Term	Abbreviation	Term	Abbreviation	Term	Abbreviation
Decibel	dB	Logarithm (common, base 10)	log	Most significant bit	MSB
Decibels referred to a level of one milliwatt	dBm	Logarithm (any base)	$\log_a$	Nano (prefix, $= 1 \times 10^{-9}$)	n
Degree (interval or change)	deg	Logarithm (natural base ε)	$\log_\varepsilon$, ln	Nanoampere	nA
Degrees Celsius	°C	Low frequency	LF	Nanofarad	nF
Degrees Fahrenheit	°F	Lowest common denominator	LCD	Nanosecond	ns
Degrees Kelvin	K	Lowest common multiple	LCM	Nanowatt	nW
Diameter	diam	Lowest significant bit	LSB	Neper	Np
Direct current	dc	Lumen	lm	Number	No. or spell
Dozen	spell	Lux	lx	Ohms	Ω
Efficiency	spell	Maximum	max	Ohms per kilometer	Ω/km
Electromotive force	emf	Mega (prefix, $= 1 \times 10^6$)	M	Ounce	oz
Equation	Eq.	Megacycles per second	MHz	Peak-to-peak	p-p
Farad	F	Megahertz	MHz	Pico (prefix, $= 1 \times 10^{-12}$)	p
Foot, feet	ft	Megavolt	MV	Picoampere	pA
Feet per minute	ft/min	Megawatt	MW	Picofarad	pF
Feet per second	ft/s	Megohm	MΩ	Picosecond	ps
Feet per second squared	ft/s²	Meter	m	Picowatt	pW
Figure	Fig.	Meter-kilogram-second system	MKS	Pound	lb
Frequency	spell	Meters per second	m/s	Power factor	PF
Frequency modulation	FM	Mho	S	Problem	Prob
Giga (prefix, $= 1 \times 10^9$)	G	Micro (prefix, $= 1 \times 10^{-6}$)	μ	Radian	. . .ʳ
Gigacycles per second	GHz	Microampere	μA	Radians per second	ʳ/s
Gigahertz	GHz	Microfarad	μF	Radio frequency	RF
Gram	g	Microhenry	μH	Radius	r, R
Henry	H	Micromho	μS	Range (distance)	R
Hertz	Hz	Micromicro (prefix, $= 1 \times 10^{-12}$)	p	Revolutions per minute	rev/min
High frequency	HF	Micromicrofarad	pF	Revolutions per second	rev/s
Highest common factor	HCF	Microsecond	μs	Root mean square	rms
Hour	h	Microsiemens	μS	Secant	sec
Hundred	spell, or × 10²	Microvolt	μV	Second	s
Inch	in	Microwatt	μW	Siemens	S
Inches per second	in/s	Mile	mi	Sine	sin
Intermediate frequency	IF	Miles per hour	mi/h	Square centimeter	cm²
Joule	J	Miles per minute	mi/min	Square foot	ft²
Kilo (prefix, $= 1 \times 10^3$)	k	Miles per second	mi/s	Square inch	in²
Kilocycles per second	kHz	Milli (prefix, $= 1 \times 10^{-3}$)	m	Square meter	m²
Kilogram	kg	Milliampere	mA	Square yard	yd²
Kilohertz	kHz	Millihenry	mH	Tangent	tan
Kilohm	kΩ	Millimeter	mm	Ultrahigh frequency	UHF
Kilometer	km	Millisecond	ms	Var (reactive voltampere)	var
Kilometers per hour	km/h	Millivolt	mV	Very high frequency	VHF
Kilovars	kvar	Milliwatt	mW	Volt	V
Kilovolt	kV	Minimum	min	Voltampere	VA
Kilovoltampere	kVA	Minute	min	Watt	W
Kilowatt	kW			Watthour	Wh
Kilowatthour	kWh			Wattsecond	Ws
Knot	kn			Webers per square meter	Wb/m²
				Yard	yd

TABLE 4   GREEK ALPHABET

Name	Capital	Lower-case	Commonly Used to Designate
Alpha	A	α	Angles, area, coefficients
Beta	B	β	Angles, flux density, coefficients
Gamma	Γ	γ	Conductivity, specific gravity
Delta	Δ	δ	Variation, density
Epsilon	E	ϵ	Base of natural logarithms
Zeta	Z	ζ	Impedance, coefficients, coordinates
Eta	H	η	Hysteresis coefficient, efficiency
Theta	Θ	θ	Temperature, phase angle
Iota	I	ι	
Kappa	K	κ	Dielectric constant, susceptibility
Lambda	Λ	λ	Wavelength
Mu	M	μ	Micro, amplification factor, permeability
Nu	N	ν	Reluctivity
Xi	Ξ	ξ	
Omicron	O	o	
Pi	Π	π	Ratio of circumference to diameter = 3.1416
Rho	P	ρ	Resistivity
Sigma	Σ	σ	Summation
Tau	T	τ	Time constant, time phase displacement
Upsilon	Υ	υ	
Phi	Φ	ϕ	Magnetic flux, angles
Chi	X	χ	
Psi	Ψ	ψ	Dielectric flux, phase difference
Omega	Ω	ω	Capital, ohms; lower case, angular velocity

TABLE 5   CONVERSION FACTORS*

Multiply	By	To Obtain
Avoirdupois pounds	0.4536	Kilograms
Circular mils	5.067×10^{-4}	Square millimeters
Coulombs	6.242×10^{18}	Electric charges
Feet	0.3048	Meters
Gallons (imperial)	4.546	Liters
Gallons (U.S. dry)	4.405	Liters
Gallons (U.S. liquid)	3.785	Liters
Horsepower	0.746	Kilowatts
Inches	25.4	Millimeters
Kilograms	2.205	Avoirdupois pounds
Kilometers	3.28×10^{3}	Feet
Meters	3.28	Feet
Meters	39.37	Inches
Mils	2.54×10^{-2}	Millimeters
Statute miles	1.609	Kilometers
Yards	0.9144	Meters

*Selected from H. F. R. Adams, *SI Metric Units: An Introduction,* McGraw-Hill Ryerson Ltd., 1974, by permission.

TABLE 6   STANDARD ANNEALED COPPER WIRE SOLID*
AMERICAN WIRE GAGE (BROWN AND SHARPE) (20° C)

Gage	Diameter, mm	Cross Section, sq mm	Ohms per Kilometer	Meters per Ohm	Kilograms per Kilometer
0000	11.68	107.2	0.160 8	6 219	953.2
000	10.40	85.01	0.202 8	4 931	755.8
00	9.266	67.43	0.255 7	3 911	599.5
0	8.252	53.49	0.322 3	3 102	475.5
1	7.348	42.41	0.406 5	2 460	377.0
2	6.543	33.62	0.512 8	1 950	298.9
3	5.827	26.67	0.646 6	1 547	237.1
4	5.189	21.15	0.815 2	1 227	188.0
5	4.620	16.77	1.028	972.4	149.0
6	4.115	13.30	1.297	771.3	118.2
7	3.665	10.55	1.634	612.0	93.80
8	3.264	8.367	2.061	485.3	74.38
9	2.906	6.631	2.600	384.6	58.95
10	2.588	5.261	3.277	305.2	46.77
11	2.30	4.17	4.14	242	37.1
12	2.05	3.31	5.21	192	29.4
13	1.83	2.63	6.56	152	23.4
14	1.63	2.08	8.28	121	18.5
15	1.45	1.65	10.4	95.8	14.7
16	1.29	1.31	13.2	75.8	11.6
17	1.15	1.04	16.6	60.3	9.24
18	1.02	0.823	21.0	47.7	7.32
19	0.912	.653	26.4	37.9	5.81
20	.813	.519	33.2	30.1	4.61
21	.724	.412	41.9	23.9	3.66
22	.643	.324	53.2	18.8	2.88
23	.574	.259	66.6	15.0	2.30
24	.511	.205	84.2	11.9	1.82
25	.455	.162	106	9.42	1.44
26	.404	.128	135	7.43	1.14
27	.361	.102	169	5.93	0.908
28	.320	.080 4	214	4.67	.715
29	.287	.064 7	266	3.75	.575
30	.254	.050 7	340	2.94	.450
31	.226	.040 1	430	2.33	.357
32	.203	.032 4	532	1.88	.288
33	.180	.025 5	675	1.48	.227
34	.160	.020 1	857	1.17	.179
35	.142	.015 9	1 090	0.922	.141

* Bureau of Standards Handbook 100, reproduced by permission.

Gage	Diameter, mm	Cross Section, sq mm	Ohms per Kilometer	Meters per Ohm	Kilograms per Kilometer
36	.127	.012 7	1 360	.735	.113
37	.114	.010 3	1 680	.595	.091 2
38	.102	.008 11	2 130	.470	.072 1
39	.089	.006 21	2 780	.360	.055 2
40	.079	.004 87	3 540	.282	.043 3
41	.071	.003 97	4 340	.230	.035 3
42	.064	.003 17	5 440	.184	.028 2
43	.056	.002 45	7 030	.142	.021 8
44	.051	.002 03	8 510	.118	.018 0
45	.0447	.001 57	11 000	.0910	.014 0
46	.0399	.001 25	13 800	.0724	.011 1
47	.0356	.000 993	17 400	.0576	.008 83
48	.0315	.000 779	22 100	.0452	.006 93
49	.0282	.000 624	27 600	.0362	.005 55
50	.0251	.000 497	34 700	.0288	.004 41
51	.0224	.000 392	43 900	.0228	.003 49
52	.0198	.000 308	55 900	.0179	.002 74
53	.0178	.000 248	69 400	.0144	.002 21
54	.0157	.000 195	88 500	.0113	.001 73
55	.0140	.000 153	112 000	.008 89	.001 36
56	.0124	.000 122	142 000	.007 06	.001 08

TABLE 7   DECIMAL MULTIPLIERS

0.000 000 000 000 000 001	$= 10^{-18}$	$=$ ten to the negative *eighteenth* power	$=$ atto	a
0.000 000 000 000 001	$= 10^{-15}$	$=$ ten to the negative *fifteenth* power	$=$ femto	f
0.000 000 000 001	$= 10^{-12}$	$=$ ten to the negative *twelfth* power	$=$ pico	p
0.000 000 001	$= 10^{-9}$	$=$ ten to the negative *ninth* power	$=$ nano	n
0.000 001	$= 10^{-6}$	$=$ ten to the negative *sixth* power	$=$ micro	μ
0.001	$= 10^{-3}$	$=$ ten to the negative *third* power	$=$ milli	m
1	$= 10^{0}$	$=$ ten to the *zero* power	$=$ unit	
1 000	$= 10^{3}$	$=$ ten to the *third* power	$=$ kilo	k
1 000 000	$= 10^{6}$	$=$ ten to the *sixth* power	$=$ Mega	M
1 000 000 000	$= 10^{9}$	$=$ ten to the *ninth* power	$=$ Giga	G
1 000 000 000 000	$= 10^{12}$	$=$ ten to the *twelfth* power	$=$ Tera	T
1 000 000 000 000 000	$= 10^{15}$	$=$ ten to the *fifteenth* power	$=$ Peta	P
1 000 000 000 000 000 000	$= 10^{18}$	$=$ ten to the *eighteenth* power	$=$ Exa	E

TABLE 8   ROUNDED VALUES OF PREFERRED NUMBERS*

Series	R5	R10	R20	R40	R6†	R12†	R24†
Approximate Ratio	1.6	1.25	1.12	1.06	1.46	1.21	1.1
	1	1	1	1	1	1	1
				1.06			1.1
			1.12	1.12		1.2	1.2
				1.18			1.3
		1.25	1.25	1.25	1.5	1.5	1.5
				1.32			1.6
			1.40	1.40		1.8	1.8
				1.50			2.0
	1.60	1.60	1.60	1.60	2.2	2.2	2.2
				1.70			2.4
			1.80	1.80		2.7	2.7
				1.90			3.0
		2.00	2.00	2.00	3.3	3.3	3.3
				2.12			3.6
			2.24	2.24		3.9	3.9
				2.36			4.3
	2.50	2.50	2.50	2.50	4.7	4.7	4.7
				2.65			5.1
			2.80	2.80		5.6	5.6
				3.00			6.2
		3.15	3.15	3.15	6.8	6.8	6.8
				3.35			7.5
			3.55	3.55		8.2	8.2
				3.75			9.1
	4.00	4.00	4.00	4.00	10	10	10
				4.25			
			4.50	4.50			
				4.75			
		5.00	5.00	5.00			
				5.30			
			5.60	5.60			
				6.00			
	6.30	6.30	6.30	6.30			
				6.70			
			7.10	7.10			
				7.50			
		8.00	8.00	8.00			
				8.50			
			9.00	9.00			
				9.50			
	10.00	10.00	10.00	10.00			

*These tables have been adapted from various international, American, and British standards.
†The R6, R12, and R24 tables are sometimes referred to as E6, E12, and E24, respectively.

Answers to Odd-Numbered Problems

Note The accuracy of answers to numerical computations is, in general, to three significant figures from a five-figure readout of a hand-held electronic calculator.

PROBLEMS 2-1

1. (a) 25 times R
(b) 6 times r
(c) 0.25 times I

3. (a) $396.00
(b) $2.75n

5. 12.5I A

7. $\frac{2}{3}C$ pF, $4C$ pF, $48C$ pF

9. (a) $16 + R\ \Omega$
(b) $v + 220$ V
(c) $i - I$ A

11. $L_2 = L_1 - 125$ mH

13. (a) 44 A
(b) 0.25 A

15. (a) 2.99 s
(b) 0.685 s

17. 0.293 m

PROBLEMS 2-2

1. (a) 72
(b) 276
(c) 1296
(d) 72
(e) 36
(f) 207

3. (a) Monomial
(b) Monomial
(c) Monomial
(d) Binomial
(e) Trinomial

(f) Binomial
(g) Trinomial
(h) Trinomial
(i) Monomial
(j) Trinomial

5. (a) $I = \dfrac{V}{R}$
(b) $V = IR$
(c) $P = RI^2$
(d) $R_1 = R_2 + R$
(e) $K = \dfrac{M}{\sqrt{L_1 L_2}}$

(f) $R_p = \dfrac{R_1 R_2}{R_1 + R_2}$
(g) $N = \dfrac{R_m}{R_s} + 1$

7. 10 197 μH Note that, all other factors remaining equal, if the number of turns is tripled, the inductance is multiplied by a factor of nine (3^2).

9. (a) Increased by a factor of 4 (b) Increased by a factor of 9
(c) Reduced to a value one-fourth the original

PROBLEMS 3-1

1. 71
3. -46
5. 28
7. -1081
9. 208.56
11. 4
13. $-10\frac{7}{32}$
15. $\frac{2}{15}$

PROBLEMS 3-2

1. 61
3. 213
5. 994
7. 3.84
9. $-10\frac{1}{16}$
11. (a) 67°
(b) 26°
(c) 159°

13. $364.80
15. 8 V

PROBLEMS 3-3

1. $11i$
3. $112IZ$
5. $4I - 5i$
7. $3IR + 13V$
9. $128\theta - 110\phi$

11. $27i^2r + 10W - 3vi + 49w$
13. $1.46vI + 3.82W + 0.75I^2r$
15. $-\frac{11}{48}\pi ft - 2\frac{1}{8}\pi Z$

17. $10\phi + 10\theta$
19. $47.6\dfrac{V^2}{R} - 16.4VI + 5.8I^2R$
21. $3.90IZ - 1.31IR - 0.41IX$

23. $6.64\psi - 7.1\lambda$

PROBLEMS 3-4

1. $3 - 7y$ **3.** $10R - 3X + 3$ **5.** $10\dfrac{V^2}{R} - 3VI$ **7.** $\alpha + 2\beta$ **9.** $17a - 10b + 6c$

PROBLEMS 3-5

1.
(a) $3X + (X_C - X_L + Z)$
(b) $\alpha + (6\beta - 3\phi + \lambda)$
(c) $5W + (6I^2R - 3VI + 7I^2Z)$
(d) $\dfrac{V^2}{R} + (-3I^2R + 7I^2Z - 4VI)$
(e) $8\lambda + 3\mu + (-7\theta - 3\phi + 6\alpha)$

3. $X^2 + R^2 - N$

7. $Z - \sqrt{r^2 + x^2}$

11. $X_C - \dfrac{1}{2\pi f C_1}$

5. $16.8 + v\,V$

9. $P - I^2R - \dfrac{V^2}{R}$

PROBLEMS 4-1

1. 12   **3.** 13.6   **5.** $-\frac{15}{256}$ **7.** 0.000 000 938

9. eit **11.** $2\pi f L_1 L_2$ **13.** $\dfrac{1}{2\pi f C_p}$ **15.** $-\dfrac{\psi\mu}{\theta\phi}$

PROBLEMS 4-2

1. x^5 **3.** $-e^{10}$ **5.** $6m^4$ **7.** $-60m^4x^3$ **9.** abm^{n+p}

11. $8p^3$ **13.** $6a^3b^4c^3d^7$ **15.** $-\dfrac{\pi M X_L}{4}$ **17.** $-0.075v^3i^4rw$ **19.** a^6

PROBLEMS 4-3

1. $18a + 30b$
3. $4I^2R_1 + 8I^2R_2$
5. $4.7\lambda^2\phi + 9.4\theta\phi - 14.1\mu\phi$
7. $2\alpha^4\beta^2 + 1.5\alpha^3\beta^3 - 2.5\alpha^2\beta^4$
9. $15a^3r_1r_2 + 6a^2r_1^2r_2 - 18ar_1^3r_2$
11. $\dfrac{iI^3RZ}{3} - \dfrac{iI^3R^2Z}{6} - \dfrac{2i^2IZ^2}{9}$
13. $3I^3PR - 6Ii^2Pr + 2IP^2$

15. $0.157V^3IZ^3 + 0.314VIZ^5 - 10.5IZ^6$
17. $15\phi - 21\theta$
19. $\theta^3 - \phi^3$
21. $0.9\pi\omega + 3\eta\pi^2 + 2.5\eta\omega^2 - 6.5\pi\omega^3$
23. $\dfrac{8\lambda E^2}{3} + 4\lambda Ee - \dfrac{\lambda e^2}{2}$
25. $0.125IR - 0.025IR_1 - 0.4125IR_2$
27. 0
29. $7s$

PROBLEMS 4-4

1. $\alpha^2 + 2\alpha + 1$ **3.** $\alpha^2 - 2\alpha + 1$ **5.** $\beta^2 - 9$
7. $p^2 + 8p + 15$ **9.** $r^2 - 8r - 33$ **11.** $m^2 + 62m + 8$
13. $3\alpha^2 + 15\alpha\beta - 42\beta^2$ **15.** $6\theta^2 - 7\theta\lambda - 5\lambda^2$ **17.** $6m^2 - 5mn - 6n^2$
19. $5R^2 - 17RZ + 6Z^2$ **21.** $6a^3 + 17a^2 + 2a - 1$ **23.** $2R^3 - 2R^2r - 2Rr^2 + 2r^3$
25. $a^3 - a^2b - ab^2 + b^3$ **27.** $\theta^3 - \theta^2\phi - \theta\phi^2 + \phi^3$ **29.** $a^3 + 3a^2b + 3ab^2 + b^3$
31. $x^2 + 2xy + y^2$ **33.** $M^2 - 2MN + N^2$ **35.** $8a^3 + 24a^2w + 24aw^2 + 8w^3$
37. $16I^4R^2 - 38I^2R + 14$ **39.** $10a^3 + 4a^2b - 16a^2 - 5ab^2 - 14ab + a$

PROBLEMS 4-5

1. 5   **3.** 5   **5.** $-\frac{4}{3}$ **7.** $-2\pi f C$ **9.** $\dfrac{V \times 10^8}{L_v}$ **11.** -6

13. -225 **15.** $-\frac{5}{2}$

PROBLEMS 4-6

1. $4x^2y^4$ **3.** $-20\phi^2\psi^3$ **5.** $-\dfrac{4X_cZ^2}{3}$ **7.** $3\eta^4\lambda^3\pi$ **9.** $\dfrac{3mn^2p}{4}$ **11.** $-9c$

13. $-4\lambda^3\psi^4$ **15.** $\dfrac{b^7d^4}{3ac^6}$ **17.** $-\dfrac{\phi^6}{4\theta^{12}\psi\Omega^3}$ **19.** $\dfrac{90\,000\alpha^2\beta^5}{\gamma^2}$

PROBLEMS 4-7

1. $4x + 5y$ **3.** $12\alpha^2 - 9\beta^2$ **5.** $3R_1 + 6R_2 - 4R_3$ **7.** $\dfrac{0.005\mu^3}{\pi} + 10\mu\pi$

9. $\dfrac{3m^3}{10} - \dfrac{7m}{5} - \dfrac{6}{5m}$ **11.** $6 + 10xz - 5x^2z^2 - 3x^4y$

13. $2(\theta + \phi) - 4(\theta + \phi)^3 + 3(\theta + \phi)^5$ **15.** $\dfrac{(VI + P)^2}{2} - 2 + \dfrac{6}{VI + P}$

17. $\dfrac{1}{I\left(\omega L - \dfrac{1}{\omega C}\right)} - 2I\left(\omega L - \dfrac{1}{\omega C}\right) - 5I^3\left(\omega L - \dfrac{1}{\omega C}\right)^3$

19. $3(\theta - \phi)^2 - 6(\theta + \phi)(\theta - \phi)^3 - \dfrac{9(\theta - \phi)}{\theta + \phi}$

PROBLEMS 4-8

1. $x + 1$ **3.** $\theta + 3$ **5.** $2E - 6$ **7.** $3R^2 - 4Z - 7$

9. $K^2 + 7K + 14 + \dfrac{6}{K - 1}$ **11.** $E + e$ **13.** $E^3 + E^2e + Ee^2 + e^3$ **15.** $V^2 + I^2R^2$

17. $X^5 - X^4Y + X^3Y^2 - X^2Y^3 + XY^4 - Y^5$ **19.** $\theta^2 + 2\theta\phi + \phi^2$ **21.** $2R_2 - 3$

23. $10E^2 - 3E - 12 + \dfrac{7E - 45}{3E^2 + 2E - 4}$ **25.** $3R + \dfrac{1}{3}$ **27.** $6x - \dfrac{y}{3} - \dfrac{1}{2}$

29. $\dfrac{3L_1^2}{8} - \dfrac{L_1}{4} - \dfrac{2}{3}$

PROBLEMS 5-1

1. $x = 4$ **3.** $k = -5$ **5.** $p = 6$ **7.** $\pi = 5$ **9.** $IR = 4$
11. $\alpha = -10$ **13.** $E = -5$ **15.** $Q = -2$ **17.** $I = -1.4$ **19.** $\beta = 1$

PROBLEMS 5-2

1. $V - 75$ V **3.** $d = rt$ km **5.** $y - t$ years **7.** $\dfrac{Z}{t}$ kilometers per minute (km/min)

9. 110 V **11.** $I = \dfrac{V}{R}$ **13.** 12 m by 6 m **15.** 4.25, 10.75, and 8.5 m

17. $h^2 = a^2 + b^2$ **19.** 63, 64, 65

PROBLEMS 5-3

1. $C = \dfrac{Q}{V}, V = \dfrac{Q}{C}$ **3.** $Z^2 = R^2 + X^2, X^2 = Z^2 - R^2$

5. $R = \dfrac{KL}{m}, K = \dfrac{Rm}{L}, m = \dfrac{KL}{R}$ **7.** $\lambda = \dfrac{v}{f}, v = f\lambda$

9. $L = \dfrac{RQ}{\omega}$, $Q = \dfrac{\omega L}{R}$, $\omega = \dfrac{RQ}{L}$

11. $X_C = \dfrac{1}{2\pi f C}$, $f = \dfrac{1}{2\pi C X_C}$

13. $\phi = HA$, $A = \dfrac{\phi}{H}$

15. $V = \dfrac{BLv}{10^8}$, $L = \dfrac{V \times 10^8}{Bv}$, $v = \dfrac{V \times 10^8}{BL}$

17. $V_s = \dfrac{V_p I_p}{I_s}$

19. $I = \dfrac{V - v}{R}$, $V = IR + v$, $v = V - IR$

21. $\theta = \omega t$, $\omega = \dfrac{\theta}{t}$

23. $V = \dfrac{V_0 + V_t}{2}$, $V_t = 2V - V_0$

25. $r^3 = \dfrac{3A}{4\pi}$

27. $Z_t = \dfrac{F(R - r)}{C}$, $F = \dfrac{CZ_t}{R - r}$, $R = \dfrac{CZ_t + Fr}{F}$, $r = \dfrac{FR - CZ_t}{F}$

29. $V_b = iR_L + v_b$, $v_b = V_b - iR_L$, $i = \dfrac{V_b - v_b}{R_L}$

31. $l = \dfrac{Rd^2}{\rho}$, $\rho = \dfrac{Rd^2}{l}$, $d^2 = \dfrac{\rho l}{R}$

33. $A = \dfrac{Cd}{0.0884K(n - 1)}$, $n = \dfrac{Cd + 0.0884KA}{0.0884KA}$

35. $L = CRZ_r$, $C = \dfrac{L}{RZ_r}$, $R = \dfrac{L}{CZ_r}$

37. $\beta = \dfrac{\gamma\omega\alpha}{\eta}$

39. $Q = \dfrac{\rho h v}{e}$

41. $C_2 = \dfrac{V_3 - V_2}{\omega^2 L V_3}$

43. $I_n = \dfrac{Q - I_p p}{n}$

45. $R_1 = \dfrac{1}{\omega_{01} C} - R_2$

47. $4000\ \Omega$

49. 5 m

PROBLEMS 5-4

1. $\frac{1}{4}$ **3.** $\frac{5}{4}$ **5.** $\frac{4}{6}, \frac{6}{9}, \frac{12}{18}$, etc. **7.** $\frac{1}{3}$ **9.** 32:1

PROBLEMS 5-5

1. 10 **3.** 200 **5.** $X = 15$ **7.** $IR = 8$ **9.** $Q = 0.0014$

PROBLEMS 5-6

1. $D \propto R$, $D = kR$ **3.** $C \propto A$, $C = kA$ **5.** $X_C \propto \dfrac{1}{C}$, $X_C = \dfrac{k}{C}$ **7.** $T \propto \sqrt{L}$, $T = k\sqrt{L}$

9. $V \propto \dfrac{1}{P}$, $V = \dfrac{k}{P}$ **11.** $L \propto \dfrac{1}{d^2}$, $L = \dfrac{k}{d^2}$ **13.** $1.74\ kV$ **15.** 5400 kg

PROBLEMS 6-1

1. 6 **3.** 6 **5.** 3 **7.** 1 **9.** 4

PROBLEMS 6-2

1. 6.43×10^5 **3.** 6.53×10^3 **5.** 9.44×10^{-9} **7.** 3.67×10^{-1}
9. 2.50×10^{-1} **11.** 3.99×10^4 **13.** 2.59×10^{-2} **15.** 2.76×10^5
17. 1.08×10^{-7} **19.** 3.00

PROBLEMS 6-3

1. 1.00×10^{-2} **3.** 3.92×10^{-1} **5.** 7.14×10^{-12} **7.** 3.11×10^{11}
9. 3.20×10 **11.** 5.65×10^0 **13.** $9.42 \times 10^4\ \Omega$ **15.** $2.20 \times 10^{-1}\ \Omega$

PROBLEMS 6-4

1. 5.00×10^{-7} **3.** 1.05×10^{-4} **5.** 1.00×10^{-13} **7.** 2.87×10^{10}
9. 2.55×10^2 **11.** 1.63×10^{-3} **13.** $6.63 \times 10^2\ \Omega$ **15.** $1.26 \times 10^{-6}\ \Omega$

PROBLEMS 6-5

1. 10^{12} **3.** 10^{20} **5.** 6.25×10^{14} **7.** 2.56 **9.** 5×10^{-3} **11.** 30
13. 1.01×10^{4} **15.** 1.50×10^{6} Hz **17.** 7.50×10^{6} Hz **19.** 1.20×10^{6} Hz

PROBLEMS 6-6

1. (*a*) 3100
(*b*) 3.10×10^{3}

3. (*a*) 4 190 000 000 000
(*b*) 4.19×10^{12}

5. (*a*) 6 279 999.841
(*b*) 6.28×10^{6}

PROBLEMS 7-1

1. (*a*) 4.30×10^{6} mV
(*b*) 4.30×10^{9} µV
(*c*) 4.30 kV

3. (*a*) 1.35×10^{-3} kV
(*b*) 1.35×10^{6} µV
(*c*) 1.35×10^{3} mV

5. (*a*) 3.30 kΩ
(*b*) 3.30×10^{-3} MΩ
(*c*) 3.03×10^{-4} S

7. (*a*) 2.00×10^{-8} F
(*b*) 2.00×10^{-2} µF

9. (*a*) 3.47×10^{-1} kW
(*b*) 3.47×10^{5} mW
(*c*) 3.47×10^{8} µW

11. (*a*) 1.32×10^{0} MHz
(*b*) 1.32×10^{6} Hz

13. (*a*) 4.00×10^{-1} W
(*b*) 4.00×10^{-4} kW

15. (*a*) 1.50×10^{-2} MHz
(*b*) 1.50×10^{4} Hz

17. (*a*) 5.50×10^{4} µA
(*b*) 5.50×10^{1} mA

19. (*a*) 2.70×10^{6} Ω
(*b*) 2.70×10^{3} kΩ

21. (*a*) 3.35×10^{6} µH
(*b*) 3.35 H

23. (*a*) 5.00×10^{2} pF
(*b*) 5.00×10^{-10} F

25. (*a*) 2.50×10^{6} µS
(*b*) 4.00×10^{-1} Ω

27. (*a*) 2.35×10^{0} mA
(*b*) 2.35×10^{-3} A

29. (*a*) 1.50×10^{8} W
(*b*) 1.50×10^{5} kW

PROBLEMS 7-2

1. (*a*) 108 in
(*b*) 274 cm
(*c*) 2.74×10^{3} mm

3. (*a*) 80.7 in
(*b*) 205 cm
(*c*) 2.24 yd

5. (*a*) 2.88 mi
(*b*) 4.63×10^{3} m
(*c*) 4.63 km

7. 1.63 mm
13. 3.22×10^{-3} Ω/cm

9. 3.74 mH/km
15. 4.72 in/min

11. 6×10^{-2} dB/100 m

PROBLEMS 7-3

3. $X_L = 2\pi f L \ \Omega$ **5.** $f = \dfrac{159}{\sqrt{LC}}$ MHz **7.** $\delta = \dfrac{6.62}{\sqrt{f}}$ cm

9. $R_{ac} = 83.2 \times 10^{-9}\, \dfrac{\sqrt{f}}{d}$ Ω/cm **13.** 435 cm **15.** 66.2 cm

17. (*a*) 72.4 cm (*b*) 75.9 cm (*c*) 30.7 cm
19. (*a*) 72.4 cm (*b*) 75.9 cm (*c*) 68.6 cm (*d*) 30.7 cm (*e*) 15.4 cm

PROBLEMS 8-1

1. 4.40 A **3.** 6.20 A **5.** 0.080 µA **7.** 3.75 A **9.** (*a*) 0.571 A (*b*) 0.635 A

PROBLEMS 8-2

1. (*a*) 5.6×10^{3} W (*b*) 5.6 kW **3.** 0.833 A **5.** 15.0 hp **7.** 1119 kW
9. 0.108 W **11.** (*a*) 0.1 mA (*b*) 0.47 mW **13.** (*a*) 20.8 pW (*b*) 0.231 µA
15. (*a*) 90.5% (*b*) $24.72 **17.** 18.4 hp **19.** 2.4 kW

PROBLEMS 8-3

1. (a) 69.6 mA 3. (a) 121 Ω 5. 179 Ω 7. 1.43 Ω 9. (a) 1.8 Ω
   (b) 47.3 V   (b) 100 W   (b) 1.5 Ω
   (c) 1.60 W   (c) 80.3 W   (c) 3.48 kW

PROBLEMS 8-4

1. (a) 500 Ω 3. (a) R_S = 3.3 kΩ 5. (a) I_C = 0.5 mA; I_E = 0.51 mA
   (b) 19.5 V   (b) I_D = 1.8 mA   (b) R_B = 1.97 MΩ
                     (c) I_T = 1.82 mA   (c) V_{CE} = +15 V
                     (d) P = 45 mW   (d) P = 7.5 mW
7. (a) 60 kΩ 9. 2.5 kΩ
   (b) 600 μW

PROBLEMS 9-1

1. (a) 4.14 Ω 3. 268 Ω 5. 1.48 Ω 7. 0.513 Ω 9. 1.31 m
   (b) 0.166 Ω

PROBLEMS 9-2

1. 13 Ω 3. 73.3 Ω 5. 0.0382 $\mu\Omega \cdot$ m 7. 2.36 km 9. 444 m

PROBLEMS 9-3

1. 4.60 Ω 3. 15.8 Ω 5. No

PROBLEMS 9-4

1. (a) 0.639 Ω, (b) 1.499 kg 3. (a) 4.08 km, (b) 21.3 Ω 5. 2112 Ω
7. No. 1 wire   9. (a) No. 5 wire, (b) 96.3%

PROBLEMS 10-1

1. x^2y^2 3. $v^3i^6Z^3$ 5. $16\pi^2\phi^2$ 7. $-8I^3R^3$ 9. $2\pi X_L^2$
11. $-\dfrac{1}{4\pi^2 f^2 C^2}$ 13. $-\dfrac{125P^6}{V^3I^3}$ 15. $-\dfrac{V^6}{8g^3}$ 17. $\dfrac{B^6A^3I^3}{512\omega^3}$ 19. $-\frac{16}{9}\pi^2R^6$ 21. $\dfrac{x^{12}y^{18}}{p^{15}}$

PROBLEMS 10-2

1. $\pm a$ 3. $\pm 3i$ 5. $-\omega$ 7. $\pm 5\lambda^2\Omega^3$ 9. $3x^2$ 11. 4
13. $\pm 13m^2np^3$ 15. $30^2\phi^4\omega$ 17. $\pm\dfrac{16\pi rx^2}{17z^3\phi^2}$ 19. $\pm\dfrac{25r^3s^2t^4}{4x^3z^5}$ 21. $\dfrac{4a\omega^2}{5x^2z^4}$ 23. $\pm\dfrac{5vt}{16a^4bx}$

PROBLEMS 10-3

1. $2(a + 3)$ 3. $\theta(3 + \phi + 4\omega)$ 5. $10i(2r - z)$ 7. $\dfrac{ay}{36}(4ay + 12a^2 - 3y^2)$
9. $2a^2bc(ab + 4c^2 + 6bc)$ 11. $36\alpha^2\beta^2\omega^2(\alpha^2\beta - 2\omega^3 + 5\beta^3)$
13. $\dfrac{1}{3648}Ii^2(57Ii + 48i^2 - 76I^2)$ 15. $120\eta\theta^2\phi\omega(6\eta^3\phi^2 + 9\eta\theta^2\omega + 5\eta^2\theta\omega - 4\theta^4)$

PROBLEMS 10-4

1. $\theta^2 + 6\theta + 9$ **3.** $m^2 - 2mR + R^2$ **5.** $\alpha^2 + 32\alpha + 256$ **7.** $9X^2 - 6XR + R^2$

9. $F^2 - 2Ff + f^2$ **11.** $25\theta^2 + 40\theta\phi + 16\phi^2$ **13.** $81r_1^2 - 54r_1r_2 + 9r_2^2$

15. $1 + 2X_L^2 + X_L^4$ **17.** $36v^4 - 24v^2t^3 + 4t^6$ **19.** $900 - 180 + 9 = 729$

21. $36\pi^2R^4 - 24\pi^2R^2r^2 + 4\pi^2r^4$ **23.** $2.25\theta^4 - 1.5\theta^2\alpha + 0.25\alpha^2$

25. $\frac{9}{16}X^4 - \frac{3}{4}X^2Z^2 + \frac{1}{4}Z^4$ **27.** $36\phi^4\omega^2 - 3\phi^2\omega\lambda^2 + \frac{1}{16}\lambda^4$ **29.** $x^2 + x + \frac{1}{4}$

31. $\frac{1}{4} - E + E^2$ **33.** $1 + 2e^3 + e^6$ **35.** $L^4 - \frac{7}{4}L^2P + \frac{49}{64}P^2$ **37.** $\frac{b^2}{9} + \frac{bm}{3} + \frac{m^2}{4}$

39. $R_1^2 - \frac{5}{4}R_1R_2 + \frac{25}{64}R_2^2$

PROBLEMS 10-5

1. $6e$ **3.** 4λ **5.** $10xy$ **7.** $14\omega\pi$ **9.** $12mp$ **11.** I^2 **13.** $16p^2$

15. $\frac{1}{3}\theta\phi\omega$ **17.** $\frac{1}{9}\pi^2$ **19.** $\pm(M + 1)$ **21.** $\pm(4q_1 + q_2)$ **23.** $\pm(3\alpha^2\beta + 9\gamma)$

25. $\pm(\frac{3}{5}\pi R^2 + \frac{2}{3})$ **27.** $\pm(\frac{5}{6}\phi + \frac{2}{7}\lambda)$

PROBLEMS 10-6

1. $3c(a + 2b)$ **3.** $2\lambda(\theta + \phi)^2$ **5.** $6\alpha^2(2\alpha + 5\beta)^2$ **7.** $\frac{20f_0}{\omega}(\omega_1 - \omega_2)^2$

9. $\frac{5r}{16e}(\lambda - 4f^2)^2$ or $\left(\frac{5}{e}\right)\left(\frac{r}{2^4}\right)(\lambda - 4f^2)^2$

PROBLEMS 10-7

1. $\theta^2 - 4$ **3.** $I^2 - i^2$ **5.** $9Q^2 - 4L^2$ **7.** $\frac{4}{9}V^2I^2 - P^2$ **9.** $\frac{4V^4}{R^2} - \frac{9I^4R^2}{P^2}$

PROBLEMS 10-8

1. $(a + b)(a - b)$ **3.** $(2\theta + 4\phi)(2\theta - 4\phi)$ **5.** $(\frac{1}{2} + \theta)(\frac{1}{2} - \theta)$

7. $(1 + 15\omega)(1 - 15\omega)$ **9.** $(9\theta\mu + 1)(9\theta\mu - 1)$

13. $9(ab - 2m - 3pq)(ab - 2m + 3pq)$ **15.** $(5a + 10cl + 12l)(5a + 10cl - 12l)$

PROBLEMS 10-9

1. $\theta^2 + 7\theta + 12$ **3.** $R^2 - R - 2$ **5.** $\theta^2 + 9\theta + 18$ **7.** $9\theta^2 - 3\theta - 2$

9. $I^2 - 7I + 12$ **11.** $\alpha^2 - \frac{5\alpha}{4} + \frac{1}{4}$ **13.** $I^2R^2 + \frac{IR}{6} - \frac{1}{6}$ **15.** $\alpha^2 + \alpha + \frac{2}{9}$

17. $\frac{1}{LC} - \frac{4f}{\sqrt{LC}} + 3f^2$ **19.** $\alpha^2\beta^4 + \frac{3\alpha\beta^2}{10} + \frac{1}{50}$

PROBLEMS 10-10

1. $(a + 1)(a + 2)$ **3.** $(R + 6)(R + 2)$ **5.** $(\beta + 6)(\beta - 4)$ **7.** $(\theta + 4)(\theta + 6)$

9. $(t + 11)(t - 2)$ **11.** $(Z^2 - 2)(Z^2 + 10)$ **13.** $(\pi + 8)(\pi - 7)$

15. $(\omega + 2f)(\omega - 3f)$ **17.** $(\theta - \frac{1}{2})(\theta - \frac{1}{3})$ **19.** $(\phi^2 + \frac{1}{5})(\phi^2 - \frac{1}{10})$

PROBLEMS 10-11

1. $x^2 - 3x - 10$ **3.** $6\phi^2 + 11\phi + 3$ **5.** $12j^2 - 2j - 4$ **7.** $6\omega^2 + 13\omega - 5$

9. $\frac{\omega^2}{4} + 2\omega - 32$ **11.** $6Z^2 + 13IRZ + 5I^2R^2$ **13.** $15X^2 - 94X - 40$

15. $15\theta^2 - 77\theta + 10$ **17.** $35 - 31\pi + 6\pi^2$ **19.** $6\alpha^2 + 31\alpha\beta + 35\beta^2$

21. $4a^2 - 24at + 35t^2$ **23.** $\omega^2 + 0.5\omega f - 0.14f^2$

25. $\frac{x^2}{4} - \frac{3x\lambda}{2} - 4\lambda^2$ **27.** $24Z^2 + \frac{4Z}{IR} + \frac{1}{6I^2R^2}$ **29.** $0.16p^2 - 0.62pq + 0.21q^2$

PROBLEMS 10-12

1. $(\omega + 2)(\omega - 5)$ **3.** $(2m - 3)(4m + 5)$ **5.** $(2x + 5)(3x - 2)$

7. $(3\phi + 4)(3\phi + 2)$ **9.** $(\alpha + 3\beta)(2\alpha - 7\beta)$ **11.** $(10m - 7)(4m + 3)$

13. $(8l + 3w)(10l - 2w)$ **15.** $(12\beta^2 - 9\gamma)(2\beta^2 - \gamma)$ or: $(6\beta^2 - 3r)(4\beta^2 - 3r)$

17. $(9lm - w)(3lm + 2w)$ **19.** $6(\psi + 2\Omega)(\psi - 2\Omega)$ **21.** $(5x + \Delta)(3x - 2\Delta)$

23. $(8\theta + \frac{1}{2})(6\theta + \frac{1}{4})$ **25.** $(0.6\theta + 2)(0.3\theta - 1)$

PROBLEMS 10-13

1. $16\omega^2L^2$ **3.** $\frac{a^{12}b^{12}c^4d^8}{p^8q^{12}r^4}$ **5.** $\pm\frac{12IR}{13FX_C^2}$ **7.** $-\frac{5lm^2}{3x^4y^5z}$ **9.** $6\theta\phi^2\omega^2$

11. $I(R + r)(R - r)$ **13.** $\frac{v^2}{8}\left(\frac{3}{r_1} + \frac{5}{r_2} - \frac{7}{r_3}\right)$ **15.** $\frac{x}{16}(7k - 3l - 9m)$

17. $R^2 + 24R + 144$ **19.** $144I^4 + 16I^2 + \frac{4}{9}$ **21.** $\frac{25\beta^2}{81} - \frac{10\beta\lambda}{3} + 9\lambda^2$ **23.** $6r$

25. $49Q^2$ **27.** $\frac{\lambda^2}{16}$ **29.** $\pm(m + 5)$ **31.** $\pm(4\alpha + 10\beta)$ **33.** $\pm\left(\frac{\phi}{6} - \frac{\lambda}{2}\right)$

35. $6R(4i + 7I)$ **37.** $3i(r + 3)^2$ **39.** $12\omega(8\theta - \phi)^2$ **41.** $\alpha^2 - 4\beta^2$

43. $Z^2 - 144$ **45.** $\frac{576V^2}{I^2R^2} - 4P^2$ **47.** $(Q + 1)(Q - 1)$

49. $\left(2\omega L + \frac{1}{4\omega C}\right)\left(2\omega L - \frac{1}{4\omega C}\right)$ **51.** $(0.05\psi + 0.6\mu)(0.05\psi - 0.6\mu)$ **53.** $\lambda - 2$

55. $\frac{1}{3}\alpha + \frac{2}{7}\beta$ **57.** $\frac{3}{5}e - \frac{4}{9}ir$ **59.** $\kappa^2 - 2\kappa - 8$ **61.** $0.2X_C^2 - 2.9X_C - 1.5$

63. $A^2 - \frac{2A}{15} - \frac{1}{15}$ **65.** $24\mu^2 + 2\mu g_m - 12g_m^2$ **67.** $0.6R^2 + 0.1Rr - 0.2r^2$

69. $24\phi^2 + 2\theta\phi - \frac{\theta^2}{3}$ **71.** $(3z + 1)(2z + 3)$ **73.** $(\lambda - 5)(\lambda - 3)$

75. $(x - 2)(x - 0.6)$ **77.** $(2R + 3X)(6R - 5X)$ **79.** $(2V - 0.5IR)(V + 0.3IR)$

81. $\left(\frac{X_C}{3} + Z\right)^2$ **83.** $(2\pi + f)(8\pi - 5f)$ **85.** $3(x + 2)(x - 2)$ **87.** $\frac{3}{2i}(V - 6v)^2$

89. $\frac{c}{144d}(8a - 9b)(9a - 8b)$

PROBLEMS 11-1

1. 8 **3.** $4\theta\phi$ **5.** $0.5a^2bc$ **7.** $3IR$ **9.** $X_L + X_C$ **11.** $E - 1$

13. $\sqrt{L_1L_2} + M$ **15.** $5\left(2I + 3\dfrac{V}{R}\right)$

PROBLEMS 11-2

1. 420 **3.** 360 **5.** $\theta^4\phi^3\lambda^3\mu\omega$ **7.** $180\ m^3n^2p^4$ **9.** $t^2 - 5t + 6$

11. $\mu(\mu + 3)(\mu + 5)$ **13.** $11(3\theta - 1)(2\theta + 1)(2\theta + 3)$

15. $\left(Q + \dfrac{\omega L}{R}\right)\left(Q - \dfrac{\omega L}{R}\right)\left(4Q - \dfrac{5\omega L}{R}\right)\left(2Q - \dfrac{7\omega L}{R}\right)$

PROBLEMS 11-3

1. 18 **3.** xy **5.** $9abd$ **7.** $t^2 - 2t + 1$ **9.** $6i + 6\alpha$ **11.** $\dfrac{12}{64}$

13. $\dfrac{\omega LR^2 - \omega LX^2}{R^3 - RX^2}$ **15.** $\dfrac{2VCQ - 3Q}{2V^2C^2 - VC - 3}$

PROBLEMS 11-4

1. $\dfrac{3}{4}$ **3.** $\dfrac{1}{13}$ **5.** $\dfrac{1}{ab^3}$ **7.** $\dfrac{5I}{R}$ **9.** $\dfrac{x}{x^2 + y^2}$ **11.** $\dfrac{a + b}{a - b}$ **13.** $\dfrac{x + y}{3}$ **15.** $\dfrac{\omega(\pi + 3\lambda)}{3\pi + \lambda}$

PROBLEMS 11-5

1. $\dfrac{a}{x}$ **3.** $\dfrac{2\pi fL}{X_C - X_L}$ **5.** $\dfrac{\omega L}{R_2 - R_1}$ **7.** $\dfrac{-IR}{V + v}$ **9.** $\dfrac{\pi R^2}{A_2 - A_1}$ **11.** -1 **13.** $-\dfrac{1}{\phi + \theta}$ **15.** $\dfrac{4 - \pi}{5 + \pi}$

PROBLEMS 11-6

1. $\dfrac{17}{8}$ **3.** $\dfrac{ac + b}{c}$ **5.** $\dfrac{4F - 5}{F}$ **7.** $\dfrac{4\pi + 6}{\pi + 1}$ **9.** $\dfrac{IR - I - V}{I}$

11. $\dfrac{(9 + 2x)(1 - x)}{x^2}$ **13.** $\dfrac{(R - 1)(R + 7)}{R^2}$ **15.** $\dfrac{9\lambda^2 - 4\lambda - 2}{(3\lambda + 1)(3\lambda - 1)}$

17. $\dfrac{-\theta^3 + 13\theta^2 + 31\theta - 45}{\theta^2(\theta - 1)}$ **19.** $\dfrac{2\alpha^4 - 7\alpha^2 - 1}{\alpha^2 - 3}$ **21.** $5\dfrac{3}{16}$ **23.** $1 + \dfrac{y^2}{x^2}$

25. $R^2 + 7R + 14 + \dfrac{6}{R - 1}$ **27.** $E^3 - E^2e + Ee^2 - e^3 - \dfrac{1}{E + e}$ **29.** $2x + 2 - \dfrac{x}{x^2 +}$

PROBLEMS 11-7

1. $\dfrac{35}{70}, \dfrac{30}{70}, \dfrac{28}{70}$ **3.** $\dfrac{36}{48}, \dfrac{21}{48}, \dfrac{20}{48}$ **5.** $\dfrac{\theta\omega}{\phi\omega}, \dfrac{\lambda\phi}{\phi\omega}$ **7.** $\dfrac{ei}{ir}, \dfrac{1}{ir}, \dfrac{ei^2r}{ir}$ **9.** $\dfrac{a + b}{a^2 - b^2}, \dfrac{a - b}{a^2 - b^2}$

11. $\dfrac{3\phi + 3\pi}{\phi^2 - \$}, \dfrac{4\phi - 4\pi}{\phi^2 - \pi^2}$ **13.** $\dfrac{ac - ad}{c^2 - d^2}, \dfrac{bc + bd}{c^2 - d^2}, \dfrac{bc + bd - ac - ad}{c^2 - d^2}$ **15.** $\dfrac{\pi^2 - \phi^2}{\pi\phi}, -\dfrac{\phi^2}{\pi\phi}$

PROBLEMS 11-8

1. $\frac{33}{70}$ **3.** $-\frac{5}{48}$ **5.** $\frac{65IR}{48}$ **7.** $\frac{\alpha\delta - \beta\gamma}{\beta\delta}$ **9.** $\frac{ayz - bxz - cxy}{xyz}$ **11.** $\frac{10R - 3I^2 + 4}{I^2R}$

13. $\frac{19I + i}{6}$ **15.** $\frac{3\alpha + \beta}{\alpha^2 - \beta^2}$ **17.** $\frac{3L_1 + 34}{L_1^2 + 4L_1 - 12}$ **19.** $\frac{7 + 25\theta}{6(1 - \theta^2)}$ **21.** $\frac{23I + 133}{I(I + 7)(I - 7)}$

23. $\frac{2\pi + 7}{3 + \pi}$ **25.** $\frac{8\theta\phi}{\theta^2 - \phi^2}$ **27.** $\frac{10 - 4E}{(E - 4)(E - 5)(E - 6)}$ **29.** $\frac{12\omega^2}{\omega^3 + 27}$

PROBLEMS 11-9

1. $\frac{1}{4}$ **3.** $-\frac{1}{50}$ **5.** 4 **7.** $4xy^3$ **9.** $\frac{4}{\theta^3\omega^2}$ **11.** $\frac{\omega}{2\pi fR}$ **13.** $\frac{4x + 4y}{(x - y)^2}$

15. $\frac{5x + y}{3x + 2}$ **17.** $\frac{1}{\phi - 2}$ **19.** ϕ^3 **21.** $14\alpha^2$ **23.** $4c$ **25.** ϕ^2 **27.** $\frac{1}{3\omega L + R}$ **29.** $2m$

PROBLEMS 11-10

1. $-\frac{7}{8}$ **3.** $-\frac{7}{40}$ **5.** $\frac{Q\omega L_1 L_2}{L_1 + L_2}$ **7.** $\frac{Ir}{Ir - V}$ **9.** $\frac{2E(E - e)}{e(E + e)}$ **11.** $\frac{l - w}{l + w}$

13. $\frac{b}{a}$ **15.** $-\frac{I^2 + i^2}{2Ii}$

PROBLEMS 12-1

1. $\phi = 8$ **3.** $\alpha = 8$ **5.** $\omega = \frac{7}{16}$ **7.** $\phi = 5$ **9.** $\lambda = 3$
11. $\omega = -12$ **13.** $\theta = -2$ **15.** $m = 12\frac{19}{24}$

PROBLEMS 12-2

1. $Q = 40$ **3.** $\theta = 4$ **5.** $r = 30$ **7.** $R = 2.5$ **9.** $b = \frac{1}{17}$
11. $a = -3$ **13.** $\lambda = 8$ **15.** $\alpha = 3$

PROBLEMS 12-3

1. $I = 2$ **3.** $q = 3$ **5.** $\phi = \frac{1}{4}$ **7.** $\omega = 2$ **9.** $\pi = 5$ **11.** $v_o = 5$
13. $x = 3$ **15.** $\alpha = 13$ **17.** $\omega = 5$ **19.** $\alpha = 3$ **23.** 42 min
25. $x = \frac{abc}{ab + ac + bc}$ days **27.** 5 h **31.** 90 kg **33.** $43.75
35. 25, 600 **37.** $\frac{1}{2}, 1\frac{1}{2}, 2\frac{1}{2}$ **39.** 8×2 m

PROBLEMS 12-4

1. $V_0 = \frac{LbV_d}{2aY_d}$ **3.** $V_b = IR + v$ **5.** $V_2 = V_3(1 - \omega^2 LC_2)$ **7.** $R_t = R_0(1 + \alpha t)$

$V_d = \frac{2aV_0Y_d}{Lb}$ $\quad v = V_b - IR$ $\quad L = \frac{V_3 - V_2}{\omega^2 C_2 V_3}$ $\quad t = \frac{R_t - R_0}{\alpha R_0}$

9. $r = \dfrac{vR}{V - v}$

$R = \dfrac{r(V - v)}{v}$

11. $R_t = \dfrac{1}{\omega^2 C_1 C_2 R_3}$

13. $\beta = \dfrac{V_0 - I_0 R_0}{\mu V_0}$

15. $a = \dfrac{b(1 + C_0)}{1 - C_0}$

$b = \dfrac{a(1 - C_0)}{1 + C_0}$

17. $Z_1 = \dfrac{Z_3(E - IZ_2)}{I(Z_2 + Z_3)}$

$Z_2 = \dfrac{Z_3(E - IZ_1)}{I(Z_1 + Z_3)}$

$Z_3 = \dfrac{IZ_1 Z_2}{E - I(Z_1 + Z_2)}$

19. $R_x = \dfrac{R_t(AV_1 - V)}{V(A + 1)}$

$R_y = \dfrac{VR_x(A + 1)}{AV_1 - V}$

$A = \dfrac{V(R_y + R_x)}{R_y V_1 - R_x V}$

21. $X_p = \dfrac{X_s^2 R + Z_{ab}^2 X_s^2 + Z_{ab}^2 R}{Z_{ab}^2}$

$R = \dfrac{Z_{ab}^2(X_p - X_s)}{X_s^2 + Z_{ab}^2}$

23. $V_n = \dfrac{I_2 R(R_1 + R_2)}{2R_1 + R_2}$

$R_t = \dfrac{R_2(V_n - I_2 R)}{I_2 R - 2V_n}$

25. $C = \tfrac{5}{9}(F - 32)$

27. $R = \dfrac{R_0(BI_0 - V_1)}{V_1}$

$R_0 = \dfrac{V_1 R}{BI_0 - V_1}$

29. $R_H = \dfrac{\mu m N}{Kg} - r$

$r = \dfrac{\mu m N}{Kg} - R_H$

31. $R = (1 - \alpha)Z_2 + (1 - \alpha + k\alpha)Z$

$Z_1 = \dfrac{R - Z_2(1 \Sigma \alpha)}{1 - \alpha + k\alpha}$

$k = \dfrac{R + (\alpha - 1)(Z_1 + Z_2)}{\alpha Z_1}$

33. $F_2 = \dfrac{2fF_s}{\alpha F_{12} - 2f}$

$f = \dfrac{\alpha F_{12} F_s}{2(F_2 + F_s)}$

35. $\alpha = \dfrac{1}{H_2 R_1} - S$

$S = \dfrac{1 - \alpha H_2 R_1}{H_2 R_2}$

37. $Z_1 = \dfrac{R_p(\mu v_g - v_1)}{v_1}$

$R_p = \dfrac{v_1 Z_2}{\mu v_g - v_1}$

39. $C_v = \dfrac{C_0(f_c - fX)}{f_c}$

$C_0 = \dfrac{C_v f_c}{f_c - fX}$

41. $v_0' = \dfrac{\omega s v_m'}{\omega s + 2\mu v_m'}$

$v_m' = \dfrac{\omega s v_0'}{\omega s - 2\mu v_0'}$

43. $v_0 = \dfrac{V i_1 R_1}{R_2(i_1 + i_2) + i_1 R_1}$

$i_1 = \dfrac{R v_0 i_2}{V R_1 - v_0(R_1 + R_2)}$

45. $G = \dfrac{\mu_1 \mu_2}{(1 - \mu_1 \beta_1)(1 - \mu_2 \beta_2)}$

$\beta_1 = \dfrac{\mu_1 \mu_2 + G(\mu_2 \beta_2 - 1)}{G\mu_1(\mu_2 \beta_2 - 1)}$

47. $I_1 = \dfrac{\pi \lambda^2 \gamma_1(2I_f + 1)}{\sigma_0(\gamma_1 + \gamma_f)} - \tfrac{1}{2}$

$I_f = \dfrac{\sigma_0(2I_1 + 1)(\gamma_1 + \gamma_f)}{4\pi \lambda^2 \gamma_1} - \tfrac{1}{2}$

49. $\lambda = \dfrac{\pi n' d_0(d_1 - d_0)}{1 - d_1}$

$d_1 = \dfrac{\lambda + \pi n' d_0^2}{\lambda + \pi n' d_0}$

51. $R_2 = \dfrac{-Z_2(Z - Z\alpha + Rk\alpha)}{Z_1 + Z_2}$

$Z = \dfrac{RkZ_2\alpha + R_2(Z_1 + Z_2)}{Z_2(\alpha - 1)}$

$\alpha = \dfrac{Z_1 R_2 + Z_2 R_2 + ZZ_2}{ZZ_2 - RkZ_2}$

53. $r_1 = \dfrac{r_2 r_3}{r_4}$

$r_3 = \dfrac{r_1 r_4}{r_2}$

$r_4 = \dfrac{r_2 r_3}{r_1}$

55. $G = \dfrac{V_{out} C_f C_{fg}}{C_{fg} Q - V_{out} C_f (C_d + C_{fg})}$

57. $p_2 = \dfrac{CNP_L p_1}{p\omega \varepsilon_2 (\tan \delta) - CNP_L}$

59. $R_p = \dfrac{R_1 R_2}{R_1 + R_2}$

$R_1 = \dfrac{R_2 R_p}{R_2 - R_p}$

$R_2 = \dfrac{R_1 R_p}{R_1 - R_p}$

61. $R_3 = \dfrac{V_0 R_a}{\mu V - V_0(\mu + 1)}$

$R_a = \dfrac{R_3}{V_0}[\mu V - V_0(\mu + 1)]$

$\mu = \dfrac{V_0}{R_3}\left(\dfrac{R_a + R_3}{V - V_0}\right)$

63. $\pi = \dfrac{MNk}{4(kH_0 - M)}$

$k = \dfrac{4\pi M}{4\pi H_0 - MN}$

65. $b = \dfrac{d(X^2 + X'^2)}{(X + X')^2}$

67. $R_a = \dfrac{\mu R_1 R_3 (V - V_0) - V_0 R_3 (R_s + R_1)}{V_0(R_1 + R_3 + R_s) - VR_3}$

$R_s = \dfrac{\mu R_1 R_3 (V - V_0) - V_0(R_1 R_3 + R_a R_3 + R_a R_1) + VR_a R_3}{V_0(R_a + R_3)}$

$\mu = \dfrac{R_a R_3 (V - V_0) - V_0(R_a R_s + R_a R_1 + R_s R_3 + R_1 R_3)}{R_1 R_3 (V_0 - V)}$

69. $R_a = \dfrac{\mu R_0 R_1 (R_s + R_1 + R_2)}{R_2(R_s + R_1) - R_0(R_s + R_1 + R_2)}$

$R_2 = \dfrac{R_0(R_s + R_1)(R_a + \mu R_1)}{R_a(R_s + R_1) - R_0(R_a + \mu R_1)}$

$\mu = \dfrac{R_a[R_2(R_s + R_1) - R_0(R_s + R_1 + R_2)]}{R_0 R_1 (R_s + R_1 + R_2)}$

71. $R_a = \dfrac{R_1 R_2 R_3 (\mu R_1 - R_1)}{(R_i - R_1)(R_1 R_2 + R_2 R_3 + R_1 R_3)}$

73. $\pi = \dfrac{\alpha^2(\beta - \alpha)}{\alpha + 2\beta}$

$\beta = \dfrac{\alpha(\alpha^2 + \pi)}{\alpha^2 - 2\pi}$

75. $A = 5.89 \times 10^{-14}$ m² **77.** $Z_2 = 6\ \Omega$ **79.** $R_2 = 100\ \Omega$

81. $F = C$ at $-40°$ **83.** $R_0 = 32\ \Omega$ **85.** $C_1 = 3$ pF **87.** $p = 133.3$

89. (a) Increased by a factor of 4, (b) Halved **91.** $R = 0.9\ \Omega$

93. $R = \dfrac{n(V - Ir)}{I}$ **95.** $g_m = \dfrac{v_d}{v_g i_d r_o} - \dfrac{1}{r_s}$ **99.** $V_0 = \dfrac{S}{t_S} - \tfrac{1}{2}gt$ **101.** $V_0 = 50.95$ m/s

$n = \dfrac{IR}{V - Ir}$ cells

103. $\Delta i_d = g_m v_{gs}$ **105.** $V_{CC} = -34.5$ V; PNP **107.** $V_1 - V_2 = 98$ V **109.** $\alpha = \dfrac{\beta}{1 + \beta}$

PROBLEMS 13-1

1. $165\ \Omega$ **3.** 37.2 kΩ **5.** (a) $50\ \Omega$ (b) 340 kΩ (c) 1.95 kΩ **7.** 440 V
9. 112 kΩ **11.** 4.47 W **13.** 14.7 W **15.** 1 kV

PROBLEMS 13-2

1. $5\ \Omega$ **3.** $4.8\ \Omega$ **5.** $6.11\ \Omega$ **7.** (a) 2.1 kΩ **9.** $R_p = \dfrac{R}{n}\ \Omega$
(b) 17 kΩ

11. 1.5 kW **13.** (a) $R_3 = 22\ \Omega$ **15.** 10 kΩ
(b) $P_t = 1.84$ kW

PROBLEMS 13-3

1. 332 mA **3.** 176 W **5.** (a) $V_G = 230$ V (d) $I_2 = 1.86$ mA
7. (a) $V_1 = 702$ V (e) $I_t = 180$ mA (b) $R_3 = 20$ kΩ (e) $I_3 = 1.4$ mA
(b) $V_2 = 298$ V (f) $I_3 = 44.6$ mA (c) $R_1 = 70.6$ kΩ
(c) $R_2 = 2.2$ kΩ (g) $P_t = 180$ W
(d) $R_3 = 6.7$ kΩ

9. 354 Ω   **11.** 600 Ω   **13.** (a) 4.2 kΩ   **15.** 730 W   **17.** 2.47 A
(b) 10 kΩ
(c) 48 W

PROBLEMS 14-1

1. 1.08 Ω   **3.** 3.18 m   **5.** (a) 0–10 mA: 6.11 Ω   **7.** R_1 = 150 Ω
(b) 0–100 mA: 0.556 Ω R_2 = 15 Ω
(c) 0–1A: 0.0551 Ω R_3 = 1.5 Ω
(d) 0–10A: 0.005 51 Ω R_4 = 0.167 Ω

PROBLEMS 14-2

1. (a) 37.5 V   **3.** R_1 = 9.6 kΩ $R_3 \cong$ 1 MΩ
(b) 25 V R_2 = 99.6 kΩ $R_4 \cong$ 10 MΩ

PROBLEMS 15-1

1. 1 = 60 V   **3.** 27 kΩ: 0.114 W   **5.** P_t = 14 W   **7.** R_1 = 13.2 W (use 20 W)
2 = 6 V   68 kΩ: 0.288 W P_1 = 6.4 W R_2 = 7.13 W (use 10 W)
3 = 0.6 V   75 kΩ: 0.318 W P_2 = 2.4 W R_3 = 3.37 W (use 5 W)
4 = 0.06 V P_3 = 5.2 W
9. R_1 = 3 kΩ R_3 = 10 kΩ   **11.** 42 W
R_2 = 3 kΩ R_4 = 500 Ω

PROBLEMS 15-2

1. (a) I_1 = 0.25 A   **3.** (a) I_1 = 103.2 A   **5.** I_2 = 4.94 A
(b) I_2 = 0.0833 A   (b) I_2 = 46.8 A

PROBLEMS 15-3

1. 0.0524 Ω   **3.** 0.0226   **5.** 35.9 km   **7.** $X = \dfrac{R_2L - R_1R_3}{R_1 + R_2}$

PROBLEMS 16-1

1. Current varies directly as the applied voltage. (Graph of current is a straight line.)
3. With velocity constant, distance varies directly as time. (Graph of distance is a straight line.)
5. (a) 2 P.M.,   (b) 480 km,   (c) 80 km   **7.** Third, sixth, ninth, and fifteenth

PROBLEMS 16-2

1. Latitude

PROBLEMS 16-5

1. $y = \frac{2}{5}x - 2$ **3.** $y = 0.113x + 1.2$ **5.** $R = -0.000\ 667T + 0.4$
7. (a) 0.021 25:1   (b) 47:1   (c)47 Ω

PROBLEMS 17-1

1. $x = 6, y = 2$ **3.** $x = 5, y = 3$ **5.** $E = 6, I = -10$
7. $\alpha = -2, \beta = -3$ **9.** $I_1 = 5, i = 5$

PROBLEMS 17-2

1. $a = 2.5, b = 4$ **3.** $R = 3, Z = 2$ **5.** $R_1 = 1, R_2 = 3$
7. $s = -2, t = 2$ **9.** $L = 1, M = 2$ **11.** $I_1 = 3, I_2 = -2$
13. $V = 2, v = 3$ **15.** $\lambda = 4, \pi = -1$ **17.** $V = -11, v = 12$
19. $I = 12, i = 9$

PROBLEMS 17-3

1. $V = 3, I = 2$ **3.** $I = -2, i = 3$ **5.** $\alpha = 8, \beta = 5$
7. $V = \frac{1}{2}, v = \frac{1}{3}$ **9.** $\theta = 16, \phi = -10$ **11.** $F = -3, f = 2$
13. $\gamma = 14, \delta = -4$ **15.** $\varepsilon = 2, \psi = 3.5$ **17.** $\theta = 3, \phi = 1$
19. $a = 1.5, b = 0.4$

PROBLEMS 17-4

1. $I = i = 1$ **3.** $\lambda = 6, \pi = -4$ **5.** $x = 3, y = 4$
7. $\alpha = 6, \beta = 1$ **9.** $p = 3, q = -7$

PROBLEMS 17-5

1. $a = 6, b = 2$ **3.** $\theta = 11, \phi = -5$ **5.** $\varepsilon = 40, \eta = 5$
7. $X_C = 2, X_L = 3$ **9.** $\theta = \frac{3}{4}, \lambda = \frac{1}{2}$

PROBLEMS 17-6

1. $R = 3, Z = 4$ **3.** $X_L = 5, X_C = 11$ **5.** $\theta = 16, \phi = 5$
7. $G = 60, Y = 33$ **9.** $L_1 = -\frac{15}{7}, M = \frac{15}{11}$

PROBLEMS 17-7

1. $\alpha = \dfrac{P + Q}{6}, \beta = \dfrac{2Q - P}{3}$ **3.** $V = \dfrac{7a - b}{4}, IR = \dfrac{b - 3a}{4}$

5. $\theta = \dfrac{3\alpha - 5\beta}{38}, \phi = \dfrac{2\alpha + 3\beta}{19}$ **7.** $X_C = 50(Z_1 - Z_2), X_L = \dfrac{20Z_2 - 10Z_1}{3}$

9. $R_1 = \dfrac{R_p R_t}{R_p - 2R_t}, R_2 = \dfrac{R_p R_t}{3R_t - R_p}$

PROBLEMS 17-8

1. $\theta = -2, \phi = 4, \pi = 1$ **3.** $R_1 = 9, R_2 = 2, R_3 = -4$
5. $R_L = 3, R_p = 5, R_1 = 8$ **7.** $r = 5, R = 6, R_L = 7$ **9.** $s = 12, t = 4, v = 8$

PROBLEMS 17-9

1. $\dfrac{I_t + I_d}{2}, \dfrac{I_t - I_d}{2}$ A   **3.** $\frac{2}{3}$ **5.** $\dfrac{90 + \alpha^\circ}{2}, \dfrac{90 - \alpha^\circ}{2}$

7. Resistors, 10¢ each, capacitors, 20¢ each   **9.** $L = 60$ km/h, $Q = 70$ km/h

11. $W = \dfrac{Q^2}{2C}$ **13.** $s = ut + \frac{1}{2}at^2$ **15.** Gain $= G_m r_d$ **17.** $R = \dfrac{L}{Cr}$

19. $Q = CV$ **21.** $H = 7.20$ **23.** $R = R_p \dfrac{V_p}{\mu V_g - V_p}$

25. $R_X = \dfrac{R_A}{R_A + R_B}(V_3 - V_2)$ **27.** $R_1 = \dfrac{R_a R_b + R_b R_c + R_a R_c}{R_c}$

$R_Y = \dfrac{R_A}{R_A + R_B}(V_2 - V_1)$ $R_2 = \dfrac{R_a R_b + R_b R_c + R_a R_c}{R_a}$

$R_T = \dfrac{R_A}{R_A + R_B}(V_3 - V_1)$ $R_3 = \dfrac{R_a R_b + R_b R_c + R_a R_c}{R_b}$

PROBLEMS 18-1

1. 2   **3.** 34   **5.** 0   **7.** -114 **9.** -0.02 **11.** 0
13. $bx - ay$ **15.** $bx - ay$

PROBLEMS 18-2

1. $a = 4$, $b = 2$ **3.** $\theta = 10$, $\pi = 2$ **5.** $I = 3$, $i = -2$ **7.** $r_p = \frac{1}{2}$, $r_L = \frac{1}{3}$
9. $R_1 = 150$, $R_2 = 1200$

PROBLEMS 18-3

1. 21   **3.** -245 **5.** -2016 **7.** $x = 5, y = 7, z = 3$
9. $\alpha = 3, \beta = 4, \gamma = 7$ **11.** $V = \frac{1}{2}, v = \frac{1}{3}, IR = \frac{1}{4}$

PROBLEMS 18-4

1. 10   **3.** -25 **5.** 23   **7.** 0

PROBLEMS 18-5

1. 297   **3.** -56 **5.** 22.1   **7.** 220   **9.** $I_1 = 1, I_2 = 3, I_3 = 5$
11. $\alpha = -2, \beta = 4, \gamma = 1$ **13.** $R_1 = 5.5, R_2 = 3.6, R_3 = 1.3$
15. $\varepsilon = 1, \eta = 2, \kappa = 3, \lambda = 4$

PROBLEMS 19-1

1. $1 \, \Omega$ **3.** $0.4 \, \Omega$ **5.** (a) 5.6 W
                                                        (b) 93.3%

PROBLEMS 19-2

1. (a) 1.3 V   (c) $0.975 \, \Omega$ **3.** (a) $0.48 \, \Omega$ (d) 370 mW   **5.** 11.1 A
    (b) 1.33 A   (b) 30 mW (e) 92.6%
                                                 (c) $5.92 \, \Omega$

7. (*a*) 8.4 Ω   **9.** (*a*) 385 mA   **11.** (*a*) 4.08 V   **13.** (*a*) 0.0733 Ω
    (*b*) 600 mW   (*b*) 12.7 V   (*b*) 1.6 V   (*b*) 200 A
    (*c*) 15 A   (*c*) 4.88 W   (*c*) 288 mW

17. $V = 1.4$ V, $r = 0.2$ Ω   **19.** $V = 2.1$ V, $R = 0.665$ Ω

PROBLEMS 20-1

1. a^7 **3.** x^3 **5.** p^{q+r} **7.** $I^{\alpha+\beta}$ **9.** x^3 **11.** X^{5y-2} **13.** $\theta^{2\beta}$ **15.** I^9

17. $x^6 y^9$ **19.** a^{4x} **21.** $x^{4l}y^{4m}z^{4p}$ **23.** $\dfrac{V^2}{R^2}$ **25.** $\dfrac{\omega^{18}}{64\pi^6 f^{12}}$ **27.** $\dfrac{-X_C^6}{X_L^3}$

29. α^{4x-4} **31.** $\dfrac{I^2}{R}$ **33.** $\dfrac{z^{3\lambda}}{y^\pi}$ **35.** $\dfrac{\theta^4}{\phi^3\lambda^{2x}}$ **37.** $\dfrac{bc}{a^3}$ **39.** $\dfrac{Ir^3}{4R^2}$

PROBLEMS 20-2

1. ±4 **3.** ±2 **5.** $-4a^2bc^4$ **7.** $\pm I^6 R^3$ **9.** $\dfrac{9\lambda^6}{\omega^8}$ **11.** $\sqrt[z]{9}$ **13.** $\sqrt[3]{8\alpha}$ or $2\sqrt[3]{\alpha}$

15. $\sqrt[4]{(\theta\lambda)^3}$ **17.** $a^{\frac{3}{2}}$ **19.** $2^{\frac{4}{3}}E^{\frac{1}{3}}$ **21.** $\theta^{\frac{2}{3}}\omega^{\frac{4}{3}}$ **23.** $(\alpha\beta)^{\frac{2}{5}}$ **25.** $4\pi f 2^{\frac{1}{3}}$

PROBLEMS 20-3

1. $\pm2\sqrt{2}$ **3.** $\pm3\sqrt{2}$ **5.** $\pm5\sqrt{2}$ **7.** $\pm4\sqrt{5}$ **9.** $\pm12\sqrt{5}$ **11.** $\pm2\theta\phi^2\sqrt{3}$

13. $\pm20I\sqrt{6R}$ **15.** $\pm18\omega f^2 FT^2\sqrt{7FT}$ **17.** $\pm33\alpha^4\beta^3\gamma^4\sqrt{2\alpha\beta}$ **19.** $\pm28\pi^2 L^2 X_L r^3\sqrt{3}$

PROBLEMS 20-4

1. $\dfrac{\sqrt{3}}{3}$ **3.** $\dfrac{\sqrt{10}}{5}$ **5.** $\pm\dfrac{\sqrt{3}}{2}$ **7.** $4\sqrt{2}$ **9.** $\dfrac{\sqrt{\lambda}}{\lambda}$ **11.** $\pm\dfrac{3\sqrt{\theta}}{4\theta}$

13. $\sqrt{\theta\lambda}$ **15.** $\pm\dfrac{\alpha\sqrt{\gamma}}{\gamma}$ **17.** $\dfrac{R^2\sqrt{\pi A}}{A}$ **19.** $\pm\dfrac{X_L\sqrt{15}}{2}$ **21.** $\pm\dfrac{4Q^2\sqrt{5}}{9}$

PROBLEMS 20-5

1. $3\sqrt{3}$ **3.** $\sqrt{5}$ **5.** $(m-p+q)\sqrt{3}$ **7.** $30\sqrt{3}$ **9.** 0 **11.** $6\sqrt{2}$

13. $\dfrac{2\sqrt{5}-\sqrt{15}}{5}$ **15.** $\dfrac{\sqrt{2\pi}}{8}$

PROBLEMS 20-6

1. $\pm\sqrt{6}$ **3.** $\pm4\sqrt{5}$ **5.** $\pm12\sqrt{10}$ **7.** 2 **9.** $A-D$

11. $2\alpha-7-2\sqrt{\alpha^2-7\alpha}$ **13.** $\theta-\phi$ **15.** $\pm6\alpha\pi\sqrt{2}$ **17.** 6 **19.** 9

PROBLEMS 20-7

1. $\pm\sqrt{5}$ **3.** $4(3-\sqrt{7})$ **5.** $-\dfrac{3(1+\sqrt{3})}{2}$ **7.** $\dfrac{x^2-2x\sqrt{y}+y}{x^2-y}$

9. $\dfrac{3+\sqrt{3}}{4}$ **11.** $-\sqrt{6}-3\sqrt{3}+2\sqrt{2}+6$

PROBLEMS 20-8

1. $j6$ **3.** $j12$ **5.** $-jZ$ **7.** jI^2x **9.** $-j35$ **11.** $j\frac{4}{11}$

13. $j\frac{4\sqrt{6}}{15}$ **15.** $-j\frac{V\sqrt{P}}{P}$

PROBLEMS 20-9

1. $5 + j20$ **3.** $41 - j2$ **5.** $172 + j5$ **7.** $20 - j2$ **9.** $1 + j4$
11. $9 + j18$ **13.** $-78 - j11$ **15.** $20 + j8$

PROBLEMS 20-10

1. $3 - j9$ **3.** $75 - j30$ **5.** $\theta^2 - \phi^2 + j2\theta\phi$ **7.** $\dfrac{1 - j1}{2}$ **9.** $j1$

11. $\dfrac{1 - j1}{2}$ **13.** $\dfrac{6(6 + jx)}{36 + x^2}$ **15.** $\dfrac{R^2 + j2R\omega X - \omega^2 X^2}{R^2 + \omega^2 X^2}$ **17.** $\dfrac{-\phi^2 + j\theta\phi}{\theta^2 + \phi^2}$

19. $\dfrac{R^2 - jR\left(\omega L - \dfrac{1}{\omega C}\right)}{R^2 + \left(\omega L - \dfrac{1}{\omega C}\right)^2}$

PROBLEMS 20-11

1. $x = 4$ **3.** $\gamma = 9$ **5.** $Z = 625$ **7.** $M = 67$ **9.** $\lambda = 1$ **11.** $\phi = 25$

13. $P_r = \dfrac{i_s^2}{2\rho^2 P_s}$ **15.** $\eta = \dfrac{S^2}{\alpha^2 N^2 \tau}$ **17.** $W = \dfrac{V^2 a(1 - \varepsilon_1)}{V^2 - C^2}$ **19.** $Q_2 = \dfrac{n^2(Y_n{}^2 - G)}{G^2(n^2 - 1)^2}$

21. $g_m{}^2 = \dfrac{G_L(G_1 - G_a{}^2)}{R_{eq}(G_a{}^2 - G_1) - G_1}$ **23.** $C = 250$ pF   **25.** $C_a = \dfrac{C_b}{(2\pi f)^2 LC_b - 1}$

PROBLEMS 21-1

1. $E = \pm 5$ **3.** $i = \pm\sqrt{189}$ **5.** $\omega = \pm 6$ **7.** $\lambda = \pm\frac{3}{11}$ **9.** $\mu = \pm\frac{4}{5}$

11. $m = \pm\sqrt{2}$ **13.** $\lambda = \pm 6$ **15.** $X_C = \pm\dfrac{\sqrt{95}}{5}$

PROBLEMS 21-2

1. $\alpha = -1$ or -4 **3.** $R = 2$ or 7 **5.** $\lambda = 1$ or -2 **7.** $E = 2$ or 20
9. $Q = 2$ or 11 **11.** $\alpha = -2$ or -25 **13.** $Z = 3$ or 6 **15.** $i = 3$ or $-\frac{7}{4}$

PROBLEMS 21-3

1. $x = 2$ or 6 **3.** $E = 6$ or 9 **5.** $i = 2$ or 25 **7.** $\theta = 1$ or 2
9. $M = -2$ or 24 **11.** $\theta = 3$ or -2 **13.** $\phi = 10$ or $G6$ **15.** $R = 5$ or $5\frac{1}{3}$

PROBLEMS 21-4

1. $\theta = 1$ or -4 **3.** $I = 7$ or -5 **5.** $q = \frac{3}{4}$ or $-\frac{5}{2}$ **7.** $Z = \dfrac{3 \pm \sqrt{129}}{12}$

9. $m = \frac{1}{4}$ or $-\frac{1}{6}$ **11.** $R_1 = -5$ or 0 **13.** $\beta = 5$ or $5\frac{1}{3}$ **15.** $i = 2$ or $-\frac{2}{15}$

PROBLEMS 21-7

1. (a) 16, roots real and unequal   (b) 0, roots are equal   (c) -80, roots are imaginary

3. 21 and 23   **5.** 140×160 m   **7.** 220 and 240

9. (a) $V = \pm \sqrt{\dfrac{PnR}{k}}$ **11.** $r = \dfrac{-PXx \pm x\sqrt{P^2X^2 + 4R^2(P-1)}}{2R(P-1)}$

(b) No change in V

$$x = \dfrac{PXr \pm r\sqrt{P^2X^2 + 4R^2(P-1)}}{2R}$$

13. $v = 1.0 \times 10^3$ m/s   **15.** $v = \sqrt{2gs}$ m/s   **17.** $h = 0.0156\, v^2$ m

19. $R = 50\ \Omega$ **25.** (a) 2 A   **27.** 20 V and 15 A or

(b) 120 V   60 V and 5 A

(c) $R_1 = 10\ \Omega$

$R_2 = 20\ \Omega$

$R_3 = 30\ \Omega$

PROBLEMS 22-1

1. 1.03 mA   **3.** 54.3 V   **5.** 267 V   **7.** (a) 0.5 A (b) 12.3 V   **9.** 2.22 Ω

PROBLEMS 22-2

1. 39.9 V   **3.** 32 V   **5.** (a) 1.19 A   **7.** (a) 1.0 A

                                       (b) 53.2 mW   (b) From a to b

PROBLEMS 22-3

1. (a) 1.27 A   **3.** (a) 1.64 A   **5.** (a) 95.5 W   **7.** (a) 86.3 W

(b) 14.6 W   (b) 2.86 A from a to b (b) 3.18 V   (b) 16 V

9. (a) 5.19 A   **11.** (a) 64.8 V   **13.** (a) 220 V

(b) 167 W   (b) 2.1 kW   (b) 313 W

PROBLEMS 22-4

1. $R_a = 4.8\ \Omega$, $R_b = 4\ \Omega$, $R_c = 6\ \Omega$ **3.** $R_a = R_b = R_c = 167\ \Omega$

5. $R_1 = 16.6$ kΩ, $R_2 = 6.36$ kΩ, $R_3 = 9.06$ kΩ **7.** 72.7 mA   **9.** 6.52 mA

11. Zero A   **13.** $I = 5$ A   **15.** 3.1 A   **17.** 14.7 A

PROBLEMS 22-5

1. (a) Constant 120-V source in series with 0.8 Ω

(b) Constant 150-A source in parallel with 0.8 Ω

3. 0.187 V in series with 18.75 Ω; $I_2 = 6.52$ mA

5. 65.6 mA in parallel with 12.6 Ω; $I_s = 30$ mA

PROBLEMS 23-1

1. (a) 22°, (b) 67°, (c) 49°, (d) −80°, (e) −165°, (f) 100°   **7.** 26   **9.** 21 600°/s

PROBLEMS 23-2

1. (a) $\frac{\pi^r}{3}$, 1.05^r, (b) $\frac{2\pi^r}{3}$, 2.09^r, (c) $\frac{11\pi^r}{12}$, 2.88^r, (d) $\frac{5\pi^r}{4}$, 3.93^r, (e) $\frac{19\pi^r}{12}$, 4.97^r, (f) $\frac{\pi^r}{36}$, 0.0873^r

3. 330π^r **5.** $\frac{5\pi^r}{2}$ **7.** $\frac{40\pi}{3}$ ^{r}/s   **9.** 30 rev/min

PROBLEMS 23-3

1. (a) 50^g, (b) 33.3^g, (c) 66.7^g, (d) 133^g, (e) 250^g, (f) 350^g
3. (a) 0.7854^r, (b) 0.3142^r, (c) 1.178^r, (d) 2.356^r, (e) 3.142^r, (f) 6.283^r

PROBLEMS 23-4

1. 4.5 m and 6 m   **3.** $a = 4$, $c = 5$, $B = 36.9°$ **5.** $a = 11.4$, $c = 11.4$, $B = 20°$
7. $b = 17.7$, $c = 13$, $A = 33.8°$ **9.** $c = 10$, $A = 48.9°$, $B = 101.8°$

PROBLEMS 23-5

1. $c = 58$, $B = 15°$ **3.** $a = 18$, $A = 13°$ **5.** 12 m   **7.** 19.9 m   **9.** 150 m

PROBLEMS 24-1

1. $\sin\theta = \frac{a}{c}$ $\sin\phi = \frac{b}{c}$

$\cos\theta = \frac{b}{c}$ $\cos\phi = \frac{a}{c}$

$\tan\theta = \frac{a}{b}$ $\tan\phi = \frac{b}{a}$

$\cot\theta = \frac{b}{a}$ $\cot\phi = \frac{a}{b}$

$\sec\theta = \frac{c}{b}$ $\sec\phi = \frac{c}{a}$

$\csc\theta = \frac{c}{a}$ $\csc\phi = \frac{c}{b}$

3. (a) $\frac{OP}{OR} = \tan\beta$

(b) $\frac{PR}{PO} = \sec\alpha$

(c) $\frac{OR}{PR} = \cos\beta$

(d) $\frac{OP}{RP} = \sin\beta$

(e) $\frac{PR}{RO} = \csc\alpha$

5. $\sin\theta = 0.707$
$\cos\theta = 0.707$
$\tan\theta = 1.00$
$\cot\theta = 1.00$
$\sec\theta = 1.41$
$\csc\theta = 1.41$

7. $\sin\theta = 0.894$
$\cos\theta = 0.447$
$\tan\theta = 2.000$

9. $\sin x = \frac{8}{10}$ $\cot x = \frac{6}{8}$

$\cos x = \frac{6}{10}$

$\tan x = \frac{8}{6}$

$\sec x = \frac{10}{6}$

$\csc x = \frac{10}{8}$

11. $\sin B = \frac{4}{5}$ $\csc B = \frac{5}{4}$

$\cos B = \frac{3}{5}$

$\tan B = \frac{4}{3}$

$\cot B = \frac{3}{4}$

$\sec B = \frac{5}{3}$

PROBLEMS 24-2

1. I or II   **3.** III or IV   **5.** II or III   **7.** I   **9.** IV   **11.** I or III   **13.** No

Q	sin	cos	tan	Q	sin	cos	tan
15.	+	+	+	**17.**	+	−	−
19.	−	−	+	**21.**	−	−	+
23.	+	+	+				

Q	sin	cos	tan	sec	csc	cot
27.	$\frac{5}{13}$	$\frac{12}{13}$	$\frac{5}{12}$	$\frac{13}{12}$	$\frac{13}{5}$	$\frac{12}{5}$
29.	$\frac{-5\sqrt{41}}{41}$	$\frac{-4\sqrt{41}}{41}$	$\frac{5}{4}$	$\frac{-\sqrt{41}}{4}$	$\frac{-\sqrt{41}}{5}$	$\frac{4}{5}$
31.	$-\frac{3}{5}$	$\frac{4}{5}$	$-\frac{3}{4}$	$\frac{5}{4}$	$-\frac{5}{3}$	$-\frac{4}{3}$
33.	$\frac{-3\sqrt{34}}{34}$	$\frac{-5\sqrt{34}}{34}$	$\frac{3}{5}$	$\frac{-\sqrt{34}}{5}$	$\frac{-\sqrt{34}}{3}$	$\frac{5}{3}$

PROBLEMS 24-3

1. 0   **3.** ∞   **5.** No   **7.** (a) 1, (b) −1, (c) −1, (d) 1

PROBLEMS 25-1

		sin	cos	tan			sin	cos	tan
1.	(a)	0.309 02	0.951 06	0.324 92	**3.**	(a)	0.033 85	0.999 43	0.033 87
	(b)	0.927 18	0.374 61	2.475 09		(b)	0.842 08	0.539 36	1.561 25
	(c)	0.161 60	0.986 86	0.163 76		(c)	0.628 10	0.778 13	0.807 19
	(d)	0.793 35	0.608 76	1.303 23		(d)	0.646 52	0.762 89	0.847 46
	(e)	0.045 36	0.998 97	0.045 41		(e)	0.822 84	0.568 27	1.447 96

PROBLEMS 25-2

1. (a) 27°, (b) 6.7°, (c) 61.5°, (d) 40.1°, (e) 2.14°
3. (a) 85.4°, (b) 0.5°, (c) 40.1°, (d) 58.8°, (e) 25.75°
5. (a) 13.16°, (b) 0.53°, (c) 74.38°, (d) 41.77°, (e) 47.11°

PROBLEMS 25-3

		sin	cos	tan			sin	cos	tan
1.	(a)	0.956 30	−0.292 37	−3.270 85	**3.**	(a)	−0.984 81	0.173 65	−5.671 28
	(b)	0.342 02	−0.939 69	−0.363 97		(b)	−0.669 13	0.743 14	−0.900 40
	(c)	0.764 92	−0.644 12	−1.187 54		(c)	−0.175 37	0.984 50	−0.178 13
	(d)	0.537 30	−0.843 39	−0.637 07		(d)	−0.865 15	0.501 51	−1.725 09
	(e)	0.066 27	−0.997 80	−0.066 42		(e)	−0.008 73	0.999 96	−0.008 73

		sin	cos	tan					
5.	(a)	−0.087 16	−0.996 19	0.087 49	**7.**	(a) ϕ = −47.1°		**9.**	1.42 m
	(b)	−0.294 04	0.955 79	−0.307 64		(b) ϕ = 91.6°		**11.**	90°
	(c)	−0.652 10	−0.758 13	0.860 14		(c) ϕ = 51.3°		**13.**	175 lx
	(d)	−0.045 36	0.998 97	−0.045 41		(d) ϕ = 167.5°		**15.**	No
	(e)	−0.003 49	−0.999 99	0.003 49		(e) ϕ = −69.9°		**17.**	40.7°

PROBLEMS 26-1

1. $Z = 26.8, X = 15.2, \phi = 55.3°$
5. $Z = 70.0, X = 29.5, \phi = 65.1°$
9. $Z = 1030, R = 557, \phi = 32.7°$
13. $Z = 0.239, X = 0.214, \phi = 26.1°$

3. $Z = 600, R = 424, \theta = 45°$
7. $Z = 1 \times 10^6, X = 4.65 \times 10^5, \phi = 62.3°$
11. $Z = 159, R = 100, \phi = 38.9°$
15. $Z = 0.378, R = 0.0500, \phi = 7.6°$

PROBLEMS 26-2

1. $R = 73.6, X = 19.7, \theta = 15°$
5. $R = 7.84 \times 10^3, X = 6.21 \times 10^3, \theta = 38.4°$
9. $R = 3.12, X = 4.04, \phi = 37.7°$

3. $R = 17.0, X = 44.5, \phi = 20.9°$
7. $R = 0.932, X = 0.171, \theta = 10.4°$

PROBLEMS 26-3

1. $\theta = 60.8°, \phi = 29.2°, R = 112$
5. $\theta = 8.4°, \phi = 81.6°, X = 0.109$
9. $\theta = 51.9°, \phi = 38.1°, X = 0.849$

3. $\theta = 69.1°, \phi = 20.9°, X = 44.5$
7. $\theta = 38.4°, \phi = 51.6°, R = 7.84 \times 10^3$

PROBLEMS 26-4

1. $\theta = 9.9°, \phi = 80.1°, Z = 36.0$
5. $\theta = 2.7°, \phi = 87.3°, Z = 430$
9. $\theta = 46°, \phi = 44°, Z = 0.403$

3. $\theta = 47.9°, \phi = 42.1°, Z = 7.14$
7. $\theta = 83.6°, \phi = 6.4°, Z = 48.7$

PROBLEMS 26-5

1. $33.7°$ **3.** $4.21°$ **5.** $62.7°$ **7.** 9.84 m **9.** 20.2 m **11.** 97.7 m

PROBLEMS 26-6

1. (a) $\phi = 68°$, (b) 278 m², (c) 37.1 m, (d) 278 m² **3.** 4.14×10^3 mm²

PROBLEMS 27-2

1. $b = 7.66, c = 9.01, \gamma = 70°$
5. $a = 33, c = 91.7, \gamma = 108°$
9. $a = 11.3, c = 63.6, \beta = 55.5°$

3. $a = 12.9, c = 18, \gamma = 75°$
7. $a = 1.14, b = 7.1, \alpha = 8°$
11. 2.53 km

PROBLEMS 27-3

1. $a = 7.3, \beta = 39.4°, \gamma = 77.6°$
5. $c = 4691, \alpha = 10.5°, \beta = 21.8°$
9. $\alpha = 17.5°, \beta = 50°, \gamma = 112.5°$

3. $c = 0.908, \alpha = 8°, \beta = 40°$
7. $\alpha = 21.8°, \beta = 38.2°, \gamma = 120°$
11. 55.7 by 146.8 mm

PROBLEMS 27-4

1. $0.366 \sin \theta + 1.366 \cos \theta$ **3.** $0.366(\cos \theta - \sin \theta)$ **5.** $\frac{33}{65}$

PROBLEMS 28-1

1. 182.4 at 28.3°   **3.** 239 at 244.7°

PROBLEMS 28-2

1. $x = 12.4$, $y = 27.3$ **3.** $x = 0.0423$, $y = 0.864$ **5.** $x = -46.3$, $y = 0$ **7.** $x = -56.7$, $y = -177$
9. $x = -28.4$, $y = 11.9$ **11.** 728 N, 234 N   **13.** 849 m   **15.** 131 N

PROBLEMS 28-3

1. $420 \underline{/81.2°}$ **3.** $1.92 \underline{/39.9°}$ **5.** $364 \underline{/15.1°}$ **7.** $183 \underline{/0°}$
9. $125 \underline{/270°}$ **11.** $7.67 \underline{/252.2°}$ **13.** $25.9 \underline{/160.8°}$ **15.** $24.4 \underline{/216.5°}$

PROBLEMS 28-4

1. $321 \underline{/55.9°}$ **3.** $111 \underline{/19.4°}$ **5.** $31.2 \underline{/167.4°}$

PROBLEMS 29-2

1. (a) $\dfrac{\pi^r}{21\,600}$/s, (b) $\dfrac{\pi^r}{1800}$/s, (c) $\dfrac{\pi^r}{30}$/s   **3.** (a) 4.5°/min, (b) $\dfrac{\pi^r}{2400}$/s
5. (a) $72\pi^r$, (b) $7.2\pi^r$, (c) $3.6\pi^r$

PROBLEMS 29-3

1. (a) 100   (b) 2π (c) 1   (d) 1   (e) 40° lead   **13.** (b) $y = 60 \sin 40\pi t$ cm,
3. 0.750   628   100   0.01   3° lead   (c) -35.3 cm,
5. V_{max} 157   25   0.04   17° lag   (d) 60 cm,
                                                      (e) $10\pi^r$

PROBLEMS 30-1

1. (a) 51 A, (b) 152 A, (c) 115 A, (d) -146 A, (e) -92.3 A   **3.** 440 V
5. -91.7 V   **7.** -1.11 A   **9.** 210° and 330°

PROBLEMS 30-2

1. (a) 400 Hz
(b) 2.5 ms
(c) $v = 314 \sin 800\pi t$ V   **3.** (a) 40 poles   **5.** 600 rev/min
(b) $v = 250 \sin 800\pi t$ V
(c) -238 V
7. 500 MHz   **9.** $i = (3 \times 10^{-5}) \sin (1000\pi \times 10^6)t$ A

PROBLEMS 30-3

1. 49 V   **3.** 16.5 V   **5.** 127 V   **7.** 21.2 A   **9.** 232 mA

PROBLEMS 30-4

1. (a) $i = 6.5 \sin (377t + 36°)$ A   3. -5.39 A   5.   (a) $i = 283 \sin (314t - 25°)$ A
(b) 6.46 A   (b) 63.7 A
7. 9.2° lag   9.   49° lead or lag

PROBLEMS 31-1

1. $20.7 + j11.3 = 23.6 \underline{/28.6°}$ 3. $1100 + j400 = 1170 \underline{/20°}$
5. $-442 + j741 = 863 \underline{/120.8°}$ 7. $11.2 - j36.4 = 38.1 \underline{/-72.9°}$
9. $2500 - j400 = 2532 \underline{/9.09°}$ 11. $10 + j9 = 13.5 \underline{/42°}$

PROBLEMS 31-2

1. $34 - j22 = 40.5 \underline{/-32.9°}$ 3. $20.9 + j25.13 = 32.7 \underline{/50.3°}$
5. $4.96 - j74.7 = 74.9 \underline{/-86.2°}$ 7. $0.0588 - j0.765 = 0.767 \underline{/-85.6°}$
9. $-0.588 - j4.35 = 4.39 \underline{/262.3°}$

PROBLEMS 31-3

1. $20.7 + j11.3 = 23.6 \underline{/28.7°}$ 3. $1104 + j400 = 1174 \underline{/19.9°}$
5. $-445 + j741 = 864 \underline{/121°}$ 7. $11.2 - j36.5 = 38.2 \underline{/-72.9°}$
9. $2478 + j798 = 2.6 \times 10^3 \underline{/17.9°}$ 11. $-300 + j400 = 500 \underline{/126.9°}$

PROBLEMS 31-4

1. $33 - j5.99 = 33.5 \underline{/-10.3°}$ 3. $-77 + j40.8 = 87.1 \underline{/152.1°}$
5. $4.43 - j74.6 = 74.7 \underline{/-86.6°}$ 7. $1.92 + j0.565 = 2 \underline{/16.4°}$
9. $-0.237 - j0.145 = 0.278 \underline{/-148.6°}$ 11. $13.9 - j2.69 = 14.1 \underline{/-11°}$
13. $\pm 12 \underline{/15°}$ 15. $2.89 \underline{/44°}$
17. $4 \underline{/90°}$ 19. $27 \underline{/33°}$

PROBLEMS 32-1

1. (a) 368 mA, (b) $v = 124 \sin 800\pi t$ V, (c) $i = 0.520 \sin 800\pi t$ A,
(d) 24.7 V, (e) 2.98 W, (f) 109 mA
3. 22.9 V   5. $i = 51.6 \sin (2 \times 10^4 \pi t)$ mA

PROBLEMS 32-2

1. 5.65 Ω   3.   94.2 kΩ   5.   2.54 kΩ   7.   297 mA   9.   200 MHz
11. -424 mA   13.   (a) X_L is doubled. (b) X_L is tripled. (c) X_L is halved.

PROBLEMS 32-3

1. 18.1 Ω   3.   72.3 mΩ   5.   138 Ω   7.   18.8 μA   9.   4 μF
11. $i = 0.639 \sin (377t + 90°)$ A   13.   153 pF   15.   137 μA   17. X_C varies inversely as C

PROBLEMS 32-4

1. (*a*) 565 Ω, (*b*) 567 Ω, (*c*) 388 mA, (*d*) $i = 549 \sin (377t - 86.5°)$ mA, (*e*) 13.6 V, (*f*) 219.6 V
3. (*a*) $f\downarrow$, $X_L\downarrow$, $Z\downarrow$, $I\uparrow$ (*b*) 12.7 kΩ, 3.9 MHz, 75° lag
5. (*a*) 121 Ω, (*b*) 351 Ω, (*c*) 342 mA, (*d*) 113 V, (*e*) 41.2 V
7. 358 Ω **9.** (*a*) 1.03 kΩ, (*b*) 583 mA, (*c*) 583 V, (*d*) 144 V

PROBLEMS 32-5

Q	Z	I	i	PF	P
1.	$528\,\underline{/67.7°}\ \Omega$	417 mA	$i = 589 \sin (377t - 67.7°)$ mA	38.0%	34.8 W
3.	$2024\,\underline{/8.88°}\ \Omega$	54.3 mA	$i = 76.9 \sin (314t - 8.88°)$ mA	99.0%	5.91 W
5.	$558\,\underline{/66.8°}\ \Omega$	2.15 A	$i = 3.04 \sin [(3.14 \times 10^7)t - 66.8°]$A	39.4%	1.02 kW
7.	$15\,\underline{/2.4°}\ \Omega$	7.79 A	$i = 11 \sin (377t - 2.4°)$A	99.9%	911 W
9.	$515\,\underline{/13.7°}\ \Omega$	3.42 A	$i = 4.84 \sin [(15.7 \times 10^6)t - 13.7°]$A	97.0%	5.88 kW

11. (*a*) 200 Ω **13.** (*a*) $55.8\,\underline{/-32.6°}\ \Omega$ **15.** (*a*) 23.5 A
    (*b*) 282 Ω (*b*) 505 μF   (*b*) 8.95 kW
    (*c*) 0.75 H

PROBLEMS 32-6

1. (*a*) 4.53 mA   (*b*) 2.56 mW   (*c*) 328 V across the capacitor, 228 V across the coil
3. 12 kHz   **5.** (*a*) 0.239 mH   (*b*) 2.3 MHz

PROBLEMS 33-1

1. 720 pF **3.** (*a*) $v = 311 \sin 377t$ V   (*b*) $i = 138 \sin (377t + 90°)$ mA   (*c*) 104 V
        (*d*) 1.82 μF   (*e*) 17.6 mA
5. (*a*) 0.47 H   (*b*) 1.41 H

PROBLEMS 33-2

1. $40.2\,\underline{/41.5°}\ \Omega$ **3.** $39.2\,\underline{/25°}\ \Omega$ **5.** $38.2\,\underline{/12.5°}\ \Omega$ **7.** $68\,\underline{/74.2°}\ \Omega$
9. $114\,\underline{/188°}\ \Omega$ **11.** $144\,\underline{/1.52°}\ \Omega$ **13.** $2.86\,\underline{/-69.5°}\ $A

PROBLEMS 33-3

1. (*a*) 5.623 MHz, (*b*) 5.627 MHz, (*c*) 25.7   **3.** (*a*) 113 kHz, (*b*) 5.66 MΩ, (*c*) 8.84 Ω
5. 500 pF   **7.** 7.08 mW   **9.** 0.6% leading   **11.** 32.2 pF

PROBLEMS 33-4

1. $Z_a = 11\,\underline{/-5.2°}\ \Omega$, $Z_b = 19.7\,\underline{/76.7°}\ \Omega$, $Z_c = 17.2\,\underline{/2°}\ \Omega$
3. $Z_1 = 73.2\,\underline{/40°}\ \Omega$, $Z_2 = 88.9\,\underline{/-73.1°}\ \Omega$, $Z_3 = 94.2\,\underline{/48°}\ \Omega$
5. $Z_{ab} = 64.0\,\underline{/2.3°}\ \Omega$ **7.** $Z_{ab} = 187\,\underline{/27.1°}\ \Omega$ **9.** 21.9 W **11.** $Z_{ab} = 89\,\underline{/-25°}\ \Omega$
13. 84.8 W   **15.** $Z_{ab} = 279\,\underline{/58.1°}\ \Omega$ **17.** $Z_{ab} = 81.5\,\underline{/67.3°}\ \Omega$

PROBLEMS 34-1

1. $2 = \log_{10} 100$ **3.** $2 = \log_7 49$ **5.** $0.5 = \log_4 2$ **7.** $1 = \log_a a$ **9.** $0 = \log_a 1$
11. $10^3 = 1000$ **13.** $5^2 = 25$ **15.** $6^0 = 1$ **17.** $5^4 = 625$ **19.** $r^s = t$ **21.** $x = 2$
23. $x = 6$ **25.** $x = 0.5$ **27.** $5 - 3 = 2$ **29.** $1, 2, 3, 4, 5, 6, 7, 8, 9$

PROBLEMS 34-2

1. 1 **3.** 2 **5.** 0 **7.** 2 **9.** $\bar{4}$ or $6 - 10$ **11.** 1 **13.** 3
15. $\bar{4}$ or $6 - 10$ **17.** $\bar{3}$ or $7 - 10$ **19.** 3 **21.** $\bar{1}$ or $9 - 10$ **23.** 1 **25.** -1.5
27. $\log 6792 + \log 20.9 - \log 176$ **29.** $\frac{1}{2}[\log 512 + \log 0.36 - (\log 2 + \log \pi + \log 177)]$
31. $3(\log 159 + \log 0.837 - \log 82.2)$ **33.** $0.437\ 12$ **35.** $\bar{2}.437\ 12$ or $8.437\ 12 - 10$
37. $4.437\ 12$ **39.** $7.437\ 12$ **41.** $\overline{11}.437\ 12$ or $9.437\ 12 - 20$ **43.** 7570 or 7.57×10^3
45. 7.57×10^6 **47.** 7.57×10^{-7} **49.** 7.57×10^{10}

PROBLEMS 34-3

1. $0.845\ 10$ **3.** $1.845\ 10$ **5.** $2.857\ 94$ **7.** $2.012\ 84$ **9.** $5.582\ 06$
11. $3.966\ 80$ **13.** $9.041\ 79 - 10$ or $-0.958\ 21$ **15.** $2.402\ 54$ **17.** $4.497\ 21 - 10$
19. $0.797\ 96$ **21.** $0.434\ 30$ **23.** $6.673\ 02 - 10$ **25.** $4.845\ 04$ **27.** $6.301\ 68$
29. $8.698\ 97 - 20$

PROBLEMS 34-4

1. 3 **3.** 30 **5.** 642 **7.** 101 **9.** 2.42×10^5 **11.** 8.425×10^3
13. 9.792×10^{-1} **15.** 3.4253×10^2 **17.** 1.493×10^{-5} **19.** 6.28 **21.** 2.718
23. 9.82×10^{-4} **25.** 7.9983×10^4 **27.** 1728 **29.** 8×10^{-12}

PROBLEMS 34-5

1. $5.951\ 49$ **3.** $\bar{2}.403\ 19$ **5.** $\bar{4}.273\ 66$
7. $2.559\ 78$ **9.** $2.347\ 87$ **11.** $3.855\ 91$

PROBLEMS 34-6

1. 2.56×10^2 **3.** 2.5×10^2 **5.** 3.78×10 **7.** -2.47×10^2 **9.** 2.37×10^3
11. 1×10^4 **13.** -9.81×10^{-1} **15.** 7.59×10^4 **17.** -1.35×10^3 **19.** 9.4

PROBLEMS 34-7

1. 3 **3.** 20 **5.** -700 **7.** -578 **9.** 4.88×10^6

PROBLEMS 34-8

1. 4.94 **3.** 2.42 **5.** -1.25×10^{-1} **7.** 2.61 **9.** 5.95×10^{-2}

PROBLEMS 34-9

1. 164 **3.** 9.59×10^{-8} **5.** 5.65 **7.** 5.93 **9.** 9.28×10^{-1} **11.** 25.3
13. 169 **15.** 1.16 **17.** 4.85×10^{-1} **19.** 8.76×10^{-1}

PROBLEMS 34-10

1. 2.563 10   **3.** $-0.186\ 97$ **5.** 1.95   **7.** 7.77×10^{-1} **9.** 4.7005

PROBLEMS 34-11

1. $x = 5.42$ **3.** $x = 31.6$ **5.** $x = 115$ **7.** $P_1 = 1.01 \times 10^4$
9. $x = 5.62 \times 10^6$ **11.** $x = 3.69$ **13.** $m = -1.84 \times 10^{-1}$ **15.** $x = 5 \times 10^{-1}$
17. $x = 4$ **19.** $L_1 = \sqrt[3]{L_2{}^2}$ **21.** $I_o = I_g \cdot 10^{5043(V_g/T)}$

23. (a) $V = i_c R \varepsilon$

     (b) $C = \dfrac{0.4343t}{R(\log V - \log i_c R)}$

     (c) $t = 2.3026RC(\log V - \log i_c R)$

25. (a) $V = \dfrac{i_L R}{1 - \varepsilon^{-Rt/L}}$

     (b) $L = \dfrac{0.4343Rt}{\log V - \log(V - i_L R)}$

     (c) $t = \dfrac{2.3026L}{R}[\log V - \log(V - i_L R)]$

27. (a) $V = \left(\dfrac{I_p + I_g}{K}\right)^{\frac{2}{3}} - \dfrac{V_p}{\mu}$

     (b) $V_p = \mu\left[-V + \left(\dfrac{I_p + I_g}{K}\right)^{\frac{2}{3}}\right]$

     (c) $\mu = \dfrac{V_p}{\left(\dfrac{I_p + I_g}{K}\right)^{\frac{2}{3}} - V}$

29. $i = 1.01$ A   **31.** 34.7 ms   **33.** 208 ms
35. 265 words/min   **39.** (a) 1.66×10^{-4} C
                     (b) 8 V
41. 296 mA   **43.** 1.26

PROBLEMS 35-1

1. R12 series of preferred values: 1.0, 1.2, 1.5, 1.8, 2.2, 2.7, 3.3, 3.9, 4.7, 5.6, 6.8, 8.2, 10.
Maximum % error: $\pm 11.1\%$
3. R10 series of preferred values: 1.0, 1.25, 1.6, 2.0, 2.5, 3.2, 4.0, 5.0, 6.4, 8.0, 10.
Probable published tolerance: $\pm 15\%$

PROBLEMS 35-2

1. (a) 13 dB   (b) 14 dB   (c) 18 dB   (d) -22.5 dB   **3.** 0.735 V   **5.** 775 mV
7. 1.94 V   **9.** (a) 12 mW, 2.68 V   (b) 60 mW, 6V   (c) 0.6 mW, 0.6 V   (d) 6×10^{-8} mW, 6 mV
11. 10^9 **13.** 10^3 W   **15.** 100 nW   **17.** 10^{-8}
19. 10^3 **21.** 100 dB   **23.** 2.92×10^6 **25.** 54 dB   **27.** 1.05 V
29. (a) 223 (b) 37 dB (c) 6.02 mW   **31.** 3.92 W   **33.** 1.1 μV   **35.** 0.48 dB/km
37. 250   **39.** 109 kW   **41.** 25.8 dB

PROBLEMS 35-3

1. 269 mH   **3.** (a) 7.13 mH (b) 22.5 nF   **5.** 356 mm   **7.** No. 2   **9.** 171.2 nF/km
11. 115 km   **13.** (a) 19 nH/cm (b) 0.0584 pF/cm (c) 0.295 mH (d) 906 pF

PROBLEMS 35-4

1. 570 Ω **3.** -1.3% **5.** No   **7.** No. 6   **9.** 138 Ω **13.** (a) 0.3 dB (b) 93.3%
15. (a) 0.7 dB (b) 85.1%   **17.** 604 pF   **19.** 5.6%

PROBLEMS 36-1

1. 5   **3.** 1   **5.** 7   **7.** 42   **9.** 35

PROBLEMS 36-2

1. 110   **3.** 10010   **5.** 11111   **7.** 1100001   **9.** 10110001

PROBLEMS 36-3

1. 2   **3.** 51   **5.** 63   **7.** 596   **9.** 3256

PROBLEMS 36-4

1. 31   **3.** 124   **5.** 245   **7.** 1467   **9.** 7530

PROBLEMS 36-5

1. 100_3 **3.** 102_5 **5.** 10000_4 **7.** 1000_{12} **9.** 40240_6 **11.** 64   **13.** 59   **15.** 3
17. 70   **19.** 6842

PROBLEMS 36-6

1. 011 110 001   **3.** 101 011 010   **5.** 001 000 110   **7.** 101 010 110 110
9. 111 111 111 111   **11.** 5   **13.** 35   **15.** 675   **17.** 632   **19.** 252

PROBLEMS 36-7

1. 111 001   **3.** 1 110 000   **5.** 111 100   **7.** 1 000 001   **9.** 110 100

PROBLEMS 36-8

1. 001001   **3.** 001001   **5.** 100111   **7.** 010111   **9.** 000101

PROBLEMS 36-9

1. 001011   **3.** 001100   **5.** 001100   **7.** 001011   **9.** 100101

PROBLEMS 36-10

1. 1011110   **3.** 10111001   **5.** 100011111   **7.** 101100000001   **9.** 101110110101

PROBLEMS 36-11

1. 011   **3.** 1001   **5.** 110   **7.** 1100   **9.** 10100

PROBLEMS 36-12

1. 0.1110011_2 99.935% accuracy   **3.** 0.10110011_2 99.888% accuracy
5. 0.11010101_2 99.884% accuracy   **7.** 0.010011001_2 99.933% accuracy
9. 0.000110011_2 99.976% accuracy

PROBLEMS 36-13

1. 0.1111_2 (100% accuracy)   **3.** 1.0011011_2 (100% accuracy)   **5.** 1.1101_2
7. 0.101101_2 (100% accuracy)   **9.** -0.1001_2 (100% accuracy)

PROBLEMS 36-14

1. $100.101001_2 = 4.640\ 625_{10}$ **3.** $100010.111_2 = 34.875_{10}$ **5.** $1001.00_2 = 9_{10}$
7. $1.1_2 = 1.5_{10}$ **9.** $10001.0111_2 = 17.4375_{10}$

PROBLEMS 36-15

	Binary	Denary	Accuracy, %
1.	1100.10011	12.59375 (12.6)	99.95
3.	10101010.10	170.5 (170.667)	99.9
5.	10	2.0	100
7.	10	2.0	100
9.	10.101010	2.65625 (2.6667)	99.6

PROBLEMS 36-16

1. 12.24_8 **3.** 11.74_8 **5.** 7.52_8 **7.** 17.74_8 **9.** 0.25_8

PROBLEMS 36-17

1. 105_{16} **3.** $53D_{16}$ **5.** $1C_{16}$ **7.** 264_{10} **9.** 400_{10}

PROBLEMS 36-18

1. 452_{10} **3.** 863_{10} **5.** 265_{10} **7.** 318_{16} **9.** $A7C_{16}$

PROBLEMS 36-19

1. 5003_{10} **3.** $0.206\ 787\ 109_{10}$ **5.** $170.101\ 562\ 500_{10}$ **7.** $B7C_{16}$ **9.** $27F_{16}$

PROBLEMS 37-1

1. sl **3.** $s + l$ **5.** sl **7.** $\bar{s} + l$ **9.** $sl + \bar{s} \cdot \bar{l}$

PROBLEMS 37-2

1.

a	b	c	ac	bc	$ac + bc$
1	1	1	1	1	1
1	1	0	0	0	0
1	0	1	1	0	1
1	0	0	0	0	0
0	1	1	0	1	1
0	1	0	0	0	0
0	0	1	0	0	0
0	0	0	0	0	0

3.

a	b	c	$a + b + c$	$\overline{a + b + c}$
1	1	1	1	0
1	1	0	1	0
1	0	1	1	0
1	0	0	1	0
0	1	1	1	0
0	1	0	1	0
0	0	1	1	0
0	0	0	0	1

PROBLEMS 37-4

1. (a) $Z_{xy} = \bar{a} + \bar{b} \cdot \bar{c}$ (b) $Y_{xy} = a(b + c)$

3. (a) $Z_{LM} = (\bar{A} + B)(A + \bar{B})$
(b) $Y_{LM} = A\bar{B} + \bar{A}B$ or $(A + B)(\overline{AB})$

5. (a) $Z_{pq} = \bar{a} + \bar{b}(ab\bar{c} + \bar{a}) + \bar{c}$ or $\bar{a} + \bar{a}\bar{b} + \bar{c}$
(b) $Y_{pq} = abc + a(\bar{a} + \bar{b} + c)c$ or ac

7.

9.

PROBLEMS 37-7

1. $\overline{\overline{a + b}} = a + b$: OR gate **3.** $\overline{\overline{a} + \overline{b}} = \overline{\bar{a} + \bar{b}} = \overline{ab}$: NAND gate

INDEX